HANDBOOK OF FINITE FIELDS

DISCRETE MATHEMATICS AND ITS APPLICATIONS

Series Editor

Kenneth H. Rosen, Ph.D.

R. B. J. T. Allenby and Alan Slomson, How to Count: An Introduction to Combinatorics, Third Edition

Craig P. Bauer, Secret History: The Story of Cryptology

Juergen Bierbrauer, Introduction to Coding Theory

Katalin Bimbó, Combinatory Logic: Pure, Applied and Typed

Donald Bindner and Martin Erickson, A Student's Guide to the Study, Practice, and Tools of Modern Mathematics

Francine Blanchet-Sadri, Algorithmic Combinatorics on Partial Words

Miklós Bóna, Combinatorics of Permutations, Second Edition

Richard A. Brualdi and Dragoš Cvetković, A Combinatorial Approach to Matrix Theory and Its Applications

Kun-Mao Chao and Bang Ye Wu, Spanning Trees and Optimization Problems

Charalambos A. Charalambides, Enumerative Combinatorics

Gary Chartrand and Ping Zhang, Chromatic Graph Theory

Henri Cohen, Gerhard Frey, et al., Handbook of Elliptic and Hyperelliptic Curve Cryptography

Charles J. Colbourn and Jeffrey H. Dinitz, Handbook of Combinatorial Designs, Second Edition

Abhijit Das, Computational Number Theory

Martin Erickson, Pearls of Discrete Mathematics

Martin Erickson and Anthony Vazzana, Introduction to Number Theory

Steven Furino, Ying Miao, and Jianxing Yin, Frames and Resolvable Designs: Uses, Constructions, and Existence

Mark S. Gockenbach, Finite-Dimensional Linear Algebra

Randy Goldberg and Lance Riek, A Practical Handbook of Speech Coders

Jacob E. Goodman and Joseph O'Rourke, Handbook of Discrete and Computational Geometry, Second Edition

Titles (continued)

Jonathan L. Gross, Combinatorial Methods with Computer Applications

Jonathan L. Gross and Jay Yellen, Graph Theory and Its Applications, Second Edition

Jonathan L. Gross and Jay Yellen, Handbook of Graph Theory

David S. Gunderson, Handbook of Mathematical Induction: Theory and Applications

Richard Hammack, Wilfried Imrich, and Sandi Klavžar, Handbook of Product Graphs, Second Edition

Darrel R. Hankerson, Greg A. Harris, and Peter D. Johnson, Introduction to Information Theory and Data Compression, Second Edition

Darel W. Hardy, Fred Richman, and Carol L. Walker, Applied Algebra: Codes, Ciphers, and Discrete Algorithms, Second Edition

Daryl D. Harms, Miroslav Kraetzl, Charles J. Colbourn, and John S. Devitt, Network Reliability: Experiments with a Symbolic Algebra Environment

Silvia Heubach and Toufik Mansour, Combinatorics of Compositions and Words

Leslie Hogben, Handbook of Linear Algebra

Derek F. Holt with Bettina Eick and Eamonn A. O'Brien, Handbook of Computational Group Theory

David M. Jackson and Terry I. Visentin, An Atlas of Smaller Maps in Orientable and Nonorientable Surfaces

Richard E. Klima, Neil P. Sigmon, and Ernest L. Stitzinger, Applications of Abstract Algebra with Maple™ and MATLAB®, Second Edition

Richard E. Klima and Neil P. Sigmon, Cryptology: Classical and Modern with Maplets

Patrick Knupp and Kambiz Salari, Verification of Computer Codes in Computational Science and Engineering

William Kocay and Donald L. Kreher, Graphs, Algorithms, and Optimization

Donald L. Kreher and Douglas R. Stinson, Combinatorial Algorithms: Generation Enumeration and Search

Hang T. Lau, A Java Library of Graph Algorithms and Optimization

C. C. Lindner and C. A. Rodger, Design Theory, Second Edition

San Ling, Huaxiong Wang, and Chaoping Xing, Algebraic Curves in Cryptography

Nicholas A. Loehr, Bijective Combinatorics

Toufik Mansour, Combinatorics of Set Partitions

Alasdair McAndrew, Introduction to Cryptography with Open-Source Software

Elliott Mendelson, Introduction to Mathematical Logic, Fifth Edition

Alfred J. Menezes, Paul C. van Oorschot, and Scott A. Vanstone, Handbook of Applied Cryptography

Stig F. Mjølsnes, A Multidisciplinary Introduction to Information Security

Jason J. Molitierno, Applications of Combinatorial Matrix Theory to Laplacian Matrices of Graphs

Titles (continued)

Richard A. Mollin, Advanced Number Theory with Applications

Richard A. Mollin, Algebraic Number Theory, Second Edition

Richard A. Mollin, Codes: The Guide to Secrecy from Ancient to Modern Times

Richard A. Mollin, Fundamental Number Theory with Applications, Second Edition

Richard A. Mollin, An Introduction to Cryptography, Second Edition

Richard A. Mollin, Quadratics

Richard A. Mollin, RSA and Public-Key Cryptography

Carlos J. Moreno and Samuel S. Wagstaff, Jr., Sums of Squares of Integers

Gary L. Mullen and Daniel Panario, Handbook of Finite Fields

Goutam Paul and Subhamoy Maitra, RC4 Stream Cipher and Its Variants

Dingyi Pei, Authentication Codes and Combinatorial Designs

Kenneth H. Rosen, Handbook of Discrete and Combinatorial Mathematics

Douglas R. Shier and K.T. Wallenius, Applied Mathematical Modeling: A Multidisciplinary Approach

Alexander Stanoyevitch, Introduction to Cryptography with Mathematical Foundations and Computer Implementations

Jörn Steuding, Diophantine Analysis

Douglas R. Stinson, Cryptography: Theory and Practice, Third Edition

Roberto Togneri and Christopher J. deSilva, Fundamentals of Information Theory and Coding Design

W. D. Wallis, Introduction to Combinatorial Designs, Second Edition

W. D. Wallis and J. C. George, Introduction to Combinatorics

Jiacun Wang, Handbook of Finite State Based Models and Applications

Lawrence C. Washington, Elliptic Curves: Number Theory and Cryptography, Second Edition

DISCRETE MATHEMATICS AND ITS APPLICATIONS

Series Editor KENNETH H. ROSEN

HANDBOOK OF FINITE FIELDS

Gary L. Mullen

Daniel Panario

CRC Press

Taylor & Francis Group

Boca Raton London New York

CRC Press is an imprint of the Taylor & Francis Group, an **informa** business

A CHAPMAN & HALL BOOK

First published in paperback 2024

First published 2013 by CRC Press
2385 NW Executive Center Drive, Suite 320, Boca Raton FL 33431

and by CRC Press
4 Park Square, Milton Park, Abingdon, Oxon, OX14 4RN

CRC Press is an imprint of Taylor & Francis Group, LLC

Publisher's Note
The publisher has gone to great lengths to ensure the quality of this reprint but points out that some imperfections in the original copies may be apparent.

ISBN: 978-1-4398-7378-6 (hbk)
ISBN: 978-1-03-291732-0 (pbk)
ISBN: 978-0-429-10519-7 (ebk)

DOI: 10.1201/b15006

Visit the Taylor & Francis Web site at
http://www.taylorandfrancis.com

and the CRC Press Web site at
http://www.crcpress.com

To Bevie Sue, with love,
Gary L. Mullen

Para Lucia, Natan, Diego y Lucas
por todo lo vivido juntos... y por lo que vendrá.
Daniel Panario

Contents

Part I: Introduction

Part II: Theoretical Properties

Part III: Applications

Preface

The CRC Handbook of Finite Fields (hereafter referred to as the *Handbook*) is a reference book for the theory and applications of finite fields. It is not intended to be an introductory textbook. Our goal is to compile in one volume the state of the art in research in finite fields and their applications. Hence, our aim is a comprehensive book, with easy-to-access references for up-to-date facts and results regarding finite fields.

The *Handbook* is organized into three parts. Part I contains just one chapter which is devoted to the history of finite fields through the 18-th and 19-th centuries.

Part II contains theoretical properties of finite fields. This part of the *Handbook* contains 12 chapters. Chapter 2 deals with basic properties of finite fields; properties that are used in various places throughout the entire *Handbook*. Near the end of Section 2.1 is a rather extensive list of recent finite field-related books; these books include textbooks, books dealing with theoretical topics as well as books dealing with various applications to such topics as combinatorics, algebraic coding theory for the error-free transmission of information, and cryptography for the secure transmission of information. Also included is a list of recent finite field-related conference proceedings volumes.

Chapter 2 also provides rather extensive tables of polynomials useful when dealing with finite field computational issues. The website `http://www.crcpress.com/product/isbn/9781439873786` provides larger and more extensive versions of the tables presented in Section 2.2.

The next two chapters deal with polynomials such as irreducible and primitive polynomials over finite fields. Chapter 5 discusses various kinds of bases over finite fields, and Chapter 6 discusses character and exponential sums over finite fields.

In Chapter 7, results on solutions of equations over finite fields are discussed. Chapter 8 covers permutation polynomials in one and several variables, as well as a discussion of value sets of polynomials, and exceptional polynomials over finite fields. Chapter 9 discusses special functions over finite fields. This discussion includes Boolean, APN, PN, bent, kappa polynomials, planar functions and Dickson polynomials, and finishes with a discussion of Schur's conjecture.

Sequences over finite fields are considered in Chapter 10. This chapter includes material on finite field transforms, LFSRs and maximal length sequences, correlation and autocorrelation and linear complexity of sequences as well as algebraic dynamical systems over finite fields.

Chapter 11 deals with various kinds of finite field algorithms including basic finite field computational techniques, formulas for polynomial counting, irreducible techniques, factorizations of polynomials in one and several variables, discrete logarithms, and standard models for finite fields.

In Chapter 12, curves over finite fields are discussed in great detail. This discussion includes elliptic and hyperelliptic curves. Rational points on curves are considered as well as towers and zeta functions over finite fields. In addition, there is a discussion of p-adic estimates of zeta and L-functions over finite fields.

Chapter 13 discusses a variety of topics over finite fields. These topics include relations between the integers and polynomials over finite fields, matrices over finite fields, linear algebra and related computational topics, as well as classical groups over finite fields, and Carlitz and Drinfeld modules.

Part III of the *Handbook*, containing four chapters, discusses various important applications, including mathematical as well as very practical applications of finite fields. Latin squares and the polynomial method, useful in various areas of combinatorics, are considered. In addition, affine and projective planes, projective spaces, block designs, and difference sets are discussed in detail. In each of these areas, since these topics contain an immense number of papers, we discuss only those techniques and topics related to finite fields. Other topics included in Chaper 14 are (t, m, s)-nets useful in numerical integration, applications of primitive polynomials over finite fields, and Ramanujan and expander graphs.

Chapter 15 is another important chapter in the *Handbook*. It discusses algebraic coding theory and includes a long introductory section dealing with basics properties of codes. This is followed by sections on special kinds of codes including LDPC codes, turbo codes, algebraic geometry codes, raptor codes, and polar codes.

Chapter 16 deals with cryptographic systems over finite fields. In the first section various basic issues dealing with cryptography are discussed. Next to be discussed are stream and block ciphers, multivariate cryptographic systems, elliptic and hyperelliptic curve cryptographic systems as well as systems arising from Abelian varieties over finite fields.

Finally, in Chapter 17 we discuss several additional applications of finite fields including finite fields in biology, quantum information theory, and various applications in engineering including topics like optimal orthogonal codes, binary sequences with small aperiodic autocorrelation, and sequences with small Hamming correlation.

In the bibliography, we have included for each reference, the pages where that reference is discussed in the *Handbook*. There is also a large index to help readers quickly locate various topics in the *Handbook*.

The *Handbook* is not meant to be read in a sequential way. Instead, each section is meant to be self-contained. Basic properties of finite fields are included in Chapter 2. Proofs are not included in the *Handbook*; instead authors have given references where proofs of the important results can be found. In an effort to help the reader locate proofs and important results for each section, at the end of each section we have provided a list of references used in that section. Those reference numbers refer to the main bibliography at the end of the *Handbook* which contains over 3,000 references. A short "See also" section is included for most sections; these are intended to provide the reader with references to other related sections and references of the *Handbook*.

The following numbering system is in effect in the *Handbook*. Within a given section, all results, theorems, corollaries, definitions, examples, etc., are numbered consecutively (with the exception of tables and figures). For example, the result numbered 2.1.5 happens to be a theorem which is the fifth listing in Section 2.1 of Chapter 2. We have also included many remarks in each section. These are also numbered as part of the same system so that for example, Remark 2.1.4 is the fourth listing in Section 2.1.

Readers are encouraged to make us aware of corrections to the material presented here. Readers should contact the author(s) of the section involved, as well as both of the Editors-in-Chief.

We would of course like to first thank the authors of the various sections for their time and effort. Without their help, the *Handbook* would, quite simply, not exist. We also greatly appreciate the authors' willingness to use our style and format so that the entire *Handbook* has a consistent and uniform style and format. While we appreciate the help of all of the authors, we would especially like to thank Ian Blake, Steve Cohen, Cary Huffman, Alfred Menezes, Harald Niederreiter, Henning Stichtenoth, and Arne Winterhof who not only wrote several sections, but who also provided the Editors-in-Chief with valuable input in numerous aspects of the *Handbook*. Every section was reviewed by at least two external reviewers, in addition to the Editors-in-Chief. We also would like to thank the many reviewers who took the time to read and send us comments on the various sections and drafts. Without their help, we would of course have ended up with a volume of considerably diminished quality. We would like to thank Brett Stevens and David Thomson for their help with various

LaTeX and file issues. Finally, we would like to thank Shashi Kumar for his invaluable help in setting up, reworking, and running the style files that define the overall look of the entire Handbook. His efforts were of tremendous help to us. We would also like to thank Bob Stern for his continued support.

Needless to say, this project has involved many, many hours. We thank Bevie Sue Mullen and Lucia Moura for their encouragement, support, love, and patience during the entire process.

Gary L. Mullen
Daniel Panario

Editors-in-Chief

Gary L. Mullen
Department of Mathematics
The Pennsylvania State University
University Park, PA 16802
U.S.A.
Email: mullen@math.psu.edu

Daniel Panario
School of Mathematics and Statistics
Carleton University
Ottawa ON K1S 5B6
Canada
Email: daniel@math.carleton.ca

Contributors

Omran Ahmadi
School of Mathematics
Institute for Research in Fundamental Sciences (IPM)
P. O. Box: 19395-5746, Tehran
Iran
Email: oahmadid@ipm.ir

Simeon Ball
Departament Matematica Aplicada IV
Universitat Politecnica Catalunya
c/Jordi Girona 1-3
08034 Barcelona
Spain
Email: simeon@ma4.upc.edu

Daniel J. Bernstein
Department of Computer Science (MC 152)
University of Illinois at Chicago
851 S. Morgan Street
Chicago, IL 60607-7053
U.S.A
Email: djb@cr.yp.to

Régis Blache
IUFM de Guadeloupe
Morne Ferret
97139 Les Abymes
French West Indies
Email: rblache@iufm.univ-ag.fr

Ian F. Blake
Department of Electrical and Computer Engineering
University of British Columbia
Vancouver, BC V6T 1Z4
Canada
Email: ifblake@ece.ubc.ca

Aart Blokhuis
Department of Mathematics and Computer Science
Eindhoven University of Technology
P.O. Box 513
5600 MB
The Netherlands
Email: aartb@win.tue.nl

Claude Carlet
University of Paris 8
Department of Mathematics
2 rue de la Liberté
93526 Saint-Denis, Cedex
France
Email: claude.carlet@univ-paris8.fr

Francis Castro
Department of Mathematics
University of Puerto Rico
Rio Piedras Campus
San Juan, 00931
Puerto Rico
Email: franciscastr@gmail.com

Pascale Charpin
INRIA-Rocquencourt
Domaine de Voluceau, B.P. 105
F-78153 Le Chesnay Cedex
France
Email: Pascale.Charpin@inria.fr

Sebastian M. Cioabă
Department of Mathematical Sciences
University of Delaware
Newark, DE 19716-2553
U.S.A.
Email: cioaba@math.udel.edu

Stephen D. Cohen
School of Mathematics & Statistics
University of Glasgow
Glasgow G12 8QW
Scotland
Email: Stephen.Cohen@glasgow.ac.uk

Charles J. Colbourn
School of CIDSE
Arizona State University
Tempe, AZ 85287-8809
U.S.A.
Email: Charles.Colbourn@asu.edu

Robert Coulter
Department of Mathematical Sciences
University of Delaware
520 Ewing Hall
Newark, DE 19716
U.S.A.
Email: coulter@math.udel.edu

Jintai Ding
Department of Mathematical Sciences
University of Cincinnati
2815 Commons Way
Cincinnati, OH 45221-0025
U.S.A.
Email: jintai.ding@gmail.com

Jeff Dinitz
Department of Mathematics and Statistics
University of Vermont
Burlington, VT 05405
U.S.A.
Email: Jeff.Dinitz@uvm.edu

Christophe Doche
Department of Computing
Macquarie University
North Ryde, NSW 2109
Australia
Email: christophe.doche@mq.edu.au

Jean-Guillaume Dumas
Université Joseph Fourier, Grenoble I
Laboratoire Jean Kuntzmann
Mathématiques Appliquées et Informatique
38041 Grenoble
France
Email: Jean-Guillaume.Dumas@imag.fr

Gary Ebert
N3691 Sylvan Isle Drive
Watersmeet, MI 49969
U.S.A.
Email: ebert@math.udel.edu

Gove Effinger
Department of Mathematics and Computer Science
Skidmore College
Saratoga Springs, NY 12866
U.S.A.
Email: effinger@skidmore.edu

Andreas Enge
INRIA Bordeaux - Sud-Ouest
Université Bordeaux 1
351 cours de la Libération
33405 Talence Cedex
France
Email: andreas.enge@inria.fr

Ron Evans
Department of Mathematics
University of California at San Diego
La Jolla, CA 92093-0112
U.S.A.
Email: revans@ucsd.edu

Haining Fan
School of Software
Tsinghua University
Beijing
China
Email: fhn@tsinghua.edu.cn

Robert Fitzgerald
Department of Mathematics
Southern Illinois University
Carbondale, IL 62901-4408
U.S.A.
Email: rfitzg@siu.edu

Michael D. Fried
3548 Prestwick Rd.
Billings, MT 59101
U.S.A.
Email: mfried@math.uci.edu

Lei Fu
Chern Institute of Mathematics
Nankai University
Tianjin 300071
P. R. China
Email: leifu@nankai.edu.cn

Shuhong Gao
Department of Mathematical Sciences
Clemson University
Clemson, SC 29634-0975
U.S.A.
Email: sgao@clemson.edu

Moubariz Z. Garaev
Centro de Ciencias Matemáticas
Universidad Nacional Autónoma de México
Morelia 58089, Michoacán
México
Email: garaev@matmor.unam.mx

Arnaldo Garcia
Instituto Nacional de Matemática Pura
e Aplicada, IMPA
Estrada Dona Castorina 110
22460-320, Rio de Janeiro, RJ
Brazil
Email: garcia@impa.br

Joachim von zur Gathen
B-IT, Universität Bonn
Dahlmannstr. 2
53179 Bonn
Germany
Email: gathen@bit.uni-bonn.de

Mark Giesbrecht
Cheriton School of Computer Science
University of Waterloo
Waterloo, Ontario, N2L 3G1
Canada
Email: mwg@uwaterloo.ca

Guang Gong
Department of Electrical and Computer
Engineering
University of Waterloo
Waterloo, Ontario N2L 3G1
Canada
Email: ggong@uwaterloo.ca

David Goss
Department of Mathematics
Ohio State University
231 West 18th Avenue
Columbus, OH 43210
U.S.A.
Email: goss@math.ohio-state.edu

Roderick Gow
School of Mathematical Sciences
University College Dublin
Belfield, Dublin 4
Ireland
Email: rod.gow@ucd.ie

Kishan Chand Gupta
Applied Statistics Unit
Indian Statistical Institute
203 B. T. Road
Kolkata 700108
India
Email: kishan@isical.ac.in

Dirk Hachenberger
Institut für Mathematik
Universität Augsburg
86135 Augsburg
Germany
Email: hachenberger@math.uni-augsburg.de

M. Anwarul Hasan
Department of Electrical and Computer Engineering
University of Waterloo
Waterloo, ON, N2L 3G1
Canada
Email: ahasan@uwaterloo.ca

Tor Helleseth
Institutt for informatikk
Universitetet i Bergen
PB. 7803, N-5020 Bergen
Norway
Email: Tor.Helleseth@ii.uib.no

Franziska Hinkelmann
Mathematical Biosciences Institute
Ohio State University
1735 Neil Ave
Columbus, OH 43210
U.S.A.
Email: hinkelmann.1@mbi.osu.edu

James W.P. Hirschfeld
Department of Mathematics
University of Sussex
Brighton BN1 9QH
United Kingdom
Email: jwph@sussex.ac.uk

Xiang-dong Hou
Department of Mathematics and Statistics
University of South Florida
4202 E. Fowler Ave
Tampa, FL 33620
U.S.A.
Email: xhou@usf.edu

W. Cary Huffman
Department of Mathematics and Statistics
Loyola University Chicago
1032 W. Sheridan Road
Chicago, IL 60660
U.S.A.
Email: whuffma@luc.edu

Michael Jacobson, Jr.
Department of Computer Science
University of Calgary
Calgary, Alberta, T2N 1N4
Canada
Email: jacobs@ucalgary.ca

Jonathan Jedwab
Department of Mathematics
Simon Fraser University
Burnaby, British Columbia V5A 1S6
Canada
Email: jed@sfu.ca

Dieter Jungnickel
Mathematical Institute
University of Augsburg
D-86135 Augsburg
Germany
Email: jungnickel@math.uni-augsburg.de

Erich Kaltofen
Department of Mathematics
Campus Box 8205
North Carolina State University
Raleigh, NC 27695-8205
U.S.A.
Email: kaltofen@math.ncsu.edu

Alexander Kholosha
Department of Informatics
University of Bergen
P.O. Box 7800
N-5020 Bergen
Norway
Email: Alexander.Kholosha@uib.no

Melsik Kyuregyan
Institute for Informatics and Automation Problems
National Academy of Sciences of Armenia
1, P. Sevak str., Yerevan 0014
Armenia
Email: melsik@ipia.sci.am

Tanja Lange
Coding Theory and Cryptology, MF6.104 B
Department of Mathematics and Computer Science
Technische Universiteit Eindhoven
P.O. Box 513
5600 MB Eindhoven
Netherlands
Email: tanja@hyperelliptic.org

Reinhard Laubenbacher
Virginia Bioinformatics Institute
Blacksburg, VA 24061
U.S.A.
Email: reinhard@vbi.vt.edu

Grégoire Lecerf
Laboratoire d'informatique (LIX, UMR 7161 CNRS)
Campus de l'École polytechnique
Bâtiment Alan Turing
91128 Palaiseau Cedex
France
Email: gregoire.lecerf@math.cnrs.fr

Hendrik Lenstra
Mathematisch Instituut
Universiteit Leiden
Postbus 9512
2300 RA Leiden
The Netherlands
Email: hwl@math.leidenuniv.nl

Antonio Rojas-León
Departamento de Álgebra - Facultad de Matemáticas
Universidad de Sevilla
Apdo. de correos 1160
41080 Sevilla
Spain
Email: arojas@us.es

Qunying Liao
Institute of Mathematics and Software Science
Sichuan Normal University
Chengdu, Sichuan Province, 610066
P. R. China
Email: qunyingliao@sicnu.edu.cn

Rudolf Lidl
7 Hill Street
West Launceston
Tasmania 7250
Australia
Email: rudi@rlidl.com.au

Simon Litsyn
School of Electrical Engineering
Tel Aviv University
Ramat Aviv 69978
Israel
Email: litsyn@eng.tau.ac.il

Gary McGuire
School of Mathematical Sciences
University College Dublin
Dublin 4
Ireland
Email: gary.mcguire@ucd.ie

Wilfried Meidl
Faculty of Engineering and Natural Sciences
Sabanci University
Orhanli 34956, Tuzla-Istanbul
Turkey
Email: wmeidl@sabanciuniv.edu

Alfred Menezes
Department of Combinatorics & Optimization
University of Waterloo
Waterloo, Ontario N2L 3G1
Canada
Email: ajmeneze@uwaterloo.ca

Gary L. Mullen
Department of Mathematics
The Pennsylvania State University
University Park, PA 16802
U.S.A.
Email: mullen@math.psu.edu

Kumar Murty
Department of Mathematics
University of Toronto
40 St. George Street
Toronto, ON M5S 2E4
Canada
Email: murty@math.toronto.edu

M. Ram Murty
Department of Mathematics and Statistics
Queen's University
Kingston, Ontario K7L 3N6
Canada
Email: murty@mast.queensu.ca

Harald Niederreiter
Radon Institute for Computational and Applied Mathematics
Austrian Academy of Sciences
Altenbergerstr. 69
A-4040 Linz
Austria
Email: ghnied@gmail.com

Andrew Odlyzko
School of Mathematics
University of Minnesota
Minneapolis, MN 55455
U.S.A.
Email: odlyzko@umn.edu

Alina Ostafe
Department of Computing
Faculty of Science
Macquarie University
North Ryde, NSW 2109
Australia
Email: alina.ostafe@mq.edu.au

Daniel Panario
School of Mathematics and Statistics
Carleton University
Ottawa ON K1S 5B6
Canada
Email: daniel@math.carleton.ca

Clément Pernet
INRIA/LIG MOAIS
51, avenue Jean Kuntzmann
F-38330 Montbonnot Saint-Martin
France
Email: clement.pernet@imag.fr

Alexander Pott
Otto-von-Guericke University Magdeburg
39106 Magdeburg
Germany
Email: alexander.pott@ovgu.de

Martin Roetteler
NEC Laboratories America, Inc.
4 Independence Way, Suite 200
Princeton, NJ 08540
U.S.A.
Email: mroetteler@nec-labs.com

Ivelisse Rubio
Department of Computer Science
University of Puerto Rico
Rio Piedras Campus
P.O. Box 70377
San Juan, PR 00936-8377
Email: iverubio@gmail.com

Renate Scheidler
Department of Mathematics and Statistics
University of Calgary
2500 University Drive NW
Calgary, Alberta, T2N 1N4
Canada
Email: rscheidl@ucalgary.ca

Kai-Uwe Schmidt
Faculty of Mathematics
Otto-von-Guericke University
Universitätsplatz 2
39106 Magdeburg
Germany
Email: kaiuwe.schmidt@ovgu.de

Igor Shparlinski
Department of Computing
Macquarie University
Sydney, NSW 2109
Australia
Email: igor.shparlinski@mq.edu.au

Joseph H. Silverman
Mathematics Department, Box 1917
Brown University
Providence, RI 02912
U.S.A.
Email: jhs@math.brown.edu

Bart de Smit
Mathematisch Institut
Universiteit Leiden
Postbus 9512
2300 RA Leiden
The Netherlands
Email: desmit@math.leidenuniv.nl

Brett Stevens
School of Mathematics and Statistics
Carleton University
Ottawa ON K1S 5B6
Canada
Email: brett@math.carleton.ca

Henning Stichtenoth
Faculty of Engineering and Natural Sciences
Sabanci University
Orhanli 34956, Tuzla-Istanbul
Turkey
Email: henning@sabanciuniv.edu

Leo Storme
Department of Mathematics
Ghent University
Krijgslaan 281, Building S22
B-9000 Ghent, Belgium
Email: ls@cage.ugent.be

Oscar Takeshita
Email: oscar_takeshita@hotmail.com

Joseph A. Thas
Department of Mathematics
Ghent University
Krijgslaan 281 - S22
9000 Gent
Belgium
Email: `jat@cage.ugent.be`

Nicolas Thériault
Departamento de Matemática
Universidad del Bio-Bio
Avda. Collao 1202
Casilla 5-C, Concepcion, 4051381
Chile
Email: `ntheriau@ubiobio.cl`

David Thomson
School of Mathematics and Statistics
Carleton University
Ottawa ON K1S 5B6
Canada
Email: `dthomson@math.carleton.ca`

José Felipe Voloch
The University of Texas at Austin
Mathematics Dept, RLM 8.100
2515 Speedway Stop C1200
Austin, Texas 78712-1202
U.S.A.
Email: `voloch@math.utexas.edu`

Daqing Wan
Department of Mathematics
University of California
Irvine, CA 92697-3875
U.S.A.
Email: `dwan@math.uci.edu`

Zhe-Xian Wan
Academy of Mathematics and Systems Science
Chinese Academy of Sciences
No 55, Zhongguancun East Road
Zhongguancun, Beijing 100190
P. R. China
Email: `wan@amss.ac.cn`

Qiang Wang
School of Mathematics and Statistics
Carleton University
Ottawa ON K1S 5B6
Canada
Email: `wang@math.carleton.ca`

Arne Winterhof
Johann Radon Institute for Computational and Applied Mathematics
Austrian Academy of Sciences
Altenbergerstr. 69
4040 Linz, Austria
Email: `arne.winterhof@oeaw.ac.at`

Joseph L. Yucas
68 Rock Springs Rd.
Makanda, IL 62958
U.S.A.
Email: `joeyucas@yahoo.com`

Michael E. Zieve
Department of Mathematics
University of Michigan
Ann Arbor, MI 48109-1043
U.S.A.
Email: `zieve@umich.edu`

I

Introduction

1

History of finite fields

1.1 Finite fields in the 18-th and 19-th centuries

Roderick Gow, *University College Dublin*

1.1.1 Introduction

While the theory of finite fields emerged as an independent discipline at the end of the 19-th century, aspects of the subject can be traced back at least to the middle of the 17-th century. It is our intention to present here a survey of highlights of finite field theory as they emerged in the 18-th and 19-th centuries, culminating in a description of Eliakim Hastings Moore's [2139], which began the study of abstract finite fields.

Leonard Eugene Dickson (1874-1954), in the first volume of his *History of the Theory of Numbers* [851] gives many references to works that can be interpreted as dealing with finite fields, although not always described explicitly as such. Chapters VII and VIII are especially relevant, and give remarkably complete listings of what had been achieved before 1918. Chapter VIII, entitled *Higher Congruences*, occasionally uses the language of finite fields, although the emphasis is largely number theoretic. Dickson had already written a textbook, entitled *Linear Groups with an Exposition of the Galois Field Theory* [850] which is probably the first work devoted exclusively to finite fields. This book remained without any serious rival until the emergence in the 1950s of more geometric, less computational, methods, such as those pioneered by Artin in his *Geometric Algebra* [135]. The first 71 pages of Dickson's work constitute a very full account of finite fields, and its exercises, partly based on the work of earlier researchers, are still a valuable source of problems and ideas.

Of course, finite fields are mentioned in general histories of algebra, such as that of van der Waerden [2848]. Furthermore, *Finite Fields* by Lidl and Niederreiter [1939] contains much historical information and a very extensive bibliography, especially of the older literature. Another brief but useful source of information is found in the historical notes scattered throughout Cox's *Galois Theory* [749].

The first use of the English expressions *field of order s* and *Galois-field of order* $s = q^n$ occurs in a paper of E. H. Moore (1862-1932), which he presented in 1893. Moore states that the term *field* was an equivalent to the German term *endlicher Körper*, used by Heinrich Weber (1842-1913). We observe that Richard Dedekind (1831-1916) had already introduced such a term as *Zahlenkörper*, which can be traced back to lectures he gave in 1858. The 1933 edition of the *Oxford English Dictionary* does not include a definition of the mathematical term *field*, although it does define *group* in its mathematical meaning, but more recent editions of the dictionary include the mathematical use of *field*, with an attribution to Moore.

The name Galois field is synonymous with finite field, and it signifies the importance to the subject of an innovatory paper by Évariste Galois (1811-1832), published in 1830 [1168], when the author was only 18. We will comment in greater detail on Galois's work later in this article, but we will briefly mention here that Galois lays the foundations of finite field theory by showing that for each prime p and positive integer n, there is a finite field of order p^n, and its multiplicative group of non-zero elements is cyclic of order $p^n - 1$.

Galois's arguments are rather sketchy, but there is no doubt that he understood the fundamental principles of the structure of a finite field, including the role of the automorphism given by raising elements to the p-th power. As has proved to be the case on a number of occasions, it seems that most of Galois's discoveries were already known to Gauss, in this case, in the late 1790s, but as Gauss never published an account of his work, Galois was unaware of Gauss's priority. (Gauss is credited with the discovery of non-Euclidean geometry before Bolyai and Lobachevsky, with the discovery of quaternions before Hamilton, and the discovery of the method of least squares before Legendre.) We will also give a sketch of Gauss's approach to finite fields, which he called the theory of *higher congruences*, as it is described in Volume 2 of his *Werke* [1259].

1.1.2 Early anticipations of finite fields

Our approach to the early history of finite fields will be largely chronological. An early occurrence of a theorem that may be interpreted in the language of finite fields is Fermat's Little Theorem, that $x^{p-1} - 1$ is divisible by p when p is a prime and x an integer not divisible by p. Dickson states that the special case when $x = 2$ was already known to the ancient Chinese around 500 BCE. As was usual with Fermat (1601-1665), he did not give a formal proof, but he communicated his conjecture that the theorem holds true, in a letter to Bernard Frénicle de Bessy (1605-1675), dated 18 October, 1640. Leonhard Euler (1707-1783) gave a complete proof of the theorem in 1736, but unpublished manuscripts of Gottfried Wilhelm Leibniz (1646-1716) show that he was in possession of a similar proof by 1680.

In the 18-th century, further theorems, expressed in terms of congruences modulo a prime, that we can see as precursors of basic facts in finite field theory were discovered by mathematicians such as Euler, Joseph-Louis Lagrange (1736-1813), and Adrien-Marie Legendre (1752-1833). However, the first complete account of that body of knowledge that relates to the finite field of prime order was presented by Carl Friedrich Gauss (1777-1855) in Sections I-IV of his *Disquisitiones Arithmeticae* [1258], which we describe in the next section.

1.1.3 Gauss's Disquisitiones Arithmeticae

Gauss's *Disquisitiones Arithmeticae* [1258] was an epoch making work in mathematics, introducing totally new ideas and demanding far higher standards of proof than had hitherto been required or expected. The book also serves as a commentary on the discoveries and shortcomings of his predecessors. For a very full account of the contents and influence of

Gauss's *magnum opus*, we refer to the book *The Shaping of Arithmetic* [1297].

Gauss introduces the concept of congruence in Article (Art.) 1, and designates congruence by means of the now familiar symbol $\equiv$. This is the first published use of this symbol, which seems to have entered into conventional use quite rapidly. It occurs for instance in C. Kramp's *Éléments d'arithmétique universelle* [1804] an elementary work much influenced by Gauss's masterpiece (the use of the exclamation mark in $n!$ makes its first appearance here). In Section 2, Gauss proves in Art. 14 that if p is a prime integer and a, b are integers not divisible by p, then p does not divide the product ab. This basic result is fundamental for the proof that the integers modulo p form a field. Gauss comments that the theorem was already in Euclid's *Elements.* Oddly enough, for a person as notoriously meticulous as Gauss, he mistakenly says that it is Proposition 32 of Book VII, when it is in fact Proposition 30. Concerning this result, Gauss wrote magisterially: *However we did not wish to omit it because many modern authors have employed vague computations in place of proof or have neglected the theorem entirely, and because by this very simple case we can more easily understand the nature of the method which will be used later for solving much more difficult problems.* He uses Art. 14 to prove Art. 16, a result often called the fundamental theorem of arithmetic: *a composite number can be resolved into prime factors in only one way.* This basic result is not in Euclid.

Gauss describes Euler's totient (or phi) function, which he denotes by the symbol ϕ (following Art. 38). (We recall that the totient function measures the number of totitives of a positive integer n, that is, the number of integers lying between 1 and n that are relatively prime to n.) This is again the first occurrence of a now familiar symbol in mathematics. Euler himself, although introducing the idea of the function in 1760, did not use such notation. Art. 43 is a proof that an integer polynomial of degree m cannot have more than m incongruent roots modulo a prime. This basic theorem on polynomial arithmetic was first published by Lagrange in 1768. Euler had shown that the congruence $x^n - 1 \equiv 0$ modulo a prime has at most n roots in 1774, and Gauss notes that Euler's method is easily generalized.

Section III, on residues of powers, contains Art. 49: *if p is a prime number that does not divide a, and a^t is the lowest power of a that is congruent to unity to the modulus p, the exponent t will either $= p - 1$ or be a factor of this number.* Gauss notes that this implies Fermat's Little Theorem, and he gives some of the history of this theorem that we described above. Art. 55 is the fundamental statement: *There always exist numbers with the property that no power less than the $p - 1$st is congruent to unity.* This of course amounts to saying that the multiplicative group of the integers modulo a prime p is cyclic of order $p - 1$. Again, it is interesting to observe the authority of Gauss's language as he describes earlier approaches to Art. 55: *This theorem furnishes an outstanding example of the need for circumspection in number theory so that we do not accept fallacies as certainties. ... No one has attempted the demonstration except Euler ... See especially his article 37 where he speaks at great length of the need for demonstration. But the demonstration which this shrewdest of men presents has two defects. ...* In Art. 57, Gauss adopts the nomenclature *primitive roots*, due originally to Euler, for the integers, or residues, described in Art. 55.

1.1.4 Gauss's Disquisitiones Generales de Congruentiis

Gauss had intended to include an eighth section of *Disquisitiones Arithmeticae*, and he even refers to this section at least twice in the published version. However, the section was omitted, possibly for reasons of saving space in an already long work. A manuscript of the missing section was found after Gauss's death, and an edited version, with notes by Dedekind, was published in volume 2 of Gauss's *Werke* [1259] in 1863, under the title *Disquisitiones Generales de Congruentiis.* A German translation followed in 1889.

Günther Frei, [1103], has given a lengthy description of the genesis and contents of the unpublished Section Eight, and we will make use of some of his analysis here, since it has considerable bearing on the early theory of finite fields. Gauss's work on finite fields can be traced back at least to 1796, as there are references to it in his *Mathematical Diary* [1257]. It is well known that Gauss was particularly fascinated by the law of quadratic reciprocity, and he gave several different proofs of this fundamental theorem, the first dating from 1796. The third and fourth of these proofs drew Gauss into the study of polynomials modulo a prime, and his surviving investigations enable us to discern much of the theory of finite extensions of a field of prime order.

In Frei's translation, Gauss wrote *But at the same time one sees that the solution of congruences constitutes only a part of a much higher investigation, namely the investigation of the decomposition of functions into factors.* Accordingly, Gauss developed a theory of factorization of polynomials whose coefficients are integers modulo a prime p, including the determination of greatest common divisors by Euclid's algorithm. He introduced the concept of a *prime* polynomial, corresponding to *irreducible* polynomial in modern terminology, and showed that arbitrary polynomials can be factored into products of prime polynomials.

Among the highlights of his discoveries, we may mention his proof that every irreducible polynomial modulo p, different from x, and of degree m, is a divisor of $x^{p^m-1}-1$. Furthermore, $x^{p^m-1}-1$ is the product of all monic irreducible polynomials of degree d dividing m, apart from x. From this fact, he obtained a formula for the number of irreducible monic polynomials of degree n with coefficients integers modulo p. Frei also notes that Gauss appreciated the importance of the *Frobenius automorphism*, and came close to discovering a form of Hensel's Lemma, significant in p-adic analysis.

The idea of using the imaginary roots of such irreducible polynomials to simplify some of his work had occurred to Gauss, and, in Frei's translation, Gauss wrote *Indeed, we could have shortened incomparably all our following investigations, had we wanted to introduce such imaginary quantities by taking the same liberty some more recent mathematicians have taken, but nevertheless, we have preferred to deduce everything from first principles.* It should be recalled that Gauss sometimes displayed a conservative approach to new concepts in mathematics, and his public aversion to using imaginary roots of congruences is akin to his disinclination to use complex numbers. Thus, for example, his thesis, published in 1799, states that every real polynomial is a product of real factors of degree one or two, rather than stating that every complex polynomial is a product of factors of degree 1.

1.1.5 Galois's Sur la théorie des nombres

We turn now to presenting a synopsis of Galois's 1830 paper [1168] *Sur la théorie des nombres* on finite fields since it is a landmark in the subject. In Frei's opinion, Galois establishes the additive and multiplicative structure of finite extensions of the field of prime order. It certainly seems that the spirit of Galois's paper is closer to the modern presentation of finite field theory than Gauss's version. It is worth noting that there are several misprints in the paper, which would have made it difficult to follow for the uninitiated, and Galois's attempts to illustrate the theory are hopelessly flawed.

Rather than translating the original French literally, we will instead try to convey some idea in modern terms of what Galois must have intended. For example, when Galois talks of a *function*, he means a polynomial in a single variable, with integer coefficients. (This convention was common among mathematicians before the twentieth century.) He notes that we usually look for integer roots of the polynomial modulo a prime p, say. We call these *real* roots of the polynomial congruence. He proceeds to generalize the notion of real roots, and begins by introducing the concept of an integer polynomial $F(x)$ being *irreducible* modulo p, meaning that it is impossible to find three integer polynomials $\phi(x)$, $\psi(x)$ and

$\chi(x)$ such that

$$F(x) + p\chi(x) = \phi(x)\psi(x).$$

(Galois does not use the term irreducible for this concept, but later in the paper speaks of an *irreducible congruence*.)

Such an irreducible polynomial $F(x)$ obviously has no integer roots modulo p, nor any *irrational roots* of degree less than that of $F(x)$ (Galois does not explain these terms). He states that we must regard the roots of the congruence $F(x) \equiv 0$ (notation he attributes to Gauss) as a type of imaginary symbols, and opines that such imaginary roots will prove to be as useful as $\sqrt{-1}$ is in conventional analysis. These imaginary roots were subsequently called *Galois imaginaries* by later writers.

Let i be a root of the congruence $F(x) \equiv 0$, where F has degree ν. (Galois does not justify why we may assume that $F(x) \equiv 0$ has roots, a point Serret attempted to rectify.) Galois then considers a general expression

$$a + a_1 i + a_2 i^2 + \cdots + a_{\nu-1} i^{\nu-1},$$

where a, a_1, ..., $a_{\nu-1}$ are integers modulo p. There are p^ν different values for these expressions.

Let α be an expression of the form above. If we raise α to the second, third, *etc*, powers, we obtain a sequence of expressions of the same form. Thus we must have $\alpha^n = 1$ for a certain positive integer n, which we choose to be as small as possible. We then have n different expressions

$$1, \alpha, \ldots, \alpha^{n-1}.$$

Galois shows that n divides $p^\nu - 1$, and thus $\alpha^{p^\nu - 1} = 1$. Galois next aims to prove that there is some α for which the corresponding n is $p^\nu - 1$. He makes an analogy at this stage with existence of primitive roots modulo p in the theory of numbers. We did not find that Galois provided a convincing argument for this key issue.

Galois then draws the remarkable conclusion that all the algebraic quantities that arise in this theory are roots of equations of the form $x^{p^\nu} = x$. Furthermore, if $F(x)$ is an integer polynomial of degree ν irreducible modulo p, there are integer polynomials $f(x)$ and $\phi(x)$ such that

$$f(x)F(x) = x^{p^\nu} - x + p\phi(x).$$

Galois also notes that if α is a root of the irreducible congruence $F(x) \equiv 0$, then the other roots are

$$\alpha^p, \alpha^{p^2}, \ldots, \alpha^{p^{\nu-1}}.$$

This is a consequence of the fact that

$$F(x)^{p^n} \equiv F(x^{p^n}).$$

We remark that this is an early indication of the role of the so-called Frobenius mapping as a generator of the associated Galois group. Galois later notes that all the roots of the congruence $x^{p^\nu} \equiv x$ depend only on the roots of a single irreducible polynomial of degree ν.

To illustrate all this theory, Galois attempts to find a primitive root of the congruence

$$x^{7^3} \equiv x \pmod{7}.$$

He aims to do this by exhibiting elements having orders 9 and 19. In fact, he makes several noteworthy errors, which may have confused any readers of this exposition of the new theory. He begins by noting that $x^3 \equiv 2 \pmod{7}$ is irreducible, and lets i be a root of the congruence. He claims that $-1-i$ has order 19, but this is false, as it has order 9×19. He then claims that $\alpha = i + i^2$ is primitive, but this is again false, as it has order 114, not $342 = 7^3 - 1$. Finally, Galois claims that his α satisfies

$$\alpha^3 + 3\alpha + 1 = 0,$$

but this is again incorrect, as in truth it satisfies

$$\alpha^3 + \alpha + 1 = 0.$$

Indeed, the polynomial $x^3 + 3x + 1$ is not even irreducible modulo 7, as 4 is a root of it. As we mentioned earlier, the paper contains several misprints, possibly because the compositor found the notation difficult to handle, but Galois's errors are not just typographical (although they are of course ultimately trivial and in no way invalidate his theory).

Joseph-Alfred Serret gives a treatment of this problem of finding a primitive root in the second edition of *Cours d'algèbre supérieure* [2596], pp. 367-370, following Galois's methods. The required primitive element Galois might have had in mind was $\beta = i - i^2$, not $i + i^2$. This element β is a root of $x^3 - x + 2$, which is certainly an irreducible primitive polynomial. While $i + i^2$ may have replaced $i - i^2$ because of a typographical error, Galois nonetheless made further mistakes which are difficult to explain. Serret himself made no comment on this strange aspect of Galois's paper.

As justification for introducing this theory, Galois explains that it is required in the theory of permutations which arise in the study of primitive (rational) polynomials which are solvable by radicals. He alludes, in effect, to what is the affine group of the finite field of order p^ν, which must be the Galois group of such a polynomial when the action on the roots is doubly transitive. He excludes degrees 9 and 25, where he must have known that there exist exceptional doubly transitive solvable permutation groups. There is another one in degree 49, which he did not mention.

1.1.6 Serret's Cours d'algèbre supérieure

The early editions of Serret's textbook mentioned above provide us with a good opportunity to gauge the progress of the project to publicize Galois's research, considered very advanced at the time. In the first edition [2596], Serret writes that his (Serret's) work was a summary of lectures given at the Sorbonne, Paris, where he had been appointed to a chair in 1848. On p. 4, he notes that the difficult problem of when an equation can be solved algebraically had been resolved, at least in the case of irreducible equations of prime degree, by Évariste Gallois (*sic*), in a memoir of 1831. This memoir had been published in 1846 by Joseph Liouville in his *Journal*, and Liouville had wanted Serret to communicate part of Galois's findings. Lesson 23 of this first edition is devoted to the theory of congruence modulo a prime, but it includes nothing that was not already known at the time of Lagrange or Euler.

The second edition of Serret's work (1854) gives a fairly complete account of Galois's theory of finite fields, as presented in his 1830 paper. Serret devotes almost 30 pages (the whole of Lesson 25) to the material that Galois had covered in six pages, with a view to making a difficult part of the writings of this great mathematician more intelligible. Thus, for example, he proves a uniqueness theorem for factoring polynomials into irreducible factors, and also gives a lengthy discussion of why primitive elements exist. On the other hand, he does not attempt to give a rigorous explanation of why roots of irreducible congruences may be taken to exist. Lesson 25 seems to be the first exposure of finite fields at the textbook level.

Serret devotes 68 pages of the third edition of his textbook (1866) to the theory of finite fields. He considered his approach to be new, and based it on a memoir he had presented in

1865. In fact, it bears many similarities to Gauss's unpublished Section 8 (itself published for the first time in Latin in 1863), and to Dedekind's 1857 paper (for details, see later in this section). Serret was presumably unaware of this material, as he made no mention of it. We can recognize several classical theorems of finite field theory described clearly in Serret's Chapter 3 of Volume 2 of the third edition. Thus for example, if the integer g is not divisible by the prime p, the polynomial (later said to be of Artin-Schreier type)

$$x^p - x - g$$

is irreducible modulo p. Art. 372 presents six theorems summarizing Galois's findings of 1830, Art. 349 gives a formula for the number of monic irrreducible polynomials modulo a prime p, and Art. 350 gives upper and lower bounds for their number.

1.1.7 Contributions of Schönemann and Dedekind

Another early contribution to finite field theory is a paper by Theodor Schönemann, *Grundzügc ciner allgemeinen Theorie der Höheren Congruenzen, deren Modul eine reele Primzahl ist* [2556] published in 1845. Schönemann is described as an *oberlehrer* (head teacher) at the Gymnasium in Brandenburg. He begins his paper with an apology, acknowledging that Gauss's unpublished Section 8 was to contain contributions to the theory of higher congruences, and that he (Schönemann) may have inadvertently rediscovered some of Gauss's results. Schönemann starts with a monic integer polynomial f of degree n which is irreducible modulo a prime p. He then takes a complex root α of f and considers in effect the quotient ring, $\mathbf{Z}[\alpha]/p\mathbf{Z}[\alpha]$, which is a finite field of order p^n. In this way, he avoids the question of whether imaginary roots of irreducible congruences may be taken to exist. This is described well in Cox [749], p. 296. One of Schönemann's main theorems is that if we allow α also to represent a root of f modulo p, then

$$f \equiv (x-\alpha)(x-\alpha^p)\cdots(x-\alpha^{p^{n-1}}) \pmod{p}.$$

He also obtained a formula for the number of monic irreducible polynomials of degree n modulo p. Schönemann's paper is long (56 pages) and not very clear. It is also written in a very formal style, each result being presented in the form of *Erklärung* and *Lehrsatz*, followed by *Beweis*, in imitation of the approach characteristic of Euclid's *Elements*. Nonetheless, Schönemann did innovatory work, which, even if anticipated by Gauss, was quoted reasonably frequently in the second half of the nineteenth century, for instance, by Kronecker.

In his paper *Abriss einer Theorie der Höheren Congruenzen in Bezug auf einen reellen Primzahl-Modulus* [793] written in late 1856, and published in 1857, Dedekind covered much of the same ground pioneered by Gauss in *Disquisitiones Generales de Congruentiis*. While we pointed out above that Dedekind was responsible for editing Gauss's manuscript for publication in 1863, Frei presents several strong reasons to suppose that, at the time he wrote, Dedekind was unacquainted with this key work, and did not see it until 1860. Frei suggests that Dedekind was more concerned to give a solid foundation to Kummer's theory of ideal numbers. In any case, Dedekind notes that there is a strong analogy between the theory of polynomials modulo a prime and elements of number theory. By way of illustrating this analogy, let p be an odd prime and let P and Q be different irreducible monic polynomials of degrees m and n, respectively. Then working modulo Q, P determines an element of the field of order p^n, and this element is either a square or a non-square. By analogy with the Legendre symbol, we set $(\frac{P}{Q})$ equal to 1 if P is a square modulo Q, and $(\frac{P}{Q})$ equal to -1 if it is a non-square. Working modulo P, we may likewise define $(\frac{Q}{P})$. Then, in complete analogy with the law of quadratic reciprocity, we have

$$\left(\frac{P}{Q}\right)\left(\frac{Q}{P}\right) = (-1)^{(\frac{p-1}{2})mn}.$$

In his proof, Dedekind uses a version of Gauss's Lemma, employed in one of Gauss's proofs of the quadratic reciprocity theorem.

1.1.8 Moore's characterization of abstract finite fields

Towards the end of the 19th century, finite fields began to assume a more prominent position in contemporary algebraic research, partly because of their importance in the construction of finite analogues of the classical linear groups. Camille Jordan, in his *Traité des substitutions* [1621] had investigated classical groups, such as the general linear group (also called the linear homogeneous group), over finite fields of prime order. E. H. Moore observed that certain of these constructions could be extended to arbitrary finite fields and he discovered the simple groups usually denoted by $\mathrm{PSL}_2(q^n)$. (He became aware by 1895 that these groups were already known to Émile Mathieu, who had come across them in 1860, [2021], pp. 38-42.) In any case, Moore was led to the investigation of abstract finite fields.

We quote from Moore's paper *A doubly infinite system of simple groups*, read on August 25, 1893, at the International Mathematical Congress held in Chicago [2139]. This paper gives the details of Moore's discoveries.

Suppose that we have a system of s distinct symbols or marks *(s being some positive integer) and suppose that these marks may be combined by the four fundamental operations of algebra–addition, subtraction, multiplication, and division–these operations being subject to the ordinary abstract* operational identities *of algebra*

$$\mu_i + \mu_j = \mu_j + \mu_i; \quad \mu_i\mu_j = \mu_j\mu_i; \quad (\mu_i + \mu_j)\mu_k = \mu_i\mu_k + \mu_j\mu_k; \ \textit{etc},$$

and that when the marks are so combined the results of these operations are in every case uniquely determined and belong to the system of marks. Such a system of marks we shall call a field of order *s, using the notation F*[*s*]. ...

We are led at once to seek [t]*o determine all such fields of order s, F*[*s*].

Moore notes that Galois had defined a field of order q^n, for each prime q and each positive integer n. Moore denotes this field by $GF[q^n]$, presumably in honor of Galois. This $GF[q^n]$ is defined via an irreducible polynomial of degree n modulo q, and is unique, in the sense that such irreducible polynomials exist for all q and n, and the $GF[q^n]$ so constructed is independent of the particular irreducible polynomial chosen. Moore's main theorem is then stated as: *Every existent field* $F[s]$ *is the abstract form of a Galois field* $GF[q^n]$; $s = q^n$. Moore remarks: *This interesting result I have not seen stated before.*

Moore's proof occupies pages 212-220 of his paper, and he derives further properties of $GF[q^n]$ in the next few pages. We feel that Moore's paper marks the beginning of the abstract theory of finite fields. In 1896, Dickson was awarded the first doctorate in mathematics at the new University of Chicago, for a thesis written under Moore's direction, the subject matter being permutation polynomials over finite fields. Dickson's 1901 book gave a streamlined proof of Moore's uniqueness theorem on pp. 13-14.

1.1.9 Later developments

Following the work of his thesis, Dickson was to extend Jordan's analysis of classical groups to their counterparts over arbitrary finite fields, and his research was the subject of his monograph of 1901. Dickson even generalized ideas of Élie Cartan on continuous groups and their Lie algebras, and published in 1901 (with later additions) details of his discovery of versions of the groups of type E_6 and G_2 over finite fields [842, 848, 843, 844]. It was not until later work of Chevalley in 1955 that further finite analogues of the exceptional continuous groups were constructed in a uniform way.

The theory of finite fields may be said to have acquired a more conceptual form in the twentieth century after Emil Artin (1898-1962) introduced the notion of a zeta function for a quadratic extension of the rational function field $\mathbf{F}_p(t)$, where p is a prime. Artin formulated a version of the Riemann hypothesis for these zeta functions, and verified the hypothesis for a number of curves in his dissertation, published in 1924. Helmut Hasse (1899-1979) subsequently proved the Riemann hypothesis for function fields of genus 1 in 1934, but the complete proof for arbitrary non-singular curves by André Weil (1906-1998) in 1948 employed sophisticated methods of algebraic geometry. The analogy between counting rational points on algebraic varieties over finite fields and the cohomology theories of complex varieties has been a powerful motivating force in the more recent theory of finite fields.

References Cited: [135, 749, 793, 842, 843, 844, 848, 850, 851, 1168, 1257, 1258, 1259, 1297, 1621, 1804, 1939, 2021, 2139, 2556, 2596, 2848]

2

Introduction to finite fields

2.1 Basic properties of finite fields

Gary L. Mullen, The Pennsylvania State University
Daniel Panario, Carleton University

Proofs for most of the results in this chapter can be found in Chapters 2 and 3 of [1939]; see also [1631, 1938, 2017, 2049, 2077, 2179, 2921]. We refer the reader to Section 2.1.8 for a comprehensive list of other finite field related books.

2.1.1 Basic definitions

2.1.1 **Definition** A *ring* $(R, +, \cdot)$ is a nonempty set R together with two operations, "+" and "·" such that:

(1) $(R, +)$ is an abelian group;
(2) $\cdot$ is associative, that is for all $a, b, c \in R, a \cdot (b \cdot c) = (a \cdot b) \cdot c$;
(3) left and right distributive laws hold: for all $a, b, c \in R$

$$a \cdot (b + c) = a \cdot b + a \cdot c \quad \text{and} \quad (b + c) \cdot a = b \cdot a + c \cdot a.$$

2.1.2 **Definition** Let R be a ring.
(1) R is a *ring with identity* if the ring has a multiplicative identity;
(2) R is *commutative* if "·" is commutative;
(3) R is an *integral domain* if it is commutative with identity and $a \cdot b = 0$ implies $a = 0$ or $b = 0$, for any $a, b \in R$;

(4) R is a *division ring* (also called a *skew field*) if the nonzero elements of R form a group under "$\cdot$";
(5) R is a *field* if it is a commutative division ring.

2.1.3 Definition The *order of a finite field* $\mathbb{F}$ is the number of distinct elements in $\mathbb{F}$.

2.1.4 Remark The following theorem is a famous result due to Wedderburn.

2.1.5 Theorem Every finite division ring is a field.

2.1.6 Definition If R is a ring and there exists a positive integer n such that $nr = 0$ for all $r \in R$, then the least such positive integer n is the *characteristic* of the ring, and R has *positive characteristic.* Otherwise, R has *characteristic zero.*

2.1.7 Theorem A ring $R \neq \{0\}$ of positive characteristic having an identity and no zero divisors must have prime characteristic.

2.1.8 Corollary A finite field has prime characteristic.

2.1.9 Proposition For a commutative ring R of characteristic p, we have

$$(a_1 + \cdots + a_s)^{p^n} = a_1^{p^n} + \cdots + a_s^{p^n}$$

for every $n \geq 1$ and $a_i \in R$.

2.1.2 Fundamental properties of finite fields

2.1.10 Lemma Suppose F is a finite field with a subfield K containing q elements. Then F is a vector space over K and $|F| = q^m$, where m is the dimension of F viewed as a vector space over K.

2.1.11 Definition A field containing no proper subfield is a *prime field.*

2.1.12 Theorem Let F be a finite field. The cardinality of F is p^n, where p is the characteristic of F and n is the dimension of F over its prime subfield.

2.1.13 Remark We denote by $\mathbb{F}_q$ a finite field with q elements. We note that by Remark 2.1.34 there is only one finite field (up to isomorphism) with q elements.

2.1.14 Remark Another common notation for a field of order q is $GF(q)$, where GF stands for Galois field. This name is used in honor of Évariste Galois (1811–1832), who in 1830 was the first person to seriously study properties of general finite fields (fields with a prime power but not necessarily a prime number of elements).

2.1.15 Remark The recent publication of *The Mathematical Writings of Evariste Galois* by Neumann [2223] will make Galois's own words available to readers.

2.1.16 Lemma If $\mathbb{F}_q$ is a finite field with q elements and $a \neq 0 \in \mathbb{F}_q$, then $a^{q-1} = 1$, and thus $a^q = a$, for all a in $\mathbb{F}_q$.

2.1.17 Remark An immediate consequence of the previous lemma is that the multiplicative inverse of any $a \neq 0$ in a field of order q is a^{q-2}, because $a^{q-2} \cdot a = a^{q-1}$.

2.1.18 Theorem The sum of all elements of a finite field is 0, except for the field $\mathbb{F}_2$.

2.1.19 Definition A polynomial f over $\mathbb{F}_q$ is an expression of the form $f(x) = \sum_{i=0}^{n} a_i x^i$, where n is a nonnegative integer, and $a_i \in \mathbb{F}_q$ for $i = 0, 1, \ldots, n$. A polynomial is *monic* if the coefficient of the highest power of x is 1. The ring formed by the polynomials over $\mathbb{F}_q$ with sum and product of polynomials is the *ring of polynomials over* $\mathbb{F}_q$ and is denoted by $\mathbb{F}_q[x]$.

2.1.20 Definition A polynomial $f \in \mathbb{F}_q[x]$ is an *irreducible polynomial* over $\mathbb{F}_q$ if f has positive degree and $f = gh$ with $g, h \in \mathbb{F}_q[x]$ implies that either g or h is a constant polynomial.

2.1.21 Remark Both $\mathbb{F}_q[x]$ and the ring of polynomials in $n \geq 1$ variables, $\mathbb{F}_q[x_1, \ldots, x_n]$, have unique factorization into irreducibles.

2.1.22 Definition The Möbius μ function is defined on the set of positive integers by

$$\mu(m) = \begin{cases} 1 & \text{if } m = 1, \\ (-1)^k & \text{if } m = m_1 m_2 \cdots m_k, \text{ where the } m_i \text{ are distinct primes,} \\ 0 & \text{otherwise, i.e., if } p^2 \text{ divides } m \text{ for some prime } p. \end{cases}$$

2.1.23 Definition The number of monic irreducible polynomials of degree n over $\mathbb{F}_q$ is denoted by $I_q(n)$.

2.1.24 Theorem For all $n \geq 1$ and any prime power q, we have

$$I_q(n) = \frac{1}{n} \sum_{d|n} \mu(d) q^{n/d}.$$

2.1.25 Remark We have that $I_q(n) > 0$ for all prime powers q and all integers $n > 1$:

$$I_q(n) = \frac{1}{n} \sum_{d|n} \mu(d) q^{n/d} \geq \frac{1}{n} \left(q^n - q^{n-1} - q^{n-2} - \cdots - q \right) > 0.$$

2.1.26 Remark For a polynomial $f \in \mathbb{F}_q[x]$, we have $(f(x))^q = f(x^q)$. This property is of great use in finite field calculations.

2.1.27 Lemma If $\mathbb{F}_q$ is a finite field with q elements then in $\mathbb{F}_q[x]$ we have

$$x^q - x = \prod_{a \in \mathbb{F}_q} (x - a).$$

2.1.28 Remark The next theorem is crucial for fast polynomial irreducibility testing and factorization algorithms over finite fields; see Sections 11.3 and 11.4.

2.1.29 Theorem Let f be an irreducible polynomial of degree n over $\mathbb{F}_q$. Then $f(x) | (x^{q^r} - x)$ if and only if $n | r$.

2.1.30 Definition Let $f \in F[x]$ be of positive degree and E an extension of F. Then f *splits* in E if f can be written as a product of linear factors in $E[x]$, that is, there exist $\alpha_1, \alpha_2, \ldots, \alpha_n \in E$ such that

$$f(x) = a(x - \alpha_1)(x - \alpha_2) \cdots (x - \alpha_n),$$

where $a \in F$ is the leading coefficient of f and E is the smallest such field. The field E is a *splitting field* of f over F if f splits in E.

2.1.31 Theorem If F is a field, and f any polynomial of positive degree in $F[x]$, then there exists a splitting field of f over F. Any two splitting fields of f over F are isomorphic under an isomorphism which keeps the elements of F fixed and maps the roots of f into each other.

2.1.32 Theorem For every prime p and positive integer $n \geq 1$ there is a finite field with p^n elements. Any finite field with p^n elements is isomorphic to the splitting field of $x^{p^n} - x$ over $\mathbb{F}_p$.

2.1.33 Remark The previous theorem shows that a finite field of a given order is unique up to field isomorphism because splitting fields are unique up to isomorphism. Thus we speak of "the" finite field of a particular order q.

2.1.34 Remark We note that when p is a prime the field $\mathbb{F}_p$ is the same as (isomorphic to) the ring $\mathbb{Z}_p$ of integers modulo p. The ring $\mathbb{Z}_p$ is also denoted by $\mathbb{Z}/p\mathbb{Z}$. When $n > 1$ the finite field $\mathbb{F}_{p^n}$ is not the same as the ring $\mathbb{Z}_{p^n}$ of integers modulo p^n. Indeed, $\mathbb{Z}_{p^n}$ is not a field if $n > 1$.

2.1.35 Theorem Let $\mathbb{F}_{p^n}$ be the finite field with p^n elements. Every subfield of $\mathbb{F}_{p^n}$ has p^m elements for some positive integer m dividing n. Conversely, for any positive integer m dividing n there is a unique subfield of $\mathbb{F}_{p^n}$ of order p^m.

2.1.36 Remark The subfields of $\mathbb{F}_{q^{36}}$ are illustrated in the following diagram:

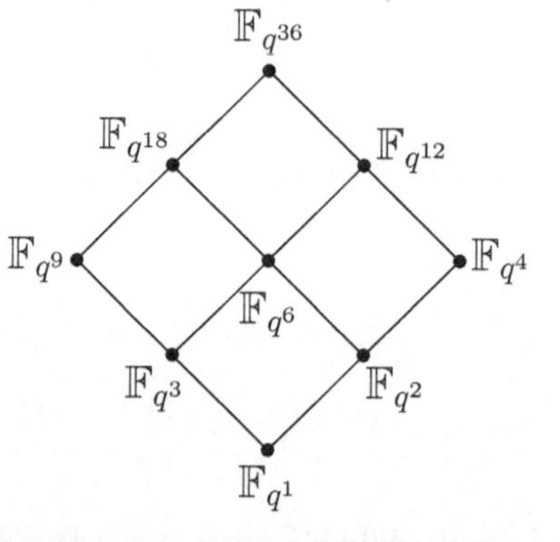

Figure 2.1.1 The subfields of $\mathbb{F}_{q^{36}}$.

2.1.37 Theorem The multiplicative group $\mathbb{F}_q^*$ of all nonzero elements of the finite field $\mathbb{F}_q$ is cyclic.

2.1.38 Definition An element $\alpha \in \mathbb{F}_q$ which multiplicatively generates the group $\mathbb{F}_q^*$ of all nonzero elements of the field $\mathbb{F}_q$ is a *primitive element*, sometimes also a *primitive root*.

2.1.39 Remark Let θ be a primitive element of a finite field $\mathbb{F}_q$. Then every nonzero element of $\mathbb{F}_q$ can be written as a power of θ. This representation makes multiplication of field elements very easy to compute. However, in general, it may not be easy to find the power s of θ such that $\theta^t + \theta^r = \theta^s$; see Subsection 2.1.7.5. Conversely, as we will see later in our discussion of bases for finite fields, representations which make exponentiation easy to compute often have a more complex multiplicative structure.

2.1.40 Definition Let $\alpha \in \mathbb{F}_q^*$. The *order* of α is the smallest positive integer n such that $\alpha^n = 1$.

2.1.41 Remark We use the notation (a, b) or $\gcd(a, b)$ to represent the *greatest common divisor (gcd)* of a and b, where a and b belong to a Euclidean domain (usually integers or polynomials).

2.1.42 Lemma If g is a primitive element of $\mathbb{F}_q$ then g^t is a primitive element of $\mathbb{F}_q$ if and only if $(t, q-1) = 1$.

2.1.43 Definition The number of positive integers $e \leq n$ such that $(n, e) = 1$ is denoted by $\phi(n)$, and is the *Euler function.*

2.1.44 Remark The Euler function is multiplicative: if $(m, n) = 1$, then $\phi(mn) = \phi(m)\phi(n)$.

2.1.45 Remark It follows from Lemma 2.1.42 that there are exactly $\phi(q-1)$ primitive elements in $\mathbb{F}_q$.

2.1.46 Definition A monic polynomial all of whose roots are primitive elements is a *primitive polynomial.*

2.1.47 Remark Primitive polynomials are treated in Chapter 4.

2.1.48 Definition The *reciprocal* f^* of a monic polynomial f of degree n is defined by $f^*(x) = x^n f(1/x)$. The polynomial f is *self-reciprocal* if $f^* = f$.

2.1.49 Remark The reciprocal polynomial of an irreducible polynomial f, $f(x) \neq x$, over $\mathbb{F}_q$ is again irreducible over $\mathbb{F}_q$. In addition, the *monic* reciprocal polynomial defined by $f(x)/f(0)$ of a primitive polynomial is also primitive.

2.1.50 Remark If f is a self-reciprocal irreducible polynomial of degree $n > 1$, in $\mathbb{F}_q[x]$, then n must be even.

2.1.51 Definition Let $f \in \mathbb{F}_q[x]$ be a nonzero polynomial. If $f(0) \neq 0$, the *order* of f is the least positive integer e such that $f | x^e - 1$. If $f(0) = 0$, let $f(x) = x^r g(x)$ for some integer $r \geq 1$ and $g \in \mathbb{F}_q[x]$ with $g(0) \neq 0$. In this case, the order of f is the order of g.

2.1.52 Remark We denote the order of f by $\mathrm{ord}(f)$. The order of a polynomial is also called the *period* or *exponent* of the polynomial.

2.1.53 Theorem Let $f \in \mathbb{F}_q[x]$ be an irreducible polynomial over $\mathbb{F}_q$ of degree n with $f(0) \neq 0$. Then $\mathrm{ord}(f)$ is equal to the order of any root of f in the multiplicative group of $\mathbb{F}_{q^n}^*$.

2.1.54 Corollary If $f \in \mathbb{F}_q[x]$ is an irreducible polynomial over $\mathbb{F}_q$ of degree n, then $\mathrm{ord}(f) | (q^n - 1)$.

2.1.55 Theorem Let $\mathbb{F}_q$ be a finite field of characteristic p, and let $f \in \mathbb{F}_q[x]$ be a polynomial of positive degree with $f(0) \neq 0$. Let $f = a f_1^{b_1} \cdots f_k^{b_k}$ be the canonical factorization of f into irreducibles in $\mathbb{F}_q[x]$, where $a \in \mathbb{F}_q, b_1, \ldots, b_k \in \mathbb{N}$, and $f_1, \ldots, f_k$ are distinct monic irreducible polynomials in $\mathbb{F}_q[x]$. Then $\mathrm{ord}(f) = ep^t$, where e is the least common multiple of $\mathrm{ord}(f_1), \ldots, \mathrm{ord}(f_k)$ and t is the smallest integer with $p^t \geq \max(b_1, \ldots, b_k)$.

2.1.3 Extension fields

2.1.56 Definition Let K be a subfield of F and let M be a subset of F. Then $K(M)$ denotes the intersection of all subfields of F containing K and M as subsets. This field is K *adjoin* M. When M is finite, say $M = \{\alpha_1, \ldots, \alpha_k\}$, we write $K(\alpha_1, \ldots, \alpha_k)$ for $K(M)$.

2.1.57 Definition Let $K \subseteq F$, $\alpha \in F$, and $f(\alpha) = 0$ where f is a monic polynomial in $K[x]$. Then f is the *minimal polynomial* of α if α is not a root of any nonzero polynomial in $K[x]$ of lower degree.

2.1.58 Proposition The minimal polynomial of any extension field element is irreducible over the base field. This result provides a method by which one can obtain irreducible polynomials.

2.1.59 Definition A field F is a *finite extension* of K if $K \subseteq F$ and F is a finite dimensional vector space over K. In this case we refer to the dimension m of F over K as the *degree* of the extension, and we write $[F : K] = m$.

2.1.60 Theorem Let F be a finite extension of K and let E be a finite extension of F. Then E is a finite extension of K. Moreover, we have $[E : K] = [E : F][F : K]$.

2.1.61 Definition Let $K \subseteq F$ and let $\alpha \in F$. Then α is *algebraic* over K if there is a nonzero polynomial $f \in K[x]$ such that $f(\alpha) = 0$ in $F[x]$. An extension field is *algebraic* if every element of the extension field is algebraic.

2.1.62 Theorem Every finite extension of a field is algebraic.

2.1.63 Theorem Let K be a subfield of F with $\alpha \in F$ algebraic of degree n over K and let g be the minimal polynomial of α over K. Then:

1. The field $K(\alpha)$ is isomorphic to the factor ring $K[x]/(g)$.
2. The dimension of $K(\alpha)$ over K is n.
3. The set $\{1, \alpha, \alpha^2, \ldots, \alpha^{n-1}\}$ is a basis for $K(\alpha)$ over K.
4. Every element of $K(\alpha)$ is algebraic over K with degree dividing n.

2.1.64 Remark An extension obtained by adjoining a single element is a *simple* extension. The next theorem gives an important property of finite fields which is not shared by infinite fields (there are finite extensions of infinite fields which are not simple).

2.1.65 Theorem Let $\mathbb{F}_q$ be a finite field and let $\mathbb{F}_r$ be a finite extension of $\mathbb{F}_q$. Then $\mathbb{F}_r$ is a simple algebraic extension of $\mathbb{F}_q$, and for any primitive element α of $\mathbb{F}_r$ the relation $\mathbb{F}_r = \mathbb{F}_q(\alpha)$ holds.

2.1.66 Corollary For any prime power q and any integer $n \geq 1$ there is an irreducible polynomial of degree n over $\mathbb{F}_q$.

2.1.67 Example Consider $q = 2^{100}$. We can identify the elements of $\mathbb{F}_q$ with polynomials of the form $a_0 + a_1\alpha + a_2\alpha^2 + \cdots + a_{99}\alpha^{99}$, where $0 \leq a_i < 2$ for each i and where α is a root of an irreducible polynomial of degree 100 over the field $\mathbb{F}_2$. Corollary 2.1.66 shows that such an irreducible polynomial always exists. Using Theorem 2.1.24 we have that there are exactly

$$\frac{1}{100}\left(2^{100} - 2^{50} - 2^{20} + 2^{10}\right)$$

irreducible polynomials of degree 100 over $\mathbb{F}_2$.

2.1.68 Remark Corollary 2.1.66 can also be derived using the formula for the number of irreducible polynomials in Theorem 2.1.24.

2.1.69 Example Consider the polynomial $f(x) = x^2+x+1$ over the field $\mathbb{F}_2$. Since f does not have a root in $\mathbb{F}_2$, f is irreducible over $\mathbb{F}_2$. Let α be a root of f so that $\alpha^2 + \alpha + 1 = 0$, that is, $\alpha^2 = -(\alpha+1) = \alpha+1$. The field $\mathbb{F}_4 = \mathbb{F}_{2^2}$ can be represented as the set $\{a\alpha+b \colon a, b \in \mathbb{F}_2\}$. We give the addition and multiplication tables for the field $\mathbb{F}_{2^2}$. We note that α is a primitive element in the field $\mathbb{F}_4$, so $\alpha^1 = \alpha, \alpha^2 = \alpha+1$ and $\alpha^3 = 1$.

$+$	0	1	α	$\alpha+1$
0	0	1	α	$\alpha+1$
1	1	0	$\alpha+1$	α
α	α	$\alpha+1$	0	1
$\alpha+1$	$\alpha+1$	α	1	0

$\times$	0	1	α	$\alpha+1$
0	0	0	0	0
1	0	1	α	$\alpha+1$
α	0	α	$\alpha+1$	1
$\alpha+1$	0	$\alpha+1$	1	α

2.1.70 Example Consider the field $\mathbb{F}_9$, which is a vector space of dimension 2 over $\mathbb{F}_3$. Consider $f(x) = x^2+x+2$ in $\mathbb{F}_3[x]$. This polynomial has no roots in $\mathbb{F}_3$ so it is irreducible over $\mathbb{F}_3$. Let α be a root of f, so $\alpha^2+\alpha+2 = 0$. Hence $\alpha^2 = -\alpha-2 = 2\alpha+1$. The field $\mathbb{F}_{3^2}$ is isomorphic to the set $\{a\alpha+b \mid a, b \in \mathbb{F}_3\}$ with its natural operations. We can compute the addition and multiplication tables by hand. For example, $2\alpha(\alpha+2) = 2\alpha^2+4\alpha = 2(2\alpha+1)+\alpha = 2\alpha+2$. The following addition and multiplication tables are obtained. We can use the multiplication table to check that the multiplicative order of α in $\mathbb{F}_9$ is 8, and thus α is a primitive element of $\mathbb{F}_9$.

$+$	0	1	2	α	$\alpha+1$	$\alpha+2$	2α	$2\alpha+1$	$2\alpha+2$
0	0	1	2	α	$\alpha+1$	$\alpha+2$	2α	$2\alpha+1$	$2\alpha+2$
1	1	2	0	$\alpha+1$	$\alpha+2$	α	$2\alpha+1$	$2\alpha+2$	2α
2	2	0	1	$\alpha+2$	α	$\alpha+1$	$2\alpha+2$	2α	$2\alpha+1$
α	α	$\alpha+1$	$\alpha+2$	2α	$2\alpha+1$	$2\alpha+2$	0	1	2
$\alpha+1$	$\alpha+1$	$\alpha+2$	α	$2\alpha+1$	$2\alpha+2$	2α	1	2	0
$\alpha+2$	$\alpha+2$	α	$\alpha+1$	$2\alpha+2$	2α	$2\alpha+1$	2	0	1
2α	2α	$2\alpha+1$	$2\alpha+2$	0	1	2	α	$\alpha+1$	$\alpha+2$
$2\alpha+1$	$2\alpha+1$	$2\alpha+2$	2α	1	2	0	$\alpha+1$	$\alpha+2$	α
$2\alpha+2$	$2\alpha+2$	2α	$2\alpha+1$	2	0	1	$\alpha+2$	α	$\alpha+1$

$\times$	0	1	2	α	$\alpha+1$	$\alpha+2$	2α	$2\alpha+1$	$2\alpha+2$
0	0	0	0	0	0	0	0	0	0
1	0	1	2	α	$\alpha+1$	$\alpha+2$	2α	$2\alpha+1$	$2\alpha+2$
2	0	2	1	2α	$2\alpha+2$	$2\alpha+1$	α	$\alpha+2$	$\alpha+1$
α	0	α	2α	$2\alpha+1$	1	$\alpha+1$	$\alpha+2$	$2\alpha+2$	2
$\alpha+1$	0	$\alpha+1$	$2\alpha+2$	1	$\alpha+2$	2α	2	α	$2\alpha+1$
$\alpha+2$	0	$\alpha+2$	$2\alpha+1$	$\alpha+1$	2α	2	$2\alpha+2$	1	α
2α	0	2α	α	$\alpha+2$	2	$2\alpha+2$	$2\alpha+1$	$\alpha+1$	1
$2\alpha+1$	0	$2\alpha+1$	$\alpha+2$	$2\alpha+2$	α	1	$\alpha+1$	2	2α
$2\alpha+2$	0	$2\alpha+2$	$\alpha+1$	2	$2\alpha+1$	α	1	2α	$\alpha+2$

2.1.71 Example Let $f(x) = x^2+1 \in \mathbb{F}_3[x]$. It is straightforward to check that f is irreducible over the field $\mathbb{F}_3$. Let α be a root of f. We compute $\alpha^2 = -1$ and $\alpha^4 = 1$. Hence no root of f can have order 8, that is, no root of f can be a primitive element. Nevertheless, the splitting field of f over $\mathbb{F}_3$ is $\mathbb{F}_9$. It can be seen that $\alpha+1$ has order 8 and is thus a primitive element for $\mathbb{F}_9$ over $\mathbb{F}_3$.

2.1.72 Remark Tables of irreducible and primitive polynomials can be found in Section 2.2. In that section is a discussion of some computer algebra packages for implementing finite field arithmetic.

2.1.73 Theorem If f is an irreducible polynomial of degree n over $\mathbb{F}_q$ then f has a root α in $\mathbb{F}_{q^n}$. Moreover all of the roots of f are simple and are given by $\alpha, \alpha^q, \alpha^{q^2}, \ldots, \alpha^{q^{n-1}}$.

2.1.74 Definition Let $\alpha \in \mathbb{F}_{q^n}$. Then $\alpha, \alpha^q, \alpha^{q^2}, \ldots, \alpha^{q^{n-1}}$ are the *conjugates* of α over $\mathbb{F}_q$.

2.1.75 Lemma Let $\alpha \in \mathbb{F}_{q^n}$ and let the minimal polynomial of α over $\mathbb{F}_q$ have degree d. Consider the set $\alpha, \alpha^q, \alpha^{q^2}, \ldots, \alpha^{q^{n-1}}$ of conjugates of α. The elements of this set are distinct if $n = d$; otherwise each distinct conjugate is repeated n/d times.

2.1.76 Theorem The distinct automorphisms of $\mathbb{F}_{q^n}$ over $\mathbb{F}_q$ are given by the functions $\sigma_0, \sigma_1, \ldots, \sigma_{n-1}$ where $\sigma_j \colon \mathbb{F}_{q^n} \to \mathbb{F}_{q^n}$ and is defined by $\sigma_j(\alpha) = \alpha^{q^j}$ for any $\alpha \in \mathbb{F}_{q^n}$.

2.1.77 Remark The set of automorphisms of $\mathbb{F}_q$ forms a group with the operation of functional composition. This group is called the *Galois group of* $\mathbb{F}_{q^n}$ *over* $\mathbb{F}_q$. It is a cyclic group with generator $\sigma_1 \colon \mathbb{F}_{q^n} \to \mathbb{F}_{q^n}$ that maps $\alpha \in \mathbb{F}_{q^n}$ to α^q, and is called the *Frobenius automorphism.* The conjugates of α are thus the elements to which α is sent by iterated applications of the Frobenius automorphism.

2.1.78 Remark The subfields of $\mathbb{F}_{q^n}$ are exactly the fields of the form $\mathbb{F}_{q^m}$ where $m|n$. The subgroups of the Galois group of $\mathbb{F}_{q^n}$ over $\mathbb{F}_q$ are exactly the groups generated by σ_1^m where $m|n$. Moreover, $\sigma_1^m(\alpha) = \alpha$ if and only if $\alpha \in \mathbb{F}_{q^m}$. Thus there is a one-to-one correspondence between the subfields of $\mathbb{F}_{q^n}$ and the subgroups of its Galois group.

2.1.79 Remark In general, if F is an extension of a field K then the set of automorphisms of F that leave K fixed pointwise is the Galois group of F over K. The field of *Galois theory* is the study of Galois groups. Thus, if K is finite and F is a finite extension of K then the Galois group is cyclic. When K is infinite, the Galois group need not be cyclic, even if F is a finite extension of K.

2.1.4 Trace and norm functions

2.1.80 Definition Let $K = \mathbb{F}_q$ and $F = \mathbb{F}_{q^n}$. For $\alpha \in F$, we define the *trace* of α over K as $\mathrm{Tr}_{F/K}(\alpha) = \alpha + \alpha^q + \cdots + \alpha^{q^{n-1}}$. Equivalently, $\mathrm{Tr}_{F/K}(\alpha)$ is the sum of the conjugates of α. If K is the prime subfield of F then the trace function is the *absolute trace.*

2.1.81 Example Let $K = \mathbb{F}_2$ and $F = \mathbb{F}_{2^4}$. Then $\mathrm{Tr}_{F/K}(\alpha) = \alpha + \alpha^2 + \alpha^4 + \alpha^8$. For $K = \mathbb{F}_4$ and $F = \mathbb{F}_{16}$ we have $\mathrm{Tr}_{F/K}(\alpha) = \alpha + \alpha^4$.

2.1.82 Remark Since $\left(\mathrm{Tr}_{F/K}(\alpha)\right)^q = \mathrm{Tr}_{F/K}(\alpha)$ the trace of an element always lies in the base field K.

2.1.83 Theorem Let $K = \mathbb{F}_q$ and $F = \mathbb{F}_{q^n}$. The trace function has the following properties:

1. for any $\alpha \in F$, $\mathrm{Tr}_{F/K}(\alpha) \in K$;
2. $\mathrm{Tr}_{F/K}(\alpha + \beta) = \mathrm{Tr}_{F/K}(\alpha) + \mathrm{Tr}_{F/K}(\beta)$ for $\alpha, \beta \in F$;
3. $\mathrm{Tr}_{F/K}(c\alpha) = c\mathrm{Tr}_{F/K}(\alpha)$ for $\alpha \in F$ and $c \in K$;
4. the trace function is a K-linear map from F onto K;

5. $\mathrm{Tr}_{F/K}(\alpha) = n\alpha$ for $\alpha \in K$;
6. $\mathrm{Tr}_{F/K}(\alpha^q) = \mathrm{Tr}_{F/K}(\alpha)$ for $\alpha \in F$;
7. for any $\alpha \in K$, we have $|\{\beta \in F \mid \mathrm{Tr}_{F/K}(\beta) = \alpha\}| = q^{n-1}$;
8. Suppose that $K \subseteq F \subseteq E$ are finite fields; then for any $\alpha \in E$

$$\mathrm{Tr}_{E/K}(\alpha) = \mathrm{Tr}_{F/K}(\mathrm{Tr}_{E/F}(\alpha)).$$

2.1.84 Theorem For $\beta \in F$ let L_β be the map $\alpha \mapsto \mathrm{Tr}_{F/K}(\beta\alpha)$. Then $L_\beta \neq L_\gamma$ if $\beta \neq \gamma$. Moreover the K-linear transformations from F to K are exactly the maps of the form L_β as β varies over the elements of the field F.

2.1.85 Remark The result in Theorem 2.1.84 provides a method to generate all of the linear transformations from the extension field F to the subfield K.

2.1.86 Definition Let $K = \mathbb{F}_q$ and $F = \mathbb{F}_{q^n}$. The *norm* over K of an element $\alpha \in F$ is defined by

$$\mathrm{Norm}_{F/K}(\alpha) = \alpha\alpha^q \cdots \alpha^{q^{n-1}} = \prod_{i=0}^{n-1} \alpha^{q^i} = \alpha^{(q^n-1)/(q-1)}.$$

2.1.87 Remark The norm of an element α is thus calculated by taking the product of all of the conjugates of α, just as the trace of α is obtained by taking the sum of all of the conjugates of α.

2.1.88 Theorem Let $K = \mathbb{F}_q$ and $F = \mathbb{F}_{q^n}$. The norm function has the following properties:

1. $\mathrm{Norm}_{F/K}(\alpha) \in K$;
2. $\mathrm{Norm}_{F/K}(\alpha\beta) = \mathrm{Norm}_{F/K}(\alpha)\,\mathrm{Norm}_{F/K}(\beta)$ for $\alpha, \beta \in F$;
3. the norm maps F onto K and F^* onto K^*;
4. $\mathrm{Norm}_{F/K}(\alpha) = \alpha^n$ if $\alpha \in K$;
5. $\mathrm{Norm}_{F/K}(\alpha^q) = \mathrm{Norm}_{F/K}(\alpha)$;
6. if $K \subseteq F \subseteq E$ are finite fields then

$$\mathrm{Norm}_{E/K}(\alpha) = \mathrm{Norm}_{F/K}(\mathrm{Norm}_{E/F}(\alpha))\,.$$

2.1.5 Bases

2.1.89 Remark Every finite field F is a vector space over each of its subfields, and thus has a vector space basis over each of its subfields. There are several different kinds of bases for finite fields. Each kind of basis facilitates certain computations. When doing computations in finite fields, there are some important operations like addition, multiplication, q-th powering and finding inverses. With some bases computing inverses and q-th powers are easy, while multiplication could be more involved. With other bases, one can calculate multiplications quickly at the cost of more complicated inverse computations or exponentiations.

2.1.90 Remark The vector space of all $n \times r$ matrices over a field $\mathbb{F}_q$ is of dimension nr over $\mathbb{F}_q$. Taking into account the order of the elements, the total number of distinct bases of $\mathbb{F}_{q^n}$ over $\mathbb{F}_q$ is given by

$$(q^n - 1)(q^n - q) \cdots (q^n - q^{n-1}),$$

which is also equal to the number of elements in the general linear group $GL_n(\mathbb{F}_q)$, the ring of nonsingular $n \times n$ matrices over $\mathbb{F}_q$.

2.1.91 Remark Consider $\mathbb{F}_{q^n}$ as a vector space over $\mathbb{F}_q$ of dimension n. We know there are many bases for this vector space. Given $B = \{\alpha_1, \ldots, \alpha_n\} \subseteq \mathbb{F}_{q^n}$, how can we tell if B is a basis for $\mathbb{F}_{q^n}$ over $\mathbb{F}_q$? We begin with a test which determines whether a set of elements of $\mathbb{F}_{q^n}$ is independent over $\mathbb{F}_q$. If this result is applied to a set containing n elements, it can thus be used to determine whether these elements form a basis of $\mathbb{F}_{q^n}$ over $\mathbb{F}_q$. We require the following notation.

2.1.92 Definition Let $K = \mathbb{F}_q$ and $F = \mathbb{F}_{q^n}$. Let $\{\alpha_1, \ldots, \alpha_n\}$ be a set of n elements of $\mathbb{F}$ viewed as a vector space over the subfield K. We define the *discriminant* $\Delta_{\mathbb{F}/K}(\alpha_1, \ldots, \alpha_n)$ with the following rule:

$$\Delta_{\mathbb{F}/K}(\alpha_1, \ldots, \alpha_n) = \begin{vmatrix} \mathrm{Tr}_{F/K}(\alpha_1\alpha_1) & \cdots & \mathrm{Tr}_{F/K}(\alpha_1\alpha_n) \\ \vdots & \ddots & \vdots \\ \mathrm{Tr}_{F/K}(\alpha_n\alpha_1) & \cdots & \mathrm{Tr}_{F/K}(\alpha_n\alpha_n) \end{vmatrix}.$$

2.1.93 Theorem If $\alpha_1, \ldots, \alpha_n \in \mathbb{F}_{q^n}$, then the set $\{\alpha_1, \ldots, \alpha_n\}$ is a basis for $\mathbb{F}_{q^n}$ over $\mathbb{F}_q$ if and only if $\Delta_{\mathbb{F}_{q^n}/\mathbb{F}_q}(\alpha_1, \ldots, \alpha_n)$ is nonzero.

2.1.94 Remark The following result provides an alternative method to determine if a given set of elements forms a basis. We note that the calculations for this method must be done in the extension field, not in the base field. Working in the extension field may have a significant computational cost. For example, if the base field is $\mathbb{F}_2$ and the extension field is $\mathbb{F}_{2^{1000}}$ then computations in the base field are much faster than computations in the extension field.

2.1.95 Corollary The set $\{\alpha_1, \ldots, \alpha_n\}$ is a basis for $\mathbb{F}_{q^n}$ over $\mathbb{F}_q$ if and only if

$$\begin{vmatrix} \alpha_1 & \cdots & \alpha_n \\ \alpha_1^q & \cdots & \alpha_n^q \\ \vdots & \ddots & \vdots \\ \alpha_1^{q^{n-1}} & \cdots & \alpha_n^{q^{n-1}} \end{vmatrix} \neq 0.$$

2.1.96 Definition Let α be a root of an irreducible polynomial of degree n over $\mathbb{F}_q$. The set $\{1, \alpha, \alpha^2, \ldots, \alpha^{n-1}\}$ is a *polynomial basis* of the field $\mathbb{F}_{q^n}$ over $\mathbb{F}_q$.

2.1.97 Remark When we use a polynomial basis for $\mathbb{F}_{q^n}$ we can regard field elements, which in reality are polynomials in α of degree at most $n-1$, as vectors. We can then easily add vectors in the usual way by adding the corresponding coefficients. Field multiplication is more complicated since we must gather terms with like powers of the basis elements when we simplify a product.

2.1.98 Definition If $\alpha \in \mathbb{F}_{q^n}$ and $\{\alpha, \alpha^q, \ldots, \alpha^{q^{n-1}}\}$ is a basis for $\mathbb{F}_{q^n}$ over $\mathbb{F}_q$, then the basis is a *normal basis* of $\mathbb{F}_{q^n}$ over $\mathbb{F}_q$, and α is a *normal element.*

2.1.99 Remark If $\beta = a_0\alpha + a_1\alpha^q + \cdots + a_{n-1}\alpha^{q^{n-1}}$ so that β is represented by the vector $(a_0, \ldots, a_{n-1})$, then α^q is simply represented by the shifted vector $(a_{n-1}, a_0, \ldots, a_{n-2})$. Thus if we have a normal basis, it is extremely easy to raise a field element to the power q. Addition is of course also still easy to compute using a normal basis. We note that multiplication of field elements is quite complicated using a normal basis. In Section 5.2 we give important properties of normal bases including their existence for any finite extension field of $\mathbb{F}_q$.

2.1.100 Definition Two ordered bases of $\mathbb{F}_{q^n}$ over $\mathbb{F}_q$ $\{\alpha_1, \ldots, \alpha_n\}$ and $\{\beta_1, \ldots, \beta_n\}$ are *complementary* (or *dual*) if $\mathrm{Tr}_{\mathbb{F}_{q^n}/\mathbb{F}_q}(\alpha_i \beta_j) = \delta_{ij}$, where $\delta_{ij} = 0$ if $j \neq i$ and $\delta_{ij} = 1$ if $i = j$. An ordered basis is *self-dual* if it is dual with itself.

2.1.101 Definition A *primitive normal basis* for an extension field $\mathbb{F}_{q^n}$ over $\mathbb{F}_q$ is a basis of the form $\{\alpha, \alpha^q, \alpha^{q^2}, \ldots, \alpha^{q^{n-1}}\}$, where α is a primitive element for $\mathbb{F}_{q^n}$ over $\mathbb{F}_q$.

2.1.102 Remark Further kinds of bases for finite fields and their properties are discussed in detail in Chapter 5. For example, we show that each basis of $\mathbb{F}_{q^n}$ has a unique dual basis. We give fundamental properties of normal bases and primitive normal bases in Section 5.2. We give there, among other results, the fundamental theorem that for any prime power q and any integer $n \geq 2$ there exists a primitive normal basis for $\mathbb{F}_{q^n}$ over $\mathbb{F}_q$.

2.1.6 Linearized polynomials

2.1.103 Definition Let $L(x) = \sum_{i=0}^{n-1} \alpha_i x^{q^i}$, where $\alpha_i \in \mathbb{F}_{q^n}$. A polynomial of this form is a *linearized polynomial* over $\mathbb{F}_{q^n}$ (also a *q-polynomial* because the exponents are all powers of q).

2.1.104 Remark These polynomials form an important class of polynomials over finite fields because they are $\mathbb{F}_q$-linear functions from $\mathbb{F}_{q^n}$ to $\mathbb{F}_{q^n}$.

2.1.105 Theorem Let $L(x)$ be a linearized polynomial. Then for all $\alpha, \beta \in \mathbb{F}_{q^n}$ and all $c \in \mathbb{F}_q$, we have

1. $L(\alpha + \beta) = L(\alpha) + L(\beta)$,
2. $L(c\alpha) = cL(\alpha)$.

2.1.106 Theorem Let L be a nonzero linearized polynomial over $\mathbb{F}_{q^n}$ and assume that the roots of L lie in the field $\mathbb{F}_{q^s}$, an extension field of $\mathbb{F}_{q^n}$. Then each root of L has the same multiplicity, which is either 1, or a positive power of q.

2.1.107 Remark The Frobenius automorphism $x \mapsto x^q$ is one such example, and the trace function $\mathrm{Tr}(x) = \sum_{i=0}^{n-1} x^{q^i}$ provides another important example of a linearized polynomial over $\mathbb{F}_q$.

2.1.108 Definition Let L be a linearized polynomial over $\mathbb{F}_{q^n}$. A polynomial of the form $A(x) = L(x) - \alpha$, for $\alpha \in \mathbb{F}_{q^n}$, is an *affine polynomial* over $\mathbb{F}_{q^n}$.

2.1.109 Theorem Let A be a nonzero affine polynomial over $\mathbb{F}_{q^n}$ and assume that the roots of A lie in the field $\mathbb{F}_{q^s}$, an extension field of $\mathbb{F}_{q^n}$. Then each root of A has the same multiplicity, which is either 1, or a positive power of q.

2.1.7 Miscellaneous results

2.1.110 Remark We collect here some concepts and results needed in later sections of the handbook.

2.1.7.1 The finite field polynomial Φ function

2.1.111 **Definition** For $f \in \mathbb{F}_q[x]$, $\Phi_q(f)$ denotes the number of polynomials over $\mathbb{F}_q$ which are of smaller degree than the degree of f and which are relatively prime to f. This is also the number of units in the ring $\mathbb{F}_q[x]/(f(x))$.

2.1.112 **Remark** Similarly to the corresponding properties for the Euler function from elementary number theory, we have the following result (see Lemma 3.69 of [1939] and Definition 2.1.43).

2.1.113 **Lemma** The function Φ_q has the following properties:

1. $\Phi_q(f) = 1$ if the degree of f is 0;
2. $\Phi_q(fg) = \Phi_q(f)\Phi_q(g)$ if f and g are relatively prime;
3. if f has degree $n \geq 1$ then

$$\Phi_q(f) = q^n(1 - q^{-n_1}) \cdots (1 - q^{-n_r}),$$

where $n_1, \ldots, n_r$ are the degrees of the distinct monic irreducible polynomials appearing in the unique factorization of f in $\mathbb{F}_q[x]$.

2.1.114 **Remark** One important consequence of Lemma 2.1.113 is that if f is irreducible of degree n over $\mathbb{F}_q$, then $\Phi_q(f^e) = q^{ne} - q^{n(e-1)}$ for any positive integer e.

2.1.7.2 Cyclotomic polynomials

2.1.115 **Remark** The following is a synopsis of properties of roots of unity and cyclotomic polynomials, which can be found in [1939, Chapters 2 and 3].

2.1.116 **Remark** Let n be a positive integer. The polynomial $x^n - 1$ has many special properties over any field. For example, $x^n - 1$ is the minimal polynomial of the Frobenius automorphism which generates the Galois group of $\mathbb{F}_{q^n}$ over $\mathbb{F}_q$, which is useful when studying normal bases over finite fields, see Section 5.2. Many of the basic properties of cyclotomic polynomials (and their roots) hold over arbitrary fields, however in this section we restrict to the finite field case.

2.1.117 **Definition** The roots $\alpha_1, \alpha_2, \ldots, \alpha_n \in \mathbb{F}_{q^n}$ of the polynomial $x^n - 1 \in \mathbb{F}_q[x]$ are the *n-th roots of unity* over $\mathbb{F}_q$.

2.1.118 **Remark** The roots of any degree n polynomial over $\mathbb{F}_q$ must be in $\mathbb{F}_{q^n}$. Thus, the n-th roots of unity of $\mathbb{F}_q$ are all contained in $\mathbb{F}_{q^n}$.

2.1.119 **Theorem** Let n be a positive integer and let $\mathbb{F}_q$ be a finite field of characteristic p. If p does not divide n, the roots of unity form a cyclic group of order n with respect to multiplication in $\mathbb{F}_q^*$. Otherwise, let $n = mp^e$, where $e > 0$ and $\gcd(m, p) = 1$. Then $x^n - 1 = (x^m - 1)^{p^e}$ and the n-th roots of unity are the m-th roots of unity with multiplicity p^e.

2.1.120 **Definition** Let $\mathbb{F}_q$ be a finite field of characteristic p which does not divide n. Denote the cyclic group of n-th roots of unity as U_n. Suppose that U_n is generated by $\alpha \in \mathbb{F}_{q^n}$, that is $U_n = \langle \alpha \rangle$. Then α is a *primitive n-th root of unity* over $\mathbb{F}_q$.

2.1.121 Definition Let $\mathbb{F}_q$ have characteristic p, not dividing n, and let ζ be a primitive n-th root of unity over $\mathbb{F}_q$. Then the polynomial

$$Q_n(x) = \prod_{s=1 \gcd(s,n)=1}^{n} (x - \zeta^s)$$

is the *n-th cyclotomic polynomial over* $\mathbb{F}_q$.

2.1.122 Remark The n-th cyclotomic polynomial does not depend on the choice of primitive root of unity chosen, since ζ^s, $\gcd(s,n) = 1$, runs over all primitive n-th roots of unity.

2.1.123 Theorem Let $\mathbb{F}_q$ be a finite field with characteristic p which does not divide n. Then

1. $\deg(Q_n) = \phi(n)$;
2. $x^n - 1 = \prod_{d|n} Q_d(x)$;
3. the coefficients of $Q_n(x)$ lay within $\mathbb{F}_p$.

2.1.124 Proposition Let r be a prime and let k be a positive integer. Then

$$Q_{r^k}(x) = 1 + x^{r^{k-1}} + x^{2r^{k-1}} + \cdots + x^{(r-1)r^{k-1}}.$$

2.1.125 Theorem Suppose $\gcd(n,q) = 1$, then Q_n factors into $\phi(n)/d$ distinct monic irreducible polynomials in $\mathbb{F}_q[x]$ of degree d, where d is the order of q modulo n. Furthermore, $\mathbb{F}_{q^d}$ is the splitting field of any such factor.

2.1.126 Corollary The cyclotomic polynomial Q_n is irreducible over $\mathbb{F}_q$ if and only if $n = 4, r^k, 2r^k$, $k \geq 0$, where r is an odd prime and q is a primitive root modulo n.

2.1.127 Proposition Let p be a prime and let m and k be positive integers. The following properties of cyclotomic polynomials hold over any field for which they are defined:

1. $Q_{mp}(x) = Q_m(x^p)/Q_m(x)$, if p does not divide m;
2. $Q_{mp}(x) = Q_m(x^p)$, if p divides m;
3. $Q_{mp^k}(x) = Q_{mp}(x^{p^{k-1}})$;
4. $Q_{2n}(x) = Q_n(-x)$ if $n \geq 3$ and n is odd;
5. $Q_n(0) = 1$ if $n \geq 2$;
6. $Q_n(x^{-1})x^{\phi(n)} = Q_n(x)$ if $n \geq 2$;
7. $$Q_n(1) = \begin{cases} 0 & \text{if } n = 1, \\ p & \text{if } n = p^e, \\ 1 & \text{if } n \text{ has at least two distinct prime factors;} \end{cases}$$
8. $$Q_n(-1) = \begin{cases} -2 & \text{if } n = 1, \\ 0 & \text{if } n = 2, \\ p & \text{if } n = 2p^e, \\ 1 & \text{otherwise.} \end{cases}$$

2.1.128 Theorem Let n be a positive integer not divisible by the characteristic of $\mathbb{F}_q$. An explicit factorization of the Q_n over $\mathbb{F}_q$ is given by

$$Q_n(x) = \prod_{d|n} (x^d - 1)^{\mu(n/d)} = \prod_{d|n} (x^{n/d} - 1)^{\mu(d)},$$

where μ is the Möbius function, see Definition 2.1.22.

2.1.129 Proposition Suppose $\gcd(n,q)=1$, then Q_n is irreducible over $\mathbb{F}_q$ if and only if the multiplicative order of q modulo n is $\phi(n)$.

2.1.130 Theorem The product of all monic irreducible polynomials of degree n over $\mathbb{F}_q$, denoted $I(q,n;x)$, is given by

$$I(q,n;x)=\prod_m Q_m(x),$$

where the product is taken over all positive divisors m of q^n-1 such that n is the multiplicative order of q modulo m.

2.1.7.3 Lagrange interpolation

2.1.131 Theorem If $f\colon \mathbb{F}_q \to \mathbb{F}_q$, there is a unique polynomial P_f with coefficients in $\mathbb{F}_q$ and of degree at most $q-1$ so that P_f represents the function f, that is, $P_f(b)=f(b)$ for all $b\in\mathbb{F}_q$. In particular,

$$P_f(x)=\sum_{a\in\mathbb{F}_q} f(a)[1-(x-a)^{q-1}].$$

2.1.132 Remark The property that every function over a finite commutative ring with identity can be represented by a polynomial with coefficients in that ring characterizes finite fields. In particular, if a finite commutative ring R with unity has the property that every function from the ring to itself can be represented by a polynomial with coefficients in the ring, then R is a finite field, and conversely.

2.1.133 Remark The Lagrange Interpolation Formula can also be stated in the following form: for $n\geq 0$, let $a_0,\ldots,a_n$ be $n+1$ distinct elements of $\mathbb{F}_q$, and let $b_0,\ldots,b_n$ be $n+1$ arbitrary elements of $\mathbb{F}_q$. Then, there exists exactly one polynomial $f\in\mathbb{F}_q[x]$ of degree less than or equal to n such that $f(a_i)=b_i$, $i=0,\ldots,n$. This polynomial is given by

$$f(x)=\sum_{i=0}^{n} b_i \prod_{k=0,k\neq i}^{n} \frac{x-a_k}{a_i-a_k}.$$

2.1.134 Theorem Let $f\colon \mathbb{F}_q^n \to \mathbb{F}_q$. The polynomial $P_f(x_1,\ldots,x_n)$ represents f, that is, $P_f(b_1,\ldots,b_n)=f(b_1,\ldots,b_n)$ for all $(b_1,\ldots,b_n)\in\mathbb{F}_q^n$, where

$$P_f(x_1,\ldots,x_n)=\sum_{(a_1,\ldots,a_n)\in\mathbb{F}_q^n} f(a_1,\ldots,a_n)[1-(x_1-a_1)^{q-1}]\cdots[1-(x_n-a_n)^{q-1}].$$

2.1.7.4 Discriminants

2.1.135 Definition Let f be a polynomial of degree n in $\mathbb{F}_q[x]$ with leading coefficient a, and with roots $\alpha_1,\alpha_2,\ldots,\alpha_n$ in its splitting field, counted with multiplicity. The *discriminant* of f is given by

$$D(f)=a^{2n-2}\prod_{1\leq i<j\leq n}(\alpha_i-\alpha_j)^2.$$

2.1.136 Example The discriminant of $ax^2+bx+c=a(x-\alpha_1)(x-\alpha_2)$ is

$$\begin{aligned} D(ax^2+bx+c) &= a^2(\alpha_1-\alpha_2)^2=a^2\left((\alpha_1+\alpha_2)^2-4\alpha_1\alpha_2\right)\\ &= a^2\left(b^2a^{-2}-4ca^{-1}\right)=b^2-4ac. \end{aligned}$$

The discriminant of $ax^3 + bx^2 + cx + d$ is

$$D(ax^3 + bx^2 + cx + d) = b^2c^2 - 4b^3d - 4ac^3 - 27a^2d^2 + 18abcd.$$

2.1.137 Remark The discriminant of f is a polynomial in the coefficients of f. It is also a symmetric function in the roots of f and thus lies in $\mathbb{F}_q$. Clearly, f has a multiple root if and only if $D(f) = 0$.

2.1.138 Proposition An alternative formula for the discriminant of a polynomial f is

$$D(f) = (-1)^{\frac{n(n-1)}{2}} \prod_{i=1}^{n} f'(\alpha_i),$$

where f' denotes the derivative of the polynomial f.

2.1.7.5 Jacobi logarithms

2.1.139 Definition If the elements of $\mathbb{F}_q^*$ are represented as powers of a fixed primitive element $b \in \mathbb{F}_q$, then addition in $\mathbb{F}_q$ can be facilitated by using *Jacobi logarithms* (sometimes also called *Zech logarithms*) $L(n)$ defined by the equation $1 + b^n = b^{L(n)}$, where the case $b^n = -1$ is excluded.

2.1.140 Remark One can show that $b^m + b^n = b^{m+L(n-m)}$ whenever this is defined. Tables of Jacobi logarithms for fields of characteristic 2 and order at most 64 can be found on Table B of [1939]. Jacobi logarithms were first studied by Jacobi [1584].

2.1.7.6 Field-like structures

2.1.141 Remark In this subsection we briefly describe several algebraic systems that have many but perhaps not all of the properties of a field. We are indebted to John Sheekey (Università di Padova) for this section.

2.1.142 Definition A *left (resp. right) prequasifield* is a set Q together with two operations, "+" and "·" such that:

(1) (Q, +) is an abelian group;

(2) for all $a, b, c \in Q$ there exist unique $x, y, z \in Q$ such that

$$a \cdot x = b \quad \text{and} \quad y \cdot a = b \quad \text{and} \quad a \cdot z = b \cdot z + c;$$

(3) left (resp. right) distributive laws hold: for all $a, b, c \in Q$

$$a \cdot (b + c) = a \cdot b + a \cdot c \quad (resp. (b + c) \cdot a = b \cdot a + c \cdot a).$$

2.1.143 Definition Let Q be a left prequasifield.

(1) A left prequasifield is a *left quasifield* if it has a multiplicative identity.

(2) A left prequasifield is a *presemifield* if it is also a right prequasifield.

(3) A presemifield is a *semifield* if it has a multiplicative identity.

(4) A semifield is *commutative* if "·" is commutative.

(5) A left quasifield is a *left nearfield* if "·" is associative.

2.1.144 Remark All left prequasifields have prime power order. Left prequasifields coordinatize translation planes. The smallest left prequasifield which is not a field has order 9. The smallest semifield which is not a field has order 16. For more on the above structures see [807] or [1560].

2.1.145 Remark The multiplicative structure of a left prequasifield is a *quasigroup*. The multiplicative structure of a semifield is a *loop*. The multiplicative structure of a nearfield is a *group*.

2.1.146 Definition Let Q be a set together with two operations, "+" and "·", containing additive identity 0 and multiplicative identity 1, such that:

(1) $(Q/\{0\}, \cdot)$ is a group;

(2) left and right distributive laws hold: for all $a, b, c \in Q$

$$a \cdot (b + c) = a \cdot b + a \cdot c \quad \text{and } (b + c) \cdot a = b \cdot a + c \cdot a;$$

(3) there exists a unique element 0 such that for all $a \in Q$

$$a + 0 = 0 + a = a.$$

(4A) Q is a *neofield* if in addition to (1), (2) and (3) it satisfies for all $a, b \in Q$ there exist unique $x, y \in Q$ such that

$$a + x = b \quad \text{and} \quad y + a = b.$$

(4B) Q is a *division semiring* if in addition to (1), (2) and (3) above it also satisfies + is associative and commutative.

2.1.147 Remark The additive structure of a neofield is a *loop*. The additive structure of a division semiring is a *commutative monoid*.

2.1.148 Remark For more properties of semirings see [1291]. Note that a division semiring in which multiplication is commutative is sometimes also referred to as a semifield, but this definition does not coincide with the previously defined structures.

2.1.149 Remark For more properties of neofields see [2343].

2.1.7.7 Galois rings

2.1.150 Remark We briefly describe Galois rings. We are indebted to Horacio Tapia-Recillas (Universidad Autónoma Metropolitana, Unidad Iztapalapa, México) for this subsection.

2.1.151 Remark Galois rings represent a natural (Galois) extension of the (local) modular ring of integers $\mathbb{Z}/p^m\mathbb{Z}$ where p is a prime and m a positive integer. Krull [1808] recognized their existence and later, Janusz [1593] and Raghavendran [2436] independently obtained additional properties of these rings. More details on Galois rings can be found in [283, 1409, 2045, 2921].

2.1.152 Definition Let $\mathbb{Z}/p^m\mathbb{Z}$ be the ring of integers modulo p^m, p a prime and $m > 1$ an integer. A monic irreducible (primitive) polynomial $f \in (\mathbb{Z}/p^m\mathbb{Z})[x]$ of degree n is a *monic basic irreducible* (*primitive*) if its reduction modulo p is *irreducible* (*primitive*) in $(\mathbb{Z}/p\mathbb{Z})[x]$.

2.1.153 Remark Monic basic irreducible (primitive) polynomials in $(\mathbb{Z}/p^m\mathbb{Z})[x]$ can be determined by means of Hensel's Lifting Lemma from monic irreducible (primitive) polynomials in $(\mathbb{Z}/p\mathbb{Z})[x]$.

2.1.154 Definition Let $f \in (\mathbb{Z}/p^m\mathbb{Z})[x]$ be a monic basic irreducible polynomial of degree n. Then the Galois ring determined by f is

$$GR(p^m, n) = (\mathbb{Z}/p^m\mathbb{Z})[x]/\langle f(x)\rangle,$$

where $\langle f(x)\rangle$ is the principal ideal of $(\mathbb{Z}/p^m\mathbb{Z})[x]$ generated by $f(x)$.

2.1.155 Remark With the above notation, an equivalent definition of a Galois ring is the following.

2.1.156 Definition Let $\mathbb{Z}$ be the ring of (rational) integers and let $f \in \mathbb{Z}[x]$ be a monic polynomial of degree n such that its reduction modulo $p\mathbb{Z}$ is irreducible, then

$$GR(p^m, n) = \mathbb{Z}[x]/\langle p^m, f\rangle.$$

2.1.157 Remark The Galois ring $GR(p^m, n)$ can also be defined by means of the p-adic numbers in the following way.

2.1.158 Definition Let p be a prime, $\mathbb{Q}_p$ be the field of p-adic (rational) numbers and $\mathbb{Z}_p$ be the ring of p-adic integers (for details see [2588]). Let n be a positive integer and let ω be a $(p^n - 1)$ root of unity. Then $\mathbb{Q}_p(\omega)$ is an unramified Galois extension of degree n of $\mathbb{Q}_p$. Let $\mathbb{Z}_p[\omega]$ be the ring of elements of $\mathbb{Q}_p(\omega)$ integral over $\mathbb{Z}_p$. Let $p\mathbb{Z}_p[\omega]$ be the (unique) maximal ideal of $\mathbb{Z}_p[\omega]$ generated by p. Then the quotient $\mathbb{Z}_p[\omega]/p\mathbb{Z}_p[\omega]$ is a field isomorphic to the Galois field $\mathbb{F}_{p^n}$.

2.1.159 Definition With the notation as above let m be a positive integer and let $p^m\mathbb{Z}_p[\omega]$ be the principal ideal of $\mathbb{Z}_p[\omega]$ generated by p^m. Then the Galois ring $GR(p^m, n)$ is defined as:

$$GR(p^m, n) = \mathbb{Z}_p[\omega]/p^m\mathbb{Z}_p[\omega].$$

2.1.160 Remark This ring contains as a subring the ring of integers modulo p^m, $\mathbb{Z}/p^m\mathbb{Z}$, and can be thought of as an extension of $\mathbb{Z}/p^m\mathbb{Z}$ by adjoining a $(p^n - 1)$ root of unity ω:

$$GR(p^m, n) = (\mathbb{Z}/p^m\mathbb{Z})[\omega].$$

2.1.161 Theorem With the notation as above, the basic properties of the Galois ring $GR(p^m, n)$ are the following [283, 1409, 2045, 2921]:

1. $GR(p^m, n)$ contains $\mathbb{Z}/p^m\mathbb{Z}$ as a subring, it has characteristic p^m and cardinality p^{mn}. The integer m is the *nilpotency* index of the Galois ring.
2. The ring $GR(p^m, n)$ is local with maximal ideal $\mathcal{M} = \langle p\rangle = pGR(p^m, n)$ generated by p, and a principal ideal ring where any ideal is of the form $\langle p^i\rangle$ for $i = 0, 1, 2, \ldots, m$. Furthermore, it is a finite chain ring:

   $$GR(p^m, n) = \langle p^0\rangle \supset \langle p\rangle \supset \cdots \supset \langle p^{m-1}\rangle \supset \langle p^m\rangle = \{0\}.$$

   The ideal $\langle p^i\rangle$ has cardinality $p^{n(m-i)}$ for $i = 0, 1, \ldots, m$.

3. Each non-zero element of the Galois ring $GR(p^m, n)$ can be written as up^k, where u is a unit and $0 \le k \le m-1$. In this representation the integer k is unique and the unit u is unique modulo the ideal $\langle p^{m-k} \rangle$.
4. The canonical homomorphism $\phi : GR(p^m, n) \longrightarrow GR(p^m, n)/\mathcal{M}$, between the Galois ring and its residue field $GR(p^m, n)/\mathcal{M}$ is such that $\phi(\xi) = \overline{\xi}$ is a root of $\phi(f(x))$. The residue field is isomorphic to the Galois field $GF(p^n) = \mathbb{F}_{p^n}$ with p^n elements. Furthermore, $GF(p^n)^* = \langle \overline{\xi} \rangle$.
5. The Galois ring is a $(\mathbb{Z}/p^m\mathbb{Z})$-module:

$$GR(p^m, n) = (\mathbb{Z}/p^m\mathbb{Z})[\xi] = (\mathbb{Z}/p^m\mathbb{Z}) + \xi(\mathbb{Z}/p^m\mathbb{Z}) + \cdots + \xi^{n-1}(\mathbb{Z}/p^m\mathbb{Z}).$$

6. The group of units $\mathcal{U}$ of the Galois ring $GR(p^m, n)$ has the following structure:

$$\mathcal{U} = C \times G,$$

where C is a cyclic group of order $p^n - 1$ generated by ξ and G is an abelian group of order $p^{(m-1)n}$. Furthermore,

   (a) if p is odd, or if $p = 2$ and $m \le 2$ then G is a direct product of n cyclic groups, each of order p^{m-1};

   (b) if $p = 2$ and $m \ge 3$ the group G is the direct product of a cyclic group of order 2, a cyclic group of order 2^{m-2} and $n-1$ cyclic groups each of order 2^{m-1}.

7. There is a subset $\mathcal{T}$ of $GR(p^m, n)$, the *Teichmüller* set of representatives of the Galois ring, such that any element $\beta \in GR(p^m, n)$ has a unique p-adic (multiplicative) representation:

$$\beta = \rho_0(\beta) + \rho_1(\beta)p + \cdots + \rho_{m-1}(\beta)p^{m-1},$$

where $\rho_i(\beta) \in \mathcal{T}$ for $0 \le i \le m-1$. The elements of the maximal ideal of the Galois ring correspond to $\rho_0(\beta) = 0$. There is a bijection, induced by the canonical homomorphism ϕ, between $\mathcal{T}$ and the residue field of the Galois ring $GR(p^m, n)$.
8. The Teichmüller set of representatives of the Galois ring can be taken as

$$\mathcal{T} = \{0, 1, \xi, \xi^2, \ldots, \xi^{q-2}\} = \{0\} \cup C,$$

where $q = p^n$.
9. Given a prime p and an integer $n > 1$, for each divisor r of n there is a unique Galois ring $GR(p^m, r)$, and any subring of the Galois ring $GR(p^m, n)$ is of this form.
10. For each positive integer t, there is a natural injective ring homomorphism $GR(p^m, n) \longrightarrow GR(p^m, nt)$.
11. There is a natural surjective ring homomorphism $GR(p^m, n) \longrightarrow GR(p^{m-1}, n)$ with kernel $\langle p^{m-1} \rangle$.
12. The group of automorphisms of the Galois ring $GR(p^m, n)$ is a cyclic group of order n.
13. The Galois ring $GR(p^m, n)$ is quasi-Frobenius.

2.1.162 Example $GR(p, n) = GF(p, n) = \mathbb{F}_{p^n}, \quad GR(p^m, 1) = (\mathbb{Z}/p^m\mathbb{Z})$.

2.1.163 Example [2045] The polynomial $f(x) = x^3 + x + 1 \in (\mathbb{Z}/2^2\mathbb{Z})[x]$ is monic basic irreducible over $(\mathbb{Z}/2^2\mathbb{Z})$. Then $GR(2^2, 3) = (\mathbb{Z}/2^2\mathbb{Z})[x]/\langle f(x) \rangle$.

2.1.164 Example [1409] The polynomial $g(x) = x^3 + 2x^2 + x - 1 \in (\mathbb{Z}/2^2\mathbb{Z})[x]$ is also monic basic irreducible over $(\mathbb{Z}/2^2\mathbb{Z})$. Then $GR(2^2, 3) = (\mathbb{Z}/2^2\mathbb{Z})[x]/\langle g(x)\rangle$.

2.1.165 Example [283] The polynomial $g(x) = x^3 - 2x^2 - x - 1 \in (\mathbb{Z}/2^3\mathbb{Z})[x]$ is monic basic irreducible over $(\mathbb{Z}/2^3\mathbb{Z})$. Then $GR(2^3, 3) = (\mathbb{Z}/2^3\mathbb{Z})[x]/\langle g(x)\rangle$.

2.1.8 Finite field related books

2.1.166 Remark We give a list of finite field related books, divided into categories and listed without duplication even though a number of these books could be listed in two or more categories.

2.1.8.1 Textbooks

2.1.167 Remark We begin by listing a number of books that could be used as textbooks. Reference [1939] by Lidl and Niederreiter is, by far, the most comprehensive. Other textbooks include Jungnickel [1631], Lidl and Niederreiter [1938], Masuda and Panario [2017], McEliece [2049], Menezes et al. [2077], Mullen and Mummert [2179], Small [2681], and Wan [2921, 2923].

2.1.8.2 Finite field theory

2.1.168 Remark We list a number of books dealing with various theoretical topics related to finite fields: [240, 398, 557, 850, 961, 1121, 1122, 1333, 1389, 1511, 1631, 1701, 1756, 1773, 1843, 1845, 1922, 1936, 1938, 1939, 2017, 2049, 2054, 2077, 2107, 2548, 2637, 2641, 2667, 2670, 2672, 2681, 2711, 2714, 2793, 2920, 2921, 2923, 2949, 2950].

2.1.8.3 Applications

2.1.169 Remark The use of finite fields in algebraic coding theory has been the focus for numerous books: [231, 270, 304, 311, 1558, 1943, 1945, 1991, 2252, 2281, 2404, 2405, 2819, 2820, 2849].

2.1.170 Remark Theoretical and applied aspects of cryptography have been treated in: [245, 312, 313, 661, 759, 762, 922, 1105, 1303, 1413, 1521, 1563, 1694, 1774, 2076, 2080, 2644, 2720].

2.1.171 Remark There have been several books on the applications of finite fields in combinatorics, especially in combinatorial design theory and finite geometries: [131, 141, 211, 260, 261, 262, 453, 484, 706, 785, 807, 819, 1509, 1510, 1515, 1560, 1875, 2445, 2719, 2781, 2851].

2.1.8.4 Algorithms

2.1.172 Remark Several books contain results on algorithmic and computational finite field topics. These include [761, 1227, 2632].

2.1.8.5 Conference proceedings

2.1.173 Remark The *Finite Fields and Applications Conferences (Fq n series)* have been held every two years (except 2005) since 1991. The proceedings from these conferences are: [663, 1636, 1870, 2052, 2181, 2182, 2184, 2185, 2187, 2197]. Other conference proceedings volumes include: [533, 535, 596, 869, 1057, 1228, 1306, 1316, 1436, 1478, 1480, 1928].

References Cited: [131, 136, 141, 211, 231, 240, 245, 260, 261, 262, 270, 304, 311, 312, 313, 398, 453, 484, 533, 535, 557, 596, 661, 663, 706, 759, 761, 762, 785, 797, 807, 819, 850,

869, 922, 961, 1057, 1105, 1121, 1122, 1227, 1228, 1291, 1303, 1306, 1316, 1333, 1389, 1413, 1436, 1478, 1480, 1509, 1510, 1511, 1515, 1521, 1558, 1560, 1563, 1570, 1584, 1631, 1636, 1694, 1701, 1756, 1773, 1774, 1843, 1845, 1848, 1875, 1922, 1928, 1936, 1938, 1939, 1943, 1945, 1991, 2017, 2049, 2052, 2054, 2076, 2077, 2080, 2107, 2144, 2179, 2181, 2182, 2184, 2185, 2187, 2197, 2223, 2252, 2280, 2281, 2343, 2404, 2405, 2445, 2548, 2632, 2637, 2641, 2644, 2667, 2670, 2672, 2681, 2711, 2714, 2719, 2720, 2781, 2793, 2819, 2820, 2849, 2851, 2920, 2921, 2923, 2949, 2950]

2.2 Tables

David Thomson, Carleton University

2.2.1 Remark Unless otherwise stated, all of the data given in this section was created by the author and, when possible, was verified with known results. Basic algorithms (for example, brute force) were preferred due to their reliability and ease of verification. Unless stated, all simulations were done in C/C++ using the NTL version 5.5.2 library [2633] for modular computations. NTL was compiled using the GMP version 4.3.2 library [2797] for multi-precision arithmetic. Extended and machine-readable versions of the tables found in this section can be found on the book's website [2180].

2.2.2 Remark Since most computer algebra packages can readily handle basic finite field computations, our aim is not to repeat tables whose purpose is to improve hand-calculations. For reference, we briefly recall the list of tables found in [1939].

Tables A and B are aids to perform fast arithmetic by hand over small finite fields. Table A is a list of all elements over small finite fields and their discrete logarithms with respect to a primitive element. Table B provides a list of *Jacobi's logarithms* $L(\cdot)$ for $\mathbb{F}_{2^n}$, $2 \leq n \leq 6$. These logarithms allow the computation of field elements by the relationship $\zeta^\alpha + \zeta^\beta = \zeta^{\alpha+L(\beta-\alpha)}$.

Table C provides a list of all monic irreducible polynomials of degree n over small prime fields. Particularly, these tables cover $p = 2$ and $n \leq 11$, $p = 3$ and $n \leq 7$, $p = 5$ and $n \leq 5$, $p = 7$ and $n \leq 4$.

Tables D, E, and F deal with primitive polynomials. Table D lists one primitive polynomial over $\mathbb{F}_2$ for degrees $n \leq 100$. Table E lists all quadratic primitive polynomials for $11 \leq p \leq 31$ and Table F lists one primitive polynomial of degree n over $\mathbb{F}_p$ for all $n \geq 2$ with $p < 50$ and $p^n < 10^9$.

2.2.1 Low-weight irreducible and primitive polynomials

2.2.3 Remark Low-weight irreducible polynomials are highly desired due to their efficiency in hardware and software implementations of finite fields. Irreducible polynomials of degree at least 2 over $\mathbb{F}_2$ must have an odd number of terms. The use of irreducible trinomials (having 3 terms) and, in their absence, irreducible pentanomials (having 5 terms) are useful; see, for example, [1413, Chapter 2]. For cryptographic use, the irreducible trinomial or pentanomial of lowest lexicographical order (for a fixed n, prefer the trinomial x^n+x^k+1 over $x^n+x^{k_1}+1$ when $k < k_1$, the analogue for pentanomials is obvious) is often preferred for transparency reasons. However, the irreducible with the optimal performance for a given implementation is not necessarily the lowest lex-order, see [2573] and Section 11.1. A list of the lowest-weight lowest-lex-order irreducible over $\mathbb{F}_2$ is given in [2582] for degree $n \leq 10000$. Table 2.2.1 gives the lowest-weight, lowest-lex-order irreducible polynomial for $n \leq 1025$. The output of the table follows the format n, k (for trinomials $x^n + x^k + 1$) or n, k_1, k_2, k_3 (for pentanomials $x^n + x^{k_1} + x^{k_2} + x^{k_3} + 1$). We have extended these tables to larger n and to larger q for small values of n. Furthermore, the computer algebra package Magma [712] contains similar tables, due to Steel (2004-2007), for the following values of q and n:

q	$n \leq$	q	$n \leq$	q	$n \leq$	q	$n \leq$
2	120,000	3	50,000	4, 5, 7	2000	$9 \leq q \leq 127$	1000 (or more).

Sections 3.4 and 4.3 give more information on weights of irreducible and primitive polynomials.

2,1	3,1	4,1	5,2	6,1	7,1	8,4,3,1	9,1
10,3	11,2	12,3	13,4,3,1	14,5	15,1	16,5,3,1	17,3
18,3	19,5,2,1	20,3	21,2	22,1	23,5	24,4,3,1	25,3
26,4,3,1	27,5,2,1	28,1	29,2	30,1	31,3	32,7,3,2	33,10
34,7	35,2	36,9	37,6,4,1	38,6,5,1	39,4	40,5,4,3	41,3
42,7	43,6,4,3	44,5	45,4,3,1	46,1	47,5	48,5,3,2	49,9
50,4,3,2	51,6,3,1	52,3	53,6,2,1	54,9	55,7	56,7,4,2	57,4
58,19	59,7,4,2	60,1	61,5,2,1	62,29	63,1	64,4,3,1	65,18
66,3	67,5,2,1	68,9	69,6,5,2	70,5,3,1	71,6	72,10,9,3	73,25
74,35	75,6,3,1	76,21	77,6,5,2	78,6,5,3	79,9	80,9,4,2	81,4
82,8,3,1	83,7,4,2	84,5	85,8,2,1	86,21	87,13	88,7,6,2	89,38
90,27	91,8,5,1	92,21	93,2	94,21	95,11	96,10,9,6	97,6
98,11	99,6,3,1	100,15	101,7,6,1	102,29	103,9	104,4,3,1	105,4
106,15	107,9,7,4	108,17	109,5,4,2	110,33	111,10	112,5,4,3	113,9
114,5,3,2	115,8,7,5	116,4,2,1	117,5,2,1	118,33	119,8	120,4,3,1	121,18
122,6,2,1	123,2	124,19	125,7,6,5	126,21	127,1	128,7,2,1	129,5
130,3	131,8,3,2	132,17	133,9,8,2	134,57	135,11	136,5,3,2	137,21
138,8,7,1	139,8,5,3	140,15	141,10,4,1	142,21	143,5,3,2	144,7,4,2	145,52
146,71	147,14	148,27	149,10,9,7	150,53	151,3	152,6,3,2	153,1
154,15	155,62	156,9	157,6,5,2	158,8,6,5	159,31	160,5,3,2	161,18
162,27	163,7,6,3	164,10,8,7	165,9,8,3	166,37	167,6	168,15,3,2	169,34
170,11	171,6,5,2	172,1	173,8,5,2	174,13	175,6	176,11,3,2	177,8
178,31	179,4,2,1	180,3	181,7,6,1	182,81	183,56	184,9,8,7	185,24
186,11	187,7,6,5	188,6,5,2	189,6,5,2	190,8,7,6	191,9	192,7,2,1	193,15
194,87	195,8,3,2	196,3	197,9,4,2	198,9	199,34	200,5,3,2	201,14
202,55	203,8,7,1	204,27	205,9,5,2	206,10,9,5	207,43	208,9,3,1	209,6
210,7	211,11,10,8	212,105	213,6,5,2	214,73	215,23	216,7,3,1	217,45
218,11	219,8,4,1	220,7	221,8,6,2	222,5,4,2	223,33	224,9,8,3	225,32
226,10,7,3	227,10,9,4	228,113	229,10,4,1	230,8,7,6	231,26	232,9,4,2	233,74
234,31	235,9,6,1	236,5	237,7,4,1	238,73	239,36	240,8,5,3	241,70
242,95	243,8,5,1	244,111	245,6,4,1	246,11,2,1	247,82	248,15,14,10	249,35
250,103	251,7,4,2	252,15	253,46	254,7,2,1	255,52	256,10,5,2	257,12
258,71	259,10,6,2	260,15	261,7,6,4	262,9,8,4	263,93	264,9,6,2	265,42
266,47	267,8,6,3	268,25	269,7,6,1	270,53	271,58	272,9,3,2	273,23
274,67	275,11,10,9	276,63	277,12,6,3	278,5	279,5	280,9,5,2	281,93
282,35	283,12,7,5	284,53	285,10,7,5	286,69	287,71	288,11,10,1	289,21
290,5,3,2	291,12,11,5	292,37	293,11,6,1	294,33	295,48	296,7,3,2	297,5
298,11,8,4	299,11,6,4	300,5	301,9,5,2	302,41	303,1	304,11,2,1	305,102
306,7,3,1	307,8,4,2	308,15	309,10,6,4	310,93	311,7,5,3	312,9,7,4	313,79
314,15	315,10,9,1	316,63	317,7,4,2	318,45	319,36	320,4,3,1	321,31
322,67	323,10,3,1	324,51	325,10,5,2	326,10,3,1	327,34	328,8,3,1	329,50
330,99	331,10,6,2	332,89	333,2	334,5,2,1	335,10,7,2	336,7,4,1	337,55
338,4,3,1	339,16,10,7	340,45	341,10,8,6	342,125	343,75	344,7,2,1	345,22
346,63	347,11,10,3	348,103	349,6,5,2	350,53	351,34	352,13,11,6	353,69
354,99	355,6,5,1	356,10,9,7	357,11,10,2	358,57	359,68	360,5,3,2	361,7,4,1
362,63	363,8,5,3	364,9	365,9,6,5	366,29	367,21	368,7,3,2	369,91
370,139	371,8,3,2	372,111	373,8,7,2	374,8,6,5	375,16	376,8,7,5	377,41
378,43	379,10,8,5	380,47	381,5,2,1	382,81	383,90	384,12,3,2	385,6
386,83	387,8,7,1	388,159	389,10,9,5	390,9	391,28	392,13,10,6	393,7
394,135	395,11,6,5	396,25	397,12,7,6	398,7,6,2	399,26	400,5,3,2	401,152
402,171	403,9,8,5	404,65	405,13,8,2	406,141	407,71	408,5,3,2	409,87
410,10,4,3	411,12,10,3	412,147	413,10,7,6	414,13	415,102	416,9,5,2	417,107
418,199	419,15,5,4	420,7	421,5,4,2	422,149	423,25	424,9,7,2	425,12
426,63	427,11,6,5	428,105	429,10,8,7	430,14,6,1	431,120	432,13,4,3	433,33
434,12,11,5	435,12,9,5	436,165	437,6,2,1	438,65	439,49	440,4,3,1	441,7
442,7,5,2	443,10,6,1	444,81	445,7,6,4	446,105	447,73	448,11,6,4	449,134
450,47	451,16,10,1	452,6,5,4	453,15,6,4	454,8,6,1	455,38	456,18,9,6	457,16
458,203	459,12,5,2	460,19	461,7,6,1	462,73	463,93	464,19,18,13	465,31
466,14,11,6	467,11,6,1	468,27	469,9,5,2	470,9	471,1	472,11,3,2	473,200
474,191	475,9,8,4	476,9	477,16,15,7	478,121	479,104	480,15,9,6	481,138

482,9,6,5	483,9,6,4	484,105	485,17,16,6	486,81	487,94	488,4,3,1	489,83
490,219	491,11,6,3	492,7	493,10,5,3	494,17	495,76	496,16,5,2	497,78
498,155	499,11,6,5	500,27	501,5,4,2	502,8,5,4	503,3	504,15,14,6	505,156
506,23	507,13,6,3	508,9	509,8,7,3	510,69	511,10	512,8,5,2	513,26
514,67	515,14,7,4	516,21	517,12,10,2	518,33	519,79	520,15,11,2	521,32
522,39	523,13,6,2	524,167	525,6,4,1	526,97	527,47	528,11,6,2	529,42
530,10,7,3	531,10,5,4	532,1	533,4,3,2	534,161	535,8,6,2	536,7,5,3	537,94
538,195	539,10,5,4	540,9	541,13,10,4	542,8,6,1	543,16	544,8,3,1	545,122
546,8,2,1	547,13,7,4	548,10,5,3	549,16,4,3	550,193	551,135	552,19,16,9	553,39
554,10,8,7	555,10,9,4	556,153	557,7,6,5	558,73	559,34	560,11,9,6	561,71
562,11,4,2	563,14,7,3	564,163	565,11,6,1	566,153	567,28	568,15,7,6	569,77
570,67	571,10,5,2	572,12,8,1	573,10,6,4	574,13	575,146	576,13,4,3	577,25
578,23,22,16	579,12,9,7	580,237	581,13,7,6	582,85	583,130	584,14,13,3	585,88
586,7,5,2	587,11,6,1	588,35	589,10,4,3	590,93	591,9,6,4	592,13,6,3	593,86
594,19	595,9,2,1	596,273	597,14,12,9	598,7,6,1	599,30	600,9,5,2	601,201
602,215	603,6,4,3	604,105	605,10,7,5	606,165	607,105	608,19,13,6	609,31
610,127	611,10,4,2	612,81	613,19,10,4	614,45	615,211	616,19,10,3	617,200
618,295	619,9,8,5	620,9	621,12,6,5	622,297	623,68	624,11,6,5	625,133
626,251	627,13,8,4	628,223	629,6,5,2	630,7,4,2	631,307	632,9,2,1	633,101
634,39	635,14,10,4	636,217	637,14,9,1	638,6,5,1	639,16	640,14,3,2	641,11
642,119	643,11,3,2	644,11,6,5	645,11,8,4	646,249	647,5	648,13,3,1	649,37
650,3	651,14	652,93	653,10,8,7	654,33	655,88	656,7,5,4	657,38
658,55	659,15,4,2	660,11	661,12,11,4	662,21	663,107	664,11,9,8	665,33
666,10,7,2	667,18,7,3	668,147	669,5,4,2	670,153	671,15	672,11,6,5	673,28
674,11,7,4	675,6,3,1	676,31	677,8,4,3	678,15,5,3	679,66	680,23,16,9	681,11,9,3
682,171	683,11,6,1	684,209	685,4,3,1	686,197	687,13	688,19,14,6	689,14
690,79	691,13,6,2	692,299	693,15,8,2	694,169	695,177	696,23,10,2	697,267
698,215	699,15,10,1	700,75	701,16,4,2	702,37	703,12,7,1	704,8,3,2	705,17
706,12,11,8	707,15,8,5	708,15	709,4,3,1	710,13,12,4	711,92	712,5,4,3	713,41
714,23	715,7,4,1	716,183	717,16,7,1	718,165	719,150	720,9,6,4	721,9
722,231	723,16,10,4	724,207	725,9,6,5	726,5	727,180	728,4,3,2	729,58
730,147	731,8,6,2	732,343	733,8,7,2	734,11,6,1	735,44	736,13,8,6	737,5
738,347	739,18,16,8	740,135	741,9,8,3	742,85	743,90	744,13,11,1	745,258
746,351	747,10,6,4	748,19	749,7,6,1	750,309	751,18	752,13,10,3	753,158
754,19	755,12,10,1	756,45	757,7,6,1	758,233	759,98	760,11,6,5	761,3
762,83	763,16,14,9	764,6,5,3	765,9,7,4	766,22,19,9	767,168	768,19,17,4	769,120
770,14,5,2	771,17,15,6	772,7	773,10,8,6	774,185	775,93	776,15,14,7	777,29
778,375	779,10,8,3	780,13	781,17,16,2	782,329	783,68	784,13,9,6	785,92
786,12,10,3	787,7,6,3	788,17,10,3	789,5,2,1	790,9,6,1	791,30	792,9,7,3	793,253
794,143	795,7,4,1	796,9,4,1	797,12,10,4	798,53	799,25	800,9,7,1	801,217
802,15,13,9	803,14,9,2	804,75	805,8,7,2	806,21	807,7	808,14,3,2	809,15
810,159	811,12,10,8	812,29	813,10,3,1	814,21	815,333	816,11,8,2	817,52
818,119	819,16,9,7	820,123	821,15,11,2	822,17	823,9	824,11,6,4	825,38
826,255	827,12,10,7	828,189	829,4,3,1	830,17,10,7	831,49	832,13,5,2	833,149
834,15	835,14,7,5	836,10,9,2	837,8,6,5	838,61	839,54	840,11,5,1	841,144
842,47	843,11,10,7	844,105	845,2	846,105	847,136	848,11,4,1	849,253
850,111	851,13,10,5	852,159	853,10,7,1	854,7,5,3	855,29	856,19,10,3	857,119
858,207	859,17,15,4	860,35	861,14	862,349	863,6,3,2	864,21,10,6	865,1
866,75	867,9,5,2	868,145	869,11,7,6	870,301	871,378	872,13,3,1	873,352
874,12,7,4	875,12,8,1	876,149	877,6,5,4	878,12,9,8	879,11	880,15,7,5	881,78
882,99	883,17,16,12	884,173	885,8,7,1	886,13,9,8	887,147	888,19,18,10	889,127
890,183	891,12,4,1	892,31	893,11,8,6	894,173	895,12	896,7,5,3	897,113
898,207	899,18,15,5	900,1	901,13,7,6	902,21	903,35	904,12,7,2	905,117
906,123	907,12,10,2	908,143	909,14,4,1	910,15,9,7	911,204	912,7,5,1	913,91
914,4,2,1	915,8,6,3	916,183	917,12,10,7	918,77	919,36	920,14,9,6	921,221
922,7,6,5	923,16,14,13	924,31	925,16,15,7	926,365	927,403	928,10,3,2	929,11,4,3
930,31	931,10,9,4	932,177	933,16,6,1	934,22,6,5	935,417	936,15,13,12	937,217
938,207	939,7,5,4	940,10,7,1	941,11,6,1	942,45	943,24	944,12,11,9	945,77
946,21,20,13	947,9,6,5	948,189	949,8,3,2	950,13,12,10	951,260	952,16,9,7	953,168
954,131	955,7,6,3	956,305	957,10,9,6	958,13,9,4	959,143	960,12,9,3	961,18
962,15,8,5	963,20,9,6	964,103	965,15,4,2	966,201	967,36	968,9,5,2	969,31
970,11,7,2	971,6,2,1	972,7	973,13,6,4	974,9,8,7	975,19	976,17,10,6	977,15
978,9,3,1	979,178	980,8,7,6	981,12,6,5	982,177	983,230	984,24,9,3	985,222
986,3	987,16,13,12	988,121	989,10,4,2	990,161	991,39	992,17,15,13	993,62
994,223	995,15,12,2	996,65	997,12,6,3	998,101	999,59	1000,5,4,3	1001,17
1002,5,3,2	1003,13,8,3	1004,10,9,7	1005,12,8,2	1006,5,4,3	1007,75	1008,19,17,8	1009,55
1010,99	1011,10,7,4	1012,115	1013,9,8,6	1014,385	1015,186	1016,15,6,3	1017,9,4,1
1018,12,10,5	1019,10,8,1	1020,135	1021,5,2,1	1022,317	1023,7	1024,19,6,1	1025,294

Table 2.2.1 Lowest weight lowest-lexicographical order irreducible polynomial of degree n over $\mathbb{F}_2$. Output: n, k (for trinomials $x^n + x^k + 1$) or n, k_1, k_2, k_3 (for pentanomials $x^n + x^{k_1} + x^{k_2} + x^{k_3} + 1$).

2.2.4 Remark Constructions of irreducible low-weight polynomials are rare; see Sections 3.4 and 3.5. Instead, conditions for reducibility are often more tractable; see Section 3.3. Swan [2753] gives conditions for when a trinomial $x^n + x^k + 1 \in \mathbb{F}_2[x]$ is reducible. In particular, the trinomial is reducible when 8 divides n. Only partial results for Swan-like conditions on pentanomials over $\mathbb{F}_2$ exist in the literature; see, for example, [1777] and Section 3.3.

2.2.5 Conjecture [2582] For every n, there exists either an irreducible trinomial of degree n over $\mathbb{F}_2$ or, in the absence of an irreducible trinomial, there exists an irreducible pentanomial of degree n over $\mathbb{F}_2$.

2.2.6 Remark A polynomial over $\mathbb{F}_q$ is *primitive* if all of its roots are generators of the (cyclic) multiplicative group $\mathbb{F}_q^*$. We give an analogous table to Table 2.2.1 but instead list the lowest-weight lowest-lexicographical order primitive polynomial of degree $n \leq 577$ over $\mathbb{F}_2$. To compute primitivity, we used the *Cunningham project* to find the factorization of $2^n - 1$; see Section 2.2.3 for more details.

2,1	3,1	4,1	5,2	6,1	7,1	8,4,3,2	9,4
10,3	11,2	12,6,4,1	13,4,3,1	14,5,3,1	15,1	16,5,3,2	17,3
18,7	19,5,2,1	20,3	21,2	22,1	23,5	24,4,3,1	25,3
26,6,2,1	27,5,2,1	28,3	29,2	30,6,4,1	31,3	32,7,6,2	33,13
34,8,4,3	35,2	36,11	37,6,4,1	38,6,5,1	39,4	40,5,4,3	41,3
42,7,4,3	43,6,4,3	44,6,5,2	45,4,3,1	46,8,7,6	47,5	48,9,7,4	49,9
50,4,3,2	51,6,3,1	52,3	53,6,2,1	54,8,6,3	55,24	56,7,4,2	57,7
58,19	59,7,4,2	60,1	61,5,2,1	62,6,5,3	63,1	64,4,3,1	65,18
66,9,8,6	67,5,2,1	68,9	69,6,5,2	70,5,3,1	71,6	72,10,9,3	73,25
74,7,4,3	75,6,3,1	76,5,4,2	77,6,5,2	78,7,2,1	79,9	80,9,4,2	81,4
82,9,6,4	83,7,4,2	84,13	85,8,2,1	86,6,5,2	87,13	88,11,9,8	89,38
90,5,3,2	91,8,5,1	92,6,5,2	93,2	94,21	95,11	96,10,9,6	97,6
98,11	99,7,5,4	100,37	101,7,6,1	102,6,5,3	103,9	104,11,10,1	105,16
106,15	107,9,7,4	108,31	109,5,4,2	110,6,4,1	111,10	112,11,6,4	113,9
114,11,2,1	115,8,7,5	116,6,5,2	117,5,2,1	118,33	119,8	120,9,6,2	121,18
122,6,2,1	123,2	124,37	125,7,6,5	126,7,4,2	127,1	128,7,2,1	129,5
130,3	131,8,3,2	132,29	133,9,8,2	134,57	135,11	136,8,3,2	137,21
138,8,7,1	139,8,5,3	140,29	141,13,6,1	142,21	143,5,3,2	144,7,4,2	145,52
146,5,3,2	147,11,4,2	148,27	149,10,9,7	150,53	151,3	152,6,3,2	153,1
154,9,5,1	155,7,5,4	156,9,5,3	157,6,5,2	158,8,6,5	159,31	160,5,3,2	161,18
162,8,7,4	163,7,6,3	164,12,6,5	165,9,8,3	166,10,3,2	167,6	168,16,9,6	169,34
170,23	171,6,5,2	172,7	173,8,5,2	174,13	175,6	176,12,11,9	177,8
178,87	179,4,2,1	180,12,10,7	181,7,6,1	182,8,6,1	183,56	184,9,8,7	185,24
186,9,8,6	187,7,6,5	188,6,5,2	189,6,5,2	190,13,6,2	191,9	192,15,11,5	193,15
194,87	195,8,3,2	196,11,9,2	197,9,4,2	198,65	199,34	200,5,3,2	201,14
202,55	203,8,7,1	204,10,4,3	205,9,5,2	206,10,9,5	207,43	208,9,3,1	209,6
210,12,4,3	211,11,10,8	212,105	213,6,5,2	214,5,3,1	215,23	216,7,3,1	217,45
218,11	219,8,4,1	220,12,10,9	221,8,6,2	222,8,5,2	223,33	224,12,7,2	225,32
226,10,7,3	227,10,9,4	228,12,11,2	229,10,4,1	230,8,7,6	231,26	232,11,9,4	233,74
234,31	235,9,6,1	236,5	237,7,4,1	238,5,2,1	239,36	240,8,5,3	241,70
242,11,6,1	243,8,5,1	244,9,4,1	245,6,4,1	246,11,2,1	247,82	248,15,14,10	249,86
250,103	251,7,4,2	252,67	253,7,3,2	254,7,2,1	255,52	256,10,5,2	257,12
258,83	259,10,6,2	260,10,8,7	261,7,6,4	262,9,8,4	263,93	264,10,9,1	265,42
266,47	267,8,6,3	268,25	269,7,6,1	270,53	271,58	272,9,6,2	273,23
274,67	275,11,10,9	276,6,3,1	277,12,6,3	278,5	279,5	280,9,5,2	281,93
282,35	283,12,7,5	284,119	285,10,7,5	286,69	287,71	288,11,10,1	289,21
290,5,3,2	291,12,11,5	292,97	293,11,6,1	294,61	295,48	296,11,9,4	297,5
298,11,8,4	299,11,6,4	300,7	301,9,5,2	302,41	303,13,12,6	304,11,2,1	305,102
306,7,3,1	307,8,4,2	308,15,9,2	309,10,6,4	310,8,5,1	311,7,5,3	312,11,10,5	313,79
314,15	315,10,9,1	316,135	317,7,4,2	318,8,6,5	319,36	320,4,3,1	321,31
322,67	323,10,3,1	324,6,4,3	325,10,5,2	326,10,3,1	327,34	328,9,7,5	329,50
330,8,7,2	331,10,6,2	332,123	333,2	334,7,4,1	335,10,7,2	336,7,4,1	337,55
338,6,3,2	339,16,10,7	340,11,4,3	341,14,11,5	342,125	343,75	344,11,10,6	345,22
346,11,7,2	347,11,10,3	348,8,7,4	349,6,5,2	350,53	351,34	352,13,11,6	353,69
354,14,13,5	355,6,5,1	356,10,9,7	357,11,10,2	358,14,8,7	359,68	360,26,25,1	361,7,4,1
362,63	363,8,5,3	364,67	365,9,6,5	366,29	367,21	368,17,9,7	369,91
370,139	371,8,3,2	372,15,7,3	373,8,7,2	374,8,6,5	375,16	376,8,7,5	377,41
378,43	379,10,8,5	380,47	381,5,2,1	382,81	383,90	384,16,15,6	385,6
386,83	387,9,8,2	388,14,3,1	389,10,9,5	390,89	391,28	392,13,10,6	393,7
394,135	395,11,6,5	396,25	397,12,7,6	398,14,6,5	399,86	400,5,3,2	401,152
402,9,4,3	403,9,8,5	404,189	405,17,8,7	406,157	407,71	408,7,5,1	409,87
410,10,4,3	411,12,10,3	412,147	413,10,7,6	414,16,13,9	415,102	416,9,5,2	417,107
418,15,3,1	419,15,5,4	420,13,10,8	421,5,4,2	422,149	423,25	424,9,7,2	425,12
426,14,12,11	427,11,6,5	428,105	429,10,8,7	430,15,13,11	431,120	432,13,4,3	433,33
434,12,11,5	435,12,9,5	436,165	437,6,2,1	438,65	439,49	440,4,3,1	441,31

442,7,5,2	443,10,6,1	444,13,12,9	445,7,6,4	446,105	447,73	448,11,6,4	449,134
450,79	451,16,10,1	452,6,5,4	453,15,6,4	454,10,9,5	455,38	456,23,11,2	457,16
458,203	459,12,5,2	460,61	461,7,6,1	462,73	463,93	464,23,9,4	465,59
466,14,11,6	467,11,6,1	468,15,9,4	469,9,5,2	470,149	471,1	472,11,3,2	473,8,6,3
474,191	475,9,8,4	476,15	477,16,15,7	478,121	479,104	480,16,13,7	481,138
482,9,6,5	483,9,6,4	484,105	485,17,16,6	486,14,8,5	487,94	488,4,3,1	489,83
490,219	491,11,6,3	492,8,7,1	493,10,5,3	494,137	495,76	496,16,5,2	497,78
498,11,9,3	499,11,6,5	500,10,6,1	501,5,4,2	502,8,5,4	503,3	504,21,14,2	505,156
506,95	507,13,6,3	508,109	509,8,7,3	510,12,10,9	511,10	512,8,5,2	513,85
514,7,5,3	515,14,7,4	516,7,5,2	517,12,10,2	518,33	519,79	520,17,13,11	521,32
522,15,13,4	523,13,6,2	524,167	525,6,4,1	526,9,5,1	527,47	528,11,6,2	529,42
530,10,7,3	531,12,6,2	532,1	533,4,3,2	534,7,5,1	535,8,6,2	536,7,5,3	537,94
538,5,2,1	539,10,5,4	540,179	541,13,10,4	542,9,3,2	543,16	544,13,9,6	545,122
546,8,2,1	547,13,7,4	548,10,5,3	549,16,4,3	550,193	551,135	552,20,5,2	553,39
554,11,8,3	555,10,9,4	556,153	557,7,6,5	558,14,9,5	559,34	560,11,9,6	561,71
562,11,4,2	563,14,7,3	564,163	565,11,6,1	566,153	567,143	568,17,11,10	569,77
570,67	571,10,5,2	572,12,8,1	573,10,6,4	574,13	575,146	576,13,4,3	577,25

Table 2.2.2 Lowest weight lowest-lexicographical order primitive polynomial of degree $n \leq 577$ over $\mathbb{F}_2$. Output: n, k (for trinomials $x^n + x^k + 1$) or n, k_1, k_2, k_3 (for pentanomials $x^n + x^{k_1} + x^{k_2} + x^{k_3} + 1$).

2.2.7 Remark Table 2.2.3 is the analogous table to Table 2.2.1, giving the lowest-weight, lowest-lexicographical order irreducible polynomial of degree $n \leq 516$ over $\mathbb{F}_3$.

2,(1)	3,1(2),(1)	4,1(1),(2)	5,1(2),(1)	6,1(1),(2)
7,2(1),(2)	8,2(1),(2)	9,4(1),(2)	10,2(2),(1)	11,2(1),(2)
12,2(1),(2)	13,1(2),(1)	14,1(1),(2)	15,2(1),(2)	16,4(1),(2)
17,1(2),(1)	18,7(1),(2)	19,2(1),(2)	20,5(1),(2)	21,5(2),(1)
22,4(2),(1)	23,3(2),(1)	24,4(1),(2)	25,3(2),(1)	26,2(2),(1)
27,7(2),(1)	28,2(1),(2)	29,4(1),(2)	30,1(1),(2)	31,5(2),(1)
32,5(1),(2)	33,5(2),(1)	34,2(2),(1)	35,2(1),(2)	36,14(1),(2)
37,6(1),(2)	38,4(2),(1)	39,7(2),(1)	40,1(1),(2)	41,1(2),(1)
42,7(1),(2)	43,17(2),(1)	44,3(1),(2)	45,17(2),(1)	46,5(1),(2)
47,15(2),(1)	48,8(1),(2)	49,3(2),2(1),(1)	50,6(2),(1)	51,1(2),(1)
52,7(1),(2)	53,13(2),(1)	54,1(1),(2)	55,11(2),(1)	56,3(1),(2)
57,7(1),2(1),(2)	58,8(2),(1)	59,17(2),(1)	60,2(1),(2)	61,7(2),(1)
62,10(2),(1)	63,26(1),(2)	64,3(1),(2)	65,5(1),3(1),(1)	66,10(2),(1)
67,2(1),(2)	68,3(1),2(1),(1)	69,17(2),(1)	70,4(2),(1)	71,20(1),(2)
72,28(1),(2)	73,1(2),(1)	74,12(2),(1)	75,5(2),4(1),(1)	76,9(1),(2)
77,16(1),(2)	78,13(1),(2)	79,26(1),(2)	80,2(1),(2)	81,40(1),(2)
82,2(2),(1)	83,27(2),(1)	84,14(1),(2)	85,16(1),(2)	86,13(1),(2)
87,26(1),(2)	88,6(1),(2)	89,13(2),(1)	90,19(1),(2)	91,17(2),(1)
92,10(1),(2)	93,23(2),(1)	94,30(2),(1)	95,47(2),(1)	96,16(1),(2)
97,12(1),(2)	98,4(1),3(1),(1)	99,19(2),(1)	100,25(1),(2)	101,31(2),(1)
102,2(2),(1)	103,47(2),(1)	104,5(1),(2)	105,6(1),2(1),(1)	106,26(2),(1)
107,3(2),(1)	108,2(1),(2)	109,9(2),(1)	110,22(2),(1)	111,2(1),(2)
112,6(1),(2)	113,19(2),(1)	114,7(1),(2)	115,32(1),(2)	116,15(1),(2)
117,52(1),(2)	118,34(2),(1)	119,2(1),(2)	120,4(1),(2)	121,1(2),(1)
122,14(2),(1)	123,7(1),4(1),(2)	124,25(1),(2)	125,52(1),(2)	126,49(1),(2)
127,8(1),(2)	128,6(1),(2)	129,3(2),2(1),(1)	130,10(1),6(1),(1)	131,27(2),(1)
132,19(1),14(1),(1)	133,15(2),(1)	134,4(2),(1)	135,44(1),(2)	136,57(1),(2)
137,1(2),(1)	138,34(2),(1)	139,59(2),(1)	140,59(1),(2)	141,5(2),(1)
142,40(2),(1)	143,35(2),(1)	144,56(1),(2)	145,24(1),(2)	146,2(2),(1)
147,8(1),(2)	148,3(1),(2)	149,11(2),10(1),(1)	150,73(1),(2)	151,2(1),(2)
152,18(1),(2)	153,59(2),(1)	154,32(2),(1)	155,12(1),(2)	156,26(1),(2)
157,22(1),(2)	158,52(2),(1)	159,32(1),(2)	160,4(1),(2)	161,9(1),5(1),(1)
162,19(1),(2)	163,59(2),(1)	164,15(1),(2)	165,22(1),(2)	166,54(2),(1)
167,71(2),(1)	168,28(1),(2)	169,24(1),(2)	170,32(2),(1)	171,20(1),(2)
172,19(1),(2)	173,7(2),(1)	174,52(2),(1)	175,10(1),8(1),(1)	176,12(1),(2)
177,52(1),(2)	178,11(1),(2)	179,59(2),(1)	180,38(1),(2)	181,37(2),(1)
182,25(1),(2)	183,2(1),(2)	184,20(1),(2)	185,64(1),(2)	186,46(2),(1)
187,8(1),(2)	188,11(1),(2)	189,9(1),7(1),(1)	190,94(2),(1)	191,71(2),(1)
192,32(1),(2)	193,12(1),(2)	194,24(2),(1)	195,26(1),(2)	196,79(1),(2)
197,9(1),7(1),(1)	198,29(1),(2)	199,35(2),(1)	200,3(1),(2)	201,88(1),(2)
202,62(2),(1)	203,3(2),(1)	204,50(1),(2)	205,9(2),(1)	206,61(1),(2)
207,11(2),8(1),(1)	208,10(1),(2)	209,40(1),(2)	210,7(1),(2)	211,89(2),(1)
212,14(1),3(1),(1)	213,17(2),4(1),(1)	214,6(2),(1)	215,36(1),(2)	216,4(1),(2)
217,85(2),(1)	218,18(2),(1)	219,25(2),(1)	220,15(1),(2)	221,12(1),2(1),(1)
222,4(2),(1)	223,8(1),5(2),(1)	224,12(1),(2)	225,16(1),(2)	226,38(2),(1)
227,11(2),(1)	228,14(1),(2)	229,72(1),(2)	230,64(2),(1)	231,8(1),7(1),(2)
232,30(1),(2)	233,6(1),2(1),(1)	234,91(1),(2)	235,26(1),(2)	236,9(1),(2)
237,70(1),(2)	238,4(2),(1)	239,5(2),(1)	240,8(1),(2)	241,88(1),(2)
242,2(2),(1)	243,121(2),(1)	244,31(1),(2)	245,97(2),(1)	246,13(1),(2)
247,122(1),(2)	248,50(1),(2)	249,59(2),(1)	250,104(2),(1)	251,9(2),(1)
252,98(1),(2)	253,7(2),(1)	254,16(2),(1)	255,26(1),(2)	256,12(1),(2)

257,22(1),(2)	258,7(1),(2)	259,65(2),(1)	260,35(1),(2)	261,119(2),(1)
262,54(2),(1)	263,69(2),(1)	264,23(1),16(1),(1)	265,61(2),(1)	266,30(2),(1)
267,9(1),2(1),(2)	268,15(1),(2)	269,7(2),(1)	270,88(2),(1)	271,50(1),(2)
272,114(1),(2)	273,46(1),(2)	274,2(2),(1)	275,12(1),(2)	276,10(1),2(1),(2)
277,24(1),(2)	278,118(2),(1)	279,7(2),(1)	280,15(1),(2)	281,10(1),7(1),(2)
282,10(2),(1)	283,23(2),(1)	284,5(1),(2)	285,89(2),(1)	286,70(2),(1)
287,101(2),(1)	288,112(1),(2)	289,73(2),(1)	290,43(1),(2)	291,25(2),(1)
292,13(1),12(1),(1)	293,7(2),(1)	294,16(2),(1)	295,83(2),(1)	296,6(1),(2)
297,3(2),2(1),(1)	298,13(1),3(1),(2)	299,51(2),(1)	300,146(1),(2)	301,30(1),(2)
302,4(2),(1)	303,8(1),2(1),(1)	304,36(1),(2)	305,46(1),(2)	306,118(2),(1)
307,17(2),(1)	308,53(1),(2)	309,3(2),2(1),(1)	310,24(2),(1)	311,13(2),12(1),(1)
312,52(1),(2)	313,93(2),(1)	314,44(2),(1)	315,127(2),(1)	316,87(1),(2)
317,7(2),(1)	318,64(2),(1)	319,10(1),9(1),(2)	320,3(1),(2)	321,83(2),(1)
322,71(1),(2)	323,9(2),(1)	324,38(1),(2)	325,157(2),(1)	326,118(2),(1)
327,7(2),(1)	328,3(1),(2)	329,52(1),(2)	330,11(1),(2)	331,2(1),(2)
332,13(1),6(1),(1)	333,94(1),(2)	334,142(2),(1)	335,8(1),(2)	336,56(1),(2)
337,3(2),(1)	338,48(2),(1)	339,49(2),(1)	340,86(1),(2)	341,25(2),(1)
342,40(2),(1)	343,12(1),10(1),(1)	344,38(1),(2)	345,101(2),(1)	346,14(2),(1)
347,18(1),(2)	348,146(1),(2)	349,54(1),(2)	350,157(1),(2)	351,20(1),(2)
352,7(1),(2)	353,142(1),(2)	354,104(2),(1)	355,41(2),(1)	356,15(1),(2)
357,71(2),(1)	358,77(1),(2)	359,15(2),(1)	360,76(1),(2)	361,157(2),(1)
362,74(2),(1)	363,26(1),(2)	364,1(1),(2)	365,88(1),(2)	366,4(2),(1)
367,107(2),(1)	368,27(1),(2)	369,11(2),(1)	370,11(1),(2)	371,27(2),(1)
372,94(1),(2)	373,25(2),(1)	374,16(2),(1)	375,67(2),(1)	376,9(1),(2)
377,160(1),(2)	378,7(1),(2)	379,44(1),(2)	380,9(1),(2)	381,143(2),(1)
382,137(1),(2)	383,80(1),(2)	384,64(1),(2)	385,22(1),(2)	386,24(2),(1)
387,152(1),(2)	388,87(1),(2)	389,76(1),(2)	390,13(1),(2)	391,22(1),21(2),(1)
392,158(1),(2)	393,185(2),(1)	394,14(1),9(1),(1)	395,23(2),(1)	396,58(1),(2)
397,12(1),5(1),(2)	398,70(2),(1)	399,181(2),(1)	400,3(1),(2)	401,11(2),10(1),(1)
402,176(2),(1)	403,161(2),(1)	404,9(1),2(1),(1)	405,25(1),18(1),(2)	406,6(2),(1)
407,48(1),(2)	408,100(1),(2)	409,99(2),(1)	410,18(2),(1)	411,8(1),2(1),(1)
412,79(1),(2)	413,22(1),(2)	414,37(1),(2)	415,13(1),3(1),(1)	416,20(1),(2)
417,40(1),(2)	418,80(2),(1)	419,26(1),(2)	420,14(1),(2)	421,13(2),(1)
422,178(2),(1)	423,68(1),(2)	424,45(1),(2)	425,61(2),(1)	426,9(1),7(1),(2)
427,167(2),(1)	428,71(1),(2)	429,65(2),(1)	430,72(2),(1)	431,66(1),(2)
432,8(1),(2)	433,120(1),(2)	434,67(1),(2)	435,8(1),2(1),(1)	436,13(1),2(1),(1)
437,14(1),3(2),(1)	438,17(1),(2)	439,16(1),3(2),(1)	440,11(1),(2)	441,7(1),6(1),(2)
442,11(1),3(1),(2)	443,188(1),(2)	444,178(1),(2)	445,141(2),(1)	446,1(1),(2)
447,157(2),(1)	448,24(1),(2)	449,52(1),(2)	450,32(2),(1)	451,17(2),(1)
452,17(1),(2)	453,17(2),4(1),(1)	454,22(2),(1)	455,32(1),(2)	456,28(1),(2)
457,67(2),(1)	458,144(2),(1)	459,13(2),6(1),(1)	460,57(1),(2)	461,13(2),(1)
462,73(1),(2)	463,15(1),13(1),(1)	464,60(1),(2)	465,41(2),(1)	466,167(1),(2)
467,48(1),(2)	468,182(1),(2)	469,166(1),(2)	470,52(2),(1)	471,8(1),(2)
472,18(1),(2)	473,73(2),(1)	474,83(1),(2)	475,17(2),(1)	476,10(1),(2)
477,101(2),(1)	478,10(2),(1)	479,221(2),(1)	480,16(1),(2)	481,22(1),(2)
482,127(1),(2)	483,26(1),(2)	484,39(1),(2)	485,1(2),(1)	486,125(1),(2)
487,29(2),(1)	488,62(1),(2)	489,7(1),5(1),(1)	490,194(2),(1)	491,11(2),(1)
492,26(1),(2)	493,4(1),(2)	494,244(2),(1)	495,7(2),(1)	496,85(1),(2)
497,7(2),6(1),(1)	498,118(2),(1)	499,20(1),(2)	500,39(1),(2)	501,88(1),(2)
502,18(2),(1)	503,35(2),(1)	504,196(1),(2)	505,61(2),(1)	506,14(2),(1)
507,80(1),(2)	508,91(1),(2)	509,151(2),(1)	510,52(2),(1)	511,215(2),(1)
512,24(1),(2)	513,14(1),10(1),(1)	514,44(2),(1)	515,8(1),(2)	516,14(1),(2)

Table 2.2.3 Lowest weight lowest lexicographical order irreducible polynomial of degree n over $\mathbb{F}_3$. Output: n, $\{$*degrees, (coefficients)*$\}$, *(constant term)*.

2.2.8 Remark Necessary and sufficient conditions for the existence of an irreducible binomial of degree n over finite fields of odd characteristic are given in [1939, Theorem 3.75]. A constructive derivation of the degrees for which there exists an irreducible binomial over $\mathbb{F}_q$, q odd, is given in [2356]. The following conjecture summarizes empirical observations of extending Tables 2.2.1 and 2.2.3 to higher characteristics.

2.2.9 Conjecture Let $q > 2$. For every n, there is an irreducible polynomial of degree n over $\mathbb{F}_q$ of weight at most 4.

2.2.2 Low-complexity normal bases

2.2.10 Remark Normal bases are often required in hardware implementations of finite fields due to the efficiency of exponentiation when the finite field is represented using a normal basis.

The *complexity* of a normal basis N, C_N, is defined in Definition 5.3.1. Normal bases with low complexity are highly preferred. An *optimal* normal basis of $\mathbb{F}_{q^n}$ over $\mathbb{F}_q$ is a normal basis attaining the minimum complexity $C_N = 2n - 1$. See Sections 5.2 and 5.3 for more details on normal bases and their complexities.

2.2.2.1 Exhaustive search for low complexity normal bases

2.2.11 Remark Table 2.2.4 is due to an exhaustive search for normal bases of $\mathbb{F}_{2^n}$ over $\mathbb{F}_2$, $n \leq 39$, originally given in [2015]. The table gives the number of normal bases, the smallest and largest complexities (m_{C_N}, M_{C_N}), the average and variance (Avg_{C_N}, Var_{C_N}) of complexities and the smallest and largest complexities for self-dual normal elements. In Table 2.2.4, we fix a typo on the minimum complexity of $n = 37$, originally noted in [130], and make some minor corrections to the calculations of the averages and variances. In the "Notes" column, "Optimal" indicates that the basis with minimal complexity is an optimal normal basis (Theorem 5.3.6), and "sd" indicates that the minimal complexity basis is self-dual.

						Self-dual		
n	# Normal bases	m_{C_N}	M_{C_N}	Avg_{C_N}	Var_{C_N}	m_{C_N}	M_{C_N}	Notes
2	1	3	3	3.00	0	3	3	Optimal, sd
3	1	5	5	5.00	0	5	5	Optimal, sd
4	2	7	9	8.00	1.00	-	-	
5	3	9	15	11.67	6.22	9	9	Optimal, sd
6	4	11	17	15.00	6.00	11	15	Optimal, sd
7	7	19	27	23.00	9.14	21	21	$m_{C_N} = 3n - 2$
8	16	21	35	29.00	11	-	-	$m_{C_N} = 3n - 3$
9	21	17	45	35.57	41.57	17	29	Optimal, sd
10	48	19	61	44.83	61.31	27	51	
11	93	21	71	55.82	57.65	21	57	Optimal, sd
12	128	23	83	64.13	107.23	-	-	
13	315	45	101	78.38	71.07	45	81	sd
14	448	27	135	91.07	108.42	27	135	Optimal, sd
15	675	45	137	105.89	127.36	45	105	sd
16	2048	85	157	115.82	114.59	-	-	
17	3825	81	177	136.83	136.67	81	171	sd
18	5376	35	243	153.51	185.12	35	243	Optimal, sd
19	13797	117	229	172.00	171.91	117	201	sd
20	24576	63	257	190.81	205.81	-	-	
21	27783	95	277	210.97	216.43	105	237	
22	95232	63	363	231.93	238.56	63	363	$m_{C_N} = 3n - 3$
23	182183	45	325	254.02	254.60	45	309	Optimal, sd
24	262144	105	375	276.89	281.01	-	-	
25	629145	93	383	301.01	300.37	93	357	sd
26	1290240	51	555	325.96	328.59	51	555	Optimal, sd
27	1835001	141	443	351.99	351.38	141	413	
28	3670016	55	517	378.98	379.12	-	-	Optimal
29	9256395	57	521	407.00	406.21	57	465	Optimal, sd
30	11059200	59	759	435.95	438.52	59	759	Optimal, sd
31	28629151	237	587	466.00	465.21	237	537	sd
32	67108864	361	621	497.00	496.07	-	-	
33	97327197	65	693	529.00	528.44	65	693	Optimal, sd
34	250675200	243	819	562.00	561.52	243	819	sd
35	352149515	69	779	596.00	595.08	69	693	Optimal, sd
36	704643060	71	1017	630.99	630.51	-	-	Optimal
37	1857283155	141	823	667	666.04	141		sd
38	3616800703	207	1131	704.00	703.18	207		
39	5282242828	77	933	742.00	741.09	77		Optimal, sd

Table 2.2.4 Statistics for normal bases of $\mathbb{F}_{2^n}$ over $\mathbb{F}_2$ obtained by exhaustive search, $n \leq 39$.

2.2.12 Remark Our first conjecture based on Table 2.2.4 appears in [3036] and elsewhere. We also summarize the conjectures found in [2015].

2.2.13 Conjecture When no optimal normal basis of $\mathbb{F}_{2^n}$ over $\mathbb{F}_2$ exists, the minimum complexity of a normal basis of $\mathbb{F}_{2^n}$ over $\mathbb{F}_2$ is $3n - 3$.

2.2.14 Remark Normal bases of $\mathbb{F}_{2^n}$ over $\mathbb{F}_2$ achieving a complexity of $3n-3$ are given in Proposition 5.3.46 and this complexity is the minimal found when $n = 8$ and $n = 22$.

2.2.15 Conjecture The number of normal bases of $\mathbb{F}_{2^n}$ over $\mathbb{F}_2$ are normally distributed with respect to their complexities. Furthermore, the average complexity of a normal basis of $\mathbb{F}_{2^n}$ over $\mathbb{F}_2$ is $(n^2-n+3)/2$ and the variance is also $n^2/2 - cn$, for a small positive constant c.

2.2.16 Remark We remark that the conspicuous wording in Conjecture 2.2.15, that normal bases are normally distributed, is mostly coincidental. Indeed, as n grows, the number of normal bases grow like $2^n/\log(n)$, see Theorem 5.2.13, so the Central Limit Theorem supports this conjecture. The precise distribution of the complexities is still an open and interesting problem.

2.2.17 Remark Self-dual normal bases are often preferred in normal basis implementations due to their highly symmetric properties; see Sections 5.1, 5.2, 5.3, 16.7 as well as [1264, 2925], for more information on self-dual normal bases and their implementations. Exhaustive searches of self-dual normal bases of $\mathbb{F}_{2^n}$ over $\mathbb{F}_2$ appear in [130, 1263, 1631, 2015] and [130] gives an exhaustive search of self-dual normal bases of $\mathbb{F}_{q^n}$ over $\mathbb{F}_q$ for larger q and odd n. Tables 2.2.5, 2.2.6, and 2.2.7 are directly from [130]; we note that we did not implement their algorithm. Table 2.2.5 gives the minimum complexity C_n of a self-dual normal basis of $\mathbb{F}_{2^n}$ over $\mathbb{F}_2$ for odd $n \leq 45$, Table 2.2.6 for q a power of 2 and small n, and Table 2.2.7 for $\mathbb{F}_{q^n}$ over $\mathbb{F}_q$ for odd $q \leq 19$ and small n.

n	3	5	7	9	11	13	15	17	19	21	23
C_n	5	9	21	17	21	45	45	81	117	105	45

n	25	27	29	31	33	35	37	39	41	43	45
C_n	93	141	57	237	65	69	141	77	81	165	153

Table 2.2.5 The lowest complexity for self-dual normal bases of $\mathbb{F}_{2^n}$ over $\mathbb{F}_2$ for odd n, $n \leq 45$.

q/n	3	5	7	9	11	13	15	17	19	21	23	25
2	5	9	21	17	21	45	45	81	117	105	45	93
4	5	9	21	17	21	45	45	81	117	105	45	93
8	9	9	21	45	21	45	81	81				
16	5	9	21	17	21	45						
32	5	19	21	17	21							
64	9	9	21	45								
128	5	9	37									
256	5	9										

Table 2.2.6 Lowest complexity for self-dual normal bases of $\mathbb{F}_{q^n}$ over $\mathbb{F}_q$ where q is a power of 2 for small odd values of n.

q/n	3	5	7	9	11	13	15	17	19	21	23	25
3	7	13	25	37	55	67	–	91	172	–	127	135
5	6	13	25	46	64	85	–	157	153	150		
7	6	16	19	41	61	96	87			–		
11	6	13	25	52	31	100	78					
13	6	13	25	51	64	37						
17	8	13	25	51	64	100		–				
19	8	13	31	51	67				–			

Table 2.2.7 Lowest complexity for self-dual normal bases of $\mathbb{F}_{q^n}$ over $\mathbb{F}_q$ for odd primes $q \leq 19$ and small odd values of n.

2.2.2.2 Minimum type of a Gauss period admitting a normal basis of $\mathbb{F}_{2^n}$ over $\mathbb{F}_2$

2.2.18 Remark We briefly recall the definition of a Gauss period (Definition 5.3.16). Let $r = nk+1$ be a prime not dividing q and let γ be a primitive r-th root of unity in $\mathbb{F}_{q^{nk}}$. Furthermore, let K be the unique subgroup of order k in $\mathbb{Z}_r^*$ and $K_i = \{a \cdot q^i : a \in K\} \subseteq \mathbb{Z}_r^*$ be cosets of $K, 0 \le i \le n-1$. The elements

$$\alpha_i = \sum_{a \in K_i} \gamma^a \in \mathbb{F}_{q^n}, \quad 0 \le i \le n-1,$$

are *Gauss periods* of type (n,k) over $\mathbb{F}_q$. Gauss periods over finite fields are highly desirable as normal bases since, when they exist, they have low complexity; see Theorem 5.3.23. Normal bases due to Gauss periods of type $(n,1)$, for all q, and of type $(n,2)$, for $q = 2$, characterize the *optimal normal bases* (Theorem 5.3.6) and have complexity $2n-1$. Gauss periods also often have high order, see Remark 5.3.49. For conditions on when Gauss periods of type (n,k) admit normal bases of $\mathbb{F}_{q^n}$ over $\mathbb{F}_q$, see Theorem 5.3.17. In particular, we note that there is no Gauss period of $\mathbb{F}_{2^n}$ over $\mathbb{F}_2$ which admits a normal basis when 8 divides n.

Table 2.2.8 gives the lowest k for which a *Gauss period* of type (n,k) admits a normal basis of $\mathbb{F}_{2^n}$ pver $\mathbb{F}_2$ for $n \le 577$. We give a similar table over $\mathbb{F}_3$ in Table 2.2.9. This range was chosen to cover degrees for common implementations of finite field arithmetic. The output of the table is in the format "n,k" where k is the minimum number admitting a type (n,k) Gauss period over $\mathbb{F}_{q^n}$, where $q = 2,3$.

2,1	3,2	4,1	5,2	6,2	7,4	9,2	10,1	11,2	12,1	13,4	14,2
15,4	17,6	18,1	19,10	20,3	21,10	22,3	23,2	25,4	26,2	27,6	28,1
29,2	30,2	31,10	33,2	34,9	35,2	36,1	37,4	38,6	39,2	41,2	42,5
43,4	44,9	45,4	46,3	47,6	49,4	50,2	51,2	52,1	53,2	54,3	55,12
57,10	58,1	59,12	60,1	61,6	62,6	63,6	65,2	66,1	67,4	68,9	69,2
70,3	71,8	73,4	74,2	75,10	76,3	77,6	78,7	79,4	81,2	82,1	83,2
84,5	85,12	86,2	87,4	89,2	90,2	91,6	92,3	93,4	94,3	95,2	97,4
98,2	99,2	100,1	101,6	102,6	103,6	105,2	106,1	107,6	108,5	109,10	110,6
111,20	113,2	114,5	115,4	116,3	117,8	118,6	119,2	121,6	122,6	123,10	124,3
125,6	126,3	127,4	129,8	130,1	131,2	132,5	133,12	134,2	135,2	137,6	138,1
139,4	140,3	141,8	142,6	143,6	145,10	146,2	147,6	148,1	149,8	150,19	151,6
153,4	154,25	155,2	156,13	157,10	158,2	159,22	161,6	162,1	163,4	164,5	165,4
166,3	167,14	169,4	170,6	171,12	172,1	173,2	174,2	175,4	177,4	178,1	179,2
180,1	181,6	182,3	183,2	185,8	186,2	187,6	188,5	189,2	190,10	191,2	193,4
194,2	195,6	196,1	197,18	198,22	199,4	201,8	202,6	203,12	204,3	205,4	206,3
207,4	209,2	210,1	211,10	212,5	213,4	214,3	215,6	217,6	218,5	219,4	220,3
221,2	222,10	223,12	225,22	226,1	227,24	228,9	229,12	230,2	231,2	233,2	234,5
235,4	236,3	237,10	238,7	239,2	241,6	242,6	243,2	244,3	245,2	246,11	247,6
249,8	250,9	251,2	252,3	253,10	254,2	255,6	257,6	258,5	259,10	260,5	261,2
262,3	263,6	265,4	266,6	267,8	268,1	269,8	270,2	271,6	273,2	274,9	275,14
276,3	277,4	278,2	279,4	281,2	282,6	283,6	284,3	285,10	286,3	287,6	289,12
290,5	291,6	292,1	293,2	294,3	295,16	297,6	298,6	299,2	300,19	301,10	302,3
303,2	305,6	306,2	307,4	308,15	309,2	310,6	311,6	313,6	314,5	315,8	316,1
317,26	318,11	319,4	321,12	322,6	323,2	324,5	325,4	326,2	327,8	329,2	330,2
331,6	332,3	333,24	334,7	335,12	337,10	338,2	339,8	340,3	341,8	342,6	343,4
345,4	346,1	347,6	348,1	349,10	350,2	351,10	353,14	354,2	355,6	356,3	357,10
358,10	359,2	361,30	362,5	363,4	364,3	365,24	366,22	367,6	369,10	370,6	371,2
372,1	373,4	374,3	375,2	377,14	378,1	379,12	380,5	381,8	382,6	383,12	385,6
386,2	387,4	388,1	389,24	390,3	391,6	393,2	394,9	395,6	396,11	397,6	398,2
399,12	401,8	402,5	403,16	404,3	405,4	406,6	407,8	409,4	410,2	411,2	412,3
413,2	414,2	415,28	417,4	418,1	419,2	420,1	421,10	422,11	423,4	425,6	426,2
427,16	428,5	429,2	430,3	431,2	433,4	434,9	435,4	436,13	437,18	438,2	439,10
441,2	442,1	443,2	444,5	445,6	446,6	447,6	449,8	450,13	451,6	452,11	453,2
454,19	455,26	457,30	458,6	459,8	460,1	461,6	462,10	463,12	465,4	466,1	467,6
468,21	469,4	470,2	471,8	473,2	474,5	475,4	476,5	477,46	478,7	479,8	481,6
482,5	483,2	484,3	485,18	486,10	487,4	489,12	490,1	491,2	492,13	493,4	494,3
495,2	497,20	498,9	499,4	500,11	501,10	502,10	503,6	505,10	506,5	507,4	508,1
509,2	510,3	511,6	513,4	514,33	515,2	516,3	517,4	518,14	519,2	521,32	522,1
523,10	524,5	525,8	526,3	527,6	529,24	530,2	531,2	532,3	533,12	534,7	535,4
537,8	538,6	539,12	540,1	541,18	542,3	543,2	545,2	546,1	547,10	548,5	549,14
550,7	551,6	553,4	554,2	555,4	556,1	557,6	558,2	559,4	561,2	562,1	563,14
564,3	565,10	566,3	567,4	569,12	570,5	571,10	572,5	573,4	574,3	575,2	577,4

Table 2.2.8 Lowest type of a Gauss period forming a normal basis for $q = 2$ and $n \le 577$.

2,2	3,2	4,1	5,2	6,1	7,4	8,2	9,2	10,3	11,2	13,4	14,2
15,2	16,1	17,6	18,1	19,10	20,5	21,2	22,3	23,2	25,4	26,2	27,4
28,1	29,2	30,1	31,10	32,8	33,6	34,3	35,2	37,4	38,15	39,2	40,7
41,2	42,1	43,4	44,2	45,4	46,3	47,6	49,4	50,2	51,8	52,1	53,2
54,3	55,6	56,2	57,4	58,4	59,12	61,6	62,21	63,2	64,4	65,2	66,3
67,4	68,2	69,2	70,3	71,8	73,4	74,2	75,8	76,10	77,6	78,1	79,4
80,5	81,2	82,9	83,2	85,16	86,2	87,4	88,1	89,2	90,7	91,10	92,5
93,4	94,3	95,2	97,4	98,2	99,2	100,1	101,6	102,11	103,6	104,5	105,2
106,10	107,6	109,10	110,3	111,2	112,1	113,2	114,5	115,4	116,2	117,8	118,9
119,2	121,6	122,3	123,6	124,13	125,2	126,1	127,4	128,2	129,8	130,4	131,2
133,16	134,2	135,4	136,1	137,6	138,1	139,4	140,2	141,2	142,4	143,6	145,10
146,2	147,10	148,1	149,8	150,5	151,6	152,5	153,14	154,3	155,2	157,10	158,2
159,34	160,4	161,6	162,1	163,4	164,5	165,2	166,3	167,14	169,4	170,8	171,12
172,1	173,2	174,9	175,4	176,2	177,4	178,15	179,2	181,6	182,14	183,4	184,7
185,8	186,15	187,6	188,5	189,2	190,3	191,2	193,4	194,2	195,10	196,1	197,18
198,1	199,4	200,2	201,10	202,3	203,12	205,4	206,3	207,4	208,10	209,2	210,1
211,10	212,5	213,6	214,3	215,6	217,6	218,15	219,4	220,4	221,2	222,1	223,12
224,2	225,8	226,15	227,24	229,12	230,2	231,2	232,1	233,2	234,5	235,4	236,8
237,6	238,4	239,2	241,6	242,3	243,2	244,4	245,24	246,3	247,6	248,11	249,8
250,3	251,2	253,4	254,2	255,12	256,1	257,6	258,5	259,10	260,2	261,6	262,3
263,6	265,4	266,8	267,4	268,1	269,8	270,3	271,6	272,5	273,10	274,3	275,12
277,4	278,2	279,10	280,1	281,2	282,1	283,6	284,2	285,2	286,3	287,6	289,12
290,20	291,6	292,1	293,2	294,5	295,12	296,2	297,8	298,4	299,2	301,10	302,3
303,2	304,4	305,6	306,7	307,4	308,2	309,4	310,15	311,6	313,6	314,14	315,2
316,1	317,26	318,17	319,4	320,2	321,18	322,3	323,2	325,4	326,2	327,10	328,7
329,2	330,1	331,6	332,8	333,6	334,15	335,6	337,10	338,2	339,10	340,4	341,8
342,13	343,4	344,5	345,2	346,3	347,6	349,10	350,2	351,22	352,1	353,14	354,3
355,12	356,11	357,4	358,4	359,2	361,30	362,3	363,4	364,7	365,18	366,5	367,6
368,11	369,2	370,4	371,2	373,4	374,3	375,2	376,7	377,14	378,1	379,12	380,5
381,20	382,10	383,12	385,6	386,2	387,14	388,1	389,24	390,5	391,6	392,8	393,10
394,9	395,6	397,6	398,2	399,12	400,1	401,8	402,5	403,4	404,2	405,2	406,21
407,8	409,4	410,2	411,2	412,19	413,2	414,9	415,30	416,5	417,6	418,15	419,2
421,10	422,21	423,4	424,4	425,12	426,3	427,4	428,2	429,2	430,3	431,2	433,4
434,3	435,4	436,13	437,18	438,9	439,10	440,2	441,6	442,3	443,2	445,6	446,15
447,4	448,1	449,8	450,33	451,6	452,8	453,2	454,28	455,2	457,30	458,15	459,8
460,1	461,6	462,1	463,12	464,2	465,10	466,3	467,6	469,10	470,2	471,8	472,4
473,2	474,3	475,4	476,2	477,14	478,4	479,8	481,28	482,3	483,10	484,7	485,2
486,1	487,4	488,2	489,12	490,15	491,2	493,4	494,3	495,30	496,13	497,14	498,11
499,4	500,8	501,16	502,9	503,6	505,10	506,2	507,18	508,1	509,2	510,7	511,6
512,23	513,4	514,3	515,2	517,4	518,9	519,2	520,1	521,32	522,3	523,10	524,2
525,20	526,3	527,6	529,24	530,2	531,2	532,4	533,12	534,27	535,4	536,8	537,8
538,4	539,12	541,18	542,3	543,2	544,10	545,6	546,5	547,10	548,2	549,18	550,21
551,2	553,4	554,2	555,6	556,1	557,6	558,5	559,4	560,5	561,2	562,9	563,14
565,6	566,3	567,28	568,1	569,12	570,1	571,10	572,5	573,4	574,3	575,2	577,4

Table 2.2.9 Lowest type of a Gauss period forming a normal basis for $q = 3$ and $n \leq 577$.

2.2.2.3 Minimum-known complexity of a normal basis of $\mathbb{F}_{2^n}$ over $\mathbb{F}_2$, $n \geq 40$

2.2.19 Remark Table 2.2.10 gives the minimum complexity of a normal basis of $\mathbb{F}_{2^n}$ over $\mathbb{F}_2$ for $40 \leq n \leq 721$ by using a combination of the exhaustive search data of Table 2.2.4 and theorems from Section 5.3. In each row, we give the degree n, the minimum complexity C_n of a normal basis of $\mathbb{F}_{2^n}$ over $\mathbb{F}_2$, the method by which the normal basis was obtained and what property or parameters were used. In the "Method" column, "Optimal" indicates existence of an optimal normal basis, "GNB" indicates the basis arises as a Gauss period and their type is given in the "Property" column. Proposition 5.3.38 constructs normal bases of $\mathbb{F}_{q^n}$ using normal bases of subfields of coprime degree. When this method wins, the values of these coprime factors are indicated in the "Property" column. Corollary 5.3.15 requires an optimal normal basis of $\mathbb{F}_{2^{kn}}$ and the type of the optimal normal basis and the value of k are indicated in the "Property" column. Finally, "sd" indicates that the basis is self-dual.

When n is a power of 2, the best result, when available, is by random search since known methods do not apply. Gauss periods cannot form normal bases when 8 divides n, see Proposition 5.3.20, and n contains no coprime factors with which to apply Proposition 5.3.38. By Conjecture 2.2.15, the complexity of these bases is likely to approach $n^2/2$.

2.2.20 Problem Find constructions of low complexity normal bases of $\mathbb{F}_{2^n}$ over $\mathbb{F}_2$ when n is a prime power, specifically a power of 2.

n	C_n	Method	Property
40	189	Prop. 5.3.38	5, 8
41	81	Optimal	Type 2, sd
42	135	Prop. 5.3.38	3, 14
43	165	GNB	$k = 4$, sd
44	147	Prop. 5.3.38	4, 11
45	153	Search	[130], sd
46	135	Prop. 5.3.38	2, 23
47	261	GNB	$k = 6$, sd
48	425	Prop. 5.3.38	3, 16
49	189	GNB	$k = 4$, sd
50	99	Optimal	Type 2, sd
51	101	Optimal	Type 2, sd
52	103	Optimal	Type 1
53	105	Optimal	Type 2, sd
54	209	GNB	$k = 3$
55	189	Prop. 5.3.38	5, 11
56	399	Prop. 5.3.38	7, 8
57	497	Search	[1263], sd
58	115	Optimal	Type 1
59	597	Search	[1263], sd
60	119	Optimal	Type 1
61	345	GNB	$k = 6$, sd
62	351	GNB	$k = 6$, sd
63	323	Prop. 5.3.38	7, 9
64	1829	Random	
65	129	Optimal	Type 2, sd
66	131	Optimal	Type 1
67	261	GNB	$k = 4$, sd
68	567	Prop. 5.3.38	4, 17
69	137	Optimal	Type 2, sd
70	207	Prop. 5.3.38	2, 35
71	567	GNB	$k = 8$, sd
72	357	Prop. 5.3.38	8, 9
73	285	GNB	$k = 4$, sd
74	147	Optimal	Type 2, sd
75	465	Prop. 5.3.38	3, 25
76	297	GNB	$k = 3$
77	399	Prop. 5.3.38	7, 11
78	231	Prop. 5.3.38	2, 39
79	309	GNB	$k = 4$, sd
80	765	Prop. 5.3.38	5, 16
81	161	Optimal	Type 2, sd
82	163	Optimal	Type 1
83	165	Optimal	Type 2, sd
84	275	Prop. 5.3.38	3, 28
85	729	Prop. 5.3.38	5, 17
86	171	Optimal	Type 2, sd
87	285	Prop. 5.3.38	3, 29
88	441	Prop. 5.3.38	8, 11
89	177	Optimal	Type 2, sd
90	179	Optimal	Type 2, sd
91	525	GNB	$k = 6$, sd
92	315	Prop. 5.3.38	4, 23
93	365	GNB	$k = 4$, sd
94	369	GNB	$k = 3$
95	189	Optimal	Type 2, sd
96	1805	Prop. 5.3.38	3, 32
97	381	GNB	$k = 4$, sd
98	195	Optimal	Type 2, sd
99	197	Optimal	Type 2, sd
100	199	Optimal	Type 1
101	585	GNB	$k = 6$, sd
102	303	Prop. 5.3.38	2, 51
103	597	GNB	$k = 6$, sd
104	945	Prop. 5.3.38	8, 13
105	209	Optimal	Type 2, sd
106	211	Optimal	Type 1
107	621	GNB	$k = 6$, sd
108	627	GNB	$k = 5$
109	1081	Cor. 5.3.15	Type 2, $k = 5$, sd
110	399	Prop. 5.3.38	10, 11
111	705	Prop. 5.3.38	3, 37
112	1615	Prop. 5.3.38	7, 16
113	225	Optimal	Type 2, sd
114	663	GNB	$k = 5$
115	405	Prop. 5.3.38	5, 23
116	399	Prop. 5.3.38	4, 29
117	765	Prop. 5.3.38	9, 13
118	687	GNB	$k = 6$, sd
119	237	Optimal	Type 2, sd
120	945	Prop. 5.3.38	3, 40
121	705	GNB	$k = 6$, sd
122	711	GNB	$k = 6$, sd
123	405	Prop. 5.3.38	3, 41
124	489	GNB	$k = 3$
125	729	GNB	$k = 6$, sd
126	459	Prop. 5.3.38	9, 14
127	501	GNB	$k = 4$, sd
128	7821	Random	
129	825	Prop. 5.3.38	3, 43
130	259	Optimal	Type 1
131	261	Optimal	Type 2, sd
132	455	Prop. 5.3.38	4, 33
133	1595	GNB	$k = 12$, sd
134	267	Optimal	Type 2, sd
135	269	Optimal	Type 2, sd
136	1701	Prop. 5.3.38	8, 17
137	801	GNB	$k = 6$, sd
138	275	Optimal	Type 1
139	549	GNB	$k = 4$, sd
140	483	Prop. 5.3.38	4, 35
141	1127	GNB	$k = 8$, sd
142	831	GNB	$k = 6$, sd
143	837	GNB	$k = 6$, sd
144	1445	Prop. 5.3.38	9, 16
145	513	Prop. 5.3.38	5, 29
146	291	Optimal	Type 2, sd
147	861	GNB	$k = 6$, sd
148	295	Optimal	Type 1
149	1191	GNB	$k = 8$, sd
150	495	Prop. 5.3.38	3, 50
151	885	GNB	$k = 6$, sd
152	2457	Prop. 5.3.38	8, 19
153	605	GNB	$k = 4$, sd
154	567	Prop. 5.3.38	11, 14
155	309	Optimal	Type 2, sd
156	515	Prop. 5.3.38	3, 52
157	1561	Cor. 5.3.15	Type 2, $k = 5$, sd
158	315	Optimal	Type 2, sd
159	525	Prop. 5.3.38	3, 53
160	3249	Prop. 5.3.38	5, 32
161	855	Prop. 5.3.38	7, 23
162	323	Optimal	Type 1
163	645	GNB	$k = 4$, sd
164	567	Prop. 5.3.38	4, 41
165	585	Prop. 5.3.38	5, 33
166	495	Prop. 5.3.38	2, 83
167	2325	Cor. 5.3.15	Type 2, $k = 7$, sd
168	1995	Prop. 5.3.38	3, 56
169	669	GNB	$k = 4$, sd
170	999	GNB	$k = 6$, sd
171	1989	Prop. 5.3.38	9, 19
172	343	Optimal	Type 1
173	345	Optimal	Type 2, sd
174	347	Optimal	Type 2, sd
175	693	GNB	$k = 4$, sd
176	1785	Prop. 5.3.38	11, 16
177	701	GNB	$k = 4$, sd
178	355	Optimal	Type 1
179	357	Optimal	Type 2, sd
180	359	Optimal	Type 1
181	1065	GNB	$k = 6$, sd
182	721	GNB	$k = 3$
183	365	Optimal	Type 2, sd
184	945	Prop. 5.3.38	8, 23
185	1269	Prop. 5.3.38	5, 37
186	371	Optimal	Type 2, sd
187	1101	GNB	$k = 6$, sd
188	1107	GNB	$k = 5$
189	377	Optimal	Type 2, sd
190	567	Prop. 5.3.38	2, 95
191	381	Optimal	Type 2, sd
192	9145	Prop. 5.3.38	3, 64
193	765	GNB	$k = 4$, sd
194	387	Optimal	Type 2, sd
195	645	Prop. 5.3.38	3, 65
196	391	Optimal	Type 1
197	3529	Cor. 5.3.15	Type 2, $k = 9$, sd
198	591	Prop. 5.3.38	2, 99
199	789	GNB	$k = 4$, sd
200	1953	Prop. 5.3.38	8, 25
201	1305	Prop. 5.3.38	3, 67
202	1191	GNB	$k = 6$, sd
203	1083	Prop. 5.3.38	7, 29
204	707	Prop. 5.3.38	4, 51
205	729	Prop. 5.3.38	5, 41
206	817	GNB	$k = 3$
207	765	Prop. 5.3.38	9, 23
208	3825	Prop. 5.3.38	13, 16
209	417	Optimal	Type 2, sd
210	419	Optimal	Type 1
211	2101	Cor. 5.3.15	Type 2, $k = 5$, sd
212	735	Prop. 5.3.38	4, 53
213	845	GNB	$k = 4$, sd
214	849	GNB	$k = 3$
215	1269	GNB	$k = 6$, sd

n	C_n	Method	Property
216	2961	Prop. 5.3.38	8, 27
217	1281	GNB	$k = 6$, sd
218	1287	GNB	$k = 5$
219	869	GNB	$k = 4$, sd
220	873	GNB	$k = 3$
221	441	Optimal	Type 2, sd
222	735	Prop. 5.3.38	3, 74
223	2665	Cor. 5.3.15	Type 2, $k = 6$, sd
224	6859	Prop. 5.3.38	7, 32
225	1581	Prop. 5.3.38	9, 25
226	451	Optimal	Type 1
227	5447	GNB	$k = 24$, sd
228	1485	Prop. 5.3.38	3, 76
229	2747	GNB	$k = 12$, sd
230	459	Optimal	Type 2, sd
231	461	Optimal	Type 2, sd
232	1197	Prop. 5.3.38	8, 29
233	465	Optimal	Type 2, sd
234	867	Prop. 5.3.38	9, 26
235	933	GNB	$k = 4$, sd
236	937	GNB	$k = 3$
237	1545	Prop. 5.3.38	3, 79
238	711	Prop. 5.3.38	2, 119
239	477	Optimal	Type 2, sd
240	3825	Prop. 5.3.38	3, 80
241	1425	GNB	$k = 6$, sd
242	1431	GNB	$k = 6$, sd
243	485	Optimal	Type 2, sd
244	969	GNB	$k = 3$
245	489	Optimal	Type 2, sd
246	815	Prop. 5.3.38	3, 82
247	1461	GNB	$k = 6$, sd
248	4977	Prop. 5.3.38	8, 31
249	825	Prop. 5.3.38	3, 83
250	2187	Prop. 5.3.38	2, 125
251	501	Optimal	Type 2, sd
252	935	Prop. 5.3.38	9, 28
253	945	Prop. 5.3.38	11, 23
254	507	Optimal	Type 2, sd
255	909	Prop. 5.3.38	5, 51
256		No data	Prime power
257	1521	GNB	$k = 6$, sd
258	855	Prop. 5.3.38	3, 86
259	2581	Cor. 5.3.15	Type 2, $k = 5$, sd
260	903	Prop. 5.3.38	4, 65
261	521	Optimal	Type 2, sd
262	783	Prop. 5.3.38	2, 131
263	1557	GNB	$k = 6$, sd
264	1365	Prop. 5.3.38	8, 33
265	945	Prop. 5.3.38	5, 53
266	1575	GNB	$k = 6$, sd
267	885	Prop. 5.3.38	3, 89
268	535	Optimal	Type 1
269	2151	GNB	$k = 8$, sd
270	539	Optimal	Type 2, sd
271	1605	GNB	$k = 6$, sd
272	6885	Prop. 5.3.38	16, 17
273	545	Optimal	Type 2, sd
274	2403	Prop. 5.3.38	2, 137
275	1953	Prop. 5.3.38	11, 25
276	959	Prop. 5.3.38	4, 69
277	1101	GNB	$k = 4$, sd
278	555	Optimal	Type 2, sd
279	1109	GNB	$k = 4$, sd
280	1449	Prop. 5.3.38	8, 35
281	561	Optimal	Type 2, sd
282	1671	GNB	$k = 6$, sd
283	1677	GNB	$k = 6$, sd
284	1129	GNB	$k = 3$
285	945	Prop. 5.3.38	3, 95
286	1071	Prop. 5.3.38	11, 26
287	1539	Prop. 5.3.38	7, 41
288	6137	Prop. 5.3.38	9, 32
289	3457	Cor. 5.3.15	Type 2, $k = 6$, sd
290	1035	Prop. 5.3.38	5, 58
291	1725	GNB	$k = 6$, sd
292	583	Optimal	Type 1
293	585	Optimal	Type 2, sd
294	975	Prop. 5.3.38	3, 98
295	4719	GNB	$k = 16$, sd
296	2961	Prop. 5.3.38	8, 37
297	1761	GNB	$k = 6$, sd
298	1767	GNB	$k = 6$, sd
299	597	Optimal	Type 2, sd
300	995	Prop. 5.3.38	3, 100
301	3001	Cor. 5.3.15	Type 2, $k = 5$, sd
302	1201	GNB	$k = 3$
303	605	Optimal	Type 2, sd

n	C_n	Method	Property
304	9945	Prop. 5.3.38	16, 19
305	1809	GNB	$k = 6$, sd
306	611	Optimal	Type 2, sd
307	1221	GNB	$k = 4$, sd
308	1155	Prop. 5.3.38	11, 28
309	617	Optimal	Type 2, sd
310	927	Prop. 5.3.38	2, 155
311	1845	GNB	$k = 6$, sd
312	1617	Prop. 5.3.38	8, 39
313	1857	GNB	$k = 6$, sd
314	1863	GNB	$k = 5$
315	1173	Prop. 5.3.38	9, 35
316	631	Optimal	Type 1
317	8217	Cor. 5.3.15	Type 2, $k = 13$, sd
318	1055	Prop. 5.3.38	3, 106
319	1197	Prop. 5.3.38	11, 29
320	16461	Prop. 5.3.38	5, 64
321	3105	Prop. 5.3.38	3, 107
322	1215	Prop. 5.3.38	14, 23
323	645	Optimal	Type 2, sd
324	1127	Prop. 5.3.38	4, 81
325	1293	GNB	$k = 4$, sd
326	651	Optimal	Type 2, sd
327	2615	GNB	$k = 8$, sd
328	1701	Prop. 5.3.38	8, 41
329	657	Optimal	Type 2, sd
330	659	Optimal	Type 2, sd
331	1965	GNB	$k = 6$, sd
332	1155	Prop. 5.3.38	4, 83
333	2397	Prop. 5.3.38	9, 37
334	2629	GNB	$k = 7$
335	2349	Prop. 5.3.38	5, 67
336	8075	Prop. 5.3.38	3, 112
337	3361	Cor. 5.3.15	Type 2, $k = 5$, sd
338	675	Optimal	Type 2, sd
339	1125	Prop. 5.3.38	3, 113
340	1353	GNB	$k = 3$
341	2727	GNB	$k = 8$, sd
342	2031	GNB	$k = 6$, sd
343	1365	GNB	$k = 4$, sd
344	3465	Prop. 5.3.38	8, 43
345	1233	Prop. 5.3.38	5, 69
346	691	Optimal	Type 1
347	2061	GNB	$k = 6$, sd
348	695	Optimal	Type 1
349	3481	Cor. 5.3.15	Type 2, $k = 5$, sd
350	699	Optimal	Type 2, sd
351	3501	Cor. 5.3.15	Type 2, $k = 5$, sd
352	7581	Prop. 5.3.38	11, 32
353	4929	Cor. 5.3.15	Type 2, $k = 7$, sd
354	707	Optimal	Type 2, sd
355	2109	GNB	$k = 6$, sd
356	1239	Prop. 5.3.38	4, 89
357	1185	Prop. 5.3.38	3, 119
358	1071	Prop. 5.3.38	2, 179
359	717	Optimal	Type 2, sd
360	3213	Prop. 5.3.38	5, 72
361	10801	Cor. 5.3.15	Type 2, $k = 15$, sd
362	2151	GNB	$k = 5$
363	1445	GNB	$k = 4$, sd
364	1449	GNB	$k = 3$
365	2565	Prop. 5.3.38	5, 73
366	1095	Prop. 5.3.38	2, 183
367	2181	GNB	$k = 6$, sd
368	3825	Prop. 5.3.38	16, 23
369	1377	Prop. 5.3.38	9, 41
370	1323	Prop. 5.3.38	5, 74
371	741	Optimal	Type 2, sd
372	743	Optimal	Type 1
373	1485	GNB	$k = 4$, sd
374	1489	GNB	$k = 3$
375	749	Optimal	Type 2, sd
376	5481	Prop. 5.3.38	8, 47
377	2565	Prop. 5.3.38	13, 29
378	755	Optimal	Type 1
379	4547	GNB	$k = 12$, sd
380	1323	Prop. 5.3.38	4, 95
381	2505	Prop. 5.3.38	3, 127
382	1143	Prop. 5.3.38	2, 191
383	4595	GNB	$k = 12$, sd
384	39105	Prop. 5.3.38	3, 128
385	1449	Prop. 5.3.38	11, 35
386	771	Optimal	Type 2, sd
387	1541	GNB	$k = 4$, sd
388	775	Optimal	Type 1
389	9335	GNB	$k = 24$, sd
390	1295	Prop. 5.3.38	3, 130
391	2325	GNB	$k = 6$, sd

n	C_n	Method	Property
392	3969	Prop. 5.3.38	8, 49
393	785	Optimal	Type 2, sd
394	3915	Cor. 5.3.15	Type 1, $k = 9$
395	2349	GNB	$k = 6$, sd
396	1379	Prop. 5.3.38	4, 99
397	2361	GNB	$k = 6$, sd
398	795	Optimal	Type 2, sd
399	4777	Cor. 5.3.15	Type 2, $k = 6$, sd
400	7905	Prop. 5.3.38	16, 25
401	3207	GNB	$k = 8$, sd
402	1335	Prop. 5.3.38	3, 134
403	6447	GNB	$k = 16$, sd
404	1609	GNB	$k = 3$
405	1449	Prop. 5.3.38	5, 81
406	1539	Prop. 5.3.38	14, 29
407	2961	Prop. 5.3.38	11, 37
408	2121	Prop. 5.3.38	8, 51
409	1629	GNB	$k = 4$, sd
410	819	Optimal	Type 2, sd
411	821	Optimal	Type 2, sd
412	1641	GNB	$k = 3$
413	825	Optimal	Type 2, sd
414	827	Optimal	Type 2, sd
415	1485	Prop. 5.3.38	5, 83
416	16245	Prop. 5.3.38	13, 32
417	1661	GNB	$k = 4$, sd
418	835	Optimal	Type 1
419	837	Optimal	Type 2, sd
420	839	Optimal	Type 1
421	4209	GNB	$k = 10$, sd
422	5053	GNB	$k = 11$
423	1685	GNB	$k = 4$, sd
424	2205	Prop. 5.3.38	8, 53
425	2529	GNB	$k = 6$, sd
426	851	Optimal	Type 2, sd
427	6555	Prop. 5.3.38	7, 61
428	2547	GNB	$k = 5$
429	857	Optimal	Type 2, sd
430	1539	Prop. 5.3.38	5, 86
431	861	Optimal	Type 2, sd
432	11985	Prop. 5.3.38	16, 27
433	1725	GNB	$k = 4$, sd
434	3843	Prop. 5.3.38	2, 217
435	1733	GNB	$k = 4$, sd
436	6091	GNB	$k = 13$
437	5265	Prop. 5.3.38	19, 23
438	875	Optimal	Type 2, sd
439	4381	Cor. 5.3.15	Type 2, $k = 5$, sd
440	3969	Prop. 5.3.38	5, 88
441	881	Optimal	Type 2, sd
442	883	Optimal	Type 1
443	885	Optimal	Type 2, sd
444	1475	Prop. 5.3.38	3, 148
445	1593	Prop. 5.3.38	5, 89
446	2655	GNB	$k = 6$, sd
447	2661	GNB	$k = 6$, sd
448	34751	Prop. 5.3.38	7, 64
449	3591	GNB	$k = 8$, sd
450	1683	Prop. 5.3.38	9, 50
451	1701	Prop. 5.3.38	11, 41
452	1575	Prop. 5.3.38	4, 113
453	905	Optimal	Type 2, sd
454	9025	Cor. 5.3.15	Type 1, $k = 19$
455	2451	Prop. 5.3.38	7, 65
456	10437	Prop. 5.3.38	8, 57
457	13681	Cor. 5.3.15	Type 2, $k = 15$, sd
458	2727	GNB	$k = 6$, sd
459	3671	GNB	$k = 8$, sd
460	919	Optimal	Type 1
461	2745	GNB	$k = 6$, sd
462	1383	Prop. 5.3.38	2, 231
463	5545	Cor. 5.3.15	Type 2, $k = 6$, sd
464	4845	Prop. 5.3.38	16, 29
465	1545	Prop. 5.3.38	3, 155
466	931	Optimal	Type 1
467	2781	GNB	$k = 6$, sd
468	1751	Prop. 5.3.38	9, 52
469	1869	GNB	$k = 4$, sd
470	939	Optimal	Type 2, sd
471	3767	GNB	$k = 8$, sd
472	12537	Prop. 5.3.38	8, 59
473	945	Optimal	Type 2, sd
474	1575	Prop. 5.3.38	3, 158
475	1893	GNB	$k = 4$, sd
476	1659	Prop. 5.3.38	4, 119
477	1785	Prop. 5.3.38	9, 53
478	1431	Prop. 5.3.38	2, 239
479	3831	GNB	$k = 8$, sd
480	16245	Prop. 5.3.38	3, 160
481	2865	GNB	$k = 6$, sd
482	2871	GNB	$k = 5$
483	965	Optimal	Type 2, sd
484	1929	GNB	$k = 3$
485	3429	Prop. 5.3.38	5, 97
486	1455	Prop. 5.3.38	2, 243
487	1941	GNB	$k = 4$, sd
488	7245	Prop. 5.3.38	8, 61
489	3225	Prop. 5.3.38	3, 163
490	979	Optimal	Type 1
491	981	Optimal	Type 2, sd
492	1863	Prop. 5.3.38	12, 41
493	1965	GNB	$k = 4$, sd
494	1969	GNB	$k = 3$
495	989	Optimal	Type 2, sd
496	20145	Prop. 5.3.38	16, 31
497	9921	Cor. 5.3.15	Type 2, $k = 10$, sd
498	1815	Prop. 5.3.38	6, 83
499	1989	GNB	$k = 4$, sd
500	5103	Prop. 5.3.38	4, 125
501	5001	Cor. 5.3.15	Type 2, $k = 5$, sd
502	1503	Prop. 5.3.38	2, 251
503	2997	GNB	$k = 6$, sd
504	6783	Prop. 5.3.38	7, 72
505	5041	Cor. 5.3.15	Type 2, $k = 5$, sd
506	2835	Prop. 5.3.38	2, 253
507	2021	GNB	$k = 4$, sd
508	1015	Optimal	Type 1
509	1017	Optimal	Type 2, sd
510	1919	Prop. 5.3.38	10, 51
511	3045	GNB	$k = 6$, sd
512		No data	Prime power
513	2045	GNB	$k = 4$, sd
514	4563	Prop. 5.3.38	2, 257
515	1029	Optimal	Type 2, sd
516	1715	Prop. 5.3.38	3, 172
517	2061	GNB	$k = 4$, sd
518	2793	Prop. 5.3.38	7, 74
519	1037	Optimal	Type 2, sd
520	2709	Prop. 5.3.38	8, 65
521	16671	GNB	$k = 32$, sd
522	1043	Optimal	Type 1
523	5221	Cor. 5.3.15	Type 2, $k = 5$, sd
524	1827	Prop. 5.3.38	4, 131
525	3465	Prop. 5.3.38	3, 175
526	2097	GNB	$k = 3$
527	3141	GNB	$k = 6$, sd
528	5525	Prop. 5.3.38	16, 33
529	12695	GNB	$k = 24$, sd
530	1059	Optimal	Type 2, sd
531	1061	Optimal	Type 2, sd
532	2121	GNB	$k = 3$
533	3645	Prop. 5.3.38	13, 41
534	1775	Prop. 5.3.38	3, 178
535	2133	GNB	$k = 4$, sd
536	5481	Prop. 5.3.38	8, 67
537	1785	Prop. 5.3.38	3, 179
538	3207	GNB	$k = 6$, sd
539	3969	Prop. 5.3.38	11, 49
540	1079	Optimal	Type 1
541	9737	GNB	$k = 18$, sd
542	2161	GNB	$k = 3$
543	1085	Optimal	Type 2, sd
544	29241	Prop. 5.3.38	17, 32
545	1089	Optimal	Type 2, sd
546	1091	Optimal	Type 1
547	5469	GNB	$k = 10$, sd
548	3267	GNB	$k = 5$
549	5865	Prop. 5.3.38	9, 61
550	2079	Prop. 5.3.38	11, 50
551	3285	GNB	$k = 6$, sd
552	2877	Prop. 5.3.38	8, 69
553	2205	GNB	$k = 4$, sd
554	1107	Optimal	Type 2, sd
555	2213	GNB	$k = 4$, sd
556	1111	Optimal	Type 1
557	3321	GNB	$k = 6$, sd
558	1115	Optimal	Type 2, sd
559	2229	GNB	$k = 4$, sd
560	5865	Prop. 5.3.38	16, 35
561	1121	Optimal	Type 2, sd
562	1123	Optimal	Type 1
563	7869	Cor. 5.3.15	Type 2, $k = 7$, sd
564	2249	GNB	$k = 3$
565	2025	Prop. 5.3.38	5, 113
566	2257	GNB	$k = 3$
567	2261	GNB	$k = 4$, sd

n	C_n	Method	Property
568	11907	Prop. 5.3.38	8, 71
569	6817	Cor. 5.3.15	Type 2, $k = 6$, sd
570	2079	Prop. 5.3.38	6, 95
571	5709	GNB	$k = 10$, sd
572	2163	Prop. 5.3.38	11, 52
573	1905	Prop. 5.3.38	3, 191
574	2187	Prop. 5.3.38	14, 41
575	1149	Optimal	Type 2, sd
576	31093	Prop. 5.3.38	9, 64
577	2301	GNB	$k = 4$, sd
578	3447	GNB	$k = 6$, sd
579	3825	Prop. 5.3.38	3, 193
580	2313	GNB	$k = 3$
581	3135	Prop. 5.3.38	7, 83
582	1935	Prop. 5.3.38	3, 194
583	2205	Prop. 5.3.38	11, 53
584	5985	Prop. 5.3.38	8, 73
585	1169	Optimal	Type 2, sd
586	1171	Optimal	Type 1
587	8205	Cor. 5.3.15	Type 2, $k = 7$, sd
588	1955	Prop. 5.3.38	3, 196
589	2349	GNB	$k = 4$, sd
590	6183	Prop. 5.3.38	5, 118
591	3525	GNB	$k = 6$, sd
592	11985	Prop. 5.3.38	16, 37
593	1185	Optimal	Type 2, sd
594	4389	Prop. 5.3.38	11, 54
595	2133	Prop. 5.3.38	5, 119
596	2377	GNB	$k = 3$
597	2381	GNB	$k = 4$, sd
598	1791	Prop. 5.3.38	2, 299
599	4791	GNB	$k = 8$, sd
600	9765	Prop. 5.3.38	3, 200
601	3585	GNB	$k = 6$, sd
602	3249	Prop. 5.3.38	7, 86
603	4437	Prop. 5.3.38	9, 67
604	4789	GNB	$k = 7$
605	3609	GNB	$k = 6$, sd
606	1211	Optimal	Type 2, sd
607	3621	GNB	$k = 6$, sd
608	42237	Prop. 5.3.38	19, 32
609	2429	GNB	$k = 4$, sd
610	5427	Prop. 5.3.38	2, 305
611	1221	Optimal	Type 2, sd
612	1223	Optimal	Type 1
613	6121	Cor. 5.3.15	Type 2, $k = 5$, sd
614	1227	Optimal	Type 2, sd
615	1229	Optimal	Type 2, sd
616	8379	Prop. 5.3.38	7, 88
617	4935	GNB	$k = 8$, sd
618	1235	Optimal	Type 1
619	2469	GNB	$k = 4$, sd
620	2163	Prop. 5.3.38	4, 155
621	3705	GNB	$k = 6$, sd
622	2481	GNB	$k = 3$
623	3363	Prop. 5.3.38	7, 89
624	6545	Prop. 5.3.38	16, 39
625	22465	Cor. 5.3.15	Type 2, $k = 18$, sd
626	5571	Prop. 5.3.38	2, 313
627	2085	Prop. 5.3.38	3, 209
628	4981	GNB	$k = 7$
629	1257	Optimal	Type 2, sd
630	2415	Prop. 5.3.38	18, 35
631	6301	Cor. 5.3.15	Type 2, $k = 5$, sd
632	6489	Prop. 5.3.38	8, 79
633	10505	Prop. 5.3.38	3, 211
634	8839	Cor. 5.3.15	Type 1, $k = 13$
635	4509	Prop. 5.3.38	5, 127
636	2415	Prop. 5.3.38	12, 53
637	2541	GNB	$k = 4$, sd
638	1275	Optimal	Type 2, sd
639	1277	Optimal	Type 2, sd
640	70389	Prop. 5.3.38	5, 128
641	1281	Optimal	Type 2, sd
642	3831	GNB	$k = 6$, sd
643	7705	Cor. 5.3.15	Type 2, $k = 6$, sd
644	2475	Prop. 5.3.38	23, 28
645	1289	Optimal	Type 2, sd
646	1935	Prop. 5.3.38	2, 323
647	9045	Cor. 5.3.15	Type 2, $k = 7$, sd
648	3381	Prop. 5.3.38	8, 81
649	6481	Cor. 5.3.15	Type 2, $k = 5$, sd
650	1299	Optimal	Type 2, sd
651	1301	Optimal	Type 2, sd
652	1303	Optimal	Type 1
653	1305	Optimal	Type 2, sd
654	6435	Prop. 5.3.38	3, 218
655	2349	Prop. 5.3.38	5, 131
656	6885	Prop. 5.3.38	16, 41
657	4845	Prop. 5.3.38	9, 73
658	1315	Optimal	Type 1
659	1317	Optimal	Type 2, sd
660	1319	Optimal	Type 1
661	3945	GNB	$k = 6$, sd
662	2641	GNB	$k = 3$
663	2205	Prop. 5.3.38	3, 221
664	3465	Prop. 5.3.38	8, 83
665	3591	Prop. 5.3.38	7, 95
666	2499	Prop. 5.3.38	9, 74
667	2565	Prop. 5.3.38	23, 29
668	7985	Cor. 5.3.15	Type 1, $k = 11$
669	2669	GNB	$k = 4$, sd
670	2403	Prop. 5.3.38	5, 134
671	4005	GNB	$k = 6$, sd
672	34295	Prop. 5.3.38	3, 224
673	2685	GNB	$k = 4$, sd
674	4023	GNB	$k = 5$
675	13113	Prop. 5.3.38	25, 27
676	1351	Optimal	Type 1
677	5415	GNB	$k = 8$, sd
678	2255	Prop. 5.3.38	3, 226
679	6789	GNB	$k = 10$, sd
680	15309	Prop. 5.3.38	5, 136
681	14961	Cor. 5.3.15	Type 2, $k = 11$, sd
682	4071	GNB	$k = 6$, sd
683	1365	Optimal	Type 2, sd
684	2729	GNB	$k = 3$
685	2733	GNB	$k = 4$, sd
686	1371	Optimal	Type 2, sd
687	6869	GNB	$k = 10$, sd
688	14025	Prop. 5.3.38	16, 43
689	4725	Prop. 5.3.38	13, 53
690	1379	Optimal	Type 2, sd
691	6901	Cor. 5.3.15	Type 2, $k = 5$, sd
692	2415	Prop. 5.3.38	4, 173
693	3743	Prop. 5.3.38	7, 99
694	2769	GNB	$k = 3$
695	4941	Prop. 5.3.38	5, 139
696	5985	Prop. 5.3.38	3, 232
697	2781	GNB	$k = 4$, sd
698	4167	GNB	$k = 5$
699	2325	Prop. 5.3.38	3, 233
700	1399	Optimal	Type 1
701	12601	Cor. 5.3.15	Type 2, $k = 9$, sd
702	7191	Prop. 5.3.38	26, 27
703	4197	GNB	$k = 6$, sd
704	38409	Prop. 5.3.38	11, 64
705	4209	GNB	$k = 6$, sd
706	14787	Prop. 5.3.38	2, 353
707	4221	GNB	$k = 6$, sd
708	1415	Optimal	Type 1
709	2829	GNB	$k = 4$, sd
710	2833	GNB	$k = 3$
711	5253	Prop. 5.3.38	9, 79
712	3717	Prop. 5.3.38	8, 89
713	1425	Optimal	Type 2, sd
714	2607	Prop. 5.3.38	6, 119
715	2709	Prop. 5.3.38	11, 65
716	2499	Prop. 5.3.38	4, 179
717	2385	Prop. 5.3.38	3, 239
718	2151	Prop. 5.3.38	2, 359
719	1437	Optimal	Type 2, sd
720	13005	Prop. 5.3.38	5, 144
721	4305	GNB	$k = 6$, sd

Table 2.2.10 Minimum found complexity of a normal basis of $\mathbb{F}_{2^n}$ over $\mathbb{F}_2$, $40 \leq n \leq 721$.

2.2.3 Resources and standards

2.2.21 Remark The *Combinatorial Object Server* (COS) [2507] allows the user to specify a type of combinatorial object with specific parameter values and COS will return a list of the objects having the desired parameters. In many cases, the format of the output can be chosen to be more machine-readable or human-readable. COS does not rely on a list, rather it generates the objects requested on-the-fly; for this reason, the output is restricted to 200 objects. Examples of the objects generated are permutations, subsets and combinations, set and integer partitions, irreducible and primitive polynomials over small finite fields and spanning trees of a graph.

2.2.22 Remark The *Cunningham project* produces a set of tables to factor the numbers $b^n \pm 1$ for $b = 2, 3, 5, 6, 7, 10, 11, 12$ for n as large as possible. The current factorization methods employed are the elliptic curve method, the multiple polynomial quadratic sieve and the number field sieve. For more information on factorization methods, see [2080, Chapter 3]. The Cunningham tables appear in published form [415] and as an electronic resource [2890].

2.2.23 Remark The *Great Internet Mersenne Prime Search* (GIMPS) [2084] is a distributed computing effort dedicated to finding and verifying Mersenne primes (that is, primes of the form $2^p - 1$, where p is also a prime). GIMPS uses a combination of trial factoring using the Sieve of Eratosthenes, followed by the Pollard $P - 1$ method and ending with the Lucas-Lehmer primality test. For more information on primality testing, see [724, Chapter 31], for example. GIMPS provides the *Prime95* software, which automates all factoring and distributed computing processes. The (currently) largest known Mersenne prime is $2^{43112609} - 1$ containing 12978189 decimal digits [2084].

The search for Mersenne primes is of particular interest in searching for primitive trinomials of large degrees. Primitive polynomials of low-weight are useful in cryptographic applications and pseudo-random number generation; see Sections 14.9 and 16.2. If p is a Mersenne prime, then any irreducible polynomial of degree p over $\mathbb{F}_2$ is primitive. Since binomials of degree at least 2 cannot be irreducible over $\mathbb{F}_2$, we consider trinomials $x^p + x^r + 1$, for some $0 \leq r \leq p - 1$. Sieving trinomials for reducibles is possible by Swan's Theorem; see Section 3.3. For more details on the algorithms and methods used in the search for primitive trinomials, see [408]. An implementation of polynomial arithmetic over $\mathbb{F}_2$ which was motivated by the GIMPS project, entitled *gf2x*, is necessarily highly optimized and is preferred in some finite field software implementations; see Table 2.2.11 for more details.

2.2.24 Remark The *On-Line Encyclopedia of Integer Sequences*^{TM} (OEIS^{TM}) [2800] is a constantly-updated, searchable database of integer sequences. Examples of famous sequences in the OEIS^{TM} are the Catalan numbers (A000108), prime numbers (A000040), and the Fibonnacci numbers (A000045). Users can search by sequence, "word" (for example, "number of irreducible polynomials" yields sequence A001037) or sequence number. Sequences are sorted lexicographically, so the sequence references may have changed since the date of publication.

2.2.25 Remark In Table 2.2.11, we present a number of software packages which are useful for finite field implementations. We distinguish between packages which are open-source and commercial. We refer the reader to the citation, which provides a current (as of the date of publication) Web URL to the most recent build of the software. We note that this is not an exhaustive list of software packages, simply a useful list of packages used or researched by the author.

Open-source software packages for computations in discrete mathematics

Name	Ver.	Description	
Fast Library for Number Theory (FLINT)	2.3	A C library for performing computations in number theory. Routines include fast algorithms, on par with the other most efficient packages listed, for arbitrary precision integers, rational numbers, modular arithmetic, and p-adic numbers. Most libraries also contain vector, polynomial and matrix methods. Multicore support to come in future versions.	[1426]
Groups, Algorithms, Programming (GAP)	4.4.x	A system for computational discrete algebra emphasizing computational group theory. Can do basic computations with arbitrary integers, rationals, finite fields, p-adic numbers, polynomials, rational functions, and more. Contains a coding theory package, combinatorial functions and prime factorization routines. Provides its own programming language, libraries of algebraic algorithms written in the GAP language as well as data libraries of algebraic objects, particularly various types of groups.	[2796]
gf2x	1.0	Library for efficient arithmetic of single-variable polynomials over $\mathbb{F}_2$. Primarily introduces fast-fourier transform (FFT) for large-degree polynomial multiplication.	[399]
The GNU Multiple Precision Arithmetic Library (GMP)	5.0.x	C/C++ library providing fast arbitrary precision arithmetic on integers, rational numbers, and floating point numbers.	[2797]
Macaulay2	1.4	Software system focusing on algebraic geometry and commutative algebra. Contains core algorithms combined with a high-level interpreted language and debugger to support package creation. Uses elements of *PARI*, *NTL*, and others in its routines.	[1355]
Number Theory Library (NTL)	5.5.x	C++ library providing data structures and routines for arbitrary length integer arithmetic, arbitrary precision floating-point arithmetic, and finite field arithmetic. Also contains lattice basis reduction algorithms and basic linear algebra packages. Interfaces with *gf2x* and *GMP* libraries for additional speedups.	[2633]
PARI/GP	2.5.x	C library designed for fast computations in number theory including integer factorization and elliptic curve compututations. Also contains useful function for use with matrices, polynomials, power series, and others. GP is a scripting language used by the gp interactive shell, which accesses the PARI functions. A subset of the GP language can be compiled as C code, resulting in a substantial speed-up.	[2801]

Open-source software packages for computations in discrete mathematics

Name	Ver.	Description	
Singular	3.1.x	A computer algebra system focusing on polynomial computations. Specializes on commutative and non-commutative algebra, algebraic geometry, and singularity theory. Provides a C-like programming language, extendable using libraries. Its core algorithms handle Gröbner bases, polynomial factorization, resultants, and root finding. Advanced libraries and third-party software provide further functionality.	[792]
SAGE	5.0	Comprehensive Python-based open-source computer algebra package. Natively contains a finite field implementation as well as wrappers for other useful packages including Flint, GAP, NTL, PARI, and Singular. Interpreted but contains the ability to compile, using Cython, as C code for a drastic improvement in speed.	[2709]

Commercial stand-alone packages containing finite field implementations

Name	Ver.	Description	
Magma	2.18-x	Computational algebra system focusing on algebra, algebraic combinatorics, algebraic geometry and number theory. Language built to closely approximate the user's mode of thought and usual notation. Major algorithms are designed to give comparable performance to specialized programs. Also contains a number of large databases of elliptic curves, linear codes, irreducible polynomials over finite fields, graphs, Cunningham factorizations, and others.	[712]
Maple	16	Comprehensive computer algebra suite, contains a full featured programming language to create scripts or full applications. A "smart" document environment allows embedding equations, visualizations, or components in the document. Can take advantage of parallelism, multi-threading and multi-process programing. Finite field arithmetic natively given by the "GF" package.	[2002]
Mathematica	8	Development platform concetrating on integrating computation into workflows. Finite field computations are performed using the "FiniteFields" package and "GF" class.	[3004]
Matlab	R2012a	Programming environment for algorithm development and data analysis. Contains arithmetic over finite fields $\mathbb{F}_{2^n}$ over $\mathbb{F}_2$ for $n \leq 16$ within the "Communications System Toolbox."	[2799]

Table 2.2.11 Software packages useful for finite field implementations.

See Also

§3.2, §3.3, §3.4	For reducibility and irreducibility of low-weight polynomials.
§5.2, §5.3	For normal bases and their complexities.
§11.1	For computational techniques over finite fields.
[1413], [2080]	For patents and standards of elliptic curve cryptography, most of which contain guidelines for finite field implementations.

References Cited: [130, 399, 408, 415, 712, 724, 792, 1263, 1264, 1355, 1413, 1426, 1631, 1777, 1939, 2002, 2015, 2080, 2084, 2180, 2356, 2507, 2573, 2582, 2633, 2709, 2753, 2796, 2797, 2799, 2800, 2801, 2890, 2925, 3004, 3036]

II

Theoretical Properties

3

Irreducible polynomials

3.1 Counting irreducible polynomials

Joseph L. Yucas, Southern Illinois University

3.1.1 Remark In this section we* are concerned with exact formulae for the number of (univariate) irreducible polynomials over finite fields possessing various properties. There is some overlap with Section 3.5 where specifically polynomials with prescribed coefficients are discussed. Formulae and asymptotic expressions for mutivariate polynomials are given in Section 3.6.

3.1.2 Theorem (Theorem 2.1.24). Denote the number of monic irreducible polynomials of degree n over $\mathbb{F}_q$ by $I_q(n)$. Then

$$I_q(n) = \frac{1}{n}\sum_{d|n} \mu(d) q^{n/d}.$$

*The author wishes to thank Stephen Cohen for a number of helpful improvements in this section.

3.1.3 Definition For a positive integer n set

$$D_n = \{r : r | q^n - 1 \text{ but } r \text{ does not divide } q^m - 1 \text{ for } m < n\}.$$

3.1.4 Theorem [3048] We have

$$I_q(n) = \frac{1}{n} \sum_{r \in D_n} \phi(r),$$

where ϕ denotes Euler's function.

3.1.1 Prescribed trace or norm

3.1.5 Definition The *trace* of a monic polynomial f of degree n over $\mathbb{F}_q$ is $-a_1$, where a_1 is the *first coefficient* of f, i.e., the coefficient of x^{n-1} in f. The *norm* of a monic polynomial f of degree n over $\mathbb{F}_q$ is $(-1)^n a_n$, where a_n is the *last coefficient* of f, i.e., the constant term in f. The trace and norm of a monic irreducible polynomial are, respectively, the trace and norm of any of its roots in $\mathbb{F}_{q^n}$ over $\mathbb{F}_q$.

3.1.6 Remark In [2508, 3048] (cited below) and sometimes in the literature, the trace of a polynomial f is taken to be the first coefficient a_1 itself.

3.1.7 Theorem [541, 2508] For a non-zero $a \in \mathbb{F}_q$ the number $I_q(n, a)$ of monic irreducible polynomials of degree n over $\mathbb{F}_q$ with trace a is

$$I_q(n, a) = \frac{1}{qn} \sum_{\substack{d|n \\ (d,q)=1}} \mu(d) q^{n/d}.$$

3.1.8 Theorem [3048] Let q be a power of the prime p. Write $n = p^k m$ with m being p-free (i.e., p does not divide m). The number $I_q(n, 0)$ of monic irreducible polynomials of degree n over $\mathbb{F}_q$ with trace 0 is

$$I_q(n, 0) = \frac{1}{qn} \sum_{d|m} \mu(d) q^{n/d} - \frac{\epsilon}{n} \sum_{d|m} \mu(d) q^{n/dp},$$

where $\epsilon = 1$ if $k > 0$ and $\epsilon = 0$ if $k = 0$.

3.1.9 Definition For $r \in D_n$, write $r = d_r m_r$ where $d_r = \left(r, \frac{q^n - 1}{q-1}\right)$.

3.1.10 Theorem [3048]

1. Let $r \in D_n$ and suppose $a \in \mathbb{F}_q$ has order m_r. Further, let $I_q(n, r, a)$ denote the number of monic irreducible polynomials over $\mathbb{F}_q$ of degree n, order r and (non-zero) norm a. Then
$$I_q(n, r, a) = \frac{\phi(r)}{n \phi(m_r)}.$$

2. Suppose $a \in \mathbb{F}_q^*$ has order m and let $I_q(n, a)$ denote the number of monic irreducible polynomials over $\mathbb{F}_q$ of degree n with (non-zero) norm a. Then
$$I_q(n, a) = \frac{1}{n \phi(m)} \sum_{\substack{r \in D_n \\ m_r = m}} \phi(r).$$

3.1.11 Remark In [541], Carlitz obtained formulae for the number of monic irreducible polynomials over $\mathbb{F}_q$, q odd, with prescribed trace and whose norm is a (non-zero) square or a non-square, respectively. These involve the quadratic character λ on $\mathbb{F}_q$; thus for $0 \neq b \in \mathbb{F}_q$, $\lambda(b) = 1$ or -1 according as b is a square or non-square in $\mathbb{F}_q$, respectively. (See also Remark 3.5.49.)

3.1.12 Theorem [541, 1783] Let $q = p^m$ be an odd prime power. For $a \in \mathbb{F}_q$ denote by $I_q(n,a,h)$, $h = 1, -1$, respectively, the number of monic irreducible polynomials of degree n over $\mathbb{F}_q$ with trace a whose norm is a (non-zero) square or a non-square. Then,

1. if $a = 0$
$$I_q(n,0,h) = \begin{cases} \frac{1}{2p}(q^{p-1} - q) & \text{for } n = p, \\ \frac{1}{2n}(q^{n-1} - 1) & \text{for } n \neq p; \end{cases}$$
2. if $a \neq 0$
$$I_q(n,a,h) = \begin{cases} \frac{1}{2p}(q^{p-1} + S) & \text{for } n = p, \\ \frac{1}{2n}(q^{n-1} + S - (-1)^h \lambda(na) - 1) & \text{for } n \neq p, \end{cases}$$

where $S = (-1)^h q^{\frac{n-1}{2}} \lambda((-1)^{\frac{n-1}{2}} a)$, and λ is the quadratic character in $\mathbb{F}_q$.

3.1.2 Prescribed coefficients over the binary field

3.1.13 Remark Subsection 3.5.4 contains various formulae for the numbers of irreducible polynomials with some prescribed coefficients in a general finite field $\mathbb{F}_q$. We list here a specialized result over the binary field $\mathbb{F}_2$ not included there.

3.1.14 Definition For $\beta \in \mathbb{F}_{2^n}$ let
$$T_j(\beta) = \sum_{0 \le i_1 < i_2 < \cdots < i_j \le n-1} \beta^{2^{i_1}} \beta^{2^{i_2}} \ldots \beta^{2^{i_j}};$$
T_j maps $\mathbb{F}_{2^n}$ to $\mathbb{F}_2$ and T_1 is the usual trace function. For $n = 2$ and $j = 3$, we define $T_3(\beta) = 0$ for all $\beta \in \mathbb{F}_4$. For an integer r with $1 \le r \le n$, define $F(n, t_1, t_2, \ldots, t_r)$ to be the number of elements $\beta \in \mathbb{F}_{2^n}$ with $T_j(\beta) = t_j$ for $j = 1, \ldots, r$ and let $(I_2(n, t_1, t_2, \ldots, t_r) =)\ I(n, t_1, t_2, \ldots, t_r)$ be the number of monic irreducible polynomials $f(x)$ over $\mathbb{F}_2$ of degree n with coefficient of $x^{n-j} = t_j$ for $j = 1, \ldots, r$.

3.1.15 Theorem [3049] We have
$$nI(n,0,0,0) = \sum_{d|n, d \text{ odd}} \mu(d) F(n/d, 0, 0, 0) - \sum_{d|n, d \text{ odd}, n/d \text{ even}} \mu(d) 2^{\frac{n}{2d}-1};$$
$$nI(n,0,0,1) = \sum_{d|n, d \text{ odd}} \mu(d) F(n/d, 0, 0, 1);$$
$$nI(n,0,1,0) = \sum_{d|n, d \text{ odd}} \mu(d) F(n/d, 0, 1, 0) - \sum_{d|n, d \text{ odd}, n/d \text{ even}} \mu(d) 2^{\frac{n}{2d}-1};$$
$$nI(n,0,1,1) = \sum_{d|n, d \text{ odd}} \mu(d) F(n/d, 0, 1, 1);$$
$$nI(n,1,0,0) = \sum_{d|n, d \equiv 1} \mu(d) F(n/d, 1, 0, 0) + \sum_{d|n, d \equiv 3} \mu(d) F(n/d, 1, 1, 1);$$
$$nI(n,1,0,1) = \sum_{d|n, d \equiv 1} \mu(d) F(n/d, 1, 0, 1) + \sum_{d|n, d \equiv 3} \mu(d) F(n/d, 1, 1, 0);$$

$$nI(n,1,1,0) = \sum_{d|n, d\equiv 1} \mu(d)F(n/d,1,1,0) + \sum_{d|n, d\equiv 3} \mu(d)F(n/d,1,0,1);$$
$$nI(n,1,1,1) = \sum_{d|n, d\equiv 1} \mu(d)F(n/d,1,1,1) + \sum_{d|n, d\equiv 3} \mu(d)F(n/d,1,0,0).$$

3.1.16 Remark Explicit formulae for $I(n,t_1,t_2,t_3)$ (the number of irreducible polynomials over $\mathbb{F}_2$ whose first three coefficients are prescribed) can be recovered from Theorem 3.1.15 via the next theorem.

3.1.17 Theorem [1076, 3049] For $n \geq 3$, $F(n,t_1,t_2,t_3) = 2^{n-3} + G(n,t_1,t_2,t_3)$, where the values of $G(n,t_1,t_2,t_3)$ are displayed in the following tables.

1. Case $n = 2m+1$ (in the first column m is calculated modulo 12):

m	000	001	010	011	100	101	110	111
0	$-3\cdot 2^{m-2}$	$3\cdot 2^{m-2}$	2^{m-2}	2^{m-2}	0	0	0	0
1 or 5	2^{m-2}	-2^{m-2}	2^{m-2}	-2^{m-2}	-2^{m-2}	-2^{m-2}	$3\cdot 2^{m-2}$	-2^{m-2}
2 or 10	0	2^{m-1}	0	-2^{m-1}	2^{m-1}	-2^{m-1}	-2^{m-1}	2^{m-1}
3	2^{m-2}	-2^{m-2}	2^{m-2}	-2^{m-2}	2^{m-1}	0	2^{m-1}	-2^{m}
4 or 8	-2^{m-1}	0	-2^{m-1}	2^{m}	0	0	0	0
6	$3\cdot 2^{m-2}$	-2^{m-2}	$-3\cdot 2^{m-2}$	2^{m-2}	2^{m-1}	-2^{m-1}	-2^{m-1}	2^{m-1}
7 or 11	2^{m-2}	-2^{m-2}	2^{m-2}	-2^{m-2}	-2^{m-2}	$3\cdot 2^{m-2}$	-2^{m-2}	-2^{m-2}
9	2^{m-2}	-2^{m-2}	2^{m-2}	-2^{m-2}	-2^{m}	2^{m-1}	0	2^{m-1}

2. Case $n = 2m$ with 3 not dividing n (in the first column m is calculated modulo 4):

m	000	001	010	011	100	101	110	111
0	$3\cdot 2^{m-2}$	-2^{m-2}	-2^{m-2}	-2^{m-2}	$3\cdot 2^{m-2}$	-2^{m-2}	-2^{m-2}	-2^{m-2}
1	$-3\cdot 2^{m-2}$	2^{m-2}	2^{m-2}	2^{m-2}	2^{m-2}	2^{m-2}	2^{m-2}	$-3\cdot 2^{m-2}$
2	$-3\cdot 2^{m-2}$	2^{m-2}	2^{m-2}	2^{m-2}	$-3\cdot 2^{m-2}$	2^{m-2}	2^{m-2}	2^{m-2}
3	$3\cdot 2^{m-2}$	-2^{m-2}	-2^{m-2}	-2^{m-2}	-2^{m-2}	-2^{m-2}	-2^{m-2}	$3\cdot 2^{m-2}$

3. Case $n = 2m$ with 3 dividing n (in the first column m is calculated modulo 4):

m	000	001	010	011	100	101	110	111
0	0	2^{m-1}	2^{m-1}	-2^{m}	0	2^{m-1}	-2^{m}	2^{m-1}
1	0	-2^{m-1}	-2^{m-1}	2^{m}	-2^{m-1}	2^{m}	-2^{m-1}	0
2	0	-2^{m-1}	-2^{m-1}	2^{m}	0	-2^{m-1}	2^{m}	-2^{m-1}
3	0	2^{m-1}	2^{m-1}	-2^{m}	2^{m-1}	-2^{m}	-2^{m-1}	0

3.1.18 Remark Formulae for $I(n,t_1,t_2)$ and $F(n,t_1,t_2)$ can be obtained by adding appropriate terms from above [565]. For the binary field these will agree with the earlier general expressions of Kuz'min (Theorems 3.5.43 and 3.5.45) from [1815].

3.1.3 Self-reciprocal polynomials

3.1.19 Remark Self-reciprocal polynomials were defined in Remark 2.1.48. For results on these polynomials see [3050]. Any monic self-reciprocal polynomial R of (even) degree $2n$ has the form $x^n f\left(\frac{x^2+1}{x}\right)$, where f is a monic polynomial of degree n in $\mathbb{F}_q[x]$. (We observe that, if R is irreducible, then necessarily f is irreducible.) Let $SRMI_q(2n)$ denote number of monic irreducible self-reciprocal polynomials of degree $2n$ over $\mathbb{F}_q$.

3.1.20 Theorem [547, 666, 2091, 2200] If q is odd, then

$$SRMI_q(2n) = \frac{1}{2n} \sum_{\substack{d|n \\ d \text{ odd}}} \mu(d)(q^{n/d} - 1),$$

and, if q is even, then

$$SRMI_q(2n) = \frac{1}{2n} \sum_{\substack{d|n \\ d \text{ odd}}} \mu(d)q^{n/d}.$$

3.1.21 Remark The notion of self-reciprocal polynomial has recently been generalized and Theorem 3.1.20 extended correspondingly [48].

3.1.22 Definition Let $g(x) = a_1x^2 + b_1x + c_1$ and $h(x) = a_2x^2 + b_2x + c_2$ be relatively prime polynomials over $\mathbb{F}_q$ with $\max(\deg f, \deg g) = 2$ (so that a_1 and a_2 are not both zero). Let $I_q(2n, g, h)$ denote the number of irreducible polynomials (not necessarily monic) of degree $2n$ that can be expressed in the form $h(x)^n f\left(\frac{g(x)}{h(x)}\right)$, where f is a monic polynomial (necessarily irreducible) of degree n.

3.1.23 Theorem [48] Suppose $n > 1$ and g, h are polynomials over $\mathbb{F}_q$ as in Definition 3.1.22. Then

$$I_q(2n, g, h) = \begin{cases} 0 & \text{if } q \text{ is even and } b_1 = b_2 = 0, \\ \frac{1}{2n}(q^n - 1) & \text{if } q \text{ is odd and } n = 2^m, \\ \frac{1}{2n} \sum_{\substack{d|n \\ d \text{ odd}}} \mu(d)q^{n/d} & \text{otherwise.} \end{cases}$$

3.1.24 Remark Theorem 3.1.20 is recovered from Theorem 3.1.23 on setting $g(x) = x^2+1$, $h(x) = x$ (using $\sum_{d|n} \mu(d) = 0$ whenever $n > 1$).

3.1.4 Compositions of powers

3.1.25 Definition The *radical* of an integer m (> 1) (denoted here by m^*) is the product of the distinct primes dividing m.

3.1.26 Definition For $t > 1$, a *t-polynomial* T over $\mathbb{F}_q$ of degree tn is one that has the form $T(x) = f(x^t)$ for some monic polynomial f of degree n.

3.1.27 Remark If T is irreducible, then f is also irreducible. Further (see Theorem 3.2.5), (i) $t^*|(q^n - 1)$, and (ii) if $4|t$, then $4|(q^n - 1)$. In fact, if (i) holds then n can be expressed as $n = klm$, where k is the order of q (mod t^*), $l^*|t$, and m and t are relatively prime [666].

3.1.28 Theorem [666] Suppose $t > 1$ and that (i) and (ii) of Remark 3.1.27 hold with $n = klm$. Let $TMI_q(tn)$ be the number of monic irreducible t-polynomials of degree tn. Then

$$TMI_q(tn) = \begin{cases} \frac{\phi(t)}{tn}(q^n - 1) & \text{for } m = 1, \\ \frac{m\phi(t)}{tn} I_{q^{n/m}}(m) & \text{for } m > 1. \end{cases}$$

3.1.29 Remark For $t > 1$, a *t-reciprocal polynomial* T over $\mathbb{F}_q$ of degree $2tn$ is one that is both a t-polynomial and a self-reciprocal polynomial. Thus it has the form $T(x) = x^{tn} f\left(\frac{x^{2t}+1}{x^t}\right)$, where f is a monic polynomial of degree n. If T is irreducible then f is irreducible. Moreover, from [666] we have (i) t is odd and (ii) $t^*|(q^n + 1)$. Also $2n = klm$, where $l^*|2t$ and m and $2t$ are relatively prime (so that $m|n$).

3.1.30 Theorem [666, 2200] Suppose $t > 1$ and that (i) and (ii) of Remark 3.1.29 hold with $2n = klm$. Let $TSRMI_q(2tn)$ be the number of monic irreducible t-reciprocal polynomials of degree $2tn$. Then

$$TSRMI_q(2tn) = \begin{cases} \frac{\phi(t)}{2tn}(q^n + 1) & \text{for } m = 1, \\ \frac{m\phi(t)}{2tn} I_{q^{n/m}}(m) & \text{for } m > 1. \end{cases}$$

3.1.5 Translation invariant polynomials

3.1.31 Definition A polynomial f over $\mathbb{F}_q$ is *translation invariant* if $f(x+a) = f(x)$ for all $a \in \mathbb{F}_q$.

3.1.32 Theorem [2200] Let $TIMI_q(qn)$ denote the number of translation invariant monic irreducible polynomials of degree qn over $\mathbb{F}_q$. Then

$$TIMI_q(qn) = \frac{q-1}{qn} \sum_{\substack{d|n \\ (q,d)=1}} \mu(d) q^{n/d}.$$

3.1.6 Normal replicators

3.1.33 Remark Many of the above formulae and other similar ones can be obtained using the general counting technique [2200] which follows.

3.1.34 Definition A rational function $r = f/g \in \mathbb{F}_q(x)$ is a *replicator* over $\mathbb{F}_q$ if for every $n \geq 1$, $x^{q^n-1} - 1$ divides $f(x)^{q^n} - f(x)g(x)^{q^n-1}$. In this case, we write

$$f(x)^{q^n} - f(x)g(x)^{q^n-1} = (x^{q^n-1} - 1)\widehat{r}(n, x)$$

for some polynomial $\widehat{r}(n, x) \in \mathbb{F}_q[x]$. The polynomial $\widehat{r}(n, x)$ is the *n-th order transform* of r.

3.1.35 Definition Let k be a positive integer. A rational function $r = f/g \in \mathbb{F}_q(x)$ is *k-normal* if for every $n \geq 1$ and for each $\lambda \in \mathbb{F}_{q^n}$, the degrees of the factors of the associated polynomial $f - \lambda g$ divide k, and for some λ, at least one factor has degree equal to k.

3.1.36 Remark For a polynomial $f(x) = x^m + a_{m-1}x^{m-1} + \cdots + a_1 x + a_0$ over $\mathbb{F}_{q^n}$ we define the following sequence of polynomials:

$$f^{(j)}(x) = x^m + a_{m-1}^{q^j} x^{m-1} + \cdots + a_1^{q^j} x + a_0^{q^j}.$$

We observe that $f^{(s)} = f$ where s is the least common multiple of the degrees of the minimal polynomials of the coefficients of f.

3.1.37 Definition Define the *spin* $S_f(x)$ of the polynomial $f(x)$ by

$$S_f(x) = \prod_{j=0}^{s-1} f^{(j)}(x).$$

3.1.38 Definition Let $r = f/g$ be a k-normal replicator over $\mathbb{F}_q$ and suppose h is an irreducible factor of $f - \lambda g$ for $\lambda \in \mathbb{F}_{q^n}$ of degree d. Then, S_h is *hard* for r if the degree of S_h does not divide n.

3.1.39 Remark Write

$$f^{q^n}(x) - f(x)g^{q^n-1}(x) = (x^{q^n-1} - 1)G(n,x)\bar{r}(n,x),$$

where $\bar{r}(n,x)$ is the factor of $f^{q^n} - fg^{q^n-1}$ of largest degree which is square-free and satisfies $(G(n,x), \bar{r}(n,x)) = 1$. Further let $g(n) = \deg(G(n,x))$. Let $HMI_q(n, r(x))$ denote the number of monic irreducible polynomials of degree n which are hard for r.

3.1.40 Theorem [2200] We have

$$\begin{aligned} HMI_q(kn, r(x)) &= \frac{1}{kn} \sum_{\substack{d \mid n \\ d \nmid (n/k)}} \mu(n/d)[(m-1)q^n - g(d) + 1] \\ &= \frac{1}{kn} \sum_{\substack{d \mid n \\ k \nmid d}} \mu(d)[(m-1)q^{n/d} - g(n/d) + 1]. \end{aligned}$$

See Also

§3.5 For further formulae and estimates for irreducible polynomials with prescribed coefficients.
§3.6 For formulae and asymptotic expressions for irreducible multivariate polynomials.

References Cited: [48, 541, 547, 565, 666, 1076, 2091, 2200, 2508, 3048, 3049, 3050]

3.2 Construction of irreducibles

Melsik Kyuregyan, Armenian National Academy of Sciences

3.2.1 Construction by composition

3.2.1 Remark Known constructions of irreducible polynomials depend on the composition of an initial irreducible polynomial with a further polynomial or rational function. Often this process can be iterated or continued recursively to produce an infinite sequence of irreducible polynomials of increasing degrees.

3.2.2 Theorem [505, 666] Let $f,\, g \in \mathbb{F}_q[x]$ be relatively prime polynomials and let $P \in \mathbb{F}_q[x]$ be an irreducible polynomial of degree n. Then the composition

$$F(x) = (g(x))^n \, P\left(f(x)/g(x)\right)$$

is irreducible over $\mathbb{F}_q$ if and only if $f - \alpha g$ is irreducible over $\mathbb{F}_{q^n}$ for any zero $\alpha \in \mathbb{F}_{q^n}$ of P.

3.2.3 Remark Theorem 3.2.2 was employed by several authors [589, 678, 685, 1172, 1820, 1819, 1821, 1822, 1823, 1824, 1939, 2091] to give iterative constructions of irreducible polynomials over finite fields. A further extension of the theorem is produced in [1825], which is also instrumental in the construction of irreducible polynomials of relatively higher degree from given ones.

3.2.4 Theorem [2077] Let $P \in \mathbb{F}_q[x]$ be irreducible of degree n. Then for any $a, b, c, d \in \mathbb{F}_q$ such that $ad - bc \neq 0$,

$$F(x) = (cx + d)^n \, P\left(\frac{ax + b}{cx + d}\right)$$

is also irreducible over $\mathbb{F}_q$.

3.2.5 Theorem [2077] Let t be a positive integer and $P \in \mathbb{F}_q[x]$ be irreducible of degree n and exponent e (equal to the order of any root of P). Then $P(x^t)$ is irreducible over $\mathbb{F}_q$ if and only if

1. $(t, (q^n - 1)/e) = 1$,
2. each prime factor of t divides e, and
3. if $4|t$ then $4|(q^n - 1)$.

3.2.6 Theorem [1939] Let $f_1, f_2, \dots, f_N$ be all the distinct monic irreducible polynomials in $\mathbb{F}_q[x]$ of degree m and order e, and let $t \geq 2$ be an integer whose prime factors divide e but not $(q^m - 1)/e$. Assume also that $q^m \equiv 1 \pmod 4$ if $t \equiv 0 \pmod 4$. Then $f_1(x^t), f_2(x^t), \dots, f_N(x^t)$ are all the distinct monic irreducible polynomials in $\mathbb{F}_q[x]$ of degree mt and order et.

3.2.7 Remark Agou [36] has established a criterion for $f(g(x))$ to be irreducible over $\mathbb{F}_q$, where $f, g \in \mathbb{F}_q[x]$ are monic and f is irreducible over $\mathbb{F}_q$. This criterion was used in Agou [36, 38, 39, 40] to characterize irreducible polynomials of special types such as $f\left(x^{p^r} - ax\right)$, $f(x^p - x - b)$, and others. Such irreducible compositions of polynomials are also studied in Cohen [666, 671], Long [1954, 1955], and Ore [2324].

Irreducibility criteria for compositions of polynomials of the form $f(x^t)$ have been established by Agou [34, 35, 36], Butler [469], Cohen [671], Pellet [2376], Petterson [2392], and Serret [2597, 2600]. Berlekamp [231, Chapter 6] and Varshamov and Ananiashvilii [2860] discussed the relationship between the orders of $f(x^t)$ and that of $f(x)$.

3.2.8 Theorem [2077] Let $q = 2^m$ and let $P(x) = \sum_{i=0}^{n} c_i x^i \in \mathbb{F}_q[x]$ be irreducible over $\mathbb{F}_q$ of degree n and $P^*(x) = x^n P\left(\frac{1}{x}\right)$. Denote $\mathbb{F}_{2^m} = F$ and $\mathbb{F}_2 = K$. Then

1. $x^n P\left(x + x^{-1}\right)$ is irreducible over F if and only if $Tr_{F/K}(c_1/c_0) = 1$;
2. $x^n P^*\left(x + x^{-1}\right)$ is irreducible over F if and only if $Tr_{F/K}(c_{n-1}/c_n) = 1$.

3.2.9 Remark Part 1 of Theorem 3.2.8 was obtained by Meyn [2091] and by Kyuregyan [1820] in the present general form; for the case $q = 2$ it was earlier obtained by Varshamov and Garakov [2861].

3.2.10 Theorem [2091] Let q be an odd prime power. If P is an irreducible polynomial of degree n over $\mathbb{F}_q$, then $x^n P\left(x + x^{-1}\right)$ is irreducible over $\mathbb{F}_q$ if and only if the element $P(2)P(-2)$ is a non-square in $\mathbb{F}_q$.

3.2.11 Theorem [1822] Let q be odd, $P(x) \neq x$ be an irreducible polynomial of degree $n \geq 1$ over $\mathbb{F}_q$, and $ax^2 + bx + c$ and $dx^2 + rx + h$ be relatively prime polynomials in $\mathbb{F}_q[x]$ with a or d being non-zero and $r^2 \neq 4dh$. Suppose

$$(ah)^2 + (cd)^2 + acr^2 + b^2dh - bcdr - abhr - 2acdh = \delta^2$$

for some $\delta \neq 0$ from $\mathbb{F}_q$. Then the polynomial

$$F(x) = \left(dx^2 + rx + h\right)^n P\left(\frac{ax^2 + bx + c}{dx^2 + rx + h}\right)$$

is irreducible over $\mathbb{F}_q$ if and only if the element

$$\left(r^2 - 4dh\right)^n P\left(\frac{br - 2\left(cd + ah - \delta\right)}{r^2 - 4hd}\right) P\left(\frac{br - 2\left(cd + ah + \delta\right)}{r^2 - 4hd}\right)$$

is a non-square in $\mathbb{F}_q$.

3.2.12 Remark The case $a = c = r = 1$ and $b = d = h = 0$ of Theorem 3.2.11 reduces to Theorem 3.2.10.

3.2.13 Remark We briefly describe some constructive aspects of irreducibility of certain types of polynomials, particularly binomials and trinomials.

3.2.14 Definition A *binomial* is a polynomial with two nonzero terms, one of them being the constant term.

3.2.15 Remark Irreducible binomials can be characterized explicitly. For this purpose it suffices to consider nonlinear, monic binomials.

3.2.16 Theorem [1939] Let $t \geq 2$ be an integer and $a \in \mathbb{F}_q^*$. Then the binomial $x^t - a$ is irreducible in $\mathbb{F}_q[x]$ if and only if the following two conditions are satisfied:

1. each prime factor of t divides the order e of a in $\mathbb{F}_q^*$, but not $(q-1)/e$;
2. $q \equiv 1 \pmod{4}$ if $t \equiv 0 \pmod{4}$.

3.2.17 Remark Theorem 3.2.16 was essentially shown by Serret [2600] for finite prime fields. Further characterizations of irreducible binomials can be found in Albert [70, Chapter 5], Cappeli [505, 506, 507], Dickson [850, Part I, Chapter 3]. Lowe and Zelinsky [1962], Rédei [2443, Chapter 11], and Schwarz [2568].

3.2.18 Theorem Let a be a nonzero element in an extension field of $\mathbb{F}_q$, $q = 2^A u - 1$, with $A \geq 2$ and u odd. Suppose e is the order of the subgroup of $\mathbb{F}_q^*$ generated by a and the condition of Part 1 in Theorem 3.2.16 is satisfied for some natural number t divisible by $2^A = 2B$. Then the binomial $x^t - a$ factors as a product of B monic irreducible polynomials in $\mathbb{F}_q[x]$ of degrees $v = t/B$, that is in $\mathbb{F}_q[x]$ we have the canonical factorization

$$x^t - a = \prod_{j=1}^{B} \left(x^v - bc_j x^{v/2} - b^2 \right),$$

where $b = a^r$, $2Br = e/2 + 1 \,(\mathrm{mod}\ (q-1))$, and the elements $c_1, \ldots, c_B$ are the roots of the polynomial

$$F(x) = \sum_{i=0}^{B/2} \frac{(B-i-1)!B}{i!(B-2i)!} x^{B-2i} \in \mathbb{F}_q[x],$$

and all c_j, $1 \le j \le B$ are in $\mathbb{F}_q$.

3.2.19 Remark The factorization in Theorem 3.2.18 is due to Serret [2600], see also Albert [70, Chapter 5] and Dickson [850, Part I, Chapter 3]. Shiva and Allard [2616] discuss a method for factoring $x^{2^k-1} + 1$ over $\mathbb{F}_2$. The factorization of $x^{q-1} - a$ over $\mathbb{F}_q$ is considered in Dickson [849], see also Agou [37]. Schwarz [2569] has a formula for the number of monic irreducible factors of fixed degree for a given binomial and Rédei [2442] gives a short proof of it; see also Agou [34], Butler [469], and Schwarz [2568]. Gay and Vélez [1261] prove a formula for the degree of the splitting field of an irreducible binomial over an arbitrary field that was shown by Darbi [769] for fields of characteristic 0. Agou [33] studied the factorization of an irreducible binomial over $\mathbb{F}_q$ in an extension field of $\mathbb{F}_q$. Beard and West [213] and McEliece [2046] tabulate factorizations of the binomials $x^n - 1$. The factorization of more general polynomials $g(x)^t - a$ over finite prime fields is considered in Ore [2323] and Petterson [2392]. Applications of factorizations of binomials are contained in Agou [34], Berlekamp [229], and Vaughan [2863].

3.2.20 Definition A *trinomial* is a polynomial with three nonzero terms, one of them being the constant term.

3.2.21 Remark The trinomials that we consider are also affine polynomials.

3.2.22 Theorem [1939] Let $a \in \mathbb{F}_q$ and let p be the characteristic of $\mathbb{F}_q$. Then the trinomial $x^p - x - a$ is irreducible in $\mathbb{F}_q[x]$ if and only if it has no root in $\mathbb{F}_q$.

3.2.23 Corollary [1939] With the notation of Theorem 3.2.22 the trinomial $x^p - x - a$ is irreducible in $\mathbb{F}_q[x]$ if and only if the absolute trace $\mathrm{Tr}_{F/K}(a) \neq 0$, where $F = \mathbb{F}_q$ and $K = \mathbb{F}_p$.

3.2.24 Remark Theorem 3.2.22 and Corollary 3.2.23 were first shown by Pellet [2378]. The fact that $x^p - x - a$ is irreducible over $\mathbb{F}_p$ if $a \in \mathbb{F}_p^*$ was already established by Serret [2597, 2600]. See also Dickson [841], [850, Part I, Chapter 3] and Albert [70, Chapter 5] for these results.

3.2.25 Remark Since for $b \in \mathbb{F}_p^*$ the polynomial $f(x)$ is irreducible over $\mathbb{F}_q$ if and only if $f(bx)$ is irreducible over $\mathbb{F}_q$, the criteria above hold also for trinomials of the form $b^p x^p - bx - a$.

3.2.26 Remark If we consider more general trinomials of the above type for which the degree is a higher power of the characteristic, then these criteria need not be valid any longer. In fact, the following decomposition formula can be established.

3.2.27 Theorem [1939] For $x^q - x - a$ with a being an element of the subfield $K = \mathbb{F}_r$ of $F = \mathbb{F}_q$ we have the decomposition

$$x^q - x - a = \prod_{j=1}^{q/r} (x^r - x - \beta_j)$$

in $\mathbb{F}_q[x]$, where the β_j are the distinct elements of $\mathbb{F}_q$ with $\mathrm{Tr}_{F/K}(\beta_j) = a$.

3.2.28 Remark Theorem 3.2.27 is due to Dickson [841], [850, Part I, Chapter 3], but in the special case $a = 0$ it was already noted by Mathieu [2022].

3.2.29 Theorem [2857, 2858] Let $P(x) = x^n + a_{n-1}x^{n-1} + \cdots + a_1x + a_0$ be an irreducible polynomial over the finite field $\mathbb{F}_q$ of characteristic p and let $b \in \mathbb{F}_q$. Then the polynomial $P(x^p - x - b)$ is irreducible over $\mathbb{F}_q$ if and only if the absolute trace $\mathrm{Tr}_{F/K}(nb - a_{n-1}) \neq 0$, where $F = \mathbb{F}_q$ and $K = \mathbb{F}_p$.

3.2.30 Remark Theorem 3.2.29 was shown in this general form by Varshamov [2857, 2858]; see also Agou [36]. The case $b = 0$ received considerable attention much earlier. The corresponding result for $b = 0$ and finite prime fields was stated by Pellet [2378] and proved in Pellet [2377]. Polynomials $f(x^p - x)$ over $\mathbb{F}_p$ with $\deg(f)$ a power of p were treated by Serret [2598, 2599]. The case $b = 0$ for arbitrary finite fields was considered in Dickson [850, Part I, Chapter 3] and Albert [70, Chapter 5]. More general types of polynomials such as $f(x^{p^r} - ax)$, $f(x^{p^{2r}} - ax^{p^r} - bx)$ and others have also been studied, see Agou [36, 37, 38, 39, 40, 41, 42], Cohen [671], Long [1953, 1954, 1955, 1956], Long and Vaughan [1957, 1958], and Ore [2324].

3.2.31 Theorem Let $f(x) = x^r - ax - b \in \mathbb{F}_q[x]$, where $r > 2$ is a power of the characteristic of $\mathbb{F}_q$, and suppose that the binomial $x^{r-1} - a$ is irreducible over $\mathbb{F}_q$. Then $f(x)$ is the product of a linear polynomial an an irreducible polynomial over $\mathbb{F}_q$ of degree $r - 1$.

3.2.32 Remark Theorem 3.2.31 generalizes results of Dickson [849] and Albert [70, Chapter 5]. See Schwarz [2571] for further results in this direction.

3.2.2 Recursive constructions

3.2.33 Theorem [1823] Let $q = p^s$ be a prime power and let $f(x) = \sum_{u=0}^{n} c_u x^u$ be a monic irreducible polynomial over $\mathbb{F}_q$. Denote $\mathbb{F}_q = F$ and $\mathbb{F}_p = K$. Suppose that there exists an element $\delta_0 \in \mathbb{F}_q$ such that $f(\delta_0) = a$ with $a \in \mathbb{F}_p^*$, and

$$\mathrm{Tr}_{F/K}(n\delta_0 + c_{n-1}) \cdot \mathrm{Tr}_{F/K}(f'(\delta_0)) \neq 0,$$

where f' is the formal derivative of f. Let $g_0(x) = x^p - x + \delta_0$ and $g_k(x) = x^p - x + \delta_k$, where $\delta_k \in \mathbb{F}_p^*$, $k \geq 1$. Define $f_0(x) = f(g_0(x))$, and $f_k(x) = f_{k-1}^*(g_k(x))$ for $k \geq 1$, where $f_{k-1}^*(x)$ is the monic reciprocal polynomial of $f_{k-1}(x)$, i.e., $f_{k-1}^*(x) = \frac{1}{f_{k-1}(0)} x^{n_{k-1}} f_{k-1}\left(\frac{1}{x}\right)$. Then for each $k \geq 0$ the polynomial $f_k(x)$ is irreducible over $\mathbb{F}_q$ of degree $n_k = n \cdot p^{k+1}$.

3.2.34 Remark The case $s = 1$ and the sequence $(\delta_k)_{k\geq 0}$ is constant, i.e., $\delta_k = \delta \in \mathbb{F}_p^*$ of Theorem 3.2.33, has been studied by Varshamov in [2859], where no proof is given. For a proof see [1172, 2077].

3.2.35 Theorem [1820, 1821, 1823] Let $\delta \in \mathbb{F}_{2^s}^*$ and $f_1(x) = \sum_{u=0}^{n} c_u x^u$ be a monic irreducible polynomial over $\mathbb{F}_{2^s}$ whose coefficients satisfy the conditions

$$Tr_{F/K}\left(\frac{c_1\delta}{c_0}\right) = 1 \text{ and } Tr_{F/K}\left(\frac{c_{n-1}}{\delta}\right) = 1,$$

where $\mathbb{F}_{2^s} = F$ and $\mathbb{F}_2 = K$. Then all the terms in the sequence $(f_k(x))_{k\geq 1}$ defined as

$$f_{k+1}(x) = x^{2^{k-1}n} f_k\left(x + \delta^2 x^{-1}\right),\ k \geq 1,$$

are irreducible polynomials over $\mathbb{F}_{2^s}$.

3.2.36 Theorem [1820, 1821, 2077] Let $f(x) = \sum_{i=0}^{n} c_i x^i$ be irreducible over $\mathbb{F}_{2^m}$ of degree n. Denote $\mathbb{F}_{2^m} = F$ and $\mathbb{F}_2 = K$. Suppose that $\mathrm{Tr}_{F/K}(c_1/c_0) \neq 0$ and $\mathrm{Tr}_{F/K}(c_{n-1}/c_n) \neq 0$. Define the polynomials $a_k(x)$ and $b_k(x)$ recursively by $a_0(x) = x$, $b_0(x) = 1$ and for $k \geq 1$

$$\begin{aligned} a_{k+1}(x) &= a_k(x)b_k(x), \\ b_{k+1}(x) &= a_k^2(x) + b_k^2(x). \end{aligned}$$

Then

$$f_k(x) = (b_k(x))^n \, f\,(a_k(x)/b_k(x))$$

is irreducible over $\mathbb{F}_{2^m}$ of degree $n2^k$ for all $k \geq 0$.

3.2.37 Remark [2077]

1. For the case $q = 2$ in Theorem 3.2.36 the trace function is the identity map on $\mathbb{F}_q$.
2. Let $f(x) = \sum_{i=0}^{n} c_i x^i$ be a monic irreducible polynomial over $\mathbb{F}_2$ of degree n with $c_1 c_{n-1} \neq 0$. Then $f_k(x) = \sum_{i=0}^{n} c_i a_k^i(x) b_k^{n-i}(x)$ is irreducible over $\mathbb{F}_2$ of degree $n2^k$ for all $k \geq 0$.
3. The irreducibility of f_k over $\mathbb{F}_2$ has been studied by several authors, including Varshamov [2859], Wiedemann [2977], Meyn [2091], Gao [1172], Menezes et al. [2077].
4. The irreducibility of f_k over $\mathbb{F}_{2^s}$ has been studied by Kyuregyan [1820, 1819, 1821] and Menezes et al. [1172, 2077].

3.2.38 Theorem [678, 2077] Let f be a monic irreducible polynomial of degree $n \geq 1$ over $\mathbb{F}_q$, q odd, where n is even if $q \equiv 3 \pmod 4$. Suppose that $f(1)f(-1)$ is a non-square in $\mathbb{F}_q$. Define

$$\begin{aligned} f_0(x) &= f(x), \\ f_k(x) &= (2x)^{t_{k-1}} f_{k-1}\left(\left(x + x^{-1}\right)/2\right), \ k \geq 1, \end{aligned}$$

where $t_k = n2^k$ denotes the degree of $f_k(x)$. Then $f_k(x)$ is an irreducible polynomial over $\mathbb{F}_q$ of degree $n2^k$ for every $k \geq 1$.

3.2.39 Remark Further constructions similar to the one from Theorem 3.2.38 can be found in [1822, 1824].

3.2.40 Theorem [1822] Let $P(x) \neq x$ be an irreducible polynomial of degree $n \geq 1$ over $\mathbb{F}_q$, where n is even if $q \equiv 3 \pmod 4$, with $r, h, \delta \in \mathbb{F}_q$ and $r \neq 0$, $\delta \neq 0$. Suppose that $P\left(\frac{2\delta - rh}{r^2}\right) P\left(-\frac{2\delta + rh}{r^2}\right)$ is a non-square in $\mathbb{F}_q$. Define

$$\begin{aligned} F_0(x) &= P(x), \\ F_k(x) &= \left(2x + \frac{2h}{r}\right)^{t_{k-1}} F_{k-1}\left(\left(x^2 + \frac{4\delta^2 - (hr)^2}{r^4}\right) \Big/ \left(2x + \frac{2h}{r}\right)\right), \ k \geq 1, \end{aligned}$$

where $t_k = n2^k$ denotes the degree of $F_k(x)$. Then $F_k(x)$ is an irreducible polynomial over $\mathbb{F}_q$ of degree $n2^k$ for every $k \geq 1$.

3.2.41 Remark For $r = \delta = 2$ and $h = 0$ Theorem 3.2.40 coincides with Theorem 3.2.38 due to Cohen [678, 685]; see also [2077, Theorem 3.24].

3.2.42 Theorem [1822] Let $P(x) \neq x$ be an irreducible polynomial of degree $n \geq 1$ over $\mathbb{F}_q$. Suppose that the elements $P(0), h^n$ and $(2r)^n$ are squares in $\mathbb{F}_q$ and the element $P\left(\frac{2h}{r}\right)$ is a non-square in $\mathbb{F}_q$. Define

$$\begin{aligned} F_0(x) &= P(x), \\ F_k(x) &= (F_{k-1}(0))^{-1}\left(\frac{(rx+2h)^2}{4h}\right)^{t_{k-1}} F_k\left(\frac{(4h)^2 x}{(rx+2h)^2}\right), \ k \geq 1, \end{aligned}$$

where $t_k = n2^k$ denotes the degree of $F_k(x)$. Then $F_k(x)$ is an irreducible polynomial over $\mathbb{F}_q$ of degree $n2^k$ for every $k \geq 1$.

3.2.43 Theorem [1822] Let P be an irreducible polynomial of degree $n \geq 1$ over $\mathbb{F}_q$, where n is even if $q \equiv 3 \pmod 4$ and $b \in \mathbb{F}_q$. Suppose that the element $P\left(-\frac{b}{2}\right)$ is a non-square in $\mathbb{F}_q$. Define

$$\begin{aligned} F_0(x) &= P(x), \\ F_k(x) &= F_{k-1}\left(x^2 + bx + \frac{b^2}{4} - \frac{b}{2}\right), \ k \geq 1. \end{aligned}$$

Then for every $k \geq 1$, $F_k(x)$ is an irreducible polynomial over $\mathbb{F}_q$ of degree $n2^k$.

3.2.44 Definition For a given polynomial $P(x) = \sum_{u=0}^{n} a_u x^u \in \mathbb{F}_q[x]$ define the polynomial g_P as

$$g_P(x) = (-1)^n \sum_{j=0}^{n} \sum_{u=0}^{2j} (-1)^u a_u a_{2j-u} x^j.$$

3.2.45 Theorem [1824] Let q be an odd prime power and $P(x) = \sum_{u=0}^{n} a_u x^u$ be an irreducible polynomial of degree $n > 1$ over $\mathbb{F}_q$ with at least one coefficient $a_{2i+1} \neq 0$ $\left(0 \leq i \leq \lfloor \frac{n}{2} \rfloor\right)$. Let $ax^2 + 2hx + ahd^{-1}$ and $dx^2 + 2ax + h$ be relatively prime, where $a, d, h, \in \mathbb{F}_q^*$ and $a^2 \neq hd$. Suppose that the element $\left(hd^{-1}\right)^n$ is a non-zero square in $\mathbb{F}_q$ and the element $\left(hd - a^2\right)^n g_{F_0}\left(\frac{h}{d}\right)$ is non-square in $\mathbb{F}_q$ (see Definition 3.2.44 for g_P). Define

$$\begin{aligned} &F_0(x) = P(x), \\ &F_k(x) = H_{k-1}(a,d)^{-1}\left(dx^2 + 2ax + h\right)^{t_{k-1}} F_{k-1}\left(\tfrac{ax^2+2hx+ahd^{-1}}{dx^2+2ax+h}\right), \ \text{for } k \geq 1, \end{aligned}$$

where $H_{k-1}(a,d) = d^{t_{k-1}} F_{k-1}\left(\frac{a}{d}\right)$, and t_k is the degree of $F_k(x)$. Then $F_k(x)$ is an irreducible polynomial over $\mathbb{F}_q$ of degree $t_k = n2^k$ for every $k \geq 1$.

3.2.46 Theorem [1824] Let q and P satisfy the hypothesis of Theorem 3.2.45. Suppose $a, c \in \mathbb{F}_q^*$, $(ac)^n$ is a square in $\mathbb{F}_q$ and the element $(-1)^n g_{F_0}\left(\frac{c}{a}\right)$ is a non square in $\mathbb{F}_q$ (see Definition 3.2.44 for g_P). Define

$$\begin{aligned} F_0(x) &= P(x), \\ F_k(x) &= (2x)^{t_{k-1}} F_{k-1}\left(\frac{ax^2+c}{2ax}\right), \ k \geq 1, \end{aligned}$$

where t_k is the degree of $F_k(x)$. Then $F_k(x)$ is an irreducible polynomial over $\mathbb{F}_q$ of degree $t_k = n2^k$ for every $k \geq 1$.

3.2.47 Remark In particular, the case $q \equiv 3 \pmod 4)$ and $F_0(x) = x^2 + 2x + c$ with $a = 1$ of Theorem 3.2.46 was considered by McNay [589]. The case $a = c = 1$ was derived by Cohen; see [678, 685, 2077].

See Also

§3.3	For composite polynomials.
§3.4	For weights of irreducible polynomials.
§3.5	For irreducible polynomials with prescribed coefficients.

References Cited: [33, 34, 35, 36, 37, 38, 39, 40, 41, 42, 70, 213, 229, 231, 469, 505, 506, 507, 589, 666, 671, 678, 685, 769, 841, 849, 850, 1172, 1261, 1819, 1820, 1821, 1822, 1823, 1824, 1825, 1939, 1953, 1954, 1955, 1956, 1957, 1958, 1962, 2022, 2046, 2077, 2091, 2323, 2324, 2376, 2377, 2378, 2392, 2442, 2443, 2568, 2569, 2571, 2597, 2598, 2599, 2600, 2616, 2857, 2858, 2859, 2860, 2861, 2863, 2977]

3.3 Conditions for reducible polynomials

Daniel Panario, Carleton University

We present qualitative results on the reducibility of univariate polynomials over finite fields. First we provide some classical work; see the comments at the end of Chapters 3 and 4 of Lidl and Niederreiter [1939] for other results published before 1983. Then we cover several reducibility results that follow from a theorem of Pellet and Stickelberger [2379, 2716].

3.3.1 Composite polynomials

3.3.1 Remark There has been substantial work showing that some classes of polynomials are irreducible using several types of composition of polynomials. Here we are interested in "if and only if" irreducibility statements that as a consequence provide reducibility results.

3.3.2 Remark Let f be a polynomial of degree m over $\mathbb{F}_q$, $q = p^k$, and let L be the linearized polynomial $L(x) = \sum_{i=0}^{n} a_i x^{p^i}$. Ore [2324] considers the irreducibility of $f(L)$. Agou in several articles [37, 39, 41] considers special types of linearized polynomials including $x^{p^r} - ax$ and $x^{p^{2r}} - ax^{p^r} - bx$.

3.3.3 Theorem [37] Let $f(x) = x^m + b_{m-1}x^{m-1} + \cdots + b_0$ be an irreducible polynomial over $\mathbb{F}_q$, $q = p^k$, with root β. Then, for any nonzero a in $\mathbb{F}_q$, $f(x^p - ax)$ is irreducible over $\mathbb{F}_q$ if and only if $a^{(q-1)\gcd(m,p-1)/(p-1)} = 1$ and $\mathrm{Tr}_{km}(\beta/A^p) \neq 0$, where $A \in \mathbb{F}_{q^m}$ satisfies $A^{p-1} = a$ and $\mathrm{Tr}_{km}(x) = x + x^p + \cdots + x^{p^{km-1}}$. In particular, if A is in $\mathbb{F}_q$, then $f(x^p - A^{p-1}x)$ is irreducible over $\mathbb{F}_q$ if and only if $\mathrm{Tr}_k(b_{m-1}/A^p) \neq 0$.

3.3.4 Remark Similar results can be found in the papers by Agou cited above.

3.3.5 Remark We are also interested in results that guarantee classes of polynomials that are reducible. The concluding reducibility result for compositions of the type $f(L)$, where f and L are as above, is given next.

3.3.6 Theorem [41] Let f be an irreducible polynomial of degree m over $\mathbb{F}_q$, $q = p^k$, and let L be the linearized polynomial $L(x) = \sum_{i=0}^{n} a_i x^{p^i}$. If $n \geq 3$, $f(L)$ is reducible.

3.3.7 Remark The cases when $n \leq 2$ in the previous theorem were studied by Agou [39, 40].

3.3.8 Remark Cohen [671, 672] gives alternative proofs for Agou's results.

3.3.9 Remark Moreno [2145] considers the irreducibility of a related composition of functions.

3.3.10 Theorem [2145] Let f and g be polynomials over $\mathbb{F}_q$, $q = p^k$, and let f be irreducible of degree m. The polynomial $f(g(x))$ is irreducible over $\mathbb{F}_q$ if and only if $g(x) + \beta$ is irreducible over $\mathbb{F}_{q^m}$ for any root β of f.

3.3.11 Remark Brawley and Carlitz [394] define root-based polynomial compositions called *composed products.*

3.3.12 Definition Let f and g be monic polynomials in $\mathbb{F}_q^*$ with factorizations in the algebraic closure of $\mathbb{F}_q$, given by

$$f(x) = \prod_{\alpha}(x - \alpha) \quad \text{and} \quad g(x) = \prod_{\beta}(x - \beta).$$

The *composed products* $f \circ g$ and $f \star g$ are defined, respectively, by

$$(f \circ g)(x) = \prod_{\alpha}\prod_{\beta}(x - \alpha\beta), \quad \text{and} \quad (f \star g)(x) = \prod_{\alpha}\prod_{\beta}(x - (\alpha + \beta)).$$

3.3.13 Theorem [394] The composed products of f and g, $f \circ g$ and $f \star g$, are irreducible if and only if f and g are irreducible with coprime degrees.

3.3.14 Remark Related results to composed products can be found in [394, 395, 2103].

3.3.2 Swan-type theorems

3.3.15 Remark Pellet [2379] and Stickelberger [2716] relate the parity of the number of irreducible factors of a squarefree polynomial with its discriminant. When the parity of the number of irreducible factors of a polynomial is even, the polynomial is reducible.

3.3.16 Remark We recall, from Section 2.1, the definition of discriminant (Definition 2.1.135).

3.3.17 Definition Let f be a polynomial of degree n in $\mathbb{F}_q[x]$ with leading coefficient a, and with roots $\alpha_1, \alpha_2, \ldots, \alpha_n$ in its splitting field, counted with multiplicity. The *discriminant* of f is given by

$$D(f) = a^{2n-2} \prod_{1 \le i < j \le n} (\alpha_i - \alpha_j)^2.$$

3.3.18 Remark The discriminant of f is zero if and only if f has multiple roots.

3.3.19 Remark Next is a result given by Stickelberger [2716] although the theorem was originally shown by Pellet [2379]; see also [846].

3.3.20 Theorem [2379, 2716] Let p be an odd prime and suppose that f is a monic polynomial of degree n with integral coefficients in a p-adic field $\mathbb{F}$. Let $\bar{f}$ be the result of reducing the coefficients of f modulo p. Assume further that $\bar{f}$ has no repeated roots. If $\bar{f}$ has r irreducible factors over the residue class field, then $r \equiv n \pmod 2$ if and only if $D(f)$ is a square in $\mathbb{F}$.

3.3.21 Proposition [2753] Let q be a power of an odd prime p and let $\mathbb{F}_q$ be the finite field with q elements. Let g be a polynomial over $\mathbb{F}_q$ of degree n with no repeated roots. Furthermore, let r be the number of irreducible factors of g over $\mathbb{F}_q$. Then $r \equiv n \pmod 2$ if and only if $D(f)$ is a square in $\mathbb{F}_q$.

3.3.22 Remark Swan extends the previous result to the case $p = 2$ by noting that a p-adic integer a coprime to p, is a p-adic square if and only if a is a square modulo $4p$.

3.3.23 Corollary [2753] Let g be a polynomial of degree n over $\mathbb{F}_2$ with $D(g) \neq 0$ and let f be a monic polynomial over the 2-adic integers such that g is the reduction of f modulo 2. Furthermore, let r be the number of irreducible factors of g over $\mathbb{F}_2$. Then $r \equiv n \pmod 2$ if and only if $D(f) \equiv 1 \pmod 8$.

3.3.24 Remark Swan also characterizes the parity of the number of irreducible factors of a trinomial over $\mathbb{F}_2$. These polynomials are of practical importance when implementing finite field extensions; for example, see [1567] and Section 3.4.

3.3.25 Theorem [2753] Let $n > k > 0$. Assume precisely one of n, k is odd. If r is the number of irreducible factors of $f(x) = x^n + x^k + 1 \in \mathbb{F}_2[x]$, then r is even (and hence f is not irreducible) in the following cases:

1. n even, k odd, $n \neq 2k$ and $nk/2 \equiv 0, 1 \pmod 4$;
2. n odd, k even, $k \nmid 2n$ and $n \equiv 3, 5 \pmod 8$;
3. n odd, k even, $k \mid 2n$ and $n \equiv 1, 7 \pmod 8$.

In other cases f has an odd number of factors.

3.3.26 Remark The case when n and k are both odd can be covered by making use of the fact that the reciprocal polynomial of f has the same number of irreducible factors as f; for reciprocal polynomials see Definition 2.1.48. If both n and k are even the trinomial is a square and has an even number of irreducible factors.

3.3.27 Corollary [2753] Let n be a positive integer divisible by 8. Then every trinomial over $\mathbb{F}_2$ of degree n has an even number of irreducible factors in $\mathbb{F}_2[x]$, and hence it is not irreducible.

3.3.28 Remark Many results have been given following Swan's technique for other types of polynomials and finite field extensions and characteristics. We state several of them starting from characteristic two results and then focusing on odd characteristic.

3.3.29 Remark Vishne [2878] considers trinomials in finite extensions of $\mathbb{F}_2$; see also Theorems 6.69 and 6.695 in [231]. Evaluating the discriminant of a trinomial (modulo $8R$, where R is the valuation ring of the corresponding extension of 2-adic numbers), Vishne's studies are a direct analogue of Swan's proof over $\mathbb{F}_2$.

3.3.30 Corollary [2878] Let $\mathbb{F}_{2^s}$ be an even degree extension of $\mathbb{F}_2$ and n an even number. Then, $g(x) = x^n + ax^k + b \in \mathbb{F}_{2^s}[x]$ has an odd number of irreducible factors only when $g(x) = x^{2d} + ax^d + b$ and $t^2 + at + b$ has no root in $\mathbb{F}_{2^s}$.

3.3.31 Remark Similar results to the one in Corollary 3.3.30 are given in [2878]. Special cases of trinomials of low degrees over extensions of $\mathbb{F}_2$ are given in [570].

3.3.32 Remark Hales and Newhart give a Swan-like result for binary tetranomials; see Theorem 2 in [1399].

3.3.33 Remark It is convenient to use irreducible trinomials over $\mathbb{F}_2$ when constructing extension fields. The usage of pentanomials (polynomials with 5 nonzero coefficients) when trinomials do not exist is in the IEEE standard specifications for public-key cryptography [1567]. However, Scott [2573] shows that some of the recommended irreducible polynomials are not optimal.

3.3.34 Remark Next we show some partial results attempting to characterize the reducibility of binary pentanomials.

3.3.35 Remark Koepf and Kim [1777] give a Swan-like result for a class of binary pentanomials of the form $x^m + x^{n+1} + x^n + x + 1$, where m is even and $1 \leq n \leq \lfloor m/2 \rfloor - 1$. Ahmadi [45] shows that there are no self-reciprocal irreducible pentanomials of degree n over $\mathbb{F}_2$ if n is a multiple of 12.

3.3.36 Problem Characterize completely the reducible pentanomials over $\mathbb{F}_2$.

3.3.37 Remark The reducibility of some classes of binary polynomials with more than five nonzero coefficients has also been studied. In all cases, the polynomials have a special form.

3.3.38 Remark Fredricksen, Hales, and Sweet [1096] give a Swan-like result for polynomials of the form $x^n f(x) + g(x)$ where $f, g \in \mathbb{F}_2[x]$.

3.3.39 Remark Let n and $v \geq 2$ be coprime positive integers, let r be the least positive residue of n modulo v, and set $d = (n-r)/v$. A polynomial of the form $x^r g(x^v) + h(x^v)$ with g monic of degree d and h of degree $\leq d$ is an *(n, v)-windmill polynomial.* These polynomials are related to a method to generate pseudorandom bit sequences by combining linear feedback shift registers in a "windmill" configuration [2686].

3.3.40 Theorem [673] Suppose that n and v are as above and f is a squarefree (n, v)-windmill polynomial over $\mathbb{F}_{2^m}$. Let $n_m(f)$ denote the number of irreducible factors of f over $\mathbb{F}_{2^m}$.

1. If $n \equiv \pm 1 \pmod 8$ or m is even, then $n_m(f)$ is odd.
2. If $n \equiv \pm 3 \pmod 8$ and m is odd, then $n_m(f)$ is even.

3.3.41 Theorem [333] Let $f(x) = x^n + \sum_{iS} x^i + 1 \in \mathbb{F}_2[x]$, where

$$S \subset \{i \colon i \text{ odd}, 0 < i < n/3\} \cup \{i : i \equiv n \pmod 4, 0 < i < n\}.$$

Then f has no repeated roots. If $n \equiv \pm 1 \pmod 8$, then f has an odd number of irreducible factors. If $n \equiv \pm 3 \pmod 8$, then f has an even number of irreducible factors.

3.3.42 Remark A short proof of Theorem 3.3.41 is given in [52].

3.3.43 Remark Swan-type results for composite, linearized, and affine polynomials over $\mathbb{F}_2$ are given in [1735, 3063]

3.3.44 Problem The complete characterization of reducible polynomials over $\mathbb{F}_2$ is a hard open problem. Provide new reducibility characterizations for classes of polynomials over $\mathbb{F}_2$.

3.3.45 Remark There have also been reducibility results over finite fields of odd characteristic. Binomials over finite fields of odd characteristic can be treated easily with a Swan-type approach [1417].

3.3.46 Remark For trinomials over finite fields of odd characteristic only partial results are known [46, 1223, 1417].

3.3.47 Remark Over $\mathbb{F}_3$, Loidreau [1952] gives the parity for the number of irreducible factors for any trinomial over $\mathbb{F}_3$ by examining the discriminant using all possible congruences of n and k modulo 12; see also [1223]. This type of analysis holds for higher characteristic, but the number of cases grows quickly with the characteristic, making a complete analysis for large q hard to achieve.

3.3.48 Problem Completely characterize the reducibility of trinomials over finite fields of characteristic different from 2 and 3.

See Also

§3.2	For results on composition of polynomials.
§3.4	For results on low weight irreducible polynomials.
§4.3	For results on low weight primitive polynomials.
§11.2	For counting polynomials with given factorization patterns.
§11.3	For irreducibility tests.
[231]	Chapter 6, for extensions of Swan's theorem.
[236]	For results on quadratic and cubic polynomials in extension fields of characteristic two.
[1300]	Chapter 5, for factorizations of binary trinomials.
[3054]	For a family of reducible polynomials related to word-oriented feedback shift registers; see [827, 828, 2818].

References Cited: [37, 39, 40, 41, 45, 46, 52, 231, 236, 333, 394, 395, 570, 671, 672, 673, 827, 828, 846, 1096, 1223, 1300, 1399, 1417, 1567, 1735, 1777, 1939, 1952, 2103, 2145, 2324, 2379, 2573, 2686, 2716, 2753, 2818, 2878, 3054, 3063]

3.4 Weights of irreducible polynomials

Omran Ahmadi, Institute for Research in Fundamental Sciences (IPM)

3.4.1 Basic definitions

3.4.1 **Definition** The weight of a polynomial $f(x) = \sum_{i=0}^{n} a_i x^i$ in $\mathbb{F}_q[x]$ is the number of its nonzero coefficients.

3.4.2 **Definition** A polynomial in $\mathbb{F}_q[x]$ is a *binomial*, *trinomial*, *tetranomial*, or a *pentanomial* if it has two, three, four, or five nonzero coefficients, respectively.

3.4.3 **Definition** A polynomial in $\mathbb{F}_q[x]$ is a *fewnomial* or a *sparse polynomial* if it has a small number of nonzero coefficients.

3.4.2 Existence results

3.4.4 **Theorem** [850] Let the order of $a \in \mathbb{F}_q^*$ be e. Then the binomial $x^n - a \in \mathbb{F}_q[x]$, $n \geq 2$, is irreducible over $\mathbb{F}_q$ if and only if the following conditions are satisfied:

1. if r is a prime number and $r|n$, then $r \mid e$ and $r \nmid (q-1)/e$;
2. $q \equiv 1 \pmod 4$ if $n \equiv 0 \pmod 4$.

3.4.5 **Remark** The study of irreducible binomials over finite fields can be traced back to the works of researchers who were trying to generalize Fermat's little theorem – see for example the expositions of Poinsot [2407] and Serret [2600]. Theorem 3.4.4 in the format given above appears in [850, 1938, 1939, 2077].

3.4.6 Corollary Let n be an odd number, and let $a \in \mathbb{F}_q$. Then $x^n - a$ is irreducible over $\mathbb{F}_q$ if and only if for every prime divisor r of n, a is not an r-th power of an element of $\mathbb{F}_q$.

3.4.7 Corollary [1938, 1939, 2077] Let the order of $a \in \mathbb{F}_q^*$ be e, and let r be a prime factor of $q-1$ which does not divide $(q-1)/e$. Assume that $q \equiv 1 \pmod 4$ if $r = 2$ and $k \geq 2$. Then for any non-negative integer k, $x^{r^k} - a$ is irreducible over $\mathbb{F}_q$.

3.4.8 Corollary [2356] Let $\mathbb{F}_q$ be a finite field of odd characteristic p, $p \geq 5$. There exists an irreducible binomial over $\mathbb{F}_q$ of degree m, $m \not\equiv 0 \pmod 4$, if and only if every prime factor of m is also a prime factor of $q-1$. For $m \equiv 0 \pmod 4$ then there exists an irreducible binomial over $\mathbb{F}_q$ of degree m if and only if $q \equiv 1 \pmod 4$ and every prime factor of m is also a prime factor of $q-1$.

3.4.9 Example

1. $x^6 - a$ is irreducible over $\mathbb{F}_q$ if and only if a is neither a quadratic nor a cubic residue in $\mathbb{F}_q$.
2. $x^{2^k} \pm 2$ are irreducible over $\mathbb{F}_5$ [2077].
3. $x^{3^k} \pm 2$ and $x^{3^k} \pm 3$ are irreducible over $\mathbb{F}_7$ [2077].
4. $x^{3^k} + \omega$ is irreducible over $\mathbb{F}_4$ where $\mathbb{F}_4 = \mathbb{F}_2(\omega)$ [2077].

3.4.10 Remark It is clear that there is no irreducible binomial of degree > 1 over $\mathbb{F}_2$. From results above, it follows that x^2+1 is the only irreducible binomial of degree > 1 over $\mathbb{F}_3$ and there are infinitely many irreducible binomials over $\mathbb{F}_q$ for $q \geq 4$. Also, it is clear that for every q there are infinitely many positive integers m, for example p the characteristic of $\mathbb{F}_q$, such that there is no irreducible binomial of degree m over $\mathbb{F}_q$. This fact motivates the following results.

3.4.11 Theorem [2378] Let $\mathbb{F}_q$ be of characteristic p. Then the trinomial $f(x) = x^p - x - b$ is irreducible over $\mathbb{F}_q$ if and only if one of the following equivalent conditions are satisfied

1. f has no root in $\mathbb{F}_q$;
2. $\mathrm{Tr}_{\mathbb{F}_q}(b) \neq 0$;
3. p does not divide $[\mathbb{F}_q : \mathbb{F}_p]$.

3.4.12 Corollary [2077] Let $\mathbb{F}_q$ be of characteristic p. For $a, b \in \mathbb{F}_q^*$, the trinomial $x^p - ax - b$ is irreducible over $\mathbb{F}_q$ if and only if $a = c^{p-1}$for some $c \in \mathbb{F}_q$ and $\mathrm{Tr}_{\mathbb{F}_q}(a/c^p) \neq 0$.

3.4.13 Corollary [2324] Let $x^p - x - a \in \mathbb{F}_q[x]$ be an irreducible trinomial over $\mathbb{F}_q$ of characteristic p, and let α be a root of this polynomial in an extension field of $\mathbb{F}_{q^p}$. Then $x^p - x - a\alpha^{p-1}$ is irreducible over $\mathbb{F}_q(\alpha)$.

3.4.14 Theorem [2077] Let $p \equiv 3 \pmod 4$ be a prime and let $p+1 = 2^r s$ with s odd. Then, for any integer $k \geq 1$, $x^{2^k} - 2ax^{2^{k-1}} - 1$ is irreducible over $\mathbb{F}_p$, and hence irreducible over any odd degree extension $\mathbb{F}_q$, where $a = a_r$ is obtained recursively as follows:

1. $a_1 = 0$;
2. for j from 2 to $r-1$, set $a_j = \left(\frac{a_{j-1}+1}{2}\right)^{(p+1)/4}$;
3. $a_r = \left(\frac{a_{r-1}-1}{2}\right)^{(p+1)/4}$.

3.4.15 Remark The following constitutes a partial result towards proving Conjecture 3.4.31.

3.4.16 Theorem [667, 2446] Let $\mathbb{F}_q$ be a finite field of characteristic p, and let $n \geq 2$ be such that p does not divide $2n(n-1)$. Let $T_n(q)$ denote the number of $a \in \mathbb{F}_q$ for which the trinomial $x^n + x + a$ is irreducible over $\mathbb{F}_q$. Then

$$T_n(q) = \frac{q}{n} + O(q^{1/2}), \tag{3.4.1}$$

where the implied constant depends only on n.

3.4.17 Remark There is not any substantial result proving the existence of irreducible fewnomials of weight at least four over finite fields.

3.4.18 Remark There are many conjectures concerning the existence of irreducible fewnomials over finite fields (see the next section) whose resolution currently seems to be out of reach. The following is a partial result towards proving the existence of irreducible fewnomials over binary fields.

3.4.19 Theorem [2634] There exists a primitive polynomial of degree n over $\mathbb{F}_2$ whose weight is $n/4 + o(n)$.

3.4.20 Remark The weight of a monic polynomial of degree n over a finite field is between 1 and $n+1$ inclusive. On one end of the weight spectrum we have the monomial x and binomials, whose irreducibility is well understood. We have some partial results concerning the irreducibility of polynomials corresponding to the other end of the weight spectrum.

3.4.21 Theorem [850] The all one polynomial $x^n + x^{n-1} + \cdots + x^2 + x + 1 \in \mathbb{F}_q[x]$, which is of weight $n+1$, is irreducible over $\mathbb{F}_q$ if and only if $n+1$ is a prime number and q is a primitive root modulo $n+1$.

3.4.22 Example

1. If $n = 2, 4, 10, 12, 18, 28, 36, 52, 58, 60, 66$, then $x^n + x^{n-1} + \cdots + x^2 + x + 1$ is irreducible over $\mathbb{F}_2$.
2. Let $m > 1$ be an integer. If m is even with $m \equiv 0$ or $6 \pmod 8$, then the number of irreducible factors of $f(x) = x^m + x^{m-1} + \cdots + x^3 + x^2 + x + 1$ is an even number and hence f is reducible over $\mathbb{F}_2$ [55].

3.4.23 Conjecture [1847] (Artin) Let a be a non-square integer different from 1 and -1. Then there are infinitely many primes p so that a is a primitive element in $\mathbb{F}_p$.

3.4.24 Remark Using Theorem 3.4.21 and Conjecture 3.4.23, naturally we have the following conjecture.

3.4.25 Conjecture There are infinitely many n for which $x^n + x^{n-1} + \cdots + x^2 + x + 1$ is irreducible over $\mathbb{F}_q$.

3.4.26 Theorem [1531] Artin's conjecture holds provided that the generalized Riemann hypothesis is true.

3.4.27 Remark The following is an immediate corollary of Theorem 3.4.21 and Theorem 3.4.26.

3.4.28 Theorem There are infinitely many n for which $x^n + x^{n-1} + \cdots + x^2 + x + 1$ is irreducible over $\mathbb{F}_q$ provided that the generalized Riemann hypothesis is true.

3.4.3 Conjectures

3.4.29 Remark The following conjectures are supported by extensive computations, but it seems that resolving them will be very hard. The first three conjectures have become part of folklore and it is very difficult to trace back their origin. Here we provide a reference for the interested readers who wish to have more information about these conjectures. Note that the following is by no means an exhaustive list of conjectures related to the weight distribution of irreducible polynomials over finite fields.

3.4.30 Conjecture [1233] For every n, there exists a polynomial of degree n and of weight at most five which is irreducible over $\mathbb{F}_2$.

3.4.31 Conjecture [1233] Let $q \geq 3$ be a prime power. For every n, there exists a polynomial of degree n and of weight at most four which is irreducible over $\mathbb{F}_q$.

3.4.32 Conjecture [2186] The set of positive integers n for which there exists an irreducible trinomial of degree n over $\mathbb{F}_2$ has a positive density in the set of positive integers.

3.4.33 Conjecture [307] The number of irreducible trinomials over $\mathbb{F}_2$ of degree at most n is $3n + o(n)$.

3.4.34 Conjecture [52] Let i, j be two positive integers such that $j < i$, and let

$$F_{i,j}(x) = \frac{x^{i+1} + 1}{x + 1} + x^j.$$

Then the number of irreducible polynomials $F_{i,j}$ of degree at most n over $\mathbb{F}_2$ is $2n + o(n)$.

3.4.35 Conjecture [1233] For every positive integer n, there exists a polynomial g of degree at most $\log_q n + 3$ such that $f(x) = x^n + g(x)$, called a *sedimentary polynomial*, is irreducible over $\mathbb{F}_q$.

See Also

§3.3	For results on reducibility of low weight polynomials.
§3.5	For results on polynomials with prescribed coefficients.
§4.2	For results on primitive polynomials with prescribed coefficients.
§4.3	For results on low weight primitive polynomials.

References Cited: [52, 55, 307, 667, 850, 1233, 1531, 1938, 1939, 2077, 2186, 2324, 2356, 2378, 2407, 2446, 2600, 2634]

3.5 Prescribed coefficients

Stephen D. Cohen, University of Glasgow

3.5.1 Definition Given a monic polynomial $f(x) = x^n + \sum_{i=1}^{n} a_i x^{n-i}$ of degree n in $\mathbb{F}_q[x]$ then a_i is the *i-th coefficient*. For $0 \leq k, m \leq n$, $\{a_1, \ldots, a_k\}$ are the *first k coefficients* and $\{a_{n-m+1}, \ldots, a_n\}$ are the *last m coefficients*. In particular $-a_1$ is the *trace of f* and $(-1)^n a_n$ is the *norm of f*.

3.5.2 Remark Results on the distribution of irreducible polynomials of degree n with certain coefficients prescribed generally fall into the categories (i) existence, (ii) asymptotic estimates, (iii) explicit estimates, (iv) exact formulae or expressions. For those (few) in category (iv) further classification under (i)–(iii) may be desirable. Historically, the distribution of irreducible polynomials with prescribed first and/or last coefficients is approachable by number-theoretic methods [134, 541, 1447, 2834] yielding asymptotic behaviour. Recently, more specialized finite field techniques have refined these into explicit estimates, whilst retaining asymptotic features. As for existence, in some cases there are theorems which yield the existence of polynomials of degree n with the stronger property of being primitive for all but a few pairs (q, n); see Section 4.2.

3.5.3 Remark To keep notation simple, the number of irreducible polynomials of any particular type will be denoted at that specific juncture by I (or, for example, $I(a, b)$). All polynomials will be *monic* polynomials in $\mathbb{F}_q[x]$ (with q a power of the characteristic p) of degree n unless mentioned otherwise.

3.5.4 Remark See Section 3.1 for further formulae for irreducible polynomials.

3.5.1 One prescribed coefficient

3.5.5 Remark For convenience, the results of Theorems 3.1.7, 3.1.8, and 3.1.10 are summarized again here.

3.5.6 Theorem Write $n = p^j n_0$, where $j \geq 0$ and $p \nmid n_0$. Then the number of irreducible polynomials with trace $a \in \mathbb{F}_q$ is

$$I = \frac{1}{nq} \sum_{d \mid n_0} \mu(d) q^{n/d} - \frac{\varepsilon}{n} \sum_{d \mid n_0} q^{n/dp},$$

where $\varepsilon = 1$ if $a = 0$ and $j > 0$; otherwise $\varepsilon = 0$.

3.5.7 Theorem As in Definition 3.1.3 set $D_n = \{r : r | q^n - 1; r \nmid q^i - 1, i < n\}$. For $r \in D_n$, write $m_r = r/\gcd(r, \frac{q^n-1}{q-1})$. Then an irreducible polynomial with norm $a \in \mathbb{F}_q^*$ is such that a has order m_r for some $r \in D_n$ and the number of irreducible polynomials of order r and norm a is

$$I = \frac{\phi(r)}{n\phi(m_r)}.$$

Thus, the total number of irreducible polynomials of norm a, having order m, is

$$I = \frac{1}{n\phi(m)} \sum_{\substack{r \in D_n \\ m_r = m}} \phi(r).$$

3.5.8 Remark An asymptotic description of the number in Theorem 3.5.7 occurs in Theorem III.5 of [512].

3.5.9 Theorem [512] The number I of irreducible polynomials with prescribed non-zero constant term satisfies

$$\frac{1}{n}\left(\frac{q^n-1}{q-1} - 2q^{n/2}\right) \leq I \leq \frac{q^n-1}{n(q-1)}.$$

3.5.10 Remark The next theorem, like Theorem 3.5.9, effectively concerns polynomials with one fixed coefficient, although it involves three coefficients a_1, a_{n-1} and a_n.

3.5.11 Theorem [310] Let q be a power of p, where $p = 2$ or 3 and $n = p^j n_0$, $p \nmid n_0$. Then the number $I(b)$ of irreducible polynomials, with $a_1 + a_{n-1}/a_n$ having prescribed value b, satisfies

$$\left| I(b) - \frac{q^{n-1}}{n} \right| \leq \frac{3q^{\frac{n}{2}}}{n}.$$

More generally, this estimate holds for $I(0)$ for any prime power q whenever $p \nmid n$.

3.5.12 Remark The general existence theorem for irreducible polynomials with a prescribed coefficient which follows had been conjectured by Hansen and Mullen [1416]. The key theoretical work (a dual approach effective for small values of m, $n - m$, respectively) on *weighted* (as opposed to *exact*) numbers of irreducibles is in [2893].

3.5.13 Theorem [1405, 2893] Given integers m, n with $0 \leq m < n$ and $a \in \mathbb{F}_q$ (non-zero if $m = 0$) there exists an irreducible polynomial with the coefficient of x^m equal to a, except when $n = 2, m = 1$ and $a = 0$.

3.5.2 Prescribed trace and norm

3.5.14 Remark By considering each monic reciprocal polynomial $f^*(x) = a_n^{-1} x^n f(1/x)$ (see Definition 2.1.48), provided $a_n \neq 0$, one sees that the number of irreducible polynomials with prescribed first coefficient a_1 and last coefficient a_n is the same as that with last two coefficients a_1/a_n and $1/a_n$, respectively. Hence it suffices to consider the first of these situations. Equivalently, examine the set of irreducible polynomials with prescribed trace and norm.

3.5.15 Theorem [512, 2893] The number of irreducible polynomials $I(a, b)$ with prescribed trace a and prescribed non-zero norm b satisfies

$$\left| I(a,b) - \frac{q^{n-1}}{n(q-1)} \right| \leq \frac{3}{n} q^{n/2}.$$

3.5.16 Theorem [2121] The number $I(a, b)$ of irreducible polynomials of degree n with trace a and non-zero norm b satisfies

$$\left| I(0,b) - \frac{q^{n-1} - 1}{n(q-1)} \right| \leq \frac{s-1}{n} q^{\frac{n-2}{2}} + \frac{q^{\frac{n}{2}} - 1}{q-1} < \frac{2}{q-1} q^{\frac{n}{2}},$$

where $s = \gcd(n, q-1)$, and, when $a \neq 0$,

$$\left| I(a,b) - \frac{q^n - 1}{nq(q-1)} \right| \leq q^{\frac{n-2}{2}} + \frac{q^{\frac{n}{2}-1}}{q(q-1)} + \frac{n}{2} q^{\frac{n-4}{4}} < \frac{2}{q-1} q^{\frac{n}{2}}.$$

3.5.17 Remark Theorem 3.5.16 is an improvement on Theorem 3.5.15 whenever $n \leq \frac{3}{2} q^{n/2}$.

3.5.18 Definition Given a pair (q, n) let P be the largest prime factor of $q^n - 1$. Then (q, n) is an *lps (largest prime survives) pair* if $P \in D_n$ as in Theorem 3.5.7.

3.5.19 Remark According to [2318], empirical evidence indicates that many pairs (q, n) are *lps* pairs although there are some "sporadic" pairs that are not.

3.5.20 Theorem [2318] Suppose $n \geq 3$ and (q, n) is an *lps* pair. Then the number $I(a, b)$ of irreducible polynomials with non-zero trace a and non-zero norm b satisfies

$$I(a,b) \geq \frac{(1 - \frac{1}{P})(q^n - 1) - q^{n-1} - 2nq^{\frac{n}{2}} + 1}{n(q-1)^2}.$$

3.5.21 Example Take $(q, n) = (9, 7)$. Since $9^7 - 1 = 2^3 \times 547 \times 1093$, then $(9, 7)$ is an *lps* pair — though $(3, 14)$ which also relates to the factorization of $9^7 - 1$ is not an *lps* pair! Theorem 3.5.15 yields a lower bound for $I(a, b)$ of 8552.73; Theorem 3.5.16 yields lower bounds of 9216.75 and 9198.47 when a is zero and non-zero, respectively, and, when $a \neq 0$, Theorem 3.5.20 yields an improved lower bound of 9411.91.

3.5.22 Remark Even when (q, n) is not an *lps* pair a classical theorem of Zsigmondy (see e.g., [2467] or Wikipedia) implies there is always a prime $l \in D_n$ except when q is a Mersenne prime and $n = 2$ and when $(q, n) = (2, 6)$. Given (q, n) with these exceptions, such a prime is a *Zsigmondy prime* and the *largest Zsigmondy prime* is the largest of such primes. The authors of Theorem 3.5.20 were unaware of Zsigmondy's theorem and it is evident that the same argument yields the following modification which is generally applicable. It appears here for the first time.

3.5.23 Theorem Suppose $n \geq 3$ and $(q, n) \neq (2, 6)$. Let P_Z be the corresponding largest Zsigmondy prime. Then the number $I(a, b)$ in Theorem 3.5.20 satisfies

$$I(a, b) \geq \frac{(1 - \frac{1}{P_Z})(q^n - 1) - q^{n-1} - 2nq^{\frac{n}{2}} + 1}{n(q-1)^2}.$$

3.5.24 Example Take $(q, n) = (47, 4)$, not an *lps* pair since $q^2 - 1 = 2^5 \times 3 \times 23$ and $q^2 + 1 = 2 \times 5 \times 13 \times 17$. Here $5, 13$, and 17 are Zsigmondy primes with $P_Z = 17$. Then Theorem 3.5.15 provides no information, whereas Theorem 3.5.16 yields lower bounds for $I(a, b)$ of 504.4 and 515.23, respectively, when a is zero and non-zero. Now, when $a \neq 0$, Theorem 3.5.23 delivers an improved lower bound for $I(a, b)$ of 528.25.

3.5.25 Remark [2318] also provides an expression for upper bounds for $I(a, b)$. These are always better than those of Theorem 3.5.16 when n is a multiple of $q - 1$.

3.5.26 Remark Existence results for (merely) irreducible polynomials with prescribed trace and norm follow from those on the existence of *primitive* polynomials (Section 4.2).

3.5.3 More prescribed coefficients

3.5.27 Remark The question of estimating the number of irreducible polynomials f with the first k coefficients $a_1, \ldots, a_k$ and the last m coefficients $a_{n-m+1}, \ldots, a_n$ of f prescribed (where $2 \leq k + m < n$) can be generalized as follows and tackled by number-theoretical methods. Given a monic polynomial $M \in \mathbb{F}_q[x]$ of degree m, where $1 \leq m < n$, let $R \in \mathbb{F}_q[x]$ be a (not necessarily monic) polynomial prime to M. Then consider polynomials f of degree n with the first k coefficients prescribed and such that $f \equiv R \pmod{M}$. The original question of choosing k first and m last coefficients is recovered by selecting $M(x) = x^m$ and $R(x) = a_{n-m+1}x^{m-1} + \cdots + a_n$, $a_n \neq 0$. In this connection, note from Lemma 2.1.113 that $\Phi_q(x^m) = q^{m-1}(q-1)$.

3.5.28 Theorem [512, 1551, 2454] Let $2 \leq k + m < n$. Suppose $a_1, \ldots, a_k \in \mathbb{F}_q$, M is a monic polynomial in $\mathbb{F}_q[x]$ of degree m, and $R \in \mathbb{F}_q[x]$ is a fixed (not necessarily monic) polynomial of degree $< m$ prime to M. Then the number I of irreducible polynomials f of degree n with prescribed first k coefficients and such that $f \equiv R \pmod{M}$ satisfies

$$\frac{1}{n}\left\{\frac{q^{n-k}}{\Phi_q(M)} - (k + m + 1)q^{\frac{n}{2}}\right\} \leq I \leq \frac{1}{n}\left\{\frac{q^{n-k}}{\Phi_q(M)} + (1 - \delta)(k + m - 1)q^{\frac{n}{2}}\right\},$$

where $0 < \delta = 1 - \dfrac{1}{q^k \Phi_q(M)} < 1$.

3.5.29 Remark Theorem 3.5.28 leads to explicit existence results such as those which follow.

3.5.30 Corollary [685] Under the conditions of Theorem 3.5.28, suppose that $k + m < n/2$ and that

$$q > \frac{n}{2} \text{ } (n \text{ even}); \quad q > \left(\frac{n+1}{2}\right)^2 \text{ } (n \text{ odd}).$$

Then there exists an irreducible polynomial $f \equiv R \pmod{M}$ of degree n with its first k coefficients prescribed. In particular, there exists an irreducible polynomial with its first k coefficients and its last m coefficients prescribed (and non-zero constant term).

3.5.31 Corollary [685] There exists an irreducible polynomial of degree n with its first k and last m coefficients prescribed whenever $2 \le k + m \le n/3$.

3.5.32 Remark In contrast to Corollaries 3.5.30 and 3.5.31 giving existence conditions for irreducible polynomials with prescribed first and last coefficients, [1209] investigates those whose middle coefficients are fixed *providing these are all zero.*

3.5.33 Theorem [1209] Suppose $1 < k \le m < n$ and $q^{2k-m-2} \ge q(n - k + 1)^4$. Then there exists an irreducible polynomial of degree n with $a_k = \cdots = a_m = 0$.

3.5.34 Corollary [1209] For any c with $0 < c < 1$ and any positive integer n such that $(1 - 3c)n \ge 2 + 8 \log_q n$, there exists an irreducible polynomial of degree n over $\mathbb{F}_q$ with any $\lfloor cn \rfloor$ consecutive coefficients (other than the first or last) equal to 0.

3.5.35 Example [1209] With $c = 1/4$ in Corollary 3.5.34, there exists an irreducible polynomial of degree n with $\lfloor n/4 \rfloor$ consecutive coefficients equal to 0 whenever $q \ge 61$ and $n \ge 37$ and for smaller prime powers for $n \ge n_q$ where $n_q \le n_2 = 266$.

3.5.36 Remark When the prescribed coefficients do not comprise first and last, or middle coefficients, the only general estimates are asymptotic. In the following two theorems $n - m \le n - 2$ coefficients a_j of f are prescribed and given their assigned values. The remaining m coefficients $A = \{a_{j_1}, \ldots, a_{j_m}\}$, say, are allowed to take any values in $\mathbb{F}_q$.

3.5.37 Theorem [2710] Let the $n - m$ coefficients of f *not* in A (as in Remark 3.5.36) be prescribed and given their assigned values in $\mathbb{F}_q$. Regard the members of A as algebraically independent indeterminates (or transcendentals). Suppose that f is absolutely irreducible in $\mathbb{F}_q[x, A]$. Then, for sufficiently large q, the number I of irreducible polynomials $f \in \mathbb{F}_q[x]$ of degree n with the coefficients not in A as prescribed satisfies

$$I = cq^m + O(q^{m/2}),$$

where c satisfies $1/n \le c < 1$.

3.5.38 Theorem [669] Suppose $m \le n - 2$ and the $n - m$ coefficients of f *not* in A are prescribed and given their assigned values (with $a_n \ne 0$, if prescribed). Suppose that there is not a $d > 1$ such that $f \in \mathbb{F}_q(A)[x^d]$. Assume also that $p > n$. Then, for sufficiently large q, the number I as in Theorem 3.5.37 satisfies

$$I = \frac{1}{n} q^m + O\left(q^{m-\frac{1}{2}}\right),$$

where the implied constant depends only on n.

3.5.39 Remark From [669], alternative versions of Theorem 3.5.38 apply even if $p \le n$. Indeed, f need not be monic, in the sense that the coefficient of x^n may be prescribed in $\mathbb{F}_q^*$ or allowed to vary in $\mathbb{F}_q$ so long as the total number of prescribed coefficients does not exceed $n - 1$.

3.5.40 Remark The final theorem in this section relates to irreducible polynomials of even degree of a specific type, namely self-reciprocal polynomials of degree $2n$ (see Definition 2.1.48). Note that, if F is a self-reciprocal polynomial of degree $2n$, then its last coefficient $a_{2n} = 1$ and its first n coefficients are the same as its last n non-constant coefficients in the sense that $a_i = a_{2n-i}$, $i = 1, \ldots, n$.

3.5.41 Theorem [1210] The number I of irreducible self-reciprocal polynomial of degree $2n$ with the first m $(< n/2)$ coefficients prescribed satisfies

$$\left| I - \frac{q^{n-m}}{2n} \right| \leq \frac{m+5}{n} q^{\frac{n}{2}+1}.$$

See also [1211].

3.5.4 Further exact expressions

3.5.42 Remark Expressions for the number of irreducible polynomials with the first two coefficients prescribed are given in [1815]. These are described in terms of the function $H(a, n)$, $a \in \mathbb{F}_q$. Here, if $p \nmid n$, then

$$H(a,n) = \begin{cases} q^{n-2} - \lambda((-1)^{l-1} la) q^{l-1} & \text{for } n = 2l, \\ q^{n-2} + \delta(a)\lambda((-1)^l n) q^{l-1} & \text{for } n = 2l+1, \end{cases}$$

where λ denotes the quadratic character on $\mathbb{F}_q$, $\delta(a) = -1$ $(a \neq 0)$ and $\delta(0) = q - 1$. If $p|n$, then

$$H(a,n) = \begin{cases} q^{n-2} - \delta(a)\lambda((-1)^l) q^{l-1} & \text{for } n = 2l, \\ q^{n-2} + \lambda((-1)^l 2a) q^l & \text{for } n = 2l+1. \end{cases}$$

Given $a_1, a_2 \in \mathbb{F}_q$, in Theorems 3.5.43 and 3.5.45 (which relate, respectively, to odd and even q), $I(a_1, a_2)$ denotes the number of irreducible polynomials of degree n over $\mathbb{F}_q$ with the indicated first two coefficients.

3.5.43 Theorem [1815] Suppose q is odd. Then, if $p \nmid n$ and $a \in \mathbb{F}_q$,

$$I(0, -a) = \frac{1}{n} \sum_{d|n} \mu(d) H(a/d, n/d).$$

Further, if $p|n$ with $n = p^j n_0$ $(j \geq 1, p \nmid n_0)$, then

$$\begin{aligned} I(0,-a) &= \frac{1}{n} \sum_{d|n_0} \mu(d) H(a/d, n/d), \quad a \neq 0, \\ I(0,0) &= \frac{1}{n} \sum_{d|n_0} \mu(d) [H(0, n/d) - q^{n/pd}], \\ I(1,0) &= \frac{1}{nq^2} \sum_{d|n_0} \mu(d) q^{n/d}. \end{aligned}$$

3.5.44 Remark The general value of $I(a_1, a_2)$ can be recovered from Theorem 3.5.43 using, if $p \nmid n$, $I(a_1, a_2) = I(0, a_2 - \frac{(n-1)}{2n} a_1^2)$. If $p|n$ and $a_1 \neq 0$, then $I(a_1, a_2) = I(1, 0)$.

3.5.45 Theorem [1815] Suppose q is even. Then for $a \in \mathbb{F}_q$,

$$I(0,a) = \begin{cases} \dfrac{1}{n} \displaystyle\sum_{d|n} \mu(d) H(a, n/d) & \text{for } n \text{ odd}, \\ \dfrac{1}{n} \displaystyle\sum_{d|n,\ d \text{ odd}} \mu(d) [H(a, n/d) - q^{\frac{n}{2d}-1}] & \text{for } n \text{ even}; \end{cases}$$

$$I(1,a) = \frac{1}{n} \sum_{d|n,\ d \text{ odd}} \mu(d) H^*\left(a + \frac{d-1}{2}, \frac{n}{d}\right),$$

where, with $q = 2^r$ and χ the canonical additive character on $\mathbb{F}_q$,

$$H^*(a,n) = \begin{cases} q^{n-2} & \text{for } n = 4l, \\ q^{n-2} + (-1)^{lr}\delta(a) q^{\frac{n-3}{2}} & \text{for } n = 4l+1, \\ q^{n-2} + (-1)^{lr}\delta(1+a) q^{\frac{n-3}{2}} & \text{for } n = 4l-1, \\ q^{n-2} - (-1)^{lr}\chi(a) q^{\frac{n-2}{2}} & \text{for } n = 4l+2. \end{cases}$$

3.5.46 Remark The general value of $I(a_1, a_2)$ can be recovered from Theorem 3.5.45 since, for $a_1 \neq 0, I(a_1, a_2) = I(1, a_2/a_1^2)$ and $I(0, a) = I(0, 1),\ a \neq 0$.

3.5.47 Remark For the binary field $\mathbb{F}_2$, [565] contains alternative expressions for the number $I(a_1, a_2)$ in Theorem 3.5.45. For expressions for the number of irreducible polynomials over $\mathbb{F}_2$ with the first three coefficients prescribed, see [1076, 3049]. The flavor of these results is caught by the following conjecture which holds for $k \leq 3$.

3.5.48 Conjecture [3049] Let $n = 2l$ be even and $I(a_1, \ldots, a_s)$ denote the number of irreducible polynomials of degree n (over $\mathbb{F}_2$) whose first s coefficients are as shown. Then

$$I(a_1, \ldots, a_s) = \sum_{2|n,\ d \text{ odd}} \mu(d) J(n/d, a_1, \ldots, a_s),$$

where

$$J(n, a_1, \ldots, a_s) = 2^{n-s} + c_{l-r+1} 2^{l-r+1} + \cdots + c_l 2^l,$$

for some r with $1 \leq r \leq l$ and $c_i \in \{-1, 0, 1\}$.

3.5.49 Remark Let γ be a primitive element of $\mathbb{F}_q$ and $s|q-1$. A coset C of the subgroup of $\mathbb{F}_q^*$ generated by γ^s takes the form $\{\gamma^{is+h} : 0 \leq i < (q-1)/s\}$, for some h with $0 \leq h < s$. In [1783] general expressions are obtained for the number of irreducible polynomials of degree n with prescribed trace and norm in a specified coset C. These involve quantities like Gauss and Jacobi sums. Such expressions are made explicit (i.e., the trigonometric sums are precisely determined) in the following cases:

1. $s = 2$ (compare with [541]), $s = 3$ and $s = 4$;
2. $q = p^{2er}$ and s (> 1) a factor of $p^e + 1$.

3.5.50 Remark For q powers of 2 or 3, [2123] contains completely explicit expressions for the number of irreducible polynomials of degrees n in the indicated ranges for polynomials with two prescribed coefficients as follows:

1. a_1 any prescribed value, $a_{n-1} = 0$, for $n \leq 10$;
2. $a_1 = 0$, a_3 any prescribed value, for $n \leq 30$.

See Also

§4.2	For discussion of primitive polynomials with prescribed coefficients.
§6.1	For evaluation of Gauss and Jacobi sums appearing in estimates.
§13.1	For arithmetical comparisons of irreducibles with primes.

References Cited: [134, 310, 512, 541, 565, 669, 685, 1076, 1209, 1210, 1211, 1405, 1416, 1447, 1551, 1783, 1815, 2121, 2123, 2318, 2454, 2467, 2710, 2834, 2893, 3049]

3.6 Multivariate polynomials

Xiang-dong Hou, University of South Florida

3.6.1 Theorem [1561] The polynomial ring $\mathbb{F}_q[x_1,\ldots,x_k]$ is a unique factorization domain. Let $f(x_1,\ldots,x_k) = a_0 + a_1x_k + \cdots + a_nx_k^n \in \mathbb{F}_q[x_1,\ldots,x_k]$, where $n > 0$, $a_i \in \mathbb{F}_q[x_1,\ldots,x_{k-1}]$, $0 \le i \le n$, $a_n \ne 0$. Then f is irreducible in $\mathbb{F}_q[x_1,\ldots,x_k]$ if and only if it is irreducible in $(\mathbb{F}_q(x_1\ldots,x_{k-1}))[x_k]$ and $\gcd_{\mathbb{F}_q[x_1,\ldots,x_{k-1}]}(a_0,\ldots,a_n) = 1$, where $\mathbb{F}_q(x_1\ldots,x_{k-1})$ is the field of rational functions in $x_1,\ldots,x_{k-1}$ over $\mathbb{F}_q$ and $\gcd_{\mathbb{F}_q[x_1,\ldots,x_{k-1}]}(a_0,\ldots,a_n)$ denotes the gcd of $a_0,\ldots,a_n$ in $\mathbb{F}_q[x_1,\ldots,x_{k-1}]$.

3.6.2 Definition For a k-tuple of indeterminates $\boldsymbol{z} = (z_1,\ldots,z_k)$ and for $\mathfrak{n} = (n_1,\ldots,n_k) \in \mathbb{N}^k$, where $\mathbb{N} = \{0,1,2,\cdots\}$, define $\boldsymbol{z}^{\mathfrak{n}} = z_1^{n_1}\cdots z_k^{n_k}$. Let $[\boldsymbol{z}^{\mathfrak{n}}]P(\boldsymbol{z})$ denote the coefficient of $\boldsymbol{z}^{\mathfrak{n}}$ in a formal power series $P(\boldsymbol{z})$.

3.6.1 Counting formulas

3.6.3 Definition Let $\mathcal{N}_k = \mathbb{F}_q[x_1,\ldots,x_k]/\sim$, where for $f,g \in \mathbb{F}_q[x_1,\ldots,x_k]$, $f \sim g$ means that $f = cg$ for some $c \in \mathbb{F}_q^*$. Elements of $\mathcal{N}_k$ are normalized polynomials in $\mathbb{F}_q[x_1,\ldots,x_k]$. Let $\mathcal{N}_k(m) = \{f \in \mathcal{N}_k : \deg f = m\}$, where $\deg f$ denotes the total degree of f. Let $N_k(m) = |\mathcal{N}_k(m)| = \frac{1}{q-1}\left[q^{\binom{m+k}{k}} - q^{\binom{m+k-1}{k}}\right]$, $I_k(m) = |\{f \in \mathcal{N}_k(m) : f \text{ is irreducible}\}|$, $P_k(m;n) = |\{(f,g) \in \mathcal{N}_k(m) \times \mathcal{N}_k(n) : \gcd(f,g) = 1\}|$.

3.6.4 Remark We have that $N_k(m)$ is the number of normalized polynomials of (total) degree m in $\mathbb{F}_q[x_1,\ldots,x_k]$, $I_k(m)$ is the number of normalized irreducible polynomials of degree m, and $P_k(m;n)$ is the number of relatively prime pairs of normalized polynomials of degrees m and n, respectively.

3.6.5 Theorem [336] We have $I_k(0) = 0$, and for $m > 0$, $I_k(m)$ is given by the recursive formula

$$I_k(m) = N_k(m) - \sum_{1a_1+2a_2+\cdots+(m-1)a_{m-1}=m} \binom{I_k(1)+a_1-1}{a_1}\cdots\binom{I_k(m-1)+a_{m-1}-1}{a_{m-1}}.$$

3.6.6 Theorem [337] Let $k \ge 1$ be a fixed integer. Define $N(z) = \sum_{n\ge0} N_k(n)z^n$ and $I(z) = \sum_{n\ge1} I_k(n)z^n$. Then

1. $I(z) = \sum_{m\ge1} \frac{\mu(m)}{m} \log N(z^m)$.
2. $I_k(n) = \sum_{m|n} \frac{\mu(m)}{m} [z^{n/m}] \log N(z)$.

3.6.7 Theorem [1546] We have

$$P_k(m;n) = \sum_{0\le d\le\min\{m,n\}} N_k(m-d)N_k(n-d)A_k(d),$$

where

$$A_k(d) = \sum_{1a_1+2a_2+\cdots+da_d=d} (-1)^{a_1+\cdots+a_d}\binom{I_k(1)}{a_1}\cdots\binom{I_k(d)}{a_d}.$$

3.6.8 Remark In Theorem 3.6.7, when $k \ge 2$, no closed formula for $A_k(d)$ is known. When $k = 1$, it is known that $A_1(0) = 1$, $A_1(1) = -q$, and $A_1(d) = 0$ for $d \ge 2$ [537, 668].

3.6.2 Asymptotic formulas

3.6.9 Theorem [1546] Let $k \geq 2$ and $t \geq 0$ be integers. Then as $m \to \infty$,

$$I_k(m) = \sum_{i=0}^{t} \alpha_i N_k(m-i) + O\Big(q^{\binom{m-t-1+k}{k}}\Big),$$

where $\alpha_0 = 1$, and for $i > 0$

$$\alpha_i = \sum_{1a_1+\cdots+ia_i=i} \frac{(a_1+\cdots+a_i)!}{a_1!\cdots a_i!}(-1)^{a_1+\cdots+a_i} N_k(1)^{a_1}\cdots N_k(i)^{a_i},$$

where the constant in the O-term depends only on q, k, t.

3.6.10 Remark There is a fundamental difference between the distribution of irreducible polynomials over $\mathbb{F}_q$ in the univariate case and in the multivariate case. For the univariate case, $\lim_{m\to\infty} \frac{I_1(m)}{N_1(m)} = 0$. For $k \geq 2$, $\lim_{m\to\infty} \frac{I_k(m)}{N_k(m)} = 1$.

3.6.11 Theorem [1546] Let $k \geq 2$. Then $\lim_{m+n\to\infty} \frac{P_k(m;n)}{N_k(m)N_k(n)} = 1$.

3.6.12 Remark [226] We have for the univariate case $\frac{P_1(m;n)}{N_1(m)N_1(n)} = 1 - \frac{1}{q}$; see Section 11.2 for more results on counting univariate polynomials.

3.6.13 Theorem [1546] Let $k \geq 2$ and $t \geq 0$ be fixed integers. Then

$$P_k(m;n) = \sum_{d=0}^{t} N_k(m-d)N_k(n-d)A_k(d) + O\big(N_k(m-t-1)N_k(n-t-1)\big),$$

where $A_k(d)$ is defined in Theorem 3.6.7. The constant in the O-term depends only on q, k, t.

3.6.3 Results for the vector degree

3.6.14 Definition Let $f \in \mathbb{F}_q[x_1, \ldots, x_k]$. The *vector degree* of f, denoted by $\mathrm{Def}\, f$, is the k-tuple $(\deg_{x_1} f, \ldots, \deg_{x_k} f)$.

3.6.15 Definition For $\mathfrak{m} = (m_1, \ldots, m_k)$ and $\mathfrak{n} = (n_1, \ldots, n_k) \in \mathbb{N}^k$, let $\mathcal{N}_k(\mathfrak{m}) = \{f \in \mathcal{N}_k : \mathrm{Deg}\, f = \mathfrak{m}\}$, $N_k(\mathfrak{m}) = |\mathcal{N}_k(\mathfrak{m})|$, $I_k(\mathfrak{m}) = |\{f \in \mathcal{N}_k(\mathfrak{m}) : f \text{ is irreducible}\}|$ and $P_k(\mathfrak{m};\mathfrak{n}) = |\{(f,g) \in \mathcal{N}_k(\mathfrak{m}) \times \mathcal{N}_k(\mathfrak{n}) : \gcd(f,g) = 1\}|$.

3.6.16 Remark We have that $N_k(\mathfrak{m})$ is the number of normalized polynomials of vector degree $\mathfrak{m}$ in $\mathbb{F}_q[x_1, \ldots, x_k]$, $I_k(\mathfrak{m})$ is the number of normalized irreducible polynomials of vector degree $\mathfrak{m}$, and $P_k(\mathfrak{m};\mathfrak{n})$ is the number of relative prime pairs of normalized polynomials of vector degrees $\mathfrak{m}$ and $\mathfrak{n}$, respectively.

3.6.17 Remark [664, 1546] For $\mathfrak{m} = (m_1, \ldots, m_k) \in \mathbb{N}^k$,

$$N_k(\mathfrak{m}) = \frac{1}{q-1} \sum_{(\delta_1,\ldots,\delta_k)\in\{0,1\}^k} (-1)^{k+\delta_1+\cdots+\delta_k} q^{(m_1+\delta_1)\cdots(m_k+\delta_k)}.$$

3.6.18 Definition Let $\mathfrak{o} = (0, \ldots, 0) \in \mathbb{N}^k$. For $\mathfrak{i} = (i_1, \ldots, i_k)$, $\mathfrak{m} = (m_1, \ldots, m_k) \in \mathbb{N}^k$, define $\mathfrak{i} \le \mathfrak{m}$ if $i_j \le m_j$ for all $1 \le j \le k$.

3.6.19 Theorem [1546] We have $I_k(\mathfrak{o}) = 0$, and for $\mathfrak{m} > \mathfrak{o}$, $I_k(\mathfrak{m})$ is given by the recursive formula

$$I_k(\mathfrak{m}) = N_k(\mathfrak{m}) - \sum_{\substack{(a_\mathfrak{i})_{\mathfrak{o}<\mathfrak{i}<\mathfrak{m}} \\ \sum_\mathfrak{i} a_\mathfrak{i} \mathfrak{i}=\mathfrak{m}}} \prod_\mathfrak{i} \binom{I_k(\mathfrak{i}) + a_\mathfrak{i} - 1}{a_\mathfrak{i}}.$$

3.6.20 Theorem [337] Let $k \ge 1$ be a fixed integer and let $\boldsymbol{z} = (z_1, \ldots, z_k)$ be a k-tuple of indeterminates. Define $N(\boldsymbol{z}) = \sum_{\mathfrak{n}\in\mathbb{N}^k} N_k(\mathfrak{n})\boldsymbol{z}^\mathfrak{n}$, $I(\boldsymbol{z}) = \sum_{\mathfrak{o}\ne\mathfrak{n}\in\mathbb{N}^k} I_k(\mathfrak{n})\boldsymbol{z}^\mathfrak{n}$. Then

1. $I(\boldsymbol{z}) = \sum_{m\ge1} \frac{\mu(m)}{m} \log N(\boldsymbol{z}^k)$.
2. $I_k(\boldsymbol{z}) = \sum_{m|\gcd(\mathfrak{n})} \frac{\mu(m)}{m} [z^{\frac{1}{m}\mathfrak{n}}] \log N(\boldsymbol{z})$.

3.6.21 Theorem [1546] Let $k \ge 2$ and $(m_1, \ldots, m_k) \in (\mathbb{Z}^+)^k$ with $m_1 = \max_{1\le i\le k-1} m_i$. Further assume that $m_1 \ge 3$ if $k = 2$ and $m_1 \ge 2$ if $k = 3$. Then

$$I_k(m_1, \ldots, m_k) = N_k(m_1, \ldots, m_k) - qN_k(m_1, \ldots, m_{k-1}, m_k - 1) + O\big(q^{m_1(m_2+1)\cdots(m_k+1)}\big).$$

3.6.22 Remark The asymptotic formula in Theorem 3.6.21 is interesting only when $m_k > m_1$ since otherwise the O-term is bigger than or comparable to the term $qN_k(m_1, \ldots, m_{k-1}, m_k - 1)$. When $m_k > m_1$, Theorem 3.6.21 indicates that most of the reducible polynomials in $\mathcal{N}_k(m_1, \ldots, m_k)$ are of the form $(x_k + \alpha)f$ for some $\alpha \in \mathbb{F}_q$, $f \in \mathcal{N}_k(m_1, \ldots, m_{k-1}, m_k - 1)$.

3.6.23 Corollary Under the assumptions of Theorem 3.6.21, we have

$$I_k(m_1, \ldots, m_k) = (1 - q^{-M_k})N(m_1, \ldots, m_k) + O\big(q^{m_1(m_2+1)\cdots(m_k+1)}\big),$$

where $M_k = (m_1 + 1)\cdots(m_{k-1} + 1) - 1$. The main term here was given by Cohen [664, Theorem 1]. For fixed $m_1, \ldots, m_{k-1}$, it is not the case that almost all polynomials in $\mathcal{N}_k(m_1, \ldots, m_k)$ are irreducible: we have $\frac{I_k(m_1,\ldots,m_k)}{N_k(m_1,\ldots,n_k)} \to 1 - q^{-M_k}$ as $m_k \to \infty$.

3.6.24 Remark [1546] For $k \ge 2$, $\frac{I_k(m_1,\ldots,m_k)}{N_k(m_1,\ldots,m_k)} \to 1$ as both $m_1, m_k \to \infty$.

3.6.25 Theorem [1546] For $\mathfrak{m} = (m_1, \ldots, m_k)$, $\mathfrak{n} = (n_1, \ldots, n_k) \in \mathbb{N}^k$, define $\min\{\mathfrak{m}, \mathfrak{n}\} = (\min\{m_1, n_1\}, \ldots, \min\{m_k, n_k\}) \in \mathbb{N}^k$. We have

$$P_k(\mathfrak{m}; \mathfrak{n}) = \sum_{\mathfrak{o}\le\mathfrak{d}\le\min\{\mathfrak{m},\mathfrak{n}\}} N_k(\mathfrak{m} - \mathfrak{d})N_k(\mathfrak{n} - \mathfrak{d})A_k(\mathfrak{d}),$$

where

$$A_k(\mathfrak{d}) = \sum_{\substack{(a_\mathfrak{i})_{\mathfrak{o}<\mathfrak{i}\le\mathfrak{d}} \\ \sum_{\mathfrak{o}<\mathfrak{i}\le\mathfrak{d}} a_\mathfrak{i}\mathfrak{i}=\mathfrak{d}}} (-1)^{\sum_{\mathfrak{o}<\mathfrak{i}\le\mathfrak{d}} a_\mathfrak{i}} \prod_{\mathfrak{o}<\mathfrak{i}\le\mathfrak{d}} \binom{I_k(\mathfrak{i})}{a_\mathfrak{i}}.$$

3.6.26 Theorem [1546] Let $k \ge 2$ and $\mathfrak{m} = (m_1, \ldots, m_k)$, $\mathfrak{n} = (n_1, \ldots, n_k) \in \mathbb{N}^k$. Then $\lim_{m_{k-1},m_k\to\infty} \frac{P_k(\mathfrak{m};\mathfrak{n})}{N_k(\mathfrak{m})N_k(\mathfrak{n})} = 1$.

3.6.27 Theorem [1546] Let $k \ge 2$, $\mathfrak{m} = (m_1, \ldots, m_k)$, $\mathfrak{n} = (n_1, \ldots, n_k)$, $\mathfrak{t} = (t_1, \ldots, t_k) \in \mathbb{N}^k$ such that $\mathfrak{t} \le \min\{\mathfrak{m}, \mathfrak{n}\}$ and $\max\{m_i, n_i\} > 2t_i + 1$ for all $1 \le i \le k$. Then

$$P_k(\mathfrak{m}; \mathfrak{n}) = \sum_{\mathfrak{o}\le\mathfrak{d}\le\mathfrak{t}} N_k(\mathfrak{m} - \mathfrak{d})N_k(\mathfrak{n} - \mathfrak{d})A_k(\mathfrak{d})$$
$$+ O\left(q^{\max_{1\le j\le k}\left(\frac{m_j-t_j}{m_j+1}\prod_{i=1}^k(m_i+1)+\frac{n_j-t_j}{n_j+1}\prod_{i=1}^k(n_i+1)\right)}\right).$$

3.6.28 Theorem [1546] Let $k \geq 2$, $\mathfrak{m} = (m_1, \ldots, m_k), \mathfrak{n} = (n_1, \ldots, n_k) \in \mathbb{N}^k$ such that $m_k > 0$, $n_k > 0$ and $\max\{m_i, n_i\} > 1$ for all $1 \leq i \leq k-1$. Then

$$P_k(\mathfrak{m}; \mathfrak{n}) = N_k(\mathfrak{m})N_k(\mathfrak{n}) - qN_k(m_1, \ldots, m_{k-1}, m_k - 1)N_k(n_1, \ldots, n_{k-1}, n_k - 1) + O\left(q^{\max_{1\leq j\leq k-1}\left(\frac{m_j}{m_j+1}\prod_{i=1}^{k}(m_i+1)+\frac{n_j}{n_j+1}\prod_{i=1}^{k}(n_i+1)\right)}\right).$$

3.6.29 Theorem [668] Let $k \geq 2$ and $\mathfrak{m} = (m_1, \ldots, m_k) \in \mathbb{N}^k$ with $\max_{1\leq i\leq k-1} m_i \geq 1$. Then as $m_k \to \infty$ (with $m_1, \ldots, m_{k-1}$ fixed), $\frac{P_k(\mathfrak{m};\mathfrak{m})}{N_k(\mathfrak{m})^2} \to 1 - q^{1-2(m_1+1)\cdots(m_{k-1}+1)}$.

3.6.4 Indecomposable polynomials and irreducible polynomials

3.6.30 Definition Let $\mathbb{F}$ be a field and $k \geq 2$. A nonconstant polynomial $f \in \mathbb{F}[x_1, \ldots, x_k]$ is *indecomposable* over $\mathbb{F}$ if there do not exist $h \in \mathbb{F}[x_1, \ldots, x_k]$ and $u \in \mathbb{F}[t]$ with $\deg u \geq 2$ such that $f = u(h(x_1, \ldots, x_k))$.

3.6.31 Theorem [2214] Let $k \geq 2$. Assume that $f \in \mathbb{F}_q[x_1, \ldots, x_k]$ is indecomposable over $\overline{\mathbb{F}}_q$, the algebraic closure of $\mathbb{F}_q$. For each $\lambda \in \overline{\mathbb{F}}_q$, let $\mathcal{I}_\lambda$ denote the set of all distinct irreducible factors of $f - \lambda$ over $\overline{\mathbb{F}}_q$. Then

$$\sum_{\lambda\in\overline{\mathbb{F}}_q} (|\mathcal{I}_\lambda| - 1) \leq \min_{\lambda\in\overline{\mathbb{F}}_q} \sum_{g\in\mathcal{I}_\lambda} \deg g - 1.$$

In particular, $f - \lambda$ is reducible over $\overline{\mathbb{F}}_q$ for at most $\deg f - 1$ values of λ.

3.6.32 Remark [338, Theorem 4.2] $f \in \mathbb{F}_q[x_1, \ldots, x_k]$ is indecomposable over $\overline{\mathbb{F}}_q$ if and only if it is indecomposable over $\mathbb{F}_q$.

3.6.33 Definition [1224] Let $\mathbb{F}$ be a field and fix a term order in $\mathbb{F}[x_1, \ldots, x_k]$ that respects the total degree. A monic polynomial $f \in \mathbb{F}[x_1, \ldots, x_k]$ with $f(0, \ldots, 0) = 0$ is *monic original.*

3.6.34 Remark [338, 1224] Let $\mathbb{F}$ be a field and let $k \geq 2$. Every monic original polynomial $f \in \mathbb{F}[x_1, \ldots, x_k]$ has a unique decomposition $f = u \circ h$, where $u \in \mathbb{F}[t]$, $h \in \mathbb{F}[x_1, \ldots, x_k]$ are both monic original and h is indecomposable.

3.6.35 Theorem [1224] Let $\mathbb{F}$ be an algebraically closed field and let $k \geq 2$ and $n \geq 2$ be integers. Denote by l the smallest prime divisor of n. Then the set of all decomposable monic original polynomials in $\mathbb{F}[x_1, \ldots, x_k]$ of degree n is an affine algebraic set of dimension $\binom{k+\frac{n}{m}}{k}+m-3$, where

$$m = \begin{cases} n & \text{if } k = 2,\ \frac{n}{l} \text{ is a prime and } \frac{n}{l} \leq 2l - 5, \\ l & \text{otherwise.} \end{cases}$$

3.6.36 Definition Let $k \geq 2$ and $n \geq 1$. Denote the number of monic original (respectively indecomposable monic original, decomposable monic original) polynomials of (total) degree n in $\mathbb{F}_q[x_1, \ldots, x_k]$ by $\mathsf{P}_{k,n}$ (respectively $\mathsf{I}_{k,n}$, $\mathsf{D}_{k,n}$).

3.6.37 Theorem [338] We have $\mathsf{I}_{k,1} = q^{k-1}$, and for $n > 1$, $\mathsf{I}_{k,n}$ is given by the recursive formula

$$\mathsf{I}_{k,n} = \mathsf{P}_{k,n} - \sum_{m|n,\, m<n} q^{\frac{n}{m}-1}\,\mathsf{I}_{k,m}.$$

3.6.38 Remark [338] Let $k \geq 2$. We have $\frac{\mathsf{I}_{k,n}}{\mathsf{P}_{k,n}} \to 1$ when $n \to \infty$ with q fixed or when $q \to \infty$ with n fixed.

3.6.39 Theorem [1224] Assume $k \geq 2$ and $n \geq 2$. Let l be the smallest prime divisor of n and let m defined in Theorem 3.6.35. Then the following hold.

1. $|\mathsf{D}_{k,n} - \alpha_{k,n}| \leq \alpha_{k,n}\beta^*_{k,n}$ where

$$\begin{aligned} \alpha_{k,n} &= q^{\binom{k+\frac{n}{m}}{k}+m-3} \cdot \frac{1-q^{-\binom{k-1+\frac{n}{m}}{k-1}}}{1-q^{-1}}, \\ \beta^*_{k,n} &= \frac{2}{1-q^{-1}}\, q^{-\frac{1}{2}\binom{k-1+\frac{n}{l}}{k-1}+1}. \end{aligned}$$

2. $\mathsf{I}_{k,n} \geq \mathsf{P}_{k,n} - 2\alpha_{k,n}$.

3.6.40 Remark For a refinement of the bound in Theorem 3.6.39, see [1224, Theorem 4.1], where $\beta^*_{k,n}$ is replaced by an expression $\beta_{k,n}$ which is of smaller magnitude but more complicated; also see [1244] for related results.

3.6.5 Algorithms for the gcd of multivariate polynomials

3.6.41 Remark Computation of the gcd of multivariate polynomials over finite fields is more difficult than that of univariate polynomials; the Extended Euclidean Algorithm alone is not sufficient to produce the gcd due to the fact that $\mathbb{F}_q[x_1, \ldots, x_k]$ with $k > 1$ is no longer a Euclidean domain. In this subsection we gather several algorithms for computing the multivariate gcd based on different approaches.

3.6.42 Definition Let R be a unique factorization domain. For $f \in R[x]$, $\mathrm{lc}(f)$ denotes the leading coefficient of f; $\mathrm{pp}(f)$ denotes the *primitive part* of f, i.e., f divided by the gcd of its coefficients. The resultant of $f, g \in R[x]$ is denoted by $\mathrm{res}(f, g)$.

3.6.43 Algorithm [1227]

Input: Primitive $f, g \in (\mathbb{F}_q[\boldsymbol{y}])[x]$, where $\boldsymbol{y} = (y_1, \ldots, y_k)$.

Output: $\gcd_{\mathbb{F}_q[x,\boldsymbol{y}]}(f, g)$.

1. Use the Extended Euclidean Algorithm to compute the monic $v = \gcd_{(\mathbb{F}_q(\boldsymbol{y}))[x]}(f, g)$.
2. Compute $b = \gcd_{\mathbb{F}_q[\boldsymbol{y}]}(\mathrm{lc}(f), \mathrm{lc}(g))$.
3. $\gcd_{\mathbb{F}_q[x,\boldsymbol{y}]}(f, g) = \mathrm{pp}(bv)$.

3.6.44 Algorithm [1227] (Modular bivariate gcd, small primes version)

Input: Primitive $f, g \in (\mathbb{F}_q[y])[x]$, $\deg_x f = n \geq \deg_x g \geq 1$, $\deg_y f, \deg_y g \leq d$ and $q \geq (4n+2)d + 2$.

Output: $\gcd_{\mathbb{F}_q[x,y]}(f, g)$.

1. Compute $b = \gcd_{\mathbb{F}_q[y]}(\mathrm{lc}_x(f), \mathrm{lc}_x(g))$. Let $l = d + 1 + \deg_y b$.
2. Repeat steps 3 – 7 until the conditions in step 7 are satisfied.
3. Choose $S \subset \mathbb{F}_q$ with $|S| = 2l$.
4. Delete the roots of b from S. For each $u \in S$, use the Extended Euclidean Algorithm to compute the monic $v_u = \gcd_{\mathbb{F}_q[x]}(f(x, u), g(x, u))$.
5. Determine $e = \min\{\deg_x v_u : u \in S\}$. Delete $\{u \in S : \deg_x v_u > e\}$ from S. If $|S| \geq l$, delete $|S| - l$ elements from S; otherwise, go to step 3.

6. Compute by interpolation the coefficients in $\mathbb{F}_q[y]$ of w, f^*, $g^* \in (\mathbb{F}_q[y])[x]$ of degrees in y less than l such that

$$w(x,u) = b(u)v_u, \quad f^*(x,u)w(x,u) = b(u)f(x,u), \quad g^*(x,u)w(x,u) = b(u)g(x,u)$$

for all $u \in S$.
7. Check if $\deg_y(f^*w) = \deg_y(bf)$ and $\deg_y(g^*w) = \deg_y(bg)$.
8. $\gcd_{\mathbb{F}_q[x,y]}(f,g) = \mathrm{pp}_x(w)$.

3.6.45 Remark Let $h = \gcd_{\mathbb{F}_q[x,y]}(f,g)$. The conditions in Step 7 of Algorithm 3.6.44 are satisfied if and only if S does not contain any root of $\mathrm{res}_x(f/h, g/h)$.

3.6.46 Algorithm [13] (Gröbner basis)
Input: $f, g \in \mathbb{F}_q[x_1, \ldots, x_k]$, $f, g \neq 0$.
Output: $\gcd(f,g)$.

1. Compute the reduced Gröbner basis G for the ideal $\langle wf, (1-w)f\rangle$ of $\mathbb{F}_q[x_1, \ldots, x_k, w]$ with respect to an elimination order with $x_1, \ldots, x_k$ smaller than w.
2. $\mathrm{lcm}(f,g)$ is the polynomial in G which does not involve w.
3. $\gcd(f,g) = \frac{fg}{\mathrm{lcm}(f,g)}$.

3.6.47 Remark For the correctness of Algorithm 3.6.43, see [1227, Theorem 6.12]. The cost of the Extended Euclidean Algorithm is given in [1227, Theorem 3.11]. For the correctness and the cost of Algorithm 3.6.44, see [1227, Theorem 6.37]. The correctness of Algorithm 3.6.46 is given in [13, p.72]. For the complexity of computing reduced Gröbner bases, see [1227, Section 21.7].

See Also

§3.1	For counting univariate irreducible polynomials.
§11.4	For factorization algorithms of univariate polynomials.
§11.5	For factorization algorithms of multivariate polynomials.

References Cited: [13, 226, 336, 337, 338, 537, 664, 668, 1224, 1227, 1244, 1546, 1561, 2214]

4

Primitive polynomials

4.1 Introduction to primitive polynomials

Gary L. Mullen, The Pennsylvania State University
Daniel Panario, Carleton University

4.1.1 Definition An element $\alpha \in \mathbb{F}_q$ is a *primitive element* if α generates the multiplicative group $\mathbb{F}_q^*$ of nonzero elements in $\mathbb{F}_q$.

4.1.2 Definition A polynomial $f \in \mathbb{F}_q[x]$ of degree $n \geq 1$ is a *primitive polynomial* if it is the minimal polynomial of a primitive element of $\mathbb{F}_{q^n}$.

4.1.3 Theorem The number of primitive polynomials of degree n over $\mathbb{F}_q$ is $\phi(q^n - 1)/n$, where ϕ denotes Euler's function.

4.1.4 Remark [1939, Section 3.1] The reciprocal polynomial f^* (with leading coefficient different from 0) of a primitive polynomial f of degree n is defined by $f^*(x) = x^n f(1/x)$; see Definition 2.1.48. The monic reciprocal polynomial of a primitive polynomial is again primitive. In general, for any polynomial f, the order of f^* is the same as the order of f. For the definition of order of a polynomial see Definition 2.1.51.

4.1.5 Theorem [1939, Theorem 3.16] A polynomial $f \in \mathbb{F}_q[x]$ of degree n is primitive if and only if f is monic, $f(0) \neq 0$, and the order of f is $q^n - 1$.

4.1.6 Theorem [1939, Theorem 3.18] A monic polynomial $f \in \mathbb{F}_q[x]$ of degree $n \geq 1$ is a primitive polynomial if and only if the smallest positive integer r for which x^r is congruent modulo f to some element of $\mathbb{F}_q$ is $r = (q^n - 1)/(q - 1)$ and $(-1)^n f(0)$ is a primitive element in $\mathbb{F}_q$. In case f is primitive over $\mathbb{F}_q$, $x^r \equiv (-1)^n f(0) \pmod{f(x)}$.

4.1.7 Theorem [1071] Let f be an irreducible polynomial of degree k over $\mathbb{F}_q$. Set $m = q^k - 1$ and define $g(x) = (x^m - 1)/((x-1)f(x))$. Then f is primitive if and only if g has exactly $(q-1)q^{k-1} - 1$ nonzero terms.

4.1.8 Theorem [1857] Let $f(x) = f_k + f_{k-1}x + \cdots + f_0x^k$ be irreducible over $\mathbb{F}_q$ of degree $k \geq 2$. Set $m = q^k - 1, t = m/(q-1), y = x^{q-1}$ and

$$g(y) = \frac{y^t - 1}{(y-1)f(y)} = h(y) + \frac{r_0 + r_1y + \cdots + r_{k-1}y^{k-1}}{f(y)},$$

where $h(y) = \epsilon_{t-k} + \epsilon_{t-k-1}y + \cdots + \epsilon_1 y^{t-k-1}$. Then f is primitive if and only if the number of nonzero terms in $h(y)$, considered as a polynomial in y over $\mathbb{F}_q$, is equal to $q^{k-1}(q-1)-1-N$, where N is the number of nonzero terms in the finite sequence $\epsilon_{t-k+1}, \epsilon_{t-k+2}, \ldots, \epsilon_m$ defined by $\epsilon_{t-n} = r_n - \sum_{i=1}^{k-n-1} f_i\epsilon_{t-n-i}, n = 0, 1, \ldots, k-1$ where the empty sum is interpreted as 0, and $\epsilon_{t+n} = 1 - \sum_{i=1}^{k} f_i\epsilon_{t+n-i}, n = 1, 2\ldots, m-t$.

4.1.9 Corollary [1857] An irreducible polynomial is primitive if and only if the finite sequence $\epsilon_1, \ldots, \epsilon_m$ defined in Theorem 4.1.8 contains no two identical periodic subsequences.

4.1.10 Definition [2186] Define $w_n(q)$ and $W_n(q)$ as the minimal weight (the number of nonzero coefficients) among all monic irreducible and primitive, respectively, polynomials of degree n over $\mathbb{F}_q$.

4.1.11 Remark It is stated in [2186] that $w_n(p) \leq W_n(p) \leq (n+1)/2$ for any sufficiently large prime p. For $p = 2$, $w_n(2) \leq W_n(2) \leq n/4 + o(n)$.

4.1.12 Problem Extend these results to $\mathbb{F}_q$ when q is a power of a prime.

4.1.13 Conjecture [2186] $W_n(2) = 3$ infinitely often.

4.1.14 Problem [2186] Find examples of fields $\mathbb{F}_{q^n}$ with $W_n(q) = o(n)$ or at least with $w_n(q) = o(n)$ for infinitely many n.

4.1.15 Remark The following conjectures require the notions of primitive normal polynomials and completely normal primitive polynomials; see Sections 5.2 and 5.4.

4.1.16 Conjecture [1416, 2156]

1. For each prime p and $n \geq 2$, $w_n(p) \leq 5$ and if $p \neq 2, 3$, then $w_n(p) \leq 4$.
2. For each prime p and $n \geq 2$ there is a primitive normal polynomial of degree n over $\mathbb{F}_p$ with weight at most 5.
3. If $p \geq 11$ weight 5 can be replaced by weight 4.

4.1.17 Definition For a prime $p \geq 3$, assume that the field $\mathbb{F}_p$ consists of the elements $0, \pm 1, \ldots, \pm(p-1)/2$. The *height* of a polynomial is the maximum absolute value of its coefficients. We define $h_n(p)$ and $H_n(p)$ as the minimal height of all monic irreducible and primitive, respectively, polynomials of degree n over $\mathbb{F}_p$.

4.1.18 Remark It is stated in [2186] that $h_n(p) = O(p^{2/3})$ and $H_n(p) = O(p^{n/(n+1)+\varepsilon})$. The bound $h_n(p) = O(p^{2/3})$ has been improved to $h_n(p) \leq p^{1/2+o(1)}$ in [2654].

4.1.19 Problem Improve the above bounds for $h_n(p)$ and $H_n(p)$.

4.1.20 Problem Extend the bounds for $\mathbb{F}_q$ when q is a power of a prime.

4.1.21 Theorem [690] For $n \geq 2$ and $a \in \mathbb{F}_q^*$, there is a primitive normal polynomial of degree n over $\mathbb{F}_q$ with trace a.

4.1.22 Theorem There is a primitive normal polynomial of degree $n \geq 3$ with given norm and nonzero trace; see [682] for $n \geq 5$, [692] for $n = 4$, and [1557] for $n = 3$.

4.1.23 Conjecture [2157] For each $n \geq 2$ there is a completely normal primitive polynomial of degree n over $\mathbb{F}_q$. (This is true if n is a prime, or $n = 4$, or if $q^n \leq 2^{31}$ with $q \leq 97$.)

4.1.24 Remark [2186] For $q \geq n \log n$ there is a completely normal primitive polynomial of $\mathbb{F}_{q^n}$ over $\mathbb{F}_q$.

4.1.25 Conjecture [2816] For any k there is a trinomial f so that $\left(f(x), x^{2^k-1}+1\right)$ is a primitive polynomial of degree k over $\mathbb{F}_2$. (This conjecture has been proved for $k \leq 500$ [307].)

4.1.26 Definition Let $f(x) = x^n + \sum_{k=0}^{t} a_k x^{n_k}$, $a_k \neq 0$, and $0 = n_0 < n_1 < \cdots < n_t < n$. We define the *excess* of f by

$$E(f) = \sum_{t \geq j \geq (t+1)/2} n_j - \sum_{t/2 \geq j \geq 1} n_j.$$

4.1.27 Remark The excess of a polynomial is related to self-dual, weakly self-dual and almost weakly self-dual bases; see Section 5.1.

4.1.28 Remark In [2186] it is proved that there are binary primitive polynomials of degree n with $E(f) \leq \frac{3}{32}n^2 + o(n)$.

4.1.29 Problem [2158] Let $E(f)$ denote the excess of the polynomial f. Prove or disprove that for p a prime, there is a primitive polynomial f for each degree $n \geq 2$ with excess $E(f)$ at most as follows:

1. $E(f) \leq 1$, if $p > 5$;
2. $E(f) \leq 2$, if $p = 5$;
3. $E(f) \leq 3$, if $p = 3$;
4. $E(f) \leq 6$, if $p = 2$.

4.1.30 Remark See [2158] for more results related to the excess of a polynomial.

4.1.31 Remark Several classes of irreducible and primitive polynomials seem to have asymptotic density δ_{irr} (as $q \to \infty$) and δ_{prim} given by

$$\delta_{\text{irr}} = \frac{1}{n} \quad \text{and} \quad \delta_{\text{prim}} = \frac{\phi(q^n - 1)}{nq^n}.$$

4.1.32 Problem [2158] Find natural examples of families of polynomials over $\mathbb{F}_q$ having densities of irreducible and/or primitive polynomials different from δ_{irr} and δ_{prim}. Some such examples are given in Table 7 of [1181] and in [673] in connection to windmill polynomials.

See Also

§2.2	For tables of primitives of various kinds and weights.
§4.2	For discussion of primitive polynomials with prescribed coefficients.
§4.3	For discussion of primitive polynomials with prescribed weights.
§5.2	For information on primitive normal polynomials.
§5.4	For discussion of completely normal primitive polynomials.
§10.2	For connections to linear feedback shift registers.
§11.3	For algorithms related to primitive polynomials.
§14.9	For combinatorial applications of primitive polynomials; see also [2204].
[588]	Considers formulas for primitive polynomials of special forms.
[2807]	Develops the notion of "typical primitive polynomials" over $\mathbb{Z}_n$.

References Cited: [307, 588, 673, 682, 690, 692, 1071, 1181, 1416, 1557, 1857, 1939, 2156, 2157, 2158, 2186, 2204, 2654, 2807, 2816]

4.2 Prescribed coefficients

Stephen D. Cohen, University of Glasgow

4.2.1 Definition Given a positive integer r define $\tau(r) = \phi(r)/r$, where ϕ is Euler's function.

4.2.2 Remark For a pair (q, n) with q (as always) a prime power, $\tau(q^n - 1)$ is the proportion of non-zero elements of $\mathbb{F}_{q^n}$ that are primitive (see Theorem 4.1.3). It also signifies the proportion of powers of irreducible polynomials of degree n that are primitive.

4.2.3 Remark For a primitive polynomial $f \in \mathbb{F}_q$ and non-zero $a \in \mathbb{F}_q$ the polynomial $F(x + a)$ need not be primitive. The arithmetical structure of the set of primitive polynomials is less marked than that of the set of irreducible polynomials.

4.2.4 Remark The monic reciprocal $f^*(x) = a_n^{-1}x^n f(1/x)$ (see Remark 3.5.14) of a primitive polynomial is also primitive (Remark 4.1.4).

4.2.5 Remark Except as mentioned, all polynomials (in particular in statements of theorems, etc) will be assumed to be *monic polynomials of degree n over $\mathbb{F}_q$* taken to have the form $f(x) = x^n + \sum_{i=1}^n a_i x^{n-i}$. Here a_i is the *i-th coefficient*. The meaning of the terms first (or last) m coefficients will be as described in Definition 3.5.1. In particular $-a_1$ is the *trace of* f and $(-1)^n a_n$ is the *norm of* f.

4.2.6 Remark The norm of a primitive polynomial has to be a primitive element of $\mathbb{F}_q$ (Theorem 4.1.6).

4.2.7 Remark Asymptotic estimates (for large n) of the number of primitive polynomials of degree n with prescribed coefficients (e.g., with prescribed first or last coefficients) in themselves do not lead to strong existence results. All the theorems in this section (except Theorem 4.2.14) are existence results that are unaccompanied by useful lower bounds on the number of primitive polynomials over a range of pairs (q, n).

4.2.8 Problem Supply non-trivial bounds for the number of primitive polynomials with prescribed coefficients in the problems in the following subsections.

4.2.9 Definition A *normal polynomial* of degree n over $\mathbb{F}_q$ is an irreducible polynomial whose roots comprise a normal basis of $\mathbb{F}_{q^n}$ over $\mathbb{F}_q$ (Definition 2.1.98, Section 5.2).

4.2.10 Remark This section will also feature results on the distribution of polynomials with prescribed coefficients that are simultaneously primitive and normal.

4.2.1 Approaches to results on prescribed coefficients

4.2.11 Remark Character sum techniques and estimates constitute the principal underlying mechanism. Specifically, one can characterize both primitivity and the prescribed coefficient requirement in terms of such sums. The primitivity condition is in terms of *multiplicative* character sums over $\mathbb{F}_{q^n}$, whereas the coefficient conditions involve lifted (multiplicative and additive) characters over $\mathbb{F}_q$.

4.2.12 Remark Theorem 4.2.14 is an illustration of a typical asymptotic type of lower bound estimate that can be attained. All existence results described derive from such estimates (perhaps "weighted"). Further examples are not given here because the best estimates for absolute existence purposes are not optimal asymptotically (see Remark 4.2.17).

4.2.13 Definition A *square-free divisor* of a positive integer r is any factor of the *radical* of r, i.e., the product of the distinct primes dividing r. Thus the number of square-free divisors or r (denoted by $W(r)$) is given by $W(r) = 2^{\omega(r)}$, where $\omega(r)$ is the number of distinct primes dividing r (with $\omega(1) = 0$).

4.2.14 Theorem [1031] The number N of primitive polynomials whose first m coefficients are arbitrarily prescribed satisfies

$$|N - \tau(q^n - 1)(q^n - 1)| \leq m\tau(q^n - 1)W(q^n - 1)(q^m - 1)q^{n/2}, \qquad (4.2.1)$$

where τ is given in Definition 4.2.1.

4.2.15 Remark Necessarily, Theorem 4.2.14 can give a positive result on the existence of a primitive polynomial with prescribed first m coefficients only if $m < n/2$, and indeed for m close to $n/2$ only asymptotically as $n \to \infty$. An improved asymptotic result on the existence of primitive normal polynomials with prescribed first m coefficients occurs in [1034].

4.2.16 Remark If the last m coefficients of a primitive polynomial are prescribed, one can obtain estimates by considering instead the number of monic primitive reciprocal polynomials $f^*(x) = a_n^{-1}x^n f(1/x)$ whose first $m-1$ coefficients and trace are prescribed.

4.2.17 Remark The effect of the factor $W(q^n - 1)$ in estimates such as (4.2.1) can be significantly reduced through a sieving technique described in many of the papers cited below. In this way $W(q^n - 1)$ is reduced to $W(k)$ where k is an essential "core," a factor of $q^n - 1$, and the sieving process generally proceeds over the remaining primes in $q^n - 1$ not in the core. This produces sharper existence results (while simultaneously blunting asymptotic quality).

4.2.18 Remark When the characteristic p of $\mathbb{F}_q$ is less than m, an equivalent process to prescribing directly the first m coefficients is to consider (irreducible) polynomials f with the first m values σ_j prescribed, where σ_j is the trace over $\mathbb{F}_q$ of γ^j for a root γ of f. By these means it suffices to prescribe m alternative parameters (or conditions).

4.2.19 Remark One can focus on polynomials with a specific coefficient prescribed (the m-th, say), [686]. For $m \leq n/2$ this can be achieved with around $m/2$ constraints (rather than the m conditions needed to fix the first m coefficients, noted in Remark 4.2.18). For m exceeding $n/2$ the reciprocal polynomial can be considered but then an extra condition (relating to fixing the norm as a selected primitive element of $\mathbb{F}_q$) is needed.

4.2.20 Remark Dealing with the situation when one (or more) of the first m coefficients are prescribed via an alternative set of prescribed values as in Remark 4.2.18 breaks down if $p \geq m$. To overcome this, a p-adic method, devised initially by Fan and Han, for example in [1030, 1031, 1032], and improved by Cohen, for example in [686], is used in many of the papers cited below. This eliminates from the situation difficulties caused by the characteristic.

4.2.21 Remark To establish a specific existence result, it is shown that it is valid for pairs (q, n) satisfying an arithmetical condition that holds for almost all pairs (q, n). Computation is required; firstly numerical checks to show that some of the finite number of remaining pairs satisfy the condition and then, for pairs which fail, direct working in the field to exhibit a polynomial with the required property.

4.2.2 Existence theorems for primitive polynomials

4.2.22 Remark A conjecture of Hansen and Mullen, [1416], has been the driver for the key theorem on the existence of a primitive polynomial with an arbitrary prescribed coefficient. Prior to its formulation the only known result was the following existence theorem for primitive polynomials with arbitrary trace [675, 700, 1637]. Paper [700] establishes a self-contained proof of the full theorem.

4.2.23 Theorem [686, 701, 702] Given m with $1 \leq m < n$ and $a \in \mathbb{F}_q$, there exists a primitive polynomial with m-th coefficient a, with (genuine) exceptions only when

$$(q, n, m, a) = (q, 2, 1, 0),\ (4, 3, 1, 0),\ (4, 3, 2, 0),\ (2, 4, 2, 1).$$

4.2.24 Remark An existence result on primitive polynomials with prescribed norm and trace can be derived by combining a theorem on primitive *normal* polynomials with prescribed norm and trace (Theorem 4.2.46) with one on the last two coefficients prescribed (Theorem 4.2.50), since these two coefficient prescriptions are equivalent for primitive polynomials, as for irreducible polynomials (Remark 3.5.14). (Note, however, that the monic reciprocal of a normal polynomial need not be normal and therefore the equivalence breaks down for primitive normal polynomials.) For the question of prescribing the norm and trace it is sensible to assume $n \geq 3$ and that the prescribed norm is a primitive element of $\mathbb{F}_q$.

4.2.25 Theorem [682, 691, 692, 1037, 1557] Let $a, b \in \mathbb{F}_q$ with b a primitive element of $\mathbb{F}_q$. Suppose $n \geq 3$ if $a \neq 0$ and $n \geq 5$ if $a = 0$. Then there exists a primitive polynomial with trace a and norm b.

4.2.26 Remark Theorem 4.2.25 is complete except for the cases in which $a = 0$ and $n = 3, 4$; these remaining cases have recently been resolved in [687]: the only exceptions occur when $n = 3$, $a = 0$, and $q = 4$ or 7.

4.2.27 Problem Similarly, other existence results on *primitive* polynomials may be derived as a consequence from those on *primitive normal* polynomials (Section 4.2.3). Enunciate these and fill gaps that arise because the trace of a normal polynomial must be non-zero.

4.2.28 Theorem [627] Suppose $n \geq 5$. Then there exists a primitive polynomial with $a_1 = a_{n-1} = 0$, except when $(q, n) = (4, 5), (2, 6), (3, 6)$.

4.2.29 Remark As an existence result, Theorem 4.2.28 is complete: the question must have a negative answer when $n \leq 4$. It can be rephrased as asserting that the existence of a primitive f for which both f and its monic reciprocal (Remark 3.5.14) have trace 0.

4.2.30 Theorem [683] Suppose $n \geq 5$ and $a, b \in \mathbb{F}_q$. Then there exists a primitive polynomial f such that f has trace a and its monic reciprocal has trace b.

4.2.31 Problem Extend Theorem 4.2.30 to cover degrees $n = 3, 4$, perhaps with some listed exceptions.

4.2.32 Theorem [198, 695, 698, 1411, 1412, 2658] Suppose $n \geq 5$ if q is odd, and $n \geq 7$ if q is even. Then there exists a primitive polynomial with its first two coefficients arbitrarily prescribed. If $n = 4$ and q is odd, then the same conclusion holds for sufficiently large q.

4.2.33 Remark In [1411] it is claimed that when q is even, then the conclusion of Theorem 4.2.32 holds (with some exceptions) when $n \geq 4$, although some of the given detail assumes $n \geq 7$.

4.2.34 Theorem [695, 1030, 1033, 2104] Suppose $n \geq 7$. Then there exists a primitive polynomial with its first three coefficients arbitrarily prescribed.

4.2.35 Remark An *asymptotic* result Theorem 4.2.52 on the existence on a *primitive normal* polynomial with its first $\lfloor \frac{n-1}{2} \rfloor$ coefficients prescribed yields a corresponding one for a *primitive* polynomial [695, 1032] (see Problem 4.2.27). What follows now is an *unconditional* result on the existence of a primitive polynomial with up to one-third of its first coefficients prescribed.

4.2.36 Theorem [684] Suppose $m \leq \frac{n}{3}$ (except that $m \leq \frac{n}{4}$ when $q = 2$). Then there exists a primitive polynomial with its first m coefficients arbitrarily prescribed, with the exception that there is no primitive cubic over $\mathbb{F}_4$ with zero first coefficient.

4.2.3 Existence theorems for primitive normal polynomials

4.2.37 Remark Underlying these results is the fundamental existence theorem of Lenstra and Schoof [1899] (see also [693]).

4.2.38 Theorem [693, 1899] For every pair (q, n) there exists a primitive normal polynomial of degree n over $\mathbb{F}_q$.

4.2.39 Remark The original proof of Theorem 4.2.38 in [1899] requires significant numerical computation to verify plus direct examination of a few fields. For the modified approach of [693] a computer is not required.

4.2.40 Remark If $n \leq 2$, then a primitive polynomial is automatically normal (theorem statements may or may not include these values).

4.2.41 Remark For normal polynomials the character sums referred to in Remark 4.2.11 involve additive characters over $\mathbb{F}_{q^n}$. The resulting estimates (corresponding to Theorem 4.2.14, for example) now feature such quantities as $W(x^n - 1)$, defined as the number of square-free polynomial divisors of the polynomial $x^n - 1 \in \mathbb{F}_q[x]$. Furthermore, the sieve (referred to in Remark 4.2.17) now can additionally relate to the factorization of the polynomial $x^n - 1$, wherein the sieving process proceeds over the irreducible factors in $x^n - 1$ not in the "core." Indeed, because the factorization of $x^n - 1$ can be examined more systematically than the corresponding numerical factorization of $q^n - 1$, theoretical existence can be treated initially more effectively by means of an additive rather than a multiplicative sieve. Naturally, the resulting arithmetic conditions are more demanding than in the case of (merely) primitive polynomials and the consequent computations more substantial.

4.2.42 Remark In the study of polynomials which are both primitive and normal, prescribed first or last coefficients cannot be so easily interchanged via use of reciprocal polynomials since the monic reciprocal of a normal polynomial is not necessarily normal. Similarly, the requirements for specifying one coefficient (see Remark 4.2.19) are more difficult. There are, however, few new difficulties associated with the characteristic (see Remark 4.2.20).

4.2.43 Theorem [1036] Suppose $n \geq 15$ and $1 \leq m < n$ and that $a \in \mathbb{F}_q$ (with $a \neq 0$ if $m = 1$). Then there exists a primitive normal polynomial with m-th coefficient $a_m = a$.

4.2.44 Remark Paper [1036] relies on some substantial computations which are only briefly summarized. Its authors note that the theory extends unchanged to polynomials of smaller degree and indeed there is a manuscript that claims an extension of Theorem 4.2.43 to degrees $n \geq 9$, though there are some problems with the details. The problem is sensible for $n \geq 3$ but the present method cannot currently be applied effectively to small degrees.

4.2.45 Conjecture [1036] Suppose $n \geq 2$ and $1 \leq m < n$ and that $a \in \mathbb{F}_q$ (with $a \neq 0$ if $m = 1$). Then there exists a primitive normal polynomial with $a_m = a$, except when (q, n, m, a) takes any of the values

$$(2,3,2,1), (2,4,2,1), (2,4,3,1), (2,6,3,1), (3,4,2,2), (5,3,4,3), (4,3,2,1+\gamma),$$

where $\mathbb{F}_4 = \mathbb{F}_2(\gamma)$ with $\gamma^2 + \gamma + 1 = 0$.

4.2.46 Theorem [682, 691, 692, 1557] (Compare with Theorem 4.1.22) Assume $n \geq 3$. Suppose $a \neq 0 \in \mathbb{F}_q$ and b is a primitive element in $\mathbb{F}_q$. Then there exists a primitive normal polynomial with trace a and norm b.

4.2.47 Remark Theorem 4.2.46 is a striking *complete* existence result. Thus, in the case of cubics, there is a primitive polynomial $x^3 + ax^2 + cx + b$ with a, b fixed appropriately and only c allowed to vary. Further results listed below are incomplete.

4.2.48 Problem The existence of primitive polynomials of the lowest degrees remains to be established.

4.2.49 Theorem [1035] Let $a \neq 0, b \in \mathbb{F}_q$. Suppose $n \geq 7$. Then there exists a primitive normal polynomial with first coefficient $a_1 = a$ and second coefficient $a_2 = b$.

4.2.50 Theorem [1037] Let $a \in \mathbb{F}_q$ and b be a primitive element in $\mathbb{F}_q$. Suppose $n \geq 5$. Then there exists a primitive normal polynomial with penultimate coefficient $a_{n-1} = a$ and last coefficient $a_n = (-1)^n b$.

4.2.51 Remark The next results (on prescribed first and last coefficients) are asymptotic, i.e., apply for sufficiently large values of q.

4.2.52 Theorem [1034] Suppose $n \geq 2$. Then there exists a constant $C(n)$ such that, for $q > C(n)$, there exists a primitive normal polynomial with its first $\lfloor \frac{n}{2} \rfloor$ coefficients arbitrarily prescribed, subject to the constraint that the first coefficient is non-zero.

4.2.53 Theorem [1029] Suppose $n \geq 2$. Then there exists a constant $C(n)$ such that, for $q > C(n)$, there exists a primitive normal polynomial with its last $\lfloor \frac{n}{2} \rfloor$ coefficients arbitrarily prescribed subject to the constraint that the last coefficient is $(-1)^n b$, where b is a given primitive element of $\mathbb{F}_q$. Moreover, if the prescribed coefficients (other than the last) are all zero, the total number of prescribed coefficients may be taken to be $\lfloor \frac{n+1}{2} \rfloor$.

4.2.54 Remark Further constraints on primitive normal polynomials with prescribed coefficients could lead to new areas of research as in the illustrations which follow.

4.2.55 Definition An element α of $\mathbb{F}_{q^n}$ is *completely normal* if it generates over any intermediate field $\mathbb{F}_{q^d}$ (where $d|n$) a normal basis of $\mathbb{F}_{q^n}$ over $\mathbb{F}_{q^d}$. The minimal polynomial of α over $\mathbb{F}_q$ is a *completely normal polynomial.*

4.2.56 Conjecture [2157] For every n there exists a primitive completely normal polynomial.

4.2.57 Remark Major contributions towards establishing this conjecture have been made by Hachenberger [1391, 1394] (see Section 5.4). Part of the construction involves trace-compatible sequences. Also the methods employed emphasize algebraic structure rather than character sums and lead to useful lower bounds on the number of constructed polynomials.

4.2.58 Problem For which pairs (q, n) does there exist a primitive completely normal polynomial with prescribed (non-zero) trace and/or (primitive) norm?

4.2.59 Remark Although the (monic) reciprocal of a primitive polynomial is primitive, the reciprocal of a normal polynomial need not be normal.

4.2.60 Definition A *strong primitive normal polynomial* f is such that both f and its monic reciprocal are primitive normal polynomials.

4.2.61 Theorem [694] For every pair (q, n) there exists a strong primitive normal polynomial, except when

$$(q, n) = (2, 3), (2, 4), (3, 4), (4, 3), (5, 4).$$

4.2.62 Problem For which pairs (q, n) does there exist a strong primitive normal polynomial with prescribed (non-zero) trace and/or (primitive) norm?

See Also

§3.5	For discussion of irreducible polynomials with prescribed coefficients.
§5.2	For information on primitive normal polynomials.
§5.4	For discussion of completely normal primitive polynomials.

References Cited: [198, 627, 675, 682, 683, 684, 686, 691, 692, 693, 694, 695, 698, 700, 701, 702, 1029, 1030, 1031, 1032, 1033, 1034, 1035, 1036, 1037, 1391, 1394, 1411, 1412, 1416, 1557, 1637, 1899, 2104, 2157, 2658]

4.3 Weights of primitive polynomials

Stephen D. Cohen, *University of Glasgow*

4.3.1 Definition As in Definition 4.1.10, denote by $W_n(q)$ the minimal weight (or number of nonzero coefficients) of a primitive polynomial of degree n over $\mathbb{F}_q$.

4.3.2 Remark Hansen and Mullen [1416] list for each prime $p < 100$ and each degree n, with $p^n < 10^{50}$, a primitive polynomial over $\mathbb{F}_p$ of weight $W_n(p)$. Earlier, Stahnke [2699] had listed a primitive polynomial over $\mathbb{F}_2$ for each degree $n \leq 168$ of weight $W_n(2)$. In every case $W_n(p) = 3$ or 5. Though Definition 4.3.1 makes sense for any prime power q, most relevant literature relates to the binary field $\mathbb{F}_2$. Because of numerous applications there is particular interest in primitive polynomials of low or minimal weight. When $n \geq 2$, $W_n(2) \geq 3$: thus primitive trinomials (failing which pentanomials) are especially sought. In this research area there are few "theorems," most work being empirical, heuristic or conjectural. References are scattered in journals in diverse fields. The citations given here are selective and incomplete.

4.3.3 Theorem [2634] (See Remark 4.1.11) For any sufficiently large prime p,

$$W_n(p) \leq \frac{n+1}{2}.$$

4.3.4 Theorem [2634] (See Remark 4.1.11) $W_n(2) \leq \frac{n}{4} + o(1)$.

4.3.5 Conjecture [1302, 2186] For all n, $W_n(2) \leq 5$.

4.3.6 Conjecture [1302, 2186] For infinitely many values of n, $W_n(2) \leq 3$.

4.3.7 Remark Progress on Conjectures 4.3.5 and 4.3.6 may be difficult. The next conjecture has a less ambitious goal.

4.3.8 Conjecture [1302] There is a positive integer m such that for infinitely many values of n, $W_n(2) \leq m$.

4.3.9 Remark When $2^n - 1$ is a (Mersenne) prime any irreducible polynomial of degree n over $\mathbb{F}_2$ is primitive. This means that polynomials of these degrees and small weight are valuable, because primitivity can be tested (or at least ruled out) with the aid of theorems such as Swan's theorem for trinomials from [2753]. See also Theorem 3.3.25.

4.3.10 Theorem [408, 2753] Let $n > s > 0$ and assume $n+s$ is odd. Then the trinomial $x^n + x^s + 1 \in \mathbb{F}_2[x]$ has an even number of irreducible factors if and only if one of the following holds:

1. n even, $n \neq 2s$, $ns/2 \equiv 0$ or $1 \pmod 4$;
2. n odd, $s \nmid 2n$, $n \equiv \pm 3 \pmod 8$;
3. n odd $s | 2n$, $n \equiv 1 \pmod 8$.

4.3.11 Remark A trinomial of degree n over $\mathbb{F}_2$ where $2^n - 1$ is a prime is a *Mersenne trinomial (of degree n)*. To date 47 Mersenne primes are known (numbered in increasing order) as $M_1 = 2$, $\ldots$, $M_{47} = 43112609$. A total of 30 of these yield primitive trinomials. Zierler [3071] lists Mersenne trinomials of degrees up to 11213 (corresponding to M_{23}) and additional primitive trinomials have been listed in [406, 1489, 1811]. The "Great Trinomial Hunt," driven by Brent and Zimmermann [408], parallels GIMPS, the "Great Internet Prime Search" (see www.mersenne.org). Note that M_{47} is the largest known prime (as of November 2011), having 12978189 digits. Table 4.3 gives a list of all known Mersenne trinomials $x^n + x^s + 1$ with $M_r = 2^n - 1$ and $s < n/2$. Their reciprocals (wherein $s > n/2$) are also primitive.

4.3.12 Remark Lists of primitive *Mersenne pentanomials* have also been produced, see for example [1811, 3011]. Those in [3011] have the form $x^n + x^{n-1} + x^m + x^{m-1} + 1$. In particular, it has been suggested [3011] that, for random number generation, the use of pentanomials might be preferable to trinomials. Lists of primitive polynomials over $\mathbb{F}_2$ of weights 5, 7, and 9 and *all* degrees between 9 and 660 occur in [2437]. These possess the quality that the difference between each pair of consecutive indices is almost the same. This property promotes the implementation of the generation of linear recurring sequences based on highly modular devices (*ring generators*) leading to enhanced performance. On the other hand, there are cryptographic reasons why primitive polynomials of high weight might be important especially if all multiples of moderate degree also have high weight [1994].

r	n	s
1	2	1
2	3	1
3	5	2
4	7	1, 3
6	17	3, 5, 6
8	31	3, 6, 7, 13
10	89	38
12	127	1, 7, 15, 30, 63
13	521	32, 48, 158, 168
14	607	105, 147, 273
15	1 279	216, 418
17	2 281	715, 915, 1029
18	3 217	67, 576
20	4 423	271, 369, 370, 649, 1393, 1419, 2098
21	9 689	84, 471, 1836, 2444, 4187
24	19 937	881, 7083, 9842
26	23 209	1530, 6619, 9739
27	44 497	8575, 21034
29	110 503	25230, 53719
30	132 049	7000, 33912, 41469, 52549, 54454
32	756 839	215747, 267428, 279695
33	859 433	170340, 288477
37	3 021 377	361604, 1010202
38	6 972 593	3037958
41	24 036 583	8412642, 8785528
42	25 964 951	880890, 4627670, 4830131, 6383880
43	30 402 457	2162059
44	32 582 657	5110722, 5552421, 7545455
46	42 643 801	55981, 3706066, 3896488, 12899278, 20150445
47	43 112 609	3569337, 4463337, 17212521, 21078848

Table 4.3.1 Mersenne trinomials.

4.3.13 Example [3011] There is no Mersenne trinomial of degree 61 (M_9) but $x^{61}+x^{60}+x^{46}+x^{45}+1$ is a primitive pentanomial.

See Also

§3.3	For Swan's theorem type results.
§3.4	For discussion of the weights of irreducible polynomials.
§14.9	For applications of primitive polynomials.

References Cited: [406, 408, 1302, 1416, 1489, 1811, 1994, 2186, 2437, 2634, 2699, 2753, 3011, 3071]

4.4 Elements of high order

José Felipe Voloch, University of Texas at Austin

4.4.1 Remark There is no known explicit construction of primitive elements (Definition 2.1.38). For some applications, it suffices to construct elements of sufficiently large order. The current state of the art on explicit constructions of elements of large order usually proves lower bounds for their orders which are much smaller than the expected actual order.

4.4.2 Definition The *(absolute) degree* of an element α of a finite field of characteristic p is $\deg\alpha = [\mathbb{F}_p(\alpha) : \mathbb{F}_p]$. The order of $\alpha \neq 0$ is defined in Definition 2.1.40 and is denoted by $\operatorname{ord}\alpha$.

4.4.3 Remark If $\alpha \neq 0, 1$ is an element of a finite field of characteristic p, then $\deg\alpha < \operatorname{ord}\alpha \leq p^{\deg\alpha} - 1$.

4.4.1 Elements of high order from elements of small orders

4.4.4 Theorem There exists an absolute constant $c > 0$ and, for every $\epsilon > 0$, there exists a $\delta > 0$, such that whenever $\alpha \neq 0, 1$ is an element of a finite field of characteristic p with $\deg\alpha = n$.

1. [1241, 1243] If $\operatorname{ord}\alpha = n + 1$ then $\operatorname{ord}(1 - \alpha) \geq \exp(c\sqrt{n})$.
2. [2884] If $\operatorname{ord}\alpha < n^{2-\epsilon}$ then $\operatorname{ord}(1 - \alpha) \geq \exp(cn^{\delta})$.

4.4.5 Remark If α is as in Theorem 4.4.4 Part 1, then $n + 1$ is prime and p is a primitive root modulo $n + 1$. Likewise, the possible values of n in Theorem 4.4.4 Part 2 are restricted.

4.4.6 Remark Similar, but weaker, results can be proved about the order of $R(\alpha), R \in \mathbb{F}_p(x)$ or even $\beta, F(\alpha, \beta) = 0, F(x, y) \in \mathbb{F}_p[x, y]$, with α as in the previous theorem [2884].

4.4.7 Remark Poonen conjectured (as part of a more general conjecture, see [2884]) that, with notation as in the previous theorem, $\max\{\operatorname{ord}\alpha, \operatorname{ord}(1 - \alpha)\} \geq \exp(cn)$. A special case was also conjectured by Cheng [611].

4.4.8 Theorem [611] Let α satisfy $\alpha^m = g$ where $m|(q - 1)$ and let g be a primitive element in $\mathbb{F}_q$. Then, $\deg\alpha = m \deg g$ and $\operatorname{ord}(1 - \alpha) \geq \exp(cm)$.

4.4.2 Gao's construction and a generalization

4.4.9 Theorem [1173] Given an integer n and a prime p, let $m = \lceil \log_p n \rceil$. If $g \in \mathbb{F}_p[x], \deg(g) \leq 2m$ is such that $x^{p^m} - g(x)$ has an irreducible factor of degree n then any root of this factor in $\mathbb{F}_{p^n}$ has order at least $\exp((\log n)^2 / \log\log n)$.

4.4.10 Remark A search for the polynomial g satisfying the hypotheses in the above theorem can be done in time polynomial in $n \log p$. It is conjectured that such a polynomial exists for all p, n.

4.4.11 Remark An improvement on the bounds of [1173] was given in [713].

4.4.3 Iterative constructions

4.4.12 Theorem [2886] Let $\alpha_0 = 1 \in \mathbb{F}_2$, α_k a root of $x^2 + \alpha_{k-1}x + 1, k > 0$. Then $\mathbb{F}_{2^{2^k}} = \mathbb{F}_2(\alpha_k), n = \deg \alpha_k = 2^k$ and $\text{ord}\alpha_k \geq \exp(n^\delta)$, for some absolute $\delta > 0$.

4.4.13 Theorem [465] Define $f(x, y) := y^2 + (6 - 8x^2)y + (9 - 8x^2)$. If $q = p^m$ is an odd prime power such that $q \equiv 1 \pmod 4$, $\alpha_0 \in \mathbb{F}_q$ is such that $\alpha_0^2 - 1$ is not a square in $\mathbb{F}_q$, define α_k by $f(\alpha_{k-1}, \alpha_k) = 0, k > 0$. Let $\delta_k = \alpha_k^2 - 1$. Then $n = \deg \delta_k = m2^k$ and $\text{ord}(\delta_k) \geq \exp(c(\log n)^2)$ for some constant $c > 0$.

See Also

[54]	For multiplicative orders of Gauss periods.
[612]	For high order elements using subspace polynomials.
[1242]	For high order elements using Gauss periods.

References Cited: [54, 465, 611, 612, 713, 1173, 1241, 1242, 1243, 2884, 2886]

5

Bases

5.1 Duality theory of bases

Dieter Jungnickel, University of Augsburg

As noted in Section 2.1, $\mathbb{F}_{q^m}$ may be viewed as a vector space of dimension m over $\mathbb{F}_q$, and therefore has a basis (in fact, many bases) over $\mathbb{F}_q$. Some fundamental definitions and results on bases were already given there. In particular, Theorem 2.1.93 and Corollary 2.1.95 provide criteria when a set $\{\alpha_1, \ldots, \alpha_m\}$ of elements in $\mathbb{F}_{q^m}$ forms a basis over $\mathbb{F}_q$.

The present section provides a more detailed treatment of the general theory of bases, the unifying theme being the notion of duality introduced in Definition 2.1.100. Proofs for most of the results in this section can be found in [1631]; however, a somewhat different notation is used there.

5.1.1 Dual bases

We restate and expand the basic Definition 2.1.100 as follows.

5.1.1 **Definition** Two ordered bases $\{\alpha_1, \ldots, \alpha_m\}$ and $\{\beta_1, \ldots, \beta_m\}$ of $F = \mathbb{F}_{q^m}$ over $K = \mathbb{F}_q$ are *dual* (or *complementary*) if $\mathrm{Tr}_{F/K}(\alpha_i\beta_j) = \delta_{ij}$, where $\delta_{ij} = 0$ if $i \neq j$ and $\delta_{ij} = 1$ if $i = j$. An ordered basis is *trace-orthogonal* if it satisfies $\mathrm{Tr}_{F/K}(\alpha_i\alpha_j) = 0$ whenever $j \neq i$; and it is *self-dual* if it is dual with itself, that is, if it additionally satisfies $\mathrm{Tr}_{F/K}(\alpha_i^2) = 1$ for all i.

5.1.2 Remark The following simple but fundamental result guarantees the existence of dual bases. Remark 5.1.4 provides a proof by giving an explicit construction using the well-known concept of dual bases in linear algebra.

5.1.3 Theorem For every ordered basis $B = \{\alpha_1, \ldots, \alpha_m\}$ of $F = \mathbb{F}_{q^m}$ over $K = \mathbb{F}_q$, there is a uniquely determined dual ordered basis $B^* = \{\beta_1, \ldots, \beta_m\}$.

5.1.4 Remark As F is a finite-dimensional vector space over K, it is isomorphic to the dual vector space F^* consisting of all linear transformations from F to K. In Theorem 2.1.84, these transformations were given in terms of the trace function $\mathrm{Tr}_{F/K}$ in the form L_β with $\beta \in F$. Then the map $L : F \to F^*$ with $\beta \mapsto L_\beta$ is an isomorphism. This allows one to obtain the following explicit description of the dual basis B^*:

$$\beta_j = L^{-1}(\alpha_j^*), \tag{5.1.1}$$

where α_j^* is the linear transformation defined by the requirement

$$\alpha_j^*(\alpha_i) = \delta_{ij} \quad \text{for } i = 1, \ldots, m. \tag{5.1.2}$$

5.1.5 Remark One reason for the importance of dual bases is the fact that they provide an easy way for determining the coordinate representation of arbitrary elements of F, using the concept of primal and dual coordinates.

5.1.6 Definition Given a dual pair of ordered bases B, B^* as above, we use the following notation. Let

$$\xi = x_1\alpha_1 + \cdots + x_m\alpha_m = (x)_1\beta_1 + \cdots + (x)_m\beta_m.$$

Then

$$r_B(\xi) = (x_1, \ldots, x_m) \quad \text{and} \quad r_{B^*}(\xi) = ((x)_1, \ldots, (x)_m)$$

are, respectively, the *primal coordinates* and the *dual coordinates* of ξ (with respect to B).

5.1.7 Lemma Let $B = \{\alpha_1, \ldots, \alpha_m\}$ and $B^* = \{\beta_1, \ldots, \beta_m\}$ be a dual pair of ordered bases of $F = \mathbb{F}_{q^m}$ over $K = \mathbb{F}_q$, let $\xi \in F$, and let $r_B(\xi)$ and $r_{B^*}(\xi)$ be as in Definition 5.1.6. Then

$$x_i = \mathrm{Tr}_{F/K}(\xi\beta_i) \quad \text{and} \quad (x_i) = \mathrm{Tr}_{F/K}(\xi\alpha_i). \tag{5.1.3}$$

5.1.8 Remark The following two results deal with the dual basis of a basis B in the important special cases where B is either a polynomial basis or a normal basis; see Definitions 2.1.96 and 2.1.98.

5.1.9 Theorem Let $\theta \in F = \mathbb{F}_{q^m}$, and assume that $B = \{\alpha_1 = \theta, \alpha_2 = \theta^q, \ldots, \alpha_m = \theta^{q^{m-1}}\}$ is a normal basis for F over $K = \mathbb{F}_q$. Then the dual basis $B^* = \{\beta_1, \ldots, \beta_m\}$ is likewise normal:

$$B^* = \{\zeta, \zeta^q, \ldots, \zeta^{q^{m-1}}\}, \quad \text{where } \zeta = \beta_1.$$

5.1.10 Remark [1172] The dual normal basis B^* can be described explicitly as follows. For $i = 0, \ldots, m-1$, put $t_i := \mathrm{Tr}_{F/K}(\theta\theta^{q^i})$. Let t be the polynomial

$$t(x) = t_{m-1}x^{m-1} + \cdots + t_1x + t_0 \in K[x],$$

and let $d(x) = d_{m-1}x^{m-1} + \cdots + d_1x + d_0$ be the unique monic polynomial of degree $< m$ in $K[x]$ satisfying

$$d(x)t(x) \equiv 1 \pmod{x^m - 1}.$$

Then B^* is the normal basis generated by the element

$$\zeta = d_0\theta + d_1\theta^q + \cdots + d_{m-1}\theta^{q^{m-1}}.$$

Normal bases will be studied in detail in Section 5.2.

5.1.11 Remark In contrast to the case of normal bases considered in Theorem 5.1.9, the dual basis of a polynomial basis is usually *not* a polynomial basis. Nevertheless, an explicit description is possible.

5.1.12 Theorem [1573] Let θ be a root of a monic irreducible polynomial f of degree m over $K = \mathbb{F}_q$, and let $B = \{1, \theta, \theta^2, \ldots, \theta^{m-1}\}$ be the corresponding polynomial basis of $F = \mathbb{F}_{q^m}$ over K. Assume that f splits over F as

$$f(x) = (x - \theta)\left(\gamma_{m-1}x^{m-1} + \cdots + \gamma_1 x + \gamma_0\right).$$

Then the dual basis $B^* = \{\beta_1, \ldots, \beta_m\}$ can be computed as follows:

$$\beta_i = \gamma_{i-1}/f'(\theta) \quad \text{for } i = 1, \ldots, m,$$

where f' denotes the formal derivative of f.

5.1.13 Theorem [1265, 1298] Let θ be a root of a monic irreducible polynomial f of degree m over $K = \mathbb{F}_q$, and let $B = \{1, \theta, \theta^2, \ldots, \theta^{m-1}\}$ be the corresponding polynomial basis of $F = \mathbb{F}_{q^m}$ over K. Then the dual basis B^* of B is likewise a polynomial basis if and only if f is a binomial and $m \equiv 1 \pmod{p}$, where q is a power of the prime p.

5.1.14 Corollary There exists a dual pair of polynomial bases of $\mathbb{F}_{q^m}$ over $\mathbb{F}_q$ if and only if the following three conditions are satisfied:

1. $m \equiv 1 \pmod{p}$;
2. every prime r dividing m also divides $q - 1$;
3. $m \equiv 0 \pmod{4}$ implies $q \equiv 1 \pmod{4}$.

5.1.15 Corollary Let $B = \{1, \theta, \theta^2, \ldots, \theta^{m-1}\}$ be a polynomial basis of of $\mathbb{F}_{2^m}$ over $\mathbb{F}_2$, where $m \geq 2$. Then the dual basis of B cannot be a polynomial basis.

5.1.16 Example There is no dual pair of polynomial bases of $\mathbb{F}_{3^m}$ over $\mathbb{F}_3$. There exists a dual pair of polynomial bases of $\mathbb{F}_{4^m}$ over $\mathbb{F}_4$ if and only if m is a power of 3. There exists a dual pair of polynomial bases of $\mathbb{F}_{5^m}$ over $\mathbb{F}_5$ if and only if m is a power of 16.

5.1.2 Self-dual bases

5.1.17 Remark Lemma 5.1.7 shows that the computation of coordinates is particularly simple when the basis used is self-dual. This also has important applications in the design of hardware implementations for the multiplication in finite (extension) fields; see, for instance, Sections 4.1, 4.4, and 5.5 of [1631]. Unfortunately, a self-dual basis does not always exist, which motivates considering a slightly weaker notion.

5.1.18 Theorem [2583] There exists a self-dual basis of $\mathbb{F}_{q^m}$ over $\mathbb{F}_q$ if and only if either q is even or both q and n are odd.

5.1.19 Definition A basis $\{\alpha_1, \ldots, \alpha_m\}$ of $F = \mathbb{F}_{q^m}$ over $K = \mathbb{F}_q$ is *almost self-dual* if it is trace-orthogonal and if it additionally satisfies $\mathrm{Tr}_{F/K}(\alpha_i^2) = 1$ for $i = 1, \ldots, m-1$, with possibly one exception.

5.1.20 Theorem [1635] There always exists an almost self-dual basis of $\mathbb{F}_{q^m}$ over $\mathbb{F}_q$.

5.1.21 Remark If there exists a self-dual basis for F over K, there are many such bases. All these bases are related by suitable transformations, namely via orthogonal matrices; see Definition 13.2.35. The number of such matrices is given in Theorem 13.2.37, which implies an explicit formula for the number of self-dual bases.

5.1.22 Lemma [1635] Let $B = \{\alpha_1, \ldots, \alpha_m\}$ be a self-dual ordered basis of $F = \mathbb{F}_{q^m}$ over $K = \mathbb{F}_q$, and let $A = (a_{ij})$ be an invertible $m \times m$ matrix over K. Then the ordered basis $B' = \{\beta_1, \ldots, \beta_m\}$ with

$$\beta_i = \sum_{i=1}^{m} a_{ij}\alpha_j \quad \text{for } i = 1, \ldots, m$$

is likewise self-dual if and only if A is an orthogonal matrix. In particular, the number of ordered self-dual bases of F over K equals the number of orthogonal $m \times m$ matrices over K, provided that one such basis exists.

5.1.23 Theorem [1635] The number of (unordered) self-dual bases of $\mathbb{F}_{q^m}$ over $\mathbb{F}_q$ is equal to

$$sd(m, q) = \frac{\gamma}{m!} \prod_{i=1}^{m-1} (q^i - \epsilon_i), \tag{5.1.4}$$

where

$$\epsilon_i = \begin{cases} 1 & \text{if } i \text{ is even,} \\ 0 & \text{if } i \text{ is odd,} \end{cases} \quad \text{and} \quad \gamma = \begin{cases} 2 & \text{if } q \text{ is even,} \\ 1 & \text{if } q \text{ and } m \text{ are odd,} \\ 0 & \text{otherwise.} \end{cases}$$

5.1.3 Weakly self-dual bases

5.1.24 Theorem [1573] For $m \geq 2$, there does not exist a self-dual polynomial basis of $\mathbb{F}_{q^m}$ over $\mathbb{F}_q$.

5.1.25 Remark In contrast to Theorem 5.1.24, self-dual *normal* bases often exist; see Section 5.2.

5.1.26 Remark For computational purposes, it would be helpful to have a self-dual polynomial basis. However, Theorem 5.1.24 excludes this possibility. Fortunately, there are weaker notions which are still useful for hardware implementations; one such notion is discussed in the present subsection.

5.1.27 Definition A basis $\{\alpha_1, \ldots, \alpha_m\}$ of $F = \mathbb{F}_{q^m}$ over $K = \mathbb{F}_q$ with associated dual basis $B^* = \{\beta_1, \ldots, \beta_m\}$ is *weakly self-dual* if there exist an element $\delta \in F^*$ and a permutation σ of $\{1, \ldots, m\}$ such that the following condition holds:

$$\beta_i = \delta\alpha_{\sigma(i)} \quad \text{for } i = 1, \ldots, m. \tag{5.1.5}$$

5.1.28 Remark For computational purposes, weakly self-dual polynomial bases are quite attractive, as they lead to rather simple transformations between dual and primal coordinates. Consider some element $\xi \in F$, with primal and dual coordinates as in Definition 5.1.6. Using Equation (5.1.5), one obtains

$$\xi\delta = \sum_{j=1}^{m} x_{\sigma(j)}\beta_j. \tag{5.1.6}$$

Thus the dual coordinates of the product $\xi\delta$ arise by simply permuting the primal coordinates of ξ according to σ.

5.1.29 **Remark** The observation in Remark 5.1.28 is of particular interest for hardware implementations of the multiplication in F if a polynomial basis B is used. Consider a further element $\eta \in F$, given in primal coordinates $(y_1, \dots, y_m)$, and write $\pi = \xi\eta$. Then one may design a simple hardware device – called a *dual basis multiplier* with respect to B – which computes the product $\pi\delta = (\xi\eta)\delta = (\xi\delta)\eta$ in dual coordinates from $\xi\delta$ in dual coordinates (which, according to Equation (5.1.6), are obtained from the primal coordinates of ξ by just applying the permutation σ) and η in primal coordinates. Using Equation (5.1.6) for π instead of ξ, the primal coordinates of the product π can then be obtained by simply permuting the dual coordinates of $\pi\delta$ according to σ^{-1}, so that all required coordinate transformations reduce to permutations. Moreover, the permutations arising are very simple ones; see Theorem 5.1.30 below. In the particularly important binary case, this allows the hardware design of efficient dual basis multipliers in many instances of practical interest; see [1631] for more details and for examples of dual basis multipliers.

5.1.30 **Theorem** [1298, 2936] A polynomial basis $B = \{1, \theta, \theta^2, \dots, \theta^{m-1}\}$ of $\mathbb{F}_{q^m}$ over $\mathbb{F}_q$ is weakly self-dual if and only if the minimal polynomial f of θ is either a trinomial with constant term -1 or a binomial. Moreover, for $i = 1, \dots, m$,

$$\delta = \beta_k = \frac{1}{\theta^k f'(\theta)} \quad \text{and} \quad \sigma(i) := k - i + 1 \pmod{m} \quad \text{if } f(x) = x^m + ax^k - 1,$$

and

$$\delta = \frac{1}{f'(\theta)} = \frac{1}{m\theta^{m-1}} \quad \text{and} \quad \sigma(i) := 1 - i \pmod{m} \quad \text{if } f(x) = x^m - a.$$

5.1.31 **Corollary** Let $B = \{1, \theta, \theta^2, \dots, \theta^{m-1}\}$ be a polynomial basis of $F = \mathbb{F}_{q^m}$ over $\mathbb{F}_q$. Then the transformation from the primal coordinates of an arbitrary element $\xi \in F$ to the dual coordinates of the element $\xi\delta$ (for a suitable constant element δ of F) is just a permutation if and only if the minimal polynomial of θ is either a trinomial with constant term -1 or a binomial.

5.1.32 **Remark** As the binary case is of particular practical importance, we state some results for this special case explicitly.

5.1.33 **Theorem** Let $f(x) = x^m + x^k + 1$ be an irreducible trinomial over $\mathbb{F}_2$, let θ be a root of f, and $B^* = \{\beta_1, \dots, \beta_m\}$ the dual basis of the polynomial basis $B = \{1, \theta, \theta^2, \dots, \theta^{m-1}\}$ generated by θ. Then B is weakly self-dual, and Equation (5.1.5) is satisfied with

$$\delta = \beta_k = \frac{1}{\theta^k f'(\theta)} \quad \text{and} \quad \sigma(i) := k - i + 1 \pmod{m} \quad \text{for } i = 1, \dots, m.$$

5.1.34 **Example** For $m = 6$, we may use a root θ of the cyclotomic polynomial $\Phi_9(x) = x^6 + x^3 + 1$ over $\mathbb{F}_2$ to generate $F = \mathbb{F}_{2^6}$. Then the polynomial basis $B = \{1, \theta, \theta^2, \theta^3, \theta^4, \theta^5\}$ is weakly self-dual with $\delta = \frac{1}{\theta^3\theta^2} = \theta^4$ and

$$\sigma = \begin{pmatrix} 1 & 2 & 3 & 4 & 5 & 6 \\ 3 & 2 & 1 & 6 & 5 & 4 \end{pmatrix}.$$

Explicitly, the dual basis of B is given by

$$B^* = \{\theta^6 = \theta^3 + 1,\ \theta^5,\ \theta^4,\ \theta^9 = 1,\ \theta^8 = \theta^5 + \theta^2,\ \theta^7 = \theta^4 + \theta\}.$$

It is easily checked that B and B^* indeed form a pair of dual bases.

5.1.35 Corollary Let θ be a root of a monic irreducible polynomial f of degree m over $K = \mathbb{F}_2$, and let $B = \{1, \theta, \theta^2, \ldots, \theta^{m-1}\}$ be the corresponding polynomial basis of $\mathbb{F}_{2^m}$ over K. Then the dual basis B^* of B is given by a permutation of B if and only if m is odd and $f(x) = x^m + x + 1$.

5.1.36 Remark As noted before, the weakly self-dual polynomial bases are those polynomial bases for which the transformation between primal and dual coordinates can be realized with a minimum of complexity. In view of Theorem 5.1.33, the irreducible polynomials one should use for designing dual basis multipliers are therefore the irreducible trinomials. Fortunately, these exist in many – though by no means all – cases; see Sections 2.2 and 3.4. For $n \leq 10,000$, this holds in about half of the cases.

5.1.4 Binary bases with small excess

5.1.37 Remark In cases where no weakly self-dual polynomial basis can exist, further generalizations of the notion of weak self-duality are useful. We begin with some results for the binary case; proofs for these results can be found in [1631]. Consider a polynomial basis $B = \{1, \theta, \theta^2, \ldots, \theta^{m-1}\}$ of $\mathbb{F}_{2^m}$ over $\mathbb{F}_2$ and a scalar multiple $C = B^*/\delta$ of its dual basis B^*. According to Definition 5.1.27, B is a weakly self-dual basis (with respect to the given value of δ) if and only if the coordinate transformation from C to B is just a permutation; equivalently, the matrix S associated with this change of basis has to be a permutation matrix. If no weakly self-dual basis exists, one wants to find a polynomial basis B and a suitable element δ for which the associated transformation matrix S is as simple as possible, meaning that S should have the smallest possible number of non-zero entries. This leads to the following definition.

5.1.38 Definition The *weight* $w(S)$ of an invertible $m \times m$ matrix S is the number of non-zero entries of S. As the minimum weight is always at least m, one defines the *excess* of S as $e(S) = w(S) - m$.

5.1.39 Remark In the hardware design of dual basis multipliers over $\mathbb{F}_2$, one wants to use an irreducible polynomial for which the matrix S associated with this change of basis has the smallest possible excess, as this turns out to be the number of XOR-gates required for computing the primal coordinates from the *generalized dual coordinates*, that is, from the coordinates with respect to $C = B^*/\delta$; see, for instance, [1631, Section 4.5]. Weakly self-dual bases correspond to the smallest possible case, namely $e(S) = 0$.

5.1.40 Example Consider the irreducible polynomial $f(x) = x^8 + x^4 + x^3 + x^2 + 1$ over $\mathbb{F}_2$, and let θ be a root of f. Then $f = (x - \theta)g$, where

$$g(x) = \theta^{-1} + \theta^{-2}x + (1 + \theta + \theta^5)x^2 + (1 + \theta^4)x^3 + \theta^3 x^4 + \theta^2 x^5 + \theta x^6 + x^7.$$

By Theorem 5.1.12, the dual basis $B^* = \{\beta_1, \ldots, \beta_m\}$ of the polynomial basis B defined by θ is obtained by dividing the coefficients of g by $f'(\theta) = \theta^2$. Choosing $\delta = (\theta^3 f'(\theta))^{-1} = 1/\theta^5$ gives $B^*/\delta = \{\gamma_1, \ldots, \gamma_m\}$, where

$$\gamma_1 = \theta^2,\ \gamma_2 = \theta,\ \gamma_3 = 1 + \theta^2,\ \gamma_4 = \theta^3 + \theta^7,\ \gamma_5 = \theta^6,\ \gamma_6 = \theta^5,\ \gamma_7 = \theta^4,\ \gamma_8 = \theta^3.$$

Thus the transformation matrix S from generalized dual to primal coordinates has excess 2 in this case, giving indeed a quite simple coordinate transformation. This heavily depends on the choice of δ; for instance, the choice $\delta = f'(\theta)^{-1} = 1/\theta^2$ would lead to a transformation matrix with excess 16.

5.1.41 Theorem [2718] Let θ be a root of an irreducible polynomial f of degree m over $K = \mathbb{F}_2$, and let $B = \{1, \theta, \theta^2, \ldots, \theta^{m-1}\}$ be the corresponding polynomial basis of $\mathbb{F}_{2^m}$ over K, and B^* the dual basis of B. Write

$$f(x) = x^m + x^{m_t} + \cdots + x^{m_1} + x^{m_0}, \quad \text{where } 0 = m_0 < m_1 < \cdots < m_t < m,$$

and put

$$\delta = \frac{1}{f'(\theta)\theta^{m_s}}, \quad \text{where } s = \lceil t/2 \rceil.$$

Then the transformation matrix S from $C = B^*/\delta$ to B has excess

$$e(S) = \sum_{j=s+1}^{t} m_j - \sum_{j=1}^{s-1} m_j.$$

5.1.42 Corollary Let θ be a root of an irreducible pentanomial f of degree m over $\mathbb{F}_2$, say

$$f(x) = x^m + x^{m_3} + x^{m_2} + x^{m_1} + 1,$$

and let B be a corresponding polynomial basis of $\mathbb{F}_{2^m}$ over $\mathbb{F}_2$, and B^* the dual basis of B. Put $\delta = (f'(\theta)\theta^{m_2})^{-1}$. Then the transformation matrix S from $C = B^*/\delta$ to B has excess $e(S) = m_3 - m_1$.

5.1.43 Corollary The binary irreducible polynomials leading to a transformation matrix of excess 2 are precisely the irreducible pentanomials of the special form $x^m + x^{k+1} + x^k + x^{k-1} + 1$.

5.1.44 Theorem A binary irreducible polynomial leads to a transformation matrix of excess 1 if and only if it has the form $x^m + x + 1$, where m is even.

5.1.45 Remark It is also of interest to investigate the possible spectra of the number of elements of trace 1 in a polynomial basis for $\mathbb{F}_{2^m}$ over $\mathbb{F}_2$. This problem was first considered in [51] where it was noted that using a polynomial basis with a small number of elements of trace 1 is desirable, since it allows a particularly efficient implementation of the trace function. For example, this is important for halving a point on an elliptic curve over $\mathbb{F}_{2^m}$, and for generating pseudo random sequences using elliptic curves. In particular, it is of interest to find trinomials and pentanomials associated with bases with a small number of elements of trace 1. We mention one striking result in this direction; more results on the trace spectra of polynomial bases can be found in [47, 51, 2646].

5.1.46 Theorem Suppose that there exists an irreducible trinomial of degree m over $\mathbb{F}_2$. Then there also exists an irreducible trinomial such that the corresponding polynomial basis for $\mathbb{F}_{2^m}$ over $\mathbb{F}_2$ contains exactly one element with trace 1.

5.1.5 Almost weakly self-dual bases

5.1.47 Remark In this subsection, we present some results on polynomial bases corresponding to matrices with small excess over a general field $\mathbb{F}_q$. These results are taken from [2158]. For $q \neq 2$, a matrix of excess 0 is not necessarily a permutation matrix. This leads to the following generalization of weakly self-dual bases.

5.1.48 Definition A basis $\{\alpha_1, \ldots, \alpha_m\}$ of $F = \mathbb{F}_{q^m}$ over $K = \mathbb{F}_q$ with associated dual basis $B^* = \{\beta_1, \ldots, \beta_m\}$ is *almost weakly self-dual* if there exist an element $\delta \in F^*$, elements $c_1, \ldots, c_n \in K^*$, and a permutation σ of $\{1, \ldots, m\}$ such that the following condition holds:

$$\beta_i = c_i \delta \alpha_{\sigma(i)} \quad \text{for } i = 1, \ldots, m. \tag{5.1.7}$$

5.1.49 Remark The almost weakly self-dual bases are precisely those bases corresponding to a transformation matrix S with $e(S) = 0$. An almost weakly self-dual basis is actually weakly self-dual if and only if $c_1 = \cdots = c_n$.

5.1.50 Theorem A polynomial basis $B = \{1, \theta, \theta^2, \ldots, \theta^{m-1}\}$ of $\mathbb{F}_{q^m}$ over $\mathbb{F}_q$ is almost weakly self-dual if and only if the minimal polynomial f of θ is either a trinomial or a binomial.

5.1.51 Remark There are no proper almost weakly self-dual bases which belong to irreducible binomials: in this case, the basis is actually weakly self-dual by Theorem 5.1.30. Hence it suffices to consider the case of irreducible trinomials in Theorem 5.1.52; in the special case where $d = 1$, one recovers Theorem 5.1.30.

5.1.52 Theorem Let $f(x) = x^m - ax^k - d$ be an irreducible trinomial over $\mathbb{F}_2$, let θ be a root of f, and $B^* = \{\beta_1, \ldots, \beta_m\}$ the dual basis of the polynomial basis $B = \{1, \theta, \theta^2, \ldots, \theta^{m-1}\}$ generated by θ. Then B is almost weakly self-dual, and Equation (5.1.7) is satisfied with

$$\delta = \beta_k = \frac{d}{\theta^k f'(\theta)}, \quad c_i = \begin{cases} 1 & \text{if } i \leq k, \\ d^{-1} & \text{if } i > k \end{cases}$$

and

$$\sigma(i) := k - i + 1 \pmod{m} \quad \text{for } i = 1, \ldots, m.$$

5.1.53 Theorem Let θ be a root of a monic irreducible polynomial f of degree m over $K = \mathbb{F}_q$, and let $B = \{1, \theta, \theta^2, \ldots, \theta^{m-1}\}$ be the corresponding polynomial basis of $\mathbb{F}_{q^m}$ over K, and B^* the dual basis of B. Write

$$f(x) = x^m + f_{m_t} x^{m_t} + \cdots + f_{m_1} x^{m_1} + f_{m_0} x^{m_0}, \quad \text{where } 0 = m_0 < m_1 < \cdots < m_t < m,$$

and put

$$\delta = \frac{1}{f'(\theta)\theta^{m_s}}, \quad \text{where } s = \lceil t/2 \rceil.$$

Then the transformation matrix S from $C = B^*/\delta$ to B has excess

$$e(S) = \begin{cases} \sum_{j=s+1}^{t} m_j - \sum_{j=1}^{s-1} m_j & \text{if } t \text{ is odd,} \\ \sum_{j=s+1}^{t} m_j - \sum_{j=1}^{s} m_j & \text{if } t \text{ is even.} \end{cases}$$

5.1.54 Corollary Let θ be a root of an irreducible polynomial f of degree m over $\mathbb{F}_q$, let $B = \{1, \theta, \theta^2, \ldots, \theta^{m-1}\}$ be the corresponding polynomial basis of $\mathbb{F}_{q^m}$ over K, and B^* the dual basis of B. Then the transformation matrix S from $C = B^*/\delta$ to B has excess

$$e(S) = \begin{cases} 0 & \text{if } f \text{ is either a trinomial or a binomial,} \\ 1 & \text{if } f \text{ is of the form } f(x) = x^m + f_k x^k + f_{k-1} x^{k_1} + f_0, \\ 2 & \text{if } f \text{ is of the form } f(x) = x^m + f_{k+1} x^{k+1} + f_k x^k + f_{k-1} x^{k-1} + f_0 \\ & \quad \text{or } f(x) = x^m + f_{k+1} x^{k+1} + f_{k-1} x^{k-1} + f_0. \end{cases}$$

5.1.6 Connections to hardware design

5.1.55 Remark Many aspects of the study of various types of bases for $\mathbb{F}_{q^m}$ over $\mathbb{F}_q$ are to a large extent motivated by the hardware design of efficient multipliers for $\mathbb{F}_{q^m}$. The seminal paper in this area is due to Berlekamp [234]. Another seminal idea – which motivated the study of optimal and low complexity normal bases, see Section 5.3 – was contained in a 1981 US patent application by Massey and Omura; see "Computational Method and Apparatus for Finite Field Arithmetic," US Patent No. 4,587,627, 1986. As already mentioned, introductory examples and some references can be found in Sections 4.1, 4.4, and 5.5 of [1631]. There is an abundance of papers in this area, largely due to its importance for hardware architectures for public key cryptography; the interested reader should consult the relevant sections of the extensive survey [208]. A more recent survey concerning the special topic of polynomial basis multipliers in the binary case is given in [982].

See Also

§3.4, §3.5	For irreducible trinomials.
§5.2	For normal bases.
§5.3	For complexities of normal bases.
§13.2	For matrices over finite fields.
§16.7	For hardware implementations.

References Cited: [47, 51, 208, 234, 982, 1172, 1265, 1298, 1573, 1631, 1635, 2158, 2583, 2646, 2718, 2936]

5.2 Normal bases

Shuhong Gao, *Clemson University*
Qunying Liao, *Sichuan Normal University*

We present basic results on normal and self-dual normal elements, and we give a unified approach following the PhD thesis [1172]. Let p be a prime and let q be a power of p. Denote by σ the Frobenius map of $\mathbb{F}_{q^n}$:

$$\sigma(\alpha) = \alpha^q, \qquad \text{for } \alpha \in \mathbb{F}_{q^n}.$$

Then

$$\sigma^i(\alpha) = \alpha^{q^i}, \qquad \text{for all } i \geq 0,$$

and $\sigma^n = 1$ as a map on $\mathbb{F}_{q^n}$ (since $\sigma^n(\alpha) = \alpha^{q^n} = \alpha$ for all $\alpha \in \mathbb{F}_{q^n}$). The Galois group of $\mathbb{F}_{q^n}$ over $\mathbb{F}_q$ consists of the n maps σ^i, $0 \leq i \leq n-1$. Recall from Definition 2.1.98 that an element $\alpha \in \mathbb{F}_{q^n}$ is a *normal element* over $\mathbb{F}_q$ if the conjugates $\sigma^i(\alpha)$, $0 \leq i \leq n-1$, are linearly independent over $\mathbb{F}_q$. Hence a normal basis for $\mathbb{F}_{q^n}$ over $\mathbb{F}_q$ is of the form $\{\alpha, \sigma(\alpha), \ldots, \sigma^{n-1}(\alpha)\}$, where $\alpha \in \mathbb{F}_{q^n}$ is normal over $\mathbb{F}_q$. Also, an irreducible polynomial $f \in \mathbb{F}_q[x]$ of degree n is an *N-polynomial* (or *normal polynomial*) if its roots are linearly independent over $\mathbb{F}_q$, that is, its roots form a normal basis of $\mathbb{F}_{q^n}$ over $\mathbb{F}_q$.

5.2.1 Basics on normal bases

5.2.1 Theorem (Normal basis theorem) For every prime power q and every integer $n \geq 1$, $\mathbb{F}_{q^n}$ has a normal basis over $\mathbb{F}_q$.

5.2.2 Remark The normal basis theorem was proved first by Hensel [1486]. A more general normal basis theorem for any finite Galois extension of an arbitrary field was proved by Noether [2297] and Deuring [824].

5.2.3 Proposition

1. For any two integers $m, n \geq 1$ and for any normal element $\alpha \in \mathbb{F}_{q^{mn}}$ over $\mathbb{F}_q$, the element
$$\beta = \mathrm{Tr}_{\mathbb{F}_{q^{mn}}/\mathbb{F}_{q^n}}(\alpha),$$
the trace of α from $\mathbb{F}_{q^{mn}}$ to $\mathbb{F}_{q^n}$, is a normal element of $\mathbb{F}_{q^n}$ over $\mathbb{F}_q$.
2. If $\gcd(m,n) = 1$, then any normal element $\alpha \in \mathbb{F}_{q^n}$ over $\mathbb{F}_q$ is still normal in $\mathbb{F}_{q^{mn}}$ over $\mathbb{F}_{q^m}$.
3. If $\gcd(m,n) = 1$, then for any normal elements $\alpha \in \mathbb{F}_{q^n}$ and $\beta \in \mathbb{F}_{q^m}$ over $\mathbb{F}_q$, the product $\alpha\beta$ is a normal element of $\mathbb{F}_{q^{mn}}$ over $\mathbb{F}_q$.

5.2.4 Definition Let $\mathbb{F}_q[\sigma]$ denote the set of all polynomials in σ with coefficients in $\mathbb{F}_q$. The ring $\mathbb{F}_q[\sigma]$ acts on $\mathbb{F}_{q^n}$ in a natural way: for any $f(\sigma) = \sum_{i=0}^{m} c_i\sigma^i \in \mathbb{F}_q[\sigma]$ and $\alpha \in \mathbb{F}_{q^n}$,
$$f \circ \alpha = \sum_{i=0}^{m} c_i\sigma^i(\alpha) \in \mathbb{F}_{q^n}.$$

5.2.5 Remark

1. An element $\alpha \in \mathbb{F}_{q^n}$ is normal over $\mathbb{F}_q$ if and only if every $\beta \in \mathbb{F}_{q^n}$ is equal to $f(\sigma) \circ \alpha$ for some $f \in \mathbb{F}_q[x]$.
2. For any $f, g \in \mathbb{F}_q[\sigma]$, we have
$$(fg) \circ \alpha = f \circ (g \circ \alpha).$$
3. For any nonzero polynomial $f(\sigma) = \sum_{i=0}^{m} c_i\sigma^i \in \mathbb{F}_q[\sigma]$ with $m < n$, f is not equal to the zero map on $\mathbb{F}_{q^n}$. In fact, if $f \circ \alpha = 0$ for all $\alpha \in \mathbb{F}_{q^n}$, then the polynomial $\sum_{i=0}^{m} a_i x^{q^i}$ has q^n zeros in $\mathbb{F}_{q^n}$, which is more than the degree q^m. Hence $x^n - 1$ is the minimal polynomial of σ and
$$\mathbb{F}_q[\sigma] \cong \mathbb{F}_q[x]/(x^n - 1)$$
as rings over $\mathbb{F}_q$. This implies that
$$\mathbb{F}_q[\sigma] = \left\{ \sum_{i=0}^{n-1} c_i\sigma^i : c_i \in \mathbb{F}_q, 0 \leq i \leq n-1 \right\}.$$
There is an obvious correspondence between polynomials of degree at most $n-1$ in $\mathbb{F}_q[x]$ and elements in $\mathbb{F}_q[\sigma]$, so we use the two notations interchangeably.
4. An element $f(\sigma) \in \mathbb{F}_q[\sigma]$ is invertible if there is $g(\sigma) \in \mathbb{F}_q[\sigma]$ so that $f(\sigma)g(\sigma) = 1$. By the ring isomorphism above, $f(\sigma)$ is invertible in $\mathbb{F}_q[\sigma]$ if and only if $\gcd(f(x), x^n - 1) = 1$ in $\mathbb{F}_q[x]$.

5. Each element $f(\sigma) = a_0 + a_1\sigma + \cdots + a_{n-1}\sigma^{n-1} \in \mathbb{F}_q[\sigma]$ corresponds to an $n \times n$ *circulant matrix* (see Definition 13.2.29)

$$C(f) = (a_{j-i})$$

whose rows are indexed by i and columns by j, where $0 \le i, j \le n-1$, with $j-i$ computed modulo n. For any $f, g \in \mathbb{F}_q[\sigma]$, we have

$$C(fg) = C(f)C(g),$$

hence f is invertible in $\mathbb{F}_q[\sigma]$ if and only if the matrix $C(f)$ is nonsingular.

5.2.6 Remark An $\mathbb{F}_q$-linear subspace V of $\mathbb{F}_{q^n}$ is *σ-invariant* (or *σ-stable*) if $\sigma(\alpha) \in V$ for every $\alpha \in V$. We can characterize normal elements in $\mathbb{F}_{q^n}$ by invariant subspaces. Let

$$n = em, \quad e = p^v,$$

where p is the characteristic of $\mathbb{F}_q$ and $\gcd(p, m) = 1$. Suppose $x^n - 1$ factors in $\mathbb{F}_q[x]$ as

$$x^n - 1 = (g_1(x)g_2(x)\cdots g_r(x))^e, \tag{5.2.1}$$

where the $g_i(x)$'s are distinct monic irreducible factors of $x^m - 1$ in $\mathbb{F}_q[x]$. Let $E_i(x) \in \mathbb{F}_q[x]$ for $1 \le i \le r$ so that, with $1 \le j \le r$,

$$E_i(x) \equiv \begin{cases} 1 \pmod{g_j^e(x)} & \text{if } j = i, \\ 0 \pmod{g_j^e(x)} & \text{if } j \ne i. \end{cases}$$

This means that, for $1 \le i, j \le r$,

$$E_i(\sigma)E_j(\sigma) = \begin{cases} E_i(\sigma) & \text{if } j = i, \\ 0 & \text{if } j \ne i. \end{cases}$$

Let $R_i = E_i(\sigma)\mathbb{F}_q[\sigma]$ for $1 \le i \le r$. Then

$$R_i \cong \mathbb{F}_q[x]/(g_i(x)^e), \qquad 1 \le i \le r,$$

and

$$\mathbb{F}_q[\sigma] = R_1 + R_2 + \cdots + R_r, \tag{5.2.2}$$

is a direct sum of subrings. For $1 \le i \le r$, let

$$V_i = R_i \circ \mathbb{F}_{q^n} = \{f \circ \alpha : f \in R_i \text{ and } \alpha \in \mathbb{F}_{q^n}\} \subseteq \mathbb{F}_{q^n}.$$

Then V_i is a σ-invariant subspace of $\mathbb{F}_{q^n}$ annihilated by $g_i(x)^e$, that is

$$g_i(\sigma)^e \circ \alpha = 0, \quad \text{for all } \alpha \in V_i.$$

Also, V_i has dimension $e \cdot \deg(g_i(x))$ over $\mathbb{F}_q$. Let W_i be the subspace of V_i annihilated by $g_i(x)^{e-1}$, $1 \le i \le r$, that is

$$W_i = \{\alpha \in V_i : g_i(\sigma)^{e-1} \circ \alpha = 0\}.$$

Then W_i is a σ-invariant subspace of V_i with dimension $(e-1)\deg(g_i(x))$ over $\mathbb{F}_q$.

5.2.7 Theorem [2395, 2580] With the notation as in the above remark, we have that

$$\mathbb{F}_{q^n} = V_1 + V_2 + \cdots + V_r$$

is a direct sum of $\mathbb{F}_q$-vector spaces. Furthermore, an arbitrary element $\alpha \in \mathbb{F}_{q^n}$, written as

$$\alpha = \alpha_1 + \alpha_2 + \cdots + \alpha_r,$$

where $\alpha_i \in V_i$, is normal over $\mathbb{F}_q$ if and only if

$$\alpha_i \notin W_i, \quad 1 \leq i \leq r.$$

5.2.8 Corollary [1486, 2324] Let $d_i = \deg(g_i(x))$ for $1 \leq i \leq r$. Then the number of normal elements in $\mathbb{F}_{q^n}$ over $\mathbb{F}_q$ is

$$\prod_{i=1}^{r} \left(q^{ed_i} - q^{(e-1)d_i}\right) = \Phi_q(x^n - 1),$$

where, Φ_q is the Euler Phi function for polynomials, see Definition 2.1.111.

5.2.9 Corollary Let n be a power of p, the characteristic of $\mathbb{F}_q$. Then

1. an element $\alpha \in \mathbb{F}_{q^n}$ is normal over $\mathbb{F}_q$ if and only if $\mathrm{Tr}_{\mathbb{F}_{q^n}/\mathbb{F}_q}(\alpha) \neq 0$;
2. an irreducible polynomial $f(x) \in \mathbb{F}_q[x]$ of degree n is an N-polynomial if and only if the coefficient of x^{n-1} in $f(x)$ is nonzero.

5.2.10 Corollary Let n be a prime such that q is primitive modulo n. Then

1. an element $\alpha \in \mathbb{F}_{q^n}$ is normal over $\mathbb{F}_q$ if and only if $\alpha \notin \mathbb{F}_q$ and $\mathrm{Tr}_{\mathbb{F}_{q^n}/\mathbb{F}_q}(\alpha) \neq 0$;
2. an irreducible polynomial $f(x) \in \mathbb{F}_q[x]$ of degree n is an N-polynomial if and only if the coefficient of x^{n-1} in $f(x)$ is nonzero.

5.2.11 Theorem For any $\alpha \in \mathbb{F}_{q^n}$, define

$$T_\alpha(x) = \sum_{i=0}^{n-1} \sigma^i(\alpha)x^i \in \mathbb{F}_{q^n}[x], \qquad t_\alpha(x) = \sum_{i=0}^{n-1} t_i x^i \in \mathbb{F}_q[x],$$

where $t_i = \mathrm{Tr}_{\mathbb{F}_{q^n}/\mathbb{F}_q}(\alpha\, \sigma^i(\alpha)) \in \mathbb{F}_q$, $0 \leq i \leq n-1$. We have the following characterizations of normal elements.

1. [1486] $\alpha \in \mathbb{F}_{q^n}$ is normal over $\mathbb{F}_q$ if and only if $\gcd(T_\alpha(x), x^n - 1) = 1$ in $\mathbb{F}_{q^n}[x]$.
2. [1172] $\alpha \in \mathbb{F}_{q^n}$ is normal over $\mathbb{F}_q$ if and only if $\gcd(t_\alpha(x), x^n - 1) = 1$ in $\mathbb{F}_q[x]$, that is, if $t_\alpha(\sigma)$ is invertible in $\mathbb{F}_q[\sigma]$.
3. [2385] Suppose that $\alpha \in \mathbb{F}_{q^n}$ is normal over $\mathbb{F}_q$. Then, for any $g(\sigma) \in \mathbb{F}_q[\sigma]$, the element $\beta = g(\sigma) \circ \alpha$ is normal over $\mathbb{F}_q$ if and only if $g(\sigma)$ is invertible in $\mathbb{F}_q[\sigma]$, that is, $\gcd(g(x), x^n - 1) = 1$ in $\mathbb{F}_q[x]$.

5.2.12 Remark Part 3 above shows again that $\Phi_q(x^n - 1)$ is equal to the number of normal elements in $\mathbb{F}_{q^n}$ over $\mathbb{F}_q$. There is a nice formula for $\Phi_q(x^n - 1)$ due to Ore [2324]. Suppose $n = p^v m$ where m is not divisible by p. For a positive integer d, let $\tau_m(d)$ denote the multiplicative order of d modulo m, and $\phi(d)$ be the Euler ϕ-function (equal to the number of integers between 1 and d that are relatively prime to d). Then

$$\Phi_q(x^n - 1) = q^n \prod_{d|m} \left(1 - \frac{1}{q^{\tau_m(d)}}\right)^{\phi(d)/\tau_m(d)}.$$

Also, for a general polynomial $f \in \mathbb{F}_q[x]$ with r distinct irreducible factors in $\mathbb{F}_q[x]$ with degrees $d_1, d_2, \ldots, d_r$ (the degrees need not be distinct), we have

$$\Phi_q(f(x)) = q^n \prod_{i=1}^{r} \left(1 - \frac{1}{q^{d_i}}\right).$$

5.2.13 Theorem [1186] For any $f(x) \in \mathbb{F}_q[x]$ of degree $n \geq 1$ with $f(0) \neq 0$, we have

$$\Phi_q(f) \geq \begin{cases} \frac{q^n}{e} & \text{if } n < q, \\ \frac{q^n}{e^{\gamma + \frac{1}{2(1+\log_q n)}}(1+\log_q n)} > \frac{q^n}{e^{0.83}(1+\log_q n)} & \text{if } n \geq q, \end{cases}$$

where $\gamma \approx 0.577216$ is Euler's constant and $e \approx 2.71828$ is Euler's number.

5.2.14 Remark In [1186], it is also proved that, for $f(x) = x^n - 1$, the lower bound above is almost tight for an infinite sequence of values of n.

5.2.15 Proposition [1631] Let α be a normal basis generator for $\mathbb{F}_{q^n}$ over $\mathbb{F}_q$ and let $C = (c_{ij})$ be an invertible matrix over $\mathbb{F}_q$. Then the basis $\{\beta_0, \beta_1, \ldots, \beta_{n-1}\}$ defined by

$$\beta_i = \sum_{j=0}^{n-1} c_{ij}\alpha^{q^i}, \quad i = 0, 1, \ldots, n-1,$$

is a normal basis for $\mathbb{F}_{q^n}$ over $\mathbb{F}_q$ if and only if C is circulant.

5.2.16 Corollary The number of normal basis generators of $\mathbb{F}_{q^n}$ over $\mathbb{F}_q$ is the order of the group $C(n,q)$ of invertible circulant $n \times n$ matrices over $\mathbb{F}_q$.

5.2.17 Proposition [2077, 2859] Define two sequences of polynomials $a_k(x), b_k(x) \in \mathbb{F}_2[x]$ as follows:

$$\begin{array}{ll} a_0(x) = x, & b_0(x) = 1 \\ a_{k+1}(x) = a_k(x)b_k(x), & b_{k+1}(x) = a_k^2(x) + b_k^2(x), \quad k \geq 0. \end{array}$$

Then, for every $k \geq 1$, the polynomial $a_k(x) + b_k(x)$ is an N-polynomial over $\mathbb{F}_2$ of degree 2^k.

5.2.18 Remark The above sequences of polynomials are used by Gao and Mateer [1185] to perform fast additive Fourier transforms over fields of characteristic two.

5.2.19 Proposition [2859] Let p be any prime. Define a sequence of polynomials $f_k(x) \in \mathbb{F}_p[x]$ as follows:

$$\begin{aligned} f_1(x) &= x^p + x^{p-1} + \cdots + x - 1, \\ f_2(x) &= f_1(x^p - x - 1), \\ f_{k+1}(x) &= f_k^*(x^p - x - 1), \quad k \geq 2, \end{aligned}$$

where $f^*(x)$ denotes the reciprocal polynomial of $f(x)$, that is, $f^*(x) = x^d f(1/x)$ where d is the degree of $f(x)$; see Definition 2.1.48. Then, for every $k \geq 1$, $f_k^*(x)$ is an N-polynomial over $\mathbb{F}_p$ of degree p^k.

5.2.20 Proposition [309]

1. Suppose $q \equiv 1 \pmod 4$ and let $a \in \mathbb{F}_q$ be a non-square. Then the polynomial

$$x^{2^k} - a(x-1)^{2^k}$$

is an N-polynomial over $\mathbb{F}_q$ for all $k \geq 1$.

2. Suppose $q \equiv 3 \pmod 4$ and $x^2 - bx - c^2 \in \mathbb{F}_q[x]$ is irreducible with $b \neq 2$. Then the polynomial

$$(x-1)^{2^{k+1}} - b(x^2 - x)^{2^k} - cx^{2^k}$$

is an N-polynomial over $\mathbb{F}_q$ for all $k \geq 0$.

3. Let q be any prime power, $r_1, \ldots, r_t$ be the distinct prime factors of $q-1$ and
$$n = r_1^{i_1} \cdots r_t^{i_t}, \quad i_1, \ldots, i_t \geq 0.$$
Assume that $q \equiv 1 \pmod 4$ if some r_i is equal to 2. Let $a \in \mathbb{F}_q$ be any element of order $q-1$. Then the polynomial $x^n - a(x-1)^n$ is an N-polynomial.

5.2.2 Self-dual normal bases

5.2.21 **Definition** For any $\alpha, \beta \in \mathbb{F}_{q^n}$, define the *trace polynomial* of α and β over $\mathbb{F}_q$ to be
$$t_{\alpha,\beta}(\sigma) = \sum_{i=0}^{n-1} \mathrm{Tr}(\alpha\,\sigma^i(\beta))\,\sigma^i \in \mathbb{F}_q[\sigma].$$
When $\alpha = \beta$, $t_{\alpha,\alpha}(\sigma)$ is denoted by $t_\alpha(\sigma)$, which agrees with the definition of t_α in Theorem 5.2.11, and is the *trace polynomial* of α over $\mathbb{F}_q$. An element α is *dual* to β in $\mathbb{F}_{q^n}$ over $\mathbb{F}_q$ if $t_{\alpha,\beta}(\sigma) = 1$, and α is *self-dual* if it is dual to itself.

5.2.22 **Remark** Note that $\alpha \in \mathbb{F}_{q^n}$ is normal over $\mathbb{F}_q$ if and only if $t_\alpha(\sigma)$ is invertible, while $\alpha \in \mathbb{F}_{q^n}$ is self-dual over $\mathbb{F}_q$ if and only if $t_\alpha(\sigma) = 1$. Hence an element $\alpha \in \mathbb{F}_{q^n}$ is self-dual over $\mathbb{F}_q$ if and only if it generates a self-dual normal basis for $\mathbb{F}_{q^n}$ over $\mathbb{F}_q$.

5.2.23 **Theorem** There is a self-dual normal basis for $\mathbb{F}_{q^n}$ over $\mathbb{F}_q$ if and only if q and n are odd, or q is even and n is not divisible by 4.

5.2.24 **Remark** Imamura and Morii [1574] proved the "only if" part of the theorem, and Lempel and Weinberger [1890] proved the "if" part. For self-dual normal bases in Galois extensions of arbitrary fields, see Bayer-Fluckier and Lenstra [212].

5.2.25 **Proposition** Suppose $\gcd(m, n) = 1$. Then

1. for any self-dual element $\alpha \in \mathbb{F}_{q^n}$ over $\mathbb{F}_q$, α remains self-dual in $\mathbb{F}_{q^{mn}}$ over $\mathbb{F}_{q^m}$;
2. for any self-dual elements $\alpha \in \mathbb{F}_{q^n}$ over $\mathbb{F}_q$ and $\beta \in \mathbb{F}_{q^m}$ over $\mathbb{F}_q$, the product $\alpha\beta$ is a self-dual element of $\mathbb{F}_{q^{mn}}$ over $\mathbb{F}_q$.

5.2.26 **Lemma** Let $\alpha \in \mathbb{F}_{q^n}$ be any normal element over $\mathbb{F}_q$. For any $\beta = f(\sigma) \circ \alpha$ and $\gamma = g(\sigma) \circ \alpha$ where $f(\sigma), g(\sigma) \in \mathbb{F}_q[\sigma]$, we have the following equation in $\mathbb{F}_q[\sigma]$:
$$t_{\beta,\gamma}(\sigma) = f(\sigma)g(\sigma^{-1})t_\alpha(\sigma).$$

5.2.27 **Proposition** Let $\alpha \in \mathbb{F}_{q^n}$ be any normal element over $\mathbb{F}_q$, and let $h_\alpha(\sigma)$ be the inverse of $t_\alpha(\sigma)$, that is, $h_\alpha(x)t_\alpha(x) \equiv 1 \pmod{x^n - 1}$.

1. The element $\beta = h_\alpha(\sigma) \circ \alpha$ is dual to α; hence, the dual basis of the normal basis generated by α is a normal basis and is generated by β.
2. An element $\beta = f(\sigma) \circ \alpha$ in $\mathbb{F}_{q^n}$ has a dual if and only if $f(\sigma)$ is invertible, and its dual is equal to $\gamma = g(\sigma) \circ \alpha$ where $g(\sigma) = (f(\sigma^{-1})t_\alpha(\sigma^{-1}))^{-1}$.
3. An element $\beta = f(\sigma) \circ \alpha$ is self-dual if and only if
$$f(\sigma)f(\sigma^{-1}) = h_\alpha(\sigma).$$
4. Suppose $\mathbb{F}_{q^n}$ has a self-dual element α over $\mathbb{F}_q$ (so $t_\alpha(\sigma) = 1$). Then $\beta = f(\sigma) \circ \alpha$ is self-dual if and only if
$$f(\sigma)f(\sigma^{-1}) = 1. \tag{5.2.3}$$

5.2.28 Remark In terms of circulant matrices, Equation (5.2.3) is equivalent to

$$C(f)C(f)^t = I,$$

that is, $C(f)$ is orthogonal. Let $\mathrm{oc}(n,q)$ denote the number of orthogonal $n \times n$ circulant matrices over $\mathbb{F}_q$. When $\mathbb{F}_{q^n}$ has a self-dual normal basis for $\mathbb{F}_q$, $\mathrm{oc}(n,q)$ is the number of self-dual elements in $\mathbb{F}_{q^n}$ over $\mathbb{F}_q$, and $\mathrm{oc}(n,q)/n$ is the number of self-dual normal bases in $\mathbb{F}_{q^n}$ over $\mathbb{F}_q$. Suppose $x^n - 1$ factors as in Equation (5.2.1) where $e = p^v$. We classify the irreducible factors $g_i(x)$ into three types:

1. $x-1$ and possibly $x+1$ (if n is even and q is odd);
2. self-reciprocal factors $g_i(x)$ with roots consisting of pairs $\{\xi, \xi^{-1}\}$ with $\xi \neq \xi^{-1}$, hence $g_i(x) = x^d g_i(1/x)/g_i(0)$ where d is the degree of $g_i(x)$. Suppose there are s such irreducible factors in Equation (5.2.1) and their degrees are $2d_1, \ldots, 2d_s$;
3. the remaining irreducible factors, which come in pairs $g_i(x)$ and $\tilde{g}_i(x)$ of the same degree so that, for each root ξ of $g_i(x)$, ξ^{-1} is a root of $\tilde{g}_i(x)$, hence $\tilde{g}_i(x) = x^d g_i(1/x)/g_i(0)$. Suppose there are t such pairs of irreducible factors in Equation (5.2.1) and their degrees are $e_1, \ldots, e_t$.

We order the components of R in Equation (5.2.2) accordingly:

$$R = R_0 + R_1 + \cdots R_s + (R_{s+1} + \tilde{R}_{s+1}) + \cdots + (R_{s+t} + \tilde{R}_{s+t}),$$

where R_0 represents the component for $x-1$ plus that for $x+1$ if it is a factor. Then the space of solutions for Equation (5.2.3) is the direct sum of solutions from each component. This leads to the following theorem.

5.2.29 Theorem [1990] Let $n = p^v m$, d_i's and e_j's be as defined in the above remark. Then

$$\mathrm{oc}(n,q) = \begin{cases} \epsilon_1 \prod_{i=1}^{s}(q^{d_i}+1)\prod_{i=1}^{t}(q^{e_i}-1) & \text{if } v = 0, \\ \epsilon_2\, q^{(p^v-1)m/2} \cdot \mathrm{oc}(m,q) & \text{if } v \geq 1, \end{cases}$$

where

$$\epsilon_1 = \begin{cases} 1 & \text{if } q \text{ is even}, \\ 2 & \text{if } q \text{ and } m \text{ are odd}, \\ 4 & \text{if } q \text{ is odd and } m \text{ is even}, \end{cases} \qquad \epsilon_2 = \begin{cases} 1 & \text{if } p \neq 2, \\ q^{1/2} & \text{if } p = 2 \text{ and } v = 1, \\ 2q^{1/2} & \text{if } p = 2 \text{ and } v \geq 2. \end{cases}$$

5.2.30 Corollary [130, 258, 1635] If the condition in Theorem 5.2.23 is satisfied, then there are $\mathrm{oc}(n,q)$ self-dual normal elements in $\mathbb{F}_{q^n}$ over $\mathbb{F}_q$.

5.2.3 Primitive normal bases

5.2.31 Definition An element $\alpha \in \mathbb{F}_{q^n}$ is *primitive normal* over $\mathbb{F}_q$ if it is normal over $\mathbb{F}_q$ and has multiplicative order $q^n - 1$. A normal basis generated by a primitive normal element is a *primitive normal basis*. A polynomial of degree n in $\mathbb{F}_q[x]$ is *primitive normal* if its roots are primitive normal in $\mathbb{F}_{q^n}$ over $\mathbb{F}_q$.

5.2.32 Theorem (Primitive normal basis theorem) For any prime power q and any integer $n \geq 1$, $\mathbb{F}_{q^n}$ has a primitive normal element over $\mathbb{F}_q$.

5.2.33 Remark The primitive normal basis theorem was proved by Carlitz [540] for q^n sufficiently large, by Davenport [775] when q is a prime and by Lenstra and Schoof [1899] in the general case. For a theoretical proof which does not require any machine calculation, see Cohen and Huczynska [692]. The next few theorems are further strengthenings of the primitive normal basis theorem.

5.2.34 Theorem [690] For any prime power q, any integer $n \geq 2$ and any nonzero $a \in \mathbb{F}_q$, there is a primitive normal polynomial $f(x) = x^n + c_1x^{n-1} + c_2x^{n-2} + \cdots + c_{n-1}x + c_n \in \mathbb{F}_q[x]$ with $c_1 = a$.

5.2.35 Theorem [1036] For any prime power q, any integer $n \geq 15$, any integer m with $1 \leq m < n$, and any $a \in \mathbb{F}_q$ (with $a \neq 0$ if $m = 1$), there is a primitive normal polynomial $f(x) = x^n + c_1x^{n-1} + c_2x^{n-2} + \cdots + c_{n-1}x + c_n \in \mathbb{F}_q[x]$ with $c_m = a$.

5.2.36 Theorem [694, 2806] For any prime power q and any integer $n \geq 2$, there is an element $\alpha \in \mathbb{F}_{q^n}$ such that both α and α^{-1} are primitive normal over $\mathbb{F}_q$ except when (q, n) is one of the pairs $(2, 3)$, $(2, 4)$, $(3, 4)$, $(4, 3)$ and $(5, 4)$.

5.2.37 Theorem [1035] For any prime power q, any integer $n \geq 7$ and any $a, b \in \mathbb{F}_q$ with $a \neq 0$, there is a primitive normal polynomial $f(x) = x^n + c_1x^{n-1} + c_2x^{n-2} + \cdots + c_{n-1}x + c_n \in \mathbb{F}_q[x]$ with $c_1 = a$ and $c_2 = b$.

5.2.38 Theorem [1034] For any integer $n \geq 2$ and sufficiently large prime power q, there is a primitive normal polynomial $f(x) = x^n + c_1x^{n-1} + c_2x^{n-2} + \cdots + c_{n-1}x + c_n \in \mathbb{F}_q[x]$ with its first $\lfloor n/2 \rfloor$ coefficients arbitrarily prescribed except that $c_1 \neq 0$.

5.2.39 Theorem [1029] For any integer $n \geq 2$ and sufficiently large prime power q, there is a primitive normal polynomial $f(x) = x^n + c_1x^{n-1} + c_2x^{n-2} + \cdots + c_{n-1}x + c_n \in \mathbb{F}_q[x]$ with its last $\lfloor n/2 \rfloor$ coefficients arbitrarily prescribed except that $(-1)^n c_n$ must be a primitive element of $\mathbb{F}_q$.

See Also

§2.2	For tables of primitive polynomials of various kinds and standards requiring normal basis arithmetic.
§5.3	For normal bases of low complexity.
§5.4	For completely normal bases.
§16.7	For hardware implementations of finite field arithmetic.

References Cited: [130, 212, 258, 309, 540, 690, 692, 694, 775, 824, 1029, 1034, 1035, 1036, 1172, 1185, 1186, 1486, 1574, 1631, 1635, 1890, 1899, 1990, 2077, 2297, 2324, 2385, 2395, 2580, 2806, 2859].

5.3 Complexity of normal bases

Shuhong Gao, Clemson University
David Thomson, Carleton University

5.3.1 Optimal and low complexity normal bases

5.3.1 **Definition** Let $\alpha \in \mathbb{F}_{q^n}$ be normal over $\mathbb{F}_q$ and let $N = (\alpha_0, \alpha_1, \ldots, \alpha_{n-1})$ be the normal basis of $\mathbb{F}_{q^n}$ over $\mathbb{F}_q$ generated by α, where

$$\alpha_i = \alpha^{q^i}, \quad 0 \le i \le n-1.$$

Denote by $T = (t_{ij})$ the $n \times n$ matrix given by

$$\alpha\alpha_i = \sum_{j=0}^{n-1} t_{ij}\alpha_j, \quad 0 \le i \le n-1,$$

where $t_{ij} \in \mathbb{F}_q$. The matrix T is the *multiplication table* of the basis N. Furthermore, the number of non-zero entries of T, denoted by C_N, is the *complexity* (also called the *density*) of the basis N.

5.3.2 **Remark** An exhaustive search for normal bases of $\mathbb{F}_{2^n}$ over $\mathbb{F}_2$ for $n < 40$ is given in [2015], extending previous tables such as those found in [1631]. Using data from the search, the authors in [2015] indicate that normal bases of $\mathbb{F}_{2^n}$ over $\mathbb{F}_2$ follow a normal distribution (with respect to their complexities) which is tightly compacted about a mean of roughly $n^2/2$. We define *low complexity* normal bases loosely to mean normal bases known to have sub-quadratic bounds, with respect to n, on their complexity.

5.3.3 **Remark** In addition, [2015] gives the minimum-known complexity of a normal basis of $\mathbb{F}_{2^n}$ over $\mathbb{F}_2$ for many values of n using a variety of constructions that appear in this section. Further tables on normal bases are provided in Section 2.2.

5.3.4 **Proposition** [2199] The complexity C_N of a normal basis N of $\mathbb{F}_{q^n}$ over $\mathbb{F}_q$ is bounded by

$$2n - 1 \le C_N \le n^2 - n + 1.$$

5.3.5 **Definition** A normal basis is *optimal normal* if it achieves the lower bound in Proposition 5.3.4.

5.3.6 **Theorem** [139, 2199]

1. (Type I optimal normal basis) Suppose $n+1$ is a prime and q is a primitive element in $\mathbb{Z}_{n+1}$. Let α be a primitive $(n+1)$-st root of unity. Then α generates an optimal normal basis of $\mathbb{F}_{q^n}$ over $\mathbb{F}_q$.
2. (Type II optimal normal basis) Suppose $2n+1$ is a prime and let γ be a primitive $(2n+1)$-st root of unity. Assume that the multiplicative group of $\mathbb{Z}_{2n+1}$ is generated by 2 and -1 (that is, either 2 is a primitive element in $\mathbb{Z}_{2n+1}$, or $2n+1 \equiv 3 \pmod 4$ and 2 generates the quadratic residues in Z_{2n+1}). Then $\alpha = \gamma + \gamma^{-1}$ generates an optimal normal basis of $\mathbb{F}_{2^n}$ over $\mathbb{F}_2$.

5.3.7 Theorem (Optimal normal basis theorem) [1184] Every optimal normal basis is equivalent to either a Type I or a Type II optimal normal basis. More precisely, suppose $\mathbb{F}_{q^n}$ has an optimal normal basis over $\mathbb{F}_q$ generated by α and let $b = \mathrm{Tr}(\alpha) \in \mathbb{F}_q$. Then one of the following must hold:

1. $n+1$ is a prime, q is primitive modulo $n+1$ and $-\alpha/b$ is a primitive $(n+1)$-st root of unity;
2. $q = 2^v$ with $\gcd(v, n) = 1$, $2n+1$ is a prime such that 2 and -1 generate the multiplicative group of $\mathbb{Z}_{2n+1}$, and $\alpha/b = \gamma + \gamma^{-1}$ for some primitive $(2n+1)$-st root of unity γ.

5.3.8 Remark Gao and Lenstra [1184] prove a more general version of the optimal normal basis theorem. They show that if a finite Galois extension L/K, where K is an arbitrary field, has an optimal normal basis, say generated by α, then there is a prime number r, an r-th root of unity γ in some algebraic extension of L and a nonzero constant $c \in K$ so that one of the following holds:

1. $\alpha = c\gamma$ and L has degree $r-1$ over K (so the polynomial $x^{r-1} + x^{r-2} + x + 1$ is irreducible over K);
2. $\alpha = c(\gamma + \gamma^{-1})$ and L has degree $(r-1)/2$ over K (so the minimal polynomial of $\gamma + \gamma^{-1}$ over K has degree $(r-1)/2$).

5.3.9 Theorem [308] Let $F(x) = x^{q+1} + dx^q - (ax+b)$ with $a, b, d \in \mathbb{F}_q$ and $b \neq ad$. Let f be an irreducible factor of F of degree $n > 1$ and let α be a root of f. Then all the roots of f are

$$\alpha_i = \alpha^{q^i} = \varphi^i(\alpha), \ i = 0, 1, \ldots, n-1,$$

where $\varphi(x) = (ax+b)/(x+d)$. If $\tau = \mathrm{Tr}_{\mathbb{F}_{q^n}/\mathbb{F}_q}(\alpha) \neq 0$, then $(\alpha_0, \alpha_1, \ldots, \alpha_{n-1})$ is a normal basis of $\mathbb{F}_{q^n}$ over $\mathbb{F}_q$ such that

$$\alpha \begin{pmatrix} \alpha_0 \\ \alpha_1 \\ \alpha_2 \\ \vdots \\ \alpha_{n-1} \end{pmatrix} = \begin{pmatrix} \tau^* & -e_{n-1} & -e_{n-2} & \cdots & -e_1 \\ e_1 & e_{n-1} & & & \\ e_2 & & e_{n-2} & & \\ \vdots & & & \ddots & \\ e_{n-1} & & & & e_1 \end{pmatrix} \begin{pmatrix} \alpha_0 \\ \alpha_1 \\ \alpha_2 \\ \vdots \\ \alpha_{n-1} \end{pmatrix} + \begin{pmatrix} b^* \\ b \\ b \\ b \\ b \end{pmatrix}, \qquad (5.3.1)$$

where $e_1 = a$, $e_{i+1} = \varphi(e_i)$ $(i \geq 1)$, $b^* = -b(n-1)$ and $\tau^* = \tau - \epsilon$ with

$$\epsilon = \sum_{i=0}^{n-1} e_i = \begin{cases} (n-1)(a-d)/2 & \text{if } p \neq 2, \\ a = d & \text{if } p = n = 2, \\ a - d & \text{if } p = 2 \text{ and } n \equiv 3 \pmod 4, \\ 0 & \text{if } p = 2 \text{ and } n \equiv 1 \pmod 4. \end{cases}$$

5.3.10 Corollary [308] The following are two special cases of the above theorem.

1. For every $a, \beta \in \mathbb{F}_q^*$ with $\mathrm{Tr}_{\mathbb{F}_q/\mathbb{F}_p}(\beta) = 1$,

$$x^p - \frac{1}{\beta} a x^{p-1} - \frac{1}{\beta} a^p$$

is irreducible over $\mathbb{F}_q$ and its roots form a normal basis of $\mathbb{F}_{q^p}$ over $\mathbb{F}_q$ of complexity at most $3p-2$. This corresponds to the case of Theorem 5.3.9 with $n = p$, $e_1 = a$, $\varphi(x) = ax/(x+a)$, $b = b^* = 0$, and $\tau^* = a/\beta$ if $p \neq 2$ and $\tau^* = a/\beta - a$ if $p = 2$.

2. Let n be any factor of $q-1$. Let $\beta \in \mathbb{F}_q$ have multiplicative order t such that $\gcd(n,(q-1)/t)=1$ and let $a=\beta^{(q-1)/n}$. Then

$$x^n - \beta(x-a+1)^n$$

is irreducible over $\mathbb{F}_q$ and its roots form a normal basis of $\mathbb{F}_{q^n}$ over $\mathbb{F}_q$ with complexity at most $3n-2$. This corresponds to the case of Theorem 5.3.9 with $e_1=a$, $\varphi(x)=ax/(x+1)$, $b=b^*=0$, and $\tau^*=-n(a-1)\beta/(1-\beta)-\epsilon$, with ϵ given as in Theorem 5.3.9 with $d=1$.

5.3.11 Conjecture [3036] If there does not exist an optimal normal basis of $\mathbb{F}_{q^n}$ over $\mathbb{F}_q$, then the complexity of a normal basis of $\mathbb{F}_{q^n}$ over $\mathbb{F}_q$ is at least $3n-3$.

5.3.12 Remark Explicit constructions of low complexity normal bases beyond the optimal normal bases and the constructions given in Theorem 5.3.10 are rare. In Section 5.3.2 we give a generalization of optimal normal bases arising from *Gauss periods*. Below, we illustrate how to construct new normal bases of low complexity arising from previously known normal bases.

5.3.13 Proposition [1172, 2578, 2580] Suppose $\gcd(m,n)=1$ and α and β generate normal bases A and B for $\mathbb{F}_{q^m}$ and $\mathbb{F}_{q^n}$ over $\mathbb{F}_q$, respectively. By Proposition 5.2.3, $\alpha\beta$ generates a normal basis N for $\mathbb{F}_{q^{mn}}$ over $\mathbb{F}_q$. Furthermore, we have $C_N=C_AC_B$ and if α and β both generate optimal normal bases, then $C_N=4mn-2m-2n+1$.

5.3.14 Proposition [634] Let $n=mk$ and suppose $\alpha \in \mathbb{F}_{q^n}$ generates a normal basis $(\alpha_0,\alpha_1,\dots,\alpha_{n-1})$ over $\mathbb{F}_q$ with multiplication table $T=(t_{ij})$ for $0\le i,j\le n-1$. Then

$$\beta = \mathrm{Tr}_{\mathbb{F}_{q^n}/\mathbb{F}_{q^m}}(\alpha) = \alpha_0+\alpha_m+\alpha_{2m}+\cdots+\alpha_{(k-1)m}$$

generates a normal basis $(\beta_0,\beta_1,\dots,\beta_{m-1})$ for $\mathbb{F}_{q^m}$ over $\mathbb{F}_q$ with

$$\beta\beta_i = \sum_{j=0}^{m-1} s_{ij}\beta_j, \quad 0\le i\le m-1,$$

where

$$s_{ij} = \sum_{0\le u,v\le k-1} t_{um+i,vm+j}, \quad 0\le i,j\le m-1.$$

5.3.15 Corollary [633, 634, 1931] Let $n=mk$. Upper-bounds on the complexity obtained from traces of optimal normal bases of $\mathbb{F}_{q^n}$ over $\mathbb{F}_q$ are given in Table 5.3.1.

	Type I (q odd):	Type I ($q=2$):	Type II ($q=2$):
m even, k odd, $p \mid k$	$km-(k+1)/2$	–	–
m even, k odd, $k\equiv 1 \pmod p$	$(k+1)m-(3k+1)/2$	$(k+1)m-3k+2^*$	$2km-2k+1$
m even, k odd, all other k	$(k+1)m-(3k+1)/2$	–	–
k even, $p \mid k$	$km-k/2^\dagger$	$km-k+1$	$2km-2k+1$
k even, $k\equiv 2 \pmod p$	$(k+1)m-3k/2+1^\dagger$	$km-k+1$	$2km-2k+1$
k even, all other k	$(k+1)m-k$	–	–

Table 5.3.1 Upper-bounds on the complexity obtained from of traces of optimal normal bases of $\mathbb{F}_{q^n}$ over $\mathbb{F}_q$, where $n=mk$. *Tight when $k=3$; †tight when $k=2,3$.

5.3.2 Gauss periods

5.3.16 **Definition** [139] Let $r = nk + 1$ be a prime not dividing q and let γ be a primitive r-th root of unity in $\mathbb{F}_{q^{nk}}$. Furthermore, let K be the unique subgroup of order k in $\mathbb{Z}_r^*$ and $K_i = \{a \cdot q^i : a \in K\} \subseteq \mathbb{Z}_r^*$ be cosets of K, $0 \le i \le n-1$. The elements

$$\alpha_i = \sum_{a \in K_i} \gamma^a \in \mathbb{F}_{q^n}, \quad 0 \le i \le n-1,$$

are *Gauss periods* of type (n, k) over $\mathbb{F}_q$.

5.3.17 **Theorem** [1180, 2951] Let $\alpha_i \in \mathbb{F}_{q^n}$ be Gauss periods of type (n, k) as defined in Definition 5.3.16. The following are equivalent:

1. $N = (\alpha_0, \alpha_1, \ldots, \alpha_{n-1})$ is a normal basis of $\mathbb{F}_{q^n}$ over $\mathbb{F}_q$;
2. $\gcd(nk/e, n) = 1$, where e is the order of q modulo r;
3. the union of $K_0, K_1, \ldots, K_{n-1}$ is $\mathbb{Z}_r^*$; equivalently, $\mathbb{Z}_r^* = \langle q, K\rangle$.

5.3.18 **Remark** Gauss periods of type $(n, 1)$ define Type I optimal normal bases and Gauss periods of type $(n, 2)$ define Type II optimal normal bases when $q = 2$.

5.3.19 **Remark** For the remainder of this section, we are concerned with Gauss periods which are *admissible* as normal bases, that is, where the properties in Theorem 5.3.17 hold. When the characteristic p does not divide n, the existence of admissible Gauss periods of type (n, k) is shown assuming the ERH in [14, 159] for any n with $k \le (cn)^3(\log(np))^2$. For any k and prime power q, assuming the GRH, there are infinitely many n such that there is an admissible Gauss period of $\mathbb{F}_{q^n}$ over $\mathbb{F}_q$ [1236]. In contrast, when p divides n, [2952] contains necessary and sufficient conditions for admissible Gauss periods, thus showing the non-existence of admissible Gauss periods in certain cases.

5.3.20 **Proposition** [1180] There is no admissible Gauss period of type (n, k) over $\mathbb{F}_2$ if 8 divides nk.

5.3.21 **Definition** Let K_i be defined as in Definition 5.3.16 for $i = 0, 1, \ldots, n-1$. The *cyclotomic numbers* are given by $c_{ij} = |(1 + K_i) \cap K_j|$.

5.3.22 **Proposition** [259, 1180] Let $N = (\alpha_0, \alpha_1, \ldots, \alpha_{n-1})$ be the normal basis arising from Gauss periods of type (n, k) for $\mathbb{F}_{q^n}$ over $\mathbb{F}_q$. Let $j_0 < n$ be the unique index such that $-1 \in \kappa_{j_0}$, and let $\delta_j = 1$ if $j = j_0$ and 0 if $j \ne j_0$. Then

$$\alpha\alpha_i = \delta_i k + \sum_{j=0}^{n-1} c_{ij}\alpha_j, \quad 0 \le i \le n-1,$$

hence $C_N \le (n-1)k + n$.

5.3.23 **Proposition** [139, 259] Let p be the characteristic of $\mathbb{F}_q$ and let $N = (\alpha_0, \alpha_1, \ldots, \alpha_{n-1})$ be the normal basis of $\mathbb{F}_{q^n}$ over $\mathbb{F}_q$ arising from Gauss periods of type (n, k).

1. If p divides k, then $C_N \le nk - 1$.
2. If $p = 2$, then

$$\begin{cases} kn - (k^2 - 3k + 3) \le C_N \le (n-1)k + 1 & \text{if } k \text{ even,} \\ (k+1)n - (k^2 - k + 1) \le C_N \le (n-2)k + n + 1 & \text{if } k \text{ odd.} \end{cases}$$

3. If $q = 2$ and $k = 2^v r$, where either $r = 1$ or both r is an odd prime and $v \leq 2$, then the lower bounds above are tight for sufficiently large n.

5.3.24 Problem Find the complexity of Gauss periods of type (n, k) over $\mathbb{F}_2$ for all n when k is not a prime, twice an odd prime, or four times an odd prime.

5.3.25 Remark [139, 634] The complexities of normal bases arising from Gauss periods of type (n, k), $2 \leq k \leq 6$, are given in Table 5.3.2 for all characteristics p when $n > p$.

Type $(n,1)$	Type $(n,2)$		Type $(n,3)$			Type $(n,4)$		
all p	$p=2$	$p>2$	$p=2$	$p=3$	$p>3$	$p=2$	$p=3$	$p>3$
$2n-1$	$2n-1$	$3n-2$	$4n-7$	$3n-2$	$4n-4$	$4n-7$	$5n-7$	$5n-6$

Type $(n,5)$				Type $(n,6)$			
$p=2$	$p=3$	$p=5$	$p>5$	$p=2$	$p=3$	$p=5$	$p>5$
$6n-21$	$6n-11$	$5n-7$	$6n-11$	$6n-21$	$6n-11$	$7n-15$	$7n-14$

Table 5.3.2 Complexities of normal bases from Gauss periods of Type (n, k), $2 \leq k \leq 6$, $n > p$.

5.3.26 Remark Let $q = 2$. Proposition 5.3.13 can be used to create normal bases of large extension degree by combining normal bases of subfields with coprime degree. By Proposition 5.3.20 Gauss periods of type (n, k) do not exist when 8 divides nk. Hence, Proposition 5.3.13 cannot be used to construct low complexity normal bases when the degree is a prime power. Thus, when n is a prime power (specifically a power of two), there are no constructions of low-complexity normal bases arising from the above propositions.

5.3.27 Problem Find explicit constructions of low-complexity normal bases of $\mathbb{F}_{2^n}$ over $\mathbb{F}_2$ when n is a power of two.

5.3.28 Remark Normal bases of low complexity are useful in fast encoding and decoding of network codes, see [2665] for more details.

5.3.3 Normal bases from elliptic periods

5.3.29 Remark Proposition 5.2.20, Theorem 5.3.9, and its corollaries show how the multiplicative group of $\mathbb{F}_q$ or $\mathbb{F}_{q^2}$ can be used to construct irreducible polynomials and normal bases for those degrees n whose prime factors divide $q - 1$ or $q + 1$. Also, Gauss periods use the multiplicative group of $\mathbb{F}_{q^{r-1}}$ for some prime r. Couveignes and Lercier [746] show how these methods can be generalized by using elliptic curve groups. The normal bases from their construction may not have low complexity, but these bases still allow a fast algorithm for multiplication. We outline their construction below; for more details on how to perform fast multiplication using elliptic periods, we refer the reader to [746]. For properties of elliptic curves, see Section 12.2.

5.3.30 Remark Let E be an elliptic curve over $\mathbb{F}_q$ defined by a Weierstrass equation

$$Y^2 + a_1XY + a_3Y = X^3 + a_2X^2 + a_4X + a_6,$$

where $a_i \in \mathbb{F}_q$. The points of E over every extension of $\mathbb{F}_q$ form an additive group with the point O at infinity as the identity. The order of the group $E(\mathbb{F}_q)$ is $q+1-t$ for some integer t with $|t| \leq 2\sqrt{q}$. Let $n > 1$ be an integer such that $E(\mathbb{F}_q)$ has a cyclic subgroup F of order n. The quotient $E' = E/F$ is also an elliptic curve over $\mathbb{F}_q$ and there is an isogeny

$$\phi : E \to E',$$

that has F as its kernel, and ϕ is defined by rational functions in $\mathbb{F}_q[X,Y]$. For any point $P \in E$, let $x(P)$ denote the x-coordinate of P and similarly denote $y(P)$, thus

$$P = (x(P), y(P)).$$

Vélu [2865] gives a formula for E' and ϕ. In fact, for $P \in E$,

$$\phi(P) = \left(x(P) + \sum_{T \in F \setminus \{O\}} (x(P+T) - x(T)),\ y(P) + \sum_{T \in F \setminus \{O\}} (y(P+T) - y(T)) \right).$$

5.3.31 Remark We describe here an explicit formula due to Kohel [1779] for E' and ϕ when E is of the form

$$E: \quad Y^2 = X^3 + aX + b.$$

We denote by D the kernel polynomial given by

$$\begin{aligned} D(X) &= \prod_{Q \in F \setminus \{O\}} (X - x(Q)) \\ &= X^n - c_1 X^{n-1} + c_2 X^{n-2} - c_3 X^{n-3} + \cdots + (-1)^n c_n \in \mathbb{F}_q[X]. \end{aligned}$$

Then, for $P = (x, y) \in E$,

$$\phi(P) = \left(\frac{N(x)}{D(x)},\ y \cdot \left(\frac{N(x)}{D(x)} \right)' \right),$$

where $N(x)$ is determined by the equation:

$$\frac{N(x)}{D(x)} = nx - c_1 - (3x^2 + a)\frac{D'(x)}{D(x)} - 2(x^3 + ax + b)\left(\frac{D'(x)}{D(x)} \right)'.$$

Furthermore, E' is defined by

$$Y^2 = X^3 + (a - 5v)X + (b - 7w),$$

where

$$v = a(n-1) + 3(c_1^2 - 2c_2), \quad w = 3ac_1 + 2b(n-1) + 5(c_1^3 - 3c_1c_2 + 3c_3).$$

5.3.32 Definition Let $T \in E(\mathbb{F}_q)$ be a point of order n and ϕ be the corresponding isogeny with its kernel generated by T. For any point $P \in E(\mathbb{F}_{q^n})$ with $\phi(P) \in E'(\mathbb{F}_q)$, let $\theta(P,T)$ denote the slope of the line passing through the two points T and $P+T$, that is

$$\theta(P,T) = \frac{y(P+T) - y(T)}{x(P+T) - x(T)} \in \mathbb{F}_{q^n}.$$

The element $\theta(P,T)$ is an *elliptic period* over $\mathbb{F}_q$.

5.3.33 Theorem [746] Let $T \in E(\mathbb{F}_q)$ be a point of order $n \geq 3$ and ϕ be the corresponding isogeny with its kernel generated by T. Suppose there is a point $P \in E(\mathbb{F}_{q^n})$ so that $nP \neq O$ in E and $\phi(P) \in E'(\mathbb{F}_q)$. Then either

1. the elliptic period $\theta(P,T)$ is a normal element of $\mathbb{F}_{q^n}$ over $\mathbb{F}_q$ if the trace of $\theta(P,T)$ from $\mathbb{F}_{q^n}$ to $\mathbb{F}_q$ is nonzero, or

2. the element $1+\theta(P,T)$ is a normal element of $\mathbb{F}_{q^n}$ over $\mathbb{F}_q$ if the trace of $\theta(P,T)$ is zero.

5.3.34 Example [746] Consider the following curve over $\mathbb{F}_7$

$$E: \quad y^2 + xy - 2y = x^3 + 3x^2 + 3x + 2.$$

The point $T = (3,1) \in E(\mathbb{F}_7)$ has order $n = 5$, so the subgroup $F = \langle T \rangle$ has order 5. By Vélu's formula, the equation for $E' = E/F$ is

$$E': \quad y^2 + xy - 2y = x^3 + 3x^2 - 3x - 1,$$

and the corresponding isogeny is

$$\phi(x,y) = \left(\frac{x^5 + 2x^2 - 2x - 1}{x^4 + 3x^2 - 3}, \frac{(x^6 - 3x^4 + 3x^3 - x^2 + 3x - 3)\, y + 3x^5 + x^4 + x^3 + 3x^2 - 3x + 1}{x^6 + x^4 - 2x^2 - 1} \right).$$

Take $A = (4,2) \in E'(\mathbb{F}_7)$. We note that the polynomial

$$f(X) = (X^5 + 2X^2 - 2X - 1) + 3(X^4 + 3X^2 - 3) = X^5 + 3X^4 - 3X^2 - 2X - 3$$

is irreducible over $\mathbb{F}_7$. Hence $\mathbb{F}_{7^5} = \mathbb{F}_7[\alpha]$, where α is a root of f. Compute β so that $\phi(\alpha,\beta) = (4,2)$. We find that

$$\beta = \alpha^{4756} = -\alpha^3 - \alpha^2 + 3\alpha + 2,$$

and $P = (\alpha,\beta) \in E(\mathbb{F}_{7^5})$. We check that $5\,P \neq O$ in E and we note that

$$P + T = (-3\alpha^4 + 3\alpha^2 + 2\alpha - 1, -\alpha^4 + \alpha^3 + \alpha^2 + 1).$$

Hence

$$\theta(P,T) = \frac{Y(P+T) - Y(T)}{X(P+T) - X(T)} = -\alpha^4 + \alpha^3 + 3\alpha^2 - 3\alpha - 3$$

is a normal element in $\mathbb{F}_{7^5}$ over $\mathbb{F}_7$.

5.3.4 Complexities of dual and self-dual normal bases

5.3.35 Remark For the definition of dual and self-dual bases, see Definition 2.1.100. Self-dual normal bases have been well studied due to their efficiency in implementation, see Section 16.7. A complete treatment of dual bases over finite fields can be found in [1631, Chapter 4], see also Sections 5.1 and 5.2.

5.3.36 Remark It is computationally easier to restrict an exhaustive search to self-dual normal bases. Geiselmann in [1263, 1631] computes the minimum complexity for a self-dual normal basis of $\mathbb{F}_{2^n}$ over $\mathbb{F}_2$ for all $n \leq 47$. These computations are repeated for odd degrees $n \leq 45$ in [130] and the authors also give tables of minimum complexity self-dual normal bases over finite fields of odd characteristic and for extensions of $\mathbb{F}_{2^\ell}, \ell > 1$. Some additional searches for self-dual normal bases can be found in [2015].

5.3.37 Proposition [1632] Let N be a normal basis with multiplication table T. Then N is self-dual if and only if T is symmetric.

5.3.38 Proposition [1632] Let $\gcd(m,n)=1$. Suppose α and β generate normal bases A and B for $\mathbb{F}_{q^m}$ and $\mathbb{F}_{q^n}$ over $\mathbb{F}_q$, respectively. Then $\gamma=\alpha\beta$ generates a self-dual normal basis N for $\mathbb{F}_{q^{mn}}$ over $\mathbb{F}_q$ if and only if both A and B are self-dual, as in Proposition 5.2.3. The complexity of the basis N is $C_N=C_AC_B$, as in Proposition 5.3.13.

5.3.39 Proposition [1632, 2422] Let n be even, $\alpha\in\mathbb{F}_{2^n}$ and $\gamma=1+\alpha$. Then,

1. the element α generates a self-dual normal basis for $\mathbb{F}_{2^n}$ over $\mathbb{F}_2$ if and only if γ does;
2. if α and $\gamma=1+\alpha$ generate self-dual normal bases B and $\bar{B}$, respectively, for $\mathbb{F}_{2^n}$ over $\mathbb{F}_2$, then the complexities of B and $\bar{B}$ are related by

$$C_{\bar{B}}=n^2-3n+8-C_B.$$

5.3.40 Corollary Suppose $n\equiv 2 \pmod 4$, then the following hold

1. the average complexity of a self-dual normal basis of $\mathbb{F}_{2^n}$ over $\mathbb{F}_2$ is $\frac{1}{2}(n^2-3n+8)$;
2. if B is a self-dual normal basis for $\mathbb{F}_{2^n}$ over $\mathbb{F}_2$, we have

$$2n-1\le C_B\le n^2-5n+9,$$

and one of the equalities holds if and only if either B or its complement $\bar{B}$ is optimal.

5.3.41 Proposition [308] Let q be a power of a prime p. For any $\beta\in\mathbb{F}_q^*$ with $\mathrm{Tr}_{\mathbb{F}_q/\mathbb{F}_p}(\beta)=1$,

$$x^p-x^{p-1}-\beta^{p-1}$$

is irreducible over $\mathbb{F}_q$ and its roots form a self-dual normal basis of $\mathbb{F}_{q^p}$ over $\mathbb{F}_q$ with complexity at most $3p-2$. The multiplication table is as in Theorem 5.3.10 with $e_1=\beta$, $e_{i+1}=\varphi(e_i)$ for $i\ge 1$, $\varphi(x)=\beta x/(x+\beta)$, $\tau^*=1$ if $p\ne 2$ and $\tau^*=1-\beta$ if $p=2$.

5.3.42 Proposition [308] Let n be an odd factor of $q-1$ and let $\xi\in\mathbb{F}_q$ have multiplicative order n. Then there exists $u\in\mathbb{F}_q$ such that $(u^2)^{(q-1)/n}=\xi$. Let $x_0=(1+u)/n$ and $x_1=(1+u)/(nu)$. Then the monic polynomial

$$\frac{1}{1+u^2}\left((x-x_0)^n-u^2(x-x_1)^n\right)$$

is irreducible over $\mathbb{F}_q$ and its roots form a self-dual normal basis of $\mathbb{F}_{q^n}$ over $\mathbb{F}_q$. The multiplication table is as in Theorem 5.3.9 with $a=(x_0-\xi x_1)/(1-\xi)$, $b=-x_0x_1$, $d=a-(x_0-x_1)$ and $\tau=1$.

5.3.43 Proposition [308] Let n be an odd factor of $q+1$ and let $\xi\in\mathbb{F}_{q^2}$ be a root of $x^{q+1}-1$ with multiplicative order n. Then there is a root u of $x^{q+1}-1$ such that $(u^2)^{q+1}/n=\xi$. Let $x_0=(1+u)/n$ and $x_1=(1+u)/(nu)$. Then

$$\frac{1}{1-u^2}\left((x-x_0)^n-u^2(x-x_1)^n\right)$$

is irreducible over $\mathbb{F}_q$ and its roots form a self-dual normal basis of $\mathbb{F}_{q^n}$ over $\mathbb{F}_q$. The multiplication table is as in Theorem 5.3.9 with $a=(x_1-\xi x_0)/(1-\xi)$, $b=-x_0x_1$, $d=a-(x_0+x_1)$ and $\tau=1$.

5.3.4.1 Duals of Gauss periods

5.3.44 Proposition [1180, 1930] Let α be a type (n, k) Gauss period generating a normal basis N and let $j_0 = 0$ if k is even and $j_0 = n/2$ if k is odd. Then the element

$$\gamma = \frac{\alpha^{q^{j_0}} - k}{nk + 1}$$

is dual to α, and hence γ generates the dual basis $\tilde{N}$ of N. Furthermore, the complexity of the dual basis $\tilde{N}$ is

$$C_{\tilde{N}} \leq \begin{cases} (k+1)n - k & \text{if } p \nmid k, \\ kn - 1 & \text{if } p \mid k. \end{cases}$$

5.3.45 Corollary [1180] For $n > 2$, a normal basis of $\mathbb{F}_{q^n}$ over $\mathbb{F}_q$ arising from Gauss periods of type (n, k) is self-dual if and only if k is even and divisible by the characteristic of $\mathbb{F}_q$. In particular, Type II optimal normal bases are self-dual.

5.3.46 Proposition [2924] The complexity of the dual of a Type I optimal normal basis is $3n - 2$ if q is odd and $3n - 3$ if q is even.

5.3.47 Remark [633] Upper bounds on the complexities of the dual basis of the $\mathbb{F}_{q^m}$-trace of optimal normal bases of $\mathbb{F}_{q^n}$ over $\mathbb{F}_q$, where $n = mk$, are given in Table 5.3.3.

	Type I (q odd)	Type I (q even)	Type II (q even)
m odd	$(k+2)m - 2$	$(k+2)(m-1)+1$	$2k(m-1)+1$
m even	$(k+3)m - k - 4$	$(k+3)m - 2k - 3$	$2k(m-1)+1$

Table 5.3.3 Upper bounds on complexities of the dual bases of the trace of optimal normal bases.

5.3.5 Fast arithmetic using normal bases

5.3.48 Remark In practical applications it is important to know how to do fast arithmetic in finite fields, for example addition, multiplication, and division; and for cryptographic applications it is also desirable to have elements of high orders and a fast algorithm for exponentiation. Details for the basic operations discussed in this section can be found in Section 11.1, see also [1227]. In hardware implementations, normal bases are often preferred, see Section 16.7 for details on hardware implementations. This subsection presents some theoretical results related to fast multiplication and exponentiation under normal bases generated by Gauss periods.

5.3.49 Remark Gao and Vanstone [1188] first observed that a Type II optimal normal basis generator has high order, which was proved later by von zur Gathen and Shparlinski [1240]; for more details see Section 4.4. Computer experiments by Gao, von zur Gathen, and Panario [1179] indicate that Gauss periods of type (n, k) with $k > 2$ also have high orders; however, it is still open whether one can prove a subexponential lower bound on their orders.

5.3.50 Problem Give tight bounds on the orders of Gauss periods of type (n, k), $k > 2$.

5.3.51 Proposition [1179, 1188] Suppose $\alpha \in \mathbb{F}_{q^n}$ is a Gauss period of type (n, k) over $\mathbb{F}_q$. Then for any integer $1 \leq t < q^n - 1$, α^t can be computed using at most $n^2 k$ operations in $\mathbb{F}_q$.

5.3.52 Theorem [1180] Suppose γ is an element of order r (not necessarily a prime) and

$$\alpha = \sum_{i \in K} \gamma^i$$

generates a normal basis N for $\mathbb{F}_{q^n}$ over $\mathbb{F}_q$, where K is a subgroup of $\mathbb{Z}_r^\times$. With $\mathbb{F}_{q^n}$ represented under the normal basis N, we have

1. addition and subtraction can be computed in $O(n)$ operations in $\mathbb{F}_q$;
2. multiplication can be computed in $O(r \log r \log\log r)$ operations in $\mathbb{F}_q$;
3. division can be computed in $O(r \log^2 r \log\log r)$ operations in $\mathbb{F}_q$;
4. exponentiation of an arbitrary element in $\mathbb{F}_{q^n}$ can be computed in $O(nr \log r \log\log r)$ operations in $\mathbb{F}_q$.

5.3.53 Remark Theorem 1.5 in [1174] tells us when and how to find such a subgroup K in the above theorem. The element α can be a Gauss period or a generalized Gauss period, see Example 5.3.54 for more information. We outline the algorithm from [1180] for fast multiplication and division. The basic idea is to convert the normal basis representation to a polynomial basis in the ring $R = \mathbb{F}_q[x]/(x^r - 1)$, do fast multiplication of polynomials in the ring, then convert the result back to the normal basis. More precisely, let γ and α be as in Theorem 5.3.52. The condition of the theorem implies that that $K, qK, \ldots, q^{n-1}K$ are disjoint subsets of $\mathbb{Z}_r$ and $q^n K = K$. For $0 \le j \le n-1$, let $K_j = q^j K \subseteq \mathbb{Z}_r$, and

$$\alpha_j = \sum_{i \in K_j} \beta^i.$$

Then $(\alpha_0, \alpha_1, \ldots, \alpha_{n-1})$ is the normal basis generated by α, with the following property:

$$\alpha_0 + \alpha_1 + \cdots + \alpha_{n-1} = -1.$$

For each element $A = a_0\alpha_0 + a_1\alpha_1 + \cdots + a_{n-1}\alpha_{n-1} \in \mathbb{F}_{q^n}$, where $a_i \in \mathbb{F}_q$, we associate a polynomial

$$A(x) = \sum_{i=0}^{r-1} u_i x^i,$$

where $u_i = a_j$ if $i \in K_j$ for some j, and $u_i = 0$ if i is not in any K_j. This can be viewed as a map from $\mathbb{F}_{q^n}$ to $R = \mathbb{F}_q[x]/(x^r - 1)$. The map is in fact a ring homomorphism.

Suppose we have two arbitrary elements $A, B \in \mathbb{F}_{q^n}$. To compute AB, we first write them as polynomials $A(x), B(x) \in \mathbb{F}_q[x]$ of degree at most $r-1$ as above. Then we use a fast algorithm to compute the product polynomial $C_1(x) = A(x)B(x)$ of degree at most $2r-2$. This step needs $O(r \log r \log\log r)$ operations in $\mathbb{F}_q$, see [1227]. Next, we reduce C_1 modulo $x^r - 1$ (just reduce the exponents of x modulo r) to get a polynomial

$$C_2(x) = c_0 + c_1 x + \cdots + c_{r-1}x^{r-1}.$$

The coefficients satisfy the property that $c_i = c_j$ whenever $i, j \in K_\ell$ for some $0 \le \ell \le n-1$. Since $\sum_{j=0}^{n-1} \alpha_j = -1$, we conclude that

$$AB = d_0\alpha_0 + d_1\alpha_1 + \cdots + d_{n-1}\alpha_{n-1}, \quad \text{where } d_i = c_j - c_0 \text{ for any } j \in K_i.$$

To compute A^{-1} (assuming $A \ne 0$), we apply a fast gcd algorithm to the two polynomials $A(x)$ and $x^r - 1$ to get a polynomial $U(x)$ of degree at most $r-1$ so that $A(x)U(x) \equiv 1 \pmod{x^r - 1}$. The element in $\mathbb{F}_{q^n}$ corresponding to the polynomial $U(x)$ is the desired inverse of A. The fast gcd step needs $O(r \log^2 r \log\log r)$ operations in $\mathbb{F}_q$, see [1227].

5.3.54 Example (Generalized Gauss Periods [1047, 1174]) For any normal basis from Gauss periods of type (n, k), we can apply Theorem 5.3.52 to perform fast arithmetic in $\mathbb{F}_{q^n}$. To obtain an admissible Gauss period of type (n, k), $r = nk + 1$ must be a prime. Here we give an example of generalized Gauss periods where r is not prime. Suppose we want to perform fast arithmetic in $\mathbb{F}_{2^{954}}$. Let $n = 954$ and note that the smallest k so that there is an admissible Gauss period of type (n, k) over $\mathbb{F}_2$ is $k = 49$. The corresponding $r = nk + 1 = 46747$ is a little big in this case. We observe that $954 = 106 \cdot 9$ and that there is an admissible Gauss period α_1 of type $(106, 1)$ over $\mathbb{F}_2$, and an admissible Gauss period α_2 of type $(9, 2)$. Then $\alpha = \alpha_1\alpha_2$ is a normal element of $\mathbb{F}_{2^n}$ over $\mathbb{F}_2$. We construct this α as follows. Let

$$r = (106 \cdot 1 + 1)(9 \cdot 2 + 1) = 2033, \quad K = \{1, 322\}.$$

Then K is a subgroup of $\mathbb{Z}_r^\times$ satisfying the condition in Theorem 5.3.52. Let γ be any primitive r-th root of unity in an extension field of $\mathbb{F}_2$. Then

$$\alpha = \gamma + \gamma^{322}$$

is a generalized Gauss period that is normal for $\mathbb{F}_{q^n}$ over $\mathbb{F}_q$. Now we can apply Theorem 5.3.52 to perform fast arithmetic in $\mathbb{F}_{q^n}$ with a much smaller r. In [1174], it is shown how to find generalized Gauss periods with minimum r and the related subgroups K; see [1174, Tables 2-4] for many more examples for which generalized Gauss periods are better than Gauss periods.

5.3.55 Example (Fast arithmetic under type II optimal normal bases) For type II optimal normal bases over $\mathbb{F}_2$, we describe below a slightly faster algorithm from [248, 1238]. Suppose $2n+1$ is a prime and the multiplicative group of $\mathbb{Z}_{2n+1}$ is generated by -1 and 2. Let $\gamma \in \mathbb{F}_{2^{2n}}$ be an element of order $2n + 1$. For any $i \geq 0$, define

$$\gamma_i = \gamma^i + \gamma^{-i}.$$

Then $N = (\gamma_1, \gamma_2, \dots, \gamma_n)$ is a permutation of the normal basis for $\mathbb{F}_{2^n}$ over $\mathbb{F}_2$ generated by

$$\alpha = \gamma_1 = \gamma + \gamma^{-1}.$$

We note that $\gamma_0 = 2$ and

$$\gamma_1 + \gamma_2 + \cdots + \gamma_n = 1.$$

To do fast multiplication and division in $\mathbb{F}_{2^n}$, we first perform a basis transition from N to the polynomial basis $P = (\alpha, \alpha^2, \dots, \alpha^n)$, then perform a fast multiplication of polynomials and finally transform the result back to the basis N. To do the basis transitions, we need the following properties:

$$\gamma_{i+j} = \gamma_i\gamma_j + \gamma_{j-i}, \quad \text{for all } i, j.$$

To see how to go from the basis N to the basis P, suppose we have an expression

$$A = a_1\gamma_1 + a_2\gamma_2 + \cdots + a_\ell\gamma_\ell,$$

where $a_i \in \mathbb{F}_2$ and $\ell \geq 1$ is arbitrary. We want to express A as a combination of $\alpha, \alpha^2, \dots, \alpha^\ell$ over $\mathbb{F}_2$. Let m be a power of 2 so that $\ell/2 \leq m < \ell$. Then

$$\gamma_m = \alpha^m.$$

We observe that

$$\begin{aligned} & a_1\gamma_1 + a_2\gamma_2 + \cdots + a_\ell\gamma_\ell \\ = \; & a_1\gamma_1 + \cdots + a_m\gamma_m + a_{m+1}(\gamma_m\gamma_1 + \gamma_{m-1}) + \cdots + a_\ell(\gamma_m\gamma_{\ell-m} + \gamma_{m-(\ell-m)}), \\ = \; & (a_1\gamma_1 + \cdots + a_m\gamma_m + a_{m+1}\gamma_{m-1} + \cdots + a_\ell\gamma_{m-(\ell-m)}) \\ & + \alpha^m (a_{m+1}\gamma_1 + a_{m+2}\gamma_2 + \cdots + a_\ell\gamma_{\ell-m}). \end{aligned}$$

We note that the first part is of the form $U = u_1\gamma_1 + \cdots + u_m\gamma_m$, where $u_i \in \mathbb{F}_2$, which can be computed using m bit operations. Let $V = a_{m+1}\gamma_1 + a_{m+2}\gamma_2 + \cdots + a_\ell\gamma_{\ell-m}$, where $\ell - m \le m$. Apply the method recursively to convert U and V into the power basis, say

$$U = b_1\alpha + \cdots + b_m\alpha^m, \quad V = b_{m+1}\alpha + \cdots + b_\ell\alpha^{\ell-m},$$

where $b_i \in \mathbb{F}_2$. Then

$$A = U + \alpha^m V = b_1\alpha + \cdots + b_m\alpha^m + b_{m+1}\alpha^{m+1} + \cdots + b_\ell\alpha^\ell.$$

This gives an algorithm for going from N to P using at most $\frac{1}{2}n\log_2(n)$ operations in $\mathbb{F}_2$, where $\log_2(n)$ is the logarithm of n in base 2. By reversing the above procedure, we get an algorithm for going from P to N using at most $\frac{1}{2}n\log_2(n)$ operations in $\mathbb{F}_2$.

This shows that multiplication in $\mathbb{F}_{2^n}$ under N can be computed by (a) two transformations from N to P, (b) one multiplication of polynomials of degree at most n in $\mathbb{F}_2[x]$, and (c) one transformation from $(\alpha, \alpha^2, \ldots, \alpha^{2n})$ to $(\gamma_1, \gamma_2, \ldots, \gamma_{2n})$ which is easily converted to N, as $\gamma_{n+1+i} = \gamma_{n-i}$. The number of bit operations used is the cost for one multiplication of polynomials of degree at most n, plus $2n\log_2 n$ bit operations for the basis transformations. Finally, for division in $\mathbb{F}_{2^n}$, we need to precompute the minimal polynomial of α and apply a fast gcd algorithm for polynomials of degree at most n in $\mathbb{F}_2[x]$.

See Also

§2.2	For standards requiring normal basis arithmetic.
§5.2	For general results on normal bases.
§11.1	For basic operations over finite fields.
§16.7	For hardware implementarions of finite fields arithmetic.
[1179, 1188, 1240]	For orders and cryptographic applications of Gauss Periods.

References cited: [14, 130, 139, 159, 248, 259, 308, 633, 634, 746, 1047, 1172, 1174, 1179, 1180, 1184, 1188, 1227, 1236, 1238, 1240, 1263, 1631, 1632, 1779, 1930, 1931, 2015, 2199, 2422, 2578, 2580, 2665, 2865, 2924, 2951, 2952, 3036].

5.4 Completely normal bases

Dirk Hachenberger, University of Augsburg

We present some theoretical results concerning algebraic extensions of finite fields. The starting point is the Complete Normal Basis Theorem, which is a strengthening of the classical Normal Basis Theorem. The search for completely normal elements leads to an interesting structure theory for finite fields comprising a generalization of the class of finite Galois field extensions to the class of cyclotomic modules.

5.4.1 The complete normal basis theorem

5.4.1 **Remark** Let $\overline{\mathbb{F}}_q$ denote an algebraic closure of the finite field $\mathbb{F}_q$. The Frobenius automorphism of $\overline{\mathbb{F}}_q/\mathbb{F}_q$ (throughout denoted by σ) is the field automorphism mapping each $\theta \in \overline{\mathbb{F}}_q$

to its q-th power θ^q. For every integer $m \geq 1$ there is a unique subfield E_m of $\overline{\mathbb{F}}_q$ such that $\mathbb{F}_q \subseteq E_m$ and $|E_m| = q^m$. As usual we write $\mathbb{F}_{q^m}$ for E_m. Given integers $d, m \geq 1$, one has $\mathbb{F}_{q^d} \subseteq \mathbb{F}_{q^m}$ if and only if $d|m$. Moreover, if d is a divisor of m, then $\mathbb{F}_{q^m}/\mathbb{F}_{q^d}$ is a Galois extension of degree $\frac{m}{d}$; its Galois group is cyclic and generated by σ^d (when restricted to $\mathbb{F}_{q^m}$). The $(\mathbb{F}_{q^m}, \mathbb{F}_{q^d})$-trace of $w \in \mathbb{F}_{q^m}$ is $\sum_{i=0}^{m/d-1} w^{q^{di}}$, while $\prod_{i=0}^{m/d-1} w^{q^{di}}$ is the $(\mathbb{F}_{q^m}, \mathbb{F}_{q^d})$-norm of w.

5.4.2 Remark Recall from Definition 2.1.98 that $\theta \in \mathbb{F}_{q^m}$ is a *normal element* of $\mathbb{F}_{q^m}$ over $\mathbb{F}_q$ provided its conjugates $\theta, \theta^q, \ldots, \theta^{q^{m-1}}$ under the Galois group of $\mathbb{F}_{q^m}/\mathbb{F}_q$ form an $\mathbb{F}_q$-basis of $\mathbb{F}_{q^m}$.

5.4.3 Definition An element $\theta \in \mathbb{F}_{q^m}$ is a *completely normal element* over $\mathbb{F}_q$, provided θ (simultaneously) is a normal element of $\mathbb{F}_{q^m}$ over $\mathbb{F}_{q^d}$ for every intermediate field $\mathbb{F}_{q^d}$ of $\mathbb{F}_{q^m}$ over $\mathbb{F}_q$, i.e., for every divisor $d \geq 1$ of m. A monic irreducible polynomial $g \in \mathbb{F}_q[x]$ of degree m is *completely normal* over $\mathbb{F}_q$ if one (and hence all) of its roots is a completely normal element of $\mathbb{F}_{q^m}/\mathbb{F}_q$.

5.4.4 Example (Based on [1387, 1388]) Let $q = 7$ and $m = 3^c$, where $c \geq 1$, and let $\eta \in \overline{\mathbb{F}}_7$ be a primitive 3^{c+1}-th root of unity. Then $\mathbb{F}_{7^{3^c}}$ is obtained by adjoining η to $\mathbb{F}_7$. For every $i = 1, \ldots, c$, let $\tau(i) := 3^{\lfloor (i-1)/2 \rfloor}$. Then

$$\theta := 1 + \sum_{i=1}^{c} \sum_{\substack{j=1,\\ \gcd(3,j)=1}}^{2\cdot\tau(i)} \eta^{j \cdot 3^{c-i}}$$

is a completely normal element of $\mathbb{F}_{7^{3^c}}$ over $\mathbb{F}_7$.

5.4.5 Theorem (Complete Normal Basis Theorem) [317] Given any finite field $\mathbb{F}_q$ and any integer $m \geq 1$ there exists a completely normal element of $\mathbb{F}_{q^m}$ over $\mathbb{F}_q$.

5.4.6 Remark The Complete Normal Basis Theorem was first proved by Blessenohl and Johnsen [317] in 1986. A simplification of the proof is given in [1386]. The analogous result for (finite dimensional) Galois extensions over infinite fields was already settled by Faith [1023] in 1957. For an outline of the general proof, see [1389, Chapter I]. Often, an element that is completely normal is also called *completely free* (or *vollständig frei* or *vollständig regulär*) [315, 317, 1386, 1387, 1388, 1389, 2090].

5.4.7 Remark Let $m = \prod_{i=1}^{k} r_i^{\nu_i}$ be the prime power decomposition of m. For every $i = 1, \ldots, k$ assume that θ_i is completely normal in $\mathbb{F}_{q^{r_i^{\nu_i}}}$ over $\mathbb{F}_q$. Then $\theta := \prod_{i=1}^{k} \theta_i$ is completely normal in $\mathbb{F}_{q^m}/\mathbb{F}_q$. This is an application of Hilfssatz 4.4 of [317] (see also the *Reduction Theorem* in Section 4 of [1389]); it reduces the existence problem for completely normal elements to extensions of prime power degree.

5.4.8 Remark Concerning the existence of completely normal elements for prime power extensions $\mathbb{F}_{q^{r^n}}/\mathbb{F}_q$ for some prime r, the cases where $r = p$ (the characteristic of $\mathbb{F}_q$) and $r \neq p$ have to be dealt with separately. The case of equal characteristic is covered by Subsection 5.4.2 on completely basic extensions. The case $r \neq p$ is much harder to establish; it is covered by the structure theory on completely normal elements to be outlined in Subsections 5.4.3-5.4.6. For explicit constructions (in the spirit of Example 5.4.4) of completely normal elements in $\mathbb{F}_{q^{r^n}}$ over $\mathbb{F}_q$ with $r \neq p$, see [1387, 1388].

5.4.9 Definition Let M be a nonempty set of positive integers such that $m \in M$ and $d|m$ imply $d \in M$, and $k, n \in M$ imply $\text{lcm}(k, n) \in M$ (where lcm denotes the least common multiple). Then M is a *Steinitz number*.

5.4.10 Remark The intermediate fields of $\overline{\mathbb{F}}_q/\mathbb{F}_q$ are in one-to-one correspondence with the Steinitz numbers (see Brawley and Schnibben [398]). The field corresponding to the Steinitz number M is the union $\mathbb{F}_{q^M} := \bigcup_{m \in M} \mathbb{F}_{q^m}$. This algebraic extension of $\mathbb{F}_q$ is infinite if and only if M is infinite. Any finite Steinitz number is of the form $\{d : d \in \mathbb{Z}, d \geq 1, d|m\}$ for some positive integer m; the corresponding field is $\mathbb{F}_{q^m}$.

5.4.11 Definition Let M be a Steinitz number. Then a sequence $(w_m)_{m \in M}$ of elements of $\overline{\mathbb{F}}_q$ is *trace-compatible*, if for all $d, m \in M$ with $d|m$ the $(\mathbb{F}_{q^m}, \mathbb{F}_{q^d})$-trace of w_m is equal to w_d. Similarly, the sequence is *norm-compatible*, if the $(\mathbb{F}_{q^m}, \mathbb{F}_{q^d})$-norm of w_m is equal to w_d for every $d, m \in M$ such that $d|m$.

5.4.12 Remark For an infinite Steinitz number M a trace-compatible sequence $(w_m)_{m \in M}$ such that every w_m is a normal element of $\mathbb{F}_{q^m}/\mathbb{F}_q$ can be interpreted as an (infinite) normal basis for $\mathbb{F}_{q^M}$ over $\mathbb{F}_q$. The existence of sequences of that kind in infinite Galois extensions over an arbitrary field is proved by Lenstra [1896]. Representations of finite fields within computer algebra systems which rely on subfield embeddings and trace-compatible sequences of normal elements are studied by Scheerhorn [2537, 2538]. For finite fields, [2537] provides an elementary proof of the existence of normal trace-compatible sequences for the entire algebraic closure $\overline{\mathbb{F}}_q/\mathbb{F}_q$ (i.e., for the Steinitz number $\mathbb{N}$).

5.4.13 Theorem [1389, Section 26] Consider the Steinitz number $\mathbb{N}$ and let $\mathbb{F}_q$ be any finite field. Then there exists a trace-compatible sequence $(w_m)_{m \in \mathbb{N}}$ in $\overline{\mathbb{F}}_q$ such that, for every m, the element w_m is *completely* normal in $\mathbb{F}_{q^m}$ over $\mathbb{F}_q$.

5.4.2 The class of completely basic extensions

5.4.14 Definition A Galois field extension $\mathbb{F}_{q^m}/\mathbb{F}_q$ is *completely basic*, if every normal element of $\mathbb{F}_{q^m}/\mathbb{F}_q$ is completely normal in $\mathbb{F}_{q^m}/\mathbb{F}_q$.

5.4.15 Remark The notion of a completely basic extension in the context of a general finite dimensional Galois extension goes back to Faith [1023]. Blessenohl and Johnsen [318] have characterized the completely basic extensions among the abelian extensions. Previously, Blessenohl [315] considered cyclic extensions of prime power degree. Meyer [2090] has extended [318] to a certain class of not necessarily abelian Galois extensions. For finite fields an elementary proof of Theorem 5.4.18 below is given in Section 15 of [1389].

5.4.16 Definition For relatively prime integers $q, \ell \geq 1$, the *order of* q *modulo* ℓ is the least integer $n \geq 1$ such that $q^n \equiv 1 \pmod{\ell}$; it is denoted by $\text{ord}_\ell(q)$.

5.4.17 Definition When considering a finite field with characteristic p, for an integer $t \geq 1$ the *p-free-part* t' of t is the largest divisor of t which is relatively prime to p.

5.4.18 Theorem [1389, Section 15] The following are equivalent:

1. $\mathbb{F}_{q^m}$ is completely basic over $\mathbb{F}_q$.

2. For every prime divisor r of m, every normal element of $\mathbb{F}_{q^m}/\mathbb{F}_q$ is normal in $\mathbb{F}_{q^m}/\mathbb{F}_{q^r}$.
3. For every prime divisor r of m, the number $\mathrm{ord}_{(m/r)'}(q)$ is not divisible by r.

5.4.19 Example Let $\mathbb{F}_q$ be any finite field. Then $\mathbb{F}_{q^m}$ is completely basic over $\mathbb{F}_q$ in all the following cases:

1. $m = r$ or $m = r^2$, where r is a prime;
2. m divides $q - 1$;
3. $m = p^b$, where p is the characteristic of $\mathbb{F}_q$ and $b \geq 0$ is any integer.

5.4.20 Remark Actually, in the latter case of Example 5.4.19, θ is a normal element for $\mathbb{F}_{q^{p^b}}/\mathbb{F}_q$ if and only if the $(\mathbb{F}_{q^{p^b}}, \mathbb{F}_q)$-trace of θ is non-zero (see [1389, Section 5]). For this class of field extensions, Blake, Gao, and Mullin [309] provide an iterative construction of completely normal elements.

5.4.3 Cyclotomic modules and complete generators

5.4.21 Remark In the present subsection the class of finite extensions of a finite field is generalized to the class of cyclotomic modules; thereby, completely normal elements are generalized to complete generators of cyclotomic modules. This generalization is necessary in order to understand how completely normal elements are additively composed. The main references are [1389, 1390].

5.4.22 Remark Consider again an algebraic closure $\overline{\mathbb{F}}_q$ of $\mathbb{F}_q$. For every integer $d \geq 1$ the additive group of $\overline{\mathbb{F}}_q$ is equipped with a module structure over the algebra $\mathbb{F}_{q^d}[x]$ of polynomials in the variable x. The corresponding scalar multiplication (of field elements by polynomials) is carried out by first evaluating a polynomial $f \in \mathbb{F}_{q^d}[x]$ at σ^d (the Frobenius automorphism over $\mathbb{F}_{q^d}$) and afterwards by applying the $\mathbb{F}_{q^d}$-endomorphism $f(\sigma^d)$ to an element $\omega \in \overline{\mathbb{F}}_q$, resulting in $f(\sigma^d)(\omega)$. In this context, $\overline{\mathbb{F}}_q$ is an $\mathbb{F}_{q^d}[x]$-module (with respect to σ^d). For $\theta \in \overline{\mathbb{F}}_q$ and an integer $d \geq 1$ let

$$\mathbb{F}_{q^d}[x]\theta := \{h(\sigma^d)(\theta) : h \in \mathbb{F}_{q^d}[x]\}$$

denote the $\mathbb{F}_{q^d}[x]$-submodule generated by θ.

5.4.23 Definition For $\theta \in \overline{\mathbb{F}}_q$ and an integer $d \geq 1$, the *q^d-order* of θ is the monic polynomial $g \in \mathbb{F}_{q^d}[x]$ of least degree such that $g(\sigma^d)(\theta) = 0$. The q^d-order of θ is denoted by $\mathrm{Ord}_{q^d}(\theta)$.

5.4.24 Theorem [1389, Section 8] Let $d \geq 1$ be an integer.

1. The finite $\mathbb{F}_{q^d}[x]$-submodules of $\overline{\mathbb{F}}_q$ correspond bijectively to the monic polynomials of $\mathbb{F}_{q^d}[x]$ that are not divisible by x: If $f \in \mathbb{F}_{q^d}[x]$ is monic and $f(0) \neq 0$, the corresponding $\mathbb{F}_{q^d}[x]$-submodule is the kernel of the $\mathbb{F}_{q^d}$-linear mapping $f(\sigma^d)$.
2. Every finite $\mathbb{F}_{q^d}[x]$-submodule of $\overline{\mathbb{F}}_q$ is cyclic. If U is the finite $\mathbb{F}_{q^d}[x]$-submodule of $\overline{\mathbb{F}}_q$ corresponding to the monic polynomial $f \in \mathbb{F}_{q^d}[x]$ (not divisible by x), then $\omega \in U$ if and only if $\mathrm{Ord}_{q^d}(\omega)$ divides f, and $\mathbb{F}_{q^d}[x]\omega = U$ if and only if $\mathrm{Ord}_{q^d}(\omega) = f$.

5.4.25 Remark For integers $m, d \geq 1$ with $d|m$, the field $\mathbb{F}_{q^m}$ is the $\mathbb{F}_{q^d}[x]$-submodule of $\overline{\mathbb{F}}_q$ corresponding to the polynomial $x^{m/d} - 1 \in \mathbb{F}_{q^d}[x]$. The generators of $\mathbb{F}_{q^m}$ as $\mathbb{F}_{q^d}[x]$-module are precisely the normal elements of $\mathbb{F}_{q^m}$ over $\mathbb{F}_{q^d}$.

5.4.26 **Definition** For an integer $k \geq 1$ which is relatively prime to the characteristic p of the underlying field $\mathbb{F}_q$, let $\Phi_k(x) \in \mathbb{F}_q[x]$ denote the k-th cyclotomic polynomial (see Definition 2.1.121). Given a further integer $t \geq 1$, the *generalized cyclotomic polynomial* corresponding to the pair (k, t) is the polynomial $\Phi_k(x^t) \in \mathbb{F}_q[x]$.

5.4.27 **Definition** For an integer $m \geq 1$, its *square-free part* $\nu(m)$ is the product of the distinct prime divisors of m. We let $\nu(1) := 1$.

5.4.28 **Proposition** [1389, Section 18] Consider $\mathbb{F}_q[x]$ and let (k, t) be as in Definition 5.4.26. Write $t = p^b t'$, where p is the characteristic of $\mathbb{F}_q$ and t' the p-free part of t. Then

1. $\Phi_k(x^t) = \Phi_k(x^{t'})^{p^b}$.
2. If d is a common divisor of k and t, then $\Phi_k(x^t) = \Phi_{kd}(x^{t/d})$. In particular, $\Phi_k(x^t) = \Phi_{\nu(k)}(x^{kt/\nu(k)})$.
3. If k and t are relatively prime, then $\Phi_k(x^t) = \prod_{e|t'} \Phi_{ke}(x)^{p^b}$. The latter is the *canonical decomposition* of $\Phi_k(x^t)$ over $\mathbb{F}_q$.

5.4.29 **Remark** Observe that different pairs (k, t) and (ℓ, s) may lead to the same generalized cyclotomic polynomial. However, $\Phi_k(x^t) = \Phi_\ell(x^s)$ if and only if $\nu(k) = \nu(\ell)$ and $\frac{kt}{\nu(k)} = \frac{\ell s}{\nu(\ell)}$.

5.4.30 **Definition** Consider an algebraic closure $\overline{\mathbb{F}}_q$ of $\mathbb{F}_q$. Given a pair (k, t) as in Definition 5.4.26, the *cyclotomic module* corresponding to (k, t) is the kernel of $\Phi_k(\sigma^t)$, where σ is the Frobenius automorphism of $\overline{\mathbb{F}}_q/\mathbb{F}_q$. Throughout, this $\mathbb{F}_q[x]$-submodule is denoted by $C_{k,t}$, hence

$$C_{k,t} = \{\theta \in \overline{\mathbb{F}}_q : \Phi_k(\sigma^t)(\theta) = 0\}.$$

The *module character* of $C_{k,t}$ is the number $\frac{kt}{\nu(k)}$.

5.4.31 **Remark** In the setting of Definition 5.4.30, one has $\lambda\omega \in C_{k,t}$ for all $\omega \in C_{k,t}$ if and only if $\lambda \in \mathbb{F}_{q^n}$ where $n = \frac{kt}{\nu(k)}$ [1389, Section 18]. Therefore, $C_{k,t}$ carries the structure of an $\mathbb{F}_{q^d}[x]$-module if and only if d is a divisor of $\frac{kt}{\nu(k)}$. This motivates the notion of the module character in Definition 5.4.30.

5.4.32 **Definition** Let $C_{k,t}$ be a cyclotomic module over $\mathbb{F}_q$. An element θ is a *complete generator* for $C_{k,t}$ over $\mathbb{F}_q$, provided θ (simultaneously) generates $C_{k,t}$ as an $\mathbb{F}_{q^d}[x]$-module for every divisor d of its module character $\frac{kt}{\nu(k)}$.

5.4.33 **Theorem** (Complete Cyclotomic Generator Theorem; [1389, Section 18]) Given any finite field $\mathbb{F}_q$ and any cyclotomic module $C_{k,t}$ over $\mathbb{F}_q$, there exists a complete generator for $C_{k,t}$ over $\mathbb{F}_q$.

5.4.34 **Remark** Theorem 5.4.33 generalizes the Complete Normal Basis Theorem 5.4.5 from the class of finite field extensions of $\mathbb{F}_q$ to the class of cyclotomic modules over $\mathbb{F}_q$: If $(k, t) = (1, m)$, then the cyclotomic module $C_{k,t}$ is equal to the extension field $\mathbb{F}_{q^m}$; over $\mathbb{F}_q$, the module character of $\mathbb{F}_{q^m}$ is equal to m and the complete generators of $\mathbb{F}_{q^m}/\mathbb{F}_q$ are exactly the completely normal elements of $\mathbb{F}_{q^m}$ over $\mathbb{F}_q$. The following theorem generalizes Remark 5.4.7.

5.4.35 **Theorem** (Cyclotomic Reduction Theorem; [1389, Section 25] and [1390, Section 3]) Consider two cyclotomic modules $C_{k,s}$ and $C_{\ell,t}$ over $\mathbb{F}_q$. Assume that ks and ℓt are relatively prime. Then, for $\theta \in C_{k,s}$ and $\omega \in C_{\ell,t}$ the following two assertions are equivalent:

1. θ and ω are complete generators for $C_{k,s}$ and $C_{\ell,t}$ over $\mathbb{F}_q$, respectively.
2. $\theta \cdot \omega$ is a complete generator for the cyclotomic module $C_{k\ell,st}$ over $\mathbb{F}_q$.

5.4.4 A decomposition theory for complete generators

5.4.36 **Definition** Consider a generalized cyclotomic polynomial $\Phi_k(x^t) \in \mathbb{F}_q[x]$. A set $\Delta \subseteq \mathbb{F}_q[x]$ of generalized cyclotomic polynomials is a *cyclotomic decomposition* of $\Phi_k(x^t)$, if the members of Δ are pairwise relatively prime and

$$\Phi_k(x^t) = \prod_{\Psi(x)\in\Delta} \Psi(x).$$

In that case, $i(\Delta)$ denotes a corresponding set of pairs (ℓ, s) with $\Phi_\ell(x^s) \in \Delta$.

5.4.37 **Proposition** [1389, Section 19] Let Δ be a cyclotomic decomposition of $\Phi_k(x^t) \in \mathbb{F}_q[x]$. Then

1. $C_{k,t} = \bigoplus_{(\ell,s)\in i(\Delta)} C_{\ell,s}$ is a decomposition into a direct sum of cyclotomic modules.
2. For $\theta \in C_{k,t}$ let $\theta = \sum_{(\ell,s)\in i(\Delta)} \theta_{\ell,s}$ be the corresponding decomposition into its Δ-components. If θ is a complete generator of $C_{k,t}$ over $\mathbb{F}_q$, then (necessarily) for every $(\ell, s) \in i(\Delta)$ the component $\theta_{\ell,s}$ is a complete generator of $C_{\ell,s}$ over $\mathbb{F}_q$.

5.4.38 **Definition** Let Δ be a cyclotomic decomposition of $\Phi_k(x^t)$ considered over $\mathbb{F}_q$. Then Δ is an *agreeable decomposition*, provided the following holds: if for every $(\ell, s) \in i(\Delta)$, one chooses any element $\theta_{\ell,s}$ that is a complete generator of $C_{\ell,s}$ over $\mathbb{F}_q$, then the sum $\theta = \sum_{(\ell,s)\in i(\Delta)} \theta_{\ell,s}$ is a complete generator of $C_{k,t}$ over $\mathbb{F}_q$.

5.4.39 **Theorem** (Complete Decomposition Theorem; [1389, Section 19] and [1390, Section 5]) Consider a generalized cyclotomic polynomial $\Phi_k(x^t)$ over the finite field $\mathbb{F}_q$ with characteristic p. Let r be a prime divisor of t. Assume that $r \neq p$ and that r does not divide k. Then

$$\Delta_r := \{\Phi_k(x^{t/r}), \Phi_{kr}(x^{t/r})\}$$

is a cyclotomic decomposition of $\Phi_k(x^t)$. Moreover, the following two statements are equivalent:

1. Δ_r is an agreeable decomposition over $\mathbb{F}_q$.
2. $\mathrm{ord}_{\nu(kt')}(q)$ is not divisible by r^a, where $a \geq 1$ is maximal such that r^a divides t (recall that t' is the p-free part of t and $\nu(kt')$ is the square-free part of kt').

5.4.40 **Remark** Consider $\Phi_k(x^t)$ and Δ_r as in Theorem 5.4.39. Then the module character of each cyclotomic module corresponding to a member of Δ_r is equal to $\frac{kt}{r\nu(k)}$ and therefore a proper divisor of the module character $\frac{kt}{\nu(k)}$ of $C_{k,t}$ over $\mathbb{F}_q$.

5.4.41 **Example** [1389, Section 19] Consider an extension $\mathbb{F}_{q^m}/\mathbb{F}_q$, where $m > 1$ is not a power of the characteristic p. Let $r|m$ be the largest prime divisor of m that is different from p. Then $\{x^{m/r} - 1, \Phi_r(x^{m/r})\}$ is an agreeable decomposition of $x^m - 1$ over $\mathbb{F}_q$. If in particular $m = r^n$, this decomposition is equal to $\{x^{r^{n-1}} - 1, \Phi_{r^n}(x)\}$. The Complete Decomposition Theorem 5.4.39 may then be applied to $x^{r^{n-1}} - 1$ if $n \geq 2$, and an induction argument shows that the canonical decomposition $\{x-1, \Phi_r(x), \Phi_{r^2}(x), \ldots, \Phi_{r^n}(x)\}$ is an agreeable decomposition of $x^{r^n}-1$ over $\mathbb{F}_q$.

5.4.42 Remark In Section 6 of [1390] a *Uniqueness Theorem* is proved: Starting from a generalized cyclotomic polynomial $\Phi_k(x^t)$ over $\mathbb{F}_q$, one obtains a unique (finest) agreeable decomposition of $\Phi_k(x^t)$ by a recursive application of the Complete Decomposition Theorem 5.4.39 independent of the order the various primes r are chosen.

5.4.43 Example [1389, Section 19] The set $\{x-1, \Phi_2(x), \Phi_4(x), \Phi_3(x^4), \Phi_9(x^4), \Phi_7(x^{36})\}$ is an agreeable decomposition of $x^{252}-1$ over $\mathbb{F}_q$, where q is any prime power relatively prime to 252. When specializing q to be equal to 5, the Complete Decomposition Theorem may be applied several further times to reach the following agreeable decomposition of $x^{252}-1$ over $\mathbb{F}_5$: $\{x-1, \Phi_2(x), \Phi_4(x), \Phi_3(x^2), \Phi_{12}(x), \Phi_9(x^2), \Phi_{36}(x), \Phi_7(x^6), \Phi_{28}(x^3), \Phi_{63}(x^2), \Phi_{252}(x)\}$.

5.4.5 The class of regular extensions

5.4.44 Remark Consider a cyclotomic module $C_{k,t}$ over $\mathbb{F}_q$, where k and t are relatively prime. Let $t = p^b t'$, where p is the characteristic of $\mathbb{F}_q$ and t' is the p-free part of t. The canonical decomposition $\{\Phi_{ke}(x)^{p^b} : e|t'\}$ is (by definition) the finest possible cyclotomic decomposition of $\Phi_k(x^t)$. By Theorem 19.10 of [1389], the canonical decomposition is agreeable provided that t' and $\mathrm{ord}_{\nu(kt')}(q)$ are relatively prime. (Recall that $\nu(kt')$ is the square-free part of kt'.)

5.4.45 Theorem [1389, Section 19] Let $\mathbb{F}_q$ be a finite field and p its characteristic. Write $m = m'p^b$ (with m' being the p-free part of m). Then the canonical decomposition $\{\Phi_d(x)^{p^b} : d|m'\}$ of $x^m - 1$ is agreeable over $\mathbb{F}_q$ if and only if m' and $\mathrm{ord}_{\nu(m')}(q)$ are relatively prime.

5.4.46 Definition Consider a cyclotomic module $C_{k,t}$ over $\mathbb{F}_q$, with k and t being relatively prime. Then $C_{k,t}$ is *regular* over $\mathbb{F}_q$, provided that $\mathrm{ord}_{\nu(kt')}(q)$ and kt are relatively prime (t' is the p-free part of t and $\nu(kt')$ is the square-free part of kt'). In particular, in the case where $(k,t) = (1,m)$, the extension $\mathbb{F}_{q^m}$ is *regular* over $\mathbb{F}_q$, provided that m and $\mathrm{ord}_{\nu(m')}(q)$ are relatively prime.

5.4.47 Example These examples (taken from [1391, Section 1]), indicate that the class of regular extensions is quite large. Given any finite field $\mathbb{F}_q$, we have that $\mathbb{F}_{q^m}$ is regular over $\mathbb{F}_q$ in all the following cases:

1. m is a power of a prime r;
2. $\nu(m)$ divides $\nu(q-1)$;
3. m is a power of a Carmichael number (a *Carmichael number* is an odd composite integer $n \geq 1$ such that $r-1$ divides $n-1$ for every prime divisor r of n; by Alford, Granville, and Pomerance [75] there are infinitely many Carmichael numbers, examples being 561, 1105, 1729, and 2465);
4. m has all its prime divisors from $\{7, 11, 13, 17, 19, 31, 41, 47, 49, 61, 73, 97, 101, 107, 109, 139, 151, 163, 167, 173, 179, 181, 193\}$, without any restriction on their multiplicity.

5.4.48 Remark Any completely basic extension (Definition 5.4.14) is regular. In order to compare the completely basic extensions with the regular ones, consider a finite set π of prime numbers, and let ν be the product of all $r \in \pi$. Let further $N(\pi) := \{m \in \mathbb{Z} : m \geq 1, \nu(m)|\nu\}$. If $\mathbb{F}_{q^\nu}$ is completely basic over $\mathbb{F}_q$, then $\mathbb{F}_{q^m}$ is regular over $\mathbb{F}_q$ for *every* $m \in N(\pi)$. In contrast, the subset of those $m \in N(\pi)$ for which $\mathbb{F}_{q^m}$ is completely basic over $\mathbb{F}_q$ is only *finite* (see [1391, Proposition 3.3]).

5.4.6 Complete generators for regular cyclotomic modules

5.4.49 Remark When considering the complete generation of a regular cyclotomic module $C_{k,t}$ over $\mathbb{F}_q$, due to Remark 5.4.44, one can independently work on the components (which are also regular) arising from the canonical decomposition of $\Phi_k(x^t)$. It is therefore sufficient to consider the subclass of (regular) cyclotomic modules of the form C_{n,p^b} (corresponding to the generalized cyclotomic polynomial $\Phi_n(x)^{p^b}$), where p is the characteristic of $\mathbb{F}_q$ and n is relatively prime to q.

5.4.50 Definition Let p be the characteristic of $\mathbb{F}_q$ and n be relatively prime to q. Assume that C_{n,p^b} is a regular cyclotomic module over $\mathbb{F}_q$. Write $n = 2^c \cdot \overline{n}$ with $\overline{n}$ odd. Then C_{n,p^b} is *exceptional* over $\mathbb{F}_q$, provided the following conditions are satisfied: $q \equiv 3 \pmod 4$ and $c \geq 3$ and the order of q modulo 2^c is equal to 2. In all other cases, C_{n,p^b} is *non-exceptional* over $\mathbb{F}_q$.

5.4.51 Remark For an integer $n \geq 1$ let $\pi(n)$ denote the set of prime divisors of n. If n and q are relatively prime, the order of q modulo n is of the form

$$\mathrm{ord}_n(q) = \mathrm{ord}_{\nu(n)}(q) \cdot \prod_{r \in \pi(n)} r^{\alpha(r)},$$

where $\alpha(r) \geq 0$ for all $r \in \pi(n)$. Assume next that C_{n,p^b} is a regular cyclotomic module over $\mathbb{F}_q$, hence $\mathrm{ord}_{\nu(n)}(q)$ is relatively prime to np^b. Let further

$$\tau = \tau(q,n) := \prod_{r \in \pi(n)} r^{\lfloor \alpha(r)/2 \rfloor}.$$

5.4.52 Theorem [1389, Section 20] Let p be the characteristic of $\mathbb{F}_q$ and n be relatively prime to q. Assume that C_{n,p^b} is a regular cyclotomic module over $\mathbb{F}_q$. Let $\tau = \tau(q,n)$ be as in Remark 5.4.51. Then τ divides $\frac{n}{\nu(n)}$ in general, and 2τ divides $\frac{n}{\nu(n)}$ if C_{n,p^b} is exceptional. Moreover the following hold:

1. If C_{n,p^b} is non-exceptional, then θ is a complete generator of C_{n,p^b} over $\mathbb{F}_q$ if and only if $\mathrm{Ord}_{q^\tau}(\theta) = \Phi_{n/\tau}(x)^{p^b}$.
2. If C_{n,p^b} is exceptional, then θ is a complete generator of C_{n,p^b} over $\mathbb{F}_q$ if and only if $\mathrm{Ord}_{q^\tau}(\theta) = \Phi_{n/\tau}(x)^{p^b}$ and $\mathrm{Ord}_{q^{2\tau}}(\theta) = \Phi_{n/(2\tau)}(x)^{p^b}$.

5.4.53 Theorem [1389, Section 21] Let p be the characteristic of $\mathbb{F}_q$ and n be relatively prime to q. Assume that C_{n,p^b} is a regular cyclotomic module over $\mathbb{F}_q$. Let $\tau = \tau(q,n)$ be as in Remark 5.4.51 and let φ denote Euler's function (see Definition 2.1.43). Then the number of complete generators of C_{n,p^b} over $\mathbb{F}_q$ is equal to

1. $\left(q^{\mathrm{ord}_n(q)/\tau} - 1\right)^{\tau\varphi(n)/\mathrm{ord}_n(q)} \cdot q^{(p^b-1)\varphi(n)}$, if C_{n,p^b} is non-exceptional;
2. $\left(q^{2\mathrm{ord}_n(q)/\tau} - 4q^{\mathrm{ord}_n(q)/\tau} + 3\right)^{\tau\varphi(n)/(2\mathrm{ord}_n(q))} \cdot q^{(p^b-1)\cdot\varphi(n)}$, if C_{n,p^b} is exceptional.

5.4.54 Theorem [1389, Section 21] Let $C_{k,t}$ be a regular cyclotomic module over $\mathbb{F}_q$ and write $t = p^b t'$, where p is the characteristic of $\mathbb{F}_q$ and t' is the p-free part of t. Then the number of complete generators of $C_{k,t}$ over $\mathbb{F}_q$ is at least

$$(q-1)^{\varphi(k)t'} \cdot q^{\varphi(k)t'(p^b-1)},$$

where φ denotes Euler's function. Moreover, equality holds if and only if kt' divides $q-1$, in which case θ is a complete generator of $C_{k,t}$ if and only if the q-order of θ is $\Phi_k(x^t)$.

5.4.55 Conjecture If $\mathbb{F}_{q^m}/\mathbb{F}_q$ is a regular extension, then the number of completely normal elements of $\mathbb{F}_{q^m}/\mathbb{F}_q$ is at least $(q-1)^{m'} \cdot q^{m'(p^b-1)}$ by Theorem 5.4.54 (where $m = p^b m'$ with m' being the p-free part of m and p the characteristic of $\mathbb{F}_q$); moreover, equality holds if and only if m' divides $q-1$, in which case $\mathbb{F}_{q^m}$ is completely basic over $\mathbb{F}_q$. We conjecture that, for *any* extension $\mathbb{F}_{q^m}/\mathbb{F}_q$, the number of completely normal elements of $\mathbb{F}_{q^m}/\mathbb{F}_q$ is at least

$$(q-1)^{m'} \cdot q^{m'(p^b-1)}.$$

5.4.56 Remark The aim of Theorem 5.4.57 below and its subsequent remarks is a construction (in the spirit of Example 5.4.4) of complete generators for regular cyclotomic modules over $\mathbb{F}_q$ that are of the form $C_{n,1}$. It is based on Theorem 5.4.52. A suitable application of the Complete Decomposition Theorem 5.4.39 in combination with the Cyclotomic Reduction Theorem 5.4.35 gives rise to a complete generator for a general cyclotomic module, in particular a completely normal element for an arbitrary extension $\mathbb{F}_{q^m}$ over $\mathbb{F}_q$. Thereby, Remark 5.4.20 on extensions of the form $\mathbb{F}_{q^{p^b}}$ may be used to cover the cases $C_{k,t}$ and $\mathbb{F}_{q^m}$, where the characteristic p of $\mathbb{F}_q$ divides t and m, respectively. Throughout, gcd denotes the greatest common divisor.

5.4.57 Theorem [1389, Chapter VI] Consider a finite field $\mathbb{F}_q$ and an integer $n \geq 1$ with $\gcd(n,q) = 1$. Let $s := \mathrm{ord}_{\nu(n)}(q)$. Assume that $C_{n,1}$ is a regular cyclotomic module over $\mathbb{F}_q$ (hence n and s are relatively prime). If n is odd, or if $q \equiv 1 \pmod 4$, or if $n \equiv 2 \pmod 4$ and $q \equiv 3 \pmod 4$, let $(Q,N) := (q^s, n)$. If $C_{n,1}$ is exceptional over $\mathbb{F}_q$, let $(Q,N) := (q^{2s}, \frac{n}{2})$. Write $Q-1 = \rho \cdot \overline{\rho}$, where $\nu(\rho) = \nu(N)$ and $\gcd(N, \overline{\rho}) = 1$ (hence ρ is the largest divisor of $Q-1$ composed from prime divisors of N), and let $a := \gcd(\rho, N)$ and $I := \{i \in \mathbb{Z} : 1 \leq i \leq a,\ \gcd(i,a) = 1\}$. On I there is defined an equivalence relation by $i \sim j$ if and only if $i \equiv q^\ell j \pmod a$ for some $\ell \in \mathbb{Z}$. Let $\mathcal{R}$ be a set of representatives of $\sim$ on I. Finally, let $y \in \overline{\mathbb{F}}_q$ be a primitive $(N\rho)$-th root of unity, and

$$w := \sum_{i \in \mathcal{R}} y^i \quad \text{and} \quad u := \sum_{i \in \mathcal{R}} (y^i + y^{iq}).$$

With Tr denoting the $(\mathbb{F}_{q^{ns}}, \mathbb{F}_{q^n})$-trace mapping, the following hold:

1. $\mathrm{Ord}_q(\mathrm{Tr}(w)) = \Phi_n(x)$ in the first case, that is, n odd, or $q \equiv 1 \pmod 4$, or $n \equiv 2 \pmod 4$ and $q \equiv 3 \pmod 4$.
2. $\mathrm{Ord}_q(\mathrm{Tr}(u)) = \Phi_n(x)$ and $\mathrm{Ord}_{q^2}(\mathrm{Tr}(u)) = \Phi_{\frac{n}{2}}(x)$ in the second case, that is, where $C_{n,1}$ is exceptional over $\mathbb{F}_q$.

5.4.58 Remark The case where $4|n$ and $q \equiv 3 \pmod 4$ and $C_{n,1}$ is regular but non-exceptional over $\mathbb{F}_q$ is missed in the formulation of Theorem 5.4.57. It can be covered, however, by applying Theorem 5.4.57 to the pair $(q^2, \frac{n}{2})$ in order to determine an element having q^2-order $\Phi_{n/2}(x)$. Any such element has q-order $\Phi_n(x)$ (Section 24 of [1389]).

5.4.59 Remark When searching for a complete generator of the regular cyclotomic module $C_{n,1}$ over $\mathbb{F}_q$, where $\gcd(n,q) = 1$, define the parameter $\tau = \tau(q,n)$ as in Theorem 5.4.52. Then (q,n) is exceptional if and only if $(q^\tau, \frac{n}{\tau})$ is exceptional. Consider $C_{n,1} = \{\theta \in \overline{\mathbb{F}}_q : \Phi_n(\sigma)(\theta) = 0\}$ as the cyclotomic module $C_{\frac{n}{\tau},1}$ over $\mathbb{F}_{q^\tau}$. Apply the construction of Theorem 5.4.57 (and Remark 5.4.58) to the pair $(q^\tau, \frac{n}{\tau})$ instead of (q,n). Then the resulting elements constitute complete generators of $C_{n,1}$ over $\mathbb{F}_q$ by Theorem 5.4.52.

5.4.7 Towards a primitive complete normal basis theorem

5.4.60 Remark Completing previous work of Carlitz [539] in 1952 and Davenport [775] in 1968, Lenstra and Schoof [1899] proved the Primitive Normal Basis Theorem (Theorem 5.2.32) in 1987. It states that for any finite field $\mathbb{F}_q$ and any integer $m \geq 1$ there exists a primitive element of $\mathbb{F}_{q^m}$ that is normal over $\mathbb{F}_q$. Recall that a primitive element of $\mathbb{F}_{q^m}$ is a generator of the (cyclic) multiplicative group of $\mathbb{F}_{q^m}$; see Theorem 2.1.37 and Definition 2.1.38. In this subsection we consider the existence of primitive elements that are completely normal.

5.4.61 Remark By means of a computer search, Morgan and Mullen [2157] calculated for every pair (p, m), with $p \leq 97$ a prime and with $p^m < 10^{50}$, a monic irreducible polynomial of degree m over $\mathbb{F}_p$ whose roots are primitive and completely normal elements for $\mathbb{F}_{p^m}$ over $\mathbb{F}_p$. It is conjectured in [2157] that for every extension $\mathbb{F}_{q^m}/\mathbb{F}_q$ there exists a primitive element of $\mathbb{F}_{q^m}$ that is completely normal over $\mathbb{F}_q$.

5.4.62 Theorem [1391] Let q be a prime power and assume that $\mathbb{F}_{q^m}$ is a regular extension over $\mathbb{F}_q$ (see Definition 5.4.46). Assume further that $q-1$ is divisible by 4 if q is odd and m is even. Then there exists a primitive element of $\mathbb{F}_{q^m}$ that is completely normal over $\mathbb{F}_q$.

5.4.63 Remark Blessenohl [316] settles the existence of primitive completely normal elements for extensions $\mathbb{F}_{q^m}/\mathbb{F}_q$, where $m = 2^\ell$ is a divisor of q^2-1 and $\ell \geq 3$ and $q \equiv 3 \pmod 4$. In [1394], the case where $q \equiv 3 \pmod 4$ and where m is a sufficiently large power of 2 is handled, giving rise to the first bound in Theorem 5.4.64 below. In a not yet published work by the author [1395], the existence of primitive completely normal elements is proved for all regular extensions, i.e., the assertion of Theorem 5.4.62 also holds without the additional assumption that $q-1$ is divisible by 4 if q is odd and m is even.

5.4.64 Theorem [1394] Let $\mathbb{F}_q$ be a finite field with characteristic p. For an integer $m \geq 1$ let $\mathrm{PCN}(q, m)$ denote the number of primitive elements of $\mathbb{F}_{q^m}$ that are completely normal over $\mathbb{F}_q$. Then (with φ denoting Euler's function):

1. $\mathrm{PCN}(q, 2^\ell) \geq 4(q-1)^{2^{\ell-2}}$, if $q \equiv 3 \pmod 4$ and $\ell \geq e+3$ (where e is maximal such that $2^e | q^2-1$), or if $q \equiv 1 \pmod 4$ and $\ell \geq 5$.
2. $\mathrm{PCN}(q, r^\ell) \geq r^2(q-1)^{r^{\ell-2}}$, if $r \neq p$ is an odd prime and $\ell \geq 2$.
3. $\mathrm{PCN}(q, r^\ell) \geq r(q-1)^{r^{\ell-1}} \cdot \varphi(q^{r^{\ell-1}}-1)$, if $r \geq 7$ and $r \neq p$ is a prime and $\ell \geq 2$.
4. $\mathrm{PCN}(q, p^\ell) \geq pq^{p^{\ell-1}-1}(q-1)$, if $\ell \geq 2$.
5. $\mathrm{PCN}(q, p^\ell) \geq pq^{p^{\ell-1}-1}(q-1) \cdot \varphi(q^{p^{\ell-1}}-1)$, if $p \geq 7$ and $\ell \geq 2$.

5.4.65 Definition Consider a finite field $\mathbb{F}_q$ and let M be a Steinitz number. Assume that $(w_m)_{m \in M}$ is a sequence in $\mathbb{F}_{q^M}$ that is both, norm-compatible and trace-compatible over $\mathbb{F}_q$ (see Definition 5.4.11). Then $(w_m)_{m \in M}$ is a *complete universal generator* of $\mathbb{F}_{q^M}$ over $\mathbb{F}_q$ if, for every $m \in M$, w_m is a primitive element of $\mathbb{F}_{q^m}$ that is completely normal over $\mathbb{F}_q$.

5.4.66 Theorem [1392] Let $q > 1$ be any prime power and $r \geq 7$ be any prime. Furthermore, let $M := \{r^n : n \in \mathbb{N}\}$. Then there exists a complete universal generator for $\mathbb{F}_{q^M}$ over $\mathbb{F}_q$.

5.4.67 Remark The conclusion of Theorem 5.4.66 could also be proved for $r = 5$ when $\mathbb{F}_q$ has characteristic 5, or when q mod 25 is not equal to 1, 7, 18 or 24, or when q is sufficiently large [1392, 1393]. For the cases $r = 2$ and $r = 3$ similar results are available only when the assumption that M is a Steinitz number is weakened, e.g., M has to consist of powers of 9 or powers of 8, respectively.

See Also

§5.1, §5.2	For self-dual, weakly self-dual, and primitive normal bases.
§5.3	For information on low-complexity normal bases.
[309], [589], [2539], [2540]	For explicit constructions of completely normal polynomials.
[1389]	Section 27 discusses aspects of completely normal polynomials and iterative constructions.
[2539], [2540]	For connections to Dickson polynomials; see also Section 9.6.

References Cited: [75, 309, 315, 316, 317, 318, 398, 539, 589, 775, 1023, 1386, 1387, 1388, 1389, 1390, 1391, 1392, 1393, 1394, 1395, 1896, 1899, 2090, 2157, 2537, 2538, 2539, 2540]

6

Exponential and character sums

6.1 Gauss, Jacobi, and Kloosterman sums

Ronald J. Evans, *University of California at San Diego*

In this section, we focus mainly on Gauss, Jacobi, and Kloosterman sums over finite fields, with brief mention of Eisenstein and Jacobsthal sums. Throughout, $\mathbb{F}_q$ is a finite field of characteristic p with $q = p^r$ elements. (In Subsection 6.1.3, the exponent r will be taken to be the order of p (mod k) for a fixed integer k.) We refer to [240] for proofs of many of the results in this section. In some cases, the proofs need modification because of differing definitions of the trivial character χ_0: in Definition 6.1.1 below, $\chi_0(0) = 0$, while in [240], $\chi_0(0) = 1$.

6.1.1 Properties of Gauss and Jacobi sums of general order

6.1.1 **Definition** A *multiplicative character* χ on $\mathbb{F}_q^*$ is a map from the cyclic group $\mathbb{F}_q^*$ into the group of complex roots of unity such that $\chi(\alpha\beta) = \chi(\alpha)\chi(\beta)$ for all $\alpha, \beta \in \mathbb{F}_q^*$. We extend χ to a function on $\mathbb{F}_q$ by setting $\chi(0) = 0$. The *trivial character* χ_0 satisfies $\chi_0(\alpha) = 1$ for every $\alpha \in \mathbb{F}_q^*$. The *order* of χ is the smallest positive integer n for which $\chi^n = \chi_0$. The unique character ρ of order 2 is the *quadratic character*.

6.1.2 Definition Write $\zeta_p = e^{2\pi i/p}$. For a character χ of order k on $\mathbb{F}_q$ and for $\beta \in \mathbb{F}_q$, the *Gauss sum* $G(\beta, \chi)$ of order k over $\mathbb{F}_q$ is defined by

$$G(\beta, \chi) = \sum_{\alpha \in \mathbb{F}_q} \chi(\alpha) \zeta_p^{\mathrm{Tr}(\alpha\beta)},$$

where $\mathrm{Tr}(\alpha)$ denotes the trace of α from $\mathbb{F}_q$ to the prime field $\mathbb{F}_p$. When $\beta = 1$, we abbreviate $G(\chi) = G(\beta, \chi)$.

6.1.3 Remark The next theorem shows that $G(\beta, \chi)$ can be evaluated in terms of $G(\chi)$.

6.1.4 Theorem [240, p. 9]. For a character χ on $\mathbb{F}_q$ and $\beta \in \mathbb{F}_q$,

$$G(\beta, \chi) = \begin{cases} q - 1 & \text{if } \beta = 0,\ \chi = \chi_0, \\ \overline{\chi}(\beta) G(\chi) & \text{otherwise.} \end{cases}$$

In particular,

$$G(\overline{\chi}) = \chi(-1)\overline{G(\chi)}.$$

6.1.5 Remark For proofs of the following two theorems, see [240, p. 10].

6.1.6 Theorem For a character χ on $\mathbb{F}_q$,

$$|G(\chi)| = \sqrt{q} \text{ if } \chi \neq \chi_0, \quad \text{and} \quad G(\chi) = -1 \text{ if } \chi = \chi_0.$$

6.1.7 Theorem For a character χ on $\mathbb{F}_q$ and $\beta \in \mathbb{F}_q$,

$$G(\beta, \chi^p) = G(\beta^p, \chi).$$

6.1.8 Remark We next present two theorems on uniform distribution of Gauss sums. The first, due to Katz and Zheng [1713], was subsequently extended by Shparlinski [2652]. For the second and some generalizations thereof, see Iwaniec and Kowalski [1581, Theorem 21.6], Katz [1701, Chapter 9], and Fu and Liu [1139].

6.1.9 Theorem Consider the collection of $(q-1)(q-2)$ normalized Gauss sums

$$G(\beta, \chi)/\sqrt{q}, \qquad \beta \in \mathbb{F}_q^*, \quad \chi \neq \chi_0.$$

As q tends to infinity, this collection is asymptotically equidistributed on the complex unit circle.

6.1.10 Theorem Consider the collection of $q - 2$ normalized Gauss sums

$$G(\chi)/\sqrt{q}, \qquad \chi \neq \chi_0.$$

As q tends to infinity, this collection is asymptotically equidistributed on the complex unit circle.

6.1.11 Definition Let $\beta \in \mathbb{F}_q^*$ and suppose that $q = kf + 1$ for some positive integer k. The *(reduced) f-nomial Gaussian periods* $g(\beta, k)$ of order k are defined by

$$g(\beta, k) = \sum_{\alpha \in \mathbb{F}_q} \zeta_p^{\mathrm{Tr}(\beta\alpha^k)}.$$

When $\beta = 1$, we abbreviate $g(k) = g(\beta, k)$. For example, if $q = p$, then

$$g(k) = \sum_{m=1}^{p} \zeta_p^{m^k}.$$

The periods $g(\beta, k)$ are also known as *Gauss sums.*

6.1.12 Remark Gauss sums and periods have been used for counting solutions to diagonal equations over $\mathbb{F}_q$ [240, Chapters 10, 12] and for counting points on more general varieties [1343, 1344, 2167]. Thaine [2791] has given an application to class groups of cyclotomic fields. For some applications to coding theory, see [1078, 2225].

6.1.13 Theorem [240, p. 11]. Let $\beta \in \mathbb{F}_q^*$. If χ is a character on $\mathbb{F}_q$ of order k, then

$$g(\beta, k) = \sum_{j=1}^{k-1} G(\beta, \chi^j).$$

In particular, $|g(\beta, k)| \leq (k-1)\sqrt{q}$.

6.1.14 Remark The inequality above has been strengthened in several different ways, depending on the relationship between p and k; see Heath-Brown and Konyagin [1454]. For further estimates for Gauss sums, see [374, 1790, 2645].

6.1.15 Definition Let $q = p^r = kf + 1$ and let γ be a generator of the cyclic group $\mathbb{F}_q^*$. The polynomial

$$R_k(x) = \prod_{s=0}^{k-1} (x - g(\gamma^s, k)),$$

whose zeros are (reduced) f-nomial Gaussian periods, is the *(reduced) period polynomial* of degree k.

6.1.16 Example $R_2(x) = x^2 - q(-1)^f$, by Theorem 6.1.86.

6.1.17 Remark The period polynomial $R_k(x)$ defined above has integral coefficients and is independent of the choice of generator γ. It is irreducible over $\mathbb{Q}$ when $r = 1$, but not necessarily irreducible when $r > 1$ [240, p. 426]. See [1538] for examples of factorizations of $R_k(x)$, particularly in the case $k = 5$. Determinations of period polynomials over $\mathbb{F}_p$ have applications to residuacity criteria mod p; see for example [999].

6.1.18 Definition For any positive integer n, write $\zeta_n = e^{2\pi i/n}$.

6.1.19 Remark The following theorem is a restatement of Theorem 6.1.13.

6.1.20 Theorem (Finite Fourier expansion of Gaussian periods.) Let γ be a generator of the group $\mathbb{F}_q^*$ and let χ be a character of order k on $\mathbb{F}_q$ such that $\chi(\gamma) = \zeta_k$. Then for each period $g(\gamma^s, k)$, we have the Fourier expansion

$$g(\gamma^s, k) = \sum_{v=1}^{k-1} G(\chi^v)\zeta_k^{-sv}.$$

6.1.21 Definition For a character χ on $\mathbb{F}_q$, define the *Eisenstein sum* $E(\chi)$ by

$$E(\chi) = \sum_{\substack{\alpha \in \mathbb{F}_q \\ \mathrm{Tr}(\alpha)=1}} \chi(\alpha).$$

6.1.22 Remark Eisenstein sums can be applied to obtain congruences for binomial coefficients [240, Section 12.9] and to evaluate Brewer sums [240, Chapter 13]. They are also useful for evaluating Gauss sums $g(k)$, via the following theorem.

6.1.23 Theorem [240, p. 421] Let χ be a character of order k on $\mathbb{F}_q$, and let χ^* denote the restriction of χ to the prime field $\mathbb{F}_p$, so that χ^* is a character on $\mathbb{F}_p$ of order

$$k^* = \frac{k}{\gcd(k, (q-1)/(p-1))}.$$

Then

$$g(k) = \sum_{\substack{j=1 \\ k^* \nmid j}}^{k-1} G(\chi^{*j}) E(\chi^j) - p \sum_{\substack{j=1 \\ k^* \mid j}}^{k-1} E(\chi^j).$$

6.1.24 Theorem [240, p. 391]. Let χ be a nontrivial character on $\mathbb{F}_q$, and let χ^* denote the restriction of χ to $\mathbb{F}_p$. Then the Eisenstein sum $E(\chi)$ can be expressed in terms of the Gauss sum $G(\chi)$ over $\mathbb{F}_q$ and the Gauss sum $G(\chi^*)$ over $\mathbb{F}_p$ as follows:

$$E(\chi) = \begin{cases} G(\chi)/G(\chi^*) & \text{if } \chi^* \text{ is nontrivial}, \\ -G(\chi)/p & \text{if } \chi^* \text{ is trivial}. \end{cases}$$

As a consequence,

$$|E(\chi)| = \begin{cases} p^{(r-1)/2} & \text{if } \chi^* \text{ is nontrivial}, \\ p^{(r-2)/2} & \text{if } \chi^* \text{ is trivial}. \end{cases}$$

6.1.25 Definition Let χ, ψ be multiplicative characters on $\mathbb{F}_q$. The *Jacobi sum* $J(\chi, \psi)$ over $\mathbb{F}_q$ is defined by

$$J(\chi, \psi) = \sum_{\alpha \in \mathbb{F}_q} \chi(\alpha)\psi(1-\alpha).$$

We say that the Jacobi sum has order k if k is the least common multiple of the orders of its arguments.

6.1.26 Remark Clearly $J(\chi, \psi) = J(\psi, \chi) \in \mathbb{Q}(\zeta_{q-1})$. The next four theorems follow easily from the results in [240, Section 2.1].

6.1.27 Theorem (Trivial Jacobi sums.) For characters χ, ψ on $\mathbb{F}_q$,

$$J(\chi, \psi) = \begin{cases} q-2 & \text{if } \psi, \chi \text{ are both trivial}, \\ -1 & \text{if exactly one of } \psi, \chi \text{ is trivial}, \\ -\chi(-1) & \text{if } \chi\psi \text{ is trivial with } \chi \text{ nontrivial}. \end{cases}$$

6.1.28 Theorem If χ and ψ are characters on $\mathbb{F}_q$ with $\chi\psi$ nontrivial, then

$$J(\chi, \psi) = G(\chi)G(\psi)/G(\chi\psi).$$

Thus if χ, ψ, and $\chi\psi$ are all nontrivial, then

$$|J(\chi,\psi)| = \sqrt{q}.$$

6.1.29 Theorem If χ and ψ are characters on $\mathbb{F}_q$ with χ nontrivial, then

$$J(\chi,\psi) = \psi(-1)G(\psi)G(\overline{\chi\psi})/G(\overline{\chi}).$$

6.1.30 Theorem If χ is a character on $\mathbb{F}_q$ of order $k > 1$, then

$$G(\chi)^k = q\chi(-1)\prod_{j=1}^{k-2} J(\chi,\chi^j).$$

Thus if f denotes the order of $p \pmod{k}$, then $G(\chi)^k$ lies in a subfield of $\mathbb{Q}(\zeta_k)$ of index f.

6.1.31 Remark Louboutin [1960] used Theorem 6.1.30 for power residue characters attached to the simplest real cyclic cubic, quartic, quintic, and sextic number fields, to efficiently compute class numbers of these fields.

6.1.32 Definition Let $\chi_1,\dots,\chi_t$ be characters on $\mathbb{F}_q$. Define the *multiple Jacobi sum* $J(\chi_1,\dots,\chi_t)$ by

$$J(\chi_1,\dots,\chi_t) = \sum_{\substack{\alpha_1,\dots,\alpha_t\in\mathbb{F}_q \\ \alpha_1+\cdots+\alpha_t=1}} \chi_1(\alpha_1)\cdots\chi_t(\alpha_t).$$

Similarly define

$$J_0(\chi_1,\dots,\chi_t) = \sum_{\substack{\alpha_1,\dots,\alpha_t\in\mathbb{F}_q \\ \alpha_1+\cdots+\alpha_t=0}} \chi_1(\alpha_1)\cdots\chi_t(\alpha_t).$$

6.1.33 Remark Jacobi sums have applications to solving diagonal equations over finite fields and to discrete log cryptosystems [240, Chapter 10]. They have been used to determine the cardinality of certain classes of irreducible polynomials over $\mathbb{F}_q$ with prescribed trace and restricted norm [1365, 1783]. See [2948, Chapter 16] for an application to primality testing, and [1146] for an application to coding theory.

6.1.34 Remark The next four theorems can be proved by a straightforward modification of the proofs in [240, Sections 10.1–10.3].

6.1.35 Theorem If $\chi_1,\dots,\chi_t$ are all trivial characters on $\mathbb{F}_q$, then

$$J(\chi_1,\dots,\chi_t) = \frac{(q-1)^t-(-1)^t}{q}, \quad J_0(\chi_1,\dots,\chi_t) = J(\chi_1,\dots,\chi_t) + (-1)^t.$$

6.1.36 Theorem Suppose that $\chi_1,\dots,\chi_t$ are characters on $\mathbb{F}_q$ which are not all trivial. Then

$$J_0(\chi_1,\dots,\chi_t) = \begin{cases} (1-q)J(\chi_1,\dots,\chi_t) & \text{if } \chi_1\cdots\chi_t \text{ is trivial,} \\ 0 & \text{otherwise.} \end{cases}$$

6.1.37 Theorem (Reduction formula) Suppose that $\chi_1,\dots,\chi_t$ are characters on $\mathbb{F}_q$ such that χ_t is nontrivial. Then

$$J(\chi_1,\dots,\chi_t) = \begin{cases} -(-1)^t & \text{if } \chi_1,\dots,\chi_{t-1} \text{ are all trivial,} \\ J(\chi_t,\chi_1\cdots\chi_{t-1})J(\chi_1,\dots,\chi_{t-1}) & \text{if } \chi_1\cdots\chi_{t-1} \text{ is nontrivial,} \\ -qJ(\chi_1,\dots,\chi_{t-1}) & \text{otherwise.} \end{cases}$$

6.1.38 Theorem Suppose that the characters $\chi_1, \ldots, \chi_t$ on $\mathbb{F}_q$ are not all trivial. Then

$$J(\chi_1, \ldots, \chi_t) = \begin{cases} G(\chi_1) \cdots G(\chi_t)/G(\chi_1 \cdots \chi_t) & \text{if } \chi_1 \cdots \chi_t \text{ is nontrivial,} \\ -G(\chi_1) \cdots G(\chi_t)/q & \text{otherwise.} \end{cases}$$

Thus if $\chi_1, \ldots, \chi_t$ are all nontrivial,

$$|J(\chi_1, \ldots, \chi_t)| = \begin{cases} q^{(t-1)/2} & \text{if } \chi_1 \cdots \chi_t \text{ is nontrivial,} \\ q^{(t-2)/2} & \text{otherwise.} \end{cases}$$

6.1.39 Remark We next present two theorems on the uniform distribution of Jacobi sums. The first is due to Katz and Zheng [1713]. For the second and generalizations thereof, see Katz [1711, Corollary 20.3]. (Katz's Corollary 20.3 for $r = 1$ is equivalent to his Theorem 17.5 with $n = 1$.)

6.1.40 Theorem Consider the collection of $(q-2)(q-3)$ normalized Jacobi sums

$$J(\chi, \psi)/\sqrt{q}, \qquad \chi, \psi, \chi\psi \text{ all nontrivial on } \mathbb{F}_q.$$

As q tends to infinity, this collection is asymptotically equidistributed on the complex unit circle.

6.1.41 Theorem For a fixed nontrivial character ψ on $\mathbb{F}_q$, consider the collection of $q-3$ normalized Jacobi sums

$$J(\chi, \psi)/\sqrt{q}, \qquad \chi \neq \chi_0,\ \chi \neq \overline{\psi}.$$

As q tends to infinity, this collection is asymptotically equidistributed on the complex unit circle.

6.1.42 Remark Suppose that in place of the collection above, one considers the more general collection of normalized *multiple* Jacobi sums

$$J(\chi_1, \ldots, \chi_m, \psi_1, \ldots, \psi_n)/q^{(n+m-1)/2},$$

where the ψ_j are fixed nontrivial characters and where the χ_i run through all nontrivial characters for which $\chi_1 \cdots \chi_m \psi_1 \cdots \psi_n$ is nontrivial. Katz [email communication, 2011] has shown that as q tends to infinity, this collection is asymptotically equidistributed on the complex unit circle, except in the "degenerate" case where both $m = 1$ and $\psi_1 \cdots \psi_n$ is trivial.

6.1.43 Definition (Lifted Gauss sums) Let χ be a character on $\mathbb{F}_q$, and let m be a positive integer. Recall that the Gauss sum $G(\chi)$ on $\mathbb{F}_q$ is defined by

$$G(\chi) = \sum_{\alpha \in \mathbb{F}_q} \chi(\alpha) \zeta_p^{\mathrm{Tr}(\alpha)}.$$

The *lift* of $G(\chi)$ to the extension field $\mathbb{F}_{q^m}$ is the Gauss sum

$$G_m(\chi') = \sum_{\delta \in \mathbb{F}_{q^m}} \chi'(\delta) \zeta_p^{\mathrm{Tr}(\delta)},$$

where Tr is the trace from $\mathbb{F}_{q^m}$ to $\mathbb{F}_p$ and χ' is the character on $\mathbb{F}_{q^m}$ defined by

$$\chi'(\delta) = \chi\left(\mathrm{Norm}_{\mathbb{F}_{q^m}/\mathbb{F}_q}(\delta)\right), \quad \delta \in \mathbb{F}_{q^m}.$$

We call χ' the *lift* of the character χ from $\mathbb{F}_q$ to $\mathbb{F}_{q^m}$.

6.1.44 Definition (Lifted Jacobi sums) Suppose that $\chi_1, \ldots, \chi_t$ are characters on $\mathbb{F}_q$, and let m be a positive integer. The *lift* of the Jacobi sum

$$J(\chi_1, \ldots, \chi_t) = \sum_{\substack{\alpha_1, \ldots, \alpha_t \in \mathbb{F}_q \\ \alpha_1 + \cdots + \alpha_t = 1}} \chi_1(\alpha_1) \cdots \chi_t(\alpha_t)$$

on $\mathbb{F}_q$ to the extension field $\mathbb{F}_{q^m}$ is the Jacobi sum

$$J_m(\chi'_1, \ldots, \chi'_t) = \sum_{\substack{\delta_1, \ldots, \delta_t \in \mathbb{F}_{q^m} \\ \delta_1 + \cdots + \delta_t = 1}} \chi'_1(\delta_1) \cdots \chi'_t(\delta_t),$$

where χ'_i is the the lift of χ_i from $\mathbb{F}_q$ to $\mathbb{F}_{q^m}$.

6.1.45 Theorem (Hasse-Davenport theorem on lifted Gauss sums [240, p. 360]) Let χ be a character on $\mathbb{F}_q$, and let m be a positive integer. Then in the notation of Definition 6.1.43,

$$G_m(\chi') = (-1)^{m-1} G(\chi)^m.$$

6.1.46 Remark The next corollary follows with the aid of Theorem 6.1.38.

6.1.47 Corollary (Hasse-Davenport theorem on lifted Jacobi sums.) Suppose that $\chi_1, \ldots, \chi_t$ are characters on $\mathbb{F}_q$ which are not all trivial, and let m be a positive integer. Then in the notation of Definition 6.1.44,

$$J_m(\chi'_1, \ldots, \chi'_t) = (-1)^{(m-1)(t-1)} J(\chi_1, \ldots, \chi_t)^m.$$

6.1.48 Definition A Gauss or Jacobi sum is *pure* if some positive integral power of it is real.

6.1.49 Example Quadratic Gauss sums are pure, by Theorem 6.1.86. Another example is given by the following theorem of Stickelberger, proved in [240, Section 11.6].

6.1.50 Theorem Let χ be a character of order $k > 2$ on $\mathbb{F}_q$, where $q = p^r$. Suppose that there is a positive integer t such that $p^t \equiv -1 \pmod{k}$, with t chosen minimal. Then $r = 2ts$ for some positive integer s, and

$$q^{-1/2} G(\chi) = \begin{cases} (-1)^{s-1} & \text{if } p = 2, \\ (-1)^{s-1+(p^t+1)s/k} & \text{if } p > 2. \end{cases}$$

6.1.51 Theorem [1011] Let χ be a character of order k on $\mathbb{F}_q$. Then the Gauss sums $G(\chi^j)$ are pure for all integers j if and only if -1 is a power of $p \pmod{k}$. In the special case that k is a prime power, $G(\chi)$ is pure if and only if -1 is a power of $p \pmod{k}$.

6.1.52 Theorem [1011] Let χ be a character of order $k > 1$ on $\mathbb{F}_q$, where $q = p^r$. If $G(\chi)$ is pure, then $2(q-1)/(k(p-1))$ is an integer with the same parity as r. In particular, if $r = 1$ and $k > 2$, then $G(\chi)$ is not pure, i.e., the normalized Gauss sum $G(\chi)/\sqrt{p}$ on $\mathbb{F}_p$ cannot equal a root of unity when χ is a character on $\mathbb{F}_p$ of order > 2. Also, if $r = 2$, then $G(\chi)$ is pure if and only if $k \mid (p+1)$.

6.1.53 Remark The next theorem, due to Aoki [112], gives further examples of pure Gauss sums. See also Aoki [113].

6.1.54 Theorem Let χ be a character of order $k > 2$ on $\mathbb{F}_q$, where $q = p^r$. For $r = 3$, if $G(\chi)$ is pure, then either (a) $k = 14$ and $p \equiv 2$ or $4 \pmod 7$; (b) $k = 42$ and $p \equiv 4$ or $16 \pmod{21}$; or (c) $k = 78$ and $p \equiv 16$ or $32 \pmod{39}$. For $r = 4$, $G(\chi)$ is pure if and only if $k \mid (p^2+1)$, except in the following four cases: (i) $k = 20$ and $p \equiv 13$ or $17 \pmod{20}$; (ii) $k = 30$ and $p \equiv 17$ or $23 \pmod{30}$; (iii) $k = 60$ and $p \equiv 17$ or $53 \pmod{60}$; (iv) $k = 120$ and $p \equiv 83$ or $107 \pmod{120}$.

6.1.55 Remark The next theorem gives examples of pure Jacobi sums over $\mathbb{F}_{p^2}$. It is due to Shiratani and Yamada, who point out an application to algebraic combinatorics [2615]. It is also due independently to Akiyama, who gave a different proof [66]. Other examples have been given by Aoki [112].

6.1.56 Theorem Let $r = 2$, i.e., $q = p^2$, and consider the Jacobi sum $J(\rho, \chi)$ on $\mathbb{F}_q$, where ρ, χ are characters on $\mathbb{F}_q$ of orders 2, k, respectively, with $k > 2$. Then $J(\rho, \chi) = \pm p$ (and is thus pure) if either (a) $k \mid (p+1)$; (b) $2(p-1)/k$ is an odd integer; (c) $k = 24$ and $p \equiv 17$ or $19 \pmod{24}$; or (d) $k = 60$ and $p \equiv 41$ or $49 \pmod{60}$. Moreover, if none of (a)–(d) hold, then $J(\rho, \chi)$ is not pure.

6.1.57 Remark The Jacobi sums satisfying one of (a)–(d) above of course generate the subfield $\mathbb{Q}$ of $\mathbb{Q}(\zeta_k)$. Aoki [111] has determined the subfield of $\mathbb{Q}(\zeta_k)$ generated by more general Jacobi sums in $\mathbb{Q}(\zeta_k)$.

6.1.58 Remark Pure multiple Jacobi sums are connected with supersingularity and ranks of certain elliptic curves; see Aoki [114] and Ulmer [2835, Section 5].

6.1.59 Theorem (Hasse-Davenport product formula for Gauss sums [240, p. 351]) Let ψ be a character on $\mathbb{F}_q$ of order $\ell > 1$. For every character χ on $\mathbb{F}_q$,

$$\prod_{i=1}^{\ell-1} G(\chi\psi^i)/G(\psi^i) = \overline{\chi}^{\ell}(\ell)G(\chi^{\ell})/G(\chi).$$

6.1.60 Remark The Hasse-Davenport formula for products of Gauss sums (Theorem 6.1.59) is the finite field analogue of the Gauss multiplication formula for gamma functions. Work of Kubert and Lichtenbaum [1809] on Jacobi sum Hecke characters led to identities involving products of Gauss sums which extended the Hasse-Davenport product formula. (For calculation of conductors of Jacobi sum Hecke characters, see [2610].) Other examples of identities involving products of Gauss sums may be found in [813, 1010, 1013, 1014, 1015, 2854]. For evaluations of special hypergeometric character sums over $\mathbb{F}_q$ in terms of products of Gauss sums, see papers of Evans and Greene [1007, 1008].

6.1.61 Example If $\ell = 2$ and ψ is the quadratic character ρ on $\mathbb{F}_q$, then the Hasse-Davenport product formula becomes

$$G(\chi\rho)/G(\rho) = \overline{\chi}(4)G(\chi^2)/G(\chi),$$

which is equivalent to the following theorem.

6.1.62 Theorem [240, p. 59] Let χ, ρ be characters on $\mathbb{F}_q$ with χ nontrivial and ρ quadratic. Then

$$J(\chi, \rho) = \chi(4)J(\chi, \chi).$$

6.1.63 Remark The next theorem gives a variant of Theorem 6.1.62.

6.1.64 Theorem [240, p. 60] Let χ, ρ be characters on $\mathbb{F}_q$ such that χ has order > 2 and ρ is quadratic. Then

$$\rho(-1)\overline{\chi}\rho(4)J(\overline{\chi}\rho, \overline{\chi}\rho) = \chi(4)J(\chi, \chi) = \chi(-1)J(\chi, \overline{\chi}\rho).$$

6.1.65 Remark We next give two congruences for Jacobi sums; see [240, pp. 60, 97]. More general congruences are given in [1001].

6.1.66 Theorem Let χ, ψ be nontrivial characters on $\mathbb{F}_q$ of orders a, b, respectively. Then

$$J(\chi, \psi) \equiv -q \pmod{(1-\zeta_a)(1-\zeta_b)}.$$

If moreover $a = b > 2$, then the right member $-q$ may be replaced by -1.

6.1.67 Theorem Let χ be a character on $\mathbb{F}_q$ of order 2ℓ, where $\ell > 1$ is odd, and let n denote an even integer not divisible by ℓ. Then

$$J(\chi, \chi^n) \equiv -\chi^n(4) \pmod{(1-\zeta_\ell)^2}.$$

6.1.68 Definition Let γ be a generator of the cyclic group $\mathbb{F}_q^*$, and let χ be a character of order k on $\mathbb{F}_q$ for which $\chi(\gamma) = \zeta_k$. For any pair of integers $a, b \pmod{k}$, define the *cyclotomic number* $C(a,b) = C(\gamma, a, b)$ of order k over $\mathbb{F}_q$ to be the number of $\alpha \in \mathbb{F}_q^*$ for which

$$\chi(\alpha/\gamma^a) = \chi((\alpha+1)/\gamma^b) = 1.$$

6.1.69 Remark Cyclotomic numbers are useful for obtaining residuacity criteria. For example, the cyclotomic numbers of order 12 can be used to prove that for a prime $p \equiv 1 \pmod{12}$, 3 is a quartic residue $(\text{mod } p)$ if and only if $a_3 \equiv -1 \pmod{4}$, where a_3 is as in Definition 6.1.74. See [240, p. 231].

6.1.70 Theorem [240, p. 365] In the notation of Definition 6.1.68, the cyclotomic numbers $C(a,b)$ are related to the Jacobi sums $J(\chi^u, \chi^v)$ by the following finite Fourier series expansions:

$$k^2 C(a,b) = \sum_{u=0}^{k-1}\sum_{v=0}^{k-1} \chi^u(-1) J(\chi^u, \chi^v) \zeta_k^{-au-bv}$$

and

$$\chi^u(-1) J(\chi^u, \chi^v) = \sum_{a=0}^{k-1}\sum_{b=0}^{k-1} C(a,b) \zeta_k^{au+bv}.$$

6.1.71 Definition Let p be an odd prime. For a positive integer k and an integer a not divisible by p, define the *Jacobsthal sums* $U_k(a)$, $V_k(a)$ over $\mathbb{F}_p$ by

$$U_k(a) = \sum_{m=0}^{p-1} \left(\frac{m}{p}\right)\left(\frac{m^k + a}{p}\right), \quad V_k(a) = \sum_{m=0}^{p-1} \left(\frac{m^k + a}{p}\right),$$

where the symbols in the summands are Legendre symbols.

6.1.72 Remark Jacobsthal sums have applications to the distribution of quadratic residues [240, Chapter 6], to evaluations of Brewer sums [240, Chapter 13], and to the evaluation of certain hypergeometric character sums [240, Equation (13.3.2)]. The next theorem expresses Jacobsthal sums in terms of Jacobi sums.

6.1.73 Theorem [240, pp. 188–189]. For a prime $p \equiv 1 \pmod{2k}$, let χ be a character on $\mathbb{F}_p$ of order $2k$. Let a be an integer not divisible by p. Then

$$U_k(a) = \chi(-1)\left(\frac{a}{p}\right) \sum_{j=0}^{k-1} \chi^{2j+1}(4a) J(\chi^{2j+1}, \chi^{2j+1})$$

and

$$V_k(a) = \left(\frac{a}{p}\right) \sum_{j=1}^{k-1} \chi^{2j}(4a) J(\chi^{2j}, \chi^{2j}).$$

6.1.2 Evaluations of Jacobi and Gauss sums of small orders

6.1.74 Definition For a prime $p = 6f + 1$, define integer parameters a_3, b_3, r_3, s_3, u_3, v_3 as follows:

$$p = a_3^2 + 3b_3^2, \quad a_3 \equiv -1 \pmod 3,$$
$$4p = r_3^2 + 3s_3^2, \quad r_3 \equiv 1 \pmod 3, \quad s_3 \equiv 0 \pmod 3,$$
$$4p = u_3^2 + 3v_3^2, \quad u_3 \equiv 1 \pmod 3, \text{ where}$$

$$\begin{cases} v_3 \equiv \pm 1 \pmod 6 & \text{if 2 is a cubic nonresidue } (\text{mod } p), \\ v_3 = s_3 \equiv 0 \pmod 3 & \text{if 2 is a cubic residue } (\text{mod } p). \end{cases}$$

6.1.75 Remark Given p, the parameters a_3, r_3, and u_3 are uniquely determined, but b_3, s_3, and v_3 are determined only up to sign. These parameters appear below in the evaluations of cubic and sextic Gauss and Jacobi sums over $\mathbb{F}_p$. In the case that 2 is a cubic nonresidue (mod p), we have $3 \nmid b_3$ and the parameters r_3, s_3, u_3, v_3 are odd. In the case that 2 is a cubic residue (mod p), we have $3 \mid b_3$ and r_3, s_3, u_3, v_3 are even; see [240, Section 3.1].

6.1.76 Theorem (Cubic Jacobi sums [240, Section 3.1]) For $q = p = 6f + 1$, let χ be a character of order 3 on $\mathbb{F}_p$. Then

$$J(\chi, \chi) = (r_3 + i s_3 \sqrt{3})/2.$$

6.1.77 Theorem (Sextic Jacobi sums [240, Section 3.1]) For $q = p = 6f + 1$, let χ be a character of order 6 on $\mathbb{F}_p$. Then

$$J(\chi, \chi) = (-1)^f (u_3 + i v_3 \sqrt{3})/2 = (-1)^f J(\chi, \chi^4),$$

$$J(\chi, \chi^2) = J(\chi^3, \chi^2) = a_3 + i b_3 \sqrt{3} = (-1)^f J(\chi, \chi^3).$$

6.1.78 Definition For a prime $p = 4f + 1$, define integer parameters a_4, b_4 by

$$p = a_4^2 + b_4^2, \quad a_4 \equiv -(-1)^f \pmod 4.$$

6.1.79 Theorem (Quartic Jacobi sums [240, Section 3.2]) For $q = p = 4f + 1$, let χ be a character of order 4 on $\mathbb{F}_p$. Then

$$J(\chi, \chi) = (-1)^f (a_4 + i b_4) = (-1)^f J(\chi, \chi^2).$$

6.1.80 Definition For a prime $p = 8f + 1$, define integer parameters a_8, b_8 by

$$p = a_8^2 + 2b_8^2, \quad a_8 \equiv -1 \pmod 4.$$

6.1.81 Theorem (Octic Jacobi sums [240, Section 3.3]) For $q = p = 8f + 1$, let χ be a character of order 8 on $\mathbb{F}_p$. Then

$$J(\chi, \chi) = \chi(4)(a_8 + ib_8\sqrt{2}) = \chi(-4)J(\chi, \chi^3), \quad J(\chi, \chi^2) = \chi(-4)(a_4 + ib_4).$$

6.1.82 Theorem (Duodecic Jacobi sums [240, Section 3.5]) For $q = p = 12f+1$, let χ be a character of order 12 on $\mathbb{F}_p$ so that

$$J(\chi^3, \chi^3) = (-1)^f(a_4 + ib_4)$$

as in Theorem 6.1.79. Then

$$\chi(-1)J(\chi, \chi^5) = \chi(4)J(\chi, \chi) = \chi(4)^5 J(\chi^5, \chi^5) = \pm(a_4 + ib_4),$$

where the plus sign is chosen if $3 \mid b_4$ and the minus sign is chosen if $3 \mid a_4$. The three additional duodecic Jacobi sums below are expressed in terms of previously evaluated Jacobi sums of orders 6, 4, 3, respectively:

$$J(\chi, \chi^2) = \chi(4)^2 J(\chi^2, \chi^2)/c_{12}, \quad J(\chi, \chi^3) = J(\chi^3, \chi^3)/c_{12}, \quad J(\chi, \chi^4) = J(\chi^4, \chi^4),$$

where

$$\begin{cases} c_{12} = \pm 1 \text{ with } c_{12} \equiv -a_4 \pmod 3 & \text{if } 3 \mid b_4, \\ c_{12} = \pm i \text{ with } c_{12} \equiv -ib_4 \pmod 3 & \text{if } 3 \mid a_4. \end{cases}$$

6.1.83 Remark Values of *all* duodecic Jacobi sums may be deduced from Theorem 6.1.82. For example,

$$J(\chi^3, \chi^5) = \sigma_5 J(\chi, \chi^3) = J(\chi, \chi^3),$$

where σ_5 is as in Definition 6.1.99 with $k = 12$. Niitsuma [2288] applied duodecic Jacobi sum evaluations to count rational points on certain hyperelliptic curves over $\mathbb{F}_p$.

6.1.84 Remark Jacobi sums of various other small orders are explicitly evaluated in [240]; e.g., quintic Jacobi sums are computed in [240, Section 3.7]. For the quintic case, see also Hoshi [1538] for an analysis of Gauss sums, Jacobi sums, and period polynomials. Values of Jacobi sums of order 16 have been applied to construct regular Hadamard matrices [1908].

6.1.85 Theorem (Quadratic multiple Jacobi sums [240, p. 299]) Suppose that $\chi_1, \ldots, \chi_t$ are all equal to the quadratic character ρ on $\mathbb{F}_q$. Then

$$J(\chi_1, \ldots, \chi_t) = \begin{cases} \rho(-1)^{(t-1)/2} q^{(t-1)/2} & \text{if } t \text{ is odd}, \\ -\rho(-1)^{t/2} q^{(t-2)/2} & \text{if } t \text{ is even}. \end{cases}$$

6.1.86 Theorem (Quadratic Gauss sums [240, p. 362]) Let $q = p^r$ for an odd prime p, and let ρ be the quadratic character on $\mathbb{F}_q$. Then

$$g(2) = G(\rho) = \begin{cases} (-1)^{r-1}\sqrt{q} & \text{if } p \equiv 1 \pmod 4, \\ (-1)^{r-1} i^r \sqrt{q} & \text{if } p \equiv 3 \pmod 4. \end{cases}$$

In particular, if $q = p$, we obtain the evaluation of Gauss:

$$\sum_{n=0}^{p-1} \zeta_p^{n^2} = \sum_{n=0}^{p-1} \left(\frac{n}{p}\right) \zeta_p^n = \begin{cases} \sqrt{p} & \text{if } p \equiv 1 \pmod 4, \\ i\sqrt{p} & \text{if } p \equiv 3 \pmod 4, \end{cases}$$

where (n/p) is the Legendre symbol.

6.1.87 Remark The Gauss sums in Theorem 6.1.86 lie in a quadratic extension of $\mathbb{Q}$. For evaluations of general Gauss sums over $\mathbb{F}_q$ lying in quadratic and multi-quadratic extensions of $\mathbb{Q}$, see [115, 2042, 3026, 3027].

6.1.88 Theorem (Cubic periods [240, Section 4.1]) Let $q = p \equiv 1 \pmod 6$, so that $4p = r_3^2 + 3s_3^2$ with $r_3 \equiv 1 \pmod 3$, $s_3 \equiv 0 \pmod 3$. Let b be a primitive root of p. Then the cubic irreducible polynomial $x^3 - 3px - pr_3$ has the three real zeros $g(3)$, $g(b, 3)$, and $g(b^2, 3)$, with one zero in each of the three intervals $(-2\sqrt{p}, -\sqrt{p})$, $(-\sqrt{p}, \sqrt{p})$, $(\sqrt{p}, 2\sqrt{p})$.

6.1.89 Remark No *simple* criterion is known for determining which of the three intervals above contains the cubic Gauss sum $g(3)$. However, $g(3)$ and the Gauss sums $G(\chi)$ for cubic characters χ have been evaluated in terms of products of values of Weierstrass $\wp$-functions; see [240, p. 158]. The next theorem, due to Heath-Brown and Patterson [1455], gives an equidistribution result for cubic Gauss character sums over $\mathbb{F}_p$.

6.1.90 Problem Determining the distribution of n-th order Gauss sums over $\mathbb{F}_p$ for a general fixed n is an open problem.

6.1.91 Theorem Consider the collection of all normalized cubic Gauss sums

$$G(\chi)/\sqrt{p}, \quad \chi \text{ cubic on } \mathbb{F}_p, \quad q = p \equiv 1 \pmod 3, \quad p < x.$$

As x tends to infinity, this collection is asymptotically equidistributed on the complex unit circle.

6.1.92 Remark The next theorem determines the sextic Gaussian period $g(6)$ unambiguously, once $g(3)$ is known.

6.1.93 Theorem (Sextic periods [1003]) Let $q = p \equiv 1 \pmod 6$, so that $4p = r_3^2 + 3s_3^2$ with $r_3 \equiv 1 \pmod 3$, $s_3 \equiv 0 \pmod 3$. In the case that 2 is a cubic residue $\pmod p$,

$$g(6) = g(3) + i^{(p-1)^2/4}\left(g(3)^2 - p\right)/\sqrt{p}.$$

In the case that 2 is a cubic nonresidue $\pmod p$, then with the sign of s_3 specified by $s_3 \equiv -r_3 \pmod 4$,

$$g(6) = g(3) + i^{(p-1)^2/4}\{4p - g(3)^2 + s_3^{-1}\left(2pg(3) + 2pr_3 - r_3 g(3)^2\right)\}/(2\sqrt{p}).$$

6.1.94 Remark For the history behind the quartic Gauss sum evaluations in the two theorems below, see [240, p. 162].

6.1.95 Theorem (Quartic Gauss sums [240, Section 4.2]) Let $q = p \equiv 1 \pmod 4$, and let χ be a quartic character on $\mathbb{F}_p$. As in Theorem 6.1.79, write $J(\chi, \chi) = a + bi$, where $p = a^2 + b^2$ with $a \equiv -1 \pmod 4$. (Note that the sign of b depends on the choice of the quartic character χ.) Define $C = \pm 1$ by

$$C \equiv (-1)^{(p-1)/4} \frac{|b|}{a} \left(\frac{p-1}{2}\right)! \pmod p.$$

Then

$$G(\chi) = \left(\frac{|b|}{|a|}\right) C (-1)^{(b^2+2b)/8} \left(\sqrt{\frac{p + a\sqrt{p}}{2}} + i\frac{|b|}{b}\sqrt{\frac{p - a\sqrt{p}}{2}}\right),$$

where the first symbol on the right is the Jacobi symbol.

6.1.96 Theorem (Quartic periods [240, Section 4.2]) In the notation of the previous theorem, if $p \equiv 1 \pmod 8$,

$$g(4) = \sqrt{p} + C\left(\frac{|b|}{|a|}\right)(-1)^{(b^2+2|b|)/8}\sqrt{2p + 2a\sqrt{p}},$$

while if $p \equiv 5 \pmod 8$,

$$g(4) = \sqrt{p} + iC\left(\frac{|b|}{|a|}\right)(-1)^{(b^2+2|b|)/8}\sqrt{2p - 2a\sqrt{p}}.$$

6.1.97 Remark The next theorem determines the Gaussian period $g(12)$ unambiguously, once $g(3)$ is known. For an extension to $\mathbb{F}_q$, see Gurak [1363].

6.1.98 Theorem (Duodecic periods [1003]) Let $q = p = 12f + 1$, so that as in Theorem 6.1.78, $p = a^2 + b^2$ with $a \equiv -(-1)^f \pmod 4$. In the case that -3 is a quartic residue (mod p),

$$g(12) = g(6) + (g(4) - \sqrt{p})\left(1 + \left(\frac{-a}{3}\right) g(3)/\sqrt{p}\right).$$

In the case that -3 is a quartic nonresidue (mod p) (which is equivalent to $3 \nmid b$ by [240, Section 7.2]) then with the sign of b specified by $b \equiv -1 \pmod 3$,

$$g(12) = g(6) + g(4) - \sqrt{p} + 2b\left(\frac{2}{p}\right) g(3)/\left(g(4) - \sqrt{p}\right).$$

6.1.3 Prime ideal divisors of Gauss and Jacobi sums

6.1.99 Definition Fix an integer $k > 1$. In this subsection, $q = p^f$, where f is the order of p (mod k). For the group $R = (\mathbb{Z}/k\mathbb{Z})^*$, let T denote a complete set of $\phi(k)/f$ coset representatives of the quotient group $R/\langle p\rangle$. For an integer a, let $\ell(a)$ denote the least nonnegative integer congruent to a (mod k). Write

$$s(a) = \sum_{i=0}^{f-1} a_i\,, \quad t(a) = \prod_{i=0}^{f-1} a_i!\,,$$

where the a_i are the digits in the base p expansion

$$\ell(a)(q-1)/k = \sum_{i=0}^{f-1} a_i p^i, \quad 0 \le a_i \le p-1.$$

Consider the cyclotomic fields

$$K = \mathbb{Q}(\zeta_k), \quad M = \mathbb{Q}(\zeta_k, \zeta_p)$$

with rings of integers $\mathcal{O}_K, \mathcal{O}_M$, respectively. Let P be a prime ideal of $\mathcal{O}_K$ above p. Since $\mathrm{Norm}(P) = q$, the finite field $\mathcal{O}_K/P$ is isomorphic to $\mathbb{F}_q$. Since P is totally ramified in $\mathcal{O}_M$, we have

$$\mathcal{O}_M P = \mathfrak{P}^{p-1} \text{ for a prime ideal } \mathfrak{P} \subset \mathcal{O}_M.$$

For $j \in R$, let σ_j denote the element in $\mathrm{Gal}(\mathbb{Q}(\zeta_k, \zeta_p)/\mathbb{Q}(\zeta_p))$ for which $\sigma_j(\zeta_k) = \zeta_k^j$, and write

$$P_j = \sigma_j(P), \quad \mathfrak{P}_j = \sigma_j(\mathfrak{P}), \quad \mathfrak{P}_{j^{-1}} = \sigma_j^{-1}(\mathfrak{P}).$$

6.1.100 Theorem [240, p. 343] Let $\pi = \zeta_p - 1$. We have the prime ideal factorizations

$$\pi\mathcal{O}_M = \prod_{j\in T} \mathfrak{P}_j, \quad p\mathcal{O}_M = \prod_{j\in T} \mathfrak{P}_j^{p-1}, \quad p\mathcal{O}_K = \prod_{j\in T} P_j.$$

6.1.101 Definition (Power residue symbol χ_P) Define the character χ_P of order k on the finite field $\mathcal{O}_K/P$ by setting $\chi_P(\alpha + P)$ equal to the unique power of ζ_k which is congruent to $\alpha^{(q-1)/k}$ (mod P), for every $\alpha \in \mathcal{O}_K$ with $\alpha \notin P$. If $\alpha \in P$, set $\chi_P(\alpha + P) = 0$.

6.1.102 Theorem (Stickelberger's congruence for Gauss sums [240, p. 344]) For any integer a, the Gauss sum $G(\chi_P^{-a})$ over the finite field $\mathcal{O}_K/P$ is an element of $\mathcal{O}_M$ satisfying the congruence

$$G(\chi_P^{-a}) \equiv \frac{-\pi^{s(a)}}{t(a)} \left(\text{mod } \mathfrak{P}^{s(a)+1}\right).$$

6.1.103 Remark For the following two corollaries, see Conrad [714].

6.1.104 Corollary If $0 \le a, b < k$ with a, b not both 0, then with $u := (q-1)/k$,

$$J(\chi_P^{-a}, \chi_P^{-b}) \equiv -(-1)^{au}\binom{q-1-bu}{au} \text{ (mod } P).$$

6.1.105 Corollary Let $k = q - 1$ (so that χ_P has order $q - 1$), and let

$$0 \le b_i < q - 1, \quad i = 1, 2, \ldots, t,$$

where not all b_i equal 0. Then

$$J(\chi_P^{-b_1}, \chi_P^{-b_2}, \ldots, \chi_P^{-b_t}) \equiv (-1)^{t+1}\frac{(b_1 + b_2 + \cdots + b_t)!}{b_1!b_2!\cdots b_t!} \text{ (mod } P).$$

6.1.106 Remark A consequence of Stickelberger's congruence is the prime ideal factorization of Gauss sums (Theorem 6.1.107). An important application of this factorization is the determination of the annihilator in $\mathbb{Z}[\text{Gal}(L/\mathbb{Q})]$ of the ideal class group of an abelian number field L; see Washington [2948, Theorem 6.10]. For applications of Theorem 6.1.107 to the theory of difference sets and coding theory, see [144, 580, 1009].

6.1.107 Theorem (Prime ideal factorization of Gauss sums [240, p. 346]) For any integer a,

$$G(\chi_P^{-a})\mathcal{O}_M = \prod_{j\in T} \mathfrak{P}_{j^{-1}}^{s(aj)}.$$

6.1.108 Theorem [240, p. 347] For any integer a, we have $G(\chi_P^{-a})^k \in \mathcal{O}_K$ and

$$G(\chi_P^{-a})^k\mathcal{O}_K = \prod_{j\in T} P_{j^{-1}}^{ks(aj)/(p-1)} = \prod_{j\in R} P_{j^{-1}}^{\ell(aj)}.$$

6.1.109 Theorem (Prime ideal factorization of Jacobi sums [240, p. 346]) Suppose that a, b are integers such that $a + b$ is not divisible by k. Then

$$J(\chi_P^{-a}, \chi_P^{-b})\mathcal{O}_K = \prod_{j\in T} P_{j^{-1}}^{v(a,b,j)}, \quad \text{with} \quad v(a,b,j) = \frac{s(aj) + s(bj) - s(aj+bj)}{p-1}.$$

In the special case $q = p$, this reduces to

$$J(\chi_P^{-a}, \chi_P^{-b})\mathcal{O}_K = \prod_{\substack{j \in T \\ \ell(aj)+\ell(bj)>k}} P_{j^{-1}}.$$

In particular, for $q = p$,

$$J(\chi_P, \chi_P)\mathcal{O}_K = \prod_{\substack{1 \le j < k/2 \\ (j,k)=1}} P_{j^{-1}}.$$

6.1.110 Definition Let $\mathbb{Q}_p$ denote the field of p-adic rationals, and let $\mathbb{Z}_p$ be its ring of p-adic integers. Consider the extension field $\mathbb{Q}_p(\zeta)$, where ζ is a primitive p-th root of the element $1 \in \mathbb{Z}_p$. For $\pi = \zeta - 1$, let λ denote the prime element in $\mathbb{Q}_p(\zeta)$ satisfying

$$\lambda^{p-1} = -p, \quad \lambda \equiv \pi \pmod{\pi^2}.$$

6.1.111 Definition (Morita's p-adic gamma function Γ_p) Define $\Gamma_p : \mathbb{Z}_p \to \mathbb{Z}_p^*$ by

$$\Gamma_p(z) = \lim_{N \to z} (-1)^N \prod_{\substack{0<j<N \\ p \nmid j}} j,$$

where N runs through any sequence of positive integers p-adically approaching z.

6.1.112 Remark The following theorem of Gross-Koblitz expresses the p-adic Gauss sum in terms of Morita's p-adic gamma functions. For a relatively elementary proof, see Robert [2463].

6.1.113 Theorem (Gross-Koblitz formula for p-adic Gauss sums) Let a be any integer. Viewing the Gauss sum $G(\chi_P^{-a}) \in \mathcal{O}_M$ as embedded in the subfield $\mathbb{Q}_p(\zeta)$ of the $\mathfrak{P}$-adic completion of M, we have the following equality in $\mathbb{Z}_p[\zeta]$:

$$G(\chi_P^{-a}) = -\lambda^{s(a)} \prod_{i=0}^{f-1} \Gamma_p\left(\frac{\ell(ap^i)}{k}\right).$$

6.1.114 Remark The next corollary follows with the aid of Theorem 6.1.38.

6.1.115 Corollary (Gross-Koblitz formula for p-adic Jacobi sums) Let $b_1, \ldots, b_t$ be integers such that $c := b_1 + \cdots + b_t$ is not divisible by k. Viewing the Jacobi sum $J(\chi_P^{-b_1}, \ldots, \chi_P^{-b_t}) \in \mathcal{O}_K$ as embedded in the subfield $\mathbb{Q}_p$ of the P-adic completion of K, we have the following equality in $\mathbb{Z}_p$:

$$J(\chi_P^{-b_1}, \ldots, \chi_P^{-b_t}) = (-1)^{t-1}(-p)^u \prod_{i=0}^{f-1} \Gamma_p\left(\frac{\ell(b_1p^i)}{k}\right) \cdots \Gamma_p\left(\frac{\ell(b_tp^i)}{k}\right) / \Gamma_p\left(\frac{\ell(cp^i)}{k}\right),$$

with $u = \{(s(b_1) + \cdots + s(b_t)) - s(c)\}/(p-1)$.

6.1.116 Remark The exponent u in Theorem 6.1.115 is an integer, since for any integer x, we have $s(x) \equiv x(q-1)/k \pmod{p-1}$.

6.1.4 Kloosterman sums

6.1.117 Remark In this subsection, we revert back to the notation $q = p^r$.

6.1.118 Definition For $u \in \mathbb{F}_q$ and a multiplicative character χ on $\mathbb{F}_q$, define the (twisted) *Kloosterman sum* $K(u, \chi)$ over $\mathbb{F}_q$ by

$$K(u, \chi) = \sum_{\alpha \in \mathbb{F}_q^*} \chi(\alpha) \zeta_p^{\mathrm{Tr}(\alpha + u/\alpha)}.$$

Note that $K(0, \chi) = G(\chi)$. When χ is trivial, we abbreviate $K(u) = K(u, \chi)$.

6.1.119 Remark Kloosterman sums occur frequently in the theory of modular forms, and they have many applications in analytic number theory [1453, 1581, 2524]. For some applications to coding theory, see [620, 1562, 1732, 2124]. For further applications, see the references in [503, p. 448]. Some congruences for Kloosterman sums may be found in [592, 1299, 1946, 2128]. In [1785], it is proved that if $K(u)$ is an integer and $p > 3$, then $K(u)$ is even. This proves in particular the nonvanishing of $1 + K(u)$ for $p > 3$. For analysis of the cases $p = 2, 3$, see [49]. The vanishing of $1 + K(u)$ has applications to bent functions, defined in Section 9.3. (See [2086] for recent work on bent and hyper-bent functions.)

6.1.120 Theorem For odd q and $u \in \mathbb{F}_q$,

$$K(u) = \sum_{y \in \mathbb{F}_q} \rho(y^2 - 4u) \zeta_p^{\mathrm{Tr}(y)},$$

where ρ is the quadratic character on $\mathbb{F}_q$.

6.1.121 Remark The theorem above follows easily from the definition of $K(u)$, by counting, for each $y \in \mathbb{F}_q$, the number of $\alpha \in \mathbb{F}_q^*$ for which $\alpha + u/\alpha = y$. For the case where q is even, see Conrad [715].

6.1.122 Definition For $u \in \mathbb{F}_q$ and characters $\chi_1, \dots, \chi_m$ on $\mathbb{F}_q$, define the (twisted) *multiple Kloosterman sum* $K(u, \chi_1, \dots, \chi_m)$ over $\mathbb{F}_q$ by

$$K(u, \chi_1, \dots, \chi_m) = \sum_{\substack{\alpha_0, \dots, \alpha_m \in \mathbb{F}_q \\ \alpha_0 \cdots \alpha_m = u}} \chi_1(\alpha_1) \cdots \chi_m(\alpha_m) \zeta_p^{\mathrm{Tr}(\alpha_0 + \cdots + \alpha_m)}.$$

This is also known as a (twisted) *hyper-Kloosterman sum*. It reduces to the sum given in Definition 6.1.118 when $m = 1$.

6.1.123 Remark Wan [2915] studied the algebraic degree of multiple Kloosterman sums. For further results on the degrees of Kloosterman sums, see [1784]. The next theorem gives an easily proved expression for multiple Kloosterman sums in terms of Gauss sums.

6.1.124 Theorem [1701, p. 47] For $u \in \mathbb{F}_q^*$ and characters $\chi_1, \dots, \chi_m$ on $\mathbb{F}_q$,

$$K(u, \chi_1, \dots, \chi_m) = \frac{1}{q-1} \sum_{\chi} \overline{\chi}(u) \prod_{i=0}^{m} G(\chi \chi_i),$$

where χ_0 is the trivial character and where the sum is over all characters χ on $\mathbb{F}_q$.

6.1.125 Theorem For a positive integer m, let ψ be a character on $\mathbb{F}_q$ of order $m+1$. Then for any $u \in \mathbb{F}_q$,

$$K(u, \psi, \psi^2, \ldots, \psi^m) = \epsilon(q, m) q^{m/2} \sum_{\substack{\alpha \in \mathbb{F}_q \\ \alpha^{m+1} = u}} \zeta_p^{\mathrm{Tr}(\alpha(m+1))},$$

where

$$\epsilon(q, m) = \begin{cases} 1 & \text{if } 2 \mid m, \\ (-1)^{r-1+(q-1)(m-1)/8} i^{(p-1)^2 r/4} & \text{if } 2 \nmid m. \end{cases}$$

6.1.126 Remark Theorem 6.1.125 is due to Duke [928], who showed it to be equivalent to the Hasse-Davenport product formula for Gauss sums (Theorem 6.1.59). For a related identity, see Ye [3034]. The Kloosterman sum $K(u, \psi, \psi^2, \ldots, \psi^m)$ is connected to Fourier expansions of certain Poincaré series [928]. In the case $m = 1$, Theorem 6.1.125 reduces to the following well-known evaluation of the *Salié sum* $K(u, \rho)$.

6.1.127 Theorem (Salié sum) Let $u \in \mathbb{F}_q$ and let ρ be the quadratic character on $\mathbb{F}_q$. Then

$$K(u, \rho) = \begin{cases} (-1)^{r-1} i^{(p-1)^2 r/4} \sqrt{q} & \text{if } u = 0, \\ 0 & \text{if } u \in \mathbb{F}_q^* \text{ is not a square}, \\ 2(-1)^{r-1} i^{(p-1)^2 r/4} \sqrt{q} \cos(4\pi \mathrm{Tr}(b)/p) & \text{if } u = b^2 \text{ for some } b \in \mathbb{F}_q^*. \end{cases}$$

6.1.128 Remark Theorem 6.1.129 below reduces to Theorem 6.1.127 when $\chi = \rho$ and reduces to Theorem 6.1.120 when χ is trivial.

6.1.129 Theorem [715, Equation (4)] Let $u \in \mathbb{F}_q$ and let ρ denote the quadratic character on $\mathbb{F}_q$. Then for any character χ on $\mathbb{F}_q$,

$$K(u, \chi) = \frac{G(\rho)\overline{\chi}(4)}{G(\chi\rho)} \sum_{y \in \mathbb{F}_q} \chi\rho(y^2 - 4u) \zeta_p^{\mathrm{Tr}(y)}.$$

6.1.130 Remark Sums closely related to the above sum on y form an orthogonal set of eigenfunctions for a collection of adjacency matrices of "finite upper half plane" Cayley graphs [1000].

6.1.131 Theorem (Upper bound for multiple Kloosterman sums) For $u \in \mathbb{F}_q$ and characters $\chi_1, \ldots, \chi_m$ on $\mathbb{F}_q$,

$$|K(u, \chi_1, \ldots, \chi_m)| \leq (m+1) q^{m/2}.$$

6.1.132 Remark The upper bound above is due to Deligne for trivial characters, and in full generality to Katz [1701, p. 49]. See Conrad [715] for a nice proof of the special case $|K(u, \chi)| \leq 2\sqrt{q}$, patterned on Weil's original proof for trivial χ. Equality in Theorem 6.1.131 can occur; for example, let $u = 1$, $(m+1) \mid (p-1)$, $p \mid r$ in Theorem 6.1.125. However, equality cannot occur in Theorem 6.1.131 in the case that all m characters are trivial, since then

$$K(u, \chi_1, \ldots, \chi_m) \equiv (-1)^m \pmod{(1 - \zeta_p)}.$$

Nor can equality occur in the case $m = 1$, $q = p$; see [1004, p. 446].

6.1.133 Definition The Weil bound $|K(u)| < 2\sqrt{q}$ is a consequence of the formula expressing $-K(u)$ as a sum of conjugate Frobenius eigenvalues:

$$-K(u) = g(u) + \overline{g}(u), \quad u \in \mathbb{F}_q^*,$$

where

$$g(u) = \sqrt{q}\exp(i\theta(q,u)), \quad \theta(q,u) \in (0,\pi).$$

We call $\theta(q,u)$ the *Kloosterman angle*, noting that

$$-K(u) = 2\sqrt{q}\cos(\theta(q,u)), \quad u \in \mathbb{F}_q^*.$$

6.1.134 Definition Let $u \in \mathbb{F}_q^*$. For a positive integer n, let $K_n(u)$ denote the Kloosterman sum over $\mathbb{F}_{q^n}$ (so that $K_1(u) = K(u)$). We call $K_n(u)$ the *lift* of the Kloosterman sum $K(u)$ from $\mathbb{F}_q$ to $\mathbb{F}_{q^n}$.

6.1.135 Remark It is shown in [1785] that $K_n(u)$ is an integer if and only if $K(u)$ is an integer.

6.1.136 Remark The following theorem, analogous to the Hasse-Davenport lifting formula for Gauss sums (Theorem 6.1.45), offers a formula of Carlitz [548] for the lift $K_n(u)$ in terms of a Dickson polynomial in $K(u)$. For an extension to $K(u,\chi)$, see [1581, p. 281]. (Dickson polynomials over $\mathbb{F}_q$ are discussed in Section 9.6.)

6.1.137 Theorem We have

$$K_n(u) = (-1)^{n-1}D_n(K(u), q),$$

where $D_n(x,a)$ is the Dickson polynomial of degree n.

6.1.138 Remark In the notation of Definition 6.1.133,

$$-K_n(u) = g(u)^n + \overline{g}(u)^n = 2q^{n/2}\cos(n\theta(q,u)), \quad u \in \mathbb{F}_q^*.$$

Since

$$g(u) = \left(-K(u) \pm \sqrt{K(u)^2 - 4q}\right)/2,$$

we have

$$-2^n K_n(u) = \left(-K(u) + \sqrt{K(u)^2 - 4q}\right)^n + \left(-K(u) - \sqrt{K(u)^2 - 4q}\right)^n.$$

This is equivalent to Theorem 6.1.137; see [240, pp. 440–441].

6.1.139 Theorem (Equidistribution of Kloosterman angles) The set of Kloosterman angles $\{\theta(q,u) : u \in \mathbb{F}_q^*\}$ is asymptotically equidistributed with respect to the Sato-Tate measure on $(0,\pi)$, as q tends to infinity. In other words, as q tends to infinity, the proportion of these $q-1$ Kloosterman angles that lie in a fixed subinterval $[x,y] \subset (0,\pi)$ approaches

$$\frac{2}{\pi}\int_x^y \sin^2 t \, dt.$$

6.1.140 Conjecture Fix a positive integer u and consider the collection of Kloosterman angles $\{\theta(p,u) : u < p < x\}$. As x tends to infinity, this collection is asymptotically equidistributed with respect to the Sato-Tate measure on $(0,\pi)$.

6.1.141 Remark Theorem 6.1.139 is due to Katz [1701, p. 240]. For a quantitative refinement and related results, see [2092, 2245, 2650]. For an extension to Kloosterman sums over rings, see [1725]. Statements of the conjecture above may be found in Katz's books [1700, Conjecture 1.2.5] and [1701, p. 5]. See also the references in Shparlinski [2650, p. 420].

6.1.142 Definition Fix an integer u which is not a perfect square. Motivated by Theorem 6.1.127 with $q = p > 2$, we define the *Salié angle* $\vartheta(p, u) \in (0, \pi)$ for each prime p with $\left(\frac{u}{p}\right) = 1$ by

$$\vartheta(p, u) = 2\pi b(u, p)/p,$$

where $b = b(u, p)$ is the smallest positive integer for which $b^2 \equiv u \pmod{p}$.

6.1.143 Remark The next theorem gives an equidistribution result for angles of Salié sums $K(u, \rho)$; see [1581, p. 496] and the references in Shparlinski [2649], where one finds further results of this type.

6.1.144 Theorem (Equidistribution of Salié angles) Fix an integer u which is not a perfect square. The set of Salié angles $\{\vartheta(p, u) : \left(\frac{u}{p}\right) = 1,\ p < x\}$ is asymptotically equidistributed in the interval $(0, \pi)$, as x tends to infinity.

6.1.145 Remark The following equidistribution result for normalized Kloosterman sums $K(-1, \chi)/\sqrt{q}$ was conjectured by Evans and proved by Katz [1711]. Note that each such sum is real and lies in the interval $[-2, 2]$ by Theorem 6.1.131.

6.1.146 Theorem (Equidistribution of $K(-1, \chi)$) Consider the collection of $q-1$ normalized Kloosterman sums $K(-1, \chi)/\sqrt{q}$, where χ runs through the characters on $\mathbb{F}_q$. As q tends to infinity, the "angles" of this collection are asymptotically equidistributed with respect to the Sato-Tate measure on $[-2, 2]$. In other words, as q tends to infinity, the proportion of the members of this collection that lie in a fixed subinterval $[v, w] \subset [-2, 2]$ approaches

$$\frac{1}{2\pi} \int_v^w \sqrt{4 - x^2}\, dx.$$

6.1.147 Definition Let n be a positive integer. Define the n-th *power moment* S_n of the Kloosterman sums $K(u)$ by

$$S_n = \sum_{u \in \mathbb{F}_q} K(u)^n.$$

6.1.148 Remark Moisio [2119, 2122] gave evaluations of S_n for $n \leq 10$ when q is a power of 2 or 3, and he related them to cyclic codes; see also [1733]. In the remainder of this subsection, we restrict our attention to the case $q = p > n$.

6.1.149 Remark Various congruences have been given for S_n, but explicit evaluations of S_n for all $q = p > n$ are known only for $n = 1, 2, 3, 4, 5, 6$ [622]. The values for $n \leq 4$ below, due to Salié, may be found in Iwaniec's book [1580, Section 4.4] (where $-p$ should be replaced by $-3p$ in Equation (4.25)).

6.1.150 Theorem (Power moments of Kloosterman sums [622]) Let $q = p > n$. The power moments S_n are integer multiples of p. We have

$$S_1 = 0, \quad S_2 = p^2 - p, \quad S_3 = \left(\frac{p}{3}\right) p^2 + 2p, \quad S_4 = 2p^3 - 3p^2 - 3p.$$

For $p > 5$,

$$S_5 = 4\left(\frac{p}{3}\right) p^3 + (a_p + 5)p^2 + 4p,$$

where a_p is the integer of absolute value $< 2p$ satisfying

$$a_p = \begin{cases} 2p - 12u^2 & \text{if } \ p = 3u^2 + 5v^2, \\ 4x^2 - 2p & \text{if } \ p = x^2 + 15y^2, \\ 0 & \text{if } \ p \equiv 7, 11, 13, \text{ or } 14 \pmod{15}. \end{cases}$$

Also for $p > 5$,

$$S_6 = 5p^4 - 10p^3 - (b_p + 9)p^2 - 5p,$$

where b_p is the integer of absolute value $< 2p^{3/2}$ defined to be the coefficient of q^p in the q-expansion of the weight 4, level 6 newform

$$\{\eta(6z)\eta(3z)\eta(2z)\eta(z)\}^2.$$

Here $\eta(z)$ is the Dedekind eta function.

6.1.151 Remark Evans [1006] has conjectured an explicit formula for S_7 in terms of the coefficient of q^p in the q-expansion of a weight 3, level 525 newform.

6.1.152 Definition Let $q = p > n \geq 1$. In the notation of Definition 6.1.133, the Kloosterman power moments S_n can be expressed as

$$S_n = (-1)^n + (-1)^n \sum_{u=1}^{p-1} (g(u) + \overline{g}(u))^n.$$

Closely related to S_n is the sum

$$T_n = \sum_{u=1}^{p-1} \sum_{j=0}^{n} g(u)^{n-j} \overline{g}(u)^j = \sum_{u=1}^{p-1} p^{n/2} U_n(2\cos(\theta(p,u))),$$

where U_n is the n-th monic Chebychev polynomial of the second kind. We normalize the sum T_n by defining

$$Y_n := (-1 - T_n)/p^2.$$

6.1.153 Remark For some twists of S_n and T_n, see Liu [1947] and Evans [1005].

6.1.154 Theorem If $q = p > n$, then Y_n is an integer. In particular,

$$Y_1 = Y_2 = 0, \quad Y_3 = \left(\frac{p}{3}\right), \quad Y_4 = 1, \quad Y_5 = a_p, \quad Y_6 = b_p,$$

where a_p, b_p are defined in Theorem 6.1.150.

6.1.155 Remark Theorem 6.1.154 can be found in Evans [1006]. There it is moreover conjectured (for $p > 7$) that

$$Y_7 = \left(\frac{p}{105}\right) (|A(p)|^2 - p^2),$$

where $A(p)$ is the p-th Fourier coefficient of a weight 3, level 525 eigenform with quartic nebentypus of conductor 105. Furthermore, Evans [1005, p. 523] conjectured (for $p > 7$) that

$$Y_8 = B(p) + p^2,$$

where $B(p)$ is the p-th Fourier coefficient of a weight 6, level 6 newform with trivial nebentypus. These conjectured values for Y_n satisfy the following upper estimate due to Katz [1701, Theorem 0.2].

6.1.156 Theorem For $q = p > n$,

$$|Y_n| \leq \left\lfloor \frac{n-1}{2} \right\rfloor p^{(n-3)/2}.$$

6.1.5 Gauss and Kloosterman sums over finite rings

6.1.157 Remark Let k be a positive integer. We briefly discuss some basic Gauss and Kloosterman sums over $\mathbb{Z}/k\mathbb{Z}$, as they are natural extensions of sums over the finite field $\mathbb{Z}/p\mathbb{Z}$.

6.1.158 Definition For integers m and $k > 0$, define the *quadratic Gauss sum* $q_k(m)$ over $\mathbb{Z}/k\mathbb{Z}$ by

$$q_k(m) = \sum_{n=0}^{k-1} \zeta_k^{mn^2}.$$

6.1.159 Remark The special case $q_p(m)$ is the quadratic Gaussian period $g(m, 2)$ over $\mathbb{F}_p$ given in Definition 6.1.11.

6.1.160 Definition For integers a, b, c with $ac \neq 0$, define a *generalized quadratic Gauss sum* $S(a, b, c)$ by

$$S(a, b, c) = \sum_{n=0}^{|c|-1} \exp(\pi i(an^2 + bn)/c).$$

6.1.161 Remark The special case $S(2m, 0, k)$ is the quadratic Gauss sum $q_k(m)$ given in Definition 6.1.158.

6.1.162 Theorem (Reciprocity theorem for Gauss sums [240, p. 13]) For integers a, b, c with $ac \neq 0$ and $ac + b$ even,

$$S(a, b, c) = |c/a|^{1/2} \exp\left(\frac{\pi i}{4}\left(\operatorname{sgn}(ac) - \frac{b^2}{ac}\right)\right) S(-c, -b, a).$$

6.1.163 Theorem [240, Section 1.5] If $(m, k) = 1$ and $k > 1$, then

$$q_k(m) = \begin{cases} \left(\frac{k}{m}\right)(1 + i^m)\sqrt{k} & \text{if } k \equiv 0 \pmod 4, \\ \left(\frac{m}{k}\right)\sqrt{k} & \text{if } k \equiv 1 \pmod 4, \\ 0 & \text{if } k \equiv 2 \pmod 4, \\ \left(\frac{m}{k}\right) i\sqrt{k} & \text{if } k \equiv 3 \pmod 4. \end{cases}$$

In particular (cf. Theorem 6.1.86),

$$q_k(1) = \frac{1}{2}(1 + i)(1 + i^{-k})\sqrt{k}.$$

6.1.164 Theorem [240, p. 47] If $(m,k)=1$ with $k>1$ odd and squarefree, then (cf. Theorem 6.1.86)

$$q_k(m) = \sum_{n=1}^{k-1} \left(\frac{n}{k}\right) \zeta_k^{mn}.$$

6.1.165 Definition (Gauss character sums over $\mathbb{Z}/k\mathbb{Z}$.) Let χ be a Dirichlet character (mod k), where $k>1$. For any integer m, define the Gauss sum $\tau_k(m,\chi)$ over $\mathbb{Z}/k\mathbb{Z}$ by

$$\tau_k(m,\chi) = \sum_{n=1}^{k-1} \chi(n)\zeta_k^{mn}.$$

6.1.166 Remark When χ is trivial, $\tau_k(m,\chi)$ is a Ramanujan sum. When $k=p$, $\tau_k(m,\chi)$ is the Gauss sum $G(m,\chi)$ over $\mathbb{F}_p$. The next two theorems deal with *primitive* Gauss sums; for proofs and generalizations, see [1581, pp. 48–49].

6.1.167 Theorem For $k>1$, let χ (mod k) be a primitive Dirichlet character [240, pp. 28–29]. Then

$$|\tau_k(1,\chi)| = \sqrt{k}.$$

If further χ is quadratic, then

$$\tau_k(1,\chi) = \begin{cases} \sqrt{k} & \text{if } \chi(-1)=1, \\ i\sqrt{k} & \text{if } \chi(-1)=-1. \end{cases}$$

6.1.168 Theorem If χ (mod k) is primitive, then for any integer m,

$$\tau_k(m,\chi) = \overline{\chi}(m)\tau_k(1,\chi).$$

6.1.169 Remark There is a reduction formula which reduces the problem of evaluating $\tau_k(m,\chi)$ to the case where χ (mod k) is primitive, $m=1$, and k is a prime power p^s; see [240, p. 29]. When $s>1$, such Gauss sums have known closed form evaluations; see [240, Section 1.6] and [659]. There are similar reduction formulae for Kloosterman sums. Evaluations and bounds for Kloosterman sums over $\mathbb{Z}/p^s\mathbb{Z}$ with $s>1$ are given in [655, 1004]. For Gauss and Kloosterman sums over rings of algebraic or p-adic integers, see [1002, 1364, 1366]. Extensions of Gaussian periods to finite rings are discussed in [1012]. For Hecke Gauss sums in quadratic number fields, see [382]. Gauss sums connected with Hecke L-functions are discussed in [1581, p. 60].

6.1.170 Remark Let $k>1$ be an odd integer. We close with an evaluation of a Salié sum over the ring $\mathbb{Z}/k\mathbb{Z}$, which for prime $k=p$ reduces to the evaluation of the Salié sum $K(u,\rho)$ over $\mathbb{F}_p$ given in Theorem 6.1.127. For a short proof, see [2815].

6.1.171 Definition Fix an odd integer $k>1$. For an integer a with $(a,k)=1$, define the *Salié sum* $S(a)$ over $\mathbb{Z}/k\mathbb{Z}$ by

$$S(a) = \sum_{x\in(\mathbb{Z}/k\mathbb{Z})^*} \left(\frac{x}{k}\right) \zeta_k^{x+a/x},$$

where (x/k) is the Jacobi symbol.

6.1.172 Theorem (Salié sums over $\mathbb{Z}/k\mathbb{Z}$) Fix an odd integer $k > 1$ and let $(a, k) = 1$. If a is not congruent to a square mod k, then $S(a) = 0$. If $a \equiv b^2 \pmod{k}$ for some integer b, then

$$S(a) = i^{(k-1)^2/4}\sqrt{k} \sum_{\substack{x \in \mathbb{Z}/k\mathbb{Z} \\ x^2=1}} \zeta_k^{2xb}.$$

See Also

§3.1, §3.5	For counting irreducible polynomials with prescribed norm or trace.
§6.2	For more general exponential and character sums.
§6.3	For further applications of character sums.
§7.3	For solutions to diagonal equations over finite fields.
§10.1	For the discrete Fourier transform and Gauss sums.
§12.2, §12.4, §12.5, §12.7, §12.8, §12.9	For curves and varieties, counting rational points, zeta and L-functions.
§14.6	For applications to difference sets.
§15.1	For cyclic codes.

References Cited: [49, 66, 111, 112, 113, 114, 115, 144, 240, 374, 382, 503, 548, 580, 592, 620, 622, 655, 659, 714, 715, 813, 928, 999, 1000, 1001, 1002, 1003, 1004, 1005, 1006, 1007, 1008, 1009, 1010, 1011, 1012, 1013, 1014, 1015, 1078, 1139, 1146, 1299, 1343, 1344, 1363, 1364, 1365, 1366, 1453, 1454, 1455, 1538, 1562, 1580, 1581, 1700, 1701, 1711, 1713, 1725, 1732, 1733, 1783, 1784, 1785, 1790, 1809, 1908, 1946, 1947, 1960, 2042, 2086, 2092, 2119, 2122, 2124, 2128, 2167, 2225, 2245, 2288, 2463, 2524, 2610, 2615, 2645, 2649, 2650, 2652, 2791, 2815, 2835, 2854, 2915, 2948, 3026, 3027, 3034]

6.2 More general exponential and character sums

Antonio Rojas-León, University of Sevilla

6.2.1 One variable character sums

6.2.1 Theorem Let $f \in \mathbb{F}_q[x]$ be a polynomial of degree $d > 0$ and $\psi : \mathbb{F}_q \to \mathbb{C}^*$ a non-trivial additive character. If d is prime to p, then

$$\left|\sum_{x \in \mathbb{F}_q} \psi(f(x))\right| \leq (d-1)\sqrt{q}.$$

If d is divisible by p and $\psi(t) = \psi(t^p)$ for every $t \in \mathbb{F}_q$ then

$$\left|\sum_{x \in \mathbb{F}_q} \psi(f(x))\right| \leq (d'-1)\sqrt{q}$$

if $d' \neq 0$, where d' is the lowest degree of a polynomial in $\mathbb{F}_q[x]$ Artin-Schreier equivalent to f (i.e., of the form $f + g^p - g$ with $g \in \mathbb{F}_q[x]$).

6.2.2 Theorem Let $f \in \mathbb{F}_q[x]$ be a polynomial of degree $d > 0$ and $\chi : \mathbb{F}_q^* \to \mathbb{C}^*$ a non-trivial multiplicative character of order m (extended by zero to $\mathbb{F}_q$). Then, if f is not an m-th power in $\overline{\mathbb{F}}_q[x]$ (where $\overline{\mathbb{F}}_q$ is the algebraic closure of $\mathbb{F}_q$),

$$\left| \sum_{x \in \mathbb{F}_q} \chi(f(x)) \right| \leq (d-1)\sqrt{q}.$$

6.2.3 Theorem Let $f, g \in \mathbb{F}_q[x]$ be polynomials of degrees $d > 0$ and $e > 0$ respectively, $\psi : \mathbb{F}_q \to \mathbb{C}^*$ a non-trivial additive character and $\chi : \mathbb{F}_q^* \to \mathbb{C}^*$ a non-trivial multiplicative character of order m (extended by zero to $\mathbb{F}_q$), such that either f is not of the form $\bar{f}^p - \bar{f}$ with $\bar{f} \in \overline{\mathbb{F}}_q[x]$ or g is not an m-th power in $\overline{\mathbb{F}}_q[x]$. Then

$$\left| \sum_{x \in \mathbb{F}_q} \psi(f(x))\chi(g(x)) \right| \leq (d+e-1)\sqrt{q}.$$

6.2.4 Remark The previous three results are a consequence of Weil's conjectures for curves, as pointed out by Hasse [1443] and Weil [2961], so they follow from Weil's proof of the conjectures [2962]. They are also particular cases of the more general higher dimensional results stated in this section.

6.2.2 Additive character sums

6.2.5 Definition Let $V \subseteq \mathbb{A}^N_{\mathbb{F}_q}$ be an affine variety, $f : V \to \mathbb{A}^1_{\mathbb{F}_q}$ a regular map and $\psi : \mathbb{F}_q \to \mathbb{C}^*$ a non-trivial additive character. We denote by $S(V, f, \psi)$ the additive character sum

$$S(V, f, \psi) = \sum_{x \in V(\mathbb{F}_q)} \psi(f(x)).$$

6.2.6 Definition The L-function associated to V, f and ψ is the power series

$$L(V, f, \psi; T) = \exp\left(\sum_{m=1}^{\infty} S_m(V, f, \psi) \frac{T^m}{m} \right) \in 1 + T\mathbb{C}[[T]],$$

where

$$S_m(V, f, \psi) = S(V \times_{\mathbb{F}_q} \mathbb{F}_{q^m}, f, \psi \circ \mathrm{Tr}_{\mathbb{F}_{q^m}/\mathbb{F}_q}).$$

6.2.7 Remark Character sums can be used to count the number of rational points on a variety. In particular, zeta functions of affine varieties are special cases of L-functions of additive character sums: it is easy to check that, if V is the affine variety defined by the vanishing of $f_1, \ldots, f_r \in \mathbb{F}_q[x_1, \ldots, x_n]$, then for any non-trivial character ψ, and variables $y_1, \ldots, y_r$,

$$S_m(\mathbb{A}^{n+r}, y_1 f_1 + \cdots + y_r f_r, \psi) = q^{mr} \cdot \#V(\mathbb{F}_{q^m}).$$

6.2.8 Definition A number $z \in \mathbb{C}$ is a *Weil integer* if it is an algebraic integer and all its conjugates over $\mathbb{Q}$ have the same absolute value. If $A \in \mathbb{R}$, $A > 0$ is fixed, z has *weight* $w \in \mathbb{R}$ (relative to A) if all its conjugates have absolute value $A^{w/2}$.

6.2.9 Theorem For every affine variety $V \subseteq \mathbb{A}^N_{\mathbb{F}_q}$ of dimension n, every regular map $f : V \to \mathbb{A}^1_{\mathbb{F}_q}$ and every non-trivial additive character $\psi : \mathbb{F}_q \to \mathbb{C}^*$, the L-function $L(V, f, \psi; T)$ is rational, and all its reciprocal roots and poles are Weil integers of weight $\leq 2n$ relative to q.

6.2.10 Remark This result is just a particular case of the more general theory of L-functions of ℓ-adic sheaves on varieties over finite fields; see Section 12.7 for the general statements and some references. For additive exponential sums, rationality was first proven by Bombieri [339] following Dwork's ideas [940].

6.2.11 Remark The L-function encodes the sequence of exponential sums $S_m(V, f, \psi)$ for $m \geq 1$. More precisely, if

$$L(V, f, \psi; T) = \frac{\prod_i (1 - \alpha_i T)}{\prod_j (1 - \beta_j T)}$$

is the decomposition into linear factors, then

$$S_m(V, f, \psi) = \sum_j \beta_j^m - \sum_i \alpha_i^m.$$

In this way, estimates about character sums can be derived from statements about the number of roots and poles of the corresponding L-function and their absolute values.

6.2.12 Remark The type of statements that one tries to prove about these sums are estimates of the form $|S_m(V, f, \psi)| \leq C q^{m(n+i)/2}$, where i is as small as possible and C is a constant that depends only on certain quantities attached to the polynomials that define the variety V and the regular map f. In some particularly nice cases, one can give geometric conditions that imply the optimal bound ($i = 0$).

6.2.13 Theorem [795, Théorème 8.4] Let $f \in \mathbb{F}_q[x_1, \ldots, x_n]$ be a polynomial of degree $d > 0$. Suppose that

1. d is prime to p.
2. The projective hypersurface defined by the highest degree homogeneous form f_d of f is smooth, that is, the polynomials $f_d, \frac{\partial f_d}{\partial x_1}, \ldots, \frac{\partial f_d}{\partial x_n}$ do not have any common zero in $\mathbb{P}^{n-1}(\overline{\mathbb{F}}_q)$.

Then for every non-trivial ψ, $L(\mathbb{A}^n_{\mathbb{F}_q}, f, \psi; T)^{(-1)^n}$ is a polynomial of degree $(d-1)^n$, all whose reciprocal roots have weight n relative to q. In particular, for every $m \geq 1$ the following estimate holds:

$$|S_m(\mathbb{A}^n, f, \psi)| \leq (d-1)^n q^{mn/2}.$$

6.2.14 Theorem [28, Theorems 1.4, 1.11][2469, Corollary 3] Let $f \in \mathbb{F}_q[x_1, \ldots, x_n]$ be a polynomial of degree d. Suppose that

1. d is divisible by p.
2. The projective hypersurface defined by the highest degree homogeneous form f_d of f is smooth.
3. The projective hypersurface defined by the homogeneous form f_{d-1} of degree $d-1$ of f does not contain any of the common roots of the partial derivatives of f_d.

Then for every non-trivial ψ, $L(\mathbb{A}^n_{\mathbb{F}_q}, f, \psi; T)^{(-1)^n}$ is a polynomial of degree $((d-1)^{n+1} - (-1)^{n+1})/d$, all whose reciprocal roots have weight n relative to q. In particular, for every $m \geq 1$ the following estimate holds:

$$|S_m(\mathbb{A}^n, f, \psi)| \leq \frac{(d-1)^{n+1} - (-1)^{n+1}}{d} q^{mn/2}.$$

6.2.15 Theorem [341, Theorem 1] Let $f \in \mathbb{F}_q[x_1, \ldots, x_n]$ be a polynomial of degree d. Then for every non-trivial ψ, the total degree of the L-function $L(\mathbb{A}^n_{\mathbb{F}_q}, f, \psi; T)$ does not exceed $(4d+5)^n$.

6.2.16 Theorem [341, Theorem 2] Let $V \subseteq \mathbb{A}^N_{\mathbb{F}_q}$ be an affine variety of dimension n and degree e and $f \in \mathbb{F}_q[x_1, \ldots, x_N]$ a polynomial of degree d. Then for every non-trivial ψ, the total degree of the L-function $L(V, f, \psi; T)$ does not exceed $(4\max(e+1, d)+5)^{2N+1}$.

6.2.17 Theorem [1532, Theorem 5] Let $V \subseteq \mathbb{A}^3_{\mathbb{F}_q}$ be the surface defined by $g = 0$, where $g \in \mathbb{F}_q[x_1, x_2, x_3]$, and let $f \in \mathbb{F}_q[x_1, x_2, x_3]$. Suppose that

1. All geometric fibres of $f : V \to \mathbb{A}^1_{\mathbb{F}_q}$ have dimension ≤ 1.
2. The generic fibre of $f : V \to \mathbb{A}^1_{\mathbb{F}_q}$ is a geometrically irreducible curve.

Then for every non-trivial $\psi : \mathbb{F}_q \to \mathbb{C}^*$ and every $m \geq 1$ the following estimate holds:

$$|S_m(V, f, \psi)| \leq C(f, g) q^m,$$

where $C(f, g)$ depends only on the degrees of f and g.

6.2.18 Theorem [342, Theorem 1] Let $V \subseteq \mathbb{A}^N_{\mathbb{F}_q}$ be a geometrically irreducible variety of dimension $n \geq 3$ and $f : V \to \mathbb{A}^1_{\mathbb{F}_q}$ a regular map. Suppose that

1. Every geometric fibre V_t of f has a unique irreducible component V_t^{d-1} of dimension $d-1$, and possibly other irreducible components of lower dimensions.
2. For all but finitely many $t \in \overline{\mathbb{F}}_q$, $V_t = V_t^{d-1}$ and the Albanese varieties of V and V_t [1844, Chapter II, Section 3] have the same dimension.

Then, if $\mathbb{F}_q$ has sufficiently large characteristic, for every non-trivial ψ all reciprocal roots and poles of $L(V, f, \psi; T)$ have weight $\leq 2n-3$ relative to q. Morover, its degree is bounded by a constant C depending only on the number and the degrees of the polynomials that define V and f. In particular, for every $m \geq 1$ the following estimate holds:

$$|S_m(V, f, \psi)| \leq C(q^m)^{n-\frac{3}{2}}.$$

6.2.19 Theorem [1700, Théorème 5.1.1] Let $X \subseteq \mathbb{P}^N_{\mathbb{F}_q}$ be a geometrically connected smooth projective variety of dimension n, $V = X \cap \mathbb{A}^N_{\mathbb{F}_q}$ its affine part, $f \in \mathbb{F}_q[x_1, \ldots, x_N]$ a polynomial of degree d and $F \in \mathbb{F}_q[x_0, x_1, \ldots, x_n]$ its homogenized with respect to x_0. Suppose that

1. d is prime to p.
2. The hyperplane at infinity $L = \mathbb{P}^N_{\mathbb{F}_q} \backslash \mathbb{A}^N_{\mathbb{F}_q}$ intersects X transversally (i.e., the scheme-theoretic intersection $X \cap L$ is smooth of dimension $n-1$).
3. The hypersurface H defined by $F = 0$ intersects $X \cap L$ transversally (i.e., the scheme-theoretic intersection $X \cap H \cap L$ is smooth of dimension $n-2$).

Then for every non-trivial ψ, $L(V, f, \psi; T)^{(-1)^n}$ is a polynomial of degree

$$C(X, d) = \left| \int_X \frac{c(X)}{(1+L)(1+dL)} \right|$$

where $c(X)$ is the total Chern class of X [799, Exposé XVII, 5.2] and L the class of a hyperplane section, all whose reciprocal roots have weight n relative to q. In particular, for every $m \geq 1$ the following estimate holds:

$$|S_m(V, f, \psi)| \leq C(X, d) q^{mn/2}.$$

6.2.20 Theorem [1704, Theorem 4] Let $X \subseteq \mathbb{P}^N_{\mathbb{F}_q}$ be a geometrically irreducible integral projective variety of dimension n, $V = X \cap \mathbb{A}^N_{\mathbb{F}_q}$ its affine part, $f \in \mathbb{F}_q[x_1, \ldots, x_N]$ a polynomial of degree d and $F \in \mathbb{F}_q[x_0, x_1, \ldots, x_n]$ its homogenized with respect to x_0. Suppose that

1. d is prime to p.
2. The scheme-theoretic intersection $X \cap H \cap L$ has dimension $n - 2$, where $L = \mathbb{P}^N_{\mathbb{F}_q} \backslash \mathbb{A}^N_{\mathbb{F}_q}$ is the hyperplane at infinity and H is the hypersurface defined by $F = 0$. Let $\delta \geq -1$ be the dimension of its singular locus.
3. The dimension of the singular locus of the scheme-theoretic intersection $X \cap L$ is smaller than or equal to δ.

Then for every non-trivial ψ, all reciprocal roots and poles of $L(V, f, \psi; T)$ have weight smaller than or equal to $n+\delta+1$ relative to q, and its total degree is bounded by a constant $C(X, d)$ depending only on number and the degrees of the forms defining X and on d. In particular, for every $m \geq 1$ the following estimate holds:

$$|S_m(V, f, \psi)| \leq C(X, d) q^{m(n+\delta+1)/2}.$$

6.2.21 Definition Let $f = \sum_{i \in \mathbb{Z}^n} a_i x^i \in \mathbb{F}_q[x_1^{\pm 1}, \ldots, x_n^{\pm 1}]$ be a Laurent polynomial. The *Newton polyhedron* $\Delta_\infty(f)$ of f at infinity is the convex hull in $\mathbb{R}^n$ of the set $\{\mathbf{0}\} \cup \{i \in \mathbb{Z}^n | a_i \neq 0\}$. A polynomial f is *non-degenerate* with respect to $\Delta_\infty(f)$ if, for every face δ of $\Delta_\infty(f)$ that does not contain the origin, the equations

$$\frac{\partial f_\delta}{\partial x_1} = \cdots = \frac{\partial f_\delta}{\partial x_n} = 0$$

do not have any common solution in $(\overline{\mathbb{F}}_q^*)^n$, where $f_\delta = \sum_{i \in \mathbb{Z}^n \cap \delta} a_i x^i$ is the restriction of f to the face δ.

6.2.22 Theorem [23, Theorem 4.2][812, Theorem 1.3] Let $\mathbb{T}^n$ be the n-dimensional split torus over $\mathbb{F}_q$, $f \in \mathbb{F}_q[x_1^{\pm 1}, \ldots, x_n^{\pm 1}]$ a Laurent polynomial and $\Delta_\infty(f)$ its Newton polyhedron at infinity. Suppose that

1. $\dim \Delta_\infty(f) = n$.
2. f is non-degenerate with respect to $\Delta_\infty(f)$.

Then for every non-trivial ψ, $L(\mathbb{T}^n(\mathbb{F}_q), f, \psi; T)^{(-1)^n}$ is a polynomial of degree $n!\mathrm{Vol}(\Delta_\infty(f))$, all whose reciprocal roots have weight smaller than or equal to n relative to q. If, in addition, the origin is an interior point of $\Delta_\infty(f)$, then all its reciprocal roots have weight n relative to q. In particular, for every $m \geq 1$ the following estimate holds:

$$|S_m(\mathbb{T}^n, f, \psi)| \leq n!\mathrm{Vol}(\Delta_\infty(f)) q^{mn/2}.$$

6.2.3 Multiplicative character sums

6.2.23 Definition Let $V \subseteq \mathbb{A}^N_{\mathbb{F}_q}$ be an affine variety, $f : V \to \mathbb{A}^1_{\mathbb{F}_q}$ a regular map and $\chi : \mathbb{F}_q^* \to \mathbb{C}^*$ a non-trivial multiplicative character, extended by zero to $\mathbb{F}_q$. We denote by $S(V, f, \chi)$ the multiplicative character sum

$$S(V, f, \chi) = \sum_{x \in V(\mathbb{F}_q)} \chi(f(x)).$$

6.2.24 Definition The L-function associated to V, f and χ is the power series

$$L(V, f, \chi; T) = \exp\left(\sum_{m=1}^{\infty} S_m(V, f, \chi)\frac{T^m}{m}\right) \in 1 + T\mathbb{C}[[T]],$$

where

$$S_m(V, f, \chi) = S(V \otimes \mathbb{F}_{q^m}, f, \chi \circ \mathrm{Norm}_{\mathbb{F}_{q^m}/\mathbb{F}_q}).$$

6.2.25 Theorem For every affine variety $V \subseteq \mathbb{A}^N_{\mathbb{F}_q}$ of dimension n, every regular map $f : V \to \mathbb{A}^1_{\mathbb{F}_q}$ and every non-trivial multiplicative character $\chi : \mathbb{F}_q^* \to \mathbb{C}^*$, the L-function $L(V, f, \chi; T)$ is rational, and all its reciprocal roots and poles are Weil integers of weight at most $2n$ relative to q.

6.2.26 Remark As in the additive case, this result is a particular case of the more general theory of L-functions of ℓ-adic sheaves on varieties over finite fields, see Section 12.7 for the general statements and some references.

6.2.27 Remark More generally, one can construct "mixed" character sums of the form

$$\sum_{x \in V(\mathbb{F}_q)} \psi(f(x))\chi(g(x))$$

where $\psi : \mathbb{F}_q \to \mathbb{C}^*$ is a non-trivial additive character, $\chi : \mathbb{F}_q^* \to \mathbb{C}^*$ a non-trivial multiplicative character and $f, g : V \to \mathbb{A}^1_{\mathbb{F}_q}$ are regular functions. The corresponding L-function is again rational, and all its reciprocal roots and poles are Weil integers of weight at most $2n$ relative to q.

6.2.28 Theorem [1707, Theorems 2.1, 2.2][2468, Theorem 4.4] Let $f \in \mathbb{F}_q[x_1, \ldots, x_n]$ be a polynomial of degree $d > 0$. Suppose that

1. The affine scheme defined by f is smooth.
2. The projective scheme defined by the highest degree homogeneous form f_d of f is smooth.

Then for every non-trivial χ, $L(\mathbb{A}^n(\mathbb{F}_q), f, \chi; T)^{(-1)^n}$ is a polynomial of degree $(d-1)^n$, all whose reciprocal roots have weight smaller than or equal to n relative to q. In particular, for every $m \geq 1$ the following estimate holds:

$$|S_m(\mathbb{A}^n, f, \chi)| \leq (d-1)^n q^{mn/2}.$$

6.2.29 Theorem [1707, Theorem 3.1, 3.2][2468, Corollary 4.3] Let $X \subseteq \mathbb{A}^N_{\mathbb{F}_q}$ be a geometrically connected smooth projective variety of dimension n, $V = X \cap \mathbb{A}^N_{\mathbb{F}_q}$ its affine part, $f \in \mathbb{F}_q[x_1, \ldots, x_N]$ a polynomial of degree d and $F \in \mathbb{F}_q[x_0, x_1, \ldots, x_n]$ its homogenized with respect to x_0. Suppose that

1. The hypersurface H defined by $F = 0$ intersects X transversally (i.e., the scheme-theoretic intersection $X \cap H$ is smooth of dimension $n-1$).
2. The hyperplane at infinity $L = \mathbb{P}^N_{\mathbb{F}_q} \backslash \mathbb{A}^N_{\mathbb{F}_q}$ intersects $X \cap H$ transversally (i.e., the scheme-theoretic intersection $X \cap H \cap L$ is smooth of dimension $n-2$).

Then for every non-trivial χ, $L(V, f, \chi; T)^{(-1)^n}$ is a polynomial of degree

$$C(X, d) = \left| \int_X \frac{c(X)}{(1+L)(1+dL)} \right|$$

where $c(X)$ is the total Chern class of X [799, Exposé XVII, 5.2] and L the class of a hyperplane section, all whose reciprocal roots have weight smaller than or equal to n relative to q. In particular, for every $m \geq 1$ the following estimate holds:

$$|S_m(V, f, \chi)| \leq C(X, d) q^{mn/2}.$$

6.2.30 Theorem [2468, Theorem 1.1] Let $X \subseteq \mathbb{P}^N_{\mathbb{F}_q}$ be a geometrically irreducible Cohen-Macaulay projective variety of dimension n, $V = X \cap \mathbb{A}^N_{\mathbb{F}_q}$ its affine part, $f \in \mathbb{F}_q[x_1, \ldots, x_N]$ a polynomial of degree d and $F \in \mathbb{F}_q[x_0, x_1, \ldots, x_n]$ its homogenized with respect to x_0. Let $\chi : (\mathbb{F}_q)^\star \to \mathbb{C}^*$ be a non-trivial multiplicative character. Suppose that

1. The scheme-theoretic intersection $X \cap H \cap L$ has dimension $n-2$ and singular locus of dimension $\delta \geq -1$, where $L = \mathbb{P}^N_{\mathbb{F}_q} \backslash \mathbb{A}^N_{\mathbb{F}_q}$ is the hyperplane at infinity and H is the hypersurface defined by $F = 0$.
2. Either the singular locus of the scheme-theoretic intersection $X \cap H$ has dimension at most δ, or else d is prime to p, χ^d is non-trivial and the singular locus of the scheme-theoretic intersection $X \cap L$ has dimension at most δ.

Then all reciprocal roots and poles of $L(V, f, \chi; T)$ have weights at most $n+\delta+1$ relative to q, and its total degree is bounded by a constant $C(X, d)$ depending only on the number and degrees of the forms that define X and on d. In particular, for every $m \geq 1$ the following estimate holds:

$$|S_m(V, f, \chi)| \leq C(X, d) q^{m(n+\delta+1)/2}.$$

6.2.4 Generic estimates

6.2.31 Theorem [1712, Theorem 5.5.1] Let $V \subseteq \mathbb{A}^N_{\mathbb{F}_q}$ be a geometrically connected smooth variety of dimension n, $f_1, \ldots, f_r \in \mathbb{F}_q[x_1, \ldots, x_N]$ such that the map $(f_1, \ldots, f_r) : V \to \mathbb{A}^r_{\mathbb{F}_q}$ is finite and $\psi : \mathbb{F}_q \to \mathbb{C}^*$ a non-trivial additive character. Then there exists a non-zero polynomial $g \in \mathbb{F}_q[x_1, \ldots, x_r]$ and a constant $C \geq 0$ such that for every $m \geq 1$ and every $(a_1, \ldots, a_r) \in \mathbb{F}^r_{q^m}$ with $g(a_1, \ldots, a_r) \neq 0$ the following estimate holds:

$$|S_m(V, a_1 f_1 + \cdots + a_r f_r, \psi)| \leq C q^{mn/2}.$$

6.2.32 Theorem [1703, Corollary 4.1.3] Let V be an affine variety of dimension n over $\mathbb{F}_q$, $f : V \to \mathbb{A}^1_{\mathbb{F}_q}$ a regular map, $\pi = (\pi_1, \ldots, \pi_N) : V \to \mathbb{A}^N_{\mathbb{F}_q}$ a quasi-finite map and $\psi : \mathbb{F}_q \to \mathbb{C}^*$ a non-trivial additive character. Then there exists a non-zero polynomial $g \in \mathbb{F}_q[x_1, \ldots, x_{N+1}]$ and a constant $C \geq 0$ such that for every $m \geq 1$ and every $(a_1, \ldots, a_N, b) \in \mathbb{F}^{N+1}_{q^m}$ with $g(a_1, \ldots, a_N, b) \neq 0$ the following estimate holds:

$$|S_m(V, f + a_1\pi_1 + \cdots + a_N\pi_N + b, \psi)| \leq C q^{mn/2}.$$

6.2.33 Theorem [1703, Corollary 4.1.3] Let V be an affine variety of dimension n over $\mathbb{F}_q$, $f : V \to \mathbb{A}^1_{\mathbb{F}_q}$ a regular map, $\pi = (\pi_1, \ldots, \pi_N) : V \to \mathbb{A}^N_{\mathbb{F}_q}$ a quasi-finite map and

$\chi : \mathbb{F}_q \to \mathbb{C}^*$ a non-trivial multiplicative character. Then there exists a non-zero polynomial $g \in \mathbb{F}_q[x_1, \dots, x_{N+1}]$ and a constant $C \geq 0$ such that for every $m \geq 1$ and every $(a_1, \dots, a_N, b) \in \mathbb{F}_{q^m}^{N+1}$ with $g(a_1, \dots, a_N, b) \neq 0$ the following estimate holds:

$$|S_m(V, f + a_1\pi_1 + \cdots + a_N\pi_N + b, \chi)| \leq Cq^{mn/2}.$$

6.2.5 More general types of character sums

6.2.34 Theorem [560, Theorem 13][2384] Let X be a smooth projective curve of genus g over $\mathbb{F}_q$, $f, h \in \mathbb{F}_q(X)$ rational functions and $\psi : \mathbb{F}_q \to \mathbb{C}^*$ (respectively $\chi : \mathbb{F}_q^* \to \mathbb{C}^*$) a non-trivial additive (resp. multiplicative) character. Suppose that f is not of the form $\bar{f}^p - \bar{f}$ with $\bar{f} \in \overline{\mathbb{F}}_q(X)$, and g is not an ord(χ)-th power in $\overline{\mathbb{F}}_q(X)$. Then for every $m \geq 1$,

$$\left| \sum_{x \in V(\mathbb{F}_{q^m})} \psi(\mathrm{Tr}_{\mathbb{F}_{q^m}/\mathbb{F}_q} f(x))\chi(\mathrm{Norm}_{\mathbb{F}_{q^m}/\mathbb{F}_q} h(x)) \right| \leq (2g - 2 + s + l + d)q^{m/2},$$

where $V \subseteq X$ is the open set where f and h are defined, l is the number of poles of f, s is the number of zeroes and poles of h and d is the degree of the polar part of the divisor (f).

6.2.35 Theorem [1709, Theorem 1.1] Let $f, g \in \mathbb{F}_q[x_1, \dots, x_n]$ be polynomials of degrees d and e, respectively, prime to p. Suppose that the projective hypersurfaces defined by the highest degree forms of f and g are smooth and intersect transversally. Then for every $m \geq 1$ the following estimate holds:

$$\left| \sum_{x \in \mathbb{F}_{q^m}} \psi(\mathrm{Tr}_{\mathbb{F}_{q^m}/\mathbb{F}_q} f(x))\chi(\mathrm{Norm}_{\mathbb{F}_{q^m}/\mathbb{F}_q} g(x)) \right| \leq C_{n,d,e}q^{mn/2},$$

where

$$C_{n,d,e} = \sum_{a+b=n} (d-1)^a(e-1)^b + \sum_{a+b=n-1} (d-1)^a(e-1)^b.$$

6.2.36 Theorem [1138, Proposition 0.1] Let $\mathbb{T}^n$ be the n-dimensional split torus over $\mathbb{F}_q$, $f \in \mathbb{F}_q[x_1^{\pm 1}, \dots, x_n^{\pm 1}]$ a Laurent polynomial, $\Delta_\infty(f)$ its Newton polyhedron at infinity and $\chi : \mathbb{T}^n(\mathbb{F}_q) \to \mathbb{C}^*$ a character. Suppose that

1. $\dim \Delta_\infty(f) = n$.
2. f is non-degenerate with respect to $\Delta_\infty(f)$.

Then for every non-trivial ψ and every $m \geq 1$ the following estimate holds:

$$\left| \sum_{x \in (\mathbb{F}_{q^m}^*)^n} \psi(\mathrm{Tr}_{\mathbb{F}_{q^m}/\mathbb{F}_q} f(x))\chi(\mathrm{Norm}_{\mathbb{F}_{q^m}/\mathbb{F}_q} x) \right| \leq n!\mathrm{Vol}(\Delta_\infty(f))q^{mn/2}.$$

6.2.37 Theorem [1702, Theorem 1][2893, Corollary 2.2] Let $a \in \mathbb{F}_{q^m}$ such that $\mathbb{F}_{q^m} = \mathbb{F}_q(a)$ and $\chi : \mathbb{F}_{q^m}^* \to \mathbb{C}^*$ a non-trivial multiplicative character. Then the following estimate holds:

$$\left| \sum_{x \in \mathbb{F}_q} \chi(x - a) \right| \leq (m-1)\sqrt{q}.$$

6.2.38 Theorem [2893, Corollary 2.8] Let $f \in \mathbb{F}_q[x]$ be a monic polynomial, and $\chi : (\mathbb{F}_q[x]/(f))^* \to \mathbb{C}^*$ a non-trivial character. Then

$$\left|\sum \chi(g \bmod f)\right| \leq \frac{1}{d}(\deg(f)+1)q^{d/2},$$

where the sum is taken over the set of monic irreducible polynomials of degree d in $\mathbb{F}_q[x]$ which are coprime to f.

6.2.39 Theorem [1140, Theorem 3.7] Let $f \in \mathbb{F}_q[x_1, \ldots, x_n, y_1, \ldots, y_{n'}]$ be a polynomial of degree d, $r \geq 1$ an integer and let g be the polynomial

$$\sum_{j=1}^{r} f(x_{1,j}, \ldots, x_{n,j}, y_1, \ldots, y_{n'})$$

in $\mathbb{F}_q[x_{1,1}, \ldots, x_{n,r}, y_1, \ldots, y_{n'}]$. Suppose that

1. d is prime to p.
2. The degree d homogeneous part of g defines a smooth hypersurface in $\mathbb{P}^{rn+n'-1}$.

Then for every non-trivial additive character ψ and every $m \geq 1$ the following estimate holds:

$$\left|\sum_{x_i \in \mathbb{F}_{q^{mr}}, y_j \in \mathbb{F}_{q^m}} \psi(\mathrm{Tr}_{\mathbb{F}_{q^{mr}}/\mathbb{F}_q} f(x_1, \ldots, x_n, y_1, \ldots, y_{n'}))\right| \leq (d-1)^{nr+n'} q^{m(nr+n')/2}.$$

The constant $(d-1)^{nr+n'}$ can be replaced by $C(p,f)r^{3(d+1)^n-1}$, where $C(p,f)$ depends only on p and f.

See Also

§6.1	For specific results about Gauss, Jacobi, and Kloosterman sums.
§12.7	For the general theory of ℓ-adic sheaves and their L-functions.
[22], [2698]	For a study of exponential sums and their L-functions using Dwork's p-adic methods.
[253]	For applications of p-adic cohomology to the study of exponential sums.
[796], [2589]	For the general theory of exponential sums using ℓ-adic cohomology.
[1700], [1867]	For a study of the total degree of the L-function associated to an exponential sum and its change with the characteristic of the base field.
[1869]	For a comprehensive survey of the main estimates for exponential sums obtained using ℓ-adic cohomology.

References Cited: [22, 23, 28, 253, 339, 341, 342, 560, 795, 796, 799, 812, 940, 1138, 1140, 1532, 1700, 1702, 1703, 1704, 1707, 1709, 1712, 1844, 1867, 1869, 2384, 2468, 2469, 2589, 2698, 2893, 2961, 2962]

6.3 Some applications of character sums

Alina Ostafe, Macquarie University
Arne Winterhof, Austrian Academy of Sciences

The main goal of this chapter is to show that character sums are very useful and friendly tools for a variety of problems in many areas such as coding theory, cryptography, and algorithms. There are so many applications of character sums that any survey will be incomplete. Here we chose a combination of some classical applications and newer, less known applications using different types of character sums.

6.3.1 Applications of a simple character sum identity

6.3.1 Proposition [1631, Lemma 7.3.7] Let χ denote a nontrivial multiplicative character of $\mathbb{F}_q$. Then we have

$$\sum_{x\in\mathbb{F}_q} \chi(x+a)\overline{\chi(x+b)} = -1, \quad a,b\in\mathbb{F}_q, a\neq b.$$

6.3.1.1 Hadamard matrices

6.3.2 Definition A *Hadamard matrix of order* n is an $n\times n$ matrix H with entries from $\{-1,+1\}$ satisfying $HH^T = nI$.

6.3.3 Construction (Paley) [2344] Let q be the power of an odd prime, η the quadratic character of $\mathbb{F}_q$ and $\mathbb{F}_q = \{\xi_1,\dots,\xi_q\}$ any fixed ordering of $\mathbb{F}_q$. For $q\equiv 3 \pmod 4$ there exists a Hadamard matrix $H=(h_{ij})$ of order $n=q+1$ defined by

$$h_{i,n} = h_{n,i} = 1, \quad i=1,\dots,n, \quad h_{j,j} = -1, \quad j=1,\dots,n-1,$$

$$h_{i,j} = \eta(\xi_j - \xi_i), \quad i,j=1,\dots,n-1,\ i\neq j.$$

6.3.4 Remark

1. The inner product of two different rows of H is zero by Proposition 6.3.1.
2. Paley also presented a similar, but slightly more complicated construction for Hadamard matrices of order $n=2(q+1)$ if $q\equiv 1 \pmod 4$.
3. The Hadamard conjecture proposes that a Hadamard matrix of order $n=4k$ exists for every positive integer k. The smallest order n for which no Hadamard matrix of order $n=4k$ has been constructed is $n=668$. The last progress known by the authors was the construction of a Hadamard matrix of order $n=428$ by [1730].
4. For a monograph on Hadamard matrices see [1536]; see also Subsection 14.6.4.
5. Hadamard matrices can be used to construct good error correcting codes [1911].
6. For relations between Hadamard matrices and designs see [260].

6.3.1.2 Cyclotomic complete mappings and check digit systems

6.3.5 **Definition** Let n be a positive divisor of $q-1$ and γ a primitive element of $\mathbb{F}_q$. Then the sets $C_i = \{\gamma^{jn+i} : j = 0, 1, \dots, (q-1)/n - 1\}$, $i = 0, 1, \dots, n-1$, are *cyclotomic cosets of order* n. For $a_0, a_1, \dots, a_{n-1} \in \mathbb{F}_q^*$ we define a *cyclotomic mapping* $f_{a_0,a_1,\dots,a_{n-1}}$ *(of index* n*)* by $f_{a_0,a_1,\dots,a_{n-1}}(0) = 0$ and

$$f_{a_0,a_1,\dots,a_{n-1}}(\xi) = a_i\xi \quad \text{if } \xi \in C_i, \quad i = 0, 1, \dots, n-1.$$

6.3.6 **Proposition** [998, Theorem 3.7] The mapping $f_{a_0,a_1,\dots,a_{n-1}}$ is a permutation of $\mathbb{F}_q$ if and only if $a_iC_i \neq a_jC_j$ for all $0 \le i < j \le n-1$.

6.3.7 **Corollary** For $n \ge 2$ let j be an integer with $0 \le j < n$, χ a multiplicative character of $\mathbb{F}_q$ of order n, and $a, b \in \mathbb{F}_q$ with $a \neq b$. If $a_j = a$ and $a_i = b$ for $i \neq j$, then $g_{a,b} = f_{a_0,a_1,\dots,a_{n-1}}$ is a permutation if and only if $\chi(a) = \chi(b)$.

6.3.8 **Definition** A permutation f of $\mathbb{F}_q$ is a *complete mapping* of $\mathbb{F}_q$ if $f(x) + x$ is also a permutation of $\mathbb{F}_q$.

6.3.9 **Remark** Complete mappings are pertinent to the problem of constructing orthogonal Latin squares [1939, Section 9.4].

6.3.10 **Remark** Substituting $b = ac$ we see that $g_{a,b}$ is a complete mapping if and only if $\chi(c) = 1$ and $\chi(a+1) = \chi(ac+1)$ and the number N of complete mappings $g_{a,ac}$ with $c \neq 1$ is

$$N = \sum_{c\in\mathbb{F}_q^*,\ \chi(c)=1,\ c\neq 1} \frac{1}{n}\sum_{i=0}^{n-1} \sum_{a\in\mathbb{F}_q^*\setminus\{1,c^{-1}\}} \chi^i(a+1)\overline{\chi^i(a-c^{-1})}.$$

Proposition 6.3.1 implies the following theorem.

6.3.11 **Theorem** [2278, Theorem 3] Let $n \ge 2$ be a divisor of $q-1$ and $j \in \{0, 1, \dots, n-1\}$. Then the number N of ordered pairs $(a, b) \in \mathbb{F}_q^* \times \mathbb{F}_q^*$ with $a \neq b$ such that the cyclotomic mapping $g_{a,b} = f_{a_0,a_1,\dots,a_{n-1}}$ with $a_j = a$ and $a_i = b$ for $i \neq j$ is a complete mapping of $\mathbb{F}_q$ equals $N = (q-n-1)(q-2n-1)/n^2$.

6.3.12 **Definition** A *check digit system* over $\mathbb{F}_q$ consists of s permutations $p_1, \dots, p_s$ of $\mathbb{F}_q$ and a symbol $c \in \mathbb{F}_q$ such that each word $a_1 \dots a_{s-1} \in \mathbb{F}_q^{s-1}$ is extended by a *check digit* a_s such that $p_1(a_1) + \cdots + p_s(a_s) = c$.

6.3.13 **Remark** All single errors are detected since all p_i are permutations. Another frequent family of errors are *adjacent transpositions* $\dots ab \dots \to \dots ba \dots$ which are all detected if $p_{i+1}(x)p_i^{-1}(x) - x$ are also permutations for $i = 1, \dots, s-1$. A permutation f such that $f(x) - x$ is also a permutation is an *orthomorphism*. Since f is an orthomorphism whenever $-f$ is a complete mapping, the number of orthomorphisms and complete mappings of the form $g_{a,b}$ is the same and the probability that a random choice of the parameters (a, b) gives an orthomorphism is asymptotically n^{-2} by Theorem 6.3.11.

6.3.14 **Example** [International Standard Book Number (ISBN-10)] An ISBN-10 consists of 10 digits $x_1 - x_2x_3x_4x_5x_6 - x_7x_8x_9 - x_{10}$. The first digit x_1 characterizes the language group, $x_2x_3x_4x_5x_6$ is the actual book number, $x_7x_8x_9$ is the number of the publisher, and x_{10} is a check digit. A correct ISBN satisfies $x_1+2x_2+3x_3+4x_4+5x_5+6x_6+7x_7+8x_8+9x_9+10x_{10}=0\in\mathbb{F}_{11}$, i.e., $p_i(x)=ix$, $i=1,\dots,10$, and $p_{i+1}p_i^{-1}(x)=(i^{-1}+1)x$, $i=1,\dots,9$, which are all orthomorphisms.

6.3.1.3 Periodic autocorrelation of cyclotomic generators

6.3.15 Definition Put $\varepsilon_n=\exp(2\pi i/n)$. Let (s_k) be a T-periodic sequence over $\mathbb{Z}_n$. The *(periodic) autocorrelation* of (s_k) is the complex-valued function defined by

$$A(t)=\frac{1}{T}\sum_{k=0}^{T-1}\varepsilon_n^{s_{k+t}-s_k},\quad 1\le t<T.$$

6.3.16 Remark Sequences with low autocorrelation have several applications in wireless communication, cryptography, and radar, see the monograph [1303].

6.3.17 Definition The p-periodic sequence (s_k) over $\mathbb{Z}_n$ defined by $s_0=0$ and $s_k=j$ if $k\in C_j$ for $0\le j<n$, $1\le k<p$, where C_j denotes the jth cyclotomic coset of order n defined by Definition 6.3.5, is a *cyclotomic generator of order n.*

6.3.18 Remark Proposition 6.3.1 implies the exact values of the autocorrelation function of the cyclotomic generator of order n; see [2068] for the proof of a generalization to arbitrary finite fields.

6.3.19 Theorem The autocorrelation function $f(t)$ of the cyclotomic generator of order n is given by $A(t)=(-1+\varepsilon_n^j+\varepsilon_n^{-j-k})/p$ if $t\in C_j$ and $-1\in C_k$.

6.3.2 Applications of Gauss and Jacobi sums

6.3.20 Definition Let χ be a multiplicative and ψ be an additive character of $\mathbb{F}_q$ and $a\in\mathbb{F}_q$ of order T. The sums

$$G(\chi,\psi)=\sum_{c\in\mathbb{F}_q^*}\chi(c)\psi(c)\quad\text{and}\quad G_a(\psi)=\sum_{n=0}^{T-1}\psi(a^n)$$

are *Gauss sums of first kind* and *second kind*, respectively. Let $\chi_1,\dots,\chi_k$ be $k\ge 2$ multiplicative characters of $\mathbb{F}_q$. The sum

$$J(\chi_1,\dots,\chi_k)=\sum_{c_1+\cdots+c_k=1}\chi_1(c_1)\dots\chi_k(c_k),$$

is a *Jacobi sum*, where the summation is extended over all $(c_1,\dots,c_k)\in\mathbb{F}_q^k$ such that $c_1+\cdots+c_k=1$.

6.3.21 Remark Gauss sums of the first and second kinds are closely related by

$$G_a(\psi)=\frac{T}{q-1}\sum_{j=0}^{(q-1)/T-1}G(\chi^j,\psi),$$

where χ is a multiplicative character of order $(q-1)/T$, and Gauss sums of first kind and Jacobi sums by

$$J(\chi_1, \ldots, \chi_k) = \frac{G(\chi_1, \psi) \cdots G(\chi_k, \psi)}{G(\chi_1 \cdots \chi_k, \psi)}$$

if all involved characters are nontrivial. For background on Gauss and Jacobi sums see Section 6.1, [240] or [1939, Chapter 5]. In particular, we have (if all characters are nontrivial)

$$|G(\chi, \psi)| = q^{1/2}, \quad |G_a(\psi)| \leq q^{1/2}, \quad \text{and } |J(\chi_1, \ldots, \chi_k)| = q^{(k-1)/2}. \tag{6.3.1}$$

We note that the bound on $|G_a(\psi)|$ is only nontrivial if $T > q^{1/2}$. If $q = p$ is a prime, bounds which are nontrivial for $T \geq p^{1/3+\varepsilon}$ and $T \geq p^{\varepsilon}$ are given in [380, 1454] (see also [371] for an improvement), respectively.

6.3.2.1 Reciprocity laws

6.3.22 Remark Gauss and Jacobi sums are involved in the proofs and statements of reciprocity laws. For example, let p and q be two distinct odd primes, let r be the order of p modulo q, ξ be a primitive q-th root of unity in $\mathbb{F}_{p^r}$, and $G = \sum_{x \in \mathbb{F}_q} \left(\frac{x}{q}\right) \xi^x$ be a Gauss sum of first kind over $\mathbb{F}_q$ in $\mathbb{F}_{p^r}$. Then from $G^2 = (-1)^{(q-1)/2} q$ we get

$$G^p = (-1)^{(p-1)(q-1)/4} \left(\frac{q}{p}\right) G.$$

On the other hand, simple calculations give $G^p = \left(\frac{p}{q}\right) G \pmod p$ and we get the following result.

6.3.23 Theorem (Gauss law of quadratic reciprocity) [1939, Theorem 5.17] For any two distinct odd primes p and q we have

$$\left(\frac{p}{q}\right)\left(\frac{q}{p}\right) = (-1)^{(p-1)(q-1)/4}.$$

6.3.24 Definition For an integer $n \geq 2$ and two distinct primes $p, q \equiv 1 \pmod n$ we denote by $\chi_{p,n}(a) = \left(\frac{a}{p}\right)_n$ the character of order n with values in $\mathbb{F}_q$ defined by

$$\left(\frac{\gamma^j}{p}\right)_n = \xi^j, \quad j = 0, \ldots, n-1,$$

where γ is a fixed primitive element of $\mathbb{F}_p$ and ξ a primitive n-th root of unity in $\mathbb{F}_q$.

6.3.25 Remark The sums $K_{p,n} = (-1)^{(p-1)(q-1)/4} \sum_{x \in \mathbb{F}_p} \chi_{p,n}(x) \chi_{p,n}(1-x)$ are up to signs Jacobi sums in $\mathbb{F}_q$ and appear in reciprocity laws of higher order.

6.3.26 Theorem (Cubic and biquadratic reciprocity laws) [463] For two distinct primes $p, q \equiv 1 \pmod 3$ we have

$$\left(\frac{p}{q}\right)_3 \left(\frac{q}{p}\right)_3 \left(\frac{K_{p,3}}{q}\right)_3 = 1.$$

For two distinct primes $p, q \equiv 1 \pmod 4$ we have

$$\left(\frac{p}{q}\right)_4 \left(\frac{p}{q}\right)_4 \left(\frac{K_{p,4}}{q}\right)_2 = 1.$$

6.3.27 Remark In terms of the decompositions of p and q in the rings of Eisenstein and Gaussian integers we have

$$K_{p,3} = \frac{ad - bc}{3} \quad \text{if } p = a^2 - ab + b^2; q = c^2 - cd + d^2; a, c \equiv 2 \pmod 3; b, d \equiv 0 \pmod 3$$

and

$$K_{p,4} = \frac{ad - bc}{2} \quad \text{if } p = a^2 + b^2; q = c^2 + d^2; a, c \equiv 1 \pmod 4; b, d \equiv 0 \pmod 2.$$

6.3.2.2 Distribution of linear congruential pseudorandom numbers

6.3.28 Definition The sequences

$$x_{n+1} = ax_n + b, \quad n \geq 0,$$

where $x_0, a, b \in \mathbb{F}_p$ with $x_0, a \neq 0$, and $a \neq 1$, are *linear congruential pseudorandom number generators.*

6.3.29 Remark If $a \neq 1$, they can also be given explicitly by

$$x_n = a^n x_0 + \frac{a^n - 1}{a - 1} b, \quad n \geq 0. \tag{6.3.2}$$

The sequence (x_n) is T-periodic if $b \neq (1 - a)x_0$, where T is the order of a.

6.3.30 Definition Let Γ be a sequence of N elements $(\gamma_n)_{n=1}^N$ in the unit interval $[0, 1)$. The *discrepancy* $D_N(\Gamma)$ is defined by

$$D_N(\Gamma) = \sup_{B \subseteq [0,1)} \left| \frac{T_\Gamma(B)}{N} - |B| \right|,$$

where the supremum is taken over all subintervals $B = [\alpha, \beta) \subseteq [0, 1)$, and $T_\Gamma(B)$ is the number of elements of Γ inside B.

6.3.31 Remark The discrepancy is a measure for the uniform distribution of Γ and a small discrepancy is a desirable feature for (quasi-)Monte Carlo integration [2248]. The problem of estimating the discrepancy can be reduced to the problem of estimating certain exponential sums.

6.3.32 Proposition (Erdős-Turan inequality) [922, Theorem 1.2.1] Let Γ be a sequence $(\gamma_n)_{n=1}^N$ in $[0, 1)$. We have for any integer $H \geq 1$,

$$D_N(\Gamma) \ll \frac{1}{H} + \frac{1}{N} \sum_{h=1}^{H} \frac{1}{h} |S_N(h)|, \quad \text{where } S_N(h) = \sum_{n=0}^{N-1} \exp(2\pi i h \gamma_n).$$

6.3.33 Remark For the sequence (x_n/p), $n = 0, 1, \ldots, T - 1$, in $[0, 1)$ derived from a linear pseudorandom number generator (x_n) (where we identify $\mathbb{F}_p$ with the integers $\{0, 1, \ldots, p-1\}$), the absolute value of the sums $S_N(h)$ equals the absolute value of Gauss sums of second kind provided that $b \neq (1 - a)x_0$ and $a \neq 1$. A discrepancy bound can be easily obtained by combining Proposition 6.3.32 with the bound $|G_a(\psi)| \leq p^{1/2}$.

6.3.34 Theorem [2230, Theorem 1] For the sequence $\Gamma = (x_n/p : n = 0, \ldots, N - 1)$, where x_n is defined by (6.3.2), $N < T$ and T is the order of a, we have $D_N(\Gamma) \ll N^{-1} p^{1/2} (\log p)^2$.

6.3.2.3 Diagonal equations, Waring's problem in finite fields, and covering radius of certain cyclic codes

6.3.35 Definition A *diagonal equation* over $\mathbb{F}_q$ is an equation of the type

$$c_1 x_1^{k_1} + \cdots + c_s x_s^{k_s} = b \qquad (6.3.3)$$

for any positive integers $k_1, \ldots, k_s$, $c_1, \ldots, c_s \in \mathbb{F}_q^*$, and $b \in \mathbb{F}_q$. We denote by N_b the number of solutions in $\mathbb{F}_q^s$ of (6.3.3); see Section 7.3.

6.3.36 Theorem [1939, Theorem 6.34] The number N_b of solutions of (6.3.3) for $b \in \mathbb{F}_q^*$ is

$$N_b = q^{s-1} + \sum_{j_1=1}^{d_1-1} \cdots \sum_{j_s=1}^{d_s-1} \chi_1^{j_1}(bc_1^{-1}) \ldots \chi_s^{j_s}(bc_s^{-1}) J(\chi_1^{j_1}, \ldots, \chi_s^{j_s}),$$

where χ_i denotes a multiplicative character of order $d_i = \gcd(k_i, q-1)$.

6.3.37 Remark Put $T_j = (q-1)/\gcd(k_j, q-1)$ and let $a_j \in \mathbb{F}_q^*$ be an element of order T_j for $j = 1, \ldots, s$. Since $\{x^{k_j} : x \in \mathbb{F}_q\} = \{a_j^n : 0 \le n < T_j\} \cup \{0\} =: G_j$, we can also express N_b in terms of Gauss sums of the second kind,

$$N_b = \frac{1}{q} \sum_{(y_1,\ldots,x_s) \in G_1 \times \ldots G_s} \sum_{\psi} \psi(c_1 y_1 + \cdots + c_s y_s - b) = \frac{1}{q} \sum_{\psi} \psi(-b) \prod_{j=1}^{s} \left(1 + G_{a_j}(\psi_{c_j})\right),$$

where the sum over ψ runs over all additive characters of $\mathbb{F}_q$ and $\psi_{c_j}(x) = \psi(c_j x)$.

6.3.38 Definition Let $g(k,q)$ be the smallest s such that every element $b \in \mathbb{F}_q$ can be written as a sum of at most s summands of k-th powers in $\mathbb{F}_q$. The problem of determining or estimating $g(k,q)$ is *Waring's problem in* $\mathbb{F}_q$.

6.3.39 Remark We note that $g(k,q) = g(d,q)$ if $d = \gcd(k, q-1)$ and we may restrict ourselves to the case that $k \mid (q-1)$. Combining Theorem 6.3.36 (with $k_1 = \cdots = k_s = k \mid q-1$ and $c_1 = \cdots = c_s$) with the result on the absolute value of Jacobi sums (6.3.1) we get immediately

$$N_b \ge q^{s-1} - (k-1)^s q^{(s-1)/2}$$

which implies the following bound.

6.3.40 Theorem [2990] For any divisor k of $q-1$ we have $g(k,q) \le s$ if $q^{s-1} > (k-1)^{2s}$.

6.3.41 Remark Theorem 6.3.40 applies only to $k < q^{1/2-\varepsilon}$ and for $q^{3/7}+1 \le k < q^{1/2}$ we have the improvement $g(k,q) \le 8$ of [650, Corollary 7]. Moreover, from [1283, Theorem 6] it follows that for any $\varepsilon > 0$, $k \le q^{1-\varepsilon}$ and if $\mathbb{F}_q = \mathbb{F}_p(x^k)$ for some $x \in \mathbb{F}_q$, there is a constant $c(\varepsilon)$ such that $g(k,q) \le c(\varepsilon)$. However, if $q = p$ is a prime, a very moderate but nontrivial bound on Gauss sums of the second kind from [1789] leads to the nontrivial bound on $g(k,p) \ll (\ln k)^{2+\varepsilon}$ if $k < p(\log\log p)^{1-\varepsilon}/\log p$.

6.3.42 Definition The *covering radius* $\rho(C)$ of a code $C \subseteq \mathbb{F}_q^n$ is

$$\rho(C) = \max_{\mathbf{x} \in \mathbb{F}_q^n} \min_{\mathbf{c} \in C} d(\mathbf{c}, \mathbf{x}),$$

where d is the *Hamming distance.*

6.3.43 Proposition [1468, Lemma 1.1] Let H be the parity check matrix of a linear $[n,k]$-code C over $\mathbb{F}_q$, i.e., $C = \{\mathbf{c} \in \mathbb{F}_q^n : H\mathbf{c}^T = \mathbf{0}\}$. The covering radius is the least integer ρ such that every $\mathbf{x} \in \mathbb{F}_q^{n-k}$ is a linear combination of at most ρ columns of H.

6.3.44 Remark Let $g \in \mathbb{F}_q[X]$ be the minimal polynomial of an element $\alpha \in \mathbb{F}_q^*$ of order n and r be the order of q modulo n, i.e., $\mathbb{F}_q(\alpha) = \mathbb{F}_{q^r}$. Then the cyclic code $C = (g)$ is the $[n, n-r]$-code with parity check matrix $H = (1, \alpha, \alpha^2, \ldots, \alpha^{n-1})$, where the elements of $\mathbb{F}_{q^r}$ are identified with r-dimensional column vectors.

Put $N = (q^r - 1)/n$. Then $\alpha = \gamma^N$ for some primitive element γ of $\mathbb{F}_{q^r}$ and the columns of H consist of the nonzero N-th powers in $\mathbb{F}_{q^r}$. By Proposition 6.3.43, $\rho(C)$ is the least integer ρ such that any $x \in \mathbb{F}_{q^r}$ can be written as a linear combination of at most ρ N-th powers in $\mathbb{F}_{q^r}$. Hence, we have $\rho(C) \le g(N,q)$, where we have equality for $q = 2$.

6.3.2.4 Hidden number problem and noisy interpolation

6.3.45 Definition ((Extended) Hidden number problem) [347, 348] Let $\mathcal{T} \subseteq \mathbb{F}_p$. Recover a number $a \in \mathbb{F}_p$ if for many known $t \in \mathcal{T}$ the l most significant bits of at are given.

6.3.46 Remark If l is of order $\log^{1/2} p$ and $\mathcal{T}$ has some uniform distribution property, a lattice reduction technique solves the hidden number problem in polynomial time. The uniform distribution property is fulfilled if the maximum over all nontrivial additive character sums of $\mathbb{F}_p$ over $\mathcal{T}$ is small,

$$\max_{\psi} \left| \sum_{t \in \mathcal{T}} \psi(t) \right| = O(\#\mathcal{T}^{1-\varepsilon}).$$

If $\mathcal{T}$ is a subgroup of $\mathbb{F}_p^*$, the sums are Gauss sums of the second kind and the desired uniform distribution property is fulfilled by the bounds of [1454] and [380, 371] if $\#T \ge p^{1/3+\varepsilon}$ and $\#T \ge p^{\varepsilon}$, respectively. The bound of [1789] and ideas reminiscent to Waring's problem solve the problem for smaller $\#T \ge \log p/(\log\log p)^{1-\varepsilon}$ using more bits, that is, l of order $\log^4 p$.

6.3.47 Definition [2655] The *sparse polynomial noisy interpolation problem* consists of finding an unknown polynomial $f \in \mathbb{F}_p[X]$ of small weight from approximate values of $f(t)$ at polynomially many points $t \in \mathbb{F}_p$ selected uniformly at random.

6.3.48 Remark

1. The case $f(X) = aX$ corresponds to the hidden number problem.
2. For more details we refer to the survey [2647] and [2644, Chapter 30].

6.3.3 Applications of the Weil bound

6.3.49 Theorem [2548, Theorem 2G] Let $g \in \mathbb{F}_q[X]$ be of degree n and $f \in \mathbb{F}_q[X]$ have d distinct roots in its splitting field over $\mathbb{F}_q$. Let χ be a multiplicative character of $\mathbb{F}_q$ of order s and let ψ be an additive character of $\mathbb{F}_q$. If either $s > 1$ and f is not, up to a multiplicative constant, an s-th power, or ψ is nontrivial and $\gcd(n,q) = 1$, we have

$$\left| \sum_{c \in \mathbb{F}_q} \chi(f(c))\psi(g(c)) \right| \le (d + n - 1)q^{1/2}.$$

6.3.50 Remark The condition $\gcd(n,q) = 1$ can be replaced by the weaker condition that $Y^q - Y - g(X)$ is absolutely irreducible. The condition on f is fulfilled if $\gcd(\deg(f), s) = 1$.

6.3.3.1 Superelliptic and Artin-Schreier equations

6.3.51 Definition For polynomials $f, g \in \mathbb{F}_q[X]$ and a divisor s of $q-1$

$$Y^s = f(X) \quad \text{and} \quad Y^q - Y = g(X)$$

are *superelliptic* and *Artin-Schreier equations* defined over $\mathbb{F}_q$, respectively. For $s = 2$ these equations are *hyperelliptic equations* and, if additionally $\deg(f) = 3$, *elliptic equations*.

6.3.52 Proposition [2711, Lemma 1 and Lemma 2] The number N_{f,s,q^r} of solutions $(x, y) \in \mathbb{F}_{q^r}^2$ of a superelliptic equation is

$$N_{f,s,q^r} = \sum_{\mathrm{ord}(\chi)|s} \sum_{x \in \mathbb{F}_{q^r}} \chi(f(x)),$$

where the outer sum runs over all multiplicative characters χ of $\mathbb{F}_{q^r}$ such that χ^s is trivial. The number N_{g,q^r} of an Artin-Schreier equation is

$$N_{g,q^r} = \sum_{\psi_r} \sum_{x \in \mathbb{F}_{q^r}} \psi_r(g(x)),$$

where the outer sum runs over all additive characters ψ of $\mathbb{F}_q$ and $\psi_r(x) = \psi(\mathrm{Tr}(x))$, and Tr denotes the trace from $\mathbb{F}_{q^r}$ to $\mathbb{F}_q$.

6.3.53 Remark After isolating the trivial characters, the Weil bound implies immediately bounds on N_{f,s,q^r} and N_{g,q^r}.

6.3.54 Theorem [2711, Chapter 1.4] If $f \in \mathbb{F}_q[X]$ with $\gcd(\deg(f), s) = 1$ and $d > 0$ different zeros (in the algebraic closure of $\mathbb{F}_q$), then the number of solutions N_{f,s,q^r} over $\mathbb{F}_{q^r}$ of the superelliptic equation $Y^s = f(X)$ satisfies

$$|N_{f,s,q^r} - q^r| \le (s-1)(d-1)q^{r/2}.$$

If $g \in \mathbb{F}_q[X]$ has degree n with $\gcd(n,q) = 1$, then the number of solutions N_{g,q^r} over $\mathbb{F}_{q^r}$ of the Artin-Schreier equation $Y^q - Y = g(X)$ satisfies

$$|N_{g,q^r} - q^r| \le (n-1)(q-1)q^{r/2}.$$

6.3.3.2 Stable quadratic polynomials

6.3.55 Definition For a polynomial $f \in \mathbb{F}_q[X]$, q odd, we define

$$f^{(0)}(X) = X, \qquad f^{(n)}(X) = f\left(f^{(n-1)}(X)\right), \quad n = 1, 2, \ldots .$$

A polynomial f is *stable* if all polynomials $f^{(n)}$ are irreducible over $\mathbb{F}_q$.

6.3.56 Definition For a quadratic polynomial $f(X) = aX^2 + bX + c \in \mathbb{F}_q[X]$, $a \neq 0$, we define $\alpha = -b/2a$ as the unique *critical point* of f (that is, the zero of the derivative f') and the *critical orbit* of f,

$$\mathrm{Orb} = \{f^{(n)}(\alpha) \ : \ n = 2, 3, \ldots\} = \{f^{(n)}(\alpha) \ : \ n = 2, \ldots, t_f\},$$

where t_f is the smallest value of t with $f^{(t)}(\alpha) = f^{(s)}(\alpha)$ for some positive integer $s < t$, i.e., the *orbit length.*

6.3.57 Proposition [1620, Proposition 3] Let η denote the quadratic character of $\mathbb{F}_q$. A quadratic polynomial $f \in \mathbb{F}_q[X]$ is stable if and only if $\eta(x) = -1$ for all elements x of the *adjusted orbit* $\overline{\mathrm{Orb}}(f) = \{-f(\alpha)\} \bigcup \mathrm{Orb}(f)$.

6.3.58 Remark If f is stable, then the elements $f^{(n)}(\alpha)$, $n = 2, \ldots, t_f - 1$, are different and Proposition 6.3.57 implies for any positive integer K,

$$t_f - 2 = \frac{1}{2^K} \sum_{n=2}^{t_f - 1} \prod_{k=1}^{K} (1 - \eta(f^{(k)}(f^{(n)}(\alpha)))) \leq \frac{1}{2^K} \sum_{x \in \mathbb{F}_q} \prod_{k=1}^{K} \left(1 - \eta\left(f^{(k)}(x)\right)\right).$$

Expanding the product on the right hand side, we obtain one trivial sum equal to q and $2^K - 1$ sums which can be bounded by the Weil bound giving

$$t_f = \frac{q}{2^K} + O(2^K q^{1/2}).$$

The optimal choice of K gives the following result.

6.3.59 Theorem [2331, Theorem 1] For any odd q and any stable quadratic polynomial $f \in \mathbb{F}_q[X]$ we have $t_f = O\left(q^{3/4}\right)$.

6.3.60 Remark Similarly, Gomez and Nicolás estimated in [1310] the number of stable quadratic polynomials over $\mathbb{F}_q$ for an odd prime power q. As for Theorem 6.3.59, this reduces to estimating the character sum

$$\sum_{(a,b,c) \in \mathbb{F}_q^* \times \mathbb{F}_q \times \mathbb{F}_q} \prod_{k=1}^{K} (1 - \eta\left(F_k(a, b, c)\right)),$$

where $F_k(a, b, c)$ is the k-th element of the critical orbit of f and K any positive integer parameter. Gomez and Nicolás [1310] have proved that there are $O(q^{5/2}(\log q)^{1/2})$ stable quadratic polynomials over $\mathbb{F}_q$ for the power q of an odd prime.

6.3.3.3 Hamming distance of dual BCH codes

6.3.61 Definition Let γ be a primitive element of $\mathbb{F}_{2^r}$, t be a positive integer with $1 \leq 2t - 1 \leq 2^{\lceil r/2 \rceil} + 1$, and G_t denote the set of polynomials

$$G_t = \left\{ \sum_{i=1}^{t} g_i X^{2i-1} \in \mathbb{F}_{2^r}[X] \right\}.$$

For each $g \in G_t$ we define a binary word c_g of length $2^r - 1$,

$$c_g = \left(\mathrm{Tr}(g(1)), \mathrm{Tr}(g(\gamma)), \ldots, \mathrm{Tr}(g(\gamma^{2^r - 2}))\right),$$

where Tr denotes the absolute trace from $\mathbb{F}_{2^r}$ to $\mathbb{F}_2$, and define a code

$$C_t = \{c_g : g \in G_t\}.$$

6.3.62 Remark The code C_t is linear and is the dual of the primitive, binary BCH code with designed distance $2t+1$ [1991].

6.3.63 Remark Following [2372], one can use the Weil bound to estimate the Hamming weight for C_t. To do this, we recall that any nonzero codeword $c_g \in C_t$ comes from a nonzero polynomial $g(X) = \sum_{i=1}^{t} g_i X^{2i-1}$. Hence, we have

$$2^r - 2\mathrm{wt}_H(c_g) = \sum_{x \in \mathbb{F}_{2^r}} (-1)^{\mathrm{Tr}(g(x))} = \sum_{x \in \mathbb{F}_{2^r}} \psi(g(x)),$$

where $\psi(x) = (-1)^{\mathrm{Tr}(x)}$ is the additive canonical character [2372, Equation (4)]. Applying now the Weil bound, we get the following result.

6.3.64 Theorem [2372, Theorem 4] For $t \geq 1$ the minimum Hamming weight of C_t is at least $2^{r-1} - (t-1)2^{r/2}$.

6.3.4 Applications of Kloosterman sums

6.3.65 Definition Let ψ be a nontrivial additive character of $\mathbb{F}_q$ and let $a, b \in \mathbb{F}_q$. We define the *Kloosterman sum* (see Section 6.1) by

$$K(\psi; a, b) = \sum_{x \in \mathbb{F}_q^*} \psi(ax + bx^{-1}).$$

6.3.66 Proposition [1939, Theorem 5.45] If $ab \neq 0$, we have $|K(\psi; a, b)| \leq 2q^{1/2}$.

6.3.4.1 Kloosterman equations and Kloosterman codes

6.3.67 Definition For $a, b, c \in \mathbb{F}_q$ with $ab \neq 0$

$$Y^q - Y = aX + bX^{-1} + c$$

is a *Kloosterman equation.*

6.3.68 Theorem The number $N_{a,b,c}$ of solutions $(x, y) \in \mathbb{F}_{q^r}^2$ of a Kloosterman equation satisfies $|N_{a,b,c} - q^r| \leq 2(q-1)q^{r/2}$.

6.3.69 Remark We note that this estimate is obtained in a similar way as the number of solutions to an Artin-Schreier equation in Theorem 6.3.54, but using the estimate in Proposition 6.3.66 instead of the Weil bound.

6.3.70 Definition Let γ be a primitive element of $\mathbb{F}_{2^r}$. The codewords of the *Kloosterman code* $C = \{\mathbf{c}_{a,b} : a, b \in \mathbb{F}_{2^r}\}$ are defined by

$$\mathbf{c}_{a,b} = (\mathrm{Tr}(a+b), \mathrm{Tr}(a\gamma + b\gamma^{-1}) + \cdots + \mathrm{Tr}(a\gamma^{2^r-2} + b\gamma^{2-2^r})), \quad a, b \in \mathbb{F}_{2^r}.$$

6.3.71 Theorem [2372] The minimum weight of the Kloosterman code is at least $2^{r-1} - 2^{r/2}$.

6.3.4.2 Distribution of inversive congruential pseudorandom numbers

6.3.72 Definition For $a \in \mathbb{F}_p^*$, $b \in \mathbb{F}_p$ we define, with the convention $0^{-1} = 0$, the sequence (u_n) by the recurrence relation

$$u_{n+1} = au_n^{-1} + b, \qquad n = 0, 1, \ldots, \tag{6.3.4}$$

with u_0 the initial value. Then the numbers u_n/p, $n = 0, 1, \ldots$, in the interval $[0, 1)$ form a sequence of *inversive congruential pseudorandom numbers.*

6.3.73 Remark The sequence (u_n) defined by (6.3.4) is purely periodic with some period $T \le p$. Using the Erdős-Turan inequality given by Theorem 6.3.32, one reduces the problem of estimating the discrepancy of the elements in the sequence (u_n/p), $n = 0, \ldots, T-1$, to estimating character sums given by

$$S_T(h) = \sum_{n=0}^{T-1} \psi(hu_n)$$

with a fixed integer $h \not\equiv 0 \pmod p$ and ψ is the additive canonical character of $\mathbb{F}_p$. Niederreiter and Shparlinski [2270] developed a method to reduce the problem of estimating $|S_T(h)|$ to estimating Kloosterman sums.

6.3.74 Theorem [2270, Theorem 2] For the sequence $\Gamma = (u_n/p : n = 0, \ldots, T-1)$, where (u_n) is defined by (6.3.4) and T is the period of the sequence (u_n), we have

$$D_T(\Gamma) \ll T^{-1/2} p^{1/4} \log p.$$

6.3.4.3 Nonlinearity of Boolean functions

6.3.75 Definition Let $\mathcal{B}_r = \{0,1\}^r$. The *Fourier coefficients* (or *Walsh-Hadamard coefficients*) $\widehat{B}(a)$ of $B(U_1, \ldots, U_r)$, where $a \in \mathcal{B}_r$, are defined as

$$\widehat{B}(a) = \sum_{u \in \mathcal{B}_r} (-1)^{B(u) + <a,u>},$$

where $< a, u >= a_1u_1 + a_2u_2 + \cdots + a_ru_r$ denotes the standard inner product. The *nonlinearity* of the Boolean function $B(U_1, \ldots, U_r)$ is defined by

$$N(B) = 2^{r-1} - \frac{1}{2} \max_{a \in \mathcal{B}_r} \left| \widehat{B}(a) \right|.$$

6.3.76 Remark Boolean functions used in cryptography must have high nonlinearity, see for example [523] and Section 9.1.

6.3.77 Remark Each Boolean function $B : \mathbb{F}_2^r \to \mathbb{F}_2$ can be represented with a polynomial $f \in \mathbb{F}_{2^r}$ and the absolute trace function as $B(x_1, \ldots, x_r) = \mathrm{Tr}(f(x_1\beta_1 + \cdots + x_r\beta_r))$ with some fixed ordered basis $\{\beta_1, \ldots, \beta_r\}$ of $\mathbb{F}_{2^r}$. Moreover, if we identify $x \in \mathbb{F}_{2^r}$ with its coordinate vector, we have that $x \to (-1)^{<a,x>}$ is an additive character of $\mathbb{F}_{2^r}$. Hence, we get

$$N(B) = 2^{r-1} - \frac{1}{2} \max_{b \in \mathbb{F}_{2^r}} \left| \sum_{u \in \mathbb{F}_{2^r}} \psi(B(u) + bu) \right|,$$

where ψ is the additive canonical character of $\mathbb{F}_{2^r}$. For example, if $f(x) = x^{-1}$ (with the convention $0^{-1} = 0$), we get Kloosterman sums on the right hand side.

6.3.5 Incomplete character sums

6.3.78 Theorem (Pòlya-Vinogradov-Weil bound) Let χ be a nontrivial multiplicative character of order s of $\mathbb{F}_p$ and $f \in \mathbb{F}_p[X]$ with $d > 0$ different zeros such that f is not, up to a constant multiple, an s-th power. Then we have

$$\left| \sum_{n=0}^{N-1} \chi(f(n)) \right| \leq dp^{1/2} \log p, \quad 1 \leq N < p.$$

6.3.79 Remark The bound of Theorem 6.3.78 can be obtained using the standard method for reducing incomplete character sums to complete ones, see for example [1581, Section 12.2], and the Weil bound. This reducing method can be traced back to Pòlya and Vinogradov; for references see [1581]. Generalizations to certain incomplete character sums over arbitrary finite fields are given in [776, 2989, 2991, 2993].

6.3.5.1 Finding deterministically linear factors of polynomials

6.3.80 Algorithm [1887] Let $f \in \mathbb{F}_p[X]$, p an odd prime, be a squarefree polynomial with $f(0) \neq 0$ which splits over $\mathbb{F}_p$. For $t = 0, 1, \ldots, N$ we compute

$$L_t(X) = \gcd((X + t)^{(p-1)/2} - 1, f(X)) = \gcd(g_t(X) - 1, f(X))$$

via the Euclidean algorithm, where N is the main parameter of the algorithm, hoping that at least one polynomial L_t is nontrivial, that is, is equal to neither 1 nor f.

For each t, the polynomial

$$g_t(X) \equiv (X + t)^{(p-1)/2} \pmod{f(X)}, \quad \deg(g_t) < \deg(f)$$

is calculated efficiently using repeated squaring.

6.3.81 Remark Since $x^{(p-1)/2} = 1$ if and only if x is a quadratic residue modulo p, L_t is trivial, i.e., either 1 or f, only if

$$\left(\frac{a+t}{p}\right) = \left(\frac{b+t}{p}\right)$$

for any two distinct roots of f. Therefore, the algorithm does not determine a nontrivial factor of f only if

$$N + 1 = \sum_{t=0}^{N} \left(\frac{(a+t)(b+t)}{p}\right) \ll p^{1/2} \log p$$

by Theorem 6.3.78.

6.3.82 Remark Any polynomial f can be factorized into squarefree polynomials by calculating $\gcd(f, f')$. The part of a squarefree polynomial f which splits over $\mathbb{F}_p$ and is not divisible by X is $\gcd(X^{p-1}-1, f)$. More details about the factorization of univariate polynomials over finite fields can be found in Section 11.4.

6.3.83 Remark Shoup [2626] extended Legendre's idea to design a deterministic factoring algorithm for all squarefree polynomials.

6.3.5.2 Measures of pseudorandomness

6.3.84 Definition [Mauduit and Sárközy [2037]] For a finite binary sequence $E_N = (e_1, \dots, e_N) \in \{-1,+1\}^N$ the *well-distribution measure* of E_N is defined by

$$W(E_N) = \max_{a,b,M} \left| \sum_{j=1}^{M} e_{a+bj} \right|,$$

where the maximum is taken over all $a, b, M \in \mathbb{Z}$ and $b, M > 0$ such that $1 \le a+b \le a+bM \le N$, and the *correlation measure of order* ℓ of E_N is defined as

$$C_\ell(E_N) = \max_{M,D} \left| \sum_{n=1}^{M} e_{n+d_1} e_{n+d_2} \cdots e_{n+d_\ell} \right|,$$

where the maximum is taken over all $D = (d_1, \dots, d_\ell)$ and M is such that $0 \le d_1 < \dots < d_\ell \le N - M$.

6.3.85 Definition Let p be an odd prime. The binary sequence $E_p = (1, e_1, \dots, e_{p-1})$ defined by

$$e_n = \left(\frac{n}{p}\right), \quad 0 \le n < p,$$

is the *Legendre sequence.*

6.3.86 Remark The sums in the definitions of well-distribution measure and correlation measure of order ℓ for the Legendre sequence are essentially sums of products of Legendre symbols which can be estimated by Theorem 6.3.78.

6.3.87 Remark We note that $W(E_N)$ and $C_\ell(E_N)$ of a "truly random" sequence are of the order of magnitude $N^{1/2}\log N$ and $N^{1/2}(\log N)^{c(\ell)}$, respectively, see [82, 554].

6.3.88 Theorem [2037] For the Legendre sequence E_p we have

$$W(E_p) \ll p^{1/2}\log p \quad \text{and} \quad C_\ell(E_p) \ll \ell p^{1/2}\log p.$$

6.3.89 Remark The linear complexity is a measure for the unpredictability and thus suitability of a sequence in cryptography. For sequences $(u_0, \dots, u_{N-1}) \in \mathbb{F}_2^N$ it is closely related to the correlation measure of order ℓ of the sequence $(e_0, \dots, e_{N-1}) \in \{-1,+1\}^N$ defined by $e_n = (-1)^{u_n}$, $n = 0, \dots, N-1$. Hence, from a suitable upper bound on $C_\ell(E_N)$ up to a sufficiently large ℓ we can derive a lower bound on the linear complexity of (u_n), see [393] and Subsection 10.4.5.

6.3.6 Other character sums

6.3.6.1 Distribution of primitive elements and powers

6.3.90 Theorem (Vinogradov's Formula) [1631, Lemma 7.5.3] For a subset $S \subseteq \mathbb{F}_q^*$, the number Q of primitive elements in S is

$$Q = \frac{\varphi(q-1)}{q-1} \sum_{d|(q-1)} \frac{\mu(d)}{\varphi(d)} \sum_{\substack{\chi \\ \mathrm{ord}(\chi)=d}} \sum_{x \in S} \chi(x),$$

where μ and φ denote Möbius' and Euler's totient function, respectively, and χ is a nontrivial multiplicative character of $\mathbb{F}_q$.

6.3.91 Theorem [1631, Lemma 7.5.3] Let S be a subset of $\mathbb{F}_q^*$. Then the number R of s-th powers in S is

$$R = \frac{1}{s} \sum_{\mathrm{ord}(\chi)|s} \sum_{x \in S} \chi(x).$$

6.3.92 Remark [1939, Theorem 5.4] For a set S of $1 \leq N \leq p$ consecutive elements in $\mathbb{F}_p$, the Burgess bound [464]

$$\sum_{n=M+1}^{M+N} \chi(n) \ll N^{1-1/r} p^{(r+1)/4r^2+\varepsilon},$$

which is nontrivial for any $N \geq p^{1/4+\varepsilon}$, implies $Q = \frac{\varphi(p-1)}{p-1}(N + O(N^{1-1/r} p^{(r+1)/4r^2+\varepsilon}))$ and $R = \frac{N}{s} + O(N^{1-1/r} p^{(r+1)/4r^2+\varepsilon})$. For generalizations of the Burgess bound see [375, 583, 584, 1791].

6.3.93 Remark For $\mathbb{F}_{p^r}$ with $2 \leq r < p^{1/2}$ and a defining element α of $\mathbb{F}_{p^r}$, the bound

$$\left| \sum_{x \in \mathbb{F}_p} \chi(\alpha + x) \right| \leq (r-1)p^{1/2}$$

of [1702] guarantees the existence of primitive elements in rather small subsets of $\mathbb{F}_{p^r}$.

6.3.6.2 Distribution of Diffie-Hellman triples

6.3.94 Remark Let $g \in \mathbb{F}_p$ be an element of order $T \mid (p-1)$. The Diffie-Hellman key exchange [859] is a way for two parties to establish a common secret key over an insecure channel. The question of studying the distribution of the Diffie-Hellman triples (g^x, g^y, g^{xy}), $x, y = 0, \ldots, T-1$, is motivated by the assumption that these triples cannot be distinguished from totally random triples in feasible computational time, see [487, 488]. In the series of papers [197, 368, 487, 488, 587, 1191] it has been shown that such triples are uniformly distributed via estimating double exponential sums with linear combinations of the entries in such triples.

6.3.95 Definition Let ψ be the additive canonical character of $\mathbb{F}_p$. For integers a, b, c, we define the exponential sum

$$S_{a,b,c}(T) = \sum_{x,y=1}^{T} \psi\left(ag^x + bg^y + cg^{xy}\right).$$

6.3.96 Remark Bounds on the sums $S_{a,b,c}(T)$ were obtained by actually estimating different sums

$$W_{a,c}(T) = \sum_{y=1}^{T} \left| \sum_{x=1}^{T} \psi(ag^x + cg^{xy}) \right|$$

since $|S_{a,b,c}(T)| \le W_{a,c}(T)$. In [487] the bound $W_{a,c}(T) \ll T^{5/3}p^{1/4}$ is obtained, which is nontrivial for $T > p^{3/4+\varepsilon}$. Moreover, Bourgain [368] gave a nontrivial estimate for $T > p^{\varepsilon}$ for any $\varepsilon > 0$ and Garaev [1191] improved not only the bound in [487] obtaining $W_{a,c}(T) \ll T^{7/4}p^{1/8+\varepsilon}$, but also the range of nontriviality $T > p^{1/2+\varepsilon}$. Furthermore, Chang and Yao [587] gave an estimate that works in a range that was not covered by any explicit bound in any other work.

6.3.6.3 Thin sets with small discrete Fourier transform

6.3.97 Definition Let ψ denote the additive canonical character of the prime field $\mathbb{F}_p$. Given a set $T = \{t_1, t_2, \ldots, t_n\}$ of n elements in $\mathbb{F}_p$, the sequence

$$f_T(k) = \sum_{j=1}^{n} \psi(t_j k), \quad k = 0, 1, \ldots, p-1,$$

is the *discrete Fourier transform* of T.

6.3.98 Remark It is well known that the discrete Fourier transform (see Section 10.1) captures a lot of information about the "random-like" behavior of sets, that is, "good" sets have a small discrete Fourier transform for all $1 \le k < p$. In [59, 1702], several constructions of *thin* sets T were given, that is, sets of size $|T| = O((\log p)^{2+\varepsilon})$ with small maximum discrete Fourier transform $\max_{1\le k\le p-1} |f_T(k)| = O(|T|(\log p)^{-\varepsilon})$.

6.3.99 Remark Constructions of thin sets have applications in graph theory, computer science [1161], and in combinatorial number theory [2510, 2996]. For example, in additive number theory, for an infinite set A of natural numbers, one defines the lower density of A as

$$d(A) = \liminf_{x\to\infty} \frac{N(x)}{x},$$

where $N(x) = \#\{a \in A \mid a \le x\}$. One problem of interest is to construct *essential components*, that are sets H with the property that $d(A+H) > d(A)$ for all such A with $0 < d(A) < 1$, and investigate how thin an essential component is. For such estimates, see [59, 2510, 2996].

6.3.6.4 Character sums over arbitrary sets

6.3.100 Theorem [1384, Corollaries 1 and 5] Let $A, B \subseteq \mathbb{F}_q$ and ψ and χ be a nontrivial additive and multiplicative character, respectively. Then we have

$$\left| \sum_{a\in A, b\in B} \psi(ab) \right| \le (|A||B|q)^{1/2} \quad \text{and} \quad \left| \sum_{a\in A, b\in B} \chi(ab+1) \right| \le (|A||B|q)^{1/2}.$$

6.3.101 Remark Based on Theorem 6.3.100, Gyarmati and Sárközy showed in [1385, Corollaries 1 and 2] that for all subsets $A, B, C, D \subseteq \mathbb{F}_q$ with $|A||B||C||D| > q^3$, the equation

$$a + b = cd, \quad a \in A, b \in B, c \in C, d \in D$$

has a solution, and with $|A||B||C||D| > 100q^3$, the equation

$$ab + 1 = cd, \quad a \in A, b \in B, c \in C, d \in D$$

has a solution.

6.3.102 Remark When $q = p$ and ψ is the additive canonical character of $\mathbb{F}_p$, Bourgain and Garaev proved in [378, Theorem 1.2] the following estimate for any subsets A, B, C of $\mathbb{F}_p^*$,

$$\left| \sum_{a \in A, b \in B, c \in C} \psi(abc) \right| < (|A||B||C|)^{13/16} p^{5/18+o(1)}.$$

See Also

§3.1, §3.5, §4.2	For estimating the number of polynomials with certain features.
§6.1, §6.2	For basics on character sums.
§6.4, §7.3	For diagonal equations and Waring's problem.
§7.1	For Kummer and Artin-Schreier curves.
§9.1, §9.3	For Boolean functions and nonlinearity.
§10.3, §10.4, §17.3	For correlation and related measures.
§10.5	For nonlinear recurrence sequences.
§11.4	For univariate factorization.
§11.6, §17.2	For discrete logarithm based cryptosystems.
§14.1	For Latin squares.
§14.5, §14.6	For Hadamard matrices and related combinatorial structures.
§15.1	For basics on coding theory.
§17.2	For applications of character sums in quantum information theory.

References Cited: [59, 82, 197, 240, 260, 347, 348, 368, 371, 375, 378, 380, 393, 463, 464, 487, 488, 523, 554, 583, 584, 587, 650, 776, 859, 922, 998, 1161, 1191, 1283, 1303, 1310, 1384, 1385, 1454, 1468, 1536, 1581, 1620, 1631, 1702, 1730, 1789, 1791, 1887, 1911, 1939, 1991, 2037, 2068, 2230, 2248, 2270, 2278, 2331, 2344, 2372, 2510, 2548, 2626, 2644, 2655, 2647, 2711, 2989, 2990, 2991, 2993, 2996]

6.4 Sum-product theorems and applications

Moubariz Z. Garaev, *Universidad Nacional Autónoma de México*

6.4.1 Notation

6.4.1 Remark The order $q = p^n$ of the field $\mathbb{F}_q$ is assumed to be sufficiently large. The elements of the prime residue field $\mathbb{F}_p$ are occasionally associated with their concrete representatives. The cardinality of a finite set X is denoted by $|X|$.

6.4.2 Definition Given nonempty sets A and B the *sum set* $A+B$ is defined by

$$A+B := \{a+b : a \in A, b \in B\}$$

and the *product set* AB is defined by

$$AB = \{ab : a \in A, b \in A\}.$$

The number of solutions of the equation

$$a_1b_1 = a_2b_2, \qquad (a_1,a_2) \in A \times A, (b_1,b_2) \in B \times B$$

is denoted by $E_\times(A,B)$ and is the *multiplicative energy* between the sets A and B.

6.4.3 Remark There is a simple and important connection between the cardinality of the product set AB and the multiplicative energy $E_\times(A,B)$:

$$|AB| \geq \frac{|A|^2|B|^2}{E_\times(A,B)}.$$

6.4.4 Remark All subsets in this section are assumed to be nonempty. For quantities U and V, the notations $U = O(V)$, $U \ll V$ and $V \gg U$ are all equivalent to the statement that the inequality $|U| \leq cV$ holds with some absolute constant $c > 0$. We use the abbreviation $e_p(x)$ to denote $e^{2\pi i x/p}$.

6.4.2 The sum-product estimate and its variants

6.4.5 Remark Bourgain, Katz, and Tao [381], with subsequent refinement by Bourgain, Glibichuk and Konyagin [380], proved the following theorem, which is called the sum-product estimate in prime fields.

6.4.6 Theorem [381] For any $\varepsilon > 0$ there exists $\delta = \delta(\varepsilon) > 0$ such that, if $A \subset \mathbb{F}_p$ and $|A| < p^{1-\varepsilon}$, then

$$\max\{|A+A|, |AA|\} \geq |A|^{1+\delta}.$$

6.4.7 Remark The proof of Theorem 6.4.6 uses results from additive combinatorics and ideas of Edgar and Miller [955]. The condition $|A| < p^{1-\varepsilon}$ is essential, because if the cardinality of $|A|$ is close to p, then $|A+A|$ and $|AA|$ have cardinalities close to $|A|$.

6.4.8 Theorem [1194] There is a positive constant c such that for any nonempty set $A \subset \mathbb{F}_p$ the following bound holds:

$$\max\{|A+A|, |AA|\} > c\min\Big\{\frac{|A|^2}{p^{1/2}}, p^{1/2}|A|^{1/2}\Big\}.$$

6.4.9 Remark Theorem 6.4.8 is meaningful for sets A of cardinality larger than $p^{1/2}$. Explicit sum-product estimates for large subsets for the first time were given by Hart, Iosevich, and Solymosi [1424]. Theorem 6.4.8 is due to Garaev [1194]. It follows that if $|A| > p^{2/3}$, then

$$\max\{|A+A|, |AA|\} > cp^{1/2}|A|^{1/2}.$$

This bound is optimal in the sense that for any positive integer $N < p$ there exists $A \subset \mathbb{F}_p$ with $|A| = N$ such that

$$\max\{|A+A|, |AA|\} < c_1p^{1/2}|A|^{1/2},$$

where c_1 is some positive constant.

6.4.10 Theorem [378] For any subset $A \subset \mathbb{F}_p$ there exists a subset $A' \subset A$ with $|A'| \geq 0.1|A|$ such that

$$E_\times(A', A')^4 \ll \left(|A - A| + \frac{|A|^3}{p}\right) \cdot |A|^3 \cdot |A - A|^7 \cdot \log^4(e|A|).$$

6.4.11 Remark Theorem 6.4.10 in this formulation is convenient for applications to explicit exponential sum estimates. It also implies that if $|A| < p^{1/2}$ then one has

$$\max\{|A - A|, |AA|\} > |A|^{13/12+o(1)}.$$

An explicit sum-product estimate for subsets of cardinality $|A| < p^{1/2}$ was given for the first time by Garaev [1192] in the form

$$\max\{|A + A|, |AA|\} > |A|^{15/14+o(1)}.$$

This estimate subsequently was improved by several authors; the most recent one is due to Rudnev [2500]: if $|A| < p^{1/2}$, then $\max\{|A + A|, |AA|\} > |A|^{12/11+o(1)}$.

6.4.12 Theorem [1195] Let $A, B \subset \mathbb{F}_p^*$, $L = \min\left\{|B|, p|A|^{-1}\right\}$. Then

$$|A - A|^2 \cdot \frac{|A|^2|B|^2}{E_\times(A, B)} \gg |A|^3 \, L^{1/9} \, (\log L)^{-1}.$$

6.4.13 Remark Theorem 6.4.12 is an explicit version of Bourgain's sum product estimate for subsets of incomparable sizes. The presence of the multiplicative energy is important in applications to multilinear exponential sum estimates. The proof can be found in [1195] and is based on ideas from [371, 1192, 1193].

6.4.14 Theorem [371] Let $A, B \subset \mathbb{F}_p^*$. Then

$$|8AB - 8AB| \geq \frac{1}{2}\min\{|A||B|, p - 1\}.$$

6.4.15 Remark Here $8AB - 8AB = \{\sum_{i=1}^{8} a_i b_i - \sum_{i=9}^{16} a_i b_i \,:\, a_i \in A, b_i \in B\}$. From the result of Glibichuk and Konyagin [1284] it was known that

$$|3AA - 3AA| \geq \frac{1}{2}\min\{|A|^2, p - 1\}.$$

6.4.16 Remark Theorem 6.4.14 is proved by Bourgain [371] with subsequent application to multilinear exponential sum estimates with nearly optimal entropy conditions.

6.4.17 Theorem [379] Let $A, B \subset \mathbb{F}_q$ such that $|A| > 1$ and B is not contained in any proper subfield of $\mathbb{F}_q$. Then

$$\max\{|A + AB|, |A - AB|\} \geq \frac{1}{2}|A|^{6/7}\min\{|A||B|, q\}^{1/7}.$$

6.4.18 Remark Theorem 6.4.17 is proved by Bourgain and Glibichuk [379] with subsequent applications to exponential sum estimates over small subgroups of $\mathbb{F}_q^*$. Because of the presence of subfields an additional condition on subsets is needed. An explicit sum-product estimate in $\mathbb{F}_q$ for the first time was obtained by Katz and Shen [1696] in the following form: suppose that A is a subset of $\mathbb{F}_q$ such that for any $A' \subset A$ with $|A'| \geq |A|^{18/19}$ and for any $G \subset \mathbb{F}_q$

a subfield (not necessarily proper) and for any elements $c, d \in \mathbb{F}_q$ if $A' \subset cG + d$ then $|A'| \leq |G|^{1/2}$. Then it must be that

$$\max\{|A+A|, |AA|\} \gg |A|^{20/19-\varepsilon},$$

where the implied constant depends only on a small positive quantity ε. This estimate subsequently has been improved in several works, the most recent one is due to Li and Rocher-Newton [1920].

6.4.19 Remark There are many interesting versions of sum-product estimates, see the works of Ahmadi and Shparlinski [53], Shparlinski [2651], Hart, Li, and Shen [1425] among others. The following useful result is due to Bukh and Tsimerman [456].

6.4.20 Theorem [456] Let f be a polynomial over $\mathbb{F}_p$ of degree $d \geq 2$. Then, for every set $A \subset \mathbb{F}_p$ of size $|A| \leq p^{1/2}$, we have

$$|A+A| + |f(A) + f(A)| \gg |A|^{1+1/(16 \cdot 6^d)},$$

where the implied constant depends only on d.

6.4.21 Theorem Let A, B be finite subsets of an additive group. Assume that $E \subset A \times B$ is such that $|E| \geq |A||B|/K$. Then there exists a subset $A' \subset A$ such that $|A'| \geq 0.1|A|/K$ and

$$|\{a - b : (a,b) \in E|^4 \geq \frac{|A' - A'||A||B|^2}{10^4 K^5}.$$

6.4.22 Remark Theorem 6.4.21 is a version of the Balog-Szemerédi-Gowers estimate, which is an important tool in additive combinatorics. It takes its origin from the works by Balog and Szemerédi [194] and Gowers [1347]. Balog-Szemerédi-Gowers type estimates are also important in multilinear exponential sum estimates over small subsets. The proof of Theorem 6.4.21 can be found in [378] and is based on ideas and results from [193, 2738, 2781].

6.4.3 Applications

6.4.23 Remark The sum-product estimate and its variants have found many spectacular applications in various areas of mathematics. Using sum-product estimates, Bourgain, Glibichuk, and Konyagin [380] obtained a new estimate of multilinear rational trigonometric sums, which has important applications to classical Gauss trigonometric sums.

6.4.24 Theorem [380] For any $\varepsilon > 0$ there exists $\delta = \delta(\varepsilon) > 0$ and a positive integer $k = k(\varepsilon)$ such that if $X \subset \mathbb{F}_p$ with $|X| > p^{\varepsilon}$, then

$$\max_{(a,p)=1} \Big| \sum_{x_1 \in X} \cdots \sum_{x_k \in X} e_p(ax_1 \ldots x_k) \Big| < |X|^k p^{-\delta}.$$

6.4.25 Corollary [380] For any $\varepsilon > 0$ there exists $\delta = \delta(\varepsilon) > 0$ such that if H is a subgroup of the multiplicative group $\mathbb{F}_p^*$ with $|H| > p^{\varepsilon}$, then

$$\max_{(a,p)=1} \Big| \sum_{h \in H} e_p(ah) \Big| < |H| p^{-\delta}.$$

6.4.26 Theorem [371] Let $0 < \delta < 1/4$ and $r \geq 2$ be an integer. There is a $\delta' > (\delta/r)^{Cr}$ such that if p is a sufficiently large prime and $X_1, X_2, \ldots, X_r \subset \mathbb{F}_p$ satisfy $|X_i| \geq p^{\delta}$ for all $1 \leq i \leq r$ and if

$$\prod_{i=1}^{r} |X_i| > p^{1+\delta},$$

then there is the exponential sum bound

$$\Big|\sum_{x_1\in X_1}\cdots\sum_{x_r\in X_r} e_p(x_1\ldots x_r)\Big| < |X_1|\cdots|X_r|p^{-\delta'}.$$

6.4.27 Remark Theorem 6.4.26 is due to Bourgain [371]. It gives a nontrivial estimate for multilinear exponential sums under nearly optimal conditions on the sizes of the sets X_i.

6.4.28 Theorem [1195] Let $3 \le r \le 1.44\log\log p$ be an integer, ε be a fixed positive constant. Let $X_1, X_2, \ldots, X_r$ be subsets of $\mathbb{F}_p^*$ such that

$$|X_1|\cdot|X_2|\cdot\Big(|X_3|\cdots|X_r|\Big)^{1/81} > p^{1+\varepsilon}.$$

Then

$$\Big|\sum_{x_1\in X_1}\cdots\sum_{x_r\in X_r} e_p(x_1\ldots x_r)\Big| < |X_1|\cdots|X_r|p^{-0.45\,\varepsilon/2^r}$$

for all sufficiently large primes $p > p_0(\varepsilon)$.

6.4.29 Remark Theorem 6.4.28 is an explicit version of Bourgain's exponential sum estimate from [371].

6.4.30 Corollary [1195] Let H be a subgroup of $\mathbb{F}_p^*$ with

$$|H| > e^{57\log p/\log\log p}.$$

Then, as $p\to\infty$, we have

$$\max_{(a,p)=1}\Big|\sum_{x\in H} e_p(ax)\Big| = o(|H|).$$

6.4.31 Corollary [1195] Let g be a generator of $\mathbb{F}_p^*$ and let $N > e^{57\log p/\log\log p}$. Then, as $p\to\infty$, we have

$$\max_{(a,p)=1}\Big|\sum_{x\le N} e_p(ag^x)\Big| = o(N).$$

6.4.32 Theorem [1195] Let $X, Y, Z \subset \mathbb{F}_p^*$ be such that $|X||Y| > \delta\, p$ for some constant $\delta > 0$. Then, for any $\varepsilon > 0$ one has the bound

$$\Big|\sum_{x\in X}\sum_{y\in Y}\sum_{z\in Z} e_p(xyz)\Big| \ll |X|\cdot|Y|\cdot|Z|^{539/540+\varepsilon},$$

where the implied constant depends only on ε and δ.

6.4.33 Theorem [368] For any $\delta > 0$ there exists $\delta' > 0$ such that if $\theta\in\mathbb{F}_p^*$ is of multiplicative order t and $t \ge t_1 > p^\delta$, $t\ge t_2 > p^\delta$, then

$$\max_{(a,b,p)=1}\sum_{x=1}^{t_1}\Big|\sum_{y=1}^{t_2} e_p(a\theta^y + b\theta^{xy})\Big| < t_1t_2p^{-\delta'}.$$

6.4.34 Remark Theorem 6.4.33 is due to Bourgain [368]. Previously known results only applied for large values of t, t_1, t_2. The exponential sum that appears in Theorem 6.4.33 for the first time was investigated by Canetti, Friedlander, and Shparlinski [488] motivated by cryptographic applications. Subsequently Banks, Conflitti, Friedlander, and Shparlinski [196] found applications to estimate exponential sums with Mersenne numbers. For further details, see Bibak [268], Chang [587] and the references therein.

6.4.35 Theorem [369] Given a positive integer r and $\varepsilon > 0$, there exists $\delta = \delta(r,\varepsilon) > 0$ satisfying the following property: if

$$f(x) = \sum_{i=1}^{r} a_i x^{k_i} \in \mathbb{Z}[x], \quad (a_i, p) = 1,$$

where the exponents k_i, $1 \le k_i < p^{1-\varepsilon}$, satisfy

$$(k_i, p-1) < p^{1-\varepsilon} \quad \text{for all } i,\ 1 \le i \le r,$$
$$(k_i - k_j, p-1) < p^{1-\varepsilon} \quad \text{for all } i, j,\ 1 \le i < j \le r,$$

then there is an exponential sum estimate

$$\left|\sum_{x=1}^{p-1} e_p(f(x))\right| < p^{1-\delta}.$$

6.4.36 Remark Theorem 6.4.35 is due to Bourgain, its proof is based on sum-product estimates for subsets of $\mathbb{F}_p \times \mathbb{F}_p$. It gives a nontrivial bound for the exponential sums under essentially optimal conditions on the exponents k_i. The exponential sum that appears in Theorem 6.4.33 was estimated by Mordell [2143] in 1932, see Cochrane, Coffelt, and Pinner [654] for more details.

6.4.37 Definition Denote $\mathrm{Tr}(x) = 1 + x^p + \cdots + x^{p^{n-1}}$ the trace of $x \in \mathbb{F}_q$. Let $\psi(x) = e_p(a\mathrm{Tr}(x))$, $a \in \mathbb{F}_q^*$, be a nontrivial additive character of $\mathbb{F}_q$.

6.4.38 Theorem [373] Let $0 < \delta, \delta_2 < 1$ and $r \ge 2$ be integer. Let $A_1, \ldots, A_r \subset \mathbb{F}_q$ satisfy

$$|A_i| > q^{\delta} \quad \text{for} \quad 1 \le i \le r,$$
$$|A_i \cap (aG + b)| < q^{-\delta_2} \quad \text{for} \quad 1 \le i \le r,$$

whenever $a, b \in \mathbb{F}_q$ and G a proper subfield, and let

$$|A_1||A_2| \prod_{i=3}^{r} |A_i|^{1/2} > q^{1+\delta}.$$

Then

$$\left|\sum_{x_1 \in A_1} \cdots \sum_{x_r \in A_r} \psi(x_1 \ldots x_r)\right| < |A_1| \cdots |A_r| q^{-\delta'},$$

where we may take $\delta' = C^{-r/\delta_2}(\delta/r)^{Cr}$ for some positive constant C.

6.4.39 Theorem [379] Let $0 < \eta \le 1$ be fixed and H be a multiplicative subgroup of $\mathbb{F}_q^*$ with

$$|H| \ge q^{\frac{\max(135/\eta, 180000)}{\log_2 \log_2 q}}.$$

Suppose that for any subfield G we have

$$|H \cap G| \le |H|^{1-\eta}.$$

Then for sufficiently large q there is an exponential sum estimate

$$\left|\sum_{h \in H} \psi(h)\right| < 2^{-0.0045(\log_2 q)^{0.1}} |H|.$$

6.4.40 Remark Using the sum-product estimate in $\mathbb{F}_p$, Bourgain, Katz, and Tao [380] obtained the following $\mathbb{F}_p$ - analogy of the Szemerédi-Trotter theorem.

6.4.41 Theorem [380] Let $\mathcal{P}$ (correspondingly $\mathcal{L}$) be a set of points (correspondingly a set of lines) in the plane $\mathbb{F}_p \times \mathbb{F}_p$. Assume that for some $\alpha < 2$ we have $\max\{|\mathcal{P}|, |\mathcal{L}|\} \le M < p^\alpha$. Then for some constant $\gamma = \gamma(\alpha) > 0$ one has

$$W(\mathcal{P},\mathcal{L}) := \#\Big\{\big((x,y),\ell\big) \in \mathcal{P} \times \mathcal{L} : (x,y) \in \ell\Big\} \ll M^{3/2-\gamma}.$$

6.4.42 Remark An explicit bound for $W(\mathcal{P},\mathcal{L})$ has been obtained by Helfgott and Rudnev [1465]. When the cardinalities of $\mathcal{P}$ and $\mathcal{L}$ are large, a sharp bound for $W(\mathcal{P},\mathcal{L})$ has been obtained by Vinh [2876]; see also Cilleruelo [643].

6.4.43 Remark The following two results are due to Bourgain [370]. They solve one of the questions of Wigderson on expander maps in two variables. Their proofs are based on the $\mathbb{F}_p$ - analogy of Szemerédi-Trotter theorem.

6.4.44 Theorem [370] Let A, B be subsets of $\mathbb{F}_p$ with $|A| = |B| = N < p^\alpha$, where $\alpha < 1$ is a positive constant. Then for some constant $\beta = \beta(\alpha) > 0$ one has

$$|\{a^2 + ab : a \in A,\, b \in B\}| \ge N^{1+\beta}.$$

6.4.45 Theorem [370] Let A, B be subsets of $\mathbb{F}_p$ with $|A| < p^{1/2}$, $|B| < p^{1/2}$. Then

$$\max_{(a,p)=1} \Big| \sum_{x\in A} \sum_{y \in B} e_p(a(xy + x^2y^2)) \Big| < p^{1-\gamma},$$

where $\gamma > 0$ is an absolute constant.

6.4.46 Remark From Theorem 6.4.45 one can deduce that there are absolute constants $\rho > 0$ and $\gamma > 0$ such that for any subsets $A, B \subset \mathbb{F}_p$ of cardinalities $|A| > p^{1/2-\rho}$, $|B| > p^{1/2-\rho}$, the following bound holds:

$$\Big| \sum_{x\in A} \sum_{y\in B} \mathrm{sign}\Big(\sin\Big(\frac{2\pi}{p}(xy + x^2y^2)\Big)\Big)\Big| < p^{-\gamma}|A||B|.$$

Thus, at the same time Bourgain constructed 2-source extractors with entropies less than 1/2, breaking the barrier 1/2. Regarding the problem of extractors and applications of sum-product results to problems of computer science, see [200, 268, 370].

6.4.47 Definition Let G be a finite group, $A \subset G$ be a set of generators of G (that is, every $g \in G$ can be expressed as a product of elements of $A \cup A^{-1}$). The *Cayley graph* $\Gamma(G, A)$ is the graph (V, E) with vertex set $V = G$ and edge set $E = \{(ag, g) : g \in G, a \in A\}$. The *diameter* of a graph $X = (V, E)$ is $\max_{v_1,v_2 \in V} d(v_1, v_2)$, where $d(v_1, v_2)$ is the length of the shortest path between v_1 and v_2 in X.

6.4.48 Theorem [1463] Let p be a prime. Let A be a set of generators of $G = SL_2(\mathbb{Z}/p\mathbb{Z})$. Then the Cayley graph $\Gamma(G, A)$ has diameter $O((\log p)^c)$, where c and the implied constant are absolute.

6.4.49 Remark Theorem 6.4.48 is due to Helfgott [1463], the sum-product estimate played a crucial role in the proof of this result. It initiated a series of very important works in the diameter problem and expansion theory of Cayley graphs of finite groups. See also [376, 377, 1464, 1466] for further references.

6.4.50 Remark The sum product estimates have found numerous applications in many other problems. For instance, in the work of Chang [583] the sum-product estimate and its versions have found a number of applications to multiplicative character sum estimates. Cochrane and Pinner [656] applied the sum-product estimates to finite field versions of the Waring problem. Ostafe and Shparlinski [2333] applied the result of Glibichuk and Rudnev [1282] to a version of the Waring problem with Dickson's function.

References Cited: [53, 193, 194, 196, 200, 268, 368, 370, 371, 373, 376, 377, 378, 379, 380, 381, 456, 488, 587, 643, 654, 656, 955, 1192, 1193, 1194, 1195, 1282, 1284, 1347, 1424, 1425, 1463, 1464, 1465, 1466, 1696, 1697, 2143, 2333, 2500, 2651, 2738, 2781, 2876]

7

Equations over finite fields

7.1 General forms

Daqing Wan, University of California Irvine

7.1.1 Remark There are lots of results on equations over finite fields. In this section, we give a collection of sample results and examples focusing on hypersurfaces. Additional results can be found in [1708, 2548, 2902] and in the references of this section.

7.1.1 Affine hypersurfaces

7.1.2 Definition For a polynomial $f(x_1,\dots,x_n) \in \mathbb{F}_q[x_1,\dots,x_n]$, let A_f denote the *affine hypersurface* in the affine n-space $\mathbb{A}^n$ defined by the equation $f = 0$. Then, $A_f(\mathbb{F}_q)$ denotes the set of $\mathbb{F}_q$-rational points $(x_1,\dots,x_n) \in \mathbb{F}_q^n$ such that

$$f(x_1,\dots,x_n) = 0.$$

The notation $\#A_f(\mathbb{F}_q)$ denotes the cardinality of the set $A_f(\mathbb{F}_q)$, namely, the number of $\mathbb{F}_q$-rational points on the affine hypersurface A_f.

7.1.3 Remark More generally, for a system of polynomials $f_1,\dots,f_m \in \mathbb{F}_q[x_1,\dots,x_n]$, let $A_{f_1,\cdots,f_m}(\mathbb{F}_q)$ denote the set of $\mathbb{F}_q$-rational points $(x_1,\dots,x_n) \in \mathbb{F}_q^n$ such that

$$f_1(x_1,\dots,x_n) = \cdots = f_m(x_1,\dots,x_n) = 0.$$

It is straightforward to check the simple relation

$$\#A_{f_1,\dots,f_m}(\mathbb{F}_q) = \frac{1}{q^m - q^{m-1}}(\#A_{y_1 f_1+\cdots+y_m f_m}(\mathbb{F}_q) - q^{m+n-1}),$$

where $y_1, \ldots, y_m$ are variables. Thus, for simplicity and without too much loss of generality, we shall focus on the hypersurface case.

7.1.4 Theorem [2548] For a non-zero polynomial f of (total) degree d in $\mathbb{F}_q[x_1, \ldots, x_n]$, we have

1. $0 \leq \#A_f(\mathbb{F}_q) \leq dq^{n-1}$.
2. If $\#A_f(\mathbb{F}_q) \geq 1$, then $\#A_f(\mathbb{F}_q) \geq q^{n-d}$.
3. If f is homogenous of degree d, then $q^{n-d} \leq \#A_f(\mathbb{F}_q) \leq d(q^{n-1} - 1) + 1$.

7.1.5 Definition Let $\Omega_{d,n}(q)$ denote the $\mathbb{F}_q$-vector space of polynomials in $\mathbb{F}_q[x_1, ..., x_n]$ with degree at most d.

7.1.6 Theorem [2548] For positive integers d and n, we have

$$\frac{1}{|\Omega_{d,n}(q)|} \sum_{f \in \Omega_{d,n}(q)} \#A_f(\mathbb{F}_q) = q^{n-1},$$

$$\frac{1}{|\Omega_{d,n}(q)|} \sum_{f \in \Omega_{d,n}(q)} (\#A_f(\mathbb{F}_q) - q^{n-1})^2 = q^{n-1} - q^{n-2}.$$

7.1.7 Remark The above average or probabilistic result suggests that for most polynomials $f \in \Omega_{d,n}(q)$, one should expect that $\#A_f(\mathbb{F}_q)$ is approximately q^{n-1} for large q. This is indeed the case, as the next few theorems show. The first one is an effective version of the Lang-Weil theorem.

7.1.8 Definition A polynomial $f \in F[x_1, \ldots, x_n]$ over a field F is *absolutely irreducible* if it is irreducible over an algebraic closure of F.

7.1.9 Theorem [473, 1849] If $f \in \mathbb{F}_q[x_1, \ldots, x_n]$ is an absolutely irreducible polynomial over $\mathbb{F}_q$ of degree $d > 0$, then

$$|\#A_f(\mathbb{F}_q) - q^{n-1}| \leq (d-1)(d-2)q^{n-\frac{3}{2}} + 5d^{13/3}q^{n-2}.$$

If in addition, $q > 15d^{13/3}$, one has

$$|\#A_f(\mathbb{F}_q) - q^{n-1}| \leq (d-1)(d-2)q^{n-\frac{3}{2}} + (5d^2 + d + 1)q^{n-2}.$$

7.1.10 Definition A polynomial $f \in \mathbb{F}_q[x_1, \ldots, x_n]$ (or the affine hypersurface it defines) is *smooth* if the following system of equations

$$\frac{\partial f}{\partial x_1} = \cdots = \frac{\partial f}{\partial x_n} = f = 0$$

has no solutions in the algebraic closure of $\mathbb{F}_q$. A homogenous polynomial f (or the projective hypersurface it defines) is *smooth* if the above system of equations has no solutions other than possibly the trivial one $(0, \ldots, 0)$.

7.1.11 Theorem Let $f \in \mathbb{F}_q[x_1, \ldots, x_n]$ be a smooth non-homogenous polynomial of degree d such that its homogenous leading form is also smooth. Then,

$$|\#A_f(\mathbb{F}_q) - q^{n-1}| \leq (d-1)^n q^{(n-1)/2}.$$

7.1.12 Remark This result follows from Deligne's well known theorem for projective hypersurfaces in the next subsection. One simply applies Deligne's theorem twice, to the projective closure of A_f (which is itself smooth) and to the infinite part of A_f.

7.1.13 Definition The *singular locus* $Sing(f)$ of a polynomial $f \in \mathbb{F}_q[x_1, \ldots, x_n]$ (or the affine hypersurface it defines) is the affine algebraic set in $\mathbb{A}^n$ defined by the following system of equations

$$\frac{\partial f}{\partial x_1} = \cdots = \frac{\partial f}{\partial x_n} = f = 0.$$

Similarly, the singular locus $Sing(f)$ of a homogeneous polynomial f (or the projective hypersurface it defines) is the projective algebraic set in the projective space $\mathbb{P}^{n-1}$ defined by the same system of equations.

7.1.14 Theorem Let $f \in \mathbb{F}_q[x_1, \ldots, x_n]$ be a polynomial of degree $d > 0$ such that the singular loci of both f and its homogenous leading form have dimension at most s for some integer $s \geq -1$. Then,

$$|\#A_f(\mathbb{F}_q) - q^{n-1}| \leq C_{d,n} q^{(n+s)/2},$$

where $C_{d,n}$ is an explicit constant depending only on d and n.

7.1.15 Remark This result is the affine version of the Hooley-Katz theorem [1533] which unifies the previous two theorems, at least in a qualitative way. If f is absolutely irreducible, then we can take $s \leq n-2$ and this recovers the Lang-Weil theorem. If f and its leading form are both smooth, then we can take $s = -1$ and this recovers Deligne's theorem.

7.1.2 Projective hypersurfaces

7.1.16 Definition For a homogeneous polynomial $f \in \mathbb{F}_q[x_0, x_1, \ldots, x_n]$ of degree d, let P_f denote the *projective hypersurface* in $\mathbb{P}^n$ defined by $f = 0$. Thus, $P_f(\mathbb{F}_q)$ denotes the set of projective solutions $(x_0, \ldots, x_n) \in \mathbb{F}_q^{n+1}$ of the equation

$$f(x_0, \ldots, x_n) = 0,$$

where two solutions are identified if they differ by a non-zero scalar multiple.

7.1.17 Theorem For a homogeneous polynomial $f \in \mathbb{F}_q[x_0, x_1, \ldots, x_n]$ of degree $d > 0$, we have

$$\frac{q^{n+1-d} - 1}{q - 1} \leq \#P_f(\mathbb{F}_q) \leq d\frac{q^n - 1}{q - 1}.$$

7.1.18 Remark This is simply a consequence of the corresponding affine theorem in the previous subsection.

7.1.19 Theorem [795] If $f \in \mathbb{F}_q[x_0, x_1, \ldots, x_n]$ is a smooth homogeneous polynomial of degree $d > 0$, then

$$\left|\#P_f(\mathbb{F}_q) - \frac{q^n - 1}{q - 1}\right| \leq \frac{d-1}{d}((d-1)^n - (-1)^n)q^{(n-1)/2}.$$

7.1.20 Theorem [1270, 1533] If $f \in \mathbb{F}_q[x_0, x_1, \ldots, x_n]$ is a homogeneous polynomial of degree $d > 0$ such that the projective singular locus $Sing(f)$ is of dimension at most s for some integer $s \geq -1$, then

$$\left|\#P_f(\mathbb{F}_q) - \frac{q^n - 1}{q - 1}\right| \leq C_{d,n} q^{(n+s)/2},$$

where $C_{d,n}$ is an explicit constant depending only on d and n.

7.1.21 Remark Similar results hold for more general complete intersections, see [1270].

7.1.3 Toric hypersurfaces

7.1.22 Definition Let $f \in \mathbb{F}_q[x_1^{\pm 1}, \ldots, x_n^{\pm 1}]$. The *Newton!polytope* $\Delta(f)$ of f is the convex closure in $\mathbb{R}^n$ of the exponents of the non-zero monomials in f. We shall assume that $\Delta(f) = \Delta$ is a fixed n-dimensional integral polytope in $\mathbb{R}^n$.

7.1.23 Definition A Laurent polynomial $f \in \mathbb{F}_q[x_1^{\pm 1}, \ldots, x_n^{\pm 1}]$ is Δ-*regular* if $\Delta(f) = \Delta$ and for every face δ (of any dimension) of Δ, the system

$$f^\delta = x_1 \frac{\partial f^\delta}{\partial x_1} = \cdots = x_n \frac{\partial f^\delta}{\partial x_n} = 0$$

has no solutions in $(\bar{\mathbb{F}}_q^*)^n$, where

$$f^\delta = \sum_{u \in \delta \cap \mathbb{Z}^n} a_u x^u$$

is the restriction of f to the face δ.

7.1.24 Definition For a Laurent polynomial $f \in \mathbb{F}_q[x_1^{\pm 1}, \ldots, x_n^{\pm 1}]$, let T_f denote the toric affine hypersurface in the n-torus $\mathbb{G}_m^n$ defined by $f = 0$. Thus, $T_f(\mathbb{F}_q)$ is the set of n-tuples $(x_1, \ldots, x_n) \in (\mathbb{F}_q^*)^n$ such that $f(x_1, \ldots, x_n) = 0$.

7.1.25 Theorem [23, 812, 2902] Assume that $f \in \mathbb{F}_q[x_1^{\pm 1}, \ldots, x_n^{\pm 1}]$ is Δ-regular. Then,

$$\left| \#T_f(\mathbb{F}_q) - \frac{(q-1)^n - (-1)^n}{q} \right| \le (n!\mathrm{Vol}(\Delta) - 1)q^{(n-1)/2}.$$

7.1.26 Example [2471] Let $f = x_1 + \cdots + x_n + \frac{1}{x_1 \cdots x_n} - \lambda$, where $\lambda \in \mathbb{F}_q$ and $n \ge 2$. Assume that the characteristic p does not divide $n+1$. If $(\lambda/(n+1))^{n+1} \ne 1$, then f is $\Delta(f)$-regular and thus

$$\left| \#T_f(\mathbb{F}_q) - \frac{(q-1)^n - (-1)^n}{q} \right| \le nq^{(n-1)/2}.$$

If $(\lambda/(n+1))^{n+1} = 1$, then f is singular and we have the improved bound

$$\left| \#T_f(\mathbb{F}_q) - \frac{(q-1)^n - (-1)^n}{q} \right| \le (n-1)q^{(n-1)/2}.$$

If, in addition, n is even, then we have the even better bound

$$\left| \#T_f(\mathbb{F}_q) - \frac{(q-1)^n - (-1)^n}{q} \right| \le (n-2)q^{(n-1)/2} + q^{(n-2)/2}.$$

These results are used to obtain improved bounds for the number of elements with given trace and norm, see [2125]. The toric hypersurface T_f in this example is a toric Calabi-Yau hypersurface. It is the most important example in arithmetic mirror symmetry, see [2900].

7.1.27 Remark For results on more general toric complete intersections, see [27].

7.1.4 Artin-Schreier hypersurfaces

7.1.28 **Definition** Let $f \in \mathbb{F}_q[x_1, \ldots, x_n]$ be a polynomial of degree $d > 0$. Let AS_f denote the *Artin-Schreier hypersurface* in $\mathbb{A}^{n+1}$ defined by the equation

$$y^q - y = f(x_1, \ldots, x_n).$$

For any positive integer k, let $\#AS_f(\mathbb{F}_{q^k})$ denote the number of $\mathbb{F}_{q^k}$-rational points on AS_f.

7.1.29 **Remark** We note that for $k = 1$, one has the relation $\#AS_f(\mathbb{F}_q) = q\#A_f(\mathbb{F}_q)$, where A_f is the affine hypersurface in $\mathbb{A}^n$ defined by $f = 0$.

7.1.30 **Theorem** Assume that d is not divisible by p and the leading form of f is smooth. Then,

$$|\#AS_f(\mathbb{F}_{q^k}) - q^{kn}| \leq (d-1)^n q^{(kn+2)/2}.$$

In particular, taking $k = 1$, we deduce the estimate

$$|\#A_f(\mathbb{F}_q) - q^{n-1}| \leq (d-1)^n q^{n/2}.$$

7.1.31 **Remark** This is simply a consequence of Deligne's estimate [795] for exponential sums, see Section 6.2 for more details. It can be improved and generalized in various ways as the next few theorems show. The next theorem is a consequence of Katz's estimate [1704] for singular exponential sums.

7.1.32 **Theorem** Assume that d is not divisible by p and the singular locus of the leading form of f has dimension at most s for some integer $s \geq -1$. Then,

$$|\#AS_f(\mathbb{F}_{q^k}) - q^{kn}| \leq C_{d,n} q^{(kn+s+3)/2},$$

where $C_{d,n}$ is an explicit constant depending only on d and n. In particular, taking $k = 1$, we deduce the estimate

$$|\#A_f(\mathbb{F}_q) - q^{n-1}| \leq C_{d,n} q^{(n+s+1)/2}.$$

7.1.33 **Remark** In the case $s = -1$, that is, the leading form of f is smooth, the above result reduces to Deligne's estimate. In this case, the subscheme defined by the Jacobian ideal $< \partial f/\partial x_1, \ldots, \partial f/\partial x_n >$ has at most a finite number of points and the above Deligne estimate can be improved further in some cases.

7.1.34 **Theorem** [2472] Assume that d is not divisible by $p > 2$ and the leading form of f is smooth. Assume further that the Jacobian subscheme in $\mathbb{A}^n$ defined by the ideal $< \partial f/\partial x_1, \ldots, \partial f/\partial x_n >$ is a zero-dimensional smooth variety and the image of its $\bar{\mathbb{F}}_q$-points under f are distinct (a Morse condition). If nk is even, suppose additionally that the hypersurface defined by

$$f(x_{11}, \ldots, x_{1n}) + \cdots + f(x_{k1}, \ldots, x_{kn}) = 0$$

is also smooth in $\mathbb{A}^{kn}$. Then,

$$|\#AS_f(\mathbb{F}_{q^k}) - q^{kn}| \leq C_{d,k,n} q^{(kn+1)/2},$$

for some explicit constant $C_{d,k,n}$.

7.1.35 Corollary Assume d is not divisible by $p > 2$, the leading form of f is smooth, and the Jacobian subscheme in $\mathbb{A}^n$ defined by the ideal $< \partial f/\partial x_1, \ldots, \partial f/\partial x_n >$ is a zero-dimensional smooth variety and the images of its $\bar{\mathbb{F}}_q$-points under f are distinct. Assume that n is odd, then

$$|\#A_f(\mathbb{F}_q) - q^{n-1}| \leq C_{d,1,n} q^{(n-1)/2}.$$

7.1.36 Definition Let $f(x, y) \in \mathbb{F}_q[x_1, \ldots, x_n, y_1, \ldots, y_{n'}]$ be a polynomial with two sets of variables, where $n, n' \geq 1$. For a positive integer k, let $N_k(f)$ denote the number of solutions of the equation

$$x_0^p - x_0 = f(x_1, \ldots, x_n, y_1, \ldots, y_{n'})$$

such that $(x_0, x_1, \ldots, x_n) \in \mathbb{F}_{q^k}^{n+1}$ and $(y_1, \ldots, y_{n'}) \in \mathbb{F}_q^{n'}$.

7.1.37 Remark We note that the two sets of coordinates in the previous definition run over different extension fields of $\mathbb{F}_q$.

7.1.38 Definition Let $f(x, y)$ be a polynomial as in the previous definition. For a positive integer k, we define the *k-th fibred sum of f along y* to be the new polynomial

$$\oplus_y^k f := f(x_{11}, \ldots, x_{1n}, y_1, \ldots, y_{n'}) + \cdots + f(x_{k1}, \ldots, x_{kn}, y_1, \ldots, y_{n'}).$$

7.1.39 Theorem [1140] Given f of degree $d > 0$ as above. Let f_d be the homogeneous leading form of f. Assume that the k-th fibred sum $\oplus_y^k f_d$ is smooth in $\mathbb{P}^{kn+n'-1}$ and assume that d is not divisible by p. Then, we have the following two estimates

$$|N_k(f) - q^{kn+n'}| \leq (p-1)(d-1)^{kn+n'} q^{(kn+n')/2},$$

$$|N_k(f) - q^{kn+n'}| \leq c(p, n, n') k^{3(d+1)^n - 1} q^{(kn+n')/2},$$

where the constant $c(p, n, n')$, depending only on p, n, n', is not known to be effective if $n' \geq 2$.

7.1.40 Remark In the case $k = 1$, the first estimate reduces to Deligne's theorem as above. More generally, for a degree d polynomial $f \in \mathbb{F}_q[x_1, \ldots, x_n]$ and positive integers $k_1, \ldots, k_n$, let $N_{k_1,\ldots,k_n}(f)$ denote the number of solutions of the equation $f(x_1, \ldots, x_n) = 0$ with $x_i \in \mathbb{F}_{q^{k_i}}$ for all $1 \leq i \leq n$. Under suitable nice conditions, one expects [2898] the estimate of the type

$$|N_{k_1,\ldots,k_n}(f) - q^{k_1+\cdots+k_n-k}| \leq C_{d,n,k} q^{(k_1+\cdots+k_n)/2},$$

where k is the least common multiple of the integers k_i. The above theorem is one example of this kind. For two additional examples, see [1705, 2471].

7.1.5 Kummer hypersurfaces

7.1.41 Definition Let $f \in \mathbb{F}_q[x_1, \ldots, x_n]$ be a polynomial of degree $d > 0$. For a positive integer m not divisible by p, let $K_{f,m}$ denote the *Kummer hypersurface* in $\mathbb{A}^{n+1}$ defined by

$$y^m = f(x_1, \ldots, x_n).$$

7.1.42 Theorem [1707, 2468] Assume that f is smooth in $\mathbb{A}^n$ and its leading form f_d is also smooth in $\mathbb{P}^{n-1}$. Then,

$$|\#K_{f,m}(\mathbb{F}_q) - q^n| \le (m-1)(d-1)^n q^{n/2}.$$

7.1.43 Remark Since $\#K_{f,m}(\mathbb{F}_q) = \#K_{f,(m,q-1)}(\mathbb{F}_q)$, we may assume that m divides $q-1$. Write $m = (q-1)/e$, then for fixed e, the above bound reduces to

$$|\#K_{f,m}(\mathbb{F}_q) - q^n| \le \frac{1}{e}(d-1)^n q^{(n+2)/2}.$$

Similar to the Artin-Schreier hypersurface case, we expect that for fixed e, under suitable hypotheses, an improved estimate of the form

$$|\#K_{f,m}(\mathbb{F}_q) - q^n| \le c(n,k,d) q^{(n+1)/2},$$

see Rojas-Leon [2470] for new results in this direction.

7.1.6 p-Adic estimates

7.1.44 Remark In this subsection, we review several results on p-divisibility of the solution number for an affine algebraic set, where p is the characteristic of the finite field $\mathbb{F}_q$. We begin with the well known Chevalley-Warning theorem.

7.1.45 Theorem [2548] Let $f_1, \dots f_m \in \mathbb{F}_q[x_1, \dots, x_n]$ be polynomials of degrees $d_1, \dots, d_m$ respectively. Assume that $n > d_1 + \cdots + d_m$. Then,

$$\#A_{f_1,\dots,f_m}(\mathbb{F}_q) \equiv 0 \pmod{p}.$$

7.1.46 Theorem Let $f_1, \dots f_m \in \mathbb{F}_q[x_1, \dots, x_n]$ be polynomials of degrees $d_1, \dots, d_m$ respectively. Let μ be the smallest integer such that

$$\mu \ge \frac{n - (d_1 + \cdots + d_m)}{\max_i d_i}.$$

Then,

$$\#A_{f_1,\dots,f_m}(\mathbb{F}_q) \equiv 0 \pmod{q^{\mu}}.$$

7.1.47 Remark This theorem is due to Ax [150] in the case $m = 1$ and to Katz [1698] for general m. Ax's proof is elementary and based on the Stickelberger relation for Gauss sums (for Gauss sums, see Section 6.1). Katz's proof uses more advanced methods from Dwork's p-adic theory. In [2907], it was noted that Katz's result can be proved by Ax's more elementary method. Later, a short elementary reduction of Katz's theorem for general m to Ax's theorem for $m = 1$ was given in [1540]. This is consistent with Remark 7.1.3 that a system of equations can often be reduced to the one equation case. Additional simpler proofs to more general results are given in [2914]. If one takes into account the actual terms (the polytope) of the polynomials $f_1, \dots, f_m$, the Ax-Katz theorem can be generalized or improved in certain cases, see Adolphson-Sperber [21] for p-divisibility of exponential sums. We next describe its consequence for equations over finite fields.

7.1.48 Theorem Let Δ be an integral convex polytope in $\mathbb{R}^{m+n}$ which contains all the exponents of the monomials in the polynomial $y_1 f_1(x_1, \dots, x_n) + \cdots + y_m f_m(x_1, \dots, x_n)$ with coefficients in $\mathbb{F}_q$. Let ω be the smallest positive integer such that the dilation $\omega\Delta$ contains a lattice point in $\mathbb{Z}^{m+n}$ with all coordinates positive. Then,

$$\#A_{f_1,\dots,f_m}(\mathbb{F}_q) \equiv 0 \pmod{q^{\omega - m}}.$$

7.1.49 Remark For non-prime fields (i.e., q is not a prime), the Ax-Katz theorem can be improved in some cases by considering the p-weights of the exponents of the polynomials f_i and the Weil descent, see Moreno-Moreno [2151]. Let $q = p^r$, and let $\{\alpha_1, \ldots, \alpha_r\}$ be an $\mathbb{F}_p$-basis of $\mathbb{F}_q$. Then, any element $x_i \in \mathbb{F}_q$ can be written uniquely as

$$x_i = \sum_{j=1}^{r} x_{ij}\alpha_j,\ x_{ij} \in \mathbb{F}_p.$$

Let $e = e_0 + e_1 p + \cdots + e_{r-1}p^{r-1}$ be the p-digit expansion of an integer $e \in [0, q-1]$. Then, one has the relation

$$x_i^e = \prod_{j=0}^{r-1} (x_{i1}\alpha_1^{p^j} + \cdots + x_{ir}\alpha_r^{p^j})^{e_j}.$$

In this way, one finds a system of mr polynomials $g_{ij} \in \mathbb{F}_p[x_{11}, \ldots, x_{mr}]$ such that

$$\#A_{f_1,\ldots,f_m}(\mathbb{F}_q) = \#A_{g_{11},\ldots,g_{mr}}(\mathbb{F}_p).$$

One can then apply the Ax-Katz theorem to the right side over the prime field $\mathbb{F}_p$.

7.1.50 Remark For a homogeneous polynomial $f \in \mathbb{F}_q[x_0, \ldots, x_n]$ of degree $d > 0$, the Ax-Katz theorem implies that the projective hypersurface P_f in $\mathbb{P}^n$ satisfies the congruence

$$\#P_f(\mathbb{F}_q) \equiv \#\mathbb{P}^{n-1}(\mathbb{F}_q) \pmod{q^u},$$

where μ is the smallest positive integer greater than or equal to $(n+1-d)/d$. This gives a non-trivial congruence only in the case $n+1 > d$ (Fano varieties). In the case $n+1 = d$ (Calabi-Yau variety) or $n+1 < d$ (varieties of general type), one cannot expect such general p-divisibility results. However, the following example suggests that one may still expect some congruences for some pair of varieties. For $\lambda \in \mathbb{F}_q$, let X_λ denote the Dwork family of Calabi-Yau hypersurfaces in $\mathbb{P}^n$ defined by the equation

$$x_0^{n+1} + \cdots + x_n^{n+1} + \lambda x_0 x_1 \cdots x_n = 0.$$

Let Y_λ be the projective closure in the projective toric variety $\mathbb{P}_\Delta$ of the toric affine hypersurface defined by

$$x_1 + \cdots + x_n + \frac{1}{x_1 \cdots x_n} + \lambda = 0,$$

where Δ is the simplex in $\mathbb{R}^n$ with vertices

$$\{(1,0,\ldots,0), \ldots, (0,\ldots,0,1), (-1,-1,\ldots,-1)\}.$$

Then, we have the mirror congruence [2900]

$$\#X_\lambda(\mathbb{F}_q) \equiv \#Y_\lambda(\mathbb{F}_q) \pmod{q}.$$

In this example, the mirror variety Y_λ is a quotient of X_λ by a finite group G, see [254, 1141] for extensions of such congruence to a pair of more general quotient varieties.

7.1.51 Remark From a zeta function point of view, see Section 12.7 for more on zeta functions, the p-adic estimate in this subsection corresponds to an estimate for the first non-trivial slope. The study of all slopes for the zeta function is significantly deeper, and the reader is refered to Section 12.8.

See Also

§6.1	For Gauss, Jacobi, and Kloosterman sums.
§6.2	For other related results to character sums.
§7.3	For diagonal equations.
§11.6	For discrete logarithms.
§12.7	For zeta functions.
§12.8	For p-adic estimates of zeta functions.

References Cited: [21, 23, 27, 150, 254, 473, 795, 812, 1140, 1141, 1270, 1533, 1540, 1698, 1704, 1705, 1707, 1708, 1849, 2125, 2151, 2468, 2470, 2471, 2472, 2548, 2898, 2900, 2902, 2907, 2914]

7.2 Quadratic forms

Robert Fitzgerald, Southern Illinois University

7.2.1 Basic definitions

7.2.1 **Definition** A *quadratic form* f over a field F is a homogeneous polynomial over F of degree two:

$$f(X) = \sum_{i \le j} a_{ij} x_i x_j \qquad a_{ij} \in F,$$

where X is the (column) vector of the variables x_i.

Let CM_f be the upper triangular matrix of coefficients. Two quadratic forms, f and g are *equivalent*, denoted $f \simeq g$, if there is an invertible matrix P over F such that $f(PX) = g(X)$.

7.2.2 **Remark** The theory is different in the characteristic 2 and odd characteristic cases.

7.2.3 **Definition** A *quadratic space* is a pair (Q, V) where V is a finite dimensional vector space over F and $Q : V \to F$ satisfies

1. $Q(\lambda v) = \lambda^2 v$, for all $\lambda \in F$ and $v \in V$, and
2. for char$(F) = 2$, we require $B_Q(v, w) = Q(v + w) + Q(v) + Q(w)$ to be a symmetric bilinear form; for char$(F) \neq 2$, we require that $B_Q(v, w) = \frac{1}{2}(Q(v+w) - Q(v) - Q(w))$ to be a symmetric bilinear form.

7.2.4 **Remark** Each choice of a basis $\{e_1, \ldots, e_n\}$ of V yields a quadratic form $f(X) = Q(\sum_i x_i e_i)$; a different choice of basis yields an equivalent form.

7.2.5 **Remark** Suppose char$(F) = 2$. The associated matrix of f is CM_f. Then $f(X) = X^T CM_f X$. There are many matrices M with $f(X) = X^T M X$ but CM_f is the only upper triangular one. The matrix for the associated symmetric bilinear form $b_f(X, Y) =$

$B_Q(\sum_i x_i e_i, \sum_j y_j e_j)$ is $CM_f + CM_f^T$. Inequivalent f may have the same associated symmetric bilinear form.

7.2.6 Definition The *radical* of a quadratic space is:

$$\mathrm{rad}(Q) = \{v \in V : B_Q(v, w) = 0 \quad \text{for all } w \in V\}.$$

If $\mathrm{rad}(Q) = 0$, Q is *non-degenerate.*

7.2.7 Remark Continue to assume $\mathrm{char}(F) = 2$. Suppose Q is non-degenerate. Then there is a *symplectic basis* $e_1, f_1, e_2, f_2, \ldots, e_n, f_n$ with associated matrix

$$\begin{pmatrix} \alpha_1 & 1 & 0 & 0 & \ldots & 0 & 0 \\ 0 & \beta_1 & 1 & 0 & \ldots & 0 & 0 \\ 0 & 0 & \alpha_2 & 1 & \ldots & 0 & 0 \\ \vdots & \vdots & \vdots & \vdots & \vdots & \vdots & \vdots \\ 0 & 0 & 0 & 0 & \ldots & \alpha_n & 1 \\ 0 & 0 & 0 & 0 & \ldots & 0 & \beta_n \end{pmatrix}.$$

Let $\mathcal{P}(F)$ be the additive subgroup $\{a^2 + a : a \in F\}$. The *Arf invariant* $\Delta(Q)$ is $\sum \alpha_i \beta_i \in F/\mathcal{P}(F)$. $\Delta(Q)$ does not depend on the choice of the symplectic basis. If Q is degenerate then write $V = \mathrm{rad}(Q) \oplus W$. Then $Q|_W$ is non-degenerate and we define $\Delta(Q) = \Delta(Q|_W)$.

Now suppose $\mathrm{char}(F) \neq 2$. The associated matrix of f is $M_f = \frac{1}{2}(CM_f + CM_f^T)$. Then $f(X) = X^T M_f X$. There are many matrices M with $f(X) = X^T M X$ but M_f is the only symmetric one. The matrix for the associated symmetric bilinear form b_f is also M_f. The function B_Q determines Q (and b_f determines f) via $Q(v) = B_Q(v, v)$. Suppose Q is non-degenerate. Then there is an orthogonal basis with the associated matrix diagonal. Let F^{*2} be the multiplicative subgroup $\{a^2 : a \in F\}$. The determinant of f is $\det M_f \in F^*/F^{*2}$ (equivalently, the product of the entries in the diagonalization) and it does not depend on the choice of basis. Again, if Q is degenerate then write $V = \mathrm{rad}(Q) \oplus W$ and set $\det(Q) = \det(Q|_W)$.

7.2.2 Quadratic forms over finite fields

7.2.8 Remark Details on the results of this section can be found in [1939, Section 6.2].

7.2.9 Theorem Let q be even and let (Q, V) be a quadratic space over $\mathbb{F}_q$. Let $n = \dim V$ and $r = \dim \mathrm{rad}(Q)$. Then:

1. $n - r = 2s$ is even.
2. $|\mathbb{F}_q/\mathcal{P}(\mathbb{F}_q)| = 2$.
3. Fix $0 \neq d \in \mathbb{F}_q/\mathcal{P}(\mathbb{F}_q)$. Then V has a symplectic basis such that the resulting quadratic form is one of the following:
   (a) $\mathcal{E}1: \quad x_1x_2 + x_3x_4 + \cdots + x_{2s-1}x_{2s}$,
   (b) $\mathcal{E}2: \quad x_1^2 + x_1x2 + dx_2^2 + x_3x_4 + \cdots + x_{2s-1}x_{2s}$,
   (c) $\mathcal{E}3: \quad x_0^2 + x_1x_2 + x_3x_4 + \cdots + x_{2s-1}x_{2s}$.
4. $\mathcal{E}1$ occurs if and only if $Q(\mathrm{rad}(Q)) = 0$ and $\Delta(Q) = 0$; $\mathcal{E}2$ occurs if and only if $Q(\mathrm{rad}(Q)) = 0$ and $\Delta(Q) = d \neq 0$; $\mathcal{E}3$ occurs if and only if $Q(\mathrm{rad}(Q)) \neq 0$.

7.2.10 Definition The *rank* of Q is the minimal number of variables in a quadratic form induced from Q.

7.2.11 Remark The rank of Q is $\dim V - \dim \mathrm{rad}(Q)$ except for the case $\mathcal{E}3$ when it is $\dim V - \dim \mathrm{rad}(Q) + 1$.

7.2.12 Theorem Let q be odd and let (Q, V) be a quadratic space over $\mathbb{F}_q$. Let $n = \dim V$ and $r = \dim \mathrm{rad}(Q)$. Write $n - r = 2s$ or $2s + 1$. Then

1. $|\mathbb{F}_q^*/\mathbb{F}_q^{*2}| = 2$.
2. Fix $1 \neq d \in \mathbb{F}_q^*/\mathbb{F}_q^{*2}$. Then V has a basis such the resulting quadratic form is one of the following:
   (a) $\mathcal{O}1:$ $\quad x_1x_2 + x_3x_4 + \cdots + x_{2s-1}x_{2s}$,
   (b) $\mathcal{O}2:$ $\quad x_1^2 - dx_2^2 + x_3x_4 + \cdots + x_{2s-1}x_{2s}$,
   (c) $\mathcal{O}3:$ $\quad x_0^2 + x_1x_2 + x_3x_4 + \cdots + x_{2s-1}x_{2s}$,
   (d) $\mathcal{O}4:$ $\quad dx_0^2 + x_1x_2 + x_3x_4 + \cdots + x_{2s-1}x_{2s}$.
3. In cases $\mathcal{O}1$, $\mathcal{O}3$, $\det Q = (-1)^s$, and in cases $\mathcal{O}2$, $\mathcal{O}4$, $\det Q = (-1)^s d$.

7.2.13 Remark When $\mathrm{char}(F) \neq 2$:

1. we always have $Q(\mathrm{rad}(Q)) = 0$;
2. the rank of Q is always $\dim V - \dim \mathrm{rad}(Q)$; and
3. $xy \simeq x^2 - y^2$ so that each form in Theorem 7.2.12 can be written as a diagonal form.

7.2.14 Definition For a quadratic space (Q, V) over a finite field, let $N(Q = 0)$ denote the number of $v \in V$ such that $Q(v) = 0$.

7.2.15 Theorem Let (Q, V) be a quadratic space over $\mathbb{F}_q$, where q is even. Let $n = \dim V$ and $r = \dim \mathrm{rad}(Q)$. Then

$$N(Q = 0) = \frac{1}{q}\left[q^n + (q-1)\Lambda(Q)\sqrt{q^{n+r}}\right],$$

where

$$\Lambda(Q) = \begin{cases} +1, & \text{if } Q \text{ is type } \mathcal{E}1, \\ -1, & \text{if } Q \text{ is type } \mathcal{E}2, \\ 0, & \text{if } Q \text{ is type } \mathcal{E}3. \end{cases}$$

7.2.16 Theorem Let (Q, V) be a quadratic space over $\mathbb{F}_q$, where q is odd. Let $n = \dim V$ and $r = \dim \mathrm{rad}(Q)$. Then

$$N(Q = 0) = \frac{1}{q}\left[q^n + (q-1)\Lambda(Q)\sqrt{q^{n+r}}\right],$$

where

$$\Lambda(Q) = \begin{cases} +1, & \text{if } Q \text{ is type } \mathcal{O}1, \\ -1, & \text{if } Q \text{ is type } \mathcal{O}2, \\ 0, & \text{if } Q \text{ is type } \mathcal{O}3 \text{ or } \mathcal{O}4. \end{cases}$$

7.2.17 Remark With a little more work [1745, 1746], one can give the number of solutions to $Q(x) = c$, for any scalar c, and even to $Q(x) + F(x) = c$, where $F(x)$ is an arbitrary linear function.

7.2.3 Trace forms

7.2.18 Remark In almost all applications the quadratic spaces arise as follows: Let $F = \mathbb{F}_q$, $K = \mathbb{F}_{q^n}$ and let $L(x) = \sum_{i=0}^{m} \alpha_i x^{q^i}$ be a linearized polynomial over K, see Definition 2.1.103.

7.2.19 Definition A *trace form* over $F = \mathbb{F}_q$ is the quadratic space (Q_L^K, K) where $K = \mathbb{F}_{q^n}$ and $Q_L^K : K \to F$ is given by $Q_L^K(x) = \mathrm{Tr}_{K/F}(x \cdot L(x))$.

7.2.20 Remark If (Q, V) is a quadratic space with $\dim V = n$ then, replacing V by K, Q is a Q_L^K for some linearized $L(x)$. In even characteristic, for $m = \lfloor n/2 \rfloor$, L is uniquely determined except that if $n = 2m$ then α_m is only determined modulo $\mathbb{F}_{q^m}$ [1073]. We set

$$L^*(x) = \sum_{i=1}^{m} \alpha_i (x^{q^{m+i}} - x^{q^{m-i}}).$$

7.2.21 Proposition For F, K and L as above:

1. [1073] $|\mathrm{rad}(Q_L^K)|$ is equal to the number of roots of $L^*(x)$ in K.
2. [1075] If all $\alpha_i \in F$ then $\dim \mathrm{rad}(Q_L^K) = \deg(L^*(x)_{dn}, x^k - 1)$. Here L^*_{dn} is the polynomial associated to L^*, namely

$$L^*(x)_{dn} = \sum_{i=1}^{m} \alpha_i (x^{m+i} - x^{m-i}).$$

7.2.22 Remark The invariants $\dim \mathrm{rad}(Q) = d$ and $\Lambda(Q)$ are completely determined only in the case of q even and L having one term. Let $v_2(k)$ demote the highest power of 2 dividing k.

7.2.23 Theorem Let $F = \mathbb{F}_q$ with q even. Let $Q(x) = \mathrm{Tr}_{K/F}(\gamma x \cdot x^{q^a})$, where $K = \mathbb{F}_{q^n}$ and $\gamma \in K$. Set $d = (n, a)$.

1. [1747] If $v_2(n) \le v_2(a)$ then $\dim \mathrm{rad}(Q) = d$ and $\Lambda(Q) = 0$.
2. [1745] If $v_2(n) = v_2(a) + 1$ then

$$(\dim \mathrm{rad}(Q), \Lambda(Q)) = \begin{cases} (2d, +1) & \text{if } \gamma \text{ is a } (q^a+1)\text{-th power in } K, \\ (0, -1) & \text{if } \gamma \text{ is not a } (q^a+1)\text{-th power in } K. \end{cases}$$

3. [1745] If $v_2(n) > v_2(a) + 1$ then

$$(\dim \mathrm{rad}(Q), \Lambda(Q)) = \begin{cases} (2d, -1) & \text{if } \gamma \text{ is a } (q^a+1)\text{-th power in } K, \\ (0, +1) & \text{if } \gamma \text{ is not a } (q^a+1)\text{-th power in } K. \end{cases}$$

7.2.24 Theorem [1746] Let $F = \mathbb{F}_q$ with q odd. Let $Q(x) = \mathrm{Tr}_{K/F}(\gamma x \cdot x^{q^a})$, where $K = \mathbb{F}_{q^n}$ and $\gamma \in K$. Set $d = (n, a)$. Let ω be a primitive element of K and write $\gamma = \omega^g$ for some $0 \le g < q^n - 1$.

1. If $v_2(n) \le v_2(a)$ then $\dim \mathrm{rad}(Q) = 0$.
2. If $v_2(n) = v_2(a) + 1$ then

$$(\dim \mathrm{rad}(Q), \Lambda(Q)) = \begin{cases} (2d, +1), & \text{if } g \equiv \frac{1}{2}(q^d + 1) \pmod{q^d + 1}, \\ (0, -1), & \text{if } g \not\equiv \frac{1}{2}(q^d + 1) \pmod{q^d + 1}. \end{cases}$$

3. If $v_2(n) > v_2(a) + 1$ then

$$(\dim \operatorname{rad}(Q), \Lambda(Q)) = \begin{cases} (2d, -1), & \text{if } g \equiv 0 \pmod{q^d + 1}, \\ (0, +1), & \text{if } g \not\equiv 0 \pmod{q^d + 1}. \end{cases}$$

7.2.25 Problem Find the value of $\Lambda(Q)$ in Case 1 above.

7.2.26 Remark In the following we use the Jacobi symbol $(\frac{a}{n})$.

7.2.27 Theorem [1074] Let $F = \mathbb{F}_2$, $K = \mathbb{F}_{2^n}$ and $Q(x) = \text{Tr}_{K/F}(x(x^{2^a} + x^{2^b}))$ with $a < b$. Set $d = (b - a, b + a)$ and $M = \max\{v_2(b - a), v_2(b + a)\}$. Then

$$\dim \operatorname{rad}(Q) = \begin{cases} (b - a, n) + (b + a, n) - (d, n) & \text{if } v_2(n) \leq M, \\ (b - a, n) + (b + a, n) & \text{if } v_2(n) > M. \end{cases}$$

We have $\Lambda(Q) = 0$ if and only if $v_2(b - a) = v_2(b + a) = v_2(n) - 1$. For an intermediate field $E = \mathbb{F}_{2^m}$ write $\Lambda(m)$ for the invariant of $\text{Tr}_{E/F}(x(x^{2^a} + x^{2^b}))$. Let $k = v_2(n)$.

1. If n is odd then $\Lambda(n) = \prod(\frac{2}{p})$, where the product is over odd prime divisors p of n with $v_p(n) + \min\{v_p(n), \max\{v_p(b - a), v_p(b + a)\}\}$ odd.
2. If n is even and $b \pm a$ is odd then $\Lambda(n) = \prod(\frac{2}{p})\Lambda(2^k)$, where the product is over odd prime divisors p of n with $\min\{v_p(n), v_p(b - a)\} + \min\{v_p(n), v_p(b + a)\}$ odd.
3. If n is even and $b \pm a$ is even then $\Lambda(n) = \Lambda(2^k)$.
4. For $n = 2^k$:
   (a) If $k \leq M$ then $\Lambda(n) = +1$.
   (b) If $k = 1 + M$ and $v_2(b - a) \neq v_2(b + a)$ then $\Lambda(n) = -1$.
   (c) If $k = 1 + M$ and $v_2(b - a) = v_2(b + a)$ then $\Lambda(n) = 0$.
   (d) If $k \geq 2 + M$ and $v_2(b - a) \neq v_2(b + a)$ then $\Lambda(n) = -1$.
   (e) If $k \geq 2 + M$ and $v_2(b - a) = v_2(b + a)$ then
      i. If $k = 2$ and one of a, b is odd and the other is equivalent to $2 \pmod 4$ then $\Lambda(n) = +1$.
      ii. If $k \geq 3$ and one of a, b is odd and the other is divisible by 4 then $\Lambda(n) = +1$.
      iii. Otherwise, $\Lambda(n) = -1$.

7.2.4 Applications

7.2.28 Remark The first appearance of trace forms seems to be Welch's Theorem [231] which is Theorem 7.2.23 for $\gamma = 1$ and $2a$ dividing n. It was used to compute weight enumerators of double-error correcting BCH codes. Other particular trace forms have been used to compute weight enumerators of second order Reed-Muller codes [1991] and minimal codes [3000]. Weights of irreducible codes have been found via counting the number of polynomials L with fixed invariants by Feng and Luo [1056, 1981].

7.2.29 Remark Quadratic forms have been used to construct Artin-Schreier curves $y^q + y = xL(x)$ with many rational points [1072, 1075, 2844]. Quadratic forms over $\mathbb{F}_2$, partitioned by their bilinear forms, were used to construct systems of linked symmetric designs in [483]. Maximal rank quadratic forms give rise to the first examples of bent functions which have cryptographic importance, see Section 9.3. Piecewise functions, with each piece a trace form, yield other bent functions and a presentation of the Kerdock code in [536]. Families of trace forms have been used to construct Gold-like sequences [1077, 1731]. Correlations of maximal and other sequences have been computed with trace forms [1475, 1736, 1745, 1746].

See Also

§9.3, §10.3, §12.6, §15.2	For more information on applications of quadratic forms.

References Cited: [231, 483, 536, 1056, 1072, 1073, 1074, 1075, 1077, 1475, 1731, 1736, 1745, 1746, 1747, 1939, 1981, 1991, 2844, 3000]

7.3 Diagonal equations

Francis Castro and Ivelisse Rubio, University of Puerto Rico

7.3.1 Preliminaries

We present a summary of results on diagonal equations. The selection gives an overview of the area as well as some of its recent developments. General references for diagonal equations are: Chapter 10 in [240], Chapter 8 in [1575], Chapters 3-6 in [1617], Chapter 6, Section 3 in [1939], and Chapter 6 in [2681].

7.3.1 **Definition** A *diagonal equation* over $\mathbb{F}_q$ is an equation of the type

$$c_1x_1^{k_1} + \cdots + c_sx_s^{k_s} = b \tag{7.3.1}$$

for any positive integers $k_1, \ldots, k_s; c_1, \ldots, c_s \in \mathbb{F}_q^*$, and $b \in \mathbb{F}_q$. We denote by N_b the number of solutions in $\mathbb{F}_q^s$ of (7.3.1). A *deformed diagonal equation* is a diagonal equation with $b = g(x_1, \ldots, x_s) \in \mathbb{F}_q[x_1, \ldots, x_s]$.

7.3.2 **Remark** The number of solutions N_b can be expressed in terms of Jacobi and Gauss sums. The relation between Jacobi and Gauss sums and other results on these sums are included in Section 6.3.

7.3.3 **Theorem** [1939, Theorem 6.33] The number N_0 of solutions of (7.3.1) for $b = 0$ is

$$N_0 = q^{s-1} + \sum_{(j_1,\ldots,j_s)\in T} \overline{\chi}_1^{j_1}(c_1)\cdots\overline{\chi}_s^{j_s}(c_s)\, J_0\left(\chi_1^{j_1},\ldots,\chi_s^{j_s}\right), \tag{7.3.2}$$

where T is the set of all $(j_1, \ldots, j_s) \in \mathbb{Z}^s$ such that $1 \le j_i \le d_i - 1$ for $1 \le i \le s$, $\chi_1^{j_1} \ldots \chi_s^{j_s}$ is trivial, χ_i is a multiplicative character of order $d_i = \gcd(k_i, q-1)$, and J_0 is a Jacobi sum.

7.3.4 **Theorem** [1939, Theorem 6.34] The number N_b of solutions of (7.3.1) for $b \in \mathbb{F}_q^*$ is

$$N_b = q^{s-1} + \sum_{j_1=1}^{d_1-1} \cdots \sum_{j_s=1}^{d_s-1} \chi_1^{j_1}\left(bc_1^{-1}\right)\cdots\chi_s^{j_s}\left(bc_s^{-1}\right) J\left(\chi_1^{j_1},\ldots,\chi_s^{j_s}\right), \tag{7.3.3}$$

where χ_i is a multiplicative character of order $d_i = \gcd(k_i, q-1)$, and J is a Jacobi sum.

7.3.5 Remark Results on N_0 and N_b can be related to the s-tuples $(j_1, \ldots, j_s) \in \mathbb{Z}^s$, $j_i \geq 1$, such that

$$\frac{j_1}{k_1} + \cdots + \frac{j_s}{k_s} \tag{7.3.4}$$

is an integer.

7.3.6 Definition Let $I(k_1, \ldots, k_s)$ be the number of s-tuples $(j_1, \ldots, j_s) \in \mathbb{Z}^s$ such that $1 \leq j_i \leq k_i - 1$ and expression (7.3.4) is an integer.

7.3.7 Remark Note that $I(k_1, \ldots, k_s) \leq (k_1 - 1) \cdots (k_s - 1)$. The number $I(k_1, \ldots, k_s)$ can be interpreted as the degree of the numerator of the zeta-function of $\sum_{i=1}^{s} c_i x_i^{k_i}$; see [2681]. This number appears in the estimates (7.3.5) and (7.3.6) for the number of solutions N_0 and N_b, $b \neq 0$, of Equation (7.3.1).

7.3.8 Theorem [1939, Theorems 6.36, 6.37] Let $d_i = \gcd(k_i, q-1)$ for $i = 1, \ldots, s$. Then,

1. For $b = 0$,

$$|N_0 - q^{s-1}| \leq I(d_1, \ldots, d_s)(q-1)q^{\frac{s-2}{2}}. \tag{7.3.5}$$

2. For $b \neq 0$,

$$|N_b - q^{s-1}| \leq \left[(d_1 - 1) \cdots (d_s - 1) - \left(1 - q^{-1/2}\right) I(d_1, \ldots, d_s)\right] q^{\frac{s-1}{2}}. \tag{7.3.6}$$

7.3.9 Remark Note that if q is sufficiently large with respect to $d_1, \ldots, d_s$ then (7.3.5) and (7.3.6) imply that Equation (7.3.1) is solvable. One can obtain an improvement of (7.3.5) and (7.3.6) if the p-weights of all the d_i's are small and the solutions are in $\mathbb{F}_{q^2}$ [1274].

7.3.2 Solutions of diagonal equations

7.3.10 Remark When dealing with any type of equation, the first question that one might ask is whether or not the equation has solutions over a given field. A classical result is Chevalley's theorem [615] that guarantees a non-trivial solution whenever the number of variables is larger than the degree of the polynomial and there is no constant term. In [2151] the authors improve Chevalley's result by considering the p-weight degree of the polynomial instead of its degree. The following results determine solvability of some families of diagonal equations.

7.3.11 Theorem [2809] Let q be any prime power and k a positive integer such that $k \neq p-1$ in the case $q = p$. The equation $\sum_{i=1}^{s} c_i x_i^k = 0$ has a nontrivial solution for $s \geq \frac{k+3}{2}$.

7.3.12 Theorem [2679, Theorems 1, 2] Let $\mathbb{F}_{q'} \subseteq \mathbb{F}_q$ be finite fields, and suppose that $c_1, \ldots, c_s \in \mathbb{F}_{q'}$.

1. If $s \geq 2$ and $q > \prod_{i=1}^{s} (k_i - 1)^{\frac{2}{s-1}}$, then Equation (7.3.1) is solvable in $\mathbb{F}_q$ for any $b \in \mathbb{F}_q$.
2. If $s \geq 3$ and

$$\frac{q^{s-1} - 1}{(q-1)q^{\frac{s}{2}-1}} > \frac{1}{D} \left| \sum_{l=0}^{D-1} \left(\prod_{k_i \mid l} (1 - k_i) \right) \right|,$$

where $D = \prod_i k_i$, then Equation (7.3.1) has a nontrivial solution over $\mathbb{F}_q$ for $b = 0$.

7.3.13 Theorem [2165] If $\sum_{i=1}^{s} \frac{1}{k_i} > 1$, then Equation (7.3.1) is solvable over $\mathbb{F}_q$ for any b. Furthermore, for $b = 0$ there is a nontrivial solution.

7.3.14 Remark It is not easy to find general formulas for the exact number of solutions of the diagonal Equation (7.3.1) and, in most cases, the formulas are rather complex. Results on the exact number of solutions of particular families of diagonal equations can be found in [199, 1548, 2120]. The case $b = 0$ has been studied separately and some general results are included below.

7.3.15 Theorem [3001, Theorem 1] Let $q = p^{2t}$, $s \geq 2$, $nk = q - 1$, and consider $\sum_{i=1}^{s} c_i x_i^k = b$. If there exists a divisor r of t such that $p^r \equiv -1 \pmod{k}$, then

$$N_0 = q^{s-1} + \frac{\epsilon^s q^{\frac{s}{2}-1}(q-1)}{k} \sum_{j=0}^{k-1} (1-k)^{\tau(j)}, \quad \text{and,}$$

$$\text{for } b \neq 0,\ N_b = q^{s-1} - \epsilon^{s+1} q^{\frac{s}{2}-1} \left[(1-k)^{\theta(b)} q^{\frac{1}{2}} - \frac{(q^{\frac{1}{2}} - \epsilon)}{k} \sum_{j=0}^{k-1} (1-k)^{\tau(j)} \right],$$

where $\epsilon = (-1)^{\frac{t}{r}}$, $\theta(b) = |\{i \mid c_i^n = (-b)^n\}|$, $\tau(j) = \left|\{i \mid c_i^n = (\alpha^j)^n\}\right|$, $1 \leq i \leq s$, and α is a primitive root in $\mathbb{F}_q$.

7.3.16 Theorem [2432] Let $k_j = 2m_j$ for $j = 1, \ldots, s-2$, $k_{s-1} = km_{s-1}$, $k_s = k^r m_s$, where $r \geq 1$, $(2k, m_1 \cdots m_s) = 1$ and $m_1, \ldots, m_s$ are pairwise coprime.

1. If $k = 1$ or if s and k are odd, then $N_0 = q^{s-1}$.
2. If s is even, then

$$N_0 = q^{s-1} + (-1)^{\frac{(s-2)(q-1)}{4}} (q-1) q^{\frac{s-2}{2}} \left[\frac{\left(1 + (-1)^{\frac{q-1}{k}}\right) k}{2} - 1 \right].$$

7.3.17 Remark Theorem 7.3.16 holds when the number of variables s is even. Explicit formulas for the case when s is odd are given in [199].

7.3.18 Theorem [2744] Let $s > 2$. Then $I(k_1, \ldots, k_s) = 0$ (and hence $N_0 = q^{s-1}$) if and only if one of the following holds:

1. For some i, $\left(k_i, \frac{k_1 \cdots k_s}{k_i}\right) = 1$.
2. If $\{k_{i_1}, \ldots, k_{i_r} \mid 1 \leq i_1 < \cdots < i_r \leq s\}$ is the set of all even integers among $\{k_1, \ldots, k_s\}$, then $2 \nmid r$, $\frac{k_{i_1}}{2}, \ldots, \frac{k_{i_r}}{2}$ are pairwise coprime, and k_{i_j} is coprime to any odd number in $\{k_1, \ldots, k_s\}$ for $j = 1, \ldots, r$ and $r < s$.

7.3.19 Remark Diagonal equations with few variables have received special attention [2144, 2714]. Hermitian curves, a particular case of "Fermat like" equations, have been studied extensively because they provide good examples of curves with maximal number of rational points [1199, 1278, 1514, 2499]. We present a recent result on Fermat equations and refer the reader to [110, 1717] for results on other specific families of "Fermat like" equations.

7.3.20 Theorem [1798, Theorem 1.1] The number of solutions of $X^k + Y^k + Z^k = 0$ over $\mathbb{F}_{q^m}$ is $N_0 = 3k + k^2(q-2) + (d-1)(d-2)$ provided that

$$p > \left(\frac{2}{\sqrt[t+1]{\sin(\frac{k\pi}{2N})}} + 1 \right)^{\frac{N(t-1)}{k-d}},$$

where $N = \frac{q^m-1}{q-1}$, $k \mid N$, $d = (\frac{N}{k}, t+1)$, and $q \equiv t \pmod{\frac{N}{k}}$ with $0 < t < \frac{N}{k}$.

7.3.21 Definition Let $L(k_1, \ldots, k_s)$ be the least positive integer represented by (7.3.4) if there is such an integer, or, otherwise, let $L(k_1, \ldots, k_s) = s - 1$.

7.3.22 Definition For k an integer, let $v_q(k)$ denote the highest power of q dividing k.

7.3.23 Remark Many results on diagonal equations give bounds for the number of solutions N_b of (7.3.1), while others give bounds for $v_p(N_b)$. Most of these bounds depend on the numbers I and L of Definitions 7.3.6 and 7.3.21. Exact values for I and L are hard to compute in general; [2743] includes their exact values for certain equations, and [1350] provides sharp general lower bounds for I.

7.3.24 Theorem [1939, 2679, 2744]

$$I(k_1, \ldots, k_s) = (-1)^s + \sum_{r=1}^{s} (-1)^{s-r} \sum_{1 \le i_1 < i_2 < \cdots < i_r \le s} \frac{k_{i_1} \cdots k_{i_r}}{\operatorname{lcm}[k_{i_1}, \ldots, k_{i_r}]}.$$

7.3.25 Theorem [2906, Theorem 1] If there is a positive integer μ such that

$$\frac{1}{k_1} + \cdots + \frac{1}{k_s} > \mu \ge 1,$$

then $v_q(N_b) \ge \mu$.

7.3.26 Theorem [2449] Let $q = p^n$, $k_i | (q-1)$, r_i be the least integer such that $k_i | (p^{r_i} - 1)$, $K_i k_i = p^{r_i} - 1$ for $i = 1, \ldots, s$, and μ_1 be the least integer such that

$$\mu_1 \ge \sum_{i=1}^{s} \frac{n(K_i, p-1)}{r_i(p-1)}.$$

If $\mu = \left[\frac{\mu_1}{n}\right] - 1$, then $v_q(N_b) \ge \mu$, where $[a]$ denotes the integer part of a.

7.3.27 Theorem [2956] Suppose $k_i | (q-1)$, $I(k_1, \ldots, k_s) > 0$ and let $w_i = \gcd(k_i, \operatorname{lcm}[k_j | i \ne j])$. Then $L(k_1, \ldots, k_s) = \left\lceil \sum_{i=1}^{s} \frac{1}{w_i} \right\rceil$ if one of the following conditions holds:

1. $\sum_{i=1}^{s} \frac{1}{w_i} \equiv 0 \pmod 1$,
2. $\operatorname{lcm}[w_1, \ldots, w_s] | w_s$,
3. $I(k_1, \ldots, k_s) \le 10$,
4. $s \le 3$.

7.3.28 Theorem [2745, Theorems 2,4] For each i, define $u_i = \gcd\left(k_i, \frac{k_1 k_2 \cdots k_s}{k_i}\right)$. Then,

1. $I(k_1, \ldots, k_s) = I(u_1, \ldots, u_s)$ and $L(k_1, \ldots, k_s) = L(u_1, \ldots, u_s)$.
2. For all $1 \le j \le s$, $I(k_1, \ldots, k_s) \le \prod_{i \ne j}(u_i - 1)$.

3. The following are equivalent:

   (a) $I(k_1, \ldots, k_s) = 1$,

   (b) s is even and $u_i = 2$ for all i except $u_j = 2^m$, $m > 0$ for one j.

7.3.29 Corollary Assume that each k_i satisfies one of the conditions of Part 3 of Theorem 7.3.28. Then $L(k_1, \ldots, k_s) = \frac{s}{2}$.

7.3.30 Theorem [2906, Theorem 3] $v_q(N_0) \geq L(k_1, \ldots, k_s) - 1$.

7.3.31 Definition Let $k = a_0 + a_1 p + \cdots + a_r p^r$, where $0 \leq a_i \leq p-1$. The *p-weight of* k is defined and denoted as $\sigma_p(k) = \sum_{i=0}^{r} a_i$. The *p-weight degree of a monomial* $x_1^{k_1} \cdots x_s^{k_s}$ is defined and denoted as $w_p(x_1^{k_1} \cdots x_s^{k_s}) = \sigma_p(k_1) + \cdots + \sigma_p(k_s)$. The *p-weight degree* $w_p(g)$ *of a polynomial* g is the largest p-weight degree of the terms in g.

7.3.32 Remark Some of the bounds for the powers of p dividing the number of solutions of diagonal equations can be improved if one considers the p-weight $\sigma_p(k_i)$ of the degrees of the terms in the equation instead of their degrees k_i.

7.3.33 Theorem [2149, Theorem 10] Let $q = p^n$. If μ is the least integer satisfying

$$\mu \geq n\left(\frac{1}{\sigma_p(k_1)} + \cdots + \frac{1}{\sigma_p(k_s)} - 1\right),$$

then $v_p(N_b) \geq \mu$.

7.3.34 Remark If $\sum_{i=1}^{s} \frac{1}{\sigma_p(k_i)} > 1$ then the equation $\sum_{i=1}^{s} c_i x_i^{k_i} = 0$ has a non-trivial solution.

7.3.35 Theorem [2149, Theorem 2] Let $\gamma = \min_{(j_1,\ldots,j_s)}\left\{\frac{\sum_{i=1}^{s} \sigma_p(j_i(q-1)/k_i)}{p-1}\right\} - 1$, where $(j_1, \ldots, j_s)$ satisfies (7.3.4). If N_0 is the number of solutions of Equation (7.3.1) over $\mathbb{F}_{q^m}$, then $v_p(N_0) \geq m\gamma$. Also, $m\gamma$ is best possible, i.e., there exists an equation $c_1 x_1^{k_1} + \cdots + c_s x_1^{k_s} = 0$ such that $v_p(N_0) = m\gamma$.

7.3.36 Remark We note that Theorem 7.3.35 uses a calculation on $\mathbb{F}_q$ to give information on $v_p(N_0)$ for any extension of $\mathbb{F}_q$.

7.3.3 Generalizations of diagonal equations

7.3.37 Remark Generalizations of diagonal equations have been considered by several authors. Some examples are systems of diagonal equations [21, 1698, 2151, 2808], deformed diagonal equations [546, 2698], and equations with terms of disjoint support [501, 502, 563].

7.3.38 Theorem [1051, Theorem 5] Let $F = c_1 x_1^k + \cdots + c_s x_s^k + g(x_1, \ldots, x_s)$ be a polynomial over $\mathbb{F}_p$, where $c_1 \cdots c_s \neq 0$ and $\deg(g) < k$. Then $F = 0$ is solvable in $\mathbb{F}_p^s$ if $s \geq \left\lceil \frac{p-1}{\lfloor \frac{p-1}{k} \rfloor} \right\rceil$.

7.3.39 Remark Theorem 7.3.38 was proved using the combinatorial nullstellensatz and it generalizes Theorem 1 in [546] to include the case $k \nmid (p-1)$. Prior to this result, Newton polyhedra were used in [23] to prove results that allows one to estimate the number of zeros of polynomials of the type in Theorem 7.3.38. Note that, even if $s > k$, Chevalley's theorem does not guarantee the solvability of the equations in Theorem 7.3.38.

7.3.40 Theorem [2746, Theorem 1.1] Let $q = p^n$ and $F = x_1^k + \cdots + x_s^k + g(x_1, \ldots, x_s)$ be a polynomial over $\mathbb{F}_q$ where $\deg(g) < k$. Then, for nonempty subsets $A_1, \ldots, A_s \subset \mathbb{F}_q$,

$$|\{F(a_1, \ldots, a_s) \mid a_i \in A_i\}| \geq \min\left\{p, \sum_{i=1}^{s} \left\lfloor \frac{|A_i| - 1}{k} \right\rfloor + 1\right\}.$$

7.3.41 Remark If $q = p$ and $A_1 = \cdots = A_s = \mathbb{F}_p^*$, $s > k$, one obtains Theorem 1.3 in [1838].

7.3.42 Theorem Let $q = p^n$ and $k_i > 1$ be positive integers satisfying $k_i \mid (p-1)$. Let $F = x_1^{k_1} + \cdots + x_s^{k_s} + g(x_1, \ldots, x_s)$ be a polynomial over $\mathbb{F}_q$, where $w_p(g) < \min_i \{k_i\}$, and N_0 be the number of solutions of $F = 0$. Then, $v_p(N_0) = n\left(\sum_{i=1}^{s} \frac{1}{k_i} - 1\right)$ whenever $\sum_{i=1}^{s} \frac{1}{k_i}$ is an integer. In particular, F is solvable over $\mathbb{F}_q$.

7.3.43 Remark The result in Theorem 7.3.42 is a special case of Theorem 15 of [562].

7.3.4 Waring's problem in finite fields

7.3.44 Remark Waring's problem is to find the smallest $s = g(k, q)$ such that every element $b \in \mathbb{F}_q$ can be written as a sum of s summands of k-th powers in $\mathbb{F}_q$. This is an active area of research and the introduction of new techniques from arithmetic combinatorics has allowed the improvement of the bounds on $g(k, q)$. Some results on Waring's problem and their application to coding theory were presented in Section 6.3.

7.3.45 Definition The smallest $s = g(k, q)$ such that the equation $x_1^k + \cdots + x_s^k = b$ has a solution for every $b \in \mathbb{F}_q$ is *Waring's number* for $\mathbb{F}_q$ with respect to k.

7.3.46 Theorem Waring's number $g(k, p^n)$ exists if and only if $\frac{p^n-1}{p^d-1} \nmid k$ for all $d | n$, $d \neq n$. Also, if $d = \gcd(k, q-1)$, then $g(k, q) = g(d, q)$.

7.3.47 Remark Because of the previous theorem it is enough to consider $k | (q-1)$. From now on we assume that $g(k, q)$ exists and $k | (q-1)$.

7.3.48 Theorem [566] We have $g(k, p) \leq k$ and equality holds if $k = 1,\ 2,\ \frac{p-1}{2},\ p-1$.

7.3.49 Theorem [2677] If $2 \leq k < q^{\frac{1}{4}} + 1$, then $g(k, q) = 2$.

7.3.50 Remark To find the exact value for $g(k, q)$ is a difficult problem and, given Theorem 7.3.49, one might ask, for each k, which is the largest q such that $g(k, q) \neq 2$ [2677]. The next table contains some of the exact values known for $g(k, p)$. These and other exact values can be found in [12, 2148, 2676, 2678].

k	$k \neq p-1, \frac{p-1}{2}$	5	6	6	7	7	8
p	$p \leq 29$	31, 41 61	31	$37 \leq p \leq 67$ 109, 139, 223	43	71, 113 127	41
$g(k,p)$	$\lfloor \frac{k}{2} \rfloor + 1$	3	4	3	4	3	4

k	8	9	10	10	10	11	11	11
p	$73 \leq p \leq 137$ 233, 257 337, 761	73, 109 127, 163 181, 199 271, 307	31	41 61	$71 \leq p \leq 491$ $521 \leq p \leq 631$ $661 \leq p \leq 881$ 641, 911	89	67	199, 331 353, 419 463, 617
$g(k,p)$	3	3	5	4	3	4	5	3

7.3.51 Remark If the value of $g(k, p)$ for any $3 \leq k \leq 11$, $k \neq p-1$, $\frac{p-1}{2}$ is not included in the table above, then $g(k, p) = 2$. Hence, the table and Theorem 7.3.48 provide all the exact values of Waring's number for $k = 1, \ldots, 11$.

7.3.52 Theorem [2149, Theorem 18] Let $l \neq 1$. If $k|(p^n+1), k \neq p^n+1$, then $g(k, p^{2nl}) = 2$.

7.3.53 Theorem [651, Theorem 2] Let $t = \frac{p-1}{k}$. If a, b are the unique positive integers with $a > b$ and $a^2+b^2+ab = p$, then, for $t = 3$, $g(k,p) = a+b-1$, and for $t = 6$, $g(k,p) = \lfloor \frac{2}{3}a + \frac{1}{3}b \rfloor$. If $t = 4$ and a, b are the unique positive integers with $a > b$ and $a^2+b^2 = p$, then $g(k,p) = a-1$.

7.3.54 Theorem [1782, Theorems 1,2] Let ϕ denote Euler's function, m be a positive integer and p,r be primes such that p is a primitive root modulo r^m. Then,

$$g\left(\frac{p^{\phi(r^m)}-1}{r^m}, p^{\phi(r^m)}\right) = \frac{(p-1)\phi(r^m)}{2}.$$

If, in addition, p and r are odd primes, then

$$g\left(\frac{p^{\phi(r^m)}-1}{2r^m}, p^{\phi(r^m)}\right) = \begin{cases} r^{m-1}\left\lfloor \frac{pr}{4} - \frac{p}{4r} \right\rfloor & \text{if } r < p, \\ r^{m-1}\left\lfloor \frac{pr}{4} - \frac{r}{4p} \right\rfloor & \text{if } r \geq p. \end{cases}$$

7.3.55 Remark Theorem 7.3.54 generalizes the results in [2995]. A good survey on bounds for $g(k,q)$ can be found in [2990].

7.3.56 Theorem [1282, Theorem 5] If $k < \sqrt{q}$, then $g(k,q) \leq 8$.

7.3.57 Theorem [2990, Theorem 2] If $d = \frac{p-1}{\gcd(\frac{p^n-1}{k}, p-1)}$, then $g(k,p^n) \leq ng(d,p)$.

7.3.58 Theorem [1206, Theorem 4] If $k \geq 2$ is a proper divisor of $p-1$ and $k \geq (p-1)^{4/7}$, then $g(k,p) \leq 170\frac{k^{7/3}}{(p-1)^{4/3}} \log p$.

7.3.59 Remark By using a result in [2030] an improvement to Theorem 7.3.58 can be obtained.

7.3.60 Theorem [1789, Theorem 2] For any $\epsilon > 0$ there exists $c_\epsilon > 0$ such that for any $k \geq 2$, $p \geq \frac{k \ln k}{(\ln(\ln k+1))^{1-\epsilon}}$, we have $g(k,p) \leq c_\epsilon (\ln k)^{2+\epsilon}$.

7.3.61 Theorem [651, Theorem 1] Let $t = \frac{p-1}{k}$ and l be a positive integer. If $\phi(t) \geq l$, then $g(k,p) \leq C(l)k^{\frac{1}{l}}$ for some constant $C(l)$, where ϕ is Euler's function.

7.3.62 Remark Theorems 7.3.60 and 7.3.61 prove conjectures made by Heilbronn in [1460]. The next theorem gives an explicit value for the constant $C(l)$ in Theorem 7.3.61 when $l = 2$. Other estimates for $C(l)$ are also given in [656].

7.3.63 Theorem [656, Theorem 1.1] For $t = \frac{p-1}{k} > 2$ we have the uniform upper bound $g(k,p) \leq 83k^{1/2}$.

7.3.64 Definition Let $A_k := \{x^k : x \in \mathbb{F}_q\}$ define the set of k-th powers of the elements in the field $\mathbb{F}_q$, and $A'_k := A_k \cap \mathbb{F}_p$.

7.3.65 Theorem [650, Theorems 1, 3, 4], [2550, Theorem 4.1]

1. $g(k,p^n) \leq 8n \left\lceil \frac{(k+1)^{1/n}-1}{|A'_k|-1} \right\rceil$. If $|A'_k| \leq 3$, then $g(k,p^n) \leq 4n \left\lceil \frac{(k+1)^{1/n}-1}{|A'_k|-1} + 2 \right\rceil$.
2. For any $\epsilon > 0$, if $|A'_k| \geq 4^{2/\epsilon n}$, then $g(k,p^n) \leq C(\epsilon)k^\epsilon$, for some constant $C(\epsilon)$.
3. $g(k,p^2) \leq 16\sqrt{k+1}$. If $n \geq 3$, then $g(k,p^n) \leq 10\sqrt{k+1}$.
4. If $|A'_k| \geq p^\epsilon$ for $\epsilon > \frac{41}{83}$, then, for all sufficiently large p, we have $g(k,p) \leq 6$.

7.3.66 Remark Part 1 of Theorem 7.3.65 improves Theorem 1 in [2992]. Parts 2 and 3 prove the extensions of Heilbronn's conjectures [1460] to arbitrary fields. See also [374].

7.3.67 Remark Waring's problem has been generalized to systems of diagonal equations [561, 2808], to general polynomials [372, 549, 658], to Dickson polynomials [1314], to factorials [1197], and to reciprocals [754, 2643].

7.3.68 Theorem [1196, Theorem 1] Any residue class λ modulo p can be represented in the form $\sum_{i=1}^{5} m_i!n_i! \equiv \lambda \pmod{p}$ for some positive integers $m_1, n_1, \ldots, m_5, n_5$ with $\max_{1\le i\le 5}\{m_i, n_i\} \le cp^{27/28}$, where c is an absolute constant.

7.3.69 Theorem [1196, Theorem 2] Let $l(p)$ be the smallest integer such that for every integer λ the congruence $n_1! + \cdots + n_l! \equiv \lambda \pmod{p}$ has a solution in positive integers. Then $l(p) \le C\log^3(p)$, for some constant C.

7.3.70 Definition Let $D_k(x, a)$ be the Dickson polynomial of degree k and parameter $a \in \mathbb{F}_q$ (see Section 8.3). The *Waring problem for Dickson polynomials over* $\mathbb{F}_q$ is to find the smallest positive integer $s = g_a(k, q)$ such that the equation

$$D_k(x_1, a) + \cdots + D_k(x_s, a) = b, \ x_1, \ldots, x_s \in \mathbb{F}_q$$

is solvable for any $b \in \mathbb{F}_q$.

7.3.71 Theorem [2333, Theorem 1] Let $g_a(k, q)$ be defined as in Definition 7.3.70. The inequality $g_a(k, q) \le 16$ holds

1. For any $a \in \mathbb{F}_q^*$ and $(k, q-1) \le 2^{-3/2}(q-1)^{1/2}$.
2. For any a that it is a square in $\mathbb{F}_q^*$ and $(k, q+1) \le 2^{-3/2}(q-1)^{1/2}$.

See Also

[168], [1917], [1918], [2137], [2146], [2147], [2150], [2697], [3066]	References on solvability/divisibility of diagonal equations.
[21], [24], [30], [150], [1540], [1698], [2151]	For solvability/divisibility of general systems of equations.
[511], [515], [538], [2883]	For studies of diagonal equations over function fields.
[416], [1757]	For results on diagonal equations over p-adic fields.
[494], [2146], [2147], [3002], [3003]	For applications of diagonal equations to coding theory.

References Cited: [12, 21, 23, 24, 30, 110, 150, 168, 199, 240, 372, 374, 416, 494, 501, 502, 511, 515, 538, 546, 549, 561, 562, 563, 566, 615, 650, 651, 656, 658, 754, 1051, 1196, 1197, 1199, 1206, 1274, 1278, 1282, 1314, 1350, 1460, 1514, 1540, 1548, 1575, 1617, 1698, 1717, 1757, 1782, 1789, 1798, 1838, 1917, 1918, 1939, 2030, 2120, 2137, 2144, 2146, 2147, 2148, 2149, 2150, 2151, 2165, 2333, 2432, 2449, 2499, 2550, 2643, 2676, 2677, 2678, 2679, 2681, 2697, 2698, 2714, 2743, 2744, 2745, 2746, 2808, 2809, 2883, 2906, 2956, 2990, 2992, 2995, 3001, 3002, 3003, 3066]

8

Permutation polynomials

8.1 One variable

Gary L. Mullen, *The Pennsylvania State University*
Qiang Wang, *Carleton University*

8.1.1 Introduction

8.1.1 Definition For q a prime power, let $\mathbb{F}_q$ denote the finite field containing q elements. A polynomial $f \in \mathbb{F}_q[x]$ is a *permutation polynomial (PP)* of $\mathbb{F}_q$ if the function $f : c \to f(c)$ from $\mathbb{F}_q$ into itself induces a permutation. Alternatively, f is a PP of $\mathbb{F}_q$ if the equation $f(x) = a$ has a unique solution for each $a \in \mathbb{F}_q$.

8.1.2 Remark The set of all PPs on $\mathbb{F}_q$ forms a group under composition modulo $x^q - x$, isomorphic to the symmetric group S_q of order $q!$. For $q > 2$, the group S_q is generated by x^{q-2} and all linear polynomials $ax + b$, and if c is a primitive element in $\mathbb{F}_q$, S_q is generated by $cx, x+1$, and x^{q-2}.

8.1.3 Remark Given a permutation g of $\mathbb{F}_q$, the unique permutation polynomial $P_g(x)$ of $\mathbb{F}_q$ of degree at most $q-1$ representing the function g can be found by the Lagrange Interpolation Formula (see Theorem 1.71 in [1939]). In particular $P_g(x) = \sum_{a \in \mathbb{F}_q} g(a)(1 - (x-a)^{q-1})$; see also Theorem 2.1.131.

8.1.4 Remark If f is a PP and $a \neq 0, b \neq 0, c \in \mathbb{F}_q$, then $f_1 = af(bx+c)$ is also a PP. By suitably choosing a, b, c we can arrange to have f_1 in *normalized form* so that f_1 is monic, $f_1(0) = 0$, and when the degree n of f_1 is not divisible by the characteristic of $\mathbb{F}_q$, the coefficient of x^{n-1} is 0.

8.1.5 Remark A few well known classes of PPs from [1939]:

Monomials: The monomial x^n is a PP of $\mathbb{F}_q$ if and only if $(n, q-1) = 1$.

Dickson: For $a \neq 0 \in \mathbb{F}_q$, the polynomial $D_n(x,a) = \sum_{i=0}^{\lfloor n/2 \rfloor} \frac{n}{n-i}\binom{n-i}{i}(-a)^i x^{n-2i}$ is a PP of $\mathbb{F}_q$ if and only if $(n, q^2-1) = 1$; see Section 9.6.

Linearized: The polynomial $L(x) = \sum_{s=0}^{n-1} a_s x^{q^s} \in \mathbb{F}_{q^n}[x]$ is a PP of $\mathbb{F}_{q^n}$ if and only if $\det(a_{i-j}^{q^j}) \neq 0$, $0 \leq i, j \leq n-1$. The set of linearized PPs forms the *Betti-Mathieu group* isomorphic to the group $GL(n, \mathbb{F}_q)$, the general linear group of all non-singular $n \times n$ matrices over $\mathbb{F}_q$ under matrix multiplication. Recently, there have been several papers devoted to explicit constructions of linearized PPs; see for example, [504, 3047, 3065].

For odd q, $f(x) = x^{(q+1)/2} + ax$ is a PP of $\mathbb{F}_q$ if and only if $a^2 - 1$ is a nonzero square in $\mathbb{F}_q$. Moreover, the polynomial $f(x) + cx$ is a PP of $\mathbb{F}_q$ for $(q-3)/2$ values of $c \in \mathbb{F}_q$ [1939].

The polynomial $x^r(f(x^d))^{(q-1)/d}$ is a PP of $\mathbb{F}_q$ if $(r, q-1) = 1$, $d \mid q-1$, and $f(x^d)$ has no nonzero root in $\mathbb{F}_q$.

8.1.6 Remark For more information, we refer to Chapter 7 of [1939], and survey papers [681, 1933, 1935, 2176, 2178].

8.1.2 Criteria

8.1.7 Theorem (Hermite) Let p be the characteristic of $\mathbb{F}_q$. A polynomial $f \in \mathbb{F}_q[x]$ is a PP if and only if

1. the polynomial f has exactly one root in $\mathbb{F}_q$;
2. for each integer t with $1 \leq t \leq q-2$ and $t \not\equiv 0 \pmod{p}$, the reduction of $(f(x))^t \pmod{x^q - x}$ has degree at most $q-2$.

8.1.8 Remark Hermite's criterion was used by Dickson to obtain all normalized PPs of degree at most 5 [1939] in the list below.

Normalized PPs of $\mathbb{F}_q$	q
x	any q
x^2	$q \equiv 0 \pmod{2}$
x^3	$q \not\equiv 1 \pmod{3}$
$x^3 - ax$ (a not a square)	$q \equiv 0 \pmod{3}$
$x^4 \pm 3x$	$q = 7$
$x^4 + a_1x^2 + a_2x$ (if its only root in $\mathbb{F}_q$ is 0)	$q \equiv 0 \pmod{2}$
x^5	$q \not\equiv 1 \pmod{5}$
$x^5 - ax$ (a not a fourth power)	$q \equiv 0 \pmod{5}$
$x^5 + ax$ ($a^2 = 2$)	$q = 9$
$x^5 \pm 2x^2$	$q = 7$
$x^5 + ax^3 \pm x^2 + 3a^2x$ (a not a square)	$q = 7$
$x^5 + ax^3 + 5^{-1}a^2x$ (a arbitrary)	$q \equiv \pm 2 \pmod{5}$
$x^5 + ax^3 + 3a^2x$ (a not a square)	$q = 13$
$x^5 - 2ax^3 + a^2x$ (a not a square)	$q \equiv 0 \pmod{5}$

A list of PPs of degree 6 over finite fields with odd characteristic can be found in [840]. A list of PPs of degree 6 and 7 over finite fields with characteristic two can be found in [1916].

A recent preprint [2605] tabulates all monic PPs of degree 6 in the normalized form.

8.1.9 Theorem [1939] The polynomial f is a PP of $\mathbb{F}_q$ if and only if $\sum_{c\in\mathbb{F}_q}\chi(f(c))=0$ for all nontrivial additive characters χ of $\mathbb{F}_q$.

8.1.10 Remark Another criterion for PPs conjectured by Mullen [2176] in terms of the size $|V_f|$ of the value set $V_f=\{f(a)\mid a\in\mathbb{F}_q\}$ of a polynomial f of degree n was proved by Wan [2911]. Namely, if $|V_f|>q-\frac{q-1}{n}$ then f is a PP of $\mathbb{F}_q$ [2911]. We refer to Section 8.3 for more information on value sets. A variation of Hermite's criterion in terms of combinatorial identities is given in [2018]. Hermite's criterion can be rewritten in terms of the invariant, $u_p(f)$, the smallest positive integer k such that $\sum_{x\in\mathbb{F}_q} f(x)^k\neq 0$. That is, f is a PP of $\mathbb{F}_q$ if and only if $u_p(f)=q-1$. In the case $q=p$, this criterion was improved in [1813] and only requires $k>\frac{p-1}{2}$. Using Teichmüller liftings, Wan et al. [2919] obtained an upper bound for $|V_f|$ and improved Hermite's criterion. Several other criteria were obtained by Turnwald [2826] in terms of invariants associated with elementary symmetric polynomials, without using Teichmüller liftings.

8.1.11 Theorem [2826] Let $f\in\mathbb{F}_q[x]$ be a polynomial of degree n with $1\le n<q$. Let u be the smallest positive integer k with $s_k\neq 0$ if such k exists and otherwise set $u=\infty$, where s_k denotes the k-th elementary symmetric polynomial of the values $f(a)$. Let v be the size of the value set $V_f=\{f(a)\mid a\in\mathbb{F}_q\}$. Let w be the smallest positive integer k with $p_k=\sum_{a\in\mathbb{F}_q} f(a)^k\neq 0$ if such k exists and otherwise set $w=\infty$. The following are equivalent:

(1) f is a PP of $\mathbb{F}_q$;	(2) $u=q-1$;	(3) $u>q-\frac{q}{n}$;
(4) $u>q-v$;	(5) $v>q-\frac{q-1}{n}$;	(6) $w=q-1$;
(7) $\frac{2q}{3}-1<w<\infty$;	(8) $q-\frac{q+1}{n}<w<\infty$;	(9) $q-u\le w<\infty$;
(10) $u>\frac{q-1}{2}$ and $w<\infty$.		

8.1.12 Remark A criterion in terms of resultants was given by von zur Gathen [1221]. Using the Euclidean algorithm to compute the resultant, von zur Gathen provided a probabilistic test to determine whether a given polynomial is a PP or not. The number of operations in $\mathbb{F}_q$ has a softly linear running time $O(n\log q(\log(n\log q))^k)$ for some k. Furthermore, Ma and von zur Gathan showed that this decision problem has a zero-error probabilistic polynomial time in [1983] and provided a random polynomial time test for rational functions over finite fields, along with several related problems in [1984]. Earlier, Shparlinski had given a deterministic superpolynomial time algorithm for testing PP [2638]. In 2005 Kayal provided a deterministic polynomial-time algorithm for testing PP [1716].

8.1.3 Enumeration and distribution of PPs

8.1.13 Problem [1935] Let $N_n(q)$ denote the number of PPs of $\mathbb{F}_q$ which have degree n. We have the trivial boundary conditions: $N_1(q)=q(q-1)$, $N_n(q)=0$ if n is a divisor of $q-1$ larger than 1, and $\sum N_n(q)=q!$ where the sum is over all $1\le n<q-1$ such that n is either 1 or is not a divisor of $q-1$. Find $N_n(q)$.

8.1.14 Remark In an invited address before the MAA in 1966, Carlitz conjectured that for each even integer n, there is a constant C_n so that for each finite field of odd order $q>C_n$, there does not exist a PP of degree n over $\mathbb{F}_q$. A polynomial f over $\mathbb{F}_q$ is *exceptional* if the only absolutely irreducible factors of $f(x)-f(y)$ in $\mathbb{F}_q[x,y]$ are scalar multiples of $x-y$. One can also characterize an exceptional polynomial as a polynomial which induces a permutation of infinitely many finite extension fields of $\mathbb{F}_q$. As first proved by Cohen in [667], any exceptional polynomial is a PP, and the converse holds if q is large compared to the degree of f. Cohen's equivalent statement of Carlitz's conjecture [677] says that there is no exceptional polynomial of even degree in odd characteristic. This was proved by Fried, Guralnick, and Saxl in [1120]; in fact an even stronger result was obtained through the use of powerful group theoretic methods, including the classification of finite simple groups.

8.1.15 Remark For the next theorem we require the concept of an exceptional cover; see Section 9.7.

8.1.16 Theorem [1120] There is no exceptional cover of nonsingular absolutely irreducible curves over $\mathbb{F}_q$ of degree $2p$ where q is a power of p and p is prime.

8.1.17 Remark Several partial results on Carlitz's conjecture can be found in [677, 2908]. Moreover, Wan generalized Carlitz's conjecture in [2909] proving that if $q > n^4$ and $(n, q-1) = 1$ then there is no PP of degree n over $\mathbb{F}_q$. Later Cohen and Fried [688] gave an elementary proof of Wan's conjecture following an argument of Lenstra and this result was stated in terms of exceptional polynomials; see Section 8.4 for more information on exceptional polynomials.

8.1.18 Theorem [688, 2909] There is no exceptional polynomial of degree n over $\mathbb{F}_q$ if $(n, q-1) > 1$.

8.1.19 Theorem [551]

1. Let $\ell > 1$. For q sufficiently large, there exists $a \in \mathbb{F}_q$ such that the polynomial $x(x^{(q-1)/\ell} + a)$ is a PP of $\mathbb{F}_q$.
2. Let $\ell > 1$, $(r, q-1) = 1$, and k be a positive integer. For q sufficiently large, there exists $a \in \mathbb{F}_q$ such that the polynomial $x^r(x^{(q-1)/\ell} + a)^k$ is a PP of $\mathbb{F}_q$.

8.1.20 Remark Any non-constant polynomial $h(x) \in \mathbb{F}_q[x]$ of degree $\leq q-1$ can be written *uniquely* as $ax^r f(x^{(q-1)/\ell}) + b$ with index ℓ [61]. Namely, write

$$h(x) = a(x^n + a_{n-i_1}x^{n-i_1} + \cdots + a_{n-i_k}x^{n-i_k}) + b,$$

where a, $a_{n-i_j} \neq 0$, $j = 1, \ldots, k$. Here we suppose that $j \geq 1$ and $n - i_k = r$. Then $h(x) = ax^r f(x^{(q-1)/\ell}) + b$, where $f(x) = x^{e_0} + a_{n-i_1}x^{e_1} + \cdots + a_{n-i_{k-1}}x^{e_{k-1}} + a_r$,

$$\ell = \frac{q-1}{(n-r, n-r-i_1, \ldots, n-r-i_{k-1}, q-1)},$$

and $(e_0, e_1, \ldots, e_{k-1}, \ell) = 1$. Clearly, h is a PP of $\mathbb{F}_q$ if and only if $g(x) = x^r f(x^{(q-1)/\ell})$ is a PP of $\mathbb{F}_q$. Then ℓ is the *index* of h.

8.1.21 Remark If $\ell = 1$ then $f(x) = 1$ so that $g(x) = x^r$. In this case $g(x)$ is a PP of $\mathbb{F}_q$ if and only if $(r, q-1) = 1$. We can assume $\ell > 1$.

8.1.22 Remark More existence and enumerative results for binomials can be found in [61, 64, 1833, 2018, 2019, 2020, 2825, 2905, 2913]. In [1833], Laigle-Chapuy proved the first assertion of Theorem 8.1.19 assuming $q > \ell^{2\ell+2}\left(1 + \frac{\ell+1}{\ell+2}\right)^2$. In [2020], Masuda and Zieve obtained a stronger result for more general binomials of the form $x^r(x^{e_1(q-1)/\ell} + a)$. More precisely they showed the truth of Part 1 of Theorem 8.1.19 for $q > \ell^{2\ell+2}$. Here we present a general result of Akbary-Ghioca-Wang (Theorem 8.1.25) which shows that there exist permutation polynomials of index ℓ for any prescribed exponents satisfying conditions (8.1.1). This result generalizes all the existence results from [551, 1833, 2020].

8.1.23 Definition [61] Let q be a prime power, and $\ell \geq 2$ be a divisor of $q-1$. Let m, r be positive integers, and $\bar{e} = (e_1, \ldots, e_m)$ be an m-tuple of integers that satisfy the following conditions:

$$0 < e_1 < e_2 < \cdots < e_m \leq \ell - 1, \ (e_1, \ldots, e_m, \ell) = 1 \text{ and } r + e_m s \leq q - 1, \qquad (8.1.1)$$

where $s := (q-1)/\ell$. For a tuple $\bar{a} := (a_1, \dots, a_m) \in (\mathbb{F}_q^*)^m$, we let

$$g_{r,\bar{e}}^{\bar{a}}(x) := x^r \left(x^{e_m s} + a_1 x^{e_{m-1} s} + \cdots + a_{m-1} x^{e_1 s} + a_m\right).$$

We define $N_{r,\bar{e}}^m(\ell, q)$ as the number of all tuples $\bar{a} \in (\mathbb{F}_q^*)^m$ such that $g_{r,\bar{e}}^{\bar{a}}(x)$ is a PP of $\mathbb{F}_q$. In other words $N_{r,\bar{e}}^m(\ell, q)$ is the number of all monic permutation $(m+1)$-nomials $g_{r,\bar{e}}^{\bar{a}}(x) = x^r f(x^{(q-1)/\ell})$ over $\mathbb{F}_q$ with vanishing order at zero equal to r, set of exponents $\bar{e}$ for $f(x)$, and index ℓ. Note that if r and $\bar{e}$ satisfy (8.1.1) then $g_{r,\bar{e}}^{\bar{a}}(x)$ has index ℓ.

8.1.24 Theorem [61] With the above notation, we have

$$\left| N_{r,\bar{e}}^m(\ell, q) - \frac{\ell!}{\ell^\ell} q^m \right| < \ell! \ell q^{m-1/2}.$$

8.1.25 Theorem [61] For any q, r, $\bar{e}$, m, ℓ that satisfy (8.1.1), $(r, s) = 1$, and $q > \ell^{2\ell+2}$, there exists an $\bar{a} \in (\mathbb{F}_q^*)^m$ such that the $(m+1)$-nomial $g_{r,\bar{e}}^{\bar{a}}(x)$ is a permutation polynomial of $\mathbb{F}_q$.

8.1.26 Remark For $q \geq 7$ we have $\ell^{2\ell+2} < q$ if $\ell < \frac{\log q}{2 \log \log q}$.

8.1.27 Theorem [771] The value $N_{p-2}(p) \sim (\varphi(p)/p)p!$ as $p \to \infty$, where φ is the Euler function. More precisely, $\left|N_{p-2}(p) - \frac{\varphi(p)}{p} p!\right| \leq \sqrt{\frac{p^{p+1}(p-2)+p^2}{p-1}}$.

8.1.28 Theorem [1787] Let q be a prime power. Then $|N_{q-2}(q) - (q-1)!| \leq \sqrt{\frac{2e}{\pi}} q^{\frac{q}{2}}$.

8.1.29 Theorem [1788] Fix j integers $k_1, \dots, k_j$ with the property that $0 < k_1 < \cdots < k_j < q-1$ and define $N(k_1, \dots, k_j; q)$ as the number of PPs of $\mathbb{F}_q$ of degree less than $q-1$ such that the coefficient of x^{k_i} equals 0, for $i = 1, \dots, j$. Then

$$\left| N(k_1, \dots, k_j; q) - \frac{q!}{q^j} \right| < \left(1 + \sqrt{\frac{1}{e}}\right)^q ((q - k_1 - 1)q)^{q/2}.$$

In particular, $N_{q-2}(q) = q! - N(q-2; q)$.

8.1.30 Remark We note that for $1 \leq t \leq q-2$ the number of PPs of degree at least $q-t-1$ is $q! - N(q-t-1, q-t, \dots, q-2; q)$. In [1788] Konyagin and Pappalardi proved that $N(q-t-1, q-t, \dots, q-2; q) \sim \frac{q!}{q^t}$ holds for $q \to \infty$ and $t \leq 0.03983\, q$. This result guarantees the existence of PPs of degree at least $q-t-1$ for $t \leq 0.03983\, q$ (as long as q is sufficiently large). However, the following theorem establishes the existence of PPs with exact degree $q-t-1$.

8.1.31 Theorem [61] Let $m \geq 1$. Let q be a prime power such that $q-1$ has a divisor ℓ with $m < \ell$ and $\ell^{2\ell+2} < q$. Then for every $1 \leq t < \frac{(\ell-m)}{\ell}(q-1)$ coprime with $(q-1)/\ell$ there exists an $(m+1)$-nomial $g_{r,\bar{e}}^{\bar{a}}(x)$ of degree $q-t-1$ which is a PP of $\mathbb{F}_q$.

8.1.32 Corollary [61] Let $m \geq 1$ be an integer, and let q be a prime power such that $(m+1) \mid (q-1)$. Then for all $n \geq 2m+4$, there exists a permutation $(m+1)$-nomial of $\mathbb{F}_{q^n}$ of degree $q-2$.

8.1.33 Definition Let $m_{[k]}(q)$ be the number of permutations of $\mathbb{F}_q$ which are k-cycles and are represented by polynomials of degree $q-k$.

8.1.34 Theorem [2967] Every transposition of $\mathbb{F}_q$ is represented by a unique polynomial of degree $q-2$. Moreover,

$$m_{[3]}(q) = \begin{cases} \frac{2}{3}q(q-1) & \text{if } q \equiv 1 \pmod 3, \\ 0 & \text{if } q \equiv 2 \pmod 3, \\ \frac{1}{3}q(q-1) & \text{if } q \equiv 0 \pmod 3. \end{cases}$$

8.1.35 Theorem [1996]

1. If $q \equiv 1 \pmod{k}$, then $m_{[k]}(q) \geq \frac{\varphi(k)}{k} q(q-1)$.
2. If $\mathrm{char}(\mathbb{F}_q) > e^{(k-3)/e}$, then $m_{[k]}(q) \leq \frac{(k-1)!}{k} q(q-1)$.

8.1.36 Remark It is conjectured in [1996] that the above upper bound for $m_{[k]}(q)$ holds for any $k < q$. Small k-cycles such as $k = 4, 5$ are also studied in [1995, 1996, 1997].

8.1.37 Theorem [2967] Let r and s be fixed positive integers, and $k_2, \ldots, k_s$ be non-negative integers such that $\sum_{i=2}^{s} i k_i = r$. Let $P(k_2, \ldots, k_s)$ be the set of permutations of $\mathbb{F}_q$ which are the disjoint products of k_2 transpositions, k_3 3-cycles, etc. Then the number of permutations in $P(k_2, \ldots, k_s)$ represented by a polynomial of degree $q-2$ is asymptotic to the number of all permutations in $P(k_2, \ldots, k_s)$ as q goes to ∞.

8.1.4 Constructions of PPs

8.1.38 Remark For the purpose of introducing the construction of PPs in the next few sections, we present the following recent result by Akbary-Ghioca-Wang (AGW).

8.1.39 Theorem (AGW's criterion, [62]) Let A, S, and $\bar{S}$ be finite sets with $\#S = \#\bar{S}$, and let $f : A \to A$, $\bar{f} : S \to \bar{S}$, $\lambda : A \to S$, and $\bar{\lambda} : A \to \bar{S}$ be maps such that $\bar{\lambda} \circ f = \bar{f} \circ \lambda$. If both λ and $\bar{\lambda}$ are surjective, then the following statements are equivalent:

1. f is a bijection (a permutation of A);
2. $\bar{f}$ is a bijection from S to $\bar{S}$ and f is injective on $\lambda^{-1}(s)$ for each $s \in S$.

8.1.40 Remark We note that this criterion does not require any restriction on the structures of the sets S and $\bar{S}$ in finding new classes of PPs of a set A. In particular, if we take A as a group and S and $\bar{S}$ as homomorphic images of A, then we obtain the following general result for finding permutations of a group.

8.1.41 Theorem [62] Let $(G, +)$ be a finite group, and let $\varphi, \psi, \bar{\psi} \in \mathrm{End}(G)$ be group endomorphisms such that $\bar{\psi} \circ \varphi = \varphi \circ \psi$ and $\#\mathrm{im}(\psi) = \#\mathrm{im}(\bar{\psi})$. Let $g : G \longrightarrow G$ be any mapping, and let $f : G \longrightarrow G$ be defined by $f(x) = \varphi(x) + g(\psi(x))$. Then,

1. f permutes G if and only if the following two conditions hold:
   a. $\ker(\varphi) \cap \ker(\psi) = \{0\}$ (or equivalently, φ induces a bijection between $\ker(\psi)$ and $\ker(\bar{\psi})$); and
   b. the function $\bar{f}(x) := \varphi(x) + \bar{\psi}(g(x))$ restricts to a bijection from $\mathrm{im}(\psi)$ to $\mathrm{im}(\bar{\psi})$.
2. For any fixed endomorphisms φ, ψ and $\bar{\psi}$ satisfying Part 1.a, there are $(\#\mathrm{im}(\psi))! \cdot (\#\ker(\bar{\psi}))^{\#\mathrm{im}(\psi)}$ such permutation functions f (when g varies).
3. Let $g : G \longrightarrow G$ be such that $(\bar{\psi} \circ g)\,|_{\mathrm{im}(\psi)} = 0$. Then $f = \varphi + g \circ \psi$ permutes G if and only if φ is a permutation of G.
4. Assume $\varphi \circ \psi = 0$ and $g : G \longrightarrow G$ is a mapping such that $g(x)$ restricted to $\mathrm{im}(\psi)$ is a permutation of $\mathrm{im}(\psi)$. Then $f(x) = \varphi(x) + g(\psi(x))$ permutes G if and only if φ and ψ satisfy Part 1.a, and $\bar{\psi}$ restricted to $\mathrm{im}(\psi)$ is a bijection from $\mathrm{im}(\psi)$ to $\mathrm{im}(\bar{\psi})$.

8.1.42 Remark One can apply this result to a multiplicative group of a finite field, an additive group of a finite field, or the group of rational points of an elliptic curve over finite fields [62]. This reduces a problem of determining whether a given polynomial over a finite field

$\mathbb{F}_q$ is a permutation polynomial to a problem of determining whether another polynomial permutes a smaller set.

8.1.43 Corollary [2359, 2940, 3076] Let $q-1=\ell s$ for some positive integers ℓ and s. Then $P(x)=x^r f(x^s)$ is a PP of $\mathbb{F}_q$ if and only if $(r,s)=1$ and $x^r f(x)^s$ permutes the set μ_ℓ of all distinct ℓ-th roots of unity.

8.1.44 Remark The above corollary is a consequence of Theorem 8.1.41 when taking multiplicative group endomorphisms x^r and x^s. There are several other equivalent descriptions of PPs of the form $x^r f(x^s)$; see for example, [65, 2359, 2916, 2940, 3076]. All of the classes of PPs in Subsection 8.1.5 are of this type. There are also many recent results on new classes of PPs when taking additive group endomorphisms in Theorem 8.1.41, see [62, 2003, 3046, 3077]. We give some classes of PPs in Subsection 8.1.6. We note that AGW's criterion does not require any restriction on the structures of subsets S and $\bar{S}$, which has even broader applications in finding new classes of PPs. Many classes of PPs in Subsection 8.1.7 can be obtained through this general construction method.

8.1.5 PPs from permutations of multiplicative groups

8.1.45 Definition [2278, 2940] Let γ be a primitive element of $\mathbb{F}_q$, $q-1=\ell s$ for some positive integers ℓ and s, and the set of all nonzero ℓ-th powers of $\mathbb{F}_q$ be $C_0=\{\gamma^{\ell j} : j = 0,1,\ldots,s-1\}$. Then C_0 is a subgroup of $\mathbb{F}_q^*$ of index ℓ. The elements of the factor group $\mathbb{F}_q^*/C_0$ are the *cyclotomic cosets*

$$C_i := \gamma^i C_0, \quad i=0,1,\ldots,\ell-1.$$

For any integer $r>0$ and any $A_0, A_1, \ldots, A_{\ell-1} \in \mathbb{F}_q$, we define an *r-th order cyclotomic mapping* $f^r_{A_0,A_1,\ldots,A_{\ell-1}}$ *of index* ℓ from $\mathbb{F}_q$ to itself by $f^r_{A_0,A_1,\ldots,A_{\ell-1}}(0)=0$ and

$$f^r_{A_0,A_1,\ldots,A_{\ell-1}}(x) = A_i x^r \quad \text{if } x\in C_i,\ i=0,1,\ldots,\ell-1.$$

Moreover, $f^r_{A_0,A_1,\ldots,A_{\ell-1}}$ is an *r-th order cyclotomic mapping of the least index* ℓ if the mapping cannot be written as a cyclotomic mapping of any smaller index.

8.1.46 Remark Cyclotomic mapping permutations were introduced in [2278] when $r=1$ and in [2940] for any positive r. Let $\zeta=\gamma^s$ be a primitive ℓ-th root of unity in $\mathbb{F}_q$ and $P(x)=x^r f(x^s)$ be a polynomial of index ℓ over $\mathbb{F}_q$ with positive integer r. Then $P(x)=x^r f(x^s)=f^r_{A_0,A_1,\ldots,A_{\ell-1}}(x)$ where $A_i=f(\zeta^i)$ for $0\le i\le \ell-1$. We note that the least index of a cyclotomic mapping is equal to the index of the corresponding polynomial. If P is a PP of $\mathbb{F}_q$ then $(r,s)=1$ and $A_i=f(\zeta^i)\ne 0$ for $0\le i\le\ell-1$. Under these two necessary conditions, $P(x)=x^r f(x^s)$ is a PP of $\mathbb{F}_q$ if and only if $f^r_{A_0,A_1,\ldots,A_{\ell-1}}$ is a PP of $\mathbb{F}_q$ [2940]. The concept of cyclotomic mapping permutations have recently been generalized in [2943] allowing each branch to take a different r_i value so that $P(x)$ has the form $\sum_{i=0}^n x^{r_i} f_i(x^s)$. More results can be found in [2943] and related piecewise constructions in [1061, 3058].

8.1.47 Remark There are several other equivalent descriptions of PPs of the form $x^r f(x^s)$, see [65, 2359, 2916, 2940, 3076] for example. In particular, in [65], it is shown that $P(x)=x^r f(x^s)$ is a PP of $\mathbb{F}_q$ if and only if $(r,s)=1$, $A_i=f(\zeta^i)\ne 0$ for $0\le i\le \ell-1$, and $\sum_{i=0}^{\ell-1}\zeta^{cri}A_i^{cs}=0$ for all $c=1,\ldots,\ell-1$. This criterion is equivalent to Hermite's criterion when the index

ℓ equals $q-1$. However, if $\ell < q-1$, this criterion in fact improves Hermite's criterion because we only need to verify that the coefficient of x^{q-1} is 0 for $\ell - 1$ different powers of P instead of all of the $q-2$ powers of P. We note that $\ell = 2$ implies that q must be odd.

8.1.48 Remark [65] For odd q, the polynomial $P(x) = x^r f(x^{(q-1)/2})$ is a PP of $\mathbb{F}_q$ if and only if $(r, (q-1)/2) = 1$ and $\eta(f(-1)f(1)) = (-1)^{r+1}$. Here η is the quadratic character of $\mathbb{F}_q$ with the standard convention $\eta(0) = 0$.

8.1.49 Remark Let $P(x) = x^k + ax^r$ be a binomial of index ℓ and $s = \frac{q-1}{\ell}$. Then $P(x) = x^r(x^{es}+a)$ for some e such that $(e, \ell) = 1$. If $a = b^s$ for some $b \in \mathbb{F}_q$, then $x^r(x^{es}+a)$ is a PP of $\mathbb{F}_q$ if and only if $x^r(x^{es}+1)$ is a PP of $\mathbb{F}_q$.

8.1.50 Remark Necessary conditions for $P(x) = x^r(x^{es}+1)$ of index ℓ to be a PP are as follows:

$$(r, s) = 1, (2e, \ell) = 1, (2r + es, \ell) = 1, \text{ and } 2^s = 1. \tag{8.1.2}$$

8.1.51 Remark For $\ell = 3$, the conditions in (8.1.2) are sufficient to determine whether P is a PP of $\mathbb{F}_q$ [2927]. However, for $\ell > 3$, it turns out not to be the case (for example, see [63, 64, 2927]). For general ℓ, a characterization of PPs of the form $x^r(x^{es}+1)$ in terms of generalized Lucas sequences of order $k := \frac{\ell-1}{2}$ is given in [2940, 2942].

8.1.52 Definition [64] For any integer $k \geq 1$ and η a fixed primitive $(4k+2)$-th root of unity, the *generalized Lucas sequence* (or *unsigned generalized Lucas sequence*) of order k is defined as $\{a_n\}_{n=0}^{\infty}$ such that

$$a_n = \sum_{\substack{t=1 \\ t \text{ odd}}}^{2k} (\eta^t + \eta^{-t})^n = \sum_{t=1}^{k} ((-1)^{t+1}(\eta^t + \eta^{-t}))^n.$$

The *characteristic polynomial* of the generalized Lucas sequence is defined by $g_0(x) = 1$, $g_1(x) = x - 1$, $g_k(x) = xg_{k-1}(x) - g_{k-2}(x)$ for $k \geq 2$.

8.1.53 Theorem [2940, 2942] Let $q = p^m$ be an odd prime power and $q - 1 = \ell s$ with odd $\ell \geq 3$ and $(e, \ell) = 1$. Let $k = \frac{\ell-1}{2}$. Then $P(x) = x^r(x^{es}+1)$ is a PP of $\mathbb{F}_q$ if and only if $(r, s) = 1$, $(2r + es, \ell) = 1$, $2^s = 1$, and

$$R_{j_c,k}(L)(a_{cs}) = -1, \text{ for all } c = 1, \ldots, \ell - 1, \tag{8.1.3}$$

where a_{cs} is the (cs)-th term of the generalized Lucas sequence $\{a_i\}_{i=0}^{\infty}$ of order k over $\mathbb{F}_p$, $j_c = c(2e^{\phi(\ell)-1}r + s) \pmod{2\ell}$, $R_{j_c,k}(x)$ is the remainder of the Dickson polynomial $D_{j_c}(x)$ of the first kind of degree j_c divided by the characteristic polynomial $g_k(x)$, and L is a left shift operator on sequences. In particular, all j_c are distinct even numbers between 2 and $2\ell - 2$.

8.1.54 Remark We note that the degree of any remainder $R_{n,k}$ is at most $k-1$ and $R_{n,k}$ is either a Dickson polynomial of degree $\leq k-1$ or the degree $k-1$ characteristic polynomial $g_{k-1}(x)$ of the generalized Lucas sequence of order k, or a negation of the above polynomials [2942]. We can extend the definition of $\{a_n\}$ to negative subscripts n using the same recurrence relations. We also remark that, for $\ell \leq 7$, the sequences used in the descriptions of permutation binomials have simple structures and are fully described [63, 2927]. We note that signed generalized Lucas sequences are defined in [2942] and they are used to compute the coefficients of the compositional inverse of permutation binomials $x^r(x^{es}+1)$. Finally we note that Equation (8.1.3) always holds if the sequence $\{a_n\}$ is *s-periodic* over $\mathbb{F}_p$, which means that $a_n \equiv a_{n+ks} \pmod{p}$ for integers k and n. We remark that these sequences are defined over prime fields and checking the s-periodicity of these sequences is a much simpler task than checking whether the polynomial is a PP over the extension field directly.

8.1.55 Theorem [64] Assume the conditions (8.1.2) on ℓ, r, e, and s hold. If $\{a_n\}$ is s-periodic over $\mathbb{F}_p$, then the binomial $P(x) = x^r(x^{es}+1)$ is a permutation binomial of $\mathbb{F}_q$.

8.1.56 Theorem [65] Let p be an odd prime and $q = p^m$ and let ℓ, r, s be positive integers satisfying that $q-1 = \ell s$, $(r,s) = 1$, $(e,\ell) = 1$, and ℓ odd. Let $p \equiv -1 \pmod{\ell}$ or $p \equiv 1 \pmod{\ell}$ and $\ell \mid m$. Then the binomial $P(x) = x^r(x^{es}+1)$ is a permutation binomial of $\mathbb{F}_q$ if and only if $(2r+es,\ell) = 1$. In particular, if $p \equiv 1 \pmod{\ell}$ and $\ell \mid m$, then the conditions $(r,s) = 1$, $(e,\ell) = 1$, and ℓ odd imply that $(2r+es,\ell) = 1$ [60].

8.1.57 Remark For $a = 1$ (equivalent to $a = b^s$ for some b), under the assumptions on q, s, ℓ, r, e, it is shown in [3076] that the s-periodicity of the generalized Lucas sequence implies that $(\eta+\eta^{-1})^s = 1$ for every (2ℓ)-th root of unity η. However, we note that these two conditions are in fact equivalent for $a = 1$. The following result extends Theorem 8.1.55 as it also deals with even characteristic.

8.1.58 Theorem [3076] For q, s, ℓ, e, r, a satisfying $q-1 = \ell s$, $(r,s) = 1$, $(e,\ell) = 1$, $r, e > 0$ and $a \in \mathbb{F}_q^*$, suppose $(-a)^\ell \neq 1$ and $(z+a/z)^s = 1$ for every (2ℓ)-th root of unity z. Then $P(x) = x^r(x^{es}+a)$ is a permutation binomial of $\mathbb{F}_q$ if and only if $(2r+es, 2\ell) \leq 2$.

8.1.59 Theorem [2020] Suppose $x^r(x^{es}+a)$ permutes $\mathbb{F}_p$, where $a \in \mathbb{F}_p^*$ and $r, e, s > 0$ such that $p-1 = \ell s$ and $(\ell, e) = 1$. Then $s \geq \sqrt{p-3/4} - 1/2 > \sqrt{p} - 1$.

8.1.60 Remark For earlier results on permutation binomials, we refer to [569, 2018, 2020, 2127, 2680, 2681, 2825, 2905, 2913].

8.1.61 Theorem [65] Let $q-1 = \ell s$. Assume that $f(\zeta^t)^s = 1$ for any $t = 0, \ldots, \ell-1$. Then $P(x) = x^r f(x^s)$ is a PP of $\mathbb{F}_q$ if and only if $(r, q-1) = 1$.

8.1.62 Corollary Let $q-1 = \ell s$ and g be any polynomial over $\mathbb{F}_q$. Then $P(x) = x^r g(x^s)^\ell$ is a PP of $\mathbb{F}_q$ if and only if $(r, q-1) = 1$ and $g(\zeta^t) \neq 0$ for all $0 \leq t \leq \ell - 1$.

8.1.63 Corollary [65, 1833] Let p be a prime, ℓ be a positive integer and v be the order of p in $\mathbb{Z}/\ell\mathbb{Z}$. For any positive integer n, let $q = p^m = p^{\ell v n}$ and $\ell s = q-1$. Assume f is a polynomial in $\mathbb{F}_{p^{vn}}[x]$. Then the polynomial $P(x) = x^r f(x^s)$ is a PP of $\mathbb{F}_q$ if and only if $(r, q-1) = 1$ and $f(\zeta^t) \neq 0$ for all $0 \leq t \leq \ell - 1$.

8.1.64 Remark Corollary 8.1.63 is reformulated as Theorem 2.3 in [3076]. Namely, let $\ell, r > 0$ satisfy $\ell s = q-1$. Suppose $q = q_0^m$ where $q_0 \equiv 1 \pmod{\ell}$ and $\ell \mid m$, and $f \in \mathbb{F}_{q_0}[x]$. Then $P(x) = x^r f(x^s)$ permutes $\mathbb{F}_q$ if and only if $(r,s) = 1$ and f has no roots in the μ_ℓ, the set of ℓ-th roots of unity.

8.1.65 Theorem [65] Let $q-1 = \ell s$, and suppose that $\overline{\mathbb{F}}_q$ (the algebraic closure of $\mathbb{F}_q$) contains a primitive $(j\ell)$-th root of unity η. Assume that $\left(\eta^{-ut} f(\eta^{jt})\right)^s = 1$ for any $t = 0, \ldots, \ell-1$ and a fixed u. Moreover assume that $j \mid us$. Then $P(x) = x^r f(x^s)$ is a PP of $\mathbb{F}_q$ if and only if $(r,s) = 1$ and $(r + \frac{us}{j}, \ell) = 1$.

8.1.66 Remark Some concrete classes of PPs satisfying the above assumptions can be found in [60, 65, 2033, 3075, 3076]. For example, let $h_k(x) := x^k + \cdots + x + 1$. Then the permutation behavior of the polynomials $x^r h_k(x^s) = x^r(x^{ks} + \cdots + x^s + 1)$ and $x^r h_k(x^{es})^t$ has been studied in detail. Moreover, for certain choices of indices ℓ and finite fields $\mathbb{F}_q$ (for example, $p \equiv -1 \pmod{2\ell}$ where $\ell > 1$ is either odd or $2\ell_1$ with ℓ_1 odd), several concrete classes of PPs can be obtained [65, 3075, 3076].

8.1.67 Remark When $\ell \leq 5$, much simpler descriptions involving congruences and gcd conditions can be found in [60, 2033]. A reformulation of these results in terms of roots of unity can be found in [3075] which also covers the case of $\ell = 7$. For larger index ℓ, one can also construct PPs of this form when ℓ is an odd prime such that $\ell < 2p+1$.

8.1.68 Theorem [60] Let ℓ be an odd prime such that $\ell < 2p+1$, then $P(x) = x^r(x^{ks} + \cdots + x^s + 1)$ is a PP of $\mathbb{F}_q$ if and only if $(r,s) = 1$, $(\ell, k+1) = 1$, $(2r+ks, \ell) = 1$, and $(k+1)^s \equiv 1 \pmod{p}$.

8.1.6 PPs from permutations of additive groups

8.1.69 Remark There are several results on new classes of PPs when using additive group endomorphisms ψ, $\bar{\psi}$, and φ in Theorem 8.1.39 [62, 2003, 3046, 3077].

8.1.70 Theorem [62] Consider any polynomial $g \in \mathbb{F}_{q^n}[x]$, any additive polynomials $\varphi, \psi \in \mathbb{F}_{q^n}[x]$, any $\mathbb{F}_q$-linear polynomial $\bar{\psi} \in \mathbb{F}_{q^n}[x]$ satisfying $\varphi \circ \psi = \bar{\psi} \circ \varphi$ and $\#\psi(\mathbb{F}_{q^n}) = \#\bar{\psi}(\mathbb{F}_{q^n})$, and any polynomial $h \in \mathbb{F}_{q^n}[x]$ such that $h(\psi(\mathbb{F}_{q^n})) \subseteq \mathbb{F}_q \setminus \{0\}$. Then

1. $f(x) := h(\psi(x))\varphi(x) + g(\psi(x))$ permutes $\mathbb{F}_{q^n}$ if and only if
   a. $\ker(\varphi) \cap \ker(\psi) = \{0\}$; and
   b. $\bar{f}(x) := h(x)\varphi(x) + \bar{\psi}(g(x))$ is a bijection between $\psi(\mathbb{F}_{q^n})$ and $\bar{\psi}(\mathbb{F}_{q^n})$.
2. For any fixed h, φ, ψ and $\bar{\psi}$ satisfying the above hypothesis and Part 1.a, there are $(\#\mathrm{im}(\psi))! \cdot (\#\ker(\bar{\psi}))^{\#\mathrm{im}(\psi)}$ such permutation functions f (when g varies) (where ψ and $\bar{\psi}$ are viewed as endomorphisms of $(\mathbb{F}_{q^n}, +)$).
3. Assume in addition that $(\bar{\psi} \circ g)\mid_{\mathrm{im}(\psi)} = 0$. Then $f(x) = h(\psi(x))\varphi(x) + g(\psi(x))$ permutes $\mathbb{F}_{q^n}$ if and only if $\ker(\varphi) \cap \ker(\psi) = \{0\}$ and $h(x)\varphi(x)$ induces a bijection from $\psi(\mathbb{F}_{q^n})$ to $\bar{\psi}(\mathbb{F}_{q^n})$.
4. Assume in addition that $\varphi \circ \psi = 0$, and that $g(x)$ restricted to $\mathrm{im}(\psi)$ is a permutation of $\mathrm{im}(\psi)$. Then $f(x) = h(\psi(x))\varphi(x) + g(\psi(x))$ permutes $\mathbb{F}_{q^n}$ if and only if $\ker(\varphi) \cap \ker(\psi) = \{0\}$ and $\bar{\psi}$ restricted to $\mathrm{im}(\psi)$ is a bijection between $\mathrm{im}(\psi)$ and $\mathrm{im}(\bar{\psi})$.

8.1.71 Theorem [62, 3046] Let $q = p^e$ for some positive integer e.

1. If k is an even integer or k is odd and q is even, then $f_{a,b,k}(x) = ax^q + bx + (x^q - x)^k$, $a, b \in \mathbb{F}_{q^2}$, permutes $\mathbb{F}_{q^2}$ if and only if $b - a^q \in \mathbb{F}_q^*$ and $a + b \neq 0$.

2. If k and q are odd positive integers, then $f_{a,k}(x) = ax^q + a^q x + (x^q - x)^k$, $a \in \mathbb{F}_{q^2}^*$ and $a + a^q \neq 0$, permutes $\mathbb{F}_{q^2}$ if and only if $(k, q-1) = 1$.

8.1.72 Remark The classes $f_{a,b,k}$ with $a, b \in \mathbb{F}_q$ and k even and $f_{a,k}$ for $a \in \mathbb{F}_q$ and p and k odd were first constructed in [62]. The remaining classes were obtained in [3046]. For other concrete classes of PPs of additive groups, we refer to [62, 325, 730, 1832, 2003, 3046, 3077].

8.1.7 Other types of PPs from the AGW criterion

8.1.73 Remark In this subsection, we give several other constructions of PPs that can be obtained by using arbitrary surjective maps λ and $\bar{\lambda}$ in Theorem 8.1.39, instead of using multiplicative or additive group homomorphism.

8.1.74 Theorem [62] Let q be a prime power, let n be a positive integer, and let L_1, L_2, L_3 be $\mathbb{F}_q$-linear polynomials over $\mathbb{F}_q$ seen as endomorphisms of $(\mathbb{F}_{q^n}, +)$. Let $g \in \mathbb{F}_{q^n}[x]$ be such that $g(L_3(\mathbb{F}_{q^n})) \subseteq \mathbb{F}_q$. Then $f(x) = L_1(x) + L_2(x)g(L_3(x))$ is a PP of $\mathbb{F}_{q^n}$ if and only if

1. $\ker(F_y) \cap \ker(L_3) = \{0\}$, for any $y \in \text{im}(L_3)$, where $F_y(x) := L_1(x) + L_2(x)g(y)$; and
2. $\bar{f}(x) := L_1(x) + L_2(x)g(x)$ permutes $L_3(\mathbb{F}_{q^n})$.

8.1.75 Remark The above result extends some constructions in [325, 730].

8.1.76 Theorem [62] Let q be any power of the prime number p, let n be any positive integer, and let S be any subset of $\mathbb{F}_{q^n}$ containing 0. Let $h, k \in \mathbb{F}_{q^n}[x]$ be polynomials such that $h(0) \neq 0$ and $k(0) = 0$, and let $B \in \mathbb{F}_{q^n}[x]$ be any polynomial satisfying $h(B(\mathbb{F}_{q^n})) \subseteq S$ and $B(a\alpha) = k(a)B(\alpha)$ for all $a \in S$ and all $\alpha \in \mathbb{F}_{q^n}$. Then the polynomial $f(x) = xh(B(x))$ is a PP of $\mathbb{F}_{q^n}$ if and only if $\bar{f}(x) = xk(h(x))$ induces a permutation of the value set $B(\mathbb{F}_{q^n})$.

8.1.77 Remark The case that $S = \mathbb{F}_q$ and $k(x) = x^2$ was considered in [2003]. Some examples of B are given in [62]. It is remarked in [62] that Theorem 8.1.76 can be generalized for $f(x) = A(x)h(B(x))$ and $\bar{f}(x) = C(x)k(h(x))$ where $A, C \in \mathbb{F}_{q^n}[x]$ are polynomials such that $B(A(x)) = C(B(x))$ with $C(0) = 0$ and A is injective on $B^{-1}(s)$ for each $s \in B(\mathbb{F}_{q^n})$, under the similar assumptions $h(B(\mathbb{F}_{q^n})) \subseteq S \setminus \{0\}$ and $B(a\alpha) = k(a)B(\alpha)$ for all $a \in S$ and all $\alpha \in \mathbb{F}_{q^n}$.

8.1.78 Definition [62] Let $S \subseteq \mathbb{F}_q$ and let $\gamma, b \in \mathbb{F}_q$. Then γ is a *b-linear translator* with respect to S for the mapping $F : \mathbb{F}_q \longrightarrow \mathbb{F}_q$, if

$$F(x + u\gamma) = F(x) + ub$$

for all $x \in \mathbb{F}_{q^n}$ and for all $u \in S$.

8.1.79 Remark The above definition is a generalization of the concept of b-linear translator studied in [593, 595, 1817], which deals with the case $q = p^{mn}$, and $S = \mathbb{F}_{p^m}$. The relaxation on the condition for S to be any subset of $\mathbb{F}_q$ provides a much richer class of functions (see examples in [62]). Using the original definition of linear translators, several classes of PPs of the form $G(x) + \gamma Tr(H(x))$ are constructed in [593, 595, 1817]. In the case that G is also a PP, it is equivalent to constructing polynomials of the form $x + \gamma Tr(H'(x))$.

8.1.80 Theorem [62] Let $S \subseteq \mathbb{F}_q$ and $F : \mathbb{F}_q \longrightarrow S$ be a surjective map. Let $\gamma \in \mathbb{F}_q$ be a b-linear translator with respect to S for the map F. Then for any $G \in \mathbb{F}_q[x]$ which maps S into S, we have that $x + \gamma G(F(x))$ is a PP of $\mathbb{F}_q$ if and only if $x + bG(x)$ permutes S.

8.1.81 Definition A *complete mapping* f of $\mathbb{F}_q$ is a permutation polynomial $f(x)$ of $\mathbb{F}_q$ such that $f(x) + x$ is also a permutation polynomial of $\mathbb{F}_q$.

8.1.82 Corollary [62, 593] Under the conditions of Theorem 8.1.80, we have

1. If $G(x) = x$ then $x + \gamma F(x)$ is a PP of $\mathbb{F}_q$ if and only if $b \neq -1$.
2. If q is odd and $2S = S$, then $x + \gamma F(x)$ is a complete mapping of $\mathbb{F}_q$ if and only if $b \notin \{-1, -2\}$.

8.1.83 Theorem [1817] Let $L : \mathbb{F}_{q^n} \longrightarrow \mathbb{F}_{q^n}$ be an $\mathbb{F}_q$-linear mapping of $\mathbb{F}_{q^n}$ with kernel $\alpha\mathbb{F}_q$, $\alpha \neq 0$. Suppose α is a b-linear translator with respect to $\mathbb{F}_q$ for the mapping $f : \mathbb{F}_{q^n} \longrightarrow \mathbb{F}_q$ and $h : \mathbb{F}_q \longrightarrow \mathbb{F}_q$ is a permutation of $\mathbb{F}_q$. Then the mapping $G(x) = L(x) + \gamma h(f(x))$ permutes $\mathbb{F}_{q^n}$ if and only if $b \neq 0$ and γ does not belong to the image set of L.

8.1.84 Corollary [1817] Let t be a positive integer with $(t, q-1) = 1$, $H \in \mathbb{F}_{q^n}[x]$ and $\gamma, \beta \in \mathbb{F}_{q^n}$. Then the mapping $G(x) = x^q - x + \gamma \left(Tr(H(x^q - x)) + \beta x\right)^t$ is a PP of $\mathbb{F}_{q^n}$ if and only if $Tr(\gamma) \neq 0$ and $Tr(\beta) \neq 0$.

8.1.85 Theorem [3046] Let A be a finite field, S and $\bar{S}$ be finite sets with $\#S = \#\bar{S}$ such that the maps $\psi : A \to S$ and $\bar{\psi} : A \to \bar{S}$ are surjective and ψ is additive, i.e., $\bar{\psi}(x+y) = \bar{\psi}(x)+\bar{\psi}(y)$, for all $x, y \in A$. Let $g : S \to A$ and $f : A \to A$ be maps such that $\bar{\psi}(f + g \circ \psi) = f \circ \psi$ and $\bar{\psi}(g(\psi(x))) = 0$ for every $x \in A$. Then the map $f(x) + g(\psi(x))$ permutes A if and only if f permutes A.

8.1.86 Corollary [3046] Let n and k be positive integers such that $(n, k) = d > 1$, and let s be any positive integer with $s(q^k-1) \equiv 0 \pmod{q^n-1}$. Let $L_1(x) = a_0x+a_1x^{q^d}+\cdots+a_{n/d-1}x^{q^{n-d}}$ be a polynomial with $L_1(1) = 0$ and let $L_2 \in \mathbb{F}_q[x]$ be a linearized polynomial and $g \in \mathbb{F}_{q^n}[x]$. Then $f(x) = (g(L_1(x)))^s + L_2(x)$ permutes $\mathbb{F}_{q^n}$ if and only if L_2 permutes $\mathbb{F}_{q^n}$.

8.1.87 Corollary [3046] Let n and k be positive integers such that $(n, k) = d > 1$, let s be any positive integer with $s(q^k - 1) \equiv 0 \pmod{q^n - 1}$. Then $h(x) = (x^{q^k} - x + \delta)^s + x$ permutes $\mathbb{F}_{q^n}$ for any $\delta \in \mathbb{F}_{q^n}$.

8.1.88 Remark More classes of PPs of the form $(x^{q^k} - x + \delta)^s + L(x)$ and their generalization can be found in [1061, 3058]. See [1481, 3043, 3044, 3057] for more classes of PPs of the form $(x^{p^k} + x + \delta)^s + L(x)$. One can also find several classes of PPs of the form $x^d + L(x)$ over $\mathbb{F}_{2^n}$ in [1929, 2363, 2362]. It is proven in [1929] that, under the assumption $\gcd(d, 2^n - 1) > 1$, if $x^d + L(x)$ is a PP of $\mathbb{F}_{2^n}$ then L must be a PP of $\mathbb{F}_{2^n}$. Hence some of these classes of PPs of the form $x^d + L(x)$ are compositional inverses of PPs of the form $L_1(x)^d + x$.

8.1.89 Remark See [872, 3045] for some explicit classes of PPs over $\mathbb{F}_{3^m}$.

8.1.8 Dickson and reversed Dickson PPs

8.1.90 Remark The permutational behavior of Dickson polynomials of the first kind is simple and classical; see Remark 8.1.5. For more information on Dickson polynomials, see Section 9.6.

8.1.91 Definition For any positive integer n, let $E_n(x, a)$ be the *Dickson polynomial of the second kind* (DPSK) defined by

$$E_n(x, a) = \sum_{i=0}^{\lfloor n/2 \rfloor} \binom{n-i}{i} (-a)^i x^{n-2i}.$$

8.1.92 Remark Matthews observed in his Ph.D. thesis [2034] that if q is a power of an odd prime p and n satisfies the system of congruences

$$\begin{aligned} n+1 &\equiv \pm 2 \pmod{p}, \\ n+1 &\equiv \pm 2 \pmod{\tfrac{1}{2}(q-1)}, \\ n+1 &\equiv \pm 2 \pmod{\tfrac{1}{2}(q+1)}, \end{aligned} \tag{8.1.4}$$

then E_n is a PP of $\mathbb{F}_q$. However, the above is not a necessary condition in general. When $p = 3$ or 5 and q is composite there are examples of DPSK E_n known which are PP for which (8.1.4) does not hold, see [1592, 2175]. Moreover, there are several papers concentrating on the permutational behavior of DPSK over finite fields with small characteristics including characteristic two and general a which is not necessarily ± 1 [733, 1482, 1483, 1484]. On the other hand, when $q = p$ or p^2 and $a = 1$, Cipu and Cohen proved that the condition (8.1.4) is also necessary [652, 653, 679, 680].

8.1.93 Definition [1547] For any positive integer n, the *reversed Dickson polynomial*, $D_n(a,x)$, is defined by

$$D_n(a,x) = \sum_{i=0}^{\lfloor n/2 \rfloor} \frac{n}{n-i}\binom{n-i}{i}(-x)^i a^{n-2i}.$$

8.1.94 Remark It is easy to check that $D_n(0,x) = 0$ if n is odd and $D_n(0,x) = 2(-x)^k$ if $n = 2k$. Hence $D_n(0,x)$ is a PP of $\mathbb{F}_q$ if and only if q is odd, $n = 2k$, and $(k, q-1) = 1$. For $a \neq 0$, we can also check that $D_n(a,x) = a^n D_n(1, x/a^2)$. Hence $D_n(a,x)$ is a PP of $\mathbb{F}_q$ where $a \neq 0$ if and only if $D_n(1,x)$ is a PP of $\mathbb{F}_q$. If $D_n(1,x)$ is a PP of $\mathbb{F}_q$, then (q,n) is a *desirable pair*.

8.1.95 Definition A mapping f from $\mathbb{F}_q$ to $\mathbb{F}_q$ is *almost perfect nonlinear* (APN) if the difference equation $f(x+a) - f(x) = b$ has at most two solutions for any fixed $a \neq 0, b \in \mathbb{F}_q$.

8.1.96 Remark We refer to Section 9.2 for more information on APN functions and their applications.

8.1.97 Theorem [1547] Let $q = p^e$ with p a prime and $e > 0$. If $p = 2$ or $p > 3$ and n is odd, then x^n is an APN on $\mathbb{F}_{q^2}$ implies that $D_n(1,x)$ is a PP of $\mathbb{F}_q$, which also implies that x^n is an APN over $\mathbb{F}_q$.

8.1.98 Theorem [1545] The pair (p^e, n) is a desirable pair in each of the following cases:

p	e	n	
2		2^k+1, $(k,2e)=1$	[1295, 1547]
2		$2^{2k}-2^k+1$, $(k,2e)=1$	[1547, 1690]
2	even	2^e+2^k+1, $k>0$, $(k-1,e)=1$	[1547]
2	$5k$	$2^{8k}+x^{6k}+2^{4k}+2^{2k}-1$	[906, 1547]
3		$(3^k+1)/2$, $(k,2e)=1$	[1547]
3	even	3^e+5	[1542]
5		$(5^k+1)/2$, $(k,2e)=1$	[1479]
≥ 3		p^k+1, $k\geq 0$, $p^k \equiv 1 \pmod 4$, $v_2(e) \leq v_2(k)$	[1547]
≥ 3		p^e+2, $k \geq 0$, $p^e \equiv 1 \pmod 3$	[1479, 1547]
≥ 5		3	

8.1.99 Remark Two pairs (q,n_1) and (q,n_2), where n_1 and n_2 are positive integers, are equivalent if n_1 and n_2 are in the same p-cyclotomic coset modulo q^2-1, i.e., $D_{n_1}(1,x) \equiv D_{n_2}(1,x) \pmod{x^q - x}$. No desirable pairs outside the ten families (up to equivalence) given in Theorem 8.1.98 are known. There are several papers on necessary conditions for a reverse Dickson polynomial to be a PP [1544, 1545]. In particular, it is proved in [1545] that (p^e,n) is a desirable pair if and only if $f_n(x) = \sum_{j\geq 0}\binom{n}{2j}x^j$ is a PP of $\mathbb{F}_{p^e}$. However, it is not known whether the above classes are the only non-equivalent desirable pairs. Several new classes of reversed Dickson polynomials can be found in [1542, 1544].

8.1.100 Remark Dickson polynomials are used to prove the following class of PPs.

8.1.101 Theorem [1527] Let $m \geq 1$ and $1 \leq k, r \leq m-1$ be positive integers satisfying that $kr \equiv 1 \pmod m$. Let $q = 2^m$, $\sigma = 2^k$, and $Tr(x) := x + x^2 + \cdots + x^{2^{m-1}}$. For α, γ in $\{0,1\}$, we define

$$H_{\alpha,\gamma}(x) := \gamma Tr(x) + \frac{(\alpha Tr(x) + \sum_{i=0}^{r-1} x^{\sigma^i})^{\sigma+1}}{x^2}.$$

Then $H_{\alpha,\gamma}(x)$ is a PP of $\mathbb{F}_{2^m}$ if and only if $r + (\alpha+\gamma)m \equiv 1 \pmod 2$.

8.1.9 Miscellaneous PPs

8.1.102 Remark Cyclic and Dickson PPs play a vital role in the Schur Conjecture from 1922 which postulated that if f is a polynomial with integer coefficients which is a PP of $\mathbb{F}_p$ (when considered modulo p) for infinitely many primes p, then f must be a composition of binomials $ax^n + b$ and Dickson polynomials. This has been shown to be true; see [1109], the notes to Chapter 7 of [1939] and [2192]. A proof without the use of complex analysis can be found in [2827]. More generally, a matrix analogue of the Schur conjecture was studied in [2178]. Let $\mathbb{F}_q^{m\times m}$ denote the ring of $m \times m$ matrices over the finite field $\mathbb{F}_q$. A polynomial $f \in \mathbb{F}_q[x]$ is a *permutation polynomial* (PP) on $\mathbb{F}_q^{m\times m}$ if it gives rise to a permutation of $\mathbb{F}_q^{m\times m}$. Using a characterization of PPs of the matrix ring over finite fields $\mathbb{F}_q$ in [396], it is shown by Mullen in [2178] that any polynomial f with integral coefficients which permutes the matrices of fixed size over a field of p elements for infinitely many p is a composition of linear polynomials and Dickson polynomials $D_n(x, a)$ with $n \neq 3$ an odd prime and $a \neq 0$ an integer. Several related questions are also addressed in [2178]; see Section 9.7 for more information on Schur's conjecture.

8.1.103 Remark Let f be an integral polynomial of degree $n \geq 2$. Cohen [676] proved one of the Chowla and Zassenhaus conjectures [632] (concerning irreducible polynomials), which postulated that if f is a PP over $\mathbb{F}_p$ of degree n modulo p for any $p > (n^2 - 3n + 4)^2$, then $f(x) + cx$ is not a PP of $\mathbb{F}_p$ unless $c = 0$. This shows that if both f and g are integral PPs of degree $n \geq 2$ over a large prime field, then their difference $h = f - q$ cannot be a linear polynomial cx where $c \neq 0$. Suppose that $t \geq 1$ denotes the degree of h. More generally, it is proved in [699] that $t \geq 3n/5$. Moreover, if $n \geq 5$ and $t \leq n - 3$ then $(t, n) > 1$. Roughly speaking, two PPs over a large prime field $\mathbb{F}_p$ cannot differ by a polynomial with degree less than $3n/5$.

8.1.104 Remark Evans [998] considers *orthomorphisms*, mappings θ with $\theta(0) = 0$ so that θ and $\theta(x) - x$ are both PPs of $\mathbb{F}_q$. He studies connections between orthomorphisms, latin squares, and affine planes. A map θ is an orthomorphism if and only if $\theta(x) - x$ is a complete mapping. Complete mappings of small degrees and existence of complete mappings (in particular, binomials) are studied in [2268]. Enumeration results for certain types of cyclotomic orthomorphisms are provided in [2278]. It is proved in [2268] for odd q and in [2904] for even q that the degree of a complete mapping is at most $q - 3$. It is known that families of permutation polynomials of the form $f(x) + cx$ can be used in the construction of maximal sets of mutually orthogonal Latin squares [997, 2918]. Let $C(f)$ be the number of c in $\mathbb{F}_q$ such that $f(x) + cx$ is a permutation polynomial over $\mathbb{F}_q$. Cohen's theorem [676] on the Chowla-Zassenhaus conjecture shows that $C(f) \leq 1$ if the degree n of f is not divisible by p and q is sufficiently large compared to n. Chou showed that $C(f) \leq q - 1 - n$ in his thesis [624]. Then Evans, Greene, and Niederreiter proved $C(f) \leq q - \frac{q-1}{n-1}$ in [1016], which also proves a conjecture of Stothers [2729] when q is prime. In the case that q is an odd prime, it gives the best possible result $C(f) \leq (q - 3)/2$ for polynomials of the form $x^{(q+1)/2} + cx$. A general bound for $C(f)$ which implies Chou's bound was obtained by Wan, Mullen, and Shiue in [2918], as well as a significant bound $C(f) \leq r$ where r is the least nonnegative residue of $q - 1$ modulo n under certain mild conditions. It is conjectured in [1016] that $f(x) - f(0)$ is a linearized p-polynomial over $\mathbb{F}_q$ if $C(f) \geq \lfloor q/2 \rfloor$; this was proved to be true for $q = p$ or any monomial $f(x) = x^e$ in [1016]. Wan observed that this conjecture holds also for $q = p^2$ from the results in [624, 1016]. Several other related results on the function $C(f)$ can be found in [997, 2826, 2912].

8.1.105 Remark Results on the cycle structure of monomials and of Dickson polynomials can be found in [44] and [1934], respectively. Cycle decomposition, in particular, decomposition of them into cycles of the same length (which are motivated by Turbo codes [2742, 2769]), are studied in [44, 2152, 2296, 2495, 2496, 2519]. Moreover, we refer readers to [68, 575, 625] for cycle structure of permutation polynomials with small Carlitz rank [575] or with full cycles.

8.1.106 Remark Finding the compositional inverse of a PP is a hard problem except for the trivial well known classes such as the inverses of linear polynomial, monomials, and Dickson polynomials. There are several papers on the explicit format of the inverses of some special classes of permutation polynomials, for example, [727, 1817, 2205, 2206, 2942]. Because the problem is equivalent to finding the inverses of PPs of the form $x^r f(x^s)$, the most general result can be found in [2941].

8.1.107 Remark PPs are related to special functions. For example, Dobbertin constructed several classes of PPs [903, 904] over finite fields of even characteristic and used them to prove several conjectures on APN monomials. The existence of APN permutations on $\mathbb{F}_{2^{2n}}$ is a long-term open problem in the study of vectorial Boolean functions. Hou [1541] proved that there are no APN permutations over $\mathbb{F}_{2^4}$ and there are no APN permutations on $\mathbb{F}_{2^{2n}}$ with coefficients in $\mathbb{F}_{2^n}$. Only recently the authors in [424] found the first APN permutation over $\mathbb{F}_{2^6}$. However, the existence of APN permutations on $\mathbb{F}_{2^{2n}}$ for $n \geq 4$ remains open. Over finite fields of odd characteristics, a function f is a *planar function* if $f(x+a) - f(x)$ is a PP for each nonzero a; see Sections 9.2 and 9.5. In [874], Ding and Yuan constructed a new family of planar functions over $\mathbb{F}_{3^m}$, where m is odd, and then obtained the first examples of skew Hadamard difference sets, which are inequivalent to classical Paley difference sets. Permutation polynomials of $\mathbb{F}_{3^{2h+1}}$ obtained from the Ree-Tits slice symplectic spreads in $\mathrm{PG}(3, 3^{2h+1})$ were studied in [192]. Later on, they were used in [871] to construct a family of skew Hadamard difference sets in the additive group of this field. For more information on these special functions and their applications, we refer the readers to Sections 9.2, 9.5, and 14.6.

8.1.108 Remark Golomb and Moreno [1305] show that PPs are useful in the construction of circular Costas arrays, which are useful in sonar and radar communications. They gave an equivalent conjecture for circular Costas arrays in terms of permutation polynomials and provided some partial results. The connection between Costas arrays and APN permutations of integer rings $\mathbb{Z}_n$ was studied in [917]. Composed with discrete logarithms, permutation polynomials of finite fields are used to produce permutations of integer rings $\mathbb{Z}_n$ with optimum ambiguity and deficiency [2353, 2355], which generate APN permutations in many cases. Earlier results on PPs of $\mathbb{Z}_n$ can be found in Section 5.6 of [870] and [2171, 2460, 3062].

See Also

§8.2	For discussion of PPs in several variables.
§8.3	For value sets of polynomials.
§8.4	For exceptional polynomials over finite fields.
§9.2	For discussion of PN and APN functions.
§9.5	For studies of planar functions.
§9.6	For Dickson polynomials over finite fields.
§9.7	For connections to Schur's Conjecture.
[902]	For connections with monomial graphs.
[2603]	For connections with check digit systems.

References Cited: [44, 60, 61, 62, 63, 64, 65, 68, 192, 325, 396, 424, 504, 551, 569, 575, 593,

595, 624, 625, 632, 652, 653, 667, 676, 677, 679, 680, 681, 688, 699, 727, 730, 733, 771, 840, 870, 871, 872, 874, 902, 903, 904, 906, 917, 997, 998, 1016, 1061, 1109, 1120, 1221, 1295, 1305, 1479, 1481, 1482, 1483, 1484, 1527, 1541, 1542, 1544, 1545, 1547, 1592, 1690, 1716, 1787, 1788, 1813, 1817, 1832, 1833, 1916, 1929, 1933, 1934, 1935, 1936, 1939, 1983, 1984, 1995, 1996, 1997, 2003, 2018, 2019, 2020, 2033, 2034, 2127, 2152, 2171, 2175, 2176, 2178, 2192, 2205, 2206, 2268, 2278, 2296, 2353, 2355, 2359, 2362, 2363, 2460, 2495, 2496, 2519, 2603, 2605, 2638, 2680, 2681, 2729, 2742, 2769, 2825, 2826, 2827, 2904, 2905, 2908, 2909, 2911, 2912, 2913, 2916, 2918, 2919, 2927, 2940, 2941, 2942, 2943, 2967, 3043, 3044, 3045, 3046, 3047, 3057, 3058, 3062, 3065, 3075, 3076, 3077]

8.2 Several variables

Rudolf Lidl, University of Tasmania
Gary L. Mullen, The Pennsylvania State University

8.2.1 Definition A polynomial $f \in \mathbb{F}_q[x_1, \ldots, x_n]$ is a *permutation polynomial in n variables over $\mathbb{F}_q$* if the equation $f(x_1, \ldots, x_n) = \alpha$ has exactly q^{n-1} solutions in $\mathbb{F}_q^n$ for each $\alpha \in \mathbb{F}_q$.

8.2.2 Remark A permutation polynomial $f(x_1, \ldots, x_n)$ induces a mapping from $\mathbb{F}_{q^n}$ to $\mathbb{F}_q$ but does not induce a $1-1$ mapping unless $n = 1$.

8.2.3 Remark [2172] A combinatorial computation shows that there are $(q^n)!/((q^{n-1})!)^q$ permutation polynomials in n variables over $\mathbb{F}_q$.

8.2.4 Definition A set of polynomials $f_i(x_1, \ldots, x_n) \in \mathbb{F}_q[x_1, \ldots, x_n], 1 \leq i \leq r$, forms an *orthogonal system in n variables over $\mathbb{F}_q$* if the system of equations $f_i(x_1, \ldots, x_n) = \alpha_i$, $1 \leq i \leq r$, has exactly q^{n-r} solutions in $\mathbb{F}_q^n$ for each $(\alpha_1, \ldots, \alpha_r) \in \mathbb{F}_q^r$.

8.2.5 Theorem [2228] For every orthogonal system $f_1, \ldots, f_m \in \mathbb{F}_q[x_1, \ldots, x_n], 1 \leq m < n$, over $\mathbb{F}_q$ and every $r, 1 \leq r \leq n - m$, there exist $f_{m+1}, \ldots, f_{m+r} \in \mathbb{F}_q[x_1, \ldots, x_n]$ so that $f_1, \ldots, f_{m+r}$ forms an orthogonal system over $\mathbb{F}_q$.

8.2.6 Theorem [542] The system $f_1, \ldots, f_m, 1 \leq m \leq n$, is orthogonal over $\mathbb{F}_q$ if and only if

$$\sum_{(c_1,\ldots,c_n)\in\mathbb{F}_q^n} \chi_{b_1}(f_1(c_1, \ldots, c_n)) \cdots \chi_{b_m}(f_m(c_1, \ldots, c_n)) = 0$$

for all additive characters $\chi_{b_1}, \ldots, \chi_{b_m}$ of $\mathbb{F}_q$ with $(b_1, \ldots, b_m) \neq (0, \ldots, 0)$.

8.2.7 Corollary [542] A polynomial $f \in \mathbb{F}_q[x_1, \ldots, x_n]$ is a permutation polynomial over $\mathbb{F}_q$ if and only if

$$\sum_{(c_1,\ldots,c_n)\in\mathbb{F}_q^n} \chi(f(c_1, \ldots, c_n)) = 0$$

for all nontrivial additive characters χ of $\mathbb{F}_q$.

8.2.8 Theorem [2228] A set of polynomials $f_1, \ldots, f_r$ in n variables over $\mathbb{F}_q$ forms an orthogonal system over $\mathbb{F}_q$ if and only if the polynomial $b_1 f_1 + \cdots + b_r f_r$ is a permutation polynomial for each $(b_1, \ldots, b_r) \neq (0, \ldots, 0) \in \mathbb{F}_q^r$.

8.2.9 Theorem [1937] Suppose $f \in \mathbb{F}_q[x_1, \ldots, x_n]$ is of the form

$$f(x_1, \ldots, x_n) = g(x_1, \ldots, x_m) + h(x_{m+1}, \ldots, x_n), \quad 1 \le m < n.$$

If at least one of g and h is a permutation polynomial over $\mathbb{F}_q$ then f is a permutation polynomial over $\mathbb{F}_q$. If q is prime, then the converse holds.

8.2.10 Theorem [2032] Let $f_1, \ldots, f_t$ be polynomials in disjoint sets of variables, where f_i is a polynomial in v_i variables, $i = 1, \ldots, t$. Then $f_1 + \cdots + f_t$ is a permutation polynomial over $\mathbb{F}_q$, $q = p^e$, if and only if, for every subgroup H of the additive group G of $\mathbb{F}_q$ of order p^{e-1}, there is an f_i which distributes $\mathbb{F}_q^{v_i}$ uniformly over the cosets of H in G.

8.2.11 Theorem [1937] If q is not prime then for $1 \le m < n$ there exist polynomials $g(x_1, \ldots, x_m)$ and $h(x_{m+1}, \ldots, x_n)$ over $\mathbb{F}_q$ such that $g(x_1, \ldots, x_m) + h(x_{m+1}, \ldots, x_n)$ is a permutation polynomial over $\mathbb{F}_q$ but neither $g(x_1, \ldots, x_m)$ nor $h(x_{m+1}, \ldots, x_n)$ is a permutation polynomial over $\mathbb{F}_q$.

8.2.12 Theorem [2228] If $n = mk$ for positive integers n, m, k, there is a 1-1 correspondence between orthogonal systems over $\mathbb{F}_q$ consisting of m polynomials over $\mathbb{F}_q$ of degree less than q in each of their n variables and permutation polynomials over $\mathbb{F}_{q^m}$ of degree less than q^m in each of their k variables.

8.2.13 Corollary [2228] There is a 1-1 correspondence between orthogonal systems over $\mathbb{F}_q$ consisting of n polynomials over $\mathbb{F}_q$ of degree less than q in each of their n variables and permutation polynomials in one variable over $\mathbb{F}_{q^n}$ of degree less than q^n.

8.2.14 Remark Dickson permutation polynomials are studied in Section 8.1 and in Section 9.6 where results related to them in both one as well as several variables are provided.

8.2.15 Theorem [1940]

1. For $a \in \mathbb{F}_q^*$ the system $g_k(a)$ (see Definition 9.6.51) is orthogonal over $\mathbb{F}_q$ if and only if $(k, q^s - 1) = 1$ for $s = 1, \ldots, n+1$.
2. The system $g_k(0)$ is orthogonal if and only if $(k, q^s - 1) = 1$ for $s = 1, \ldots, n$.

8.2.16 Definition Two polynomials are *equivalent* if one can be transformed into the other by a transformation of the form $x_i = \sum_{j=1}^{n} a_{ij} y_j + b_i, 1 \le i \le n$, where $a_{ij}, b_i \in \mathbb{F}_q$ and the matrix (a_{ij}) is nonsingular.

8.2.17 Theorem [2227] Let $f \in \mathbb{F}_q[x_1, \ldots, x_n]$ with the degree of f at most two and $n \ge 2$. For q odd, f is a permutation polynomial over $\mathbb{F}_q$ if and only if f is equivalent to a polynomial of the form $g(x_1, \ldots, x_{n-1}) + x_n$ for some g. For q even, f is a permutation polynomial over $\mathbb{F}_q$ if and only if f is equivalent to a polynomial of the form $g(x_1, \ldots, x_{n-1}) + x_n$ or $g(x_1, \ldots, x_{n-1}) + x_n^2$.

8.2.18 Theorem [2174] If q is odd and $n \ge 2$, then $f(x_1, \ldots, x_n)$ of degree at most two is a nonsingular feedback function if and only if $f(x_1, \ldots, x_n) = cx_1 + f_0(x_2, \ldots, x_n)$, $c \in \mathbb{F}_q$ where $f_0(x_2, \ldots x_n)$ is any polynomial in the variables $x_2, \ldots, x_n$ of degree at most two over $\mathbb{F}_q$.

8.2.19 Remark Niederreiter [2229] provides criteria for quadratic polynomials over $\mathbb{Z}$ to be permutation polynomials modulo p, i.e., permutation polynomials over $\mathbb{F}_p$, that involve the rank of a matrix of coefficients. A result similar to Theorem 8.2.18 is obtained in [2174] for q even; see the corresponding result for quadratic forms in Theorem 8.2.17. These results provide an application of several variable permutation polynomials.

8.2.20 Remark [2736] Results are given on permutation polynomials in several variables over a finite field and orthogonal systems of polynomials whose image spaces are allowed to be arbitrary subfields of the finite field. The results developed in this paper are used to construct additional complete sets of frequency squares, rectangles and hyperrectangles, and orthogonal arrays. Some of the theorems present a relationship between permutation polynomials of a finite field and field permutation functions, an implicit bound on the possible number of functions in an orthogonal field system, and results related to the generation of orthogonal field systems.

8.2.21 Remark For finite rings there are two concepts to distinguish: permutation polynomials (as above) and strong permutation polynomials. The latter are defined via the cardinality of the inverse image. Polynomials are *strong permutation polynomials* (or *strong orthogonal systems*) in n variables if they can be completed to an orthogonal system of n polynomials in n variables.

8.2.22 Theorem [1132] If R is a local ring whose maximal ideal has a minimal number m of generators, then for every $n > m$ there exists a permutation polynomial in n variables that is not a strong permutation polynomial.

8.2.23 Corollary [1132] Every permutation polynomial in any number of variables over a local ring R is strong if and only if R is a finite field.

8.2.24 Remark Wei and Zhang showed [2957] that if $n \leq m$ in the setting of Theorem 8.2.22, then every orthogonal system of k polynomials in n variables can be completed to an orthogonal system of n polynomials (and in particular, every permutation polynomial is strong).

See Also

§9.4	Considers κ-polynomials used for constructions of semifields.
§14.1	Discusses orthogonal latin squares and hypercubes.
[1939]	Section 7.5 discusses permutations and orthogonal systems in several variables.
[2188]	Considers bounds for value sets of polynomial vectors in several variables.
[2326]	Considers an application of permutation polynomials and orthogonal systems to pseudorandom number generation.
[3062]	Considers permutation polynomials over finite commutative rings.

References Cited: [542, 1132, 1937, 1939, 1940, 2032, 2172, 2174, 2227, 2228, 2229, 2326, 2736, 2957, 3062]

8.3 Value sets of polynomials

Gary L. Mullen, The Pennsylvania State University
Michael E. Zieve, University of Michigan

8.3.1 Definition For $f \in \mathbb{F}_q[x]$, the *value set of* f is the set $V_f = \{f(a) | a \in \mathbb{F}_q\}$; the cardinality of V_f is denoted by $\#V_f$.

8.3.2 Remark Every subset of $\mathbb{F}_q$ occurs as V_f for some $f \in \mathbb{F}_q[x]$ of degree at most $q-1$ (by the Lagrange Interpolation Formula); see Theorem 2.1.131.

8.3.1 Large value sets

8.3.3 Remark Any $f \in \mathbb{F}_q[x]$ satisfies $\#V_f \le q$; equality occurs precisely when f is a permutation polynomial; see Section 8.1.

8.3.4 Theorem Suppose $f \in \mathbb{F}_q[x]$ of degree n is not a permutation polynomial. Then:

1. [2826, 2911, 2982] $\#V_f \le q - \lceil (q-1)/n \rceil$.
2. [1086, 1367, 1368] If $\#V_f \ne (1-1/n)q$ and $n > 5$ then $\#V_f \le (1-2/n)q + O_n(\sqrt{q})$.
3. [1367] If $\gcd(n, q) = 1$ then $\#V_f \le (5/6)q + O_n(\sqrt{q})$.

8.3.5 Example [760] Let $q = r^k$ where r is a prime power and k is a positive integer. Then $f(x) := x^r + x^{r-1}$ satisfies $\#V_f = q - q/r$, and hence achieves equality in (1).

8.3.2 Small value sets

8.3.6 Remark If $f \in \mathbb{F}_q[x]$ has degree n, then $\#V_f \ge \lceil q/n \rceil$ (since each $\alpha \in \mathbb{F}_q$ has at most n preimages under f).

8.3.7 Definition A polynomial $f \in \mathbb{F}_q[x]$ of degree n is a *minimal value set polynomial* if $\#V_f = \lceil q/n \rceil$.

8.3.8 Theorem [549] Let $f \in \mathbb{F}_p[x]$ have degree n, where p is prime. If $n < p$ and $\#V_f = \lceil p/n \rceil \ge 3$, then n divides $p-1$ and $f(x) = a(x+b)^n + c$ with $a, b, c \in \mathbb{F}_p$.

8.3.9 Theorem [2105] Let $f \in \mathbb{F}_q[x]$ be monic of degree n, where $\gcd(n, q) = 1$ and $n \le \sqrt{q}$. If $\#V_f = \lceil q/n \rceil$, then n divides $q-1$ and $f(x) = (x+b)^n + c$ with $b, c \in \mathbb{F}_q$.

8.3.10 Problem Determine all minimal value set polynomials over $\mathbb{F}_{p^k}$. This is done for $k \le 2$ in [2105].

8.3.11 Remark Minimal value set polynomials whose values form a subfield are characterized in [351]. A connection between minimal value set polynomials and Frobenius non-classical curves is given in [350].

8.3.12 Theorem [628, 1308] If $f(x) \in \mathbb{F}_q[x]$ is monic of degree $n > 15$, where $n^4 < q$ and $\#V_f < 2q/n$, then $f(x)$ has one of the forms:

1. $(x+a)^n + b$, where $n \mid (q-1)$;
2. $((x+a)^{n/2} + b)^2 + c$, where $n \mid (q^2-1)$;
3. $((x+a)^2 + b)^{n/2} + c$, where $n \mid (q^2-1)$.

8.3.13 Theorem [289] Let $f \in \mathbb{F}_p[x]$ have degree less than $\frac{3}{4}(p-1)$, where p is prime. If $\#f(\mathbb{F}_p^*) = 2$ then f is a polynomial in $x^{(p-1)/d}$ for some $d \in \{2, 3\}$.

8.3.14 Remark This result indicates that some phenomena become apparent only when one considers $\#f(\mathbb{F}_p^*)$ rather than $\#V_f$.

8.3.3 General polynomials

8.3.15 Theorem [285, 2981] Fix n, and let $e_n := \sum_{j=1}^{n}(-1)^{j-1}/j!$. There is a constant a_n such that, for each q, there are $q^n + O_n(q^{n-1})$ monic polynomials $f \in \mathbb{F}_q[x]$ of degree n satisfying $|\#V_f - e_n q| \leq a_n\sqrt{q}$.

8.3.16 Remark The previous result says that if q is large compared to n then most polynomials over $\mathbb{F}_q$ of degree n take approximately $e_n q$ values. Note that $e_n \to 1 - 1/e$ as $n \to \infty$.

8.3.17 Theorem [667] For fixed n, there is a finite set T_n of rational numbers such that: for any q, and any $f \in \mathbb{F}_q[x]$ of degree n, there is an element $c_f \in T_n$ such that $\#V_f = c_f q + O_n(\sqrt{q})$.

8.3.18 Remark The set T_n may be chosen to be $\{a/n!\colon (n-1)! \leq a \leq n!\}$.

8.3.19 Remark For fixed n, if q is large and $f \in \mathbb{F}_q[x]$ has degree n, then $\#V_f/q$ lies in a tiny interval around a member of a finite set; crucially, this finite set depends only on n, and not on q.

8.3.20 Theorem [667] Let $f \in \mathbb{F}_q[x]$ have degree n, and write $f = g(x^{p^j})$ where $j \geq 0$ and $g \in \mathbb{F}_q[x] \setminus \mathbb{F}_q[x^p]$; here p is the characteristic of $\mathbb{F}_q$. Let t be transcendental over $\mathbb{F}_q$, and let A and G be the Galois groups of $g(x) - t$ over $\mathbb{F}_q(t)$ and $\bar{\mathbb{F}}_q(t)$, respectively, where $\bar{\mathbb{F}}_q$ denotes an algebraic closure of $\mathbb{F}_q$. Then G is a normal subgroup of A, and A/G is cyclic. The quantity c_f in Theorem 8.3.17 may be taken to be the proportion of elements in a generating coset of A/G which fix at least one of the roots of $g(x) - t$.

8.3.21 Example Let $f \in \mathbb{F}_q[x]$ have degree n.

1. If $n = 2$ then $\#V_f \in \{q/2,\ (q+1)/2,\ q\}$.
2. If $n = 3$ then $\#V_f \in \{q/3,\ (q+2)/3,\ (2q-1)/3,\ 2q/3,\ (2q+1)/3,\ q\}$.
3. [2043] If $n = 4$ and q is an odd prime then $\#V_f$ is either $(q+3)/4$, $(q+1)/2$, $(3q+4+i)/8$ with $\pm i \in \{1,3,5\}$, or $5q/8 + O(\sqrt{q})$.

8.3.4 Lower bounds

8.3.22 Theorem [2919] If $\mu_q(f)$ is the smallest positive integer i so that $\sum_{a \in F_q}(f(a))^i \neq 0$, then $\#V_f \geq \mu_q(f) + 1$.

8.3.23 Remark Assume that for a polynomial f the degree n of f satisfies $n < q - 1$. Write $(f(x))^i = \sum_{j=0}^{q-1} a_{ij} \bmod (x^q - x)$. Let A_f be the matrix $A_f = (a_{ij}^{q-1})$, for $1 \leq i, j \leq q-1$ so that the (i,j)-th entry of A_f is 1 if the coefficient of x^j in $(f(x))^i \bmod (x^q - x)$ is nonzero. If f is not the zero polynomial, then A_f has at least one nonzero column. If the j-th column of A_f consists entirely of 0s or entirely of 1s, set $l_j = 0$. Otherwise, for a nonzero j-th column of A_f, arrange the entries in a circle and define l_j to be the maximum number of consecutive zeros appearing in this circular arrangement. Let L_f be the maximum of the values of l_j, where the maximum is taken over all of the $q-1$ columns of the matrix A_f.

8.3.24 Theorem [774] With notation as above, $|V_f| \geq L_f + 2$.

8.3.25 Remark [630] If f is a polynomial over $\mathbb{F}_q$ and A_f' is the matrix from Remark 8.3.23 without the $(q-1)$-st powers, i.e., the matrix $A_f' = (a_{ij})$, then $\#V_f = 1 + \mathrm{rank}(A_f')$.

8.3.26 Corollary Since $|V_f| \geq l_{q-1} + 2$, we have the result of Theorem 2.1 of [2919].

8.3.27 Remark The Hermite/Dickson criterion from Section 8.1 is essentially equivalent to the first $q-2$ consecutive elements of the last column of the matrix A_f being 0, with the last element being 1. Thus f is a permutation polynomial if and only if $L_f = q-2$.

8.3.28 Remark See [2826] for further inequalities.

8.3.5 Examples

8.3.29 Theorem $\#V_{x^n} = 1 + (q-1)/(n, q-1)$.

8.3.30 Definition The *Dickson polynomial* of degree n and parameter $a \in \mathbb{F}_q$ is defined by

$$D_n(x,a) = \sum_{i=0}^{\lfloor n/2 \rfloor} \frac{n}{n-i} \binom{n-i}{i} (-a)^i x^{n-2i}.$$

8.3.31 Remark See Section 9.6 for a discussion of Dickson polynomials over $\mathbb{F}_q$.

8.3.32 Theorem [628] Suppose q is odd with $2^r \parallel (q^2-1)$. Then for each $n \geq 1$, and each $a \in \mathbb{F}_q^*$,

$$\#V_{D_n(x,a)} = \frac{q-1}{2(n,q-1)} + \frac{q+1}{2(n,q+1)} + \alpha,$$

where $\alpha = 1$ if $2^{r-1} \parallel n$; $\alpha = 1/2$ if $2^t \parallel n$ and $1 \leq t \leq r-2$; $\alpha = 0$ otherwise. Here η denotes the quadratic character defined by $\eta(0) = 0$, $\eta(a) = 1$ if a is a square in $\mathbb{F}_q$ and $\eta(a) = -1$ if a is a nonsquare in $\mathbb{F}_q$.

8.3.33 Remark If $(n_1, q^2-1) = (n_2, q^2-1)$, then $\#V_{D_{n_1}(x,a)} = \#V_{D_{n_2}(x,a)}$.

8.3.34 Theorem [628] Suppose q is even. Then for each $n \geq 1$, and each $a \in \mathbb{F}_q^*$,

$$\#V_{D_n(x,a)} = \frac{q-1}{2(n,q-1)} + \frac{q+1}{2(n,q+1)}$$

8.3.35 Remark For further examples, see for instance [757, 758].

8.3.6 Further value set papers

8.3.36 Remark There are many other papers describing results concerning value sets of polynomials over finite fields; however, for lack of space, we are unable to precisely state them. Pages 379-381 of [1939] provide a wealth of descriptions of older papers dealing with value sets; pages 388-389 of [1939] provide summaries of value set results for polynomials in several variables. The paper [2826] presents the state of knowledge about value sets as of 1995.

8.3.37 Remark Since the publication of [1939], in [772] are given formulas for the number of polynomials of degree $q-1$ with a value set of cardinality k. Paper [773] describes values sets of diagonal equations over finite fields by giving a new proof of the Cauchy-Davenport theorem. See [757] and [758] for a discussion of polynomials over $\mathbb{F}_{2^n}$ which take on each nonzero value only a small number of times (at most six).

8.3.38 Remark Paper [1222] shows that if f is not a permutation polynomial over $\mathbb{F}_q$ and $q \geq n^4$, then $\#V_f < q - q/(2n)$, while [760] shows that by using the polynomial $(x+1)x^{q-1}$, Wan's bound is sharp for every extension of the base field. The paper [57] discusses results concerning the size of the intersection of the value sets of two nonconstant polynomials and [1458] discusses lower bounds for the size of the value set for the polynomial $(x^m + b)^n$ improving the bound given in [1307]. Paper [2746] discusses cardinalities of value sets for diagonal kinds of polynomials in several variables where the preimage points come from subsets of the field rather than the entire field.

See Also

§8.1	Discusses permutation polynomials in one variable.
§8.2	Discusses permutation polynomials in several variables.
§8.4	Considers exceptional polynomials.
[629]	Considers polynomials whose value sets lie in a subfield.
[734]	Studies value sets as they relate to Dembowski-Ostrom and planar polynomials.
[2188]	Considers bounds for value sets of polynomial vectors in several variables.

References Cited: [57, 285, 289, 350, 351, 549, 628, 629, 630, 667, 734, 757, 758, 760, 772, 773, 774, 1086, 1222, 1307, 1308, 1367, 1368, 1458, 1939, 2043, 2105, 2188, 2746, 2826, 2911, 2919, 2981, 2982]

8.4 Exceptional polynomials

Michael E. Zieve, University of Michigan

8.4.1 Fundamental properties

8.4.1 Definition An *exceptional polynomial* over $\mathbb{F}_q$ is a polynomial $f \in \mathbb{F}_q[x]$ which is a permutation polynomial on $\mathbb{F}_{q^m}$ for infinitely many m.

8.4.2 Remark If $f \in \mathbb{F}_q[x]$ is exceptional over $\mathbb{F}_{q^k}$ for some k, then f is exceptional over $\mathbb{F}_q$.

8.4.3 Definition A polynomial $F(x, y) \in \mathbb{F}_q[x, y]$ is *absolutely irreducible* if it is irreducible in $\bar{\mathbb{F}}_q[x, y]$, where $\bar{\mathbb{F}}_q$ is an algebraic closure of $\mathbb{F}_q$.

8.4.4 Theorem [667] A polynomial $f \in \mathbb{F}_q[x]$ is exceptional over $\mathbb{F}_q$ if and only if every absolutely irreducible factor of $f(x) - f(y)$ in $\mathbb{F}_q[x, y]$ is a constant times $x - y$.

8.4.5 Corollary If $f \in \mathbb{F}_q[x]$ is exceptional, then there are integers $1 < e_1 < e_2 < \cdots < e_k$ such that: f is exceptional over $\mathbb{F}_{q^n}$ if and only if n is not divisible by any e_i.

8.4.6 Corollary If $f \in \mathbb{F}_q[x]$ is exceptional, then there is an integer $M > 1$ such that f permutes each field $\mathbb{F}_{q^m}$ for which m is coprime to M.

8.4.7 Corollary For $g, h \in \mathbb{F}_q[x]$, the composition $g \circ h$ is exceptional if and only if both g and h are exceptional.

8.4.8 Definition A polynomial $f \in \mathbb{F}_q[x]$ is *indecomposable* if it cannot be written as the composition $f = g \circ h$ of two nonlinear polynomials $g, h \in \mathbb{F}_q[x]$.

8.4.9 Corollary A polynomial $f \in \mathbb{F}_q[x]$ is exceptional if and only if it is the composition of indecomposable exceptional polynomials.

8.4.2 Indecomposable exceptional polynomials

8.4.10 Theorem [1120] Let f be an indecomposable exceptional polynomial over $\mathbb{F}_q$ of degree n, and let p be the characteristic of $\mathbb{F}_q$. Then either

1. n is coprime to p, or
2. n is a power of p, or
3. $n = \frac{p^r(p^r-1)}{2}$ where $r > 1$ is odd and $p \in \{2, 3\}$.

8.4.11 Theorem [1755, 2192] The indecomposable exceptional polynomials over $\mathbb{F}_q$ of degree coprime to q are precisely the polynomials of the form $\ell_1 \circ f \circ \ell_2$ where $\ell_1, \ell_2 \in \mathbb{F}_q[x]$ are linear and either

1. $f(x) = ax + b$ with $a \in \mathbb{F}_q^*$ and $b \in \mathbb{F}_q$, or
2. $f(x) = x^n$ where n is a prime which does not divide $q - 1$, or
3. $f(x) = D_n(x, a)$ (a Dickson polynomial) where $a \in \mathbb{F}_q^*$ and n is a prime which does not divide $q^2 - 1$.

8.4.12 Theorem [1372, 1374] The indecomposable exceptional polynomials over $\mathbb{F}_q$ of degree $s(s-1)/2$, where $s = p^r > 3$ and $q = p^m$ with p prime, are precisely the polynomials of the form $\ell_1 \circ f \circ \ell_2$ where $\ell_1, \ell_2 \in \mathbb{F}_q[x]$ are linear, $r > 1$ is coprime to $2m$, and f is one of the following polynomials:

1. $x^{-s}T(ax^e)^{(s+1)/e}$ where $p = 2$, $T(x) = x^{s/2} + x^{s/4} + \cdots + x$, $e \mid (s+1)$, and $a \in \mathbb{F}_q^*$,
2. $\left(\frac{T(x)+a}{x}\right)^s \cdot \left(T(x) + \frac{T(x)+a}{a+1} \cdot T\left(\frac{x(a^2+a)}{(T(x)+a)^2}\right)\right)$ where $p = 2$, $a \in \mathbb{F}_q \setminus \mathbb{F}_2$, and $T(x) = x^{s/2} + x^{s/4} + \cdots + x$,
3. $x(x^{2e} - a)^{(s+1)/(4e)} \left(\frac{(x^{2e}-a)^{(s-1)/2} + a^{(s-1)/2}}{x^{2e}}\right)^{(s+1)/(2e)}$ where $p = 3$, $e \,\middle|\, \frac{s+1}{4}$, and $a \in \mathbb{F}_q^*$ is an element whose image in $\mathbb{F}_q^*/(\mathbb{F}_q^*)^{2e}$ has even order.

8.4.13 Remark The proofs of Theorems 8.4.10 and 8.4.12 rely on the classification of finite simple groups.

8.4.14 Theorem [1120, 1755] For prime p, the degree-p exceptional polynomials over $\mathbb{F}_{p^m}$ are precisely the polynomials $\ell_1 \circ f \circ \ell_2$ where $\ell_1, \ell_2 \in \mathbb{F}_{p^m}[x]$ are linear and $f(x) = x(x^{(p-1)/r} - a)^r$ with $r \mid (p-1)$ and $a \in \mathbb{F}_{p^m}$ such that $a^{r(p^m-1)/(p-1)} \neq 1$.

8.4.15 Proposition [674, 840] Let L be a *linearized polynomial* (i.e., $L(x) = \sum_{i=0}^{d} a_i x^{p^i}$ with $a_i \in \mathbb{F}_{p^m}$), and let $S(x) = x^j H(x)^k$ where $H \in \mathbb{F}_{p^m}[x]$ satisfies $L(x) = x^j H(x^k)$. Then S is exceptional over $\mathbb{F}_{p^m}$ if and only if S has no nonzero roots in $\mathbb{F}_{p^m}$.

8.4.16 Proposition [1369] Let $s = p^r$ where p is an odd prime. If $a \in \mathbb{F}_{p^m}$ is not an $(s-1)$-th power, then

$$\frac{(x^s - ax - a)\cdot(x^s - ax + a)^s + \left((x^s - ax + a)^2 + 4a^2x\right)^{(s+1)/2}}{2x^s}$$

is an indecomposable exceptional polynomial over $\mathbb{F}_{p^m}$.

8.4.17 Proposition [1371] Let $s = 2^r$. If $a \in \mathbb{F}_{2^m}$ is not an $(s-1)$-th power, then

$$\frac{(x^s + ax + a)^{s+1}}{x^s}\cdot\left(\frac{x^s + ax}{x^s + ax + a} + T\left(\frac{a^2x}{(x^s + ax + a)^2}\right)\right)$$

is an indecomposable exceptional polynomial over $\mathbb{F}_{2^m}$, where $T(x) = x^{s/2} + x^{s/4} + \cdots + x$.

8.4.18 Remark The previous three Propositions describe all known indecomposable exceptional polynomials over $\mathbb{F}_{p^m}$ of degree p^r with $r > 0$, up to composing on both sides with linear polynomials. It is expected that there are no further examples. Theorem 8.4.14 shows this when $r = 1$, and [1371, 2126] show it under different hypotheses.

8.4.3 Exceptional polynomials and permutation polynomials

8.4.19 Theorem A permutation polynomial over $\mathbb{F}_q$ of degree at most $q^{1/4}$ is exceptional over $\mathbb{F}_q$.

8.4.20 Remark A weaker version of Theorem 8.4.19 was proved in [777]; the stated result is obtained from the same proof by using the fact that an absolutely irreducible degree-d bivariate polynomial over $\mathbb{F}_q$ has at least $q + 1 - (d-1)(d-2)\sqrt{q}$ roots in $\mathbb{F}_q \times \mathbb{F}_q$. For proofs of this estimate, see [145, 1122, 1886]. A stronger (but false) version of this estimate was stated in [1939], and [1222] deduced Theorem 8.4.19 from this false estimate. Finally, [145] states a stronger version of Theorem 8.4.19, but the proof is flawed and when fixed it yields Theorem 8.4.19.

8.4.21 Remark Up to composing with linears on both sides, the only known non-exceptional permutation polynomials over $\mathbb{F}_q$ of degree less than $\sqrt{q}$ are $x^{10} + 3x$ over $\mathbb{F}_{343}$ and $\frac{(x+1)^N + 1}{x}$ over $\mathbb{F}_{2^{4r-1}}$, where $r \geq 3$ and $N = (4^r + 2)/3$.

8.4.22 Remark Heuristics predict that "at random" there would be no permutation polynomials over $\mathbb{F}_q$ of degree less than $\frac{q}{2\log q}$.

8.4.23 Remark There are no known examples of non-exceptional permutation polynomials over $\mathbb{F}_q$ of degree less than $\frac{q}{2\log q}$ when q is prime.

8.4.24 Remark Nearly all known examples of permutation polynomials over $\mathbb{F}_q$ of degree less than $\frac{q}{2\log q}$ can be written as the restriction to $\mathbb{F}_q$ of a permutation π of an infinite algebraic extension K of $\mathbb{F}_q$, where π is induced by a rational function in the symbols $\sigma^i(x)$, with σ being a fixed automorphism of K. Such a permutation π may be viewed as an exceptional rational function over the *difference field* (K, σ); see [703, 1912, 1913].

8.4.4 Miscellany

8.4.25 Theorem [543, 3074] Every permutation of $\mathbb{F}_q$ is induced by an exceptional polynomial.

8.4.26 Theorem [688, 1369, 1898] Exceptional polynomials over $\mathbb{F}_q$ have degree coprime to $q-1$.

8.4.27 Remark Theorem 8.4.26 is called the Carlitz–Wan conjecture. It follows from Theorems 8.4.10, 8.4.11, and 8.4.12. However, the known proofs of Theorems 8.4.10 and 8.4.12 rely on the classification of finite simple groups, whereas [688, 1369, 1898] present short self-contained proofs of Theorem 8.4.26.

8.4.28 Theorem If $f \in \mathbb{Z}[x]$ is a permutation polynomial over $\mathbb{F}_p$ for infinitely many primes p, then f is the composition of linear and Dickson polynomials.

8.4.29 Remark Theorem 8.4.28 was proved in [2563] when f has prime degree. It was shown in [2192] (confirming an assertion in [2563]) that the full Theorem 8.4.28 follows quickly from the main lemma in [2563] together with a group-theoretic result from [2564]. A different proof of Theorem 8.4.28 appears in [1109, 1936, 2827], which combines this group-theoretic result with Weil's bound on the number of $\mathbb{F}_q$-rational points on a genus-g curve over $\mathbb{F}_q$.

8.4.30 Remark Theorem 8.4.28 is called the Schur conjecture, although Schur did not pose this conjecture. The paper [1109] made the incorrect assertion that Schur had conjectured Theorem 8.4.28 in [2563], and this assertion has become widely accepted despite its falsehood.

8.4.31 Remark The concept of exceptionality can be extended to rational functions or more general maps between varieties [1373]. In particular, many exceptional rational functions arise as coordinate projections of isogenies of elliptic curves [1115, 1370, 2193].

8.4.5 Applications

8.4.32 Remark Exceptional polynomials were used in [2785] to produce families of hyperelliptic curves whose Jacobians have an unusually large endomorphism ring. These curves were used in [770] to realize certain groups $\mathrm{PSL}_2(q)$ as Galois groups of extensions of certain cyclotomic fields.

8.4.33 Remark Exceptional polynomials were used in [516, 2341] to produce curves whose Jacobian is isogenous to a power of an elliptic curve, and in particular to produce maximal curves (see Section 12.5).

8.4.34 Lemma [862] We have $(x+1)^N + x^N + 1 = f(x^2+x)$ in $\mathbb{F}_2[x]$, where $N = 4^r - 2^r + 1$ and $f(x) = T(x)^{2^r+1}/x^{2^r}$ with $T(x) = x^{2^{r-1}} + x^{2^{r-2}} + \cdots + x$. This polynomial $f(x)$ is obtained from case 1 of Theorem 8.4.12 by putting $a = 1$ and $e = 1$.

8.4.35 Remark This result (together with exceptionality of f) has been used to produce new examples of binary sequences with ideal autocorrelation [862], cyclic difference sets with Singer parameters [864], almost perfect nonlinear functions [863], and bent functions [864, 3012]. See Sections 10.3, 14.6, 9.2, and 9.3, respectively.

8.4.36 Remark For further results about the polynomials f from Lemma 8.4.34, including formulas for a polynomial inducing the inverse of the permutation induced by f on $\mathbb{F}_{2^m}$, see [905]. These polynomials are shown to be exceptional in [696, 697, 864, 905, 3073].

8.4.37 Remark The polynomials in cases 1 and 3 of Theorem 8.4.12 have been used to produce branched coverings of the projective line in positive characteristic whose Galois group is either symplectic [9] or orthogonal [8].

See Also

§8.1	For discussion of permutation polynomials in one variable.
§8.3	For value sets of polynomials.
[58], [1368]	For *Davenport pairs*, which are pairs (f, g) of polynomials in $\mathbb{F}_q[x]$ such that $f(\mathbb{F}_{q^m}) = g(\mathbb{F}_{q^m})$ for infinitely many m. This notion generalizes exceptionality, since $f \in \mathbb{F}_q[x]$ is exceptional if and only if (f, x) is a Davenport pair.
[696], [697], [3073]	For the factorization of $f(x) - f(y)$ where $f(x)$ is a polynomial from case 1 or 3 of Theorem 8.4.12.
[696], [1900], [2190]	For the discovery of some of the polynomials in Theorem 8.4.12.
[840]	For a thorough study of exceptional polynomials using only the Hermite–Dickson criterion, and the discovery of the polynomials in Theorem 8.4.11 and Proposition 8.4.15.

References Cited: [8, 9, 58, 145, 516, 543, 667, 674, 688, 696, 697, 703, 770, 777, 840, 862, 863, 864, 905, 1109, 1115, 1120, 1122, 1222, 1368, 1369, 1370, 1371, 1372, 1373, 1374, 1755, 1886, 1898, 1900, 1912, 1913, 1936, 1939, 2126, 2190, 2192, 2193, 2341, 2563, 2564, 2785, 2827, 3012, 3073, 3074]

9

Special functions over finite fields

9.1 Boolean functions

Claude Carlet, *Université Paris 8, LAGA*

9.1.1 Definition Let n be a positive integer.

1. An *n-variable Boolean function* (or *Boolean function in dimension n*) is a function defined over the vector space $\mathbb{F}_2^n$ and valued in $\mathbb{F}_2$. *Notation*: $f(x)$, where $x = (x_1, \ldots, x_n) \in \mathbb{F}_2^n$.
2. The domain $\mathbb{F}_2^n$ can be endowed with the structure of the field $\mathbb{F}_{2^n}$. *Same notation*: $f(x)$, where $x \in \mathbb{F}_{2^n}$.

9.1.2 Remark In this section, we study *single-output Boolean functions*. Multi-output (or vectorial) Boolean functions are studied in Section 9.2.

9.1.3 Remark Endowing $\mathbb{F}_2^n$ with the structure of $\mathbb{F}_{2^n}$ allows taking advantage of the field structure for designing Boolean functions (this is more true for multi-output Boolean functions, however, see Section 9.2), despite the fact that the important parameters in applications are more related to the vector space structure of $\mathbb{F}_2^n$ than to the field structure of $\mathbb{F}_{2^n}$.

9.1.4 Remark We emphasize that in all the definitions and propositions of this section, "$x \in \mathbb{F}_2^n$" can be replaced by "$x \in \mathbb{F}_{2^n}$," except when the coordinates of x are specifically involved or when the Hamming weight of x plays a role.

9.1.1 Representation of Boolean functions

9.1.5 Remark The simplest way of representing a Boolean function is its truth table. But this representation gives little insight on the function. Another well-known way is with the disjunctive and conjunctive normal forms. But these representations, which do not allow uniqueness, are not well adapted to coding, cryptography, and sequences for communications, which are main applications of Boolean functions.

9.1.1.1 Algebraic normal form

9.1.6 Proposition Every n-variable Boolean function f can be represented by a multivariate polynomial (*i.e.*, is a polynomial mapping) over $\mathbb{F}_2$ of the form

$$f(x) = \sum_{I \subseteq \{1,\ldots,n\}} a_I \left(\prod_{i \in I} x_i \right) \in \mathbb{F}_2[x_1, \ldots, x_n]/(x_1^2 + x_1, \ldots, x_n^2 + x_n). \tag{9.1.1}$$

This representation is unique.

9.1.7 Remark The sum in (9.1.1) is in $\mathbb{F}_2$. Every coordinate x_i appears in this polynomial with exponents at most 1 (more precisely, we consider the polynomial f modulo the ideal generated by $x_1^2 + x_1, \ldots, x_n^2 + x_n$) because x_i represents a bit and every bit in $\mathbb{F}_2$ equals its own square.

9.1.8 Definition Representation (9.1.1) is the *algebraic normal form* of f (in brief, ANF). The terms $\prod_{i \in I} x_i$ are *monomials*.

9.1.9 Remark The ANF can also be represented in the form

$$f(x) = \sum_{u \in \mathbb{F}_2^n} a_u \left(\prod_{j=1}^{n} x_j{}^{u_j} \right) = \sum_{u \in \mathbb{F}_2^n} a_u x^u, \tag{9.1.2}$$

with the convention $0^0 = 1$.

9.1.10 Example The 3-variable function $f(x) = x_1x_2x_3 + x_1x_3 + x_2 + 1 = x^{111} + x^{101} + x^{010} + x^{000}$ takes at input $(1, 0, 1)$, for instance, the value $0 + 1 + 0 + 1 = 0$.

9.1.11 Proposition Let the Boolean function f be given by (9.1.1) or equivalently by (9.1.2). We have

$$f(x) = \sum_{I \subseteq \text{supp}(x)} a_I = \sum_{u \preceq x} a_u, \tag{9.1.3}$$

where supp(x) denotes the support $\{i \in \{1, \ldots, n\} \mid x_i = 1\}$ of x and $u \preceq x$ means that supp$(u) \subseteq$ supp(x). Conversely, for every $I \subseteq \{1, \ldots, n\}$, respectively, $u \in \mathbb{F}_2^n$

$$a_I = \sum_{x \in \mathbb{F}_2^n \mid \text{supp}(x) \subseteq I} f(x), \quad \text{respectively,} \quad a_u = \sum_{x \in \mathbb{F}_2^n \mid x \preceq u} f(x). \tag{9.1.4}$$

9.1.12 Remark The transform giving the expression of the coefficients of the ANF by means of the values of f, and *vice versa*, is the *binary Möbius transform.*

9.1.13 Remark Proposition 9.1.11 results in an algorithm for calculating the ANF from the truth table of the function and *vice versa*, with complexity $O(n2^n)$ (so, a little higher than linear, since the size of the input f is 2^n), see for example [523].

9.1.14 Remark A multivariate representation similar to the ANF but over $\mathbb{Z}$ also exists, called *numerical normal form*, with similar formulas involving the *Möbius transform over* $\mathbb{Z}$ [530],[523, Section 8.2.1].

9.1.1.2 Trace representation

9.1.15 Proposition [1303] Let $\mathbb{F}_2^n$ be identified with $\mathbb{F}_{2^n}$ and let f be an n-variable Boolean function of even weight (i.e., of algebraic degree at most $n-1$). There exists a unique representation of f as a univariate polynomial mapping of the form

$$f(x) = \sum_{j \in \Gamma_n} \text{Tr}_{\mathbb{F}_{2^{o(j)}}/\mathbb{F}_2}(A_j x^j), \quad x \in \mathbb{F}_{2^n}, \tag{9.1.5}$$

where Γ_n is the set of integers obtained by choosing one element in each cyclotomic coset of 2 (mod $2^n - 1$), $o(j)$ is the size of the cyclotomic coset containing j, $A_j \in \mathbb{F}_{2^{o(j)}}$ and $\text{Tr}_{\mathbb{F}_{2^{o(j)}}/\mathbb{F}_2}$ is the trace function from $\mathbb{F}_{2^{o(j)}}$ to $\mathbb{F}_2$.

9.1.16 Remark The A_j's can be calculated by using the *Mattson-Solomon polynomial* [523, 1992].

9.1.17 Definition Representation (9.1.5) is the *trace representation* (or *univariate representation*) of f.

9.1.18 Example Let n be even. Then we can define the function $f(x) = \text{Tr}_{\mathbb{F}_{2^n}/\mathbb{F}_2}(x^3) + \text{Tr}_{\mathbb{F}_{2^{n/2}}/\mathbb{F}_2}(x^{2^{n/2}+1})$.

9.1.19 Remark Any Boolean function f can be simply represented in the form $\text{Tr}_{\mathbb{F}_{2^n}/\mathbb{F}_2}(P(x))$ where P is a polynomial over $\mathbb{F}_{2^n}$, but there is no uniqueness of such a representation, unless $o(j) = n$ for every j such that $A_j \neq 0$.

9.1.2 The Walsh transform

9.1.20 **Definition** The *Walsh transform* W_f of an n-variable Boolean function f is the discrete Fourier transform (or Hadamard transform) of the *sign function* $(-1)^{f(x)}$. Given an inner product $x \cdot y$ in $\mathbb{F}_2^n$ (for instance $x \cdot y = \sum_{i=1}^n x_i y_i$ over $\mathbb{F}_2^n$, or $x \cdot y = \mathrm{Tr}_{\mathbb{F}_{2^n}/\mathbb{F}_2}(xy)$ over $\mathbb{F}_{2^n}$), the value of the Walsh transform of f at $u \in \mathbb{F}_2^n$ is given by:

$$W_f(u) = \sum_{x \in \mathbb{F}_2^n} (-1)^{f(x)+x \cdot u},$$

where the sum is over the integers. The set $\{u \in \mathbb{F}_2^n \mid W_f(u) \neq 0\}$ is the *Walsh support* of f.

9.1.21 **Remark** There exists also a normalized version, in which the value $W_f(u)$ above is divided by $\sqrt{2^n}$, which simplifies the inverse Walsh transform formula (see below) but gives a non-integer value.

9.1.22 **Remark** There exists an algorithm for calculating the Walsh transform whose complexity is $O(n2^n)$, see for example [523]. A more general definition and an example are given in Section 9.3.

9.1.23 **Proposition** (Inverse Walsh transform) [1992], [523, Section 8.2.2] For every $x \in \mathbb{F}_2^n$ and every n-variable Boolean function f, we have

$$\sum_{u \in \mathbb{F}_2^n} W_f(u)(-1)^{u \cdot x} = 2^n (-1)^{f(x)}.$$

9.1.24 **Proposition** (Parseval's relation) [1992], [523, Section 8.2.2] For every n-variable Boolean function f, we have

$$\sum_{u \in \mathbb{F}_2^n} W_f^2(u) = 2^{2n}.$$

9.1.25 **Proposition** (Poisson summation formula) [523, Section 8.2.2] For every n-variable Boolean function f, for every vector subspace E of $\mathbb{F}_2^n$, and for all elements a and b of $\mathbb{F}_2^n$, we have

$$\sum_{u \in a+E^\perp} (-1)^{b \cdot u} W_f(u) = |E^\perp| (-1)^{a \cdot b} \sum_{x \in b+E} (-1)^{f(x)+a \cdot x},$$

where $E^\perp$ is the orthogonal subspace of E (that is, $E^\perp = \{u \in \mathbb{F}_2^n \mid u \cdot x = 0, \text{ for all } x \in E\}$) and $|E^\perp|$ denotes the cardinality of $E^\perp$.

9.1.26 **Proposition** [491] Let E and E' be two supplementary vector subspaces of $\mathbb{F}_2^n$. Then, for every element a of $\mathbb{F}_2^n$, we have

$$\sum_{u \in a+E^\perp} W_f^2(u) = |E^\perp| \sum_{b \in E'} \left(\sum_{x \in b+E} (-1)^{f(x)+a \cdot x} \right)^2.$$

9.1.3 Parameters of Boolean functions

9.1.27 **Definition** The degree $d^\circ f$ of the ANF is the *algebraic degree* of the function: if f is given by (9.1.1), respectively, by (9.1.2), then $d^\circ f = \max\{|I| \mid a_I \neq 0\} = \max\{\mathrm{wt}(u) \mid a_u \neq 0\}$.

9.1.28 Remark A function is *affine* (respectively, *quadratic*) if it has algebraic degree at most 1 (respectively, 2). Affine functions are those functions of the form $x \mapsto u \cdot x + \epsilon$, $u \in \mathbb{F}_2^n$, $\epsilon \in \mathbb{F}_2$.

9.1.29 Example The function of Example 9.1.10 has algebraic degree 3.

9.1.30 Proposition [523, Section 8.2.1] Let f be represented by its trace representation (9.1.5). Then f has algebraic degree $\max_{j \in \Gamma_n \mid A_j \neq 0} w_2(j)$, where $w_2(j)$ is the Hamming weight of the binary expansion of j (called the 2-weight of j).

9.1.31 Example The function of Example 9.1.18 has algebraic degree 2 (*i.e.*, is quadratic).

9.1.32 Proposition [523, Section 8.2.1] The algebraic degree of any n-variable Boolean function f equals the maximum dimension of the subspaces $\{x \in \mathbb{F}_2^n \mid \mathrm{supp}(x) \subseteq I\}$, where I is any subset of $\{1, \ldots, n\}$ (equivalently, of all affine subspaces of $\mathbb{F}_2^n$), on which f takes value 1 an odd number of times.

9.1.33 Definition The *Hamming weight* of an n-variable Boolean function f is the integer $\mathrm{wt}(f) = |\{x \in \mathbb{F}_2^n \mid f(x) = 1\}|$. The function is *balanced* if it has Hamming weight 2^{n-1} (that is, if its output is uniformly distributed over $\mathbb{F}_2$).

9.1.34 Proposition An n-variable Boolean function has algebraic degree at most $n-1$ if and only if it has even Hamming weight.

9.1.35 Proposition (McEliece's theorem) Let f be an n-variable Boolean function of algebraic degree at most r with $0 < r < n$. Then the Hamming weight of f is divisible by $2^{\lceil \frac{n}{r} \rceil - 1} = 2^{\lfloor \frac{n-1}{r} \rfloor}$ (equivalently, the Walsh transform of f takes values divisible by $2^{\lceil \frac{n}{r} \rceil} = 2^{\lfloor \frac{n-1}{r} \rfloor + 1}$).

9.1.36 Proposition [1855], [523, Section 8.2.2] For $n \geq 2$ and $1 \leq k \leq n$, if the Walsh transform of f takes values divisible by 2^k, then f has algebraic degree at most $n - k + 1$.

9.1.37 Proposition A quadratic Boolean function f is balanced if and only if its restriction to the vector subspace $\mathcal{E}_f = \{a \in \mathbb{F}_2^n \mid D_a f(x) := f(x) + f(x+a) \equiv cst\} = \{a \in \mathbb{F}_2^n \mid D_a f(x) = D_a f(0), \text{ for all } x \in \mathbb{F}_2^n\}$ (the *linear kernel* of f) is not constant. If it is not balanced, then its Hamming weight equals $2^{n-1} \pm 2^{\frac{n+k}{2}-1}$ where k is the dimension of $\mathcal{E}_f$.

9.1.38 Definition The *Hamming distance* between two n-variable Boolean functions f and g equals the size of the set $\{x \in \mathbb{F}_2^n \mid f(x) \neq g(x)\}$, that is, equals $\mathrm{wt}(f + g)$.

The *nonlinearity* $\mathcal{NL}(f)$ of a Boolean function f is its minimal distance to affine functions.

9.1.39 Proposition [1992], [523, Section 8.4.1] For every n-variable Boolean function, we have

$$\mathcal{NL}(f) = 2^{n-1} - \frac{1}{2} \max_{u \in \mathbb{F}_2^n} |W_f(u)|. \tag{9.1.6}$$

9.1.40 Proposition [1992], [523, Section 8.4.1] Parseval's relation implies the *covering radius bound*

$$\mathcal{NL}(f) \leq 2^{n-1} - 2^{n/2-1}.$$

9.1.41 Definition The Boolean functions achieving the covering radius bound with equality are *bent* (see Section 9.3 and the references therein).

9.1.42 Proposition [1992] An n-variable Boolean function is bent if and only if its Walsh transform takes values $\pm 2^{n/2}$ only (n even).

9.1.43 Definition For every $0 \leq r \leq n$, the *r-th order nonlinearity* $\mathcal{NL}_r(f)$ of a Boolean function f equals its minimum Hamming distance to Boolean functions of algebraic degrees at most r.

9.1.44 Remark This notion is related to the *Gowers norm* [267].

9.1.45 Proposition [522] Let f be any n-variable function and r a positive integer smaller than n. Denoting (again) by $D_a f(x) = f(x) + f(x+a)$ the first-order derivatives of f, we have

$$\mathcal{NL}_r(f) \geq \max\left(\frac{1}{2}\max_{a\in\mathbb{F}_2^n}\mathcal{NL}_{r-1}(D_a f);\ 2^{n-1} - \frac{1}{2}\sqrt{2^{2n} - 2\sum_{a\in\mathbb{F}_2^n}\mathcal{NL}_{r-1}(D_a f)}\right).$$

9.1.46 Remark Proposition 9.1.45, iteratively applied, allows deducing lower bounds on the higher order nonlinearity of a function from lower bounds on the nonlinearities (*i.e.*, the first-order nonlinearities) of its higher-order derivatives.

9.1.4 Equivalence of Boolean functions

9.1.47 Remark The parameters above (algebraic degree, Hamming weight, nonlinearity) are invariant under composition of the Boolean function on the right by any affine automorphism $x \mapsto L(x) + a$ (where L is linear bijective). The algebraic degree and the nonlinearity are invariant under addition of any affine Boolean function (in the case of the algebraic degree, though, the invariance needs the function to be non-affine) [1992].

9.1.48 Definition Two n-variable Boolean functions f and g are *extended-affine equivalent* (in brief, EA-equivalent) if there exists a linear automorphism L, an affine Boolean function ℓ and a vector a such that $g(x) = f(L(x) + a) + \ell(x)$. A parameter is *EA-invariant* if it is preserved by EA-equivalence.

9.1.49 Remark If x is in $\mathbb{F}_2^n$ (and is viewed as a $1\times n$ matrix over $\mathbb{F}_2$), then $L(x) = x \times A$ where A is a non-singular $n\times n$ matrix over $\mathbb{F}_2$. If x lives in $\mathbb{F}_{2^n}$, then $L(x) = a_1x + a_2x^2 + \cdots + a_{n-1}x^{2^{n-1}}$ where $a_1, a_2, \ldots, a_{n-1}$ are chosen in $\mathbb{F}_{2^n}$ such that the kernel of L is reduced to $\{0\}$.

9.1.50 Remark Little is known on the number of n-variable Boolean functions up to EA-equivalence, except of course that it is larger than the number 2^{2^n} of Boolean functions divided by the number $2^n(2^n-1)(2^n-2)(2^n-2^2)\cdots(2^n-2^{n-1})$ of affine automorphisms of $\mathbb{F}_2^n$ and by the number 2^{n+1} of affine Boolean functions.

9.1.51 Remark There exists another notion of equivalence: the CCZ-equivalence; for vectorial functions, it is more general than EA-equivalence, but for Boolean functions, it coincides with EA-equivalence [444].

9.1.5 Boolean functions and cryptography

9.1.52 Remark Boolean functions are used in pseudo-random generators, in stream ciphers (in conventional cryptography), to ensure a sufficient "nonlinearity" (since linear ciphers are weak [2011]), see Section 16.2. They must then be balanced to avoid distinguishing attacks.

9.1.53 Remark A high algebraic degree of a Boolean function f ensures a good resistance of the stream ciphers (using it as the nonlinear part) against the Berlekamp-Massey attack [2011] and the Rønjom-Helleseth attack [2474] and its variants.

9.1.54 Remark A large nonlinearity (that is, a nonlinearity near the covering radius bound) is an important cryptographic characteristic which ensures resistance to the fast correlation attack [2075].

9.1.55 Remark Vectorial Boolean functions are also used in cryptography (in block ciphers); they are addressed in Section 9.2.

9.1.56 Definition Let n be a positive integer and $m < n$ a non-negative integer. A Boolean function over $\mathbb{F}_2^n$ is *m-resilient* (respectively, m-th order *correlation immune*) if any of its restrictions obtained by fixing at most m of its input coordinates x_i is balanced (respectively, has same output distribution as f itself). The *resiliency order* of f is the largest value of m such that f is m-resilient.

9.1.57 Remark The notions of resilient and correlation immune functions have a sense for $x \in \mathbb{F}_2^n$ only (that is, not for $x \in \mathbb{F}_{2^n}$).

9.1.58 Remark A function is m-resilient if and only if it is balanced and m-th order correlation immune. Since Boolean functions used in stream ciphers must be balanced, we shall address only resiliency in the sequel.

9.1.59 Remark The resiliency order quantifies the resistance to the Siegenthaler *correlation attack* [2663] of a stream cipher using f as a combiner (that is, combining the outputs to linear feedback registers by applying f, see Section 16.2).

9.1.60 Remark The resiliency order is not EA-invariant. It is invariant under permutations of the coordinates of $x \in \mathbb{F}_2^n$.

9.1.61 Proposition (Siegenthaler's bound) [2663], [523, Section 8.7] Any m-resilient n-variable Boolean function has algebraic degree smaller than or equal to $n - m - 1$ if $0 \leq m < n - 1$ and is affine if $m = n - 1$.

9.1.62 Proposition (Xiao-Massey's characterization) [3015] Any n-variable Boolean function f is m-resilient if and only if $W_f(u) = 0$ for all $u \in \mathbb{F}_2^n$ such that $0 \leq \mathrm{wt}(u) \leq m$.

9.1.63 Proposition (Improved Sarkar-Maitra's bound) [520, 534, 2522], [523, Section 8.7] Let f be any n-variable m-resilient function ($m \leq n-2$) and let d be its algebraic degree. The values of the Walsh transform of f are divisible by $2^{m+2+\lfloor \frac{n-m-2}{d} \rfloor}$. Hence, according to Equation (9.1.6), the nonlinearity of f is divisible by $2^{m+1+\lfloor \frac{n-m-2}{d} \rfloor}$ and is therefore bounded above by the largest number divisible by $2^{m+1+\lfloor \frac{n-m-2}{d} \rfloor}$ and smaller than $2^{n-1} - 2^{n/2-1}$. In particular, if $m + 1 + \lfloor \frac{n-m-2}{d} \rfloor < \frac{n}{2}$ and n is even, we have $\mathcal{NL}(f) \leq 2^{n-1} - 2^{n/2-1} - 2^{m+1+\lfloor \frac{n-m-2}{d} \rfloor}$.

9.1.64 Remark The divisibility property in Proposition 9.1.63, combined with known bounds on the nonlinearity, implies improvements of these bounds (as shown in Proposition 9.1.63 for the covering radius bound).

9.1.65 Definition [2073] A function g such that $fg = 0$ is an *annihilator* of f. The minimum algebraic degree of the nonzero functions g such that $fg = 0$ or $(f + 1)g = 0$ is the *algebraic immunity* of f. It is denoted by $AI(f)$.

9.1.66 Example The function $f(x_1, x_2, x_3) = x_1x_2x_3 + x_2 + 1$ has algebraic immunity 1 since, being non-constant, it does not have algebraic immunity 0, and since $f + 1$ has annihilator $x_2 + 1$.

9.1.67 Remark The algebraic immunity quantifies the resistance to the *algebraic attack* [741] of the ciphers using f as a combiner, or as a filter (taking as input n bits at fixed positions in an LFSR, see Section 16.2).

9.1.68 Remark The set of annihilators of f equals the ideal of all multiples of $f+1$.

9.1.69 Remark Let g be any function of algebraic degree at most d, then expressing that g is an annihilator of f (that is, $f(x) = 1$ implies $g(x) = 0$, for every $x \in \mathbb{F}_2^n$) by means of the (unknown) coefficients of the ANF of g results in a system of homogeneous linear equations. In this system, we have $\sum_{i=0}^{d} \binom{n}{i}$ number of variables (the coefficients of the monomials of degrees at most d) and $\mathrm{wt}(f)$ many equations.

9.1.70 Proposition [741] For every n-variable Boolean function, we have: $AI(f) \leq \lceil \frac{n}{2} \rceil$.

9.1.71 Remark Let f filter an LFSR. Let us assume for instance that the LFSR has length 256 (it can then be initialized for instance with a key of length 128 and an IV of length 128 as well). Then Proposition 9.1.70 requires $n \geq 16$ to ensure a complexity of the algebraic attack larger than exhaustive search, see *e.g.*, [490]. A minimal security margin would be $n \geq 18$. Algebraic attacks have therefore changed the number of variables of the Boolean functions used in practice (before them, for reasons of efficiency, the number of variables was rarely more than 10).

9.1.72 Proposition (Bounds on the weight, Lobanov's bound on the nonlinearity) [523, Section 8.9] For every n and every n-variable Boolean function f, we have $\sum_{i=0}^{AI(f)-1} \binom{n}{i} \leq \mathrm{wt}(f) \leq \sum_{i=0}^{n-AI(f)} \binom{n}{i}$ (in particular, if n is odd and f has optimal algebraic immunity $\frac{n+1}{2}$, then f is balanced) and $\mathcal{NL}(f) \geq 2\sum_{i=0}^{AI(f)-2} \binom{n-1}{i}$.

9.1.73 Remark Bounds also exist for the higher order nonlinearities [523, Section 8.9], [2085].

9.1.74 Proposition [490] If an n-variable balanced Boolean function f, with n odd, has no non-zero annihilator of algebraic degree at most $\frac{n-1}{2}$, then it has optimal algebraic immunity.

9.1.75 Remark

1. Stream ciphers must also resist fast algebraic attacks, which work if one can find $g \neq 0$ of low degree and h of algebraic degree not much larger than $\lceil n/2 \rceil$, such that $fg = h$ [735, 1446]. These attacks need more data (of a particular shape) than standard algebraic attacks but can be faster if g has lower degree due to the relaxation of the condition on h.
2. Stream ciphers must also resist algebraic attacks on the augmented function [1069].
3. Finally, they must resist the already mentioned attack by Rønjom and Helleseth (designed for the filter generator and later generalized) which requires f to have an algebraic degree close to n.

9.1.76 Definition Let $1 \leq l \leq n$. An n-variable Boolean function f satisfies the *propagation criterion* of order l, denoted by $PC(l)$, if for every $a \in \mathbb{F}_2^n$ such that $1 \leq \mathrm{wt}(a) \leq l$, the derivative $D_a f$ is balanced. Property $PC(1)$ is the *strict avalanche criterion.*

9.1.77 Remark This characteristic is related to the diffusion of the cipher in which f is involved. It is less important than the characteristics seen above, but it has attracted some attention.

9.1.78 Remark Other cryptographic characteristics (less essential than the algebraic degree, the nonlinearity and the algebraic immunity) exist [523]: the non-existence of nonzero linear structure, the global avalanche criterion, the maximum correlation to subsets, the algebraic thickness, the nonhomomorphicity.

9.1.6 Constructions of cryptographic Boolean functions

9.1.79 Remark Boolean functions being tools for the design of cryptosystems, an important aspect of the research in this domain is to design constructions of Boolean functions having the necessary cryptographic features (contrary to other domains of the study of Boolean functions, which mainly study their properties).

9.1.80 Remark A Boolean function obtained by some construction and satisfying a given cryptographic criterion, or several criteria, will be considered as new if it is EA-inequivalent to all previously found functions satisfying the same criteria.

9.1.81 Remark We call *secondary* the constructions which use already defined functions satisfying a given property, to build a new one satisfying the same property. A construction from first principles will be *primary*.

9.1.6.1 Primary constructions of resilient functions

9.1.82 Proposition (Maiorana-McFarland's construction) [485, 519], [523, Section 8.7] Let r and s be positive integers; let g be any s-variable Boolean function and let ϕ be a mapping from $\mathbb{F}_2^s$ to $\mathbb{F}_2^r$. Then the function

$$f_{\phi,g}(x,y) = x \cdot \phi(y) + g(y),\ x \in \mathbb{F}_2^r,\ y \in \mathbb{F}_2^s, \tag{9.1.7}$$

where "$\cdot$" is an inner product in $\mathbb{F}_2^r$, is (at least) k-resilient where $k = \min\{\mathrm{wt}(\phi(y)),\ y \in \mathbb{F}_2^s\} - 1$, and has nonlinearity satisfying:

$$2^{n-1} - 2^{r-1} \max_{a \in \mathbb{F}_2^r} |\phi^{-1}(a)| \le \mathcal{NL}(f_{\phi,g}) \le 2^{n-1} - 2^{r-1} \sqrt{\max_{a \in \mathbb{F}_2^r} |\phi^{-1}(a)|}.$$

9.1.83 Remark This construction was originally developed for bent functions (see Section 9.3) and has been later adapted to resilient functions.

9.1.84 Remark The resiliency order of $f_{\phi,g}$ can be larger, for some well-chosen functions g [519].

9.1.85 Remark The Maiorana-McFarland construction has been generalized in several ways [523].

9.1.6.2 Secondary constructions of resilient functions

9.1.86 Proposition (Indirect sum) [521], [523, Section 8.7] Let r and s be positive integers and let $0 \le t < r$ and $0 \le m < s$. Let f_1 and f_2 be two r-variable t-resilient functions. Let g_1 and g_2 be two s-variable m-resilient functions. Then the function

$$h(x,y) = f_1(x) + g_1(y) + (f_1 + f_2)(x)\,(g_1 + g_2)(y); \quad x \in \mathbb{F}_2^r, y \in \mathbb{F}_2^s$$

is an $(r+s)$-variable $(t+m+1)$-resilient function. The Walsh transform of h takes the value

$$W_h(a,b) = \frac{1}{2} W_{f_1}(a)\left[W_{g_1}(b) + W_{g_2}(b)\right] + \frac{1}{2} W_{f_2}(a)\left[W_{g_1}(b) - W_{g_2}(b)\right]. \tag{9.1.8}$$

If the Walsh transforms of f_1 and f_2 have disjoint supports and if the Walsh transforms of g_1 and g_2 have disjoint supports, then

$$\mathcal{NL}(h) = \min_{i,j \in \{1,2\}} \left(2^{r+s-2} + 2^{r-1}\mathcal{NL}(g_j) + 2^{s-1}\mathcal{NL}(f_i) - \mathcal{NL}(f_i)\mathcal{NL}(g_j)\right). \tag{9.1.9}$$

In particular, if f_1 and f_2 have nonlinearity $2^{r-1} - 2^{t+1}$ and disjoint Walsh supports, if g_1 and g_2 have nonlinearity $2^{s-1} - 2^{m+1}$ and disjoint Walsh supports, and if $f_1 + f_2$ has algebraic degree $r - t - 1$ and $g_1 + g_2$ has algebraic degree $s - m - 1$, then h achieves Siegenthaler's and (improved) Sarkar-Maitra's bounds with equality.

9.1.87 Remark Some particular choices of functions f_1, f_2, g_1, g_2 give secondary constructions previously introduced by several authors (Siegenthaler, Tarannikov) [523, Section 8.7]; in particular the well-known direct sum $h(x, y) = f(x) + g(y)$ is obtained by putting $g_1 = g_2$ and/or $f_1 = f_2$.

9.1.88 Remark The indirect sum gives also a secondary construction of bent functions [523].

9.1.89 Proposition (Secondary construction without extension of the number of variables) [523, Section 8.7] Let $0 \leq k \leq n$. Let f_1, f_2 and f_3 be three k-resilient n-variable functions. Then the function $s_1 = f_1 + f_2 + f_3$ is k-resilient if and only if the function $s_2 = f_1 f_2 + f_1 f_3 + f_2 f_3$ is k-resilient. Moreover

$$\mathcal{NL}(s_2) \geq \frac{1}{2}\left(\mathcal{NL}(s_1) + \sum_{i=1}^{3} \mathcal{NL}(f_i)\right) - 2^{n-1}, \tag{9.1.10}$$

and if the Walsh supports of f_1, f_2 and f_3 are pairwise disjoint, then

$$\mathcal{NL}(s_2) \geq \frac{1}{2}\left(\mathcal{NL}(s_1) + \min_{1 \leq i \leq 3} \mathcal{NL}(f_i)\right). \tag{9.1.11}$$

9.1.90 Remark It has been impossible until now to obtain resilient functions of sufficient orders with good algebraic immunity. For this reason, the filter model is preferred to the combiner model (the correlation attack works on the latter).

9.1.6.3 Constructions of highly nonlinear functions with optimal algebraic immunity

9.1.91 Proposition [528] Let n be any positive integer and α a primitive element of the field $\mathbb{F}_{2^n}$. Let f be the balanced Boolean function on $\mathbb{F}_{2^n}$ whose support equals $\{0, \alpha^s, \ldots, \alpha^{s+2^{n-1}-2}\}$ for some s. Then f has optimum algebraic immunity $\lceil n/2 \rceil$. Moreover, f has algebraic degree $n - 1$ and nonlinearity $\mathcal{NL}(f) \geq 2^{n-1} - n \cdot \ln 2 \cdot 2^{\frac{n}{2}} - 1$.

9.1.92 Remark This bound on the nonlinearity, which has been recently slightly improved, is not enough for ensuring that the function allows resisting the fast correlation attacks, but it has been checked, for $n \leq 26$, that the exact value of $\mathcal{NL}(f)$ is much better than this lower bound. Improving significantly the bound of Proposition 9.1.91 is an open problem.

9.1.93 Remark The function in Proposition 9.1.91 shows also good immunity against fast algebraic attacks as shown by Liu et al. [1951].

9.1.94 Remark Despite the fact that complexity for computing the output is roughly the same as for computing the discrete logarithm, the function can be efficiently computed because n is small; the Pohlig-Hellman method is efficient, at least for some values ($n = 18$ or 20).

9.1.95 Remark We note that a similar function had been previously studied by Brandstätter, Lange, and Winterhof in [391] but the algebraic immunity was not addressed by these authors.

9.1.96 Remark Other infinite classes of balanced functions with optimal algebraic immunity have been found [2777, 3056] which have good nonlinearity and good resistance to fast algebraic attacks (checked by computer), at least for small n. More classes exist but, either their optimal algebraic immunity is not completely proved, or they are closely related to the classes mentioned above, or they do not have good nonlinearity, or they have bad resistance to fast algebraic attacks.

9.1.7 Boolean functions and error correcting codes

9.1.97 Remark Boolean functions are used to design error correcting codes whose lengths are powers of 2, in particular Reed-Muller codes (see Section 15.1).

9.1.7.1 Reed-Muller codes

9.1.98 Definition Let n be any positive integer and $0 \le r \le n$. The *Reed-Muller code* of length 2^n and order r is the set of binary words of length 2^n corresponding to the output columns of the truth-tables of all the n-variable Boolean functions of algebraic degrees at most r.

9.1.99 Proposition [1992] The Reed-Muller code of length 2^n and order r is a linear code over $\mathbb{F}_2$ (*i.e.*, is a vector subspace of $\mathbb{F}_2^{2^n}$) of dimension $1+n+\binom{n}{2}+\cdots+\binom{n}{r}$ and minimum distance 2^{n-r}.

9.1.100 Remark The Reed-Muller code of order 1 is optimal according to the Griesmer bound [1992].

9.1.7.2 Kerdock codes

9.1.101 Remark The Reed-Muller code of length 2^n and of order 2 contains a nonlinear optimal code: the Kerdock code of the same length, introduced in [1728] (not as in the definition below, though). The Kerdock code of length 2^n is the union of cosets of the first order Reed-Muller code, chosen such that the sum of two elements from different cosets is a bent function.

9.1.102 Definition The Kerdock code of length 2^n is the set of functions of the form $(x,\epsilon) \in \mathbb{F}_{2^{n-1}} \times \mathbb{F}_2 \mapsto f(ux,\epsilon) + \mathrm{Tr}_{\mathbb{F}_{2^{n-1}}/\mathbb{F}_2}(ax) + \eta\epsilon + \tau$ where $f(x,\epsilon) = \mathrm{Tr}_{\mathbb{F}_{2^{n-1}}/\mathbb{F}_2}\left(\sum_{i=1}^{\frac{n}{2}-1} x^{2^i+1}\right) + \epsilon\,\mathrm{Tr}_{\mathbb{F}_{2^{n-1}}/\mathbb{F}_2}(x)$, with $u, a \in \mathbb{F}_{2^{n-1}}$, and $\eta, \tau \in \mathbb{F}_2$.

9.1.103 Proposition [1992] The Kerdock code of length 2^n has size 2^{2n} and minimum distance $2^{n-1} - 2^{n/2-1}$.

9.1.104 Remark The Kerdock code of length 2^n is optimal, as proved by Delsarte [800].

9.1.105 Remark Other codes with the same parameters exist, called generalized Kerdock codes [1673].

9.1.106 Remark The Kerdock code of length 2^n is not linear but it is the image of a linear code over $\mathbb{Z}/4\mathbb{Z}$ by a distance preserving mapping called the Gray map [1409].

9.1.8 Boolean functions and sequences

9.1.107 Remark Boolean functions are related to sequences for communications: see Chapter 10.

9.1.8.1 Boolean functions and cross correlation of m-sequences

9.1.108 Definition A sequence $(s_i)_{i\geq 0}$ over $\mathbb{F}_2$ satisfying an order n linear homogeneous recurrence relation (with constant coefficients), is an *m-sequence* if it has (optimal) period 2^n-1.

9.1.109 Proposition [1303] The m-sequences of period 2^n-1 are the sequences of the form $s_i = \mathrm{Tr}_{\mathbb{F}_{2^n}/\mathbb{F}_2}(\lambda\alpha^i)$, where $\lambda \in \mathbb{F}_{2^n}^*$ and α is a primitive element of $\mathbb{F}_{2^n}$. Given such an m-sequence, any other m-sequence of the same period differs with $(s_i)_{i\geq 0}$, up to a cyclic shift, by a decimation d such that $\gcd(d, 2^n-1) = 1$ (that is, the second sequence equals $(s_{di+t})_{i\geq 0}$, where t is some integer).

9.1.110 Remark Since sequences are viewed up to cyclic shifts, we shall, in the sequel, take λ equal to 1 and the integer t equal to 0.

9.1.111 Remark According to Proposition 9.1.109, the crosscorrelation $C_d(\tau) = \sum_{t=0}^{n-1}(-1)^{s_{t+\tau}+s_{dt}}$ between two m-sequences equals

$$\sum_{x\in\mathbb{F}_{2^n}^*}(-1)^{\mathrm{Tr}_{\mathbb{F}_{2^n}/\mathbb{F}_2}(cx+x^d)}$$

where $c = \alpha^\tau$; hence, $1 + C_d(\tau)$ is the value at c of the Walsh transform of the monomial function $f(x) = \mathrm{Tr}_{\mathbb{F}_{2^n}/\mathbb{F}_2}(x^d)$. The nonlinearity of $f(x) = \mathrm{Tr}_{\mathbb{F}_{2^n}/\mathbb{F}_2}(x^d)$ is therefore

$$\mathcal{NL}(f) = 2^{n-1} - \frac{1}{2}\max_{\tau}|1 + C_d(\tau)|. \tag{9.1.12}$$

9.1.112 Remark A family of balanced sequences of period 2^n-1 with good correlation properties and large linear complexity can be constructed from a bent function. These sequences are *bent sequences* [2317]. Let $n = 0 \pmod 4$, $k = n/2$ and $h : \mathbb{F}_{2^k} \to \mathbb{F}_2$ be a bent function. Let $(e_1, e_2, \ldots, e_k)$ be a basis of the vector space $\mathbb{F}_{2^k}$ over $\mathbb{F}_2$ and let $\sigma \in \mathbb{F}_{2^n} \setminus \mathbb{F}_{2^k}$. Let

$$h_i(y_1, y_2, \ldots, y_k) = h(y_1, y_2, \ldots, y_k) + a_1y_1 + a_2y_2 + \cdots + a_ky_k$$

for some ordering of the 2^k possible choices of $(a_1, \ldots, a_k)$ in $\mathbb{F}_2^k$. Let

$$f_i(x) = h_i(\mathrm{Tr}_{\mathbb{F}_{2^n}/\mathbb{F}_2}(e_1x), \mathrm{Tr}_{\mathbb{F}_{2^n}/\mathbb{F}_2}(e_2x), \ldots, \mathrm{Tr}_{\mathbb{F}_{2^n}/\mathbb{F}_2}(e_kx)) + \mathrm{Tr}_{\mathbb{F}_{2^n}/\mathbb{F}_2}(\sigma x)$$

for $i = 1, 2, \ldots, 2^k$. The family is composed of the sequences $\left(s_t^{(i)}\right)_{0\leq t\leq 2^n-2}$ for $1 \leq i \leq 2^k$, where $s_t^{(i)} = f_i(\alpha^t)$ for some fixed primitive element α in $\mathbb{F}_{2^n}$.

See Also

§15.1	For properties of Reed-Muller codes and their sub-codes; see also [1992].
[523]	For a survey on Boolean functions for coding and cryptography (the chapter which follows it in the same monograph deals with vectorial functions, which are the subject of Section 9.2 in the present handbook). This survey includes binary bent functions (see also Section 9.3 in the present handbook).
[1303]	For a recent survey on sequences.

References Cited: [267, 391, 444, 485, 490, 491, 519, 520, 521, 522, 523, 528, 530, 534, 735, 741, 800, 1069, 1303, 1409, 1446, 1673, 1728, 1855, 1951, 1992, 2011, 2073, 2075, 2085, 2317, 2474, 2522, 2663, 2777, 3015, 3056]

9.2 PN and APN functions

Pascale Charpin, INRIA

9.2.1 Remark Around 1992, two cryptanalysis methods had been introduced in the literature devoted to symmetric cryptosystems, the differential cryptanalysis [278], and the linear cryptanalysis [2026]. It was shown later that these methods are basically linked [577]. In order to resist these attacks the round function used in an iterated block cipher must satisfy some mathematical properties. These properties are mainly covered by the concepts of nonlinearity and of differential uniformity for functions on extension fields.

9.2.1 Functions from $\mathbb{F}_{2^n}$ into $\mathbb{F}_{2^m}$

9.2.2 Definition Any function from $\mathbb{F}_{2^n}$ into $\mathbb{F}_{2^m}$ is an *(n, m)-function*. It is a Boolean function when $m = 1$ and a function on $\mathbb{F}_{2^n}$ when $m = n$. Here we assume that $n \geq m$. Basic properties on Boolean functions can be found in Section 9.1.

9.2.3 Definition Let F be an (n, m)-function. The *component* functions of F are the Boolean functions

$$f_\lambda : x \in \mathbb{F}_{2^n} \mapsto \mathrm{Tr}(\lambda F(x)), \quad \lambda \in \mathbb{F}_{2^m}^*,$$

where Tr is the absolute trace on $\mathbb{F}_{2^m}$ (see Definition 2.1.80).

9.2.4 Definition An (n, m)-function F is *balanced* when it is uniformly distributed, i.e., F takes every value of $\mathbb{F}_{2^m}$ each 2^{n-m} times.

9.2.5 Proposition [524] An (n, m)-function is balanced if and only if all its component functions are balanced.

9.2.6 Remark When $m = n$, a balanced function is a permutation of $\mathbb{F}_{2^n}$, as it is shown in a more general context in [1939, Theorem 7.7].

9.2.7 Remark The nonlinearity of a Boolean function f is usually computed by means of the highest magnitude of its Walsh spectrum W_f. These quantities are denoted by $\mathcal{NL}(f)$ and $\mathcal{L}(f)$ respectively (see the definitions in Section 9.1).

9.2.8 Definition A Boolean function f is *plateaued* if its Walsh coefficients take at most three values, namely $0, \pm\mathcal{L}(f)$. Then, $\mathcal{L}(f) = 2^s$ with $s \geq n/2$.
If $s = n/2$ (and n even) then f is *bent* and its Walsh coefficients take two values only, namely $\pm 2^{\frac{n}{2}}$.
Also, f is *semi-bent* if $s = (n+1)/2$ for odd n and $s = (n+2)/2$ for even n.
An (n, m)-function is *plateaued* when its components are plateaued.

9.2.9 Definition Let F be an (n, m)-function with components f_λ. Let $\mathcal{NL}(f_\lambda)$ be the nonlinearity of f_λ. The *nonlinearity* of F, say $\mathcal{NL}(F)$, is the lowest nonlinearity achieved by one of its components:

$$\mathcal{NL}(F) = \min_{\lambda \in \mathbb{F}_{2^m}} (\mathcal{NL}(f_\lambda)) = 2^{n-1} - \frac{\mathcal{L}(F)}{2},$$

where $\mathcal{L}(F)$ is the highest magnitude appearing in the Walsh spectrum of all f_λ.

9.2.10 Remark By definition the nonlinearity of an (n, m)-function F satisfies

$$\mathcal{NL}(F) \leq 2^{n-1} - 2^{n/2-1}.$$

Indeed, this holds for any f_λ (see the covering radius bound in Definition 9.1.41).

9.2.11 Definition Let an (n, m)-function F be expressed as a univariate polynomial of degree less than 2^n. The *algebraic degree* of F is the maximal Hamming weight of its exponents, considering the 2-ary expansion of exponents.

9.2.12 Definition Let F be an (n, m)-function. For any $a \in \mathbb{F}_{2^n}$, the *derivative of F with respect to a* is the (n, m)-function D_aF defined for all $x \in \mathbb{F}_{2^n}$ by

$$D_aF(x) = F(x + a) + F(x).$$

9.2.2 Perfect Nonlinear (PN) functions

9.2.13 Definition An (n, m)-function F is a *bent* function, also called a *perfect nonlinear* (PN) function, if and only if $\mathcal{NL}(F) = 2^{n-1} - 2^{(n/2-1)}$, i.e., all its components are Boolean bent functions.

9.2.14 Theorem [2302] An (n, m)-function can be PN only if n is even and $m \leq n/2$.

9.2.15 Remark To construct a PN function is exactly to characterize a subspace of Boolean bent functions. This PN function is an (n,m)-function if this subspace is of dimension m composed of bent functions on $\mathbb{F}_{2^n}$. *Primary* constructions, by means of the main classes of bent functions are explained in [524, Section 3.1.1]. To find subspaces of bent functions is an important research problem (see recent results in [264, 495]).

9.2.16 Proposition [2302] An (n, m)-function F is PN if and only if all of its derivatives are balanced. In other terms, let

$$\delta_{a,b}(F) = |\{x \in \mathbb{F}_{2^n} | D_aF(x) = b\}|, \ a \in \mathbb{F}_{2^n}^*, \ b \in \mathbb{F}_{2^m}. \tag{9.2.1}$$

Then F is PN if and only $\delta_{a,b}(F) = 2^{n-m}$ for any a and any b.

9.2.17 Remark From Theorem 9.2.14, PN functions do not exist for $m = n$. However they do exist for odd characteristics. They are *planar* functions and were introduced in [808]. They are functions on $\mathbb{F}_{p^n}$, p odd, such that for every non-zero $\alpha \in \mathbb{F}$ the *difference mapping* $x \mapsto F(x + \alpha) - F(x)$ is a permutation of $\mathbb{F}_{p^n}$ (see Section 9.5).

9.2.18 Remark In cryptology, PN functions were introduced in 1990-95 as functions which provide the optimal resistance to linear attacks and to differential attacks (see [2302] and [2304]). They have the best nonlinearity, by Definition 9.2.13, and the $\delta_{a,b}$ are as small as possible, by Proposition 9.2.16. A major drawback is that these optimal functions are not balanced; also, they presuppose the use of non-invertible round functions (see [489], [496, Chapter 3]).

9.2.3 Almost Perfect Nonlinear (APN) and Almost Bent (AB) functions

9.2.19 Remark In cryptography, most works focused on optimal functions with respect to differential attacks. The aim is to exhibit such functions that, moreover, are bijective and oppose a good resistance to linear attacks. According to Remark 9.2.18, functions on $\mathbb{F}_{2^n}$, i.e., (n,n)-functions) are generally considered. Algebraic properties of almost perfect nonlinear (resp. almost bent) functions and their links with error-correcting codes are introduced in [525].

9.2.20 Definition A function F on $\mathbb{F}_{2^n}$ is *almost perfect nonlinear* (APN) if and only if all the equations

$$F(x) + F(x+a) = b,\ a \in \mathbb{F}_{2^n},\ a \neq 0,\ b \in \mathbb{F}_{2^n}, \tag{9.2.2}$$

have at most two solutions. The function F is *almost bent* (AB) if and only if the value of

$$W_F(\beta, \lambda) = \sum_{x \in \mathbb{F}_{2^n}} (-1)^{\mathrm{Tr}(\lambda F(x) + \beta x)} \tag{9.2.3}$$

is equal either to 0 or to $\pm 2^{\frac{n+1}{2}}$, for any β and λ in $\mathbb{F}_{2^n}$, $\beta \neq 0$.

9.2.21 Theorem [577] AB functions exist for n odd only. Any AB function is APN.

9.2.22 Remark A function F on $\mathbb{F}_{2^n}$ is APN if and only if all its derivatives are 2-to-1. This can be derived from the definition.

9.2.23 Proposition [525, Theorem 1] Let F be an AB function on $\mathbb{F}_{2^n}$. Then the algebraic degree of F is less than or equal to $(n+1)/2$.

9.2.24 Remark According to (9.2.3), a function F on $\mathbb{F}_{2^n}$ is an AB function if and only if its components f_λ are *semi-bent*. Thus

$$\mathcal{NL}(f_\lambda) = 2^{n-1} - 2^{(n-1)/2}, \quad \text{i.e.,} \quad \mathcal{L}(f_\lambda) = 2^{(n+1)/2}, \quad \text{for all } \lambda \in \mathbb{F}_{2^n}^*.$$

Consequently, $\mathcal{L}(F) = 2^{(n+1)/2}$.

9.2.25 Proposition [493] Let n be odd. Then F is an AB function on $\mathbb{F}_{2^n}$ if and only if F is APN and all Walsh coefficients of all components f_λ are divisible by $2^{(n+1)/2}$.

9.2.26 Remark APN functions are optimal concerning the resistance of a cipher to differential attacks and to its variants. More generally, this resistance is related to the following quantities, introduced in [2304].

9.2.27 Definition Let F be a function on $\mathbb{F}_{2^n}$. For any $a \in \mathbb{F}_{2^n}^*$ and $b \in \mathbb{F}_{2^n}$, we denote

$$\delta(a,b) = |\{x \in \mathbb{F}_{2^n} \mid D_a F(x) = b\}|.$$

Then, the *differential uniformity* of F is

$$\delta(F) = \max_{a \neq 0,\ b \in \mathbb{F}_{2^n}} \delta(a,b).$$

9.2.28 Remark Those functions for which $\delta(F) = 2$ are APN. It is worth noticing that such functions are rare and hard to find. From a recent result of Voloch [2885], it follows that these functions asymptotically have density zero in the set of all functions. Few infinite families are known. The monomial APN functions are related with exceptional objects (see Subsection 9.2.8). The known infinites families of non-monomial APN functions are listed in [388]. They are all quadratic. Edel and Pott propose in [954] an original construction providing a sporadic non-quadratic non-monomial APN function.

9.2.29 Problem AB functions on $\mathbb{F}_{2^n}$, with n odd, provide an optimal resistance to both differential attacks and linear attacks. There are several classes of AB permutations. The situation is different for even n: there are APN functions F such that $\mathcal{L}(F) = 2^{(n+2)/2}$ and it is conjectured that this value is the minimum; also the existence of APN permutations, for $n > 6$, is not resolved.

9.2.30 Example The *inverse function**, $F(x) = x^{-1}$, is an APN (not AB) permutation on $\mathbb{F}_{2^n}$ when n is odd. For even n, it is a permutation too, but $\delta(a, b)$ takes three values, namely 0, 2 and 4 so that $\delta(F) = 4$. This function has the highest degree and satisfies $\mathcal{L}(F) = 2^{(n+2)/2}$ for n even. The inverse function on $\mathbb{F}_{2^8}$ is used in the *AES S-boxes* (see Section 16.2).

9.2.31 Remark Regarding the PN property, the APN property can be extended on fields of odd characteristic (see [908, 1479], for instance).

9.2.4 APN permutations

9.2.32 Remark The first APN permutation of $\mathbb{F}_{2^6}$ ($n = 6$) was presented by Dillon at the conference *Finite Fields and their Applications* (Fq 9) in 2009. Thus, a long-standing (and famous) conjecture stating that *there is no APN permutation of* $\mathbb{F}_{2^n}$ *when* n *is even* was disproved. The method, explained in [424], uses mostly the representation of APN functions by codes (see Theorem 9.2.47). However the existence of APN permutations of an even number n (with $n \geq 8$) of variables remains a research problem of great interest. The next theorem, is due to several authors [228, 1541, 2303].

9.2.33 Conjecture APN permutations on $\mathbb{F}_{2^n}$ exist for any $n \geq 6$.

9.2.34 Theorem [228, Theorem 3] Let F be any permutation polynomial on $\mathbb{F}_{2^n}$, $n = 2t$. Then F is not APN when one of the statements below is satisfied:

1. $n = 4$;
2. $F \in \mathbb{F}_4[x]$;
3. $F \in \mathbb{F}_{2^t}[x]$;
4. F is plateaued, i.e., all its components f_λ are plateaued.

9.2.35 Remark When n is odd, any APN monomial is bijective; little is known about the other APN functions. Clearly, they are not generally bijective. For example, in the case where n is even the APN monomials are 3-to-1.

9.2.36 Proposition [228, Proposition 3] Let r be a divisor of n. Let F be any function on $\mathbb{F}_{2^n}$ such that $F \in \mathbb{F}_{2^r}[x]$. If F satisfies for some $a \in \mathbb{F}_{2^r}$

$$F(y) + F(y + a) = \beta, \quad \beta \in \mathbb{F}_{2^r},$$

for some $y \notin \mathbb{F}_{2^r}$ such that $y^{2^r} + y + a \neq 0$, then F is not APN.
Consequently, if F is an APN monomial, $F(x) = x^d$, then $\gcd(d, 2^n - 1) = 1$ for odd n and $\gcd(d, 2^n - 1) = 3$ for even n.

---

*Note that, as a function on $\mathbb{F}_{2^n}$, $F(x) = x^{2^n-2}$ so that $F(0) = 0$.

9.2.37 Remark It was conjectured in [525] that for any AB function F, there exists a linear function L such that $F+L$ is a permutation. A counterexample for this conjecture is given in [449, Remark 4].

9.2.38 Example [447, Proposition 1] Let $n=3k$ where k is odd and not divisible by 3. Let u be any element in $\mathbb{F}_{2^n}^*$ of order $2^{2k}+2^k+1$. Then any function

$$F(x)=x^{2^s+1}+ux^{2^{ik}+2^{tk+s}}$$

with $\gcd(s,3k)=1, i\equiv sk \pmod 3, t=3-i$, is an AB permutation.

9.2.5 Properties of stability

9.2.39 Definition Let F and F' be two functions on $\mathbb{F}_{2^n}$. They are *equivalent* if F' is obtained from F by compositions of 1 and/or 2:

1. $F\mapsto A_1\circ F\circ A_2+A$, where A_1 and A_2 are affine permutations and A is any function which is affine or constant;
2. $F\mapsto F^{-1}$, the inverse function of F when F is a permutation.

They are *extended affine equivalent* (EA-equivalent) when F' is obtained from F by transformations of type 1.

9.2.40 Proposition [525] Let F be a function on $\mathbb{F}_{2^n}$ which is APN (resp. AB). Then any function which is equivalent to F is an APN (resp. AB) function too.

9.2.41 Remark The definition of *Carlet-Charpin-Zinoviev (CCZ)-equivalence*, naturally derived from [525, Proposition 3], was proposed in [449]. It necessitates to introduce the *graph* of any function F on $\mathbb{F}_{2^n}$:

$$G_F=\{\ (x,F(x))\mid x\in\mathbb{F}_{2^n}\}.$$

9.2.42 Definition Let F and F' be two functions on $\mathbb{F}_{2^n}$. They are *CCZ-equivalent* if their graphs are affine equivalent, i.e., if there exists an affine automorphism A of $\mathbb{F}_{2^n}\times\mathbb{F}_{2^n}$ such that $A(G_F)=G_{F'}$.

9.2.43 Theorem [449, Proposition 2] Let F, F' be CCZ-equivalent functions. Then F is APN (resp. AB) if and only if F' is APN (resp. AB). In this case, F and F' have the same nonlinearity but may have different algebraic degrees.

9.2.44 Remark In [449], Budaghyan, Carlet, and Pott proved that CCZ-equivalence is more general than equivalence. They are then able to present APN functions which are EA-inequivalent to all known APN functions.

9.2.45 Remark The important question whether affinely inequivalent functions are CZZ-equivalent was first investigated by Edel, Kyureghyan, and Pott [953]. They exhibited an APN binomial, which is new since it cannot be obtained from another known APN function (see example below). A number of constructions of APN functions followed the definition of CCZ-equivalence (see, for instance: [386, 423, 443, 447]). A classification up to CCZ-equivalence, for small dimensions, is proposed in [417].

9.2.46 Example [953, Theorem 2] Let the function F on $\mathbb{F}_{2^{10}}$, $F(x)=x^3+ux^{36}$. Let $\omega\in\mathbb{F}_4\setminus\{0,1\}$. Then F is APN for any $u=\omega^i y$, $i\in\{1,2\}$ and $y\in\mathbb{F}_{2^5}^*$. Moreover, these APN functions are not CCZ-equivalent to any APN monomial on $\mathbb{F}_{2^{10}}$.

9.2.6 Coding theory point of view

9.2.47 Theorem [525] Let F be a function on $\mathbb{F}_{2^n}$ such that $F(0) = 0$. Let α be a primitive element of $\mathbb{F}_{2^n}$. Let us denote by C_F the linear binary code of length $2^n - 1$ defined by its parity check matrix

$$\mathcal{H}_F = \begin{bmatrix} 1 & \alpha & \alpha^2 & \dots & \alpha^{2^n-2} \\ F(1) & F(\alpha) & F(\alpha^2) & \dots & F(\alpha^{2^n-2}) \end{bmatrix}$$

where each entry is viewed as a binary vector. The dual code is denoted by $(C_F)^\perp$. Then we have

1. The function F is APN if and only if the code C_F has minimum distance five.
2. The function F is AB if and only if the weights of the non zero codewords of the code $(C_F)^\perp$ form the following set: $\{2^{n-1},\ 2^{n-1} \pm 2^{(n-1)/2}\}$.

9.2.48 Corollary [525] Let F be a function on $\mathbb{F}_{2^n}$ such that $F(0) = 0$. Then we have

1. If F is APN then the dimension of C_F is equal to $2^n - 2n - 1$.
2. If F is APN then $(C_F)^\perp$ does not contain the all-one vector.
3. If F is AB then the weight distribution of $(C_F)^\perp$ is unique and the same as the weight distribution of the dual of the 2-error-correcting BCH code.

9.2.49 Remark The weight distribution of $(C_F)^\perp$ exactly corresponds to the *Walsh spectrum* of F, that is the multiset of the values $W_F(\beta, \lambda)$ given by (9.2.3). Thus, if this weight distribution is known, the nonlinearity of F is also known.

9.2.50 Definition A binary code $\mathcal{C}$ is 2^ℓ*-divisible* if the weight of any of its codewords is divisible by 2^ℓ. Moreover $\mathcal{C}$ is *exactly* 2^ℓ*-divisible* if, additionally, it contains at least one codeword whose weight is not divisible by $2^{\ell+1}$.

9.2.51 Theorem [494] Let F be a function on $\mathbb{F}_{2^n}$, with $n = 2\ell + 1$. Then F is AB if and only if F is APN and the code $(C_F)^\perp$, defined in Theorem 9.2.47, is 2^ℓ-divisible.

9.2.52 Remark The determination of the weight distributions of codes of type C_F remains an open problem except when the code is optimal in a certain sense. The work of Kasami remains fundamental, proving notably the uniqueness of the weight enumerator of these optimal codes ([1688, 1689]; see also an extensive study in [590, Section 3.4.2]).

9.2.53 Example The functions $F(x) = x^d$, $d = 2^{2i} - 2^i + 1$ with $\gcd(i, n) = 1$ are APN for any n. Such exponents d are called the *Kasami exponents.* When n is odd, the code C_F is equivalent to the cyclic code with two zeros α^{2^i+1} and $\alpha^{2^{3i}+1}$ (α being a primitive root of $\mathbb{F}_{2^n}$) and F is AB. The proof is due to Kasami [1688] for odd n (see also [590, Theorem 3.32]) and to Janwa and Wilson [1595] for even n.

9.2.54 Remark The code C_F always contains a subcode which is a *cyclic code.* This is of most interest when F is a monomial as we will see later (see [525] for more details).

9.2.7 Quadratic APN functions

9.2.55 Definition A function F on $\mathbb{F}_{2^n}$ is *quadratic* when it has algebraic degree 2.

9.2.56 Proposition [525, Section 3.4] Let F be a quadratic function on $\mathbb{F}_{2^n}$ with n odd. Then F is AB if and only if F is APN.

9.2.57 Proposition [525] Let F be a quadratic function. Then F is APN if and only if the code C_F does not contain any codeword of weight three.

9.2.58 Remark When F is quadratic, $C_F^\perp$ is contained in the punctured Reed-Muller code of order 2 (see [1991, Chapter 15]).

9.2.59 Example The functions $F(x) = x^{2^i+1}$ with $\gcd(i, n) = 1$ are APN for any n (see [1689, 2303]). Such quadratic exponents are called *Gold exponents.*

9.2.60 Remark There are several constructions of quadratic non-monomial APN functions; notably, there are infinite families of such functions (see [272, 386, 388, 443, 447, 449] and Remark 9.2.28). There is only one infinite family of binomials, which was introduced and extensively studied in [447, Section II] by Budaghyan, Carlet, and Leander. Their Walsh spectrum was computed in [387]. Some such binomials are bijective (see Example 9.2.38).

9.2.61 Problem The Walsh spectrum of a non-monomial quadratic function is generally not known. Concerning APN such functions, the problem is discussed in [387]. For the computation of some spectrum, see [384, 385, 387].

9.2.62 Example The function $F(x) = x^3 + Tr(x^9)$, was recently discovered [448]. It is APN for any n and inequivalent (in any sense) to the Gold functions, while it has the same Walsh spectrum [385].

9.2.63 Theorem [389, Theorem 3] Let F be a quadratic APN function and G be a Gold function, i.e., $G(x) = x^{2^k+1}$ with $\gcd(k, n) = 1$, on $\mathbb{F}_{2^n}$. If F and G are CCZ-equivalent then they are EA-equivalent.

9.2.64 Theorem [228, Theorem 6] There are no APN quadratic mappings on $\mathbb{F}_{2^n}$ of the form

$$F(x) = \sum_{i=1}^{n-1} c_i x^{2^i+1}, \quad c_i \in \mathbb{F}_{2^n} \tag{9.2.4}$$

unless F is a monomial: $F(x) = cx^{2^k+1}$ with $\gcd(k, n) = 1$, $c \in \mathbb{F}_{2^n}$.

9.2.65 Definition The APN function F, on $\mathbb{F}_{2^n}$, is *crooked* if for every $a \in \mathbb{F}_{2^n}^*$ the image of its derivative D_aF is a hyperplane or the complement of a hyperplane.

9.2.66 Proposition Any APN quadratic function is crooked.

9.2.67 Remark Definition 9.2.65 was introduced in [1816], as an extension of the original definition where crooked functions are implicitly bijective (see [224, 2838] for crooked AB functions). The next conjecture is strengthened by the work of Bierbrauer and Kyureghyan (Theorem 9.2.69).

9.2.68 Conjecture Any APN function, which is crooked, is quadratic.

9.2.69 Theorem [277, 1816] Let $F(x) = x^d + \beta x^e$ be an APN function on $\mathbb{F}_{2^n}$, where $\beta \in \mathbb{F}_{2^n}$ and d, e are integers modulo $2^n - 1$. If F is crooked then F is quadratic.

9.2.70 Proposition [2838] Let F be an APN permutation on $\mathbb{F}_{2^n}$ which is crooked. Then n is odd and F is an AB function.

9.2.71 Proposition [1816, 2838] Any APN function, which is is crooked, is plateaued.

9.2.8 APN monomials

9.2.72 Remark In cryptography, monomials are usually called *power functions*. They were intensively studied, since they have a lower implementation cost in hardware. Moreover, their properties regarding differential attacks can be studied more easily, since they are related to the weight enumerators of some cyclic codes with two zeros [590, Section 3.4.2].

9.2.73 Proposition [525] Let $F(x) = x^d$ be a function on $\mathbb{F}_{2^n}$. Then C_F, defined as in Theorem 9.2.47, is the binary cyclic code of length $2^n - 1$ whose zeros are α, α^d and their conjugates; C_F *is a cyclic code with two zeros.*

9.2.74 Remark According to Corollary 9.2.48, if F is APN then the cyclotomic coset of d modulo $2^n - 1$ has size n (with $F(x) = x^d$). In this case, C_F has minimum distance 5 and dimension $2^n - 2n - 1$.

9.2.75 Example The Melas code is the cyclic code of length $2^n - 1$ with two zeros α and α^{-1}. Its minimum distance is 5 for odd n and 3 for even n. It is the code C_F corresponding to the inverse function $F(x) = x^{-1}$ (see Example 9.2.30). Lachaud and Wolfmann described in [1828] the set of weights of $C_F^{\perp}$.

9.2.76 Proposition [494, 1467] Let $F(x) = x^d$ with $\gcd(d, 2^n - 1) = 1$. Then $C_F^{\perp}$ is exactly 2-divisible if and only if C_F is the Melas code, i.e., $d = -1$.

9.2.77 Remark Let $F(x) = x^d$ with $\gcd(d, 2^n - 1) = 1$. The study of the weights of the code $C_F^{\perp}$ corresponds to the study of the crosscorrelation of a pair of maximal length linear sequences, called *m-sequences* (see Section 10.3).

9.2.78 Remark Janwa, McGuire, and Wilson characterized several classes of cyclic codes with two zeros whose minimum distance is at most four. More precisely, by applying a form of Weil's theorem, they showed that, for a large class of such codes, only a finite number could be *good*, i.e., have minimum distance five [1594, 1595].

9.2.79 Theorem [1594] For any fixed d satisfying $d \equiv 3 \pmod 4$ and $d > 3$, the cyclic codes C_F, $F(x) = x^d$, of length $2^n - 1$, with two zeros α and α^d, have codewords of weight 4 for all but finitely many values of n.

9.2.80 Remark By their work [1594], the authors strengthened the conjecture that APN (a fortiori AB) functions are *exceptional*. Their main conjecture was solved recently by Hernando and McGuire (see Theorem 9.2.83 below).

9.2.81 Definition Let $F(x) = x^d$. The exponent d is *exceptional* if F is APN on infinitely many extension fields of $\mathbb{F}_2$.

9.2.82 Remark The epithet *exceptional* is due to Dillon who observed that Gold and Kasami exponents can be defined by means of *exceptional polynomials* (see Section 8.4). Dillon extensively describes the links between these exceptional codes, maps, difference sets, and polynomials in [863], a review of great interest. More can be found in [862, 864]. Moreover, reversed permutation Dickson polynomials are closely related with APN monomials [1547].

9.2.83 Theorem [1490] Only the Gold and the Kasami numbers are exceptional exponents.

9.2.84 Problem Gold functions and Kasami functions, presented in Examples 9.2.59 and 9.2.53, form the exceptional classes of APN power functions. The other classes, known as *sporadic*, are listed in Example 9.2.85. It is currently conjectured that any APN power function, which is not exceptional, belongs to one of the known "sporadic" classes (up to equivalence).

9.2.85 Example The "sporadic" classes, which are known, are the following classes of APN power functions. Such a function F on $\mathbb{F}_{2^n}$, $F(x) = x^d$, is designed by its exponent d.

1. The Welch functions: $d = 2^t + 3$ with $n = 2t + 1$. Any Welch function is AB, proved by Canteaut, Charpin, and Dobbertin [493] (see also [492, 494]). This was conjectured by Welch around 1968, concerning the crosscorrelation function of binary m-sequences. Dobbertin previously proved that F is APN [904].
2. The Niho functions: for $n = 2t + 1$,

$$d = \begin{cases} 2^t + 2^{t/2} - 1 & \text{if } t \text{ is even,} \\ 2^t + 2^{(3t+1)/2} - 1 & \text{otherwise} \end{cases}$$

   (proposed by Niho [2287], also concerning properties of sequences). The APN property was proved by Dobbertin [903]. Hollmann and Xiang proved that Niho functions are AB in [1526], where they also give another proof of Welch's conjecture.
3. The Dobbertin functions: $d = 2^{4k} + 2^{3k} + 2^{2k} + 2^k - 1$ with $n = 5k$. Dobbertin introduced these functions as potential APN and in [906] proved that these functions are APN. For odd n, the Dobbertin functions are not AB [494, Proposition 7.13].
4. The inverse function: $d = 2^{n-1} - 1$ for odd n (see Examples 9.2.30 and 9.2.75).

9.2.86 Remark The nonlinearity of APN functions is generally not known. Our last general result concerns monomials: Let $n = 2t$ and $F(x) = x^d$ with $s = \gcd(d, 2^n - 1)$, $s > 1$. Then $\mathcal{L}(F) \geq 2^{t+1}$ and, if equality holds, then $s = 3$ (see [228] and [489, Theorem 8.14]). We note that if F is APN then $s = 3$ (Proposition 9.2.36).

9.2.87 Conjecture If $\gcd(d, 2^n - 1) = 1$, n even, then $\mathcal{L}(x^d) \geq 2^{(n+2)/2}$.

9.2.88 Remark APN monomials appear as specific objects with very particular properties. Although these are rare they are not completely classified, as explained in Problem 9.2.84. Very little is known about differential properties of other power functions. It is important to explore these properties, in view of future possible applications [331].

9.2.89 Remark Monomial APN functions in odd characteristic are investigated in [908, 1479]. We note that there exist *planar monomial* PN functions (see Section 9.5).

See Also

§8.4	For exceptional polynomials.
§9.1	For basic properties on Boolean functions.
§9.5	For planar functions, that is, PN functions in odd characteristics.
§10.3	For connections to m-sequences.
§16.2	For the design of block ciphers.

References Cited: [224, 228, 264, 272, 277, 278, 331, 384, 385, 386, 387, 388, 389, 417, 423, 424, 443, 447, 448, 449, 489, 492, 493, 494, 495, 496, 524, 525, 577, 590, 808, 862, 863, 864, 903, 904, 906, 908, 953, 954, 1467, 1479, 1490, 1526, 1541, 1547, 1594, 1595, 1688, 1689, 1816, 1828, 1939, 1991, 2026, 2287, 2302, 2303, 2304, 2838, 2885]

9.3 Bent and related functions

Alexander Kholosha, University of Bergen
Alexander Pott, Otto-von-Guericke-Universität Magdeburg

9.3.1 Remark In this section, we give basics and mainly focus on recent results on Boolean bent functions and also provide a rather comprehensive survey of generalized bent functions. For other results on Boolean bent functions, together with most of the proofs, the reader is referred to [523].

9.3.1 Definitions and examples

9.3.2 Definition (Walsh transform) Let $f : \mathbb{F}_p^n \to \mathbb{C}$ be a complex-valued function, where p is a prime. Let $\zeta_p = e^{\frac{2\pi i}{p}}$ be a primitive complex p-th root of unity. Then the *Walsh transform* of f is the function $\hat{f} : \mathbb{F}_p^n \to \mathbb{C}$ such that $\hat{f}(y) = \sum_{x\in\mathbb{F}_p^n} f(x)\cdot\zeta_p^{y\cdot x}$, where "$\cdot$" denotes some non-degenerate symmetric bilinear form on $\mathbb{F}_p^n$, sometimes also called an *inner product.* If $f : \mathbb{F}_p^n \to \mathbb{F}_p$, we may define a corresponding complex-valued function f_c by $f_c(x) = \zeta_p^{f(x)}$. The Walsh transform of f is by definition the Walsh transform of f_c. The *normalized Walsh coefficients* are the numbers $p^{-n/2}\hat{f}(y)$.

9.3.3 Remark

1. When we compute $\zeta_p^{y\cdot x}$, we interpret the exponent as an integer in $\{0,\dots,p-1\}$.
2. The Walsh transform is a special case of the discrete Fourier transform of Abelian groups.
3. More information about the Walsh transform of Boolean functions is contained in Section 9.1, where $\hat{f}(y)$ is denoted $W_f(y)$, instead.
4. If we identify the vector space $\mathbb{F}_p^n$ with the additive group of $\mathbb{F}_{p^n}$, we may use the trace function to define the bilinear form $x\cdot y$ that occurs in Definition 9.3.2. We take the bilinear form $x\cdot y := \mathrm{Tr}_n(xy)$, where $x, y \in \mathbb{F}_{p^n}$.
5. The standard inner product on $\mathbb{F}_p^n$ is $(x_1,\dots,x_n)\cdot(y_1,\dots,y_n) = \sum_{i=1}^n x_i y_i$.

9.3.4 Remark

1. If we identify $\mathbb{F}_p^n$ with $\mathbb{F}_{p^n}$, all p-ary functions can be described by $\mathrm{Tr}_n(F(x))$ for some function $F : \mathbb{F}_{p^n} \to \mathbb{F}_{p^n}$ of degree at most $p^n - 1$. This is the *univariate representation.* If we do not identify $\mathbb{F}_p^n$ with $\mathbb{F}_{p^n}$, the p-ary function has a representation as a multinomial in $x_1,\dots,x_n$, where the variables x_i occur with exponent at most $p-1$. This is the *multivariate representation.*
2. The multivariate representation is unique.
3. The univariate representation is not unique. Unique univariate form of a Boolean function, *trace representation,* is discussed in Section 9.1. The case of odd p is similar.

9.3.5 Definition The *algebraic degree of a p-ary function* is the degree of the polynomial giving its multivariate representation.

9.3.6 Remark The algebraic degree of a p-ary function is not the degree of the polynomial F in the univariate representation, see Section 9.1. Throughout this section, degree always means the algebraic degree.

9.3.7 Remark The Walsh transform can also be defined for functions $F : \mathbb{F}_p^n \to \mathbb{F}_p^m$. In this case, one considers all functions $F_a : \mathbb{F}_{p^n} \to \mathbb{F}_p$ of the form $F_a(x) = a \cdot F(x)$, where $a \cdot x$ denotes an inner product on $\mathbb{F}_p^m$. In this section, we mostly consider functions $\mathbb{F}_p^n \to \mathbb{F}_p$, where p is prime. Interesting classes of functions $\mathbb{F}_p^n \to \mathbb{F}_p^n$ include planar functions, APN functions and permutation polynomials, see Chapter 8 and Sections 9.2, 9.5.

9.3.8 Definition [1812] A function $f : \mathbb{F}_p^n \to \mathbb{F}_p$ is a *p-ary bent function* (or *generalized bent function*) if all its Walsh coefficients satisfy $|\hat{f}(y)|^2 = p^n$. A bent function f is *regular* if for every $y \in \mathbb{F}_p^n$, the normalized Walsh coefficient $p^{-n/2}\hat{f}(y)$ is equal to a complex p-th root of unity, i.e., $p^{-n/2}\hat{f}(y) = \zeta_p^{f^*(y)}$ for some function f^* mapping $\mathbb{F}_p^n$ into $\mathbb{F}_p$. A bent function f is *weakly regular* if there exists a complex number u having unit magnitude such that $up^{-n/2}\hat{f}(y) = \zeta_p^{f^*(y)}$ for all $y \in \mathbb{F}_p^n$. Hereafter, $p^{n/2}$ with odd n stands for the *positive* square root of p^n. If f is a weakly regular p-ary bent function (in particular, a Boolean bent function) then f^* is the *dual* of f.

9.3.9 Definition [2486] A Boolean function $f : \mathbb{F}_2^n \to \mathbb{F}_2$ is *bent* if

$$|\hat{f}(y)| = |\sum_{x \in \mathbb{F}_2^n} (-1)^{f(x)+y\cdot x}| = 2^{n/2}$$

for all $y \in \mathbb{F}_2^n$.

9.3.10 Remark Definition 9.3.9 is a special case of Definition 9.3.8. We include it here since most papers on bent functions just deal with the Boolean case. In the Boolean case, the dimension n must obviously be even. Boolean bent functions are regular.

9.3.11 Remark Boolean bent functions that are equal to their dual are *self-dual* and those whose dual is the complement of the function are *anti self-dual* [526, 1543]. The class of self-dual bent functions can be extended by defining *formally self-dual* Boolean functions [1565].

9.3.12 Example

1. The function

$$f(x_1, \ldots, x_{2m}) := x_1x_2 + x_3x_4 + \cdots + x_{2m-1}x_{2m}$$

is bent on $\mathbb{F}_2^{2m}$.

2. If we identify $\mathbb{F}_p^n$ with the additive group of $\mathbb{F}_{p^n}$, the function $f(x) = \mathrm{Tr}_n(x^2)$ is p-ary bent for all odd primes p. This example shows that in the odd characteristic case, the dimension n need not be even.

9.3.13 Definition (EA-equivalence) Functions $f, g : \mathbb{F}_p^n \to \mathbb{F}_p$ are *extended-affine equivalent* (in brief, EA-equivalent) if there exists an affine permutation L of $\mathbb{F}_p^n$, an affine function $l : \mathbb{F}_p^n \to \mathbb{F}_p$ and $a \in \mathbb{F}_p^*$ such that $g(x) = a(f \circ L)(x) + l(x)$.

9.3.14 Definition A class of functions is *complete* if it is a union of EA-equivalence classes. The *completed class* is the smallest possible complete class that contains the original one.

9.3.15 Remark

1. EA-equivalence of Boolean functions from Definition 9.1.48 is the case $p = 2$ in Definition 9.3.13.
2. If a function f is not affine then any function EA-equivalent to it has the same algebraic degree as f. On the other hand, the function f and its dual f^* do not necessarily have the same degree (this can be seen in many examples below).
3. Any function which is EA-equivalent to a bent function is bent.
4. For bent functions, CCZ equivalence coincides with EA equivalence (see [451, Theorem 3]).
5. Each Boolean quadratic bent function over $\mathbb{F}_2^{2m}$ is EA-equivalent to the function in Example 9.3.12 (1). Thus, the class of Boolean quadratic bent functions is complete and consists of a single equivalence class. For the case of odd p, see Remark 9.3.36.

9.3.16 Definition [3038] A Boolean function f over $\mathbb{F}_{2^n}$ is *hyper bent* if $f(x^k)$ is bent for any k coprime with $2^n - 1$ [529, 591].

9.3.2 Basic properties of bent functions

9.3.17 Theorem [1539, 2486] If f is a p-ary bent function on $\mathbb{F}_p^n$ then

$$2 \le \deg f \le \frac{(p-1)n}{2} + 1.$$

Moreover, if f is a weakly regular bent function (that includes the case of Boolean bent functions) with $(p-1)n \ge 4$ then

$$2 \le \deg f \le \frac{(p-1)n}{2}.$$

9.3.18 Remark The Maiorana-McFarland and Dillon classes of bent functions (see Construction 9.3.37, Remark 9.3.39, and Theorem 9.3.59) contain examples of regular functions where the upper bound on the degree is achieved when n is even. Other examples of ternary (weakly) regular bent functions in even dimension that achieve the maximal degree come from Theorem 9.3.62 and are given by Coulter-Matthews exponents with $k = n - 1$ (see Remark 9.3.65). Infinite classes of ternary bent functions in odd dimension (both weakly and non weakly regular) attaining the bound are obtained from plateaued functions using Construction 9.3.69. This construction also gives examples of weakly regular functions in odd dimension with $p = 5$ that have maximal degree.

9.3.19 Problem It is not known whether the bound for the degree of non weakly regular bent functions is sharp for general p. The same question remains open for weakly regular bent functions with odd n and $p > 5$.

9.3.20 Theorem [1812, 2486] The dual of a weakly regular bent function is again a weakly regular bent function.

9.3.21 Theorem [1812, Property 8] The Walsh transform coefficients of a p-ary bent function f with odd p satisfy

$$p^{-n/2}\hat{f}(y) = \begin{cases} \pm\zeta_p^{f^*(y)} & \text{if } n \text{ is even or } n \text{ is odd and } p \equiv 1 \pmod 4, \\ \pm i\,\zeta_p^{f^*(y)} & \text{if } n \text{ is odd and } p \equiv 3 \pmod 4, \end{cases}$$

where i is a complex primitive 4-th root of unity. Regular bent functions can only be found for even n and for odd n with $p \equiv 1 \pmod 4$. Also, for a weakly regular bent function, the constant u in Definition 9.3.8 can only be equal to ± 1 and $\pm i$.

9.3.22 Theorem [1812, 2486] (Propagation criterion) A function $f : \mathbb{F}_p^n \to \mathbb{F}_p$ is bent if and only if the mappings $D_a(f) : \mathbb{F}_p^n \to \mathbb{F}_p$ with $D_a(f)(x) = f(x+a) - f(x)$ are balanced for all $a \neq 0$, i.e., the number of solutions $f(x+a) - f(x) = b$ is p^{n-1} for all $a \neq 0$ and all $b \in \mathbb{F}_p$.

9.3.23 Example If $f(x) = \mathrm{Tr}_n(x^{p+1})$, defined on $\mathbb{F}_{p^n}$ with p odd and n odd, then $D_a(f)(x) = f(x+a) - f(x) = \mathrm{Tr}_n(x^p a + a^p x + a^{p+1})$. The polynomial $x^p a + a^p x$ is linearized and $x^p a + a^p x = 0$ if and only if $x = 0$ or $(x/a)^{p-1} + 1 = 0$. But the latter expression has no solution if n is odd, hence $x^p a + a^p x$ is a permutation polynomial and $\mathrm{Tr}_n(x^p a + a^p x + a^{p+1})$ is balanced, hence f is bent.

9.3.24 Theorem Let $n = 2m$ and $f : \mathbb{F}_{p^n} \to \mathbb{F}_p$ be a p-ary bent function. Then

$$N(0) = p^{n-1} - \mu(p^m - p^{m-1}), \quad N(1) = \cdots = N(p-1) = p^{n-1} + \mu p^{m-1} ,$$

where $\mu = \pm 1$ and $N(j) = \#\{x \in \mathbb{F}_{p^n} : f(x) = j\}$. In particular, $\hat{f}(0) = \pm p^m$.

9.3.3 Bent functions and other combinatorial objects

9.3.25 Remark Weakly regular bent functions are useful for constructing certain combinatorial objects such as partial difference sets, strongly regular graphs, and association schemes (see [598, 1058, 2424, 2775]).

9.3.26 Theorem [861] A function $f : \mathbb{F}_2^n \to \mathbb{F}_2$ is bent if and only if the set $\{x \in \mathbb{F}_2^n : f(x) = 1\}$ is a difference set in $\mathbb{F}_2^n$ with parameters $(2^n, 2^{n-1} \pm 2^{(n/2)-1}, 2^{n-2} \pm 2^{(n/2)-1}; 2^{n-2})$ (see Section 14.6.4). Equivalent difference sets give rise to EA-equivalent bent functions, but EA-equivalent bent functions need not necessarily give rise to equivalent difference sets, see Example 9.3.28.

9.3.27 Remark Theorem 9.3.26 does not hold for p-ary bent functions with p odd. For general p, the set $R_f := \{(x, f(x)) : x \in \mathbb{F}_p^n\} \subset \mathbb{F}_p^{n+1}$ is a *relative difference set* [2423].

9.3.28 Example If f and g are EA-equivalent Boolean bent functions, the corresponding difference sets need not be equivalent. One reason is that the complemented function f, i.e., $f+1$ describes the complementary difference set. But, more seriously, adding a linear mapping l to f may not preserve equivalence of the corresponding difference sets. Using the bivariate notation introduced in Remark 9.3.52, the functions $f(x,y) = \mathrm{Tr}_4(x \cdot y^7)$ and $g(x,y) = f(x,y) + \mathrm{Tr}_4(x)$ with $x, y \in \mathbb{F}_{2^4}$, are EA equivalent. Both functions describe difference sets with parameters $(256, 120, 56; 64)$. These difference sets are inequivalent since the corresponding designs are not isomorphic.

9.3.29 Theorem [1475] The function $f : \mathbb{F}_2^n \to \mathbb{F}_2$ is bent if and only if

$$\#\{x \in \mathbb{F}_2^n : f(x) \neq l(x) + \epsilon\} = 2^{n-1} \pm 2^{(n/2)-1}$$

for all linear mappings $l : \mathbb{F}_2^n \to \mathbb{F}_2$ and all $\epsilon \in \{0,1\}$.

9.3.30 Remark Theorem 9.3.29 shows that f has the largest distance to all affine functions, hence bent functions solve the covering radius problem for first order Reed-Muller codes of length 2^n with n even. In other words, bent functions are *maximum nonlinear functions* $\mathbb{F}_2^n \to \mathbb{F}_2$ if n is even.

9.3.4 Fundamental classes of bent functions

9.3.31 Example [2486] The following is a complete list of Boolean bent functions on $\mathbb{F}_2^{2m}$ for $1 \le m \le 3$ (up to equivalence). For the functions F_3, F_4, F_5, and F_6, we have $m = 3$.

1. x_1x_2 for $m = 1$,
2. $x_1x_2 + x_3x_4$ for $m = 2$,
3. $x_1x_4 + x_2x_5 + x_3x_6 = F_3$,
4. $F_3 + x_1x_2x_3 = F_4$,
5. $F_4 + x_2x_4x_6 + x_1x_2 + x_4x_6 = F_5$,
6. $F_5 + x_3x_4x_5 + x_1x_2 + x_3x_5 + x_4x_5 = F_6$.

9.3.32 Remark The corresponding difference sets are inequivalent, but the design corresponding to the difference set from F_3 is equivalent to the design corresponding to F_4.

9.3.33 Proposition [1991, Chapter 15] Let f be a quadratic function $\sum_{i,j} a_{i,j}x_ix_j$ in $2m$ variables over $\mathbb{F}_2$. Then f is bent if and only if one of the following equivalent conditions is satisfied:

1. The matrix $(a_{i,j} + a_{j,i})_{i,j=1,\ldots,2m} \in \mathbb{F}_2^{(2m,2m)}$ is invertible.
2. The symplectic form $B(x,y) := f(x+y) + f(x) + f(y)$ is nondegenerate.

9.3.34 Example The matrix corresponding to the Boolean function $f(x) = x_1x_2 + x_2x_3 + x_3x_4$ is $\begin{pmatrix} 0 & 1 & 0 & 0 \\ 1 & 0 & 1 & 0 \\ 0 & 1 & 0 & 1 \\ 0 & 0 & 1 & 0 \end{pmatrix}$ and it is invertible, hence f is bent. Similarly, the function $x_1x_2 + x_2x_3 + x_3x_4 + x_1x_4$ is not bent, since the corresponding matrix $\begin{pmatrix} 0 & 1 & 0 & 1 \\ 1 & 0 & 1 & 0 \\ 0 & 1 & 0 & 1 \\ 1 & 0 & 1 & 0 \end{pmatrix}$ is singular.

9.3.35 Proposition [1470] Any quadratic p-ary function f mapping $\mathbb{F}_{p^n}$ to $\mathbb{F}_p$ is bent if and only if the bilinear form $B(x,y) := f(x+y) - f(x) - f(y) + f(0)$ associated with f is nondegenerate. Moreover, all quadratic p-ary bent functions are (weakly) regular.

9.3.36 Remark There are exactly two inequivalent nondegenerate quadratic forms on $\mathbb{F}_{p^n}$ with p odd:

$$f_1 := x_1^2 + x_2^2 + \cdots + x_n^2,$$
$$f_2 := x_1^2 + x_2^2 + \cdots + g \cdot x_n^2,$$

where g is a nonsquare in $\mathbb{F}_p$. Here "equivalence" of quadratic forms is the usual linear algebra equivalence. Both functions are bent. If n is odd, then $g \cdot f_2$ is EA-equivalent to f_1, hence there is only one quadratic bent function if n is odd. If n is even, the two quadratic bent functions are EA-inequivalent.

9.3.37 Construction [861, 1812, 2051] (Maiorana-McFarland) Take any permutation π of $\mathbb{F}_p^m$ and any function $\sigma : \mathbb{F}_p^m \to \mathbb{F}_p$. Then $f : \mathbb{F}_p^m \times \mathbb{F}_p^m \to \mathbb{F}_p$ with $f(x,y) := x \cdot \pi(y) + \sigma(y)$ is a bent function. Moreover, the bijectiveness of π is necessary and sufficient for f being bent. Such bent functions are regular and the dual function is equal to $f^*(x,y) = y \cdot \pi^{-1}(x) + \sigma(\pi^{-1}(x))$. Boolean quadratic bent functions all belong to the completed Maiorana-McFarland class. On the other hand, for odd p and even n, quadratic bent function f_2 in Remark 9.3.36 does not belong to the completed Maiorana-McFarland class [445].

9.3.38 Example The Boolean function $x_1x_4 + x_2x_5 + x_3x_6 + x_4x_5x_6$ is a Maiorana-McFarland bent function on 6 variables.

9.3.39 Remark The Maiorana-McFarland construction can be used to construct bent functions of degree $(p-1)m$ by choosing a function σ of degree $(p-1)m$.

9.3.40 Construction [861] (Partial spread) Let $V = \mathbb{F}_2^n$ with $n = 2m$. Let U_i, $i \in I$, be a collection of subspaces of dimension m with $U_i \cap U_j = \{0\}$ for all $i \neq j$. If $|I| = 2^{m-1}$, the set $D_I^- = \bigcup_{i \in I} U_i \setminus \{0\}$ is a difference set with parameters $(2^{2m}, 2^{2m-1} - 2^{m-1}, 2^{2m-2} - 2^{m-1}; 2^{2m-2})$. Similarly, if $|I| = 2^{m-1} + 1$, the union $D_I^+ = \bigcup_{i \in I} U_i$ is a difference set with parameters $(2^{2m}, 2^{2m-1} + 2^{m-1}, 2^{2m-2} + 2^{m-1}; 2^{2m-2})$. The functions $f^- : V \to \mathbb{F}_2$ (or f^+) with $f^-(x) = 1$ if and only if $x \in D_I^-$ (or $f^+(x) = 1$ if and only if $x \in D_I^+$) are bent functions (see Theorem 9.3.26). For more information on difference sets see Section 14.6.

9.3.41 Remark These functions are bent functions of *partial spread type*, since a collection of subspaces of dimension m with pairwise trivial intersection is a partial spread. The functions f^+ are of type $\mathcal{PS}^+$, the others of type $\mathcal{PS}^-$.

9.3.42 Example Let $q = 2^m$, and view $\mathbb{F}_{q^2}$ as a 2-dimensional space over $\mathbb{F}_q$, but also as a $2m$-dimensional vector space over $\mathbb{F}_2$. Then the 1-dimensional subspaces of $\mathbb{F}_{q^2}$ (viewed as a 2-dimensional vector space) are m-dimensional subspaces over $\mathbb{F}_2$. The $q+1$ subspaces of dimension 1 over $\mathbb{F}_q$ are

$$U_\alpha := \{\alpha \cdot x : x \in \mathbb{F}_q\}$$

where $\alpha^{q+1} = 1$, i.e., α is in the multiplicative group of order $q+1$ in $\mathbb{F}_{q^2}$. These subspaces intersect pairwise trivially, hence we may take any 2^{m-1} or $2^{m-1}+1$ of these subspaces to construct f^- or f^+.

9.3.43 Remark

1. The parameters of D_I^+ are complementary to the parameters of D_I^-, but the difference sets are, in general, not complements of each other.
2. All functions in $\mathcal{PS}^-$ have algebraic degree $n/2$. On the contrary, if $n/2$ is even then class $\mathcal{PS}^+$ contains all quadratic bent functions [861].
3. A partial spread where the union of the subspaces covers the entire vector space is called a *spread*.
4. The spread constructed in Example 9.3.42 is the *regular spread*.
5. Spreads of subspaces of dimension m in $2m$-dimensional subspaces can be used to describe translation planes.
6. There are numerous spreads, hence many partial spreads. Many partial spreads are not contained in spreads. For partial spreads contained in spreads, the two constructions in Construction 9.3.40 are complements of each other, i.e., for any f^-, there is another partial spread such that $f^- = f^+ + 1$.

9.3.44 Definition The class $\mathcal{PS}_{ap}$ is a subclass of $\mathcal{PS}^-$ where a subspread of the regular spread is used.

9.3.45 Remark [523] All of the functions in the $\mathcal{PS}_{ap}$ class are hyper bent (see Definition 9.3.16).

9.3.46 Remark Some of the known constructions of bent functions are direct, that is, do not use as building blocks previously constructed bent functions. We call these *primary constructions.* The others, sometimes leading to recursive constructions, will be called *secondary constructions.* Most of the secondary constructions of Boolean bent functions are not explained here and can be found in [523, Section 6.4.2].

9.3.47 Construction (Recursive construction) Let $f_1 : \mathbb{F}_p^{m_1} \to \mathbb{F}_p$ and $f_2 : \mathbb{F}_p^{m_2} \to \mathbb{F}_p$ be p-ary bent functions, then $f_1 * f_2 : \mathbb{F}_p^{m_1+m_2}$ with $(f_1 * f_2)(x_1, x_2) = f_1(x_1) + f_2(x_2)$ is p-ary bent.

9.3.48 Remark This construction was formulated originally in a much more general form using relative difference sets [2423]. Since the function $f : \mathbb{F}_p \to \mathbb{F}_p$ with $f(x) = x^2$ is a p-ary bent function if p is odd, Construction 9.3.47 yields bent functions $\mathbb{F}_p^n \to \mathbb{F}_p$ for all n (p odd).

9.3.5 Boolean monomial and Niho bent functions

9.3.49 Example The following monomial functions $f(x) = \mathrm{Tr}_n(\alpha x^d)$ are bent on $\mathbb{F}_{2^n}$ with $n = 2m$:

1. $d = 2^k + 1$ with $n/\gcd(k, n)$ being even and $\alpha \notin \{y^d : y \in \mathbb{F}_{2^n}\}$ [1295];
2. $d = r(2^m - 1)$ with $\gcd(r, 2^m + 1) = 1$ and $\alpha \in \mathbb{F}_{2^m}$ being -1 of the Kloosterman sum (see Remark 9.3.61) [591, 861];
3. $d = 2^{2k} - 2^k + 1$ with $\gcd(k, n) = 1$ and $\alpha \notin \{y^3 : y \in \mathbb{F}_{2^n}\}$ [864, 1856];
4. $d = (2^k + 1)^2$ with $n = 4k$ and k odd, $\alpha \in \omega\mathbb{F}_{2^k}$ with $\omega \in \mathbb{F}_4 \setminus \mathbb{F}_2$ [594, 1879];
5. $d = 2^{2k} + 2^k + 1$ with $n = 6k$ and $k > 1$, $\alpha \in \mathbb{F}_{2^{3k}}$ with $\mathrm{Tr}_{\mathbb{F}_{2^{3k}}/\mathbb{F}_{2^k}}(\alpha) = 0$ [495].

9.3.50 Remark These functions are *monomial bent functions.* Functions in Part 1 are quadratic, and those in Parts 4 and 5 belong to the Maiorana-McFarland class. Functions in Part 2 are in the class $\mathcal{PS}_{ap}$. An exhaustive search shows that there are no other monomial bent functions for $n \leq 20$.

9.3.51 Example [531, 532, 907, 1474, 1878] (Niho bent functions) A positive integer d (always understood modulo $2^n - 1$ with $n = 2m$) is a *Niho exponent* if $d \equiv 2^j \pmod{2^m - 1}$ for some $j < n$. As we consider $\mathrm{Tr}_n(ax^d)$ with $a \in \mathbb{F}_{2^n}$, without loss of generality, we can assume that d is in normalized form, i.e., with $j = 0$. Then we have a unique representation $d = (2^m - 1)s + 1$ with $2 \leq s \leq 2^m$. Following are examples of bent functions consisting of one or more Niho exponents:

1. Quadratic function $\mathrm{Tr}_m(ax^{2^m+1})$ with $a \in \mathbb{F}_{2^m}^*$ (here $s = 2^{m-1} + 1$).
2. Binomials of the form $f(x) = \mathrm{Tr}_n(\alpha_1 x^{d_1} + \alpha_2 x^{d_2})$, where $2d_1 \equiv 2^m + 1 \pmod{2^n - 1}$ and $\alpha_1, \alpha_2 \in \mathbb{F}_{2^n}^*$ are such that $(\alpha_1 + \alpha_1^{2^m})^2 = \alpha_2^{2^m+1}$. Equivalently, denoting $a = (\alpha_1 + \alpha_1^{2^m})^2$ and $b = \alpha_2$ we have $a = b^{2^m+1} \in \mathbb{F}_{2^m}^*$ and

$$f(x) = \mathrm{Tr}_m(ax^{2^m+1}) + \mathrm{Tr}_n(bx^{d_2}).$$

We note that if $b = 0$ and $a \neq 0$ then f is a bent function listed under number 1.

The possible values of d_2 are:

$$d_2 = (2^m - 1)3 + 1,$$
$$6d_2 = (2^m - 1) + 6 \quad \text{(with the condition that } m \text{ is even).}$$

These functions have algebraic degree m and do not belong to the completed Maiorana-McFarland class [446].

3. Take $r > 1$ with $\gcd(r, m) = 1$ and define

$$f(x) = \mathrm{Tr}_n\left(ax^{2^m+1} + \sum_{i=1}^{2^{r-1}-1} x^{d_i}\right),$$

where $2^r d_i = (2^m - 1)i + 2^r$ and $a \in \mathbb{F}_{2^n}$ is such that $a + a^{2^m} = 1$. The dual of f, calculated using Proposition 9.3.56, is equal to

$$f^*(y) = \mathrm{Tr}_m\Big((u(1 + y + y^{2^m}) + u^{2^{n-r}} + y^{2^m})(1 + y + y^{2^m})^{1/(2^r-1)}\Big),$$

where $u \in \mathbb{F}_{2^n}$ is arbitrary with $u + u^{2^m} = 1$. Moreover, if $d < m$ is a positive integer defined uniquely by $dr \equiv 1 \pmod{m}$ then the algebraic degree of f^* is equal to $d+1$. Both functions f and its dual belong to the completed Maiorana-McFarland class. On the other hand, f^* is not a Niho bent function.

4. Bent functions in a bivariate representation obtained from the known o-polynomials (see Remarks 9.3.52 and 9.3.54).

9.3.52 Remark The *bivariate representation* of a Boolean function is defined as follows: we identify $\mathbb{F}_2^n$ with $\mathbb{F}_{2^m} \times \mathbb{F}_{2^m}$ and consider the argument of f as an ordered pair (x, y) of elements in $\mathbb{F}_{2^m}$. There exists a unique bivariate polynomial $\sum_{0 \le i,j \le 2^m-1} a_{i,j} x^i y^j$ over $\mathbb{F}_{2^m}$ that represents f. The algebraic degree of f is equal to $\max_{(i,j)\,|\,a_{i,j} \neq 0}(w_2(i) + w_2(j))$ ($w_2(j)$ is the Hamming weight of the binary expansion of j). Since f is Boolean the bivariate representation can be written in the form of $f(x, y) = \mathrm{Tr}_m(P(x, y))$, where $P(x, y)$ is some polynomial of two variables over $\mathbb{F}_{2^m}$.

9.3.53 Construction [532, 861] Define class $\mathcal{H}$ of functions in their bivariate representation as follows

$$g(x, y) = \begin{cases} \mathrm{Tr}_m\left(xH(\frac{y}{x})\right) & \text{if } x \neq 0, \\ \mathrm{Tr}_m(\mu y) & \text{if } x = 0, \end{cases} \tag{9.3.1}$$

where $\mu \in \mathbb{F}_{2^m}$ and H is a mapping from $\mathbb{F}_{2^m}$ to itself with $G(z) := H(z) + \mu z$ satisfying

$$z \mapsto G(z) + \beta z \quad \text{is 2-to-1 on } \mathbb{F}_{2^m} \text{ for any } \beta \in \mathbb{F}_{2^m}^*. \tag{9.3.2}$$

Condition (9.3.2) is necessary and sufficient for g to be bent.

9.3.54 Remark

1. Any mapping G on $\mathbb{F}_{2^m}$ that satisfies (9.3.2) is an *o-polynomial.*
2. Any o-polynomial defines a hyperoval in $\mathrm{PG}(2, 2^m)$. Using Construction 9.3.53, every o-polynomial results in a bent function in class $\mathcal{H}$. For a list of known, up to equivalence, o-polynomials, see [532].

9.3.55 Remark A bent function $\mathbb{F}_{2^n} \to \mathbb{F}_2$ with $n = 2m$ belongs to class $\mathcal{H}$ if and only if its restriction to each coset $u\mathbb{F}_{2^m}$ with $u \in \mathbb{F}_{2^n}^*$ is linear. Thus, Niho bent functions from

Example 9.3.51 are just functions of class $\mathcal{H}$ viewed in their univariate representation. In particular, binomial Niho bent functions with $d_2 = (2^m - 1)3 + 1$ correspond to Subiaco hyperovals [1474], functions with $6d_2 = (2^m - 1) + 6$ correspond to Adelaide hyperovals and functions listed under number 3 (consisting of 2^r terms) are obtained from Frobenius map $G(z) = z^{2^{m-r}}$ (i.e., translation hyperovals) [531].

9.3.56 Proposition [532] Let g be a bent function having the form of (9.3.1). Then its dual function g^*, represented in its bivariate form, satisfies $g^*(\alpha, \beta) = 1$ if and only if the equation $H(z) + \beta z = \alpha$ has no solutions in $\mathbb{F}_{2^m}$.

9.3.6 p-ary bent functions in univariate form

9.3.57 Theorem [1470] Consider an odd prime p and nonzero $a \in \mathbb{F}_{p^n}$. Then for any $j \in \{1, \ldots, n\}$, the quadratic p-ary function f mapping $\mathbb{F}_{p^n}$ to $\mathbb{F}_p$ and given by

$$f(x) = \mathrm{Tr}_n\big(ax^{p^j+1}\big)$$

is bent if and only if

$$p^{\gcd(2j,n)} - 1 \nmid \frac{p^n - 1}{2} - \mathrm{ind}(a)(p^j - 1)\,, \qquad (9.3.3)$$

where $\mathrm{ind}(a)$ denotes the unique integer $t \in \{0, \ldots, p^n - 1\}$ with $a = \xi^t$ and ξ being a primitive element of $\mathbb{F}_{p^n}$.

9.3.58 Remark The functions in Theorem 9.3.57 are quadratic, hence they are (weakly) regular. In [1471], the dual function has been determined explicitly.

9.3.59 Theorem [1470] Let $n = 2m$ and t be an arbitrary positive integer with $\gcd(t, p^m + 1) = 1$ for an odd prime p. For any nonzero $a \in \mathbb{F}_{p^n}$, define the following p-ary function mapping $\mathbb{F}_{p^n}$ to $\mathbb{F}_p$

$$f(x) = \mathrm{Tr}_n\big(ax^{t(p^m-1)}\big). \qquad (9.3.4)$$

Let K denote the Kloosterman sum (see Definition 6.1.118). Then for any $y \in \mathbb{F}_{p^n}^*$, the corresponding Walsh transform coefficient of f is equal to

$$\hat{f}(y) = 1 + K\big(a^{p^m+1}\big) + p^m \zeta_p^{-\mathrm{Tr}_n(a^{p^m} y^{t(p^m-1)})} \quad \text{and}$$
$$\hat{f}(0) = 1 - (p^m - 1)K\big(a^{p^m+1}\big).$$

Assuming $p^m > 3$, then f is bent if and only if $K\big(a^{p^m+1}\big) = -1$. Moreover, if the latter holds then f is a regular bent function of degree $(p-1)m$.

9.3.60 Remark According to the result of Katz and Livné [1695] (see also [2121, Theorem 6.4]), with c running over $\mathbb{F}_{3^m}^*$, the Kloosterman sum $K(c)$ takes on *all* the integer values in the range $(-2\sqrt{3^m}, 2\sqrt{3^m})$ that are equal to -1 modulo 3. In particular, there exists at least one $a \in \mathbb{F}_{3^n}$ such that $K\big(a^{3^m+1}\big) = -1$. This means that in the ternary case (i.e., when $p = 3$) and given the conditions of Theorem 9.3.59, there exists at least one $a \in \mathbb{F}_{3^n}$ such that the function (9.3.4) is bent. Moreover, there are no bent functions having the form of (9.3.4) when $p > 3$ since in this case, the Kloosterman sum never takes on the value -1 as shown in [1785]. More on Kloosterman sums is contained in Section 6.1.

9.3.61 Remark Let $p = 2$ and without loss of generality assume $a \in \mathbb{F}_{2^m}$. Then exactly the same result as in Theorem 9.3.59 holds for any m in the binary case giving the *Dillon class* of bent functions [591, 861, 1879] (see Example 9.3.49 Part 2). Moreover, the Kloosterman sum over $\mathbb{F}_{2^m}$ takes on *all* the integer values in the closed range $[-2\sqrt{2^m}, 2\sqrt{2^m}]$ that are equal to -1 modulo 4 [1828]. This means that binary Dillon bent functions exist.

9.3.62 Theorem [1315, 1469, 1470] Let $n = 2m$ with m odd and $a = \xi^{\frac{3^m+1}{4}}$, where ξ is a primitive element of $\mathbb{F}_{3^n}$. Then the ternary function f mapping $\mathbb{F}_{3^n}$ to $\mathbb{F}_3$ and given by

$$f(x) = \mathrm{Tr}_n\left(ax^{\frac{3^n-1}{4}+3^m+1}\right)$$

is a weakly regular bent function of degree n.

For any $y \in \mathbb{F}_{3^n}$ the Walsh transform coefficient of f is equal to $\hat{f}(y) = -3^m\zeta_3^{g(y)}$ with

$$g(y) = -\mathrm{Tr}_n\left(a^{-1}y^{\frac{3^n-1}{4}+3^m+1}\right)\sum_{t\in I_m}\mathrm{Tr}_{o(t)}\left((ay^{-2})^{t(3^m+1)}\right),$$

where $o(t)$ is the size of the cyclotomic coset modulo $3^m - 1$ that contains t and the set I_m is defined as follows. Select all such integers in the range $\{0, \ldots, 3^m - 1\}$ that do not contain 2-digits in the ternary expansion and none of 1-digits are adjacent (the least significant digit is cyclically linked with the most significant). Split this set into cyclotomic cosets modulo $3^m - 1$, take coset leaders and denote this subset I_m.

9.3.63 Theorem [1472] Let $n = 4k$. Then the p-ary function f mapping $\mathbb{F}_{p^n}$ to $\mathbb{F}_p$ and given by

$$f(x) = \mathrm{Tr}_n\left(x^{p^{3k}+p^{2k}-p^k+1} + x^2\right)$$

is a weakly regular bent function of degree $(p-1)k+2$. Moreover, for any $y \in \mathbb{F}_{p^n}$ the corresponding Walsh transform coefficient of f is equal to

$$\hat{f}(y) = -p^{2k}\zeta_p^{\mathrm{Tr}_k(x_0)/4},$$

where x_0 is a unique root in $\mathbb{F}_{p^k}$ of the polynomial

$$y^{p^{2k}+1} + (y^2+X)^{(p^{2k}+1)/2} + y^{p^k(p^{2k}+1)} + (y^2+X)^{p^k(p^{2k}+1)/2}.$$

In particular, if $y^2 \in \mathbb{F}_{p^{2k}}$ then $x_0 = -\mathrm{Tr}_{\mathbb{F}_{p^{2k}}/\mathbb{F}_{p^k}}(y^2)$.

9.3.64 Remark Some sporadic examples of ternary (non) weakly regular bent functions consisting of one or two terms in the univariate representation can be found in [1473].

9.3.7 Constructions using planar and s-plateaued functions

9.3.65 Remark Definition and other information on planar functions can be found in Section 9.5. In particular, except for one class, all known planar functions are quadratic [451] which means that they can be represented by so called Dembowski-Ostrom polynomials (see Definition 9.5.17 and [732]). The only known example of a nonquadratic planar function is $F(x) = x^{\frac{3^k+1}{2}}$ over $\mathbb{F}_{3^n}$ with $\gcd(k,n) = 1$ and odd k, known as Coulter-Matthews function [732]. The following theorem shows that every planar function gives a family of generalized bent functions.

9.3.66 Theorem [527] A function F mapping $\mathbb{F}_{p^n}$ to itself is planar if and only if for every nonzero $a \in \mathbb{F}_{p^n}$ the function $\mathrm{Tr}_n(aF)$ is generalized bent.

9.3.67 Remark The generalized bent functions $\mathrm{Tr}_n(aF)$ obtained from Dembowski-Ostrom polynomials are quadratic, hence they are weakly regular (see Proposition 9.3.35). It was shown in [1055, 3042] that the bent functions coming from the Coulter-Matthews planar functions are also weakly regular.

9.3.68 Definition A function f mapping $\mathbb{F}_{p^n}$ to $\mathbb{F}_p$ is *s-plateaued* if its Walsh coefficients either equal zero or satisfy $|\hat{f}(y)|^2 = p^{n+s}$. The case $s = 0$ corresponds to bent functions.

9.3.69 Construction [571, 572] For every $\boldsymbol{a} = (a_1, a_2, \ldots, a_s) \in \mathbb{F}_p^s$ let $f_{\boldsymbol{a}}$ be an s-plateaued function from $\mathbb{F}_{p^n}$ to $\mathbb{F}_p$ such that $\hat{f}_{\boldsymbol{a}}(t) \cdot \hat{f}_{\boldsymbol{b}}(t) \equiv 0$ for any $\boldsymbol{a}, \boldsymbol{b} \in \mathbb{F}_p^s$ with $\boldsymbol{a} \neq \boldsymbol{b}$ and $t \in \mathbb{F}_{p^n}$. Then function $f(x, y_1, y_2, \ldots, y_s)$ from $\mathbb{F}_{p^n} \times \mathbb{F}_p^s$ to $\mathbb{F}_p$ defined by

$$f(x, y_1, y_2, \ldots, y_s) = (-1)^s \sum_{\boldsymbol{a} \in \mathbb{F}_p^s} \frac{\prod_{i=1}^{s} y_i(y_i - 1) \cdots (y_i - (p-1))}{(y_1 - a_1) \cdots (y_s - a_s)} f_{\boldsymbol{a}}(x)$$

is bent. Moreover, for any $t \in \mathbb{F}_{p^n}$ and $\boldsymbol{a} \in \mathbb{F}_p^s$, the corresponding Walsh transform coefficient of f is equal to

$$\hat{f}(t, \boldsymbol{a}) = \zeta_p^{-\boldsymbol{a} \cdot \boldsymbol{y}} \hat{f}_{\boldsymbol{y}}(t),$$

where $\boldsymbol{y} \in \mathbb{F}_p^s$ is unique with $\hat{f}_{\boldsymbol{y}}(t) \neq 0$.

9.3.70 Remark Quadratic functions are always plateaued. Then one can construct s-plateaued functions with the prescribed properties by adding suitably chosen linear functions to such quadratic s-plateaued functions. In this case, the constructed bent function f has degree $(p-1)s + 2$ (resp. $(p-1)s + 1$) if $\sum_{\boldsymbol{a} \in \mathbb{F}_p^s} f_{\boldsymbol{a}}$ is quadratic (resp. affine). It is possible to construct specific families of quadratic s-plateaued functions that lead to weakly regular bent functions when $n - s$ is even. Similar constructions with $n - s$ odd can lead both to weakly and non weakly regular bent functions. On the other hand, one can take a suitable family of $(n-1)$-plateaued quadratic functions (maximal order achievable by a non affine function). Then in the case when the degree of the corresponding bent function exceeds $(p-1)n/2$ being the maximum for a weakly regular function (see Theorem 9.3.17), the obtained functions are non weakly regular.

9.3.8 Vectorial bent functions and Kerdock codes

9.3.71 Definition A mapping $f : \mathbb{F}_p^n \to \mathbb{F}_p^m$ is a *vectorial bent function* if all its component functions $f_a : \mathbb{F}_p^n \to \mathbb{F}_p$ defined by $f_a(x) = a \cdot x$ are bent for all $a \neq 0$. Here "$\cdot$" denotes a symmetric bilinear form (see Definition 9.3.2).

9.3.72 Example We identify the vector spaces $\mathbb{F}_p^n$ and $\mathbb{F}_p^m$ with the finite fields $\mathbb{F}_{p^n}$ and $\mathbb{F}_{p^m}$.

1. Any planar function is a vectorial bent function $\mathbb{F}_{p^n} \to \mathbb{F}_{p^n}$.
2. If $n = 2m$, then $f(x, y) = xy$ is a vectorial bent function $\mathbb{F}_{p^m} \times \mathbb{F}_{p^m} \to \mathbb{F}_{p^m}$. All component functions are quadratic.
3. If $n = 2m$, then $f(x, y) = xy^{p^m - 2}$ is a vectorial bent function $\mathbb{F}_{p^m} \times \mathbb{F}_{p^m} \to \mathbb{F}_{p^m}$.

9.3.73 Theorem [2302] If $f : \mathbb{F}_2^{2m} \to \mathbb{F}_2^n$ is vectorial bent, then $n \leq m$ and equality can be attained (see Example 9.3.72).

9.3.74 Remark A collection of $2^{2m-1} - 1$ quadratic Boolean bent functions in $2m$ variables such that the sum of any two of them is bent again, gives rise to a *Kerdock code* of length 2^{2m}. The Boolean functions in Example 9.3.72 Part 2 only give rise to $2^m - 1$ such bent functions.

See Also

§6.1	Kloosterman sums are related to some classes of bent functions.
§7.2	The classical bent functions are related to quadratic forms.
§9.1	Boolean bent functions are special types of Boolean functions.
§9.2	Via the trace mapping, PN and some APN functions give rise to bent functions.
§9.5	Any planar function is a vectorial bent function.
§14.6	Boolean bent functions are special types of difference sets.
[524]	A very good source for vectorial functions.
[809]	This paper contains a lot of information about the symmetries of bent functions.
[1409]	A description of Kerdock codes which initiated work on codes over rings.
[1842]	Lander's book includes information about bent functions and difference sets.
[2224]	A thesis about bent functions has been written by Timo Neumann.

References Cited: [445, 446, 451, 495, 523, 524, 526, 527, 529, 531, 532, 571, 572, 591, 594, 598, 732, 809, 861, 864, 907, 1055, 1058, 1295, 1315, 1409, 1469, 1470, 1471, 1472, 1473, 1474, 1475, 1539, 1543, 1565, 1695, 1785, 1812, 1828, 1842, 1856, 1878, 1879, 1991, 2051, 2121, 2224, 2302, 2423, 2424, 2486, 2775, 3038, 3042]

9.4 κ-polynomials and related algebraic objects

Robert Coulter, The University of Delaware

9.4.1 Definitions and preliminaries

9.4.1 **Definition** A polynomial $f(x_1,\dots,x_n)$ is a *κ-polynomial* over $\mathbb{F}_q$ if

$$k_a = \#\{(x_1,\dots,x_n) \in \mathbb{F}_q^n : f(x_1,\dots,x_n) = a\},$$

is independent of a for $a \in \mathbb{F}_q^*$.

9.4.2 **Theorem** [2034] The polynomial $f(x_1,\dots,x_n)$ is a κ-polynomial over $\mathbb{F}_q$ if and only if

$$\sum_{\mathbf{a}\in\mathbb{F}_q^n} \chi(f(\mathbf{a})) = 0$$

for all non-trivial multiplicative characters χ of $\mathbb{F}_q$.

9.4.3 **Remark** This should be compared to Corollary 8.2.7 of Section 8.2. There are almost no results in the literature directly discussing κ-polynomials. This seems altogether surprising since the specified regularity on preimages of all non-zero elements of the field suggests such polynomials must almost certainly appear in many guises other than those outlined here.

9.4.4 **Lemma** [2034] Let $f(x_1,\dots,x_n)$ be a κ-polynomial over $\mathbb{F}_q$ and g be any polynomial in variables disjoint from $x_1,\dots,x_n$. Then fg is a κ-polynomial.

9.4.5 Definition Let $\mathcal{S} = \langle \mathcal{S}, +, \star \rangle$ be a set with two binary operations, addition + and multiplication $\star$ and where $\langle \mathcal{S}, + \rangle$ forms a group.

1. $\mathcal{S}$ is a *Cartesian group* if it has no zero divisors and $\langle \mathcal{S}^*, \star \rangle$ forms a loop.
2. $\mathcal{S}$ is a *left (resp. right) quasifield* if, in addition to being a Cartesian group, $\mathcal{S}$ also has a left (resp. right) distributive law.
3. $\mathcal{S}$ is a *semifield* if it is a non-associative division ring – that is, a Cartesian group with both distributive laws.

If we do not insist on a multiplicative identity, so that $\langle \mathcal{S}^*, \star \rangle$ forms a quasigroup instead of a loop, then we may speak of a *pre-Cartesian group, prequasifield, or pre-semifield.*

9.4.6 Remark Each algebraic object corresponds with a type of projective plane under the Lenz-Barlotti classification. Specifically, Cartesian groups correspond to type II, quasifields to type IV, and semifields to type V. For more information along these lines see Section 14.3, as well as the classical texts of Dembowski [807] and Hughes and Piper [1560]. See also Subsection 2.1.7.6.

9.4.7 Theorem

1. The additive group of a quasifield is necessarily an elementary abelian p-group.
2. [1764] Any semifield of prime or prime squared order is necessarily a finite field.
3. [1764] Proper semifields, i.e., semifields which are not fields, exist for all prime power orders $p^e \geq 16$ with $e \geq 3$.

9.4.8 Theorem Let $M(x, y)$ be a κ-polynomial over $\mathbb{F}_q$ satisfying $M(x, y) = 0$ if and only if $xy = 0$. Define the sets $\mathcal{L}$ and $\mathcal{R}$ as follows:

$$\mathcal{L} = \{M(x, a) \in \mathbb{F}_q[x] \, : \, a \in \mathbb{F}_q^*\}; \quad \mathcal{R} = \{M(a, x) \in \mathbb{F}_q[x] \, : \, a \in \mathbb{F}_q^*\}.$$

Consider the algebraic object $\mathcal{S} = \langle \mathbb{F}_q, +, \star \rangle$ where $x \star y = M(x, y)$ for all $x, y \in \mathbb{F}_q$.

1. If $\mathcal{L} \cup \mathcal{R}$ contains only permutation polynomials over $\mathbb{F}_q$, then $\mathcal{S}$ is a pre-Cartesian group.
2. If in addition to Part 1, precisely one of $\mathcal{L}$ or $\mathcal{R}$ contains only linearized polynomials, then $\mathcal{S}$ is a prequasifield.
3. If in addition to Part 1, $\mathcal{L} \cup \mathcal{R}$ contains only linearized polynomials, then $\mathcal{S}$ is a presemifield.

9.4.9 Remark Theorem 9.4.8 shows how κ-polynomials underpin all of these algebraic structures. Indeed, it follows from Theorem 9.4.7 Part 1 that there is a one-to-one correspondence between either quasifields or semifields and κ-polynomials satisfying the corresponding part of Theorem 9.4.8. Hence, when we talk of a quasifield or semifield $\mathcal{S} = \langle \mathbb{F}_q, +, \star \rangle$, we may refer to the κ-polynomial corresponding to $\mathcal{S}$, by which we mean the reduced bivariate polynomial $M \in \mathbb{F}_q[x, y]$ satisfying $M(x, y) = x \star y$ for all $x, y \in \mathbb{F}_q$. It is this point that underlines the philosophy behind the presentation of material in this section, which is essentially to provide a polynomial-based or, if one prefers, algebraic approach to this area, thematically linking the material with the chapter in which it resides. That said, it needs to be underlined that the literature on quasifields and especially semifields is vast, and that κ-polynomials represent only one approach to this subject. An excellent and detailed survey of what might be termed the geometric approach to semifields was recently given by Lavrauw and Polverino [1871]; there one will find many more results, along with discussions on several topics reluctantly omitted here, such as the Knuth operations.

9.4.10 Remark There are many examples of quasifield planes – for example, the classification of translation planes of order 49 by Mathon and Royle [2025] reveals there are 1347 non-isomorphic such planes. However, there does not appear to have been any attempt to study the corresponding κ-polynomials, nor the non-linearized permutation polynomials in $\mathcal{L}$ or $\mathcal{R}$ that arise from quasifields (through the known examples or by attempting to prove restrictions on the permissible sets of permutation polynomials which can represent the non-distributive side of a quasifield).

9.4.11 Remark Unsurprisingly, given the significant amount of extra structure, much more is known about semifields than quasifields or Cartesian groups. Consequently, the remainder of this section focuses on the semifield case.

9.4.2 Pre-semifields, semifields, and isotopy

9.4.12 Definition Let $\mathcal{S}_1 = \langle \mathbb{F}_q, +, \star \rangle$ and $\mathcal{S}_2 = \langle \mathbb{F}_q, +, * \rangle$ be two presemifields. Then $\mathcal{S}_1$ and $\mathcal{S}_2$ are *isotopic* if there exists three non-singular linear transformations $L, M, N \in \mathbb{F}_q[x]$ of $\mathbb{F}_q$ over $\mathbb{F}_p$ such that

$$\text{for all } x, y \in \mathbb{F}_q \ : \ M(x) \star N(y) = L(x * y).$$

The triple (M, N, L) is an *isotopism* between $\mathcal{S}_1$ and $\mathcal{S}_2$. An isotopism of the form (N, N, L) is a *strong isotopism* and two presemifields are *strongly isotopic* if there exists a strong isotopism between them.

9.4.13 Remark This definition of equivalence, which is clearly much weaker than the standard ring isomorphism, arises from projective geometry, where its importance is underlined by the following result of Albert.

9.4.14 Theorem [71] Two presemifields coordinatize isomorphic planes if and only if they are isotopic.

9.4.15 Remark Any presemifield $\mathcal{S} = \langle \mathbb{F}_q, +, \circ \rangle$ is isotopic to a semifield via the following transformation: choose any $\alpha \in \mathbb{F}_q^*$ and define a new multiplication $\star$ by

$$(x \circ \alpha) \star (\alpha \circ y) = x \circ y$$

for all $x, y \in \mathbb{F}_q$. Then $\mathcal{S}' = \langle \mathbb{F}_q, +, \star \rangle$ is a semifield with identity $\alpha \circ \alpha$, isotopic to $\mathcal{S}$. Any such $\mathcal{S}'$ ia a semifield corresponding to $\mathcal{S}$. If $\mathcal{S}$ is commutative, then $\mathcal{S}'$ is strongly isotopic to $\mathcal{S}$.

9.4.16 Remark There is no general classification result for semifields, in direct contrast to finite fields, which were classified in the 1890s. However, some computational results on the classification problem do exist. In light of Menichetti's results – see Theorem 9.4.24 below – the problem remains open for semifields of order p^e with $e \geq 4$. Semifields of order N have been classified up to isotopism for the following N: $N = 16$ [1751], $N = 32$ [2891], $N = 64$ [2492], $N = 81$ [810], and $N = 243$ [2491].

9.4.3 Semifield constructions

9.4.17 Proposition (Dickson's commutative semifields) Let $q = p^e$ with p an odd prime, $e > 1$ and let $\{1, \lambda\}$ be basis for $\mathbb{F}_{q^2}$ over $\mathbb{F}_q$. For any j a non-square of $\mathbb{F}_q$ and any non-trivial automorphism σ of $\mathbb{F}_q$, we define a binary operation $\star$ by for all $a, b, c, d \in \mathbb{F}_q$, $(a + \lambda b) \star (c + \lambda d) = ac + j(bd)^\sigma + \lambda(ad + bc)$.

1. [847] $\mathcal{D}_{j,\sigma} = \langle \mathbb{F}_{q^2}, +, \star \rangle$ is a commutative semifield.
2. [845] For j, k distinct non-squares of $\mathbb{F}_q$, $\mathcal{D}_{j,\sigma}$ and $\mathcal{D}_{k,\sigma}$ are isotopic.
3. [466, 2521] $\mathcal{D}_{j,\sigma}$ and $\mathcal{D}_{j,\tau}$ are isotopic if and only if $\sigma = \tau$ or $\sigma = \tau^{-1}$. In such cases, they are strongly isotopic.

9.4.18 Proposition (Albert's generalized twisted fields) [72, 73] For any prime power q, select $\sigma, \tau \in \mathrm{Aut}(\mathbb{F}_q)$ and let $j \in \mathbb{F}_q$ be any element satisfying $(xy)^{-1} x^\sigma y^\tau = j$ has no solution for $x, y \in \mathbb{F}_q^*$. Define a binary operation $\star$ by

$$\text{for all } x, y \in \mathbb{F}_q,\ x \star y = xy - j x^\sigma y^\tau.$$

1. $\mathcal{A}(\sigma, \tau, j) = \langle \mathbb{F}_q, +, \star \rangle$ is a presemifield.
2. $\mathcal{A}(\sigma, \tau, j)$ is isotopic to a commutative semifield if and only if $\sigma = \tau^{-1}$ and $j = -1$.
3. $\mathcal{A}(\sigma, \tau, j)$ is isomorphic to a finite field if and only if $\sigma = \tau$.

9.4.19 Remark The Dickson semifields of Proposition 9.4.17 were the first published examples of non-associative finite division rings, while Albert's construction is both historically important and fundamental, as it has played a significant role in subsequent classification results, see Theorem 9.4.24 below. These two constructions are the only ones needed for our discussion. To highlight the variety of construction techniques developed, even for isotopic semifields, we direct the interested reader to the following not exhaustable list:

1. the paper [1764] of Knuth which contains several constructions;
2. the Cohen-Ganley and Ganley commutative semifields [689, 1170];
3. the Jha-Johnson semifields [1608, 1680];
4. the Hiramine-Matsumoto-Oyama quasifield construction [1508, 1679];
5. the Kantor-Williams semifields [1682];
6. the semifield construction using spread sets viewed as linear maps [518, 952].

9.4.4 Semifields and nuclei

9.4.20 Definition Let $\mathcal{S} = \langle \mathbb{F}_q, +, \star \rangle$ be a finite semifield. We define the *left, middle, and right nucleus* of $\mathcal{S}$, denoted by $\mathcal{N}_l, \mathcal{N}_m$ and $\mathcal{N}_r$, respectively, as follows:

$$\mathcal{N}_l(\mathcal{S}) = \{\alpha \in \mathcal{S} \mid (\alpha \star x) \star y = \alpha \star (x \star y) \text{ for all } x, y \in \mathcal{S}\},$$
$$\mathcal{N}_m(\mathcal{S}) = \{\alpha \in \mathcal{S} \mid (x \star \alpha) \star y = x \star (\alpha \star y) \text{ for all } x, y \in \mathcal{S}\},$$
$$\mathcal{N}_r(\mathcal{S}) = \{\alpha \in \mathcal{S} \mid (x \star y) \star \alpha = x \star (y \star \alpha) \text{ for all } x, y \in \mathcal{S}\}.$$

The set $\mathcal{N}(\mathcal{S}) = \mathcal{N}_l \cap \mathcal{N}_m \cap \mathcal{N}_r$ is the *nucleus* of $\mathcal{S}$.

9.4.21 Remark It is easy to show that all nuclei are finite fields. The nuclei measure how far $\mathcal{S}$ is from being associative. Moreover, the orders of the nuclei are invariants of $\mathcal{S}$ under isotopy and so act as a coarse signature of the semifield. We note that if $\mathcal{S}$ is commutative, then $\mathcal{N}_l = \mathcal{N}_r \subset \mathcal{N}_m$.

9.4.22 Theorem (Restricting the corresponding κ-polynomial) [729] Let n and e be natural numbers. Set $q = p^e$ for some odd prime p and $t(x) = x^q - x$. Let $\mathcal{S}$ be a semifield of order q^n with middle nucleus containing $\mathbb{F}_q$. Then there exists a semifield $\mathcal{S}' = \langle \mathbb{F}_{q^n}, +, \star \rangle$, isotopic to

$\mathcal{S}$, with corresponding κ-polynomial $M \in \mathbb{F}_{q^n}[x,y]$ satisfying $M(x,y) = K(t(x),t(y)) + \frac{1}{2}xy$ and where $K(x,y)$ is of the shape

$$K(x,y) = \sum_{i,j=0}^{(n-1)e-1} a_{ij} x^{p^i} y^{p^j}.$$

Furthermore, if $\mathcal{S}'$ is commutative and $\mathcal{N} = \mathbb{F}_{p^k}$, then K is symmetric in x and y and the only non-zero terms of K are of the form $x^{p^{ki}} y^{p^{kj}}$.

9.4.23 Remark Any semifield $\mathcal{S}$ can be represented as a right vector space over $\mathcal{N}_l$, a left vector space over $\mathcal{N}_r$ and both a left or right vector space over $\mathcal{N}_m$. The concept of dimension therefore naturally arises in the study of semifields.

9.4.24 Theorem (Menichetti's classification results based on dimension)

1. [2081] Any three dimensional semifield over $\mathcal{N}$ is necessarily either a finite field or a twisted field.
2. [2082] Fix d to be a prime. For sufficiently large q, any semifield of dimension d over $\mathcal{N} = \mathbb{F}_q$ is necessarily either a finite field or a twisted field.

9.4.25 Theorem (Dimension two results for commutative semifields) Let $\mathcal{S}$ be a commutative semifield of order q^{2n} with $[\mathcal{S} : \mathcal{N}_m] = 2$.

1. [689] If q is even, then $\mathcal{S}$ is a finite field.
2. [327] If q is odd and $q \geq 4n^2 - 8n + 2$, then $\mathcal{S}$ is necessarily a finite field or a Dickson semifield.

9.4.26 Theorem (Strong isotopy for commutative semifields) [728] Let $\mathcal{S}_1 = \langle \mathbb{F}_q, +, \star \rangle$ and $\mathcal{S}_2 = \langle \mathbb{F}_q, +, * \rangle$ be isotopic commutative presemifields and let $\mathcal{S}_1'$ be any commutative semifield corresponding to $\mathcal{S}_1$. Set $d = [\mathcal{N}_m(\mathcal{S}_1') : \mathcal{N}(\mathcal{S}_1')]$.

1. If d is odd, then $\mathcal{S}_1$ and $\mathcal{S}_2$ are strongly isotopic.
2. If d is even, then either $\mathcal{S}_1$ and $\mathcal{S}_2$ are strongly isotopic or the only isotopisms between any two corresponding commutative semifields $\mathcal{S}_1'$ and $\mathcal{S}_2'$ are of the form $(\alpha \star N, N, L)$ where α is a non-square element of $\mathcal{N}_m(\mathcal{S}_1')$.

In particular, if $q = p^e$ with $p = 2$ or e odd, then $\mathcal{S}_1$ and $\mathcal{S}_2$ are strongly isotopic.

See Also

[138]	For a recent attempt to resolve the problem of establishing inequivalence between projective planes. Whether dealing with Cartesian groups, quasifields or semifields, present methods for establishing the inequivalence of examples are technical and generally unwieldy. Of major interest would be a new and efficient method for doing so.
[787]	Determines many ovals in semifields, as well as in the planes generated by the planar functions of Proposition 9.5.11 Part 2 of the next section.
[1560]	Gives a good discussion on the coordinatization method for projective planes.
[1678]	Gives a conjecture concerning the asymptotic number of pairwise non-isotopic semifields of fixed characteristic. The conjecture is proved for characteristic two by Kantor [1677]; see also [1679, 1682].
[1871]	For a recent survey emphasizing the geometric approach to semifields by two of the leading authors in that field.
[2969]	Conjectures the existence of left or right primitive elements (suitably defined) in finite semifields. This was proved by Gow and Sheekey [1346] for semifields of sufficiently large order relative to the characteristic. See [1487, 2490] for counterexamples of small order in characteristic two.

References Cited: [71, 138, 327, 466, 518, 689, 728, 729, 787, 807, 810, 845, 847, 952, 1170, 1346, 1487, 1508, 1560, 1608, 1677, 1678, 1679, 1680, 1682, 1751, 1764, 1871, 2025, 2034, 2081, 2082, 2490, 2491, 2492, 2521, 2891, 2969]

9.5 Planar functions and commutative semifields

Robert Coulter, The University of Delaware

9.5.1 Definitions and preliminaries

9.5.1 **Definition** Let G and H be arbitrary finite groups, written additively, but not necessarily abelian. A function $f : G \to H$ is a *planar function* if for every non-identity $a \in G$ the functions $\Delta_{f,a} : x \mapsto f(a+x) - f(x)$ and $\nabla_{f,a} : x \mapsto -f(x) + f(x+a)$ are bijections. A polynomial $f \in \mathbb{F}_q[x]$ is *planar* over $\mathbb{F}_q$ if the function induced by f on $\mathbb{F}_q$ is a planar function (on $\langle \mathbb{F}_q, + \rangle$).

9.5.2 **Remark** A few points should be made clear from the outset.

1. Planar functions do not exist over groups of even order.
2. If $f : G \to H$ is planar with G and H abelian, then so is $f + \phi + c$ where ϕ is any homomorphism from G to H and $c \in H$ a constant. In particular, if $f \in \mathbb{F}_q[x]$ is planar, then so is $f + L$ for any linearized polynomial $L \in \mathbb{F}_q[x]$.
3. At the time of writing, all known planar functions can be defined over finite fields; that is, when $G = H = \langle \mathbb{F}_q, + \rangle$ with q odd.

4. The polynomial x^2 is planar over any finite field of odd characteristic.
5. A planar function is a perfect non-linear function in one variable, see Section 9.2.

9.5.3 Theorem [326] Suppose $f : G \to H$ is a planar function with G and H abelian. Then $G \times H$ is a p-group of order p^{2b} for some natural number b and the minimum number of generators of $G \times H$ is at least $b + 1$.

9.5.4 Remark In a letter to the author, Jill C.D.S. Yaqub claimed to have a proof of "about 8 hand-written pages" in length which showed that if a planar function existed mapping G to H, then G and H were necessarily abelian. Along with the above result, this would go a long way to proving that G and H are necessarily elementary abelian. Unfortunately, Yaqub passed away before she received my return correspondence urging her to publish. In her memory, we record this as a conjecture.

9.5.5 Conjecture (Yaqub's Conjecture) If $f : G \to H$ is a planar function, then G and H are abelian.

9.5.2 Constructing affine planes using planar functions

9.5.6 Remark Given groups G and H as in Definition 9.5.1 and a function $f : G \to H$, an incidence structure $I(G, H; f)$ may be defined as follows: "Points" are the elements of $G \times H$; "Lines" are the symbols $\mathcal{L}(a, b)$ with $(a, b) \in G \times H$, together with the symbols $\mathcal{L}(c)$ with $c \in G$; incidence is defined by

$$(x, y) \in \mathcal{L}(a, b) \text{ if and only if } y = f(x - a) + b; \text{ and}$$
$$(x, y) \in \mathcal{L}(c) \text{ if and only if } x = c.$$

9.5.7 Theorem [808] Let G and H be groups and $f : G \to H$ be a function.

1. The function f is planar if and only if $I(G, H; f)$ is an affine plane.
2. Suppose $G \times H$ is abelian and f is planar. Then $I(G, H; f)$ allows a collineation group, isomorphic to $G \times H$, which acts transitively on the affine points as well as on the set of lines $\{\mathcal{L}(a, b) : (a, b) \in G \times H\}$.

9.5.8 Remark Theorem 9.5.7 constitutes Dembowski and Ostrom's original motivation for studying planar functions. In the case where f is planar over $\mathbb{F}_q$, we use $I(f)$ to denote the corresponding affine plane and $P(f)$ its projective closure.

9.5.9 Lemma [2393] If $f \in \mathbb{F}_q[x]$ is a planar polynomial, then $P(f)$ can only be of Lenz-Barlotti type II.1, V.1, or VII.2.

9.5.10 Lemma [811] Let $\Gamma = \Gamma L(1, q)$ denote the group of all semilinear automorphisms of $\mathbb{F}_q$, viewed as a vector space over its prime field, and set $G = \langle \mathbb{F}_q, + \rangle$. If x^n is planar over $\mathbb{F}_q$ and $P(x^n)$ is Lenz-Barlotti type II.1, then $\mathrm{Aut}(P(x^n)) \approx \Gamma \cdot (G \times G)$.

9.5.3 Examples, constructions, and equivalence

9.5.11 Proposition

1. [732] $x^{p^\alpha+1}$ is planar over $\mathbb{F}_{p^e}$ if and only if $e/\gcd(\alpha, e)$ is odd.
2. [732] $x^{(3^\alpha+1)/2}$ is planar over $\mathbb{F}_{3^e}$ if and only if $\gcd(\alpha, 2e) = 1$.
3. [732, 874] For any $a \in \mathbb{F}_{3^e}$, $x^{10} + ax^6 - a^2x^2$ is planar over $\mathbb{F}_{3^e}$ if and only if e is odd or $e = 2$.

9.5.12 Remark Many more examples can be generated from commutative semifields of odd order, see Section 9.5.5. The examples of Proposition 9.5.11 Part 2 do not generalize to larger characteristic – apply Proposition 9.5.24 Part 2. When e is odd, these examples generate the only Lenz-Barlotti type II planes of non-square order known [732]; see [263, 787, 1758] for more concerning these planes.

9.5.13 Definition Let $f, g \in \mathbb{F}_q[x]$ be two planar polynomials over $\mathbb{F}_q$. Set f_1 (respectively g_1) to be the polynomial which results when f (respectively g) is stripped of all linearized and constant terms. Then f and g are *planar equivalent* if there exist two linearized permutation polynomials $L, M \in \mathbb{F}_q[x]$ satisfying $L(f_1(x)) \equiv g_1(M(x)) \bmod (x^q - x)$.

9.5.14 Theorem [732] Let $f \in \mathbb{F}_q[x]$ and let $L \in \mathbb{F}_q[x]$ be a linearized polynomial. Then the following are equivalent.

1. $f(L)$ is a planar polynomial.
2. $L(f)$ is a planar polynomial.
3. f is a planar polynomial and L is a permutation polynomial.

9.5.15 Theorem (Isomorphic planes and planar equivalence)

1. [732] Let $f, g \in \mathbb{F}_q[x]$ be planar polynomials. If f and g are planar equivalent, then $I(f)$ and $I(g)$ are isomorphic.
2. [811] Let x^m and x^n be planar over $\mathbb{F}_{p^e}$. Then $I(x^m)$ and $I(x^n)$ are isomorphic if and only if $m \equiv np^i \bmod (p^e - 1)$ for some suitable choice of integer i.

9.5.16 Remark The converse of Theorem 9.5.15 Part 1 is false as both situations of Theorem 9.4.26 Part 2 do actually occur; see Theorem 9.5.25 below.

9.5.4 Classification results, necessary conditions, and the Dembowski-Ostrom Conjecture

9.5.17 Definition A *Dembowski-Ostrom (or DO)* polynomial over a field of characteristic p is any polynomial of the shape

$$\sum_{i,j} a_{ij} x^{p^i + p^j}.$$

9.5.18 Theorem [732] Let $f \in \mathbb{F}_q[x]$ with $\deg(f) < q$. The following are equivalent.

1. $f = D + L + c$, where D is a Dembowski-Ostrom polynomial, L is a linearized polynomial and $c \in \mathbb{F}_q$ is a constant.
2. For each $a \in \mathbb{F}_q^*$, $\Delta_{f,a} = L_a + c_a$ where L_a is a linearized polynomial and $c_a \in \mathbb{F}_q$ is a constant (both depending on a).

9.5.19 Conjecture (The Dembowski-Ostrom Conjecture) Let q be a power of a prime $p \geq 5$. If $f \in \mathbb{F}_q[x]$ is a planar polynomial, then $f(x) = D(x) + L(x) + c$, where D is a DO polynomial, L is a linearized polynomial and c is a constant.

9.5.20 Remark Though it was only formally made a conjecture by Rónyai and Szönyi [2481], the conjecture stems from a question posed by Dembowski and Ostrom in [808], hence the attribution. Originally stated for all odd characteristics, Proposition 9.5.11 Part 2 shows the conjecture is false in characteristic 3. However, the class of all polynomials planar equivalent to those examples remain the only known counterexamples in any characteristic and it is eminently possible that no other non-DO examples exist.

9.5.21 Theorem (Classification results)

1. [1286, 1506, 2481] The polynomial $f \in \mathbb{F}_p[x]$ is planar over $\mathbb{F}_p$ if and only if the reduced form of f is a quadratic.
2. [726] The polynomial x^n is planar over $\mathbb{F}_{p^2}$ if and only if $n \equiv 2p^i \bmod (p^2-1)$ for some integer $i \in \{0,1\}$.
3. [731] For prime $p \geq 5$, the polynomial x^n is planar over $\mathbb{F}_{p^4}$ if and only if $n \equiv 2p^i \bmod (p^4-1)$ for some integer $i \in \{0,1,2,3\}$.

9.5.22 Remark These results represent the only general classification results on planar polynomials so far obtained.

9.5.23 Proposition (General necessary conditions) [734, 871, 1818] Let $f \in \mathbb{F}_q[x]$ and let $V(f) = \{f(a) : a \in \mathbb{F}_q\}$. A necessary condition for f to be planar over $\mathbb{F}_q$ is $\#V(f) \geq (q+1)/2$. If f is a DO polynomial, then this condition is also sufficient.

9.5.24 Proposition (Necessary conditions for monomials) Let x^n be planar over $\mathbb{F}_{p^e}$. The following statements hold.

1. [808] $\gcd(n, p^e-1) = 2$.
2. [1611] $n \equiv 2 \bmod (p-1)$.
3. [726] If $2|e$, then $n \equiv 2p^i \bmod (p^2-1)$.
4. [731] If $4|e$, then $n \equiv 2p^i \bmod (p^4-1)$.

9.5.5 Planar DO polynomials and commutative semifields of odd order

9.5.25 Theorem (Correspondence) If $\mathcal{S} = \langle \mathbb{F}_q, +, \star \rangle$ is a commutative presemifield of odd order, then $f(x) = \frac{1}{2}(x \star x)$ describes a planar DO polynomial. Conversely, let $f \in \mathbb{F}_q[x]$ be a planar DO polynomial and define the κ-polynomial $M \in \mathbb{F}_q[x,y]$ by $M(x,y) = f(x+y) - f(x) - f(y)$. Then $\mathcal{S}_f = \langle \mathbb{F}_q, +, \star \rangle$, where $x \star y = M(x,y)$ for all $x, y \in \mathbb{F}_q$, defines a commutative presemifield, the presemifield *corresponding to* f.

9.5.26 Remark More information on κ-polynomials can be found in Section 9.4.

9.5.27 Remark By Theorem 9.5.25, classifying commutative semifields of odd order is equivalent to determining the "isotopism classes" of planar DO polynomials. Moreover, Theorem 9.4.26 shows that any strong isotopism class of commutative semifields can split into at most two isotopism classes, so that there are sound reasons for considering the strong isotopism problem on planar DO polynomials instead of the more difficult isotopism problem.

9.5.28 Theorem (Strong isotopy and planar equivalence) [728] Let $f, g \in \mathbb{F}_q[x]$ be planar DO polynomials with corresponding commutative presemifields $\mathcal{S}_f$ and $\mathcal{S}_g$. There is a strong isotopism (N, N, L) between $\mathcal{S}_f$ and $\mathcal{S}_g$ if and only if $f(N(x)) \equiv L(g(x)) \bmod (x^q - x)$.

See Also

§9.2	For APN functions that are closely related to planar functions.
§9.3	For bent functions that are closely related to planar functions.
§14.3	For affine and projective planes; the seminal paper [808] clearly outlines the main properties of the planes constructed via planar functions.
§14.6	Discusses difference sets. Ding and Yuan [874] used the examples of Proposition 9.5.11 Part 3 to disprove a long-standing conjecture on skew Hadamard difference sets; see also [871, 2972].

[273]	Construct further classes of planar DO polynomials; see also [274], [450], [451], [2383], [3059], [3060]. The problem of planar (in)equivalence between these constructions is not completely resolved at the time of writing. An incredible new class, which combines Albert's twisted fields with Dickson's semifields, was very recently discovered by Pott and Zhou [2425].
[728]	Classifies planar DO polynomials over fields of order p^2 and p^3. This does not constitute a classification of planar polynomials over fields of these orders.
[729]	Applies Theorem 9.4.22 to commutative presemifields of odd order to restrict both the form of the DO polynomials and the isotopisms that need to be considered; see also [1799]. A promising alternative approach (which applies also to APN functions, see Section 9.2) is outlined in [274], while a third approach was given recently in [2973].
[1507]	For results on possible forms of planar functions not defined over finite fields.
[1799]	Gives specific forms for planar DO polynomials corresponding to the Dickson semifields [847], the Cohen-Ganley semifields [689], the Ganley semifields [1170], and the Penttila-Williams semifield [2383].

References Cited: [263, 273, 274, 326, 450, 451, 689, 726, 728, 729, 731, 732, 734, 787, 808, 811, 847, 871, 874, 1170, 1286, 1506, 1507, 1611, 1758, 1799, 1818, 2383, 2393, 2425, 2481, 2972, 2973, 3059, 3060]

9.6 Dickson polynomials

Qiang Wang, Carleton University
Joseph L. Yucas, Southern Illinois University

9.6.1 Basics

9.6.1 **Definition** Let n be a positive integer. For $a \in \mathbb{F}_q$, we define the *n-th Dickson polynomial of the first kind* $D_n(x,a)$ over $\mathbb{F}_q$ by

$$D_n(x,a) = \sum_{i=0}^{\lfloor n/2 \rfloor} \frac{n}{n-i} \binom{n-i}{i} (-a)^i x^{n-2i}.$$

9.6.2 Theorem (Waring's formula, [1939, Theorem 1.76]) Let $\sigma_1, \ldots, \sigma_k$ be elementary symmetric polynomials in the variables $x_1, \ldots, x_k$ over a ring R and $s_n = s_n(x_1, \ldots, x_k) = x_1^n + \cdots + x_k^n \in R[x_1, \ldots, x_k]$ for $n \geq 1$. Then we have

$$s_n = \sum (-1)^{i_2+i_4+i_6+\cdots} \frac{(i_1 + i_2 + \cdots + i_k - 1)!n}{i_1!i_2! \cdots i_k!} \sigma_1^{i_1} \sigma_2^{i_2} \cdots \sigma_k^{i_k},$$

for $n \geq 1$, where the summation is extended over all tuples $(i_1, i_2, \ldots, i_n)$ of nonnegative integers with $i_1 + 2i_2 + \cdots + ki_k = n$. The coefficients of the $\sigma_1^{i_1} \sigma_2^{i_2} \cdots \sigma_k^{i_k}$ are integers.

9.6.3 Theorem Dickson polynomials of the first kind are the unique monic polynomials satisfying the functional equation

$$D_n\left(y + \frac{a}{y}, a\right) = y^n + \frac{a^n}{y^n},$$

where $a \in \mathbb{F}_q$ and $y \in \mathbb{F}_{q^2}$. Moreover, they satisfy the recurrence relation

$$D_n(x, a) = xD_{n-1}(x, a) - aD_{n-2}(x, a),$$

for $n \geq 2$ with initial values $D_0(x, a) = 2$ and $D_1(x, a) = x$.

9.6.4 Remark The Dickson polynomial $D_n(x, a)$ of the first kind satisfies a commutative type of relation under composition. That is, $D_{mn}(x, a) = D_m(D_n(x, a), a^n)$. Hence the set of all Dickson polynomials $D_n(x, a)$ of even degree over $\mathbb{F}_q$ are closed under composition if and only if $a = 0$ or $a = 1$. In particular, if $a = 0$ or 1 then $D_m(x, a) \circ D_n(x, a) = D_n(x, a) \circ D_m(x, a)$. Moreover, the set of all Dickson polynomials $D_n(x, a)$ of odd degree over $\mathbb{F}_q$ is closed under composition if and only if $a = 0$, $a = 1$ or $a = -1$. See [1936, 1939] for more details.

9.6.5 Definition For $a \in \mathbb{F}_q$, we define the *n-th Dickson polynomial of the second kind* $E_n(x, a)$ over $\mathbb{F}_q$ by

$$E_n(x, a) = \sum_{i=0}^{\lfloor n/2 \rfloor} \binom{n-i}{i} (-a)^i x^{n-2i}.$$

9.6.6 Theorem Dickson polynomials of the second kind have a functional equation

$$E_n(x, a) = E_n\left(y + \frac{a}{y}, a\right) = \frac{y^{n+1} - a^{n+1}/y^{n+1}}{y - a/y},$$

for $y \neq \pm\sqrt{a}$; for $y = \pm\sqrt{a}$, we have $E_n(2\sqrt{a}, a) = (n + 1)(\sqrt{a})^n$ and $E_n(-2\sqrt{a}, a) = (n + 1)(-\sqrt{a})^n$; here $a \in \mathbb{F}_q$ and $y \in \mathbb{F}_{q^2}$. Moreover, they satisfy the recurrence relation

$$E_n(x, a) = xE_{n-1}(x, a) - aE_{n-2}(x, a),$$

for $n \geq 2$ with initial values $E_0(x, a) = 1$ and $E_1(x, a) = x$.

9.6.7 Remark In the case $a = 1$, denote Dickson polynomials of degree n of the first and the second kinds by $D_n(x)$ and $E_n(x)$, respectively. These Dickson polynomials are closely related over the complex numbers to the Chebychev polynomials through the connections $D_n(2x) = 2T_n(x)$ and $E_n(2x) = U_n(x)$, where $T_n(x)$ and $U_n(x)$ are Chebychev polynomials of degree n of the first and the second kind, respectively. In recent years these polynomials have received an extensive examination. The book [1936] is devoted to a survey of the algebraic and number theoretic properties of Dickson polynomials.

9.6.8 Remark Suppose q is odd and a is a nonsquare in $\mathbb{F}_q$. Then

$$(x+\sqrt{a})^n = r_n(x) + s_n(x)\sqrt{a},$$

where

$$r_n(x) = \sum_{i=0}^{\lfloor n/2 \rfloor} \binom{n}{2i} a^i x^{n-2i},\ s_n(x) = \sum_{i=0}^{\lfloor n/2 \rfloor} \binom{n}{2i+1} a^i x^{n-2i-1}.$$

The *Rédei function* is the rational function $R_n(x) = \frac{r_n(x)}{s_n(x)}$. It is shown in [1116] that $2r_n(x) = D_n(2x, x^2 - a)$.

9.6.9 Remark Permutation properties of Dickson polynomials are important; see Section 8.1. The famous Schur conjecture postulating that every integral polynomial that is a permutation polynomial for infinitely many primes is a composition of linear polynomials and Dickson polynomials was proved by Fried [1109]. We refer readers to Section 9.7.

9.6.2 Factorization

9.6.10 Remark The factorization of the Dickson polynomials of the first kind over $\mathbb{F}_q$ was given [626] and simplified in [266].

9.6.11 Theorem [266, 626] If q is even and $a \in \mathbb{F}_q^*$ then $D_n(x,a)$ is the product of squares of irreducible polynomials over $\mathbb{F}_q$ which occur in cliques corresponding to the divisors d of n, $d > 1$. Let k_d be the least positive integer such that $q^{k_d} \equiv \pm 1 \pmod{d}$. To each such d there corresponds $\phi(d)/(2k_d)$ irreducible factors of degree k_d, each of which has the form

$$\prod_{i=0}^{k_d - 1} (x - \sqrt{a}(\zeta^{q^i} + \zeta^{-q^i}))$$

where ζ is a d-th root of unity and ϕ is Euler's totient function.

9.6.12 Theorem [266, 626] If q is odd and $a \in \mathbb{F}_q^*$ then $D_n(x,a)$ is the product of irreducible polynomials over $\mathbb{F}_q$ which occur in cliques corresponding to the divisors d of n for which n/d is odd. Let m_d is the least positive integer satisfying $q^{m_d} \equiv \pm 1 \pmod{4d}$. To each such d there corresponds $\phi(4d)/(2N_d)$ irreducible factors of degree N_d, each of which has the form

$$\prod_{i=0}^{N_d - 1} (x - \sqrt{a^{q^i}}(\zeta^{q^i} + \zeta^{-q^i}))$$

where ζ is a $4d$-th root of unity and

$$N_d = \begin{cases} m_d/2 & \text{if } \sqrt{a} \notin \mathbb{F}_q, m_d \equiv 2 \pmod{4} \text{ and } q^{m_d/2} \equiv 2d \pm 1 \pmod{4d}, \\ 2m_d & \text{if } \sqrt{a} \notin \mathbb{F}_q \text{ and } m_d \text{ is odd}, \\ m_d & \text{otherwise}. \end{cases}.$$

9.6.13 Example Let $(q,n) = (5,12)$. Then $D_{12}(x,2) = x^{12} + x^{10} + x^8 + 4x^6 + 3x^2 + 3$ is the product of irreducible polynomials over $\mathbb{F}_5$ which occur in cliques corresponding to the divisors $d = 4$ and $d = 12$ of $n = 12$. By direct computation, $m_4 = N_4 = 4$ and $m_{12} = N_{12} = 4$. For $d = 4$, there corresponds one irreducible factor of degree 4, while there are two irreducible factors of degree 4 for $d = 12$, each of which has the form $\prod_{i=0}^{N_d - 1}(x - \sqrt{a^{q^i}}(\zeta^{q^i} + \zeta^{-q^i}))$, where ζ is a $4d$-th root of unity.

9.6.14 Remark Similar results hold for Dickson polynomials of the second kind and they can be found in [266] and [626]. Dickson polynomials of other kinds are defined in [2945] and the factorization of the Dickson polynomial of the third kind is obtained similarly in [2945]. We note that the factors appearing in the above results are over $\mathbb{F}_q$, although their description uses elements in an extension field of $\mathbb{F}_q$. In [1079] Fitzgerald and Yucas showed that these factors can be obtained from the factors of certain cyclotomic polynomials. This in turn gives a relationship between a-self-reciprocal polynomials and these Dickson factors. In the subsequent subsections we explain how this works. These results come mainly from [1079].

9.6.2.1 a-reciprocals of polynomials

9.6.15 Definition Let q be an odd prime power and fix $a \in \mathbb{F}_q^*$. For a monic polynomial f over $\mathbb{F}_q$ of degree n, with $f(0) \neq 0$, define the *a-reciprocal* of f by

$$\hat{f}_a(x) = \frac{x^n}{f(0)} f(a/x).$$

The polynomial f is *a-self-reciprocal* if $f(x) = \hat{f}_a(x)$.

9.6.16 Remark We note that the notion of a 1-self-reciprocal is the usual notion of a self-reciprocal.

9.6.17 Lemma

1. If α is a root of f then a/α is a root of $\hat{f}_a$.
2. The polynomial f is irreducible over $\mathbb{F}_q$ if and only if $\hat{f}_a$ is irreducible over $\mathbb{F}_q$.

9.6.18 Remark The a-reciprocal of an irreducible polynomial f may not have the same order as f. For example, consider $f(x) = x^3 + 3$ when $q = 7$. Then f has order 9 while $\hat{f}_3(x) = x^3 + 2$ has order 18.

9.6.19 Theorem [1079] Suppose f is a polynomial of even degree $n = 2m$ over $\mathbb{F}_q$. The following statements are equivalent:

1. f is a-self-reciprocal;
2. $n = 2m$ and f has the form

$$f(x) = b_m x^m + \sum_{i=0}^{m-1} b_{2m-i}(x^{2m-i} + a^{m-i}x^i)$$

for some $b_j \in \mathbb{F}_q$.

9.6.20 Definition Let n be an even integer. Define

$$D_n = \{r : r \text{ divides } q^n - 1 \text{ but } r \text{ does not divide } q^s - 1 \text{ for } s < n\}.$$

For $r \in D_n$ and even n, we write $r = d_r t_r$ where $d_r = (r, q^m + 1)$ and $m = n/2$.

9.6.21 Theorem [1079] Suppose f is an irreducible polynomial of degree $n = 2m$ over $\mathbb{F}_q$. The following statements are equivalent:

1. f is a-self-reciprocal for some $a \in \mathbb{F}_q^*$ with $\text{ord}(a) = t$,
2. f has order $r \in D_n$ and $t_r = t$.

9.6.22 Theorem [1079] Let $r \in D_n$ and suppose t_r divides $q - 1$. Then the cyclotomic polynomial $Q(r, x)$ factors into all a-self-reciprocal monic irreducible polynomials of degree n and order r where a ranges over all elements of $\mathbb{F}_q$ of order t_r.

9.6.2.2 The maps Φ_a and Ψ_a

9.6.23 Definition Define the mapping $\Phi_a : P_m \to S_n$ from the polynomials over $\mathbb{F}_q$ of degree m to the a-self-reciprocal polynomials over $\mathbb{F}_q$ of degree $n = 2m$ by

$$\Phi_a(f(x)) = x^m f(x + a/x).$$

9.6.24 Remark In the case $a = 1$ this transformation has appeared often in the literature. The first occurrence is Carlitz [547]. Other authors writing about Φ are Chapman [589], Cohen [678], Fitzgerald-Yucas [1079], Kyuregyan [1821], Miller [2100], Meyn [2091], and Scheerhorn [2538].

9.6.25 Definition Define $\Psi_a : S_n \to P_m$ by

$$\Psi_a\left(b_m x^m + \sum_{i=0}^{m-1} b_{2m-i}(x^{2m-i} + a^{m-i}x^i)\right) = b_m + \sum_{i=0}^{m-1} b_{2m-i} D_{m-i}(x, a).$$

9.6.26 Theorem Maps Φ_a and Ψ_a are multiplicative and are inverses of each other.

9.6.2.3 Factors of Dickson polynomials

9.6.27 Theorem [1079] The polynomial $D_n(x, a)$ is mapped to $x^{2n} + a^n$ by the above defined Φ_a, namely, $\Phi_a(D_n(x, a)) = x^{2n} + a^n$.

9.6.28 Theorem [1079] The polynomial $x^{2n} + a^n$ factors over $\mathbb{F}_q$ as

$$x^{2n} + a^n = \prod f(x),$$

where each f is either an irreducible a-self-reciprocal polynomial or a product of an irreducible polynomial and its a-reciprocal over $\mathbb{F}_q$.

9.6.29 Theorem [1079] The polynomial $D_n(x, a)$ factors over $\mathbb{F}_q$ as

$$D_n(x, a) = \prod \Psi_a(f(x)),$$

where $x^{2n} + a^n = \prod f(x)$ such that f is either an irreducible a-self-reciprocal polynomial or a product of an irreducible polynomial and its a-reciprocal.

9.6.30 Theorem The following is an algorithm for factoring $D_n(x, a)$ over $\mathbb{F}_q$.

1. Factor $x^{2n} + a^n$.
2. For each factor f of $x^{2n} + a^n$ which is not a-self-reciprocal, multiply f with $\hat{f}_a$.
3. Apply Ψ_a.

9.6.31 Example We factor $D_{12}(x, 2) = x^{12} + x^{10} + x^8 + 4x^6 + 3x^2 + 3$ when $q = 5$.

$$\begin{aligned} x^{24} + 2^{12} &= [(x^4 + x^2 + 2)(x^4 + 2x^2 + 3)][(x^4 + 3)(x^4 + 2)][(x^4 + 4x^2 + 2)(x^4 + 3x^2 + 3)] \\ &= (x^8 + 3x^6 + 2x^4 + 2x^2 + 1)(x^8 + 1)(x^8 + 2x^6 + 2x^4 + 3x^2 + 1). \end{aligned}$$

Then apply Ψ_2 to obtain

$$\begin{aligned} D_{12}(x,2) &= (D_4(x,2)+3D_2(x,2)+2)D_4(x,2)(D_4(x,2)+2D_2(x,2)+2) \\ &= (x^4+3)(x^4+2x^2+3)(x^4+4x^2+2). \end{aligned}$$

9.6.32 Definition For $a \in \mathbb{F}_q^*$, define $\eta(n,a)$ by

$$\eta(n,a) = \begin{cases} n \cdot \mathrm{ord}(a^n) & \text{if } n \text{ is odd and } a \text{ is a non-square,} \\ 4n \cdot \mathrm{ord}(a^n) & \text{otherwise.} \end{cases}$$

9.6.33 Theorem [1079] For a monic irreducible polynomial f over $\mathbb{F}_q$ and $a \in \mathbb{F}_q^*$, the following statements are equivalent:

1. f divides $D_n(x,a)$.
2. There exists a divisor d of n with n/d odd and $\mathrm{ord}(\Phi_a(f)) = \eta(d,a)$, where Φ_a is defined in Definition 9.6.23.

9.6.2.4 *a*-cyclotomic polynomials

9.6.34 Definition For $a \in \mathbb{F}_q^*$, define the *a-cyclotomic polynomial* $Q_a(n,x)$ over $\mathbb{F}_q$ by

$$Q_a(n,x) = \prod_{\substack{d|n \\ d \text{ even}}} (x^d - a^{d/2})^{\mu(n/d)}.$$

9.6.35 Remark When $n \equiv 0 \pmod 4$, we have $Q_1(n,x) = Q(n,x)$, the n-th cyclotomic polynomial over $\mathbb{F}_q$. When $n \equiv 2 \pmod 4$, we have $Q_1(n,x) = Q(n/2,-x^2)$. Similar to the factorization of $x^n - 1 = \prod_{d|n} Q(d,x)$ [1939], we can reduce the factorization of $x^{2n} \pm a^n$ to the factorization of a-cyclotomic polynomials.

9.6.36 Theorem [1079] We have

1. $x^{2n} - a^n = \prod_{d|n} Q_a(2d,x)$;
2. $x^{2n} + a^n = \prod_{\substack{d|n \\ d \text{ odd}}} Q_a(4d,x)$.

9.6.37 Remark A factorization of these a-cyclotomic polynomials $Q_a(m,x)$ is also given in [1079].

9.6.3 Dickson polynomials of the $(k+1)$-th kind

9.6.38 Definition [2945] For $a \in \mathbb{F}_q$, any integers $n \geq 0$ and $0 \leq k < p$, we define the *n-th Dickson polynomial of the $(k+1)$-th kind* $D_{n,k}(x,a)$ over $\mathbb{F}_q$ by $D_{0,k}(x,a) = 2-k$ and

$$D_{n,k}(x,a) = \sum_{i=0}^{\lfloor n/2 \rfloor} \frac{n-ki}{n-i} \binom{n-i}{i} (-a)^i x^{n-2i}.$$

9.6.39 Definition [2945] For $a \in \mathbb{F}_q$, any integers $n \geq 0$ and $0 \leq k < p$, we define the *n-th reversed Dickson polynomial of the $(k+1)s$-th kind* $D_{n,k}(a,x)$ over $\mathbb{F}_q$ by $D_{0,k}(a,x) = 2-k$ and

$$D_{n,k}(a,x) = \sum_{i=0}^{\lfloor n/2 \rfloor} \frac{n-ki}{n-i} \binom{n-i}{i} (-1)^i a^{n-2i} x^i.$$

9.6.40 Remark [2945] It is easy to see that $D_{n,0}(x,a) = D_n(x,a)$ and $D_{n,1}(x,a) = E_n(x,a)$. Moreover, if $\mathrm{char}(\mathbb{F}_q) = 2$, then $D_{n,k}(x,a) = D_n(x,a)$ if k is even and $D_{n,k}(x,a) = E_n(x,a)$ if k is odd.

9.6.41 Theorem [2945] For any integer $k \geq 1$, we have

$$D_{n,k}(x,a) = kD_{n,1}(x,a) - (k-1)D_{n,0}(x,a) = kE_n(x,a) - (k-1)D_n(x,a).$$

9.6.42 Theorem [2945] The fundamental functional equation for $D_{n,k}$ is

$$\begin{aligned} D_{n,k}(y+ay^{-1},a) &= \frac{y^{2n} + kay^{2n-2} + \cdots + ka^{n-1}y^2 + a^n}{y^n} \\ &= \frac{y^{2n}+a^n}{y^n} + \frac{ka}{y^n}\frac{y^{2n} - a^{n-1}y^2}{y^2-a}, \quad \text{for } y \neq 0, \pm\sqrt{a}. \end{aligned}$$

9.6.43 Theorem [2945] The Dickson polynomial of the $(k+1)$-th kind satisfies the following recurrence relation

$$D_{n,k}(x,a) = xD_{n-1,k}(x,a) - aD_{n-2,k}(x,a),$$

for $n \geq 2$ with initial values $D_{0,k}(x,a) = 2-k$ and $D_{1,k}(x,a) = x$.

9.6.44 Theorem [2945] The generating function of $D_{n,k}(x,a)$ is

$$\sum_{n=0}^{\infty} D_{n,k}(x,a)z^n = \frac{2-k+(k-1)xz}{1-xz+az^2}.$$

9.6.45 Remark The Dickson polynomial $D_{n,k}(x,a)$ of the $(k+1)$-th kind satisfies a second order differential equation; see [2723, 2945] for more details.

9.6.46 Theorem [2945] Suppose ab is a square in $\mathbb{F}_q^*$. Then $D_{n,k}(x,a)$ is a PP of $\mathbb{F}_q$ if and only if $D_{n,k}(x,b)$ is a PP of $\mathbb{F}_q$. Furthermore,

$$D_{n,k}(\alpha,a) = (\sqrt{a/b})^n D_{n,k}((\sqrt{b/a})\alpha, b).$$

9.6.47 Definition Define S_{q-1}, S_{q+1}, and S_p by

$$S_{q-1} = \{\alpha \in \mathbb{F}_q : u_\alpha^{q-1} = 1\}, \quad S_{q+1} = \{\alpha \in \mathbb{F}_q : u_\alpha^{q+1} = 1\}, \quad S_p = \{\pm 2\},$$

where $u_\alpha \in \mathbb{F}_{q^2}$ satisfies $\alpha = u_\alpha + \frac{1}{u_\alpha} \in \mathbb{F}_q$.

9.6.48 Theorem [2945] As functions on $\mathbb{F}_q$, we have

$$D_{n,k}(\alpha) = \begin{cases} D_{(n)_{2p},k}(\alpha) & \text{if } \alpha \in S_p, \\ D_{(n)_{q-1},k}(\alpha) & \text{if } \alpha \in S_{q-1}, \\ D_{(n)_{q+1},k}(\alpha) & \text{if } \alpha \in S_{q+1}, \end{cases}$$

where for positive integers n and r we use the notation $(n)_r$ to denote $n \pmod r$, the smallest positive integer congruent to n modulo r.

9.6.49 Theorem [2945] Let $\alpha = u_\alpha + \frac{1}{u_\alpha}$ where $u_\alpha \in \mathbb{F}_{q^2}$ and $\alpha \in \mathbb{F}_q$. Let $\epsilon_\alpha = u_\alpha^c \in \{\pm 1\}$ where $c = \frac{p(q^2-1)}{4}$. As functions on $\mathbb{F}_q$ we have

$$D_{c+n,k}(\alpha) = \epsilon_\alpha D_{n,k}(\alpha).$$

Moreover, $D_{n,k}(x)$ is a PP of $\mathbb{F}_q$ if and only if $D_{c+n,k}(x)$ is a PP of $\mathbb{F}_q$.

9.6.50 Theorem [2945] For $k \neq 1$, let $k' = \frac{k}{k-1} \pmod p$ and $\epsilon_\alpha = u_\alpha^c \in \{\pm 1\}$ where $c = \frac{p(q^2-1)}{4}$. For $n < c$, as functions on $\mathbb{F}_q$ we have

$$D_{c-n,k'}(\alpha) = \frac{-\epsilon_\alpha}{k-1} D_{n,k}(\alpha).$$

Moreover, $D_{n,k}(x)$ is a PP of $\mathbb{F}_q$ if and only if $D_{c-n,k'}(x)$ is a PP of $\mathbb{F}_q$.

9.6.4 Multivariate Dickson polynomials

9.6.51 Definition [1936] The Dickson polynomial of the first kind $D_n^{(i)}(x_1, \ldots, x_t, a)$, $1 \le i \le t$, is given by the functional equations

$$D_n^{(i)}(x_1, \ldots, x_t, a) = s_i(u_1^n, \ldots, u_{t+1}^n),\ 1 \le i \le t,$$

where $x_i = s_i(u_1, \ldots, u_{t+1})$ are elementary symmetric functions and $u_1 \cdots u_{t+1} = a$. The vector $D(t,n,a) = (D_n^{(1)}, \ldots, D_n^{(t)})$ of the t Dickson polynomials is a *Dickson polynomial vector.*

9.6.52 Remark Let $r(c_1, \ldots, c_t, z) = z^{t+1} - c_1 z^t + c_2 z^{t-1} + \cdots + (-1)^t c_t z + (-1)^{t+1} a$ be a polynomial over $\mathbb{F}_q$ and $\beta_1, \ldots, \beta_{t+1}$ be the roots (not necessarily distinct) in a suitable extension of $\mathbb{F}_q$. For any positive integer n, we let $r_n(c_1, \ldots, c_t, z) = (z - \beta_1^n) \ldots (z - \beta_{t+1}^n)$. Then $r_n(c_1, \ldots, c_t, z) = z^{t+1} - D_n^{(1)}(c_1, \ldots, c_t, a) z^t + D_n^{(2)}(c_1, \ldots, c_t, a) z^{t-1} + \cdots + (-1)^t D_n^{(t)}(c_1, \ldots, c_t, a) z + (-1)^{t+1} a^t$.

9.6.53 Remark For the Dickson polynomial $D_n^{(1)}(x_1, \ldots, x_t, a)$, an explicit expression, a generating function, a recurrence relation, and a differential equation satisfied by $D_n^{(1)}(x_1, \ldots, x_t, a)$ can be found in [1936]. Here we only give the generating function and recurrence relation.

9.6.54 Theorem The Dickson polynomial of the first kind $D_n^{(1)}(x_1, \ldots, x_t, a)$ satisfies the generating function

$$\sum_{n=0}^{\infty} D_n^{(1)}(x_1, \ldots, x_t, a) z^n = \frac{\sum_{i=0}^{t} (t+1-i)(-1)^i x_i z^i}{\sum_{i=0}^{t+1} (-1)^i x_i z^i}, \text{ for } n \ge 0,$$

and the recurrence relation

$$D_{n+t+1}^{(1)} - x_1 D_{n+t}^{(1)} + \cdots + (-1)^t x_k D_{n+1}^{(1)} + (-1)^{t+1} a D_n^{(1)} = 0,$$

with the $t+1$ initial values

$$D_0^{(1)} = t+1, D_j^{(1)} = \sum_{r=1}^{j} (-1)^{r-1} x_r D_{j-r}^{(1)} + (-1)^j (t+1-j) x_j, \text{ for } 0 < j \le t.$$

9.6.55 Remark Much less is known for the multivariate Dickson polynomials of the second kind. The same recurrence relation of $D_n^{(1)}(x_1,\ldots,x_t,a)$ is used to define the multivariate Dickson polynomials of the second kind $E_n^{(1)}(x_1,\ldots,x_t,a)$ with the initial condition $E_0 = 1$, $E_j = \sum_{r=1}^{j}(-1)^{r-1}x_rE_{j-r}$ for $1 \le j \le t$. The generating function is $\sum_{n=0}^{\infty} E_nz^n = \frac{1}{\sum_{i=0}^{t+1}(-1)^i x_i z^i}$; see [1936] for more details.

See Also

§8.1	For permutation polynomials with one variable.
§8.3	For value sets of polynomials over fnite fields.
§9.7	For Schur's conjecture and exceptional covers.
[1936]	For a comprehensive book on Dickson polynomials.

References Cited: [266, 547, 589, 626, 678, 1079, 1109, 1116, 1821, 1936, 1939, 2091, 2100, 2538, 2723, 2945]

9.7 Schur's conjecture and exceptional covers

Michael D. Fried, *Irvine campus of the University of California*

9.7.1 Rational function definitions

9.7.1 Remark (Extend values) The historical functions of this section are polynomials and rational functions: $f(x) = N_f(x)/D_f(x)$ with N_f and D_f relatively prime (nonzero) polynomials, denoted $f \in F(x)$, F a field (almost always $\mathbb{F}_q$ or a number field). The subject takes off by including functions f – *covers* – where the domain and range are varieties of the same dimension. Still, we emphasize functions between projective algebraic curves (nonsingular), often where the target and domain are projective 1-space.

9.7.2 Definition The *degree of* $f \in F(x)$, $\deg(f)$, is the maximum of $\deg(N_f)$ and $\deg(D_f)$. Add a point at ∞ to F, $F \cup \{\infty\} = \mathbb{P}_x^1(F)$, to get *the* F *points of projective 1-space.*

9.7.3 Remark (Plug in ∞) Using Definition 9.7.2 requires plugging in and getting out ∞. We sometimes use the notion of value sets V_f and their cardinality $\#V_f$ (Section 8.3).

1. The value of $f(x')$ for $x' \in F$ is ∞ if x' is a zero of $D_f(x)$.
2. The value of $f(\infty)$ is respectively $\infty, 0$, or the ratio of the N_f and D_f leading coefficients, if the degree of N_f is greater, less than, or equal to the degree of D_f.

If z is a variable indicating the range, this gives f as a function from $\mathbb{P}_x^1(F)$ to $\mathbb{P}_z^1(F)$. We abbreviate this as $f : \mathbb{P}_x^1 \to \mathbb{P}_z^1$.

9.7.4 Definition (Möbius equivalence) Denote the group – under composition – of *Möbius transformations* $x \mapsto \frac{ax+b}{cx+d}$ with $ad - bc \neq 0$, $a, b, c, d \in F$ by $\mathrm{PGL}(F)$. Refer to $f_1, f_2 \in F(x)$ as Möbius equivalent if $f_2 = \alpha \circ f_1 \circ \beta$ for $\alpha, \beta \in \mathrm{PGL}(F)$.

9.7.5 Example If $f(x) = x^n$, with $\gcd(n, q-1) = 1$, then $\#V_f = q^k + 1$ on $\mathbb{P}^1_x(\mathbb{F}_{q^k})$ exactly for those infinitely many k with $\gcd(n, q^k - 1) = 1$.

9.7.6 Remark Initial motivation came from *Schur's Conjecture* Theorem 9.7.32, which starts over a *number field* K – a finite extension of $\mathbb{Q}$, the rational numbers – with its ring of integers $\mathcal{O}_K$. That asks about V_f over residue class fields, $\mathcal{O}_K/\boldsymbol{p}$ of prime ideals $\boldsymbol{p}$, denoting this $V_f(\mathcal{O}/\boldsymbol{p})$ ($V_f(\mathbb{F}_p)$ if $\mathcal{O} = \mathbb{Z}$). Assume N_f and D_f have coefficients in $\mathcal{O}_K$. Avoid $\boldsymbol{p}$ – it is a *bad* prime – if it contains the leading coefficient of either N_f or D_f.

9.7.7 Definition For $f \in F(x)$, if $f = f_1 \circ f_2$ with $f_1, f_2 \in F(x)$, $\deg(f_i) > 1$, $i = 1, 2$, we say f *decomposes* over F. Then, the f_is are *composition factors* of f.

9.7.8 Definition (Cofinite) For B a subset of A, we say B is *cofinite* in A if $A \setminus B$ is finite.

9.7.9 Proposition [2203, p. 390] Consider $X'_h = \{(x, y) \mid h(x, y) = 0\}$, an algebraic curve, defined by $h \in K[x, y]$. Then, there is a unique nonsingular curve X_h – the *normalization* of X'_h – and a morphism $\mu_h : X'_h \to X_h$ that is an isomorphism on the complement of a finite subset of points in X_h. Indeed, every variety X'_h has such a unique normalization, but in higher dimensions it may be singular, and μ_h is an isomorphism off a codimension 1 set.

9.7.10 Definition (Components) A *definition field* for an algebraic set W is a field containing all coefficients of all polynomials defining W. *Components* of W over F are algebraic subsets which are not the union of two closed non-empty proper algebraic subsets over F [1427, p. 3]. We say W is a *variety* if it has just one component. It is *absolutely irreducible* if it has just one component over $\bar{F}$, an algebraic closure of F.

9.7.11 Remark (Points on varieties) [1427, Chapters 1 and 2] and [2203, Section 2] introduce affine and projective algebraic sets, and their components (Definition 9.7.10), except they are over an algebraically closed field. For perfect fields F (including finite fields and number fields) this extends for normal varieties. Since their components do not meet, taking any disjoint union of distinct varieties under the action of the absolute Galois group of F defines components in general. Points on an algebraic set X over F refers here to *geometric* points: points with coordinates in $\bar{F}$. It is an F point if its coordinates are in F.

9.7.12 Definition A general $f : X \to Z$ is a *cover* means it is a finite, flat morphism (see Definition 9.7.25) of quasi-projective varieties [2203, p. 432, Proposition 2].

9.7.13 Lemma Definition 9.7.12 simplifies for curves, because all our varieties will be normal, and so for curves, nonsingular. Then, any nonconstant morphism is a cover: That includes any nonconstant rational function $f : \mathbb{P}^1_x \to \mathbb{P}^1_z$.

9.7.14 Example If $f : X \to Z$ is finite and X and Z are nonsingular, generalizing what happens for curves, and no matter their dimension, then f is automatically flat [1427, p. 266, 9.3a)]. This does not extend to weakening nonsingular to normal varieties. [2203, p. 434] has a finite morphism, where X is nonsingular (it is affine 2-space), and Z is normal. But, the fiber degree is 2 over each $z \in Z$, excluding one point where it is 3.

9.7.15 Remark (Assuming normality) Starting with Subsection 9.7.2 all results assume that the algebraic sets are normal. Some constructions (especially Definition 9.7.45) momentarily produce nonnormal sets, that we immediately replace with their normalizations.

9.7.2 MacCluer's Theorem and Schur's Conjecture

9.7.16 Definition An $f \in \mathbb{F}_q(x)$ is *exceptional* if it maps one-one on $\mathbb{P}^1(\mathbb{F}_{q^k})$ for infinitely many k. Similarly, with K a number field, $f \in K(x)$ is *exceptional* if it is exceptional $\bmod \boldsymbol{p}$ for infinitely many primes $\boldsymbol{p}$.

9.7.17 Remark We use K, allowing decoration, for a number field. Section 8.3 refers to the splitting field, Ω_f (respectively $\Omega_{f,\bar{F}}$), of $f(x) - z$ over $F(z)$ (respectively over $\bar{F}(z)$). The automorphism group of the extension $\Omega_f/F(z)$ (respectively $\Omega_{f,\bar{F}}/\bar{F}(z)$) is the *arithmetic* (respectively*geometric*) *monodromy* group A (respectively G) of a separable function (Definition 9.7.25) $f \in F(x)$. When there are several functions, we denote these A_f and G_f. They act on the zeros, $\{x_1, \ldots, x_n\}$ (often denoted $\{1, \ldots, n\}$), of $f(x) - z$, giving a natural permutation representation on n symbols.

9.7.18 Definition Every cover $f : X \to Z$ over a field F with X irreducible has an associated extension of *function fields* that determines the cover up to birational morphisms (see Lemma 9.7.43).

9.7.19 Remark Essentially all the Galois theory of fields translates to useful statements about a cover $f : X \to Z$ (over F) of an irreducible variety Z. It does this by corresponding to f the composite of the *function field* extensions $F(X')/F(Z)$ where X' runs over the components of X [2203, p. 396]. Several papers in our references (say, [1114, Section 0.C]) give oft-used examples, with Lemma 9.7.20 a simple archetype.

9.7.20 Lemma (See Remark 9.7.21) Any separable cover $f : X \to Y$ over F has a Galois closure cover $\hat{f} : \hat{X} \to Z$ over F. Then, A_f is the group of $\hat{f}$ with its natural permutation representation T_{A_f} (of degree the degree of f). Do this over $\bar{F}$ to get the geometric monodromy G_f. Then, X is irreducible (resp. absolutely irreducible) if and only if T_{A_f} (resp. T_{G_f}) is transitive. For f a rational function it is automatic that T_{G_f} (and so T_{A_f}) is transitive.

9.7.21 Remark [1118, Section 2.1] explains how to form the Galois closure cover of a cover using fiber products (see Remark 9.7.54). This shows how to form the Galois closure cover of any collection of covers as in Lemma 9.7.50.

9.7.22 Remark Normalization gives a nearly invertible process to Remark 9.7.19: going from field extensions of $F(Z)$ to covers of Z. While this does not translate *all* arithmetic cover problems to Galois theory, we apply the phrase "monodromy precision" (Remark 9.7.26) to when it does. Example: It does in the topic of exceptional covers, as in Proposition 9.7.28.

9.7.23 Definition Denote the elements of a group G, under a representation T_G, that fix 1 by $G(1)$. When T_G is transitive, refer to T_G as *primitive* (respectively *doubly transitive*) if there is no group properly between $G(1)$ and G (respectively $G(1)$ is transitive on $\{2, \ldots, n\}$).

9.7.24 Theorem [1112, Theorem 1]: An $f \in \mathbb{F}_q(x)$ is exceptional if and only if the following holds for each orbit O of $A_f(1)$ on $\{2, \ldots, n\}$:

$$O \text{ breaks into strictly smaller orbits under } G_f(1). \tag{9.7.1}$$

Denote the projective normalization of $\{(x,y) \mid \frac{f(x)-f(y)}{x-y} = 0\}$ by $X_{f,f} \setminus \Delta$. Also equivalent to (9.7.1): Each $\mathbb{F}_q$ component of $X_{f,f} \setminus \Delta$ has at least 2 components over $\bar{\mathbb{F}}_q$.

Similarly, an $f \in K(x)$, K a number field, is exceptional if and only if (9.7.1) holds for f mod $\boldsymbol{p}$ for infinitely many primes $\boldsymbol{p}$.

9.7.25 Definition (Covers) Let $f \in \mathbb{F}_q(x)$ be nonconstant and *separable*: not $g(x^p)$ for some $g \in \mathbb{F}_q(x)$. Then, $f : \mathbb{P}^1_x(\bar{\mathbb{F}}_q) \to \mathbb{P}^1_z(\bar{\mathbb{F}}_q)$ by $x \mapsto f(x)$ has these *cover* properties.

1. Excluding a finite set $\{z_1, \ldots, z_r\} \subset \mathbb{P}^1_z(\bar{\mathbb{F}}_q)$, *branch points of* f, there are exactly $n = \deg(f)$ points over z'.
2. For z' a branch point, counting zeros, x' of $f(x) - z'$ with multiplicity, the sum at all x's over z' is still n. An $x' \in \mathbb{P}^1_x$ with multiplicity > 1 is a *ramified point*.

For K a number field, the same properties hold, without any separable condition.

9.7.26 Remark (MacCluer's Theorem) Theorem 9.7.24 has a surprise: (9.7.1) implies exceptionality over $\mathbb{F}_q$. An error term in applying *Chebotarev's density theorem* with branch points (as in Section 8.3, in Section 8.3.3) vanishes. A ramified point with p not dividing its multiplicity is *tame*.

Macluer's thesis [1986] responded to a Davenport-Lewis conjecture [777] by showing Theorem 9.7.24 for a polynomial tame at every point. We say: *MacCluer's Theorem* shows tame polynomial exceptional covers exhibit *monodromy precision* [1119, Section 3.2.1]. Proposition 9.7.28 shows monodromy precision holds for general exceptional covers.

9.7.27 Example A polynomial f over $\mathbb{F}_q$ for which $p|\deg(f)$ is not tame at ∞.

9.7.28 Proposition [1112] combined with [1118, Principle 3.1]: Let $f : X \to Z$ be any cover (Definition 9.7.12) over $\mathbb{F}_q$ with X absolutely irreducible. Then [1118, Corollary 2.5]:

1. the extended meaning of (9.7.1) is that the 2-fold fiber product (Section 9.7.3) of f minus the diagonal has no absolutely irreducible $\mathbb{F}_q$ components; and
2. (9.7.1) is equivalent to f being exceptional: $X(\mathbb{F}_{q^k}) \to Y(\mathbb{F}_{q^k})$ is one-one (and onto) for infinitely many k.

9.7.29 Remark As noted in [1118, Comments on Principle 3.1], the proof of [1112] applies without change to give Proposition 9.7.28 Part 2 when X and Z are non-singular; indeed, it applies to pr-exceptionality (Definition 9.7.93). Without, however, this nonsingularity assumption, there are complications considered in [1119, Section A.4.1] (see Example 9.7.14).

9.7.30 Definition Let f in Proposition 9.7.28 over $\mathbb{F}_q$ be an exceptional cover. Denote values k where (9.7.1) holds with $\mathbb{F}_{q^k}$ replacing $\mathbb{F}_q$, by $E_{f,q}$: the *exceptionality set* of f.

Similarly, for f satisfying the hypotheses of Proposition 9.7.28 over a number field K, denote those primes $\boldsymbol{p}$ where f mod $\boldsymbol{p}$ has $E_{f,\mathcal{O}/\boldsymbol{p}}$ infinite, by $E_{f,K}$.

9.7.31 Definition The equation $T_u(\cos(\theta)) = \cos(u\theta)$ defines the u-th Chebychev polynomial, T_u. From it define a *Chebychev conjugate*: $\alpha \circ T_u \circ \alpha^{-1}$ with $\alpha(x) = \alpha_{z'}(x) = z'x$ and either $z' = 1$, or z' and $-z'$ are conjugate in a quadratic extension of K.

9.7.32 Theorem (Schur's Conjecture) [1109, Theorem 2]: With K a number field, the $f \in \mathcal{O}[x]$ for which $E_{f,K}$ is infinite are compositions with maps $a \mapsto ax + b$ (affine) over K with polynomials of the following form for some odd prime u:

$$x^u \text{ (\textit{cyclic}) or, } \alpha \circ T_u \circ \alpha^{-1}, u > 3, \text{ a Chebychev conjugate.} \qquad (9.7.2)$$

9.7.33 Remark Many still refer to Theorem 9.7.32 as *Schur's Conjecture*, though Schur conjectured it only over $\mathbb{Q}$. Paper [1109] refers to all Chebychev conjugates as Chebychev polynomials, rather than *Dickson* as in Remark 9.7.34. Reference [1936] assiduously distinguishes Dickson polynomials.

Here is a simple branch point Chebychev Conjugate characterization: f has two finite ($\neq \infty$) branch points, $\pm z' \in \mathbb{P}^1_z(\bar{\mathbb{Q}})$, which identify with the *unique* unramified points (in $\mathbb{P}^1_x(\bar{\mathbb{Q}})$) over the branch points, as in [1109, Proof of Lemma 9].

1. A corollary of [1124, Theorem 3.5] is that *any* cover with a unique totally and tamely ramified point decomposes over F if and only if it decomposes over $\bar{F}$. This applies if $f \in F[x]$ has $\deg(f)$ prime to the characteristic of F.
2. If f from Part 1 is indecomposable, then G_f is primitive (see Definition 9.7.23) and it contains an n-cycle.
3. If $f \in K[x]$ is exceptional, since (9.7.1) says G_f cannot be doubly transitive, up to composing with K affine maps, f from Part 2 is in (9.7.2).

9.7.34 Remark (Dickson doppelgangers, see Section 9.6) Each Chebychev conjugate is a constant times a Dickson polynomial [1118, Proposition 5.3]. The Remark 9.7.33 characterization – by locating their branch points – avoids using equations. That is the distinction at the last step between the proof of Theorem 9.7.32 and [1936, Chapter 6].

9.7.35 Remark Use the notation in Theorem 9.7.32. Suppose $f \in \mathcal{O}_K[x]$ is an exceptional polynomial. Define $n_{f,c}$ (resp. $n_{f,C}$) to be the product of distinct primes s for which f has a degree s cyclic (resp. Chebychev conjugate) composition factor. One can check that Corollary 9.7.36 follows from 9.7.28 combined with 9.7.32.

9.7.36 Corollary For $f \in \mathcal{O}_K[x]$ an exceptional polynomial, one can determine $E_{f,K}$ (excluding bad primes, Remark 9.7.6) from $n_{c,f}$ and $n_{f,C}$ by congruences. When $\mathcal{O}_K = \mathbb{Z}$, then $p \in E_{f,\mathbb{Q}}$ if and only if $\gcd(p-1, s) = 1$ for each $s | n_{f,c}$ and $\gcd(p^2-1, s) = 1$ for each $s | n_{f,C}$.

9.7.37 Example (Infinite $E_{f,\mathbb{Q}}$) It is necessary that $\gcd(2, n_c) = 1$ and $\gcd(6, n_C) = 1$ for there to be infinitely many p that satisfy the conclusion of Corollary 9.7.36. But it is sufficient, too. Without loss, assume $\gcd(n_c, n_C) = 1$. If $3 \not| n_c$, then Dirichlet's Theorem on primes in arithmetic progressions gives an infinite set of $p \equiv 3 \pmod{n_c n_C}$. They are in $E_{f,\mathbb{Q}}$. If $3 | n_c$, the Chinese remainder theorem gives an arithmetic progression of p satisfying $p \equiv 3 \pmod{n_C}$ and $p \equiv -1 \pmod{n_c}$. So, $E_{f,\mathbb{Q}}$ is infinite whenever it has a chance to be.

9.7.38 Remark Combine [1113, Lemma 1] with monodromy precision in Proposition 9.7.39. This shows, the Proposition 9.7.28 fiber product statement is equivalent to $f \in K(x)$ being exceptional, and therefore permutation, mod $\boldsymbol{p}$. If $\mathcal{O}_K/\boldsymbol{p}$ is sufficiently large, the fiber product statement is also necessary for f to be permutation (well-known, for example [1109, proof of Theorem 2, last paragraph]).

9.7.39 Proposition (Permutation functions) From Remark 9.7.38, for $f \in \mathbb{F}_q(x)$, those k where f *permutes* $\mathbb{P}^1(\mathbb{F}_{q^k})$ contains $E_{f,\mathbb{F}_q}$ as a cofinite subset. Similarly, for K a number field, those $\boldsymbol{p}$ where f functionally permutes $\mathbb{P}^1(\mathcal{O}/\boldsymbol{p})$ contains $E_{f,K}$ as a cofinite subset.

9.7.40 Remark Section 8.1 shows permutation polynomials are abundant. Exceptional polynomials satisfy a much stronger property, but Corollary 9.7.36 shows they are abundant, too. One difference: Section 9.7.3 combines them in ways with no analog for permutation polynomials.

9.7.41 Corollary An analog of Theorem 9.7.32 holds over $\mathbb{F}_q$ to characterize exceptional polynomials of degree prime to p ([1120, Introduction to Section 5] or [1118, Proposition 5.1]). There, z' in $\alpha_{z'}$ is either 1 or in the unique quadratic extension of $\mathbb{F}_q$. Consider a Chebychev conjugate $\alpha_{z'} \circ T_n \circ \alpha_{z'}^{-1}$ as a permutation polynomial on $\mathbb{F}_{q^k}$ with $\gcd(q^{2k}-1, n) = 1$. Then, when $n \cdot m \equiv 1 \pmod{q^{2k}-1}$, $\alpha_{z'} \circ T_m \circ \alpha_{z'}^{-1}$ is its functional inverse.

9.7.3 Fiber product of covers

9.7.42 **Definition** For any field extension F_1/F_2 containing $\mathbb{F}_p$, there is the notion of being *separable* [1121, p. 111]. For $f \in \mathbb{F}_q(x)$, the extension $\bar{\mathbb{F}}_q(x)/\bar{\mathbb{F}}_q(f(x))$ being separable is equivalent to f is separable (Definition 9.7.25). Many of our examples inherit separableness from this special case.

9.7.43 **Lemma** (Curve covering maps [1427, Chapter I, Section 6]) Any nonsingular projective algebraic curve X over a perfect field F has a field of functions $F(X)$ that uniquely determines X up to isomorphism over F.

Each non-constant element $f \in F(X)$ determines a finite map $X \to \mathbb{P}_z^1$ over F [1427, Chapter I, Exercise 6.4]. If $F(X)/F(f)$ is separable, then f has the covering properties of (9.7.25): finite number of branch points, and uniform count of points in a fiber over $\bar{F}$ (including multiplicity in the fiber) [1427, Chapter IV, Proposition 2.2].

9.7.44 **Definition** Refer to any f in the conclusion of Lemma 9.7.43 as a *nonsingular cover* of $\mathbb{P}_z^1$.

9.7.45 **Definition** (Fiber product) Let $f_i : X_i \to \mathbb{P}_z^1$, $i = 1, 2$, be two nonsingular covers of $\mathbb{P}_z^1$. The *set theoretic* fiber product consists of the algebraic curve

$$\{(x_1, x_2) \in X_1 \times X_2 | f_1(x_1) = f_2(x_2)\}.$$

Denote this $X_1 \times_{\mathbb{P}_z^1}^{\text{set}} X_2$. Its normalization (Proposition 9.7.9), $X_1 \times_{\mathbb{P}_z^1} X_2$, is the *fiber product* of f_1 and f_2.

9.7.46 **Remark** Definition 9.7.45 works equally for any covers $X_i \to Z$, $i = 1, 2$, with Z a normal projective variety. Then, $X_1 \times_Z X_2$ is normal and projective (possibly with several components) with natural maps $\text{pr}_i : X_1 \times_Z X_2 \to X_i$, $i = 1, 2$, given by its projection on each factor. The functions $f_i \circ \text{pr}_i$, $i = 1, 2$ are identical, giving a well-defined map:

$$(f_1, f_2) : X_1 \times_Z X_2 \to Z. \tag{9.7.3}$$

9.7.47 **Remark** (Fiber equations) Consider $x' \in X_1 \times_Z X_2$ that is simultaneously over $x_i' \in X_i$, $i = 1, 2$, where both x_i's ramify over $\text{pr}_1(x_1') = \text{pr}_2(x_2')$. Then, $f_1(x_1) = f_2(x_2)$, with x_i in a neighborhood of x_i', is *not* a correct local description around x'.

There is another complication when Z is not a curve (dimension 1). The fiber product might be singular even when the X_is are not. So (f_1, f_2) in (9.7.3) may not be a cover because it is not flat (Remark 9.7.67).

9.7.48 **Example** Consider two polynomials, $f_1, f_2 \in K[x]$, of the same degree n. They define $f_j : \mathbb{P}_{x_j}^1 \to \mathbb{P}_z^1$, $j = 1, 2$. Then, there are n points over $z = \infty$ on $\mathbb{P}_{y_1}^1 \times_{\mathbb{P}_z^1} \mathbb{P}_{y_2}^1$, but only one point on the set theoretic fiber product over ∞. Proposition 1 of [1111] gives the generalization of this, showing – when the covers are tame – how to compute the genus of the fiber product components from the covers f_j, $j = 1, 2$.

9.7.49 **Definition** The fiber product $X_{f,f} = X \times_Z X$ for a cover $f : X \to Z$ of degree exceeding 1 has at least two components. One is the *diagonal*: the set $\Delta(X) = \{(x, x) \mid x \in X\}$. The normal variety $X \times_Z X \setminus \Delta(X)$ generalizes the set in Theorem 9.7.24.

9.7.50 Lemma (Fiber product monodromy [1118, Section 2.1.3]) Consider the covers in Definition (9.7.45). To each f_j there is an arithmetic (resp. geometric) monodromy group A_{f_j} (resp. G_{f_j}), $j = 1, 2$. Similarly, for (f_1, f_2) in (9.7.3). Then, $A_{(f_1,f_2)}$ maps naturally, surjectively, to A_{f_j} by homomorphisms pr_j^*, $j = 1, 2$. There is a largest simultaneous quotient, H, of both A_{f_j} s given by homomorphisms $m_i : A_{f_j} \to H$, $j = 1, 2$, so that

$$A_{(f_1,f_2)} = \{(\sigma_1, \sigma_2) \in A_{f_1} \times A_{f_2} \mid m_1(\sigma_1) = m_2(\sigma_2)\}.$$

Similarly with geometric replacing arithmetic monodromy.

9.7.51 Corollary (Components) With the hypotheses of Lemma 9.7.50, let $\{1_j, \ldots, n_j\}$, be integers on which A_j acts, $j = 1, 2$. Then, $A_{(f_1,f_2)}$ acts on the pairs (i_1, i_2) and on each of the sets $\{1_j, \ldots, n_j\}$ separately. If X_1 is absolutely irreducible, then the components of $X_1 \times_Z X_2$ over F (resp. $\bar{F}$) correspond to the orbits of $A_{(f_1,f_2)}(1_1)$ (resp. $G_{(f_1,f_2)}(1_1)$; see Definition 9.7.23) on $\{1_2, \ldots, n_2\}$. We note that the degrees n_1 and n_2 may be different.

9.7.52 Definition (Absolute components) Given $X_1 \times_Z X_2$ in Corollary 9.7.51, denote the union of its absolutely irreducible F components by $X_1 \times_Z^{\mathrm{abs}} X_2$. Denote the *complementary set*, $X_1 \times_Z X_2 \setminus X_1 \times_Z^{\mathrm{abs}} X_2$, of components by $X_1 \times_Z^{\mathrm{cp}} X_2$.

9.7.53 Theorem (Explicit $E_{f,q}$ – see Remark 9.7.54) Let $f : X \to Z$ (as in Proposition 9.7.28) be an exceptional cover over $\mathbb{F}_q$. For X_i', an $\mathbb{F}_q$ component of $X \times_Z^{\mathrm{cp}} X$, denote the number of components in its breakup over $\bar{\mathbb{F}}_q$ by s_i, $i = 1, \ldots, u$.

$$\text{With } s_{\mathrm{exc}} = \mathrm{lcm}(s_1, \ldots, s_u), E_{f,q} = \{k \pmod{s_{\mathrm{exc}}} \mid \gcd(k, s_i) < s_i, i = 1, \ldots, u\}.$$

The group $G(\mathbb{F}_{q^{s_{\mathrm{exc}}}}/\mathbb{F}_q)$ is naturally a quotient of A_f/G_f. We can interpret all quantities using A_f and G_f.

9.7.54 Remark All but the last sentence of Theorem 9.7.53 is [1118, Corollary 2.8]. The last sentence is from [1118, Lemma 2.6], using that the Galois closure cover of f is a(ny) component (over $\mathbb{F}_q$) of the $\deg(f) = n$-fold fiber product of f with itself. Project that fiber product onto the 2-fold fiber product of f over $\mathbb{F}_q$ to finish. Corollary 9.7.51 shows the orbit lengths of $A_f(1)$ on $\{2, \ldots, n\}$ divided by the corresponding orbit lengths of $G_f(1)$, give the s_i s.

9.7.55 Theorem (Explicit $E_{f,K}$ – see Remark 9.7.56) Now change $\mathbb{F}_q$ to K (number field) in the first sentence of Theorem 9.7.53. For each cyclic subgroup $C \le A_f/G_f$ denote those $\sigma \in A_f$ that map to C by A_C. As previously, denote the stabilizers of 1 in the representation by $A_C(1)$ and $G_C(1)$. Consider the set, $\mathcal{C}_{f,K}$, of cyclic C (as in (9.7.1)):

$$\{C \mid \text{each orbit of } A_C(1) \text{ on } \{2, \ldots, n\} \text{ breaks into strictly smaller orbits under } G_C(1)\}.$$

Then, f is exceptional over K if and only if $\mathcal{C}_{f,K}$ is nonempty. Further, $E_{f,K}$ consists of those primes $\boldsymbol{p}$ for which the Frobenius attached to $\boldsymbol{p}$ is a generator of some $C \in \mathcal{C}_{f,K}$.

9.7.56 Remark Theorem 9.7.55 comes from applying [1113, Section 2] exactly as in Remark 9.7.28. If $E_{f,K}$ is infinite, then $X \times_Z X \setminus \Delta(X)$ has no absolutely irreducible component. The converse, however, does not hold.

9.7.4 Combining exceptional covers; the $(\mathbb{F}_q, Z)$ exceptional tower

9.7.57 Definition (Category of exceptional covers) For Z absolutely irreducible over $\mathbb{F}_q$, denote the collection of exceptional covers of Z over $\mathbb{F}_q$ by $\mathcal{T}_{Z,\mathbb{F}_q}$.

9.7.58 Theorem [1118, Section 4.1] Given $(f_i, X_i) \in \mathcal{T}_{Z,\mathbb{F}_q}$, $i = 1, 2$, $X_1 \times_Z^{\text{abs}} X_2$ (Definition 9.7.52) has one component. We conclude that:

$$(f_1 \circ \text{pr}_1, X_1 \times_Z^{\text{abs}} X_2) \in \mathcal{T}_{Z,\mathbb{F}_q}.$$

Also, there is at most one morphism between any two objects in $\mathcal{T}_{Z,\mathbb{F}_q}$.

9.7.59 Remark (When $f_1 = f_2$ in Theorem 9.7.58) We definitely include the fiber product of a cover in $\mathcal{T}_{Z,\mathbb{F}_q}$ with itself. Then, the only absolutely irreducible component of the fiber product is the diagonal (Definition 9.7.49), which is equivalent to the original cover.

9.7.60 Definition We call $X_1 \times_Z^{\text{abs}} X_2$ the fiber product of f_1 and f_2 in $\mathcal{T}_{Z,\mathbb{F}_q}$, and continue to denote its morphism to Z by (f_1, f_2). This defines $\mathcal{T}_{Z,\mathbb{F}_q}$ as a *category with fiber products.* Theorem 9.7.53 shows $E_{f_1,q} \cap E_{f_2,q} = E_{(f_1,f_2),\mathbb{F}_q}$ is infinite.

9.7.61 Remark Consider $(f_i, X_i) \in \mathcal{T}_{Z,\mathbb{F}_q}$, $i = 1, 2$, for which there exists $\psi : X_1 \to X_2$ over $\mathbb{F}_q$ that factors through f_2: $f_2 \circ \psi = f_1$. Then, Theorem 9.7.58 says ψ is unique.

9.7.62 Corollary For $(f, X) \in \mathcal{T}_{Z,\mathbb{F}_q}$, denote the group of the Galois closure cover of f over X by $A_f(1)$. Then, A_f has the representation T_f by acting on cosets of $A_f(1)$. If $(f_i, X_i) \in \mathcal{T}_{Z,\mathbb{F}_q}$, $i = 1, 2$, we write $(f_1, X_1) > (f_2, X_2)$ if f_1 factors through X_2. [1118, Proposition 4.3] produces from these pairs a canonical group $A_{Z,\mathbb{F}_q}$ with a profinite permutation representation $T_{Z,\mathbb{F}_q}$.

9.7.63 Remark (A projective limit) Given $(f_i, X_i) \in \mathcal{T}_{Z,\mathbb{F}_q}$, $i = 1, 2$, there is a 3rd $(f, X) \in \mathcal{T}_{Z,\mathbb{F}_q}$, given by the fiber product, that factors through both. This is the condition defining a projective sequence. So, $A_{Z,\mathbb{F}_q}$ in Corollary 9.7.62 is a projective limit.

9.7.64 Definition $(A_{Z,\mathbb{F}_q}, T_{Z,\mathbb{F}_q})$ is the (arithmetic) monodromy group, in its natural permutation representation, of the exceptional tower $\mathcal{T}_{Z,\mathbb{F}_q}$.

9.7.65 Theorem Let $f_i : X_i \to Z$, $i = 1, 2$, be exceptional covers over K: $E_{f_i,K}$ is infinite, $i = 1, 2$. Then, $X_1 \times_Z X_2$ is exceptional in the sense that

$$X_1 \bmod \boldsymbol{p} \times_{Z \bmod \boldsymbol{p}}^{\text{abs}} X_2 \bmod \boldsymbol{p} \text{ is exceptional for infinitely } \boldsymbol{p}$$

if and only if $E_{f_1,K} \cap E_{f_2,K}$ is infinite.

9.7.66 Remark Theorem 9.7.65 forces considering if there is an infinite intersection of two exceptionality sets over K. As Theorem 9.7.53 shows, this is automatic over $\mathbb{F}_q$. Example 9.7.37 shows it is not automatic over a number field. Subsections 9.7.5 and 9.7.6 have examples along these lines: If both f_i s, $i = 1, 2$, are exceptional rational functions, then their composition is again exceptional over K if and only if $E_{f_1,K} \cap E_{f_2,K}$ is infinite. Beyond cyclic and Chebychev situations, it is very difficult to decide when this intersection is infinite.

9.7.67 Remark The same definition for exceptional works for any finite, surjective, map of normal varieties over $\mathbb{F}_q$. Such maps may not be flat (say, when Remark 9.7.14 does not apply), so they may not be covers. Normalization of any projective variety is projective: Segre's Embedding [2203, Theorem 4, p. 400].

For irreducible X, flatness says the multiplicity sum of points in the fiber over z is constant in z: the function field extension degree, $[K(X) : K(Z)]$ [2203, Proposition 2, p. 432]. That is, Definition 9.7.25, Part 2, holds. For finite morphisms that characterizes flatness [2203, Corollary p. 432]. With normality, but not flatness, this may hold only outside a codimension 2 set in the target. Appendix A.4 of [1119] has a liesurely discussion; see Example 9.7.14.

9.7.5 Exceptional rational functions; Serre's Open Image Theorem

9.7.68 Definition Definition 9.7.31 explains Chebychev conjugates. Consider $l_{z'} : x \mapsto \frac{x-z'}{x+z'}$, mapping $\pm z'$ to $0, \infty$, with $a = (z')^2 \in K, z' \notin K$. Then, for n odd, characterize $R_{n,a} = (l_{z'})^{-1} \circ (l_{z'}(x))^n$, a *cyclic conjugate*, by these conditions:

$$\pm z' \text{ are its sole ramified points, } R_{n,a}(\pm z') = \pm z' \text{ and it maps } \infty \mapsto \infty. \tag{9.7.4}$$

9.7.69 Remark According to [1936, Chapter 2, Section 5]), the function $R_{n,a}$ in Definition 9.7.68 is a *Redei function.* From [1936, Theorem 3.11], under the hypotheses on z', the exceptionality set $E_{R_{n,a},K}$ is

$$\begin{array}{ll} \{\boldsymbol{p} \mid (|\mathcal{O}_K/\boldsymbol{p}| - 1, n) = 1\} & \text{if } z' \text{ is a quadratic residue modulo } \boldsymbol{p}, \text{ and} \\ \{\boldsymbol{p} \mid (|\mathcal{O}_K/\boldsymbol{p}| + 1, n) = 1\} & \text{if not.} \end{array}$$

9.7.70 Remark (Addendum Remark 9.7.69) Quadratic reciprocity determines nonempty arithmetic progressions for which z' is a quadratic residue and those for which it is not. If z' in Definition 9.7.68 were in K, then – of course – the exceptional set is the same as for x^n. Whether or not $z' \in K$, we refer to $R_{n,a}$ as a *cyclic conjugate.*

9.7.71 Definition [1118, Section 4.2] Suppose a collection $\mathcal{C}$ of covers from an exceptional tower $\mathcal{T}_{Y,\mathbb{F}_q}$ is closed under the categorical fiber product. We say $\mathcal{C}$ is a *subtower.* We also speak of the (minimal) subtower any collection generates under fiber product.

9.7.72 Remark Section 4.3 of [1118] uses that the fiber product of two unramified covers is unramified to create *cryptographic* exceptional subtowers. Section 5.2.3 of [1118] computes the arithmetic monodromy attached to the *Dickson subtower* generated by all the exceptional Chebychev conjugates over $\mathbb{F}_q$. The analog of Remark 9.7.69 over $\mathbb{F}_q$ gives a similar – *Redei* – subtower of $\mathcal{T}_{\mathbb{P}^1_z,\mathbb{F}_q}$ generated by exceptional cyclic conjugates.

9.7.73 Remark Theorem 9.7.65 requires common exceptional intersection (Remark 9.7.66) to form fiber products in $\mathcal{T}_{Z,K}$, Z absolutely irreducible over a number field K. For fiber products (or composites) of Chebychev and cyclic conjugates, we easily decide if exceptional sets have infinite intersection. Exceptional rational functions from Serre's O(pen) I(mage) T(heorem) give much harder versions of such problems.

9.7.74 Definition (j-line $\mathbb{P}^1_j$) A special copy of projective 1-space, the *j-line,* occurs in the study of *modular curves* (see Theorem 9.7.76). Each $j \in \mathbb{P}^1_j \setminus \{\infty\}(\bar{\mathbb{Q}}) = \mathbb{A}^1_j(\bar{\mathbb{Q}})$ has an attached isomorphism class of elliptic curves E_j. For each integer $n > 0$, consider a special case of a modular curve, $\mu_0(n) : X_0(n) \to \mathbb{P}^1_j$, with its cover of $\mathbb{P}^1_j$. Denote the points of $X_0(n)$ not lying over $j = \infty$ by $Y_0(n)$.

9.7.75 Definition For E an elliptic curve, denote by $E \to E/C$ an *isogeny* from quotienting E by a (finite) torsion subgroup C of E. When C is a cyclic, generated by $e' \in E$ (resp. all torsion points killed by multiplication by n), write $C = \langle e' \rangle$ (resp. C_n).

9.7.76 Theorem There are two approaches to giving "meaning" to each algebraic point $y \in Y_0(n)$, whose image in $\mathbb{P}^1_j$ is j_y

1. [2311, p. 108] or [1115, p. 158]: $y \mapsto [E_{j_y} \to E_{j_y}/\langle e'_y \rangle]$ with $e'_y \in E_{j_y}$ of order n where brackets, [ ], indicate an isomorphism class of isogenies.
2. [1115, Lemma 2.1]: $y \mapsto f_y \in \bar{\mathbb{Q}}(x)$ (up to Möbius equivalence) of degree n.

9.7.77 Theorem [1115, Theorem 2.1] Suppose $f \in K(x)$ is exceptional and of prime degree u. Then, f is Möbius equivalent over K to either:

1. a cyclic (Remark 9.7.70) or a Chebychev (Remark 9.7.34) conjugate; or
2. to some f_y ($u = n$) in Theorem 9.7.76, Part 2.

9.7.78 Definition For a dense set of $j' \in \mathbb{A}^1_j$, the corresponding $E_{j'}$ is of *CM*-type if its ring of isogenies, tensored by $\mathbb{Q}$, has dimension 2 over $\mathbb{Q}$. Such isogenies form a complex quadratic extension of $\mathbb{Q}$ (containing j', which is an algebraic integer; [2586, II-28] or [2614, Chapter 2, Section 5.2]). Otherwise, j' is of GL_2-type.

9.7.79 Theorem [1115, (2.10)] Continue the notation of Theorem 9.7.77. Except for the two cases where j_y is one of the two finite branch points of $\mu_0(u)$, the geometric monodromy G_{f_y} is the order $2u$ dihedral group D_u, and f_y has four branch points (Definition 9.7.25). For u in Theorem 9.7.77, Part 2, for which $E_{j'}$ has good reduction, the coordinates of e'_y generate a constant extension of K with group A_{f_y}/G_{f_y} (explained in Theorem 9.7.85).

9.7.80 Theorem [1115, Section 2.B] For j' of CM-type, complex multiplication theory gives (an infinite) $E_{f_y,K}$. Computing this would use [1370, Sections 6.3.1-6.3.2].

9.7.81 Remark (Addendum to Theorem 9.7.80) Using adelic (modular) arithmetic gives analogs of Corollary 9.7.36; and Corollary 9.7.41 for explicitly finding the functional inverse of a CM-type reduced modulo a prime in the exceptional set $E_{f_y,K}$. If $K = \mathbb{Q}(j')$, then $E_{f_y,K}$ depends on the congruence defining the Frobenius in the (cyclic of degree $u-1$ over K) constant field. Only finitely many j' in $\mathbb{Q}$ have CM-type, corresponding to class number 1 for complex quadratic extensions.

9.7.82 Problem Take one of the CM-type j s in $\mathbb{Q}$. Then, consider two allowed values of u, u_i, $i = 1, 2$, denoting the corresponding f_y s by f_i, $i = 1, 2$. Test for explicitness in Remark 9.7.81 as to whether $E_{f_1,\mathbb{Q}} \cap E_{f_2,\mathbb{Q}}$ is infinite.

9.7.83 Definition (Composition factor definition field) For $f \in F(x)$ consider a minimal field $F_f(\mathrm{ind})$ over which f decomposes into composition factors indecomposable over $\bar{F}$. Similarly, denote the minimal field over which $X_{f,f} \setminus \Delta$ in Theorem 9.7.24 breaks into absolutely irreducible components by $\hat{F}_f(2)$.

9.7.84 Proposition [1118, Proposition 6.5] If $f : X \to Z$ is a cover over F, then $F_f(\mathrm{ind}) \subset \hat{F}_f(2)$.

9.7.85 Theorem (See Remark 9.7.87) Assume $j' \in \mathbb{A}^1_j$ is of GL_2-type. For $K = \mathbb{Q}(j')$, consider $C = C_u$ in Definition 9.7.75 with u a prime. The corresponding $f_y \in K(x)$, y over j', has degree u^2. Use the monodromy groups of Definition 9.7.17.

There is a constant $M_{1,j'}$ so that if $u > M_{1,j'}$, then the arithmetic/geometric monodromy quotient A_{f_y}/G_{f_y} is $\mathrm{GL}_2(\mathbb{Z}/u)/\{\pm 1\}$. Further, f_y decomposes into two degree u rational functions over $K_f(\mathrm{ind})$, but it is indecomposable over K.

9.7.86 Theorem [1118, Proposition 6.6] Continue Theorem 9.7.85 hypotheses. For a second constant $M_{2,j'}$, and for any prime $\boldsymbol{p}$ of $\mathcal{O}_K$ with $|\mathcal{O}_K/\boldsymbol{p}| > M_{2,j'}$ assume $A_{\boldsymbol{p}} \in \mathrm{GL}_2(\mathbb{Z}/u)/\langle \pm 1 \rangle$ represents the conjugacy class of the Frobenius for $\boldsymbol{p}$. Then, f_y mod $\boldsymbol{p}$ is an exceptional indecomposable rational function, and it decomposes over the algebraic closure of $\mathcal{O}_K/\boldsymbol{p}$, precisely when $\langle A_{\boldsymbol{p}}\rangle$ acts irreducibly on $(\mathbb{Z}/u)^2 = V_u$. This holds for infinitely many primes $\boldsymbol{p}$. In particular, f_y is exceptional over K (Definition 9.7.30).

9.7.87 Remark (Using Serre's OIT) [2586] lays the groundwork for [2587]. The latter has the existence of the constant $M_{1,j'}$. Appendix A.1 and Section 3.2 of [2586] proves that it exists when $j' \in \mathbb{A}^1_j(\bar{\mathbb{Q}})$ is *not* an algebraic integer. Then, the computation of $M_{i,j'}$, $i = 1, 2$, in Theorems 9.7.85 and 9.7.86 is effective. Even after all these years, there is no effective computation of these constants when j is not CM-type, but is an algebraic integer. Section 2 of [1115] gets Theorem 9.7.85 from the OIT using the relation between Parts 1 and 2 in Theorem 9.7.76.

9.7.88 Remark (More elementary, but less precise than Theorem 9.7.86) Theorem 2.2 of [1115] shows, for every K and any prime $u > 3$, the $j' \in K$, with f_y satisfying the exceptionality and decomposability conclusions of Theorem 9.7.86, are dense. Applying the [1113, Theorem 3] (or [1121, Theorem 12.7]) version of Hilbert's Irreducibility Theorem to $X_0(u)$ gives the corresponding $M_{2,j'}$ explicitly.

9.7.89 Example ($M_{1,j'}$ effectiveness?) Appendix A.1 and Section 3.3 of [2586] gives Ogg's example [2310] with $j' \in \mathbb{Q}$. Section 6.2.2 of [1118] reviews this case, where $M_{2,j'} = 6$, to show how to pick an $A_{\boldsymbol{p}}$ acting irreducibly on V_u as in Theorem 9.7.86 (for infinitely many $\boldsymbol{p}$), assuring that $E_{f_y,\mathbb{Q}}$ is infinite for $u > M_{2,j'}$.

Section 6.3.2 of [1118] – still Ogg's case – aims at finding an automorphic function, a la Langland's Program, that would characterize the primes in $E_{f_y,\mathbb{Q}}$. This is akin to the unrelated examples of [2595], but uses results on automorphic functions in [2590, Theorem 22]. Primes of $E_{f_y,\mathbb{Q}}$ *do not* lie in arithmetic progressions. So, Problem 9.7.90 is much harder than Problem 9.7.82.

9.7.90 Problem (Analog of Problem 9.7.82) For the Ogg curve in Example 9.7.89, consider two allowed values of u, u_i, $i = 1, 2$, denoting the corresponding f_ys by f_i, $i = 1, 2$. Test for explicitness in Remark 9.7.81 as to whether $E_{f_1,\mathbb{Q}} \cap E_{f_2,\mathbb{Q}}$ is infinite.

9.7.91 Remark Paper [1123] connects "variables separated factors" of $X_{f,f}$ and composition factors of f. Reference [84] used this to effectively test for composition factors (and primitivity) of covers.

9.7.92 Theorem [1370, Chapter 3] Excluding finitely many degrees, all indecomposable exceptional $f \in K(x)$ (K a number field) are Möbius equivalent to a cyclic or Chebychev conjugate, or to a CM function from Theorem 9.7.77 of prime degree; or they are from Theorem 9.7.86 and of prime degree squared.

9.7.6 Davenport pairs and Poincaré series

9.7.93 Definition [1118, Definition 2.2] Consider $f : X \to Z$, a cover of normal varieties over $\mathbb{F}_q$, with Z absolutely irreducible, but X *possibly reducible*. Then f is *pr-exceptional* if it is surjective on $\mathbb{F}_{q^k}$ points for infinitely many k. There is a similar definition extending Definition 9.7.30 over a number field, and for both a notation for exceptional sets.

9.7.94 Definition Use the value set notation of Remark 9.7.3. We say $f_i \in \mathbb{F}_q(x)$, $i = 1, 2$, is a *Davenport pair* over $\mathbb{F}_q$ if $V_{f_1}(\mathbb{P}^1(\mathbb{F}_{q^k})) = V_{f_2}(\mathbb{P}^1(\mathbb{F}_{q^k}))$ for infinitely many k. So, take $f_2(x) = x$ to see Davenport pairs generalize exceptional functions. The notion applies to any pair of covers $f_i : X_i \to Z$, $i = 1, 2$. For K a number field, this similarly generalizes Definition 9.7.30: $f_1, f_2 \in K(x)$ are a Davenport pair if they are a Davenport pair for infinitely many residue class fields.

9.7.95 Theorem [1118, Corollary 3.6] Monodromy precision (Definition 9.7.26) applies to pr-exceptional covers and so to Davenport pairs. That is, generalizing Theorems 9.7.53 and 9.7.55, a precise monodromy statement generalizes MacCluer's Theorem (Proposition 9.7.28) to pr-exceptional covers and to Davenport pairs.

9.7.96 Theorem [1118, Section 3.1.2] With the notation of Definition 9.7.93, a pr-exceptional cover over $\mathbb{F}_q$ is exceptional if and only if X is absolutely irreducible.

9.7.97 Remark The proof of Schur's Conjecture began the solution of Davenport's problem for polynomial pairs (f_1, f_2) over a number field, the main result of [1110]. Section 3.2 in [1118] shows the exceptional set characterization for Davenport pairs in general is given by the intersection of exceptionality sets for *pr-exceptionality correspondences.* A full description of many authors' results that came from the solution of Davenport's problem – especially the study of general zeta functions attached to diophantine problems – is in [1119, Section 7.3].

9.7.98 Remark (The Genus 0 Problem) Geometric monodromy groups of rational functions are severely limited. The mildest statement for $f \in \bar{\mathbb{Q}}(x)$ is that excluding cyclic and alternating groups the composition factors of G_f fall among a finite set of simple groups. That is the original genus 0 problem.

There is a large literature distinguishing between geometric monodromy of $f \in \bar{\mathbb{Q}}(x)$ and those in $\bar{\mathbb{F}}_q(x)$, because of *wild* (not tame; Remark 9.7.26) ramification. The contrast starts from the [2191, Section 8.1.2, Guralnick's Optimistic Conjecture] list of all primitive monodromy groups of indecomposable $f \in \bar{\mathbb{Q}}[x]$.

9.7.99 Example (Davenport pairs) A significant part of the exceptional primitive monodromy groups (Remark 9.7.98), without cyclic or alternating group composition factors, came from the finitely many possible degrees of Davenport pairs $f_1, f_2 \in K[x]$ (polynomials) over number fields, with f_1 indecomposable and $V_{f_1}(\mathcal{O}_K/\boldsymbol{p}) = V_{f_2}(\mathcal{O}_K/\boldsymbol{p})$.

Important hints about what to expect for primitive monodromy groups of $f \in \bar{\mathbb{F}}_q(x)$ came also from Davenport pairs. Section 3.3.3 in [1118] (explicitly in [332]): Over every $\mathbb{F}_q$, there are infinitely many degrees of Davenport pairs, where $(\deg(f_1), p) = 1$, f_1 is indecomposable, and $V_{f_1}(\mathbb{F}_{q^k}) = V_{f_2}(\mathbb{F}_{q^k})$ for all k.

9.7.100 Example Theorem 14.1 in [696] described the geometric monodromy ($\mathrm{PSL}_2(p^a)$, $p = 2, 3$, a odd) of the only possible exceptional polynomials over $\mathbb{F}_p$ whose degrees were neither prime to p or a power of p. Then, [1120] produced these: the first exceptional polynomials over finite fields with nonsolvable monodromy.

9.7.101 Remark (Zeta functions attached to problems) Chapters 25 and 26 in [1121] details how Davenport pairs led to attaching Poincaré series – based on the *Galois stratification* procedure of [1117] – to counting the values of parameters for any diophantine problem interpretable over all extensions of $\mathbb{F}_q$, or for infinitely many primes $\boldsymbol{p}$ of K.

9.7.102 Example Denote $w_1, \ldots, w_u$ by $\boldsymbol{w}$. Suppose $f(\boldsymbol{w}, x), g(\boldsymbol{w}, y) \in \mathbb{F}_q[\boldsymbol{w}, x, y]$. Denote the cardinality of $\boldsymbol{w}' \in \mathbb{A}^u(\mathbb{F}_{q^k})$ with

$$V(f(\boldsymbol{w}', x))(\mathbb{P}^1(\mathbb{F}_{q^k})) = V(g(\boldsymbol{w}', x))(\mathbb{P}^1(\mathbb{F}_{q^k})) \tag{9.7.5}$$

by $N_{f,g,k}$. Define $\mathrm{P}_{f,g,\mathbb{F}_q}(t)$ to be the Poincaré series $\sum_{i=1}^{\infty} N_{f,g,k} t^k$.

9.7.103 Example With notation over $\mathbb{Z}$, as in Example 9.7.102, suppose $f(\boldsymbol{w}, x), g(\boldsymbol{w}, y) \in \mathbb{Z}[\boldsymbol{w}, x, y]$. Denote the cardinality of $\boldsymbol{w}' \in \mathbb{A}^u(\mathbb{F}_{p^k})$ with (9.7.5) holding over $\mathbb{F}_{p^k}$ by $N_{f,g,\mathbb{Z}/p,k}$. Define $\mathrm{P}_{f,g,\mathbb{Z}/p}(t)$ to be $\sum_{i=1}^{\infty} N_{f,g,\mathbb{Z}/p,k} t^k$.

9.7.104 Theorem [1121, Chapter 25], [1119, Section 7.3.3] For any diophantine problem over $\mathbb{F}_q$ expressed in a first order language, the attached Poincaré series is a rational function. Further, there is an effective computation of the coefficients of its numerator and denominator based on expressing those coefficients in p-adic Dwork cohomology.

9.7.105 Theorem [Theorem 9.7.104 continued] Given a diophantine problem D over $\mathbb{Z}$ (or $\mathcal{O}_K$) expressed in a first order language, there is an effective split of the primes of $\mathbb{Q}$ (or over K) into two sets: $L_{D,1}$ and $L_{D,2}$, with $L_{D,2}$ finite. Further, there is a set of varieties $V_1, \ldots, V_s$ over $\mathbb{Z}$, from which we produce linear equations in variables $Y_1, \ldots, Y_{s'}$ that serve as the coefficients of the numerator and denominator of a rational function $P_D(t)$. To each (p, Y_i), $p \in L_{D,1}$ there is a universal attachment of a p-adic Dwork cohomology group, $H(p, Y_i)$, computed in the category of such Dwork cohomology attached to $V_1, \ldots, V_s$.

The corresponding Poincaré series $P_{D,p}$ at $p \in L_{D,1}$ comes by substituting $H(p, Y_i)$ for each $Y_1, \ldots, Y_{s'}$ in $P_D(t)$. Then apply the Frobenius operator at p to these coefficients.

9.7.106 Remark [814] In Theorem 9.7.105 it is possible to take $V_1, \ldots, V_s$ to be nonsingular projective varieties with Y_i representing a *Chow motive* (over $\mathbb{Q}$). Applying the Frobenius operator at p is meaningful as Chow motives are formed from étale cohomology groups of $V_1, \ldots, V_s$.

9.7.107 Remark The effectiveness of Theorem 9.7.104 is based on Dwork cohomology [943], and the explicit calculations of [341]. Theorem 9.7.105 and Remark 9.7.106 both rest on the Galois stratification procedure of [1117] or [1121, Chapter 24].

On the plus side, the uniform use of étale cohomology from characteristic 0 produces wonderful invariants – like, Euler characteristics – attached to diophantine problems. On the negative, all the effectiveness disappears. In particular, the relation between the sets denoted $L_{D,1}$ in the two results is a mystery.

9.7.108 Remark Relating exceptional covers (and Davenport pairs) and other problems about algebraic equations is a running theme in [1118] and [1119]. Detecting these relations comes from *pr-exceptional correspondences* [1118, Section 3.2]. We catch the possible appearance of such correspondences when two Poincaré series have infinitely many identical coefficients.

9.7.109 Example An exceptional cover, $X \to \mathbb{P}^1_z$, over $\mathbb{Q}$, will be a curve whose Poincaré series is the same as that of $\mathbb{P}^1_z$ at infinitely many primes. The systematic use of such characterizations combines *monodromy precision* (where it applies) and Theorem 9.7.110.

9.7.110 Theorem ([1119, Proposition 7.17], based on [1021]) The zero support of the difference of two Poincaré series consists of the union of arithmetic progressions.

See Also

§8.1	Discusses the large literature on permutation polynomials (as in Proposition 9.7.39). This contrasts with the use of a cover given by an exceptional polynomial, where one fixed polynomial works for infinitely many finite fields.
§8.3	Mentions several explicit Chebotarev density theorem error terms. Such error terms have improved over time, but, like Proposition 9.7.28, this sections' results exhibit monodromy precision: the error term vanishes.
§9.6	Discusses Dickson polynomials in detail, including their various combinatorial formulas. This contrasts with Remark 9.7.34 which provides a formula free characterization.

References Cited: [84, 332, 341, 696, 777, 814, 943, 1021, 1109, 1110, 1111, 1112, 1113, 1114, 1115, 1117, 1118, 1119, 1120, 1121, 1123, 1124, 1370, 1427, 1936, 1986, 2031, 2191, 2203, 2310, 2311, 2586, 2587, 2590, 2595, 2614]

10

Sequences over finite fields

10.1 Finite field transforms

Gary McGuire, University College Dublin

10.1.1 Basic definitions and important examples

A finite field contains two important finite abelian groups, the additive group and the multiplicative group. We shall first define the Fourier transform for an arbitrary finite abelian group, and then we shall focus on the two groups in a finite field. The definition involves the characters of a finite abelian group.

10.1.1 **Definition** Let G be a finite abelian group. A *character* of G is a group homomorphism $G \longrightarrow \mathbb{C}^\times$, where $\mathbb{C}^\times$ is the multiplicative group of nonzero complex numbers.

10.1.2 Remark Let $\widehat{G}$ denote the group of characters of G, which is a group under the operation $(\chi\chi')(x) = \chi(x)\chi'(x)$. Sometimes $\widehat{G}$ is called the dual group of G. The dual group $\widehat{G}$ is isomorphic to G, although this isomorphism is not canonical.

10.1.3 Definition Let G be a finite abelian group. The *Fourier transform* of any function $f : G \longrightarrow \mathbb{C}$ is the function $\widehat{f} : \widehat{G} \longrightarrow \mathbb{C}$ defined by

$$\widehat{f}(\chi) = \sum_{x \in G} f(x)\overline{\chi(x)}$$

where χ is a character of G.

10.1.4 Remark The Fourier transform can also be defined without the complex conjugation of $\chi(x)$.

10.1.5 Remark It is common to choose an identification of G and $\widehat{G}$. If χ_α denotes the image of $\alpha \in G$ under some isomorphism from G to $\widehat{G}$, we write the Fourier transform as

$$\widehat{f}(\alpha) = \sum_{x \in G} f(x)\overline{\chi_\alpha(x)}$$

and as a result we may consider $\widehat{f}$ to be defined on G.

10.1.6 Example The characters of the additive group of $\mathbb{F}_q$ have the form

$$\mu_\alpha(x) = \zeta_p^{\langle \alpha, x \rangle}$$

where ζ_p is a fixed primitive p-th root of unity, and $\langle , \rangle$ is any $\mathbb{F}_p$-valued inner product on $\mathbb{F}_q$. If we take $\langle \alpha, x \rangle$ to be $\mathrm{Tr}(\alpha x)$ (absolute trace) then the Fourier transform of a function $f : \mathbb{F}_q \longrightarrow \mathbb{C}$ becomes

$$\widehat{f}(\alpha) = \sum_{x \in \mathbb{F}_q} f(x)\zeta_p^{-\mathrm{Tr}(\alpha x)}.$$

10.1.7 Example The characters of $\mathbb{F}_q^*$ (the multiplicative group of nonzero elements of $\mathbb{F}_q$) have the form

$$\chi_j(\gamma^k) = \zeta_{q-1}^{jk}$$

where $0 \le j \le q-2$, ζ_{q-1} is a fixed primitive complex $(q-1)$-th root of unity, and γ is a fixed generator of $\mathbb{F}_q^*$. For example, if q is odd and $j = (q-1)/2$ then χ_j is the quadratic character. The Fourier transform of a function $f : \mathbb{F}_q^* \longrightarrow \mathbb{C}$ can then be written

$$\widehat{f}(j) = \sum_{k=0}^{q-2} f(\gamma^k)\zeta_{q-1}^{-jk}$$

10.1.8 Example Let $n > 1$ be a positive integer. If $G = \mathbb{Z}/n\mathbb{Z}$, the characters are the functions $\chi_j : \mathbb{Z}/n\mathbb{Z} \to \mathbb{C}$, for $j \in \mathbb{Z}/n\mathbb{Z}$, where

$$\chi_j(k) = \zeta_n^{jk}$$

where ζ_n is a complex primitive n-th root of unity. The Fourier transform of a function $f : \mathbb{Z}/n\mathbb{Z} \longrightarrow \mathbb{C}$ can then be written

$$\widehat{f}(j) = \sum_{k=0}^{n-1} f(k)\zeta_n^{-jk}.$$

See Subsection 10.1.3 for a discussion of the Discrete Fourier Transform.

10.1.9 Remark The space $\mathbb{C}^G$ of all complex-valued functions defined on G is a Hermitian inner product space via

$$\langle f, g \rangle = \frac{1}{|G|} \sum_{x \in G} f(x)\overline{g(x)}.$$

This vector space has dimension $|G|$.

10.1.10 Theorem The characters form an orthonormal basis of $\mathbb{C}^G$.

10.1.11 Remark We note that

$$\widehat{f}(\chi) = \langle f, \chi \rangle.$$

10.1.12 Remark The Fourier transform can be inverted, in the sense that

$$f(x) = \frac{1}{|G|} \sum_{\chi \in \widehat{G}} \langle f, \chi \rangle \chi(x)$$

for all $x \in G$. This expresses f as a linear combination of the basis of characters. Note that it also expresses each value of f as a linear combination of roots of unity. This expression explains why the values $\widehat{f}(\chi)$ are sometimes called the Fourier coefficients of f.

10.1.13 Remark Next we present some of the fundamental theorems in Fourier analysis. Proofs can be found in [2789].

10.1.14 Theorem Plancherel's Theorem states that

$$\langle f, g \rangle = \frac{1}{|G|} \langle \widehat{f}, \widehat{g} \rangle.$$

10.1.15 Theorem Parseval's Theorem states that

$$\langle f, f \rangle = \frac{1}{|G|} \langle \widehat{f}, \widehat{f} \rangle.$$

10.1.16 Theorem The Poisson Summation Formula states that, for any subgroup H of G,

$$\frac{1}{|H|} \sum_{x \in H} f(x) = \frac{1}{|G|} \sum_{\chi \in H^\perp} \widehat{f}(\chi)$$

where $H^\perp$ consists of all the characters of G that are trivial on H.

10.1.17 Remark This is a discrete version of the Poisson Summation Formula. There are many other versions, and many applications, see [2789] for further details.

10.1.18 Definition The *convolution* of f and g is defined by

$$(f * g)(a) = \sum_{x \in G} f(a - x) g(x)$$

for all $a \in G$.

10.1.19 Theorem The convolution theorem of Fourier analysis states that the Fourier transform of a convolution of two functions is equal to the ordinary product of their Fourier transforms:

$$\widehat{f * g} = \widehat{f} \cdot \widehat{g}.$$

10.1.2 Functions between two groups

10.1.20 Remark We have defined above the Fourier transform of a function $G \longrightarrow \mathbb{C}$. The definition can be extended to functions defined on G taking values in another abelian group, B. The definition involves characters of B as well as characters of G.

10.1.21 Definition Let $f : G \longrightarrow B$ be a function between finite abelian groups. The *Fourier transform of* f is the function $\widehat{f} : \widehat{G} \times \widehat{B} \longrightarrow \mathbb{C}$ defined by

$$\widehat{f}(\chi, \psi) = \sum_{x \in G} \psi(f(x)) \overline{\chi(x)}. \tag{10.1.1}$$

10.1.22 Remark We observe that if $\psi \in \widehat{B}$ is the principal character, the Fourier transform evaluates to 0 if χ is not principal, and evaluates to $|G|$ if χ is principal.

10.1.23 Remark This Fourier transform of $f : G \longrightarrow B$ can also be viewed as the ordinary Fourier transform (Definition 10.1.3) on the group $G \times B$ of the characteristic function of the set $\{(x, f(x)) : x \in G\}$.

10.1.24 Remark If we use isomorphisms $\alpha \mapsto \chi_\alpha$ from G to $\widehat{G}$ and $\beta \mapsto \psi_\beta$ from B to $\widehat{B}$, we may consider $\widehat{f}$ to be defined on $G \times B$ and we write the Fourier transform of f at $(\alpha, \beta) \in G \times B$ as

$$\widehat{f}(\alpha, \beta) = \sum_{x \in G} \psi_\beta(f(x)) \overline{\chi_\alpha(x)}. \tag{10.1.2}$$

10.1.25 Example The Fourier transform of a function $f : \mathbb{F}_q \longrightarrow \mathbb{F}_q$ is

$$\widehat{f}(\alpha, \beta) = \sum_{x \in \mathbb{F}_q} \zeta^{\mathrm{Tr}(\beta f(x) - \alpha x)}$$

because the additive characters (see Example 10.1.6) have the form $\mathrm{Tr}(\alpha x)$.

10.1.26 Remark We note that in the important special case where $f(x) = x^d$ and $\gcd(d, q-1) = 1$, we may write any $\beta \in \mathbb{F}_q$ as c^d, and then

$$\widehat{f}(\alpha, \beta) = \sum_{x \in \mathbb{F}_q} \zeta^{\mathrm{Tr}(c^d x^d - \alpha x)} = \sum_{x \in \mathbb{F}_q} \zeta^{\mathrm{Tr}(x^d - \alpha c^{-1} x)} = \widehat{f}(\alpha c^{-1}, 1).$$

It follows that, when $f(x) = x^d$ and $\gcd(d, q-1) = 1$, we may often assume without loss of generality that $\beta = 1$.

10.1.27 Example For a Boolean function $f : \mathbb{F}_2^n \longrightarrow \mathbb{F}_2$, the characters of $\mathbb{F}_2^n$ have the form

$$\chi_\alpha(x) = (-1)^{\langle \alpha, x \rangle}$$

where $\langle , \rangle$ is any inner product on $\mathbb{F}_2^n$. Also, the only nonzero β in $\mathbb{F}_2$ is $\beta = 1$, so we may drop the dependence on β and the Fourier transform is written

$$\widehat{f}(\alpha) = \sum_{x \in \mathbb{F}_2^n} (-1)^{f(x) + \langle \alpha, x \rangle}.$$

This is the Walsh transform of the Boolean function f, or the Hadamard transform of the ± 1 valued function $(-1)^f$.

10.1.28 Remark One often uses the finite field $\mathbb{F}_{2^n}$ for the vector space $\mathbb{F}_2^n$, and the trace inner product $\langle \alpha, x \rangle = \mathrm{Tr}(\alpha x)$. In this case the Walsh transform of a Boolean function f is

$$\widehat{f}(\alpha) = \sum_{x \in \mathbb{F}_2^n} (-1)^{f(x)+\mathrm{Tr}(\alpha x)}.$$

See Section 9.1 for further details on Boolean functions.

10.1.29 Example Let $n > 1$ be a positive integer. If $G = B = \mathbb{Z}/n\mathbb{Z}$, the characters are the functions $\chi_j(k) = \zeta_n^{jk}$ where ζ_n is a complex primitive n-th root of unity (see Example 10.1.8). For a function $f : \mathbb{Z}/n\mathbb{Z} \longrightarrow \mathbb{Z}/n\mathbb{Z}$ it follows that

$$\widehat{f}(\alpha, \beta) = \sum_{x \in \mathbb{Z}/n\mathbb{Z}} \zeta_n^{\beta f(x) - \alpha x}. \tag{10.1.3}$$

10.1.3 Discrete Fourier Transform

10.1.30 Remark A function $f : G \longrightarrow B$ can also be considered as a vector or sequence

$$(f(g_1), \ldots, f(g_n)) \in B^n$$

when $G = \{g_1, \ldots, g_n\}$. Through this correspondence, transforms of functions are sometimes considered as transforms of vectors and sequences.

10.1.31 Definition If $G = \{g_1, \ldots, g_n\}$ and $\widehat{G} = \{\chi_1, \ldots, \chi_n\}$, the *character table of* G is the matrix with (i, j) entry $\chi_i(g_j)$.

10.1.32 Remark If we identify a function $f : G \longrightarrow \mathbb{C}$ with its vector $(f(g_1), \ldots, f(g_n))$, the Fourier Transform of f is the vector obtained by multiplying the vector f by the character table of G:

$$\widehat{f} = \overline{X} f$$

where X is the character table of G.

10.1.33 Remark We have already presented a case of the Discrete Fourier Transform in Example 10.1.8, however we shall give the matrix formulation here, which is more common. In fact, the field K below does not need to be a finite field. Further information can be found in many places, see [2789] for example.

10.1.34 Definition Let K be a field containing all the n-th roots of unity. Let ζ_n be a primitive n-th root of unity in K. Let F_n be the $n \times n$ matrix whose (i, j) entry is ζ_n^{ij}, where $0 \le i, j \le n - 1$. The matrix F_n is the *n-th Fourier matrix.*

10.1.35 Remark When $K = \mathbb{C}$ the Fourier matrix F_n is an important example of a Butson-Hadamard matrix. We note that F_n is also a Vandermonde matrix.

10.1.36 Remark We modify the matrix F_n to the matrix D_n whose (i, j) entry is ζ_n^{-ij} (we complex conjugate each entry of F_n). We remark that it does not matter whether we use F_n or D_n in the definition of Discrete Fourier Transform, but we shall use D_n to be consistent with our earlier definitions.

10.1.37 **Definition** Let $f = (f_0, f_1, \ldots, f_{n-1})$ be in K^n. The *Discrete Fourier Transform* of f is the vector

$$\widehat{f} = D_n f = (\widehat{f_0}, \widehat{f_1}, \ldots, \widehat{f}_{n-1})$$

where $\widehat{f_j} = \sum_{k=0}^{n-1} f_k \zeta_n^{-jk}$.

10.1.38 **Remark** The Fourier matrix F_n is the character table of $G = \mathbb{Z}/n\mathbb{Z}$, and this definition is a case of Remark 10.1.32.

10.1.39 **Remark** If $K = \mathbb{C}$ the Discrete Fourier Transform (DFT) is the same as Example 10.1.8. If $K = \mathbb{F}_q$ and n is a divisor of $q-1$, the DFT is known as the Discrete Fourier Transform over a finite field.

10.1.40 **Remark** The inverse of F_n has (i,j) entry $\frac{1}{n}\zeta_n^{-ij}$, i.e., $D_n F_n = nI_n$. Therefore the Discrete Fourier Transform has an inverse, which is almost a Discrete Fourier Transform itself, except for the factor of $1/n$.

Sometimes authors include a factor of $1/\sqrt{n}$ in the definition, which then also appears in the inverse, so with this definition the inverse DFT is also a DFT.

If K is a field of characteristic p, we assume here that n is relatively prime to p so that $1/n$ exists in K. If p divides n, a generalized DFT has been defined in [2013] using the values of the Hasse derivatives.

10.1.41 **Remark** The DFT as defined here may also be viewed as the DFT on the group ring $K[G]$ where G is a cyclic group. This definition can be generalized to a DFT on other group rings. Rings which support Fourier transforms are characterized in [418].

10.1.42 **Remark** Sometimes we identify the vector $f = (f_0, f_1, \ldots, f_{n-1})$ with the polynomial $f(x) = f_0 + f_1 x + \cdots + f_{n-1}x^{n-1}$, and then

$$\widehat{f} = (f(1), f(\overline{\zeta_n}), f(\overline{\zeta_n^2}), \ldots, f(\overline{\zeta_n^{n-1}})).$$

Thus the Discrete Fourier Transform of f is the vector of values of the polynomial f at the n-th roots of unity.

10.1.43 **Remark** The Fast Fourier Transform is a computationally efficient way of computing the Discrete Fourier Transform, and has many applications. Traditionally, n is a power of 2 and one uses polynomial evaluations as in Remark 10.1.42. There is a huge literature on this topic, see [2789] for example, so we do not go into this here.

10.1.44 **Remark** Given a vector $f = (f_0, f_1, \ldots, f_{n-1})$ in K^n, we construct a circulant matrix F with f as its top row. Similarly we construct a circulant matrix $\widehat{F}$ with $\widehat{f} = (\widehat{f_0}, \widehat{f_1}, \ldots, \widehat{f}_{n-1})$ as its top row. There is a result sometimes known as Blahut's theorem [141] stating that the weight of f is equal to the rank of this circulant $\widehat{F}$, and the weight of $\widehat{f}$ is equal to the rank of F. This is true because if we let D be the diagonal matrix $\mathrm{diag}(f_0, f_1, \ldots, f_{n-1})$, we observe that $F_n D F_n = \widehat{F}$, and because F_n is invertible, the rank of D (which is the weight of f) is equal to the rank of $\widehat{F}$. Because the linear complexity of f is equal to the rank of F, this result can be useful for linear complexities of sequences.

10.1.4 Further topics

10.1.45 **Remark** In this subsection, as before, G and B denote finite abelian groups.

10.1.4.1 Fourier spectrum

10.1.46 Remark The Fourier spectrum is the set of values of the Fourier transform. Most authors take the Fourier spectrum to be the multi-set of values, i.e., the values including their multiplicities. The Fourier spectrum is an important invariant in many applications.

10.1.47 Example The Fourier spectrum of the function x^3 on $\mathbb{F}_{2^n}$ is $\{0, \pm 2^{(n+1)/2}\}$ if n is odd, and $\{0, \pm 2^{n/2}, \pm 2^{(n+2)/2}\}$ if n is even.

10.1.48 Remark There are many papers calculating the Fourier spectrum of specific functions that are of particular interest, see [384] or [387] for example. There are also general papers on the structure of the Fourier spectrum of arbitrary functions. The Weil bound gives an upper bound on the absolute value of any Fourier coefficient, a result that we state here, and that has many variations.

10.1.49 Definition The *Weil sum* associated to an additive character μ of $\mathbb{F}_q$ and a polynomial $f \in \mathbb{F}_q[x]$ is

$$\sum_{x \in \mathbb{F}_q} \mu(f(x)).$$

10.1.50 Theorem (Weil bound) If the Weil sum is non-degenerate then

$$\left| \sum_{x \in \mathbb{F}_q} \mu(f(x)) \right| \leq (\deg(f) - 1)\sqrt{q}.$$

10.1.4.2 Nonlinearity

10.1.51 Definition The linearity of a function $f : G \longrightarrow B$ is defined to be the maximum of the absolute values of the numbers in the Fourier spectrum, i.e., we define the *linearity* of f by

$$\mathbb{L}(f) = \max_{\alpha \in G, \beta \in B^*} |\widehat{f}(\alpha, \beta)|.$$

10.1.52 Definition The *nonlinearity* of $f : G \longrightarrow B$ is defined by $\mathbb{NL}(f) = (|G| - \mathbb{L}(f))/|B|$.

10.1.53 Remark For a Boolean function $f : \mathbb{F}_2^n \longrightarrow \mathbb{F}_2$, the nonlinearity of f is $(2^n - \mathbb{L}(f))/2$. The nonlinearity of a Boolean function is often a useful measure in cryptography; see Section 9.1.

10.1.4.3 Characteristic functions

10.1.54 Definition If A is a subset of $\mathbb{F}_q$, the *characteristic function of A* is

$$\chi_A(x) = \begin{cases} 1 & \text{if } x \in A \\ 0 & \text{if } x \notin A. \end{cases}$$

10.1.55 Remark The Fourier Transform of χ_A can be used to obtain information about A. There are some general principles, such as: if all the Fourier coefficients of χ_A are small, relatively speaking, then A is usually a fairly "random" subset; see [155] for further details. In a different direction, the paper [865] is an example of utilizing the Fourier Transform of the characteristic function of the support of a specific function of interest.

10.1.4.4 Gauss sums

10.1.56 Definition The *Gauss sum* associated to a multiplicative character χ (i.e., a character of $\mathbb{F}_q^*$) and an additive character μ (i.e., a character of $\mathbb{F}_q$) is

$$S(\chi, \mu) = \sum_{x \in \mathbb{F}_q^*} \chi(x)\mu(x). \tag{10.1.4}$$

See Section 6.1 for further details.

10.1.57 Remark Here is one interpretation of Gauss sums. Working in the space of functions $\mathbb{F}_q^* \longrightarrow \mathbb{C}$ we consider μ as a function on $\mathbb{F}_q^*$ by restriction. Using Fourier inversion on (10.1.4), we can express an additive character μ in terms of the basis of multiplicative characters; and the coefficients in this expansion are Gauss sums.

10.1.4.5 Uncertainty principle

10.1.58 Remark If G is any finite abelian group, the uncertainty principle [2789] states that

$$|\text{supp } (f)| \cdot |\text{supp } (\widehat{f})| \geq |G|$$

for any nonzero complex-valued function f defined on G. In particular, if $G = \mathbb{F}_q$ we have

$$|\text{supp } (f)| \cdot |\text{supp } (\widehat{f})| \geq q.$$

Thus, a function and its Fourier transform cannot both have "small" support. Tao [2780] recently showed that, in the case $q = p$, the uncertainty principle can be improved as follows:

$$|\text{supp } (f)| + |\text{supp } (\widehat{f})| \geq p + 1.$$

See Also

§6.2	For discussion of character sums.
§6.3	For applications of character sums.
§9.1	For discussion of Boolean functions.
§9.2	For discussion of PN and APN functions.
§9.3	For a study of bent functions.

References Cited: [141, 155, 384, 387, 418, 865, 2013, 2780, 2789]

10.2 LFSR sequences and maximal period sequences

Harald Niederreiter, KFUPM

10.2.1 General properties of LFSR sequences

10.2.1 **Definition** Let k be a positive integer and let $a_0, a_1, \ldots, a_{k-1}$ be fixed elements of the finite field $\mathbb{F}_q$. A sequence $s_0, s_1, \ldots$ of elements of $\mathbb{F}_q$ satisfying the linear recurrence relation

$$s_{n+k} = \sum_{i=0}^{k-1} a_i s_{n+i} \qquad \text{for } n = 0, 1, \ldots$$

is an *LFSR sequence* (or a *linear feedback shift register sequence*, also a *linear recurring sequence*) in $\mathbb{F}_q$. The integer k is the *order* of the LFSR sequence or of the linear recurrence relation.

10.2.2 **Remark** We usually abbreviate the sequence $s_0, s_1, \ldots$ by (s_n). The sequence (s_n) in Definition 10.2.1 is uniquely determined by the linear recurrence relation and by the *initial values* $s_0, s_1, \ldots, s_{k-1}$.

10.2.3 **Remark** In electrical engineering, LFSR sequences in $\mathbb{F}_q$ are generated by special switching circuits called linear feedback shift registers. A linear feedback shift register consists of adders and multipliers for arithmetic in $\mathbb{F}_q$ as well as delay elements. In the binary case $q = 2$, multipliers are not needed.

10.2.4 **Theorem** Any LFSR sequence (s_n) in $\mathbb{F}_q$ of order k is ultimately periodic with least period at most $q^k - 1$. A sufficient condition for (s_n) to be (purely) periodic is that the coefficient a_0 in the linear recurrence relation in Definition 10.2.1 is nonzero.

10.2.5 **Definition** Let (s_n) be an LFSR sequence in $\mathbb{F}_q$ of order k satisfying the linear recurrence relation in Definition 10.2.1. Then the polynomial

$$f(x) = x^k - \sum_{i=0}^{k-1} a_i x^i \in \mathbb{F}_q[x]$$

is a *characteristic polynomial* of (s_n) and also a characteristic polynomial of the linear recurrence relation. The reciprocal polynomial $f^*(x) = x^k f(1/x)$ of f is a *connection polynomial* of (s_n) and also a connection polynomial of the linear recurrence relation.

10.2.6 **Definition** Let (s_n) be an LFSR sequence in $\mathbb{F}_q$ of order k. Then for $n = 0, 1, \ldots$, the vector

$$\mathbf{s}_n = (s_n, s_{n+1}, \ldots, s_{n+k-1}) \in \mathbb{F}_q^k$$

is the n-th *state vector* of (s_n).

10.2.7 **Remark** Let (s_n) be an LFSR sequence in $\mathbb{F}_q$ with characteristic polynomial $f \in \mathbb{F}_q[x]$ and let A be the companion matrix of f. Then the linear recurrence relation for (s_n) in Definition 10.2.1 can be written as the identity $\mathbf{s}_{n+1} = \mathbf{s}_n A$, $n = 0, 1, \ldots$, for the state vectors. Consequently, we get $\mathbf{s}_n = \mathbf{s}_0 A^n$ for $n = 0, 1, \ldots$. Since A^n can be calculated by $O(\log n)$ matrix multiplications using the standard square-and-multiply technique, this identity leads to an efficient algorithm for computing remote terms of the LFSR sequence (s_n).

10.2.8 Remark Let $\mathbb{F}_q^\infty$ be the sequence space over $\mathbb{F}_q$, viewed as a vector space over $\mathbb{F}_q$ under termwise operations for sequences. Let T be the shift operator

$$T\omega = (w_{n+1}) \qquad \text{for all } \omega = (w_n) \in \mathbb{F}_q^\infty.$$

Then for a characteristic polynomial $f \in \mathbb{F}_q[x]$ of an LFSR sequence $\sigma = (s_n)$ in $\mathbb{F}_q$ we have $f(T)(\sigma) = (0)$, where (0) denotes the zero sequence. The set

$$I(\sigma) = \{g \in \mathbb{F}_q[x] : g(T)(\sigma) = (0)\}$$

of annihilating polynomials is a nonzero ideal in $\mathbb{F}_q[x]$.

10.2.9 Definition The uniquely determined monic polynomial over $\mathbb{F}_q$ generating the ideal $I(\sigma)$ in Remark 10.2.8 is the *minimal polynomial* of the LFSR sequence σ in $\mathbb{F}_q$.

10.2.10 Remark If σ is the zero sequence, then its minimal polynomial is the constant polynomial 1. If σ is a nonzero LFSR sequence, then its minimal polynomial has positive degree and is the characteristic polynomial of the linear recurrence relation of least possible order satisfied by σ.

10.2.11 Theorem The minimal polynomial of an LFSR sequence (s_n) in $\mathbb{F}_q$ divides any characteristic polynomial of (s_n). A characteristic polynomial of (s_n) of degree $k \geq 1$ is the minimal polynomial of (s_n) if and only if the corresponding state vectors $\mathbf{s}_0, \mathbf{s}_1, \ldots, \mathbf{s}_{k-1}$ are linearly independent over $\mathbb{F}_q$.

10.2.12 Example Let (s_n) be an LFSR sequence in $\mathbb{F}_q$ of order k with initial values $s_0 = s_1 = \cdots = s_{k-2} = 0$, $s_{k-1} = 1$ ($s_0 = 1$ if $k = 1$). Then the linear independence property in the second part of Theorem 10.2.11 is clearly satisfied, and so the characteristic polynomial of (s_n) of degree k is also the minimal polynomial of (s_n). An LFSR sequence with these special initial values is an *impulse response sequence.*

10.2.13 Theorem If $m \in \mathbb{F}_q[x]$ is the minimal polynomial of the LFSR sequence σ in $\mathbb{F}_q$, then the least period of σ is equal to the order of m and the least preperiod of σ is equal to the multiplicity of 0 as a root of m.

10.2.14 Corollary An LFSR sequence in $\mathbb{F}_q$ is periodic if and only if its minimal polynomial $m \in \mathbb{F}_q[x]$ satisfies $m(0) \neq 0$.

10.2.15 Remark The basic theory of LFSR sequences in finite fields, as presented above, has quite a long history. An important early paper is Zierler [3070]. Other milestones in the history of LFSR sequences in finite fields are the lecture notes of Selmer [2579] and the book of Golomb [1300]. A treatment of LFSR sequences in the wider context of general feedback shift register sequences in finite fields, i.e., those including also nonlinear feedback functions, is given in the monograph of Ronse [2475]. The proofs of many results in this section can be found in Chapters 6 and 7 of the book [1938].

10.2.16 Theorem [1063] Let (s_n) be an LFSR sequence in $\mathbb{F}_q$ with characteristic polynomial $f \in \mathbb{F}_q[x]$. Let e_0 be the multiplicity of 0 as a root of f, where we can have $e_0 = 0$, and let $\alpha_1, \ldots, \alpha_h$ be the distinct nonzero roots of f (in its splitting field F over $\mathbb{F}_q$) with multiplicities $e_1, \ldots, e_h$, respectively. Then

$$s_n = t_n + \sum_{i=1}^{h} \sum_{j=0}^{e_i-1} \binom{n+j}{j} \beta_{ij} \alpha_i^n \qquad \text{for } n = 0, 1, \ldots,$$

where all $t_n \in \mathbb{F}_q$, $t_n = 0$ for $n \geq e_0$, and all $\beta_{ij} \in F$.

10.2.17 Corollary If a characteristic polynomial f of (s_n) is irreducible over $\mathbb{F}_q$ and α is a root of f in the splitting field F of f over $\mathbb{F}_q$, then there exists a uniquely determined $\beta \in F$ such that

$$s_n = \mathrm{Tr}_{F/\mathbb{F}_q}(\beta\alpha^n) \qquad \text{for } n = 0, 1, \ldots.$$

10.2.2 Operations with LFSR sequences and characterizations

10.2.18 Theorem For each $i = 1, \ldots, h$, let σ_i be an LFSR sequence in $\mathbb{F}_q$ with minimal polynomial $m_i \in \mathbb{F}_q[x]$ and least period r_i. If $m_1, \ldots, m_h$ are pairwise coprime, then the minimal polynomial of the (termwise) sum sequence $\sigma_1 + \cdots + \sigma_h$ is equal to the product $m_1 \cdots m_h$ and the least period of $\sigma_1 + \cdots + \sigma_h$ is equal to the least common multiple of $r_1, \ldots, r_h$.

10.2.19 Remark In general, operations with LFSR sequences are treated in terms of the spaces $S(f)$, where for a monic $f \in \mathbb{F}_q[x]$ we let $S(f)$ be the kernel of the linear operator $f(T)$ on $\mathbb{F}_q^\infty$ (compare with Remark 10.2.8). Any $S(f)$ is a linear subspace of $\mathbb{F}_q^\infty$ of dimension $\deg(f)$. The following result characterizes the spaces $S(f)$.

10.2.20 Theorem A subset E of $\mathbb{F}_q^\infty$ is equal to $S(f)$ for some monic $f \in \mathbb{F}_q[x]$ if and only if E is a finite-dimensional subspace of $\mathbb{F}_q^\infty$ which is closed under the shift operator T.

10.2.21 Theorem For any monic $f_1, \ldots, f_h \in \mathbb{F}_q[x]$, we have

$$\begin{aligned} S(f_1) \cap \cdots \cap S(f_h) &= S(\gcd(f_1, \ldots, f_h)), \\ S(f_1) + \cdots + S(f_h) &= S(\mathrm{lcm}(f_1, \ldots, f_h)). \end{aligned}$$

10.2.22 Remark The (termwise) product sequence $\sigma_1 \cdots \sigma_h$ of LFSR sequences $\sigma_1, \ldots, \sigma_h$ in $\mathbb{F}_q$ is more difficult to analyze. For monic polynomials $f_1, \ldots, f_h \in \mathbb{F}_q[x]$, let $S(f_1) \cdots S(f_h)$ be the subspace of $\mathbb{F}_q^\infty$ spanned by all product sequences $\sigma_1 \cdots \sigma_h$ with $\sigma_i \in S(f_i)$ for $1 \leq i \leq h$. It follows from Theorem 10.2.20 that

$$S(f_1) \cdots S(f_h) = S(g)$$

for some monic $g \in \mathbb{F}_q[x]$. If each f_i, $1 \leq i \leq h$, is nonconstant and has only simple roots, then g is the monic polynomial whose roots are the distinct elements of the form $\alpha_1 \cdots \alpha_h$, where each α_i is a root of f_i in the splitting field of $f_1 \cdots f_h$ over $\mathbb{F}_q$. For the general case, a procedure to determine g can be found in Zierler and Mills [3072].

10.2.23 Definition If $\sigma = (s_n)$ is a sequence of elements of $\mathbb{F}_q$ and d is a positive integer, then the operation of *decimation* produces the decimated sequence $\sigma^{(d)} = (s_{nd})$. Thus, $\sigma^{(d)}$ is obtained by taking every d-th term of σ, starting from s_0.

10.2.24 Theorem [939, 2237] Let σ be an LFSR sequence in $\mathbb{F}_q$. Then so is $\sigma^{(d)}$ for any positive integer d. If f is a characteristic polynomial of σ and $f(x) = \prod_{j=1}^{k}(x - \beta_j)$ is the factorization of f in its splitting field over $\mathbb{F}_q$, then $g_d(x) = \prod_{j=1}^{k}(x - \beta_j^d)$ is a characteristic polynomial of $\sigma^{(d)}$. Furthermore, if f is the minimal polynomial of σ and d is coprime to the least period of σ, then g_d is the minimal polynomial of $\sigma^{(d)}$.

10.2.25 Definition [1294] An LFSR sequence σ in $\mathbb{F}_q$ which has a characteristic polynomial $f \in \mathbb{F}_q[x]$ and satisfies $\sigma^{(q)} = \sigma$ is a *characteristic sequence* for f.

10.2.26 Remark Characteristic sequences play an important role in the Niederreiter algorithm for factoring polynomials over finite fields [2249, 2260]. Characteristic sequences can be described explicitly in terms of their generating functions (see Theorem 10.2.35 below).

10.2.27 Remark In the case $q = 2$, an interesting operation on sequences is that of binary complementation. If σ is a sequence of elements of $\mathbb{F}_2$, then its *binary complement* $\overline{\sigma}$ is obtained by replacing each term 0 in σ by 1 and each term 1 in σ by 0.

10.2.28 Theorem Let σ be an LFSR sequence in $\mathbb{F}_2$ with minimal polynomial $m \in \mathbb{F}_2[x]$. Write m in the form $m(x) = (x+1)^h m_1(x)$ with an integer $h \geq 0$ and $m_1 \in \mathbb{F}_2[x]$ satisfying $m_1(1) = 1$. Then the minimal polynomial $\overline{m}$ of the binary complement $\overline{\sigma}$ is given by $\overline{m}(x) = (x+1)m(x)$ if $h = 0$, $\overline{m}(x) = m_1(x)$ if $h = 1$, and $\overline{m}(x) = m(x)$ if $h \geq 2$.

10.2.29 Remark LFSR sequences in $\mathbb{F}_q$ can be characterized in terms of Hankel determinants. For an arbitrary sequence (s_n) of elements of $\mathbb{F}_q$ and for integers $n \geq 0$ and $b \geq 1$, define the Hankel determinant

$$D_n^{(b)} = \det((s_{n+i+j})_{0 \leq i,j \leq b-1}).$$

10.2.30 Theorem The sequence (s_n) of elements of $\mathbb{F}_q$ is an LFSR sequence in $\mathbb{F}_q$ if and only if there exists an integer $b \geq 1$ such that $D_n^{(b)} = 0$ for all sufficiently large n. Furthermore, (s_n) is an LFSR sequence in $\mathbb{F}_q$ with minimal polynomial of degree k if and only if $D_0^{(b)} = 0$ for all $b \geq k+1$ and $k+1$ is the least positive integer for which this holds.

10.2.31 Remark If an LFSR sequence in $\mathbb{F}_q$ is known to have a minimal polynomial of degree at most k for some integer $k \geq 1$, then the Berlekamp-Massey algorithm produces the minimal polynomial from the first $2k$ terms of the sequence [231, 2011].

10.2.32 Remark LFSR sequences in $\mathbb{F}_q$ can also be characterized in terms of their generating functions. There are two different characterizations, depending on whether the generating function is a formal power series in x or in x^{-1}.

10.2.33 Theorem The sequence (s_n) of elements of $\mathbb{F}_q$ is an LFSR sequence in $\mathbb{F}_q$ of order k with connection polynomial $c \in \mathbb{F}_q[x]$ if and only if

$$\sum_{n=0}^{\infty} s_n x^n = \frac{g(x)}{c(x)}$$

with $g \in \mathbb{F}_q[x]$ and $\deg(g) < k$.

10.2.34 Theorem The sequence (s_n) of elements of $\mathbb{F}_q$ is an LFSR sequence in $\mathbb{F}_q$ with minimal polynomial $m \in \mathbb{F}_q[x]$ if and only if

$$\sum_{n=0}^{\infty} s_n x^{-n-1} = \frac{g(x)}{m(x)}$$

with $g \in \mathbb{F}_q[x]$ and $\gcd(g, m) = 1$.

10.2.35 Theorem [2260] Let $f \in \mathbb{F}_q[x]$ be a monic polynomial of positive degree. Then the sequence (s_n) of elements of $\mathbb{F}_q$ is a characteristic sequence for f if and only if

$$\sum_{n=0}^{\infty} s_n x^{-n-1} = \sum_{i=1}^{h} b_i \frac{p_i'(x)}{p_i(x)} \quad \text{with } b_1, \ldots, b_h \in \mathbb{F}_q,$$

where $p_1, \ldots, p_h \in \mathbb{F}_q[x]$ are the distinct monic irreducible factors of f and p_i' is the first derivative of p_i.

10.2.3 Maximal period sequences

10.2.36 **Definition** An LFSR sequence in $\mathbb{F}_q$ whose minimal polynomial is a primitive polynomial over $\mathbb{F}_q$ is a *maximal period sequence* (or an *m-sequence*) in $\mathbb{F}_q$.

10.2.37 **Theorem** Any maximal period sequence σ in $\mathbb{F}_q$ is periodic with least period $q^k - 1$, where k is the degree of the minimal polynomial of σ.

10.2.38 **Remark** The terminology "maximal period sequence" stems from the fact that, by Theorems 10.2.4 and 10.2.37, $q^k - 1$ is the largest value that can be achieved by the least period of an LFSR sequence in $\mathbb{F}_q$ of order k.

10.2.39 **Theorem** Let (s_n) be a maximal period sequence in $\mathbb{F}_q$ with minimal polynomial of degree k. Then the state vectors $\mathbf{s}_0, \mathbf{s}_1, \ldots, \mathbf{s}_{q^k-2}$ of (s_n) run exactly through all nonzero vectors in $\mathbb{F}_q^k$.

10.2.40 **Theorem** A nonzero periodic sequence σ of elements of $\mathbb{F}_q$ is a maximal period sequence in $\mathbb{F}_q$ if and only if all its shifted sequences $T^h\sigma$, $h = 0, 1, \ldots$, together with the zero sequence form an $\mathbb{F}_q$-linear subspace of $\mathbb{F}_q^\infty$.

10.2.41 **Theorem** Let σ be a maximal period sequence in $\mathbb{F}_q$ with minimal polynomial of degree k. Then every LFSR sequence in $\mathbb{F}_q$ having an irreducible minimal polynomial g with $g(0) \neq 0$ and $\deg(g)$ dividing k can be obtained from σ by applying a shift and then a decimation.

10.2.42 **Remark** For maximal period sequences, the autocorrelation function has a simple form. Let (s_n) be a maximal period sequence in $\mathbb{F}_q$ with least period r. Then its *autocorrelation function* C_r is defined by

$$C_r(h) = \sum_{n=0}^{r-1} \chi(s_n - s_{n+h})$$

for all positive integers h, where χ is a fixed nontrivial additive character of $\mathbb{F}_q$.

10.2.43 **Theorem** For any maximal period sequence in $\mathbb{F}_q$ with least period r, its autocorrelation function C_r satisfies $C_r(h) = r$ if $h \equiv 0 \pmod{r}$ and $C_r(h) = -1$ if $h \not\equiv 0 \pmod{r}$.

10.2.4 Distribution properties of LFSR sequences

10.2.44 **Remark** For LFSR sequences in $\mathbb{F}_q$ for which the least period is sufficiently large compared to the order, the terms in the full period are almost evenly distributed over $\mathbb{F}_q$. Without loss of generality, it suffices to consider periodic LFSR sequences in $\mathbb{F}_q$. For such a sequence $\sigma = (s_n)$ with least period r and for $b \in \mathbb{F}_q$, let $Z(b;\sigma)$ be the number of integers n with $0 \leq n \leq r-1$ such that $s_n = b$. In other words, $Z(b;\sigma)$ is the number of occurrences of b in a full period of σ.

10.2.45 **Theorem** Let σ be a periodic LFSR sequence in $\mathbb{F}_q$ of order k and with least period r. Then for any $b \in \mathbb{F}_q$ we have

$$\left| Z(b;\sigma) - \frac{r}{q} \right| \leq \left(1 - \frac{1}{q}\right) q^{k/2}.$$

10.2.46 **Remark** If σ is a maximal period sequence in $\mathbb{F}_q$ with minimal polynomial of degree k, then it follows from Theorem 10.2.39 that $Z(b;\sigma) = q^{k-1}$ for $b \neq 0$ and $Z(0;\sigma) = q^{k-1} - 1$.

10.2.47 **Remark** Let the sequence $\sigma = (s_n)$ be as in Remark 10.2.44. For $b \in \mathbb{F}_q$ and a positive integer N, let $Z(b;N;\sigma)$ be the number of integers n with $0 \leq n \leq N-1$ such that $s_n = b$.

10.2.48 Theorem [2231] Let σ be a periodic LFSR sequence in $\mathbb{F}_q$ of order k and with least period r. Then for any $b \in \mathbb{F}_q$ and any integer N with $1 \leq N \leq r$ we have

$$\left|Z(b; N; \sigma) - \frac{N}{q}\right| \leq \left(1 - \frac{1}{q}\right) q^{k/2} \left(\frac{2}{\pi} \log r + \frac{7}{5}\right).$$

10.2.49 Remark The distribution of blocks of elements in a periodic LFSR sequence $\sigma = (s_n)$ in $\mathbb{F}_q$ has also been investigated. Let r be the least period of σ. For $\mathbf{b} = (b_1, \ldots, b_t) \in \mathbb{F}_q^t$ with a positive integer t, let $Z(\mathbf{b}; \sigma)$ be the number of integers n with $0 \leq n \leq r-1$ such that $s_{n+i-1} = b_i$ for $1 \leq i \leq t$. The simplest case is that of a maximal period sequence σ in $\mathbb{F}_q$. If k is the degree of the minimal polynomial of σ and $1 \leq t \leq k$, then $Z(\mathbf{b}; \sigma) = q^{k-t}$ for $\mathbf{b} \in \mathbb{F}_q^t$ with $\mathbf{b} \neq \mathbf{0}$, whereas $Z(\mathbf{0}; \sigma) = q^{k-t} - 1$.

10.2.50 Theorem [2233] Let σ be a periodic LFSR sequence in $\mathbb{F}_q$ with least period r. Let $m \in \mathbb{F}_q[x]$ be the minimal polynomial of σ and put $k = \deg(m)$. Suppose that the positive integer t is less than or equal to the degree of any irreducible factor of m in $\mathbb{F}_q[x]$. Then for any $\mathbf{b} \in \mathbb{F}_q^t$ we have

$$\left|Z(\mathbf{b}; \sigma) - \frac{r}{q^t}\right| \leq \left(1 - \frac{1}{q^t}\right) q^{k/2}.$$

10.2.51 Remark More general results on distribution properties of LFSR sequences are available in [2233]. For instance, there is an analog of Theorem 10.2.48 for the distribution of blocks of elements in parts of the full period. Furthermore, one can consider not only blocks of successive terms of an LFSR sequence as in Remark 10.2.49, but also blocks of terms with arbitrary lags. Refined results for various special cases can be found in [2635, 2662].

10.2.5 Applications of LFSR sequences

10.2.52 Remark LFSR sequences in $\mathbb{F}_q$ have numerous applications. In this subsection, we mention some typical applications. We start with an application to combinatorics.

10.2.53 Definition An (h, k) *de Bruijn sequence* is a finite sequence $d_0, d_1, \ldots, d_{N-1}$ with $N = h^k$ terms from a nonempty finite set of h elements such that the k-tuples $(d_n, d_{n+1}, \ldots, d_{n+k-1})$, $n = 0, 1, \ldots, N-1$, with subscripts considered modulo N are all different.

10.2.54 Remark Let $s_0, s_1, \ldots$ be an impulse response sequence of order k (see Example 10.2.12) which is also a maximal period sequence in $\mathbb{F}_q$ with minimal polynomial of degree k. Then $0, s_0, s_1, \ldots, s_{q^k-2}$ is a (q, k) de Bruijn sequence. This follows immediately from Theorem 10.2.39.

10.2.55 Remark Any periodic sequence $\sigma = (s_n)$ of elements of $\mathbb{F}_q$, say with period r, satisfies the linear recurrence relation $s_{n+r} = s_n$ for $n = 0, 1, \ldots$ and is thus an LFSR sequence in $\mathbb{F}_q$. The *linear complexity* (or the *linear span*) of σ is defined to be the degree of the minimal polynomial of σ. The linear complexity is an important complexity measure in the theory of stream ciphers in cryptology. For details on the linear complexity, the reader is referred to Section 10.4.

10.2.56 Remark There is a family of cryptosystems which are based on LFSR sequences in $\mathbb{F}_q$ and the operation of decimation (see Definition 10.2.23). These cryptosystems were introduced in [2241] and are *FSR cryptosystems*.

10.2.57 Remark LFSR sequences in $\mathbb{F}_q$ can be used for the encoding of cyclic codes. We refer to Section 8.7 in the book of Peterson and Weldon [2390] for an account of this application.

This connection between LFSR sequences and cyclic codes, when combined with results of the type stated in Theorem 10.2.45, yields information on the weight distribution of cyclic codes [2232].

10.2.58 Remark Maximal period sequences in finite prime fields are used in methods for generating uniform pseudorandom numbers in the interval $[0, 1]$. Well-known methods of this type are the digital multistep method and the generalized feedback shift register (GFSR) method. We refer to Chapter 9 in the book [2248] for a detailed discussion of these methods.

10.2.59 Remark LFSR sequences in $\mathbb{F}_q$ have important applications in digital communication systems. A celebrated example is code division multiple access (CDMA) in wireless communication. The book of Viterbi [2880] is the standard reference for CDMA.

See Also

§10.3	For applications to correlation of sequences.
§10.4	For connections with linear complexity and cryptology.

References Cited: [231, 939, 1063, 1294, 1300, 1938, 2011, 2231, 2232, 2233, 2237, 2241, 2248, 2249, 2260, 2390, 2475, 2579, 2635, 2662, 2880, 3070, 3072]

10.3 Correlation and autocorrelation of sequences

Tor Helleseth, University of Bergen

10.3.1 Basic definitions

10.3.1 Definition Let $\{u(t)\}$ and $\{v(t)\}$ be two complex-valued sequences of period n. The *periodic correlation* of $\{u(t)\}$ and $\{v(t)\}$ at shift τ is the inner product

$$\theta_{u,v}(\tau) = \sum_{t=0}^{n-1} u(t+\tau)\overline{v}(t), \quad 0 \le \tau < n,$$

where $\overline{a}$ denotes the complex conjugation of a and $t + \tau$ is calculated modulo n. If the sequences $\{u(t)\}$ and $\{v(t)\}$ are the same, the correlation $\theta_{u,u}(\tau)$ is denoted by $\theta_u(\tau)$ and is the *autocorrelation*. When they are distinct $\theta_{u,v}(\tau)$ is the *crosscorrelation*.

10.3.2 Remark Sequences with good correlation properties have numerous applications in communication systems and lead to many challenging problems in finite fields. The main problems from an application point of view are to find single sequences with low autocorrelation for all nonzero shifts and families of sequences where the maximum nontrivial auto- and crosscorrelation values between any two sequences in the family is low. For a more detailed survey on the design and analysis of sequences with low correlation the reader is referred to [1475]. Other related references are [1215, 1303, 1476, 2526, 2673] and Chapter V (Section 7) in [706].

10.3.3 Remark The crosscorrelation between two sequences $\{a(t)\}$ and $\{b(t)\}$ that take on values in $\mathbb{Z}_q = \{0, 1, \ldots, q-1\}$ is defined using Definition 10.3.1 where $u(t) = \omega^{a(t)}$, $v(t) = \omega^{b(t)}$ and ω is a complex q-th root of unity, i.e.,

$$\theta_{a,b}(\tau) = \sum_{t=0}^{n-1} \omega^{a(t+\tau)-b(t)}, \quad 0 \leq \tau < n.$$

10.3.2 Autocorrelation of sequences

10.3.4 Definition A sequence $\{s(t)\}$ has *ideal autocorrelation* if $\theta_s(\tau) = 0$ for all $\tau \not\equiv 0 \pmod{n}$.

10.3.5 Remark Sequences with ideal autocorrelation do not always exist so a sequence has *optimal autocorrelation* if the maximal value of its autocorrelation is as small as it can be for a sequence of the given period and symbol alphabet. Many optimal sequences have constant autocorrelation of -1 for all out-of-phase shifts, i.e., when $\tau \not\equiv 0 \pmod{n}$.

10.3.6 Definition A q-ary *maximal-length linear sequence* $\{s(t)\}$ (or *m-sequence*) is a sequence of elements with symbols from $\mathbb{F}_q$ of period $q^m - 1$ generated from a nonzero initial state $(s(0), s(1), \ldots, s(m-1))$ and a linear recursion of degree m given by

$$\sum_{i=0}^{m} f_i s(t+i) = 0,$$

where the characteristic polynomial of the recursion, defined by $f(x) = \sum_{i=0}^{m} f_i x^i$, is a primitive polynomial in $\mathbb{F}_q[x]$ of degree m.

10.3.7 Theorem [1939, Theorem 8.24] An m-sequence, after a suitable cyclic shift, can be described using the trace function from $F = \mathbb{F}_{q^m}$ to $K = \mathbb{F}_q$ as

$$s(t) = \mathrm{Tr}_{F/K}(\alpha^t),$$

where α is a zero of the primitive characteristic polynomial of the m-sequence.

10.3.8 Remark Maximal-length linear sequences are balanced (with each nonzero element occurring q^{m-1} times and 0 occurring $q^{m-1} - 1$ times). Furthermore, for any $\tau \not\equiv 0 \pmod{q^m - 1}$ there is a δ such that $s(t+\tau) - s(t) = s(t+\delta)$. In combination with Remark 10.3.3 this leads to the following well-known theorem.

10.3.9 Theorem Let q be a prime. The autocorrelation of the q-ary m-sequence $\{s(t)\}$ is two-valued,

$$\theta_s(\tau) = \begin{cases} q^m - 1 & \text{if } \tau = 0 \pmod{q^m - 1}, \\ -1 & \text{if } \tau \not\equiv 0 \pmod{q^m - 1}. \end{cases}$$

10.3.10 Construction (GMW sequences) [1324, 2555] Let k, m be integers with $k \mid m$, $k \geq 1$ and let q be a prime power. Let also $\gcd(r, q^k - 1) = 1$, $1 \leq r \leq q^k - 2$. Let $F = \mathbb{F}_{q^m}$, $M = \mathbb{F}_{q^k}$, and $K = \mathbb{F}_q$. Let α be a primitive element of F and $c \in F$. A *GMW sequence* of period $q^m - 1$ is defined by

$$s(t) = \mathrm{Tr}_{M/K}\left(\left(\mathrm{Tr}_{F/M}\left(c\alpha^t\right)\right)^r\right).$$

10.3.11 Theorem [2555] A GMW sequence is balanced and, if q is a prime, has a two-valued autocorrelation with value -1 for all out-of-phase shifts.

10.3.12 Remark There is a close connection between a balanced binary sequence $\{s(t)\}$ of period $2^m - 1$ with autocorrelation -1 for all shifts $\tau \neq 0 \pmod{2^m - 1}$ and difference sets with *Singer* parameters $(2^m - 1, 2^{m-1} - 1, 2^{m-2} - 1)$. The connection is that $\{s(t)\}$ gives a Singer difference set $\pmod{2^m - 1}$ defined by $D = \{t \mid s(t) = 0\}$ (see Section 14.6).

10.3.13 Theorem [864] Let $1 \leq k < \frac{m}{2}$ and $\gcd(k, m) = 1$ and let α be a primitive element in $\mathbb{F}_{2^m}$. Define the subset of $\mathbb{F}_{2^m}$ by

$$U_k = \left\{(x+1)^{2^{2k}-2^k+1} + x^{2^{2k}-2^k+1} + 1 \mid x \in \mathbb{F}_{2^m}\right\}.$$

The binary sequence defined by $s_k(t) = 1$ if $\alpha^t \in \mathbb{F}_{2^m} \setminus U_k$ and $s_k(t) = 0$ otherwise, is balanced with $2^{m-1} - 1$ zeros and 2^{m-1} ones, and has two-valued autocorrelation with out-of-phase correlation value -1.

10.3.14 Construction (Legendre sequence) The *Legendre sequence* is a binary sequence $\{s(t)\}$ of a prime period p, and defined by

$$s(t) = \begin{cases} 1 & \text{if } t = 0 \text{ or } t \text{ is a nonsquare} \pmod{p}, \\ 0 & \text{otherwise.} \end{cases}$$

10.3.15 Theorem [2388] If $p \equiv 3 \pmod 4$ then the Legendre sequence has a two-valued autocorrelation with values $\theta_s(0) = p$ and $\theta_s(\tau) = -1$ for $\tau \neq 0 \pmod p$. If $p \equiv 1 \pmod 4$ the autocorrelation is inferior and takes on values 1 and -3 when $\tau \neq 0 \pmod p$ in addition to $\theta_s(0) = p$.

10.3.16 Construction (Binary Sidelnikov sequences) Let α be a primitive element in $\mathbb{F}_{p^m}$. The *binary Sidelnikov sequence* has period $p^m - 1$ and is defined by

$$s(t) = \begin{cases} 1 & \text{if } \alpha^t + 1 \text{ is a nonsquare in } \mathbb{F}_{p^m}, \\ 0 & \text{otherwise.} \end{cases}$$

10.3.17 Theorem [2659] Binary Sidelnikov sequences are balanced with three-valued optimal autocorrelation with out-of-phase values 0 or -4 when $n \equiv 0 \pmod 4$ and 2 and -2 when $n \equiv 2 \pmod 4$.

10.3.18 Theorem [1475] Let h be a function from $\mathbb{Z}_q$ to $\mathbb{Z}_q$. The sequence $\{h(t)\}$ of period q with symbols from $\mathbb{Z}_q$ has ideal autocorrelation if and only if h is a bent function.

10.3.19 Remark Information on bent functions can be found in Section 9.3.

10.3.3 Sequence families with low correlation

10.3.20 Remark In code-division multiple-access (CDMA) systems it is important to find large families of sequences where the (nontrivial) auto- and crosscorrelation between all pairs of sequences in the family are low.

10.3.21 Definition Let $\mathcal{F} = \{\{s_i(t)\} \mid i = 1, 2, \ldots, M\}$ be a family of M sequences of period n with symbols from the alphabet $\mathbb{Z}_q = \{0, 1, \ldots, q-1\}$. Let $\theta_{i,j}(\tau)$ denote the crosscorrelation between the sequences $\{s_i(t)\}$ and $\{s_j(t)\}$ at shift τ. The parameters of family $\mathcal{F}$ are (n, M, θ_{max}) where

$$\theta_{max} = \max\{|\theta_{i,j}(\tau)| : \text{ either } i \neq j \text{ or } \tau \neq 0\}.$$

10.3.22 Remark There are three well-known bounds on θ_{max} for a family of sequences with given period n and family size M. These bounds are due to Welch [2964], and Sidelnikov and Levenshtein [1475]. The Welch and Sidelnikov bounds are based on bounds on the inner products between complex vectors. In the Welch (resp. Sidelnikov) bound the sequences are considered as complex vectors of norm $\sqrt{n}$ (resp. complex q-th roots of unity).

10.3.23 Theorem (The Welch bound) Let $k \geq 1$ be an integer and $\mathcal{F}$ a family of M cyclically distinct sequences of period n. Then

$$(\theta_{\max})^{2k} \geq \frac{1}{Mn-1}\left(\frac{Mn^{2k+1}}{\binom{k+n-1}{n-1}} - n^{2k}\right).$$

10.3.24 Corollary (The Welch bound for $k = 1$)

$$\theta_{max} \geq n\sqrt{\frac{M-1}{Mn-1}}.$$

10.3.25 Remark For even moderate values of M this implies that $\theta_{max} \geq \sqrt{n}$.

10.3.26 Theorem (The Sidelnikov bound)

1. In the case $q = 2$, then

$$(\theta_{\max})^2 > (2k+1)(n-k) + \frac{k(k+1)}{2} - \frac{2^k n^{2k+1}}{M(2k)!\binom{n}{k}}, \quad 0 \leq k < \frac{2n}{5}.$$

2. In the case $q > 2$, then

$$(\theta_{\max})^2 > \frac{k+1}{2}(2n-k) - \frac{2^k n^{2k+1}}{M(k!)^2\binom{2n}{k}}, \quad k \geq 0.$$

10.3.27 Remark The crosscorrelation between two m-sequences of the same period and with symbols from $K = \mathbb{F}_p$, p prime, is equivalent to calculating the following exponential sum for all nonzero $c \in F = \mathbb{F}_{p^m}$,

$$\theta_{a,b}(\tau) = -1 + \sum_{x \in F} \omega^{\mathrm{Tr}_{F/K}(cx - x^d)},$$

where two zeros α and β of the two characteristic polynomials are related by $\beta = \alpha^d$ where $\gcd(d, p^m - 1) = 1$ and $c = \alpha^\tau$. General results and open problems can be found in [1467, 1475].

10.3.28 Remark Many binary families of sequences with excellent correlation properties are constructed from m-sequences. The most well-known is the family of Gold sequences.

10.3.29 Construction (Gold sequences) Let m be odd, $d = 2^k + 1$ and $\gcd(k, m) = 1$. Let $\{s(t)\}$ be an m-sequence of period $n = 2^m - 1$. The *family of Gold sequences* is defined by

$$\mathcal{G} = \{s(t)\} \cup \{s(dt)\} \cup \{\{s(t+\tau) + s(dt)\} \mid 0 \leq \tau \leq n-1\}.$$

10.3.30 Theorem [1295] The parameters of the Gold family are $n = 2^m - 1$, $M = 2^m + 1$, and $\theta_{max} = 2^{\frac{m+1}{2}} + 1$.

10.3.31 Remark [2660] The Gold sequence family is optimal with respect to the Sidelnikov bound and has the smallest possible θ_{max} for the given period, alphabet, and family size.

10.3.32 Construction (The small family of Kasami sequences) Let $m = 2k, k \geq 2$, and α be a primitive element of $F = \mathbb{F}_{2^m}$. Let also $M = \mathbb{F}_{2^k}$, $K = \mathbb{F}_2$, and

$$s_a(t) = \mathrm{Tr}_{F/K}(\alpha^t) + \mathrm{Tr}_{M/K}\left(a\alpha^{(2^k+1)t}\right)$$

where $a \in M$. The *small family of Kasami sequences* is

$$\mathcal{K} = \{\{s_a(t)\} \mid a \in M\}.$$

10.3.33 Theorem [1688] The parameters of the small Kasami family are $n = 2^m - 1$, $M = 2^k$, and $\theta_{max} = 2^k + 1$.

10.3.34 Remark [1475] The small Kasami family is optimal with respect to the Welch bound and has the smallest possible θ_{max} for the given period, alphabet, and family size.

10.3.35 Construction (No sequences) Let $m = 2k, k \geq 2$, and α be a primitive element of $F = \mathbb{F}_{2^m}$. Let also $1 \leq r \leq 2^k - 1$, $r \neq 2^i$ for any i and $\gcd(r, 2^k - 1) = 1$. Let $M = \mathbb{F}_{2^k}$, $K = \mathbb{F}_2$, and define

$$s_a(t) = \mathrm{Tr}_{M/K}\left(\left(\mathrm{Tr}_{F/M}\left(\alpha^t + a\alpha^{(2^k+1)t}\right)\right)^r\right)$$

where $a \in F$. The *No sequence family* is defined by

$$\mathcal{N} = \{\{s_a(t)\} \mid a \in F\}.$$

10.3.36 Theorem [2295] The parameters of the No sequence family are the same as for the small Kasami family. No sequences have higher linear complexity than the Kasami sequences.

10.3.37 Theorem [2661] (Sidelnikov sequence family) Let p be a prime, $0 < d < p$ and α be an element of order $p^m - 1$. Let also $F = \mathbb{F}_{p^m}$ and $K = \mathbb{F}_p$. Let $\{s(t)\}$ be a sequence over $\mathbb{F}_p$ of the form

$$s(t) = \mathrm{Tr}_{F/K}\left(\sum_{k=1}^{d} a_k \alpha^{kt}\right),$$

where $a_k \in F$ for $1 \leq k \leq d$. The parameters of the *Sidelnikov sequence family* are $n = p^m - 1$, $M \geq p^{m(d-1)}$ and $\theta_{max} \leq (d-1)p^{\frac{m}{2}} + 1$.

10.3.38 Remark The condition on d can be relaxed but leads to a more complicated bound on the parameters. The main idea is that θ_{max} is bounded by the Carlitz-Uchiyama bound [550] by the maximal degree of the polynomials involved in the correlation computations.

10.3.4 Quaternary sequences

10.3.39 Remark Families of sequences with symbols from $\mathbb{Z}_4$ can have correlation properties that are superior to binary sequences. Family $\mathcal{A}$ in Construction 10.3.42 is an important example.

10.3.40 Definition (Lifting of $f(x)$) The *lifting of a binary polynomial* $f(x)$ is the quaternary polynomial $g(x) = \sum_{i=0}^{m} g_i x^i$ where $g(x^2) \equiv (-1)^m f(x)f(-x) \pmod 4$.

10.3.41 Example The lifting of the primitive binary polynomial $f(x) = x^3 + x + 1$ is $g(x) \equiv x^3 + 2x^2 + x + 3 \pmod 4$ since $g(x^2) = (-1)^3 f(x)f(-x) \equiv x^6 + 2x^4 + x^2 + 3 \pmod 4$.

10.3.42 Construction (Family $\mathcal{A}$) Let $g(x)$ be a lifting of a binary primitive polynomial of degree m. The recursion $\sum_{i=0}^{m} g_i s(t+i) \equiv 0 \pmod 4$ generates $4^m - 1$ quaternary nonzero sequences corresponding to all nonzero initial states $(s(0), s(1), \ldots, s(m-1))$. These sequences are known to have period $2^m - 1$. *Family* $\mathcal{A}$ is constructed by selecting $2^m + 1$ cyclically distinct sequences from this set.

10.3.43 Theorem [383, 2692] Family $\mathcal{A}$ has parameters $n = 2^m - 1$, $M = 2^m + 1$ and $\theta_{max} \leq 2^{\frac{m}{2}} + 1$.

10.3.44 Remark Compared with the family of binary Gold sequences, family $\mathcal{A}$ has the same period $n = 2^m - 1$ and family size $M = 2^m + 1$ but θ_{max} is a factor of $\sqrt{2}$ lower than for the Gold sequence family.

10.3.45 Definition Let $s = a + 2b \in \mathbb{Z}_4$ where $a, b \in \mathbb{Z}_2$. The *most significant bit map* π is defined by $\pi(s) = b$.

10.3.46 Remark Any quaternary sequence $\{s(t)\}$ of odd period n defines a quaternary sequence $\{(-1)^t s(t)\}$ of period $2n$. Thus a binary sequence $\{b(t)\}$ of period $2n$ is obtained by $b(t) = \pi((-1)^t s(t))$.

10.3.47 Construction (Family of binary Kerdock sequences) Let $g(x)$ be the lifting of a primitive binary polynomial of odd degree m and define $h(x) = -g(-x)$. The recursion with characteristic polynomial $h(x)$ is applied to all initial states that are nonzero modulo 2. This generates $4^m - 2^m$ quaternary sequences of period $2(2^m - 1)$. Selecting cyclically distinct sequences from this set and using the most significant bit map π to these sequences leads to the *Kerdock family* $\mathcal{K}$ of 2^{m-1} binary sequences of period $2(2^m - 1)$.

10.3.48 Theorem [1475] The parameters of family $\mathcal{K}$ are $n = 2(2^m - 1), M = 2^{m-1}$ and $\theta_{max} \leq 2^{\frac{m+1}{2}} + 2$.

10.3.49 Remark The sequence family $\mathcal{K}$ is superior to the small Kasami set. The size of family $\mathcal{K}$ can be increased by a factor of 2 without increasing θ_{max}. For further details see [1475].

10.3.5 Other correlation measures

10.3.50 Definition The *aperiodic correlation* of two complex-valued sequences $\{u(t)\}$ and $\{v(t)\}$ for $t = 0, 1, \cdots, n-1$ is defined by

$$\rho_{u,v}(\tau) = \sum_{t=\max\{0,-\tau\}}^{\min\{n-1,n-1-\tau\}} u(t+\tau)\overline{v}(t), \quad -(n-1) \leq \tau \leq n-1.$$

10.3.51 Remark If $\{s(t)\}$ is a binary $\{+1, -1\}$ sequence then $\rho_{s,s}(\tau) = \rho_{s.s}(-\tau)$ and the *aperiodic autocorrelation* is determined by the values

$$\rho_s(\tau) = \sum_{t=0}^{n-1-\tau} s(t+\tau)s(t), \quad 0 \leq \tau \leq n-1.$$

10.3.52 Definition *A Barker sequence* is a binary $\{-1, +1\}$ sequence of length n if the aperiodic values $\rho_s(\tau)$ satisfy $|\rho_s(\tau)| \leq 1$ for all τ, $1 \leq \tau \leq n-1$.

10.3.53 Remark Barker sequences are only known for the following lengths $n = 2, 3, 4, 5, 7, 11, 13$. For example the sequence $(+1+1+1+1+1-1-1+1+1-1+1-1+1)$ of length $n = 13$ is the longest known Barker sequence; see also Section 17.3.

10.3.54 Remark It has been shown by Turyn and Storer [2828] that there are no Barker sequences of odd length $n > 13$ and if they exist then $n = 0 \pmod 4$. There is an overwhelming evidence that no Barker sequence of length $n > 13$ exists.

10.3.55 Definition The *merit factor* of a binary $\{-1, +1\}$ sequence $\{s(t)\}$ of length n is defined by

$$F = \frac{n^2}{2\sum_{\tau=1}^{n-1} \rho_s^2(\tau)}.$$

10.3.56 Remark The highest known merit factor for a sequence is $F = 14.08$ coming from a Barker sequence of length 13. For a long time the largest proven asymptotical value of the merit factor for a family of arbitrarily long sequences was 6 [1523]. In [352] and [1806] new constructions were presented of families of sequences with asymptotic merit factor believed to be greater than 6.34. Recently, this claim has been proved [1605].

10.3.57 Remark *Low correlation zone sequences* (LCZ) are designed with small auto- and crosscorrelation values for small values of their relative time shifts. The parameters of a family of LCZ sequences are the period n of the sequences, the number M of sequences, the length L of the low correlation zone, and the upper bound δ on the correlation value in the low correlation zone. For further details see [1215].

10.3.58 Definition A *family of low correlation zone sequences* is defined by the parameters (n, M, L, δ), where

$$|\theta_{i,j}(\tau)| \leq \delta \text{ when } 0 \leq |\tau| < L, i \neq j \text{ and for } 1 \leq |\tau| < L \text{ when } i = j.$$

10.3.59 Theorem [2778] For an (n, M, L, δ) LCZ sequence family it holds that

$$ML - 1 \leq \frac{n-1}{1 - \frac{\delta^2}{n}}.$$

10.3.60 Definition The *periodic Hamming correlation* between a pair of binary sequences $\{s_1(t)\}$ and $\{s_2(t)\}$ of period n is the integer

$$\theta_{1,2}(\tau) = \sum_{t=0}^{n-1} s_1(t+\tau)s_2(t), \quad 0 \leq \tau < n.$$

10.3.61 Remark The periodic Hamming correlation is important in evaluating optical communication systems where 0s and 1s indicate presence or absence of pulses of transmitted light.

10.3.62 Definition An (n, w, λ) *optical orthogonal code* (OOC) is a family

$$\mathcal{F} = \{\{s_i(t)\} \mid i = 1, 2, \ldots, M\}$$

of M binary sequences of period n and constant Hamming weight w where $1 \leq w \leq n-1$ and $\theta_{i,j}(\tau) \leq \lambda$ when either $i \neq j$ or $\tau \neq 0$.

10.3.63 Remark The close relations between (n, w, λ) OOCs and constant weight codes provides good bounds on OOC from known bounds on constant weight codes. For further information see [1475].

10.3.64 Remark Other correlation measures include the *partial-period correlation* between two very long sequences where the correlation is calculated over a partial period. In practice, there is also some interest in the *mean-square correlation* of a sequence family rather than in $\theta_{\max}$. For the evaluation of these correlation measures coding theory sometimes plays an important role. For more information the reader is referred to [1475].

See Also

§6.1	Exponential sums are crucial in calculating the correlation of sequences.
§9.1	Boolean functions are closely related to sequences.
§9.3	Bent functions can be used in sequence constructions and vice versa.
§10.2	Linear Feedback Shift Registers (LFSRs) are important in constructing sequences with low correlation.
§14.6	Cyclic difference sets are related to sequences with two-level autocorrelation.
§17.3	Describes some applications of sequences and results on aperiodic correlation.
[1215]	An overview of some recent advances of low correlation sequences.
[1303]	This textbook by Golomb and Gong gives important information on sequences and their applications to signal design and cryptography.
[1475]	Provides an extensive survey of sequences with low correlation and their connections to coding theory.
[1476]	An elementary introduction to pseudonoise sequences.
[2526]	A classical paper on the crosscorrelation of pseudorandom sequences.

References Cited: [352, 383, 550, 706, 864, 1215, 1295, 1303, 1324, 1467, 1475, 1476, 1523, 1605, 1688, 1806, 1939, 2295, 2388, 2526, 2555, 2659, 2660, 2661, 2673, 2692, 2778, 2828, 2964]

10.4 Linear complexity of sequences and multisequences

Wilfried Meidl, Sabanci University
Arne Winterhof, Austrian Academy of Sciences

10.4.1 Linear complexity measures

10.4.1 **Definition** A sequence $S = s_0, s_1, \ldots$ over the finite field $\mathbb{F}_q$ is called a *(homogeneous) linear recurring sequence* over $\mathbb{F}_q$ with *characteristic polynomial*

$$f(x) = \sum_{i=0}^{l} c_i x^i \in \mathbb{F}_q[x]$$

of degree l, if S satisfies the *linear recurrence relation*

$$\sum_{i=0}^{l} c_i s_{n+i} = 0 \quad \text{for } n = 0, 1, \ldots. \tag{10.4.1}$$

10.4.2 Definition The *minimal polynomial* of a linear recurring sequence S is the uniquely defined monic polynomial $M \in \mathbb{F}_q[x]$ of smallest degree for which S is a linear recurring sequence with characteristic polynomial M. The *linear complexity* $L(S)$ of S is the degree of the minimal polynomial M.

10.4.3 Remark Without loss of generality one can assume that f is monic, i.e., $c_l = 1$. A sequence S over $\mathbb{F}_q$ is a linear recurring sequence if and only if S is ultimately periodic, if c_0 in (10.4.1) is nonzero then S is purely periodic, see [1939, Chapter 8]. Consequently Definition 10.4.1 is only meaningful for (ultimately) periodic sequences. Using the notation of [1134, 2064], we let $\mathcal{M}_q^{(1)}(f)$ be the set of sequences over $\mathbb{F}_q$ with characteristic polynomial f. The set of sequences with a fixed period N is then $\mathcal{M}_q^{(1)}(f)$ with $f(x) = x^N - 1$. The minimal polynomial M of a sequence $S \in \mathcal{M}_q^{(1)}(f)$ is always a divisor of f. For an N-periodic sequence S we have $L(S) \leq N$; see Section 10.2.

10.4.4 Remark The linear complexity of a sequence S can alternatively be defined as the length of the shortest linear recurrence relation satisfied by S. In engineering terms, $L(S)$ is 0 if S is the zero sequence and otherwise it is the length of the shortest linear feedback shift register (Section 10.2) that can generate S [1631, 1939, 2502, 2503].

10.4.5 Definition For $n \geq 1$ the *n-th linear complexity* $L(S, n)$ of a sequence S over $\mathbb{F}_q$ is the length L of a shortest linear recurrence relation

$$s_{j+L} = c_{L-1}s_{j+L-1} + \cdots + c_0 s_j, \quad 0 \leq j \leq n - L - 1,$$

over $\mathbb{F}_q$ satisfied by the first n terms of the sequence. The polynomial $\sum_{i=0}^{L} c_i x^i \in \mathbb{F}_q[x]$ is an *n-th minimal polynomial* of S. The *linear complexity* $L(S)$ of a periodic sequence can then be defined by

$$L(S) := \sup_{n \geq 1} L(S, n).$$

10.4.6 Remark Again one may assume that the n-th minimal polynomial is monic. Then it is unique whenever $L \leq n/2$. Definition 10.4.5 is also applicable for finite sequences, i.e., strings of elements of $\mathbb{F}_q$ of length n.

10.4.7 Definition For an infinite sequence S, the non-decreasing integer sequence $L(S, 1), L(S, 2), \ldots$ is the *linear complexity profile* of S.

10.4.8 Remark Linear complexity and linear complexity profile of a given sequence (as well as the linear recurrence defining it) can be determined by using the Berlekamp-Massey algorithm; see Section 15.1 or [1631, Section 6.7], and [2011]. The algorithm is efficient for sequences with low linear complexity and hence such sequences can easily be predicted.

10.4.9 Remark A sequence used as a keystream in stream ciphers must consequently have a large linear complexity, but also altering a few terms of the sequence should not cause a significant decrease of the linear complexity. An introduction to the stability theory of stream ciphers is the monograph [873]. For a general comprehensive survey on the theory of stream ciphers we refer to [2502, 2503].

10.4.10 **Definition** The *k-error linear complexity* $L_k(S,n)$ of a sequence S of length n is defined by

$$L_k(S,n) = \min_T L(T,n),$$

where the minimum is taken over all sequences T of length n with Hamming distance $d(T,S)$ from S at most k. For an N-periodic sequence S over $\mathbb{F}_q$ the *k-error linear complexity* is defined by [2700]

$$L_k(S) = \min_T L(T),$$

where the minimum is taken over all N-periodic sequences T over $\mathbb{F}_q$ for which the first N terms differ in at most k positions from the corresponding terms of S.

10.4.11 **Remark** The concept of the k-error linear complexity is based on the *sphere complexity* introduced in [873].

10.4.12 **Remark** Recent developments in stream ciphers point toward an increasing interest in word-based or vectorized stream ciphers (see for example [784, 1445]), which requires the study of multisequences.

10.4.13 **Definition** For an arbitrary positive integer m, an *m-fold multisequence* $\mathbf{S} = (S_1, \dots, S_m)$ over $\mathbb{F}_q$ (of finite or infinite length) is a string of m parallel sequences $S_1, \dots, S_m$ over $\mathbb{F}_q$ (of finite or infinite length, respectively).

Let $f_1, \dots, f_m \in \mathbb{F}_q[x]$ be arbitrary monic polynomials with $\deg(f_i) \geq 1$, $1 \leq i \leq m$. The set $\mathcal{M}_q(f_1, \dots, f_m)$ is defined to be the set of m-fold multisequences $(S_1, \dots, S_m)$ over $\mathbb{F}_q$ such that for each $1 \leq i \leq m$, S_i is a linear recurring sequence with characteristic polynomial f_i.

10.4.14 **Definition** The *joint minimal polynomial* of an m-fold multisequence $\mathbf{S} \in \mathcal{M}_q(f_1, \dots, f_m)$ is the (uniquely determined) monic polynomial $M \in \mathbb{F}_q[x]$ of smallest degree which is a characteristic polynomial of S_i for all $1 \leq i \leq m$. The *joint linear complexity* of $\mathbf{S}$ is the degree of the joint minimal polynomial M.

10.4.15 **Remark** The set of N-periodic m-fold multisequences is $\mathcal{M}_q(f_1, \dots, f_m)$ with $f_1 = \dots = f_m = x^N - 1$, alternatively denoted by $\mathcal{M}_q^{(m)}(f)$ with $f(x) = x^N - 1$. The joint linear complexity of an m-fold multisequence can also be defined as the length of the shortest linear recurrence relation the m parallel sequences satisfy simultaneously. The joint minimal polynomial M of $\mathbf{S} \in \mathcal{M}_q(f_1, \dots, f_m)$ is always a divisor of $\mathrm{lcm}(f_1, \dots, f_m)$.

10.4.16 **Definition** For an integer $n \geq 1$ the *n-th joint linear complexity* $L(\mathbf{S}, n)$ of an m-fold multisequence $\mathbf{S} = (S_1, \dots, S_m)$ is the length of the shortest linear recurrence relation the first n terms of the m parallel sequences $S_1, \dots, S_m$ satisfy simultaneously. The *joint linear complexity profile* of $\mathbf{S}$ is the non-decreasing integer sequence $L(\mathbf{S}, 1), L(\mathbf{S}, 2), \dots$.

10.4.17 **Remark** As the $\mathbb{F}_q$-linear spaces $\mathbb{F}_q^m$ and $\mathbb{F}_{q^m}$ are isomorphic, an m-fold multisequence $\mathbf{S}$ can also be identified with a single sequence $\mathcal{S}$ having its terms in the extension field $\mathbb{F}_{q^m}$. If $s_j^{(i)}$ denotes the j-th term of the i-th sequence S_i, $1 \leq i \leq m$, and $\{\beta_1, \dots, \beta_m\}$ is a basis of $\mathbb{F}_{q^m}$ over $\mathbb{F}_q$, then the j-th term of $\mathcal{S}$ is $\sigma_j = \sum_{i=1}^m \beta_i s_j^{(i)}$. The ($n$-th) joint linear complexity of $\mathcal{S}$ coincides then with the *$\mathbb{F}_q$-linear complexity* of $\mathcal{S}$, which is the length of the shortest linear recurrence relation with coefficients exclusively in $\mathbb{F}_q$ (the first n terms of) $\mathcal{S}$ satisfies (see [759, pp. 83–85]).

10.4.18 **Definition** We identify an m-fold multisequence $\mathbf{S}$ of length n (or period n) with an $m \times n$ matrix and write $\mathbf{S} \in \mathbb{F}_q^{m\times n}$ ($\mathbf{S} \in (\mathbb{F}_q^{m\times n})^\infty$). For two m-fold multisequences $\mathbf{S} = (S_1, \ldots, S_m), \mathbf{T} = (T_1, \ldots, T_m) \in \mathbb{F}_q^{m\times n}$ the *term distance* $d_T(\mathbf{S}, \mathbf{T})$ between $\mathbf{S}$ and $\mathbf{T}$ is the number of terms in the matrix for $\mathbf{S}$ that are different from the corresponding terms in the matrix for $\mathbf{T}$.

The *column distance* $d_C(\mathbf{S}, \mathbf{T})$ between $\mathbf{S}$ and $\mathbf{T}$ is the number of columns in which the matrices of $\mathbf{S}$ and $\mathbf{T}$ differ.

The *individual distances vector* for $\mathbf{S}, \mathbf{T}$ is defined by $d_V(\mathbf{S}, \mathbf{T}) = (d_H(S_1, T_1), \ldots, d_H(S_m, T_m))$, where d_H denotes the Hamming distance.

10.4.19 **Example** For $q = 2$, $m = 2$, $n = 5$, and

$$\mathbf{S} = \begin{pmatrix} 1 & 1 & 0 & 0 & 1 \\ 0 & 1 & 0 & 1 & 1 \end{pmatrix}, \quad \mathbf{T} = \begin{pmatrix} 1 & 1 & 0 & 1 & 1 \\ 1 & 1 & 0 & 0 & 1 \end{pmatrix},$$

we have $d_T(\mathbf{S}, \mathbf{T}) = 3$, $d_C(\mathbf{S}, \mathbf{T}) = 2$ and $d_V(\mathbf{S}, \mathbf{T}) = (1, 2)$.

10.4.20 **Definition** For an integer k with $0 \le k \le mn$, the *(n-th) k-error joint linear complexity* $L_k(\mathbf{S}, n)$ of an m-fold multisequence $\mathbf{S}$ over $\mathbb{F}_q$ is defined by

$$L_k(\mathbf{S}, n) = \min_{\mathbf{T} \in \mathbb{F}_q^{m\times n}, d_T(\mathbf{S},\mathbf{T}) \le k} L(\mathbf{T}, n).$$

For an integer $0 \le k \le n$ the *(n-th) k-error $\mathbb{F}_q$-linear complexity* $L_k^q(\mathbf{S}, n)$ of $\mathbf{S}$ is defined by

$$L_k^q(\mathbf{S}, n) = \min_{\mathbf{T} \in \mathbb{F}_q^{m\times n}, d_C(\mathbf{S},\mathbf{T}) \le k} L(\mathbf{T}, n).$$

We define a partial order on $\mathbb{Z}^m$ by $\mathbf{k} = (k_1, \ldots, k_m) \le \mathbf{k}' = (k_1', \ldots, k_m')$ if $k_i \le k_i'$, $1 \le i \le m$. For $\mathbf{k} = (k_1, \ldots, k_m) \in \mathbb{Z}^m$ such that $0 \le k_i \le n$ for $1 \le i \le m$, the *(n-th)* **k**-*error joint linear complexity* $L_{\mathbf{k}}(\mathbf{S}, n)$ of $\mathbf{S}$ is

$$L_{\mathbf{k}}(\mathbf{S}, n) = \min_{\mathbf{T} \in \mathbb{F}_q^{m\times n}, d_V(\mathbf{S},\mathbf{T}) \le \mathbf{k}} L(\mathbf{T}, n),$$

i.e., the minimum is taken over all m-fold length n multisequences $\mathbf{T} = (T_1, \ldots, T_m)$ over $\mathbb{F}_q$ with Hamming distances $d_H(S_i, T_i) \le k_i$, $1 \le i \le m$.
The definitions for periodic multisequences are analogous.

10.4.2 Analysis of the linear complexity

10.4.21 **Proposition** Let $f \in \mathbb{F}_q[x]$ be a nonconstant monic polynomial.

1. [1631, Theorem 6.1.2], [1939, Chapter 8] For a sequence $S = s_0, s_1, \ldots$ over $\mathbb{F}_q$ consider the element $\sum_{i=0}^{\infty} s_i x^i$ in the ring $\mathbb{F}_q[[x]]$ of formal power series over $\mathbb{F}_q$. Then S is a linear recurring sequence with characteristic polynomial f if and only if $\sum_{i=0}^{\infty} s_i x^i = g(x)/f^*(x)$ with $g \in \mathbb{F}_q[x]$, $\deg(g) < \deg(f)$ and $f^*(x) = x^{\deg(f)} f(1/x)$ is the reciprocal polynomial of $f(x)$.
2. [2240, Lemma 1] For a sequence $S = s_1, s_2, \ldots$ over $\mathbb{F}_q$ consider the element $\sum_{i=1}^{\infty} s_i x^{-i}$ in the field $\mathbb{F}_q((x^{-1}))$ of formal Laurent series in x^{-1} over $\mathbb{F}_q$. Then S is a linear recurring sequence with characteristic polynomial f if and only if $\sum_{i=1}^{\infty} s_i x^{-i} = g(x)/f(x)$ with $g \in \mathbb{F}_q[x]$ and $\deg(g) < \deg(f)$.

10.4.22 Remark For more information and discussion of linear recurring sequences, we refer to Section 10.2.

10.4.23 Remark The reciprocal of a characteristic polynomial of a sequence S is also called a *feedback polynomial* of S.

10.4.24 Remark Proposition 10.4.21 implies a one-to-one correspondence between sequences in $\mathcal{M}_q^{(1)}(f)$ and rational functions g/f with $\deg(g) < \deg(f)$ (when the approach via Laurent series is used), and more generally between m-fold multisequences in $\mathcal{M}_q(f_1, \ldots, f_m)$ and m-tuples of rational functions $(g_1/f_1, \ldots, g_m/f_m)$ with $\deg(g_i) < \deg(f_i)$, $1 \le i \le m$. We note that in Proposition 10.4.21 Part 2 it is more convenient to start the indices for the sequence elements s_i with $i = 1$.

10.4.25 Proposition [1135] Let $(g_1/f_1, \ldots, g_m/f_m)$ be the m-tuple of rational functions corresponding to $\mathbf{S} \in \mathcal{M}_q(f_1, \ldots, f_m)$. The joint minimal polynomial of $\mathbf{S}$ is the unique monic polynomial $M \in \mathbb{F}_q[x]$ such that $\frac{g_1}{f_1} = \frac{h_1}{M}, \ldots, \frac{g_m}{f_m} = \frac{h_m}{M}$ for some (unique) polynomials $h_1, \ldots, h_m \in \mathbb{F}_q[x]$ with $\gcd(M, h_1, \ldots, h_m) = 1$.

10.4.26 Remark For an N-periodic sequence $S = s_0, s_1, \ldots$, let $S^N(x)$ be the polynomial $S^N(x) = s_0 + s_1 x + \cdots + s_{N-1} x^{N-1}$ of degree at most $N - 1$. Then $\sum_{i=0}^{\infty} s_i x^i = S^N(x)/(1 - x^N)$, which gives rise to the following theorem.

10.4.27 Theorem [759, Lemma 8.2.1], [2061] The joint linear complexity of an N-periodic m-fold multisequence $\mathbf{S} = (S_1, \ldots, S_m)$ is given by

$$L(\mathbf{S}) = N - \deg(\gcd(x^N - 1, S_1^N(x), \ldots, S_m^N(x))).$$

10.4.28 Remark Theorem 10.4.27 implies the famous Blahut theorem [303, 2503], [1631, Theorem 6.8.2] for the linear complexity of N-periodic sequences over $\mathbb{F}_q$, $\gcd(N, q) = 1$, which we state in 3 commonly used different versions.

10.4.29 Theorem (Blahut's Theorem) Let S be an N-periodic sequence over $\mathbb{F}_q$, let $\gcd(N, q) = 1$, and let α be a primitive N-th root of unity in an extension field of $\mathbb{F}_q$. Then

$$L(S) = N - |\{j \ : \ S^N(\alpha^j) = 0, 0 \le j \le N - 1\}|.$$

10.4.30 Theorem (Blahut's Theorem) Let $\gcd(N, q) = 1$, α be a primitive N-th root of unity in an extension field of $\mathbb{F}_q$ and let $A = (a_{ij})$ be the $N \times N$ Vandermonde matrix with $a_{ij} = \alpha^{ij}$, $0 \le i, j \le N - 1$. Let $\mathbf{s} = (s_0, s_1, \ldots, s_{N-1})$ be the vector corresponding to one period of an N-periodic sequence S over $\mathbb{F}_q$. The linear complexity $L(S)$ of S is the Hamming weight of the vector $A\mathbf{s}^T$.

10.4.31 Remark The vector $\mathbf{a} = A\mathbf{s}^T$ is called the *discrete Fourier transform* of $\mathbf{s}$. Several generalizations of the discrete Fourier transform have been suggested in the literature that can be used to determine the linear complexity of periodic sequences and multisequences with period not relatively prime to the characteristic of the field. We refer to [297, 2013, 2061].

10.4.32 Theorem (Blahut's Theorem) Let $S = s_0, s_1, \ldots$ be a sequence over $\mathbb{F}_q$ with period N dividing $q - 1$, and let $g \in \mathbb{F}_q[x]$ be the unique polynomial of degree at most $N - 1$ satisfying $g(\alpha^j) = s_j$, $j = 0, 1, \ldots$, where α is a fixed element of $\mathbb{F}_q$ of order N. Then $L(S) = w(g)$, where $w(g)$ denotes the weight of g, i.e., the number of nonzero coefficients of g.

10.4.33 Theorem [298, Theorem 8] Let f be a polynomial over a prime field $\mathbb{F}_p$ with degree of f at most $p - 1$ and let $S = s_0, s_1, \ldots$ be the p-periodic sequence over $\mathbb{F}_p$ defined by $s_j = f(j)$, $j = 0, 1, \ldots$. Then $L(S) = \deg(f) + 1$.

10.4.34 Remark Theorem 8 in [298] more generally describes the linear complexity of p^r-periodic sequences over $\mathbb{F}_p$. A generalization of Theorem 10.4.33 to arbitrary finite fields is given in Theorem 1 of [2066].

10.4.35 Remark The linear complexity of an N-periodic sequence over $\mathbb{F}_q$ can be determined by the Berlekamp-Massey algorithm in $O(N^2)$ elementary operations. For some classes of period lengths, faster algorithms (of complexity $O(N)$) are known, the earliest being the Games-Chan algorithm [1169] for binary sequences with period $N = 2^v$. A collection of algorithms for several period lengths can be found in [3013] (see also [2958, 2959, 2960, 3014]). Some techniques to establish fast algorithms for arbitrary periods are presented in [85, 599, 600, 2057]. Stamp and Martin [2700] established a fast algorithm for the k-error linear complexity for binary sequences with period $N = 2^v$. Generalizations are presented in [1642, 1864, 2520], and for odd characteristic in [2056, 3013].

10.4.36 Remark In contrast to the faster algorithms introduced in the literature for certain period lengths, the Berlekamp-Massey algorithm also can determine the linear complexity profile of a (single) sequence. As an application, the general behavior of linear complexity profiles can be analyzed.

10.4.37 Theorem [1631, Theorem 6.7.4],[2502] Let $S = s_1, s_2, \ldots$ be a sequence over $\mathbb{F}_q$. If $L(S,n) > n/2$ then $L(S,n+1) = L(S,n)$. If $L(S,n) \leq n/2$, then $L(S,n+1) = L(S,n)$ for exactly one choice of $s_{n+1} \in \mathbb{F}_q$ and $L(S,n+1) = n+1-L(S,n)$ for the remaining $q-1$ choices of $s_{n+1} \in \mathbb{F}_q$.

10.4.38 Remark The linear complexity profile is uniquely described by the *increment sequence* of S, i.e., by the sequence of the positive integers among $L(S,1), L(S,2) - L(S,1)$, $L(S,3) - L(S,2)$, $\ldots$ [2246, 2935, 2937]. Another tool for the analysis of the linear complexity profile arises from a connection to the continued fraction expansion of Laurent series [2239, 2240].

10.4.39 Theorem [2240] Let $S = s_1, s_2, \ldots$ be a sequence over $\mathbb{F}_q$, let $S(x) = \sum_{i=1}^{\infty} s_i x^{-i} \in \mathbb{F}_q((x^{-1}))$ be the corresponding formal Laurent series, and let $A_1, A_2, \ldots$ be the polynomials in the continued fraction expansion of $S(x)$, i.e., $S(x) = 1/(A_1 + 1/(A_2 + \cdots))$ where $A_j \in \mathbb{F}_q[x]$, $\deg(A_j) \geq 1$, $j \geq 1$. Let $Q_{-1} = 0, Q_0 = 1$ and $Q_j = A_j Q_{j-1} + Q_{j-2}$ for $j \geq 1$. Then $L(S,n) = \deg(Q_j)$ where j is determined by

$$\deg(Q_{j-1}) + \deg(Q_j) \leq n < \deg(Q_j) + \deg(Q_{j+1}).$$

The n-th minimal polynomials are all (monic) polynomials of the form $M = aQ_j + gQ_{j-1}$, $a \in \mathbb{F}_q^*$, $g \in \mathbb{F}_q[x]$ with $\deg(g) \leq 2\deg(Q_j) - n - 1$. In particular, the increment sequence of S is $\deg(A_1), \deg(A_2), \ldots$.

10.4.40 Remark Generalizations of the Berlekamp-Massey algorithm and of continued fraction analysis for the linear complexity of multisequences can be found in [127, 763, 764, 765, 766, 873, 1053, 1671, 2516, 2517, 2944].

10.4.3 Average behavior of the linear complexity

10.4.41 Remark We use the notation $N_n^{(m)}(L)$ and $E_n^{(m)}$ for the number of m-fold multisequences over $\mathbb{F}_q$ with length n and joint linear complexity L and the expected value for the joint linear complexity of a random m-fold multisequence over $\mathbb{F}_q$ of length n.

10.4.42 Theorem [1378, 2502, 2685] For $1 \leq L \leq n$

$$N_n^{(1)}(L) = (q-1)q^{\min(2L-1, 2n-2L)}.$$

The expected value for $L(S,n)$ for a random sequence S over $\mathbb{F}_q$ is

$$E_n^{(1)} = \frac{1}{q^n}\sum_{S\in\mathbb{F}_q^n} L(S,n) = \begin{cases} \frac{n}{2} + \frac{q}{(q+1)^2} - q^{-n}\frac{n(q+1)+q}{(q+1)^2} & \text{for even } n,\\ \frac{n}{2} + \frac{q^2+1}{2(q+1)^2} - q^{-n}\frac{n(q+1)+q}{(q+1)^2} & \text{for odd } n.\end{cases}$$

10.4.43 Remark Theorem 10.4.42 was obtained by an analysis of the Berlekamp-Massey algorithm. Rueppel and Smeets [2502, 2685] provide closed formulas for the variance, showing that the variance is small. A detailed analysis of the linear complexity profile of sequences over $\mathbb{F}_q$ is given by Niederreiter in the series of papers [2235, 2239, 2240, 2243, 2246]. As a main tool, the continued fraction expansion of formal Laurent series is used. For a more elementary combinatorial approach, see [2242].

10.4.44 Theorem [2239] The linear complexity profile of a random sequence follows closely but irregularly the $n/2$-line, deviations from $n/2$ of the order of magnitude $\log n$ must appear for infinitely many n.

10.4.45 Remark The asymptotic behavior of the joint linear complexity is investigated by Niederreiter and Wang in the series of papers [2275, 2276, 2933] using a sophisticated multisequence linear feedback shift-register synthesis algorithm based on a lattice basis reduction algorithm in function fields [2549, 2928, 2934].

10.4.46 Theorem [2253, 2275, 2276]

$$\begin{aligned} N_n^{(m)}(L) &= (q^m-1)q^{(m+1)L-m}, \quad 1\le L\le n/2,\\ N_n^{(m)}(L) &\le C(q,m)L^m q^{2mn-(m+1)L}, \quad 1\le L\le n,\end{aligned}$$

where $C(q,m)$ is a constant only depending on q and m. We have $N_n^{(m)}(L) \le q^{(m+1)L}$.

10.4.47 Remark In [2933] a method to determine $N_n^{(m)}(L)$ is presented and a closed formula for $N_n^{(2)}(L)$ is given. A closed formula for $N_n^{(3)}(L)$ is presented in [2276]. In [2275, 2276] it is shown that the joint linear complexity profile of a random m-fold multisequence follows closely the $mn/(m+1)$-line, generalizing Theorem 10.4.44 for $m=1$.

10.4.48 Theorem [2275, 2276]

$$E_n^{(m)} = \frac{mn}{m+1} + o(n) \quad \text{as } n\to\infty.$$

For $m = 2,3$ [1059, 2276, 2933]

$$E_n^{(m)} = \frac{mn}{m+1} + O(1), \quad \text{as } n\to\infty.$$

10.4.49 Remark Feng and Dai [1059] obtained their result with different methods, namely with multi-dimensional continued fractions.

10.4.50 Conjecture [2276]

$$E_n^{(m)} = \frac{mn}{m+1} + O(1) \quad \text{as } n\to\infty.$$

10.4.51 Remark For a detailed survey on recent developments in the theory of the n-th joint linear complexity of m-fold multisequences we refer to [2257].

10.4.52 Theorem [1134, 1136] For a monic polynomial $f\in\mathbb{F}_q[x]$ with $\deg(f)\ge 1$, let

$$f = r_1^{e_1} r_2^{e_2}\cdots r_k^{e_k}$$

be the canonical factorization of f into monic irreducible polynomials over $\mathbb{F}_q$. For $1 \le i \le k$, let $\alpha_i = q^{m \deg(r_i)}$. Then for an arbitrary positive integer m the expected value $E^{(m)}(f)$ of the joint linear complexity of a random m-fold multisequence from $\mathcal{M}_q^{(m)}(f)$ is

$$E^{(m)}(f) = \deg(f) - \sum_{i=1}^{k} \frac{1-\alpha_i^{-e_i}}{\alpha_i - 1} \deg(r_i).$$

10.4.53 Remark In [1134, 1136] an explicit formula for the variance $\mathrm{Var}^{(m)}(f)$ of the joint linear complexity of random multisequences of $\mathcal{M}_q^{(m)}(f)$ is given. In [1135, 1136] it is shown how to obtain from Theorem 10.4.52 closed formulas for the more general case of m-fold multisequences in $\mathcal{M}_q(f_1, \ldots, f_m)$.

10.4.54 Remark Since for $f(x) = x^N - 1$ the set $\mathcal{M}^{(m)}(f)$ is the set of N-periodic sequences, earlier formulas on expectation (and variance) of the (joint) linear complexity of periodic (multi)sequences can be obtained as a corollary of Theorem 10.4.52: [2059, Theorem 3.2], [2060, Theorem 1], [3025, Theorem 1] on $E^{(1)}(x^N - 1)$, and [1137, Theorem 1], [2061, Theorem 1] on $E^{(m)}(x^N - 1)$ for arbitrary m.

10.4.55 Remark In [1137, 2061] lower bounds on the expected joint linear complexity for periodic multisequences are presented, estimating the magnitude of the formula for $E^{(m)}(x^N - 1)$ in Theorem 10.4.52. In [1137] it is also noted that the variance $\mathrm{Var}^{(m)}(x^N - 1)$ is small, showing that for random N-periodic multisequences over $\mathbb{F}_q$ the joint linear complexity is close to N (the trivial upper bound), with a small variance.

10.4.56 Remark Lower bounds for the expected n-th k-error joint linear complexity, the expected n-th k-error $\mathbb{F}_q$-linear complexity and the expected n-th $\mathbf{k}$-error joint linear complexity for an integer vector $\mathbf{k} = (k_1, \ldots, k_m)$ for a random m-fold multisequence over $\mathbb{F}_q$ are established in [2063]. These results generalize earlier bounds for the case $m = 1$ presented in [2058].

10.4.57 Remark For periodic sequences, lower bounds on the expected k-error linear complexity have been established in [2059, 2060]. For periodic multisequences (with prime period N different from the characteristic), lower bounds for the expected error linear complexity are presented in [2063] for all 3 multisequence error linear complexity measures.

10.4.58 Remark In the papers [2062, 2254, 2273, 2274, 2866] the question is addressed if linear complexity and k-error linear complexity can be large simultaneously. Among others, the existence of N-periodic sequences attaining the upper bounds N and $N - 1$ for linear and k-error linear complexity is shown for infinitely many period lengths (and a certain range for k depending on the period length), and it is shown that for several classes of period length a large number of N-periodic (multi)sequences with (joint) linear complexity N also exhibits a large k-error linear complexity.

10.4.59 Remark In [3016] methods from function fields are used to construct periodic multisequences with large linear complexity and k-error linear complexity simultaneously for various period lengths.

10.4.4 Some sequences with large n-th linear complexity

10.4.4.1 Explicit sequences

10.4.60 Definition For $a, b \in \mathbb{F}_p$ with $a \neq 0$ the *explicit inversive congruential sequence* $Z = z_0, z_1, \ldots$ is

$$z_j = (aj + b)^{p-2}, \quad j \ge 0. \tag{10.4.2}$$

10.4.61 Theorem [2067] We have

$$L(Z,n) \geq \begin{cases} (n-1)/3 & \text{for } 1 \leq n \leq (3p-7)/2, \\ n-p+2 & \text{for } (3p-5)/2 \leq n \leq 2p-3, \\ p-1 & \text{for } n \geq 2p-2. \end{cases}$$

10.4.62 Remark We note that $j^{p-2} = j^{-1}$ for $j \in \mathbb{F}_p^*$. Since inversion is a fast operation this sequence is, despite its high n-th linear complexity, still highly predictable.

10.4.63 Remark Analogous sequences of (10.4.2) over arbitrary finite fields $\mathbb{F}_q$ are studied in [2067]. Multisequences of this form are investigated in [2070]. Explicit inversive sequences and multisequences can also be defined using the multiplicative structure of $\mathbb{F}_q$.

10.4.64 Definition For $m \geq 1$, $\alpha_i, \beta_i \in \mathbb{F}_q^*$, $1 \leq i \leq m$, and an element $\gamma \in \mathbb{F}_q$ of order N, the *explicit inversive congruential sequence of period* N, $\mathbf{Z} = (Z_1, \ldots, Z_m)$, with $Z_i = \sigma_0^{(i)}, \sigma_1^{(i)}, \ldots$ is

$$\sigma_j^{(i)} = (\alpha_i \gamma^j + \beta_i)^{q-2}, \quad j \geq 0. \tag{10.4.3}$$

10.4.65 Remark Sequences of the form (10.4.3) are analyzed in [2069, 2070]. With an appropriate choice of the parameters one can obtain (multi)sequences with *perfect linear complexity profile*, i.e., $L(\mathbf{Z}, n) \geq mn/(m+1)$.

10.4.66 Theorem [2070] Let $m < (q-1)/N$ and let $C_1, \ldots, C_m$ be different cosets of the group $\langle\gamma\rangle$ generated by γ, such that none of them contains the element -1. For $1 \leq i \leq m$ choose α_i, β_i such that $\alpha_i \beta_i^{-1} \in C_i$, then

$$L(\mathbf{Z}, n) \geq \min\left\{\frac{mn}{m+1}, N\right\}, \quad n \geq 1.$$

10.4.67 Definition Given an element $\vartheta \in \mathbb{F}_q^*$, the *quadratic exponential sequence* $Q = q_0, q_1, \ldots$ is

$$q_j = \vartheta^{j^2}, \quad j \geq 0.$$

10.4.68 Theorem [1382] We have

$$L(Q, n) \geq \frac{\min\{n, N\}}{2}, \quad n \geq 1.$$

10.4.69 Remark The period N of Q is at least half of the multiplicative order of ϑ.

10.4.4.2 Recursive nonlinear sequences

10.4.70 Definition Given a polynomial $f \in \mathbb{F}_p[x]$ of degree $d \geq 2$, the *nonlinear congruential sequence* $U = u_0, u_1, \ldots$ is defined by the recurrence relation

$$u_{j+1} = f(u_j), \quad j \geq 0, \tag{10.4.4}$$

with some initial value $u_0 \in \mathbb{F}_p$ such that U is purely periodic with some period $N \leq p$.

10.4.71 Theorem [1382] Let U be as in (10.4.4), where $f \in \mathbb{F}_p[x]$ is of degree $d \geq 2$, then

$$L(U,n) \geq \min\left\{\log_d(n - \lfloor \log_d n \rfloor), \log_d N\right\}, \quad n \geq 1.$$

10.4.72 Remark For some special classes of polynomials much better results are available, see [1359, 1382, 2642]. For instance, in case of the largest possible period $N = p$ we have

$$L(U,n) \geq \min\{n - p + 1, p/d\}, \quad n \geq 1.$$

10.4.73 Theorem [1382] The *inversive (congruential) sequence* $Y = y_0, y_1, \ldots$ defined by

$$y_{j+1} = a y_j^{p-2} + b, \quad j \geq 0,$$

with $a, b, y_0 \in \mathbb{F}_p$, $a \neq 0$, has linear complexity profile

$$L(Y,n) \geq \min\left\{\frac{n-1}{3}, \frac{N-1}{2}\right\}, \quad n \geq 1.$$

10.4.74 Theorem [1359, 2642] The *power sequence* $P = p_0, p_1, \ldots,$ defined as

$$p_{j+1} = p_j^e, \quad j \geq 0,$$

with some integer $e \geq 2$ and initial value $0 \neq p_0 \in \mathbb{F}_p$ satisfies

$$L(P,n) \geq \min\left\{\frac{n^2}{4(p-1)}, \frac{N^2}{p-1}\right\}, \quad n \geq 1.$$

10.4.75 Remark Two more classes of nonlinear sequences provide much better results than in the general case, nonlinear sequences with *Dickson polynomials* [87] and *Rédei functions* [2072]. See Section 9.6 and [1936] for the definitions.

10.4.4.3 Legendre sequence and related bit sequences

10.4.76 Definition Let $p > 2$ be a prime. The *Legendre sequence* $\Lambda = l_0, l_1, \ldots,$ for $j \geq 0$, is

$$l_j = \begin{cases} 1 & \text{if } \left(\frac{j}{p}\right) = -1, \\ 0 & \text{otherwise,} \end{cases}$$

where $\left(\frac{\cdot}{p}\right)$ is the Legendre symbol.

10.4.77 Theorem [759, 2829] The linear complexity of the Legendre sequence is

$$L(\Lambda) = \begin{cases} (p-1)/2 & \text{if } p \equiv 1 \pmod 8, \\ p & \text{if } p \equiv 3 \pmod 8, \\ p-1 & \text{if } p \equiv 5 \pmod 8, \\ (p+1)/2 & \text{if } p \equiv 7 \pmod 8. \end{cases}$$

10.4.78 Theorem [2644, Theorem 9.2] The linear complexity profile of the Legendre sequence satisfies

$$L(\Lambda,n) > \frac{\min\{n,p\}}{1 + p^{1/2}(1+\log p)} - 1, \quad n \geq 1.$$

10.4.79 Remark For similar sequences, that are defined by the use of the quadratic character of arbitrary finite fields and the study of their linear complexity profiles, see [1786, 2065, 2994].

10.4.80 **Definition** Let γ be a primitive element and η be the quadratic character of the finite field $\mathbb{F}_q$ of odd characteristic. The *Sidelnikov sequence* $\sigma = \sigma_0, \sigma_1, \ldots$ for $j \geq 0$, is

$$\sigma_j = \begin{cases} 1 & \text{if } \eta(\gamma^j + 1) = -1, \\ 0 & \text{otherwise.} \end{cases}$$

10.4.81 **Remark** In many cases one is able to determine the linear complexity $L(\sigma)$ over $\mathbb{F}_2$ exactly, see Meidl and Winterhof [2071]. For example, if $(q-1)/2$ is an odd prime such that 2 is a primitive root modulo $(q-1)/2$, then σ attains the largest possible linear complexity $L(\sigma) = q - 1$. Moreover we have the lower bound [2071]

$$L(\sigma, n) \gg \frac{\min\{n, q\}}{q^{1/2} \log q}, \quad n \geq 1.$$

The k-error linear complexity of the Sidelnikov sequence seen as a sequence over $\mathbb{F}_p$ has been estimated in [86, 641, 1198]. For results on similar sequences with composite modulus see [392] and [759, Chapter 8.2].

10.4.4.4 Elliptic curve sequences

10.4.82 **Definition** Let $p > 3$ be a prime and E be an elliptic curve over $\mathbb{F}_p$ of the form

$$Y^2 = X^3 + aX + b$$

with coefficients $a, b \in \mathbb{F}_p$ such that $4a^3 + 27b^2 \neq 0$. For a given initial point $W_0 \in E(\mathbb{F}_p)$, a fixed point $G \in E(\mathbb{F}_p)$ of order N and a rational function $f \in \mathbb{F}_p(E)$ the *elliptic curve congruential sequence* $W = w_0, w_1, \ldots$ (with respect to f) is

$$w_j = f(W_j), \quad j \geq 0, \quad \text{where} \quad W_j = G \oplus W_{j-1} = jG \oplus W_0, \quad j \geq 1.$$

10.4.83 **Remark** Obviously, W is N-periodic.

10.4.84 **Remark** For example, choosing the function $f(x, y) = x$, the work of Hess and Shparlinski [1493] gives the lower bound

$$L(W, n) \geq \min\{n/3, N/2\}, \quad n \geq 2.$$

10.4.5 Related measures

10.4.5.1 Kolmogorov complexity

10.4.85 **Remark** The *Kolmogorov complexity* is a central topic in *algorithmic information theory.* The Kolmogorov complexity of a binary sequence is, roughly speaking, the length of the shortest computer program that generates the sequence. The relationship between linear complexity and Kolmogorov complexity was studied in [257, 2946]. The Kolmogorov complexity is twice the linear complexity for almost all sequences over $\mathbb{F}_2$ of sufficiently (but only moderately) large length. In contrast to the linear complexity the Kolmogorov complexity is in general not computable and so of no practical significance.

10.4.5.2 Lattice test

10.4.86 **Definition** Let $S = s_0, s_1, \ldots$ be a sequence over $\mathbb{F}_q$, and for $s \geq 1$ let $V(S,s)$ be the subspace of $\mathbb{F}_q^s$ spanned by the vectors $\mathbf{s}_j - \mathbf{s}_0$, $j = 1, 2, \ldots$, where

$$\mathbf{s}_j = (s_j, s_{j+1}, \ldots, s_{j+s-1}), \quad j \geq 0.$$

The sequence S passes the *s-dimensional lattice test* for some $s \geq 1$, if $V(S,s) = \mathbb{F}_q^s$. For given $s \geq 1$ and $n \geq 2$ we say that S passes the *s-dimensional n-lattice test* if the subspace spanned by the vectors $\mathbf{s}_j - \mathbf{s}_0$, $1 \leq j \leq n - s$, is $\mathbb{F}_q^s$. The largest s for which S passes the s-dimensional n-lattice test is the *lattice profile at* n and is denoted by $\mathcal{S}(S,n)$.

10.4.87 **Theorem** [914] We have either

$$\begin{aligned} \mathcal{S}(S,n) &= \min\{L(S,n), n+1-L(S,n)\} \quad \text{or} \\ \mathcal{S}(S,n) &= \min\{L(S,n), n+1-L(S,n)\} - 1. \end{aligned}$$

10.4.88 **Remark** The results of [913] on the expected value of the lattice profile show that a "random" sequence should have $\mathcal{S}(S,n)$ close to $\min\{n/2, N\}$.

10.4.5.3 Correlation measure of order k

10.4.89 **Definition** The *correlation measure of order* k of a binary sequence S is

$$C_k(S) = \max_{M,D} \left| \sum_{n=0}^{M-1} (-1)^{s_{n+d_1}} \cdots (-1)^{s_{n+d_k}} \right|, \quad k \geq 1,$$

where the maximum is taken over all $D = (d_1, d_2, \ldots, d_k)$ with non-negative integers $d_1 < d_2 < \cdots < d_k$ and M such that $M - 1 + d_k \leq T - 1$. Obviously, $C_2(S)$ is bounded by the maximal absolute value of the aperiodic autocorrelation of S.

10.4.90 **Remark** The correlation measure of order k was introduced by Mauduit and Sárközy in [2037]. The linear complexity profile of a given N-periodic sequence can be estimated in terms of its correlation measure and a lower bound on $L(S,n)$ can be obtained whenever an appropriate bound on $\max C_k(S)$ is known.

10.4.91 **Theorem** [393] We have

$$L(S,n) \geq n - \max_{1 \leq k \leq L(S,n)+1} C_k(S), \quad 1 \leq n \leq N-1.$$

10.4.5.4 FCSR and p-adic span

10.4.92 **Remark** In [1748] an alternative feedback shift register architecture was presented, *feedback with carry shift registers (FCSR)*. For binary sequences the procedure is as follows: Differently to linear recurring sequences the bits are added as integers (again following a linear recurrence relation). The result is added to the content of a memory, which is a nonnegative integer m, to obtain an integer σ. The parity bit $\sigma \pmod 2$, of σ is then the next term of the sequence, and the higher order bits $\lfloor \sigma/2 \rfloor$ are the new content of the memory.

FCSR-sequences share many properties with linear recurring sequences, but for their analysis instead of arithmetics in finite fields, arithmetics in the 2-adic numbers is used - or in the more general case of sequences modulo p in the p-adic numbers.

An FCSR-equivalent to the linear complexity is the 2-*adic span*, respectively the *p*-*adic span* of a sequence, which measures the size of the smallest FCSR that generates the sequence.

Since their introduction, FCSR-sequences attracted a lot of attention. We refer to [129, 1330, 1331, 1749, 2774] and the references therein.

10.4.5.5 Discrepancy

10.4.93 **Definition** Let $X = x_0, x_1, \ldots$ be a sequence in the unit interval $[0, 1)$. For $0 \le d_1 < \cdots < d_k < n$ we put

$$\mathbf{x}_j = \mathbf{x}_j(d_1, \ldots, d_k) = (x_{j+d_1}, \ldots, x_{j+d_k}), \quad 1 \le j \le n - d_k.$$

The *discrepancy* of the vectors $\mathbf{x}_1(d_1, \ldots, d_k), \ldots, \mathbf{x}_{n-d_k}(d_1, \ldots, d_k)$ is

$$\sup_I \left| \frac{A(I, \mathbf{x}_1, \ldots, \mathbf{x}_{n-d_k})}{n - d_k} - V(I) \right|,$$

where the supremum is taken over all subintervals of $[0, 1)^k$, $V(I)$ is the volume of I and $A(I, \mathbf{x}_1, \ldots, \mathbf{x}_{n-d_k})$ is the number of points $\mathbf{x}_j$, $j = 1, \ldots, n - d_k$, in the interval I.

10.4.94 **Remark** We can derive a binary sequence $B = e_0, e_1, \ldots$ from X by $e_j = 1$ if $0 \le x_j < 1/2$ and $e_j = 0$ otherwise.

10.4.95 **Remark** In [2036, Theorem 1] the correlation measure of order k of B is estimated in terms of the above discrepancy of vectors derived from the sequence X. Hence, using the relation between linear complexity profile and correlation measure of B we can obtain (weak) linear complexity profile lower bounds for B from discrepancy upper bounds for X.

See Also

§6.3, §10.2, §17.3	For related measures.
§9.1, §9.3	For Boolean functions and nonlinearity.
§10.2	For LFSR.
§10.5, §17.1	For nonlinear recurrence sequences.
§12.2, §12.3, §16.4	For elliptic curves.
§15.1	For basics on coding theory and the Berlekamp-Massey algorithm.
§16.2	For stream ciphers.

References Cited: [85, 86, 87, 127, 129, 257, 297, 298, 303, 392, 393, 599, 600, 641, 759, 763, 764, 765, 766, 784, 873, 913, 914, 1053, 1059, 1134, 1135, 1136, 1137, 1169, 1198, 1330, 1331, 1359, 1378, 1382, 1445, 1493, 1631, 1642, 1671, 1748, 1749, 1786, 1864, 1936, 1939, 2011, 2013, 2036, 2037, 2056, 2057, 2058, 2059, 2060, 2061, 2062, 2063, 2064, 2065, 2066, 2067, 2069, 2070, 2071, 2072, 2235, 2239, 2240, 2242, 2243, 2246, 2253, 2254, 2257, 2273, 2274, 2275, 2276, 2502, 2503, 2516, 2517, 2520, 2549, 2642, 2644, 2685, 2700, 2774, 2829, 2866, 2928, 2933, 2934, 2935, 2937, 2944, 2946, 2958, 2959, 2960, 2994, 3013, 3014, 3016, 3025]

10.5 Algebraic dynamical systems over finite fields

Igor Shparlinski, Macquarie University

10.5.1 Introduction

10.5.1 **Definition** Let $F_1, \ldots, F_m \in \mathbb{F}_q(X_1, \ldots, X_m)$ be m rational functions in m variables over the finite field $\mathbb{F}_q$ of q elements. The *algebraic dynamical system (ADS)* generated by $\mathcal{F} = \{F_1, \ldots, F_m\}$ is the dynamical system formed by the iterations

$$F_i^{(k)} = F_i(F_1^{(k-1)}, \ldots, F_m^{(k-1)}), \quad k = 1, 2, \ldots, \quad i = 1, \ldots, m,$$

where $F_i^{(0)} = X_i$.

10.5.2 **Remark** ADSs have proved to be exciting and challenging mathematical objects. They have very interesting algebraic and number theoretic properties and also exhibit a very complex behavior; see [91, 964, 1020, 1087, 1313, 1618, 1619, 2421, 2547, 2668] for the foundations of the theory. This makes them an invaluable building block for various applications including pseudorandom number generators (PRNGs), which are of crucial value in quasi-Monte Carlo methods and cryptography; see [2271, 2648, 2814]. Recently very surprisingly links with other natural sciences such as biology [1596, 1860, 2717] and physics [106, 154, 216, 222, 1348, 1498, 2464, 2870, 2871] have emerged.

10.5.2 Background and main definitions

10.5.3 **Definition** Given an *initial vector* $\mathbf{u}_0 = (u_{0,1}, \ldots, u_{0,m}) \in \mathbb{F}_q^m$, the *trajectory* of an ADS (see Definition 10.5.1) originating at this vector, is the sequence of vectors $\mathbf{u}_n = (u_{n,1}, \ldots, u_{n,m}) \in \mathbb{F}_p^m$ defined by the recurrence relation

$$u_{n+1,i} = F_i(u_{n,1}, \ldots, u_{n,m}), \quad n = 0, 1, \ldots, \quad i = 1, \ldots, m.$$

10.5.4 **Remark** Using the following vector notation

$$\mathbf{F}(X_1, \ldots, X_m) = (F_1(X_1, \ldots, X_m), \ldots, F_m(X_1, \ldots, X_m)),$$

we have the recurrence relation

$$\mathbf{u}_{n+1} = \mathbf{F}(\mathbf{u}_n), \quad n = 0, 1, \ldots. \tag{10.5.1}$$

In particular, for any $n, k \geq 0$ and $i = 1, \ldots, m$ we have

$$u_{n+k,i} = F_i^{(k)}(\mathbf{u}_n) = F_i^{(k)}(u_{n,1}, \ldots, u_{n,m})$$

or

$$\mathbf{u}_{n+k} = \mathbf{F}^{(k)}(\mathbf{u}_n).$$

10.5.5 **Remark** If $F_1, \ldots, F_m$ are rational functions, then one has to decide what to do if $\mathbf{u}_n$ is a pole of some of them. A canonical way to resolve this is to define these functions on the set of their poles separately (for example, define $0^{-1} = 0$). Certainly this problem does not occur if all functions $F_1, \ldots, F_m$ are polynomials.

10.5.6 Remark Recently, new applications have emerged to cryptography and quasi-Monte Carlo methods, where ADSs have been shown to provide a very attractive alternative to the classical linear congruential PRNG.

10.5.7 Definition An *attack* on a PRNG is an algorithm that observes several outputs of a PRNG and then is able to continue to generate the same sequence with a nontrivial probability.

10.5.8 Remark The interest in nonlinear dynamical systems as sources of pseudorandom numbers [2271, 2814] has been driven by a series of devastating attacks on traditional linear constructions which have made them useless for cryptographic purposes; see [716, 1831] and references therein.

10.5.9 Remark Although nonlinear PRNGs are believed to be cryptographically stronger; see, however [210, 299, 300, 301, 1309, 1380] for some attacks. Yet, obtaining concrete efficient constructions with good rigorously proven estimates on their statistical and other properties, has also been a challenging task [1358, 1379, 2329].

10.5.3 Degree growth

10.5.10 Definition The *algebraic entropy* of the ADS generated by $\mathcal{F} = \{F_1, \ldots, F_m\}$ is

$$\delta(\mathbf{F}) = \lim_{n \to \infty} \frac{\log D_n(\mathbf{F})}{n},$$

where $D_k(\mathbf{F})$ is the degree of $\mathbf{F}^{(k)}$, defined as the largest degree of the components $F_1^{(k)}, \ldots, F_m^{(k)}$.

10.5.11 Remark The existence of the limit follows immediately from the inequality $D_{k+m}(\mathbf{F}) \leq D_k(\mathbf{F}) D_m(\mathbf{F})$.

10.5.12 Remark Studying how the iterates of an ADS grow is a classical research direction with a rich history and a variety of results. In the case of ADSs over the complex and p-adic numbers there is a well-studied measure of "size" called the *height*, [91, 964, 1020, 2668]. Unfortunately this measure does not apply to ADSs over finite fields. However, in this case the degree $D_k(\mathbf{F})$ provides a very natural and adequate substitute for the notion of height. Thus, the algebraic entropy plays a very essential role in the theory of ADSs over finite fields [215, 216, 222, 1498, 2817, 2870, 2871].

10.5.13 Remark The degree growth is important for many applications of ADSs. In the univariate case, it is obvious that the n-th iterate of a polynomial of degree d is a polynomial of degree d^n. This however is not true anymore in the multidimensional case. Yet, the exponential growth is expected for a "typical" ADS. For example in [1358, 1379] the exponential growth is shown for some very special classes of polynomial systems, corresponding to nonlinear recurrence sequences of order m, that is, sequences satisfying a recurrence relation of the form

$$w_{n+m} = F(w_{n+m-1}, \ldots, w_n), \qquad n = 0, 1, \ldots,$$

with $F \in \mathbb{F}_q(X_1, \ldots, X_m)$. An alternative approach with a combinatorial flavor to studying such sequences has been suggested in [2329].

10.5.14 Remark Recently, a family of multivariate ADSs with polynomial degree growth of their iterates has been constructed in [1312, 2328, 2330, 2332, 2334]. These ADSs are formed by systems of rational functions of the form:

$$\begin{aligned} F_1(X_1,\dots,X_m) &= X_1^{e_1}G_1(X_2,\dots,X_m)+H_1(X_2,\dots,X_m),\\ &\ \vdots \\ F_{m-1}(X_1,\dots,X_m) &= X_{m-1}^{e_{m-1}}G_{m-1}(X_m)+H_{m-1}(X_m),\\ F_m(X_1,\dots,X_m) &= g_mX_m^{e_m}+h_m, \end{aligned} \tag{10.5.2}$$

where $e_i \in \{-1,1\}$, $i=1,\dots,m$,

$$g_m, h_m \in \mathbb{F}_q, \qquad g_m \neq 0,$$

and G_i has a *unique leading monomial*:

$$G_i(X_{i+1},\dots,X_m) = g_iX_{i+1}^{s_{i,i+1}}\cdots X_m^{s_{i,m}} + \widetilde{G_i}(X_{i+1},\dots,X_m),$$

which "dominates" all other terms:

$$g_i \neq 0, \quad \deg_{X_j}\widetilde{G_i} < s_{i,j}, \quad \deg_{X_j} H_i \le s_{i,j},$$

for $1 \le i < j \le m$.

10.5.15 Remark The structure and the degree of iterates of ADSs in 10.5.14 is given by Theorem 10.5.16 which is essentially [2332, Lemma 1].

10.5.16 Theorem [2332, Lemma 1] In the case of $e_i = 1$, $i = 1,\dots,m$, for any ADS of the form (10.5.2), we have

$$F_i^{(k)} = X_iG_{i,k}(X_{i+1},\dots,X_m)+H_{i,k}(X_{i+1},\dots,X_m), \quad i=1,\dots,m,\ k=0,1,\dots,$$

where

$$G_{i,k}, H_{i,k} \in \mathbb{F}_q[X_{i+1},\dots,X_m], \quad i=1,\dots,m-1$$

and

$$\deg G_{i,k} = \frac{1}{(m-i)!}k^{m-i}s_{i,i+1}\dots s_{m-1,m}+\psi_i(k),$$

with $\psi_i(T)\in\mathbb{Q}[T]$, $\deg\psi_i < m-i$, $i=1,\dots,m-1$, and $G_{m,k}=g_m^k\in\mathbb{F}_q^*$.

10.5.17 Remark In [2328, 2334] a more general (but also more technically involved) form of Theorem 10.5.16 is given which applies to exponents $e_i = \pm 1$, $i=1,\dots,m$.

10.5.18 Problem Find new constructions of ADSs with polynomial degree grows of the iterates.

10.5.19 Remark In [2335] yet another new class of ADSs is given that also leads to good pseudorandom number generators, namely, given ADSs that are formed by systems of multivariate polynomials $F_1,\dots,F_m \in \mathbb{F}_p[X_1,\dots,X_m]$ of the form:

$$\begin{aligned} F_1 &= (X_1-h_1)^{e_1}G_1+h_1,\\ &\ \vdots \\ F_{m-1} &= (X_{m-1}-h_{m-1})^{e_{m-1}}G_{m-1}+h_{m-1},\\ F_m &= g_m(X_m-h_m)^{e_m}+h_m, \end{aligned} \tag{10.5.3}$$

where $G_i \in \mathbb{F}_p[X_{i+1}, \ldots, X_m]$, $i = 1, \ldots, m-1$, e_j are positive integers and $g_m, h_j \in \mathbb{F}_p$, $j = 1, \ldots, m$. The corresponding pseudorandom number generator generalizes the classical power generator (see Section 10.4) but is free of some of its undesirable features such as homogeneity. Although the degree of the iterates of (10.5.3) grows exponentially, under some additional condition using a recent result of Cochrane and Pinner [657], one can get a reasonably good estimate of the discrepancy of the corresponding sequence; see [2335, Theorem 8] for details. This direction has been further developed in [1312].

10.5.20 Problem Find new constructions of ADSs with sparse iterates (that is, having a constant or slowly growing with the number of iterations number of monomials).

10.5.4 Linear independence and other algebraic properties of iterates

10.5.21 Remark Surprisingly enough, investigating the distribution and other number theoretic properties of PRNGs ultimately requires to study additive character sums with their elements. In turn, this leads to investigation of their algebraic properties such as the degree growth or absolute irreducibility of certain linear combinations of the iterates. Deeper algebraic properties such as the dimension of the singularity locus of certain associated algebraic varieties are also of interest. Below we explain how these properties become ultimately related to the rather analytic question of the uniformity of distribution.

10.5.22 Remark An application of the celebrated *Koksma–Szüsz inequality* (see the original works [1780, 2759] and also Section 6.3) reduces the question of studying the distribution of the vectors (10.5.1) to the question of estimating the following additive character sums

$$S_{\mathbf{a}}(N) = \sum_{n=0}^{N-1} \psi\left(\sum_{i=1}^{m} a_i u_{n,i}\right),$$

where ψ is a fixed additive character of $\mathbb{F}_q$ and $\mathbf{a} = (a_1, \ldots, a_m) \in \mathbb{F}_q^m$. The standard methodology, suggested in [2269, 2270] and recently improved in [2279], after a certain chain of standard transformations used in estimating character sums, leads to additive character sums with

$$L_{\mathbf{a},k,l}(\mathbf{X}) = \sum_{i=1}^{m} a_i \left(F_i^{(k)}(\mathbf{X}) - F_i^{(l)}(\mathbf{X})\right),$$

where $\mathbf{X} = (X_1, \ldots, X_m)$. So a natural next step is to use the Weil bound, see Section 6.3. However, for this the rational function $L_{\mathbf{a},k,l}(\mathbf{X})$ has to be "exponential sums friendly," that is,

1. nontrivial (that is, with a nontrivial trace);
2. of small degree (so that the Weil bound is nontrivial for this function);
3, if possible, linear in some of the variables (thus to avoid using the Weil bound at all);
4. if possible, have a low dimensional locus of singularity (to apply the Deligne bound, see Section 6.3 or Deligne-like bounds by Katz [1704]).

10.5.23 Problem Find general (necessary and sufficient) conditions under which, the linear forms $L_{\mathbf{a},k,l}(\mathbf{X})$ are not constant for any non-zero vector $\mathbf{a} \in \mathbb{F}_q^m$ and $k \neq l$.

10.5.24 Remark Over $\mathbb{F}_q$, for a special class of iterations, some sufficient conditions are given in [1358, 1379]. In [2329], in some special case, the above condition of on $L_{\mathbf{a},k,l}(\mathbf{X})$ has been replaced by some combinatorial argument, which, however, does not seem to generalize any further.

10.5.25 Remark For ADSs of the shape (10.5.2) the degree of the iterates grows polynomially. Furthermore the iterates are linear in one of the variables. Both properties together, have led to rather strong results about the distribution of the vectors (10.5.1); see [2330, 2332]. This idea and construction has been further developed in [2326, 2327, 2336, 2337]. Furthermore, in [2332] a new construction of *hash functions* is suggested that is based on the above ADSs.

10.5.26 Remark It is clear that the systems of the shape (10.5.2) with $e_i = 1$, $i = 1, \ldots, m$ (or for arbitrary $e_i = \pm 1$, if one defines $0^{-1} = 0$) define a permutation of $\mathbb{F}_q^m$ if and only if the "coefficients" G_i, $i = 1, \ldots, m-1$, have no zeros over $\mathbb{F}_q$. For such permutation systems, Ostafe [2326] has established rather strong results about the distribution of elements in trajectories on average over all initial values; see also [2334, 2653].

10.5.27 Remark At the same time it has become clear that in order to improve the results of [2330, 2332] one needs to obtain more detailed information about the *algebraic structure* of polynomial iterates, in particular about the number of absolutely irreducible components of the polynomials $G_{i,k_1} - G_{i,k_2}$ where $0 \leq k_2 < k_1$ and $G_{i,k}$ is as in Theorem 10.5.16.

10.5.28 Remark Sums of multiplicative characters along trajectories of ADSs have also been considered in the literature [2272, 2277, 2337]. Such sums can be estimated within the same lines that have been used for additive character sums, however instead of linear combinations $L_{\mathbf{a},k,l}(\mathbf{X})$ one has to study some other algebraic expressions including the iterates. For instance, the argument of [2337] is based on studying the bilinear combinations $G_{i,k}H_{i,l} - G_{i,l}H_{i,k}$ (in the notation of Theorem 10.5.16) and showing that they are not constant.

10.5.5 Multiplicative independence of iterates

10.5.29 Theorem [1173, Theorem 1.4] Suppose that a polynomial $f \in \mathbb{F}_q[X]$ is not a monomial or a binomial of the form $ax^{p^\ell} + b$ where p is the characteristic of $\mathbb{F}_q$. Then for any integer $N \geq 1$, the polynomials $f, f^{(1)}, \ldots, f^{(N)}$ are multiplicative independent.

10.5.30 Remark Theorem 10.5.29 is used in an construction of elements of large multiplicative order over finite fields; see [1173, Theorem 1.1].

10.5.31 Remark There are several possible interpretations of what a multivariate analogue of Theorem 10.5.29 may look like. All of them are interesting, however no results in this direction have been obtained so far.

10.5.6 Trajectory length

10.5.32 Remark Clearly the sequence of vectors $\{\mathbf{u}_n\}$, given by (10.5.1) is eventually periodic with some period τ. That is, for some integer $s \geq 0$ we have $\mathbf{u}_{n+\tau} = \mathbf{u}_n$ for $n \geq s$.

10.5.33 Definition If s is the smallest integer with this above property of Remark 10.5.32, then $T = s + \tau$ is the *trajectory length*.

10.5.34 Remark If we work in a finite field of q elements, then the trajectory length T satisfies $T \leq q^m$, where m is the number of variables.

10.5.35 Remark No general lower bounds on the trajectory length of sequences generated by ADSs are known. Furthermore, assuming that the map generated by $\mathbf{F}$ behaves as a random map (and for a generic polynomial system this seems to be a natural and well tested assumption), one should expect that in fact T is of order $q^{m/2}$. On the other hand, most of the results about the distribution of the sequence (10.5.1) are nontrivial only if the trajectory length T is close to its largest possible value [2326, 2327, 2330, 2332, 2336, 2337]. Hence we see that "generic" ADSs are not likely to satisfy this property. Thus one needs constructions of special ADSs tailored for these applications.

10.5.36 Remark Very little is known about constructing ADSs with guaranteed large trajectory length. The only known rigorous results are those of Ostafe [2328] and their generalizations in [2334], that gives a complete characterization (which in turn leads to explicit constructions) of ADSs of the type (10.5.2) which achieve the largest possible value of the trajectory length $T = q^m$.

10.5.37 Remark A result of [2669] gives a nontrivial (albeit very weak) lower bound on the length of a reduction of a trajectory of an ADS over the rationals modulo a prime p that holds for almost all p (in fact the result applies to more general settings).

10.5.7 Irreducibility of iterates

10.5.38 Remark As we have seen in Section 10.5.4, the algebraic structure of iterates becomes very important for studying the distribution of trajectories. Questions of this type are also of great intrinsic interest for the theory of ADSs. Unfortunately, they are notoriously hard, and most of the few known results apply only to iterates of univariate quadratic polynomials; see [76, 152, 153, 1310, 1313, 1618, 1619, 1620, 2331] and references therein.

10.5.39 Definition A polynomial f over a field $\mathbb{K}$ is *stable* if all its iterations $f^{(n)}$ are irreducible over $\mathbb{K}$.

10.5.40 Definition For a quadratic polynomial $f(X) = aX^2 + bX + c \in \mathbb{K}[X]$, over a field $\mathbb{K}$ of characteristic $p \neq 2$, we define the *critical orbit* of f as the set of consecutive iterations

$$\mathrm{Orb}(f) = \{f^{(n)}(\gamma) \ : \ n = 2, 3, \ldots\},$$

where $\gamma = -b/2a$.

10.5.41 Remark Clearly γ in Definition 10.5.40 is the unique critical point of f (that is, the zero of the derivative f').

10.5.42 Remark It is shown in [1618, 1619, 1620] that critical orbits play a very important role in the dynamics of polynomial iterations.

10.5.43 Remark Capelli's Lemma, describing the conditions on the irreducibility of polynomial compositions, plays a prominent role in this area.

10.5.44 Theorem [1620, Proposition 3] A quadratic polynomial $f \in \mathbb{K}[X]$ is stable if the set $\{-f(\gamma)\} \cup \mathrm{Orb}(f)$ contains no squares. If $\mathbb{K} = \mathbb{F}_q$ is a finite field of odd characteristic, this property is also necessary.

10.5.45 Remark If $\mathbb{K} = \mathbb{F}_q$ is a finite field, there is some integer t such that $f^{(t)}(\gamma) = f^{(s)}(\gamma)$ for some positive integer $s < t$. Then $f^{(n+t)}(\gamma) = f^{(n+s)}(\gamma)$ for any $n \geq 0$. Accordingly, for the smallest value of t with the above condition denoted by t_f, we have

$$\mathrm{Orb}(f) = \{f^{(n)}(\gamma) \ : \ n = 2, \ldots, t_f\}$$

and $\#\mathrm{Orb}(f) = t_f - 1$ or $\#\mathrm{Orb}(f) = t_f - 2$ (depending whether $s = 1$ or $s \geq 2$ in the above).

10.5.46 Remark Remark 10.5.45 immediately implies that a quadratic polynomial $f \in \mathbb{F}_q[X]$ can be tested for stability in $q^{1+o(1)}$ arithmetic operations over $\mathbb{F}_q$. Using bounds of character sums, it has been shown in [2331] that this can be improved.

10.5.47 Theorem [2331] A quadratic polynomial over $\mathbb{F}_q$ can be tested for stability in $q^{3/4+o(1)}$ arithmetic operations over $\mathbb{F}_q$.

10.5.48 Remark In [50], a generalization of Theorem 10.5.47 is given to polynomials $g(f)$ where f is quadratic and g is an arbitrary polynomial over $\mathbb{F}_q$.

10.5.49 Remark Since a random polynomial of degree d is irreducible over $\mathbb{F}_q$ with probability about $1/d$ and the degree of the iterations $f^{(n)}$ grows exponentially with n, it is natural to expect that there are only very few stable polynomials over $\mathbb{F}_q$. For quadratic polynomials this has been confirmed in a quantitative form by Gomez and Nicolás [1310]. With several ingenious extensions of the method of [1310], Gomez, Nicolás, Ostafe, and Sadornil [1311] have shown this for polynomials of arbitrary degree $d \geq 2$.

10.5.50 Theorem [1310] For any odd prime power q, there are at most $q^{5/2+o(1)}$ stable quadratic polynomials over $\mathbb{F}_q$.

10.5.51 Remark The number of irreducible divisors and other arithmetic properties of iterations of polynomials over large finite fields have been studied in [1313].

10.5.52 Problem Give explicit constructions of polynomial systems, over some "interesting" fields such as $\mathbb{Q}$ and $\mathbb{F}_q$, such that all polynomials $F_i^{(k)}$, $i = 1, \ldots, m$, $k = 1, 2, \ldots$, are

1. irreducible over $\mathbb{F}_q$;
2. absolutely irreducible over $\mathbb{F}_q$.

10.5.53 Remark Over the field of rational numbers $\mathbb{Q}$ we define the class $\mathcal{E}_{p,m}$ (where p is an arbitrary prime) of m-variate analogues of the *Eisenstein polynomials*: We say that $F \in \mathbb{Z}[X_1, \ldots, X_m]$ belongs to $\mathcal{E}_{p,m}$ if

$$F(X_1, \ldots, X_m) = A_0 X_1^{d_1} \cdots X_m^{d_m} + pf(X_1, \ldots, X_m)$$

for some $f \in \mathbb{Z}[X_1, \ldots, X_m]$, where $A_0 \not\equiv 0 \pmod p$, $d_1 + \cdots + d_m > \deg f$ and such that $F(\mathbf{0}) \not\equiv 0 \pmod{p^2}$, where $\mathbf{0} = (0, \ldots, 0)$. Clearly if $F = GH$, where $G, H \in \mathbb{Z}[X_1, \ldots, X_m]$, is reducible then $F \equiv GH \pmod p$. Since if F is a monomial modulo p then so are G and H. As $\deg F = d_1 + \cdots + d_m$, we conclude that G and H are nonconstant monomials modulo p. Therefore $G(\mathbf{0}) \equiv H(\mathbf{0}) \equiv 0 \pmod p$ which implies that $F(\mathbf{0}) = G(\mathbf{0})H(\mathbf{0}) \equiv 0 \pmod{p^2}$, contradicting $F \in \mathcal{E}_{m,p}$. If $F, G_1, \ldots, G_m \in \mathcal{E}_{p,m}$ then because

$$F(G_1(\mathbf{0}), \ldots, G_m(\mathbf{0})) \equiv F(\mathbf{0}) \not\equiv 0 \pmod{p^2},$$

we also have $F(G_1, \ldots, G_m) \in \mathcal{E}_{p,m}$. Therefore if $F_1, \ldots, F_m \in \mathcal{E}_{m,p}$ for some prime p then their iterations $F_1^{(k)}, \ldots, F_m^{(k)}$, $k = 1, 2, \ldots$ are all irreducible over $\mathbb{Q}$. This however does not lead to a construction of absolutely irreducible iterates. There is no obvious finite field analogue of this construction.

10.5.8 Diameter of partial trajectories

10.5.54 Definition The *diameter* $D_{\mathcal{F},\mathbf{u}_0}(N)$ of the sequence of the first N vectors (10.5.1) over $\mathbb{F}_p$ is defined as

$$D_{\mathcal{F},\mathbf{u}_0}(N) = \max_{0 \leq k,n \leq N-1} \|\mathbf{u}_k - \mathbf{u}_n\|,$$

where $\|\mathbf{u}\|$ is the Euclidean norm of a vector $\mathbf{u}$ and we assume that the elements of $\mathbb{F}_p$ are represented by the set $\{0, \ldots, p-1\}$.

10.5.55 Remark Certainly results about the asymptotically uniform distribution of the first N vectors (10.5.1) (mentioned in Section 10.5.4) immediately imply that $D_{\mathcal{F},\mathbf{u}_0}(N)$ is close to the largest possible value $n^{1/2}p$ in this case. However such results are known only for very long segments of the trajectories.

10.5.56 Remark The notion of the diameter (under a slightly different name) is introduced in [1381] and then has also been studied in [585, 586, 644] where a wide variety of methods has been used. However, the only known results about the diameter are in the univariate case, that is, when $\mathcal{F} = \{f\} \subseteq \mathbb{F}_p[X]$ and $\mathbf{u}_0 = u_0 \in \mathbb{F}_p$.

10.5.57 Theorem [1381, Theorem 6] For any fixed $\varepsilon > 0$ and $T_{f,u_0} \geq N \geq p^{1/2+\varepsilon}$ where T_{f,u_0} is the trajectory length corresponding to iterations of $f \in \mathbb{F}_p[X]$ originating at u_0, we have

$$D_{f,u_0}(N) = p^{1+o(1)}$$

as $p \to \infty$.

10.5.58 Remark Theorem 10.5.57, when it applies, provides the asymptotically best possible bound. For smaller values of N, one can use a variety of the results from [585, 586, 644] that are nontrivial in essentially the best possible range $T_{f,u_0} \geq N \geq p^{\varepsilon}$ for any fixed $\varepsilon > 0$.

10.5.59 Problem Let $\mathbb{F}_{q^s}$ be an extension of degree $s \geq 2$ of $\mathbb{F}_q$. Obtain a lower bound on the smallest dimension of an affine space over $\mathbb{F}_q$ containing the first N elements of the trajectory of iterations of $f \in \mathbb{F}_{q^s}[X]$ originating at $u_0 \in \mathbb{F}_{q^s}$.

10.5.60 Problem In the multidimensional case, besides estimating $D_{\mathcal{F},\mathbf{u}_0}(N)$, one can also study other geometric characteristics, such as

1. the volume and the number of vertices of the convex hull of the set $\{\mathbf{u}_0, \ldots, \mathbf{u}_{N-1}\}$;
2. the number of directions and the number of distances defined by pairs of vectors $(\mathbf{u}_k, \mathbf{u}_n)$, $0 \leq k, n \leq N-1$;
3. the number of directions and the number of distances defined by pairs of consecutive vectors $(\mathbf{u}_n, \mathbf{u}_{n+1})$, $0 \leq n \leq N-1$.

References Cited: [50, 76, 91, 106, 152, 153, 154, 210, 215, 216, 222, 299, 300, 301, 585, 586, 644, 657, 716, 964, 1020, 1087, 1309, 1310, 1311, 1312, 1313, 1348, 1358, 1379, 1380, 1381, 1498, 1596, 1618, 1619, 1620, 1704, 1780, 1831, 1860, 2269, 2270, 2271, 2272, 2277, 2279, 2326, 2327, 2328, 2329, 2330, 2331, 2332, 2334, 2335, 2336, 2337, 2421, 2464, 2547, 2648, 2653, 2668, 2669, 2717, 2759, 2814, 2817, 2870, 2871]

11

Algorithms

11.1 Computational techniques

Christophe Doche, Macquarie University

This section presents algorithmic methods for finite fields. It deals with concrete implementations and describes how some fundamental operations on the elements are performed.

Finite fields play a central role in many areas, such as cryptography, coding theory, and random number generation, where the speed of the computations is paramount. Therefore we discuss algorithms with a particular attention to efficiency and we address implementation techniques and complexity aspects as often as possible. In many instances, the complexity of the algorithms that we describe depends on the field multiplication method that is implemented. In the following, $\mathrm{M}(\log p)$ denotes the complexity to multiply two positive integers less than p. Similarly, $\mathrm{M}_q(n)$ represents the complexity to multiply two polynomials in $\mathbb{F}_q[x]$ of degree less than n. We note that many software tools or libraries implement some of the algorithms that are presented next. A non-exhaustive list includes Magma [2798], Pari [209], Sage [2709], Mathematica [3004], Maple [2002], NTL [2633], GMP [1102], MPFQ [1255], FLINT [1426], and ZEN [576]. Detailed algorithms and complexity analyses can be found in [660, 661, 751, 1227, 1413, 1768, 2080, 2632].

11.1.1 Preliminaries

11.1.1.1 Prime field generation

11.1.1 Remark For many applications, a random prime field of a given size is needed. This implies finding a prime number of a given bit length that is random in the following sense: the probability for any particular prime of that size to be selected is sufficiently small so that it is impossible for anyone to take advantage of this event. There are two different ways to find such a random prime number. The first technique is to identify a prime number among integers of a prescribed size successively picked at random. The second technique is to construct integers with special properties so that it is easy to prove that they are prime and whose distribution over the set of all the primes of the desired size is close to uniform.

11.1.2 Algorithm (Prime number random search)

Input: An integer $\ell \geq 2$.

Output: An ℓ-bit prime number.

1. **repeat**
2. Pick an ℓ-bit random integer n
3. **until** n is not composite
4. **return** n

11.1.3 Remark The prime number theorem [2080, Section 4.1] ensures that it takes $O(\ell)$ random draws among ℓ-bit integers before picking an integer that is actually prime.

11.1.4 Remark The approach used at Line 3 of Algorithm 11.1.2 to test if n is composite or not determines the quality of the prime number that is returned. In practice, Algorithm 11.1.2 relies on trial division, to quickly eliminate most composite numbers, and on the Miller–Rabin compositeness test, properly set up depending on the size of the desired prime. See Remark 11.1.5 and [2080, Section 4.4] for implementation details. In the end, Algorithm 11.1.2 returns a probable prime, i.e., not certified but prime with very high probability.

11.1.5 Remark The first compositeness test of real practical significance is due to Solovay and Strassen [2694]. It was superseded by the Miller–Rabin test [2099, 2435], which is faster and easier to implement. The error probability in the Miller–Rabin test can be adjusted using a security parameter t that also drives the complexity of the algorithm. The Miller–Rabin test is always correct when it declares that an integer n is composite, but it may be wrong with probability at most 4^{-t} when it declares that n is prime. Taking into account the distribution of prime numbers and using advanced probabilistic analysis techniques, it can be shown that Algorithm 11.1.2 coupled with the Miller–Rabin test requires a very small t in order to return a high quality probable prime. For instance, $t = 5$ is enough to produce a 500-bit integer that is prime with probability greater than $1 - 2^{-85}$; see [2080, Note 4.47 and Fact 4.48] and [2632, Section 10.3] for more examples.

11.1.6 Remark Given the parameter t and the factorization of $n-1$ as $2^s m$ with m odd, the Miller–Rabin test computes at most t modular exponentiations to determine if n is composite. Each exponentiation is of the form a^k, where a is a $O(n)$ random value and $k = 2^r m$ for some $r < s$. Its complexity is therefore $O(t \log^3 n)$. As noted in [660], the algorithm is significantly faster when single precision integers a are used instead of random values in $[0, n-1]$. Although the error bound may no longer hold in that case.

11.1.7 Remark An interesting variant of Algorithm 11.1.2 consists in testing integers in an arithmetic progression. For instance, considering the odd numbers in some interval may reduce the complexity of the process, see [2080, Section 4.4].

11.1.8 Remark When a provable prime is required, we can run a more expensive primality test on the integer n returned by Algorithm 11.1.2. There are mainly two primality tests used in practice. APRCL [660, 662], named after its inventors, relies on Jacobi sums and is a simplified version of the test presented in [19]. Its complexity is $O(\log^{c \log\log\log n} n)$, for some effective constant c. The Elliptic Curve Primality Proving test (ECPP) [143] uses elliptic curves and its complexity is $\tilde{O}(\log^5 n)$. The fast version of ECPP, called `fastECPP` runs in $\tilde{O}(\log^4 n)$, see [2141] for implementation details. Both ECPP and `fastECPP` offer the advantage of producing a certificate that can be used to prove that n is prime much quicker than running the test again. In 2002, Agrawal, Kayal, and Saxena [43] announced the first deterministic polynomial time primality testing algorithm. For an input n, the complexity of the so-called AKS algorithm is $O(\log^{10.5} n)$. One variant of AKS [610] runs in time $\tilde{O}(\log^4 n)$. Although this algorithm has the same complexity as `fastECPP`, the constants are much larger and as a result `fastECPP` is still the method of choice to prove the primality of general integers.

11.1.9 Remark As hinted in Remark 11.1.1, a radically different approach to find a prime number is to construct it from scratch. There are two popular provable prime generation methods in the literature, respectively due to Mihăilescu [2095] and Maurer [2038, 2080]. They both use Pocklington's lemma recursively, but Maurer's method generates random provable primes whose distribution is close to uniform over the set of all primes of a given size, whereas Mihăilescu's approach is more efficient but slightly reduces the set of primes that may be produced.

11.1.1.2 Extension field generation

11.1.10 Remark To define $\mathbb{F}_{q^n}$, it is enough to construct a basis of the vector space $\mathbb{F}_{q^n}$ over $\mathbb{F}_q$. A normal basis offers several advantages, such as a cheap evaluation of the Frobenius automorphism, see Section 5.2. Alternatively, any irreducible polynomial of degree n with coefficients in $\mathbb{F}_q$ can be used to represent $\mathbb{F}_{q^n}$. This gives rise to a polynomial basis; see Definition 2.1.96.

11.1.11 Remark Quite similarly to the prime field case, there are two different ways to find an irreducible polynomial of a given degree. We can identify an irreducible polynomial among many polynomials generated at random. We may also construct an irreducible polynomial directly. This is well illustrated in [1187].

11.1.12 Remark It follows from Theorem 2.1.24 that a random monic polynomial of degree n with coefficients in $\mathbb{F}_q$ is irreducible with probability close to $\frac{1}{n}$. So it takes an expected $O(n)$ attempts to find an irreducible polynomial of degree n at random; see Subsection 11.3.2, for a description of some irreducibility testing algorithms.

11.1.13 Remark It is well known that different irreducible polynomials of degree n generate isomorphic finite fields. The isomorphism can even be computed explicitly [77, 1895]. So it may seem that the choice of the irreducible polynomial used to generate $\mathbb{F}_{q^n}$ is irrelevant. However, certain polynomials are more suitable than others when it comes to the efficiency of the operations in $\mathbb{F}_{q^n}$. In particular, polynomials with a low number of nonzero terms have a clear advantage, as they provide a faster modular reduction, see Subsection 11.1.3.2.

11.1.14 Definition Let $f \in \mathbb{F}_q[x]$. The polynomial f is *s-sparse*, if f has s nonzero terms. The terms *binomial*, *trinomial*, *quadrinomial*, and *pentanomial* are frequently used to refer to 2-sparse, 3-sparse, 4-sparse, and 5-sparse polynomials, respectively. Furthermore, f is *t-sedimentary* if $f = x^n + h$ where $\deg h = t$.

11.1.15 Remark According to Definition 11.1.14, any polynomial is s-sparse. Similarly any polynomial is t-sedimentary. However only those with sufficiently small parameters s or t are considered in practice. The relevance of s-sparse polynomials is known for a long time, whereas the interest for t-sedimentary polynomials is more recent [717, 2306].

11.1.16 Definition We denote by $\sigma_q(n)$ the minimal s such that there exists an irreducible s-sparse polynomial of degree n in $\mathbb{F}_q[x]$. Similarly, $\tau_q(n)$ denotes the minimal t such that there exists an irreducible polynomial of degree n in $\mathbb{F}_q[x]$ of the form $x^n + h$ with $\deg h = t$.

11.1.17 Conjecture Let n be a positive integer and let q be a power of a prime greater than 2, then we have $\sigma_q(n) \leq 4$. If $q = 2$, then $\sigma_q(n) \leq 5$ [1233].

11.1.18 Remark Conjecture 11.1.17 states that except for certain extensions of degree n of $\mathbb{F}_2$ where a pentanomial is required because there is no irreducible trinomial of degree n, it is always possible to find an irreducible binomial, trinomial, or quadrinomial to define $\mathbb{F}_{q^n}$ over $\mathbb{F}_q$. A similar conjecture exists for primitive polynomials, see Section 4.1.

11.1.19 Example We have $\sigma_2(8) = 5$. An exhaustive search shows that there is no irreducible trinomial of degree 8 over $\mathbb{F}_2$ but it is easy to see that $x^8 + x^4 + x^3 + x + 1$ is irreducible over $\mathbb{F}_2$. The field $\mathbb{F}_{2^8}$ is the smallest extension of $\mathbb{F}_2$ that requires a pentanomial. Similarly, $\sigma_3(49) = 4$. Again, an exhaustive search shows that there is no irreducible binomial or trinomial of degree 49 over $\mathbb{F}_3$ but $x^{49} + 2x^3 + x^2 + 1$ is irreducible over $\mathbb{F}_3$. The field $\mathbb{F}_{3^{49}}$ is the smallest extension of $\mathbb{F}_3$ that requires a quadrinomial.

11.1.20 Conjecture Let n be a positive integer and let $q \geq 2$ be a prime power, then we have $\tau_q(n) \leq 3 + \log_q n$ [1233].

11.1.21 Remark Conjectures 11.1.17 and 11.1.20 are supported by extensive computations [1181, 1187, 1233, 1327].

11.1.22 Remark We refer to [2582] for a table of irreducible trinomials and pentanomials in $\mathbb{F}_2[x]$ of degree ranging from 2 to 10,000; see also Section 2.2.

11.1.23 Remark See [403] for a dedicated algorithm with reduced space complexity designed to test the irreducibility of trinomials of large degree in $\mathbb{F}_2[x]$. See [1223] for additional algorithms specifically designed for trinomials.

11.1.24 Remark Finally, we refer to Subsection 11.3.2 for methods to construct an irreducible polynomial of degree n over $\mathbb{F}_q$. We note that there exist infinite families of irreducible polynomials. For instance, the polynomials $x^{2.3^k} + x^{3^k} + 1$ with $k \geq 0$ are all irreducible in $\mathbb{F}_2[x]$ [1187].

11.1.1.3 Primitive elements

11.1.25 Remark Finding a primitive element of a finite field is an interesting problem both from a theoretical and a practical point of view [439, 927, 2627, 2628, 2640]. Given the complete factorization of $q-1$, Algorithm 11.1.26 returns a generator of $\mathbb{F}_q^*$ in polynomial time in $\log q$. No efficient method is available when the factorization of $q-1$ is not known.

11.1.26 Algorithm (Primitive element random search)

Input: A prime power q and the complete factorization of $q-1$ as $p_1^{d_1} \cdots p_k^{d_k}$.
Output: A generator γ of $\mathbb{F}_q^*$.

for $i = 1$ **to** k **do**
    repeat
        Choose $\alpha \in \mathbb{F}_q^*$ at random and compute $\beta = \alpha^{(q-1)/p_i}$
    until $\beta \neq 1$
    $\gamma_i = \alpha^{(q-1)/p_i^{d_i}}$
end for
return $\gamma = \prod_{i=1}^{k} \gamma_i$

11.1.27 Remark Algorithm 11.1.26 is more efficient than the naïve approach, which consists in searching for an element γ such that $\gamma^{(q-1)/p_i} \neq 1$, for all i. This is especially true as the number of prime factors of $q-1$ grows. The complexity of finding a generator of the multiplicative group of a prime field $\mathbb{F}_p^*$ with Algorithm 11.1.26 is $O(\log^4 p)$ [2632].

11.1.28 Remark To generate a random prime field $\mathbb{F}_p$ together with a primitive element, simply modify Algorithm 11.1.2 so that the random integer n produced at Line 2 is of the form $n = m+1$, where all the prime factors of m are known. An algorithm to generate a random factored number is given in [2632, Section 9.6]. Then use Algorithm 11.1.26 to return a generator of $\mathbb{F}_p^*$.

11.1.1.4 Order of an irreducible polynomial and primitive polynomials

11.1.29 Remark The order of a polynomial is introduced in Definition 2.1.51. The order of an irreducible polynomial f is equal to the order of any of its roots. It can be found with Algorithm 11.1.30 [2473].

11.1.30 Algorithm (Order of an irreducible polynomial)

Input: An irreducible polynomial f of degree n with coefficients in $\mathbb{F}_q$ and the complete factorization of $q^n - 1$ as $q_1^{e_1} \cdots q_\ell^{e_\ell}$.
Output: The order of f.

for $i = 1$ **to** ℓ **do**
    Find the smallest nonnegative integer f_i such that $f \mid \left(x^{{q_1}^{e_1} \cdots {q_i}^{f_i} \cdots {q_\ell}^{e_\ell}} - 1\right)$
end for
return ${q_1}^{f_1} \cdots {q_\ell}^{f_\ell}$

11.1.31 Remark If f is monic of order $q^n - 1$ such that $f(0) \neq 0$ then f is a primitive polynomial. In fact, the conditions are equivalent [1939, Theorem 3.16]; see also Section 4.1 for more details on primitive polynomials. Algorithm 11.1.32 [1416] is specifically designed to quickly test if a random polynomial is primitive or not.

11.1.32 Algorithm (Primitive polynomial testing)

Input: An irreducible polynomial f of degree n with coefficients in $\mathbb{F}_q$ and the list of all the prime factors of $q^n - 1$.
Output: **true** if f is primitive and **false** otherwise.

```
1. if (−1)^n f(0) is not a primitive element of F_q then
2.     return false
3. end if
4. if x^((q^n−1)/(q−1)) ≢ (−1)^n f(0) (mod f) then
5.     return false
6. end if
7. for each prime factor r_i of (q^n−1)/(q−1) that does not divide q − 1 do
8.     if x^((q^n−1)/((q−1)r_i)) (mod f) ∈ F_q then
9.         return false
10.    end if
11. end for
12. return true
```

11.1.33 Remark The prime factors of $q-1$ are needed at Line 1 to test if $(-1)^n f(0)$ is a primitive element or not. If q is small enough, it is worth evaluating $f(\alpha)$ for all $\alpha \in \mathbb{F}_q^*$ in order to detect linear factors. Another useful test, if n is not too large, is to form the Berlekamp matrix Q (see Subsection 11.3.2) and compute the rank of $Q - I$. If the rank is strictly less than $n-1$ then the polynomial is not irreducible and hence not primitive. Those two optional tests should be implemented after the initial one that ends at Line 3.

11.1.34 Remark Given a single primitive polynomial f of degree n over $\mathbb{F}_q$, it is quite easy to find them all. Simply consider γ, the primitive element whose minimal polynomial is f and form the minimal polynomial of γ^k, for each k coprime with $q-1$. A more efficient algorithm is presented in [2420].

11.1.35 Remark If $n = 2^k - 1$ is a Mersenne prime, then an irreducible polynomial of degree k is necessarily primitive in $\mathbb{F}_2[x]$. We refer to [403, 404, 406, 408] for a search of binary primitive trinomials by Brent and Zimmermann.

11.1.1.5 Minimal polynomial of an element

11.1.36 Remark The simplest approach to find the minimal polynomial of $\alpha \in \mathbb{F}_{q^n}$ is to compute the different conjugates α^{q^i} of α until we have $\alpha^{q^k} = \alpha$. The minimal polynomial of α is then equal to the product $(x-\alpha)(x-\alpha^q)\cdots(x-\alpha^{q^{k-1}})$. The worst case complexity of this algorithm is $O(\mathrm{M}_q(n) n \log q)$ in time and $O(n)$ elements in space. This technique is refined in [1328] for the special case $\mathbb{F}_q = \mathbb{F}_2$.

11.1.37 Remark In [2631], Shoup proposes a more general algorithm with a better time complexity but also with an increased space complexity. Indeed, given the ring $\mathbb{F}_q[\alpha][\beta]$ of dimension n over $\mathbb{F}_q$, Shoup's algorithm finds the minimal polynomial of an element in that ring with complexity $O(\mathrm{M}_q(n) n^{1/2} + n^2)$ in time and $O(n^{3/2})$ elements in space. It relies on several subroutines, including Wiedemann's projection method [2976], a polynomial evaluation technique by Brent–Kung [401], and finally, the Berlekamp/Massey algorithm [2011] applied to recover the minimal polynomial from a sequence of projected points. Kedlaya and Umans [1721] give an algorithm for constructing a minimal polynomial in $n^{1+o(1)} \log^{1+o(1)} q$ bit operations. It relies on a fast modular composition method via multivariate multipoint evaluation, see Remark 11.1.86.

11.1.2 Representation of finite fields

11.1.38 Remark There are essentially three different ways to represent the elements of a finite field. Small finite fields can be represented with Jacobi logarithms, also known as Zech's logarithms, see Subsection 2.1.7.5. Extension fields $\mathbb{F}_{q^n}$ can always be represented with a normal basis over $\mathbb{F}_q$, see Section 5.2. Finally, prime fields and extension fields can be seen as a quotient set, respectively modulo a prime number and an irreducible polynomial.

11.1.39 Remark Elements in $\mathbb{F}_p$ are usually represented as integers in $[0, p-1]$ or in $[-\lfloor p/2 \rfloor, \lfloor p/2 \rfloor]$. We can also use alternative systems, such as the Montgomery representation, see Remark 11.1.46, redundant systems [3030], or even floating point numbers [933].

11.1.40 Remark When $\mathbb{F}_{q^n}$ is defined as $\mathbb{F}_q[x]/(f)$, where f is an irreducible polynomial of degree n in $\mathbb{F}_q[x]$, elements are usually represented as polynomials in $\mathbb{F}_q[x]$ of degree strictly less than n. However, a generalization of Montgomery representation, see Remark 11.1.57, and various redundant systems [405, 909, 3009] exist for extension fields as well.

11.1.41 Remark The *Kronecker substitution* [2111] allows to represent a polynomial in $\mathbb{F}_q[x]$ as an integer by formally replacing x by a suitable integer, usually a sufficiently large power of 2. Polynomial multiplication can be done faster with this system [930, 1430]. This technique is implemented in the FLINT library [1426].

11.1.3 Modular reduction

11.1.42 Remark The reduction of an integer u modulo a prime number p and the reduction of a polynomial g modulo an irreducible polynomial f are denoted by $u \pmod{p}$ and $g \pmod{f}$, respectively. In any case, an efficient reduction method is crucial for fast finite field arithmetic. Indeed, a reduction is usually performed after each operation to ensure that the size of the operands, i.e., the bit length or the degree, remains under control. Assuming that the operands are always reduced, an addition requires at worst one straightforward reduction. On the contrary, reducing the result of a multiplication is more involved even if we know that the size of the product is at most twice the size of the modulus.

11.1.3.1 Prime fields

11.1.43 Remark We assume that integers are represented using radix b. For most architectures, we have $b = 2^w$, for a fixed $w \geq 2$. A *word* then corresponds to w bits and we assume that multiplications and divisions by b, which correspond respectively to word left shift and word right shift instructions, are free. The following is presented in [2080, Algorithm 14.42].

11.1.44 Algorithm (Barrett reduction)

Input: A $2n$-word integer u, the n-word integer p, and the value $\mu = \lfloor b^{2n}/p \rfloor$.
Output: The n-word integer r such that $u \equiv r \pmod{p}$.

$\hat{q} \leftarrow \lfloor \lfloor (u/b^{n-1}) \rfloor \mu / b^{n+1} \rfloor$
$r_1 \leftarrow u \pmod{b^{n+1}}$
$r_2 \leftarrow (\hat{q}p) \pmod{b^{n+1}}$
$r \leftarrow r_1 - r_2$
if $r < 0$ **then**
    $r \leftarrow r + b^{n+1}$
end if
while $r \geq p$ **do**
    $r \leftarrow r - p$
end while
return r

11.1.45 Remark The quantity $\hat{q}$ computed at Line 1 of Algorithm 11.1.44 is an approximation of $q = \lfloor u/p \rfloor$. It satisfies $q - 2 \le \hat{q} \le q$ but it is obtained with reduced efforts thanks to the precomputed value μ. When b is large enough with respect to k, it is possible to implement Algorithm 11.1.44 so that it requires at most $n(n+4)+1$ single-precision multiplications; see [2080, Note 14.45] for a precise analysis.

11.1.46 Remark Given R some fixed power of the radix b larger than p, Montgomery developed in [2132] a very interesting arithmetic system modulo p where an integer $x \in [0, p-1]$ is represented by $xR \pmod{p}$. The *Montgomery representation* of x is denoted by $[x]$. The advantage of this system lies in a notion of reduction that can be evaluated very efficiently. Namely, the *Montgomery reduction* of $u \in [0, Rp-1]$ denoted by $\textsc{Redc}(u)$ is defined as $uR^{-1} \pmod{p}$. Assuming that $p' = -p^{-1} \pmod{R}$ has been precomputed, simply determine $k \equiv up' \pmod{R}$ with $k \in [0, p-1]$ and let t be the result of the exact division of $(u+kp)$ by R. Then it is easy to see that $t = \textsc{Redc}(u)$ or $t = \textsc{Redc}(u) + p$. Also, since only the bottom half of up' and the upper half of $(u+kp)$ are relevant, the overall complexity of $\textsc{Redc}$ is close to one multiprecision multiplication of size n. Algorithm 11.1.47 is a multiprecision variant with a reduction at each step [2080, Subsection 14.3.2].

11.1.47 Algorithm (Montgomery reduction)

Input: An n-word integer p, $R = b^n$, $p' = -p^{-1} \pmod{b}$ and a $2n$-word integer $u = (u_{2n-1} \dots u_0)_b < Rp$.

Output: The n-word integer $t = uR^{-1} \pmod{p}$.

1. $(t_{2n-1} \dots t_0)_b \leftarrow (u_{2n-1} \dots u_0)_b$
2. **for** $i = 0$ **to** $n-1$ **do**
3. $\quad k_i \leftarrow t_i p' \pmod{b}$
4. $\quad t \leftarrow t + k_i p b^i$
5. **end for**
6. $t \leftarrow t/R$
7. **if** $t \geqslant p$ **then**
8. $\quad t \leftarrow t - p$
9. **end if**
10. **return** t

11.1.48 Remark The computation of $\textsc{Redc}(u)$ does not require any division, only n word right shifts at Line 6. A precise analysis shows that Algorithm 11.1.47 needs $n(n+1)$ single-precision multiplications [2080, Note 14.34]. A classical reduction modulo p of a $2n$-word integer relying on a Euclidean division also requires approximately n^2 single-precision multiplications but it also needs n single-precision divisions on top.

11.1.49 Remark The conversion to the Montgomery representation is not particularly cheap. However, once the conversion is done, it is possible to efficiently add, multiply, and even invert values within this system, see Remarks 11.1.67 and 11.1.101. At the end of the whole computation, for instance a modular exponentiation, the result is then converted back to its normal representation using the relation $\textsc{Redc}([z]) = z$.

11.1.50 Remark Other representation systems, such as the residue number system [163, 2754], the modular number system [166], and the polynomial modular number system [165] offer interesting features for prime field arithmetic. All the computations can easily be parallelized for increased performance.

11.1.51 Remark If working with a random prime is not crucial, choosing a modulus of a special form can lead to substantial savings. An extreme example is illustrated by a Mersenne prime $p = 2^k - 1$, for which a reduction modulo p is extremely cheap. Indeed, since a shift by a full word is free, reducing $x < p^2$ modulo p only requires to shift x by exactly $r = k$

(mod w) bits, i.e., less than w bits. Unfortunately, Mersenne primes are too scarce to be of any use for practical applications. This explains the introduction of generalizations of the form $p = 2^k + c$ with c small [750] and later $p = 2^{n_k w} \pm 2^{n_{k-1} w} \pm \cdots \pm 2^{n_1 w} \pm 1$ where $w = 16$, 32, or 64 [2693]. For instance, $p = 2^{192} - 2^{64} - 1$ and $p = 2^{256} - 2^{224} + 2^{192} + 2^{96} - 1$ are prime and their low Hamming weight allows a fast reduction. Those integers are part of a list of recommended primes of various sizes known as NIST primes [1066]. A further generalization is to consider prime numbers of the form $p = g(t)$, where g is a sparse polynomial and t is an integer not necessarily equal to 2 [640].

11.1.3.2 Extension fields

11.1.52 Remark The remainder $g \pmod f$ can always be computed with the Euclidean division algorithm, but if f is a sparse polynomial, the following algorithm [1233] is particularly well suited and more efficient.

11.1.53 Algorithm (Polynomial modular reduction)

Input: Two polynomials f and g with coefficients in $\mathbb{F}_q$, where $f = x^n + \sum_{i=1}^{s-1} a_i x^{b_i}$ with $0 = b_1 < b_2 < \cdots < b_{s-1} < n$.

Output: The polynomial r such that $g \equiv r \pmod f$ with $\deg r < n$.

1. $r \leftarrow g$
2. **while** $\deg(r) \geq n$ **do**
3. $\quad k \leftarrow \max\{n, \deg r - n + b_{s-1} + 1\}$
4. $\quad$ Write r as $r_1 x^k + r_2$ with $\deg r_2 < k$
5. $\quad r \leftarrow r_2 - r_1 (f - x^n) x^{k-n}$
6. **end while**
7. **return** r

11.1.54 Remark For an s-sparse polynomial f of degree n, the reduction of g modulo f requires at most $2(s-1)(\deg g - n + 1)$ operations in $\mathbb{F}_q$. The impact of s on the overall complexity is obvious in this complexity analysis and justifies the interest for low weight irreducible polynomials; see Subsection 11.1.1.2.

11.1.55 Remark The concept of an *almost irreducible trinomial*, i.e., a trinomial f defined over $\mathbb{F}_2$ and having an irreducible factor of degree n, is introduced in [405] and [909]. The arithmetic is then performed in the ring $\mathbb{F}_2[x]/(f)$ containing the field $\mathbb{F}_{2^n}$, using a redundant set of representatives. In [2573], Scott observes that for a given architecture, some well chosen irreducible pentanomials over $\mathbb{F}_2$ may provide a faster arithmetic than trinomials, including irreducible trinomials.

11.1.56 Remark Dedicated reduction methods have been developed for specific polynomials. For instance, [1413, Algorithms 2.41 and 2.42] gives highly optimized reduction methods modulo $x^{163} + x^7 + x^6 + x^3 + 1$ and $x^{233} + x^{74} + 1$.

11.1.57 Remark A generalization of the Montgomery representation for polynomials over finite fields of characteristic 2 is described in [1775].

11.1.58 Remark For cryptographic applications, especially for applications running on embedded devices, a new type of finite field, called an *optimal extension field*, has been recently introduced [161, 2096].

11.1.59 Definition An *optimal extension field* is one of the form $\mathbb{F}_{p^n}$ where p is a generalized Mersenne prime of the form $2^k + c$ that fits in a word and such that the irreducible polynomial f used to define $\mathbb{F}_{p^n}$ is the binomial $x^n - w$, where $w \in \mathbb{F}_p$. The field is of *Type I* if $c = \pm 1$ and it is of *Type II* when $w = 2$.

11.1.60 Remark Considering finite fields of similar cardinality, optimal extension fields compare favorably against prime fields thanks to an inversion that is usually faster. Also, due to the lack of dedicated instructions in some old processors to multiply polynomials in $\mathbb{F}_2[x]$, a multiplication in an optimal extension field is usually faster than in characteristic 2 on those platforms.

11.1.4 Addition

11.1.61 Remark Adding, or subtracting, elements in a finite field is in general pretty straightforward; see for instance [1413, Algorithms 2.7 and 2.32]. This is not the case when nonzero elements are expressed as a power of a generator of the multiplicative group. The notion of Jacobi logarithm, see Subsection 2.1.7.5, gives a way to add elements easily. However, Jacobi logarithms must be precomputed and stored for each nonzero element [661, Subsection 2.3.3] and this explains why they are only used for relatively small fields.

11.1.62 Remark For prime fields represented as $\mathbb{Z}/p\mathbb{Z}$, a reduction is sometimes needed after adding two integers modulo a prime number p and this might lead to branch mispredictions [1428]. In [3078], Zimmermann discuss specifically designed algorithms, which do not need any adjustment step to perform additions, subtractions, and multiplications under certain conditions.

11.1.5 Multiplication

11.1.63 Remark Multiplying two elements is a task significantly more complicated than adding them. There is a wide range of multiplication methods whose efficiency and level of sophistication increase with the size of the operands.

11.1.5.1 Prime fields

11.1.64 Remark One approach to perform a modular multiplication is to compute the product first and then reduce it independently. This is especially effective for large values where it is worth using advanced multiplication techniques, such as Karatsuba [1684], Toom-Cook [1768], or fast Fourier transform [751]. For smaller values, Algorithm 11.1.65, which is based on the schoolbook method, reduces the result while it is computed for increased performance.

11.1.65 Algorithm (Interleaved multiplication-reduction)

Input: The n-word prime p and two n-word integers $u = (u_{n-1} \dots u_0)_b$ and v.

Output: An integer r such that $r \equiv uv \pmod{p}$.

1. $r \leftarrow 0$
2. **for** $i = 0$ **to** $n-1$ **do**
3. $\quad r \leftarrow rb + u_{n-i-1}v$
4. $\quad$ Approximate $q = \lfloor r/p \rfloor$ by $\hat{q}$
5. $\quad r \leftarrow r - \hat{q}p$
6. **end for**
7. **while** $r \geq p$ **do**
8. $\quad r \leftarrow r - p$
9. **end while**
10. **return** r

11.1.66 Remark Although r is relatively small, different techniques, some of them quite involved, exist to determine $\hat{q}$ at Line 4 of Algorithm 11.1.65; see [829, Section 2.2] and [661, Subsection 11.1.2].

11.1.67 Remark Remark 11.1.46 gives a presentation of the notions of Montgomery representation and of Montgomery reduction. Montgomery multiplication consists in multiplying elements in Montgomery representation, before applying Montgomery reduction in order to have the result of the product, again, in Montgomery representation. Formally, we have the relation $\text{REDC}([x][y]) = [xy]$.

11.1.68 Remark Karatsuba's method relies on a clever use of the divide and conquer strategy. While the schoolbook multiplication has complexity $O(n^2)$ to compute the product of two n-word integers, Karatsuba multiplication [1683, 1684] has asymptotic complexity $O(n^{\log_2 3}) = O(n^{1.585})$. Karatsuba multiplication is thus faster than the naïve approach, but due to a certain overhead this occurs only for integers of size larger than some threshold n_0, which depends on the platform. It is reported in [409] that this threshold can vary from 10 up to more than 100 words. Note that GMP [1102] uses specifically optimized values for a wide range of architectures.

11.1.69 Algorithm (Karatsuba multiplication)

Input: Two n-word integers $u = (u_{n-1} \dots u_0)_b$, $v = (v_{n-1} \dots v_0)_b$, and $n_0 \geq 2$.
Output: The $2n$-word integer uv.

if $n \leq n_0$ **then**
  return uv computed with the schoolbook method
else
  $k \leftarrow \lceil n/2 \rceil$
  Split u and v in two parts:
    $U_1 \leftarrow (u_{n-1} \dots u_k)_b$ and $U_0 \leftarrow (u_{k-1} \dots u_0)_b$
    $V_1 \leftarrow (v_{n-1} \dots v_k)_b$ and $V_0 \leftarrow (v_{k-1} \dots v_0)_b$
  $U_s \leftarrow U_0 + U_1$ and $V_s \leftarrow V_0 + V_1$
  Compute recursively U_0V_0, U_1V_1, and U_sV_s
  return $U_1V_1b^{2k} + (U_sV_s - U_1V_1 - U_0V_0)b^k + U_0V_0$
end if

11.1.70 Remark Since multiplications by b^q are free, multiplying two n-word integers with Algorithm 11.1.69 requires only three multiplications of size $n/2$ and a few additions of size n. This observation applied recursively justifies the complexity $O(n^{\log_2 3})$ of Karatsuba's method.

11.1.71 Remark Using polynomial evaluation and interpolation techniques, Toom 3-ways reduces one multiplication of size n to five multiplications of size $n/3$ and thus runs in $O(n^{\log_3 5}) = O(n^{1.465})$ [1768, Subsection 4.3.3.A]. Generalizing this idea, Schönhage–Strassen multiplication [2559] using the fast Fourier transform runs in $O(n \log n \log\log n)$ [751, Section 9.5]. The overhead is such that this method is only worth using for integers having several thousand digits [1102].

11.1.5.2 Extension fields

11.1.72 Remark For finite fields $\mathbb{F}_{q^n}$ defined via an irreducible polynomial $f \in \mathbb{F}_q[x]$ of degree n, all the techniques presented for prime fields apply in this context as well. In particular, there is a generalization of Algorithm 11.1.65 where polynomial multiplications and reductions modulo f are interleaved. Also, more advanced multiplication methods, such as Karatsuba or based on the fast Fourier transform are available as well, with similar complexities as in the integer case.

11.1.73 Remark Many practical applications use finite fields of characteristic two and specific multiplication methods, such as the right-to-left comb, left-to-right comb, or even window methods involving some precomputations, have been developed in this context; see [1413] for some explicit algorithms and [400] for a discussion focused on the implementation of Karatsuba, Toom-Cook, Schönhage, and Cantor methods.

11.1.74 Remark We refer to Section 5.3 for multiplication in a normal basis, where the complexity is thoroughly discussed and the concept of optimal normal basis is introduced.

11.1.6 Squaring

11.1.6.1 Finite fields of odd characteristic

11.1.75 Remark The formula

$$\left(\sum_{i=0}^{n-1} u_i b^i\right)^2 = \sum_{i=0}^{n-1} u_i^2 b^{2i} + 2\sum_{i<j} u_i u_j b^{i+j}$$

suggests that it takes less effort to compute u^2 than to compute uv, where u and v are distinct integers of the same size. In practice, a dedicated modular squaring algorithm can be up to 20% faster than its general counterpart [661].

11.1.76 Remark For each multiplication algorithm, there exists a specific variant exclusively designed to square elements; see [661, Subsection 10.3.3] for a description of the schoolbook and Karatsuba squaring methods.

11.1.6.2 Finite fields of characteristic two

11.1.77 Remark For the binary field $\mathbb{F}_{2^n}$ defined over $\mathbb{F}_2$ with a normal basis, a squaring corresponds to a circular shift of the coordinates. With a polynomial basis modulo f, the squaring of an element requires slightly more work, as we have

$$\left(\sum_{i=0}^{n-1} a_i x^i\right)^2 = \sum_{i=0}^{n-1} a_i x^{2i}.$$

Thus, only a reduction modulo f is necessary to find the result.

11.1.7 Exponentiation

11.1.78 Remark The exponentiation operation consists in computing α^m, for a nonnegative integer m and $\alpha \in \mathbb{F}_q$. It should be optimized as much as possible as it is a crucial subroutine in many algorithms. For instance, exponentiation is key for finding a generator of the multiplicative group or a primitive polynomial, for computing an inverse or the square root of an element, or for factoring a polynomial; see [1232] for a very complete survey dedicated to exponentiation methods in finite fields.

11.1.7.1 Prime fields

11.1.79 Remark When the exponent m is fixed, it may be worth searching for an addition chain with low complexity computing m; see [661, Section 9.2]. If the element $\alpha \in \mathbb{F}_p$ is fixed while the exponent varies, precomputing some powers of α can lead to considerable speedups. See in particular Yao's method [1767, 3032] also known as BGMW [414] and fixed-base comb algorithm [242, 2396] presented in [1941] as well. More details are available in [661, Section 9.3]. In the general case, where both the element and the exponent vary, Algorithm 11.1.80 is the most efficient exponentiation method that is known.

11.1.80 Algorithm (Sliding window exponentiation)

Input: An element $\alpha \in \mathbb{F}_p$, a nonnegative exponent $m = \sum_{i=0}^{\ell-1} m_i 2^i$, a parameter $k \geq 1$, and the stored values $\alpha, \alpha^3, \ldots, \alpha^{2^k-1}$.

Output: The element $\alpha^m \in \mathbb{F}_p$.

$\beta \leftarrow 1$ and $i \leftarrow \ell - 1$
while $i \geq 0$ **do**
    if $m_i = 0$ **then**
        $\beta \leftarrow \beta^2 \pmod p$ and $i \leftarrow i - 1$
    else
        $s \leftarrow \max\{i - k + 1, 0\}$
        while $m_s = 0$ **do**
            $s \leftarrow s + 1$
        end while
        for $j = 1$ **to** $i - s + 1$ **do**
            $\beta \leftarrow \beta^2 \pmod p$
        end for
        Form $t = (m_i \ldots m_s)_2$
        $\beta \leftarrow \beta \times \alpha^t \pmod p$
        $i \leftarrow s - 1$
    end if
end while
return β

11.1.81 Remark The parameter k controls the size of the window used to scan the bits of n. For $k = 1$, Algorithm 11.1.80 coincides with the well-known square and multiply method. For $k > 1$, Algorithm 11.1.80 relies on $2^{k-1} - 1$ precomputed values obtained at a cost of 2^{k-1} multiplications. Given an ℓ-bit exponent m, it requires $\ell/(k+1)$ extra field multiplications, on average [2305]. The size of the window k should therefore be selected to minimize the quantity $2^{k-1} + \ell/(k+1)$. The number of squarings is independent of k and is equal to ℓ in any case. Algorithm 11.1.80 is implemented in NTL [2633].

11.1.7.2 Extension fields

11.1.82 Remark The main difference with the prime field case is the existence of the Frobenius automorphism that can be used in an extension field to speed up the computation of an exponentiation.

11.1.83 Algorithm (Fast exponentiation in polynomial basis)

Input: Two polynomials f and g with coefficients in $\mathbb{F}_q$ such that $\deg f = n$ and $\deg g < n$, an exponent $0 < m < q^n$, and a positive integer r.

Output: The polynomial $g^m \pmod f$.

Write m in base q^r as $m = (m_{\ell-1} \ldots m_0)_{q^r}$
for $i = 0$ **to** $\ell - 1$ **do**
    Compute and store $g^{m_i} \pmod f$
end for
$h \leftarrow x^{q^r} \pmod f$ and $t \leftarrow 1$
for $i = \ell - 1$ **down to** 0 **do**

7. $t \leftarrow t(h) \pmod{f}$
8. $t \leftarrow tg^{m_i} \pmod{f}$
9. **end for**
10. **return** t

11.1.84 Remark Algorithm 11.1.83 relies on a generalization of the square and multiply method in base q^r. It is discussed in [1180] where r is set to be $\lceil n/\log_q n\rceil$ and where the computation of the residues $g^{m_i} \pmod{f}$ at Line 3 is done with the BGMW method [414]. The computation $t(h) \pmod{f}$ at Line 7 is a modular composition; see Remarks 11.1.85 and 11.1.86.

11.1.85 Remark The first nontrivial method to compute $t(h) \pmod{f}$ is due to Brent and Kung [401]. Assuming that $\deg h,\ \deg t < \deg f = n$ and that square matrices of dimension n can be multiplied with $O(n^w)$ field multiplications, the Brent–Kung approach takes $O(\mathrm{M}_q(n)n^{1/2} + n^{(w+1)/2})$ field operations. With $\mathrm{M}_q(n) = O(n^{\log_3 2})$ corresponding to Karatsuba's multiplication (Subsection 11.1.5), the overall complexity becomes $O(n^{\log_3 2+1/2}) = O(n^{2.085})$. If instead fast integer multiplication methods *à la* Schönhage–Strassen [2559] are used, the complexity is $O(n^{(w+1)/2})$. The natural bound $w = 3$ was improved by Strassen who introduced a method with $w = \log_2 7 < 2.8074$ [2731]. The best known upper bound is $w < 2.3727$ [2984, 2985].

11.1.86 Remark Umans proposed a completely different approach to perform a modular composition, based on multivariate multipoint evaluation [2836]. Initially, this work only addressed characteristic at most $n^{o(1)}$ but was later generalized to any characteristic by Umans and Kedlaya [1721, 1722]. In any case, the asymptotic complexity is $n^{1+o(1)} \log^{1+o(1)} q$ bit operations, which is optimal up to lower order terms.

11.1.87 Remark For the particular case $\mathbb{F}_q = \mathbb{F}_{2^n}$ and for $r = \lceil n/\log_2 n\rceil$, the complexity of Algorithm 11.1.83 becomes $O(\mathrm{M}_2(n)n/\log n)$. This complexity does not depend on the method, either [401] or [1722], used to perform the modular composition.

11.1.88 Remark To compute $\alpha^m \in \mathbb{F}_{q^n}$, where $\mathbb{F}_{q^n}$ is represented using a normal basis over $\mathbb{F}_q$, the exponent m is again expressed in base q^k, for some fixed k, in order to take advantage of the Frobenius automorphism that allows one to compute α^q with just a cyclic shift of the coordinates of α. In this case, only Line 7 of Algorithm 11.1.83 needs to be modified and replaced by $t \leftarrow t^{q^k}$; see also Subsection 5.3.5.

11.1.89 Remark When $\mathbb{F}_q$ is an optimal extension field (Definition 11.1.59), the action of the Frobenius automorphism can also be computed extremely efficiently, leading to very fast exponentiation techniques; see [2097] and [661, Subsection 11.3.3].

11.1.8 Inversion

11.1.90 Remark There are two ways to compute the inverse of a field element $\alpha \in \mathbb{F}_{q^n}$. If the field is defined as a quotient set, we can use the extended Euclidean algorithm, see Algorithm 11.1.92. Alternatively, Lagrange's theorem implies that we have $\alpha^{q^n-2} = 1/\alpha$. This last method is totally general but it is particularly adapted to finite fields defined with a normal basis or when the action of the Frobenius automorphism $\alpha \mapsto \alpha^q$ can be computed efficiently.

11.1.91 Remark Let R be a Euclidean ring with Euclidean function φ. For elements $a, b \in R$, we can always write $a = bq + r$ with $\varphi(r) < \varphi(b)$. In practice, $R = \mathbb{Z}$ with natural order $<$ or $R = \mathbb{F}_q[x]$ with the degree function. Assuming that b is a prime number or an irreducible polynomial and a is an element such that $\varphi(a) < \varphi(b)$, we then have $\gcd(a, b) = 1$. The Bézout identity $au + bv = 1$, whose coefficients u and v are returned by Algorithm 11.1.92, implies that $au \equiv 1 \pmod{b}$, i.e., u is the inverse of a modulo b.

11.1.92 Algorithm (Extended Euclidean Algorithm)

Input: Two elements a, b in a Euclidean ring R such that $\varphi(a) < \varphi(b)$.

Output: Elements (u, v, d) in R such that $au + bv = d$ with $d = \gcd(a, b)$.

1. $t \leftarrow 0$, $u \leftarrow 1$, $c \leftarrow b$, and $d \leftarrow a$
2. **while** $c \neq 0$ **do**
3.   Write $d = cq + r$ with $\varphi(r) < \varphi(c)$
4. $s \leftarrow u - tq$, $u \leftarrow t$, $t \leftarrow s$, $d \leftarrow c$, and $c \leftarrow r$
5. **end while**
6. $v \leftarrow (d - au)/b$
7. **return** (u, v, d)

11.1.93 Remark The division at Line 6 of Algorithm 11.1.92 is exact. Dedicated methods to compute the quotient of an exact division are given in [1601, 1805].

11.1.94 Remark In a finite field, an inversion is in general considerably more expensive than a multiplication. When k elements $a_1, \ldots, a_k$ need to be inverted, a trick due to Montgomery [2133] allows one to replace k inversions by one inversion and $3k - 3$ field multiplications. The principle is to compute the inverse $(a_1 \cdots a_k)^{-1}$ and then multiply it by suitable precomputed terms in order to recover successively each inverse individually. It was introduced initially to speed up the elliptic curve factorization method and is therefore described for integers defined modulo a composite number. However, this trick can be applied to any structure where the notion of inverse exists. It is also presented in [660] and [661].

11.1.95 Remark In order to compute the division e/a, where $e, a \in \mathbb{F}_{q^n}$, one could invert a and multiply the result by e. However, e/a can be obtained directly with Algorithm 11.1.92. Simply replace $u \leftarrow 1$ by $u \leftarrow e$ at Line 1.

11.1.8.1 Prime fields

11.1.96 Remark Algorithm 11.1.92 is usually preferred over Lagrange's method to compute the inverse of α modulo p. There are at least two reasons for that. Computing the exponentiation $\alpha^{p-2} \pmod p$ requires on average twice as many arithmetic operations as the extended Euclidean method [709]. Also, there are a number of improvements and variants of Algorithm 11.1.92 that can be implemented to speed up its execution on different platforms.

11.1.97 Remark Algorithm 11.1.92 returns the inverse modulo p after $O(\log p)$ steps. When suitably implemented, in particular if the precision of the Euclidean division is adjusted and decreases with the size of its arguments d and c, then the overall time complexity is $O(\log^2 p)$ [660, Section 1.3]. In [1888], Lehmer suggests to replace the exact computation of the quotient q at Line 3 of Algorithm 11.1.92 by an approximation obtained by dividing the most significant digits of d and c. This remark has been explored further by many authors [710, 1602, 1903].

11.1.98 Remark In a variant introduced by Brent and Kung [402, 661], divisions are eliminated and replaced by shifts, additions, and subtractions. This approach, known as the *binary method* or the *plus-minus method*, is well-suited for architectures where a division is expensive. Not surprisingly, the number of steps needed is still $O(\log p)$. Indeed, the quotient q at each step of Algorithm 11.1.92 is small most of the time. The probability that $q = 1$ is close to 0.415 and $q \leq 5$ in more than 77% of the cases [660].

11.1.99 Remark Schönhage [2557], improving on Knuth's work [1766], showed how the extended Euclidean algorithm, and hence inversion, can be done asymptotically in time $O(M(\log p) \log\log p)$; see also [3033] and [56] for a description of the method. Stehlé and Zimmermann [2705] developed a binary recursive gcd method. Although it does not improve on the $O(M(\log p) \log\log p)$ asymptotic complexity, its description, implementation, and proof of correctness are simpler than Schönhage's method.

11.1.100 Remark A very simple approach presented in [2803] allows one to find an inverse modulo p. Interestingly, it is not related to the extended Euclidean gcd method nor Lagrange's method. It is particularly efficient for certain types of primes, such as Mersenne primes. The method is recalled in [661, Subsection 11.1.3].

11.1.101 Remark There is a notion of Montgomery inverse [2132], which completes the other operations already existing in Montgomery representation; see Remarks 11.1.46 and 11.1.67, and [1644, 2536] for additional improvements regarding the Montgomery inverse.

11.1.8.2 Extension fields

11.1.102 Remark As in the integer case, there is a binary version of Algorithm 11.1.92 that does not require any division; see [436] for an efficient version in even characteristic.

11.1.103 Remark A notion of reduction in a finite field defined by a normal basis is discussed in [2748]. It is then possible to compute the inverse of an element with a variant of Algorithm 11.1.92. However, since the Frobenius automorphism can be evaluated for free in a normal basis, Lagrange's method, which computes α^{-1} as α^{q^n-2} is preferred. Algorithm 11.1.104 follows this idea with an additional improvement, i.e., the use of addition chains computing $q-2$ and $n-1$ [1576]. See [661, Section 9.2] for a definition of addition chains and related techniques to find short chains.

11.1.104 Algorithm (Inversion using Lagrange's theorem)

Input: An element $\alpha \in \mathbb{F}_{q^n}^*$, two addition chains, namely $(a_0, a_1, \ldots, a_{s_1})$ computing $q-2$ and $(b_0, b_1, \ldots, b_{s_2})$ computing $n-1$.

Output: The inverse of α, i.e., $\alpha^{-1} = \alpha^{q^n-2}$.

1. Compute $\beta \leftarrow \alpha^{q-2}$ using the addition chain $(a_0, a_1, \ldots, a_{s_1})$
2. $T[0] \leftarrow \alpha \times \beta$
3. **for** $i = 1$ **to** s_2 **do**
4. $\quad \gamma \leftarrow T[k]^{q^{b_j}}$ where $b_i = b_k + b_j$
5. $\quad T[i] \leftarrow \gamma \times T[j]$
6. **end for**
7. $\gamma \leftarrow T[s_2]$
8. **return** $\beta \times \gamma^q$

11.1.105 Remark Algorithm 11.1.104 needs $s_1 + s_2 + 2$ multiplications in $\mathbb{F}_{q^n}$ and $1 + b_1 + \cdots + b_{s_2}$ applications of the Frobenius automorphism to compute α^{-1}.

11.1.106 Remark Algorithm 11.1.104 is particularly well suited for $q = 2$. For $q > 2$, set $r = (q^n-1)/(q-1)$ and decompose α^{-1} as $\alpha^{-r}\alpha^{r-1}$. Since $r = q^{n-1}+\cdots+q+1$, it follows that α^r is the norm of α. Thus, it is in $\mathbb{F}_q$ and can be easily inverted. Also, α^{r-1} can be obtained with at most $n-1$ evaluations of the Frobenius automorphism and at most $n-2$ multiplications in $\mathbb{F}_q$. This is the standard way to compute an inversion in an optimal extension field. Further optimizations exist for specific extension degrees; see [661, Subsections 11.3.4 and 11.3.6] for more details and concrete examples.

11.1.9 Squares and square roots

11.1.107 Remark Computing square roots in $\mathbb{F}_{p^n}$ efficiently is important in many applications and is the subject of active research; see for instance [204, 1410, 1781, 2194]. In a finite field of characteristic 2, every element is a square and it is straightforward to compute a square root. In odd characteristic, not every element is a square. When it exists, a square root can be computed in most cases with a closed formula. If that is not the case, and in fact in any case, there are algorithms returning the result fairly efficiently.

11.1.9.1 Finite fields of odd characteristic

11.1.108 Remark Let us assume that p is an odd prime number and that q is some power of p. We know that $\mathbb{F}_q^*$ is a cyclic group, generated by, say, γ. Because the cardinality of $\mathbb{F}_q^*$ is even, all the square elements in $\mathbb{F}_q^*$ must be even powers of γ and all the nonsquare elements correspond to the odd powers of γ. Thus, there are $(q-1)/2$ squares and just as many nonsquare elements in $\mathbb{F}_q^*$.

11.1.109 Remark To check if a nonzero element α is a square or not, it is enough to perform an exponentiation. Indeed, α is a square if and only if $\alpha^{(q-1)/2} = 1$. The outcome is -1 when α is not a square.

11.1.110 Remark For the prime number p, we introduce the Legendre symbol denoted by $\left(\frac{\alpha}{p}\right)$ such that $\left(\frac{0}{p}\right) = 0$ and $\left(\frac{\alpha}{p}\right) = \alpha^{(p-1)/2} \pmod{p}$ otherwise; see Algorithm 11.1.111 for an efficient way to compute $\left(\frac{\alpha}{p}\right)$ and thus decide if α is a square or not in $\mathbb{F}_p$.

11.1.111 Algorithm (Legendre symbol)

Input: An integer α and an odd prime number p.
Output: The Legendre symbol $\left(\frac{\alpha}{p}\right)$.

$k \leftarrow 1$
while $p \neq 1$ **do**
  if $\alpha = 0$ **then**
    return 0
  end if
  $v \leftarrow 0$
  while $\alpha \equiv 0 \pmod{2}$ **do**
    $v \leftarrow v + 1$ and $\alpha \leftarrow \alpha/2$
  end while
  if $v \equiv 1 \pmod{2}$ **and** $p \equiv \pm 3 \pmod{8}$ **then**
    $k \leftarrow -k$
  end if
  if $\alpha \equiv 3 \pmod{4}$ **and** $p \equiv 3 \pmod{4}$ **then**
    $k \leftarrow -k$
  end if
  $r \leftarrow \alpha$, $\alpha \leftarrow p \pmod{r}$, and $p \leftarrow r$
end while
return k

11.1.112 Remark Algorithm 11.1.111 relies on a quadratic reciprocity law that allows one to reduce the size of the operands in a way that is similar to the computation of the gcd with Euclid's algorithm.

11.1.113 Remark The evaluation of $\left(\frac{\alpha}{p}\right)$ via the exponentiation $\alpha^{(p-1)/2} \pmod{p}$ has complexity $O(\log^3 p)$ with the square and multiply method. By contrast, Algorithm 11.1.111 has complexity $O(\log^2 p)$. In [407], Brent and Zimmermann describe a new algorithm to compute $\left(\frac{\alpha}{p}\right)$ with complexity $O(\mathrm{M}(\log p) \log\log p)$.

11.1.114 Remark There is a generalization of the Legendre symbol for the elements of $\mathbb{F}_{p^n}$. Let $f \in \mathbb{F}_p[x]$ be an irreducible polynomial of degree n and let $g \in \mathbb{F}_p[x]$. The Legendre symbol for polynomials, again denoted by $\left(\frac{g}{f}\right)$ satisfies $\left(\frac{0}{f}\right) = 0$ and $\left(\frac{g}{f}\right) = g^{(q-1)/2} \pmod{f}$ for nonzero g. Algorithm 11.1.115 relies also on a quadratic reciprocity law, quite similarly to Algorithm 11.1.111, and allows to efficiently determine if g is a square or not modulo f.

11.1.115 Algorithm (Legendre symbol for polynomials)

Input: A polynomial $g \in \mathbb{F}_p[x]$ and an irreducible polynomial $f \in \mathbb{F}_p[x]$.
Output: The Legendre symbol for polynomials $\left(\frac{g}{f}\right)$.

$k \leftarrow 1$
repeat
  if $g = 0$ **then**
    return 0
  end if
  $a \leftarrow$ the leading coefficient of g
  $g \leftarrow g/a$
  if $\deg f \equiv 1 \pmod 2$ **then**
    $k \leftarrow \left(\frac{a}{p}\right)k$
  end if
  if $p^{\deg f} \equiv 3 \pmod 4$ **and** $\deg f \deg g \equiv 1 \pmod 2$ **then**
    $k \leftarrow -k$
  end if
  $r \leftarrow g$, $g \leftarrow f \pmod r$, and $f \leftarrow r$
until $\deg f = 0$
return k

11.1.116 Remark Once we know that a nonzero element $\alpha \in \mathbb{F}_p$ is a square, it may be necessary to compute a square root of α, i.e., an element $\rho \in \mathbb{F}_p$, satisfying $\rho^2 = \alpha$. If ρ is one square root, then $-\rho$ is the other one and there exist closed formulas to compute ρ in most cases [660, Section 1.5]. Indeed, we have

1. $\rho = \pm\alpha^{(q+1)/4}$, if $q \equiv 3 \pmod 4$;
2. $\rho = \pm\alpha^{(q+3)/8}$, if $q \equiv 5 \pmod 8$ and $\alpha^{(q-1)/4} = 1$;
3. $\rho = \pm 2\alpha(4\alpha)^{(q-5)/8} \pmod p$, if $q \equiv 5 \pmod 8$, $\alpha^{(q-1)/4} = -1$, and q is an odd power of p.

11.1.117 Remark For the other cases, i.e., when $q \equiv 1 \pmod 8$ or when $q \equiv 5 \pmod 8$, $\alpha^{(q-1)/4} = -1$, and q is an even power of p, there exist several polynomial time methods to compute a square root. The most efficient factorization methods, applied to the polynomial $x^2 - \alpha$, return a square root of α in $\mathbb{F}_q$ using $O(\log q)$ field operations; see Remark 11.4.3. There are also dedicated methods, such as Tonelli and Shanks algorithm [660], which returns a square root in the prime field $\mathbb{F}_p$ in time $O(\log^4 p)$. Another example is Algorithm 11.1.118, which is a generalization of Cipolla's method [751, Subsection 2.3.9]. Algorithm 11.1.118 is remarkably simple and easy to implement. It works in the quadratic extension $\mathbb{F}_{q^2}$ and also requires $O(\log q)$ field operations.

11.1.118 Algorithm (Square root computation)

Input: A square α in $\mathbb{F}_q$.
Output: A square root $\rho \in \mathbb{F}_q$ of α.

repeat
  Choose $\beta \in \mathbb{F}_q$ at random
until $\beta^2 - 4\alpha$ is not a square in $\mathbb{F}_q$
$T \leftarrow x^2 - \beta x + \alpha$
$\rho \leftarrow x^{(q+1)/2} \pmod T$
return ρ

11.1.9.2 Finite fields of even characteristic

11.1.119 Remark Every element $\alpha \in \mathbb{F}_{2^n}$ is a square and the square root ρ of α can be easily obtained thanks to the multiplicative structure of $\mathbb{F}_{2^n}^*$, which implies that $\rho = \alpha^{2^{n-1}}$. With a normal basis, the computation is immediate. Using a polynomial representation, modulo an irreducible polynomial f, there is a different approach. If α is represented by $\sum_{i=0}^{n-1} g_i x^i$, then observe that

$$\sqrt{\alpha} = \sum_{i \text{ even}} g_i x^{i/2} + \sqrt{x} \sum_{i \text{ odd}} g_i x^{\frac{i-1}{2}}$$

where $\sqrt{x}$ has been precomputed modulo f. We note that $\sqrt{x}$ can be obtained very easily when f is a trinomial of odd degree; see [661, Subsection 11.2.6].

11.1.120 Remark Unlike for finite fields of odd characteristic, solving a quadratic equation in $\mathbb{F}_{2^n}$ is more involved than just extracting square roots. Considering the polynomial $x^2 + x + \beta$ in $\mathbb{F}_{2^n}[x]$, we can show that it has a root in $\mathbb{F}_{2^n}$ if and only if the trace of β is zero. In that case, a solution τ is given by

$$\tau = \sum_{i=0}^{(n-3)/2} \beta^{2^{2i+1}}$$

when n is odd. When n is even, τ satisfies

$$\tau = \sum_{i=0}^{n-1} \left(\sum_{j=0}^{i} \beta^{2^j} \right) \omega^{2^i}$$

where $\omega \in \mathbb{F}_{2^n}$ is any element of trace 1. In any case, the other solution is $\tau + 1$. Reference [1088] gives techniques to speed up the computation of a square root in characteristic two at the expense of extra storage.

See Also

§3.4	For discussion on the weights of irreducible polynomials.
§5.3	For discussion on optimal and low complexity normal bases.
§11.6	For discussion on discrete logarithms.
§16.4	For discussion on elliptic cryptographic systems.
§16.5	For discussion on hyperelliptic cryptographic systems.

References Cited: [19, 43, 56, 77, 143, 161, 163, 165, 166, 204, 209, 242, 400, 401, 402, 403, 404, 405, 406, 407, 408, 409, 414, 436, 439, 576, 610, 640, 660, 661, 662, 709, 710, 717, 750, 751, 829, 909, 927, 930, 933, 1066, 1088, 1102, 1180, 1181, 1187, 1223, 1227, 1232, 1233, 1255, 1327, 1328, 1410, 1413, 1416, 1426, 1428, 1430, 1576, 1601, 1602, 1644, 1683, 1684, 1721, 1722, 1766, 1767, 1768, 1775, 1781, 1805, 1888, 1895, 1903, 1939, 1941, 2002, 2011, 2038, 2080, 2095, 2096, 2097, 2099, 2111, 2132, 2133, 2141, 2194, 2305, 2306, 2396, 2420, 2435, 2473, 2536, 2557, 2559, 2573, 2582, 2627, 2628, 2631, 2632, 2633, 2640, 2693, 2694, 2705, 2709, 2731, 2748, 2754, 2798, 2803, 2836, 2976, 2984, 2985, 3004, 3009, 3030, 3032, 3033, 3078]

11.2 Univariate polynomial counting and algorithms

Daniel Panario, Carleton University

We* give basic counting estimates for univariate polynomials over finite fields. First, we provide some classical counting results. Then we focus on a methodology based on analytic combinatorics that allows the derivation of many nontrivial counting results. A series of counting results for univariate polynomials over finite fields, some of them linked to the analysis of algorithms, is then provided.

11.2.1 Classical counting results

11.2.1 Remark The most classical counting estimate is for the number I_n of monic irreducible polynomials of degree n over $\mathbb{F}_q$; see Theorem 2.1.24. (We observe that in the results of this section there is only one finite field involved, and so we drop the notation $I_q(n)$ for the simpler one I_n. Sometimes, however, we let q go to infinity.)

11.2.2 Theorem For all $n \geq 1$ and any prime power q, we have

$$I_n = \frac{1}{n}\sum_{d|n} \mu(d) q^{n/d} = \frac{q^n}{n} + O\left(\frac{q^{n/2}}{n}\right).$$

11.2.3 Remark Since $I_n > 0$ for any prime power q and integer $n > 1$ and the number of monic polynomials of degree n over $\mathbb{F}_q$ is q^n, then the probability of a polynomial being irreducible is close to $1/n$. This probability tends to zero with $n \to \infty$.

11.2.4 Remark A polynomial is squarefree if it has no repeated factors. The number of squarefree polynomials over $\mathbb{F}_q$ was first given by Carlitz [537].

11.2.5 Theorem [537] Let Q_n be the number of squarefree polynomials of degree n over $\mathbb{F}_q$. Then,

$$Q_n = \begin{cases} q^n - q^{n-1} & \text{for } n \geq 2, \\ q^n & \text{for } n = 0, 1. \end{cases}$$

11.2.6 Remark Theorem 11.2.5 implies, for $n \geq 2$, that the proportion of squarefree polynomials is $1 - 1/q$. As a consequence, for large finite fields $\mathbb{F}_q$ most polynomials are squarefree.

11.2.7 Remark Let us consider the number of irreducible factors of fixed degree d in a random polynomial of degree n over $\mathbb{F}_q$. Zsigmondy [3083] concentrates on the prime field case $\mathbb{F}_p$ and gives results for the number of monic polynomials having no irreducible factors of degree

*Originally, this section was to be written by Philippe Flajolet and the author, but sadly, Philippe passed away before we started working on it. This section is dedicated to the memory of my friend Philippe Flajolet, for all his many lessons and guidance.

d, $1 \le d \le n$, and for the number of monic polynomials having a given number of distinct roots. We will return to this problem in Subsection 11.2.3.2.

11.2.8 Remark The notes at the end of Chapter 4 of Lidl and Niederreiter's book [1939] point to several classical counting references published before 1983.

11.2.9 Remark Other classical results related to polynomials with prescribed trace or norm and to self-reciprocal polynomials are given in Sections 3.1 and 3.5, respectively.

11.2.10 Remark As a final classical estimate we consider the probability that two polynomials are coprime. It has been known since at least the 1960s (see Berlekamp [231] and Knuth [1765]) but most likely for a long time before then, that with probability $1 - 1/q$, the greatest common divisor of two polynomials over a finite field is 1, independently of the degrees of the polynomials. This result has been reinvented a large number of times. In Subsection 11.2.3.4, a much more precise refinement of this classical result is given.

11.2.2 Analytic combinatorics approach

11.2.11 Remark Flajolet has made major methodological contributions to the research area known as *analytic combinatorics*. Among other things, analytic combinatorics provides a general methodology that can be successfully applied to the analysis of algorithms from many diverse areas. Its main and classical reference is the book by Flajolet and Sedgewick [1083]. In this section this general framework is presented only in relation to polynomials over a finite field $\mathbb{F}_q$ although it can be used in much more general settings. For a longer introduction to this methodology and its application to the analysis of polynomial factorization algorithms see Flajolet, Gourdon, and Panario [1080].

11.2.12 Remark This framework has two basic components: generating functions to express, combinatorially, properties of interest, and asymptotic analysis for the derivation of estimates when exact extraction of coefficients is not possible. The studied properties could be for pure mathematical interest or for their application to the analysis of algorithms for polynomials over finite fields [1080, 2347].

11.2.13 Remark Generating functions for counting some properties of polynomials over finite fields have been previously used in some specific cases by Berlekamp [231, Chapter 3], Knuth [1765, Subsection 4.6.2], and Odlyzko [2306]. The global usage of this technique to count many interesting expressions, and its usage in the analysis of polynomial over finite fields algorithms, only became possible after the establishment of analytic combinatorics [1083].

11.2.14 Definition Let I_n be the number of monic irreducible polynomials of degree n in $\mathbb{F}_q$. The generating functions of monic irreducible polynomials and monic polynomials are, respectively,

$$I(z) = \sum_{j \ge 1} I_j z^j, \quad \text{and} \quad P(z) = \sum_{j \ge 0} P_j z^j.$$

We denote by $[z^n]T(z)$ the coefficient of z^n in the generating function $T(z)$.

11.2.15 Proposition We have

$$P(z) = \prod_{k\geq 1}(1 + z^k + z^{2k} + \cdots)^{I_k} = \prod_{k\geq 1}(1 - z^k)^{-I_k}.$$

11.2.16 Remark Since $[z^n]P(z)$ is q^n, it follows that $P(z) = (1 - qz)^{-1}$. This expression and the one in Proposition 11.2.15 implicitly determine that I_n satisfies

$$I_n = \frac{1}{n}\sum_{k|n}\mu(k)q^{n/k}.$$

11.2.17 Remark Carlitz's result [537] for squarefree polynomials (Theorem 11.2.5) can be easily recovered under this framework. Indeed, the generating function for squarefree polynomials is

$$Q(z) = \prod_{k\geq 1}\left(1 + z^k\right)^{I_k}.$$

Moreover, considering the multiplicity of its irreducible factors, each polynomial f factors as $f = st^2$, where s is squarefree and t is an arbitrary polynomial. We thus have $P(z) = Q(z)P(z^2)$, and therefore,

$$Q(z) = \frac{P(z)}{P(z^2)} = \frac{1 - qz^2}{1 - qz}.$$

Carlitz's estimate, given in Theorem 11.2.5, can be recovered after extracting coefficients from $Q(z)$.

11.2.18 Remark Generating functions encode exact counting information in their coefficients. However, their extraction from a given generating function is in general a difficult task. Nevertheless, when considering generating functions as analytic functions, their behavior near their dominant singularities (those with smallest modulus) is an important source of information to extract coefficient asymptotics as their index tends to infinity.

11.2.19 Remark Most of the generating functions $f(z)$ of interest in this section are singular at $z = 1/q$ with an isolated singularity of the algebraic-logarithmic type. In that case, we can apply the following important result due to Flajolet and Odlyzko [1081].

11.2.20 Theorem [1081] Let $f(z)$ be a function analytic in a domain

$$\mathcal{D} = \{z\colon |z| \leq z_1, |\text{Arg}(z - 1/q)| > \pi/2 - \varepsilon\},$$

where $z_1 > 1/q$ and ε are positive real numbers. Let $k \geq 0$ be any integer, and α a real number with $\alpha \neq 0, -1, -2, \ldots$. If in a neighborhood of $z = 1/q$, $f(z)$ has an expansion of the form

$$f(z) = \frac{1}{(1 - qz)^\alpha}\left(\log\frac{1}{1 - qz}\right)^k (1 + o(1)),$$

then the coefficients satisfy, asymptotically as $n \to \infty$,

$$[z^n]f(z) = q^n\frac{n^{\alpha-1}}{\Gamma(\alpha)}(\log n)^k (1 + o(1)).$$

11.2.21 Remark This theorem requires analytic continuation of $f(z)$ outside its circle of convergence. However, there are some situations in which generating functions do not satisfy this hypothesis. For instance in some of the generating functions related to *smooth polynomials* (Subsection 11.2.3.3) analytic continuation is not possible. Saddle point methods are used in these cases. All asymptotic enumeration methods required in this section are explained in detail in the excellent presentations by Flajolet and Sedgewick [1083] and Odlyzko [2307].

11.2.3 Some illustrations of polynomial counting

11.2.22 Remark We list several results that are mostly derived in the framework of analytic combinatorics.

11.2.23 Remark Bivariate generating functions are used to study important parameters of interest. The exact counting problem is now refined with two parameters, namely, the degree of the polynomial and an additional property to be studied (for example, the number of its irreducible factors). With an appropriate normalization, successive differentiation of the bivariate generating function with respect to the additional parameter (evaluated at 1) gives the factorial moments of interest. The classical book by Flajolet and Sedgewick [1083] presents a comprehensive explanation of this methodology. The complete study of the *number* of irreducible factors of a random polynomial is presented below as an example.

11.2.3.1 Number of irreducible factors of a polynomial

11.2.24 Remark Let us consider the bivariate generating function

$$P(u,z) = \prod_{j\geq 1}(1 + uz^j + u^2z^{2j} + \cdots)^{I_j} = \prod_{j\geq 1}(1 - uz^j)^{-I_j},$$

where $[u^k z^n]P(u,z)$ is the number of polynomials of degree n with k irreducible factors. Successive differentiation of $P(u,z)$ with respect to u (evaluated at $u = 1$) give univariate generating functions for the factorial moments of this parameter. Asymptotic analysis of these univariate generating functions gives an expectation of $\log n$ and standard deviation $\sqrt{\log n}$. More can be said for this problem as the next theorem states. Flajolet and Soria [1084] prove that the number of irreducible factors in a random polynomial over a finite field satisfies a central limit theorem with mean and variance asymptotic to $\log n$. We note that this result is equivalent to the Erdös-Kac theorem stating that the number of prime factors in a random integer at most n satisfies a central limit theorem with mean and variance asymptotic to $\log\log n$. We refer to Subsection 11.2.3.5 for analogies among irreducible decompositions of polynomials over finite fields, prime decompositions of integers, and cycle decompositions of permutations.

11.2.25 Theorem Let Ω_n be a random variable counting the number of irreducible factors of a random polynomial of degree n over $\mathbb{F}_q$, where each factor is counted with its order of multiplicity.

1. The mean value of Ω_n is asymptotic to $\log n$ [231, 1765].
2. The variance of Ω_n is asymptotic to $\log n$ [1084, 1765].

3. For any two real constants $\lambda < \mu$, we have [1084]

$$\Pr\left\{\log n + \lambda\sqrt{\log n} < \Omega_n < \log n + \mu\sqrt{\log n}\right\} \to \frac{1}{\sqrt{2\pi}}\int_{\lambda}^{\mu} e^{-t^2/2}dt.$$

4. The distribution of Ω_n admits exponential tails [1085].
5. A local limit theorem holds [1190].
6. The behavior of $\Pr\{\Omega_n = m\}$ for all m is known [509, 665, 1564].

11.2.3.2 Factorization patterns

11.2.26 Remark As a first example of a factorization pattern, let us consider the number of irreducible factors in a random polynomial of a given fixed degree. As it is indicated in Remark 11.2.7, the number of roots was studied by Zsigmondy [3083] for prime fields. Knopfmacher and Knopfmacher [1760] present a detailed analysis, for any finite field, including variance.

11.2.27 Remark The case of polynomials with no roots is interesting when studying the distinct values that a polynomial can take. This is related to permutation polynomials (Section 8.1) and was studied by Uchiyama [2833]; see also [670].

11.2.28 Remark The generating function of polynomials with no linear factors is

$$\prod_{k\geq 2}\left(\frac{1}{1-z^k}\right)^{I_k} = \frac{1}{1-qz}(1-z)^{I_1} = \frac{1}{1-qz}(1-z)^q.$$

As a consequence, the number of polynomials of degree n with no irreducible factors of degree 1 is asymptotic, as $n \to \infty$, to $q^n(1-1/q)^q$. That is, the probability of a polynomial with no linear factor tends to $1/e = 0.3678\ldots$ when q grows. This implies that "most" polynomials are reducible and have at least one irreducible linear factor.

11.2.29 Remark The number of irreducible factors of a given degree d in a polynomial of degree n was studied by Williams [2983]. In [1761] can be found a detailed analysis of this problem as well as a determination of the variance, in both the cases where repetitions are allowed, and where they are not allowed.

11.2.30 Remark Knopfmacher, Knopfmacher, and Warlimont [1762] provide the mean and variance of what they call the "length" of a general factorization pattern by studying the number of polynomial factorizations into exactly k factors.

11.2.31 Remark When factoring univariate polynomials (Section 11.4) using the method based on the squarefree, distinct-degree, and equal-degree factorizations, it is relevant to determine if a polynomial has all its irreducible factors of different degrees. In this case, the third stage, the "equal-degree factorization," is not required.

11.2.32 Theorem [1080, 1763] The probability that all irreducible factors of a random polynomial of degree n over $\mathbb{F}_q$ have different degrees (but with single factors possibly repeated) is asymptotic to

$$c_q = \prod_{k\geq 1}\left(1+\frac{I_k}{q^k-1}\right)(1-q^{-k})^{I_k},$$

where $c_2 = 0.6656\ldots$, $c_{257} = 0.5618\ldots$, $c_\infty = e^{-\gamma} = 0.5614\ldots$, where γ is Euler's constant.

11.2.33 Remark Several related results used in the analysis of polynomial factorization algorithms can be found in [1080]. Recent fast factorization algorithms like [1226, 1239, 1667] require a study of the distribution of irreducible factors in parts of interval partitions of $[1, n]$. See von zur Gathen, Panario, and Richmond [1235] for first steps towards an understanding of these advanced algorithms.

11.2.3.3 Largest and smallest degree irreducibles

11.2.34 Remark Information on the degrees of the largest irreducible factors is crucial to measure stopping rules for factorization algorithms [1080]. Information on the degrees of the smallest irreducible factors helps in the analysis of irreducible test algorithms [2349, 2350].

11.2.35 Remark A random polynomial of degree n has with high probability several irreducible factors whose degrees sum to near n; see Theorem 11.2.37 and Remark 11.2.38. Let $D_n^{[j]}$ be the j-th largest degree of the factors of a random polynomial of degree n in $\mathbb{F}_q$. Car [510] obtained an asymptotic expression for the cumulative distribution function of $D_n^{[1]}$ in terms of the Dickman function. This number-theoretic function was originally introduced to describe the distribution of the largest prime divisor of a random integer [2788].

11.2.36 Definition The *Dickman function* is defined as the unique continuous solution of the difference-differential equation

$$\rho(u) = 1 \qquad (0 \leq u \leq 1), \qquad\qquad u\rho'(u) = -\rho(u-1) \qquad (u > 1).$$

11.2.37 Theorem The distribution of the largest degree $D_n^{[1]}$ satisfies for all $x \in (0,1)$:

$$\lim_{n\to\infty} \Pr\{D_n^{[1]} \leq x\} = F_1(x),$$

where $F_1(x) = \rho(1/x)$ and ρ denotes the Dickman function. In particular, one has $E(D_n^{[1]}) \sim gn$, where $g = 0.62432\ldots$ is known as the Golomb-Dickman constant.

11.2.38 Remark The most complete results about $D_n^{[j]}$, for any fixed positive integer j, are due to Gourdon [1341]; see also [1080]. For instance, we also have $E(D_n^{[2]}) \sim 0.20958\ldots n$, and $E(D_n^{[3]}) \sim 0.08831\ldots n$.

11.2.39 Remark Information on the relation between the first and second largest degree irreducible factors is used to compute the average-case analysis of the classical factorization method based on squarefree, distinct-degree, and equal-degree factorization under the "early-abort" stopping strategy [1080]. This also requires information on the joint distribution of the first two largest degree irreducible factors.

11.2.40 Remark Estimates for the largest degree of the irreducible factors are related to the study of *smooth polynomials* that play an important role in the discrete logarithm problem in finite fields, especially in the index calculus method; see Section 11.6.

11.2.41 Definition A polynomial of degree n over $\mathbb{F}_q$ is m-smooth if all its irreducible factors have degree at most m.

11.2.42 Remark In the index calculus method a search is repeated until an m-smooth polynomial is found. Hence, the analysis of the index calculus method requires information on the number of polynomials that are m-smooth. The generating function for the number $N_q(m;n)$ of monic polynomials over $\mathbb{F}_q$ of degree n which are m-smooth is

$$S_m(z) = \sum_{n\geq 0} N_q(m;n)\, z^n = \prod_{k=1}^{m} \left(\frac{1}{1-z^k}\right)^{I_k}.$$

For the cryptographic applications, m tends to infinity with n. More precisely, we have $m = \sqrt{n \log n}/\sqrt{2 \log 2}$; see [2306]. Hence, singularity analysis does not apply since analytic continuation is not possible. Odlyzko [2306] uses the saddle point method for deriving an asymptotic estimate for the numbers $N_q(m;n)$ as $n \to \infty$, uniformly for m in the range $n^{1/100} \leq m \leq n^{99/100}$. (Actually, his results hold for $n^\delta \leq m \leq n^{1-\delta}$, where $\delta > 0$.)

11.2.43 Remark A variant of the index calculus method over $\mathbb{F}_{2^n}$ (the *Waterloo algorithm*) was introduced in [306]; see also [2306]. It improves the running time of the method but it does not improve its asymptotic order. The running time was proven rigorously by Drmota and Panario [921] using a bivariate saddle point analysis that follows closely Odlyzko's estimates in [2306].

11.2.44 Remark For related estimates for the index calculus method without using smooth polynomials see [1212, 1213]. The fastest variant of the index calculus method for $\mathbb{F}_{2^n}$ is still due to Coppersmith [717]; see Section 11.6 for more details and references.

11.2.45 Remark We focus now on the smallest degrees of the irreducible factors of a polynomial. These estimates are useful, for example, to analyze Ben-Or's irreducible test; see Section 11.3. This analysis requires the study of the probability that a random polynomial of degree n contains no irreducible factors of degree up to a certain value m (such polynomials are sometimes called *m-rough*) and are related to the Buchstab function.

11.2.46 Definition The *Buchstab* function is the unique continuous solution of the difference-differential equation

$$u\omega(u) = 1 \qquad 1 \leq u \leq 2, \qquad\qquad (u\omega(u))' = \omega(u-1) \qquad u > 2.$$

11.2.47 Remark This function was introduced by Buchstab [442] when studying the analogous problem for integer numbers, that is, numbers with no small prime factors. This function has been largely studied [2788]. It is known that it tends quickly to $e^{-\gamma} = 0.56416\ldots$, where γ is Euler's constant.

11.2.48 Remark Car [510] gives estimates for m-roughness that depend on the Buchstab function for m large with respect to n, say $m > c_1 n \log\log n / \log n$. Gao and Panario [1187] show that for m small with respect to n, say $m < c_2 \log n$, the estimate $e^{-\gamma}/m$ holds; see also [2351].

11.2.49 Remark The study of the probability that a random polynomial is m-rough for the complete range $1 \leq m \leq n$, is given by Panario and Richmond [2352]. The estimates are in terms of the Buchstab function when $m \to \infty$. When m is fixed Flajolet and Odlyzko singularity analysis is applied.

11.2.50 Theorem [2352] The smallest degree S_n among the irreducible factors of a random polynomial of degree n over $\mathbb{F}_q$ satisfies

$$Pr(S_n \geq m) = \frac{1}{m}\,\omega\left(\frac{n}{m}\right) + O\left(\max\left\{\frac{1}{m^2}, \frac{\log n}{mn}\right\}\right),$$

when m tends to infinity with n.

11.2.51 Remark Using Theorem 11.2.50 it is not difficult to prove that the expected smallest degree among the irreducible factors of a random polynomial is asymptotic to $e^{-\gamma}\log n$ [2352]. More generally, the expected r-th smallest degree among the irreducible factors of a random polynomial is asymptotic to $e^{-\gamma}\log^r n/r!$.

11.2.3.4 Greatest common divisor of polynomials

11.2.52 Remark As it is pointed out in Remark 11.2.10 two polynomials over $\mathbb{F}_q$ are coprime with probability $1 - 1/q$. Much more can be said about the distribution of the degrees of the irreducible factors in the gcd of several polynomials. Indeed, the limiting distribution of a random variable counting the total degree of the greatest common divisor of two or more random univariate polynomials over the finite field $\mathbb{F}_q$ is geometric, and the distributions of random variables counting the number of common factors (with and without repetitions) are very close to Poisson distributions when q is large. The main reference for these results, from where we extract a couple of main theorems, is [1189]. For simplicity we state the results for two polynomials but they immediately generalize to several polynomials.

11.2.53 Theorem [1189] Let us consider two polynomials over $\mathbb{F}_q$ of degrees n_1 and n_2, respectively, and the random variables $Z_d(n_1, n_2)$ for the number of distinct irreducible factors in the gcd, $Z_r(n_1, n_2)$ for the number of irreducible factors in the gcd counting repetitions, and $Z_t(n_1, n_2)$ for the total degree of the gcd of the two polynomials. Then, as $n_1 \to \infty$ and $n_2 \to \infty$ and for $I(z)$ the generating function of irreducible polynomials, we have

1. the probability generating function for Z_d is

$$PD(u) = \exp\left(-\sum_{m\geq 1} \frac{(1-u)^m}{m} I\left(q^{-2m}\right)\right);$$

2. the probability generating function for Z_r is
$$PR(u) = \exp\left(\sum_{m\geq 1} \frac{u^m - 1}{m} I(q^{-2m})\right);$$
3. the probability generating function for Z_t is
$$PT(u) = \frac{q-1}{q-u}.$$

11.2.54 Remark As a corollary we obtain, for example, the probability that the gcd has zero (coprime polynomials), one or two irreducible factors:
$$\begin{aligned} P(Z_r = 0) &= 1 - 1/q, \\ P(Z_r = 1) &= (1 - 1/q)I(1/q^2), \text{ and} \\ P(Z_r = 2) &= (1/2)(1 - 1/q)\left(I^2(1/q^2) + I(1/q^4)\right). \end{aligned}$$
We also obtain, from the total degree results, that the probability that the gcd has degree k is asymptotic to $q^{-k}(1 - q^{-1})$ as the degrees grow.

11.2.55 Remark Exact values of these probabilities can be computed for small values of the degrees n_1 and n_2 and of field size q. For tables of probabilities for a few common irreducible factors (counted with, or without repetitions), and for the mean and variance of Z_d and Z_r for small values of q, see [1189].

11.2.3.5 Relations to permutations and integers

11.2.56 Remark There are several analogies among the irreducible decomposition of polynomials over finite fields, the prime decomposition of integers, and the cycle decomposition of permutations. We exemplify below with several specific results and then give a heuristic argument to justify these analogies.

11.2.57 Remark We focus first on the splitting degree of a random polynomial of degree n over $\mathbb{F}_q$. Let λ be a partition of n (denoted $\lambda \vdash n$) and write λ in the form $\left[1^{k_1} 2^{k_2} \cdots n^{k_n}\right]$ where λ has k_s parts of size s. We say that a polynomial is of *shape* λ if it has k_s irreducible factors of degree s for each s, $1 \leq s \leq n$. Let $w(\lambda, q)$ be the proportion of polynomials of degree n over $\mathbb{F}_q$ which have shape λ. Let $m(\lambda)$ be the least common multiple of the sizes of the parts of λ. Then the degree of the splitting field over $\mathbb{F}_q$ of a polynomial of shape λ is $m(\lambda)$. The average degree of a splitting field is then given by
$$E_n(q) := \sum_{\lambda \vdash n} w(\lambda, q) m(\lambda).$$

11.2.58 Remark [901] Consider the following classes of polynomials:

1. $\mathcal{M}_1(q)$: the class of all monic polynomials over $\mathbb{F}_q$. In this class the number of polynomials of degree n is q^n.
2. $\mathcal{M}_2(q)$: the class of all monic square-free polynomials over $\mathbb{F}_q$. In this class the number of polynomials of degree n is $(1 - q^{-1})q^n$.

3. $\mathcal{M}_3(q)$: the class of all monic square-free polynomials over $\mathbb{F}_q$ whose irreducible factors have distinct degrees. In this class the number of polynomials of degree n is $a(n,q)q^n$ where, $a(n,q) \to a(q) := \prod_{k\geq 1}(1+I_k q^{-k})\exp(-1/k)$ as $n \to \infty$.

Let us define, for $x > 0$,

$$\Phi_n(x) := \left\{\lambda \vdash n\colon \left|\log m(\lambda) - \frac{1}{2}(\log n)^2\right| > \frac{x}{\sqrt{3}}(\log n)^{3/2}\right\}.$$

11.2.59 Theorem [901] Fix one of the classes $\mathcal{M}_i(q)$ described above. For each $\lambda \vdash n$, let $w(\lambda, q)$ denote the proportion of polynomials in this class whose factorizations have shape λ. Then there exists a constant $c_0 > 0$ (independent of the class) such that for each $x \geq 1$ there exists $n_0(x)$ such that

$$\sum_{\lambda\in\Phi_n(x)} w_i(\lambda, q) \leq c_0 e^{-x/4} \text{ for all } q \text{ and all } n \geq n_0(x).$$

In particular, almost all polynomials of degree n over $\mathbb{F}_q$ in $\mathcal{M}_i(q)$ have splitting fields of degree $\exp((\frac{1}{2}+o(1))(\log n)^2)$, as $n \to \infty$.

11.2.60 Theorem [901] In each of the classes described above the average degree $E_n(q)$ of a splitting field of a polynomial of degree n in that class satisfies

$$\log E_n(q) = C\sqrt{\frac{n}{\log n}} + \mathrm{O}\left(\frac{\sqrt{n}\log\log n}{\log n}\right),$$

uniformly in q, and for $C = 2.99047...$ an explicitly defined constant.

11.2.61 Remark The constant C in the above theorem was obtained by Goh and Schmutz [1290] and Stong [2724] in their study of the analogous problem in the symmetric group S_n. A permutation in S_n is of type $\lambda = \left[1^{k_1}2^{k_2}\cdots n^{k_n}\right]$ if it has exactly k_s cycles of length s for each s, and its order is then equal to $m(\lambda)$. If $w(\lambda)$ denotes the proportion of permutations in S_n which are of type λ, then the average order of a permutation in S_n is equal to $E_n := \sum_{\lambda\vdash n} w(\lambda)m(\lambda)$. We can think of $m(\lambda)$ as a random variable where λ ranges over the partitions of n and the probability of λ is $w(\lambda)$. Properties of the random variable $m(\lambda)$ (and related random variables) under the distribution $w(\lambda)$ have been studied by Erdös and Turán [983, 984, 985, 986]. In particular, the distribution of $\log m(\lambda)$ is approximated by a normal distribution with mean $\frac{1}{2}(\log n)^2$ and variance $\frac{1}{3}(\log n)^3$ in a precise sense [985].

11.2.62 Remark There is a general heuristic that Flajolet, Gourdon, and Panario [1080] call *the permutation model.* Probabilistic properties of the decomposition of polynomials into irreducible factors are expected to have a shape resembling (as $q \to \infty$) that of the corresponding properties of the cycle decomposition of permutations. As the cardinality q of the finite field $\mathbb{F}_q$ goes to infinity (with n staying fixed!), the joint distribution of the degrees of the irreducible factors in a random polynomial of degree n converges to the joint distribution of the lengths of cycles in a random permutation of size n. As stated in [1080]: *This property is visible at the generating functions level when any generating function of polynomials taken at z/q converges (as $q \to \infty$) to the corresponding exponential generating function of permutations.* For example, the generating function of monic polynomials, when normalized with the change of variable $z \mapsto z/q$, is the exponential generating function of permutations:

$$P\left(\frac{z}{q}\right) = \frac{1}{1-z} = \sum_{n=1}^{\infty} n!\frac{z^n}{n!}.$$

Similarly, we have

$$I\left(\frac{z}{q}\right) \underset{(q\to\infty)}{\to} \log\frac{1}{1-z} = \sum_{n=1}^{\infty}(n-1)!\frac{z^n}{n!},$$

the exponential generating function for the number of cycles in permutations.

11.2.63 Remark Several of the results for polynomials and permutations have been generalized to problems in the *exp-log* combinatorial class; see for instance [1084, 1085, 1341, 1342, 2351]. The exp-log class includes problems such as 2-regular graphs, several types of permutations, random mappings (functional digraphs), polynomials over finite fields, random mappings patterns for unlabelled objects, and arithmetical semigroups.

11.2.64 Remark Relations between the irreducible decomposition of polynomials over finite fields and the prime decomposition of integers are discussed in detail in Section 13.1.

See Also

§3.1	For formulas for self-reciprocal polynomials.
§3.5	For formulas for irreducible polynomials with prescribed coefficients.
§11.3	For irreducible test algorithms.
§11.4	For polynomial factorization algorithms.
§11.6	For the index calculus method for discrete logarithms.
§13.1	For relations between polynomials over finite fields and integers.

References Cited: [231, 306, 442, 509, 510, 537, 670, 665, 717, 901, 921, 983, 984, 985, 986, 1080, 1081, 1083, 1084, 1085, 1187, 1189, 1190, 1212, 1213, 1226, 1235, 1239, 1290, 1341, 1342, 1564, 1667, 1760, 1761, 1762, 1763, 1765, 1939, 2306, 2307, 2347, 2349, 2350, 2351, 2352, 2724, 2788, 2833, 2983, 3083]

11.3 Algorithms for irreducibility testing and for constructing irreducible polynomials

Mark Giesbrecht, University of Waterloo

11.3.1 Introduction

We consider the problem of testing polynomials for irreducibility and constructing irreducible polynomials of prescribed degree. Such constructions are essential in many computations with finite fields, including cryptosystems, error-correcting codes, random number generators, combinatorial designs, complexity theory, and many other mathematical computations. In particular, they allow us to construct finite fields of specified order. We confine ourselves to univariate polynomials in this section. Factorization algorithms and irreducibility tests for multivariate polynomials are discussed in Section 11.5.

11.3.1 Definition The following notation will be used in the analyses of algorithms.

$\mathrm{M}(n)\colon \mathbb{N} \to \mathbb{N}$ is defined such that two polynomials over a field $\mathbb{F}$ of degree at most n can be multiplied with $O(\mathrm{M}(n))$ operations in $\mathbb{F}$. $\mathrm{M}(n) = n^2$ using the "school" method, while $\mathrm{M}(n) = n \log n \log\log n$ for FFT-based methods [498].

$\mathrm{MM}(n)\colon \mathbb{N} \to \mathbb{N}$ is defined such that two $n \times n$ matrices over a field $\mathbb{F}$ can be multiplied with $O(\mathrm{MM}(n))$ operations in $\mathbb{F}$. Using the standard algorithm, $\mathrm{MM}(n) = n^3$ and $\mathrm{MM}(n) = n^{2.3727}$ for the best currently known algorithm [2985].

For $f, g\colon \mathbb{R} \to \mathbb{R}$, $f = \tilde{O}(g)$ if $f = O(g(\log|g|)^c)$ for some absolute constant $c \geq 0$.

11.3.2 Early irreducibility tests of univariate polynomials

11.3.2 Remark An essential early construction for certifying irreducibility, as well as for factoring, was established by Petr in 1937 [2391], though it was not presented in algorithmic terms.

11.3.3 Definition Let $f \in \mathbb{F}_q[x]$ be a squarefree polynomial of degree n over a finite field $\mathbb{F}_q$. The *Petr/Berklekamp matrix* $Q \in \mathbb{F}_q^{n\times n}$ of f is defined such that

$$x^{iq} \equiv \sum_{0 \leq j < n} Q_{ij} x^j \pmod{f}.$$

This is the matrix representation of the Frobenius map $(a \mapsto a^q \pmod{f})$ on the basis $\langle 1, x, x^2, \ldots, x^{n-1}\rangle$ for $\mathbb{F}_q[x]/(f)$.

11.3.4 Theorem [2391, 2567, 2570] Let $f = f_1^{e_1} \cdots f_k^{e_k} \in \mathbb{F}_q[x]$, for irreducible $f_1, \ldots, f_k \in \mathbb{F}_q[x]$, have Petr/Berlekamp matrix $Q \in \mathbb{F}_q^{n\times n}$. The characteristic polynomial $\det(Q - \lambda I) \in \mathbb{F}_q[x]$ of Q satisfies

$$\det(Q - \lambda I) = (-1)^n \left(\prod_{1 \leq i \leq k} (\lambda^{e_i} - 1) \right) \cdot \lambda^{(e_1-1)\cdots(e_k-1)}.$$

11.3.5 Remark Theorem 11.3.4 could, in principle, have been cast as an algorithm for testing irreducibility with the technology of the day, using the method for computing the characteristic polynomial of a matrix by Danilevsky [768] from 1937. The characteristic polynomial of Q has the form $x^n - 1$ for an irreducible polynomial.

11.3.6 Remark In 1954 Butler [468] presented an explicit method for determining the number of irreducible factors of a polynomial based on the following theorem.

11.3.7 Theorem [468] Let $f \in \mathbb{F}_q[x]$ have degree n, with Petr/Berlekamp matrix $Q \in \mathbb{F}_q^{n\times n}$. Then $\mathrm{rank}(Q - I) = n - k$, where k is the number of distinct irreducible factors of f. A squarefree f is irreducible if and only if $\mathrm{rank}(Q - I) = n - 1$.

11.3.8 Remark The Petr/Butler approach is developed into a complete algorithm for polynomial factorization by Berlekamp [230] in 1967. While the cost of Butler's method was not analyzed in the modern sense, it is straightforward that constructing Q and taking the rank of $Q - I$ would cost $O(\mathrm{M}(n)(n + \log q) + \mathrm{MM}(n))$ or $\tilde{O}(n \log q + \mathrm{MM}(n))$ operations in $\mathbb{F}_q$.

11.3.3 Rabin's irreducibility test

11.3.9 Remark In 1980, Rabin [2434] based a more efficient test around the following theorem. It was originally presented for polynomials over prime fields, but works for polynomials over any finite field $\mathbb{F}_q$.

11.3.10 Theorem A polynomial $f \in \mathbb{F}_q[x]$ of degree $n \geq 1$ is irreducible if and only if

1. $f \mid x^{q^n} - x$, and
2. $\gcd(x^{q^m} - x, f) = 1$ for all divisors m of n.

11.3.11 Algorithm: Rabin's Irreducibility Test [2434]

Input: $f \in \mathbb{F}_q[x]$ of degree n

Output: "Irreducible" if f is irreducible in $\mathbb{F}_q[x]$; "Reducible" otherwise

1. If $(x^{q^n} \operatorname{rem} f) \neq x$ then return "Reducible"
2. For all prime factors d of n do
3. $\quad h \leftarrow x^{q^{n/d}} \operatorname{rem} f$
4. $\quad$ If $\gcd(h - x, f) \neq 1$ return "Reducible"
5. Return "Irreducible"

11.3.12 Remark For $h \in \mathbb{F}_q[x]$, $h \operatorname{rem} f$ is defined as the unique polynomial of degree less than that of f which is equivalent to h modulo f.

11.3.13 Remark Using repeated squaring to compute powers (see Section 11.1), and the fact that n has at most $\log_2 n$ prime factors, Rabin [2434] provides a cost estimate of $O(n^2 \log^2(n) \log\log(n) \log(q))$ operations in $\mathbb{F}_q$ to test irreducibility of a polynomial. A similar, alternative algorithm based on the Petr/Berlekamp matrix is presented by Calmet and Loos [482].

11.3.14 Remark Panario and Gao [1187] show that a slight variant of Rabin's algorithm requires $O(n\mathrm{M}(n) \log q + \mathrm{M}(n) \log^2 n)$ operations in $\mathbb{F}_q$, deterministically.

11.3.15 Remark The fast modular composition algorithm of von zur Gathen and Shoup [1239] allows us to compute x^{q^m} from x^q using $O((\mathrm{MM}(n^{1/2})n^{1/2} + n^{1/2}\mathrm{M}(n)) \log m)$ operations in $\mathbb{F}_q$. Employing this in Rabin's algorithm gives an algorithm for testing irreducibility which requires $O(\mathrm{M}(n) \log q + (\mathrm{MM}(n^{1/2})n^{1/2} + n^{1/2}\mathrm{M}(n) \log^2 n))$ or $O\tilde{~}(\mathrm{MM}(n^{1/2})n^{1/2} + n \log q)$ operations in $\mathbb{F}_q$.

11.3.16 Remark The algorithm for modular composition of Kedlaya and Umans [1722] requires $n^{1+o(1)} \cdot (\log q)^{1+o(1)}$ *bit operations*, for prime q. This yields an algorithm for testing irreducibility which requires $n^{1+o(1)} \cdot (\log q)^{1+o(1)}$ bit operations. Note that this algorithm is not in the algebraic model of all operations in $\mathbb{F}_q$, and counts bit operations instead.

11.3.17 Remark A more precise average-case analysis of the cost of Rabin's algorithm is given by Panario et al. [2357, 2349]. This involves the study of the probability that a polynomial has an irreducible factor of degree dividing a maximal divisor of n. They give an exact expression for this probability when n is prime or a product of two primes, as well as an asymptotic analysis. They also present analyses of variants of Rabin's algorithm suggested in [1187, 1227].

11.3.18 Remark Any modern factoring algorithm can be used to test irreducibility, possibly at the cost of randomization. Asymptotically, the fastest currently known algorithms are those of von zur Gathen and Shoup [1239] and Kaltofen and Shoup [1667], with improvements in the bit complexity model by Kedlaya and Umans [1722]. These are slower than the above irreducibility tests; see Section 11.4.

11.3.4 Constructing irreducible polynomials: randomized algorithms

11.3.19 Remark Density estimates for irreducible polynomials easily reduce the problem of finding irreducible polynomials to that of certifying a polynomial is irreducible, assuming we have a way to generate random field elements. Such an approach was even suggested by Galois in 1830 [1168]. However, one can do considerably better than the most naïve reduction, and both the asymptotic complexity of this problem and more practical concerns hold considerable interest.

11.3.20 Remark Gauss gives an explicit formula for the number $I_q(n)$ of irreducible polynomials of degree n, from which it is easily derived that $q^n/(2n) < I(n,q) < q^n/n$ (see [1939], Exercises 3.26 and 3.27). Thus, given a way to randomly generate elements of $\mathbb{F}_q$, any algorithm for testing irreducibility also yields a probabilistic algorithm for finding irreducible polynomials of any specified degree. The expected number of operations is n times the cost of the chosen irreducibility test.

11.3.5 Ben-Or's algorithm for construction of irreducible polynomials

11.3.21 Remark In 1981, Ben-Or [223] described a simple probabilistic algorithm for constructing an irreducible polynomial with a better expected cost than the straightforward approach suggested in Remark 11.3.20.

11.3.22 Algorithm: Ben-Or's irreducible polynomial constructor
Input: $n \in \mathbb{Z}_{>0}$
Output: A uniformly random monic irreducible polynomial of degree n in $\mathbb{F}_q[x]$

1. Randomly choose a monic $f \in \mathbb{F}_q[x]$ of degree n
2. For i from 1 to $\lfloor n/2 \rfloor$ do
3. If $\gcd(x^{q^i} - x, f) \neq 1$ goto Step 1.
4. Return f

11.3.23 Theorem [223]. The expected value of the degree of the smallest factor of a randomly and uniformly chosen polynomial of degree n in $\mathbb{F}_q[x]$ is $O(\log n)$.

11.3.24 Theorem [223]. Using Ben-Or's algorithm we can construct an irreducible polynomial of degree n over $\mathbb{F}_q$ with an expected number of $O(n\mathrm{M}(n)\log n \log(nq))$ or $\tilde{O}(n^2 \log q)$ operations in $\mathbb{F}_q$.

11.3.25 Remark A more precise analysis is provided by Panario and Gao [1187], who consider the probability that a polynomial is *rough*, i.e., has all irreducible factors of degree greater than some bound. Reducible polynomials with no small degree factors cause Ben-Or's algorithm to perform more operations.

11.3.26 Theorem [1187] Let $P_q^I(n,m)$ be the probability that a polynomial in $\mathbb{F}_q[x]$ of degree n is irreducible if it has no factors less than or equal to $m = O(\log n)$. Then $P_q^I(n,m) \sim e^{\gamma} m/n$ as n, m, and q approach infinity, where γ is Euler's constant.

11.3.27 Remark Panario and Richmond [2350] give a much more precise average-case analysis of Ben-Or's algorithm. They study the probability that the smallest degree irreducible factor of a degree n polynomial is greater than some m. As well, they study the expectation and variance of the smallest degree among the irreducible factors of a random polynomial of degree n. Results are stated in terms of the Buchstab function, a classical number-theoretic function used in the study of numbers with no prime factors smaller than some bound; see [2350] for definitions and details. They also show that the expected average-case cost of an irreducible test in Ben-Or's algorithm is $O(\mathrm{M}(n)\log(n)(\log n + \log q))$ operations in $\mathbb{F}_q$, and provide an asymptotic analysis with explicit constants for this complexity. The expected cost of actually finding an irreducible polynomial is n times this cost.

11.3.28 Remark Von zur Gathen and Gerhard [1227], Section 14.9 suggest that a reasonable approach to finding irreducible polynomials might be a hybrid of Ben-Or's algorithm, with a switch to the irreducibility test of Rabin when testing for factors of degrees over some prescribed bound such as $\log_2 n$.

11.3.6 Shoup's algorithm for construction of irreducible polynomials

11.3.29 Remark Shoup [2629] gives an asymptotically faster probabilistic algorithm for constructing an irreducible polynomial of degree n, which requires $O((n^2 \log n + n \log q) \log n \log\log n)$ operations in $\mathbb{F}_q$. It also has the benefit of reducing use of randomness to $O(n)$ random elements from $\mathbb{F}_q$ (the algorithms described above all require an expected $O(n^2)$ random elements). Shoup's algorithm is very different from the Rabin's and Ben-Or's, and more closely resembles the deterministic constructions discussed in Section 11.3.7 below. The algorithm proceeds by looking at each prime power r^e dividing n (where r is prime). There are a number of special cases. In particular, if r is the characteristic of $\mathbb{F}_q$, then Artin-Schreier type polynomials can be employed (see Remark 11.3.33 below), and when $r = 2$ a simple construction suffices. In the remaining cases, for each prime-power-factor r^e of n, it first factors the r-th cyclotomic polynomial $\Phi_r \in \mathbb{F}_q[x]$. This can be done quickly. It then finds an r-th power non-residue in $\xi \in \mathbb{F}_q[\theta]$, where θ is an adjoined root of Φ_r; $x^{r^e} - \xi$ is irreducible of degree r^e in $\mathbb{F}_q[\theta][x]$, from which an irreducible polynomial of degree r^e over $\mathbb{F}_q[x]$ is constructed via traces. The good asymptotic cost is maintained through a fast algorithm for finding minimal polynomials in algebraic extensions, employing a very elegant use of the "Transposition Principle" (also known as Tellegen's theorem). Despite the intricate construction, the polynomial produced is uniformly selected from the set of all irreducible polynomials of degree n in $\mathbb{F}_q[x]$.

11.3.30 Remark An algorithm for finding irreducible polynomials of degree n, along the lines of Shoup's algorithm above, is exhibited by Couveignes and Lercier [747] and has an expected cost of $n^{1+o(1)}(\log q)^{5+o(1)}$. The key innovation is the fast construction of irreducible polynomials (for all but some exceptional degrees) using isogenies of a random elliptic curve. This is combined with the recent fast modular composition algorithm of Kedlaya and Umans [1722], along with the fast algorithm of Bostan et al. [363] to compute "composed sums" of polynomials.

11.3.7 Constructing irreducible polynomials: deterministic algorithms

11.3.31 Remark Deterministic algorithms for constructing univariate polynomials of specified degree over a finite field of characteristic p are considerably more difficult. The goal is algorithms which require time polynomial in n and $\log p$. Given such an algorithm it is straightforward to construct an irreducible polynomial over an extension field $\mathbb{F}_q$ of $\mathbb{F}_p$, at least up to polynomial time in n and $\log q$. We thus concern ourselves with constructing irreducible polynomials over prime fields $\mathbb{F}_p$. A more efficient algorithm for non-prime fields is established by Shoup [2625].

11.3.32 Remark The algorithms of Chistov from 1984 [621] and Semaev [2580] are the first to require time $(np)^{O(1)}$, and hence are effective over small finite fields; see also [2856].

11.3.33 Remark In 1986 Adleman and Lenstra [14] exhibited an algorithm for constructing irreducible polynomials which runs in polynomial time in n and $\log p$, assuming that the Extended Riemann Hypothesis is true. Their algorithm proceeds by first factoring $n = p^e n_0$, where $\gcd(n_0, p) = 1$. An extension of degree n_0 of $\mathbb{F}_p$ is built, and then an extension of degree p^e is constructed on top of that. A minimal polynomial of a generating element is irreducible of degree n. The degree n_0 extension requires finding a prime $r \equiv 1 \pmod{n_0}$ such that p is inert in the unique subfield K of the r-th cyclotomic field such that $[\mathsf{K} : \mathbb{Q}] = n$. From this an extension of $\mathbb{F}_p$ of degree n_0 is constructed via Gauss periods; see Section 5.3. To find the inert primes, they simply test numbers of the form $n_0 k + 1$ for $k = 1, 2, \ldots$ in sequence for primality and inertness. The Extended Riemann Hypothesis guarantees that a $k = O(n_0^4 (\log np)^2)$ exists, but without this assumption (or randomization) there is no way known to find such a p. To find an extension of degree p^e, Artin-Schreier polynomials of the form $x^p - x - \alpha \in \mathbb{F}_q[x]$ are employed, which are irreducible when $\alpha \neq \beta^p - \beta$ for some $\beta \in \mathsf{K}$, where K is any extension of $\mathbb{F}_p$. A tower of fields is constructed, starting with $\mathbb{F}_{p^{n_0}}$, and ending with $\mathbb{F}_{p^n} = \mathbb{F}_{p^{n_0 p^e}}$.

11.3.34 Remark Evdokimov [1017], in 1986, independently presented a similar result to Adelman and Lenstra's. In this construction n is factored as a product $r_1^{e_1} \cdots r_\ell^{e_\ell}$ for distinct primes r_i, and an irreducible polynomial of each degree $r_i^{e_i}$ is constructed. When $r_i = p$ the Artin-Schreier polynomials are again employed. When $r_i \neq p$ he employs a reducibility theorem for the polynomial $x^{r_i^{e_i}} - a$ (where a lies in a small cyclotomic extension) which requires finding r_i-th non-residues, small examples of which are ensured by the Extended Riemann Hypothesis. Irreducible polynomials of different prime power degrees are combined by computing the minimal polynomial of the sum of the roots of the two polynomials. This is easily accomplished with basic linear algebra, though the "composed sum" algorithm of [363] can now accomplish this in quasi-linear time; see Remark 11.3.30 above.

11.3.35 Remark In 1989 Shoup [2624, 2625] gave the currently fastest completely unconditional deterministic algorithm for finding irreducible polynomials, along similar lines to [1017]. He demonstrates a polynomial-time (in n and $\log p$) reduction from the problem of finding irreducible polynomials of degree n to the problem of factoring polynomials in $\mathbb{F}_p[x]$. Shoup shows that if, for every prime $r \mid n$, we are given a representation of a splitting field K of $x^r - 1$ (i.e., a non-trivial irreducible factor of $x^r - 1$), and an r-th nonresidue in K, then we can construct an irreducible polynomial of degree n. Both finding the factor of $x^r - 1$ and the non-residue (by repeatedly taking r-th roots in K) can be accomplished through factorization. Shoup [2624, 2625] also provides the fastest (unconditional) deterministic algorithm for factoring polynomials over $\mathbb{F}_p[x]$. Employing this in his irreducible polynomial construction method yields a deterministic algorithm requiring $O(p^{1/2}(\log p)^3 n^3 (\log n)^c + (\log p)^2 n^4 (\log n)^c)$ or $O\tilde{~}(p^{1/2} n^3 + n^4)$ operations in $\mathbb{F}_p$, for some absolute constant $c > 0$.

11.3.8 Construction of irreducible polynomials of approximate degree

11.3.36 Remark If the requirement that the degree of the constructed polynomial in $\mathbb{F}_p[x]$ be exactly n is relaxed, and only an approximation of this degree is required, some unconditional algorithms are known.

11.3.37 Remark In 1986, von zur Gathen [1219] gave a polynomial-time algorithm (in n and $\log p$) which finds an irreducible polynomial of degree at least n, assuming we can do preprocessing only with respect to p (and requiring time polynomial in p).

11.3.38 Remark Adleman and Lenstra [14] present an algorithm which outputs an irreducible polynomial $f \in \mathbb{F}_p[x]$ with $n/(c \log p) < \deg f \leq n$, for some absolute constant c, and requires time polynomial in n and $\log p$.

11.3.39 Remark For a fixed prime p and $\tilde{n} \in \mathbb{N}$, Shparlinski [2639, Theorem 1], gives a deterministic, unconditional algorithm that finds an irreducible polynomial of degree $\tilde{n} \cdot (1 + \exp(-(\log\log\tilde{n})^{1/2-o(1)}))$, and requires time polynomial in p and $\log\tilde{n}$.

11.3.40 Remark For a sufficiently large $\tilde{q}$, Shparlinski [2639, Theorem 2], provides a deterministic, unconditional algorithm that constructs a prime p, an integer $n \geq 0$, and an irreducible polynomial $f \in \mathbb{F}_p[x]$ of degree n, such that $p^n = \tilde{q} + o(\tilde{q})$. The algorithm requires time polynomial in $\log\tilde{q}$; see also [2636, Section 2].

See Also

[1227, Section 14.9]	For an exposition of Rabin's algorithm for testing irreducible polynomials, and Ben-Or's algorithm for generating irreducible polynomials, and their complexity analyses.
[2636, Section 2]	For early deterministic algorithms for constructing irreducible polynomials; see also the introduction of [2625].

References Cited: [14, 223, 230, 363, 468, 482, 498, 621, 747, 768, 1017, 1168, 1187, 1219, 1227, 1239, 1667, 1722, 1939, 2349, 2350, 2357, 2391, 2434, 2567, 2570, 2580, 2624, 2625, 2629, 2636, 2639, 2856, 2985]

11.4 Factorization of univariate polynomials

Joachim von zur Gathen, Universität Bonn

11.4.1 Remark The pioneering algorithms for the factorization of univariate polynomials over a finite field are due to Berlekamp [230, 232]. Subsequent improvements of the asymptotic running time were presented by Cantor and Zassenhaus [499], von zur Gathen and Shoup [1239], Huang and Pan [1554], and Kaltofen and Shoup [1667]. The currently fastest method yields the following main result of this section.

11.4.2 Theorem [1722] The factorization of a univariate polynomial of degree n over $\mathbb{F}_q$ can be computed with an expected number $(n \log q)^{1+o(1)} \cdot (n^{1/2} + \log q)$ of bit operations.

11.4.3 Remark Ignoring asymptotically small factors, this running time corresponds to $n^{3/2} + n \log q$ field operations. These methods rely on a long line of algorithmic developments. The most basic nontrivial task is that of multiplying polynomials. At degree n, its cost $\mathsf{M}(n)$ is discussed in Definition 11.3.1. The current record of Fürer [1148] for integer multiplication stands at $n \log n \, 2^{O(\log^* n)}$ bit operations, where $\log^* n$ is the smallest k for which the k-fold iterated logarithm $\log\log\cdots\log n$ is less than 1. Fürer's algorithm can be adapted to multiplication of polynomials. For a k-bit prime p and two polynomials in $\mathbb{F}_p[x]$ of degree at most n it takes $O(\mathsf{M}(n(k + \log n)))$ bit operations. Next come division with remainder at $O(\mathsf{M}(n))$ (Sieveking [2664], Brent and Kung [401]), and univariate gcd, multipoint evaluation, and interpolation at $O(\mathsf{M}(n)\log n)$ field operations (Strassen [2734]). Further tasks include squarefree factorization, with $O(\mathsf{M}(n)\log n)$ operations (Yun [3051]). Furthermore, the matrix multiplication exponent ω is such that $\mathsf{MM}(n) \leq n^{\omega}$; see Definition 11.3.1. It appears in an algorithm of Brent and Kung [401] for modular composition, which computes $f \circ g \pmod h$ from f, g, and h with $O(n^{1.687})$ operations.

Multivariate multipoint evaluation asks for the evaluation of a multivariate polynomial, of degree at most d in each of its r variables, at n points in $\mathbb{F}_q^r$. One can achieve this with $O(d^{(\omega+1)(r-1)+1})$ field operations (Nüsken and Ziegler [2301]). A major advance of Kedlaya and Umans [1722], at the heart of their subsequent results, was a modular approach to this problem. They consider it as a problem over the integers, say for a prime field $\mathbb{F}_q$, and solve it in a standard modular fashion, modulo various small primes. This leads to a method using $r(d^r + q^r + n)(\log q)^{o(1)}$ bit operations. It also yields a univariate modular composition method with $(n \log q)^{1+o(1)}$ bit operations, roughly corresponding to only $O(n)$ field operations. The more efficient methods use a "polynomial representation of the Frobenius automorphism," suggested by Erich Kaltofen, fast algorithms for it, and an interval blocking strategy. This appeared first in von zur Gathen and Shoup [1239].

11.4.4 Remark Many other works have contributed to the factoring problem. We only mention Serret [2600], Arwin [137], Petr [2391], Kempfert [1726], Niederreiter [2249, 2250], Kaltofen and Lobo [1663], Gao and von zur Gathen [1178], Kaltofen and Shoup [1667], von zur Gathen and Gerhard [1226].

11.4.5 Remark Berlekamp [230] presented a deterministic algorithm to factor in $\mathbb{F}_p[x]$ with $O(n^{\omega} + pn^{2+o(1)})$ field operations; see [1220] for this estimate. It was improved by Shoup [2624] to $O(p^{1/2}\log^2 p + \log p \cdot n^{2+o(1)})$ operations. Berlekamp [232] introduced the fundamental idea of probabilistic algorithms, although it became widely accepted in computer science only after the polynomial-time primality test of Solovay and Strassen [2694]. The factorization algorithms mentioned are all probabilistic, in the Las Vegas sense where the output can be verified and thus guaranteed to be correct, but the running time is a random variable (with exponentially decaying tails). Removing randomness while conserving polynomial time is still an open question, of great theoretical interest but presumably no practical import. Already Berlekamp [232] observed that it boils down to the following.

11.4.6 Open Question Given the coefficients of a polynomial which is a product of linear factors over a prime field $\mathbb{F}_p$, can one find a nontrivial factor deterministically in polynomial time?

11.4.7 Remark Several papers provide steps in this direction, often assuming the Extended Riemann Hypothesis: Moenck [2112], Adleman, Manders, and Miller [15], Huang [1552, 1553], von zur Gathen [1220], Rónyai [2476, 2477, 2478, 2479], Mignotte and Schnorr [2094], Evdokimov [1019, 1018], Bach, von zur Gathen, H. Lenstra [158], Rónyai and Szántó [2480], Gao [1176], Ivanyos, Karpinski, and Saxena [1579], Ivanyos, Karpinski, Rónyai, and Saxena [1578], Arora, Ivanyos, Karpinski, and Saxena [132].

11.4.8 Remark Computations just using field operations can be modeled by arithmetic circuits (Strassen [2732]). All algorithms discussed here, except those of Kedlaya and Umans [1722] and Fürer's multiplication, are of this type. Their running time is $\Omega(n \log q)$ field operations.

11.4.9 Open Question Do all arithmetic circuits over $\mathbb{F}_q$ that factor univariate polynomials of degree n require $\Omega(n \log q)$ arithmetic operations?

11.4.10 Remark A positive answer is known only for $n = 2$ (von zur Gathen and Seroussi [1237]).

11.4.11 Remark The survey articles of Kaltofen [1646, 1655], von zur Gathen and Panario [1234] and the textbooks by Shparlinski [2637, 2641] and von zur Gathen and Gerhard [1227] present the details of most of these algorithms, more references, and historical information.

It is intriguing that the basics of most modern algorithms go back to Legendre, Gauß, and Galois.

11.4.12 Remark The average cost of factoring algorithms, for polynomials picked uniformly at random from those of a fixed degree, is analyzed in Flajolet, Gourdon, and Panario [1080] and von zur Gathen, Panario, and Richmond [1235].

See Also

§11.3	For univariate irreducibility testing.
§11.5	For multivariate factorization methods.

References Cited: [15, 132, 137, 158, 230, 232, 401, 498, 499, 723, 1018, 1019, 1080, 1148, 1176, 1178, 1220, 1225, 1226, 1227, 1234, 1235, 1237, 1239, 1552, 1553, 1554, 1578, 1579, 1646, 1655, 1663, 1667, 1684, 1722, 1726, 2094, 2112, 2249, 2250, 2301, 2391, 2476, 2477, 2478, 2479, 2480, 2558, 2559, 2600, 2624, 2637, 2641, 2664, 2694, 2732, 2734, 2984, 3051]

11.5 Factorization of multivariate polynomials

Erich Kaltofen, *North Carolina State University*
Grégoire Lecerf, *CNRS & École polytechnique*

We extend the univariate factorization techniques of the previous section to several variables. Two major ingredients are the reduction from the bivariate case to the univariate one, and the reduction from any number to two variables. We present most of the known techniques according to the representation of the input polynomial.

11.5.1 Factoring dense multivariate polynomials

11.5.1 Remark In this subsection we are concerned with different kinds of factorizations of a multivariate polynomial $f \in \mathbb{F}_q[x_1, \ldots, x_n]$ stored in dense representation.

11.5.2 Definition Let R be any ring. A *dense representation* of a polynomial $f \in R[x_1, \ldots, x_n]$ is the data of the vector $(d_1, \ldots, d_n)$ of the partial degrees of f, and the vector of the coefficients of the monomials $x_1^{e_1} \cdots x_n^{e_n}$ for all $0 \le e_1 \le d_1, \ldots, 0 \le e_n \le d_n$, sorted in reverse lexicographical ordering on the exponents $(e_1, \ldots, e_n)$, which means that $(e_1, \ldots, e_n) < (e'_1, \ldots, e'_n)$ if, and only if, there exists j such that $(e_n = e'_n, \ldots, e_{j+1} = e'_{j+1}$, and $e_j < e'_j)$.

11.5.3 Remark The representation of multivariate polynomials is an important issue, which has been discussed from the early ages of computer algebra [761, 778, 1519, 1616, 2129, 2130, 2131, 2730, 3021].

11.5.1.1 Separable factorization

11.5.4 Remark Separable factorization can be seen as a preprocess to the other factorizations (squarefree, irreducible, and absolutely irreducible, as defined below), which allows to reduce to considering separable polynomials.

11.5.5 **Definition** Let R be an integral domain. A polynomial $f \in R[x]$ is *primitive* if the common divisors in R of all the coefficients of f are invertible in R.

11.5.6 **Definition** Let R be a unique factorization domain of characteristic p, and let E_p represent $\{1\}$ if $p = 0$ and $\{1, p, p^2, p^3, \ldots\}$ otherwise. If f is a primitive polynomial in $R[y]$ of degree $d \geq 1$, then the *separable decomposition* of f, written $\operatorname{Sep}(f)$, is defined to be the set $\operatorname{Sep}(f) := \{(f_1, q_1, m_1), \ldots, (f_s, q_s, m_s)\} \subseteq (R[y] \setminus R) \times E_p \times \mathbb{N}$, satisfying the following properties:

1. $f(y) = \prod_{i=1}^{s} f_i(y^{q_i})^{m_i}$,
2. for all $i \neq j$ in $\{1, \ldots, s\}$, $f_i(y^{q_i})$ and $f_j(y^{q_j})$ are coprime,
3. for all $i \in \{1, \ldots, s\}$, $m_i \pmod{p} \neq 0$,
4. for all $i \in \{1, \ldots, s\}$, f_i is separable and primitive,
5. for all $i \neq j$ in $\{1, \ldots, s\}$, $(q_i, m_i) \neq (q_j, m_j)$.

The process of computing the separable decomposition is the *separable factorization*.

11.5.7 **Example** With $R := \mathbb{F}_3$ and $f := y^2(y+1)^3(y+2)^4 = y^9 + 2y^8 + 2y^3 + y^2$, we have that $\operatorname{Sep}(f) = \{(y, 1, 2), (y+1, 3, 1), (y+2, 1, 4)\}$.

11.5.8 **Example** With $R := \mathbb{F}_3[x]$ and $f := (y+2x)^7(y^3+2x)^3(y^6+x)$, we have that $\operatorname{Sep}(f) = \{(y+2x, 1, 7), (y+2x^3, 9, 1), (y^2+x, 3, 1)\}$.

11.5.9 **Theorem** [2108, Chap. VI, Theorem 6.3] Any primitive polynomial $f \in R[y]$ admits a unique (up to permutations and units in R) separable decomposition, which only depends on the coefficients of f.

11.5.10 **Remark** Roughly speaking, the separable decomposition corresponds to sorting the roots of the given polynomial according to their multiplicity. A constructive proof of Theorem 11.5.9 can be found in [2108, Chap. VI, Theorem 6.3], and another proof using the irreducible factorization in [1882, Proposition 4].

11.5.11 **Remark** Since the separable decomposition only depends on the coefficients of f it can be computed in any extension of R.

11.5.12 **Theorem** [1882, Proposition 5] If F is a field then the separable decomposition of a polynomial $f \in F[y]$ of degree d can be computed with $O(\mathsf{M}(d) \log d)$ arithmetic operations in F. Let us recall that $\mathsf{M}(d)$ represents a bound for the complexity of multiplying two polynomials of degree at most d with coefficients in a commutative ring with unity, in terms of the number of arithmetic operations in the latter ring.

11.5.13 **Theorem** [1882, Propositions 8 and 9] Let $R = F[x]$, where F is a field, and let $f \in F[x][y]$ be a primitive polynomial of degree d_x in x and d_y in y.

1. If F has cardinality at least $d_x(2d_y+1)+1$ then $\operatorname{Sep}(f)$ can be computed (deterministically) with $O(d_y(d_y\mathsf{M}(d_x)\log d_x + d_x\mathsf{M}(d_y)\log d_y))$ or $\tilde{O}(d_x d_y^2)$ operations in F.
2. If F has cardinality at least $4d_xd_y$ then $\operatorname{Sep}(f)$ can be computed with an *expected* number of $O(d_y\mathsf{M}(d_x)\log d_x + d_x\mathsf{M}(d_y)\log d_y))$ or $\tilde{O}(d_xd_y)$ operations in F.

Let us recall that $f(d) \in \tilde{O}(g(d))$ means that $f(d) \in g(d)(\log_2(3+g(d)))^{O(1)}$. With the second randomized algorithm, the ouput is always correct, and the cost estimate is the average of the number of operations in F taken over all the possible executions.

11.5.1.2 Squarefree factorization

11.5.14 **Definition** If R is a unique factorization domain then the *squarefree decomposition* of $a \in R$, written $\mathrm{Sqr}(a)$, is the set of pairs (a_m, m), where a_m represents the product of all the irreducible factors of a of multiplicity m. The process of computing the squarefree decomposition is the *squarefree factorization.*

11.5.15 **Definition** For convenience, we say that a polynomial $f \in R[x_1, \dots, x_n]$ is *primitive* (resp. *separable*) in x_i if it is so when seen in $R[x_1, \dots, x_{i-1}, x_{i+1}, \dots, x_n][x_i]$.

11.5.16 **Algorithm** (Sketch of the algorithm squarefree factorization)

Input: a polynomial $f \in \mathbb{F}_q[x_1, \dots, x_n]$, primitive in $x_1, \dots, x_n$.
Output: the squarefree decomposition $\mathrm{Sqr}(f)$ of f.

1. First compute the separable decomposition of f seen in $\mathbb{F}_q[x_1, \dots, x_{n-1}][x_n]$. Then for each separable factor g of $\mathrm{Sep}(f)$ compute the separable decomposition of g seen in $\mathbb{F}_q[x_1, \dots, x_{n-2}, x_n][x_{n-1}]$. Then for each separable factor h of $\mathrm{Sep}(g)$ compute the separable decomposition of h seen in $\mathbb{F}_q[x_1, \dots, x_{n-3}, x_{n-1}, x_n][x_{n-2}]$, etc. At the end rewrite f as the product of polynomials of the form $f_i(x_1^{q_{i,1}}, \dots, x_n^{q_{i,n}})^{m_i}$, where the f_i are separable in $x_1, \dots, x_n$, and where the $q_{i,j}$ are powers of p.
2. The squarefree factorization of each $f_i(x_1^{q_{i,1}}, \dots, x_n^{q_{i,n}})^{m_i}$ is simply obtained by extracting the $\min_{j \in \{1,\dots,n\}} q_{i,j}$-th root of $f_i(x_1^{q_{i,1}}, \dots, x_n^{q_{i,n}})$.

11.5.17 **Theorem** [1882, Proposition 12] Let $f \in \mathbb{F}_q[x, y]$ be a polynomial of degree d_x in x and d_y in y. If $q \geq 4(3d_y + 1)d_x$ then $\mathrm{Sqr}(f)$ can be computed with an *expected* number of $O(d_y \mathsf{M}(d_x) \log d_x + d_x \mathsf{M}(d_y) \log d_y))$ or $\tilde{O}(d_x d_y)$ operations in $\mathbb{F}_q$.

11.5.18 **Remark** Practical multivariate squarefree factorization algorithms have been designed in [237] to be specifically efficient in small and medium sizes, when M does not behave as softly linear. Algorithms for deducing the squarefree decomposition from the separable one were proposed in [1271] and then improved in [1882] in particular cases.

11.5.1.3 Bivariate irreducible factorization

11.5.19 **Definition** If R is a unique factorization domain then the *irreducible decomposition* of $a \in R$, written $\mathrm{Irr}(a)$, is the set of pairs (a_i, m_i), where a_i is an irreducible factor of a of multiplicity m_i. The process of computing the irreducible decomposition is the *irreducible factorization.*

11.5.20 **Definition** If F is a field then the *absolutely irreducible decomposition* of $f \in F[x_1, \dots, x_n]$ is the irreducible decomposition of f in $\bar{F}[x_1, \dots, x_n]$, where $\bar{F}$ represents the algebraic closure of F. The process of computing the absolutely irreducible decomposition is the *absolutely irreducible factorization*, or *absolute factorization.*

11.5.21 **Remark** We do not discuss specific algorithms for computing the absolute factorization. In fact, whenever F is a finite field, the absolutely irreducible decomposition of $f \in F[x_1, \dots, x_n]$ can be obtained from the irreducible decomposition over the algebraic extension of F of degree $\deg f$. For more details and advanced algorithms we refer the reader to [617].

11.5.22 Theorem [1883, Theorem 2] Let $q = p^k$, and let $f \in \mathbb{F}_q[x, y]$ be a polynomial of degree d_x in x and d_y in y. If $q \geq 10 d_x d_y$ then $\mathrm{Irr}(f)$ can be computed with factoring several polynomials in $\mathbb{F}_q[y]$ whose degree sum does not exceed $d_x + d_y$, plus an *expected* number of $\tilde{O}(k(d_x d_y)^{1.5})$ operations in $\mathbb{F}_p$.

11.5.23 Remark If q is not sufficiently large to apply Theorem 11.5.22 then one can compute the irreducible factorization of f over a slightly larger finite field, and then recover the factorization over $\mathbb{F}_q$ by computing the norm of the factors.

The algorithm underlying Theorem 11.5.22 is summarized in the following.

11.5.24 Algorithm (Sketch of the lifting and recombination technique)

Input: a primitive and separable polynomial $f \in \mathbb{F}_q[x][y]$, of partial degrees d_x in x and d_y in y.

Output: the irreducible decomposition $\mathrm{Irr}(f)$ of f.

1. *Normalization.* If the cardinality of $\mathbb{F}_q$ is sufficiently large then a suitable shift of the variable x reduces the problem to the *normalized* case defined as follows:

$$\deg f(0, y) = d_y \text{ and } \mathrm{Res}\left(f(0, y), \frac{\partial f}{\partial y}(0, y)\right) \neq 0.$$

2. *Univariate factorization.* Compute $\mathrm{Irr}(f(0, y))$ in $\mathbb{F}_q[y]$.
3. *Lifting.* Use the classical *Hensel lifting* from the previously computed irreducible factors $\mathfrak{f}_1(0, y), \ldots, \mathfrak{f}_s(0, y)$ of $f(0, y)$ in order to deduce the irreducible *analytic decomposition* $\mathfrak{f}_1, \ldots, \mathfrak{f}_s$ of f in $\mathbb{F}_q[[x]][y]$ to a certain finite precision σ in x.
4. *Recombination.* Discover how the latter analytic factors $\mathfrak{f}_1, \ldots, \mathfrak{f}_s$ *recombine* into the irreducible factors.

11.5.25 Remark Since any proper factor g of f is the product of a subset of the analytic factors, the precision $\sigma = d_x$ is sufficient in Algorithm 11.5.24 to discover $\mathrm{Irr}(f)$ by means of exhaustive search. To be precise, it suffices to run over all the subsets S of $\{1, \ldots, s\}$ of cardinality at most $s/2$ and test whether the truncated polynomial of $\prod_{i \in S} \mathfrak{f}_i$ to precision d_x in $\mathbb{F}_q[x][y]$ divides f or not. This approach was originally popularized in computer algebra by Zassenhaus in [3052] in the context of factoring in $\mathbb{Q}[y]$ *via* the p-adic completion of $\mathbb{Q}$. The adaptation to two and several variables was first pioneered in [2209, 2938, 2939]. In particular, [2209] introduced coefficient field abstractions that marked the beginning of generic programming. Von zur Gathen adopted Musser's approach to valuation rings [1217]. The cost of this approach is, of course, exponential in s. However, as proved in [1183] the cost of the recombination process behaves in softly linear time in average over finite fields, which explains the practical efficiency of this approach.

11.5.26 Remark For details concerning Hensel lifting, we refer the reader to [1227, Chap. 15], that contains a variant of the multifactor Hensel lifting first designed by Shoup for his C++ library NTL (`http://www.shoup.net`). An improvement obtained thanks to the transposition principle is proposed in [365]. Parallelization has been studied in [238].

11.5.27 Remark The first attempt to reduce the recombination stage to linear algebra seems to be due to Sasaki et al. [2528, 2529, 2530], with a method called the *trace recombination.* But the first successes in the design and proofs of complete algorithms are due to van Hoeij [1518] for the factorization in $\mathbb{Z}[x]$, and then to Belabas et al. [219] for $F(x)[y]$, with the *logarithmic derivative recombination* method, where the precision $\sigma = \deg f(\deg f - 1) + 1$ is shown to be sufficient in general. Then a precision linear in $\deg f$ in characteristic 0 or large enough characteristic has been shown to suffice in [365, 1880].

11.5.28 Remark In [1177], Gao designed the first softly quadratic time probabilistic reduction of the factorization problem from two to one variable whenever the characteristic of the coefficient field is zero or sufficiently large. His algorithm makes use of the first algebraic de Rham cohomology group of $F[x, y, 1/f(x, y)]$, as previously used by Ruppert [2504, 2505] for testing the absolute irreducibility. In fact, if f factors into $f_1 \cdots f_r$ over the algebraic closure of F then

$$\left(\frac{\hat{f}_i \frac{\partial f_i}{\partial x}}{f} dx + \frac{\hat{f}_i \frac{\partial f_i}{\partial y}}{f} dy\right)_{i \in \{1,\ldots,r\}}$$

is a basis of the latter group, where $\hat{f}_i := f/f_i$ [2504, Satz 2]. In consequence, this group can be obtained by searching for closed differential 1-forms with denominators f and numerators of degrees at most $\deg f - 1$, which can be easily done by solving a linear system. A nice presentation of Ruppert's results is made in Schinzel's book [2542, Chapter 3]. The algorithm underlying Theorem 11.5.22 makes use of these ideas in order to show that a precision $\sigma = d_x + 1$ of the series in the Hensel lifting suffices.

11.5.1.4 Reduction from any number to two variables

11.5.29 Remark Let $f \in F[x_1, \ldots, x_n]$ continue to denote a polynomial in n variables over a field F of total degree d. For any points $(\alpha_1, \ldots, \alpha_n)$, $(\beta_1, \ldots, \beta_n)$ and $(\gamma_1, \ldots, \gamma_n)$ in F^n, we define the bivariate polynomial $f_{\alpha,\beta,\gamma}$ in the variables x and y by $f_{\alpha,\beta,\gamma} := f(\alpha_1 x + \beta_1 y + \gamma_1, \ldots, \alpha_n x + \beta_n y + \gamma_n)$.

11.5.30 Theorem (Bertini's theorem) [2602, Chapter II, Section 6.1] If f is irreducible, then there exists a proper Zariski open subset of $(F^n)^3$ such that $f_{\alpha,\beta,\gamma}$ is irreducible for any triple $(\alpha_1, \ldots, \alpha_n), (\beta_1, \ldots, \beta_n), (\gamma_1, \ldots, \gamma_n)$ in this subset.

11.5.31 Definition For any irreducible factor g of f, a triple $(\alpha_1, \ldots, \alpha_n), (\beta_1, \ldots, \beta_n), (\gamma_1, \ldots, \gamma_n)$ in $(F^n)^3$ is a *Bertinian good point* for g if $g(\alpha_1 x + \beta_1 y + \gamma_1, \ldots, \alpha_n x + \beta_n y + \gamma_n)$ is irreducible with the same total degree as g. In other words, the irreducible factors of f are in one-to-one correspondence with those of $f_{\alpha,\beta,\gamma}$. The complementary set of Bertinian good points is written $\mathcal{B}(f)$ and is the set of *Bertinian bad points.*

11.5.32 Remark For algorithmic purposes, the entries of $(\alpha_1, \ldots, \alpha_n)$, $(\beta_1, \ldots, \beta_n)$ and $(\gamma_1, \ldots, \gamma_n)$ must be taken in a finite subset S of F, so that we are naturally interested in upper bounding the number of Bertinian bad points in $(S^n)^3$. We refer to such a bound as a *quantitative Bertini theorem.* The density of Bertinian bad points with entries in a non-empty finite subset S of F is

$$\mathcal{B}(f, S) := \frac{|\mathcal{B}(f) \cap (S^n)^3|}{|S|^{3n}},$$

where $|S|$ represents the cardinality of S.

11.5.33 Theorem (Quantitative Bertini theorem) [1658, Corollary 2] and [1881, Corollary 8] If F is a perfect field of characteristic p, and according to the above notation, we have that:

1. $\mathcal{B}(f, S) \leq (3d(d-1) + 1)/|S|$ if $p \geq d(d-1) + 1$;
2. $\mathcal{B}(f, S) \leq 2d^4/|S|$ otherwise.

11.5.34 Remark What we call "Bertini's theorem" here is a particular but central case of more general theorems such as in [2602, Chapter II, Section 6.1]. As pointed out by Kaltofen [1658], the special application of Bertini's theorem to reduce the factorization problem from several to two variables was already known by Hilbert [1499, p. 117]. This is why Kaltofen and some authors say "(effective) Hilbert Irreducibility Theorem" instead of "Bertini's theorem." For more historical details about Bertini's work, we refer the reader to [1623, 1750].

11.5.35 Remark Bertini's theorem was introduced in complexity theory by Heintz and Sieveking [1462], and Kaltofen [1645]. It quickly became a cornerstone of many randomized factorization or reduction techniques including [1218, 1229, 1647, 1648, 1649]. Over the field of complex numbers, Bajaj et al. [162] obtained the bound $\mathcal{B}(f,S) \leq (d^4 - 2d^3 + d^2 + d + 1)/|S|$ by following Mumford's proof [2202, Theorem 4.17] of Bertini's theorem. Gao [1177] proved the bound $\mathcal{B}(f,S) \leq 2d^3/|S|$ whenever F has characteristic 0 or larger than $2d^2$. Then Chèze pointed out [616, Chapter 1] that the latter bound can be refined to $\mathcal{B}(f,S) \leq d(d^2-1)/|S|$ by using directly [2504, Satz C]. The paper [1218] contains a version for non-perfect fields with a bound that is exponential in d. If the cardinality $|F|$ is too small, one can switch to an extension (see Remark 11.5.67 below).

11.5.36 Corollary Let $\mathsf{S}(n,d)$ represent a cost function for the product of two power series over a field F in n variables truncated to precision d. Let $f \in \mathbb{F}_q[x_1,\ldots,x_n]$ be a polynomial of total degree d. If $q \geq 4d^4$ then $\mathrm{Irr}(F)$ can be computed with an *expected* number of $O(1)$ factorizations of polynomials in $\mathbb{F}_q[x,y]$ of total degree d, plus an *expected* number of $\tilde{O}(d\mathsf{S}(n-1,d))$ operations in $\mathbb{F}_q$.

11.5.37 Remark Softly optimal series products exist in particular cases [1519], for which the factorization thus reduces to the univariate case in expected softly linear time as soon as $n \geq 3$.

11.5.38 Remark The first deterministic polynomial time multivariate factorization algorithms are due to Kaltofen [1645, 1646]. Kaltofen constructed polynomial-time reductions to bi- (in 1981) and univariate (in 1982) factorization over an abstract field, which were discovered independently of the 1982 univariate factorization algorithm over the rationals by A. K. Lenstra, H. W. Lenstra, and Lovász [1893]. Kaltofen's reduction to univariate factorization, however, was inspired by Zassenhaus's algorithm [3053]. For more references to work by others (Chistov, von zur Gathen, Grigoriev, A. K. Lenstra) that immediately followed, we refer the reader to Kaltofen's surveys [1654, 1655, 1659], and to [1227].

11.5.39 Remark Polynomial factorization over finite fields has been implemented in MAPLE by Bernardin and Monagan [239]. Other practical techniques have been reported in [2299]. At the present time, the most general algorithm is due to Steel [2703]: it handles all coefficient fields being explicitly finitely generated over their prime field, and it has been implemented within the MAGMA computer algebra system [360]. Steel's algorithm actually completes and improves a previous approach investigated by Davenport and Trager [779].

11.5.40 Remark It is possible, via the rank of the Petr matrix or the distinct degree factorization algorithm, to count the number of irreducible factors of a univariate polynomial over a field $\mathbb{F}_q$ of characteristic p in deterministic polynomial time in $\log p$. The same remains true for multivariate polynomials [1182, 1651], but the algorithms are not straightforward. In [1182] a multivariate deterministic distinct degree factorization is presented. There "distinct degree" is with respect to any degree order.

11.5.2 Factoring sparse multivariate polynomials

11.5.41 Remark Let F be a field. A polynomial f in $F[x_1,\ldots,x_n]$ is made of a sum of terms, with each term composed of a coefficient and an exponent seen as a vector in $\mathbb{N}^n$. For any $e = (e_1,\ldots,e_n) \in \mathbb{N}^n$, we let f_e denote the coefficient of the monomial $x_1^{e_1}\cdots x_n^{e_n}$ in f. If a polynomial has only a few of nonzero terms in its dense representation, one prefers to use the following representation.

11.5.42 **Definition** A *sparse representation* of a multivariate polynomial stores the sequence of the nonzero terms as pairs of monomials and coefficients, sorted for instance in reverse lexicographical order.

11.5.43 **Definition** The *support* of f is $\mathrm{Supp}(f) := \{e \in \mathbb{N}^n \mid f_e \neq 0\}$.

11.5.2.1 Ostrowski's theorem

11.5.44 **Definition** The *Minkowski sum* of two subsets Q and R of $\mathbb{R}^n$, written $Q+R$, is $Q+R := \{e+f \mid (e,f) \in Q \times R\}$.

11.5.45 **Definition** A polytope in $\mathbb{R}^n$ is *integral* if all of its vertices are in $\mathbb{Z}^n$. An integral polytope P is *integrally decomposable* if there exists two integral polytopes Q and R such that $P = Q + R$, where both Q and R have at least two points. Otherwise, P is *integrally indecomposable.*

11.5.46 **Definition** The *Newton polytope* of f, written $N(f)$, is the convex hull in $\mathbb{R}^n$ of $\mathrm{Supp}(f)$. The *integral convex hull* of f is the subset of points in $\mathbb{Z}^n$ lying in $N(f)$.

11.5.47 **Theorem** (Ostrowski's theorem) [2338], translated in [2339] If f factors into gh then we have $N(f) = N(g) + N(h)$.

11.5.2.2 Irreducibility tests based on indecomposability of polytopes

11.5.48 **Remark** The previous theorem leads to the following irreducibility test.

11.5.49 **Corollary** (Irreducibility criterion) [1175, p. 507] If $f \in F[x_1, \ldots, x_n]$ is a nonzero polynomial not divisible by any x_i, and if $N(f)$ is integrally indecomposable, then f is irreducible over any algebraic extension of F.

11.5.50 **Theorem** [1175, Theorem 4.2] Let P be an integral polytope in $\mathbb{R}^n$ contained in a hyperplane H and let $e \in \mathbb{Z}^n$ be a point lying outside of H. If $e_1, \ldots, e_k$ are all the vertices of P, then the convex hull of P and e is integrally indecomposable if, and only if, all the entries of $e - e_1$, $e - e_2$, …, $e - e_k$ are coprime.

11.5.51 **Theorem** [1175, Theorem 4.11] Let P be an indecomposable integral polytope in $\mathbb{R}^n$ with at least two points, that is contained in a hyperplane H, and let $e \in \mathbb{R}^n$ be a point outside of H. Let S be any subset of points in $\mathbb{Z}^n$ contained in the convex hull of e and P. Then the convex hull of S and Q is integrally indecomposable.

11.5.2.3 Sparse bivariate Hensel lifting driven by polytopes

11.5.52 **Remark** Let $f \in \mathbb{F}_p[x, y]$ be a polynomial with t nonzero terms and of total degree d such that $t < d$. Let r be a vector in $\mathbb{R}^2$, and let Γ be a subset of edges of $N(f)$ satisfying the following properties:

1. $N(f) \subseteq \Gamma + r\mathbb{R}_{\geq 0}$,
2. each of the two infinite edges of $\Gamma + r\mathbb{R}_{\geq 0}$ contains exactly one point of $N(f)$,

3. no proper subset of Γ satisfies the previous two conditions.

Assume furthermore that:

1. f factorizes into $f = gh$ for two proper factors g and h in $\mathbb{F}_p[x, y]$ with t_g and t_h terms respectively, such that $\max(t_g, t_h) \leq t^\lambda$ for some constant λ satisfying $1/2 \leq \lambda < 1$.
2. For each edge $\gamma \in \Gamma$ we are given polynomials g_γ and h_γ supported by γ_g and γ_h respectively, where γ_g and γ_h are the unique vertices or edges of $N(g)$ and $N(h)$ respectively such that $\gamma = \gamma_g + \gamma_h$.
3. For each edge $\gamma \in \Gamma$ the given polynomials g_γ and h_γ are coprime up to monomial factors.

11.5.53 Theorem [11, Theorem 28] Under the above assumptions, there exists an integral decomposition $N(f) = N(g) + N(h)$ such that $N(g)$ is not a single point or a line segment parallel to $r\mathbb{R}_{\geq 0}$. There exists at most one full factorization of f which extends the boundary factorization defined by the given $(g_\gamma)_{\gamma \in \Gamma}$ and $(h_\gamma)_{\gamma \in \Gamma}$. Assuming that d and p fit a machine word, this factorization can be computed, or shown not to exist, using $O(t^\lambda d^2 + t^{2\lambda} d \log d \log\log d + t^{4\lambda} d)$ bit-operations, and $O(t^\lambda d)$ bits of memory.

11.5.54 Remark Theorem 11.5.53 extends previous results from [10]. Although it does not provide a complete factoring algorithm, it proves to be very efficient in practice for large particular problems.

11.5.2.4 Convex-dense bivariate factorization

11.5.55 Remark In the worst case, the size of the irreducible factorization is exponential in the sparse size of the polynomial f to be factored. However Theorem 11.5.47 ensures that the size of the output is upper bounded by the number π, called the *convex size*, of points in $\mathbb{Z}^n$ lying inside of $N(f)$. The next theorem to be presented reduces the bivariate sparse factorization to the usual dense case.

11.5.56 Definition The *affine group* over $\mathbb{Z}^2$, written $\mathrm{Aff}(\mathbb{Z}^2)$, is the set of the maps U

$$U : (i, j) \mapsto \begin{pmatrix} \alpha & \beta \\ \alpha' & \beta' \end{pmatrix} \begin{pmatrix} i \\ j \end{pmatrix} + \begin{pmatrix} \gamma \\ \gamma' \end{pmatrix}, \qquad (11.5.1)$$

with α, β, γ, α', β', and γ' in $\mathbb{Z}$, such that $\alpha\beta' - \alpha'\beta = \pm 1$.

11.5.57 Definition Let S be a finite subset of $\mathbb{Z}^2$. The set S is *normalized* if it belongs to $\mathbb{N}^2$ and if it contains at least one point in $\{0\} \times \mathbb{N}$, and also at least one point in $\mathbb{N} \times \{0\}$.

11.5.58 Theorem [256, Theorem 1.2] For any normalized finite subset S of $\mathbb{Z}^2$, of cardinality σ, convex size π, and included in $[0, d_x] \times [0, d_y]$, one can compute an affine map $U \in \mathrm{Aff}(\mathbb{Z}^2)$ as in (11.5.1), together with $U(S)$, with $O(\sigma \log^2((d_x+1)(d_y+1)))$ bit-operations, such that $U(S)$ is normalized and contained in a block $[0, d'_x] \times [0, d'_y]$ satisfying $(d'_x+1)(d'_y+1) \leq 9\pi$.

11.5.59 Lemma For any field F, for any $f \in F[x, y]$ not divisible by x and y, for any U as in Equation (11.5.1), the polynomial

$$U(f) := \sum_{(e_x, e_y) \in \mathrm{Supp}(f)} f_{(e_x, e_y)} x^{\alpha e_x + \beta e_y + \gamma} y^{\alpha' e_x + \beta' e_y + \gamma'}$$

is irreducible in $F[x, y, x^{-1}, y^{-1}]$ if and only if f is irreducible.

11.5.60 Remark In order to compute the irreducible factorization of F, we can compute a reduction map U as in Theorem 11.5.58 for Supp(f), then compute the irreducible factorization of $U(f)$, and finally apply U^{-1} to each factor. In this way we benefit from complexity bounds that only depend on the convex size π of f instead of its dense size $(d_x + 1)(d_y + 1)$.

11.5.3 Factoring straight-line programs and black boxes

11.5.61 Remark The sparse representation (see Definition 11.5.42) of a polynomial allows for space efficient storage of polynomials of very high degree, since the degree of the term $x^{2^{500}}$ can be represented by a 501 bit integer. Polynomials f whose sparse representation occupies $(\log(\deg f))^{O(1)}$ bit space are *supersparse* (*lacunary*) [1661, 1897]. While computing small degree factors of such polynomials over the rational numbers can be accomplished in bit time that is polynomial in the input size, over finite fields such tasks are NP- or co-NP-hard [1739]. Here we have a situation where factoring over the rational numbers is provably easier than factoring over a sufficiently large finite field.

11.5.62 Remark In [1660] it is shown, by transferring the construction in [2400], that several other operations on univariate and bivariate supersparse polynomials over a sufficiently large finite field are NP- or co-NP-hard. For instance, in [1660] the following is proven.

11.5.63 Theorem Suppose we have a Monte Carlo polynomial-time irreducibility test for supersparse polynomials in $\mathbb{F}_{2^m}[X, Y]$ for sufficiently large m. Then large integers can be factored in Las Vegas polynomial-time.

11.5.64 Remark A polynomial in n variables of (total) degree d can have $\binom{n+d}{n}$ terms, i.e., exponentially many terms in the number of variables. Sparse polynomials are those that have $(n + d)^{O(1)}$ non-zero terms. Note that, as in all asymptotic analysis, one considers not a single polynomial but an infinite set of sparse input polynomials that a given algorithm processes, now in polynomial time in the sparse size. By using the factorization $x^d - 1 = (x - 1)(x^{d-1} + \cdots + 1)$ one can easily generate examples where the sparse size of one irreducible factor is super-polynomially larger than the input size [1230, Example 5.1]. Motivated by algebraic computation models, straight-line programs were adopted as an alternate polynomial representation, first only for inputs [1218], but ultimately and importantly as a representation of the irreducible factors themselves [1650, 1653]. Here is an example of a division-free straight-line program (single assignment program), where $\mathbb{F}$ is the field generated by those operands $c_1, c_2, \ldots$ which are constants, while $x_1, x_2, \ldots$ are input variables:

$$
\begin{aligned}
v_1 &\leftarrow c_1 \times x_1;\\
v_2 &\leftarrow x_2 - c_2;\\
v_3 &\leftarrow v_2 \times v_2;\\
v_4 &\leftarrow v_3 + v_1;\\
v_5 &\leftarrow v_4 \times x_3;\\
&\vdots\\
v_{101} &\leftarrow v_{100} + v_{51};
\end{aligned}
$$

The variable v_{101} represents a polynomial in $\mathbb{F}[x_1, x_2, \ldots]$, which can be evaluated by use of the straight-line program. For instance, the determinant of an $n \times n$ matrix whose entries are n^2 variables can be represented, via Gaussian elimination, by a straight-line program with divisions of length $O(n^3)$. Because those divisions can cause divisions by zero on evaluation at certain points, it is desirable to remove them from such programs [2733]: the shortest division-free straight-line program for the determinant that is known today has length $O(n^{2.7})$ and uses no constants other than 1 and -1 in $\mathbb{F}$ [1670]. In any case, divisions can be removed by increasing the length by a factor $O((\deg f)^{1+\epsilon})$ for any $\epsilon > 0$. The 1986 algorithm in [1650, 1653] produces from a straight-line program of length l for a polynomial of degree d in Monte Carlo random polynomial-time straight-line programs for the irreducible factors (and their multiplicities). The factor programs themselves have length $O(d^2 l + d^{3+\epsilon})$. Over finite fields of characteristic p, for an irreducible factor g of multiplicity $p^m m'$, where $\gcd(p, m') = 1$, a straight-line program for g^{p^m} is returned; see Remark 11.5.68 below. The algorithm is implemented in the Dagwood system [1100] and can factor matrix determinants. A shortcoming of the straight-line representations, which later were adopted by the TERA project, was exposed by the Dagwood program: the lengths, while polynomial in the input lengths, become quite large (over a million assignments). The construction, however, plays a key role in complexity theory [1640].

11.5.65 Remark Since polynomials represented by straight-line programs can be converted to sparse polynomials in polynomial-time in their sparse size by the algorithm in [3080], the straight-line factorization algorithm brought to a successful conclusion the search for polynomial-time sparse factorizers. Previous attempts based on sparse Hensel lifting [1216, 1230, 1649, 3080, 3081], retained an exponential substep for many factors, namely the computation of the so-called right-side Hensel correction coefficients. The problem of computing the coefficient of a given term in a sparse product is in general #P-hard. Nonetheless, if a polynomial has only a few sparse factors, such sparse lifting can be quite efficient, in practice.

11.5.66 Remark Instead of straight-line programs, one can use a full-fledged programmed procedure that evaluates the input polynomial. The irreducible factors are then evaluated at values for the variables by another procedure that makes ("oracle") calls to the input evaluation procedure. Thus is the genesis of algorithms for *black box* polynomials [1668, 1669]..

The idea is the following: Suppose one can call a black evaluation box for the polynomial $f(x_1, \ldots, x_n) \in \mathbb{F}[x_1, \ldots, x_n]$. First, uniformly randomly select from a sufficiently large finite set field elements a_i, c_i $(2 \le i \le n)$ and b_j $(1 \le j \le n)$ and interpolate and factor the bivariate image

$$\hat{f}(X, Y) = f(X + b_1, c_2 Y + a_2 X + b_2, \ldots, c_n Y + a_n X + b_n) = \prod_{k=1}^{r} \hat{g}_k(X, Y)^{e_k}.$$

By the effective Hilbert Irreducibility Theorem 11.5.33 above, the irreducible polynomials $\hat{g}_k$ are with high probability bivariate images of the irreducible factors $h_k(x_1, \ldots, x_n)$ of f. For small coefficient fields we shall assume that the black box can evaluate f at elements in a finite algebraic extension E of $\mathbb{F}$. Already the bivariate interpolation algorithm may require such an extension in order to have sufficiently many distinct points.

11.5.67 Remark If one selects an extension E of degree $[E : \mathbb{F}] > \deg(f)$ that is a prime number, all h_k remain irreducible over that extension. Indeed, the Frobenius norm $\mathrm{Norm}_{E/\mathbb{F}}(\tilde{h}) \in \mathbb{F}[x_1, \ldots, x_n]$ of a possible non-trivial irreducible factor $\tilde{h} \in E[x_1, \ldots, x_n]$ of an h_k must be a power of an irreducible polynomial over $\mathbb{F}$, hence a power of h_k itself. For otherwise $\gcd(h_k, \mathrm{Norm}_{E/\mathbb{F}}(\tilde{h}))$ would constitute a non-trivial factor of h_k over $\mathbb{F}$. But then $\deg(\tilde{h}) \cdot [E : \mathbb{F}] = \deg(h_k) \cdot m$, where m is the exponent of that power, and because $[E : \mathbb{F}]$ is a prime $> \deg(f) \ge \deg(h_k)$, we obtain the contradiction $\deg(\tilde{h}) = \deg(h_k) \cdot (m/[E : \mathbb{F}]) \ge \deg(h_k)$.

Remark 11.5.66 continued. Now the black box for evaluating all $h_k(\xi_1, \ldots, \xi_n)$ at field elements $\xi_i \in \mathbb{F}$ stores ("hard-wires") the a_i, b_j and the factors $g_k(X) = \hat{g}_k(X, 0)$ in its constant pool. We note that the g_k are not necessarily irreducible, but with high probability they are pairwise relatively prime [1669, Section 2, Step 3], and their leading terms only depend on the variable X. The black box first interpolates

$$\begin{aligned}\bar{f}(X, Y) = f(&X + b_1, Y(\xi_2 - a_2(\xi_1 - b_1) - b_2) + a_2 X + b_2,\\ &\ldots, Y(\xi_n - a_n(\xi_1 - b_1) - b_n) + a_n X + b_n)\end{aligned} \tag{11.5.2}$$

and then factors $\bar{f}$ such that

$$\bar{f}(X,Y) = \prod_{k=1}^{r} \bar{h}_k(X,Y)^{e_k} \quad \text{with} \quad \bar{h}_k(X,0) = g_k(X). \tag{11.5.3}$$

We note that again the $\bar{h}_k$ are not necessarily irreducible. One may Hensel-lift the factorization

$$f(X + b_1, a_2X + b_2, \ldots, a_nX + b_n) = \prod_{k=1}^{r} g_k(X)^{e_k} \tag{11.5.4}$$

provided none of the multiplicities e_k is divisible by p. Otherwise, one can fully factor $\bar{f}(X,Y)$ and lump (multiply) those irreducible factors $\bar{h}_\kappa(X,Y)$ together where $\bar{h}_\kappa(X,0)$ divide one and the same $g_k(X)$. Alternatively, if p^m divides e_k one could lift the p^m-th power of g_k and take a p^m-th root of the lifted factor. We have $\bar{f}(\xi_1 - b_1, 1) = f(\xi_1, \ldots, \xi_n)$, and for all k we obtain $\bar{h}_k(\xi_1 - b_1, 1) = h_k(\xi_1, \ldots, \xi_n)$. We observe that the scalar multiple of h_k is fixed in all evaluations by the choice of g_k.

11.5.68 Remark Over finite coefficient fields, there is no restriction on the multiplicities e_k. One does not obtain a pure straight-line program for the polynomial h_k because a bivariate factorization of $\bar{f}$ or a p^m-th root of the lifted factor, which depend on the evaluation points ξ_i, are performed on each evaluation. One can obtain straight-line polynomials that equal the irreducible factors modulo $(x_1^q - x_1, \ldots, x_n^q - x_n)$ by powering by q/p^m, where $q < \infty$ is the cardinality of the coefficient field. Those straight-line programs produce correct evaluations of the irreducible factors.

11.5.69 Remark The blackbox factorization algorithm is implemented in the FoxBox system [830]. The size blowup experienced in the straight-line factorization algorithm does not occur. In fact, the factor evaluation black box makes $O(\deg(f)^2)$ calls to the black box for f and factors a bivariate polynomial, either by lifting (11.5.4) or, if multiplicities are divisible by the characteristic, by factoring $\bar{f}$. The program is fixed except for the constants a_i, b_j and the polynomials g_k.

11.5.70 Remark We conclude that the sparse representations of the factors can be recovered by sparse interpolation over a finite field; see [1662] and the literature cited there. Dense factors can be identified to have more than a given number of terms and skipped.

See Also

§3.6	For irreducible multivariate polynomials.
§8.1	Where absolute factorization intervenes for testing if a univariate rational function generates a permutation of a finite field as in the algorithms of [1716, 1983].

References Cited: [10, 11, 162, 219, 237, 238, 239, 256, 360, 365, 616, 617, 761, 778, 779, 830, 1100, 1175, 1177, 1182, 1183, 1216, 1217, 1218, 1227, 1229, 1230, 1271, 1462, 1499, 1518, 1519, 1616, 1623, 1640, 1645, 1646, 1647, 1648, 1649, 1650, 1651, 1653, 1654, 1655, 1658, 1659, 1660, 1661, 1662, 1668, 1669, 1670, 1716, 1739, 1750, 1880, 1881, 1882, 1883, 1893, 1897, 1983, 2108, 2129, 2130, 2131, 2202, 2209, 2299, 2338, 2339, 2400, 2504, 2505, 2528, 2529, 2530, 2542, 2602, 2703, 2730, 2733, 2938, 2939, 3021, 3052, 3053, 3080, 3081]

11.6 Discrete logarithms over finite fields

Andrew Odlyzko, *University of Minnesota*

Surveys and detailed expositions with proofs can be found in [661, 1939, 1938, 2080, 2308, 2306, 2720].

11.6.1 Basic definitions

11.6.1 Remark Discrete exponentiation in a finite field is a direct analog of ordinary exponentiation. The exponent can only be an integer, say n, but for w in a field F, w^n is defined except when $w = 0$ and $n \leq 0$, and satisfies the usual properties, in particular $w^{m+n} = w^m w^n$ and (for u and v in F) $(uv)^m = u^m v^m$. The discrete logarithm is the inverse function, in analogy with the ordinary logarithm for real numbers. If F is a finite field, then it has at least one primitive element g; i.e., all nonzero elements of F are expressible as powers of g, see Chapter 2.

11.6.2 Definition Given a finite field F, a primitive element g of F, and a nonzero element w of F, the *discrete logarithm* of w to base g, written as $\log_g(w)$, is the least non-negative integer n such that $w = g^n$.

11.6.3 Remark The value $\log_g(w)$ is unique modulo $q-1$, and $0 \leq \log_g(w) \leq q-2$. It is often convenient to allow it to be represented by any integer n such that $w = g^n$.

11.6.4 Remark The discrete logarithm of w to base g is often called the *index* of w with respect to the base g. More generally, we can define discrete logarithms in groups. They are commonly called *generic discrete logarithms*.

11.6.5 Definition If G is a group (with multiplication as group operation), and g is an element of G of finite order m, then for any element h of $\langle g \rangle$, the cyclic subgroup of G generated by g, the *discrete logarithm* of h to base g, written as $\log_g(h)$, is the least non-negative integer n such that $h = g^n$ (and therefore $0 \leq \log_g(h) \leq m-1$).

11.6.6 Remark The definition of a group discrete logarithm allows for consideration of discrete logarithms in finite fields when the base g is not primitive, provided the argument is in the group $\langle g \rangle$. This situation arises in some important applications, in particular in the U.S. government standard for the Digital Signature Algorithm (DSA). DSA operations are performed in a field $\mathbb{F}_p$ with p a prime (nowadays recommended to be at least 2048 bits). This prime p is selected so that $p-1$ is divisible by a much smaller prime r (specified in the standard to be of 160, 224, or 256 bits), and an element h of $\mathbb{F}_p$ is chosen to have multiplicative order r (say by finding a primitive element g of $\mathbb{F}_p$ and setting $h = g^{(p-1)/r}$). The main element of the signature is of the form h^s for an integer s, and ability to compute s would break DSA. DSA can be attacked either by using generic finite group discrete logarithm algorithms in the group $\langle h \rangle$ or finite field algorithms in the field $\mathbb{F}_p$ (which can then easily yield a solution in $\langle h \rangle$).

11.6.7 Remark The basic properties of discrete logarithms given below, such as the change of base formula, apply universally. On the other hand, many of the discrete logarithm algorithms described later are valid only in finite fields. Generally speaking, discrete logarithms are comparatively easy to compute in finite fields, since they have a rich algebraic structure that can be exploited for cryptanalytic purposes. Much of the research on discrete logarithms in other settings has been devoted to embedding the relevant groups inside finite fields in order to apply finite field discrete logarithm algorithms.

11.6.8 Remark This section is devoted to finite field discrete logarithms, and only gives a few references to other ones. For elliptic curve discrete logarithms, the most prominent collection, see Section 16.4. However, other groups have also been used, for example class groups of number fields [2044].

11.6.2 Modern computer implementations

11.6.9 Remark Most popular symbolic algebra systems contain some implementations of discrete logarithm algorithms. For example, Maple has the *mlog* function, while Mathematica has *FieldInd*. More specialized systems for number theoretic and algebraic computations, such as Magma, PARI, and Sage, also have implementations, and typically can handle larger problems. Thus for all but the largest problems that are at the edge of computability with modern methods, widely available and easy to use programs are sufficient. Tables of finite fields, such as those in [1939], are now seldomly printed in books.

11.6.3 Historical remarks

11.6.10 Remark Until the mid-1970s, the main applications for discrete logarithms were similar to those of ordinary logarithms, namely in routine computations, but this time in finite fields. They allowed replacement of relatively hard multiplications by easier additions. What was frequently used was *Zech's logarithm* (also called *Jacobi's logarithm*, cf. [1939]), which is a modification of the ordinary discrete logarithm. In a finite field F with primitive element g, Zech's logarithm of an integer n is defined as the integer $Z(n) \bmod (q-1)$ which satisfies $g^{Z(n)} = 1 + g^n$. This provides a quick way to add elements given in terms of their discrete logarithms: aside from boundary cases, $g^m + g^n = g^m(1 + g^{n-m}) = g^{m+Z(n-m)}$.

11.6.11 Remark As with ordinary logarithms, where slide rules and logarithm tables have been replaced by calculators, such routine applications of discrete logarithms in small or moderately large fields now rely on computer algebra systems.

11.6.12 Remark Interest in discrete logarithms jumped dramatically in the mid-1970s with the invention of public key cryptography, see Chapter 16. While discrete exponentiation is easy, the discrete logarithm, its inverse, appeared hard, and this motivated the invention of the Diffie–Hellman key exchange protocol, the first practical public key cryptosystem. Efficient algorithms for discrete logarithms in the field over which this protocol is implemented would make it insecure.

11.6.13 Remark The Diffie–Hellman problem is to compute g^{xy}, the key that the two parties to the Diffie–Hellman protocol obtain, from the g^x and g^y that are visible to the eavesdropper. Although this problem has attracted extensive attention, it has not been solved, and for the most important cases of finite field and elliptic curve discrete logarithms, it is still unknown whether the Diffie–Hellman problem is as hard as the discrete logarithm one; see [410] for recent results and references.

11.6.14 Remark It is known that single bits of discrete logarithms are about as hard to compute as the entire discrete logarithms [1444].

11.6.15 Remark There are some rigorous lower bounds on discrete log problems, but only for groups given in ways that restrict what can be done in them [2219, 2630].

11.6.16 Remark Various cryptosystems other than the Diffie-Hellman one have been proposed whose security similarly depends on the intractability of the discrete logarithm problem; see Section 16.1. Many of them can be used in settings other than finite fields.

11.6.17 Remark There are close analogies between integer factorization and discrete logarithms in finite fields, and most (but not all) of the algorithms in one area have similar ones in the other. This will be seen from some of the references later. In general, considerably less attention has been devoted to discrete logarithms than to integer factorization. Hence the smaller sizes of discrete logarithm problems that have been solved result both from the greater technical difficulty of this problem as compared to integer factorization and from less effort being devoted to it.

11.6.18 Remark Shor's 1994 result [2623] shows that if quantum computers become practical, discrete logarithms will become easy to compute. Therefore cryptosystems based on discrete logarithms may all become suddenly insecure.

11.6.4 Basic properties of discrete logarithms

11.6.19 Remark Suppose that G is a group, and g an element of finite order m in G. If u and v are two elements of $\langle g \rangle$, then

$$\log_g(uv) \equiv \log_g(u) + \log_g(v) \pmod{m},$$

$$\log_g(u^{-1}) \equiv -\log_g(u) \pmod{m}.$$

11.6.20 Remark (Change of base formula) Suppose that G is a group, and that g and h are two elements of G that generate the same cyclic subgroup $\langle g \rangle = \langle h \rangle$ of order m. If u is an element of $\langle g \rangle$, then

$$\log_g(u) \equiv \log_h(u) * \log_g(h) \pmod{m},$$

and therefore

$$\log_g(h) \equiv 1/\log_h(g) \pmod{m}.$$

These formulas mean that one can choose the most convenient primitive element to work with in many applications. For example, in finite fields $\mathbb{F}_{2^k}$, elements are usually represented as polynomials with binary coefficients, and one can find (as verified by experiment and inspired by heuristics, but not proved rigorously) primitive elements that are represented as polynomials of very low degree. This can offer substantial efficiencies in implementations. However, it does not affect the security of the system. If discrete logarithms are easy to compute in one base, they are easy to compute in other bases. Similarly, the change of the irreducible polynomial that defines the field has little effect on the difficulty of the discrete logarithm problem.

11.6.5 Chinese Remainder Theorem reduction: The Silver–Pohlig–Hellman algorithm

11.6.21 Remark If the order of the element g can be factored even partially, the discrete logarithm problem reduces to easier ones. This is the Silver–Pohlig–Hellman technique [2406]. Suppose that g is an element of finite order m in a group G, and m is written as $m = m_1 m_2$ with $\gcd(m_1, m_2) = 1$. Then the cyclic group $\langle g \rangle$ is the direct product of the cyclic groups $\langle g^{m_2} \rangle$ and $\langle g^{m_1} \rangle$ of orders m_1 and m_2, respectively. If we determine $a = \log_{g^{m_2}}(w^{m_2})$ and

$b = \log_{g^{m_1}}(w^{m_1})$, the Chinese Remainder Theorem tells us that $\log_g(w)$ is determined completely, and in fact we obtain

$$\log_g(w) \equiv b * x * m_1 + a * y * m_2 \pmod{m},$$

where x and y come from the Euclidean algorithm computation of $\gcd(m_1, m_2)$, namely $1 = xm_1 + ym_2$. This procedure extends easily to more than two relatively prime factors.

11.6.22 Remark When m, the order of g, is a prime power, say $m = p^k$, the computation of $\log_g(w)$ reduces to k discrete logarithm computations in a cyclic group of p elements. For example, if $r = p^{k-1}$ and $h = g^r$, $u = w^r$, then h has order p, and computing $\log_h(u)$ yields the reduction of $\log_g(w) \pmod{p}$. This process can then be iterated to obtain the reduction modulo p^2, and so on.

11.6.23 Remark The above remarks, combined with results of the next section, show that when the complete factorization of the order of g can be obtained, discrete logarithms can be computed in not much more than $r^{1/2}$ operations in the group, where r is the largest prime in the factorization.

11.6.24 Remark In a finite field, any function can be represented by a polynomial. For the discrete logarithm, such polynomials do turn out to have some aesthetically pleasing properties, see [2189, 2244, 2922, 2968]. However, so far they have turned out to be of no practical use whatsoever.

11.6.6 Baby steps–giant steps algorithm

11.6.25 Remark We next consider some algorithms for discrete logarithms that work in very general groups. The basic one is the baby steps–giant steps method that combines time and space, due to Shanks [2607].

11.6.26 Algorithm Baby steps–giant steps algorithm: Suppose that G is a group and g is an element of G of finite order m. If $h \in \langle g \rangle$, $h = g^k$, and $w = \lceil m^{1/2} \rceil$, then k can be written as $k = aw+b$ for some (often non-unique) a, b with $0 \le a,\ b < w$. To find such a representation, compute the set $A = \{g^{jw} : 0 \le j < w\}$ and sort it. This takes $m^{1/2} + O(\log(m))$ group operations and $O(m^{1/2} \log(m))$ sorting steps, which are usually very easy, since they can be performed on bit strings, or even initial segments of bit strings. Next, for $0 \le i < w$, compute hg^{-i} and check whether it is present in A. When it is, we obtain the desired representation $k = jw+i$.

11.6.27 Remark The baby steps–giant steps technique has the advantage of being fully deterministic. Its principal disadvantage is that it requires storage of approximately $m^{1/2}$ group elements. A space-time tradeoff is available, in that one can store a smaller list (the set A in the notation above, with fewer but larger "giant steps") but then have to do more computing (more "baby steps").

11.6.28 Remark The baby steps–giant steps algorithm extends easily to many cases where the discrete logarithm is restricted in some way. For example, if it is known that $\log_g(w)$ lies in an interval of length n, the basic approach sketched above can be modified to find it in $O(n^{1/2})$ group operations (plus the usual sorting steps). Similarly, if the discrete logarithm k is allowed to have only small digits when represented in some base (say binary digits in base 10), then the running time will be about the square root of the number of possibilities for k. For some other recent results, see [2708].

11.6.7 Pollard rho and kangaroo methods for discrete logarithms

11.6.29 Remark In 1978, Pollard invented two randomized methods for computing discrete logarithms in any group, the rho method, and the kangaroo (or lambda) technique [2413]. Just like Pollard's earlier rho method for integer factorization, they depend on the birthday paradox, which says that if one takes a (pseudo) random walk on a completely connected graph of n vertices, one is very likely to revisit the same vertex in about $n^{1/2}$ steps. These discrete logarithm algorithms also depend, just as the original rho method does, on the Floyd algorithm (Section 3.1 of [1765]) for detecting cycles with little memory at some cost in running time, in that they compare x_{2i} to x_i, where x_i is the position of the random walk at time i.

11.6.30 Remark Since the rho and kangaroo methods for discrete logarithms are probabilistic, they cannot guarantee a solution, but heuristics suggest, and experiments confirm, that both run in expected time $O(m^{1/2})$, where m is the order of the group. This is the same computational effort as for the baby steps–giant steps algorithm. However, the rho and kangaroo methods have two advantages. One is that they use very little memory. Another one is that, as was first shown by van Oorschot and Wiener [2852], they can be parallelized, with essentially linear speedup, so that k processors find a solution about k times faster than a single one. We sketch just the standard version of the rho method, and only briefly.

11.6.31 Algorithm Rho algorithm for discrete logarithms: Partition the group $\langle g \rangle$ of order m into three roughly equal sets S_1, S_2, and S_3, using some property that is easy to test, such as the first few bits of a canonical representation of the elements of G. To compute $\log_g(h)$, define a sequence $w_0, w_1, \ldots$ by $w_0 = g$ and for $i > 0$, $w_{i+1} = w_i{}^2, w_i g$, or $w_i h$, depending on whether $w_i \in S_1, S_2$, or S_3. Then each w_i is of the form

$$w_i = g^{a_i} h^{b_i}$$

for some integers a_i, b_i. If the procedure of moving from w_i to w_{i+1} behaves like a random walk (as is expected), then in $O(m^{1/2})$ steps we will find i such that $w_i = w_{2i}$, and this will give a congruence

$$a_i + b_i \log_g(h) \equiv a_{2i} + b_{2i} \log_g(h) \pmod{m}.$$

Depending on the greatest common divisor of m and $b_i - b_{2i}$ this congruence will typically either yield $\log_g(h)$ completely, or give some stringent congruence conditions, which with the help of additional runs of the algorithm will provide a complete solution.

11.6.32 Remark The low memory requirements and parallelizability of the rho and kangaroo algorithms have made them the methods of choice for solving general discrete logarithm problems. There is a substantial literature on various modifications, although they do not improve too much on the original parallelization observations of [2852]. Some references are [613, 1734, 2414, 2790].

11.6.33 Remark The rho method, as outlined above, requires knowledge of the exact order m of the group. The kangaroo method only requires an approximation to m. The kangaroo algorithm can also be applied effectively when the discrete logarithm is known to lie in a restricted range.

11.6.8 Index calculus algorithms for discrete logarithms in finite fields

11.6.34 Remark The rest of this section is devoted to a brief overview of index calculus algorithms for discrete logarithms. Unlike the Shanks and Pollard methods of the previous two subsections, which take exponential time, about $m^{1/2}$ for a group of order m, the index calculus techniques are subexponential, with running times closer to $\exp((\log(m))^{1/2})$ and even $\exp((\log(m))^{1/3})$. However, they apply directly only to finite fields. That is why much of the research on discrete logarithms in other groups of cryptographic interest, such as on elliptic curves, is devoted to finding ways to reduce those problems to ones in finite fields.

11.6.35 Remark In the case of DSA mentioned at the beginning of this section, the recommended size of the modulus p has increased very substantially, from 512 to 1024 bits when DSA was first adopted, to the range of 2048 to 3072 bits more recently. The FIPS 186-3 standard specifies bit lengths for the two primes p and r of (1024, 160), (2048, 224), (2048, 256), and (3072, 256). The relative sizes of p and r were selected to offer approximately equal levels of security against index calculus algorithms (p) and generic discrete logarithm attacks (r). The reason for the much faster growth in the size of p is that with the subexponential running time estimates, the effect of growing computing power is far more pronounced on the p side than on the r side. In addition, while there has been no substantial theoretical advance in index calculus algorithms in the last two decades, there have been numerous small incremental improvements, several cited later in more detailed discussions. On the other hand, there has been practically no progress in generic discrete logarithm algorithms, except for parallelization.

11.6.36 Remark The basic idea of index calculus algorithms dates back to Kraitchik, and is also key to all fast integer factorization algorithms. In a finite field $\mathbb{F}_q$ with primitive element g, if we find some elements $x_i, y_j \in \mathbb{F}_q$ such that

$$\prod_{i=1}^{r} x_i = \prod_{j=1}^{s} y_j,$$

then

$$\sum_{i=1}^{r} \log_g x_i \equiv \sum_{j=1}^{s} \log_g y_j \pmod{q-1}.$$

If enough equations are collected, this linear system can be solved for the $\log_g x_i$ and $\log_g y_j$. Singular systems are not a problem in practice, since typically computations generate considerably more equations than unknowns, and one can arrange for g itself to appear in the multiplicative relations.

11.6.37 Remark To compute $\log_g w$ for some particular $w \in F$ with index calculus algorithms, it is often necessary to run a second stage that produces a relation involving w and the previously computed discrete logarithms. In some algorithms the second stage is far easier than the initial computation, in others it is of comparable difficulty.

11.6.38 Remark For a long time (see [2306] for references), the best index calculus algorithms for both integer factorization and discrete logarithms had running times of the form

$$\exp((c + o(1))(\log q)^{1/2}(\log\log q)^{1/2}) \quad \text{as} \quad p \to \infty$$

for various constants $c > 0$, where q denotes the integer being factored or the size of the finite field. The first practical method that broke through this running time barrier was Coppersmith's algorithm [717] for discrete logarithms in fields of size $q = 2^k$ (and more generally, of size $q = p^k$ where p is a small prime and k is large). It had running time of approximately

$$\exp(C(\log q)^{1/3}(\log\log q)^{2/3}),$$

where the C varied slightly, depending on the distance from k to the nearest power of p, and in the limit as $k \to \infty$ it oscillated between two bounds [2306]. The function field sieve of

Adleman [16], which also applies to fields with $q = p^k$ where p is relatively small, improves on the Coppersmith method, but has similar asymptotic running time estimate. For the latest results on its developments, see [1626, 1628, 2543].

11.6.39 Remark The running time of Coppersmith's algorithm turned out to also apply to the number field sieve. This method, which uses algebraic integers, was developed for integer factorization by Pollard and H. Lenstra, with subsequent contributions by many others. It was adopted for discrete log computations in prime fields by Gordon [1325], with substantial improvements by other reseachers. For the latest estimates and references, see [711, 1627, 2544, 2545].

11.6.9 Smooth integers and smooth polynomials

11.6.40 Remark The index calculus algorithms depend on a multiplicative splitting of some elements, such as integers or polynomials, into such elements drawn from a smaller collection. This smaller collection usually is made up of elements that by some measure (norm) are small. The essence of index calculus algorithms is to select general elements from the large set at random, but as intelligently as possible in order to maximize the chances they will have the desired type of splitting. Usually elements that do have such splittings are called "smooth."

11.6.41 Remark There are rigorous analyses that provide estimates of how often elements in various domains are "smooth." For ordinary integers, there are the estimates of [1501]. For algebraic integers, we can use [441, 2574]. For polynomials over finite fields, recent results are [2348].

11.6.10 Sparse linear systems of equations

11.6.42 Remark Index calculus algorithms for discrete logarithms require the solution of linear equations modulo $q - 1$, where q is the size of the field. As in the Silver–Pohlig–Hellman method, the Chinese Remainder Theorem (and an easy reduction of the case of a power of a prime to that of the prime itself) reduces the problem to that of solving the system modulo primes r that divide $q - 1$. (For more extensive discussion of linear algebra over finite fields, see Section 13.4.)

11.6.43 Remark The linear algebra problems that arise in index calculus algorithms for integer factorization are very similar, but simpler, in that they are all just modulo 2. For discrete logarithm problems to be hard, they have to be resistant to the Silver–Pohlig–Hellman attack. Hence $q - 1$ has to have at least one large prime factor r, and so the linear system has to be solved modulo a large prime. That increases the complexity of the linear solution computation, and thus provides slightly higher security for discrete logarithm cryptosystems.

11.6.44 Remark A key factor that enables the solution of the very large linear systems that arise in index calculus algorithms is that these systems are very sparse. (Those "smooth" elements do not involve too many of the "small" elements in the multiplicative relations.) Usually the structured Gaussian elimination method (proposed in [2306] and called there intelligent gaussian elimination, afterwards renamed in the first practical demonstration of it [1839], now sometimes called filtering) is applied first. It combines the relations in ways that reduce the system to be solved and do not destroy the sparsity too far. Then the conjugate gradient, the Lanczos, or the Wiedemann methods (developed in [721, 2976], the first two demonstrated in practice in [1839]) that exploit sparsity are used to obtain the final solution.

11.6.45 Remark For the extremely large linear systems that are involved in record-setting computations, distributed computation is required. The methods of choice, once structured Gaussian elimination is applied, are the block Lanczos and block Wiedemann methods [718, 719, 2134].

11.6.46 Remark Some symbolic algebra systems incorporate implementations of the sparse linear system solvers mentioned above.

11.6.47 Remark As a demonstration of the effectiveness of the sparse methods, the record factorization of RSA768 [1753], mentioned below, produced 64 billion linear relations. These were reduced, using structured gaussian elimination, to a system of almost 200 million equations in about that many unknowns. This system was still sparse, with the average equation involving about 150 unknowns. The block Wiedemann method was then used to solve the resulting system.

11.6.11 Current discrete logarithm records

11.6.48 Remark Extreme caution should be exercised when drawing any inferences about relative performance of various integer factorization and discrete logarithm algorithms from the record results listed here. The computing resources, as well as effort involved in programming, varied widely among the various projects.

11.6.49 Remark As of the time of writing (early 2012), the largest cryptographically hard integer (i.e., one that was chosen specifically to resist all known factoring attacks, and is a product of two roughly equal primes) that has been factored is RSA768, a 768-bit (232 decimal digit) integer from the RSA challenge list [1753]. This was the result of a large collaboration across the globe stretching over more than two years, and used the general number field sieve.

11.6.50 Remark The largest discrete logarithm case for a prime field $\mathbb{F}_p$ (with p chosen to resist simple attacks) that has been solved is for a 530-bit (160 decimal digit) prime p. This was accomplished by Kleinjung in 2007 [1752]. The number field sieve was used.

11.6.51 Remark In fields of characteristic two, the largest case that has been solved is that of $\mathbb{F}_q$ with $q = 2^{613}$, using the function field sieve. (An earlier record was for $q = 2^{607}$ using the Coppersmith algorithm.) This computation took several weeks on a handful of processors, and was carried out by Joux and Lercier in 2005 [1625].

11.6.52 Remark The largest generic discrete logarithm problem that has been solved in a hard case is that of discrete logarithms over an elliptic curve modulo a 112-bit prime, thus a group of size about 2^{112}. This is due to Bos and Kaihara [353], and was done in 2009. Right now, a large multi-year collaborative effort is under way to break the Certicom ECC2K-130 challenge, which involves computing discrete logarithms on an elliptic curve over a field with 2^{131} elements [160]. All these efforts rely on parallelized versions of the Pollard rho method.

See Also

Chapter 2	For basic properties of finite fields.
§11.1	For basic computational techniques in finite fields.
§13.4	For linear algebra over finite fields.
Chapter 16	For public key cryptographic systems.

References Cited: [16, 160, 353, 410, 441, 613, 661, 711, 717, 718, 719, 721, 1325, 1444, 1501, 1625, 1626, 1627, 1628, 1734, 1752, 1753, 1765, 1839, 1938, 1939, 2044, 2080, 2134, 2189, 2219, 2244, 2306, 2308, 2348, 2406, 2413, 2414, 2543, 2544, 2545, 2574, 2607, 2623, 2630, 2708, 2720, 2790, 2852, 2922, 2968, 2976]

11.7 Standard models for finite fields

Bart de Smit, Universiteit Leiden
Hendrik Lenstra, Universiteit Leiden

11.7.1 **Definition** Let p be a prime number and let n be a positive integer. An *explicit model* for a finite field of size p^n is a field whose underlying additive group is $\mathbb{F}_p^n = \mathbb{F}_p \times \mathbb{F}_p \times \cdots \times \mathbb{F}_p$.

11.7.2 **Remark** Let $e_0, \ldots, e_{n-1}$ be the standard $\mathbb{F}_p$-basis of $\mathbb{F}_p^n$. Then a field structure on $\mathbb{F}_p^n$ is uniquely determined by the n^3 elements $a_{ijk} \in \mathbb{F}_p$ such that

$$e_i \cdot e_j = \sum_{k=0}^{n-1} a_{ijk} e_k.$$

Thus, one can specify an explicit model using $O(n^3 \log p)$ bits. When we say that for an algorithm an explicit model is input or output we assume that the explicit data consisting of p and (a_{ijk}) are given as input or output. There is a deterministic polynomial time algorithm that given such data first decides whether p is prime [43], and then decides whether it defines an explicit model for a finite field [1895, Section 2].

11.7.3 **Remark** Let A be a field of characteristic $p > 0$ and size p^n and let $b_0, \ldots, b_{n-1}$ be a basis of A as a vector space over $\mathbb{F}_p$. We then obtain an explicit model as follows. Write ψ for the unique $\mathbb{F}_p$-vector space isomorphism $\mathbb{F}_p^n \to A$ sending e_i to b_i for $0 \le i < n$. Define a multiplication map on $\mathbb{F}_p^n$ by $v \cdot w = \psi^{-1}(\psi(v) \cdot \psi(w))$, for $v, w \in \mathbb{F}_p^n$. Together with vector addition, this multiplication makes $\mathbb{F}_p^n$ into a field.

11.7.4 **Remark** An alternative space-efficient way to give explicit models is to give an irreducible polynomial $f \in \mathbb{F}_p[x]$ of degree n over $\mathbb{F}_p$, which we can encode with $O(n \log p)$ bits. Using the $\mathbb{F}_p$-basis $1, x, \ldots, x^{n-1}$ of the field $\mathbb{F}_p[x]/(f)$ we then obtain an explicit model of a field of size p^n as in Remark 11.7.3. One can convert between this representation and the representation with n^3 elements by deterministic polynomial time algorithms [1895, Theorem 1.1]. Since our concern in this section is only about whether an algorithm runs in polynomial time or not, and not about the degree of the polynomial if it does, we use the more flexible setup of the explicit data consisting of the n^3 elements a_{ijk} in $\mathbb{F}_p$.

11.7.5 **Theorem** [789] There is a deterministic polynomial time algorithm such that

1. on input two explicit models for finite fields A, B of the same cardinality, it produces a field isomorphism $\phi_{A,B}\colon A \to B$;
2. for any three explicit models A, B, C for finite fields of the same size we have $\phi_{B,C} \circ \phi_{A,B} = \phi_{A,C}$.

11.7.6 **Remark** The isomorphism that the algorithm produces is given as explicit output by listing the entries of the square matrix associated to the underlying linear map over the prime field; see [1895, Section 2] for a proof of this theorem without Property 2. By Property 2 this algorithm can be used for "coercion" in computer algebra [361].

11.7.7 Definition An *algorithmic model for finite fields* is a sequence $(A_q)_q$, where q runs over all prime powers and A_q is an explicit model for a finite field of size q, such that there is an algorithm that on input q produces A_q.

11.7.8 Example For each prime p and $n \geq 1$ one can take the explicit model A_{p^n} to be given as in Remark 11.7.4 by the first irreducible polynomial of degree n over $\mathbb{F}_p$ with respect to a lexicographic ordering of polynomials of degree n over $\mathbb{F}_p$.

11.7.9 Example Conway polynomials [1452, 1966] provide an algorithmic model for finite fields that has some additional properties. However computing Conway polynomials is laborious and there is only a rather limited table of known Conway polynomials.

11.7.10 Theorem [789] There is an algorithmic model $(S_q)_q$ such that there is a deterministic polynomial time algorithm that on input an explicit model A for a field of size q

1. computes the model S_q;
2. computes an isomorphism of fields $A \to S_q$.

11.7.11 Remark The theorem in fact determines $(S_q)_q$ uniquely up to "polynomial time base change." More precisely, suppose that for every prime power $q = p^n$ we have an invertible $n \times n$-matrix M_q over $\mathbb{F}_p$ and suppose that there is a deterministic polynomial time algorithm that on input an explicit model for a field of size q produces M_q. Given $(S_q)_q$ as in the theorem we let $(S'_q)_q$ be the algorithmic model that one gets in the manner of Remark 11.7.3 by considering the columns of M_q as an $\mathbb{F}_p$-basis of S_q for every q. Then Theorem 11.7.10 also holds for $(S'_q)_q$. Moreover, every algorithmic model $(S'_q)_q$ with the Properties 1 and 2 of the theorem arises in this way from $(S_q)_q$.

11.7.12 Remark Theorem 11.7.10 implies Theorem 11.7.5. To prove Theorem 11.7.10 one can show [789] that it holds for the algorithmic model of finite fields $(S_q)_q$, where S_q is the *standard model* for a finite field of size q defined below.

11.7.13 Definition (Construction of the standard model) [790]

Step 1: cyclotomic rings. Let r be a prime number and write $\mathbf{r} = r \cdot \gcd(r, 2)$. We write $\mathbb{Z}_r$ for the ring of r-adic integers, $\mathbb{Z}_r^*$ for its group of units, and Δ_r for the torsion subgroup of $\mathbb{Z}_r^*$; the group Δ_r is cyclic of order $\varphi(\mathbf{r})$, where φ denotes the Euler φ-function.

The ring A_r is the polynomial ring $\mathbb{Z}[x_0, x_1, x_2, \ldots]$ modulo the ideal generated by $\{\sum_{j=0}^{r-1} x_0^{j\mathbf{r}/r},\ x_{k+1}^r - x_k : k \geq 0\}$. For $k \in \mathbb{Z}_{\geq 0}$, we write $\zeta_{\mathbf{r}r^k}$ for the residue class of x_k in A_r, which is a unit of multiplicative order $\mathbf{r}r^k$. For each $u \in \mathbb{Z}_r^*$ there is a unique ring automorphism of A_r that maps each $\zeta_{\mathbf{r}r^k}$ to $\zeta_{\mathbf{r}r^k}^{\bar{u}}$, where $\bar{u} = (u \bmod \mathbf{r}r^k)$; we denote this ring automorphism by σ_u.

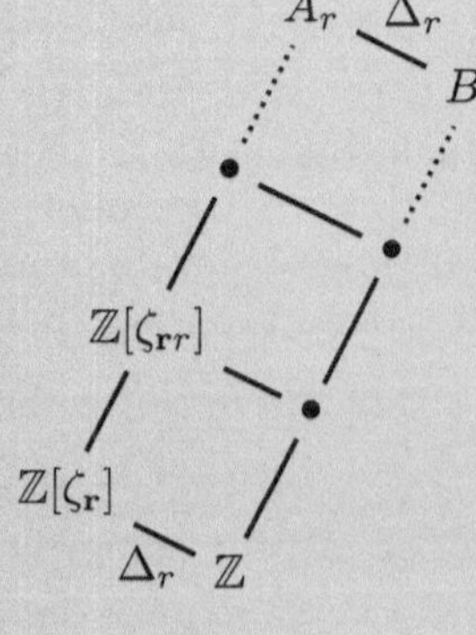

The ring B_r is defined by $B_r = \{a \in A_r : \sigma_u(a) = a \text{ for all } u \in \Delta_r\}$. For $k \in \mathbb{Z}_{>0}$, $i \in \{0, 1, \ldots, r-1\}$ the element $\eta_{r,k,i} \in B_r$ is defined by $\eta_{r,k,i} = \sum_{u \in \Delta_r} \sigma_u(\zeta_{\mathbf{r}r^k}^{1+i\mathbf{r}r^{k-1}})$.

Step 2: prime ideals. Let p, r be prime numbers with $p \neq r$, and let l be the number of factors r in the integer $(p^{\varphi(\mathbf{r})} - 1)/(\mathbf{r}^2/r)$. Denote by $S_{p,r}$ the set of prime ideals $\mathfrak{p}$ of B_r that satisfy $p \in \mathfrak{p}$. This set is finite of cardinality r^l, and for each $\mathfrak{p} \in S_{p,r}$ there exists a unique system $(a_{\mathfrak{p},j})_{0 \leq j < lr}$ of integers $a_{\mathfrak{p},j} \in \{0, 1, \ldots, p-1\}$ such that $\mathfrak{p}$ is generated by p together with $\{\eta_{r,k+1,i} - a_{\mathfrak{p},i+kr} : 0 \leq k < l, 0 \leq i < r\}$. We define a total ordering on $S_{p,r}$ by putting $\mathfrak{p} < \mathfrak{q}$ if there exists $h \in \{0, 1, \ldots, lr-1\}$ such that $a_{\mathfrak{p},j} = a_{\mathfrak{q},j}$ for all $j < h$ and $a_{\mathfrak{p},h} < a_{\mathfrak{q},h}$. The smallest element of $S_{p,r}$ in this ordering is denoted by $\mathfrak{p}_{p,r}$.

We define $\mathbb{F}_{p,r}$ to be the ring $B_r/\mathfrak{p}_{p,r}$, and for $k \in \mathbb{Z}_{>0}$ we define $\alpha_{p,r,k} \in \mathbb{F}_{p,r}$ to be the residue class of $\eta_{r,k+l,0}$ modulo $\mathfrak{p}_{p,r}$.

Step 3: equal characteristic. Let p be a prime number and put $\mathbb{F}_p = \mathbb{Z}/p\mathbb{Z}$. Let the element $f = f(x, y)$ of the polynomial ring $\mathbb{F}_p[x, y]$ be defined by $f = x^p - 1 - y \cdot \sum_{i=1}^{p-1} x^i$. We define $\mathbb{F}_{p,p}$ to be the polynomial ring $\mathbb{F}_p[x_1, x_2, x_3, \ldots]$ modulo the ideal generated by $\{f(x_1, 1), f(x_{k+1}, x_k) : k > 0\}$. For $k \in \mathbb{Z}_{>0}$ we denote the image of x_k in $\mathbb{F}_{p,p}$ by $\alpha_{p,p,k}$.

Step 4: an algebraic closure. Let p be a prime number. Then for any prime number r it is true that the ring $\mathbb{F}_{p,r}$ is a field containing $\mathbb{F}_p$; that for each $k \in \mathbb{Z}_{>0}$, the element $\alpha_{p,r,k}$ of $\mathbb{F}_{p,r}$ is algebraic of degree r^k over $\mathbb{F}_p$; and that one has $\mathbb{F}_{p,r} = \mathbb{F}_p(\alpha_{p,r,1}, \alpha_{p,r,2}, \ldots)$.

We write $\bar{\mathbb{F}}_p$ for the tensor product, over $\mathbb{F}_p$, of the rings $\mathbb{F}_{p,r}$, with r ranging over the set of all prime numbers. For any prime number r and $k \in \mathbb{Z}_{>0}$, the image of $\alpha_{p,r,k}$ under the natural ring homomorphism $\mathbb{F}_{p,r} \to \bar{\mathbb{F}}_p$ is again denoted by $\alpha_{p,r,k}$.

The ring $\bar{\mathbb{F}}_p$ is a field containing $\mathbb{F}_p$, and it is an algebraic closure of $\mathbb{F}_p$. We have $\bar{\mathbb{F}}_p = \mathbb{F}_p(\alpha_{p,r,k} : r \text{ prime}, k \in \mathbb{Z}_{>0})$, each $\alpha_{p,r,k}$ being algebraic of degree r^k over $\mathbb{F}_p$.

Step 5: a vector space basis. Let p be a prime number. For each $s \in \mathbb{Q}/\mathbb{Z}$, the element $\epsilon_s \in \bar{\mathbb{F}}_p$ is defined as follows. There exists a unique system of integers $(c_{r,k})_{r,k}$, with r ranging over the set of prime numbers and k over $\mathbb{Z}_{>0}$, such that each $c_{r,k}$ belongs to $\{0, 1, \ldots, r-1\}$ and s equals the residue class of $\sum_{r,k} c_{r,k}/r^k$ modulo $\mathbb{Z}$, the sum being finite in the sense that $c_{r,k} = 0$ for all but finitely many pairs r, k. With that notation, ϵ_s is defined to be the finite product $\prod_{r,k} \alpha_{p,r,k}^{c_{r,k}}$.

The system $(\epsilon_s)_{s \in \mathbb{Q}/\mathbb{Z}}$ is a vector space basis of $\bar{\mathbb{F}}_p$ over $\mathbb{F}_p$. In addition, for each $s \in \mathbb{Q}/\mathbb{Z}$ the degree of ϵ_s over $\mathbb{F}_p$ equals the order of s in the additive group $\mathbb{Q}/\mathbb{Z}$.

For $n \geq 1$ the *standard model* S_{p^n} *for a field of size* p^n is the explicit model that one obtains in the manner of Remark 11.7.3 by considering the $\mathbb{F}_p$-basis $\epsilon_0, \epsilon_{1/n}, \epsilon_{2/n}, \ldots, \epsilon_{(n-1)/n}$ of the unique subfield $\mathbb{F}_{p^n}$ of $\bar{\mathbb{F}}_p$ of size p^n.

11.7.14 Example Suppose that p is an odd prime number. The standard models for fields of size p^n where n is a power of 2 can be computed as follows. Put $l = \mathrm{ord}_2((p^2 - 1)/8)$ and for each $k \geq -l$ consider the image $\alpha_{p,2,k}$ of $\eta_{2,k+l,0} = \zeta_{2^{k+l+2}} + \zeta_{2^{k+l+2}}^{-1}$ in the field $\mathbb{F}_{p,2} = B_2/\mathfrak{p}_{p,2}$. We have $\alpha_{p,2,-l} = 0$, and for each $k \geq -l$ the element $\alpha_{p,2,k+1}$ is a root of the quadratic polynomial $f_k = x^2 - 2 - \alpha_{p,2,k} \in \mathbb{F}_p(\alpha_{p,2,k})[x]$.

If $k < 0$ then f_k has two roots in $\mathbb{F}_p$ and by the choice of the prime $\mathfrak{p}_{p,2}$ in Definition 11.7.13, the element $\alpha_{p,2,k+1}$ is the smallest of the two roots if we order $\mathbb{F}_p$ as $0, 1, \ldots, p-1$. Starting from $\alpha_{p,2,-l} = 0$, this enables us to find $\alpha_{p,2,-l+1}, \ldots, \alpha_{p,2,0} \in \mathbb{F}_p$.

One can show that f_k is irreducible in $\mathbb{F}_p(\alpha_{p,2,k})[x]$ for all $k \geq 0$. In particular, the polynomial $f_0 = x^2 - 2 - \alpha_{p,2,0} \in \mathbb{F}_p[x]$ gives the standard model for a field of size p^2. Moreover, for every $k \geq 1$ the irreducible polynomial over $\mathbb{F}_p$ of the generator $\alpha_{p,2,k}$ of $\mathbb{F}_{p^{2^k}}$ over $\mathbb{F}_p$ is $(x^2 - 2)^{\circ k} - \alpha_{p,2,0}$, where for a polynomial $f \in \mathbb{F}_p[x]$ we let $f^{\circ k}$ denote the k-fold composition $f(f(\cdots f(x) \cdots))$.

11.7.15 Remark Note that the definition above also provides a *standard embedding* $S_q \to S_{q^d}$ for every prime power q and every integer $d \geq 1$, which is induced by the inclusion $\mathbb{F}_q \subset \mathbb{F}_{q^d}$. A composition of standard embeddings is again a standard embedding.

11.7.16 Theorem [789] There is a probabilistic algorithm that on input a prime p and $n \geq 1$ computes S_{p^n} in expected time bounded by a polynomial in $\log p$ and n.

11.7.17 Remark This theorem is an easy consequence of the fact that explicit models can be made in probabilistic polynomial time [1892], and Theorem 11.7.10 above. Under the assumption of the generalized Riemann hypothesis, this can also be achieved with a deterministic polynomial time algorithm [14].

11.7.18 Remark An algorithm as in Theorem 11.7.16 is *semi-deterministic*: it is a probabilistic algorithm that gives the same output when it runs twice on the same input. Thus, the output does not depend on the random numbers drawn by the algorithm.

11.7.19 Example One way to find a non-square in $\mathbb{F}_p$ for a given odd prime p in semi-deterministic polynomial time is as follows. Start with the value -1. If it is a square, find the two square roots with the probabilistic method of Tonelli-Shanks [660, 1.5.1] and select the root that has the smallest representative in the set $\{1, \ldots, p-1\}$. If this is a square in $\mathbb{F}_p$, repeat. Within $O(\log p)$ iterations we find a non-square u_p in $\mathbb{F}_p$. By considering the $\mathbb{F}_p$-basis $1, \sqrt{u_p}$ of the field $\mathbb{F}_p(\sqrt{u_p})$ we find a semi-deterministic algorithm that given p produces an explicit model for a field of size p^2 in polynomial time. This proves a special case of Theorem 11.7.16.

The models produced in this way are not the standard models, which as we saw in Example 11.7.14 are obtained by taking inverse images of 0 under iterates of the map $x \mapsto x^2 - 2$ rather than inverse images of -1 under iterates of the map $x \mapsto x^2$. In both cases the algorithm produces quadratic polynomials until it encounters an irreducible one, but only the method of Example 11.7.14 has the advantage that for all p all subsequent quadratic polynomials are also irreducible.

11.7.20 Remark In a similar way, one can prove that there is a semi-deterministic algorithm that given a prime power q, and a prime l and $r \geq 1$ with $l^r \mid q-1$, computes a root of unity of order l^r in S_q in expected time polynomial in l and $\log q$.

See Also

Chapter 2	For basic properties of finite fields.
§11.1	For basic computational techniques in finite fields.
§11.3	For constructing irreducible polynomials.
§13.4	For linear algebra over finite fields.

References Cited: [14, 43, 361, 660, 789, 790, 1452, 1892, 1895, 1966]

12

Curves over finite fields

12.1 Introduction to function fields and curves

Arnaldo Garcia, IMPA
Henning Stichtenoth, Sabanci University

The theory of algebraic curves is essentially equivalent to the theory of algebraic function fields. The latter requires less background and is closer to the theory of finite fields; therefore we present here the theory of function fields. At the end of the section, we give a brief introduction to the language of algebraic curves. Our exposition follows mainly the book [2714]*, other references are [1147, 1296, 1511, 2280, 2281, 2872].

Throughout this section, K denotes a *finite field*. However, almost all results of this section hold for arbitrary perfect fields.

12.1.1 Valuations and places

12.1.1 Definition An *algebraic function field over* K is an extension field F/K with the following properties:

1. There is an element $x \in F$ such that x is transcendental over K and the extension $F/K(x)$ has finite degree.
2. No element $z \in F \setminus K$ is algebraic over K.

The field K is the *constant field* of F.

12.1.2 Remark

1. We often use the term *function field* rather than *algebraic function field.*
2. Property 2 in Definition 12.1.1 is often referred to as: K is *algebraically closed in* F, or K is the *full constant field of* F.
3. If F/K is a function field, then the degree $[F : K(z)]$ is finite for *every* $z \in F\setminus K$.
4. Every function field F/K can be generated by *two* elements, $F = K(x,y)$, where the extension $F/K(x)$ is finite and separable.

Throughout this section, F/K always means a function field over K.

12.1.3 Example (Rational function fields) The simplest example of a function field over K is the *rational function field* $F = K(x)$, with x being transcendental over K. The elements of $K(x)$ are the *rational functions* $z = f(x)/g(x)$ where f, g are polynomials over K and g is not the zero polynomial.

12.1.4 Example (Elliptic and hyperelliptic function fields) Let F be an extension of the rational function field $K(x)$ of degree $[F : K(x)] = 2$. For simplicity we assume that $\mathrm{char} K \neq 2$. Then there exists an element $y \in F$ such that $F = K(x,y)$, and y satisfies an equation over $K(x)$ of the form

$$y^2 = f(x), \quad \text{with } f \in K[x] \text{ square-free}$$

(i.e., f is not divisible by the square of a polynomial $h \in K[x]$ of degree ≥ 1). One shows that F is rational if $\deg(f) = 1$ or 2. F is an *elliptic function field* if $\deg(f) = 3$ or 4, and

*The authors thank Springer Science+Business Media for their permission to use in Section 12.1 parts from H. Stichtenoth's book *Algebraic Function Fields and Codes*, GTM 254, 2009.

it is a *hyperelliptic function field* if $\deg(f) \geq 5$. See also Definition 12.1.108 and Example 12.1.109. A detailed exposition of elliptic and hyperelliptic function fields is given in Sections 12.2 and 12.4.

12.1.5 Remark In case of $\mathrm{char} K = 2$, the definition of elliptic and hyperelliptic function fields requires some modification, see [2714, Chapters 6.1, 6.2].

12.1.6 Definition A *valuation* of F/K is a map $\nu : F \to \mathbb{Z} \cup \{\infty\}$ with the following properties:

1. $\nu(x) = \infty$ if and only if $x = 0$.
2. $\nu(xy) = \nu(x) + \nu(y)$ for all $x, y \in F$.
3. $\nu(x + y) \geq \min\{\nu(x), \nu(y)\}$ for all $x, y \in F$.
4. There exists an element $z \in F$ such that $\nu(z) = 1$.
5. $\nu(a) = 0$ for all $a \in K \setminus \{0\}$.

12.1.7 Remark The symbol ∞ denotes an element not in $\mathbb{Z}$ such that $\infty + \infty = \infty + n = n + \infty = \infty$ and $\infty > m$ for all $m, n \in \mathbb{Z}$. It follows that $\nu(x^{-1}) = -\nu(x)$ for every nonzero element $x \in F$. Property 3 above is the *Triangle Inequality.* The following proposition is often useful.

12.1.8 Proposition (Strict triangle inequality) Let ν be a valuation of the function field F/K and let $x, y \in F$ such that $\nu(x) \neq \nu(y)$. Then $\nu(x + y) = \min\{\nu(x), \nu(y)\}$.

12.1.9 Remark For a valuation ν of F/K, consider the following subsets $\mathcal{O}, \mathcal{O}^*, P$ of F:

$$\mathcal{O} := \{z \in F \mid \nu(z) \geq 0\}, \;\; \mathcal{O}^* := \{z \in F \mid \nu(z) = 0\}, \;\; P := \{z \in F \mid \nu(z) > 0\}.$$

Then $\mathcal{O}$ is a ring, $\mathcal{O}^*$ is the group of invertible elements (*units*) of $\mathcal{O}$, and P is a maximal ideal of $\mathcal{O}$. In fact, P is the *unique* maximal ideal of $\mathcal{O}$, which means that $\mathcal{O}$ is a *local ring.* The ideal P is a *principal ideal,* which is generated by every element $t \in F$ with $\nu(t) = 1$.

For distinct valuations ν_1, ν_2, the corresponding ideals $P_1 = \{z \in F \mid \nu_1(z) > 0\}$ and $P_2 = \{z \in F \mid \nu_2(z) > 0\}$ are distinct.

12.1.10 Definition

1. A subset $P \subseteq F$ is a *place* of F/K if there exists a valuation ν of F/K such that $P = \{z \in F \mid \nu(z) > 0\}$. The valuation ν is uniquely determined by the place P. Therefore we write $\nu =: \nu_P$ and say that ν_P is the *valuation corresponding to the place* P.
2. If P is a place of F/K and ν_P is the corresponding valuation, then the ring $\mathcal{O}_P := \{z \in F \mid \nu_P(z) \geq 0\}$ is the *valuation ring of* F corresponding to P.
3. An element $t \in F$ with $\nu_P(t) = 1$ is a *prime element at the place* P.
4. Let $\mathbb{P}_F := \{P \mid P \text{ is a place of } F\}$.

12.1.11 Remark Since P is a maximal ideal of its valuation ring $\mathcal{O}_P$, the residue class ring $\mathcal{O}_P/P$ is a field. The constant field K is contained in $\mathcal{O}_P$, and $P \cap K = \{0\}$. Hence one has a canonical embedding $K \hookrightarrow \mathcal{O}_P/P$. We always consider K as a subfield of $\mathcal{O}_P/P$ via this embedding.

12.1.12 Definition Let P be a place of F/K.

1. The field $F_P := \mathcal{O}_P/P$ is the *residue class field of* P.

2. The degree of the field extension F_P/K is finite and is the *degree of the place* P. We write $\deg P := [F_P : K]$.
3. A place $P \in \mathbb{P}_F$ is *rational* if $\deg P = 1$. This means that $F_P = K$.
4. For $z \in \mathcal{O}_P$, denote by $z(P) \in F_P$ the residue class of z in F_P. For $z \in F \setminus \mathcal{O}_P$, set $z(P) := \infty$. The map from F to $F_P \cup \{\infty\}$ given by $z \mapsto z(P)$ is the *residue class map* at P.

12.1.13 Remark For a *rational* place $P \in \mathbb{P}_F$ and an element $z \in \mathcal{O}_P$, the residue class $z(P)$ is the (unique) element $a \in K$ such that $\nu_P(z - a) > 0$. In this case, one calls the map $z \mapsto z(P)$ from $\mathcal{O}_P$ to K the *evaluation map* at the place P. We note that the evaluation map is K-linear. This map plays an important role in the theory of *algebraic–geometry codes*, see Section 15.2.

12.1.14 Example We want to describe all places of the rational function field $K(x)/K$.

1. Let $h \in K[x]$ be an irreducible monic polynomial. Every nonzero element $z \in K(x)$ can be written as
$$z = h(x)^r \cdot \frac{f(x)}{g(x)}$$
with polynomials $f, g \in K[x]$ which are relatively prime to h, and $r \in \mathbb{Z}$. Then the map $\nu_P : K(x) \to \mathbb{Z} \cup \{\infty\}$ with $\nu_P(z) := r$ (and $\nu_P(0) := \infty$) defines a valuation of $K(x)/K$. The corresponding place P is
$$P = \left\{ \frac{u(x)}{v(x)} \mid u, v \in K[x],\ h \text{ divides } u \text{ but not } v \right\}.$$
The residue class field of this place is isomorphic to $K[x]/(h)$ and therefore we have $\deg P = \deg(h)$.
2. Another valuation of $K(x)/K$ is defined by $\nu(z) = \deg(g) - \deg(f)$ for $z = f(x)/g(x) \neq 0$. The corresponding place is called the *place at infinity* and is denoted by P_∞ or $(x = \infty)$. It follows from the definition that
$$P_\infty = \left\{ \frac{f(x)}{g(x)} \mid \deg(f) < \deg(g) \right\}.$$
The place P_∞ has degree one, that is, it is a rational place.
3. There are no places of $K(x)/K$ other than those described in Parts 1 and 2.
4. For $a \in K$, the polynomial $x - a$ is irreducible of degree 1 and defines a place P of degree one. We sometimes denote this place as $P = (x = a)$. The set $K \cup \{\infty\}$ is therefore in 1–1 correspondence with the set of rational places of $K(x)/K$ via $a \longleftrightarrow (x = a)$.
5. The residue class map corresponding to a place $P = (x = a)$ with $a \in K$ is given as follows: If $z = f(x)/g(x) \in \mathcal{O}_P$ then $g(a) \neq 0$ and
$$z(P) =: z(a) = f(a)/g(a) \in K.$$
In order to determine $z(\infty) := z(P)$ at the infinite place $P = P_\infty$, we write $f(x) = a_n x^n + \cdots + a_0$ and $g(x) = b_m x^m + \cdots + b_0$ with $a_n b_m \neq 0$. Then $z(\infty) = 0$ if $n < m$, $z(\infty) = \infty$ if $n > m$, and $z(\infty) = a_n/b_n$ if $n = m$.

12.1.2 Divisors and Riemann–Roch theorem

12.1.15 Remark [2714, Corollary 1.3.2] Every function field F/K has infinitely many places.

12.1.16 Remark The following theorem states that distinct valuations of F/K are *independent* of each other.

12.1.17 Theorem (Approximation theorem) [2714, Theorem 1.3.1] Let $P_1, \ldots, P_n \in \mathbb{P}_F$ be pairwise distinct places of F. Let $x_1, \ldots, x_n \in F$ and $r_1, \ldots, r_n \in \mathbb{Z}$. Then there exists an element $z \in F$ such that

$$\nu_{P_i}(z - x_i) = r_i \quad \text{for } i = 1, \ldots, n.$$

12.1.18 Definition Let F/K be a function field, $x \in F$ and $P \in \mathbb{P}_F$.

1. P is a *zero of* x if $\nu_P(x) > 0$, and the integer $\nu_P(x)$ is the *zero order* of x at P.
2. P is a *pole of* x if $\nu_P(x) < 0$. The integer $-\nu_P(x)$ is the *pole order* of x at P.

12.1.19 Remark

1. A nonzero element $a \in K$ has neither zeros nor poles.
2. For all $x \neq 0$ and $P \in \mathbb{P}_F$, P is a pole of x if and only if P is a zero of x^{-1}.

12.1.20 Theorem [2714, Theorem 1.4.11] For $x \in F \setminus K$ the following hold:

1. x has at least one zero and one pole.
2. The number of zeros and poles of x is finite.
3. Let $P_1, \ldots, P_r$ and $Q_1, \ldots, Q_s$ be all zeros and poles of x, respectively. Then

$$\sum_{i=1}^{r} \nu_{P_i}(x) \deg P_i = \sum_{j=1}^{s} -\nu_{Q_j}(x) \deg Q_j = [F : K(x)].$$

12.1.21 Definition

1. The *divisor group* of F/K is the free abelian group generated by the set of places of F/K. It is denoted by $\mathrm{Div}(F)$. The elements of $\mathrm{Div}(F)$ are *divisors* of F. That means, a divisor of F is a formal sum

$$D = \sum_{P \in \mathbb{P}_F} n_P P \quad \text{with } n_P \in \mathbb{Z} \text{ and } n_P \neq 0 \text{ for at most finitely many } P.$$

The set of places with $n_P \neq 0$ is the *support* of D and denoted as $\operatorname{supp} D$. If $\operatorname{supp} D \subseteq \{P_1, \ldots, P_k\}$ then D is also written as

$$D = n_1 P_1 + \cdots + n_k P_k \quad \text{where } n_i = n_{P_i}.$$

Two divisors $D = \sum_P n_P P$ and $E = \sum_P m_P P$ are added coefficientwise, that is $D + E = \sum_P (n_P + m_P) P$. The *zero divisor* is the divisor $0 = \sum_P r_P P$ where all $r_P = 0$.

2. A divisor of the form $D = P$ with $P \in \mathbb{P}_F$ is a *prime divisor*.
3. The *degree* of the divisor $D = \sum_P n_P P$ is

$$\deg D := \sum_{P \in \mathbb{P}_F} n_P \cdot \deg P.$$

We note that this is a finite sum since $n_P \neq 0$ only for finitely many P.

4. A partial order on $\mathrm{Div}(F)$ is defined as follows: if $D = \sum_P n_P P$ and $E = \sum_P m_P P$, then

$$D \leq E \text{ if and only if } n_P \leq m_P \text{ for all } P \in \mathbb{P}_F.$$

A divisor $D \geq 0$ is *positive* (or *effective*).

12.1.22 Remark Since every nonzero element $x \in F$ has only finitely many zeros and poles, the following definitions are meaningful.

12.1.23 Definition For a nonzero element $x \in F$, let Z and N denote the set of zeros and poles of x, respectively.

1. The divisor $(x)_0 := \sum_{P\in Z} \nu_P(x)P$ is the *zero divisor* of x.
2. The divisor $(x)_\infty := -\sum_{P\in N} \nu_P(x)P$ is the *divisor of poles* of x.
3. The divisor $\mathrm{div}(x) := \sum_{P\in\mathbb{P}_F} \nu_P(x)P = (x)_0 - (x)_\infty$ is the *principal divisor* of x.

12.1.24 Remark

1. We note that both divisors $(x)_0$ and $(x)_\infty$ are positive divisors. By Theorem 12.1.20, $\deg(x)_0 = \deg(x)_\infty$ and hence $\deg(\mathrm{div}(x)) = 0$.
2. For $x \in F \setminus K$ we have $\deg(x)_0 = \deg(x)_\infty = [F : K(x)]$. The principal divisor of a nonzero element $a \in K$ is the zero divisor. We observe that for the element $0 \in K$, no principal divisor is defined.
3. The sum of two principal divisors and the negative of a principal divisor are principal, since $\mathrm{div}(xy) = \mathrm{div}(x)+\mathrm{div}(y)$ and $\mathrm{div}(x^{-1}) = -\mathrm{div}(x)$. Therefore the principal divisors form a subgroup of the divisor group of F.

12.1.25 Example We consider again the *rational function field* $F = K(x)$. Let $f \in K[x]$ be a nonzero polynomial and write f as a product of irreducible polynomials,

$$f(x) = a \cdot p_1(x)^{r_1} \cdots p_n(x)^{r_n},$$

where $0 \neq a \in K$ and $p_1, \ldots, p_n$ are pairwise distinct, monic, irreducible polynomials. Let P_i be the place of $K(x)$ corresponding to the polynomial p_i (see Example 12.1.14), and P_∞ be the place at infinity. Then the principal divisor of f in $\mathrm{Div}(K(x))$ is

$$\mathrm{div}(f) = r_1P_1 + \cdots + r_nP_n - dP_\infty \quad \text{where} \quad d = \deg(f).$$

As every element of $K(x)$ is a quotient of two polynomials, we thus obtain the principal divisor for any nonzero element $z \in K(x)$ in this way.

12.1.26 Definition

1. Two divisors $D, E \in \mathrm{Div}(F)$ are *equivalent* if $E = D + \mathrm{div}(x)$ for some $x \in F$. This is an equivalence relation on the divisor group of F/K. We write

$$D \sim E \text{ if } D \text{ and } E \text{ are equivalent.}$$

2. $\mathrm{Princ}(F) := \{A \in \mathrm{Div}(F) \mid A \text{ is principal}\}$ is the *group of principal divisors* of F.

3. The factor group $\mathrm{Cl}(F) := \mathrm{Div}(F)/\mathrm{Princ}(F)$ is the *divisor class group* of F.
4. For a divisor $D \in \mathrm{Div}(F)$ we denote by $[D] \in \mathrm{Cl}(F)$ its class in the divisor class group.

12.1.27 Remark The equivalence relation $\sim$ as defined in Definition 12.1.26 is often denoted as *linear equivalence* of divisors.

12.1.28 Remark

1. It follows from the definitions that $D \sim E$ if and only if $[D] = [E]$.
2. $D \sim E$ implies $\deg D = \deg E$.
3. In a *rational* function field $K(x)$, the converse of Part 2 also holds. If F/K is *non-rational*, then there exist, in general, divisors of the same degree which are not equivalent.

12.1.29 Definition Let F/K be a function field and let $A \in \mathrm{Div}(F)$ be a divisor of F. Then the set

$$\mathcal{L}(A) := \{x \in F \mid \mathrm{div}(x) \geq -A\} \cup \{0\}$$

is the *Riemann–Roch space associated to the divisor A*.

12.1.30 Proposition $\mathcal{L}(A)$ is a finite-dimensional vector space over K.

12.1.31 Definition For a divisor A, the integer

$$\ell(A) := \dim \mathcal{L}(A)$$

is the *dimension* of A. We point out that $\dim \mathcal{L}(A)$ denotes here the dimension as a vector space over K.

12.1.32 Remark

1. If $A \sim B$ then the spaces $\mathcal{L}(A)$ and $\mathcal{L}(B)$ are isomorphic (as K-vector spaces). Hence $A \sim B$ implies $\ell(A) = \ell(B)$.
2. $A \leq B$ implies $\mathcal{L}(A) \subseteq \mathcal{L}(B)$ and hence $\ell(A) \leq \ell(B)$.
3. $\deg A < 0$ implies $\ell(A) = 0$.
4. $\mathcal{L}(0) = K$ and hence $\ell(0) = 1$.

12.1.33 Remark The following theorem is one of the main results of the theory of function fields.

12.1.34 Theorem (Riemann–Roch theorem) [2714, Theorem 1.5.15] Let F/K be a funcion field. Then there exist an integer $g \geq 0$ and a divisor $W \in \mathrm{Div}(F)$ with the following property: for all divisors $A \in \mathrm{Div}(F)$,

$$\ell(A) = \deg A + 1 - g + \ell(W - A).$$

12.1.35 Definition The integer $g =: g(F)$ is the *genus* of F, and the divisor W is a *canonical divisor* of F.

12.1.36 Remark [2714, Proposition 1.6.1]

1. If $W' \sim W$, then the equation above also holds when W is replaced by W'.

2. Suppose that $g_1, g_2 \in \mathbb{Z}$ and $W_1, W_2 \in \mathrm{Div}(F)$ satisfy the equations $\ell(A) = \deg A + 1 - g_1 + \ell(W_1 - A) = \deg A + 1 - g_2 + \ell(W_2 - A)$ for all divisors A. Then $g_1 = g_2$ and $W_1 \sim W_2$.
3. As a consequence of 1 and 2, the canonical divisors of F/K form a uniquely determined divisor class $[W] \in \mathrm{Cl}(F)$, the *canonical class* of F.

12.1.37 Corollary [2714, Corollary 1.5.16] Let W be a canonical divisor and $g = g(F)$ the genus of F. Then

$$\deg W = 2g - 2 \quad \text{and} \quad \ell(W) = g.$$

Conversely, every divisor C with $\deg C = 2g - 2$ and $\ell(C) = g$ is canonical.

12.1.38 Remark A slightly weaker version of the Riemann–Roch theorem is often sufficient.

12.1.39 Theorem (Riemann's theorem) [2714, Theorem 1.4.17] Let F/K be a function field of genus g. Then for all divisors $A \in \mathrm{Div}(F)$,

$$\ell(A) \geq \deg A + 1 - g.$$

Equality holds for all divisors A with $\deg A > 2g - 2$.

12.1.40 Example Consider the *rational function field* $F = K(x)$. The following hold:

1. The genus of $K(x)$ is 0.
2. Let P_∞ be the infinite place of $K(x)$, see Example 12.1.14. For every $k \geq 0$ we obtain

$$\mathcal{L}(kP_\infty) = \{f \in K[x] \mid \deg(f) \leq k\}.$$

This shows that Riemann–Roch spaces are natural generalizations of spaces of polynomials.
3. The divisor $W = -2P_\infty$ is canonical.

12.1.41 Remark Conversely, if F/K is a function field of genus $g(F) = 0$, then there exists an element $x \in F$ such that $F = K(x)$. (This does not hold in general if K is not a finite field.)

12.1.42 Remark For divisors of degree $\deg A > 2g - 2$, Riemann's Theorem gives a precise formula for $\ell(A)$. On the other hand, $\ell(A) = 0$ if $\deg A < 0$. For the interval $0 \leq \deg A \leq 2g - 2$, there is no exact formula for $\ell(A)$ in terms of $\deg A$.

12.1.43 Theorem (Clifford's theorem) [2714, Theorem 1.6.13] For all divisors $A \in \mathrm{Div}(F)$ with $0 \leq \deg A \leq 2g - 2$,

$$\ell(A) \leq 1 + \frac{1}{2} \cdot \deg A.$$

12.1.44 Remark The genus $g(F)$ of a function field F is its most important numerical invariant. In general it is a difficult task to determine $g(F)$. Some methods are discussed in Subsection 12.1.3. Here we give upper bounds for $g(F)$ in some special cases.

12.1.45 Remark Assume that $F = K(x, y)$ is a function field over K, where x, y satisfy an equation $\varphi(x, y) = 0$ with an *irreducible* polynomial $\varphi(X, Y) \in K[X, Y]$ of degree d. Then

$$g(F) \leq \frac{(d-1)(d-2)}{2}.$$

Equality holds if and only if the plane projective curve which is defined by the affine equation $\varphi(X, Y) = 0$, is nonsingular. (These terms are explained in Subsection 12.1.5.)

12.1.46 Remark (Riemann's inequality) [2714, Corollary 3.11.4] Suppose that $F = K(x, y)$. Then

$$g(F) \leq ([F : K(x)] - 1)([F : K(y)] - 1).$$

12.1.3 Extensions of function fields

12.1.47 **Remark** In this subsection we consider the following situation: F/K and F'/K' are function fields with $F \subseteq F'$ and $K \subseteq K'$. We always assume that K (respectively K') is algebraically closed in F (respectively in F') and that the degree $[F : F']$ is finite. As before, K is a *finite* field.

12.1.48 **Remark** The extension degree $[K' : K]$ divides $[F' : F]$.

12.1.49 **Definition** Let $P \in \mathbb{P}_F$ and $P' \in \mathbb{P}_{F'}$. The place P' is an *extension* of P (equivalently, P' *lies over* P, or P *lies under* P') if one of the following equivalent conditions holds:

1. $P \subseteq P'$,
2. $\mathcal{O}_P \subseteq \mathcal{O}_{P'}$,
3. $P' \cap F = P$,
4. $\mathcal{O}_{P'} \cap F = \mathcal{O}_P$.

We write $P'|P$ to indicate that P' is an extension of P.

12.1.50 **Remark** If P' lies over P then the inclusion $\mathcal{O}_P \hookrightarrow \mathcal{O}_{P'}$ induces a natural embedding of the residue class fields $F_P \hookrightarrow F'_{P'}$. We therefore consider F_P as a subfield of $F'_{P'}$ via this embedding.

12.1.51 **Definition** Let P' be a place of F' lying above P.

1. There exists an integer $e \geq 1$ such that $\nu_{P'}(z) = e \cdot \nu_P(z)$ for all $z \in F$. This integer $e =: e(P'|P)$ is the *ramification index of* $P'|P$.
2. The degree $f(P'|P) := [F'_{P'} : F_P]$ is finite and is the *relative degree* of $P'|P$.

12.1.52 **Remark** Suppose that F''/K'' is another finite extension of F'/K'. Let P, P', P'' be places of F, F', F'' such that $P'|P$ and $P''|P'$. Then

$$e(P''|P) = e(P''|P') \cdot e(P'|P) \quad \text{and} \quad f(P''|P) = f(P''|P') \cdot f(P'|P).$$

12.1.53 **Theorem** (Fundamental equality) [2714, Theorem 3.1.11] Let P be a place of F/K. Then there exists at least one but only finitely many places of F' lying above P. If $P_1, \ldots, P_m$ are *all* extensions of P in F' then

$$\sum_{i=1}^{m} e(P_i|P) f(P_i|P) = [F' : F].$$

12.1.54 **Corollary** Let F'/F be an extension of degree $[F' : F] = n$, and let $P \in \mathbb{P}_F$. Then

1. For every place $P' \in \mathbb{P}_{F'}$ lying over P, $e(P'|P) \leq n$ and $f(P'|P) \leq n$.
2. There are at most n distinct places of F' lying over P.

12.1.55 **Definition** Let F'/F be an extension of degree $[F' : F] = n$, and let $P \in \mathbb{P}_F$.

1. A place $P' \in \mathbb{P}_{F'}$ over P is *ramified* if $e(P'|P) > 1$, and it is *unramified* if $e(P'|P) = 1$.
2. P is *ramified in* F'/F if there exists an extension of P in F' that is ramified. Otherwise, P is *unramified* in F'.
3. P is *totally ramified* in F'/F if there is a place P' of F' lying over P with $e(P'|P) = n$. It is clear that P' is then the *only* extension of P in F'.
4. P *splits completely* in F'/F if P has n distinct extensions $P_1, \ldots, P_n$ in F'. It is clear that P is then *unramified* in F'.

12.1.56 Theorem [2714, Corollary 3.5.5] If F'/F is a finite *separable* extension of function fields, then at most *finitely many* places of F are ramified in F'/F.

12.1.57 Remark More precise information about the ramified places in F'/F is given in Theorem 12.1.72.

12.1.58 Definition For $P \in \mathbb{P}_F$ one defines the *conorm* of P in F'/F as

$$\mathrm{Con}_{F'/F}(P) := \sum_{P'|P} e(P'|P) \cdot P'.$$

For an arbitrary divisor of F we define its conorm as

$$\mathrm{Con}_{F'/F}\left(\sum_P n_P P\right) := \sum_P n_P \cdot \mathrm{Con}_{F'/F}(P).$$

12.1.59 Remark $\mathrm{Con}_{F'/F}$ is a homomorphism from the divisor group of F to the divisor group of F', which sends principal divisors of F to principal divisors of F'.

12.1.60 Remark For every divisor $A \in \mathrm{Div}(F)$, one has

$$\deg \mathrm{Con}_{F'/F}(A) = \frac{[F' : F]}{[K' : K]} \cdot \deg A.$$

In particular, if $K' = K$ then $\deg \mathrm{Con}_{F'/F}(A) = [F' : F] \cdot \deg A$.

12.1.61 Definition Let F'/K' be a finite extension of F/K, let $P \in \mathbb{P}_F$ and $\mathcal{O}_P$ its valuation ring.

1. An element $z \in F'$ is *integral over* $\mathcal{O}_P$ if there exist elements $u_0, \ldots, u_{m-1} \in \mathcal{O}_P$ such that $z^m + u_{m-1}z^{m-1} + \cdots + u_1 z + u_0 = 0$. Such an equation is an *integral equation for* z over $\mathcal{O}_P$.
2. The set $\mathcal{O}'_P := \{z \in F' \mid z \text{ is integral over } \mathcal{O}_P\}$ is a subring of F'. It is the *integral closure of* $\mathcal{O}_P$ *in* F'.

12.1.62 Proposition [2714, Chapter 3.2, 3.3] With notation as in Definition 12.1.61, the following hold:

1. $z \in F'$ is integral over $\mathcal{O}_P$ if and only if the coefficients of the minimal polynomial of z over F are in $\mathcal{O}_P$.
2. $\mathcal{O}'_P = \bigcap_{P'|P} \mathcal{O}_{P'}$.
3. There exists a basis $(z_1, \ldots, z_n)$ of F'/F such that $\mathcal{O}'_P = \sum_{i=1}^n z_i \mathcal{O}_P$, that is, every element $z \in F'$ which is integral over $\mathcal{O}_P$, has a unique representation $z = \sum x_i z_i$ with $x_i \in \mathcal{O}_P$. Such a basis $(z_1, \ldots, z_n)$ is an *integral basis* at the place P.
4. Every basis $(y_1, \ldots, y_n)$ of F'/F is an integral basis for *almost all* places $P \in \mathbb{P}_F$ (that is, for all P with only finitely many exceptions). In particular, if $F' = F(y)$ then $(1, y, \ldots, y^{n-1})$ is an integral basis for almost all P.

12.1.63 Remark Using integral bases one can often determine all extensions of a place $P \in \mathbb{P}_F$ in F'. In the following theorem, denote by $\bar{u} := u(P) \in F_P$ the residue class of an element $u \in \mathcal{O}_P$ in the residue class field $F_P = \mathcal{O}_P/P$. For a polynomial $\psi(T) = \sum u_i T^i \in \mathcal{O}_P[T]$ we set $\bar{\psi}(T) := \sum \bar{u}_i T^i \in F_P[T]$.

12.1.64 Theorem (Kummer's theorem) [2714, Theorem 3.3.7] Suppose that $F' = F(y)$ with y integral over $\mathcal{O}_P$. Let $\varphi \in \mathcal{O}_P[T]$ be the minimal polynomial of y over F and decompose $\bar{\varphi}$ into irreducible factors over F_P

$$\bar{\varphi}(T) = \gamma_1(T)^{\epsilon_1} \cdots \gamma_r(T)^{\epsilon_r}$$

with distinct irreducible monic polynomials $\gamma_i \in F_P[T]$ and $\epsilon_i \geq 1$. Choose monic polynomials $\varphi_i \in \mathcal{O}_P[T]$ such that $\bar{\varphi}_i = \gamma_i$. Then the following hold:

1. For each $i \in \{1, \dots, r\}$ there exists a place $P_i|P$ such that $\varphi_i(y) \in P_i$. The relative degree of $P_i|P$ satisfies $f(P_i|P) \geq \deg(\gamma_i)$.
2. If $(1, y, \dots, y^{n-1})$ is an *integral basis* at P, then there exists for each $i \in \{1, \dots, r\}$ a *unique* place $P_i|P$ with $\varphi_i(y) \in P_i$, and we have $e(P_i|P) = \epsilon_i$ and $f(P_i|P) = \deg(\gamma_i)$.
3. If $\bar{\varphi}(T) = \prod_{i=1}^{n}(T - a_i)$ with distinct elements $a_1, \dots, a_n \in K$, then P splits completely in F'/F.

12.1.65 Example Consider a field K with char$K \neq 2$ and a function field $F = K(x, y)$, where y satisfies an equation $y^2 = f(x)$ with a polynomial $f(x) \in K[x]$ of odd degree. Then $[F : K(x)] = 2$, and $\varphi(T) = T^2 - f(x)$ is the minimal polynomial of y over $K(x)$. Let $a \in K$.

1. If $f(a)$ is a nonzero square in K (that is, $f(a) = c^2$ with $0 \neq c \in K$), then the place $(x = a)$ of $K(x)$ (see Example 12.1.14) splits into two rational places of F.
2. If $f(a)$ is a non-square in K, then the place $(x = a)$ has exactly one extension Q in F, and $\deg Q = 2$.
3. If $a \in K$ is a simple root of the equation $f(x) = 0$, then the place $(x = a)$ of $K(x)$ is totally ramified in $F/K(x)$, and its unique extension $P \in \mathbb{P}_F$ is rational.

For more examples see Section 12.5.

12.1.66 Remark In what follows, we assume that F'/F is a *separable* extension of function fields of degree $[F' : F] = n$. As before, P denotes a place of F and $\mathcal{O}'_P$ is the integral closure of $\mathcal{O}_P$ in F'. By $\mathrm{Tr}_{F'/F} : F' \to F$ we denote the *trace mapping*. For information about separable extensions and the trace map, see any standard textbook on algebra, e.g., [1846].

12.1.67 Definition

1. For $P \in \mathbb{P}_F$, the set

$$\mathcal{C}_P := \{z \in F' \mid \mathrm{Tr}_{F'/F}(z\mathcal{O}'_P) \subseteq \mathcal{O}_P\}$$

is the *complementary module* of P in F'.

2. There is an element $t_P \in F'$ such that $\mathcal{C}_P = t_P\mathcal{O}'_P$, and we define for $P' \in \mathbb{P}_{F'}$ with $P'|P$ the *different exponent of P' over P* as

$$d(P'|P) := -\nu_{P'}(t_P).$$

We observe that the element t_P is not unique, but the different exponent is well-defined (independent of the choice of t_P).

12.1.68 Lemma [2714, Definition 3.4.3]

1. For all $P'|P$, $d(P'|P) \geq 0$.
2. For almost all $P \in \mathbb{P}_F$, $d(P'|P) = 0$ holds for all extensions $P'|P$ in F'.

12.1.69 Definition The *different* of a finite separable extension of function fields F'/F is the divisor of the function field F' defined as

$$\mathrm{Diff}(F'/F) := \sum_{P \in \mathbb{P}_F} \sum_{P'|P} d(P'|P)P'.$$

12.1.70 Theorem [2714, Theorems 3.4.6, 3.4.13] Let F'/K' be a finite separable extension of F/K.

1. If W is a canonical divisor of F/K, then the divisor

$$W' := \mathrm{Con}_{F'/F}(W) + \mathrm{Diff}(F'/F)$$

is a canonical divisor of F'/K'.

2. (Hurwitz genus formula) The genera of F' and F satisfy the equation

$$2g(F') - 2 = \frac{[F' : F]}{[K' : K]}(2g(F) - 2) + \deg \mathrm{Diff}(F'/F).$$

12.1.71 Remark We note that Part 2 is an immediate consequence of Part 1 since the degree of a canonical divisor of F is $2g(F) - 2$. Next we give some results that help to compute the different exponents $d(P'|P)$.

12.1.72 Theorem (Dedekind's different theorem) [2714, Theorem 3.5.1] Let F'/F be a finite separable extension of function fields, let $P \in \mathbb{P}_F$ and $P' \in \mathbb{P}_{F'}$ with $P'|P$. Then

1. $d(P'|P) \geq e(P'|P) - 1 \geq 0$.
2. $d(P'|P) = e(P'|P) - 1$ if and only if the characteristic of F does not divide $e(P'|P)$.

12.1.73 Remark In other words, the different of F'/F contains exactly the places of F' which are ramified in F'/F. In particular it follows that only finitely many places are ramified. The following definition is motivated by Dedekind's Different Theorem.

12.1.74 Definition Assume that $P'|P$ is ramified.

1. $P'|P$ is *tame* if the characteristic of F does not divide $e(P'|P)$.
2. $P'|P$ is *wild* if the characteristic of F divides $e(P'|P)$.

12.1.75 Lemma In a tower of separable extensions $F'' \supseteq F' \supseteq F$, the different is *transitive*, that is:

$$d(P''|P) = d(P''|P') + e(P''|P') \cdot d(P'|P) \quad \text{for} \quad P'' \supseteq P' \supseteq P, \text{ and hence}$$
$$\mathrm{Diff}(F''/F) = \mathrm{Diff}(F''/F') + \mathrm{Con}_{F''/F'}(\mathrm{Diff}(F'/F)).$$

12.1.76 Proposition [2714, Theorem 3.5.10] Let $F' = F(y)$ be a separable extension of degree $[F' : F] = n$. Let $P \in \mathbb{P}_F$ and assume that the minimal polynomial φ of y has all of its coefficients in $\mathcal{O}_P$. Let $P_1, \ldots, P_r$ be all extensions of P in F'. Then one has:

1. $0 \le d(P_i|P) \le \nu_{P_i}(\varphi'(y))$ for $i = 1, \ldots, r$.
2. $\{1, y, \ldots, y^{n-1}\}$ is an integral basis at P if and only if $d(P_i|P) = \nu_{P_i}(\varphi'(y))$ for $i = 1, \ldots, r$.

Here φ' denotes the derivative of φ in the polynomial ring $F[T]$.

12.1.77 Remark Recall that a finite field extension F'/F is *Galois* if the automorphism group $G := \{\sigma : F' \to F' \mid \sigma$ is an automorphism of F' which is the identity on $F\}$ has order $\operatorname{ord} G = [F' : F]$. In this case, $\mathrm{Gal}(F'/F) := G$ is the *Galois group* of F'/F.

12.1.78 Remark If F'/F is Galois and P is a place of F, the Galois group $\mathrm{Gal}(F'/F)$ acts on the set of extensions of P via $\sigma(P') = \{\sigma(z) \mid z \in P'\}$.

12.1.79 Proposition [2714, Theorem 3.7.1] Suppose that F'/F is a *Galois* extension, and let $P \in \mathbb{P}_F$.

1. The Galois group acts *transitively* on the set of extensions of P in F'. That is, for any two extensions P_1, P_2 of P in F', there is an automorphism $\sigma \in \mathrm{Gal}(F'/F)$ such that $P_2 = \sigma(P_1)$.
2. If $P_1, \ldots, P_r$ are *all* extensions of P in F', then

$$e(P_i|P) = e(P_j|P), \quad f(P_i|P) = f(P_j|P), \text{ and } d(P_i|P) = d(P_j|P)$$

holds for all $i, j = 1, \ldots, r$.
3. Setting $e(P) := e(P_i|P)$ and $f(P) := f(P_i|P)$, we have the equality

$$e(P) \cdot f(P) \cdot r = [F' : F].$$

12.1.80 Proposition (Kummer extensions) [2714, Proposition 3.7.3] Let $F' = F(y)$ be an extension of function fields of degree $[F' : F] = n$, where the constant field of F is the finite field $\mathbb{F}_q$. Assume that

$$y^n = u \in F \text{ and } n \text{ divides } (q - 1).$$

Then F'/F is Galois, and the Galois group $\mathrm{Gal}(F'/F)$ is cyclic of order n.

1. For $P \in \mathbb{P}_F$ define $r_P := \gcd(n, \nu_P(u))$, the greatest common divisor of n and $\nu_P(u)$. Then

$$e(P'|P) = \frac{n}{r_P} \text{ and } d(P'|P) = \frac{n}{r_P} - 1 \text{ for all } P'|P.$$

2. Denote by K (K', respectively) the constant field of F (F', respectively). Then

$$g(F') = 1 + \frac{n}{[K' : K]}\left(g(F) - 1 + \frac{1}{2}\sum_{P \in \mathbb{P}_F}\left(1 - \frac{r_P}{n}\right)\deg P\right).$$

3. If $K = K'$ and $F = K(x)$ is a rational function field, then

$$g(F') = -n + 1 + \frac{1}{2}\sum_{P \in \mathbb{P}_F}(n - \gcd(n, \nu_P(u)))\deg P.$$

12.1.81 Remark Let F'/F be a Galois extension of function fields of degree $[F' : F] = n$ whose Galois group is cyclic. Suppose that n divides $q - 1$ (where the constant field of F is $\mathbb{F}_q$). Then $F' = F(y)$ with some element y satisfying $y^n \in F$. So Proposition 12.1.80 applies.

12.1.82 Example Assume that the characteristic of K is *odd.* Let $F = K(x,y)$ with $y^2 = f(x)$, where $f \in K[x]$ is a square-free polynomial of degree $\deg(f) = 2m+1$. This means that $f = f_1 \cdots f_s$ with pairwise distinct irreducible polynomials $f_i \in K[x]$. Let $P_i \in \mathbb{P}_{K(x)}$ be the place corresponding to f_i, $i = 1,\ldots,s$, and P_∞ be the pole of x in $K(x)$. For $P \in \{P_1,\ldots,P_s,P_\infty\}$ we have $\gcd(2,\nu_P(f)) = 1$, and for all other places $Q \in \mathbb{P}_{K(x)}$ we have $\nu_Q(f) = 0$. Then Part 3 of the Proposition above yields $g(F) = (\deg(f)-1)/2 = m$. Hence for every integer $m \geq 0$ there exist function fields F/K of genus $g(F) = m$.

12.1.83 Proposition (Artin–Schreier extensions) [2714, Proposition 3.7.8] Let F/K be a function field, where K is a finite field of characteristic p. Let $F' = F(y)$ with $y^p - y = u \in F$. We assume that *for all poles P of u in F, p does not divide $\nu_P(u)$*, and that $u \notin K$. Then the following hold:

1. F'/F is Galois of degree $[F' : F] = p$, and F, F' have the same constant field.
2. Exactly the poles of u are ramified in F'/F (in fact, they are *totally ramified*); all other places of F are unramified.
3. Let P be a pole of u in F and let P' be the unique place of F' lying over P. Then the different exponent of $P'|P$ is $d(P'|P) = (p-1)(-\nu_P(u)+1)$.
4. The genus of F' is given by the formula

$$g(F') = p \cdot g(F) + \frac{p-1}{2}\left(-2 + \sum_{P:\ \nu_P(u)<0} (-\nu_P(u)+1) \cdot \deg P\right).$$

12.1.84 Remark Every Galois extension F'/F of degree $[F' : F] = p = \mathrm{char}K$ can be written as $F' = F(y)$, where y satisfies an equation of the form $y^p - y = u \in F$. If moreover $F = K(x)$ is a *rational* function field, one can choose u in such a way that for all poles P of u in $K(x)$, p does not divide $\nu_P(u)$.

12.1.85 Example Suppose that $\mathrm{char}K = p$ and $F = K(x,y)$, where $y^p - y = f(x) \in K[x]$, $\deg(f) = m$ and m is not divisible by p. Then $F/K(x)$ is Galois of degree p and K is algebraically closed in F. The pole of x is the only place of $K(x)$ that is ramified in $F/K(x)$, and the genus of F is $g(F) = (p-1)(m-1)/2$.

12.1.86 Definition The function field F'/K' is a *constant field extension* of F/K, if $F' = FK'$ (that is, if $K' = K(\alpha)$ then $F' = F(\alpha)$).

12.1.87 Remark If E/K' is a finite extension of F/K (meaning that E/F is a finite extension and K' is the constant field of E), we consider the intermediate field $F \subseteq F' := FK' \subseteq E$. Then F'/K' is a constant field extension of F/K, and E/F' is an extension of function fields having the same constant field K'.

12.1.88 Theorem [2714, Chapter 3.6] Let $F' = FK'$ be a constant field extension of F. Then the following hold:

1. $[F' : F] = [K' : K]$, and K' is algebraically closed in F'.
2. F'/F is unramified, that is, all $P \in \mathbb{P}_F$ are unramified in F'/F.
3. $g(F') = g(F)$.
4. For every divisor $A \in \mathrm{Div}(F)$, $\deg \mathrm{Con}_{F'/F}(A) = \deg A$ and $\ell(\mathrm{Con}_{F'/F}(A)) = \ell(A)$.

12.1.4 Differentials

12.1.89 Remark In this subsection we consider a function field F/K where $K = \mathbb{F}_q$ is a finite field of characteristic p. The aim is to give an interpretation of the *canonical divisors* of F.

12.1.90 Remark The set $F^p := \{z^p \mid z \in F\}$ is a subfield of F which contains K. The extension F/F^p has degree $[F : F^p] = p$ and is *purely inseparable*. An element $z \in F \setminus F^p$ is called a *separating element* for F/K. For every separating element z, the extension $F/K(z)$ is finite and separable.

12.1.91 Remark Recall that a *module* over a field L is just a vector space over L.

12.1.92 Definition Let M be a module over F. A *derivation* of F into M is a map $\delta : F \to M$, which is K-linear and satisfies the *product rule*

$$\delta(u \cdot v) = u \cdot \delta(v) + v \cdot \delta(u) \quad \text{for all} \quad u, v \in F.$$

12.1.93 Remark Let $\delta : F \to M$ be a derivation of F, $z \in F$ and $n \geq 0$. Then $\delta(z^n) = nz^{n-1} \cdot \delta(z)$. In particular, $\delta(z^p) = 0$ for all $z \in F$.

12.1.94 Proposition [2714, Proposition 4.1.4] Let x be a separating element for F/K. Then there exists a unique derivation $\delta_x : F \to F$ with the property $\delta_x(x) = 1$. We call δ_x the *derivation of F with respect to x.*

12.1.95 Proposition [2714, Chapter 4.1] There is a one–dimensional F-module Ω_F and a derivation $d : F \to \Omega_F$ (written as $z \mapsto dz$) with the following properties:

1. $dz \neq 0$ for every separating element $z \in F$.
2. $dz = \delta_x(z) \cdot dx$ for every $z \in F$ and $x \in F \setminus F^p$.

The pair (Ω_F, d) is the *differential module* of F/K, the elements of Ω_F are *differentials* of F/K.

12.1.96 Remark

1. If $z \in F$ is not separating then $dz = 0$.
2. Given a separating element $x \in F$, every differential $\omega \in \Omega_F$ has a unique representation $\omega = udx$ with $u \in F$, since Ω_F is a one-dimensional F-module.
3. Suppose that $\omega, \eta \in \Omega_F$ and $\omega \neq 0$. Then there is a unique element $u \in F$ such that $\eta = u\omega$. We write then $u = \eta/\omega$.
4. Item 2 in Proposition 12.1.95 indicates that, for a separating element $x \in F$,

$$\delta_x(z) = \frac{dz}{dx} \quad \text{for all } z \in F.$$

12.1.97 Remark One can attach a divisor to every nonzero differential $\omega \in \Omega_F$ as follows.

12.1.98 Definition [2714, Theorem 4.3.2(e)] Let $\omega \in \Omega_F$, $\omega \neq 0$.

1. Let $P \in \mathbb{P}_F$ and let t be a P-prime element (that is, $\nu_P(t) = 1$). Then t is a separating element of F/K, and we can write $\omega = u \cdot dt$ with $u \in F$. We define

$$\nu_P(\omega) := \nu_P(u).$$

This definition is independent of the choice of the prime element t, and one can show that $\nu_P(\omega) = 0$ for almost all $P \in \mathbb{P}_F$.

2. The *divisor of* ω is

$$\operatorname{div}(\omega) := \sum_{P \in \mathbb{P}_F} \nu_P(\omega)P.$$

12.1.99 Remark Divisors have the property $\operatorname{div}(u\omega) = \operatorname{div}(u) + \operatorname{div}(\omega)$ for $u \in F \setminus \{0\}$ and $\omega \in \Omega_F \setminus \{0\}$. Therefore $\operatorname{div}(\omega) \sim \operatorname{div}(\eta)$ for any two nonzero differentials $\omega, \eta \in \Omega_F$.

12.1.100 Remark Recall that the divisor of poles of an element $0 \neq x \in F$ is denoted by $(x)_\infty$.

12.1.101 Proposition [2714, Chapter 4.3] Let $x \in F$ be a separating element for F/K. Then

$$\operatorname{div}(dx) = -2(x)_\infty + \operatorname{Diff}(F/K(x)).$$

12.1.102 Theorem [2714, Chapter 4.3] Let $\omega \in \Omega_F$ be a nonzero differential of F/K. Then the divisor $W := \operatorname{div}(\omega)$ is a *canonical divisor* of F. In particular,

$$2g(F) - 2 = \deg(\operatorname{div}(\omega)).$$

12.1.103 Definition For every divisor $A \in \operatorname{Div}(F)$, we define the set

$$\Omega_F(A) := \{\omega \in \Omega_F \mid \operatorname{div}(\omega) \geq A\}.$$

12.1.104 Remark $\Omega_F(A)$ is a finite-dimensional K-vector space.

12.1.105 Theorem (Riemann–Roch theorem, 2nd version) For every divisor $A \in \operatorname{Div}(F)$,

$$\ell(A) = \deg A + 1 - g(F) + \dim \Omega_F(A),$$

where $\dim \Omega_F(A)$ means the dimension as a K-vector space.

12.1.106 Corollary We have $\dim \Omega_F(0) = g(F)$.

12.1.107 Remark We finish this subsection with examples of function fields that will be discussed in detail in Sections 12.2 and 12.4.

12.1.108 Definition

1. A function field F/K of genus $g(F) = 1$ is an *elliptic function field.*
2. A function field F/K is *hyperelliptic* if $g(F) \geq 2$, and there exists an element $x \in F$ such that $[F : K(x)] = 2$.

12.1.109 Example [2714, Chapters 6.1, 6.2] Let K be a finite field of characteristic $\neq 2$, and let F/K be an elliptic or hyperelliptic function field of genus g. Assume that F has at least one rational place P. Then there exist $x, y \in F$ such that $F = K(x, y)$ and $y^2 = f(x)$ with a square-free polynomial $f \in K[x]$ of degree $2g + 1$. The differentials

$$\omega_i := \frac{x^i}{y} dx \ , \quad i = 0, \ldots, g-1$$

form a basis of $\Omega_F(0)$.

12.1.5 Function fields and curves

12.1.110 Remark There is an alternative *geometric* approach to function fields via *algebraic curves.* We give here only a very brief (and incomplete) introduction. For more information we refer to [1147, 1427, 2281].

12.1.111 Remark Let K be a finite field, and denote by $\bar{K}$ the algebraic closure of K. Let $K[X_1, \ldots, X_n]$ be the ring of polynomials in n variables over K.

12.1.112 Definition

1. The *n-dimensional affine space* $\mathbf{A}^n = \mathbf{A}^n(\bar{K})$ *over* $\bar{K}$ is the set of all n-tuples of elements of $\bar{K}$. An element $P = (a_1, \ldots, a_n) \in \mathbf{A}^n$ is a *point*, and $a_1, \ldots, a_n$ are its *coordinates.*
2. Let $f_1, \ldots, f_m \in K[X_1, \ldots, X_n]$ be polynomials. Then the set $V := \{P \in \mathbf{A}^n \mid f_1(P) = \cdots = f_m(P) = 0\}$ is the *affine algebraic set defined by* $f_1 = \cdots = f_m = 0$. We say that V is *defined over* K since the polynomials $f_1, \ldots, f_m$ have coefficients in K.
3. Let V be as in 2. The set $I(V) := \{f \in \bar{K}[X_1, \ldots, X_n] \mid f(P) = 0 \text{ for all } P \in V\}$ is an ideal of $\bar{K}[X_1, \ldots, X_n]$, which is the *ideal of* V.
4. The algebraic set V is *absolutely irreducible* if $I(V)$ is a prime ideal of $\bar{K}[X_1, \ldots, X_n]$. Then the residue class ring $\Gamma(V) := \bar{K}[X_1, \ldots, X_n]/I(V)$ is an integral domain, and its quotient field $\bar{K}(V) := \mathrm{Quot}(\Gamma(V))$ is the *field of rational functions on* V. The residue class of X_i in $\bar{K}(V)$ is the *i-th coordinate function on* V and is denoted by x_i. The subfield $K(V) := K(x_1, \ldots, x_n) \subseteq \bar{K}(V)$ is the field of *K-rational functions on* V.
5. An absolutely irreducible affine algebraic set V is an *absolutely irreducible affine algebraic curve over* K (briefly, an *affine curve over* K), if the field $K(V)$ as defined in Part 4 has transcendence degree one over K. This means that $K(V)$ is an *algebraic function field* over K, as in Definition 12.1.1. The curve V is a *plane affine curve* if $V \subseteq \mathbf{A}^2$.
6. Let V be an affine curve over K. A point $P \in V$ is *K-rational* if all its coordinates are in K. We set $V(K) := \{P \in V \mid P \text{ is } K\text{-rational}\}$.
7. Two affine curves V_1 and V_2 are *birationally equivalent* if their function fields $K(V_1)$ and $K(V_2)$ are isomorphic.

12.1.113 Example Let F/K be an algebraic function field. Then there exist elements $x, y \in F$ such that $F = K(x, y)$, and there is an irreducible polynomial $f \in K[X, Y]$ such that $f(x, y) = 0$. Let $V \subseteq \mathbf{A}^2$ be the plane affine curve defined by $f = 0$. Then $K(V) = F$.

12.1.114 Definition

1. Let V be an affine curve as in Definition 12.1.112, and let $P \in V$. A rational function $\varphi \in \bar{K}(V)$ is *defined at* P if $\varphi = g(x_1, \ldots, x_n)/h(x_1, \ldots, x_n)$ with $g, h \in \bar{K}[x_1, \ldots, x_n]$ and $h(P) \neq 0$. The set $\mathcal{O}_P(V)$ of all rational functions on V which are defined at P, is a ring and it is the *local ring of* V *at* P.
2. The point P is *non-singular* if its local ring is *integrally closed.* This means, by definition, that every $z \in \bar{K}(V)$ which satisfies an *integral equation* over $\mathcal{O}_P(V)$, is in $\mathcal{O}_P(V)$; see Definition 12.1.61.
3. The curve V is *non-singular* if all of its points are non-singular.

12.1.115 Remark Let $f \in K[X,Y]$ be an *absolutely irreducible polynomial* (that is, f is irreducible in $\bar{K}[X,Y]$). Then the equation $f = 0$ defines a plane affine curve $\mathcal{C} \subseteq \mathbf{A}^2(\bar{K})$. A point $P \in \mathcal{C}$ is non-singular if and only $f_X(P) \neq 0$ or $f_Y(P) \neq 0$, where $f_X(X,Y)$ and $f_Y(X,Y)$ denote the partial derivatives with respect to X and Y, respectively.

12.1.116 Example Let $n > 0$ be relatively prime to the characteristic of K. Then the *Fermat curve* $\mathcal{C}$ which is defined by the equation $f(X,Y) = X^n + Y^n - 1 = 0$, is non-singular.

12.1.117 Remark In a sense, affine curves are not "complete," one has to add a finite number of points "at infinity." To be precise, one introduces the projective space $\mathbf{P}^n$ over $\bar{K}$ and the "projective closure" of an affine curve in $\mathbf{P}^n$. This leads to the concept of *projective curves.* We do not give details here and refer to textbooks on algebraic geometry, for example [1147, 1427, 2281].

12.1.118 Remark

1. Two projective curves are *birationally equivalent* if their function fields are isomorphic.
2. For every projective curve $\mathcal{C}$ there exists a *non-singular* projective curve $\mathcal{X}$ which is birationally equivalent to $\mathcal{C}$. The curve $\mathcal{X}$ is uniquely determined up to isomorphism and it is *the non-singular model of* $\mathcal{C}$.

12.1.119 Remark There is a 1–1 correspondence between {algebraic function fields F/K, up to isomorhism} and {absolutely irreducible, non-singular, projective curves $\mathcal{X}$ defined over K, up to isomorphism}. Under this correspondence, extensions F'/F of function fields correspond to coverings $\mathcal{X}' \to \mathcal{X}$ of curves, composites of function fields $E = F_1F_2$ correspond to fibre products of curves, etc. What corresponds to a place P of a function field F/K? If P is rational, then it corresponds to a K-rational point of the associated projective curve. Now let $K = \mathbb{F}_q$ and let P be a place of F with $\deg P = n$. Then P corresponds to exactly n points on the associated projective curve, with coordinates in the field $\mathbb{F}_{q^n}$. These points form an orbit under the Frobenius map, which is the map that raises the coordinates of points to the q-th power. For details, see [2281].

See Also

§12.2	For more material on function fields of genus 1 (elliptic curves).
§12.4	For more material on hyperelliptic function fields.
§12.5	For rational places of function fields (rational points on curves).
§12.6	For towers of function fields.
§15.2	For applications of function fields to coding theory.

References Cited: [1147, 1296, 1427, 1511, 1846, 2280, 2281, 2714, 2872]

12.2 Elliptic curves

Joseph Silverman, Brown University

12.2.1 Remark Most of the background material for this section may be found in [2670, Chapters III, V, XI]. Other standard references for the theory of elliptic curves include the books [557, 1563, 1756, 1773, 1843, 1845, 2054, 2107, 2667, 2672, 2950] and survey articles [556, 2784].

12.2.1 Weierstrass equations

12.2.2 Definition A *Weierstrass equation* over a field K is an equation of the form

$$y^2 + a_1xy + a_3y = x^3 + a_2x^2 + a_4x + a_6 \qquad \text{with } a_1, a_2, a_3, a_4, a_6 \in K. \tag{12.2.1}$$

Associated to a Weierstrass equation are the following quantities:

$$\begin{aligned}
&b_2 = a_1^2 + 4a_4, \qquad b_4 = 2a_4 + a_1a_3, \qquad b_6 = a_3^2 + 4a_6,\\
&b_8 = a_1^2a_6 + 4a_2a_6 - a_1a_3a_4 + a_2a_3^2 - a_4^2,\\
&c_4 = b_2^2 - 24b_4, \qquad c_6 = -b_2^3 + 36b_2b_4 - 216b_6,\\
&\Delta = -b_2^2b_8 - 8b_4^3 - 27b_6^2 + 9b_2b_4b_6, \qquad j = c_4^3/\Delta.
\end{aligned}$$

The quantity Δ is the *discriminant* and the quantity j is the *j-invariant.*

12.2.3 Remark The quantities in Definition 12.2.2 satisfy the relations

$$4b_8 = b_2b_6 - b_4^2 \qquad \text{and} \qquad 1728\Delta = c_4^3 - c_6^2.$$

12.2.4 Remark The discriminant vanishes if and only if the curve defined by the Weierstrass equation has a singular point, i.e, a point (x_0, y_0) on the curve where both partial derivatives vanish:

$$2y_0 + a_1x_0 + a_3 = 0 \quad \text{and} \quad 3x_0^2 + 2a_2x_0 + a_4 - a_1y_0 = 0.$$

12.2.5 Remark If $\mathrm{char}(K) \neq 2$, then the substitution $y \longmapsto \frac{1}{2}(y - a_1x - a_3)$ transforms the Weierstrass equation into the simpler form

$$y^2 = 4x^3 + b_2x^2 + 2b_4x + b_6.$$

If also $\mathrm{char}(K) \neq 3$, then the substitution $(x, y) \longmapsto \left(\frac{x-3b_2}{36}, \frac{y}{108}\right)$ yields the further simplification

$$y^2 = x^3 - 27c_4x - 54c_6.$$

12.2.6 Definition An *elliptic curve E defined over a field K* is a Weierstrass equation over K that is nonsingular, i.e., $\Delta \neq 0$, together with an extra point "at infinity" which is denoted $\mathcal{O}$. For any extension field L of K, the *set of points of E defined over L* is the set

$$E(L) = \left\{ \begin{matrix} \text{solutions } (x_0, y_0) \in L^2 \text{ to the Weier-} \\ \text{strass equation (12.2.1) defining } E \end{matrix} \right\} \cup \{\mathcal{O}\}.$$

12.2.7 Remark A fancier definition of an elliptic curve over K is a nonsingular projective curve of genus one defined over K with a marked point whose coordinates are in K. Using the Riemann–Roch theorem, one can show that every such curve is given by a Weierstrass equation, with the marked point being the point $\mathcal{O}$, which is the unique point at infinity; see [2670, III.3.1].

12.2.8 Example The real points on an elliptic curve defined over $\mathbb{R}$ may have one or two components, as illustrated in Figure 12.2.8*.

---

*The author thanks Springer Science+Business Media for their permission to include Figures 3.1 and 3.3 from his book *The Arithmetic of Elliptic Curves*, GTM 106, 2009.

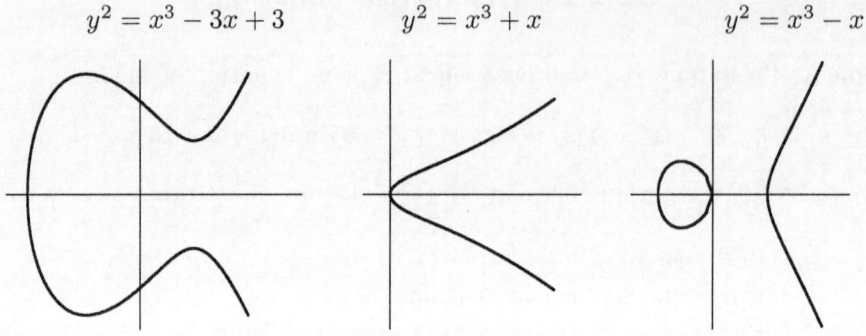

Figure 12.2.1 Three elliptic curves.

12.2.9 Proposition
The substitution

$$x \longrightarrow u^2x + r, \qquad y \longrightarrow u^3y + u^2sx + t, \tag{12.2.2}$$

transforms the Weierstrass equation (12.2.1) into a Weierstrass equation

$$y^2 + a_1'xy + a_3'y = x^3 + a_2'x^2 + a_4'x + a_6'$$

whose coefficients and associated quantities satisfy

$ua_1' = a_1 + 2s$
$u^2a_2' = a_2 - sa_1 + 3r - s^2$
$u^3a_3' = a_3 + ra_1 + 2t$
$u^4a_4' = a_4 - sa_3 + 2ra_2 - (t + rs)a_1 + 3r^2 - 2st$
$u^6a_6' = a_6 + ra_4 + r^2a_2 + r^3 - ta_3 - t^2 - rta_1$
$u^2b_2' = b_2 + 12r$
$u^4b_4' = b_4 + rb_2 + 6r^2$
$u^6b_6' = b_6 + 2rb_4 + r^2b_2 + 4r^3$
$u^8b_8' = b_8 + 3rb_6 + 3r^2b_4 + r^3b_2 + 3r^4$
$u^4c_4' = c_4 \quad u^6c_6' = c_6 \quad u^{12}\Delta' = \Delta \quad j' = j$

12.2.10 **Definition** Two elliptic curves E/K and E'/K are *isomorphic over* K if there is a substitution (12.2.2) with $u \in K^*$ and $r, s, t \in K$ that transforms the Weierstrass equation of E into the Weierstrass equation of E'.

12.2.11 **Theorem** [2670, III.1.4] Let K be an algebraically closed field. Then E/K and E'/K are isomorphic over K if and only if they have the same j-invariant, i.e., if and only if $j(E) = j(E')$.

12.2.12 **Example** According to Theorem 12.2.11, there are two $\overline{\mathbb{F}}_2$-isomorphism classes of elliptic curves defined over $\mathbb{F}_2$, namely those with j-invariant 0 and those with j-invariant 1. However, there are five $\mathbb{F}_2$-isomorphism classes of elliptic curves defined over $\mathbb{F}_2$. An example from each isomorphism class is listed in the following table.

curve	j
$y^2 + y = x^3$	0
$y^2 + y = x^3 + x$	0
$y^2 + y = x^3 + x + 1$	0
$y^2 + xy = x^3 + 1$	1
$y^2 + xy + y = x^3 + 1$	1

12.2.13 Theorem [2670, X.5.4.1] Let $p \geq 5$, let q be a power of p, and let $E_1/\mathbb{F}_q$ and $E_2/\mathbb{F}_q$ be elliptic curves. Then E_1 and E_2 are isomorphic over $\mathbb{F}_q$ if and only if $j(E_1) = j(E_2)$ and there exists a $u \in \mathbb{F}_q^*$ such that

$$\begin{cases} c_4(E)c_6(E) = u^2 c_4(E')c_6(E') & \text{if } j(E_1) \neq 0 \text{ and } j(E_1) \neq 1728, \\ c_4(E) = u^4 c_4(E') & \text{if } j(E_1) = 1728, \\ c_6(E) = u^6 c_6(E') & \text{if } j(E_1) = 0. \end{cases}$$

12.2.14 Definition Let $\mathcal{E}_q$ be the set of $\mathbb{F}_q$-isomorphism classes of elliptic curves defined over $\mathbb{F}_q$.

12.2.15 Remark The set $\mathcal{E}_2$ is described in Example 12.2.12; it has five elements. For $p \geq 5$ and q a power of p, we have

$$\#\mathcal{E}_q = 2(q-2) + \#\mathbb{F}_q^*/\mathbb{F}_q^{*4} + \#\mathbb{F}_q^*/\mathbb{F}_q^{*6}.$$

12.2.16 Remark There are curves over finite fields $\mathbb{F}_q$ that have no points with coordinates in $\mathbb{F}_q$, but a theorem of Lang says that this does not happen for curves of genus one.

12.2.17 Theorem [2670, Exercise 10.6] Let $C/\mathbb{F}_q$ be a smooth projective curve of genus one. Then $C(\mathbb{F}_q)$ is not empty. More generally, if $V/\mathbb{F}_q$ is a variety that is isomorphic over $\overline{\mathbb{F}}_q$ to an abelian variety, then $V(\mathbb{F}_q)$ is not empty.

12.2.18 Remark In particular, if $F(X, Y, Z) \in \mathbb{F}_q[X, Y, Z]$ is homogeneous of degree 3 and if the associated curve $F = 0$ is nonsingular, then there are values $x, y, z \in \mathbb{F}_q$, not all zero, such that $F(x, y, z) = 0$.

12.2.2 The group law

12.2.19 Definition Let E/K be an elliptic curve defined over a field K, and let L/K be an extension field. The set of points $E(L)$ forms an abelian group using the following rules:

1. The point $\mathcal{O}$ is the identity element. In what follows, we use the convention that $\mathcal{O}$ is on every vertical line.
2. The negative of the point $P = (x, y)$ is the point $-P = (x, -y-a_1x-a_3)$. If E is given by a simpler Weierstrass equation as in Remark 12.2.5, then $a_1 = a_3 = 0$, and $-P = (x, -y)$ is the reflection of P about the x-axis.
3. The sum of distinct points P and Q is obtained by intersecting the line through P and Q with E. This yields three points P, Q, and R (counted with appropriate multiplicities). Then $P + Q$ equals $-R$.
4. The sum of P with itself is obtained similarly, using the tangent line to E at P.

The geometric definition of the group law is illustrated in Figure 12.2.2.

12.2.20 Remark All of the group axioms are easy to verify except for associativity; see [2670, III.3.4] for a proof of associativity.

12.2.21 Algorithm Let E/K be an elliptic curve defined over a field K, and let L/K be an extension field. The following *addition algorithm* gives the group structure on the set of points $E(L)$.

1. The point $\mathcal{O}$ is the identity element, so

$$P + \mathcal{O} = \mathcal{O} + P = P \quad \text{for all } P \in E(L).$$

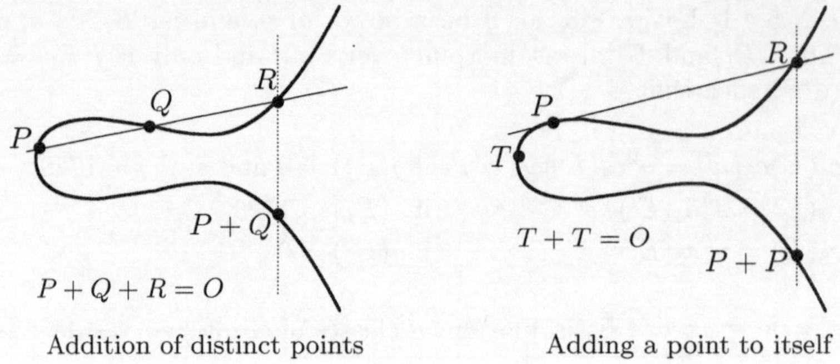

Figure 12.2.2 Examples of the addition law on an elliptic curve.

2. Let $P_0 = (x_0, y_0)$. Then

$$-P_0 = (x_0, -y_0 - a_1x_0 - a_3).$$

3. Let $P_1 = (x_1, y_1)$ and $P_2 = (x_2, y_2)$. If $x_1 = x_2$ and $y_1 + y_2 + a_1x_2 + a_3 = 0$, then

$$P_1 + P_2 = \mathcal{O}.$$

4. Otherwise, define λ and ν by the following formulas:

	λ	ν
$x_1 \neq x_2$	$\dfrac{y_2 - y_1}{x_2 - x_1}$	$\dfrac{y_1x_2 - y_2x_1}{x_2 - x_1}$
$x_1 = x_2$	$\dfrac{3x_1^2 + 2a_2x_1 + a_4 - a_1y_1}{2y_1 + a_1x_1 + a_3}$	$\dfrac{-x_1^3 + a_4x_1 + 2a_6 - a_3y_1}{2y_1 + a_1x_1 + a_3}$

Then the sum $P_3 = P_1 + P_2$ is given by

$$P_3 = (x_3, y_3) \quad \text{with} \quad x_3 = \lambda^2 + a_1\lambda - a_2 - x_1 - x_2 \quad \text{and} \quad y_3 = -(\lambda + a_1)x_3 - \nu - a_3.$$

12.2.22 Remark A special case of the addition algorithm is the *duplication formula*. Let $P = (x, y) \in E(L)$. Then the x-coordinate of $2P$ is

$$x(2P) = \frac{x^4 - b_4x^2 - 2b_6x - b_8}{4x^3 + b_2x^2 + 2b_4x + b_6}.$$

12.2.23 Remark Algorithm 12.2.21 explains how to add and double points on an elliptic curve. For cryptographic applications, it is important to do these operations as efficiently as possible. There are tradeoffs between using affine versus projective coordinates. It may also be possible to use the Frobenius endomorphism in place of the doubling map for "double-and-add" algorithms, and alternative equations for elliptic curves may also allow for more efficient operations. For details, see Sections 12.3 and 16.4.

12.2.24 Definition For a positive integer m, the *multiplication-by-m map* on $E(L)$ is the map

$$[m] : E(L) \longrightarrow E(L), \qquad [m](P) = \underbrace{P + P + \cdots + P}_{m \text{ terms}}.$$

For $m < 0$ define $[m](P) = -[-m](P)$, and set $[0](P) = \mathcal{O}$. The *kernel of multiplication-by-m* is the subgroup

$$E(L)[m] = \{P \in E(L) : [m](P) = \mathcal{O}\}.$$

12.2.25 Example The elliptic curve $E : y^2 = x^3 + x + 1$ over the field $\mathbb{F}_{11}$ has discriminant $\Delta = 10$ and j-invariant $j = 9$. The points $P = (2, 0)$, $Q = (6, 5)$, and $R = (8, 9)$ are in $E(\mathbb{F}_{11})$. The addition algorithm (Algorithm 12.2.21) allows us to compute quantities such as

$$P + Q = (8, 9), \quad Q + R = (4, 4), \quad [2]Q = (0, 1), \quad [3]P + [4]R = (4, 5).$$

The group $E(\mathbb{F}_{11})$ has 14 elements. The orders of the elements P, Q, and R are given by

$$[2]P = [7]Q = [14]R = \mathcal{O}.$$

So for example, the kernel of multiplication by 7 in $E(\mathbb{F}_{11})$ consists of the seven points

$$E(\mathbb{F}_{11})[7] = \big\{[n](Q) : 0 \le n \le 6\big\}, \quad \text{where } Q = (6, 5).$$

However, going to an extension field, one finds that $E(\overline{\mathbb{F}}_{11})[7]$ contains 49 points; see Theorem 12.2.60.

12.2.3 Isogenies and endomorphisms

12.2.26 Definition Let E_1/K and E_2/K be elliptic curves. An *isogeny* from E_1 to to E_2 is a map $\phi : E_1 \to E_2$ that is defined using rational functions and that sends the zero point on E_1 to the zero point on E_2. If the coefficients of the rational functions are in K, the isogeny is *defined over K*. If there is an isogeny from E_1 to E_2, then E_1 and E_2 are *isogenous.*

12.2.27 Definition Let E/K be an elliptic curve. An *endomorphism* is an isogeny from E to itself. An *automorphism* of E is an endomorphism of E that has an inverse, i.e., an isomorphism of E with itself.

12.2.28 Definition The set of isogenies from E_1 to E_2 is denoted $\mathrm{Hom}(E_1, E_2)$, the set of endomorphisms of E is denoted $\mathrm{End}(E)$, and the set of automorphisms of E is denoted $\mathrm{Aut}(E)$. A subscript K indicates that we take only maps that are defined over K, thus $\mathrm{Hom}_K(E_1, E_2)$, $\mathrm{End}_K(E)$, and $\mathrm{Aut}_K(E)$.

12.2.29 Example The multiplication-by-m map $[m] : E \to E$ is an endomorphism of E.

12.2.30 Example Let K be a field with $\mathrm{char}(K) \neq 2$, and let $a, b \in K$ satisfy $b(a^2 - 4b) \neq 0$. Then the elliptic curves

$$E_1 : y^2 = x^3 + ax^2 + bx, \qquad E_2 : Y^2 = X^3 - 2aX^2 + (a^2 - 4b)X,$$

are isogenous via the map

$$\phi : E_1 \longrightarrow E_2, \qquad (x, y) \longmapsto \left(\frac{y^2}{x^2}, \frac{y(b - x^2)}{x^2}\right).$$

There is also an isogeny going the other direction,

$$\hat{\phi} : E_2 \longrightarrow E_1, \qquad (X,Y) \longmapsto \left(\frac{Y^2}{4X^2}, \frac{Y(a^2 - 4b - X^2)}{8X^2} \right).$$

A direct computation shows that $\hat{\phi} \circ \phi = [2]$ on E_1 and $\phi \circ \hat{\phi} = [2]$ on E_2. The maps ϕ and $\hat{\phi}$ are examples of *dual isogenies* as described in Theorem 12.2.41.

12.2.31 Example Let K be a field of characteristic $p > 0$, let q be a power of p, let E/K be an elliptic curve given by a Weierstrass equation (12.2.1), and define an elliptic curve $E^{(q)}/K$ using the Weierstrass equation

$$E^{(q)} : y^2 + a_1^q xy + a_3^q y = x^3 + a_2^q x^2 + a_4^q x + a_6^q.$$

Then the *Frobenius map*

$$\phi_q : E(\overline{K}) \longrightarrow E^{(q)}(\overline{K}), \qquad \phi_q(x,y) = (x^q, y^q),$$

is an isogeny from E to $E^{(q)}$. Note that if $K = \mathbb{F}_q$, then $E = E^{(q)}$, so in this case ϕ_q is an endomorphism of E. Further,

$$\{P \in E(\overline{\mathbb{F}}_q) : \phi_q(P) = P\} = E(\mathbb{F}_q),$$

and more generally, since $\phi_{q^n} = \phi_q^n$,

$$\{P \in E(\overline{\mathbb{F}}_q) : \phi_q^n(P) = P\} = E(\mathbb{F}_{q^n}).$$

12.2.32 Proposition [2670, II.2.3] Let E_1/K and E_2/K be elliptic curves and let $\phi : E_1 \to E_2$ be an isogeny. Then either $\phi(P) = \mathcal{O}$ for all $P \in E_1(\overline{K})$, or else $\phi(E_1(\overline{K})) = E_2(\overline{K})$. (The constant map $\phi(P) = \mathcal{O}$ is the *zero isogeny*.)

12.2.33 Definition In general, the *degree* of a finite map $\phi : C_1 \to C_2$ between algebraic curves is the degree of the extension of function fields $K(C_1)/\phi^* K(C_2)$. The map is *separable* if the field extension $K(C_1)/\phi^* K(C_2)$ is separable, and otherwise it is inseparable (which can only happen in finite characteristic). The *inseparability degree of* ϕ, denoted $\deg_i(\phi)$, is the inseparability degree of the extension $K(E_1)/\phi^* K(E_2)$. Thus ϕ is separable if and only if $\deg_i(\phi) = 1$, and ϕ is *purely inseparable* if $\deg_i(\phi) = \deg(\phi)$.

12.2.34 Example The Frobenius map ϕ_q defined in Example 12.2.31 is purely inseparable. In general, for integers m and n, the map

$$[m + n\phi_q] : E(\overline{K}) \longrightarrow E^{(q)}(\overline{K}), \qquad [m + n\phi_q](P) = [m](P) + [n](\phi_q(P)),$$

is separable if and only if $\gcd(m, q) = 1$.

12.2.35 Proposition [2670, III.4.10] Let $\phi : E_1 \to E_2$ be a nonconstant isogeny of elliptic curves defined over K. If ϕ is separable, then ϕ is unramified, and for every point $Q \in E_2(\overline{K})$ we have

$$\#\{P \in E_1(\overline{K}) : \phi(P) = Q\} = \deg(\phi).$$

More generally, for every nonconstant isogeny ϕ,

$$\#\{P \in E_1(\overline{K}) : \phi(P) = Q\} = \frac{\deg(\phi)}{\deg_i(\phi)}.$$

12.2.36 **Proposition** [2670, II.2.12] Let $\phi : E_1 \to E_2$ be a nonconstant isogeny of elliptic curves defined over a field of characterstic p, and let $q = \deg_i(\phi)$ be the inseparability degree of ϕ. Then ϕ factors as

$$E_1 \xrightarrow{\phi_q} E_1^{(q)} \xrightarrow{\psi} E_2,$$

where ϕ_q is the Frobenius map (Example 12.2.31) and ψ is separable.

12.2.37 **Remark** An isogeny is only required to send zero to zero, but it turns out that this suffices to force it to be a homomorphism, as described in the next result.

12.2.38 **Theorem** [2670, III.4.8] Let E_1/K and E_2/K be elliptic curves and let $\phi : E_1 \to E_2$ be an isogeny. Then the map

$$\phi : E_1(\overline{K}) \longrightarrow E_2(\overline{K})$$

is a homomorphism, i.e., $\phi(P + Q) = \phi(P) + \phi(Q)$ for all $P, Q \in E_1(\overline{K})$.

12.2.39 **Definition** The *sum* of two isogenies $\phi, \psi \in \operatorname{Hom}(E_1, E_2)$ is the map

$$(\phi + \psi) : E_1(\overline{K}) \longrightarrow E_2(\overline{K}), \qquad (\phi + \psi)(P) = \phi(P) + \psi(P).$$

In this way, $\operatorname{Hom}(E_1, E_2)$ becomes a group. The *product* of two endomorphisms $\phi, \psi \in \operatorname{End}(E)$ is the composition

$$(\phi\psi) : E(\overline{K}) \longrightarrow E(\overline{K}), \qquad (\phi\psi)(P) = \phi(\psi(P)).$$

Addition and multiplication of endomorphisms give $\operatorname{End}(E)$ the structure of a (not necessarily commutative) ring. Similarly, multiplication (composition) gives $\operatorname{Aut}(E)$ the structure of a group; it is the group of units in the ring $\operatorname{End}(E)$.

12.2.40 **Remark** For every isogeny $E_1 \to E_2$ there is an isogeny going in the opposite direction, as described in the next theorem.

12.2.41 **Theorem** [2670, III.6.1, III.6.2] Let E_1/K and E_2/K be a elliptic curves, and let $\phi : E_1 \to E_2$ be an isogeny of degree n defined over K.

1 Then there is a unique isogeny $\hat{\phi} : E_2 \to E_1$ satisfying $\hat{\phi} \circ \phi = [n]$ on E_1. It is the *dual isogeny* to ϕ. It is defined over K and has the same degree as ϕ.
2. The dual isogeny satisfies $\phi \circ \hat{\phi} = [n]$ on E_2.
3. Let $\lambda : E_2 \to E_3$ be another isogeny. Then $\widehat{\lambda \circ \phi} = \hat{\phi} \circ \hat{\lambda}$.
4. Let $\psi : E_1 \to E_2$ be another isogeny. Then $\widehat{\phi + \psi} = \hat{\phi} + \hat{\psi}$.
5. Let $m \in \mathbb{Z}$. Then $\widehat{[m]} = [m]$.
6. $\hat{\hat{\phi}} = \phi$.

12.2.42 **Remark** Let E/K be an elliptic curve. Recall that $\operatorname{Div}(E)$ is the group of formal sums $\sum_{P \in E(\overline{K})} n_P(P)$, where the n_P are integers and only finitely many of them are nonzero. Also $\operatorname{Pic}(E)$, the Picard group of E, is the quotient of $\operatorname{Div}(E)$ by the subgroup of principal divisors, i.e., divisors of functions. So there is an exact sequence

$$1 \longrightarrow \overline{K}^* \longrightarrow \overline{K}(E)^* \longrightarrow \operatorname{Div}(E) \longrightarrow \operatorname{Pic}(E) \longrightarrow 0.$$

(For background on divisors, see Section 12.1.) On an elliptic curve, the group law can be used to describe the principal divisors, as in the next result.

12.2.43 Proposition [2670, III.3.5] Let E/K be an elliptic curve. A divisor $D = \sum n_P(P) \in \mathrm{Div}(E)$ is principal if and only if

$$\sum_{P \in E(\overline{K})} n_P = 0 \qquad \text{and} \qquad \sum_{P \in E(\overline{K})} [n_P](P) = \mathcal{O}.$$

(We note that the first sum is a sum of integers, while the second sum is a sum of points on the elliptic curve E.)

12.2.44 Proposition [2670, III.3.4] Let E/K be an elliptic curve, let $\mathrm{Div}^0(E)$ be the group of divisors of degree 0, and let $\mathrm{Pic}^0(E)$ be the corresponding group of divisor classes. Then there is an isomorphism

$$E(\overline{K}) \longrightarrow \mathrm{Pic}^0(E), \qquad P \longmapsto \text{divisor class of } (P) - (\mathcal{O}).$$

The inverse is the map that sends the divisor class of $\sum n_P(P)$ to the sum of points $\sum [n_P](P)$.

12.2.4 The number of points in $E(\mathbb{F}_q)$

12.2.45 Remark If $E/\mathbb{F}_q$ is an elliptic curve defined over a finite field, then $E(\mathbb{F}_q)$ is a finite (abelian) group.

12.2.46 Theorem (Hasse–Weil estimate) [2670, V.3.1] Let $E/\mathbb{F}_q$ be an elliptic curve defined over a finite field. Then

$$\left|q + 1 - \#E(\mathbb{F}_q)\right| \leq 2\sqrt{q}.$$

12.2.47 Definition Let $E/\mathbb{F}_q$ be an elliptic curve. The quantity

$$a_q(E) = q + 1 - \#E(\mathbb{F}_q)$$

is the *trace of the Frobenius map*; see Theorem 12.2.66.

12.2.48 Remark The following theorem of Birch describes how $a_q(E)$ is distributed as E ranges over all isomorphism classes of elliptic curves defined over $\mathbb{F}_q$.

12.2.49 Theorem [284], [2101, Appendix B]. For each $E \in \mathcal{E}_q$ (see Definition 12.2.14), write $a_q(E) = 2\sqrt{q}\cos\theta_q(E)$ with $0 \leq \theta_q(E) \leq \pi$. Then for all $0 \leq \alpha \leq \beta \leq \pi$,

$$\lim_{q \to \infty} \frac{\#\{E \in \mathcal{E}_q : \alpha \leq \theta_q(E) \leq \beta\}}{\#\mathcal{E}_q} = \frac{2}{\pi}\int_\alpha^\beta \sin^2(t)\, dt.$$

12.2.50 Remark The number of points in $E(\mathbb{F}_q)$ is constrained by Theorem 12.2.46. The exact orders that occur are given in the following theorem.

12.2.51 Theorem [2954] Let $q = p^n$ be a prime power, and let b be an integer with $|b| \leq 2\sqrt{q}$. Then there exists an elliptic curve $E/\mathbb{F}_q$ with $\#E(\mathbb{F}_q) = q + 1 - b$ if and only if b satisfies one of the following conditions:

1. $\gcd(b, p) = 1$;
2. n is even and $b = \pm 2\sqrt{q}$;
3. n is even and $p \not\equiv 1 \pmod 3$ and $b = \pm\sqrt{q}$;
4. n is odd and p equals 2 or 3 and $b = \pm p^{(n+1)/2}$;
5. n is odd and $b = 0$;

6. n is even and $p \not\equiv 1 \pmod 4$ and $b = 0$.

12.2.52 Remark Waterhouse [2954] proved a generalization of Theorem 12.2.51 for abelian varieties of arbitrary dimension. Deuring [825] had earlier proven the theorem in the case that q is prime, in which case every possible value of $\#E(\mathbb{F}_p)$ allowed by the Hasse–Weil constraint (Theorem 12.2.46) occurs. See also Rück's description [2497] of all possible structures for the group $E(\mathbb{F}_q)$ subject to the constraints provided by Theorem 12.2.51.

12.2.53 Remark For elliptic curves over finite fields, the number of points on the curve determines its isogeny class, as in the following result.

12.2.54 Theorem [2783] Let $E_1/\mathbb{F}_q$ and $E_2/\mathbb{F}_q$ be elliptic curves. Then the following are equivalent:

1. $\#E_1(\mathbb{F}_q) = \#E_2(\mathbb{F}_q)$;
2. E_1 and E_2 are $\mathbb{F}_q$-isogenous;
3. $\text{End}(E_1) \otimes \mathbb{Q} = \text{End}(E_2) \otimes \mathbb{Q}$.

12.2.5 Twists

12.2.55 Definition Let E/K be an elliptic curve. A *twist of* E is an elliptic curve E'/K such that E' is isomorphic to E over $\overline{K}$, but not necessarily isomorphic over K. Two twists E'/K and E''/K are *equivalent* if they are isomorphic over K.

12.2.56 Remark If E' and E'' are equivalent, then clearly $E'(K) \cong E''(K)$, but if E' and E'' are inequivalent twists of E, then $E'(K)$ and $E''(K)$ may differ.

12.2.57 Proposition [2670, X.5.4] Assume that $\text{char}(K) \neq 2, 3$. Let E/K be an elliptic curve. The set of equivalence classes of twists of E/K is isomorphic to $K^*/(K^*)^d$, where $d = 2$ if $j(E) \notin \{0, 1728\}$, $d = 4$ if $j(E) = 1728$, and $d = 6$ if $j(E) = 0$. (Equivalently, $d = \#\,\text{Aut}(E)$; see Theorem 12.2.83.) Precisely, for $D \in K^*/(K^*)^d$, the corresponding twist E' and isomorphism $E' \to E$ defined over $K(\sqrt[d]{D})$ are as follows:

$$\begin{array}{llll} [\mathbf{d=2}] & E : y^2 = x^3 + Ax + B, & E' : y^2 = x^3 + D^2Ax + D^3B, & (x, y) \mapsto (D^{-1}x, D^{-3/2}y). \\ [\mathbf{d=4}] & E : y^2 = x^3 + Ax, & E' : y^2 = x^3 + DAx, & (x, y) \mapsto (D^{-1/2}x, D^{-3/4}y). \\ [\mathbf{d=6}] & E : y^2 = x^3 + B, & E' : y^2 = x^3 + DB, & (x, y) \mapsto (D^{-1/3}x, D^{-1/2}y). \end{array}$$

12.2.58 Remark Let E/K and E'/K be quadratic twists ($d = 2$) as in Proposition 12.2.57,

$$E : y^2 = x^3 + Ax + B, \quad E' : y^2 = x^3 + D^2Ax + D^3B, \quad (x, y) \mapsto (D^{-1}x, D^{-3/2}y).$$

Let $\text{Gal}\big(K(\sqrt{D})/K\big) = \{1, \sigma\}$. Then the group $E'(K)$ may be identified with a subgroup of $E\big(K(\sqrt{D})\big)$ via

$$E'(K) \cong \big\{P \in E\big(K(\sqrt{D})\big) : \sigma(P) = -P\big\},$$

just as $E(K)$ is the subgroup of $E\big(K(\sqrt{D})\big)$ defined by the relation $\sigma(P) = P$. With this identification, the kernel and cokernel of the natural map

$$E(K) \oplus E'(K) \longrightarrow E\big(K(\sqrt{D})\big), \qquad (P, Q) \longmapsto P + Q,$$

are finite 2-groups.

12.2.59 Proposition Let $p \geq 5$, let q be a power of p, let $E/\mathbb{F}_q$ be an elliptic curve, and let $d > 1$ divide $\#\,\text{Aut}(E)$. Then Proposition 12.2.57 says that for each $D \in \mathbb{F}_q^*/(\mathbb{F}_q^*)^d$, the curve $E/\mathbb{F}_q$ has a twist $E_D/\mathbb{F}_q$. The orders of the groups of points on the twists satisfy the following relations:

1. $\sum_{D\in\mathbb{F}_q^*/(\mathbb{F}_q^*)^d} \#E_D(\mathbb{F}_q) = d(q-1)$;
2. $\prod_{D\in\mathbb{F}_q^*/(\mathbb{F}_q^*)^d} \#E_D(\mathbb{F}_q) = \#E(\mathbb{F}_{q^d})$.

12.2.6 The torsion subgroup and the Tate module

12.2.60 Theorem [2670, III.6.4] Let E/K be an elliptic curve, and let $\overline{K}$ be an algebraic closure of K.

1. Let $p = \mathrm{char}(K)$. Then
$$E(\overline{K})[m] \cong \mathbb{Z}/m\mathbb{Z} \times \mathbb{Z}/m\mathbb{Z} \quad \text{for all } m \geq 1 \text{ with } p \nmid m.$$
(If $\mathrm{char}(K) = 0$, then this holds for all $m \geq 1$.)
2. If $\mathrm{char}(K) = p > 0$, then one of the following is true
$$\begin{aligned} E(\overline{K})[p^r] &\cong \mathbb{Z}/p^r\mathbb{Z} && \text{for all } r \geq 1, \\ E(\overline{K})[p^r] &= 0 && \text{for all } r \geq 1. \end{aligned}$$

12.2.61 Example Let E be an elliptic curve defined by a Weierstrass equation (12.2.1). One can determine conditions on the coordinates of a point $P = (x, y) \in E$ for it to be a torsion point. For example
$$\begin{aligned} [2](P) = \mathcal{O} \quad &\text{if and only if} \quad 2y + a_1x + a_3 = 0 \\ &\text{if and only if} \quad 4x^4 - b_4x^2 - 2b_6x - b_8 = 0, \end{aligned}$$
and
$$[3](P) = \mathcal{O} \quad \text{if and only if} \quad 3x^4 + b_2x^3 + 3b_4x^2 + 3b_6x + b_8 = 0.$$
In general, there is a *division polynomial* $\psi_n(x, y) \in K[x, y]$ with the property that $[n](P) = \mathcal{O}$ if and only if $\psi_n(x, y) = 0$; see [2670, Exercise 3.7]. These polynomials can be computed recursively using the formula
$$\psi_{n+m}\psi_{n-m}\psi_r^2 = \psi_{n+r}\psi_{n-r}\psi_m^2 - \psi_{m+r}\psi_{m-r}\psi_n^2, \qquad \text{valid for all } n > m > r.$$

12.2.62 Remark Letting $m = \ell^i$ run over larger and larger powers of a prime ℓ, we obtain a module over the ring of ℓ-adic integers $\mathbb{Z}_\ell$. This is convenient because $\mathbb{Z}_\ell$ is a ring of characteristic zero.

12.2.63 Definition Let E/K be an elliptic curve. The *ℓ-adic Tate module of* E is the inverse limit
$$T_\ell(E) = \varprojlim E(\overline{K})[\ell^n].$$
If the characteristic of K is different from ℓ, then $T_\ell(E) \cong \mathbb{Z}_\ell \times \mathbb{Z}_\ell$, although the isomorphism depends on choosing a basis.

12.2.64 Remark Let $E/\mathbb{F}_q$ be an elliptic curve, where q is a power of p. The Frobenius map $\mathbb{F}_q$: $E(\overline{\mathbb{F}}_q) \longrightarrow E(\overline{\mathbb{F}}_q)$ (Example 12.2.31) maps $E(\overline{\mathbb{F}}_q)[m]$ to $E(\overline{\mathbb{F}}_q)[m]$. If $p \nmid m$, then $E(\overline{\mathbb{F}}_q)[m]$ is a free $\mathbb{Z}/m\mathbb{Z}$-module of rank two, so choosing a $\mathbb{Z}/m\mathbb{Z}$-basis T_1, T_2 for $E(\overline{\mathbb{F}}_q)[m]$, the Frobenius map satisfies
$$\phi_q(T_1) = aT_1 + bT_2 \quad \text{and} \quad \phi_q(T_2) = cT_1 + dT_2 \quad \text{for some } a, b, c, d \in \mathbb{Z}/m\mathbb{Z}.$$
Thus the action of ϕ_q on $E(\overline{\mathbb{F}}_q)[m]$ is represented by the matrix $\tilde{\phi}_{q,m} = \left(\begin{smallmatrix} a & b \\ c & d \end{smallmatrix}\right)$. The matrix depends on the choice of the basis for $E(\overline{\mathbb{F}}_q)[m]$, but its trace and determinant do not.

The maps $\tilde{\phi}_{q,\ell^i} : E(\overline{\mathbb{F}}_q)[\ell^i] \to E(\overline{\mathbb{F}}_q)[\ell^i]$ fit together to give a map $\phi_{q,\ell} : T_\ell(E) \to T_\ell(E)$. Choosing a basis for $T_\ell(E) \cong \mathbb{Z}_\ell^2$ allows us to evaluate the trace and the determinant of $\phi_{q,\ell}$ as ℓ-adic numbers.

12.2.65 Remark The following theorem explains why the quantity $a_q(E)$ in Definition 12.2.47 is the trace of the Frobenius map.

12.2.66 Theorem [2670, V.2.6] Let $E/\mathbb{F}_q$ be an elliptic curve with q a power of the prime p. Then for every integer m with $p \nmid m$,

$$\operatorname{Tr}(\tilde{\phi}_{q,m}) \equiv q + 1 - \#E(\mathbb{F}_q) \pmod{m} \qquad \text{and} \qquad \det(\tilde{\phi}_{q,m}) \equiv q \pmod{m},$$

and for every prime $\ell \neq p$,

$$\operatorname{Tr}(\phi_{q,\ell}) = q + 1 - \#E(\mathbb{F}_q) \qquad \text{and} \qquad \det(\phi_{q,\ell}) = q.$$

12.2.67 Remark More generally, any isogeny $\phi : E_1 \to E_2$ induces a homomorphism of the associated Tate modules $T_\ell(E_1) \to T_\ell(E_2)$, and if ϕ is defined over K, then the induced map is $\operatorname{Gal}(\overline{K}/K)$-invariant.

12.2.68 Theorem [2783] Let $E_1/\mathbb{F}_q$ and $E_2/\mathbb{F}_q$ be elliptic curves defined over a finite field, and let ℓ be a prime different from the characteristic of $\mathbb{F}_q$. Then the natural map

$$\operatorname{Hom}_K(E_1, E_2) \otimes \mathbb{Z}_\ell \longrightarrow \operatorname{Hom}_{\mathbb{Z}_\ell}\big(T_\ell(E_1), T_\ell(E_2)\big)^{\operatorname{Gal}(\overline{K}/K)}$$

is an isomorphism.

12.2.69 Remark Theorem 12.2.68 is also true for elliptic curves defined over number fields. This was conjectured by Tate and proven by Faltings [1024].

12.2.7 The Weil pairing and the Tate pairing

12.2.70 Theorem (Weil pairing) [2670, III.8.1, III.8.2] Let E/K be an elliptic curve, let $m \geq 1$ be an integer that is prime to the characteristic of K, and let $\boldsymbol{\mu}_m \subset \overline{K}^*$ denote the group of m-th roots of unity. There is a pairing

$$e_m : E(\overline{K})[m] \times E(\overline{K})[m] \longrightarrow \boldsymbol{\mu}_m$$

with the following properties:

1. It is bilinear, i.e., $e_m(P+Q, R) = e_m(P, R)e_m(Q, R)$ and $e_m(P, Q+R) = e_m(P, Q)e_m(P, R)$.
2. It is alternating, i.e., $e_m(P, P) = 1$, so in particular $e_m(Q, P) = e_m(P, Q)^{-1}$.
3. It is non-degenerate, i.e., if $e_m(P, Q) = 1$ for all $Q \in E(\overline{K})[m]$, then $P = 0$.
4. It is Galois invariant, i.e., $e_m\big(\sigma(P), \sigma(Q)\big) = \sigma\big(e_m(P, Q)\big)$ for all $\sigma \in \operatorname{Gal}(\overline{K}/K)$.
5. It is compatible in towers, i.e., $e_{mn}(P, Q) = e_m([n](P), Q)$ for $P \in E(\overline{K})[mn]$ and $Q \in E(\overline{K})[n]$.
6. It respects duality, i.e., let $\phi : E_1 \to E_2$ be an isogeny, then $e_m(P, \hat{\phi}(Q)) = e_m(\phi(P), Q)$ for $P \in E_1(\overline{K})[m]$ and $Q \in E_2(\overline{K})[m]$.

12.2.71 Remark The non-degeneracy and Galois invariance of the Weil pairing imply that there exist points $P, Q \in E(\overline{K})[m]$ such that $e_m(P, Q)$ is a primitive m-th root of unity. In particular, if $E(\overline{K})[m] \subset E(K)$, then $\boldsymbol{\mu}_m \subset K$.

12.2.72 Remark The Weil pairings on $E(\overline{K})[\ell^i]$ fit together to give a bilinear, alternating, non-degenerate, Galois invariant pairing

$$e_{\ell^\infty} : T_\ell(E) \times T_\ell(E) \longrightarrow T_\ell(\boldsymbol{\mu}),$$

where $T_\ell(\boldsymbol{\mu})$ is the inverse limit of $\boldsymbol{\mu}_{\ell^i}$. Further, $e_{\ell^\infty}(P, \hat{\phi}(Q)) = e_{\ell^\infty}(\phi(P), Q)$.

12.2.73 Theorem Each of the following recipes computes the Weil pairing $e_m(P,Q)$.

1. Choose a function $f_P \in \overline{K}(E)$ whose divisor satisfies $\operatorname{div}(f_P) = m(P) - m(\mathcal{O})$. Choose a function $g_P \in \overline{K}(E)$ satisfying $f_P \circ [m] = g_P^m$. Then the quantity $g_P(S+Q)/g_P(S)$ does not depend on the point $S \in E(\overline{K})$, and its value is $e_m(P,Q)$.
2. Choose arbitrary points $S, T \in E(\overline{K})$ and choose functions F_P and F_Q in $\overline{K}(E)$ whose divisors satisfy $\operatorname{div}(F_P) = m(S+P) - m(S)$ and $\operatorname{div}(F_Q) = m(T+Q) - m(T)$. Then the quantity
$$\frac{F_Q(T+P)}{F_Q(T)} \Big/ \frac{F_P(S+Q)}{F_P(S)}$$
does not depend on S or T and is equal to $e_m(P,Q)$.
3. Let $f_P \in \overline{K}(E)$ have divisor $\operatorname{div}(f_P) = m(P) - m(\mathcal{O})$ as in Part 1. This determines f_P up to multiplication by a constant. The function f_P has a pole of order m at $\mathcal{O}$, so the function $(x/y)^m f_P$ is defined and nonzero at $\mathcal{O}$. We adjust the constant so that $\big((x/y)^m f_P\big)(\mathcal{O}) = 1$. Similarly, let $f_Q \in \overline{K}(E)$ have divisor $\operatorname{div}(f_Q) = m(Q) - m(\mathcal{O})$, and normalize f_Q in the same way. Then $e_m(P,Q) = (-1)^m f_P(Q)/f_Q(P)$.

12.2.74 Remark The fact that functions with the stated properties used in Theorem 12.2.73 exist follows from Proposition 12.2.43, which explains when a divisor is the divisor of a function. However, it is far from obvious that the formulas in Theorem 12.2.73 yield the same value; see [2670, page 462] for the equality of formulas of Parts 1 and 2, but note that the last line of [2670, page 462] should be replaced by $= e(Q,P) = e(P,Q)^{-1}$.

12.2.75 Theorem (Tate Pairing) [2670, XI §9] Let E/K be an elliptic curve and let $m \geq 1$ be an integer that is prime. There is a bilinear pairing

$$\mathsf{T} : E(K)[m] \times E(K)/mE(K) \longrightarrow K^*/(K^*)^m$$

defined as follows: Let $T \in E(K)[m]$ and $P \in E(K)$. Choose a point $Q \in E(\overline{K})$ with $[m](Q) = P$. Then there exists an $\alpha \in K^*$ such that

$$e_m(\sigma(Q) - Q, T) = (\sqrt[m]{\alpha})^\sigma / \sqrt[m]{\alpha} \quad \text{for all } \sigma \in \operatorname{Gal}(\overline{K}/K),$$

and we set $\mathsf{T}(T,P) = \alpha \bmod (K^*)^m$. The Tate pairing may be computed by choosing a function $f_T \in \overline{K}(E)$ with divisor $\operatorname{div}(f_T) = m(T) - m(\mathcal{O})$, and then $\mathsf{T}(T,P) = f_T(P+Q)/f_T(Q)$ for any $Q \in E(\overline{K})$ such that the functions are defined and nonzero.

12.2.76 Remark If m is large, it is not clear in practice how to compute the functions used to evaluate the Weil pairing (Theorem 12.2.73) and the Tate pairing (Theorem 12.2.75). A double-and-add algorithm due to Miller allows these pairings to be computed quite efficiently; see Theorem 16.4.38.

12.2.77 Remark There are functorial definitions of the Weil and Tate pairings from which our rather ad hoc definitions may be derived. Briefly, the Weil pairing is a pairing between the m torsion on an abelian variety A and the m-torsion on its dual $\hat{A} \cong \operatorname{Ext}(A, \mathbb{G}_m)$, combined with an identification of $\hat{A}$ with the Picard group $\mathrm{Pic}^0(A)$. The Tate pairing is a cup product pairing on Galois cohomology that uses the Weil pairing to map $A[m] \otimes \hat{A}[m]$ to $\boldsymbol{\mu}_m$. For details, see for example [2671].

12.2.8 The endomorphism ring and automorphism group

12.2.78 **Definition** Let $\mathcal{A}$ be a finite $\mathbb{Q}$-algebra, i.e., $\mathcal{A}$ is a ring that contains $\mathbb{Q}$ as a subring and that is a finite dimensional $\mathbb{Q}$-vector space. An *order* of $\mathcal{A}$ is a subring of $\mathcal{A}$ that is finitely generated as a $\mathbb{Z}$-module and that contains a $\mathbb{Q}$-basis of $\mathcal{A}$. A *maximal order* is an order that is contained in no other orders.

12.2.79 **Example** Let $D \in \mathbb{Z}$ be a positive integer that is not a perfect square and let $\mathbb{Z}[\delta]$ be the ring of integers of $\mathbb{Q}(\sqrt{-D})$. For example, we can take $\delta = \frac{-D+\sqrt{-D}}{2}$. Then every order in $\mathbb{Q}(\sqrt{-D})$ has the form $\mathbb{Z} + f\mathbb{Z}[\delta]$ for some integer $f \geq 1$. The integer f is the *conductor* of the order $\mathbb{Z} + f\mathbb{Z}[\delta]$.

12.2.80 **Definition** A (definite) *quaternion algebra* over $\mathbb{Q}$ is a non-commutative ring of the form $\mathbb{Q}+\mathbb{Q}\sqrt{a}\boldsymbol{i}+\mathbb{Q}\sqrt{b}\boldsymbol{j}+\mathbb{Q}\sqrt{ab}\boldsymbol{k}$, where $a, b \in \mathbb{Q}$ are positive numbers and $\boldsymbol{i}, \boldsymbol{j}, \boldsymbol{k}$ are quantities satisfying the multiplication rules $\boldsymbol{i}^2 = \boldsymbol{j}^2 = \boldsymbol{k}^2 = \boldsymbol{ijk} = -1$.

12.2.81 **Remark** A number field has a unique maximal order, its ring of integers. Quaternion algebras may have many maximal orders.

12.2.82 **Theorem** [2670, III.9.4] Let E/K be an elliptic curve. The endomorphism ring $\mathrm{End}(E)$ of E has one of the following forms:

1. $\mathrm{End}(E) = \mathbb{Z}$;
2. $\mathrm{End}(E)$ is an order in a quadratic imaginary field $\mathbb{Q}(\sqrt{-D})$;
3. $\mathrm{End}(E)$ is an order in a quaternion algebra. (This form is only possible if K is a finite field.)

12.2.83 **Theorem** [2670, III.10.1] Let E/K be an elliptic curve. Then its automorphism group $\mathrm{Aut}(E)$ is a finite group whose order is given in the following table, where $j(E)$ is the j-invariant of E:

$\#\mathrm{Aut}(E)$	$j(E)$	$\mathrm{char}(K)$
2	$j(E) \neq 0, 1728$	—
4	$j(E) = 1728$	$\mathrm{char}(K) \neq 2, 3$
6	$j(E) = 0$	$\mathrm{char}(K) \neq 2, 3$
12	$j(E) = 0 = 1728$	$\mathrm{char}(K) = 3$
24	$j(E) = 0 = 1728$	$\mathrm{char}(K) = 2$

12.2.84 **Example** Let K be a field whose characteristic is neither 2 nor 3, and let $A, B \in K^*$ with $4A^3 + 27B^2 \neq 0$. Then

$$E_{A,B} : y^2 = x^3 + Ax + B$$

is an elliptic curve whose automorphism group is as follows:

1. If $AB \neq 0$, then $\mathrm{Aut}(E_{A,B}) = \boldsymbol{\mu}_2$.
2. If $B = 0$, then $j(E) = 1728$ and $\mathrm{Aut}(E_{A,0}) = \boldsymbol{\mu}_4$, where $[\zeta](x, y) = (\zeta^2 x, \zeta y)$ for $\zeta \in \boldsymbol{\mu}_4$.
3. If $A = 0$, then $j(E) = 0$ and $\mathrm{Aut}(E_{0,B}) = \boldsymbol{\mu}_6$, where $[\zeta](x, y) = (\zeta^2 x, \zeta^3 y)$ for $\zeta \in \boldsymbol{\mu}_6$.

12.2.85 Remark Theorem 12.2.51 lists the possible values of $\#E(\mathbb{F}_q)$. These determine the associated endomorphism rings as described in the following theorem.

12.2.86 Theorem [2954] Let $E/\mathbb{F}_q$ be an elliptic curve, let $\phi_q \in \text{End}(E)$ be the q-power Frobenius map, and let $\mathcal{A} = \text{End}(E) \otimes \mathbb{Q}$. Referring to the classification in Theorem 12.2.51:

In Case 1, $\mathcal{A} = \mathbb{Q}(\phi_q)$ is a quadratic imaginary field and $\text{End}(E)$ is an arbitrary order in $\mathcal{A}$.

In Case 2, $\mathcal{A}$ is a quaternion algebra, $\phi_q \in \mathbb{Z}$, and $\text{End}(E)$ is a maximal order in $\mathcal{A}$.

In Cases 3–6, $\mathcal{A} = \mathbb{Q}(\phi_q)$ is a quadratic imaginary field and $\text{End}(E)$ is an order in $\mathcal{A}$ whose conductor is not divisible by p.

12.2.9 Ordinary and supersingular elliptic curves

12.2.87 Theorem [2670, V.3.1] Let E/K be an elliptic curve defined over a field of characteristic $p > 0$. The following are equivalent:

1. $E[p^n] = 0$ for some $n \geq 1$, equivalently for all $n \geq 1$.
2. The dual of the Frobenius map, $\hat{\phi}_{p^n} : E^{(p^n)} \to E$, is purely inseparable for some $n \geq 1$, equivalently for all $n \geq 1$.
3. The multiplication-by-p^n map $[p^n] : E \to E$ is purely inseparable for some $n \geq 1$, equivalently for all $n \geq 1$.
4. The endomorphism ring $\text{End}(E)$ is an order in a quaternion algebra.
5. The formal group associated to E has height 2. (See [2670, IV §1] for the construction of the formal group associated to an elliptic curve and [2670, IV §7] for the definition the height of a formal group.)

12.2.88 Definition Let K be a field of positive characteristic p. An elliptic curve E/K is *supersingular* if one (equivalently all) of the conditions in Theorem 12.2.87 are true. Otherwise E is *ordinary*.

12.2.89 Theorem [2670, V.3.1] If E is supersingular, then $j(E) \in \mathbb{F}_{p^2}$.

12.2.90 Remark We comment that despite their name, supersingular elliptic curves are nonsingular curves, since they are elliptic curves, which are nonsingular by definition.

12.2.91 Remark In characteristic 2 one can check that $E/\overline{\mathbb{F}}_2$ is supersingular if and only if $j(E) = 0$. There is thus only one $\overline{\mathbb{F}}_2$-isomorphism class of such curves, a representative being $y^2 + y = x^3$. However, there are three $\mathbb{F}_2$-isomorphism classes of supersingular elliptic curves; see Example 12.2.12.

12.2.92 Remark The following theorem describes all supersingular curves in characteristic $p \geq 3$.

12.2.93 Theorem [2670, V.4.1] Let $q = p^n$ with $p \geq 3$.

1. Let $E/\mathbb{F}_q$ be an elliptic curve given by a Weierstrass equation of the form $y^2 = f(x)$ with $f \in \mathbb{F}_q[x]$ a monic cubic polynomial. Then E is supersingular if and only if the coefficient of x^{p-1} in $f(x)^{(p-1)/2}$ is zero.
2. Let $\lambda \in \overline{\mathbb{F}}_p$. Then the elliptic curve $y^2 = x(x-1)(x-\lambda)$ is supersingular if and only if λ is a root of the polynomial

$$H_p(T) = \sum_{i=0}^{\frac{p-1}{2}} \binom{\frac{p-1}{2}}{i}^2 T^i.$$

3. The polynomial $H_p(T)$ has distinct roots in $\overline{\mathbb{F}}_p$. There is one supersingular elliptic curve in characteristic 3. For $p \geq 5$, the number of $\overline{\mathbb{F}}_p$-isomorphism classes of supersingular elliptic curves is

$$\frac{p-1}{12} + \frac{1}{2}A_p + \frac{1}{2}B_p,$$

where

$$A_p = \begin{cases} 0 & \text{if } p \equiv 1 \pmod 3, \\ 1 & \text{if } p \equiv 2 \pmod 3, \end{cases} \qquad \text{and} \qquad B_p = \begin{cases} 0 & \text{if } p \equiv 1 \pmod 4, \\ 1 & \text{if } p \equiv 3 \pmod 4. \end{cases}$$

Equivalently, this number is equal to

$$\left\lfloor \frac{p}{12} \right\rfloor + \begin{cases} 0 & \text{if } p \equiv 1 \pmod{12}, \\ 1 & \text{if } p \equiv 5 \text{ or } 7 \pmod{12}, \\ 2 & \text{if } p \equiv 11 \pmod{12}. \end{cases}$$

12.2.94 Example The factorization of the first few polynomials $H_p(T)$ modulo p are listed in the following table.

p	$H_p(T) \pmod p$
3	$T+1$
5	T^2+4T+1
7	$(T+1)(T+3)(T+5)$
11	$(T+1)(T+5)(T+9)(T^2+10T+1)$
13	$(T^2+4T+9)(T^2+7T+1)(T^2+12T+3)$
17	$(T^2+T+16)(T^2+14T+1)(T^2+16T+1)(T^2+16T+16)$
19	$(T+1)(T+9)(T+17)(T^2+4T+1)(T^2+13T+6)(T^2+18T+16)$

12.2.95 Example The elliptic curve $y^2 = x^3 + 1$ is supersingular over $\mathbb{F}_p$ if $p \equiv 2 \pmod 3$ and ordinary if $p \equiv 1 \pmod 3$. Similarly, the curve $y^2 = x^3 + x$ is supersingular over $\mathbb{F}_p$ if $p \equiv 3 \pmod 4$ and ordinary if $p \equiv 1 \pmod 4$.

12.2.96 Remark [2670, V.4.2] Associated to the elliptic curve $y^2 = x(x-1)(x-T)$ is the Picard–Fuchs differential operator

$$\mathcal{D} = 4T(1-T)\frac{d^2}{dT^2} + 4(1-2T)\frac{d}{dT} - 1.$$

The polynomial $H_p(T)$ in Theorem 12.2.93, Part 3 satisfies

$$\mathcal{D}H_p(T) = p\sum_{i=0}^{\frac{p-1}{2}} (p-2-4i)\binom{\frac{p-1}{2}}{i}^2 T^i.$$

In particular, $\mathcal{D}H_p(T) = 0$ in characteristic p, which shows that $H_p(T)$ has simple roots in $\overline{\mathbb{F}}_p$, since $H_p(0) = 1$ and $H_p(1) = (-1)^{(p-1)/2}$ in $\mathbb{F}_p$.

12.2.97 Theorem [2670, Exercise 5.9] The following mass formula for supersingular elliptic curves is due to Eichler and Deuring:

$$\sum_{\substack{E/\overline{\mathbb{F}}_p \\ \text{supersingular}}} \frac{1}{\#\operatorname{Aut}(E)} = \frac{p-1}{24}.$$

12.2.98 Remark Let $E/\mathbb{Q}$ be an elliptic curve given by a Weierstrass equation with integer coefficients. Reducing the coefficients modulo p gives an elliptic curve $\tilde{E}_p/\mathbb{F}_p$ for all primes $p \nmid \Delta$. If $\text{End}(E)$ is an order in a quadratic imaginary field k, then $\tilde{E}_p$ is supersingular if p is inert in k and ordinary if p is split in k; see Definition 12.4.3.

12.2.99 Remark The situation if $\text{End}(E) = \mathbb{Z}$ is more complicated. Serre and Elkies [969, 2590] have proven that $\mathcal{SS}(X) = \#\{p < X : \tilde{E}_p \text{ is supersingular}\}$ is smaller that $X^{3/4+\epsilon}$ as $X \to \infty$. Lang and Trotter have conjectured [1848] that $\mathcal{SS}(X)$ is asymptotic to $C\sqrt{X}/\log(X)$ for a certain positive constant C. In the opposite direction, Elkies [968] has proven that $\tilde{E}_p$ is supersingular for infinitely many p, i.e., $\mathcal{SS}(X) \to \infty$ as $X \to \infty$.

12.2.10 The zeta function of an elliptic curve

12.2.100 Remark This section describes the zeta function of an elliptic curve. See Sections 11.6 and 12.7 for zeta functions of arbitrary curves and higher dimensional algebraic varieties.

12.2.101 Definition Let $E/\mathbb{F}_q$ be an elliptic curve. The *zeta function of* $E/\mathbb{F}_q$ is the formal power series

$$Z(E/\mathbb{F}_q, T) = \exp\left(\sum_{n=1}^{\infty} \frac{\#E(\mathbb{F}_{q^n})}{n} T^n\right)$$

12.2.102 Remark It might seem more natural to use the series $\sum_{n=1}^{\infty} \#E(\mathbb{F}_{q^n})T^n$, but the series defining $Z(E/\mathbb{F}_q, T)$ has better transformation properties.

12.2.103 Theorem [2670, V.2.4] Let $E/\mathbb{F}_q$ be an elliptic curve, and let $a = a_q(E)$ be the trace of Frobenius for E (Definition 12.2.47). Then

$$Z(E/\mathbb{F}_q, T) = \frac{1 - aT + qT^2}{(1-T)(1-qT)}. \tag{12.2.3}$$

12.2.104 Remark The zeta function satisfies

$$Z\left(E/\mathbb{F}_q, \frac{1}{qT}\right) = Z(E/\mathbb{F}_q, T).$$

This is an instance of Poincaré duality. For the general statement of Poincaré duality and the associated functional equation for zeta functions of varieties over finite fields, see Theorems 12.7.18 and 12.7.20.

12.2.105 Remark The formula (12.2.3) for $Z(E/\mathbb{F}_q, T)$ is equivalent to the statement that if we factor $1 - aT + qT^2 = (1-\alpha T)(1-\beta T)$ over the complex numbers, then

$$\#E(\mathbb{F}_{q^n}) = q^n + 1 - \alpha^n - \beta^n \qquad \text{for all } n \geq 1.$$

Further, the Hasse–Weil estimate (Theorem 12.2.46) is equivalent to the statement that α and β are complex conjugates satisfying $|\alpha| = |\beta| = \sqrt{q}$. So in particular,

$$\left|\#E(\mathbb{F}_{q^n}) - q^n - 1\right| = |\alpha^n + \beta^n| \leq |\alpha|^n + |\beta|^n = 2q^{n/2}.$$

12.2.106 Example Consider the curve $E/\mathbb{F}_2$ defined by the Weierstrass equation

$$E : y^2 + xy = x^3 + 1.$$

Then $\#E(\mathbb{F}_2) = 4$ and $a_2(E) = -1$. The roots of $1 + T + 2T^2$ are $\frac{-1\pm\sqrt{-7}}{2}$, so for all $n \geq 1$,

$$\#E(\mathbb{F}_{2^n}) = 2^n + 1 - \left(\frac{-1+\sqrt{-7}}{2}\right)^n - \left(\frac{-1-\sqrt{-7}}{2}\right)^n.$$

Using this formula, it is easy to compute $\#E(\mathbb{F}_{2^n})$ for large values of n, for example,

$$\#E(\mathbb{F}_{2^{173}}) = 2^{173} + 1 - 6787078360394475405304229.$$

12.2.11 The elliptic curve discrete logarithm problem

2.2.107 **Remark** The elliptic curve discrete logarithm problem (ECDLP) underlies the use of elliptic curves in cryptography. In this section we discuss ECDLP and some related problems. For applications to cryptography, see Section 16.4.

2.2.108 **Definition** Let $E/\mathbb{F}_q$ be an elliptic curve and let $P, Q \in E(\mathbb{F}_q)$. A *(discrete) logarithm of Q to the base P* is an integer N such that $Q = [N](P)$. The discrete logarithm, which is denoted by $\log_P(Q)$, is well-defined modulo the order m_P of the element P in the group $E(\mathbb{F}_q)$, so one may view $\log_P$ as a group homomorphism

$$\log_P : \{Q \in E(\mathbb{F}_q) : Q \text{ is a multiple of } P\} \longrightarrow \mathbb{Z}/m_P\mathbb{Z}.$$

The *elliptic curve discrete logarithm problem (ECDLP)* is the problem of computing $\log_P(Q)$ for given points P and Q. (Note the analogy with the classical discrete logarithm problem for the multiplicative group $\mathbb{F}_q^*$; see Section 11.6.)

2.2.109 **Remark** If the order m_P of P is prime, then the fastest known general algorithm for solving the ECDLP (as of 2012) has running time on the order of $\sqrt{m_P}$. This may be compared to the DLP in $\mathbb{F}_q^*$, for which there are algorithms with running times that are subexponential in $\log q$.

2.2.110 **Definition** Let $E/\mathbb{F}_q$ be an elliptic curve and let $P \in E(\mathbb{F}_q)$. The *(computational) elliptic curve Diffie–Hellman problem (ECDHP-comp)* is the following: Given the values of P, $[M](P)$, and $[N](P)$, compute the value of $[MN](P)$.

2.2.111 **Definition** The *(decisional) elliptic curve Diffie–Hellman problem (ECDHP-dec)* is the following: Given the values of P, $[M](P)$, and $[N](P)$, with better than equal probability, distinguish between the points $[MN](P)$ and a randomly chosen point Q.

2.2.112 **Definition** The *embedding degree of the integer m in the field $\mathbb{F}_q$* is the smallest integer k such that $\boldsymbol{\mu}_m \subset \mathbb{F}_{q^k}^*$, where $\boldsymbol{\mu}_m$ is the group of m-th roots of unity. Equivalently, it is the smallest integer k such that $q^k \equiv 1 \pmod{m}$.

2.2.113 **Remark** The importance of the embedding degree is that it describes the degree of an extension field over which the Weil and Tate pairings are defined.

2.2.114 **Remark** [2079], [2670, XI.6.1] Let $E/\mathbb{F}_q$, $m \geq 1$, and $T \in E(\mathbb{F}_q)[m]$ be as in Theorem 12.2.75, and let T be the Tate pairing. Then Menezes, Okamoto, and Vanstone have noted that the ECDLP in $E(\mathbb{F}_q)$ can be reduced to the DLP in $\mathbb{F}_q$, since if $N = \log_P(Q)$, then $\mathsf{T}(P, Q) = \mathsf{T}(P, P)^N$. Similarly, the decisional ECDHP is as easy to solve as computing the Tate pairing, since (with rare exceptions) $\mathsf{T}([M]P, [N]P) = \mathsf{T}(P, Q)$ if and only if $Q = [MN]P$.

12.2.115 **Definition** An elliptic curve $E/\mathbb{F}_p$ is *anomalous* if $a_p(E) = 1$, that is, if $\#E(\mathbb{F}_p) = p$.

12.2.116 **Remark** Let $p \geq 3$ and let $E/\mathbb{F}_p$ be an anomalous elliptic curve. Then Araki, Satoh, Semaev, and Smart [2535, 2581, 2682] observed that the ECDLP in $E(\mathbb{F}_p)$ can be solved in essentially linear time [2670, XI.6.5].

See Also

§10.5	For elliptic curves and Lattès dynamical systems.
§11.6	For the discrete logarithm problem for the multiplicative group $\mathbb{F}_q^*$.
§12.3	For efficient addition formulas on elliptic curves.
§12.7, §12.9	For zeta functions of higher genus curves and higher dimensional algebraic varieties.
§16.4, §16.5, §16.6	For cryptographic applications of elliptic curves, hyperelliptic curves, and abelian varieties.

References Cited: [284, 556, 557, 825, 968, 969, 1024, 1563, 1756, 1773, 1843, 1845, 1848, 2054, 2079, 2101, 2107, 2497, 2535, 2581, 2590, 2667, 2670, 2671, 2672, 2682, 2783, 2784, 2950, 2954]

12.3 Addition formulas for elliptic curves

Daniel J. Bernstein, University of Illinois at Chicago
Tanja Lange, Technische Universiteit Eindhoven

12.3.1 Curve shapes

12.3.1 **Remark** Section 12.2 defined elliptic curves using Weierstrass equations (12.2.1). The following definitions present other curve shapes which have algorithmic advantages.

12.3.2 **Definition** A *short Weierstrass curve* over a field K of characteristic not equal to 2 is a curve of the form $y^2 = x^3 + ax + b$ with $a, b \in K$ and $4a^3 + 27b^2 \neq 0$.

12.3.3 **Definition** A *short Weierstrass curve* over a field K of characteristic 2 is a curve of the form $y^2 + xy = x^3 + ax^2 + b$ with $a, b \in K$ and $b \neq 0$.

12.3.4 **Remark** The curve shape defined in Definition 12.3.3 covers only ordinary curves; supersingular curves over a field K of characteristic 2 have short Weierstrass equations of the form $y^2 + cy = x^3 + ax + b$ with $a, b, c \in K$ and $c \neq 0$.

12.3.5 **Definition** A *Montgomery curve* over a field K of characteristic not equal to 2 is a curve of the form $by^2 = x^3 + ax^2 + x$ with $a \in K \setminus \{-2, 2\}$ and $b \in K \setminus \{0\}$.

12.3.6 Definition An *Edwards curve* over a field K of characteristic not equal to 2 is a curve of the form $x^2 + y^2 = 1 + dx^2y^2$ with $d \in K \setminus \{0, 1\}$.

A *twisted Edwards curve* over a field K of characteristic not equal to 2 is a curve of the form $ax^2 + y^2 = 1 + dx^2y^2$ with $a, d \in K \setminus \{0\}$ and $a \neq d$.

12.3.7 Definition A *Hessian curve* over a field K is a curve of the form $x^3 + y^3 + 1 = dxy$ with $d \in K$ and $d^3 \neq 27$.

A *twisted Hessian curve* over a field K is a curve of the form $ax^3 + y^3 + 1 = dxy$ with $a, d \in K$, $a \neq 0$, and $d^3 \neq 27a$.

12.3.8 Remark Other curve shapes studied in the literature include binary Edwards curves, Jacobi quartics, Jacobi intersections, and Doche-Icart-Kohel curves. Chudnovsky and Chudnovsky's work [636] studied addition formulas on Hessian curves, Jacobi quartics, Jacobi intersections, and Weierstrass curves. Montgomery curves were proposed by Montgomery in [2133] in the context of the elliptic-curve method to factor integers. Edwards curves were introduced by Edwards in [956] in the form $x^2 + y^2 = a^2(1 + x^2y^2)$, and generalized to $x^2 + y^2 = 1 + dx^2y^2$ by Bernstein and Lange in [247]. Twisted Edwards curves were introduced by Bernstein, Birkner, Joye, Lange, and Peters in [244].

12.3.2 Addition

12.3.9 Remark Given a nonsingular projective genus one curve with a specified neutral element O, one can abstractly define addition of points P on the curve to correspond to addition of divisors $P - O$ modulo principal divisors. However, computations use more concrete addition laws specified as rational functions of the input coordinates.

12.3.10 Remark Addition on Weierstrass curves is described concretely in Definition 12.2.19 and Algorithm 12.2.21. It is necessary to distinguish generic additions from doublings and from computations involving $\mathcal{O}$ as input or output; the generic addition formulas fail for those cases.

There are other addition formulas that do not require so many distinctions. An addition law is *strongly unified* if it can be used to double generic points. It is *K-complete* (abbreviated *complete* when K is clear from context) if it can be used to add each pair of points defined over K.

12.3.11 Example Consider the twisted Edwards curve $ax^2 + y^2 = 1 + dx^2y^2$ over K with neutral element $(0, 1)$. The negative of (x, y) on this curve is $(-x, y)$. The Edwards addition law, valid for almost all points $P = (x_1, y_1)$ and $Q = (x_2, y_2)$ on the curve, states that the sum $P + Q$ is

$$\left(\frac{x_1y_2 + y_1x_2}{1 + dx_1x_2y_1y_2}, \frac{y_1y_2 - ax_1x_2}{1 - dx_1x_2y_1y_2}\right).$$

The Edwards addition law is strongly unified: it does not make an exception for $P = Q$. If a is a square in K and d is not a square in K then the Edwards addition law is K-complete.

12.3.12 Example Consider the twisted Hessian curve $ax^3 + y^3 + 1 = dxy$ over K with neutral element $(0, -1)$. The negative of (x, y) on this curve is $(x/y, 1/y)$. The rotated Hessian addition law, valid for almost all points $P = (x_1, y_1)$ and $Q = (x_2, y_2)$ on the curve, states that the sum $P + Q$ is

$$\left(\frac{x_1 - y_1^2x_2y_2}{ax_1y_1x_2^2 - y_2}, \frac{y_1y_2^2 - ax_1^2x_2}{ax_1y_1x_2^2 - y_2}\right).$$

This addition law is strongly unified. It is K-complete if a is not a cube in K.

12.3.13 Remark The completeness of the Edwards addition law was proven by Bernstein and Lange in [247]. Earlier addition laws, including the Weierstrass addition law and the usual (non-rotated) Hessian addition law, were not complete.

12.3.3 Coordinate systems

12.3.14 Remark Additions in affine coordinates require inversions in the underlying field K. Inversions are computationally expensive compared to multiplications and additions. Implementations thus work with fractions, delaying the inversions for as long as possible. Mathematically this means working in projective coordinates.

12.3.15 Definition The *projective representations* over a field K of a vector $(x, y) \in K^2$ are the vectors $(X, Y, Z) \in K^3$ such that $Z \neq 0$, $X/Z = x$, and $Y/Z = y$.

12.3.16 Remark The conditions $Z \neq 0$, $X/Z = x$, and $Y/Z = y$ are equivalent to $(X : Y : Z) \in \mathbb{P}^2(K)$ mapping to $(x, y) \in \mathbb{A}^2(K)$ under the natural rational map from $\mathbb{P}^2(K)$ to $\mathbb{A}^2(K)$. For computational purposes, however, projective representations of points are elements $(X, Y, Z) \in K^3$, not equivalence classes $(X : Y : Z) \in \mathbb{P}^2(K)$.

12.3.17 Remark Weighted projective coordinates sometimes lead to more efficient formulas. In $(a, b, 1)$-weighted coordinates, the point $(X, Y, Z) \in K^3$ with $Z \neq 0$ represents $(X/Z^a, Y/Z^b) \in K^2$; any scaled point $(X\lambda^a, Y\lambda^b, Z\lambda)$ with $\lambda \in K^*$ represents the same point in K^2; one defines $(X : Y : Z)$ as the set of such scaled points. For example, *Jacobian coordinates* for Weierstrass curves are $(2, 3, 1)$-weighted coordinates. Standard projective coordinates are $(1, 1, 1)$-weighted coordinates.

12.3.18 Remark Extra coordinates sometimes lead to more efficient formulas. For example, *extended coordinates* for Edwards curves represent an affine point (x, y) as any element of the equivalence class $(x : y : 1 : xy)$ in $(1, 1, 1, 1)$-weighted coordinates; i.e., (X, Y, Z, T) with $Z \neq 0$ and $XY = ZT$ represents $(X/Z, Y/Z)$.

12.3.19 Example The twisted Edwards curve $ax^2 + y^2 = 1 + dx^2y^2$ is naturally embedded into $\mathbb{P}^1 \times \mathbb{P}^1$ as $aX^2T^2 + Y^2Z^2 = Z^2T^2 + dX^2Y^2$ with $x = X/Z$ and $y = Y/T$ for $Z, T \neq 0$. In this representation addition can be written as follows:

$$((X_1 : Z_1), (Y_1 : T_1)) + ((X_2 : Z_2), (Y_2 : T_2)) =$$
$$\begin{cases} ((X_1Y_2Z_2T_1 + X_2Y_1Z_1T_2 : Z_1Z_2T_1T_2 + dX_1X_2Y_1Y_2), \\ \quad (Y_1Y_2Z_1Z_2 - aX_1X_2T_1T_2 : Z_1Z_2T_1T_2 - dX_1X_2Y_1Y_2)) & \text{if defined,} \\ ((X_1Y_1Z_2T_2 + X_2Y_2Z_1T_1 : aX_1X_2T_1T_2 + Y_1Y_2Z_1Z_2), \\ \quad (X_1Y_1Z_2T_2 - X_2Y_2Z_1T_1 : X_1Y_2Z_2T_1 - X_2Y_1Z_1T_2)) & \text{if defined.} \end{cases}$$

Bernstein and Lange showed in [249] that these two addition laws cover all possible pairs of curve points; the outputs coincide if they are both defined; each defined output is on the curve; and this addition turns the set of curve points into a group.

An analogous 56-monomial pair of addition laws for Weierstrass curves was presented by Bosma and Lenstra in [362] before Edwards curves were introduced.

12.3.4 Explicit formulas

12.3.20 Remark One can compute the scalar multiple nP for any positive integer n by writing n as $\sum_i n_i 2^i$ with $n_i \in \{0,1\}$ and computing nP as $\sum_i n_i 2^i P$. This uses about $\log_2 n$ doublings and, typically, about $(1/2)\log_2 n$ additions. The number of group operations to compute nP is reduced by more advanced scalar-multiplication methods such as windows, sliding windows, and fractional windows.

12.3.21 Remark Algorithms to compute elliptic-curve group operations using field operations are referred to as *explicit formulas.* These algorithms try to minimize the cost of group operations and of higher-level operations such as scalar multiplication. These algorithms translate addition formulas such as in Algorithm 12.2.21, and Examples 12.3.11, and 12.3.12 into sequences of field operations; they reduce cost by reusing intermediate results and applying various optimization techniques. Explicit formulas depend on the curve shape.

A complete addition algorithm is typically built as a combination of an incomplete addition formula with various special-case formulas, glued together by appropriate comparisons between P and Q, as in Algorithm 12.2.21.

12.3.22 Definition Explicit formulas to compute $P, Q \mapsto P + Q$ for generic P, Q are *addition formulas.* Explicit formulas to compute $P \mapsto 2P$ for generic P are *doubling formulas.* Explicit formulas to compute $P, Q, P - Q \mapsto P + Q$ for generic P, Q are *differential addition formulas.*

12.3.23 Remark One way to compute a scalar multiplication $P \mapsto nP$ using differential additions, where $n = \sum_{i=0}^{l} n_i 2^i$, is to compute the pairs $(L_j, R_j) = (\sum_{i=j}^{l} n_i 2^{i-j} P, P + \sum_{i=j}^{l} n_i 2^{i-j} P)$ for $l \geq j \geq 0$ recursively as follows: if $n_{j-1} = 0$ then $(L_{j-1}, R_{j-1}) = (2L_j, L_j + R_j)$; otherwise $(L_{j-1}, R_{j-1}) = (L_j + R_j, 2R_j)$. Note that the differences in the addition are equal to P and that the doublings are additions of points with difference $\mathcal{O}$. This scalar-multiplication method was introduced for Montgomery curves in [2133] and is called the *Montgomery ladder.*

12.3.24 Definition Addition formulas with inputs and output in projective coordinates are *projective addition formulas.* Addition formulas with one input and output in projective coordinates but the other input in affine coordinates are *mixed addition formulas.*

12.3.25 Remark Affine coordinates are equivalent to projective coordinates with $Z = 1$; mixed addition formulas eliminate multiplications by 1 and sometimes save time in other ways. The literature also contains various speedups for other types of restricted representations, such as additions of projective points having $X = 1$ or of two points having the same Z-coordinate.

12.3.26 Remark Often an addition formula involves some computations that depend only on one input point. A *readdition* of the same input point reuses those computations.

12.3.27 Remark The following subsections state the most efficient explicit formulas known for the most popular curve shapes used in computations. Cost is reported as squarings (**S**), multiplications by curve constants (**D**), and general multiplications (**M**); multiplications by curve constants are counted separately because often these constants can be chosen small. Additions and subtractions are ignored. Coordinate systems here are chosen primarily to minimize doubling cost and secondarily to minimize addition cost, where cost counts multiplications and a fraction of squarings; the fraction is 0.8 in characteristic $\neq 2$ and 0.2 in characteristic 2. See the Explicit Formulas Database [246] (`http://hyperelliptic.org/EFD/`)

for a much more comprehensive collection of curve shapes, coordinate systems, computer-verified addition formulas, and references.

12.3.5 Short Weierstrass curves, large characteristic: $y^2 = x^3 - 3x + b$

12.3.28 Remark The following algorithms choose weights $(2, 3, 1)$ and $a = -3$. These choices minimize the cost of known doubling algorithms on short Weierstrass curves, except for a few curves (at most 6 isomorphism classes) having $a = 0$. All of the standard NIST curves over prime fields have $a = -3$, and almost all curves over prime fields have low-degree isogenies to curves with $a = -3$.

12.3.29 Algorithm: Doubling.
Input: $P_1 = (X_1 : Y_1 : Z_1)$.
Output: $P_3 = (X_3 : Y_3 : Z_3) = 2P_1$.
$\delta = Z_1^2$; $\gamma = Y_1^2$; $\beta = X_1\gamma$; $\alpha = 3(X_1 - \delta)(X_1 + \delta)$; $X_3 = \alpha^2 - 8\beta$; $Z_3 = (Y_1 + Z_1)^2 - \gamma - \delta$; $Y_3 = \alpha(4\beta - X_3) - 8\gamma^2$.

12.3.30 Algorithm: Addition.
Input: $P_1 = (X_1 : Y_1 : Z_1)$, $P_2 = (X_2 : Y_2 : Z_2)$.
Output: $P_3 = (X_3 : Y_3 : Z_3) = P_1 + P_2$.
$A = Z_1^2$; $B = Z_2^2$; $U_1 = X_1 \cdot B$; $U_2 = X_2 \cdot A$; $S_1 = Y_1 \cdot Z_2 \cdot B$; $S_2 = Y_2 \cdot Z_1 \cdot A$; $H = U_2 - U_1$; $I = (2H)^2$; $J = H \cdot I$; $r = 2(S_2 - S_1)$; $V = U_1 \cdot I$; $X_3 = r^2 - J - 2V$; $Y_3 = r \cdot (V - X_3) - 2S_1 \cdot J$; $Z_3 = ((Z_1 + Z_2)^2 - A - B) \cdot H$.

12.3.31 Remark Doubling takes 3**M**+5**S**. Addition takes 11**M**+5**S**. Readdition of P_2 saves 1**M**+1**S** by caching B and $Z_2 \cdot B$. Mixed addition with $Z_2 = 1$ takes 7**M**+4**S**: it skips all operations involving Z_2 and computes $K = H^2; I = 4K; Z_3 = (Z_1 + H)^2 - A - K$.

12.3.6 Short Weierstrass curves, characteristic 2, ordinary case: $y^2 + xy = x^3 + a_2x^2 + a_6$

12.3.32 Remark The following algorithms for short binary Weierstrass curves choose weights $(1, 2, 1)$, called *Lopez-Dahab coordinates.* The formulas use $\sqrt{a_6}$ as a curve constant, assuming implicitly that a_6 is a square (which is automatic for characteristic-2 finite fields and other characteristic-2 perfect fields).

For fields $\mathbb{F}_{2^n}$ with n odd each isomorphism class contains a curve with $a_2 = 1$ or one with $a_2 = 0$. Optimizations are different, as the following algorithms show. Lopez-Dahab coordinates $(X : Y : Z)$ are extended to include $T = Z^2$, and further extended to include $W = XZ$ for $a_2 = 0$.

Differential-addition formulas represent a point by its x-coordinate, and represent x in turn as a fraction X/Z. These formulas do not differ between $a_2 = 0$ and $a_2 = 1$.

12.3.33 Algorithm: Doubling, $a_2 = 1$.
Input: $P_1 = (X_1 : Y_1 : Z_1 : T_1)$.
Output: $P_3 = (X_3 : Y_3 : Z_3 : T_3) = 2P_1$.
$A = X_1^2$; $B = Y_1^2$; $Z_3 = T_1 \cdot A$; $T_3 = Z_3^2$; $X_3 = (A + \sqrt{a_6}T_1)^2$; $Y_3 = B \cdot (B + X_3 + Z_3) + a_6T_3 + T_3$.

12.3.34 Algorithm: Addition, $a_2 = 1$.
Input: $P_1 = (X_1 : Y_1 : Z_1 : T_1)$, $P_2 = (X_2 : Y_2 : Z_2 : T_2)$.
Output: $P_3 = (X_3 : Y_3 : Z_3 : T_3) = P_1 + P_2$.
$A = X_1 \cdot Z_2$; $B = X_2 \cdot Z_1$; $C = A^2$; $D = B^2$; $E = A + B$; $F = C + D$; $G = Y_1 \cdot T_2$;

$H = Y_2 \cdot T_1$; $I = G + H$; $J = I \cdot E$; $Z_3 = F \cdot Z_1 \cdot Z_2$; $T_3 = Z_3^2$; $X_3 = A \cdot (H + D) + B \cdot (C + G)$; $Y_3 = (A \cdot J + F \cdot G) \cdot F + (J + Z_3) \cdot X_3$.

12.3.35 Algorithm: Mixed addition, $a_2 = 1$.
Input: $P_1 = (X_1 : Y_1 : Z_1 : T_1)$, $P_2 = (X_2 : Y_2 : 1 : 1)$.
Output: $P_3 = (X_3 : Y_3 : Z_3 : T_3) = P_1 + P_2$.
$A = X_1 + X_2 \cdot Z_1$; $B = Y_1 + Y_2 \cdot T_1$; $C = A \cdot Z_1$; $D = C \cdot (B + C)$; $Z_3 = C^2$; $T_3 = Z_3^2$; $X_3 = B^2 + C \cdot A^2 + D$; $Y_3 = (X_3 + X_2 \cdot Z_3) \cdot D + (X_2 + Y_2) \cdot T_3$.

12.3.36 Remark For $a_2 = 1$, doubling takes $2\mathbf{M} + 4\mathbf{S} + 2\mathbf{D}$; addition takes $13\mathbf{M} + 3\mathbf{S}$; readdition does not save anything; mixed addition with $Z_2 = 1$ takes $8\mathbf{M} + 4\mathbf{S}$.

12.3.37 Algorithm: Doubling, $a_2 = 0$.
Input: $P_1 = (X_1 : Y_1 : Z_1 : T_1 : W_1)$.
Output: $P_3 = (X_3 : Y_3 : Z_3 : T_3 : W_3) = 2P_1$.
$A = X_1^2$; $B = Y_1^2$; $Z_3 = W_1^2$; $X_3 = (A + \sqrt{a_6} T_1)^2$; $T_3 = Z_3^2$; $W_3 = X_3 \cdot Z_3$; $Y_3 = B \cdot (B + X_3 + Z_3) + a_6 T_3 + W_3$.

12.3.38 Algorithm: Addition, $a_2 = 0$.
Input: $P_1 = (X_1 : Y_1 : Z_1 : T_1 : W_1)$, $P_2 = (X_2 : Y_2 : Z_2 : T_2 : W_2)$.
Output: $P_3 = (X_3 : Y_3 : Z_3 : T_3 : W_3) = P_1 + P_2$.
$A = X_1 \cdot Z_2$; $B = X_2 \cdot Z_1$; $C = A^2$; $D = B^2$; $E = A + B$; $F = C + D$; $G = Y_1 \cdot T_2$; $H = Y_2 \cdot T_1$; $I = G + H$; $J = I \cdot E$; $Z_3 = F \cdot Z_1 \cdot Z_2$; $X_3 = A \cdot (H + D) + B \cdot (C + G)$; $Y_3 = (A \cdot J + F \cdot G) \cdot F + (J + Z_3) \cdot X_3$; $T_3 = Z_3^2$; $W_3 = X_3 \cdot Z_3$.

12.3.39 Algorithm: Mixed addition, $a_2 = 0$.
Input: $P_1 = (X_1 : Y_1 : Z_1 : T_1 : W_1)$, $P_2 = (X_2 : Y_2 : 1 : 1 : 1)$.
Output: $P_3 = (X_3 : Y_3 : Z_3 : T_3 : W_3) = P_1 + P_2$.
$A = Y_1 + Y_2 \cdot T_1$; $B = X_1 + X_2 \cdot Z_1$; $C = B \cdot Z_1$; $Z_3 = C^2$; $T_3 = Z_3^2$; $D = X_2 \cdot Z_3$; $X_3 = A^2 + C \cdot (A + B^2 + a_2 C)$; $Y_3 = (D + X_3) \cdot (A \cdot C + Z_3) + (Y_2 + X_2) \cdot T_3$; $W_3 = X_3 \cdot Z_3$.

12.3.40 Remark For $a_2 = 0$, doubling takes $2\mathbf{M} + 5\mathbf{S} + 2\mathbf{D}$; addition takes $14\mathbf{M} + 3\mathbf{S}$; readdition does not save anything; mixed addition with $Z_2 = 1$ takes $8\mathbf{M} + 4\mathbf{S} + 1\mathbf{D}$.

12.3.41 Algorithm: Differential addition and doubling.
Input: $P_2 = (X_2 : Z_2)$, $P_3 = (X_3 : Z_3)$ with $x(P_3 - P_2) = x_1$.
Output: $P_4 = (X_4 : Z_4) = 2P_2, P_5 = (X_5 : Z_5) = P_2 + P_3$.
$A = X_2 \cdot Z_3$; $B = X_3 \cdot Z_2$; $C = X_2^2$; $D = Z_2^2$; $Z_5 = (A + B)^2$; $X_5 = x_1 \cdot Z_5 + A \cdot B$; $X_4 = (C + \sqrt{a_6} D)^2$; $Z_4 = C \cdot D$.

12.3.42 Remark The combined operation of differential addition and doubling takes $5\mathbf{M} + 4\mathbf{S} + 1\mathbf{D}$. The Montgomery ladder thus takes $5\mathbf{M} + 4\mathbf{S} + 1\mathbf{D}$ per bit of the scalar.

12.3.7 Montgomery curves: $by^2 = x^3 + ax^2 + x$

12.3.43 Remark Montgomery curves are of interest primarily for their efficient differential addition. Points are represented via their x-coordinate $x(P) = X/Z$. The formulas depend on the curve constant $c = (a + 2)/4$.

12.3.44 Algorithm: Differential addition and doubling.
Input: $P_2 = (X_2 : Z_2)$, $P_3 = (X_3 : Z_3)$ with $x(P_3 - P_2) = x_1$.
Output: $P_4 = (X_4 : Z_4) = 2P_2, P_5 = (X_5 : Z_5) = P_2 + P_3$.
$A = X_2 + Z_2$; $\bar{A} = A^2$; $B = X_2 - Z_2$; $\bar{B} = B^2$; $E = \bar{A} - \bar{B}$; $C = X_3 + Z_3$; $D = X_3 - Z_3$; $F = D \cdot A$; $G = C \cdot B$; $X_5 = (F + G)^2$; $Z_5 = x_1 \cdot (F - G)^2$; $X_4 = \bar{A} \cdot \bar{B}$; $Z_4 = E \cdot (\bar{B} + cE)$.

12.3.45 Remark The combined operation of differential addition and doubling takes $5\mathbf{M}+4\mathbf{S}+1\mathbf{D}$. A Montgomery curve is birationally equivalent to a twisted Edwards curve. Differential addition formulas on twisted Edwards curves trade $1\mathbf{M}$ for $1\mathbf{D}$; if the curve constant can be chosen small those formulas are more efficient.

12.3.8 Twisted Edwards curves: $ax^2 + y^2 = 1 + dx^2y^2$

12.3.46 Remark The following formulas use extended coordinates: $x = X/Z, y = Y/Z, xy = T/Z$. Three different formulas are stated below: a complete addition formula for the case that a is a square and d is not a square; an incomplete but faster addition formula for the case $a = -1$; and a faster doubling formula.

12.3.47 Algorithm: Complete addition.
Input: $P_1 = (X_1 : Y_1 : Z_1 : T_1)$, $P_2 = (X_2 : Y_2 : Z_2 : T_2)$.
Output: $P_3 = (X_3 : Y_3 : Z_3 : T_3) = P_1 + P_2$.
$A = X_1 \cdot X_2$; $B = Y_1 \cdot Y_2$; $C = dT_1 \cdot T_2$; $D = Z_1 \cdot Z_2$; $E = (X_1 + Y_1) \cdot (X_2 + Y_2) - A - B$; $F = D - C$; $G = D + C$; $H = B - aA$; $X_3 = E \cdot F$; $Y_3 = G \cdot H$; $T_3 = E \cdot H$; $Z_3 = F \cdot G$.

12.3.48 Algorithm: Addition, $a = -1$.
Input: $P_1 = (X_1 : Y_1 : Z_1 : T_1)$, $P_2 = (X_2 : Y_2 : Z_2 : T_2)$.
Output: $P_3 = (X_3 : Y_3 : Z_3 : T_3) = P_1 + P_2$.
$A = (Y_1 - X_1) \cdot (Y_2 + X_2)$; $B = (Y_1 + X_1) \cdot (Y_2 - X_2)$; $C = Z_1 \cdot 2 \cdot T_2$; $D = T_1 \cdot 2 \cdot Z_2$; $E = D + C$; $F = B - A$; $G = B + A$; $H = D - C$; $X_3 = E \cdot F$; $Y_3 = G \cdot H$; $T_3 = E \cdot H$; $Z_3 = F \cdot G$.

12.3.49 Algorithm: Doubling.
Input: $P_1 = (X_1 : Y_1 : Z_1 : T_1)$.
Output: $P_3 = (X_3 : Y_3 : Z_3 : T_3) = 2P_1$.
$A = X_1^2$; $B = Y_1^2$; $C = 2Z_1^2$; $D = aA$; $E = (X_1 + Y_1)^2 - A - B$; $G = D + B$; $F = G - C$; $H = D - B$; $X_3 = E \cdot F$; $Y_3 = G \cdot H$; $T_3 = E \cdot H$; $Z_3 = F \cdot G$.

12.3.50 Remark Doubling takes $4\mathbf{M} + 4\mathbf{S} + 1\mathbf{D}$. Doubling does not use the input T; suppressing the computation of T from the previous addition or doubling reduces the effective cost of doubling to $3\mathbf{M} + 4\mathbf{S} + 1\mathbf{D}$ in typical addition chains.

A complete addition takes $9\mathbf{M} + 2\mathbf{D}$. Readdition saves $1\mathbf{D}$. Mixed addition with $Z_2 = 1$ saves $1\mathbf{M}$. The incomplete addition formulas for $a = -1$ take $8\mathbf{M}$. Mixed addition saves $1\mathbf{M}$. There are also complete addition formulas for $a = -1$ taking $8\mathbf{M} + 1\mathbf{D}$, where the $\mathbf{D}$ is a multiplication by $2d$.

See Also

§12.2	For more material on function fields of genus 1 (elliptic curves).
§12.4	For more material on hyperelliptic function fields.
§12.5	For rational places of function fields (rational points on curves).
§12.6	For towers of function fields.
§15.2	For applications of function fields to coding theory.
§16.4	For applications in cryptography.

References Cited: [244, 246, 247, 249, 362, 636, 956, 2133]

12.4 Hyperelliptic curves

Michael John Jacobson, Jr., University of Calgary
Renate Scheidler, University of Calgary

Recall that by Remark 12.1.119 every K-rational point on a projective curve over a field K, and hence also every K-rational point on an affine curve over K, corresponds to a degree one place of the associated function field. This section predominantly uses the language of curves and points rather than function fields and places.

12.4.1 Hyperelliptic equations

12.4.1 **Definition** A *hyperelliptic equation* over a field K is an equation of the form

$$y^2 + h(x)y = f(x) \qquad \text{with } h, f \in K[x]. \tag{12.4.1}$$

12.4.2 **Remark** The curve defined by a hyperelliptic equation (12.4.1) is non-singular if and only if for no point (x_0, y_0) on the curve, both partial derivatives vanish, i.e.,

$$2y_0 + h(x_0) = 0 \quad \text{and} \quad h'(x_0)y_0 = f'(x_0).$$

12.4.3 **Definition** A *hyperelliptic curve C of genus g defined over a field K* is given by a hyperelliptic equation over K that is irreducible in $K(x, y)$, non-singular, and satisfies one of the following three conditions:

1. $\deg(f) = 2g + 1$ and $\deg(h) \le g$;
2. $\deg(f) \le 2g + 1$ and h is monic of degree $g + 1$, or $\deg(f) = 2g + 2$ and
   a. either K has characteristic different from 2, $\deg(h) \le g$ and the leading coefficient of f is a square in K,
   b. or K has characteristic 2, or h is monic of degree $g + 1$ and the leading coefficient of f is of the form $s^2 + s$ for some $s \in K^*$;
3. $\deg(f) = 2g + 2$ and
   a. either K has characteristic different from 2, $\deg(h) \le g$, and the leading coefficient of f is not a square in K,
   b. or K has characteristic 2, or h is monic of degree $g + 1$ and the leading coefficient of f is not of the form $s^2 + s$ for any $s \in K^*$.

A curve C is *imaginary* or *ramified* in case (1), *real* or *split* in case (2), and *unusual* or *inert* in case (3).

12.4.4 **Remark** An elliptic curve can be thought of as an imaginary hyperelliptic curve of genus $g = 1$.

12.4.5 **Remark** Every unusual hyperelliptic curve is a real curve over a quadratic extension of K.

12.4.6 **Remark** It is customary, although not necessary, to take h to be identically zero if K does not have characteristic 2. This is always possible by completing the square, i.e., adding $h^2/4$ to both sides of (12.4.1) and replacing y by $y - h/2$.

12.4.7 Definition The *infinite places* of a hyperelliptic curve C/K as given in (12.4.1) are exactly the poles of x (see Definition 12.1.18).

12.4.8 Definition Let C/K be a hyperelliptic curve. For any extension field L of K, the set of *finite points of C defined over L* is the set of solutions $(x_0, y_0) \in L \times L$ to the hyperelliptic equation (12.4.1) defining C. The set of *points at infinity of C defined over L* is the set

$$S = \begin{cases} \{\infty\} & \text{if } C/L \text{ is imaginary,} \\ \{\infty, \overline{\infty}\} & \text{if } C/L \text{ is real,} \\ \emptyset & \text{if } C/L \text{ is unusual.} \end{cases}$$

The finite points and points at infinity defined over L together form the set of *points of C defined over L* or the set of *L-rational points of C*, denoted by $C(L)$.

12.4.9 Definition The *hyperelliptic involution* of a hyperelliptic curve C/K is the map $\iota : C(\overline{K}) \to C(\overline{K})$ that sends a finite point $P = (x_0, y_0)$ of C to the point $\overline{P} = \iota(P) = (x_0, -y_0 - h(x_0))$ of C. If C is imaginary, then $\iota(\infty) = \infty$. If C is real, then $\iota(\infty) = \overline{\infty}$ and $\iota(\overline{\infty}) = \infty$.

12.4.10 Remark A point on a hyperelliptic curve is ramified (when viewed as a place) if it is fixed by ι, and unramified otherwise.

12.4.11 Remark If C is imaginary, then the infinite place of $K(x)$ is a ramified K-rational point on C; if C is real, then it is an unramified K-rational point on C, and if C is unusual, then it is not a point on C, but rather a place of degree 2.

12.4.12 Theorem (Generalization of [2373, Section 5] and [1587, Proposition 2.1]) Let C be a hyperelliptic curve of genus g over a perfect field K defined by (12.4.1), and let $x_0, y_0 \in K$. Substituting

$$x = t^{-1} + x_0, \qquad y = \frac{bz}{t^{g+1}} + a$$

into (12.4.1), with

$$a = \begin{cases} y_0 & \text{if char}(K) = 2, \\ -h(x_0)/2 & \text{otherwise,} \end{cases}$$

and

$$b = \begin{cases} 1 & \text{if } h(x_0) + 2y_0 = 0, \\ h(x_0) + 2y_0 & \text{otherwise,} \end{cases}$$

yields a hyperelliptic curve $C' : z^2 + H(t)z = F(t)$ of genus g over K where

$$H(t) = b^{-1}t^{g+1}(h(t^{-1} + x_0) + 2a), \quad F(t) = b^{-2}t^{2g+2}(f(t^{-1} + x_0) - ah(t^{-1} + x_0) - a^2),$$

and the following conditions hold:

1. If $P = (x_0, y_0)$ is a finite ramified point on C, then C' is an imaginary hyperelliptic curve.
2. If $P = (x_0, y_0)$ is a finite unramified point on C, then C' is a real hyperelliptic curve.
3. If no finite point on C has x-coordinate x_0, then C' is an unusual hyperelliptic curve.

12.4.13 Example A hyperelliptic curve $y^2 = f(x)$ of genus g defined over $\mathbb{R}$ may have as few as one and as many as $g+1$ connected components, as illustrated in Figure 12.4.13, depending on the number of real roots of f.

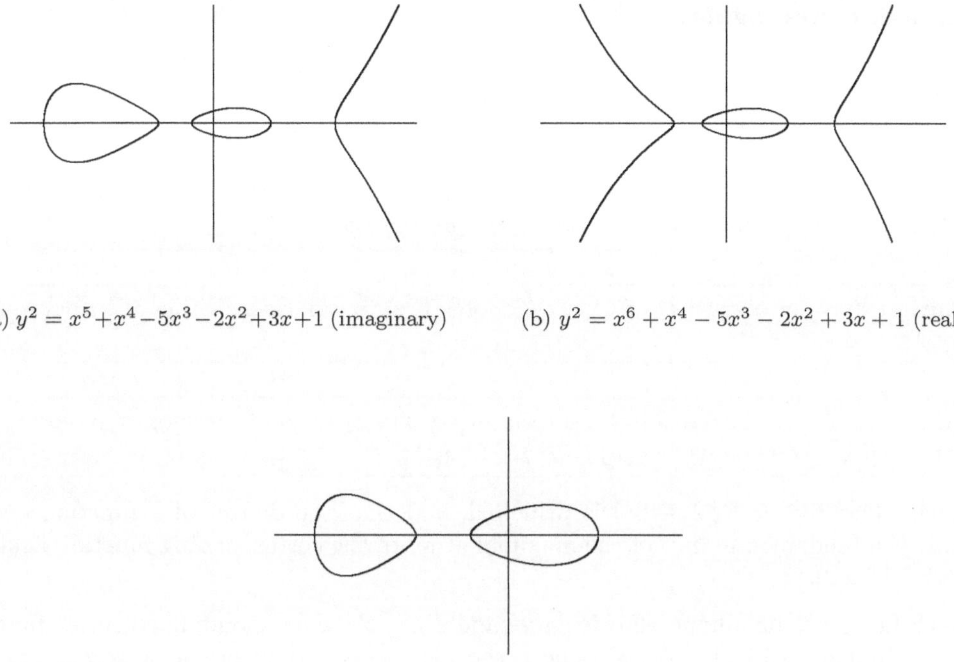

(a) $y^2 = x^5 + x^4 - 5x^3 - 2x^2 + 3x + 1$ (imaginary)

(b) $y^2 = x^6 + x^4 - 5x^3 - 2x^2 + 3x + 1$ (real)

(c) $y^2 = -x^6 + x^4 - 5x^3 - 2x^2 + 3x + 1$ (unusual)

Figure 12.4.1 Examples of imaginary, real, and unusual hyperelliptic curves of genus 2 defined over $\mathbb{R}$.

12.4.2 The degree zero divisor class group

12.4.14 Remark Unlike the case of elliptic curves, the set of points of C defined over any extension field of K does not form an abelian group. Instead, one needs to use *divisors*, as in Definition 12.1.21.

12.4.15 Definition Let C/K be a hyperelliptic curve and L an extension field of K. A divisor of C is *defined over* L if $D^\sigma = \sum_P n_P P^\sigma = D$ for all L-automorphisms σ on $\overline{K}$, where $P^\sigma = (\sigma(x), \sigma(y))$ if $P = (x, y)$, $\infty^\sigma = \infty$ and (in case C is real) $\overline{\infty}^\sigma = \overline{\infty}$.

12.4.16 Definition [661, Definition 14.3] Let C/K be a hyperelliptic curve. The set of degree zero divisors of C defined over K, denoted by $\mathrm{Div}_K^0(C)$, is a subgroup of the divisor group of C/K (see Definition 12.1.21). The set of principal divisors defined over K, denoted by $\mathrm{Princ}_K(C)$, is a subgroup of $\mathrm{Div}_K^0(C)$ (see Definition 12.1.26).

12.4.17 Definition Let C/K be a hyperelliptic curve. The *degree zero divisor class group defined over* K is given by $\mathrm{Pic}_K^0(C) = \mathrm{Div}_K^0(C)/\mathrm{Princ}_K(C)$. We denote $\mathrm{Pic}^0(C) = \mathrm{Pic}_{\overline{K}}^0(C)$.

12.4.18 Remark In the case of genus 1, i.e., elliptic curves, $\mathrm{Pic}_K^0(C)$ is isomorphic to the group of points defined over K on the elliptic curve; see Proposition 12.2.44. This is *not* true for hyperelliptic curves of genus $g > 1$.

12.4.19 Remark The group $\mathrm{Pic}_K^0(C)$ is isomorphic to the group of K-rational points on the Jacobian variety $J_C(K)$ of C; see [661, Section 4.4.6.a]. As a result, the two terminologies and notations are sometimes used interchangeably in the literature.

12.4.20 Definition Let $K = \mathbb{F}_q$. Then $\mathrm{Pic}_{\mathbb{F}_q}^0(C)$ is a finite abelian group. Its order, denoted by h, is the *degree zero divisor class number*, or *class number*, of $C/\mathbb{F}_q$.

12.4.21 Definition For any divisor D of C, $[D]$ denotes the divisor class of $\mathrm{Pic}_K^0(C)$ represented by D.

12.4.22 Definition Let $C/\mathbb{F}_q$ be a real hyperelliptic curve. The order R of the subgroup of $\mathrm{Pic}_K^0(C)$ generated by $[\overline{\infty} - \infty]$ is the *regulator* of C/K.

12.4.23 Remark The divisor $R(\overline{\infty} - \infty)$ is principal, and thus the divisor of a function ϵ. This function is a fundamental unit of the maximal order of the corresponding function field; see Remark 12.1.119.

12.4.24 Remark Let $C/\mathbb{F}_q$ be a hyperelliptic curve and $F/\mathbb{F}_q$ the corresponding algebraic function field; see Remark 12.1.119. Let $\mathcal{C}$ denote the ideal class group of the maximal order of $F/\mathbb{F}_q(x)$. Then the following relationships hold between $\mathrm{Pic}_K^0(C)$ and $\mathcal{C}$.

1. If C/K is imaginary, then $h = |\mathcal{C}|$, and the groups $\mathrm{Pic}_K^0(C)$ and $\mathcal{C}$ are isomorphic.
2. If C/K is real, then $h = R|\mathcal{C}|$.
3. If C/K is unusual, then $h = |\mathcal{C}|/2$.

12.4.25 Remark The ideal class number $|\mathcal{C}|$ of a real hyperelliptic curve is expected to be small (frequently 1), so we expect that $R \approx h$ [1125]; see also [1129, 1130, 1131] for the genus 1 case.

12.4.3 Divisor class arithmetic over finite fields

12.4.26 Remark In the case of imaginary and real hyperelliptic curves, and to a lesser extent for unusual curves, there exist algorithms for efficient arithmetic in $\mathrm{Pic}_K^0(C)$. We restrict our attention to $K = \mathbb{F}_q$.

12.4.27 Definition A divisor $D = \sum_P n_P P$ of a hyperelliptic curve $C/\mathbb{F}_q$ is *finitely effective* if $n_P \geq 0$ for all finite points P of C.

12.4.28 Remark Every finitely effective degree zero divisor D of a hyperelliptic curve $C/\mathbb{F}_q$ can be represented uniquely as

$$D = \begin{cases} D_S - \deg(D_S)\infty & \text{if } C/\mathbb{F}_q \text{ is imaginary,} \\ D_S - \deg(D_S)\infty + n_{\overline{\infty}}(\overline{\infty} - \infty) & \text{if } C/\mathbb{F}_q \text{ is real,} \\ D_S - (\deg(D_S)/2)\infty & \text{if } C/\mathbb{F}_q \text{ is unusual,} \end{cases}$$

where $D_S = \sum_P n_P P$ is only supported at finite points on C and $n_{\overline{\infty}} \in \mathbb{Z}$; see [1588, Section 3].

12.4.29 Definition A degree zero divisor D of a hyperelliptic curve $C/\mathbb{F}_q$ with a representation as given in Remark 12.4.28 is *semi-reduced* if for all $P \in \text{supp}(D_S)$ we have $\overline{P} \notin \text{supp}(D_S)$, unless $P = \overline{P}$, in which case $n_P = 1$.

12.4.30 Definition A semi-reduced divisor D is *reduced* if $\deg(D_S) \leq g$.

12.4.31 Remark If $C/\mathbb{F}_q$ is imaginary, then every divisor class of $\text{Pic}^0_{\mathbb{F}_q}(C)$ contains a unique reduced divisor.

12.4.32 Remark If $C/\mathbb{F}_q$ is real, then divisor classes of $\text{Pic}^0_{\mathbb{F}_q}(C)$ do not generically contain unique reduced divisors. However, each class does contain a unique reduced divisor D such that $n_{\overline{\infty}}$ lies in a specified range of length $g + 1 - \deg(D_S)$. Paulus and Rück [2373] proposed the interval $[0, g - \deg(D_S)]$. Galbraith, Harrison, and Mireles Morales [1155] proposed a *balanced* divisor representation, using the interval centered around $\lceil \deg(D_S)/2 \rceil$.

12.4.33 Remark If $C/\mathbb{F}_q$ is unusual, then every divisor class of $\text{Pic}^0_{\mathbb{F}_q}(C)$ contains at most one reduced divisor. If the class contains a divisor of the form given in Remark 12.4.28, then it contains either a unique reduced divisor or $q + 1$ divisors of the form as in Remark 12.4.28 with $\deg(D_S) = g+1$. Arithmetic in $\text{Pic}^0_{\mathbb{F}_q}(C)$ when $C/\mathbb{F}_q$ is unusual is not as well developed as for imaginary and real hyperelliptic curves $C/\mathbb{F}_q$. For details, see [133].

12.4.34 Theorem [1774, Theorem 5.1] Let $C/\mathbb{F}_q$ be a hyperelliptic curve. If D is a semi-reduced divisor defined over $\mathbb{F}_q$ as given in Remark 12.4.28, then D_S can be represented uniquely by a pair of polynomials $u, v \in \mathbb{F}_q[x]$ where

$$u(x) = \prod_{P \in \text{supp}(D_S)} (x - x_P)^{n_P}$$

is monic and v is the unique polynomial such that $\deg(v) < \deg(u)$, $v(x_P) = -y_P$ for all $P = (x_P, y_P) \in \text{supp}(D_S)$, and $u \mid v^2 + hv - f$.

12.4.35 Definition The pair $[u, v]$ is the *Mumford representation* of D.

12.4.36 Remark In some sources such as [1774], the condition $v(x_P) = -y_P$ is replaced by $v(x_P) = y_P$. This also describes a unique representation of semi-reduced divisors in which D_S from Theorem 12.4.34 is replaced by $\overline{D}_S = \sum_P n_P \overline{P}$.

12.4.37 Remark Alternative representations to $[u, v]$ can be obtained by taking $[u, v']$ where v' is any polynomial with $v' \equiv v \pmod u$ that satisfies the same interpolation condition. In the real case, it is sometimes computationally advantageous to use an alternative with $\deg(v') = g + 1$ [988].

12.4.38 Remark If $C/\mathbb{F}_q$ is imaginary, then semi-reduced divisors are uniquely determined by their Mumford representation. Thus, the Mumford representation gives an explicit, efficient representation of divisor classes of $\text{Pic}^0_{\mathbb{F}_q}(C)$ via reduced representatives D with $\deg(v) < \deg(u) \leq g$. The identity divisor class, $\text{Princ}_{\mathbb{F}_q}(C)$, is represented by $[1, 0]$, and the inverse of $[u, v]$ is given by $[u, -h - v]$. Cantor's algorithm [497] (see also [661, Algorithm 14.7]) describes how to compute the reduced sum of two divisors in Mumford representation in polynomial time using only arithmetic with polynomials in $\mathbb{F}_q[x]$.

12.4.39 Remark If $C/\mathbb{F}_q$ is real, then semi-reduced divisors are uniquely determined by their Mumford representation and $n_{\overline{\infty}}$-value. Thus, the Mumford representation, together with the integer $n_{\overline{\infty}}$, can be used to represent divisor classes in $\mathrm{Pic}^0_{\mathbb{F}_q}(C)$. The identity class is represented by $u = 1$, $v = 0$, and $n_{\overline{\infty}} = 0$. The inverse of the divisor class $[u, v, n_{\overline{\infty}}]$ is given by $[u, -h - v, -n_{\overline{\infty}} - \deg(u)]$, after which additional adjustment steps are performed to obtain a value $n_{\overline{\infty}}$ in the required range, as described in [1155, 2373]. A modification of Cantor's algorithm (see, for example, [1586]) can be used to compute the reduced sum of two reduced divisors, after which additional operations are performed to obtain a value $n_{\overline{\infty}}$ in the required range.

12.4.40 Remark A generalization of Shanks' NUCOMP algorithm is more efficient than Cantor's algorithm for moderate and large genus [1588], and can be adapted for use in both the imaginary and real models. For $g \le 4$, optimized explicit formulas exist in the imaginary case that describe the addition and reduction algorithm in terms of operations in $\mathbb{F}_q$; see [661, Chapter 14] for a survey. Explicit formulas for genus 2 also exist in the real case [987, 988].

12.4.41 Remark The multiplication-by-m map on elliptic curves (see Definition 12.2.24) generalizes naturally to $\mathrm{Pic}^0_K(C)$. The double-and-add method, as well as more advanced methods for scalar multiplication, can also be applied to compute this map efficiently; see [661, Chapter 9] for a survey.

12.4.42 Definition Let $C/\mathbb{F}_q$ be a real hyperelliptic curve. The *infrastructure* of $C/\mathbb{F}_q$, denoted by $\mathcal{R}$, is the finite set of all reduced principal divisors D with $0 \ge n_{\overline{\infty}} > -R$.

12.4.43 Remark The infrastructure is often described in terms of reduced principal ideals of the maximal order of the function field associated to $C/\mathbb{F}_q$; see, for example [2707].

12.4.44 Definition Let $C/\mathbb{F}_q$ be a real hyperelliptic curve, and $D \in \mathcal{R}$. The *distance* of D is $\delta(D) = -n_{\overline{\infty}}$.

12.4.45 Remark Divisors in the infrastructure can be represented using a combination of the Mumford representation and distance.

12.4.46 Remark Distance imposes a natural ordering on the set $\mathcal{R}$. The *baby step* operation moves cyclically from one divisor to the next in this ordering [2706]. The distance obtained by traversing one entire cycle is exactly the regulator R.

12.4.47 Remark A modification of Cantor's algorithm applied to two divisors in the infrastructure, where the reduction process terminates as soon as a reduced divisor is obtained, produces another infrastructure divisor [2706]. NUCOMP [1588] and explicit formulas for genus 2 [987, 988] can also be used for this purpose.

12.4.48 Definition The operation of computing the reduced sum of two divisors in $\mathcal{R}$, as described in the previous remark, is a *giant step*.

12.4.49 Remark The divisor $[1, 0]$ with distance 0 acts as the identity with respect to the giant step operation. The inverse of $D = [u, v] \neq [1, 0]$ with respect to the giant step operation is $[u, -h - v]$ and has distance $R + \deg(u) - \delta(D)$.

12.4.50 Remark Giant steps move through $\mathcal{R}$ in larger steps than baby steps, because the distance of a giant step applied to inputs D and D' is $\delta(D) + \delta(D') - d$, where $0 \le d \le 2g$; see, for example [1587]. Distances are not exactly additive due to the adjustments required to achieve reduction. Thus, although $\mathcal{R}$ is structurally similar to a cyclic group of order R under the giant step operation, it is not a group as associativity does not hold [1587].

12.4.51 Remark The analogue of the multiplication-by-m map in the infrastructure is computing a divisor with distance as close to m as possible but not exceeding m. This divisor can be computed in $O(\log m)$ giant steps using methods similar to the double-and-add method, and in some cases, can be sped up using the fact that computing baby steps is faster than giant steps; see [1587] for details.

12.4.4 Endomorphisms and supersingularity

12.4.52 Definition Let C/K be a hyperelliptic curve. An *endomorphism* of $\mathrm{Pic}^0(C)$ is a group homomorphism of $\mathrm{Pic}^0(C)$. An endomorphism of $\mathrm{Pic}^0(C)$ is *defined over* K if it is a group homomorphism of $\mathrm{Pic}^0_K(C)$. The set of endomorphisms of $\mathrm{Pic}^0(C)$ is denoted by $\mathrm{End}(\mathrm{Pic}^0(C))$, and the set of endomorphisms defined over K is denoted by $\mathrm{End}_K(\mathrm{Pic}^0(C))$.

12.4.53 Remark If C has genus 1, then Definition 12.4.52 agrees with the definition of an endomorphism of an elliptic curve as given in Definition 12.2.27.

12.4.54 Remark As in the elliptic curve case (see Definition 12.2.39), $\mathrm{End}(\mathrm{Pic}^0(C))$ and $\mathrm{End}_K(\mathrm{Pic}^0(C))$ are rings.

12.4.55 Example The multiplication-by-m map $[m] : \mathrm{Pic}^0(C) \longrightarrow \mathrm{Pic}^0(C)$ is an endomorphism of $\mathrm{Pic}^0(C)$ that is defined over K. Thus, $\mathrm{End}(\mathrm{Pic}^0(C))$ and $\mathrm{End}_K(\mathrm{Pic}^0(C))$ always contain $\mathbb{Z}$.

12.4.56 Definition [661, Definition 14.13] Let C/K be a hyperelliptic curve. If $\mathrm{End}(\mathrm{Pic}^0(C))$ contains an order of a number field of degree $2g$ over $\mathbb{Q}$, then $\mathrm{End}(\mathrm{Pic}^0(C))$ has *complex multiplication*.

12.4.57 Example Let $C/\mathbb{F}_q$ be hyperelliptic curve. As in the elliptic curve case (see Example 12.2.31), the Frobenius automorphism of $\overline{\mathbb{F}}_q$ that sends an element a to a^q extends to an endomorphism of $\mathrm{Pic}^0(C)$ that is defined over K and is different from $[m]$ for all $m \in \mathbb{Z}$.

12.4.58 Definition Let $C/\mathbb{F}_q$ be a hyperelliptic curve. The group $\mathrm{Pic}^0(C)$ (or, more properly, the Jacobian J_C) is *supersingular* if it is isogenous to the product of supersingular elliptic curves.

12.4.59 Remark If $C/\mathbb{F}_q$ is a hyperelliptic curve that is not supersingular, then C may have complex multiplication. The Frobenius endomorphism satisfies a monic polynomial equation of degree $2g$ with integer coefficients (its characteristic polynomial, see Remark 12.4.60). If that polynomial is irreducible, then the Frobenius corresponds to an algebraic integer of degree $2g$ and $C/\mathbb{F}_q$ has complex multiplication.

12.4.5 Class number computation

12.4.60 Remark The zeta function of a hyperelliptic curve $C/\mathbb{F}_q$ of genus g (Definition 12.5.12) is of the form

$$Z(C/\mathbb{F}_q, t) = \frac{L(t)}{(1-t)(1-qt)},$$

where $L(t)$ is the L-polynomial of $C/\mathbb{F}_q$ (Theorem 12.5.13). The reciprocal polynomial $P(t) = t^{2g}L(1/t)$ is the characteristic polynomial of the Frobenius endomorphism (Remark 12.5.15), a polynomial of degree $2g$ with integer coefficients whose roots have absolute value $\sqrt{q}$ (Theorem 12.5.17).

12.4.61 Theorem (Special case of Theorem 12.5.13) Let $C/\mathbb{F}_q$ be a hyperelliptic curve of genus g and class number h. Then $h = L(1)$.

12.4.62 Remark Theorem 12.4.61 and Remark 12.4.60 immediately imply the bounds $(\sqrt{q}-1)^{2g} \leq h \leq (\sqrt{q}+1)^{2g}$. The interval $[(\sqrt{q}-1)^{2g}, (\sqrt{q}+1)^{2g}]$ is called the *Hasse–Weil interval* for hyperelliptic curves of genus g over $\mathbb{F}_q$.

12.4.63 Remark There is an extensive body of literature on algorithms for computing class numbers of hyperelliptic curves over finite fields; see [815, 817, 1249, 1555, 1556, 1718, 1719, 1865, 1866, 1905, 2089, 2867] For genus 2 curves, see also [2399]. An overview of the main methods can be found in [661, Section 17.3].

12.4.6 The Tate-Lichtenbaum pairing

12.4.64 Definition The *kernel* of the multiplication-by-m map on $\text{Pic}^0(C)$ is the subgroup

$$\text{Pic}^0(C)[m] = \{[D] \in \text{Pic}^0(C) \ : \ m[D] = \text{Princ}_K(C)\}.$$

12.4.65 Definition Let $C/\mathbb{F}_q$ be a hyperelliptic curve. Let $m \geq 1$ a prime integer with embedding degree k in $\mathbb{F}_q$ (see Definition 12.2.112). The *Tate-Lichtenbaum pairing* is defined as

$$T : \text{Pic}^0_{\mathbb{F}_q}(C)[m] \times \text{Pic}^0_{\mathbb{F}_q}(C)/m\,\text{Pic}^0_{\mathbb{F}_q}(C) \to \mathbb{F}^*_{q^k}/(\mathbb{F}^*_{q^k})^m$$

$$([D_1],[D_2]) \mapsto f_{m,D_1}(D_2) = \prod_P f_{m,D_1}(P)^{n_P},$$

where f_{m,D_1} is a function with divisor $m[D_1]$ and $D_2 = \sum_P n_P(P)$.

12.4.66 Remark The Tate-Lichtenbaum pairing is bilinear, i.e.,

$$T([D_1]+[D_2],[D_3]) = T([D_1],[D_3])T([D_2],[D_3])$$

and

$$T([D_1],[D_2]+[D_3]) = T([D_1],[D_2])T([D_1],[D_3]),$$

non-degenerate (if $T([D_1],[D_2]) = 1$ for all $[D_2] \in \text{Pic}^0_K(C)[m]$ then $[D_1] = \text{Princ}_K(C)$), and the result is independent of the divisor class representatives used.

12.4.67 Remark The Tate-Lichtenbaum pairing can be computed using an analogue of Miller's algorithm for elliptic curves [182, Section 5].

12.4.68 Remark Hyperelliptic curves with small embedding degree exist, i.e., for which computing the Tate-Lichtenbaum pairing is efficient. For example, Galbraith [1154] proved that supersingular hyperelliptic curves of genus g have embedding degree bounded by an integer $k(g)$. For $g \leq 6$, Rubin and Silverberg [2493] show that $k(g) \leq 7.5g$. Various constructive methods for non-supersingular hyperelliptic curves also exist; see [182] for a recent survey.

12.4.69 Definition Let $C/\mathbb{F}_q$ be a hyperelliptic curve. Let $m \geq 1$ a prime integer with embedding degree k in $\mathbb{F}_q$. Suppose that $\mathrm{Pic}^0_{\mathbb{F}_q}(C)$ contains no elements of order m^2. The *modified Tate-Lichtenbaum pairing* is defined as

$$T' : \mathrm{Pic}^0_{\mathbb{F}_q}(C)[m] \times \mathrm{Pic}^0_{\mathbb{F}_q}(C)[m] \to \boldsymbol{\mu}_m$$
$$([D_1],[D_2]) \mapsto T([D_1],[D_2])^{(q^k-1)/m},$$

where $\boldsymbol{\mu}_m$ is the group of m-th roots of unity.

12.4.70 Remark The main advantage of the modified Tate-Lichtenbaum pairing over the Tate-Lichtenbaum pairing is that it takes specific values in $\boldsymbol{\mu}_m$ as opposed to equivalence classes.

12.4.71 Remark There are other types of pairings and algorithms to compute them, many designed to have computational advantages over the Tate-Lichtenbaum pairing. For a recent survey, see [182].

12.4.7 The hyperelliptic curve discrete logarithm problem

12.4.72 Remark Similar to the elliptic curve discrete logarithm problem (see Section 12.2.11), the hyperelliptic curve discrete logarithm problem (HCDLP) is the basis of many hyperelliptic curve cryptosystems. In this section we discuss the HCDLP and some related problems. For applications to cryptography, see Section 16.5.

12.4.73 Definition Let $C/\mathbb{F}_q$ be a hyperelliptic curve and $[D_1],[D_2] \in \mathrm{Pic}^0_{\mathbb{F}_q}(C)$. The *hyperelliptic curve discrete logarithm problem (HCDLP)* is the problem of computing $n \in \mathbb{Z}$ such that $[D_1] = n[D_2]$, if it exists.

12.4.74 Remark If the order l of $[D_2]$ is prime, then the fastest known *general* algorithm for solving the HCDLP (as of 2011) has running time on the order of $\sqrt{l}$. As with the ECDLP, there are a number of cases where the problem can be solved more easily; see [661, Part V] and [1586] for recent surveys.

12.4.75 Remark If the genus g is sufficiently large compared to the finite field order q, then the HCDLP can be solved in expected time subexponential in q^g using the index-calculus method [17, 977, 980]. The current state-of-the-art is Enge and Gaudry's result [980] that if $g/\log_g q > \vartheta$, the expected bit complexity is $L_{q^g}[1/2, \sqrt{2}((1+1/2\vartheta)^{1/2} + (1/2\vartheta)^{1/2})]$ where

$$L_n[\beta, c] = e^{\left((c+o(1))(\log n)^\beta (\log\log n)^{1-\beta}\right)}.$$

12.4.76 Remark Index-calculus can also be used to solve the HCDLP faster than the generic methods for smaller genera [1245, 1256, 2802]. The current state-of-the-art is Gaudry, Thomé, Thériault, and Diem's result [1256] that the HCDLP can be solved in expected time $O(g^5 q^{2-\frac{2}{g}+\epsilon})$ if $q > g!$. This is asymptotically faster than the generic algorithms for $g \geq 3$.

12.4.77 Remark Frey, Müller, and Rück [1106, 1413] showed how the modified Tate-Lichtenbaum pairing can be used to reduce the HCDLP to the DLP in the group of m-th roots of unity $\boldsymbol{\mu}_m \subset \mathbb{F}_{q^k}$, where k is the embedding degree of m in the field $\mathbb{F}_q$ (see Definition 12.2.112). If k is sufficiently small, for example if $C/\mathbb{F}_q$ is supersingular, this is more efficient than the generic algorithms.

12.4.78 Remark If $m = p^n$ where p is the characteristic of $\mathbb{F}_q$, then an algorithm of Rück [2498] can be used to solve the HCDLP in time $O(n^2 \log p)$.

12.4.79 Remark If $\mathbb{F}_q = \mathbb{F}_{p^n}$ where n is composite, the Weil descent methodology [661, Section 22.3] can in certain cases be used to reduce the HCDLP to another instance of the HCDLP on a curve of higher genus over a smaller finite field, where faster non-generic algorithms may apply.

12.4.80 Remark The Diffie-Hellman and decisional Diffie-Hellman problems also generalize to $\mathrm{Pic}^0_{\mathbb{F}_q}(C)$, and are also sometimes used as the underlying security assumption of certain cryptographic protocols (see Section 16.5). Both reduce to the HCDLP, but equivalence is not known.

12.4.81 Definition Let $C/\mathbb{F}_q$ be a real hyperelliptic curve and let $D \in \mathcal{R}$. The *infrastructure discrete logarithm problem (IDLP)* is the problem of computing $\delta(D)$.

12.4.82 Remark This problem has also been used as the underlying security assumption of cryptographic protocols [1587]. It is computationally easy to compute a divisor in the infrastructure close to a given distance [1587], but solving the IDLP is believed to be difficult. In fact, the IDLP can be reduced to the DLP in the subgroup of $\mathrm{Pic}^0_{\mathbb{F}_q}(C)$ generated by $[\overline{\infty} - \infty]$ [1089, 2142].

See Also

§12.1	For analagous material for general function fields and curves.
§12.2	For analagous material for the genus 1 case (i.e., elliptic curves).
§16.5	For applications of hyperelliptic curves to cryptography.
[1105]	A general reference to hyperelliptic curves and their cryptographic applications.
[1774]	The appendix of this reference consists of an excellent introduction to the arithmetic of hyperelliptic curves.

References Cited: [17, 133, 182, 497, 661, 815, 817, 977, 980, 987, 988, 1089, 1105, 1106, 1125, 1129, 1130, 1131, 1154, 1155, 1245, 1249, 1256, 1413, 1555, 1556, 1586, 1587, 1588, 1718, 1719, 1774, 1865, 1866, 1905, 2089, 2142, 2373, 2399, 2493, 2498, 2706, 2707, 2802, 2867]

12.5 Rational points on curves

Arnaldo Garcia, IMPA
Henning Stichtenoth, Sabanci University

12.5.1 Remark In this section we use the language of function fields rather than algebraic curves, see Section 12.1. A simple way for switching from function fields to algebraic curves is as follows.

A *function field $F/\mathbb{F}_q$ of genus g* corresponds to a *curve $\mathcal{X}$ of genus g over* $\mathbb{F}_q$, that is an absolutely irreducible, non-singular, projective curve which is defined over $\mathbb{F}_q$. If $F = \mathbb{F}_q(x, y)$ and x, y satisfy the equation $\varphi(x, y) = 0$ for an irreducible polynomial $\varphi(X, Y) \in \mathbb{F}_q[X, Y]$, then $\mathcal{X}$ is a non-singular, projective model of the plane curve which is defined by

$\varphi(X, Y) = 0$. By abuse of notation, we say briefly that the curve $\mathcal{X}$ is given by $\varphi(x, y) = 0$. *Rational places* of the function field correspond to $\mathbb{F}_q$*-rational points* of $\mathcal{X}$.

12.5.1 Rational places

12.5.2 **Remark** Let F be a function field over $\mathbb{F}_q$. Then F has only finitely many rational places.

12.5.3 **Definition** Define $N(F) := |\{P \mid P \text{ is a rational place of } F\}|$.

12.5.4 **Example** For the *rational function field* $F = \mathbb{F}_q(x)$ we have $N(F) = q + 1$. The rational places are the zeros of $x - a$ with $a \in \mathbb{F}_q$, and the pole P_∞ of x.

12.5.5 **Lemma** [2714, Lemma 5.1] Let F'/F be a finite extension of function fields having the same constant field $\mathbb{F}_q$. Then the following hold.

1. Let P be a place of F and P' a place of F' lying above P. If P' is rational, then P is rational.
2. $N(F') \leq [F' : F] \cdot N(F)$.

12.5.6 **Remark** The following special case of Kummer's Theorem [2714, Theorem 3.3.7] is often useful to determine rational places of a function field.

12.5.7 **Lemma** Let P be a rational place of F and let $\mathcal{O}_P$ be its valuation ring. Consider a finite extension $E = F(y)$ of F such that $\mathbb{F}_q$ is also the full constant field of E. Assume that the minimal polynomial $\varphi(T)$ of y over F has all its coefficients in $\mathcal{O}_P$ (that is, y is integral over $\mathcal{O}_P$). Suppose that the reduction $\bar{\varphi}(T)$ of $\varphi(T)$ modulo P (which is a polynomial over the residue class field $\mathcal{O}_P/P = \mathbb{F}_q$) splits over $\mathbb{F}_q$ as follows:

$$\bar{\varphi}(T) = (T - a_1) \cdots (T - a_s) \cdot p_1(T) \cdots p_r(T)$$

with distinct elements $a_1, \ldots, a_s \in \mathbb{F}_q$ and distinct irreducible polynomials $p_1, \ldots, p_r \in \mathbb{F}_q[T]$ of degree > 1. Then there are exactly s rational places $P_1, \ldots, P_s$ of E lying over P.

12.5.8 **Example** Assume that $q = 2^m$ with $m \geq 2$, and consider the function field $F = \mathbb{F}_q(x, y)$ with

$$y^2 + y = x^{q-1}.$$

The pole P_∞ of x is totally ramified in the extension $F/\mathbb{F}_q(x)$; this gives one rational place of F. Next we consider the place $P = (x = a)$ of $\mathbb{F}_q(x)$ which is the zero of $x - a$ with $a \in \mathbb{F}_q$. The reduction of the minimal polynomial $\varphi(T) = T^2 + T + x^{q-1}$ modulo P is then

$$\bar{\varphi}(T) = \begin{cases} T^2 + T + 1 & \text{if } a \neq 0, \\ T^2 + T & \text{if } a = 0. \end{cases}$$

The polynomial $T^2 + T = T(T + 1)$ splits over $\mathbb{F}_q$ into linear factors. If m is odd, then $T^2 + T + 1$ is irreducible over $\mathbb{F}_q$, and for m even, $T^2 + T + 1$ splits into two distinct linear polynomials over $\mathbb{F}_q$. Therefore

$$N(F) = \begin{cases} 3 & \text{if } m \text{ is odd}, \\ 2q + 1 & \text{if } m \text{ is even}. \end{cases}$$

12.5.2 The Zeta function of a function field

12.5.9 Definition Throughout this subsection we use the following notations:

1. F is an algebraic function field over $\mathbb{F}_q$ of genus $g(F) = g$, and $\mathbb{F}_q$ is algebraically closed in F;
2. $\mathbb{P}_F$ is the set of places of $F/\mathbb{F}_q$;
3. $N(F)$ is the number of rational places of F;
4. $\mathrm{Div}(F)$ is the divisor group of F;
5. $\mathrm{Div}^0(F) := \{A \in \mathrm{Div}(F) \mid \deg A = 0\}$ is the *group of divisors of degree zero*, and $\mathrm{Princ}(F) \subseteq \mathrm{Div}^0(F)$ is the *group of principal divisors* of F;
6. $\mathrm{Cl}^0(F) := \mathrm{Div}^0(F)/\mathrm{Princ}(F)$ is the *class group* of F. In terms of algebraic curves $\mathcal{X}$, the class group corresponds to the rational points of the *Jacobian of* $\mathcal{X}$ and is then denoted as $\mathrm{Jac}(\mathcal{X})(\mathbb{F}_q)$.

12.5.10 Lemma [2714, Proposition 5.1.3]

1. For every $n \geq 0$, there are only finitely many divisors $A \geq 0$ with $\deg A = n$.
2. The class group $\mathrm{Cl}^0(F)$ is a finite group.

12.5.11 Definition The number $h := h_F := \mathrm{ord}(\mathrm{Cl}^0(F))$ is the *class number of* F.

12.5.12 Definition The *Zeta function* of F is defined by the power series in $\mathbb{C}[[t]]$ below (here $\mathbb{C}$ is the complex number field):

$$Z(t) := \sum_{n=0}^{\infty} A_n t^n,$$

where A_n denotes the number of *positive* divisors $D \in \mathrm{Div}(F)$ of degree n.

12.5.13 Theorem [2714, Theorem 5.1.15]

1. The power series $Z(t)$ converges for all $t \in \mathbb{C}$ with $|t| < q^{-1}$.
2. $Z(t)$ can be written as

$$Z(t) = \frac{L(t)}{(1-t)(1-qt)}$$

with a polynomial $L(t) = a_0 + a_1 t + \cdots + a_{2g} t^{2g} \in \mathbb{Z}[t]$ of degree $2g$. This polynomial is the *L-polynomial of F*.
3. (Functional equation of the L-polynomial) The coefficients of the L-polynomial of F satisfy
   a. $a_0 = 1$ and $a_{2g} = q^g$,
   b. $a_{2g-i} = q^{g-i} a_i$ for $0 \leq i \leq g$.
4. $N(F) = a_1 + q + 1$.
5. $L(1) = h_F$ is the class number of F.

12.5.14 Lemma [2714, Theorem 5.1.15]

1. The L-polynomial factors into linear factors over $\mathbb{C}$ as follows:

$$L(t) = \prod_{j=1}^{2g}(1 - \omega_j t)$$

with *algebraic integers* $\omega_j \in \mathbb{C}$. As $L(\omega_j^{-1}) = 0$, the complex numbers ω_j are the *reciprocals of the roots of* $L(t)$.

2. One can arrange $\omega_1, \ldots, \omega_{2g}$ in such a way that $\omega_j \cdot \omega_{g+j} = q$ for $1 \leq j \leq g$.

12.5.15 Remark The reciprocal polynomial $P(t) := t^{2g} \cdot L(1/t)$ has an interpretation as the *characteristic polynomial of the Frobenius endomorphism* acting on the Tate module T_ℓ; see [1959, 2783]. The roots of $P(t)$ are just the reciprocals of the roots of $L(t)$. Therefore, the complex numbers ω_j in Lemma 12.5.14 are also called the *eigenvalues of the Frobenius endomorphism.*

12.5.16 Remark The following theorem is fundamental for the theory of function fields over finite fields. It was first proved by Hasse for $g = 1$; the generalization to all $g \geq 1$ is due to Weil.

12.5.17 Theorem (Hasse–Weil theorem) [2714, Theorem 5.2.1] The reciprocals of the roots of the L-polynomial satisfy

$$|\omega_j| = q^{1/2} \quad \text{for} \quad 1 \leq j \leq 2g \; .$$

12.5.18 Remark The Hasse–Weil theorem is often referred to as the *Riemann Hypothesis for function fields over finite fields.*

12.5.3 Bounds for the number of rational places

12.5.19 Remark The next result is an easy consequence of the Hasse–Weil theorem 12.5.17.

12.5.20 Theorem (Hasse–Weil bound) [2714, Theorem 5.2.3] The number $N = N(F)$ of rational places of a function field $F/\mathbb{F}_q$ of genus g satisfies the inequality

$$|N - (q + 1)| \leq 2gq^{1/2}.$$

12.5.21 Remark If q is not a square, this bound can be improved as follows.

12.5.22 Theorem (Serre bound) [2591], [2714, Theorem 5.3.1]

$$|N - (q + 1)| \leq g \cdot \left\lfloor 2q^{1/2} \right\rfloor ,$$

where $\lfloor \alpha \rfloor$ means the integer part of the real number α.

12.5.23 Definition For every $g \geq 0$, we define

$$N_q(g) := \max\{N \in \mathbb{N} \mid \text{there is a function field } F/\mathbb{F}_q \text{ of genus } g \text{ with } N(F) = N\}.$$

12.5.24 Remark Clearly $N_q(g) \leq q + 1 + g \cdot \lfloor 2q^{1/2} \rfloor$. Further improvements of this bound can be obtained.

12.5.25 Proposition (Serre's explicit formulas) [2592], [2714, Proposition 5.3.4] Suppose that $u_1, \ldots, u_m$ are non-negative real numbers, not all of them equal to zero, satisfying

$1 + \sum_{n=1}^{m} u_n \cos n\theta \geq 0$ for all $\theta \in \mathbb{R}$. Then

$$N_q(g) \leq 1 + \frac{2g + \sum_{n=1}^{m} u_n q^{n/2}}{\sum_{n=1}^{m} u_n q^{-n/2}}.$$

12.5.26 Remark The results of the examples and tables below are proved in the following way. First one derives upper bounds for $N_q(g)$ using Serre's explicit formulas. In some cases, these upper bounds can be improved slightly by rather subtle arguments [1549]. Lower bounds for $N_q(g)$ are usually obtained by providing explicit examples of function fields having that number of rational places. Many methods of construction have been proposed, see [1550, 2280, 2846] for some of them.

12.5.27 Example (The case $g = 1$) [2591] Let $q = p^e$ with a prime number p.

1. If e is odd, $e \geq 3$ and p divides $\lfloor 2q^{1/2} \rfloor$, then $N_q(1) = q + \lfloor 2q^{1/2} \rfloor$.
2. $N_q(1) = q + 1 + \lfloor 2q^{1/2} \rfloor$, otherwise.

12.5.28 Example (The case $g = 2$) For all prime powers q,

$$q - 2 + 2 \cdot \left\lfloor 2q^{1/2} \right\rfloor \leq N_q(2) \leq q + 1 + 2 \cdot \left\lfloor 2q^{1/2} \right\rfloor.$$

In fact, the exact value of $N_q(2)$ is known in all cases [2591].

12.5.29 Example (The case $g = 3$) The value of $N_q(3)$ is known for many but not for all q. For instance, one knows $N_q(3)$ for all $q \leq 169$ and for all $q = 2^k$ with $k \leq 20$. For details we refer to [2459].

12.5.30 Remark The following tables show $N_q(g)$ for some small values of q and g. Updated tables can be found on the website http://www.manypoints.org/; see [1550].

12.5.31 Example (Values of $N_q(g)$ for $q = 2, 4, 8$ and small g) In the tables below, an entry like $21-24$ means that the exact value of $N_4(8)$ is not known; one knows only that $21 \leq N_4(8) \leq 24$ (at the time of printing).

g	0	1	2	3	4	5	6	7	8	9	10	20
$N_2(g)$	3	5	6	7	8	9	10	10	11	12	13	19-21
$N_4(g)$	5	9	10	14	15	17	20	21	21-24	26	27	40-45
$N_8(g)$	9	14	18	24	25	29	33-34	34-38	35-42	45	42-49	76-83

12.5.32 Example (Values of $N_q(g)$ for $1 \leq g \leq 4$ and prime numbers $q \leq 43$) (at the time of printing)

q	2	3	5	7	11	13	17	19	23	29	31	37	41	43
$N_q(1)$	5	7	10	13	18	21	26	28	33	40	43	50	54	57
$N_q(2)$	6	8	12	16	24	26	32	36	42	50	52	60	66	68
$N_q(3)$	7	10	16	20	28	32	40	44	48	60	62	72	78	80
$N_q(4)$	8	12	18	24	33	38	46	48-50	57	67-70	72	82	88-90	92

12.5.33 Remark If the genus $g(F)$ is large with respect to q, the Hasse–Weil bound can be improved considerably.

12.5.34 Proposition (Ihara's bound) [1569], [2714, Proposition 5.3.3] Suppose that $N_q(g) = q+1+2gq^{1/2}$. Then $g \leq q^{1/2}(q^{1/2} - 1)/2$.

12.5.35 Example Let q be a square. Then there exists a function field of genus $g = q^{1/2}(q^{1/2}-1)/2$ having $q + 1 + 2gq^{1/2}$ rational places. For more about function fields which attain the Hasse–Weil upper bound, see Subsection 12.5.4.

12.5.36 Example

1. For $q = 2^{2m+1}$ with $m \geq 1$, and $g = 2^{3m+1} - 2^m$, one knows that $N_q(g) = q^2 + 1$.
2. Similarly, for $q = 3^{2m+1}$ with $m \geq 1$ and $g = 3^{m+1}(3^{4m+2} + 3^{3m+1} - 3^m - 1)/2$ one has $N_q(g) = q^3 + 1$.

The function fields which attain the values $N_q(g)$ in this example, correspond to the *Deligne–Lusztig curves* associated to the *Suzuki group* and to the *Ree group*, respectively [474, 1414, 2593].

12.5.4 Maximal function fields

12.5.37 Definition A function field $F/\mathbb{F}_q$ is *maximal* if $g(F) > 0$ and $N(F)$ attains the Hasse-Weil upper bound $N(F) = q + 1 + 2gq^{1/2}$.

12.5.38 Remark It is clear that q must be the square of a prime power, if there exists a maximal function field $F/\mathbb{F}_q$. Therefore we assume in this subsection that $q = \ell^2$ is a square. By Ihara's bound 12.5.34, the genus of a maximal function field F over $\mathbb{F}_{\ell^2}$ satisfies $1 \leq g(F) \leq \ell(\ell-1)/2$.

12.5.39 Example [2714, Lemma 6.4.4] Let $H := \mathbb{F}_{\ell^2}(x, y)$ where x, y satisfy the equation $y^\ell + y = x^{\ell+1}$. Then H is a maximal function field over $\mathbb{F}_{\ell^2}$ with $g(H) = \ell(\ell-1)/2$ and $N(H) = \ell^3 + 1 = \ell^2 + 1 + 2g(H)\ell$. The field H is called the *Hermitian function field over* $\mathbb{F}_{\ell^2}$.

12.5.40 Remark The rational places of the Hermitian function field H are the following: there is a unique common pole of x and y, and for any $\alpha, \beta \in \mathbb{F}_{\ell^2}$ with $\alpha^\ell + \alpha = \beta^{\ell+1}$ there is a unique common zero of $y - \alpha$ and $x - \beta$. In this way one obtains all $1 + \ell^3$ rational places of H.

12.5.41 Remark There are generators u, v of the Hermitian function field H which satisfy the equation $u^{\ell+1} + v^{\ell+1} = 1$. Hence the Hermitian function field is a special case of a *Fermat function field*, which is defined by an equation $u^n + v^n = 1$ with $\gcd(n, q) = 1$.

12.5.42 Proposition

1. Suppose that $F/\mathbb{F}_{\ell^2}$ is a maximal function field of genus $g(F) = \ell(\ell-1)/2$. Then F is isomorphic to the Hermitian function field H [2499].
2. There is no maximal function field $E/\mathbb{F}_{\ell^2}$ whose genus satisfies $\frac{1}{4}(\ell-1)^2 < g(E) < \frac{1}{2}\ell(\ell-1)$ for ℓ odd (and $\frac{1}{4}\ell(\ell-2) < g(E) < \frac{1}{2}\ell(\ell-1)$ for ℓ even) [1145].
3. Up to isomorphism there is a unique maximal function field $E/\mathbb{F}_{\ell^2}$ of genus $g(E) = \frac{1}{4}(\ell-1)^2$ for ℓ odd (and $g(E) = \frac{1}{4}\ell(\ell-2)$ for ℓ even) [3, 1144].

12.5.43 Proposition [1826]. Let F be a maximal function field over $\mathbb{F}_q$. Then every function field E of positive genus with $\mathbb{F}_q \subset E \subseteq F$ is also maximal over $\mathbb{F}_q$.

12.5.44 Remark The Hermitian function field $H/\mathbb{F}_{\ell^2}$ has a large automorphism group G. Every subgroup $U \subseteq G$ whose fixed field is not rational, provides then an example of a maximal function field H^U over $\mathbb{F}_{\ell^2}$. Most known examples of maximal function fields over $\mathbb{F}_{\ell^2}$ have been constructed in this way [474, 1205, 1280], and [1511, Chapter 10].

12.5.45 Example [1279] Over the field $\mathbb{F}_q$ with $q = r^6$, consider the function field $F = \mathbb{F}_q(x, y, z)$ which is defined by the equations

$$x^r + x = y^{r+1} \quad \text{and} \quad y \cdot \frac{x^{r^2} - x}{x^r + x} = z^{\frac{r^3+1}{r+1}}.$$

Here F is the *Giulietti–Korchmáros function field*; it is maximal over $\mathbb{F}_q$ of genus $g(F) = (r-1)(r^4+r^3-r^2)/2$. It is (at the time of printing) the only known example of a maximal function field over $\mathbb{F}_q$ which is *not* a subfield of the Hermitian function field $H/\mathbb{F}_q$.

12.5.46 Remark [2722] An important ingredient in many proofs of results on maximal function fields (for example, Parts 2 and 3 of Proposition 12.5.42) is the *Stöhr–Voloch theory* which sometimes gives an improvement of the Hasse–Weil upper bound. The method of Stöhr–Voloch involves the construction of an auxiliary function which has zeros of high order at the $\mathbb{F}_q$-rational points of the corresponding non-singular curve. We illustrate this method in the case of plane curves. Let $f(X,Y) \in \mathbb{F}_q[X,Y]$ be an absolutely irreducible polynomial that defines a non-singular projective plane curve. Recall that an affine point (a,b) with $f(a,b) = 0$ is non-singular if at least one of the partial derivatives $f_X(X,Y)$ or $f_Y(X,Y)$ does not vanish at the point (a,b). The auxiliary function $h(X,Y)$ in this case is obtained from the equation of the tangent line as $h(X,Y) = (X - X^q)f_X(X,Y) + (Y - Y^q)f_Y(X,Y)$. Suppose now that $f(X,Y)$ does not divide $h(X,Y)$. Then

$$N(F) \le d(d+q-1)/2 ,$$

where $F = \mathbb{F}_q(x,y)$ with $f(x,y) = 0$ is the corresponding function field, and d denotes the degree of the polynomial $f(X,Y)$. As an example consider the case $d = 4$. The genus of F is $g(F) = (d-1)(d-2)/2 = 3$. The bound above gives $N(F) \le 2q+6$ which is better than the Hasse–Weil upper bound for all $q \le 23$. We note that $N_q(3) = 2q+6$ for $q = 5, 7, 11, 13, 17$ and 19; see Example 12.5.32.

12.5.5 Asymptotic bounds

12.5.47 Remark In this subsection we give some results about the asymptotic growth of the numbers $N_q(g)$, see 12.5.23. As was mentioned in Proposition 12.5.34, the Hasse-Weil upper bound $N_q(g) \le q + 1 + 2gq^{1/2}$ cannot be attained if the genus is large with respect to q.

12.5.48 Definition The real number $A(q) := \limsup_{g\to\infty} N_q(g)/g$ is *Ihara's quantity.*

12.5.49 Remark As follows from the Hasse–Weil bound, $A(q) \le 2q^{1/2}$. The following bound is a significant improvement of this estimate.

12.5.50 Theorem (Drinfeld–Vlăduţ bound) [2714, Theorem 7.1.3], [2882]

$$A(q) \le q^{1/2} - 1.$$

12.5.51 Remark The proof of the Drinfeld–Vlăduţ bound is a clever application of Serre's explicit formulas 12.5.25. If q is a square, the Drinfeld–Vlăduţ bound is sharp.

12.5.52 Theorem [1569, 2821]

$$A(q) = q^{1/2} - 1 \text{ if } q \text{ is a square.}$$

12.5.53 Remark If q is a non-square, the exact value of $A(q)$ is not known. The lower bounds for $A(q)$, given below, are proved by providing specific sequences of function fields $F_n/\mathbb{F}_q$ such that $\lim_{n\to\infty} N(F_n)/g(F_n) > 0$. Every such sequence gives then a lower bound for $A(q)$. For details, see Section 12.6.

12.5.54 Theorem

1. [2280, Theorem 5.2.9], [2592] There is an absolute constant $c > 0$ such that $A(q) > c \cdot \log q$ for all prime powers q.
2. [265, 3079]
$$A(q^3) \geq 2(q^2 - 1)/(q + 2).$$

12.5.55 Remark Recall that $\lfloor \alpha \rfloor$ and $\lceil \alpha \rceil$ denote the floor and the ceiling of a real number α. The harmonic mean of two positive real numbers α, β is given by the formula $H(\alpha, \beta) = 2\alpha\beta/(\alpha+\beta)$. The following result contains Theorem 12.5.52 and Part 2 of Theorem 12.5.54 as special cases.

12.5.56 Theorem [1203] (see also Example 12.6.29) For every prime number p and every $n \geq 2$,

$$A(p^n) \geq H(p^{\lfloor n/2 \rfloor} - 1, p^{\lceil n/2 \rceil} - 1).$$

For example, one has for $p = 2$ and all sufficiently large odd integers n,

$$0.9428 \times (2^{n/2} - 1) \leq A(2^n) \leq 2^{n/2} - 1.$$

12.5.57 Example [105, 938] The best known lower bounds for $A(q)$ for $q = 2, 3, 5$ were obtained from class field towers:

$$\begin{aligned} A(2) &\geq 0.316999\ldots, \\ A(3) &\geq 0.492876\ldots, \\ A(5) &\geq 0.727272\ldots. \end{aligned}$$

12.5.58 Remark A counterpart to Ihara's quantity $A(q)$ is the following quantity.

12.5.59 Definition We set $A^-(q) := \liminf_{g\to\infty} N_q(g)/g$.

12.5.60 Proposition [973] $A^-(q) > 0$ for all q. More precisely,

1. $A^-(q) \geq (q^{1/2} - 1)/4$, if q is a square.
2. There is an absolute constant $d > 0$ such that $A^-(q) \geq d \cdot \log q$ for all q.

See Also

§12.2	For more material on function fields of genus 1 (elliptic curves).
§12.4	For more material on hyperelliptic function fields.
§12.6	For towers of function fields.
§12.7	For zeta functions for curves.
§15.2	For applications of function fields to coding theory.

References Cited: [3, 105, 265, 474, 938, 973, 1144, 1145, 1203, 1205, 1279, 1280, 1414, 1511, 1549, 1550, 1569, 1826, 1959, 2280, 2459, 2499, 2591, 2592, 2593, 2714, 2722, 2783, 2821, 2846, 2882, 3079]

12.6 Towers

Arnaldo Garcia, IMPA
Henning Stichtenoth, Sabanci University

We use terminology as in Sections 12.1 and 12.5, see also [2714]. Some methods are discussed how to obtain *lower bounds* for Ihara's quantity $A(q)$, see Definition 12.5.48. Such bounds have a great impact in applications, for instance in coding theory, see Section 15.2.

12.6.1 Introduction to towers

12.6.1 Remark Lower bounds for $A(q)$ are usually obtained in the following way: one constructs a sequence of function fields $(F_i/\mathbb{F}_q)_{i\geq 0}$ with $g(F_i) \to \infty$ such that the limit $\lim_{i\to\infty} N(F_i)/g(F_i)$ exists. If this limit is positive, then it provides a non-trivial lower bound for $A(q)$.

12.6.2 Remark Essentially three methods are known for constructing such sequences of function fields: *modular towers, class field towers*, and *explicit towers*. In the following two remarks we give a very brief description of the first two methods.

12.6.3 Remark (Modular towers) [217, 971, 972, 1569, 2821] Modular towers were introduced by Ihara, and independently by Tsfasman, Vlăduţ, and Zink. Let N be a positive integer and p a prime number not dividing N. There exists an affine algebraic curve $Y_0(N)$ defined over $\mathbb{F}_p$ such that, for any field K of characteristic p, $Y_0(N)$ parametrizes the set of isomorphy classes of pairs (E, C), where E is an elliptic curve (see Section 12.2) and C is a cyclic subgroup of E of order N, defined over K, in a functorial way. The construction of $Y_0(N)$ is independent of p and can be done in characteristic zero also. The complete curve obtained from $Y_0(N)$ is denoted $X_0(N)$. If $\ell \neq p$ is another prime, then the curves $X_0(\ell^n)$, $n = 1, 2, \ldots$ form a tower with the maps sending (E, C) to (E, C') where C' is the unique subgroup of C of index ℓ. Over $\mathbb{F}_{p^2}$, the supersingular elliptic curves (see Subsection 12.2.9) together with all their cyclic subgroups of order ℓ^n give rational points on $X_0(\ell^n)(\mathbb{F}_{p^2})$, because Frobenius is multiplication by $-p$ on those curves. This gives a tower of curves over $\mathbb{F}_{p^2}$ which attains the Drinfel'd–Vlăduţ bound.

For $\mathbb{F}_{q^2}$, with q arbitrary, a similar construction can be made using Shimura curves which parametrize abelian varieties of higher dimension with additional structure.

12.6.4 Remark (Class field towers) [938, 2280, 2561, 2592] Starting with any function field F_0 of genus $g_0 \geq 2$ and a set S_0 of rational places of F_0, one defines inductively the field F_{n+1} to be the maximal abelian unramified extension of F_n in which all places of S_n split completely, and S_{n+1} to be the set of all places of F_{n+1} which lie over S_n. If $F_n \subsetneq F_{n+1}$ for all n (which is not always the case), the tower thus obtained is called a *class field tower*, and its limit (see Definition 12.6.8) is at least $|S_0|/(g_0 - 1)$. The hard part is to choose F_0, S_0 so that the tower is infinite. This is analogous to the corresponding problem in the number field case of infinite class field towers which was solved by Golod and Shafarevich. A choice of F_0, S_0 then can be used to show that $A(p) \geq c \cdot \log p$, for p prime, with an absolute constant $c > 0$. This approach which is due to Serre [2592], is so far the only way to prove that $A(p) > 0$ holds for prime numbers p.

12.6.5 Remark (Explicit towers of function fields) These towers were introduced by Garcia and Stichtenoth [1200, 2714]. The method, which is more elementary than modular towers and class field towers, is presented below in some detail.

12.6.6 Definition A *tower* $\mathcal{F}$ *over* $\mathbb{F}_q$ is an infinite sequence $\mathcal{F} = (F_0, F_1, F_2, \ldots)$ of function fields $F_i/\mathbb{F}_q$ (with $\mathbb{F}_q$ algebraically closed in all F_i) such that

1. $F_0 \subsetneqq F_1 \subsetneqq F_2 \subsetneqq \cdots \subsetneqq F_n \subsetneqq \cdots$;
2. each extension F_{n+1}/F_n is finite and separable;
3. for some $n \geq 0$, the genus $g(F_n)$ is ≥ 2.

12.6.7 Remark Items 2 and 3 imply that $g(F_i) \to \infty$ as $i \to \infty$. The following limit exists for every tower over $\mathbb{F}_q$ [2714, Lemma 7.2.3].

12.6.8 Definition Let $\mathcal{F} = (F_0, F_1, \ldots)$ be a tower of function fields over $\mathbb{F}_q$. The limit $\lambda(\mathcal{F}) := \lim_{i\to\infty} N(F_i)/g(F_i)$ is the *limit of the tower* $\mathcal{F}$.

12.6.9 Remark We note that the inequalities $0 \leq \lambda(\mathcal{F}) \leq A(q)$ hold for every tower over $\mathbb{F}_q$.

12.6.10 Definition A tower $\mathcal{F}/\mathbb{F}_q$ is *asymptotically good* if $\lambda(\mathcal{F}) > 0$. It is *asymptotically bad* if $\lambda(\mathcal{F}) = 0$.

12.6.11 Remark The notion of asymptotically good (bad) towers is related to the notion of asymptotically good (bad) sequences of codes; see Section 15.2. The remark below follows immediately from the definitions.

12.6.12 Remark As $A(q) \geq \lambda(\mathcal{F})$, every asymptotically good tower $\mathcal{F}$ over $\mathbb{F}_q$ provides a non-trivial lower bound for Ihara's quantity.

12.6.13 Remark Most towers turn out to be asymptotically bad and some effort is needed to find asymptotically good ones. We discuss now some criteria which ensure that a tower is good.

12.6.14 Definition Let $\mathcal{F} = (F_0, F_1, \ldots)$ be a tower over $\mathbb{F}_q$.

1. A place P of F_0 is *ramified in* $\mathcal{F}/F_0$, if there is some $n \geq 1$ and some place Q of F_n lying over P with ramification index $e(Q|P) > 1$. Otherwise, P is *unramified* in $\mathcal{F}$.
2. A *rational* place P of F_0 *splits completely in* $\mathcal{F}/F_0$, if P splits completely in the extensions F_n/F_0, for all $n \geq 1$.
3. The set $\mathrm{Ram}(\mathcal{F}/F_0) := \{P \mid P$ is a place of F_0 which is ramified in $\mathcal{F}/F_0\}$ is the *ramification locus* of $\mathcal{F}$ over F_0.
4. The set $\mathrm{Split}(\mathcal{F}/F_0) := \{P \mid P$ is a rational place of F_0 splitting completely in $\mathcal{F}/F_0\}$ is the *splitting locus* of $\mathcal{F}$ over F_0.

12.6.15 Remark The splitting locus is always finite (it may be empty). The ramification locus can be finite or infinite.

12.6.16 Theorem [2714, Theorem 7.2.10] Assume that the tower $\mathcal{F} = (F_0, F_1, \ldots)$ over $\mathbb{F}_q$ has the following properties.

1. The splitting locus $\mathrm{Split}(\mathcal{F}/F_0)$ is non-empty.
2. The ramification locus $\mathrm{Ram}(\mathcal{F}/F_0)$ is finite.

3. For every $P \in \mathrm{Ram}(\mathcal{F}/F_0)$ there is a constant $c_P \in \mathbb{R}$ such that for all $n \geq 0$ and all places Q of F_n lying over P, the different exponent $d(Q|P)$ is bounded by
$$d(Q|P) \leq c_P \cdot (e(Q|P) - 1).$$

Then the tower $\mathcal{F}$ is asymptotically good, and its limit satisfies the inequality
$$\lambda(\mathcal{F}) \geq \frac{s}{g(F_0) - 1 + r},$$
where
$$s := \left|\ \mathrm{Split}(\mathcal{F}/F_0)\ \right| \quad \text{and} \quad r := \frac{1}{2} \sum_{P \in \mathrm{Ram}(\mathcal{F}/F_0)} c_P \cdot \deg P.$$

12.6.17 Remark Of course, one should choose the constant c_P as small as possible (if it exists). In general it is a difficult task to prove its existence in towers having wild ramification.

12.6.18 Remark A tower $\mathcal{F}/F_0$ is *tame* if all places $P \in \mathrm{Ram}(\mathcal{F}/F_0)$ are tame in all extensions F_n/F_0; that is, the ramification index $e(Q|P)$ is relatively prime to q for all places Q of F_n lying over P.

12.6.19 Remark For a tame tower, the constants c_P in Theorem 12.6.16 can be chosen as $c_P = 1$. Hence a tame tower with finite ramification locus and non-empty splitting locus is asymptotically good, and the inequality for $\lambda(\mathcal{F})$ given in Theorem 12.6.16 holds with
$$r := \frac{1}{2} \sum_{P \in \mathrm{Ram}(\mathcal{F}/F_0)} \deg P.$$

12.6.20 Remark All *known* asymptotically good towers of function fields have properties 1, 2, and 3 of Theorem 12.6.16.

12.6.2 Examples of towers

12.6.21 Definition Let $f(Y) \in \mathbb{F}_q(Y)$ and $h(X) \in \mathbb{F}_q(X)$ be non-constant rational functions, and let $\mathcal{F} = (F_0, F_1, \ldots)$ be a tower of function fields over $\mathbb{F}_q$. The tower $\mathcal{F}$ is *recursively defined by the equation* $f(Y) = h(X)$, if there exist elements $x_i \in F_i$ $(i = 0, 1, \ldots)$ such that

1. $F_0 = \mathbb{F}_q(x_0)$ is a rational function field;
2. $F_i = F_{i-1}(x_i)$ for all $i \geq 1$;
3. for all $i \geq 1$, the elements x_{i-1}, x_i satisfy the equation $f(x_i) = h(x_{i-1})$.

12.6.22 Example [2714, Proposition 7.3.2] Let $q = \ell^2$ be a square, $\ell > 2$. Then the equation
$$Y^{\ell-1} = 1 - (X+1)^{\ell-1}$$
defines an asymtotically good tame tower $\mathcal{F}$ over $\mathbb{F}_q$. The ramification locus of this tower is the set of all places $(x_0 = \alpha)$ with $\alpha \in \mathbb{F}_\ell$, and the place $(x_0 = \infty)$ splits completely. By Theorem 12.6.16 the limit satisfies the inequality
$$\lambda(\mathcal{F}) \geq 2/(\ell - 2).$$

For $q = 9$ this limit attains the Drinfeld–Vlăduţ bound $\lambda(\mathcal{F}) = 2 = \sqrt{9} - 1$.

12.6.23 Example [2714, Proposition 7.3.3] Let $q = \ell^e$ with $e \geq 2$ and set $m := (q-1)/(\ell-1)$. Then the equation

$$Y^m = 1 - (X+1)^m$$

defines an asymptotically good tame tower $\mathcal{F}$ over $\mathbb{F}_q$ with limit

$$\lambda(\mathcal{F}) \geq 2/(q-2).$$

This gives a simple proof that $A(q) > 0$ for all non-prime values of q. For $q = 4$ the tower attains the Drinfeld–Vlăduţ bound $\lambda(\mathcal{F}) = 1 = \sqrt{4} - 1$.

12.6.24 Example [1204] Let $q = p^2$ where p is an *odd prime.* Then the equation

$$Y^2 = \frac{X^2+1}{2X}$$

defines a tame tower $\mathcal{F}$ over $\mathbb{F}_q$. Its ramification locus is

$$\mathrm{Ram}(\mathcal{F}/F_0) = \{(x_0 = \alpha) \mid \alpha^4 = 1 \text{ or } \alpha = 0 \text{ or } \alpha = \infty\}.$$

There are $2(p-1)$ rational places of F_0 which split completely in the tower. The inequality in Theorem 12.6.16 gives $\lambda(\mathcal{F}) \geq p - 1$ which coincides with the Drinfeld–Vlăduţ bound. So,

$$\lambda(\mathcal{F}) = p - 1.$$

The fact that the splitting locus of this tower has cardinality $2(p-1)$ is not easy to prove. For $p = 3, 5$ one can check directly that the places $(x_0 = \alpha)$ with $\alpha^4 + 1 = 0$ (for $p = 3$) and $\alpha^8 - \alpha^4 + 1 = 0$ (for $p = 5$) split completely in $\mathcal{F}$.

12.6.25 Remark Now we give some examples of *wild towers*, that is, there are some places of F_0 whose ramification index in some extension F_n/F_0 is divisible by the characteristic of $\mathbb{F}_q$.

In wild towers, it is usually difficult to find a bound, if it exists, for the different exponents in terms of ramification indices (see Theorem 12.6.16).

12.6.26 Example [1200] Let $q = \ell^2$ be a square and define the tower $\mathcal{F} = (F_0, F_1, \ldots)$ over $\mathbb{F}_q$ as follows: $F_0 := \mathbb{F}_q(x_0)$ is the rational function field, and for all $n \geq 0$, set $F_{n+1} := F_n(x_{n+1})$ with

$$(x_{n+1}x_n)^\ell + x_{n+1}x_n = x_n^{\ell+1}.$$

The ramification locus of $\mathcal{F}$ is $\mathrm{Ram}(\mathcal{F}/F_0) = \{\ (x_0 = 0),\ (x_0 = \infty)\ \}$, and all other rational places of F_0 split completely in the tower. We note however that Theorem 12.6.16 is not directly applicable to determine the limit $\lambda(\mathcal{F})$. One can show that

$$\lambda(\mathcal{F}) = \ell - 1,$$

so this tower attains the Drinfeld–Vlăduţ bound.

12.6.27 Example [1201] The equation

$$Y^\ell + Y = \frac{X^\ell}{X^{\ell-1}+1}$$

defines a tower over $\mathbb{F}_q$ with $q = \ell^2$, whose limit attains the Drinfeld–Vlăduţ bound $\lambda(\mathcal{F}) = \ell - 1$. The determination of the splitting locus and the ramification locus for this tower is easy. The hard part is to show that $c_P = 2$ for all ramified places (for the definition of c_P see Theorem 12.6.16).

12.6.28 Example [265, 2845] Over the field $\mathbb{F}_q$ with $q = \ell^3$, the equation

$$Y^\ell - Y^{\ell-1} = 1 - X + X^{-(\ell-1)}$$

defines an asymptotically good tower $\mathcal{F}$ with limit

$$\lambda(\mathcal{F}) \geq \frac{2(\ell^2-1)}{\ell+2}.$$

It follows that

$$A(\ell^3) \geq \frac{2(\ell^2-1)}{\ell+2},$$

for all prime powers ℓ (see Theorem 12.5.54).

12.6.29 Example [1203] Let $q = \ell^n$ with $n \geq 2$. For every partition of n into relatively prime parts,

$$n = j + k \text{ with } j \geq 1, k \geq 1 \text{ and } \gcd(j,k) = 1,$$

a tower $\mathcal{F}$ over $\mathbb{F}_q$ is recursively defined by the equation

$$\mathrm{Tr}_j\Big(\frac{Y}{X^{\ell^k}}\Big) + \mathrm{Tr}_k\Big(\frac{Y^{\ell^j}}{X}\Big) = 1,$$

where

$$\mathrm{Tr}_a(T) = T + T^{\ell} + T^{\ell^2} + \cdots + T^{\ell^{a-1}} \quad \text{for any} \quad a \in \mathbb{N}.$$

The limit of this tower satisfies the inequality

$$\lambda(\mathcal{F}) \geq 2\Big(\frac{1}{\ell^j-1} + \frac{1}{\ell^k-1}\Big)^{-1},$$

which is the harmonic mean of $\ell^j - 1$ and $\ell^k - 1$. This tower gives the best known lower bound for Ihara's quantity $A(q)$, for all non-prime fields $\mathbb{F}_q$ (at the time of printing).

12.6.30 Remark None of the towers in Examples 12.6.22 - 12.6.24 or 12.6.26 - 12.6.29 is Galois over F_0, that is, not all of the extensions F_n/F_0, $n \geq 0$ are Galois extensions. In some special cases however, one can prove that the tower $\hat{\mathcal{F}} := (\hat{F}_0, \hat{F}_1, \ldots)$, where $\hat{F}_n$ is the Galois closure of F_n/F_0, is also asymptotically good [1202, 2713].

12.6.31 Remark There are examples of function fields with many rational points which are *abelian extensions* of a rational function field (for instance, the Hermitian function field H; see Example 12.5.39). Other abelian extensions over $\mathbb{F}_q(x)$ having many rational places can be obtained via the method of *cyclotomic function fields* [2280]. However, abelian extensions $F/\mathbb{F}_q(x)$ of *large* genus have only few rational places. More precisely, if $(F_i)_{i\geq 0}$ is a sequence of abelian extensions of a rational function field with $g(F_i) \to \infty$, then $\lim_{i\to\infty} N(F_i)/g(F_i) = 0$ [1107].

12.6.32 Remark We conclude this section with a warning: not every irreducible equation $f(Y) = h(X)$ defines a recursive tower. For instance, if one replaces $X+1$ by X in Examples 12.6.22 and 12.6.23, one just gets a finite extension $\mathcal{F}/F_0$ but not a tower. Also, one has to show that $\mathbb{F}_q$ is algebraically closed in each field F_i of the tower. In most of the examples above this follows from the fact that there is some place which is totally ramified in all extensions F_i/F_0.

See Also

§13.5	For discussion of Drinfeld modules.
§15.2	For applications of towers to coding theory.

References Cited: [217, 265, 938, 971, 972, 1107, 1200, 1201, 1202, 1203, 1204, 1569, 2280, 2561, 2592, 2713, 2714, 2821, 2845]

12.7 Zeta functions and L-functions

Lei Fu, *Nankai University*

We use the terminology in [1427]. For the definitions of schemes, morphisms between schemes, and the affine Spec A for a commutative ring A, sec [1427, II 2]. For the definitions of schemes or morphisms of finite type, see [1427, II 3]. For the definitions of separated, proper or projective schemes or morphisms, see [1427, II 4]. For the definition of smooth morphisms, see [1427, III 10].

Throughout this section, we assume our schemes are separated. Let X be a scheme of finite type over $\mathbb{Z}$. Denote the set of Zariski closed points in X by $|X|$ (observe that in the rest of the handbook this notation indicates the cardinality of the set X). For any $x \in |X|$, the residue field $k(x)$ of X at x is a finite field. Let $N(x)$ be the number of elements of $k(x)$.

12.7.1 Zeta functions

12.7.1 **Definition** The *Zeta function* $\zeta_X(s)$ of X is

$$\zeta_X(s) = \prod_{x \in |X|} \frac{1}{1 - N(x)^{-s}}.$$

12.7.2 **Remark** When X is the affine scheme Spec $\mathbb{Z}$, $\zeta_X(s)$ is just the Riemann zeta function

$$\zeta(s) = \prod_p \frac{1}{1 - p^{-s}} = \sum_{n=1}^{\infty} \frac{1}{n^s}.$$

12.7.3 **Remark** We are concerned with the case where X is a scheme of finite type over a finite field $\mathbb{F}_q$ with q elements of characteristic p. For any $x \in |X|$, $k(x)$ is a finite extension of $\mathbb{F}_q$. Set $\deg(x) = [k(x) : \mathbb{F}_q]$. Then we have $N(x) = q^{\deg(x)}$.

12.7.4 **Definition** Suppose X is a scheme of finite type over $\mathbb{F}_q$. We define

$$Z(X, t) = \prod_{x \in |X|} \frac{1}{1 - t^{\deg(x)}}.$$

12.7.5 Remark The relation between $\zeta_X(s)$ and $Z(X,t)$ is

$$\zeta_X(s) = Z(X, q^{-s}).$$

We have

$$Z(X,t) = \sum_{\alpha} t^{\deg(\alpha)},$$

where α goes over the set of effective 0-cycles in X.

12.7.6 Theorem [795, Equation (1.5.2)] For any positive integer m, let $X(\mathbb{F}_{q^m})$ be the set of $\mathbb{F}_{q^m}$-rational points in X. We have an equation of formal power series

$$t\frac{d}{dt}\ln Z(X,t) = \sum_{m=1}^{\infty} \#X(\mathbb{F}_{q^m})t^m.$$

12.7.7 Remark For the n-dimensional affine space $\mathbb{A}^n_{\mathbb{F}_q}$, we have

$$\#\mathbb{A}^n_{\mathbb{F}_q}(\mathbb{F}_{q^m}) = q^{mn},$$
$$Z(\mathbb{A}^n_{\mathbb{F}_q}, t) = \frac{1}{1-q^n t}.$$

For the n-dimensional projective space $\mathbb{P}^n_{\mathbb{F}_q}$, we have

$$\#\mathbb{P}^n_{\mathbb{F}_q}(\mathbb{F}_{q^m}) = 1 + q^m + \cdots + q^{mn},$$
$$Z(\mathbb{P}^n_{\mathbb{F}_q}, t) = \frac{1}{(1-t)(1-qt)\cdots(1-q^n t)}.$$

12.7.8 Theorem Let X be a smooth projective scheme over $\mathbb{F}_q$ of dimension n.

1. *Rationality*: $Z(X,t)$ is a rational function of t, that is, a quotient of polynomials with rational coefficients.
2. *Functional equation*: $Z(X,t)$ satisfies a functional equation of the form

$$Z\left(X, \frac{1}{q^n t}\right) = \epsilon q^{\frac{n\chi(X)}{2}} t^{\chi(X)} Z(X,t)$$

   for some constants $\chi(X)$ and $\epsilon = \pm 1$.
3. *Riemann hypothesis*: $Z(X,t)$ can be written in the form

$$Z(X,t) = \prod_{i=0}^{2n} P_i(X,t)^{(-1)^{i+1}}$$

   such that $P_i(X,t) \in \mathbb{Z}[t]$ and all reciprocal zeros of $P_i(X,t)$ lie on the circle $|t| = q^{\frac{i}{2}}$.

12.7.9 Remark The above theorem is usually called the Weil conjecture. It was proved to be true by Dwork, Grothendieck, and Deligne. Weil points out that to prove his conjecture, one needs to construct a cohomology theory for schemes over an abstract field. One such theory is the ℓ-adic cohomology theory constructed by Grothendieck, where ℓ is a prime number distinct from the characteristic of the ground field. Results on ℓ-adic cohomology theory used in this section can be found in [136, 797, 1570].

12.7.10 Remark In the Riemann hypothesis, we consider the Archimedean absolute values of the zeros and poles of $Z(X,t)$. One can show that for any prime number $\ell \neq p$, all the zeros and poles of $Z(X,t)$ are ℓ-adic units. (This follows from ℓ-adic cohomology theory.) It is interesting to study the p-adic absolute values of the zeros and poles of $Z(X,t)$; see Section 12.8.

12.7.11 Definition [1570, Exp. XV] Let $\mathrm{Fr}_X : X \to X$ be the morphism of schemes $(\mathrm{Fr}_X, \mathrm{Fr}_X^\#) : (X, \mathcal{O}_X) \to (X, \mathcal{O}_X)$ such that on the underlying topological space, $\mathrm{Fr}_X : X \to X$ is the identity, and $\mathrm{Fr}_X^\# : \mathcal{O}_X \to \mathcal{O}_X$ maps a section s of $\mathcal{O}_X$ to s^q. Fix an algebraic closure $\mathbb{F}$ of $\mathbb{F}_q$, and let $\overline{X} = X \otimes_{\mathbb{F}_q} \mathbb{F}$. The *geometric Frobenius correspondence* on X is the base change $F_X : \overline{X} \to \overline{X}$ of Fr_X from $\mathbb{F}_q$ to $\mathbb{F}$.

12.7.12 Remark There is a canonical one-to-one correspondence between the set $X(\mathbb{F}_{q^m})$ of $\mathbb{F}_{q^m}$-rational points in X, and the set of fixed points of F_X^m on $\overline{X}$.

12.7.13 Remark Let $H^i(\overline{X}, \mathbb{Q}_\ell)$ and $H^i_c(\overline{X}, \mathbb{Q}_\ell)$ be the ℓ-adic cohomology groups and the ℓ-adic cohomology groups with compact support of $\overline{X}$, respectively. They are finite dimensional vector spaces, and they vanish if $i \notin [0, 2\dim X]$. If X is proper, we have $H^i(\overline{X}, \mathbb{Q}_\ell) \cong H^i_c(\overline{X}, \mathbb{Q}_\ell)$.

12.7.14 Theorem [797, Rapport 3.2] (Lefschetz fixed point theorem) We have

$$\#X(\mathbb{F}_{q^m}) = \sum_{i=0}^{2\dim X} (-1)^i \mathrm{Tr}\Big(F_X^m, H^i_c(\overline{X}, \mathbb{Q}_\ell)\Big).$$

12.7.15 Remark The following theorem follows from Theorems 12.7.6 and 12.7.14. It proves the rationality of the function $Z(X,t)$.

12.7.16 Theorem [797, Rapport 3.1] (Grothendieck's formula) Let X be a scheme of finite type over $\mathbb{F}_q$. We have

$$Z(X,t) = \prod_{i=0}^{2\dim X} \det\Big(1 - F_X t, H^i_c(\overline{X}, \mathbb{Q}_\ell)\Big)^{(-1)^{i+1}}.$$

12.7.17 Remark Suppose a finite group G acts on X. Then each $H^i_c(\overline{X}, \mathbb{Q}_\ell)$ is a representation G. Let

$$H^i_c(\overline{X}, \mathbb{Q}_\ell) = \bigoplus_{j \in I_i} V_{ij}$$

be the isotypic decomposition of this representation. Then each V_{ij} is invariant under the action of F_X, and we have a further factorization for the formula of $Z(X,t)$ in Theorem 12.7.16:

$$Z(X,t) = \prod_{i=0}^{2\dim X} \prod_{j \in I_i} \det(1 - F_X t, V_{ij})^{(-1)^{i+1}}.$$

In this way, we can get further information about $Z(X,t)$. As an example, let X_λ be the Dwork hypersurface in $\mathbb{P}^{n-1}_{\mathbb{F}_q}$ defined by the equation

$$x_1^n + \cdots + x_n^n - n\lambda x_1 \cdots x_n = 0,$$

where λ is a parameter. The group

$$A = \{(\zeta_1, \ldots, \zeta_n) | \zeta_i \in \mathbb{F}_q, \zeta_i^n = \prod_i \zeta_i = 1\} / \{(\zeta, \ldots, \zeta) | \zeta \in \mathbb{F}_q, \ \zeta^n = 1\}$$

acts on X_λ by coordinatewise multiplication, and the symmetric group $\mathfrak{S}_n$ acts on X_λ by permuting coordinates. Thus the finite group $G = A \rtimes \mathfrak{S}_n$ acts on X_λ. Goutet [1345] studies the factorization of $Z(X_\lambda, t)$ with respect to the group action of G on $H^i(\overline{X}_\lambda, \mathbb{Q}_\ell)$. The function $Z(X_\lambda, t)$ has also been studied by Candelas, de la Ossa, and Rodriguez-Villegas [486] for the case $n = 5$, by Katz [1710], and by Brünjes [435] for the case $\lambda = 0$.

12.7.18 Theorem [136, Exp. XVIII, Paragraph 3.2.6] (Poincaré duality) Suppose X is proper smooth over $\mathbb{F}_q$ and pure of dimension n. Then we have a perfect pairing

$$(\ ,\ ) : H^i(\overline{X}, \mathbb{Q}_\ell) \times H^{2n-i}(\overline{X}, \mathbb{Q}_\ell) \to \mathbb{Q}_\ell$$

such that for any $s \in H^i(\overline{X}, \mathbb{Q}_\ell)$ and $t \in H^{2n-i}(\overline{X}, \mathbb{Q}_\ell)$, we have

$$(F_X(s), F_X(t)) = q^n(s, t).$$

12.7.19 Remark Poincaré duality implies the functional equation for $Z(X, t)$.

12.7.20 Theorem [795, Paragraph 2.6] Suppose X is proper smooth over $\mathbb{F}_q$ and pure of dimension n. We have

$$Z\left(X, \frac{1}{q^n t}\right) = \epsilon q^{\frac{n\chi(\overline{X})}{2}} t^{\chi(\overline{X})} Z(X, t),$$

where $\chi(\overline{X}) = \sum_{i=0}^{2n} (-1)^i \dim H^i(\overline{X}, \mathbb{Q}_\ell)$ is the Euler characteristic of $\overline{X}$, and if N is the multiplicity of the eigenvalue $q^{\frac{n}{2}}$ of F_X acting on $H^n(\overline{X}, \mathbb{Q}_\ell)$, then we have

$$\epsilon = \begin{cases} 1 & \text{if } n \text{ is odd,} \\ (-1)^N & \text{if } n \text{ is even.} \end{cases}$$

12.7.21 Remark Together with Theorem 12.7.16, the following theorem of Deligne proves the Riemann hypothesis for $Z(X, t)$.

12.7.22 Theorem [795], [798, Corollaires 3.3.4-3.3.5] Suppose X is a scheme of finite type over $\mathbb{F}_q$. For any eigenvalue α of F_X on $H_c^i(\overline{X}, \mathbb{Q}_\ell)$, α is an algebraic integer, and all the Galois conjugates of α have Archimedean absolute value $q^{\frac{w}{2}}$ for some integer $w \le i$. The equality $w = i$ holds if X is proper smooth over $\mathbb{F}_q$.

12.7.23 Remark For each $0 \le i \le 2\dim X$, let $b_i = \dim H_c^i(\overline{X}, \mathbb{Q}_\ell)$, and let α_{ij} $(j = 1, \ldots, b_i)$ be all the eigenvalues of F_X on $H_c^i(\overline{X}, \mathbb{Q}_\ell)$. By Theorem 12.7.14, we have

$$\#X(\mathbb{F}_q) = \sum_{i=0}^{2\dim X} \sum_{j=1}^{b_i} (-1)^i \alpha_{ij}.$$

Theorem 12.7.22 provides bounds for $|\alpha_{ij}|$. To get a bound for $\#X(\mathbb{F}_q)$, it suffices to find a bound for $\sum_i b_i$.

12.7.24 Remark Suppose X is proper smooth over $\mathbb{F}_q$, pure of dimension n, and geometrically connected (i.e., $\overline{X}$ is connected). Then $H^0(\overline{X}, \mathbb{Q}_\ell)$ and $H^{2n}(\overline{X}, \mathbb{Q}_\ell)$ are one dimensional, and F_X acts on them by scalar multiplications 1 and q^n, respectively.

12.7.25 Corollary Suppose X is proper smooth over $\mathbb{F}_q$, pure of dimension n, and geometrically connected. Then we have

$$|\#X(\mathbb{F}_q) - (1+q^n)| \leq \sum_{i=1}^{2n-1} b_i q^{\frac{i}{2}}.$$

12.7.26 Proposition [795, Théorème 8.1] Let $X \subset \mathbb{P}^{n+r}_{\mathbb{F}_q}$ be a nonsingular complete intersection over $\mathbb{F}_q$ of dimension n and multi-degree $(d_1, \ldots, d_r)$. Let $b' = \dim H^n(\overline{X}, \mathbb{Q}_\ell)$. Set $b = b'$ if n is odd, and set $b = b' - 1$ if n is even. Then we have

$$|\#X(\mathbb{F}_q) - \#\mathbb{P}^n(\mathbb{F}_q)| \leq bq^{\frac{n}{2}}.$$

12.7.27 Remark By Theorems 12.7.16 and 12.7.22, we can write

$$Z(X,t) = \frac{Q(t)}{P(t)},$$

where $P(t) = (1-\lambda_1)\cdots(1-\lambda_s)$ and $Q(t) = (1-\mu_1)\cdots(1-\mu_t)$ are relatively prime polynomials, and λ_i, μ_i are algebraic integers satisfying $|\lambda_i|$, $|\mu_j| \leq q^{\dim X}$. By Theorem 12.7.6, we have

$$\#X(\mathbb{F}_q) = \lambda_1 + \cdots + \lambda_s - \mu_1 - \cdots - \mu_t.$$

To get a bound for $\#X(\mathbb{F}_q)$, it suffices to find a bound for $s+t$. We call $s+t$ the *total degree* of $Z(X,t)$, and denote it by $\operatorname{totdeg} Z(X,t)$. Note that $t-s$ is the degree of $Z(X,t)$, and we have

$$s - t = \sum_{i=0}^{2\dim X} (-1)^i \dim H^i_c(\overline{X}, \mathbb{Q}_\ell).$$

12.7.28 Remark Using Dwork's theory and Theorem 12.7.22, Bombieri obtains the following bound for $\operatorname{totdeg} Z(X,t)$; see [341, Theorems 1, 2, and Proposition in IV] and [339, Theorem 1].

12.7.29 Theorem Let X be a closed affine subvariety in $\mathbb{A}^N_{\mathbb{F}_q}$ defined by the vanishing of r polynomials $f_1, \ldots, f_r \in \mathbb{F}_q[t_1, \ldots, t_N]$ of degrees $\leq d$. Then we have

$$\operatorname{totdeg} Z(X,t) \leq (4(d+1)+5)^{N+r}.$$

12.7.30 Remark Adolphson and Sperber generalize Bombieri's result as follows.

12.7.31 Proposition [22, Theorem 5.27, Corollary 6.13] Let X be a closed affine subvariety in $\mathbb{A}^N_{\mathbb{F}_q}$ defined by the vanishing of r polynomials $f_1, \ldots, f_r \in \mathbb{F}_q[t_1, \ldots, t_N]$ of degrees $d_1, \ldots, d_r$, respectively. Set

$$D_N(x_0, x_1, \ldots, x_r) = \sum_{i_0+i_1+\cdots+i_r=N} x_0^{i_0} x_1^{i_1} \cdots x_r^{i_r}.$$

We have

$$\left|\sum_{i=0}^{2\dim X} (-1)^i \dim H^i_c(\overline{X}, \mathbb{Q}_\ell)\right| \leq 2^r D_N(1, d_1+1, \ldots, d_r+1),$$

$$\operatorname{totdeg} Z(X,t) \leq (2e^3)^N (2e^3+1)^N (5\max\{d_1, \ldots, d_r\}+1)^N.$$

12.7.32 Remark Starting from a universal bound $|\sum_{i=0}^{2\dim X}(-1)^i \dim H^i_c(\overline{X}, \mathbb{Q}_\ell)|$, Katz [1706] deduces a bound for $\sum_{i=0}^{2\dim X} \dim H^i_c(\overline{X}, \mathbb{Q}_\ell)$, and hence a bound for $\operatorname{totdeg} Z(X,t)$. In particular, Katz gets the following estimate from those of Bombieri and Adolphson-Sperber.

12.7.33 Proposition [1706, Corollary of Theorem 1] Let X be a closed affine subvariety in $\mathbb{A}^N_{\mathbb{F}_q}$ defined by the vanishing or r polynomials $f_1, \dots, f_r \in \mathbb{F}_q[t_1, \dots, t_N]$ of degrees $\leq d$. Then we have

$$\sum_{i=0}^{2\dim X} \dim H^i_c(\overline{X}, \mathbb{Q}_\ell) \leq 2^{r-2} \cdot 5 \cdot (4rd + 13)^{N+2},$$

$$\sum_{i=0}^{2\dim X} \dim H^i_c(\overline{X}, \mathbb{Q}_\ell) \leq 2^{r+1} \cdot 3 \cdot (rd + 3)^{N+1}.$$

12.7.2 L-functions

12.7.34 Remark For any scheme X of finite type over $\mathbb{F}_q$ and any ℓ-adic sheaf $\mathcal{F}$, we can associate an L-function $L(X, \mathcal{F}, t)$. For the definition of an ℓ-adic sheaf on a scheme, we refer to [1570, Exp. VI]. We simply mention that in the case where $X = \operatorname{Spec} F$ for a field F, giving an ℓ-adic sheaf on X is equivalent to giving a continuous ℓ-adic Galois representation

$$\mathrm{Gal}(\overline{F}/F) \to \mathrm{GL}(n, \overline{\mathbb{Q}}_\ell).$$

Suppose X is a scheme of finite type over $\mathbb{F}_q$. An $\mathbb{F}_{q^m}$*-rational point* $x \in X(\mathbb{F}_{q^m})$ is an $\mathbb{F}_q$-morphism $\operatorname{Spec} \mathbb{F}_{q^m} \to X$. Let $\mathcal{F}$ be an ℓ-adic sheaf on X. Then the inverse image of $\mathcal{F}$ on $\operatorname{Spec} \mathbb{F}_{q^m}$ defines a Galois representation which we denote by

$$\mathrm{Gal}(\mathbb{F}/\mathbb{F}_{q^m}) \to \mathrm{GL}(\mathcal{F}_{\bar{x}}).$$

Here $\mathcal{F}_{\bar{x}}$ is the *stalk* of $\mathcal{F}$ at the geometric point $\operatorname{Spec} \mathbb{F} \to X$ over x. The Galois group $\mathrm{Gal}(\mathbb{F}/\mathbb{F}_{q^m})$ has a special element, the *Frobenius substitution*

$$\phi_x : \mathbb{F}_{q^m} \to \mathbb{F}_{q^m}, \quad \phi_x(\alpha) = \alpha^{q^m}.$$

Denote by F_x the inverse of ϕ_x and call it the *geometric Frobenius* at x. Let $x \in |X|$ be a Zariski closed point in X. Then we have a closed immersion $\operatorname{Spec} k(x) \to X$, and hence x defines a $k(x)$-rational point in X. We denote the corresponding geometric Frobenius also by F_x.

12.7.35 Definition The *L-function* $L(X, \mathcal{F}, t)$ is the formal power series with variable t and with coefficients in $\overline{\mathbb{Q}}_\ell$ defined by

$$L(X, \mathcal{F}, s) = \prod_{x \in |X|} \frac{1}{\det(1 - F_x t^{\deg(x)}, \mathcal{F}_{\bar{x}})}.$$

12.7.36 Remark When $\mathcal{F}$ is the constant ℓ-adic sheaf $\overline{\mathbb{Q}}_\ell$, $L(X, \mathcal{F}, t)$ coincides with $Z(X, t)$.

12.7.37 Theorem [797, Rapport 3] For any positive integer m, let

$$S_m(X, \mathcal{F}) = \sum_{x \in X(\mathbb{F}_{q^m})} \mathrm{Tr}(F_x, \mathcal{F}_{\bar{x}}).$$

We have an equation of formal power series

$$t \frac{d}{dt} \ln L(X, \mathcal{F}, t) = \sum_{n=1}^{\infty} S_m(X, \mathcal{F}) t^m.$$

12.7.38 Remark [797, Sommes trig.] Let $\psi : \mathbb{F}_q \to \overline{\mathbb{Q}}_\ell^*$ be a nontrivial additive character. One can construct an ℓ-adic sheaf $\mathcal{L}_\psi$ on $\mathbb{A}^1_{\mathbb{F}_q}$ such that for any $\mathbb{F}_{q^m}$-rational point $x \in \mathbb{A}^1_{\mathbb{F}_q}(\mathbb{F}_{q^m}) = \mathbb{F}_{q^m}$, we have

$$\mathrm{Tr}(F_x, (\mathcal{L}_\psi)_{\bar{x}}) = \psi(\mathrm{Tr}_{\mathbb{F}_{q^m}/\mathbb{F}_q}(x)).$$

Let $f \in \mathbb{F}_q[t_1, \ldots, t_N]$ be a polynomial. It defines a morphism $f : \mathbb{A}^N_{\mathbb{F}_q} \to \mathbb{A}^1_{\mathbb{F}_q}$. For any $\mathbb{F}_{q^m}$-rational point $x \in \mathbb{A}^N_{\mathbb{F}_q}(\mathbb{F}_{q^m}) = \mathbb{F}^N_{q^m}$ with coordinates $x = (x_1, \ldots, x_N)$, the ℓ-adic sheaf $f^*\mathcal{L}_\psi$ has the property

$$\mathrm{Tr}(F_x, (f^*\mathcal{L}_\psi)_{\bar{x}}) = \psi(\mathrm{Tr}_{\mathbb{F}_{q^m}/\mathbb{F}_q}(f(x_1, \ldots, x_N))).$$

We note that

$$S_m(\mathbb{A}^N_{\mathbb{F}_q}, f^*\mathcal{L}_\psi) = \sum_{x_1,\ldots,x_N \in \mathbb{F}_{q^m}} \psi(\mathrm{Tr}_{\mathbb{F}_{q^m}/\mathbb{F}_q}(f(x_1, \ldots, x_N)))$$

is the classical exponential sum associated to the polynomial f.

12.7.39 Remark Let $\overline{X} = X \otimes_{\mathbb{F}_q} \mathbb{F}$ and let $\overline{\mathcal{F}}$ be the inverse image of $\mathcal{F}$ on $\overline{X}$. We have the cohomology groups $H^i(\overline{X}, \overline{\mathcal{F}})$ and the cohomology groups with compact support $H^i_c(\overline{X}, \overline{\mathcal{F}})$. They are finite dimensional vector spaces, and they vanish if $i \notin [0, 2\dim X]$. If X is proper, we have $H^i(\overline{X}, \overline{\mathcal{F}}) \cong H^i_c(\overline{X}, \overline{\mathcal{F}})$. Moreover, we have a morphism of sheaves $F^* : F^*_X\overline{\mathcal{F}} \to \overline{\mathcal{F}}$. The pair (F_X, F^*) is the *geometric Frobenius correspondence* for $\mathcal{F}$. Denote the homomorphisms induced by this pair on cohomology groups with compact support by $F : H^i_c(\overline{X}, \overline{\mathcal{F}}) \to H^i_c(\overline{X}, \overline{\mathcal{F}})$.

12.7.40 Theorem [797, Rapport 3.2] (Grothendieck trace formula) We have

$$\sum_{x \in X(\mathbb{F}_{q^m})} \mathrm{Tr}(F_x, \mathcal{F}_{\bar{x}}) = \sum_{i=0}^{2\dim X} (-1)^i \mathrm{Tr}\Big(F^m, H^i_c(\overline{X}, \overline{\mathcal{F}})\Big).$$

12.7.41 Remark The following theorem follows from Theorems 12.7.37 and 12.7.40. It proves the rationality of the function $L(X, \mathcal{F}, t)$.

12.7.42 Theorem [797, Rapport 3.1] (Grothendieck's formula) Let X be a scheme of finite type over $\mathbb{F}_q$. We have

$$L(X, \mathcal{F}, t) = \prod_{i=0}^{2\dim X} \det\Big(1 - Ft, H^i_c(\overline{X}, \overline{\mathcal{F}})\Big)^{(-1)^{i+1}}.$$

12.7.43 Definition A number α in $\overline{\mathbb{Q}}_\ell$ is *pure of weight* w (relative to q) if it is an algebraic number and all its Galois conjugates have Archimedean absolute value $q^{\frac{w}{2}}$. An ℓ-adic sheaf $\mathcal{F}$ on X is *punctually pure of weight* w if for any $x \in |X|$, the eigenvalues of F_x on the stalk $\mathcal{F}_{\bar{x}}$ are pure of weight w (relative to $q^{\deg(x)}$). The sheaf $\mathcal{F}$ is *mixed of weights* $\leq w$ if there exists a finite filtration of $\mathcal{F}$ such that the successive quotients are punctually pure of weights $\leq w$.

12.7.44 Remark Together with 12.7.42, the following theorem of Deligne proves the Riemann hypothesis for $L(X, \mathcal{F}, t)$.

12.7.45 Theorem [798, Corollaire 3.3.4] Suppose X is a scheme of finite type over $\mathbb{F}_q$ and $\mathcal{F}$ is a mixed ℓ-adic sheaf of weights $\leq w$ on X. Then any eigenvalue of F on $H^i_c(\overline{X}, \overline{\mathcal{F}})$ is pure of weight $\leq i + w$ relative to q.

12.7.46 Remark For each $0 \leq i \leq 2\dim X$, let $b_i = \dim H^i_c(\overline{X}, \overline{\mathcal{F}})$, and let α_{ij} $(j = 1, \ldots, b_i)$ be all the eigenvalues of F on $H^i_c(\overline{X}, \overline{\mathcal{F}})$. By 12.7.40, we have

$$S_m(X, \mathcal{F}) = \sum_{x \in X(\mathbb{F}_{q^m})} \mathrm{Tr}(F_x, \mathcal{F}_{\bar{x}}) = \sum_{i=0}^{2\dim X} \sum_{j=1}^{b_i} (-1)^i \alpha_{ij}^m.$$

Theorem 12.7.45 provides bounds for $|\alpha_{ij}|$. To get a bound for $S_m(X, \mathcal{F})$, it suffices to find bounds for $\sum_i b_i$.

12.7.47 Remark Applying Theorem 12.7.45 to the ℓ-adic sheaf $f^*\mathcal{L}_\psi$ in Remark 12.7.38, we can get a bound for the exponential sum $|\sum_{x_1,\ldots,x_N \in \mathbb{F}_{q^m}} \psi(\mathrm{Tr}_{\mathbb{F}_{q^m}/\mathbb{F}_q}(f(x_1, \ldots, x_N)))|$. We have the following results.

12.7.48 Theorem [795, Théorème 8.4], [798, Paragraphs 3.7.2-3.7.4] Let $f \in F_q[t_1, \ldots, t_N]$ be a polynomial of degree d, and let f_d be the homogeneous part of f of degree d. Suppose f_d defines a smooth hypersurface in $\mathbb{P}^{N-1}_{\mathbb{F}_q}$ and d is relatively prime to p.

1. $H^i_c(\mathbb{A}^N_{\overline{\mathbb{F}}}, f^*\mathcal{L}_\psi) = 0$ for $i \neq N$.
2. $\dim H^N_c(\mathbb{A}^N_{\overline{\mathbb{F}}}, f^*\mathcal{L}_\psi) = (d-1)^N$.
3. All eigenvalues of F on $H^N_c(\mathbb{A}^N_{\overline{\mathbb{F}}}, f^*\mathcal{L}_\psi)$ are pure of weight N.
4. $L(\mathbb{A}^N_{\mathbb{F}_q}, f^*\mathcal{L}_\psi) = P(t)^{(-1)^{N+1}}$ for a polynomial $P(t)$ of degree $(d-1)^N$ so that all reciprocal roots of $P(t)$ have Archimedean absolute value q^N.
5. $|\sum_{x_1,\ldots,x_N \in \mathbb{F}_{q^m}} \psi(\mathrm{Tr}_{\mathbb{F}_{q^m}/\mathbb{F}_q}(f(x_1, \ldots, x_N)))| \leq (d-1)^N q^{\frac{Nm}{2}}$.

12.7.49 Theorem [23, Theorem 4.2], [812, Theorem 1.3] Let $f \in F_q[t_1, \ldots, t_N, 1/t_1, \ldots, 1/t_N]$ be a Laurent polynomial. Write

$$f = \sum_{i_1,\ldots,i_N} c_{i_1\ldots i_N} t_1^{i_1} \cdots t_N^{i_N}.$$

Let $\Delta_\infty(f)$ be the convex hull in $\mathbb{Q}^N$ of the set $\{(i_1, \ldots, i_N) | c_{i_1\ldots i_N} \neq 0\} \cup \{0\}$. For any face τ of $\Delta_\infty(f)$, let $f_\tau = \sum_{(i_1,\ldots,i_N)\in\tau} c_{i_1\ldots i_N} t_1^{i_1} \cdots t_N^{i_N}$. Suppose f is *nondegenerate with respect to* $\Delta_\infty(f)$ in the sense that for any face τ of $\Delta_\infty(f)$ that does not contain the origin, the subscheme of $\mathbb{G}^N_{m,\mathbb{F}_q} = (\mathbb{A}^1_{\mathbb{F}_q} - \{0\})^N$ defined by

$$\frac{\partial f_\tau}{\partial t_1} = \cdots = \frac{\partial f_\tau}{\partial t_N} = 0$$

is empty. Suppose furthermore that $\dim \Delta_\infty(f) = N$.

1. $H^i_c(\mathbb{G}^N_{m,\overline{\mathbb{F}}}, f^*\mathcal{L}_\psi) = 0$ for $i \neq N$.
2. $\dim H^N_c(\mathbb{G}^N_{m,\overline{\mathbb{F}}}, f^*\mathcal{L}_\psi) = N!\mathrm{Vol}(\Delta_\infty(f))$.
3. If, in addition, the origin is an interior point of $\Delta_\infty(f)$, then all eigenvalues of F on $H^N_c(\mathbb{G}^N_{m,\overline{\mathbb{F}}}, f^*\mathcal{L}_\psi)$ are pure of weight N.
4. $L(\mathbb{G}^N_{m,\mathbb{F}_q}, f^*\mathcal{L}_\psi) = P(t)^{(-1)^{N+1}}$ for a polynomial $P(t)$ of degree $N!\mathrm{Vol}(\Delta_\infty(f))$. If, in addition, the origin is an interior point of $\Delta_\infty(f)$, then all reciprocal roots of $P(t)$ have Archimedean absolute value q^N.
5. $|\sum_{x_1,\ldots,x_N \in \mathbb{F}^*_{q^m}} \psi(\mathrm{Tr}_{\mathbb{F}_{q^m}/\mathbb{F}_q}(f(x_1, \ldots, x_N)))| \leq N!\mathrm{Vol}(\Delta_\infty(f)) q^{\frac{Nm}{2}}$.

12.7.50 Remark The estimates in Theorem 12.7.29, and Propositions 12.7.31 and 12.7.33 can also be extended to the L-functions associated to exponential sums [22, 339, 341, 1706].

12.7.3 The case of curves

12.7.51 Remark Suppose X is a geometrically connected smooth projective curve over $\mathbb{F}_q$ of genus g. Then $H^1(\overline{X}, \mathbb{Q}_\ell)$ can be identified with $T_\ell(J_{\overline{X}}) \otimes_{\mathbb{Z}_\ell} \mathbb{Q}_\ell$, where $T_\ell(J_{\overline{X}})$ is the Tate module of the Jacobian $J_{\overline{X}}$ of $\overline{X}$, and $\dim H^1(\overline{X}, \mathbb{Q}_\ell) = 2g$. By Theorems 12.7.16 and 12.7.22 and Corollary 12.7.25, we have the following.

12.7.52 Theorem Suppose X is a geometrically connected smooth projective curve over $\mathbb{F}_q$ of genus g. We have

$$Z(X,t) = \frac{P(t)}{(1-t)(1-qt)},$$

where $P(t) = \det(1 - Ft, H^1(\overline{X}, \mathbb{Q}_\ell))$ is a polynomial with integer coefficients, and all its reciprocal roots have Archimedean absolute value $\sqrt{q}$. Moreover, we have

$$|\#X(\mathbb{F}_q) - (1+q)| \leq 2g\sqrt{q}.$$

12.7.53 Remark The above theorem was proved by Hasse for elliptic curves and by Weil for curves of higher genus. An elementary proof was given by Stepanov, Schmidt, and Bombieri [340].

12.7.54 Definition Let $K(X)$ be the function field of X, and let $\rho : \mathrm{Gal}(\overline{K(X)}/K(X)) \to \mathrm{GL}(V)$ be a continuous Galois representation, where V is a finite dimensional vector space over $\overline{\mathbb{Q}}_\ell$. Suppose there exists a finite subset S of $|X|$ such that ρ is unramified everywhere on $X - S$. We define the *L-function* $L(X, \rho, t)$ to be

$$L(X,\rho,t) = \prod_{x \in |X|} \frac{1}{\det(1 - F_x t^{\deg(x)}, V^{I_x})},$$

where I_x is the inertia subgroup.

12.7.55 Remark The Galois representation ρ defines an ℓ-adic sheaf $\mathcal{F}_V$ on $X - S$ such that for any $x \in |X - S|$, the Galois representation $\mathrm{Gal}(\overline{k(x)}/k(x)) \to \mathrm{GL}(\mathcal{F}_{\bar{x}})$ coincides with the Galois representation $\rho|_{\mathrm{Gal}(\overline{k(x)}/k(x))}$. Let $j : X - S \hookrightarrow X$ be the open immersion. Then we have

$$\det(1 - F_x t, V^{I_x}) = \det(1 - F_x t, (j_*\mathcal{F}_V)_{\bar{x}})$$

for any $x \in |X|$. It follows that

$$L(X, \rho, t) = L(X, j_*\mathcal{F}_V, t).$$

We have

$$\begin{aligned} H^0(\overline{X}, \overline{j_*\mathcal{F}_V}) &\cong V^{\mathrm{Gal}(\overline{K(X)}/K(\overline{X}))}, \\ H^2(\overline{X}, \overline{j_*\mathcal{F}_V}) &\cong V_{\mathrm{Gal}(\overline{K(X)}/K(\overline{X}))}, \end{aligned}$$

where $K(\overline{X})$ is the function field of $\overline{X}$, and $\overline{j_*\mathcal{F}_V}$ is the inverse image of $j_*\mathcal{F}_V$ on $\overline{X}$. It follows from 12.7.42 that we have

$$L(X,\rho,t) = \frac{\det(1 - Ft, H^1(\overline{X}, \overline{j_*\mathcal{F}_V}))}{\det\left(1 - Ft, V^{\mathrm{Gal}(\overline{K(X)}/K(\overline{X}))}\right) \det\left(1 - qFt, V_{\mathrm{Gal}(\overline{K(X)}/K(\overline{X}))}\right)}.$$

12.7.56 Theorem [136, Exp. XVIII. Paragraph 3.2.6], [797, Dualité 1.3] (Poincaré duality) We have a perfect pairing

$$(\ ,\ ) : H^i(\overline{X}, \overline{j_*\mathcal{F}_V}) \times H^{2-i}(\overline{X}, \overline{j_*\mathcal{F}_{V^*}}) \to \overline{\mathbb{Q}}_\ell,$$

where V^* is the dual representation $\rho^* : \mathrm{Gal}(\overline{K(X)}/K(X)) \to \mathrm{GL}(V^*)$ of ρ. For any $s \in H^i(\overline{X}, \overline{j_*\mathcal{F}_V})$ and $t \in H^{2-i}(\overline{X}, \overline{j_*\mathcal{F}_{V^*}})$, we have

$$(F_X(s), F_X(t)) = q(s,t).$$

12.7.57 Remark By Poincaré duality and Theorem 12.7.42, we have the following functional equation for L-functions.

12.7.58 Theorem [1868, Equation (3.1.1.8)] We have

$$L\left(X, \rho, \frac{1}{qt}\right) = \epsilon(X,\rho)(qt)^{\chi(X,\rho)} L(X, \rho^*, t),$$

where

$$\begin{aligned}\chi(X,\rho) &= \sum_{i=0}^{2}(-1)^i \dim H^i(\overline{X}, \overline{j_*\mathcal{F}_V}),\\ \epsilon(X,\rho) &= \prod_{i=0}^{2}\det\Big(-F, H^i(\overline{X}, \overline{j_*\mathcal{F}_V})\Big)^{(-1)^{i+1}}.\end{aligned}$$

12.7.59 Remark Using Theorem 12.7.45 and Poincaré duality, one can prove the following, which gives the Riemann hypothesis for $L(X, \rho, t)$ by Remark 12.7.55.

12.7.60 Theorem Suppose for any $x \in |X - S|$, all eigenvalues of F_x on V are pure of weight w relative to $q^{\deg(x)}$. Then any eigenvalue of F on $H^i(\overline{X}, \overline{j_*\mathcal{F}})$ is pure of weight $i + w$ relative to q.

12.7.61 Remark For any point $x \in |X|$, let K_x be the completion of $K(X)$ with respect to the valuation corresponding to x, and let $\rho_x : \mathrm{Gal}(\overline{K_x}/K_x) \to \mathrm{GL}(V)$ be the restriction of the representation ρ. In Theorem 12.7.58, the Euler characteristic $\chi(X,\rho)$ and the constant $\epsilon(X,\rho)$ in the functional equation can be expressed in terms of the invariants of the Galois representations ρ_x ($x \in |X|$) of the local fields K_x.

12.7.62 Theorem [1570, Exp. X, Théorème 7.1] (Grothendieck-Ogg-Shafarevich formula) We have

$$\chi(X,\rho) = (2-2g)\dim V - \sum_{x\in|X|} \deg(x) a_x(\rho_x),$$

where $a_x(\rho_x)$ is the Artin conductor of ρ_x.

12.7.63 Theorem [1868, Théorème 3.2.1.1] (Laumon's product formula) Let ω be any nonzero meromorphic differential 1-form on X. Then we have

$$\epsilon(X,\rho) = q^{(1-g)\dim(V)} \prod_{x\in|X|} \epsilon(K_x, \rho_x, \omega|_{\mathrm{Spec}\, K_x}),$$

where $\epsilon(K_x, \rho_x, \omega|_{\mathrm{Spec}\, K_x})$ are the epsilon-factors defined by Deligne [794].

See Also

§6.2	For information on estimating exponential and character sums.
§12.8	For p-adic estimates of zeta-functions and L-functions.

References Cited: [22, 23, 136, 339, 340, 341, 435, 486, 794, 795, 797, 798, 812, 1345, 1427, 1570, 1706, 1710, 1868]

12.8 p-adic estimates of zeta functions and L-functions

Régis Blache, *IUFM de Guadeloupe*

12.8.1 Introduction

12.8.1 Remark We know from the preceding section that the reciprocal roots and poles of zeta and L-functions defined over the finite field $\mathbb{F}_q$ are algebraic integers, which are units at all primes except the Archimedean ones and those lying over p. Many Archimedean estimates (the Riemann hypothesis over finite fields) have been given in Section 12.7; in this section we are interested in p-adic estimates, i.e., the p-adic Riemann hypothesis. We often refer to notations and results from Section 12.9.

12.8.2 Remark The first result in this direction seems to be Stickelberger's congruence which gives the valuation of Gauss sums and Jacobi sums (see Section 6.1). The modern results come from the work of Dwork [940], who was the first to prove the rationality of zeta and L-functions, by p-adic means. From his pioneering work, many p-adic cohomology theories originated, such as Monsky-Washnitzer, crystalline, or rigid cohomology [1572]; most of the results described below follow from the explicit description of these cohomologies. Note also they proved very useful for explicit calculations.

12.8.3 Remark One can describe the variation of the p-adic cohomology spaces from differential equations, such as the Picard-Fuchs one; they sometimes allow one to give an analytic expression for roots or poles of some zeta and L-functions. The best known example is the family of ordinary elliptic curves in Legendre form, which is linked to a hypergeometric function ${}_2F_1$ [941]; the Gross-Koblitz formula (Theorem 6.1.113) links Gauss sums with the p-adic gamma function. As a consequence one gets a p-adic expression for Jacobi sums and the zeta function of a diagonal hypersurface. Other examples are cubic sums linked to the solutions of the Airy differential equation [1397], Kloostermann sums to the Bessel differential equation [942], or the zeta function of a monomial deformation of a diagonal hypersurface which can be expressed from hypergeometric functions ${}_nF_{n-1}$ [1754, 3041].

12.8.4 Remark Here we shall be concerned with p-adic valuations; we only mention briefly these subjects; we neither speak about unit root functions, the reader interested in this subject should refer to [2895, 2896, 2897] and the references therein.

12.8.5 Remark [798] Let $L(T)$ be an L-function as in Definition 12.7.35, coming from a scheme over $\mathbb{F}_q$ of dimension n and a punctually pure sheaf of weight 0. From Deligne's integrality theorem and Poincaré duality (see Section 12.7), the q-adic valuations of its reciprocal roots and poles are rational numbers lying in the interval $[0, n]$.

12.8.6 Definition Let F be a p-adic field, O_F its ring of integers, π a uniformizing parameter, and v_q the valuation on F normalized by $v_q(q) = 1$. If $P = \sum_{i=0}^{d} a_i T^i \in F[T]$ is a one variable polynomial, its *q-adic Newton polygon*, denoted $\mathrm{NP}_q(P)$, is the lower convex hull of the set of points $\{(i, v_q(a_i)),\ 0 \le i \le d\}$.

12.8.7 Theorem [1770] Assume that $P(0) = 1$. Let $s_1, \dots, s_r$ be the slopes of $\mathrm{NP}_q(P)$, of respective multiplicity l_i (i.e., each s_i is the slope of a segment of horizontal length l_i); then the polynomial P has exactly l_i reciprocal roots of q-adic valuation s_i for any $1 \le i \le r$.

12.8.8 Remark One can give a more general statement about the valuations of the roots of P, removing the hypothesis $P(0) = 1$. However, the theorem above is sufficient in the following results.

12.8.2 Lower bounds for the first slope

12.8.9 Remark We give lower bounds for the first slope of the Newton polygon of L-functions attached to families of (additive) exponential sums over affine space, first uniform, then depending on the characteristic. We end the subsection with an (incomplete) historical account on these questions.

12.8.10 Proposition [150] Let X be a scheme of finite type over $\mathbb{F}_q$ and $L(X, \mathcal{F}, T)$ be an L-function as in Definition 12.7.35; for $\mu \in \mathbb{R}_+$, the following statements are equivalent:

1. The q-adic valuations of the reciprocal roots and poles of $L(X, \mathcal{F}, T)$ are greater than or equal to μ.
2. For any m, we have $v_{q^m}(S_m(X, \mathcal{F})) \ge \mu$.
3. All slopes of the q-adic Newton polygons of the factors of $L(X, \mathcal{F}, T)$ are greater than or equal to μ.

12.8.11 Theorem [21, Theorem 1.2] Let $f \in \mathbb{F}_q[x_1, \dots, x_N]$ be a polynomial, and $\Delta := \Delta_\infty(f)$ its Newton polytope at infinity (see Theorem 12.7.49); denote by $\omega(\Delta)$ the smallest positive rational number such that $\omega(\Delta)\Delta$, the dilation of Δ by the factor $\omega(\Delta)$, contains a lattice point with all coordinates positive (a point in $\mathbb{Z}^N_{>0}$). Then every reciprocal root or pole of $L(\mathbb{A}^N, f^*\mathcal{L}_\psi, T)$ has q-adic valuation greater than or equal to $\omega(\Delta)$.

12.8.12 Remark Note that we make no assumption about the polynomial being non-degenerate with respect to its Newton polytope here.

12.8.13 Remark (see Section 7.1) One can deduce divisibility results on the numbers of points of algebraic varieties via the orthogonality relation on additive characters. Actually the Chevalley-Warning, Ax and Katz theorems are all consequences of the theorem above.

12.8.14 Definition Let $\mathcal{D} \subset (\mathbb{N}\backslash\{0\})^N$ be a finite subset, which is not contained in some $\mathbb{N}^k$, $k < N$; for any $m \ge 1$, define the subset $E_{\mathcal{D},p}(m)$ of $\{0, \dots, p^m - 1\}^{\#\mathcal{D}}$ consisting of all $(u_d)_{d\in\mathcal{D}}$ such that $\sum_{d\in\mathcal{D}} du_d \equiv 0 \pmod{p^m - 1}$ and $\sum_{d\in\mathcal{D}} du_d$ has all coordinates positive. We set

$$\sigma_{\mathcal{D},p}(m) := \min\left\{\sum_{d\in\mathcal{D}} \sigma_p(u_d),\ (u_d) \in E_{\mathcal{D},p}(m)\right\}.$$

12.8.15 Proposition [293] The set $\left\{\frac{\sigma_{\mathcal{D},p}(m)}{m}\right\}_{m\ge1}$ has a minimum.

12.8.16 Definition The *p-density* of the set $\mathcal{D}$ is the rational number $\pi_p(\mathcal{D}) := \frac{1}{p-1}\min_{m\geq 1}\left\{\frac{\sigma_{\mathcal{D},p}(m)}{m}\right\}$.

12.8.17 Theorem Let $\mathbb{F}_q[x_1,\ldots,x_N]_{\mathcal{D}}$ be the vector space of polynomials whose monomials have their exponents in $\mathcal{D}$.

1. For any $f\in\mathbb{F}_q[x_1,\ldots,x_N]_{\mathcal{D}}$, the reciprocal roots and poles of $L(\mathbb{A}^n, f^*\mathcal{L}_\psi, T)$ have q-adic valuation greater than or equal to $\pi_p(\mathcal{D})$.

2. Moreover, this bound is optimal in the sense that there exists a polynomial f in $\mathbb{F}[x_1,\ldots,x_N]_{\mathcal{D}}$ such that a reciprocal root or pole of $L(\mathbb{A}^N, f^*\mathcal{L}_\psi, T)$ has q-adic valuation equal to $\pi_p(\mathcal{D})$.

12.8.18 Remark For $f\in\mathbb{F}_q[x_1,\ldots,x_N]$ a degree d polynomial, there are many lower bounds in the literature for the q-adic valuation of the exponential sum

$$S(f) := \sum_{(x_1,\ldots,x_N)\in\mathbb{F}_q^N}\psi(f(x_1,\ldots,x_N)),$$

giving in turn lower bounds for the valuations or the reciprocal roots and poles of the associated L-function. We give a brief account of these results here.

1. Sperber [2698] proves the uniform bound $v_q(S(f))\geq\frac{N}{d}$; then with Adolphson they give the bound in Theorem 12.8.11. This last bound is the best possible uniform one, since it is attained for some large enough p.

2. Later on, Moreno and Moreno take into account the characteristic [2151] via Weil descent; this leads to generally better bounds, less uniform however. Recently, Moreno, Shum, Castro, and Kumar [2153] give a bound depending on the exponents effectively appearing in the polynomial, and on the cardinality of the field (namely $\frac{\sigma_{\mathcal{D},p}(m)}{m(p-1)}$, with $q=p^m$). Note this last bound depends on too many parameters to say anything about the valuations of the reciprocal roots and poles of the L-function.

12.8.3 Uniform lower bounds for Newton polygons

12.8.19 Remark In this subsection, we describe lower bounds for the Newton polygons associated to zeta functions of smooth projective varieties, to L-functions associated to an additive character and a Laurent polynomial (toric exponential sums) or a rational function of one variable. In the case of zeta functions, these bounds come from the Hodge numbers of related varieties in characteristic 0; for this reason we shall call these bounds Hodge polygons.

12.8.20 Definition Let X be a smooth projective variety of dimension n, defined over $\mathbb{F}_q$. For any $0\leq m\leq 2n$, define $\mathrm{NP}_m(X)$ as the q-adic Newton polygon of the characteristic polynomial of the action of Frobenius on the m-th etale cohomology space $\det(1-F_Xt, H^m(\overline{X},\mathbb{Q}_\ell))$.

Define the *Hodge polygon in degree m* of X as the polygon $\mathrm{HP}_m(X)$ having slope i with multiplicity $h^{i,m-i} := \dim H^{m-i}(X,\Omega^i)$ for $0\leq i\leq m$.

12.8.21 Theorem [29, 255, 2041] For any $0\leq m\leq 2n$, the polygon $\mathrm{NP}_m(X)$ lies on or above the polygon $\mathrm{HP}_m(X)$, and they have the same endpoints.

12.8.22 Remark There is an analogous result in the case of a smooth complete intersection in $\mathbb{G}^m\times\mathbb{A}^n$ [27].

12.8.23 **Definition** Let X be a smooth projective variety of dimension n, defined over $\mathbb{F}_q$. The variety X is *ordinary* when we have $\mathrm{NP}_m(X) = \mathrm{HP}_m(X)$ for any $0 \le m \le 2n$.

12.8.24 **Definition** Notations and assumptions are as in Theorem 12.7.49. We set $\Delta := \Delta_\infty(f)$, and denote by $\mathrm{NP}_q(f)$ the q-adic Newton polygon of the polynomial $L(\mathbb{G}_m^N, f^*\mathcal{L}_\psi, T)^{(-1)^{N+1}}$.

Denote by $C(\Delta) := \mathbb{R}_+\Delta$ the cone of Δ in $\mathbb{R}^N$, $M_\Delta := C(\Delta) \cap \mathbb{Z}^N$ the monoid associated to this cone, and $\mathcal{A}_\Delta$ the algebra $k[\mathbf{x}^{M_\Delta}]$. One can define a map from $C(\Delta)$ to $\mathbb{R}_+$, the *weight associated to* Δ, by

$$w_\Delta(\mathbf{u}) = \min\{\rho \in \mathbb{R}_+,\ \mathbf{u} \in \rho\Delta\}.$$

The vertices of Δ lie in $\mathbb{Z}^N$, thus the image of M_Δ by w_Δ lies in $\mathbb{Q}_+$; more precisely there is a positive integer $D(\Delta)$ such that $\mathrm{Im} w_\Delta \subseteq \frac{1}{D(\Delta)}\mathbb{N}$. The least integer $D := D(\Delta)$ having this property is the *denominator of* Δ. The weight w_Δ turns the algebra $\mathcal{A}_\Delta$ into a graded algebra

$$\mathcal{A}_\Delta = \oplus_{i \ge 0} \mathcal{A}_{\Delta, \frac{i}{D}},\ \mathcal{A}_{\Delta, \frac{i}{D}} = \mathrm{Vect}\left\{\mathbf{x}^{\mathbf{u}},\ w_\Delta(\mathbf{u}) = \frac{i}{D}\right\}$$

to which we associate the Poincaré series $P_{\mathcal{A}_\Delta}(t) := \sum_{i \ge 0} \dim \mathcal{A}_{\Delta, \frac{i}{D}} t^i$.

12.8.25 **Proposition** [1802, Lemme 2.9] The series $P_{\mathcal{A}_\Delta}(t)$ is a rational function. Precisely, the series $P_\Delta(t) := (1 - t^D)^N P_{\mathcal{A}_\Delta}(t)$ is a polynomial with degree less than or equal to ND, such that $P_\Delta(1) = N!\mathrm{Vol}(\Delta)$.

12.8.26 **Example** Let f be a polynomial of degree d in the N variables $x_1, \ldots, x_N$, containing the monomials $x_1^d, \ldots, x_N^d$ with non-zero coefficients; its Newton polytope at infinity is the simplex with vertices $(0, \ldots, 0), (d, \ldots, 0), \ldots, (0, \ldots, d)$. The associated cone is $\mathbb{R}_+^N$, the weight is $w_\Delta(u_1, \ldots, u_N) = \frac{\sum u_i}{d}$, and the denominator is d. In this case the Poincaré series can be written $P_{\mathcal{A}_\Delta}(t) = \frac{1}{(1-t)^N}$.

12.8.27 **Definition** Set $P_\Delta(t) := \sum \ell_i t^{s_i}$. The *Hodge polygon of* Δ, $\mathrm{HP}(\Delta)$, is the polygon starting at the origin, and formed by the slopes $\frac{s_i}{D}$ with multiplicity ℓ_i.

12.8.28 **Theorem** [23, Theorem 3.10] For any polynomial f in $\mathbb{F}_q[t_1, \ldots, t_N, \frac{1}{t_1 \cdots t_N}]$, non-degenerate with respect to its Newton polytope Δ, the polygon $\mathrm{NP}_q(f)$ lies on or above $\mathrm{HP}(\Delta)$, and they have the same endpoints.

12.8.29 **Remark** One can show that the function $L(\mathbb{G}_m^N, f^*\mathcal{L}_\psi, T)^{(-1)^{N+1}}$ has exactly one unit root. Adolphson and Sperber [31] give an analytic expression for this root, as an eigenvalue for the action of an operator on a p-adic Banach space.

12.8.30 **Remark** Adolphson and Sperber [25, 26] also consider twisted L-functions of the form $L(\mathbb{G}_m^N, f^*\mathcal{L}_\psi \otimes \mathcal{L}_\chi, T)$, associated to the product of an additive character evaluated at a Laurent polynomial, and of a multiplicative character χ of $\mathbb{G}_m^N$. They construct a Hodge polygon which is a lower bound for the Newton polygon as above; this Hodge polygon can be written in terms of the Hodge polygon $\mathrm{HP}(\Delta)$ and the valuations of the Gauss sums associated to the multiplicative characters appearing, as given by Stickelberger's theorem.

12.8.31 **Definition** Let $d_1, \ldots, d_s$ denote positive integers. We define the *Hodge polygon* $\mathrm{HP}(d_1, \ldots, d_s)$ as the polygon with slopes 0 and 1 with multiplicity $s-1$, $\frac{1}{d_1}, \ldots, \frac{d_1-1}{d_1}, \ldots, \frac{1}{d_s}, \ldots, \frac{d_s-1}{d_s}$, each with multiplicity 1.

12.8.32 **Theorem** [3069, Theorem 1.1] Let $f \in \mathbb{F}_q(x)$ be a rational function having s poles of prime to p orders $d_1, \ldots, d_s$, and X denote the projective line with the poles removed. Then the q-adic Newton polygon of the function $L(X, f^*\mathcal{L}_\psi, T)$ lies on or above the Hodge polygon $\mathrm{HP}(d_1, \ldots, d_s)$, and they have the same endpoints.

12.8.33 **Remark** Lower bounds for Newton polygons of L-functions associated to a character of order p^l, evaluated at a Witt vector of functions, pure or twisted by a multiplicative character can be found in [1948, 1950]. See also [1949], in which T-adic exponential sums are introduced, giving a framework in which to unify the study of the p-adic properties of these L-functions when l varies.

12.8.34 **Remark** There are also some results and conjectures on the p-adic theory of L-functions associated to multiplicative characters; see [20] and the references therein.

12.8.4 Variation of Newton polygons in a family

12.8.35 **Remark** It is in general very hard to determine the exact Newton polygon of a given L-function. Instead we consider L-functions associated to data varying in a family; Grothendieck's specialization Theorem 12.8.37 asserts that in such a family, most L-functions share the same Newton polygon, the generic Newton polygon; for instance our first result states that smooth complete intersections are generically ordinary. We describe this last polygon (or parts of it) in the known cases (toric sums, affine sums in one variable). Then we deal with its asymptotic variation with the characteristic.

12.8.36 **Theorem** [1571, 2910] Let $\mathcal{S}$ denote the scheme parametrizing smooth complete intersections of dimension n and multi-degree $(d_1, \ldots, d_r)$ in $\mathbb{P}^{n+r}_{\mathbb{F}_q}$, and $X \to \mathcal{S}$ the universal family. There exists a Zariski dense open subset $\mathcal{U}$ such that for any $s \in \mathcal{U}$, X_s is ordinary.

12.8.37 **Theorem** [1699] (Grothendieck's specialization theorem) Assume $f_t : X \to \mathbb{A}^1$ belongs to a family parametrized by t varying in an affine scheme $\mathcal{S}/\mathbb{F}_p$. We assume the L-function $L(X, f_t^*\mathcal{L}_\psi, T)$ (or its inverse) to be a polynomial of constant degree when t varies, and we denote by $\mathrm{NP}_q(f_t)$ its q-adic Newton polygon (the q-adic Newton polygon of its inverse). There is a Zariski dense open subset $\mathcal{U}$ in $\mathcal{S}$ (the *open stratum*) and a polygon $\mathrm{GNP}(\mathcal{S}, p)$ such that

1. For any $t \in \mathcal{U}(\mathbb{F}_q)$, $\mathrm{NP}_q(f_t) = \mathrm{GNP}(\mathcal{S}, p)$.
2. For any $t \in \mathcal{S}(\mathbb{F}_q)$, $\mathrm{NP}_q(f_t)$ lies above $\mathrm{GNP}(\mathcal{S}, p)$.

12.8.38 **Definition** The polygon $\mathrm{GNP}(\mathcal{S}, p)$ defined above is the *generic Newton polygon* of the family $f_t, t \in \mathcal{S}$. When it exists, the *Hasse polynomial for the generic Newton polygon* of this family is the polynomial generating the ideal defining the Zariski closed subset $\mathcal{S} \backslash \mathcal{U}$.

12.8.39 **Remark** One can define more general Hasse polynomials, for instance for the first vertex, or for the first m vertices.

12.8.40 **Example** Set $f_\lambda(x, y, z) := z(y^2 - x(x-1)(x-\lambda))$ for $\lambda \in \mathbb{A}^1_{\mathbb{F}_q}(\overline{\mathbb{F}}_p) \backslash \{0, 1\}$. The generic Newton polygon has vertices $(0,0), (1,1), (2,3)$, and the Hasse polynomial for the generic Newton polygon (actually for the vertex $(1,1)$) is the polynomial $F(\lambda) = \sum_{i=0}^{\frac{p-1}{2}} \binom{\frac{p-1}{2}}{i}^2 \lambda^i$. In other words, the supersingular elliptic curves in Legendre form are those for which $F(\lambda) = 0$.

12.8.41 Theorem [2899, 2910] Let $\mathcal{S}_\Delta$ parametrize the space of Laurent polynomials over $\overline{\mathbb{F}}_p$ with Newton polytope $\Delta \subset \mathbb{R}^N$, non-degenerate with respect to it.

1. If $N \leq 3$, then $\mathrm{GNP}(\mathcal{S}_\Delta, p) = \mathrm{HP}(\Delta)$ for any $p \equiv 1 \pmod{D(\Delta)}$.

2. If $N \geq 4$, there exists an integer $D'(\Delta)$ (in general strictly greater than $D(\Delta)$) depending only on Δ such that $\mathrm{GNP}(\mathcal{S}_\Delta, p) = \mathrm{HP}(\Delta)$ for any large enough prime p such that $p \equiv 1 \pmod{D'(\Delta)}$.

12.8.42 Remark The L-function associated to f has its coefficients in $\mathbb{Q}_p(\zeta_p)$, which is a totally ramified extension of $\mathbb{Q}_p$ of degree $p-1$. Since the ordinates for the vertices of the Hodge polygon are in $\frac{1}{D(\Delta)}\mathbb{N}$, a necessary condition in order to have $\mathrm{GNP}(\mathcal{S}_\Delta, p) = \mathrm{HP}(\Delta)$ is $p \equiv 1 \pmod{D(\Delta)}$. The theorem above shows that it is not a sufficient condition for $N \geq 4$.

12.8.43 Remark When $N = 1$, something stronger is true: if $p \equiv 1 \pmod{\mathrm{lcm}(d, d')}$, then for any $f(x) = \sum_{i=-d'}^{d} a_i x^i \in \mathbb{F}_q[x, x^{-1}]$ with $a_d a_{-d'} \neq 0$, we have $\mathrm{NP}(f) = \mathrm{HP}([-d', d])$.

12.8.44 Remark [3069] More generally, this result remains true for rational functions of one variable with poles of orders $d_1, \ldots, d_s$ when we have $p \equiv 1 \pmod{\mathrm{lcm}(d_1, \ldots, d_s)}$.

12.8.45 Theorem [2553] Let $\mathcal{D} = \{1, \ldots, d\}$; if $p > 2d$, the first vertex of the generic Newton polygon $\mathrm{GNP}(\mathcal{S}_\mathcal{D}, p)$ is $(1, \frac{1}{p-1}\lceil \frac{p-1}{d} \rceil)$, with Hasse polynomial $\left\{ f^{\lceil \frac{p-1}{d} \rceil} \right\}_{p-1}$, the polynomial associating to the coefficients $a_1, \ldots, a_d$ of f the degree $p-1$ coefficient of the $\lceil \frac{p-1}{d} \rceil$-th power of f.

12.8.46 Theorem [293] Let $\mathcal{D} \subset \mathbb{N}$ be finite, $d = \max \mathcal{D}$, and consider $\mathcal{S}_\mathcal{D}$, the affine variety parametrizing degree d polynomials whose monomials have their exponents in $\mathcal{D}$. The first slope of the generic Newton polygon $\mathrm{GNP}(\mathcal{S}_\mathcal{D}, p)$ is equal to the p-density $\pi_p(\mathcal{D})$ of the set $\mathcal{D}$.

12.8.47 Theorem [291] Let $p = 2$, and $\mathcal{D} = \{1 \leq i \leq d,\ 2 \nmid i\}$; set $\mathcal{S}_\mathcal{D} = \mathrm{Spec}\mathbb{F}_p[\{a_i\}_{i \in \mathcal{D}}, a_d^{-1}]$ (we parametrize the polynomials by their coefficients). The first vertex of the generic Newton polygon $\mathrm{GNP}(\mathcal{S}_\mathcal{D}, p)$ is

1. $(n, 1)$ if $2^n - 1 \leq d < 2^{n+1} - 3$, with Hasse polynomial $a_{2^n - 1}$;

2. $(2n, 2)$ if $d = 2^{n+1} - 3$, with Hasse polynomial $a_{3 \cdot 2^{n-1} - 1}$;

3. in the second case, if we consider $\mathcal{D}' = \mathcal{D} \backslash \{3 \cdot 2^{n-1} - 1\}$, the first vertex of the generic Newton polygon $\mathrm{GNP}(\mathcal{S}_{\mathcal{D}'}, p)$ is $(n, 1)$, with Hasse polynomial $a_{2^n - 1}$.

12.8.48 Remark The first result in the direction of Theorem 12.8.47 can be found in [2552] where the authors determine the first slope and necessary conditions to get it. There are also results for the first vertex in any characteristic p [291], but only dealing with the cases $p^n - 1 \leq d \leq 2p^n - 2$.

12.8.49 Theorem [295] Let $d \geq 2$ be an integer prime to p, and $\mathcal{S}_d = \mathrm{Spec}\ \mathbb{F}_p[a_1, \ldots, a_d, a_d^{-1}]$ parametrize the degree d polynomials. If $p \geq 3d$, the generic Newton polygon $\mathrm{GNP}(\mathcal{S}_d, p)$ has vertices

$$\left(i, \frac{Y_i}{p-1}\right)_{0 \leq i \leq d-1}, \quad Y_i = \sum_{j=1}^{i} \left\lceil \frac{pj - i}{d} \right\rceil,$$

and the Hasse polynomial for the i-th vertex is a polynomial in $\mathbb{F}_p[a_1, \ldots, a_d]$, homogeneous of degree Y_i.

12.8.50 Remark The Newton polygons corresponding to degree 4 [1528], degree 6 [1529] polynomials, and to the family $x^d + \lambda x$ when p is large enough and $p \equiv -1 \pmod{d}$ [3028] are completely determined.

12.8.51 Remark The generic Newton polygons associated to twisted one variable sums (coming from the product of an additive character evaluated at a Laurent polynomial and of a multiplicative character evaluated at the variable) are given in [296] for p large enough. Using the Poisson formula, they give the generic Newton polygons attached to families of polynomials $P(x^s)$, $\deg P = d$.

12.8.52 Theorem [3067, 3068]

1. We have the limit $\lim_{p\to\infty} \mathrm{GNP}(\mathcal{S}_d, p) = \mathrm{HP}(d)$.

2. There is a Zariski dense open subset $\mathcal{U}$ in $\mathrm{Spec}\mathbb{Q}[a_1, \ldots, a_d, a_d^{-1}]$ (parametrizing the degree d polynomials defined over $\overline{\mathbb{Q}}$) such that for any $f \in \mathcal{U}$ we have $\lim_{p\to\infty} \mathrm{NP}(f \bmod p) = \mathrm{HP}(d)$.

12.8.53 Remark In the case of twisted exponential sums, by a multiplicative character of order s (or for the family of polynomials $P(x^s)$, $\deg P = d$), the limit no longer exists as in the first assertion above. Actually the limit exists if we restrict to the primes in a fixed residue class modulo s [296], and there is a result similar to the second assertion in this case.

12.8.54 Remark For L-functions associated to a one variable rational function of fixed pole orders $d_1, \ldots, d_s$, the generic Newton polygon tends to the Hodge polygon $\mathrm{HP}(d_1, \ldots, d_s)$ when p tends to infinity [1915].

12.8.55 Problem There are conjectures by Wan [2899] asserting that under certain additional hypotheses, Theorem 12.8.52 remains true for the space $\mathcal{S}_\Delta$ of polynomials over $\overline{\mathbb{F}}_p$ with Newton polytope at infinity $\Delta \subset \mathbb{R}^N$, non-degenerate with respect to it. Actually a consequence of Theorem 12.8.41 is that $\liminf_{p\to\infty} \mathrm{GNP}(\mathcal{S}_\Delta, p) = \mathrm{HP}(\Delta)$. Some special cases of these conjectures are proved in [292].

12.8.5 The case of curves and abelian varieties

12.8.56 Remark We consider the Newton polygons of curves. To each curve one can associate its Jacobian variety, and more generally we consider the Newton polygons of abelian varieties. They encode useful invariants, such as the p-rank. We give the stratification of the space of principally polarized abelian varieties by their Newton polygon as described in the work of Oort and others. Then we focus on curves; even if the situation is less well-known than in the case of abelian varieties, the subject has drawn much attention and we give the principal results. We end the section with some remarks about Artin-Schreier curves.

12.8.57 Definition The *Newton polygon of a curve* C defined over $\mathbb{F}_q$ is the q-adic Newton polygon of the numerator $L(C,T)$ of its zeta function.

Let A be an abelian variety of dimension g defined over $\mathbb{F}_q$; for any prime $\ell \neq p$, the inverse limit of the ℓ-th power torsion subgroups of A is the ℓ-adic Tate module of A, a $\mathbb{Z}_\ell$-module of rank $2g$. Let $P_A(T)$ denote the characteristic polynomial of the action of Frobenius (q-th power) on the $\mathbb{Q}_\ell$-vector space $T_\ell(A) \otimes \mathbb{Q}_\ell$. The *Newton polygon of the abelian variety* A defined over $\mathbb{F}_q$ is the q-adic Newton polygon of the polynomial $T^{2g} P_A(\frac{1}{T})$.

12.8.58 Remark The Newton polygon of a curve C coincides with one of its Jacobian variety J_C, as defined above.

12.8.59 Remark For a curve of genus g, or an abelian variety of dimension g, the Newton polygon starts at $(0,0)$ and ends at $(2g, g)$. Moreover, it follows from Poincaré duality that it is symmetric in the sense that if it contains the slope s with multiplicity m, it also contains the slope $1 - s$ with the same multiplicity.

12.8.60 Remark Some authors consider the Newton polygon of the polynomial $P_A(T)$; this Newton polygon is symmetric to the one we consider here, with respect to the line $x = g$.

12.8.61 Definition A genus g curve (an abelian variety of dimension g) is *ordinary* when its Newton polygon has the slopes 0 and 1 each with multiplicity g; it is *supersingular* when it has the slope $\frac{1}{2}$ with multiplicity $2g$.

A polygon satisfying the requirements of Remark 12.8.59 is *admissible.*

12.8.62 Theorem [1769] Curves of genus g (*resp.* abelian varieties of dimension g) are generically ordinary.

12.8.63 Theorem [2322] Let $\mathcal{A}_{g,1} \otimes \mathbb{F}_p$ denote the space parametrizing principally polarized abelian varieties of dimension g defined over $\overline{\mathbb{F}}_p$. It has dimension $\frac{g(g+1)}{2}$.

Let ζ be an admissible polygon; the space W_ζ of principally polarized abelian varieties having their Newton polygon lying on or above ζ is closed, and has dimension

$$\dim W_\zeta = \#\{(x, y) \in \mathbb{N}^2,\ y < x \leq g,\ (x, y) \text{ above } \zeta\}.$$

12.8.64 Remark There are more precise results by Li and Oort on the supersingular stratum [1919].

12.8.65 Problem As a consequence, every admissible polygon is the Newton polygon of an abelian variety. It is not known whether this is true for curves. Van der Geer and Van der Vlugt have shown that for $p = 2$, there are supersingular curves of every genus [2843], but this is not even known in odd characteristic.

12.8.66 Definition The *p-rank* of the curve C (*resp.* of the abelian variety A) is the integer in $\{0, \ldots, g\}$ defined as either the length of the horizontal segment of its Newton polygon, or the dimension of the $\mathbb{F}_p$-vector space $J_C[p]$ (*resp.* $A[p]$).

12.8.67 Remark From Theorem 12.8.63, the space of principally polarized abelian varieties having p-rank f has codimension $g - f$ in $\mathcal{A}_{g,1} \otimes \mathbb{F}_p$; this is also true in the space $\mathcal{M}_g$ of genus g curves [1022] (which has dimension $3g - 3$) and in the space $\mathcal{H}_g$ of genus g hyperelliptic curves [1281] (of dimension $2g - 1$).

12.8.68 Definition The *Hasse-Witt matrix* of a non-singular curve C of genus g is the matrix of the Frobenius (p-th power) mapping on the g dimensional space $H^1(C, \mathcal{O}_C)$.

12.8.69 Remark Via Serre's duality, the Hasse-Witt matrix is the transpose of the matrix of the Cartier-Manin operator [553] on the space of differentials of the first kind.

12.8.70 Theorem [1999] Let H denote the Hasse Witt matrix of C, a curve defined over $\mathbb{F}_q$, $q = p^m$, and $H^{(p^i)}$ denote the matrix obtained by raising all coefficients of H to the power p^i. Let $H_a := HH^{(p)} \cdots H^{(p^{a-1})}$. Then

1. The rank of H_g is the p-rank of the curve C.
2. We have the congruence $L(C, T) \equiv \det(\mathbf{I}_g - TH_m) \pmod{p}$.

12.8.71 Example With respect to the basis dual to the one given in Example 12.1.85, the Hasse-Witt matrix of the hyperelliptic curve $y^2 = f(x)$ is the matrix $(\{f^{\frac{p-1}{2}}\}_{pi-j})_{1 \leq i,j \leq g}$.

12.8.72 Remark There are more general congruences for the factor of the zeta function of an hypersurface coming from the primitive middle cohomology space [799].

12.8.73 Theorem [752, Corollary 1.8] (Deuring-Shafarevich formula) Let C and C' be two curves with function fields E and F; assume that the extension E/F is Galois with Galois group G a p-group. If f and f' denote the respective p-ranks of C and C', then the relation

$$1 - f = \#G(1 - f') + \sum_{x \in C(\overline{k})} (e_x - 1),$$

holds, where e_x is the ramification index of the place x in the extension E/F.

12.8.74 Example Let $E = k(x, y)$ be the extension of the rational function field $k(x)$ defined by the (Artin-Schreier) equation $y^p - y = f(x)$; assume that f has s poles with orders prime to p. Then $E/k(x)$ is a Galois extension with Galois group $\mathbb{Z}/p\mathbb{Z}$, and the p-rank of E is $f = s - 1$.

12.8.75 Remark One can deduce a stratification by the p-rank of the space of Artin-Schreier curves (i.e., Artin-Schreier coverings of the projective line) [2430].

12.8.76 Remark [3068] We have an expression for the numerator of the zeta function of the Artin-Schreier curve $C : y^p - y = f(x)$ from the L-functions $L(af, T)$ associated to the sums $\sum'_{x \in \mathbb{P}^1} \psi(af(x))$, where the sum is taken over the points which are not poles of f. Precisely we have $L(C, T) = \prod_{a \in \mathbb{F}_p} L(af, T)$. The L-functions on the right are conjugated under the action of $\text{Gal}(\mathbb{Q}(\zeta_p)/\mathbb{Q})$; thus they all have the same Newton polygon, and $\text{NP}_q(C)$ is the dilation by the factor $p-1$ of $\text{NP}_q(f)$. As a consequence, the above determination of (parts of) Newton polygons of L-functions associated to one variable exponential sums translate to results on the Newton polygons of Artin-Schreier curves.

12.8.77 Remark The same argument gives the Newton polygon for curves with equation $A(y) = f(x)$, A an additive polynomial, as a dilation of the Newton polygon $\text{NP}_q(f)$.

12.8.78 Remark We focus on the case $p = 2$: here Artin-Schreier curves are hyperelliptic curves. From the remark above, we have $\text{NP}_q(C) = \text{NP}_q(f)$ for $C : y^2 + y = f(x)$. The stratification of $\mathcal{H}_g$ by the 2-rank is described in [2430]: the irreducible components of the stratum of curves with 2-rank f, $\mathcal{H}_{g,f}$ are in bijection with the partitions of $g + 1$ in $f + 1$ positive integers. Inside $\mathcal{H}_{g,0}$, one can reduce to f a polynomial, and Theorem 12.8.47 gives the first vertex for the generic Newton polygon in this space. One can also deduce from these results a theorem originally proved by Scholten and Zhu: there is no supersingular hyperelliptic curve of genus $g = 2^n - 1$, $n \geq 2$ in characteristic 2 [2552].

12.8.79 Remark This result stands in striking contrast with the situation of genus $g = 2^n$ [2842]. In this case the dimension of the space of hyperelliptic supersingular curves is greater than or equal to n; it can be as large as possible.

12.8.80 Remark For $g \leq 8$, the supersingular hyperelliptic curves are completely determined when $p = 2$ in [2551], and when $p = 3$ in [291]. When $p = 2$ one can also give the first vertices occuring for Newton polygons of curves in $\mathcal{H}_{g,0}$ for $g \leq 9$ [291].

12.8.81 Remark There are results asserting the non-existence of supersingular Artin-Schreier curves in odd characteristic for some infinite families of genera [291].

See Also

§7.1	For discussion of equations over finite fields.
§12.7	For discussion of zeta-functions and L-functions.
§12.9	For discussion of rational points and zeta-functions.

References Cited: [20, 21, 23, 25, 26, 27, 29, 31, 150, 255, 291, 292, 293, 295, 296, 553, 752, 798, 799, 940, 941, 942, 1022, 1281, 1397, 1528, 1529, 1571, 1699, 1754, 1769, 1770, 1802, 1915, 1919, 1948, 1949, 1950, 1999, 2041, 2151, 2153, 2322, 2430, 2551, 2552, 2553, 2698, 2842, 2843, 2895, 2896, 2897, 2899, 2910, 3028, 3041, 3067, 3068, 3069]

12.9 Computing the number of rational points and zeta functions

Daqing Wan, *University of California Irvine*

12.9.1 Remark As in Section 7.1, we shall restrict to the case of hypersurfaces. We focus on theoretical and deterministic results. Probabilistic algorithms and improvements are not discussed. As always, $\mathbb{F}_q$ denotes a finite field of characteristic p. The time for algorithms means the number of field operations.

12.9.1 Point counting: sparse input

12.9.2 Definition For a polynomial $f \in \mathbb{F}_q[x_1, \ldots, x_n]$, the sparse representation of f is the sum of its non-zero terms

$$f(x_1, \ldots, x_n) = \sum_{j=1}^{m} a_j x^{V_j},\ a_j \in \mathbb{F}_q^*,$$

where

$$V_j = (v_{1j}, \ldots, v_{nj}),\ x^{V_j} = x_1^{v_{1j}} \cdots x_n^{v_{nj}}.$$

The *point counting problem* is to compute the number $\#A_f(\mathbb{F}_q)$ of $\mathbb{F}_q$-rational points of the equation $f = 0$.

12.9.3 Remark For this problem, we may replace x_i^q by x_i and assume that the degree of f in each variable is at most $q-1$. The sparse input size of f is then $mn\log(q)$.

12.9.4 Example For non-zero elements $a_i \in \mathbb{F}_q$ and $b \in \mathbb{F}_q$, let

$$f(x) = a_1 x_1^{q-1} + \cdots + a_n x_n^{q-1} + b,$$

be a diagonal polynomial. Deciding if $\#A_f(\mathbb{F}_q) > 0$ is equivalent to deciding if there is a subset $\{a_{i_1}, \ldots, a_{i_k}\}$ of the set $\{a_1, \ldots, a_n\}$ such that

$$a_{i_1} + \cdots + a_{i_k} + b = 0.$$

The latter problem is the subset sum problem over $\mathbb{F}_q$, which is well known to be **NP**-complete. The fastest known deterministic algorithm for deciding if $\#A_f(\mathbb{F}_q) > 0$ in this case is the baby-step-giant-step method, which runs in time $O(n2^{n/2}\log(q))$.

12.9.5 Theorem [1231] Computing $\#A_f(\mathbb{F}_q)$ is **NP**-hard, even in the case $n = 2$ or $\deg(f) = 3$.

12.9.6 Remark For a positive integer $r > 1$, the modular counting problem is to compute the residue class of $\#A_f(\mathbb{F}_q)$ modulo r. It is clear that $0 \le \#A_f(\mathbb{F}_q) \le q^n$. Thus, if one can compute $\#A_f(\mathbb{F}_q)$ modulo r for a single large $r > q^n$ or for many small r, the Chinese remainder theorem implies that one can compute $\#A_f(\mathbb{F}_q)$ as well. This suggests that even for small r, the modular counting problem is not going to be much easier than the full counting problem.

12.9.7 Theorem [1318] Let r be a positive integer. Let $q = p^h$. If r is not a power of p, then computing $\#A_f(\mathbb{F}_q)$ modulo r is **NP**-hard. If $r = p^b$ is a power of p, computing $\#A_f(\mathbb{F}_q)$ modulo r is also **NP**-hard, if either $p \ge 2n$ or $h \ge 2n$ or $b > nh$, that is, $r = p^b > q^n$.

12.9.8 Remark This complexity result shows that if r is not a power of p, one cannot expect a fast algorithm to compute $\#A_f(\mathbb{F}_q)$ modulo r. Even in the case $r = p^b$ is a power of p, any general algorithm computing $\#A_f(\mathbb{F}_q)$ modulo p^b is expected to be fully exponential in each of the three parameters $\{p, b, h\}$. The next two results provide non-trivial algorithms in this direction.

12.9.9 Theorem [1318] Let $q = p^h$ and $r = p^b$. The number $\#A_f(\mathbb{F}_q)$ modulo p^b can be computed in time $O(nm^{2qb}) = O(nm^{2p^h b})$, where m is the number of monomials of f.

12.9.10 Theorem [2903] Let $q = p^h$ and $r = p^b$. The number $\#A_f(\mathbb{F}_q)$ modulo p^b can be computed in time $O(n(8m)^{p(h+b)})$, where m is the number of monomials of f.

12.9.11 Problem Improve the exponent $p(h + b)$ to $O(p + h + b)$ if possible.

12.9.2 Point counting: dense input

12.9.12 Definition For a polynomial $f \in \mathbb{F}_q[x_1, \ldots, x_n]$ of degree at most $d > 1$, the *dense representation of f* is the sum of all terms of degree at most d:

$$f(x_1, \ldots, x_n) = \sum_{i_1+\cdots+i_n \le d} a_{i_1,\ldots,i_n} x_1^{i_1} \cdots x_n^{i_n},\ a_{i_1,\ldots,i_n} \in \mathbb{F}_q.$$

12.9.13 Remark Replacing x_i^q by x_i and we may again assume that the degree of f in each variable is at most $q - 1$. The dense input size of f is then $(d + 1)^n \log(q)$.

12.9.14 Remark There is no known complexity result for the general point counting problem with dense input. On the contrary, there are polynomial time algorithms in various special cases. This suggests that the dense input point counting problem may have polynomial time algorithms in much greater generality. We shall describe some of these positive results below. In the special case that both n and d are fixed, the sparse input size agrees with the dense input size. This is the case for elliptic curves for instance.

12.9.15 Theorem [1866] There is a p-adic algorithm which computes the number $\#A_f(\mathbb{F}_q)$ in time $O(p^{2n+4}(dn \log_p q)^{3n+7})$.

12.9.16 Remark This is a general purpose algorithm, which runs in polynomial time if p is small and n is fixed. It assumes no conditions on the affine hypersurface A_f. If one assumes additional conditions on f, significant improvements can be made. In the following, we give several such examples.

12.9.17 Theorem [1862] Assume that both the affine hypersurface A_f and its infinite part are smooth of degree d not divisible by $p > 2$. Then, the number $\#A_f(\mathbb{F}_q)$ can be computed in time $O_\epsilon(p^{2+\epsilon}(d^n \log_p q)^{O(1)})$.

12.9.18 Remark It may be possible to improve the factor $p^{2+\epsilon}$ to $p^{0.5+\epsilon}$. This has been done in the special case of hyperelliptic curves and more generally superelliptic curves [1429, 2109].

12.9.19 Theorem [18, 2394, 2560] Let $n = 2$ and assume that the projective curve defined by f is smooth. Then, the number $\#A_f(\mathbb{F}_q)$ can be computed in time $O((\log q)^{c_d})$, where c_d is a constant depending only on d.

12.9.20 Remark The constant c_d is in general exponential in d. For hyperelliptic curves, the constant c_d can be taken to be a polynomial in d. It is not clear if the same theorem is true for singular plane curves.

12.9.21 Remark In the case of the diagonal hypersurface

$$f(x_1, \dots, x_n) = a_1 x_1^{d_1} + \cdots + a_n x_n^{d_n} + b,$$

the number $\#A_f(\mathbb{F}_q)$ has a compact expression in terms of Jacobi sums, which can be computed in polynomial time using LLL lattice basis reduction. This is worked out in some cases in [455].

12.9.3 Computing zeta functions: general case

12.9.22 Remark For an affine hypersurface A_f defined by a polynomial $f(x_1, \dots, x_n)$ over $\mathbb{F}_q$, the zeta function

$$Z(A_f, T) = \exp\left(\sum_{k=1}^{\infty} \frac{\#A_f(\mathbb{F}_{q^k})}{k} T^k\right)$$

is a rational function in T; see Section 12.8 for more details. The degrees of its numerator and the denominator can be bounded by a function depending only on the degree d of the polynomial f and the number n of variables. The output size for the zeta function is comparable to the dense input size $O(d^n \log q)$ of f. Thus, in computing the zeta function, we always use the dense input size.

12.9.23 Theorem [2894] There is an algorithm, that given $f \in \mathbb{F}_q[x_1, \dots, x_n]$ of degree d, computes the reduction of the zeta function $Z(A_f, T)$ modulo p in time bounded by a polynomial in $p^{\binom{d}{n}} \log q$.

12.9.24 Remark This is a polynomial time algorithm if p is small. However, it only gives the modulo p reduction of the zeta function. For any other prime $\ell \neq p$, no nontrivial general algorithm is known which computes the reduction of the zeta function modulo ℓ, except when $n \leq 2$. By the Chinese remainder theorem, this is not much easier than computing the full zeta function in general.

12.9.25 Theorem [1866] There is an algorithm, that given $f \in \mathbb{F}_q[x_1, \dots, x_n]$ of degree d, computes the zeta function $Z(A_f, T)$ in time bounded by a polynomial in $(d^n p \log q)^n$.

12.9.26 Remark This is a polynomial time algorithm if p is small and n is fixed. In the case that f is sufficiently smooth, this result can be greatly improved as follows.

12.9.27 Theorem [1862, 1863] There is an algorithm, that given $f \in \mathbb{F}_q[x_1, \dots, x_n]$ of degree d, computes the zeta function $Z(A_f, T)$ in time bounded by a polynomial in $d^n p \log q$, provided that the affine hypersurface A_f and its infinite part are both smooth and d is not divisible by $p > 2$.

12.9.28 Remark In various special cases, one can expect significantly better results. For instance, when $n = 1$, there is always a polynomial time algorithm which computes the zeta function of the zero-dimensional hypersurface $Z(A_f, T)$ [2894]. The case $n = 2$ (the curve case) has been studied most extensively. We state two such results in the next subsection. For a diagonal hypersurface

$$f(x_1, \ldots, x_n) = a_1 x_1^{d_1} + \cdots + a_n x_n^{d_n} + b,$$

the zeta function has an explicit expression in terms of Jacobi sums, which can then be computed in polynomial time using LLL basis reduction; see [455] for the main ideas.

12.9.4 Computing zeta functions: curve case

12.9.29 Remark In this subsection, we restrict to the smooth plane curve C_f in $\mathbb{P}^2$ defined by a smooth homogenous polynomial $f(x_1, x_2, x_3)$ of degree d over $\mathbb{F}_q$. The genus of the curve is then $g = (d-1)(d-2)/2$. The zeta function of C_f is of the form

$$Z(C_f, T) = \frac{P_f(T)}{(1-T)(1-qT)},$$

where $P_f(T) \in 1 + T\mathbb{Z}[T]$ is a polynomial of degree $2g$. The special value $P_f(1)$ is the order of the Jacobian variety of C_f. Thus, any algorithm for computing the zeta function gives an algorithm for computing the order of the Jacobian. We state two general results for curves. The first one is ℓ-adic in nature, and the second one is p-adic in nature.

12.9.30 Theorem [18, 2394, 2560] There is an algorithm which computes the zeta function $Z(C_f, T)$ of the curve C_f in time $O((\log q)^{c_g})$, where c_g is a constant which in general depends exponentially on g.

12.9.31 Remark This is a polynomial time algorithm for fixed genus g. For hyperelliptic curves, the constant c_g can be taken to be a polynomial in g.

12.9.32 Theorem [564, 1718] There is an algorithm which for fixed p, computes the zeta function $Z(C_f, T)$ of the curve C_f in time $\tilde{O}(g^6(\log_p q)^3 + g^{6.5}(\log_p q)^2)$.

12.9.33 Remark This is a polynomial time algorithm if p is fixed. The results in [564] work for more general non-degenerate toric curves.

12.9.34 Remark There are a lot more specialized and more precise results on computing the zeta function in various special cases, such as elliptic curves, hyperelliptic curves, superelliptic curve, C_{ab}-curves, Kummer curves, Artin-Scheirer curves, Fermat curves, etc. We refer to [552, 816, 817, 1248, 1556, 1861, 1865, 1905, 2533] for more details and for further references. For general survey, see [579, 1720, 2901].

See Also

§12.5	Discusses rational points on curves.
§12.7	Discusses zeta-functions and L-functions.
§12.8	Discusses p-adic estimates of zeta-functions.

References Cited: [18, 455, 552, 564, 579, 816, 817, 1231, 1248, 1318, 1429, 1556, 1718, 1720, 1861, 1862, 1863, 1865, 1866, 1905, 2109, 2394, 2533, 2560, 2894, 2901, 2903].

13

Miscellaneous theoretical topics

13.1 Relations between integers and polynomials over finite fields

Gove Effinger, Skidmore College

The arithmetic structures of the ring of integers $\mathbb{Z}$ and the ring of polynomials $\mathbb{F}_q[x]$, where q is a prime power, are strikingly similar. In particular, the densities of irreducible elements in these rings are virtually identical, leading to very closely analogous theorems (and conjectures) in the two settings. For a general exposition on the ideas contained in this section, see, for example, [963]. Here we state some of these analogous definitions, theorems, and conjectures, listing first the item involving $\mathbb{Z}$ and second its analogue in $\mathbb{F}_q[x]$. The latter of these first two definitions will be used throughout this section.

13.1.1 Definition For $m\ (\neq 0) \in \mathbb{Z}$, the *absolute value* of m, denoted $|m|$, is $|\mathbb{Z}/\langle m\rangle|$.

13.1.2 Definition For $f\ (\neq 0) \in \mathbb{F}_q[x]$, the *absolute value* of f, denoted $|f|$, is $|\mathbb{F}_q[x]/\langle f\rangle|$.

13.1.3 Remark Suppose the degree of f in the above definition is n. Since the quotient ring consists of all polynomials of degree less than n, we see that in fact $|f| = q^n$. We note also that $n = \log_q(|f|)$. These facts will be relevant frequently in what follows.

13.1.4 Remark Throughout this section we shall use the notation $f(k) \sim g(k)$ to mean that $\lim_{k\to\infty} f(k)/g(k) = 1$. In addition, the notation log denotes the natural logarithm.

13.1.1 The density of primes and irreducibles

13.1.5 Theorem (The Prime Number Theorem) Let $\pi(m)$ be the number of prime numbers less than or equal to m. Then

$$\pi(m) \sim \frac{m}{\log(m)}.$$

13.1.6 Remark For the polynomial case, we employ the notation of Definition 2.1.23 but, in analogy with integers, add the notation $\pi_q(n)$ to mean the number of monic irreducible polynomials over $\mathbb{F}_q$ of degree *less than or equal to* n. The next result then follows immediately from Theorem 2.1.24.

13.1.7 Theorem (The Polynomial Prime Number Theorem) If $I_q(n)$ is as in Definition 2.1.23, we have

$$I_q(n) \sim \frac{q^n}{n}.$$

13.1.8 Remark We note that if $f \in \mathbb{F}_q[x]$ is of degree n, this theorem says that

$$I_q(n) \sim \frac{|f|}{\log_q(|f|)},$$

in exact analogy to the Prime Number Theorem.

13.1.9 Remark The two Prime Number Theorems (integer and polynomial) both say that "near" an element, the density of primes is 1 out of the log of the absolute value of the element. In the integer case the log is natural; in the polynomial case the log is base q.

13.1.10 Remark The next result of Lenskoi [1891] gives asymptotic information for the case of counting monic irreducibles of degree less than or equal to n (see also [2412]).

13.1.11 Theorem Let $\pi_q(n)$ be as defined in 13.1.6 above. Then we have

$$\pi_q(n) \sim \frac{q}{q-1}\frac{q^n}{n}.$$

13.1.2 Primes and irreducibles in arithmetic progression

13.1.12 Theorem (Primes in Arithmetic Progression) Let the Euler function ϕ be as in Definition 2.1.43 and suppose a and d are relatively prime positive integers. By $\pi_{a,d}(m)$ we mean the number of primes less than or equal to m which are congruent to a modulo d. Then

$$\pi_{a,d}(m) \sim \frac{m}{\log(m)\phi(m)}.$$

13.1.13 Remark This then says that for any such pair $\{a, d\}$ there are infinitely many primes which are congruent to $a \pmod d$, and moreover that the primes are "ultimately uniformly distributed" among the eligible congruence classes of d. Artin [134] proved the following analogous result for polynomials.

13.1.14 Theorem [134] Let the "polynomial Euler function" Φ_q be as in Definition 2.1.111 and suppose A and D are relatively prime polynomials in $\mathbb{F}_q[x]$. By $I_{q;A,D}(n)$ we mean the number of monic irreducible polynomials of degree n which are congruent to A modulo D. Then

$$I_{q;A,D}(n) \sim \frac{q^n}{n\Phi_q(D)}.$$

13.1.15 Remark Hayes [1447] generalizes this result to a much broader class of congruence relations which he calls "arithmetically distributed" relations. As an example, he shows that the relation of two monic polynomials having the same first k and last m coefficients (see Definition 3.5.1 in [1447]) is arithmetically distributed, and so the following theorem holds.

13.1.16 Theorem [1447] Let $I_{q;k,m}(n)$ be the number of monic irreducible polynomials of degree n for which the first k and last m coefficients are prescribed. Then

$$I_{q;k,m}(n) \sim \frac{q}{q-1}\frac{q^{n-k-m}}{n}.$$

13.1.17 Remark This says that monic irreducibles are "ultimately uniformly distributed" with respect to their first k and last m coefficients, provided of course that the constant term is not 0. For much more information on irreducible polynomials with prescribed coefficients, see Section 3.5.

13.1.3 Twin primes and irreducibles

13.1.18 Definition Two odd prime numbers are *twin primes* if the absolute value of their difference is as small as possible, i.e., is 2.

13.1.19 Definition Two monic irreducible polynomials over $\mathbb{F}_q$ are *twin irreducibles* if the absolute value of their difference is as small as possible, i.e., is 1 if $q > 2$ and is 4 if $q = 2$.

13.1.20 Remark This last definition implies that for $q > 2$, two monic irreducibles are twins provided they are identical except in their constant terms (so their difference has degree 0). For $q = 2$, however, all irreducible have 1 as their constant coefficient and all have an odd number of terms (otherwise they are divisible by $x + 1$), and so in this case twins will differ in their linear and quadratic terms (so their difference has degree 2).

13.1.21 Conjecture [1418] Let $\pi_2(m)$ be the number of twin prime pairs less than or equal to m. Then

$$\pi_2(m) \sim 2\frac{m}{(\log m)^2} \prod_{\text{odd } p} \left(1 - \frac{1}{(p-1)^2}\right).$$

13.1.22 Remark The product over odd primes in the above conjecture is the "twin primes constant" and has value approximately 0.66016. If this conjecture were proved, it then implies the "Twin Primes Conjecture," i.e., the existence of infinitely many twin prime pairs. We have an analogous conjecture in the polynomial setting.

13.1.23 Conjecture [962] Let $I_{2,q}(n)$ be the number of twin irreducible polynomials of degree n. Then

$$I_{2,q}(n) \sim \delta\left(\frac{q-1}{2}\right)\frac{q^n}{n^2}\prod_P\left(1-\frac{1}{(|P|-1)^2}\right),$$

where either $\delta = 1$ and the product is over all monic irreducibles P provided $q > 2$, or $\delta = 4$ and the product excludes linear irreducibles when $q = 2$.

13.1.24 Remark Though unproven, these two conjectures are strongly supported by numerical evidence. For example, for polynomials of degree 16 over $\mathbb{F}_3$, the above formula accurately predicts the 66606 twin irreducible pairs. For more information on what is currently known along the lines of Conjecture 13.1.23, especially for the "fixed n, q going to ∞" case, see Pollack [2410] and [2411].

13.1.25 Remark Though the Twin Primes Conjecture remains unproven, it was first observed by Hall [1402] and then further explored by Pollack [2409] and Effinger [959], that the "Polynomial Twin Primes Conjecture" holds provided only that $q > 2$. The proof technique makes use of the fact that unlike prime numbers, irreducible polynomials have "internal structure." Specifically, "irreducibility preserving substitutions" can be exploited to guarantee the following theorem.

13.1.26 Theorem [959] (Polynomial Twin Primes Theorem) Over every $\mathbb{F}_q$ with $q > 2$, there exist infinitely many twin irreducible pairs.

13.1.27 Remark In fact one can prove the existence of "t-tuplets" (twins being 2-tuplets) in the polynomial case. For example, over $\mathbb{F}_7$ we can guarantee the existence of infinitely many 4-tuplets. This follows from the next result.

13.1.28 Theorem [959] Let $\mathbb{F}_q$ satisfy that $q \geq 4$. If $q \equiv 0 \pmod 4$ or $q \equiv 1 \pmod 4$ and if p is any prime dividing $q-1$, then there exist exactly $t = \frac{(p-1)(q-1)}{p}$ irreducible polynomials of the form $x^{p^k} - a$ over $\mathbb{F}_q$ for every $k \geq 1$. If $q \equiv 3 \pmod 4$, the conclusion holds for all *odd* primes p, but *no* irreducibles of the form $x^{2^k} - a$ exist provided $k \geq 2$.

13.1.29 Problem The polynomial twin primes conjecture remains unsolved over $\mathbb{F}_2$.

13.1.4 The generalized Riemann hypothesis

13.1.30 Definition If χ is a Dirichlet character (see, for example, Section 1.4.3 of [751]), then the corresponding *Dirichlet L-function* is

$$L(s,\chi) = \prod_p \left(1 - \frac{\chi(p)}{p^s}\right)^{-1},$$

where $s \in \mathbb{C}$ and p ranges over all primes.

13.1.31 Remark The function $L(s,\chi)$, which converges for all Re(s) > 1, can, like the Riemann zeta function, be extended analytically to a meromorphic function on the whole complex plane.

13.1.32 Conjecture (The Generalized Riemann Hypothesis) For every Dirichlet character χ, if $\text{Re}(s) > 0$ and if $L(s, \chi) = 0$, then $\text{Re}(s) = 1/2$.

13.1.33 Definition If χ is a character "of Dirichlet type" on $\mathbb{F}_q[x]$ (see, for example, Chapter 5 of [961]), then the corresponding *polynomial Dirichlet L-function* is

$$L_q(s, \chi) = \prod_P \left(1 - \frac{\chi_q(P)}{|P|^s}\right)^{-1},$$

where $s \in \mathbb{C}$ and P ranges over all monic irreducibles over $\mathbb{F}_q$.

13.1.34 Remark The following polynomial analogue of the Generalized Riemann Hypothesis follows from the deep results of André Weil [2962] and is a key ingredient of the results of our next subsection and of many other results in the number theory of polynomials over finite fields. For example, see [1606] for an exposition of the polynomial analogue of Artin's conjecture on primitive roots.

13.1.35 Theorem [2962] (A Polynomial Generalized Riemann Hypothesis) The function $L_q(s, \chi)$ is a complex polynomial F_χ in q^{-s} and when factored into

$$F_\chi(q^{-s}) = \prod (1 - \gamma_i q^{-s}),$$

each γ_i satisfies $|\gamma_i| = q^{1/2}$.

13.1.5 The Goldbach problem over finite fields

13.1.36 Remark In both the integer and polynomial cases, the "two-primes" Goldbach problem of writing an appropriate element of the ring $\mathbb{Z}$ or $\mathbb{F}_q[x]$ as a sum of two irreducible elements remains unsolved. However, if one moves to the "three-primes" case, a great deal more can be said. The first giant step forward was the pioneering work of Hardy and Littlewood in the 1920s. In the following result, *Hypothesis R* (a Weak Generalized Riemann Hypothesis) replaces the 1/2 in Conjecture 13.1.32 with Θ, where $1/2 \le \Theta < 3/4$.

13.1.37 Theorem [1420] If Hypothesis R holds, then, as $m \to \infty$, the number $N_3(m)$ of representations of the odd integer m as a sum of three odd primes satisfies:

$$N_3(m) \sim \frac{m^2}{(\log m)^3} \prod_{p>2} \left(1 + \frac{1}{(p-1)^3}\right) \prod_{p|m} \left(1 - \frac{1}{p^2 - 3p + 3}\right),$$

where p runs over prime numbers as specified.

13.1.38 Remark Vinogradov [2877] succeeded in removing Hypothesis R from the above Hardy-Littlewood result, hence proving unconditionally that every sufficiently large odd number is a sum of three primes. It has since been established using refinements of his analysis that "sufficiently large" can be assumed to mean above 10^{43000} [602]. It has also been established that *if* one assumes Conjecture 13.1.32, then every odd number greater than 5 is a sum of three primes (the so-called "Complete 3-Primes Theorem under GRH"; see, for example, [823]).

13.1.39 Remark Turning to the polynomial case, Hayes [1448] adapted the Hardy-Littlewood "Circle Method" to the function field setting, using the completion of $\mathbb{F}_q(x)$ at the infinite place as the analogue of the unit circle in the complex plane and employing an appropriate version of Theorem 13.1.35, to obtain an asymptotic "3-primes" result for polynomials. However, the irreducibles in his analysis were not monic and hence the result was not a perfect analogue of the integer result. Later, using adelic analysis (wherein $\mathbb{F}_q(x)$ is completed at all its places, not just the infinite one) and again an appropriate version of Theorem 13.1.35, Hayes and Effinger [961] obtained the theorem below. First we need a definition.

13.1.40 **Definition** A polynomial f over $\mathbb{F}_q$ is *even* if f is divisible by an irreducible P with $|P| = 2$. Otherwise f is *odd.*

13.1.41 **Remark** Obviously, even polynomials only exist over $\mathbb{F}_2$ and are then ones which are divisible by x or $x + 1$.

13.1.42 **Theorem** [961] Let f be an odd monic polynomial of degree n over $\mathbb{F}_q$. As $q \to \infty$ *or* $n \to \infty$, the number $M_3(f)$ of representations of $f = P_1 + P_2 + P_3$, where $\deg(P_1) = n$, $\deg(P_2) < n$ and $\deg(P_3) < n$, satisfies:

$$M_3(f) \sim \frac{q^{2n}}{n^3} \prod_{|P|>2} \left(1 + \frac{1}{(|P|-1)^3}\right) \prod_{P|f} \left(1 - \frac{1}{|P|^2 - 3|P| + 3}\right),$$

where P runs over irreducible polynomials over $\mathbb{F}_q$ as specified.

13.1.43 **Remark** Because this result is asymptotic in both q and n, careful analysis of the error terms and further investigation, both theoretical and numerical, of the low degree and small field cases [957, 958, 960] yield the following "best possible" result.

13.1.44 **Theorem** (A Complete Polynomial 3-Primes Theorem) Every odd monic polynomial of degree $n \geq 2$ over every finite field $\mathbb{F}_q$ (except the case of $x^2 + \alpha$ when q is even) is a sum of three monic irreducible polynomials, one of degree n and the others of lesser degrees.

13.1.45 **Remark** Hence, we get "complete 3-primes theorems" in both the integer and polynomial cases provided that we have a *proven* Generalized Riemann Hypothesis. In the polynomial case, thanks to Weil, we do; in the integer case we still do not.

13.1.6 The Waring problem over finite fields

13.1.46 **Remark** The general problem of writing an arbitrary positive integer as a sum of a limited number of k-th powers, known as the Waring Problem, was first settled by Hilbert in 1909.

13.1.47 **Theorem** (The Hilbert-Waring Theorem) [1500] For every positive integer k, there exists an integer $s(k)$ such that every positive integer m can be written as a sum of at most $s(k)$, k-th powers.

13.1.48 **Remark** Tremendous effort over the years has been put toward the question: Given k, what is $s(k)$? To that end we have following definitions, the latter having first been introduced by Hardy and Littlewood [1419].

13.1.49 **Definition** Let $g(k)$ be smallest s such that *every* positive integer m is a sum of at most s, k-th powers. Let $G(k)$ be the smallest s for which every *sufficiently large* positive integer is a sum of at most s, k-th powers.

13.1.50 **Remark** It is known that $g(2) = G(2) = 4$, that $g(3) = 9$, and that $4 \leq G(3) \leq 7$. Exact formulas are now known for $g(k)$ for all k. Although exact values of $G(k)$ are known only for $k = 2$ and $k = 4$ ($G(4) = 16$), considerable progress has been made on good lower and upper bounds for $G(k)$. For excellent surveys of the Waring Problem, see Ellison [974] and the more current Vaughan and Wooley [2862].

13.1.51 Remark Turning to the polynomial case, one must first observe that no cancellation occurs in the integer Waring Problem, and so the most appropriate analogue in the polynomial case will allow as little cancellation as possible. This is achieved as follows:

13.1.52 Definition The representation $f = X_1^k + \cdots + X_s^k$ of $f \in \mathbb{F}_q[x]$ as a sum of k-th powers of polynomials in $\mathbb{F}_q[x]$ is a *strict sum* provided that for every $1 \le i \le s$, $\deg(X_i) \le \lceil \deg(f)/k \rceil$ (equivalently, provided that $k\deg(X_i) < k + \deg(f)$).

13.1.53 Remark The following result, first proved independently by Car [508], Webb [2955], and Kubota [1810], is the best analogue to the Hilbert-Waring Theorem.

13.1.54 Theorem (Polynomial Waring Theorem) Suppose $k < p = \text{char}(\mathbb{F}_q)$. Then there exists an integer $s(k)$, independent of q, such that every $f \in \mathbb{F}_q[x]$ is a strict sum of $s(k)$ k-th powers of polynomials in $\mathbb{F}_q[x]$.

13.1.55 Remark Just as in the integer case, the question becomes: Given k, what is $s(k)$? The following definition is from Section 1.1 of [961].

13.1.56 Definition Given $k \ge 2$, let $g_{\text{poly}}(k)$ be the smallest value of s such that for every q with $p = \text{char}(\mathbb{F}_q) > k$, every $f \in \mathbb{F}_q[x]$ is a strict sum of s k-th powers in $\mathbb{F}_q[x]$. Let $G_{\text{poly}}(k)$ be the smallest s such that this same condition holds except possibly for a finite number of polynomials in the collection of all $\mathbb{F}_q[x]$ with $p > k$.

13.1.57 Remark As discussed in Section 1.2 of [961], the case of $k = 2$ is settled by Serre, with Webb showing that the only exceptions are two polynomials of degree 3 and six polynomials of degree 4 over $\mathbb{F}_3$ which require four squares.

13.1.58 Theorem [961] We have $g_{\text{poly}}(2) = 4$ and $G_{\text{poly}}(2) = 3$.

13.1.59 Remark To date no exact values of $g_{\text{poly}}(k)$ or $G_{\text{poly}}(k)$ are known for $k > 2$. The cases of $k = 3$ and $k = 4$ have been extensively studied. See the Introduction of [513] for a good summary of what is currently known, including cases not satisfying the hypothesis of Theorem 13.1.54 (i.e., that $k < p$). Following our Definition 13.1.56, however, it is known that $g_{\text{poly}}(3) \le 9$, $G_{\text{poly}}(3) \le 7$, and that $G_{\text{poly}}(4) \le 11$; see [514, 515]. For results on upper bounds for $g_{\text{poly}}(k)$, especially for large k, see [1167]. Finally, in [513], the following upper bounds for general k are established for both parameters. Note that in the case of $g_{\text{poly}}(k)$, there is dependence on q.

13.1.60 Theorem [513] We have

1. $G_{\text{poly}}(k) \le k(\log(k-1) + 3) + 3$;
2. If $k \ge 4$, then $g_{\text{poly}}(k) \le \gcd(q-1, k)(k^3 - 2k^2 - k + 1)$.

See Also

§3.1	For discussion of counting irreducible polynomials.
§3.3	For discussion of reducibility criteria for polynomials.
§3.5	For discussion of prescribed coefficients of irreducible polynomials.
§11.2	For discussion of polynomial counting algorithms.
§11.4	For discussion of univariate polynomial factoring algorithms.
[1606]	For discussion of polynomial analogies of Artin's primitive root conjecture.
[2016]	For discussion of polynomial analogies of sequences of smooth numbers.
[1497]	For discussion of polynomial analogies of the $3n + 1$ problem.
[1166]	For discussion of polynomial analogies of perfect numbers.

References Cited: [134, 508, 513, 514, 515, 602, 751, 823, 957, 958, 959, 960, 961, 962, 963, 974, 1166, 1167, 1402, 1418, 1419, 1420, 1447, 1448, 1497, 1500, 1606, 1810, 1891, 2016, 2409, 2410, 2411, 2412, 2862, 2877, 2955, 2962]

13.2 Matrices over finite fields

Dieter Jungnickel, University of Augsburg

On one hand we collect results about the numbers of matrices of various types over $\mathbb{F}_q$; on the other hand, we shall also discuss matrix representations of the field $\mathbb{F}_{q^m}$ over $\mathbb{F}_q$ and give a few results concerning determinants.

13.2.1 Matrices of specified rank

13.2.1 Remark As noted in Remark 2.1.90, the vector space of all $m \times n$ matrices over a field $F = \mathbb{F}_q$ has dimension mn, and thus the number of $m \times n$ matrices is q^{mn}. We shall mainly concentrate on square matrices. Clearly, the $m \times m$ matrices over $\mathbb{F}_q$ constitute a ring $R = \mathbb{F}_q^{(m,m)}$, and the invertible matrices in R form a group $G = GL(m, q)$, the *general linear group.* The elements of G are exactly those matrices in R which transform the elements of a fixed ordered basis $B = \{\alpha_1, \ldots, \alpha_m\}$ for $\mathbb{F}_{q^m}$ over $\mathbb{F}_q$ into an ordered basis again. Hence the order of G agrees with the total number of distinct ordered bases:

$$|GL(m, q)| = q^{m(m-1)/2}(q^m - 1)(q^{m-1} - 1) \cdots (q - 1). \qquad (13.2.1)$$

Trivially, the determinants of the elements of G are uniformly distributed over F^*. In particular, the matrices in G with determinant 1 form a group $SL(m, q)$ of order

$$|SL(m, q)| = q^{m(m-1)/2}(q^m - 1)(q^{m-1} - 1) \cdots (q^2 - 1),$$

called the *special linear group.* These two groups are the most elementary instances of the *classical groups*, which are discussed in detail in Section 13.4.

13.2.2 Remark Generalizing Equation (13.2.1), we give a formula for the number of all $m \times n$ matrices with prescribed rank k over $\mathbb{F}_q$. This is closely related to the number of subspaces of prescribed dimension in a vector space over $\mathbb{F}_q$.

13.2.3 **Definition** The number of k-dimensional subspaces of an m-dimensional vector space over $\mathbb{F}_q$ is denoted by $\begin{bmatrix} m \\ k \end{bmatrix}_q$. These numbers are the *Gaussian coefficients*; they constitute a q-analogue of the binomial coefficients.

13.2.4 **Lemma** The Gaussian coefficient $\begin{bmatrix} m \\ k \end{bmatrix}_q$ is given explicitly as follows:

$$\begin{bmatrix} m \\ k \end{bmatrix}_q = \frac{(q^m - 1)(q^{m-1} - 1) \cdots (q^{m-k+1} - 1)}{(q^k - 1)(q^{k-1} - 1) \cdots (q - 1)}.$$

13.2.5 **Theorem** [1070] Let $V = \mathbb{F}_q^{(m,n)}$ be the vector space of all $m \times n$ matrices over the field $F = \mathbb{F}_q$. Then the number $f(m, n, k)$ of matrices with rank $k \leq \min\{m, n\}$ in V is

$$\begin{aligned} f(m,n,k) &= \begin{bmatrix} n \\ k \end{bmatrix}_q q^{k(k-1)/2} (q^m - 1)(q^{m-1} - 1) \cdots (q^{m-k+1} - 1) \\ &= \begin{bmatrix} m \\ k \end{bmatrix}_q q^{k(k-1)/2} (q^n - 1)(q^{n-1} - 1) \cdots (q^{n-k+1} - 1) \\ &= q^{k(k-1)/2} \prod_{i=0}^{k-1} \left(q^{m-i} - 1\right) \left(q^{n-i} - 1\right) \left(q^{i+1} - 1\right)^{-1}. \end{aligned}$$

13.2.6 **Remark** Clearly, the probability that a given entry of an $m \times n$ matrix of rank k over $\mathbb{F}_q$ is $\neq 0$ does not depend on the position of the entry. In [2093] this probability is determined to be

$$\frac{\left(1 - \frac{1}{q}\right)\left(1 - \frac{1}{q^k}\right)}{\left(1 - \frac{1}{q^m}\right)\left(1 - \frac{1}{q^n}\right)}.$$

In [412], functions counting matrices of given rank over a finite field with specified positions equal to 0 are studied. Such matrices may be viewed as q-analogues of permutations with certain restricted values. In particular, a simple closed formula for the number of invertible matrices with zero diagonal is obtained (see below), as well as recursions to enumerate matrices with zero diagonal by rank.

13.2.7 **Theorem** [2093] The number of invertible $m \times m$ matrices over $\mathbb{F}_q$ whose diagonal consists entirely of zeros is

$$q^{\binom{m-1}{2}} (q-1)^m \left(q^{-1} \sum_{i=0}^{m} (-1)^i \binom{m}{i} ([m-i]_q)! \right),$$

where $[k]_q = q^{k-1} + \cdots + q + 1$.

13.2.2 Matrices of specified order

13.2.8 **Remark** By basic group theory, the order of an invertible $m \times m$ matrix over $\mathbb{F}_q$ has to divide the order of $GL(m, q)$ given in Equation (13.2.1). This elementary result can be strengthened in two ways.

13.2.9 **Theorem** [2293] The least common multiple of all orders of matrices in $GL(m, q)$ is $p^e M$, where q is a power of the prime p, e is the least integer satisfying $m \leq p^e$, and M is the least common multiple of $q - 1$, $q^2 - 1, \ldots, q^m - 1$. In particular, the order $o(A)$ of $A \in GL(m, q)$ divides $p^e M$. Moreover, $o(A)$ is always bounded by $q^m - 1$.

13.2.10 Example Consider $F = \mathbb{F}_{q^m}$ as a vector space over $K = \mathbb{F}_q$, and let γ be any element of F^*. Then γ defines a K-linear mapping $T_\gamma : F \to F$ via

$$T_\gamma : \xi \mapsto \gamma\xi \quad \text{for } \xi \in F,$$

and the order of T_γ equals the order $o(\gamma)$ of γ in F^*. Now represent T_γ with respect to some basis B of F over K; then the associated matrix $A_\gamma(B)$ is a matrix of order $o(\gamma)$ in $GL(m, q)$. In particular, choosing γ as a primitive element ω for F gives a matrix $A_\omega(B)$ of the maximum possible order $q^m - 1$.

13.2.11 Definition An invertible $m \times m$ matrix A of the maximum possible order $q^m - 1$ over $\mathbb{F}_q$ is a *Singer cycle*, and the group generated by A is a *Singer subgroup* of $GL(m, q)$. Moreover, an *involutory* matrix is simply a matrix of order two.

13.2.12 Theorem [1269] Any two Singer subgroups of $G = GL(m, q)$ are conjugate in G. The number $S(m, q)$ of Singer subgroups of G equals

$$S(m, q) = \frac{|GL(m, q)|}{m(q^m - 1)},$$

and the number of Singer cycles in G is given by

$$S(m, q)\phi(q^m - 1) = \frac{\phi(q^m - 1)}{m} \cdot \prod_{i=1}^{m-1} \left(q^m - q^i\right),$$

where ϕ is the Euler function given in Definition 2.1.43.

13.2.13 Theorem [1517] The number $i(m, q)$ of involutory $m \times m$ matrices over $\mathbb{F}_q$ equals

$$i(m, q) = \begin{cases} g_m \cdot \sum_{t=0}^{m} g_t^{-1} g_{m-t}^{-1} & \text{if } q \text{ is odd,} \\ g_m \cdot \sum_{t=0}^{\lceil m/2 \rceil} g_t^{-1} g_{m-2t}^{-1} q^{-t(2m-3t)} & \text{if } q \text{ is even,} \end{cases}$$

where g_t denotes the number of invertible $t \times t$ matrices over $\mathbb{F}_q$ as given in (13.2.1) (for $t \neq 0$) and $g_0 = 1$.

13.2.14 Remark [2166] More generally, there is a (rather involved) formula for the number of $m \times m$ matrices of order k over $\mathbb{F}_q$. Clearly, the probability that a randomly chosen $m \times m$ matrix over $\mathbb{F}_q$ is invertible (and therefore *has* an order) is g_m/q^{m^2}, where g_m denotes the number of invertible $m \times m$ matrices over $\mathbb{F}_q$ as given in (13.2.1). For fixed q, this probability has a limit:

$$\lim_{m\to\infty} \frac{g_m}{q^{m^2}} = \prod_{r\geq 1} \left(1 - \frac{1}{q^r}\right).$$

For $q = 2$, this limit is $0.28878\ldots$; for $q = 3$, the limit is $0.56012\ldots$; and, as $q \to \infty$, the probability of being invertible goes to 1.

13.2.15 Remark [1064] On the other end of the spectrum, the number of *nilpotent* $m \times m$ matrices over $\mathbb{F}_q$ – that is, matrices A satisfying $A^k = 0$ for some k – is given by a very simple formula: it equals $q^{m(m-1)}$.

13.2.3 Matrix representations of finite fields

13.2.16 **Definition** Any subring R of the full matrix ring $K^{(m,m)}$ over a field K which is itself a field is a *matrix field* (of degree m) over K. If K is a finite field and if $R \cong \mathbb{F}_q$, R is a *matrix representation* of degree m for $\mathbb{F}_q$ (over K).

13.2.17 **Remark** The following result concerning the existence of matrix representations for $\mathbb{F}_q$ is an obvious consequence of Theorem 13.2.9 and Example 13.2.10. For details, see Example 13.2.19.

13.2.18 **Theorem** Any matrix representation for $\mathbb{F}_{q^m}$ over $\mathbb{F}_q$ has degree at least m. Moreover, there always exists a representation of degree m.

13.2.19 **Example** With the notation of Example 13.2.10, the $q^m - 1$ matrices $A_\gamma(B)$ with $\gamma \in F^*$ together with the zero matrix form a matrix representation $R(B)$ of degree m for $\mathbb{F}_{q^m}$ over $\mathbb{F}_q$. Clearly, one may write $R(B)$ as the $q^m - 1$ powers of the Singer cycle $A_\omega(B)$ together with the zero matrix. Now let $f(x) = x^m + f_{m-1}x^{m-1} + \cdots + f_1 x + f_0$ be the minimal polynomial of a primitive element ω (so that f is a primitive polynomial of degree m over K), and let $B = \{1, \omega, \omega^2, \ldots, \omega^{m-1}\}$ be the associated polynomial basis. Then $A_\omega(B)$ is the *companion matrix* of f, that is,

$$A_\omega(B) = \begin{pmatrix} 0 & 0 & \cdots & 0 & -f_0 \\ 1 & 0 & \cdots & 0 & -f_1 \\ 0 & 1 & \ddots & \vdots & \vdots \\ \vdots & \vdots & \ddots & 0 & -f_{m-2} \\ 0 & 0 & \cdots & 1 & -f_{m-1} \end{pmatrix}.$$

As this example shows, Singer cycles in $GL(m, q)$ give rise to matrix representations for $\mathbb{F}_{q^m}$ over $\mathbb{F}_q$ of the smallest possible degree. Essentially, all minimal degree representations may be obtained in this way.

13.2.20 **Theorem** [2980] Let $R \subset \mathbb{F}_q^{(m,m)}$ be a matrix representation for $F = \mathbb{F}_{q^m}$ over $K = \mathbb{F}_q$. Then there exists a matrix $A \in R$ such that

$$R = \{A^k : k = 1, \ldots, q^m - 1\} \cup \{0\}.$$

Moreover, A is similar to the companion matrix of a primitive polynomial of degree m over K, and $\det A$ is a primitive element of K.

13.2.21 **Remark** Theorem 13.2.20 characterizes the matrix representations for $\mathbb{F}_{q^m}$ of smallest degree. There are considerably more general results due to Willett [2980] who classified all matrix fields of degree m over $\mathbb{F}_q$ in terms of primitive polynomials over the underlying prime field $\mathbb{F}_p$. The general result is rather technical, so we only give the special case $q = p$ here. Proofs for all these results can also be found in Section 1.5 of [1631].

13.2.22 **Theorem** [2980] Let p be a prime, and let R be any representation of $\mathbb{F}_{p^m}$ as a matrix field of degree n over $\mathbb{F}_p$. Up to similarity, R has the form

$$R = \{\operatorname{diag}(0, \ldots, 0, A^k, \ldots, A^k) : k = 1, \ldots, p^m - 1\} \cup \{0\},$$

where A is the companion matrix of a primitive polynomial of degree m over $\mathbb{F}_p$.

13.2.23 **Remark** It is of interest to study representations for $\mathbb{F}_{q^m}$ consisting of special types of matrices. We present some results due to Seroussi and Lempel [2584] concerning *symmetric* *matrix representations*, that is, matrix representations consisting of symmetric matrices only. This turns out to be closely related to the duality theory for bases, see Section 5.1. With the exception of Theorem 13.2.27, proofs for the subsequent results can be found in Section 4.6 of [1631].

13.2.24 Lemma [2584] Let B be a basis for $F = \mathbb{F}_{q^m}$ over $\mathbb{F}_q$, and let $R(B)$ be the associated matrix representation. Then $R(B)$ is symmetric if and only if there exists an element $\lambda \in F^*$ such that the dual basis B^* of B satisfies $B^* = \lambda B$, where λB consists of all elements $\lambda\beta$ with $\beta \in B$.

13.2.25 Theorem [2584] Every finite field $F = \mathbb{F}_{q^m}$ admits a basis B over $\mathbb{F}_q$ such that the associated matrix representation $R(B)$ is symmetric.

13.2.26 Theorem [2584] Let B be a basis of $F = \mathbb{F}_{q^m}$ over $\mathbb{F}_q$ with associated matrix representation $R(B)$, and assume that q is even or that q and m are both odd. Then the dual basis B^* of B satisfies $B^* = \lambda B$ for some $\lambda \in F^*$ (so that $R(B)$ is symmetric) if and only if λ is a square, say $\lambda = \mu^2$, and the basis μB is self-dual.

13.2.27 Theorem [2584] Let B be a basis of $F = \mathbb{F}_{q^m}$ over $\mathbb{F}_q$, and assume that the associated matrix representation $R(B)$ is closed under transposition of matrices. Then $R(B)$ is symmetric provided that either q is even or q and m are both odd. If q is odd and m is even, then $R(B)$ is *not* symmetric if and only if $B^* = (\lambda B)^{q^{m/2}}$ for some $\lambda \in F^*$.

13.2.4 Circulant and orthogonal matrices

13.2.28 Remark There are many results on the number of matrices of special types over $\mathbb{F}_q$. In the remaining subsections, we present a selection of such results. We begin with those cases which are related to the enumeration of various types of bases as discussed in Chapter 5. Let us summarize these connections as follows:

Type of basis	Associated transformation matrices	Reference
normal bases	circulant matrices	5.2.15
self-dual bases	orthogonal matrices	5.1.22
self-dual normal bases	orthogonal circulant matrices	5.2.28

13.2.29 Definition An $m \times m$ matrix $C = (c_{ij})_{i,j=1,\ldots,m}$ is *circulant* if its rows are generated by successive cyclic shifts of its first row, that is

$$c_{i+1,j+1} = c_{ij} \quad \text{for all } i, j = 1, \ldots, m, \tag{13.2.2}$$

where indices are computed modulo m. Thus C is specified by the entries $c_j = c_{1,j}$ in its first row:

$$C = \begin{pmatrix} c_1 & c_2 & \cdots & c_{m-1} & c_m \\ c_m & c_1 & \cdots & c_{m-2} & c_{m-1} \\ c_{m-1} & c_m & \cdots & c_{m-3} & c_{m-2} \\ \vdots & \vdots & \ddots & \vdots & \vdots \\ c_2 & c_3 & \cdots & c_m & c_1 \end{pmatrix}.$$

Such a matrix C is denoted as $\mathrm{circ}(c_1, \ldots, c_m)$.

13.2.30 Remark The following three results are well-known. They may be found, for example, in [1631].

13.2.31 Lemma Let F be a field. Mapping the matrix $C = \mathrm{circ}(c_1, \dots, c_m)$ to the coset of the polynomial $c(x) = c_0 + c_1 x + \cdots + c_{m-1}x^{m-1}$ modulo $x^m - 1$ gives an isomorphism between the ring of all circulant $m \times m$ matrices over F and the ring $R = F[x]/(x^m - 1)$. In particular, C is invertible if and only if the associated polynomial $c(x)$ is a unit in R.

13.2.32 Corollary Let $C(m,q)$ denote the multiplicative group of all invertible circulant $m \times m$ matrices over $\mathbb{F}_q$. Then the order of $C(m,q)$ equals $\Phi_q(x^m - 1)$, where Φ_q is the function introduced in Definition 2.1.111.

13.2.33 Theorem Let q be a power of the prime p, let m be a positive integer, and write $m = p^b n$, where p does not divide n. Then

$$|C(m,q)| = q^m \prod_{d|n} \left(1 - q^{o_d(q)}\right)^{\phi(d)/o_d(q)}, \tag{13.2.3}$$

where ϕ is the Euler function given in Definition 2.1.43 and where $o_d(q)$ denotes the multiplicative order of d modulo q.

13.2.34 Corollary Assume that q and m are co-prime. Then $|C(m,q)| = \prod_{j=1}^{r} (q^{m_j} - 1)$, where $m_1, \dots, m_r$ are the degrees of the irreducible factors of $x^m - 1$ over $\mathbb{F}_q$.

13.2.35 Definition An $m \times m$ matrix A is *orthogonal* if it satisfies the condition $AA^T = I$. We denote the multiplicative group of all orthogonal $m \times m$ matrices over $\mathbb{F}_q$ by $O(m,q)$.

13.2.36 Remark The preceding terminology is somewhat ambiguous, as the term *orthogonal group* usually refers to the group of isometries of an orthogonal geometry, that is, of a vector space equipped with a quadratic form. In the case of finite fields, no ambiguity arises if both q and m are odd. However, if q is odd and m is even, then there are two distinct orthogonal groups (of different orders); and if q is even, an orthogonal geometry is by definition a symplectic geometry refined by an additional quadratic form, which forces m to be even. In particular, the standard text books do not contain the order of $O(m,q)$ for even values of q. Nevertheless, the definition given in 13.2.35 makes sense for all cases. See Section 13.4 for the "classical" orthogonal groups.

13.2.37 Theorem [1989] Assume that either q is even or both q and m are odd. Then

$$|O(m,q)| = \gamma \prod_{i=1}^{m-1} \left(q^i - \epsilon_i\right), \tag{13.2.4}$$

where

$$\epsilon_i = \begin{cases} 1 & \text{if } i \text{ is even,} \\ 0 & \text{if } i \text{ is odd,} \end{cases} \quad \text{and} \quad \gamma = \begin{cases} 2 & \text{if } q \text{ is even,} \\ 1 & \text{if } q \text{ and } n \text{ are odd.} \end{cases}$$

Now let q be odd and m even, say $m = 2s$. Then

$$|O(m,q)| = q^{s(s-1)/2} \left(q^s + \epsilon\right) \prod_{i=1}^{s-1} \left(q^{2i} - 1\right), \tag{13.2.5}$$

where

$$\epsilon = \begin{cases} 1 & \text{if } s \text{ is odd and } q \equiv 3 \pmod 4, \\ -1 & \text{otherwise.} \end{cases}$$

13.2.38 Lemma Let F be a field, $C = \text{circ}(c_1, \dots, c_m)$ a circulant matrix, and $c(x) = c_0 + c_1 x + \dots + c_{m-1}x^{m-1}$ the associated polynomial $c(x)$ in the ring $R = F[x]/(x^m - 1)$, as in Lemma 13.2.31. Then C is an orthogonal matrix if and only if $c(x)c^T(x) = 1$ in R, where $c^T(x) = c_0 + c_{m-1}x + \dots + c_2 x^{m-2} + c_1 x^{m-1}$ is the polynomial corresponding to the transpose C^T of C.

13.2.39 Remark We require some notation to state the number of orthogonal circulant $m \times m$ matrices over $\mathbb{F}_q$; clearly all these matrices form a group which is denoted by $OC(m, q)$. First assume that q and m are co-prime. (The general case is reduced recursively to this special case.) Then let $x - 1$, possibly $x + 1$ (this factor arises only if q is odd and m is even) and $f_1, \dots, f_r$ be the monic irreducible factors of $x^m - 1$. Some of the f_i may be self-reciprocal (that is, $f_i = f_i^*$, see Definition 2.1.48), say $f_1, \dots, f_s$. Then the remaining f_j split into pairs of the form $\{f, f^\wedge\}$ with $f \neq f^\wedge$, where $f^\wedge = f^*/f_0$ and where f_0 is the constant term of f; say $r = s + 2t$, and $(f_{s+j})^\wedge = f_{s+t+j}$ for $j = 1, \dots, t$.

13.2.40 Theorem [258, 471, 1633, 1990] Assume that q and m are co-prime, and write, using the notation introduced in Remark 13.2.39,

$$\deg(f_i) = 2d_i \text{ for } i = 1, \dots, s \quad \text{and} \quad \deg(f_{s+j}) = \deg(f_{s+t+j}) = e_j \text{ for } j = 1, \dots, t.$$

Then the order of $OC(m, q)$ is given by

$$|OC(m, q)| = \gamma \prod_{i=1}^{s} (q^{d_i} + 1) \prod_{i=1}^{t} (q^{e_j} - 1), \tag{13.2.6}$$

where

$$\gamma = \begin{cases} 1 & \text{if } q \text{ is even}, \\ 2 & \text{if } q \text{ and } m \text{ are odd}, \\ 4 & \text{if } q \text{ is odd and } m \text{ is even}. \end{cases}$$

Now let q be a power of the prime p. If $p \neq 2$, then

$$|OC(pn, q)| = q^{n(p-1)/2} \, |OC(n, q)| \tag{13.2.7}$$

for every positive integer n. Finally,

$$|OC(2n, q)| = \begin{cases} q^{(n+1)/2} \, |OC(n, q)| & \text{if } n \text{ is odd}, \\ 2q^{n/2} \, |OC(n, q)| & \text{if } n \equiv 2 \pmod 4, \\ q^{n/2} \, |OC(n, q)| & \text{if } n \equiv 0 \pmod 4, \end{cases} \tag{13.2.8}$$

for every positive integer n.

13.2.41 Remark With the exception of the special case of Theorem 13.2.37 where q is odd and m is even, proofs for all the preceding results can also be found in [1631].

13.2.5 Symmetric and skew-symmetric matrices

13.2.42 Remark In order to prove the results on (circulant) orthogonal matrices presented in the previous subsection, one needs a connection with and enumeration results for two other interesting classes of matrices, which are the topic of the present subsection. Again, proofs for these results are in [1631].

13.2.43 **Definition** A matrix A over $\mathbb{F}_q$ is *symmetric* if it satisfies the condition $A = A^T$, and *skew-symmetric* if it satisfies $A = -A^T$ and has diagonal entries 0 only. (This extra condition has to be added in view of the special case where q is a power of 2.)

13.2.44 **Definition** Let A be a symmetric invertible matrix over $\mathbb{F}_q$. Then any (invertible) $m \times m$ matrix M satisfying $A = MM^T$ is a *factor* of A.

13.2.45 **Lemma** [1989] The invertible symmetric $m \times m$ matrices over $\mathbb{F}_q$ admitting a factor are in a 1-to-1 correspondence with the cosets of $O(m,q)$ in $GL(m,q)$. Hence

$$|O(m,q)| = \frac{|GL(m,q)|}{sf(m,q)},$$

where $sf(m,q)$ denotes the number of invertible symmetric matrices over $\mathbb{F}_q$ admitting a factor.

13.2.46 **Lemma** [69] Let A be a symmetric invertible matrix over $\mathbb{F}_q$. If q is odd, A admits a factor if and only if $\det A$ is a square; and if q is even, A admits a factor if and only if A has at least one non-zero diagonal entry.

13.2.47 **Theorem** [544, 1989] The number $N(m,r)$ of symmetric $m \times m$ matrices of rank r over $\mathbb{F}_q$ equals

$$N(m,r) = \prod_{i=1}^{s} \frac{q^{2i}}{q^{2i}-1} \cdot \prod_{i=0}^{r-1} \left(q^{m-i}-1\right),$$

where $r \le m$ and $s = \lfloor r/2 \rfloor$. In particular, the number of invertible symmetric $m \times m$ matrices over $\mathbb{F}_q$ is given by

$$N(m,m) = \prod_{i=1}^{m} \left(q^i - \delta_i\right), \quad \text{where } \delta_i = \begin{cases} 0 & \text{if } i \text{ is even,} \\ 1 & \text{otherwise.} \end{cases}$$

13.2.48 **Theorem** [545, 1989] The number $N_0(m,r)$ of skew-symmetric $m \times m$ matrices of rank r over $\mathbb{F}_q$ is given by

$$N_0(m,2s) = \prod_{i=1}^{s} \frac{q^{2i-2}}{q^{2i}-1} \cdot \prod_{i=0}^{2s-1} \left(q^{m-i}-1\right) \quad \text{and} \quad N_0(m,2s+1) = 0,$$

where $2s \le m$. In particular, there are no invertible skew-symmetric $m \times m$ matrices over $\mathbb{F}_q$ if m is odd; if m is even, the number of invertible skew-symmetric $m \times m$ matrices over $\mathbb{F}_q$ equals

$$N_0(m,m) = \prod_{i=1}^{m-1} \left(q^i - \delta_i\right), \quad \text{where } \delta_i = \begin{cases} 0 & \text{if } i \text{ is even,} \\ 1 & \text{otherwise.} \end{cases}$$

13.2.6 Hankel and Toeplitz matrices

13.2.49 **Remark** (Infinite) Toeplitz and Hankel matrices with complex entries play a prominent role in classical linear algebra and have many important applications; see, for instance, [1354] for an introductory treatment or [367] for a monograph on large finite Toeplitz matrices. In this subsection, we discuss such matrices over $\mathbb{F}_q$. Both classes of matrices are closely related, as a Hankel matrix may be viewed as an "upside-down" Toeplitz matrix, so it suffices to concentrate on one of these two classes.

13.2.50 **Definition** An $m \times m$ matrix $A = (a_{ij})_{i,j=1,\ldots,m}$ is a *Toeplitz matrix* if

$$a_{ij} = a_{kl} \quad \text{whenever } i - j = k - l. \tag{13.2.9}$$

Thus A is specified by the entries in its first row and column:

$$A = \begin{pmatrix} a_m & a_{m-1} & \cdots & a_2 & a_1 \\ a_{m+1} & a_m & \cdots & a_3 & a_2 \\ a_{m+2} & a_{m+1} & \cdots & a_4 & a_3 \\ \vdots & \vdots & \ddots & \vdots & \vdots \\ a_{2m-1} & a_{2m-2} & \cdots & a_{m+1} & a_m \end{pmatrix}.$$

13.2.51 **Remark** Note that indices are *not* computed modulo m in the defining condition (13.2.9) for a Toeplitz matrix. If we would do so, we get further restrictions and arrive at an equivalent formulation for the defining condition (13.2.2) for circulant matrices. Thus circulant matrices constitute a special class of Toeplitz matrices.

13.2.52 **Definition** An $m \times m$ matrix $A = (a_{ij})_{i,j=1,\ldots,m}$ is a *Hankel matrix* if

$$a_{ij} = a_{kl} \quad \text{whenever } i + j = k + l.$$

Thus A is specified by the entries in its first row and column:

$$A = \begin{pmatrix} a_1 & a_2 & \cdots & a_{m-1} & a_m \\ a_2 & a_3 & \cdots & a_m & a_{m+1} \\ a_3 & a_4 & \cdots & a_{m+1} & a_{m+2} \\ \vdots & \vdots & \ddots & \vdots & \vdots \\ a_m & a_{m+1} & \cdots & a_{2m-2} & a_{2m-1} \end{pmatrix}. \tag{13.2.10}$$

13.2.53 **Remark** "Rotating" a Toeplitz matrix by 90 degrees (counter-clockwise) transforms it into a Hankel matrix. More formally, let P be the $m \times m$ matrix with entries 1 on the antidiagonal and 0 elsewhere, that is,

$$P = \left(\delta_{i,m-j+1}\right)_{i,j=1,\ldots,m},$$

where $\delta_{ik} = 0$ if $i \neq k$ and $\delta_{ik} = 1$ if $i = k$. Then the bijection defined by $A \mapsto AP$ transforms the set of all $m \times m$ Toeplitz matrices into the set of all $m \times m$ Hankel matrices. Note that this bijection preserves the rank of matrices, so that the following result applies for Hankel matrices as well.

13.2.54 **Theorem** [1207] The number $t(m, r, q)$ of $m \times m$ Toeplitz matrices over $\mathbb{F}_q$ with rank r is given by

$$t(m, r, q) = \begin{cases} q^{2m-2}(q-1) & \text{if } r = m, \\ q^{2r-2}(q^2-1) & \text{if } 1 \leq r \leq m-1, \\ 1 & \text{if } r = 0. \end{cases}$$

13.2.55 **Corollary** For any positive integer r, the number of $m \times m$ Toeplitz (or Hankel) matrices over $\mathbb{F}_q$ with rank r is constant for every $m \geq r + 1$.

13.2.56 **Remark** We note that Toeplitz and Hankel matrices over finite fields have some interesting applications. For instance, Toeplitz matrices are used as pre-conditioners in the process of solving linear systems having unstructured coefficient matrices [1665]. In [1207], an explicit surjective map σ from the set of all ordered pairs of coprime monic polynomials of degree m over $\mathbb{F}_q$ to the set of all invertible $m \times m$ Hankel matrices over $\mathbb{F}_q$ is constructed; this map has the additional property that, for any such matrix A, the pre-image $\sigma^{-1}(A)$ is in a one-to-one correspondence with $\mathbb{F}_q$. Therefore Theorem 13.2.54 gives a proof for the following result.

13.2.57 Theorem [226] The number of ordered pairs of coprime monic polynomials of degree m over $\mathbb{F}_q$ equals $q^{2m-1}(q-1)$.

13.2.58 Corollary The probability that two randomly chosen monic polynomials of the same positive degree with coefficients in $\mathbb{F}_q$ are coprime is $1-(1/q)$.

13.2.59 Remark Further results on the greatest common divisor of polynomials are given in Section 11.2.

13.2.7 Determinants

13.2.60 Definition Let $\{\alpha_1, \ldots, \alpha_m\}$ bc a set of m elements of $\mathbb{F}_{q^m}$. Then the determinant

$$\begin{vmatrix} \alpha_1 & \cdots & \alpha_m \\ \alpha_1^q & \cdots & \alpha_m^q \\ \vdots & \ddots & \vdots \\ \alpha_1^{q^{m-1}} & \cdots & \alpha_m^{q^{m-1}} \end{vmatrix}$$

is the *Moore determinant* of $\{\alpha_1, \ldots, \alpha_m\}$.

13.2.61 Remark By Corollary 2.1.95, the set $\{\alpha_1, \ldots, \alpha_m\}$ is a basis for $\mathbb{F}_{q^m}$ over $\mathbb{F}_q$ if and only if its Moore determinant is nonzero. The Moore determinant may be viewed as a finite field analogue of the well-known Vandermonde determinant. It is used extensively in the theory of Drinfeld modules, see Section 13.3. Moore [2140] proved the following formula for his determinant, which immediately implies the validity of Corollary 2.1.95.

13.2.62 Theorem Let $\{\alpha_1, \ldots, \alpha_m\}$ be a set of m elements of $\mathbb{F}_{q^m}$. Then its Moore determinant $\det M(\alpha_1, \ldots, \alpha_m)$ is given by

$$\det M(\alpha_1, \ldots, \alpha_m) = \alpha_1 \prod_{i=1}^{m-1} \prod_{c_1, \ldots, c_i \in \mathbb{F}_q} \left(\alpha_{i+1} - \sum_{j=1}^{i} c_j \alpha_j \right).$$

13.2.63 Remark [910] Equation (13.2.1) immediately gives an exact formula for the probabilty that the determinant of a random $m \times m$ matrix A over $\mathbb{F}_q$ equals 0. From this one easily sees that the probability in question is in the order of magnitude $\text{prob}(\det A = 0) = \frac{1}{q} + O(\frac{1}{q^2})$. It is remarkable that one actually has the following much more general result for both determinants and permanents:

$$\text{prob}(\det A = \alpha) = \text{prob}(\text{per } A = \alpha) = \frac{1}{q} + O\left(\frac{1}{q^2}\right)$$

for every $\alpha \in \mathbb{F}_q$. This result is obtained in [910] as a byproduct of the authors' study of the Polya permanent problem for matrices over finite fields.

13.2.64 Remark Several further results concerning the enumeration of various types of matrices over finite fields can be found in the nice survey article [2166]. We have considered the number of invertible $m \times m$ matrices over $\mathbb{F}_q$ with the smallest (viz. 2) and the largest (viz. $q^m - 1$) possible orders in Subsection 13.2.2. Matrices of a specified order k are, of course, closely related to the solutions of the matrix equation $f(X) = 0$, where $f(x) = x^k - 1$. There are many papers concerning matrix equations over finite fields; see the notes to Section 6.2 in [1939] for a collection of references.

See Also

§5.1	For bases of finite fields.
§5.2	For a discussion of normal bases.
§13.3	For a discussion of classical groups over finite fields.

References Cited: [69, 226, 258, 367, 412, 471, 544, 545, 910, 1064, 1070, 1207, 1269, 1354, 1517, 1631, 1633, 1665, 1989, 1990, 2093, 2140, 2166, 2293, 2584, 2980]

13.3 Classical groups over finite fields

Zhe-Xian Wan, *Chinese Academy of Sciences*

13.3.1 Linear groups over finite fields

13.3.1 Definition Let $\mathbb{F}_q$ be a finite field with q elements and n an integer > 1. The set of $n \times n$ nonsingular matrices over $\mathbb{F}_q$ forms a group with respect to matrix multiplication, called the *general linear group* of degree n over $\mathbb{F}_q$ and denoted by $GL_n(q)$. The set of $n \times n$ matrices over $\mathbb{F}_q$ of determinant 1 forms a subgroup of $GL_n(q)$, called the *special linear group* of degree n over $\mathbb{F}_q$ and denoted by $SL_n(q)$.

The group $GL_n(q)$ can also be defined as the group consisting of nonsingular linear transformations of an n-dimensional vector space over $\mathbb{F}_q$.

13.3.2 Remark In literatures of group theory the groups $GL_n(q)$ and $SL_n(q)$ are sometimes written as $GL(n, q)$ instead of $GL_n(q)$ and $SL(n, q)$ instead of $SL_n(q)$.

13.3.3 Theorem [1589] The order of $GL_n(q)$ is

$$|GL_n(q)| = (q^n - 1)(q^n - q) \cdots (q^n - q^{n-1});$$

$SL_n(q)$ is a normal subgroup of $GL_n(q)$ and is of order

$$|SL_n(q)| = (q^n - 1)(q^n - q) \cdots (q^n - q^{n-2})q^{n-1}.$$

13.3.4 Theorem [1589] The center Z_n of $GL_n(q)$ consists of those matrices λI_n, where $\lambda \in \mathbb{F}_q^*$ and I_n is the $n \times n$ identity matrix. The center of $SL_n(q)$ is $SL_n(q) \cap Z_n$, which consists of those matrices λI_n, where $\lambda \in \mathbb{F}_q^*$ and $\lambda^n = 1$.

13.3.5 **Definition** The factor group $GL_n(q)/Z_n$ is the *projective general linear group* of degree n over $\mathbb{F}_q$ and denoted by $PGL_n(q)$. The factor group $SL_n(q)/SL_n(q)\cap Z_n$ is the *projective special linear group* of degree n over $\mathbb{F}_q$ and is denoted by $PSL_n(q)$.

13.3.6 **Theorem** [1589] The order of $PGL_n(q)$ is

$$|PGL_n(q)| = (q^n - 1)(q^n - q)\cdots(q^n - q^{n-2})q^{n-1};$$

and the order of $PSL_n(q)$ is

$$|PSL_n(q)| = (q^n - 1)(q^n - q)\cdots(q^n - q^{n-2})q^{n-1}/d,$$

where $d = \gcd\ (n, q-1)$.

13.3.7 **Theorem** [1589] The group $SL_n(q)$ is generated by the elementary matrices

$$T_{ij}(b) = I_n + bE_{ij},\quad 1 \le i,j \le n \text{ and } i \ne j,$$

where I_n is the $n \times n$ identity matrix, $b \in \mathbb{F}_q^*$ and E_{ij} is the $n \times n$ matrix with 1 in the (i,j) position and 0 elsewhere.

13.3.8 **Lemma** Every element A of $GL_n(q)$ can be expressed in the form

$$A = BD(\mu)$$

where $B \in SL_n(q)$ and $D(\mu)$ is a diagonal matrix with $1, 1, \ldots, 1, \mu$ along its main diagonal.

13.3.9 **Theorem** [1589] The group $SL_n(q)$ is the commutator subgroup of $GL_n(q)$ unless $n = 2$ and $q = 2$. Moreover, $SL_n(q)$ is its own commutator subgroup unless $n = 2$ and $q = 2$ or 3. When $n = 2$ and $q = 2$, $SL_2(2) = GL_2(2) \simeq S_3$, where S_3 denotes the symmetric group on three letters. When $n = 2$ and $q = 3$, $|SL_2(3)| = 24$ and the commutator subgroup of $SL_2(3)$ is of order 8.

13.3.10 **Theorem** (Dickson) [1589] The group $PSL_n(q)$ is simple except in the cases $n = 2$, $q = 2$ or 3.

13.3.11 **Remark** Except for the cases $n = 2$, $q = 2$ or 3, $GL_n(q)$ has the normal series

$$GL_n(q) \supset SL_n(q) \supset Z_n \cap SL_n(q) \supset \{I\},$$

where $GL_n(q)/SL_n(q)$ and $Z_n \cap SL_n(q)$ are cyclic and $SL_n(q)/(Z_n \cap SL_n(q)) \simeq PSL_n(q)$ is simple.

When $n = 2$ and $q = 2$, the group $GL_2(2) = SL_2(2) \simeq S_3$ has composition factors 2 and 3 and is not simple.

When $n = 2$ and $q = 3$, the group $GL_2(3)$ has generators

$$A = \begin{pmatrix} -1 & 0 \\ 1 & 1 \end{pmatrix},\ B = \begin{pmatrix} 0 & 1 \\ -1 & 1 \end{pmatrix},\ C = \begin{pmatrix} 0 & -1 \\ 1 & 0 \end{pmatrix},$$

$$D = \begin{pmatrix} 1 & 1 \\ 1 & -1 \end{pmatrix},\ E = \begin{pmatrix} -1 & 0 \\ 0 & -1 \end{pmatrix}$$

and defining relations

$$\begin{aligned}
&E^2 = I_2,\\
&D^2 = E,\ DE = ED,\\
&C^2 = E,\ CE = EC,\ CD = EDC,\\
&B^3 = E,\ BE = EB,\ BD = EDBC,\ BC = ECDB,\\
&A^2 = I_2,\ AE = EA,\ AD = CA,\ AC = DA,\ AB = ECB^2AC.
\end{aligned}$$

Then $GL_2(3)$ has the composition series

$$\begin{aligned} GL_2(3) &= \{E, D, C, B, A\} \supset SL_2(3) \\ &= \{E, D, C, B\} \supset \{E, D, C\} \supset \{E, D\} \supset Z_2 = \{E\} \supset \{I_2\} \end{aligned}$$

and the orders of these groups are $48, 24, 8, 4, 2$, respectively. Thus $PSL_2(3)$ is not simple.

13.3.12 Theorem [858] The only isomorphisms between the groups $PSL_n(q)$ are

1. $PSL_2(4) \simeq PSL_2(5)$, and
2. $PSL_2(7) \simeq PSL_3(2)$.

13.3.13 Theorem [858] The only isomorphisms between the groups $PSL_n(q)$ and A_m are

1. $PSL_2(3) \simeq A_4$,
2. $PSL_2(4) \simeq PSL_2(5) \simeq A_5$,
3. $PSL_2(9) \simeq A_6$, and
4. $PSL_4(2) \simeq A_8$.

13.3.14 Definition Let $\mathbb{F}_q^n$ be the n-dimensional row vector space over $\mathbb{F}_q$ consisting of all n-dimensional row vectors $(x_1, x_2, \ldots, x_n)$, $x_i \in \mathbb{F}_q$, $1 \le i \le n$, over $\mathbb{F}_q$. Let P be an m-dimensional subspace of $\mathbb{F}_q^n$ and let $v_1, v_2, \ldots, v_m$ be a basis of P. Then the $m \times n$ matrix

$$\begin{pmatrix} v_1 \\ v_2 \\ \vdots \\ v_m \end{pmatrix}$$

is a *matrix representation* of P and is denoted by P also. Two matrix representations of P differ from an $m \times m$ nonsingular matrix multiplied on the left. Every element $A \in GL_n(q)$ acts on $\mathbb{F}_q^n$ in the following way:

$$(x_1, x_2, \ldots, x_n) \mapsto (x_1, x_2, \ldots, x_n)A,$$

which induces a transformation on the set of subspaces of $\mathbb{F}_q^n$:

$$P \mapsto PA.$$

The subset of subspaces of the same dimension forms an *orbit.* The cardinality of the orbit of m-dimensional subspaces ($0 \le m \le n$) is denoted by $N(m, n)$.

13.3.15 Theorem [2920] We have $N(m, n) = \begin{bmatrix} n \\ m \end{bmatrix}_q$, where $\begin{bmatrix} n \\ m \end{bmatrix}_q$ is the Gaussian coefficient

$$\begin{bmatrix} n \\ m \end{bmatrix}_q = \frac{(q^n - 1)(q^{n-1} - 1) \cdots (q^{n-m+1} - 1)}{(q^m - 1)(q^{m-1} - 1) \cdots (q - 1)}.$$

13.3.16 Remark For more details on Gaussian coefficients, see Section 13.2.

13.3.2 Symplectic groups over finite fields

13.3.17 Definition Let n be an integer > 0. An $n \times n$ matrix K over $\mathbb{F}_q$ is *alternate*, if ${}^tK = -K$ and the diagonal elements are 0. When q is even, alternate matrices are simply skew-symmetric matrices. Let $n = 2\nu$ be even and K be a $2\nu \times 2\nu$ nonsingular alternate matrix over $\mathbb{F}_q$. The set of $2\nu \times 2\nu$ matrices T satisfying $TK\,{}^tT = K$ forms a group with respect to matrix multiplication, the *symplectic group* of degree 2ν with respect to K over $\mathbb{F}_q$ and denoted by $Sp_{2\nu}(q, K)$.

Let $K = (k_{ij})$ be an $n \times n$ alternate matrix over $\mathbb{F}_q$. The bilinear form $K(x, y)$ defined by $K(x, y) = \sum k_{ij}x_iy_j$ is an alternate form, which is nonsingular if $\det K \neq 0$. Now Let $n = 2\nu$ be even and $K(x, y)$ be a nonsingular alternate form on a 2ν-dimensional vector space V over $\mathbb{F}_q$. Then $Sp_{2\nu}(q, K)$ can also be defined as the group of linear transformations T of V satisfying $K(xT, yT) = K(x, y)$ for all $x, y \in V$.

13.3.18 Theorem [857] All elements of $Sp_{2\nu}(q, K)$ are nonsingular matrices with determinant 1 and, hence, $Sp_{2\nu}(q, K)$ is a subgroup of $SL_{2\nu}(q)$. Moreover, $Sp_2(q, K) = SL_2(q)$.

13.3.19 Definition Two $n \times n$ matrices A and B over $\mathbb{F}_q$ are *cogredient*, if there is an $n \times n$ nonsingular matrix P over $\mathbb{F}_q$ such that $A = PB\,{}^tP$.

13.3.20 Theorem [857] Let K_1 and K_2 be two cogredient $2\nu \times 2\nu$ nonsingular alternate matrices, then $Sp_{2\nu}(q, K_1) \simeq Sp_{2\nu}(q, K_2)$.

13.3.21 Remark By the previous theorem it is sufficient to consider the symplectic group $Sp_{2\nu}(q, K_0)$ where

$$K_0 = \begin{pmatrix} 0 & I_\nu \\ -I_\nu & 0 \end{pmatrix}.$$

The group $Sp_{2\nu}(q, K_0)$ is simply the symplectic group of degree 2ν over $\mathbb{F}_q$ and denoted by $Sp_{2\nu}(q)$.

13.3.22 Theorem [1589] The order of $Sp_{2\nu}(q)$ is

$$|Sp_{2\nu}(q)| = q^{\nu^2} \prod_{i=1}^{\nu} (q^{2i} - 1).$$

13.3.23 Definition Let $T \in Sp_{2\nu}(q)$. If $I_{2\nu} - T$ is of rank 1, then T is a *symplectic transvection.*

13.3.24 Lemma Every symplectic transvection can be expressed in the form

$$T \begin{pmatrix} I_\nu & 0 \\ \operatorname{diag}(\lambda, 0, \ldots, 0) & I_\nu \end{pmatrix} T^{-1}$$

where $T \in Sp_{2\nu}(q)$, and $\operatorname{diag}(\lambda, 0, \ldots, 0)$ is a diagonal matrix with $\lambda, 0, \ldots, 0$ along the main diagonal and $\lambda \in \mathbb{F}_q^*$.

13.3.25 Theorem [1589] The group $Sp_{2\nu}(q)$ is generated by symplectic transvections.

13.3.26 Theorem [1589] The group $Sp_{2\nu}(q)$ is its own commutator subgroup in all cases except $\nu = 1$, $q = 2$ or 3 and $\nu = 2$, $q = 2$.

13.3.27 Remark The group $Sp_2(2) = SL_2(2) \simeq S_3$ and its commutator subgroup is A_3. The group $Sp_2(3) = SL_2(3)$ is of order 24 and is solvable.

13.3.28 Theorem [2786] The group $Sp_4(2) \cong S_6$.

13.3.29 Theorem [2786] For $\nu \geq 2$, the symmetric group $S_{2\nu+2}$ is a subgroup of $Sp_{2\nu}(2)$.

13.3.30 Theorem [1589] The center of $Sp_{2\nu}(q)$ consists of $I_{2\nu}$ and $-I_{2\nu}$.

13.3.31 **Definition** The factor group $Sp_{2\nu}(q)/\{\pm I_{2\nu}\}$ is the *projective symplectic group* of degree 2ν over $\mathbb{F}_q$ and denoted by $PSp_{2\nu}(\mathbb{F}_q)$.

13.3.32 **Theorem** (Dickson) [1589] The group $PSp_{2\nu}(q)$ is a simple group except for the cases $PSp_2(2)$, $PSp_2(3)$, and $PSp_4(2)$.

13.3.33 **Definition** Let P be an m-dimensional subspace of $\mathbb{F}_q^{2\nu}$. Then $PK_0{}^tP$ is an $m \times m$ alternate matrix. Suppose $PK_0{}^tP$ is cogredient to

$$\begin{pmatrix} 0 & I_s & 0 \\ -I_s & 0 & 0 \\ 0 & 0 & 0^{(2\nu-2s)} \end{pmatrix},$$

then P is a subspace of *type* (m, s). Clearly, $0 \le 2s \le m \le 2\nu$.

13.3.34 **Theorem** [2920] The action of $GL_{2\nu}(q)$ on the subspaces of $\mathbb{F}_q^{2\nu}$ induces an action of $Sp_{2\nu}(q)$ on the subspaces of $\mathbb{F}_q^{2\nu}$. Two subspaces P and Q are in the same orbit of $Sp_{2\nu}(q)$ if and only if they are of the same type.

13.3.35 **Theorem** [2920] Denote the cardinality of the orbit of subspaces of type (m, s) by $N(m, s; 2\nu)$. Then

$$N(m, s; 2\nu) = q^{2s(\nu+s-m)} \frac{\prod_{i=\nu+s-m+1}^{\nu}(q^{2\nu}-1)}{\prod_{i=1}^{s}(q^{2i}-1)\prod_{i=1}^{m-2s}(q^i-1)}.$$

13.3.3 Unitary groups over finite fields

13.3.36 **Definition** Let $\mathbb{F}_{q^2}$ be a finite field with q^2 elements, where q is a prime power. The Frobenius automorphism (refer to Remark 2.1.77)

$$a \mapsto a^q$$

of $\mathbb{F}_{q^2}$ is denoted by $-$, i.e., $\overline{a} = a^q$ for all $a \in \mathbb{F}_{q^2}$, and is the *involution* of $\mathbb{F}_{q^2}$.

13.3.37 **Definition** Let n be an integer > 1, and $H = (h_{ij})$ be an $n \times n$ matrix over $\mathbb{F}_{q^2}$. The matrix $(\overline{h}_{ij})$ is denoted by $\overline{H}$. If ${}^t\overline{H} = H$, H is a *Hermitian matrix*. Let H be an $n \times n$ nonsingular Hermitian matrix over $\mathbb{F}_{q^2}$. The set of $n \times n$ matrices T satisfying $TH\,{}^t\overline{T} = H$ forms a group with respect to matrix multiplication, the *unitary group* of degree n with respect to H over $\mathbb{F}_{q^2}$ and denoted by $U_n(q^2, H)$. The subgroup of $U_n(q^2, H)$ consisting of those $T \in U_n(q^2, H)$ with determinant 1 is the *special unitary group* and denoted by $SU_n(q^2, H)$.

Let $H = (h_{ij})$ be an $n \times n$ matrix over $\mathbb{F}_{q^2}$ and $H(x, y) = \sum h_{ij} x_i \overline{y}_j$ be the corresponding *Hermitian form*, which is nonsingular if $\det H \neq 0$. Let $H(x, y)$ be a nonsingular Hermitian form on an n-dimensional vector space V over $\mathbb{F}_{q^2}$. Then $U_n(q^2, H)$ can also be defined as the group of linear transformations T of V satisfying $H(xT, yT) = H(x, y)$ for $x, y \in V$.

13.3.38 **Theorem** [857] All elements of $U_n(q^2, H)$ are nonsingular matrices and, hence, $U_n(q^2, H)$ is a subgroup of $GL_n(q^2)$ and $SU_n(q^2, H)$ is a subgroup of $SL_n(q^2)$.

13.3.39 **Definition** Two $n \times n$ Hermitian matrices H_1 and H_2 over $\mathbb{F}_{q^2}$ are *cogredient* if there is an $n \times n$ nonsingular matrix P over $\mathbb{F}_{q^2}$ such that $H_1 = PH_2{}^t\overline{P}$.

13.3.40 **Theorem** [857] Let H_1 and H_2 be two cogredient nonsingular Hermitian matrices, then $U_n(q^2, H_1) \simeq U_n(q^2, H_2)$, and also $SU_n(q^2, H_1) \simeq SU_n(q^2, H_2)$.

13.3.41 **Remark** When n is even, write n as $n = 2\nu$; and when n is odd, write n as $n = 2\nu + 1$. By the above theorem, it is sufficient to consider the unitary groups $U_n(q^2, H_0)$, $SU_n(q^2, H_0)$ and $U_n(q^2, H_1)$, $SU_n(q^2, H_1)$, where

$$H_0 = \begin{pmatrix} 0 & I_\nu \\ I_\nu & 0 \end{pmatrix} \text{ and } H_1 = \begin{pmatrix} 0 & I_\nu & 0 \\ I_\nu & 0 & 0 \\ 0 & 0 & 1 \end{pmatrix}.$$

Here $U_n(q^2, H_0)$ and $U_n(q^2, H_1)$ are the unitary groups of degree n over $\mathbb{F}_{q^2}$, and are denoted by $U_{2\nu}(q^2)$ and $U_{2\nu+1}(q^2)$, respectively. Similarly, we have $SU_{2\nu}(q^2)$ and $SU_{2\nu+1}(q^2)$. We use the symbols $U_n(q^2)$ and $SU_n(q^2)$ to cover these two cases. In the literatures of group theory these groups are sometimes written as $U_n(q)$ instead of $U_n(q^2)$ and $SU_n(q)$ instead of $SU_n(q^2)$.

13.3.42 **Theorem** [2786] The orders of $U_n(q^2)$ and $SU_n(q^2)$ are, respectively,

$$|U_n(q^2)| = q^{\frac{1}{2}n(n-1)} \prod_{i=1}^{n} (q^i - (-1)^i),$$

and

$$|SU_n(q^2)| = q^{\frac{1}{2}n(n-1)} \prod_{i=2}^{n} (q^i - (-1)^i).$$

13.3.43 **Theorem** [857] We have $SU_2(q^2) \simeq SL_2(q)$.

13.3.44 **Lemma** [857] The center W_n of $U_n(q^2)$ is

$$W_n = U_n(q^2) \cap Z_n = \{aI_n : a\overline{a} = 1\}$$

and the center of $SU_n(q^2)$ is

$$SU_n(q^2) \cap Z_n = \{aI_n : a\overline{a} = 1 \text{ and } a^n = 1\}.$$

(Recall that Z_n is the subgroup of $n \times n$ nonsingular scalar matrices over $\mathbb{F}_{q^2}$, i.e., the center of $GL_n(q^2)$.)

13.3.45 **Definition** The factor group $U_n(q^2)/W_n$ is the *projective unitary group* of degree n over $\mathbb{F}_{q^2}$ and denoted by $PU_n(q^2)$. Similarly, the factor group $SU_n(q^2)/(SU_n(q^2) \cap Z_n)$ is the *projective special unitary group* of degree n over $\mathbb{F}_{q^2}$ and denoted by $PSU_n(q^2)$.

13.3.46 **Theorem** [2786] The order of $PU_n(q^2)$ and $PSU_n(q^2)$ are, respectively,

$$|PU_n(q^2)| = q^{\frac{1}{2}n(n-1)} \prod_{i=2}^{n} (q^i - (-1)^i)$$

and

$$|PSU_n(q^2)| = (\gcd(n, q+1))^{-1} q^{\frac{1}{2}n(n-1)} \prod_{i=2}^{n} (q^i - (-1)^i).$$

13.3.47 Definition Let $T \in U_n(q^2)$. If $I_n - T$ is of rank 1, then T is a *unitary transvection.*

13.3.48 Lemma [857] Every unitary transvection can be expressed in the form

$$T\begin{pmatrix} I_\nu & 0 \\ \operatorname{diag}(\lambda, 0, \ldots, 0) & I_\nu \end{pmatrix} T^{-1} \text{ or } T\begin{pmatrix} I_\nu & 0 & 0 \\ \operatorname{diag}(\lambda, 0, \ldots, 0) & I_\nu & 0 \\ 0 & 0 & 1 \end{pmatrix} T^{-1}$$

corresponding to $n = 2\nu$ or $n = 2\nu+1$, respectively, where $T \in U_n(q^2)$ and $\lambda \in \mathbb{F}_q^*$ satisfying $\lambda + \overline{\lambda} = 0$.

13.3.49 Theorem [857] For $n \geq 2$, the group $SU_n(q^2)$ is generated by unitary transvections except $SU_3(2^2)$.

13.3.50 Theorem (Dickson) [857] For $n \geq 2$, the group $PSU_n(q^2)$ is simple except for $PSU_2(2^2)$, $PSU_2(3^2)$ and $PSU_3(2^2)$.

13.3.51 Remark [857, 858] For the exceptional cases in Theorem 13.3.50, the group

$$PSU_2(2^2) \simeq PSL_2(2) \simeq SL_2(2) \simeq S_3,$$

the group

$$PSU_2(3^2) \simeq PSL_2(3) \simeq A_4,$$

and the group $PSU_3(2^2)$ is of order $2^3 \cdot 3^2 = 72$ and is solvable.

13.3.52 Theorem [858] We have $PSU_4(2^2) \simeq PSp_4(3)$.

13.3.53 Definition Let P be an m-dimensional subspace of $\mathbb{F}_{q^2}^n$ and $H = H_0$ or H_1 according to $n = 2\nu$ or $n = 2\nu + 1$, respectively. The matrix $PH\,{}^t\overline{P}$ is an $m \times m$ Hermitian matrix. Suppose $PH\,{}^t\overline{P}$ is of rank r, then P is a *subspace of type* (m, r). Clearly, $0 < r \leq m \leq n$.

13.3.54 Theorem [2920] The action of $GL_n(q^2)$ on the subspaces of $\mathbb{F}_{q^2}^n$ induces an action of $U_n(q^2)$ on the subspaces of $\mathbb{F}_{q^2}^n$. Two subspaces P and Q are in the same orbit of $U_n(q^2)$ if and only if they are of the same type.

13.3.55 Theorem [2920] Denote the cardinality of the orbit of subspaces of type (m, r) by $N(m, r; n)$. Then

$$N(m, r; n) = q^{r(n+r-2m)} \frac{\prod_{i=n+r-2m+1}^{n} (q^i - (-1)^i)}{\prod_{i=1}^{r} (q^i - (-1)^i) \prod_{i=1}^{m-r} (q^{2i} - 1)}.$$

13.3.4 Orthogonal groups over finite fields of characteristic not two

13.3.56 Definition Let n be an integer > 1, $\mathbb{F}_q$ be a finite field with q elements where q is an odd prime power; and S be an $n \times n$ nonsingular symmetric matrix over $\mathbb{F}_q$. The set of $n \times n$ matrices T satisfying $TS\,{}^tT = S$ forms a group with respect to matrix multiplication, the *orthogonal group* of degree n with respect to S over $\mathbb{F}_q$ and denoted by $O_n(q, S)$.

Let $S = (s_{ij})$ be an $n \times n$ symmetric matrix over $\mathbb{F}_q$ and $Q(x) = \sum s_{ij} x_i x_j$ be the corresponding quadratic form, which is nonsingular if S is nonsingular. Let $Q(x)$ be a nonsingular quadratic form on an n-dimensional vector space V over $\mathbb{F}_q$. Then $O_n(q, S)$ can also be defined as the group of linear transformations T of V such that $Q(xT) = Q(x)$ for all $x \in V$.

13.3.57 Theorem [1589] All elements of $O_n(q, S)$ are nonsingular matrices of determinant ± 1 and, hence, $O_n(q, S)$ is a subgroup of $GL_n(q)$.

13.3.58 Theorem [1589] Let S_1 and S_2 be two cogredient $n \times n$ nonsingular symmetric matrices, then $O_n(q, S_1) \simeq O_n(q, S_2)$. Moreover, for any $n \times n$ nonsingular symmetric matrix S over $\mathbb{F}_q$ and any $\lambda \in \mathbb{F}_q^*$, $O_n(q, S) = O_n(q, \lambda S)$.

13.3.59 Remark Choose a fixed non-square element z of $\mathbb{F}_q^*$. By the previous theorem it is sufficient to consider the four orthogonal groups with respect to the following four $n \times n$ nonsingular symmetric matrices

$$S_{2\nu} = \begin{pmatrix} 0 & I_\nu \\ I_\nu & 0 \end{pmatrix},$$

$$S_{2\nu+1,1} = \begin{pmatrix} 0 & I_\nu & 0 \\ I_\nu & 0 & 0 \\ 0 & 0 & 1 \end{pmatrix}, \quad S_{2\nu+1,z} = \begin{pmatrix} 0 & I_\nu & 0 \\ I_\nu & 0 & 0 \\ 0 & 0 & z \end{pmatrix}$$

$$S_{2\nu+2} = \begin{pmatrix} 0 & I_\nu & 0 & 0 \\ I_\nu & 0 & 0 & 0 \\ 0 & 0 & 1 & 0 \\ 0 & 0 & 0 & -z \end{pmatrix},$$

where $n = 2\nu, 2\nu+1, 2\nu+1$ and $2\nu+2$, respectively. In order to cover these four cases, we introduce the notation $S_{2\nu+\delta,\Delta}$, where $\delta = 0, 1$ or 2 and Δ denotes its definite part, i.e.,

$$\Delta = \begin{cases} \phi & \text{if } \delta = 0, \\ 1 \text{ or } z & \text{if } \delta = 1, \\ \begin{pmatrix} 1 & 0 \\ 0 & -z \end{pmatrix} & \text{if } \delta = 2. \end{cases}$$

The orthogonal group of degree $2\nu + \delta$ with respect to $S_{2\nu+\delta,\Delta}$ over $\mathbb{F}_q$ will be denoted by $O_{2\nu+\delta}(q, S_{2\nu+\delta,\Delta})$. Clearly, $O_{2\nu+1}(q, S_{2\nu+1,z}) = O_{2\nu+1}(q, zS_{2\nu+1,z})$. Since $zS_{2\nu+1,z}$ and $S_{2\nu+1,1}$ are cogredient, the groups $O_{2\nu+1}(q, zS_{2\nu+1,z})$ and $O_{2\nu+1}(q, S_{2\nu+1,1})$ are isomorphic. It follows that $O_{2\nu+1}(q, S_{2\nu+1,z})$ and $O_{2\nu+1}(q, S_{2\nu+1,1})$ are isomorphic. Therefore actually only three types of orthogonal groups need to be considered; they are $O_{2\nu}(q, S_{2\nu}), O_{2\nu+1}(q, S_{2\nu+1,1})$, and $O_{2\nu+2}(q, S_{2\nu+2,\Delta})$, which simply denoted by $O_{2\nu}(q), O_{2\nu+1}(q)$ and $O_{2\nu+2}(q)$, respectively. We use $O_{2\nu+\delta}(q)$ to cover these three cases.

13.3.60 Remark For odd n, there is only one type of orthogonal groups, and for even $n = 2\nu$ there are two types of orthogonal groups: $O_{2\nu}(q)$ and $O_{2(\nu-1)+2}(q)$. In literatures of group theory they are sometimes called the plus type and the minus type, and denoted by $O_n^+(q)$ and $O_n^-(q)$, respectively.

13.3.61 Remark Throughout the remainder of this section we assume $\nu \geq 1$.

13.3.62 Theorem [1589] The order of $O_{2\nu+\delta}(q)$ is

$$|O_{2\nu+\delta}(q)| = q^{\nu(\nu+\delta-1)} \prod_{i=1}^{\nu} (q^i - 1) \prod_{i=0}^{\nu+\delta-1} (q^i + 1).$$

13.3.63 Definition Let $T \in O_{2\nu+\delta}(q)$. If $T^2 = I$ and $T - I$ is of rank 1, T is a *symmetry.*

13.3.64 Theorem [857] Every element of $O_{2\nu+\delta}(q)$ is a product of at most $2\nu + \delta$ symmetries.

13.3.65 Definition Elements of $O_{2\nu+\delta}(q)$ are *orthogonal matrices* and those of determinant 1 are *proper orthogonal matrices*. All proper orthogonal matrices form a subgroup of $O_{2\nu+\delta}(q)$, the *proper orthogonal group* also referred to as the *special orthogonal group*, and denoted by $SO_{2\nu+\delta}(q)$. The commutator subgroup of $O_{2\nu+\delta}(q)$ is denoted by $\Omega_{2\nu+\delta}(q)$.

13.3.66 Lemma The center of $O_{2\nu+\delta}(q)$ is $\{I_{2\nu+\delta}, -I_{2\nu+\delta}\}$.

13.3.67 Definition The factor group $O_{2\nu+\delta}(q)/\{\pm I_{2\nu+\delta}\}$ over $\mathbb{F}_q$ is the *projective orthogonal group* of degree $2\nu+\delta$ with respect to $S_{2\nu+\delta}$ over $\mathbb{F}_q$ and is denoted by $PO_{2\nu+\delta}(q)$. Similarly, the factor group $SO_{2\nu+\delta}(q)/(SO_{2\nu+\delta}(q)\cap\{\pm I_{2\nu+\delta}\})$ is the *projective proper orthogonal group* of degree $2\nu+\delta$ with respect to $S_{2\nu+\delta}$ over $\mathbb{F}_q$ and is denoted by $PSO_{2\nu+\delta}(q)$. We also define $P\Omega_{2\nu+\delta}(q) = \Omega_{2\nu+\delta}(q)/(\Omega_{2\nu+\delta}(q)\cap\{\pm I_{2\nu+\delta}\})$.

13.3.68 Remark We have the normal series of $O_{2\nu+\delta}(q)$

$$O_{2\nu+\delta}(q) \supset SO_{2\nu+\delta}(q) \supset \Omega_{2\nu+\delta}(q) \supset \Omega_{2\nu+\delta}(q)\cap\{\pm I_{2\nu+\delta}\} \supset \{I\}.$$

Clearly, $O_{2\nu+\delta}(q) : SO_{2\nu+\delta}(q) = 2$.

13.3.69 Theorem [857] We have $SO_{2\nu+\delta}(q)/\Omega_{2\nu+\delta}(q) \simeq \mathbb{F}_q^*/\mathbb{F}_q^{*2}$.

13.3.70 Theorem (Dickson) [1589] The group $P\Omega_{2\nu+\delta}(q)$ is a simple group except the following cases:

1. $\nu = 2, \delta = 0$,
2. $\nu = 1, \delta = 1$ and $q = 3$.

13.3.71 Remark For the exceptional cases in the above Theorem, we have

$$P\Omega_{2\cdot 2}(q) \simeq PSL_2(q) \times PSL_2(q),$$

$$P\Omega_{2\cdot 1+1}(3) \simeq PSL_2(3).$$

13.3.72 Theorem [2786] When $\nu \geq 2$, $P\Omega_{2\nu+1}(q)$ $(= \Omega_{2\nu+1}(q))$ and $PSp_{2\nu}(q)$ are non-isomorphic simple groups of the same order.

13.3.73 Remark The action of $GL_{2\nu+\delta}(q)$ on the subspaces of $\mathbb{F}_q^{2\nu+\delta}$ induces an action of $O_{2\nu+\delta}(q)$ on the subspaces. Let P be an m-dimensional subspace of $\mathbb{F}_q^{2\nu+\delta}$. $PS_{2\nu+\delta}\,{}^tP$ is cogredient to one of the following normal forms

$$M(m, 2s, s) = \begin{pmatrix} 0 & I_s & 0 \\ I_s & 0 & 0 \\ 0 & 0 & 0^{(m-2s)} \end{pmatrix},$$

$$M(m, 2s+1, s, 1) = \begin{pmatrix} 0 & I_s & 0 & 0 \\ I_s & 0 & 0 & 0 \\ 0 & 0 & 1 & 0 \\ 0 & 0 & 0 & 0^{(m-2s-1)} \end{pmatrix},$$

$$M(m, 2s+1, s, z) = \begin{pmatrix} 0 & I_s & 0 & 0 \\ I_s & 0 & 0 & 0 \\ 0 & 0 & z & 0 \\ 0 & 0 & 0 & 0^{(m-2s-1)} \end{pmatrix},$$

and

$$M(m, 2s+2, s) = \begin{pmatrix} 0 & I_s & 0 & 0 & 0 \\ I_s & 0 & 0 & 0 & 0 \\ 0 & 0 & 1 & 0 & 0 \\ 0 & 0 & 0 & -z & 0 \\ \\ 0 & 0 & 0 & 0 & 0^{(m-2s-2)} \end{pmatrix}.$$

Then P is a *subspace of type* $(m, 2s, s), (m, 2s+1, s, 1), (m, 2s+1, s, z)$, and $(m, 2s+2, s, \mathrm{diag}(1, -z))$, respectively.

13.3.74 Theorem [2920] Two subspaces P and Q of $\mathbb{F}_q^{2\nu+\delta}$ belong to the same orbit of $O_{2\nu+\delta}(q)$ if and only if they are of the same type.

13.3.75 Remark The cardinality of any orbit of $O_{2\nu+\delta}(q)$ has already been determined [2920].

13.3.5 Orthogonal groups over finite fields of characteristic two

13.3.76 Definition Let q be a power of 2 and n be an integer > 1. Let G be an $n \times n$ regular matrix over $\mathbb{F}_q$. An $n \times n$ matrix T over $\mathbb{F}_q$ is *orthogonal* with respect to G, if $TG\,{}^tT + G$ is an alternate matrix. The set of $n \times n$ orthogonal matrices with respect to G over $\mathbb{F}_q$ forms a group with respect to matrix multiplication, the *orthogonal group* of degree n with respect to G over $\mathbb{F}_q$ and denoted by $O_n(q, G)$.

Let $G(x)$ be a nonsingular quadratic form on an n-dimensional vector space V over $\mathbb{F}_q$. Then $O_n(q, G)$ can also be defined as the group of linear transformations T of V such that $G(vT) = G(v)$ for all $v \in V$.

13.3.77 Theorem [857] All elements of $O_n(q, G)$ are nonsingular matrices of determinant 1 and, hence, $O_n(q, G)$ is a subgroup of $SL_n(q)$.

13.3.78 Theorem [857] Let G_1 and G_2 be two cogredient $n \times n$ regular matrices, then $O(q, G_1) \simeq O(q, G_2)$.

13.3.79 Remark Choose a fixed element α such that α cannot be expressed in the form $x^2 + x$ where $x \in \mathbb{F}_q$. Write $n = 2\nu + \delta$ where $\delta = 0, 1$ or 2. Assume $\nu > 0$. By the previous theorem it is sufficient to consider the orthogonal groups $O_{2\nu+\delta}(q, G)$, where G is an $n \times n$ matrix of one of the following forms

$$\begin{pmatrix} 0 & I_\nu \\ 0 & 0 \end{pmatrix}, \begin{pmatrix} 0 & I_\nu & 0 \\ 0 & 0 & 0 \\ 0 & 0 & 1 \end{pmatrix}, \begin{pmatrix} 0 & I_\nu & 0 & 0 \\ 0 & 0 & 0 & 0 \\ 0 & 0 & \alpha & 1 \\ 0 & 0 & 0 & \alpha \end{pmatrix},$$

corresponding to the cases $n = 2\nu, n = 2\nu + 1$ and $n = 2\nu + 2$, respectively. The orthogonal group with respect to G over $\mathbb{F}_q$ will be denoted by $O_{2\nu+\delta}(q)$.

13.3.80 Theorem [857] We have $O_{2\nu+1}(q) \simeq Sp_{2\nu}(q)$.

13.3.81 Remark In the following we consider only the groups $O_{2\nu}(q)$ and $O_{2\nu+2}(q)$. In the literature on group theory, $O_{2\nu}(q)$ and $O_{2\nu+2}(q)$ are sometimes denoted by $O_{2\nu}^+(q)$ and $O_{2\nu+2}^-(q)$, and called the plus type and the minus type, respectively.

13.3.82 Theorem [2786] The order of $O_{2\nu+\delta}(q)$, where $\delta = 0$ or 2, is

$$|O_{2\nu+\delta}(q)| = q^{\nu(\nu+\delta-1)} \prod_{i=1}^{\nu}(q^i - 1) \prod_{i=0}^{\nu+\delta-1}(q^i + 1).$$

13.3.83 Definition Let $T \in O_{2\nu+\delta}(q)$. If $I_{2\nu+\delta} - T$ is of rank 1, T is an *orthogonal transvection.*

13.3.84 Theorem [857] Every element of $O_{2\nu+\delta}(q)$ is a product of at most $2\nu+\delta$ orthogonal transvections except for the case $n = 4, \nu = 2$ and $q = 2$; if an element of $O_{2\nu+\delta}(q)$ is expressed as a product of an even number of orthogonal transvections, so is every such expression.

13.3.85 Definition Except for the case $n = 4, \nu = 2$ and $q = 2$, an orthogonal matrix $T \in O_{2\nu+\delta}(q)$ which is a product of an even number of orthogonal transvections is a *rotation*. The set of rotations forms a subgroup of $O_{2\nu+\delta}(q)$, the group of rotations and denoted by $SO_{2\nu+\delta}(q)$. The commutator subgroup of $O_{2\nu+\delta}(q)$ is denoted by $\Omega_{2\nu+\delta}(q)$.

13.3.86 Lemma For $n \geq 4$ the center of $\Omega_{2\nu+\delta}(q)$ consists of the identity element only.

13.3.87 Remark Except for the case $n = 4, \nu = 2$, and $q = 2$, we have the normal series of $O_{2\nu+\delta}(q)$

$$O_{2\nu+\delta}(q) \supset SO_{2\nu+\delta}(q) \supset \Omega_{2\nu+\delta}(q) \supset \{I_{2\nu+\delta}\}.$$

13.3.88 Theorem [857, 858] Except for the case $n = 4, \nu = 2$ and $q = 2$,

$$|O_{2\nu+\delta}(q)/SO_{2\nu+\delta}(q)| = |SO_{2\nu+\delta}(q)/\Omega_{2\nu+\delta}(q)| = 2.$$

13.3.89 Theorem (Dickson) [857] The group $\Omega_{2\nu+\delta}(q)$ is a simple group except for the case $n = 4,\ \nu = 2$.

13.3.90 Theorem [858] If $q \neq 2$ and $n = 4, \nu = 2$, then

$$\Omega_{2\cdot 2}(q) \simeq SL_2(q) \times SL_2(q).$$

If $q = 2$ and $n = 4, \nu = 2$, then $SO_{2\cdot 2}(2) = SL_2(2) \times SL_2(2)$ and $\Omega_{2\cdot 2}(2)$ is a direct product of two cyclic groups of order 3.

13.3.91 Remark The action of $GL_{2\nu+\delta}(q)$ on the subspaces of $\mathbb{F}_q^{2\nu+\delta}$ induces an action of $O_{2\nu+\delta}(q)$ on the subspaces of $\mathbb{F}_q^{2\nu+\delta}$. The cardinality of any orbit of $O_{2\nu+\delta}(q)$ has already been determined [2920].

See Also

§7.2	For a discussion of quadratic forms.
§13.2	For a discussion of orthogonal, symmetric, and skew-symmetric matrices.

References Cited: [135, 857, 858, 1589, 2786, 2920]

13.4 Computational linear algebra over finite fields

Jean-Guillaume Dumas, Université de Grenoble
Clément Pernet, Université de Grenoble

We present algorithms for efficient computation of linear algebra problems over finite fields. Implementations* of the proposed algorithms are available through the MAGMA, MAPLE (within the `LinearAlgebra[Modular]` subpackage) and SAGE systems; some parts can also be found within the C/C++ libraries NTL, FLINT, IML, M4RI, and the special purpose LINBOX template library for exact, high-performance linear algebra computation with dense, sparse, and structured matrices over the integers and over finite fields [931].

13.4.1 Dense matrix multiplication

13.4.1 **Definition** For $A \in \mathbb{F}_q^{m\times k}$ and $B \in \mathbb{F}_q^{k\times n}$ with elements $A_{i,j}$ and $B_{i,j}$, the matrix $C = A\times B$ has $C_{i,j} = \sum_{l=1}^{k} A_{i,l}B_{l,j}$. We denote by $\mathtt{MM}(m,k,n)$ a time complexity bound on the number of field operations necessary to compute C.

13.4.2 **Remark** Classical triple loop implementation of matrix multiplication makes $\mathtt{MM}(m,k,n) \leq 2mkn$. The best published estimates to date gives $\mathtt{MM}(n,n,n) \leq \mathcal{O}(n^{\omega})$ with $\omega \approx 2.3755$ [723], though improvements to 2.3737 and 2.3727 are now claimed [2728, 2985]. For very rectangular matrices one also have astonishing results like $\mathtt{MM}(n,n,n^{\alpha}) \leq \mathcal{O}\left(n^{2+\epsilon}\right)$ for a constant $\alpha > 0.294$ and any $\epsilon > 0$ [720]. Nowadays practical implementations mostly use Strassen-Winograd's algorithm, see Subsection 13.4.1.4, with an intermediate complexity and $\omega \approx 2.8074$.

13.4.1.1 Tiny finite fields

13.4.3 **Remark** The practical efficiency of matrix multiplication depends highly on the representation of field elements. We thus present three kinds of compact representations for elements of a finite field with very small cardinality: bitpacking (for $\mathbb{F}_2$), bit-slicing (for say $\mathbb{F}_3, \mathbb{F}_5, \mathbb{F}_7, \mathbb{F}_{2^3}$, or $\mathbb{F}_{3^2}$), and Kronecker substitution. These representations are designed to allow efficient linear algebra operations, including matrix multiplication.

13.4.4 **Algorithm** (Greasing) Over $\mathbb{F}_2$, the method of *the four Russians* [124], also called *Greasing*, can be used as follows:

1. A 64 bit machine word can be used to represent a row vector of dimension 64.
2. Matrix multiplication of a $m \times k$ matrix A by a $k \times n$ matrix B can be done by first storing all 2^k k-dimensional linear combinations of rows of B in a table. Then the i-th row of the product is copied from the row of the table indexed by the i-th row of A.
3. By ordering indices of the table according to a binary Gray Code, each row of the table can be deduced from the previous one, using only one row addition. This brings the bit operation count to build the table from $k2^kn$ to 2^kn.
4. Choosing $k = \log_2 n$ in the above method implies $\mathtt{MM}(n) = \mathcal{O}\left(n^3/\log n\right)$ over $\mathbb{F}_2$.

*`http://magma.maths.usyd.edu.au`, `http://www.maplesoft.com`, `http://sagemath.org`, `http://www.shoup.net/ntl`, `http://www.flintlib.org`, `http://www.cs.uwaterloo.ca/~astorjoh/iml.html`, `http://m4ri.sagemath.org`, `http://linalg.org`

13.4.5 Definition [349] *Bit slicing* consists in representing an n-dimensional vector of k-bit sized coefficients using k binary vectors of dimension n. In particular, one can use Boolean word instruction to perform arithmetic on 64 dimensional vectors.

1. Over $\mathbb{F}_3$, the binary representation $0 \equiv [0, 0], 1 \equiv [1, 0], -1 \equiv [11]$ allows to add and subtract two elements in 6 Boolean operations:

$$\begin{aligned} \text{Add}([x_0, x_1], [y_0, y_1]) : \quad & s \leftarrow x_0 \oplus y_1, t \leftarrow x_1 \oplus y_0 \\ & \text{Return}(s \wedge t, (s \oplus x_1) \vee (t \oplus y_1)) \\ \text{Sub}([x_0, x_1], [y_0, y_1]) : \quad & t \leftarrow x_0 \oplus y_0 \\ & \text{Return}(t \vee (x_1 \oplus y_1), (t \oplus y_1) \wedge (y_0 \oplus x_1)) \end{aligned}$$

2. Over $\mathbb{F}_5$ (resp. $\mathbb{F}_7$), a redundant representation $x = x_0 + 2x_1 + 4x_2 \equiv [x_0, x_1, x_2]$ allows to add two elements in 20 (resp. 17) Boolean operations, negate in 3 (resp. 6) Boolean operations, and double in 0 (resp. 5) Boolean operations.

	$\mathbb{F}_3$	$\mathbb{F}_5$	$\mathbb{F}_7$
Addition	6	20	17
Negation	1	5	3
Double		5	0

Table 13.4.1 Boolean operation counts for basic arithmetic using bit slicing

13.4.6 Definition *Bit packing* consists in representing a vector of field elements as an integer fitting in a single machine word using a 2^k-adic representation:

$$(x_0, \ldots, x_{n-1}) \in \mathbb{F}_q^n \equiv X = x_0 + 2^k x_1 + \cdots + (2^k)^{n-1} x_{n-1} \in \mathbb{Z}_{2^{64}}.$$

Elements of extension fields are viewed as polynomials and stored as the evaluation of this polynomial at the characteristic of the field. The latter evaluation is known as *Kronecker substitution.*

13.4.7 Remark We first need a way to simultaneously reduce coefficients modulo the characteristic, see [929].

13.4.8 Algorithm (REDQ: Q-adic REDuction)

Require: three integers p, q and $\tilde{r} = \sum_{i=0}^{d} \widetilde{\mu_i} q^i \in \mathbb{Z}$
Ensure: $\rho \in \mathbb{Z}$, with $\rho = \sum_{i=0}^{d} \mu_i q^i$ where $\mu_i = \widetilde{\mu_i} \bmod p$

<u>REDQ COMPRESSION</u>

$s = \left\lfloor \frac{\tilde{r}}{p} \right\rfloor$;
for $i = 0$ to d **do**
  $u_i = \left\lfloor \frac{\tilde{r}}{q^i} \right\rfloor - p \left\lfloor \frac{s}{q^i} \right\rfloor$;
end for

<u>REDQ CORRECTION</u> {only when $p \nmid q$, otherwise $\mu_i = u_i$ is correct}

$\mu_d = u_d$;
for $i = 0$ to $d - 1$ **do**
  $\mu_i = u_i - q u_{i+1} \bmod p$;
end for

9. Return $\rho = \sum_{i=0}^{d} \mu_i q^i$;

13.4.9 Remark Once we can pack and simultaneously reduce coefficients of finite field in a single machine word, the obtained parallelism can be used for matrix multiplication. Depending on the respective sizes of the matrix in the multiplication one can pack only the left operand or only the right one or both [930]. We give here only a generic algorithm for packed matrices, which use multiplication of a right packed matrix by a non-packed left matrix.

13.4.10 Algorithm (Right packed matrix multiplication)

Require: a prime p and $A_c \in \mathbb{F}_p^{m\times k}$ and $B_c \in \mathbb{F}_p^{k\times n}$, stored with several field elements per machine word

Ensure: $C_c = A_c \times B_c \in \mathbb{F}_p^{m\times n}$

1. $A = \text{Uncompress}(A_c)$; {extract the coefficients}
2. $C_c = A \times B_c$; {Using e.g., Algorithm 13.4.14}
3. Return REDQ(C_c);

13.4.11 Remark Then, over extensions, fast floating point operations can be used on the Kronecker substitution of the elements. Indeed, it is very often desirable to use floating point arithmetic, *exactly*. For instance floating point routines can more easily use large hardware registers, they can more easily optimize the memory hierarchy usage [1336, 2974] and portable implementations are more widely available. We present next the dot product and the matrix multiplication is then straightforward [929, 930, 932].

13.4.12 Algorithm (Compressed dot product over extension fields)

Require: a field $\mathbb{F}_{p^k}$ with elements represented as exponents of a generator of the field

Require: two vectors v_1 and v_2 of elements of $\mathbb{F}_{p^k}$

Require: a sufficiently large integer q

Ensure: $R \in \mathbb{F}_{p^k}$, with $R = v_1^T \cdot v_2$

{Tabulated conversion: uses tables from exponent to floating point evaluation}

1. Set $\widetilde{v_1}$ and $\widetilde{v_2}$ to the floating point Kronecker substitution of the elements of v_1 and v_2.
2. Compute $\tilde{r} = \widetilde{v_1}^T \cdot \widetilde{v_2}$; {The floating point computation}
3. $r = \mathit{REDQ\_COMPRESSION}(\tilde{r}, p, q)$; {Computing a radix decomposition}
   {Variant of REDQ_CORRECTION: $\mu_i = \widetilde{\mu_i} \bmod p$ for $\tilde{r} = \sum_{i=0}^{2k-2} \widetilde{\mu_i} q^i$}
4. Set $L = \mathit{representation}(\sum_{i=0}^{k-2} \mu_i X^i)$;
5. Set $H = \mathit{representation}(X^{k-1} \times \sum_{i=k-1}^{2k-2} \mu_i X^{i-k+1})$;
6. Return $R = H + L \in \mathbb{F}_{p^k}$; {Reduction in the field}

13.4.1.2 Word size prime fields

13.4.13 Remark Over word-size prime fields one can also use the reduction to floating point routines of algorithm 13.4.12. The main point is to be able to perform efficiently the matrix multiplication of blocks of the initial matrices without modular reduction. Thus delaying the reduction as much as possible, depending on the algorithm and internal representations, in order to amortize its cost. We present next such a delaying with the classical matrix multiplication algorithm and a centered representation [933].

13.4.14 Algorithm (`fgemm`: Finite Field GEneric Matrix Multiplication)

Require: an odd prime p of size smaller than the floating point mantissa β and F_p elements stored by values between $\frac{1-p}{2}$ and $\frac{p-1}{2}$

Require: $A \in \mathbb{F}_p^{m\times k}$ and $B \in \mathbb{F}_p^{k\times n}$

Ensure: $C = A \times B \in \mathbb{F}_p^{m\times n}$

if $n(p-1)^2 < 2^{\beta+1}$ **then**
    Convert A and B to floating point matrices A_f and B_f;
    Use floating point routines to compute $C_f = A_f \times B_f$;
    $C = C_f \mod p$;
else
    Cut A and B into smaller blocks;
    Call the algorithm recursively for the block multiplications;
    Perform the block additions modulo p;
end if

13.4.1.3 Large finite fields

13.4.15 Remark If the field is too large for the strategy 13.4.14 over machine words, then two main approaches would have to be considered:

1. Use extended arithmetic, either arbitrary of fixed precision, if the characteristic is large, and a polynomial representation for extension fields. The difficulty here is to preserve an optimized memory management and to have an almost linear time extended precision polynomial arithmetic.
2. Use a residue number system and an evaluation/interpolation scheme: one can use Algorithm 13.4.14 for each prime in the residue number system (RNS) and each evaluation point. For $\mathbb{F}_{p^k}$, the number of needed primes is roughly $2\log_{2^\beta}(p)$ and the number of evaluations points is $2k-1$.

13.4.1.4 Large matrices: subcubic time complexity

13.4.16 Remark With matrices of large dimension, sub-cubic time complexity algorithms, such as Strassen-Winograd's [2987] can be used to decrease the number of operations. Algorithm 13.4.17 describes how to compute one recursive level of the algorithm, using seven recursive calls and 15 block additions.

13.4.17 Algorithm (Strassen-Winograd)

$$A = \begin{bmatrix} A_{11} & A_{12} \\ A_{21} & A_{22} \end{bmatrix}; B = \begin{bmatrix} B_{11} & B_{12} \\ B_{21} & B_{22} \end{bmatrix}; C = \begin{bmatrix} C_{11} & C_{12} \\ C_{21} & C_{22} \end{bmatrix};$$

$$\begin{array}{llll} S_1 \leftarrow A_{21} + A_{22}; & T_1 \leftarrow B_{12} - B_{11}; & P_1 \leftarrow A_{11} \times B_{11}; & P_2 \leftarrow A_{12} \times B_{21}; \\ S_2 \leftarrow S_1 - A_{11}; & T_2 \leftarrow B_{22} - T_1; & P_3 \leftarrow S_4 \times B_{22}; & P_4 \leftarrow A_{22} \times T_4; \\ S_3 \leftarrow A_{11} - A_{21}; & T_3 \leftarrow B_{22} - B_{12}; & P_5 \leftarrow S_1 \times T_1; & P_6 \leftarrow S_2 \times T_2; \\ S_4 \leftarrow A_{12} - S_2; & T_4 \leftarrow T_2 - B_{21}; & P_7 \leftarrow S_3 \times T_3; & \end{array}$$

$$\begin{array}{llll} C_{11} \leftarrow P_1 + P_2; & U_2 \leftarrow P_1 + P_6; & U_3 \leftarrow U_2 + P_7; & U_4 \leftarrow U_2 + P_5; \\ C_{12} \leftarrow U_4 + P_3; & C_{21} \leftarrow U_3 - P_4; & C_{22} \leftarrow U_3 + P_5; & \end{array}$$

13.4.18 Remark In practice, one uses a threshold in the matrix dimension to switch to a base case algorithm, that can be any of the previously described ones. Following Subsection 13.4.1.2, one can again delay the modular reductions, but the intermediate computations of Strassen-Winograd's algorithm impose a tighter bound.

13.4.19 Theorem [933] Let $A \in \mathbb{Z}^{m \times k}$, $B \in \mathbb{Z}^{k \times n}$ $C \in \mathbb{Z}^{m \times n}$ and $\beta \in \mathbb{Z}$ with $a_{i,j}, b_{i,j}, c_{i,j}, \beta \in \{0, \ldots, p-1\}$. Then every intermediate value z involved in the computation of $A \times B + \beta C$

with l ($l \geq 1$) recursive levels of Algorithm 13.4.17 satisfy

$$|z| \leq \left(\frac{1+3^l}{2}\right)^2 \left\lfloor \frac{k}{2^l} \right\rfloor (p-1)^2.$$

Moreover, this bound is tight.

13.4.20 Remark For instance, on a single Xeon 2.8GHz core with gcc-4.6.3, Strassen-Winograd's variant implemented with LinBox-1.2.1 and GotoBLAS2-1.13 can be 37% faster for the multiplication of $10\,000 \times 10\,000$ matrices over $\mathbb{F}_{2^{19}-1}$, in less than 1′49".

13.4.2 Dense Gaussian elimination and echelon forms

13.4.21 Remark We present algorithms computing the determinant and inverse of square matrices; the rank, rank profile, nullspace, and system solving for arbitrary shape and rank matrices. All these problems are solved a la Gaussian elimination, but recursively in order to effectively incorporate matrix multiplication. The latter is denoted generically `gemm` and, depending on the underlying field, can be implemented using any of the techniques of Subsections 13.4.1.1, 13.4.1.2, or 13.4.1.3.

13.4.22 Remark A special care is given to the asymptotic time complexities: the exponent is reduced to that of matrix multiplication using block recursive algorithms, and the constants are also carefully compared. Meanwhile, this approach is also effective for implementations: grouping arithmetic operations into matrix-matrix products allow to better optimize cache accesses.

13.4.2.1 Building blocks

13.4.23 Remark Algorithms 13.4.24, 13.4.25, 13.4.26, and 13.4.27 show how to reduce the computation of triangular matrix systems, triangular matrix multiplications, and triangular matrix inversions to matrix-matrix multiplication. Note that they do not require any temporary storage other than the input and output arguments.

13.4.24 Algorithm [`trsm`: Triangular System Solve with Matrix right hand side)
Require: $A \in \mathbb{F}_q^{m\times m}$ non-singular upper triangular, $B \in \mathbb{F}_q^{m\times n}$
Ensure: $X \in \mathbb{F}_q^{m\times n}$ s.t. $AX = B$

Using the conformal block decomposition: $\begin{bmatrix} A_1 & A_2 \\ & A_3 \end{bmatrix} \begin{bmatrix} X_1 \\ X_2 \end{bmatrix} = \begin{bmatrix} B_1 \\ B_1 \end{bmatrix}$

1. **if** m=1 **then return** $X = A_{1,1}^{-1} \times B$ **end if**
2. $X_2 =$`trsm`(A_3, B_2);
3. $B_1 = B_1 - A_2X_2$; {using `gemm`, e.g., via Algorithm 13.4.14}
4. $X_1 =$`trsm`(A_1, B_1);
5. **return** $X = \begin{bmatrix} X_1 \\ X_2 \end{bmatrix}$;

13.4.25 Algorithm (`trmm`: Triangular Matrix Multiplication)
Require: $A \in \mathbb{F}_q^{m\times m}$ upper triangular, $B \in \mathbb{F}_q^{m\times n}$
Ensure: $C \in \mathbb{F}_q^{m\times n}$ s.t. $AB = C$

Using the conformal block decomposition: $\begin{bmatrix} A_1 & A_2 \\ & A_3 \end{bmatrix} \begin{bmatrix} B_1 \\ B_2 \end{bmatrix} = \begin{bmatrix} C_1 \\ C_2 \end{bmatrix}$

1. **if** m=1 **then return** $C = A_{1,1} \times B$ **end if**
2. $C_1 =$`trmm`(A_1, B_1);
3. $C_1 = C_1 + A_2B_2$; {using `gemm`}
4. $C_2 =$`trmm`(A_3, B_2);
5. **return** $C = \begin{bmatrix} C_1 \\ C_2 \end{bmatrix}$;

13.4.26 Algorithm (`trtri`: Triangular Matrix Inversion)
Require: $A \in \mathbb{F}_q^{n\times n}$ upper triangular and non-singular
Ensure: $C = A^{-1}$

if m=1 **then return** $C = A_{1,1}^{-1}$ **end if**
$C_1 = A_1^{-1}$; {using `trtri` recursively}
$C_3 = A_3^{-1}$; {using `trtri` recursively}
$C_2 = A_2C_3$; {using `trmm` }
$C_2 = -C_1C_2$; {using `trmm` }
return $C = \begin{bmatrix} C_1 & C_2 \\ & C_3 \end{bmatrix}$;

Using the conformal block decomposition:
$\begin{bmatrix} A_1 & A_2 \\ & A_3 \end{bmatrix}, \begin{bmatrix} C_1 & C_2 \\ & C_3 \end{bmatrix}$

13.4.27 Algorithm (`trtrm`: Upper-Lower Triangular Matrix Multiplication)
Require: $L \in \mathbb{F}_q^{n\times n}$ lower triangular
Require: $U \in \mathbb{F}_q^{n\times n}$ upper triangular
Ensure: $A = UL$

if m=1 **then return** $A = U_{1,1}L_{1,1}$ **end if**
$A_1 = U_1L_1$; {using `trtrm` recursively}
$A_1 = A_1 + U_2L_2$; {using `gemm`}
$A_2 = U_2L_3$; {using `trmm` }
$A_3 = U_3L_2$; {using `trmm` }
$A_4 = U_3L_3$; {using `trtrm` recursively}
return $A = \begin{bmatrix} A_1 & A_2 \\ A_3 & A_4 \end{bmatrix}$;

Using the conformal block decomposition:
$\begin{bmatrix} L_1 & \\ L_2 & L_3 \end{bmatrix}, \begin{bmatrix} U_1 & U_2 \\ & U_3 \end{bmatrix}, \begin{bmatrix} A_1 & A_2 \\ A_3 & A_4 \end{bmatrix}$

13.4.2.2 PLE decomposition

13.4.28 Remark Dense Gaussian elimination over finite fields can be reduced to matrix multiplication, using the usual techniques for the LU decomposition of numerical linear algebra [457]. However, in applications over a finite field, the input matrix often has non-generic rank profile and special care needs to be taken about linear dependencies and rank deficiencies. The PLE decomposition is thus a generalization of the PLU decomposition for matrices with any rank profile.

13.4.29 Definition A matrix is in *row-echelon form* if all its zero rows occupy the last row positions and the leading coefficient of any non-zero row except the first one is strictly to the right of the leading coefficient of the previous row. Moreover, it is in *reduced row-echelon form* if all coefficients above a leading coefficient are zeros.

13.4.30 Definition For any matrix $A \in F_q^{m\times n}$ of rank r, there is a *PLE decomposition* $A = PLE$ where P is a permutation matrix, L is a $m \times r$ lower triangular matrix and E is a $r \times n$ matrix in row-echelon form, with unit leading coefficients.

13.4.31 Remark Algorithm 13.4.32 shows how to compute such a decomposition by a block recursive algorithm, thus reducing the complexity to that of matrix multiplication.

13.4.32 Algorithm (PLE decomposition)
Require: $A \in \mathbb{F}_q^{m\times n}$
Ensure: (P, L, E) a PLE decomposition of A

if $n = 1$ **then**
  if $A = 0_{m\times 1}$ **then return** (I_m, I_0, A); **end if**

3. Let j be the column index of the first non-zero entry of A and $P = T_{1,j}$ the transposition between indices 1 and j;
4. **return** $(P, PA, [1])$;
5. **else**
6. $(P_1, L_1, E_1) = \texttt{PLE}(A_1)$; {recursively}
7. $A_2 = P_1 A_2$;
8. $A_3 = L_{1,1}^{-1} A_3$; {using **trsm**}
9. $A_4 = A_4 - L_{1,2} A_3$; {using **gemm**}
10. $(P_2, L_2, E_2) = \texttt{PLE}(A_4)$; {recursively}
11. **return** $\left(P_1 \begin{bmatrix} I_{r_1} & \\ & P_2 \end{bmatrix}, \begin{bmatrix} L_{1,1} & \\ P_2 L_{1,2} & L_2 \end{bmatrix}, \begin{bmatrix} E_1 & A_3 \\ & E_2 \end{bmatrix}\right)$;
12. **end if**

Split A columnwise in halves: $A = \begin{bmatrix} A_1 & A_2 \end{bmatrix}$

Split $A_2 = \begin{bmatrix} A_3 \\ A_4 \end{bmatrix}$, $L_1 = \begin{bmatrix} L_{1,1} \\ L_{1,2} \end{bmatrix}$ where A_3 and $L_{1,1}$ have r_1 rows.

13.4.2.3 Echelon forms

13.4.33 Remark The row-echelon and reduced row-echelon forms can be obtained from the PLE decomposition, using additional operations: **trsm**, **trtri**, and **trtrm**, as shown in Algorithms 13.4.34 and 13.4.35.

13.4.34 Algorithm (**RowEchelon**)

Require: $A \in \mathbb{F}_q^{m \times n}$

Ensure: (X, E) such that $XA = E$, X is non-singular and E is in row-echelon form

1. $(P, L, E) = \texttt{PLE}(A)$;
2. $X_1 = L_1^{-1}$; {using **trtri**}
3. $X_2 = -L_2 X_1$; {using **trmm**}
4. **return** $\left(X = \begin{bmatrix} X_1 & \\ X_2 & I_{m-r} \end{bmatrix} P^T, E\right)$;

Split $L = \begin{bmatrix} L_1 \\ L_2 \end{bmatrix}$, $L_1 : r \times r$.

13.4.35 Algorithm (**ReducedRowEchelon**)

Require: $A \in \mathbb{F}_q^{m \times n}$

Ensure: (Y, R) such that $YA = R$, Y is non-singular and R is in reduced row-echelon form

1. $(X, E) = \texttt{RowEchelon}(A)$;
2. Let Q be the permutation matrix that brings the leading row coefficients of E to the diagonal;
3. Set $EQ = \begin{bmatrix} U_1 & U_2 \end{bmatrix}$; {where U_1 is $r \times r$ upper triangular}
4. $Y_1 = U_1^{-1}$; {using **trtri**}
5. $Y_1 = Y_1 X_1$; {using **trtrm**}
6. $R = \begin{bmatrix} I_r & U_1^{-1} U_2 \end{bmatrix} Q^T$; {using **trsm**}
7. **return** $\left(Y = \begin{bmatrix} Y_1 & \\ U_2 & I_{n-r} \end{bmatrix} P^T, R\right)$;

13.4.36 Remark Figure 13.4.2 shows the various steps between the classical Gaussian elimination (LU decomposition), the computation of the echelon form and of the reduced echelon form, together with the various problems that each of them solve. Table 13.4.3 shows the leading constant K_ω in the asymptotic time complexity of these algorithms, assuming that two $n \times n$ matrices can be multiplied in $C_\omega n^\omega + o(n^\omega)$.

13.4.37 Remark If the rank r is very small compared to the dimensions $m \times n$ of the matrix, a system $Ax = b$ can be solved in time bounded by $\mathcal{O}\left((m+n)r^2\right)$ [2170, Theorem 1].

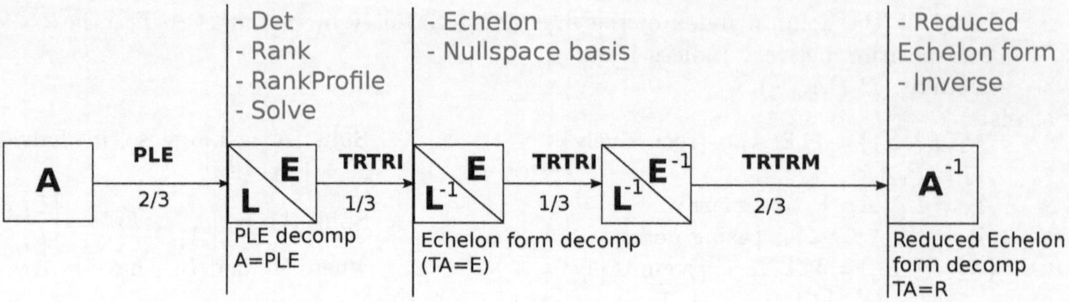

Figure 13.4.2 Reductions from PLE decomposition to reduced echelon form.

Algorithm	Constant K_ω	K_3	$K_{\log_2 7}$
gemm	C_ω	2	6
trsm	$\frac{C_\omega}{2^{\omega-1}-2}$	1	4
trtri	$\frac{C_\omega}{(2^{\omega-1}-2)(2^{\omega-1}-1)}$	$\frac{1}{3} \approx 0.33$	$\frac{8}{5} = 1.6$
trtrm, PLE	$\frac{C_\omega}{2^{\omega-1}-2} - \frac{C_\omega}{2^{\omega}-2}$	$\frac{2}{3} \approx 0.66$	$\frac{14}{5} = 2.8$
Echelon	$\frac{C_\omega}{2^{\omega-2}-1} - \frac{3C_\omega}{2^{\omega}-2}$	1	$\frac{22}{5} \approx 4.4$
RedEchelon	$\frac{C_\omega(2^{\omega-1}+2)}{(2^{\omega-1}-2)(2^{\omega-1}-1)}$	2	$\frac{44}{5} = 8.8$

Table 13.4.3 Complexity of elimination algorithms

13.4.3 Minimal and characteristic polynomial of a dense matrix

13.4.38 Definition

1. A *Las-Vegas algorithm* is a randomized algorithm which is always correct. Its expected running time is always finite.
2. A *Monte-Carlo algorithm* is a randomized algorithm which is correct with a certain probability. Its running time is deterministic.

13.4.39 Remark The computation of the minimal and characteristic polynomials is closely related to that of the Frobenius normal form.

13.4.40 Definition Any matrix $A \in \mathbb{F}_q^{n\times n}$ is similar to a unique block diagonal matrix $F = P^{-1}AP = diag(C_{f_1}, \ldots, C_{f_t})$ where the blocks C_{f_i} are companion matrices of the polynomials f_i, which satisfy $f_{i+1} | f_i$. The f_i are the *invariant factors* of A and F is the *Frobenius normal form* of A.

13.4.41 Remark Most algorithms computing the minimal and characteristic polynomial or the Frobenius normal form rely on Krylov basis computations.

13.4.42 Definition

1. The *Krylov matrix* of order d for a vector v with respect to a matrix A is the matrix $K_{A,v,d} = \begin{bmatrix} v & Av & \ldots & A^{d-1}v \end{bmatrix} \in \mathbb{F}_q^{n\times d}$.
2. The *minimal polynomial* $P_{\min}^{A,v}$ of A and v is the least degree monic polynomial P such that $P(A)v = 0$.

13.4.43 Theorem

1. $AK_{A,v,d} = K_{A,v,d}C_{P_{\min}^{A,v}}$, where $d = \deg(P_{\min}^{A,v})$.
2. For linearly independent vectors $(v_1, \ldots, v_k)$, if $K = \begin{bmatrix} K_{A,v_1,d_1} & \ldots & K_{A,v_k,d_k} \end{bmatrix}$ is non-singular. Then $AK = K \begin{bmatrix} C_{P_{\min}^{A,v_1}} & B_{1,2} & \ldots & B_{1,k} \\ B_{2,1} & C_{P_{\min}^{A,v_1}} & \ldots & B_{2,k} \\ \vdots & \vdots & \ddots & \vdots \\ B_{k,1} & B_{k,2} & & C_{P_{\min}^{A,v_k}} \end{bmatrix}$, where the blocks $B_{i,j}$ are zero except on the last column.
3. For linearly independent vectors $(v_1, \ldots, v_k)$, let $(d_1, \ldots d_k)$ be the lexicographically largest sequence of degrees such that $K = \begin{bmatrix} K_{A,v_1,d_1} & \ldots & K_{A,v_k,d_k} \end{bmatrix}$ is non-singular. Then

$$K^{-1}AK = \begin{bmatrix} C_{P_{\min}^{A,v_1}} & B_{1,2} & \ldots & B_{1,k} \\ & C_{P_{\min}^{A,v_1}} & \ldots & B_{2,k} \\ & & \ddots & \vdots \\ & & & C_{P_{\min}^{A,v_k}} \end{bmatrix} = H. \tag{13.4.1}$$

13.4.44 Remark

1. Some choice of vectors $v_1, \ldots, v_k$ lead to a matrix H block diagonal: this is the Frobenius normal form [1171].
2. The matrix obtained from Equation (13.4.1) is a *Hessenberg form*. It suffices to compute the characteristic polynomial from its diagonal blocks.

13.4.45 Theorem The Frobenius normal form can be computed:

1. by a deterministic algorithm [2727] in $6n^3 + \mathcal{O}\left(n^2 \log^2 n\right)$ field operations, (only $(2 + \frac{2}{3})n^3 + \mathcal{O}\left(n^2\right)$ for the characteristic polynomial [934]);
2. by a deterministic algorithm [2726] in $\mathcal{O}\left(n^\omega \log n \log\log n\right)$, together with a transformation matrix (only $\mathcal{O}\left(n^\omega \log n\right)$ for the characteristic polynomial [1723]);
3. by a Las-Vegas algorithm [944] in $\mathcal{O}\left(n^\omega \log n\right)$ field operations for any field, together with a transformation matrix;
4. by a Las-Vegas algorithm [2386] in $\mathcal{O}\left(n^\omega\right)$ for $q > 2n^2$, without a transformation matrix.

13.4.46 Remark The minimal and characteristic polynomials, obtained as the first invariant factor and the product of all invariant factors, can be computed with the same complexities.

13.4.47 Remark These algorithms are all based upon Krylov bases. The algorithm in Part 1 iteratively computes the Krylov iterates one after the other. Their cubic time complexity with a small leading constant makes them comparable to Gaussian elimination. A fast exponentiation scheme by Keller-Gehrig [1723] achieves a sub-cubic time complexity for the characteristic polynomial, off by a logarithmic factor of n from the matrix multiplication. The choice for the appropriate vectors that will generate the Frobenius normal form can be done either probabilistically (Las Vegas) or deterministically with an extra $\log\log n$ factor. The algorithm in Part 4 uses a different iteration where the size of the Krylov increases according to an arithmetic progression rather than geometric (as all others) and the transformation matrix is not computed. This allows the algorithm to match the complexity of matrix multiplication. This reduction is practical and is implemented as in LinBox.

13.4.48 Remark These probabilistic algorithms depend on the ability to sample uniformly from a large set of coefficients from the field. Over small fields, it is always possible to embed the problem into an extension field, in order to make the random sampling set sufficiently large. In the worst case, this could add a $\mathcal{O}(\log(n))$ factor to the arithmetic cost and prevent most of the bit-packing techniques. Instead, the effort of [944] is to handle cleanly the small finite field case.

13.4.4 Blackbox iterative methods

13.4.49 Remark We consider now the case where the input matrix is sparse, i.e., has many zero elements, or has a structure which enables fast matrix-vector products. Gaussian elimination would fill-in the sparse matrix or modify the interesting structure. Therefore one can use iterative methods instead which only use matrix-vector iterations (*blackbox methods* [1669]). There are two major differences with numerical iterative routines: over finite fields there exists isotropic vectors and there is no notion of convergence, hence the iteration must proceed until exactness of the result [1840]. Probabilistic early termination can nonetheless be applied when the degree of the minimal polynomial is smaller than the dimension of the matrix [935, 945, 1664]. More generally the probabilistic nature of the algorithms presented in this section is subtle: e.g., the computation of the minimal polynomial is Monte-Carlo, but that of system solving, using the minimal polynomial, is Las Vegas (by checking consistency of the produced solution with the system). Making some of the Monte-Carlo solutions Las Vegas is a key open problem in this area.

13.4.4.1 Minimal polynomial and the Wiedemann algorithm

13.4.50 Remark The first iterative algorithm and its analysis are due to Wiedemann [2976]. The algorithm computes the minimal polynomial in the Monte-Carlo probabilistic fashion.

13.4.51 Definition For a linearly recurring sequence $S = (S_i)$, its minimal polynomial is denoted by Π_S.

1. The minimal polynomial of a matrix is denoted $\Pi_A = \Pi_{(A^i)}$.
2. For a matrix A and a vector b, we note $\Pi_{A,b} = \Pi_{(A^i \cdot b)}$.
3. For another vector u, we note $\Pi_{u,A,b} = \Pi_{(u^T \cdot A^i \cdot b)}$.

13.4.52 Algorithm (Wiedemann minimal polynomial)

Require: $A \in \mathbb{F}_q^{n \times n}$, $u, b \in \mathbb{F}_q^n$
Ensure: $\Pi_{u,A,b}$

1. Compute $S = (u^T A^i b)$ for $i \leq 2n$;
2. Use the Berlekamp-Massey algorithm to compute the minimal polynomial of the scalar sequence S;

13.4.53 Definition [See Definition 2.1.111] We extend Euler's totient function by $\Phi_{q,k}(f) = \prod(1 - q^{-kd_i})$, where d_i are the degrees of the distinct monic irreducible factors of the polynomial f.

13.4.54 Theorem For $u_1, \ldots, u_j$ selected uniformly at random, the probability that $\mathrm{lcm}(\Pi_{u_j,A,b}) = \Pi_{A,b}$ is at least $\Phi_{q,k}(\Pi_{A,b})$.

13.4.55 Theorem For $b_1, \ldots, b_k$ selected uniformly at random, the probability that $\mathrm{lcm}(\Pi_{A,b_i}) = \Pi_A$ is at least $\Phi_{q,k}(\Pi_A)$.

13.4.4.2 Rank, determinant, and characteristic polynomial

13.4.56 Remark It is possible to compute the rank, determinant, and characteristic polynomial of a matrix from its minimal polynomial. All these reductions require to precondition the matrix so that the minimal polynomial of the obtained matrix will reveal the information sought, while keeping a low cost for the matrix-vector product [607, 935, 947, 1666, 2824, 2874, 2875].

13.4.57 Theorem [947] Let S be a finite subset of a field $\mathbb{F}$ that does not include 0. Let $A \in \mathbb{F}^{m\times n}$ having rank r. Let $D_1 \in S^{n\times n}$ and $D_2 \in S^{m\times m}$ be two random diagonal matrices then $\deg(\text{minpoly}(D_1 \times A^t \times D_2 \times A \times D_1)) = r$, with probability at least $1 - \frac{11n^2-n}{2|S|}$.

13.4.58 Theorem [2824] Let S be a finite subset of a field $\mathbb{F}$ that does not include 0. Let $U \in S^{n\times n}$ be a unit upper bi-diagonal matrix where the second diagonal elements $u_1, \ldots, u_{n-1}$ are randomly selected in S. For $A \in \mathbb{F}^{n\times n}$, the term of degree 0 of the minimal polynomial of UA is the determinant of A with probability at least $1 - \frac{n^2-n}{2|S|}$.

13.4.59 Remark If A is known to be non-singular the algorithm can be repeated with different matrices U until the obtained minimal polynomial is of degree n. Then it is the characteristic polynomial of UA and the determinant is certified. Alternatively if the matrix is singular then X divides the minimal polynomial. As Wiedemann's algorithm always returns a factor of the true minimal polynomial, and U is invertible, the algorithm can be repeated on UA until either the obtained polynomial is of degree n or it is divisible by X. Overall the determinant has a Las-Vegas blackbox solution.

13.4.60 Theorem [2874, 2875] Let S be a finite subset of a field $\mathbb{F}$ that does not include 0 and $A \in \mathbb{F}^{n\times n}$ with $s_1, \ldots, s_t$ as invariant factors. Let $U \in S^{n\times k}$ and $V \in S^{k\times n}$ be randomly chosen rank k matrices in $\mathbb{F}$. Then $\gcd(\Pi_A, \Pi_{A+UV}) = s_{k+1}$ with probability at least $1 - \frac{nk+n+1}{|S|}$.

13.4.61 Remark Using the divisibility of the invariant factors and the fact that their product is of degree n, one can see that the number of degree changes between successive invariant factors is of order $\mathcal{O}(\sqrt{n})$ [2874]. Thus by a binary search over successive applications of Theorem 13.4.60 one can recover all of the invariant factors and thus the characteristic polynomial of the matrix in a Monte-Carlo fashion.

13.4.4.3 System solving and the Lanczos algorithm

13.4.62 Remark For the solution of a linear system $Ax = b$, one could compute the minimal polynomial $\Pi_{A,b}$ and then derive a solution of the system as a linear combination of the A^ib. The following Lanczos approach is more efficient for system solving as it avoids recomputing (or storing) the latter vectors [947, 1273].

13.4.63 Algorithm (Lanczos system solving)

Require: $A \in \mathbb{F}^{m\times n}$, $b \in \mathbb{F}^m$

Ensure: $x \in \mathbb{F}^n$ such that $Ax = b$ or *failure*

1. Let $\tilde{A} = D_1A^TD_2AD_1$ and $\tilde{b} = D_1A^TD_2b + \tilde{A}v$ with D_1 and D_2 random diagonal matrices and v a random vector;
2. $w_0 = \tilde{b}$; $v_1 = \tilde{A}w_0$; $t_0 = v_1^Tw_0$; $\gamma = \tilde{b}^tw_0t_0^{-1}$; $x_0 = \gamma w_0$;
3. **repeat**
4. $\quad \alpha = v_{i+1}^Tv_{i+1}t_i^{-1}$; $\beta = v_{i+1}^Tv_it_{i-1}^{-1}$; $w_{i+1} = v_{i+1} - \alpha w_i - \beta w_{i-1}$;
5. $\quad v_{i+2} = \tilde{A}w_{i+1}$; $t_{i+1} = w_{i+1}^Tv_{i+2}$;
6. $\quad \gamma = \tilde{b}^tw_{i+1}t_{i+1}^{-1}$; $x_{i+1} = x_i + \gamma w_{i+1}$;
7. **until** $w_{i+1} = 0$ or $t_{i+1} = 0$;
8. Return $x = D_1(x_{i+1} - v)$;

13.4.64 Remark The probability of success of Algorithm 13.4.63 follows from Theorem 13.4.57.

13.4.65 Remark Over small fields, if the rank of the matrix is known, the diagonal matrices of line 1 can be replaced by sparse preconditioners with $\mathcal{O}(n\log(n))$ non-zero coefficients to avoid the need of field extensions [607, Corollary 7.3].

13.4.66 Remark If the system with A and b is known to have a solution then the algorithm can be turned Las Vegas by checking that the output x indeed satisfies $Ax = b$. In general, we do not know if this algorithm returns failure because of bad random choices or because the system is inconsistent. However, Giesbrecht, Lobo, and Saunders have shown that when the system is inconsistent, it is possible to produce a certificate vector u such that $u^T A = 0$ together with $u^T b \neq 0$ within the same complexity [1273, Theorem 2.4]. Overall, system solving can be performed by blackbox algorithms in a Las-Vegas fashion.

13.4.5 Sparse and structured methods

13.4.67 Remark Another approach to sparse linear system is to use Gaussian elimination with pivoting, taking into account the zero coefficients. This algorithm modifies the structure of the matrix and might suffer from fill-in. Consequently the available memory is usually the bottleneck. From a triangularization one can naturally derive the rank, determinant, system solving, and nullspace. Comparisons with the blackbox approaches above can be found, e.g., in [935].

13.4.5.1 Reordering

13.4.68 Algorithm (Gaussian elimination with linear pivoting)

Require: a matrix $A \in \mathbb{F}^{m\times n}$

Ensure: An upper triangular matrix U such that there exists a unitary lower-triangular matrix L and permutations matrices P and Q over $\mathbb{F}$, with $A = P{\cdot}L{\cdot}U{\cdot}Q$

1. **for all** elimination steps **do**
2. Choose as pivot row the sparsest remaining row;
3. In this row choose the non-zero pivot with lowest number of non-zero elements in its column;
4. Eliminate using this pivot;
5. **end for**

13.4.69 Remark Yannakakis showed that finding the minimal fill-in (or equivalently the best pivots) during Gaussian elimination is an NP-complete task [3031]. In numerical algorithms, heuristics have been developed and comprise minimal degree ordering, cost functions, or nested dissection; see for example [89, 1485, 3082]. These heuristics for reducing fill-in in the numerical setting, often assume symmetric and invertible matrices, and do not take into account that new zeros may be produced by elimination operations ($a_{ij} = a_{ij} + \delta_i * a_{kj}$), as is the case with matrices over finite fields. Thus, [935] proposes the heuristic 13.4.68 to take those new zeros into account, using a local optimization of a cost function at each elimination step.

13.4.5.2 Structured matrices and displacement rank

13.4.70 Remark Originating from the seminal paper [1643] most of the algorithms dealing with structured matrices use the displacement rank approach [2346].

13.4.71 Definition For $A \in \mathbb{F}^{m\times m}$ and $B \in \mathbb{F}^{n\times n}$, the *Sylvester (respectively Stein) linear displacement operator* $\nabla_{A,B}$ *(respectively* $\triangle_{A,B}$*)* satisfies for $M \in \mathbb{F}^{m\times n}$:

$$\nabla_{A,B}(M) = AM - MB,$$
$$\triangle_{A,B}(M) = M - AMB.$$

A pair of matrices $(Y, Z) \in \mathbb{F}^{m\times\alpha} \times \mathbb{F}^{n\times\alpha}$ is an *(A, B)-Sylvester-generator of length* α (respectively Stein) for M if $\nabla_{A,B}(M) = YZ^T$ (respectively $\triangle_{A,B}(M) = YZ^T$).

13.4.72 Remark The main idea behind algorithms for structured matrices is to use such generators as a compact data structure, in cases where the displacement has low rank.

13.4.73 Remark Usual choices of matrices A and B are diagonal matrices and cyclic down shift matrices.

13.4.74 Definition We denote the diagonal matrix whose (i, i) entry is x_i by $\mathbb{D}_x, x \in \mathbb{F}^n$ and by $\mathbb{Z}_{n,\varphi}, \varphi \in \mathbb{F}$ the $n \times n$ unit circulant matrix having φ at position $(1, n)$, ones in the subdiagonal $(i + 1, i)$ and zeros elsewhere.

operator matrices A	B	class of structured matrices M	rank of $\nabla_{A,B}(M)$	number of flops for computing $M \cdot v$
$\mathbb{Z}_{n,1}$	$\mathbb{Z}_{n,0}$	Toeplitz and its inverse	≤ 2	$\mathcal{O}((m+n)\log(m+n))$
$\mathbb{Z}_{n,1}$	$\mathbb{Z}_{n,0}^T$	Hankel and its inverse	≤ 2	$\mathcal{O}((m+n)\log(m+n))$
$\mathbb{Z}_{n,0} + \mathbb{Z}_{n,0}^T$	$\mathbb{Z}_{n,0} + \mathbb{Z}_{n,0}^T$	Toeplitz + Hankel	≤ 4	$\mathcal{O}((m+n)\log(m+n))$
$\mathbb{D}_x$	$\mathbb{Z}_{n,0}$	Vandermonde	≤ 1	$\mathcal{O}((m+n)\log^2(m+n))$
$\mathbb{Z}_{n,0}$	$\mathbb{D}_x$	inverse of Vandermonde	≤ 1	$\mathcal{O}((m+n)\log^2(m+n))$
$\mathbb{Z}_{n,0}^T$	$\mathbb{D}_x$	transposed of Vandermonde	≤ 1	$\mathcal{O}((m+n)\log^2(m+n))$
$\mathbb{D}_x$	$\mathbb{D}_y$	Cauchy and its inverse	≤ 1	$\mathcal{O}((m+n)\log^2(m+n))$

Table 13.4.4 Complexity of the matrix-vector product for some structured matrices

13.4.75 Remark As computing matrix vector products with such structured matrices have close algorithmic correlation to computations with polynomials and rational functions, these matrices can be quickly multiplied by vectors, in nearly linear time as shown on Table 13.4.4. Therefore the algorithms of Subsection 13.4.4 can naturally be applied to structured matrices, to yield almost $\mathcal{O}(n^2)$ time linear algebra.

13.4.76 Remark If the displacement rank is small there exist algorithms quasilinear in n, the dimension of the matrices, which over finite fields are essentially variations or extensions of the Morf/Bitmead-Anderson divide-and-conquer [290, 2154] or Cardinal's [517] approaches. The method is based on dividing the original problem repeatedly into two subproblems with one leading principal submatrix and the related Schur complement. This leads to $\mathcal{O}(\alpha^2 n^{1+o(1)})$ system solvers, which complexity bound have recently been reduced to $\mathcal{O}(\alpha^{\omega-1} n^{1+o(1)})$ [364, 1600]. With few exceptions, all algorithms thus need matrices in generic rank profile. Over finite fields this can be achieved using Kaltofen and Saunders unit upper triangular Toeplitz preconditioners [1666] and by controlling the displacement rank growth and non-singularity issues [1656].

13.4.6 Hybrid methods

13.4.6.1 Hybrid sparse-dense methods

13.4.77 **Remark** Overall, as long as the matrix fits into memory, Gaussian elimination methods over finite fields are usually faster than iterative methods [935]. There are heuristics trying to take advantage of both strategies. Among those we briefly mention the most widely used:

1. Perform the Gaussian elimination with reordering 13.4.68 until the matrix is almost filled up. If the remaining non-eliminated part would fit as a dense matrix, switch to the dense methods of Subsection 13.4.2.
2. Maintain two sets of rows (or columns), sparse and dense. Favor elimination on the sparse set. This is particularly adapted to index calculus [1839].
3. Perform a preliminary reordering in order to cut the matrix into four quadrants, the upper left one being triangular. This, together with the above strategies has proven effective on matrices which are already quasi-triangular, e.g., Gröbner bases computations in finite fields [1043].
4. If the rank is very small compared to the dimension of the matrix, one can use left and right highly rectangular projections to manipulate smaller structures [2040].
5. The arithmetic cost and thus timing predictions are easier on iterative methods than on elimination methods. On the other hand the number of non-zero elements at a given point of the elimination is usually increasing during an elimination, thus providing a lower bound on the remaining time to triangularize. Thus a heuristic is to perform one matrix-vector product with the original matrix and then eliminate using Gaussian elimination. If at one point the lower bound for elimination time surpasses the predicted iterative one or if the algorithm runs out of memory, stop the elimination and switch to the iterative methods [937].

13.4.6.2 Block-iterative methods

13.4.78 **Remark** Iterative methods based on one-dimensional projections, such as Wiedmann and Lanczos algorithm can be generalized with block projections. Via efficient preconditioning [607] these extensions to the scalar iterative methods can present enhanced properties:

1. Usage of dense sub-blocks, after multiplications of blocks of vectors with the sparse matrix or the blackboxes, allows for a better locality and optimization of memory accesses, via the application of the methods of Subsection 13.4.1.
2. Applying the matrix to several vectors simultaneously introduces more parallelism [718, 719, 1664].
3. The probability of success augments with the size of the considered blocks, especially over small fields [1657, 2873].

13.4.79 **Definition** Let $X \in \mathbb{F}_q^{k\times n}$, $Y \in \mathbb{F}_q^{n\times k}$ and $H_i = XA^iY$ for $i = 0, \ldots, n/k$. The *matrix minimal polynomial* of the sequence H_i is the matrix polynomial $F_{X,A,Y} \in \mathbb{F}_q[X]^{k\times k}$ of least degree, with its leading degree matrix column-reduced, that annihilates the sequence (H_i).

13.4.80 **Theorem** The degree d matrix minimal polynomial of a block sequence $(H_i) \in (F_q^{k\times k})^{\mathbb{Z}}$ can be computed in $\mathcal{O}\left(k^3d^2\right)$ using block versions of Hermite-Pade approximation and extended Euclidean algorithm [214] or Berlkamp-Massey algorithm [719, 1657, 2873]. Further improvement by [214, 1275, 1670, 2804] bring this complexity down to $\mathcal{O}\left((k^\omega d)^{1+o(1)}\right)$, using a matrix extended Euclidean algorithm.

13.4.81 Algorithm (Nullspace vector)
Require: $A \in \mathbb{F}_q^{n\times n}$
Ensure: $\omega \in \mathbb{F}_q^n$ a vector in the nullspace of A
1. Pick $X \in \mathbb{F}_q^{k\times n}, Y \in \mathbb{F}_q^{n\times k}$ uniformly at random;
2. Compute the sequence $H_i = XA^iY$;
3. Compute $F_{X,A,Y}$ the matrix minimal polynomial;
4. Let $f = f_r x^r + \cdots + f_d x^d$ be a column of $F_{X,A,Y}$;
5. Return $\omega = Yf_r + AYf_{r+1} + \cdots + A^{d-r}Yf_d$;

13.4.82 Remark These block-Krylov techniques are used to achieve the best known time complexities for several computations with black-box matrices over a finite field or the ring of integers: computing the determinant, the characteristic polynomial [1670], and the solution of a linear system of equations [946].

See Also

Chapter 10	For generating sequences and their links to Wiedemann's algorithm via Berlekamp-Massey's shift register synthesis.
§11.1, §11.3	For effective finite field constructions.
§13.2	For structured matrices over finite fields.

References Cited: [89, 124, 214, 290, 349, 364, 457, 517, 607, 718, 719, 720, 723, 929, 930, 931, 932, 933, 934, 935, 937, 944, 945, 946, 947, 1043, 1171, 1273, 1275, 1336, 1485, 1600, 1643, 1656, 1657, 1664, 1666, 1669, 1670, 1723, 1839, 1840, 2040, 2154, 2170, 2346, 2386, 2726, 2727, 2728, 2804, 2824, 2873, 2874, 2875, 2974, 2976, 2985, 2987, 3031, 3082]

13.5 Carlitz and Drinfeld modules

David Goss, Ohio State University

13.5.1 Remark Much of the theory presented here will be familiar to the reader from the theory of elliptic curves. Moreover, the reader may profit on first reading this by only focusing on the basic case $A = \mathbb{F}_q[t]$. We have given a very rapid introduction to Drinfeld modules and the like and have naturally omitted many topics; the subject is quite active and changing rapidly. For much of the elided material please consult [919, 920, 1333, 1451, 2793].*

13.5.2 Remark One of the salient points about Drinfeld modules is that they allow one to go as far as possible in replacing the integers $\mathbb{Z}$ as the fundamental object in arithmetic. In other words, while all of the fields involved obviously lie over Spec($\mathbb{Z}$), the theory allows one to study invariants coming from characteristic 0 (such as class groups, etc.) *as well*

*This survey is dedicated to the memory of my friend, and function field pioneer, David Hayes.

as analogous invariants arising only in finite characteristic (such as the "class module" of Subsection 13.5.6 below).

13.5.1 Quick review

13.5.3 Remark It is well known that the most important function in classical analysis is the *exponential function* e^z; in analysis in characteristic p, it is the q-th power mapping $\tau_q(z) := z^q$ ($q = p^{n_0}$, n_0 a positive integer) and functions created out of it. Note that τ_q is clearly an $\mathbb{F}_q$-linear mapping by the Binomial Theorem and separable polynomials in τ_q are characterized among all separable polynomials as those having their zeroes form an $\mathbb{F}_q$ vector space. For any $\mathbb{F}_q$-field E we let $E\{\tau_q\}$ be the algebra of polynomials in τ_q (which is noncommutative in general); the ring $E\{\tau_q\}$ is the ring of algebraic $\mathbb{F}_q$-linear endomorphisms of the additive group over E. We also let $\bar{E}$ be a fixed algebraic closure of E.

13.5.4 Remark Let E be an arbitrary field (of arbitrary characteristic).

13.5.5 Definition A *non-Archimedean absolute value* on E is a mapping $|\ |\colon E \to \mathbb{R}$ satisfying:

1. $|x| \geq 0$,
2. $|x| = 0$ if and only if $x = 0$,
3. $|xy| = |x||y|$, and,
4. $|x + y| \leq \max\{|x|, |y|\}$.

13.5.6 Example Let $E = \mathbb{F}_q(t)$ where t is an indeterminate. Let $g \in E$ and suppose that $g(t)$ has a zero of order j at ∞; set $|g| := q^{-j}$.

13.5.7 Remark The field E is *complete* if every Cauchy sequence converges to an element of E. If the field is not complete one can *complete* it by mimicking the construction of the real numbers (the field E of Example 13.5.6 completes to the formal Laurent series field $\mathbb{F}_q((1/t))$ in $1/t$). We assume that E is complete for the rest of this subsection.

13.5.8 Remark Property 4 in Definition 13.5.5 immediately implies that a series with coefficients in E converges if and only if the n-th term goes to 0.

13.5.9 Remark Let F be a finite extension of E. It is well-known that $|\cdot|$ extends uniquely to F (and thus to any algebraic closure $\bar{E}$ of E).

13.5.10 Proposition [1846] Let F be a normal extension of E. Let $\sigma\colon F \to F$ be an E-automorphism. Then $|\sigma(x)| = |x|$ for all $x \in F$.

13.5.11 Corollary [1846] Let F/E be finite of degree d and let N_E^F be the norm. Then $|x| = |N_E^F(x)|^{1/d}$ for all $x \in E$.

13.5.12 Remark The absolute value $|\cdot|$ is also readily seen to extend uniquely to the completion of $\bar{E}$ which remains algebraically closed.

13.5.13 Definition A power series $\sum_{i=0}^{\infty} a_i x^i$ with coefficients in E is *entire* if it converges for all $x \in E$.

13.5.14 Theorem [1333] Let $f(x)$ be an entire power series with coefficients in E. Then there exists a nonnegative integer j, an element $c \in E$ and a (possibly empty or finite) sequence $\{0 \neq \lambda_i\} \subset \bar{E}$, where $|\lambda_i| \to \infty$ as $i \to \infty$, with $f(x) = cx^j \prod_i (1 - x/\lambda_i)$.

13.5.15 Remark Conversely a product, as in Theorem 13.5.14, defines an E-entire power series if $\{\lambda_i\}$ is stable under E-automorphisms of $\bar{E}$ and inseparability indices are taken into account. Note that Theorem 13.5.14 immediately implies that the power series for e^z is *never* entire as an E-power series for non-Archimedean E of characteristic 0.

13.5.16 Remark The notion of analytic continuation in complex analysis is essential; it is what allows analytic functions to be defined locally. In non-Archimedean analysis, as just described, there are *far* too many open sets for any analogous theory. Following ideas from algebraic geometry (and Grothendieck), Tate had the fantastic idea to use a "Grothendieck topology" to very seriously cut down on the number of such open sets. This theory is called "rigid analysis" and it does in fact allow for analytic continuation, etc. More importantly, via this theory one is able to pass between (rigid) analytic sheaves and algebraic sheaves in the manner of Serre's famous G.A.G.A. paper [2585]. In particular, one is then able to construct algebraic functions via analysis. For details on all of this, see for example [354].

13.5.2 Drinfeld modules: definition and analytic theory

13.5.17 Remark Let X be a smooth, complete, geometrically irreducible curve over $\mathbb{F}_q$ with function field k. Note that k is precisely a *global field* of characteristic p. We let $\infty \in X$ be a fixed closed point of degree d_∞ over $\mathbb{F}_q$, and A the ring of functions holomorphic away from ∞; set $K := k_\infty =$ the completion of k under the absolute value $|x|_\infty = q^{-v(x)}$ where $v(x)$ is the order of zero of x at ∞. We set $\mathbb{C}_\infty$ to be the completion of a fixed algebraic closure $\bar{K}$ of K. Let $K_1 \subset \mathbb{C}_\infty$ be a finite extension of K (automatically also complete) with separable closure $K_1^{\text{sep}} \subset \mathbb{C}_\infty$.

13.5.18 Definition 1. A *K_1-lattice* is a discrete, finitely generated, $\text{Gal}(K_1^{\text{sep}}/K_1)$-stable A-submodule M of K_1^{sep}. 2. A *morphism* between two lattices M_1 and M_2 of the same A-rank is a scalar c such $cM_1 \subseteq M_2$. 3. To a lattice M we attach the *exponential function* $e_M(z) := z\prod_{0\neq\beta\in M}(1 - z/\beta)$.

13.5.19 Remark By Theorem 13.5.14, $e_M(z)$ is entire with K_1 coefficients. As M can be exhausted by finite dimensional $\mathbb{F}_q$-vector spaces, one deduces that $e_M(z)$ is a limit of $\mathbb{F}_q$-linear polynomials and so is, itself, an $\mathbb{F}_q$-linear surjection from $\mathbb{C}_\infty$ to itself. We obtain an exact sequence

$$0 \longrightarrow M \longrightarrow \mathbb{C}_\infty \overset{e_M(z)}{\longrightarrow} \mathbb{C}_\infty \longrightarrow 0\,. \tag{13.5.1}$$

Let $a \in A$; by transport of structure (and matching divisors) we now obtain an exotic A-module structure $\phi_a^M(z)$ on $\mathbb{C}_\infty$ via the *functional equation*

$$e_M(az) := \phi_a^M(e_M(z)) = a \cdot e_M(z) \prod_{0\neq\alpha\in a^{-1}M/M} (1 - e_M(z)/e_M(\alpha))\,. \tag{13.5.2}$$

The *essential* observation is that $a \mapsto \phi_a^M(z)$ embeds A into $K_1\{\tau_q\}$ as $\mathbb{F}_q$-algebras; that is we are representing an element $a \in A$ by the nontrivial $\mathbb{F}_q$-linear polynomial $\phi_a^M(z)$ with $\phi_{a+b}^M(z) = \phi_a^M(z) + \phi_b^M(z)$ *and* $\phi_{ab}^M(z) = \phi_a^M(\phi_b^M(z))$.

13.5.20 Definition Let E be a field with an $\mathbb{F}_q$-linear homomorphism $\imath\colon A \to E$. Then a *Drinfeld module* over $(E, \imath)$ is a homomorphism $\phi\colon A \to E\{\tau_q\}$, $a \mapsto \phi_a$, such that $\phi_a = \imath(a)\tau_q^0 + \{\text{higher terms}\}$ and such that there exists $a \in A$ such that $\phi_a \neq \imath(a)\tau_q^0$. Let ϕ and ψ be two Drinfeld modules over E; then a *morphism* from ϕ to ψ is a polynomial $P \in E\{\tau_q\}$ such that $P\phi_a = \psi_a P$ for all a. Nonzero morphisms are *isogenies*.

13.5.21 Remark Via ϕ, E and $\bar{E}$ become A-modules with $a*z := \phi_a(z)$ for $a \in A$, $z \in \bar{E}$. Let $I \subseteq A$ be an ideal and let $\phi[I] \subset \bar{E}$ be those points annihilated by all $i \in I$; commutativity of A implies that $\phi[I]$ is a *finite* A-module. Now let $a \in A \notin \ker \imath$. As A is a Dedekind domain, simple counting arguments show the existence of an integer d, independent of a, such that $\phi[(a)] \simeq A/(a)^d$; d is the *rank of* ϕ. Isogenies are only possible between Drinfeld modules of the *same* rank and isomorphisms correspond to those P of the form $c\tau_q^0$ where $c \neq 0$.

We then have the following fundamental result of Drinfeld.

13.5.22 Theorem [919] The exponential construction gives an equivalence of categories between Drinfeld modules of rank d over K_1 and rank d K_1-lattices.

13.5.23 Example Let $A = \mathbb{F}_q[t]$. The *Carlitz module* C is the rank 1 Drinfeld module over $\mathbb{F}_q(t)$ given by $C_t := t\tau_q^0 + \tau_q$. The associated *Carlitz exponential* is $e_C(z) := \sum_{j=0}^{\infty} z^{q^j}/D_j$, where D_j is the product of all monic polynomials of degree j, and the *Carlitz lattice* is $A\xi$ where $\xi := (-t)^{q/(q-1)} \prod_{j=1}^{\infty}(1 - t^{1-q^j})^{-1}$ is the *Carlitz period.* For general A, Hayes has constructed generalizations of C which are defined over a Hilbert class field of k.

13.5.24 Remark It was a fundamental observation of Carlitz that adding the elements of $C[a]$ to k gave an abelian extension. These extensions were then shown to provide the abelian closure of k which is tamely ramified at ∞ by Hayes [1449] and generalized by him to arbitrary A in [1450] in analogy with cyclotomic fields. Drinfeld gave a modular approach to the construction of these class fields in his paper [919]; in [920] he obtained the full abelian closure. This full abelian closure has also recently been explicitly constructed by Zywina [3084].

13.5.25 Remark Carlitz originally constructed his module very concretely based on the famous combinatorial formula called the *Moore Determinant* (a finite characteristic analog of the Wronskian as well as the Vandermonde determinant) as given in the next definition (see also Section 13.2).

13.5.26 Definition We define $\Delta_q(w_1, \ldots, w_n)$ to be the determinant of

$$\begin{pmatrix} w_1 & \ldots & w_n \\ w_1^q & \ldots & w_n^q \\ \vdots & & \vdots \\ w_1^{q^{n-1}} & \ldots & w_n^{q^{n-1}} \end{pmatrix} = \begin{pmatrix} \tau_q^0(w_1) & \ldots & \tau_q^0(w_n) \\ \vdots & & \vdots \\ \tau_q^{n-1}(w_1) & \ldots & \tau_q^{n-1}(w_n) \end{pmatrix}.$$

13.5.27 Remark Moore then shows the following basic result analogous to Abel's calculation of the Wronskian.

13.5.28 Proposition [1333] $\Delta_q(w_1, \ldots, w_d)$ equals

$$\prod_{i=1}^{d} \prod_{k_{i-1} \in \mathbb{F}_q} \cdots \prod_{k_1 \in \mathbb{F}_q} (w_i + k_{i-1}w_{i-1} + \cdots + k_1 w_1).$$

13.5.29 Remark Proposition 13.5.28 allowed Carlitz to calculate various products of polynomials over finite fields, and thus to explicitly construct his exponential, lattice, and module.

13.5.30 Remark Returning to a general Drinfeld modules ϕ, let $\mathfrak{p}$ now be a prime of A with completions $k_\mathfrak{p}$, $A_\mathfrak{p}$. Let $\phi[\mathfrak{p}^\infty]$ be the set of all $\mathfrak{p}$-power torsion points.

13.5.31 Definition We set $T_{\mathfrak{p}}(\phi) := \mathrm{Hom}_A(k_{\mathfrak{p}}/A_{\mathfrak{p}}, \phi[\mathfrak{p}^\infty])$; this is the *$\mathfrak{p}$-adic Tate module of* ϕ.

13.5.32 Remark If $\mathfrak{p} \neq \ker \imath$, then $T_{\mathfrak{p}}(\phi)$ is isomorphic to $A_{\mathfrak{p}}^d$ as A-module; the construction of Tate modules is functorial.

13.5.3 Drinfeld modules over finite fields

13.5.33 Remark Let E be as in Definition 13.5.20 now also be *finite*; let ϕ be a Drinfeld module over E of rank d. Let $F_E := \tau_q^t$ which is an *endomorphism* of ϕ. Set $\mathfrak{b} := \ker \imath$ (which is a maximal ideal of A) and let $\mathfrak{p} \neq \mathfrak{b}$ be another nontrivial prime of A; set $f_\phi(u) := \det(1 - uF_E \mid T_{\mathfrak{p}}(\phi))$. Very clever use of central simple arithmetic allows one to establish the following results.

13.5.34 Theorem [920] The polynomial $f_\phi(u)$ depends only on the isogeny class of $f_\phi(u)$ and has coefficients in A which are independent of the choice of $\mathfrak{p}$. The reciprocal roots of $f_\phi(u)$ in $\mathbb{C}_\infty$ have absolute value $q^{t/d}$ and their product generates the ideal $\mathfrak{b}^{[E:\, A/\mathfrak{b}]}$.

13.5.35 Definition An element $\alpha \in \mathbb{C}_\infty$ is a *Weil number of rank* d for E if and only if

1. α is integral over A,
2. there is only place in $k(\alpha)$ which is a zero of α and there is only one place of $k(\alpha)$ above ∞,
3. $|\alpha|_\infty = q^{t/d}$, and,
4. $[k(\alpha) : k] \mid d$.

13.5.36 Remark Clearly the set of Weil numbers of rank d for E is acted on by the group of k-automorphism of $\bar{k} \subset \mathbb{C}_\infty$ and we let $W_d(E)$ be the set of orbits under this action.

13.5.37 Theorem [920] The mapping from the set of isogeny classes of Drinfeld modules over E of rank d to $W_d(E)$ given by Theorem 13.5.34 is a bijection.

13.5.4 The reduction theory of Drinfeld modules

13.5.38 Remark Let E now be a finite extension of k equipped with a Drinfeld module ϕ of rank d; let $\mathfrak{O}$ be the A-integers of E. As A is a finitely generated $\mathbb{F}_q$-algebra, for almost all $\mathfrak{O}$-primes $\mathfrak{P}$ we can reduce the coefficients of ϕ modulo $\mathfrak{P}$ to obtain a Drinfeld module of rank d over $\mathfrak{O}/\mathfrak{P}$. For the other primes, we now localize and assume that E is equipped with a nontrivial discrete valuation v with $v(A)$ nonnegative; let $\mathfrak{O}_v \subset E$ be the associated valuation ring of v-integers and let $M_v \subset \mathfrak{O}_v$ be the maximal ideal.

13.5.39 Definition 1. The Drinfeld module ϕ has *stable reduction at* v if there exists ψ which is isomorphic to ϕ such that ψ has coefficients in $\mathfrak{O}_v$ and such that the reduction ψ^v of ψ is a Drinfeld module (of rank $d_1 \leq d$). 2. The module ϕ has *good reduction at* v if it has stable reduction and $d_1 = d$. 3. The module ϕ has *potential stable* (resp. *potential good*) reduction if there is an extension (E_1, w) of (E, v) such that ϕ has stable (resp. good) reduction at w.

13.5.40 Remark Drinfeld [919] established that every Drinfeld module has potential stable reduction. Let $\mathfrak{p}$ be a prime of A not contained in M_v and view $T_{\mathfrak{p}}(\phi)$ as a $\mathrm{Gal}(E^{\mathrm{sep}}/E)$-module.

13.5.41 Theorem [2767] The Drinfeld module ϕ has good reduction at v if and only if $T_{\mathfrak{p}}(\phi)$ is unramified at v as a $\mathrm{Gal}(E^{\mathrm{sep}}/E)$-module.

13.5.42 Remark Theorem 13.5.41 is an obvious analog of the theorem of *Ogg-Néron-Shafarevich* and is due to Takahashi. Taguchi [2765] has established that $T_{\mathfrak{p}}(\phi)$ is a *semisimple* $\mathrm{Gal}(E^{\mathrm{sep}}/E)$-module; this was also independently established by Tamagawa [2773],

13.5.5 The A-module of rational points

13.5.43 Remark Let E continue to be a finite extension of k equipped with a Drinfeld module ϕ of rank d; using ϕ, E becomes a natural A-module which is denoted "$\phi(E)$".

13.5.44 Theorem [2419] The A-module $\phi(E)$ is isomorphic to $T \oplus N$ where T is a finite torsion module and N is a free A-module of rank $\aleph_0$.

13.5.45 Remark The above result, due to Poonen, applies more generally to other modules of rational points (such as a ring $\mathfrak{O}_S$ of S-integers which contains the coefficients of ϕ). The main technique in the proof is to establish the *tameness* of $\phi(L)$; that is, let $M \subseteq \phi(L)$ be a submodule such that $k \otimes_A M$ is finite dimensional, then M is itself finitely generated. This result is established via the use of *heights* and is analogous to the theorem of *Mordell-Weil* for elliptic curves/abelian varieties *except* that the rank is always infinite.

13.5.6 The invariants of a Drinfeld module

13.5.46 Remark We now present very recent work of Taelman [2760, 2761] that establishes an analog of the classical class group/Tate-Shafarevich group *and* the group of units/Mordell-Weil group in the theory of Drinfeld modules.

13.5.47 Remark As above, let E be a finite extension of k with A-integers $\mathfrak{O}$; set $E_\infty := E \otimes_k K$. Let ϕ be a Drinfeld A-module of rank d over E *whose coefficients will be assumed to lie in* $\mathfrak{O}$ (N.B., this does *not* imply that ϕ has good reduction at all primes of $\mathfrak{O}$); let $e_\phi(z)$ be the associated exponential function. The functional equation satisfied by $e_\phi(z)$ implies that the coefficients of ϕ lie in E. Thus $e_\phi(z)$ induces a natural $\mathbb{F}_q$-linear, continuous endomorphism of E_∞ (also denoted e_ϕ); it is an *open map* as the derivative of $e_\phi(z)$ is identically 1. Let $\hat{M}(\phi/\mathfrak{O}) := e_\phi^{-1}(\mathfrak{O}) \subset E_\infty$. The openness of $e_\phi(z)$ immediately implies the next result.

13.5.48 Proposition [2760, 2761] The A-module $\hat{M}(\phi/\mathfrak{O})$ is discrete and cocompact in E_∞.

13.5.49 Definition We set $M(\phi/\mathfrak{O}) \subset \phi(\mathfrak{O})$ to be the image of $\hat{M}(\phi/\mathfrak{O})$ under $e_\phi(z)$.

13.5.50 Corollary [2760, 2761] $M(\phi/\mathfrak{O})$ is a finitely generated A-module under the Drinfeld action.

13.5.51 Remark $M(\phi/\mathfrak{O})$ is an analog of both the group of units of a number field *and* the Mordell-Weil group of an elliptic curve.

13.5.52 Definition We set $H(\phi/\mathfrak{O}) := \phi(E_\infty)/(\phi(\mathfrak{O}) + e_\phi(E_\infty))$.

13.5.53 Remark As $\mathfrak{O}$ is cocompact in E_∞ and $e_\phi(E_\infty)$ is open in E_∞, we deduce the next result.

13.5.54 Proposition [2760, 2761] $H(\phi, \mathfrak{O})$ is a finite A-module.

13.5.55 Remark $H(\phi, \mathfrak{O})$ is the *class module* or the *Tate-Shafarevich module* of ϕ over $\mathfrak{O}$.

13.5.56 Example Let $A = \mathbb{F}_q[t]$ and let $\phi = C$ be the Carlitz module.

1. Suppose first of all that $E = k$. In this case, one finds that $H(C/A) = \{0\}$ and that $e_C^{-1}(A)$ has rank one with generator the logarithm $e_C^{-1}(1)$.

2. For general E, all torsion elements of $C(\mathfrak{O})$ are contained in $M(C/\mathfrak{O})$.
3. When E is obtained by adjoining torsion elements of C, Anderson [95] has produced an A-module of *special points* (analogous to *cyclotomic units*) which has maximal rank inside $M(C/\mathfrak{O})$.

13.5.57 Remark One can reformulate the invariants of the Carlitz module in terms of Zariski sheaves and shtuka (see Remark 13.5.116) thereby reinterpreting them à la *motivic cohomology*, see [2762].

13.5.7 The L-series of a Drinfeld module

13.5.58 Remark Let $\mathbb{F}_\infty \subset K$ be the field of constants and let $\pi \in K$ be a *fixed* element of order 1 at ∞. Every element $x \in K^*$ has a decomposition $x = \zeta_{x,\pi}\pi^{v_\infty(x)}\langle x\rangle_\pi$ with $\zeta_{x,\pi} \in \mathbb{F}_\infty^*$ and $\langle x\rangle_\pi \equiv 1 \pmod{\pi}$; x is *positive* if $\zeta_{x,\pi} = 1$ (generalizing the notion of "monic polynomial"). Let $\mathcal{I}$ be the group of A-fractional ideals with subgroups $\mathcal{P}^+ \subseteq \mathcal{P}$ where $\mathcal{P}$ (resp. $\mathcal{P}^+$) is the group of principal (resp. and positively) generated ideals; $\mathcal{I}/\mathcal{P}^+$ is finite, etc.

13.5.59 Remark Let $U_1 \subset \mathbb{C}_\infty$ be the group of elements $1 + w$ where $|w|_\infty < 1$. The binomial theorem implies that U_1 is a $\mathbb{Z}_p$-module in the usual exponential fashion; as we can also take p-power roots uniquely in U_1, it is in fact a $\mathbb{Q}_p$-vector space. Let $I = (i) \in \mathcal{P}^+$ where i is positive and define $\langle I\rangle_\pi := \langle i\rangle_\pi$. As U_1 is divisible the next result follows directly.

13.5.60 Proposition [1333] The map $I \mapsto \langle I\rangle_\pi$ extends uniquely to a homomorphism $I \mapsto \langle I\rangle_\pi$ of $\mathcal{I}$ to U_1.

13.5.61 Remark Let $\deg(I)$ denote the degree over $\mathbb{F}_q$ of the associated divisor on Spec(A).

13.5.62 Definition We set $\mathbb{S}_\infty := \mathbb{C}_\infty \times \mathbb{Z}_p$. For a fractional ideal I and $s = (x, y) \in \mathbb{S}_\infty$, we set $I^s := x^{\deg(I)}\langle I\rangle_\pi^y$.

13.5.63 Remark Let $\pi_* \in \mathbb{C}_\infty$ be a fixed d_∞-th root of π and let $j \in \mathbb{Z}$; set $s_j := (\pi_*^{-j}, j) \in \mathbb{S}_\infty$. For positive $a \in A$, we have $(a)^{s_j} = a^j$ (with the standard meaning).

13.5.64 Remark Let $E, \mathfrak{O}$ be as above and suppose ϕ is a Drinfeld module over E. Let N_k^E be the norm mapping on $\mathfrak{O}$-ideals. For each $\mathfrak{O}$-prime $\mathfrak{p}$ of good reduction, we let $f_\mathfrak{p}(u) \in A[u]$ be the characteristic polynomial of the reduction as in Subsection 13.5.3. The next definition builds on classical theory.

13.5.65 Definition We formally put $L(\phi, s) := \prod_{\mathfrak{p}\text{ good}} f_\mathfrak{p}(N_k^E\mathfrak{p}^{-s})^{-1}$ for $s \in \mathbb{S}_\infty$.

13.5.66 Remark Theorem 13.5.34 immediately implies that $L(\phi, s)$ converges on a "half-plane" of $\mathbb{S}_\infty$ consisting of (x, y) with $|x|_\infty$ bounded below (see [1208] for factors at the finitely many bad primes).

13.5.67 Theorem [335] The function $L(\phi, s)$ analytically continues to a continuous (in $y \in \mathbb{Z}_p$) family of entire power series in x^{-1} which is also continuous on $\mathbb{S}_\infty$.

13.5.68 Remark Note that these entire power series are obtained by expanding the Euler products and "summing by degree."

13.5.69 Corollary [335] Let j be a nonnegative integer. Then $L(\phi, (x, -j))$ is a *polynomial* in x^{-1}.

13.5.70 Remark These *special polynomials* play an essential role in the theory. By [335], their degree grows *logarithmically* in j. They interpolate continuously to entire families at *all* places of

k. Moreover, this logarithmic growth allows one to analogously handle *all* associated *partial* L-series obtained by summing only over a residue class [1334].

13.5.71 Definition We set $\zeta_A(s) := \sum_I I^{-1} = \prod_{\mathfrak{p},\,\text{prime}}(1-\mathfrak{p}^{-s})^{-1}$ where I runs over the nonzero ideals of A, etc. In a similar fashion one defines the zeta function of the A-integers in a finite extension of k.

13.5.72 Example Let $A = \mathbb{F}_q[t]$ and let "positive=monic." Then $L(C, s) = \zeta_A(s-1)$.

13.5.8 Special values

13.5.73 Remark Let ξ be a *Hayes-period* (as in Example 13.5.23). Judicious use of the associated exponential function gives the next result.

13.5.74 Theorem [1333] Let j be a positive integer divisible by $q^{d_\infty} - 1$. Then $0 \neq \zeta_A(j)/\xi^j$ is algebraic over k.

13.5.75 Example In the case $A = \mathbb{F}_q[t]$, Carlitz in the 1930's defined a suitable "factorial" element and thus analogs of Bernoulli numbers; these are called "Bernoulli-Carlitz" elements.

13.5.76 Theorem [1333] Let j be as in Theorem 13.5.74. Then $\zeta_A(-j) = 0$.

13.5.77 Remark Clearly, for all $j \in \mathbb{Z}$, we have $\zeta_A(jp) = \zeta_A(j)^p$. For the rest of this subsection, we let $A = \mathbb{F}_q[t]$, etc. In this case, one knows that the order of zero (i.e., of the polynomial $\zeta_A(x, -j)$ at $x = 1$) in Theorem 13.5.76 is exactly 1.

13.5.78 Theorem [582] The *only* algebraic relations on $\{\zeta_A(j)\}$, $j > 0$, arise from Theorem 13.5.74 and the p-th power mapping.

13.5.79 Remark The obvious analog of Theorem 13.5.78 *without* the p-th power mapping is *conjectured* for the Riemann zeta function.

13.5.80 Remark Let $E, \mathfrak{O}$ be as in Subsection 13.5.6. One lets $|H(C/\mathfrak{O})|$ be the monic generator of the associated *Fitting ideal* and one defines a natural *regulator* $R(C/\mathfrak{O})$ using $\hat{M}(C/\mathfrak{O})$. Taelman then establishes the next fundamental result.

13.5.81 Theorem [2764] We have $\zeta_{\mathfrak{O}}(1) = |H(C/\mathfrak{O})|R(C/\mathfrak{O})$.

13.5.82 Remark Let ϕ be as in Subsection 13.5.6. Using the *dual* representation of the Tate module of ϕ, one defines the L-series $L(\phi^\vee, s)$. Theorem 13.5.81 then generalizes to a formula for $L(\phi^\vee, 0)$.

13.5.83 Remark In [2792], Thakur extends Theorem 13.5.81 to some basic examples involving certain factorial, non-rational A thus indicating that Theorem 13.5.81 should hold in general.

13.5.84 Remark In [2763] Taelman establishes the analog for the Bernoulli-Carlitz elements and the class module of the classical *Theorem of Herbrand-Ribet.*

13.5.9 Measures and symmetries

13.5.85 Remark Let $A = \mathbb{F}_q[t]$, $\mathfrak{p}$ a prime of A of degree d, and let $\pi_\mathfrak{p} \in A$ have order 1 at $\mathfrak{p}$; thus the completion $A_\mathfrak{p}$ is isomorphic to $\mathbb{F}[[\pi_\mathfrak{p}]]$ where $\mathbb{F} \simeq \mathbb{F}_{q^d}$. Set $R = A_\mathfrak{p}$ equipped with the canonical topology.

13.5.86 **Definition** An *R-valued measure on R* is a finitely additive R-valued function on the compact opens of R.

13.5.87 **Remark** Let $f\colon R \to R$ be a continuous function and μ a measure; it is easy to see that the associated "Riemann sums" will converge to an element denoted "$\int_R f(t)\,d\mu(t)$." Let μ and ν be two R-valued measures; it is clear that their sum is also an R-valued measure. Their product is defined, as usual, by convolution as below.

13.5.88 **Definition** Let μ, ν be as above. Their *convolution* $\mu * \nu$ is the R-valued measure on R given by $\int_R f(t)\,d\mu * \nu(t) := \int_R \int_R f(x+y)\,d\mu(x)d\nu(y)$.

13.5.89 **Remark** It is easy to check that the space of measures forms a commutative R-algebra under convolution. Now let $\mathfrak{D}_j$ be the hyperdifferential operator given by $\mathfrak{D}_j x^i := \binom{i}{j} x^{i-j}$. Notice that $\mathfrak{D}_i \mathfrak{D}_j = \binom{i+j}{i} \mathfrak{D}_{i+j}$. Let $R\{\{\mathfrak{D}\}\}$ be the algebra of formal power-series in the $\mathfrak{D}_i$ with the above multiplication rule. Building certain bases for the continuous functions out of additive polynomials (following Carlitz) one can establish the next result.

13.5.90 **Theorem** [1333] There is an isomorphism (which depends on the basis constructed) of the R-algebra of measures and $R\{\{\mathfrak{D}\}\}$.

13.5.91 **Remark** Let $y \in \mathbb{Z}_p$ be written as $y = \sum_{i=0}^{\infty} c_i q^i$, where $0 \le c_i < q$. Let ρ be a permutation of the set $\{0, 1, 2, \ldots\}$.

13.5.92 **Definition** We define $\rho_*(y)$, $y \in \mathbb{Z}_p$, by $\rho_*(y) := \sum_{i=0}^{\infty} c_i q^{\rho(i)}$. We let $S_{(p)}$ denote the induced group of bijections of $\mathbb{Z}_p$.

13.5.93 **Theorem** [1335] The map ρ_* is a *homeomorphism* of $\mathbb{Z}_p$ and $\rho(y_0 + y_1) = \rho(y_0) + \rho(y_1)$ if there is *no* carry-over of q-adic digits in the sum of y_0 and y_1. Furthermore, ρ_* stabilizes both the positive and negative integers. Finally, $n \equiv \rho_*(n) \pmod{q-1}$ for all integers n.

13.5.94 **Remark** It is remarkable that $S_{(q)}$ *appears* to act as a group of symmetries of the L-series of Drinfeld modules. We present some evidence of this here and refer the reader to [1335] for more details. Our first result is completely general and is an application of *Lucas' formula* for the reduction modulo p of binomial coefficients.

13.5.95 **Proposition** [1335] Let $\rho_* \in S_{(p)}$, $y \in \mathbb{Z}_p$, and j a nonnegative integer. Then we have $\binom{\rho_*(y)}{\rho_*(j)} \equiv \binom{y}{j} \pmod{p}$.

13.5.96 **Theorem** [1335] Let j be a nonnegative integer. Then, as polynomials in x^{-1}, $\zeta_A(x, -j)$ and $\zeta_A(x, -\rho_*(j))$ have the same degree.

13.5.97 **Remark** Theorem 13.5.96 also depends on *another* old formula of Carlitz (resurrected by Thakur) that allows one to compute the relevant degrees; these degrees then turn out to be *invariant* of the ρ_* action as a corollary of Theorem 13.5.93. Using Lucas' formula again, one also establishes the next result.

13.5.98 **Theorem** [1335] Let ρ_* be as above. Then the map $\mathfrak{D}_i \to \mathfrak{D}_{\rho_*(i)}$ gives rise to an *automorphism* of $R\{\{\mathfrak{D}\}\}$.

13.5.99 **Remark** Thus, of course, $S_{(p)}$ also acts as automorphisms of measure algebras. Moreover, long ago Carlitz computed the denominators of his Bernoulli-Carlitz elements (as in Example 13.5.75) in analogy with the classical result of von Staudt-Clausen. It is quite remarkable that the condition that a prime $f \in A$ of degree d divides this denominator is *invariant* under $S_{(q^d)}$.

13.5.100 Remark We finish this subsection by noting that, *in examples* including $A = \mathbb{F}_q[t]$, the zeroes of $\zeta_A(s)$ *"lie on a line"* (with perhaps finitely many exceptions for nonrational A) with orders agreeing with classical predictions. For this, see [2611, 2892].

13.5.10 Multizeta

13.5.101 Remark L-series of Drinfeld modules give rise to entire power series upon summing by degree. If one takes these sums and intermixes them, one obtains "multi-L-series" whose study was initiated by Thakur (see for example [2792]). One obtains a remarkably rich edifice with analogs of many essential classical results such as shuffle relations and the realization of multizeta values as periods, etc. For instance, Anderson and Thakur [97] have shown the analog of a result of Terasoma giving an interpretation of multizeta values as *periods* of "mixed Carlitz-Tate t-modules."

13.5.11 Modular theory

13.5.102 Remark Let A again be a general base ring. Let M be a rank d lattice as in Definition 13.5.18. Let $\{m_1, \ldots, m_d\} \in M$ be chosen so that $M = Am_1 + \cdots + Am_{d-1} + Im_d$ where I is a nonzero ideal of A (as can always be done since A is a Dedekind domain). The discreteness of M implies that $(m_1, \ldots, m_d)$ does *not* belong to any hyperplane defined over $K = k_\infty$.

13.5.103 Definition We define $\Omega^d := \mathbb{P}^{d-1}(\mathbb{C}_\infty) \setminus Y_d$, where Y_d is the subset of all points lying in K-hyperplanes.

13.5.104 Remark The space Ω^d comes equipped with the structure of a rigid analytic space. It is *connected* (in the appropriate rigid sense) [2847]; in particular, functions are determined by their local expansions.

13.5.105 Example Let $d = 2$. Then $\Omega^2 = \mathbb{P}^1(\mathbb{C}_\infty) - \mathbb{P}^1(K)$. This space is analogous to the classical upper *and* lower half-planes.

13.5.106 Remark Let M, $\{m_1, \ldots, m_d\}$, be as above; clearly M is isomorphic (Definition 13.5.18) to the lattice $\hat{M} := A + A\frac{m_2}{m_1} + \cdots + A\frac{m_{d-1}}{m_1} + I\frac{m_d}{m_1}$ which is, itself, associated to the element $(m_2/m_1, \ldots, m_d/m_1) \in \Omega^d$. It is therefore reasonable that *modular spaces* associated to Drinfeld modules (i.e., spaces whose points parametrize Drinfeld modules with, perhaps, some added structure on their torsion points) can be described using Ω^d in a manner exactly analogous to elliptic modular curves and the classical upper half plane. Indeed, this was accomplished by Drinfeld in his original paper [919].

13.5.107 Remark Drinfeld's construction can be quite roughly sketched as follows: For simplicity, assume that $I = A$ and let $J \subset A$ be a nontrivial ideal. Let $\Gamma_J \subset \mathrm{GL}_d(A)$ be the principal congruence subgroup. Then Γ_J acts on Ω^d in complete analogy with the standard action of congruence subgroups of $\mathrm{SL}_2(\mathbb{Z})$. This action of Γ_J is *rigid analytic* and the quotient $\Gamma_J \backslash \Omega^d$ also exists as a rigid space which is denoted M_J^d.

13.5.108 Theorem [919] The space M_J^d is a regular affine variety of dimension d. It is equipped with a natural morphism to $\mathrm{Spec}(A)$ which is flat and smooth away from the primes dividing J.

13.5.109 Remark The space M_J^2 is a relative curve highly analogous to elliptic modular curves. Indeed, it also can be compactified via "Tate objects" at a finite number of cusps.

13.5.110 Example Let $I = A$ before and let $\Gamma = \mathrm{SL}_2(A)$. Let $e_A(z)$ be the exponential function associated to A considered as a rank 1 lattice (Definition 13.5.18). Set $E_A(z) := e_A(z)^{-1}$.

Then one shows [1332] that $E_A(z)$ is a uniformizing parameter at the infinite cusp for Γ acting on Ω^2 in analogy with $e^{2\pi i z}$ and the upper half plane classically. (We note that our notation "$E_A(z)$" is highly nonstandard but necessary as the commonly used symbols are already taken here.)

13.5.111 Remark Given the analogy between the spaces Ω^d (and especially Ω^2) and the classically upper half plane, it makes sense to discuss modular forms in this context [1332]. For simplicity, let $\Gamma = \mathrm{GL}_2(A)$.

13.5.112 Definition [1267] A rigid analytic function f on Ω^2 is a *modular form of weight* k *and type* m (for a nonnegative integer k and class m of $\mathbb{Z}/(q-1)$) if and only if

1. for $\gamma = \begin{pmatrix} a & b \\ c & d \end{pmatrix} \in \Gamma$ one has $f\left(\frac{az+b}{cz+d}\right) = \det \gamma^{-m}(cz+d)^k f(z)$,
2. f is holomorphic at the cusps.

13.5.113 Remark Property 1 of Definition 13.5.112 implies that $f(z)$ has a Laurent-series expansion in terms of $E_A(z)$ and Property 2 is the requirement that this expansion has no negative terms; similar expansions are mandated at the other cusps. Modular forms of a given weight and type then comprise *finite dimensional* vector spaces.

13.5.114 Example Let $A = \mathbb{F}_q[t]$. A Drinfeld module ρ of rank 2 is determined by $\rho_t = t\tau^0 + g\tau + \Delta\tau^2$. Here g is a modular form of weight $q-1$ and type 0 and Δ is a form of weight q^2-1 and type 0.

13.5.115 Remark One can readily equip such modular forms with an action of the *Hecke operators* which are related to Galois representations. For this see, for example, [125, 126, 334, 1267].

13.5.116 Remark The modular theory has been highly important in a number of ways. In analogy with work on the Korteweg-de Vries equation, Drinfeld found a way to sheafify his modules, (see, for example, [2201] for a very nice account). These sheaves, called "shtuka," have been essential to the work of Lafforgue completing the *Langlands program* for function fields and the general linear group [1829, 1830].

13.5.117 Remark Building on the notion of shtuka, in a fundamental paper Anderson [94] generalized Drinfeld modules to "t-modules" by replacing the additive group with additive n-space. Such t-modules have a tensor product as well as associated exponential functions and Anderson presents an extremely useful criterion for the surjectivity of these functions. The paper [94] has played a key role in many developments.

13.5.118 Remark Another application of the Drinfeld modular curves has been in the algebraic-geometric construction of codes due to Goppa. For example, the reader may consult [2280, 2881].

13.5.119 Remark Finally, we mention the very recent result of Pellarin [2375] which establishes a *two-variable* deformation of Theorem 13.5.74. In particular, one obtains deformations of Bernoulli-Carlitz elements.

13.5.12 Transcendency results

13.5.120 Remark There is a vast theory related to the transcendency of elements arising in the theory of Drinfeld modules and the like. For instance, the Carlitz period (Example 13.5.23) was shown to be transcendental by Wade [2888]. Since then a vast number of other results and techniques were introduced into the theory including ideas from *automata theory.* For a *sampling* of what can be found, we refer the reader to [78, 79, 96, 581, 582, 3039, 3040].

References Cited: [78, 79, 94, 96, 97, 125, 126, 334, 335, 354, 581, 582, 919, 920, 1208, 1267, 1332, 1333, 1334, 1335, 1449, 1450, 1451, 1829, 1830, 1846, 2201, 2280, 2375, 2419, 2585, 2611, 2760, 2761, 2762, 2763, 2764, 2765, 2767, 2773, 2792, 2793, 2847, 2881, 2888, 2892, 3039, 3040, 3084]

III

Applications

III

Applications

14

Combinatorial

14.1 Latin squares

Gary L. Mullen, The Pennsylvania State University

14.1.1 Definition A *latin square of order* n is an $n \times n$ array based upon n distinct symbols with the property that each row and each column contains each of the n symbols exactly once.

14.1.2 Example The following are latin squares of orders 3 and 5

$$\begin{array}{ccc} 0 & 1 & 2 \\ 1 & 2 & 0 \\ 2 & 0 & 1 \end{array}, \qquad \begin{array}{ccccc} 1 & 2 & 3 & 4 & 0 \\ 3 & 4 & 0 & 1 & 2 \\ 0 & 1 & 2 & 3 & 4 \\ 2 & 3 & 4 & 0 & 1 \\ 4 & 0 & 1 & 2 & 3 \end{array}.$$

14.1.3 Remark Given any latin square of prime power order q (with symbols from $\mathbb{F}_q$, the finite field of order q), using the Lagrange Interpolation Formula from Theorem 2.1.131, we can construct a polynomial $P(x, y)$ of degree at most $q - 1$ in both x and y which represents the given latin square. The field element $P(a, b)$ is placed at the intersection of row a and column b. For example, the two squares given in the previous example can be represented by the polynomials $x + y$ over $\mathbb{F}_3$ and $2x + y + 1$ over $\mathbb{F}_5$.

14.1.4 Definition Assume that a latin square of order n is based upon the n distinct symbols $1, 2, \ldots, n$. Such a latin square of order n is *reduced* if the first row and first column are in the standard order $1, 2, \ldots, n$. Let l_n denote the number of reduced latin squares of order n. Let L_n denote the total number of distinct latin squares of order n.

14.1.5 Theorem [706] For each $n \geq 1$, $L_n = n!(n-1)!l_n$.

14.1.6 Remark Using the addition table of the ring $\mathbb{Z}_n$ of integers modulo n, it is easy to see that $l_n \geq 1$ and hence $L_n \geq n!(n-1)!$ for each $n \geq 2$. The total number L_n of latin squares of order n is unknown if $n > 11$ [2053]. The table from [706, p. 142], gives the values of l_n for $n \leq 11$.

14.1.7 Definition Two latin squares of order n are *orthogonal* if upon placing one of the squares on top of the other, we obtain each of the possible n^2 distinct ordered pairs exactly once. In addition, a set $\{L_1, \ldots, L_t\}$ of latin squares all of the same order is *orthogonal* if each distinct pair of squares is orthogonal, i.e., if L_i is orthogonal to L_j whenever $i \neq j$. Such a set of squares is a set of *mutually orthogonal latin squares (MOLS)*.

14.1.8 Remark There are numerous combinatorial objects which are equivalent to sets of MOLS. These include transversal designs, orthogonal arrays, edge-partitions of a complete bipartite graph, and (k, n)-nets. We refer to Chapter III, Theorem 3.18 of [706] for a more detailed discussion of these topics; see also Sections 14.5 and 14.7.

14.1.9 Definition Let $N(n)$ denote the maximum number of mutually orthogonal latin squares (MOLS) of order n.

14.1.10 Theorem [706] For $n \geq 2$, $N(n) \leq n - 1$.

14.1.11 Definition A set $\{L_1, \ldots, L_t\}$ of MOLS of order n is *complete* if $t = n - 1$.

14.1.1 Prime powers

14.1.12 Theorem [355] For any prime power q, the polynomials $ax + y$ with $a \neq 0 \in \mathbb{F}_q$ represent a complete set of $q-1$ MOLS of order q by placing the field element $ax+y$ at the intersection of row x and column y of the a-th square.

14.1.13 Remark In Subsection 14.1.5 we discuss connections of complete sets of MOLS with other combinatorial objects; in particular with affine and projective planes where it is stated that the existence of a complete set of MOLS of order n is equivalent to the existence of an affine, or projective, plane of order n.

14.1.14 Example The following gives a complete set of 4 MOLS of order 5, arising from the polynomials $x + y, 2x + y, 3x + y, 4x + y$ over the field $\mathbb{F}_5$

$$\begin{matrix} 0 & 1 & 2 & 3 & 4 \\ 1 & 2 & 3 & 4 & 0 \\ 2 & 3 & 4 & 0 & 1 \\ 3 & 4 & 0 & 1 & 2 \\ 4 & 0 & 1 & 2 & 3 \end{matrix}, \quad \begin{matrix} 0 & 1 & 2 & 3 & 4 \\ 2 & 3 & 4 & 0 & 1 \\ 4 & 0 & 1 & 2 & 3 \\ 1 & 2 & 3 & 4 & 0 \\ 3 & 4 & 0 & 1 & 2 \end{matrix}, \quad \begin{matrix} 0 & 1 & 2 & 3 & 4 \\ 3 & 4 & 0 & 1 & 2 \\ 1 & 2 & 3 & 4 & 0 \\ 4 & 0 & 1 & 2 & 3 \\ 2 & 3 & 4 & 0 & 1 \end{matrix}, \quad \begin{matrix} 0 & 1 & 2 & 3 & 4 \\ 4 & 0 & 1 & 2 & 3 \\ 3 & 4 & 0 & 1 & 2 \\ 2 & 3 & 4 & 0 & 1 \\ 1 & 2 & 3 & 4 & 0 \end{matrix}.$$

14.1.15 Theorem For $q \geq 5$ an odd prime power, the polynomials $ax + y, a \neq 0, 1, -1 \in \mathbb{F}_q$ give a (maximal) set of $q - 3$ MOLS of order q, each of which is diagonal, i.e., which has distinct elements on both of the main diagonals. When $q \geq 4$ is even the same construction with $a \neq 0, 1 \in \mathbb{F}_q$ gives a (maximal) set of $q - 2$ diagonal MOLS of order q.

14.1.16 Remark The construction of sets of infinite latin squares containing nested sets of mutually orthogonal finite latin squares is discussed in [397, 398]. The construction involves use of polynomials of the form $ax + y$ over infinite algebraic extensions of finite fields.

14.1.17 Remark If q is odd, a latin square of order $q - 1$ which is the multiplication table of the group $\mathbb{F}_q^*$, is mateless; i.e., there is no latin square which is orthogonal to the given square. In fact, a latin square arising from the multiplication table of a cyclic group of even order is mateless; $\mathbb{F}_q$ with q odd is such an example.

14.1.18 Conjecture [1875] A complete set of $n - 1$ MOLS of order n exists if and only if n is a prime power.

14.1.19 Remark The above conjecture is the *prime power conjecture*, and is discussed in many articles. In [2177] this conjecture is referred to as the *next Fermat problem.*

14.1.2 Non-prime powers

14.1.20 Definition If A is a latin square of order m and B is a latin square of order n, denote the entry at row i and column j of A by a_{ij}. Similarly we denote the (i, j) entry of B by b_{ij}. Then the *Kronecker product* of A and B is the $mn \times mn$ square $A \otimes B$, given by

$$A \otimes B = \begin{matrix} (a_{11}, B) & (a_{12}, B) & \cdots & (a_{1m}, B) \\ (a_{21}, B) & (a_{22}, B) & \cdots & (a_{2m}, B) \\ \vdots & \vdots & & \vdots \\ (a_{m1}, B) & (a_{m2}, B) & \cdots & (a_{mm}, B) \end{matrix}$$

where for each entry a of A, (a, B) is the $n \times n$ matrix

$$(a, B) = \begin{matrix} (a, b_{11}) & (a, b_{12}) & \cdots & (a, b_{1n}) \\ (a, b_{21}) & (a, b_{22}) & \cdots & (a, b_{2n}) \\ \vdots & \vdots & & \vdots \\ (a, b_{n1}) & (a, b_{n2}) & \cdots & (a, b_{nn}) \end{matrix}.$$

14.1.21 Example As an illustration of this Kronecker product construction, for $m = 2, n = 3$ let

$$A = \begin{matrix} 0 & 1 \\ 1 & 0 \end{matrix}, \qquad B = \begin{matrix} 0 & 1 & 2 \\ 1 & 2 & 0 \\ 2 & 0 & 1 \end{matrix}.$$

Then the Kronecker product construction using A and B yields the following 6×6 square whose elements are ordered pairs:

00	01	02	10	11	12
01	02	00	11	12	10
02	00	01	12	10	11
10	11	12	00	01	02
11	12	10	01	02	00
12	10	11	02	00	01

14.1.22 Lemma If H and K are latin squares of orders n_1 and n_2, then $H \otimes K$ is a latin square of order $n_1 n_2$.

14.1.23 Lemma If H_1 and H_2 are orthogonal latin squares of order n_1 and K_1 and K_2 are orthogonal latin squares of order n_2, then $H_1 \otimes K_1$ and $H_2 \otimes K_2$ are orthogonal latin squares of order $n_1 n_2$.

14.1.24 Corollary If there is a pair of MOLS of order n and a pair of MOLS of order m, then there is a pair of MOLS of order mn.

14.1.25 Theorem If $n = q_1 \cdots q_r$, where the q_i are distinct prime powers with $q_1 < \cdots < q_r$, then $N(n) \geq q_1 - 1$.

14.1.26 Remark In 1922, MacNeish [1988] conjectured that $N(n) = q_1 - 1$. This has been shown to be false for all non-prime power values of $n \leq 62$; it is in fact conjectured in [1876] that this conjecture is false at all values of n other than 6 and prime powers.

14.1.3 Frequency squares

14.1.27 Definition Let $n = \lambda m$. An $F(n;\lambda)$ *frequency square* is an $n \times n$ square based upon m distinct symbols so that each of the m symbols occurs exactly λ times in each row and column. Thus an $F(n;1)$ frequency square is a latin square of order n. Two $F(n;\lambda)$ frequency squares are *orthogonal* if when one square is placed on top of the other, each of the m^2 possible distinct ordered pairs occurs exactly λ^2 times [2173]. A set of such squares is *orthogonal* if any two distinct squares are orthogonal. Such a set of mutually orthogonal squares is a set of *MOFS*.

14.1.28 Theorem [1456] The maximum number of MOFS of the form $F(n;\lambda)$ is bounded above by $(n-1)^2/(m-1)$.

14.1.29 Theorem [2173] If q is a prime power and $i \geq 1$ is an integer, a complete set of $(q^i-1)^2/(q-1)$, $F(q^i;q^{i-1})$ MOFS can be constructed using the linear polynomials $a_1x_1 + \cdots + a_{2i}x_{2i}$ over the field $\mathbb{F}_q$ where

1. The vector $(a_1,\ldots,a_i) \neq (0,\ldots,0)$,
2. The vector $(a_{i+1},\ldots,a_{2i}) \neq (0,\ldots,0)$,
3. The vector $(a'_1,\ldots,a'_{2i}) \neq e(a_1,\ldots,a_{2i})$ for any $e \neq 0 \in \mathbb{F}_q$.

14.1.4 Hypercubes

14.1.30 Definition A *d-dimensional hypercube* of order n is an $n \times \cdots \times n$ array with n^d points based on n distinct symbols with the property that if any single coordinate is fixed, each of the n symbols occurs exactly n^{d-2} times in that subarray. Such a hypercube is of *type* j, $0 \leq j \leq d-1$ if whenever any j of the coordinates are fixed, each of the n symbols appears n^{d-j-1} times in that subarray. Note that the definition implies that a hypercube of type j is also of types $0,1,\ldots,j-1$.

14.1.31 Definition Two hypercubes are *orthogonal* if, when superimposed, each of the n^2 ordered pairs appears n^{d-2} times. Again the $d=2$ case reduces to that of latin squares. A set of $t \geq 2$ hypercubes is *orthogonal* if every pair of distinct hypercubes is orthogonal.

14.1.32 Theorem [1876] The maximum number of mutually orthogonal hypercubes of order $n \geq 2$, dimension $d \geq 2$, and fixed type j with $0 \leq j \leq d-1$ is bounded above by

$$\frac{1}{n-1}\left(n^d - 1 - \binom{d}{1}(n-1) - \binom{d}{2}(n-1)^2 - \cdots - \binom{d}{j}(n-1)^j\right).$$

14.1.33 Corollary The maximum number of order n, dimension d, and type 1 hypercubes is bounded above by

$$N_d(n) \leq \frac{n^d-1}{n-1} - d.$$

14.1.34 Remark In the case that $d=2$, $N_d(n)$ reduces to the familiar bound of $n-1$ for sets of MOLS of order n. As was the case for $d=2$, the bound for $d>2$ can always be realized when n is a prime power.

14.1.35 Corollary There are at most

$$(n-1)^{d-1} + \binom{d}{d-1}(n-1)^{d-2} + \cdots + \binom{d}{j+1}(n-1)^j$$

cube 1	cube 2	cube 3	cube 4	cube 5	cube 6	cube 7	cube 8	cube 9	cube 10
012	012	012	012	000	000	012	012	012	012
120	201	012	012	111	111	120	201	201	120
201	120	012	012	222	222	201	120	120	201
012	012	120	201	111	222	120	201	120	201
120	201	120	201	222	000	201	120	012	012
201	120	120	201	000	111	012	012	201	120
012	012	201	120	222	111	201	120	201	120
120	201	201	120	000	222	012	012	120	201
201	120	201	120	111	000	120	201	012	012
$x+y$	$2x+y$	$y+z$	$y+2z$	$x+z$	$x+2z$	$x+y+z$	$2x+y+2z$	$2x+y+z$	$x+y+2z$

Figure 14.1.1 A complete set of mutually orthogonal cubes of order 3.

hypercubes of order n, type j, and dimension d.

14.1.36 Theorem The polynomials $a_1x_1 + \cdots + a_dx_d$ with

1. the elements $a_i \in \mathbb{F}_q$ for $i = 1, \ldots, d$ with at least $j+1$ of the $a_i \neq 0$,
2. and $(a'_1, \ldots, a'_d) \neq e(a_1, \ldots, a_d)$ for any $e \neq 0 \in \mathbb{F}_q$, represent a complete set of mutually orthogonal hypercubes of dimension d, order q, and type j.

14.1.37 Remark In [2155] another definition of orthogonality for hypercubes, called *equi-orthogonality* is studied. In [992, 993] sets of hypercubes using various other definitions of orthogonality are considered. Such stronger definitions of orthogonality turn out to be useful in the study of MDS codes (see Section 15.1). In one definition, not only does one keep track of the total number of times that ordered pairs occur, but their locations are also taken into account. In other definitions, one studies various notions of orthogonality involving more than the usual two hypercubes at a time. In all of these definitions, polynomials over finite fields are used to construct complete sets of such orthogonal hypercubes of prime power orders.

14.1.38 Remark In [2737] sets of very general mutually orthogonal frequency hyperrectangles of prime power orders are constructed using linear polynomials over finite fields.

14.1.5 Connections to affine and projective planes

14.1.39 Remark Affine and projective planes are discussed in Section 14.3. We first state the following fundamental result; and then discuss a few other related results.

14.1.40 Theorem [355], [706, Theorem III.3.20] There exists a projective plane (or an affine plane) of order n if and only if there exists a complete set of MOLS of order n.

14.1.41 Definition Two complete sets of MOLS of order n are *isomorphic* if after permuting the rows, permuting the columns with a possibly different permutation, and permuting the symbols with a third possibly different permutation of each square of the first set, we obtain the second set of MOLS. See Part III of [706] for further discussion of non-isomorphic sets of MOLS, affine, and projective planes.

14.1.42 Conjecture If p is a prime, any two complete sets of MOLS of order p are isomorphic.

14.1.43 Remark The above conjecture is only known to be true for $p = 3, 5, 7$. Truth of the conjecture would imply that all planes of prime order are Desarguesian.

14.1.44 Theorem [2917, 2918] For $q = p^n$, let $0 \le k < n$, $N = (q-1)/(q-1, p^k-1)$ and set $e = q - N$. Let u be a primitive N-th root of unity in $\mathbb{F}_q$. Assume that $x^{p^k} + c_i x$ is a permutation polynomial for e elements $c_1, \ldots, c_e \in \mathbb{F}_q$, where one can assume that $c_1 - 1 = c_2$. Let $a \neq 0$ and c_1 be such that $f(x) = ax^{p^k} + c_1 x$ is an orthomorphism of $\mathbb{F}_q$ (so f is a permutation polynomial with $f(0) = 0$, and $f(x) - x$ is also a permutation). Let $d_i = c_1 - c_i$. Then the polynomials $au^j x^{p^k} + c_1 x + y, j = 1, \ldots, N; d_i x + y, i = 3, \ldots, e; x + y$ represent a complete set of $q - 1$ MOLS of order q.

14.1.45 Corollary For each $n \ge 2$ and any odd prime p, the above construction gives $\tau(n) \ge 2$, non-isomorphic complete sets of MOLS of order p^n, where $\tau(n)$ denotes the number of positive divisors of n.

14.1.46 Example For any odd prime p, this construction gives an example of a non-Desarguesian affine translation plane of order p^2, constructed without the use of a right quasifield as used in [818].

14.1.47 Remark For $q = 9$, let $\mathbb{F}_9$ be generated by the primitive polynomial $f(x) = x^2 + 2x + 2$ over $\mathbb{F}_3$. Let α be a root of $f(x)$. The Desarguesian plane of order 9 may be constructed by using the polynomials $\alpha^i x + y, i = 0, \ldots, 7$. Since $u = \alpha^2$ is a primitive 4-th root of unity, the construction from the above corollary leads to the polynomials $\alpha x^3 + y, \alpha^3 x^3 + y, \alpha^5 x^3 + y, \alpha^7 x^3 + y$ which represent four MOLS of order 9. To extend these four MOLS to a complete set of 8 MOLS of order 9, we consider the polynomials $x + y, \alpha^2 x + y, \alpha^4 x + y, \alpha^6 x + y$. Thus four of the latin squares are the same in both the Desarguesian and non-Desarguesian constructions.

14.1.6 Other finite field constructions for MOLS

14.1.48 Remark There are other finite field constructions for sets of MOLS; here we briefly allude to a few of them which are described in much more detail in [706]. Quasi-difference matrices and $V(m, t)$ vectors are discussed in Section VI.17.4; self-orthogonal latin squares are considered in Section III.5.6; MOLS with holes are considered in Section III.1.7; starters are studied in VI.55.; and atomic latin squares are studied in Section III.1.6.

See Also

§14.3	Discusses affine and projective planes.
§14.5	Discusses block designs.
[706]	Part III discusses latin squares.
[706]	Part III, Section 3 discusses sets of MOLS.
[1875]	Discusses topics in discrete mathematics with topics motivated by latin squares.

References Cited: [355, 397, 398, 706, 818, 992, 993, 1456, 1875, 1876, 1988, 2053, 2155, 2173, 2177, 2737, 2917, 2918]

14.2 Lacunary polynomials over finite fields

Simeon Ball, Universitat Politècnica de Catalunya
Aart Blokhuis, Eindhoven University of Technology

14.2.1 Introduction

14.2.1 Remark In 1970 Rédei published his treatise *Lückenhafte Polynome über endlichen Körpern* [2444], soon followed by the English translation *Lacunary Polynomials over Finite Fields* [2445], the title of this chapter. One of the important applications of his theory is to give information about the following two sets.

14.2.2 Definition For $f\colon \mathbb{F}_q \to \mathbb{F}_q$, or $f \in \mathbb{F}_q[X]$ define the set of directions (slopes of secants of the graph):

$$D(f) := \left\{ \frac{f(x)-f(y)}{x-y} \,|\, x \neq y \in \mathbb{F}_q \right\}.$$

14.2.3 Definition For $f \in \mathbb{F}_q[X]$ let

$$P(f) := \{m \in \mathbb{F}_q \,|\, f(X) + mX \text{ is a permutation polynomial}\}.$$

14.2.4 Remark The sets $P(f)$ and $D(f)$ partition $\mathbb{F}_q$. If $(f(x) - f(y))/(x - y) = m$ then the polynomial $f(x) + mx = f(y) + my$, so m is a direction determined by f precisely when $f(X) + mX$ is *not* a permutation polynomial (on $\mathbb{F}_q$).

14.2.2 Lacunary polynomials

14.2.5 Definition Let K be a (finite) field. A polynomial $f \in K[x]$ is *fully reducible* if K is a splitting field for f, that is, if f factors completely into linear factors in $K[X]$.

14.2.6 Definition Denote by f° the degree of f, and by $f^{\circ\circ}$ the second degree, the degree of the polynomial we obtain by removing the leading term.

14.2.7 Definition If $f^{\circ\circ} < f^\circ - 1$ then f is *lacunary* and the difference $f^\circ - f^{\circ\circ}$ is the *gap.*

14.2.8 Remark We want to survey what is known about lacunary polynomials (with a large gap) that are fully reducible. In many applications however the gap is not between the degree and the second degree, so instead of being of the form $f(X) = X^n + h(X)$, where $h^\circ \leq n-2$, it is of the more general form $f(X) = g(X)X^n + h(X)$, where $h^\circ \leq n-2$, for some polynomial g.

14.2.9 Example For $d \mid (q-1)$ the field $K = \mathbb{F}_q$ contains the d-th roots of unity, so the polynomial $X^d - a^d$ is fully reducible.

14.2.10 Remark In many applications the degree $f^\circ = q$, as is the case in the following examples.

14.2.11 Example The lacunary polynomials $X^q + c$, $X^q - X$, and if q is odd then $X^q \pm X^{(q+1)/2}$ and $X^q \pm 2X^{(q+1)/2} + X$, are fully reducible in $\mathbb{F}_q[X]$.

14.2.12 Theorem [2445] Let $f(X) = X^p + g(X)$, with $g^\circ = f^{\circ\circ} < p$, be fully reducible in $\mathbb{F}_p[X]$, p prime. Then either g is constant, or $g = -X$ or g° is at least $(p+1)/2$.

14.2.13 Remark Let $s(X)$ be the zeros polynomial of f, that is the polynomial with the same set of zeros as f, but each with multiplicity one. So $s = \gcd(f, X^p - X)$. It follows that

$$s \mid f - (X^p - X) = X + g.$$

We may write $f = s \cdot r$, where r is the fully reducible polynomial that has the zeroes of f with multiplicity one less. Hence r divides the derivative $f' = g'$. So we conclude that

$$f = s \cdot r \mid (X + g)g'.$$

If the right hand side is zero, then either $g = -X$, corresponding to the fully reducible polynomial $f(X) = X^q - X$, or $g' = 0$ which (since $g^\circ < p$) implies $g(X) = c$ for some $c \in K$ and $f(X) = X^p + c = (X+c)^p$. If the right hand side is nonzero, then, being divisible by f, it has degree at least p, so $g^\circ + g^\circ - 1 \geq p$ which gives $g^\circ \geq (p+1)/2$.

14.2.14 Remark In the next section we see how this result can be applied to obtain information about the number of directions determined by a function.

14.2.3 Directions and Rédei polynomials

14.2.15 Definition Let $AG(2, q)$ be the Desarguesian affine plane of order q, where points of $AG(2, q)$ are denoted by pairs (a, b), $a, b \in \mathbb{F}_q$.

14.2.16 Definition Let $PG(2, q)$ be the Desarguesian projective plane of order q with homogeneous point coordinates $(a : b : c)$ and line coordinates $[u : v : w]$. The point $(a : b : c)$ is *incident* with the line $[u : v : w]$ precisely when $au + bv + cw = 0$. The equation of the line $[u : v : w]$ is then $uX + vY + wZ = 0$.

14.2.17 Remark We consider $AG(2, q)$ as part of the projective plane $PG(2, q)$ where $[0 : 0 : 1]$ is the line at infinity, the line with equation $Z = 0$. The affine point (a, b) corresponds to the projective point $(a : b : 1)$.

14.2.18 Definition Let $u = (u_1, u_2)$ and $v = (v_1, v_2)$ be two affine points. The pair u, v determines the *direction* m if the line joining them has slope m, or equivalently, if $(u_2 - v_2)/(u_1 - v_1) = m$.

14.2.19 Definition Let R be a set of q points in $AG(2,q)$. We define $D_R \subseteq \mathbb{F}_q \cup \{\infty\}$ to be the set of directions determined by the pairs of points in R.

14.2.20 Remark The reason we take R to have size q is two-fold. Firstly, in Rédei's formulation of the problem R is the graph of a function f and $D_R = D_f$. Secondly, any set with more than q points determines all directions, by the pigeon hole principle: there are exactly q lines in every parallel class, so if $|R| > q$, then there is a line with at least two points of R in each parallel class. For results concerning the case $|R| < q$, see [2756].

14.2.21 Definition With R we associate its *Rédei polynomial*

$$F(U,W) = \prod_{(a,b)\in R} (W + aU + b).$$

14.2.22 Lemma If the direction $m \notin D_R$ then $F(m,W) = W^q - W$.

14.2.23 Lemma If the direction $m \in D_R$, then $F(m,W)$ is a fully reducible lacunary polynomial of degree q, and second degree at most $|D_R| - 1$.

14.2.24 Theorem [322, Theorem 1] Let R be a set of q points in $AG(2,q)$, and let $N = |D_R|$. Then either $N = 1$, or $N \geq (q+3)/2$, or $2 + (q-1)/(p^e+1) \leq N \leq (q-1)/(p^e-1)$ for some e, $1 \leq e \leq \lfloor n/2 \rfloor$.

14.2.4 Sets of points determining few directions

14.2.25 Remark The third case in Theorem 14.2.24, $2+(q-1)/(p^e+1) \leq N \leq (q-1)/(p^e-1)$ for some e satisfying $1 \leq e \leq \lfloor n/2 \rfloor$, is not sharp. The following are some examples of functions that determine few directions.

14.2.26 Example The function $f(X) = X^{(q+1)/2}$, where q is odd, determines $(q+3)/2$ directions.

14.2.27 Example The function $f(X) = X^s$, where $s = p^e$ is the order of a subfield of $\mathbb{F}_q$, determines $(q-1)/(s-1)$ directions.

14.2.28 Example The function $f(X) = \mathrm{Tr}_{\mathbb{F}_q/\mathbb{F}_s}(X)$, the trace from $\mathbb{F}_q$ to the subfield $\mathbb{F}_s$, determines $(q/s)+1$ directions.

14.2.29 Example If $f(X) \in \mathbb{F}_q[X^s]$, where s is the order of a subfield of $\mathbb{F}_q$ and is chosen maximal with this property, in other words, f is $\mathbb{F}_s$-linear (apart from the constant term) but not linear over a larger subfield, then $(q/s)+1 \leq N \leq (q-1)/(s-1)$.

14.2.30 Remark Motivated by the form of the examples the following theorem was obtained (in a number of steps) by Ball, Blokhuis, Brouwer, Storme, and Szőnyi. Initial results are in [322], then the classification was all but obtained in [321], and completed in [185].

14.2.31 Theorem [185] If, for $f\colon \mathbb{F}_q \to \mathbb{F}_q$, with $f(0) = 0$, the number $N = |D(f)| > 1$ of directions determined by f is less than $(q+3)/2$, then for a subfield $\mathbb{F}_s$ of $\mathbb{F}_q$

$$\frac{q}{s} + 1 \leq N \leq \frac{q-1}{s-1},$$

and if $s > 2$ then f is $\mathbb{F}_s$-linear.

14.2.32 Remark This result is obtained using several lemmas about fully reducible lacunary polynomials which are of independent interest.

14.2.33 Lemma [2445, Satz 18] Let $s = p^e$ be a power of p with $1 \leq s < q$. If

$$X^{q/s} + g(X) \in \mathbb{F}_q[X] \setminus \mathbb{F}_q[X^p]$$

is fully reducible over $\mathbb{F}_q$ then either $s = 1$ and $g(X) = -X$ or

$$g^\circ \geq ((q/s) + 1)/(s + 1).$$

14.2.34 Lemma [185] Let s be a power of p with $1 \leq s < q$ and suppose that

$$X^{q/s} + g(X) \in \mathbb{F}_q[X] \setminus \mathbb{F}_q[X^p]$$

is fully reducible over $\mathbb{F}_q$. If $s > 2$, $g^\circ = q/s^2$ and $2(g')^\circ < g^\circ$ then $X^{q/s} + g(X)$ is $\mathbb{F}_s$-linear.

14.2.35 Remark Theorem 14.2.31 completely characterizes the case in which the number of directions is *small*, that is less than $(q + 3)/2$. In the case that $q = p$ is prime, $N < (p + 3)/2$ implies $N = 1$, and the characterization of $N = (p + 3)/2$ directions was given by Lovász and Schrijver [1961].

14.2.36 Theorem [1961] If $f \in \mathbb{F}_p[X]$, p prime, determines $(p + 3)/2$ directions, then $f(X) = X^{(p+1)/2}$ up to affine equivalence.

14.2.37 Remark Much more can be said in this case, the following surprising theorem by Gács [1151] shows that there is a huge gap in the spectrum of possible number of directions.

14.2.38 Theorem [1151] If the number of directions determined by $f \in \mathbb{F}_p[X]$, p prime, is more than $(p + 3)/2$, then it is at least $\lceil \frac{2}{3}(p - 1) \rceil + 1$.

14.2.39 Remark This bound is almost tight, there are examples that determine $\frac{2}{3}(p-1)+2$ directions if $p \equiv 1 \pmod 3$. Progress was made using Gács' approach in [190] indicating that a further gap is possible from $2p/3$ to $3p/4$. If there is an example with less than $3p/4$ directions then lines meet the graph of f in at most 3 points or at least $p/4$. Futhermore, if there are 3 lines meeting the graph of f in more than 3 points then the graph of f is contained in these 3 lines. There are examples that determine $\frac{3}{4}(p - 1) + 2$ directions if $p \equiv 1 \pmod 4$ and some constructions where $|D(f)| \approx 7p/9$ can be found in [1423].

14.2.40 Remark For results concerning the case $q = p^2$, see [1152]. For related results on functions $f \colon \mathbb{F}_q^k \to \mathbb{F}_q$, with $k \geq 2$, that determine few directions, see [186], and for results on functions $f, g \colon \mathbb{F}_q \to \mathbb{F}_q$, where the set

$$P(f, g) = \{(r, s) \in \mathbb{F}_q^2 \mid X + rf(X) + sg(X) \text{ is a permutation polynomial}\}$$

is large [191].

14.2.5 Lacunary polynomials and blocking sets

14.2.41 Remark Let R be a subset of $AG(2, q)$ of size q. The set of points of $PG(2, q)$

$$B = \{(a : b : 1) \mid (a, b) \in R\} \cup \{(1 : m : 0) \mid m \in D_R\}$$

has the property that every line of $PG(2, q)$ intersects B.

14.2.42 Definition A *blocking set* of $PG(2, q)$ is a set of points B of $PG(2, q)$ with the property that every line of is incident with a point of B.

14.2.43 Lemma [358] A blocking set of $PG(2,q)$ has at least $q+1$ points and equality can only be obtained if these points all are on a line.

14.2.44 Definition A blocking set of $PG(2,q)$ that contains a line is *trivial.*

14.2.45 Remark We tacitly assume that all blocking sets under consideration are *minimal*, so they do not contain a proper subset that is also a blocking set. For blocking sets of non-Desarguesian planes and for further reading on blocking sets see [320, 329, 430, 432, 433, 1150, 1153] and for more recent references, see Remark 14.2.54.

14.2.46 Lemma [319] Suppose that B is a blocking set of size $q+k+1$ and that $(1:0:0) \in B$ and assume that the line with equation $Z=0$, that is $[0:0:1]$ is a tangent to B. Then the non-horizontal lines $[1:u:v]$ are blocked by the affine points of B and the Rédei polynomial of the affine part of B can be written as

$$F(V,W) = (V^q - V)G(V,W) + (W^q - W)H(V,W),$$

where G and H are of total degree k in the variables V and W.

14.2.47 Lemma [319] Let F_0 denote the part of F that is homogeneous of degree $q+k$, and let G_0 and H_0 be the parts of G and H that are homogeneous of total degree k. Restricting to the terms of total degree $q+k$ we obtain the homogeneous equation

$$F_0 = V^q G_0 + W^q H_0,$$

with

$$F_0(V,W) = \prod_{(a:b:1)\in B} (bV + W).$$

Writing $f(W) = F_0(1,W)$ and defining g and h analogously, we obtain a one-variable fully reducible lacunary polynomial in $\mathbb{F}_q[W]$,

$$f(W) = g(W) + W^q h(W).$$

14.2.48 Lemma [320] Let $f \in \mathbb{F}_q[X]$ be fully reducible, and suppose that $f(X) = X^q g(X) + h(X)$, where g and h have no common factor. Let k be the maximum of the degrees of g and h. Then $k=0$, or $k=1$ and $f(X) = a(X^q - X)$ for some $a \in \mathbb{F}_q^*$, or q is prime and $k \geq (q+1)/2$, or q is a square and $k \geq \sqrt{q}$, or $q = p^{2e+1}$ for some prime p and $k \geq p^{e+1}$.

14.2.49 Theorem [430] A non-trivial blocking set B in $PG(2,q)$, q square, has at least $q+\sqrt{q}+1$ points. If equality holds then B consists of the points of a subplane of order $\sqrt{q}$.

14.2.50 Theorem [320] A non-trivial blocking set B in $PG(2,q)$, $q = p^{2e+1}$, p prime, $q \neq p$, has at least $q + p^{e+1} + 1$ points. This bound is sharp only in the case $e=1$.

14.2.51 Theorem [319] A non-trivial blocking set B in $PG(2,p)$, p prime, has at least $\frac{3}{2}(p+1)$ points. If equality holds then every point of B is on precisely $\frac{1}{2}(p-1)$ tangents.

14.2.52 Remark The bound in Theorem 14.2.51 was conjectured in [831].

14.2.53 Remark The proof of Lemma 14.2.48 leads to the following divisibility condition

$$f \mid (Xg+h)(h'g - g'h).$$

It would be good (and probably not infeasible) to characterize the case of equality in the case p is prime, that is find all f, g, and h with f of degree $q+(q+1)/2$, g and h of degree at most $(q+1)/2$ and $f \hat{=} (Xg+h)(h'g-g'h)$, where $a \hat{=} b$ means there exists a scalar $c \in \mathbb{F}_q$ such that $a = cb$. This is the subject of the next section.

14.2.54 Remark Blocking sets in $PG(2,p^n)$, p prime, of size less than $3(p^n+1)/2$ have been classified for $n=2$ [2757] and $n=3$ [2417] and they come from the construction in Remark 14.2.41. However, for $n \geq 4$, there are examples known which are not of this form. These examples, called linear blocking sets, include those obtained by the construction in Remark 14.2.41. It is conjectured that all small blocking sets are linear blocking sets. More precisely, we have the following conjecture which is called the linearity conjecture. For recent articles concerning this conjecture see [1872, 1977, 1978, 2408, 2755, 2757].

14.2.55 Conjecture [2755] If B is a blocking set in $PG(2,p^n)$, p prime, of less than $3(p^n+1)/2$ points then there exists an n-dimensional subspace U of $PG(3n-1,p)$ with the property that every point of B, when viewed as an $(n-1)$-dimensional subspace of $PG(3n-1,p)$, has non-trivial intersection with U.

14.2.6 Lacunary polynomials and blocking sets in planes of prime order

14.2.56 Remark The blocking set problem in $PG(2,p)$, p prime, leads one to search for polynomials $f(X)$, $g(X)$, $h(X)$, where $f = X^p g + h$ factors completely into linear factors and g and h have degree at most $\frac{1}{2}(p+1)$. More precisely, given a blocking set B of size $\frac{3}{2}(p+1)$, for each point $P \in B$, and each tangent ℓ passing through P, there is a polynomial f with the above property. A factor of f of multiplicity e corresponds to a line incident with P distinct from ℓ meeting B in $e+1$ points.

14.2.57 Remark The equation $f \hat{=} (Xg+h)(h'g-g'h)$ has several infinite families of solutions, and some sporadic ones, not all of them necessarily corresponding to blocking sets.

14.2.58 Theorem [323] The following list contains all non-equivalent solutions for $f = X^p g + h$, where f factors completely into linear factors and g and h have degree at most $\frac{1}{2}(p+1)$, for $p < 41$.

1. (For odd p, say $p = 2r+1$.) Take $f(X) = X\prod(X-a)^3$ where the product is over the nonzero squares a. Then f satisfies $f(X) = X(X^r-1)^3 = X^p g + h$ with $g(X) = X^r - 3$, $h(X) = 3X^{r+1} - X$. This would correspond to line intersections of the lines incident with P (with frequencies written as exponents) 1^r, 2^2, 4^r. For $p = 7$ this is the function for the blocking set $\{(1:0:0),(0:1:0),(0:0:1)\} \cup \{(a:b:1) \mid a,b \in \{1,2,4\}\}$.
2. (For $p = 4t+1$.) Take $f(X) = X\prod(X-a)\prod(X-b)^4$ where the product is over the nonzero squares a and fourth powers b. Here $f(X) = X(X^{2t}-1)(X^t-1)^4 = X^p g + h$ with $g(X) = X^{2t} - 4X^t + 5$ and $h(X) = -5X^{2t+1} + 4X^{t+1} - X$. This would correspond to line intersections 1^{2t}, 2^{t+2}, 6^t.
3. (For $p = 4t+1$.) Take $f(X) = X^{t+1}\prod(X-a)\prod(X-b)^2$ where the product is over the nonzero squares a and fourth powers b. Here $f(X) = X^{t+1}(X^{2t}-1)(X^t-1)^2 = X^p g + h$ with $g(X) = X^t - 2$ and $h(X) = 2X^{2t+1} - X^{t+1}$. This would correspond to line intersections 1^{2t}, 2^t, 4^t $(t+2)^2$. For $p = 13$ this is a function for the blocking set $\{(1:0:0),(0:1:0),(0:0:1)\} \cup \{(1:a:0),(0:1:a),(a:0:1) \mid a^3 = -1\} \cup \{(b:c:1) \mid b^3 = c^3 = 1\}$.
4. (For $p = 13$.) Take $f(X) = X\prod(X-a)^4\prod(X-\frac{1}{2}a)$ where the product is over the values a with $a^3 = 1$. Here $f(X) = X(X^3-1)^4(X^3-\frac{1}{8}) = X^p g + h$ with $g(X) = X^3 + 4$ and $h(X) = 5X^7 - 5X^4 - 5X$. This corresponds to line

intersections 1^6, 2^4, 5^4, and indeed occurs.

5. Take $f(X) = X^p - X^{(p+1)/2} = X^{(p+1)/2}\prod(X-a)$ where the product is over the nonzero squares a.
6. Take $f(X) = X^p - 2X^{(p+1)/2} + X = X\prod(X-a)^2$ where the product is over the nonzero squares a.

14.2.59 Remark These lacunary polynomials are just weighted subsets of the projective line, so equivalence means that $f = X^p g + h$ is equivalent to those polynomials obtained under the maps $f(X) \mapsto (cX+d)^{3(p+1)/2} f((aX+b)/(cX+d))$, for some $a, b, c, d \in \mathbb{F}_p$, where $ad \neq bc$.

14.2.60 Theorem [323] Let B be a non-trivial blocking set in $PG(2,p)$ of size $\frac{3}{2}(p+1)$, where p is a prime less than 41. Then either there is a line incident with $(p+3)/2$ points of B (and hence is the example characterized in Theorem 14.2.36) or $p \in \{7, 13\}$ and there is a unique other example in both cases.

14.2.61 Conjecture [323] The restriction $p < 41$ is unnecessary in the above theorem.

14.2.7 Lacunary polynomials and multiple blocking sets

14.2.62 Definition A t-fold blocking set B of $PG(2,q)$ is a set of points such that every line is incident with at least t points of B.

14.2.63 Theorem [328] Let B be a t-fold blocking set in $PG(2,q)$, $q = p^h$, p prime, of size $t(q+1)+c$. Let $c_2 = c_3 = 2^{-1/3}$ and $c_p = 1$ for $p > 3$.

1. If $q = p^{2d+1}$ and $t < q/2 - c_p q^{2/3}/2$, then $c \geq c_p q^{2/3}$, unless $t = 1$ in which case B, with $|B| < q + 1 + c_p q^{2/3}$, contains a line.
2. If $4 < q$ is a square, $t < q^{1/4}/2$ and $c < c_p q^{2/3}$, then $c \geq t\sqrt{q}$ and B contains the union of t disjoint Baer subplanes, except for $t = 1$ in which case B contains a line or a Baer subplane.
3. If $q = p^2$, p prime, and $t < q^{1/4}/2$ and $c < p\left\lceil \frac{1}{4} + \sqrt{\frac{p+1}{2}} \right\rceil$, then $c \geq t\sqrt{q}$ and B contains the union of t disjoint Baer subplanes, except for $t = 1$ in which case B contains a line or a Baer subplane.

14.2.64 Remark For more precise results in the case $t = 2$ see [188]; for $t = 3$ see [184]; for $q = p^3$ see [2416, 2417, 2418]; for $q = p^{6n+3}$ see [328]; and for $q = p^{6n}$ see [328, 2418].

14.2.65 Remark The proof of Theorem 14.2.63 starts with the main theorem of [330] on fully reducible lacunary polynomials.

14.2.66 Theorem [330] Let $f \in \mathbb{F}_q[X]$, $q = p^n$, p prime, be fully reducible, $f(X) = X^q g(X) + h(X)$, where $(g, h) = 1$. Let $k = \max(g^\circ, h^\circ) < q$. Let e be maximal such that f is a p^e-th power. Then we have one of the following:

1. $e = n$ and $k = 0$;
2. $e \geq 2n/3$ and $k \geq p^e$;
3. $2n/3 > e > n/2$ and $k \geq p^{n-e/2} - (3/2)p^{n-e}$;
4. $e = n/2$ and $k = p^e$ and $f(X) = a\mathrm{Tr}(bX+c)+d$ or $f(X) = a\,\mathrm{Norm}(bX+c)+d$ for suitable constants a, b, c, d. Here Tr and Norm respectively denote the trace and norm function from $\mathbb{F}_q$ to $\mathbb{F}_{\sqrt{q}}$;
5. $e = n/2$ and $k \geq p^e\left\lceil \frac{1}{4} + \sqrt{(p^e+1)/2} \right\rceil$;

6. $n/2 > e > n/3$ and $k \geq p^{n/2+e/2} - p^{n-e} - p^e/2$, or if $3e = n + 1$ and $p \leq 3$, then $k \geq p^e(p^e + 1)/2$;
7. $n/3 \geq e > 0$ and $k \geq p^e\lceil (p^{n-e} + 1)/(p^e + 1)\rceil$;
8. $e = 0$ and $k \geq (q + 1)/2$;
9. $e = 0$, $k = 1$ and $f(X) = a(X^q - X)$.

14.2.67 Remark Lacunary polynomials over finite fields and in particular Redei's theorem, Theorem 14.2.12, and Blokhuis' theorem, Theorem 14.2.51, have also been used in algebra, algebraic number theory, group theory, and group factorization. For a survey of these applications, see [2758].

Polynomials over finite fields have been used to tackle a variety of problems associated with incidence geometries. Various extensions of the ideas first used for lacunary polynomials have been studied. This has led to some interesting techniques involving field extensions, algebraic curves which in turn have led to classification, non-existence, and stability results concerning subsets of points of a finite projective spaces with a certain given property. For a recent survey, see [187].

See Also

Chapter 8	For more on permutation polynomials over finite fields.
§14.3	For more on affine and projective planes over finite fields.
§14.4	For more on higher dimensional spaces over finite fields.
§14.9	For more on polynomials over finite fields with restricted weights.

References Cited: [184, 185, 186, 187, 188, 190, 191, 319, 320, 321, 322, 323, 328, 329, 330, 358, 430, 432, 433, 831, 1150, 1151, 1152, 1153, 1423, 1872, 1961, 1977, 1978, 2408, 2416, 2417, 2418, 2444, 2445, 2755, 2756, 2757, 2758]

14.3 Affine and projective planes

Gary Ebert, University of Delaware
Leo Storme, Universiteit Gent

All structures in this section are finite. Reference [1560] is an excellent introduction to projective and affine planes. See Section VII.2 of [706] for a concise description of the Hall, André, Hughes, and Figueroa planes.

14.3.1 Projective planes

14.3.1 Definition A *finite projective plane* is a finite incidence structure of points and lines such that

1. every two distinct points together lie on a unique line;
2. every two distinct lines meet in a unique point;
3. there exists a quadrangle (four points with no three collinear).

14.3.2 Remark If π is a finite projective plane, then there is a positive integer n such that any line of π has exactly $n+1$ points, every point lies on exactly $n+1$ lines, the total number of points is n^2+n+1, and the total number of lines is n^2+n+1. This number n is called the *order* of π.

14.3.3 Construction [1510] The classical examples of finite projective planes are constructed as follows. Let V be a 3-dimensional vector space over the finite field $\mathbb{F}_q$ of order q. Take as *points* the 1-dimensional subspaces of V and as *lines* the 2-dimensional subspaces of V, and let incidence be given by containment. The resulting incidence structure is a finite projective plane of order q, denoted by PG$(2,q)$. These projective planes are *Desarguesian* since they satisfy the classical configurational theorem of Desargues (for instance, see [555]). Note that this construction shows that there exists a finite projective plane of order q for any prime power q. Alternatively, one may use *homogeneous coordinates* $(x:y:z)=\{(fx,fy,fz)\colon f\in \mathbb{F}_q\backslash\{0\}\}$ for the points of PG$(2,q)$, and $[a:b:c]=\{[fa,fb,fc]\colon f\in\mathbb{F}_q\backslash\{0\}\}$ for the lines of PG$(2,q)$, where the point $(x:y:z)$ is incident with the line $[a:b:c]$ if and only if $ax+by+cz=0$.

14.3.4 Remark Some non-classical finite projective planes are discussed in Subsections 14.3.3 to 14.3.5. Many other constructions can be found in [807]. One of the most difficult problems in finite geometry is determining the spectrum of possible orders for finite projective planes. All known examples have prime power order, but it is unknown if this must be true in general. The Bruck-Ryser-Chowla Theorem (Section 14.5) excludes an infinite number of positive integers as possible orders. In addition, order 10 has been excluded via a computer search [1837]. There are precisely four different (*non-isomorphic*, as defined in Subsection 14.3.3) projective planes of order 9, the smallest order for which non-classical examples exist [1836].

An overview of the state of knowledge concerning small projective planes follows:

order n	2	3	4	5	6	7	8	9	10	11	12	13	14	15	16
existence	y	y	y	y	n	y	y	y	n	y	?	y	n	?	y
number	1	1	1	1	0	1	1	4	0	≥ 1	?	≥ 1	0	?	≥ 22

14.3.2 Affine planes

14.3.5 Definition A *finite affine plane* is a finite incidence structure of points and lines such that

1. any two distinct points together lie on a unique line;
2. for any point P and any line ℓ not containing P, there exists a unique line m through P that has no point in common with ℓ (the "parallel axiom");
3. there exists a triangle (three points not on a common line).

14.3.6 Remark If one defines a *parallelism* on the lines of an affine plane by saying that two lines are parallel if they are equal or have no point in common, then parallelism is an equivalence relation whose equivalence classes are called *parallel classes*. Each parallel class of lines is a partition of the point set, and every line belongs to exactly one parallel class.

14.3.7 Remark If one removes a line ℓ together with all its points from a projective plane π, then one obtains an affine plane $\pi_0=\pi^\ell$. Two lines of the affine plane π^ℓ are parallel if and only if the projective lines containing them meet the line ℓ in the same point. We call ℓ the *line at infinity* of π_0, and the points of ℓ are called the *points at infinity*. Conversely, to construct a projective plane from an affine plane π_0, create a new point for each parallel class of π_0 and adjoin this new point to each line in that parallel class. Also adjoin a new line that contains all the new points and no other points. The resulting incidence structure is a projective plane π, called the *projective completion* of π_0. The *order* of π_0 is the order of its projective completion.

14.3.8 Construction The classical way to construct finite affine planes is as follows. Take as points the ordered pairs (a, b), with $a, b \in \mathbb{F}_q$, and as lines the sets of points (x, y) satisfying an equation of the form $Y = mX + b$ for some $m, b \in \mathbb{F}_q$ or an equation of the form $X = c$ for some $c \in \mathbb{F}_q$. The resulting structure is an affine plane of order q, denoted by $\mathrm{AG}(2, q)$. Such an affine plane is also *Desarguesian* since the projective completion of $\mathrm{AG}(2, q)$ is (isomorphic to) $\mathrm{PG}(2, q)$. Alternatively, $\mathrm{AG}(2, q)$ may be constructed from a 2-dimensional vector space V over $\mathbb{F}_q$ by taking as points all vectors in V and as lines all cosets of 1-dimensional subspaces, where incidence is then given by containment.

14.3.3 Translation planes and spreads

14.3.9 Definition Let π be a projective plane. A *collineation* (*automorphism*) of π is a bijective map ϕ on the point set of π that preserves collinearity. All collineations of π form the *automorphism group* $\mathrm{Aut}(\pi)$ of π under composition of maps. A *collineation group* of π is any subgroup of $\mathrm{Aut}(\pi)$. Two projective planes are *isomorphic* if there is a bijective map from the point set of one plane to the point set of the other plane that sends collinear points to collinear points.

14.3.10 Definition If ϕ is a collineation of a projective plane π, and ϕ fixes all lines through a point P and all points on a line ℓ, then ϕ is a *(P, ℓ)-perspectivity*. In particular, it is a *(P, ℓ)-elation* if $P \in \ell$.

14.3.11 Definition A projective plane π is *(P, ℓ)-transitive* if for any distinct points A, B not on ℓ and collinear with P $(A \neq P \neq B)$, there is a (P, ℓ)-perspectivity ϕ in $\mathrm{Aut}(\pi)$ such that $A^\phi = B$. Similarly, π is *(m, ℓ)-transitive* if it is (P, ℓ)-transitive for all points P on m. If π is (ℓ, ℓ)-transitive for some line ℓ, then ℓ is a *translation line* of π and π is a *translation plane* with respect to ℓ.

14.3.12 Remark If π is a translation plane with respect to a line ℓ, then the affine plane $\pi^\ell = \pi \setminus \ell$ is also a *translation plane*. Most often a translation plane is considered an affine plane, with its line at infinity the translation line. The *translation group* of such an affine plane is the group of all (ℓ, ℓ)-elations, which acts sharply transitively on the points of the affine plane π^ℓ. References [279, 1613] provide extensive information on translation planes.

14.3.13 Remark Translation planes are coordinatized by algebraic structures called *quasifields* (see Section 2.1). Every quasifield has an algebraic substructure called its *kernel*, which in the finite setting is necessarily a finite field. The quasifield is then a finite dimensional vector space over its kernel, and the *dimension* of the translation plane is the dimension of this vector space.

14.3.14 Definition Let $\Sigma = \mathrm{PG}(2t+1, q)$ be a $(2t+1)$-dimensional projective space for some non-negative integer t (see Section 14.4 for the definition of projective space). A *spread* of Σ is a set S of t-subspaces of Σ such that any point of Σ belongs to exactly one element of S. The set-wise stabilizer of S in $\mathrm{Aut}(\Sigma)$ is the *automorphism group* $\mathrm{Aut}(S)$ of the spread.

14.3.15 Construction View the finite field $F = \mathbb{F}_{q^{2t+2}}$ as a $(2t+2)$-dimensional vector space V over its subfield $\mathbb{F}_q$, and let $\Sigma = \mathrm{PG}(2t+1, q)$ be the associated $(2t+1)$-dimensional projective space. If θ is a primitive element of F and $L = \mathbb{F}_{q^{t+1}}$ is the subfield of order q^{t+1}, then for each positive integer i, $\theta^i L$ is a $(t+1)$-dimensional vector subspace of V that represents a t-subspace of Σ. Moreover, $S = \{L, \theta L, \theta^2 L, \ldots, \theta^{q^{t+1}} L\}$ is a spread of Σ. The spreads obtained in this way are *regular* as defined below.

14.3.16 Definition A *t-regulus* of $\mathrm{PG}(2t+1, q)$ is a set R of $q+1$ mutually disjoint t-subspaces such that any line intersecting three elements of R intersects all elements of R. These lines are the *transversals* of R.

14.3.17 Proposition [1515, p. 200] Any three mutually skew t-subspaces of the projective space $\mathrm{PG}(2t+1, q)$ determine a unique t-regulus containing them.

14.3.18 Remark The points covered by a 1-regulus R in $\mathrm{PG}(3, q)$ are the points of a hyperbolic quadric. The transversals to R form another 1-regulus covering the same hyperbolic quadric. This 1-regulus is the *opposite regulus* R^{opp} to R.

14.3.19 Definition Let $q > 2$ be a prime power. A spread S in $\mathrm{PG}(2t+1, q)$ is *regular* if for any three elements of S, the t-regulus determined by them is contained in S. (See [1515] for an alternative definition valid for $q = 2$.)

14.3.20 Construction (Bruck-Bose [428]) Let $\Sigma \cong \mathrm{PG}(2t+1, q)$ be a hyperplane of $\overline{\Sigma} = \mathrm{PG}(2t+2, q)$, for some integer $t \geq 0$, and let S be a spread of Σ. Define $A(S)$ to be the geometry whose points are the points of $\overline{\Sigma} \backslash \Sigma$, and whose lines are the $(t+1)$-subspaces of $\overline{\Sigma}$ that intersect Σ precisely in an element of S.

14.3.21 Theorem [428] The structure $A(S)$ is an affine translation plane of order q^{t+1} which is at most $(t+1)$-dimensional over its kernel. Conversely, any finite affine translation plane can be constructed in this way for an appropriate choice of t. In particular, every finite translation plane has prime power order. In addition, $A(S)$ is isomorphic to $\mathrm{AG}(2, q^{t+1})$ if and only if S is regular.

14.3.22 Remark The automorphism group of an affine translation plane $A(S)$ is isomorphic to the semidirect product of the translation group with the group of all nonsingular semilinear mappings of the underlying vector space which fix the spread S [104]. The affine translation plane $A(S)$ is completed to a projective plane $P(S)$ by adding the members of the spread S as the points at infinity. Projective planes $P(S_1)$ and $P(S_2)$ are isomorphic if and only if there is a collineation of Σ mapping S_1 to S_2 [1980].

14.3.23 Remark Let $t = 1$ above. Replacing a regulus R by its opposite regulus R^{opp} in a regular spread S_0 produces a new spread S which is not regular, provided $q > 2$. The associated planes $A(S)$ are the *Hall planes.* Simultaneously replacing mutually disjoint reguli by their opposite reguli in a regular spread S_0 produces a *subregular spread* S, whose associated translation planes $A(S)$ are also called *subregular.* If the set of mutually disjoint reguli in S_0

is "linear" in a well-defined way [427], then the resulting subregular planes are the *André planes* which are two-dimensional over their kernels. Thus Hall planes are André planes, but not necessarily vice versa.

14.3.24 Remark In [1508], a method is given for obtaining a spread of $PG(3,q^2)$ from a spread of $PG(3,q)$ for every odd prime power q.

14.3.4 Nest planes

14.3.25 Definition Let S_0 be a regular spread of $PG(3,q)$. A *nest* N in S_0 is a set of reguli in S_0 such that every line of S_0 belongs to 0 or 2 reguli of N. Thus a nest is a 2-cover of the lines of S_0 which are contained in the nest.

14.3.26 Remark Counting arguments show that the number t of reguli in a nest must satisfy $(q+3)/2 \leq t \leq 2(q-1)$. In particular, we note that q must be odd for nests to exist. If a nest contains t reguli, it is a *t-nest.* If U denotes the $t(q+1)/2$ lines of S_0 contained in the reguli of a t-nest N, there is a natural potential replacement set for U. Namely, if $(q+1)/2$ lines can be found in the opposite regulus to each regulus of N such that the resulting set W of $t(q+1)/2$ lines are mutually disjoint, then $S = (S_0 \setminus U) \cup W$ is a non-regular spread of $PG(3,q)$ and hence $A(S)$ is a non-Desarguesian translation plane. In this case, the nest N is *replaceable*, and the resulting plane $A(S)$ is a *nest plane.*

14.3.27 Definition An *inversive plane* is a $3-(n^2+1, n+1, 1)$ design (see Section 14.5), for some integer $n \geq 2$. That is, an inversive plane is an incidence structure of n^2+1 points and $n(n^2+1)$ blocks, each block of size $n+1$, such that every three points lie in a unique block. The blocks are the *circles* of the inversive plane.

14.3.28 Construction Let q be any prime power. Take as points the elements of $\mathbb{F}_{q^2}$ together with the symbol ∞. Take as circles the images of $\mathbb{F}_q \cup \{\infty\}$ under the nonsingular linear fractional mappings on $\mathbb{F}_{q^2}$, with the usual conventions on ∞. If incidence is given by containment, this produces an inversive plane with q^2+1 points whose circles have size $q+1$. This inversive plane is *Miquelian* because it satisfies the classical configurational result of Miquel, and is denoted by $M(q)$ [807].

14.3.29 Theorem [427] There is a one-to-one correspondence between the points and circles of $M(q)$, and the lines and reguli of a regular spread of $PG(3,q)$. There is an associated homomorphism from the stabilizer of the regular spread to the automorphism group of $M(q)$, whose kernel is a cyclic group of order $q+1$.

14.3.30 Remark Using the above correspondence, it is usually easier to search for nests in $M(q)$ rather than directly in a regular spread S_0 of $PG(3,q)$. Such nests can often be constructed by taking the orbit of some carefully chosen "base" circle under a natural cyclic or elementary abelian subgroup of $\mathrm{Aut}(M(q))$. However, to check if the resulting nest is replaceable, one must pull back to S_0 and work in $PG(3,q)$. Some nests are replaceable and some are not. Computations involving finite field arithmetic lead to the following results.

14.3.31 Theorem [172, 173, 949, 950, 2374] For any odd prime power $q \geq 5$, there exist replaceable t-nests for $t = q-1, q, q+1, 2(q-1)$. The resulting spreads determine non-Desarguesian translation planes of order q^2 which are two-dimensional over their kernels.

14.3.32 Remark The nesting technique for constructing two-dimensional translation planes is quite robust. In addition to the above examples, replaceable t-nests have been constructed for many values of t in the range $3(q+1)/4 - \sqrt{q}/2 \leq t \leq 3(q+1)/4 + \sqrt{q}/2$; see [174]. Moreover, the translation planes associated with nests often can be characterized by the action of certain collineation groups [1609, 1612, 1614, 1615].

14.3.33 Remark Circle geometries and the notion of subregularity can be extended to higher dimensions. Using algebraic pencils of Sherk surfaces, in [756] several infinite families of non-André subregular translation planes are constructed which are 3-dimensional over their kernels. Proofs use intricate finite field computations involving the trace, norm, and bitrace.

14.3.5 Flag-transitive affine planes

14.3.34 Definition An affine plane is *flag-transitive* if it admits a collineation group which acts transitively on incident point-line pairs.

14.3.35 Remark A straightforward counting argument shows that transitivity on lines implies transitivity on flags for affine planes.

14.3.36 Remark By a celebrated result of Wagner [2889], every finite flag-transitive affine plane is necessarily a translation plane, and hence arises from a spread S of $\mathrm{PG}(2t+1,q)$, for some positive integer t, according to Theorem 14.3.21. The affine plane $A(S)$ is flag-transitive if and only if the spread S admits a transitive collineation group.

14.3.37 Construction Let $F = \mathbb{F}_{q^{2t+2}}$ be treated as a $(2t+2)$-dimensional vector space over its subfield $\mathbb{F}_q$, thus serving as the underlying vector space for $\Sigma = \mathrm{PG}(2t+1,q)$. If θ is a primitive element of $\mathbb{F}$, the collineation $\overline{\theta}$ induced by multiplication by θ is a Singer cycle of Σ; that is, the cyclic group $\langle\overline{\theta}\rangle$ acts sharply transitively on the points and hyperplanes of Σ. If G denotes the Singer subgroup of order $q^{t+1}+1$, let $\mathcal{O}$ denote the partition of the points of Σ into $(q^{t+1}-1)/(q-1)$ G-orbits of size $q^{t+1}+1$ each. As shown in [948], these point orbits are caps when t is odd (see Section 14.4 for the definition of a cap). For future reference we let H denote the index two subgroup of G.

14.3.38 Example [1674] Let q be an odd prime power, and let t be an odd integer. Using the above model, choose $b \in F$ such that $b^{q^{t+1}-1} = -1$. Let $\sigma : F \to F$ via $\sigma : x \mapsto x^q$, and let E denote the subfield of F whose order is q^{t+1}. Then $A_1 = \{x + bx^\sigma : x \in E\}$ represents a t-space Γ_1 of Σ that meets half the G-orbits of $\mathcal{O}$ in two points each (from different H-orbits) and is disjoint from the rest. Similarly, $A_2 = \{bx + b^{\sigma+1}x^\sigma : x \in E\}$ represents a t-space Γ_2 of Σ that meets the G-orbits of $\mathcal{O}$ which are disjoint from Γ_1 in two points each (from different H-orbits). Moreover, $S = \Gamma_1^H \cup \Gamma_2^H$ is a spread of Σ admitting a transitive collineation group, which yields a non-Desarguesian flag-transitive affine plane $A(S)$ of order q^{t+1} with $\mathbb{F}_q$ in its kernel.

14.3.39 Example [1674, 1681] Let q be an odd prime power, and let t be an even integer. Using the notation of Example 14.3.38, Γ_1 now meets every G-orbit of $\mathcal{O}$ in one point each, and hence $S_1 = \Gamma_1^G$ is a spread of Σ which admits a transitive (cyclic) collineation group. The resulting affine plane $A(S_1)$ is a non-Desarguesian flag-transitive affine plane of order q^{t+1} with $\mathbb{F}_q$ in its kernel. If $q \equiv 1 \pmod 4$, then Γ_2 also meets each G-orbit of $\mathcal{O}$ in one point each, and these points lie in H-orbits that are disjoint from Γ_1. Moreover, $S_2 = \Gamma_1^H \cup \Gamma_2^H$ is a spread of Σ admitting a transitive (non-cyclic) collineation group, thereby yielding another non-Desarguesian flag-transitive affine plane $A(S_2)$ of order q^{t+1} with $\mathbb{F}_q$ in its kernel. This plane does not admit a cyclic collineation group acting transitively on the line at infinity. For $q \equiv 3 \pmod 4$, one may obtain such a spread S_2 by replacing Γ_2 with the t-space of Σ represented by $\{\mu x^{q^{t+1}} + \mu b^{q^{t+1}}(x^\sigma)^{q^{t+1}} : x \in E\}$, where $\mu = \theta^{(q^{t+1}-1)/(q-1)}$.

14.3.40 Remark The field automorphism σ in the above examples may be replaced by any element of $\text{Gal}(\mathbb{F}_{q^{2t+2}}/\mathbb{F}_q)$. The resulting planes are non-Desarguesian provided σ does not induce the identity map on the subfield E. Lower bounds are given in [1674, 1681] for the number of mutually non-isomorphic planes obtained as b and σ vary.

14.3.41 Example [1675] Let q be a power of 2, and let $t \geq 2$ be an even integer. Using the notation of Example 14.3.38, let Tr denote the trace from E to $\mathbb{F}_q$, and choose some element $r \in \mathbb{F}_{q^2} \backslash \mathbb{F}_q$. Let Γ' be the t-space of Σ represented by $\{\text{Tr}\,(x)+rx : x \in E\}$. Then Γ' meets every G-orbit of $\mathcal{O}$ in one point each, and hence $S' = (\Gamma')^G$ is a spread of Σ which admits a transitive (cyclic) collineation group. The resulting flag-transitive affine plane $A(S')$ of order q^{t+1} with $\mathbb{F}_q$ in its kernel is non-Desarguesian provided $q^{t+1} > 8$.

14.3.42 Remark Other than the Hering plane [1488] of order 27 and the Lüneburg planes [1979] of order 2^{2d} for odd $d \geq 3$, all known finite flag-transitive affine planes arise from spreads consisting of a single G-orbit or the union of two H-orbits, where G and H are the Singer subgroups defined in Construction 14.3.37. It is shown in [951] that if $q = p^e$ for some odd prime p and some positive integer e and if $\gcd((q^{t+1}+1)/2, (t+1)e) = 1$, then any flag-transitive affine plane of order q^{t+1} with $\mathbb{F}_q$ in its kernel (other than the above Hering plane) must arise in this way. More can be said for $t = 1, 2$.

14.3.43 Theorem [175] If $q = p^e$ is an odd prime power such that $\gcd((q^2+1)/2, e) = 1$, then any two-dimensional flag-transitive affine plane of order q^2 is isomorphic to one of the planes constructed in Example 14.3.38 with $t = 1$. The number of such isomorphism classes can be determined by Möbius inversion. For $e = 1$ (hence $q = p$ prime), the above gcd condition is necessarily satisfied and the number of isomorphism classes is precisely $(q-1)/2$.

14.3.44 Theorem [169, 176] If $q = p^e$ is an odd prime power such that $\gcd((q^3+1)/2, 3e) = 1$, then any three-dimensional flag-transitive affine plane of order q^3, other than Hering's plane of order 27, is isomorphic to one of the planes constructed in Example 14.3.39 with $t = 2$. For $e = 1$ (hence $q = p$ prime), the number of isomorphism classes of each type arising from Example 14.3.39 is precisely $(q-1)/2$.

14.3.45 Problem For q even, the classification and complete enumeration of finite flag-transitive affine planes of dimension two or three over their kernel remains an open problem. The only known two-dimensional examples are the Lüneburg planes.

14.3.46 Problem The classification of finite flag-transitive affine planes is one of the few open cases in the program announced in [454] to classify all finite flag-transitive linear spaces. For arbitrary dimension over the kernel, it is not known if there exist examples of finite flag-transitive affine planes other than the ones listed above, and the classification seems to be quite difficult. In the projective setting, it is believed that the only flag-transitive projective plane is the Desarguesian one, although this remains an open problem.

14.3.6 Subplanes

14.3.47 Definition Let π be a projective plane with point set $\mathcal{P}$ and line set $\mathcal{L}$. A projective plane π' with point set $\mathcal{P}'$ and line set $\mathcal{L}'$ is a *subplane* of π if $\mathcal{P}' \subseteq \mathcal{P}$ and $\mathcal{L}' \subseteq \mathcal{L}$, and π' inherits its incidence relation from π.

14.3.48 Theorem [425] Let π be a finite projective plane of order n, and let π' be a subplane of π with order $m < n$. Then $n = m^2$ or $m^2 + m \leq n$.

14.3.49 Remark It is unknown whether equality can hold in the above inequality; if so, this would imply that the order $n = m^2+m$ of π is not a prime power. The case $n = m^2$ is of particular interest. In this case, every point of $\pi \setminus \pi'$ is incident with a unique line of π', and dually every line of $\pi \setminus \pi'$ is incident with a unique point of π'.

14.3.50 Definition A subplane π' of order m in a projective plane π of order $n = m^2$ is a *Baer subplane* of π.

14.3.51 Remark In the classical setting, the lattice of subplanes follows directly from the lattice of subfields. Namely, if $q = p^e$ for some prime p and some positive integer e, then the subplanes of $\mathrm{PG}(2,q)$, up to isomorphism, are precisely $\mathrm{PG}(2,p^k)$ as k varies over all positive divisors of e. So $\mathrm{PG}(2,q)$ has a Baer subplane if and only if q is a square. Moreover, one can easily count the number of subplanes of a given order in this classical (Desarguesian) setting.

14.3.52 Theorem [1510, Lemma 4.20] If q is any prime power and $n \geq 2$ is any integer, then the number of subplanes of order q in $\mathrm{PG}(2,q^n)$, all of which are isomorphic to $\mathrm{PG}(2,q)$, is

$$\frac{q^{3(n-1)}(q^{3n}-1)(q^{2n}-1)}{(q^3-1)(q^2-1)}.$$

In particular, the number of Baer subplanes in $\mathrm{PG}(2,q^2)$ is $q^3(q^3+1)(q^2+1)$.

14.3.53 Remark It is currently unknown if $\mathrm{PG}(2,q^2)$ has the greatest number of Baer subplanes among all projective planes of order q^2. No counter-examples have been found. Amazingly, there are affine planes of order q^2 which contain more affine subplanes of order q than does $\mathrm{AG}(2,q^2)$ [1099].

14.3.54 Definition A *Baer subplane partition*, or BSP for short, of $\mathrm{PG}(2,q^2)$ is a partition of the points of $\mathrm{PG}(2,q^2)$ into subplanes, each isomorphic to $\mathrm{PG}(2,q)$.

14.3.55 Example [426] Consider the full Singer group of order q^4+q^2+1 acting sharply transitively on the points and lines of $\pi = \mathrm{PG}(2,q^2)$. Then the orbits under the Singer subgroup of order q^2+q+1 are Baer subplanes, and the orbit of any one of these Baer subplanes under the complementary Singer subgroup of order q^2-q+1 forms a BSP of π. Such a BSP is *classical*.

14.3.56 Remark It is shown in [171] that any spread of $\mathrm{PG}(5,q)$ admitting a linear cyclic sharply transitive action corresponds to a "perfect" BSP of $\mathrm{PG}(2,q^2)$, and this spread is regular if and only if the BSP is classical. By definition, a BSP is *perfect* if and only if it is an orbit of some Baer subplane under an appropriate Singer subgroup, although the Baer subplane itself need not be a point orbit under a Singer subgroup. Examples 14.3.39 and 14.3.41 for $t = 2$ yield the following result.

14.3.57 Theorem [171] Let $q \neq 2$ be a prime power. Then there exist non-classical BSPs of $\mathrm{PG}(2,q^2)$.

14.3.58 Remark Relatively little is known about the subplane structure of non-Desarguesian planes. There is no known example of a square order projective plane which has been shown not to contain a Baer subplane. However, it is not known if every square order projective plane must contain a Baer subplane. At the other extreme, the Hall planes, the Hughes planes, the Figueroa planes, and many two-dimensional subregular translation planes have been proven to contain subplanes of order two (that is, Fano subplanes). It has been conjectured that every finite non-Desarguesian plane must contain a subplane of order two. More surprisingly, it is shown in [480] that the Hughes plane of order q^2 (q odd) has a subplane of order 3 when $q \equiv 2 \pmod 3$. Extensive, but not exhaustive, computer searches for small q have found no subplanes of order 3 in this plane when $q \equiv 1 \pmod 3$. Very recently [481], subplanes of order 3 have been proven to exist in all odd order Figueroa planes.

14.3.7 Embedded unitals

14.3.59 Remark Reference [205] provides an excellent introduction to the topic of unitals. Proofs and precise statements of most results in this subsection may be found in the above reference.

14.3.60 Definition A *unital* is a $2\text{--}(n^3+1, n+1, 1)$ design for some integer $n \geq 3$ (that is, a geometry having n^3+1 points, with $n+1$ points on each line such that any two distinct points are on exactly one line).

14.3.61 Remark Here the interest is not in unitals as designs, but in unitals embedded in a projective plane of order n^2. The lines (blocks) of the unital are then the lines of the ambient projective plane which meet the unital in more than one point (and hence in $n+1$ points).

14.3.62 Example Let $\mathrm{PG}(2, q^2)$ be represented using homogeneous coordinates. Then the points $(x : y : z)$ for which $xx^q + yy^q + zz^q = 0$ form a unital. This unital is a *Hermitian curve.*

14.3.63 Construction (Buekenhout [452]) Using the Bruck-Bose representation of Construction 14.3.20, with $t = 1$ for a 2-dimensional translation plane, let S be any spread of Σ and let $\overline{U}$ be an ovoidal cone of $\overline{\Sigma}$ (that is, the point cone over some 3-dimensional ovoid as defined in Section 14.4) that meets Σ in a line of S. Then $\overline{U}$ corresponds to a unital U in $P(S)$ which is tangent to the line at infinity. Also, if $\overline{U}$ is a nonsingular (parabolic) quadric in $\overline{\Sigma}$ that meets Σ in a regulus of the spread S, then $\overline{U}$ corresponds to a unital U in $P(S)$ which meets the line at infinity in $q+1$ points. Of course, the second construction is valid only for those 2-dimensional translation planes of order q^2 whose associated spread contains at least one regulus.

14.3.64 Remark If the ovoidal cone above is an orthogonal cone (with an elliptic quadric as base), the resulting unital in $P(S)$ is an *orthogonal Buekenhout unital.* Units embedded in $P(S)$ which arise from the nonsingular quadric construction are *nonsingular Buekenhout unitals.*

14.3.65 Remark Orthogonal Buekenhout unitals embedded in $\mathrm{PG}(2, q^2)$ have been completely enumerated. In particular, if q is an odd prime, then the number of mutually inequivalent orthogonal Buekenhout unitals in $\mathrm{PG}(2, q^2)$ is $\frac{1}{2}(q+1)$, one of which is the Hermitian curve. The only nonsingular Buekenhout unital embedded in $\mathrm{PG}(2, q^2)$ is the Hermitian curve. Exhaustive computer searches in [195, 2382] show that there are precisely two inequivalent unitals embedded in each of $\mathrm{PG}(2, 9)$ and $\mathrm{PG}(2, 16)$, the Hermitian curve and one other orthogonal Buekenhout unital. The enumeration of orthogonal and nonsingular Buekenhout unitals in various non-Desarguesian translation planes may be found in [179, 180]. For instance, if $q \geq 5$ is a prime, then, up to equivalence, the Hall plane of order q^2 has $\frac{1}{2}(q+1)$ nonsingular Buekenhout unitals and $1 + \lfloor \frac{3q}{4} \rfloor$ orthogonal Buekenhout unitals.

14.3.66 Remark In [915], it is shown that the Hall planes contain unitals which are not obtainable from any Buekenhout construction. This is the only infinite family of unitals embedded in translation planes which has been proven to be non-Buekenhout. There are also square-order non-translation planes which are known to contain unitals, necessarily non-Buekenhout. For instance, the Hughes planes of order q^2 are known to contain unitals for all odd prime powers q [2482, 2947]. In [788], the Figueroa planes of order q^6 are shown to contain unitals for any prime power q. In fact, there is no known example of a square-order projective plane which has been shown not to contain a unital.

14.3.8 Maximal arcs

14.3.67 Remark Proofs of almost all results in this subsection may be found in Chapter 12 of [1510].

14.3.68 Definition A $\{k;r\}$-*arc* in $\mathrm{PG}(2,q)$ is a set K of k points such that r is the maximum number of points in K that are collinear. A $\{k;2\}$-arc is a k-*arc.*

14.3.69 Theorem Let K be a $\{k;r\}$-arc in $\mathrm{PG}(2,q)$. Then $k \le (q+1)(r-1)+1$.

14.3.70 Definition The $\{k;r\}$-arcs in $\mathrm{PG}(2,q)$ with $k=(q+1)(r-1)+1$ are *maximal* $\{k;\ r\}$-*arcs.*

14.3.71 Example Singleton points ($r=1$), the whole plane ($r=q+1$), and the complement of a line ($r=q$) are *trivial* maximal $\{k;r\}$-arcs. The $\{q+2;2\}$-arcs in $\mathrm{PG}(2,q)$ for q even, also called *hyperovals* (see Section VII.2.9 of [706]), are examples of non-trivial maximal $\{k;r\}$-arcs, and have been objects of intense interest for many years.

14.3.72 Lemma If K is a non-trivial maximal $\{k;r\}$-arc in $\mathrm{PG}(2,q)$, then r must be a (proper) divisor of q.

14.3.73 Theorem [189] If $\mathrm{PG}(2,q)$ contains a non-trivial maximal $\{k;r\}$-arc, then q must be even.

14.3.74 Construction [820] Let $X^2+\beta X+1$ be an irreducible quadratic polynomial over $\mathbb{F}_q$, q even. Consider the algebraic pencil in $\mathrm{PG}(2,q)$ consisting of the conics $C_\lambda : X_0^2+\beta X_0X_1+X_1^2+\lambda X_2^2=0$ for $\lambda \in \mathbb{F}_q \cup \{\infty\}$. Let A be an additive subgroup of $(\mathbb{F}_q,+)$ of order r, and let K be the set of points which is the union of the conics C_λ for $\lambda \in A$. Then K is a maximal $\{k;r\}$-arc of $\mathrm{PG}(2,q)$.

14.3.75 Construction [2024] Let q be even, and let Tr denote the absolute trace from $\mathbb{F}_q$ to $\mathbb{F}_2$. In $\mathrm{PG}(2,q)$, consider a set $\mathcal{C}$ consisting of conics $C_{\alpha,\beta,\lambda} : \alpha X_0^2+X_0X_1+\beta X_1^2+\lambda X_2^2=0$, where $\alpha,\beta \in \mathbb{F}_q$ with $\mathrm{Tr}\,(\alpha\beta)=1$ and $\lambda \in \mathbb{F}_q \cup \{\infty\}$. Define the "composition" of two distinct conics from $\mathcal{C}$ in the following way:

$$C_{\alpha,\beta,\lambda} \oplus C_{\alpha',\beta',\lambda'} = C_{\alpha\oplus\alpha',\beta\oplus\beta',\lambda\oplus\lambda'},$$

where

$$\alpha\oplus\alpha' = \frac{\alpha\lambda+\alpha'\lambda'}{\lambda+\lambda'}, \quad \beta\oplus\beta' = \frac{\beta\lambda+\beta'\lambda'}{\lambda+\lambda'}, \quad \lambda\oplus\lambda'=\lambda+\lambda'.$$

Let $\mathcal{F} \subset \mathcal{C}$ be a set of 2^d-1 non-singular conics with common nucleus $(0,0,1)$, which is closed under the composition of distinct conics. Then the points of the conics in $\mathcal{F}$, together with $(0,0,1)$, form a maximal $\{k;2^d\}$-arc in $\mathrm{PG}(2,q)$. To obtain one such set $\mathcal{F}$, assume $q=2^{4m+2}$ and let $\epsilon \in \mathbb{F}_{2^{4m+2}}$ be such that $\mathrm{Tr}\,(\epsilon)=1$. Let $A=\{x \in \mathbb{F}_{2^{4m+2}} : x^2+x \in \mathbb{F}_{2^{2m+1}}\}$, and let $r(\lambda)=\lambda^3+\epsilon$ for all $\lambda \in A$. Then $|A|=2^{2m+2}$ and

$$\mathcal{F} = \{C_{1,r(\lambda),\lambda} : \lambda \in A\setminus\{0\}\}$$

is such a subset of $\mathcal{C}$ which determines, as indicated above, a maximal $\{k;2^{2m+2}\}$-arc in $\mathrm{PG}(2,2^{4m+2})$. These arcs do not arise from Construction 14.3.74. Other possibilities for $\mathcal{F}$ may be found in [1062, 1406, 1407].

14.3.9 Other results

14.3.76 Remark Semifields (see Section 2.1) are algebraic structures that may be used to coordinatize certain translation planes, called *semifield planes.* These are the only translation planes which are also dual translation planes. Many new examples have recently been found. Chapter 6 of [785] is an excellent source for many of these new developments.

14.3.77 Problem As previously mentioned, all known finite projective (and affine) planes have prime power order, although it is certainly unclear whether this must be true in general. However, it is now known that if a projective plane of order n admits an abelian collineation group of order n^2 or $n^2 - n$, then n must be a prime power [326, 1634]. An equally important open problem is whether any finite projective (or affine) plane of prime order p must be Desarguesian. This appears to be a very difficult problem; the smallest open case is $p = 11$.

14.3.78 Remark A *hyperbolic fibration* of $\mathrm{PG}(3, q)$ is a collection of $q - 1$ hyperbolic quadrics and two lines that partition the points of $\mathrm{PG}(3, q)$. By selecting one of the two reguli ruling each hyperbolic quadric in the fibration, one obtains 2^{q-1} spreads of $\mathrm{PG}(3, q)$, which in turn give rise to 2^{q-1} translation planes of order q^2 which are at most two-dimensional over their kernels. Although there may be some isomorphisms among these planes, this is a very robust method for constructing two-dimensional translation planes (see [177] for isomorphism counts). An easy example of a hyperbolic fibration is the *hyperbolic pencil*, which is an algebraic pencil of quadrics of the appropriate types. Other examples of hyperbolic fibrations may be found in [170, 178]. All known hyperbolic fibrations have the property that the two lines in the fibration form a conjugate (skew) pair with respect to each of the polarities associated with the $q - 1$ hyperbolic quadrics (such hyperbolic fibrations are called *regular* in the literature), and also have the property that all $q - 1$ hyperbolic quadrics intersect one of the two skew lines in the same pair of conjugate points with respect to the quadratic extension $\mathbb{F}_{q^2}$ of $\mathbb{F}_q$ (such hyperbolic fibrations are said to *agree* on one of the two skew lines). In [422], it is shown that all hyperbolic fibrations are necessarily regular if q is even (the problem is still open for q odd), and it is also shown for any q that a hyperbolic fibration which agrees on one of its two skew lines is necessarily regular. In [177], it is shown that there is a bijection between regular hyperbolic fibrations which agree on one of their two lines and flocks of a quadratic cone, once a conic of the flock is specified. This further leads to a correspondence with normalized q-clans and certain types of generalized quadrangles.

14.3.79 Remark There are other survey articles on substructures in projective planes. The section on Finite Geometry in *The Handbook of Combinatorial Designs* [706] and the survey article [1513] state the main results on arcs, $\{k; r\}$-arcs, caps, unitals, and blocking sets in $\mathrm{PG}(2, q)$, where exact definitions, tables, and supplementary results are provided. In addition, the collected work [785] contains a great variety of results on substructures in $\mathrm{PG}(2, q)$, techniques for investigating these substructures, and important open problems in this area. The *linearity conjecture* on small minimal blocking sets in $\mathrm{PG}(2, q)$ is one of the most important such open problems (see Chapter 3 of [785] for an explicit statement of this conjecture). Proving this conjecture would imply several new results on various substructures in $\mathrm{PG}(2, q)$ as well as in $\mathrm{PG}(n, q)$, for $n > 2$. In particular, this would include new results on maximal partial spreads, minihypers, extendability of linear codes, tight sets, and Cameron-Liebler line classes. The investigation of maximal arcs in $\mathrm{PG}(2, q)$ for q even, inspired by the new construction method of Mathon described in Construction 14.3.75, and the investigation of embedded unitals as discussed in Section 14.3.7 are central problems on substructures in projective planes which also merit further research.

See Also

§2.1	For a discussion of traces, norms, and linearized polynomials.
§9.4	For some semifield constructions.
§9.5	For some examples of semifield planes constructed from planar functions.
§13.3	For a discussion of the classical projective groups over finite fields.
§14.4	For a discussion of projective spaces of higher dimensions.
[807]	Develops the notion of inversive planes.
[1560]	Develops the notion of coordinatizing non-Desarguesian projective planes.

References Cited: [104, 169, 170, 171, 172, 173, 174, 175, 176, 177, 178, 179, 180, 189, 195, 205, 279, 326, 422, 425, 426, 427, 428, 452, 454, 480, 481, 555, 706, 756, 785, 788, 807, 820, 915, 948, 949, 950, 951, 1062, 1099, 1406, 1407, 1488, 1508, 1510, 1513, 1515, 1560, 1609, 1612, 1613, 1614, 1615, 1634, 1674, 1675, 1681, 1836, 1837, 1979, 1980, 2024, 2374, 2382, 2482, 2889, 2947]

14.4 Projective spaces

James W.P. Hirschfeld, University of Sussex
Joseph A. Thas, Ghent University

14.4.1 Projective and affine spaces

14.4.1 **Definition** Let $V = V(n+1, F)$, with $n \geq 1$, be an $(n+1)$-dimensional vector space over the field F with zero element 0. Consider the equivalence relation on the elements of $V\backslash\{0\}$ whose equivalence classes are the one-dimensional subspaces of V with the zero deleted. Thus, if $X, Y \in V\backslash\{0\}$, then X is equivalent to Y if $Y = tX$ for some t in $F_0 = F\backslash\{0\}$.

1. The set of equivalence classes is the *n-dimensional projective space over F* and is denoted by $\mathrm{PG}(n, F)$ or, when $F = \mathbb{F}_q$, by $\mathrm{PG}(n, q)$.
2. The elements of $\mathrm{PG}(n, F)$ are *points*; the equivalence class of the vector X is the point $\mathbf{P}(X)$. The vector X is a *coordinate vector* for $\mathbf{P}(X)$ or X is a vector *representing* $\mathbf{P}(X)$. In this case, tX with t in F_0 also represents $\mathbf{P}(X)$; that is, by definition, $\mathbf{P}(tX) = \mathbf{P}(X)$.
3. If $X = (x_0, \ldots, x_n)$ for some basis, then the x_i are the *coordinates* of the point $\mathbf{P}(X)$.
4. The points $\mathbf{P}(X_1), \ldots, \mathbf{P}(X_r)$ are *linearly independent* if a set of vectors $X_1, \ldots, X_r$ representing them is linearly independent.

14.4.2 Definition

1. For any $m = -1, 0, 1, 2, \ldots, n$, a *subspace of dimension* m, or m*-space*, of $\mathrm{PG}(n, F)$ is a set of points all of whose representing vectors form, together with the zero, a subspace of dimension $m+1$ of $V = V(n+1, F)$; it is denoted by Π_m.
2. A subspace of dimension zero is a point; a subspace of dimension -1 is the empty set. A subspace of dimension one is a *line*, of dimension two is a *plane*, of dimension three is a *solid.* A subspace of dimension $n-1$ is a *hyperplane.* A subspace of dimension $n-r$ is a subspace of *codimension* r.

14.4.3 Definition

1. The set of m-spaces of $\mathrm{PG}(n, F)$ is denoted $\mathrm{PG}^{(m)}(n, F)$ or, when $F = \mathbb{F}_q$, by $\mathrm{PG}^{(m)}(n, q)$.
2. For $r, s, m, n \in \mathbb{N}$, let
   (a) $\theta(n, q) = (q^{n+1} - 1)/(q - 1)$, also denoted by $\theta(n)$;
   (b) $|\mathrm{PG}^{(m)}(n, q)| = \phi(m; n, q)$;
   (c) $[r, s]_- = \prod_{i=r}^{s}(q^i - 1)$, for $s \geq r$.

14.4.4 Theorem [1510, Chapter 3]
For $n \geq 1$, $m \geq 0$, and q any prime power,

1. $|\mathrm{PG}(n, q)| = \theta(n, q)$;
2. $\phi(m; n, q) = [n - m + 1, n + 1]_- / [1, m + 1]_-$.

14.4.5 Theorem [1510, Chapter 2]

1. A hyperplane is the set of points $\mathbf{P}(X)$ whose vectors $X = (x_0, \ldots, x_n)$ satisfy a linear equation
$$u_0x_0 + u_1x_1 + \cdots + u_nx_n = 0,$$
with $U = (u_0, \ldots, u_n)$ in $F^{n+1} \backslash \{(0, \ldots, 0)\}$; it is denoted $\pi(U) = \Pi_{n-1}$.
2. An m-space Π_m is the set of points whose representing vectors $X = (x_0, \ldots, x_n)$ satisfy the equations $XA = 0$, where A is an $(n+1) \times (n-m)$ matrix of rank $n - m$ with coefficients in F.

14.4.6 Remark [1510, Chapter 2] The vector U in the theorem is a *coordinate vector* of the hyperplane; the u_i are *hyperplane* or *tangential coordinates.*

14.4.7 Definition

1. If a point P lies in a subspace Π_m, then P is *incident* with Π_m or, equally well, Π_m is *incident* with P.
2. If Π_r and Π_s are subspaces of $\mathrm{PG}(n, F)$, then the *meet* or *intersection* of Π_r and Π_s, written $\Pi_r \cap \Pi_s$, is the set of points common to Π_r and Π_s; it is also a subspace.
3. The *join* of Π_r and Π_s, written $\Pi_r\Pi_s$, is the smallest subspace containing Π_r and Π_s.

14.4.8 Theorem [1510, Chapter 2] Subspaces have the following properties.

1. If Π_r and Π'_r are both r-spaces in $\mathrm{PG}(n, F)$ and $\Pi'_r \subset \Pi_r$, then $\Pi'_r = \Pi_r$.
2. (Grassmann Identity) If $\Pi_r \cap \Pi_s = \Pi_t$ and $\Pi_r \Pi_s = \Pi_m$, then $r + s = m + t$.
3. A subspace Π_m is the join of $m + 1$ linearly independent points; it is also the intersection of $n - m$ linearly independent hyperplanes.
4. Equivalently, the set of all representing vectors of the points of Π_m, together with the zero vector, is the intersection of $n - m$ hyperplanes of the vector space V, which define $n - m$ linearly independent vectors $U = (u_0, \dots, u_n)$.

14.4.9 Theorem (The principle of duality) [1510, Chapter 2] For any space $S = \mathrm{PG}(n, F)$, there is a *dual* space S^*, whose *points* and *hyperplanes* are respectively the hyperplanes and points of S. For any theorem true in S, there is an equivalent theorem true in S^*. In particular, if T is a theorem in S stated in terms of points, hyperplanes, and incidence, the same theorem is true in S^* and gives a dual theorem T^* in S by substituting "hyperplane" for "point" and "point" for "hyperplane." Thus "join" and "meet" are dual. Hence the dual of an r-space in $\mathrm{PG}(n, F)$ is an $(n - r - 1)$-space.

14.4.10 Remark For small dimensions, in $\mathrm{PG}(2, F)$, point and line are dual; in $\mathrm{PG}(3, F)$, point and plane are dual, whereas the dual of a line is a line.

14.4.11 Definition

1. If $\mathcal{H}_\infty$ is any hyperplane in $\mathrm{PG}(n, F)$, then $\mathrm{AG}(n, F) = \mathrm{PG}(n, F) \backslash \mathcal{H}_\infty$ is an *affine space of n dimensions over F*. When $F = \mathbb{F}_q$, write $\mathrm{AG}(n, F) = \mathrm{AG}(n, q)$.
2. The *subspaces* of $\mathrm{AG}(n, F)$ are the subspaces of $\mathrm{PG}(n, F)$, apart from $\mathcal{H}_\infty$, with the points of $\mathcal{H}_\infty$ deleted in each case.
3. This hyperplane $\mathcal{H}_\infty$ is the *hyperplane at infinity* of $\mathrm{AG}(n, F)$.

14.4.2 Collineations, correlations, and coordinate frames

14.4.12 Definition

1. If S and S' are two spaces $\mathrm{PG}(n, F)$, $n \geq 2$, then a *collineation* $\alpha : S \longrightarrow S'$ is a bijection which preserves incidence; that is, if $\Pi_r \subset \Pi_s$, then $\Pi_r^\alpha \subset \Pi_s^\alpha$.
2. It is sufficient that α is a bijection such that, if $\Pi_0 \subset \Pi_1$, then $\Pi_0^\alpha \subset \Pi_1^\alpha$.
3. When $n = 1$, consider the lines S and S' embedded in planes over F; then a collineation $\alpha : S \longrightarrow S'$ is a transformation induced by a collineation of the planes; that is, if S_0 and S'_0 are planes with $S \subset S_0$ and $S' \subset S'_0$, and $\alpha_0 : S_0 \longrightarrow S'_0$ is a collineation mapping S onto S', then let α be the restriction of α_0 to S.

14.4.13 Definition A *projectivity* $\alpha : S \longrightarrow S'$ is a bijection given by a matrix T, necessarily non-singular, such that $\mathbf{P}(X') = \mathbf{P}(X)^\alpha$ if and only if $tX' = XT$, where $t \in F_0$. Write $\alpha = \mathbf{M}(T)$; then $\alpha = \mathbf{M}(\lambda T)$ for any λ in F_0.

14.4.14 Remark A projectivity is a collineation. Mostly the case to be considered is when $S = S'$.

14.4.15 **Definition** With respect to a fixed basis of $V(n+1, F)$, an automorphism σ of F defines an *automorphic collineation* σ of $S = \mathrm{PG}(n, F)$; in coordinates, this is given by $\mathbf{P}(X)^{\sigma} = \mathbf{P}(X^{\sigma})$, where $X^{\sigma} = (x_0^{\sigma}, x_1^{\sigma}, \ldots, x_n^{\sigma})$.

14.4.16 **Theorem** (The fundamental theorem of projective geometry) [1510, Chapter 2]

1. If $\alpha' : S \longrightarrow S'$ is a collineation, then $\alpha' = \sigma\alpha$, where σ is an automorphic collineation, given by a field automorphism σ, and α is a projectivity. In particular, if $K = \mathbb{F}_q$ with $q = p^h$, p prime, and $\mathbf{P}(X') = \mathbf{P}(X)^{\alpha'}$, then there exists m in $\{1, 2, \ldots, h\}$, $t_{ij} \in F$ for $i, j \in \{0, 1, \ldots, n\}$, and $t \in F_0$ such that
$$tX' = X^{p^m}T,$$
where
$$\begin{aligned} X^{p^m} &= (x_0^{p^m}, \ldots, x_n^{p^m}), \\ T &= (t_{ij}); \end{aligned}$$
that is,
$$tx_i' = x_0^{p^m} t_{0i} + \cdots + x_n^{p^m} t_{ni},$$
for $i = 0, 1, \ldots, n$.
2. If $\{P_0, \ldots, P_{n+1}\}$ and $\{P_0', \ldots, P_{n+1}'\}$ are both subsets of $\mathrm{PG}(n, F)$ of cardinality $n+2$ such that no $n+1$ points chosen from the same set lie in a hyperplane, then there exists a unique projectivity α such that $P_i' = P_i^{\alpha}$, for all $i \in \{0, 1, \ldots, n+1\}$.

14.4.17 **Remark** There are cases where Theorem 14.4.16 simplifies.

1. For $n = 1$, there is a unique projectivity transforming any three distinct points on a line to any other three.
2. When $F = \mathbb{F}_2$, it suffices to give the images of $P_0, \ldots, P_n$ to determine a projectivity. For $n = 1$, the images of two points determine the projectivity.

14.4.18 **Remark** Part 2 of Theorem 14.4.16 emphasizes a difference between the spaces $V(n+1, F)$ and $\mathrm{PG}(n, F)$. In the former, linear transformations are determined by the images of $n+1$ vectors; in the latter, projectivities are determined by the images of $n+2$ points.

14.4.19 **Definition** Let $\{P_0, \ldots, P_{n+1}\}$ be any set of $n+2$ points in $\mathrm{PG}(n, F)$, no $n+1$ in a hyperplane. If P is any other point of the space, then a coordinate vector for P is determined in the following manner. Let P_i be represented by the vector X_i for some vector X_i in $V(n+1, F)$. For any given t in F_0 there exist a_i in F for all $i \in \{0, 1, \ldots, n\}$ such that
$$tX_{n+1} = a_0 X_0 + \cdots + a_n X_n.$$
So, for any t, the ratios a_i/a_j remain fixed. Thus, if P is any point with $P = \mathbf{P}(X)$, then
$$X = t_0 a_0 X_0 + \cdots + t_n a_n X_n.$$
Hence, with respect to $\{P_0, \ldots, P_{n+1}\}$, the point P is given by $(t_0, \ldots, t_n)$ where the t_i are determined up to a common factor. Then $\{P_0, \ldots, P_n\}$ is the *simplex of reference* and P_{n+1} the *unit point*. Together the $n+2$ points form a (*coordinate*) *frame*.

14.4.20 Remark In $V(n+1,F)$, a basis is a set of $n+1$ linearly independent vectors and, in $\mathrm{PG}(n,F)$, a frame is a set of $n+2$ points, no $n+1$ in a hyperplane; that is, every set of $n+1$ points is linearly independent.

14.4.21 Theorem [1510, Chapter 2] Again from Theorem 14.4.16, if two coordinate frames are given by the vectors $X=(x_0,\dots,x_n)$ and $Y=(y_0,\dots,y_n)$, then a change from one frame to the other is given by $Y=XA$, where A is an $(n+1)\times(n+1)$ non-singular matrix over F. If a projectivity α in the one frame is given by $X'=XT$, then, since $Y'=X'A$, in the other frame it is given by $Y'=X'A=XTA=YA^{-1}TA$.

14.4.22 Definition Let S be a space $\mathrm{PG}(n,F)$ and S' its dual space superimposed on S; that is, the points of S' are the hyperplanes of S and the hyperplanes of S' are the points of S. Consider a function $\alpha: S\longrightarrow S'$. If α is a collineation, it is a *correlation* of S and induces a collineation, also named α, of S' to S; that is, as the points of S are transformed to hyperplanes, then hyperplanes are transformed to points since α preserves incidence. If α is a projectivity, then it is a *reciprocity* of S. In either case, if α is involutory, that is $\alpha^2=1$, where 1 is the identity, then α is a *polarity* of S.

14.4.23 Remark If P and P' are points and π is a hyperplane such that $P^\alpha=\pi$ and $\pi^\alpha=P'$, then, in a polarity, $P=P'$.

14.4.24 Definition Let $\mathrm{AG}(n,F)=\mathrm{PG}(n,F)\backslash\mathcal{H}_\infty$ be an affine space over F. Then, in a given coordinate frame where $\mathcal{H}_\infty$ has equation $x_0=0$, a point of $\mathrm{AG}(n,F)$ can be written $\mathbf{P}(1,x_1,\dots,x_n)$ and hence as $(x_1,\dots,x_n)$. So the points of $\mathrm{AG}(n,F)$ are the elements of $V(n,F)$. The x_i are the *affine* or *non-homogeneous coordinates* of the given point.

14.4.25 Remark It is assumed that, for any $\mathrm{AG}(n,F)$, the coordinate frame is this one.

14.4.26 Theorem [1510, Chapter 2]

1. The subspaces of $\mathrm{AG}(n,F)$ have the form $X+S$, where X is any vector and S is any subspace of $V(n,F)$.
2. Three points X,Y,Z of $\mathrm{AG}(n,F)$ are collinear if and only if there exists λ in $F\backslash\{0,1\}$ such that $X=\lambda Y+(1-\lambda)Z$.

14.4.27 Theorem [1510, Chapter 2]

1. If $\mathbb{F}_{q^n}=\mathbb{F}_q(\beta)$, then the map
$$X=(x_1,\dots,x_n)\mapsto x=x_1+x_2\beta+\cdots+x_n\beta^{n-1}$$
gives a bijection between $\mathrm{AG}(n,q)$ and $\mathbb{F}_{q^n}$.
2. Distinct points X,Y,Z in $\mathrm{AG}(n,F)$ are collinear if and only if, in $\mathbb{F}_{q^n}$,
$$(x-y)^{q-1}=(x-z)^{q-1}.$$

14.4.3 Polarities

14.4.28 Theorem [1510, Chapter 2] Suppose that α is a correlation of $\mathrm{PG}(n,F)$; then it is the product of an automorphic collineation σ and a projectivity of $\mathrm{PG}(n,F)$ to its dual space given by the matrix T. Then α is a polarity if and only if
$$\sigma^2=1 \quad\text{and}\quad T^\sigma T^{*^{-1}}=tI,$$

where T^* denotes the transpose of T and $t \in F_0$.

14.4.29 Theorem [1510, Chapter 2] If α is a polarity of $\mathrm{PG}(n, F)$, then, with σ and T as in Theorem 14.4.28, there are the following possibilities.

1. If $\sigma = 1$ and $\operatorname{char} F \neq 2$, then $T = T^*$ or $T = -T^*$.
2. If $\sigma = 1$ and $\operatorname{char} F = 2$, then $T = T^*$.
3. If $\sigma^2 = 1$, $\sigma \neq 1$, then $T^\sigma = tT^*$ with $t^{\sigma+1} = 1$. For a given frame, T can be chosen so that $t = 1$; that is, $T^\sigma = T^*$.

14.4.30 Remark [1560, Chapter 2] If α is a polarity of $\mathrm{PG}(n, F)$ with n even, then, for a given frame, T can be chosen so that $T^\sigma = T^*$ with $\sigma^2 = 1$.

14.4.31 Definition Let α be a polarity of $\mathrm{PG}(n, F)$.

1. If $\sigma = 1$, $\operatorname{char} F \neq 2$, and $T = T^*$, then α is an *orthogonal* or *ordinary polarity* or a *polarity with respect to a quadric.*
2. If $\sigma = 1$, $\operatorname{char} F \neq 2$, and $T = -T^*$, then α is a *null polarity* or *symplectic polarity* or a *polarity with respect to a linear complex.* Since T is non-singular and skew-symmetric, n is odd.
3. If $\sigma = 1$, $\operatorname{char} F = 2$, $T = T^*$, and all elements on the main diagonal of T are zero, then α is a *null polarity* or *symplectic polarity* or a *polarity with respect to a linear complex.* This only occurs for n odd.
4. If $\sigma = 1$, $\operatorname{char} F = 2$, $T = T^*$, and some element on the main diagonal of T is not zero, then α is a *pseudo-polarity.*
5. If $\sigma \neq 1$, then α is a *Hermitian* or *unitary polarity.*

14.4.32 Definition

1. In a polarity α, if $P^\alpha = \pi$ and $\pi'^\alpha = P'$, with P, P' points and π, π' hyperplanes, then π is the *polar* (*hyperplane*) of P and P' is the *pole* of π'. Since $\alpha^2 = 1$, the converse is also true.
2. If P' lies in $\pi = P^\alpha$, then P lies in $\pi' = P'^\alpha$. In this case, P and P' are *conjugate* points, and π and π' are *conjugate* hyperplanes. The point P is *self-conjugate* if it lies in its own polar hyperplane; the hyperplane π is *self-conjugate* if it contains its own pole.

14.4.33 Remark [1510, Chapter 2] The self-conjugate points $\mathbf{P}(X)$ of α are given by $X^\sigma T X^* = 0$.

14.4.34 Remark For the definition of a linear complex see [1509, Section 15.2].

14.4.35 Definition

1. A *quadric* $\mathcal{Q}$ (or $\mathcal{Q}_n$) in $\mathrm{PG}(n, F)$, $n \geq 1$, is the set of points $\mathbf{P}(x_0, \ldots, x_n)$ satisfying a quadratic equation

$$\sum_{\substack{i,j=0 \\ i \leq j}}^{n} a_{ij} x_i x_j = 0,$$

with a_{ij} in F and not all zero. For $n = 2$, a quadric is a *conic*; for $n = 3$, a quadric is a *quadric surface.*

2. A *Hermitian variety* $\mathcal{H}$ (or $\mathcal{H}_n$) in $\mathrm{PG}(n,F)$, $n \geq 1$, is the set of points $\mathbf{P}(x_0,\ldots,x_n)$ satisfying an equation

$$\sum_{i,j=0}^{n} a_{ij}x_i x_j^{\sigma} = 0,$$

with a_{ij} in F and not all zero, with σ an automorphism of F of order 2, and with $a_{ij}^{\sigma} = a_{ji}$. For $n=2$, a Hermitian variety is a *Hermitian curve*; for $n=3$, a Hermitian variety is a *Hermitian surface*.

14.4.36 Definition Let $\mathcal{F}$ be a quadric or Hermitian variety in $\mathrm{PG}(n,F)$.

1. The point P of $\mathrm{PG}(n,F)$ is *singular* for $\mathcal{F}$ if $\ell \cap \mathcal{F} = \{P\}$ or $\ell \cap \mathcal{F} = \ell$ for every line ℓ through P.
2. If $\mathcal{F}$ has a singular point, then $\mathcal{F}$ is *singular*; otherwise, it is *non-singular*.

14.4.37 Theorem [1510, Chapters 2,5]

1. If α is an orthogonal polarity of $\mathrm{PG}(n,F)$, then the set of self-conjugate points of α is a non-singular quadric $\mathcal{Q}$ of $\mathrm{PG}(n,F)$.
2. If α is a symplectic polarity of $\mathrm{PG}(n,F)$, then every point of $\mathrm{PG}(n,F)$ is self-conjugate.
3. If α is a pseudo-polarity of $\mathrm{PG}(n,F)$, then the set of self-conjugate points of α is a subspace of $\mathrm{PG}(n,F)$.
4. If α is a Hermitian polarity of $\mathrm{PG}(n,F)$, then the set of self-conjugate points of α is a non-singular Hermitian variety $\mathcal{H}$ of $\mathrm{PG}(n,F)$.

14.4.38 Theorem [1510, Chapter 5] If α is a symplectic polarity of $\mathrm{PG}(n,F)$, then, in a suitable coordinate frame, the matrix of α is

$$T = \begin{bmatrix} 0 & 1 & 0 & 0 & \cdots & 0 & 0 \\ -1 & 0 & 0 & 0 & \cdots & 0 & 0 \\ 0 & 0 & 0 & 1 & \cdots & 0 & 0 \\ 0 & 0 & -1 & 0 & \cdots & 0 & 0 \\ \vdots & \vdots & \vdots & \vdots & & \vdots & \vdots \\ 0 & 0 & 0 & 0 & \cdots & 0 & 1 \\ 0 & 0 & 0 & 0 & \cdots & -1 & 0 \end{bmatrix}.$$

14.4.39 Theorem [1510, Chapter 5] If α is a pseudo-polarity of $\mathrm{PG}(n,q)$, q even, then, in a suitable coordinate frame, the matrix of α is as follows:

1. for n even,

$$T = \begin{bmatrix} 1 & 0 & 0 & 0 & 0 & \cdots & 0 & 0 \\ 0 & 0 & 1 & 0 & 0 & \cdots & 0 & 0 \\ 0 & 1 & 0 & 0 & 0 & \cdots & 0 & 0 \\ 0 & 0 & 0 & 0 & 1 & \cdots & 0 & 0 \\ 0 & 0 & 0 & 1 & 0 & \cdots & 0 & 0 \\ \vdots & \vdots & \vdots & \vdots & \vdots & & \vdots & \vdots \\ 0 & 0 & 0 & 0 & 0 & \cdots & 0 & 1 \\ 0 & 0 & 0 & 0 & 0 & \cdots & 1 & 0 \end{bmatrix};$$

2. for n odd,

$$T = \begin{bmatrix} 1 & 1 & 0 & 0 & \cdots & 0 & 0 \\ 1 & 0 & 0 & 0 & \cdots & 0 & 0 \\ 0 & 0 & 0 & 1 & \cdots & 0 & 0 \\ 0 & 0 & 1 & 0 & \cdots & 0 & 0 \\ \vdots & \vdots & \vdots & \vdots & & \vdots & \vdots \\ 0 & 0 & 0 & 0 & \cdots & 0 & 1 \\ 0 & 0 & 0 & 0 & \cdots & 1 & 0 \end{bmatrix}.$$

14.4.40 Remark [1510, Chapter 5] Theorem 14.4.39 holds for every field F of characteristic two with the property that $\{x^2 \mid x \in F\} = F$.

14.4.41 Theorem [1510, Chapter 5] Let $\mathcal{Q}$ be a non-singular quadric of $\mathrm{PG}(n, q)$. The coordinate frame can be chosen so that $\mathcal{Q}$ has the following equation:

1. n even,

$$x_0^2 + x_1x_2 + x_3x_4 + \cdots + x_{n-1}x_n = 0;$$

2. n odd,

   a.

$$x_0x_1 + x_2x_3 + \cdots + x_{n-1}x_n = 0;$$

   b.

$$f(x_0, x_1) + x_2x_3 + \cdots + x_{n-1}x_n = 0,$$

   where f is any chosen irreducible quadratic form.

Hence, up to a projectivity, there is a unique non-singular quadric in $\mathrm{PG}(n, q)$ with n even; for n odd, there are two types of non-singular quadric.

14.4.42 Definition

1. In case 1 of Theorem 14.4.41, the quadric is *parabolic*. In case 2.a, it is *hyperbolic*; in case 2.b, it is *elliptic*;
2. The *character* w of a non-singular quadric in $\mathrm{PG}(n, q)$ is 0 if it is elliptic, 1 if it is parabolic, and 2 if it is hyperbolic.

14.4.43 Theorem [1510, Chapter 5] If $\mathcal{Q}_n$ is a non-singular quadric of $\mathrm{PG}(n, q)$ with character w, then

$$|\mathcal{Q}_n| = (q^n - 1)/(q - 1) + (w - 1)q^{(n-1)/2}.$$

14.4.44 Theorem [1510, Chapter 5] Let $\mathcal{H}$ be a non-singular Hermitian variety of $\mathrm{PG}(n, q^2)$. The coordinate frame can be chosen so that $\mathcal{H}$ has the following equation:

$$x_0^{q+1} + x_1^{q+1} + \cdots + x_n^{q+1} = 0.$$

14.4.45 Theorem [1510, Chapter 5] If $\mathcal{H}_n$ is a non-singular Hermitian variety of $\mathrm{PG}(n, q^2)$, then

$$|\mathcal{H}_n| = [q^{n+1} + (-1)^n][q^n - (-1)^n]/(q^2 - 1).$$

14.4.4 Partitions and cyclic projectivities

14.4.46 Definition A *spread* $\mathcal{S}$ of $\mathrm{PG}(n,q)$ by r-spaces is a set of r-spaces which partitions $\mathrm{PG}(n,q)$; that is, every point of $\mathrm{PG}(n,q)$ lies in exactly one r-space of $\mathcal{S}$. Hence any two r-spaces of $\mathcal{S}$ are disjoint.

14.4.47 Theorem [1510, Chapter 4] The following are equivalent:

1. there exists a spread $\mathcal{S}$ of r-spaces of $\mathrm{PG}(n,q)$;
2. $\theta(r,q) \mid \theta(n,q)$;
3. $(r+1) \mid (n+1)$.

14.4.48 Remark Spreads of $\mathrm{PG}(2r+1,q)$ by r-spaces have been much studied, particularly for their application to non-Desarguesian planes. The latter are considered in more detail in Section 14.3.

14.4.49 Definition Since $\mathbb{F}_q$ is a subfield of $\mathbb{F}_{q^k}$ for $k \in \mathbb{N}\backslash\{0\}$, so $\mathrm{PG}(n,q)$ is naturally embedded in $\mathrm{PG}(n,q^k)$ once the coordinate frame is fixed. Any $\mathrm{PG}(n,q)$ embedded in $\mathrm{PG}(n,q^k)$ is a *subgeometry* of $\mathrm{PG}(n,q^k)$.

14.4.50 Theorem [1510, Chapter 4] If $s(n,q,q^k)$ is the number of subgeometries $\mathrm{PG}(n,q)$ embedded in $\mathrm{PG}(n,q^k)$, then

$$\begin{aligned} s(n,q,q^k) &= |\mathrm{PGL}(n+1,q^k)|/|\mathrm{PGL}(n+1,q)| \\ &= q^{n(n+1)(k-1)/2}\prod_{i=2}^{n+1}[(q^{ki}-1)/(q^i-1)]. \end{aligned}$$

14.4.51 Corollary [1510, Chapter 4] On the line $\mathrm{PG}(1,q^k)$,

1. $s(1,q,q^k) = q^{k-1}(q^{2k}-1)/(q^2-1)$;
2. $s(1,q,q^2) = q(q^2+1)$.

14.4.52 Corollary [1510, Chapter 4] In the plane $\mathrm{PG}(2,q^2)$,

$$s(2,q,q^2) = q^3(q^2+1)(q^3+1).$$

14.4.53 Theorem [1510, Chapter 4] The following are equivalent:

1. there exists a partition of $\mathrm{PG}(n,q^k)$ into subgeometries $\mathrm{PG}(n,q)$;
2. $\theta(n,q) \mid \theta(n,q^k)$;
3. $(k,n+1) = 1$.

14.4.54 Corollary [1510, Chapter 4] The line $\mathrm{PG}(1,q^k)$ can be partitioned into sublines $\mathrm{PG}(1,q)$ if and only if k is odd.

14.4.55 Corollary [1510, Chapter 4] The plane $\mathrm{PG}(2,q^k)$ can be partitioned into subplanes $\mathrm{PG}(2,q)$ if and only if $(k,3) = 1$.

14.4.56 Corollary [1510, Chapter 4] The plane $\mathrm{PG}(2,q^2)$ can be partitioned into q^2-q+1 subplanes $\mathrm{PG}(2,q)$.

14.4.57 **Definition** A projectivity α which permutes the $\theta(n)$ points of $\mathrm{PG}(n,q)$ in a single cycle is a *cyclic projectivity*; it is a *Singer cycle* and the group it generates a *Singer group.*

14.4.58 **Theorem** [1510, Chapter 4] A projectivity α of $\mathrm{PG}(n,q)$ is cyclic if and only if the characteristic polynomial of an associated matrix is *subprimitive*; that is, the smallest power m of a characteristic root that lies in $\mathbb{F}_q$ is $m = \theta(n)$.

14.4.59 **Corollary** [1510, Chapter 4] A cyclic projectivity permutes the hyperplanes of $\mathrm{PG}(n,q)$ in a single cycle.

14.4.60 **Corollary** [1510, Chapter 4] The number of cyclic projectivities in $\mathrm{PG}(n,q)$ is given by

$$\sigma(n,q) = q^{n(n+1)/2}\prod_{i=1}^{n}(q^i-1)\varphi(\theta(n))/(n+1),$$

where φ is the Euler function.

14.4.61 **Corollary** [1510, Chapter 4] The number of conjugacy classes of $\mathrm{PGL}(n+1,q)$ consisting of cyclic projectivities is $\varphi(\theta(n))/(n+1)$.

14.4.62 **Corollary** [1510, Chapter 4] In $\mathrm{PGL}(n+1,q)$ there is just one conjugacy class of Singer groups.

14.4.63 **Theorem** [1510, Chapter 4] If $(k, n+1) = 1$ and α is a projectivity of $\mathrm{PG}(n,q^k)$ which acts as a cyclic projectivity on some $\mathrm{PG}(n,q)$ embedded in $\mathrm{PG}(n,q^k)$, then, letting

$$u = (q^{k(n+1)}-1)(q-1)/[(q^k-1)(q^{n+1}-1)],$$

1. there exists a cyclic projectivity ρ of $\mathrm{PG}(n,q^k)$ such that $\rho^u = \alpha$;
2. every orbit of α is a subgeometry $\mathrm{PG}(n,q)$;
3. if γ is any cyclic projectivity of $\mathrm{PG}(n,q^k)$, then the orbits of γ^u are subgeometries $\mathrm{PG}(n,q)$.

14.4.64 **Theorem** [1510, Chapter 4] If $(k, n+1) = 1$, then the number of projectivities α of $\mathrm{PG}(n,q^k)$ that act cyclically on at least one $\mathrm{PG}(n,q)$ of $\mathrm{PG}(n,q^k)$ is

$$q^{kn(n+1)/2}\prod_{i=1}^{n}(q^{ki}-1)\varphi(\theta(n,q))/(n+1).$$

14.4.5 *k*-Arcs

14.4.65 **Definition**

1. A *k-arc* in $\mathrm{PG}(n,q)$, $n \geq 2$, is a set $\mathcal{K}$ of k points, with $k \geq n+1$, such that no $n+1$ of its points lie in a hyperplane.
2. An arc $\mathcal{K}$ is *complete* if it is not properly contained in a larger arc.
3. Otherwise, if $\mathcal{K}\cup\{P\}$ is an arc for some point P of $\mathrm{PG}(n,q)$, the point P *extends* $\mathcal{K}$.

14.4.66 Definition A *normal rational curve* in $\mathrm{PG}(n,q)$, $n \geq 2$, is any set of points of $\mathrm{PG}(n,q)$ which is projectively equivalent to

$$\{\mathbf{P}(t^n, t^{n-1}, \ldots, t, 1) \mid t \in \mathbb{F}_q\} \cup \{\mathbf{P}(1, 0, \ldots, 0, 0)\}.$$

14.4.67 Remark [1515, Chapter 27] Any normal rational curve contains $q+1$ points. For $n=2$, it is a non-singular conic; for $n=3$, it is a *twisted cubic.* Any $(n+3)$-arc in $\mathrm{PG}(n,q)$ is contained in a unique normal rational curve of this space.

14.4.68 Problem (The three problems of Segre)

I. For given n and q, what is the maximum value of k such that a k-arc exists in $\mathrm{PG}(n,q)$?

II. For what values of n and q with $q > n+1$ is every $(q+1)$-arc of $\mathrm{PG}(n,q)$ a normal rational curve?

III. For given n and q with $q > n+1$, what are the values of k such that each k-arc of $\mathrm{PG}(n,q)$ is contained in a normal rational curve of this space?

14.4.69 Remark For a survey of solutions to Problems I, II, III, see [1512, 1513] and [1511, Chapter 13].

14.4.70 Theorem [1510, Chapter 8] Let $\mathcal{K}$ be a k-arc of $\mathrm{PG}(2,q)$. Then

1. $k \leq q+2$;
2. for q odd, $k \leq q+1$;
3. any non-singular conic of $\mathrm{PG}(2,q)$ is a $(q+1)$-arc;
4. each $(q+1)$-arc of $\mathrm{PG}(2,q)$, q even, extends to a $(q+2)$-arc.

14.4.71 Definition

1. The $(q+1)$-arcs of $\mathrm{PG}(2,q)$ are *ovals.*
2. The $(q+2)$-arcs of $\mathrm{PG}(2,q)$, q even, are *complete ovals* or *hyperovals.*

14.4.72 Theorem [1510, Chapter 8] In $\mathrm{PG}(2,q)$, q odd, every oval is a non-singular conic.

14.4.73 Remark Theorem 14.4.72 is a celebrated result due to Segre [2575]. For more details on k-arcs in $\mathrm{PG}(2,q)$, ovals and hyperovals see [1510, Chapters 8–10]. The fundamental Theorem 14.4.76, also due to Segre [2577], relates k-arcs of $\mathrm{PG}(2,q)$ to plane algebraic curves.

14.4.74 Definition Let $\mathcal{K}$ be a k-arc of $\mathrm{PG}(2,q)$.

1. A *tangent* of $\mathcal{K}$ is a line of $\mathrm{PG}(2,q)$ meeting $\mathcal{K}$ in a unique point.
2. A *secant* of $\mathcal{K}$ is a line meeting $\mathcal{K}$ in two points.

14.4.75 Remark At each point, $\mathcal{K}$ has $t = q+2-k$ tangents; the total number of tangents is tk.

14.4.76 Theorem [1510, Chapter 10]

1. Let $\mathcal{K}$ be a k-arc in $\mathrm{PG}(2,q)$, with q even. Then the tk tangents of $\mathcal{K}$ belong to an algebraic envelope Γ_t of class t, that is, the dual of a plane algebraic curve of order t, with the following properties:

a. Γ_t is unique if $k > (q+2)/2$;

b. Γ_t contains no secant of $\mathcal{K}$ and so no pencil with vertex P in $\mathcal{K}$, where a *pencil* is the set of lines through a point;

c. if Δ_P is the pencil of lines with vertex P in $\mathcal{K}$ and ℓ is a tangent at P, then the intersection multiplicity of Δ_P and Γ_t at ℓ is one.

2. Let $\mathcal{K}$ be a k-arc in $\mathrm{PG}(2,q)$, with q odd. Then the tk tangents of $\mathcal{K}$ belong to an algebraic envelope Γ_{2t} of class $2t$ with the following properties:

a. Γ_{2t} is unique if $k > (2q+4)/3$;

b. Γ_{2t} contains no secant of $\mathcal{K}$ and so no pencil with vertex P in $\mathcal{K}$;

c. if Δ_P is the pencil with vertex P in $\mathcal{K}$ and ℓ is a tangent at P, then the intersection multiplicity of Δ_P and Γ_{2t} at ℓ is two;

d. Γ_{2t} may contain components of multiplicity at most two, but does not consist entirely of double components.

14.4.77 Corollary [1510, Chapter 10]

1. If q is even and $k > (q+2)/2$, then $\mathcal{K}$ is contained in a unique complete arc of $\mathrm{PG}(2,q)$.
2. If q is odd and $k > (2q+4)/3$, then $\mathcal{K}$ is contained in a unique complete arc of $\mathrm{PG}(2,q)$.

14.4.78 Remark

1. For a survey on k-arcs in $\mathrm{PG}(n,q)$, $n > 2$, see [1512, 1513].
2. For q odd, the main results were obtained by induction and projection of the k-arc $\mathcal{K}$ from one of its points P onto a hyperplane not containing P; see Thas [2794] and [1515, Chapter 27].
3. For any q, a theorem of Bruen, Thas, and Blokhuis [434] relates k-arcs in $\mathrm{PG}(3,q)$ to dual algebraic surfaces. For q even, this result enables estimates to be made for the three problems of Segre. A generalisation by Blokhuis, Bruen, and Thas [324] now follows.

14.4.79 Theorem [1515, Chapter 27] Let $\mathcal{K} = \{P_1, P_2, \ldots, P_k\}$ be a k-arc of $\mathrm{PG}(n,q)$. For distinct $i_1, i_2, \ldots, i_{n-1} \in \{1, 2, \ldots, k\}$, let $Z_{\{i_1,i_2,\ldots,i_{n-1}\}}$ be the set of $t = q+n-k$ hyperplanes through the $(n-2)$-dimensional subspace Π_{n-2}, generated by $P_{i_1}, P_{i_2}, \ldots, P_{i_{n-1}}$, that contains no other point of $\mathcal{K}$.

1. a. For q even, there exists a dual algebraic hypersurface Φ_t of class t in $\mathrm{PG}(n,q)$ which contains the hyperplanes of each set $Z_{\{i_1,i_2,\ldots,i_{n-1}\}}$.

   b. This dual hypersurface is unique when $k > (q+2n-2)/2$.

2. a. For q odd, there exists a dual algebraic hypersurface Φ_{2t} of class $2t$ in $\mathrm{PG}(n,q)$ which contains the hyperplanes of each set $Z_{\{i_1,i_2,\ldots,i_{n-1}\}}$.

   b. The intersection multiplicity of Φ_{2t} and the pencil of hyperplanes with vertex Π_{n-2} at each hyperplane of $Z_{\{i_1,i_2,\ldots,i_{n-1}\}}$ is two.

   c. This dual hypersurface is unique when $k > (2q+3n-2)/3$.

14.4.80 Corollary [1515, Chapter 27]

1. If q is even and $k > (q+2n-2)/2$, then a k-arc $\mathcal{K}$ of $\mathrm{PG}(n,q)$ is contained in a unique complete arc.

2. If q is odd and $k > (2q + 3n - 2)/3$, then a k-arc $\mathcal{K}$ of $\mathrm{PG}(n, q)$ is contained in a unique complete arc.

14.4.81 Theorem (The duality principle for k-arcs) [1515, Chapter 27] A k-arc of $\mathrm{PG}(n, q)$, with $n \geq 2$ and $k \geq n + 4$, exists if and only if a k-arc of $\mathrm{PG}(k - n - 2, q)$ exists.

14.4.82 Corollary [1515, Chapter 27] In $\mathrm{PG}(q - 2, q)$, q even and $q \geq 4$, there exist $(q + 2)$-arcs.

14.4.83 Conjecture [1511, Chapter 13]

1. If $\mathcal{K}$ is a k-arc in $\mathrm{PG}(n, q)$, with $q - 1 \geq n \geq 2$, then $k \leq q + 1$ for q odd and $k \leq q + 2$ for q even.
2. In $\mathrm{PG}(n, q)$, with $q \geq n \geq 2$ and q even, there exist $(q + 2)$-arcs if and only if $n \in \{2, q - 2\}$.

14.4.6 k-Arcs and linear MDS codes

14.4.84 Definition

1. Let C be a *code* of *length* k over an alphabet A of size q with $q \geq 2$. In other words, C is a set of *(code)words*, where each word is an ordered k-tuple over A.
2. For a given m with $2 \leq m \leq k$, impose the following condition: no two words in C agree in as many as m positions. Then $|C| \leq q^m$. If $|C| = q^m$, then C is a *maximum distance separable* (*MDS*) code.

14.4.85 Remark MacWilliams and Sloane [1991, Chapter 11] introduce the chapter on MDS codes as "one of the most fascinating in all of coding theory."

14.4.86 Definition

1. The (*Hamming*) *distance* between two codewords

$$X = (x_1, \ldots, x_k) \text{ and } Y = (y_1, \ldots, y_k)$$

is the number of indices i for which $x_i \neq y_i$; it is denoted $d(X, Y)$.
2. The *minimum distance* of a code C, with $|C| > 1$, is

$$d(C) = \min\{d(X, Y) \mid X, Y \in C,\ X \neq Y\}.$$

14.4.87 Theorem [2849, Chapter 5] For an MDS code, $d(C) = k - m + 1$; see Section 15.1.

14.4.88 Remark One of the main problems concerning MDS codes is to maximize $d(C)$, and so k, for given m and q. Another problem is to determine the structure of C in the optimal case.

14.4.89 Theorem [434] For an MDS code, $k \leq q + m - 1$.

14.4.90 Remark For $m = 2$, the MDS code C gives a set of q^2 codewords of length k, no two of which agree in more than one position. This is equivalent to the existence of a net of order q and degree k; see also Section 14.1. It follows that $k \leq q + 1$, the case of equality corresponding to an affine plane of order q; see Section 14.3. Theorem 14.4.89 follows by an inductive argument.

14.4.91 Remark

1. The case $m = 3$ and $k = q + 2$ is equivalent to the existence of an affine plane of order q, with q even, containing an appropriate system of $(q+2)$-arcs. For all known examples the plane is an affine plane $\mathrm{AG}(2, q)$ with $q = 2^h$; see Willems and Thas [2979].
2. For $m = 4$ and $k = q + 3$, it has been shown that either $q = 2$ or 36 divides q; no example other than for $q = 2$ is known to exist; see Bruen and Silverman [431].

14.4.92 Remark Henceforth, only *linear* MDS codes are considered; that is C is an m-dimensional subspace of the vector space $V(k, q)$, which is $\mathbb{F}_q^k$ with the usual addition and scalar multiplication.

14.4.93 Theorem [1511, Chapter 13] For $m \geq 3$, linear MDS codes and arcs are equivalent objects.

14.4.94 Remark Let C be an m-dimensional subspace of $V(k, q)$ and let G be an $m \times k$ generator matrix for C; that is, the rows of G are a basis for C. Then C is MDS if and only if any m columns of G are linearly independent; this property is preserved under multiplication of the columns by non-zero scalars. So consider the columns of G as points $P_1, \ldots, P_k$ of $\mathrm{PG}(m-1, q)$. It follows that C is MDS if and only if $\{P_1, \ldots, P_k\}$ is a k-arc of $\mathrm{PG}(m-1, q)$. This gives the relation between linear MDS codes and arcs.

14.4.95 Theorem [1511, Chapter 13] For $2 \leq m \leq k - 2$, the dual of a linear MDS code is again a linear MDS code.

14.4.96 Remark For $3 \leq m \leq k - 3$, Theorem 14.4.95 is the translation of Theorem 14.4.81 from geometry to coding theory.

14.4.7 k-Caps

14.4.97 Definition

1. In $\mathrm{PG}(n, q)$, $n \geq 3$, a set $\mathcal{K}$ of k points no three of which are collinear is a *k-cap.*
2. A k-cap is *complete* if it is not contained in a $(k + 1)$-cap.
3. A line of $\mathrm{PG}(n, q)$ is a *secant, tangent,* or *external line* as it meets $\mathcal{K}$ in $2, 1$ or 0 points.

14.4.98 Theorem [1509, Chapter 16]

1. For any k-cap $\mathcal{K}$ in $\mathrm{PG}(3, q)$ with $q \neq 2$, the cardinality k satisfies $k \leq q^2 + 1$.
2. In $\mathrm{PG}(3, 2)$, a k-cap satisfies $k \leq 8$; an 8-cap is the complement of a plane.

14.4.99 Definition A $(q^2 + 1)$-cap of $\mathrm{PG}(3, q)$, $q \neq 2$ is an *ovoid*; the *ovoids* of $\mathrm{PG}(3, 2)$ are its elliptic quadrics.

14.4.100 Theorem [1509, Chapter 16] At each point P of an ovoid $\mathcal{O}$ of $\mathrm{PG}(3, q)$, there is a unique *tangent plane* π such that $\pi \cap \mathcal{O} = \{P\}$.

14.4.101 Theorem [1509, Chapter 16]

1. Apart from the tangent planes, every plane meets an ovoid $\mathcal{O}$ in a $(q + 1)$-arc.
2. When q is even, the $(q^2 + 1)(q + 1)$ tangents of $\mathcal{O}$ are the totally isotropic lines of a symplectic polarity α of $\mathrm{PG}(3, q)$, that is, the lines ℓ for which $\ell^\alpha = \ell$.

14.4.102 Theorem [1509, Chapter 16] In PG$(3, q)$, q odd, every ovoid is an elliptic quadric.

14.4.103 Remark Theorem 14.4.102 is a celebrated result, due independently to Barlotti [202] and Panella [2358]. Both proofs rely on Theorem 14.4.72.

14.4.104 Theorem [421] In PG$(3, q)$, q even, every ovoid containing at least one conic section is an elliptic quadric.

14.4.105 Theorem [1509, Chapter 16] In PG$(3, q)$, let $W(q)$ be the incidence structure formed by all the points and the totally isotropic lines of a symplectic polarity α. Then $W(q)$ admits a polarity α' if and only if $q = 2^{2e+1}$; in that case, the absolute points of α', namely the points lying in their image lines, form an ovoid of PG$(3, q)$. Such an ovoid is an elliptic quadric if and only if $q = 2$.

14.4.106 Definition For $q = 2^{2e+1}$, with $e \geq 1$, the ovoids in Theorem 14.4.105 are *Tits ovoids.*

14.4.107 Theorem [1509, Chapter 16] With $q = 2^{2e+1}$, the canonical form of a Tits ovoid is

$$\mathcal{O} = \{\mathbf{P}(1, z, y, x) \mid z = xy + x^{\sigma+2} + y^{\sigma}\} \cup \{\mathbf{P}(0, 1, 0, 0)\},$$

where σ is the automorphism $t \mapsto t^{2^{e+1}}$ of $\mathbb{F}_q$.

14.4.108 Theorem [1509, Chapter 16] For $q = 2^{2e+1}$, $e \geq 1$, the group of all projectivities of PG$(3, q)$ fixing the Tits ovoid $\mathcal{O}$ is the Suzuki group Sz(q), which acts doubly transitively on $\mathcal{O}$.

14.4.109 Remark The case $q = 4$ is the same as q odd; that is, an ovoid of PG$(3, 4)$ is an elliptic quadric; see Barlotti [202] or [1509, Chapter 16]. For $q = 8$, Segre [2576] found an ovoid other than an elliptic quadric; Fellegara [1050] showed that this example is a Tits ovoid. She also showed, using a computer program, that every ovoid in PG$(3, 8)$ is either an elliptic quadric or a Tits ovoid. O'Keefe and Penttila [2313, 2314] showed, also using a computer program, that in PG$(3, 16)$ every ovoid is an elliptic quadric. O'Keefe, Penttila, and Royle [2315], also using a computer program, showed that in PG$(3, 32)$ every ovoid is an elliptic quadric or a Tits ovoid.

14.4.110 Definition Let $\mathcal{O}$ be an ovoid of PG$(3, q)$ and let $\mathcal{B}$ be the set of all intersections $\pi \cap \mathcal{O}$, with π a non-tangent plane of $\mathcal{O}$. Then the incidence structure formed by the triple $\mathcal{I}(\mathcal{O}) = (\mathcal{O}, \mathcal{B}, \in)$ is a $3-(q^2+1, q+1, 1)$ design. A $3-(n^2+1, n+1, 1)$ design $\mathcal{I} = (\mathcal{P}, \mathcal{B}, \in)$ is an *inversive plane of order* n and the elements of $\mathcal{B}$ are *circles*; the inversive planes arising from ovoids are *egglike.*

14.4.111 Theorem [807, Chapter 6] Every inversive plane of even order is egglike.

14.4.112 Definition If the ovoid $\mathcal{O}$ is an elliptic quadric, then the inversive plane $\mathcal{I}(\mathcal{O})$, and any inversive plane isomorphic to it, is *classical* or *Miquelian.*

14.4.113 Remark By Theorem 14.4.102, an egglike inversive plane of odd order is Miquelian. For odd order, no other inversive planes are known.

14.4.114 Definition Let $\mathcal{I}$ be an inversive plane of order n. For any point P of $\mathcal{I}$, the points of $\mathcal{I}$ other than P, together with the circles containing P with P removed, form a $2-(n^2, n, 1)$ design, that is, an affine plane of order n. This plane is denoted $\mathcal{I}_P$ and is the *internal plane* or *derived plane* of $\mathcal{I}$ at P.

14.4.115 Remark [807, Chapter 6] For an egglike inversive plane $\mathcal{I}(\mathcal{O})$ of order q, each internal plane is Desarguesian, that is, the affine plane $\mathrm{AG}(2,q)$ over $\mathbb{F}_q$.

14.4.116 Theorem [2795] Let $\mathcal{I}$ be an inversive plane of odd order n. If, for at least one point P of $\mathcal{I}$, the internal plane $\mathcal{I}_P$ is Desarguesian, then $\mathcal{I}$ is Miquelian.

14.4.117 Remark Up to isomorphism, there is a unique inversive plane of order n for the values $n = 2, 3, 4, 5, 7$; see Chen [608], Denniston [821, 822], Witt [2997]. As a corollary of Theorem 14.4.116 and the uniqueness of the projective plane of order n for $n = 3, 5, 7$, a computer-free proof of the uniqueness of the inversive plane of order n is obtained for these n.

14.4.118 Remark For more information about designs, see Section 14.5. For more information about projective spaces, see [1509, 1510, 1515] and [1511, Chapter 13].

See Also

§12.5	For results on curves which impinge on k-arcs.
§14.2	For a technique to resolve problems on blocking sets.
§14.3	For other aspects of Desarguesian planes.

References Cited: [202, 324, 421, 431, 434, 608, 807, 821, 822, 1050, 1509, 1510, 1511, 1512, 1513, 1515, 1560, 1991, 2313, 2314, 2315, 2358, 2575, 2576, 2577, 2794, 2795, 2849, 2979, 2997]

14.5 Block designs

Charles J. Colbourn, Arizona State University
Jeffrey H. Dinitz, University of Vermont

14.5.1 Basics

14.5.1 Definition A *balanced incomplete block design* (BIBD) is a pair $(V, \mathcal{B})$ where V is a v-set and $\mathcal{B}$ is a collection of b k-subsets of V (*blocks*) such that each element of V is contained in exactly r blocks and any 2-subset of V is contained in exactly λ blocks. The numbers v, b, r, k, and λ are *parameters* of the BIBD. Its *order* is v; its *replication number* is r; its *blocksize* is k; and its *index* is λ.

14.5.2 Proposition Trivial necessary conditions for the existence of a BIBD(v, b, r, k, λ) are

1. $vr = bk$, and
2. $r(k-1) = \lambda(v-1)$.

Parameter sets that satisfy conditions 1 and 2 are *admissible*.

14.5.3 Remark The three parameters v, k, and λ determine the remaining two as $r = \frac{\lambda(v-1)}{k-1}$ and $b = \frac{vr}{k}$. Hence one often writes (v, k, λ) *design* to denote a BIBD(v, b, r, k, λ).

14.5.4 Example The unique $(6, 3, 2)$ design and the unique $(7, 3, 1)$ design have blocks shown below as columns:

```
0000011122        0001123
1123423433        1242534
2345554545        3654656
```

14.5.5 Definition A BIBD $(V, \mathcal{B})$ with parameters v, b, r, k, λ is

simple	if it has no repeated blocks;
complete or *full*	if it is simple and contains $\binom{v}{k}$ blocks;
decomposable	if $\mathcal{B}$ can be partitioned into two nonempty collections $\mathcal{B}_1$ and $\mathcal{B}_2$ so that $(V, \mathcal{B}_i)$ is a (v, k, λ_i) design for $i = 1, 2$;
Hadamard	if $v = 4n - 1$, $k = 2n - 1$, and $\lambda = n - 1$ for some integer $n \geq 2$;
m-multiple	if $v, \frac{b}{m}, \frac{r}{m}, k, \frac{\lambda}{m}$ are the parameters of a BIBD;
nontrivial	if $3 \leq k < v$;
quasi-symmetric	if every two distinct blocks intersect in either μ_1 or μ_2 elements;
resolvable (an *RBIBD*)	if there exists a partition R of its set of blocks B into *parallel classes*, each of which in turn partitions the set V (R is a *resolution*);
a *Steiner 2-design* $S(2, k, v)$	if $\lambda = 1$;
a *Steiner triple system* $\mathrm{STS}(v)$	if $k = 3$ and $\lambda = 1$;
symmetric	if $v = b$, or equivalently $k = r$;
a *triple system* $\mathrm{TS}(v, \lambda)$	if $k = 3$.

14.5.2 Triple systems

14.5.6 Remark A Steiner triple system of order v can exist only when $v - 1$ is even because every element occurs with $v - 1$ others, and in each block in which it occurs it appears with two other elements. Moreover, every block contains three pairs and hence $\binom{v}{2}$ must be a multiple of 3. Thus, it is necessary that $v \equiv 1, 3 \pmod 6$. This condition was shown to be sufficient in 1847.

14.5.7 Theorem [1741] A Steiner triple system of order v exists if and only if $v \equiv 1, 3 \pmod 6$.

14.5.8 Theorem [708] A $\mathrm{TS}(v, \lambda)$ exists if and only if $v \neq 2$ and $\lambda \equiv 0 \pmod{\gcd(v - 2, 6)}$.

14.5.9 Remark Existence theorems such as Theorems 14.5.7 and 14.5.8 are typically established by a combination of *direct* constructions to make designs for specific values of v, and *recursive* constructions to make solutions for large values of v from solutions with smaller values of v. Finite fields are most often used in providing direct constructions, both to provide ingredients for recursive constructions, and to produce solutions with specific properties. Examples for triple systems are developed to demonstrate these; see [708].

14.5.10 Construction [2222] Let p be a prime, $n \geq 1$, and $p^n \equiv 1 \pmod 6$. Then there is an $\mathrm{STS}(p^n)$. To construct one, let $\mathbb{F}_{p^n}$ be a finite field on a set X of size $p^n = 6t + 1$ with 0 as its zero element, and ω a primitive root of unity. Then

$$\{\{\omega^i + j, \omega^{2t+i} + j, \omega^{4t+i} + j\} : 0 \leq i < t, j \in X\}$$

(with computations in $\mathbb{F}_{p^n}$) is the set of blocks of an STS(p^n) on X.

14.5.11 Remark In order to verify Construction 14.5.10, consider two distinct elements $x, y \in \mathbb{F}_{p^n}$. Let $d = x - y$ (arithmetic in $\mathbb{F}_{p^n}$). Now since $d \neq 0$ and $\omega^{2t} - 1 \neq 0$, d can be uniquely written in the form $(\omega^{2t} - 1)\omega^{2jt}\omega^i$ for $j \in \{0, 1, 2\}$ and $0 \leq i < 2t$. Since $\omega^{3t} = -1$, if $i \geq t$, we may write

$$-d = y - x = (\omega^{2t} - 1)\omega^{2(j+1)t}\omega^{i-t}.$$

Thus we suppose without loss of generality that $d = x - y$ and $i < t$. Then $\{x, y\}$ appears in the triple $\{\omega^i, \omega^{2t+i}, \omega^{4t+i}\} + (x - \omega^{2(j+1)t+i})$. Consequently, every pair of distinct elements in $\mathbb{F}_{p^n}$ appears in at least one of the triples defined. Because the total number of pairs in the triples defined is precisely $\binom{p^n}{2}$, every pair occurs in exactly one triple.

14.5.12 Definition Two BIBDs (V_1, B_1), (V_2, B_2) are *isomorphic* if there exists a bijection $\alpha : V_1 \to V_2$ such that $B_1\alpha = B_2$. An *automorphism* is an isomorphism of a design with itself. The set of all automorphisms of a design forms a group, the *(full) automorphism group*. An *automorphism group* is any subgroup of the full automorphism group.

14.5.13 Remark If $(V, \mathcal{B})$ is a BIBD(v, b, r, k, λ) with automorphism group G, the action of G partitions $\mathcal{B}$ into classes (*orbits*). A set of orbit representatives is a set of *starter blocks* or *base blocks*. Applying the action of G to a set of base blocks yields a design, the *development*.

14.5.14 Remark In Construction 14.5.10, we can treat $\mathcal{D} = \{\{\omega^i, \omega^{2t+i}, \omega^{4t+i}\} : 0 \leq i < t\}$ as the *base* or *starter triples* of the design. Their *development* is the result of applying the action of the elementary abelian group of order p^n to the base triples. The verification requires that for every *difference* $d \in \mathbb{F}_{p^n} \setminus \{0\}$, there is exactly one way to choose $x, y \in D \in \mathcal{D}$ so that $d = x - y$, with arithmetic in the elementary abelian group of order p^n.

14.5.15 Construction [807] Let p be a prime, $n \geq 1$, and $p^n \equiv 7 \pmod{12}$. Let $\mathbb{F}_{p^n}$ be a finite field on a set X of size $p^n = 6t + 1 = 12s + 7$ with 0 as its zero element and ω a primitive root of unity. Then

$$\{\{\omega^{2i} + j, \omega^{2t+2i} + j, \omega^{4t+2i} + j\} : 0 \leq i < t, j \in X\}$$

forms the blocks of an STS(p^n) on X. (These are the *Netto triple systems*.)

14.5.16 Remark The Netto triple systems provide examples of STS(v)s that admit 2-homogeneous automorphism groups but (for $v > 7$) do not admit 2-transitive groups. We give another construction of the Netto triple systems. Let

$$\Gamma = \{a^2 x\sigma + b \ : \ a, b \in \mathbb{F}_{p^n}, \ \sigma \in \text{Aut}(\mathbb{F}_{p^n})\}.$$

Let ε be a primitive sixth root of unity in $\mathbb{F}_{p^n}$. Then the orbit of $\{0, 1, \varepsilon\}$ under the action of Γ is the Netto triple system of order p^n. This illustrates one of the principal reasons for using large automorphism groups, and in particular for using the additive and multiplicative structure of the finite field – a single triple represents the entire triple system.

14.5.17 Construction [1459] Let $p = 2t + 1$ be an odd prime. Let ω be a primitive root of unity in $\mathbb{Z}_p$ satisfying $\omega \equiv 1 \pmod 3$. Then

$$\{\{\omega^i + j, \omega^{2t+i} + j, \omega^{4t+i} + j\} : 0 \leq i < t, j \in \mathbb{Z}_{3p}\} \cup \{\{j, j + p, j + 2p\} : j \in \mathbb{Z}_p\}$$

(with computations modulo $3p$) is the set of blocks of an STS($3p$).

14.5.18 **Definition** A set of blocks is a *partial parallel class (PPC)* if no two blocks in the set share an element. A PPC is an *almost parallel class* if it contains $\frac{v-1}{3}$ blocks; when it contains $\frac{v}{3}$ blocks, it is a *parallel class* or *resolution class*. A partition of all blocks of a $\mathrm{TS}(v,\lambda)$ into parallel classes is a *resolution* and the STS is *resolvable*. An $\mathrm{STS}(v)$ together with a resolution of its blocks is a *Kirkman triple system*, $\mathrm{KTS}(v)$.

14.5.19 **Remark** In Construction 14.5.17, the triples $\{\{j, j+p, j+2p\} : j \in \mathbb{Z}_p\}$ form a parallel class. Indeed we can say much more in certain cases. The method of "pure and mixed differences" [356] is applied, using a set of elements $\mathbb{F}_q \times X$, for X a finite set; a *pure(x) difference* is the difference $d = a - b$ associated with the pair $\{(a, x), (b, x)\}$ and a *mixed(x,y) difference* is the difference $d = a - b$ associated with the pair $\{(a, x), (b, y)\}$.

14.5.20 **Construction** [2440] If $q = p^\alpha \equiv 1 \pmod 6$ is a prime power, then there exists a $\mathrm{KTS}(3q)$. Let $t = (q-1)/6$. To construct a $\mathrm{KTS}(3q)$, take as elements $\mathbb{F}_q \times \{1, 2, 3\}$, writing a_i for (a, i). Let ω be a primitive element in $\mathbb{F}_q$, and let $\mathcal{B}$ consist of triples:

1. $C = \{0_1, 0_2, 0_3\}$;
2. $B_{ij} = \{\omega^i_j, \omega^{i+2t}_j, \omega^{i+4t}_j\}$, $0 \le i < t$, $j \in \{1, 2, 3\}$;
3. $A_i = \{\omega^i_1, \omega^{i+2t}_2, \omega^{i+4t}_3\}$, $0 \le i < t$.

Each of the (nonzero) pure and mixed differences occurs exactly once in triples of $\mathcal{B}$, and thus $\mathcal{B}$ is the set of starter triples for an $\mathrm{STS}(3q)$. This $\mathrm{STS}(3q)$ is resolvable. Indeed, $R_0 = C \cup \{B_{ij} : 0 \le i < t, j \in \{1, 2, 3\}\} \cup \{A_{jt+i} : j \in \{1, 3, 5\}, 0 \le i < t\}$ forms a parallel class; when developed modulo $6t+1$, it yields a further $6t$ parallel classes. Each A_i, when developed modulo $6t+1$, also yields a parallel class; taking those parallel classes only from A_{jt+i} with $j \in \{0, 2, 4\}$ and $0 \le i < t\}$ thus yields a further $3t$ parallel classes, for a total of $9t+1$ forming the resolution.

14.5.21 **Construction** [2440] If $q = p^\alpha \equiv 1 \pmod 6$ is a prime power, then there exists a $\mathrm{KTS}(2q+1)$. Let $t = (q-1)/6$. To construct a $\mathrm{KTS}(2q+1)$, take as elements $(\mathbb{F}_q \times \{1, 2\}) \cup \{\infty\}$. Let ω be a primitive element of $\mathbb{F}_q$, and let m satisfy $2\omega^m = \omega^t + 1$. Let $\mathcal{B}$ consist of triples

1. $C = \{0_1, 0_2, \infty\}$;
2. $B_i = \{\omega^i_1, \omega^{i+t}_1, \omega^{i+m}_2\}$, $0 \le i < t$, $2t \le i < 3t$, or $4t \le i < 5t$;
3. $A_i = \{\omega^{i+m}_2, \omega^{i+m+3t}_2, \omega^{i+m+5t}_2\}$, $0 \le i < t$.

Every pure and every mixed difference occurs exactly once and hence $\mathcal{B}$ is a set of starter triples for an $\mathrm{STS}(2q+1)$. But $\mathcal{B}$ itself forms a parallel class, whose development modulo q yields the required q parallel classes for the $\mathrm{KTS}(2q+1)$.

14.5.3 Difference families and balanced incomplete block designs

14.5.22 **Definition** Let $B = \{b_1, ..., b_k\}$ be a subset of an additive group G. The *G-stabilizer* of B is the subgroup G_B of G consisting of all elements $g \in G$ such that $B + g = B$. B is *full* or *short* according to whether G_B is or is not trivial. The *G-orbit* of B is the set $Orb_G B$ of all distinct right translates of B, namely, $Orb_G B = \{B + s \mid s \in S\}$ where S is a complete system of representatives for the right cosets of G_B in G.

14.5.23 **Definition** The multiset $\Delta B = \{b_i - b_j \mid i, j = 1, \ldots, k, i \ne j\}$ is the *list of differences from B*. The multiplicity in ΔB of an element $g \in G$ is of the form $\mu_g |G_B|$ for some integer μ_g. The *list of partial differences from* B is the multiset ∂B where each $g \in G$ appears exactly μ_g times. ($\Delta B = \partial B$ if and only if B is a full block.)

14.5.24 Definition Let G be a group of order v. A collection $\{B_1, ..., B_t\}$ of k-subsets of G forms a (v, k, λ) *difference family* (or *difference system*) if every nonidentity element of G occurs λ times in $\partial B_1 \cup \cdots \cup \partial B_t$. The sets B_i are *base blocks.* A difference family having at least one short block is *partial.*

14.5.25 Remark

1. All definitions given can be extended to a multiplicative group by replacing $B+g$ with $B \cdot g$ and $b_i - b_j$ with $b_i b_j^{-1}$.
2. If $t = 1$, then B_1 is a (v, k, λ) difference set; see Section 14.6.
3. If $\{B_1, \ldots, B_t\}$ is a (v, k, λ) difference family over G, $Orb_G(B_1) \cup \cdots \cup Orb_G(B_t)$ is the collection of blocks of a BIBD(v, k, λ) admitting G as a sharply point-transitive automorphism group. This BIBD is *cyclic (abelian, nonabelian, dihedral, and so on)* if the group G has the respective property. In this case the difference family is a *cyclic (abelian, nonabelian, dihedral, respectively) difference family.*
4. A BIBD(v, k, λ) with an automorphism group G acting sharply transitively on the points is (up to isomorphism) generated by a suitable (v, k, λ) difference family.
5. Every short block of a $(v, k, 1)$ difference family over an abelian group G is a coset of a suitable subgroup of G.

14.5.26 Theorem [261] The set of order p subgroups of $\mathbb{F}_{p^n}$ forms a $(p^n, p, 1)$ difference family generating the point-line design associated with the affine geometry $AG(n, p)$.

14.5.27 Definition

1. $C = \{c_1, \ldots, c_k\}$ is a *multiple* of $B = \{b_1, \ldots, b_k\}$ if, for some w, $c_i = w \cdot b_i$ for all i.
2. w is a *multiplier* of order n of $B = \{b_1, \ldots, b_k\}$ if $w^n = 1$ but $w^i \neq 1$ for $0 < i < n$, and for some $g \in G$, $B = w \cdot B + g = \{w \cdot b_1 + g, w \cdot b_2 + g, \ldots, w \cdot b_k + g\}$.
3. w is a *multiplier* of a difference family D if, for each base block $B \in D$, there exists $C \in D$ and $g \in G$ for which $w \cdot B + g = C$.
4. If q is a prime power and D is a (q, k, λ) difference family over $\mathbb{F}_q$ in which one base block, B, has a multiplier of order k or $k-1$ and all other base blocks are multiples of B, then D is *radical.*

14.5.28 Theorem [5] Suppose $q \equiv 7 \pmod{12}$ is a prime power and there exists a cube root of unity ω in $\mathbb{F}_q$ such that $x = \omega - 1$ is a primitive root. Then the following base blocks form a $(7q,4,1)$ difference family over $\mathbb{Z}_7 \times \mathbb{F}_q$:

1. $\{(0,0), (0, (x-1)x^{2t-1}), (0, \omega(x-1)x^{2t-1}), (0, \omega^2(x-1)x^{2t-1})\}$ for $1 \leq t \leq (q-7)/12$,
2. $\{(0,0), (1, x^{2t}), (2, \omega x^{2t}), (4, \omega^2 x^{2t})\}$ for $1 \leq t \leq (q-3)/2$, $x^{2t} \neq \omega$, and
3. $\{(0,0), (2^t, \omega^t), (2^t, x \cdot \omega^t), (2^{t+2}, 0)\}$ for $0 \leq t \leq 2$.

14.5.29 Remark The $(7q,4,1)$ difference families are obtainable by Theorem 14.5.28 for $q =$ 7, 19, 31, 43, 67, 79, 103, 127, 151, 163, 199, 211, 367, 379, 439, 463, 487, 571, but not for $q =$

139, 223, 271, 283, 307, 331, 523, 547. A more general construction for (7q,4,1) difference families with q a prime power $\equiv 7 \pmod{12}$ can be found in [5].

14.5.30 Theorem [2986] Suppose q is a prime power, and $\lambda(q-1) \equiv 0 \pmod{k(k-1)}$. Then a (q, k, λ) difference family over $\mathbb{F}_q$ exists if

1. λ is a multiple of $k/2$ or $(k-1)/2$;
2. $\lambda \geq k(k-1)$; or
3. $q > \binom{k}{2}^{k(k-1)}$.

14.5.31 Theorem [459] Suppose q is an odd prime power. Then there exists a (q, k, λ) radical difference family if either:

1. λ is a multiple of $k/2$ and $q \equiv 1 \pmod{k-1}$, or
2. λ is a multiple of $(k-1)/2$ and $q \equiv 1 \pmod{k}$.

14.5.32 Remark For radical difference families with $\lambda = 1$, the multiplier must have odd order (that is, order k if k is odd, or order $k-1$ if k is even).

14.5.33 Theorem [459] Let $q = 12t + 1$ be a prime power and 2^e be the largest power of 2 dividing t. Then a $(q, 4, 1)$ radical difference family in $\mathbb{F}_q$ exists if and only if -3 is not a 2^{e+2}-th power in $\mathbb{F}_q$. (This condition holds for $q =$ 13, 25, 73, 97, 109, 121, 169, 181, 193, 229, 241, 277, 289, 313, 337, 409, 421, 433, 457, 529, 541, 577, 601, 625, 673, 709, 733, 757, 769, 829, 841.)

14.5.34 Theorem [459] Let $q = 20t + 1$ be a prime power, and let 2^e be the largest power of 2 dividing t. Then a $(q, 5, 1)$ radical difference family in $\mathbb{F}_q$ exists if and only if $(11 + 5\sqrt{5})/2$ is not a 2^{e+1}-th power in $\mathbb{F}_q$. (This condition holds for $q =$ 41, 61, 81, 241, 281, 401, 421, 601, 641, 661, 701, 761, 821, 881.)

14.5.35 Remark In [460], necessary and sufficient conditions are given for a $(q, k, 1)$ radical difference family with $k \in \{6, 7\}$ to exist over $\mathbb{F}_q$; a sufficient condition is also given for $k \geq 8$.

14.5.36 Theorem [460, 1356, 2986] Among others, $(q, k, 1)$ radical difference families exist for the following values of q and k:

$k = 6$ $q \in \{181, 211, 241, 631, 691, 1531, 1831, 1861, 2791, 2851, 3061\}$;
$k = 7$ $q \in \{337, 421, 463, 883, 1723, 3067, 3319, 3823, 3907, 4621, 4957, 5167, 5419, 5881, 6133, 8233, 8527, 8821, 9619, 9787, 9829\}$;
$k = 8$ $q \in \{449, 1009, 3137, 3697, 6217, 6329, 8233, 9869\}$;
$k = 9$ $q \in \{73, 1153, 1873, 2017, 6481, 7489, 7561, 8359\}$.

14.5.37 Theorem [1268] If there exists a $(p, k, 1)$ radical difference family with p a prime and k odd, there exists a cyclic RBIBD$(kp, k, 1)$ whose resolution is invariant under the action of $\mathbb{Z}_{kp}$.

14.5.4 Nested designs

14.5.38 Theorem [708] Let p be a prime, $n \geq 1$, $p^n \equiv 1 \pmod 6$, and ω be a primitive root of $\mathbb{F}_{p^n}$, $p^n = 6t + 1$. Let $\mathcal{S} = \{\omega^0, \omega^{2t}, \omega^{4t}\}$, and $\mathcal{S}_i = \omega^i \mathcal{S}$.

1. For $0 \leq c < t$, the development of $\{0\} \cup \mathcal{S}_c$ under the addition and multiplication of $\mathbb{F}_{p^n}$ forms a $(p^n, 4, 2)$ design in which the omission of the first element in each block yields an STS(p^n).
2. For $0 \leq c < d < t$, the development of $\mathcal{S}_c \cup \mathcal{S}_d$ under the addition and multiplication of $\mathbb{F}_{p^n}$ forms a $(p^n, 6, 5)$ design.

14.5.39 Remark The STSs in Theorem 14.5.38 Part 1 have been called *nested Steiner triple systems*, but the standard statistical notion of nested design is different – see Definition 14.5.40 and [2159].

14.5.40 Definition If the blocks of a BIBD $(\mathcal{V}, \mathcal{D}_1)$ with v symbols in b_1 blocks of size k_1 are each partitioned into sub-blocks of size k_2, and the $b_2 = b_1k_1/k_2$ sub-blocks themselves constitute a BIBD $(V, \mathcal{D}_2)$, then the system of blocks, sub-blocks, and symbols is a *nested balanced incomplete block design* (nested BIBD or NBIBD) with parameters $(v, b_1, b_2, r, k_1, k_2)$, r denoting the common replication. Also $(V, \mathcal{D}_1)$ and $(V, \mathcal{D}_2)$ are the *component BIBDs* of the NBIBD.

14.5.41 Remark A resolvable BIBD (RBIBD) $(V, \mathcal{D})$ is a nested block design $(V, \mathcal{D}_1, \mathcal{D}_2)$ where the blocks of $\mathcal{D}_1$, of size $k_1 = v$, are the resolution classes of $\mathcal{D}$, and $\mathcal{D}_2 = \mathcal{D}$.

14.5.42 Remark Nested block designs may have more than two blocking systems and consequently more than one level of nesting. A *doubly nested* block design is a system $(\mathcal{V}, \mathcal{D}_1, \mathcal{D}_2, \mathcal{D}_3)$ where both $(\mathcal{V}, \mathcal{D}_1, \mathcal{D}_2)$ and $(\mathcal{V}, \mathcal{D}_2, \mathcal{D}_3)$ are nested block designs. A resolvable NBIBD is a doubly nested block design.

14.5.43 Definition A *multiply nested* BIBD (MNBIBD) is a nested block design $(\mathcal{V}, \mathcal{D}_1, \mathcal{D}_2, \ldots, \mathcal{D}_s)$ with parameters $(v, b_1, \ldots, b_s, r, k_1, \ldots, k_s)$ for which the systems $(\mathcal{V}, \mathcal{D}_j, \mathcal{D}_{j+1})$ are NBIBDs for $j = 1, \ldots, s-1$.

14.5.44 Theorem [2159] Let v be a prime power of the form $v = a_0a_1a_2 \cdots a_n + 1$ ($a_0 \geq 1$, $a_n \geq 1$ and $a_i \geq 2$ for $1 \leq i \leq n-1$ are integers). Then there is an MNBIBD with n component designs having $k_1 = ua_1a_2 \cdots a_n$, $k_2 = ua_2a_3 \cdots a_n, \ldots,$ $k_n = ua_n$, and with a_0v blocks of size k_1, for any integer u with $1 \leq u \leq a_0$ and $u > 1$ if $a_n = 1$. If integer $t \geq 2$ is chosen so that $2 \leq tu \leq a_0$, then there is an MNBIBD with $n+1$ component designs, with the same number of big blocks but of size $k_0 = tk_1$, and with its n other block sizes being $k_1, \ldots, k_n$ as given. Moreover, if a_0 is even and a_i is odd for $i \geq 1$, then MNBIBDs can be constructed with the same block sizes but with $a_0v/2$ blocks of size k_1.

14.5.45 Definition A *nested row-column design* is a system $(\mathcal{V}, \mathcal{D}_1, \mathcal{D}_2, \mathcal{D}_3)$ for which (1) each of $(\mathcal{V}, \mathcal{D}_1, \mathcal{D}_2)$ and $(\mathcal{V}, \mathcal{D}_1, \mathcal{D}_3)$ is a nested block design, (2) each block of $\mathcal{D}_1$ may be displayed as a $k_2 \times k_3$ row-column array, one member of the block at each position in the array, so that the columns are the $\mathcal{D}_2$ sub-blocks in that block, and the rows are the $\mathcal{D}_3$ sub-blocks in that block.

14.5.46 Definition A *(completely balanced) balanced incomplete block design with nested rows and columns*, BIBRC(v, b_1, k_2, k_3), is a nested row-column design $(\mathcal{V}, \mathcal{D}_1, \mathcal{D}_2, \mathcal{D}_3)$ for which each of $(\mathcal{V}, \mathcal{D}_1, \mathcal{D}_2)$ and $(\mathcal{V}, \mathcal{D}_1, \mathcal{D}_3)$ is a NBIBD.

14.5.47 Theorem [2159] If $v = mpq + 1$ is a prime power and p and q are relatively prime, then initial nesting blocks for a BIBRC(v, mv, sp, tq) are $A_l = x^{l-1}L \otimes M$ for $l = 1, \ldots, m$, where $L_{s \times t} = (x^{i+j-2})_{i,j}$, $M_{p \times q} = (x^{[(i-1)q+(j-1)p]m})_{i,j}$, s and t are integers with $st \leq m$, and x is a primitive element of $\mathbb{F}_v$. (Here $\otimes$ is the Kronecker product.) If m is even and pq is odd, $A_1, \ldots, A_{m/2}$ are intial nesting blocks for BIBRC$(v, mv/2, sp, tq)$;

14.5.48 Theorem [2159] Write $x^{u_i} = 1 - x^{2mi}$ where x is a primitive element of $\mathbb{F}_v$ and $v = 4tm + 1$ is a prime power. Let A be the addition table with row margin $(x^0, x^{2m}, \ldots, x^{(4t-2)m})$ and column margin $(x^m, x^{3m}, \ldots, x^{(4t-1)m})$, and set $A_l = x^{l-1}A$. If $u_i - u_j \not\equiv m \pmod{2m}$ for $i, j = 1, \ldots, t$, then $A_1, \ldots, A_m$ are initial nesting blocks for BIBRC$(v, mv, 2t, 2t)$. Including 0 in each margin for A, if further $u_i \not\equiv m \pmod{2m}$ for $i = 1, \ldots, t$, then $A_1, \ldots, A_m$ are initial nesting blocks for BIBRC$(v, mv, 2t+1, 2t+1)$.

14.5.5 Pairwise balanced designs

14.5.49 **Definition** Let K be a subset of positive integers and let λ be a positive integer. A *pairwise balanced design* (PBD(v, K, λ) or (K, λ)-PBD) of order v with block sizes from K is a pair $(\mathcal{V}, \mathcal{B})$, where $\mathcal{V}$ is a finite set (the *point set*) of cardinality v and $\mathcal{B}$ is a family of subsets (*blocks*) of $\mathcal{V}$ that satisfy (1) if $B \in \mathcal{B}$, then $|B| \in K$ and (2) every pair of distinct elements of $\mathcal{V}$ occurs in exactly λ blocks of $\mathcal{B}$. The integer λ is the *index* of the PBD. The notations PBD(v, K) and K-PBD of order v are often used when $\lambda = 1$.

14.5.50 **Example** A PBD$(10, \{3, 4\})$ is given below where the blocks are listed columnwise.

1	1	1	2	2	2	3	3	3	4	4	4
2	5	8	5	6	7	5	6	7	5	6	7
3	6	9	8	9	10	10	8	9	9	10	8
4	7	10									

14.5.51 **Remark** Many constructions of pairwise balanced designs employ sub-structures in balanced incomplete block designs. In a (v, k, λ)-design $(V, \mathcal{B})$, useful sub-structures include those specified by a set $S \subset V$ so that for every $B \in \mathcal{B}$, $|B \cap S| \in L$; then setting $K = \{k - \ell : \ell \in L\}$, a $(|V \setminus S|, K, \lambda)$-PBD arises by removing all points of S. When the BIBD is made by a finite field construction, such sub-structures may arise from algebraic properties of the field. Other useful sub-structures arise from the presence of parallel classes; when a parallel class of blocks is present, a new element can be added and adjoined to each block in the parallel class to increase the size of some blocks by one. Example 14.5.50 is produced in this way from a (9,3,1)-design. This can be applied to more than one parallel class, when present [706, §IV].

14.5.6 Group divisible designs

14.5.52 **Definition** Let K and G be sets of positive integers and let λ be a positive integer. A *group divisible design of index λ and order v* ((K, λ)-GDD) is a triple $(\mathcal{V}, \mathcal{G}, \mathcal{B})$, where $\mathcal{V}$ is a finite set of cardinality v, $\mathcal{G}$ is a partition of $\mathcal{V}$ into parts (*groups*) whose sizes lie in G, and $\mathcal{B}$ is a family of subsets (*blocks*) of $\mathcal{V}$ that satisfy (1) if $B \in \mathcal{B}$ then $|B| \in K$, (2) every pair of distinct elements of $\mathcal{V}$ occurs in exactly λ blocks or one group, but not both, and (3) $|\mathcal{G}| > 1$. If $v = a_1 g_1 + a_2 g_2 + \cdots + a_s g_s$, and if there are a_i groups of size g_i, $i = 1, 2, \ldots, s$, then the (K, λ)-GDD is of type $g_1^{a_1} g_2^{a_2} \ldots g_s^{a_s}$. This is *exponential notation* for the group type. Alternatively, if the GDD has groups $G_1, G_2, \ldots, G_t$, then the list $T = [|G_i| : i = 1, 2, \ldots, t]$ is the *type* of the GDD when more convenient. If $K = \{k\}$, then the (K, λ)-GDD is a (k, λ)-GDD. If $\lambda = 1$, the GDD is a K-GDD. Furthermore, a $(\{k\}, 1)$-GDD is a k-GDD.

14.5.53 **Definition** Let H be a subgroup of order h of a group G of order v. A collection $\{B_1, ..., B_t\}$ of k-subsets of G forms a (v, h, k, λ) *difference family over G and relative to H* if $\partial B_1 \cup \cdots \cup \partial B_t$ covers each element of $G - H$ exactly λ times and covers no element in H.

14.5.54 Remark

1. A $(v, 1, k, \lambda)$ difference family is a (v, k, λ) difference family.
2. If $\{B_1, \ldots, B_t\}$ is a (v, h, k, λ) difference family over G and relative to H, then $Orb_G(B_1) \cup \cdots \cup Orb_G(B_t)$ is the collection of blocks of a (k, λ) GDD of type $h^{v/h}$ where the groups are the right cosets of H in G. This GDD admits G as a sharply point-transitive automorphism group.
3. A (k, λ) GDD of type $h^{v/h}$ with an automorphism group G acting sharply transitively on the points is, up to isomorphisms, generated by a suitable (v, h, k, λ) difference family.
4. If $\{B_1, \ldots, B_t\}$ is a (v, k, k, λ) difference family over G and relative to H, then $\{B_1, \ldots, B_t\} \cup \underbrace{\{H, \ldots, H\}}_{\lambda \text{ times}}$ is a (v, k, λ) difference family.

14.5.55 Theorem [4, 1357] Suppose $q \equiv 1 \pmod{k-1}$ is a prime power. Then a $(kq, k, k, 1)$ relative difference family over $\mathbb{F}_k \times \mathbb{F}_q$ exists if one of the following holds:

1. $k \in \{3, 5\}$ (for $k = 5$, the initial block B can be taken as $\{(0, 0), (1, 1), (1, -1), (4, x), (4, -x)\}$ where x is any nonsquare in $\mathbb{F}_q$ such that exactly one of $x-1, x+1$ is a square);
2. $k = 7$ and $q \neq 19$;
3. $k = 9$ and $q \notin \{17, 25, 41, 97, 113\}$;
4. $k = 11, q < 1202$, q is prime, and $q \notin$ [30,192], [240,312] or [490,492].

14.5.56 Theorem [461]

1. Let $q = 12t + 1$ be a prime power, and let 3^e be the largest power of 3 dividing t. If, in $\mathbb{F}_q$, 3 and $2 + \sqrt{3}$ are both 3^e-th powers but not 3^{e+1}-th powers and 6 is not a 3^{e+1}-th power, then a $(13q, 13, 13, 1)$ relative difference family exists.
2. If p and q are odd prime powers with $q > p$, then a $(pq, p, p, (p-1)/2)$ relative difference family exists. If further $p \equiv 1 \pmod 4$ and $q \equiv 1 \pmod{p-1}$, then a $(pq, p, p, (p-1)/4)$ relative difference family exists.

14.5.7 *t*-designs

14.5.57 Definition A t-(v, k, λ) *design*, a t-design in short, is a pair $(X, \mathcal{B})$ where X is a v-set of *points* and $\mathcal{B}$ is a collection of k-subsets of X (*blocks*) with the property that every t-subset of X is contained in exactly λ blocks. The parameter λ is the *index* of the design.

14.5.58 Example Let q be a prime power and $n > 0$ be an integer. Then $G = \mathrm{PGL}(2, q^n)$ acts sharply 3-transitively on $X = \mathbb{F}_{q^n} \cup \{\infty\}$. If $S \subseteq X$ is the natural inclusion of $\mathbb{F}_q \cup \{\infty\}$, then the orbit of S under G is a 3-$(q^n + 1, q + 1, 1)$ design. These designs are *spherical geometries*; when $n = 2$, they are *inversive planes* or *Möbius planes.*

14.5.59 Theorem [1559] Let B be a subgroup of the multiplicative group of nonzero elements of $\mathbb{F}_q$. Then the orbit of $S = B \cup \{0, \infty\}$ under the action of $\mathrm{PGL}(2, q)$ is the block set of one of the designs:

1. 3-$(q^n+1, q+1, 1)$ design where q is a prime power and $n \geq 2$ (a spherical geometry);
2. 3-$(q+1, k+1, k(k+1)/2)$ design if $(k-1)|(q-1)$, $k \nmid q$, and $k \notin \{3,5\}$;
3. 3-$(q+1, 4, 3)$ design for $q \equiv 1, 5 \pmod 6$;
4. 3-$(q+1, 6, 5)$ design for $q \equiv 1, 9, 13, 17 \pmod{20}$.

14.5.8 Packing and covering

14.5.60 Remark Packings and coverings relax the conditions on block designs, and have been extensively studied; see [2106] for a more detailed exposition.

14.5.61 Definition Let $v \geq k \geq t$. A t-(v,k,λ) *covering* is a pair $(X, \mathcal{B})$, where X is a v-set of elements (*points*) and $\mathcal{B}$ is a collection of k-subsets (*blocks*) of X, such that every t-subset of points occurs in at least λ blocks in $\mathcal{B}$. Repeated blocks in $\mathcal{B}$ are permitted.

14.5.62 Theorem (Schönheim bound) [2106] $C_\lambda(v,k,t) \geq \lceil v\,C_\lambda(v-1,k-1,t-1)/k \rceil$. Iterating this bound yields $C_\lambda(v,k,t) \geq L_\lambda(v,k,t)$, where

$$L_\lambda(v,k,t) = \left\lceil \frac{v}{k} \left\lceil \frac{v-1}{k-1} \cdots \left\lceil \frac{\lambda(v-t+1)}{k-t+1} \right\rceil \right\rceil \right\rceil.$$

14.5.63 Definition Let $v \geq k \geq t$. A t-(v,k,λ) *packing* is a pair $(X, \mathcal{B})$, where X is a v-set of elements (*points*) and $\mathcal{B}$ is a collection of k-subsets of X (*blocks*), such that every t-subset of points occurs in at most λ blocks in $\mathcal{B}$. If $\lambda > 1$, then $\mathcal{B}$ is allowed to contain repeated blocks.

14.5.64 Remark A t-$(v,k,1)$ packing with b blocks is equivalent to a binary code of length v, size b, constant weight k, and minimum Hamming distance at least $2(k-t+1)$; see Section 15.1.

14.5.65 Theorem (First Johnson bound) [2106] $D_\lambda(v,k,t) \leq \left\lfloor \frac{vD_\lambda(v-1,k-1,t-1)}{k} \right\rfloor$. Iterating this bound yields $D_\lambda(v,k,t) \leq U_\lambda(v,k,t)$, where

$$U_\lambda(v,k,t) = \left\lfloor \frac{v}{k} \left\lfloor \frac{v-1}{k-1} \cdots \left\lfloor \frac{\lambda(v-t+1)}{k-t+1} \right\rfloor \right\rfloor \right\rfloor.$$

Further, if $\lambda(v-1) \equiv 0 \pmod{k-1}$ and $\frac{\delta(k-1)}{2} > \lambda\frac{\delta(\delta-1)}{2}$, where $\delta = \frac{\lambda v(v-1)}{k-1} - kU_\lambda(v,k,2)$, then $D_\lambda(v,k,2) \leq U_\lambda(v,k,2) - 1$.

14.5.66 Theorem (Second Johnson bound) [2106] Suppose $d = D_1(v,k,t) = qv + r$, where $0 \leq r \leq v-1$. Then $q(q-1)v + 2qr \leq (t-1)d(d-1)$, and hence $D_1(v,k,t) \leq \left\lfloor \frac{v(k+1-t)}{k^2 - v(t-1)} \right\rfloor$.

See Also

§14.1	For latin squares, MOLS, and transversal designs.
§14.3	For affine and projective planes.
§14.4	For projective spaces.
§14.6	For difference sets.
§14.7	For other combinatorial structures.
[261]	A textbook on combinatorial designs.
[262]	Another textbook on combinatorial designs.
[706]	For triple systems (§II.2); balanced incomplete block designs (§II.1,II.3,II.5); t-designs (§II.4,II.5); symmetric designs (§II.6); Hadamard designs and matrices(§V.1); difference sets (§VI.18); resolvable designs (§II.7); pairwise balanced designs and group divisible designs (§IV); coverings (§VI.11); packings (§VI.40); nested designs (§VI.36); connections with MOLS and transversal designs (§III.1,III.3).
[2159]	For nested designs.
[2719]	For an introductory textbook on combinatorial designs.

References Cited: [4, 5, 261, 262, 356, 459, 460, 461, 706, 708, 807, 1268, 1356, 1357, 1459, 1559, 1741, 2106, 2159, 2222, 2440, 2719, 2986]

14.6 Difference sets

Alexander Pott, *Otto-von-Guericke-Universität Magdeburg*

14.6.1 Basics

14.6.1 **Definition** Let G be an additively written group of order v. A k-subset D of G is a $(v,k,\lambda;n)$-*difference set* of *order* $n=k-\lambda$ if every nonzero element of G has exactly λ representations as a difference $d-d'$ with elements from D. The difference set is *abelian*, *cyclic*, etc., if the group G has the respective property. The redundant parameter n is sometimes omitted, therefore the notion of (v,k,λ)-difference sets is also used.

14.6.2 Example

1. The group G itself and $G\backslash\{g\}$ for an arbitrary $g\in G$ are $(v,v,v,0)$- and $(v,v-1,v-2;1)$-difference sets.
2. The set $\{1,3,4,5,9\}$ is a cyclic $(11,5,2;3)$-difference set in the group $(\mathbb{Z}/11\mathbb{Z},+)$. These are the squares modulo 11.
3. The set $\{1,2,4\}\subset\mathbb{Z}/7\mathbb{Z}$ is a cyclic $(7,3,1;2)$-difference set.
4. The set $\{(0,1),(0,2),(0,3),(1,0),(2,0),(3,0)\}\subset\mathbb{Z}/4\mathbb{Z}\times\mathbb{Z}/4\mathbb{Z}$ is a $(16,6,2;4)$ difference set.

5. There is a non-abelian example with the same parameters: Let $G = Q \times \mathbb{Z}/2\mathbb{Z}$, where $Q = \{\pm 1, \pm i, \pm j, \pm k\}$ is the quaternion group, whose generators satisfy $i^2 = j^2 = k^2 = 1$, $ij = jk = ki = -1$. Then the set $D = \{(1,0), (i,0), (j,0), (k,0), (1,1), (-1,1)\}$ is a non-abelian $(16, 6, 2; 4)$-difference set.

14.6.3 Remark A thorough investigation of difference sets is contained in [261], see also the tables in [262]. A short summary, including a list of small examples, is contained in [706]. Classical textbooks are [211] and [1842]. A recent book including a modern treatment of necessary conditions for the existence of difference sets is [2546].

14.6.4 Remark The complement of a $(v, k, \lambda; n)$-difference set is again a difference set but with parameters $(v, v-k, v-2k+\lambda; n)$. Therefore, we may assume $k \leq v/2$ (the case $k = v/2$ is actually impossible).

14.6.5 Remark Many constructions of difference sets are closely related to the connection between the additive and the multiplicative group of a finite field:

1. The set of nonzero squares in a field $\mathbb{F}_q$, $q \equiv 3 \pmod 4$, which is a multiplicative subgroup, is a difference set in the additive group of the field, see Theorem 14.6.38. Case 2 in Example 14.6.2 is such a difference set in $(\mathbb{F}_{11}, +)$.
2. The set of elements of trace 0 in $\mathbb{F}_{2^n}$, which is an additive subgroup, forms a difference set in the multiplicative group of $\mathbb{F}_{2^n}$; see Theorem 14.6.22. Case 3 in Example 14.6.2 is such a difference set in $(\mathbb{F}_8^*, \cdot) \cong \mathbb{Z}/7\mathbb{Z}$.

14.6.6 Lemma The parameters v, k and λ of a difference set satisfy

$$\lambda \cdot (v-1) = k \cdot (k-1).$$

14.6.7 Remark Lemma 14.6.6 can be proved by counting differences. It also follows from Theorem 14.6.9 which shows that difference sets are the same objects as symmetric designs with a sharply transitive (regular) automorphism group. We refer the reader to Section 14.5 for the definition of symmetric designs and to [261] for a proof of the important Theorem 14.6.9.

14.6.8 Definition The *development* of a difference set D is the incidence structure $dev(D)$ whose points are the elements of G and whose blocks are the translates $g+D := \{g+d : d \in D\}$.

14.6.9 Theorem [262] The existence of a $(v, k, \lambda; n)$-difference set is equivalent to the existence of a symmetric (v, k, λ)-design $\mathcal{D}$ admitting G as a point regular automorphism group; i.e., for any two points P and Q, there is a unique group element g which maps P to Q. The design $\mathcal{D}$ is isomorphic with $dev(D)$.

14.6.10 Remark Necessary conditions on the parameters v, k and λ of a symmetric design are also necessary conditions for the parameters of a difference set. In particular, the following two theorems hold. We emphasize that these are necessary conditions for symmetric designs, even if the designs are not constructed from difference sets.

14.6.11 Theorem [2566] If D is a $(v, k, \lambda; n)$ difference set with v even, then $n = u^2$ is a square.

14.6.12 Theorem [429, 631] If D is a $(v, k, \lambda; n)$ difference set with v odd, then the equation $nx^2 + (-1)^{(v-1)/2}\lambda y^2 = z^2$ must have an integral solution $(x, y, z) \neq (0, 0, 0)$.

14.6.13 Example Not all symmetric designs can be constructed from difference sets. There are, for instance, no difference sets with parameters $(25, 9, 3; 6)$ or $(31, 10, 3; 7)$, but symmetric designs with these parameters exist, see the tables in [262].

14.6.14 Remark The main problem about difference sets is to give necessary and sufficient conditions for their existence. These conditions sometimes depend only on the parameters $(v, k, \lambda; n)$, sometimes also on the structure of the ambient group G. Another problem is to classify all $(v, k, \lambda; n)$-difference sets up to equivalence or isomorphism. In many nonexistence theorems, the exponent of a group plays an important role.

14.6.15 Definition The *exponent* $\exp(G)$ of a (multiplicatively written) finite group G is the smallest integer v^* such that $g^{v^*} = 1$.

14.6.16 Remark There are many necessary conditions that the parameters of a difference set have to satisfy. An important condition is Theorem 14.6.62. Many more restrictions are in [2546].

14.6.17 Definition Two difference sets D_1 (in G_1) and D_2 (in G_2) are *equivalent* if there is a group isomorphism φ between G_1 and G_2 such that $D_1^\varphi = \{d^\varphi : d \in D_1\} = g + D_2$ for a suitable $g \in G_2$. The difference sets are *isomorphic* if the designs $dev(D_1)$ and $dev(D_2)$ are isomorphic.

14.6.18 Remark Equivalent difference sets yield isomorphic designs, but a design may give rise to several inequivalent difference sets, as the following example shows.

14.6.19 Example The following three difference sets in $\mathbb{Z}/4\mathbb{Z} \times \mathbb{Z}/4\mathbb{Z}$ with parameters $(16, 6, 2; 4)$ are pairwise inequivalent, but the designs are all isomorphic:

$$\begin{aligned}
D_1 &= \{(0,0), (1,0), (2,0), (0,1), (1,2), (2,3)\},\\
D_2 &= \{(0,0), (1,0), (2,0), (0,1), (0,3), (3,2)\},\\
D_3 &= \{(0,0), (1,0), (0,1), (2,1), (1,2), (2,3)\}.
\end{aligned}$$

14.6.20 Definition Some types of $(v, k, \lambda; n)$-difference sets have special names. A difference set with $\lambda = 1$ is *planar*. The parameters can be written in terms of the order n as $(n^2 + n + 1, n + 1, 1; n)$. The corresponding design is a projective plane. Difference sets with $v = 4n$ are *Hadamard* difference sets, in which case $n = u^2$ must be a square, and the parameters are $(4u^2, 2u^2 - u, u^2 - u; u^2)$. Difference sets with $v = 4n - 1$, hence with parameters $(4n - 1, 2n - 1, n - 1; n)$, are of *Paley* type. Both Hadamard and Paley type difference sets are closely related to Hadamard matrices; see Constructions 14.6.44 and 14.6.52. The parameters $\left(\frac{q^n-1}{q-1}, \frac{q^{n-1}-1}{q-1}, \frac{q^{n-2}-1}{q-1}; q^{n-2}\right)$ are the *Singer* parameters; see Theorem 14.6.22.

14.6.21 Remark The parameters of a symmetric design, hence also the parameters of a $(v, k, \lambda; n)$-difference set satisfy $4n - 1 \le v \le n^2 + n + 1$. The extremal cases are Paley type difference sets ($v = 4n - 1$) and planar difference sets.

14.6.2 Difference sets in cyclic groups

14.6.22 Theorem [2674] Let α be a generator of the multiplicative group of $\mathbb{F}_{q^n}$, where q is a prime power. Then the set of integers $\{i : 0 \le i < (q^n - 1)/(q - 1), \mathrm{Tr}\,(\alpha^i) = 0\}$ modulo $(q^n - 1)/(q - 1)$ form a (cyclic) difference set with Singer parameters

$$\left(\frac{q^n - 1}{q - 1}, \frac{q^{n-1} - 1}{q - 1}, \frac{q^{n-2} - 1}{q - 1}; q^{n-2}\right).$$

These difference sets are *Singer difference sets*.

14.6.23 Remark

1. If $q = 2$, the Singer difference set is simply the set of nonzero elements in an additive subgroup of order 2^{n-1}, interpreted as a subset of the multiplicative group of $\mathbb{F}_{2^n}$. These difference sets are of Paley type.
2. In the general case, a Singer difference set can be viewed as follows. Let $U := \{\beta \in \mathbb{F}_{q^n}^* : \mathrm{Tr}(\beta) = 0\}$. This is a subgroup of the additive group of $\mathbb{F}_{q^n}$ fixed by multiplication with elements from $\mathbb{F}_q$, hence it is a hyperplane of the vector space $\mathbb{F}_q^n$. The Singer difference set is the image of this hyperplane under the canonical projection $\mathbb{F}_{q^n}^* \to \mathbb{F}_{q^n}^*/\mathbb{F}_q^*$.
3. The design corresponding to a Singer difference set is the classical point-hyperplane design of the projective geometry $\mathrm{PG}(n-1, q)$.

14.6.24 Conjecture If $n = 3$, the Singer parameters are $(q^2+q+1, q+1, 1; q)$. It is conjectured that the only abelian difference sets (up to equivalence) with these parameters are the Singer difference sets. Moreover, it is conjectured that planar difference sets exist only if the order q is a prime power. This holds for all orders $q \le 2,000,000$; see [1326].

14.6.25 Construction The construction of Singer difference sets is easy if a primitive polynomial (see Section 4.1) $f(x) = x^n + \sum_{i=1}^{n} a_i x^{n-i}$ of degree n in $\mathbb{F}_{q^n}$ is known. Consider the recurrence relation $\gamma_m = -\sum_{i=1}^{n} a_i \gamma_{m-i}$. Take arbitrary initial values, for instance $\gamma_0 = 1, \gamma_1 = \gamma_2 = \cdots = \gamma_{n-1} = 0$. Then the set of integers $\{0 \le i < (q^n-1)/(q-1) : \gamma_i = 0\}$ is a Singer difference set. For instance, $x^4 + x^3 + 2$ is a primitive polynomial over $\mathbb{F}_3$. The recurrence relation $\gamma_m = 2\gamma_{m-1} + \gamma_{m-4}$ yields the sequence

$$1000121220111222202021120102100221202200 2000\ldots$$

which gives the cyclic $(40, 13, 4; 9)$-difference set

$$\{1, 2, 3, 9, 17, 19, 24, 26, 29, 30, 35, 38, 39\}.$$

14.6.26 Remark In general, there are many difference sets with Singer parameters inequivalent to Singer difference sets (Theorems 14.6.27 and 14.6.29). There are even non-abelian difference sets with Singer parameters; see Example 14.6.70.

14.6.27 Theorem [1324] Let D be an arbitrary cyclic difference set with parameters $\left(\frac{q^s-1}{q-1}, \frac{q^{s-1}-1}{q-1}, \frac{q^{s-2}-1}{q-1}; q^{s-2}\right)$ in a group G. Let G be embedded into $\mathbb{F}_{q^n}^*/\mathbb{F}_q^*$ which is possible if $s|n$. Let α be a primitive element in $\mathbb{F}_{q^n}$. Then the set of integers $\{0 \le i < (q^n-1)(q-1) : \mathrm{Tr}_{\mathbb{F}_{q^n}/\mathbb{F}_{q^s}}(\alpha^i) \in D\}$ is a difference set with classical Singer parameters. The difference sets are *Gordon-Mills-Welch difference sets* corresponding to D. Note that different embeddings of the same difference set D may result in inequivalent difference sets [211].

14.6.28 Remark If D is a Singer difference set, the above construction may be reformulated as follows: if s divides n and if r is relatively prime to $q^s - 1$, then the set of integers $\{0 \le i < (q^n-1)/(q-1) : \mathrm{Tr}_{\mathbb{F}_{q^s}/\mathbb{F}_q}[(\mathrm{Tr}_{\mathbb{F}_{q^n}/\mathbb{F}_{q^s}}(\alpha^i))^r] = 0\}$ is a Gordon-Mills-Welch difference set.

14.6.29 Theorem [862, 864] Let α be a generator of the multiplicative group of $\mathbb{F}_{2^n}$, and let $t < n/2$ be an integer relatively prime to n, and $k = 4^t - 2^t + 1$. Then the set $D = \{(x+1)^k + x^k + 1 : x \in \mathbb{F}_{2^n}, x \ne 0, 1\} \subseteq \mathbb{F}_{2^n}^*$ is a *Dillon-Dobbertin difference set* with parameters $(2^n-1, 2^{n-1}-1, 2^{n-2}-1; 2^{n-2})$. If $n = 3t \pm 1$, then the set $D = \mathbb{F}_{2^n}^* \setminus \{(x+1)^k + x^k : x \in \mathbb{F}_{2^n}\} \subseteq \mathbb{F}_{2^n}^*$ is a difference set with parameters $(2^n-1, 2^{n-1}-1, 2^{n-2}-1; 2^{n-2})$.

14.6.30 Remark The series of Gordon-Mills-Welch difference sets [1324] and Dillon-Dobbertin difference sets [862, 864] show that the number of inequivalent difference sets grows rapidly. In these two series, inequivalent difference sets are in general also non-isomorphic; see [1676] for the Gordon-Mills-Welch case and [905] for the Dillon-Dobbertin case.

14.6.31 Construction [1303] Cyclic difference sets can be used to construct binary sequences with 2-level autocorrelation function. Let α be the generator of the cyclic group $\mathbb{Z}/v\mathbb{Z}$, and let D be a $(v, k, \lambda; n)$ difference set in $\mathbb{Z}/v\mathbb{Z}$. Define a sequence (a_i) by $a_i = 1$ if $\alpha^i \in D$, otherwise $a_i = 0$. This sequence has period v, and $C_t(a) := \sum_{i=0}^{n-1}(-1)^{a_i+a_{i+t}} = v - 4(k - \lambda)$ for $t = 1, \ldots, v-1$, which are the *off-phase autocorrelation coefficients*; see also Section 10.3.

14.6.32 Remark Cyclic Paley type difference sets yield sequences with constant off-phase autocorrelation -1. These sequences (difference sets) have numerous applications since the autocorrelation is small (in absolute value) [1303]. It is conjectured that no sequences with constant off-phase autocorrelation 0 exist if $v > 4$; see Remark 14.6.45.

14.6.33 Conjecture (Ryser's Conjecture) [1842, 1909] If $\gcd(v, n) \neq 1$, then there is no cyclic $(v, k, \lambda; n)$ difference set in a cyclic group. A strengthening of this conjecture is due to Lander: if D is a $(v, k, \lambda; n)$-difference set in an abelian group of order v, and p is a prime dividing v and n, then the Sylow p-subgroup of G is not cyclic.

14.6.34 Theorem [1907] Lander's conjecture is true for all abelian difference sets of order $n = p^k$, where $p > 3$ is prime.

14.6.35 Remark

1. The smallest open case for Lander's conjecture is a cyclic $(465, 145, 45; 100)$ difference set.
2. More restrictions on putative counterexamples to Lander's conjecture are contained in [1909].

14.6.36 Theorem [116] Let $R = \{a \in \mathbb{F}_{3^m}^* : a = x + x^6$ has 4 solutions with $x \in \mathbb{F}_{3^m}\}$ with $m > 1$. Then the set $\rho(R)$ is a difference set with Singer parameters $((3^m - 1)/2, (3^{m-1} - 1)/2, (3^{m-2} - 1)/2; 3^{m-2})$, where ρ is the canonical epimorphism $\mathbb{F}_{3^m}^* \to \mathbb{F}_{3^m}^*/\mathbb{F}_3^*$.

14.6.37 Theorem [1477] Let $q = 3^e$, $e \geq 1$, $m = 3k$, $d = q^{2k} - q^k + 1$. If $R = \{x \in \mathbb{F}_{q^m} : \mathrm{Tr}_{\mathbb{F}_{q^m}/\mathbb{F}_q}(x + x^d) = 1\}$, then $\rho(R)$ is a difference set with parameters $((q^m - 1)/(q - 1), q^{m-1}, q^{m-1} - q^{m-2}; q^{m-2})$, where ρ is the canonical epimorphism $\mathbb{F}_{q^m}^* \to \mathbb{F}_q^*$.

14.6.3 Difference sets in the additive groups of finite fields

14.6.38 Theorem [262] The following subsets of $\mathbb{F}_q$ are difference sets in the additive group of $\mathbb{F}_q$. They are *cyclotomic difference sets.* Some of these difference sets may have Singer parameters.

1. $\mathbb{F}_q^{(2)} := \{x^2 : x \in \mathbb{F}_q \backslash \{0\}\}$, $q \equiv 3 \pmod 4$ (quadratic residues, *Paley difference sets*);
2. $\mathbb{F}_q^{(4)} := \{x^4 : x \in \mathbb{F}_q \backslash \{0\}\}$, $q = 4t^2 + 1, t$ odd;
3. $\mathbb{F}_q^{(4)} \cup \{0\}$, $q = 4t^2 + 9, t$ odd;
4. $\mathbb{F}_q^{(8)} = \{x^8 : x \in \mathbb{F}_q \backslash \{0\}\}$, $q = 8t^2 + 1 = 64u^2 + 9, t, u$ odd;
5. $\mathbb{F}_q^{(8)} \cup \{0\}$, $q = 8t^2 + 49 = 64u^2 + 441, t$ odd, u even;
6. $H(q) = \{x^i : x \in \mathbb{F}_q \backslash \{0\}, i \equiv 0, 1 \text{ or } 3 \pmod 6\}$, $q = 4t^2 + 27, q \equiv 1 \pmod 6$ (*Hall difference sets*).

14.6.39 Remark The proofs of the statements in Theorem 14.6.38 use *cyclotomic numbers* [2725].

14.6.40 Theorem Let q and $q+2$ be prime powers. Then the set $D = \{(x, y) : x, y$ are both nonzero squares or both non-squares or $y = 0\}$ is a *twin prime power difference set* with parameters

$$\left(q^2+2q, \frac{q^2+2q-1}{2}, \frac{q^2+2q-3}{4}; \frac{q^2+2q+1}{4}\right)$$

in the group $(\mathbb{F}_q, +) \times (\mathbb{F}_{q+2}, +)$; see [2725].

14.6.41 Definition A difference set D in the group G is *skew symmetric* if D is of Paley type and $\{0, d, -d : d \in D\} = G$, hence $D \cap -D = \emptyset$.

14.6.42 Theorem The following sets are skew symmetric difference sets in the additive group of $\mathbb{F}_q$, $q \equiv 3 \pmod 4$:

1. $\{x^2 : x \in \mathbb{F}_q, x \neq 0\}$ (Paley difference sets);
2. $\{x^{10} \pm x^6 - x^2 : x \in \mathbb{F}_q, x \neq 0\}$ where $q = 3^h$, h odd [874];
3. $\{x^{4a+6} \pm x^{2a} - x^2 : x \in \mathbb{F}_q, x \neq 0\}$ where $q = 3^h$, h odd, $a = 3^{\frac{h+1}{2}}$ [871].

14.6.43 Remark A large class of skew Hadamard difference sets in elementary abelian groups of order q^3 (q prime power) has been recently constructed [2210].

14.6.4 Difference sets and Hadamard matrices

14.6.44 Construction A Hadamard difference set D in a group G of order $4u^2$ (see Definition 14.6.20) gives rise to a Hadamard matrix (Section 14.5) as follows: Label the rows and columns of a matrix $H = (h_{x,y})$ by the elements of G, and put $h_{x,y} = 1$ if $x - y \in D$, otherwise $h_{x,y} = -1$. This matrix is a Hadamard matrix, see Section 14.5.

14.6.45 Remark A special case of Rysers's Conjecture 14.6.33 is that there are no cyclic Hadamard difference sets with $v > 4$ (if $v = 4$, there is a trivial cyclic $(4, 1, 0; 1)$-difference set). This is also known as the *circulant Hadamard matrix conjecture.* The smallest open case for which one cannot prove the nonexistence of a cyclic Hadamard difference set with $v = 4u^2$ so far is $u = 11715 = 3 \cdot 5 \cdot 11 \cdot 71$; see [1910] and also [2168] for the connection to the Barker sequence conjecture (Section 10.3).

14.6.46 Remark [262] Hadamard difference sets in elementary abelian groups are equivalent to bent functions (Section 9.3). The bent function is the characteristic function of the Hadamard difference set. The following theorem gives an explicit construction.

14.6.47 Theorem The set

$$\{(x_1, \ldots, x_{2m}) \in \mathbb{F}_2^{2m} : x_1x_2 + x_3x_4 + \cdots + x_{2m-1}x_{2m} = 1\} \subset \mathbb{F}_2^{2m}$$

is a Hadamard difference set with parameters $(2^{2m}, 2^{2m-1} - 2^{m-1}, 2^{2m-2} - 2^{m-1}; 2^{2m-2})$.

14.6.48 Remark There are several other constructions of difference sets with these parameters, also in other groups. Two major construction methods are the Maiorana-McFarland method (Sections 9.1 and 9.3) and the use of partial spreads (Section 9.3).

14.6.49 Theorem [781, 1803, 2830] An abelian Hadamard difference set in a group G of order 2^{2m+2} exists if and only if $\exp(G) \leq 2^{m+2}$.

14.6.50 Remark In the following theorem, we combine knowledge about the existence of abelian Hadamard difference sets. Many authors contributed to this theorem.

14.6.51 Theorem [261] Let $G \cong H \times EA(w^2)$ be an abelian group of order $4u^2$ with $u = 2^a 3^b w^2$ where w is the product of not necessarily distinct primes $p \equiv 3 \pmod 4$ and $EA(w^2)$ denotes the group of order w^2 which is the direct product of groups of prime order. If H is of type $(2^{a_1})(2^{a_2})\cdots(2^{a_s})(3^{b_1})^2\cdots(3^{b_r})^2$ with $\sum a_i = 2a+2$ ($a \geq 0$, $a_i \leq a+2$), $\sum b_i = 2b$ ($b \geq 0$), then G contains a Hadamard difference set of order u^2.

14.6.52 Construction [262] Difference sets with parameters $(4n-1, 2n-1, n-1; n)$ (hence of Paley type) in G can be used to construct Hadamard matrices: Label the rows and columns of a matrix H by the elements of $G \cup \{\infty\}$. The matrix $H = (h_{u,v})$ such that $h_{u,v} = 1$ if $u = \infty$ or $v = \infty$ or $u - v \in D$ is a Hadamard matrix of order $4n$.

14.6.53 Theorem For the following orders n, Paley type difference sets exist in groups of order $v = 4n - 1$:

1. $4n-1$ is a prime power (Theorem 14.6.38);
2. $4n-1$ is the product $q(q+2)$ of two prime powers q and $q+2$ (Theorem 14.6.40);
3. $4n-1 = 2^m - 1$ (Theorem 14.6.22).

14.6.54 Problem It is an open question whether Paley type difference sets exist for other values.

14.6.5 Further families of difference sets

14.6.55 Theorem [2051] Let q be a prime power and d a positive integer. Let G be a group of order $v = q^{d+1}(q^d + \cdots + q^2 + q + 2)$ which contains an elementary abelian subgroup E of order q^{d+1} in its center. View E as the additive group of $\mathbb{F}_q^{d+1}$. Put $r = (q^{d+1}-1)/(q-1)$ and let $H_1, \ldots, H_r$ be the hyperplanes of order q^d of E. If $g_0, \ldots, g_r$ are distinct coset representatives of E in G, then $D = (g_1 + H_1) \cup (g_2 + H_2) \cup \cdots \cup (g_r + H_r)$ is a *McFarland difference set* with parameters

$$\left(q^{d+1}(1 + \tfrac{q^{d+1}-1}{q-1}), q^d \cdot \tfrac{q^{d+1}-1}{q-1}, q^d \cdot \tfrac{q^d-1}{q-1}; q^{2d}\right).$$

14.6.56 Remark

1. If $q = 2$, the McFarland construction gives Hadamard difference sets.
2. If $q = 2$ and G is elementary abelian, this construction is known as the Maiorana-McFarland construction of bent functions; see Section 9.3.

14.6.57 Theorem [609, 782] Let q be a prime power, and let t be any positive integer. Difference sets with parameters

$$\left(4q^{2t}\frac{q^{2t}-1}{q^2-1}, q^{2t-1}\frac{2q^{2t}+q-1}{q+1}, q^{2t-1}(q-1)\frac{q^{2t-1}+1}{q+1}; q^{4t-2}\right)$$

exist in abelian groups G in the following cases:

1. $q = 3^f$, the Sylow 3-subgroup of G is elementary abelian;
2. $q = p^{2f}$, p odd, the Sylow p-subgroup of G is elementary abelian;
3. $q = 2^f$, the Sylow 2-subgroup of G has rank $\geq 2f + 1$;
4. $q = 2$, $\exp(G) \leq 4$.

If $t = 1$, these difference sets are Hadamard difference sets.

14.6.6 Difference sets and character sums

14.6.58 Remark The existence of difference sets is closely related to character sums. Most necessary conditions on the existence of difference sets are derived from character sums and number theoretic conditions.

14.6.59 Theorem [262, 2830] Let D be a $(v, k, \lambda; n)$ difference set in G, and let χ be a homomorphism from G into the multiplicative group of a field. If $\chi(g) = 1$ for all $g \in G$ (in which case the homomorphism is denoted χ_0), then $\sum_{g \in D} \chi_0(d) = k$. If $\chi \neq \chi_0$, then

$$\left(\sum_{d \in D} \chi(d)\right) \cdot \left(\sum_{d \in D} \chi(d^{-1})\right) = n.$$

14.6.60 Remark Theorem 14.6.59 is very useful if χ is complex-valued. In this case, the sum $\chi(D) := \sum_{d \in D} \chi(d)$ is an element in the ring $\mathbb{Z}[\zeta_{v^*}]$, where $\zeta_{v^*} = e^{2\pi i/v^*}$ is a primitive v^*-th root of unity, and v^* is the exponent of G.

14.6.61 Remark If χ is complex-valued, Theorem 14.6.59 may be also viewed as an equation about the ideal generated by $\chi(D)$: For $\chi \neq \chi_0$, we have $(\chi(D))(\overline{\chi(D)}) = (n)$, where $(\cdot)$ denotes an ideal generated in $\mathbb{Z}[\zeta_{v^*}]$ and $\overline{(\ )}$ is complex conjugation. Using results from algebraic number theory, many necessary conditions can be obtained, for instance Theorem 14.6.62.

14.6.62 Theorem [2830] Let D be an abelian $(v, k, \lambda; n)$ difference set in G, and let w be a divisor of v. If p is prime, $p|n$ and $p^j \equiv -1 \pmod{w}$, then an integer i exists such that $p^{2i}|n$, but p^{2i+1} is not a divisor of n. If w is the exponent v^* of G, then p does not divide n.

14.6.63 Example [262, 1842] There is no $(40, 13, 4; 9)$-difference set in $\mathbb{Z}/2\mathbb{Z} \times \mathbb{Z}/2\mathbb{Z} \times \mathbb{Z}/2\mathbb{Z} \times \mathbb{Z}/5\mathbb{Z}$ (use $v^* = 10$ and $p = 3$ in Theorem 14.6.62). Using a different (though similar) theorem, one can also rule out the existence of a $(40, 13, 4; 9)$-difference set in $\mathbb{Z}/2\mathbb{Z} \times \mathbb{Z}/4\mathbb{Z} \times \mathbb{Z}/5\mathbb{Z}$. Note that a cyclic difference set with these parameters exists (Construction 14.6.25).

14.6.7 Multipliers

14.6.64 Definition Let D be a difference set in G. Then $\varphi \in \mathrm{Aut}(G)$ is a *multiplier* of D if $D^\varphi := \{\varphi(D) : d \in D\} = g+D$ for some $g \in G$. If G is abelian and φ is the automorphism that maps h to $t \cdot h$, then t is a *numerical* multiplier.

14.6.65 Theorem If φ is a multiplier of the difference set D, then there is at least one translate $g + D$ of D which is fixed by φ. If D is abelian and $\gcd(v, k) = 1$, then there is a translate fixed by all multipliers [261].

14.6.66 Remark

1. Multipliers play an important role, in particular in the theory of abelian difference sets.
2. The content of a multiplier theorem is the assertion that certain automorphisms (integers) have to be (numerical) multipliers of an abelian difference set depending only on the parameters v, k, and λ. Theorem 14.6.67 is the *first multiplier theorem* [261].

14.6.67 Theorem Let D be an abelian $(v, k, \lambda; n)$-difference set. If p is a prime which satisfies $\gcd(p, v) = 1$, $p|n$ and $p > \lambda$, then p is a numerical multiplier.

14.6.68 Conjecture Every prime divisor p of n which is relatively prime to v is a multiplier of a $(v,k,\lambda;n)$-difference set, i.e., the condition $p>\lambda$ in Theorem 14.6.67 is not necessary.

14.6.69 Remark

1. Multipliers are quite useful in constructing difference sets and in proofs of nonexistence.
2. Several attempts have been made to weaken the assumption "$p>\lambda$" in Theorem 14.6.67 (second multiplier theorem, McFarland's multiplier theorem) [261].

14.6.70 Example Multipliers may be used to construct non-abelian difference sets: The set $D=\{3,6,7,12,14\}$ is a $(21,5,1;4)$-difference set in $(\mathbb{Z}/21\mathbb{Z},+)$ with multiplier 4. Denote the automorphism $x\mapsto x+3$ by a, and the automorphism $x\mapsto 4x+1$ by b. Then $G=\langle a,b\colon a^7=b^3=1, b^{-1}ab=a^4\rangle$ acts regularly on the points of $dev(D)$. A difference set D' in G corresponding to this action is $D'=\{a,a^2,a^4,a^4b,a^5b\}$.

See Also

§9.2	Relative difference sets are a generalization of difference sets. An important class of relative difference sets can be described by planar functions (PN functions).
§9.3	Bent functions are equivalent to elementary abelian Hadamard difference sets.
§10.3	Cyclic difference sets are binary sequences with two-level autocorrelation function.
§14.5	Difference sets are an important tool to construct combinatorial designs.

References Cited: [116, 211, 261, 262, 429, 609, 631, 706, 781, 782, 862, 864, 871, 874, 905, 1303, 1324, 1326, 1477, 1676, 1803, 1842, 1907, 1909, 1910, 2051, 2168, 2210, 2546, 2566, 2674, 2725, 2830]

14.7 Other combinatorial structures

Jeffrey H. Dinitz, *University of Vermont*
Charles J. Colbourn, *Arizona State University*

14.7.1 Association schemes

14.7.1 Definition Let d denote a positive integer, and let X be a nonempty finite set. A *d-class symmetric association scheme* on X is a sequence $R_0,R_1,\ldots,R_d$ of nonempty subsets of the Cartesian product $X\times X$, satisfying

1. $R_0=\{(x,x)\mid x\in X\}$,
2. $X\times X=R_0\cup R_1\cup\cdots\cup R_d$ and $R_i\cap R_j=\emptyset$ for $i\neq j$,
3. for all $i\in\{0,1,\ldots,d\}$, $R_i^T=R_i$ where $R_i^T:=\{(y,x)\mid(x,y)\in R_i\}$,
4. for all integers $h,i,j\in\{0,1,\ldots,d\}$, and for all $x,y\in X$ such that $(x,y)\in R_h$, the number $p_{ij}^h:=|\{z\in X\mid(x,z)\in R_i,\ (z,y)\in R_j\}|$ depends only on h, i, j, and not on x or y.

14.7.2 Example The *Hamming scheme* $H(n,q)$ has the set F^n of all words of length n over an alphabet F of q symbols as its vertex set. Two words are *i-th* associates if and only if the Hamming distance between them is i. Generally the alphabet F is $\mathbb{F}_2$, but other finite fields are also used.

14.7.3 Example The *cyclotomic schemes* are obtained as follows. Let q be a prime power and k a divisor of $q-1$. Let C_1 be the subgroup of the multiplicative subgroup of $\mathbb{F}_q$ of index k, and let $C_i, i = 1, 2, \ldots, k$ be the cosets of C_1 (the *cyclotomic classes*). The points of the scheme are the elements of $\mathbb{F}_q$, and two points x, y are *i-th* associates if $x - y \in C_i$ (zero associates if $x - y = 0$). In order for this to be an association scheme one must have $-1 \in C_1$ or equivalently $2k$ must divide $q-1$ if q is odd.

14.7.2 Costas arrays

14.7.4 Definition A *Costas array* of *order* n is an $n \times n$ array of dots and blanks that satisfies:

1. There are n dots and $n(n-1)$ blanks, with exactly one dot in each row and column.
2. All the segments between pairs of dots differ in length or in slope.

$C(n)$ denotes the number of distinct $n \times n$ Costas arrays.

14.7.5 Construction (Welch construction) Let p be prime and α be a primitive element in the field $\mathbb{F}_p$. Let $n = p - 1$. A Costas array of order n is obtained by placing a dot at (i, j) if and only if $i = \alpha^j$, for $a \le j < n + a$, a a nonnegative integer, and $i = 1, \ldots, n$.

14.7.6 Construction [1301] Let α and β be primitive elements in the field $\mathbb{F}_q$ for q a prime power. Let $n = q - 2$. Costas arrays of order n are obtained by

1. *Lempel construction:* Put a dot at (i, j) if and only if $\alpha^i + \alpha^j = 1$, $1 \le i, j \le n$.
2. *Golomb construction:* Put a dot at (i, j) if and only if $\alpha^i + \beta^j = 1$, $1 \le i, j \le n$.

14.7.7 Remark Using Constructions 14.7.5 and 14.7.6, $C(p-1) > 1$ and $C(q-2) > 1$. Also, if a corner dot is present in a Costas array of order n, it can be removed along with its row and column to obtain a Costas array of order $n-1$.

14.7.8 Theorem [1304] If $q > 2$ is a prime power, then there exist primitive elements α and β in $\mathbb{F}_q$ such that $\alpha + \beta = 1$.

14.7.9 Corollary Removing the corner dot at $(1, 1)$ in the Costas array of order $q-2$ from Construction 14.7.6 Part 2 yields $C(q-3) \ge 1$.

14.7.10 Example If there exist primitive elements α and β satisfying the conditions stated, then a Costas array of order n can be obtained by removing one or more corner dots.

Conditions	n
$\alpha^1 = 2$	$q-3$
$\alpha^1 + \beta^1 = 1$ and $\alpha^2 + \beta^2 = 1$	$q-4$
$\alpha^2 + \alpha^1 = 1$	$q-4$
$\alpha^1 + \beta^1 = 1$ and $\alpha^2 + \beta^{-1} = 1$	$q-4$
and necessarily $\alpha^{-1} + \beta^2 = 1$	$q-5$

14.7.11 Remark All Costas arrays of order 28 are accounted for by the Golomb and Welch construction methods [918], making 28 the first order (larger than 5) for which no sporadic Costas array exists.

14.7.12 Remark For $n \geq 30$, $n \notin \{31, 53\}$, the only orders for which Costas arrays are known are orders $n = p - 1$ or $n = q - 2$ or orders for which some algebraic condition exists that guarantees corner dots whose removal leaves a smaller Costas array.

14.7.13 Remark The properties of a Costas array make it an ideal discrete waveform for Doppler sonar. Having one dot in each row and column minimizes reverberation. Distinct segments between pairs of dots give it a thumbtack ambiguity function because, shifted left-right in time and up-down in frequency, copies of the pattern can only agree with the original in one dot, no dots, or all n dots at once. Thus, the spike of the thumbtack makes a sharp distinction between the actual shift and all the near misses. See [916] for a survey on Costas arrays and `http://www.costasarrays.org/` for up-to-date information on Costas arrays.

14.7.3 Conference matrices

14.7.14 Definition A *conference matrix* of order n is an $n \times n$ $(0, \pm 1)$-matrix C with zero diagonal satisfying $CC^T = (n-1)I$. A conference matrix is *normalized* if all entries in its first row and first column are 1 (except the (1,1) entry which is 0). A square matrix A is *symmetric* if $A = A^T$ and *skew-symmetric* if $A = -A^T$. The *core* of a normalized conference matrix C consists of all the rows and columns of C except the first row and column.

14.7.15 Theorem [2851, page 360] If there exists a conference matrix of order n, then n is even; furthermore, if $n \equiv 2 \pmod 4$, then, for any prime $p \equiv 3 \pmod 4$, the highest power of p dividing $n - 1$ is even.

14.7.16 Theorem [2345] Let q be an odd prime power.

1. If $q \equiv 1 \pmod 4$, then there is a symmetric conference matrix of order $q + 1$.
2. If $q \equiv 3 \pmod 4$, then there is a skew-symmetric conference matrix of order $q+1$.

14.7.17 Construction In the construction for Theorem 14.7.16, let q be an odd prime power and let χ denote the quadratic character on the finite field $\mathbb{F}_q$ (i.e., $\chi(x) = 0$ if $x = 0$, $\chi(x) = 1$ if x is a square and $\chi(x) = -1$ if x is a nonsquare). Number the elements of $\mathbb{F}_q : 0 = a_0, a_1, \ldots, a_{q-1}$ and define a $q \times q$ matrix Q by $q_{i,j} := \chi(a_i - a_j)$ for $0 \leq i, j < q - 1$. It follows that Q is symmetric if $q \equiv 1 \pmod 4$ and skew-symmetric if $q \equiv 3 \pmod 4$. Define the $(q + 1) \times (q + 1)$ matrix C by

$$\begin{pmatrix} 0 & 1 \ 1 \ \cdots \ 1 \\ \pm 1 & \\ \vdots & Q \\ \pm 1 & \end{pmatrix}$$

where the terms ± 1 are $+1$ when $q \equiv 1 \pmod 4$ and -1 when $q \equiv 3 \pmod 4$. It follows that C is a conference matrix of order $q + 1$. In the special case when q is prime, Q is circulant.

14.7.18 Lemma 1. If C is a skew-symmetric conference matrix, then $I + C$ is a Hadamard matrix. 2. If C is a symmetric conference matrix of order n, then

$$\begin{pmatrix} I + C & -I + C \\ -I + C & -I - C \end{pmatrix}$$

is a Hadamard matrix of order $2n$.

14.7.19 Remark Theorem 14.7.20 follows from Theorem 14.7.16 and Lemma 14.7.18.

14.7.20 Theorem If q is a power of an odd prime, then a Hadamard matrix of order $q+1$ exists if $q \equiv 3 \pmod 4$, and a Hadamard matrix of order $2(q+1)$ exists if $q \equiv 1 \pmod 4$.

14.7.21 Construction Let A be a $(0,\pm1)$-matrix of order n and B a ±1-matrix of order n such that $AB = BA$ and $AA^T + BB^T = (2n-1)I$. Then the matrix $C = \begin{pmatrix} A & B \\ B^T & -A^T \end{pmatrix}$ is a conference matrix of order $2n$.

14.7.22 Definition When A and B are circulant matrices, the conference matrix C in Construction 14.7.21 is *constructible from two circulant matrices* or for short, *two circulants type*.

14.7.23 Theorem [1289, 2831, 2975] If $q \equiv 1 \pmod 4$ is a prime power, then there is a symmetric conference matrix C of order $q+1$ of two circulants type.

14.7.24 Theorem [2023] There is a symmetric conference matrix of order $q^2(q+2)+1$ whenever q is a prime power, $q \equiv 3 \pmod 4$, and $q+3$ is the order of a conference matrix.

14.7.4 Covering arrays

14.7.25 Definition A *covering array* $\mathrm{CA}_\lambda(N;t,k,v)$ is an $N \times k$ array containing v different symbols. In every $N \times t$ subarray, each t-tuple occurs at least λ times. Then t is the *strength* of the coverage of interactions, k is the number of components (*degree*), λ is the *index*, and v is the number of symbols for each component (*order*). Only the case when $\lambda = 1$ is treated; the subscript is then omitted in the notation.

14.7.26 Definition The size of a covering array is the *covering array number* $\mathrm{CAN}(t,k,v)$. The covering array is *optimal* if it has the minimum possible number of rows.

14.7.27 Construction [1457] Let q be a prime power and $q \geq s \geq 2$. Over the finite field $\mathbb{F}_q$, let $\mathcal{F} = \{f_1, \ldots, f_{q^s}\}$ be the set of all polynomials of degree less than s. Let $\mathcal{A}$ be a subset of $\mathbb{F}_q \cup \{\infty\}$. Define an $q^s \times |\mathcal{A}|$ array in which the entry in cell (j,a) is $f_j(a)$ when $a \in \mathbb{F}_q$, and is the coefficient of the term of degree $s-1$ when $a = \infty$. The result is a $\mathrm{CA}(q^s; s, |\mathcal{A}|, q)$. Because every t-tuple is covered exactly once, it is in fact an *orthogonal array* of index one and strength s.

14.7.28 Remark Covering arrays are typically constructed by a combination of computational, direct, and recursive constructions [705]. Finite fields arise most frequently in the direct construction of covering arrays. One example is the use of *permutation vectors* to construct covering arrays [2612]. A second, outlined next, uses Weil's theorem and character theoretic arguments to establish that certain cyclotomic matrices form covering arrays.

14.7.29 Construction [704] Let ω be a primitive element of $\mathbb{F}_q$, with $q \equiv 1 \pmod v$. For each q and ω, form a *cyclotomic vector* $\mathbf{x}_{q,v,\omega} = (x_i : i \in \mathbb{F}_q) \in \mathbb{F}_q^q$ by setting $x_0 = 0$ and $x_i \equiv j \pmod v$ when $i = \omega^j$ for $i \in \mathbb{F}_q^*$. Choosing a different primitive element of $\mathbb{F}_q$ can lead to the same vector $\mathbf{x}_{q,v,\omega}$, or, for some number m that is coprime to v, to a vector in which each element is multiplied by m and reduced modulo v. For our purposes, the vectors produced are equivalent, so henceforth let $\mathbf{x}_{q,v}$ denote any vector so obtained. From

$\mathbf{x}_{q,v} = (x_i : i \in \mathbb{F}_q)$, form a $q \times q$ matrix $A_{q,v} = (a_{ij})$ with rows and columns indexed by $\mathbb{F}_q$, by setting $a_{ij} = x_{j-i}$ (computing the subscript in $\mathbb{F}_q$).

14.7.30 Theorem [704] When $q > t^2 v^{2t}$, $A_{q,v}$ from Construction 14.7.29 is a covering array of strength t.

14.7.5 Hall triple systems

14.7.31 Definition [1404] A *Hall triple system* (HTS) is a pair $(S, \mathcal{L})$ where S is a set of elements (*points*) and $\mathcal{L}$ a set of *lines* satisfying:

1. every line is a 3-subset of S,
2. any two distinct points lie in exactly one line, and
3. for any two intersecting lines, the smallest subsystem containing them is isomorphic to the affine plane of nine points, AG(2,3).

14.7.32 Example Let S be some $(n+1)$-dimensional vector space over $\mathbb{F}_3$, with $n \geq 3$. Let $\{e_0, e_1, \ldots, e_n\}$ be a basis for S. For any two points $x = \sum \alpha_i e_i$ and $y = \sum \beta_i e_i$ set $z = x \circ y$ when $x + y + z = (\alpha_1 - \beta_1)(\alpha_2\beta_3 - \alpha_3\beta_2)e_0$. This defines a binary operation on S. One either has $x = y = z$, or the three points x, y, z are pairwise distinct. The 3-subsets of the form $\{x, y, z\}$ such that $z = x \circ y$ provide S with a structure of an HTS. This HTS is referred to as $H(n)$.

14.7.33 Example Any affine space AG$(n, 3)$ over $\mathbb{F}_3$ with the usual lines may be viewed as an HTS. Such an HTS is an *affine* HTS.

14.7.34 Theorem [1403, 3037] The cardinality of any HTS is 3^m for some integer $m \geq 2$. Nonaffine HTS of order 3^m exist for any $m \geq 4$ and do not exist for $m \in \{2, 3\}$.

14.7.35 Remark When $m > 3$, the existence of a nonaffine HTS of order 3^m is provided by $H(m-1)$ in Example 14.7.32. For the orders 3^4 and 3^5, there is a unique nonaffine HTS, namely, $H(3)$ and $H(4)$, respectively.

14.7.6 Ordered designs and perpendicular arrays

14.7.36 Definition An *ordered design* OD$_\lambda(t, k, v)$ is a $k \times \lambda \cdot \binom{v}{t} \cdot t!$ array with v entries such that

1. each column has k distinct entries, and
2. each tuple of t rows contains each column tuple of t distinct entries precisely λ times.

14.7.37 Definition A *perpendicular array* PA$_\lambda(t, k, v)$ is a $k \times \lambda \cdot \binom{v}{t}$ array with v entries such that

1. each column has k distinct entries, and
2. each set of t rows contains each set of t distinct entries as a column precisely λ times.

14.7.38 Definition For $0 \le s \le t$, a $\mathrm{PA}_\lambda(t,k,v)$ is an s-$\mathrm{PA}_\lambda(t,k,v)$ if, for each $w \le t$ and $u \le \min(s,w)$, the following holds. Let E_1, E_2 be disjoint sets of entries, $E_1 \cap E_2 = \emptyset$ with $|E_1| = u$ and $|E_2| = w - u$. Then the number of columns containing $E_1 \cup E_2$ and having E_2 in a given set U of $w-u$ rows is a constant, independent of the choice of E_1, E_2, and U. *Authentication perpendicular arrays* (APA) are 1-PA.

14.7.39 Definition A set $S \subseteq S_n$ of permutations is (uniformly) *t-homogeneous* if it is an $\mathrm{APA}_\lambda(t,n,n)$; it is *t-transitive* if it is an $\mathrm{OD}_\lambda(t,n,n)$.

14.7.40 Theorem [275] Permutation *groups* yield special cases of t-transitive or t-homogeneous sets.

1. The groups $\mathrm{PGL}_2(q), q$ a prime power, form $\mathrm{OD}_1(3,q+1,q+1)$; the groups $\mathrm{PSL}_2(q), q \equiv 3 \pmod 4$ are $\mathrm{APA}_3(3,q+1,q+1)$. The special cases of this last family when the prime power $q \equiv 3, 11 \pmod{12}$ form the only known infinite family of $\mathrm{APA}_\lambda(t,n,n)$ with $t > 2$ and minimal λ.
2. The groups $\mathrm{AGL}_1(q)$, (q a prime power), of order $q \cdot (q-1)$ form an $\mathrm{OD}_1(2,q,q)$; the groups $\mathrm{ASL}_1(q)$, (q a prime power $\equiv 3 \pmod 4$) of order $q \cdot (q-1)/2$ form $\mathrm{APA}_1(2,q,q)$.

14.7.41 Definition Let $q \equiv 3 \pmod 4$ be a prime power, k odd. An $\mathrm{APAV}(q,k)$ (V stands for *vector*) is a tuple $(x_1,\ldots,x_k)$ where $x_i \in \mathbb{F}_q$ and such that for each i the $x_i - x_j, j \neq i$ are evenly distributed on squares and nonsquares [1262].

14.7.42 Remark An $\mathrm{APAV}(q,k)$ implies the existence of $\mathrm{APA}_1(2,k,q)$. In [606] a theorem on character sums based on the Hasse–Weil inequality is used to prove existence of an $\mathrm{APAV}(q,k)$ when q is large enough with respect to k.

14.7.43 Theorem [606] The following exist, for a prime power q with $q \equiv 3 \pmod 4$,

1. $\mathrm{APAV}(q,7)$ for $q \ge 7, q \notin \{11,19\}$,
2. $\mathrm{APAV}(q,9)$ for $q \ge 19$,
3. $\mathrm{APAV}(q,11)$ for $q \ge 11, q \notin \{19,27\}$,
4. $\mathrm{APAV}(q,13)$ for $q \ge 13, q \notin \{19,23,31\}$, and
5. $\mathrm{APAV}(q,15)$ for $q \ge 31$.

14.7.7 Perfect hash families

14.7.44 Definition Let n, q, t, and s be positive integers and suppose (to avoid trivialities) that $n > q \ge t \ge 2$. Let V be a set of cardinality n and let W be a set of cardinality q. A function $f : V \to W$ *separates* a subset X of V if f is an injection when restricted to X. An (n,q,t)*-perfect hash family* of *size* s is a collection $\mathcal{F} = \{f_1, f_2, \ldots, f_s\}$ of functions from V to W with property that for all sets $X \subseteq V$ such that $|X| = t$, at least one of the functions $f_1, f_2, \ldots, f_s$ separates X. The notation $\mathrm{PHF}(s;n,q,t)$ is used for an (n,q,t)-perfect hash family of size s. A perfect hash family is *optimal* if s is as small as possible, given n, q, t.

14.7.45 Theorem A $\mathrm{PHF}(s;n,q,t)$ is equivalent to an $s \times n$ array A of elements from a q-set F, such that, for any t columns of A, there exists a row of A, say r, such that the entries in the t given columns of row r of A are distinct.

14.7.46 Theorem [2721] Suppose that there exists a q-ary code $\mathcal{C}$ of length K, with N codewords, having minimum distance D. Then there exists a PHF$(N; K, q, t)$, where $(N-D)\binom{t}{2} < N$.

14.7.47 Corollary [2721] Suppose N and v are given, with v a prime power and $N \leq v+1$. Then there exists a PHF$(N; v^{\lceil N/\binom{t}{2} \rceil}, v, t)$ based on a Reed–Solomon code.

14.7.48 Theorem [2853] A PHF$((i+1)^2; v^{i+1}, v, 3)$ exists whenever v is a prime power, $v \geq 3$, and $i \geq 1$. A PHF$(\frac{5}{6}(2i^3+3i^2+i)+i+1; v^{i+1}, v, 4)$ exists whenever v is a prime power, $v \geq 4$, and $i \geq 1$.

14.7.49 Theorem [2721] For any prime power q and for any positive integers n, m, i such that $n \geq m$ and $2 \leq i \leq q^n$, there exists a PHF$(q^n; q^{m+(i-1)n}, q^m, t)$ when $\binom{t}{2} < \frac{q^m}{i-1}$.

14.7.50 Definition A PHF$(N; q^s, q, t)$ is *linear* if it is an $N \times q^s$ array with rows indexed by elements of $\mathbb{F}_q \cup \{\infty\}$ and columns indexed by the polynomials of degree less than s over $\mathbb{F}_q$; each entry of the array is the evaluation of the polynomial corresponding to the column on the row index, when that index is in $\mathbb{F}_q$; otherwise it is the coefficient of the term of degree $s-1$ in the polynomial.

14.7.51 Remark In a linear PHF, columns correspond to polynomials of degree less than s over $\mathbb{F}_q$. It follows directly that two columns agree in at most $s-1$ entries, and hence that if the linear PHF has more than $(s-1)\binom{t}{2}$ rows, it has strength at least t. By judicious selection of the particular rows (i.e., a subset $\mathcal{A}$ of $\mathbb{F}_q \cup \{\infty\}$), fewer rows can often be employed. The key observation, developed in [206, 302, 707], is that when $\mathcal{A}$ is chosen properly, a system of equations over $\mathbb{F}_q$ for each set of t chosen columns never admits a solution. This is developed in an algebraic setting in [302], in a geometric setting in [206], and in a graph-theoretic setting in [707]. The results to follow all employ this basic strategy.

14.7.52 Theorem [302] Let $s \geq 2$ and $t \geq 2$. When q is a sufficiently large prime power, there is an optimal linear PHF$(s(t-1); q^s, q, t)$.

14.7.53 Theorem [206, 207, 302]

1. An optimal linear PHF$(6; q^2, q, 4)$ exists if and only if $q \geq 11$ is a prime power and $q \neq 13$.
2. An optimal linear PHF$(6; q^3, q, 3)$ exists if and only if $q \geq 11$ is a prime power.

14.7.54 Theorem [707] Let p be a prime.

1. A PHF$(9; p^4, p, 3)$ exists when $p \geq 17$.
2. A PHF$(8; p^4, p, 3)$ exists when $p \geq 19$.
3. A PHF$(12; p^3, p, 4)$ exists when $p \geq 17$.
4. A PHF$(11; p^3, p, 4)$ exists when $p \geq 29$.
5. A PHF$(10; p^3, p, 4)$ exists when $p \geq 251$ and $p \notin \{257, 263\}$.
6. A PHF$(10; p^2, p, 5)$ exists when $p \geq 19$.
7. A PHF$(9; p^2, p, 5)$ exists when $p \geq 41$.
8. A PHF$(8; p^2, p, 5)$ exists when $p \geq 241$ and $p \notin \{251, 257\}$.
9. A PHF$(15; p^2, p, 6)$ exists when $p \geq 29$.
10. A PHF$(14; p^2, p, 6)$ exists when $p \geq 41$.
11. A PHF$(13; p^2, p, 6)$ exists when $p \geq 73$.

14.7.8 Room squares and starters

14.7.55 Definition A *starter* in the odd order abelian group G (written additively), where $|G| = g$ is a set of unordered pairs $S = \{\{s_i, t_i\} : 1 \leq i \leq (g-1)/2\}$ that satisfies:

1. $\{s_i : 1 \leq i \leq (g-1)/2\} \cup \{t_i : 1 \leq i \leq (g-1)/2\} = G\backslash\{0\}$, and
2. $\{\pm(s_i - t_i) : 1 \leq i \leq (g-1)/2\} = G\backslash\{0\}$.

14.7.56 Definition A *strong starter* is a starter $S = \{\{s_i, t_i\}\}$ in the abelian group G with the additional property that $s_i + t_i = s_j + t_j$ implies $i = j$, and for any i, $s_i + t_i \neq 0$.

14.7.57 Definition A *skew starter* is a starter $S = \{\{s_i, t_i\}\}$ in the abelian group G with the additional property that $s_i + t_i = \pm(s_j + t_j)$ implies $i = j$, and for any i, $s_i + t_i \neq 0$.

14.7.58 Example A strong starter in $\mathbb{Z}_{17}$ is

$$\{9, 10\}, \{3, 5\}, \{13, 16\}, \{11, 15\}, \{1, 6\}, \{2, 8\}, \{7, 14\}, \{4, 12\}.$$

14.7.59 Definition Let $S = \{\{s_i, t_i\} : 1 \leq i \leq (g-1)/2\}$ and $T = \{\{u_i, v_i\} : 1 \leq i \leq (g-1)/2\}$ be two starters in G. Without loss of generality, assume that $s_i - t_i = u_i - v_i$, for all i. Then S and T are *orthogonal* starters if $u_i - s_i = u_j - s_j$ implies $i = j$, and if $u_i \neq s_i$ for all i.

14.7.60 Definition Let q be a prime power that can be written in the form $q = 2^k t + 1$, where $t > 1$ is odd and let ω be a primitive element in the field $\mathbb{F}_q$. Then define

1. C_0 to be the multiplicative subgroup of $\mathbb{F}_q\backslash\{0\}$ of order t,
2. $C_i = \omega^i C_0$, $0 \leq i \leq 2^k - 1$ to be the cosets of C_0 (*cyclotomic classes*), and
3. $\Delta = 2^{k-1}$, $H = \cup_{i=0}^{\Delta-1} C_i$ and $C_i^a = (1/(a-1))C_i$.

14.7.61 Theorem [2198] Let $T = \{\{x, \omega^\Delta x\} : x \in H\}$. Then T is a skew starter (the *Mullin–Nemeth starter*) in the additive subgroup of $\mathbb{F}_q$.

14.7.62 Theorem [897] For each $a \in C_\Delta$, let $S_a = \{\{x, ax\} : x \in \cup_{i=0}^{\Delta-1} C_i^a\}$. Then for any $a \in C_\Delta$, S_a is a strong starter in the additive group of $\mathbb{F}_q$. Further, S_a and S_b are orthogonal if $a, b \in C_\Delta$ with $a \neq b$. Hence, the set $\{S_a | a \in C_\Delta\}$ is a set of t pairwise orthogonal starters of order q.

14.7.63 Theorem [623] Let $p = 2^{2^n}$ be a Fermat prime with $n \geq 2$. There exists a strong starter in the additive group of $\mathbb{F}_p$.

14.7.64 Remark [1537] No strong starter in the additive groups of $\mathbb{F}_3$, $\mathbb{F}_5$, or $\mathbb{F}_9$ exists.

14.7.65 Definition Let G be an additive abelian group of order g, and let H be a subgroup of order h of G, where $g - h$ is even. A *Room frame starter* in $G\backslash H$ is a set of unordered pairs $S = \{\{s_i, t_i\} : 1 \leq i \leq (g-h)/2\}$ such that

1. $\{s_i : 1 \leq i \leq (g-h)/2\} \cup \{t_i : 1 \leq i \leq (g-h)/2\} = G\backslash H$, and
2. $\{\pm(s_i - t_i) : 1 \leq i \leq (g-h)/2\} = G\backslash H$.

14.7.66 Remark A starter is the special case of a frame starter when $H = \{0\}$. The concepts of strong, skew, and orthogonal for Room frame starters are as for starters replacing $\{0\}$ by H and $(g-1)/2$ by $(g-h)/2$.

14.7.67 Theorem [898] Let $q \equiv 1 \pmod 4$ be a prime power such that $q = 2^k t + 1$, where $t > 1$ is odd. Then there exist t orthogonal Room frame starters in $(\mathbb{F}_q \times (\mathbb{Z}_2)^n) \setminus (\{0\} \times (\mathbb{Z}_2)^n)$ for all $n \geq 1$.

14.7.68 Theorem [93] If $p \equiv 1 \pmod 6$ is a prime and $p \geq 19$, then there exist three orthogonal frame starters in $(\mathbb{F}_p \times (\mathbb{Z}_3)) \setminus (\{0\} \times (\mathbb{Z}_3))$.

14.7.69 Definition Let S be a set of $n+1$ elements (*symbols*). A *Room square* of side n (on symbol set S), RS(n), is an $n \times n$ array, F, that satisfies the following properties:

1. every cell of F either is empty or contains an unordered pair of symbols from S,
2. each symbol of S occurs once in each row and column of F,
3. every unordered pair of symbols occurs in precisely one cell of F.

14.7.70 Definition A Room square of side n is *standardized* (with respect to the symbol ∞) if the cell (i, i) contains the pair $\{\infty, i\}$.

14.7.71 Definition A standardized Room square of side n is *skew* if for every pair of cells (i, j) and (j, i) (with $i \neq j$) exactly one is filled.

14.7.72 Definition A standardized Room square of side n is *cyclic* if $S = \mathbb{Z}_n \cup \{\infty\}$ and if whenever $\{a, b\}$ occurs in the cell (i, j), then $\{a+1, b+1\}$ occurs in cell $(i+1, j+1)$ where all arithmetic is performed in $\mathbb{Z}_n$ (and $\infty + 1 = \infty$).

14.7.73 Example Below are skew Room squares of sides 7 and 9; the Room square of side 7 is cyclic.

∞0			15		46	23
34	∞1			26		50
61	45	∞2			30	
	02	56	∞3			41
52		13	60	∞4		
	63		24	01	∞5	
		04		35	12	∞6

∞1		49	37	28		56		
89	∞2				57	34		16
	58	∞3		69	24		17	
	36	78	∞4		19		25	
	79		12	∞5	38		46	
45					∞6	18	39	27
		26	59	13		∞7		48
67	14					29	∞8	35
23		15	68	47				∞9

14.7.74 Theorem [899] The existence of d pairwise orthogonal starters in an abelian group of order n implies the existence of a Room d-cube of side n.

14.7.75 Remark Construction 14.7.77 is used to establish Theorem 14.7.74 when $d = 2$. It is easily extended when $d > 2$.

14.7.76 Definition An *adder* for the starter $S = \{\{s_i, t_i\} : 1 \le i \le (g-1)/2\}$ is an ordered set $A_S = \{a_1, a_2, \ldots, a_{(g-1)/2}\}$ of $(g-1)/2$ distinct nonzero elements from G such that the set $T = \{\{s_i + a_i, t_i + a_i\} : 1 \le i \le (g-1)/2\}$ is also a starter in the group G.

14.7.77 Construction Let $S = \{\{s_i, t_i\} : 1 \le i \le (n-1)/2\}$ and $T = \{\{u_i, v_i\} : 1 \le i \le (n-1)/2\}$ be two orthogonal starters in G (usually $\mathbb{Z}_n$), (n odd) with $A_S = \{a_i : 1 \le i \le (n-1)/2\}$ the associated adder. Let R be an $n \times n$ array indexed by the elements of G. Each $\{s_i, t_i\} \in S$ is placed in the first row in cell $R(0, -a_i)$. This is cycled so that the pair $\{s_i + x, t_i + x\}$ is in the cell $R(x, -a_i + x)$, where all arithmetic is performed in G. Finally, for each $x \in G$, place the pair $\{\infty, x\}$ in cell $R(x, x)$. Then R is a Room square of side n.

14.7.78 Remark Construction 14.7.77 in conjunction with Theorem 14.7.61 yields skew Room squares of prime power orders $q \equiv 3 \pmod 4$. This is useful in proving Theorem 14.7.79.

14.7.79 Theorem [899] A (skew) Room square of side n exists if and only if n is odd and $n \notin \{3, 5\}$.

14.7.80 Theorem [1932] A cyclic skew Room square of side n exists if $n = \prod {p_i}^{\alpha_i}$ where each p_i is a non-Fermat prime or if $n = pq$ with p, q distinct Fermat primes.

14.7.81 Definition If $\{S_1, \ldots, S_n\}$ is a partition of a set S, an $\{S_1, \ldots, S_n\}$-*Room frame* is an $|S| \times |S|$ array, F, indexed by S, satisfying:

1. every cell of F either is empty or contains an unordered pair of symbols of S,
2. the subarrays $S_i \times S_i$ are empty, for $1 \le i \le n$ (these subarrays are *holes*),
3. each symbol $x \notin S_i$ occurs once in row (or column) s for any $s \in S_i$, and
4. the pairs occurring in F are those $\{s, t\}$, where $(s, t) \in (S \times S) \setminus \bigcup_{i=1}^{n} (S_i \times S_i)$.

The *type* of a Room frame F is the multiset $\{|S_i| : 1 \le i \le n\}$. An "exponential" notation is used to describe types; a Room frame has type ${t_1}^{u_1} {t_2}^{u_2} \cdots {t_k}^{u_k}$ if there are u_i S_js of cardinality t_i, $1 \le i \le k$.

14.7.82 Remark Theorem 14.7.83 gives the connection between frame starters and Room frames. The construction for a Room frame from a pair of orthogonal frame starters is a generalization of Construction 14.7.77.

14.7.83 Theorem [898] Suppose a pair of orthogonal frame starters in $G \setminus H$ exists, where $|G| = g$ and $|H| = h$. Then there exists a Room frame of type $h^{g/h}$.

14.7.84 Remark Theorems 14.7.67 and 14.7.68 in conjunction with Theorem 14.7.83 yield Corollary 14.7.85. Theorem 14.7.86 details the existence of Room frames of type t^u.

14.7.85 Corollary a) Let $q \equiv 1 \pmod 4$ be a prime power such that $q = 2^k t + 1$, where $t > 1$ is odd. Then there exist a Room frame of type $(2^n)^q$ for all $n \ge 1$. b) If $p \equiv 1 \pmod 6$ is a prime and $p \ge 19$, then there exists a Room frame of type 3^p.

14.7.86 Theorem (Existence theorems for uniform Room frames) [706, §VI.50] and [900]

1. There does not exist a Room frame of type t^u if any of the following conditions hold: (i) $u = 2$ or 3; (ii) $u = 4$ and $t = 2$; (iii) $u = 5$ and $t = 1$; (iv) $t(u-1)$ is odd.
2. Suppose t and u are positive integers, $u \ge 4$ and $(t, u) \ne (1, 5), (2, 4)$. Then there exists a uniform Room frame of type t^u if and only if $t(u-1)$ is even.

14.7.9 Strongly regular graphs

14.7.87 Definition A *strongly regular graph* with parameters (v, k, λ, μ) is a finite graph on v vertices, without loops or multiple edges, regular of degree k (with $0 < k < v - 1$, so that there are both edges and nonedges), and such that any two distinct vertices have λ common neighbors when they are adjacent, and μ common neighbors when they are nonadjacent.

14.7.88 Remark There are many constructions for strongly regular graphs. Example 14.7.89 gives several that use finite fields. For a table of the existence of strongly regular graphs with $v \leq 280$ see [706, pp. 852–866].

14.7.89 Example [419]

1. *Paley(q):* For prime powers $q = 4t + 1$, the graph with vertex set $\mathbb{F}_q$ where two vertices are adjacent when they differ by a square. This strongly regular graph has parameters $(q, \frac{1}{2}q - 1, \frac{1}{4}(q - 5), \frac{1}{4}(q - 1))$.
2. *van Lint–Schrijver(u):* a graph constructed by the cyclotomic construction in [2850], by taking the union of u classes.
3. $A_{n-1,2}(q)$ or $\left[{n \atop 2}\right]_q$: the graph on the lines in $\mathrm{PG}(n-1, q)$, adjacent when they have a point in common.
4. $Bilin_{2\times d}(q)$: the graph on the $2 \times d$ matrices over $\mathbb{F}_q$, adjacent when their difference has rank 1.
5. $O^{\varepsilon}_{2d}(q)$: the graph on the isotropic points on a nondegenerate quadric in $\mathrm{PG}(2d-1, q)$, where two points are joined when the connecting line is totally singular.
6. $Sp_{2d}(q)$: the graph on the points of $\mathrm{PG}(2d-1, q)$ provided with a nondegenerate symplectic form, where two points are joined when the connecting line is totally isotropic.
7. $U_d(q)$: the graph on the isotropic points of $\mathrm{PG}(d - 1, q^2)$ provided with a nondegenerate Hermitian form, where two points are joined when the connecting line is totally isotropic.
8. *Affine difference sets* [479]: Let V be an n-dimensional vector space over $\mathbb{F}_q$ and let X be a set of directions (a subset of the projective space PV). Two vectors are *adjacent* when the line joining them has a direction in X. Then $v = q^n$ and $k = (q-1)|X|$. This graph is strongly regular if and only if there are two integers w_1, w_2 such that all hyperplanes of PV miss either w_1 or w_2 points of X. If this is the case, then $r = k - qw_1$, $s = k - qw_2$ (assuming $w_1 < w_2$), and hence $\mu = k + (k - qw_1)(k - qw_2)$, $\lambda = k - 1 + (k - qw_1 + 1)(k - qw_2 + 1)$.

14.7.10 Whist tournaments

14.7.90 Definition A *whist tournament* Wh($4n$) for $4n$ players is a schedule of games each involving two players opposing two others, such that

1. the games are arranged into $4n - 1$ rounds, each of n games;
2. each player plays in exactly one game in each round;
3. each player partners every other player exactly once;
4. each player opposes every other player exactly twice.

14.7.91 **Definition** Each game is denoted by an ordered 4-tuple (a, b, c, d) in which the pairs $\{a, c\}$, $\{b, d\}$ are *partner pairs*; $\{a, c\}$ is a *partner pair of the first kind*, and $\{b, d\}$ is a *partner pair of the second kind*. The other pairs are *opponent pairs*; in particular $\{a, b\}$, $\{c, d\}$ are *opponent pairs of the first kind*, and $\{a, d\}$, $\{b, c\}$ are *opponent pairs of the second kind*.

14.7.92 **Definition** A *whist tournament* $\mathrm{Wh}(4n+1)$ for $4n+1$ players is defined as for $4n$, except that Conditions 1, 2 are replaced by

1′. the games are arranged into $4n+1$ rounds each of n games;
2′. each player plays in one game in each of $4n$ rounds, but does not play in the remaining round.

14.7.93 **Definition** A $\mathrm{Wh}(4n)$ is $\mathbb{Z}$-cyclic if the players are $\infty, 0, 1, \ldots, 4n-2$ and each round is obtained from the previous one by adding 1 modulo $4n-1$ to each non-∞ entry. A $\mathrm{Wh}(4n+1)$ is $\mathbb{Z}$-cyclic if the players are $0, 1, \ldots, 4n$ and the rounds are similarly developed modulo $4n+1$.

14.7.94 **Theorem** [167] If $p = 4n+1$ is prime and w is a primitive root of p, then the games $(w^i, w^{i+n}, w^{i+2n}, w^{i+3n})$, $0 \le i \le n-1$, form the initial round of $\mathbb{Z}$-cyclic $\mathrm{Wh}(4n+1)$.

14.7.95 **Theorem** Let P denote any product of primes p with each $p \equiv 1 \pmod 4$, and let q, r denote primes with both $q, r \equiv 3 \pmod 4$.
A $\mathbb{Z}$-cyclic $\mathrm{Wh}(4n)$ is known to exist when:

1. $4n \le 132$ (see [7, 98]);
2. $4n = 2^\alpha (\alpha \ge 2)$ (see [100]);
3. $4n = qP + 1$, $q \in \{3, 7, 11, 19, 23, 31, 43, 47, 59, 67, 71, 79, 83, 103, 107, 127\}$ (see [7]);
4. $4n = 3P + 1$ (see [100]);
5. $4n = 3^{2m+1} + 1$, $m \ge 0$ (see [101]);
6. $4n = qr^2P + 1$, q and r distinct, $q < 60$, $r < 100$ (see [7]).

A $\mathbb{Z}$-cyclic $\mathrm{Wh}(4n+1)$ is known to exist when:

1. $4n+1 = P$ or r^2P and $r \le 100$ (see [7, 102]);
2. $4n+1 \le 149$ (see [7]);
3. $4n+1 = 3^{2m}$ or $3^{2m}P$ (see [101]);
4. $4n+1 = 3sP$, $s \in \{7, 11, 15, 19, 23, 27, 31, 35, 39, 43, 47\}$ (see [7]).

14.7.96 **Definition** A *triplewhist tournament* $\mathrm{TWh}(v)$ is a $\mathrm{Wh}(v)$ with Condition 4 replaced by (4″) each player has every other player once as an opponent of the first kind and once as an opponent of the second kind.

14.7.97 **Theorem** A $\mathrm{TWh}(4n+1)$ exists for all $n \ge 5$, and possibly for $n = 4$. A $\mathrm{TWh}(4n)$ exists for all $n \ge 1$ except for $n = 3$; see [7, 1965].

14.7.98 **Theorem** A $\mathbb{Z}$-cyclic $\mathrm{TWh}(4n+1)$ exists when:

1. $4n+1$ is a prime $p \equiv 1 \pmod 4$, $p \ge 29$ (see [462]);

2. $4n + 1 = q^2$ where q is prime, $q \equiv 3 \pmod 4$, $7 \le q \le 83$ (see [6]);
3. $4n + 1 = 5^n$ or 13^n ($n \ge 2$) (see [6]);
4. $4n + 1$ is the product of any values in the above three items (see [102]).

14.7.99 Theorem A $\mathbb{Z}$-cyclic TWh($4n$) exists when:

1. $4n = qP + 1$, where P denotes any product of primes p with each $p \equiv 1 \pmod 4$ and $q \in \{3, 7, 19, 23, 31, 43, 47, 59, 67, 71, 79, 83, 103, 107, 127, 131\}$ (see [7]);
2. $4n = 2^n$ ($n \ge 2$) (see [100]);
3. $4n = 3^m + 1$ (m odd) (see [101]).

See Also

§14.1	For latin squares, MOLS, and transversal designs.
§14.3	For affine and projective planes.
§14.4	For projective spaces.
§14.5	For block designs.
§14.6	For difference sets.
[99]	Textbook on combinatorial designs.
[261]	Advanced textbook on combinatorial designs.
[262]	Another advanced textbook on combinatorial designs.
[706]	For association schemes (§VI.1), Costas arrays (§VI.9), conference matrices (§V.6), covering arrays (§VI.10), Hadamard designs and matrices(§V.1); Hall triple systems (§VI.28), ordered designs and perpendicular arrays (§VI.38), perfect hash families (§VI.43), Room squares (§VI.50), strongly regular graphs (§VII.11), whist tournaments (§VI.64).
[2719]	Another textbook on combinatorial designs.

References Cited: [6, 7, 93, 98, 99, 100, 101, 102, 167, 206, 207, 261, 262, 275, 302, 419, 462, 479, 606, 623, 704, 705, 706, 707, 897, 898, 899, 900, 916, 918, 1262, 1289, 1301, 1304, 1403, 1404, 1457, 1537, 1932, 1965, 2023, 2198, 2345, 2612, 2719, 2721, 2831, 2850, 2851, 2853, 2975, 3037]

14.8 (t, m, s)-nets and (t, s)-sequences

Harald Niederreiter, KFUPM

14.8.1 (t, m, s)-nets

14.8.1 Remark The theory of (t, m, s)-nets and (t, s)-sequences is significant for quasi-Monte Carlo methods in scientific computing (see the books [839] and [2248] and the recent survey article [2259]). For both (t, m, s)-nets and (t, s)-sequences, the idea is to sample the s-dimensional unit cube $[0, 1]^s$ in a uniform and equitable manner. In a nutshell, (t, m, s)-nets are finite samples (or point sets) and (t, s)-sequences are infinite sequences with special uniformity properties. The definition of a (t, m, s)-net (see Definition 14.8.2 below) has *a priori* no connection with finite fields, but it turns out that most of the interesting constructions of (t, m, s)-nets use finite fields as a tool. By a *point set* we mean a multiset in the sense of combinatorics, i.e., a set in which multiplicities of elements are allowed and taken into account.

14.8.2 Definition [2236, 2690] For integers $b \geq 2$ and $0 \leq t \leq m$ and a given dimension $s \geq 1$, a *(t, m, s)-net in base b* is a point set P consisting of b^m points in $[0, 1]^s$ such that every subinterval of $[0, 1]^s$ of volume b^{t-m} which has the form

$$\prod_{i=1}^{s} [a_i b^{-d_i}, (a_i + 1) b^{-d_i})$$

with integers $d_i \geq 0$ and $0 \leq a_i < b^{d_i}$ for $1 \leq i \leq s$ contains exactly b^t points of P.

14.8.3 Remark It is easily seen that a (t, m, s)-net in base b is also a (u, m, s)-net in base b for all integers u with $t \leq u \leq m$. Any point set consisting of b^m points in $[0, 1)^s$ is a (t, m, s)-net in base b with $t = m$. Smaller values of t mean stronger uniformity properties of a (t, m, s)-net in base b. The number t is the *quality parameter* of a (t, m, s)-net in base b.

14.8.4 Definition A (t, m, s)-net P in base b is a *strict (t, m, s)-net in base b* if t is the least integer u such that P is a (u, m, s)-net in base b.

14.8.5 Example Let $s = 2$ and let $b \geq 2$ and $m \geq 1$ be given integers. For any integer n with $0 \leq n < b^m$, let $n = \sum_{r=0}^{m-1} a_r(n) b^r$ with $a_r(n) \in \mathbb{Z}_b = \{0, 1, \ldots, b-1\}$ be the digit expansion of n in base b and put $\phi_b(n) = \sum_{r=0}^{m-1} a_r(n) b^{-r-1}$. Then the point set consisting of the points $(nb^{-m}, \phi_b(n)) \in [0, 1]^2$, $n = 0, 1, \ldots, b^m - 1$, is a $(0, m, 2)$-net in base b. This point set is the *Hammersley net* in base b.

14.8.6 Example Let $b \geq 2$, $s \geq 1$, and $t \geq 0$ be given integers. Then the point set consisting of the points $(nb^{-1}, \ldots, nb^{-1}) \in [0, 1]^s$, $n = 0, 1, \ldots, b - 1$, each taken with multiplicity b^t, is a $(t, t + 1, s)$-net in base b.

14.8.7 Remark According to Remark 14.8.3 and Example 14.8.6, a (t, m, s)-net in base b always exists for $m = t$ and $m = t + 1$. For $m \geq t + 2$ there are combinatorial obstructions to the general existence of (t, m, s)-nets in base b. This was first observed in [2236]. Later, a combinatorial equivalence between (t, m, s)-nets in base b and ordered orthogonal arrays as defined in Definition 14.8.8 below was established.

14.8.8 Definition Let b, s, k, T, λ be positive integers with $b \geq 2$ and $sT \geq k$. An *ordered orthogonal array* $\mathrm{OOA}_b(s, k, T, \lambda)$ is a $(\lambda b^k) \times (sT)$ matrix with entries from $\mathbb{Z}_b$ and column labels (i, j) for $1 \leq i \leq s$ and $1 \leq j \leq T$ such that, for any integers $0 \leq d_1, \ldots, d_s \leq T$ with $\sum_{i=1}^{s} d_i = k$, the $(\lambda b^k) \times k$ submatrix obtained by restricting to the columns (i, j), $1 \leq j \leq d_i$, $1 \leq i \leq s$, contains among its rows every element of $\mathbb{Z}_b^k$ with the same frequency λ.

14.8.9 Theorem [1873, 2183] Let $b \geq 2$, $s \geq 2$, $k \geq 2$, and $t \geq 0$ be integers. Then there exists a $(t, t+k, s)$-net in base b if and only if there exists an ordered orthogonal array $\mathrm{OOA}_b(s, k, k-1, b^t)$.

14.8.10 Corollary [2236] There exists a $(0, 2, s)$-net in base b if and only if there exist $s-2$ mutually orthogonal latin squares of order b.

14.8.11 Corollary [2236] For $m \geq 2$, a $(0, m, s)$-net in base b can exist only if $s \leq M(b) + 2$, where $M(b)$ is the maximum cardinality of a set of mutually orthogonal latin squares of order b. In particular, if $m \geq 2$, then a necessary condition for the existence of a $(0, m, s)$-net in base b is $s \leq b + 1$.

14.8.12 Remark The equivalence between nets and ordered orthogonal arrays enunciated in Theorem 14.8.9, when combined with extensions of standard parameter bounds for orthogonal arrays to the case of ordered orthogonal arrays, leads to lower bounds on the quality parameter for nets. Examples of such bounds are the linear programming bound [2008], the Rao bound [2007], and the dual Plotkin bound [271, 2009]. Extensive numerical data on these bounds are available at `http://mint.sbg.ac.at`.

14.8.2 Digital (t, m, s)-nets

14.8.13 Remark Most of the known constructions of nets are based on the *digital method* which goes back to [2236]. In order to describe the digital method for the construction of (t, m, s)-nets in base b, we need the following ingredients. First of all, let integers $b \geq 2$, $m \geq 1$, and $s \geq 1$ be given. Then we choose:

1. a commutative ring R with identity and of cardinality b;
2. bijections $\eta_j^{(i)} : R \to \mathbb{Z}_b$ for $1 \leq i \leq s$ and $1 \leq j \leq m$;
3. $m \times m$ matrices $C^{(1)}, \ldots, C^{(s)}$ over R.

Now let $\mathbf{r} \in R^m$ be an m-tuple of elements of R and define

$$\pi_j^{(i)}(\mathbf{r}) = \eta_j^{(i)}(\mathbf{c}_j^{(i)} \cdot \mathbf{r}) \in \mathbb{Z}_b \qquad \text{for } 1 \leq i \leq s,\ 1 \leq j \leq m,$$

where $\mathbf{c}_j^{(i)}$ is the j-th row of the matrix $C^{(i)}$ and $\cdot$ denotes the standard inner product. Next we put

$$\pi^{(i)}(\mathbf{r}) = \sum_{j=1}^{m} \pi_j^{(i)}(\mathbf{r}) b^{-j} \in [0, 1] \qquad \text{for } 1 \leq i \leq s$$

and

$$P(\mathbf{r}) = (\pi^{(1)}(\mathbf{r}), \ldots, \pi^{(s)}(\mathbf{r})) \in [0, 1]^s.$$

By letting $\mathbf{r}$ range over all b^m elements of R^m, we arrive at a point set P consisting of b^m points in $[0, 1]^s$.

14.8.14 Definition If the point set P constructed in Remark 14.8.13 forms a (t, m, s)-net in base b, then P is a *digital (t, m, s)-net in base b*. If we want to emphasize that the construction uses the ring R, then we speak also of a *digital (t, m, s)-net over R*. If P is a strict (t, m, s)-net in base b, then P is a *digital strict (t, m, s)-net in base b* (or *over R*).

14.8.15 Remark The matrices $C^{(1)}, \ldots, C^{(s)}$ in Remark 14.8.13 are *generating matrices* of the digital net. The quality parameter t of the digital net depends only on the generating matrices. For a convenient algebraic condition on the generating matrices to guarantee a certain value of t, we refer to Theorem 4.26 in [2248]. In the important case where the ring R is a finite field, an even simpler condition is given in Theorem 14.8.18 below.

14.8.16 Example Let $s = 2$ and let $b \geq 2$ and $m \geq 1$ be given integers. Choose $R = \mathbb{Z}_b$ and let the bijections $\eta_j^{(i)}$ in Remark 14.8.13 be identity maps. Let $C^{(1)}$ be the $m \times m$ identity matrix over $\mathbb{Z}_b$ and let $C^{(2)} = (c_{i,j})_{1 \leq i,j \leq m}$ be the $m \times m$ antidiagonal matrix over $\mathbb{Z}_b$ with $c_{i,j} = 1$

if $i+j=m+1$ and $c_{i,j}=0$ otherwise. Then the Hammersley net in Example 14.8.5 is easily seen to be a digital $(0,m,2)$-net over $\mathbb{Z}_b$ with generating matrices $C^{(1)}$ and $C^{(2)}$.

14.8.17 Definition Let $C^{(1)},\ldots,C^{(s)}$ be $m\times m$ matrices over the finite field $\mathbb{F}_q$ and for $1\le i\le s$ and $1\le j\le m$ let $\mathbf{c}_j^{(i)}$ denote the j-th row of the matrix $C^{(i)}$. Then $\varrho(C^{(1)},\ldots,C^{(s)})$ is defined to be the largest nonnegative integer d such that, for any integers $0\le d_1,\ldots,d_s\le m$ with $\sum_{i=1}^s d_i=d$, the vectors $\mathbf{c}_j^{(i)}$, $1\le j\le d_i$, $1\le i\le s$, are linearly independent over $\mathbb{F}_q$ (this property is assumed to be vacuously satisfied for $d=0$).

14.8.18 Theorem [2236] The point set P constructed in Remark 14.8.13 with $R=\mathbb{F}_q$ and $m\times m$ generating matrices $C^{(1)},\ldots,C^{(s)}$ over $\mathbb{F}_q$ is a digital strict (t,m,s)-net over $\mathbb{F}_q$ with $t=m-\varrho(C^{(1)},\ldots,C^{(s)})$.

14.8.19 Example Let $s=2$, let $b=p$ be a prime, and let the $m\times m$ matrices $C^{(1)}$ and $C^{(2)}$ over $\mathbb{F}_p$ be as in Example 14.8.16. Then it is easily seen that $\varrho(C^{(1)},C^{(2)})=m$. Using Theorem 14.8.18, this shows again that the Hammersley net in base $b=p$ is a digital $(0,m,2)$-net over $\mathbb{F}_p$.

14.8.20 Remark The equivalence between nets and ordered orthogonal arrays stated in Theorem 14.8.9 has an analog for digital nets. The special family of linear ordered orthogonal arrays was introduced in [276] and it was shown that these arrays correspond to digital nets.

14.8.21 Remark There is a very useful duality theory for digital nets which facilitates many constructions of good digital nets. A crucial ingredient is the weight function V_m on $\mathbb{F}_q^{ms}$ introduced in Definition 14.8.22 below. The main result of this duality theory is Theorem 14.8.26 below.

14.8.22 Definition Let $m\ge 1$ and $s\ge 1$ be integers. Put $v_m(\mathbf{a})=0$ if $\mathbf{a}=\mathbf{0}\in\mathbb{F}_q^m$, and for $\mathbf{a}=(a_1,\ldots,a_m)\in\mathbb{F}_q^m$ with $\mathbf{a}\ne\mathbf{0}$ let $v_m(\mathbf{a})$ be the largest value of j such that $a_j\ne 0$. Write a vector $\mathbf{A}\in\mathbb{F}_q^{ms}$ as the concatenation of s vectors of length m, i.e., $\mathbf{A}=(\mathbf{a}^{(1)},\ldots,\mathbf{a}^{(s)})\in\mathbb{F}_q^{ms}$ with $\mathbf{a}^{(i)}\in\mathbb{F}_q^m$ for $1\le i\le s$. Then the *NRT weight* of $\mathbf{A}$ is defined by

$$V_m(\mathbf{A})=\sum_{i=1}^{s} v_m(\mathbf{a}^{(i)}).$$

14.8.23 Remark The NRT weight is named after the work of Niederreiter [2234] and Rosenbloom and Tsfasman [2483]. The *NRT space* is $\mathbb{F}_q^{ms}$ with the metric $d_m(\mathbf{A},\mathbf{B})=V_m(\mathbf{A}-\mathbf{B})$ for $\mathbf{A},\mathbf{B}\in\mathbb{F}_q^{ms}$. For $m=1$ the NRT space reduces to the Hamming space in coding theory.

14.8.24 Definition The *minimum distance* $\delta_m(\mathcal{N})$ of a nonzero $\mathbb{F}_q$-linear subspace $\mathcal{N}$ of $\mathbb{F}_q^{ms}$ is given by

$$\delta_m(\mathcal{N})=\min_{\mathbf{A}\in\mathcal{N}\setminus\{\mathbf{0}\}} V_m(\mathbf{A}).$$

14.8.25 Remark Let the $m\times m$ matrices $C^{(1)},\ldots,C^{(s)}$ over $\mathbb{F}_q$ be generating matrices of a digital net P. Set up an $m\times ms$ matrix M over $\mathbb{F}_q$ as follows: for $1\le j\le m$, the j-th row of M is obtained by concatenating the transposes of the j-th columns of $C^{(1)},\ldots,C^{(s)}$. Let $\mathcal{M}\subseteq\mathbb{F}_q^{ms}$ be the row space of M and let $\mathcal{M}^\perp$ be its dual space, i.e.,

$$\mathcal{M}^\perp=\{\mathbf{A}\in\mathbb{F}_q^{ms}:\mathbf{A}\cdot\mathbf{M}=0 \text{ for all } \mathbf{M}\in\mathcal{M}\},$$

where $\cdot$ is the standard inner product in $\mathbb{F}_q^{ms}$.

14.8.26 Theorem [2267] Let $m \geq 1$ and $s \geq 2$ be integers. Then the point set P in Remark 14.8.25 is a digital strict (t, m, s)-net over $\mathbb{F}_q$ with $t = m + 1 - \delta_m(\mathcal{M}^\perp)$.

14.8.27 Corollary [2267] Let $m \geq 1$ and $s \geq 2$ be integers. Then from any $\mathbb{F}_q$-linear subspace $\mathcal{N}$ of $\mathbb{F}_q^{ms}$ with $\dim(\mathcal{N}) \geq ms - m$ we can construct a digital strict (t, m, s)-net over $\mathbb{F}_q$ with $t = m + 1 - \delta_m(\mathcal{N})$.

14.8.28 Remark There are digital nets for which a property analogous to that in Definition 14.8.2 holds for a wider range of subintervals of $[0, 1]^s$. Such generalized digital nets were introduced in [835] and are also studied in detail in Chapter 15 of [839].

14.8.29 Remark There is a generalization of the digital method which can be viewed as a nonlinear analog of the construction in Remark 14.8.13. For simplicity we consider only the case where $R = \mathbb{F}_q$ (see [2258] for a general ring R). Compared to Remark 14.8.13, the only change is that instead of linear forms $\mathbf{c}_j^{(i)} \cdot \mathbf{r}$ we now use polynomial functions, that is, for $1 \leq i \leq s$ and $1 \leq j \leq m$ we choose polynomials $f_j^{(i)}$ over $\mathbb{F}_q$ in m variables and then we replace $\mathbf{c}_j^{(i)} \cdot \mathbf{r}$ by $f_j^{(i)}(\mathbf{r})$ for $\mathbf{r} \in \mathbb{F}_q^m$. The following criterion uses the concept of permutation polynomial in several variables (see Section 8.2).

14.8.30 Theorem [2258] The point set constructed in Remark 14.8.29 is a (t, m, s)-net in base q if and only if, for any integers $d_1, \ldots, d_s \geq 0$ with $\sum_{i=1}^{s} d_i = m - t$, the polynomials $f_j^{(i)}$, $1 \leq j \leq d_i$, $1 \leq i \leq s$, have the property that all of their nontrivial linear combinations with coefficients from $\mathbb{F}_q$ are permutation polynomials over $\mathbb{F}_q$ in m variables.

14.8.3 Constructions of (t, m, s)-nets

14.8.31 Remark A general principle for the construction of (t, m, s)-nets with $s \geq 2$ is based on the use of Proposition 14.8.50 below in conjunction with the constructions of $(t, s-1)$-sequences in Subsection 14.8.6. In the present subsection, we describe constructions of (t, m, s)-nets that are not derived from this principle. One of the first constructions of this type was that of *polynomial lattices* in [2247]. Choose $f \in \mathbb{F}_q[x]$ with $\deg(f) = m \geq 1$ and an s-tuple $\mathbf{g} = (g_1, \ldots, g_s) \in \mathbb{F}_q[x]^s$ with $\deg(g_i) < m$ for $1 \leq i \leq s$. Consider the Laurent series expansions

$$\frac{g_i(x)}{f(x)} = \sum_{k=1}^{\infty} u_k^{(i)} x^{-k} \in \mathbb{F}_q((x^{-1})) \qquad \text{for } 1 \leq i \leq s.$$

Then for $1 \leq i \leq s$ the generating matrix $C^{(i)} = (c_{j,r})$ is the Hankel matrix given by $c_{j,r} = u_{j+r}^{(i)} \in \mathbb{F}_q$ for $1 \leq j \leq m$, $0 \leq r \leq m-1$. The bijections $\eta_j^{(i)}$ in Remark 14.8.13 are chosen arbitrarily. The resulting digital net over $\mathbb{F}_q$ is denoted by $P(\mathbf{g}, f)$.

14.8.32 Definition Let $s \geq 2$ and let f and $\mathbf{g}$ be as in Remark 14.8.31. Then the *figure of merit* $\varrho(\mathbf{g}, f)$ is defined by

$$\varrho(\mathbf{g}, f) = s - 1 + \min \sum_{i=1}^{s} \deg(h_i),$$

where the minimum is over all nonzero s-tuples $(h_1, \ldots, h_s) \in \mathbb{F}_q[x]^s$ with $\deg(h_i) < m$ for $1 \leq i \leq s$ and f dividing $\sum_{i=1}^{s} h_i g_i$. Here we use the convention $\deg(0) = -1$.

14.8.33 Theorem [2247] For $s \geq 2$, the point set $P(\mathbf{g}, f)$ in Remark 14.8.31 is a digital strict (t, m, s)-net over $\mathbb{F}_q$ with $t = m - \varrho(\mathbf{g}, f)$.

14.8.34 Remark It is clear from Theorem 14.8.33 that in order to obtain a good (t, m, s)-net by this construction, i.e., a net with a small value of t, we need to find $\mathbf{g}$ and f with a large figure of merit $\varrho(\mathbf{g}, f)$. A systematic method for the explicit construction of good polynomial lattices is the component-by-component algorithm in [836]; see also Chapter 10 in [839].

14.8.35 Remark Several constructions of digital nets are based on Corollary 14.8.27. A powerful construction of this type uses algebraic function fields (see Section 12.1 for background on algebraic function fields). We present only a simple version of this construction; more refined versions can be found in [2261]. Let F be an algebraic function field (of one variable) with full constant field $\mathbb{F}_q$, that is, $\mathbb{F}_q$ is algebraically closed in F. Let $N(F)$ denote the number of rational places of F. For a given dimension $s \geq 2$, we assume that $N(F) \geq s$ and let $P_1, \dots, P_s$ be s distinct rational places of F. Let G be a divisor of F. For each $i = 1, \dots, s$, let $t_i \in F$ be a prime element at P_i and let $n_i \in \mathbb{Z}$ be the coefficient of P_i in G. For f in the Riemann-Roch space $\mathcal{L}(G)$ and a given integer $m \geq 1$, let $\theta^{(i)}(f) \in \mathbb{F}_q^m$ be the vector whose coordinates are, in descending order, the coefficients of t_i^j, $j = -n_i + m - 1, -n_i + m - 2, \dots, -n_i$, in the local expansion of f at P_i. Now define the $\mathbb{F}_q$-linear map $\theta : \mathcal{L}(G) \to \mathbb{F}_q^{ms}$ by

$$\theta(f) = (\theta^{(1)}(f), \dots, \theta^{(s)}(f)) \qquad \text{for all } f \in \mathcal{L}(G).$$

A digital net over $\mathbb{F}_q$ is then obtained by applying Corollary 14.8.27 with $\mathcal{N}$ being the image of the map θ. A suitable choice of the divisor G leads to the following result.

14.8.36 Theorem [2261] Let $s \geq 2$ be an integer and let F be an algebraic function field with full constant field $\mathbb{F}_q$, genus $g \geq 1$, and $N(F) \geq s$. If k and m are integers with $0 \leq k \leq g - 1$ and $m \geq \max(1, g - k - 1)$, then there exists a digital $(g - k - 1, m, s)$-net over $\mathbb{F}_q$ provided that

$$\binom{s + m + k - g}{s - 1} A_k(F) < h(F),$$

where $A_k(F)$ is the number of positive divisors of F of degree k and $h(F)$ is the divisor class number of F.

14.8.37 Example Let $q = 9$ and let F be the Hermitian function field over $\mathbb{F}_9$, that is, $F = \mathbb{F}_9(x, y)$ with $y^3 + y = x^4$. Then $g = 3$, $N(F) = 28$, and $h(F) = 4096$. We apply Theorem 14.8.36 with $s = 28$, $k = 0$, $m = 5$, and we obtain a digital $(2, 5, 28)$-net over $\mathbb{F}_9$. The value $t = 2$ is the currently best value of the quality parameter for a $(t, 5, 28)$-net in base 9, according to the website `http://mint.sbg.ac.at` which contains an extensive database for parameters of (t, m, s)-nets.

14.8.38 Remark Another construction based on Corollary 14.8.27 was introduced in [2255]. For integers $m \geq 1$ and $s \geq 2$, consider the $\mathbb{F}_q$-linear space $\mathcal{P} = \{f \in \mathbb{F}_{q^m}[x] : \deg(f) < s\}$. Fix $\alpha \in \mathbb{F}_{q^m}$ and define the $\mathbb{F}_q$-linear subspace $\mathcal{P}_\alpha = \{f \in \mathcal{P} : f(\alpha) = 0\}$ of $\mathcal{P}$. Set up a map $\tau : \mathcal{P} \to \mathbb{F}_q^{ms}$ as follows. Write $f \in \mathcal{P}$ explicitly as $f(x) = \sum_{i=1}^{s} \gamma_i x^{i-1}$ with $\gamma_i \in \mathbb{F}_{q^m}$ for $1 \leq i \leq s$. For each $i = 1, \dots, s$, choose an ordered basis B_i of $\mathbb{F}_{q^m}$ over $\mathbb{F}_q$ and let $\mathbf{c}_i(f) \in \mathbb{F}_q^m$ be the coordinate vector of γ_i with respect to B_i. Then define

$$\tau(f) = (\mathbf{c}_1(f), \dots, \mathbf{c}_s(f)) \in \mathbb{F}_q^{ms} \qquad \text{for all } f \in \mathcal{P}.$$

A digital net over $\mathbb{F}_q$ is now obtained by applying Corollary 14.8.27 with $\mathcal{N}$ being the image of the subspace $\mathcal{P}_\alpha$ under τ. The resulting digital net is a *cyclic digital net* over $\mathbb{F}_q$ relative to the bases $B_1, \dots, B_s$.

14.8.39 Remark A generalization of the construction in Remark 14.8.38 was presented in [2398]. For integers $m \geq 1$ and $s \geq 2$, consider $\mathcal{Q} = \mathbb{F}_{q^m}^s$ as a vector space over $\mathbb{F}_q$. Fix $\boldsymbol{\alpha} \in \mathcal{Q}$ with $\boldsymbol{\alpha} \neq \mathbf{0}$ and put $\mathcal{Q}_{\boldsymbol{\alpha}} = \{\boldsymbol{\beta} \in \mathcal{Q} : \boldsymbol{\alpha} \cdot \boldsymbol{\beta} = 0\}$. Then $\mathcal{Q}_{\boldsymbol{\alpha}}$ is an $\mathbb{F}_q$-linear subspace of $\mathcal{Q}$. Let

$\sigma : \mathcal{Q} \to \mathbb{F}_q^{ms}$ be an isomorphism between vector spaces over $\mathbb{F}_q$. A digital net over $\mathbb{F}_q$ is now obtained by applying Corollary 14.8.27 with $\mathcal{N}$ being the image of the subspace $\mathcal{Q}_{\boldsymbol{\alpha}}$ under σ. The resulting digital net is a *hyperplane net* over $\mathbb{F}_q$. Detailed information on hyperplane nets and cyclic digital nets can be found in Chapter 11 of [839].

14.8.40 Theorem [1874] Given a linear code over $\mathbb{F}_q$ with length n, dimension k, and minimum distance $d \geq 3$, we can construct a digital $(n-k-d+1, n-k, s)$-net over $\mathbb{F}_q$ with $s = \lfloor (n-1)/h \rfloor$ if $d = 2h+2$ is even and $s = \lfloor n/h \rfloor$ if $d = 2h+1$ is odd.

14.8.41 Remark Further applications of coding theory to the construction of digital nets are discussed in the survey articles [2251] and [2256]. We specifically mention some principles of combining several digital nets to obtain a new digital net that are inspired by coding theory. For instance, the well-known Kronecker-product construction in coding theory has an analog for digital nets [276]. The following result is an analog of the matrix-product construction of linear codes.

14.8.42 Theorem [2263] Let h be an integer with $2 \leq h \leq q$. If for $k = 1, \ldots, h$ a digital (t_k, m_k, s_k)-net over $\mathbb{F}_q$ is given and if $s_1 \leq s_2 \leq \cdots \leq s_h$, then we can construct a digital $(t, \sum_{k=1}^{h} m_k, \sum_{k=1}^{h} s_k)$-net over $\mathbb{F}_q$ with

$$t = 1 + \sum_{k=1}^{h} m_k - \min_{1 \leq k \leq h} (h-k+1)(m_k - t_k + 1).$$

14.8.43 Proposition Given a (t, m, s)-net in base b, we can construct:

1. a (t, u, s)-net in base b for $t \leq u \leq m$;
2. a (t, m, r)-net in base b for $1 \leq r \leq s$;
3. a $(t+u, m+u, s)$-net in base b for any integer $u \geq 0$.

14.8.44 Remark A result of the type appearing in Proposition 14.8.43 is called a *propagation rule* for nets. There are also propagation rules for digital nets, in the sense that the input net and the output net are both digital nets. Furthermore, there are propagation rules that involve a base change, typically moving from a base b to a base that is a power b^k with $k \geq 2$ or vice versa. A detailed discussion of propagation rules is presented in Chapter 9 of [839].

14.8.4 (t, s)-sequences and $(\mathbf{T}, s)$-sequences

14.8.45 Remark There is an analog of (t, m, s)-nets for sequences of points in $[0,1]^s$, given in Definition 14.8.46 below. First we need some notation. For an integer $b \geq 2$ and a real number $x \in [0,1]$, let $x = \sum_{j=1}^{\infty} y_j b^{-j}$ with all $y_j \in \mathbb{Z}_b$ be a b-adic expansion of x, where the case $y_j = b-1$ for all sufficiently large j is allowed. For any integer $m \geq 1$, we define the truncation $[x]_{b,m} = \sum_{j=1}^{m} y_j b^{-j}$. Note that this truncation operates on the expansion of x and not on x itself, since it may yield different results depending on which b-adic expansion of x is used. If $\mathbf{x} = (x^{(1)}, \ldots, x^{(s)}) \in [0,1]^s$ and the $x^{(i)}$, $1 \leq i \leq s$, are given by prescribed b-adic expansions, then we define

$$[\mathbf{x}]_{b,m} = ([x^{(1)}]_{b,m}, \ldots, [x^{(s)}]_{b,m}).$$

14.8.46 Definition [2236, 2690] Let $b \geq 2$, $s \geq 1$, and $t \geq 0$ be integers. A sequence $\mathbf{x}_0, \mathbf{x}_1, \ldots$ of points in $[0,1]^s$ is a *(t, s)-sequence in base b* if for all integers $k \geq 0$ and $m > t$ the points $[\mathbf{x}_n]_{b,m}$ with $kb^m \leq n < (k+1)b^m$ form a (t, m, s)-net in base b. Here the coordinates of all points $\mathbf{x}_n$, $n = 0, 1, \ldots$, are given by prescribed b-adic expansions.

14.8.47 Remark It is easily seen that a (t, s)-sequence in base b is also a (u, s)-sequence in base b for all integers $u \geq t$. Smaller values of t mean stronger uniformity properties of a (t, s)-sequence in base b. The number t is the *quality parameter* of a (t, s)-sequence in base b.

14.8.48 Definition A (t, s)-sequence S in base b is a *strict* (t, s)*-sequence in base* b if t is the least integer u such that S is a (u, s)-sequence in base b.

14.8.49 Example Let $s = 1$ and let $b \geq 2$ be an integer. For $n = 0, 1, \ldots$, let $n = \sum_{r=0}^{\infty} a_r(n) b^r$ with all $a_r(n) \in \mathbb{Z}_b$ and $a_r(n) = 0$ for all sufficiently large r be the digit expansion of n in base b. Put $\phi_b(n) = \sum_{r=0}^{\infty} a_r(n) b^{-r-1}$. Then the sequence $\phi_b(0), \phi_b(1), \ldots$ is a $(0, 1)$-sequence in base b. This sequence is the *van der Corput sequence* in base b.

14.8.50 Proposition [2236] Given a (t, s)-sequence in base b, we can construct a $(t, m, s+1)$-net in base b for any integer $m \geq t$.

14.8.51 Remark The following result is obtained by combining Corollary 14.8.11 and Proposition 14.8.50.

14.8.52 Corollary [2236] A $(0, s)$-sequence in base b can exist only if $s \leq M(b) + 1$. In particular, a necessary condition for the existence of a $(0, s)$-sequence in base b is $s \leq b$.

14.8.53 Remark It was shown in [2238] that for any integers $b \geq 2$ and $s \geq 1$ there exists a (t, s)-sequence in base b for some value of t. Therefore it is meaningful to define $t_b(s)$ as the least value of t for which there exists a (t, s)-sequence in base b.

14.8.54 Theorem [2283, 2565] For any integers $b \geq 2$ and $s \geq 1$, we have

$$t_b(s) \geq \frac{s}{b-1} - c_b \log(s+1)$$

with a constant $c_b > 0$ depending only on b.

14.8.55 Definition [1858] Let $b \geq 2$ and $s \geq 1$ be integers and let $\mathbb{N}_0$ denote the set of nonnegative integers. Let $\mathbf{T} : \mathbb{N} \to \mathbb{N}_0$ be a function with $\mathbf{T}(m) \leq m$ for all $m \in \mathbb{N}$. Then a sequence $\mathbf{x}_0, \mathbf{x}_1, \ldots$ of points in $[0, 1]^s$ is a $(\mathbf{T}, s)$*-sequence in base* b if for all $k \in \mathbb{N}_0$ and $m \in \mathbb{N}$, the points $[\mathbf{x}_n]_{b,m}$ with $kb^m \leq n < (k+1)b^m$ form a $(\mathbf{T}(m), m, s)$-net in base b. Here the coordinates of all points $\mathbf{x}_n$, $n = 0, 1, \ldots$, are given by prescribed b-adic expansions. A $(\mathbf{T}, s)$-sequence S in base b is a *strict* $(\mathbf{T}, s)$*-sequence in base* b if there is no function $\mathbf{U} : \mathbb{N} \to \mathbb{N}_0$ with $\mathbf{U}(m) \leq m$ for all $m \in \mathbb{N}$ and $\mathbf{U}(m) < \mathbf{T}(m)$ for at least one $m \in \mathbb{N}$ such that S is a $(\mathbf{U}, s)$-sequence in base b.

14.8.56 Remark If the function $\mathbf{T}$ in Definition 14.8.55 is such that for some integer $t \geq 0$ we have $\mathbf{T}(m) = m$ for $m \leq t$ and $\mathbf{T}(m) = t$ for $m > t$, then the concept of a $(\mathbf{T}, s)$-sequence in base b reduces to that of a (t, s)-sequence in base b.

14.8.5 Digital (t, s)-sequences and digital $(\mathbf{T}, s)$-sequences

14.8.57 Remark There is an analog of the digital method in Remark 14.8.13 for the construction of sequences. Let integers $b \geq 2$ and $s \geq 1$ be given. Then we choose:

1. a commutative ring R with identity and of cardinality b;
2. bijections $\psi_r : \mathbb{Z}_b \to R$ for $r = 0, 1, \ldots$, with $\psi_r(0) = 0$ for all sufficiently large r;
3. bijections $\eta_j^{(i)} : R \to \mathbb{Z}_b$ for $1 \leq i \leq s$ and $j \geq 1$;
4. $\infty \times \infty$ matrices $C^{(1)}, \ldots, C^{(s)}$ over R.

For $n = 0, 1, \ldots$, let $n = \sum_{r=0}^{\infty} a_r(n) b^r$ with all $a_r(n) \in \mathbb{Z}_b$ and $a_r(n) = 0$ for all sufficiently large r be the digit expansion of n in base b. We put $\mathbf{n} = (\psi_r(a_r(n)))_{r=0}^{\infty} \in R^{\infty}$. Next we define

$$y_{n,j}^{(i)} = \eta_j^{(i)}(\mathbf{c}_j^{(i)} \cdot \mathbf{n}) \in \mathbb{Z}_b \qquad \text{for } n \geq 0,\ 1 \leq i \leq s, \text{ and } j \geq 1,$$

where $\mathbf{c}_j^{(i)}$ is the j-th row of the matrix $C^{(i)}$. Note that the inner product $\mathbf{c}_j^{(i)} \cdot \mathbf{n}$ is meaningful since $\mathbf{n}$ has only finitely many nonzero coordinates. Then we put

$$x_n^{(i)} = \sum_{j=1}^{\infty} y_{n,j}^{(i)} b^{-j} \qquad \text{for } n \geq 0 \text{ and } 1 \leq i \leq s.$$

Finally, we define the sequence S consisting of the points

$$\mathbf{x}_n = (x_n^{(1)}, \ldots, x_n^{(s)}) \in [0,1]^s \qquad \text{for } n = 0, 1, \ldots.$$

14.8.58 **Definition** If the sequence S constructed in Remark 14.8.57 forms a (t, s)-sequence in base b, then S is a *digital (t, s)-sequence in base b*. If we want to emphasize that the construction uses the ring R, then we speak also of a *digital (t, s)-sequence over R*. If S is a strict (t, s)-sequence in base b, then S is a *digital strict (t, s)-sequence in base b* (or *over R*).

14.8.59 **Definition** If the sequence S constructed in Remark 14.8.57 forms a (strict) $(\mathbf{T}, s)$-sequence in base b, then S is a *digital (strict) $(\mathbf{T}, s)$-sequence in base b* (or *over R*).

14.8.60 **Remark** The matrices $C^{(1)}, \ldots, C^{(s)}$ in Remark 14.8.57 are *generating matrices* of the digital sequence. The value of t for a digital (t, s)-sequence and the function $\mathbf{T}$ for a digital $(\mathbf{T}, s)$-sequence depend only on the generating matrices. For the determination of t in the general case, we refer to Theorem 4.35 in [2248]. For the case $R = \mathbb{F}_q$, see Theorem 14.8.62 below.

14.8.61 **Example** Let $s = 1$ and let $b \geq 2$ be an integer. Choose $R = \mathbb{Z}_b$ and let the bijections ψ_r and $\eta_j^{(i)}$ in Remark 14.8.57 be identity maps. Let $C^{(1)}$ be the $\infty \times \infty$ identity matrix over $\mathbb{Z}_b$. Then the van der Corput sequence in Example 14.8.49 is easily seen to be a digital $(0, 1)$-sequence over $\mathbb{Z}_b$ with generating matrix $C^{(1)}$.

14.8.62 **Theorem** Let S be the sequence constructed in Remark 14.8.57 with $R = \mathbb{F}_q$ and $\infty \times \infty$ generating matrices $C^{(1)}, \ldots, C^{(s)}$ over $\mathbb{F}_q$. For $1 \leq i \leq s$ and $m \in \mathbb{N}$, let $C_m^{(i)}$ denote the left upper $m \times m$ submatrix of $C^{(i)}$. Then S is a digital strict $(\mathbf{T}, s)$-sequence over $\mathbb{F}_q$ with $\mathbf{T}(m) = m - \varrho(C_m^{(1)}, \ldots, C_m^{(s)})$ for all $m \in \mathbb{N}$, where $\varrho(C_m^{(1)}, \ldots, C_m^{(s)})$ is given by Definition 14.8.17.

14.8.63 **Remark** It was shown in [2238] that for any prime power q and any integer $s \geq 1$, there exists a digital (t, s)-sequence over $\mathbb{F}_q$ for some value of t. In analogy with $t_b(s)$ in Remark 14.8.53, we define $d_q(s)$ as the least value of t for which there exists a digital (t, s)-sequence over $\mathbb{F}_q$. It is trivial that $t_q(s) \leq d_q(s)$, and so Theorem 14.8.54 provides also a lower bound on $d_q(s)$.

14.8.64 **Problem** With the previous notation, it is an open problem whether we can ever have $t_q(s) < d_q(s)$.

14.8.65 **Remark** An analog of the duality theory for digital nets described in Subsection 14.8.2 was developed in [838] for the case of digital $(\mathbf{T}, s)$-sequences. Let the weight function v_m on

$\mathbb{F}_q^m$ be as in Definition 14.8.22. For $\mathbf{A} = (\mathbf{a}^{(1)}, \ldots, \mathbf{a}^{(s)}) \in \mathbb{F}_q^{ms}$ with $\mathbf{a}^{(i)} \in \mathbb{F}_q^m$ for $1 \le i \le s$, we put

$$U_m(\mathbf{A}) = \max_{1 \le i \le s} v_m(\mathbf{a}^{(i)}).$$

14.8.66 Definition Let $s \ge 2$ be an integer. For each integer $m \ge 1$, let $\mathcal{N}_m$ be an $\mathbb{F}_q$-linear subspace of $\mathbb{F}_q^{ms}$ with $\dim(\mathcal{N}_m) \ge ms - m$. Let $\mathcal{N}_{m+1,m}$ be the projection of the set $\{\mathbf{A} \in \mathcal{N}_{m+1} : U_{m+1}(\mathbf{A}) \le m\}$, where $\mathbf{A} = (\mathbf{a}^{(1)}, \ldots, \mathbf{a}^{(s)})$ with all $\mathbf{a}^{(i)} \in \mathbb{F}_q^{m+1}$, on the first m coordinates of each $\mathbf{a}^{(i)}$ for $1 \le i \le s$. Suppose that $\mathcal{N}_{m+1,m}$ is an $\mathbb{F}_q$-linear subspace of $\mathcal{N}_m$ with $\dim(\mathcal{N}_{m+1,m}) \ge \dim(\mathcal{N}_m) - 1$ for all $m \ge 1$. Then the sequence $(\mathcal{N}_m)_{m \ge 1}$ of spaces is a *dual space chain* over $\mathbb{F}_q$.

14.8.67 Theorem [838] Let $s \ge 2$ be an integer. Then from any dual space chain $(\mathcal{N}_m)_{m \ge 1}$ over $\mathbb{F}_q$ we can construct a digital strict $(\mathbf{T}, s)$-sequence over $\mathbb{F}_q$ with $\mathbf{T}(m) = m + 1 - \delta_m(\mathcal{N}_m)$ for all $m \ge 1$.

14.8.68 Remark In analogy with the generalized digital nets mentioned in Remark 14.8.28, there are generalized digital sequences as introduced in [835] and also studied in Chapter 15 of [839].

14.8.69 Remark The nonlinear digital method described in Remark 14.8.29 can be used also for the construction of (t, s)-sequences [2258].

14.8.6 Constructions of (t, s)-sequences and $(\mathbf{T}, s)$-sequences

14.8.70 Remark A general family of digital (t, s)-sequences is formed by *Niederreiter sequences* [2238]. We describe only the simplest case of this construction. For a given dimension $s \ge 1$, let $p_1, \ldots, p_s \in \mathbb{F}_q[x]$ be pairwise coprime polynomials over $\mathbb{F}_q$. Let $e_i = \deg(p_i) \ge 1$ for $1 \le i \le s$. For $1 \le i \le s$ and integers $u \ge 1$ and $0 \le k < e_i$, consider the Laurent series expansion

$$\frac{x^k}{p_i(x)^u} = \sum_{r=0}^{\infty} a^{(i)}(u, k, r) x^{-r-1} \in \mathbb{F}_q((x^{-1})).$$

Then define $c_{j,r}^{(i)} = a^{(i)}(Q+1, k, r) \in \mathbb{F}_q$ for $1 \le i \le s$, $j \ge 1$, and $r \ge 0$, where $j-1 = Qe_i + k$ with integers $Q = Q(i,j)$ and $k = k(i,j)$ satisfying $0 \le k < e_i$. The generating matrices of the Niederreiter sequence are now given by $C^{(i)} = (c_{j,r}^{(i)})_{j \ge 1, r \ge 0}$ for $1 \le i \le s$. The bijections ψ_r and $\eta_j^{(i)}$ in Remark 14.8.57 are chosen arbitrarily.

14.8.71 Theorem [837, 2238] The Niederreiter sequence based on the pairwise coprime nonconstant polynomials $p_1, \ldots, p_s \in \mathbb{F}_q[x]$ is a digital strict (t, s)-sequence over $\mathbb{F}_q$ with $t = \sum_{i=1}^{s} (\deg(p_i) - 1)$.

14.8.72 Remark If q is a prime, $1 \le s \le q$, and $p_i(x) = x - i + 1 \in \mathbb{F}_q[x]$ for $1 \le i \le s$, then we obtain the digital $(0, s)$-sequences over $\mathbb{F}_q$ called *Faure sequences* [1045]. An analogous construction of digital $(0, s)$-sequences over $\mathbb{F}_q$ for arbitrary prime powers q and dimensions $1 \le s \le q$ was given in [2236]. Note that in view of Corollary 14.8.52, $s \le q$ is also a necessary condition for the existence of a $(0, s)$-sequence in base q. If $q = 2$, $s \ge 1$ is an arbitrary dimension, $p_1(x) = x \in \mathbb{F}_2[x]$, and $p_2, \ldots, p_s$ are distinct primitive polynomials over $\mathbb{F}_2$, then we obtain *Sobol' sequences* [2690].

14.8.73 Remark The construction of Niederreiter sequences in Remark 14.8.70 is optimized by letting $p_1, \ldots, p_s$ be s distinct monic irreducible polynomials over $\mathbb{F}_q$ of least degrees. If with this choice we put $T_q(s) = \sum_{i=1}^{s} (\deg(p_i) - 1)$, then for fixed q the quantity $T_q(s)$ is of the order of magnitude $s \log s$ as $s \to \infty$. Let $U(s)$ denote the least value of t that is known to be achievable by Sobol' sequences for given s. Then $T_2(s) = U(s)$ for $1 \le s \le 7$ and $T_2(s) < U(s)$ for all $s \ge 8$.

14.8.74 Remark Substantial improvements on the construction of Niederreiter sequences in Remark 14.8.70 can be obtained by using tools from the theory of algebraic function fields over finite fields (see Section 12.1 for this theory). This leads to the family of *Niederreiter-Xing sequences* [2282, 3018]. Let F be an algebraic function field with full constant field $\mathbb{F}_q$ and genus g. Assume that F contains at least one rational place P_∞ and let D be a divisor of F with $\deg(D) = 2g$ and $P_\infty \notin \operatorname{supp}(D)$. Furthermore, we choose s distinct places $P_1, \ldots, P_s$ of F with $P_i \ne P_\infty$ for $1 \le i \le s$. There exist integers $0 = n_0 < n_1 < \cdots < n_g \le 2g$ such that

$$\ell(D - n_u P_\infty) = \ell(D - (n_u + 1)P_\infty) + 1 \qquad \text{for } 0 \le u \le g.$$

We choose

$$w_u \in \mathcal{L}(D - n_u P_\infty) \setminus \mathcal{L}(D - (n_u + 1)P_\infty) \qquad \text{for } 0 \le u \le g.$$

For each $i = 1, \ldots, s$, we consider the chain $\mathcal{L}(D) \subset \mathcal{L}(D + P_i) \subset \mathcal{L}(D + 2P_i) \subset \cdots$ of vector spaces over $\mathbb{F}_q$. By starting from the basis $\{w_0, w_1, \ldots, w_g\}$ of $\mathcal{L}(D)$ and successively adding basis vectors at each step of the chain, we obtain for each $n \ge 1$ a basis

$$\{w_0, w_1, \ldots, w_g, f_1^{(i)}, f_2^{(i)}, \ldots, f_{n \deg(P_i)}^{(i)}\}$$

of $\mathcal{L}(D + nP_i)$. For a prime element z at P_∞ and for $r = 0, 1, \ldots$, we put $z_r = z^r$ if $r \notin \{n_0, n_1, \ldots, n_g\}$ and $z_r = w_u$ if $r = n_u$ for some $u \in \{0, 1, \ldots, g\}$. Each $f_j^{(i)}$ with $1 \le i \le s$ and $j \ge 1$ has then a local expansion at P_∞ of the form $f_j^{(i)} = \sum_{r=0}^{\infty} a_{j,r}^{(i)} z_r$ with all $a_{j,r}^{(i)} \in \mathbb{F}_q$. Let $\mathbf{c}_j^{(i)}$ be the sequence obtained from the sequence $a_{j,r}^{(i)}$, $r = 0, 1, \ldots$, by deleting the terms with $r = n_u$ for some $u \in \{0, 1, \ldots, g\}$. For $1 \le i \le s$, the generating matrix $C^{(i)}$ of the Niederreiter-Xing sequence is now the matrix whose j-th row is $\mathbf{c}_j^{(i)}$ for $j \ge 1$. The bijections ψ_r and $\eta_j^{(i)}$ in Remark 14.8.57 are chosen arbitrarily.

14.8.75 Theorem [3018] Let F be an algebraic function field with full constant field $\mathbb{F}_q$ and genus g which contains at least one rational place P_∞. Let D be a divisor of F with $\deg(D) = 2g$ and $P_\infty \notin \operatorname{supp}(D)$ and let $P_1, \ldots, P_s$ be distinct places of F with $P_i \ne P_\infty$ for $1 \le i \le s$. Then the corresponding Niederreiter-Xing sequence is a digital (t, s)-sequence over $\mathbb{F}_q$ with $t = g + \sum_{i=1}^{s}(\deg(P_i) - 1)$.

14.8.76 Corollary [2282] For every prime power q and every dimension $s \ge 1$, there exists a digital $(V_q(s), s)$-sequence over $\mathbb{F}_q$, where $V_q(s) = \min\{g \ge 0 : N_q(g) \ge s + 1\}$ and $N_q(g)$ is the maximum number of rational places that an algebraic function field with full constant field $\mathbb{F}_q$ and genus g can have.

14.8.77 Remark It was shown in [2282] that $V_q(s) = O(s)$ as $s \to \infty$. Since $t_q(s) \le d_q(s) \le V_q(s)$ by Remark 14.8.63 and Corollary 14.8.76, we obtain $t_q(s) = O(s)$ and $d_q(s) = O(s)$ as $s \to \infty$. In view of Theorem 14.8.54, these asymptotic bounds are best possible.

14.8.78 Remark The only improvements on Niederreiter-Xing sequences were obtained, in some special cases, in the more recent paper [2266]. For instance, let q be an arbitrary prime power and let $s = q+1$. Then $t_q(q+1) = d_q(q+1) = 1$. On the other hand, the construction in [2266] yields a digital $(\mathbf{T}, q+1)$-sequence over $\mathbb{F}_q$ with $\mathbf{T}(m) = 0$ for even $m \ge 2$ and $\mathbf{T}(m) = 1$ for odd $m \ge 1$.

See Also

§14.1	For orthogonal arrays and related combinatorial designs.
§15.1	For related constructions of linear codes.

References Cited: [271, 276, 835, 836, 837, 838, 839, 1045, 1858, 1873, 1874, 2007, 2008, 2009, 2183, 2234, 2236, 2238, 2247, 2248, 2251, 2255, 2256, 2258, 2259, 2261, 2263, 2266, 2267, 2282, 2283, 2398, 2483, 2565, 2690, 3018]

14.9 Applications and weights of multiples of primitive and other polynomials

Brett Stevens, Carleton University

14.9.1 Applications where weights of multiples of a base polynomial are relevant

14.9.1 Remark The performance of several applications of polynomials, frequently primitive, depend on the weights of multiples of the base polynomial. Many of these applications are discussed in this Handbook.

14.9.1.1 Applications from other Handbook sections

14.9.2 Remark The multiples of a polynomial f with weight w influence the statistical bias of the linear feedback shift register sequence generated from f. Fewer multiples with a given weight, w reduces the w-th moment of the Hamming weight [1622, 1944]. For more information on bias and randomness of linear feedback shift register sequences see Section 10.2.

14.9.3 Remark In Section 15.1 the use of primitive polynomials f, to generate *cyclic redundancy check codes* is discussed. The undetectable error patterns of these codes are precisely those whose errors correspond to multiples of f. This has the consequences that burst errors of length up to $\deg(f)$ are always detectable and that to understand how many arbitrary errors can be detected requires having knowledge of the weights of multiples of f.

14.9.4 Remark In Section 15.4, turbo codes are discussed. Turbo codes use feedback polynomials that are often primitive. The bit error rate (BER) of the turbo code's interleaver design depends on the weights of polynomials divisible by the feedback polynomial [2513].

14.9.5 Remark Low weight multiples of a public polynomial compromise the private key for the $\mathcal{TCH}o$ cryptosystem and its security therefore rests on the difficulty of finding low weight multiples [146, 1491]. The weight of polynomials and their multiples is important in linear feedback shift register cryptanalysis and certain attacks depend on the sparsity of the feedback polynomial or one of its multiples [2074]. Chapter 16 discusses the many connections between finite fields and cryptography.

14.9.1.2 Application of polynomials to the construction of orthogonal arrays

14.9.6 Remark We present a discussion of applications of polynomials and the weights of their multiples to the construction and strength of orthogonal arrays.

14.9.7 Definition An *orthogonal array* of *size* N, with k constraints (or k *factors* or of *degree* k), s *levels* (or of *order* s), and *strength* t, denoted $\mathrm{OA}(N,k,s,t)$, is a $k \times N$ array (sometimes $N \times k$) with entries from a set of $s \geq 2$ symbols, having the property that in every $t \times N$ submatrix, every $t \times 1$ column vector appears the same number $\lambda = \frac{N}{s^t}$ of times. The parameter λ is the *index* of the orthogonal array. An $\mathrm{OA}(N,k,s,t)$ is also denoted by $\mathrm{OA}_\lambda(t,k,s)$.

14.9.8 Remark From the definition, an orthogonal array of strength t is also an orthogonal array of strength t' for all $1 \leq t' \leq t$.

14.9.9 Theorem [357, 1457] Let C be a linear code over $\mathbb{F}_q$ with words of length n. Then the $n \times |C|$ array formed with the words of C as the columns is a (linear) orthogonal array of maximal strength t if and only if $C^\perp$, its dual code, has minimum weight $t+1$.

14.9.10 Remark The half of Theorem 14.9.9 that gives the strength of the orthogonal array from the minimum weight of the dual code was known as early as 1947 [1457, 1727]. Delsarte was able to generalize Theorem 14.9.9 to the case where the code and the orthogonal array are not required to be linear [801]. We can extend Theorem 14.9.9 and exactly determine the number of times each vector appears in any $(t+1) \times n$ submatrix of the orthogonal array.

14.9.11 Theorem [2204] Let C be a linear code of length n over $\mathbb{F}_q$ and assume that the words of C form the columns of an orthogonal array of strength t. Then for any $t+1$-subset $T = \{i_1, \dots i_t\} \subset \{1, \dots, n\}$ and for any $t+1$-tuple $b \in \mathbb{F}_q^{t+1}$, the number of times that b appears as a column of the $(t+1) \times n$ submatrix determined by T, $\lambda_b^T(C)$, is

$$\lambda_b^T(C) = \begin{cases} |C|/q^t & \text{if there is no } u \in C^\perp \text{ with support } T; \\ |C|/q^{t-1} & \text{if there exists a } u \in C^\perp \text{ with support } T \text{ and } u_{i_j} = b_j \text{ for } i_j \in T; \\ 0 & \text{otherwise.} \end{cases}$$

14.9.12 Theorem [2204] Let f be a primitive polynomial of degree m over $\mathbb{F}_q$ and let C_n^f be the set of all subintervals of the shift-register sequence with length n generated by f, together with the zero vector of length n. The dual code of C_n^f is given by

$$(C_n^f)^\perp = \{(b_1, \dots, b_n) : \sum_{i=0}^{n-1} b_{i+1} x^i \text{ is divisible by } f\}.$$

14.9.13 Remark [2204] Munemasa only proves Theorem 14.9.12 over $\mathbb{F}_2$ but the proof works more generally for any finite field.

14.9.14 Remark [2354] The primitivity condition in Theorem 14.9.12 can be substantially relaxed to polynomials with distinct roots.

14.9.15 Remark The combined effect of Theorems 14.9.9 and 14.9.12 is that to know the strength of the orthogonal array derived from a polynomial f, and its shift register sequences, it is essential to know about the weights of multiples of f.

14.9.1.3 Application of polynomials to a card trick

14.9.16 Remark Although the weight of multiples of primitive $f = x^5 + x^2 + 1 \in \mathbb{F}_2[x]$ is not relevant to this application to a card trick, the low weight of f itself facilitates the mental arithmetic so we include this application in this subsection.

14.9.17 Remark The polynomial $f(x) = x^5 + x^2 + 1 \in \mathbb{F}_2[x]$ is primitive and generates the binary shift register sequence with the property that $a_{k+5} = a_k + a_{k+2}$

$$0000100101100111110001101110101.$$

14.9.18 Remark The set of cards from a standard deck which contains the Ace, 2, 3, 4, 5, 6, and 7 of each suit and the 8 of spades, clubs, and hearts can be encoded uniquely with the non-zero binary words of length 5. The first digit encodes the cards color, 0 for red and 1 for black. The second digit encodes whether the suit is major or minor in bridge: 0 for clubs or diamonds; 1 for hearts or spades. The remaining three digits encode the value of the card via the last three digits in the binary representation of the card's value: 000 for 8, 001 for Ace, 010 for 2, 011 for 3, 100 for 4, 101 for 5, 110 for 6, and 111 for 7. This encoding has the property that the first digit in a card's code corresponds to the color of that card. Other encodings have the required properties as well [833].

14.9.19 Remark Using the shift register sequence from Remark 14.9.17 and the card encoding from Remark 14.9.18 we obtain the following sequence of cards:

$$A\diamondsuit, 2\diamondsuit, 4\diamondsuit, A\heartsuit, 2\clubsuit, 5\diamondsuit, 3\heartsuit, 6\clubsuit, 4\heartsuit, A\spadesuit, 3\clubsuit, 7\diamondsuit, 7\heartsuit, 7\spadesuit, 6\spadesuit, 4\spadesuit,$$
$$8\spadesuit, A\clubsuit, 3\diamondsuit, 6\diamondsuit, 5\heartsuit, 3\spadesuit, 7\clubsuit, 6\heartsuit, 5\spadesuit, 2\spadesuit, 5\clubsuit, 2\heartsuit, 4\clubsuit, 8\heartsuit, 8\clubsuit$$

14.9.20 Remark A deck of these 31 cards arranged in this order looks upon casual inspection to be randomly ordered. The deck can be cut arbitrarily many times (since the shift register sequence is cyclic) before removing five cards in sequence from the top of the deck. With the knowledge which cards are black, the identity of all five chosen cards can be determined [832].

14.9.21 Remark Due to the low weight of primitive $f = x^5 + x^2 + 1$, the encoding scheme and the generating polynomial are simple enough to be quickly calculated mentally which is important for the appearance of the trick [832].

14.9.22 Remark Much can be done to augment the impression this trick makes on an audience. For ideas see [832, 833, 2169]. Two sets of these 31 cards with identical backs can be placed in this order repeated to give the impression of a more normal sized deck.

14.9.23 Remark The $8\diamondsuit$, corresponding to the binary string 00000, can be added to the deck between the $8\clubsuit$ and $A\diamondsuit$. This deviation from the linear shift register can simply be memorized ad-hoc or a new, nonlinear shift register sequence memorized:

$$a_{k+5} = (1 + \overline{a_{k+1}} \cdot \overline{a_{k+2}} \cdot \overline{a_{k+3}} \cdot \overline{a_{k+4}})(a_k + a_{k+2}) + (\overline{a_k} \cdot \overline{a_{k+1}} \cdot \overline{a_{k+2}} \cdot \overline{a_{k+3}} \cdot \overline{a_{k+4}}),$$

where $\overline{a_i}$ is the complement of a_i.

14.9.24 Remark For other mathematical card tricks see [218, 832, 833, 2169].

14.9.25 Remark The applications discussed in Remarks 14.9.2 through 14.9.15 strongly motivate researching the distributions and patterns of weights of multiples of polynomials f over finite fields. Subsection 14.9.2 gives a summary of the knowledge in this area to date.

14.9.2 Weights of multiples of polynomials

14.9.26 Remark Polynomials in $\mathbb{F}_2[x]$ which have large weight or large degree will sometimes be given in hexadecimal notation. For example the polynomial

$$f = x^8 + x^7 + x^6 + x^5 + x + 1 = 1x^8 + 1x^7 + 1x^6 + 1x^5 + 0x^4 + 0x^3 + 0x^2 + 1x^1 + 1x^0$$

is 111100011 in binary notation and, grouping these from the right into fours, is 1E3 in hexadecimal notation. The use of the two notations for polynomials will always be clear in the context. Be aware that some authors in the literature denote binary polynomials in hexidecimal *after deleting* the rightmost 1, since most polynomials used in applications have a constant term 1, so it can be assumed present in many contexts.

14.9.27 Definition The set of polynomials of degree d in $\mathbb{F}_q[x]$ is denoted by $\mathbf{P}_{q,d}$. For $f \in \mathbb{F}[x]$, the dual code of length n, $(C_n^f)^\perp$, defined in Theorem 14.9.12 can be identified with all polynomials divisible by f of degree less than n. The minimum weight of a polynomial from this set is denoted by $d((C_n^f)^\perp)$. This is also the minimum weight of the code $(C_n^f)^\perp$.

14.9.28 Remark We begin with some general bounds on $d((C_n^f)^\perp)$, followed by results for polynomials f of specific degree and end with results for polynomials f of specific weights.

14.9.2.1 General bounds on $d((C_n^f)^\perp)$

14.9.29 Proposition An application of Theorem 14.9.12 with bounds on the period of polynomials gives that if $f \in \mathbf{P}_{q,m}$, then $d((C_n^f)^\perp) = 2$ for all $n \geq q^m - 1$.

14.9.30 Theorem [2083] Let $f \in \mathbf{P}_{2,m}$ and $0 \leq t \leq (m-1)/2$. Let $n_1(t)$ be the smallest positive integer such that

$$\sum_{j=0}^{t+1} \binom{n_1(t)}{j} > 2^m.$$

Set $n_2(0) = \infty$ and for $t > 0$, let $n_2(t) = 2^{\lfloor (m-1)/t \rfloor} - 1$. If $n_1(t) < n_2(t)$, then for all $n_1(t) \leq n \leq n_2(t)$, $d((C_n^f)^\perp) \leq 2t+2$. In other words, for such n, there will always be a multiple of f of weight less than $2t+3$ and degree less than n.

14.9.31 Theorem [2083] Let $e = \lfloor (m-1)/t \rfloor$ and $n_2(t) = 2^e - 1$. Let α be a primitive element in $\mathbb{F}_{2^e}$ and $M^{(i)}(x)$ be the minimal polynomial of α^i. Let

$$g = \text{lcm}\{M^{(i)}(x) | 0 \leq i \leq 2t\},$$

then $d((C_{2^e-1}^g)^\perp) \geq 2t+2$ and the BCH code (see Section 15.1) generated by g can be truncated to a code meeting the bound in Theorem 14.9.30 for any admissible $n_1(t) \leq n \leq n_2(t)$.

14.9.32 Proposition [2083] In Theorem 14.9.30, $n_1(t) \leq t + 2^{m/(t+1)}(t+1)!^{1/(t+1)}$, and whenever $m > (t+1)^2 + t(t+1)\log_2(t+1)$, we have $n_1(t) < n_2(t)$.

14.9.33 Theorem [1591] If $f \in \mathbf{P}_{2,m}$ is primitive and if $g = x^n + x^k + 1$ is the trinomial multiple of f with minimum degree then

$$n \leq \frac{2^m + 2}{3}.$$

14.9.34 Proposition If $x+1$ is a factor of $f \in \mathbb{F}_2[x]$ then f does not divide any polynomials of odd weight.

14.9.35 Lemma [558] Let $f \in \mathbb{F}_2[x]$ have simple roots, period n and suppose $(1+x)$ is a factor of f. If $d((C_n^f)^\perp) = d$ then $d((C_{n+1}^{(x+1)f})^\perp) = d$ and $d((C_j^{(x+1)f})^\perp) = 4$ for $n+2 \le j \le 2n$.

14.9.36 Theorem [1591] Let $f \in \mathbb{F}_2[x]$ be an irreducible polynomial with period ρ. Then f divides a trinomial if and only if $\gcd(x^\rho + 1, (x+1)^\rho + 1) \ne 1$.

14.9.37 Theorem [1591] If $f \in \mathbf{P}_{2,m}$ is primitive and if $g = x^n + x^k + 1$ is a trinomial divisible by f then n and k belong to the same-length cyclotomic coset modulo $2^m - 1$.

14.9.38 Theorem [1591] All primitive $f \in \mathbf{P}_{2,m}$ divide some 4-nomial of degree no bigger than

$$\left\lfloor \frac{1 + \sqrt{1 + 4.2^{m+1}}}{2} \right\rfloor.$$

14.9.39 Theorem [1591] For a given $t \ge 2$ and $s \ge 1$, if m is such that

$$1.548^m - 1 \ge (t-1)\binom{m^s+1}{t}$$

then there exists at least one primitive polynomial of degree m which does not divide any t-nomial of degree less than or equal to m^s.

14.9.40 Theorem [1994] Let $f \in \mathbf{P}_{2,m}$ be a primitive t-nomial. Then there exists a primitive $g \in \mathbf{P}_{2,m}$ which divides some t-nomial of degree sm (s odd) when $\gcd(s, 2^m - 1) = 1$. Moreover $g = \gcd(f(x^s), x^{2^m-1} - 1)$ is such a polynomial.

14.9.41 Remark The previous results give information about multiples of f that can have small degree relative to the period of f. The following gives information about multiples of f that have relatively large degree.

14.9.42 Remark Let $f \in \mathbb{F}_q[x]$ be primitive of degree m. Let $n = q^m - 1$ and T_2 be the set of binomials of the form $x^i - x^j$, satisfying $0 \le i < j$ with $i \equiv j \pmod{n}$. Let T_i be the set of all linear combinations of binomials from T_2 which have weight i. Finally define $\mu : \mathbb{F}_q[x] \to \mathbb{F}_q[x]$ by $\mu(\sum_{i=0}^d a_i x^i) = \sum_{i=0}^d a_i x^{i \pmod{n}}$.

14.9.43 Theorem [2513] Suppose that g is a polynomial of weight w and write

$$g = g_1 + g_2$$

where $g_1 \in T_i$, $g_2 \in \mathbb{F}_q[x]$ has weight j, no two exponents of g_2 are congruent modulo n, and the terms of g_1 and g_2 are disjoint (i.e., $w = i + j$). Then g is divisible by f if and only if $\mu(g_2)$ is divisible by f.

14.9.44 Remark In [2513], Theorem 14.9.43 is only stated and proved over $\mathbb{F}_2$. It is true for all finite fields $\mathbb{F}_q$.

14.9.45 Remark Using the fact that f divides $x^{n'} + a$, with $n' = (q^m - 1)/(q-1)$ for some unique $a \in \mathbb{F}_q$, and letting T being the set of corresponding binomials, Theorem 14.9.43 can be further generalized with an increase in the complexity of the statement.

14.9.46 Remark There have also been some interesting results on enumeration and probability of multiples with given weights. We discuss this next.

14.9.47 Theorem [1591] Given $t \ge 2$ and $s \ge 1$, if m is such that $\phi(2^m - 1) > (t-1)\binom{m^s+1}{t}$, then the probability that a randomly chosen primitive polynomial of degree m does not divide any t-nomial of degree less than or equal to m^s is at least

$$1 - \frac{(t-1)(\binom{m^s+1}{t})}{\phi(2^m - 1)},$$

where ϕ denotes Euler's function.

14.9.48 Theorem [1591] There exist primitive polynomials of degree m which divide a trinomial of degree $3m$ and a 4-nomial of degree less than $6m$.

14.9.49 Theorem [1362] Let $N_{m,t}$ denote the number of t-nomial multiples with degree no more than $2^m - 2$ of a primitive polynomial of degree m. Then $N_{m,2} = N_{m,1} = 0$ and

$$N_{m,t} = \frac{\binom{2^m-2}{t-2} - N_{m,t-1} - \frac{t-1}{t-2}(2^m - t + 1)N_{m,t-2}}{t-1}.$$

14.9.50 Remark See [1362] for discussion and results for solving this recurrence, and [1834] for an alternative presentation.

14.9.51 Theorem [1362] Given any primitive polynomial of degree m, the sum of the degrees of all its t-nomial multiples is

$$\frac{t-1}{t}(2^m - 1)N_{m,t}.$$

14.9.52 Theorem [1362] Given any primitive polynomial f of degree m, the average degree of its t-nomial multiples with degree no more than $2^m - 2$ is equal to the average of the maximum of all the distinct $(t-1)$-tuples from 1 to $2^m - 2$.

14.9.53 Theorem [1362] Given a primitive polynomial f of degree m, under the assumption that t-nomial multiples of f are distributed as $(t-1)$-tuples, there exists a t-nomial multiple g of f such that the degree of g is less than or equal to

$$2^{\frac{m}{t-1} + \log_2(t-1) + 1}.$$

14.9.54 Remark The assumption in Theorem 14.9.53 is motivated by Theorem 14.9.52 and empirical evidence. See [1362] for precise definition of the assumption and detailed discussion.

14.9.55 Remark Theorem 14.9.53 implies that it is highly likely to get a trinomial multiple with degree no more than $2^{m/2+2}$. This is in contrast to the bound of $(2^m + 2)/3$ from Theorem 14.9.33. In general Theorem 14.9.53 suggests that to avoid sparse multiples, f must be picked with very large degree.

14.9.56 Remark In [1994], Maitra, Gupta, and Venkateswarlu extend this enumerative and probabilistic analysis to include the product of primitive polynomials.

14.9.2.2 Bounds on $d((C_n^f)^\perp)$ for polynomials of specific degree

14.9.57 Proposition The bounds on weights of multiples of *all* polynomials from degree 4 to degree 16 and degrees 24 and 32 in $\mathbb{F}_2[x]$ are given in Table 14.9.2.2. For degrees $4 \leq m \leq 16$, Koopman and Chakravarty exhaustively searched all polynomials of degree m and all their multiples of degrees $m + 8 \leq n \leq m + 2048$ [1793]. The $m = 16$ results are from the theoretical work of Merkey and Posner [2083] and exhaustive searches by Castagnoli, Ganz, and Graber [559]. The bounds on weights of multiples of degree 24 polynomials, which are less complete than those for smaller m, are the work of Merkey and Posner [2083] and searches by Castagnoli, Ganz, and Graber [559] and Ray and Koopman [2439]. In all cases for $m = 24$ polynomials attaining the bounds are reported to be known although the specific polynomials have not been published [559, 2083, 2439]. The even more incomplete results for $m = 32$ are reported in [559, 2083].

14.9.58 Example Table 14.9.2.2 gives bounds that apply to *every* polynomial with the given degree. To aid the reading of Table 14.9.2.2, we give an example from it. The information from the

deg(f)	degree range of multiples of f	upper bound on $d(C_n^f)^\perp$	polynomial (in hexadecimal notation) attaining the bound
4	$12 \le n \le 15$	3	$f = 13$
5	$13 \le n \le 15$	4	$f = 2B$
	$16 \le n \le 31$	3	$f = 25$
6	$14 \le n \le 31$	4	$f = 59$
	$32 \le n \le 63$	3	$f = 43$
7	$15 \le n \le 63$	4	$f = B7$
	$64 \le n \le 127$	3	$f = 91$
8	$16 \le n \le 17$	5	$f = 139$
	$18 \le n \le 127$	4	$f = 12F$
	$128 \le n \le 255$	3	$f = 14D$
9	$n = 17$	6	$f = 13C$
	$18 \le n \le 22$	5	$f = 30B$
	$23 \le n \le 255$	4	$f = 297$
	$256 \le n \le 511$	3	$f = 2CF$
10	$18 \le n \le 22$	6	$f = 51D$
	$23 \le n \le 31$	5	$f = 573$
	$32 \le n \le 511$	4	$f = 633$
	$512 \le n \le 1023$	3	$f = 64F$
11	$19 \le n \le 23$	7	$f = AE1$
	$24 \le n \le 33$	6	$f = A65$
	$34 \le n \le 36$	5	$f = BAF$
	$37 \le n \le 1023$	4	$f = B07$
	$1024 \le n \le 2047$	3	$f = C9B$
12	$20 \le n \le 23$	8	$f = 149F$
	$24 \le n \le 39$	6	$f = 1683$
	$40 \le n \le 65$	5	$f = 11F1$
	$66 \le n \le 2047$	4	$f = 180F$
	$2048 \le n \le 2060$	3	$f = 16EB$
13	$21 \le n \le 24$	8	$f = 216F$
	$n = 25$	7	$f = 254B$
	$26 \le n \le 65$	6	$f = 3213$
	$66 \le n \le 2061$	4	$f = 2055$
14	$22 \le n \le 25$	8	$f = 46E3$
	$26 \le n \le 27$	7	$f = 5153$
	$28 \le n \le 71$	6	$f = 6E57$
	$72 \le n \le 127$	5	$f = 425B$
	$128 \le n \le 2062$	4	$f = 43D1$
15	$23 \le n \le 27$	8	$f = C617$
	$28 \le n \le 31$	7	$f = B7AB$
	$32 \le n \le 129$	6	$f = AE75$
	$128 \le n \le 191$	5	$f = D51B$
	$192 \le n \le 2063$	4	$f = 92ED$
16	$n = 18$	12	$f = 15BED$
	$19 \le n \le 21$	10	$f = 1D22F$
	$n = 22$	9	$f = 18F57$
	$23 \le n \le 31$	8	$f = 11FB7$
	$32 \le n \le 35$	7	$f = 126B5$
	$36 \le n \le 151$	6	$f = 13D65$
	$152 \le n \le 257$	5	$f = 15935$
	$258 \le n \le 32767$	4	$f = 1A2EB$
	$32768 \le n \le 65535$	3	$f = 1002D$
24	$18 \le n \le 47$	12	
	$48 \le n \le 50$	10	
	$51 \le n \le 63$	9	
	$64 \le n \le 129$	8	
	$130 \le n \le 255$	7	
	$466 \le n \le 2^{11} - 1$	6	
	$5793 \le n \le 2^{23} - 1$	4	
32	$n = 18$	12	
	$568 \le n \le 2^{10} - 1$	8	
	$2954 \le n \le 2^{15} - 1$	6	
	$92682 \le n \le 2^{31} - 1$	4	

Table 14.9.1 Bounds on weights of multiples of degree n polynomials for $4 \le n \le 16$.

			minimum distance					
f	standard	$d(C_n^f)^\perp$:	12	11	10	9	8	7
13D65	IEC TC57 after 1990	ranges of n:			[17,20]		[21,22]	
1F29F		ranges of n:	17				[18,22]	
15B93	IEC TC57 before 1990	ranges of n:			[17,19]		[20,25]	
15935		ranges of n:				[17,19]	[20,24]	[25,26]
16F63	IEEE WG77.1	ranges of n:		17	18			[19,29]
1A2EB		ranges of n:			[17,18]		[19,27]	
1011B		ranges of n:						
1A097	IBM SDLC	ranges of n:					[17,24]	
11021	CRC-CCITT	ranges of n:						
18005	CRC-ANSI	ranges of n:						

			minimum distance			
f	standard	$d(C_n^f)^\perp$:	6	5	4	2
13D65	IEC TC57 after 1990	ranges of n:	[23,151]			[151,∞]
1F29F		ranges of n:	[23,130]		[131,258]	[259,∞]
15B93	IEC TC57 before 1990	ranges of n:	[26,128]	[129,254]		[255,∞]
15935		ranges of n:	[27,51]	[52,257]		[258,∞]
16F63	IEEE WG77.1	ranges of n:	30	[31,255]		[256,∞]
1A2EB		ranges of n:	[28,109]		[110,32767]	[32768,∞]
1011B		ranges of n:	[17,115]		[116,28658]	[28659,∞]
1A097	IBM SDLC	ranges of n:	[25,83]		[84,32766]	[32767,∞]
11021	CRC-CCITT	ranges of n:			[17,32767]	[32768,∞]
18005	CRC-ANSI	ranges of n:			[17,32767]	[32768,∞]

Table 14.9.2 Distance profiles of specific degree 16 polynomials.

third line of the section for polynomials of degree 11, indicates that for every binary degree 11 polynomial $f \in \mathbf{P}_{2,11}$, there exists multiples of f which have degrees 34, 35, and 36 and weight less than or equal to 5. The polynomial cited in the last column, $f(x) = BAF = x^{11}+x^9+x^8+x^7+x^5+x^3+x^2+x+1$, meets this bound tightly; that is, all of its multiples of degree 34, 35, or 36 have weight 5 or above.

14.9.59 Remark In Table 14.9.2.2, three of the degree 16 polynomials meeting the bounds are known to be unique. For $d(C_n^f)^\perp = 6$, $f = 13D65$ and for $d(C_n^f)^\perp = 4$, $f = 1A2EB$ are the unique tight polynomials, up to reciprocal. For $d(C_n^f)^\perp = 5$, $f = 15935$ is unique [559].

14.9.60 Remark In contrast to Table 14.9.2.2, Tables 14.9.2 through 14.9.4 give the distance distributions of multiples of a few, specific polynomials for degrees 16, 24, and 32.

14.9.61 Remark [559] Table 14.9.2 gives the distance profiles of ten specific polynomials in $\mathbf{P}_{2,16}$ found by Castagnoli, Ganz, and Graber. They exhaustively searched all degree 16 polynomials for those with optimum profiles. The polynomial $f = 1F29F$ is the unique polynomial with $d(C_{130}^f)^\perp = 6$ and $d(C_{258}^f)^\perp = 4$. Up to reciprocal, $f = 1011B$ is the unique polynomial with $d(C_{28658}^f)^\perp = 4$ and $d(C_{115}^f)^\perp = 6$. The authors of [559] suggest that any cyclic redundancy check polynomials of degree 16 should be chosen only from the list $\{13D65, 1F29F, 15935, 1A2EB, 1011B\}$.

14.9.62 Example The third polynomial in Table 14.9.2 gives the distance distribution for the polynomial $f(x) = 15B93 = x^{16}+x^{14}+x^{12}+x^{11}+x^9+x^8+x^7+x^4+x+1$, which was the IEC TC57 standard cyclic redundancy check polynomial until 1990. All its multiples of degrees 17–19 have weight 10 or more. All its multiples of degrees 20–25 have weight 8 or more. All its multiples of degrees 26–128 have weight 6 or more. All its multiples of degree 129–254 have weight 5 or more. All its multiples of degrees 255 and higher have weight at least 2. For each degree there exist specific multiples that attain these lower bounds; for example there is a degree 17 multiple of f with weight 10.

14.9.63 Remark Table 14.9.3 gives the distance distribution for some specific polynomials of degree 24. All were constructed via the generalized BCH code method: multiplying together

		minimum distance							
f	$d(C_n^f)^\perp$:	16	15	14	12	11	10	9	ref.
1323009	ranges of n:								[558, 2083]
1401607	ranges of n:								[558, 2083]
1805101	ranges of n:								[2083]
15D6DCB	ranges of n:			25	26		[27,36]		[558]
17B01BD	ranges of n:	[25,26]					[27,41]		[558]
131FF19	ranges of n:		25				[26,33]		[558]
15BC4F5	ranges of n:		[25,26]		[27,28]	[29,31]	[32,33]	[34,35]	[558]
1328B63	ranges of n:				[25,30]		[31,36]		[558]

		minimum distance						
f	$d(C_n^f)^\perp$:	8	7	6	5	4	2	ref.
1323009	ranges of n:	[25,68]		[69,2048]		[2049,4094]	[4095,∞]	[558, 2083]
1401607	ranges of n:	[25,55]		[56,2048]		[2049,4094]	[4095,∞]	[558, 2083]
1805101	ranges of n:			[25,1023]				[2083]
15D6DCB	ranges of n:	[37,83]		[84,2050]		[2051,4098]	[4099,∞]	[558]
17B01BD	ranges of n:	[42,95]		[96,2048]		[2049,4094]	[4095,∞]	[558]
131FF19	ranges of n:		[34,37]	[38,252]	[253,4097]		[4098,∞]	[558]
15BC4F5	ranges of n:	[36,41]	[42,47]	[78,217]	[218,4095]		[4096,∞]	[558]
1328B63	ranges of n:	[37,61]		[62,846]		[847,$2^{23}-1$]	[2^{23},∞]	[558]

Table 14.9.3 Distance profiles of specific degree 24 polynomials.

minimal polynomials of elements from $\mathbb{F}_{2^e}$ and small factors, x and $x+1$ [558, 2083]. For discussion of BCH codes, see Section 15.1.

14.9.64 Remark Table 14.9.4 gives the distance distribution for some specific polynomials of degree 32. All were obtained via the generalized BCH code method: multiplying together minimal polynomials of elements from $\mathbb{F}_{2^e}$ and small factors, x and $x+1$ [558, 2083].

14.9.65 Remark For the third polynomial in Table 14.9.4, used in many standards, Jain [1590] has determined and published many of the minimum degree polynomials that establish the ranges given in Table 14.9.4. The actual polynomials are given in Table 14.9.5. Jain has determined all the polynomials that f divides which have the pattern of at most three burst errors of length 4 each and several other specific patterns of errors.

14.9.66 Remark Koopman has performed an exhaustive search over all $f \in \mathbf{P}_{2,32}$ for $40 \leq n \leq 131104$. His primary concern was finding cyclic redundancy check polynomials which were simultaneously good at typical Ethernet *maximum transmission unit* (MTU) lengths, $n = 12112$, and much longer lengths $n \geq 64,000$, so although his search has in principle solved the $d(C_n^f)^\perp$ problem for all n in this range he did not specifically publish these, rather he highlights the last three polynomials given in Table 14.9.4 and compares them to the others [1792]. Discussion of the benefits and costs of using these various polynomials in different scenarios appear in [558, 1590, 1792, 2083].

14.9.2.3 Bounds on $d((C_n^f)^\perp)$ for polynomials of specific weight

14.9.67 Remark We now present divisibility results that are organized by the weight of the base polynomial.

14.9.68 Theorem [2204] Let $f(x) = x^m + x^l + 1$ be a trinomial over $\mathbb{F}_2$ such that $\gcd(m, l) = 1$. If g is a trinomial multiple of f of degree at most $2m$, then

1. $g(x) = x^{\deg g - m} f(x)$;
2. $g(x) = f(x)^2$;
3. $g(x) = x^5 + x^4 + 1 = (x^2+x+1)(x^3+x+1)$ or; its reciprocal,

		minimum distance												
f	$d(C_n^f)^\perp$:	20	18	17	16	15	14	13	12	11	10	9	8	ref.
1404098E2	ranges of n:										[33,78]		[79,1023]	[558, 2083]
10884C912	ranges of n:										[33,79]		[80,1023]	[558, 2083]
104C11DB7*	ranges of n:					[33,42]			[43,44]	[45,53]	[54,66]	[67,89]	[90,123]	[558, 1590]
1F1922815	ranges of n:						[33,44]		[45,48]		[49,98]		[99,1024]	[558]
1F4ACFB13	ranges of n:	33	[34,35]		36		37		[38,43]		[44,56]		[57,306]	[558]
1A833982B	ranges of n:					[33,35]			[36,49]	[50,53]	[54,59]		[60,90]	[558]
1572D7285	ranges of n:			[33,34]			35	[36,38]		[39,52]	[53,68]	[69,80]	[81,110]	[558]
11EDC6F41	ranges of n:		33		[34,38]		[39,40]		[41,52]		[53,79]		[80,209]	[558]
1741B8CD7	ranges of n:								[40,48]		[49,50]		[51,184]	[1792]
132583499	ranges of n:								[40,48]		[49,58]		[59,166]	[1792]
120044009	ranges of n:													[1792]
100210801	ranges of n:													[1792]

		minimum distance						
f	$d(C_n^f)^\perp$:	7	6	5	4	3	2	ref.
1404098E2	ranges of n:						[1024,∞]	[558, 2083]
10884C912	ranges of n:						[1024,∞]	[558, 2083]
104C11DB7*	ranges of n:	[124, 203]	[204,300]	[301,3006]	[3007,91639]	[91640,$2^{32}-1$]	[2^{32},∞]	[558, 1590]
1F1922815	ranges of n:				[1025,2046]		[2047,∞]	[558]
1F4ACFB13	ranges of n:		[307,32768]		[32769,65534]		[65535,∞]	[558]
1A833982B	ranges of n:	[91,113]	[114,1092]	[1093,65537]			[65538,∞]	[558]
1572D7285	ranges of n:	[111,266]	[267,1029]	[1030,65535]			[65536,∞]	[558]
11EDC6F41	ranges of n:		[210,5275]		[5276,$2^{31}-1$]		[2^{31},∞]	[558]
1741B8CD7	ranges of n:		[185,16392]		[16393,114695]		[114696,∞]	[1792]
132583499	ranges of n:		[167,32769]		[32770,65538]		[65539,∞]	[1792]
120044009	ranges of n:		[40,32770]		[32771,65538]		[65539,∞]	[1792]
100210801	ranges of n:			[40,65537]			[65538,∞]	[1792]

Table 14.9.4 Distance profiles of degree 32 polynomials.

*The third polynomial $f = 104C11DB7$ is used in the FDDI, IEEE 802, AUTODIN-II standards.

t	smallest degree t-nomial divisible by f
3	$x^{91639} + x^{41678} + 1$
4	$x^{3006} + x^{2846} + x^{2215} + 1$
5	$x^{300} + x^{155} + x^{117} + x^{89} + 1$
6	$x^{203} + x^{196} + x^{123} + x^{85} + x^{79} + 1$
7	$x^{123} + x^{120} + x^{80} + x^{74} + x^{58} + x^{46} + 1$
8	$x^{89} + x^{88} + x^{41} + x^{36} + x^{16} + x^{3} + x^{2} + 1$
9	$x^{66} + x^{57} + x^{37} + x^{32} + x^{31} + x^{16} + x^{7} + x^{5} + 1$
10	$x^{53} + x^{38} + x^{36} + x^{33} + x^{30} + x^{27} + x^{26} + x^{7} + x^{3} + 1$
11	$x^{44} + x^{43} + x^{41} + x^{37} + x^{35} + x^{32} + x^{31} + x^{16} + x^{7} + x^{5} + 1$
12	$x^{42} + x^{30} + x^{26} + x^{24} + x^{21} + x^{18} + x^{13} + x^{8} + x^{7} + x^{5} + x^{3} + 1$
13	$x^{42} + x^{40} + x^{37} + x^{35} + x^{33} + x^{29} + x^{28} + x^{20} + x^{18} + x^{15} + x^{6} + x^{1} + 1.$

Table 14.9.5 Smallest t-nomial divisors of $f = 104C11DB7$.

4. $g(x) = x^5 + x + 1 = (x^2 + x + 1)(x^3 + x^2 + 1)$.

14.9.69 Theorem [826] Let $f(x) = x^m + x^l + x^k + x^j + 1$ be a pentanomial over $\mathbb{F}_2$ such that $\gcd(m, l, k, j) = 1$. If g is a trinomial of degree at most $2m$ divisible by f, with $g = fh$, then

1. f is one of the twenty-five polynomials given in Table 14.9.6 with the corresponding h;
2. $m \equiv 1 \pmod 3$ and f, g, h are as follows
$$\begin{aligned} f(x) &= 1 + x + x^2 + x^{m-3} + x^m \\ &= (1 + x + x^2)(1 + x^{m-3} + x^{m-2}), \\ h(x) &= (1 + x) + (x^3 + x^4) + \cdots + (x^{m-7} + x^{m-6}) + x^{m-4}, \\ f(x)h(x) = g(x) &= 1 + x^{2m-6} + x^{2m-4}; \text{ or} \end{aligned}$$
3. f is the reciprocal of one of the polynomials listed in the previous items.

No.	$f(x)$	$h(x)$	type
1	$x^5 + x^4 + x^3 + x^2 + 1$	$x^3 + x^2 + 1$	p
2	$x^5 + x^3 + x^2 + x + 1$	$x^3 + x + 1$	p
3	$x^5 + x^3 + x^2 + x + 1$	$x^4 + x + 1$	p
4	$x^5 + x^4 + x^3 + x + 1$	$x^2 + x + 1$	p
5	$x^6 + x^5 + x^4 + x^3 + 1$	$x^4 + x^3 + 1$	r
6	$x^6 + x^4 + x^2 + x + 1$	$x^3 + x + 1$	i
7	$x^6 + x^4 + x^3 + x + 1$	$x^2 + x + 1$	p
8	$x^6 + x^5 + x^2 + x + 1$	$x^5 + x^4 + x^3 + x + 1$	p
9	$x^6 + x^5 + x^3 + x + 1$	$x^2 + x + 1$	r
10	$x^7 + x^4 + x^2 + x + 1$	$x^3 + x + 1$	r
11	$x^7 + x^4 + x^3 + x^2 + 1$	$x^3 + x^2 + 1$	p
12	$x^7 + x^5 + x^2 + x + 1$	$x^7 + x^6 + x^5 + x^4 + x^3 + x + 1$	p
13	$x^7 + x^5 + x^3 + x^2 + 1$	$x^5 + x^4 + x^3 + x^2 + 1$	r
14	$x^8 + x^5 + x^3 + x + 1$	$x^5 + x^4 + x^2 + x + 1$	p
15	$x^8 + x^5 + x^3 + x^2 + 1$	$x^8 + x^7 + x^5 + x^4 + x^3 + x^2 + 1$	p
16	$x^8 + x^6 + x^3 + x + 1$	$x^6 + x^4 + x^2 + x + 1$	r
17	$x^8 + x^7 + x^5 + x^2 + 1$	$x^6 + x^5 + x^4 + x^2 + 1$	r
18	$x^9 + x^6 + x^5 + x^2 + 1$	$x^8 + x^5 + x^4 + x^2 + 1$	i
19	$x^9 + x^7 + x^4 + x^3 + 1$	$x^8 + x^6 + x^4 + x^3 + 1$	i
20	$x^9 + x^8 + x^5 + x^2 + 1$	$x^6 + x^5 + x^4 + x^2 + 1$	r
21	$x^{10} + x^4 + x^3 + x^2 + 1$	$x^8 + x^7 + x^4 + x^2 + 1$	i
22	$x^{10} + x^7 + x^2 + x + 1$	$x^6 + x^4 + x^3 + x + 1$	r
23	$x^{11} + x^7 + x^6 + x^2 + 1$	$x^8 + x^7 + x^4 + x^2 + 1$	r
24	$x^{13} + x^{10} + x^2 + x + 1$	$x^9 + x^7 + x^6 + x^4 + x^3 + x + 1$	r
25	$x^{13} + x^{10} + x^9 + x^2 + 1$	$x^{12} + x^9 + x^8 + x^6 + x^4 + x^2 + 1$	p

Table 14.9.6 Table of pentanomials which divide trinomials: "p" in *type* indicates that the given polynomial f is primitive, "i" indicates that f is irreducible, and "r" indicates that f is reducible.

14.9.70 Remark All primitive polynomials satisfy the gcd condition of Theorems 14.9.68 and 14.9.69, and thus, in particular, Theorems 14.9.68 and 14.9.69 hold for all primitive trinomials and pentanomials over $\mathbb{F}_2$.

14.9.71 Corollary If $f(x) = x^m + x^l + x^k + x^j + 1$ is primitive over $\mathbb{F}_2$ and not one of the exceptions given in Table 14.9.6 or their reciprocals, then, for $m < n \leq 2m$,

1. C_n^f is an orthogonal array of strength at least 3; or equivalently,
2. $(C_n^f)^\perp$, the dual code of C_n^f, has minimum weight at least 4;
3. the cyclic redundancy check code derived from f, of length n, can detect all errors in three or fewer positions;
4. the bias from the third moment of the Hamming weight in the linear feedback shift register sequence generated from f is small.

14.9.72 Theorem [826] Let $\mathbb{F}$ be any field and $f, g, h \in \mathbb{F}[x]$, $fh = g$, $w(f) = n > 1$ and $w(g) = m$. If there exists an $f_0 \in \mathbb{F}[x]$ such that $f(x) = f_0(x^k)$ for $k > 1$ then there exist $g_i \in \mathbb{F}[x]$, $w(g_i) = m_i$ for $0 \leq i < k$ such that

$$g(x) = \sum_{i=0}^{k-1} g_i(x^k)x^i \tag{14.9.1}$$

$$m = \sum_{i=0}^{k-1} m_i \text{ and } m_i \neq 1. \tag{14.9.2}$$

14.9.73 Remark Theorem 14.9.72 can be used to simplify the analysis of multiples of f. An example used in [2354] is given in Corollary 14.9.74 and was used in the proofs of Theorems 14.9.75 and 14.9.76.

14.9.74 Corollary [826] Let $\mathbb{F}$ be any field and $f, g, h \in \mathbb{F}[x]$, $fh = g$, $w(f) = n$ and $w(g) \leq 3$. If there exists $f_0 \in \mathbb{F}[x]$ such that $f(x) = f_0(x^k)$ for $k > 1$ then there exists $g_0 \in \mathbb{F}[x]$ such that $g(x) = g_0(x^k)$.

14.9.75 Theorem [826] Let $f(x) = a + bx^k + x^m$ $(a, b \neq 0)$ be a monic trinomial over $\mathbb{F}_3$. If $g(x) = c + x^n$ $(c \neq 0)$ is a monic binomial over $\mathbb{F}_3$ with degree at most $3m$ divisible by f with $g = fh$, then f and g are as given in Table 14.9.7.

Case	$f(x)$	$g(x)$
1.1	$1 + bx^{m/2} + x^m$	$-b + x^{3m/2}$
1.2	$-1 + bx^{m/2} + x^m$	$1 + x^{2m}$
1.3	$1 + bx^{m/2} + x^m$	$-1 + x^{3m}$
1.4	$a + x^{m/3} + x^m$	$-1 + x^{8m/3}$
1.5	$b + bx^{2m/3} + x^m$	$-1 + x^{8m/3}$

Table 14.9.7 Polynomials over $\mathbb{F}_3$ such that $g = fh$ for monic trinomial f and monic binomial g.

14.9.76 Theorem [826] Let $f(x) = a + bx^k + x^m$ $(a, b \neq 0)$ be a monic trinomial over $\mathbb{F}_3$. If $g(x) = c + dx^l + x^n$ $(c, d \neq 0)$ is a monic trinomial over $\mathbb{F}_3$ with degree at most $3m$ divisible by f with $g = fh$, then

1. $g = f^3$;
2. f and g are as in the Table 14.9.8;
3. f and g are reciprocals of polynomials listed in Table 14.9.8.

Case	$f(x)$	$g(x)$
1.1	$-1 + bx^{m/2} + x^m$	$1 - bx^{m/2} + x^{3m}$
1.2	$1 + bx^{m/2} + x^m$	$b + x^{m/2} + x^{5m/2}$
1.3	$-1 + bx^{m/2} + x^m$	$b - bx^m + x^{5m/2}$
1.4	$-1 + bx^{m/2} + x^m$	$-b - x^{3m/2} + x^{5m/2}$
1.5	$1 + bx^{m/2} + x^m$	$b + bx^{4m/2} + x^{5m/2}$
1.6	$1 + bx^{m/2} + x^m$	$1 + x^m + x^{2m}$
1.7	$-1 + bx^{m/2} + x^m$	$b + x^m + x^{3m/2}$
1.8	$-1 + bx^{m/2} + x^m$	$-b - bx^m + x^{3m/2}$
1.9	$a - x^{m/3} + x^m$	$-a - x^{m/3} + x^{3m}$
1.10	$a - x^{m/3} + x^m$	$1 + x^{2m/3} + x^{8m/3}$
1.11	$a + x^{m/3} + x^m$	$a + ax^{2m/3} + x^{7m/3}$
1.12	$a - x^{m/3} + x^m$	$a - ax^{4m/3} + x^{7m/3}$
1.13	$a - x^{m/3} + x^m$	$-a + x^{5m/3} + x^{7m/3}$
1.14	$a + x^{m/3} + x^m$	$1 + ax^{5m/3} + x^{2m}$
1.15	$a - x^{m/3} + x^m$	$a + ax^{4m/3} + x^{5m/3}$
1.16	$-1 + bx^{m/4} + x^m$	$-b + bx^{6m/4} + x^{11m/4}$
1.17	$1 + bx^{m/4} + x^m$	$1 + bx^{9m/4} + x^{10m/4}$

Table 14.9.8 Table of polynomials such that $g = fh$ with f and g monic trinomials over $\mathbb{F}_3$.

See Also

§2.2	For tables of primitives of various kinds and weights.
§3.4	For results on the weights of irreducible polynomials.
§4.3	For results on the weights of primitive polynomials.
§10.2	For discussion on bias and randomness of linear feedback shift register sequences.
§14.1	For results of latin squares which are strongly related to orthogonal arrays.
§14.5	For a discussion of block designs, which include orthogonal arrays.
§15.1	Uses primitive polynomials to generate cyclic redundancy check and BCH codes.
§15.4	For results on turbo codes.
§17.3	For more discussion of applications of finite fields and polynomials.

References Cited: [146, 218, 357, 558, 559, 801, 826, 832, 833, 1362, 1457, 1491, 1590, 1591, 1622, 1727, 1792, 1793, 1834, 1944, 1994, 2074, 2083, 2169, 2204, 2354, 2439, 2513]

14.10 Ramanujan and expander graphs

M. Ram Murty, Queen's University
Sebastian M. Cioabă, University of Delaware

In the last two decades, the theory of Ramanujan graphs has gained prominence primarily for two reasons. First, from a practical viewpoint, they resolve an extremal problem in communication network theory (see for example [269, 1535]). Second, from a more aesthetic viewpoint, they fuse diverse branches of pure mathematics, namely, number theory, representation theory, and algebraic geometry. The purpose of this survey is to unify some of the recent developments and expose certain open problems in the area. This survey is an expanded version of [2208] and is by no means an exhaustive one and demonstrates a highly number-theoretic bias. For other surveys, we refer the reader to [1535, 1922, 1967, 1968, 2525, 2837]. For a more up-to-date survey highlighting the connection between graph theory and automorphic representations, we refer the reader to Li's recent survey article [1924].

14.10.1 Graphs, adjacency matrices, and eigenvalues

14.10.1 **Definition** A *graph* X is a pair (V, E) consisting of a *vertex set* $V = V(X)$ and an *edge set* $E = E(X)$ which is a multiset of unordered pairs of (not necessarily distinct) vertices. Each edge consists of two vertices that are called its *endpoints.* A *loop* is an edge whose endpoints are equal. Multiple edges are edges having the same pair of endpoints. A *simple graph* is a graph having no loops nor multiple edges. If a graph has loops or multiple edges, we will call it a *multigraph.* When two vertices u and v are endpoints of an edge, they are *adjacent* and write $u \sim v$ to indicate this fact. A *directed graph* Y is a pair (W, F) consisting of a set of vertices W and a multiset F of ordered pairs of vertices which are called *arcs.*

14.10.2 **Remark** All the graphs in this chapter are undirected unless stated explicitly otherwise.

14.10.3 **Definition** The *degree* of a vertex v of a graph X, denoted by $\deg(v)$, is the number of edges incident with v, where we count a loop with multiplicity 1. A graph X is k-regular if every vertex has degree k.

14.10.4 **Proposition** (Handshaking Lemma) For any simple graph X, $\sum_{v \in V(X)} \deg(v) = 2|E(X)|$. If X is a k-regular graph with n vertices, then $|E(X)| = kn/2$.

14.10.5 **Definition** An *adjacency matrix* $A = A(X)$ of a graph X with n vertices is an $n \times n$ matrix with rows and columns indexed by the vertices of X, where the (x, y)-th entry equals the number of edges between vertex x and vertex y. As X is an undirected graph with n vertices, the matrix $A(X)$ is symmetric and therefore, its eigenvalues $\lambda_1 \geq \lambda_2 \geq \cdots \geq \lambda_n$ are real. The multiset of eigenvalues of X is the *spectrum* of X.

14.10.6 **Remark** We remark that the adjacency matrix defined above depends on the labeling of the vertices of X. Different labelings of the vertices of a graph X may possibly yield different adjacency matrices. However, all these adjacency matrices are similar to each other (by permutation matrices) and thus, their spectrum is the same.

14.10.7 **Remark** One can define an adjacency matrix of a directed graph $Y = (W, A)$ similarly. Given a labeling of the vertices W of Y, the (x, y)-th entry of the adjacency matrix corresponding to this labeling equals the number of arcs from x to y. Adjacency matrices of directed graphs may be non-symmetric.

14.10.8 **Example** The spectrum of the *complete graph* K_n on n vertices is $(n-1)^{(1)}, (-1)^{(n-1)}$, where the exponent signifies the multiplicity of the respective eigenvalue. The Petersen graph has spectrum $3^{(1)}, 1^{(5)}, -2^{(4)}$.

14.10.9 **Theorem** Let X be a graph on n vertices with maximum degree Δ and average degree $\overline{d}$. Then $\overline{d} \leq \lambda_1 \leq \Delta$ and $|\lambda_i| \leq \Delta$ for every $2 \leq i \leq n$.

14.10.10 Definition For a multigraph X, a *walk* of *length* r from x to y is a sequence $x = v_0, v_1, \ldots, v_r = y$ of vertices of X such that v_iv_{i+1} is an edge of X for any $i = 0, 1, \ldots, r-1$. The *length* of this walk is r. A *closed walk* is a walk where the starting vertex x is the same as the last vertex y.

14.10.11 Definition A *path* is a walk with no repeated vertices. A *cycle* is a closed walk with no repeated vertices except the starting vertex.

14.10.12 Remark A word of caution must be inserted here. In graph theory literature, the distinction between a walk and a path is as we have defined it above. However, in number theory circles, the finer distinction is not made and one uses the word "path" to mean a "walk"; see for example, [2523, 2789].

14.10.13 Definition A graph X is *connected* if for every two distinct vertices x and y, there is a path from x to y.

14.10.14 Proposition For every graph X with adjacency matrix A and any integer $r \geq 1$, the (x, y)-th entry of A^r equals the number of walks of length r between x and y.

14.10.15 Proposition The number of closed walks of length r in a graph X with n vertices equals $\lambda_1^r + \lambda_2^r + \cdots + \lambda_n^r$.

14.10.16 Definition An *independent set* in a graph X is a subset of vertices that are pairwise non-adjacent. A graph X is *bipartite* if its vertex set can be partitioned into two independent sets A and B; X is *complete bipartite* and denoted by $K_{|A|,|B|}$ if it contains all the edges between A and B.

14.10.17 Proposition A graph is bipartite if and only if it does not contain any cycles of odd length.

14.10.18 Theorem If X is a k-regular and connected graph with n vertices, then $\lambda_1 = k$ and the multiplicity of k is 1 with the eigenspace of k spanned by the all 1 vector of dimension n. If X is a k-regular and connected graph, then X is bipartite if and only if $\lambda_n = -k$.

14.10.19 Definition If X is a k-regular and connected graph, then the eigenvalue k of X is *trivial.* All other eigenvalues of X are *non-trivial.* Let $\lambda(X) = \max |\lambda_i|$, where the maximum is taken over all non-trivial eigenvalues of X. The parameter $\lambda(X)$ is the *second eigenvalue* of X by some authors. The *second largest eigenvalue* of X is $\lambda_2(X)$ and $\lambda(X) \geq \lambda_2(X)$.

14.10.20 Definition The *distance* $d(x, y)$ between two distinct vertices x and y of a connected graph X is the length of a shortest path between x and y. The *diameter* D of a connected graph X is the maximum of $d(x, y)$, where the maximum is taken over all pairs of distinct vertices $x \neq y$ of X.

14.10.21 Remark When $k \geq 3$, if X is a k-regular and connected graph with n vertices and diameter D, then $n \leq 1 + k + k(k-1) + \cdots + k(k-1)^{D-1} = 1 + k \cdot \frac{(k-1)^D - 1}{k-2}$ and consequently, $D \geq \log_{k-1}\left(\frac{(n-1)(k-2)}{k} + 1\right) > \frac{\log(n-1)}{\log(k-1)} - \frac{\log(k/(k-2))}{\log(k-1)}$. Thus, the diameter of any connected k-regular graph is at least logarithmic in the order of the graph. The next theorem implies that when the non-trivial eigenvalues of a k-regular connected graph are small, then the above inequality is tight up to a multiplicative constant.

14.10.22 Theorem [637] If X is a connected non-bipartite k-regular graph with n vertices and diameter D, then:

$$D \leq \frac{\log(n-1)}{\log(k/\lambda(X))} + 1.$$

14.10.23 Remark Kahale [1641] obtained an upper bound on the minimum distance between i subsets of the same size of a regular graph in terms of the i-th largest eigenvalue in absolute value. Kahale also constructed examples of k-regular graphs on n vertices having $\lambda(X) = (1+o(1))2\sqrt{k-1}$ and $D = 2(1+o(1))\log_{k-1} n$ showing the previous result is asymptotically best possible. Here the $o(1)$ term tends to 0 as n goes to infinity.

14.10.24 Remark A similar result can be derived for k-regular bipartite graphs; if X is a bipartite k-regular and connected graph of diameter D, we have (see Quenell [2433])

$$D \leq \frac{\log(n-2)/2}{\log(k/\lambda'(X))} + 2,$$

where $\lambda'(X)$ is the maximum absolute value of the eigenvalues of X that are not k nor $-k$.

14.10.25 Remark Chung, Faber, and Manteuffel [638] and independently, Van Dam and Haemers [2839] obtained slight improvements of the previous diameter bounds.

14.10.26 Definition The *chromatic number* $\chi(X)$ of a graph X is the minimum number of colors that can be assigned to the vertices of a graph such that any two adjacent vertices have different colors. The largest order of an independent set of vertices of X is the *independence number* of X and is denoted by $\alpha(X)$.

14.10.27 Remark The chromatic number of X is the minimum number of independent sets that partition the vertex set of X and consequently, $\chi(X) \geq \frac{|V(X)|}{\alpha(X)}$.

14.10.28 Theorem [1287] If X is a k-regular non-empty graph, then

$$\alpha(X) \leq \frac{n(-\lambda_n)}{k-\lambda_n}$$

and so

$$\chi(X) \geq 1 + \frac{k}{-\lambda_n}.$$

14.10.29 Remark An immediate consequence of the previous result is that $\alpha(X) \leq \frac{n\lambda(X)}{k+\lambda(X)}$ and $\chi(X) \geq 1 + \frac{k}{\lambda(X)}$ for any non-bipartite connected k-regular graph X. These facts show that a good upper bound for the absolute values of the non-trivial eigenvalues of a regular graph will yield non-trivial bounds for the independence and chromatic number.

14.10.30 Remark The following theorem shows that the eigenvalues of a regular graph are closely related to its edge distribution.

14.10.31 Theorem [81] If X is a k-regular connected graph with eigenvalues $k = \lambda_1 > \lambda_2 \geq \cdots \geq \lambda_n \geq -k$, let $\lambda := \max(|\lambda_2|, |\lambda_n|)$. For $S, T \subset V(X)$, denote by $e(S,T)$ the number of edges with one endpoint in S and another in T. Then for all $S, T \subset V(X)$

$$\left| e(S,T) - \frac{k|S||T|}{n} \right| \leq \lambda \sqrt{|S||T|\left(1 - \frac{|S|}{n}\right)\left(1 - \frac{|T|}{n}\right)} < \lambda\sqrt{|S||T|}.$$

14.10.32 Remark The previous theorem states that k-regular graphs X with small non-trivial eigenvalues (compared to k) have their edges uniformly distributed (similar to random k-regular graphs). Such graphs are called *pseudorandom graphs* and are important in many situations (see Krivelevich-Sudakov's survey [1807]). Bilu and Linial [282] have obtained a converse of the previous result of Alon and Chung.

14.10.2 Ramanujan graphs

14.10.33 **Definition** A k-regular connected multigraph X is a *Ramanujan multigraph* if $|\lambda_i| \leq 2\sqrt{k-1}$ for every eigenvalue $\lambda_i \neq k$. A *Ramanujan graph* is a Ramanujan multigraph having no loops nor multiple edges.

14.10.34 **Remark** We mention that the definition of a Ramanujan graph used by other authors is slightly weaker. For example, Sarnak in [2525] calls a k-regular graph Ramanujan if $\lambda_2(X) \leq 2\sqrt{k-1}$.

14.10.35 **Example** The complete graph K_n is an $(n-1)$-regular Ramanujan graph as its eigenvalues are $(n-1)^{(1)}, (-1)^{(n-1)}$, where the exponents denote the multiplicities of the eigenvalues. The complete bipartite graph $K_{n,n}$ has eigenvalues $n^{(1)}, 0^{(2n-2)}, -n^{(1)}$ and is an n-regular Ramanujan graph.

14.10.36 **Remark** In [80, p.95], Alon announced a proof with Boppana of the fact that for any k-regular graph X of order n, $\lambda_2(X) \geq 2\sqrt{k-1} - O((\log_k n)^{-1})$, where the constant in the O term depends only on k. Many researchers refer to this result as the Alon-Boppana Theorem. Other researchers refer to the following statement proved by Nilli (pseudonym for Alon) in [2289] as the Alon-Boppana Theorem.

14.10.37 **Theorem** [2289] If X is a k-regular and connected graph with diameter $D \geq 2b+2$, then

$$\lambda_2(X) \geq 2\sqrt{k-1} - \frac{2\sqrt{k-1}-1}{b+1}.$$

14.10.38 **Remark** The Alon-Boppana Theorem and Remark 14.10.21 imply that if $(X_i)_{i\geq 1}$ is a sequence of k-regular and connected graphs with $\lim_{i\to+\infty} |V(X_i)| = +\infty$, then

$$\liminf_{i\to\infty} \lambda_2(X_i) \geq 2\sqrt{k-1}.$$

14.10.39 **Remark** The best lower bound for the second largest eigenvalue $\lambda_2(X)$ of a k-regular graph of diameter D is due to Friedman [1126] who showed that

$$\lambda_2(X) \geq 2\sqrt{k-1}\cos\theta_{k,t} \geq 2\sqrt{k-1}\cos\frac{\pi}{t+1} \qquad (14.10.1)$$

where t is the largest integer such that $D \geq 2t$ and $\theta_{k,t} \in \left[\frac{\pi}{t+5}, \frac{\pi}{t+1}\right]$ is the smallest positive solution of the equation $\frac{k}{2k-2} = \frac{\sin(t+1)\theta\cos\theta}{\sin t\theta}$. The number $2\sqrt{k-1}\cos\theta_{k,t}$ is the largest eigenvalue of the k-regular tree $T_{k,t}$ of depth t; this tree has a root vertex x and exactly $k(k-1)^{i-1}$ vertices at distance i from x for each $1 \leq i \leq t$. Friedman used analytic tools involving Dirichlet and Neumann eigenvalues for graphs with boundaries to prove (14.10.1). Later, Nilli [2290] gave an elementary proof of a slightly weaker bound.

14.10.40 **Remark** We outline here an elementary proof of the inequality $\lambda_2(X) \geq 2\sqrt{k-1}\cos\frac{\pi}{t+1}$ for every connected k-regular graph X of diameter $D \geq 2t+2$. The first ingredient of the proof is that the largest eigenvalue of any subgraph induced by a ball of radius t of X is larger than the largest eigenvalue of $T_{k,t}$. The second is that if u and v are vertices at distance at least $2t+2$ in X, then the subgraph induced by the vertices at distance at most t from u or v

has exactly two components $X(u)$ and $X(v)$. By Cauchy eigenvalue interlacing, the second largest eigenvalue of X is greater than the minimum of the largest eigenvalue of $X(u)$ and $X(v)$ which by the previous argument is at least $2\sqrt{k-1}\cos\theta_{k,t} \geq 2\sqrt{k-1}\cos\frac{\pi}{t+1}$.

14.10.41 Remark At this point, it is worth stating that Friedman [1126] (see also Nilli [2290]) proved the stronger statement that if X is a k-regular graphs containing a subset of r points each of distance at least $2t$ from one another, then $\lambda_r(X) \geq 2\sqrt{k-1}\cos\theta_{k,t} \geq 2\sqrt{k-1}\cos\frac{\pi}{t+1}$. This implies that the r-th largest eigenvalue $\lambda_r(X)$ of any connected k-regular graph X is at least $2\sqrt{k-1}\left(1-\frac{\pi^2}{2f^2}+O(f^{-4})\right)$, where $f = \frac{\log_{k-1}(n/r)}{2}$.

14.10.42 Remark One might wonder if the behavior of the negative eigenvalues of a connected k-regular graph X is similar to the behavior of the negatives of the positive eigenvalues of X. If X is bipartite, then the spectrum of X is symmetric with respect to 0 and this settles the previous question. In general, it turns out that additional conditions are needed in order to obtain similar results for the negative eigenvalues. This is because there are regular graphs with increasing order whose eigenvalues are bounded from below by an absolute constant. For example, the eigenvalues of a line graph are at least -2. It turns out that the number of odd cycles plays a role in the behavior of the negative eigenvalues of regular graphs. The *odd girth* of a graph X is the smallest length of a cycle of odd length.

14.10.43 Theorem [1126, 2290] If X is a connected k-regular graph of order n with a subset of r points each of distance at least $2t$ from one another, and odd girth at least $2t$, then

$$\lambda_{n-r}(X) \leq -2\sqrt{k-1}\cos\theta_{k,t} = -2\sqrt{k-1}\cos\frac{\pi}{t+1}. \tag{14.10.2}$$

14.10.44 Corollary [1923] If $(X_i)_{i\geq 0}$ is a sequence of k-regular graphs of increasing orders such that the odd girth of X_i tends to infinity as $i \to \infty$, then $\limsup_{i\to\infty} \mu_l(X_i) \leq -2\sqrt{k-1}$, for each $l \geq 1$, where $\mu_l(X)$ denotes the l-th smallest eigenvalue of X.

14.10.45 Theorem [646, 648] For an integer $r \geq 3$, let $c_r(X)$ denote the number of cycles of length r of a graph X. If $(X_i)_{i\geq 0}$ is a sequence of k-regular graphs of increasing orders such that $\lim_{i\to\infty} \frac{c_{2r+1}(X_i)}{|V(X_i)|} = 0$ for each $r \geq 1$, then $\limsup_{i\to\infty} \mu_l(X_i) \leq -2\sqrt{k-1}$, for each $l \geq 1$.

14.10.46 Remark The difficulty of constructing infinite families of Ramanujan graphs is also illustrated by the following result of Serre.

14.10.47 Theorem [2594] For any $\epsilon > 0$, there exists a positive constant $c = c(\epsilon, k)$ such that for every k-regular graph X on n vertices, the number of eigenvalues λ_i of X such that $\lambda_i > (2-\epsilon)\sqrt{k-1}$ is at least $c \cdot n$.

14.10.48 Remark Different short and elementary proofs of Serre's theorem were found independently by Nilli [2290] and Cioabă [646, 648]. Nilli's proof is similar to Friedman's argument from [1126] while Cioabă's proof uses the fact that the trace of A^l is the number of closed walks of length l. See also [646, 648] for a similar theorem to Theorem 14.10.47 for the smallest eigenvalues of regular graphs. These proofs, as well as extensions of the Alon-Bopanna theorem (see recent work of Mohar [2118]) rely on the notion of the universal cover of a graph; see Definitions 14.10.108 and 14.10.109, and Theorem 14.10.110 for more details.

14.10.49 Remark The idea behind all the proofs of Serre's theorem indicated above is that the universal cover of a finite k-regular graph is the rooted infinite k-regular tree T_k. This implies that the number of closed walks of even length starting at some vertex of a finite k-regular graph is at least the number of closed walks of the same length starting at the root of the infinite k-regular tree. The number $2\sqrt{k-1}$ is the spectral radius of the adjacency operator of the infinite k-regular tree (for more details see [1126, 1535]). In some circumstances, the lower bound for the second eigenvalue of a k-regular graph can be improved beyond $2\sqrt{k-1}$ [1927, 2118].

14.10.50 Remark Greenberg and Lubotzky (see Chapter 4 of [1967] or [646, 647] for a short elementary proof) extended the Alon-Bopanna bound to any family of general graphs with isomorphic universal cover. If $(X_i)_{i\geq 1}$ is a family of finite connected graphs with universal cover $\tilde{X}$ and ρ is the spectral radius of the adjacency operator of $\tilde{X}$, then $\liminf \lambda_2(X_i) \geq \rho$ as $i \to +\infty$. For extensions of Alon-Boppana theorem and Serre's theorem for irregular graphs, see [646, 647, 1534, 2118].

14.10.51 Remark The notion of Ramanujan graph has been extended to hypergraphs and studied in this setting. Again, these notions lead to the use of the Ramanujan conjecture formulated for higher GL_n in the Langlands program.

14.10.52 Definition A *hypergraph* $X = (V, E)$ is a pair consisting of a vertex set V and a set of *hyperedges* E consisting of subsets of V. If all the edges are of the same size r, X is an r-uniform hypergraph or r-graph. In the familiar setting of a graph, an edge is viewed as a 2-element subset of V and is thus a 2-uniform hypergraph. One class of hypergraphs that are studied are the (k, r)-regular hypergraphs in which each edge contains r elements and each vertex is contained in k edges. For an ordinary graph, $r = 2$ and this generalizes the notion of a k-regular graph. In this special setting, the adjacency matrix A is a $|V| \times |V|$ matrix with zero diagonal entries and the (i, j)-th entry is the number of hyperedges that contain $\{i, j\}$.

14.10.53 Remark One can show easily that $k(r-1)$ is an eigenvalue of A and this is the trivial eigenvalue. With this definition in place, a *Ramanujan hypergraph* is defined as a finite connected (k, r)-regular hypergraph such that every eigenvalue λ of A with $|\lambda| \neq k(r-1)$ satisfies

$$|\lambda - (r-2)| \leq 2\sqrt{(k-1)(r-1)}.$$

We refer the reader to the important work of Li [1925] for further details.

14.10.3 Expander graphs

14.10.54 Definition For any subset A of vertices of a graph X, the *edge boundary* of A, denoted ∂A, is

$$\partial A = \{xy \in E(X) : x \in A, y \notin A\}.$$

That is, the edge boundary of A consists of edges with one endpoint in A and another outside A.

14.10.55 Definition The *edge-expansion constant* of X, denoted by $h(X)$, is defined as

$$h(X) = \min\left\{\frac{|\partial A|}{|A|} : A \subset X, |A| \leq \frac{|V(X)|}{2}\right\}.$$

14.10.56 Definition A family of k-regular graphs $(X_i)_{i\geq 1}$ with $|V(X_i)|$ increasing with i, is a *family of expanders* if there exists a positive absolute constant c such that $h(X_i) > c$ for every $i \geq 1$.

14.10.57 Remark Informally, a family of k-regular expanders is a family of sparse (k fixed and $|V(X_i)| \to +\infty$ as $i \to +\infty$ imply that the number of edges of X_i is linear in its number of vertices), but highly connected graphs ($h(X_i) > c$ means that in order to disconnect X_i, one must remove many edges).

14.10.58 Example [2005] Let X_m denote the following 8-regular graph on m^2 vertices. The vertex set of X_m is $\mathbb{Z}/m\mathbb{Z} \times \mathbb{Z}/m\mathbb{Z}$. The neighbors of a vertex (x, y) are $(x \pm y, y), (x, y \pm x), (x \pm y + 1, y), (x, y \pm x + 1)$. The family $(X_m)_{m \geq 4}$ is the first explicit family of expanders. Margulis [2005] proved that $(X_m)_{m \geq 4}$ are expanders using representation theory. Margulis [2005] used the fact that the group $SL_3(\mathbb{Z})$ has Kazhdan property T. Groups having this property or the weaker property τ can be used to construct infinite families of constant-degree Cayley graphs expanders. We refer the reader to [1535, 1967, 1968] for nice descriptions and explanations of these properties and their relation to expanders.

14.10.59 Remark Expander graphs play an important role in computer science, mathematics, and the theory of communication networks; see [269, 1535]. These graphs arise in questions about designing networks that connect many users while using only a small number of switches.

14.10.60 Theorem [80, 2117] If X is a connected k-regular graph, then

$$\sqrt{k^2 - \lambda_2^2} \geq h(X) \geq \frac{k - \lambda_2}{2}.$$

14.10.61 Remark The previous theorem shows that constructing an infinite family of k-regular expanders $(X_i)_{i \geq 1}$ is equivalent to constructing an infinite family of k-regular graphs $(X_i)_{i \geq 1}$ such that $k - \lambda_2(X_i)$ is bounded away from zero.

14.10.4 Cayley graphs

14.10.62 Definition Let G be a group written in multiplicative notation and let S be a subset of elements of G that is closed under taking inverses and does not contain the identity. The *Cayley graph of G with respect to S* (denoted by $X(G, S)$) is the graph whose vertex set is G where $x \sim y$ if and only if $x^{-1}y \in S$. If G is abelian, then it is common to use the additive notation in the definition of $X(G, S)$: $x \sim y$ if and only if $y - x \in S$.

14.10.63 Remark In general, if S is an arbitrary multiset of G, denote by $X(G, S)$ the directed graph with vertex set G and arc set $\{(x, y) : x^{-1}y \in S\}$. If S is inverse-closed and does not contain the identity, then this graph is undirected and has no loops.

14.10.64 Theorem Let G be a finite abelian group and S a symmetric subset of G of size k. The eigenvalues of the adjacency matrix of $X(G, S)$ are given by

$$\lambda_\chi = \sum_{s \in S} \chi(s)$$

where χ ranges over all irreducible characters of G.

4.10.65 Remark For each irreducible character of G, let v_χ denote the column vector $(\chi(g) : g \in G)$. The proof of Theorem 14.10.64 follows by showing that v_χ is an eigenvector of the adjacency matrix of $X(G, S)$ corresponding to eigenvalue $\lambda_\chi = \sum_{s \in S} \chi(s)$.

4.10.66 Remark Notice that for the trivial character $\chi = 1$, we have $\lambda_1 = k$. If we have for all $\chi \neq 1$

$$\left| \sum_{s \in S} \chi(s) \right| < k$$

then the graph is connected by our earlier remarks. Thus, to construct Ramanujan graphs, we require

$$\left|\sum_{s\in S}\chi(s)\right| \leq 2\sqrt{k-1}$$

for every non-trivial irreducible character χ of G. This is the strategy employed in many of the explicit constructions of Ramanujan graphs.

14.10.67 Example A simple example can be given using Gauss sums. If $p \equiv 1 \pmod 4$ is a prime, let $G = \mathbb{Z}/p\mathbb{Z}$ and $S = \{x^2 : x \in \mathbb{Z}/p\mathbb{Z}\}$ be the multiset of squares. The multigraph $X(G, S)$ is easily seen to be Ramanujan in view of the fact (see for example [2207, p. 81])

$$\left|\sum_{x\in\mathbb{Z}/p\mathbb{Z}} e^{2\pi i a x^2/p}\right| = \sqrt{p}$$

for any $a \neq 0$. By our convention in the computation of degree of a vertex, we see that $X(G, S)$ is a p-regular graph.; see [1807] for other related examples.

14.10.68 Example When $q \equiv 1 \pmod 4$ is a prime power, the *Paley graph of order* q is the Cayley graph $X(G, S)$ of the additive group of a finite field $G = \mathbb{F}_q$ with respect to the set S of non-zero squares. This simple and undirected graph has q vertices, is connected and regular of degree $\frac{q-1}{2}$ and its non-trivial eigenvalues are $\frac{-1-\sqrt{q}}{2}$ and $\frac{-1+\sqrt{q}}{2}$, each of multiplicity $\frac{q-1}{2}$. The Paley graph is Ramanujan when $q \geq 9$.

14.10.69 Remark The proof of Theorem 14.10.64 is reminiscent of the Dedekind determinant formula in number theory. This formula computes $\det A$, where A is the matrix whose (i, j)-th entry is $f(ij^{-1})$ for any function f defined on the finite abelian group G of order n. The determinant is

$$\prod_{\chi}\left(\sum_{g\in G} f(g)\chi(g)\right).$$

14.10.70 Definition Let G be an abelian group written in the additive notation and $S \subset G$. The *sum graph of* G *with respect to* S (denoted by $Y(G, S)$) has G as vertex set and $x \sim y$ if and only if $x + y \in S$.

14.10.71 Theorem [1922, p. 197] Let G be an abelian group. The eigenvalues of $Y(G, S)$ are given as follows. For each irreducible character χ of G, define

$$e_\chi = \sum_{s\in S}\chi(s).$$

If $e_\chi = 0$, then v_χ and $v_{\chi^{-1}}$ are both eigenvectors with eigenvalues 0. If $e_\chi \neq 0$, then

$$|e_\chi| v_\chi \pm e_\chi v_{\chi^{-1}}$$

are two eigenvectors with eigenvalues $\pm|e_\chi|$.

14.10.72 Example Using Theorem 14.10.71, Li [1921] constructed Ramanujan graphs in the following way. Let $\mathbb{F}_q$ denote the finite field of q elements. Let $G = \mathbb{F}_{q^2}$ and take for S the elements of G of norm 1. This is a symmetric subset of G and the Cayley graph $X(G, S)$ turns out to be Ramanujan. The latter is a consequence of a theorem of Weil estimating Kloosterman sums [2589].

14.10.73 Theorem [2208] Let $G = \mathbb{F}_q$ be a finite field of $q = p^m$ elements and $f(x)$ a polynomial with coefficients in $\mathbb{F}_q$ and of degree 2 or 3. Let S be the multiset

$$\{f(x) : x \in \mathbb{F}_q\}.$$

Suppose S is symmetric. Then $X(G, S)$ is a Ramanujan graph if the degree of f is 2 and is almost Ramanujan if the degree of f is 3.

14.10.74 Remark The required character sum estimates in Theorem 14.10.73 come from Weil's proof of the Riemann hypothesis for the zeta functions of curves over finite fields. In particular, we have for all $a \in \mathbb{F}_q$, $a \neq 0$,

$$\left| \sum_{x \in \mathbb{F}_q} \exp(2\pi i \mathrm{tr}_{\mathbb{F}_q/\mathbb{F}_p}(af(x))/p) \right| \leq (\deg f - 1)\sqrt{q}$$

provided f is not identically zero; see [1922, p. 94]. In particular, if f has degree 3, we get the estimate of $2\sqrt{q}$ for the exponential sum. For example, if $u \in \mathbb{Z}/p\mathbb{Z}$ and we take

$$S = \{x^3 + ux : x \in \mathbb{Z}/p\mathbb{Z}\},$$

then S is symmetric and, according to our convention, $X(G, S)$ is a p-regular graph. In addition, it is an almost Ramanujan graph since

$$\left| \sum_{x \in \mathbb{Z}/p\mathbb{Z}} \exp(2\pi i a(x^3 + ux)/p) \right| \leq 2\sqrt{p}$$

by virtue of the Riemann hypothesis for curves (proved by Weil).

14.10.75 Remark We observe that even though there are many constructions of Ramanujan graphs that are abelian Cayley graphs, it is actually impossible to construct an infinite family of constant-degree abelian Cayley graphs that are Ramanujan. There are several proofs of this fact in the literature, see [83, 645, 1128]. Friedman, Murty, and Tillich [1128] proved that if X is a k-regular abelian Cayley graph of order n, then $\lambda_2(X) \geq k - cn^{-4/k}$, where c is some absolute positive constant. Cioabă [645] proved that for fixed $k \geq 3$ and $\epsilon > 0$, there is a positive constant $C = C(\epsilon, k)$ such that any k-regular abelian Cayley graph on n vertices has at least Cn eigenvalues that are larger than $k - \epsilon$.

14.10.76 Remark Lubotzky and Weiss [1970] proved the stronger result that it is impossible to construct infinite families of constant-degree expanders that are Cayley graphs of solvable groups of bounded derived length.

14.10.77 Remark The eigenvalues of Cayley graphs can be calculated even in the case of non-abelian groups. This is essentially contained in a paper by Diaconis and Shahshahani [834]. Using their results, one can easily generalize the Dedekind determinant formula as follows (and which does not seem to be widely known). Let G be a finite group and f a class function on G. Then the determinant of the matrix A whose rows (and columns) are indexed by the elements of G and whose (i, j)-th entry $f(i^{-1}j)$ is given by

$$\prod_{\chi} \left(\frac{1}{\chi(1)} \sum_{g \in G} f(g)\chi(g) \right)^{\chi(1)}$$

where the product is taken over the distinct irreducible characters of G.

14.10.78 Theorem [834] Let G be a finite group and S a symmetric subset which is stable under conjugation. The eigenvalues of the Cayley graph $X(G, S)$ are given by

$$\lambda_\chi = \frac{1}{\chi(1)} \sum_{s \in S} \chi(s)$$

as χ ranges over all irreducible characters of G. Moreover, the multiplicity of λ_χ is $\chi(1)^2$.

14.10.79 Remark We remark that the λ_χ in the above theorem need not be all distinct. For example, if there is a non-trivial character χ which is trivial on S, then the multiplicity of the eigenvalue $|S|$ is at least $1 + \chi(1)^2$. We refer the reader to Babai [156] for a more detailed proof of the above result in a slightly more general context.

14.10.80 Remark The intriguing question of what groups can be used to construct infinite families of constant-degree Cayley graphs expanders was formulated as a conjecture by Babai, Kantor, and Lubotzky [157] in 1989.

14.10.81 Conjecture [157] Let $(G_i)_{i \geq 0}$ be a family of non-abelian simple groups. There exist generating sets S_i of constant size such that $(X(G_i, S_i))_{i \geq 0}$ form a family of expanders.

14.10.82 Remark As a supporting fact of this conjecture, we mention the result of Babai, Kantor, and Lubotzky [157] who proved constructively that any simple non-abelian group G contains a set S of at most 7 generators such that the diameter of the Cayley graph $X(G, S)$ is at most $c \log |G|$, where $c > 0$ is some absolute constant.

14.10.83 Remark The previous conjecture of Babai, Kantor, and Lubotzky is true and its recent resolution has been possible due to the effort of several researchers. We refer the reader to the works of Kassabov, Lubotzky, and Nikolov [1693], Breuillard, Green, and Tao [411], and the recent survey by Lubotzky [1968] for a thorough account of the solution of this conjecture.

14.10.5 Explicit constructions of Ramanujan graphs

14.10.84 Definition Let X be a graph. A *non-backtracking walk* of length r in X is a sequence $x_0, x_1, \ldots, x_r$ of vertices of X such that x_i is adjacent to x_{i+1} for each $0 \leq i \leq r-1$ and $x_{i-1} \neq x_{i+1}$ for each $1 \leq i \leq r-1$. For $r \in \mathbb{N}$, define the matrix A_r as follows: $A_r(x, y)$ equals the number of non-backtracking walks of length r that start at x and end at y.

14.10.85 Proposition If X is a k-regular graph with n vertices and adjacency matrix A, then

1. $A_0 = I_n, A_1 = A$;
2. $A_2 = A_1^2 - kI_n$;
3. $A_{r+1} = A_1 A_r - (k-1)A_{r-1}$ for every $r \geq 2$.

14.10.86 Proposition [780, 1969] Let U_m denote the Chebychev polynomial of the second kind defined by expressing $\frac{\sin(m+1)\theta}{\sin\theta}$ as a polynomial of degree m in $\cos\theta$:

$$U_m(\cos\theta) = \frac{\sin(m+1)\theta}{\sin\theta}.$$

Then

$$\sum_{r=0}^{\lfloor \frac{m}{2} \rfloor} A_{m-2r} = (k-1)^{\frac{m}{2}} U_m\left(\frac{A}{2\sqrt{k-1}}\right).$$

14.10.87 **Definition** A graph X is *vertex-transitive* if the automorphism group of X acts transitively on its vertex set which means that for any $x, y \in V(X)$ there exists an automorphism σ of X such that $\sigma(x) = y$.

14.10.88 **Proposition** [780, 1969] If X is a k-regular graph with n vertices and eigenvalues $k = \lambda_1 \geq \lambda_2 \geq \cdots \geq \lambda_n$, then

$$\sum_{x \in V} \sum_{r=0}^{\lfloor \frac{l}{2} \rfloor} (A_{l-2r})(x,x) = (k-1)^{\frac{l}{2}} \sum_{j=1}^{n} \frac{\sin(l+1)\theta_j}{\sin \theta_j}.$$

where $\cos\theta_j = \frac{\lambda_j}{2\sqrt{k-1}}$ for each $1 \leq j \leq n$. If X is vertex-transitive of degree k, then $(A_j)(x,x) = (A_j)(y,y)$ for any j and $x, y \in V(X)$ and thus,

$$n \sum_{r=0}^{\lfloor \frac{l}{2} \rfloor} (A_{l-2r})(x,x) = (k-1)^{\frac{l}{2}} \sum_{j=1}^{n} \frac{\sin(l+1)\theta_j}{\sin \theta_j}.$$

for every vertex $x \in V(X)$.

14.10.89 **Remark** Note that $\theta_1 = i \log \sqrt{k-1}$ as $\lambda_1 = k$ and $\theta_n = \pi + i \log \sqrt{k-1}$ if $\lambda_n = -k$. Also, it is important to observe that θ_j is real if $|\lambda_j| = |2\sqrt{k-1}\cos\theta_j| \leq 2\sqrt{k-1}$; otherwise, $\theta_j = iIm(\theta_j)$ is purely imaginary if $\lambda_j > 2\sqrt{k-1}$ and $\theta_j = \pi + iIm(\theta_j)$ if $\lambda_j < 2\sqrt{k-1}$.

14.10.90 **Remark** The general idea of using quaternions (see Lubotzky, Phillips, and Sarnak [1969] or Margulis [2006]) to construct infinite families of k-regular Ramanujan graphs can be summarized in the following two steps:

1. The first step consists of constructing the infinite k-regular tree T_k as the free group of some group $\mathcal{G}$ of quaternions integers with some suitable set of k generators $\mathcal{S}_k$. Thus, T_k will be identified with the Cayley graph $X(\mathcal{G}, \mathcal{S}_k)$.
2. Finite k-regular graphs are constructed from the infinite k-regular tree T_k by taking suitable finite quotients of it. More precisely, by choosing appropriate normal subgroups $\mathcal{H}$ of $\mathcal{G}$ of finite index, one can construct finite k-regular graphs which are the Cayley graphs of the quotient group $\mathcal{G}/\mathcal{H}$ with the set of generators being formed by the cosets of the form $\alpha\mathcal{H}$ where $\alpha \in \mathcal{S}_k$.

14.10.91 **Construction** [1969, 2006] Let p and q be unequal primes $p, q \equiv 1 \,(\text{mod } 4)$. Let u be an integer so that $u^2 \equiv -1 \,(\text{mod } q)$. By a classical formula of Jacobi, we know that there are $8(p+1)$ solutions $v = (a,b,c,d)$ such that $a^2 + b^2 + c^2 + d^2 = p$. Among these, there are exactly $p+1$ with $a > 0$ and b, c, d even, as is easily shown. To each such v we associate the matrix

$$\tilde{v} = \begin{pmatrix} a + ub & c + ud \\ -c + ud & a - ub \end{pmatrix}$$

which gives $p+1$ matrices in $PGL_2(\mathbb{Z}/q\mathbb{Z})$. We let S be the set of these matrices $\tilde{v}$ and define

$$X^{p,q} = \begin{cases} X(PGL_2(\mathbb{Z}/q\mathbb{Z}), S), \text{ if } \left(\frac{p}{q}\right) = -1, \\ X(PSL_2(\mathbb{Z}/q\mathbb{Z}), S), \text{ if } \left(\frac{p}{q}\right) = 1, \end{cases} \tag{14.10.3}$$

where $\left(\frac{p}{q}\right)$ is the Legendre symbol that equals 1 if p is a square modulo q and -1 if p is not a square modulo q. In [1969], it is shown that the Cayley graphs $X^{p,q}$ are Ramanujan graphs. As we vary q, we get an infinite family of such graphs, all $(p+1)$-regular.

14.10.92 Remark The integer quaternion algebra is

$$\mathrm{H}(\mathbb{Z}) = \{a_0 + a_1\mathbf{i} + a_2\mathbf{j} + a_3\mathbf{k} : a_0, a_1, a_2, a_3 \in \mathbb{Z}\},$$

where $\mathbf{i}^2 = \mathbf{j}^2 = \mathbf{k}^2 = -1$ and $\mathbf{ij} = -\mathbf{ji} = k, \mathbf{jk} = -\mathbf{kj} = \mathbf{i}, \mathbf{ki} = -\mathbf{ik} = \mathbf{j}$. If $\alpha = a_0 + a_1\mathbf{i} + a_2\mathbf{j} + a_3\mathbf{k}$, then $\overline{\alpha} = a_0 - a_1\mathbf{i} - a_2\mathbf{j} - a_3\mathbf{k}$ and $N(\alpha) = \alpha\overline{\alpha} = a_0^2 + a_1^2 + a_2^2 + a_3^2$. The units of $\mathrm{H}(\mathbb{Z})$ are $\pm 1, \pm\mathbf{i}, \pm\mathbf{j}, \pm\mathbf{k}$. Let p be a prime with $p \equiv 1 \pmod 4$. By Jacobi's Theorem, there are $8(p+1)$ integer quaternions $\alpha = a_0 + a_1\mathbf{i} + a_2\mathbf{j} + a_3\mathbf{k}$ in $\mathrm{H}(\mathbb{Z})$ such that $N(\alpha) = p$. As $p \equiv 1 \pmod 4$, only one of the integers a_i will be odd. Let S be the set of those $p+1$ elements of $\mathrm{H}(\mathbb{Z})$ such that $N(\alpha) = p$, $a_0 > 0$ is odd and a_1, a_2, a_3 even (this last fact is denoted by $\alpha \equiv 1 \pmod 2$ from now on). As $N(\alpha) = N(\overline{\alpha})$, the set S consists of $s = \frac{p+1}{2}$ conjugate pairs $S = \{\alpha_1, \overline{\alpha_1}, \ldots, \alpha_s, \overline{\alpha_s}\}$. A reduced word of length m with letters in S is defined to be a word of length m in the elements of S which does not contain subwords of the form $\alpha_j\overline{\alpha_j}$ nor $\overline{\alpha_j}\alpha_j$.

14.10.93 Construction (Different construction of the graphs $X^{p,q}$) Define $\Lambda'(2) = \{\alpha : \alpha \in \mathbb{Z}, \alpha \equiv 1 \pmod 2), N(\alpha) = p^l, l \in \mathbb{Z}\}$. As $N(\alpha\beta) = N(\alpha)N(\beta)$ and the properties of quaternion multiplication, it follows that $\Lambda'(2)$ is closed under multiplication. Define $\alpha \sim \beta$ for $\alpha, \beta \in \Lambda'(2)$ whenever $\pm p^{v_1}\alpha = p^{v_2}\beta$ for some $v_1, v_2 \in \mathbb{Z}$. This is an equivalence relation and $[\alpha]$ will denote the equivalence class of $\alpha \in \Lambda'(2)$. The set of equivalence classes $\Lambda(2) = \{[\alpha] : \alpha \in \Lambda'(2)\}$ forms a group with the multiplication $[\alpha][\beta] = [\alpha\beta]$ and $[\alpha][\overline{\alpha}] = [1]$. One of the key observations at this point is that, by previous results, the group $\Lambda(2)$ is free on $[\alpha_1], [\alpha_2], \ldots, [\alpha_s]$. This means that the Cayley graph of $\Lambda(2)$ with respect to the set $[S] = \{[\alpha_1], [\overline{\alpha_1}], \ldots, [\alpha_s], [\overline{\alpha_s}]\}$ will be the infinite $(p+1)$-regular tree.

For m coprime with p, let

$$\Lambda(2m) = \{[\alpha] : \alpha = a_0 + a_1\mathbf{i} + a_2\mathbf{j} + a_3\mathbf{k} \in \Lambda'(2), 2m | a_j, 1 \le j \le 3\}.$$

It can be shown that $\Lambda(2m)$ is a normal subgroup of $\Lambda(2)$ of finite index. Let q be a prime. The graphs $X^{p,q}$ and the Cayley graph of $\Lambda(2)/\Lambda(2q)$ with respect to the set of generators $\alpha_1\Lambda(2q), \overline{\alpha_1}\Lambda(2q), \ldots, \alpha_s\Lambda(2q), \overline{\alpha_s}\Lambda(2q)$ are isomorphic as shown by the next result.

14.10.94 Proposition Let $\phi : \Lambda(2) \to PGL(2, \mathbb{Z}/q\mathbb{Z})$ defined as follows:

$$\phi([a_0 + a_1\mathbf{i} + a_2\mathbf{j} + a_3\mathbf{k}]) = \begin{bmatrix} a_0 + ua_1 & a_2 + ua_3 \\ -a_2 + ua_3 & a_0 - ua_1 \end{bmatrix}$$

where $u^2 \equiv -1 \pmod q$. Then ϕ is a group homomorphism whose kernel is $\Lambda(2q)$ and whose image is $PGL(2, \mathbb{Z}/q\mathbb{Z})$ if $\left(\frac{p}{q}\right) = -1$ and $PSL(2, \mathbb{Z}/q\mathbb{Z})$ if $\left(\frac{p}{q}\right) = 1$.

14.10.95 Definition Let $Q = Q(x_0, x_1, x_2, x_3)$ denote the quadratic form

$$Q(x_0, x_1, x_2, x_3) = x_0^2 + (2q)^2x_1^2 + (2q)^2x_2^2 + (2q)^2x_3^2.$$

Denote by $r_Q(n)$ the number of integer solutions of $Q(x_0, x_1, x_2, x_3) = n$ which is the same as the number of $\alpha = a_0 + a_1\mathbf{i} + a_2\mathbf{j} + a_3\mathbf{k} \in \mathrm{H}(\mathbb{Z})$ such that $2q | \alpha - a_0$ and $N(\alpha) = n$.

14.10.96 Remark Estimating $r_Q(n)$ is an important and difficult problem in number theory. According to [1969], there is no simple explicit formula for $r_Q(n)$ as Jacobi's formula because of additional cusp forms that appear at the higher level. The Ramanujan conjecture for weight 2 cusp forms and its proof by Eichler and Igusa yields a good approximation for $r_Q(n)$. More precisely, if p is a prime and $l \ge 0$, then

$$r_Q(p^l) = C(p^l) + O_\epsilon(p^{\frac{l}{2}+\epsilon})$$

for any $\epsilon > 0$ as $l \to \infty$. Here

$$C(p^l) = \begin{cases} 2\sum_{d|p^l} d = 2\frac{p^{l+1}-1}{p-1} \text{ if } \left(\frac{p}{q}\right) = 1, \\ 4\sum_{d|p^l} d = 4\frac{p^{l+1}-1}{p-1} \text{ if } \left(\frac{p}{q}\right) = -1 \text{ and } l \text{ even}, \\ 0 \text{ if } \left(\frac{p}{q}\right) = -1 \text{ and } l \text{ odd}. \end{cases}$$

14.10.97 Remark The number theoretic facts above and the connection between the eigenvalues and the number of closed nonbacktracking walks in a regular graph were used by Lubotzky, Phillips, and Sarnak to prove the following result.

14.10.98 Theorem [1969, 2006] The graphs $X^{p,q}$ are Ramanujan.

14.10.99 Remark If $\left(\frac{q}{p}\right) = -1$, then $X^{p,q}$ is bipartite of high girth; its girth is at least $4\log_p q - \log_p 4 \approx \frac{4}{3}\log_p |V(X^{p,q})|$. If $\left(\frac{q}{p}\right) = 1$, then $X^{p,q}$ also has high girth; its girth is at least $2\log_p q \approx \frac{2}{3}\log_p |V(X^{p,q})|$. From the results of Hoffman, it also follows that these graphs have large chromatic number (at least $1 + \frac{p+1}{2\sqrt{p}}$).

14.10.100 Remark Morgenstern [2160] generalized Lubotzky, Phillips, and Sarnak's construction and constructed infinite families of $(q+1)$-regular Ramanujan graphs for every prime power q.

14.10.6 Combinatorial constructions of expanders

14.10.101 Construction Reingold, Vadhan, and Wigderson [2448] introduced a new graph product called the *zig-zag product* which they used to construct infinite families of constant-degree expanders.

14.10.102 Definition Let X be a k-regular graph with vertex set $[n] = \{1, \ldots, n\}$. Suppose the edges incident to each vertex of X are labeled from 1 to k in some arbitrary, but fixed way. The rotation map $Rot_X : [n] \times [k] \to [n] \times [k]$ is defined as follows: $Rot_X(u, i) = (v, j)$ if the i-th edge incident to u is the j-th edge incident to v.

14.10.103 Definition Let G_1 be a D_1-regular graph with vertex set $[N_1]$ with rotation map Rot_{G_1} and G_2 be a D_2-regular graph with vertex set $[D_1]$ with rotation map Rot_{G_2}. The *zig-zag product* $G_1 z G_2$ is the D_2^2-regular graph with vertex set $[N_1] \times [D_1]$ whose rotation map $Rot_{G_1 z G_2}$ is:

1. let $(k', i') = Rot_{G_2}(k, i)$;
2. let $(w, l') = Rot_{G_1}(v, k')$;
3. let $(l, j') = Rot_{G_2}(l', j)$;
4. define $Rot_{G_1 z G_2}((v, k), (i, j)) = ((w, l), (i', j'))$.

14.10.104 Definition A graph G is an (n, d, λ)-graph if G has n vertices, is d-regular, and the absolute value of any non-trivial eigenvalue of G is at most λd.

14.10.105 Theorem [2448] If G_1 is an (N_1, D_1, μ_1)-graph and G_2 is an (D_1, D_2, μ_2)-graph, then $G_1 z G_2$ is an $(N_1 D_1, D_2^2, \mu_1 + \mu_2 + \mu_2^2)$-graph.

14.10.106 Construction Using the previous theorem, Reingold, Vadhan, and Wigderson [2448] constructed infinite families of constant-degree expanders.

4.10.107 Construction Bilu and Linial [282] have used graph lifts to construct infinite families of d-regular graphs whose non-trivial eigenvalues have absolute value at most $C\sqrt{d \log^3 d}$, where C is some positive absolute constant. We outline their method below.

4.10.108 Definition Given two graphs G_1 and G_2, a *graph homomorphism* from G_1 to G_2 is a function $f : V(G_1) \to V(G_2)$ which preserves adjacency, namely if xy is an edge of G_1, then $f(x)f(y)$ is an edge of G_2; the function f is a *graph isomorphism* if it is bijective and preserves both adjacency and non-adjacency, namely xy is an edge of G_1 if and only if $f(x)f(y)$ is an edge of G_2.

4.10.109 Definition A surjective homomorphism $f : V(G_1) \to V(G_2)$ is a *covering map* (see [282, 1126]) if for each vertex x of G_1, the restriction of f to x and its neighbors is bijective. Given a graph G, a cover of G is a pair (H, f), where $f : V(H) \to V(G)$ is a covering map. If in addition G is connected and finite and H is finite, then for each vertex $y \in V(G)$, the preimage $f^{-1}(y)$ has the same cardinality. If $|f^{-1}(y)| = t$ for each $y \in V(G)$, then (H, f) is a *t-cover* or H is a *t-lift* of G.

4.10.110 Remark For every finite graph G, there is a universal cover or a *largest cover* $\tilde{G}$ which is an infinite tree whose vertices can be identified with the set of nonbacktracking walks from a fixed vertex $x \in V(G)$. For example, the universal cover of any finite k-regular graph is the infinite k-regular tree T_k.

4.10.111 Remark An important property of a t-cover (H, f) of a finite graph G is that the graph H inherits the eigenvalues of G. This is because the vertex set of H can be thought as $V(G) \times \{1, \ldots, t\}$ with the preimage (also called the fiber of y) $f^{-1}(y) = \{x : x \in V(H), f(x) = y\} = \{(y, i) : 1 \le i \le t\}$. The edges of H are related to the edges of G as follows: each fiber $f^{-1}(y)$ induces an independent set in H; if $yz \in E(G)$, then the subgraph of H induced by $f^{-1}(y) \cup f^{-1}(z) = \{(y, i), (z, i) : 1 \le i \le t\}$ is a perfect matching (meaning that there exists a permutation $\sigma \in S_t$ such that (y, i) is adjacent to $(z, \sigma(i))$ for each $1 \le i \le t$); if $yz \notin E(G)$, then there are no edges between $f^{-1}(y)$ and $f^{-1}(z)$. The partition of the vertex set of H as $V(H) = \cup_{y \in V(G)} f^{-1}(y)$ is equitable (see [1287]) and its quotient matrix is the same as the adjacency matrix of G. This implies the eigenvalues of $A(G)$ are the eigenvalues of $A(H)$. These eigenvalues of $A(H)$ are *old* and the remaining eigenvalues of $A(H)$ are *new*.

4.10.112 Remark In the case of a 2-lift, the new eigenvalues can be interpreted as eigenvalues of a *signed* adjacency matrix as follows. If H is a 2-lift of G, then for each edge yz of G, the subgraph induced by $f^{-1}(y) \cup f^{-1}(z) = \{(y, 0), (y, 1), (z, 0), (z, 1)\}$ in H has either $(y, 0)$ adjacent to $(z, 0)$ and $(y, 1)$ adjacent to $(z, 1)$ (in which case set $s(y, z) = s(z, y) = 1$) or $(y, 0)$ adjacent to $(z, 1)$ and $(y, 1)$ adjacent to $(z, 0)$ (in which case set $s(y, z) = s(z, y) - 1$). Let $s(y, z) = 0$ for all other $y, z \in V(G)$. The symmetric $\{0, -1, 1\}$ matrix A_s whose (y, z)-th entry is $s(y, z)$ is the *signed adjacency matrix* of the G with respect to the cover H. It is known that the eigenvalues of H are union of the eigenvalues of the adjacency matrix of G and the eigenvalues of the signed adjacency matrix of G. Bilu and Linial [282] proved that every graph G with maximum degree d has a signed adjacency matrix (which can be found efficiently) whose eigenvalues have absolute value at most $C\sqrt{d \log^3 d}$ where C is some positive absolute constant.

4.10.113 Construction Bilu and Linial's idea to construct almost Ramanujan graphs is the following: start with a k-regular Ramanujan graph G_0 (for example, the complete graph K_{k+1}) and then construct a 2-lift of G_i (denoted by G_{i+1}) such that the new eigenvalues of G_{i+1} are small in absolute value for $i \ge 0$. Bilu and Linial [282] prove that every k-regular graph G

has a 2-lift H such that the new eigenvalues of H have absolute value at most $C\sqrt{k \log^3 k}$ where C is some positive absolute constant. In this way the sequence of k-regular graphs G_i has non-trivial eigenvalues bounded from above by $C\sqrt{k \log^3 k}$.

4.10.114 **Remark** Bilu and Linial [282] make the following conjecture which if true, would imply the existence of infinite sequences of k-regular Ramanujan graphs for every $k \geq 3$.

4.10.115 **Conjecture** [282] Every k-regular graph has a signed adjacency matrix whose eigenvalues have absolute value at most $2\sqrt{k-1}$.

4.10.116 **Construction** A different combinatorial construction of almost Ramanujan graphs was proposed by de la Harpe and Musitelli [786], and independently by Cioabă and Murty [646, 649]. The idea of these constructions is that perturbing Ramanujan graphs by adding or removing perfect matchings will yield graphs with small non-trivial eigenvalues. The linear algebraic reason for this fact follows from a theorem of Weyl which bounds the eigenvalues of a sum of two Hermitian matrices in terms of the eigenvalues of the summands. De la Harpe and Musitelli [786] note that adding a perfect matching to any 6-regular Ramanujan graph will yield a 7-regular graph whose 2nd largest eigenvalue is at most $2\sqrt{5}+1 \cong 5.47$ which is larger than the Ramanujan bound of $2\sqrt{6} \cong 4.89$, but strictly less than 7. Cioabă and Murty [646, 649] use known results regarding gaps between consecutive primes to observe that by adding or removing perfect matching from Ramanujan graphs, one can construct k-regular almost Ramanujan graphs for almost all k. More precisely, their result is that given $\epsilon > 0$, for almost all $k \geq 3$, one can construct infinite families of k-regular graphs whose 2nd largest eigenvalue is at most $(2+\epsilon)\sqrt{k-1}$.

4.10.117 **Remark** The following conjecture was made in [646]; if true, this conjecture would imply the existence of infinite families of k-regular Ramanujan graphs for any $k \geq 3$.

4.10.118 **Conjecture** [646] Let X be a k-regular Ramanujan graph with an even number of vertices. Then there exists a perfect matching P with $V(P) = V(X)$ such that the $(k+1)$-regular graph obtained from the union of the edges of X and P is Ramanujan.

4.10.119 **Remark** In a recent outstanding work, Friedman [1127] solved a long-standing conjecture of Alon from the 1980s and proved that almost all regular graphs are almost Ramanujan.

4.10.120 **Theorem** [1127] Given $\epsilon > 0$ and $k \geq 3$, the probability that a random k-regular graph on n vertices has all non-trivial eigenvalues at most $(2+\epsilon)\sqrt{k-1}$ goes to 1 as n goes to infinity.

14.10.7 Zeta functions of graphs

4.10.121 **Definition** A walk in a graph X is *non-backtracking* if no edge is traversed and then immediately backtracked upon. A non-backtracking walk whose endpoints are equal is a *closed geodesic.* If γ is a closed geodesic, we denote by γ^r the closed geodesic obtained by repeating the walk γ r times. A walk is *tailless* if it is non-backtracking under any cyclic permutation of vertices. A closed geodesic which is not the power of another one and is tailless is a *prime geodesic.* We define an equivalence relation on the closed geodesics as follows: $(x_0, ..., x_n)$ and $(y_0, ..., y_m)$ are equivalent if and only if $m = n$ and there is a d such that $y_i = x_{i+d}$ for all i (and the subscripts are interpreted modulo n). An equivalence class of a prime geodesic is a *prime geodesic cycle.*

4.10.122 **Definition** Let X be a k-regular graph and denote $q = k - 1$. The *Ihara zeta function* is

$$Z_X(s) = \prod_p \left(1 - q^{-s\ell(p)}\right)^{-1}$$

where the product is over all prime geodesic cycles p and $\ell(p)$ is the length of p.

4.10.123 **Theorem** [1568] For $g = (q-1)|X|/2$, we have

$$Z_X(s) = (1 - u^2)^{-g} \det(I - Au + qu^2 I)^{-1}, \quad u = q^{-s}.$$

Moreover, $Z_X(s)$ satisfies the *Riemann hypothesis'* (that is, all the singular points in the region $0 < \Re(s) < 1$ lie on $Re(s) = 1/2$) if and only if X is a Ramanujan graph.

4.10.124 **Remark** Hashimoto [1441], as well as Stark and Terras [2701] have defined a zeta function for an arbitrary graph and established its rationality. The definition of this zeta function is simple enough. Let N_r be the number of closed walks γ of length r so that neither γ nor γ^2 have backtracking. Then, the *zeta function* of the graph X is defined as

$$\mathcal{Z}_X(t) = \exp\left(\sum_{r=1}^{\infty} \frac{N_r t^r}{r}\right).$$

This definition is very similar to the zeta function of an algebraic variety.

4.10.125 **Remark** It would be interesting to interpret the singularities of $\mathcal{Z}_X(t)$ in terms of properties of the graph. For instance, these zeta functions have a pole at $t = 1$ and Hashimoto [1441] has shown that the residue at $t = 1$ is related to the number of spanning trees of the graph X. Thus, this number is the graph-theoretic analogue of the class number of an algebraic number field. These constructions raise the intriguing question of whether there is a generalization of the notion of a graph to that of a "supergraph" whose zeta function would (in some cases) coincide with those higher dimensional zeta functions of varieties. Work in this direction has started [1926].

See Also

§6.1, §6.2	For details on Gauss sums and other character sums.
§12.7	For discussions on zeta functions and L-functions of curves.
§15.1, §15.3	For details on algebraic, LDPC, and expander codes.

References Cited: [80, 81, 83, 156, 157, 269, 282, 411, 637, 638, 645, 646, 647, 648, 649, 780, 786, 834, 1126, 1127, 1128, 1287, 1441, 1534, 1535, 1568, 1641, 1693, 1807, 1921, 1922, 1923, 1924, 1925, 1926, 1927, 1967, 1968, 1969, 1970, 2005, 2006, 2117, 2118, 2160, 2207, 2208, 2289, 2290, 2433, 2448, 2523, 2525, 2589, 2594, 2701, 2789, 2837, 2839]

15

Algebraic coding theory

15.1 Basic coding properties and bounds

Ian Blake, *University of British Columbia*
W. Cary Huffman, *Loyola University Chicago*

15.1.1 Channel models and error correction

15.1.1 **Definition** A *discrete memoryless channel (DMC)* is a set of possible inputs $\{u_1, u_2, \ldots, u_K\}$ and outputs $\{v_1, v_2, \ldots, v_L\}$ and probabilities p_{ij} such that if the input on a given channel use is u_i, then the probability that the channel output is v_j is $P(v_j \mid u_i) = p_{ij}$, and the channel behavior is independent from one usage to the next. Note that

$$\sum_{j=1}^{L} p_{ij} = 1 \quad \text{for } 1 \le i \le K.$$

Thus for a given input vector $\mathbf{x} = (x_1, x_2, \ldots, x_n)$ and output vector $\mathbf{y} = (y_1, y_2, \ldots, y_n)$

$$P(\mathbf{y} \mid \mathbf{x}) = \prod_{i=1}^{n} P(y_i \mid x_i).$$

15.1.2 Example A channel with two inputs $\{0,1\}$ and two outputs $\{0,1\}$ is a *binary symmetric channel (BSC)* if $p_{00} = p_{11} = 1-p$, $p_{01} = p_{10} = p$; p is the *crossover probability*.

15.1.3 Example A channel with two inputs $\{0,1\}$ and three outputs $\{0,E,1\}$ is a *binary erasure channel (BEC)* if $p_{00} = p_{11} = 1-\epsilon$, $p_{0E} = p_{1E} = \epsilon$, and $p_{01} = p_{10} = 0$. An *erasure* in the received word is a position containing the symbol E.

15.1.4 Remark There are many other types of channels, including q-ary channels (q inputs), burst noise channels, and the *additive white Gaussian noise (AWGN)* channel, which is a continuous channel that adds white Gaussian noise to the transmitted signal.

15.1.5 Definition A *code* is a subset of $\mathbb{F}_q^n$, the set of n-tuples over the finite field $\mathbb{F}_q$ of order q. Elements (vectors) of $\mathbb{F}_q^n$ are denoted by boldface lowercase letters.

15.1.6 Definition For $\mathbf{x} \in \mathbb{F}_q^n$, denote by $\omega(\mathbf{x})$ the *Hamming weight* of $\mathbf{x}$ as the number of nonzero coordinate positions of $\mathbf{x}$. For $\mathbf{x}$, $\mathbf{y} \in \mathbb{F}_q^n$, denote by $d(\mathbf{x},\mathbf{y})$ the *Hamming distance* between $\mathbf{x}$ and $\mathbf{y}$ as the number of coordinate positions where $\mathbf{x}$ and $\mathbf{y}$ differ.

15.1.7 Remark Note that $d(\mathbf{x},\mathbf{y}) = \omega(\mathbf{x}-\mathbf{y})$.

15.1.8 Definition A code C is an $(n,M,d)_q$ *code* if it has M distinct codewords of length n over the finite field $\mathbb{F}_q$ such that the *minimum distance* between any two distinct codewords is d where

$$d = \min_{\substack{\mathbf{c}_1,\mathbf{c}_2 \in C \\ \mathbf{c}_1 \neq \mathbf{c}_2}} d(\mathbf{c}_1,\mathbf{c}_2).$$

The *rate* r of such a code is the ratio of the number of information symbols transmitted per codeword to the number of symbols per codeword, assuming the codewords are chosen equally likely, i.e., $r = \log_q(M)/n$.

15.1.9 Remark An important problem of coding theory is to design codes that have a large minimum distance for a given code size (or large size for a given minimum distance) and to have encoding and decoding algorithms that can be implemented efficiently. Many types of encoders and decoders are considered in the literature.

15.1.10 Remark Unless specified otherwise, a received word is the sum of a codeword transmitted over a discrete memoryless channel and an error word. All such words are vectors over a finite field.

15.1.11 Definition A *maximum likelihood decoder (MLD)* [1164] is one that, on receiving a word of channel output symbols $\mathbf{r}$, chooses the codeword $\hat{\mathbf{c}} \in C$ that maximizes the probability of $\mathbf{r}$ being received, i.e., maximizes the conditional probability $P(\mathbf{r} \mid \mathbf{c})$:

$$\hat{\mathbf{c}} = \arg\max_{\mathbf{c} \in C} P(\mathbf{r} \mid \mathbf{c}).$$

15.1.12 Definition A *maximum a posteriori (MAP) decoder* [1164] is one that, on receiving the channel output word $\mathbf{r}$, chooses the codeword $\hat{\mathbf{c}} \in C$ that maximizes the a posteriori probability $P(\mathbf{c} \mid \mathbf{r})$, i.e.,

$$\hat{\mathbf{c}} = \arg\max_{\mathbf{c} \in C} P(\mathbf{c} \mid \mathbf{r}).$$

15.1.13 Definition A *minimum distance decoder* [1164] is one that, on receiving the word $\mathbf{r}$, chooses a codeword $\hat{\mathbf{c}} \in C$ that is at minimum distance from $\mathbf{r}$, i.e.,

$$\hat{\mathbf{c}} = \arg\min_{\mathbf{c}\in C} d(\mathbf{c}, \mathbf{r}).$$

A *bounded distance decoder* attempts to find the codeword, if it exists, within distance $\lfloor (d-1)/2 \rfloor$ of the received word.

15.1.14 Remark The output from these decoders may not be unique. In this case, in an equally likely scenario, the decoder chooses one of the codewords satisfying the criteria. In the case of choosing codewords equally likely, MAP is equivalent to MLD.

15.1.15 Lemma On a BSC with crossover probability $p < 1/2$, the minimum distance decoder is equivalent to the MLD [304, 1943, 2389].

15.1.16 Remark If C is an $(n, M, d)_q$ code, one may think of the codewords being surrounded with "spheres" of radius $e = \lfloor (d-1)/2 \rfloor$, i.e. the set of q-ary n-tuples at distance e or less from the center. The spheres are then nonintersecting.

15.1.17 Lemma If C is an $(n, M, d)_q$ code and a codeword $\mathbf{c}$ is transmitted and word $\mathbf{r}$ is received, with r errors and s erasures, then $\mathbf{c}$ is the unique codeword in C closest to the received $\mathbf{r}$ provided that $2r + s < d$.

15.1.18 Remark Shannon [2608] showed that with every discrete memoryless channel with a finite number of input and output symbols, one may define the notion of *channel capacity* in bits per channel use. His celebrated channel coding theorem and its converse then state there exists a code that allows essentially error-free transfer of data over the channel as long as the rate does not exceed channel capacity. Conversely, he showed that such transmission is not possible if the channel rate exceeds capacity. More specifically, Gallager [1164] showed that if C is the channel capacity, there is an *error-rate exponent* function $E(R)$ that is positive for rates $0 \leq R < C$ and zero elsewhere, and there exists a code of rate $R < C$ and length $n > N$ for which the probability P_w of word error on the channel satisfies

$$P_w < \exp(-NE(R)) \quad \text{for } 0 \leq R < C.$$

The importance of the result is that there exists a sequence of codes of increasing lengths $n > N$ and of rates less than the channel capacity for which the probability of word error, after maximum likelihood decoding, decreases exponentially to zero.

15.1.19 Remark Algebraic coding theory arose in response to the challenge of the Shannon theorems to define codes, as well as efficient encoding and decoding algorithms, that achieve asymptotically low error rate at code rates arbitrarily close to capacity. There are numerous excellent books available on the subject. The following were particularly useful references in preparing this section: [231, 304, 311, 1558, 1639, 1943, 1945, 1991, 2047, 2136, 2389, 2390, 2405, 2484, 2511, 2849].

15.1.2 Linear codes

15.1.20 Definition A *linear code* C of length n over $\mathbb{F}_q$ is a subspace of $\mathbb{F}_q^n$. If the dimension of C is k and the minimum distance is d, it is referred to as an $(n, k, d)_q$ code and the number of codewords is $M = q^k$.

15.1.21 Remark The anomaly in notation from the nonlinear case with a designation of $(n, M, d)_q$ code, where the number of codewords is M, will be clear from the context.

15.1.22 Remark To determine the minimum distance of a nonlinear $(n, M, d)_q$ code $\mathcal{C}$ will in general require a search over the distances between all $\binom{M}{2}$ pairs of codewords. For a linear code this can be reduced to determining the weight of a minimum weight codeword since

$$d = \min_{\substack{\mathbf{x},\mathbf{y}\in\mathcal{C}\\ \mathbf{x}\neq\mathbf{y}}} d(\mathbf{x},\mathbf{y}) = \min_{\substack{\mathbf{x}\in\mathcal{C}\\ \mathbf{x}\neq\mathbf{0}}} w(\mathbf{x}).$$

15.1.23 Definition The *scalar (or inner) product* of two vectors $\mathbf{u} = (u_1, u_2, \dots, u_n)$, $\mathbf{v} = (v_1, v_2, \dots, v_n)$ from $\mathbb{F}_q^n$ is

$$(\mathbf{u}, \mathbf{v}) = u_1v_1 + u_2v_2 + \cdots + u_nv_n \in \mathbb{F}_q.$$

If $(\mathbf{u}, \mathbf{v}) = 0$, the vectors are *orthogonal.*

15.1.24 Definition For a linear code $\mathcal{C} \subseteq \mathbb{F}_q^n$, the *orthogonal code* or *dual code* to $\mathcal{C}$ is defined as

$$\mathcal{C}^\perp = \{\mathbf{v} \in \mathbb{F}_q^n \mid (\mathbf{v}, \mathbf{c}) = 0 \text{ for all } \mathbf{c} \in \mathcal{C}\}.$$

15.1.25 Theorem If $\mathcal{C}$ is a linear code of dimension k in $\mathbb{F}_q^n$, then $\mathcal{C}^\perp$ is a linear code of dimension $n - k$ in $\mathbb{F}_q^n$.

15.1.26 Definition Let $\mathcal{C}$ be a linear code in $\mathbb{F}_q^n$. If $\mathcal{C} \subseteq \mathcal{C}^\perp$, the code $\mathcal{C}$ is *self-orthogonal.* If $\mathcal{C} = \mathcal{C}^\perp$, $\mathcal{C}$ is *self-dual.*

15.1.27 Remark In $\mathbb{R}^n$, the only vector orthogonal to itself is the zero vector. So nonzero linear subspaces of $\mathbb{R}^n$ cannot be self-orthogonal. However, the situation is markedly different when considering subspaces of vector spaces over fields of prime characteristic; in that case, self-orthogonal and self-dual subspaces do exist. For example, $\mathcal{C} = \{(0,0), (1,1)\}$ is a self-dual $(2,1,2)_2$ code in $\mathbb{F}_2^2$.

15.1.28 Remark If $\mathcal{C}$ is a self-dual code, its dimension is $n/2$ and its length n is even.

15.1.29 Definition For $\mathcal{C}$ an $(n, k, d)_q$ linear code, a $k \times n$ matrix G is a *generator matrix* of $\mathcal{C}$ if its row space equals $\mathcal{C}$. Similarly an $(n - k) \times n$ matrix H is a *parity check matrix* for $\mathcal{C}$ if its row space equals $\mathcal{C}^\perp$.

15.1.30 Remark Note that some authors use the term parity cheek matrix only for the binary case, as in this case the inner product of a codeword and row of the matrix involves an even number of ones in the sum. We retain the term parity check matrix for all sizes of fields.

15.1.31 Remark As any codeword of $\mathcal{C}$ is orthogonal to any word in $\mathcal{C}^\perp$, it follows that

$$GH^T = [0] \quad (\text{the } k \times (n-k) \text{ all zero matrix}),$$

where superscript T denotes matrix transpose. Thus $\mathbf{c} \in \mathcal{C}$ if and only if $H\mathbf{c}^T = \mathbf{0}^T$ where $\mathbf{0}$ is the zero row vector of length $n - k$.

15.1.32 Remark By using elementary row operations (which preserve the row space of a matrix) and possibly coordinate position permutations, any generator matrix of a code $\mathcal{C}$ can be put in the form

$$G = [I_k \mid A],$$

where I_k is the $k \times k$ identity matrix and A is a $k \times (n-k)$ matrix over $\mathbb{F}_q$. A corresponding parity check matrix is one in the form

$$H = \left[-A^T \mid I_{n-k}\right].$$

15.1.33 Definition Generator and parity check matrices of a code $\mathcal{C}$ in the form of Remark 15.1.32 are in *systematic* form.

15.1.34 Remark Some authors (e.g. [1558]) refer to the form of matrices in Remark 15.1.32 as being in *standard form* and reserve "systematic" for any form where the information bits appear explicitly in the codewords.

15.1.35 Remark From the definitions it is clear that a parity check matrix of a linear code $\mathcal{C}$ is a generator matrix of the code $\mathcal{C}^\perp$, and a generator matrix of $\mathcal{C}$ is a parity check matrix of the code $\mathcal{C}^\perp$.

15.1.36 Remark If $\mathcal{C}$ is an $(n, k, d)_q$ code with parity check matrix H, then $\mathbf{c} \in \mathcal{C}$ if and only if $H\mathbf{c}^T = \mathbf{0}^T$. Thus if $\mathbf{c}$ is a minimum weight codeword, then $H\mathbf{c}^T = \mathbf{0}^T$ implies a linear combination of d columns of H is the all zero vector. It follows that $\mathcal{C}$ has minimum distance d if every subset of $d-1$ columns of H is linearly independent, and there is a dependent set of d columns.

15.1.37 Lemma In an $(n, k, d)_q$ code $\mathcal{C}$, any $d-1$ columns of a parity check matrix for $\mathcal{C}$ are linearly independent, and there is at least one set of d columns that is dependent.

15.1.38 Definition Let $\mathcal{C}$ be an $(n, M, d)_q$ code and let A_i (resp. B_i, if $\mathcal{C}$ is linear) be the number of codewords in $\mathcal{C}$ (resp. $\mathcal{C}^\perp$, which exists if $\mathcal{C}$ is linear) of weight i. Define the *weight enumerators* of these codes as

$$W_{\mathcal{C}}(x, y) = \sum_{i=0}^{n} A_i x^{n-i} y^i \quad \text{and} \quad W_{\mathcal{C}^\perp}(x, y) = \sum_{i=0}^{n} B_i x^{n-i} y^i.$$

The sequence $\{A_0, A_1, \ldots, A_n\}$ (resp. $\{B_0, B_1, \ldots, B_n\}$) is the *weight distribution* of the code $\mathcal{C}$ (resp. $\mathcal{C}^\perp$).

15.1.39 Theorem (MacWilliams identities) [304, 1558, 2849] Let $W_{\mathcal{C}}(x, y)$ (resp. $W_{\mathcal{C}^\perp}(x, y)$) be the weight enumerator of an $(n, k, d)_q$ linear code $\mathcal{C}$ (resp. of the $(n, n-k, d')_q$ dual code $\mathcal{C}^\perp$). Then

$$W_{\mathcal{C}^\perp}(x, y) = \frac{1}{q^k} W_{\mathcal{C}}(x + (q-1)y, x - y). \tag{15.1.1}$$

15.1.40 Remark An alternative version of the MacWilliams identities is obtained by expanding the terms in (15.1.1) to obtain

$$\sum_{i=0}^{n-j} \binom{n-i}{j} A_i = q^{k-j} \sum_{i=0}^{j} \binom{n-i}{n-j} B_i \quad \text{for } 0 \le j \le n.$$

15.1.41 Remark In [801] Delsarte proposed an interesting method to examine code properties and in particular their combinatorial aspects. It gives fundamental insight to the problem. Only a brief look at this approach will be given.

15.1.42 Definition [801] For positive integers n and λ the *Krawtchouk polynomial* $P_k(x)$ is the polynomial of degree k over the rational numbers

$$P_k(x) = \sum_{j=0}^{k} (-1)^j \lambda^{k-j} \binom{x}{j} \binom{n-x}{k-j} \quad \text{for } 0 \le k \le n$$

where

$$\binom{x}{j} = x(x-1)(x-2)\cdots(x-j+1)/j!.$$

These polynomials form a set of *orthogonal polynomials* and have many interesting properties such as

1. $\sum_{i=0}^{n} \binom{n}{i} P_k(i) P_\ell(i) = \delta_{kl} \binom{n}{k}$,
2. $\sum_{i=0}^{n} P_\ell(i) P_i(k) = \delta_{kl} (\lambda+1)^k$,
3. $\sum_{k=0}^{n} \binom{n-k}{n-j} P_k(x) = (\lambda+1)^j \binom{n-x}{j}$.

Later, when required, the parameter λ will be specified.

15.1.43 Definition [801] Let $A = (A_0, A_1, \ldots, A_n)$ be an $(n+1)$-tuple of rational numbers. The *MacWilliams transform of* A is given by $A' = (A'_0, A'_1, \ldots, A'_n)$ where

$$A'_k = \sum_{i=0}^{n} A_i P_k(i) \quad \text{for } 0 \le k \le n.$$

15.1.44 Remark If the (i,k) entry of a matrix P is defined as $P_k(i), 0 \le i, k \le n$, the MacWilliams transform can be expressed as $A' = AP$.

15.1.45 Remark It can be shown that $(A')' = (\lambda+1)^n A$ for any $(n+1)$-tuple A. Likewise $P^2 = (\lambda+1)^n I_{n+1}$.

15.1.46 Definition [801] Let $\mathcal{C}$ be an $(n, M, d)_q$ code. Define the *distance distribution* of $\mathcal{C}$ as

$$A_i(\mathcal{C}) = \frac{1}{M} \left| \{(\mathbf{x}, \mathbf{y}) \in \mathcal{C}^2 \mid (\mathbf{x}, \mathbf{y}) = i\} \right| \quad \text{for } 0 \le i \le n,$$

i.e., the average number of codewords at distance i from a fixed codeword. In the case $\mathcal{C}$ is linear, this is just the weight distribution of $\mathcal{C}$.

15.1.47 Definition [801] Define an $(n+1)$-tuple A as *positive* if $A_0 = 1$, $A_i \ge 0$, and $A'_i \ge 0$ for $0 \le i \le n$.

15.1.48 Remark The following lemma gives motivation for the appearance of Krawtchouk polynomials.

15.1.49 Lemma [801] Let W_k be the set of vectors of weight k in $\mathbb{F}_q^n$, and let $\mathbf{u}$ be a vector of weight j in $\mathbb{F}_q^n$. Then with $\lambda = q-1$

$$\sum_{\mathbf{x} \in W_k} (\mathbf{u}, \mathbf{x}) = P_k(j).$$

15.1.50 Lemma [801] The distance distribution of any code over $\mathbb{F}_q$ is a positive $(n+1)$-tuple for $\lambda = q-1$.

15.1.51 **Definition** Let $A(\mathcal{C})$ be the distance distribution of an $(n, M, d)_q$ code $\mathcal{C}$, and define the *four fundamental parameters* of $\mathcal{C}$ (where the parameter of the Krawtchouk polynomials is $\lambda = q - 1$) as

1. $d = d(A(\mathcal{C}))$ is the *minimum distance*,
2. $s = s(A(\mathcal{C}))$ is the *number of nonzero distances* (of $\mathcal{C}$),
3. $d' = d(A'(\mathcal{C}))$ is the *dual distance*,
4. $s' = s(A'(\mathcal{C}))$ is the *external distance*.

In the case the code $\mathcal{C}$ is linear, $A'(\mathcal{C})$ is a constant multiple of the weight distribution of $\mathcal{C}^\perp$.

15.1.52 **Remark** It is important to note these four parameters are defined for *any* code $\mathcal{C}$. The parameter d' is the smallest i such that $A'_i \neq 0, i > 0$. The parameter s' is the number of nonzero entries in the set $\{A'_1, i = 1, \ldots, n\}$. In the case the code is linear, the four parameters are the minimum distance and number of nonzero codeword weights of $\mathcal{C}$ and the dual code $\mathcal{C}^\perp$, respectively. The following three theorems from [801] (the first theorem was first proved by MacWilliams) are representative of the effectiveness of this approach.

15.1.53 **Theorem** [801] Let A be an $(n+1)$-tuple with $A_0 \neq 0$ and let A' be its MacWilliams transform. Then $s' \geq \lfloor (d-1)/2 \rfloor$.

15.1.54 **Definition** A code $\mathcal{C}$ over $\mathbb{F}_q$ is *distance invariant* if the weight distribution of $\mathbf{c} + \mathcal{C}$ is the same for all $\mathbf{c} \in \mathcal{C}$, i.e., the number of codewords in $\mathcal{C}$ at distance i from $\mathbf{c} \in \mathcal{C}$ is independent of $\mathbf{c}$.

15.1.55 **Remark** A linear code over $\mathbb{F}_q$ is automatically distance invariant because $\mathbf{c} + \mathcal{C} = \mathcal{C}$ for all $\mathbf{c} \in \mathcal{C}$.

15.1.56 **Theorem** [801] Let $\mathcal{C}$ be a code for which $s \leq d'$. Then $\mathcal{C}$ is distance invariant.

15.1.57 **Remark** The following theorem justifies the name *external distance* given to the parameter s'.

15.1.58 **Theorem** [801] Let $\mathcal{C}$ be a code with external distance s'. Then each point of $\mathbb{F}_q^n$ is at distance at most s' from at least one codeword.

15.1.2.1 Standard array decoding of linear codes

15.1.59 **Definition** Let $\mathcal{C}$ be an $(n, k, d)_q$ code. An *encoding* is an injective mapping from the set $\mathbb{F}_q^k$ of all messages to the set $\mathcal{C}$ of codewords. A *minimum distance decoding* is a map from the set $\mathbb{F}_q^n$ of all possible received vectors to the set $\mathcal{C}$ which sends the received vector $\mathbf{r}$ to a codeword closest to $\mathbf{r}$. The codeword is unique if the number of errors in transmission is at most $e = \lfloor \frac{d-1}{2} \rfloor$.

15.1.60 **Remark** A particular decoding algorithm that works for any linear code over $\mathbb{F}_q$ is the *coset* or *standard array decoding algorithm.* It is generally too inefficient for practical consideration. Since a linear $(n, k, d)_q$ code $\mathcal{C}$ is a vector subspace of $\mathbb{F}_q^n$, it can be viewed as an (additive) subgroup of $\mathbb{F}_q^n$. Thus $\mathbb{F}_q^n$ can be decomposed into the cosets of $\mathcal{C}$. Such a decomposition can be used to derive a decoding algorithm.

15.1.61 **Definition** Let $\mathcal{C}$ be a linear $(n,k,d)_q$ code. A *coset* of $\mathcal{C}$ with representative $\mathbf{u} \in \mathbb{F}_q^n$ is the set $\mathbf{u}+\mathcal{C} = \{\mathbf{u}+\mathbf{c} \mid \mathbf{c} \in \mathcal{C}\}$. For $\mathbf{u}_1,\ \mathbf{u}_2 \in \mathbb{F}_q^n$, as sets $\mathbf{u}_1+\mathcal{C}$ and $\mathbf{u}_2+\mathcal{C}$ are either disjoint or identical. Since $\mathcal{C}$ is closed under addition, it follows that $\mathbf{u}+\mathcal{C}$ equals $\mathbf{u}+\mathbf{c}+\mathcal{C}$ if $\mathbf{c} \in \mathcal{C}$. There are q^{n-k} distinct cosets of $\mathcal{C}$ in $\mathbb{F}_q^n$.

15.1.62 **Definition** The *standard array* for an $(n,k,d)_q$ code $\mathcal{C}$ is a $q^{n-k} \times q^k$ array of vectors of $\mathbb{F}_q^n$ formed in the following manner:

1. In the first row of the array, place the codewords of $\mathcal{C}$ in some ordering, with the all-zero codeword first.
2. Choose a vector $\mathbf{u}_1$ of minimum weight in $\mathbb{F}_q^n \setminus \mathcal{C}$ and add this to each vector in the first row to form the second row.
3. Choose a vector $\mathbf{u}_2$ of minimum weight in $\mathbb{F}_q^n \setminus \{\mathcal{C} \cup \{\mathbf{u}_1 + \mathcal{C}\}\}$ and add this to each vector in the first row to form the third row.
4. Continue the process until the vectors of $\mathbb{F}_q^n$ are exhausted.

Vectors in the first column of the array are *coset leaders.* Every vector of $\mathbb{F}_q^n$ appears exactly once in the array.

15.1.63 **Remark** To determine a codeword closest to a received word $\mathbf{r} \in \mathbb{F}_q^n$ proceed as follows:

1. Locate the received word $\mathbf{r}$ in the standard array.
2. If the coset containing $\mathbf{r}$ has coset leader $\mathbf{e}$, decode $\mathbf{r}$ to the codeword $\mathbf{r} - \mathbf{e}$.

This procedure is referred to as *standard array decoding.* The coset leaders are in fact the set of correctable error patterns. If $e = \lfloor \frac{d-1}{2} \rfloor$, then every vector of weight less than or equal to e will appear as a coset leader. The coset leaders of weight greater than e are then not "complete."

15.1.64 **Lemma** The standard array decoding algorithm of Remark 15.1.63 finds a codeword $\hat{\mathbf{c}}$ in $\mathcal{C}$ at minimum distance from the received word $\mathbf{r}$. This codeword is the transmitted one if at most $e = \lfloor \frac{d-1}{2} \rfloor$ errors were made in transmission.

15.1.65 **Remark** In what follows a large number of codes are introduced. For notation, where appropriate, a linear code of length n over $\mathbb{F}_q$ that has some parameter (or set of parameters) m will be denoted $ABC_{n,q}(m)$ for some sequence of Roman letters ABC, descriptive for that particular code. The duplication of notation with $(n,M,d)_q$ and $(n,k,d)_q$ will be clear from the context.

15.1.66 **Remark** A useful quantity to define is

$$\pi_{m,q} = \frac{q^m - 1}{q-1} = q^{m-1} + q^{m-2} + \cdots + 1.$$

This is the number of nonzero *projective m-tuples* where scalar multiples are identified.

15.1.2.2 Hamming codes

15.1.67 **Definition** The *m-th order Hamming code* over $\mathbb{F}_q$, denoted $H_{\pi_{m,q},q}(m)$, has a parity check matrix formed by constructing the $m \times \pi_{m,q}$ matrix whose columns are the set of distinct nonzero projective q-ary m-tuples. The code $H_{\pi_{m,q},q}(m)$ is a $(\pi_{m,q}, \pi_{m,q} - m, 3)_q$ code.

15.1.68 Remark The dimension of $H_{\pi_{m,q},q}(m)$ is easy to establish. That the code has minimum distance 3 is seen by finding some linear combination of three columns of the parity check matrix that equals zero. That there is no linear combination of two columns equaling zero follows from the columns being projectively distinct.

15.1.69 Remark The *binary Hamming code* $H_{\pi_{m,2},2}(m)$ is a $(2^m - 1, 2^m - 1 - m, 3)_2$ code. The dual of this code has all nonzero words of weight 2^{m-1}, and hence the weight enumerator of $H^{\perp}_{\pi_{m,2},2}(m)$ is

$$W_{H^{\perp}_{\pi_{m,2},2}}(x, y) = x^n + nx^{(n-1)/2}y^{(n+1)/2}, \quad n = 2^m - 1.$$

Using the MacWilliams identities it is seen that the weight enumerator of $H_{\pi_{m,2},2}(m)$ is

$$W_{H_{\pi_{m,2},2}}(x, y) = 2^{-m}\left((x + y)^n + n(x + y)^{(n-1)/2}(x - y)^{(n+1)/2}\right).$$

In a similar manner it can be established that the dual of the code $H_{\pi_{m,q},q}(m)$ over $\mathbb{F}_q$ has all nonzero weights equal to q^{m-1}. The dual codes to Hamming codes are *simplex* codes.

15.1.2.3 Reed-Muller codes

15.1.70 Remark For vectors $\mathbf{u}, \mathbf{v} \in \mathbb{F}_2^n$ define the product as

$$\mathbf{u} \otimes \mathbf{v} = (u_1v_1, u_2v_2, \ldots, u_nv_n).$$

This is also the coordinate-wise "AND" operation. Similarly the usual coordinate-wise addition over $\mathbb{F}_2$ corresponds to "XOR."

15.1.71 Definition Define the *Reed-Muller generator matrix* $G(r, m)$, $0 \le r \le m$, as follows.

1. The zero-th row $\mathbf{v}_0$ is all ones.
2. The rows $\mathbf{v}_i$, $i = 1, 2, \ldots, m$, are formed by alternating 2^{m-i} 0's with 2^{m-i} 1s to fill the vector of length 2^m.
3. The matrix $G(r, m)$ consists of the above $m+1$ rows together with all products of the vectors $\mathbf{v}_i$, $i = 1, 2, \ldots, m$, taken j at a time for all $j \le r$. $G(r, m)$ is a $(\sum_{j=0}^{r} \binom{m}{j}) \times 2^m$ matrix.

15.1.72 Definition The matrix $G(r, m)$ is a generator matrix for the *order r and degree m binary Reed-Muller code $RM_{2^m,2}(r, m)$*.

15.1.73 Lemma The parameters of $RM_{2^m,2}(r, m)$ for $0 \le r \le m$ are

$$n = 2^m, \quad k = \sum_{j=0}^{r} \binom{m}{j}, \quad d = 2^{m-r}.$$

The dual of $RM_{2^m,2}(r, m)$ is $RM_{2^m,2}(m - r - 1, m)$ for $0 \le r < m$.

15.1.74 Remark That the matrix $G(r, m)$ has linearly independent rows can be established by showing the $2^m \times 2^m$ matrix $G(m, m)$ is nonsingular. Equivalently this can be shown by noting there exists a Boolean function, which can be realized by Boolean operations on the first $m + 1$ rows, that will generate a vector with a single 1 in a given place. The codes are in fact Euclidean geometry codes discussed later.

15.1.75 Remark [2849] The generator matrix $G(r, m)$ of $RM_{2^m,2}(r, m)$ can be written as

$$G(r,m) = \begin{bmatrix} G(r,m-1) & G(r,m-1) \\ 0 & G(r-1,m-1) \end{bmatrix}.$$

It follows that

$$RM_{2^m,2}(r,m) = \{(\mathbf{u}, \mathbf{u}+\mathbf{v}) \mid \mathbf{u} \in RM_{2^{m-1},2}(r,m-1),\ \mathbf{v} \in RM_{2^{m-1},2}(r-1,m-1)\}.$$

The minimum distance of $RM_{2^m,2}(r, m)$ can be established using Lemma 15.1.93 below.

15.1.2.4 Subfield and trace codes

15.1.76 Remark Given an $(n, k, d)_{q^m}$ code $\mathcal{C}$ over $\mathbb{F}_{q^m}$, there are several natural ways to obtain codes over subfields. Two methods are of particular interest: subfield subcodes and trace codes, introduced below.

15.1.77 Definition The *subfield subcode* of an $(n, k, d)_{q^m}$ code $\mathcal{C}$ is the set of codewords of $\mathcal{C}$ with all elements in $\mathbb{F}_q$, i.e.,

$$\mathcal{C}^{sf}_{q^m|q} = \{(c_1, c_2, \dots, c_n) \in \mathcal{C} \mid c_i \in \mathbb{F}_q \text{ for } 1 \le i \le n\} = \mathcal{C} \cap \mathbb{F}_q^n.$$

15.1.78 Lemma The subfield subcode $\mathcal{C}^{sf}_{q^m|q}$ obtained from the $(n, k, d)_{q^m}$ code $\mathcal{C}$ is a linear $(n, \ge mk - (m-1)n, \ge d)_q$ code.

15.1.79 Remark The subfield subcode of an $(n, M, d)_{q^m}$ nonlinear code is, as in the linear case, the set of codewords with all coordinates in $\mathbb{F}_q$. The minimum distance of the subfield subcode is at least d. In the linear case, the bound on the dimension of the subfield subcode is obtained from considering the dimension of the subspaces involved. Thus replacing elements of the parity check matrix of $\mathcal{C}$ by m-tuple columns over $\mathbb{F}_q$, the subfield subcode is the set of n-tuples over $\mathbb{F}_q$ orthogonal to all $(n - k)m$ rows and has dimension at least $n - (n - k)m = mk - (m - 1)n$.

15.1.80 Definition The *trace* of an element $\eta \in \mathbb{F}_{q^m}$ is defined as

$$\mathrm{Tr}_{q^m|q}(\eta) = \eta + \eta^q + \cdots + \eta^{q^{m-1}}$$

and is an $\mathbb{F}_q$-linear map from $\mathbb{F}_{q^m}$ to $\mathbb{F}_q$; see also Section 2.1.

15.1.81 Definition Given an $(n, k, d)_{q^m}$ code $\mathcal{C}$, the *trace code of* $\mathcal{C}$ is defined as

$$\mathcal{C}^{tr}_{q^m|q} = \{(\mathrm{Tr}_{q^m|q}(c_1), \mathrm{Tr}_{q^m|q}(c_2), \dots, \mathrm{Tr}_{q^m|q}(c_n)) \mid (c_1, c_2, \dots, c_n) \in \mathcal{C}\}.$$

15.1.82 Lemma The trace code of an $(n, k, d)_{q^m}$ code $\mathcal{C}$ is an $(n, \le mk, \le d)_q$ code.

15.1.83 Remark The dimension bound of Lemma 15.1.82 is the trivial one; since the parent code has $(q^m)^k$ codewords, the trace code over $\mathbb{F}_q$ can have at most q^{mk} codewords. The bound is achieved if and only if $\mathrm{Tr}_{q^m|q}(\mathbf{c}_1) \ne \mathrm{Tr}_{q^m|q}(\mathbf{c}_2)$ for any two distinct codewords $\mathbf{c}_1, \mathbf{c}_2 \in \mathcal{C}$.

15.1.84 Remark The following result, due to Delsarte [802], shows the relation between subfield subcodes and trace codes.

15.1.85 Lemma [802] Let $\mathcal{C}$ be a linear code over $\mathbb{F}_{q^m}$. Then the dual of the subfield subcode of $\mathcal{C}$ is the trace code of the dual code of $\mathcal{C}$ over $\mathbb{F}_{q^m}$, i.e.,

$$\left(\mathcal{C}^{sf}_{q^m|q}\right)^{\perp} = \left(\mathcal{C}^{\perp}\right)^{tr}_{q^m|q}.$$

15.1.86 Remark The following commutative diagram may be of value in visualizing these relationships:

$$\begin{array}{ccc} \mathcal{C} & \xrightarrow{\text{dual}} & \mathcal{C}^{\perp} \\ {\scriptstyle \text{subfield}}\downarrow & & \downarrow{\scriptstyle \text{trace}} \\ \mathcal{C}^{sf}_{q^m|q} & \xrightarrow{\text{dual}} & \left(\mathcal{C}^{sf}_{q^m|q}\right)^{\perp} = \left(\mathcal{C}^{\perp}\right)^{tr}_{q^m|q}. \end{array}$$

15.1.2.5 Modifying linear codes

15.1.87 Remark There are many ways of modifying a given code. The definitions given below are typical, although the terminology is not uniform in the literature.

15.1.88 Definition An $(n, k, d)_q$ code $\mathcal{C}$ can be modified in the following ways:

1. *Expurgation:* Certain codewords are deleted (no change of length).
2. *Augmentation:* The number of codewords is increased (no change of length).
3. *Puncturing:* The length of the code is reduced by deleting a coordinate position, generally without reducing the size of the code.
4. *Extending:* The length of the code is increased without increasing the size of the code.
5. *Shortening:* The length and dimension of the code are reduced.
6. *Lengthening:* The length and dimension of the code are increased.

15.1.89 Example Examples of the above operations on an $(n, k, d)_q$ code are given below.

1. If a row of a generator matrix is deleted, the result is an $(n, k-1, d')_q$ code where $d' \geq d$. Similarly a row could be added to a parity check matrix, which is linearly independent of the existing rows, to give the same result.
2. A row could be added to a generator matrix that is linearly independent of the existing rows to give an $(n, k+1, d')_q$ code where $d' \leq d$.
3. If we delete a coordinate position contained in the nonzero positions of a codeword of minimum weight (assuming it is not of weight 1), the result is an $(n-1, k, d-1)_q$ code.
4. Adding an overall parity check produces an $(n+1, k, d')_q$ code where $d' = d$ or $d+1$ depending on the code structure. For example, if $q = 2$ and the original code contains minimum weight codewords of odd weight (in which case half the words are of odd weight and half of even weight), the extended code has minimum distance $d+1$. This is equivalent to adding a column of zeroes to the parity check matrix and then a row of all ones.
5. If a generator matrix of the code is in systematic form and the first row and column are deleted, the result is an $(n-1, k-1, d')_q$ code where $d' \geq d$.
6. Adding a row to a generator matrix, linearly independent of the existing rows, and then adding a column gives an $(n+1, k+1, d')_q$ code where $d' \leq d+1$.

15.1.90 Remark The trace code and subfield subcode of the previous subsection may also be thought of as ways of modifying a given code.

15.1.91 Remark Given two codes, an $(n_1, k_1, d_1)_q$ code $\mathcal{C}_1$ with generator matrix G_1 and an $(n_2, k_2, d_2)_q$ code $\mathcal{C}_2$ with generator matrix G_2, methods to produce a third code $\mathcal{C}$ are considered.

15.1.92 Remark The following construction proves useful in some situations, such as the construction of the Reed-Muller codes of Subsection 15.1.2.3, already noted.

15.1.93 Lemma Let $\mathcal{C}_i$ be an $(n, k_i, d_i)_q$ code, $i = 1, 2$. Then the code

$$\mathcal{C} = \{(\mathbf{u}, \mathbf{u} + \mathbf{v}) \mid \mathbf{u} \in \mathcal{C}_1, \ \mathbf{v} \in \mathcal{C}_2\}$$

is a $(2n, k_1 + k_2, d)_q$ code where $d = \min\{2d_1, d_2\}$.

15.1.94 Definition The *direct sum* code is the $(n_1 + n_2, k_1 + k_2, d)_q$ code where $d = \min\{d_1, d_2\}$ with generator matrix $G_1 \oplus G_2$ (block diagonal sum of matrices).

15.1.95 Definition The *product code* $\mathcal{C}_1 \otimes \mathcal{C}_2$ is the $(n_1 n_2, k_1 k_2, d_1 d_2)_q$ code with generator matrix $G_1 \otimes G_2$, the tensor product of the two component generator matrices.

15.1.96 Remark One can think of the codewords of the product code in the following manner. Consider an array of $k_1 \times k_2$ information symbols from $\mathbb{F}_q$. The array is first extended to a $k_1 \times n_2$ array by completing each row to codewords in $\mathcal{C}_2$. The columns of this array are completed to an $n_1 \times n_2$ array by completing each column to a codeword in $\mathcal{C}_1$. Notice the bottom right $(n_1 - k_1) \times (n_2 - k_2)$ array consists of checks on checks. The resulting $n_1 \times n_2$ matrices are the codewords of $\mathcal{C}_1 \otimes \mathcal{C}_2$. The process could have been done by first completing the columns of the information symbols with codewords in $\mathcal{C}_1$ and then completing rows. It is easily verified the same final array is obtained.

15.1.97 Definition Let $\mathcal{C}_1$ be a linear $(N, K, D)_{q^m}$ code and $\mathcal{C}_2$ a linear $(n, m, d)_q$ code. Fix a basis of $\mathbb{F}_{q^m}$ over $\mathbb{F}_q$ and represent elements of $\mathbb{F}_{q^m}$ as m-tuples over $\mathbb{F}_q$. These m-tuples represent information positions in $\mathcal{C}_2$ which therefore give a one-to-one correspondence between elements of $\mathbb{F}_{q^m}$ and codewords of $\mathcal{C}_2$. The *concatenated code* of $\mathcal{C}_1$ by $\mathcal{C}_2$ is obtained by replacing the $\mathbb{F}_{q^m}$-elements of codewords of $\mathcal{C}_1$ by codewords of $\mathcal{C}_2$ corresponding to the information m-tuples of the $\mathbb{F}_{q^m}$-elements.

15.1.98 Lemma The concatenated code of $\mathcal{C}_1$ by $\mathcal{C}_2$ is a linear $(nN, kK, \geq dD)_q$ code.

15.1.2.6 Bounds on codes

15.1.99 Remark The central problem of coding theory is to construct codes with as many codewords as possible for a given minimum distance (or as large a minimum distance as possible for a given code size), in such a way that the code can be efficiently encoded and decoded. This section reviews many of the bounds available. While interest is largely in linear codes, a version of a few of the bounds holds true for nonlinear codes as well. Most of the books on coding have a treatment similar to the one here. The treatment of bounds in [2849] is particularly elegant and succinct.

15.1.100 Definition Define $A_q(n,d)$ to be the size of the largest code of length n over $\mathbb{F}_q$ with minimum distance d, i.e.,

$$A_q(n,d) = \max\{M \mid \text{an } (n,M,d)_q \text{ code exists}\}.$$

15.1.101 Definition The *sphere* of radius r and center $\mathbf{x} \in \mathbb{F}_q^n$ is

$$S_q(r,n,\mathbf{x}) = \{\mathbf{y} \in \mathbb{F}_q^n \mid d(\mathbf{y},\mathbf{x}) \le r\}.$$

The cardinality of $S_q(r,n,\mathbf{x})$ is

$$s_q(r,n) = \mid S_q(r,n,\mathbf{x}) \mid = \sum_{i=0}^{r} \binom{n}{i} (q-1)^i.$$

15.1.102 Remark The following bound is a lower bound on the size of a maximal code.

15.1.103 Theorem (Sphere covering bound, Varshamov-Gilbert bound) [2849, 2855]

$$A_q(n,d) \ge \frac{q^n}{s_q(d-1,n)}.$$

15.1.104 Remark Theorem 15.1.103 is shown by surrounding each codeword of a code of maximum size with a sphere of radius $d-1$ and considering the union of such spheres. If the space is not exhausted, it would be possible to add a word at distance at least d from every codeword, implying the code was not optimal.

For linear codes a different argument leads to a similar result. For a given n and k, let $r = n-k$, and attempt to construct an $r \times n$ parity check matrix, with the property that any $d-1$ columns are linearly independent, by adding columns sequentially. The process may be started with the $r \times r$ identity matrix. Suppose $j-1$ such columns have been found. A j-th column can be added if

$$\sum_{i=1}^{d-2} \binom{j-1}{i} (q-1)^i < q^r - 1.$$

If n is the largest value of j for which this inequality holds, then an $(n,k,d)_q$ code exists. Notice that n is also the smallest value of n for which

$$\sum_{i=1}^{d-2} \binom{n}{i} (q-1)^i \ge q^{n-k} - 1,$$

an expression which can be rewritten as

$$q^k \ge q^n / s_q(d-2,n).$$

This result can be compared to the result of Theorem 15.1.103 and is also referred to as the Varshamov-Gilbert bound.

15.1.105 Remark The following bounds are upper bounds on the size of codes. In many cases it is possible to find codes that meet the bounds with equality.

15.1.106 Theorem (Sphere packing bound, Hamming bound) [2849] For a positive integer d with $1 \le d \le n$, let $e = \lfloor \frac{d-1}{2} \rfloor$. Then

$$A_q(n,d) \le \frac{q^n}{\sum_{i=0}^{e} \binom{n}{i}(q-1)^i} = \frac{q^n}{s_q(e,n)}.$$

15.1.107 Remark In an optimal code with $A_q(n,d)$ codewords, it is possible to surround codewords with nonintersecting spheres of radius e and the bound follows.

15.1.108 Definition A code whose size meets the bound of Theorem 15.1.106 with equality is *perfect.*

15.1.109 Remark In a perfect code the spheres of radius e around the codewords are nonintersecting and exhaust the space. The Hamming codes, both binary and nonbinary, are perfect codes. Besides the linear Hamming codes there are nonlinear perfect codes with the same parameters. The $(23,12,7)_2$ binary Golay code and the $(11,6,5)_3$ ternary Golay code (to be discussed later) are perfect. There are also trivial perfect codes: (i) the $(n,n,1)_q$ code (the complete space), $e=0$, (ii) the $(n,1,n)_2$ binary repetition code for n odd, $e=(n-1)/2$, (iii) a code of length n over $\mathbb{F}_q$ consisting of a single codeword, $e=n$, and (iv) a binary code of odd length containing only two words $\mathbf{c}$ and its complement $\mathbf{c}+\mathbf{1}$, $e=(n-1)/2$ [1558, 2810, 2811].

15.1.110 Theorem (Plotkin bound) [2849] If an $(n,M,d)_q$ code exists, then

$$d \le \frac{nM(q-1)}{(M-1)q}.$$

15.1.111 Remark In the following it will be convenient to define $\theta = (q-1)/q$. By rearranging terms in the above theorem, another expression for the Plotkin bound is

$$A_q(n,d) \le \left\lfloor \frac{d}{d-\theta n} \right\rfloor \quad \text{for } d > \theta n.$$

15.1.112 Theorem (Griesmer bound) [2849] If an $(n,k,d)_q$ code exists, then

$$n \ge \sum_{i=0}^{k-1} \left\lceil \frac{d}{q^i} \right\rceil.$$

15.1.113 Theorem (Singleton bound) [2849] For a positive integer d, $1 \le d \le n$,

$$A_q(n,d) \le q^{n-d+1}.$$

15.1.114 Remark Puncturing the code on $d-1$ nonzero coordinate positions of a minimum weight codeword implies all remaining codewords are distinct and the Singleton bound follows.

15.1.115 Remark For a linear $(n,k,d)_q$ code, the Singleton bound reduces to

$$d \le n-k+1.$$

15.1.116 Definition A code (either linear or nonlinear) that meets the Singleton bound with equality is a *maximum distance separable (MDS)* code.

15.1.117 Remark

1. For a linear $(n,k,d)_q$ code, the Singleton bound is simply a reflection that every set of $d-1$ columns of the parity check matrix (an $(n-k)\times n$ matrix) is linearly independent, and hence the largest value of $d-1$ is $n-k$.
2. The dual of an $(n,k,d=n-k+1)_q$ MDS code is an $(n,n-k,k+1)_q$ MDS code.
3. Important examples of MDS codes are the Reed-Solomon codes discussed in the next section. Trivial examples include the $(n,1,n)_q$ repetition code and the $(n,n,1)_q$ full code.

15.1.118 Theorem (Elias bound) [2390, 2849] Assume that $r \leq \theta n$ and $r^2 - 2\theta nr + \theta nd > 0$. Then

$$A_q(n,d) \leq \frac{\theta nd}{r^2 - 2\theta nr + \theta nd} \cdot \frac{q^n}{s_q(r,n)}.$$

15.1.2.7 Asymptotic bounds

15.1.119 Remark It is of interest to consider the constraints on codes and how they are reflected in the bounds discussed as the length of the code increases to infinity. Notice that for a linear $(n,k,d)_q$ code, the rate is k/n, the ratio of the number of information symbols to code symbols. Asymptotically as the length increases, this rate is denoted $\alpha_q(\delta)$ where δ is the asymptotic normalized distance d/n. The following definition is valid for both linear and nonlinear codes.

15.1.120 Definition The *asymptotic normalized rate* is

$$\alpha_q(\delta) = \limsup_{n\to\infty} \frac{1}{n} \log_q A_q(n,\delta n).$$

15.1.121 Remark To discuss asymptotic versions of the bounds, the following definition of the q-ary entropy function is needed.

15.1.122 Definition The *q-ary entropy function* is defined as

$$H_q(x) = \begin{cases} x\log_q(q-1) - x\log_q(x) - (1-x)\log_q(1-x) & \text{for } 0 < x \leq \theta, \\ 0 & \text{elsewhere.} \end{cases}$$

15.1.123 Remark The work of Delsarte [801] on distance distributions for codes and their MacWilliams transforms (Definition 15.1.43) suggests the following linear programming approach to upper bounding the size of M for an $(n,M,d)_q$ code [2050, 2849]. Let $A = (A_0 = 1, A_1, \ldots, A_n)$ be the distance distribution of a putative $(n,M,d)_q$ code, where $A_1 = \cdots = A_{d-1} = 0$ and $\sum_{i=0}^n A_i = M$. The linear program consists of maximizing $\sum_{i=0}^n A_i$ subject to $A_0 = 1$, $A_1 = \cdots = A_{d-1} = 0$, $A_i \geq 0$ for $i = d, \ldots, n$, and the MacWilliams transform constraints

$$\sum_{i=0}^{n} A_i P_k(i) \geq 0, \quad k = 0,1,\ldots,n.$$

This bound, often referred to as the *linear programming* or *LP* bound, is not very explicit, but it leads to the following two theorems [2050] which are quite strong upper bounds for binary codes. Noting that $H_2(x) = -x\log_2(x) - (1-x)\log_2(1-x)$, define $g(x)$ by

$$g(x) = H_2((1-\sqrt{1-x})/2).$$

15.1.124 Theorem [2050] For binary codes, if $0 < \delta < 1/2$,

$$\alpha_2(\delta) \le \min_{0 \le u \le 1-2\delta} \{1 + g(u^2) - g(u^2 + 2\delta u + 2\delta)\}.$$

15.1.125 Remark The above bound is referred to as the *MRRW bound* (after the initials of the four authors of [2050]), and it implies the following one - the bounds are actually the same over a range of values of δ.

15.1.126 Theorem For binary codes, if $0 < \delta < 1/2$,

$$\alpha_2(\delta) \le g((1 - 2\delta)^2).$$

15.1.127 Remark The following relationship is crucial to considering the asymptotic bounds.

15.1.128 Theorem [1558, 2389, 2849] For $0 \le \lambda \le \theta,\ q \ge 2$

$$\lim_{n \to \infty} \frac{1}{n} \log_q s_q(\lfloor \lambda n \rfloor, n) = H_q(\lambda).$$

15.1.129 Remark The above theorem implies that, asymptotically, $s_q(r, n)$ tends to

$$q^{nH_q(r/n)}\ .$$

Apart from the MRRW bound, the following asymptotic versions of the bounds follow directly from their finite counterparts.

15.1.130 Theorem (Asymptotic bounds) [1558, 2390, 2849]

1. Asymptotic Varshamov-Gilbert bound:

$$\alpha_q(\delta) \ge 1 - H_q(\delta), \quad 0 \le \delta \le \theta.$$

2. Asymptotic Singleton bound:

$$\alpha_q(\delta) \le 1 - \delta, \quad 0 \le \delta \le 1.$$

3. Asymptotic Plotkin bound:

$$\alpha_q(\delta) \le 1 - \delta/\theta, \quad 0 \le \delta < \theta.$$

4. Asymptotic Hamming bound:

$$\alpha_q(\delta) \le 1 - H_q(\delta/2), \quad 0 \le \delta \le \theta.$$

5. Asymptotic Elias bound:

$$\alpha_q(\delta) \le 1 - H_q(\theta - \sqrt{\theta(\theta - \delta)}), \quad 0 \le \delta < \theta.$$

15.1.3 Cyclic codes

15.1.131 Definition Let $\mathbf{x} = (x_0, x_1, \ldots, x_{n-1}) \in \mathbb{F}_{q^n}$. A cyclic (right) shift of $\mathbf{x}$ (with wraparound) is $(x_{n-1}, x_0, x_1, \ldots, x_{n-2})$. Let $\mathcal{C}$ be a linear code. Then $\mathcal{C}$ is a *cyclic* code if every cyclic shift of a codeword in $\mathcal{C}$ is also a codeword in $\mathcal{C}$.

15.1.132 Remark Some benefits to assuming a code is linear have been seen. Restricting attention further to cyclic codes allows the formulation of efficient algorithms for the construction, encoding, and decoding of them. The simple addition of requiring cyclic shifts of codewords to be codewords introduces a strong algebraic structure into the picture that allows these benefits.

15.1.3.1 Algebraic prerequisites

15.1.133 **Definition** Let R be a commutative ring with identity. A subset I of R is an *ideal* of R if for all $a, b \in I$ and $r \in R$, then $a - b \in I$ and $ra \in I$. The ideal is a *principal ideal* if it has a single generator a where $I = \langle a \rangle = \{ra \mid r \in R\}$. The ring R is an *integral domain* if whenever $a, b \in R$ and $ab = 0$, either $a = 0$ or $b = 0$. The ring R is a *principal ideal domain (PID)* if it is an integral domain and all its ideals are principal. The quotient ring of R by the ideal I is denoted by R/I with addition and multiplication of cosets given by $(r + I) + (s + I) = (r + s) + I$ and $(r + I)(s + I) = rs + I$.

15.1.134 **Example** The ring $\mathbb{F}_q[x]$ is the set of polynomials in the indeterminate x with coefficients in $\mathbb{F}_q$. The rings $\mathbb{Z}$, $\mathbb{F}_q[x]$, and $\mathbb{F}_q[x]/\langle x^n - 1\rangle$ are PIDs. In $\mathbb{Z}$ the ideal $\langle 2\rangle$ is $\{2k \mid k \in \mathbb{Z}\}$, i.e., the set of even integers. In $\mathbb{F}_q[x]$

$$\langle g(x)\rangle = \{a(x)g(x) \mid a(x) \in \mathbb{F}_q[x]\}$$

is the ideal generated by the polynomial $g(x)$.

15.1.135 **Remark** In the quotient ring $\mathbb{F}_q[x]/\langle x^n-1\rangle$, each coset has a unique coset representative that is either the zero polynomial or has degree less than n. For simplicity, the coset $a(x)+\langle x^n-1\rangle$ with $a(x) = a_0 + a_1x + \cdots + a_{n-1}x^{n-1}$ will be denoted $a(x)$. Therefore the quotient ring $\mathbb{F}_q[x]/\langle x^n - 1\rangle$ is the set of polynomials

$$\{a_0 + a_1x + \cdots + a_{n-1}x^{n-1} \mid a_i \in \mathbb{F}_q,\ i = 0, 1, \ldots, n-1\}$$

with addition and multiplication modulo $x^n - 1$.

There is a natural map between n-tuples over $\mathbb{F}_q$ and polynomials in $\mathbb{F}_q[x]/\langle x^n - 1\rangle$, namely

$$\begin{array}{ccc} \mathbb{F}_q^n & \longrightarrow & \mathbb{F}_q[x]/\langle x^n - 1\rangle \\ (a_0, a_1, \ldots, a_{n-1}) & \mapsto & a_0 + a_1x + \cdots + a_{n-1}x^{n-1}. \end{array}$$

Thus a linear code $\mathcal{C}$ can be viewed equivalently as a subspace of $\mathbb{F}_q^n$ and an $\mathbb{F}_q$-subspace of $\mathbb{F}_q[x]/\langle x^n - 1\rangle$. Notice that if $(a_0, a_1, \ldots, a_{n-1}) \mapsto a_0 + a_1x + \cdots + a_{n-1}x^{n-1}$, then the cyclic shift $(a_{n-1}, a_0, a_1, \ldots, a_{n-2})$ corresponds to the polynomial $x(a_0 + a_1x + \cdots + a_{n-1}x^{n-1})$ $(\text{mod } x^n - 1)$ which gives the reason for interest in the quotient ring $\mathbb{F}_q[x]/\langle x^n - 1\rangle$.

15.1.136 **Lemma** A linear code $\mathcal{C}$ is a cyclic code if and only if it is an ideal in $\mathbb{F}_q[x]/\langle x^n - 1\rangle$.

15.1.137 **Remark** Since $\mathbb{F}_q[x]/\langle x^n - 1\rangle$ is a PID, every nonzero ideal $\mathcal{C}$ has a generator polynomial $g(x)$, i.e., $\mathcal{C} = \langle g(x)\rangle$, and one such generator polynomial is the unique monic codeword polynomial of least degree. It also follows that $g(x)$ divides $x^n - 1$ (written $g(x) \mid (x^n - 1)$) as otherwise $x^n - 1 = a(x)g(x) + r(x)$ for some $a(x)$ and $r(x)$ where $r(x)$ has degree strictly less than that of $g(x)$. Since by definition $r(x)$ would be in $\mathcal{C}$, $r(x)$ nonzero contradicts the fact that $g(x)$ was of least degree. For $g(x) \mid (x^n-1)$, $g(x)$ generates an ideal in $\mathbb{F}_q[x]/\langle x^n-1\rangle$ that is the set of polynomials divisible by $g(x)$. More specifically

$$\langle g(x)\rangle = \{a(x)g(x) \mid a(x) \in \mathbb{F}_q[x],\ \deg a(x) < n - \deg g(x)\}.$$

The term *generator polynomial* of a cyclic code is reserved for the unique monic polynomial generating the code and dividing x^n-1. Thus cyclic codes are determined by factors of x^n-1. Indeed, if $\gcd(n, q) = 1$, $x^n - 1$ factors into distinct irreducible polynomials $f_i(x)$ over $\mathbb{F}_q$, $i = 1, 2, \ldots, t$, and then there are 2^t cyclic codes of length n over $\mathbb{F}_q$. If $\gcd(n, q) \neq 1$, $x^n - 1$ has repeated irreducible factors. For example if $\gcd(n, q) = p$ where $\mathbb{F}_q$ has characteristic p, then $n = n_1p$ and $x^n - 1 = (x^{n_1} - 1)^p$. More generally, if $x^n - 1 = \prod_{1=1}^{t} f_i^{e_i}(x)$ is the factorization, the possible number of cyclic codes is $\prod_{i=1}^{t}(e_i + 1)$; see [1945]. Henceforth it is assumed that $\gcd(n, q) = 1$.

15.1.3.2 Properties of cyclic codes

15.1.138 Remark A cyclic $(n,k,d)_q$ code $\mathcal{C}$ has a generator polynomial $g(x)$ of degree $n-k$. If $g(x) = g_0+g_1x+\cdots+g_{n-k}x^{n-k}$, the code is viewed equivalently as the ideal of $\mathbb{F}_q[x]/\langle x^n-1\rangle$

$$\mathcal{C} = \{a(x)g(x) \mid a(x) \in \mathbb{F}_q[x],\ \deg a(x) < k\}$$

and the row space of the matrix

$$G = \begin{bmatrix} g_0 & g_1 & g_2 & \cdots & g_{n-k} & 0 & 0 & \cdots & 0 \\ 0 & g_0 & g_1 & \cdots & g_{n-k-1} & g_{n-k} & 0 & \cdots & 0 \\ 0 & 0 & g_0 & \cdots & g_{n-k-2} & g_{n-k-1} & g_{n-k} & \cdots & 0 \\ & & & \vdots & & & & \vdots & \\ 0 & 0 & 0 & \cdots & g_0 & g_1 & g_2 & \cdots & g_{n-k} \end{bmatrix}, \qquad (15.1.2)$$

i.e., G is a generator matrix for the code $\mathcal{C}$.

15.1.139 Remark As the generator polynomial $g(x) \mid (x^n-1)$, $\deg g(x) = n-k$, there is a polynomial $h(x) = h_0 + h_1x + \cdots + h_kx^k$ such that $g(x)h(x) = x^n - 1$. Since $\deg h(x) = k$ and $h(x) \mid (x^n-1)$, it generates a cyclic $(n, n-k, d')_q$ code $\mathcal{C}'$ over $\mathbb{F}_q$ for some minimum distance d'. However $\mathcal{C}'$ is not $\mathcal{C}^\perp$. Define the reciprocal polynomial $h^*(x) = x^kh(1/x)$, which is the polynomial obtained by reversing the coefficients of $h(x)$. Then $(1/h_0)h^*(x)$ is the generator polynomial of $\mathcal{C}^\perp$, which is $\mathcal{C}'$ with coordinates reversed and has the same parameters as $\mathcal{C}'$. A generator matrix of $\mathcal{C}^\perp$ (and parity check matrix of $\mathcal{C}$) is then given by

$$H = \begin{bmatrix} h_k & h_{k-1} & h_{k-2} & \cdots & h_0 & 0 & 0 & \cdots & 0 \\ 0 & h_k & h_{k-1} & \cdots & h_1 & h_0 & 0 & \cdots & 0 \\ 0 & 0 & h_k & \cdots & h_2 & h_1 & h_0 & \cdots & 0 \\ & & & \vdots & & & & \vdots & \\ 0 & 0 & 0 & \cdots & h_k & h_{k-1} & h_{k-2} & \cdots & h_0 \end{bmatrix}.$$

15.1.140 Remark Let $\mathcal{C}$ be an $(n,k,d)_q$ cyclic code with generator polynomial $g(x)$ whose roots are $\alpha_1, \alpha_2, \ldots, \alpha_{n-k}$. The roots will in general be in an extension field of $\mathbb{F}_q$, say $\mathbb{F}_{q^m}$ where $n \mid (q^m-1)$. Another way of defining the code is to note that $c(x) = a(x)g(x)$ is a codeword polynomial if and only if

$$c(\alpha_i) = 0, \quad i = 1, 2, \ldots, n-k.$$

Thus $\mathbf{c} = (c_0, c_1, \ldots, c_{n-1}) \in \mathcal{C}$ if and only if $H\mathbf{c}^T = \mathbf{0}^T$ where

$$H = \begin{bmatrix} 1 & \alpha_1 & \alpha_1^2 & \cdots & \alpha_1^{n-1} \\ 1 & \alpha_2 & \alpha_2^2 & \cdots & \alpha_2^{n-1} \\ & & & \vdots & \\ 1 & \alpha_{n-k} & \alpha_{n-k}^2 & \cdots & \alpha_{n-k}^{n-1} \end{bmatrix}.$$

15.1.141 Remark Let $\mathcal{C}_1$ and $\mathcal{C}_2$ be cyclic codes with generator polynomials $g_1(x)$ and $g_2(x)$, respectively. Then $\mathcal{C}_1 \subseteq \mathcal{C}_2$ if and only if $g_2(x) \mid g_1(x)$.

15.1.142 Remark In practice there is often a preference to use systematic codes where the information symbols appear explicitly in the codeword. To achieve this one might row-reduce the generator matrix of (15.1.2) (with possible column permutations - see Remark 15.1.32) to be of the form

$$G' = [I_k \mid A]$$

as can be done for any linear code. One could then encode the information word **m** (of length k, recalling that the message length is the same as code dimension) as $\mathbf{m}G'$. However if column permutations are required to obtain G', the resulting code with generator matrix G' may not be cyclic. For an $(n,k,d)_q$ cyclic code with generator polynomial $g(x)$ of degree $n-k$, one might also do the following encoding. Divide $x^{n-k}m(x)$ by $g(x)$, where $m(x)$ is the information polynomial, to obtain

$$x^{n-k}m(x) = a(x)g(x) + r(x), \quad \deg r(x) < n-k.$$

Thus $x^{n-k}m(x) - r(x) = a(x)g(x)$ is a codeword with the k information symbols in the "high end" and $n-k$ parity check symbols in the "low end."

15.1.143 Theorem [1558] Let $\mathcal{C}$ be a cyclic code of length n over $\mathbb{F}_q$ and let $v(x) \in \mathbb{F}_q[x]$. Then $\mathcal{C} = \langle v(x) \rangle$ if and only if $\gcd(v(x), x^n - 1) = g(x)$. Equivalently $v(x)$ generates $\mathcal{C}$ if and only if the n-th roots of unity that are zeros of $v(x)$ are precisely the zeros of $g(x)$.

15.1.144 Theorem Let $\mathcal{C}_i$, $i = 1, 2$, be cyclic codes with generator polynomials $g_i(x)$, $i = 1, 2$. Then $\mathcal{C}_1 \cap \mathcal{C}_2$ has generator polynomial $\mathrm{lcm}(g_1(x), g_2(x))$ and $\mathcal{C}_1 + \mathcal{C}_2$ has generator polynomial $\gcd(g_1(x), g_2(x))$.

15.1.3.3 Classes of cyclic codes

15.1.145 Remark The additional constraint of cyclic codes (over linear) allows considerably more information on the minimum distance of the code to be obtained. The *minimal polynomial* over $\mathbb{F}_q$ of an element β in some extension field of $\mathbb{F}_q$ is the monic irreducible polynomial, denoted $M_\beta(x)$, of least degree in $\mathbb{F}_q[x]$ that has β as a zero.

15.1.146 Remark Recall that the q-ary Hamming code $H_{n,q}(m)$ is a $(\pi_{m,q}, \pi_{m,q} - m, 3)_q$ code where $n = \pi_{m,q} = (q^m - 1)/(q-1)$. When $\gcd(n, q-1) = 1$, $H_{n,q}(m)$ can be made cyclic in the following manner. Let α be a primitive element in $\mathbb{F}_{q^m}$. Then $\beta = \alpha^{q-1}$ is a primitive n-th root of unity. The parity check matrix of $H_{n,q}(m)$ can be taken as

$$H = \begin{bmatrix} 1 & \beta & \beta^2 & \cdots & \beta^{n-1} \end{bmatrix}.$$

Under the stated conditions the minimal polynomial of β over $\mathbb{F}_q$ is of degree m. It remains to verify that no two columns of H are multiples of each other over $\mathbb{F}_q$ to ensure a minimum distance of 3. Since the dual of this code has all codewords of weight q^{m-1} the weight enumerator of the q-ary Hamming code can be obtained via the MacWilliams identities, as in the binary case. This is an example of the first class of cyclic codes to be considered.

15.1.147 Definition Let n be a positive integer relatively prime to q. For i an integer, the *q-cyclotomic coset modulo n containing i* is $C_i = \{i, iq, \ldots, iq^{r-1}\} \pmod{n}$ where r is the smallest positive integer such that $iq^r \equiv i \pmod{n}$.

15.1.148 Remark The smallest extension field of $\mathbb{F}_q$ that contains a primitive n-th root of unity is $\mathbb{F}_{q^m}$ where m is the size $|C_1|$ of the q-cylotomic coset modulo n containing 1. If β is a primitive n-th root of unity in $\mathbb{F}_{q^m}$, then the minimal polynomial of β^a over $\mathbb{F}_q$ is

$$M_{\beta^a}(x) = \prod_{j \in C_a} (x - \beta^j).$$

There is a one-to-one correspondence between the monic irreducible factors of $x^n - 1$ and the q-cyclotomic cosets modulo n. The factorization of $x^n - 1$ into irreducible factors over $\mathbb{F}_q$ is given by

$$x^n - 1 = \prod_s M_{\beta^s}(x)$$

where s runs through a set of representatives of the distinct q-cyclotomic cosets modulo n.

15.1.149 **Definition** Let β be a primitive n-th root of unity in an extension field of $\mathbb{F}_q$. Let $\mathcal{C}$ be a cyclic code over $\mathbb{F}_q$ of length n with generator polynomial $g(x) \in \mathbb{F}_q[x]$. Then there is a set $T \subseteq \{0, 1, \ldots, n-1\}$ such that the roots of $g(x)$ are $\{\beta^t \mid t \in T\}$. T is a *defining set* (with respect to β) of $\mathcal{C}$.

15.1.150 **Remark** Changing β will change T. For a given β, knowing $g(x)$ is equivalent to knowing T because $g(x) = \prod_{t \in T}(x - \beta^t)$. If $g(\beta^t) = 0$, then $g(\beta^{tq}) = 0$ also, implying that T is a union of q-cyclotomic cosets modulo n. Conversely, any set $T \subseteq \{0, 1, \ldots, n-1\}$ that is a union of q-cyclotomic cosets modulo n has the property that $\prod_{t \in T}(x - \beta^t) \in \mathbb{F}_q[x]$ and hence is the defining set of a cyclic code of length n over $\mathbb{F}_q$.

BCH codes:

15.1.151 **Definition** [359, 1329, 1516, 1558, 2035] The cyclic code $BCH_{n,q}(\delta)$ of length n and *designed distance* δ has a generator polynomial of the form

$$g(x) = \operatorname{lcm}\{M_{\beta^a}(x), M_{\beta^{a+1}}(x), \ldots, M_{\beta^{a+\delta-2}}(x)\}$$

for some sequence of elements $\beta^a, \beta^{a+1}, \ldots, \beta^{a+\delta-2}$, where β is a primitive n-th root of unity in some extension field of $\mathbb{F}_q$. Alternately, $BCH_{n,q}(\delta)$ has defining set $C_a \cup C_{a+1} \cup \cdots \cup C_{a+\delta-2}$ relative to β. If $n = q^m - 1$ for some positive integer m, the code is *primitive*, and if $a = 1$, it is *narrow sense*.

15.1.152 **Remark** These codes are referred to as $BCH_{n,q}(\delta)$ codes, where BCH stands for the code name (Bose-Chaudhuri-Hocquenghem after the code originators [359, 1516]), and the subscripts on BCH are the length and field of definition, and δ is the designed distance of the code. The relationship between the designed distance and the true minimum distance of $BCH_{n,q}(\delta)$ is discussed below.

15.1.153 **Remark** The reason for requiring the generator polynomial of the BCH code to have a consecutive sequence of roots derives from properties of *Vandermonde* matrices as follows. Let $\beta_1, \beta_2, \ldots, \beta_\ell$ be *distinct* elements in some extension field of $\mathbb{F}_q$. The determinant of the matrix

$$\begin{bmatrix} \beta_1 & \beta_2 & \cdots & \beta_\ell \\ \beta_1^2 & \beta_2^2 & \cdots & \beta_\ell^2 \\ \beta_1^3 & \beta_2^3 & \cdots & \beta_\ell^3 \\ & & \vdots & \\ \beta_1^\ell & \beta_2^\ell & \cdots & \beta_\ell^\ell \end{bmatrix} \tag{15.1.3}$$

is $\prod_{i=1}^{\ell} \beta_i$ times the determinant of the matrix

$$\begin{bmatrix} 1 & 1 & \cdots & 1 \\ \beta_1 & \beta_2 & \cdots & \beta_\ell \\ \beta_1^2 & \beta_2^2 & \cdots & \beta_\ell^2 \\ & & \vdots & \\ \beta_1^{\ell-1} & \beta_2^{\ell-1} & \cdots & \beta_\ell^{\ell-1} \end{bmatrix}. \tag{15.1.4}$$

The determinant of this matrix is $\prod_{i>j}(\beta_i - \beta_j)$. If $\beta_i = \beta_j$ for some $i \neq j$, the determinant of (15.1.4) is zero and the matrix is singular. If the entries of the second row are distinct, the determinant is nonzero and the matrix is nonsingular. The determinant of the matrix in (15.1.3) is

$$\left(\prod_{i=1}^{\ell} \beta_i\right) \prod_{i>j} (\beta_i - \beta_j).$$

15.1.154 Remark To complete the discussion, it is straightforward to find the inverse of a Vandermonde matrix of the form (15.1.4) in the following manner. Define the polynomials

$$f_i(x) = \sum_{j=0}^{\ell-1} f_{ij} x^j = \prod_{j=1, j \neq i}^{\ell} \frac{x - \beta_j}{\beta_i - \beta_j}$$

that take on the values of 0 for $x = \beta_j,\ j \neq i$ and 1 for $x = \beta_i$. The inverse of the Vandermonde matrix (15.1.4) is then the matrix $[f_{ij}]$.

15.1.155 Remark To consider the dimension and minimum distance of the BCH code of Definition 15.1.151, note that the parity check matrix may be written in the form

$$H = \begin{bmatrix} 1 & \beta^a & \beta^{2a} & \cdots & \beta^{(n-1)a} \\ 1 & \beta^{a+1} & \beta^{2(a+1)} & \cdots & \beta^{(n-1)(a+1)} \\ \vdots & \vdots & \vdots & \vdots & \vdots \\ 1 & \beta^{a+\delta-2} & \beta^{2(a+\delta-2)} & \cdots & \beta^{(n-1)(a+\delta-2)} \end{bmatrix}.$$

The word $\mathbf{c} \in \mathbb{F}_q^n$ is in the BCH code if and only if $H\mathbf{c}^T = \mathbf{0}^T$. Since by the Vandermonde argument, any $\delta - 1$ columns of H are linearly independent, the code has minimum distance at least δ. In addition, as the entries of H are in $\mathbb{F}_{q^m}$, they can be expressed as column m-tuples over $\mathbb{F}_q$. Therefore the rank of H over $\mathbb{F}_q$ is at most $m(\delta - 1)$. Thus the BCH code of Definition 15.1.151 is an $(n, \geq (n - m(\delta - 1)), \geq \delta)_q$ cyclic code.

15.1.156 Remark There are numerous techniques to improve the bound on the actual minimum distance of a BCH code, i.e., improve on the lower bound of the designed distance.

Reed-Solomon (RS) codes:

15.1.157 Definition [2447] A q-ary *Reed-Solomon* code $RS_{q-1,q}(k)$ of dimension k is a BCH code of length $n = q - 1$ and designed distance $\delta = n - k + 1 = q - k$ over $\mathbb{F}_q$, i.e., for $\alpha \in \mathbb{F}_q$ a primitive element, the code has a generator polynomial of the form

$$g(x) = \prod_{i=a}^{a+\delta-2} (x - \alpha^i).$$

15.1.158 Remark The RS code defined above is a $(q-1, k, d = q-k)_q$ MDS code. That the minimum distance d is exactly $q-k$ is shown as follows. By the Singleton bound $d \leq n-k+1 = q-k$. Since a BCH code has minimum distance at least its designed distance δ, $d \geq \delta = q - k$. Hence $d = q - k$. If 1 is not a root of $g(x)$, then the extended narrow sense RS code, obtained by adding an overall parity check, is a $(q, k, d+1)_q$ code and hence is also MDS. With appropriate care one can define RS codes of length less than $q - 1$.

Notice also that the code defined as

$$\mathcal{C} = \{(f(1), f(\alpha), \ldots, f(\alpha^{q-2})) \mid f(x) \in \mathbb{F}_q[x],\ \deg f(x) \leq k - 1\}$$

is the $(q-1, k, q-k)_q$ narrow sense RS code, as now outlined. The code $\mathcal{C}$ is certainly linear and k-dimensional as distinct polynomials of degree at most $k - 1$ cannot produce equal vectors in $\mathcal{C}$ as otherwise the difference of these polynomials has at least $q-1>k-1$ roots. Let $\mathcal{C}_1$ be the $(q-1,k,q-k)_q$ narrow sense RS code, which has defining set $T=\{1,2,\dots,q-k-1\}$ relative to α. Since $\mathcal{C}$ and $\mathcal{C}_1$ have equal dimensions, $\mathcal{C}=\mathcal{C}_1$ if $\mathcal{C}\subseteq\mathcal{C}_1$. Let $c(x)=\sum_{j=0}^{q-2} f(\alpha^j)x^j\in\mathcal{C}$ for some $f(x)=\sum_{m=0}^{k-1} f_m x^m\in\mathbb{F}_q[x]$. For $i\in T$,

$$c(\alpha^i)=\sum_{j=0}^{q-2}\left(\sum_{m=0}^{k-1} f_m\alpha^{jm}\right)\alpha^{ij}=\sum_{m=0}^{k-1} f_m\sum_{j=0}^{q-2}\alpha^{(i+m)j}=\sum_{m=0}^{k-1} f_m\frac{\alpha^{(i+m)(q-1)}-1}{\alpha^{i+m}-1}$$

noting that $\alpha^{i+m}\neq 1$ as $1\le i+m\le q-2$. Since $\alpha^{(i+m)(q-1)}=(\alpha^{q-1})^{i+m}=1$, $c(\alpha^i)=0$ implying $c(x)\in\mathcal{C}_1$, and so $\mathcal{C}=\mathcal{C}_1$. Furthermore, in this realization, the extended code (adding a simple parity check) is realized by adding the coordinate position containing $f(0)$.

15.1.159 Remark More generally one could define an $(n,k,d=n-k+1)_q$ code $\mathcal{C}$ by choosing n distinct elements $\alpha_1,\alpha_2,\dots,\alpha_n$ in a field $\mathbb{F}_q$ and letting

$$\mathcal{C}=\{(f(\alpha_1),f(\alpha_2),\dots,f(\alpha_n))\mid f(x)\in\mathbb{F}_q[x],\ \deg f(x)\le k-1\}.$$

The code is not, in general, cyclic but is MDS.

15.1.160 Definition A set of k column positions of a linear $(n,k,d)_q$ code is an *information set* if the corresponding columns of a code generator matrix are linearly independent.

15.1.161 Lemma If $\mathcal{C}$ is an $(n,k,n-k+1)_q$ MDS code, then $\mathcal{C}^\perp$ is an $(n,n-k,k+1)_q$ MDS code. Any k columns of a generator matrix for an $(n,k,d)_q$ MDS code $\mathcal{C}$ are linearly independent and hence these column positions form an information set. Similarly any $n-k$ columns of the parity check matrix for $\mathcal{C}$ are linearly independent and any such set of columns forms an information set for $\mathcal{C}^\perp$.

15.1.162 Remark If $\mathbb{F}_q$ has characteristic 2 ($q=2^s$ for some s), then by fixing a basis of $\mathbb{F}_q$ over $\mathbb{F}_2$ one could expand the elements of $\mathbb{F}_q$ in each coordinate position of each codeword of a $(q-1,k,q-k)_q$ RS code to obtain an $(s(q-1),sk,\ge(n-k+1))_2$ code that is effective in correcting burst errors. This is simply a concatenated code with trivial inner code.

15.1.163 Remark It is easy to see that a BCH code can be viewed as a subfield subcode of an RS code.

15.1.164 Remark [304, 1945, 2484] In order to study the structure of other codes, the notion of a *generalized RS (GRS)* code is defined. Let $\alpha_1,\alpha_2,\dots,\alpha_n$ be $n\le q$ distinct elements of $\mathbb{F}_q$ and $\mathbf{v}=(v_1,v_2,\dots,v_n)\in(\mathbb{F}_q^*)^n$ where $\mathbb{F}_q^*=\mathbb{F}_q\setminus\{0\}$. Let $k\le n$. The GRS code $GRS_{n,q}(\boldsymbol{\alpha},\mathbf{v},k)$ is defined as

$$GRS_{n,q}(\boldsymbol{\alpha},\mathbf{v},k)=\{(v_1f(\alpha_1),v_2f(\alpha_2),\dots,v_nf(\alpha_n))\mid f(x)\in\mathbb{F}_q[x],\ \deg f(x)\le k-1\}.$$

15.1.165 Theorem The above code $GRS_{n,q}(\boldsymbol{\alpha},\mathbf{v},k)$ is an $(n,k,n-k+1)_q$ MDS code. Further, there is a vector $\mathbf{w}\in(\mathbb{F}_q^*)^n$ such that $\mathcal{C}^\perp$ is $GRS_{n,q}(\boldsymbol{\alpha},\mathbf{w},n-k)$, i.e., the dual of a GRS code is a GRS code, and the code $GRS_{n,q}(\boldsymbol{\alpha},\mathbf{v},k)^\perp$ is the code $GRS_{n,q}(\boldsymbol{\alpha},\mathbf{w},n-k)$ for some $\mathbf{w}\in(\mathbb{F}_q^*)^n$.

15.1.166 Remark The proof of the above theorem is a simple generalization of the RS/BCH argument. Notice that the vector of nonzero elements $\mathbf{v}$ has the effect of multiplying each coordinate position by a nonzero element which has little effect on code parameters. However when subfield subcodes are considered, the effect can be more significant.

15.1.167 Remark A code being MDS is a strong condition. In particular, as the following theorem shows, its weight enumerator is uniquely determined.

15.1.168 Theorem [2390] If $\mathcal{C}$ is an $(n,k,d)_q$ MDS code, then its weight distribution $\{A_i,\ i = 0,1,\ldots,n\}$ is given by $A_0 = 1$, $A_i = 0$ for $i = 1,2,\ldots,d-1$, and

$$A_i = \binom{n}{i}(q-1)\sum_{j=0}^{i-d}(-1)^j\binom{i-1}{j}q^{i-d-j} \quad \text{for } i = d, d+1, \ldots, n.$$

15.1.169 Remark The existence of MDS codes is of interest. Only linear codes are considered here. As noted, there are the trivial $(n,1,n)_q$ codes (for any alphabet of size q) and parity check $(n,n-1,2)_q$ codes for any length n. If one considers the parity check matrix of a $(q-1,k,q-k)_q$ RS (MDS) code, it is always possible to add columns $(1,0,\ldots,0)^T$ and $(0,0,\ldots,0,1)^T$ to obtain a $(q+1,k,q+2-k)_q$ code for $2 \le k \le q-1$. This "doubly extended" construction also works for BCH codes [2999]. When q is even, it is possible to obtain a triply extended $(q+2,q-1,4)_q$ MDS code and its dual $(q+2,3,q)_q$ code.

15.1.170 Conjecture [1558] It is postulated that if there is a nontrivial $(n,k,n-k+1)_q$ MDS code over $\mathbb{F}_q$, then $n \le q+1$, except when q is even and $k = 3$ or $k = q-1$ in which case $n \le q+2$.

15.1.171 Remark Reed-Solomon codes are ubiquitous in communication and storage system standards of all kinds. Only a few of these are mentioned as examples: digital video broadcasting (DVB) (EN 300-421 (satellite) and 429 (cable)); ADSL (asymmetric digital subscriber loops) for low rate communications over telephone lines (ANSI T1.413); Internet high speed (gigabit) optical networks (ITU-T G.795 and G.709 and OC-192); wireless broadband access networks (including metropolitan (known commercially as WiMax) and local) in IEEE 802.16 (which is a family of standards covering the many types of networks in such applications); deep space missions (*Voyager* and *Mariner*, among others, although more recent missions have tended to favor LDPC and turbo codes); satellite communication systems (IESS 308); two-dimensional bar codes; data storage including CDs, CD-ROMs, DRAMs, and DVDs and RAID (random arrays of inexpensive disks) systems. In each application the alphabet size and code parameters are carefully chosen for given "channel" conditions. For example ordinary music CDs use $(32,28,5)_{256}$ and $(28,24,5)_{256}$ codes with 8 bit symbols, obtained by shortening a $(255,251,5)_{256}$ Reed-Solomon code, and then cross interleaved in a clever way to enable the system to withstand bursts of lengths up to 4,000 bits, caused by a scratch of up 2.5 mm on the disk surface, without replay error. The optical network standard OC-192 uses symbols from 3 to 12 bits (and RS codes of length up to 4095 symbols with a maximum of 256 parity symbols).

Duadic codes:

15.1.172 Definition A vector $\mathbf{v} = (v_0, v_1, \ldots, v_{n-1}) \in \mathbb{F}_q^n$ is *even-like* if $\sum_{i=0}^{n-1} v_i = 0$; otherwise $\mathbf{v}$ is *odd-like*. A code $\mathcal{C}$ is *even-like* if all of its codewords are even-like; $\mathcal{C}$ is *odd-like* if it is not even-like.

15.1.173 Remark The terms even-like and odd-like generalize the notion of even and odd weight for binary vectors; i.e., if $\mathbf{v} \in \mathbb{F}_2^n$, $\mathbf{v}$ is even-like if and only if $\mathbf{v}$ has even weight. If $\mathcal{C}$ is a cyclic code of length n over $\mathbb{F}_q$ with defining set T, then $\mathcal{C}$ is even-like if and only if $0 \in T$.

15.1.174 Definition Let S_1 and S_2 be subsets of $\{1,2,\ldots,n-1\}$ that are unions of q-cyclotomic cosets modulo n. Assume further that $S_1 \cap S_2 = \emptyset$, $S_1 \cup S_2 = \{1,2,\ldots,n-1\}$, and that there exists b relatively prime to n such that $S_2 = \{sb \pmod{n} \mid s \in S_1\}$ and $S_1 = \{sb \pmod{n} \mid s \in S_2\}$. The pair of sets S_1 and S_2 form a *splitting of n given by b over $\mathbb{F}_q$*. For $i \in \{1,2\}$, let $\mathcal{D}_i$ be the cyclic code of length n over $\mathbb{F}_q$ with defining set S_i. The pair $\mathcal{D}_1$ and $\mathcal{D}_2$ is a pair of *odd-like duadic codes.* For $i \in \{1,2\}$, let $T_i = \{0\} \cup S_i$, and let $\mathcal{C}_i$ be the cyclic code of length n over $\mathbb{F}_q$ with defining set T_i. The pair $\mathcal{C}_1$ and $\mathcal{C}_2$ is a pair of *even-like duadic codes.*

15.1.175 Remark Binary duadic codes were first defined in [1901]. They were later generalized to other fields in [2402, 2403, 2506, 2687]. Duadic codes include quadratic residue codes described later; quadratic residue codes exist only for prime lengths, but duadic codes can exist for composite lengths.

15.1.176 Theorem With the notation of Definition 15.1.174, the following hold.

1. Duadic codes of length n over $\mathbb{F}_q$ exist if and only if n is odd and q is a square modulo n.
2. For $i \in \{1,2\}$, $\mathcal{C}_i$ has dimension $(n-1)/2$, and there is a permutation of coordinates that sends $\mathcal{C}_1$ to $\mathcal{C}_2$.
3. For $i \in \{1,2\}$, $\mathcal{C}_i \subseteq \mathcal{D}_i$, $\mathcal{D}_i$ has dimension $(n+1)/2$, and there is a permutation of coordinates that sends $\mathcal{D}_1$ to $\mathcal{D}_2$.

15.1.177 Remark Part 1 of Theorem 15.1.176 can be used to find precisely the values of n such that duadic codes of length n exist, in a manner similar to the following. Assume $n = p_1^{a_1} p_2^{a_2} \cdots p_r^{a_r}$ where $p_1, p_2, \ldots, p_r$ are distinct odd primes. Binary duadic codes exist if and only if $p_i \equiv \pm 1 \pmod{8}$ for $1 \le i \le r$. Duadic codes over $\mathbb{F}_3$ exist if and only if $p_i \equiv \pm 1 \pmod{12}$ for $1 \le i \le r$. Duadic codes over $\mathbb{F}_4$ exist if and only if n is odd.

15.1.178 Remark Much is known about the structure of duadic codes as illustrated by the next two theorems; see Chapter 6 of [1558].

15.1.179 Theorem Let $\mathcal{C}$ be a cyclic $(n,(n-1)/2,d)_q$ code over $\mathbb{F}_q$. Then $\mathcal{C}$ is self-orthogonal if and only if $\mathcal{C}$ is an even-like duadic code where -1 gives the splitting of n over $\mathbb{F}_q$. In that case, if $\mathcal{C}$ is one of the duadic pair $\mathcal{C}_1$ and $\mathcal{C}_2$, $\mathcal{C}_i^{\perp} = \mathcal{D}_i$ for $i \in \{1,2\}$. Furthermore, if $n = p$ is a prime with $p \equiv -1 \pmod{8}$, every splitting of p over $\mathbb{F}_2$ is given by -1 and every binary duadic code of length p is self-orthogonal.

15.1.180 Theorem Let $\mathcal{D}_1$ and $\mathcal{D}_2$ be a pair of odd-like binary duadic codes of length n where the splitting is given by -1 over $\mathbb{F}_2$. Then for $i \in \{1,2\}$,

1. the weight of every even weight codeword of $\mathcal{D}_i$ is divisible by 4, and the weight of every odd weight codeword is congruent to $n \pmod{4}$; furthermore,
2. the extended codes $\widehat{\mathcal{D}}_i$ are self-dual. In $\widehat{\mathcal{D}}_i$ if $n \equiv -1 \pmod{8}$, all codewords have weights divisible by 4, and if $n \equiv 1 \pmod{8}$, all codewords have even weights but some codewords have weights not divisible by 4.

15.1.181 Remark Quadratic residue codes, considered next, are special cases of duadic codes.

Quadratic residue (QR) codes:

15.1.182 Definition For an odd prime p define $\mathcal{R}$ to be the set of integers modulo p where

$$\mathcal{R} = \{a^2 \in \mathbb{Z}_p \mid a \in \mathbb{Z}_p \text{ and } a \not\equiv 0 \pmod{p}\}.$$

The set $\mathcal{R}$ is the set of *quadratic residues modulo p*. The set $\mathcal{N}$ of nonzero elements of $\mathbb{Z}_p$ that are not in $\mathcal{R}$ are the *quadratic nonresidues modulo p*.

15.1.183 Remark Clearly $|\mathcal{R}| = |\mathcal{N}| = (p-1)/2$. The sets observe a parity of sorts since a residue times a residue modulo p is a residue, a residue times a nonresidue is a nonresidue, and a nonresidue times a nonresidue is a residue. This implies that, whenever q is a quadratic residue modulo p, the pair of sets $\mathcal{R}$ and $\mathcal{N}$ form a splitting of p given by any nonresidue over $\mathbb{F}_q$. For ν a primitive element of $\mathbb{Z}_p$ (i.e., ν is a generator of the multiplicative group $\mathbb{Z}_p^* = \mathbb{Z}_p \setminus \{0\}$), it is clear that $\mathcal{R} = \{\nu^{2i} \mid i = 1, 2, \ldots, (p-1)/2\}$, independent of the primitive element chosen. The theory of quadratic residues is a fundamental part of number theory.

15.1.184 Definition [311, 2405] Let p be an odd prime and q a prime power with $\gcd(p, q) = 1$. Assume further that $q^{(p-1)/2} \equiv 1 \pmod{p}$; hence q is a quadratic residue modulo p. Let $\mathcal{R}$ and $\mathcal{N}$ be the quadratic residues and quadratic nonresidues modulo p. Relative to a primitive p-th root of unity β in some extension field of $\mathbb{F}_q$, denote by $\mathcal{D}_\mathcal{R}$ and $\mathcal{D}_\mathcal{N}$ the $(p, (p+1)/2, d)_q$ odd-like duadic codes over $\mathbb{F}_q$ with defining sets $\mathcal{R}$ and $\mathcal{N}$, respectively. Denote by $\mathcal{C}_\mathcal{R}$ and $\mathcal{C}_\mathcal{N}$ the $(p, (p-1)/2, d')_q$ even-like duadic codes over $\mathbb{F}_q$ with defining sets $\{0\} \cup \mathcal{R}$ and $\{0\} \cup \mathcal{N}$, respectively. The four codes $\mathcal{D}_\mathcal{R}$, $\mathcal{D}_\mathcal{N}$, $\mathcal{C}_\mathcal{R}$, and $\mathcal{C}_\mathcal{N}$ are *quadratic residue (QR) codes.*

15.1.185 Remark Much is known about QR codes and their relatives, their minimum distances, and automorphism groups. As noted previously, the only two perfect codes with minimum distance greater than 3 are cosets of two linear Golay codes, which are quadratic residue codes treated in the next two examples.

15.1.186 Example The binary $(23, 12, 7)_2$ Golay code [2405, 2849]: Let $p = 23$ and $q = 2$, and note that 2 is a quadratic residue modulo 23 as $5^2 \equiv 2 \pmod{23}$. The smallest extension field of $\mathbb{F}_2$ that contains a primitive 23-rd root of unity is $\mathbb{F}_{2^{11}}$. If α is a primitive element of $\mathbb{F}_{2^{11}}$, then $\beta = \alpha^{89}$ is a primitive 23-rd root of unity. With the appropriate choice of primitive element, the polynomial

$$g_\mathcal{R}(x) = \prod_{r \in \mathcal{R}} (x - \beta^r) = x^{11} + x^9 + x^7 + x^6 + x^5 + x + 1 \in \mathbb{F}_2[x]$$

generates the $(23, 12, d)_2$ quadratic residue code $\mathcal{D}_\mathcal{R}$. Theorem 15.1.180 implies $d \equiv 3 \pmod{4}$, and the minimum distance d can be determined to be 7. Notice that since $\sum_{i=0}^{3} \binom{23}{i} = 2^{11}$, this code is perfect.

15.1.187 Example The ternary $(11, 6, 5)_3$ Golay code [2405, 2849]: For $p = 11$ and $q = 3$, note that $6^2 \equiv 3 \pmod{11}$, and hence 3 is a quadratic residue modulo 11. The smallest extension field of $\mathbb{F}_3$ containing a primitive 11-th root of unity is $\mathbb{F}_{3^5}$, and if α is a primitive root in $\mathbb{F}_{3^5}$, then $\beta = \alpha^{22}$ is a primitive 11-th root of unity. With the appropriate choice of primitive element,

$$g_\mathcal{R}(x) = \prod_{r \in \mathcal{R}} (x - \beta^r) = x^5 + x^4 + 2x^3 + x^2 + 2 \in \mathbb{F}_3[x]$$

is a generator polynomial of the $(11, 6, d)_3$ quadratic residue code $\mathcal{D}_\mathcal{R}$; it can be shown that $d = 5$. Since $\sum_{i=0}^{2} \binom{11}{i} 2^i = 3^5$, this code is also perfect. In fact, as noted earlier, this code and the binary $(23, 12, 7)_2$ Golay code are the only two possible perfect linear codes with minimum distance greater than 3. This ternary perfect code was actually discovered before Golay by Virtakallio [1522] in connection with a football pool problem.

15.1.188 Remark It is interesting to note that the Golay codes were discovered very early in the history of coding [1292] and took only half a page in the *Proceedings of the Institute of Radio Engineers*, (IRE, precursor to the IEEE), appearing prior to the paper of Hamming

[1408]. The original paper of Shannon [2608] actually contained an example of a Hamming code, prior to the appearance of the Hamming paper. The parameters of the Golay codes were found by examining Pascal's triangle for sequential summations along a row that add to an appropriate power. Once the required relationship was found, Golay was able to find generator matrices for the two (linear) codes. He also noted that

$$\sum_{i=0}^{2} \binom{90}{i} = 2^{12}$$

raising the possibility of a $(90, 78, 5)_2$ perfect binary code, but no such code can exist [2810, 2849].

Alternant codes:

15.1.189 Remark Alternant and GRS codes bear a similar relationship as BCH and RS codes in that a BCH code of block length $q^m - 1$ over $\mathbb{F}_q$ is a subfield subcode of an RS code of this length over $\mathbb{F}_{q^m}$.

15.1.190 Definition An *alternant code* $A_{n,q}(\boldsymbol{\alpha}, \mathbf{v})$ is the subfield subcode of $GRS_{n,q^m}(\boldsymbol{\alpha}, \mathbf{v}, k)$, i.e.,

$$A_{n,q}(\boldsymbol{\alpha}, \mathbf{v}) = GRS_{n,q^m}(\boldsymbol{\alpha}, \mathbf{v}, k) \,|_{q^m|q}^{sf} .$$

15.1.191 Lemma [1991, 2484] If $A_{n,q}(\boldsymbol{\alpha}, \mathbf{v}) = GRS_{n,q^m}(\boldsymbol{\alpha}, \mathbf{v}, k) \,|_{q^m|q}^{sf}$ is an $(n, k', d')_q$ code, then $k' \leq k$ and $d' \geq n - k + 1$.

15.1.192 Remark The class of alternant codes is quite large, containing all narrow sense BCH and Goppa codes. The duals of alternant codes are also of interest for which we have the following diagram [802]:

$$\begin{array}{ccc}
GRS_{n,q^m}(\boldsymbol{\alpha}, \mathbf{v}, k) & \xrightarrow{\text{dual}} & GRS_{n,q^m}(\boldsymbol{\alpha}, \mathbf{w}, n-k) \\
\text{subfield}\big\downarrow & & \big\downarrow\text{trace} \\
A_{n,q}(\boldsymbol{\alpha}, \mathbf{v}) & \xrightarrow{\text{dual}} & A_{n,q}(\boldsymbol{\alpha}, \mathbf{v})^{\perp} \\
\left(= GRS_{n,q^m}(\boldsymbol{\alpha}, \mathbf{v}, k)_{q^m|q}^{sf}\right) & & \left(= GRS_{n,q^m}(\boldsymbol{\alpha}, \mathbf{w}, n-k)_{q^m|q}^{tr}\right)
\end{array} .$$

Goppa codes:

15.1.193 Remark [1319, 1320, 2047, 2849] To discuss Goppa codes it is instructive to consider a natural transition from BCH codes in the following manner. For a variable x and α a nonzero element in a field, note that

$$\frac{1}{1 - \alpha^{-1}x} = 1 + \alpha^{-1}x + \alpha^{-2}x^2 + \cdots + \alpha^{-(2t-1)}x^{2t-1} + \alpha^{-2t}x^{2t} + \alpha^{-(2t+1)}x^{2t+1} + \cdots$$

as can be verified by multiplying each side by $(1 - \alpha^{-1}x)$. The effect of taking this equation modulo x^{2t} is to truncate the series on the right hand side to terms of degree $2t$ and higher. Consider the primitive narrow sense BCH code of length $n = q^m - 1$ and designed distance $2t + 1$ with defining set $C_1 \cup C_2 \cup \cdots \cup C_{2t}$ relative to α, a primitive element of $\mathbb{F}_{q^m}$. For convenience denote the nonzero field elements α^i as α_i^{-1}. Then $\mathbf{c} = (c_0, c_1, \ldots, c_{n-1})$ is a codeword of the BCH code if and only if

$$\sum_{i=0}^{n-1} c_i \alpha_i^{-j} = 0 \quad \text{for } j = 1, 2, \ldots, 2t.$$

For a received word $\mathbf{r} = (r_0, r_1, \ldots, r_{n-1})$, the sum of a codeword and error word (presumed of weight not more than t), define the *syndromes* as

$$S_j = \sum_{i=0}^{n-1} r_i \alpha_i^{-j} \quad \text{for } j = 1, 2, \ldots, 2t,$$

and define the *syndrome polynomial* as $\sum_{j=1}^{2t} S_j x^{j-1}$. Note that in the absence of errors (the received word is a codeword), this is the zero polynomial. However this polynomial can be expressed as

$$\begin{aligned} S(x) &= \sum_{j=1}^{2t} S_j x^{j-1} = \sum_{j=1}^{2t} x^{j-1} \left(\sum_{i=0}^{n-1} r_i \alpha_i^{-j} \right) = \sum_{i=0}^{n-1} r_i \left(\sum_{j=1}^{2t} \alpha_i^{-j} x^{j-1} \right) \\ &= \sum_{i=0}^{n-1} r_i \frac{1}{\alpha_i} \sum_{j=1}^{2t} (\alpha_i^{-1} x)^{j-1} = -\sum_{i=0}^{n-1} \frac{r_i}{x - \alpha_i} \pmod{x^{2t}}. \end{aligned}$$

It follows that $\mathbf{c} = (c_0, c_1, \ldots, c_{n-1})$ is a codeword if and only if

$$\sum_{i=0}^{n-1} \frac{c_i}{x - \alpha_i} \equiv 0 \pmod{x^{2t}}.$$

In this guise the transition from BCH to Goppa codes is more intuitive.

15.1.194 **Definition** [1945, 2047, 2849] Let $g(x) \in \mathbb{F}_{q^m}[x]$ be monic and let $L = \{\alpha_0, \alpha_1, \ldots, \alpha_{n-1}\}$ be a subset of $\mathbb{F}_{q^m}$ such that $g(\alpha_i) \neq 0$ for $0 \leq i \leq n-1$. Then $(c_0, c_1, \ldots, c_{n-1}) \in \mathbb{F}_q^n$ is in the *Goppa code* $G_{n,q}(L, g)$ if and only if

$$\sum_{i=0}^{n-1} \frac{c_i}{x - \alpha_i} \equiv 0 \pmod{g(x)}.$$

The Goppa code is more often denoted $\Gamma(L, g)$ in the literature, but the notation used here is consistent with our earlier code designations.

15.1.195 **Remark** If the Goppa polynomial is irreducible over $\mathbb{F}_{q^m}$, the code is called *irreducible.* By choosing the Goppa polynomial as $g(x) = x^{d-1}$ and L as the set of n-th roots of unity in $\mathbb{F}_{q^m}$, the Goppa code is a BCH code with designed distance d.

15.1.196 **Theorem** [2047, 2849] For the Goppa code $G_{n,q}(L, g)$, where $|L| = n$ and $\deg g(x) = s$, we have:

1. The minimum distance is bounded by $d \geq s + 1$.
2. The code dimension is bounded by $k \geq n - ms$.

15.1.197 **Remark** [1945, 2047] For any monic polynomial $g(x)$ of degree s such that $g(\alpha) \neq 0$, it is clear from previous arguments that

$$\frac{1}{x - \alpha} \equiv -\frac{1}{g(\alpha)} \left(\frac{g(x) - g(\alpha)}{x - \alpha} \right) \pmod{g(x)}$$

is a polynomial of degree less than s. (Note that this follows as $x - \alpha$ is a factor of the numerator since the numerator has α as a zero.) Thus $\mathbf{c}$ is a codeword in $G_{n,q}(L, g)$ if and only if its inner product with

$$\left(\frac{1}{g(\alpha_0)}\frac{g(x)-g(\alpha_0)}{(x-\alpha_0)},\ldots,\frac{1}{g(\alpha_{n-1})}\frac{g(x)-g(\alpha_{n-1})}{(x-\alpha_{n-1})}\right)$$

is zero. To simplify this expression, let $g(x)=\sum_{i=0}^{s}g_ix^i$ and substitute this into the above equation. By expanding the fractions and considering like powers of x, one obtains a matrix which can be row reduced to

$$\begin{bmatrix} h_0 & h_1 & \cdots & h_{n-1} \\ h_0\alpha_0 & h_1\alpha_1 & \cdots & h_{n-1}\alpha_{n-1} \\ & & \vdots & \\ h_0\alpha_0^{s-1} & h_1\alpha_1^{s-1} & \cdots & h_{n-1}\alpha_{n-1}^{s-1} \end{bmatrix}$$

where $h_i=1/g(\alpha_i)$ for $i=0,1,\ldots,n-1$; the Goppa code $G_{n,q}(L,g)$ is the set of words over $\mathbb{F}_q$ orthogonal to the rows of this matrix. Notice this form of parity check matrix implies the Goppa code is an alternant code.

15.1.198 Remark To show the class of Goppa codes contains *asymptotically good codes*, it suffices to show there exists a sequence of Goppa polynomials, of increasing degrees, for which the normalized rate and distance approach the Varshamov-Gilbert bound; see Theorem 15.1.130. While not constructive, such an argument shows the class of Goppa codes includes some asymptotically good codes. The issue is discussed in [2849].

15.1.199 Remark There are many other classes of cyclic codes not discussed here. To round out the discussion above, the generalized versions of Reed-Muller codes are mentioned, as well as codes obtained from finite Euclidean and projective geometries [2390] that will be of interest in discussing majority logic decoding.

Generalized Reed-Muller (GRM) codes:

15.1.200 Remark The binary RM codes are generalized in the following manner. Construct the $(m+1)\times(q^m-1)$ matrix over $\mathbb{F}_q$ as follows. The first row consists of all ones and is labeled $\mathbf{v}_0$. Among the m-tuples of the q^m-1 columns in rows $\mathbf{v}_1$ through $\mathbf{v}_m$, all nonzero m-tuples over $\mathbb{F}_q$ occur. A simple way of viewing this is to choose row $\mathbf{v}_1$ as a *pseudo-noise (pn) sequence over* $\mathbb{F}_q$, a sequence generated by a linear feedback shift register whose feedback coefficients are the coefficients of a primitive polynomial of degree m over $\mathbb{F}_q$. Row $\mathbf{v}_{i+1}$ is row 1 shifted by i positions for $1\le i\le m-1$. Let $\mathbb{F}_q[X_1,\ldots,X_m]_\nu=\mathbb{F}_q[X]_\nu$ denote the set of polynomials of degree at most ν over $\mathbb{F}_q$ in m variables. Define the matrix $G_{\nu,m}$ over $\mathbb{F}_q$ whose first row is $\mathbf{v}_0$ and whose remaining rows are generated by all monomials in $\mathbb{F}_q[X]_\nu$ acting on rows $\mathbf{v}_1,\mathbf{v}_2,\ldots,\mathbf{v}_m$. Let $GRM_{q^m-1,q}(\nu,m)$ be the code with generator matrix $G_{\nu,m}$; let $GRM_{q^m,q}(\nu,m)$ be the extended code obtained from $GRM_{q^m-1,q}(\nu,m)$ by adding an overall parity check. The exponent of a monomial is limited to degree at most $q-1$ in any variable since $x_i^q=x_i$ in $\mathbb{F}_q$; thus we only consider values of ν with $\nu\le m(q-1)$.

15.1.201 Theorem [805, 1691] The parameters of $GRM_{q^m-1,q}(\nu,m)$ are

$$n=q^m-1,\quad k=\sum_{t=0}^{\nu}\sum_{j=0}^{m}(-1)^j\binom{m}{j}\binom{t-jq+m-1}{t-jq},\quad d=(q-s)q^{m-r-1}-1,$$

where $\nu=r(q-1)+s,\ 0\le s<q-1,\ 0\le r\le m$. The extended code $GRM_{q^m,q}(\nu,m)$ has length and minimum distance one greater than the above, with the same dimension. The dual code to $GRM_{q^m,q}(\nu,m)$ is $GRM_{q^m,q}(\mu,m)$ where $\nu+\mu+1=m(q-1)$.

15.1.202 Remark The expression for the code dimension in the previous theorem is the number of monomials of total degree at most ν. This is also the number of ways of placing ν balls in m cells, no cell containing more than $q-1$ balls. When $\nu < q$, this expression reduces to $\binom{\nu+m}{m}$, and the minimum distance is $(q-\nu)q^{m-1}-1$.

15.1.203 Remark To show the GRM codes are cyclic requires the following notion. The *radix-q weight* of an integer j is defined as

$$w_q(j) = j_0 + j_1 + \cdots \quad \text{where} \quad j = j_0 + j_1 q + j_2 q^2 + \cdots \quad \text{with} \quad 0 \leq j_i < q.$$

15.1.204 Theorem Let α be a primitive element of $\mathbb{F}_{q^m}$. Then the code $GRM_{q^m-1,q}(\nu, m)$, for $\nu \leq m(q-1)$, is cyclic with defining set relative to α given by

$$\{j \mid 0 < j < q^m - 1,\ 0 < w_q(j) \leq m(q-1) - \nu - 1\}.$$

The code is a subcode of the primitive narrow sense BCH code with designed distance $(q-s)q^{m-r-1} - 1$ where $\nu = r(q-1) + s$ as in Theorem 15.1.201.

15.1.205 Remark To discuss *projective GRM codes* denote by $\mathbb{F}_q[X_0, \ldots, X_m]_\nu^0 = \mathbb{F}_q[X]_\nu^0$ the set of homogeneous polynomials in $m+1$ variables of degree at most ν where $\nu \leq (m+1)(q-1)$. (The difference in the use of X, compared to the nonprojective case, is resolved by context.) There are $\pi_{m+1,q}$ projective $(m+1)$-tuples, $\mathbb{P}_q^m$, which coordinatize the positions of the projective code. The projective codes are defined as

$$PGRM_{\pi_{m+1,q},q}(\nu, m) = \{(f(x)) \mid f(x) \in \mathbb{F}_q[X]_\nu^0,\ x \in \mathbb{P}_q^m\}.$$

The basic parameters of the code are given in the following theorem.

15.1.206 Theorem [1827, 2696] The code $PGRM_{\pi_{m+1,q},q}(\nu, m)$ has the following parameters:

$$n = \pi_{m+1,q} = \frac{q^{m+1}-1}{q-1}, \quad k = \sum_{\substack{t=\nu \pmod{q-1} \\ 0<t\leq \nu}} \left(\sum_{j=0}^{m+1} (-1)^j \binom{m+1}{j} \binom{t-jq+1}{t-jq} \right)$$

and

$$d = (q-s)q^{m-r-1},$$

where $\nu - 1 = r(q-1) + s,\ 0 \leq s < q-1$. In the case $\nu < q$ these expressions reduce to

$$n = \pi_{m+1,q}\,, \quad k = \binom{r+m}{m}, \quad d = (q-r+1)q^{m-1}.$$

Finite geometry codes:

15.1.207 Remark The binary (extended) RM codes have an interesting interpretation in terms of finite geometries for which some terminology is needed. Denote the n-dimensional projective geometry over $\mathbb{F}_q$ by $PG(n,q)$ and the Euclidean geometry by $EG(n,q)$. (Some authors denote $EG(n,q)$ by $AG(n,q)$; see Chapter 14.)

15.1.208 Remark [311] Projective geometries are discussed first; see also Section 14.4. Let α denote a primitive element of $\mathbb{F}_{q^{n+1}}$ and $v = (q^{n+1}-1)/(q-1) = \pi_{n+1,q}$. Then $\beta = \alpha^v$ is a primitive element of $\mathbb{F}_q$. No two points α^j with $j = 0, 1, \ldots, v-1$ are $\mathbb{F}_q$-multiples of each other, and hence these first v powers of α can be taken as the points of $PG(n,q)$. Let $\alpha^{j_1}, \alpha^{j_2}, \ldots, \alpha^{j_{m+1}}$, $m \geq 1$, be $m+1$ linearly independent points in $PG(n,q)$, i.e., there is no linear relationship between them over $\mathbb{F}_q$. The points

$$\{a_{j_1}\alpha^{j_1} + a_{j_2}\alpha^{j_2} + \cdots + a_{j_{m+1}}\alpha^{j_{m+1}} \mid a_{j_i} \in \mathbb{F}_q\},$$

with multiples over $\mathbb{F}_q$ identified, give the $\pi_{m+1,q} = (q^{m+1}-1)/(q-1)$ points of an m-flat in $PG(n,q)$. The flat is a $PG(m,q)$. In particular, the number of points on a line in $PG(n,q)$ is $q+1$. In $PG(n,q)$ any two lines intersect in a single point, and the number of lines through a fixed point in $PG(n,q)$ is given by $(q^n-1)/(q-1)$.

To construct an $EG(n,q)$ from a $PG(n,q)$, subtract an $(n-1)$-flat from $PG(n,q)$. An $(n-1)$-flat can be thought of as the set of points in $PG(n,q)$ orthogonal to a given line. Alternatively the points of $PG(n,q)$ can be divided into the two groups

$$S_1 = \{(1,\gamma_2,\ldots,\gamma_{n+1}) \mid \gamma_i \in \mathbb{F}_q\} \quad \text{and} \quad S_2 = \{(0,\eta_2,\ldots,\eta_{n+1}) \mid \eta_j \in \mathbb{F}_q\}.$$

The points of S_2 form an $(n-1)$-flat and those of S_1 an $EG(n,q)$. The points of this geometry can be represented as the elements of $\mathbb{F}_{q^n}$. The points of an m-flat in $EG(n,q)$ through a given point α^i is the set of points

$$\{\alpha^i + \gamma_1\alpha^{j_1} + \cdots + \gamma_m\alpha^{j_m} \mid \gamma_j \in \mathbb{F}_q\}$$

where the α^{j_k}'s are independent. Lines can be parallel in $EG(n,q)$. In $EG(2,q)$ the $(q+1)q$ lines can be divided into $q+1$ parallel classes, where the lines in each class are parallel, each class containing q lines. Lines in different classes intersect. In $EG(n,q)$ there are $q^{n-1}(q^n-1)/(q-1)$ lines and $(q^n-1)/(q-1)$ parallel classes, each class containing q^{n-1} lines and each line containing q points.

15.1.209 Remark The generator matrix $G(r,m)$ of the r-th order binary Reed-Muller code $RM_{2^m,2}(r,m)$ (Definitions 15.1.71 and 15.1.72) has the following interpretation. The zero-th row is the incidence vector of the Euclidean space $EG(m,2)$. Rows 1 through m are interpreted as incidence vectors for $(m-1)$-flats. The product of two such rows is the incidence vector of the intersection of two such flats (an $(m-2)$-flat) etc. One aim in pursuing this geometric view for coding is the consideration of codes derived from such finite geometries and their majority logic decoding. For the remainder of this section assume $q = p^s$ for some prime p, the characteristic of $\mathbb{F}_q$, and integer s.

15.1.210 Definition [1691, 2390] The *r-th order Euclidean geometry code* of length q^m over $\mathbb{F}_p$, denoted $EG_{q^m,p}(r)$, is the largest linear code over $\mathbb{F}_p$ that contains in its null space the incidence vectors of all $(r+1)$-flats in $EG(m,q)$.

15.1.211 Remark Note that the incidence vectors are, by definition, binary and so the codes could be defined over any finite field. However, not choosing the field $\mathbb{F}_p$ would complicate the analysis considerably. It turns out that these codes are in fact extended cyclic codes as the following theorem shows.

15.1.212 Theorem [304, 1691] Let α be a primitive element of $\mathbb{F}_{q^m} = \mathbb{F}_{p^{sm}}$. Then $EG_{q^m,p}(r)$ is the code obtained by extending the cyclic code whose defining set relative to α is

$$\{j \mid 0 < j < q^m - 1,\ 0 < \max_{0\le i<s} w_q(jp^i) \le (q-1)(m-r-1)\}.$$

15.1.213 Definition The *r-th order projective geometry code*, denoted $PG_{(q^m-1)/(q-1),p}(r)$, is the largest linear code containing the incidence vectors of all r-flats in its null space.

15.1.214 Theorem [1692] Let $\beta \in \mathbb{F}_{q^m} = \mathbb{F}_{p^{sm}}$ be a primitive $(q^m-1)/(q-1)$ root of unity. The code $PG_{(q^m-1)/(q-1),p}(r)$ is a cyclic code whose defining set relative to β is

$$\{j \mid 0 < j < (q^m-1)/(q-1),\ 0 < \max_{0\le i<s} w_q(j(q-1)p^i) \le (q-1)(m-r+1)\}.$$

15.1.215 Remark The EG and PG codes are defined as the largest codes having the appropriate flats in their dual codes. In the binary case, the EG codes are extended PG codes, and the PG codes can be made cyclic. The codes were investigated extensively in [1691, 1692, 1942, 2966]. Lower bounds on their minimum distances can be obtained from the number of errors they are capable of correcting with majority logic decoding; see Section 15.1.6.6. The *RM*, *GRM*, *EG*, and *PG* codes, as well as many others, can be discussed under a very general class of *polynomial* codes [1692].

Justesen codes:

15.1.216 Remark There have been many attempts to explicitly construct codes for which the normalized rate and distance functions do not both tend to zero with increasing block length. It is known that the class of BCH codes cannot achieve this. While the class of Goppa codes can be used for such a purpose, the construction is not explicit in that it calls for the construction of a sequence of suitable polynomials which are known to exist but are not given [2849]. Justesen [1638] provided the first explicit construction. An outline of that construction is given. Let $N = 2^m - 1$, and let α be a primitive element in $\mathbb{F}_{2^m}$. Let $\mathcal{C}_{m,K}$ be the $(N, K, D)_{2^m}$ RS code given by

$$\{(f(1), f(\alpha), \ldots, f(\alpha^{N-1})) \mid f(x) \in \mathbb{F}_{2^m}[x],\ \deg f(x) \le K - 1\}.$$

Let $\mathcal{C}'_{m,K}$ be the $(2N, K, 2D)_{2^m}$ code given by

$$\mathcal{C}'_{m,K} = \{(a_0, a_1, \ldots, a_{N-1}, a_0, \alpha a_1, \ldots, \alpha^{N-1} a_{N-1}) \mid (a_0, a_1, \ldots, a_{N-1}) \in \mathcal{C}_{m,K}\}.$$

The *Justesen* code $\mathcal{C}''_{m,K}$ is found by expanding the components of each codeword, which are in $\mathbb{F}_{2^m}$, into binary m-tuples with respect to some fixed basis.

15.1.217 Theorem [1638] For any given code rate $R < 1/2$ and given $m = 1, 2, \ldots$, choose K_m to be the smallest integer K such that $K/2N \ge R$ where $N = 2^m - 1$. The code $\mathcal{C}''_{m,K_m}$ is linear over $\mathbb{F}_2$ with length $n = 2mN$, dimension mK_m, rate $K_m/2N \ge R$, and minimum distance d_n asymptotically bounded by

$$\liminf_{n\to\infty} d_n/n \ge (1 - 2R)H^{-1}(1/2) \sim 0.11(1 - 2R).$$

15.1.218 Remark The normalized rate and distance of the code for a given rate R are both clearly nonzero. The construction calls for the expansion of elements of $\mathbb{F}_{2^m}$ into binary m-tuples as well as a primitive element $\alpha \in \mathbb{F}_{2^m}$. To make the procedure entirely constructive (i.e., to explicitly give such an expansion) an irreducible polynomial of each degree, as the degrees tend to infinity, is needed. The issue is discussed in [1638] where it is noted that an irreducible polynomial of the form $x^{2\cdot 3^\ell} + x^{3^\ell} + 1$ can be used, $\ell = 1, 2, \ldots$. Given such an irreducible polynomial for a given degree and known order, a primitive element can be determined.

15.1.4 A spectral approach to coding

15.1.219 Remark Suppose $\mathbf{a} = (a_0, a_1, \ldots, a_{n-1}) \mapsto a(x) = a_0 + a_1 x + \cdots + a_{n-1}x^{n-1} \in \mathbb{F}_q[x]$ and $n \mid (q^m - 1)$, i.e., $\mathbb{F}_{q^m}$ contains a primitive n-th root of unity α. The *Mattson-Solomon polynomial* of $\mathbf{a}$ is defined as

$$A(x) = \sum_{i=1}^{n} A_{n-i} x^{n-i} = \sum_{i=1}^{n} a(\alpha^i) x^{n-i}$$

and the correspondence is given by

$$A_i = a(\alpha^{n-i}) \longleftrightarrow a_i = \frac{1}{n}A(\alpha^i).$$

The utility of such an approach stems from the following theorem.

15.1.220 Theorem [2849] If the Mattson-Solomon polynomial A of a vector $\mathbf{a}$ has r n-th roots of unity as zeros, then the weight of $\mathbf{a}$ must be at least $n - r$.

15.1.221 Remark This theorem follows as $a_i = A(\alpha^i)/n$ for $0 \leq i < n$. The theorem allows a *spectral approach* to coding to be taken (as for example in [304]) where the correspondence between weights of vectors and zeros of Mattson-Solomon polynomials can be exploited. The coefficients of the Mattson-Solomon polynomial $A(x)$ can be viewed as a discrete Fourier transform of the coefficients of the original polynomial $a(x)$ and hence enjoy many useful properties, similar to those of a discrete Fourier transform.

15.1.5 Codes and combinatorics

15.1.222 Remark The relationships between codes and certain combinatorial structures are deep and of great interest. They include connections between codes and association schemes, difference sets, finite geometries and designs, among many others. Only a basic relationship between the weight classes of certain codes as incidence vectors for designs will be noted here. The reader is referred to many chapters in the handbook by Pless et al. [2405] for a recent view of the subject and to Chapter 14 of this Handbook.

15.1.223 Definition A t-(v, k, λ) *design* is a pair $(\mathcal{P}, \mathcal{B})$ where $\mathcal{P}$ is a collection of v distinct points and $\mathcal{B}$ is a collection of subsets of $\mathcal{P}$, called *blocks*, each of size k, with the property that any subset of $\mathcal{P}$ of size t is contained in exactly λ blocks. The number of blocks of the design is given by $b = \lambda\binom{v}{t}/\binom{k}{t}$.

15.1.224 Remark An interesting technique for obtaining classes of t-designs is to consider the codewords of fixed weight in a binary code and to determine under what conditions they might form incidence vectors for the blocks of a t-design. Such investigations have often illuminated the structure of a code and produced interesting classes of designs for combinatorial use. An important tool in this investigation has been the Assmus-Mattson theorem below. While quite technical, it is straightforward to apply if sufficient information is known about the code and its dual. The *support* of a vector is the list of coordinate positions where the vector is nonzero. The codewords of some fixed weight *hold* a design if the supports of the codewords form a t-design for some t.

15.1.225 Theorem [142] Let $\mathcal{C}$ be an $(n, k, d)_q$ code with dual code $\mathcal{C}^\perp$ having minimum distance $d^\perp$. If $q = 2$, let $w = n$; for $q > 2$, let w be the largest integer such that

$$w - \left\lfloor \frac{w + q - 2}{q - 1} \right\rfloor < d.$$

Define $w^\perp$ analogously for $\mathcal{C}^\perp$. Suppose $\{A_i\}$ and $\{B_i\}$ are the weight distributions for $\mathcal{C}$ and $\mathcal{C}^\perp$ respectively. Let s be the number of i with $B_i \neq 0$ for $0 < i \leq n-t$ for some integer t. Suppose $t < d$ and $s \leq d-t$. Then the codewords of weight i in $\mathcal{C}$ hold a t-design provided $A_i \neq 0$ and $d \leq i \leq w$. The words of weight i in $\mathcal{C}^\perp$ hold a t-design provided $B_i \neq 0$ and $d^\perp \leq i \leq \min\{n - t, w^\perp\}$.

15.1.226 Example The extended binary $(24, 12, 8)_2$ Golay code (the extension of the quadratic residue $(23, 12, 7)_2$ code) has weight distribution $\{A_0 = 1,\ A_8 = 759,\ A_{12} = 2576,\ A_{16} = 759,\ A_{24} = 1\}$ and is self-dual. For $t = 5$, the number of nonzero weights less than $24-5 = 19$ is $s = 3$. Thus the conditions and each weight class of the code satisfies the theorem, i.e., each weight class of the code holds a 5-design. In particular the 759 codewords of weight 8 hold a Steiner design ($\lambda = 1$), in fact a 5-$(24, 8, 1)$ design, one of the more interesting designs known. This design is also related to the Leech lattice which leads to a particularly dense sphere packing in 24-dimensional Euclidean space. This connection is explored in detail in [2805].

15.1.227 Example In a similar manner, the extended ternary $(12, 6, 6)_3$ Golay code is of interest. This code has nonzero codewords only of weights 6, 9, and 12, and is self-dual. Taking $t = 5$, the number of nonzero code weights less than $12 - 5 = 7$ is $s = 1$ and so the weight classes of the code each hold 5-designs. This code has 264 codewords of weight 6, 440 of weight 9 and 24 of weight 12.

15.1.228 Remark The above discussion suggests that self-dual codes with few weights and large distance might be of interest in terms of applying the Assmus-Mattson theorem and such has been the case. Particular interest is in the case of binary and ternary self-dual codes. If $\mathcal{C}$ is a binary self-dual code, its length is even and every codeword has even weight. If $\mathcal{C}$ is a ternary self-orthogonal code, every weight is divisible by 3. A binary code is *even* if every codeword has even weight. A binary code is *doubly-even* if every weight is divisible by 4. *Formally self-dual* codes are those for which $W_{\mathcal{C}}(x, y) = W_{\mathcal{C}^\perp}(x, y)$ and all self-dual codes are formally self-dual. Many results are known about such codes and a few are touched on here [1991, 2405].

15.1.229 Lemma [2405] Let $\mathcal{C}$ be a code over $\mathbb{F}_q$ with every weight divisible by Δ. If $q = 2$ and $\Delta = 4$ or if $q = 3$ and $\Delta = 3$, then $\mathcal{C}$ is self-orthogonal.

15.1.230 Lemma [2405] Binary self-dual doubly-even codes of length n exist if and only if $8 \mid n$. Ternary self-dual codes exist if and only if $4 \mid n$.

15.1.231 Theorem [2405] Let $\mathcal{C}$ be an $(n, n/2, d)_q$ code for $q = 2$ or $q = 3$.

1. If $q = 2$ and $\mathcal{C}$ is formally self-dual and even, then $d \leq 2\lfloor n/8 \rfloor + 2$.
2. If $q = 2$ and $\mathcal{C}$ is self-dual and doubly-even, then $d \leq 4\lfloor n/24 \rfloor + 4$.
3. If $q = 3$ and $\mathcal{C}$ is self-dual, then $d \leq 3\lfloor n/12 \rfloor + 3$.

15.1.232 Remark Codes which achieve the upper bounds in the above theorem are referred to as *extremal* and often have weight classes of codewords that hold designs. The binary and ternary Golay codes are extremal. The existence of other extremal codes has long been a matter of interest.

15.1.233 Remark As noted, the work of Delsarte [801, 806] has been influential in the relationship between codes and combinatorics, particularly t-designs, orthogonal arrays and graphs. The result below is but one such example.

15.1.234 Theorem [801] Let $\mathcal{C}$ be a binary code of minimum distance d and external distance s', with $d \geq s'$. Then $\mathcal{C}$ is distance invariant and the codewords of a given weight hold a t-design with $t = d - s'$.

15.1.6 Decoding

15.1.6.1 Decoding BCH codes

15.1.235 Remark Decoding a primitive narrow sense BCH code is considered. The modifications needed for a nonnarrow sense or nonprimitive code, as well as for RS and other cyclic codes, will be clear. Also, in most cases, such errors-only decoding algorithms can be extended to errors-and-erasures algorithms. Assume the code has length $n = q^m - 1$ and designed distance $d = 2t+1$ and every codeword polynomial has roots α^i for $i = 1, 2, \ldots, 2t$ where α is a primitive element of $\mathbb{F}_{q^m}$. The development below is fairly standard [231, 304, 1329, 1943, 2136, 2390, 2484]. It is convenient to think of codewords and received words as polynomials. Assume a codeword polynomial $c(x)$ is sent and the polynomial $r(x) = c(x) + e(x)$ is received, where $e(x)$ is the error polynomial, assumed to be of weight at most t. If more than t errors occur, the algorithm may fail. For example, in the extreme case, if a certain set of d errors occurs, it may happen that an incorrect codeword is received and the algorithm will assume no errors occurred in transmission.

15.1.6.2 The Peterson-Gorenstein-Zierler decoder

15.1.236 Remark The algorithm discussed in this section was first described by Peterson [2389] for binary codes and by Gorenstein and Zierler [1329] for nonbinary codes. Denote the i-th position of a codeword by α^i for $i = 0, 1, \ldots, q^m - 2$. Assume the actual number of errors that occurred (unknown) is $\nu \le t = \lfloor (d-1)/2 \rfloor$. Let the errors be in coordinate positions $X_i = \alpha^{j_i}$ with error values Y_i for $i = 1, 2, \ldots, \nu$. Noting that $c(\alpha^j) = 0$ for $j = 1, 2, \ldots, 2t$, the information of value in the received codeword is from its *syndromes* evaluated as

$$S_j = r(\alpha^j) = c(\alpha^j) + e(\alpha^j) = e(\alpha^j) = \sum_{i=1}^{\nu} Y_i X_i^j \quad \text{for } j = 1, 2, \ldots, 2t.$$

In matrix form this is

$$\begin{bmatrix} Y_1X_1 & + & Y_2X_2 & + & \cdots & + & Y_\nu X_\nu \\ Y_1X_1^2 & + & Y_2X_2^2 & + & \cdots & + & Y_\nu X_\nu^2 \\ & & & & \vdots & & \\ Y_1X_1^{2t} & + & Y_2X_2^{2t} & + & \cdots & + & Y_\nu X_\nu^{2t} \end{bmatrix} = \begin{bmatrix} S_1 \\ S_2 \\ \vdots \\ S_{2t} \end{bmatrix}. \tag{15.1.5}$$

Denote the *error locator polynomial* by

$$\sigma(x) = \prod_{i=1}^{\nu} (1 - X_i x) = 1 + \sigma_1 x + \cdots + \sigma_{\nu-1} x^{\nu-1} + \sigma_\nu x^\nu,$$

i.e., the errors occur at the inverses of the zeros of $\sigma(x)$. It follows that

$$S_{j+\nu} + \sigma_1 S_{j+\nu-1} + \cdots + \sigma_\nu S_j = 0 \quad \text{for } j = 1, 2, \ldots, 2t - \nu \tag{15.1.6}$$

which leads to the matrix equation

$$\begin{bmatrix} S_1 & S_2 & \cdots & S_{\nu-1} & S_\nu \\ S_2 & S_3 & \cdots & S_\nu & S_{\nu+1} \\ & & \vdots & & \\ S_\nu & S_{\nu+1} & \cdots & S_{2\nu-2} & S_{2\nu-1} \end{bmatrix} \begin{bmatrix} \sigma_\nu \\ \sigma_{\nu-1} \\ \vdots \\ \sigma_1 \end{bmatrix} = \begin{bmatrix} -S_{\nu+1} \\ -S_{\nu+2} \\ \vdots \\ -S_{2\nu} \end{bmatrix}.$$

Since $\nu \leq t$, all syndromes are known. If the number of errors made, ν, was known, the equations could be solved for the coefficients σ_i and error locations found from the roots of $\sigma(x)$. The following lemma suggests a method to find ν.

15.1.237 Lemma The matrix

$$\begin{bmatrix} S_1 & S_2 & \cdots & S_\ell \\ S_2 & S_3 & \cdots & S_{\ell+1} \\ & & \vdots & \\ S_\ell & S_{\ell+1} & \cdots & S_{2\ell-1} \end{bmatrix}$$

is nonsingular if $\ell \leq \nu$ and singular otherwise.

15.1.238 Remark To find the actual number of errors that occurred in transmission, first form the matrix in Lemma 15.1.237 for $\ell = t$. If nonsingular, assume t errors occurred. If singular, set ℓ to $t-1$ and repeat until the matrix is nonsingular. The algorithm is then:

1. Compute the syndromes S_i for $i = 1, 2, \ldots, 2t$.
2. Determine the actual number of errors that occurred in transmission from Lemma 15.1.237.
3. Find the error locations by finding the roots of $\sigma(x)$ (by trying all nonzero field elements if necessary).
4. Find the error values using (15.1.5).

Several steps of this algorithm are computationally expensive, such as finding successive determinants. More efficient methods are introduced next.

15.1.239 Remark The above decoding technique is easily generalized to RS, GRS, alternant, and Goppa codes.

15.1.6.3 Berlekamp-Massey decoding

15.1.240 Remark [231, 2011] Equation (15.1.6) suggests the following interpretation for determining the error locator polynomial: Given the sequence of $2t$ syndromes, S_i for $i = 1, 2, \ldots, 2t$, determine the linear feedback shift register of minimum length such that if $S_1, S_2, \ldots$ are initially loaded into it, it will generate all $2t$ syndromes. Clearly the feedback coefficients of such a shift register will be the coefficients of the error locator polynomial. An efficient algorithm to determine these coefficients was given in Berlekamp [231] and this algorithm was interpreted in the above manner by Massey [2011]. The details of the development are intricate and omitted here. The following lemma from [2011] gives the flavor of the argument. Denote the feedback connection polynomial at the r-th stage by $\sigma^{(r)}(x)$ and the length of the register by ℓ_r so that the shift register generates $S_1, S_2, \ldots, S_r$. The lemma decides on the length of the minimum length register in going from the $(r-1)$-st stage to the r-th stage.

15.1.241 Lemma [2011] Let $\{\ell_i, \sigma^{(i)}(x)\}$ be a sequence of minimum-length shift registers such that $(\ell_i, \sigma^{(i)}(x))$ generates the sequence up to S_i, for $i = 1, 2, \ldots, r-1$. Then if $\sigma^{(r)}(x) \neq \sigma^{(r-1)}(x)$, the length of the r-th shift register is

$$\ell_r = \max\{\ell_{r-1}, r - \ell_{r-1}\}.$$

15.1.242 Remark Once the length of the shift register for the r-th stage is determined, an efficient procedure to update the coefficients of $\sigma^{(r)}(x)$ is known ([304, Theorem 7.4.1], [2390, Theorem 9.10], and [231, Algorithm 7.4]). The final step of the improvements for the Berlekamp-Massey algorithm involves determining the error values, once their locations are known, due to Forney [1090]. Let $\omega(x)$ be the *error evaluator polynomial* given by

$$\omega(x) \equiv S(x)\sigma(x) \pmod{x^{2t}}. \tag{15.1.7}$$

This is the *key equation* for decoding.

15.1.243 Theorem [1090] The error values may be determined by the equations

$$\omega(x) = x \sum_{i=1}^{\nu} Y_i X_i \prod_{j \neq i} (1 - X_j x)$$

and

$$Y_i = \frac{\omega(X_i^{-1})}{X_i^{-1}\omega'(X_i^{-1})}$$

where ω' is the formal derivative of ω.

15.1.244 Remark Once the error locations are known, $\omega(x)$ can be used to compute the error values (rather than invert a $\nu \times \nu$ matrix over $\mathbb{F}_q$). The degree of $\omega(x)$ is less than that of $\sigma(x)$. Notice that this last step can be omitted for binary codes, since the error values at error locations are 1.

15.1.6.4 Extended Euclidean algorithm decoding

15.1.245 Remark Sugiyama et al. [2741] showed how the *extended Euclidean algorithm (EEA)* can be used to decode Goppa (and hence RS and BCH) codes. The technique was further developed in [2047]. If $\mathbb{F}$ is any field, the EEA is used to determine the greatest common divisor $d(x) \in \mathbb{F}[x]$ of two polynomials $a(x), b(x) \in \mathbb{F}[x]$ and to also determine two polynomials $s(x), t(x) \in \mathbb{F}[x]$ such that

$$s(x)a(x) + t(x)b(x) = d(x).$$

The application of this algorithm to solving the key equation for decoding will be seen later. To discuss this, some properties of the algorithm are required. The treatment of [2047] is followed.

Three sequences of polynomials are derived: $\{r_i(x),\ s_i(x),\ t_i(x)\}$ with the initial conditions

$$\begin{array}{lll} r_{-1}(x) = a(x) & s_{-1}(x) = 1 & t_{-1}(x) = 0 \\ r_0(x) = b(x) & s_0(x) = 0 & t_0(x) = 1 \end{array}$$

where it is assumed $\deg a(x) \geq \deg b(x)$. The sequence of polynomials $\{r_i(x)\}$ and $\{q_i(x)\}$ are derived via the equation

$$r_{i-2}(x) = q_i(x)r_{i-1}(x) + r_i(x) \quad \text{for } i \geq 1,$$

where $\deg r_i(x) < \deg r_{i-1}(x)$; thus the degrees of the $r_i(x)$'s are strictly decreasing. Using the polynomials $q_i(x)$, two auxiliary polynomial sequences are defined:

$$s_i(x) = s_{i-2}(x) - q_i(x)s_{i-1}(x) \quad \text{and} \quad t_i(x) = t_{i-2}(x) - q_i(x)t_{i-1}(x).$$

These polynomial sequences have many properties. In particular for $i \geq -1$

$$s_i(x)a(x) + t_i(x)b(x) = r_i(x).$$

Also $\deg t_i(x) + \deg r_{i-1}(x) = \deg a(x)$ for all $i > 0$, and since the degrees of the $r_i(x)$'s are strictly decreasing, the degrees of the $t_i(x)$'s are strictly increasing and in particular

$\deg t_i(x) + \deg r_i(x) < \deg a(x)$. If the last nonzero remainder polynomial $r_i(x)$ is at step n, i.e., $r_n(x) \neq 0$ but $r_{n+1}(x) = 0$, then

$$d(x) = r_n(x) = s_n(x)a(x) + t_n(x)b(x).$$

15.1.246 Remark Applying the EEA to the polynomials of the key Equation (15.1.7) with $a(x) = x^{2t}$ and $b(x) = S(x)$ leads to three sequences of polynomials $\{r_i(x)\}$, $\{s_i(x)\}$, and $\{t_i(x)\}$. The degrees of the $t_i(x)$'s are increasing and those of the $r_i(x)$'s are decreasing. Stopping the algorithm at the point where $\deg t_j(x)$ first exceeds $\deg r_j(x)$ yields the polynomials $\sigma(x) = t_j(x)$ and $\omega(x) = r_j(x)$; also

$$s_j(x)x^{2t} + t_j(x)S(x) = r_j(x).$$

Interpreting this equation modulo x^{2t} then solves the key Equation (15.1.7).

15.1.6.5 Welch-Berlekamp decoding of GRS codes

15.1.247 Remark The decoding method for GRS codes considered in [2965] relies on forming the syndrome polynomial where the syndromes are a form of Fourier transform of the received word. *Welch-Berlekamp* decoding operates directly on the received word. There are many variants of the algorithm in the literature and only the basic ideas are given here. For simplicity we consider $\mathcal{C} = GRS_{n,q}(\mathbf{x}, \mathbf{1}, k+1)$ where

$$\mathcal{C} = \{(f(x_1), f(x_2), \dots, f(x_n)) \mid f(x) \in \mathbb{F}_q[x], \deg f(x) \leq k\},$$

with distinct code positions x_i in $\mathbb{F}_q$, and let $t = \lfloor (n-k)/2 \rfloor$. Suppose the codeword corresponding to the polynomial $a(x)$ is sent, i.e., $\mathbf{c} = (c_1, c_2, \dots, c_n) = (a(x_1), a(x_2), \dots, a(x_n))$, and the word $\mathbf{r} = (r_1, r_2, \dots, r_n)$ is received. For decoding, it suffices to determine the codeword polynomial $a(x)$ from $\mathbf{r}$. Suppose an unknown number $e \leq t$ of errors has occurred in transmission and $r_i = c_i + e_i$ for $i = 1, 2, \dots, n$. Two polynomials $D(x)$, $N(x) \in \mathbb{F}_q[x]$ are sought with the properties

$$\begin{array}{ll} \text{a.} & \deg D(x) \leq t, \\ \text{b.} & \deg N(x) \leq t + k - 1, \\ \text{c.} & N(x_i) = r_i D(x_i) \text{ for } i = 1, 2, \dots, n. \end{array} \qquad (15.1.8)$$

15.1.248 Theorem There exist polynomials $N(x)$, $D(x)$ which satisfy the above conditions and can be efficiently computed. Furthermore, the ratio $N(x)/D(x)$ is the unique polynomial that gives the closest codeword to $\mathbf{r}$ if fewer than t errors were made in transmission.

15.1.249 Remark The thinking behind this theorem is straightforward. Let $E = \{i \mid r_i \neq c_i\}$, $e = |E|$, i.e., the (unknown) error positions and their number. The polynomial

$$D(x) = \prod_{j \in E} (x - x_j)$$

is the error locator polynomial, i.e., its zeros are the error locations. The polynomial corresponding to the codeword is $a(x) = N(x)/D(x)$. It can be shown that the ratio of any pair of solutions of the system in Part (c) of (15.1.8), under the conditions given in Part (a) for $D(x)$ and Part (b) for $N(x)$, gives this polynomial. Note that $N(x_i) = r_i D(x_i)$ for $i = 1, 2, \dots, n$ as follows. If $x_i \in E$, $N(x_i) = 0 = r_i D(x_i)$ since $N(x) = a(x)D(x)$ and $D(x_i) = 0$; if $x_i \notin E$, $N(x_i) = a(x_i)D(x_i) = c_i D(x_i) = r_i D(x_i)$.

15.1.250 Remark When the code is presented in one of its alternative forms, for example in terms of a generator polynomial, the Welch-Berlekamp equations above are slightly modified. Such an approach is given in [578, 2164].

15.1.6.6 Majority logic decoding

15.1.251 **Remark** Majority logic decoding involves taking a majority vote on the evaluations of certain parity check equations. It is a very simple (both to discuss and implement) decoding technique that is effective when the codes possess certain structure. The Euclidean and projective geometry codes are perhaps the prime examples, but by no means the only codes in this class. Early work on this subject is due to Massey [2010].

15.1.252 **Definition** If a code has J check sums (equations) which each check coordinate position j and checks other coordinate positions at most once, the checks are *orthogonal* on position j.

15.1.253 **Lemma** If it is possible to construct J parity checks orthogonal on each code coordinate position, the code can correct $\lfloor J/2 \rfloor$ errors and has minimum distance at least $J+1$.

15.1.254 **Definition** A set of check sums is *orthogonal on a set* S if each check sum contains all coordinate positions of S and checks each coordinate position not in S at most once.

15.1.255 **Remark** If J checks, orthogonal on a set S, can be constructed, it will be possible to determine the correct value of the sum of checks on coordinate positions for S if fewer than $J/2$ errors occur, by majority vote. Similarly, if checks orthogonal on certain smaller subsets can be constructed, one may be able to determine correct check sums on those smaller subsets. Proceeding iteratively until the subsets are of size one will decode the code. *L-step majority logic decoding* then proceeds in L steps, in this manner. Decoding to the full error correcting capability of the code with majority logic decoding may not be possible for a particular code.

15.1.256 **Theorem** [1942, 2390, 2966] The code $EG_{q^m,p}(r)$ can be $(r+1)$-step majority logic decoded up to $\lfloor (d_{ML}-1)/2 \rfloor$ errors where

$$d_{ML} = \frac{q^{m-r}-1}{q-1}$$

and the minimum distance of the code is at least this large.

The code $PG_{(q^m-1)/(q-1),p}(r)$ can be r-step majority logic decoded up to $\lfloor (d_{ML}-1)/2 \rfloor$ errors where

$$d_{ML} = \frac{q^{m-r+1}-1}{q-1} + 1.$$

15.1.6.7 Generalized minimum distance decoding

15.1.257 **Remark** In a practical communication scheme, codeword symbols are modulated in a manner suitable for channel transmission in that the bandwidth of the scheme must be within the bandwidth assigned. The receiver typically observes the demodulated waveform for the period of transmission for a symbol and makes a decision as to the symbol transmitted. This decision often has associated with it a reliability or confidence measure as to how likely the decision made is to being correct. For example if the output of a matched filter is quantized at the decision time to two levels, this is equivalent to making a hard decision. If more levels are used, it would correspond to a soft decision. Thus in a *hard decision* receiver this *soft information* is discarded and only the hard decisions on the symbols are used. This leads to a decrease in the error performance of the system and techniques to incorporate the soft information on the received symbols to be used in the error control process have been investigated. The *generalized minimum distance (GMD)* technique, due to Forney [1092], is perhaps the simplest such technique. It is discussed here for binary codes only. Assume the $\{0,1\}$ code has Hamming distance d. The generalization to the nonbinary case is straightforward.

15.1.258 Remark For the binary case assume the code is over the alphabet $\{-1,+1\}$ by changing 0s in the code over $\mathbb{F}_2$ to -1 and 1s to $+1$ and assume d is the minimum Hamming distance of the code. Assume further the received word is of the form $\mathbf{y} = (y_0, y_1, \ldots, y_{n-1})$ and define the reliability word $\mathbf{r} = (r_0, r_1, \ldots, r_{n-1})$ where r_i represents the reliability of the i-th symbol. The larger r_i is in magnitude (positive or negative), the more reliable the transmitted symbol is (in the $\{\pm 1\}$ code). For maximum likelihood decoding this is the log likelihood ratio of the two possibilities, i.e.,

$$r_i = \log \frac{P(y_i \mid 1)}{P(y_i \mid 0)} .$$

Finally, for some $T > 0$, the vector $\mathbf{r}' = (r'_0, r'_1, \ldots, r'_{n-1})$ is formed according to the equations

$$r'_i = \begin{cases} +1, & r_i \geq T, \\ r_i/T, & -T \leq r_i \leq T, \\ -1, & r_i \leq -T. \end{cases}$$

Thus $\mathbf{r}'$ is the magnitude limited version of the reliability measure formed on each coordinate position of the reliability word. The Euclidean distance from $\mathbf{r}'$ to any $\{\pm 1\}$ codeword $\mathbf{c}$ is then

$$\mid \mathbf{r}' \mid^2 -2(\mathbf{r}', \mathbf{c}) + \mid \mathbf{c} \mid^2 .$$

15.1.259 Theorem [1092] There is at most one codeword $\mathbf{c}$ in the binary $\{\pm 1\}$ linear code with Hamming distance d from the received word such that

$$(\mathbf{r}', \mathbf{c}) > n - d$$

where $\mathbf{r}'$ has components in the range $[-1,+1]$.

15.1.260 Remark In the above theorem, the case where all components of $\mathbf{r}'$ are set to $+1$ if $r_i > 0$ and -1 otherwise corresponds to hard decisions. One could also choose a threshold $\delta > 0$ and set r_i to $+1$ if $r_i > \delta > 0$ and -1 if $r_i < -\delta < 0$ and 0 otherwise. This corresponds to the case of errors-and-erasures decoding.

The $(n,k,d)_2$ code is capable of correcting s erasures and r errors if $2r + s < d$ and many of the errors-only algorithms can be adapted to errors-and-erasures algorithms. GMD utilizes the fact that a code is capable of correcting more erasures than errors. For the received word $\mathbf{r}'$ formulated above, where $-1 \leq r'_i \leq 1$, denote by $\mathbf{r}^\ell$ the codeword $\mathbf{r}'$ with the least reliable ℓ positions erased (i.e., the ℓ positions with smallest $\mid r'_i \mid$) and the remaining positions quantized to hard decision (either $+1$ or -1).

15.1.261 Theorem [1092, 2390] If $(\mathbf{r}', \mathbf{c}) > n - d$, then at least one of the $\mathbf{r}^\ell$ satisfies

$$(\mathbf{r}^\ell, \mathbf{c}) > n - d \quad \text{for} \quad \ell = 0, 1, \ldots, d-1.$$

15.1.262 Remark Successively increasing the number of erasures (but not beyond $(d-1)/2$) and using errors-and-erasures decoding will find the codeword closest to the quantized received word $\mathbf{r}'$ under the conditions noted.

15.1.6.8 List decoding - decoding beyond the minimum distance bound

15.1.263 Remark It is assumed that $\mathcal{C}$ is a *GRS* code, although the results are applicable to other code classes such as algebraic geometry codes. For simplicity we consider $\mathcal{C} = GRS_{n,q}(\mathbf{x}, 1, k+1)$ where

$$\mathcal{C} = \{(f(x_1), f(x_2), \ldots, f(x_n)) \mid f(x) \in \mathbb{F}_q[x], \deg f(x) \leq k\},$$

with distinct code positions x_i and code rate $(k+1)/n$. For such a code if the number of errors made in transmission on the channel is at most $e = \lfloor (d-1)/2 \rfloor$, the bounded distance decoding algorithm produces the unique correct codeword from the received word $\mathbf{r}$. A *list* ℓ decoder with decoding radius τ produces a list of at most ℓ codewords for *any* received word $\mathbf{r} \in \mathbb{F}_q^n$ within a radius τ of $\mathbf{r}$. The decoding is successful if the transmitted codeword is in the list and fails otherwise. The relationship between the decoding radius τ and the size of the list is of interest.

15.1.264 Remark The list decoding problem is closely related to the following polynomial interpolation problem [2739]; for the Lagrange Interpolation Formula, see Theorem 2.1.131.

15.1.265 Definition The *polynomial interpolation problem* is defined as follows. For a set of pairs of elements $(x_i, y_i) \in \mathbb{F}_q \times \mathbb{F}_q$, the x_i distinct, and for positive integers k and τ, determine a list of all polynomials $f(x) \in \mathbb{F}_q[x]$ of degree at most k such that

$$|\{i \mid f(x_i) = y_i\}| \geq \tau.$$

15.1.266 Remark In other words, the desire is to find the set of all polynomials of degree at most k for which $f(x_i) = y_i$ in at least τ out of the n places. The relationship of this problem to the list decoding problem is immediate.

15.1.267 Remark The technique of Sudan [2739] is to determine a nonzero bivariate polynomial $Q(x, y)$, of a certain degree, that is zero on the set $\{(x_i, y_i) \mid i = 1, 2, \ldots, n\}$ and show that factors of this polynomial of the form $(y - f(x))$ with $\deg f(x) \leq k$ correspond to codewords within the required distance of the received word. In the above formulation we would like τ to be as small as possible (the number of errors as large as possible). The concept of *weighted degree* of $Q(x, y)$ is of interest: the $(1, k)$ degree of a monomial $x^i y^j$ is $i + jk$, and the $(1, k)$ degree of $Q(x, y)$ is $m + \ell k$ where this is the maximum over all monomials $x^m y^\ell$ of $Q(x, y)$. This arises as we need the x-degree of $Q(x, y)$ when a polynomial of degree k in x is substituted for y.

15.1.268 Remark The algorithm proceeds as follows:

1. Find a nonzero bivariate polynomial $Q(x, y)$ of weighted degree at most $m + \ell k$ (chosen later) such that $Q(x_i, y_i) = 0$ for $i = 1, 2, \ldots, n$.
2. Factor $Q(x, y)$ into irreducible factors and output all factors of the form $y - f(x)$ with $\deg f(x) \leq k$ where $f(x_i) = y_i$ for at least τ values of i.

The number of variables q_{ab} in the linear system $\sum_{b=0}^{\ell} \sum_{a=0}^{m+(\ell-b)k} q_{ab} x_i^a y_i^b = 0$ where $i = 1, 2, \ldots, n$ is

$$(m+1)(\ell+1) + k \binom{\ell+1}{2}.$$

If this quantity is greater than n, the system of n linear equations has a nontrivial solution since the number of unknowns exceeds the number of equations. The polynomial $Q(x, y)$ is given by $Q(x, y) = \sum_{b=0}^{\ell} \sum_{a=0}^{m+(\ell-b)k} q_{ab} x^a y^b$.

15.1.269 Lemma [2739] If $Q(x, y)$ is as in Part 1 above and $f(x)$ is such that $\deg f(x) \leq k$ with $f(x_i) = y_i$ for at least τ values where $\tau > m + \ell k$, then $y - f(x)$ divides $Q(x, y)$.

15.1.270 Remark [2739] In the above development if one chooses $m = \lceil k/2 \rceil - 1$ and $\ell = \sqrt{2(n+1)/k} - 1$, it can be shown that

$$\tau \geq d\lceil \sqrt{2(n+1)/k} \rceil - \lfloor k/2 \rfloor \quad \text{or} \quad \tau > \sqrt{2kn}.$$

Thus if fewer than $n - \sqrt{2kn} \approx n(1 - \sqrt{2R})$ errors are made in transmission, the transmitted codeword will be in the list. Notice this implies the results are valid only for fairly low rate codes.

15.1.271 Remark There are a number of items to be noted.

1. As τ decreases (number of errors increases), the size of the list tends to increase although not always strictly monotonically. For a given decoding radius there are estimates of the list size which are not discussed here.
2. A significant improvement in the above bound was achieved by Guruswami and Sudan [1377] who increased the relative fraction of errors to $\sim 1 - \sqrt{R}$. This was achieved by allowing the bivariate polynomial $Q(x, y)$ to have zeros of multiplicity $m > 1$ and optimizing the results over this parameter. Parvaresh and Vardy [2361] described a class of codes that could be list-decoded to this radius.
3. Define the entropy function $h_q(p) = -p \log_q(p) - (1-p)\log_q(1-p)$ and note this is different than the q-ary entropy function of Definition 15.1.122. Denote a (p, L) *list-decodable code* as one capable of correcting a fraction p of errors with a list of size L. One can show [2501] that for code rates $R \leq 1 - h_q(p) - \epsilon$ there exists a $(p, O(1/\epsilon))$-list decodable code while for $R > 1 - h_q(p) + \epsilon$ every (p, L)-list decodable code has L exponential in q.
4. The work is extended in [1376, 1377].

15.1.7 Codes over $\mathbb{Z}_4$

15.1.272 Definition A $\mathbb{Z}_4$*-linear code* $\mathcal{C}$ of length n is an additive subgroup of $\mathbb{Z}_4^n$. Associated to $\mathcal{C}$ are two binary linear codes of length n, the *residue code* $\text{Res}(\mathcal{C}) = \{\mu(\mathbf{c}) \mid \mathbf{c} \in \mathcal{C}\}$, where $\mu : \mathbb{Z}_4 \to \mathbb{F}_2$ is given by $\mu(z) = z \pmod 2$, and the *torsion code* $\text{Tor}(\mathcal{C}) = \{\mathbf{b} \in \mathbb{F}_2^n \mid 2\mathbf{b} \in \mathcal{C}\}$. Let $\mathfrak{G} : \mathbb{Z}_4 \to \mathbb{F}_2^2$ be defined by $\mathfrak{G}(0) = 00$, $\mathfrak{G}(1) = 01$, $\mathfrak{G}(2) = 11$, and $\mathfrak{G}(3) = 10$; $\mathfrak{G}$ is the *Gray map*. The binary code $\mathfrak{G}(\mathcal{C}) = \{(\mathfrak{G}(c_1), \mathfrak{G}(c_2), \ldots, \mathfrak{G}(c_n)) \mid (c_1, c_2, \ldots, c_n) \in \mathcal{C}\}$ of length $2n$ is the *Gray image* of $\mathcal{C}$; $\mathfrak{G}(\mathcal{C})$ may not be linear over $\mathbb{F}_2$.

15.1.273 Remark The study of codes over $\mathbb{Z}_4$ began in earnest after the publication of [1409]. In that paper, a relationship, via the Gray map, was discovered between certain families of binary nonlinear codes and $\mathbb{Z}_4$-linear codes. That paper heightened interest in studying codes over $\mathbb{Z}_r$ and eventually codes over other rings; see Section 2.1.7.7 for a discussion of Galois rings.

15.1.274 Remark Since a subspace of $\mathbb{F}_q^n$ is an $\mathbb{F}_q$-submodule of $\mathbb{F}_q^n$, a linear code of length n over $\mathbb{F}_q$ can be defined as an $\mathbb{F}_q$-submodule of $\mathbb{F}_q^n$. Analogously, a $\mathbb{Z}_4$-linear code of length n is in fact a $\mathbb{Z}_4$-submodule of $\mathbb{Z}_4^n$. In general, for any ring R, an R-linear code of length n is an R-submodule of R^n.

15.1.275 Remark Like a linear code over a field, a $\mathbb{Z}_4$-linear code $\mathcal{C}$ of length n has a $k \times n$ generator matrix G, where k is chosen to be minimal so that the $\mathbb{Z}_4$-span of the rows of G is the code

$\mathcal{C}$. By permuting coordinates, the generator matrix for $\mathcal{C}$ can be put in the form

$$G = \begin{bmatrix} I_{k_1} & A & B_1 + 2B_2 \\ O & 2I_{k_2} & 2C \end{bmatrix}, \qquad (15.1.9)$$

where A, B_1, B_2, and C are matrices with entries from $\{0,1\}$, I_{k_1} and I_{k_2} are $k_1 \times k_1$ and $k_2 \times k_2$ identity matrices, and O is the $k_2 \times k_1$ zero matrix. The number of codewords in $\mathcal{C}$, called the *type* of $\mathcal{C}$, is $4^{k_1}2^{k_2}$. If $\mathcal{C}$ is a $\mathbb{Z}_4$-linear code with generator matrix (15.1.9), the generator matrices for $\mathrm{Res}(\mathcal{C})$ and $\mathrm{Tor}(\mathcal{C})$ are, respectively,

$$G_{\mathrm{Res}} = \begin{bmatrix} I_{k_1} & A & B_1 \end{bmatrix} \quad \text{and} \quad G_{\mathrm{Tor}} = \begin{bmatrix} I_{k_1} & A & B_1 \\ O & I_{k_2} & C \end{bmatrix}.$$

So $\mathrm{Res}(\mathcal{C}) \subseteq \mathrm{Tor}(\mathcal{C})$, $\mathrm{Res}(\mathcal{C})$ is an $(n, k_1, d_1)_2$ code for some d_1, and $\mathrm{Tor}(\mathcal{C})$ is an $(n, k_1 + k_2, d_2)_2$ code for some $d_2 \le d_1$.

15.1.276 **Definition** Analogous to the scalar product on $\mathbb{F}_q^n$, there is a natural *scalar product* on $\mathbb{Z}_4^n$ defined by

$$(\mathbf{x}, \mathbf{y}) = x_1y_1 + x_2y_2 + \cdots + x_ny_n \pmod 4,$$

where $\mathbf{x} = (x_1, x_2, \ldots, x_n)$ and $\mathbf{y} = (y_1, y_2, \ldots, y_n)$ are in $\mathbb{Z}_4^n$. If $\mathcal{C}$ is a $\mathbb{Z}_4$-linear code of length n, define $\mathcal{C}^\perp = \{\mathbf{v} \in \mathbb{Z}_4^n \mid (\mathbf{v}, \mathbf{c}) = 0 \text{ for all } \mathbf{c} \in \mathcal{C}\}$, the *dual* of $\mathcal{C}$. If $\mathcal{C} \subseteq \mathcal{C}^\perp$, the code $\mathcal{C}$ is *self-orthogonal*. If $\mathcal{C} = \mathcal{C}^\perp$, it is *self-dual*.

15.1.277 **Theorem** If $\mathcal{C}$ is a self-dual $\mathbb{Z}_4$-linear code, then $\mathrm{Res}(\mathcal{C})$ is doubly-even and $\mathrm{Res}(\mathcal{C}) = \mathrm{Tor}(\mathcal{C})^\perp$.

15.1.278 **Remark** If $\mathcal{C}$ is a $\mathbb{Z}_4$-linear code of type $4^{k_1}2^{k_2}$, then $\mathcal{C}^\perp$ is a $\mathbb{Z}_4$-linear code of type $4^{n-k_1-k_2}2^{k_2}$. Furthermore, if $\mathcal{C}$ has generator matrix (15.1.9), $\mathcal{C}^\perp$ has generator matrix

$$G^\perp = \begin{bmatrix} -(B_1 + 2B_2)^T - C^TA^T & C^T & I_{n-k_1-k_2} \\ 2A^T & 2I_{k_2} & O \end{bmatrix},$$

where O is the $k_2 \times (n - k_1 - k_2)$ zero matrix.

15.1.279 **Definition** The concepts of weight and distance in $\mathbb{F}_q^n$ have analogies in $\mathbb{Z}_4^n$.

1. For $\mathbf{x} \in \mathbb{Z}_4^n$ and $i \in \mathbb{Z}_4$, let $n_i(\mathbf{x})$ be the number of components of $\mathbf{x}$ equal to i. The *Hamming weight* $\omega(\mathbf{x})$ of $\mathbf{x}$ and the *Hamming distance* $d(\mathbf{x}, \mathbf{y})$ between $\mathbf{x}$ and $\mathbf{y}$ are defined exactly as in Definition 15.1.6 for $\mathbb{F}_q^n$: $\omega(\mathbf{x}) = n_1(\mathbf{x}) + n_2(\mathbf{x}) + n_3(\mathbf{x})$ and $d(\mathbf{x}, \mathbf{y}) = \omega(\mathbf{x} - \mathbf{y})$. The *Lee weight* of $\mathbf{x}$ is $\omega_L(\mathbf{x}) = n_1(\mathbf{x}) + 2n_2(\mathbf{x}) + n_3(\mathbf{x})$, and the *Lee distance* between $\mathbf{x}$ and $\mathbf{y}$ is $d_L(\mathbf{x}, \mathbf{y}) = \omega_L(\mathbf{x} - \mathbf{y})$.
2. Analogous to Definition 15.1.38, the *Hamming weight distribution* of a $\mathbb{Z}_4$-linear code $\mathcal{C}$ of length n is the sequence $\{A_0, A_1, \ldots, A_n\}$ where A_i is the number of codewords in $\mathcal{C}$ of Hamming weight i. The *Lee weight distribution* of $\mathcal{C}$ is the sequence $\{L_0, L_1, \ldots, L_{2n}\}$ where L_i is the number of codewords in $\mathcal{C}$ of Lee weight i; note that Lee weights in $\mathbb{Z}_4^n$ range from 0 to $2n$. The *minimum Hamming*, respectively *Lee*, *weight* of $\mathcal{C}$ is the smallest Hamming, respectively Lee, weight of a nonzero codeword of $\mathcal{C}$.

15.1.280 Theorem The following hold.

1. The Gray map $\mathfrak{G}$ is a distance preserving map from $\mathbb{Z}_4^n$ with Lee distance to $\mathbb{F}_2^{2n}$ with Hamming distance.
2. If $\mathcal{C}$ is a $\mathbb{Z}_4$-linear code, then $\mathfrak{G}(\mathcal{C})$ is a distance invariant binary code.
3. If $\mathcal{C}$ is a $\mathbb{Z}_4$-linear code, then the weight distribution of the binary code $\mathfrak{G}(\mathcal{C})$ is the same as the Lee weight distribution of $\mathcal{C}$.
4. If $\mathcal{C}$ is a $\mathbb{Z}_4$-linear code, then $\mathfrak{G}(\mathcal{C})$ is linear if and only if whenever $\mathbf{v}$ and $\mathbf{w}$ are in $\mathcal{C}$, so is $2(\mathbf{v} * \mathbf{w})$, where $\mathbf{v} * \mathbf{w}$ is the componentwise product of $\mathbf{v}$ and $\mathbf{w}$ in $\mathbb{Z}_4^n$.

15.1.281 Example Let o_8 be the $\mathbb{Z}_4$-linear code, called the *octacode*, with generator matrix

$$G = \begin{bmatrix} 1 & 0 & 0 & 0 & 3 & 1 & 2 & 1 \\ 0 & 1 & 0 & 0 & 1 & 2 & 3 & 1 \\ 0 & 0 & 1 & 0 & 3 & 3 & 3 & 2 \\ 0 & 0 & 0 & 1 & 2 & 3 & 1 & 1 \end{bmatrix}.$$

The octacode is self-dual, has type 4^4, and has minimum Lee weight 6. $\text{Res}(o_8) = \text{Tor}(o_8)$ is equivalent to the $[8, 4, 4]$ extended binary Hamming code. By Part 3 of Theorem 15.1.280, the Gray image $\mathfrak{G}(o_8)$ is a $(16, 256, 6)_2$ code. By [1093, 2298, 2689], $\mathfrak{G}(o_8)$ is the Nordstrom-Robinson code, a nonlinear binary code that is the largest possible binary code of length 16 and minimum Hamming distance 6. This $(16, 256, 6)_2$ code is the extension, using an overall parity check, of a $(15, 256, 5)_2$ code originally defined by Nordstrom and Robinson in [2298]. The latter code consisted of length 15 binary vectors with the first 8 coordinate positions arbitrary and the last 7 positions Boolean combinations of the first 8 coordinates.

15.1.282 Remark In 1968, Preparata [2427] defined a family of nonlinear $(2^{m+1}, 2^{2^{m+1}-2m-2}, 6)_2$ binary codes when m is odd. These codes have twice as many codewords as a linear $(2^{m+1}, 2^{m+1} - 2m - 3, 6)_2$ extended binary BCH code. Each Preparata code has the largest number of codewords of any binary code of its length 2^{m+1} and minimum distance 6; see Chapter 17, Section 3 of [1991]. These codes lie between $RM_{2^{m+1},2}(m-2, m+1)$ and $RM_{2^{m+1},2}(m-1, m+1)$. In 1972, Kerdock [1728] defined a family of nonlinear $(2^{m+1}, 4^{m+1}, 2^m - 2^{(m-1)/2})_2$ binary codes when m is odd. These codes lie between $RM_{2^{m+1},2}(1, m+1)$ and $RM_{2^{m+1},2}(2, m+1)$. Amazingly, the weight distribution of the $(2^{m+1}, 2^{2^{m+1}-2m-2}, 6)_2$ Preparata code is the MacWilliams transform of the weight distribution of the $(2^{m+1}, 4^{m+1}, 2^m - 2^{(m-1)/2})_2$ Kerdock code. When $m = 3$, both codes are equivalent to the Nordstrom–Robinson code of Example 15.1.281. In [1409], an extended cyclic $\mathbb{Z}_4$-linear code of length 2^m and type 4^{m+1}, denoted $\widehat{K}(m+1)$, was defined; the Gray image $\mathfrak{G}(\widehat{K}(m+1))$ is equivalent, by permuting coordinates, to the $(2^{m+1}, 4^{m+1}, 2^m - 2^{(m-1)/2})_2$ binary code defined by Kerdock. The Lee weight distribution $\{L_0, L_1, \ldots, L_{2^{m+1}}\}$ of $\widehat{K}(m+1)$, and hence the weight distribution of $\mathfrak{G}(\widehat{K}(m+1))$, is

$$L_i = \begin{cases} 1 & \text{if } i = 0 \text{ or } i = 2^{m+1}, \\ 2^{m+1}(2^m - 1) & \text{if } i = 2^m \pm 2^{(m-1)/2}, \\ 2^{m+2} - 2 & \text{if } i = 2^m, \\ 0 & \text{otherwise.} \end{cases}$$

Letting $P(m+1) = \widehat{K}(m+1)^{\perp}$, the Gray image $\mathfrak{G}(P(m+1))$, although generally not the same as the original Preparata code, has the same weight distribution as the $(2^{m+1}, 2^{2^{m+1}-2m-2}, 6)_2$ binary Preparata code; this weight distribution can be obtained by using the MacWilliams transform on the weight distribution of $\mathfrak{G}(\widehat{K}(m+1))$. Preparata's original construction begins with a single-error correcting binary BCH code and a double-error correcting subcode of this code. A linear code is created using a variation of the construction in Lemma 15.1.93, and then cosets of this code are adjoined to give a $(2^{m+1}-1, 2^{2^{m+1}-2m-2}, 5)_2$ code for m odd. Adding an overall parity check yields a $(2^{m+1}, 2^{2^{m+1}-2m-2}, 6)_2$ code. There is a similar connection [1409] between the nonlinear binary codes of minimum distance 8 of Goethals [1288] and the nonlinear binary codes of high minimum distance of Delsarte and Goethals [804].

15.1.8 Conclusion

15.1.283 **Remark** This section has outlined certain aspects of algebraic coding theory including much of the early work in the subject. Two books [233, 305] compile and comment on initial papers from the first 25 years of coding theory that greatly influenced the development of the discipline. The references cited in this section found in [233] include [359, 1090, 1092, 1292, 1319, 1320, 1329, 1408, 1516, 1638, 2035, 2298, 2447, 2811]. References cited in this section found in [305] include [359, 1292, 1329, 1408, 1516, 1638, 1942, 2011, 2035, 2298, 2427, 2447, 2855].

See Also

Chapter 3	Irreducible polynomials are essential for the construction of generator polynomials for cyclic codes.
Chapter 6	Character sums are used for various codes including perfect codes.
§6.3.3.3	Examines the distance properties of the duals of BCH codes.
§6.3.4	Kloosterman sums are used to examine the distance properties of codes.
§9.1.7	Discusses the application of Boolean functions to codes including the linear Reed-Muller codes and the nonlinear Kerdock codes.
§9.2.6	Examines almost perfect nonlinear (APN) and almost bent (AB) functions from a coding perspective.
§9.3.8	Relates bent functions to Kerdock codes.
§10.1	The Mattson-Solomon polynomial can be viewed as a discrete Fourier transform. The Gauss sum is used in the study of quadratic and generalized quadratic residue codes.
§10.2	Duals of binary Hamming codes are generated by a primitive polynomial. The nonzero codewords in such a code form LFSR sequences.
§11.1, §11.2	The computational techniques discussed in these subsections are useful for code implementations.
Chapter 12	Is useful in the study of algebraic geometry codes.
§14.4	Projective and affine geometries are strongly related to RM, GRM, and polynomial codes as well as majority logic decodable codes.
§14.5	Block designs can arise as the supports of codewords of fixed weight in codes, particularly those satisfying the Assmus-Mattson Theorem.
§14.7.1	Association schemes can be used to connect the distance distribution of a code and the dual distribution.
§17.2.3	This material is useful in discussing quantum error-correcting codes.
§17.3	Includes coding theory found in engineering applications.
[13]	Gröbner bases are useful in the theory of algebraic geometry codes.
[936]	Describes the essential properties of concatenated codes.
[1525]	Develops the theory of Goppa and algebraic geometry codes.
[1530]	For a proof of the classification of perfect codes.

References Cited: [13, 142, 231, 233, 304, 305, 311, 359, 578, 801, 802, 804, 805, 806, 936, 1090, 1092, 1093, 1164, 1288, 1292, 1319, 1320, 1329, 1376, 1377, 1408, 1409, 1516, 1522, 1525, 1530, 1558, 1638, 1639, 1691, 1692, 1728, 1827, 1901, 1942, 1943, 1945, 1991, 2010, 2011, 2035, 2047, 2050, 2136, 2164, 2298, 2361, 2389, 2390, 2402, 2403, 2405, 2427, 2447, 2484, 2501, 2506, 2511, 2608, 2687, 2689, 2696, 2739, 2741, 2805, 2810, 2811, 2849, 2855, 2965, 2966, 2999]

15.2 Algebraic-geometry codes

Harald Niederreiter, KFUPM

15.2.1 Classical algebraic-geometry codes

15.2.1 Remark Algebraic-geometry codes constitute a powerful family of codes which were introduced in their classical form by Goppa in the years 1977 to 1982 [1321, 1322, 1323]. Goppa used the language of algebraic curves over finite fields to define algebraic-geometry codes. Modern expositions prefer a description in terms of algebraic function fields over finite fields. In order to be consistent with the recent literature, we follow this practice in the present section.

15.2.2 Remark We follow the terminology for algebraic function fields in Chapter 12. Let F be an algebraic function field (of one variable) with full constant field $\mathbb{F}_q$, that is, $\mathbb{F}_q$ is algebraically closed in F. A *divisor* of F is a finite $\mathbb{Z}$-linear combination of places of F. If $\mathcal{P}$ denotes the set of all places of F, then a divisor G of F can be uniquely written in the form

$$G = \sum_{P\in\mathcal{P}} m_P\, P,$$

where $m_P \in \mathbb{Z}$ for all $P \in \mathcal{P}$ and $m_P \neq 0$ for only finitely many $P \in \mathcal{P}$. The *support* of G is the set of all $P \in \mathcal{P}$ with $m_P \neq 0$. Consequently, the support of G is a finite set. The *degree* $\deg(G)$ of G is defined by

$$\deg(G) = \sum_{P\in\mathcal{P}} m_P \deg(P),$$

where $\deg(P)$ denotes the degree of the place P; see Section 12.1. The divisor G is *positive* (written $G \geq 0$) if $m_P \geq 0$ for all $P \in \mathcal{P}$. The divisors of F form an abelian group with respect to addition, where two divisors of F are added by adding the corresponding integer coefficients in the above unique representation of divisors.

15.2.3 Remark A basic object in the construction of an algebraic-geometry code is a Riemann-Roch space. For any $f \in F^*$, the *principal divisor* $\mathrm{div}(f)$ of f is given by

$$\mathrm{div}(f) = \sum_{P\in\mathcal{P}} \nu_P(f)\, P,$$

where ν_P denotes the normalized discrete valuation of F associated with the place P. Note that the image of ν_P as a map is $\mathbb{Z} \cup \{\infty\}$. If $f \in F^*$ is given, then $\nu_P(f) \neq 0$ for at most finitely many $P \in \mathcal{P}$, and so $\text{div}(f)$ is indeed a divisor of F. Given an arbitrary divisor G of F, we now define the *Riemann-Roch space*

$$\mathcal{L}(G) = \{f \in F^* : \text{div}(f) + G \geq 0\} \cup \{0\}.$$

It is an important fact that $\mathcal{L}(G)$ is a finite-dimensional vector space over $\mathbb{F}_q$. We write $\ell(G)$ for the dimension of this vector space. The Riemann-Roch theorem (Chapter 12) provides important information on $\ell(G)$.

15.2.4 **Definition** Let n be a positive integer and let q be an arbitrary prime power. Let F be an algebraic function field with full constant field $\mathbb{F}_q$, genus g, and at least n rational places. Choose n distinct rational places $P_1, \ldots, P_n$ of F and a divisor G of F such that none of the P_i, $1 \leq i \leq n$, is in the support of G. The *algebraic-geometry code* $C(P_1, \ldots, P_n; G)$ is defined as the image of the $\mathbb{F}_q$-linear map $\psi : \mathcal{L}(G) \to \mathbb{F}_q^n$ given by

$$\psi(f) = (f(P_1), \ldots, f(P_n)) \qquad \text{for all } f \in \mathcal{L}(G).$$

15.2.5 **Remark** Note that $\nu_{P_i}(f) \geq 0$ for $1 \leq i \leq n$ and all $f \in \mathcal{L}(G)$, since P_i is not in the support of G. Thus, f belongs to the valuation ring $\mathcal{O}_{P_i}$ of P_i for $1 \leq i \leq n$, and so the residue class $f(P_i)$ of f, that is, the image of f under the residue class map of the place P_i, is well defined. Since P_i is a rational place, that is, a place of degree 1, we can identify $f(P_i)$ with an element of $\mathbb{F}_q$. Thus, $C(P_1, \ldots, P_n; G)$ is indeed a subset of $\mathbb{F}_q^n$.

15.2.6 **Theorem** With the notation and assumptions in Definition 15.2.4, suppose that the divisor G of F satisfies also $g \leq \deg(G) < n$. Then the algebraic-geometry code $C(P_1, \ldots, P_n; G)$ is a linear code over $\mathbb{F}_q$ with length n, dimension

$$k = \ell(G) \geq \deg(G) + 1 - g,$$

and minimum distance

$$d \geq n - \deg(G).$$

Moreover, if $\deg(G) \geq 2g - 1$, then $k = \deg(G) + 1 - g$.

15.2.7 **Remark** The condition $\deg(G) < n$ in Theorem 15.2.6 guarantees that the map ψ in Definition 15.2.4 is injective. Therefore $k = \ell(G)$ and the remaining information on k is obtained from the Riemann-Roch theorem (Chapter 12).

15.2.8 **Remark** Theorem 15.2.6 implies that $k + d \geq n + 1 - g$. This should be compared with the Singleton bound $k + d \leq n + 1$ in Section 15.1. Thus, the genus g of F controls, in a sense, the deviation of $k + d$ from the Singleton bound.

15.2.9 **Example** Let $F = \mathbb{F}_q(x)$ be the rational function field over $\mathbb{F}_q$. Let $\mathbb{F}_q = \{b_1, \ldots, b_q\}$ and for $1 \leq i \leq q$ let P_i be the rational place $x - b_i$ of $\mathbb{F}_q(x)$. Put $G = (k-1)P_\infty$, where k is an integer with $1 \leq k \leq q$ and P_∞ denotes the infinite place of $\mathbb{F}_q(x)$. Then $C(P_1, \ldots, P_q; G)$ is a *generalized Reed-Solomon code* with length q, dimension k, and minimum distance $q - k + 1$. Thus, algebraic-geometry codes can be considered as vast generalizations of generalized Reed-Solomon codes.

15.2.10 **Example** Let F be the Hermitian function field over $\mathbb{F}_{q^2}$, that is, $F = \mathbb{F}_{q^2}(x, y)$ with $y^q + y = x^{q+1}$. Then F has genus $g = (q^2 - q)/2$ and exactly $q^3 + 1$ rational places. Let Q be the rational place of F lying over the infinite place of $\mathbb{F}_{q^2}(x)$ and let $P_1, \ldots, P_n$ with $n = q^3$ be the remaining rational places of F. Put $G = mQ$ with an integer m satisfying $q^2 - q - 1 \leq m < q^3$. Then $C(P_1, \ldots, P_n; G)$ is a linear code over $\mathbb{F}_{q^2}$ with length $n = q^3$, dimension $k = m + 1 - (q^2 - q)/2$, and minimum distance $d \geq q^3 - m$. Such a code is a *Hermitian code.*

15.2.11 Example Take $q = 2$ and $m = 4$ in Example 15.2.10, so that $G = 4Q$. Then the Hermitian code $C(P_1, \ldots, P_8; G)$ is a linear code over $\mathbb{F}_4$ with length 8, dimension 4, and minimum distance $d \geq 4$. It can be shown that actually $d = 4$. The code $C(P_1, \ldots, P_8; G)$ is optimal in the sense that there is no linear code over $\mathbb{F}_4$ with length 8, dimension 4, and minimum distance at least 5.

15.2.12 Remark The condition in Definition 15.2.4 that none of the P_i, $1 \leq i \leq n$, is in the support of G is a conventional one in the area. However, one can get rid of this condition by replacing G by the divisor $G' = G - \text{div}(u)$, where $u \in F$ is such that $\nu_{P_i}(u)$ is equal to the coefficient m_{P_i} of P_i in G for $1 \leq i \leq n$. Such an element u exists by the approximation theorem for valuations. Note that by construction none of the P_i, $1 \leq i \leq n$, is in the support of G', but $\deg(G') = \deg(G)$ and $\ell(G') = \ell(G)$. Thus, the linear code $C(P_1, \ldots, P_n; G')$ satisfies the properties stated in Theorem 15.2.6.

15.2.13 Example Let F be the Hermitian function field over $\mathbb{F}_4$; see Example 15.2.10. Then F has genus 1 and exactly 9 rational places. Let $P_1, \ldots, P_9$ be the rational places of F and choose a divisor G of F with $\deg(G) = 4$. In view of Remark 15.2.12, if we replace G by a suitable divisor G', then we need not check whether the P_i, $1 \leq i \leq 9$, are in the support of G or not. Thus, $C(P_1, \ldots, P_9; G')$ is a linear code over $\mathbb{F}_4$ with length 9, dimension 4, and minimum distance $d \geq 5$. It can be shown that actually $d = 5$. The linear code $C(P_1, \ldots, P_9; G')$ is optimal.

15.2.14 Remark The dual codes of algebraic-geometry codes can be described in terms of differentials and residues for algebraic function fields [2714].

15.2.15 Remark Certain algebraic-geometry codes can be constructed without the use of algebraic geometry or algebraic function fields, but rather by using so-called order domains [92].

15.2.16 Remark There exist efficient decoding algorithms for algebraic-geometry codes. A survey of such decoding algorithms is presented in [1524], and for a more recent contribution see [1375].

15.2.17 Remark Interesting codes can be derived not only from algebraic function fields over finite fields, or equivalently from algebraic curves over finite fields, but also from higher-dimensional algebraic varieties over finite fields. A rather general approach to the construction of such codes is described in [1415].

15.2.2 Generalized algebraic-geometry codes

15.2.18 Remark The condition $g \leq \deg(G) < n$ in Theorem 15.2.6 implies that the number N of rational places of the algebraic function field F must be greater than the genus g of F, in order for Theorem 15.2.6 to be applicable. However, for small values of q such as $q = 2$ and $q = 3$, the condition $N > g$ can be satisfied only for small values of g. Consequently, for small values of q the construction of classical algebraic-geometry codes in Definition 15.2.4 can yield good codes only for small lengths. A possible remedy is to devise constructions that use not only rational places, but also places of higher degree. Such constructions will be discussed in this subsection.

15.2.19 Remark The construction of NXL codes due to Niederreiter, Xing, and Lam [2285] uses places of arbitrary degree. We present this construction in the more general form given in Chapter 5 of [2281]. Let F be an algebraic function field with full constant field $\mathbb{F}_q$ and genus g. Let $G_1, \dots, G_r$ be positive divisors $\neq 0$ of F with pairwise disjoint supports. Put $n = \sum_{i=1}^{r} s_i$, where $s_i = \deg(G_i) \geq 1$ for $1 \leq i \leq r$, and assume that $n > g$. Let E be a positive divisor of F for which the support is disjoint from the support of G_i for $1 \leq i \leq r$. Furthermore, let D be a divisor of F with $\ell(D) = \deg(D) + 1 - g$, e.g., this holds if $\deg(D) \geq 2g - 1$. Assume also that $1 \leq \deg(E - D) \leq n - g$. We observe that $\ell(D + G_i) = \ell(D) + s_i$ for $1 \leq i \leq r$. For each $i = 1, \dots, r$, we choose an $\mathbb{F}_q$-basis

$$\{f_{i,j} + \mathcal{L}(D) : 1 \leq j \leq s_i\}$$

of the factor space $\mathcal{L}(D+G_i)/\mathcal{L}(D)$. The n-dimensional factor space $\mathcal{L}(D+\sum_{i=1}^{r} G_i)/\mathcal{L}(D)$ has then the $\mathbb{F}_q$-basis

$$\{f_{i,j} + \mathcal{L}(D) : 1 \leq j \leq s_i,\ 1 \leq i \leq r\}$$

which we order in a lexicographic manner. We note further that every

$$f \in \mathcal{L}(D + \sum_{i=1}^{r} G_i - E) \subseteq \mathcal{L}(D + \sum_{i=1}^{r} G_i)$$

has a unique representation

$$f = \sum_{i=1}^{r} \sum_{j=1}^{s_i} c_{i,j} f_{i,j} + u$$

with all $c_{i,j} \in \mathbb{F}_q$ and $u \in \mathcal{L}(D)$.

15.2.20 **Definition** The *NXL code* $C(G_1, \dots, G_r; D, E)$ is defined as the image of the $\mathbb{F}_q$-linear map

$$\eta : \mathcal{L}(D + \sum_{i=1}^{r} G_i - E) \to \mathbb{F}_q^n$$

given by

$$\eta(f) = (c_{1,1}, \dots, c_{1,s_1}, \dots, c_{r,1}, \dots, c_{r,s_r})$$

for all $f \in \mathcal{L}(D + \sum_{i=1}^{r} G_i - E)$, where the $c_{i,j}$ are as in Remark 15.2.19.

15.2.21 **Theorem** Assume that the hypotheses in Remark 15.2.19 are satisfied and put $m = \deg(E - D)$. Then the NXL code $C(G_1, \dots, G_r; D, E)$ is a linear code over $\mathbb{F}_q$ with length $n = \sum_{i=1}^{r} s_i$, dimension

$$k = \ell(D + \sum_{i=1}^{r} G_i - E) \geq n - m - g + 1,$$

and minimum distance $d \geq d_0$, where d_0 is the least cardinality of a subset R of $\{1, \dots, r\}$ for which $\sum_{i \in R} s_i \geq m$. Moreover, if $n - m \geq 2g - 1$, then $k = n - m - g + 1$.

15.2.22 **Example** A simple choice for the divisors $G_1, \dots, G_r$ in the construction of NXL codes is to take distinct places $P_1, \dots, P_r$ of F. Note that these places need not be rational, but can have arbitrary degrees. This special case was considered in [2285] and is used also in the present example. Let $q = 3$, let F be the rational function field over $\mathbb{F}_3$, and put $r = 13$. For $P_1, \dots, P_{13}$ we choose four rational places, three places of degree 2, and six places of degree 3 of F. Let D be the zero divisor and E a place of degree 7 of F. Then $C(P_1, \dots, P_{13}; D, E)$ is a linear code over $\mathbb{F}_3$ with length 28, dimension $k = 22$, and minimum distance $d \geq 3$. Hence $k + d \geq 25$. By comparison, the best lower bound on $k + d$ for a classical algebraic-geometry code over $\mathbb{F}_3$ of length 28 is obtained by taking $g = 15$, and then $k + d \geq 14$ by Remark 15.2.8.

15.2.23 Remark The following construction of XNL codes is due to Xing, Niederreiter, and Lam [3019]. It is a powerful method of combining data from algebraic function fields over finite fields with (short) linear codes in order to produce a longer linear code as the output. We present the slightly more general version of XNL codes in Chapter 5 of [2281].

15.2.24 Definition Let F be an algebraic function field with full constant field $\mathbb{F}_q$. Let $P_1, \dots, P_r$ be distinct places of F which can have arbitrary degrees. Let G be a divisor of F such that none of the P_i, $1 \le i \le r$, is in the support of G. For each $i = 1, \dots, r$, let C_i be a linear code over $\mathbb{F}_q$ with length n_i, dimension $k_i \ge \deg(P_i)$, and minimum distance d_i, and let ϕ_i be an injective $\mathbb{F}_q$-linear map from the residue class field of P_i into C_i. Put $n = \sum_{i=1}^r n_i$. Then the *XNL code* $C(P_1, \dots, P_r; G; C_1, \dots, C_r)$ is defined as the image of the $\mathbb{F}_q$-linear map $\beta : \mathcal{L}(G) \to \mathbb{F}_q^n$ given by

$$\beta(f) = (\phi_1(f(P_1)), \dots, \phi_r(f(P_r))) \qquad \text{for all } f \in \mathcal{L}(G),$$

where on the right-hand side we use concatenation of vectors.

15.2.25 Theorem With the notation and assumptions in Definition 15.2.24, suppose that the divisor G of F satisfies also $g \le \deg(G) < \sum_{i=1}^r \deg(P_i)$, where g is the genus of F. Then the XNL code $C(P_1, \dots, P_r; G; C_1, \dots, C_r)$ is a linear code over $\mathbb{F}_q$ with length $n = \sum_{i=1}^r n_i$, dimension

$$k = \ell(G) \ge \deg(G) + 1 - g,$$

and minimum distance $d \ge d'$, where d' is the minimum of $\sum_{i \in \overline{M}} d_i$ taken over all subsets M of $\{1, \dots, r\}$ for which $\sum_{i \in M} \deg(P_i) \le \deg(G)$, with $\overline{M}$ denoting the complement of M in $\{1, \dots, r\}$. Moreover, if $\deg(G) \ge 2g - 1$, then $k = \deg(G) + 1 - g$.

15.2.26 Corollary If in addition $\deg(P_i) \ge d_i$ for $1 \le i \le r$, then the minimum distance d of the XNL code $C(P_1, \dots, P_r; G; C_1, \dots, C_r)$ satisfies

$$d \ge \sum_{i=1}^{r} d_i - \deg(G).$$

15.2.27 Remark If $P_1, \dots, P_r$ are distinct rational places of F and for each $i = 1, \dots, r$ we choose C_i to be the trivial linear code over $\mathbb{F}_q$ with $n_i = k_i = d_i = 1$ and ϕ_i the identity map on $\mathbb{F}_q$, then the construction of XNL codes reduces to that of classical algebraic-geometry codes. Theorem 15.2.6 is thus a special case of Theorem 15.2.25 and Corollary 15.2.26.

15.2.28 Example Many excellent examples of XNL codes were found in [875] and [3020]. The following simple example is typical. Let $q = 2$ and let $F = \mathbb{F}_2(x, y)$ be the elliptic function field defined by $y^2 + y = x + x^{-1}$. We choose $r = 6$ and let P_1, P_2, P_3, P_4 be the four rational places and P_5 and P_6 the two places of degree 2 of F. The linear codes C_i have the following parameters: for $1 \le i \le 4$ we let $n_i = k_i = d_i = 1$ and for $i = 5, 6$ we let $n_i = 3$, $k_i = 2$, $d_i = 2$. Then for $m = \deg(G) = 1, \dots, 7$ the corresponding XNL code is a linear code over $\mathbb{F}_2$ with parameters $n = 10$, $k = m$, and $d \ge 8 - m$. The linear codes with $m = 2, 3, 4$ and $d = 8 - m$ are optimal.

15.2.29 Example Let $q = 3$ and let $F = \mathbb{F}_3(x, y)$ be the elliptic function field defined by $y^2 = x(x^2 + x - 1)$. We choose $r = 9$ and let $P_1, P_2, P_3, P_4, P_5, P_6$ be the six rational places and P_7, P_8, P_9 the three places of degree 2 of F. The linear codes C_i have the following parameters: for $1 \le i \le 6$ we let $n_i = k_i = d_i = 1$ and for $7 \le i \le 9$ we let $n_i = 3$, $k_i = 2$, $d_i = 2$. Then for $m = \deg(G) = 1, \dots, 11$ the corresponding XNL code is a linear code over $\mathbb{F}_3$ with parameters $n = 15$, $k = m$, and $d \ge 12 - m$. The linear code with $m = 3$ and $d = 9$ is optimal.

15.2.30 Remark By the same argument as in Remark 15.2.12, the condition in Definition 15.2.24 that none of the P_i, $1 \le i \le r$, is in the support of G can be dropped if we replace G by a suitable divisor G'.

15.2.31 Remark Decoding algorithms for generalized algebraic-geometry codes can be found in [1496].

15.2.32 Remark As for classical algebraic-geometry codes (Remark 15.2.14), the dual codes of XNL codes can be described in terms of differentials and residues for algebraic function fields [912].

15.2.3 Function-field codes

15.2.33 Remark A very general perspective on the construction of algebraic-geometry codes is that of function-field codes. A function-field code is a special type of subspace of an algebraic function field over a finite field from which linear codes can be derived. Function-field codes were introduced in Chapter 6 of [2280] and studied in detail by Hachenberger, Niederreiter, and Xing [1396].

15.2.34 Remark Let F be an algebraic function field with full constant field $\mathbb{F}_q$. For a place P of F, let $\mathcal{O}_P$ denote the valuation ring of P and let M_P be the unique maximal ideal of $\mathcal{O}_P$. For a finite nonempty set $\mathcal{Q}$ of places of F, we write

$$\mathcal{O}_\mathcal{Q} = \bigcap_{P \in \mathcal{Q}} \mathcal{O}_P, \quad \mathsf{M}_\mathcal{Q} = \bigcap_{P \in \mathcal{Q}} \mathsf{M}_P.$$

15.2.35 Definition Let F be an algebraic function field with full constant field $\mathbb{F}_q$ and let $\mathcal{Q}$ be a finite nonempty set of places of F. A *function-field code* (in F with respect to $\mathcal{Q}$) is a nonzero finite-dimensional $\mathbb{F}_q$-linear subspace V of F which satisfies the two conditions

$$V \subseteq \mathcal{O}_\mathcal{Q} \quad \text{and} \quad V \cap \mathsf{M}_\mathcal{Q} = \{0\}.$$

15.2.36 Theorem Let $\mathcal{Q}$ be a finite nonempty set of places of F and let G be a divisor of F such that $\ell(G) \ge 1$, none of the places in $\mathcal{Q}$ is in the support of G, and

$$\deg(G) < \sum_{P \in \mathcal{Q}} \deg(P).$$

Then the Riemann-Roch space $\mathcal{L}(G)$ is a function-field code in F with respect to $\mathcal{Q}$.

15.2.37 Theorem Let $\mathcal{Q} = \{P_1, \dots, P_r\}$ be a finite nonempty set of places of F. Let $f_1, \dots, f_k \in \mathcal{O}_\mathcal{Q}$ be such that the k vectors

$$(f_j(P_1), \dots, f_j(P_r)) \in \overline{\mathbb{F}_q}^r, \quad 1 \le j \le k,$$

are linearly independent over $\mathbb{F}_q$, where $\overline{\mathbb{F}_q}$ is an algebraic closure of $\mathbb{F}_q$. Then the $\mathbb{F}_q$-linear subspace of F spanned by $f_1, \dots, f_k$ is a function-field code in F with respect to $\mathcal{Q}$.

15.2.38 Remark For suitable $\mathcal{Q}$ and k in Theorem 15.2.37, appropriate elements $f_1, \dots, f_k$ can be constructed by the approximation theorem for valuations.

15.2.39 Remark Any nonzero $\mathbb{F}_q$-linear subspace of a function-field code in F with respect to $\mathcal{Q}$ is again a function-field code in F with respect to $\mathcal{Q}$.

15.2.40 Remark The most powerful method of deriving linear codes from function-field codes is described in Definition 15.2.41 below and generalizes the construction of XNL codes in Definition 15.2.24.

15.2.41 Definition Let V be a function-field code in F with respect to $\mathcal{Q} = \{P_1, \ldots, P_r\}$, where $P_1, \ldots, P_r$ are distinct places of F. For each $i = 1, \ldots, r$, let C_i be a linear code over $\mathbb{F}_q$ with length n_i, dimension $k_i \geq \deg(P_i)$, and minimum distance d_i, and let ϕ_i be an injective $\mathbb{F}_q$-linear map from the residue class field of P_i into C_i. Put $n = \sum_{i=1}^{r} n_i$ and let the $\mathbb{F}_q$-linear map $\gamma : V \to \mathbb{F}_q^n$ be given by

$$\gamma(f) = (\phi_1(f(P_1)), \ldots, \phi_r(f(P_r))) \qquad \text{for all } f \in V,$$

where on the right-hand side we use concatenation of vectors. Then the linear code $C_{\mathcal{Q}}(V; C_1, \ldots, C_r)$ over $\mathbb{F}_q$ is defined as the image of V under γ.

15.2.42 Remark The following notation is convenient. For $I \subseteq \{1, \ldots, r\}$ we write $\overline{I}$ for the complement of I in $\{1, \ldots, r\}$ and $\mathcal{Q}(\overline{I}) = \{P_i : i \in \overline{I}\} \subseteq \mathcal{Q}$. We put

$$d' = \min_{I} \sum_{i \in I} d_i,$$

where the minimum is extended over all $I \subseteq \{1, \ldots, r\}$ for which $V \cap \mathsf{M}_{\mathcal{Q}(\overline{I})} \neq \{0\}$. The last condition is always assumed to be satisfied for $I = \{1, \ldots, r\}$. The condition $V \cap \mathsf{M}_{\mathcal{Q}} = \{0\}$ in Definition 15.2.35 implies that $d' \geq 1$.

15.2.43 Theorem The code $C_{\mathcal{Q}}(V; C_1, \ldots, C_r)$ in Definition 15.2.41 is a linear code over $\mathbb{F}_q$ with length n, dimension k, and minimum distance d, where

$$n = \sum_{i=1}^{r} n_i, \quad k = \dim(V), \quad d \geq d',$$

and where d' is as in Remark 15.2.42.

15.2.44 Definition Let V be a function-field code in F with respect to $\mathcal{Q} = \{P_1, \ldots, P_r\}$. For $f \in V$, the *block weight* of f (with respect to $\mathcal{Q}$) is defined to be

$$\vartheta_{\mathcal{Q}}(f) = |\{1 \leq i \leq r : f(P_i) \neq 0\}|.$$

The number

$$\vartheta_{\mathcal{Q}}(V) = \min \{\vartheta_{\mathcal{Q}}(f) : f \in V,\ f \neq 0\}$$

is the *minimum block weight* of V (with respect to $\mathcal{Q}$).

15.2.45 Theorem If the notation is arranged in such a way that $d_1 \leq d_2 \leq \cdots \leq d_r$, then the minimum distance d of the code $C_{\mathcal{Q}}(V; C_1, \ldots, C_r)$ in Definition 15.2.41 satisfies

$$d \geq \sum_{i=1}^{\vartheta_{\mathcal{Q}}(V)} d_i,$$

where $\vartheta_{\mathcal{Q}}(V)$ is the minimum block weight of V.

15.2.46 Remark The codes $C_{\mathcal{Q}}(V; C_1, \ldots, C_r)$ derived from function-field codes form a universal family of linear codes, in the sense that any linear code can be obtained by this construction.

In fact, as the following theorem from Chapter 6 in [2280] shows, a special family of these codes suffices to represent any linear code.

15.2.47 Theorem Let C be an arbitrary linear code over $\mathbb{F}_q$ with length n and dimension k. Then there exists an algebraic function field F with full constant field $\mathbb{F}_q$, a set $\mathcal{Q}$ of n distinct rational places of F, and a k-dimensional function-field code V in F with respect to $\mathcal{Q}$ such that C is equal to the code $C_{\mathcal{Q}}(V; C_1, \dots, C_n)$, where C_i is the trivial linear code over $\mathbb{F}_q$ of length 1 and dimension 1 for $1 \le i \le n$.

15.2.48 Remark A stronger result than Theorem 15.2.47, according to which an even smaller family of codes can represent any linear code, was proved in [2380].

15.2.4 Asymptotic bounds

15.2.49 Remark The asymptotic theory of codes studies the set of ordered pairs of asymptotic relative minimum distances and asymptotic information rates as the length of codes goes to ∞. This theory considers general (including nonlinear) codes as well as the special case of linear codes. Algebraic-geometry codes play a decisive role in this theory.

15.2.50 Remark We write $n(C)$ for the length of a code C and $d(C)$ for the minimum distance of C. Furthermore, we use $\log_q$ to denote the logarithm to the base q.

15.2.51 Definition For a given prime power q, let U_q (respectively U_q^{lin}) be the set of all ordered pairs $(\delta, R) \in [0,1]^2$ for which there exists a sequence $C_1, C_2, \dots$ of general (respectively linear) codes over $\mathbb{F}_q$ such that $n(C_i) \to \infty$ as $i \to \infty$ and

$$\delta = \lim_{i\to\infty} \frac{d(C_i)}{n(C_i)}, \qquad R = \lim_{i\to\infty} \frac{\log_q |C_i|}{n(C_i)}.$$

15.2.52 Proposition [2820] There exists a function α_q (respectively α_q^{lin}) on $[0,1]$ such that

$$U_q = \{(\delta, R) : 0 \le R \le \alpha_q(\delta),\ 0 \le \delta \le 1\}$$

and

$$U_q^{\text{lin}} = \{(\delta, R) : 0 \le R \le \alpha_q^{\text{lin}}(\delta),\ 0 \le \delta \le 1\}.$$

15.2.53 Proposition [2820] The functions α_q and α_q^{lin} have the following properties:

1. $\alpha_q(\delta) \ge \alpha_q^{\text{lin}}(\delta)$ for $0 \le \delta \le 1$;
2. α_q and α_q^{lin} are nonincreasing and continuous on $[0,1]$;
3. $\alpha_q(0) = \alpha_q^{\text{lin}}(0) = 1$;
4. $\alpha_q(\delta) = \alpha_q^{\text{lin}}(\delta) = 0$ for $(q-1)/q \le \delta \le 1$.

15.2.54 Remark Values of $\alpha_q(\delta)$ and $\alpha_q^{\text{lin}}(\delta)$ are not known for $0 < \delta < (q-1)/q$. Thus, the best one can do at present from a practical point of view is to find lower bounds on $\alpha_q(\delta)$ and $\alpha_q^{\text{lin}}(\delta)$ for $0 < \delta < (q-1)/q$. Such lower bounds guarantee an asymptotic information rate R that can be achieved for a given q and a given asymptotic relative minimum distance δ.

15.2.55 Definition For $0 < \delta < 1$, the *q-ary entropy function* H_q is defined by (Definition 15.1.122)

$$H_q(\delta) = \delta \log_q(q-1) - \delta \log_q \delta - (1-\delta)\log_q(1-\delta).$$

15.2.56 Remark The benchmark bound in the asymptotic theory of codes is the *asymptotic Gilbert-Varshamov bound* in Theorem 15.2.57 below which dates from the 1950s.

15.2.57 Theorem For any prime power q, we have

$$\alpha_q(\delta) \geq \alpha_q^{\text{lin}}(\delta) \geq R_{\text{GV}}(q,\delta) := 1 - H_q(\delta) \qquad \text{for } 0 < \delta < (q-1)/q.$$

15.2.58 Theorem [3017] For any prime power q and any real number δ with $0 < \delta < (q-1)/q$, there exists a sequence $C_1, C_2, \ldots$ of classical algebraic-geometry codes over $\mathbb{F}_q$ with $n(C_i) \to \infty$ as $i \to \infty$ which yields

$$\alpha_q^{\text{lin}}(\delta) \geq R_{\text{GV}}(q,\delta).$$

15.2.59 Remark Since no improvement on Theorem 15.2.57 was obtained for a long time, there was speculation that maybe $\alpha_q^{\text{lin}}(\delta) = R_{\text{GV}}(q,\delta)$ for $0 < \delta < (q-1)/q$. However, this conjecture was disproved by the use of algebraic-geometry codes. The crucial result in this context is the *TVZ bound* (named after Tsfasman, Vlăduţ, and Zink [2821]) in Theorem 15.2.60 below. For the formulation of this bound, we need the quantity

$$A(q) = \limsup_{g\to\infty} \frac{N_q(g)}{g},$$

where q is an arbitrary prime power and $N_q(g)$ is the maximum number of rational places that an algebraic function field with full constant field $\mathbb{F}_q$ and genus g can have. We note that $A(q) > 0$ for all prime powers q and that $A(q) = q^{1/2} - 1$ if q is a square. The proof of the TVZ bound uses appropriate sequences of classical algebraic-geometry codes for which the length goes to ∞.

15.2.60 Theorem [2821] For any prime power q, we have

$$\alpha_q(\delta) \geq \alpha_q^{\text{lin}}(\delta) \geq R_{\text{TVZ}}(q,\delta) := 1 - \frac{1}{A(q)} - \delta \qquad \text{for } 0 \leq \delta \leq 1.$$

15.2.61 Remark The following two theorems show that for certain sufficiently large values of the prime power q, we can beat the asymptotic Gilbert-Varshamov bound in Theorem 15.2.57 by using suitable sequences of classical algebraic-geometry codes. Both theorems are simple consequences of the TVZ bound in Theorem 15.2.60 and information about the quantity $A(q)$ defined in Remark 15.2.59.

15.2.62 Theorem Let $q \geq 49$ be the square of a prime power. Then there exists an open subinterval (δ_1, δ_2) of $(0, (q-1)/q)$ containing $(q-1)/(2q-1)$ such that

$$R_{\text{TVZ}}(q,\delta) > R_{\text{GV}}(q,\delta) \qquad \text{for all } \delta \in (\delta_1, \delta_2).$$

15.2.63 Theorem Let $q \geq 343$ be the cube of a prime power. Then there exists an open subinterval (δ_1, δ_2) of $(0, (q-1)/q)$ containing $(q-1)/(2q-1)$ such that

$$R_{\text{TVZ}}(q,\delta) > R_{\text{GV}}(q,\delta) \qquad \text{for all } \delta \in (\delta_1, \delta_2).$$

15.2.64 Remark More generally, it was shown in [2284] that a result like Theorem 15.2.62 holds for all sufficiently large composite nonsquare prime powers q. It is not known whether such a result holds also for all sufficiently large primes q.

15.2.65 Remark If one considers general (i.e., not necessarily linear) codes, then global improvements on the TVZ bound for $\alpha_q(\delta)$ in Theorem 15.2.60 can be obtained. By a global improvement we mean an improvement for any q and δ by a positive quantity independent of δ. The currently best global improvement on the TVZ bound is the *Niederreiter-Özbudak bound* in Theorem 15.2.66 below.

15.2.66 Theorem [2262] For any prime power q, we have

$$\alpha_q(\delta) \geq R_{\mathrm{NO}}(q,\delta) := 1 - \frac{1}{A(q)} - \delta + \log_q\left(1 + \frac{1}{q^3}\right) \qquad \text{for } 0 \leq \delta \leq 1.$$

15.2.67 Remark The proof of the Niederreiter-Özbudak bound in [2262] is quite involved. A simpler proof using a new variant of the construction of algebraic-geometry codes was presented in [2715], but this proof works only for a slightly restricted range of the parameter δ. The new construction proceeds as follows. Let n be a positive integer (which will be the length of the code) and let F be an algebraic function field with full constant field $\mathbb{F}_q$ such that F has at least $n+1$ rational places. Choose $n+1$ distinct rational places $P_0, P_1, \ldots, P_n$ of F and put $\mathcal{Q} = \{P_1, \ldots, P_n\}$. Furthermore, let m, r, and s be integers with $m \geq 1$ and $1 \leq r \leq \min(n, s)$. We let $\mathcal{G}(\mathcal{Q}; r, s)$ be the set of all positive divisors G of F of degree s such that the support of G is a subset of $\mathcal{Q}$ of cardinality r. Next we define

$$S(mP_0; \mathcal{Q}; r, s) = \bigcup_{G \in \mathcal{G}(\mathcal{Q};r,s)} \{f \in \mathcal{L}(mP_0 + G) : \nu_{P_0}(f) = -m\}.$$

Note that the union above is a disjoint union. A map

$$\phi : S(mP_0; \mathcal{Q}; r, s) \to \mathbb{F}_q^n$$

is now defined in the following way. Take $f \in S(mP_0; \mathcal{Q}; r, s)$ and let $G \in \mathcal{G}(\mathcal{Q}; r, s)$ be the unique divisor such that $f \in \mathcal{L}(mP_0 + G)$ and $\nu_{P_0}(f) = -m$. We define

$$\phi(f) = (c_1(f), \ldots, c_n(f)) \in \mathbb{F}_q^n,$$

where for $i = 1, \ldots, n$ we put $c_i(f) = f(P_i)$ if P_i is not in the support of G and $c_i(f) = 0$ if P_i is in the support of G. The desired code $C(mP_0; \mathcal{Q}; r, s)$ over $\mathbb{F}_q$ is the image of $S(mP_0; \mathcal{Q}; r, s)$ under ϕ. Note that $C(mP_0; \mathcal{Q}; r, s)$ is in general a nonlinear code. The proof of the bound $\alpha_q(\delta) \geq R_{\mathrm{NO}}(q,\delta)$ for a restricted range of δ in [2715] is obtained by using a suitable sequence of codes of this type with increasing length.

15.2.68 Remark Recently, some local improvements on the Niederreiter-Özbudak bound in Theorem 15.2.66 have been obtained, that is, improvements that are valid only for a restricted range of the parameter δ (this range may depend on the value of q). We refer to the papers [2264, 2265, 3029] for these improvements, and we note that [2265] contains also improved lower bounds on $\alpha_q^{\mathrm{lin}}(\delta)$. An important role in this work is played by distinguished divisors in the construction of algebraic-geometry codes, that is, divisors G such that $\mathcal{L}(G - D) = \{0\}$ for a (large) finite family of positive divisors D. The point of a distinguished divisor is that it yields an improvement on the lower bound for the minimum distance in Theorem 15.2.6.

See Also

§15.1 For general background on coding theory.

References Cited: [92, 875, 912, 1321, 1322, 1323, 1375, 1396, 1415, 1496, 1524, 2262, 2264, 2265, 2280, 2281, 2284, 2285, 2380, 2714, 2715, 2820, 2821, 3017, 3019, 3020, 3029]

15.3 LDPC and Gallager codes over finite fields

Ian Blake, *University of British Columbia*
W. Cary Huffman, *Loyola University Chicago*

15.3.1 **Definition** [1987] A binary linear code of length n is a *Gallager code* provided it has an $m \times n$ parity check matrix H where every column has fixed weight c and every row has weight as close to nc/m as possible. This code will be denoted $Gal_{n,2}(c, H)$. If $r = nc/m$ is an integer and every row of H has weight r, the Gallager code is called *regular*.

15.3.2 **Definition** A binary linear code of length n is a *low density parity check (LDPC) code* provided it is a regular Gallager code $Gal_{n,2}(c, H)$. This code will be denoted by $LDPC_{n,2}(c, r, H)$ where c is the weight of each column of H and r is the weight of each row of H.

15.3.3 **Remark** Note that the $m \times n$ matrix H used to define a Gallager or LDPC code $\mathcal{C}$ is not assumed to have independent rows and so is technically not a parity check matrix in the sense often used. However, H can be row reduced and the zero rows removed to form a parity check matrix for $\mathcal{C}$ with independent rows. Thus the dimension of $\mathcal{C}$ is at least $n - m$. Generally the column and row weights are chosen to be relatively small compared to n, and thus the density of 1s in H is low. A natural generalization for Gallager and LDPC codes is to allow the row weights and column weights of H to both vary, in some controlled manner. The resulting codes are sometimes termed *irregular* LDPC codes. Unless specific parameters are given, for the remainder of this section, the term "LDPC" will refer to either regular or irregular LDPC codes.

15.3.4 **Remark** Gallager and LDPC codes were developed by R. G. Gallager in [1162, 1163]. Gallager also presented two iterative decoding algorithms designed to decode these codes of long length, several thousand bits for example. These algorithms are presented in Remarks 15.3.8 and 15.3.12.

15.3.5 **Remark** The reason Gallager codes are useful is because they achieve close to optimal properties. Roughly speaking, for any $c \geq 3$ and any $\lambda > 1$, there exist Gallager codes of long enough length and rates up to $1 - 1/\lambda$ such that virtually error-free transmission occurs; the specific codes that achieve this property are dependent upon the characteristics of the communication channel being used. Additionally, for an appropriate choice of c, there exist Gallager codes $Gal_{n,2}(c, H)$, where H is $m \times n$, of rates arbitrarily close to channel capacity and c/m arbitrarily small. The precise formulation of these statements can be found in [1987]. Furthermore, there exist irregular LDPC codes whose performance very closely approaches the Shannon channel capacity for the binary additive white Gaussian noise (AWGN) channel [642, 2457]. These codes compare favorably and can even surpass the performance of turbo codes [1987, 2457].

15.3.6 **Remark** LDPC codes are part of several standards used in digital television, optical communication, and mobile wireless communication, including DVB-T2, DVB-S2, WiMAX-IEEE 802.16e, and IEEE 802.11n [1046]. Both the DVB-T2 standard [995], used in digital television, and the DVB-S2 standard [994], used in a variety of satellite applications, employ LDPC codes concatenated with BCH codes. In addition, a 2007 paper from the Consultative Committee for Space Data Systems (CCSDS) outlines the experimental specifications for the use of LDPC codes for near-Earth and deep space communications [573].

15.3.7 Remark Gallager and LDPC codes can be defined over other alphabets. Such codes were examined in Gallager's original work [1163]; see also [2455]. Only binary codes are considered here.

15.3.8 Remark Gallager's first decoding algorithm is an iterative algorithm that involves flipping bits in the received vector. This algorithm works best on a binary symmetric channel (BSC) when the code rate is well below channel capacity. Assume that the codeword $\mathbf{c}$ is transmitted and $\mathbf{y} = (y_1, y_2, \ldots, y_n)$ is received using $Gal_{n,2}(c, H)$. In the computation of the syndrome $H\mathbf{y}^T$, each received bit y_i affects at most c components of that syndrome, as the i-th bit is in c parity check equations. Let $\mathcal{S}_i$ denote the set of bits involved in these c parity check equations. If among all the bits $\mathcal{S}_i$ involved in these c parity check equations only the i-th is in error, then the c components of $H\mathbf{y}^T$, arising from these c parity check equations, will equal 1 indicating the parity check equations are not satisfied. Even if there are some other errors among the bits $\mathcal{S}_i$, one expects that several of these c components of $H\mathbf{y}^T$ will equal 1. This is the basis of the *Gallager Bit-Flipping Decoding Algorithm*.

I. Compute $H\mathbf{y}^T$ and determine the unsatisfied parity check equations, i.e., the parity check equations where the components of $H\mathbf{y}^T$ equal 1.

II. For each of the n bits, compute the number of these unsatisfied parity check equations involving that bit.

III. Flip the bits of $\mathbf{y}$, from 0 to 1 or 1 to 0, that are involved in the largest number of these unsatisfied parity check equations; call the resulting vector $\mathbf{y}$ again.

IV. Iteratively repeat I, II, and III until either $H\mathbf{y}^T = \mathbf{0}^T$, in which case the received vector is decoded as this latest $\mathbf{y}$, or until a certain number of iterations is reached, in which case the received vector is not decoded and a decoding failure is declared.

15.3.9 Remark In order to describe the second of Gallager's decoding algorithms, we need the concept of a Tanner graph [2779], which can be defined for any code.

15.3.10 Definition The *Tanner graph* is a bipartite graph constructed from an $m \times n$ parity check matrix H for a binary code. The Tanner graph has two types of vertices: *variable* and *check nodes*. There are n variable nodes, one corresponding to each coordinate or column of H. There are m check nodes, one for each parity check equation or row of H. The Tanner graph has only edges between variable nodes and check nodes; a given check node is connected to precisely those variable nodes where there is 1 in the corresponding column of H. Since a code can have many different parity check matrices, there are many different Tanner graphs for a code. If the code is $Gal_{n,2}(c, H)$, the degree of each variable node is c; if the code is $LDPC_{n,2}(c, r, H)$, the degree of each variable node is c, and the degree of each check node is r.

15.3.11 Remark The second of Gallager's algorithms is an example of an iterative "message passing" decoding algorithm. Message passing decoding algorithms have the following general characteristics.

1. Message passing algorithms are performed in iterations, also called rounds.
2. Messages are passed from a variable node to a check node and from a check node to a variable node along the edges of the Tanner graph of a code. Outgoing messages sent out from a node along an adjacent edge ϵ depend on all the incoming messages to that node except the incoming message along ϵ.
3. There is an initial passing of messages from the variable nodes to the check nodes (or perhaps from the check nodes to the variable nodes).

4. One iteration, or round, consists of passing messages from all check nodes to all adjacent variable nodes followed by the passing of messages from all variable nodes to all adjacent check nodes (or perhaps from variable nodes to check nodes and then back to variable nodes).
5. At the end of each round, a computation is done that will either end the algorithm or indicate that another iteration should be performed.
6. There is a preset maximum number of rounds that the algorithm will be allowed to run.

15.3.12 Remark The description of Gallager's message passing decoding algorithm comes from [1987] and applies to binary channels in which noise bits are independent, such as the memoryless BSC or the binary AWGN channel. The algorithm falls in a class of algorithms called "sum-product algorithms" and is implemented using message passing.

Let $\mathcal{C} = Gal_{n,2}(c, H)$ where H is an $m \times n$ matrix. In the Tanner graph $\mathcal{T}$ for $\mathcal{C}$, number the variable nodes $1, 2, \ldots, n$ and the check nodes $1, 2, \ldots, m$. Let $V(j)$ denote the variable nodes connected to the j-th check node, and let $C(k)$ denote the c check nodes connected to the k-th variable node.

Suppose the codeword $\mathbf{c}$ is transmitted and $\mathbf{y} = \mathbf{c} + \mathbf{e}$ is received where $\mathbf{e}$ is the unknown error vector. Given the syndrome $H\mathbf{y}^T = H\mathbf{e}^T = \mathbf{z}^T$, the object of the decoder is to compute the conditional probabilities $P(e_k = 1 \mid \mathbf{z})$ for $1 \leq k \leq n$. From there, the most likely error vector and hence most likely codeword can be found. The algorithm is an iterative message passing algorithm in which each message is a pair of probabilities. For $1 \leq k \leq n$, the initial message passed from the variable node k to the check node $j \in C(k)$ (in Step I of the algorithm) is the pair $(q_{j,k}^0, q_{j,k}^1) = (p_k^0, p_k^1)$, where $p_k^0 = P(e_k = 0)$ and $p_k^1 = P(e_k = 1)$ come directly from the channel statistics. For example, if communication is over a BSC with crossover probability p, then $p_k^0 = 1 - p$ and $p_k^1 = p$. Each iteration consists of first passing messages from check nodes to variable nodes (in Step II of the algorithm) and then passing messages from variable nodes back to check nodes (in Step III of the algorithm). The message from the check node j to the variable node k for $k \in V(j)$ is the pair $(r_{j,k}^0, r_{j,k}^1)$, where $r_{j,k}^e$ for $e \in \{0,1\}$ is the probability that the j-th check equation is satisfied given that $e_k = e$ and that the other bits e_i for $i \in V(j) \setminus \{k\}$ have probability distribution given by $\{q_{j,i}^0, q_{j,i}^1\}$. The message from the variable node k to the check node $j \in C(k)$ is the pair $(q_{j,k}^0, q_{j,k}^1)$, where $q_{j,k}^e$ for $e \in \{0,1\}$ is the probability that $e_k = e$ given information obtained from checks $C(k) \setminus \{j\}$. The *Gallager Message Passing Sum-Product Decoding Algorithm* for binary Gallager codes is the following.

I. For $1 \leq k \leq n$, pass the message $(q_{j,k}^0, q_{j,k}^1) = (p_k^0, p_k^1)$ from variable node k to check node $j \in C(k)$.

II. Update the values of $r_{j,k}^e$ for $k \in V(j)$ and $e \in \{0,1\}$ according to the equation

$$r_{j,k}^e = \sum_{i \in V(j) \setminus \{k\},\, e_i \in \{0,1\}} P(z_j \mid e_k = e,\ \{e_i \mid i \in V(j) \setminus \{k\}\}) \prod_{i \in V(j) \setminus \{k\}} q_{j,i}^{e_i}.$$

For $1 \leq j \leq m$, pass the message $(r_{j,k}^0, r_{j,k}^1)$ from check node j to variable node $k \in V(j)$.

III. Update the values of $q_{j,k}^e$ for $j \in C(k)$ and $e \in \{0,1\}$ according to the equation

$$q_{j,k}^e = \alpha_{j,k} p_k^e \prod_{i \in C(k) \setminus \{j\}} r_{i,k}^e$$

where $\alpha_{j,k}$ is chosen so that $q_{j,k}^0 + q_{j,k}^1 = 1$. For $1 \leq k \leq n$, pass the message $(q_{j,k}^0, q_{j,k}^1)$ from variable node k to check node $j \in C(k)$.

After each pass through Step III, compute

$$q_k^e = p_k^e \prod_{j \in C(k)} r_{j,k}^e$$

for $1 \leq k \leq n$ and $e \in \{0,1\}$. Then for $1 \leq k \leq n$, set $\widehat{e}_k = 0$ if $q_k^0 > q_k^1$ and $\widehat{e}_k = 1$ if $q_k^1 > q_k^0$. Let $\widehat{\mathbf{e}} = (\widehat{e}_1, \widehat{e}_2, \ldots, \widehat{e}_n)$. If $H\widehat{\mathbf{e}}^T = \mathbf{z}^T$, decode by setting $\mathbf{e} = \widehat{\mathbf{e}}$ and stop the algorithm; otherwise repeat Steps II and III unless the predetermined maximum number of iterations has been reached. If the maximum number of iterations has been reached and if $H\widehat{\mathbf{e}}^T$ never equals $\mathbf{z}^T$, stop the algorithm and declare a decoding failure.

15.3.13 Remark If the Tanner graph $\mathcal{T}$ is without cycles and the algorithm successfully halts, then the probabilities $P(e_k = e \mid \mathbf{z})$ equal $\alpha_k q_k^e$ for $1 \leq k \leq n$ where α_k is chosen so that $\alpha_k q_k^0 + \alpha_k q_k^1 = 1$. If the graph has cycles, $\alpha_k q_k^e$ are approximations to $P(e_k = e \mid \mathbf{z})$. In the end, one does not care exactly what these probabilities are; one only cares about obtaining a solution to $H\mathbf{e}^T = \mathbf{z}^T$. Thus the algorithm can be used effectively even when there are long cycles present. The algorithm has been successful for codes of length a few thousand, say $n = 10,000$, particularly with c small, say $c = 3$. Analysis of the algorithm can be found in [1987].

15.3.14 Remark In Step II of the algorithm, the conditional probabilities $P(z_j \mid e_k = e, \ \{e_i \mid i \in V(j) \setminus \{k\}\})$ for $k \in V(j)$ and $e \in \{0,1\}$ are required. These probabilities are either 0 or 1. Notice that the j-th check equation is the modulo 2 sum of the values e_k for $k \in V(j)$. Thus

$$P(z_j \mid e_k = e, \ \{e_i \mid i \in V(j) \setminus \{k\}\}) = \begin{cases} 1 & \text{if } z_j \equiv e + \displaystyle\sum_{i \in V(j)\setminus\{k\}} e_i \pmod 2, \\ 0 & \text{otherwise.} \end{cases}$$

15.3.15 Remark There is a message passing algorithm for Gallager codes on the binary erasure channel (BEC). For the BEC, there are two inputs $\{0,1\}$ to the channel, three possible outputs $\{0, E, 1\}$ from channel transmission, and an associated probability ϵ. In a BEC an input symbol x is received as an output symbol y with the following probabilities $P(y \mid x)$: $P(0 \mid 1) = P(1 \mid 0) = 0$, $P(0 \mid 0) = P(1 \mid 1) = 1 - \epsilon$, and $P(E \mid 0) = P(E \mid 1) = \epsilon$. So the output that is received is either an erasure, denoted E, or is the original input symbol; 0 is never received as 1 and 1 is never received as 0. The message passing algorithm for the BEC, which is described and thoroughly analyzed in [2455], employs the notation from Remark 15.3.12. Suppose that $\mathbf{c}$ is sent and $\mathbf{y} = (y_1, y_2, \cdots, y_n)$ is received. The messages on each edge of the Tanner graph are elements of $\{0, E, 1\}$. For binary Gallager codes the *Message Passing Decoding Algorithm for a BEC* is the following.

I. For $1 \leq j \leq m$, send E from every check node j to every variable node $k \in V(j)$.

II. For $1 \leq k \leq n$, determine the message to send from the variable node k to the check node $j \in C(k)$ as follows. If y_k and all of the incoming messages to the variable node k from the check nodes $i \in C(k) \setminus \{j\}$ equal E, the outgoing message from the variable node k to the check node $j \in C(k)$ is E. Otherwise, if one of y_k or the incoming messages to the variable node k from the check nodes $i \in C(k) \setminus \{j\}$ equals 0, respectively 1, send 0, respectively 1.

III. For $1 \leq j \leq m$, determine the message to send from the check node j to the variable node $k \in V(j)$ as follows. If any of the incoming messages to the check node j from the variable nodes $i \in V(j) \setminus \{k\}$ is E, the outgoing message from the check node j to the variable node $k \in V(j)$ is E. Otherwise, if none of the incoming messages to the check node j from the variable nodes $i \in V(j) \setminus \{k\}$ is

E, the outgoing message from the check node j to the variable node $k \in V(j)$ is the modulo 2 sum of the incoming messages to the check node k from the variable nodes $i \in V(j) \setminus \{k\}$. In addition, if $y_k = E$, update y_k to be the value of the incoming message.

Repeat Steps II and III until all y_k's have been determined, in which case the updated $\mathbf{y}$ is declared to equal the transmitted codeword $\mathbf{c}$, or until the number of iterations has reached a predetermined value. In the latter case, declare a decoding failure.

15.3.16 Remark In Step II of the algorithm, there is no ambiguity in assigning the value to the message to be sent. It is not possible for one of y_k or the incoming messages to have the value 0 and another of y_k or the incoming messages to have the value 1. In Step III, the only time the value 0 or 1 is passed as a message from the check node j to the variable node k is when there is no erasure from any other variable node sent to check node j and hence the value of the k-th variable is determined uniquely by the j-th parity check equation.

15.3.17 Remark When applying the message passing algorithm of Remark 15.3.15, the iterations may stabilize so that there are erasures left but no further iteration will remove any of these remaining erasures. This will occur because of the presence of stopping sets.

15.3.18 Definition Let $\mathcal{V}$ be the set of variable nodes in the Tanner graph of $Gal_{n,2}(c, H)$. A subset $\mathcal{S}$ of $\mathcal{V}$ is a *stopping set* if all the check nodes connected to a variable node in $\mathcal{S}$ are connected to at least two variable nodes in $\mathcal{S}$.

15.3.19 Theorem [2455] Suppose that a codeword of $Gal_{n,2}(c, H)$ is sent over a BEC and that the received vector has erasures on the set $\mathcal{E}$ of variable nodes. Assume that the message passing decoder of Remark 15.3.15 is allowed to run on the received vector until either all erasures are corrected or the process fails to make progress. If the algorithm fails to make progress and erasures remain, then the set of variable nodes that still contain erasures is the unique maximal stopping set inside $\mathcal{E}$.

15.3.20 Remark So far, the focus of this section has been on decoding Gallager codes. The algorithms are efficient because they take advantage of the sparsity of the parity check matrix for such codes. One might ask if this sparseness of the parity check matrix can be used to efficiently encode Gallager codes. The answer is yes, as described in [2455]. Begin with the $m \times n$ parity check matrix H for $Gal_{n,2}(c, H)$, which we assume has rank m. By using row and column permutations, which do not affect the column or row weights, H can be transformed into an approximate upper triangular form

$$\begin{bmatrix} T & A & B \\ E & C & D \end{bmatrix}.$$

The matrix T is an upper triangular $(n-k-g) \times (n-k-g)$ matrix with 1s on the diagonal; A is $(n-k-g) \times g$, B is $(n-k-g) \times k$, and E, C, and D have g rows with the appropriate number of columns. There is an algorithm with $O(n+g^2)$ operations that encodes any message.

15.3.21 Remark Constructing "good" irregular LDPC codes is an important area of research where "good" means optimal or near optimal according to some measure. One approach to the construction of good LDPC codes is to use combinatorial objects, such as finite geometries, permutation matrices, and Latin squares to derive incidence matrices of variable versus check nodes. The properties of the combinatorial objects typically allow observations on certain structural properties of the resulting graph. The objects used often give quasi-cyclic codes which have desirable encoding and decoding properties. As with most

LDPC code constructions, performance has to be obtained through simulation. The references [1801, 1841, 3055, 3061] are representative of this approach. There are numerous decoding algorithms, some variations of those given here, and their performance analysis is also an important topic of research. When using iterative decoding, one needs to know if the algorithm will converge to a codeword and if that codeword is the original codeword transmitted. The presence of "pseudocodewords" play a role in describing convergence of an iterative decoder; see [151, 1724, 1778] and the references in these papers. The error performance of a maximum likelihood decoder is determined by the distance distribution of the codewords in the code; in the case of an iterative decoder, error performance is determined by the distribution of pseudocodewords [1724]. There are three common notions of pseudocodewords: computation tree pseudocodewords, graph cover pseudocodewords, and linear programming pseudocodewords; see [151] for connections between the three types. For instance, computation tree pseudocodewords arise as follows [1724]. From the Tanner graph of an LDPC code, the *computation tree* can be constructed starting from a fixed variable root node. The actual structure of the tree is determined by the scheduling used by the message passing algorithm and the number of iterations used. In general a variable node from the Tanner graph will appear multiple times in the computation tree, as will the check nodes. A codeword in the LDPC code will be a binary assignment of variable nodes in the Tanner graph so that the binary sum of the neighbors of each check node is 0. The same assignment made in the computation tree will also yield, for the neighbors of each check node, a binary sum equal to 0. However, it is possible to make an assignment of the variable nodes in the computation tree so that the binary sum of the neighbors of each check node is 0, but the assignment gave different values to some nodes of the computation tree that actually represented the same variable node of the Tanner graph. This assignment of values is a *computation tree pseudocodeword.*

15.3.22 Remark The special case of linear codes that can be represented by Tanner graphs with no cycles has been investigated. Such a code can be decoded with maximum likelihood by using the min-sum algorithm with complexity $O(n^2)$. However, it can be shown that such cycle-free Tanner graphs cannot support good codes, in terms of either trade-off between rate and minimum distance or performance. Aspects of cycle-free Tanner graphs are explored further in [996, 1401, 2518].

15.3.23 Remark The analysis of LDPC codes, particularly for finite block lengths, poses challenges. Interest is in performance under a class of *belief propagation* decoding algorithms, a subclass of message passing algorithms, where the messages sent between variable and check nodes represent the probability or belief a given variable node has a particular value. This usually involves either likelihoods or log likelihood ratios, conditioned on values received in the previous round. While such algorithms are suboptimal, they are more efficient than maximum likelihood algorithms and generally yield good performance. In the first round of decoding, the message sent from a variable node v to its check neighbors is the log likelihood ratio of the received data, i.e., the log of the ratio of the probability the sent bit was a 0 to that it was a 1, given the received value (which may be discrete or continuous). In subsequent rounds, messages from a check node c to a neighbor variable node v is a log likelihood of the message arriving at v, involving information sent from all neighbor variable nodes other than v, in the previous round. After the first round, messages from a variable node v to a neighbor check node c is a log likelihood of the value of the message sent from v conditioned on values received in the previous round from neighbor check nodes other than c. The analysis requires evaluation of probability density functions of data received by a node that involves convolutions on the order of the degree of the node, either check or variable. It can be shown [2458] that the behavior of these densities is narrowly concentrated around their expected values and thus it suffices to consider only the expected values of the densities. The updating formulae for these expected densities, referred to as *density evolution*, gives a method of evaluating the performance of the algorithm. The equations assume the variables involved in the convolutions are independent which is only true to the extent that the graph, grown out from a given variable node, is a tree to a certain level. At some point this ceases to be true. However, the analysis given with this assumption yields results which are accurate for most codes. The technique of density evolution introduced in [2458] has been a cornerstone of the analysis of LDPC codes.

15.3.24 Remark Many LDPC codes exhibit probability of error behavior that, as the signal to noise ratio (SNR) is increased, the error curve flattens out rather than continuing as a typical "waterfall" curve. This is termed an *error floor* and is an undesirable feature for applications requiring very small probabilities of error, in the range of 10^{-12} for many standards. It is difficult to simulate curves down to such a low value - even with special hardware, simulations can typically be done only to about 10^{-9} [2456]. Much effort has been expended on finding analytical techniques to determine codes whose error floors are below such levels. The key reference for such work is [2456] where the notion of *trapping sets* is introduced. An (a, b) trapping set is a subgraph of the Tanner graph of the code induced by a set of variable nodes of size a with the property that the subgraph has b check nodes of odd degree. Although the sizes of the trapping sets are not the only relevant parameters for predicting error performance of the code [2456], it can be shown they are a major influence for error floors in performance curves. In particular such floors tend to be caused by overlapping clusters of fairly small trapping sets. The determination of such trapping sets tends to be a feasible computational task, either analytically or via simulation [2456], for many codes of interest. The design of codes without such trapping sets is a major goal of LDPC coding theorists. Trapping sets can be thought of as an analog of stopping sets for erasure codes.

References Cited: [151, 573, 642, 994, 995, 996, 1046, 1162, 1163, 1401, 1724, 1778, 1801, 1841, 1987, 2455, 2456, 2457, 2458, 2518, 2779, 3055, 3061]

15.4 Turbo codes over finite fields

Oscar Takeshita, *Silvus Technologies*

15.4.1 Introduction

15.4.1.1 Historical background

Turbo codes are a class of error control codes that was first introduced by Berrou, Glavieux, and Thitimajshima in 1993 [251]. Turbo coding was a paradigm shift in the design of error control codes that enabled them to achieve very good performance.

15.4.1.2 Terminology

15.4.1 Remark Basic coding notation and code properties are covered in Section 15.1 of this handbook. We shall need to add some terminology in order to properly address turbo codes.

15.4.2 Remark In many practical communication systems, we are often interested in encoding binary (the alphabet is restricted to two symbols) messages of a given manageable length k, i.e., messages are transmitted in packets of a predetermined size. This is because of a fixed amount of memory or other hardware limitations. However, the actual amount of data to be transmitted may be much more than k bits. In this case, long data streams are simply split into multiple messages and the encoding process is repeated as necessary. Those messages may be represented by a k-bit vector $\mathbf{u}$. In this section, we work only over $\mathbb{F}_2(= GF(2))$. However, many of these ideas can be considered over general finite fields.

15.4.3 Definition Let $\mathbf{u} = (u_0, u_1, u_2, \ldots, u_{k-1})$ be a message vector of length k, where $u_i \in \mathbb{F}_2$.

15.4.4 Example An example of a message of length $k = 5$:

$$\mathbf{u} = (u_0, u_1, u_2, u_3, u_4) = (0, 1, 1, 0, 1).$$

15.4.5 Definition The set of distinct messages of length k is $\mathbf{U} = \mathbb{F}_2^k$, where $\mathbb{F}_2^k$ is the Cartesian product of $\mathbb{F}_2$ taken k times.

15.4.6 Remark Let $\mathbf{t}$ and $\mathbf{s}$ be two vectors. One is often interested in chaining them together to form a single vector $\mathbf{r}$.

15.4.7 Definition Let $\mathbf{t} = (t_0, t_1, t_2, \ldots, t_{k_1-1})$ and $\mathbf{s} = (s_0, s_1, s_2, \ldots, s_{k_2-1})$ be two vectors. Define the *chained vector* $\mathbf{r} = \mathbf{t}|\mathbf{s} = (t_0, t_1, t_2, \ldots, t_{k_1-1}, s_0, s_1, s_2, \ldots, s_{k_2-1}) = (r_0, r_1, r_2, \ldots, r_{k_1+k_2-1})$.

15.4.8 Remark Order matters with the chaining operator, i.e., $\mathbf{t}|\mathbf{s}$ is not the same as $\mathbf{s}|\mathbf{t}$ in general. The chaining of more than two vectors follows in a similar way.

15.4.9 Example Let three vectors $\mathbf{r}$, $\mathbf{s}$, and $\mathbf{t}$ be chained together in this order. Their chaining is denoted $\mathbf{r}|\mathbf{s}|\mathbf{t}$.

15.4.10 Definition Let $\mathcal{C}$ be an (n, k) linear code. An *encoder* $\mathcal{C}^e$ for $\mathcal{C}$ is a one-to-one and onto function $\mathcal{C}^e : \mathbb{F}_2^k \to \mathbb{F}_2^n$, $\mathcal{C}^e : \mathbf{u} \mapsto \mathbf{v}$, where $n \geq k$, $\mathbf{u} \in \mathbb{F}_2^k$, $\mathbf{v} \in \mathbb{F}_2^n$, and the set of codewords $\mathcal{C} = \{\mathbf{v} \in \mathbb{F}_2^n | \mathbf{v} = \mathcal{C}^e(\mathbf{u})\}$ forms a linear subspace of $\mathbb{F}_2^n$.

15.4.11 Remark A few methods for deriving codes from existing ones are explained in Section 15.1. The methods therein modify one code in order to derive a new code. Let us denote by the term compound derivation of a code any technique that derives a new code by combining two or more codes. An important classical method of compound derivation of a code by combining two codes is code concatenation [1091].

15.4.12 Definition Let $\mathcal{C}_1$ be an (n_1, k_1) code over $\mathbb{F}_2^{n_1}$ and $\mathcal{C}_2$ be an $(n_2, k_2 = n_1)$ code over $\mathbb{F}_2^{n_2}$. A *classical* (n_2, k_1) *concatenated code* $\mathcal{C}_3$ over $\mathbb{F}_2^{n_2}$ is defined by an encoder function $\mathcal{C}_3^e$ generated by the function composition $\mathcal{C}_3^e = \mathcal{C}_2^e \circ \mathcal{C}_1^e$; $\mathcal{C}_1$ and $\mathcal{C}_2$ are the *constituent codes* of $\mathcal{C}_3$. Further, $\mathcal{C}_1$ is the *outer code* and $\mathcal{C}_2$ the *inner code*.

15.4.13 Remark In the definition above, the symbols for both $\mathcal{C}_1$ and $\mathcal{C}_2$ are over the same field $\mathbb{F}_2$. The interested reader may refer to [1091] and learn that when the above codes are defined with appropriate lengths and symbols over extension fields of different lengths, the minimum distance of the resulting code $\mathcal{C}_3$ is at least the product of the minimum distances of $\mathcal{C}_1$ and $\mathcal{C}_2$.

15.4.14 Remark The above definition of a concatenated code is precisely what is known as a serially-concatenated turbo code [225]. The novelty in turbo coding is a proper choice of the constituent codes for the encoder and a decoding method for these codes. The following type of compound derived code is used to form the classical parallel-concatenated turbo code.

15.4.15 Definition Let $\mathcal{C}_1$ be an (n_1, k) code over $\mathbb{F}_2^{n_1}$ and $\mathcal{C}_2$ be an (n_2, k) code over $\mathbb{F}_2^{n_2}$. Let $\mathbf{v}_1$ and $\mathbf{v}_2$ be the code vectors corresponding to a message vector $\mathbf{u}$ for codes $\mathcal{C}_1$ and $\mathcal{C}_2$, respectively. An $(n_2 + n_1, k)$ *parallel-concatenated* code $\mathcal{C}_3$ over $\mathbb{F}_2^{n_2+n_1}$ is formed by vector chaining these code vectors, i.e., encoding a message vector $\mathbf{u}$ with $\mathcal{C}_3$ produces $\mathbf{v}_3 = \mathbf{v}_1|\mathbf{v}_2$. The codes $\mathcal{C}_1$ and $\mathcal{C}_2$ are the *constituent* codes of $\mathcal{C}_3$.

15.4.16 Remark Several variations of a turbo encoder have been studied, but we focus on the classical parallel-concatenated version. A parallel-concatenated turbo code, which is a compound derived code, has a recursive convolutional code as one of its constituent codes. We shall see that the second constituent code itself is also a compound code derived from a recursive convolutional code and a permutation code. A permutation code is implemented in practice with a device called an interleaver.

15.4.2 Convolutional codes

15.4.17 Remark Convolutional codes are a popular type of error control code because of the simplicity of their encoding and the practicality of optimal decoding for many channels. They have been used in a variety of applications from home wireless networks to deep-space communication systems. Encoding is performed by polynomial operations over the ring of polynomials with coefficients over $\mathbb{F}_2$. The Viterbi algorithm [2879] is widely used to decode convolutional codes in practice because it is known to be optimal (maximum likelihood decoding) for any memoryless channel; see Section 15.1 for a discussion of these concepts.

15.4.2.1 Non-recursive convolutional codes

15.4.18 Remark There are two major classes of convolutional codes: non-recursive convolutional codes and recursive convolutional codes. Many of the traditional applications use non-recursive convolutional codes, but turbo codes require recursive convolutional codes for reasons briefly explained in Section 15.4.5.

15.4.19 Remark In convolutional coding, a message vector $\mathbf{u}$ is traditionally represented in polynomial form on the variable D.

15.4.20 Definition The *polynomial representation of a message vector* $\mathbf{u} = (u_0, u_1, u_2, \ldots, u_{k-1})$ is $U(D) = u_0 + u_1 D + u_2 D^2 + \cdots + u_{k-1} D^{k-1}$.

15.4.21 Example The polynomial representation of the information vector in Example 15.4.4 is

$$U(D) = D + D^2 + D^4.$$

15.4.22 Definition The set of all polynomials in the polynomial ring on the variable D with coefficients in $\mathbb{F}_2$ and whose degree is smaller than k is denoted $\mathbb{F}_2[D]_{<k}$.

15.4.23 Definition Typically convolutional codes are encoded in "non-recursive" form. The encoder is defined by a generator matrix $G(D)$ whose entries are polynomials in $\mathbb{F}_2[D]_{<d}$, where $d_{max} = d - 1$ is the maximum degree among the polynomials in the matrix.

15.4.24 Example An example generator matrix with $d_{max} = 2$ and elements in $\mathbb{F}_2[D]_{<3}$ is

$$G(D) = [g_1(D) \quad g_2(D)] = \left[1 + D + D^2 \quad 1 + D^2\right].$$

15.4.25 Definition The encoding of a message $U(D)$ is performed by multiplying it by the generator matrix. The resulting matrix $V(D) = U(D)G(D)$ is the codeword in polynomial form.

15.4.26 Remark It is worthwhile noting that the codeword length n of convolutional codes is a function of the message space $\mathbb{F}_2[D]_{<k}$. This means that a generator matrix defines a family of codes. However, in practice k is chosen to be a fixed value.

15.4.27 Example We illustrate the encoding operation with an example by encoding the message in Example 15.4.21 with the generator matrix in Example 15.4.24

$$V(D) = U(D)G(D) = U(D)\,[g_1(D) \quad g_2(D)] = [v_1(D) \quad v_2(D)]\,.$$

15.4.28 Remark The encoding operation may be generalized to multi-dimensional message matrices and generator polynomials. However, many practical systems have a $U(D)$ with a single message polynomial ($M = N = 1$) and $G(D)$ is a simple one-by-two matrix ($P = 2$). This implies $V(D)$ is typically a one-by-two matrix.

15.4.29 Definition The *Hamming weight* $w(P(D))$ *of a polynomial* $P(D) \in \mathbb{F}_2[D]_{<k}$ is the number of non-zero monomials in $P(D)$.

15.4.30 Remark The Hamming weight of a vector as defined in Section 15.1 of the corresponding vector $\mathbf{p}$ gives the same value, i.e., $w(\mathbf{p}) = w(P(D))$.

15.4.31 Definition The *Hamming weight* $w(M(D))$ *of a matrix* $M(D)$ whose entries are polynomials in $\mathbb{F}_2[D]_{<k}$ is the sum of the Hamming weights of each of the polynomials.

15.4.2.2 Distance properties of non-recursive convolutional codes

15.4.32 Remark We examine the non-recursive convolutional code with a fixed maximum degree $d_{max} = 2$ in Example 15.4.24. It is clear that its minimum distance is no larger than the weight of the generator matrix $w(G(D)) = 5$ regardless of the length n of the code since, by letting the message polynomial be $U(D) = 1$, we obtain a code polynomial $V(D) = G(D)$. Therefore, a convolutional code with a fixed generator matrix is asymptotically bad.

15.4.33 Remark A message of the form $U(D) = D^x$, $0 \leq x < k$, generates a code polynomial whose weight is $w(G(D))$.

15.4.34 Remark In order to improve the minimum distance of a convolutional code, one may increase the maximum degree d_{max} of the generator matrix; however, it is easy to see that the minimum distance is expected to only increase linearly with d_{max}. The decoding of such codes, however, unfortunately becomes exponentially complex in d_{max} when Viterbi decoding is used. A typical value for d_{max} is six to keep decoding complexity manageable.

15.4.35 Theorem Non-recursive convolutional codes $\mathcal{C}$ with a fixed generator matrix $G(D)$ are asymptotically bad.

15.4.36 Theorem Any low-weight message polynomial encoded by a non-recursive convolutional code $\mathcal{C}$ produces a low-weight codeword polynomial.

15.4.2.3 Recursive convolutional codes

15.4.37 Remark From the definition of encoding, it follows that $v_1(D)$ and $v_2(D)$ have $g_1(D)$ and $g_2(D)$ as their factors, respectively.

15.4.38 Theorem There is a one-to-one and onto mapping between each $U(D)$ and $v_1(D)$ as well as $U(D)$ and $v_2(D)$.

15.4.39 Remark We examine $v_1(D) = U(D)g_1(D)$ more carefully. Let the degree of $g_1(D)$ be $d_{max} = d - 1$. We may write $v_1(D)$ as

$$v_1(D) = U(D)g_1(D) = D^{d-1}(U'(D)) + T(D)$$

for some $U'(D) \in \mathbb{F}_2[D]_{<k}$ and $T(D) \in \mathbb{F}_2[D]_{<d-1}$.

15.4.40 Theorem $T(D)$ is the negative of the remainder $R(D)$ of the division of $D^{d-1}(U'(D))$ by $g_1(D)$.

15.4.41 Remark In characteristic two, $T(D) = R(D)$.

15.4.42 Remark We shall now see how a recursive convolutional code may be derived from a non-recursive convolutional code. The set of codewords turns out to be identical following such a derivation. The difference is how message vectors are encoded (mapped) into codewords.

15.4.43 Definition Let a generator matrix for a non-recursive convolutional code be $G(D) = [g_1(D) \quad g_2(D)]$. The corresponding generator matrix for an equivalent recursive convolutional code is $G_R(D) = \begin{bmatrix} 1 & \frac{g_2(D)}{g_1(D)} \end{bmatrix}$.

15.4.44 Remark We illustrate how to encode a message when using recursive convolutional codes. While for a non-recursive convolutional code we simply multiplied the message polynomial by the generator matrix, for a recursive convolutional code there is an additional step.

15.4.45 Remark Let $U'(D) \in \mathbb{F}_2[D]_{<k}$ be a message to be encoded. We first form a polynomial $Z(D) = D^{d-1}(U'(D)) + T(D)$, where $T(D)$ is the negative of the remainder of the division of $D^{d-1}(U'(D))$ by $g_1(D)$. Naturally, $Z(D)$ is divisible by $g_1(D)$ by construction. Next we simply multiply $Z(D)$ by the generator matrix $G_R(D)$. The resulting matrix $V(D) = Z(D)G_R(D)$ is the codeword in polynomial form.

15.4.46 Example We illustrate the encoding operation with an example by encoding the message $U'(D) = D^3 + D^4$. The remainder of the division of $D^2U'(D) = D^2(D^3 + D^4) = D^5 + D^6$ by $g_1(D) = 1 + D + D^2$ is $R(D) = D$. The polynomial $Z(D) = D + D^5 + D^6$ is then multiplied by the generator matrix $G_R(D)$

$$V(D) = Z(D)G_R(D) = Z(D)\begin{bmatrix} 1 & \dfrac{g_2(D)}{g_1(D)} \end{bmatrix} = [v_1(D) \quad v_2(D)].$$

15.4.47 Remark We note that the codeword in vector format is $\mathbf{v} = \mathbf{v_1}|\mathbf{v_2} = \mathbf{t}|\mathbf{u}'|\mathbf{v_2}$.

15.4.48 Remark While non-recursive convolutional code encoding involves multiplication of polynomials, recursive convolutional encoding involves long-division.

15.4.2.4 Distance properties of recursive convolutional codes

15.4.49 Remark Recursive convolutional codes are asymptotically bad because the set of codeword polynomials is identical to an equivalent non-recursive convolutional code.

15.4.50 Theorem Recursive convolutional codes $\mathcal{C}$ with a fixed generator matrix $G_R(D)$ are asymptotically bad.

15.4.51 Remark Despite the fact that the weight distribution of the codewords in a non-recursive convolutional code and an equivalent recursive convolutional code are the same, there are fundamental differences in the relationship between the Hamming weights of the message vectors and the Hamming weights of the corresponding codewords.

15.4.52 Remark A low-weight message polynomial encoded by a recursive convolutional code $\mathcal{C}$ does not necessarily produce a low-weight codeword polynomial.

15.4.53 Remark A message polynomial of the form $U'(D) = D^x$, $0 \leq x < k$, produces codeword polynomials whose weights are lower bounded by αx for some positive constant α. Compare this with Remark 15.4.33. This is a fundamental property of recursive convolutional codes that makes them suitable for turbo coding due to the long-division encoding process.

15.4.3 Permutations and interleavers

15.4.54 Remark Permutations have long been used along with error control codes in the area of digital communications. Their main use is to reorder the elements of a vector. A device that permutes the elements of a vector is an interleaver.

15.4.55 Definition A permutation function Π on k letters is a one-to-one and onto function $\Pi : \mathbb{Z}_k \to \mathbb{Z}_k$, where $\mathbb{Z}_k$ is the set of integers modulo k.

15.4.56 Definition Let a vector $\mathbf{u} = (u_0, u_1, u_2, \ldots, u_{k-1}) \in \mathbb{F}_2^k$. A permuted version $\mathbf{u}^{\Pi}$ of $\mathbf{u}$ under the permutation function Π is $\mathbf{u}^{\Pi} = (u_{\Pi(0)}, u_{\Pi(1)}, u_{\Pi(2)}, \ldots, u_{\Pi(k-1)})$.

15.4.57 Definition The design requirements for an interleaver in typical applications are minimal. A simple *block interleaver* may in general suffice. A block interleaver writes the elements of the vector on a matrix row-wise and then reads them back column-wise to permute the elements of the vector.

15.4.58 Example Let a vector $\mathbf{a} = (a_0, a_1, a_2, a_3, a_4, a_5)$. A 2 by 3 block interleaver first writes the elements of $\mathbf{a}$ on a 2 by 3 matrix row-wise:

$$\begin{bmatrix} a_0 & a_1 & a_2 \\ a_3 & a_4 & a_5 \end{bmatrix}.$$

The permuted vector becomes $\mathbf{a}^{\Pi} = (a_0, a_3, a_1, a_4, a_2, a_5)$ by reading the matrix column-wise.

15.4.59 Remark Algebraic formulations of interleavers are not only elegant, but an algebraic structure typically implies efficient implementations. We shall describe next one of the simplest known permutation functions.

15.4.60 Definition A *linear polynomial function (LPF)* over $\mathbb{Z}_n$ is defined by the function $\mathcal{L} : \mathbb{Z}_n \to \mathbb{Z}_n$, $\mathcal{L} : x \mapsto f_1 x + f_0$, where $f_1, f_0 \in \mathbb{Z}_n$.

15.4.61 Remark Let f_1 be co-prime to n over $\mathbb{Z}$. The linear polynomial function function $\mathcal{L} : \mathbb{Z}_n \to \mathbb{Z}_n$, $\mathcal{L} : x \mapsto f_1 x + f_0$ is a linear permutation polynomial (LPP).

15.4.62 Definition A *quadratic polynomial function (QPF)* over $\mathbb{Z}_n$ is defined by the function $\mathcal{Q} : \mathbb{Z}_n \to \mathbb{Z}_n$, $\mathcal{Q} : x \mapsto f_2x^2 + f_1x + f_0$, where $f_2, f_1, f_0 \in \mathbb{Z}_n$.

15.4.63 Proposition Let n be a power of 2. Let f_1 be odd and f_2 even. The QPF function $\mathcal{Q} : \mathbb{Z}_n \to \mathbb{Z}_n, \mathcal{Q} : x \mapsto f_2x^2 + f_1x + f_0$ is a quadratic permutation polynomial (QPP).

15.4.64 Remark Necessary and sufficient conditions to obtain for QPPs for arbitrary n and other properties have been investigated in [2742, 2768].

15.4.65 Definition Let Π be a permutation function on k symbols. There is an associated $(n, n = k)$ permutation code $\mathcal{C}_\Pi = \{\mathbf{u}^\Pi | \mathbf{u} \in \mathbf{U}\}$.

15.4.4 Encoding and decoding

15.4.66 Definition We define a *parallel-concatenated* turbo code $\mathcal{C}_T$. Recall Definition 15.4.15. We must simply specify the constituent codes $\mathcal{C}_1$ and $\mathcal{C}_2$. Let $\mathcal{C}_1$ be an (n, k) recursive convolutional code with generator matrix $G_R(D) = [1 \;\; \frac{g_2(D)}{g_1(D)}]$, and let $\mathcal{C}_2$ be a compound $(n - k, k)$ derived code. The derivation of $\mathcal{C}_2$ is in two steps. The first step is the derivation of an $(n_2 = n, k_1 = k)$ compound code $\mathcal{C}_s$ by the serial concatenation of an $(n_1, n_1 = k_1 = k)$ permutation outer code $\mathcal{C}_\Pi$ with an $(n_2 = n, k_2 = n_1)$ recursive convolutional inner code identical to $\mathcal{C}_1$. The second step is to puncture the k message positions of $\mathcal{C}_s$.

15.4.67 Example Let $\mathbf{u}$ be a message vector. First we encode $\mathbf{u}$ with the recursive convolutional code $\mathcal{C}_1$ and obtain the code vector $\mathbf{v}^{(1)} = \mathbf{t}^{(1)}|\mathbf{u}|\mathbf{v}_2^{(1)}$. Next we form the code vector for $\mathcal{C}_2$. We first encode $\mathbf{u}$ with the permutation code $\mathcal{C}_\Pi$ to obtain $\mathbf{u}^\Pi = \Pi(\mathbf{u})$. Then $\mathbf{u}^\Pi$ is encoded with a recursive convolutional code $\mathcal{C}_1$ to obtain the code vector $\mathbf{v}^{(\mathbf{s})} = \mathbf{t}^{(2)}|\mathbf{u}^\Pi|\mathbf{v}_2^{(2)}$ for $\mathcal{C}_s$. The code vector for $\mathcal{C}_2$ becomes $\mathbf{v}^{(2)} = \mathbf{t}^{(2)}|\mathbf{v}_2^{(2)}$ by puncturing the message symbols $\mathbf{u}^\Pi$ of $\mathcal{C}_s$. Finally, the code vector for the turbo code becomes $\mathbf{v}^{(\mathbf{T})} = \mathbf{t}^{(1)}|\mathbf{u}|\mathbf{v}_2^{(1)}|\mathbf{t}^{(2)}|\mathbf{v}_2^{(2)}$.

15.4.68 Remark The decoding of turbo codes is performed using iterative algorithms [1398]; for iterative algorithms see Section 15.1.

15.4.5 Design of turbo codes

15.4.69 Remark We focus on the design of parallel-concatenated turbo codes. Naturally, the design narrows down to an understanding of the overall turbo code principle and to the selection of the constituent recursive convolutional codes and an interleaver.

15.4.70 Remark Recall the definition of a linear code $\mathcal{C}$ of length n as a linear subspace of $\mathbb{F}_2^n$ from Section 15.1. Let $\mathcal{C}$ be a code over $\mathbb{F}_2^n$ (not necessarily linear). The Holy Grail design for $\mathcal{C}$ has been to maximize its minimum distance as defined in Section 15.1. For a linear code, this translates to maximizing its minimum Hamming weight. This task becomes even more challenging when we require a design for a family of codes that are asymptotically good, i.e., a family of codes with a fixed code-rate such that the minimum distance to code length ratio does not vanish as the code length goes to infinity.

15.4.71 Remark The error correction capability of a code is mostly dictated by its minimum distance, with its weight distribution a secondary factor. In turbo coding, the relationship between the weight of message vectors and the weight of codeword vectors becomes crucial.

When an error control code is used, we ultimately need not only a set of codewords $\mathcal{C}$ but an encoder to map messages from $\mathbf{U}$ to $\mathcal{C}$ and a decoder to map codewords (possibly with errors) back to messages.

15.4.72 Remark We observe again the codeword of a parallel-concatenated turbo code:

$$\mathbf{v}^{(\mathbf{T})} = \mathbf{t}^{(1)}|\mathbf{u}|\mathbf{v}_{\mathbf{2}}^{(1)}|\mathbf{t}^{(2)}|\mathbf{v}_{\mathbf{2}}^{(2)}. \tag{15.4.1}$$

The design principles for a turbo code are related to the following question: How do we minimize the chances of producing simultaneously low weight codewords for $\mathcal{C}_1$ $(\mathbf{t}^{(1)}|\mathbf{u}|\mathbf{v}_{\mathbf{2}}^{(1)})$ and $\mathcal{C}_2$ $(\mathbf{t}^{(2)}|\mathbf{v}_{\mathbf{2}}^{(2)})$?

15.4.73 Remark The first design principle is that, since the codeword includes the message vector $\mathbf{u}$, it is more important to focus on the case when $w(\mathbf{u})$ is small.

15.4.74 Remark The second design principle is, given that $w(\mathbf{u})$ is small, identify a good permutation function that minimizes the chances of producing simultaneously low weight codewords for $\mathcal{C}_1$ and $\mathcal{C}_2$.

15.4.5.1 Design of the recursive convolutional code

15.4.75 Remark The main design parameters of constituent recursive convolutional codes are the maximum degree d_{max} of the generator polynomials and the choice of the generator matrices. Surprisingly, it has been found that $d_{max} = 3$ or 4 is sufficient to achieve very good performance with turbo coding. Typically, the polynomial $g_1(D)$ is chosen to be a primitive polynomial.

15.4.76 Remark The reason for choosing $g_1(D)$ to be primitive is that it maximizes the degree of polynomials of the form $U(D) = 1 + D^\tau$ such that $U(D)$ is divisible by $g_1(D)$; this is known to improve the codeword weight of recursive convolutional codes with respect to message vectors with $w(U(D)) = 2$. One of the latest communication systems using turbo codes is the fourth generation wireless cellular *4G LTE* standard.

15.4.77 Example The *4G LTE* standard uses the generator matrix with $d_{max} = 3$:

$$G_R(D) = \begin{bmatrix} 1 & \dfrac{1 + D + D^3}{1 + D^2 + D^3} \end{bmatrix}.$$

15.4.5.2 Design of interleavers

15.4.78 Remark Interleavers or permutation functions have been extensively investigated in turbo coding applications [252, 725, 755, 767, 911, 1277, 2291, 2513, 2742, 2770, 2771]. Shannon's random coding bound theorem [2608] naturally led researchers to experiment with different pseudo-random or modified pseudo-random constructions [911]. There were two major drawbacks to early pseudo-random constructions: first, practical implementation calls for interleavers that require little storage and computation; and second, pseudo-random constructions often generate turbo codes that suffer from a performance deficiency known as an "error-floor"; see Remark 15.3.24.

15.4.79 Remark Some researchers attempted to use simple interleavers with turbo coding such as block interleavers. However, a block interleaver has been shown to have too much regularity; it has also been shown that interleavers based on LPPs behave similarly to block interleavers [2770]. A good compromise between interleaver complexity and turbo code performance can be achieved with QPP interleavers; the second degree polynomial brings a "non-linear" feature that is very desirable for turbo coding [2768, 2771]. The simplicity of QPP interleavers due to their algebraic structure and their very good performance resulted in them becoming part of the *4G LTE* standard.

See Also

§15.1	For discussion of basic coding theory properties.
§15.3	For discussion of LDPC and Gallager codes.
[2495], [2496], [2519]	For discussions of cycle structure of permutation functions over finite fields and their applications to turbo codes.

References Cited: [225, 251, 252, 725, 755, 767, 911, 1091, 1277, 1398, 2291, 2495, 2496, 2513, 2519, 2608, 2742, 2768, 2770, 2771, 2879]

15.5 Raptor codes

Ian Blake, *University of British Columbia*
W. Cary Huffman, *Loyola University Chicago*

15.5.1 Remark This section is concerned with coding for the *binary erasure channel* (BEC) (see Section 15.1), except for the last subsection which includes comments on the performance of raptor codes on the *binary symmetric channel* (BSC). Recall that for the BEC a nonerased symbol is correct and the goal is to design codes which can be encoded and decoded efficiently. The capacity of the BEC with erasure probability ϵ is $1-\epsilon$. If $\mathcal{C}$ is a linear $(n,k,d)_q$ code with parity check matrix H, it is capable of correcting up to $d-1$ erasures. If a codeword $\mathbf{c}$ is transmitted and word $\mathbf{r}$ received (whose components are in the set $\{0,1,E\}$), the problem of correcting erasures is the problem of solving a set of linear equations of the form $H'\mathbf{x}^T = \mathbf{y}^T$ where H' is the matrix of columns of the original parity check matrix H of the code corresponding to erased positions of $\mathbf{r}$, $\mathbf{x}$ is a vector of variables corresponding to the erased positions of $\mathbf{r}$, whose solutions are required, and $\mathbf{y}$ the sum of columns of the parity check matrix corresponding to the positions of $\mathbf{r}$ containing 1s. This follows from the fact that a codeword $\mathbf{c}$ satisfies the equation $H\mathbf{c}^T = \mathbf{0}^T$, the vector of all zeros. The complexity of decoding is that of matrix reduction, assumed to be $O(d^3)$ when the maximum number of erasures $d-1$ has occurred (unless a more sophisticated algorithm is used). It is quite remarkable that the codes in this section achieve essentially linear complexity for both encoding and decoding while achieving capacity. This is accomplished by constructing a code by means of a bipartite graph and interpreting the decoding operation as a graph algorithm. The forerunners of raptor codes are tornado and LT codes, considered in the following sections. It should be noted that packet loss on the internet, due to traffic conditions or node buffer overflow, give a natural model of an erasure channel and the results of this section yield an excellent solution for Internet multicast data transmission with no feedback channel to request missing packets. If a block code is used for such a purpose, to be effective, one would have to know the erasure probability for effective code design. This is usually difficult to estimate and is often time varying. The LT and raptor codes described here are rateless and such an estimate is not required. The tornado codes are block codes, unlike LT and raptor codes, which are rateless (fountain) codes. The tornado codes are considered for the important ideas they introduced that led, in a sense, to the fountain and LT codes. However tornado codes are block codes while LT and raptor codes are rateless, having no concept of block length or dimension. All codes considered are binary.

15.5.1 Tornado codes

15.5.2 **Remark** The most complete description of tornado codes is given in [1975] (although not referred to as tornado codes there). Earlier work includes the papers [1972, 1974, 1973, 2619]. The paper [1975] has been influential in the last decade of progress on the problem of coding with irregular bipartite graphs.

15.5.3 **Remark** Consider a bipartite graph B with k information nodes on the left and βk check nodes on the right. This is a version of the Tanner graph of a code; see Section 15.3. The information nodes are associated with information symbols. Since only XOR operations are used in the encoding, the information symbols can be taken as either bits or strings of bits of the same length (packets). The manner of choosing edges in the graph is critical to code performance and is discussed later. Given the graph, the complexity of encoding and decoding is proportional to the number of edges in the graph. For the bipartite graph B, denote the associated code by $\mathcal{C}(B)$. The decoding is considered next. A codeword consists of the set of information and check symbols.

15.5.4 **Remark** *Decoding process:* For the code $\mathcal{C}(B)$ and its associated bipartite graph B, consider a received vector $\mathbf{y}$, the result of transmitting a codeword through the BEC. The positions of $\mathbf{y}$ contain symbols from the alphabet $\{0, 1, E\}$ or strings of such symbols. Associate the received symbols with the appropriate information and check graph nodes. The decoding proceeds as follows: given the correct value of a check symbol and all but one of the information symbol neighbors, set the missing information symbol to the XOR value of the check and the known information symbols. The process continues until all information nodes are retrieved or decoding failure. The algorithm is a version of the *belief propagation (BP)* or *message passing* algorithms used for noisy channels.

15.5.5 **Remark** The success of the decoding procedure of $\mathcal{C}(B)$ depends on the existence of check symbols with the required property to the completion of decoding. Before considering this condition, the construction of the codes is given. The term *tornado* derives from the observation that [470] one can initiate decoding on such graphs as coded symbols arrive, and the decoder stalls until, typically, the arrival of a single further symbol allows it to proceed quickly to completion.

15.5.6 **Remark** The bipartite graph B is constructed randomly as follows. Refer to an edge as one of degree i on the left (right) if the left (right) node it is connected to has degree i. The following approach is taken. The graph is to have e edges. Two distributions (to be discussed later) are defined: a left distribution (on edges, not nodes) $(\lambda_1, \ldots, \lambda_n)$ and a right distribution $(\rho_1, \ldots, \rho_n)$, where λ_i (resp. ρ_i) is the fraction of edges in the graph of left degree i (resp. right degree i), for some appropriate integer n. The number of left nodes of degree i is $e\lambda_i/i$ and $k = e\sum_i \lambda_i/i$ and similarly for the right nodes. The average left degree of the information nodes is $1/(\sum_i \lambda_i/i)$, denoted by a_ℓ and similarly for right degrees $a_r = 1/(\sum_i \rho_i/i)$. If the graph has e edges, k left information nodes and βk right check nodes, then $a_\ell = a_r\beta$. As discussed below, it is possible to choose the left and right distributions to ensure complete decoding with high probability.

15.5.7 **Remark** *Code construction:* With the terminology of the previous remark, the construction of the bipartite graph, B, is described, given the left and right distributions. Consider a sequence of four columns of nodes. The first column has k information nodes, the second and third e nodes each, and the fourth βk check nodes. Edges appear only between adjacent columns. A fraction λ_i of the e edges have left degree i. For each i, connect i of the nodes of the second column to a node of the first column, $e\lambda_i/i$ times (truncated to form integers), to give this number of left nodes of degree i. Similarly, connect $e\rho_i/i$ of the nodes in the third column to a node in the fourth column to give this number of right vertices of this degree. Nodes in the second and third columns are all of degree 1. To complete the process connect the e nodes of the second and third columns by a random matching (permutation). The second and third columns of nodes are removed with edges now connecting the information and check nodes. The process may yield a small number of multiple edges between an information and check node. These are replaced with a single edge with negligible effect. The graph B is the random bipartite graph.

15.5.8 Remark The two distributions are chosen from the analysis of the probability of successful decoding of the algorithm of Remark 15.5.4. The decoding algorithm is a Markov process that leads to a certain differential equation. It is convenient for the analysis to define the two distribution polynomials

$$\lambda(x) = \sum_i \lambda_i x^{i-1}, \quad \rho(x) = \sum_i \rho_i x^{i-1}.$$

The use of x^{i-1} rather than x^i is for convenience in the analysis. The analysis leads to the following important results.

15.5.9 Theorem [1975, Proposition 2] Let B be a bipartite graph with k information nodes that is chosen at random with edge degree distributions $\lambda(x)$ and $\rho(x)$, and suppose that δ is fixed such that

$$\rho(1 - \delta\lambda(x)) > 1 - x \text{ for all } 0 < x \leq 1. \tag{15.5.1}$$

For all $\eta > 0$ there is some k_0 such that for all $k > k_0$, if the message symbols of $\mathcal{C}(B)$ are erased independently with probability δ, then with probability at least $1-k^{2/3}\exp(-k^{3/2}/2)$ the decoding algorithm terminates with at most ηk information symbols erased.

15.5.10 Remark While not quite enough to show the decoding terminates successfully, with a slight modification of the conditions [1975] it can be shown:

15.5.11 Lemma [1975, Lemma 1] Let B be a bipartite graph with k left information nodes, chosen at random according to the distributions $\lambda(x)$ and $\rho(x)$ satisfying Condition (15.5.1) such that $\lambda_1 = \lambda_2 = 0$. Then there is some $\eta > 0$ such that, with probability $1 - O(k^{-3/2})$, the decoding process restricted to the subgraph induced by any η-fraction of the left nodes terminates successfully.

15.5.12 Remark Condition (15.5.1) was relaxed in [1972] to:

$$\delta\lambda(1 - \rho(1 - x)) < x \quad \text{for} \quad x \in (0, \delta].$$

It remains to show that distributions satisfying (15.5.1) exist. Distributions that satisfy this condition are referred to as *capacity achieving* (CA). Numerous works have addressed the problem of finding CA sequences of distributions including [1976, 2340, 2457, 2618, 2620]. A variety of techniques are used. One such example is given in the following.

15.5.13 Example This example is taken from [1975]. Let D be a positive integer which is chosen to trade off average left degree with how well the decoding process works. Let $H(D)$ be the truncated Harmonic series: $H(D) = \sum_{i=1}^{D} 1/i \approx \ln(D)$. Let

$$\lambda_i = \frac{1}{(H(D)(i-1))}, \quad i = 2, 3, \ldots, D+1, \quad \text{and hence} \quad \lambda_D(x) = \frac{1}{H(D)} \sum_{i=1}^{D} \frac{x^i}{i}.$$

The average left degree is then $a_\ell = H(D)(D+1)/D$. For the right distribution choose

$$\rho_i = \frac{e^{-\alpha}\alpha^{i-1}}{(i-1)!}, \quad i = 1, 2, \ldots, \quad \rho_D(x) = e^{\alpha(x-1)},$$

i.e., the Poisson distribution, where α is chosen so that $a_r = \alpha e^{\alpha}/(e^{\alpha} - 1) = a_\ell/\beta$, i.e., to make the fractions of nodes on the left and the right the same. It can be shown [1975] that the two distributions satisfy (15.5.1) for $\delta \le \beta/(1 + 1/D)$ and hence are CA. These distributions are referred to as *heavy tail/Poisson* or *tornado* [1975, 2619].

15.5.14 Remark It is to be noted that CA sequences lead to a graph having a high probability of successfully completing decoding on the graph.

15.5.15 Theorem [1975, Theorem 3] For any rate R, $0 < R < 1$, and any ϵ with $0 < \epsilon < 1$, and sufficiently large block length n, there is a linear code and decoding algorithm that, with probability $1 - O(n^{-3/4})$, is able to correct a random $(1 - R)(1 - \epsilon)$ fraction of errors in time proportional to $n \ln(1/\epsilon)$.

15.5.16 Remark The original construction of tornado codes involved a cascade of $m + 1$ sections of random bipartite graphs, the i-th section having $\beta^i k$ left nodes and $\beta^{i+1}k$ check nodes, with a final code C as a good erasure correcting code. Works subsequent to [1975] typically used only a single section (and no final erasure correcting code) and perhaps this is the important impact of the work on tornado codes. The ideas of tornado codes (although not always referred to as tornado codes) first appeared in [1974]. Reference [1972] simplified the analysis. The notion of the graph construction and CA distribution sequences has proved important and a rather large set of papers on this subject now exists, many giving new techniques, including linear programming, for finding CA distributions. These include [2457, 2618, 2620]. The influence of the paper [1975] is evident in subsequent work on the problem. The ability to encode and decode codes on the BEC in linear time is a remarkable achievement (recall the initial comments on the equivalence of this problem to matrix reduction and Gaussian elimination).

15.5.17 Definition Define a bipartite graph B_0 with k information nodes and βk check nodes, $0 < \beta < 1$. Recursively form the bipartite graph B_i whose left nodes are the $\beta^i k$ right nodes of B_{i-1}, with $\beta^{i+1}k$ right nodes, $1 \le i \le m$, for an integer m chosen so that $\beta^{m+1}k$ is roughly $\sqrt{k}$. Choose a final code C as a good erasure correcting code of rate $1-\beta$ with $\beta^{m+1}k$ information nodes. The resulting cascaded code, $\mathcal{C}(B_0, B_1, \cdots, B_m, C)$ has k information symbols (the leftmost k symbols as input to the graph B_0) and

$$\sum_{i=1}^{m+1} \beta^i k + \beta^{m+2}k/(1-\beta) = k\beta/(1-\beta)$$

check nodes. The code $\mathcal{C}(B_0, B_1, \ldots, B_m, C)$ is a *tornado code.*

15.5.2 LT and fountain codes

15.5.18 Remark The notion of fountain codes first appears in the work [470]. The term “fountain” refers to the process whereby k information symbols (either bits or packets) are encoded into coded symbols by XOR’ing subsets of information symbols in such a way that a receiver may collect *any* $k(1+\epsilon)$ coded symbols, for some ϵ appropriate to the system, to recover all k information symbols. The image is of a coder producing a digital fountain of coded symbols and a receiver collects a sufficient number of such symbols to retrieve the information. Codes with such a property can be described as *universal rateless erasure codes.* The codes are rateless since there is no concept of code dimension. LT codes are the first realization of such erasure codes. The term "LT" refers to "Luby transform." The codes were devised for a situation where packets are lost, or dropped, on the Internet and there was no efficient way a receiver could request a retransmission for such a missing packet. This is a typical situation for a multicast internet protocol where a single information source transmits a large amount of data to a large user community. For a typical block erasure correcting code, each user would have to receive a specific set of codeword symbols in order to decode the remaining erasures. In the LT scenario, *any* sufficiently large collection of coded symbols will do. While often thought of as erasure correcting codes, there are no erasures in this situation - in effect erased packets are those that are lost.

15.5.19 Remark As a matter of notation, the terms used in this section are information symbols and coded symbols. Many papers use the terms input and output symbols which are also natural but in some situations, where more than one level of graphs is considered, they might be ambiguous.

15.5.20 Remark In constructing bipartite graphs for coding, two forms appear in the literature. In the first form, there are n nodes on the left, and $n-k$ check nodes on the right. This is often called the *Tanner graph* of the code. The other form is the one used above with k information nodes on the left and $n-k$ check nodes on the right.

15.5.21 Definition (The LT encoding process) To encode k information symbols a probability distribution $\{\rho_i, i = 1, 2, \ldots, k\}$ is chosen. For each coded symbol, the distribution is sampled. If the integer d is produced, this number of information symbols, chosen uniformly at random from among the k, are XOR'ed together to produce a coded symbol. Information as to how the symbol was produced is included in a symbol header. Such information is a negligible fraction for practical systems and is ignored in the analysis. Note that for the remainder of the section $\{\rho_i\}$ denotes a node degree distribution, as opposed to the edge distribution of the previous section.

15.5.22 Definition (The LT decoding rule) [1971] From a collection of K (slightly larger than k) received coded symbols, a bipartite graph is formed with k left information nodes (initially of unknown value) and K right received code nodes. Edge connections are formed from the header information of the received coded symbols. All coded symbols of degree one are said to *cover* their unique information neighbors and this set of *covered* information symbols is the *ripple.* Information symbols in the ripple are determined as they are the same as the coded symbols covering them. At the first stage an information symbol in the ripple is XOR'ed with all of its coded neighbors and all edges to the coded neighbors removed. This might produce coded nodes of degree one and their unique information symbols are added to the ripple. The process continues iteratively until either all information symbols are recovered or there are no coded symbols of degree one. If at the end some information symbols remain uncovered, decoding failure is declared.

15.5.23 Remark The decoding process above has a complexity proportional to the number of edges in the graph. The number K of coded symbols required to have a reasonably high probability of decoding success is typically in the range of 1.01 to 1.05 times k, for sufficiently large k. From a *balls in bins* analysis it can be shown that for such a coding process it is necessary that at least $k \ln(k/\delta)$ balls must be thrown into the k (information) bins in order to have a probability of δ that each of the bins has at least one ball (i.e., that each information symbol is used (covered) by at least one of the coded symbols - for if one is not covered it could not be recovered). Thus in the encoding process the average degree of an information symbol must be at least $\ln(k/\delta)$ no matter what distribution is used for producing the coded symbols.

15.5.24 Remark The random behavior of the LT decoding process is determined entirely by the distribution $\{\rho_i\}$, the number of information symbols k, and the number of coded symbols K obtained, which are typically slightly larger than k. The desirable properties for the distribution are

1. to ensure the fewest possible number of coded symbols to be able to complete the decoding with high probability and
2. the average degree of the coded symbol nodes is as small as possible (although, as noted, at least $\ln(k)$).

15.5.25 Remark In analyzing the LT decoding process, a desirable feature is to have the rate at which information symbols are added to the ripple to be about the same as the rate they are processed. The ripple size should be large enough to ensure the decoding process can continue to completion, but not so large that a coded symbol has too high a probability of covering an information node already in the ripple. A distribution that has desirable properties in terms of this and other properties is the *soliton distribution* given by:

$$\rho_i = \begin{cases} 1/k & i = 1, \\ 1/(i(i-1)) & i = 2, 3, \ldots, k, \end{cases} \tag{15.5.2}$$

and $\rho_S(x) = \sum_i \rho_i x^i$. (The name derives from physics where it arises from a similar property in a refraction problem.) The expected value of this distribution is $H(k) = \sum_{i=1}^{k} 1/i$, the harmonic sum to k. In [2621], a different analysis of the decoding process of Definition 15.5.22 is given. It is shown there that if, at each step of the decoding process one wants an expected number of degree one coded nodes, the degree distribution must satisfy the (asymptotic) equation

$$(1-x)\eta''(x) = 1, \quad 0 < x < 1$$

where $\eta(x) = \sum_i \eta_i x^i$. The solution of this equation is almost the soliton distribution - except that $\eta_1 = 0$ and the range is infinite. The fact that such a distribution yields no coded symbols of degree one is a problem since the decoding algorithm cannot start. While this distribution is ideal in the sense that the expected number of coded symbols needed to recover the information in the ripple is one, in practice it performs poorly as the probability the ripple disappears before completion is high. The *robust soliton distribution* $\{\mu_i\}$ is proposed to correct these defects. It is defined in the following manner. Let δ be the probability of decoder failure for k information symbols and K coded symbols and $R = c\ln(k\delta)\sqrt{k}$ for some suitable constant c:

$$\mu_i = (\rho_i + \tau_i)/\beta, \quad \text{where } \beta = \sum_{i=1}^{k} (\rho_i + \tau_i) \tag{15.5.3}$$

and where $\{\rho_i\}$ is the ideal soliton distribution (15.5.2), $\{\mu_i\}$ is the robust soliton distribution, and $\mu_{RS}(x) = \sum_i \mu_i x^i$, with

$$\tau_i = \begin{cases} R/(ik) & i = 1, \ldots, k/R - 1, \\ R\ln(R/\delta)/k & i = k/R, \\ 0 & i = k/R + 1, \ldots, k. \end{cases}$$

The intuition is that the probability a random walk of length k deviates by more than $\ln(k/\delta)\sqrt{k}$ from its mean is at most δ [1971]. With this background it is possible to show the following theorem.

15.5.26 Theorem [1971, Theorems 12, 13 and 17] With the above notation, the average degree of a coded symbol is $D = O(\ln(k/\delta))$, and the number of coded symbols required to achieve a decoding failure probability of at most δ is $K = k + O(\sqrt{k}\ln^2(k/\delta))$.

15.5.3 Raptor codes

15.5.27 Remark While the LT codes represent a remarkable step forward for coding on an erasure channel, such as the Internet, its complexity or running time is not linear in the number of input information symbols. The following result emphasizes this. For a code with k information symbols, we say a decoding algorithm is *reliable* if it fails to decode with a probability at most $1/k^c$ for some positive constant c. The *overhead* of the code and decoding algorithm is ϵ if the decoder needs $(1+\epsilon)k$ coded symbols in order for the decoder to succeed with high probability. The term *space complexity* refers to the amount of memory required to implement decoding. The reference [2621] is followed closely here, although the terms information and coded symbols are used rather than input and output symbols. The term raptor derives from *RAP*id *TOR*nado although the tornado and LT codes have very different constructions.

15.5.28 Lemma [2621, Proposition 1] If an LT code with k information symbols possesses a reliable decoding algorithm, then there is a constant c such that the graph associated to the decoder has at least $ck\log(k)$ edges.

15.5.29 Remark Recall that if K coded symbols are gathered to decode for the k information symbols, then Gaussian elimination is maximum likelihood and has a complexity of $O(Kk^2)$, i.e., from the K coded symbols a $K \times k$ matrix equation can be set up to be solved for the k information symbols. The decoding algorithm for LT codes is able to improve on this complexity considerably by choosing an appropriate distribution $\{\rho_i\}$ and decoding algorithm, as noted above. An LT code generated in this manner is referred to as a $(k, \rho(x))$ LT code.

15.5.30 Remark The idea behind the raptor codes is to first add parity check symbols to the information symbols, by use of an efficient linear block erasure correcting code, to form the set of *intermediate symbols* and then LT encoding this set. This relieves the LT decoder from having to correct *all* the erasures. A few can be left to the block code to correct and this eases the burden of the LT decoder considerably and allows a linear decoding time, for a good choice of code. Let $\mathcal{C}_n$ be a linear code of dimension k and block length n. It is called the *precode* of the raptor code. The intermediate symbols are then the k information symbols and $n-k$ parity checks of the precode $\mathcal{C}_n$. For the LT code, a modification of the above soliton like distribution is suggested [2621]:

$$\rho_D(x) = \frac{1}{\mu+1}\left(\mu x + \sum_{i=2}^{D}\frac{x^i}{(i-1)i} + \frac{x^{D+1}}{D}\right)$$

where $D = \lceil 4(1+\epsilon)/\epsilon \rceil$ for a given real number ϵ and $\mu = (\epsilon/2) + (\epsilon/2)^2$. Thus this is a soliton-like distribution with a positive probability of degree 1 and truncated at $D+1$. The LT code is designated a $(n, \rho_D(x))$ code. The entire code is the raptor code and designated a $(k, \mathcal{C}_n, \rho_D(x))$ raptor code. A key result on the way to the final one is the following lemma.

15.5.31 Lemma [2621, Lemma 4] There exists a positive number c (depending on ϵ) such that with an error probability of at most e^{-cn} any set of $(1+\epsilon/2)n+1$ coded symbols of the LT code with parameters $(n, \rho_D(x))$ are sufficent to recover at least $(1-\delta)n$ information symbols, where $\delta = (\epsilon/4)(1+\epsilon)$.

15.5.32 Remark Two extreme cases of raptor codes are noted. If $\rho_w = \binom{k}{w}/2^k$, corresponding to the distribution polynomial $\rho(x) = (1+x)^k/2^k$, it leads to the probability that any particular binary k-tuple being chosen as $1/2^k$, i.e., a uniform distribution over $\mathbb{F}_2^k$ since a vector of weight w is chosen with probability $\binom{k}{w}/2^k$ and each of the $\binom{k}{w}$ are chosen equally likely. If such a distribution was used in the LT process, performance would be very poor. The degree distribution is too large to permit the process to succeed. At the other extreme, suppose the LT process of the raptor code $(k, \mathcal{C}_n, \rho(x))$ uses a trivial distribution $\rho_1 = 1$, i.e., after the precoding, the coder chooses an intermediate symbol at random and declares it a coded symbol. Such a code is referred to as a *precode only (PCO)* raptor code. One can see that such a code can achieve a low overhead only for very low rate codes $\mathcal{C}_n$.

15.5.33 Remark A suitable precode has the properties:

1. The rate R of $\mathcal{C}_n$ is $(1+\epsilon/2)(1+\epsilon)$.
2. The BP decoder can decode $\mathcal{C}_n$ on a BEC with erasure probability

$$\delta = (\epsilon/4)/(1+\epsilon) = (1-R)/2,$$

in $O(n \log(1/\epsilon))$ operations. (Note this is half of capacity for the rate of the code.)

It is suggested that several types of codes meet these criteria, such as tornado codes and right regular codes (coded symbol nodes have the same degree).

15.5.34 Theorem [2621, Theorem 5] Let ϵ be a positive real number, k the number of information symbols, $D = \lceil 4(1+\epsilon)/\epsilon \rceil$, $R = (1+\epsilon/2)/(1+\epsilon)$, $n = \lceil k/(1-R) \rceil$, and let $\mathcal{C}_n$ be a code with properties described above. Then the raptor code $(k, \mathcal{C}_n, \rho_D(x))$ has space complexity $1/R$, overhead ϵ and a cost of $O(\log(1/\epsilon))$ with respect to BP decoding of both the precode and LT code.

15.5.35 Remark A problem with raptor codes is that they are not systematic. A technique is given in [2621] to generate systematic raptor codes but is not discussed here. Many applications of raptor codes in practice prefer systematic codes. Raptor codes have been incorporated into numerous standards for the reliable delivery of data objects. The codes are described in IETF RFC 5053 and IETF RFC 6330 for such applications as the DVB-H IPDC (IP datacast to handheld devices) and 3GPP TS for multimedia broadcast/multicast service (MBMS) and future standards of IEEE P2220 (a draft standard protocol for stream management in media client devices) and 3GPP eMBMS, among others. The latter standards use RaptorQ codes defined over the finite field $\mathbb{F}_{2^8}$. All of these standards use systematic raptor codes. The monograph [2622] has an extensive discussion of these codes for standards.

15.5.36 Remark A good erasure correcting (linear) code should have a good minimum distance. Thus a reasonable question to ask is how such a code would perform on a noisy channel, such as a binary symmetric channel (BSC). The performance of raptor codes on such a channel is considered in [991]. Many of the results for the erasure channel are generalized there. It builds on the landmark paper [2458] which considered more general classes of low density parity check codes and introduced such fundamental concepts as density evolution. The application of these ideas to raptor codes generalizes many of the results for the erasure channel.

References Cited: [470, 991, 1971, 1972, 1973, 1974, 1975, 1976, 2340, 2457, 2458, 2618, 2619, 2620, 2621, 2622]

15.6 Polar codes

Simon Litsyn, Tel Aviv University

15.6.1 Space decomposition

15.6.1 **Definition** Let $\mathbb{F} = \mathbb{F}_q$ be the field of cardinality q. For a positive integer ℓ decomposition of $\mathbb{F}^\ell$ is defined recursively. Let $T_0 = \mathbb{F}^\ell, |T_0| = q^\ell$, and

$$T_0 = \bigcup_{a_0\in\mathbb{F}} T_1^{(a_0)} = \bigcup_{a_0\in\mathbb{F}}\bigcup_{a_1\in\mathbb{F}} T_2^{(a_0,a_1)} = \cdots = \bigcup_{a_0\in\mathbb{F}}\bigcup_{a_1\in\mathbb{F}}\cdots\bigcup_{a_{\ell-1}\in\mathbb{F}} T_\ell^{(a_0,a_1,\ldots,a_{\ell-1})},$$

where

$$\left|T_i^{(a_0,a_1,\ldots,a_{i-1})}\right| = q^{\ell-i},\ i = 1,\ldots,\ell.$$

15.6.2 **Remark** Each of the sets $T_\ell^{(a_0,a_1,\ldots,a_{\ell-1})}$ of the last level consists of a single vector from $\mathbb{F}^\ell$.

15.6.3 **Example** Let $q = 2$ and $\ell = 2$. Then

$$T_0 = \{00, 01, 10, 11\},$$

$$T_1^{(0)} = \{00, 11\},\ T_1^{(1)} = \{01, 10\},$$

$$T_2^{(0,0)} = \{00\},\ T_2^{(0,1)} = \{11\},\ T_2^{(1,0)} = \{01\},\ T_2^{(1,1)} = \{10\}.$$

15.6.4 **Remark** A decomposition may be defined by a linear transform, with $T_i^{(a_0,\ldots,a_{i-1})}$ being cosets of a linear code $T_i = T_i^{(0,\ldots,0)}$ spanned by the first i rows of a full-rank $\ell \times \ell$ matrix over $\mathbb{F}$.

15.6.5 **Definition** Let G be an $\ell \times \ell$ matrix with rows $\mathbf{g}_i \in \mathbb{F}^\ell$, $i = 0, 1, \ldots, \ell - 1$. Define

$$T_i^{(a_0,\ldots,a_{i-1})} = (a_0\mathbf{g}_0 + \cdots + a_{i-1}\mathbf{g}_{i-1}) + \bigcup_{a_i\in\mathbb{F}}\cdots\bigcup_{a_\ell\in\mathbb{F}} (a_i\mathbf{g}_i + \cdots + a_{\ell-1}\mathbf{g}_{\ell-1}).$$

15.6.6 **Example** The decomposition of Example 15.6.3 is linear, use

$$G = \begin{pmatrix} 0 & 1 \\ 1 & 1 \end{pmatrix}.$$

15.6.7 **Remark** Cosets of extended Reed-Solomon codes can be used for space decomposition when $\ell = q$ [2161].

15.6.8 **Example** Let α be a primitive element of $\mathbb{F}$, then a decomposition of $\mathbb{F}^q$ can be defined using the following matrix

$$G = \begin{pmatrix} 0 & 1 & 1 & 1 & \ldots & 1 \\ 0 & 1 & \alpha^{q-2} & \alpha^{2(q-2)} & \ldots & \alpha^{(q-2)(q-2)} \\ \vdots & \vdots & \vdots & \vdots & \vdots & \vdots \\ 0 & 1 & \alpha & \alpha^2 & \ldots & \alpha^{q-2} \\ 1 & 1 & 1 & 1 & \ldots & 1 \end{pmatrix}.$$

15.6.9 Definition Let

$$d_i = \min d\left(T_i^{(a_0,\ldots,a_{i-1})}\right),$$

where the minimum is taken over all possible choices of $a_0, \ldots, a_{i-1}$, and $d(\cdot)$ is the minimum Hamming distance of the code. Then the *distance hierarchy* of a space decomposition is the vector

$$\mathbf{d} = (d_0 = 1, d_1, \ldots, d_{\ell-1}).$$

15.6.10 Remark The distance hierarchy of a decomposition is a parameter used to determine the decay rate of decoding error probability.

15.6.11 Example The distance hierarchy of the decomposition from Example 15.6.3 is $(1, 2)$. The distance hierarchy of the decomposition from Example 15.6.8 is $(1, 2, 3, \ldots, q)$.

15.6.12 Definition A space decomposition is *proper* if for at least one vector $(a_0, \ldots, a_{\ell-1}) \in \mathbb{F}^\ell$,

$$d\left(T^{(a_0,\ldots,a_{\ell-1})}\right) \geq 2.$$

15.6.2 Vector transformation

15.6.13 Remark Given a space decomposition and a vector $\mathbf{a} = (a_0, \ldots, a_{\ell-1}) \in \mathbb{F}^\ell$, one may define a transform $g : \mathbb{F}^\ell \to \mathbb{F}^\ell$ associated to the decomposition as

$$g(\mathbf{a}) = T_\ell^{(a_0,a_1,\ldots,a_{\ell-1})}.$$

Recall that $T_\ell^{(a_0,a_1,\ldots,a_{\ell-1})}$ here is a vector from $\mathbb{F}^\ell$. An extended transform of vectors of lengths greater than ℓ can be defined as follows.

15.6.14 Definition Let $\mathbf{b} = (b_0, \ldots, b_{\ell s-1}) \in \mathbb{F}^{\ell s}$ be a vector, and B be the matrix of size $\ell \times s$,

$$B = \begin{pmatrix} b_0 & b_1 & \ldots & b_{s-1} \\ b_s & b_{s+1} & \ldots & b_{2(s-1)} \\ \vdots & \vdots & \vdots & \vdots \\ b_{(\ell-1)s} & b_{(\ell-1)s+1} & \ldots & b_{\ell s-1} \end{pmatrix} = (\mathbf{b}_0, \ldots, \mathbf{b}_{s-1}),$$

where $\mathbf{b}_0, \ldots, \mathbf{b}_{s-1}$ are the columns of B. Then

$$\hat{g}(\mathbf{b}) = (g(\mathbf{b}'_0), \ldots, g(\mathbf{b}'_{s-1})),$$

where $\mathbf{b}'_i$ is the transposition of $\mathbf{b}_i$.

15.6.15 Example Using the decomposition from Example 15.6.3 with $s = 4$ we have for $\mathbf{b} = (01011100)$,

$$B = \begin{pmatrix} 0 & 1 & 0 & 1 \\ 1 & 1 & 0 & 0 \end{pmatrix},$$

and

$$\hat{g}(\mathbf{b}) = (g(01), g(11), g(00), g(10)) = (11100001).$$

15.6.16 **Definition** An *ℓ-step transform* $\mathcal{G} : \mathbb{F}^{\ell^m} \to \mathbb{F}^{\ell^m}$ is defined as follows. Let $\mathbf{b}^{(0)} = \mathbf{b} \in \mathbb{F}^{\ell^m}$. At the i-th step of the transform, $i = 0, \dots, m-1$, the vector $\mathbf{b}^{(i)}$ is partitioned to the successive sub-vectors of length ℓ^{i+1},

$$\mathbf{b}^{(i)} = \left(\mathbf{b}_0^{(i)}, \mathbf{b}_1^{(i)}, \dots, \mathbf{b}_{\ell^{m-i-1}-1}^{(i)}\right),$$

where

$$\mathbf{b}_j^{(i)} = (b_{j\ell}^{(i)}, b_{j\ell+1}^{(i)}, \dots, b_{(j+1)\ell^{i+1}-1}^{(i)}),\ j = 0, \dots, \ell^{m-i-1} - 1.$$

Then

$$\mathbf{b}^{(i+1)} = \left(\hat{g}(\mathbf{b}_0^{(i)}), \hat{g}(\mathbf{b}_1^{(i)}), \dots, \hat{g}(\mathbf{b}_{\ell^{m-i-1}-1}^{(i)})\right).$$

The resulting vector of the transform is

$$\mathbf{c} = \mathcal{G}(\mathbf{b}) = \mathbf{b}^{(m)}.$$

15.6.17 **Example** Let $q = 2$, $\ell = 2$, $m = 3$ and the space decomposition is defined by Example 15.6.3. Then for $\mathbf{b} = \mathbf{b}^{(0)} = (01011100)$ we have the following sequence of results:

$$\mathbf{b}^{(1)} = (g(01), g(01), g(11), g(00)) = (11111000),$$

$$\mathbf{b}^{(2)} = (g(11), g(11), g(10), g(00)) = (10100100),$$

$$\mathbf{b}^{(3)} = (g(10), g(01), g(10), g(00)) = (01110100).$$

Thus,

$$\mathcal{G}(01011100) = (01110100).$$

15.6.18 **Definition** Let $n = \ell^m$, and $J \subseteq \{0, 1, \dots, n-1\}$, $|J| = n - k$. The *polar code* $\mathcal{C} \subseteq \mathbb{F}^{\ell^m}$ of size q^k is defined as

$$\mathcal{C} = \bigcup \mathcal{G}(\mathbf{b}),$$

where the union is taken over all q^k choices of $\mathbf{b} \in \mathbb{F}^n$ such that the components of $\mathbf{b}$ are set to zero in the coordinates having index belonging to J.

15.6.19 **Remark** The way to choose the set J will be discussed later.

15.6.20 **Lemma** If the space decomposition is linear, the polar code is a linear code.

15.6.3 Decoding

15.6.21 **Lemma** Encoding of a polar code requires m steps each having complexity linear in the length of the code. The complexity of encoding a polar code is $O(n \log n)$.

15.6.22 **Remark** Let $\mathbf{c} = \mathcal{G}(\mathbf{b})$ be transmitted over a memoryless channel. At the output of the channel we obtain for each of the n coordinates a set of q probabilities, one for each of the field elements, that has been transmitted at this position. Based on this we have to conclude what is the most likely code vector that had been transmitted.

15.6.23 **Remark** Decoding polar codes can be done recursively. The algorithm is called *Successive Cancelation*. The defined transform $\mathcal{G}$ implies a natural order of the elements of $\mathbf{b}$. The elements of $\mathbf{b}$ are processed in this order under the assumption that the previous elements of $\mathbf{b}$ have been determined. It can be shown that asymptotically in the length of the code a subset J of the elements have negligible probability of being wrongly decoded, while the rest of elements are correctly decoded only with probability tending to $1/q$. This allows to employ the following encoding: place the encoded information on the positions of J, while the values of the rest of the symbols are fixed to some prescribed value, e.g., to 0. The code rate, equal to the proportion $|J|/n$, asymptotically achieves the symmetric capacity of memoryless channels. A detailed description of the algorithm is given below.

15.6.24 Remark Noticing that the last step of the transform $\mathcal{G}$ is just a concatenation of transformations $\mathbf{g}$ of the transposed columns of a matrix of size $\ell \times \ell^{m-1}$, we may reconstruct the rows of the matrix row by row. Moreover, we use knowledge from the previously decoded rows.

15.6.25 Remark The last step of transform $\mathcal{G}$ is $\hat{g}(\mathbf{b}_0^{m-1})$, i.e., if

$$\mathbf{b}_0^{m-1} = (b_0^{m-1}, b_1^{m-1}, \ldots, b_{n-1}^{m-1}),$$

then it is a concatenation of the transformations of columns of the matrix

$$B^{m-1} = \begin{pmatrix} b_0^{m-1} & b_1^{m-1} & \cdots & b_{\ell^{m-1}-1}^{m-1} \\ b_{\ell^{m-1}}^{m-1} & b_{\ell^{m-1}+1}^{m-1} & \cdots & b_{2\ell^{m-1}-1}^{m-1} \\ \vdots & \vdots & \vdots & \vdots \\ b_{\ell^m-\ell}^{m-1} & b_{\ell^m-\ell+1}^{m-1} & \cdots & b_{\ell^m-1}^{m-1} \end{pmatrix} = \begin{pmatrix} a_{0,0} & a_{0,1} & \cdots & a_{0,n/\ell-1} \\ a_{1,0} & a_{1,1} & \cdots & a_{1,n/\ell-1} \\ \vdots & \vdots & \vdots & \vdots \\ a_{\ell-1,0} & a_{\ell-1,1} & \cdots & a_{\ell-1,n/\ell-1} \end{pmatrix}.$$

Given that we managed to decode the first $(i-1)$ rows, we may compute the q probabilities for the entries of the i-th row from the channel output as follows: the probability of $a_{i,j}$, $i = 0, \ldots, \ell-1$, $j = 0, \ldots, n/\ell-1$, to be $\beta \in \mathbb{F}$, is just the probability that in the j-th segment of length ℓ in the transmitted code word was a vector belonging to $T^{(a_{0,j},\ldots,a_{(i-2),j},\beta)}$, where $a_{0,j}, \ldots, a_{(i-2),j}$ are known from the previous decoded rows.

15.6.26 Remark The problem of decoding a polar code of length ℓ^m is thus reduced to ℓ decodings of polar codes of length ℓ^{m-1}, encoding the rows of the matrix. Decoding of each row could be split into ℓ decodings of codes of length ℓ^{m-1}, etc. Finally, we arrive at decodings of single symbols, being the entries of the initial vector $\mathbf{b}$. If this entry has index belonging to the set J, then it is zero, otherwise we may choose the most probable element of $\mathbb{F}$ as our decision. It was shown in [119] that when the rate of the code is less than the channel capacity, there exists a choice of set J allowing negligible probabilities of errors in the entries where we make a choice.

15.6.27 Theorem The complexity of the successive cancellation decoding is $O(n \log n)$.

15.6.28 Theorem Let the polar code be based on a proper space decomposition and g be the corresponding transform. Let $\mathbf{d} = (d_0, d_1, \ldots, d_{\ell-1})$ be its distance hierarchy. Then for any rate less than the capacity of the channel and growing code length n, the probability of decoding error decays as $O(q^{-n^{E(g)}})$, where the decomposition exponent $E(g)$ satisfies

$$E(g) \geq \frac{1}{\ell} \sum_{i=0}^{\ell-1} \log_\ell d_i.$$

15.6.29 Remark The theorem is a consequence of [122, 2161].

15.6.30 Example For the binary decomposition from Example 15.6.3, $E(g) = 0.5$. For the quaternary ($q = 4$, $\ell = 4$) decomposition from Example 15.6.8, $E(g) = 0.573120\ldots$.

15.6.4 Historical notes and other results

15.6.31 Remark Polar codes were proposed by Arikan [119] and provided a scheme for achieving the symmetric capacity of binary memoryless channels (BMC) with polynomial encoding and decoding complexity. The original construction by Arikan yields a binary code of block length $n = 2^m$, and a flexible rate.

15.6.32 Remark Different decompositions were considered in [1796, 2161, 2163, 2428, 2429, 2531, 2776]. For the binary case the decomposition from Example 15.6.3 gives the best exponent for all lengths up to 13, see [1796]. In [2428] non-linear decompositions of lengths 14, 15, and 16 based on partitions of a Hamming code to cosets of the Nordstrom-Robinson code, which in turn is partitioned to cosets of the first-order Reed-Muller codes, are described. These decompositions provide a better exponent than any linear ones. However, for linear decompositions the smallest ℓ for which the exponent is greater than 0.5 is 16, for which $E(g) = 0.51828$; see [1796]. In [2163], along with extended Reed-Solomon codes, nested families of algebraic-geometric codes are used to construct non-binary decompositions. In [1796] it was suggested to use nested families of BCH codes and codes achieving the Gilbert-Varshamov bound to construct efficient decompositions. It was shown that when ℓ increases, the best error exponent tends to 1. A construction of polar codes using several decompositions over different fields is proposed in [2429].

15.6.33 Remark As mentioned in [119, Section I.D] the notion of polar coding is strongly related to previous ideas in coding theory, such as multi-level coding and Reed-Muller codes. Another strong origin of polar coding is a previous paper by Arikan [117] where the channel combining and splitting were used to demonstrate that improvements can be obtained for the sum cutoff rate of some appropriate channels.

15.6.34 Remark In [118, 119, 1442, 2162, 2531] the problem of optimizing the choice of the information subset was considered for different channels.

15.6.35 Remark The tradeoff between the block length, the gap to capacity and the asymptotic decoding error probability was considered in [1795]. Decoding implementations were considered in [119, 120, 1906] A list decoding for polar codes was introduced in [2772]. Using polar codes in concatenated schemes was discussed in [181, 1814].

15.6.36 Remark Use of polar codes in other areas of information theory was considered in [2, 103, 121, 314, 753, 1520, 1685, 1794, 1797, 1993, 2532].

References Cited: [2, 103, 117, 118, 119, 120, 121, 122, 181, 314, 753, 1442, 1520, 1685, 1794, 1795, 1796, 1797, 1814, 1906, 1993, 2161, 2162, 2163, 2428, 2429, 2531, 2532, 2772, 2776]

16

Cryptography

16.1 Introduction to cryptography

Alfred Menezes, *University of Waterloo*

16.1.1 Goals of cryptography

The fundamental goals of cryptography are to provide the following information security services:

1. *Confidentiality*: Keeping data secret from all but those who are authorized to see it.
2. *Authentication*: Corroborating the source of data, and that the data has not been altered by unauthorized means.
3. *Non-repudiation*: Preventing the denial of a previous commitment to being the source of some data.

Cryptography has been used throughout the ages to *encrypt* (scramble) data, so that only the intended recipient can *decrypt* (descramble) thereby recovering the original data. Traditional encryption schemes can be classified as *symmetric-key* because the sender and intended recipient share secret keying material that can be used to encrypt as well as decrypt.

In 1975, Diffie, Hellman, and Merkle invented the concept of *public-key cryptography*, wherein each party A has a pair of keys — a *public key* that is available to everyone and a *private key* that is kept secret. Any party can use A's public key to encrypt a message for A in such a way that decryption requires knowledge of the corresponding private key, thus ensuring that only A can decrypt. Furthermore, A can use her private key to generate a message-dependent *signature* on a message in such a way that any user in possession of A's public key can verify that A indeed signed the message; this facilitates provision of non-repudiation services. Note that unlike the case with symmetric-key cryptography where communicating parties must possess shared secret keys, users of a public-key cryptosystem only need to possess authentic copies of each other's public keys thus simplifying the management and distribution of keying material.

Subsection 16.1.2 provides some examples of symmetric-key encryption schemes; a more extensive treatment can be found in Section 16.2. Subsection 16.1.3 introduces the RSA public-key encryption and signature schemes and some fundamental discrete logarithm-based cryptosystems. Discrete logarithm cryptosystems based on elliptic curves, hyperelliptic curves, and abelian varieties are covered in Sections 16.4, 16.5, and 16.6, respectively. Subsection 16.1.4 introduces the relatively new field of pairing-based cryptography. Finally, Subsection 16.1.5 describes a public-key encryption scheme that may be resistant to attacks by a quantum computer.

16.1.2 Symmetric-key cryptography

16.1.1 Definition Symmetric-key encryption schemes, also called *ciphers*, are usually classified as being a *stream cipher* in which encryption is performed one character (or bit) at a time, or a *block cipher* in which encryption is performed on a block of characters (or bits).

16.1.2 Remark Stream ciphers are generally preferred over block ciphers in applications where buffering is limited and message characters must be individually processed as they are received.

16.1.2.1 Stream ciphers

16.1.3 Example A classical example of a stream cipher is the *simple substitution cipher*. The secret key is a randomly selected permutation π of the English alphabet. An English *plaintext* message $m = m_1, m_2, m_3, \ldots$ (where each m_i is a letter of the English alphabet) is encrypted one letter at a time, to produce the *ciphertext* $c = c_1, c_2, c_3, \ldots$ where $c_i = \pi(m_i)$. Decryption is performed by applying the inverse permutation to the ciphertext letters: $m_i = \pi^{-1}(c_i)$. The total number of possible keys is $26! \approx 2^{88.4}$, which is sufficiently large that exhaustively searching through the set of all keys for the secret key is computationally infeasible. However, an adversary who has captured a sufficiently long ciphertext (say, of length 500) can easily use statistical knowledge of the English language (such as the relative frequency of letters) to determine the secret key π.

16.1.4 Example The *one-time pad* is a stream cipher that perfectly hides all statistical information that may be present in the plaintext message. The secret key is a randomly selected binary string $k = k_1, k_2, k_3, \ldots$. The plaintext is written as a binary string $m = m_1, m_2, m_3, \ldots$. Encryption is performed one bit at a time — the ciphertext is $c = c_1, c_2, c_3, \ldots$, where $c_i = m_i \oplus k_i$. Here, $\oplus$ denotes bitwise addition modulo 2. Decryption is similarly performed: $m_i = c_i \oplus k_i$. As shown by Shannon [2609], the one-time pad achieves *perfect secrecy* in the sense that an adversary — even one having infinite computational resources — who intercepts a ciphertext c is unable to determine anything whatsoever about the plaintext m except for its length.

16.1.5 Remark While perfectly secure, the one-time pad has a serious deficiency that the secret key must have the same length as the plaintext message. Note that the secret key cannot be reused to encrypt a second message. Indeed, if k is used to encrypt two plaintext messages m and m', then the resulting ciphertexts $c = m \oplus k$ and $c' = m' \oplus k'$ can be added together to obtain $c \oplus c' = m \oplus m'$, from which information about m and m' might be deduced.

16.1.6 Remark To overcome the aforementioned deficiency, stream ciphers have been designed that take an initial secret key k that is relatively small (e.g., 128 bits), and use a deterministic algorithm called a *keystream generator* KG to produce a much longer binary string $KG(k)$. The bits of $KG(k)$ are then added to the plaintext bits to produce ciphertext, in this way simulating the one-time pad. Since $KG(k)$ is no longer a truly random string, the stream cipher does not achieve perfect secrecy. However, the hope is that the bits of $KG(k)$ are "sufficiently random" so that a computationally bounded adversary will be unable to deduce any information about the plaintext from the ciphertext.

16.1.7 Remark Stream ciphers are further studied in Section 16.2.

16.1.2.2 Block ciphers

16.1.8 Definition A *block cipher* consists of a family of encryption functions $E_k : \{0,1\}^n \rightarrow \{0,1\}^n$ parameterized by an ℓ-bit key k. Each function in the family is invertible. The inverse of E_k is the decryption function D_k.

16.1.9 Remark If two parties wish to communicate securely, they first agree upon a secret key $k \in \{0,1\}^\ell$. Then, to transmit a message $m \in \{0,1\}^n$, a party computes $c = E_k(m)$ and sends c. The recipient computes $m = D_k(c)$.

16.1.10 Example *Feistel ciphers* [1048] are a general class of block ciphers. The parameters of a Feistel cipher are n (the block length), ℓ (the key length), and h (the number of rounds). The ingredients are a *key scheduling algorithm* that determines *subkeys* $k_1, k_2, \ldots, k_h$ from k, and *component functions* $f_i : \{0,1\}^{n/2} \rightarrow \{0,1\}^{n/2}$ for $1 \leq i \leq h$, where f_i depends on k_i.

A plaintext block $m \in \{0,1\}^n$ is encrypted as follows:

1. Write $m = (m_0, m_1)$, where $m_0, m_1 \in \{0,1\}^{n/2}$.

2. Compute $m_{i+1} = m_{i-1} \oplus f_i(m_i)$ for $i = 1, 2, \ldots, h$.
3. The ciphertext is $c = (m_h, m_{h+1})$.

The ciphertext block c is decrypted as follows:

1. Write $c = (m_h, m_{h+1})$.
2. Compute $m_{i-1} = m_{i+1} \oplus f_i(m_i)$ for $i = h, h-1, \ldots, 1$.
3. The plaintext is $m = (m_0, m_1)$.

16.1.11 Example The *Data Encryption Standard (DES)* is the most well-known example of a Feistel cipher [1068]. DES was developed in the mid 1970s and has been widely deployed in commercial applications. The DES parameters are $n = 64$, $\ell = 56$, and $h = 16$. The set of all possible keys has cardinality only 2^{56}, whereby exhaustive key search can be performed in a few hours on a network of workstations. Hence, DES is considered to be insecure today.

16.1.12 Example Although DES is no longer recommended in practice, *Triple-DES* is considered secure and widely deployed. In Triple-DES, the secret key is comprised of a triple $k = (k', k'', k''')$ of DES keys. A plaintext block $m \in \{0,1\}^n$ is encrypted as

$$c = \mathrm{DES}_{k'''}(\mathrm{DES}_{k''}(\mathrm{DES}_{k'}(m))),$$

where DES denotes the DES encryption function. The ciphertext block c can be decrypted as follows:

$$m = \mathrm{DES}_{k'}^{-1}(\mathrm{DES}_{k''}^{-1}(\mathrm{DES}_{k'''}^{-1}(c))),$$

where DES^{-1} denotes the DES decryption function. The secret key has bitlength 168 rendering exhaustive key search infeasible. Given a few plaintext-ciphertext pairs, the secret key can be recovered by a meet-in-the-middle attack that has running time approximately 2^{112} steps; however, this attack is considered infeasible in practice.

16.1.13 Example Block ciphers encrypt a long message n bits at a time. The drawback of this method is that identical plaintext blocks result in identical ciphertext blocks, and hence the ciphertext may leak information about the plaintext. To circumvent this weakness, long messages can be encrypted using the *cipher-block-chaining (CBC) mode* of encryption. A long message m is first broken up into blocks $m_1, m_2, \ldots, m_t$, each of bitlength n. Then, a random initialization vector $c_0 \in \{0,1\}^n$ is chosen, and $c_i = E_k(m_i \oplus c_{i-1})$ is computed for $i = 1, 2, \ldots, t$. The ciphertext is $c = (c_0, c_1, \ldots, c_t)$. Decryption is accomplished by computing $m_i = \mathrm{DES}_k^{-1}(c_i) \oplus c_{i-1}$ for $i = 1, 2, \ldots, t$.

16.1.3 Public-key cryptography

16.1.3.1 RSA

16.1.14 Remark The RSA encryption and signature schemes were introduced in a 1978 paper by Rivest, Shamir, and Adleman [2462].

16.1.15 Algorithm [RSA key generation] Each party does the following:

1. Randomly select two (distinct) primes p and q of the same bitlength.
2. Compute $n = pq$ and $\phi = (p-1)(q-1)$.
3. Select an arbitrary integer e, $1 < e < \phi$, with $\gcd(e, \phi) = 1$.
4. Compute the integer d, $1 < d < \phi$, with $ed \equiv 1 \pmod{\phi}$.
5. The party's public key is (n, e); her private key is d.

16.1.16 Remark The adversary's task of computing the private key d corresponding to a public key (n, e) can be shown to be equivalent to the problem of factoring n. As of 2012, factoring 1024-bit RSA moduli n is out of reach of the fastest integer factorization algorithms. However, 2048-bit moduli and 3072-bit moduli are recommended for medium- and long-term security.

16.1.17 Remark Presented next are the basic versions of the RSA encryption and signature schemes. In the signature scheme $H : \{0,1\}^* \to [0, n-1]$ is a cryptographic hash function (Remark 16.1.23).

16.1.18 Remark In what follows, $a = b \bmod n$ is understood to mean that a is the reminder of b when divided by n, and as usual, $a \equiv b \pmod{n}$ means a and b are congruent modulo n.

16.1.19 Algorithm (RSA encryption scheme) To encrypt a message $m \in [0, n-1]$ for party A, do the following:

1. Obtain an authentic copy of A's public key (n, e).
2. Compute $c = m^e \bmod n$.
3. Send c to A.

To decrypt c, party A does the following:

1. Compute $m = c^d \bmod n$.

16.1.20 Algorithm (RSA signature scheme) To sign a message m, party A with public key (n, e) and private key d does the following:

1. Compute $h = H(m)$.
2. Compute $s = h^d \bmod n$.
3. A's signature on m is the integer s.

To verify A's signature s on the message m, do the following:

1. Obtain an authentic copy of A's public key (n, e).
2. Compute $h = H(m)$.
3. Accept the signature if and only if $s^e \equiv h \pmod{n}$.

16.1.21 Remark The RSA encryption and signature schemes work because

$$(m^e)^d \equiv m \pmod{n}$$

for all $m \in [0, n-1]$, a property that can easily be verified using Fermat's little theorem.

16.1.22 Remark Security is based on the intractability of the problem of computing e-th roots modulo n. While it is clear that this problem is no harder than that of factoring n, the equivalence of the two problems has not been proven.

16.1.23 Remark A *hash function* $H : \{0,1\}^* \to \{0,1\}^\ell$ is an efficiently-computable function that meets some cryptographic requirements such as *one-wayness* given a randomly chosen element $h \in \{0,1\}^\ell$ it is computationally infeasible to find any $m \in \{0,1\}^*$ with $H(m) = h$) and *collision resistance* (it is computationaly infeasible to find distinct $m_1, m_2 \in \{0,1\}^*$ with $H(m_1) = H(m_2)$). Examples of commonly-used hash functions are SHA-1 and SHA-256 [1065].

16.1.3.2 Discrete logarithm cryptosystems

16.1.24 **Definition** Let G be a multiplicatively-written group of prime order n, and let g be a generator of G. The *discrete logarithm problem* in G to the base g is the following problem: Given $y \in G$, find the integer $x \in [0, n-1]$ such that $y = g^x$.

16.1.25 **Remark** The fastest generic algorithm known for solving the discrete logarithm problem is Pollard's rho algorithm [2413] which has a running time of $O(\sqrt{n})$ and its parallelization by van Oorschot and Wiener [2852].

16.1.26 **Example** The first example of a discrete logarithm cryptographic system was the *Diffie-Hellman key agreement protocol* [859]. The purpose of this protocol is to enable two parties A and B to agree upon a shared secret by exchanging messages over a communications channel whose contents are authenticated but not secret. Given a cyclic group $G = \langle g \rangle$ of order n, party A randomly selects an integer $a \in [1, n-1]$ and sends g^a to B. Similarly, party B randomly selects an integer $b \in [1, n-1]$ and sends g^b to A. Both parties can compute the shared secret $k = g^{ab}$. An eavesdropper is faced with the task of determining g^{ab} given g, g^a and g^b. This is the *Diffie-Hellman* problem, whose intractability is assumed to be equal to that of the discrete logarithm problem in G [2039].

16.1.27 **Example** ElGamal designed a closely-related scheme for public-key encryption [965]. In this scheme, party A's private key is a randomly selected integer $a \in [1, n-1]$ and her public key is the group element g^a. To encrypt a message $m \in G$ for A, a party selects a random integer $k \in [1, n-1]$, computes $c_1 = g^k$ and $c_2 = m(g^a)^k$, and sends (c_1, c_2) to A. Party A decrypts by computing $m = c_2/c_1^a$. The basic security requirement is that an adversary should be unable to compute m given the public key g^a and ciphertext (c_1, c_2). It is easy to see that the adversary's task is equivalent to solving an instance of the Diffie-Hellman problem.

16.1.28 **Remark** The main criteria for selecting a suitable group G for implementing a discrete logarithm cryptosystem are that (i) the group operation can be efficiently computed (so that cryptographic operations can be efficiently performed); and (ii) the discrete logarithm problem should be intractable. Over the years, several families of groups have been proposed for cryptographic use including subgroups of:

1. The multiplicative group of a finite field.
2. The group $E(\mathbb{F}_q)$ of $\mathbb{F}_q$-rational points on an elliptic curve E defined over a finite field $\mathbb{F}_q$ [1771, 2102].
3. The divisor class group of a genus-g hyperelliptic curve defined over a finite field $\mathbb{F}_q$ [1772].
4. The group of $\mathbb{F}_q$-rational points on an abelian variety defined over a finite field $\mathbb{F}_q$ (Section 16.6).
5. The class group of an imaginary quadratic number field [440].

16.1.29 **Remark** Public-key cryptosystems designed using elliptic curves, hyperelliptic curves, and abelian varieties are studied in Sections 16.4, 16.5, and 16.6, respectively. An example of a discrete logarithm cryptographic scheme that employs a multiplicative subgroup of a finite field is presented next.

16.1.3.3 DSA

16.1.30 **Remark** The digital signature algorithm (DSA) was proposed by the U.S. government's National Institute of Standards and Technology in 1991 [1067]. It was the first digital signature scheme to be recognized by a government.

16.1.31 Algorithm (DSA key generation) System parameters are (p, q, g), where p is a prime, q is a prime divisor of $p-1$, and $g \in \mathbb{Z}/p\mathbb{Z}$ has multiplicative order q. To generate a key pair, each party does the following:

1. Randomly select an integer a from $[1, q-1]$.
2. Compute $y = g^a \bmod p$.
3. The party's public key is y; her private key is a.

16.1.32 Algorithm (DSA signature scheme) To sign a message m, party A with system parameters (p, q, g), public key y, and private key a does the following:

1. Randomly select an integer $k \in [1, q-1]$.
2. Compute $r = (g^k \bmod p) \bmod q$.
3. Compute $h = H(m)$ and $s = k^{-1}(h + ar) \bmod q$.
4. A's signature on m is the pair of integers (r, s).

To verify A's signature (r, s) on the message m, do the following:

1. Obtain an authentic copy of the system parameters (p, q, g) and A's public key y.
2. Verify that $r \in [1, q-1]$ and $s \in [1, q-1]$.
3. Compute $h = H(m)$ and $w = s^{-1} \bmod q$.
4. Compute $u_1 = w \cdot h \bmod q$ and $u_2 = r \cdot w \bmod q$.
5. Compute $v = (g^{u_1} y^{u_2} \bmod p) \bmod q$.
6. Accept the signature if and only if $v = r$.

16.1.33 Remark A correctly-generated signature (r, s) on a message m will always be accepted because

$$g^{u_1} y^{u_2} \equiv g^{wh+war} \equiv g^{s^{-1}(h+ar)} \equiv g^k \pmod{p}.$$

16.1.34 Remark Security of DSA is based on the intractability of the discrete logarithm problem in the order-q multiplicative subgroup of $\mathbb{Z}/p\mathbb{Z}$. The fastest algorithms known for solving this problem are summarized in Section 11.6. As of 2012, DSA is considered to be secure if the bitlengths of the primes p and q are 1024 and 160, respectively. However, 2048-bit p and 224-bit q are recommended for medium-term security, and 3072-bit p and 256-bit q are recommended for long-term security.

16.1.4 Pairing-based cryptography

16.1.35 Definition Let $G_1 = \langle g_1 \rangle$, $G_2 = \langle g_2 \rangle$ and G_3 be multiplicatively-written groups of prime order n. A *pairing* on (G_1, G_2, G_3) is a map $e : G_1 \times G_2 \to G_3$ that satisfies the following three conditions:

1. *Bilinearity*: For all $r_1, r_2 \in G_1$ and $s_1, s_2 \in G_2$, $e(r_1 r_2, s_1) = e(r_1, s_1) \cdot e(r_2, s_1)$ and $e(r_1, s_1 s_2) = e(r_1, s_1) \cdot e(r_1, s_2)$.
2. *Non-degeneracy*: $e(g_1, g_2) \neq 1$.
3. *Computability*: $e(r, s)$ can be efficiently computed for all $r \in G_1$ and $s \in G_2$.

16.1.36 Remark Since 2000, pairings have been widely used to design cryptographic protocols that attain objectives not known to be achievable using conventional methods. The first such protocol was a one-round three-party key agreement scheme due to Joux [1624]. Recall that the Diffie-Hellman key agreement scheme is a two-party protocol where the two exchanged messages are independent of each other, and therefore can be simultaneously exchanged. Joux showed how pairings can be used to construct an analogous one-round key agreement scheme for three parties.

16.1.37 Example Suppose that e is a *symmetric pairing*, i.e., $G_1 = G_2$. In Joux's protocol, each of the three communicating parties A, B, C, randomly selects integers $a, b, c \in [1, n-1]$, and simultaneously broadcasts the group elements g_1^a, g_1^b, g_1^c, respectively. The shared secret is $k = e(g_1, g_1)^{abc}$ which party A, for example, can compute as $k = e(g_1^b, g_1^c)^a$. A passive adversary's task is to compute k given g_1, g_1^a, g_1^b and g_1^c. This problem is the *bilinear Diffie-Hellman problem*, and is assumed to be no easier than the discrete logarithm problems in G_1 and G_3.

16.1.38 Example A fundamental pairing-based protocol is the Boneh-Franklin identity-based encryption scheme [344]. The scheme has the feature that a party B can encrypt a message for a second party A using only A's identifier (such as A's email address) and some publically-available system parameters. Party A decrypts the message using a secret key that it must obtain from a trusted third party (TTP). Unlike symmetric-key cryptography, A and B do not have to share secret keying material. Also, unlike public-key cryptography, it is not necessary for A to have a public key before B can encrypt a message for A.

16.1.39 Remark A basic version of the Boneh-Franklin scheme is described next using symmetric pairings. The scheme uses a bilinear pairing e on (G_1, G_1, G_3) and two cryptographic hash functions $H_1 : \{0,1\}^* \to G_1$ and $H_2 : G_3 \to \{0,1\}^\ell$, where ℓ is the length of the message to be encrypted.

16.1.40 Algorithm (Boneh-Franklin identity based encryption) In the setup stage, a trusted third party (TTP) generates keying material for itself.

1. The TTP randomly selects an integer $t \in [1, n-1]$.
2. The TTP's public key is $T = g_1^t$ and its private key is t.

At any time, a party A with identifier ID_A can request its private key d_A from the TTP:

1. The TTP computes $d_A = H_1(ID_A)^t$ and securely delivers d_A to A.

To encrypt a message $m \in \{0,1\}^\ell$ for A, do the following:

1. Randomly select an integer $r \in [0, n-1]$.
2. Compute $R = g_1^r$ and $C = m \oplus H_2(e(H_1(ID_A), T)^r)$.
3. Send (R, C) to A.

To decrypt (R, C), party A does the following:

1. Obtain the private key d_A from the TTP.
2. Compute $m = C \oplus H_2(e(d_A, R))$.

16.1.41 Remark Decryption works because

$$e(d_A, R) = e(H_1(ID_A)^t, g_1^r) = e(H_1(ID_A), g_1^t)^r = e(H_1(ID_A), T)^r.$$

16.1.42 Remark Security is based on the hardness of the bilinear Diffie-Hellman problem.

16.1.43 Example Another fundamental pairing-based protocol is the Boneh-Lynn-Shacham (BLS) short signature scheme [346]. Unlike the DSA signature scheme (Subsection 16.1.3.3), BLS signatures have only one component and thus can be shorter if suitable parameters are chosen.

16.1.44 Remark The BLS signature scheme is described next using asymmetric pairings. The scheme uses a bilinear pairing e on (G_1, G_2, G_3) and a cryptographic hash function $H : \{0,1\}^* \to G_1^*$.

16.1.45 Algorithm (BLS key generation) To generate a key pair, each party does the following:

1. Randomly select an integer x from $[1, n-1]$.
2. Compute $X = g_2^x$.
3. The party's public key is X; her private key is x.

16.1.46 Algorithm (BLS signature scheme) To sign a message m, party A with public key X and private key x does the following:

1. Compute $M = H(m)$ and $S = M^x$.
2. A's signature on m is the group element S.

To verify A's signature S on the message m, do the following:

1. Obtain an authentic copy of A's public key X.
2. Verify that $S \in G_1^*$.
3. Compute $M = H(m)$.
4. Accept the signature if and only if $e(S, g_2) = e(M, X)$.

16.1.47 Remark A correctly-generated signature S on a message m will always be accepted because

$$e(S, g_2) = e(M^x, g_2) = e(M, g_2^x) = e(M, X).$$

16.1.48 Remark Security is based on the hardness of the following variant of the Diffie-Hellman problem: given $M \in G_1$ and $X \in G_2$, compute M^x where $X = g_2^x$.

16.1.49 Remark Pairings that are suitable for implementing Joux's key agreement scheme, the Boneh-Franklin identity-based encryption scheme, and the BLS short signature scheme can be constructed from the Weil and Tate pairings defined on certain elliptic curves over finite fields. For further details, see Section 16.4.

16.1.5 Post-quantum cryptography

16.1.50 Remark Shor [2623] showed that integer factorization and discrete logarithm problems can be efficiently solved on a quantum computer, thus rendering RSA and all discrete logarithm cryptosystems insecure. As of 2012, the feasibility of building large-scale quantum computers is far from certain. Nonetheless, cryptographers have been designing and analyzing public-key cryptosystems that potentially resist attacks by quantum computers, and which could serve as replacements to RSA and discrete logarithm cryptosystems in the event that large-scale quantum computers become a reality. Among these *post-quantum cryptosystems* are quantum key distribution [227] and conventional cryptosystems based on hash functions, error-correcting codes, lattices, and multivariate quadratic equations [245]. Cryptosystems based on multivariate quadratic equations are examined in Section 16.3. A code-based cryptosystem is described next.

16.1.51 Remark In 1978, McEliece introduced a public-key encryption scheme based on error correcting codes [2048]. The security of McEliece's scheme is based on the hardness of the general decoding problem, a problem that is known to be NP-hard [235].

16.1.52 Algorithm (McEliece key generation) Each party does the following:

1. Select a $k \times n$ generator matrix G for a t-error correcting binary (n,k)-code for which there is an efficient decoding algorithm.
2. Randomly select a $k \times k$ binary invertible matrix S.
3. Randomly select a $n \times n$ permutation matrix P.
4. Compute the $k \times n$ matrix $\widehat{G} = SGP$.
5. The party's public key is $(n,k,t,\widehat{G})$; her private key is (S,G,P).

16.1.53 Algorithm (McEliece encryption scheme) To encrypt a message $m \in \{0,1\}^k$ for party A, do the following:

1. Obtain an authentic copy of A's public key $(n,k,t,\widehat{G})$.
2. Randomly select an error vector z of length n and Hamming weight t.
3. Compute $c = m\widehat{G} + z$.
4. Send c to A.

To decrypt c, party A does the following:

1. Compute $\widehat{c} = cP^{-1}$.
2. Use the decoding algorithm for the code generated by G to decode $\widehat{c}$ to $\widehat{m}$.
3. Compute $m = \widehat{m}S^{-1}$.

16.1.54 Remark Decryption works because

$$\widehat{c} = cP^{-1} = (m\widehat{G} + z)P^{-1} = (mSGP + z)P^{-1} = (mS)G + zP^{-1}.$$

Since zP^{-1} is a vector of Hamming weight t, the decoding algorithm for the code generated by G decodes $\widehat{c}$ to $\widehat{m} = mS$, whence $\widehat{m}S^{-1} = m$.

16.1.55 Remark McEliece's original paper [2048] proposed using a Goppa code with parameters $n = 1024$, $n = 524$, and $t = 50$. However, it has recently been shown that the McEliece encryption scheme with these parameters is insecure [250]. Research is ongoing to determine parameter sets for which one can have a high confidence that the McEliece encryption scheme will remain resistant to both classical and quantum attacks for the forseeable future.

See Also

[1521], [1694], [2080], [2720]	Textbooks on cryptography.

References Cited: [227, 235, 245, 250, 344, 346, 440, 965, 1048, 1065, 1067, 1068, 1521, 1624, 1694, 1771, 1772, 2039, 2048, 2080, 2102, 2413, 2462, 2609, 2623, 2720, 2852]

16.2 Stream and block ciphers

Guang Gong, University of Waterloo
Kishan Chand Gupta, Indian Statistical Institute

16.2.1 Remark We present some algorithms for stream ciphers and block ciphers. In *stream cipher cryptography* a pseudorandom sequence of bits of length equal to the message length is generated by a pseudorandom sequence generator (PSG). This sequence is then bitwise XOR-ed (addition modulo 2) with the message sequence and the resulting sequence is transmitted. At the receiving end, deciphering is done by generating the same pseudorandom sequence and bitwise XOR-ing the cipher bits with this sequence. The seed for the pseudorandom bit generator is the secret key. A general algorithm for a pseudorandom sequence generator is based on a recursive relation over a finite field, a finite state machine in general and a feedback shift register sequence in particular, with a filtering function or some control units.

16.2.2 Remark In *block cipher cryptography*, the message bits are divided into blocks and each block is separately provided as an input to a permutation, i.e., encryption, using the same key and transmitted. A block cipher basically is a permutation of a finite field (or a finite ring), which is a composition of multiple permutations in a subfield (or subring) of the finite field (or the finite ring). Most modern day block ciphers are iterated ciphers and use substitution boxes (S-boxes) (i.e., permutations) as the nonlinear part in the scheme.

16.2.3 Remark We consider stream ciphers like RC4 [2001] and the WG stream cipher [2217], and block ciphers like RC6 [2461] and AES [762]. The aim is to explain the underlying ideas rather than describing complete solutions.

16.2.1 Basic concepts of stream ciphers

16.2.4 Remark Stream ciphers are a very important class of cryptographic primitives for encryption, authentication, and key derivation. The basic principle behind stream cipher encryption is simple.

16.2.5 Definition (One-time pad) Let z_t, for $t \geq 0$, be a *random* key bit sequence which is known to both the sender and the receiver. Suppose the sender wants to send a message bit sequence m_t. The cipher bit sequence is computed as $c_t = m_t \oplus z_t$, and transmitted to the receiver. The receiver knowing z_t, computes $m_t = c_t \oplus z_t$.

16.2.6 Remark This simple scheme provides the highest level of security, called *perfect secrecy.* It is unbreakable under the assumption that each key is used only for one encryption.

16.2.7 Remark The main problem with the one-time pad is that the key sequence is as long as the message sequence and for each encryption we need a new random key sequence which has to be shared by sender and receiver. This creates serious key management and key distribution problems. One remedy is to use a *pseudorandom sequence generator* (PSG) also known as *keystream generator.* A PSG is a deterministic algorithm which starts with a reasonably short random bit string (called a *seed*) and expands it into a very long bit string which is used as the keystream. The seed is the secret key shared between sender and receiver. The security of the stream cipher depends on the security of the PSG. Informally a PSG is *secure* if given a segment of the generated key bits it is hard to predict the next bit. Equivalently, it must be computationally very hard to distinguish the generated pseudorandom sequence from a random sequence.

16.2.8 Remark Most modern stream ciphers use an initialization vector (IV) which is not secret. The PRG is seeded by the (key, IV) pair. The same key may be used with distinct IVs and the constraint on the protocol usage is that a (key, IV) pair should not be repeated. Current stream ciphers have a similar structure which can be described by a finite state machine

(FSM). The Ecrypt home page [989] contains thorough information and may be referred to by anybody who is interested in the design and analysis of stream ciphers.

16.2.9 Definition (Security assumptions) It is assumed that the adversary knows everything except the secret key. This is known as *Kerckhoff's principle.* A few details are:

1. Algorithms in a stream cipher are public.
2. The only secret information in the system is the pre-shared key.
3. An attacker can intercept communications (ciphertext) among communicating entities.

16.2.10 Remark From Assumption 3, attackers can always obtain ciphertext. If an attacker manages to obtain a certain amount of the corresponding plaintext, then this portion of keystream is exposed. This is referred to as a *known plaintext attack.* Thus the security of stream cipher is reduced to randomness of PSG. The attacker's goal may be to recover the secret key or partial information about the secret key using a portion of the known keystream, i.e., using the known portion of the output of PSG.

16.2.11 Definition (The two phases in stream cipher) A stream cipher consists of two phases: one is the *key initialization phase*, for which the algorithm is *key initialization algorithm (KIA)*, and the other is *the PSG running phase* and the algorithm is PSG. Usually, the algorithms used in these two phases are similar.

16.2.12 Remark More specifically, KIA is the same as PSG without outputs or it may be a slightly different function. Figure 16.2.1 shows a general model of a stream cipher.

16.2.13 Remark In the initialization phase, a key initialization algorithm (KIA) is employed, which has two inputs, one is an initial vector (IV), which is public information, and the other is a secret key, k, which is a pre-shared encryption key. The goal of KIA is to scramble key bits with IV in order to get a complex nonlinear function of k and IV. The output of KIA is provided as an initial value to the PSG. KIA only executes once for each encryption session. After the key initialization, PSG starts to output a keystream which is used in encryption.

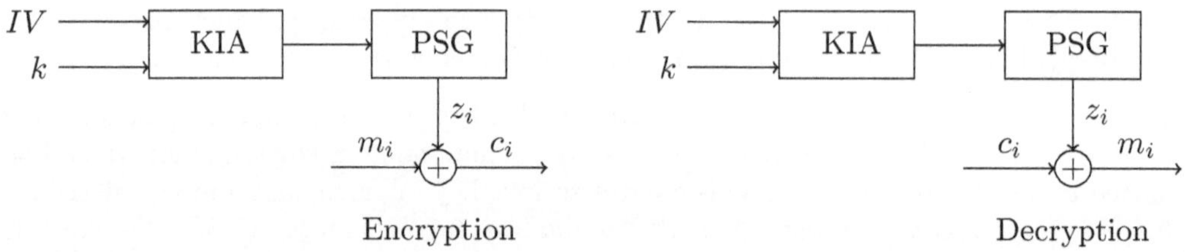

Figure 16.2.1 Two algorithms in stream ciphers: KIA and PSG.

16.2.14 Definition (Stream cipher modeled as a Finite State Machine (FSM)) In general, any PSG can be considered as a finite state machine (FSM) or some variant of it which may be defined as follows. Suppose $\mathcal{Y}$ and $\mathcal{Z}$ are finite fields (or finite rings) and the elements of $\mathcal{Z}$ are represented by m bits. An FSM is a 5-tuple (S_0, F, G, n, m) where $S_0 \in \mathcal{Y}^n$ is the initial state, $F : \mathcal{Y}^n \times \mathcal{Z} \to \mathcal{Y}^n$ is the state update function and $G : \mathcal{Y}^n \times \mathcal{Z} \to \mathcal{Z}$

is the output function. At time instant $t \geq 0$, state S_t is represented by an n-tuple $(s_t, s_{t+1}, \ldots, s_{t+n-1})$ where $s_i \in \mathcal{Y}$ and the output $z_t \in \mathcal{Z}$. The state update function and output function are given respectively by

$$S_{t+1} = F(S_t, z_t) \quad \text{and} \quad z_{t+1} = G(S_t, z_t).$$

16.2.15 Remark In modeling practical stream ciphers, in many cases it is seen that, for any S_t, $F(S_t, z_t)$ and $G(S_t, z_t)$ do not depend on z_t. In this situation we write

$$S_{t+1} = F(S_t) \quad \text{and} \quad z_{t+1} = G(S_t).$$

16.2.2 (Alleged) RC4 algorithm

16.2.16 Remark RC4 was designed by Rivest in 1987 and kept as a trade secret until it was leaked in 1994. It is widely used in Internet communications. In the open literature, RC4 is one of the very few proposed keystream generators that are not based on shift registers. A design approach of RC4 which has originated from the exchange-shuffle paradigm, is to use a relatively big array/table that slowly changes with time under the control of itself. For a detailed discussion on RC4 see the Master's thesis of Mantin [2001].

16.2.17 Definition RC4 has an N-stage register S, which holds a permutation of all $N = 2^n$ possible n-bit integers, where n is typically chosen as 8. The initial state is derived from a key (whose typical size is between 40 and 256 bits) by a Key-Scheduling Algorithm (KSA), i.e., Key initialization algorithm (KIA). The PSG is referred to as the *Pseudo-Random Generation Algorithm* (PRGA) in RC4.

16.2.18 Remark In what follows, $a = b \pmod{n}$ is understood to mean that b is the reminder of a when divided by n.

16.2.19 Definition (KSA of RC4) FSM for KSA for a given secret key K of length l bytes is a 4-tuple (Q_0, F_K, r, m). The secret key is used to scramble S by shuffling the words in S. Suppose $S = (x_0, x_1, \ldots, x_{N-1})$; we denote a state of RC4 by (i, j, S) or equivalently by $(i, j, x_0, x_1, \ldots, x_{N-1})$. Let $(i, j, x_0, x_1, \ldots, x_{N-1}) \in R$ be a state at some time instant and let $(e, d, y_0, y_1, \ldots, y_{N-1}) = F_K(i, j, x_0, x_1, \ldots, x_{N-1}) \in R$ be the next state. In the initialization process, i and j are initialized to 0, the identity permutation $(0, 1, \ldots, N-1)$ is loaded in the array S. Thus we have the initial state $Q_0 = (0, 0, 0, 1, 2, \ldots, 255)$ of KSA.

FSM for KSA:	(Q_0, F_K, r, m)
Parameters:	$N = 2^8 = 256$, $m = 8$, $r = N + 2 = 258$ and $R = \mathbb{Z}_N^{N+2}$.
Initial State:	$Q_0 = (0, 0, 0, 1, 2, \ldots, 255) \in R$
Input:	$K = (k_0, \ldots, k_{l-1})$, $k_s \in \mathbb{Z}_2^8$ for $s = 0, \ldots, l-1$.
State update function:	$F_K : R \to R$, given by $e = i + 1 \pmod{N}$, $d = (j + x_e + k_{e \pmod{l}}) \pmod{N}$, $y_d = x_e$, $y_e = x_d$ and $y_v = x_v$, for all $v \neq e$ or d.
Final State:	$F_K^{256}(Q_0) := (i, j, S)$ which will provide the initial state $I_0 = (0, 0, S)$ of PRGA.

16.2.20 Definition (PRGA of RC4.) RC4 keystream generator (PRGA) can also be represented as a finite state machine (I_0, F, G, r, m) where F is the state updating function and G is the output function. Let $(i, j, x_0, x_1, \dots, x_{N-1}) \in R$ be a state at some time instant and let $(e, d, y_0, y_1, \dots, y_{N-1}) = F(i, j, x_0, x_1, \dots, x_{N-1}) \in R$ be the next state. Figure 16.2.2 shows the state transition of PRGA.

FSM for PRGA:	(I_0, F, G, r, m)
Parameters:	$N = 2^8 = 256$, $m = 8$, $r = N + 2 = 258$ and $R = \mathbb{Z}_N^{N+2}$.
Initial state:	$I_0 \in R$ from KSA
State update function:	$F : R \to R$, given by $e = i + 1 \pmod N$, $d = j + x_e \pmod N$, $y_d = x_e$, $y_e = x_d$ and $y_v = x_v$, for all $v \neq e$ or d.
Output:	The output function is $G : R \to \mathbb{Z}_N$ and output is given by x_t where $t = x_e + x_d \pmod N$.

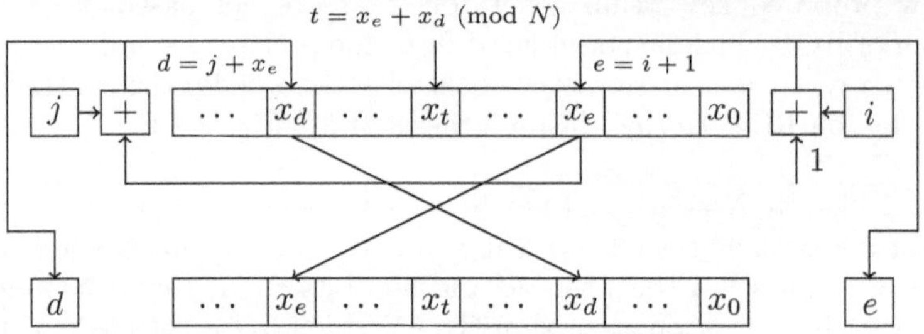

Figure 16.2.2 State transition of PRGA.

16.2.21 Remark (Attacks on RC4) RC4 has a huge internal state of $8 \times 258 = 2064$ bits. We observe that in RC4, the state update function is invertible. If the size of the internal state is s in bits ($s = (N + 2)(\log N) = 2064$ in RC4) and the next state update function is randomly chosen, then the average cycle length is about 2^{s-1} [1082]. However, it is hard theoretically to determine any randomness properties for RC4. Cryptanalysis of RC4 attracted a lot of attention in the cryptographic community after it was made public in 1994. Numerous significant weaknesses were discovered and notable weakness include weak initialization vectors, classes of weak keys, patterns that appear twice the expected number of times (the second byte bias), and biased distribution of RC4 initial permutation. Weaknesses in the key scheduling algorithm in RC4 led to a practical attack on the security protocol WEP. Currently, it has been proposed to use AES in WEP due to these weaknesses of RC4.

16.2.3 WG stream cipher

16.2.22 Remark We now introduce the WG stream cipher which was submitted to the eSTREAM project in 2005 by Nawaz and Gong [989]. The cipher is based on WG (Welch-Gong) transformations. WG cipher has desired randomness properties, like long periods, large linear complexity, two level autocorrelation and ideal t-tuple distribution. It is resistant to Time/Memory/Data tradeoff attacks, algebraic attacks, and correlation attacks. The cipher can be implemented with a small amount of hardware [1835].

16.2.23 Definition A WG cipher can be regarded as a nonlinear filter generator over an extension field, filtered by a WG transformation. As shown in Figure 16.2.3, it consists of a linear feedback shift register, followed by a WG permutation transform. The LFSR is based on an l degree primitive polynomial p over the finite field $\mathbb{F}_{2^m}$ given by, $p(x) = \sum_{i=0}^{l} c_i x^i, c_i \in \mathbb{F}_{2^m}$. The LFSR generates a maximal-length sequence (an m-sequence) over $\mathbb{F}_{2^m}$. This simple design generates a keystream whose period is $2^n - 1$, where $n = lm$, and it is easy to analyze various cryptographic properties of the generated keystream. The feedback signal Init is used only in the key initialization phase. In PSG running phase, the feedback is only from the LFSR. The output of the cipher is one bit. We denote a WG cipher with an LFSR of l stages over $\mathbb{F}_{2^m}$ as $WG(m, l)$.

16.2.24 Remark The version of the WG submitted to eSTREAM [989] is denoted by $WG(29, 11)$.

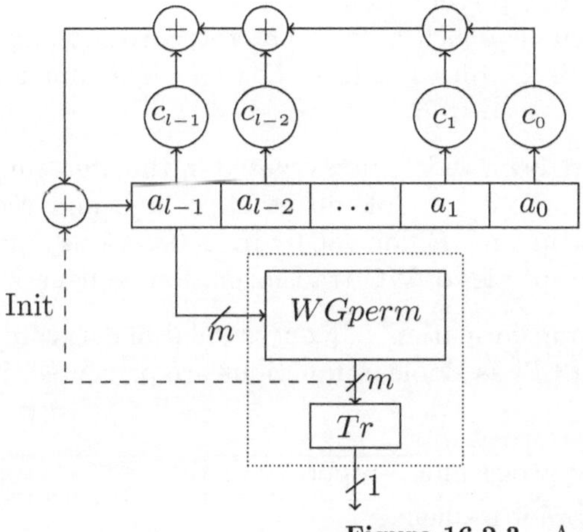

Update of LFSR

$$a_{k+l} = \begin{cases} \sum_{i=0}^{l-1} c_i a_{i+k} + WGperm(a_{k+l-1}), \\ \quad 0 \le k < 2l \text{ (in KIA phase)} \\ \\ \sum_{i=0}^{l-1} c_i a_{i+k}, k \ge 2l \text{ (in PSG)} \end{cases}$$

$$\text{Output} : s_k = WG(a_{k+l-1}), k \ge 2l$$

Figure 16.2.3 A diagram for WG ciphers.

16.2.25 Definition For $x \in \mathbb{F}_{2^m}$, $WGperm(x)$ and $WG(x)$ are defined by

$$\begin{aligned} WGperm(x) &= t(x+1)+1 \\ t(x) &= x + x^{r_1} + x^{r_2} + x^{r_3} + x^{r_4} \\ WG(x) &= Tr(WGperm(x)) \end{aligned}$$

where $r_1 = 2^k + 1, r_2 = 2^{2k} + 2^k + 1, r_3 = 2^{2k} - 2^k + 1$, and $r_4 = 2^{2k} + 2^k - 1$ where $3k \equiv 1 \pmod{m}$.

16.2.26 Remark Note that a WG transformation exists only if $m \not\equiv 0 \pmod 3$ (see [864]). In practice, we consider a value of m to be a reasonable choice for a WG cipher where $m \not\equiv 0 \pmod 3$ and either of the following holds: m is small enough to allow an efficient lookup table implementation of the permutation ($m \le 11$), or $m \equiv 2 \pmod 3$ and m has an *optimal normal basis* for efficient implementation in hardware; see Sections 5.3 and 16.7. The suitable values of m for $7 \le m \le 29$ are $7, 8, 10, 11, 23$, and 29.

16.2.27 Remark Here we compute exponents, i.e., r_i's from [864] so that t is a permutation polynomial. The exponents used in [2217] which are taken from [2294] are different. But WG sequences are identical for both representations.

16.2.28 Theorem [1303] The linear span of the WG cipher can be determined by the following formula

$$LS = m \times \sum_{i \in I} l^{w(i)} \text{ where } w(i) \text{ is the Hamming weight of } i,$$

$I = I_1 \cup I_2$ for $m = 3k-1$ where $I_1 = \{2^{2k-1} + 2^{k-1} + 2 + i : 0 \le i \le 2^{k-1} - 3\}$ and $I_2 = \{2^{2k} + 3 + 2i : 0 \le i \le 2^{k-1} - 2\}$; and $I = \{1\} \cup I_3 \cup I_4$ for $m = 3k-2$ where $I_3 = \{2^{k-1} + 2 + i : 0 \le i \le 2^{k-1} - 3\}$ and $I_4 = \{2^{2k-1} + 2^{k-1} + 2 + i : 0 \le i \le 2^{k-1} - 3\}$.

16.2.29 Remark The linear span of $WG(7, 23)$, $WG(8, 20)$, $WG(11, 16)$, and $WG(29, 11)$ are $2^{30}, 2^{33}, 2^{24}$, and 2^{45}, respectively. Here the linear spans for $m = 7$ and $m = 8$ are computed by using $WG(x^{-1})$ instead of $WG(x)$, because they have the same autocorrelation, but possess the highest algebraic degree, resulting in larger linear spans.

16.2.30 Remark (Resilient basis) The WG transformation form $\mathbb{F}_{2^m} \to \mathbb{F}_2$ can be regarded as a Boolean function in m variables. The exact Boolean representation depends on the basis used for computation in $\mathbb{F}_{2^m}$. The basis can be selected in such a way that the corresponding Boolean representation of WG transformation is 1-order resilient. For an algorithm for finding resilient bases, see [1317].

16.2.31 Remark The WG transformation sequences have been widely investigated in the literature on sequence design. These sequences were discovered by Golomb et al. [2294], and the randomness properties were proved by Dobbertin and Dillon [864]. In 2002, Gong and Youssef [1317] presented several cryptographic properties of WG transformation sequences.

16.2.32 Theorem The randomness properties of WG transformation sequences [864] and cryptographic properties of the WG transformations [1317] as Boolean functions are presented in Table 16.2.4.

Randomness Properties of Keystreams	Cryptographic Properties of WG
Period is $2^n - 1$	1-order resilient
Balanced	Algebraic degree $\lceil m/3 \rceil + 1$
Ideal 2-level autocorrelation	Nonlinearity $= 2^{m-1} - 2^{(m-1)/2}$ for odd m
Linear span increases exponentially with m, which can be determined exactly	Additive autocorrelation between $f(x+a)$ and $f(x)$ has three values: $0, \pm 2^{\frac{m+1}{2}}$
Ideal t-tuple distribution ($1 \le t \le l$)	1-order propagation property

Table 16.2.4 Randomness and cryptographic properties of WG.

16.2.33 Example (A concrete design of $WG(29, 11)$)

1. General description [2217]:
   a. synchronous stream cipher submitted in Profile 2 of eSTREAM (for hardware applications);
   b. key lengths of 80, 96, 112, and 128 bits;
   c. IVs of 32, 64 bits and also the same lengths as the key are allowed;
   d. estimated strength 2^{128} (exhaustive key search)
2. Parameters of $WG(29, 11)$ [2217]: The LFSR is of degree 11 and generates an m-sequence over the extension field $\mathbb{F}_{2^{29}}$. Then the elements of the m-sequence are filtered by a WG transformation: $\mathbb{F}_{2^{29}} \to \mathbb{F}_2$, to produce a keystream sequence. WG transformation operations can be implemented using the optimal normal

basis in the finite field $\mathbb{F}_{2^{29}}$. The parameters for implementation are listed in Table 2.

m	l	WG and Polynomials
29	11	$WG(x) = Tr(t(x+1)+1)$ where $t(x) = x + x^{r_1} + x^{r_2} + x^{r_3} + x^{r_4}$ and $r_1 = 2^{10}+1$, $r_2 = 2^{20}+2^{10}+1$, $r_3 = 2^{20}-2^{10}+1$, $r_4 = 2^{20}+2^{10}-1$
		$g(x) = x^{29}+x^{28}+x^{24}+x^{21}+x^{20}+x^{19}+x^{18}+x^{17}+x^{14}+x^{12}+x^{11}+x^{10}+x^7+x^6+x^4+x+1$ $p(x) = x^{11}+x^{10}+x^9+x^6+x^3+x+\gamma$ where $\gamma = \alpha^{464730077}$ and $g(\alpha)=0$

Table 16.2.5 Parameters for WG(29, 11) implementation using a optimal normal basis.

3. Key Initialization Phase for $WG(29,11)$: An initial state of the LFSR contains 319 bits where each register holds 29 bits. For a 128-bit key and 128-bit IV (initial vector), the rule for loading the LFSR is shown in Table 3. Here $x||y = (x_0,\ldots,x_{r-1},y_0,\ldots,y_{s-1})$ is the concatenation of two vectors $x = (x_0,\cdots,x_{r-1})$ and $y = (y_0,\ldots,y_{s-1})$. Once the LFSR has been loaded with the key and IV, the key stream generator is run for 22 clock cycles. This is the key-initialization phase of the cipher operation. During this phase the 29 bit vector of the output of the WG permutation is added to the feedback of the LFSR which is then used to update the LFSR.

Registers	29-bit Format
0,2,4,6,8	16 bits from the key||8 bits from IV||padding zeros
1,3,5,7,9	8 bits from the key||16 bits from IV||padding zeros
10	8 bits from the key||8 bits from IV||padding zeros

Table 16.2.6 Initializing the LFSR.

4. PSG phase: After running the KIA for $2l = 22$ clock cycles, PSG starts to give 1 bit output for each clock cycle. At this phase, the feedback to LFSR from the output of the WG permutation stops.

16.2.34 Remark Security against known attacks [2217].

1. Randomness and cryptographic properties of $WG(29,11)$ can be computed from Table 16.2.4.
2. Time/memory/data tradeoff attacks: Size of the internal state is 2^{319}. Thus this attack is not applicable.
3. Algebraic attacks: the number of linear equations is approximately $\binom{319}{11}$ and the attack has complexity approximately 2^{182}.
4. Correlation attacks: WG transformation is 1-order resilient, and nonlinearity is very high $2^{28}-2^{14}$. Thus the correlation between the WG transformation and any linear or affine function is very small.

16.2.35 Remark We note that the WG Stream Cipher was not selected for Phase III of the e-Stream Project since, although no attacks against WG were reported, the cipher is compromised if a relaxation of at most 2^{45} bits are generated from a single key. Also the hardware implementation seems to be larger than desirable; see [990]. However, the linear span can be increased up to at least $29 \times 11^{28} = 2^{101.722}$ using the decimation method as it is done for WG7 in [1982]. A hardware implementation has recently been reported in [1835], which shows different implementation methods and optimizations.

16.2.4 Basic structures of block ciphers

16.2.36 **Definition** A *block cipher* consists of a pair of encryption and decryption operators E and D. For a fixed key K, E maps an n bit plaintext $\mathbf{m} = (m_0, \ldots, m_{n-1})$ to an n bit ciphertext $\mathbf{c} = (c_0, \ldots, c_{n-1})$, namely *encryption*, and D maps the ciphertext back to the plaintext, i.e., *decryption.* In other words,

$$E : \mathcal{K} \times \mathbb{F}_2^n \to \mathbb{F}_2^n$$
$$(D \circ E)(\mathbf{m}) = \mathbf{m}$$

where $\mathbf{m}$ is an n-bit message, $\mathcal{K}$ is the set of possible keys and $\mathbb{F}_2^n$ is the set consisting of all n-bit vectors.

16.2.37 **Remark** E is an invertible function, i.e., E is a permutation of $\mathbb{F}_2^n$, and D is the inverse of E. The encryption operator E consists of several rounds which are applied to the plaintext one after another. The secret key K is expanded using a key scheduling algorithm into a set of *round keys* $K_1, \ldots, K_r$. Each round function takes as input the round key and the output of the previous round and produces an output. For a fixed round key, the encryption function is a bijective map.

16.2.38 **Remark** For a plaintext M, let $M_0 = M$ and M_{i-1} denote the input to the i-th round and let $M_r = C$ be the final output of $E_K(M)$. If we denote by R_i the i-th round function, then we have $M_i = R_i(K_i, M_{i-1})$. This reduces the task of designing a block cipher to the task of designing the round functions and the key scheduling algorithm. Usually the round functions are identical or very similar. Two standard methods for designing round functions are Feistel Structure and Substitution-Permutation Network (SPN).

16.2.39 **Definition** (Feistel Structure) [2080] The *Feistel Structure* is a feedback shift register with time varying feedback function. In other words, the round function is a time varying feedback function. For example, DES is of a Feistel structure, the input M_{i-1} to the i-th round is divided into two equal halves L_{i-1} and R_{i-1}, i.e., $M_{i-1} = L_{i-1}||R_{i-1}$. The output $M_i = (L_i, R_i)$ is defined as follows

$$L_i = R_{i-1}, \text{ and } R_i = L_{i-1} \oplus f(R_{i-1}, K_i).$$

16.2.40 **Remark** Note that for the invertibility of the round function, $f(\cdot, \cdot)$ need not be invertible. The security of the encryption algorithm depends on the design of $f(\cdot, \cdot)$ and the key scheduling algorithm. The block cipher RC6 and many other block ciphers have this structure. Here we consider RC6 as an example of a block cipher based on the Feistel structure.

16.2.41 **Definition** (Substitution-Permutation Network (SPN)) [2080] In an SPN, each round function consists of a few successive layers. The input to a substitution layer is divided into small blocks of bits say blocks of eight bits each. An S-box is applied to each block. Each S-box is a bijective map, so that entire substitution layer is also a bijective map.

The effect of a substitution layer is local in the sense that an output bit in a particular position depends only on a few of the input bits in its nearby positions. This local effect is compensated by having a permutation layer which permutes its input bits. The round key is usually incorporated at the beginning or at the end of the round function.

16.2.42 Remark We consider AES Rijndael as an example of a block cipher based on the SPN structure; see Subsection 16.2.6.

16.2.5 RC6

16.2.43 Remark This section is mainly from [2461]. RC6 is a symmetric key block cipher which encrypts 128-bit plaintext blocks to 128-bit ciphertext blocks and supports key sizes of 128, 192, and 256 bits. It was designed by Rivest, Robshaw, Sidney, and Yin to meet the requirements of the Advanced Encryption Standard (AES) competition [2461].

16.2.44 Remark In general RC6 is specified as RC6-$w/r/l$ where the word size is w bits, encryption consists of r rounds (generally r is 20), and l denotes the length of the encryption key in bytes.

16.2.45 Definition The encryption process involves three types of operations. Let $x = (x_0, \ldots, x_{w-1}) \in \mathbb{F}_2^w$ and $y = (y_0, \ldots, y_{w-1}) \in \mathbb{F}_2^w$ (note that x and y can be treated as binary vectors or binary numbers depending on the context).

1. Integer operations, i.e., $(x+y) \pmod{2^w}$, $(x-y) \pmod{2^w}$ and $(x \cdot y) \pmod{2^w}$.
2. Bitwise exclusive-or $x \oplus y$.
3. The rotation to the left and rotation to the right. Let L and R be the circular left and right shift operators respectively, i.e., $L(x) = (x_1, \ldots, x_{w-1}, x_0)$ and $R(x) = (x_{w-1}, x_0, \ldots, x_{w-2})$. We denote

$$
\begin{aligned}
(x \lll y) = L^y(x) &= (x_t, x_{t+1}, \ldots, x_{w-1}, x_0, x_1, \ldots, x_{t-1}), \\
(x \ggg y) = R^y(x) &= (x_{w-t}, \ldots, x_{w-1}, x_0, x_1, \ldots, x_{w-t-1}),
\end{aligned}
$$

where $t = \sigma(y_0, \ldots, y_{(\lg w)-1})$, $\lg w$ is the logarithm of w to the base 2, and $\sigma(y_0, \ldots, y_{(\lg w)-1})$ is the integer representation of binary string $(y_0, \ldots, y_{(\lg w)-1})$.

16.2.46 Remark The block cipher RC6 has the following features [2461]. It is fast and simple and the best attack on RC6 appears to be exhaustive key search.

16.2.47 Definition (Encryption and decryption) RC6 is of the Feistel structure. In details, RC6 is a feedback shift register with time varying feedback and four w-bit registers which contain the input plaintext (a_0, a_1, a_2, a_3) and the output ciphertext $(a_r, a_{r+1}, a_{r+2}, a_{r+3})$ at the end of encryption. Round keys $S_0, \ldots, S_{2r+3}$ are obtained from the key schedule algorithm, where each array element S_i is of w bits (see Definition 16.2.49 and Figure 16.2.7). The state updating is done as follows:

$$(a_i, a_{i+1}, a_{i+2}, a_{i+3}) = g_i(a_{i-1}, a_{(i-1)+1}, a_{(i-1)+2}, a_{(i-1)+3}), \text{ for } i = 1, \ldots, r,$$

where g_i is defined in the *for loop* of the Algorithm *Enc*(); see Figure 16.2.7.

16.2.48 Remark From the encryption algorithm, it can be easily seen that the process is invertible. The decryption algorithm is very similar to the encryption algorithm. It is a good exercise to write the decryption algorithm from the encryption algorithm. The input is $(a_r, a_{r+1}, a_{r+2}, a_{r+3})$, the round keys are in the reversed order $(S_{2r+3}, \ldots, S_0)$, $\lll$ is replaced by $\ggg$ and $+$ is replaced by $-$ in proper places of the encryption algorithm. See [2461] for a detailed description of the decryption algorithm.

Algorithm *Enc*() for RC6-$w/r/l$

Inputs : Plaintext (a_0, a_1, a_2, a_3),
number of rounds r,
round keys $S_0, \ldots, S_{2r+3}$.

Output : Ciphertext $(a_r, a_{r+1}, a_{r+2}, a_{r+3})$.

Procedure :
$a_1 = a_1 + S_0$
$a_3 = a_3 + S_1$
For $i = 1$ to r do
  $t = f(a_{(i-1)+1})$
  $u = f(a_{(i-1)+3})$
  $a_{i-1} = ((a_{i-1} \oplus t) \lll u) + S_{2i}$
  $a_{(i-1)+2} = ((a_{(i-1)+2} \oplus u) \lll t) + S_{2i+1}$
  The i-th state: $(a_i, a_{i+1}, a_{i+2}, a_{i+3})$
  $= (a_{(i-1)+1}, a_{(i-1)+2}, a_{(i-1)+3}, a_{i-1})$
End For
$a_r = a_r + S_{2r+2}$
$a_{r+2} = a_{r+2} + S_{2r+3}$
End Algorithm.
(Here $f(x) = x(2x+1) \lll \lg w$)

Key Schedule for RC6-$w/r/l$

Inputs : Secret key loaded in the array
$L_0, \ldots, L_{c-1}$, number of rounds r.

Output : Round keys $S_0, \ldots, S_{2r+3}$.

Procedure :
$S_0 = P_w$.
For $i = 1$ to $2r + 3$ do
  $S_i = S_{i-1} + Q_w$
End For
$A = B = i = j = 0$
$v = 3 \times \max\{c, 2r+4\}$
For $s = 1$ to v do
  $A = S_i = (S_i + A + B) \lll 3$
  $B = L_j = (L_j + A + B) \lll (A + B)$
  $i = (i + 1) \pmod{(2r + 4)}$
  $j = (j + 1) \pmod{c}$
End For
End Algorithm.

Figure 16.2.7 Encryption and key schedule algorithm for RC6 [2461].

16.2.49 Definition (Key schedule) The user supplies a key of l bytes, where $0 \le l \le 255$. Sufficient zero bytes are appended to give a key length equal to an integral number (say c) of words, and it is stored in $L_0, \ldots, L_{c-1}$. From this key, $2r + 4$ words are derived and stored in the array $S_0, \ldots, S_{2r+3}$. The constants $P_{32} = B7E15163$ and $Q_{32} = 9E3779B9$ (hexadecimal) are derived from the binary expansion of $e - 2$ (e is the base of natural logarithm) and $\phi - 1$ (ϕ is the Golden Ratio).

16.2.50 Remark Figure 16.2.7 gives a description of the key schedule algorithm.

16.2.6 Advanced Encryption Standard (AES) RIJNDAEL

16.2.51 Remark Some of the features of AES are as follows; see [762] for a more detailed description of AES. There are some differences between Rijndael and AES. Rijndael provides for several choices of block and key sizes. AES adopted only a subset of these parameter choices. Here we are ignoring these differences.

1. There are three allowable block lengths: 128, 192, and 256 bits.
2. There are three allowable key lengths (independent of selected block length): 128, 192, and 256 bits.

3. The number of rounds is 10, 12, or 14, depending on the key length.
4. Each round consists of three functions, which are in four "layers" as
   a. 8-bit inverse permutation (sub-byte transform),
   b. 32-bit linear transformation (mix columns operation),
   c. 128-bit permutation (shift rows operation) and
   d. round key addition.

16.2.52 Definition (High level description of AES) A plaintext m of 128 bits is the initial state which is represented as a four by four array of bytes (see Figure 16.2.8).

1. For a given plaintext m, the initial state is m. Perform an AddRoundKey operation which is XOR of the **RoundKey** with the initial **State**.
2. For each of the $r-1$ rounds perform a SubByte operation on **State** using an S-box; perform a ShiftRows on **State**; perform an operation called MixColumns on **State**; and perform an AddRoundKey operation.
3. For the r-th round, perform SubByte; perform ShiftRows and perform AddRoundKey.
4. The final **State** y is the ciphertext.

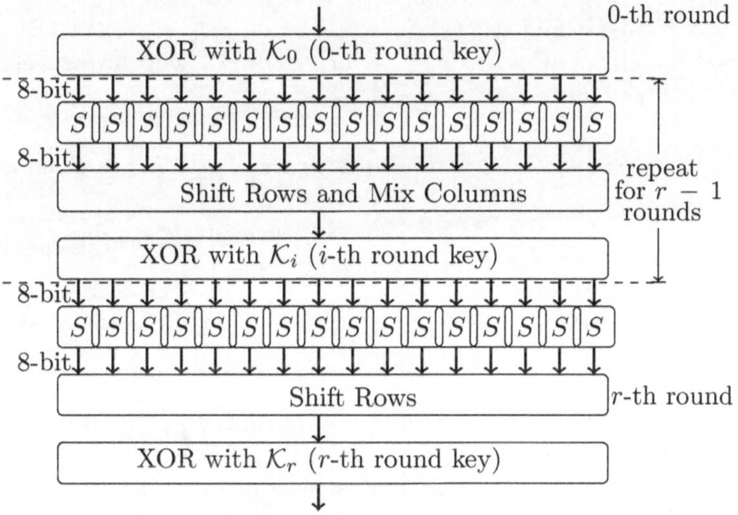

Figure 16.2.8 Round operations for AES Rijndael.

16.2.53 Remark (Algebraic structure of AES Rijndael) [762] Rijndael uses a finite field $\mathbb{F}_{2^8}$ defined by the primitive polynomial $p(x) = x^8 + x^4 + x^3 + x + 1$. Let α be a root of p, i.e., $p(\alpha) = 0$ in $\mathbb{F}_{2^8}$. We use classical polynomial representation and the elements of $\mathbb{F}_{2^8}$ are considered as a set consisting of all polynomials of degree less than or equal to 7 with coefficients from $\mathbb{F}_2$. So we can identify an element of $\mathbb{F}_{2^8}$ by an 8-bit vector.

16.2.54 Remark We introduce the following ring of matrices:

$$\mathcal{M}_4(\mathbb{F}_{2^8}) = \{X = (x_{ij})_{4\times 4} | x_{ij} \in \mathbb{F}_{2^8}\}.$$

In other words, for each matrix in $\mathcal{M}_4(\mathbb{F}_{2^8})$, the entries are taken from $\mathbb{F}_{2^8}$, i.e., each element of the matrix has 8-bit or one byte representation, and each row or column can be considered as a 32-bit word.

16.2.55 Remark For 128-bit version of Rijndael block cipher, a message M of 128 bits is parsed as 16 bytes and then further parsed as a 4 by 4 matrix:

$$M = (m_0, m_1, m_2, m_3, m_4, m_5, m_6, m_7, m_8, m_9, m_{10}, m_{11}, m_{12}, m_{13}, m_{14}, m_{15}),$$

where $m_i \in \mathbb{F}_{2^8}$, and the initial state $M_0 \in \mathcal{M}_4(\mathbb{F}_{2^8})$ is given as

$$M_0 = \begin{pmatrix} m_0 & m_4 & m_8 & m_{12} \\ m_1 & m_5 & m_9 & m_{13} \\ m_2 & m_6 & m_{10} & m_{14} \\ m_3 & m_7 & m_{11} & m_{15} \end{pmatrix}.$$

16.2.56 Remark Three basic operators for AES are SubByte, ShiftRow, and MixColumn.

16.2.57 Definition SubByte is a map $S : \mathcal{M}_4(\mathbb{F}_{2^8}) \to \mathcal{M}_4(\mathbb{F}_{2^8})$. Let $c = (1,1,0,0,0,1,1,0)$ and

$$A = \begin{pmatrix} 1 & 0 & 0 & 0 & 1 & 1 & 1 & 1 \\ 1 & 1 & 0 & 0 & 0 & 1 & 1 & 1 \\ 1 & 1 & 1 & 0 & 0 & 0 & 1 & 1 \\ 1 & 1 & 1 & 1 & 0 & 0 & 0 & 1 \\ 1 & 1 & 1 & 1 & 1 & 0 & 0 & 0 \\ 0 & 1 & 1 & 1 & 1 & 1 & 0 & 0 \\ 0 & 0 & 1 & 1 & 1 & 1 & 1 & 0 \\ 0 & 0 & 0 & 1 & 1 & 1 & 1 & 1 \end{pmatrix}.$$

For $y = (y_0, y_1 \ldots, y_7) \in \mathbb{F}_{2^8}$, define $T(y) = Ay^t + c^t$ and $\sigma(y) = y^{-1}$. So we have $T^{-1}(y) = A^{-1}(y^t + c^t)$ and $\sigma^{-1}(y) = y^{-1}$. For $X = (x_{ij}) \in \mathcal{M}_4(\mathbb{F}_{2^8})$, S is defined as $S(X) = (\mathcal{S}(x_{ij}))_{4\times 4}$ where $\mathcal{S}(x_{ij}) = (T \circ \sigma)(x_{ij})$. The inverse of S is given by $S^{-1}(X) = (\mathcal{S}^{-1}(x_{ij}))_{4\times 4} = ((\sigma^{-1} \circ T^{-1})(x_{ij}))_{4\times 4}$.

16.2.58 Definition ShiftRow transform R and its inverse on a state X are given as follows

$$R(X) = \begin{pmatrix} x_{00} & x_{01} & x_{02} & x_{03} \\ x_{11} & x_{12} & x_{13} & x_{10} \\ x_{22} & x_{23} & x_{20} & x_{21} \\ x_{33} & x_{30} & x_{31} & x_{32} \end{pmatrix}, \quad R^{-1}(X) = \begin{pmatrix} x_{00} & x_{01} & x_{02} & x_{03} \\ x_{13} & x_{10} & x_{11} & x_{12} \\ x_{22} & x_{23} & x_{20} & x_{21} \\ x_{31} & x_{32} & x_{33} & x_{30} \end{pmatrix}.$$

16.2.59 Definition MixColumn transform L is a linear transform on $\mathbb{F}_{2^8}^4$. Recall that α is a root of $p(x)$ in $\mathbb{F}_{2^8}$. Given a state X, $L(X) = LX$ where LX is the matrix multiplication of L and X over $\mathbb{F}_{2^8}$. The linear transform L and its inverse are given as follows

$$L = \begin{pmatrix} \alpha & 1+\alpha & 1 & 1 \\ 1 & \alpha & 1+\alpha & 1 \\ 1 & 1 & \alpha & 1+\alpha \\ 1+\alpha & 1 & 1 & \alpha \end{pmatrix}, \quad L^{-1} = \begin{pmatrix} \beta_0 & \beta_3 & \beta_2 & \beta_1 \\ \beta_1 & \beta_0 & \beta_3 & \beta_2 \\ \beta_2 & \beta_1 & \beta_0 & \beta_3 \\ \beta_3 & \beta_2 & \beta_1 & \beta_0 \end{pmatrix},$$

where $\beta_0 = \alpha^3 + \alpha^2 + \alpha$, $\beta_1 = \alpha^3 + 1$, $\beta_2 = \alpha^3 + \alpha^2 + 1$, and $\beta_3 = \alpha^3 + \alpha + 1$.

16.2.60 Definition Composition of operators is defined as follows:

$G(X) = (R \circ S)(X)$	$G^{-1}(X) = (S^{-1} \circ R^{-1})(X)$
$H(X) = (L \circ G)(X)$	$H^{-1}(X) = (G^{-1} \circ L^{-1})(X)$

16.2.61 Remark The total number of round key bits is equal to the block length times the number of rounds plus 1. The 128-bit version needs 10 rounds. Thus 1408 bits, or 44 words of round key bits are needed. Thus, the key schedule should extend a 128-bit key to round keys, a total of 1408 key bits. Let $\{\mathbf{k}_i\}_{i=0}^{43}$ be a sequence of words, $\mathbf{k}_i \in \mathbb{F}_{2^8}^4$, which consists of 4 bytes. Let $(\mathbf{k}_0, \mathbf{k}_1, \mathbf{k}_2, \mathbf{k}_3)$ be the 128-bit session key. The sequence $\{\mathbf{k}_i\}$ is used as the round keys. The expansion of the key is shown in Table 16.2.9.

Input:	$(\mathbf{k}_0, \mathbf{k}_1, \mathbf{k}_2, \mathbf{k}_3), \mathbf{k}_j \in \mathbb{F}_{2^8}^4$, 128-bit key.
Output	$\mathcal{K}_i = (\mathbf{k}_{4i}, \mathbf{k}_{4i+1}, \mathbf{k}_{4i+2}, \mathbf{k}_{4i+3}), i = 0, 1, \ldots, 10$, the i-th round key where each $\mathbf{k}_j$ is a 32-bit word.
Procedure.	For $i = 1, 2, \ldots, 10$, compute $\mathbf{k}_{4i} = \mathbf{k}_{4i-4} + (S(k_{4i-1,3}) + \alpha^{i-1}, S(k_{4i-1,0}), S(k_{4i-1,1}), S(k_{4i-1,2}));$ $\mathbf{k}_{4i+j} = \mathbf{k}_{4i+j-4} + \mathbf{k}_{4i+j-1}$, in $F_{2^8}^4$, $j = 1, 2, 3$.
Return	$\mathcal{K}_i = (\mathbf{k}_{4i}, \mathbf{k}_{4i+1}, \mathbf{k}_{4i+2}, \mathbf{k}_{4i+3}), i = 0, 1, \ldots, 10$

Table 16.2.9 Key scheduling, where $\mathbf{k}_i = (k_{i0}, k_{i1}, k_{i2}, k_{i3}), i = 0, 1, \ldots, 43$ and $k_{ij} \in \mathbb{F}_{2^8}$.

16.2.62 Definition (Rijndael encryption and decryption) For 128-bit version of Rijndael, a message M of 128 bits is parsed as a 4 by 4 matrix. The number of rounds is equal to 10. The process of computation of the cipher C (again written as a 4 by 4 matrix) is shown in Table 16.2.10. When viewed as an FSM, the initial state of the Rijndael block cipher is M_0, the final state is M_{10} (which is the output, i.e., ciphertext) and the state update function is a map : $\mathbb{F}_{2^8}^{16} \to \mathbb{F}_{2^8}^{16}$ as shown in Table 16.2.10.

Encryption	Decryption
$M_0 = M + \mathcal{K}_0$, in $\mathcal{M}_4(\mathbb{F}_{2^8})$	$C_0 = C + \mathcal{K}_{10}$,
$M_i = H(M_{i-1}) + \mathcal{K}_i, 1 \leq i \leq 9$	$C_1 = G^{-1}(C_0) + \mathcal{K}_9$
$M_{10} = G(M_9) + \mathcal{K}_{10}$	$C_i = H^{-1}(C_{i-1}) + \mathcal{K}_{10-i}, 2 \leq i \leq 10$
The ciphertext is $C = M_{10}$.	The plaintext is $M = C_{10}$.

Table 16.2.10 Encryption and decryption processes of Rijndael.

See Also

§16.1	For goals of cryptography and symmetric-key cryptography.
[762]	For a thorough description of the design of Rijndael: AES.
[864], [1303], [1317]	For cryptographic properties of the WG stream cipher.
[989]	Contains information about the design and analysis of stream ciphers.
[1835]	For hardware implementations of the WG stream cipher.
[2001]	For information on RC4.
[2217]	Develops the theory of WG stream cipher.
[2461]	For information on the RC6 block cipher.

References Cited: [762, 864, 989, 990, 1082, 1303, 1317, 1835, 1982, 2001, 2080, 2217, 2294, 2461]

16.3 Multivariate cryptographic systems

Jintai Ding, University of Cincinnati

16.3.1 Remark Due to limited space only a few key areas more directly related to the theory of finite fields are covered. For a more complete reference, readers should consult [894, 882]. This section grows out of [894] but with new materials from the last two years added.

16.3.2 Remark Multivariate public key cryptosystems are motivated by the need to develop new cryptosystems that have the potential to resist future quantum computer attacks. In addition, multivariate public key cryptosystems are also motivated by the need to develop efficient public key cryptosystems that could be used in small computing devices with limited computing and memory capacities like sensors, radio-frequency identification (RFID) tags, and other similar small devices.

16.3.3 Remark The foundation of any public key cryptosystem is a class of "trapdoor one-way functions." The fundamental mathematical structure of such a class of functions determines all the basic characteristics of a public key cryptosystem. In the case of multivariate (public-key) cryptosystems (MPKCs), the trapdoor one-way function is usually in the form of a multivariate quadratic polynomial map over a finite field.

16.3.4 Definition For a MPKC, the public key is, in general, given by a set of quadratic polynomials:

$$\mathcal{P}(x_1,...,x_n) = (p_1(x_1,\dots,x_n),\dots,p_m(x_1,\dots,x_n)),$$

where each p_i is a (usually quadratic) nonlinear polynomial in $X=(x_1,\dots,x_n)$:

$$y_k = p_k(X) := \sum_i P_{ik}x_i + \sum_i Q_{ik}x_i^2 + \sum_{i>j} R_{ijk}x_ix_j + S_k. \tag{16.3.1}$$

with all coefficients and variables in $\mathbb{F}_q$.

16.3.5 Remark The evaluation of these polynomials corresponds to either the encryption procedure or the verification procedure.

16.3.6 Remark Most of the constructions are quadratic constructions due to the consideration of the efficiency of encryption determined by the key size.

16.3.7 Remark Inverting a multivariate quadratic map is generally equivalent to solving a set of quadratic equations over a finite field, or the following multivariate quadratic (MQ) problem.

16.3.8 Definition (MQ problem) Solve a system $p_1(X)=p_2(X)=\cdots=p_m(X)=0$, where each p_i is a quadratic polynomial in $X=(x_1,\dots,x_n)$. All coefficients and variables are in $\mathbb{F}_q$.

16.3.9 Remark MQ is an NP-complete problem, which are believed to be hard generically. A set of random quadratic polynomials cannot be a trapdoor. Since one does not deal with "random" or "generic" systems, but systems using specific trapdoors, the security MPKCs is then not guaranteed by the NP-hardness of the MQ problem, and effective attacks may exist for any chosen trapdoor. The history of MPKCs therefore evolves as we understand better about how to design efficient secure multivariate trapdoors.

16.3.10 Remark The mathematical structure behind a system of polynomial equations is the ideal generated by those polynomials. Multivariate cryptography is based mainly on mathematics that handles polynomial ideals, namely algebraic geometry but over a finite field. In contrast, the security of RSA-type cryptosystems relies on the hardness of integer factorization and is based on number theory developed in the 17th and 18th centuries. Elliptic curve cryptosystems employ the mathematical theory developed in the 19th century. This is a remark from Whitfield Diffie at the RSA Europe conference in Paris in 2002. Algebraic geometry, the mathematics that MPKCs depend on, was developed in the 20-th century.

16.3.11 Remark This section is organized as follows: Subsection 16.3.1 provides a sketch of how MPKCs work in general; Subsection 16.3.2 describes the known trapdoor constructions in more detail; Subsection 16.3.3 describes the most important modes of attacks; the last subsection is about future research directions in this area.

16.3.1 The basics of multivariate PKCs

16.3.12 Remark After Diffie-Hellman [860], cryptographers proposed many trapdoor functions. The earliest published proposal of MPKC schemes seemed to have arisen in Japan [2029, 2822, 2823] in the early 1980s. These papers were published in Japanese, and remained largely unknown outside Japan.

16.3.13 Remark The first article in English describing a public key cryptosystem with more than one independent variable may be the one from Ong *et al* [2320], and the first use of more than one equation is by Fell and Diffie [1049]. The earliest attempt bearing some resemblance to today's MPKCs (with 4 variables) seems to be [2029]. In 1988, the first MPKC in the current form appeared in [2028], and the basic construction described below (Subsection 16.3.1.1) has not really changed much since.

16.3.1.1 The standard (bipolar) construction of MPKCs

16.3.14 Definition A usual MPKC has a private map $\mathcal{Q}$, which is the *central map* and it belongs to a certain class of quadratic maps each of which can be efficiently inverted.

16.3.15 Definition The basic construction of the public key is derived via composition with two affine maps S, T.

$$\mathcal{P} = T \circ \mathcal{Q} \circ S : \mathbb{F}_q^n \to \mathbb{F}_q^m,$$

where the maps S, T are affine (sometimes linear) invertible maps on $\mathbb{F}_q^n$ and $\mathbb{F}_q^m$, respectively.

16.3.16 Remark The purpose of T and S is to hide the trap door $\mathcal{Q}$. The key of a MPKC is indeed the design of the central map.

16.3.17 Remark

The public key consists of the polynomials in $\mathcal{P}$.
The secret key consists of the information in S, T, and $\mathcal{Q}$.
To verify a signature or to encrypt a block, one simply computes

$$Y = \mathcal{P}(X),$$

where $Y = (y_1, \dots, y_m)$ and $X = (x_1, \dots, x_n)$.
To sign or to decrypt a block, one computes

$$X = P^{-1}(Y) = S^{-1} \circ \mathcal{Q}^{-1} \circ T^{-1}(Y),$$

which is computed via the composition factors in turn. Notice that by the inverse of a map here, we mean finding one of possibly many pre-images, not necessarily an inverse function in the strict mathematical sense.

16.3.18 Remark The basics of MPKCs are provided below so that the reader has a basic sense about how these schemes can work in practice:

Cipher block or message digest size: m elements of $\mathbb{F}_q$;
Plaintext block or signature size: n elements of $\mathbb{F}_q$;
Public key size: $mn(n+3)/2$ elements of $\mathbb{F}_q$;
Secret key size: Usually $(n^2 + m^2 + [\text{size of } \mathcal{P}])$ elements of $\mathbb{F}_q$;
Secret map time complexity: $(n^2 + m^2)$ $\mathbb{F}_q$-multiplications, plus the time it is needed to invert $\mathcal{Q}$;
Public map time complexity: About $mn^2/2$ $\mathbb{F}_q$-multiplications.

16.3.19 Remark In terms of computational complexity, MPKCs usually have strong advantages as we shall see below. But a disadvantage with MPKCs is that their keys are large compared to number-theory-based systems like RSA or ECC. For example, the public key size of RSA-2048 is not much more than 2048 bits, but a current version of the Rainbow signature scheme has $n = 42$, $m = 24$, $q = 256$, i.e., the size of the public key is 22,680 bytes.

16.3.20 Remark There are other alternative forms in which multivariate polynomials can be used for public key cryptosystems, as we discuss next.

16.3.1.2 Implicit form MPKCs

16.3.21 Definition The public key of an *implicit form* MPKC is a system of l equations:

$$P(W, Z) = P(w_1, \ldots, w_n, z_1, \ldots, z_m) = (p_1(W, Z), \ldots, p_l(W, Z)) = (0, \ldots, 0), \quad (16.3.2)$$

where each p_i is a polynomial in $W = (w_1, \ldots, w_n)$ and $Z = (z_1, \ldots, z_m)$. This P is built from the secret $\mathcal{Q}$:

$$\mathcal{Q}(X, Y) = q(x_1, \ldots, x_n, y_1, \ldots, y_m) = (q_1(X, Y), \ldots, q_l(X, Y)) = (0, \ldots, 0),$$

where $q_i(X, Y)$ is polynomial in $X = (x_1, \ldots, x_n)$, $Y = (y_1, \ldots, y_m)$ such that

1. for any given specific element X', we can easily solve the equation
$$\mathcal{Q}(X', Y) = (0, \ldots, 0), \quad (16.3.3)$$
2. for any given specific element Y', we can easily solve the equation
$$\mathcal{Q}(X, Y') = (0, \ldots, 0), \quad (16.3.4)$$
3. Equation (16.3.3) is linear and Equation (16.3.4) is nonlinear but can be solved efficiently.

16.3.22 Remark The public key is built as

$$P = L \circ \mathcal{Q}(S(W), T^{-1}(Z)) = (0, \ldots, 0),$$

where S, T are invertible affine maps and L is linear.

16.3.23 Remark To verify a signature W with the digest Z, one checks that $P(W,Z) = 0$. If one wants to use P to encrypt the plaintext W, one would solve $P(W,Z) = (0,\dots,0)$, and find the ciphertext Z. To invert (i.e., to decrypt or to sign) Z, one first calculates $Y' = T^{-1}(Z)$, then substitutes Y' into the Equation (16.3.4) and solves for X. The final plaintext or signature is given by $W = S^{-1}(X)$.

16.3.24 Remark In an *implicit-form MPKC*, the public key consists of the l polynomial components of P and the field structure of $\mathbb{F}$. The secret key mainly consists of L, S, and T. Depending on the case, the equation $\mathcal{Q}(X,Y) = (0,\dots,0)$ is either known or has parameters which are a part of the secret key.

16.3.25 Remark The maps S, T, L serve to hide the equation $\mathcal{Q}(X,Y) = 0$, which otherwise could be easily solved for any Y. Mixed schemes are relatively rare, one example being Patarin's Dragon [2365].

16.3.1.3 Isomorphism of polynomials

16.3.26 Remark The isomorphism of polynomials problem originated from trying to attack MPKCs by finding the secret keys.

16.3.27 Definition Let $\bar{F}_1, \bar{F}_2$ with

$$\bar{F}_i(x_1,\dots,x_n) = (\bar{f}_{i1},\dots,\bar{f}_{im}), \tag{16.3.5}$$

be two polynomial maps from $\mathbb{F}^n$ to $\mathbb{F}^m$. The *Isomorphism of Polynomials (IP) problem* is to find two invertible affine linear transformations S on $\mathbb{F}^n$ and T over $\mathbb{F}^m$ (if they exist) such that

$$\bar{F}_1(x_1,\dots,x_n) = T \circ \bar{F}_2 \circ S(x_1,\dots,x_n). \tag{16.3.6}$$

16.3.28 Remark The system based on the IP problem was first proposed by Patarin [2366], where the verification process is performed by showing the equivalence (or isomorphism) of two different maps. A simplified version is the isomorphism of polynomials with one secret (IP1s) problem, where we only need to find the map S (if it exists), while the map T is known to be the identity map. This problem is used to build identification schemes [1044, 1266, 1914, 2371, 2387].

16.3.29 Remark Mathematically, this problem can be viewed from the perspective of the problem of classification of quadratic maps from $\mathbb{F}_q^n$ to $\mathbb{F}_q^m$ under the action of the group $GL_n(\mathbb{F}_q) \times GL_m(\mathbb{F}_q)$, namely to describe precisely the orbit space of all the quadratic maps from $\mathbb{F}_q^n$ to $\mathbb{F}_q^m$ under the action of the group $GL_n(\mathbb{F}_q) \times GL_m(\mathbb{F}_q)$; a very hard mathematical problem we do not know much about, except the case when $m = 1$, which is the classification of quadratic (or bilinear) forms, a problem we know very well.

16.3.30 Remark Since the vast majority of MPKCs are in standard form, we deal with mainly such systems in the rest of this section.

16.3.2 Main constructions and variations

16.3.2.1 Historical constructions

16.3.31 Remark The first attempt to construct a multivariate signature [2320, 2321] utilizes a quadratic equation with two variables.

$$y \equiv x_1^2 + \alpha x_2^2 \pmod{n}, \tag{16.3.7}$$

where $n = pq$ is an RSA modulus, a product of two large primes. The public key is essentially the integer n and Equation (16.3.7). Since the security is supposed to be based on the factorization of n, this system can really be viewed as a derivative of RSA, though it indeed initiated the idea of multivariate cryptosystems. This system was broken by Pollard and Schnorr in [2415], where they gave a probabilistic algorithm to solve Equation (16.3.7) for any y without even knowing the factors of n. Assuming the generalized Riemann hypothesis, a solution can be found with a computational complexity of $O((\log n)^2 \log\log |k|)$ in $O(\log n)$-bit integer operations.

16.3.32 Remark Diffie and Fell [1049] tried to build a cryptosystem using the composition of invertible linear maps and simple tame maps of the form

$$T(x_1, x_2) = (x_1 + g(x_2), x_2),$$

where g is a polynomial. Tame maps, well known in algebraic geometry, are easily invertible but hard to hide when composed with each other, [1049] used only two variables and equations; not surprisingly, the authors concluded that it appeared very difficult to build such a cryptosystem with any real practical value that is both secure and has a public key of practical size, therefore practically useful.

16.3.33 Remark An attempt to build a true multivariate (with four variables) public key cryptosystem was also made by Matsumoto, Imai, Harashima, and Miyagawa [2029], where the public keys are given by quadratic polynomials. However it was soon broken [2312]. People soon realized that more than 4 variables are needed and new mathematical ideas are needed to make MPKCs work.

16.3.2.2 Triangular constructions

16.3.34 Remark The tame maps used in [1049] are a special case of the "triangular" or *de Jonquières* maps from algebraic geometry.

16.3.35 Definition A *de Jonquières* map is a polynomial map in the following form:

$$J(x_1, \ldots, x_n) = (x_1 + g_1(x_2, \ldots, x_n), \ldots, x_{n-1} + g_{n-1}(x_n), x_n), \tag{16.3.8}$$

where the g_i are arbitrary polynomial functions.

16.3.36 Remark A *de Jonquières* map J can be efficiently inverted as long as g_i is not too complicated. The invertible affine linear maps over $\mathbb{F}_q^n$ together with the de Jonquières maps belong to the family of tame transformations from algebraic geometry, including all transformations that are in the form of a composition of elements of these two types of transformations. Tame transformations are elements of the group of automorphisms of the polynomial ring $\mathbb{F}_q[x_1, \ldots, x_n]$. Elements in this automorphism group that are not tame are *wild.* Given a polynomial map, it is in general very difficult to decide whether or not the map is tame, or even if there is indeed any wild map [2213], a question closely related to the famous Jacobian conjecture. This problem was solved in 2003 when [2613] proves that the Nagata map is indeed wild.

16.3.37 Remark The first attempt in the English literature with a clear triangular form is the birational permutations construction by Shamir [2606]. However, triangular constructions were earlier pursued in Japan under the name "sequential solution type systems" [1440, 2822, 2823]. Their construction is actually even more general in the sense that they use rational functions instead of just polynomials. These works in Japanese are not so well-known.

16.3.38 Remark Triangular maps are extremely fast to evaluate and to invert if g_i are very simple functions. However, they do have certain strong definitive characteristics. On the small end of a triangular system, so to speak, the variable x_n is mapped to the simple function of itself. On the bigger end, the variable x_i appears only once in a single equation. The other equations involve successively more variables.

16.3.39 Proposition If we write the quadratic portion of the central polynomials $y_i = q_i(X)$ as bilinear forms, or take the symmetric matrix denoting the symmetric differential of the central polynomials as in

$$q_i(X+B) - q_i(X) - q_i(B) + q_i(0) := B^T M_i X, \tag{16.3.9}$$

then the rank of the matrix M_i in general increases monotonically as i increases. If $q = 2^k$, the equation dealing with x_1 always has rank zero. Let $\ker M_i$ be the kernel of the linear map associated with M_i. Then $\ker M_1 \subset \ker M_2 \subset \cdots$, which is a *chain of kernels* [722].

16.3.40 Remark This matrix rank and the chain structure of kernels is invariant under composition with an invertible map, S. That is, considering y_i as a function of X, the corresponding differential is $B^T \left(M_S^T M_i M_S\right) X$. For the most part, M_S is full-rank, and hence $\operatorname{rank}\left(M_S^T M_i M_S\right) = \operatorname{rank} M_i$. This leads to what is known as rank attacks based on linear algebra [722, 1339]. Therefore triangular/tame constructions cannot be used alone. Some ways to design around this problem are *lock polynomials* (Subsection 16.3.2.9), *solvable segments* (Subsection 16.3.2.4), and *plus-minus* (Subsection 16.3.2.6).

16.3.2.3 Big-field families: Matsumoto-Imai (C^*) and HFE

16.3.41 Remark Triangular (and Oil-and-Vinegar, and variants thereof) systems are usually called "single-field" or "small-field" approaches to MPKC design, in contrast to the approach taken by Matsumoto and Imai in 1988 [2028]. In the "big-field" constructions, a totally new type of mathematical construction, the central map is really a map in a larger field $\mathbb{L}$, a degree n extension of a finite field $\mathbb{K}$. One builds an invertible map $\overline{\mathcal{Q}} : \mathbb{L} \to \mathbb{L}$, and picks a $\mathbb{K}$-linear bijection $\phi : \mathbb{L} \to \mathbb{K}^n$. Then we have the following multivariate polynomial map, which should presumably be quadratic in general:

$$\mathcal{Q} = \phi \circ \overline{\mathcal{Q}} \circ \phi^{-1}, \tag{16.3.10}$$

and, we "hide" this map $\mathcal{Q}$ by composing from both sides by two invertible affine linear maps S and T in $\mathbb{K}^n$.

16.3.42 Definition Matsumoto and Imai built a scheme C^* by choosing a field $\mathbb{K}$ of characteristic 2 and the map $\overline{\mathcal{Q}}$

$$\overline{\mathcal{Q}} : X \longmapsto Y = X^{1+q^\alpha}, \tag{16.3.11}$$

where q is the number of elements in $\mathbb{K}$, X is an element in $\mathbb{L}$, and $\gcd(1+q^\alpha, q^n - 1) = 1$.

16.3.43 Theorem We have

$$\overline{\mathcal{Q}}^{-1}(X) = X^h, \tag{16.3.12}$$

where $h(1+q^\alpha) \equiv 1 (\bmod (q^n - 1))$.

16.3.44 Remark In the rest of this section, for the purpose of simplicity, we simply identify the vector space $\mathbb{K}^n$ with the large field $\mathbb{L}$, and $\mathcal{Q}$ with $\overline{\mathcal{Q}}$, omitting the isomorphism ϕ from formulas. When necessary to distinguish the inner product in a vector space over $\mathbb{K}$ and the larger field $\mathbb{L}$, the former is denoted by a dot $(\cdot)$ and the latter by an asterisk $(*)$.

16.3.45 Remark The map $\mathcal{Q}$ is always quadratic due to the linearity of the Frobenius map $F_\alpha(X) = X^{q^\alpha}$.

16.3.46 Remark A significant algebraic implication of C^* and Equation (16.3.11) is $Y^{q^\alpha - 1} = X^{q^{2\alpha}-1}$ or

$$XY^{q^\alpha} = X^{q^{2\alpha}}Y. \tag{16.3.13}$$

Patarin [2364] used this bilinear relation to cryptanalyze the original C^* (see Subsection 16.3.3.1). Though the original idea of C^* failed, it has inspired many new designs.

16.3.47 Definition An HFE (Hidden Field Equations) system, as the most significant of the C^* derivatives, is constructed by replacing $\mathcal{Q}$, the monomial used by C^*, by the *extended Dembowski-Ostrom polynomial map*:

$$\mathcal{Q} : X \in \mathbb{L} = \mathbb{F}_{q^n} \longmapsto Y = \sum_{0\le i\le j<r} a_{ij}X^{q^i+q^j} + \sum_{0\le i<r} b_i X^{q^i} + c. \tag{16.3.14}$$

16.3.48 Remark This map is, in general, *not* one-to-one and we need additional structure to identify the real inverse from one of a number of possible candidates for decryption.

16.3.49 Remark Inverting $\mathcal{Q}$ is equivalent to solving a univariate equation of a certain degree in $\mathbb{L}$. It is well-studied and straightforward to implement but depends very much on the degree of the polynomial, using some version of the Berlekamp (or Cantor-Zassenhaus) algorithm [230, 499]; see Section 11.4. Typically, the cost of this solution is $O(nd^2 \log d + d^3)$, where d is the maximum degree of $\mathcal{Q}$.

16.3.50 Remark For practical applications, one might conclude right away that we should have as small a d as possible, or as small an r as possible, since usually $d = 2q^r$ or $q^r + 1$. But the situation actually becomes very subtle. Just as Equation (16.3.11) intrinsically meant that the C^* map in some form has a rank of 2 and leads to Equation (16.3.13) and all the known cryptanalysis of C^* related systems, Equation (16.3.14) fundamentally is responsible for all the algebraic properties of the HFE.

16.3.51 Remark A critical fact is that the intrinsic rank of the map is bounded by r, and usually achieves that value for randomly chosen parameters. This rank essentially determines the complexity of current attacks [737, 1042]. For example, the HFE Challenge 1 solved by Faugère and Joux [1042] has an intrinsic rank of 4.

16.3.52 Remark An HFE with a high d is unbroken, although it can be really slow to decrypt/invert. Quartz [2369] probably sets a record for the slowest cryptographic algorithm when submitted to NESSIE — on a Pentium III 500MHz, it took half a minute to do a signature, but has been improved substantially since.

16.3.53 Remark Recent research progress in [878, 884, 890] shows that the HFE still has great potential if we explore the cases using finite fields with odd characteristics.

16.3.54 Remark Schemes C^* and HFE each can be modified by techniques mentioned later (Plus-Minus, vinegar variables, and internal perturbation). Also related are the ℓIC system and probabilistic big-field based MPKCs [1340].

16.3.2.4 Oil and vinegar (unbalanced and balanced) and variations

16.3.55 Remark The Oil and Vinegar (OV) and later unbalanced Oil and Vinegar (UOV) schemes [1737, 2367] are designed only for signatures. This construction is inspired by the idea of linearization equations (Subsection 16.3.2.3). In some sense, this construction uses a method where ones transforms an attacking method into a designing method.

16.3.56 **Definition** Let $v < n$ and $m = o = n - v$. The variables $x_1, \ldots, x_v$ are *vinegar* variables and $x_{v+1}, \ldots, x_n$ *oil* variables. An *oil-vinegar map* $\mathcal{Q} : \mathbb{K}^n \to \mathbb{K}^m$ is a map in the form $Y = \mathcal{Q}(X) = (q_1(X), \ldots, q_o(X))$, where

$$q_l(X) = \sum_{i=1}^{v} \sum_{j=i}^{n} \alpha_{ij}^{(l)} x_i x_j, \; l = 1, \cdots, o$$

and all coefficients are randomly chosen from the base field $\mathbb{K}$.

16.3.57 **Remark** In each q_i, there are no quadratic terms of oil variables, which means the oil variables and vinegar variables are not fully mixed (like oil and vinegar in a salad dressing), which is the origin of the name of this scheme.

16.3.58 **Definition** The *public key map* $\mathcal{P}$ for an Oil-Vinegar scheme is constructed as

$$\mathcal{P} = \mathcal{Q} \circ S,$$

where S is an invertible linear map.

16.3.59 **Remark** The change of basis by the transformation S is a process to "mix" fully oil and vinegar, so one cannot tell what are oil variables and what are the vinegar variables. With OV and UOV constructions, there is no need to compose another affine map T.

16.3.60 **Remark** The original Oil and Vinegar signature scheme has $m = o = v = n/2$. When $o < v$, it becomes the unbalanced Oil and Vinegar signature scheme.

16.3.61 **Remark** The public key for an OV or an UOV is $\mathcal{P} = (p_1, \ldots, p_o)$, the polynomial components of $\mathcal{P}$. The secret key consists of the linear map S and the map $\mathcal{Q}$.

16.3.62 **Remark** Given a message $Y = (y_1, \ldots, y_o)$, to sign it, one needs to find a vector $W = (w_1, \ldots, w_n)$ such that $\mathcal{P}(W) = Y$. With the secret key, this can be done efficiently. First, one guesses values for each vinegar variable $x_1, \ldots, x_v$, and obtains a set of o linear equations with the o oil variables $x_{v+1}, \ldots, x_n$. With high probability, it has a solution. If the linear system does not have a solution, one may repeatedly assign random values to the vinegar variables until one finds a pre-image of a given element in $\mathbb{K}^o$. Then one applies S^{-1}.

16.3.63 **Remark** To check if W is indeed a legitimate signature for Y, one only needs to get the public map $\mathcal{P}$ and check if indeed $\mathcal{P}(W) = Y$.

16.3.64 **Remark** The algebraic property that is most significant in an unbalanced Oil-and-Vinegar system is the absence of pure oil cross-terms. Equivalently, if we have an UOV polynomial, then the quadratic part of each component q_i in the central map from X to Y, when viewed as a bilinear form using a matrix, see Equation (16.3.9), looks like

$$M_i := \left[\begin{array}{ccc|ccc} \alpha_{11}^{(i)} & \cdots & \alpha_{1v}^{(i)} & \alpha_{1,v+1,}^{(i)} & \cdots & \alpha_{1n}^{(i)} \\ \vdots & \ddots & \vdots & \vdots & \ddots & \vdots \\ \alpha_{v1}^{(i)} & \cdots & \alpha_{vv}^{(i)} & \alpha_{v,v+1,}^{(i)} & \cdots & \alpha_{vn}^{(i)} \\ \hline \alpha_{v+1,1,}^{(i)} & \cdots & \alpha_{v+1,v,}^{(i)} & 0 & \cdots & 0 \\ \vdots & \ddots & \vdots & \vdots & \ddots & \vdots \\ \alpha_{n1}^{(i)} & \cdots & \alpha_{nv}^{(i)} & 0 & \cdots & 0 \end{array}\right], \qquad (16.3.15)$$

or in the block form: $\left[\begin{array}{c|c} * & * \\ \hline * & 0 \end{array}\right]$.

16.3.2.5 UOV as a booster stage

16.3.65 Remark There have been different attempts to make UOV more efficient such as [1687, 1686], which were promptly broken [2998].

16.3.66 Remark A new way to make the UOV type construction more efficient is the Rainbow construction by stacking several layers of Unbalanced Oil-Vinegar systems together for an easily invertible central map [889].

16.3.67 Definition For a *u-stage Rainbow* $0 < v_1 < v_2 < \cdots < v_{u+1} = n$, the construction of central map over any finite field is given by

$$y_k = q_k(X) = \sum_{i=1}^{v_l} \sum_{j=i}^{n} \alpha_{ij}^{(k)} x_i x_j + \sum_{i < v_{l+1}} \beta_i^{(k)} x_i, \quad \text{if } v_l < k \leq v_{l+1}. \tag{16.3.16}$$

16.3.68 Remark In the signing process, we first choose randomly the values for the vinegar variables $x_1, \ldots, x_{v_1}$ in the first layer, and solve for the oil variables $x_{v_1+1}, \ldots, x_{v_2}$. Then we use the known values of x_i in the second layer and find the values for the oil variables in the second layer. We continue like this layer by layer until we have all the x_i's.

16.3.69 Remark The components of Y in a Rainbow-type construction are typically written to have indices $v_1 + 1, \ldots, n$. In the pure Rainbow scheme, S and T and the coefficients α and β are totally randomly chosen. The essential structure of the Rainbow instance is determined by $0 < v_1 < v_2 < \cdots < v_{u+1} = n$ or the "Rainbow structure sequence" $(v_1, o_1, o_2, \ldots, o_u)$, where $o_i := v_{i+1} - v_i$.

16.3.70 Remark The Rainbow construction is a special case of the UOV; however, the structure of the system, consist of o_u equations with the associated bilinear maps of the form $\left[\begin{array}{c|c} * & * \\ \hline * & 0 \end{array}\right]$ following $m - o_u$ equations with the associated bilinear maps of the form $\left[\begin{array}{c|c} * & 0 \\ \hline 0 & 0 \end{array}\right]$ leads to a different attack; see Subsection 16.3.3.10).

16.3.71 Remark Aside from attacks peculiar to the UOV and Rainbow systems, the Rainbow-type constructions also share certain characteristics of triangular schemes, therefore there is a need to account for rank-based attacks (Subsection 16.3.3.8), such as the two improved attacks in [280, 895]. None of these attacks are considered essentially effective.

16.3.72 Remark If one wants to make the computation of the central map and its inverse fast, another direct way is to make the Oil-vinegar polynomials sparse, such as in the case of the TTS (Tame Transformation Signatures) schemes, [604, 3023, 3024]. The TRMS [2931] of Wang *et al.* are also a TTS instance. Due to the sparsity, there also exist certain extra possibilities of linear algebra and related vulnerabilities, principally UOV-type vulnerabilities such as described in [891].

16.3.2.6 Plus-Minus variations

16.3.73 Remark *Minus* and *Plus* are simple but useful ideas, earliest mentioned by Matsumoto, Patarin, and Shamir (probably found independently [2370, 2606]).

16.3.74 Remark For the Minus method [2370], several (r) polynomials are removed from the public keys. When inverting the public map, the legitimate users take random values for the missing variables. Minus is very suitable for signature schemes without any performance loss, since a document need not have a unique signature.

16.3.75 Remark For encryption, Minus causes significant slowdown, since the missing coordinates must be searched. In theory the public map of an encryption method should be injective. If we have to search through r variables in $\mathbb{F}_q$, we effectively have q^{n+r} results, only q^n of which should represent valid ciphertexts, hence the expected number of guesses taken per decryption is q^r. Thus, decryption is slowed by that same factor of q^r.

16.3.76 Remark *Minus* or removing some public equations makes a C^*-based system much harder to solve. SFLASH [67, 739, 2368], a C^{*-} instance with $(q, n, r) = (2^7, 37, 11)$, was accepted as an European security standard for low-cost smart cards by the New European Schemes for Signatures, Integrity, and Encryption [2220].

16.3.77 Remark In 2007, the SFLASH family of cryptosystems [924, 925] was defeated. The attack uses the symmetry and the invariants of the differential of the public map $\mathcal{P}$ (Subsection 16.3.3.4) inspired by the attack on internal perturbation. Making a secure C^*-based signature scheme may require a new modifier called *Projection* [880].

16.3.78 Remark *Plus* is the opposite of *Minus*: add random central equations to the original central map, and this can be used to mask the high-end of the triangle system. For encryption methods, this again does not affect performance much; for digital signatures there is a slowdown as the extra variables again need to be guessed. Regardless, *Plus-Minus* variations defend against attacks that are predicated on the rank of equations.

16.3.79 Remark *Plus-Minus* alone does not make triangular constructions secure. Paper [1339] discusses this in detail and concludes exactly the opposite: Triangle-Plus-Minus constructions can be broken by very straightforward attacks using simple linear algebra. Therefore one must use more elaborate variations [280, 895, 3023].

16.3.2.7 Internal perturbation

16.3.80 Remark Internal perturbation is a general method of improving the security of MPKCs by adding some perturbation or controlled noise. Internal perturbation was first applied to Matsumoto-Imai systems, which produce the variation [876].

16.3.81 Remark Take $V = (v_1, \ldots, v_r)$ to be an r-tuple of random affine forms in the variables X. Let $\mathbf{f} = (f_1, \ldots, f_n)$ be a random r-tuple of quadratic functions in V. Let our new $\mathcal{Q}$ be defined by

$$X \mapsto Y = (X)^{q^\alpha+1} + \mathbf{f}(V(X))$$

where the power operation assumes the vector space to represent a field. The number of Patarin relations decreases quickly down to 0 as r increases. For every Y, we may find $\mathcal{Q}^{-1}(Y)$ by guessing at $V(X) = B$, finding a candidate $X = (Y + B)^h$ and checking the initial assumption that $V(X) = X$. Since we repeat the procedure q^r times, we are almost forced to let $q = 2$ and make r as small as possible.

16.3.82 Remark In this system, there are possible extraneous solutions just as in the system HFE. Therefore, we must manufacture some redundancy in the form of a hash segment or checksum. The original perturbation system was broken [1094] via a surprising *differential cryptanalysis* (cf. Sec. 16.3.3.4). Internal perturbation is usually coupled with the *plus* variation, and this is one of the best multivariate encryption schemes.

16.3.2.8 Vinegar as an external perturbation and projection

16.3.83 Remark The idea of *vinegar* variables had been introduced earlier with UOV, and was used as a defense in Quartz. The idea is to use an auxillary variable that occupies only a small subspace of the input space. It was pointed out [888] that internal perturbation is almost exactly equal to both vinegar variables and projection, or fixing the input to an affine subspace. We basically set one, two, or more variables of the public key to be zero to create the new public key. In the case of signature schemes, each projected dimension will slow down the signing process by a factor of q.

16.3.84 Remark Projection is a very useful simple method. Since (Sec. 16.3.3.4) structural attacks usually start by looking for an invariant or a symmetry, it is a good idea that we should try to remove both. Restricting to a subspace of the original space breaks the symmetry. Something like the *Minus* modifier destroys an invariant. Hence the use of projection by itself prevents some attacks, such as [924, 925, 1095]. The differential attack against C^* (and ℓIC) derivatives uses the structure of the big field $\mathbb{L}$. Hence projection is expected to prevent such an attack [880].

16.3.2.9 TTM and related schemes: "lock" or repeated triangular

16.3.85 Remark MPKCs based only on triangular constructions were not pursued again until a much more complex defense against rank attacks was proposed, with the tame transformation method (TTM) of Moh [2113].

16.3.86 Remark de Jonquières maps can be viewed either as upper triangular or lower triangular. Moh [2113] suggested a construction where the central map $\mathcal{Q}$ is given by

$$\mathcal{Q} = J_u \circ J_l \circ I(x_1, \ldots, x_n), \tag{16.3.17}$$

where J_u is a $\mathbb{K}^m$ upper triangular de Jonquières map and J_l is a $\mathbb{K}^m$ lower triangular de Jonquières map and the linear map I is the embedding of $\mathbb{K}^n$ into $\mathbb{K}^m$: $I(x_1, \ldots, x_n) = (x_1, \ldots, x_n, 0, 0, \ldots, 0)$. The main achievement of such a construction is how to make $\mathcal{Q}$ quadratic. Moh's trick is actually in using the map I. One can see that

$$\begin{aligned} J_l \circ I(x_1, \ldots, x_n) &= (x_1, x_2 + g_1(x_1), \ldots, x_n + g_{n-1}(x_1, \ldots, x_{n-1}), \\ &\quad g_n(x_1, \ldots, x_n), \ldots, g_{m-1}(x_1, \ldots, x_n)), \end{aligned}$$

which presents the freedom to choose any g_i, $i = n, \ldots, m-1$. When decrypting, one evaluates the de Jonquières maps from x_1 up.

16.3.87 Remark The extremely low rank of central polynomials presented in published TTM instances [603, 2113] is the main source of weakness and therefore of effective attacks.

16.3.88 Remark Courtois and Goubin [1339] used the MinRank method (Subsection 16.3.3.8) to attack this system.

16.3.89 Definition A *MinRank problem* is to look for non-zero matrices with minimum rank in a space of matrices.

16.3.90 Remark The Minrank problem is NP-hard in general but can be easy for special cases, in particular, when the minimum rank is very low.

16.3.91 Remark The idea of sequentially solvable equations (or stages) can also be used in conjunction with other ideas. Some of the more notable attempts are from Wang, who had written about a series of schemes called "Tractable Rational Map Cryptosystems" (TRMC). TRMC v1 is essentially no different from early TTM [603]. The central map of TRMC v2 [2929] has a small random overdetermined block on one end and the rest of the variables are determined in the triangular (tame) style. Versions 3 and 4 [2930, 2932] use a similar trick as 3IC [892].

16.3.92 Remark The TTM construction is a truly original and very intriguing idea. So far existing constructions of the TTM cryptosystem and related schemes do not work for public-key encryption [883, 886, 887, 2226]. Most of the schemes proposed are not presented in any systematic way, and no explanation is yet given why and how they work. More sophistication is needed and we suspect that to create a successful TTM-like scheme will probably require deep insight from algebraic geometry.

16.3.93 Remark A TTS (tamed transformation signature) scheme can be viewed as a similar but simpler construction. This system is essentially the result of an application of the Minus method in [2606] for a tame transformation. A few of them were suggested mainly by Chen and Yang [604, 605]. These systems can also be defeated by the method used by Stern and Vaudenay [722, 896].

16.3.2.10 Intermediate fields: MFE and ℓIC

16.3.94 Remark In C^* and HFE, we use one big field $\mathbb{L} = \mathbb{K}^n$. In Rainbow/TTS or similar schemes, each component is as small as the base field. We can use something in between, as seen in MFE (Medium Field Encryption) and ℓIC (ℓ-Invertible Cycles). These two constructions use a standard Cremona transformation from algebraic geometry.

16.3.95 Definition A *standard Cremona transformation* is defined as: $\mathbb{L}^* := \mathbb{L}\backslash\{0\}$ for some field $\mathbb{L}$:

$$(X_1, X_2, X_3) \in (\mathbb{L}^*)^3 \longmapsto (Y_1, Y_2, Y_3) := (X_1X_2, X_1X_3, X_2X_3) \in (\mathbb{L}^*)^3. \quad (16.3.18)$$

16.3.96 Proposition This transformation is a bijection for any field $\mathbb{L}$, and inverts via

$$X_1 := \sqrt{Y_1Y_2/Y_3}.$$

16.3.97 Remark MFE's central map uses structure related to matrix multiplications to defend against linearization relations, but it does not avoid all the problems, as can be seen in Subsection 16.3.3.3.

16.3.98 Remark The ℓ-invertible cycle [892] also uses an intermediate field $\mathbb{L} = \mathbb{K}^k$ and extends C^* by using the following central map from $(\mathbb{L}^*)^\ell$ to itself:

$$\begin{aligned} \mathcal{Q} : (X_1, \ldots, X_\ell) &\mapsto (Y_1, \ldots, Y_\ell) \\ &:= (X_1X_2,\, X_2X_3, \ldots, X_{\ell-1}X_\ell,\, \underline{X_\ell X_1^{q^\alpha}}). \end{aligned} \quad (16.3.19)$$

16.3.99 Remark This is much faster computationally than computing the inverse of C^*. But these have so much in common with C^* that we need the same variations. In other words, we need to do 3IC$^-$p (with minus and projection) and 2IC$^+$i (with internal perturbation and plus), paralleling C^{*-}p and C^{*+}i (also known as PMI+) [881].

16.3.2.11 Odd characteristic

16.3.100 **Remark** Initially all the MPKCs were constructed over fields of characteristic 2. But in the work of [890], they notice that a very good idea would be the usage of fields of odd characteristic. The rationale is that in the case of relatively large odd characteristic, the field equations in the form of $x_i^q - x_i = 0$, cannot be effectively used, therefore it forces one to find all the solutions in the algebraic closure instead of the small field itself. This would make the MPKCs much more secure in terms of direct attacks by polynomial solvers. Recent results on the degree of regularity [878, 884] further confirms this notion.

16.3.2.12 Other constructions

16.3.101 **Remark** There are also other new constructions using different new ideas. One interesting one is a construction using Diophantine equations over certain function rings [1461]. Another is the MQQ construction using quasi-groups, although it has been defeated [1285, 2115].

16.3.3 Standard attacks

16.3.102 **Remark** To attack an MPKC directly as an MQ problem instance is usually not very effective, and cryptanalysts generally try to attack MPKCs utilizing the structure of the central map by either finding the private key directly, or finding extra polynomial relations to enhance the polynomial solver.

16.3.3.1 Linearization equations

16.3.103 **Definition** For a given MPKC, a *linearization equation* is a relation between the components of the ciphertext Y and plaintext X in the following form:

$$\sum a_{ij}x_iy_j + \sum b_ix_i + \sum c_jy_j + d = 0. \qquad (16.3.20)$$

16.3.104 **Remark** The key property of these equations is that when substituted with the actual values of Y, we get an affine (linear) relation between the x_i's. In general, each equation should effectively eliminate one variable from the system.

16.3.105 **Remark** The key and first example is the attack against C^* by Patarin [2364]. For any C^* public key, we can compute Y from X, and substitute enough (X, Y) pairs and solve for a_{ij}, b_i, c_j, and d. A basis for the solution space gives us all the linearization relations.

16.3.106 **Remark** If we are given any ciphertext, i.e., the values of y_i, these n bilinear relations will produce linear equations satisfied by components of the plaintext X, and in the case of C^*, it gives us enough linearly independent linear equations to help us to find the plaintext with the help of the public equations. In similar systems like 3IC, for example, linearization equations are also present in large numbers, which need to be eliminated to make the system secure.

16.3.3.2 Critical bilinear relations

16.3.107 **Remark** If the number of linearization equations is high enough, we can defeat the system efficiently. However, it is shown in [883, 886, 887] that even when the number of linearization equations (or special form of bilinear relations) is not so large, their existence can be lethal.

16.3.108 Remark Ding and Schmidt [887] found that the low-rank central polynomials — often rank 2 — in currently existing implementation schemes for the TTM cryptosystems make it possible to extend the linearization method by Patarin [2364] to attack all current TTM implementation schemes (Subsection 16.3.3.1).

16.3.109 Remark For the Ding-Schmidt attack [887], the number of linearization equations is not that high, but they manage to eliminate the "lock polynomial" that defends a TTM instance against a simple rank attack.

16.3.3.3 HOLEs (higher-order linearization equations)

16.3.110 Definition A *Higher-Order Linearization Equation* (HOLE) is a linearization relation that is higher degree in the components of Y only. In particular, a SOLE (*second order linearization equation*) would look like

$$\sum_{i<j} a_{ijk}\, y_i y_j x_k + \sum_{i\leq j} b_{ij}\, y_i y_j + \sum c_{ij}\, y_i x_j + \sum d_i\, y_i + \sum e_j\, x_j + f = 0.$$

16.3.111 Remark This is the key to break the MFE systems. There are at least $8k$ linear dependencies derived from the SOLE out of a total of $12k$ variables in an MFE system, which makes the cryptanalyst's task much easier. Paper [885] used another trick – the fact that squaring is linear in a characteristic two field – to get it down to $2k$ remaining variables at most and concluded that solving for the remaining variables is easy.

16.3.112 Remark The existence of linearization relations of higher degrees shows multivariate encryption schemes designed in the triangular style are full of traps and to design a secure system is very difficult without an intrinsically sophisticated algebraic structure.

16.3.3.4 Differential attacks

16.3.113 Remark Structural attacks on MPKC systems to recover private keys (or equivalently useful keys) are of two related types:

Invariants: invariants (mostly subspaces) that can be tracked.

Symmetries: transformations that leave certain structures invariant and hence can be computed by a system of equations.

These two methods are closely related since invariants are defined according to symmetry. Earlier designers were not yet fully aware of the importance of symmetry. We will present the symmetry or invariants used in the new differential attacks on the C^* family of cryptosystems as exemplified by the differential attacks developed by Stern and collaborators in [1094].

16.3.3.5 Attacking internal perturbations

16.3.114 Remark The cryptanalysis of the PMI (perturbed Matsumoto-Imai) [1094] was a true novelty for a technique usually associated with symmetric key cryptography.

16.3.115 Remark Using the notation in a PMI system, we know that for a randomly chosen vector $\mathbf{b}$, the probability is q^{-r} that it lies in the kernel $\mathcal{K}$ of the linear part of V. When that happens, $V(X + \mathbf{b}) = V(X)$ for any X. Since q^{-r} is not too small, if we can use this to distinguish between a vector $\mathbf{b} \in T^{-1}\mathcal{K}$ (back-mapped into the original space) and $\mathbf{b} \notin T^{-1}\mathcal{K}$, we can bypass the protection of the perturbation, find our bilinear relations and accomplish the cryptanalysis.

16.3.116 Remark In [1094], Fouque, Granboulan, and Stern built a *distinguisher* using a test on the kernel of the symmetric difference

$$D\mathcal{P}(W, \mathbf{b}) = \mathcal{P}(\mathbf{b} + W) - \mathcal{P}(\mathbf{b}) - \mathcal{P}(W).$$

We say that $t(\mathbf{b}) = 1$ if $\dim \ker_W D\mathcal{P}(\mathbf{b}, W) = 2^{\gcd(n,\alpha)} - 1$, and $t(\mathbf{b}) = 0$ otherwise. If $\mathbf{b} \in \mathcal{K}$, then $t(\mathbf{b}) = 1$ with probability one, otherwise it is less than one. If $\gcd(n, \alpha) > 1$, it is a nearly perfect distinguisher. If not, we can employ two other tricks. For one of them, we observe $\mathcal{K}$ is a vector space, so $\Pr(t(\mathbf{b} + \mathbf{b}') = 0 | t(\mathbf{b}') = 0)$ will be relatively high if $\mathbf{b} \in \mathcal{K}$ and relatively low otherwise.

16.3.117 Remark There is a surprisingly simple defense dating back to [2370] (which introduced SFLASH). By using the "plus" variant, i.e., appending a random quadratic polynomial to $\mathcal{P}$, enough false positives are generated to overwhelm the distinguishing test of [1094]. The extra equations also serve as a distinguisher when there are extraneous solutions. Basically, the more "plus" equations, the less discriminating power of the above mentioned test. Based on empirical results of Ding and Gower [881], when $r = 6$, $a = 12$ should be sufficient, and $a = 14$ would be a rather conservative estimate for the amount of "plus" needed to mask the PMI structure.

16.3.3.6 The skew symmetric transformation

16.3.118 Remark The symmetry found in [1094] can be explained by considering the case of the C^* cryptosystem. The symmetric differential of any function G, defined formally just like in Equation (16.3.21):

$$DG(\mathbf{a}, X) := G(X + \mathbf{a}) - G(X) - G(\mathbf{a}) + G(0),$$

is bilinear and symmetric in its variables $\mathbf{a}$ and X. In the first version of this attack [925], we look at the differential of the public map $\mathcal{P}$, and look for skew-symmetric maps with respect to this bilinear function, namely, the linear maps M such that

$$D\mathcal{P}(\mathbf{c}, M(W)) + D\mathcal{P}(M(\mathbf{c}), W) = 0.$$

16.3.119 Remark The reason that this works is that the central map $\mathcal{Q}$ and the public key, which encapsulates the vital information in the central map, unfortunately have very strong symmetry in the sense that all the differentials from these maps share some common nontrivial skew-symmetric map M. Since $\mathcal{Q}(X) = X^{1+q^\alpha}$, its differential is

$$D\mathcal{Q}(\mathbf{a}, X) = \mathbf{a}^{q^\alpha} X + \mathbf{a} X^{q^\alpha}.$$

16.3.120 Theorem The maps M skew-symmetric with respect to this $D\mathcal{Q}(\mathbf{a}, X)$ [925] are precisely those induced from multiplication by some element ζ satisfying the condition

$$\zeta^{q^\alpha} + \zeta = 0.$$

16.3.121 Remark This skew-symmetry survives under change of basis. It can be seen that the skew-symmetry continues to hold even when we remove some components of $\mathcal{P}$. In terms of the public key, this means that if we write

$$D\mathcal{P}(\mathbf{c}, W) := (\mathbf{c}^T H_1 W, \mathbf{c}^T H_2 W, \ldots, \mathbf{c}^T H_m W)$$

and try to solve $M^T H_i + H_i M = 0$ for all $i = 1, \ldots, m$ simultaneously, we should find just k-multiples of the identity if n and α are co-prime, and a d-dimensional subspace in the space of linear maps if $d = \gcd(n, \alpha) > 1$.

16.3.122 **Remark** For a randomly chosen map G, it is expected that only trivial solutions $M = u1_n$, where $u \in \mathbb{K}$, will satisfy this condition. This means that there is a very strong condition on C^{*-} cryptosystems. This symmetry can be utilized to break C^{*-} systems for which $d = \gcd(n, \alpha) > 1$.

16.3.3.7 Multiplicative symmetry

16.3.123 **Remark** The second symmetry is multiplicative symmetry, which comes also from the differential $D\mathcal{P}(\mathbf{c}, W)$ [924].

16.3.124 **Proposition** Let ζ be an element in the big field $\mathbb{L}$. Then we have

$$D\mathcal{Q}(\zeta \cdot a, x) + D\mathcal{Q}(a, \zeta \cdot x) = (\zeta^{q^\alpha} + \zeta) D\mathcal{Q}(a, x).$$

16.3.125 **Theorem** Let

$$\mathrm{M}_\zeta = \mathrm{M}_S^{-1} \circ (X \mapsto \zeta X) \circ \mathrm{M}_S$$

be the linear map in $\mathbb{K}^n$ corresponding to multiplication by ζ, then

$$\mathrm{span}\{\mathrm{M}_\zeta^T H_i + H_i \mathrm{M}_\zeta : i = 1, \ldots, n\} = \mathrm{span}\{H_i : i = 1, \ldots, n\},$$

i.e., the space spanned by the quadratic polynomials from the central map is invariant under the skew-symmetric action.

16.3.126 **Remark** The public key of C^{*-} inherits some of that symmetry. Note that not every skew-symmetric action by a matrix M_ζ corresponding to an $\mathbb{L}$-multiplication results in $\mathrm{M}_\zeta^T H_i + H_i \mathrm{M}_\zeta$ being in the span of the public-key differential matrices, because $S := \mathrm{span}\{H_i : i = 1, \ldots, n - r\}$ as compared to $\mathrm{span}\{H_i : i = 1, \ldots, n\}$ is missing r of the basis matrices. However, as the authors of [924] argued heuristically and backed up with empirical evidence, if we just pick the first three $\mathrm{M}_\zeta^T H_i + H_i \mathrm{M}_\zeta$ matrices, or any three random linear combinations of the form $\sum_{i=1}^{n-r} b_i(\mathrm{M}_\zeta^T H_i + H_i \mathrm{M}_\zeta)$ and demand that they fall in S, then there is a good chance to find a nontrivial M_ζ corresponding to a multiplication by ζ, which can be used to break the C^{*-} scheme.

16.3.127 **Remark** For a set of public keys from C^*, tests [924] show that the above strategy almost surely eventually recovers the missing r equations and breaks the scheme. The only known attempted defense is [880].

16.3.3.8 Rank attacks

16.3.128 **Remark** Given a quadratic polynomial, we can always associate it with a symmetric matrix. By a rank attack, we mean an attack using the rank of those matrices. There are two types of rank attacks, attacks that specifically target high rank or low rank. Let H_i be the symmetric matrix corresponding to the quadratic part of public polynomials.

16.3.129 **Remark** Since rank attack usually means attacking via finding *low rank* matrices, some also call the high rank attack the *dual rank attack*. The high rank attack first appeared with [722] where Coppersmith et al. defeated a triangular construction.

16.3.130 **Remark** The high rank attacks of Goubin-Courtois and Yang-Chen [1339, 3023] work for "plus"-modified triangular systems; it is also easier to understand than the formulation in [722]. Against UOV, we might possibly do even better on this attack with differentials [895].

16.3.131 Remark The first Minrank attack is the Goubin-Courtois version against TTM. Let r be the smallest rank in linear combinations of central equations, which without loss of generality we take to be the first central equation itself in TTM. Goubin and Courtois outline how to find the smallest ranked combination (and hence break the Triangle-Plus-Minus system) in expected time $O(q^{\lceil \frac{m}{n} \rceil r} m^3)$. Yang and Chen have extended the effectiveness of this attack [3023] related to the number of distinct kernels of the same rank.

16.3.3.9 MinRank attacks on big-field schemes

16.3.132 Remark The defeat of the HFE Challenge 1 by Faugère and Joux [1042], a direct solution of the 80 equations in 80 variables, is not the first serious attempt on HFE systems. That credit goes to a rank-based attack by Kipnis and Shamir [1739]. The attack proceeds by moving the problem back to the extension field, where all the underlying structure can be seen. This is a very natural approach if we intend to exploit the design structure of HFE in the attack. They transform the problem of finding the secret key into a problem of finding the minimum rank of linear combinations of certain matrices, which is exactly r (as in Subsection 16.3.2.3). This is the MinRank problem [467] and is in general exponential, but can be easy if r is small.

16.3.133 Remark Kipnis and Shamir [1739] suggested using determinants of all $(r+1) \times (r+1)$ sub-matrices to derive a huge assortment of equations to solve the problem. To solve this system, they introduce an idea which they call *relinearization*, which led to the well-known XL paper [740]. It has been argued that using a Lazard-Faugère solver on this system of equations is effective [737] and equally effective as the direct attack, however the situation is still not very clear due to a recent observation in [1610].

16.3.3.10 Distilling oil from vinegar and other attacks on UOV

16.3.134 Remark To forge a signature for a UOV scheme as in Subsection 16.3.2.4, one needs to solve the equation $\mathcal{P}(W) = Y$.

16.3.135 Remark When $o = v$ as with the original Oil-and-Vinegar, this is fairly easy due to the attack by Kipnis and Shamir [1738]. The basic idea is to treat each component $y_i = p_i(W)$ of the public key $\mathcal{P}$ as a bilinear form. Namely, take their associated symmetric matrices via the symmetric differential as follows:

$$Dp_i(W, \mathbf{c}) := p_i(W + \mathbf{c}) - p_i(W) - p_i(\mathbf{c}) + p_i(0) := \mathbf{c}^T H_i W. \tag{16.3.21}$$

A basic fact of OV: each matrix M_i, see Equation (16.3.9) is in the rough form of $\begin{bmatrix} * & * \\ * & 0 \end{bmatrix}$ but not the matrices H_i associated to the public key. This reduces a cryptanalysis to the algebraic problem of finding a basis change for a set of bilinear forms into a common form.

16.3.136 Remark Assume $v = o = n/2 = m$. The vectors X that have all vinegar coordinates $x_1, \ldots, x_v$ equal to zero, form the Oil Space $\mathcal{O}$, i.e., the collection of X-vectors looking like $\begin{bmatrix} 0 \\ * \end{bmatrix}$, and similarly an X-vector in the Vinegar Space $\mathcal{V}$ has all oil coordinates $x_{v+1}, \ldots, x_n$ equal to zero and looks like $\begin{bmatrix} * \\ 0 \end{bmatrix}$. Clearly, if each M_i is nonsingular, we have

$$\begin{bmatrix} * & * \\ * & 0 \end{bmatrix} \begin{bmatrix} 0 \\ * \end{bmatrix} = \begin{bmatrix} * \\ 0 \end{bmatrix}, \text{ or } M_i \mathcal{O} = \mathcal{V} \text{ for all } i.$$

16.3.137 Proposition If M_j, H_j are invertible,

$$\left(M_j^{-1}M_i\right)\mathcal{O} = \mathcal{O};$$

$$\left(H_j^{-1}H_i\right)\left(S^{-1}\mathcal{O}\right) = \left(S^{-1}\mathcal{O}\right).$$

16.3.138 Remark This proposition states that any $H_j^{-1}H_i$ has the common invariant subspace of $S^{-1}\mathcal{O}$. Knowing $S^{-1}\mathcal{O}$ is sufficient to find an equivalent form for S. Kipnis *et al.* [1737] claim that the same argument works if $v < o$; even if $v > o$ it can be done in time directly proportional to q^{v-o}, and hence $v - o$ cannot be too small. However the situation about this claim is also not very clear due to recent observations in [500]. When there are two or three times more vinegar variables than oil variables the method appears to be secure, despite the claims of [390].

16.3.3.11 Reconciliation

16.3.139 Remark We could also attempt to find a sequence of change of basis that would lead to the inversion of the public map as an improvement to a brute force attack [895]. In the case of an attack on the UOV scheme with o oil and $v = n - o$ vinegar variables, the attack becomes a problem of solving m equations in v variables, which could be much easier than solving m equations in n variables in a direct attack. The reconciliation attack fails with a probability of approximately $\frac{1}{q-1}$.

16.3.140 Remark Reconciliation attack can be applied to Rainbow systems, which have multiple layer structure (Subsection 16.3.2.4). This attack works for all constructions with a UOV construction in the final stage, including all Rainbow and TTS constructions. This affects how the current proposed parameters of Rainbow [895] are selected.

16.3.3.12 Direct attacks using polynomial solvers

16.3.141 Remark To mount a direct attack, we try to solve the m equations $P(W) = Z$ in the n variables $w_1, \ldots, w_n$. If $m \geq n$, we are (over-)determined. If $m < n$, we are underdetermined. For most cases (n cannot be too large compared to m), we cannot do much more than to fix values for $m - n$ variables randomly and continue with $m = n$ [738].

16.3.142 Remark Due to the NP-hardness the the MQ problem [1214], the difficulty of solving "generic" or randomly chosen systems of nonlinear equations is generally conceded. But, it is very hard to quantify exactly how hard it is to solve a non-generic system. Often many techniques in algebraic cryptanalysis require solving a system of polynomial equations at the end for more or less generic systems. So we must solve the system $p_1 = p_2 = \cdots = p_m = 0$, where each p_i is a quadratic polynomial. Coefficients and variables are in the field $\mathbb{K} = \mathbb{F}_q$.

16.3.143 Remark The standard methods for solving equations are Buchberger's algorithm [437] to compute a Gröbner basis, and its descendents investigated by Lazard's group [1877]. Macaulay generalized Sylvester's matrix to multivariate polynomials [1985]. The idea is to construct a matrix whose rows are from the coefficients of the multiples of the polynomials of the original system, the columns representing all the monomials up to a given degree. Lazard [1877] observed that for a large enough degree, ordering the columns according to a monomial ordering and performing usual row reduction on the matrix is equivalent to Buchberger's algorithm.

16.3.144 Remark Faugère proposed an improved Gröbner bases algorithm called $\mathbf{F_4}$ [1040]. A later version, $\mathbf{F_5}$ [1041] made headlines [1042] when it was used for solving HFE Challenge 1 in

2002, but we do not really know what the real $\mathbf{F_5}$ is, since no one else, as far as we know, has repeated any of the many results claimed by $\mathbf{F_5}$. A version of $\mathbf{F_4}$ is implemented in the computer algebra system MAGMA [2798] and is publicly available.

16.3.145 **Remark** Lazard's idea was rediscovered in 1999 by Courtois, Klimov, Patarin, and Shamir [740] as **XL**. Courtois *et al.* proposed several adjuncts [736, 742, 743] to XL.

16.3.146 **Remark** Recently based on the concept of mutant, much work has been done to improve the XL algorithms, which produced a family of mutant XL algorithms [438, 877, 879, 2114, 2116].

16.3.147 **Remark** For a system of equations $p_1(x_1, \ldots, x_n) = \cdots = p_m(x_1, \ldots, x_n) = 0$, if we look at the ideal generated by p_i, then each element f of the ideal can be expressed in the form:

$$f = \sum_1^n f_i p_i,$$

which is not unique.

16.3.148 **Definition** For each such expression, we define the *level of this expression* to be the $\max(\deg(g_i) + \deg(p_i), 1 = 1, \ldots, m)$. If $\deg(f)$ is lower than any of the levels of f, f is a *mutant.*

16.3.149 **Remark** The key idea of mutants [877] is that we try to mathematically describe the degeneration of the systems, which is critical for the solving process to work. Currently, the best mutant algorithms beat all other algebraic solvers, including $\mathbf{F_4}$.

16.3.150 **Remark** A key question for understanding the security of the MPKCs is the complexity of those algebraic solvers. Using generating functions, there are heuristic estimates on the complexity of solving generic systems [201, 3022]. These works are used to estimate the complexity of algebraic attack on the HFE systems [1349] based on the concept of degree of regularity. Recently new breakthroughs have led us to mathematically prove new estimates for the degree of regularity of HFE [878, 884, 926], which provide theoretical support for the apparent security benefits of using fields of odd characteristics.

16.3.151 **Theorem** Let P be a HFE polynomial of degree D. If $\text{Rank}(P) > 1$, the degree of regularity of the associated HFE system is bounded by

$$\frac{(q-1)\text{Rank}(P)}{2} + 2.$$

In particular, this is less than or equal to

$$\frac{(q-1)(\lfloor \log_q(D-1) \rfloor + 1)}{2} + 2.$$

If $\text{Rank}(P) = 1$, then the degree of regularity is less than or equal to q. Here Rank is the rank of the quadratic form associated to P.

16.3.152 **Remark** The concept of degree of regularity and mutant are also closely related. Mutants can appear only at a degree equal to or higher than the degree of regularity [472].

16.3.4 The future

16.3.153 **Remark** What really drives the development of the designs in MPKCs are indeed new mathematical ideas that bring new mathematical structures and insights in the construction of MPKCs. The mathematical ideas we have used are just some of the very basic ideas developed in mathematics and there is great potential in advancing this idea further with some of the more sophisticated mathematical constructions in algebraic geometry. One particularly interesting problem would be to make the TTM cryptosystems [2113] work with the establishment of some new systematic approach. This definitely demands some deep insights and the usage of some intrinsically combinatorial structures from algebraic geometry.

16.3.154 Remark Though a lot has been done in studying the efficiency of different attacks, we still do not fully understand the full potential or the limitations of some of the attack algorithms. We still need to understand both the theory and practice of how efficiently general attack algorithms work and how to implement them efficiently. From the theoretical point of view, to answer these problems, the foundation again lies in modern algebraic geometry. One critical step would be to prove the maximum rank conjecture postulated in [853], the theoretical basis used to estimate the complexity of the polynomial solving algorithms, and the $\mathbf{F_4}$ algorithm. Another interesting problem is to mathematically prove some of the commonly used complexity estimate formulas in [3022].

16.3.155 Remark MPKCs interact more and more with other topics like algebraic attacks. Algebraic attacks are a very popular research topic in breaking symmetric block ciphers like AES [743] and stream ciphers [128] and analyzing hash functions [2740]. The origin of such an idea is from MPKCs, and in particular Patarin's linearization equation attack method. New ideas in MPKCs will have much more broad applications in the area of algebraic attacks. The idea of multivariate constructions was also applied to the symmetric constructions [281, 893]. Similar ideas may have further applications in designing stream ciphers and block ciphers. The theory of functions on a space over a finite field (multivariate functions) will play an increasingly important role in the unification of the research in all these related areas.

16.3.156 Remark Research in MPKCs has already developed new challenges that need new methods and ideas. A mutually beneficial interaction between MPKCs and algebraic geometry [748] will grow rapidly. MPKCs will provide excellent motivation and critical problems for the development of the theory of functions over finite fields, and new mathematical tools and insights are critical for MPKCs' future development.

See Also

§11.4	For algorithms to compute the roots of HFE polynomials.
§13.4	For the fast linear algebra used in multivariate polynomial solving algorithms.
§16.1	For public key cryptography and post-quantum cryptography.

References Cited: [67, 128, 201, 230, 280, 281, 390, 437, 438, 467, 472, 499, 500, 603, 604, 605, 722, 736, 737, 738, 739, 740, 742, 743, 748, 853, 860, 876, 877, 878, 879, 880, 881, 882, 883, 884, 885, 886, 887, 888, 889, 890, 891, 892, 894, 893, 895, 896, 924, 925, 926, 1040, 1041, 1042, 1044, 1049, 1094, 1095, 1214, 1266, 1285, 1339, 1340, 1349, 1440, 1461, 1610, 1686, 1687, 1737, 1738, 1739, 1877, 1914, 1985, 2028, 2029, 2113, 2114, 2115, 2116, 2213, 2220, 2226, 2312, 2320, 2321, 2367, 2364, 2365, 2366, 2368, 2369, 2370, 2371, 2387, 2415, 2606, 2613, 2740, 2798, 2822, 2823, 2929, 2930, 2931, 2932, 2998, 3022, 3023, 3024]

16.4 Elliptic curve cryptographic systems

Andreas Enge, INRIA Bordeaux–Sud-Ouest

16.4.1 Cryptosystems based on elliptic curve discrete logarithms

16.4.1 Remark The $\mathbb{F}_q$-rational points on an elliptic curve E defined over a finite field $\mathbb{F}_q$ form a finite abelian group according to Subsection 12.2.2; its group order is close to q by Theorem 12.2.46. This group can be used to implement the discrete logarithm based cryptosystems introduced in Subsection 16.1.3.2, as first observed in [1771, 2102].

16.4.2 Remark For reasons of efficiency, elliptic curve cryptosystems are usually implemented over prime fields $\mathbb{F}_p$ or fields $\mathbb{F}_{2^m}$ of characteristic two. Supersingular curves over fields $\mathbb{F}_{3^m}$ of characteristic three have attracted some attention in the context of pairing based cryptography, see Section 16.4.2.

16.4.1.1 Key sizes

16.4.3 Remark To resist generic attacks on the discrete logarithm problem, elliptic curve cryptosystems are implemented in the prime order cyclic subgroup of maximal cardinality n inside $E(\mathbb{F}_q)$. For representing group elements with the minimum number of bits, it is desirable that the curve order itself be prime. Except for special cases (see Section 16.4.1.3 and [2535, 2581, 2682]), only generic attacks are known on the elliptic curve discrete logarithm problem (ECDLP), with a running time on the order of $\sqrt{n}$. A security level of m bits, corresponding to a symmetric-key cryptosystem (see Section 16.1.2) with 2^m keys, thus requires an order n of $2m$ bits. Extrapolating the theoretical subexponential complexity of Remark 11.6.38 for factoring or the DLP in finite fields allows to derive heuristic security estimates for the corresponding public key cryptosystems of Sections 16.1.3.1 and 16.1.3.2. Several studies have been carried out in the literature, taking added heuristics on technological progress into account, see [1276]. They are summarized in the following table; the figures for the factorization based RSA system essentially carry over to systems based on discrete logarithms in finite fields; see Remark 11.6.38. The 80 bit security level is a historic figure.

security (bits)	symmetric	ECC	RSA [1276, 1894]	RSA [2426, §7.2.2.3]	RSA [2684, Table 7.2]
80	—	160	1513	1536	1248
112	Triple DES	224	4509	4096	2432
128	AES-128	256	6669	6000	3248
192	AES-192	384	22089	—	7936
256	AES-256	512	49562	—	15424

16.4.1.2 Cryptographic primitives

16.4.4 Remark Some cryptographic primitives (encryption, signatures, etc.) have been adapted and standardized specifically for elliptic curves. As other discrete logarithm based systems (see Section 16.1.3.2), they require a setup of public *domain parameters*, a cyclic subgroup G of prime order n of some curve $E(\mathbb{F}_q)$, with a fixed base point P such that $G = \langle P \rangle$. Moreover, the bit patterns representing elements of $\mathbb{F}_q$ and $E(\mathbb{F}_q)$ need to be agreed upon.

16.4.5 Example (Elliptic Curve Integrated Encryption Scheme, ECIES) This cryptosystem is essentially the same as ElGamal's, see Example 16.1.27; but the encryption of elements of G is replaced by symmetric key encryption of arbitrary bit strings with a derived secret key. So the scheme is *hybrid*, using symmetric key and public key elements. An additional *message authentication code (MAC)* prevents alterations of the encrypted message during transmission and authenticates its sender. (A MAC is essentially a hash function, see Remark 16.1.23, depending additionally on a symmetric key, and can indeed be constructed from hash functions; for more details, see [2080, Subsection 9.5.2].)

Besides the domain parameters for the elliptic curve group, the setup comprises a symmetric key scheme with an encryption function E_{k_1} and inverse decryption function D_{k_1}, using keys k_1 of length ℓ_1 bits; and a message authentication code M_{k_2} using keys k_2 of length ℓ_2. Party A has the private key $a \in [0, n-1]$ and the related public key $Q = aP$.

To encrypt a message $m \in \{0,1\}^*$, party B selects a random integer $r \in [0, n-1]$, computes $R = kP$, $S = kQ$ and $(k_1, k_2) = f(S)$, where $f : G \to \{0,1\}^{\ell_1} \times \{0,1\}^{\ell_2}$ is a *key derivation function* (for instance, a cryptographic hash function; see Remark 16.1.23). He computes $c_1 = E_{k_1}(m)$ and $c_2 = M_{k_2}(c_1)$; the ciphertext is (R, c_1, c_2).

To decrypt such a ciphertext, party A recovers $S = aR$ and $(k_1, k_2) = f(S)$. If $M_{k_1}(c_1) \neq c_2$, she rejects the ciphertext as invalid; otherwise, she obtains the clear text as $m = D_{k_1}(c_1)$.

16.4.6 Remark The scheme has been first described in a generic discrete logarithm setting (and in a slightly different form) in [221], and standardized under the name *Elliptic Curve Augmented Encryption Scheme* in [109]. For arguments supporting its security under suitable assumptions on the underlying primitives, see [221, 2683] and [313, Chapter III].

16.4.7 Example (Elliptic Curve Digital Signature Algorithm, ECDSA) The algorithm is a simple transposition of the DSA of Section 16.1.3.3 to the elliptic curve setting.

Besides the domain parameters for the elliptic curve group, the setup comprises a hash function $H : \{0,1\}^* \to [0, n-1]$ and the reduction function $f : G \to [0, n-1]$, $(x, y) \mapsto x \pmod{n}$.

Party A has the private key $a \in [0, n-1]$ and the related public key $Q = aP$.

To sign a message m, party A randomly selects an integer $k \in [1, n-1]$, computes $R = kP$, $r = f(R)$, $h = H(m)$, and $s \equiv k^{-1}(h + ar) \pmod{n}$. The signature is the pair (r, s).

To verify such a signature, party B computes $h = H(m)$, $w \equiv s^{-1} \pmod{n}$, $u_1 \equiv wh \pmod{n}$, $u_2 \equiv wr \pmod{n}$, and $R = u_1 P + u_2 Q$. He accepts the signature as valid if and only if $r = f(R)$.

16.4.8 Remark The scheme has been standardized in [108], see also [1133], [1567, Subsections 7.2.7–7.2.8], and [2292, Section 6]. For arguments supporting its security under suitable assumptions on the underlying primitives, see [313, Chapter II] and [420]. The fact that the function f depends only on the x-coordinate of its argument has raised doubts about the security of the scheme [2712]; in particular, it implies weak malleability: From a signature (r, s) on a given message, another signature $(r, -s)$ on the same message may be obtained.

16.4.1.3 Special curves

16.4.9 Remark A necessary condition for the security of an elliptic curve cryptosystem is that the order of $E(\mathbb{F}_q)$ be prime, or a prime multiplied by a small cofactor. Some special curves for which this condition is easily tested have been suggested in the literature. These are more and more deprecated in favor of random curves (see Section 16.4.1.4) in conventional discrete logarithm settings, [458]. Supersingular and especially CM curves are still needed, however, in pairing based cryptography; see Section 16.4.2.

16.4.10 Example (Supersingular curves) The orders of supersingular elliptic curves are known by Theorem 12.2.51: Over $\mathbb{F}_p$, the only occurring order is $p+1$. Over $\mathbb{F}_{p^m}$ with $p \in \{2,3\}$, the orders $p^m + 1 - t$ with $t \in \{0, \pm p^{m/2}, \pm p^{(m+1)/2}, \pm 2p^{m/2}\}$ may occur depending on the parity of m. The ECDLP on supersingular curves over $\mathbb{F}_{p^m}$ may be reduced to the DLP in the multiplicative group of $\mathbb{F}_{p^2}$ for curves over $\mathbb{F}_p$; of $\mathbb{F}_p$, $\mathbb{F}_{p^2}$, $\mathbb{F}_{p^3}$ or $\mathbb{F}_{p^4}$ for curves over $\mathbb{F}_{2^m}$; and of $\mathbb{F}_p$, $\mathbb{F}_{p^2}$, $\mathbb{F}_{p^3}$ or $\mathbb{F}_{p^6}$ for curves over $\mathbb{F}_{3^m}$. Thus, supersingular curves are deprecated except for low security pairing based cryptosystems.

16.4.11 Example (Curves over extension fields) If E is defined over a finite field $\mathbb{F}_q$ with q small, then $|E(\mathbb{F}_q)|$ can be obtained by exhaustively enumerating all points; and $|E(\mathbb{F}_{q^m})|$ is easily computed by Remark 12.2.105. In particular, the case $q = 2$ has been suggested in the literature. However, since $E(\mathbb{F}_{q^m})$ contains the subgroup $E(\mathbb{F}_q)$ (and further subgroups if m is not prime), the group order cannot be prime anymore.

16.4.12 Remark The existence of the additional Frobenius automorphism of order m, together with the negation automorphism of order 2, may be used to speed up the generic algorithms of Sections 11.6.6 and 11.6.7 by a factor of $\sqrt{2m}$ [1165, 2978], which reduces the effective security level.

16.4.13 Remark (Weil descent) If E is defined over an extension field $\mathbb{F}_{q^m}$, then $E(\mathbb{F}_{q^m})$ can be embedded into $A(\mathbb{F}_q)$, where A is an abelian variety of dimension m, called the *Weil restriction* or *restriction of scalars* of E. There is reason to believe that the discrete logarithm problem in $A(\mathbb{F}_q)$ may be easier to solve than by a generic algorithm, relying on an approach of representing the group $A(\mathbb{F}_q)$ by a set of generators (called the *factor base*) and relations which are solved by linear algebra, cf. Section 11.6.8, leading to a potential attack described first in [1104, Section 3.2]. Cases where A contains the Jacobian of a hyperelliptic curve of genus close to m have been worked out for curves over fields of characteristic 2 in [1160, 1250], and fields of odd characteristic in [852]. So far, the attack has been made effective for certain curves with prime $m \leq 7$.

Another algorithm for discrete logarithms, working directly with curves over $\mathbb{F}_{q^m}$ and specially adapted factor bases, is described in [1247]; heuristically, it is faster than the generic algorithms for $m \geq 3$ fixed and $q \to \infty$. Since it involves expensive Gröbner basis computations, it has been made effective only for $m \leq 3$.

Combinations of these approaches are also possible and have led to an attack on curves of close to cryptographic size over $\mathbb{F}_{p^6}$ [1629]. Moreover, isogenies may be used to transport the discrete logarithm problem from a seemingly secure curve to one that may be attacked by Weil descent [1156].

It thus appears cautious to prefer for cryptographic applications curves over prime fields $\mathbb{F}_p$ or, if even characteristic leads to significant performance improvements, fields $\mathbb{F}_{2^m}$ of prime extension degree m.

16.4.14 Example (Complex multiplication curves) All ordinary elliptic curves over a finite field $\mathbb{F}_q = \mathbb{F}_{p^m}$ have complex multiplication by some order $\mathcal{O}_D = \left[1, \frac{D+\sqrt{D}}{2}\right]_{\mathbb{Z}}$ of discriminant $D < 0$ in the imaginary-quadratic field $\mathbb{Q}(\sqrt{D})$; see Remark 12.2.98. For small $|D|$, this can be exploited to explicitly construct curves with a known number of points as follows.

1. Let $D < 0$, $D \equiv 0$ or $1 \pmod 4$, p prime and m minimal such that $4p^m = t^2 - v^2 D$ has a solution in integers t, v.
2. Compute the *class polynomial* $H_D \in \mathbb{Z}[X]$, the minimal polynomial of $j\left(\frac{D+\sqrt{D}}{2}\right)$, where j is the absolute elliptic modular invariant function.

3. H_D splits completely over $\mathbb{F}_{p^m}$ (and no subfield), and its roots are the j-invariants of the elliptic curves defined over $\mathbb{F}_{p^m}$ with complex multiplication by $\mathcal{O}_D$. For each such j-invariant, one easily writes down a curve with $p^m + 1 - t$ points by solving the expression of j in Definition 12.2.2 for the curve coefficients (up to isomorphisms and twists, see Section 12.2.5, the solution is unique).

16.4.15 Remark It is easy to see that a prime number of points is only possible for $D \equiv 5 \pmod 8$.

16.4.16 Remark The degree of the class polynomial is the class number of $\mathcal{O}_D$, and its total bit size is of the order of $O(|D|^{1+\epsilon})$ under GRH. Several quasi-linear algorithms of complexity $O(|D|^{1+\epsilon})$ for computing class polynomials have been described in the literature, by floating point approximations of its roots [978], lifting to a local field [744] or Chinese remaindering [220]. Nevertheless, the algorithms are restricted to small values of $|D|$, while random curves correspond to $|D|$ of the order of q, so that only a negligible fraction of curves may be reached by the CM approach.

16.4.17 Remark While no attack on this particular fraction of curves has been devised so far, random curves are generally preferred where possible; note, however, that pairing-based cryptosystems require the use of either supersingular curves or ordinary curves obtained with the CM approach; see Section 16.4.2.4.

16.4.18 Example (NIST curves) The USA standard [2292] suggests a prime field $\mathbb{F}_p$ and a pseudorandom curve (assuming that the hash function SHA-1 is secure) of prime order over $\mathbb{F}_p$ for p of 192, 224, 256, 384, and 521 bits. (The largest example is for the Mersenne prime $p = 2^{521} - 1$.) For the binary fields $\mathbb{F}_{2^{163}}$, $\mathbb{F}_{2^{233}}$, $\mathbb{F}_{2^{283}}$, $\mathbb{F}_{2^{409}}$ and $\mathbb{F}_{2^{571}}$, a pseudo-random curve (of order twice a prime) and a curve defined over $\mathbb{F}_2$ (of order twice or four times a prime) are given. As recommended in Remark 16.4.13, the extension degrees are prime for curves defined over $\mathbb{F}_2$.

16.4.19 Remark We note that the generic discrete logarithm algorithms of Subsections 11.6.6 and 11.6.7 allow for a trade-off between precomputations and the breaking of a given discrete logarithm: In a group of size about 2^m, a precomputation of 2^k group elements yields additional logarithms in time 2^{m-k}. As a precaution, one may thus wish to avoid predetermined curves, especially at lower security levels.

16.4.1.4 Random curves: point counting

16.4.20 Remark Algorithms for counting points on random elliptic curves currently come in two flavours. The first algorithm, SEA, is of polynomial complexity; for curves over extension fields $\mathbb{F}_{p^m}$, there are a variety of algorithms using p-adic numbers, with a much better polynomial exponent in m, but which are exponential in $\log p$.

16.4.21 Algorithm (Schoof) In [2560], Schoof describes the first algorithm of complexity polynomial in $\log q$ for counting the number of points on an arbitrary elliptic curve $E(\mathbb{F}_q)$. The algorithm is deterministic and computes the trace of Frobenius a_q of Definition 12.2.47 and thus the zeta function of Section 12.2.10. Given a prime ℓ not dividing q, by Theorem 12.2.66 the value of a_q modulo ℓ can be determined by checking for all possible values whether the numerator of the zeta function annihilates the ℓ-torsion points. Chinese remaindering for sufficiently many primes yields the exact value of a_q, which is bounded by Theorem 12.2.46. The algorithm has a complexity of $O\left((\log q)^{5+\epsilon}\right)$, due in part to the fact that the ℓ-torsion points generate an $\mathbb{F}_q$-algebra of dimension $O(\ell^2)$.

16.4.22 Algorithm (Schoof–Elkies–Atkin, SEA) Improvements are due to Atkin and Elkies [970]. When there is an $\mathbb{F}_q$-rational separable isogeny (see Definition 12.2.26) of degree ℓ from

$E(\mathbb{F}_q)$ to another curve, then the ℓ-torsion points may be replaced by the kernel of the isogeny, generating an algebra of dimension $O(\ell)$ over $\mathbb{F}_q$. By the complex multiplication theory of Example 16.4.14, this happens when ℓ is coprime to the conductor of the ring of endomorphisms $\mathcal{O}_D$ of E and ℓ is not inert in the quadratic number field $\mathbb{Q}(\sqrt{D})$, which holds for about half of the primes. The complexity of the algorithm becomes $O\left((\log q)^{4+\epsilon}\right)$ [312, Chapter VII], [661, Section 17.2].

16.4.23 Remark The practical bottleneck of the algorithm used to be the computation of bivariate modular polynomials, of size $O(\ell^{3+\epsilon})$, needed to derive isogenies of degree ℓ. A quasi-linear algorithm is described in [979]; eventually limited by space, it has been used for ℓ up to around 10000. A more recent algorithm [2752] computes the polynomial, reduced modulo the characteristic p of $\mathbb{F}_q$ and instantiated in one variable by an element of $\mathbb{F}_q$, also in time $O(\ell^{3+\epsilon})$, but in space $O(\ell(\ell + \log q))$; it has been used for ℓ up to about 100000. Further building blocks of the SEA algorithm have also been optimized [366, 1251, 2098]. The current record is for a prime field $\mathbb{F}_p$ with p having about 5000 decimal digits [2751].

16.4.24 Remark The SEA algorithm is implemented in several major computer algebra systems, and random elliptic curves of cryptographic size with a prime number of points are easily found, be it as domain parameters, be it in a setting where each user has his own elliptic curve as part of his public key.

16.4.25 Algorithm (p-adic point counting) For an elliptic curve E over an extension field $\mathbb{F}_{p^m}$, Satoh [2533] introduced an algorithm computing its *canonical lift* to a curve $\hat{E}$ over $\mathbb{Q}_{p^m}$, the unramified extension of degree m of the p-adic numbers $\mathbb{Q}_p$. The curve $\hat{E}$ has the same endomorphism ring $\mathcal{O}_D$ (Example 16.4.14) as E and reduces modulo the maximal ideal of $\mathbb{Q}_{p^m}$ to E. More precisely, an approximation to $\hat{E}$ may be computed by Newton iterations on a function derived from the modular polynomial of level p, Algorithm 16.4.21, at arbitrary p-adic precision. In a second step, the trace of the Frobenius map is computed in this characteristic 0 setting by the action of its dual isogeny (the reduction of which is separable) on a holomorphic differential; for this, the isogenies are computed explicitly. After a precomputation of $O(p^{3+\epsilon})$ for the p-th modular polynomial (see Algorithm 16.4.21), the complexity of the algorithm is $O(p^2 m^{3+\epsilon})$.

16.4.26 Remark Satoh's algorithm is not immediately applicable in characteristic two. Mestre suggests in [2088] to use arithmetic-geometric mean (AGM) iterations, a sequence of isogenies of degree 2, to obtain the canonical lift and the trace of the Frobenius map, also in time $O(m^{3+\epsilon})$.

16.4.27 Remark Later work concentrates on lowering the complexity in m: to quasi-quadratic for finite fields $\mathbb{F}_{q^m}$ with a Gaussian normal basis [1904] or in the general case [1422]; or on lowering the complexity in p: to quasi-linear [1248] or even quasi-square root [1429]. The record in [1904] for a curve over $\mathbb{F}_{2^{100002}}$ goes beyond all practical cryptographic needs.

16.4.28 Remark For a more thorough account, see [313, Chapter VI] or [661, Section 17.3].

16.4.2 Pairing based cryptosystems

16.4.29 Remark While conventional elliptic curve cryptography relies on the map $x \mapsto xP$, which is a group homomorphism or, equivalently, a linear map of $\mathbb{Z}/n\mathbb{Z}$-modules, pairing based cryptography requires a bilinear map $e : G_1 \times G_2 \to G_3$, see Definition 16.1.35. This introduces an additional degree of freedom and a wealth of new cryptographic primitives following the first applications described in Section 16.1.4. Since 2007, a series of conferences, *Pairing-Based Cryptography — Pairing*, has been devoted to the topic [1159, 1630, 2601, 2766].

16.4.2.1 Cryptographic pairings

16.4.30 **Definition** Let E be an elliptic curve defined over a finite field $\mathbb{F}_q$, and let n be the largest prime divisor of the cardinality of $E(\mathbb{F}_q)$. Assume that n does not divide q. (This is required in a cryptographic setting due to anomalous curves.) Then the *embedding degree* of E is the smallest integer k such that $E(\mathbb{F}_{q^k})$ contains $E(\overline{\mathbb{F}}_q)[n]$, the n^2 points of n-torsion of $E(\overline{\mathbb{F}}_q)$; see Theorem 12.2.60, i.e., k is minimal such that $E(\mathbb{F}_{q^k})[n] = E(\overline{\mathbb{F}}_q)[n]$.

16.4.31 **Theorem** [183, Theorem 1] If n does not divide $q-1$, then the embedding degree is the smallest integer k such that n divides $q^k - 1$.

16.4.32 **Definition** A *cryptographic elliptic pairing* is a map $e : G_1 \times G_2 \to G_3$ satisfying the conditions of Definition 16.1.35, where E is an elliptic curve defined over $\mathbb{F}_q$, n is the largest prime factor of $|E(\mathbb{F}_q)|$, n divides neither $q-1$ nor q, k is the embedding degree of E, and $G_1 \subseteq E(\mathbb{F}_q)$ and $G_2 \subseteq E(\mathbb{F}_{q^k})$ (denoted additively) and $G_3 \subseteq \mathbb{F}_{q^k}^*$ (denoted multiplicatively) are subgroups of order n.

16.4.33 **Remark** In this setting, $G_1 = E(\mathbb{F}_q)[n]$ and G_3 are in fact fixed, while there are $n+1$ possible choices for G_2; see Subsection 16.4.2.3. Diagonalizing the matrix of the Frobenius endomorphism on $E(\overline{\mathbb{F}}_q)[n]$ by Theorem 12.2.66 yields a mathematically canonical choice also for G_2, which is given by the following theorems.

16.4.34 **Theorem** G_1 is the subgroup of $E(\mathbb{F}_{q^k})[n]$ generated by the points having eigenvalue 1 under the Frobenius endomorphism ϕ_q of Example 12.2.31. There is a unique subgroup $\overline{G}_2 \subseteq E(\mathbb{F}_{q^k})$ of order n generated by the points having eigenvalue q under the Frobenius endomorphism.

16.4.35 **Theorem** Let $\mathrm{Tr} : E(\overline{\mathbb{F}}_q) \to E(\overline{\mathbb{F}}_q)$, $P \mapsto \sum_{i=0}^{k-1} \phi_q^i(P)$, denote the *trace endomorphism* of level k on E. Then the endomorphisms Tr and $\pi_2 = \mathrm{id} - \phi_q$, restricted to $E(\mathbb{F}_{q^k})[n]$, yield surjective group homomorphisms $\mathrm{Tr} : E(\mathbb{F}_{q^k})[n] \to G_1$ with kernel $\overline{G}_2$ and $\pi_2 : E(\mathbb{F}_{q^k})[n] \to \overline{G}_2$ with kernel G_1.

16.4.36 **Definition** For a point P on E defined over some extension field $\mathbb{F}_{q^m}$ and an integer r, let $f_{r,P}$ be the function with divisor $r(P) - (rP) - (r-1)(\mathcal{O})$ that is defined over $\mathbb{F}_{q^m}$ and has leading coefficient 1 in $\mathcal{O}$, see Definitions 12.1.21 and 12.1.23.

For finite points R and $S \neq -R$, denote by $v_R = x - x(R)$ the line with divisor $(R) + (-R) - 2(\mathcal{O})$ and by $\ell_{R,S} = (y - y(R)) - \lambda_{R,S}(x - x(R))$ the line with divisor $(R) + (S) + (-R-S) - 3(\mathcal{O})$, where

$$\lambda_{R,S} = \begin{cases} \frac{y(S)-y(R)}{x(S)-x(R)} & \text{if } R \neq S, \\ \frac{3x(R)^2+2a_2x(R)+a_4-a_1y(R)}{2y(R)+a_1x(R)+a_3} & \text{if } R = S, \end{cases}$$

see Algorithm 12.2.21; additionally, $\ell_{R,-R} = v_R$ and $v_{\mathcal{O}} = 1$.

16.4.37 **Definition** An *addition-negation chain* for an integer r is a sequence $r_1, \ldots, r_s$ such that $r_1 = 1$, $r_s = r$ and each element r_i is either

1. the negative of a previously encountered one: there is $1 \leq j(i) < i$ such that $r_i = -r_{j(i)}$; or
2. the sum of two previously encountered ones: there are $1 \leq j(i) \leq k(i) < i$ such that $r_i = r_{j(i)} + r_{k(i)}$.

16.4.38 Algorithm

Require: A point P on E and an integer r
Ensure: $f_{r,P} = \frac{L}{V}$, where L and V are given as products of lines
Compute an addition-negation chain $r_1, \ldots, r_s$ for r.
$P_1 \leftarrow P$, $L_1 \leftarrow 1$, $V_1 \leftarrow 1$
for $i = 2, \ldots, s$ **do**
  $j \leftarrow j(i), k \leftarrow k(i)$
  if $r_i = -r_j$ **then**
    $P_i \leftarrow -P_j$
    $L_i \leftarrow V_j$
    $V_i \leftarrow L_j v_{P_i}$
  else
    $P_i \leftarrow P_j + P_k$
    $L_i \leftarrow L_j L_k \ell_{P_{j(i)},P_{k(i)}}$
    $V_i \leftarrow V_j V_k v_{P_i}$
  end if
end for
return $L = L_s$, $V = V_s$

16.4.39 Example The *Weil pairing* e_n of Theorem 12.2.70 is a cryptographic pairing as long as $G_2 \neq G_1$. If P, $Q \in E(\mathbb{F}_{q^k})[n]$, then $e_n(P, Q) = (-1)^n \frac{f_{n,P}(Q)}{f_{n,Q}(P)}$.

16.4.40 Example Assume that $E(\mathbb{F}_{q^k})$ does not contain a point of order n^2, or, equivalently, that n^3 does not divide $|E(\mathbb{F}_{q^k})|$. Then the map $E(\mathbb{F}_{q^k})[n] \to E(\mathbb{F}_{q^k})/nE(\mathbb{F}_{q^k})$, $Q \mapsto Q + nE(\mathbb{F}_{q^k})$, is a group isomorphism, and the *Tate pairing* T of Theorem 12.2.75 yields a non-degenerate pairing

$$e'_T : E(\mathbb{F}_{q^k})[n] \times E(\mathbb{F}_{q^k})[n] \to \mathbb{F}_{q^k}^*/(\mathbb{F}_{q^k}^*)^n, \quad (P, Q) \mapsto \mathsf{T}(P, Q + nE(\mathbb{F}_{q^k})).$$

Since $e'_T|_{G_1 \times G_1}$ takes values in $\mathbb{F}_q^*/(\mathbb{F}_q^*)^n = \{1\}$, the restriction $e'_T|_{G_1 \times G_2}$ is non-degenerate for any $G_2 \neq G_1$.

The *reduced Tate pairing*

$$e_T : G_1 \times G_2 \to G_3, \quad (P, Q) \mapsto \big(\mathsf{T}(P, Q + nE(\mathbb{F}_{q^k}))\big)^{(q^k-1)/n},$$

is a cryptographic pairing for any $G_2 \neq G_1$. It is computed as

$$e_T(P, Q) = (f_{n,P}(Q))^{(q^k-1)/n}.$$

16.4.41 Remark We observe that during the computation of the reduced Tate pairing by Algorithm 16.4.38, all factors lying in a subfield of $\mathbb{F}_{q^k}$ may be omitted due to the final exponentiation. In particular, if the x-coordinate of Q lies in a subfield, then all v_{P_i} may be dropped, a technique known as *denominator elimination*; see Remark 16.4.51.

16.4.42 Definition A *distortion map* is an effectively computable endomorphism $\psi : E(\overline{\mathbb{F}}_q) \to E(\overline{\mathbb{F}}_q)$ that restricts to an isomorphism $\psi : G_1 \to G_2$ for some subgroup $G_2 \neq G_1$.

16.4.43 Remark Since G_1 is invariant under the Frobenius ϕ_q, but G_2 is not, the endomorphisms ψ and ϕ_q cannot commute. So the existence of ψ implies that E is supersingular; see Theorems 12.2.82 and 12.2.87. Conversely, for supersingular curves, there are distortion maps $\psi : G_1 \to G_2$ for any $G_2 \neq G_1$ [2869, Theorem 5].

16.4.44 Example Let E be a supersingular curve with distortion map ψ and $G_2 = \psi(G_1)$. Let $e : G_1 \times G_2 \to G_3$ be a cryptographic pairing. Then

$$e' : G_1 \times G_1 \to G_3, \quad (P, Q) \mapsto e(P, \psi(Q)),$$

is a cryptographic pairing in which both arguments come from the same group G_1. This setting is sometimes called a *symmetric pairing* in the literature, although it does not in general satisfy $e(P, Q) = e(Q, P)$; see also Section 16.4.2.3.

16.4.45 Remark Further work has produced a variety of pairings with a shorter loop in Algorithm 16.4.38, that is, defined by some function $f_{r,P}$ with $r < n$. In general, this is obtained by choosing special curves and restricting to the subgroups G_1 and $\overline{G}_2$ of Theorem 16.4.34. Since all involved groups are cyclic, such pairings are necessarily powers of the Tate pairing.

16.4.46 Example (Eta pairing) Let E be a supersingular curve with even $k = 2a$ and distortion map ψ as in Definition 16.4.42. Let $T = t - 1$, where t is the trace of the Frobenius map, see Definition 12.2.47. Then $T \equiv q \pmod{n}$ and $n \mid (T^a + 1)$. Assume that $n^2 \nmid (T^a + 1)$. Then the map

$$G_1 \times G_1 \to G_3, \quad (P, Q) \mapsto f_{T,P}(\psi(Q))^{aT^{a-1}\frac{q^k-1}{n}},$$

is a cryptographic pairing. For a proof, see [203, Section 4] and [1494, Section III]. By Example 16.4.62, only curves over fields of characteristic two or three may satisfy the assumptions of the theorem. Notice that T is of order $\sqrt{q}$ by Theorem 12.2.46, so that in the best case $\rho \approx 1$ (see Definition 16.4.66) the loop length in Algorithm 16.4.38 is reduced by a factor of about 2, while the final exponentiation becomes more expensive.

16.4.47 Example (Ate pairing) Let $T = t - 1$, where t is the trace of the Frobenius map, and assume that $n^2 \nmid (T^k - 1)$. Then the map

$$\overline{G}_2 \times G_1 \to G_3, \quad (Q, P) \mapsto f_{T,Q}(P)^{\frac{q^k-1}{n}},$$

is a cryptographic pairing [1494]. Notice that the roles of G_1 and $\overline{G}_2$ are inverted compared to the reduced Tate pairing of Example 16.4.40. Thus, as a price to pay for the loop shortening in Algorithm 16.4.38, the number of operations in $\overline{G}_2$ and thus $\mathbb{F}_{q^k}$ increases.

16.4.48 Conjecture (Optimal ate pairing) A loop length of essentially $\frac{\log_2 n}{\varphi(k)}$, where φ is Euler's function, may be obtained for a pairing of the previous type, for instance via a product of functions $f_{c_i,Q}(P)^{q^i \frac{q^k-1}{n}}$ with $\sum \log_2 c_i$ of the desired magnitude; concrete instances have been obtained using lattice reduction [1492, 2868].

16.4.2.2 Pairings and twists

16.4.49 Theorem Assume that E is defined over the field $\mathbb{F}_q$ of characteristic at least 5 and that $d \in \{2, 3, 4, 6\}$ is such that $d \mid \gcd(k, \# \operatorname{Aut}(E))$. By Proposition 12.2.59, there is, besides E itself and up to equivalence, precisely one twist E' of degree d such that $n \mid \#E'(\mathbb{F}_{q^{k/d}})$. As can be seen from Proposition 12.2.57, there is an isomorphism $\varphi : E' \to E$ which is defined over $\mathbb{F}_{q^d}$. The subgroup G_2' of order n of $E'(\mathbb{F}_{q^{k/d}})$ satisfies $\varphi(G_2') = \overline{G}_2$.

16.4.50 Remark Theorem 16.4.49 implies that in the presence of twists, elements of $\overline{G}_2$ are more compactly represented by elements of G_2'; or otherwise said, any cryptographic pairing $e : G_1 \times \overline{G}_2 \to G_3$ yields an equivalent cryptographic pairing $e' : G_1 \times G_2' \to G_3$, $(P, Q') \mapsto e(P, \varphi(Q'))$.

16.4.51 Remark Theorem 16.4.49 and the explicit form of φ given in Proposition 12.2.57 show that the x-coordinates of elements in $\overline{G}_2$ lie in $\mathbb{F}_{q^{k/2}}$ for d even and that the y-coordinates lie in $\mathbb{F}_{q^{k/3}}$ when $3 \mid d$. This may allow for simplifications of Algorithm 16.4.38 in conjunction with the final exponentiation; see Remark 16.4.41.

16.4.52 Example (Twisted ate pairing) Under the hypotheses of Theorem 16.4.49, let $T = t - 1$, where t is the trace of the Frobenius map, and assume that $n^2 \nmid (T^k - 1)$. Then the map

$$G_1 \times \overline{G}_2 \to G_3, \quad (P, Q) \mapsto f_{T^{k/d},P}(Q)^{\frac{q^k-1}{n}},$$

is a cryptographic pairing [1494]. Here, the roles of G_1 and $\overline{G}_2$ are again as in the reduced Tate pairing of Example 16.4.40. However, compared to the ate pairing of Example 16.4.47, the loop length in Algorithm 16.4.38 is increased by a factor of $\frac{k}{d}$. Unless t is smaller than generically expected, the twisted ate pairing is in fact less efficient to compute than the reduced Tate pairing.

16.4.2.3 Explicit isomorphisms

16.4.53 Remark For the sake of giving security arguments for pairing based systems, the cryptologic literature has taken to distinguishing pairings according to the possibility of moving efficiently between the groups G_1 and G_2. For instance, if $G_1 = G_2$, then the decisional Diffie-Hellman problem is easy in G_1: Given P, aP, bP and $R \in G_1$, one has $R = abP$ if and only if $e(P, R) = e(aP, bP)$.

16.4.54 Definition Let e be a cryptographic pairing in the sense of Definition 16.4.32. It is of

1. *type 1* if $G_1 = G_2$;
2. *type 2* if there is an efficiently computable isomorphism $\psi : G_2 \to G_1$, but no such isomorphism $G_1 \to G_2$ is known;
3. *type 3* if no efficiently computable isomorphisms $G_1 \to G_2$ or $G_2 \to G_1$ are known.

16.4.55 Remark We note that since G_1 and G_2 are cyclic of the same order n, they are trivially isomorphic; but exhibiting an effective isomorphism may require to compute discrete logarithms. In general, an efficiently computable isomorphism will be given by an endomorphism of the elliptic curve.

16.4.56 Example The pairing of Example 16.4.44 on supersingular curves with distortion map is of type 1. Any pairing with $G_2 \neq G_1, \overline{G}_2$ is of type 2: The isomorphism is given by the trace map Tr of Theorem 16.4.35. To the best of our knowledge, pairings with $G_2 = \overline{G}_2$ are of type 3; at least the trace is trivial on $\overline{G}_2$.

16.4.57 Remark The terminology *type 4* has been used for pairings in which the second argument comes from the full n-torsion group; in this case, G_2 can be seen as the group generated by this argument, which may vary with each use of the cryptographic primitive. As it is then unlikely that $G_2 = G_1$ or $\overline{G}_2$, a type 4 pairing essentially behaves as a type 2 pairing.

16.4.58 Remark Type 1 pairings, being restricted to supersingular curves, offer a very limited choice of embedding degrees, see Example 16.4.62. Type 2 pairings are sometimes preferred in the cryptographic literature since they appear to facilitate certain security arguments. On the other hand, the existence of ψ implies that the decisional Diffie-Hellman problem is easy in G_2, and it is apparently not possible to hash into any subgroup G_2 different from G_1 and $\overline{G}_2$; see Subsection 16.4.2.5. Recent work introduces a heuristic construction to transform a cryptographic primitive in the type 2 setting, together with its security argument, into an equivalent type 3 primitive [597].

16.4.59 Remark Some cryptographic primitives have been formulated with a pairing on subgroups of composite order n. More precisely, n is the product of two primes that are unknown to the general public, but form part of the private key as in the RSA system of Section 16.1.3.1. Such pairings can be realized either with supersingular curves [345, Section 2.1] or using Algorithm 16.4.64; the former leads to a ρ-value (Definition 16.4.66) at least 2, the latter to a ρ-value close to 2. There is a heuristic approach to transform such cryptosystems, together with their security proofs, into the setting of prime order subgroups [1098].

16.4.2.4 Curve constructions

16.4.60 Remark The existence of a cryptographic pairing $e : G_1 \times G_2 \to G_3$ reduces the discrete logarithm problem in G_1 or G_2 to that of G_3: For instance, given a point $P \in G_1$ and a multiple xP, choose a point $Q \in G_2$ such that $\zeta = e(P,Q) \neq 1$. Then $\xi = e(P, xQ) = e(P,Q)^x$, and x is the discrete logarithm of $\xi \in G_3$ to the base ζ [1108, 2079].

16.4.61 Remark Thus to balance the difficulty of the discrete logarithm problems in the elliptic curve groups G_1 and G_2 over $\mathbb{F}_q$ and $G_3 \subseteq \mathbb{F}_{q^k}^*$, the embedding degree k should be chosen according to the security equivalences in Section 16.4.1.1. For instance, if one follows the recommendations of [2684], for a system of 256 bit security one would choose $n \approx 2^{512}$ and thus $q \approx 2^{512}$, and $k \approx \frac{15425}{512} \approx 30$. Since by Theorem 16.4.31 the embedding degree k equals the order of n in $\mathbb{F}_q$, it will be close to q for random curves. Hence one needs special constructions to obtain *pairing-friendly curves*, curves with a prescribed, small value of k. For a comprehensive survey, see [1101].

16.4.62 Example (Supersingular curves) As first noticed in [2079], the embedding degree is always exceptionally small for supersingular curves. The following table gives the possible cardinalities according to Theorem 12.2.51, the maximal size n of a cyclic subgroup by [2497] and the embedding degree k with respect to n.

$\|E(\mathbb{F}_q)\|$	n	k
$q+1$	$q+1$	2
$q+1\pm\sqrt{q}$	$q+1\pm\sqrt{q}$	3
$q+1\pm\sqrt{2q}$	$q+1\pm\sqrt{2q}$	4
$q+1\pm\sqrt{3q}$	$q+1\pm\sqrt{3q}$	6
$q+1\pm 2\sqrt{q}$	$\sqrt{q}\pm 1$	1

16.4.63 Remark All algorithms for finding ordinary pairing-friendly curves rely on complex multiplication constructions, cf. Example 16.4.14, and construct curves over prime fields only.

16.4.64 Algorithm A very general method is due to Cocks and Pinch [1101, Section 4.1]. It allows to fix the desired group order n beforehand; choosing a low Hamming weight in the binary decomposition of n or more generally a value of n with a short addition-subtraction chain speeds up Algorithm 16.4.38.

Require: An integer $k \geq 2$, a quadratic discriminant $D < 0$ and a prime n such that $k \mid (n-1)$ and the Legendre symbol $\left(\frac{D}{n}\right) = 1$
Ensure: A prime p and an elliptic curve $E(\mathbb{F}_p)$ (with complex multiplication by $\mathcal{O}_D$) having a subgroup of order n and embedding degree k
repeat
$\zeta \leftarrow$ an integer such that ζ modulo n is a primitive k-th root of unity in $\mathbb{F}_n^*$
$t \leftarrow \zeta + 1$

$v \leftarrow$ an integer such that $v \equiv \frac{t-2}{\sqrt{D}} \pmod{n}$
$p \leftarrow \frac{t^2-v^2D}{4}$
until p is an integer and prime
Then $p \equiv t-1 \pmod{n}$
Construct the curve E over $\mathbb{F}_p$ with $p+1-t$ points as in Example 16.4.14

16.4.65 Remark Generically, in this construction t and v will be close to n, so that p will be close to n^2. This motivates the following definition.

16.4.66 Definition The *ρ-value* of a pairing-friendly curve is given by

$$\rho = \frac{\log p}{\log n}.$$

16.4.67 Remark By Theorem 12.2.46, the superior limit of ρ is at least 1 for $p \to \infty$. Values of ρ larger than 1 result in a loss of bandwidth when transmitting elements of G_1, which is a $\log_2 n$-bit subgroup embedded into a $\rho \log_2 n$-bit group, and a less efficient arithmetic in the elliptic curve. The security equivalences of Section 16.4.1.1 do in fact not fix the value of k, but that of ρk; so different values of k may lead to comparable security levels.

16.4.68 Remark Further research has concentrated on finding families of pairing-friendly curves, the parameters of which are given by values of polynomials.

16.4.69 Algorithm [413] The following is a direct transcription of Algorithm 16.4.64 to polynomials.
Require: An integer $k \geq 2$ and a quadratic discriminant $D < 0$
Ensure: Polynomials p and $n \in \mathbb{Q}(x)$ such that if the values $p(x_0)$ and $n(x_0)$ are simultaneously prime integers, then there is an elliptic curve $E(\mathbb{F}_{p(x_0)})$ (with complex multiplication by $\mathcal{O}_D$) having a subgroup of order $n(x_0)$ and embedding degree k
$n \leftarrow$ an irreducible polynomial in $\mathbb{Q}[x]$ such that the number field $K = \mathbb{Q}[x]/(n)$ contains $\sqrt{D}$ and a primitive k-th root of unity
$z \leftarrow$ a polynomial in $\mathbb{Q}[x]$ that reduces to a primitive k-th root of unity ζ in K
$t \leftarrow z+1$
$v \leftarrow$ a polynomial in $\mathbb{Q}[x]$ that reduces to the element $\frac{\zeta-1}{\sqrt{D}}$ in K
$s \leftarrow$ a polynomial in $\mathbb{Q}[x]$ that reduces to $\sqrt{D}$ in K
$v \leftarrow \frac{(z-1)s}{D} \bmod n$
$p \leftarrow \frac{t^2-Dv^2}{4}$

16.4.70 Remark The polynomials p and n need not represent primes or even integers; choosing small values of $|D|$, and n such that z, $s \in \mathbb{Z}[X]$ may help. Let $d = \deg(n)$ be the degree of K. While it is always possible to choose n such that either z or v is of low degree (as low as 1 if n is the minimal polynomial of the corresponding algebraic number), it is a priori not clear whether *both* can be chosen of low degree. Generically, p is of degree $2(d-1)$, and the asymptotic ρ-value of the family is $2 - \frac{2}{d}$, a small improvement over Algorithm 16.4.64. In many cases, however, actual ρ-values are much closer to 1, as demonstrated by the following example.

16.4.71 Example [413, p. 137] Let k be odd, $D = -4$ and $K = \mathbb{Q}(\zeta, \sqrt{-1}) = \mathbb{Q}[x]/(\Phi_{4k}(x))$ where $\Phi_{4k}(x) = \Phi_k(-x^2)$ is the $4k$-th cyclotomic polynomial. Choose $\zeta(x) = -x^2$, $t(x) = -x^2+1$, $s(x) = 2x^k$, $v(x) = \frac{1}{2}(x^{k+2}+x^k)$, $p(x) = \frac{1}{4}(x^{2k+4}+2x^{2k+2}+x^{2k}+x^4-2x^2+1)$. The polynomial p takes integral values in odd arguments and, conjecturally, represents primes if it is irreducible (since $p(1) = 1$, there is no local obstruction to representing primes).

Asymptotically for $p \to \infty$, $\rho \to \frac{k+2}{\varphi(k)}$, and $\rho \to 1$ if furthermore $k \to \infty$ with a fixed number of prime factors.

16.4.72 Remark Similar results hold for even k, and for $D = -3$ since $\sqrt{-3} \in \mathbb{Q}[x]/(\Phi_3(x))$.

16.4.73 Remark The following table, taken from [1101], gives the current best values of $\overline{\rho} = \frac{\deg p}{\deg n}$ for polynomial families of pairing-friendly curves for $k \geq 4$. (Smaller values of k may be obtained for prime fields using supersingular curves; see Example 16.4.62.) For the constructions behind each family, see [1101].

k	$\deg p$	$\deg n$	$\overline{\rho}$	$k\overline{\rho}$
4	2	2	1.00	4.0
5	14	8	1.75	8.8
6	2	2	1.00	6.0
7	16	12	1.33	9.3
8	10	8	1.25	10.0
9	8	6	1.33	12.0
10	4	4	1.00	10.0
11	24	20	1.20	13.2
12	4	4	1.00	12.0
13	28	24	1.17	15.2
14	16	12	1.33	18.7
15	12	8	1.50	22.5
16	10	8	1.25	20.0
17	36	32	1.12	13.8
18	8	6	1.33	24.0
19	40	36	1.11	21.1
20	22	16	1.38	27.5
21	16	12	1.33	28.0
22	26	20	1.30	28.6
23	48	44	1.09	25.1
24	10	8	1.25	30.0
25	52	40	1.30	32.5
26	28	24	1.17	30.3
27	20	18	1.11	30.0

k	$\deg p$	$\deg n$	$\overline{\rho}$	$k\overline{\rho}$
28	16	12	1.33	37.3
29	60	56	1.07	31.1
30	12	8	1.50	45.0
31	64	60	1.07	33.1
32	34	32	1.06	34.0
33	24	20	1.20	39.6
34	36	32	1.12	38.2
35	72	48	1.50	52.5
36	14	12	1.17	42.0
37	76	72	1.06	39.1
38	40	36	1.11	42.2
39	28	24	1.17	45.5
40	22	16	1.38	55.0
41	84	80	1.05	43.0
42	16	12	1.33	56.0
43	88	84	1.05	45.0
44	46	40	1.15	50.6
45	32	24	1.33	60.0
46	50	44	1.14	52.3
47	96	92	1.04	49.0
48	18	16	1.12	54.0
49	100	84	1.19	58.3
50	52	40	1.30	65.0

16.4.2.5 Hashing into elliptic curves

16.4.74 Remark Hashing into elliptic curve groups is often required for pairing-based cryptosystems, see for instance Algorithms 16.1.40 and 16.1.46. Standard cryptographic hash functions (see Remark 16.1.23) $\{0,1\}^* \to \{0,1\}^\ell$ of sufficient length ℓ are easily modified to yield values in $\mathbb{Z}/nZ$ (by reduction modulo n) and to arbitrary finite fields (by hashing to coefficients with respect to a fixed basis). One would like to extend such constructions to elliptic curves.

16.4.75 Remark In the setting of Definition 16.4.32, if $H : \{0,1\}^* \to \mathbb{Z}/n\mathbb{Z}$ is a collision-resistant hash function and G_1 is generated by a point P of order n, then the function $\{0,1\}^* \to G_1$, $m \mapsto H(m)P$, is trivially collision-resistant. However, this simple construction reveals the discrete logarithm of the hash value, which in general renders the cryptosystem totally insecure.

16.4.76 Remark In the following, let E be an elliptic curve defined over $\mathbb{F}_q$ as in Definition 16.4.32, and assume that $n^3 \nmid |E(\mathbb{F}_{q^k})|$. If one can hash into $E(\mathbb{F}_q)$, then one can also hash into $G_1 = E(\mathbb{F}_q)[n]$: It suffices for that to multiply the result by the cofactor $\frac{|E(\mathbb{F}_q)|}{n}$. The same argument holds for $E(\mathbb{F}_{q^k})[n]$. However, it is then in general not possible to project into an arbitrary group G_2. For G_2 as in Theorem 16.4.35, that is, type 3 pairings as in

Definition 16.4.54, the trace $\mathrm{Tr} : E(\mathbb{F}_{q^k})[n] \to \overline{G}_2$ can be used to obtain a hash function with values in $\overline{G}_2$. Alternatively, in the presence of twists as described in Theorem 16.4.49, one may more efficiently hash into the subgroup G_2' on the twisted curve, for which the cofactor is smaller.

To hash into $E(k)$ where $k = \mathbb{F}_q$ or $k = \mathbb{F}_{q^k}$, one may use a hash function $H : \{0,1\} \to k$ to obtain the x-coordinate of a point. As not all elements of k occur as x-coordinates, one may need several trials. A possibility is to concatenate the message m with a counter i, denoted by $m||i$, and to increase the counter until $H(m||i)$ is the x-coordinate of a point on E. An additional hash bit may be used to determine one of the generically two points with the given x-coordinate. The algorithm is deterministic and, if H is modeled as a random function, it needs an expected number of two trials averaged over all input values. However, for $|k| \to \infty$, there is a doubly exponentially small fraction of the input values that will take exponential time. Several recent results exhibit special cases in which polynomial time hashing is possible uniformly for all input values.

16.4.77 Example [312, Section 4.1] If $q \equiv 2 \pmod 3$, then $E : y^2 = x^3 + 1$ is a supersingular curve over $\mathbb{F}_q$ with $q+1$ points and $k = 2$. Precisely, since third powering is a bijection on $\mathbb{F}_q$ with inverse $z \mapsto z^{1/3} = z^{(2q-1)/3}$, the map $\mathbb{F}_q \to E(\mathbb{F}_q)\backslash\{\mathcal{O}\}$, $y \mapsto \left((y^2-1)^{(2q-1)/3}, y\right)$, is a bijection.

16.4.78 Example [2604] Let $E : y^2 = f(x) = x^3 + a_2x^2 + a_4x + a_6$ over $\mathbb{F}_q$ of characteristic at least 3. There are explicit rational functions $u_1(t)$, $u_2(t)$, $u_3(t)$, and $v(t)$ such that $v(t)^2 = f(u_1(t^2))f(u_2(t^2))f(u_3(t^2))$ [2675]. So for any t there is at least one $i(t)$ such that $u_{i(t)}(t^2)$ is a square in $\mathbb{F}_q$, which yields a map $\mathbb{F}_q \to E(\mathbb{F}_q)$, $t \mapsto \left(u_{i(t)}(t^2), f(u_{i(t)}(t^2))^{1/2}\right)$. In a cryptographic context, we may assume that a non-square in $\mathbb{F}_q^*$ is part of the input, and then Tonelli-Shanks's algorithm computes square roots in deterministic polynomial time; see [660, Section 1.5.1] and [2813]. The argument is refined in [2604] to give a deterministic procedure for computing points on the curve without knowing a non-square and to show that at least $\frac{q-4}{8}$ different points may be reached. The case of characteristic two is also handled.

16.4.79 Example [1566] Let $\mathbb{F}_q$ with $q \equiv 2 \pmod 3$ be of characteristic at least 5, and let $E : y^2 = x^3 + ax + b$ be an elliptic curve over $\mathbb{F}_q$. Let $v(t) = \frac{3a-t^4}{6t}$ and $x(t) = \left(v(t)^2 - b - \frac{t^6}{27}\right)^{1/3} + \frac{t^2}{3}$. Then $0 \mapsto \mathcal{O}$, $0 \neq t \mapsto (x(t), tx(t) + v(t))$ is a map $\mathbb{F}_q \to E(\mathbb{F}_q)$ with image size at least $\frac{q}{4}$, and conjecturally close to $\frac{5q}{8}$. A similar result holds for curves over $\mathbb{F}_{2^m}$ with odd m.

16.4.80 Remark Alternative encodings for elliptic curves in Hessian form over $\mathbb{F}_q$ with $q \equiv 2 \pmod 3$ and odd are given in [1039, 1672]; see also [745]. They have an image of proven size about $q/2$.

See Also

[312], [313], [661]	Give comprehensive accounts of elliptic curve cryptography.

References Cited: [108, 109, 183, 203, 220, 221, 312, 313, 345, 366, 413, 420, 458, 597, 660, 661, 744, 745, 852, 970, 978, 979, 1039, 1098, 1101, 1104, 1108, 1133, 1156, 1159, 1160, 1165, 1247, 1250, 1251, 1276, 1422, 1429, 1492, 1494, 1566, 1567, 1629, 1630, 1672, 1771, 1894, 1904, 2079, 2080, 2088, 2098, 2102, 2292, 2426, 2497, 2533, 2560, 2601, 2604, 2675, 2683, 2684, 2712, 2751, 2752, 2766, 2813, 2868, 2869, 2978]

16.5 Hyperelliptic curve cryptographic systems

Nicolas Thériault, Universidad del Bio-Bio

16.5.1 Cryptosystems based on hyperelliptic curve discrete logarithms

16.5.1 Remark As stated in Definition 12.4.17, the set of $\mathbb{F}_q$-rational reduced divisors of degree zero of a hyperelliptic curve C defined over a finite field $\mathbb{F}_q$ form a finite abelian group, the Picard group $Pic^0_{\mathbb{F}_q}(C)$.

16.5.2 Remark The Picard group can be used as a substitute for the group of points on an elliptic curve to implement cryptosystems based on the difficulty of the discrete logarithm problem [1772, 661]. For efficiency reasons, imaginary curves are usually preferred over real curves since their group operation is slightly faster in practice. We note that for real hyperelliptic curves, the cryptosystems rely on the infrastructure discrete logarithm problem which can be reduced to a discrete logarithm problem in the Picard group (see Definition 12.4.81 and Remark 12.4.82).

16.5.3 Remark For an imaginary hyperelliptic curve defined over the finite field $\mathbb{F}_q$, the Picard group is isomorphic to the *ideal class group* of the curve (the quotient group of the ideals in the polynomial ring of the curve modulo the principal ideals). For a curve of genus g, this group has order close to q^g (to be precise the group order is between $(\sqrt{q}-1)^{2g}$ and $(\sqrt{q}+1)^{2g}$, see Remark 12.4.62). In practice, Mumford's representation (Theorem 12.4.34) is used to construct divisors (or rather the corresponding ideals).

16.5.2 Curves of genus 2

16.5.4 Remark As is the case for groups obtained from elliptic curves, the fastest known attacks on the discrete logarithm problem for hyperelliptic curves of genus two are generic attacks which require $O(\sqrt{n})$ group operations to compute the discrete logarithm. Because the group order is $n \approx q^2$, solving the discrete logarithm problem requires $O(q)$ group operations.

16.5.5 Remark Following the ideas of Harley [1421], efficient implementations of hyperelliptic curve group operations are usually done via *explicit fomulae* which replace the polynomial operations of Cantor's algorithm [497] with a sequence of field operations on the coefficients of these polynomials [147, 1246, 1854].

16.5.6 Remark For efficiency reasons, for curves in odd characteristic the curve is usually assumed to be reduced (via isomorphisms) to the form $y^2 = x^5 + f_3x^3 + f_2x^2 + f_1x + f_0$. For curves in characteristic two, the curve is usually assumed to be reduced to either $y^2 + h_1xy = x^5 + f_3x^3 + f_2x^2 + f_0$ (with $h_1 = 1$ in extensions of odd degree) or $y^2 + (x^2 + h_1x + h_0)y = x^5 + f_4x^4 + f_1x + f_0$.

16.5.7 Remark Genus two curves of the form $y^2 + y = f(x)$ over a field of characteristic two are supersingular. The discrete logarithm in these curves is much easier to compute (via the Weil or Tate pairings) than for other curves of genus two. For this reason, these curves are avoided for cryptographic applications.

16.5.8 Remark For curves in characteristic two, the binary field structure can be used to easily solve quadratic equations. It becomes possible to compute "halving" operations, giving performance advantages over doubling operations in some cases [286, 287, 1038].

16.5.9 Remark As for elliptic curves, there are various coordinate systems to represent divisors, and more precisely to represent reduced divisors of weight two (divisors of the most common case). In particular, projective coordinates are sometimes used to obtain inversion-free explicit formulae. Two basic types of projective coordinates are used in practice:

1. "standard" projective coordinates [2110, 661], $[U_1, U_0, V_1, V_0, Z]$, which correspond to the Mumford representation (in affine coordinates) $[x^2 + (U_1/Z)x + (U_0/Z), (V_1/Z)x + (V_0/Z)]$;
2. *new* projective coordinates [1853], $[U_1, U_0, V_1, V_0, Z_1, Z_2]$, which correspond to the Mumford representation (in affine coordinates) $[x^2 + (U_1/Z_1^2)x + (U_0/Z_1^2), (V_1/Z_1^3 Z_2)x + (V_0/Z_1^3 Z_2)]$.

16.5.10 Remark In some cases, extended (or redundant) coordinates are used since they can save some field operations in further group operations. For example, the *recent* coordinates of Lange [1852] use the coordinates $[U_1, U_0, V_1, V_0, Z, Z^2]$ for curves over fields of characteristic two. Similarly, new coordinates are usually used in the form $[U_1, U_0, V_1, V_0, Z_1, Z_2, Z_1^2, Z_2^2]$ (in odd characteristic) or $[U_1, U_0, V_1, V_0, Z_1, Z_2, Z_1Z_2, Z_1^2, Z_2^2, Z_1^2Z_2]$ (in characteristic two).

16.5.3 Curves of genus 3

16.5.11 Remark Just as curves of genus two, hyperelliptic curves of genus three can be used to construct cryptosystems based on the discrete logarithm problems, however, evaluating the security of these cryptosystems is more complicated. There are two algorithms that must be considered when evaluating the difficulty of the discrete logarithm problem on a specific hyperelliptic curve: the index calculus algorithm and Smith's trigonal mapping to non-hyperelliptic curves.

16.5.12 Remark Index calculus algorithms are used to map the discrete logarithm problem to computing non-trivial solutions of a system of linear equations. The algorithm proceeds in two steps: a relation search (to build the system) and a (sparse) linear algebra solver.

Under the right conditions, the overall complexity of the relation search and linear algebra solver is lower than generic attacks. For curves of genus one and two, index calculus attacks appear to be slower than generic attacks, but for genus three and higher, index calculus does indeed reduce the cost of computing discrete logarithms.

16.5.13 Remark A number of variants of the index calculus relation search have been proposed [1245, 1256, 2212, 2802]. Currently, the most efficient algorithms are those of Gaudry et al. [1256] and Nagao [2212]. Both of these algorithm require the equivalent of $O(q^{4/3+\epsilon})$ group operations to compute the discrete logarithm. It should be noted that index calculus algorithms have the same (estimated) running time for all genus three curves over the field $\mathbb{F}_q$.

16.5.14 Remark For curves of genus three over fields of odd characteristic, there exists a specialized attack against the discrete logarithm problem. This attack is due to Smith [2688] and uses a *trigonal* map to send the Picard group of a hyperelliptic curve of genus three to the Picard group of a non-hyperelliptic curve (also of genus three).

As was shown by Diem [854, 856], it is easier to solve the discrete logarithm in the Picard group of a non-hyperelliptic curve than a hyperelliptic one, the running time decreasing to an equivalent of $O(q^{1+\epsilon})$ group operations for genus three curves. If the trigonal map in Smith's attack is successful, the security of the curve is then no better than that of a genus two curve (but with a higher cost for the group operation).

Smith's attack depends heavily on the structure of the 2-torsion group of the curve (over the algebraic closure of the field $\mathbb{F}_q$). More specifically, it depends on the extension degree of the field $\mathbb{F}_{q^k}$ in which each of the Weierstrass points of the curve is defined. As a result, not all curves over $\mathbb{F}_q$ are vulnerable to Smith's attack.

16.5.15 Example For curves of the form $y^2 = f(x)$ for which $f(x)$ splits into linear factors, Smith's attack has a $1 - 2^{-105}$ probability of success (i.e., roughly only 1 in 2^{105} curves will remain safe from the attack).

However, the probability of success is much lower for other types of factorization of $f(x)$, and can even reach zero. For imaginary curves, the following factorization types of $f(x)$ (degrees of the irreducible factors of $f(x)$) are completely secure against Smith's attack: $[7]$, $[5,2]$, $[5,1,1]$, $[4,3]$, $[3,2,2]$, $[3,2,1,1]$, $[3,1,1,1,1]$. For real curves, there is one more factorization type of $f(x)$ which is completely secure against the attack: $[5,3]$.

16.5.16 Remark There are a number of open problems coming from Smith's algorithm. First of all, it is not known how to adapt the trigonal map to curves over fields of characteristic two. Secondly, the trigonal map exists only for curves of genus three, but similar ideas could be important to the security of hyperelliptic curves of higher genus.

16.5.17 Remark As with genus two curves, the group operations are performed via explicit formulae rather than Cantor's algorithm [149, 1383, 2211]. Once again, curve isomorphisms are used to reduce the curve equation and improve efficiency. In odd characteristic the preferred form of the equation is $y^2 = x^7 + f_5x^5 + f_4x^4 + f_3x^3 + f_2x^2 + f_1x + f_0$. For curves in characteristic two, similar reductions can be performed for each form of $h(x)$.

Unlike elliptic curves and curves of genus two, genus three curves of the form $y^2 + y = f(x)$ over $\mathbb{F}_{2^n}$ are never supersingular [2552] and can be used in cryptographic applications. These curves allow for much faster doubling operations, giving efficiency advantages.

16.5.18 Remark As for genus two curves, for genus three curves in characteristic two, the binary field structure can be used to easily solve quadratic (and some quartic) equations. It becomes possible to compute "halving" operations, giving performance advantages over doubling operations, especially when the degree of $h(x)$ increases [288].

16.5.4 Curves of higher genus

16.5.19 Remark The highest genus that is sometimes considered for cryptographic applications is four [149, 2381]. The running time for solving the discrete logarithm problem in a hyperelliptic curve of genus four (using the algorithm of Gaudry et al. [1256]) is equivalent to $O(q^{3/2+\epsilon})$ group operations.

16.5.20 Remark For curves of small degree, the cost of computing discrete logarithm grows as $O(q^{2-2/g+\epsilon})$ (for a fixed genus and an increasing field size) [1256, 2212]. However, the complexity of the group operation is at least linear with respect to the genus of the curve (in practice this growth is closer to quadratic), in practice limiting the range of "small" genera which are of interest for cryptosystems. At higher genera, the situation is even more difficult for cryptosystems based on the discrete logarithm. This situation is discussed in Section 16.6.10.

16.5.5 Key sizes

16.5.21 Remark The security arguments for hyperelliptic curves of genus two are very similar to those of elliptic curves, with the only distinction that the group order is close to q^2 (rather than q for elliptic curves). The required bit-sizes for the finite fields are therefore half of what they are for elliptic curves with the same security level. Once again, the group order should be a prime or a prime with a small co-factor. Similarly, the secret key should be of size similar to the group order, which means that secret key sizes for hyperelliptic curves of genus two are in fact identical to those of elliptic curves.

16.5.22 Remark The following table summarizes these arguments, giving comparisons with symmetric key cryptosystems and as in Remark 16.4.3, the 80 bit security level is now considered a historic figure.

security level	ECC field size	genus two field size	secret key size
80	160	80	160
112	224	112	224
128	256	128	256
192	384	192	384
256	512	256	512

16.5.23 Remark To choose the field size for genus three hyperelliptic curve, the first concern is the possibility of an index calculus attack. To have m-bits of security level, we ask that $q^{4/3} \approx 2^m$ (using the complexity in Remark 16.5.12), hence $\log q \approx 3m/4$.

It may be noted that index calculus attacks do not obtain any significant speedups from restrictions on the key size or factorizations of the group order. In some situations it may be acceptable to allow the group order to factor into a prime with a (relatively) large cofactor, as long as the largest prime factor is of such size that the subgroup attack of Pohlig and Hellman [2406] cannot be effective. The cofactor could then take up to one quarter of the bit size of the group order.

Having large subgroups can still pose a problem for the security of a cryptosystem unless there is a mechanism in place to ensure the group element is indeed in the desired subgroup, otherwise an attacker could provide a group element which is outside of the (large-)prime-order subgroup, and use a subgroup attack to obtain partial information about the scalar, thus reducing the effective security level. Unfortunately, verifying in which subgroup a given group element is located is quite expensive (of cost similar to the scalar multiplication itself), and has a significant impact on the efficiency of the cryptosystem. For this reason, it is usually preferred to use groups whose order is (close to) a prime, even though larger cofactors may not directly decrease the security level.

16.5.24 Remark It is possible to choose keys which are significantly smaller than the group size without necessarily weakening the cryptosystem. This can be of great importance for the efficiency of the cryptosystem since the cost of the scalar multiplication (the cryptographic primitive) is directly proportional to the number of bits of the secret key. However, if the secret key is too small, it may become possible to compute the secret key using a generic attack. The bit-size of the secret must therefore be equivalent to those used for elliptic curves at the same security level.

16.5.25 Remark For curves in odd characteristic, it is recommended to ask that $f(x)$ factors into one of the factorization types given in Example 16.5.15 (to insure no trigonal mapping can be used to mount a successful attack). This final condition (factorization of the defining polynomial) can be seen as equivalent to the requirement that the group order be close to a prime for protection against generic attacks (factorization of the group order).

16.5.26 Remark The following table gives bit sizes for the field of definition and the secret key for genus three hyperelliptic curve cryptosystems, comparing with symmetric key sizes and ECC sizes. If large cofactors are allowed in the group order, its largest prime factor should be at least of the secret key size.

security level	ECC field size	genus three field size	secret key size
80	160	60	160
112	224	84	224
128	256	96	256
192	384	144	384
256	512	192	512

16.5.6 Special curves

16.5.27 Definition A *hyperelliptic Koblitz curve* is a hyperelliptic curve defined over $\mathbb{F}_2$ that is used over $\mathbb{F}_{2^n}$ for n prime (to avoid having too many subgroups).

16.5.28 Remark The following table lists all (isomorphism classes of) non-supersingular hyperelliptic Koblitz curves of genus 2 and their characteristic polynomials.

Curve C (reduced)	Characteristic polynomial
$y^2 + xy = x^5 + 1$	$t^4 + t^3 + 2t + 4$
$y^2 + xy = x^5 + x^2 + 1$	$t^4 - t^3 - 2t + 4$
$y^2 + (x^2 + x)y = x^5 + 1$	$t^4 - t^2 + 4$
$y^2 + (x^2 + x + 1)y = x^5 + 1$	$t^4 + t^2 + 4$
$y^2 + (x^2 + x + 1)y = x^5 + x$	$t^4 + 2t^3 + 3t^2 + 4t + 4$
$y^2 + (x^2 + x + 1)y = x^5 + x + 1$	$t^4 - 2t^3 + 3t^2 - 4t + 4$

16.5.29 Remark Besides an easier computation of the group order (although computing the group order of a random hyperelliptic curve over a binary field is quite efficient), the main advantage of Koblitz curves comes from the Frobenius map over $\mathbb{F}_2$.

16.5.30 Remark A similar table for genus three can be found in [1850].

16.5.31 Definition The map defined by $\tau(x, y) = (x^2, y^2)$ for points on the curve gives a degree n endomorphism on the Picard group when applied to (the support of) divisors defined over $\mathbb{F}_{2^n}$. The application of τ on the Mumford representation of a divisor consists of squaring all the coefficients of the polynomials. A τ-adic expansion of the scalar [2195] then gives a reduction in the cost of the scalar multiplication (as for Koblitz elliptic curves).

16.5.32 Remark Since the τ map has order n for divisors over $\mathbb{F}_{2^n}$, the security level of the curve must be adjusted accordingly (generic attacks can be accelerated by a factor of $O(\sqrt{n})$).

16.5.33 Example The curve

$$y^2 + (x^2 + x + 1)y = x^5 + x$$

is a genus two Koblitz curve with characteristic polynomial over $\mathbb{F}_2$ given by $t^4 + 2t^3 + 3t^2 + 4t + 4 = 0$. Over the field $\mathbb{F}_{2^{113}}$, its Picard group has order $2 \cdot 7 \cdot 1583 \cdot 476183 \cdot 102187125502054743104177319847474471863139915547642198344 09$ (i.e., 10553167646 times a prime). Taking into account the subgroups and the degree 113 endomorphism, this curve gives equivalent security to a random elliptic curve over a field of 186 bits.

16.5.34 Definition A *subfield curve* is a hyperelliptic curve of genus g defined over a field $\mathbb{F}_q$, but used as a curve over $\mathbb{F}_{q^n}$ [1850]. These curves are generalizations of Koblitz curves.

16.5.35 Remark There are two possible advantages of using subfield curves. First, computing the group order is easier than for a general curve over the same field: it can be obtained from knowledge of the characteristic polynomial of the curve over $\mathbb{F}_q$ [1850]. This is of particular interest in odd characteristic since computing the group order is much more expensive for these fields than for binary fields. Second, the field structure (of $\mathbb{F}_{q^n}$ as an extension of $\mathbb{F}_q$) allows for more efficient arithmetic than for a prime field $\mathbb{F}_p$ with $p \approx q^n$.

16.5.36 Remark One of the main drawbacks of using curves which are defined over a subfield (in this case of $\mathbb{F}_{q^n}$) is that the Picard group contains at least one large subgroup, namely the Picard group of the curve over the subfield $\mathbb{F}_q$. This opens the way for attacks on the discrete logarithm problem based on the algorithm of Pohlig and Hellman [2406]. To avoid these attacks, it is necessary to implement some techniques to ensure the group elements used are always in the larger (prime order) subgroup. Furthermore, these curves are at risk of Weil descent based attacks, in particular the variant of Gaudry [1247].

16.5.37 Definition Given a curve C defined over $\mathbb{F}_q$, the *trace-zero subvariety* of the curve is the quotient group $T = Pic^0_{\mathbb{F}_{q^n}}(C)/Pic^0_{\mathbb{F}_q}(C)$. This was proposed by Lange for genus two and $n = 3$ [1851] as a method to construct a cryptographically viable group on a field extension without having to deal with subgroups. If the order of the group T is prime and close to q^4 ($q^{g(n-1)}$ in general) this group can be used for cryptosystems based on the discrete logarithm problem.

16.5.38 Remark As with subfield curves, the group order of trace-zero subvarieties is computed from the group order of the Picard group of the curve over $\mathbb{F}_q$. Similarly, these curves take advantage of the subfield structure to make the field arithmetic more efficient. Finally, the group operations can also be tailored to work on T rather than on the general Picard group.

16.5.39 Remark Just as with any curve on a field extension, trace-zero subvarieties can be subjected to Weil descent based attacks. For curves of genus greater than two and for $n > 3$ in genus two, a Weil descent attack on the whole Picard group with the simplest form of index calculus attack [1245] is sufficient to solve the discrete logarithm in time $O(q^2)$, significantly faster than through generic attacks [855].

16.5.40 Remark For $n = 3$, Gaudry's variant would allow the attack to handle the Picard group as if it were the Picard group of a genus six curve over $\mathbb{F}_q$, and the discrete logarithm problem in the trace-zero subvariety can be lifted to a discrete logarithm problem in the whole Picard group. This attack has running time $O(q^{5/3})$, instead of the desired $O(q^2)$ for generic attacks on the group T.

Furthermore, it may be possible to adapt Gaudry's variant to work directly on the trace-zero subvariety (by selecting a different factor base, more appropriate to this context), treating it as a genus four curve over $\mathbb{F}_q$. If such an attack is possible, it would have running time $O(q^{3/2})$, making it particularly effective. The construction of an appropriate factor base is an open problem.

16.5.41 Remark Because of these attacks, field sizes for secure cryptosystems would have to be increased to preserve the security level, which in turns decreases the efficiency of the field arithmetic. As a result, groups coming from trace-zero subvarieties are now mostly of theoretical interest.

16.5.7 Random curves: point counting

16.5.42 Remark Computing the Picard group order of hyperelliptic curves in even characteristic can be done quite efficiently via p-adic (in this case 2-adic) algorithms. Currently, the fastest algorithms are due to Satoh [2534] and Streng [2735]. These algorithms can handle in reasonable time both curves of high genus and (more interesting for practical applications) curves of small genus over a large field.

16.5.43 Remark Compared with the characteristic two case, computing the group order of curves in odd characteristic is mostly an open problem. In curves over prime fields, most algorithms are based on Pila's [2394] extension of Schoof's algorithm. For genus two curves (where such works are concentrated), the fastest algorithms are due to Gaudry and Schost [1252, 1253], and Gaudry, Kohel, and Smith for curves with complex multiplication [1254].

16.5.44 Remark At this time, finding a random hyperelliptic curve in characteristic two that corresponds to required security requirements – having a given genus and field size, with a Picard group whose order is a large prime with a small co-factor – is considered quite practical.

For curves in odd characteristic, this is considerably more difficult. For genus two curves, such a search is feasible although expensive ([1253] reports such a search at the 128-bit security level taking over one million CPU-hours). For genus three (or higher), there are currently no examples of practical searches for such curves.

16.5.8 Pairings in hyperelliptic curves

16.5.45 Remark Since the Tate-Lichtenbaum pairing comes from the divisor class group structure of elliptic curves, it naturally generalizes to hyperelliptic curves, although some care has to be taken to properly adapt the definition as well as Miller's algorithm; see Subsection 12.4.6. A survey of the different techniques available for pairings on hyperelliptic curves can be found in [1157]. Actual implementations of cryptosystems using hyperelliptic curve pairings are much less common than elliptic curve ones, but [2309] demonstrates the potential interest of hyperelliptic curves for pairing-based cryptography.

16.5.46 Remark For curves of genus two over prime fields, it is possible to construct curves with a given group order using complex multiplication methods [2087, 2970, 2971]. Such curves may be of interest for pairing based cryptosystems on hyperelliptic curves [1157]. For cryptosystems based on the discrete logarithm problem, more random curves are usually preferred for fear the structure coming from complex multiplication could eventually open the way to new attacks that significantly decreases the difficulty of the discrete logarithm problem in these curves (although no such attack is currently known).

See Also

[661], [313]	Give comprehensive accounts of hyperelliptic curve cryptography.

References Cited: [1, 147, 149, 286, 287, 288, 313, 497, 661, 854, 855, 856, 1038, 1157, 1245, 1246, 1247, 1252, 1253, 1254, 1256, 1383, 1421, 1772, 1850, 1851, 1852, 1853, 1854, 2087, 2110, 2195, 2211, 2212, 2309, 2381, 2394, 2406, 2534, 2552, 2688, 2735, 2802, 2970, 2971]

16.6 Cryptosystems arising from Abelian varieties

Kumar Murty, *University of Toronto*

16.6.1 Definitions

16.6.1 Definition An *Abelian variety* A over a finite field $\mathbb{F}$ is a smooth projective algebraic variety defined over $\mathbb{F}$ on which there is an algebraic group operation, also defined over $\mathbb{F}$. In particular, the identity element O of the group is an $\mathbb{F}$-rational point.

16.6.2 Definition An *Abelian subvariety* of an Abelian variety A is an Abelian variety that is contained in A. Trivial Abelian subvarieties are $\{0\}$ and A itself. A non-trivial Abelian subvariety is *proper*.

16.6.3 Definition An Abelian variety is *simple* over $\mathbb{F}$ if it contains no proper Abelian subvarieties defined over $\mathbb{F}$. It is *absolutely simple* if it is simple over the algebraic closure $\overline{\mathbb{F}}$ of $\mathbb{F}$.

16.6.4 Definition Two Abelian varieties A_1 and A_2 are *isogenous* if there is a surjective morphism $A_1 \longrightarrow A_2$ with finite kernel.

16.6.2 Examples

16.6.5 Example Abelian varieties of dimension one are elliptic curves.

16.6.6 Remark Elliptic curves are the only curves that are also Abelian varieties. This follows from the result below which is a consequence of the Riemann-Roch theorem.

16.6.7 Proposition If C is an algebraic curve of genus g which is also an Abelian variety, then necessarily $g = 1$ and C is an elliptic curve.

16.6.8 Example Given two Abelian varieties A_1 and A_2, there is a natural structure of Abelian variety on the product $A_1 \times A_2$.

16.6.9 Example In particular, one can take a power of an elliptic curve $E^n = E \times \cdots \times E$ and products of powers of different elliptic curves $E_1^{n_1} \times \cdots \times E_k^{n_k}$.

16.6.10 Theorem An Abelian variety over $\mathbb{F}$ is isogenous to a product of simple Abelian varieties defined over $\mathbb{F}$.

16.6.3 Jacobians of curves

16.6.11 Theorem Given a curve C defined over $\mathbb{F}$, there is an Abelian variety $J(C)$, the Jacobian of C, which is also defined over $\mathbb{F}$. It has the following properties:

1. The dimension of $J(C)$ is the genus of C.
2. If g is the genus of C, there is a surjective map $C^g = C \times \cdots \times C \longrightarrow J(C)$.

16.6.12 Theorem The $\overline{\mathbb{F}}$ points on $J(C)$ are given by D^0/P where D^0 is the Abelian group of divisors of degree zero and P is the subgroup of divisors of degree zero which are divisors of functions.

16.6.4 Restriction of scalars

16.6.13 Remark Let A be an Abelian variety defined over an extension field $\mathbb{F}'$ over $\mathbb{F}$. Suppose that $[\mathbb{F}' : \mathbb{F}] = r$ and that A is of dimension d. Then there is an Abelian variety $R_{\mathbb{F}'/\mathbb{F}}A$

(called the restriction of scalars of A) defined over $\mathbb{F}$ of dimension rd with the property that $R_{\mathbb{F}'/\mathbb{F}}A(\mathbb{F}) = A(\mathbb{F}')$.

16.6.5 Endomorphisms

16.6.14 Remark The fundamental theorem is due to Tate.

16.6.15 Theorem Let A be an Abelian variety of dimension g defined over a finite field $\mathbb{F}$. Let π be the Frobenius endomorphism of A relative to $\mathbb{F}$ and P its characteristic polynomial.

1. The algebra $F = \mathbb{Q}[\pi]$ is the center of the semisimple algebra $E = \mathrm{End}_{\mathbb{F}}(A) \otimes \mathbb{Q}$.
2. E contains a semisimple $\mathbb{Q}$-subalgebra M of rank $2g$ which is maximal and commutative.
3. The following are equivalent:
   a. $[E : \mathbb{Q}] = 2g$;
   b. P has no multiple roots;
   c. $E = F$;
   d. E is commutative.
4. The following are equivalent:
   a. $[E : \mathbb{Q}] = (2g)^2$;
   b. P is a power of a linear polynomial;
   c. $F = \mathbb{Q}$;
   d. E is isomorphic to the algebra of $g \times g$ matrices over the unique quaternion division algebra D_p over $\mathbb{Q}$ which splits at all primes $\ell \neq p, \infty$;
   e. A is $\mathbb{F}$-isogenous to the g-th power of a supersingular elliptic curve, all of whose endomorphisms are defined over $\mathbb{F}$.
5. A is $\mathbb{F}$-isogenous to a power of an $\mathbb{F}$-simple Abelian variety if and only if P is a power of a $\mathbb{Q}$-irreducible polynomial. When this is the case, E is a central simple algebra over $\mathbb{F}$ which splits at all finite primes v of F not dividing p, but does not split at any real primes of F.

16.6.6 The characteristic polynomial of an endomorphism

16.6.16 Definition Let $\phi : A \longrightarrow A$ be an endomorphism. Then, there is a polynomial $P_\phi(T)$ with the property that for every integer $n \geq 1$, we have $P_\phi(n) = \deg(n - \phi)$. This polynomial is the *characteristic polynomial* of ϕ.

16.6.17 Remark An important case of this is the Frobenius endomorphism $x \mapsto x^q$ on $\overline{\mathbb{F}}$. It induces an endomorphism of A.

16.6.7 Zeta functions

16.6.18 Definition Denote by $\mathbb{F}_n$ the unique extension of $\mathbb{F}$ of degree n contained in $\overline{\mathbb{F}}$. The *zeta function* $Z(A, T)$ of an Abelian variety A defined over a finite field $\mathbb{F}$ is the power series

$$Z(A,T) = \exp\left\{\sum_{n\geq 1} |A(\mathbb{F}_n)| T^n/n\right\}.$$

16.6.19 Theorem There are polynomials $P_0(T), \ldots, P_{2d}(T)$ (where $d = \dim(A)$) such that

$$Z(A,T) = \prod_{i=0}^{2d} P_i(T)^{(-1)^{i+1}}.$$

16.6.20 Example The zeta function of an elliptic curve E is of the form

$$Z(E,T) = P_1(T)/(1-T)(1-qT).$$

Here, $P_1(T)$ is a polynomial of degree 2 and is of the form $1 - a_qT + qT^2$ with $|E(\mathbb{F})| = 1 - a_q + q$. Moreover, a_q is an integer with $a_q = \pi + \overline{\pi}$ and π is a complex number of absolute value $q^{1/2}$. We have $|E(\mathbb{F}_n)| = 1 - (\pi^n + \overline{\pi}^n) + q^n$.

16.6.21 Remark The single integer a_q determines the number of points on E over $\mathbb{F}_n$ for every n.

16.6.22 Theorem The polynomials $P_i(T)$ of the previous result have the following properties:

1. We have $P_i(T) \in \mathbb{Z}[T]$.
2. The degree of P_i is $\binom{2d}{i}$.
3. The polynomial $P_1(T)$ is the characteristic polynomial of the Frobenius endomorphism.
4. The factorization of P_1 is of the form $\prod_{j=1}^{2d}(1 - \pi_jT)$ where for $1 \le j \le d$, $\pi_{j+d} = \overline{\pi_j}$ and $\pi_j\pi_{j+d} = q$.
5. The roots of P_i are i-fold products of the π_j.
6. We have $P_0(T) = 1 - T$ and $P_{2d}(T) = 1 - q^dT$.

16.6.23 Theorem [2783] The polynomial $P_1(T)$ determines the isogeny class of A.

16.6.24 Remark The above theorem is a fundamental result of Tate.

16.6.25 Theorem Given a set of $2d$ complex numbers $\{\pi_j : 1 \le j \le 2d\}$, with $\pi_{j+d} = \overline{\pi_j}$ and $\pi_j\pi_{j+d} = q$ for $1 \le j \le d$, there is an Abelian variety A defined over $\overline{\mathbb{F}}$ for which $P_1(T) = \prod_{j=1}^{2d}(1 - \pi_jT)$.

16.6.26 Remark This is a fundamental result of Honda and Tate.

16.6.27 Definition The *Newton polygon of an Abelian variety* A is the Newton polygon of $P_1(T)$. In particular, if we write $P_1(T) = 1 + a_1T + \cdots + a_{2d}T^{2d}$, then the Newton polygon is the convex hull of the points $(j, \mathrm{ord}_p a_j)$ for $1 \le j \le 2d$. A slope of the Newton polygon is $(d-b)/(c-a)$ where (a,b) and (c,d) are two vertices. The length of the slope is $c-a$.

16.6.28 Proposition [2438, Theorem 9.1] The slopes of the Newton polygon are (counting multiplicities) the p-adic ordinals of the reciprocal roots of $P_1(T)$. More precisely, if λ is a slope of length m, then precisely m of the numbers $ord_p\pi_i$ are equal to λ.

16.6.29 Remark Given the condition that for every j, we have $\pi_j\pi_{j+d} = q$, at most d of the π_j can be prime to q (that is, have $\mathrm{ord}_p\pi_j = 0$).

16.6.30 Definition An Abelian variety A is *ordinary* if exactly d of the π_j satisfy $\mathrm{ord}_p\pi_j = 0$. An Abelian variety is *supersingular* if none of the π_j satisfy $\mathrm{ord}_p\pi_j = 0$.

16.6.31 Theorem A supersingular Abelian variety A over $\mathbb{F}$ is isogenous to a power of a supersingular elliptic curve over $\overline{\mathbb{F}}$.

16.6.8 Arithmetic on an Abelian variety

16.6.32 Remark There are efficient algorithms to compute on elliptic curves as well as on Jacobians of hyperelliptic curves. These have been described in earlier sections. There is an interesting class of curves (called $C_{a,b}$ curves) which generalizes these. We describe them below.

16.6.33 Definition Let a and b be positive integers. A $C_{a,b}$ *curve* over $\mathbb{F}$ is one that is defined by an equation of the form

$$F(x,y) = \alpha_{b,0}x^b + \alpha_{0,a}y^a + \sum_{ia+jb<ab} \alpha_{i,j}x^i y^j$$

where the coefficients $\alpha_{i,j}$ are elements of $\mathbb{F}$ and the leading coefficients $\alpha_{b,o}$ and $\alpha_{0,a}$ are nonzero.

16.6.34 Remark Such a curve has the property that it is has exactly one $\mathbb{F}$-rational point (Q say) at infinity and the polar divisors of the functions x and y are aQ and bQ, respectively.

16.6.35 Example An elliptic curve is a $C_{2,3}$ curve and more generally, a hyperelliptic curve given by

$$y^2 = f(x)$$

with f a polynomial of degree $2g+1$ is a $C_{2,2g+1}$ curve.

16.6.36 Example A superelliptic curve, in other words one of the form

$$y^a = f(x)$$

where f has degree g (say) is a $C_{a,g}$ curve.

16.6.37 Proposition Let C be a $C_{a,b}$ curve. Any element of the Jacobian $J(C)$ can be expressed (not necessarily uniquely) by a divisor of the form $E - nQ$ where E is a positive divisor whose support is disjoint from Q and $0 \leq n \leq g$. Here g is the genus of C.

16.6.38 Remark A divisor as in the proposition is called *semi normal*. It is possible that two semi normal divisors are equivalent (that is, represent the same point in the Jacobian). To get a unique representation, start with a semi normal divisor $D = E - nQ$ with E as above. Find a nonzero function f whose polar divisor is supported at Q, whose zero divisor satisfies $(f)_0 \geq E$, and which has the smallest possible order pole at Q. Then consider the divisor $-D + (f)$. Applying this process once more yields a divisor equivalent to D which is unique. Every divisor class contains a divisor of this form, henceforth called the *normal representative of the class.*

16.6.39 Proposition Let C be a $C_{a,b}$ curve. Every point in the Jacobian $J(C)$ (and hence every divisor class) can be represented uniquely by a normal divisor.

16.6.40 Remark With a canonical representative of each divisor class, arithmetic is now explicit.

16.6.41 Remark [1582] By expressing the above normal form using Gröbner bases, the complexity of addition is estimated to be $O(g^3(\log q)^2)$.

16.6.42 Remark A special case of $C_{a,b}$ curves is the family of diagonal curves. They are represented by an equation of the form $y^a = cx^b + d$ with $c, d \in \mathbb{F}_q^\times$. This curve has genus $\frac{1}{2}((a-1)(b-1) - \gcd(a,b) + 1)$. By explicitly computing the zeta function of $J(C)$ in terms of Jacobi sums, Blache, Cherdieu, and Sarlabous [294] show that for $\mathbb{F} = \mathbb{F}_p$, the Jacobian $J(C)$ is simple if $p \equiv 1, 2, 4, 7, 8, 11, 13 \pmod{15}$. Moreover, it is supersingular if $p \equiv 14 \pmod{15}$. Moreover, they develop an analogue of Cantor's addition algorithm in $J(C)$.

16.6.43 Remark Addition on the Jacobian of a general curve has been described by Volcheck but this method does not seem to be practical over fields of cryptographic size.

16.6.44 Remark Arita, Miura, and Sekiguchi [123] have shown how to use a singular plane model of a general curve to obtain an algorithm for addition on the Jacobian. However, the complexity of this algorithm has not been analyzed.

16.6.9 The group order

16.6.45 Remark For cryptographic applications, we want the group $A(\mathbb{F})$ to be "nearly" of prime order. In particular, we want A to be simple (and perhaps even absolutely simple). This rules out supersingular Abelian varieties as (by a result stated above) they are isogenous to a power of a supersingular elliptic curve.

16.6.10 The discrete logarithm problem

16.6.46 Definition Let A be an Abelian variety over the finite field $\mathbb{F}$. Let $P \in A(\mathbb{F})$ and Q an element of the subgroup of $A(\mathbb{F})$ generated by P. The *discrete logarithm problem* for this subgroup is to determine an integer n so that $Q = nP$.

16.6.47 Remark In the case of hyperelliptic curves, we have the following result of Enge and Gaudry based on an index calculus approach.

16.6.48 Theorem [980] The discrete logarithm problem on $J(C)$ (where C is a hyperelliptic curve of genus g defined over a finite field of q elements) is of complexity

$$O(\exp\{(\sqrt{2} + o(1))\sqrt{(\log q^g)(\log\log q^g)}\}).$$

16.6.49 Remark This result assumes that the group order is known and that the group itself can be computed in polynomial time. Several authors have studied analogues of this result for the Jacobian of a non-hyperelliptic curve. In particular, one has the following results.

16.6.50 Theorem [854] Fix positive integers d (the degree) and $g \le (d-1)(d-2)/2$ (the genus). Denote by $S(q)$ the set of all instances of the discrete logarithm problem in curves of genus g over $\mathbb{F}_q$ represented by plane models of degree d. Then there is a subset $S_1(q)$ with $|S_1(q)|/|S(q)| \to 1$ (as $q \to \infty$) such that the instances in $S_1(q)$ can be solved in an "expected" time of $O(q^{(2-\frac{2}{d-2})}(\log q)^A)$ for some integer $A \ge 1$.

16.6.51 Remark The analysis in [854] involves several heuristic assumptions (including assumptions on the group order and structure) and the power of the logarithm is not specified. The result, in particular, applies to non-hyperelliptic curves of genus 3 (which by the canonical embedding, may be represented by a plane quartic ($d = 4$)). This particular case is further analyzed by Diem and Thomé [856].

16.6.52 Remark In the case studied by Diem and Thomé, the heuristic assumptions can be made explicit in graph-theoretic terms. Choose a subset $\mathcal{F} \subset C(\mathbb{F})$ (called the *factor base*) of cardinality $O(\sqrt{q})$ and let $\mathcal{L}$ denote the complement of $\mathcal{F}$ in $C(\mathbb{F})$. A graph $\mathcal{G}$ is constructed with vertices $\mathcal{L} \cup \{*\}$ where $*$ is a special root vertex. The edges in this graph are determined as follows. For each pair $F_1, F_2 \in \mathcal{F}$, compute the line L through these points. Consider the divisor $D = F_1 + F_2 + D_{1,2}$ representing the intersection of C with L. If P, Q are points in the support of $D_{1,2}$, at least one of which is in $\mathcal{L}$, construct an edge joining P and Q (or P and $*$ or Q and $*$).

16.6.53 Theorem [856] Suppose that C is a non-hyperelliptic curve of genus 3 with the property that $J(C)(\mathbb{F})$ is cyclic. Suppose also that the graph $\mathcal{G}$ constructed above has a tree rooted at $*$ of depth $O((\log q)^2)$ and having at least $O(q^{5/6})$ elements. Then the discrete logarithm problem in $J(C)$ can be solved in $O(q(\log q)^A)$ steps.

16.6.54 Remark We say that the discrete logarithm problem for the pair (A, P) consisting of the Abelian variety A and the subgroup generated by a point P on A is difficult if it is computationally infeasible to solve the problem.

16.6.55 Remark Let A and P be as above and suppose that the discrete logarithm problem for (A, P) is difficult. Then we can perform a key exchange using the Diffie Hellman protocol with the group G generated by P.

16.6.56 Remark In order for such a key exchange scheme to be useful and secure against known attacks, we require:

1. arithmetic on A should be efficient;
2. the group order $A(\mathbb{F})$ should be nearly prime.

16.6.57 Remark There is a vast literature on the discrete logarithm problem. The reader is referred to the surveys [1495, 1586] for some results that are not mentioned here.

16.6.58 Remark The motivation for considering higher dimensional Abelian varieties as the basis of a cryptographic scheme is based on the fact that if A has dimension d, the number of points $A(\mathbb{F})$ is of order q^d. Thus, we should expect the difficulty of the discrete logarithm problem to be of the order $q^{d/2}$. In particular, a 2-dimensional Abelian variety has a discrete logarithm problem of difficulty $O(q)$ whereas an elliptic curve has a discrete logarithm problem of difficulty $O(q^{1/2})$. This means that there is the possibility of achieving the same level of security that an elliptic curve over a field of 2^{163} offers by using a two dimensional Abelian variety over a field of 2^{82}. The fact that we can work over a field of smaller size may mean that there is a reduction in the overhead of time and memory required to do computations. Whether this is actually the case is a matter of current research.

16.6.11 Weil descent attack

16.6.59 Remark Frey [1104] suggested that it might be possible to use Weil descent to attack the discrete logarithm problem on elliptic curves defined over $\mathbb{F}_{p^n}$ where n is composite. This idea was explicitly developed and analyzed by Gaudry, Hess, and Smart [1250] in the case $p = 2$.

16.6.60 Definition Let ℓ and n be positive integers. Set $q = 2^\ell$, $k = \mathbb{F}_q$, and $K = \mathbb{F}_{q^n}$. Let E be a non-supersingular elliptic curve defined over K by the equation

$$y^2 + xy = x^3 + ax + b$$

with $a, b \in K$ and $b \neq 0$. Suppose that $|E(K)| = dr$ where d is small and r is prime. Let $b_0 = b, b_1, \ldots, b_{n-1}$ be the conjugates of b (under the Frobenius automorphism $\sigma(x) = x^q$).

Let $m(b)$ be the dimension of the vector space over $\mathbb{F}_2$ spanned by the set

$$\{(1, b_0^{1/2}), \ldots, (1, b_{n-1}^{1/2})\}.$$

16.6.61 Theorem [1250] There is a hyperelliptic curve C of genus $g = 2^{m(b)-1}$ or $2^{m(b)-1} - 1$ defined over k so that the discrete logarithm problem on $E(K)$ can be solved in $J(C)(k)$. In particular, the complexity of the discrete logarithm problem is $O(q^{g/2})$.

16.6.62 Theorem [2078] Let n be an odd prime and t the multiplicative order of 2 modulo n. Set $s = (n-1)/t$.

1. The polynomial $x^n + 1$ factors over $\mathbb{F}_2$ as $(x+1)f_1 f_2 \cdots f_s$ where the f_i are distinct irreducible polynomials of degree t.
2. The set
$$B = \{b \in \mathbb{F}_{q^n} \backslash \mathbb{F}_q : (\sigma + 1) f_i(\sigma)(b) = 0 \text{ for some } 1 \le i \le s\}$$
has cardinality $qs(q^t - 1)$.
3. For all $b \in B$, the curves $y^2 + xy = x^3 + ax^2 + b$ has $m(b) = t + 1$.

16.6.63 Example

1. If $n = 3$ we have $s = 1$ and $t = 2$. Then B has cardinality $q(q^2-1)$ so $B = \mathbb{F}_{q^3} \backslash \mathbb{F}_q$ and $m(b) = 3$.
2. If $n = 5$, we have $s = 1$ and $t = 4$. Then again, $B = \mathbb{F}_{q^5} \backslash \mathbb{F}_q$ and $m(b) = 5$.
3. If $n = 31$ we have $t = 5$ and $s = 6$. Then B has cardinality $6q(q^5 - 1)$ and $m(b) = 6$.

16.6.64 Example The following special case was analyzed in detail by Jacobson, Menezes, and Stein [1585]. For the elliptic curve
$$y^2 + xy = x^3 + x + b$$
defined over the field $\mathbb{F}_2[z]/(z^{155} + z^{62} + 1)$ of 2^{155} elements, and with $b = z^{16} + z^2 + z$, instances of the discrete logarithm problem can be reduced to the discrete logarithm problem on the Jacobian $J(C)$ of a hyperelliptic curve C of genus 31 over the field $\mathbb{F}_{2^5}$.

16.6.65 Remark The case of odd characteristic was developed by Diem [852]. In particular, he showed that the Weil descent attack is not effective for elliptic curves defined over $\mathbb{F}_{p^n}$ when $n \ge 11$ is prime. On the other hand, if $n = 5, 7$ and p is replaced by $q = p^m$, in general, the discrete logarithm problem on an elliptic curve E over $\mathbb{F}_{q^n}$ can be reduced to a discrete logarithm problem in the Jacobian of a curve of genus 5 or 7 over $\mathbb{F}_q$.

16.6.66 Remark Weil descent attacks can also be mounted on the discrete logarithm problem for Jacobians of hyperelliptic curves. See the discussion in [148].

16.6.12 Pairings based cryptosystems

16.6.67 Remark Besides the Weil pairing described above, Frey and Rück [1108] introduced the Lichtenbaum-Tate pairing. Suppose that A is the Jacobian of the curve C defined over $\mathbb{F}$. If ℓ is a prime not dividing q (the cardinality of $\mathbb{F}$) and k is the order of q modulo ℓ, there is a pairing
$$A[\ell](\mathbb{F}) \times A(\mathbb{F})/\ell A(\mathbb{F}) \longrightarrow \mathbb{F}_{q^k}^{\times}/(\mathbb{F}_{q^k}^{\times})^{\ell}.$$

16.6.68 Theorem [1108] This pairing is non-degenerate. Using it, the discrete logarithm problem in $A[\ell](F)$ can be reduced to the corresponding problem in $\mathbb{F}_{q^k}^{\times}$ in probabilistic polynomial time in $\log q$.

16.6.69 **Definition** The number k above is the *embedding degree.*

16.6.70 **Remark** If the embedding degree is small, then solving the discrete logarithm problem in $\mathbb{F}_{q^k}^{\times}$ is practical. Several authors have found examples of Abelian varieties for which the embedding degree is small.

16.6.71 **Theorem** [1097] Given a CM field K of degree $2g \geq 4$, a primitive CM-type Φ of K, a positive integer k and a prime $r \equiv 1 \pmod{k}$ that splits completely in K, there exists a prime p and a simple, ordinary Abelian variety A defined over $\mathbb{F}_p$ with embedding degree k with respect to r, and an element $\pi \in K$ so that $|A(\mathbb{F}_p)| = N_{K/\mathbb{Q}}(\pi - 1)$.

16.6.72 **Remark** The construction of A is not explicit. Rather the polynomial $P_1(T)$ is constructed explicitly and an appeal is made to the theorem of Honda and Tate. Moreover, heuristic analysis suggests that $P_1(T)$ can be found in $O(\log r)$ steps.

16.6.73 **Remark** Galbraith, McKee, and Valenca [1158] produce examples of ordinary Abelian surfaces with embedding degrees 5, 10, and 12. Again, the construction appeals to the Honda-Tate theorem.

16.6.74 **Remark** The embedding degree and security parameters in the case of supersingular Abelian varieties is analyzed carefully by Rubin and Silverberg [2494].

See Also

§16.4	For elliptic curve cryptosystems.
§16.5	For hyperelliptic curve cryptosystems.

References Cited: [123, 148, 294, 852, 854, 856, 980, 1097, 1104, 1108, 1158, 1250, 1495, 1582, 1585, 1586, 2078, 2438, 2494, 2783]

16.7 Binary extension field arithmetic for hardware implementations

M. Anwarul Hasan, *University of Waterloo*
Haining Fan, *Tsinghua University*

We consider algorithms and architectures for hardware realization of arithmetic operations over binary extension fields $\mathbb{F}_{2^n}$. In particular, we focus on arithmetic algorithms and architectures suitable for implementation in hardware for today's cryptographic applications. For representation of elements of $\mathbb{F}_{2^n}$, we consider polynomial and normal bases over $\mathbb{F}_2$.

16.7.1 Preamble and basic terminologies

16.7.1 **Remark** Common choices for hardware are configurable semiconductor devices, such as *field programmable gate arrays* (FPGAs) and *application specific integrated circuits* (ASICs). FPGAs have a very low non-recurring engineering cost and can generally be re-configured many times. On the other hand, ASICs require a high non-recurring engineering cost, but the unit cost is relatively lower for large quantity production, and are considered to be more suitable for applications requiring high-speed, low area, or low power designs.

16.7.2 Remark An arithmetic algorithm for hardware can be evaluated based on one or more of the following: (i) number of arithmetic operations over the underlying field (hence it is essentially *arithmetic complexity* of the algorithm), (ii) amount of storage, e.g., flip-flops for temporary storage and memory for pre-computed values, and (iii) number of accesses to memories used.

16.7.3 Remark An arithmetic algorithm can be mapped onto different types of hardware architectures, e.g., serial, parallel, pipelined, systolic, etc. Two important figures of merit to characterize the performance of an architecture are space and time complexities. Generally, trade-offs between space and time are possible. Main components of a hardware architecture include: logic gates, temporary storage (flip-flops, registers, etc.), and memories.

16.7.4 Definition The *space complexity* of an architecture is its number of logic gates (e.g., AND and XOR) and the amount of storage (i.e., flip-flops and memories). For simplicity, only two-input logic gates are assumed and interconnections are ignored.

16.7.5 Definition The *critical path* of an architecture is the path that represents the longest time delay. We approximate the delay as the delay caused by the logic gates in the critical path. In actual hardware realization, other factors such as interconnections contribute to the delay.

16.7.6 Definition The *time complexity* of an architecture is the amount of time needed by the architecture to complete the required arithmetic operation upon receiving any portion of the input. For a fully bit-parallel architecture, the time complexity is simply the delay in the critical path. For architecture that requires multiple clock cycles for the arithmetic operation, the time complexity is approximated as the product of the number of clock cycles and the gate delays in the critical path.

16.7.7 Remark At the architectural level, various design components can be used, for example, a two-input XOR gate for an addition over $\mathbb{F}_2$ and a two-input AND gate for a multiplication over $\mathbb{F}_2$. The time delay of a component is denoted as T along with a suitable subscript. For example, we denote the time delay for a two-input XOR gate as T_X, and that of a two-input AND gate as T_A.

16.7.2 Arithmetic using polynomial operations

16.7.8 Definition The elements of $\mathbb{F}_{2^n}$ can be represented as polynomials over $\mathbb{F}_2$ of degree $n-1$ or less. Thus, for $a \in \mathbb{F}_{2^n}$ we can write $a = \sum_{i=0}^{n-1} a_i x^i$, where $a_i \in \mathbb{F}_2$ and $0 \le i \le n-1$. The set $\{1, x, \ldots, x^{n-1}\}$ is a *polynomial basis* of $\mathbb{F}_{2^n}$ over $\mathbb{F}_2$; see Section 2.1.

16.7.9 Definition Let $a = \sum_{i=0}^{n-1} a_i x^i$ and $b = \sum_{i=0}^{n-1} b_i x^i$ be two elements of $\mathbb{F}_{2^n}$. Then the *addition of* a *and* b is $a + b := \sum_{i=0}^{n-1} (a_i + b_i) x^i$, where $a_i + b_i$ is over $\mathbb{F}_2$.

16.7.10 Proposition Using n XOR gates in parallel, $a+b$ can be computed with a delay of one T_X only. On the other hand, $a+b$ can be be computed in bit serial fashion using only one XOR gate and a delay of nT_X.

16.7.11 Definition Let a and b be two elements of $\mathbb{F}_{2^n}$ represented using polynomials as stated above. Then the *product of a and b* is $ab := a(x)b(x) \pmod{f(x)}$, where f is an irreducible polynomial of degree n over $\mathbb{F}_2$ and defines the representation of the field.

16.7.12 Remark Various algorithms exist for computing ab. One class of algorithms involves the following two steps: first the multiplication of the two n-term polynomials, a and b, is performed, and then the resulting $(2n-1)$-term polynomial is reduced modulo f. In the straightforward or schoolbook method, the multiplication ab is performed by repeated shift-and-add operations and requires $\mathcal{O}(n^2)$ additions and multiplications over $\mathbb{F}_2$. A more efficient method is known as the Karatsuba algorithm and it is based on the following.

16.7.13 Theorem [1684] Assume that n is even and the n-term polynomial a is split as $a_H x^{n/2} + a_L$, where a_H and a_L are each $(n/2)$-term polynomials in x over $\mathbb{F}_2$. Similarly, b is split as $b_H x^{n/2} + b_L$. Then

$$ab = a_H b_H x^n + [(a_H + a_L)(b_H + b_L) - a_H b_H - a_L b_L] x^{n/2} + a_L b_L. \tag{16.7.1}$$

16.7.14 Remark We note that addition and subtraction are the same in fields of characteristic two. In case n is not even, a zero coefficient can be padded at the higher degree end of a and b allowing an even splitting. This technique can be applied recursively and the following can be proven.

16.7.15 Lemma [2342] Assuming that n is a power of 2, then a recursive application of Theorem 16.7.13 for the multiplication of a and b requires no more than $n^{\log_2 3}$ multiplications and $6n^{\log_2 3} - 8n + 2$ additions over $\mathbb{F}_2$.

16.7.16 Corollary [1026, 2342] A fully bit parallel implementation of the Karatsuba algorithm for the multiplication of two n-term $\mathbb{F}_2$ polynomials requires $6n^{\log_2 3} - 8n + 2$ XOR and $n^{\log_2 3}$ AND gates, and its time complexity is $(3\log_2 n - 1)T_X + T_A$.

16.7.17 Remark Since $\log_2 3 \approx 1.58 < 2$, the Karatsuba algorithm, which first appeared in [1684] for integer multiplication, has a subquadratic arithmetic complexity. In the context of polynomial multiplication, a generalization of the Karatsuba algorithm can be found in [2963].

16.7.18 Remark The main rationale behind designing a fully parallel multiplier is to achieve a higher speed. For multiplication of polynomials a and b, one way to achieve a time complexity that is lower than that of the Karatsuba algorithm is to split the polynomials according to the parity of the x's exponent [1028], i.e., $a(x) = a_e(y) + x a_o(y)$, where $y = x^2$, $a_e(y) = \sum_{i=0}^{n/2-1} a_{2i} y^i$ and $a_o(y) = \sum_{i=0}^{n/2-1} a_{2i+1} y^i$. Similarly, $b(x) = b_e(y) + x b_o(y)$. Then the product $a(x)b(x)$ is computed using the following Karatsuba-like formula

$$\begin{aligned} a(x)b(x) &= \{[a_e(y)b_e(y) + y a_o(y)b_o(y)]\} + \\ &\quad x\{[(a_e(y) + a_o(y))(b_e(y) + b_o(y))] - [a_e(y)b_e(y) + a_o(y)b_o(y)]\}. \end{aligned}$$

If n is a power of 2, then a recursive application of the above splitting can lead to a fully bit parallel polynomial multiplication hardware that has the same space complexity as the Karatsuba algorithm, but is about 30% faster - the main gate delays are $2\log_2 n T_X$ vs. $(3\log_2 n - 1)T_X$.

16.7.19 Remark An improvement to the original $\mathbb{F}_2[x]$ Karatsuba algorithm in terms of space complexity is given in [3064]. The main idea of this method can also be described using the following *refined* Karatsuba identity [243, p. 325]:

$$\begin{aligned} ab &= a_H b_H x^{2n} + \{[(a_H + a_L)(b_H + b_L)] - [a_H b_H + a_L b_L]\}x^n + a_L b_L \\ &= (a_H b_H x^n - a_L b_L)(x^n - 1) + (a_H + a_L)(b_H + b_L)x^n. \end{aligned}$$

This identity is in fact closer to Karatsuba's original squaring identity ([1684, p. 595]) than identity 16.7.1. Table 16.7.1 summarizes complexities of the Karatsuba algorithm and its two variants.

Algorithm	#AND	#XOR	Gate delay
Karatsuba [32, 2342]	$n^{\log_2 3}$	$6n^{\log_2 3} - 8n + 2$	$(3\log_2 n - 1)T_X + T_A$
Overlap-free Karatsuba [1028]	$n^{\log_2 3}$	$6n^{\log_2 3} - 8n + 2$	$(2\log_2 n)T_X + T_A$
Refined Karatsuba [243, 3064]	$n^{\log_2 3}$	$5.5n^{\log_2 3} - 7n + 1.5$	$(3\log_2 n - 1)T_X + T_A$

Table 16.7.1 Asymptotic complexities of three Karatsuba algorithms.

16.7.20 Remark The Winograd short convolution algorithm may be viewed as a generalization of the original Karatsuba-based algorithm [241, 2747, 2988]: while the original Karatsuba algorithm performs evaluation and interpolation using linear factors $x - \infty$, x and $x - 1$, the Winograd method also uses nonlinear factors, i.e., irreducible polynomials of degrees greater than 1. For $n = 3^i$ ($i > 0$), the algorithm yields an asymptotic time complexity of $(4\log_3 n - 1)T_X + T_A \approx (2.52\log_2 n - 1)T_X + T_A$, which is better than that of the Karatsuba algorithm. However, the Winograd method has a higher asymptotic space complexity than the Karatsuba algorithm.

16.7.21 Remark Other subquadratic arithmetic complexity algorithms exist, e.g., [164, 243]. However, when mapped onto hardware architectures, they yield a higher time complexity, e.g., [243] is linear to n for a fully bit parallel implementation [574].

16.7.22 Theorem [3006] Let d be a binary polynomial of degree at most $2n - 2$ (i.e., d could be the product of two binary polynomials - each of degree at most $n - 1$). Let W_f be the number of nonzero coefficients of f of degree n. Then $d \pmod f$ can be computed with at most $(W_f - 1)(n - 1)$ bit operations.

16.7.23 Remark In addition to the class of algorithms mentioned in Remark 16.7.12 which performs polynomial multiplication ab and reduction modulo f in two separate steps, another class exists that combines these two steps. The latter class of algorithms is often used with sequential architectures. To this end, we first look at the *shift-and-reduce* operation for hardware.

16.7.24 Lemma Let a and f be as defined above. Then the i-th coordinate of xa modulo f is given as follows:

$$(xa)_i = \begin{cases} a_{n-1} & i = 0, \\ a_{n-1} + f_i a_{i-1} & 1 \le i \le n-1. \end{cases} \tag{16.7.2}$$

16.7.25 Remark Equation (16.7.2) can be realized using an n-stage *linear feedback shift register* (LFSR) that has feedback connections corresponding to the coefficients of f. If the LFSR is initialized with the coordinates of a then after one shift the LFSR will contain the coordinates of xa modulo f; see Section 10.2.

16.7.26 Lemma Let a polynomial f be as defined above. Let $a, b \in \mathbb{F}_{2^n}$ be represented as polynomials of degree at most $n-1$ over $\mathbb{F}_2$. Then $ab = \sum_{i=0}^{n-1} b_i a^{(i)}$, where $a^{(i)} \in \mathbb{F}_{2^n}$ and $a^{(i)} = x^i a \pmod{f}$.

16.7.27 Remark Note that $a^{(0)} = a$ and, for $1 \le i \le n-1$, $a^{(i)} = xa^{(i-1)} \pmod{f}$. Thus, an LFSR can be used to obtain $a^{(i)}$ from $a^{(i-1)}$ as stated in Lemma 16.7.24. This along with Lemma 16.7.26 leads to the following algorithm for multiplication of two elements a and b of $\mathbb{F}_{2^n}$. The algorithm scans the coordinates of b from the low to the high end.

16.7.28 Algorithm (Bit-level $\mathbb{F}_{2^n}$ multiplication - low to high)

Input: $a,\ b \in \mathbb{F}_{2^n}$ and reduction polynomial f

Output: $ab \pmod{f}$

1. $g \leftarrow a,\ c \leftarrow 0$
2. For i from 0 to $n-1$ do
3. $\quad c \leftarrow c + b_i g$
4. $\quad g \leftarrow xg \pmod{f}$
5. Return c

16.7.29 Corollary Algorithm 16.7.28 requires $\mathcal{O}(n^2)$ arithmetic operations over $\mathbb{F}_2$. For a bit-serial architecture of the algorithm, where the computations of one iteration are performed in one clock cycle, the space complexity is $2n$ flip-flops, n two-input AND gates, and $n + W_f - 2$ two-input XOR gates. The gate delay in the critical path of the architecture is T_X and the computation time is nT_X.

16.7.30 Remark A multiplication algorithm that scans the coordinates of b from the high to the low can be easily obtained by writing out ab using Horner's rule, i.e., $ab = (\cdots(ab_{n-1}x + ab_{n-2})x + \cdots)x + ab_0 \pmod{f}$.

16.7.31 Definition Algorithm 16.7.28 can be modified so that in each iteration a digit is processed, reducing the number of iterations to $\lceil n/d \rceil$, where d is the number of bits in each digit. We refer to the modified algorithms as *digit-level multiplication algorithms*.

16.7.32 Remark Compared to bit-level algorithms (e.g., Algorithm 16.7.28), the number of arithmetic operations in each iteration of a digit-level algorithm is higher and so is the space complexity of the corresponding digit-serial architecture, typically by a factor of d. This is because the x multiplication operation of Algorithm 16.7.28 is replaced by a multiplication with x^d, and each AND operation by a multiplication of two polynomials of degree $d-1$ over $\mathbb{F}_2$. A number of digit-serial multiplier architectures have been proposed, see for example, [107] and [2695]. For resource constrained applications, digit-serial multipliers may offer suitable trade-offs between space and time requirements.

16.7.33 Remark For applications that demand high *throughput* (in terms of number of operations per second), multipliers can be designed based on *pipeline* or *systolic array* architecture. Examples of such multipliers include [2055, 3035]. Pipeline and systolic array multipliers usually require extra flip-flips for storing intermediate results.

16.7.34 Remark Using a polynomial basis, squaring of $a \in \mathbb{F}_{2^n}$ is $a^2 = \sum_{i=0}^{n-1} a_i x^{2i} \pmod{f}$. Thus, unlike a normal basis (see Subsection 16.7.4), the use of a polynomial basis to implement an $\mathbb{F}_{2^n}$ squaring requires bit operations or logic gates. However, the number of gates required can be quite low. For example, if the reduction polynomial f is a trinomial in some special form, then a bit parallel squaring unit can be implemented with about $n/2$ XOR gates [3007]. Like squaring, a square root in polynomial basis is also very efficient.

16.7.35 Remark Let a be a nonzero element of $\mathbb{F}_{2^n}$ and b be the multiplicative inverse of a. Then in polynomial notation we have $ab \equiv 1 \pmod{f}$, where f is as defined earlier. Since $\gcd(f, a) = 1$, the extended Euclidean algorithm can be used to obtain b. By simply changing the initialization, the extended Euclidean algorithm can also be used for computing the division $b = c/a$ in $\mathbb{F}_{2^n}$, which in polynomial notation can be written as $ab \equiv c \pmod{f}$.

16.7.36 Algorithm ($\mathbb{F}_{2^n}^*$ inversion using the extended Euclidean algorithm)

Input: $a \in \mathbb{F}_{2^n}^*$ and reduction polynomial f

Output: $b \in \mathbb{F}_{2^n}^*$ such that $ab \equiv 1 \pmod{f}$

1. $r \leftarrow f$, $s \leftarrow a$, $u \leftarrow 0$, $v \leftarrow 1$
2. while $s \neq 0$ do
3. $\quad q \leftarrow \lfloor r/s \rfloor$
4. $\quad (r, s) \leftarrow (s, r - q \cdot s)$
5. $\quad (u, v) \leftarrow (v, u - q \cdot v)$
6. $b \leftarrow u$
7. Return b

16.7.37 Remark Other algorithms for computing inverses that use polynomial operations and are suitable for hardware realization include the binary GCD [2953] and the Berlekamp-Massey algorithm [1431, 2316]. Additional inversion algorithms that rely on matrix operations over subfields or multiplications and squaring operations over $\mathbb{F}_{2^n}$ (e.g., the Itoh-Tsujii algorithm) are mentioned in the following two subsections.

16.7.38 Remark For arithmetic over $\mathbb{F}_{2^n}$, the algorithms stated earlier use operations of polynomials over $\mathbb{F}_2$. In the following subsection, we consider arithmetic algorithms that use matrix operations over $\mathbb{F}_2$.

16.7.3 Arithmetic using matrix operations

16.7.39 Remark Let a polynomial f be as defined above and an element $a \in \mathbb{F}_{2^n}$ in polynomial form be represented as $a = \sum_{i=0}^{n-1} a_i x^i$. Using the coordinates of a, we write the corresponding column vector as $A = (a_0, a_1, \ldots, a_{n-1})^T$.

16.7.40 Theorem [2014] Let a, b, and $c \in \mathbb{F}_{2^n}$ and $c = ab$. For $0 \leq i \leq n-1$, let Z_i be the column vector corresponding to ax^i. Denote the column vectors corresponding to b and c by B and C, respectively. Then, the following holds:

$$C = ZB, \tag{16.7.3}$$

where Z is an $n \times n$ matrix over $\mathbb{F}_2$ and is given by $Z = (Z_0,\ Z_1, \ldots, Z_{n-1})$.

16.7.41 Remark Theorem 16.7.40 is an algorithm for multiplication of two elements of $\mathbb{F}_{2^n}$ using matrix operations over $\mathbb{F}_2$ and has two main steps: first form the matrix Z, which is a function of input a and f, and then compute a matrix-vector product (MVP) over $\mathbb{F}_2$.

16.7.42 Remark The matrix Z is the *Mastrovito matrix* [2014]. Given Z_{i-1}, one can obtain Z_i using an LFSR operation and hence it requires $W_f - 2$ additions over $\mathbb{F}_2$. Thus, the formation of matrix Z requires no more than $(n-1)(W_f - 2)$ additions over $\mathbb{F}_2$ [1434]. For arbitrary a and b, a straightforward approach to compute the matrix vector product ZB requires $\mathcal{O}(n^2)$ arithmetic operations over $\mathbb{F}_2$.

16.7.43 **Definition** Consider an $n \times n$ matrix $T = [t_{i,j}]_{i,j=0}^{n-1}$. If $t_{i,j} = t_{i+1,j+1}$ for $0 \leq i < n-1$, then T is a *Toeplitz matrix.* Thus a Toeplitz matrix can be uniquely defined by its $2n-1$ entries that are in the top row and the left most column.

16.7.44 **Lemma** [1026] The sum of two $n \times n$ Toeplitz matrices is also an $n \times n$ Toeplitz matrix. If the matrix entries belong to $\mathbb{F}_2$, then the sum requires no more than $2n-1$ additions over $\mathbb{F}_2$.

16.7.45 **Lemma** [1026, 2988] Let $n = 2^i$ $(i > 0)$, T be an $n \times n$ Toeplitz matrix and V be a column vector over $\mathbb{F}_2$. Then the Toeplitz matrix-vector product (TMVP) TV can be computed with $n^{\log_2 3}$ multiplications and $5.5n^{\log_2 3} - 6n + 0.5$ additions over $\mathbb{F}_2$.

16.7.46 **Proposition** [2988] The matrix T and vector V can be split as follows:

$$T = \begin{pmatrix} T_1 & T_0 \\ T_2 & T_1 \end{pmatrix} \text{ and } V = \begin{pmatrix} V_0 \\ V_1 \end{pmatrix},$$

where T_0, T_1 and T_2 are $(n/2) \times (n/2)$ matrices and are individually in Toeplitz form, and V_0 and V_1 are $(n/2) \times 1$ column vectors. Now the following noncommutative formula can be used to compute the TMVP TV recursively:

$$TV = \begin{pmatrix} T_1 & T_0 \\ T_2 & T_1 \end{pmatrix} \begin{pmatrix} V_0 \\ V_1 \end{pmatrix} = \begin{pmatrix} P_0 + P_2 \\ P_1 + P_2 \end{pmatrix}, \qquad (16.7.4)$$

where $P_0 = (T_0 + T_1)V_1, P_1 = (T_1 + T_2)V_0$ and $P_2 = T_1(V_0 + V_1)$.

16.7.47 **Theorem** [1432] Let matrix Z be as defined in Theorem 16.7.40. Then there exists an $n \times n$ nonsingular matrix U over $\mathbb{F}_2$ such that UZ is an $n \times n$ Toeplitz matrix. Furthermore, let $D = UC$ and $T = UZ$. Then we have

$$D = TB. \qquad (16.7.5)$$

16.7.48 **Remark** A class of transformation matrices U is given in [1434]. Special cases exist for which $T = UZ$ and $C = U^{-1}D$ can be computed efficiently. In particular, when the reduction polynomial is a trinomial, these transformations can be done by simple permutations. For such special cases, the cost of multiplication of two elements of $\mathbb{F}_{2^n}$ is essentially the cost of the multiplication of a Toeplitz matrix and a vector over $\mathbb{F}_2$ [1026].

16.7.49 **Remark** Let $a, b, c \in \mathbb{F}_{2^n}$ and $a \neq 0$. Then the division $b = c/a$ over $\mathbb{F}_{2^n}$ can be performed by solving the matrix equation (16.7.3) over $\mathbb{F}_2$ for B, which is the column vector corresponding to the coordinates of b [1434]. Hardware architectures are available for solving such matrix equations [1433].

16.7.50 **Remark** The division $b = c/a$ over $\mathbb{F}_{2^n}$ can also be performed by solving equation (16.7.5) over $\mathbb{F}_2$ for B. Since the matrix in (16.7.5) is Toeplitz, algorithms for solving (16.7.5) are more efficient that those for (16.7.3). The use of (16.7.5) however requires pre- and post-processing. As stated in Remark 16.7.48, such processing can be as simple as a permutation when f is a trinomial.

16.7.4 Arithmetic using normal bases

16.7.51 **Remark** Below we consider arithmetic in binary extension finite fields $\mathbb{F}_{2^n}$ using normal bases. Definition and various properties of normal bases can be found in Section 5.2.

16.7.52 Remark Let $\gamma \in \mathbb{F}_{2^n}$ and $N = \{\gamma, \gamma^2, \gamma^{2^2}, \ldots, \gamma^{2^{n-1}}\}$ be a normal basis of $\mathbb{F}_{2^n}$ over $\mathbb{F}_2$. An element $a \in \mathbb{F}_{2^n}$ can be represented as $a = a_0\gamma + a_1\gamma^2 + \cdots + a_{n-1}\gamma^{2^{n-1}}$, where $a_i \in \mathbb{F}_2$ and $0 \le i \le n-1$. It is easy to see that $a^2 = a_{n-1}\gamma + a_0\gamma^2 + \cdots + a_{n-2}\gamma^{2^{n-1}}$. Thus the coordinates of a^2 are obtained by a simple cyclic shift of the coordinates of a. In other words, $a_i = (a^2)_{(i+1)} = (a^{2^j})_{(i+j)}$, where subscripts are evaluated modulo n.

16.7.53 Remark Let a be a nonzero element of $\mathbb{F}_{2^n}$. Then we can write $a^{-1} = a^{2^n-2} = a^{2(2^{n-1}-1)}$. Noting that since $2^{n-1}-1 = 1+2+2^2+\cdots+2^{n-2}$, the inverse can be computed with $\mathcal{O}(n)$ multiplication and squaring operations. Using a normal basis, squaring does not require any gates. For practical applications, this method however requires a high speed multiplier.

16.7.54 Remark Consider an integer $m > 1$. Denote $\tilde{m} = \lfloor m/2 \rfloor$ and $m_0 = m \pmod 2$. Then $2^m - 1 = 2^{m_0}(2^{\tilde{m}}-1)(2^{\tilde{m}}+1) + m_0 = 2^{m_0}\{(2^{\tilde{m}}-1)2^{\tilde{m}} + (2^{\tilde{m}}-1)\} + m_0$. Thus, given $a^{2^{\tilde{m}}-1}$, one can compute a^{2^m-1} using $1 + m_0$ multiplications (disregarding other operations). By recursively applying this technique, the inverse of a nonzero $a \in \mathbb{F}_{2^n}$ can be computed with $\lfloor \log_2(n-1) \rfloor + W(n-1) - 1$ multiplications over $\mathbb{F}_{2^n}$, where $W(n-1)$ is the number of nonzero bits in the binary representation of $n-1$. Inversion algorithms incurring only these many multiplications have been independently proposed by Itoh-Tsujii [1576] and Feng [1052], and have been subsequently implemented in hardware for cryptographic applications, see for example [1361, 2441, 2466, 2489].

16.7.55 Lemma [2196] Let $a, b, c \in \mathbb{F}_{2^n}$ be represented with respect to a normal basis N as defined above and $c = ab$. Let $A = (a_0, a_1, \ldots, a_{n-1})^T$ denote the column vector representing the coordinates of $a = a_0\gamma + a_1\gamma^2 + \cdots + a_{n-1}\gamma^{2^{n-1}}$. Similarly, let B be the column vector corresponding to b. Then

$$c = \left(\sum_{i-0}^{n-1} a_i\gamma^{2^i}\right)\left(\sum_{i-0}^{n-1} b_i\gamma^{2^i}\right) = A^T G B,$$

where $G = [g_{i,j}]_{i,j=0}^{n-1}$ is an $n \times n$ matrix over $\mathbb{F}_{2^n}$ and $g_{i,j} = \gamma^{2^i+2^j}$.

16.7.56 Remark Note that G depends only on N and is fixed or constant for a given basis. As each entry of G can be represented with respect to N, one can write $G = \sum_{k=0}^{n-1} G_k\gamma^{2^k}$, where G_k is an $n \times n$ matrix over $\mathbb{F}_2$. Using Lemma 16.7.55, for $0 \le k \le n-1$ we can write $c_k = A^T G_k B$. Consider G_{n-1} and denote $A^T G_{n-1} B$ as $g(a, b)$. Since $c^{2^k} = (ab)^{2^k}$, for $0 \le k \le n-1$ we have $c_{n-1-k} = g(a^{2^k}, b^{2^k})$. This leads to the following algorithm for multiplication using normal basis N.

16.7.57 Algorithm (Multiplication over $\mathbb{F}_{2^n}$ using a normal basis)

Input: a, $b \in \mathbb{F}_{2^n}$ represented with respect to a normal basis N

Output: $c = ab$

1. $s \leftarrow a$, $t \leftarrow b$
2. For i from 0 to $n-1$ do
3. $\quad c_{n-1-i} \leftarrow g(s, t)$
4. $\quad s \leftarrow s^2$, $t \leftarrow t^2$
5. Return c

16.7.58 Remark The multiplication scheme described above is due to Massey and Omura [2012]. For an arbitrary normal basis, half of the entries of G_{n-1} are expected to be nonzero. Thus, up to $\mathcal{O}(n^2)$ arithmetic operations over $\mathbb{F}_2$ are needed for each iteration of Algorithm 16.7.57 or $\mathcal{O}(n^3)$ for the entire algorithm.

16.7.59 Remark The number of nonzero entries in G_{n-1}, which is denoted by C_N, determines the complexity of the normal basis multiplier. It has been shown by Mullin et al. that the number of non-zeros in G_{n-1} is $\geq 2n-1$ [2196]. The normal bases that satisfy the equality are known as *optimal normal bases* (ONB). There are two types of ONB (Types I and II). Section 5.3 presents more details on ONB.

16.7.60 Remark If Algorithm 16.7.57 is mapped onto a sequential architecture that requires n clock cycles, then its space complexity is $\mathcal{O}(n^2)$ logic gates and the critical path has a gate delay of $\mathcal{O}(\log_2 n)$. The algorithm can be easily unrolled and mapped onto a fully parallel architecture, especially since the squaring operation in N does not require any logic gates. The space and the time complexities for such a fully parallel realization are $\mathcal{O}(n^3)$ gates and $\mathcal{O}(\log_2 n)$ gate delays, respectively.

16.7.61 Remark The Massey-Omura multiplication scheme has some redundancy in the sense that there are common terms in the expressions of product coordinates c_k, $0 \leq k \leq n-1$ [2451]. By removing the redundancy, a bit parallel multiplier (referred to as Reyhani-Hasan-1) that offers reduced space complexity and is applicable to any arbitrary normal basis has been reported in [2451]. By exchanging one AND gate for one XOR gate, the multiplier (referred to as Reyhani-Hasan-2) of [2452] reduces the number of AND gates to $n(n+1)/2$. Because the complexity of subfield multiplication is higher than that of subfield addition, this technique is shown to be quite effective for composite field multiplications [2452].

Scheme	#AND	#XOR	Gate delay
Wang et al. 1985 [2926]	nC_N	$n(C_N-1)$	$\lceil \log_2 C_N \rceil T_X + T_A$
Reyhani-Hasan-1 [2451]	n^2	$n(C_N+n-2)/2$	$\lceil \log_2 C_N \rceil T_X + T_A$
Reyhani-Hasan-2 [2452]	$n(n+1)/2$	$n(C_N+2n-3)/2$	$\lceil \log_2 C_N \rceil T_X + T_A$

Table 16.7.2 Complexities of quadratic $\mathbb{F}_{2^n}$ general normal basis parallel multipliers.

16.7.5 Multiplication using optimal normal bases

16.7.62 Proposition [2196] Let $\hat{X} = \{x^{2^0}, x^{2^1}, \ldots, x^{2^{n-1}}\}$ be a Type I optimal normal basis of $\mathbb{F}_{2^n}$ over $\mathbb{F}_2$. Because 2 is a primitive root of prime $n+1$, the following two sets are equal

$$\{2^0, 2^1, \ldots, 2^{n-1}\} = \{1, 2, \ldots, n\}. \tag{16.7.6}$$

Therefore, $X = \{x^1, x^2, \ldots, x^n\}$ is also a basis of $\mathbb{F}_{2^n}$ over $\mathbb{F}_2$.

16.7.63 Remark Given a field element a represented in the above two bases, i.e., $a = \sum_{i=0}^{n-1} \hat{a}_i x^{2^i}$ and $a = \sum_{i=1}^{n} a_i x^i$, it is easy to obtain the following coordinate transformation formula:

$$a_{2^i} = \hat{a}_i, \tag{16.7.7}$$

where $0 \leq i \leq n-1$ and the subscript 2^i is reduced modulo $n+1$. Equations (16.7.6) and (16.7.7) reveal that the basis conversion operation between X and $\hat{X}$ is simply a permutation and can be performed in hardware without using any logic gates.

16.7.64 Lemma [1027] Let $a, b, c \in \mathbb{F}_{2^n}$ be represented with respect to basis X, $c = ab$ and $B = (b_1, b_2, \ldots, b_n)^T$ be the column vector corresponding to coordinates of b with respect to X. Similarly, $C = (c_1, c_2, \ldots, c_n)^T$. Then $C = (Z_1, \ldots, Z_n)\, B = ZB$, where Z_i $(1 \leq i \leq n)$ is the column vector corresponding to the coordinates of field element $x^i a$ with respect to basis X, and Z is an $n \times n$ matrix over $\mathbb{F}_2$.

16.7.65 Proposition [1027] Using the identity $x^{n+1} = 1 = \sum_{j=1}^{n} x^j$, we have the following decomposition of the matrix $Z = Z_1 + Z_2$:

$$Z = \begin{pmatrix} 0 & a_n & a_{n-1} & \cdots & a_3 & a_2 \\ a_1 & 0 & a_n & \cdots & a_4 & a_3 \\ a_2 & a_1 & 0 & \cdots & a_5 & a_4 \\ \vdots & \vdots & \vdots & \ddots & \vdots & \vdots \\ a_{n-2} & a_{n-3} & a_{n-4} & \cdots & 0 & a_n \\ a_{n-1} & a_{n-2} & a_{n-3} & \cdots & a_1 & 0 \end{pmatrix} + \begin{pmatrix} a_n & a_{n-1} & a_{n-2} & \cdots & a_2 & a_1 \\ a_n & a_{n-1} & a_{n-2} & \cdots & a_2 & a_1 \\ a_n & a_{n-1} & a_{n-2} & \cdots & a_2 & a_1 \\ \vdots & \vdots & \vdots & \ddots & \vdots & \vdots \\ a_n & a_{n-1} & a_{n-2} & \cdots & a_2 & a_1 \\ a_n & a_{n-1} & a_{n-2} & \cdots & a_2 & a_1 \end{pmatrix}. \quad (16.7.8)$$

Therefore, MVP ZB may be computed via $ZB = Z_1B + Z_2B$.

16.7.66 Remark The straightforward computation of the TMVP Z_1B requires $n(n-1)$ AND gates and $n(n-2)$ XOR gates. Clearly, computing Z_2B requires only n AND gates and $n-1$ XOR gates since all the rows in Z_2 are the same. However, this fact was not noticed in the original Massey-Omura [2012] normal basis multiplier, where n AND gates and $n \times (n-1)$ XOR gates were used to compute MVP Z_2B [2926]. After removing the above redundancy in Z_2B, Hasan et al. presented a multiplier with the following complexities: $n(n-1)+n = n^2$ AND gates, $n(n-2)+(n-1)+n = n^2-1$ XOR gates, and a gate delay of $\lceil 1+\log_2 n \rceil T_X + T_A$ [1439]. The structure of Sunar-Koç's Type I optimal normal basis multiplier is based on their polynomial basis multiplier [1776] and its space complexity is the same as that in [1439], but its gate delay is $\lceil 2+\log_2 n \rceil T_X + T_A$. Another design that has the same complexities as the multiplier in [1439] is Reyhani-Hasan-1 multiplier [2451]. These multipliers all belong to the class of quadratic parallel multipliers, and their complexities are summarized in Table 16.7.3.

Scheme	#AND	#XOR	Gate delay
Wang et al. [2926]	n^2	$2n^2 - 2n$	$\lceil 1+\log_2 n \rceil T_X + T_A$
Hasan et al. [1439]	n^2	$n^2 - 1$	$\lceil 1+\log_2 n \rceil T_X + T_A$
Sunar-Koç [1776]	n^2	$n^2 - 1$	$\lceil 2+\log_2 n \rceil T_X + T_A$
Reyhani-Hasan-1 [2451]	n^2	$n^2 - 1$	$\lceil 1+\log_2 n \rceil T_X + T_A$
Reyhani-Hasan-2 [2452]	$n(n+1)/2$	$1.5n^2 - 0.5n - 1$	$\lceil 1+\log_2 n \rceil T_X + T_A$

Table 16.7.3 Complexities of quadratic $\mathbb{F}_{2^n}$ Type I optimal normal basis parallel multipliers.

16.7.67 Remark By exchanging one AND gate for one XOR gate, Reyhani-Hasan-2 Type I optimal normal basis multiplier reduces the number of AND gates to $n(n+1)/2$ [2452], while keeping the total number of AND and XOR gates unchanged. This technique is also used in the Elia-Leone-2 Type II optimal normal basis parallel multiplier listed in Table 16.7.4 [966].

16.7.68 Proposition [1188] Let $x = y + y^{-1}$ generate a Type II optimal normal basis of $\mathbb{F}_{2^n}$ over $\mathbb{F}_2$, where y is a primitive $(2n+1)$-st root of unity in $\mathbb{F}_{2^{2n}}$. Define $x_i = y^i + y^{-i}$ for $0 \le i \le n$, we have

$$\{x^{2^0}, x^{2^1}, \ldots, x^{2^{n-1}}\} = \{x_1, x_2, \ldots, x_n\}. \quad (16.7.9)$$

Therefore, $X = \{x_1, x_2, \ldots, x_n\}$ is also a basis of $\mathbb{F}_{2^n}$ over $\mathbb{F}_2$.

16.7.69 Proposition [1188] Given a field element a represented with respect to the above two bases, i.e., $a = \sum_{i=0}^{n-1} \hat{a}_i x^{2^i}$ and $a = \sum_{i=1}^{n} a_i x_i$, the coordinate transformation formula between these two bases is given as follows:

$$a_{s(2^i)} = \hat{a}_i, \quad (16.7.10)$$

where $0 \le i \le n-1$ and $s(j)$ is defined as the unique integer such that $0 \le s(j) \le n$ and $j \equiv s(j) \pmod{2n+1}$ or $j \equiv -s(j) \pmod{2n+1}$.

16.7.70 Remark From (16.7.9) and (16.7.10), it follows that the basis conversion operation between X and $\hat{X}$ is simply a permutation and hence may be performed in hardware without using any logic gates.

16.7.71 Proposition [1884, 2750] The product ab can be computed via an MVP ZB using basis X, and the matrix Z can be decomposed as the summation of two matrices i.e., $Z = Z_1 + Z_2$:

$$Z = \begin{pmatrix} a_2 & a_3 & \cdots & a_n & a_n \\ a_3 & a_4 & \cdots & a_n & a_{n-1} \\ \vdots & \vdots & \ddots & \vdots & \vdots \\ a_n & a_n & \cdots & a_3 & a_2 \\ a_n & a_{n-1} & \cdots & a_2 & a_1 \end{pmatrix} + \begin{pmatrix} 0 & a_1 & \cdots & a_{n-2} & a_{n-1} \\ a_1 & 0 & \cdots & a_{n-3} & a_{n-2} \\ \vdots & \vdots & \ddots & \vdots & \vdots \\ a_{n-2} & a_{n-3} & \cdots & 0 & a_1 \\ a_{n-1} & a_{n-2} & \cdots & a_1 & 0 \end{pmatrix}. \quad (16.7.11)$$

Here, Z_1 is a Hankel matrix, i.e., entries at (i, j) and $(i - 1, j + 1)$ are equal, and Z_2 is a circulant matrix.

16.7.72 Remark The straightforward computation of the above matrix-vector product ZB results in the following quadratic parallel multipliers: Sunar-Koç multiplier [2750], Elia-Leone-1 multiplier [966] and Reyhani-Hasan-1 multiplier [2451]. Their gate counts and delay complexities are equal, and are summarized in Table 16.7.4.

Scheme	#AND	#XOR	Gate delay
Sunar-Koç [2750]	n^2	$3n(n-1)/2$	$\lceil 1 + \log_2 n \rceil T_X + T_A$
Elia-Leone-1 [966]	n^2	$3n(n-1)/2$	$\lceil 1 + \log_2 n \rceil T_X + T_A$
Elia-Leone-2 [966]	$n(n+1)/2$	$2n(n-1)$	$\lceil 1 + \log_2 n \rceil T_X + T_A$
Reyhani-Hasan-1 [2451]	n^2	$3n(n-1)/2$	$\lceil 1 + \log_2 n \rceil T_X + T_A$

Table 16.7.4 Complexities of quadratic $\mathbb{F}_{2^n}$ Type II optimal normal basis parallel multipliers.

16.7.73 Remark After being converted from a Type I optimal normal basis into basis X using (16.7.6), the computation of ab becomes a modular multiplication operation, which can be divided into two steps. The first step, i.e., the polynomial multiplication operation step, can be performed using the Karatsuba algorithm. This results in Leone's subquadratic multiplier [1902]. Fan-Hasan's Type I optimal normal basis subquadratic multiplier is based on (16.7.8). For Type II optimal normal basis multiplication, Fan and Hasan use the TMVP formula (16.7.4) and the decomposition (16.7.11). By exploiting vector and matrix symmetry properties that exist in the matrix vector expressions of Types I and II optimal normal basis multiplications, Hasan et al. use the block recombination technique to design subquadratic parallel multipliers in [1437]. Table 16.7.5 gives gate counts and gate delays for the above-mentioned optimal normal basis parallel multipliers of subquadratic space complexity.

ONB	Scheme	#AND	#XOR	Gate delay
Type I	Leone [1902]	$n^{\log_2 3}$	$6n^{\log_2 3} - 8n + 2$	$\lceil 3\log_2 n - 1 \rceil T_X + T_A$
	Fan-Hasan [1027]	$n^{\log_2 3} + n$	$5.5n^{\log_2 3} - 4n - 0.5$	$\lceil 1 + 2\log_2 n \rceil T_X + T_A$
	Hasan et al. [1437]	$\frac{4}{3}n^{\log_2 3} + 1.5n$	$\frac{14}{3}n^{\log_2 3} - 4n + 1$	$\lceil 2 + 2\log_2 n \rceil T_X + T_A$
Type II	Fan-Hasan [1027]	$2n^{\log_2 3}$	$11n^{\log_2 3} - 12n + 1$	$\lceil 1 + 2\log_2 n \rceil T_X + T_A$
	Hasan et al. [1437]	$2n^{\log_2 3} + 0.5n$	$\frac{41}{6}n^{\log_2 3} - 7.5n + 1.5$	$\lceil 1 + 2\log_2 n \rceil T_X + T_A$

Table 16.7.5 Complexities of subquadratic $\mathbb{F}_{2^n}$ Types I and II optimal normal basis parallel multipliers.

16.7.74 Remark For Type II optimal normal bases, multiplication over $\mathbb{F}_{2^n}$ can be expressed in terms of one or more multiplications of n-term polynomials over $\mathbb{F}_2$ [1180, 1238]. Then one can use any suitable method for polynomial multiplication such as the Karatsuba algorithm. The scheme presented in [1180] uses two polynomial multiplications and that in [1238] uses only one. Both schemes however incur computational overhead to express the Type II optimal normal basis multiplication into polynomial multiplication(s). The overheads are $\mathcal{O}(n)$ for [1180] and $\mathcal{O}(n \log_2 n)$ for [1238]. In [248], the computational overhead of [1238] has been reduced by a factor of about two.

16.7.6 Additional notes

16.7.75 **Remark** For efficient hardware realization of field multiplication using polynomial bases, examples of architectures that use low weight field defining polynomials, i.e., trinomials and pentanomials, include [2453, 2465, 2749]. Other types of polynomials that have been used in multiplier architectures include *all-one*, *nearly all-one*, and *equally-spaced* polynomials.

16.7.76 **Remark** The field $\mathbb{F}_{2^n}$ can be embedded into a *cyclotomic* ring. This embedding technique leads to a redundant representation, i.e., each field element is represented using more that n bits. Several multiplier architectures have been reported using such redundant representation [923, 2216, 3009]. The best scenario for redundant representation is when only one extra bit is needed and it occurs where a Type I optimal normal basis exists, the conditions for which are the same as those for an irreducible *all-one* polynomial.

16.7.77 **Remark** Optimal normal bases do not exist for every field, so that alternatives that allow efficient squaring operations include the use of Gaussian normal bases and Dickson polynomials for the representation of field elements. Examples of multiplier architectures that use such representation include [1438, 1885].

16.7.78 **Remark** Besides polynomial and normal bases, hardware multiplier architectures have been reported using bases like *shifted polynomial*, *dual*, and *triangular*; see for example [1025, 1060, 1435, 3008].

16.7.79 **Remark** The Montgomery multiplication algorithm [2132] and Residue Number System (RNS) [2691] have been extensively studied for hardware implementations of arithmetic over *prime fields*; see for example [614, 2221, 2325, 2541]. These schemes have also been applied to arithmetic over binary extension fields [164, 1775, 2787]. For multiplication over $\mathbb{F}_{2^n}$, the main cost of a straightforward realization of the Montgomery algorithm is a multiplication of two binary polynomials of degree $n-1$ [1775], and the RNS based multiplication scheme has been shown to have $\mathcal{O}(n^{1.6})$ bit operations for a special form of reduction polynomials [164].

16.7.80 **Remark** Some cryptographic systems use exponentiation a^e, where $a \in \mathbb{F}_{2^n}$ and e is a nonzero positive integer of up to n bits long. A straightforward method for exponentiation is to use the well-known *square-and-multiply* algorithm [2080]. This method requires $\lfloor \log_2 e \rfloor$ squaring operations and $W(e) - 1$ multiplications over $\mathbb{F}_{2^n}$, where $W(e)$ is the number of nonzero bits in the binary representation of e. Many improvements have been proposed that require some re-coding of the exponent e and/or creation of look-up tables based on a [2080].

See Also

§5.2	For definition and properties of normal bases.
§5.3	For definition and properties of Types I and II optimal normal bases.

References Cited: [32, 107, 164, 241, 243, 248, 574, 614, 923, 966, 1025, 1026, 1027, 1028, 1052, 1060, 1180, 1188, 1238, 1361, 1431, 1432, 1433, 1434, 1435, 1437, 1438, 1439, 1576, 1684, 1775, 1776, 1884, 1885, 1902, 2012, 2014, 2055, 2080, 2132, 2196, 2216, 2221, 2316, 2325, 2342, 2441, 2451, 2452, 2453, 2465, 2466, 2489, 2541, 2691, 2695, 2747, 2749, 2750, 2787, 2926, 2953, 2963, 2988, 3006, 3007, 3008, 3009, 3035, 3064]

17

Miscellaneous applications

17.1 Finite fields in biology

Franziska Hinkelmann, Virginia Tech
Reinhard Laubenbacher, Virginia Tech

17.1.1 Polynomial dynamical systems as framework for discrete models in systems biology

The goal of molecular systems biology is to understand the functionality of biological systems as a whole by studying interactions among the components, e.g., genes, proteins, and metabolites. Modeling is a vital tool in achieving this goal. It allows the simulation of different types of interactions within the system, leading to testable hypotheses, which can then be experimentally validated. This process leads to a better understanding of the underlying biological system.

Discrete models are increasingly used in systems biology [74, 2514, 2515, 3010]. They can reveal valuable insight about the qualitative behavior of a system when it is infeasible to estimate enough parameters accurately to build a quantitative model. Many discrete models can be formulated as polynomial dynamical systems, that is, state and time discrete dynamical systems described by polynomial functions over a finite field. This provides access to the algorithmic theory of computational algebra and the theoretical foundation of algebraic geometry, which help with all aspects of the modeling process. Polynomial dynamical systems provide a unifying framework for many discrete modeling types. The algebraic framework allows for efficient construction and analysis of discrete models.

17.1.2 Polynomial dynamical systems

17.1.1 Remark Definitions and theorems can be found in [2864].

17.1.2 Definition A *Polynomial Dynamical System (PDS)* is a time-discrete dynamical system $f = (f_1, \ldots, f_n) : \mathbb{F}^n \to \mathbb{F}^n$ where each coordinate function f_i is a function of the n variables $x_1, \ldots, x_n$, each of which takes on values in a finite field $\mathbb{F}$, and each f_i can be assumed polynomial.

17.1.3 Remark Two directed graphs are usually assigned to each such system.

17.1.4 Definition The *dependency graph* (or *wiring diagram*) $\mathcal{D}(f)$ of f has n vertices $1, \ldots, n$, corresponding to the variables $x_1, \ldots, x_n$ of f. There is a directed edge $i \to j$ if there exists $x = (x_1, \ldots, x_i, \ldots, x_n) \in \mathbb{F}^n$ such that $f_j(x) \neq f_j(x_1, \ldots, x_i + 1, \ldots, x_n)$. That is, $\mathcal{D}(f)$ encodes the variable dependencies in f.

17.1.5 Definition The dynamics of f is encoded by its *phase space* (or *state space*), denoted by $\mathcal{S}(f)$. It is the directed graph with vertex set $\mathbb{F}^n$ and a directed edge from u to v if $f(u) = v$.

17.1.6 Definition For each $u \in \mathbb{F}^n$, the *orbit* of u is the sequence $\{u, f(u), f^2(u), \ldots\}$, where f^k means k-th composition of f. The sequence $\{u, f(u), f^2(u), \ldots, f^{t-1}(u)\}$ is a *limit cycle* of length t and u is a *periodic point* of period t if $u = f^t(u)$ and t is the smallest such number. Since $\mathbb{F}^n$ is finite, every orbit must include a limit cycle.

17.1.7 Definition The point u is a *fixed point* (or *steady state*) if $f(u) = u$.

17.1.8 Lemma [748] (Limit cycle analysis) For a polynomial dynamical system $f = (f_1, \ldots, f_n) : \mathbb{F}^n \to \mathbb{F}^n$, states in a limit cycle of length t are elements of the algebraic variety $V(f_1^t - x_1, \ldots, f_n^t - x_n)$, defined by the polynomials $f_1^t - x_1, \ldots, f_n^t - x_n$, but not of the variety $V(f_1^s - x_1, \ldots, f_n^s - x_n)$ for any $s < t$. In particular, fixed points are the points in the variety $V(f_1 - x_1, \ldots, f_n - x_n)$.

17.1.9 Definition A *component* of the phase space $\mathcal{S}(f)$ consists of a limit cycle and all orbits of f that contain it. Hence, the phase space is a disjoint union of components.

17.1.10 Definition A polynomial dynamical system can be iterated using different *update schedules*:

- *synchronous*: all variables are updated at the same time;

- *sequential*: variables are updated sequentially according to the order specified by an update schedule. A sequential system with a given update schedule can be translated into a synchronous system with the same dynamics;
- *asynchronous*: at every time step, one variable is chosen according to a probability distribution (usually uniform) to be updated. Asynchronous systems are non-deterministic;
- *delays*.

17.1.11 Remark Synchronous and asynchronous systems have the same fixed points.

17.1.12 Remark For visualization and analysis of PDS, the Web-based tool ADAM is available, see Section 17.1.5 [1502]. PDS can be used to represent dynamic biological systems.

17.1.13 Example The *lac(tose)* operon is a functional unit of three genes, LacZ, LacY, and LacA, transcribed together, responsible for the metabolism of lactose in the absence of glucose in bacteria. In the presence of lactose this genetic machinery is disabled by a repressor protein. The genes in the *lac* operon encode several proteins involved in this process. Lactose permease transports extracellular lactose into the cell, where the protein β-galactosidase breaks the lactose down into glucose, galactose, and allolactose. The allolactose binds to the repressor protein and deactivates it, which results in the transcription of the three *lac* genes, resulting in the production of permease and β-galactosidase. This positive feedback loop allows for a rapid increase of lactose when needed [1583]. The result is a bistable dynamical system. This process can be modeled by a polynomial dynamical system. Each variable can take on two states: 0 denotes the absence (or low concentration) of a substrate or the inactive state of a variable, and 1 denotes presence or activity. As there are only two states, the system is modeled over the finite field $\mathbb{F}_2$. Permease (x_P) transports external lactose (x_{eL}) inside the cell, and the update function for intracellular lactose (x_L) is $x_L(t+1) = x_{eL}(t)$ AND $x_P(t)$, or as a polynomial over $\mathbb{F}_2$, $f_L = x_{eL}x_P$. Using x_B for β-galactosidase, x_M for mRNA, x_A for allolactose, the functionality of the *lac* operon can be modeled by the polynomial dynamical system $f = (f_L, f_A, f_M, f_P, f_B) : \mathbb{F}_2^5 \to F_2^5$:

$$\begin{aligned} f_L &= x_{eL}x_P, \\ f_A &= x_L x_B, \\ f_M &= x_A, \\ f_P &= x_M, \\ f_B &= x_M. \end{aligned}$$

Figure 17.1.1 denotes the dependency graph and phase space of the above model [1504].

17.1.3 Discrete model types and their translation into PDS

17.1.14 Remark Continuous models of biological systems, such as ordinary or partial differential equation models, rely on exact rate parameters, for which it is oftentimes impossible to obtain exact measurements. When not enough information is available to build a quantitative model, discrete models can give valuable insight about the qualitative behavior of the system. Such models are state- and time-discrete. For example, in the most simple case, one distinguishes only between two states, ON and OFF, or active and inactive, present and absent, etc.

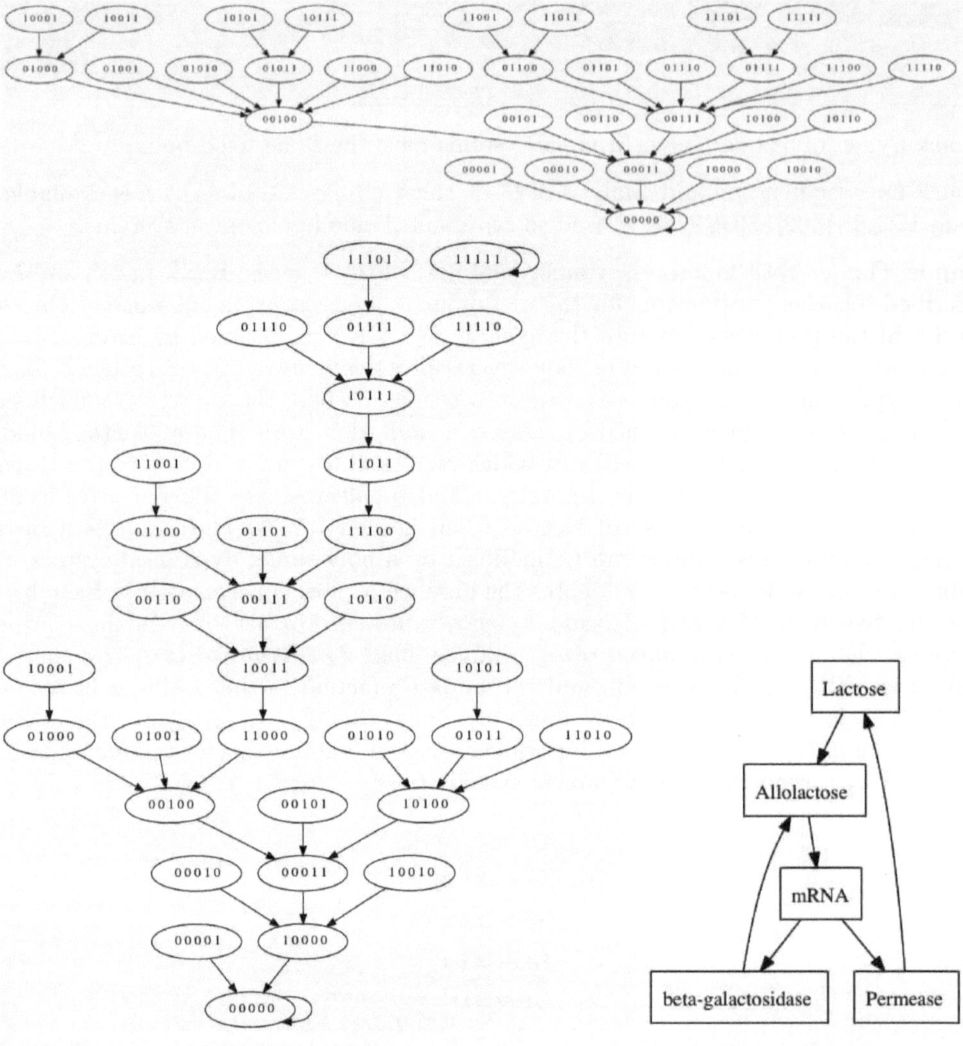

Figure 17.1.1 PDS for *lac* operon: state space in the absence (top) and presence (bottom left) of glucose, and dependency graph (bottom right). Each 5-tuple represents the states of lactose, allolactose, mRNA, permease, and β-galactosidase, (L, A, M, P, B).

17.1.15 Remark Model types include (probabilistic) Boolean networks, logical networks, Petri nets, cellular automata, and agent-based (individual-based) models, to name the most commonly found ones [343, 1859, 2215, 2512, 2617, 2704]. All these model types can be translated into PDS [1505, 2864].

17.1.3.1 Boolean network models

17.1.16 Remark In a *Boolean network model* every variable is either ON or OFF, and the state of each variable at time $t+1$ is determined by a Boolean expression that involves some or all of the variables at time t. Boolean models were first introduced in 1969 by Kauffman for gene regulatory networks, in which each gene (variable) is either expressed (ON) or not expressed (OFF) at every time step [1715].

17.1.17 Algorithm For *Boolean network models*, where there are two states (TRUE and FALSE), $\mathbb{F} = \mathbb{F}_2$, 0 denotes FALSE and 1 TRUE. Table 17.1.17 lists the Boolean expressions and the corresponding polynomials. All Boolean expressions can be translated to polynomials using this correspondence.

Boolean expression	polynomial
X	x
NOT X	$x+1$
X AND Y	xy
X OR Y	$xy+x+y$

Table 17.1.2 Correspondence of Boolean expressions and polynomials over $\mathbb{F}_2$.

17.1.3.2 Logical models

17.1.18 Remark *Logical models* are a generalization of Boolean models, in which variables can take on more than two states, e.g., to represent the three states low, medium, and high concentration of a substrate. The rules governing the temporal evolution are switch-like logical rules for the different states of each variable. Updates in logical models are specified via parameters rather than by a (Boolean) expression.

17.1.19 Definition A *logical model* is a triple $(\mathcal{V}, \mathcal{E}, \mathcal{K})$, where:

1. $V = \{v_1, \ldots, v_n\}$ is the set of vertices or nodes. Each v_i has a maximum expression level, m_i. The set $S = \{0, \ldots, m_1\} \times \cdots \times \{0, \ldots, m_n\}$ (Cartesian product) is the *state space* of the logical model and its elements are *states*.
2. $\mathcal{E}$ is the set of arcs. The elements of $\mathcal{E}$ have the form (v_i, v_j, θ), where $1 \le \theta \le m_i$ is a *threshold* for (v_i, v_j). Let $\mathcal{I}(j) = \{v : (v, v_j, \theta) \in \mathcal{E}\}$, the input of v_j, be the set of vertices that have an edge ending in v_j. Notice that an arc (v_i, v_j) is allowed to have multiplicity corresponding to different thresholds; the number of thresholds is denoted by $m_{i,j}$. These thresholds are indexed in increasing order, $1 \le \theta_{i,j,1} < \cdots < \theta_{i,j,m_{i,j}} \le m_i$; $\theta_{i,j,k}$ is the k-th threshold for the arc (v_i, v_j). By convention, we define $\theta_{i,j,m_{i,j}+1} = m_i + 1$ (actually, $\theta_{i,j,m_{i,j}+1}$ can be defined as any number greater than m_i) and $\theta_{i,j,0} = 0$. Let $x = (x_1, \ldots, x_n)$ be a state; we say that the k-th interaction for input v_i of v_j is active if $\theta_{i,j,k} \le x_i < \theta_{i,j,k+1}$; we denote this by $\Theta^{i,j}(x_i) = k$.
3. $\mathcal{K} = \{K_i : \prod_{v_j \in \mathcal{I}(i)} \{0, \ldots, m_{i,j}\} \to \{0, \ldots, m_i\}, i = 1, \ldots, n\}$ is the set of parameters.

17.1.20 Example (Logical model of lambda phage [2431]) Lambda phage is a virus that injects its DNA into a bacterial host. Once injected, it enters either a lytic cycle or a lysogenic cycle. In the lytic cycle, its DNA is replicated, the host cell lyses, and new viruses are released. In the lysogenic cycle, the virus DNA is copied into the hosts DNA where it remains without any apparent harm to the host. A number of bacterial and viral genes takes part in the decision process between lysis and lysogenisation. The core of the regulatory network that controls the life cycle consists of two regulatory genes, cI and cro. Lysogeny is maintained if cI proteins dominate, the lytic cycle if cro proteins dominate. CI inhibits cro, and vice versa. At high concentrations, cro downregulates its own production, see Figure 17.1.3. This regulatory network can be encoded in the following logical model: $\mathcal{V} = \{\text{c1}, \text{cro}\}$,

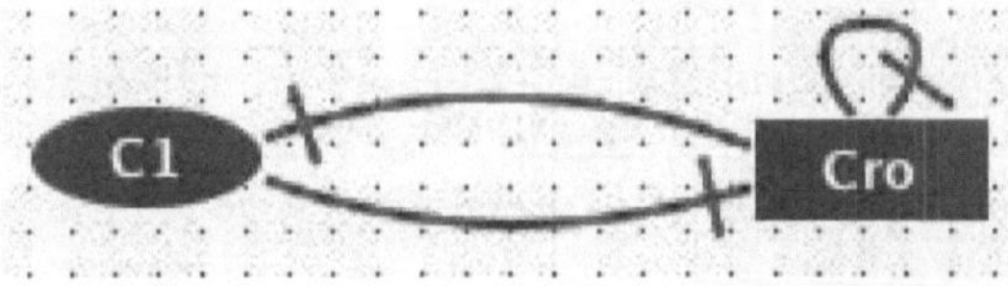

Figure 17.1.3 Logical model of lambda phage, blunt ended arrows indicate an inhibitory effect.

$\mathcal{E} = \{(\text{c1}, \text{cro}, 1), (\text{cro}, \text{c1}, 1), (\text{cro}, \text{cro}, 2)\}$, and $\mathcal{K} = \{K_{\text{c1}}, K_{\text{cro}}\}$, where $K_{\text{c1}} : \{0,1,2\} \to \{0,1\}$ is defined as $K_{\text{c1}}(0) = 1$, $K_{\text{c1}}(1) = K_{\text{c1}}(2) = 0$ and $K_{\text{cro}} : \{0,1\} \times \{0,1,2\} \to \{0,1,2\}$ as $K_{\text{cro}}(0,2) = K_{\text{cro}}(1,2) = 1$, $K_{\text{cro}}(1,0) = K_{\text{cro}}(1,1) = 0$, $K_{\text{cro}}(0,0) = 1$, and $K_{\text{cro}}(0,1) = 2$.

17.1.21 Algorithm [2864] *Logical models* are translated into a PDS by the following algorithm. Let $(\mathcal{V}, \mathcal{E}, \mathcal{K})$ be a logical model as in Definition 17.1.19. Choose a prime number p such that $p \geq m_i + 1$ for all $1 \leq i \leq n$ ($\{0, \ldots, m_i\}$ has $m_i + 1$ elements), and let $\mathbb{F} = \mathbb{F}_p = \{0, 1, \ldots, p-1\}$ be the field with p elements. Note that we may consider the set S to be a subset of $\mathbb{F}^n$. Consider a vertex v_i and let g_i be its coordinate function. Our goal is to represent g_i as a polynomial in terms of its inputs, say $x_{i_1}, \ldots, x_{i_r}$. That is, we need a polynomial function defined on $\mathbb{F}^r$ with values in $\mathbb{F}$. Denote $a \wedge b = \min\{a, b\}$, using the natural order on the set $\mathbb{F}$, viewed as integers. To extend the domain of g from $\prod_{v_j \in \mathcal{I}(i)} \{0, \ldots, m_{j,i}\}$ to $\mathbb{F}^r$, we define $g(x_{i_1}, \ldots, x_{i_r}) = g_i(x_{i_1} \wedge m_{i_1}, \ldots, x_{i_r} \wedge m_{i_r})$ for $(x_{i_1}, \ldots, x_{i_r}) \in \mathbb{F}^r$. The polynomial form of $g_i : \mathbb{F}^r \to \mathbb{F}$ is then

$$g_i(x) = \sum_{(c_{i_1}, \ldots, c_{i_r}) \in \mathbb{F}^r} g_i(c_{i_1}, \ldots, c_{i_r}) \prod_{v_j \in \mathcal{I}(i)} (1 - (x_j - c_j)^{p-1}),$$

where the right-hand side is computed modulo p; see also the Lagrange Interpolation Formula (Theorem 2.1.131).

17.1.22 Example The logical model of the lambda phage presented in Example 17.1.22 corresponds to the PDS over $\mathbb{F}_3$: $\text{c1} = x_1$, $\text{cro} = x_2$, and the polynomials are

$$f_1 = -x_2^2 + 1,$$
$$f_2 = -x_1^2 x_2^2 + x_1^2 x_2 + x_1^2 + x_2^2 - x_2 - 1.$$

17.1.23 Remark The logical model can be converted manually or with the software package ADAM [1502]. Instead of analyzing the logical model, the corresponding PDS can be analyzed for its dynamic features.

17.1.3.3 Petri nets and agent-based models

17.1.24 Remark *Petri nets* are bipartite graphs, consisting of places and transitions. Places can be marked with tokens, usually representing concentration levels or number of molecules present. Transitions fire in a non-deterministic way and move tokens between places. Petri nets have been used extensively to model chemical reaction networks, where places represent species and transitions interactions. Analysis of the Petri net can reveal which species are persistent, i.e., which species can become extinct if all species are present at the initial time. For the translation algorithm of Petri nets into polynomial dynamical systems, see [2864].

17.1.25 Remark *Agent-based models (ABM)* (or *individual-based models*) are computational models consisting of individual agents, each agent having a set of rules that defines how it interacts with other agents and the environment. Simulation is used to assess the evolution of the system as a whole. Sophisticated agent-based models have been published that simulate biological systems including tumor growth and the immune system [90, 976, 2000]. For conversion of agent-based models into polynomial dynamical systems, see [1505].

17.1.4 Reverse engineering and parameter estimation

17.1.26 Remark A central problem in systems biology is the construction of models based on system-level experimental data and biological input. When the wiring diagram is known but not the combinatorial effect of the different regulatory inputs, the process of identifying models that fit the data is analogous to parameter estimation for continuous models: estimate a function that fits the experimental data and satisfies some optimality criterion.

17.1.27 Remark Typically, the set of possible models is very large. Tools from computer algebra allow the identification of all models that fit a given experimental data set, and furthermore allow one to identify those models that satisfy a given optimality criterion [867, 1860]. Subsection 17.1.4.1 introduces an algorithm that identifies possible wiring diagrams, with the optimality criterion that every variable is affected by a minimal number of other variables, and Subsections 17.1.4.2 and 17.1.6.2 introduce parameter estimation algorithms.

17.1.4.1 The minimal-sets algorithm

17.1.28 Remark The *minimal-sets algorithm* [1597] constructs the set of all "minimal" wiring diagrams such that a model that fits the experimental data exists for each wiring diagram. Given m observations stating that the inputs t_i result in the state s_i for a variable, i.e., $(s_1, t_1), \dots, (s_m, t_m)$, where $s_i \in \mathbb{F}^n$ and $t_i \in \mathbb{F}$, the minimal-sets algorithm identifies all inclusion minimal sets $S \in \{1, \dots, n\}$ such that there exists a polynomial function $f \in \mathbb{F}[\{x_i | i \in S\}]$ with $f(s_i) = t_i$. The algorithm is based on the fact that one can define a simplicial complex Δ associated to the experimental data, such that the face ideal for the Alexander dual of Δ is a square-free monomial ideal M, and the generating sets for the minimal primes in the primary decomposition of the ideal M are exactly the desired minimal sets.

17.1.4.2 Parameter estimation using the Gröbner fan of an ideal

17.1.29 Remark Typically, there are many models that fit experimental data, even when restricting the model space to minimal models. The *minimal sampling algorithm* is used to sample the subspace of minimal models; it returns a set of weighted functions per node, assigning a higher weight to functions that are candidates for several monomial orders [868]. The algorithm is based on the Gröbner fan of an ideal.

17.1.5 Software for biologists and computer algebra software

17.1.30 **Remark** The methods described in this section heavily rely on computer algebra systems, e.g., Macaulay2 [1355]. A major advantage of translating other discrete modeling types into polynomial systems is that efficient implementations of algorithms such as Gröbner basis calculations or primary decomposition are already implemented, and can be used independently of the underlying model type. On the other hand, an algorithm implemented to analyze a Petri net cannot be re-used to analyze a logical model. As many biologists are not familiar with computer algebra systems and the mathematical theory, software packages have been developed that allow the construction and analysis of discrete models using methods described in this section, without requiring understanding of the underlying mathematics [866, 1502].

17.1.6 Specific polynomial dynamical systems

17.1.31 **Remark** Sections 17.1.6.1 to 17.1.6.4 describe specific classes of polynomial dynamical systems, and theorems that relate the structure of the PDS to its dynamics.

17.1.6.1 Nested canalyzing functions

17.1.32 **Remark** Certain polynomial functions are very unlikely to represent an interaction in a biological system. For example, in a Boolean system, $x + y$, i.e., the exclusive OR, is unlikely to represent an actual biological process. In addition, there are classes of functions that are biologically more relevant than other functions. One such class consists of nested canalyzing functions, named after the genetic concept of canalization, identified by the geneticist Waddington in the 1940s. Networks consisting of nested canalyzing functions are robust and stable [1714, 2887].

17.1.33 **Definition** A Boolean function $f(x_1, \dots, x_n)$ is *canalyzing* if there exists an index i and a Boolean value a for x_i such that $f(x_1, \dots, x_{i-1}, a, x_{i+1}, \dots, x_n) = b$ is constant. That is, the variable x_i, when given the *canalyzing value* a, determines the value of the function f, regardless of the other inputs. The output value b is the *canalyzed value*.

17.1.34 **Definition** Let f be a Boolean function in n variables.

- Let σ be a permutation on $\{1, \dots, n\}$. The function f is a *nested canalyzing function* (NCF) in the variable order $x_{\sigma(1)}, \dots, x_{\sigma(n)}$ with canalyzing input values $a_1, \dots, a_n$ and canalyzed output values $b_1, \dots, b_n$, respectively, if it can be represented in the form

$$f(x_1, x_2, \dots, x_n) = \begin{cases} b_1 & \text{if } x_{\sigma(1)} = a_1, \\ b_2 & \text{if } x_{\sigma(1)} \neq a_1 \text{ and } x_{\sigma(2)} = a_2, \\ b_3 & \text{if } x_{\sigma(1)} \neq a_1 \text{ and } x_{\sigma(2)} \neq a_2 \text{ and } x_{\sigma(3)} = a_3, \\ \vdots & \vdots \\ b_n & \text{if } x_{\sigma(1)} \neq a_1 \text{ and } \cdots \text{ and } x_{\sigma(n-1)} \neq a_{n-1} \\ & \text{and } x_{\sigma(n)} = a_n, \\ b_n + 1 & \text{if } x_{\sigma(1)} \neq a_1 \text{ and } \cdots \text{ and } x_{\sigma(n)} \neq a_n. \end{cases}$$

- The function f is nested canalyzing if f is nested canalyzing in the variable order $x_{\sigma(1)}, \dots, x_{\sigma(n)}$ for some permutation σ.

17.1.35 Remark Any Boolean function in n variables is a map $f : \{0,1\}^n \to \{0,1\}$. Denote the set of all such maps by B_n. For any Boolean function $f \in B_n$, there is a unique polynomial $g \in \mathbb{F}_2[x_1, \ldots, x_n]$ such that $g(a_1, \ldots, a_n) = f(a_1, \ldots, a_n)$ for all $(a_1, \ldots, a_n) \in \mathbb{F}_2^n$ and such that the degree of each variable appearing in g is equal to 1. Namely,

$$g(x_1, \ldots, x_n) = \sum_{(a_1,\ldots,a_n)\in\mathbb{F}_2^n} f(a_1, \ldots, a_n) \prod_{i=1}^{n} (1 - (x_i - a_i)).$$

17.1.36 Definition Let S be a a non-empty set whose highest element is r_S. The *completion* of S, which is denoted by $[r_S]$, is the set $[r_S] := \{1, 2, \ldots, r_S\}$. For $S = \emptyset$, let $[r_\emptyset] := \emptyset$.

17.1.37 Theorem [1599] Let f be a Boolean polynomial in n variables, given by

$$f(x_1, x_2, \ldots, x_n) = \sum_{S\subseteq[n]} c_S \prod_{i\in S} x_i. \tag{17.1.1}$$

The polynomial f is a nested canalyzing function in the order $x_1, x_2, \ldots, x_n$ if and only if $c_{[n]} = 1$, and for any subset $S \subseteq [n]$,

$$c_S = c_{[r_S]} \prod_{i\in[r_S]\setminus S} c_{[n]\setminus\{i\}}.$$

17.1.38 Corollary The set of points in $\mathbb{F}_2^{2^n}$ corresponding to coefficient vectors of nested canalyzing functions in the variable order $x_1, \ldots, x_n$, denoted by $V_{\text{id}}^{\text{ncf}}$, is given by

$$V_{\text{id}}^{\text{ncf}} = \left\{ (c_\emptyset, \ldots, c_{[n]}) \in \mathbb{F}_2^{2^n} : c_{[n]} = 1, c_S = c_{[r_S]} \prod_{i\in[r_S]\setminus S} c_{[n]\setminus\{i\}}, \text{ for } S \subseteq [n] \right\}.$$

17.1.39 Definition Let σ be a permutation on the elements of the set $[n]$. We define a new order relation $<_\sigma$ on the elements of $[n]$ as follows: $i <_\sigma j$ if and only if $\sigma^{-1}(i) < \sigma^{-1}(j)$.

Let S be a nonempty subset of $[n]$, say $S = \{i_1, \ldots, i_t\}$. Let $r_S^\sigma := \max\{\sigma^{-1}(i_1), \ldots, \sigma^{-1}(i_t)\}$. The *completion of S with respect to the permutation σ*, denoted by $[r_S^\sigma]_\sigma$ is the set $[r_S^\sigma]_\sigma := \{\sigma(1), \ldots, \sigma(r_S^\sigma)\}$.

17.1.40 Corollary Let $f \in B_n$ and let σ be a permutation of the set $[n]$. The polynomial f is a nested canalyzing function in the order $x_{\sigma(1)}, \ldots, x_{\sigma(n)}$, with input values $a_{\sigma(i)}$ and corresponding output values $b_{\sigma(i)}$, $1 \le i \le n$, if and only if $c[n] = 1$ and, for any subset $S \in [n]$,

$$c_S = c_{[r_S^\sigma]_\sigma} \prod_{w\in[r_S^\sigma]_\sigma\setminus S} c_{[n]\setminus\{w\}}.$$

17.1.41 Corollary Let σ be a permutation on $[n]$. The set of points in $\mathbb{F}_2^{2^n}$ corresponding to nested canalyzing functions in the variable order $x_{\sigma(1)}, \ldots, x_{\sigma(n)}$, denoted by V_{id}^σ, is defined by

$$V_{\text{id}}^\sigma = \left\{ (c_\emptyset, \ldots, c_{[n]}) \in \mathbb{F}_2^{2^n} : c_{[n]} = 1, c_S = c_{[r_S^\sigma]_\sigma} \prod_{w\in[r_S^\sigma]_\sigma\setminus S} c_{[n]\setminus\{w\}}, \text{ for } S \subseteq [n] \right\}.$$

17.1.42 Corollary Let $f \in R$ and let σ be a permutation of the elements of the set $[n]$. If f is a nested canalyzing function in the order $x_{\sigma(1)}, \ldots, x_{\sigma(n)}$, with input values a_i and corresponding output values b_i, $1 \le i \le n$, then

$$\begin{aligned} a_i &= c_{[n]\setminus\{\sigma(i)\}}, \text{ for } 1 \le i \le n-1, \\ b_1 &= c_\emptyset + c_{\sigma(1)} c_{[n]\setminus\{\sigma(1)\}}, \\ b_{i+1} - b_i &= c_{[i+1]_\sigma} c_{[n]\setminus\{\sigma(i+1)\}} + c_{[i]_\sigma}, \text{ for } 1 \le i < n-1 \text{ and} \\ b_n - a_n &= b_{n-1} + c_{[n-1]_\sigma}. \end{aligned}$$

17.1.43 Remark Thus, the family of nested canalyzing polynomials in a given number of variables can be described as an algebraic variety defined by a collection of binomials. This description has several useful implications.

17.1.6.2 Parameter estimation resulting in nested canalyzing functions

17.1.44 Remark For given time course data and a wiring diagram, one can identify all nested canalyzing functions (17.1.6.1) that fit these information. Nested canalyzing functions can be parameterized by an ideal, and intersecting the variety of this ideal with the variety of the ideal of all functions that fit the data results in the desired set of models [1503].

17.1.6.3 Linear polynomial dynamical systems

17.1.45 Remark For linear systems, i.e., all polynomials are linear functions, the dynamics of a system can be determined completely from the structure of the polynomials [975, 2812].

17.1.6.4 Conjunctive/disjunctive networks

17.1.46 Remark Conjunctive (respectively disjunctive) Boolean network models consist of functions constructed using only the AND (respectively OR) operator. For conjunctive or disjunctive networks with strongly connected dependency graph, the cycle structure is completely determined by a formula that depends on the *loop number*, the greatest common divisor of the lengths of the dependency graph's simple (no repeated vertices) directed cycles. For general Boolean conjunctive or disjunctive networks, an upper and lower bound for the number of cycles of a particular length can be calculated [1598].

See Also

§2.1	For the Lagrange Interpolation Formula.
§10.5	For dynamical systems over finite fields.

References Cited: [74, 90, 343, 748, 866, 867, 868, 975, 976, 1355, 1502, 1503, 1504, 1505, 1583, 1597, 1598, 1599, 1714, 1715, 1859, 1860, 2000, 2215, 2431, 2512, 2514, 2515, 2617, 2704, 2812, 2864, 2887, 3010]

17.2 Finite fields in quantum information theory

Martin Roetteler, NEC Laboratories America
Arne Winterhof, Austrian Academy of Sciences

In this chapter we mention some topics of the theory of finite fields related to quantum information theory. However, we will not give any background on quantum information theory and just refer to the monographs [1742, 2286, 2360]. For a more detailed treatment of quantum algorithms for algebraic problems we refer to the survey [619].

17.2.1 Mutually unbiased bases

17.2.1 **Definition** A maximal set of *mutually unbiased bases*, for short *MUBs*, is given by a set of n^2+n vectors in $\mathbb{C}^n$ which are the elements of $n+1$ orthonormal bases of $\mathbb{C}^n$:

$$\mathcal{B}_h = \{\mathbf{w}_{h,1}, \ldots, \mathbf{w}_{h,n}\}, \quad h = 0, \ldots, n.$$

Hence,

$$\langle \mathbf{w}_{h,i}, \mathbf{w}_{h,j} \rangle = \delta_{i,j},$$

where $\delta_{i,j} = \begin{cases} 1, & i = j, \\ 0, & i \neq j, \end{cases}$ and the defining property is the mutual unbiasedness, given by

$$|\langle \mathbf{w}_{f,i}, \mathbf{w}_{g,j} \rangle| = \frac{1}{\sqrt{n}} \tag{17.2.1}$$

for $0 \le f, g \le n$, $f \neq g$, and $1 \le i, j \le n$, where $\langle \mathbf{a}, \mathbf{b} \rangle = \sum_{u=1}^{n} \overline{a_u} b_u$ denotes the standard inner product of two vectors $\mathbf{a} = (a_1, \ldots, a_n)$, $\mathbf{b} = (b_1, \ldots, b_n) \in \mathbb{C}^n$.

17.2.2 **Remark** Mutually unbiased bases were introduced by Schwinger [2572]. They have applications in quantum state determination [1577, 3005], quantum cryptography [2401], quantum error-correcting codes [475, 476, 1337], and the mean king's problem [981].

17.2.3 **Theorem** [1743, 3005] Let $n = p^r$ be the power of a prime $p > 2$ and ψ be the additive canonical character of $\mathbb{F}_n = \{\xi_1, \ldots, \xi_n\}$. Then

$$\mathbf{w}_{h,k} = \frac{1}{\sqrt{n}} \left(\psi(\xi_h \xi_u^2 + \xi_k \xi_u) \right)_{u=1,\ldots,n}, \qquad h, k = 1, \ldots, n,$$

and $\mathbf{w}_{0,j} = (\delta_{j,u})_{u=1}^{n}$ is a maximal set of MUBs.

17.2.4 **Remark** Maximal sets of $n+1$ MUBs in dimension n are only known to exist in any dimension $n = p^r$ which is a power of a prime p. For $n = 2^r$ a construction based on Galois rings is given in [1743].

If we relax (17.2.1) to

$$|\langle \mathbf{w}_{f,i}, \mathbf{w}_{g,j} \rangle| = O\left(n^{-1/2} (\log n)^{1/2} \right)$$

we can construct sets of $n+1$ orthonormal bases for any dimension n [2656]. Elliptic curves over finite fields can be used to construct sets of $n+1$ orthonormal bases with

$$|\langle \mathbf{w}_{f,i}, \mathbf{w}_{g,j} \rangle| = O\left(n^{-1/2} \right)$$

which applies to almost all dimensions n and under some widely believed conjectures about the gaps between primes to all n.

17.2.5 **Theorem** [2656] Let E be an elliptic curve over a finite field $\mathbb{F}_p$ of prime order $p > 3$ with n points. For $2 \le d \le n-1$ denote by A_d the set of bivariate polynomials over E of degree at most d with $f(0,0) = 0$. Let X denote the character group of E. For $f \in \mathbb{F}_p[E]$ we define the set

$$B_f = \{\mathbf{v}_{f,\chi} : \chi \in X\},$$

where for a character $\chi \in X$, the vector $\mathbf{v}_{f,\chi}$ is given by

$$\mathbf{v}_{f,\chi} = \frac{1}{\sqrt{n}}(\psi(f(P))\chi(P))_{P\in E}$$

where ψ denotes the additive canonical character of $\mathbb{F}_p$.

For $2 \le d \le n-1$ the standard basis and the p^{d-1} sets $B_f = \{\mathbf{v}_{f,\chi} : \chi \in X\}$, with $f \in A_d$, are orthonormal and satisfy

$$|\langle \mathbf{v}_{f,\chi}, \mathbf{v}_{g,\psi}\rangle| \le \frac{2d + (2d+1)n^{-1/2}}{n^{1/2}},$$

where $f, g \in A_d$, $f \ne g$, and $\chi, \psi \in X$.

17.2.6 **Remark** Maximal sets of MUBs and geometries over finite fields were used to define a discrete analog of quantum mechanical phase space and the corresponding notion of Wigner transform [1272, 1360]. The latter is a real valued function that uniquely characterizes a quantum state and allows one to compute measurement statistics by performing summation along lines of the underlying finite geometry.

17.2.2 Positive operator-valued measures

17.2.7 **Definition** For a positive integer n let $\mathcal{A} = \{\mathbf{v}_i = (v_{i1}, \ldots, v_{in}) : i = 1, \ldots, n^2\}$ be a set of n^2 vectors in $\mathbb{C}^n$. The set of $n^2 \times n^2$ matrices

$$\mathcal{E} = \{E_i = (v_{ij}\overline{v_{ik}}/n)_{i,k=1}^{n^2} : i = 1, \ldots, n^2\}$$

is an *approximately symmetric informationally complete positive operator-valued measure (ASIC-POVM)* if it satisfies the following conditions:

1. $\sum_{i=1}^{n^2} E_i = I_n$ (completeness/POVM condition),
2. the E_i are linearly independent as elements of $\mathbb{C}^{n\times n}$ (informational completeness),
3. $|\langle \mathbf{v}_i, \mathbf{v}_j\rangle|^2 \le (1 + o(1))n^{-1}$, $1 \le i < j \le n^2$ (approximate symmetry).

17.2.8 **Remark** If $\mathcal{E}$ satisfies instead of Condition 3 the stronger condition

$$|\langle \mathbf{v}_i, \mathbf{v}_j\rangle|^2 = \frac{1}{n+1}, \quad 1 \le i < j \le n^2,$$

it is a *SIC-POVM*.

SIC-POVMs have several very desirable properties, see [568] for a discussion in the context of the quantum de Finetti theorem and more generally in their Bayesian approach to quantum mechanics and its interpretation [567]. Furthermore, see [1142, 1143] for their role in establishing the quantumness of a Hilbert space and the related question about optimal intercept-resend eavesdropping attacks on quantum cryptographic schemes. Explicit analytical constructions of SIC-POVMs have been given for dimensions $n = 2, 3, 4, 5, 6, 7, 8, 19$, see [2450] and references therein. While it has been conjectured that SIC-POVMs exist in all dimensions and numerical evidence exists for dimensions up to 45, see [2450], it is a difficult task to explicitly construct SIC-POVMs. There are no known infinite families and in fact it is not even clear if there are SIC-POVMs for infinitely many n. However, ASIC-POVMS can be constructed for any power of an odd prime.

17.2.9 Theorem [1744] Let q be a power of a prime $p \geq 3$ and ψ denote the additive canonical character of $\mathbb{F}_q$. Let

$$\mathbf{v}_{a,b} = q^{-1/2}\left(\psi(ax^2+bx)\right)\right)_{x\in\mathbb{F}_q} \in \mathbb{C}^q$$

for all $(a,b) \in \mathbb{F}_q \times \mathbb{F}_q^*$ and $\mathbf{v}_{a,0} = (\delta_{a,x})_{x\in\mathbb{F}_q}$ for all $a \in \mathbb{F}_q$. We define $E_{a,b} = (v_{a,b,x}\overline{v_{a,b,y}})_{x,y\in\mathbb{F}_q}$ and

$$G = \sum_{a\in\mathbb{F}_q}\sum_{b\in\mathbb{F}_q} E_{a,b}.$$

Then the set $\{F_{a,b} \;:\; a,b\in\mathbb{F}_q\}$, with $F_{a,b} = G^{-1/2}E_{a,b}G^{-1/2}$, is an ASIC-POVM.

17.2.3 Quantum error-correcting codes

17.2.10 Remark Recall that for a matrix $A \in \mathbb{C}^{d\times d}$ the Hermitian conjugate is defined as $A^\dagger = (A^t)^*$, where A^t denotes transposition and * denotes entry-wise complex conjugation. For two matrices A, B we denote by $A\otimes B$ their (Kronecker) tensor product.

17.2.11 Definition (QECC characterization [1759]) Let $\mathcal{C} \leq \mathbb{C}^d$ be a subspace and let $\mathcal{E} \subseteq C^{d\times d}$ be a set of error operators. Then $\mathcal{C}$ is a *quantum error-correcting code for* $\mathcal{E}$ if the projector $P_\mathcal{C}$ onto the code space $\mathcal{C}$ satisfies the identities $P_\mathcal{C}E^\dagger FP_\mathcal{C} = \alpha_{E,F}P_\mathcal{C}$ for all $E,F\in\mathcal{E}$ and for some $\alpha_{E,F}\in\mathbb{C}$. In this case $\mathcal{C}$ can detect all errors in $\mathcal{E}$.

17.2.12 Remark We next give a brief description of the stabilizer formalism which allows connecting the problem of finding quantum error-correcting codes to the problem of finding isotropic subspaces over finite fields with respect to a certain symplectic inner product.

17.2.13 Definition Let $\mathbb{F}_q$ be the finite field of order $q = p^r$ where p is prime. Denote by ψ an additive character of $\mathbb{F}_q$. For $\alpha\in\mathbb{F}_q$ denote by $\mathbf{e}_\alpha$ the corresponding standard basis vector in $\mathbb{C}^q$, i.e., the vector that has 1 in position α and is 0 elsewhere. We define the following unitary error operators: $X_\alpha := \sum_{x\in\mathbb{F}_q}\mathbf{e}_{x+\alpha}\mathbf{e}_x{}^t$ for $\alpha\in\mathbb{F}_q$ and $Z_\beta := \sum_{z\in\mathbb{F}_q}\psi(\beta z)\mathbf{e}_z\mathbf{e}_z{}^t$ for $\beta\in\mathbb{F}_q$.

17.2.14 Definition For each $\gamma\in\mathbb{F}_q$, $\alpha = (\alpha_1,\alpha_2,\ldots,\alpha_n)\in\mathbb{F}_q^n$, and $\beta = (\beta_1,\beta_2,\ldots,\beta_n)\in\mathbb{F}_q^n$ we define the corresponding n *qudit error operator*

$$E_{\gamma,\alpha,\beta} = \psi(\gamma)(X_{\alpha_1}Z_{\beta_1})\otimes(X_{\alpha_2}Z_{\beta_2})\otimes\ldots\otimes(X_{\alpha_n}Z_{\beta_n}).$$

The *weight of* $E_{\gamma,\alpha,\beta}$ is the number of indices i for which not both α_i and β_i are zero. The group $\mathcal{G}_n$ of all $E_{\gamma,\alpha,\beta}$ has size pq^{2n} and is the *Pauli group*.

17.2.15 Lemma For $(\alpha,\beta),(\alpha',\beta')\in\mathbb{F}_q^n\times\mathbb{F}_q^n$ we have that

$$(X_\alpha Z_\beta)(X_{\alpha'}Z_{\beta'}) = \psi((\alpha,\beta)*(\alpha',\beta'))(X_{\alpha'}Z_{\beta'})(X_\alpha Z_\beta),$$

where the symplectic inner product $*$ is defined by

$$(\alpha,\beta)*(\alpha',\beta') := \sum_{i=1}^{n}(\alpha_i'\beta_i - \alpha_i\beta_i'). \tag{17.2.2}$$

17.2.16 Definition [140, 477, 1337] A *stabilizer code* is a quantum code $\mathcal{C}$ with parameters $[[n,k,d]]$ that is the joint eigenspace of a set of $n-k$ commuting Pauli operators $S_1,\ldots,S_{n-k}$, where $0 \le k \le n$. The distance d is the weight of the smallest error that cannot be detected by the code.

17.2.17 Remark A stabilizer code corresponds to an additively closed subset C of $\mathbb{F}_q^n \times \mathbb{F}_q^n$ that is contained in its dual $C^* := \{v \colon v \in \mathbb{F}_q^n \mid \text{for all } c \in C\colon \operatorname{Tr}_{\mathbb{F}_q/\mathbb{F}_p}(v * c) = 0\}$.

17.2.18 Remark In order to characterize the error-correcting properties of the code, it is useful to introduce the normalizer $\mathcal{N}$ of $\mathcal{S}$ in the Pauli group $\mathcal{G}_n$ of n qudits. The elements of the normalizer can be seen as encoded logical operations on the code. On the other hand, they also correspond to undetectable errors.

17.2.19 Definition Let $\mathcal{S}$ be an Abelian subgroup of $\mathcal{G}_n$ which has trivial intersection with the center of $\mathcal{G}_n$. Furthermore, let $\{g_1, g_2, \ldots, g_\ell\}$ where $g_i = \psi(\gamma_i)X_{\alpha_i}Z_{\beta_i}$ with $\gamma_i \in \{0,\ldots,p-1\}$ and $(\alpha_i,\beta_i) \in \mathbb{F}_q^n \times \mathbb{F}_q^n$ be a minimal set of generators for $\mathcal{S}$. Then a *stabilizer matrix* of the corresponding stabilizer code $\mathcal{C}$ is a generator matrix of the (classical) additive code $C \subseteq F_q^n \times F_q^n$ generated by (α_i,β_i). The corresponding stabilizer matrix is defined as

$$\left(\begin{array}{c|c} \alpha_1 & \beta_1 \\ \vdots & \vdots \\ \alpha_\ell & \beta_\ell \end{array}\right) \in \mathbb{F}_q^{\ell\times 2n}.$$

17.2.20 Remark A special class of stabilizer codes are *CSS codes* [478, 2702]. These codes are obtained from a pair of classical linear codes $C_1 = [n,k_1,d_1]_q$ and $C_2 = [n,k_2,d_2]_q$ over $\mathbb{F}_q$. The codes have to satisfy the condition that $C_2^\perp \subseteq C_1$, where $C_2^\perp$ denotes the dual code of C_2 with respect to the standard inner product on $\mathbb{F}_q^n$. The basis states of the code space $\mathcal{C}$ are defined as $\mathbf{c}_w := \frac{1}{\sqrt{|C_2^\perp|}} \sum_{c\in C_2^\perp} \mathbf{e}_{c+w}$, where $w \in C_1$. Two states $\mathbf{c}_w$ and $\mathbf{c}_{w'}$ are identical if and only if $w - w' \in C_2^\perp$ and otherwise they are orthogonal. The dimension of the code space $\mathcal{C}$ is q^k, where $k = k_1 + k_2 - n$ and the minimum distance is $d \ge \min(d_1,d_2)$.

17.2.21 Remark It can be shown [1338, 1353] that any stabilizer code can be encoded efficiently by using quantum operations from the Clifford group. A generating set for this group is $\sum_{y\in\mathbb{F}_q} \mathbf{e}_{\gamma y}\mathbf{e}_y{}^t$ for $\gamma \in \mathbb{F}_q\setminus\{0\}$, $\frac{1}{\sqrt{q}}\sum_{x,z\in\mathbb{F}_q}\psi(xz)\mathbf{e}_z\mathbf{e}_x{}^t$, and $\sum_{x,y\in\mathbb{F}_q}\mathbf{e}_x\mathbf{e}_x{}^t \otimes \mathbf{e}_{x+y}\mathbf{e}_y{}^t$, where these gates can be applied to any (pair) of the n qudits.

17.2.22 Remark Several constructions of families of quantum error-correcting codes based on finite fields are known. Starting from classical Reed-Solomon codes (see Section 15.1) over $\mathbb{F}_{2^n}$, in [1352] a construction of codes with parameters $[[n(2^n-1), n(2^n-1-2K), d \ge K+1]]$ was given, where $K = 2^n - \delta$ and $\delta > (2^n-1)/2+1$ is the designed distance of the classical Reed-Solomon code $[2^n-1, K, \delta]$ which is underlying the construction.

For a construction of nonbinary stabilizer codes using arbitrary finite fields see [1729].

17.2.23 Remark Various bounds on the parameters of quantum error-correcting codes are known, see [476]. Similar to the classical Singleton bound, a *quantum Singleton bound* of $k + 2d \le n + 2$ for any code with parameters $[[n,k,d]]$ can be shown. Codes meeting this bound with equality are *quantum MDS codes* and several constructions based on finite fields $\mathbb{F}_q$ for large q are known [1351].

17.2.24 Remark Classical Goppa codes have been used to construct quantum error-correcting codes. A construction in [2027] is based on towers of Garcia-Stichtenoth function fields (see Section 12.6) $F_i = \mathbb{F}_{q^2}(x_0, z_1, \ldots, x_i, z_{i+1})$, defined by equations $z_i^q + z_i - x_{i-1}^{q+1} = 0$ and $x_i = z_i/x_{i-1}$,

$i = 0, 1, \ldots$ These quantum codes have been shown to be asymptotically good, i.e., their parameters $[[n_i, k_i, d_i]]$ satisfy $\lim_{i\to\infty} n_i \to \infty$, $\lim\inf_{i\to\infty} k_i/n_i > 0$, and $\lim\inf_{i\to\infty} d_i/n_i > 0$, see also [601].

17.2.25 Remark The webpage http://www.codetables.de provides tables of the best known quantum error-correcting codes for small parameters. The codes are specified by their stabilizer matrices and where applicable it is noted that a code is optimal, i.e., achieving the highest possible d for fixed n and k or the highest possible k for fixed n and d. The table also contains known bounds on the parameters and contains information about the construction of the codes. Some of these tables and constructions have also been made available in the computer algebra system MAGMA.

17.2.26 Remark For a more detailed survey on quantum error-correcting codes see [1054].

17.2.4 Period finding

17.2.27 Theorem [2623] Let (e_n) be a periodic sequence over $\mathbb{F}_p$ of (unknown) period T. If the mapping $n \mapsto e_n$, $0 \le n < T$, is injective, T can be recovered on a quantum computer in polynomial time.

17.2.28 Remark No classical polynomial time algorithm is known for period finding.

If $g \in \mathbb{F}_p^*$ is an element of (unknown) order T, Shor's algorithm [2623] determines T in quantum polynomial time.

Let (f_n) be another sequence over $\mathbb{F}_p$ of period T such that $n \mapsto f_n$ is injective and the (unknown) pair of positive integers (t_1, t_2) satisfies $e_{n+t_1} f_{n+t_2} = e_n f_n$ for all $n \ge 0$. Then Shor's algorithm also determines (t_1, t_2). Let $a, b \in \mathbb{F}_p^*$ be such that the order of b is t and $a = b^x$ with $0 \le x < t$. Then x is the *discrete logarithm* of a to the base b. If we choose $e_n = b^n$ and $f_n = a^{-n}$, we get $e_{n+x} f_{n+1} = b^{n+x} a^{-n-1} = e_n f_n$ for all n and we have $(t_1, t_2) = (x, 1)$. Hence, Shor's algorithm finds the discrete logarithm in quantum polynomial time.

However, Shor's algorithm does not provide the period of (e_n) if $n \mapsto e_n$ is not injective.

17.2.29 Theorem [1400] Given two periodic sequences (e_n) and (f_n) with periods T and t, respectively, we denote by $D(e_n, f_n)$ the number of integers $n \in [0, Tt - 1]$ with $e_n \ne f_n$.

For any constant $c > 0$, there is a quantum algorithm which computes in polynomial time, with probability at least 3/4, the period of any sequence (e_n) of period T satisfying

$$D(e_n, f_n) \ge \frac{Tt}{(\log T)^c}, \tag{17.2.3}$$

for any sequence (f_n) of period $t < T$.

17.2.30 Remark In [2397], for binary sequences (e_n) with moderately small autocorrelation (which includes all cryptographically interesting sequences) it was proved that (17.2.3) is fulfilled and the algorithm of Hales and Hallgren [1400] determines their period in quantum polynomial time. In [2657] a more general problem was studied of determining the period of a sequence over an unknown finite field $\mathbb{F}_p$, given a black-box which returns only a few most significant bits of the sequence elements. A moderately small autocorrelation is again a sufficient condition such that the condition (17.2.3) is satisfied.

17.2.5 Quantum function reconstruction

17.2.31 Theorem [2509] Any monic, square-free polynomial $f \in \mathbb{F}_p[X]$ of degree d can be reconstructed from an oracle $\mathcal{O}_f : \mathbb{F}_p \to \{-1, 0, 1\}$ given by the Legendre symbol $\mathcal{O}_f(a) = \left(\frac{f(a)}{p}\right)$ of $f(a)$ after $O(d)$ quantum-queries with probability $1 + O(p^{-1})$.

17.2.32 Remark The case $f(X) = X + s$ with an unknown $s \in \mathbb{F}_p$ is a special case of the *hidden shift problem*. In this case in [2840] an efficient quantum algorithm was presented that finds s with one query to the oracle $\mathcal{O}_f$. This algorithm uses the fact that the Legendre symbol is, essentially, equal to its discrete Fourier transform.

17.2.33 Remark The hidden shift problem has also been studied for functions other than the Legendre symbol. In [2488] the *hidden shift problem for Boolean functions* was studied and it was shown that for any bent function (see Section 9.3) $B : \mathbb{F}_2^n \to \mathbb{F}_2$ the hidden shift s can be found by making $O(n)$ queries to $\mathcal{O}_B(x) = (-1)^{B(x)}$. If furthermore an efficient circuit is known to implement the dual bent function B^*, then this can be reduced to a single query. These results were recently extended to the case of random Boolean functions [1260] where the shift s can be determined from solving a system of linear equations over $\mathbb{F}_2$ that is obtained from a sampling procedure and to the case of functions that are close to quadratic bent functions, where closeness is measured using the Gowers U_3 norm [2487].

17.2.6 Further connections

17.2.34 Remark Some additive character sums over a finite field (*twisted Kloosterman sums*) are closely connected to quantum algorithms for finding hidden nonlinear structures over finite fields [618].

17.2.35 Remark Schumacher and Westmoreland [2562] investigate a discrete version of quantum mechanics called *modal quantum computing* based on a finite field instead of the complex numbers. Some characteristics of actual quantum mechanics are retained including the notions of superposition, interference, and entanglement.

17.2.36 Remark Exponential sums over finite fields are used to construct *quantum finite automata* [88].

17.2.37 Remark Classical and quantum algorithms for solvability testing and finding integer solutions x, y of equations of the form $ag^x + bh^y = c$ over a finite field $\mathbb{F}_q$ are studied in [2841].

17.2.38 Remark Given a group G, a subgroup $H \leq G$, and a set X, a function $f : G \mapsto X$ hides the subgroup H if for all $g_1, g_2 \in G$, $f(g_1) = f(g_2)$ if and only if $g_1H = g_2H$ for the cosets of H. The function f is given via an oracle. The *hidden subgroup problem* is the problem of finding H using information gained from evaluations of f via its oracle. In the case of an Abelian group G an efficient quantum algorithm is known for solving the hidden subgroup problem. In particular, Shor's algorithm relies on this fact.

In [2138] the authors showed that the hidden subgroup problem can be also efficiently solved by a quantum computer in the case of the semi-direct product of the additive groups of $\mathbb{F}_{p_1}$ and $\mathbb{F}_{p_2}$ where p_1 and p_2 are primes with $p_2|(p_1 - 1)$ and $(p_1 - 1)/p_2$ is polynomial in $\log p_2$.

Another related problem involving finite fields is the hidden polynomial function graph polynomial studied in [791].

See Also

§6.1, §6.3	For basics on character sums and applications.
§9.1	For Boolean functions.
§10.3, §17.3	For autocorrelation of sequences.
§11.6, §16.1.3.2	For the discrete logarithm problem and public-key cryptography.
§12.2, §12.3, §16.4	For elliptic curves.
§15.1, §15.2	For classical and algebraic codes.

References Cited: [88, 140, 475, 476, 477, 478, 567, 568, 601, 618, 619, 791, 981, 1054, 1142, 1143, 1260, 1272, 1337, 1338, 1351, 1352, 1353, 1360, 1400, 1577, 1729, 1742, 1743, 1744, 1759, 2027, 2138, 2286, 2360, 2397, 2401, 2450, 2487, 2488, 2509, 2562, 2572, 2623, 2656, 2657, 2702, 2840, 2841, 3005]

17.3 Finite fields in engineering

Jonathan Jedwab, Simon Fraser University
Kai-Uwe Schmidt, Otto-von-Guericke University

17.3.1 Binary sequences with small aperiodic autocorrelation

17.3.1 **Definition** A *sequence* of length n over an alphabet $\mathcal{A}$ is an n-tuple $(a_j) = (a_0, a_1, \dots, a_{n-1})$, where each a_j is an element of the set $\mathcal{A}$.

17.3.2 **Definition** Let $A = (a_j)$ be a sequence over $\mathbb{C}$ of length n. For integer u satisfying $0 \le u < n$, the *aperiodic autocorrelation* of A at shift u is $C_u(A) = \sum_{j=0}^{n-u-1} a_j \overline{a_{j+u}}$.

17.3.3 **Remark** The most important case, from a practical and historical viewpoint, occurs when the sequence is *binary*, which means that its alphabet is $\{1, -1\}$. A binary sequence for which all aperiodic autocorrelations at nonzero shifts are small in magnitude relative to the sequence length is intrinsically suited for the separation of signals from noise, and therefore has natural applications in digital communications, including radar, synchronization, and steganography.

17.3.4 **Remark** The overall goal is to find binary sequences A having the property that, for each $u \ne 0$ independently, $|C_u(A)|$ takes its smallest possible value. An ideal binary sequence A from this viewpoint therefore satisfies $|C_u(A)| = 0$ or 1 for each $u \ne 0$, which is known as a *Barker sequence.* The longest Barker sequence currently known has length 13 and there is overwhelming evidence that no longer Barker sequence exists (see [1603] for a survey).

17.3.5 **Definition** Write $F = \mathbb{F}_2$ and $E = \mathbb{F}_{2^m}$. An m-sequence (y_j) of length $n = 2^m - 1$ satisfies $y_j = (-1)^{\mathrm{Tr}_{E/F}(c\alpha^j)}$ for $0 \le j < n$ and for some primitive element α in E and some nonzero c in E.

17.3.6 Theorem [2527] Every m-sequence Y of length $n = 2^m - 1$ satisfies $|C_u(Y)| < 1 + (2/\pi)\sqrt{n+1}\,\log(4n/\pi)$ for each u satisfying $0 < u < n$.

17.3.7 Remark The above theorem shows the existence of an infinite family of binary sequences for which the magnitude of the aperiodic autocorrelation (at nonzero shifts) grows no faster than order $\sqrt{n}\log n$. The only known improvement of this result uses probabilistic methods and guarantees a growth rate of at most order $\sqrt{n\log n}$ [2135].

17.3.8 Remark For the following definition we need quadratic characters, see Section 6.1.

17.3.9 Definition For an odd prime p, define $\lambda : \mathbb{F}_p \to \{1, -1\}$ to be the quadratic character on $\mathbb{F}_p^*$ and $\lambda(0) = 1$. For real r, the *r-shifted Legendre sequence* of length p is the sequence (x_j) of length p satisfying $x_j = \lambda(j + \lfloor rp \rfloor)$ for $0 \le j < p$.

17.3.10 Theorem [2037] For all real r, the r-shifted Legendre sequence X of length p satisfies $|C_u(X)| < 1 + 18\sqrt{p}\log p$ for each u satisfying $0 < u < n$.

17.3.11 Definition The *merit factor* of a binary sequence A of length $n > 1$ is $F(A) = n^2/(2\sum_{u=1}^{n-1}[C_u(A)]^2)$.

17.3.12 Theorem [1607] Let Y_n be an m-sequence of length $n = 2^m - 1$. Then $F(Y_n) \to 3$ as $n \to \infty$.

17.3.13 Theorem [1523] For real r satisfying $|r| \le 1/2$, let X_p be the r-shifted Legendre sequence of length p. Then $1/F(X_p) \to 1/6 + 8\big(|r| - 1/4\big)^2$ as $p \to \infty$.

17.3.14 Remark The largest asymptotic merit factor occurring in the above theorem is 6. However, by modifying the construction as shown below, an asymptotic merit factor of approximately 6.34 can be achieved, which is the largest known asymptotic merit factor for binary sequences.

17.3.15 Theorem [1604, 1605] Let $t = 1.0578\ldots$ be the middle root of $4x^3 - 30x + 27$ and write $r = 3/4 - t/2$. Let (x_j) be the r-shifted Legendre sequence of length p. Define W_p to be the sequence (w_j) of length $\lfloor tp \rfloor$ given by $w_j = x_j$ for $0 \le j < p$ and $w_j = x_{j-p}$ for $p \le j < \lfloor tp \rfloor$. Then $F(W_p) \to 6.3420\ldots$, which is the largest root of $29x^3 - 249x^2 + 417x - 27$.

17.3.2 Sequence sets with small aperiodic auto- and crosscorrelation

17.3.16 Definition Let $A = (a_j)$ and $B = (b_j)$ be sequences over $\mathbb{C}$ of length n. For integer u satisfying $0 \le u < n$, the *aperiodic crosscorrelation* of A and B at shift u is $C_u(A, B) = \sum_{j=0}^{n-u-1} a_j\overline{b_{j+u}}$ and the *periodic crosscorrelation* of A and B at shift u is $R_u(A, B) = \sum_{j=0}^{n-1} a_j\overline{b_{j+u}}$, where indices are reduced modulo n.

17.3.17 Definition A set $\mathcal{S}$ of sequences has *maximum aperiodic correlation θ* if $|C_u(A, B)| \le \theta$ for all $A, B \in \mathcal{S}$ when either $A \ne B$ or $u \ne 0$.

17.3.18 Remark [2554] Code-division multiple access (CDMA) is a technique that allows multiple users to communicate over the same medium. For example, direct-sequence CDMA employs a set $\mathcal{S}$ of sequences over $\mathbb{C}$ of the same length. Each user is assigned a sequence of $\mathcal{S}$, and encodes information by sending a modulated version of the assigned sequence in which each sequence element is multiplied by a complex number drawn from some alphabet. In order to allow synchronization at the receiver and to minimize interference between different users, it is necessary to minimize the maximum aperiodic correlation of $\mathcal{S}$.

17.3.19 Remark It is a notoriously difficult problem to design sequence sets with small maximum aperiodic correlation directly. The usual approach is therefore to design sequence sets that have good periodic crosscorrelation properties (see Section 10.3 for constructions using finite fields), and then analyze their aperiodic crosscorrelation properties using either separate methods or numerical computation.

17.3.3 Binary Golay sequence pairs

17.3.20 Definition A *binary Golay sequence pair* is a pair of binary sequences A, B of equal length n whose aperiodic autocorrelations satisfy $C_u(A) + C_u(B) = 0$ for $0 < u < n$. A *binary Golay sequence* is a member of a binary Golay sequence pair.

17.3.21 Remark Binary Golay sequence pairs were introduced to solve a problem in infrared multislit spectrometry [1293], and have since been used in many other digital information processing applications such as optical time domain reflectometry [2218] and medical ultrasound [2300]. The defining aperiodic autocorrelation property can be exploited to allow very efficient energy use when transmitting information, or to remove unwanted components from received signals.

17.3.22 Theorem [967] If there exists a binary Golay sequence pair of length n and p is an odd prime factor of n, then -1 is a square in $\mathbb{F}_p$ and so $p \equiv 1 \pmod 4$.

17.3.23 Theorem [2832] If there exist binary Golay sequence pairs of length n_1 and n_2, then there exists a binary Golay sequence pair of length $n_1 n_2$.

17.3.24 Remark Starting from "seed" binary Golay sequence pairs of length 2, 10, and 26, the above theorem produces a binary Golay sequence pair of length $2^k 10^\ell 26^m$ for all nonnegative integers k, ℓ, m. Once it is known that a binary Golay sequence of a particular length exists, it is then important in some applications to find as many such sequences of this length as possible.

17.3.25 Definition Let $A = ((-1)^{a_j})$ be a binary sequence of length 2^m. The *algebraic normal form* of A is the unique function $f_A(x_1, \ldots, x_m) : \mathbb{F}_2^m \to \mathbb{F}_2$ satisfying

$$f_A(x_1, \ldots, x_m) = a_{x_m + 2x_{m-1} + \cdots + 2^{m-1}x_1} \quad \text{for all } (x_1, \ldots, x_m) \in \mathbb{F}_2^m,$$

where indices are calculated in $\mathbb{Z}$.

17.3.26 Theorem [783] Let π be a permutation of $\{1, \ldots, m\}$ and let $e_0', e_0, e_1, \ldots, e_m \in \mathbb{F}_2$. The binary sequences A and B of length 2^m having algebraic normal form $f_A(x_1, \ldots, x_m) = \sum_{k=1}^{m-1} x_{\pi(k)} x_{\pi(k+1)} + \sum_{k=1}^{m} e_k x_{\pi(k)} + e_0$ and $f_B(x_1, \ldots, x_m) = \sum_{k=1}^{m-1} x_{\pi(k)} x_{\pi(k+1)} + \sum_{k=1}^{m} e_k x_{\pi(k)} + e_0' + x_{\pi(1)}$ form a binary Golay sequence pair.

17.3.27 Example Take $m = 3$, $(\pi(1), \pi(2), \pi(3)) = (2, 1, 3)$, and $(e_0', e_0, e_1, e_2, e_3) = (0, 1, 1, 0, 1)$. Then $f_A(x_1, x_2, x_3) = x_2 x_1 + x_1 x_3 + x_2 + x_3 + 1$ is the algebraic normal form of $A = (- + + - - - - -)$ (writing $+$ for 1, and $-$ for -1), and $f_B(x_1, x_2, x_3) = x_2 x_1 + x_1 x_3 + x_3$ is the algebraic normal form of $B = (+ - + - + + - -)$. The sequences A and B form a binary Golay sequence pair of length 8.

17.3.28 Corollary For $m > 1$, there are at least $2^m m!$ binary Golay sequences of length 2^m.

17.3.29 Remark The $2^m m!$ binary Golay sequences arising from the above theorem form $m!/2$ cosets of the first-order Reed-Muller code $\mathrm{RM}(1,m)$ in the second-order Reed-Muller code $\mathrm{RM}(2,m)$; see Section 15.1. When these sequences are used in multicarrier transmission, the Golay property tightly controls variations in the transmitted power while the code structure allows powerful error correction [783].

17.3.4 Optical orthogonal codes

17.3.30 Definition An (n, w, λ) *optical orthogonal code* is a set $\mathcal{C}$ of sequences over $\{0,1\}$ of length n and Hamming weight w such that the periodic crosscorrelation satisfies $R_u(A,B) \leq \lambda$ for all $A, B \in \mathcal{C}$ when either $A \neq B$ or $u \neq 0$.

17.3.31 Remark [639] Optical orthogonal codes are used in optical CDMA. Each user is assigned a sequence of the code, and a 1 in the sequence corresponds to a time instant when the user is allowed to transmit a light pulse. The correlation constraint enables self-synchronization of the system and controls the interference between different users.

17.3.32 Definition An (n, w, λ) optical orthogonal code of size M is *optimal* if there is no (n, w, λ) optical orthogonal code of size greater than M.

17.3.33 Remark Optical orthogonal codes are closely related to other combinatorial objects such as constant-weight codes and cyclic difference families [639]. There are numerous constructions of optimal optical orthogonal codes. Two important general constructions using finite geometries are given here.

17.3.34 Construction [2319] Write $F = \mathbb{F}_q$ and $E = \mathbb{F}_{q^m}$, and let α be primitive in E. The points in the *affine geometry* $\mathrm{AG}(m,q)$ are the elements of E; see Section 14.3. A *d-flat* in $\mathrm{AG}(m,q)$ is a translate of a d-dimensional subspace of E over F. Then the lines in $\mathrm{AG}(m,q)$ are precisely the 1-flats. Two d-flats A and B are *equivalent* if there exists $a \in E^*$ such that $A = \{ax : x \in B\}$. This partitions the d-flats into orbits. Let $\mathcal{C}$ be a set containing exactly one representative from each orbit of a d-flat that does not contain 0. With a d-flat $S \in \mathcal{C}$ we associate a sequence over $\{0,1\}$ of length $q^m - 1$ by placing a 1 at position i precisely when $\alpha^i \in S$.

17.3.35 Remark The *q-binomial coefficient* $\left[{m \atop k}\right]_q = \prod_{i=1}^{k} \frac{q^{m-i+1}-1}{q^i-1}$ equals the number of k-dimensional subspaces of $\mathbb{F}_{q^m}$ over $\mathbb{F}_q$; see Section 13.2.

17.3.36 Theorem [2319] The above code is a $(q^m - 1, q^d, q^{d-1})$ optical orthogonal code of size $\left[{m-1 \atop d}\right]_q$. If $d = 1$ or $d = m - 1$, the code is optimal.

17.3.37 Example Take $q = 2$, $m = 3$, $d = 1$, and let α be primitive in $\mathbb{F}_8$. There are 21 lines in $\mathrm{AG}(3,2)$ that do not contain 0. These are of the form $\{x, y\}$, where $x, y \in \mathbb{F}_8^*$ and $x \neq y$. A list of representatives of the orbits is $\{1, \alpha\}, \{1, \alpha^2\}, \{1, \alpha^3\}$ and $\{(1100000), (1010000), (1001000)\}$ is an optimal $(7, 2, 1)$ optical orthogonal code.

17.3.38 Construction [639] Write $F = \mathbb{F}_q$, $E = \mathbb{F}_{q^{m+1}}$, and $n = \frac{q^{m+1}-1}{q-1}$. The points in the *projective geometry* $\mathrm{PG}(m,q)$ are the elements of E^*/F^* (which is isomorphic to $\mathbb{Z}/n\mathbb{Z}$); see Section 14.4. Let α be primitive in E and let $[a]$ denote the coset of F^* in E^* containing a. A *d-space* in $\mathrm{PG}(m,q)$ is the image under the mapping $x \mapsto [x]$ of the nonzero elements of a $(d+1)$-dimensional subspace of E over F. Then the lines in $\mathrm{PG}(m,q)$ are precisely the 1-spaces. Two d-spaces A and B are *equivalent* if there exists $a \in E^*/F^*$ such that

$A = \{ax : x \in B\}$. This partitions the d-spaces into orbits, whose sizes divide n. Let $\mathcal{C}$ be a set containing exactly one representative from each orbit of size exactly n. With a d-space $S \in \mathcal{C}$ we associate a sequence over $\{0, 1\}$ of length n by placing a 1 at position i precisely when $[\alpha^i] \in S$.

17.3.39 Theorem [639] The above code is a $\left(\frac{q^{m+1}-1}{q-1}, \frac{q^{d+1}-1}{q-1}, \frac{q^d-1}{q-1}\right)$ optical orthogonal code. For $d = 1$, the code is optimal and has size $\frac{q^m-1}{q^2-1}$ for even m and $\frac{q^m-q}{q^2-1}$ for odd m.

17.3.40 Example Take $q = 2$, $m = 3$, $d = 1$, and let α be the primitive element in $\mathbb{F}_{16}$ that satisfies $\alpha^4 = \alpha + 1$. Then the 2-dimensional subspaces $\{0, 1, \alpha, \alpha^4\}$, $\{0, 1, \alpha^2, \alpha^8\}$, $\{0, 1, \alpha^5, \alpha^{10}\}$ of $\mathbb{F}_{16}$ map to the lines $\{[1], [\alpha], [\alpha^4]\}$, $\{[1], [\alpha^2], [\alpha^8]\}$, $\{[1], [\alpha^5], [\alpha^{10}]\}$ in $\mathrm{PG}(3, 2)$. Their orbits have size 15, 15, and 5, respectively, and exhaust all 35 lines in $\mathrm{PG}(3, 2)$. Hence, $\{(110010000000000), (101000001000000)\}$ is an optimal $(15, 3, 1)$ optical orthogonal code.

17.3.5 Sequences with small Hamming correlation

17.3.41 Definition Let $A = (a_j)$ and $B = (b_j)$ be sequences of length n. For integer u satisfying $0 \le u < n$, the *Hamming correlation* between A and B at shift u is $H_u(A, B) = \sum_{j=0}^{n-1} h(a_j, b_{j+u})$, where $h(x, y) = 1$ if $x = y$ and $h(x, y) = 0$ otherwise and indices are reduced modulo n.

17.3.42 Remark [2554] In frequency-hopping spread spectrum communications, several frequencies are shared by different users. Let q be the number of such frequencies and let M be the number of users. The system employs a family of M "frequency-hopping" sequences of the same length, each over the same alphabet of size q. Each user is assigned a sequence of the family and switches between the q frequencies according to the elements of this sequence. In order to minimize interference between different users, as well as to support self-synchronization of the system, the Hamming correlation between every two sequences in the family should be as small as possible.

17.3.43 Definition Let $\mathcal{S}$ be the set of all sequences of length n over some fixed alphabet, and write

$$H(A) = \max_{0<u<n} H_u(A, A),$$
$$H(A, B) = \max_{0\le u<n} H_u(A, B),$$
$$M(A, B) = \max\{H(A), H(B), H(A, B)\}.$$

1. $X \in \mathcal{S}$ is *optimal* if $H(X) \le H(X')$ for all $X' \in \mathcal{S}$.
2. $X, Y \in \mathcal{S}$ is an *optimal pair* if $M(X, Y) \le M(X', Y')$ for all $X', Y' \in \mathcal{S}$.
3. A subset $\mathcal{F} \subset \mathcal{S}$ is an *optimal family* if every pair of distinct sequences in $\mathcal{F}$ is an optimal pair.

17.3.44 Remark There are numerous constructions of optimal families. Two constructions based on finite fields are given here.

17.3.45 Construction [1889] Let k and m be integers satisfying $1 \le k \le m$. Let α be primitive in $\mathbb{F}_{p^m}$ and let β be primitive in $\mathbb{F}_{p^k}$. Write $F = \mathbb{F}_p$ and $E = \mathbb{F}_{p^m}$. For $v \in \mathbb{F}_{p^k}$, define the sequence $X_v = (x_j)$ of length $p^m - 1$ over $\mathbb{F}_{p^k}$ by $x_j = v + \sum_{\ell=0}^{k-1} \mathrm{Tr}_{E/F}(\alpha^{j+\ell})\, \beta^\ell$ for $0 \le j < p^m - 1$. Call $\{X_v : v \in \mathbb{F}_{p^k}\}$ the *Lempel-Greenberger family*.

17.3.46 Remark The sequence X_0 is obtained by mapping k consecutive elements of an m-sequence over $\mathbb{F}_p$ (as defined in Section 10.3) to an element in $\mathbb{F}_{p^k}$. The other sequences are then translates of X_0.

17.3.47 Theorem [1889] Each sequence in the Lempel-Greenberger family $\mathcal{F}$ is optimal and $\mathcal{F}$ is an optimal family. Specifically, $H(X) = p^{n-k} - 1$ and $H(X, Y) = p^{n-k}$ for all distinct $X, Y \in \mathcal{F}$.

17.3.48 Construction [635] Let $p = em + 1$ be an odd prime and let α be primitive in $\mathbb{F}_p$. For integer i, define the i-th cyclotomic class in $\mathbb{F}_p$ to be $C_i = \{\alpha^{te+i} : 0 \le t < m\}$. For integer v, define the sequence $X_v = (x_j)$ of length p over $\{\infty, 0, 1, \ldots, e-1\}$ by $x_0 = \infty$ and $x_j = i$ when $j \in C_{v+i}$ for $0 \le j < p$. Call $\{X_v : 0 \le v < e\}$ the *Chu-Colbourn family.*

17.3.49 Theorem [635] The Chu-Colbourn family $\mathcal{F}$ satisfies $H(X) = m - 1$ and $H(X, Y) = m$ for all distinct $X, Y \in \mathcal{F}$. When $e \ge 3m$ and $m \ge 2$, then each sequence in $\mathcal{F}$ is optimal and $\mathcal{F}$ is an optimal family.

17.3.50 Example Take $e = 3$, $m = 2$, so that $p = 7$. For $\alpha = 3$, we have $C_0 = \{1, 6\}$, $C_1 = \{3, 4\}$, $C_2 = \{2, 5\}$, giving $X_0 = (\infty 021120)$, $X_1 = (\infty 210012)$, $X_2 = (\infty 102201)$.

17.3.6 Rank distance codes

17.3.51 Definition An (m, n, d) *rank distance code* over $\mathbb{F}_q$ is a subset $\mathcal{C}$ of $\mathbb{F}_q^{m\times n}$ (the set of $m \times n$ matrices over $\mathbb{F}_q$) such that the rank of $X - Y$ is at least d for all distinct $X, Y \in \mathcal{C}$. Without loss of generality, it is assumed that $m \le n$.

17.3.52 Remark Rank distance codes were introduced independently in [803, 1149, 2485]. Such codes find applications in correcting crisscross errors in arrays, for example in memory chip arrays or magnetic tape recording [2485]. Further applications are given in the following two subsections. Motivated by these applications, the goal is to find (m, n, d) rank distance codes that have as many elements as possible.

17.3.53 Theorem [803] The size of an (m, n, d) rank distance code over $\mathbb{F}_q$ is at most $q^{n(m-d+1)}$.

17.3.54 Definition An (m, n, d) rank distance code over $\mathbb{F}_q$ of size $q^{n(m-d+1)}$ is a *maximum rank distance code.*

17.3.55 Definition The *rank distribution* of a rank distance code $\mathcal{C}$ is (a_i), where a_i is the number of elements in $\mathcal{C}$ of rank i, and its *distance distribution* is (b_i), where $b_i = |\{(X, Y) \in \mathcal{C} \times \mathcal{C} : \operatorname{rank}(X - Y) = i\}|/|\mathcal{C}|$.

17.3.56 Theorem [803] The rank distribution (a_i) and the distance distribution (b_i) of an (m, n, d) maximum rank distance code over $\mathbb{F}_q$ satisfy $a_0 = b_0 = 1$ and

$$a_i = b_i = \begin{bmatrix} m \\ i \end{bmatrix}_q \sum_{j=0}^{i-d} (-1)^j q^{\binom{j}{2}} \begin{bmatrix} i \\ j \end{bmatrix}_q (q^{n(i-j-d+1)} - 1) \quad \text{for } i \in \{1, 2, \ldots, m\},$$

where $\begin{bmatrix} m \\ k \end{bmatrix}_q$ is the q-binomial coefficient given in Remark 17.3.35.

17.3.57 Theorem [803] Let m, n, d be positive integers satisfying $d \le m \le n$. Write $F = \mathbb{F}_q$ and $E = \mathbb{F}_{q^n}$, and let α be primitive in E. For $\lambda = (\lambda_0, \lambda_1, \ldots, \lambda_{m-d}) \in E^{m-d+1}$, let $X_\lambda = (x_{ij})$ be the $m \times n$ matrix given by $x_{ij} = \operatorname{Tr}_{E/F}(\alpha^j \sum_{k=0}^{m-d} \lambda_k \alpha^{iq^k})$ for $i \in \{1, 2, \ldots, m\}$ and

$j \in \{1, 2, \ldots, n\}$. Then $\{X_\lambda : \lambda \in E^{m-d+1}\}$ is an (m, n, d) maximum rank distance code over $\mathbb{F}_q$.

17.3.7 Space-time coding

17.3.58 **Definition** An (m, n, d) *space-time code* $\mathcal{S}$ over a (finite) alphabet $\mathcal{A} \subset \mathbb{C}$ is a set of $m \times n$ matrices with entries in $\mathcal{A}$ such that, for all distinct $A, B \in \mathcal{S}$, the rank of $A - B$ is at least d, which is the *diversity* of $\mathcal{S}$. Without loss of generality, it is assumed that $m \leq n$.

17.3.59 **Remark** Space-time codes are used in wireless digital communications to transmit a block of n data symbols over m transmit antennas. The motivation is that, with careful code design, the transmitter introduces spatial diversity and so can prevent signal loss caused by shadowing [2782]. The usual goal is to design space-time codes whose diversity equals the number of transmit antennas m.

17.3.60 **Theorem** [1963] The size of an (m, n, d) space-time code over an alphabet of size q is at most $q^{n(m-d+1)}$.

17.3.61 **Definition** A space-time code that achieves equality above is *optimal.*

17.3.62 **Remark** There are different approaches for constructing space-time codes. Algebraic constructions based on rank distance codes over finite fields are presented in the following. The principal difficulty in this approach is to find some "rank-preserving" mapping from a finite field to a finite subset of $\mathbb{C}$.

17.3.63 **Theorem** [1964] Let $\mathcal{C}$ be an (m, n, d) maximum rank distance code over $\mathbb{F}_2$, where we interpret the elements of $\mathcal{C}$ to be in $\{0, 1\}$, as a subset of $\mathbb{Z}$. Let h and ℓ be positive integers. Write $\theta = e^{2\pi\sqrt{-1}/2^h}$, let η be a nonzero element in $2\mathbb{Z}[\theta]$, and let κ be a nonzero element in $\mathbb{C}$. Define

$$\mathcal{S} = \left\{ \kappa \sum_{u=0}^{\ell-1} \eta^u \theta^{\sum_{v=0}^{h-1} 2^v X_{u,v}} : X_{u,v} \in \mathcal{C} \right\},$$

where the exponentiation of matrices is understood to be element-wise. Then $\mathcal{S}$ is an optimal (m, n, d) space-time code over an alphabet of size $2^{h\ell}$.

17.3.64 **Remark** The above construction admits space-time codes over alphabets commonly used in digital communications. When $h = 2$, $\kappa = 1 + \sqrt{-1}$, and $\eta = 2$, the resulting alphabet is known as 4^ℓ-QAM. When $\ell = 1$ and $\kappa = 1$, and $\eta > 0$, the resulting alphabet is known as 2^h-PSK.

17.3.65 **Remark** Let K be a number field (a finite extension of the field of rationals $\mathbb{Q}$), let R be the ring of algebraic integers in K, and let p be a prime number. The ideal pR factors uniquely into prime ideals over R. If P is such a factor, then R/P is a finite field of size p^k, where k is the inertial degree of P over $p\mathbb{Z}$ (see [2004] for background on number fields and prime ideal factorization). This is usually applied with $K = \mathbb{Q}$, in which case $R = \mathbb{Z}$ and $\mathbb{Z}/p\mathbb{Z}$ is a finite field of size p.

17.3.66 **Definition** Let K be a number field and R the ring of algebraic integers in K. Suppose that R contains a prime ideal P such that R/P is isomorphic to $\mathbb{F}_q$. Let $\mathcal{A} \subset \mathbb{C}$ be a set containing exactly one representative from each of the q elements of R/P. Define $\phi : \mathbb{F}_q \to \mathcal{A}$ to be the mapping induced by the isomorphism from $\mathbb{F}_q$ to R/P, and extend ϕ to act element-wise on matrices over $\mathbb{F}_q$.

17.3.67 Theorem [1740] For each matrix X over $\mathbb{F}_q$, we have $\text{rank}(\phi(X)) \geq \text{rank}(X)$.

17.3.68 Corollary [1740] Let $\mathcal{C}$ be an (m, n, d) maximum rank distance code over $\mathbb{F}_q$. Then $\phi(\mathcal{C})$ is an optimal (m, n, d) space-time code over $\mathcal{A}$.

17.3.69 Remark The set $\mathcal{A}$ is not uniquely defined for a given K, but from a practical viewpoint it is advantageous to choose $\mathcal{A}$ such that the energy $\sum_{a \in \mathcal{A}} |a|^2$ is minimized.

17.3.70 Example Write $i = \sqrt{-1}$ and take $K = \mathbb{Q}[i]$, so that $R = \mathbb{Z}[i]$. Denote the ideal aR by (a) and let p be a prime satisfying $p \equiv 1 \pmod 4$. Then $(p) = (\pi)(\overline{\pi})$, where $\pi = a + bi$ for some $a, b \in \mathbb{Z}$ satisfying $a^2 + b^2 = p$, and $R/(\pi)$ is isomorphic to $\mathbb{F}_p$. Identify $\mathbb{F}_p$ with $\mathbb{Z}/p\mathbb{Z}$ and define $\phi : \mathbb{F}_p \to \mathbb{C}$ by $\phi(x + p\mathbb{Z}) = x - [\frac{x}{\pi}]\pi$, where $[\,\cdot\,]$ rounds to the nearest element (in the Euclidean metric) in $\mathbb{Z}[i]$. Take $\mathcal{A}$ to be the image of ϕ. Then $\mathcal{A}$ minimizes the energy. Specifically for $p = 5$, we have $(5) = (2 + i)(2 - i)$ and $\mathcal{A} = \{0, 1, -1, i, -i\}$.

17.3.8 Coding over networks

17.3.71 Definition [1800] Let $P(\mathbb{F}_q^m)$ be the set of subspaces of $\mathbb{F}_q^m$ over $\mathbb{F}_q$. The *subspace distance* d_S on $P(\mathbb{F}_q^m)$ is defined by $d_S(U, V) = \dim(U) + \dim(V) - 2\dim(U \cap V)$.

17.3.72 Definition [1800] Let $P_k(\mathbb{F}_q^m)$ be the set of k-dimensional subspaces of $\mathbb{F}_q^m$ over $\mathbb{F}_q$. An *$(m, k, 2d)$ constant-dimension code* Ω over $\mathbb{F}_q$ is a subset of $P_k(\mathbb{F}_q^m)$ such that $d_S(U, V) \geq 2d$ for all distinct $U, V \in \Omega$.

17.3.73 Remark Random linear network coding is a technique for communicating efficiently over networks. The transmitting node injects a basis for a k-dimensional subspace of $\mathbb{F}_q^m$ into the network. Each intermediate node relays a randomly chosen linear combination of its incoming vectors. The receiving node waits until enough linearly independent vectors are received and then tries to reconstruct the transmitted subspace. If the transmitted subspaces are taken from an $(m, k, 2d)$ constant-dimension code with $d > 1$, then the receiver can reconstruct the transmitted space even if there are (a limited number of) lost or erroneously inserted vectors. Specifically, a successful reconstruction is always possible if the transmitted subspace U and the received subspace V satisfy $d_S(U, V) < d$ [1800].

17.3.74 Definition [2666] The mapping $\Lambda : \mathbb{F}_q^{k \times (m-k)} \to P_k(\mathbb{F}_q^m)$ takes $X \in \mathbb{F}_q^{k \times (m-k)}$ to the rowspace of $[I \ \ X]$, where I is the identity matrix of order k.

17.3.75 Theorem [2666] We have $d_S(\Lambda(X), \Lambda(Y)) = 2\,\text{rank}(X - Y)$ for all $X, Y \in \mathbb{F}_q^{k \times (m-k)}$.

17.3.76 Corollary [2666] Let $\mathcal{C}$ be a $(k, m - k, d)$ rank distance code over $\mathbb{F}_q$. Then $\Lambda(\mathcal{C})$ is an $(m, k, 2d)$ constant-dimension code over $\mathbb{F}_q$.

17.3.77 Remark If $\mathcal{C}$ is a $(k, m - k, d)$ maximum rank distance code, the "lifted" code $\Lambda(\mathcal{C})$ has size $q^{k(m-k)-(d-1)\max\{k,m-k\}}$. This code is almost optimal in the sense that every $(m, k, 2d)$ constant-dimension code over $\mathbb{F}_q$ has fewer than four times as many codewords as $\Lambda(\mathcal{C})$ [2666]. In some special cases $\Lambda(\mathcal{C})$ can be augmented to give an optimal code, as shown below.

17.3.78 Theorem [1998] Let k and m be positive integers satisfying $m = rk$ for integer r. Let I and Z be the identity and all-zero matrix over $\mathbb{F}_q$ of size $k \times k$, respectively. Let $\Omega_0 \subset P_k(\mathbb{F}_q^m)$ contain the rowspace of $[Z \ \cdots \ Z \ I]$, and for $\ell \in \{1, 2, \ldots, r-1\}$, let $\Omega_\ell \subset P_k(\mathbb{F}_q^m)$ be the set of rowspaces corresponding to $\{[Z \ \cdots \ Z \ I \ A] : A \in \mathcal{C}_\ell\}$, where $\mathcal{C}_\ell$ is a $(k, \ell k, k)$ maximum rank distance code over $\mathbb{F}_q$. Then $\bigcup_\ell \Omega_\ell$ is an $(rk, k, 2k)$ constant-dimension code of size $1 + q^k + q^{2k} + \cdots + q^{(r-1)k}$.

17.3.79 Remark The above code is a k-spread of $\mathbb{F}_{q^m}$, namely a set of k-dimensional subspaces of $\mathbb{F}_{q^m}$ having the property that each nonzero element of $\mathbb{F}_{q^m}$ belongs to exactly one such subspace. The code therefore has largest possible size.

See Also

§10.3	For more on correlation of sequences.
§15.1	For background on coding theory.

References Cited: [635, 639, 783, 803, 967, 1149, 1293, 1523, 1603, 1604, 1605, 1607, 1740, 1800, 1889, 1963, 1964, 1998, 2004, 2037, 2135, 2218, 2300, 2319, 2485, 2527, 2554, 2666, 2782, 2832]

Bibliography

[1] R. Abarzúa, N. Thériault, R. Avanzi, I. Soto, and M. Alfaro, Optimization of the arithmetic of the ideal class group for genus 4 hyperelliptic curves over projective coordinates, *Advances in Mathematics of Communications* 4 (2010) 115–139. <803>

[2] E. Abbe, Randomness and dependencies extraction via polarization, In *Proc. Information Theory and Applications Workshop (ITA)*, 1–7, 2011. <739>

[3] M. Abdón and F. Torres, On maximal curves in characteristic two, *Manuscripta Math.* 99 (1999) 39–53. <461, 463>

[4] R. J. R. Abel, Some new BIBDs with block size 7, *J. Combin. Des.* 8 (2000) 146–150. <597, 599>

[5] R. J. R. Abel and M. Buratti, Some progress on $(v, 4, 1)$ difference families and optical orthogonal codes, *J. Combin. Theory, Ser. A* 106 (2004) 59–75. <593, 594, 599>

[6] R. J. R. Abel, N. J. Finizio, G. Ge, and M. Greig, New Z-cyclic triplewhist frames and triplewhist tournament designs, *Discrete Appl. Math.* 154 (2006) 1649–1673. <619>

[7] R. J. R. Abel and G. Ge, Some difference matrix constructions and an almost completion for the existence of triplewhist tournaments $\mathrm{TWh}(v)$, *European J. Combin.* 26 (2005) 1094–1104. <618, 619>

[8] S. S. Abhyankar, Resolution of singularities and modular Galois theory, *Bull. Amer. Math. Soc. (New Ser.)* 38 (2001) 131–169. <239, 240>

[9] S. S. Abhyankar, Symplectic groups and permutation polynomials. II, *Finite Fields Appl.* 8 (2002) 233–255. <239, 240>

[10] F. Abu Salem, S. Gao, and A. G. B. Lauder, Factoring polynomials via polytopes, In *ISSAC 2004*, 4–11, ACM, New York, 2004. <389, 392>

[11] F. K. Abu Salem, An efficient sparse adaptation of the polytope method over $\mathbb{F}_p$ and a record-high binary bivariate factorisation, *J. Symbolic Comput.* 43 (2008) 311–341. <389, 392>

[12] J.-K. Accetta, Z. Mejías, and A. Santos, Número de waring en cuerpos finitos, Preprint, 2011. <211, 213>

[13] W. W. Adams and P. Loustaunau, *An Introduction to Gröbner Bases*, American Mathematical Society, Providence, RI, first edition, 1994. <85, 702, 703>

[14] L. Adleman and H. Lenstra, Finding irreducible polynomials over finite fields, In *Proceedings of the Eighteenth Annual ACM Symposium on Theory of Computing*, 350–355, ACM, New York, NY, USA, 1986. <120, 128, 379, 380, 404>

[15] L. Adleman, K. Manders, and G. Miller, On taking roots in finite fields, In *Proceedings of the Eighteenth Annual Symposium on Foundations of Computer Science*, 175–178, IEEE Computer Society, Washington, DC, USA, 1977. <381, 382>

[16] L. M. Adleman, The function field sieve, In *Algorithmic Number Theory*, volume 877 of *Lecture Notes in Comput. Sci.*, 108–121, Springer, Berlin, 1994. <399, 401>

[17] L. M. Adleman, J. DeMarrais, and M.-D. Huang, A subexponential algorithm for discrete logarithms over the rational subgroup of the Jacobians of large genus hyperelliptic curves over finite fields, In *Algorithmic Number Theory*, volume 877 of *Lecture Notes in Comput. Sci.*, 28–40, Springer, Berlin, 1994. <455, 456>

[18] L. M. Adleman and M.-D. Huang, Counting points on curves and Abelian varieties over finite fields, *J. Symbolic Comput.* 32 (2001) 171–189. <490, 491>

[19] L. M. Adleman, C. Pomerance, and R. S. Rumely, On distinguishing prime numbers from composite numbers, *Ann. of Math., 2nd Ser.* 117 (1983) 173–206. <347, 363>

[20] A. Adolphson and S. Sperber, p-adic estimates for multiplicative character sums, Preprint, http://arxiv.org/abs/1103.5513. <483, 488>

[21] A. Adolphson and S. Sperber, p-adic estimates for exponential sums and the theorem of Chevalley-Warning, *Ann. Sci. École Norm. Sup., IVe Ser.* 20 (1987) 545–556. <199, 201, 210, 213, 480, 488>

[22] A. Adolphson and S. Sperber, On the degree of the L-function associated with an exponential sum, *Compositio Math.* 68 (1988) 125–159. <169, 473, 476, 479>

[23] A. Adolphson and S. Sperber, Exponential sums and Newton polyhedra: cohomology and estimates, *Ann. of Math., 2nd Ser.* 130 (1989) 367–406. <165, 169, 196, 201, 210, 213, 476, 479, 482, 488>

[24] A. Adolphson and S. Sperber, p-adic estimates for exponential sums, In *p-adic Analysis*, volume 1454 of *Lecture Notes in Math.*, 11–22, Springer, Berlin, 1990. <213>

[25] A. Adolphson and S. Sperber, On twisted exponential sums, *Math. Ann.* 290 (1991) 713–726. <482, 488>

[26] A. Adolphson and S. Sperber, Twisted exponential sums and Newton polyhedra, *J. Reine Angew. Math.* 443 (1993) 151–177. <482, 488>

[27] A. Adolphson and S. Sperber, On the zeta function of a complete intersection, *Ann. Sci. École Norm. Sup., IVe Ser.* 29 (1996) 287–328. <196, 201, 481, 488>

[28] A. Adolphson and S. Sperber, Exponential sums on A^n. III, *Manuscripta Math.* 102 (2000) 429–446. <163, 169>

[29] A. Adolphson and S. Sperber, On the zeta function of a projective complete intersection, *Illinois J. Math.* 52 (2008) 389–417. <481, 488>

[30] A. Adolphson and S. Sperber, Exponential sums nondegenerate relative to a lattice, *Alg. Num. Th.* 3 (2009) 881–906. <213>

[31] A. Adolphson and S. Sperber, On unit root formulas for toric exponential sums, *Alg. Num. Th.* 6 (2012) 573–585. <482, 488>

[32] V. B. Afanasyev, Complexity of VLSI implementation of finite field arithmetic, In *Proc. II. Intern. Workshop on Algebraic and Combinatorial Coding Theory, USSR*, 6–7, 1990. <814, 823>

[33] S. Agou, Sur l'irréducibilité des polynômes à coefficients dans un corps fini, *C. R. Acad. Sci. Paris, Sér. A-B* 272 (1971) A576–A577. <62, 66>

[34] S. Agou, Factorisation sur un corps fini $\mathbb{F}_{p^n}$ des polynômes composés $f(X^s)$ lorsque $f(X)$ est un polynôme irréductibile de $\mathbb{F}_{p^n}[X]$, *L' Enseignement Math., IIe Ser.* 22 (1976) 305–312. <60, 62, 66>

[35] S. Agou, Factorisation sur un corps fini K des polynômes composés $f(X^s)$ lorsque $f(X)$ est polynôme irréductibile de $K[X]$, *C. R. Acad. Sci. Paris, Sér. A-B* 282 (1976) Ai, A1067–A1068. <60, 66>

[36] S. Agou, Critères d'irréductibilité des polynômes composés à coefficients dans un corps fini, *Acta Arith.* 30 (1976/77) 213–223. <60, 63, 66>

[37] S. Agou, Factorisation sur un corps fini $\mathbb{F}_{p^n}$ des polynômes composés $f(X^{p^r} - aX)$ lorsque $f(X)$ est un polynôme irréductible de $\mathbb{F}_{p^n}(X)$, *J. Number Theory* 9 (1977) 229–239. <62, 63, 66, 70>

[38] S. Agou, Irréductibilité des polynômes $f(X^{p^r} - aX)$ sur un corps fini $\mathbb{F}_{p^s}$, *J. Reine Angew. Math.* 292 (1977) 191–195. <60, 63, 66>

[39] S. Agou, Irréductibilité des polynômes $f(X^{p^{2r}} - aX^{p^r} - bX)$ sur un corps fini $\mathbb{F}_{p^s}$, *J. Number Theory* 10 (1978) 64–69. <60, 63, 66, 70>

[40] S. Agou, Irréductibilité des polynômes $f(X^{p^{2r}} - aX^{p^r} - bX)$ sur un corps fini $\mathbb{F}_{p^s}$, *J. Number Theory* 11 (1979) 20. <60, 63, 66, 70>

[41] S. Agou, Irréductibilité des polynômes $f(\sum_{i=0}^{m} a_i X^{p^{ri}})$ sur un corps fini $\mathbb{F}_{p^s}$, *Canad. Math. Bull.* 23 (1980) 207–212. <63, 66, 70>

[42] S. Agou, Sur la factorisation des polynômes $f(X^{p^{2r}} - aX^{p^r} - bX)$ sur un corps fini $\mathbb{F}_{p^s}$, *J. Number Theory* 12 (1980) 447–459. <63, 66>

[43] M. Agrawal, N. Kayal, and N. Saxena, PRIMES is in P, *Ann. of Math., 2nd Ser.* 160 (2004) 781–793. <347, 363, 401, 404>

[44] S. Ahmad, Cycle structure of automorphisms of finite cyclic groups, *J. Combin. Theory* 6 (1969) 370–374. <228, 229>

[45] O. Ahmadi, Self-reciprocal irreducible pentanomials over $\mathbb{F}_2$, *Des. Codes Cryptogr.* 38 (2006) 395–397. <69, 70>

[46] O. Ahmadi, On the distribution of irreducible trinomials over $\mathbb{F}_3$, *Finite Fields Appl.* 13 (2007) 659–664. <69, 70>

[47] O. Ahmadi, The trace spectra of polynomial bases for $\mathbb{F}_{2^n}$, *Appl. Algebra Engrg. Comm. Comput.* 18 (2007) 391–396. <107, 109>

[48] O. Ahmadi, Generalization of a theorem of Carlitz, *Finite Fields Appl.* 17 (2011) 473–480. <57, 59>

[49] O. Ahmadi and R. Granger, An efficient deterministic test for Kloosterman sum zeros, 2012, to appear in Math. Comp. <154, 161>

[50] O. Ahmadi, F. Luca, A. Ostafe, and I. E. Shparlinski, On stable quadratic polynomials, *Glasg. Math. J.* 54 (2012) 359–369. <343, 344>

[51] O. Ahmadi and A. Menezes, On the number of trace-one elements in polynomial bases for $\mathbb{F}_{2^n}$, *Des. Codes Cryptogr.* 37 (2005) 493–507. <107, 109>

[52] O. Ahmadi and A. Menezes, Irreducible polynomials of maximum weight, *Util. Math.* 72 (2007) 111–123. <69, 70, 73>

[53] O. Ahmadi and I. E. Shparlinski, Bilinear character sums and sum-product problems on elliptic curves, *Proc. Edinb. Math. Soc., 2nd Ser.* 53 (2010) 1–12. <188, 192>

[54] O. Ahmadi, I. E. Shparlinski, and J. F. Voloch, Multiplicative order of Gauss periods, *Int. J. Number Theory* 6 (2010) 877–882. <99>

[55] O. Ahmadi and G. Vega, On the parity of the number of irreducible factors of self-reciprocal polynomials over finite fields, *Finite Fields Appl.* 14 (2008) 124–131. <72, 73>

[56] A. V. Aho, J. E. Hopcroft, and J. D. Ullman, *The Design and Analysis of Computer Algorithms*, Addison-Wesley Publishing Co., Reading, Mass.-London-Amsterdam, 1975, Second printing, Addison-Wesley Series in Computer Science and Information Processing. <359, 363>

[57] W. Aitken, On value sets of polynomials over a finite field, *Finite Fields Appl.* 4 (1998) 441–449. <235, 236>

[58] W. Aitken, M. D. Fried, and L. M. Holt, Davenport pairs over finite fields, *Pacific J. Math.* 216 (2004) 1–38. <240>

[59] M. Ajtai, H. Iwaniec, J. Komlós, J. Pintz, and E. Szemerédi, Construction of a thin set with small Fourier coefficients, *Bull. London Math. Soc.* 22 (1990) 583–590.

<184, 185>

[60] A. Akbary, S. Alaric, and Q. Wang, On some classes of permutation polynomials, *Int. J. Number Theory* 4 (2008) 121–133. <223, 224, 229>

[61] A. Akbary, D. Ghioca, and Q. Wang, On permutation polynomials of prescribed shape, *Finite Fields Appl.* 15 (2009) 195–206. <218, 219, 229>

[62] A. Akbary, D. Ghioca, and Q. Wang, On constructing permutations of finite fields, *Finite Fields Appl.* 17 (2011) 51–67. <220, 221, 224, 225, 229>

[63] A. Akbary and Q. Wang, On some permutation polynomials over finite fields, *Int. J. Math. Math. Sci.* 16 (2005) 2631–2640. <222, 229>

[64] A. Akbary and Q. Wang, A generalized Lucas sequence and permutation binomials, *Proc. Amer. Math. Soc.* 134 (2006) 15–22. <218, 222, 223, 229>

[65] A. Akbary and Q. Wang, On polynomials of the form $x^r f(x^{(q-1)/l})$, *Int. J. Math. Math. Sci.* (2007) Art. ID 23408, 7. <221, 222, 223, 229>

[66] S. Akiyama, On the pure Jacobi sums, *Acta Arith.* 75 (1996) 97–104. <146, 161>

[67] M.-L. Akkar, N. T. Courtois, R. Duteuil, and L. Goubin, A fast and secure implementation of Sflash, In *Public Key Cryptography—PKC 2003*, volume 2567 of *Lecture Notes in Comput. Sci.*, 267–278, Springer, Berlin, 2002. <773, 783>

[68] E. Aksoy, A. Çeşmelioğlu, W. Meidl, and A. Topuzoğlu, On the Carlitz rank of permutation polynomials, *Finite Fields Appl.* 15 (2009) 428–440. <229>

[69] A. A. Albert, Symmetric and alternate matrices in an arbitrary field. I, *Trans. Amer. Math. Soc.* 43 (1938) 386–436. <507, 510>

[70] A. A. Albert, *Fundamental Concepts of Higher Algebra*, University of Chicago Press, Chicago, IL, 1958. <61, 62, 63, 66>

[71] A. A. Albert, Finite division algebras and finite planes, In *Proc. Sympos. Appl. Math., Vol. 10*, 53–70, American Mathematical Society, Providence, RI, 1960. <275, 278>

[72] A. A. Albert, Generalized twisted fields, *Pacific J. Math.* 11 (1961) 1–8. <276>

[73] A. A. Albert, Isotopy for generalized twisted fields, *An. Acad. Brasil. Ci.* 33 (1961) 265–275. <276>

[74] R. Albert and H. G. Othmer, The topology of the regulatory interactions predicts the expression pattern of the segment polarity genes in *drosophila melanogaster*, *J. Theoret. Biol.* 223 (2003) 1–18. <825, 834>

[75] W. R. Alford, A. Granville, and C. Pomerance, There are infinitely many Carmichael numbers, *Ann. of Math., 2nd Ser.* 139 (1994) 703–722. <134, 138>

[76] N. Ali, Stabilité des polynômes, *Acta Arith.* 119 (2005) 53–63. <342, 344>

[77] B. Allombert, Explicit computation of isomorphisms between finite fields, *Finite Fields Appl.* 8 (2002) 332–342. <348, 363>

[78] J.-P. Allouche and J. Shallit, *Automatic Sequences: Theory, Applications, Generalizations*, Cambridge University Press, Cambridge, 2003. <546>

[79] J.-P. Allouche and D. S. Thakur, Automata and transcendence of the Tate period in finite characteristic, *Proc. Amer. Math. Soc.* 127 (1999) 1309–1312. <546>

[80] N. Alon, Eigenvalues and expanders, *Combinatorica* 6 (1986) 83–96. <646, 649, 658>

[81] N. Alon and F. R. K. Chung, Explicit construction of linear sized tolerant networks, *Discrete Math.* 72 (1988) 15–19. <645, 658>

[82] N. Alon, Y. Kohayakawa, C. Mauduit, C. G. Moreira, and V. Rödl, Measures of pseudorandomness for finite sequences: typical values, *Proc. Lond. Math. Soc., 3rd Ser.* 95 (2007) 778–812. <182, 185>

[83] N. Alon and Y. Roichman, Random Cayley graphs and expanders, *Random Structures Algorithms* 5 (1994) 271–284. <651, 658>

[84] C. Alonso, J. Gutierrez, and T. Recio, A rational function decomposition algorithm by near-separated polynomials, *J. Symbolic Comput.* 19 (1995) 527–544. <300, 302>

[85] H. Aly, R. Marzouk, and W. Meidl, On the calculation of the linear complexity of periodic sequences, In *Finite Fields: Theory and Applications*, volume 518 of *Contemp. Math.*, 11–22, Amer. Math. Soc., Providence, RI, 2010. <329, 336>

[86] H. Aly and W. Meidl, On the linear complexity and k-error linear complexity over $\mathbb{F}_p$ of the d-ary Sidel′nikov sequence, *IEEE Trans. Inform. Theory* 53 (2007) 4755–4761. <334, 336>

[87] H. Aly and A. Winterhof, On the linear complexity profile of nonlinear congruential pseudorandom number generators with Dickson polynomials, *Des. Codes Cryptogr.* 39 (2006) 155–162. <333, 336>

[88] A. Ambainis and N. Nahimovs, Improved constructions of quantum automata, *Theoret. Comput. Sci.* 410 (2009) 1916–1922. <840, 841>

[89] P. R. Amestoy, T. A. Davis, and I. S. Duff, Algorithm 837: AMD, an approximate minimum degree ordering algorithm, *ACM Trans. Math. Software* 30 (2004) 381–388. <532, 535>

[90] G. An, In silico experiments of existing and hypothetical cytokine-directed clinical trials using agent-based modeling, *Crit Care Med* 32 (2004) 2050–2060. <831, 834>

[91] V. Anashin and A. Khrennikov, *Applied Algebraic Dynamics*, volume 49 of *de Gruyter Expositions in Mathematics*, de Gruyter, Berlin, 2009. <337, 338, 344>

[92] H. E. Andersen and O. Geil, Evaluation codes from order domain theory, *Finite Fields Appl.* 14 (2008) 92–123. <705, 712>

[93] B. A. Anderson and K. B. Gross, A partial starter construction, *Congress. Numer.* 21 (1978) 57–64. <615, 619>

[94] G. W. Anderson, t-motives, *Duke Math. J.* 53 (1986) 457–502. <545, 546>

[95] G. W. Anderson, Log-algebraicity of twisted A-harmonic series and special values of L-series in characteristic p, *J. Number Theory* 60 (1996) 165–209. <541>

[96] G. W. Anderson, W. D. Brownawell, and M. A. Papanikolas, Determination of the algebraic relations among special Γ-values in positive characteristic, *Ann. of Math, 2nd Ser.* 160 (2004) 237–313. <546>

[97] G. W. Anderson and D. S. Thakur, Multizeta values for $\mathbb{F}_q[t]$, their period interpretation, and relations between them, *Int. Math. Res. Not. IMRN* (2009) 2038–2055. <544, 546>

[98] I. Anderson, A hundred years of whist tournaments, *J. Combin. Math. Combin. Comput.* 19 (1995) 129–150. <618, 619>

[99] I. Anderson, *Combinatorial Designs and Tournaments*, volume 6 of *Oxford Lecture Series in Mathematics and its Applications*, The Clarendon Press, Oxford University Press, New York, 1997. <619>

[100] I. Anderson, Some cyclic and 1-rotational designs, In *Surveys in Combinatorics*, volume 288 of *London Math. Soc. Lecture Note Ser.*, 47–73, Cambridge Univ. Press, Cambridge, 2001. <618, 619>

[101] I. Anderson and N. J. Finizio, Some new z-cyclic whist tournament designs, *Discrete Math.* 293 (2005) 19–28. <618, 619>

[102] I. Anderson, N. J. Finizio, and P. A. Leonard, New product theorems for Z-cyclic

whist tournaments, *J. Combin. Theory, Ser. A* 88 (1999) 162–166. <618, 619>

[103] M. Andersson, V. Rathi, R. Thobaben, J. Kliewer, and M. Skoglund, Nested polar codes for wiretap and relay channels, *IEEE Comm. Letters* 14 (2010) 752–754. <739>

[104] J. André, Über nicht-Desarguessche Ebenen mit transitiver Translationsgruppe, *Math. Z.* 60 (1954) 156–186. <566, 574>

[105] B. Angles and C. Maire, A note on tamely ramified towers of global function fields, *Finite Fields Appl.* 8 (2002) 207–215. <463>

[106] J.-C. Anglès d'Auriac, J.-M. Maillard, and C. M. Viallet, On the complexity of some birational transformations, *J. Phys. A* 39 (2006) 3641–3654. <337, 344>

[107] B. Ansari and M. A. Hasan, High-performance architecture of elliptic curve scalar multiplication, *IEEE Trans. Comput.* 57 (2008) 1443–1453. <815, 823>

[108] ANSI, The elliptic curve digital signature algorithm (ECDSA), Working Draft American National Standard: Public Key Cryptography for the Financial Services Industry X9.62-1998, American National Standards Institute, 1998, Available at `http://grouper.ieee.org/groups/1363/private/x9-62-09-20-98.zip`. <785, 796>

[109] ANSI, Key agreement and key transport using elliptic curve cryptography, Working Draft American National Standard: Public Key Cryptography for the Financial Services Industry X9.63-199x, American National Standards Institute, 1999, Available at `http://grouper.ieee.org/groups/1363/private/x9-63-01-08-99.zip`. <785, 796>

[110] N. Anuradha and S. A. Katre, Number of points on the projective curves $aY^l = bX^l + cZ^l$ and $aY^{2l} = bX^{2l} + cZ^{2l}$ defined over finite fields, l an odd prime, *J. Number Theory* 77 (1999) 288–313. <208, 213>

[111] N. Aoki, Abelian fields generated by a Jacobi sum, *Comment. Math. Univ. St. Paul.* 45 (1996) 1–21. <146, 161>

[112] N. Aoki, On the purity problem of Gauss sums and Jacobi sums over finite fields, *Comment. Math. Univ. St. Paul.* 46 (1997) 223–233. <145, 146, 161>

[113] N. Aoki, A finiteness theorem on pure Gauss sums, *Comment. Math. Univ. St. Pauli* 53 (2004) 145–168. <145, 161>

[114] N. Aoki, On the zeta function of some cyclic quotients of Fermat curves, *Comment. Math. Univ. St. Pauli* 57 (2008) 163–185. <146, 161>

[115] N. Aoki, On multi-quadratic Gauss sums, *Comment. Math. Univ. St. Pauli* 59 (2010) 97–117. <150, 161>

[116] K. T. Arasu and K. J. Player, A new family of cyclic difference sets with Singer parameters in characteristic three, *Des. Codes Cryptogr.* 28 (2003) 75–91. <603, 607>

[117] E. Arikan, Channel combining and splitting for cutoff rate improvement, *IEEE Trans. Inform. Theory* 52 (2006) 628–639. <739>

[118] E. Arikan, A performance comparison of polar codes and Reed-Muller codes, *IEEE Comm. Letters* 12 (2008) 447–449. <739>

[119] E. Arikan, Channel polarization: A method for constructing capacity-achieving codes for symmetric binary-input memoryless channels, *IEEE Trans. Inform. Theory* 55 (2009) 3051–3073. <738, 739>

[120] E. Arikan, Polar codes: A pipelined implementation, In Proc. Int. Symp. Broadband Communication (ISBC2010), 2010. <739>

[121] E. Arikan, Source polarization, preprint available, `http://arxiv.org/abs/1001.`

3087, 2010. <739>

[122] E. Arikan and E. Telatar, On the rate of channel polarization, preprint available, `http://arxiv.com/abs/0807.3806`, 2008. <738, 739>

[123] S. Arita, S. Miura, and T. Sekiguchi, An addition algorithm on the Jacobian varieties of curves, *J. Ramanujan Math. Soc.* 19 (2004) 235–251. <808, 811>

[124] V. L. Arlazarov, E. A. Dinic, M. A. Kronrod, and I. A. Faradžev, The economical construction of the transitive closure of an oriented graph, *Dokl. Akad. Nauk SSSR* 194 (1970) 487–488. <521, 535>

[125] C. Armana, Torsion des modules de Drinfeld de rang 2 et formes modulaires de Drinfeld, *C. R. Math. Acad. Sci. Paris, Ser. I* 347 (2009) 705–708. <545, 546>

[126] C. Armana, Coefficients of Drinfeld modular forms and Hecke operators, *J. Number Theory* 131 (2011) 1435–1460. <545, 546>

[127] M. A. Armand, Multisequence shift register synthesis over commutative rings with identity with applications to decoding cyclic codes over integer residue rings, *IEEE Trans. Inform. Theory* 50 (2004) 220–229. <329, 336>

[128] F. Armknecht and M. Krause, Algebraic attacks on combiners with memory, In *Advances in Cryptology—CRYPTO 2003*, volume 2729 of *Lecture Notes in Comput. Sci.*, 162–175, Springer, Berlin, 2003. <783>

[129] F. Arnault, T. P. Berger, and M. Minier, Some results on FCSR automata with applications to the security of FCSR-based pseudorandom generators, *IEEE Trans. Inform. Theory* 54 (2008) 836–840. <336>

[130] F. Arnault, E. J. Pickett, and S. Vinatier, Construction of self-dual normal bases and their complexity, *Finite Fields Appl.* 18 (2012) 458–472. <38, 39, 42, 49, 115, 116, 123, 128>

[131] V. I. Arnold, *Dynamics, Statistics and Projective Geometry of Galois Fields*, Cambridge University Press, Cambridge, 2011. <31>

[132] M. Arora, G. Ivanyos, M. Karpinski, and N. Saxena, Deterministic polynomial factoring and association schemes, Technical Report 68, ECCC, Electronic Colloquium on Computational Complexity, 2012. <381, 382>

[133] E. Artin, Quadratische Körper im Gebiete der höheren Kongruenzen. I, *Math. Z.* 19 (1924) 153–206. <451, 456>

[134] E. Artin, Quadratische Körper im Gebiete der höheren Kongruenzen. II., *Math. Z.* 19 (1924) 207–246. <74, 79, 495, 500>

[135] E. Artin, *Geometric Algebra*, Interscience Publishers, Inc., New York-London, 1957. <3, 11, 520>

[136] M. Artin, A. Grothendieck, and J. L. Verdier, *Théorie des Topos et Cohomologie Étale des Schémas. Tome 3*, Lecture Notes in Mathematics, Vol. 305. Springer-Verlag, Berlin, 1973. <31, 470, 472, 477, 479>

[137] A. Arwin, Über Kongruenzen von dem fünften und höheren Graden nach einem Primzahlmodulus, *Ark. Mat. Astr. Fys.* 14 (1918) 1–46. <381, 382>

[138] M. Aschbacher, Isotopy and geotopy for ternary rings of projective planes, *J. Algebra* 319 (2008) 868–892. <278>

[139] D. W. Ash, I. F. Blake, and S. A. Vanstone, Low complexity normal bases, *Discrete Appl. Math.* 25 (1989) 191–210. <117, 120, 121, 128>

[140] A. Ashikhmin and E. Knill, Nonbinary quantum stabilizer codes, *IEEE Trans. Inform. Theory* 47 (2001) 3065–3072. <838, 841>

[141] E. F. Assmus, Jr. and J. D. Key, *Designs and Their Codes*, volume 103 of *Cambridge*

Tracts in Mathematics, Cambridge University Press, Cambridge, 1992. <31, 308, 310>

[142] E. F. Assmus, Jr. and H. F. Mattson, Jr., New 5-designs, *J. Combin. Theory* 6 (1969) 122–151. <690, 703>

[143] A. O. L. Atkin and F. Morain, Elliptic curves and primality proving, *Math. Comp.* 61 (1993) 29–68. <347, 363>

[144] Y. Aubry and P. Langevin, On the weights of binary irreducible cyclic codes, In *Coding and Cryptography*, volume 3969 of *Lecture Notes in Comput. Sci.*, 46–54, Springer, Berlin, 2006. <152, 161>

[145] Y. Aubry and M. Perret, A Weil theorem for singular curves, In *Arithmetic, Geometry and Coding Theory*, 1–7, de Gruyter, Berlin, 1996. <238, 240>

[146] J.-P. Aumasson, M. Finiasz, W. Meier, and S. Vaudenay, A hardware-oriented trapdoor cipher, In J. Pieprzyk, H. Ghodosi, and E. Dawson, editors, *Information Security and Privacy*, volume 4586 of *Lecture Notes in Computer Science*, 184–199, Springer Berlin / Heidelberg, 2007. <630, 642>

[147] R. Avanzi, Aspects of hyperelliptic curves over large prime fields in software implementations, In *Proceedings of the Sixth International Workshop on Cryptographic Hardware and Embedded Systems (CHES)*, volume 3156 of *Lecture Notes in Comput. Sci.*, 148–162, Springer, Berlin, 2004. <797, 803>

[148] R. Avanzi and N. Thériault, Hyperelliptic curve security, In H. C. van Tilborg and S. Jajodia, editors, *Encyclopedia of Cryptography and Security*, volume 2, 573–576, Springer, Berlin, 2011. <810, 811>

[149] R. Avanzi, N. Thériault, and Z. Wang, Rethinking low genus hyperelliptic Jacobian arithmetic over binary fields: interplay of field arithmetic and explicit formulae, *Journal Mathematical Cryptology* 2 (2008) 227–255. <799, 803>

[150] J. Ax, Zeroes of polynomials over finite fields, *Amer. J. Math.* 86 (1964) 255–261. <199, 201, 213, 480, 488>

[151] N. Axvig, D. Dreher, K. Morrison, E. Psota, L. C. Pérez, and J. L. Walker, Analysis of connections between pseudocodewords, *IEEE Trans. Inform. Theory* 55 (2009) 4099–4107. <718, 719>

[152] M. Ayad and D. L. McQuillan, Irreducibility of the iterates of a quadratic polynomial over a field, *Acta Arith.* 93 (2000) 87–97. <342, 344>

[153] M. Ayad and D. L. McQuillan, Corrections to: "Irreducibility of the iterates of a quadratic polynomial over a field" [Acta Arith. **93** (2000), 87–97; MR1760091 (2001c:11031)], *Acta Arith.* 99 (2001) 97. <342, 344>

[154] M. Baake, J. A. G. Roberts, and A. Weiss, Periodic orbits of linear endomorphisms on the 2-torus and its lattices, *Nonlinearity* 21 (2008) 2427–2446. <337, 344>

[155] L. Babai, The Fourier transform and equations over finite abelian groups, Private Communication. <310>

[156] L. Babai, Spectra of Cayley graphs, *J. Combin. Theory, Ser. B* 27 (1979) 180–189. <652, 658>

[157] L. Babai, W. M. Kantor, and A. Lubotsky, Small-diameter Cayley graphs for finite simple groups, *European J. Combin.* 10 (1989) 507–522. <652, 658>

[158] E. Bach, J. von zur Gathen, and H. W. Lenstra, Jr., Factoring polynomials over special finite fields, *Finite Fields and Their Applications* 7 (2001) 5–28. <381, 382>

[159] E. Bach and J. Shallit, Factoring with cyclotomic polynomials, *Math. Comp.* 52 (1989) 201–219. <120, 128>

[160] D. V. Bailey, L. Batina, D. J. Bernstein, P. Birkner, J. W. Bos, H.-C. Chen, C.-M. Cheng, G. van Damme, G. de Meulenaer, L. J. Dominguez Perez, J. Fan, T. Güneysu, F. Gurkaynak, T. Kleinjung, T. Lange, N. Mentens, R. Niederhagen, C. Paar, F. Regazzoni, P. Schwabe, L. Uhsadel, A. Van Herrewege, and B.-Y. Yang, Breaking ECC2K-130, preprint, http://eprint.iacr.org/2009/541, 2009. <400, 401>

[161] D. V. Bailey and C. Paar, Optimal extension fields for fast arithmetic in public key algorithms, In *Advances in Cryptology—CRYPTO 1998*, volume 1462 of *Lecture Notes in Comput. Sci.*, Springer, Berlin, 1998. <353, 363>

[162] C. Bajaj, J. Canny, T. Garrity, and J. Warren, Factoring rational polynomials over the complex numbers, *SIAM J. Comput.* 22 (1993) 318–331. <387, 392>

[163] J.-C. Bajard, L.-S. Didier, and P. Kornerup, An RNS Montgomery modular multiplication algorithm, *IEEE Trans. Comput.* 47 (1998) 766–776. <352, 363>

[164] J.-C. Bajard, L. Imbert, and G. A. Jullien, Parallel Montgomery multiplication in $GF(2^k)$ using trinomial residue arithmetic, In *Proc. Seventeenth IEEE Symposium on Computer Arithmetic (ARITH-17)*, 164–171, 2005. <814, 822, 823>

[165] J.-C. Bajard, L. Imbert, and T. Plantard, Arithmetic operations in the polynomial modular number system, In *IEEE Symposium on Computer Arithmetic*, 206–213, 2005. <352, 363>

[166] J.-C. Bajard, L. Imbert, and T. Plantard, Modular number systems: beyond the Mersenne family, In *Selected Areas in Cryptography*, volume 3357 of *Lecture Notes in Comput. Sci.*, 159–169, Springer, Berlin, 2005. <352, 363>

[167] R. Baker, Whist tournaments, *Congr. Numer.* 14 (1975) 89–100. <618, 619>

[168] R. C. Baker, Small solutions of congruences, *Mathematika* 30 (1983) 164–188. <213>

[169] R. D. Baker, C. Culbert, G. L. Ebert, and K. E. Mellinger, Odd order flag-transitive affine planes of dimension three over their kernel, *Adv. Geom.* (2003) S215–S223. <569, 574>

[170] R. D. Baker, J. M. Dover, G. L. Ebert, and K. L. Wantz, Hyperbolic fibrations of $PG(3, q)$, *European J. Combin.* 20 (1999) 1–16. <573, 574>

[171] R. D. Baker, J. M. Dover, G. L. Ebert, and K. L. Wantz, Baer subgeometry partitions, *J. Geom.* 67 (2000) 23–34, Second Pythagorean Conference (Pythagoreion, 1999). <570, 574>

[172] R. D. Baker and G. L. Ebert, Nests of size $q - 1$ and another family of translation planes, *J. London Math. Soc., 2nd Ser.* 38 (1988) 341–355. <567, 574>

[173] R. D. Baker and G. L. Ebert, A new class of translation planes, In *Combinatorics '86*, volume 37 of *Ann. Discrete Math.*, 7–20, North-Holland, Amsterdam, 1988. <567, 574>

[174] R. D. Baker and G. L. Ebert, Filling the nest gaps, *Finite Fields Appl.* 2 (1996) 42–61. <568, 574>

[175] R. D. Baker and G. L. Ebert, Two-dimensional flag-transitive planes revisited, *Geom. Dedicata* 63 (1996) 1–15. <569, 574>

[176] R. D. Baker, G. L. Ebert, K. H. Leung, and Q. Xiang, A trace conjecture and flag-transitive affine planes, *J. Combin. Theory, Ser. A* 95 (2001) 158–168. <569, 574>

[177] R. D. Baker, G. L. Ebert, and T. Penttila, Hyperbolic fibrations and q-clans, *Des. Codes Cryptogr.* 34 (2005) 295–305. <573, 574>

[178] R. D. Baker, G. L. Ebert, and K. L. Wantz, Regular hyperbolic fibrations, *Adv. Geom.* 1 (2001) 119–144. <573, 574>

[179] R. D. Baker, G. L. Ebert, and K. L. Wantz, Enumeration of nonsingular Buekenhout unitals, *Note Mat.* 29 (2009) 69–90. <571, 574>

[180] R. D. Baker, G. L. Ebert, and K. L. Wantz, Enumeration of orthogonal Buekenhout unitals, *Des. Codes Cryptogr.* 55 (2010) 261–283. <571, 574>

[181] M. Bakshi, S. Jaggi, and M. Effros, Concatenated polar codes, In *2010 IEEE International Symposium on Information Theory Proceedings (ISIT)*, 918–922, 2010. <739>

[182] J. Balakrishnan, J. Belding, S. Chisholm, K. Eisenträger, K. E. Stange, and E. Teske, Pairings on hyperelliptic curves, *Fields Inst. Commun.* 58 (2010) 1–34. <454, 455, 456>

[183] R. Balasubramanian and N. Koblitz, The improbability that an elliptic curve has subexponential discrete log problem under the Menezes-Okamoto-Vanstone algorithm, *J. Cryptology* 11 (1998) 141–145. <789, 796>

[184] S. Ball, On the size of a triple blocking set in $\mathrm{PG}(2,q)$, *European J. Combin.* 17 (1996) 427–435. <562, 563>

[185] S. Ball, The number of directions determined by a function over a finite field, *J. Combin. Theory, Ser. A* 104 (2003) 341–350. <558, 559, 563>

[186] S. Ball, On the graph of a function in many variables over a finite field, *Des. Codes Cryptogr.* 47 (2008) 159–164. <559, 563>

[187] S. Ball, The polynomial method in Galois geometries, in *Current Research Topics in Galois Geometry*, Chapter 5, Nova Sci. Publ., New York, (2012) 105–130. <563>

[188] S. Ball and A. Blokhuis, On the size of a double blocking set in $\mathrm{PG}(2,q)$, *Finite Fields Appl.* 2 (1996) 125–137. <562, 563>

[189] S. Ball, A. Blokhuis, and F. Mazzocca, Maximal arcs in Desarguesian planes of odd order do not exist, *Combinatorica* 17 (1997) 31–41. <572, 574>

[190] S. Ball and A. Gács, On the graph of a function over a prime field whose small powers have bounded degree, *European J. Combin.* 30 (2009) 1575–1584. <559, 563>

[191] S. Ball, A. Gács, and P. Sziklai, On the number of directions determined by a pair of functions over a prime field, *J. Combin. Theory, Ser. A* 115 (2008) 505–516. <559, 563>

[192] S. Ball and M. Zieve, Symplectic spreads and permutation polynomials, In *Finite Fields and Applications*, volume 2948 of *Lecture Notes in Comput. Sci.*, 79–88, Springer, Berlin, 2004. <229>

[193] A. Balog, Many additive quadruples, In *Additive Combinatorics*, volume 43 of *CRM Proc. Lecture Notes*, 39–49, Amer. Math. Soc., Providence, RI, 2007. <188, 192>

[194] A. Balog and E. Szemerédi, A statistical theorem of set addition, *Combinatorica* 14 (1994) 263–268. <188, 192>

[195] J. Bamberg, A. Betten, C. Praeger, and A. Wassermann, Unitals in the Desarguesian projective plane of order sixteen, International Conference on Design of Experiments (ICODOE, 2011). <571, 574>

[196] W. D. Banks, A. Conflitti, J. B. Friedlander, and I. E. Shparlinski, Exponential sums over Mersenne numbers, *Compos. Math.* 140 (2004) 15–30. <189, 192>

[197] W. D. Banks, J. B. Friedlander, S. V. Konyagin, and I. E. Shparlinski, Incomplete exponential sums and Diffie-Hellman triples, *Math. Proc. Cambridge Philos. Soc.* 140 (2006) 193–206. <183, 185>

[198] H. W. Bao, On two exponential sums and their applications, *Finite Fields Appl.* 3 (1997) 115–130. <93, 95>

[199] I. Baoulina, On the number of solutions to certain diagonal equations over finite fields, *Int. J. Number Theory* 6 (2010) 1–14. <208, 213>

[200] B. Barak, G. Kindler, R. Shaltiel, B. Sudakov, and A. Wigderson, Simulating independence: new constructions of condensers, Ramsey graphs, dispersers, and extractors, *J. ACM* 57 (2010) Art. 20, 52. <191, 192>

[201] M. Bardet, J.-C. Faugère, and B. Salvy, On the complexity of Gröbner basis computation of semi-regular overdetermined algebraic equations, In *Proceedings of the International Conference on Polynomial System Solving*, 71–74, 2004. <782, 783>

[202] A. Barlotti, Un'estensione del teorema di Segre-Kustaanheimo, *Boll. Un. Mat. Ital.* 10 (1955) 498–506. <588, 589>

[203] P. S. L. M. Barreto, S. D. Galbraith, C. Ó'hÉigeartaigh, and M. Scott, Efficient pairing computation on supersingular abelian varieties, *Designs, Codes and Cryptography* 42 (2007) 239–271. <791, 796>

[204] P. S. L. M. Barreto and J. F. Voloch, Efficient computation of roots in finite fields, *Des. Codes Cryptogr.* 39 (2006) 275–280. <360, 363>

[205] S. Barwick and G. Ebert, *Unitals in Projective Planes*, Springer Monographs in Mathematics. Springer, New York, 2008. <571, 574>

[206] S. G. Barwick and W.-A. Jackson, Geometric constructions of optimal linear perfect hash families, *Finite Fields Appl.* 14 (2008) 1–13. <613, 619>

[207] S. G. Barwick, W.-A. Jackson, and C. T. Quinn, Optimal linear perfect hash families with small parameters, *J. Combin. Des.* 12 (2004) 311–324. <613, 619>

[208] L. Batina, S. B. Örs, B. Preneel, and J. Vandewalle, Hardware architectures for public key cryptography, *Integration, the VLSI Journal* 34 (2003) 1 - 64. <109>

[209] C. Batut, K. Belabas, D. Bernardi, H. Cohen, and M. Olivier, PARI/GP, version 2.5.0, `http:pari.math.u-bordeaux.fr/`, as viewed in July, 2012. <346, 363>

[210] A. Bauer, D. Vergnaud, and J.-C. Zapalowicz, Inferring sequences produced by nonlinear pseudorandom number generators using coppersmith's methods, In *Proc. 15th Intern. Conf. on Practice and Theory in Public-Key Cryptography, PKC 2012*, volume 7293 of *Lecture Notes in Comput. Sci.*, 609–626, Springer, Berlin, 2012. <338, 344>

[211] L. D. Baumert, *Cyclic Difference Sets*, Lecture Notes in Mathematics, Vol. 182. Springer-Verlag, Berlin, 1971. <31, 600, 602, 607>

[212] E. Bayer-Fluckiger and H. W. Lenstra, Jr., Forms in odd degree extensions and self-dual normal bases, *Amer. J. Math.* 112 (1990) 359–373. <114, 116>

[213] J. T. Beard, Jr. and K. I. West, Factorization tables for $x^n - 1$ over GF(q), *Math. Comp.* 28 (1974) 1167–1168. <62, 66>

[214] B. Beckermann and G. Labahn, Fraction-free computation of matrix rational interpolants and matrix GCDs, *SIAM J. Matrix Anal. Appl.* 22 (2000) 114–144. <534, 535>

[215] E. Bedford and K. Kim, Continuous families of rational surface automorphisms with positive entropy, *Math. Ann.* 348 (2010) 667–688. <338, 344>

[216] E. Bedford and T. T. Truong, Degree complexity of birational maps related to matrix inversion, *Comm. Math. Phys.* 298 (2010) 357–368. <337, 338, 344>

[217] P. Beelen and I. I. Bouw, Asymptotically good towers and differential equations, *Compos. Math.* 141 (2005) 1405–1424. <464, 469>

[218] D. Behr, Searchable magic book contents, main site: `http://archive.denisbehr.de`, `http://archive.denisbehr.de/archive/route/entries.php?url=10`,

50,1036. <632, 642>

[219] K. Belabas, M. van Hoeij, J. Klüners, and A. Steel, Factoring polynomials over global fields, *J. Théor. Nombres Bordeaux* 21 (2009) 15–39. <385, 392>

[220] J. Belding, R. Bröker, A. Enge, and K. Lauter, Computing Hilbert class polynomials, In *Algorithmic Number Theory*, volume 5011 of *Lecture Notes in Comput. Sci.*, 282–295, Springer, Berlin, 2008. <787, 796>

[221] M. Bellare and P. Rogaway, Minimizing the use of random oracles in authenticated encryption schemes, In Y. Han, T. Okamoto, and S. Qing, editors, *Information and Communications Security*, volume 1334 of *Lecture Notes in Comput. Sci.*, 1–16, Springer-Verlag, Berlin, 1997. <785, 796>

[222] M. P. Bellon and C.-M. Viallet, Algebraic entropy, *Comm. Math. Phys.* 204 (1999) 425–437. <337, 338, 344>

[223] M. Ben-Or, Probabilistic algorithms in finite fields, In *Proc. Twenty Second IEEE Symp. Foundations Computer Science*, 394–398, 1981. <377, 380>

[224] T. D. Bending and D. Fon-Der-Flaass, Crooked functions, bent functions, and distance regular graphs, *Electron. J. Combin.* 5 (1998) Research Paper 34, 14 pp. <259, 261>

[225] S. Benedetto, D. Divsalar, G. Montorsi, and F. Pollara, Serial concatenation of interleaved codes: performance analysis, design, and iterative decoding, *IEEE Trans. Inform. Theory* 44 (1998) 909–926. <721, 727>

[226] A. T. Benjamin and C. D. Bennett, The probability of relatively prime polynomials, *Math. Mag.* 80 (2007) 196–202. <81, 85, 509, 510>

[227] C. H. Bennett and G. Brassard, Quantum cryptography: Public key distribution and coin tossing, In *International Conference on Computers, Systems & Signal Processing*, 1984. <749, 750>

[228] T. P. Berger, A. Canteaut, P. Charpin, and Y. Laigle-Chapuy, On almost perfect nonlinear functions over F_2^n, *IEEE Trans. Inform. Theory* 52 (2006) 4160–4170. <256, 259, 261>

[229] E. R. Berlekamp, Distribution of cyclic matrices in a finite field, *Duke Math. J.* 33 (1966) 45–48. <62, 66>

[230] E. R. Berlekamp, Factoring polynomials over finite fields, *Bell System Tech. J.* 46 (1967) 1853–1859. <375, 380, 381, 382, 770, 783>

[231] E. R. Berlekamp, *Algebraic Coding Theory*, McGraw-Hill Book Co., New York, 1968. <31, 60, 66, 68, 70, 205, 206, 314, 317, 365, 367, 374, 661, 692, 693, 703>

[232] E. R. Berlekamp, Factoring polynomials over large finite fields, *Math. Comp.* 24 (1970) 713–735. <380, 381, 382>

[233] E. R. Berlekamp, editor, *Key Papers in the Development of Coding Theory*, IEEE Press Sel. Rep. Ser., New York, 1974. <702, 703>

[234] E. R. Berlekamp, Bit-serial Reed-Solomon encoders., *IEEE Trans. Inf. Theory* 28 (1982) 869–874. <109>

[235] E. R. Berlekamp, R. J. McEliece, and H. C. A. van Tilborg, On the inherent intractability of certain coding problems, *IEEE Trans. Inform. Theory* IT-24 (1978) 384–386. <749, 750>

[236] E. R. Berlekamp, H. Rumsey, and G. Solomon, On the solution of algebraic equations over finite fields, *Information and Control* 10 (1967) 553–564. <70>

[237] L. Bernardin, On square-free factorization of multivariate polynomials over a finite field, *Theoret. Comput. Sci.* 187 (1997) 105–116. <384, 392>

[238] L. Bernardin, On bivariate Hensel lifting and its parallelization, In *ISSAC '98: Proceedings of the 1998 International Symposium on Symbolic and Algebraic Computation*, 96–100, New York, 1998, ACM. <385, 392>

[239] L. Bernardin and M. B. Monagan, Efficient multivariate factorization over finite fields, In *Applied Algebra, Algebraic Algorithms and Error-Correcting Codes*, volume 1255 of *Lecture Notes in Comput. Sci.*, 15–28, Springer-Verlag, 1997. <387, 392>

[240] B. C. Berndt, R. J. Evans, and K. S. Williams, *Gauss and Jacobi Sums*, Canadian Mathematical Society Series of Monographs and Advanced Texts, John Wiley & Sons Inc., New York, 1998. <31, 139, 140, 141, 142, 143, 145, 146, 147, 148, 149, 150, 151, 152, 156, 159, 160, 161, 173, 185, 206, 213>

[241] D. J. Bernstein, Multiplication for mathematicians, 2001, preprint available at `http://cr.yp.to/papers.html#m3`. <814, 823>

[242] D. J. Bernstein, Pippenger's exponentiation algorithm, 2002, preprint available at `http://cr.yp.to/papers/pippenger.pdf`. <357, 363>

[243] D. J. Bernstein, Batch binary Edwards, In *Advances in Cryptology—CRYPTO 2009*, volume 5677 of *Lecture Notes in Comput. Sci.*, 317–336, Springer, Berlin, 2009. <814, 823>

[244] D. J. Bernstein, P. Birkner, M. Joye, T. Lange, and C. Peters, Twisted Edwards curves, In *Progress in Cryptology—AFRICACRYPT 2008*, volume 5023 of *Lecture Notes in Comput. Sci.*, 389–405, Springer, Berlin, 2008. <441, 446>

[245] D. J. Bernstein, J. Buchmann, and E. Dahmen, editors, *Post-Quantum Cryptography*, Springer-Verlag, Berlin, 2009. <31, 749, 750>

[246] D. J. Bernstein and T. Lange, Explicit-formulas database, `http://hyperelliptic.org/EFD/`. <443, 446>

[247] D. J. Bernstein and T. Lange, Faster addition and doubling on elliptic curves, In *Advances in Cryptology—ASIACRYPT 2007*, volume 4833 of *Lecture Notes in Comput. Sci.*, 29–50, Springer, Berlin, 2007. <441, 442, 446>

[248] D. J. Bernstein and T. Lange, Type-II optimal polynomial bases, In *Arithmetic of Finite Fields*, volume 6087 of *Lecture Notes in Comput. Sci.*, 41–61, Springer, Berlin, 2010. <127, 128, 822, 823>

[249] D. J. Bernstein and T. Lange, A complete set of addition laws for incomplete Edwards curves, *J. Number Theory* 131 (2011) 858–872. <442, 446>

[250] D. J. Bernstein, T. Lange, and C. Peters, Attacking and defending the McEliece cryptosystem, In *Post-Quantum Cryptography*, volume 5299 of *Lecture Notes in Comput. Sci.*, 31–46, Springer, Berlin, 2008. <750>

[251] C. Berrou, A. Glavieux, and P. Thitimajshima, Near Shannon limit error-correcting coding and decoding: turbo-codes, In *Proc. IEEE Int. Conf. on Commun.*, 1064–1070, Geneva, Switzerland, 1993. <719, 727>

[252] C. Berrou, S. K. Y. Saouter, C. Douillard, and M. Jézéquel, Designing good permutations for turbo codes: towards a single model, In *Proc. International Conference on Communications*, volume 1, 341–345, Paris, France, 2004. <726, 727>

[253] P. Berthelot, Cohomologie rigide et théorie de Dwork: le cas des sommes exponentielles, *Astérisque* (1984) 3, 17–49, p-adic cohomology. <169>

[254] P. Berthelot, S. Bloch, and H. Esnault, On Witt vector cohomology for singular varieties, *Compos. Math.* 143 (2007) 363–392. <200, 201>

[255] P. Berthelot and A. Ogus, *Notes on Crystalline Cohomology*, Princeton University Press, Princeton, NJ, 1978. <481, 488>

[256] J. Berthomieu and G. Lecerf, Reduction of bivariate polynomials from convex-dense to dense, with application to factorizations, *Math. Comp.* 81 (2012) 1799–1821. <389, 392>

[257] T. Beth and Z. D. Dai, On the complexity of pseudo-random sequences—or: If you can describe a sequence it can't be random, In *Advances in Cryptology—EUROCRYPT '89*, volume 434 of *Lecture Notes in Comput. Sci.*, 533–543, Springer, Berlin, 1990. <334, 336>

[258] T. Beth and W. Geiselmann, Selbstduale Normalbasen über GF(q), *Arch. Math. (Basel)* 55 (1990) 44–48. <115, 116, 506, 510>

[259] T. Beth, W. Geiselmann, and F. Meyer, Finding (good) normal bases in finite fields, In *Proceedings of the 1991 International Symposium on Symbolic and Algebraic Computation*, ISSAC '91, 173–178, ACM, New York, NY, USA, 1991. <120, 128>

[260] T. Beth, D. Jungnickel, and H. Lenz, *Design Theory*, Cambridge University Press, Cambridge, 1986. <31, 170, 185>

[261] T. Beth, D. Jungnickel, and H. Lenz, *Design Theory. Vol. I*, volume 69 of *Encyclopedia of Mathematics and its Applications*, Cambridge University Press, Cambridge, second edition, 1999. <31, 593, 599, 600, 605, 606, 607, 619>

[262] T. Beth, D. Jungnickel, and H. Lenz, *Design Theory. Vol. II*, volume 78 of *Encyclopedia of Mathematics and its Applications*, Cambridge University Press, Cambridge, second edition, 1999. <31, 599, 600, 603, 604, 605, 606, 607, 619>

[263] D. Betten and D. G. Glynn, Über endliche planare Funktionen, ihre zugehörenden Schiebebenen, und ihre abgeleiteten Translationsebenen, *Results Math.* 42 (2002) 32–36. <280, 282>

[264] C. Bey and G. M. Kyureghyan, On Boolean functions with the sum of every two of them being bent, *Des. Codes Cryptogr.* 49 (2008) 341–346. <254, 261>

[265] J. Bezerra, A. Garcia, and H. Stichtenoth, An explicit tower of function fields over cubic finite fields and Zink's lower bound, *J. Reine Angew. Math.* 589 (2005) 159–199. <463, 467, 469>

[266] M. Bhargava and M. E. Zieve, Factoring Dickson polynomials over finite fields, *Finite Fields Appl.* 5 (1999) 103–111. <284, 290>

[267] A. Bhattacharyya, S. Kopparty, G. Schoenebeck, M. Sudan, and D. Zuckerman, Optimal testing of Reed-Muller codes (report no. 86), In *Proceedings of Electronic Colloquium on Computational Complexity (2009)*, volume 3690 of *Lecture Notes in Comput. Sci.*, 269–275, Springer, 2011. <246, 252>

[268] K. Bibak, Additive combinatorics with a view towards computer science and cryptography: An exposition, *arXiv:1108.3790*. <189, 191, 192>

[269] F. Bien, Constructions of telephone networks by group representations, *Notices Amer. Math. Soc.* 36 (1989) 5–22. <642, 649, 658>

[270] J. Bierbrauer, *Introduction to Coding Theory*, Discrete Mathematics and its Applications. Chapman & Hall/CRC, Boca Raton, FL, 2005. <31>

[271] J. Bierbrauer, A direct approach to linear programming bounds for codes and (t, m, s)-nets, *Des. Codes Cryptogr.* 42 (2007) 127–143. <621, 630>

[272] J. Bierbrauer, A family of crooked functions, *Des. Codes Cryptogr.* 50 (2009) 235–241. <259, 261>

[273] J. Bierbrauer, New commutative semifields and their nuclei, In *Applied Algebra, Algebraic Algorithms, and Error-Correcting Codes*, volume 5527 of *Lecture Notes in Comput. Sci.*, 179–185, Springer, Berlin, 2009. <282>

[274] J. Bierbrauer, New semifields, PN and APN functions, *Des. Codes Cryptogr.* 54 (2010) 189–200. <282>

[275] J. Bierbrauer and Y. Edel, Theory of perpendicular arrays, *J. Combin. Des.* 2 (1994) 375–406. <612, 619>

[276] J. Bierbrauer, Y. Edel, and W. C. Schmid, Coding-theoretic constructions for (t, m, s)-nets and ordered orthogonal arrays, *J. Combin. Des.* 10 (2002) 403–418. <622, 625, 630>

[277] J. Bierbrauer and G. M. Kyureghyan, Crooked binomials, *Des. Codes Cryptogr.* 46 (2008) 269–301. <259, 261>

[278] E. Biham and A. Shamir, Differential cryptanalysis of DES-like cryptosystems, *J. Cryptology* 4 (1991) 3–72. <253, 261>

[279] M. Biliotti, V. Jha, and N. L. Johnson, *Foundations of Translation Planes*, volume 243 of *Monographs and Textbooks in Pure and Applied Mathematics*, Marcel Dekker Inc., New York, 2001. <565, 574>

[280] O. Billet and H. Gilbert, Cryptanalysis of Rainbow, In *Security and Cryptography for Networks*, volume 4116 of *Lecture Notes in Comput. Sci.*, 336–347, Springer, 2006. <772, 773, 783>

[281] O. Billet, M. J. B. Robshaw, and T. Peyrin, On building hash functions from multivariate quadratic equations, In J. Pieprzyk, H. Ghodosi, and E. Dawson, editors, *ACISP*, volume 4586 of *Lecture Notes in Computer Science*, 82–95, Springer, 2007. <783>

[282] Y. Bilu and N. Linial, Lifts, discrepancy and nearly optimal spectral gap, *Combinatorica* 26 (2006) 495–519. <646, 656, 657, 658>

[283] G. Bini and F. Flamini, *Finite Commutative Rings and their Applications*, The Kluwer International Series in Engineering and Computer Science, 680, Kluwer Academic Publishers, Boston, MA, 2002. <28, 29, 31>

[284] B. J. Birch, How the number of points of an elliptic curve over a fixed prime field varies, *J. London Math. Soc., 2nd Ser.* 43 (1968) 57–60. <430, 440>

[285] B. J. Birch and H. P. F. Swinnerton-Dyer, Note on a problem of Chowla, *Acta Arith.* 5 (1959) 417–423 (1959). <234, 236>

[286] P. Birkner, Efficient divisor class halving on genus two curves, In *Thirteenth International Workshop on Selected Areas in Cryptography*, volume 4356 of *Lecture Notes in Comput. Sci.*, 317–326, Springer, Berlin, 2007. <797, 803>

[287] P. Birkner and N. Thériault, Faster halvings in genus 2, In *Fifteenth International Workshop on Selected Areas in Cryptography*, volume 5381 of *Lecture Notes in Comput. Sci.*, 1–17, Springer, Berlin, 2009. <797, 803>

[288] P. Birkner and N. Thériault, Efficient halving for genus 3 curves over binary fields, *Advances in Mathematics of Communications* 4 (2010) 23–47. <799, 803>

[289] A. Biró, On polynomials over prime fields taking only two values on the multiplicative group, *Finite Fields Appl.* 6 (2000) 302–308. <233, 236>

[290] R. R. Bitmead and B. D. O. Anderson, Asymptotically fast solution of Toeplitz and related systems of linear equations, *Linear Algebra Appl.* 34 (1980) 103–116. <533, 535>

[291] R. Blache, First vertices for generic Newton polygons, and p-cyclic coverings of the projective line, preprint available, `http://arxiv.org/abs/0912.2051`, 2009. <484, 487, 488>

[292] R. Blache, Newton polygons for character sums and Poincaré series, *Int. J. Number Theory* 7 (2011) 1519–1542. <485, 488>

[293] R. Blache, Valuations of exponential sums and the generic first slope of artin-schreier curves, *J. Number Theory* 132 (2012) 2336–2352. <480, 484, 488>

[294] R. Blache, J.-P. Cherdieu, and J. Estrada Sarlabous, Some computational aspects of Jacobians of curves in the family $y^3 = \gamma x^5 + \delta$ over $\mathbb{F}_p$, *Finite Fields Appl.* 13 (2007) 348–365. <807, 811>

[295] R. Blache and É. Férard, Newton stratification for polynomials: the open stratum, *J. Number Theory* 123 (2007) 456–472. <484, 488>

[296] R. Blache, É. Férard, and H. J. Zhu, Hodge-Stickelberger polygons for L-functions of exponential sums of $P(x^s)$, *Math. Res. Lett.* 15 (2008) 1053–1071. <485, 488>

[297] S. R. Blackburn, A generalisation of the discrete Fourier transform: determining the minimal polynomial of a periodic sequence, *IEEE Trans. Inform. Theory* 40 (1994) 1702–1704. <328, 336>

[298] S. R. Blackburn, T. Etzion, and K. G. Paterson, Permutation polynomials, de Bruijn sequences, and linear complexity, *J. Combin. Theory, Ser. A* 76 (1996) 55–82. <328, 329, 336>

[299] S. R. Blackburn, D. Gómez-Pérez, J. Gutierrez, and I. E. Shparlinski, Predicting the inversive generator, In *Cryptography and Coding*, volume 2898 of *Lecture Notes in Comput. Sci.*, 264–275, Springer, Berlin, 2003. <338, 344>

[300] S. R. Blackburn, D. Gómez-Pérez, J. Gutierrez, and I. E. Shparlinski, Predicting nonlinear pseudorandom number generators, *Math. Comp.* 74 (2005) 1471–1494. <338, 344>

[301] S. R. Blackburn, D. Gómez-Pérez, J. Gutierrez, and I. E. Shparlinski, Reconstructing noisy polynomial evaluation in residue rings, *J. Algorithms* 61 (2006) 47–59. <338, 344>

[302] S. R. Blackburn and P. R. Wild, Optimal linear perfect hash families, *J. Combin. Theory, Ser. A* 83 (1998) 233–250. <613, 619>

[303] R. E. Blahut, Transform techniques for error control codes, *IBM J. Res. Develop.* 23 (1979) 299–315. <328, 336>

[304] R. E. Blahut, *Theory and Practice of Error Control Codes*, Addison-Wesley Publishing Company Advanced Book Program, Reading, MA, 1983. <31, 661, 663, 680, 688, 690, 692, 693, 703>

[305] I. F. Blake, editor, *Algebraic Coding Theory: History and Development*, Dowden Hutchinson & Ross Inc., Stroudsburg, Pa., 1973, Benchmark Papers in Electrical Engineering and Computer Science. <702, 703>

[306] I. F. Blake, R. Fuji-Hara, R. C. Mullin, and S. A. Vanstone, Computing logarithms in finite fields of characteristic two, *SIAM J. Algebraic Discrete Methods* 5 (1984) 276–285. <370, 374>

[307] I. F. Blake, S. Gao, and R. J. Lambert, Construction and distribution problems for irreducible trinomials over finite fields, In *Applications of Finite Fields*, volume 59 of *Inst. Math. Appl. Conf. Ser. (New Ser.)*, 19–32, Oxford Univ. Press, New York, 1996. <73, 89, 90>

[308] I. F. Blake, S. Gao, and R. C. Mullin, Normal and self-dual normal bases from factorization of $cx^{q+1} + dx^q - ax - b$, *SIAM J. Discrete Math.* 7 (1994) 499–512. <118, 124, 128>

[309] I. F. Blake, S. Gao, and R. C. Mullin, Specific irreducible polynomials with linearly independent roots over finite fields, *Linear Algebra Appl.* 253 (1997) 227–249. <113, 116, 131, 138>

[310] I. F. Blake and T. Garefalakis, A transform property of Kloosterman sums, *Discrete*

Appl. Math. 158 (2010) 1064–1072. <75, 79>

[311] I. F. Blake and R. C. Mullin, *The Mathematical Theory of Coding*, Academic Press, New York-London, 1975. <31, 661, 683, 687, 703>

[312] I. F. Blake, G. Seroussi, and N. P. Smart, *Elliptic Curves in Cryptography*, volume 265 of *London Mathematical Society Lecture Note Series*, Cambridge University Press, Cambridge, 2000, Reprint of the 1999 original. <31, 788, 796>

[313] I. F. Blake, G. Seroussi, and N. P. Smart, *Advances in Elliptic Curve Cryptography*, volume 317 of *London Mathematical Society Lecture Note Series*, Cambridge University Press, Cambridge, 2005. <31, 785, 788, 796, 803>

[314] R. Blasco-Serrano, R. Thobaben, V. Rathi, and M. Skoglund, Polar codes for compress-and-forward in binary relay channels, In *Proc. Conf Signals, Systems and Computers (ASILOMAR) Record of the Forty Fourth Asilomar Conf*, 1743–1747. <739>

[315] D. Blessenohl, Abelsche Erweiterungen, in denen jedes reguläre Element vollständig regulär ist, *Arch. Math. (Basel)* 54 (1990) 146–156. <129, 130, 138>

[316] D. Blessenohl, Zu einer Vermutung von Morgan und Mullen, *Berichtsreihe des Mathematischen Seminars der Universität Kiel* 05-21 (2005). <137, 138>

[317] D. Blessenohl and K. Johnsen, Eine Verschärfung des Satzes von der Normalbasis, *J. Algebra* 103 (1986) 141–159. <129, 138>

[318] D. Blessenohl and K. Johnsen, Stabile Teilkörper galoisscher Erweiterungen und ein Problem von C. Faith, *Arch. Math. (Basel)* 56 (1991) 245–253. <130, 138>

[319] A. Blokhuis, On the size of a blocking set in $\mathrm{PG}(2,p)$, *Combinatorica* 14 (1994) 111–114. <560, 563>

[320] A. Blokhuis, Blocking sets in Desarguesian planes, In *Combinatorics, Paul Erdős is Eighty, Vol. 2*, volume 2 of *Bolyai Soc. Math. Stud.*, 133–155, János Bolyai Math. Soc., Budapest, 1996. <560, 563>

[321] A. Blokhuis, S. Ball, A. E. Brouwer, L. Storme, and T. Szőnyi, On the number of slopes of the graph of a function defined on a finite field, *J. Combin. Theory, Ser. A* 86 (1999) 187–196. <558, 563>

[322] A. Blokhuis, A. E. Brouwer, and T. Szőnyi, The number of directions determined by a function f on a finite field, *J. Combin. Theory, Ser. A* 70 (1995) 349–353. <558, 563>

[323] A. Blokhuis, A. E. Brouwer, and H. A. Wilbrink, Blocking sets in $\mathrm{PG}(2,p)$ for small p, and partial spreads in $\mathrm{PG}(3,7)$, *Adv. Geom.* (2003) S245–S253. <561, 562, 563>

[324] A. Blokhuis, A. A. Bruen, and J. A. Thas, Arcs in $\mathrm{PG}(n,q)$, MDS-codes and three fundamental problems of B. Segre—some extensions, *Geom. Dedicata* 35 (1990) 1–11. <585, 589>

[325] A. Blokhuis, R. S. Coulter, M. Henderson, and C. M. O'Keefe, Permutations amongst the Dembowski-Ostrom polynomials, In *Finite Fields and Applications*, 37–42, Springer, Berlin, 2001. <224, 225, 229>

[326] A. Blokhuis, D. Jungnickel, and B. Schmidt, Proof of the prime power conjecture for projective planes of order n with abelian collineation groups of order n^2, *Proc. Amer. Math. Soc.* 130 (2002) 1473–1476. <279, 282, 573, 574>

[327] A. Blokhuis, M. Lavrauw, and S. Ball, On the classification of semifield flocks, *Adv. Math.* 180 (2003) 104–111. <277, 278>

[328] A. Blokhuis, L. Lovász, L. Storme, and T. Szőnyi, On multiple blocking sets in Galois planes, *Adv. Geom.* 7 (2007) 39–53. <562, 563>

[329] A. Blokhuis, R. Pellikaan, and T. Szőnyi, Blocking sets of almost Rédei type, *J. Combin. Theory, Ser. A* 78 (1997) 141–150. <560, 563>

[330] A. Blokhuis, L. Storme, and T. Szőnyi, Lacunary polynomials, multiple blocking sets and Baer subplanes, *J. London Math. Soc., 2nd Ser.* 60 (1999) 321–332. <562, 563>

[331] C. Blondeau, A. Canteaut, and P. Charpin, Differential properties of power functions, *Int. J. Inf. Coding Theory* 1 (2010) 149–170. <261>

[332] A. W. Bluher, Explicit formulas for strong Davenport pairs, *Acta Arith.* 112 (2004) 397–403. <301, 302>

[333] A. W. Bluher, A Swan-like theorem, *Finite Fields Appl.* 12 (2006) 128–138. <69, 70>

[334] G. Böckle, An eichler-shimura isomorphism over function fields between Drinfeld modular forms and cohomology classes of crystals, *preprint* (2002). <545, 546>

[335] G. Böckle, Global L-functions over function fields, *Math. Ann.* 323 (2002) 737–795. <541, 546>

[336] A. Bodin, Number of irreducible polynomials in several variables over finite fields, *Amer. Math. Monthly* 115 (2008) 653–660. <80, 85>

[337] A. Bodin, Generating series for irreducible polynomials over finite fields, *Finite Fields Appl.* 16 (2010) 116–125. <80, 82, 85>

[338] A. Bodin, P. Dèbes, and S. Najib, Indecomposable polynomials and their spectrum, *Acta Arith.* 139 (2009) 79–100. <83, 84, 85>

[339] E. Bombieri, On exponential sums in finite fields, *Amer. J. Math.* 88 (1966) 71–105. <163, 169, 473, 476, 479>

[340] E. Bombieri, Counting points on curves over finite fields (d'après S. A. Stepanov), In *Séminaire Bourbaki, 25ème année (1972/1973), Exp. No. 430*, 234–241. Lecture Notes in Math., Vol. 383, Springer, Berlin, 1974. <477, 479>

[341] E. Bombieri, On exponential sums in finite fields. II, *Invent. Math.* 47 (1978) 29–39. <164, 169, 302, 473, 476, 479>

[342] E. Bombieri and S. Sperber, On the estimation of certain exponential sums, *Acta Arith.* 69 (1995) 329–358. <164, 169>

[343] D. Bonchev, S. Thomas, A. Apte, and L. B. Kier, Cellular automata modelling of biomolecular networks dynamics, *SAR and QSAR in Environmental Research* 21 (2010) 77–102. <829, 834>

[344] D. Boneh and M. Franklin, Identity-based encryption from the Weil pairing, *SIAM J. Comput.* 32 (2003) 586–615. <748, 750>

[345] D. Boneh, E.-J. Goh, and K. Nissim, Evaluating 2-DNF formulas on ciphertexts, In *Theory of cryptography*, volume 3378 of *Lecture Notes in Comput. Sci.*, 325–341, Springer, Berlin, 2005. <793, 796>

[346] D. Boneh, B. Lynn, and H. Shacham, Short signatures from the Weil pairing, *J. Cryptology* 17 (2004) 297–319. <748, 750>

[347] D. Boneh and R. Venkatesan, Rounding in lattices and its cryptographic applications, In *Proceedings of the Eighth Annual ACM-SIAM Symposium on Discrete Algorithms*, 675–681, ACM, New York, 1997. <176, 185>

[348] D. Boneh and R. Venkatesan, Breaking RSA may not be equivalent to factoring (extended abstract), In *Advances in Cryptology—EUROCRYPT '98*, volume 1403 of *Lecture Notes in Comput. Sci.*, 59–71, Springer, Berlin, 1998. <176, 185>

[349] T. J. Boothby and R. W. Bradshaw, Bitslicing and the method of four Russians over larger finite fields, preprint available, `http://arxiv.org/abs/0901.1413`, 2009. <522, 535>

[350] H. Borges, Frobenius non-classical curves of type $g(y) = f(x)$, preprint, 2012. <233, 236>

[351] H. Borges and F. Conceição, On the characterization of minimal value set polynomials, preprint, 2012. <233, 236>

[352] P. Borwein, K.-K. S. Choi, and J. Jedwab, Binary sequences with merit factor greater than 6.34, *IEEE Trans. Inform. Theory* 50 (2004) 3234–3249. <323, 324>

[353] J. Bos and M. E. Kaihara, Playstation 3 computing breaks 2^{60} barrier: 112-bit prime ECDLP solved, online annoucement, http://lacal.epfl.ch/112bit_prime, 2009. <400, 401>

[354] S. Bosch, U. Güntzer, and R. Remmert, *Non-Archimedean Analysis: A Systematic Approach to Rigid Analytic Geometry*, volume 261 of *Grundlehren der Mathematischen Wissenschaften [Fundamental Principles of Mathematical Sciences]*, Springer-Verlag, Berlin, 1984. <537, 546>

[355] R. C. Bose, On the application of the properties of Galois fields to the construction of hyper-graeco-latin squares, *Sankhyā* 3 (1938) 323–338. <551, 554, 556>

[356] R. C. Bose, On the construction of balanced incomplete block designs, *Ann. Eugenics* 9 (1939) 353–399. <592, 599>

[357] R. C. Bose, On some connections between the design of experiments and information theory, *Bull. Inst. Internat. Statist.* 38 (1961) 257–271. <631, 642>

[358] R. C. Bose and R. C. Burton, A characterization of flat spaces in a finite geometry and the uniqueness of the Hamming and the MacDonald codes, *J. Combin. Theory* 1 (1966) 96–104. <560, 563>

[359] R. C. Bose and D. K. Ray-Chaudhuri, On a class of error correcting binary group codes, *Information and Control* 3 (1960) 68–79. <678, 702, 703>

[360] W. Bosma, J. Cannon, and C. Playoust, The Magma algebra system I: The user language, *J. Symbolic Comput.* 24 (1997) 235–265. <387, 392>

[361] W. Bosma, J. Cannon, and A. Steel, Lattices of compatibly embedded finite fields, *J. Symbolic Comput.* 24 (1997) 351–369. <402, 404>

[362] W. Bosma and H. W. Lenstra, Jr., Complete systems of two addition laws for elliptic curves, *J. Number Theory* 53 (1995) 229–240. <442, 446>

[363] A. Bostan, P. Flajolet, B. Salvy, and E. Schost, Fast computation of special resultants, *J. Symbolic Comput.* 41 (2006) 1–29. <378, 379, 380>

[364] A. Bostan, C.-P. Jeannerod, and É. Schost, Solving structured linear systems with large displacement rank, *Theoret. Comput. Sci.* 407 (2008) 155–181. <533, 535>

[365] A. Bostan, G. Lecerf, B. Salvy, É. Schost, and B. Wiebelt, Complexity issues in bivariate polynomial factorization, In *ISSAC 2004*, 42–49, ACM, New York, 2004. <385, 392>

[366] A. Bostan, F. Morain, B. Salvy, and E. Schost, Fast algorithms for computing isogenies between elliptic curves, *Math. Comp.* 77 (2008) 1755–1778. <788, 796>

[367] A. Böttcher and B. Silbermann, *Introduction to Large Truncated Toeplitz Matrices*, Universitext. Springer-Verlag, New York, 1999. <507, 510>

[368] J. Bourgain, Estimates on exponential sums related to the Diffie-Hellman distributions, *Geom. Funct. Anal.* 15 (2005) 1–34. <183, 184, 185, 189, 192>

[369] J. Bourgain, Mordell's exponential sum estimate revisited, *J. Amer. Math. Soc.* 18

(2005) 477–499. <190>

[370] J. Bourgain, More on the sum-product phenomenon in prime fields and its applications, *Int. J. Number Theory* 1 (2005) 1–32. <191, 192>

[371] J. Bourgain, Multilinear exponential sums in prime fields under optimal entropy condition on the sources, *Geom. Funct. Anal.* 18 (2009) 1477–1502. <173, 176, 185, 187, 188, 189, 192>

[372] J. Bourgain, Estimates on polynomial exponential sums, *Israel J. Math.* 176 (2010) 221–240. <213>

[373] J. Bourgain, On exponential sums in finite fields, In *An Irregular Mind: Szemerédi is 70*, 219–242, Springer, 2010. <190, 192>

[374] J. Bourgain and M.-C. Chang, A Gauss sum estimate in arbitrary finite fields, *C. R. Math. Acad. Sci. Paris* 342 (2006) 643–646. <141, 161, 213>

[375] J. Bourgain and M.-C. Chang, On a multilinear character sum of Burgess, *C. R. Math. Acad. Sci. Paris* 348 (2010) 115–120. <183, 185>

[376] J. Bourgain and A. Gamburd, Uniform expansion bounds for Cayley graphs of $SL_2(\mathbb{F}_p)$, *Ann. of Math., 2nd Ser.* 167 (2008) 625–642. <191, 192>

[377] J. Bourgain, A. Gamburd, and P. Sarnak, Affine linear sieve, expanders, and sum-product, *Invent. Math.* 179 (2010) 559–644. <191, 192>

[378] J. Bourgain and M. Z. Garaev, On a variant of sum-product estimates and explicit exponential sum bounds in prime fields, *Math. Proc. Cambridge Philos. Soc.* 146 (2009) 1–21. <185, 187, 188, 192>

[379] J. Bourgain and A. Glibichuk, Exponential sum estimates over a subgroup in an arbitrary field, *J. Analyse Math.* 115 (2011) 51–70. <187, 190, 192>

[380] J. Bourgain, A. A. Glibichuk, and S. V. Konyagin, Estimates for the number of sums and products and for exponential sums in fields of prime order, *J. London Math. Soc., 2nd Ser.* 73 (2006) 380–398. <173, 176, 185, 186, 188, 191, 192>

[381] J. Bourgain, N. Katz, and T. Tao, A sum-product estimate in finite fields, and applications, *Geom. Funct. Anal.* 14 (2004) 27–57. <186, 192>

[382] H. Boylan and N.-P. Skoruppa, Explicit formulas for Hecke Gauss sums in quadratic number fields, *Abh. Math. Semin. Univ. Hambg.* 80 (2010) 213–226. <160, 161>

[383] S. Boztas, R. Hammons, and P. Kumar, 4-phase sequences with near-optimum correlation properties, *IEEE Trans. Inform. Theory* IT-38 (1992) 1101–1113. <322, 324>

[384] C. Bracken, E. Byrne, N. Markin, and G. McGuire, Determining the nonlinearity of a new family of APN functions, In *Applied Algebra, Algebraic Algorithms and Error-Correcting Codes*, volume 4851 of *Lecture Notes in Comput. Sci.*, 72–79, Springer, Berlin, 2007. <259, 261, 309, 310>

[385] C. Bracken, E. Byrne, N. Markin, and G. McGuire, On the Walsh spectrum of a new APN function, In *Cryptography and Coding*, volume 4887 of *Lecture Notes in Comput. Sci.*, 92–98, Springer, Berlin, 2007. <259, 261>

[386] C. Bracken, E. Byrne, N. Markin, and G. McGuire, New families of quadratic almost perfect nonlinear trinomials and multinomials, *Finite Fields Appl.* 14 (2008) 703–714. <257, 259, 261>

[387] C. Bracken, E. Byrne, N. Markin, and G. McGuire, Fourier spectra of binomial APN functions, *SIAM J. Discrete Math.* 23 (2009) 596–608. <259, 261, 309, 310>

[388] C. Bracken, E. Byrne, N. Markin, and G. McGuire, A few more quadratic APN functions, *Cryptogr. Commun.* 3 (2011) 43–53. <256, 259, 261>

[389] C. Bracken, E. Byrne, G. McGuire, and G. Nebe, On the equivalence of quadratic APN functions, *Des. Codes Cryptogr.* 61 (2011) 261–272. <259, 261>

[390] A. Braeken, C. Wolf, and B. Preneel, A study of the security of unbalanced oil and vinegar signature schemes, In *Topics in Cryptology—CT-RSA 2005*, volume 3376 of *Lecture Notes in Comput. Sci.*, 29–43, Springer, Berlin, 2005. <781, 783>

[391] N. Brandstätter, T. Lange, and A. Winterhof, On the non-linearity and sparsity of Boolean functions related to the discrete logarithm in finite fields of characteristic two, In *Coding and Cryptography*, volume 3969 of *Lecture Notes in Comput. Sci.*, 135–143, Springer, Berlin, 2006. <250, 252>

[392] N. Brandstätter and A. Winterhof, Some notes on the two-prime generator of order 2, *IEEE Trans. Inform. Theory* 51 (2005) 3654–3657. <334, 336>

[393] N. Brandstätter and A. Winterhof, Linear complexity profile of binary sequences with small correlation measure, *Period. Math. Hungar.* 52 (2006) 1–8. <182, 185, 335, 336>

[394] J. V. Brawley and L. Carlitz, Irreducibles and the composed product for polynomials over a finite field, *Discrete Math.* 65 (1987) 115–139. <67, 70>

[395] J. V. Brawley and L. Carlitz, A test for additive decomposability of irreducibles over a finite field, *Discrete Math.* 76 (1989) 61–65. <67, 70>

[396] J. V. Brawley, L. Carlitz, and J. Levine, Scalar polynomial functions on the $n \times n$ matrices over a finite field, *Linear Algebra and Appl.* 10 (1975) 199–217. <228, 229>

[397] J. V. Brawley and G. L. Mullen, Infinite Latin squares containing nested sets of mutually orthogonal finite Latin squares, *Publ. Math. Debrecen* 39 (1991) 135–141. <551, 556>

[398] J. V. Brawley and G. E. Schnibben, *Infinite Algebraic Extensions of Finite Fields*, volume 95 of *Contemp. Math.*, American Mathematical Society, Providence, RI, 1989. <31, 130, 138, 551, 556>

[399] R. Brent, P. Gaudry, E. Thomé, and P. Zimmerman, The software *gf2x*, `http://gf2x.gforge.inria.fr/`, as viewed in July, 2012. <47, 49>

[400] R. P. Brent, P. Gaudry, E. Thomé, and P. Zimmermann, Faster multiplication in $\mathrm{GF}(2)[x]$, In *Algorithmic Number Theory*, volume 5011 of *Lecture Notes in Comput. Sci.*, 153–166, Springer, Berlin, 2008. <356, 363>

[401] R. P. Brent and H. T. Kung, Fast algorithms for manipulating formal power series, *J. Assoc. Comput. Mach.* 25 (1978) 581–595. <350, 358, 363, 380, 381, 382>

[402] R. P. Brent and H. T. Kung, Systolic VLSI arrays for linear-time GCD computation, In *VLSI 1983*, 145–154, Elsevier Science Publishers B. V., 1983. <359, 363>

[403] R. P. Brent, S. Larvala, and P. Zimmermann, A fast algorithm for testing reducibility of trinomials mod 2 and some new primitive trinomials of degree 3021377, *Math. Comp.* 72 (2003) 1443–1452. <348, 350, 363>

[404] R. P. Brent, S. Larvala, and P. Zimmermann, A primitive trinomial of degree 6972593, *Math. Comp.* 74 (2005) 1001–1002. <350, 363>

[405] R. P. Brent and P. Zimmermann, Algorithms for finding almost irreducible and almost primitive trinomials, In *High Primes and Misdemeanours: Lectures in Honour of the 60th Birthday of Hugh Cowie Williams*, volume 41 of *Fields Inst. Commun.*, 91–102, Amer. Math. Soc., Providence, RI, 2004. <351, 353, 363>

[406] R. P. Brent and P. Zimmermann, Ten new primitive binary trinomials, *Math. Comp.* 78 (2009) 1197–1199. <96, 97, 350, 363>

[407] R. P. Brent and P. Zimmermann, An $O(M(n) \log n)$ algorithm for the Jacobi symbol,

In *Algorithmic Number Theory*, volume 6197 of *Lecture Notes in Comput. Sci.*, 83–95, Springer, Berlin, 2010. <361, 363>

[408] R. P. Brent and P. Zimmermann, The great trinomial hunt, *Notices Amer. Math. Soc.* 58 (2011) 233–239. <46, 49, 96, 97, 350, 363>

[409] R. P. Brent and P. Zimmermann, *Modern Computer Arithmetic*, volume 18 of *Cambridge Monographs on Applied and Computational Mathematics*, Cambridge University Press, Cambridge, 2011. <355, 363>

[410] E. Bresson, O. Chevassut, and D. Pointcheval, The group Diffie-Hellman problems, In *Selected Areas in Cryptography*, volume 2595 of *Lecture Notes in Comput. Sci.*, 325–338, Springer, Berlin, 2003. <394, 401>

[411] E. Breuillard, B. Green, and T. Tao, Suzuki groups as expanders, *ArXiv e-prints* (2010). <652, 658>

[412] J. Brewster Lewis, R. Ini Liu, A. H. Morales, G. Panova, S. V. Sam, and Y. Zhang, Matrices with restricted entries and q-analogues of permutations, *ArXiv e-prints* (2010). <501, 510>

[413] F. Brezing and A. Weng, Elliptic curves suitable for pairing based cryptography, *Des. Codes Cryptogr.* 37 (2005) 133–141. <794, 796>

[414] E. F. Brickell, D. M. Gordon, K. S. McCurley, and D. B. Wilson, Fast exponentiation with precomputation, In *Advances in Cryptology—EUROCRYPT 1992*, volume 658 of *Lecture Notes in Comput. Sci.*, 200–207, Springer, Berlin, 1993. <357, 358, 363>

[415] J. Brillhart, D. H. Lehmer, J. L. Selfridge, B. Tuckerman, and S. S. Wagstaff, Jr., *Factorizations of* $b^n \pm 1$, volume 22 of *Contemp. Math.*, American Mathematical Society, Providence, RI, second edition, 1988, $b = 2, 3, 5, 6, 7, 10, 11, 12$ up to high powers. <46, 49>

[416] D. Brink, H. Godinho, and P. H. A. Rodrigues, Simultaneous diagonal equations over p-adic fields, *Acta Arith.* 132 (2008) 393–399. <213>

[417] M. Brinkmann and G. Leander, On the classification of APN functions up to dimension five, *Des. Codes Cryptogr.* 49 (2008) 273–288. <257, 261>

[418] D. J. Britten and F. W. Lemire, A structure theorem for rings supporting a discrete Fourier transform, *SIAM J. Appl. Math.* 41 (1981) 222–226. <308, 310>

[419] A. E. Brouwer and J. H. van Lint, Strongly regular graphs and partial geometries, In *Enumeration and Design*, 85–122, Academic Press, Toronto, ON, 1984. <617, 619>

[420] D. R. L. Brown, Generic groups, collision resistance, and ECDSA, *Designs, Codes and Cryptography* 35 (2005) 119–152. <785, 796>

[421] M. R. Brown, Ovoids of PG(3,q), q even, with a conic section, *J. London Math. Soc.* 62 (2000) 569–582. <588, 589>

[422] M. R. Brown, G. L. Ebert, and D. Luyckx, On the geometry of regular hyperbolic fibrations, *European J. Combin.* 28 (2007) 1626–1636. <573, 574>

[423] K. A. Browning, J. F. Dillon, R. E. Kibler, and M. T. McQuistan, APN polynomials and related codes, *Journal of Combinatorics Information and System Sciences* 34 (2009) 135–159. <257, 261>

[424] K. A. Browning, J. F. Dillon, M. T. McQuistan, and A. J. Wolfe, An APN permutation in dimension six, In *Finite Fields: Theory and Applications*, volume 518 of *Contemp. Math.*, 33–42, Amer. Math. Soc., Providence, RI, 2010. <229, 256, 261>

[425] R. H. Bruck, Difference sets in a finite group, *Trans. Amer. Math. Soc.* 78 (1955)

464–481. <569, 574>

[426] R. H. Bruck, Quadratic extensions of cyclic planes, In *Proc. Sympos. Appl. Math., Vol. 10*, 15–44, American Mathematical Society, Providence, RI, 1960. <570, 574>

[427] R. H. Bruck, Construction problems of finite projective planes, In *Combinatorial Mathematics and its Applications*, 426–514, Univ. North Carolina Press, Chapel Hill, N.C., 1969. <567, 574>

[428] R. H. Bruck and R. C. Bose, The construction of translation planes from projective spaces, *J. Algebra* 1 (1964) 85–102. <566, 574>

[429] R. H. Bruck and H. J. Ryser, The nonexistence of certain finite projective planes, *Canadian J. Math.* 1 (1949) 88–93. <600, 607>

[430] A. Bruen, Blocking sets in finite projective planes, *SIAM J. Appl. Math.* 21 (1971) 380–392. <560, 563>

[431] A. A. Bruen and R. Silverman, On the nonexistence of certain MDS codes and projective planes, *Math. Z.* 183 (1983) 171–175. <587, 589>

[432] A. A. Bruen and R. Silverman, Arcs and blocking sets. II, *European J. Combin.* 8 (1987) 351–356. <560, 563>

[433] A. A. Bruen and J. A. Thas, Blocking sets, *Geometriae Dedicata* 6 (1977) 193–203. <560, 563>

[434] A. A. Bruen, J. A. Thas, and A. Blokhuis, On M.D.S. codes, arcs in $\mathrm{PG}(n, q)$ with q even, and a solution of three fundamental problems of B. Segre, *Invent. Math.* 92 (1988) 441–459. <585, 586, 589>

[435] L. Brünjes, *Forms of Fermat Equations and their Zeta Functions*, World Scientific Publishing Co. Pte. Ltd., Hackensack, NJ, 2004. <472, 479>

[436] H. Brunner, A. Curiger, and M. Hofstetter, On computing multiplicative inverses in $\mathrm{GF}(2^m)$, *IEEE Trans. Comput.* 42 (1993) 1010–1015. <360, 363>

[437] B. Buchberger, *Ein Algorithmus zum Auffinden der Basiselemente des Restklassenringes nach einem nulldimensionalen Polynomideal*, PhD thesis, Innsbruck, 1965. <781, 783>

[438] J. Buchmann, D. Cabarcas, J. Ding, and M. S. E. Mohamed, Flexible partial enlargement to accelerate Gröbner basis computation over $\mathbb{F}_2$, In *AFRICACRYPT*, volume 6055 of *Lecture Notes in Comput. Sci.*, 69–81, Springer, 2010. <782, 783>

[439] J. Buchmann and V. Shoup, Constructing nonresidues in finite fields and the extended Riemann hypothesis, *Math. Comp.* 65 (1996) 1311–1326. <349, 363>

[440] J. Buchmann and H. C. Williams, A key-exchange system based on imaginary quadratic fields, *J. Cryptology* 1 (1988) 107–118. <746, 750>

[441] J. A. Buchmann and C. S. Hollinger, On smooth ideals in number fields, *J. Number Theory* 59 (1996) 82–87. <399, 401>

[442] A. A. Buchstab, Asymptotic estimates of a general number theoretic function, *Rec. Math. (Mat. Sbornik), New Ser.* 44 (1937) 1239–1246. <370, 374>

[443] L. Budaghyan and C. Carlet, Classes of quadratic APN trinomials and hexanomials and related structures, *IEEE Trans. Inform. Theory* 54 (2008) 2354–2357. <257, 259, 261>

[444] L. Budaghyan and C. Carlet, CCZ-equivalence of single and multi output Boolean functions, In *Finite Fields: Theory and Applications*, volume 518 of *Contemp. Math.*, 43–54, Amer. Math. Soc., Providence, RI, 2010. <246, 252>

[445] L. Budaghyan, C. Carlet, T. Helleseth, and A. Kholosha, Generalized bent functions and their relation to Maiorana-Mcfarland class, In *Proceedings of the 2012 IEEE International Symposium on Information Theory*. IEEE, 2012. <267, 273>

[446] L. Budaghyan, C. Carlet, T. Helleseth, A. Kholosha, and S. Mesnager, Further results on Niho bent functions, *IEEE Trans. Inform. Theory* 58 (2012) 6979–6985. <269, 273>

[447] L. Budaghyan, C. Carlet, and G. Leander, Two classes of quadratic APN binomials inequivalent to power functions, *IEEE Trans. Inform. Theory* 54 (2008) 4218–4229. <257, 259, 261>

[448] L. Budaghyan, C. Carlet, and G. Leander, Constructing new APN functions from known ones, *Finite Fields Appl.* 15 (2009) 150–159. <259, 261>

[449] L. Budaghyan, C. Carlet, and A. Pott, New classes of almost bent and almost perfect nonlinear polynomials, *IEEE Trans. Inform. Theory* 52 (2006) 1141–1152. <257, 259, 261>

[450] L. Budaghyan and T. Helleseth, New perfect nonlinear multinomials over $\mathbf{F}_{p^{2k}}$ for any odd prime p, In *Sequences and Their Applications—SETA 2008*, volume 5203 of *Lecture Notes in Comput. Sci.*, 403–414, Springer, Berlin, 2008. <282>

[451] L. Budaghyan and T. Helleseth, New commutative semifields defined by new PN multinomials, *Cryptogr. Commun.* 3 (2011) 1–16. <264, 271, 273, 282>

[452] F. Buekenhout, Existence of unitals in finite translation planes of order q^2 with a kernel of order q, *Geom. Dedicata* 5 (1976) 189–194. <571, 574>

[453] F. Buekenhout, An introduction to incidence geometry, In *Handbook of Incidence Geometry*, 1–25, North-Holland, Amsterdam, 1995. <31>

[454] F. Buekenhout, A. Delandtsheer, J. Doyen, P. B. Kleidman, M. W. Liebeck, and J. Saxl, Linear spaces with flag-transitive automorphism groups, *Geom. Dedicata* 36 (1990) 89–94. <569, 574>

[455] J. Buhler and N. Koblitz, Lattice basis reduction, Jacobi sums and hyperelliptic cryptosystems, *Bull. Austral. Math. Soc.* 58 (1998) 147–154. <490, 491>

[456] B. Bukh and J. Tsimerman, Sum-product estimates for rational functions, *Proc. London Math. Soc.* <188, 192>

[457] J. R. Bunch and J. E. Hopcroft, Triangular factorization and inversion by fast matrix multiplication, *Math. Comp.* 28 (1974) 231–236. <526, 535>

[458] Bundesnetzagentur für Elektrizität, Gas, Telekommunikation, Post und Eisenbahnen, Bekanntmachung zur elektronischen Signatur nach dem Signaturgesetz und der Signaturverordnung (Übersicht über geeignete Algorithmen), *Bundesanzeiger* 85, June 7 (2011) 2034. <785, 796>

[459] M. Buratti, Improving two theorems of Bose on difference families, *J. Combin. Des.* 3 (1995) 15–24. <594, 599>

[460] M. Buratti, On simple radical difference families, *J. Combin. Des.* 3 (1995) 161–168. <594, 599>

[461] M. Buratti, Old and new designs via difference multisets and strong difference families, *J. Combin. Des.* 7 (1999) 406–425. <597, 599>

[462] M. Buratti, Existence of Z-cyclic triplewhist tournaments for a prime number of players, *J. Combin. Theory, Ser. A* 90 (2000) 315–325. <618, 619>

[463] K. Burde, Zur Herleitung von Reziprozitätsgesetzen unter Benutzung von endlichen Körpern, *J. Reine Angew. Math.* 293/294 (1977) 418–427. <173, 185>

[464] D. A. Burgess, On character sums and primitive roots, *Proc. London Math. Soc., 3rd Ser.* 12 (1962) 179–192. <183, 185>

[465] J. F. Burkhart, N. J. Calkin, S. Gao, J. C. Hyde-Volpe, K. James, H. Maharaj, S. Manber, J. Ruiz, and E. Smith, Finite field elements of high order arising from modular curves, *Des. Codes Cryptogr.* 51 (2009) 301–314. <99>

[466] M. V. D. Burmester, On the commutative non-associative division algebras of even order of L. E. Dickson, *Rend. Mat. e Appl. Ser. V* 21 (1962) 143–166. <276, 278>

[467] J. F. Buss, G. S. Frandsen, and J. O. Shallit, The computational complexity of some problems of linear algebra, In *STACS 97*, volume 1200 of *Lecture Notes in Comput. Sci.*, 451–462, Springer, Berlin, 1997. <780, 783>

[468] M. Butler, On the reducibility of polynomials over a finite field, *Quart. J. Math. Oxford* 5 (1954) 102–107. <375, 380>

[469] M. C. R. Butler, The irreducible factors of $f(x^m)$ over a finite field, *J. London Math. Soc., 2nd Ser.* 30 (1955) 480–482. <60, 62, 66>

[470] J. W. Byers, M. Luby, M. Mitzenmacher, and A. Rege, A digital fountain approach to reliable distribution of bulk data, In *Proceedings of the ACM SIGCOMM '98*, 56–67, ACM, New York, 1998. <728, 730, 734>

[471] K. A. Byrd and T. P. Vaughan, Counting and constructing orthogonal circulants, *J. Combin. Theory, Ser. A* 24 (1978) 34–49. <506, 510>

[472] D. Cabarcas and J. Ding, Linear algebra to compute syzygies and Gröbner bases, In *ISSAC 2011—Proceedings of the Thirty Sixth International Symposium on Symbolic and Algebraic Computation*, 67–74, ACM, New York, 2011. <782, 783>

[473] A. Cafure and G. Matera, Improved explicit estimates on the number of solutions of equations over a finite field, *Finite Fields Appl.* 12 (2006) 155–185. <194, 201>

[474] E. Çakçak and F. Özbudak, Subfields of the function field of the Deligne-Lusztig curve of Ree type, *Acta Arith.* 115 (2004) 133–180. <461, 463>

[475] A. R. Calderbank, P. J. Cameron, W. M. Kantor, and J. J. Seidel, Z_4-Kerdock codes, orthogonal spreads, and extremal Euclidean line-sets, *Proc. London Math. Soc., 3rd Ser.* 75 (1997) 436–480. <835, 841>

[476] A. R. Calderbank, E. M. Rains, P. W. Shor, and N. J. A. Sloane, Quantum error correction and orthogonal geometry, *Phys. Rev. Lett.* 78 (1997) 405–408. <835, 838, 841>

[477] A. R. Calderbank, E. M. Rains, P. W. Shor, and N. J. A. Sloane, Quantum error correction via codes over GF(4), *IEEE Trans. Inform. Theory* 44 (1998) 1369–1387. <838, 841>

[478] A. R. Calderbank and P. W. Shor, Good quantum error-correcting codes exist, *Phys. Rev. A* 54 (1996) 1098–1105. <838, 841>

[479] R. Calderbank and W. M. Kantor, The geometry of two-weight codes, *Bull. London Math. Soc.* 18 (1986) 97–122. <617, 619>

[480] C. Caliskan and G. E. Moorhouse, Subplanes of order 3 in Hughes planes, *Electron. J. Combin.* 18 (2011) Paper 2, 8. <570, 574>

[481] C. Caliskan and B. Petrak, Subplanes of order 3 in Figueroa planes, *Finite Fields Appl.* 20 (2013) 24–29. <570, 574>

[482] J. Calmet and R. Loos, An improvement of Rabin's probabilistic algorithm for generating irreducible polynomials over $GF(p)$, *Information Processing Letters* 11 (1980) 94–95. <376, 380>

[483] P. J. Cameron and J. J. Seidel, Quadratic forms over $GF(2)$, *Proc. Kon. Nederl. Akad. Wetensch. Ser. A* 76 (1973) 1–8. <205, 206>

[484] P. J. Cameron and J. H. van Lint, *Designs, Graphs, Codes and Their Links*, volume 22

of *London Mathematical Society Student Texts*, Cambridge University Press, Cambridge, 1991. <31>

[485] P. Camion, C. Carlet, P. Charpin, and N. Sendrier, On correlation-immune functions, In *Advances in Cryptology—CRYPTO '91*, volume 576 of *Lecture Notes in Comput. Sci.*, 86–100, Springer, Berlin, 1992. <249, 252>

[486] P. Candelas, X. de la Ossa, and F. Rodriguez-Villegas, Calabi-Yau manifolds over finite fields. II, In *Calabi-Yau Varieties and Mirror Symmetry*, volume 38 of *Fields Inst. Commun.*, 121–157, Amer. Math. Soc., Providence, RI, 2003. <472, 479>

[487] R. Canetti, J. Friedlander, S. Konyagin, M. Larsen, D. Lieman, and I. E. Shparlinski, On the statistical properties of Diffie-Hellman distributions, *Israel J. Math.* 120 (2000) 23–46. <183, 184, 185>

[488] R. Canetti, J. Friedlander, and I. E. Shparlinski, On certain exponential sums and the distribution of Diffie-Hellman triples, *J. London Math. Soc., 2nd Ser.* 59 (1999) 799–812. <183, 185, 189, 192>

[489] A. Canteaut, *Analyse et Conception de Chiffrements à Clef Secrète*, Mémoire d'habilitation à diriger des recherches, Université Paris 6, 2006. <254, 261>

[490] A. Canteaut, Open problems related to algebraic attacks on stream ciphers, In *Coding and Cryptography*, volume 3969 of *Lecture Notes in Comput. Sci.*, 120–134, Springer, Berlin, 2006. <248, 252>

[491] A. Canteaut, C. Carlet, P. Charpin, and C. Fontaine, On cryptographic properties of the cosets of $R(1,m)$, *IEEE Trans. Inform. Theory* 47 (2001) 1494–1513. <244, 252>

[492] A. Canteaut, P. Charpin, and H. Dobbertin, A new characterization of almost bent functions, In *Fast Software Encryption 99*, volume 1636 of *Lecture Notes Comput. Sci.*, 186–200. Springer-Verlag, 1999. <261>

[493] A. Canteaut, P. Charpin, and H. Dobbertin, Binary m-sequences with three-valued crosscorrelation: a proof of Welch's conjecture, *IEEE Trans. Inform. Theory* 46 (2000) 4–8. <255, 261>

[494] A. Canteaut, P. Charpin, and H. Dobbertin, Weight divisibility of cyclic codes, highly nonlinear functions on $\mathbf{F}_{2^m}$, and crosscorrelation of maximum-length sequences, *SIAM J. Discrete Math.* 13 (2000) 105–138. <213, 258, 260, 261>

[495] A. Canteaut, P. Charpin, and G. M. Kyureghyan, A new class of monomial bent functions, *Finite Fields Appl.* 14 (2008) 221–241. <254, 261, 268, 273>

[496] A. Canteaut (ed.), D. Augot, C. Cid, H. Englund, H. Gilbert, M. Hell, T. Johansson, M. Parker, T. Pornin, B. Preneel, C. Rechberger, and M. Robshaw, D.STVL.9 - ongoing research areas in symmetric cryptography, ECRYPT - European NoE in Cryptology, 2008. <254, 261>

[497] D. G. Cantor, Computing in the Jacobian of a hyperelliptic curve, *Math. Comp.* 48 (1987) 95–101. <451, 456, 797, 803>

[498] D. G. Cantor and E. Kaltofen, On fast multiplication of polynomials over arbitrary algebras, *Acta Infor.* 28 (1991) 693–701. <375, 380, 382>

[499] D. G. Cantor and H. Zassenhaus, A new algorithm for factoring polynomials over finite fields, *Math. Comp.* 36 (1981) 587–592. <380, 382, 770, 783>

[500] W. Cao, L. Hu, J. Ding, and Z. Yin, Kipnis-shamir attack on unbalanced oil-vinegar scheme, In F. Bao and J. Weng, editors, *ISPEC*, volume 6672 of *Lecture Notes in Comput. Sci.*, 168–180, Springer, 2011. <781, 783>

[501] W. Cao and Q. Sun, A reduction for counting the number of zeros of general diagonal

equations over finite fields, *Finite Fields Appl.* 12 (2006) 681–692. <210, 213>

[502] W. Cao and Q. Sun, Factorization formulae on counting zeros of diagonal equations over finite fields, *Proc. Amer. Math. Soc.* 135 (2007) 1283–1291. <210, 213>

[503] X. Cao, A note on the moments of Kloosterman sums, *Appl. Algebra Engrg. Comm. Comput.* 20 (2009) 447–457. <154, 161>

[504] X. Cao and L. Hu, New methods for generating permutation polynomials over finite fields, *Finite Fields Appl.* 17 (2011) 493–503. <216, 229>

[505] A. Capelli, Sulla rudittibilità delle equazioni algebriche I, *Rend. Acad. Sci. Fis. Mat. Napoli* 3 (1897) 243–252. <60, 61, 66>

[506] A. Capelli, Sulla rudittibilità delle equazioni algebriche II, *Rend. Acad. Sci. Fis. Mat. Napoli* 4 (1898) 243–252. <61, 66>

[507] A. Capelli, Sulla redutibilita delle funzione $x^n - A$ in un campo qualunque di razionalità, *Math. Ann.* 54 (1901) 602–603. <61, 66>

[508] M. Car, Le probléme de Waring pour l'anneau des polynômes sur un corps fini, *C. R. Acad. Sci. Paris, Sér. A-B* 273 (1971) A141–A144. <499, 500>

[509] M. Car, Factorisation dans $F_q[X]$, *C. R. Acad. Sci. Paris, Sér. I, Math.* 294 (1982) 147–150. <368, 374>

[510] M. Car, Théorèmes de densité dans $\mathbf{F}_q[X]$, *Acta Arith.* 48 (1987) 145–165. <369, 371, 374>

[511] M. Car, Waring's problem in function fields, *Proc. London Math. Soc., 3rd Ser.* 68 (1994) 1–30. <213>

[512] M. Car, Distribution des polynômes irréductibles dans $\mathbf{F}_q[T]$, *Acta Arith.* 88 (1999) 141–153. <74, 75, 76, 79>

[513] M. Car, New bounds on some parameters in the Waring problem for polynomials over a finite field, In *Finite Fields and Applications*, volume 461 of *Contemp. Math.*, 59–77, Amer. Math. Soc., Providence, 2008. <499, 500>

[514] M. Car and L. Gallardo, Sums of cubes of polynomials, *Acta Arith.* 112 (2004) 41–50. <499, 500>

[515] M. Car and L. Gallardo, Waring's problem for polynomial biquadrates over a finite field of odd characteristic, *Funct. Approx. Comment. Math.* 37 (2007) 39–50. <213, 499, 500>

[516] P. Carbonne and T. Henocq, Décomposition de la jacobienne sur les corps finis, *Bull. Polish Acad. Sci. Math.* 42 (1994) 207–215. <239, 240>

[517] J.-P. Cardinal, On a property of Cauchy-like matrices, *C. R. Acad. Sci. Paris, Sér. I, Math.* 328 (1999) 1089–1093. <533, 535>

[518] I. Cardinali, O. Polverino, and R. Trombetti, Semifield planes of order q^4 with kernel F_{q^2} and center F_q, *European J. Combin.* 27 (2006) 940–961. <276, 278>

[519] C. Carlet, A larger class of cryptographic Boolean functions via a study of the Maiorana-McFarland construction, In *Advances in Cryptology—CRYPTO 2002*, volume 2442 of *Lecture Notes in Comput. Sci.*, 549–564, Springer, Berlin, 2002. <249, 252>

[520] C. Carlet, On the coset weight divisibility and nonlinearity of resilient and correlation-immune functions, In *Sequences and Their Applications*, Discrete Math. Theor. Comput. Sci. (Lond.), 131–144, Springer, London, 2002. <247, 252>

[521] C. Carlet, On the secondary constructions of resilient and bent functions, In *Coding, Cryptography and Combinatorics*, volume 23 of *Progr. Comput. Sci. Appl. Logic*, 3–28, Birkhäuser, Basel, 2004. <249, 252>

[522] C. Carlet, Recursive lower bounds on the nonlinearity profile of Boolean functions and their applications, *IEEE Trans. Inform. Theory* 54 (2008) 1262–1272. <246, 252>

[523] C. Carlet, Boolean Functions for Cryptography and Error Correcting Codes (Chapter 8), In Y. Crama and P. L. Hammer, editors, *Boolean Models and Methods in Mathematics, Computer Science, and Engineering*, 257–397, Cambridge University Press, 2010. <180, 185, 243, 244, 245, 247, 248, 249, 250, 252, 262, 268, 273>

[524] C. Carlet, Vectorial Boolean functions for cryptography, In Y. Crama and P. L. Hammer, editors, *Boolean Models and Methods in Mathematics, Computer Science, and Engineering*, 398–469, Cambridge University Press, 2010. <253, 254, 261, 273>

[525] C. Carlet, P. Charpin, and V. Zinoviev, Codes, bent functions and permutations suitable for DES-like cryptosystems, *Des. Codes Cryptogr.* 15 (1998) 125–156. <255, 257, 258, 259, 260, 261>

[526] C. Carlet, L. E. Danielsen, M. G. Parker, and P. Solé, Self-dual bent functions, *Int. J. Inf. Coding Theory* 1 (2010) 384–399. <263, 273>

[527] C. Carlet and S. Dubuc, On generalized bent and q-ary perfect nonlinear functions, In *Finite Fields and Applications*, 81–94, Springer, Berlin, 2001. <271, 273>

[528] C. Carlet and K. Feng, An infinite class of balanced functions with optimal algebraic immunity, good immunity to fast algebraic attacks and good nonlinearity, In *Advances in Cryptology—ASIACRYPT 2008*, volume 5350 of *Lecture Notes in Comput. Sci.*, 425–440, Springer, Berlin, 2008. <250, 252>

[529] C. Carlet and P. Gaborit, Hyper-bent functions and cyclic codes, *J. Combin. Theory, Ser. A* 113 (2006) 466–482. <264, 273>

[530] C. Carlet and P. Guillot, A new representation of Boolean functions, In *Applied Algebra, Algebraic Algorithms and Error-Correcting Codes*, volume 1719 of *Lecture Notes in Comput. Sci.*, 94–103, Springer, Berlin, 1999. <243, 252>

[531] C. Carlet, T. Helleseth, A. Kholosha, and S. Mesnager, On the dual of bent functions with 2^r Niho exponents, In *Proceedings of the 2011 IEEE International Symposium on Information Theory*, 657–661, IEEE, 2011. <268, 270, 273>

[532] C. Carlet and S. Mesnager, On Dillon's class H of bent functions, Niho bent functions and o-polynomials, *J. Combin. Theory, Ser. A* 118 (2011) 2392–2410. <268, 269, 270, 273>

[533] C. Carlet and A. Pott, editors, *Sequences and Their Applications*, volume 6338 of *Lecture Notes in Comput. Sci.*, Springer, Berlin, 2010. <31>

[534] C. Carlet and P. Sarkar, Spectral domain analysis of correlation immune and resilient Boolean functions, *Finite Fields Appl.* 8 (2002) 120–130. <247, 252>

[535] C. Carlet and B. Sunar, editors, *Arithmetic of Finite Fields*, volume 4547 of *Lecture Notes in Comput. Sci.*, Springer, Berlin, 2007. <31>

[536] C. Carlet and J. L. Yucas, Piecewise constructions of bent and almost optimal Boolean functions, *Des. Codes Cryptogr.* 37 (2005) 449–464. <206>

[537] L. Carlitz, The arithmetic of polynomials in a Galois field, *Amer. J. Math.* 54 (1932) 39–50. <80, 85, 364, 366, 374>

[538] L. Carlitz, Some applications of a theorem of Chevalley, *Duke Math. J.* 18 (1951) 811–819. <213>

[539] L. Carlitz, Primitive roots in a finite field, *Trans. Amer. Math. Soc.* 73 (1952) 373–382. <137, 138>

[540] L. Carlitz, Some problems involving primitive roots in a finite field, *Proc. Nat. Acad. Sci. U.S.A.* 38 (1952) 314–318; errata, 618. <115, 116>

[541] L. Carlitz, A theorem of Dickson on irreducible polynomials, *Proc. Amer. Math. Soc.* 3 (1952) 693–700. <54, 55, 59, 74, 79>

[542] L. Carlitz, Invariantive theory of equations in a finite field, *Trans. Amer. Math. Soc.* 75 (1953) 405–427. <230, 232>

[543] L. Carlitz, Permutations in a finite field, *Proc. Amer. Math. Soc.* 4 (1953) 538. <238, 240>

[544] L. Carlitz, Representations by quadratic forms in a finite field, *Duke Math. J.* 21 (1954) 123–137. <507, 510>

[545] L. Carlitz, Representations by skew forms in a finite field, *Arch. Math. (Basel)* 5 (1954) 19–31. <507, 510>

[546] L. Carlitz, Solvability of certain equations in a finite field, *Quart. J. Math. Oxford, 2nd Ser.* 7 (1956) 3–4. <210, 213>

[547] L. Carlitz, Some theorems on irreducible reciprocal polynomials over a finite field, *J. Reine Angew. Math.* 227 (1967) 212–220. <57, 59, 286, 290>

[548] L. Carlitz, Kloosterman sums and finite field cxtensions, *Acta Arith.* 16 (1969/1970) 179–193. <156, 161>

[549] L. Carlitz, D. J. Lewis, W. H. Mills, and E. G. Straus, Polynomials over finite fields with minimal value sets, *Mathematika* 8 (1961) 121–130. <213, 233, 236>

[550] L. Carlitz and S. Uchiyama, Bounds for exponential sums, *Duke Math. J.* 24 (1957) 37–41. <321, 324>

[551] L. Carlitz and C. Wells, The number of solutions of a special system of equations in a finite field, *Acta Arith* 12 (1966/1967) 77–84. <218, 229>

[552] R. Carls and D. Lubicz, A p-adic quasi-quadratic time point counting algorithm, *Int. Math. Res. Not. IMRN* (2009) 698–735. <491>

[553] P. Cartier, Une nouvelle opération sur les formes différentielles, *C. R. Acad. Sci. Paris* 244 (1957) 426–428. <486, 488>

[554] J. Cassaigne, C. Mauduit, and A. Sárközy, On finite pseudorandom binary sequences VII: The measures of pseudorandomness, *Acta Arith.* 103 (2002) 97–118. <182, 185>

[555] R. Casse, *Projective Geometry: an Introduction*, Oxford University Press, Oxford, 2006. <564, 574>

[556] J. W. S. Cassels, Diophantine equations with special reference to elliptic curves, *J. London Math. Soc., 2nd Ser.* 41 (1966) 193–291. <422, 440>

[557] J. W. S. Cassels, *Lectures on Elliptic Curves*, volume 24 of *London Mathematical Society Student Texts*, Cambridge University Press, Cambridge, 1991. <31, 422, 440>

[558] G. Castagnoli, S. Bräuer, and M. Herrmann, Optimization of cyclic redundancy-check codes with 24 and 32 parity bits, *IEEE Transactions on Communications* 41 (1993) 883–892. <634, 638, 639, 642>

[559] G. Castagnoli, J. Ganz, and P. Graber, Optimum cycle redundancy-check codes with 16-bit redundancy, *IEEE Transactions on Communications* 38 (1990) 111–114. <635, 637, 642>

[560] F. N. Castro and C. J. Moreno, Mixed exponential sums over finite fields, *Proc. Amer. Math. Soc.* 128 (2000) 2529–2537. <168, 169>

[561] F. N. Castro, I. Rubio, P. Guan, and R. Figueroa, On systems of linear and diagonal

equation of degree $p^i + 1$ over finite fields of characteristic p, *Finite Fields Appl.* 14 (2008) 648–657. <213>

[562] F. N. Castro, I. Rubio, and J. M. Vega, Divisibility of exponential sums and solvability of certain equations over finite fields, *Q. J. Math.* 60 (2009) 169–181. <211, 213>

[563] F. N. Castro and I. M. Rubio, Solvability of systems of polynomial equations with some prescribed monomials, In *Finite Fields: Theory and Applications*, volume 518 of *Contemp. Math.*, 73–81, Amer. Math. Soc., Providence, RI, 2010. <210, 213>

[564] W. Castryck, J. Denef, and F. Vercauteren, Computing zeta functions of nondegenerate curves, *IMRP Int. Math. Res. Pap.* (2006) Art. ID 72017, 57. <491>

[565] K. Cattell, C. R. Miers, F. Ruskey, J. Sawada, and M. Serra, The number of irreducible polynomials over GF(2) with given trace and subtrace, *J. Combin. Math. Combin. Comput.* 47 (2003) 31–64. <56, 59, 79>

[566] A. Cauchy, Recherches sur les nombres, *Ecole Polytechnique* 9 (1813) 99–116. <211, 213>

[567] C. M. Caves, C. A. Fuchs, and R. Schack, Quantum probabilities as Bayesian probabilities, *Phys. Rev. A., 3rd Ser.* 65 (2002) 6 pages. <836, 841>

[568] C. M. Caves, C. A. Fuchs, and R. Schack, Unknown quantum states: the quantum de Finetti representation, *J. Math. Phys.* 43 (2002) 4537–4559. <836, 841>

[569] S. R. Cavior, A note on octic permutation polynomials, *Math. Comp.* 17 (1963) 450–452. <223, 229>

[570] C. Cazacu and D. Simovici, A new approach of some problems concerning polynomials over finite fields, *Information and Control* 22 (1973) 503–511. <68, 70>

[571] A. Çeşmelioğlu, G. McGuire, and W. Meidl, A construction of weakly and non-weakly regular bent functions, *J. Combin. Theory, Ser. A* 119 (2012) 420–429. <272, 273>

[572] A. Çeşmelioğlu and W. Meidl, Bent functions of maximal degree, *IEEE Trans. Inform. Theory* 58 (2012) 1186–1190. <272, 273>

[573] CCSDS, Low density parity check codes for use in near-Earth and deep space applications, 2007. <713, 719>

[574] M. Cenk, C. Negre, and M. A. Hasan, Improved three-way split formulas for binary polynomial and Toeplitz matrix vector products, Technical Report cacr2011-30, University of Waterloo, Waterloo, 2011. <814, 823>

[575] A. Çeşmelioğlu, W. Meidl, and A. Topuzoğlu, On the cycle structure of permutation polynomials, *Finite Fields Appl.* 14 (2008) 593–614. <229>

[576] F. Chabaud and R. Lercier, ZEN, version 3.0, 2001, available at `http://zenfact.sourceforge.net/`. <346, 363>

[577] F. Chabaud and S. Vaudenay, Links between differential and linear cryptanalysis, In *Advances in Cryptology—EUROCRYPT '94*, volume 950 of *Lecture Notes in Comput. Sci.*, 356–365, Springer, Berlin, 1995. <253, 255, 261>

[578] W. Chambers, Solution of Welch-Berlekamp key equation by Euclidean algorithm, *Electronics Letters* 29 (1993) 1031. <695, 703>

[579] A. Chambert-Loir, Compter (rapidement) le nombre de solutions d'équations dans les corps finis, *Astérisque* (2008) Exp. No. 968, vii, 39–90, Séminaire Bourbaki. Vol. 2006/2007. <491>

[580] D. B. Chandler and Q. Xiang, The invariant factors of some cyclic difference sets, *J. Combin. Theory, Ser. A* 101 (2003) 131–146. <152, 161>

[581] C.-Y. Chang, M. A. Papanikolas, D. S. Thakur, and J. Yu, Algebraic independence of arithmetic gamma values and Carlitz zeta values, *Adv. Math.* 223 (2010) 1137–1154. <546>

[582] C.-Y. Chang and J. Yu, Determination of algebraic relations among special zeta values in positive characteristic, *Adv. Math.* 216 (2007) 321–345. <542, 546>

[583] M.-C. Chang, On a question of Davenport and Lewis and new character sum bounds in finite fields, *Duke Math. J.* 145 (2008) 409–442. <183, 185, 192>

[584] M.-C. Chang, Burgess inequality in $\mathbb{F}_{p^2}$, *Geom. Funct. Anal.* 19 (2009) 1001–1016. <183, 185>

[585] M.-C. Chang, Expansions of quadratic maps in prime fields, *Proc. Amer. Math. Soc.* (2012), to appear. <344>

[586] M.-C. Chang, J. Cilleruelo, M. Z. Garaev, J. Hernández, I. E. Shparlinski, and A. Zumalacárregui, Points on curves in small boxes and applications, Preprint available from `http://arxiv.org/abs/1111.1543`, 2012. <344>

[587] M.-C. Chang and C. Z. Yao, An explicit bound on double exponential sums related to Diffie-Hellman distributions, *SIAM J. Discrete Math.* 22 (2008) 348–359. <183, 184, 185, 189, 192>

[588] Y. Chang, W.-S. Chou, and P. J.-S. Shiue, On the number of primitive polynomials over finite fields, *Finite Fields Appl.* 11 (2005) 156–163. <90>

[589] R. Chapman, Completely normal elements in iterated quadratic extensions of finite fields, *Finite Fields Appl.* 3 (1997) 1–10. <60, 65, 66, 138, 286, 290>

[590] P. Charpin, Open problems on cyclic codes, In *Handbook of Coding Theory, Vol. I, II*, 963–1063, North-Holland, Amsterdam, 1998. <258, 260, 261>

[591] P. Charpin and G. Gong, Hyperbent functions, Kloosterman sums, and Dickson polynomials, *IEEE Trans. Inform. Theory* 54 (2008) 4230–4238. <264, 268, 271, 273>

[592] P. Charpin, T. Helleseth, and V. Zinoviev, Divisibility properties of classical binary Kloosterman sums, *Discrete Math.* 309 (2009) 3975–3984. <154, 161>

[593] P. Charpin and G. Kyureghyan, When does $G(x) + \gamma\mathrm{Tr}(H(x))$ permute $\mathbb{F}_{p^n}$?, *Finite Fields Appl.* 15 (2009) 615–632. <225, 229>

[594] P. Charpin and G. M. Kyureghyan, Cubic monomial bent functions: a subclass of $\mathcal{M}$, *SIAM J. Discrete Math.* 22 (2008) 650–665. <268, 273>

[595] P. Charpin and G. M. Kyureghyan, On a class of permutation polynomials over $\mathbb{F}_{2^n}$, In *Sequences and Their Applications—SETA 2008*, volume 5203 of *Lecture Notes in Comput. Sci.*, 368–376, Springer, Berlin, 2008. <225, 230>

[596] P. Charpin, A. Pott, and A. Winterhof, Finite Fields and Applications: Character Sums and Polynomials, Radon Series in Computational and Applied Mathematics, 11, to appear. <31>

[597] S. Chatterjee and A. Menezes, On cryptographic protocols employing asymmetric pairings—the role of Ψ revisited, *Discrete Appl. Math.* 159 (2011) 1311–1322. <793, 796>

[598] Y. M. Chee, Y. Tan, and X. D. Zhang, Strongly regular graphs constructed from p-ary bent functions, *J. Algebraic Combin.* 34 (2011) 251–266. <265, 273>

[599] H. Chen, Fast algorithms for determining the linear complexity of sequences over $\mathrm{GF}(p^m)$ with period $2^t n$, *IEEE Trans. Inform. Theory* 51 (2005) 1854–1856. <329, 336>

[600] H. Chen, Reducing the computation of linear complexities of periodic sequences over $\mathrm{GF}(p^m)$, *IEEE Trans. Inform. Theory* 52 (2006) 5537–5539. <329, 336>

[601] H. Chen, S. Ling, and C. Xing, Asymptotically good quantum codes exceeding the Ashikhmin-Litsyn-Tsfasman bound, *IEEE Trans. Inform. Theory* 47 (2001) 2055–2058. <839, 841>

[602] J. Chen and T. Wang, On the Goldbach problem, *Acta Math. Sinica* 32 (1989) 702–718. <497, 500>

[603] J.-M. Chen and T.-T. Moh, On the Goubin-Courtois attack on TTM, Cryptology ePrint Archive, 2001, http://eprint.iacr.org/2001/072. <774, 775, 783>

[604] J.-M. Chen and B.-Y. Yang, A more secure and efficacious TTS signature scheme, In *Information Security and Cryptology—ICISC 2003*, volume 2971 of *Lecture Notes in Comput. Sci.*, 320–338, Springer, Berlin, 2004. <772, 775, 783>

[605] J.-M. Chen, B.-Y. Yang, and B.-Y. Peng, Tame transformation signatures with Topsy-Yurvy Hashes, In *IWAP'02*, 1–8, 2002, http://dsns.csie.nctu.edu.tw/iwap/proceedings/proceedings/sessionD/7.pdf. <775, 783>

[606] K. Chen and L. Zhu, Existence of APAV(q, k) with q a prime power $\equiv 3 \pmod 4$ and k odd > 1, *J. Combin. Des.* 7 (1999) 57–68. <612, 619>

[607] L. Chen, W. Eberly, E. Kaltofen, B. D. Saunders, W. J. Turner, and G. Villard, Efficient matrix preconditioners for black box linear algebra, *Linear Algebra Appl.* 343/344 (2002) 119–146. <531, 532, 534, 535>

[608] Y. Chen, The Steiner system $S(3, 6, 26)$, *J Geometry* 2 (1972) 7–28. <589>

[609] Y. Q. Chen, A construction of difference sets, *Des. Codes Cryptogr.* 13 (1998) 247–250. <605, 607>

[610] Q. Cheng, Primality proving via one round in ECPP and one iteration in AKS, In *Advances in cryptology—CRYPTO 2003*, volume 2729 of *Lecture Notes in Comput. Sci.*, 338–348, Springer, Berlin, 2003. <347, 363>

[611] Q. Cheng, Constructing finite field extensions with large order elements, *SIAM J. Discrete Math.* 21 (2007) 726–730 (electronic). <98, 99>

[612] Q. Cheng, S. Gao, and D. Wan, Constructing high order elements through subspace polynomials, In *Proceedings of the Twenty-Third Annual ACM-SIAM Symposium on Discrete Algorithms*, 1457–1463, 2012. <99>

[613] J. H. Cheon, J. Hong, and M. Kim, Accelerating Pollard's rho algorithm in finite fields, *Journal of Cryptology* 25 (2012) 185–242. <397, 401>

[614] R. C. C. Cheung, S. Duquesne, J. Fan, N. Guillermin, I. Verbauwhede, and G. X. Yao, FPGA implementation of pairings using residue number system and lazy reduction, In *Proceedings of the 2011 Workshop on Cryptographic Hardware and Embedded Systems*, 421–441, 2011. <822, 823>

[615] C. Chevalley, Démonstration d'une hypothèse de m. artin, *Abhand. Math. Sem. Hamburg* 11 (1936) 73–75. <207, 213>

[616] G. Chèze, *Des méthodes symboliques-numériques et exactes pour la factorisation absolue des polynômes en deux variables*, PhD thesis, Université de Nice-Sophia Antipolis (France), 2004. <387, 392>

[617] G. Chèze and G. Lecerf, Lifting and recombination techniques for absolute factorization, *J. Complexity* 23 (2007) 380–420. <384, 392>

[618] A. M. Childs, L. J. Schulman, and U. V. Vazirani, Quantum algorithms for hidden nonlinear structures, In *Forty Eighth Annual IEEE Symposium on Foundations of Computer Science*, 395–404, 2007. <840, 841>

[619] A. M. Childs and W. van Dam, Quantum algorithms for algebraic problems, *Rev. Modern Phys.* 82 (2010) 1–52. <835, 841>

[620] K. Chinen and T. Hiramatsu, Hyper-Kloosterman sums and their applications to the coding theory, *Appl. Algebra Engrg. Comm. Comput.* 12 (2001) 381–390. <154, 161>

[621] A. Chistov, Polynomial time construction of a finite field, In *Abstracts of Lectures at Seventh All-Union Conference in Mathematical Logic*, 196, Novosibirsk, USSR, 1984, In Russian. <379, 380>

[622] H. T. Choi and R. Evans, Congruences for sums of powers of Kloosterman sums, *Int. J. Number Theory* 3 (2007) 105–117. <157, 161>

[623] B. C. Chong and K. M. Chan, On the existence of normalized room squares, *Nanta Math.* 7 (1974) 8–17. <614, 619>

[624] W. S. Chou, *Permutation Polynomials on Finite Fields and their Combinatorial Applications*, PhD thesis, Penn. State Univ., University Park, PA, 1990. <228, 230>

[625] W. S. Chou, The period lengths of inversive pseudorandom vector generations, *Finite Fields Appl.* 1 (1995) 126–132. <229, 230>

[626] W.-S. Chou, The factorization of Dickson polynomials over finite fields, *Finite Fields Appl.* 3 (1997) 84–96. <284, 290>

[627] W.-S. Chou and S. D. Cohen, Primitive elements with zero traces, *Finite Fields Appl.* 7 (2001) 125–141. <92, 95>

[628] W. S. Chou, J. Gómez-Calderón, and G. L. Mullen, Value sets of Dickson polynomials over finite fields, *J. Number Theory* 30 (1988) 334–344. <233, 235, 236>

[629] W.-S. Chou, J. Gómez-Calderón, G. L. Mullen, D. Panario, and D. Thomson, Subfield value sets of polynomials over finite fields, *Funct. Approx. Comment. Math.* 48 (2013) 147–165. <236>

[630] W.-S. Chou and G. L. Mullen, A note on value sets of polynomials over finite fields, preprint, 2012. <234, 236>

[631] S. Chowla and H. J. Ryser, Combinatorial problems, *Canadian J. Math.* 2 (1950) 93–99. <600, 607>

[632] S. Chowla and H. Zassenhaus, Some conjectures concerning finite fields, *Norske Vid. Selsk. Forh. (Trondheim)* 41 (1968) 34–35. <228, 230>

[633] M. Christopoulou, T. Garefalakis, D. Panario, and D. Thomson, The trace of an optimal normal element and low complexity normal bases, *Des. Codes Cryptogr.* 49 (2008) 199–215. <119, 125, 128>

[634] M. Christopoulou, T. Garefalakis, D. Panario, and D. Thomson, Gauss periods as constructions of low complexity normal bases, *Des. Codes Cryptogr.* 62 (2012) 43–62. <119, 121, 128>

[635] W. Chu and C. J. Colbourn, Optimal frequency-hopping sequences via cyclotomy, *IEEE Trans. Inform. Theory* 51 (2005) 1139–1141. <846, 849>

[636] D. V. Chudnovsky and G. V. Chudnovsky, Sequences of numbers generated by addition in formal groups and new primality and factorization tests, *Adv. in Appl. Math.* 7 (1986) 385–434. <441, 446>

[637] F. R. K. Chung, Diameters and eigenvalues, *J. Amer. Math. Soc.* 2 (1989) 187–196. <645, 658>

[638] F. R. K. Chung, V. Faber, and T. A. Manteuffel, An upper bound on the diameter of a graph from eigenvalues associated with its Laplacian, *SIAM J. Discrete Math.* 7 (1994) 443–457. <645, 658>

[639] F. R. K. Chung, J. A. Salehi, and V. K. Wei, Optical orthogonal codes: design, analysis, and applications, *IEEE Trans. Inform. Theory* 35 (1989) 595–604.

<844, 845, 849>

[640] J. Chung and A. Hasan, More generalized Mersenne numbers, In *Selected Areas in Cryptography*, volume 3006 of *Lecture Notes in Comput. Sci.*, 335–347, Springer, Berlin, 2004. <353, 363>

[641] J.-H. Chung and K. Yang, Bounds on the linear complexity and the 1-error linear complexity over F_p of M-ary Sidel′nikov sequences, In *Sequences and Their Applications—SETA 2006*, volume 4086 of *Lecture Notes in Comput. Sci.*, 74–87, Springer, Berlin, 2006. <334, 336>

[642] S.-Y. Chung, G. D. Forney, T. J. Richardson, and R. L. Urbanke, On the design of low-density parity-check codes within 0.0045 db of the Shannon limit, *IEEE Commun. Lett.* 5 (2001) 58–60. <713, 719>

[643] J. Cilleruelo, Combinatorial problems in finite fields and Sidon sets, Combinatorica, 32 (2012), 497–511. <191, 192>

[644] J. Cilleruelo, M. Z. Garaev, A. Ostafe, and I. E. Shparlinski, On the concentration of points of polynomial maps and applications, *Math. Z.* 272 (2012) 825–837. <344>

[645] S. M. Cioabă, Closed walks and eigenvalues of abelian Cayley graphs, *C. R. Math. Acad. Sci. Paris* 342 (2006) 635–638. <651, 658>

[646] S. M. Cioabă, *Eigenvalues, Expanders and Gaps Between Primes*, ProQuest LLC, Ann Arbor, MI, 2006, Thesis (Ph.D.)–Queen's University (Canada). <647, 648, 657, 658>

[647] S. M. Cioabă, Eigenvalues of graphs and a simple proof of a theorem of Greenberg, *Linear Algebra Appl.* 416 (2006) 776–782. <648, 658>

[648] S. M. Cioabă, On the extreme eigenvalues of regular graphs, *J. Combin. Theory, Ser. B* 96 (2006) 367–373. <647, 658>

[649] S. M. Cioabă and M. R. Murty, Expander graphs and gaps between primes, *Forum Math.* 20 (2008) 745–756. <657, 658>

[650] J. A. Cipra, Waring's number in a finite field, *Integers* 9 (2009) A34, 435–440. <175, 185, 212, 213>

[651] J. A. Cipra, T. Cochrane, and C. Pinner, Heilbronn's conjecture on Waring's number (mod p), *J. Number Theory* 125 (2007) 289–297. <212, 213>

[652] M. Cipu, Dickson polynomials that are permutations, *Serdica Math. J.* 30 (2004) 177–194. <226, 230>

[653] M. Cipu and S. D. Cohen, Dickson polynomial permutations, In *Finite Fields and Applications*, volume 461 of *Contemp. Math.*, 79–90, Amer. Math. Soc., Providence, RI, 2008. <226, 230>

[654] T. Cochrane, J. Coffelt, and C. Pinner, A further refinement of Mordell's bound on exponential sums, *Acta Arith.* 116 (2005) 35–41. <190, 192>

[655] T. Cochrane, M.-C. Liu, and Z. Zheng, Upper bounds on n-dimensional Kloosterman sums, *J. Number Theory* 106 (2004) 259–274. <160, 161>

[656] T. Cochrane and C. Pinner, Sum-product estimates applied to Waring's problem mod p, *Integers* 8 (2008) A46, 18. <192, 212, 213>

[657] T. Cochrane and C. Pinner, Explicit bounds on monomial and binomial exponential sums, *Q. J. Math.* 62 (2011) 323–349. <340, 344>

[658] T. Cochrane, C. Pinner, and J. Rosenhouse, Bounds on exponential sums and the polynomial Waring problem mod p, *J. London Math. Soc., 2nd Ser.* 67 (2003) 319–336. <213>

[659] T. Cochrane and Z. Zheng, A survey on pure and mixed exponential sums modulo prime powers, In *Number Theory for the Millennium I*, 273–300, A. K. Peters, Natick, MA, 2002. <160, 161>

[660] H. Cohen, *A Course in Computational Algebraic Number Theory*, volume 138 of *Graduate Texts in Mathematics*, Springer-Verlag, Berlin, 1993. <346, 347, 359, 362, 363, 404, 796>

[661] H. Cohen, G. Frey, R. Avanzi, C. Doche, T. Lange, K. Nguyen, and F. Vercauteren, editors, *Handbook of Elliptic and Hyperelliptic Curve Cryptography*, Discrete Mathematics and Its Applications. Chapman & Hall/CRC, Boca Raton, FL, 2006. <31, 346, 354, 355, 356, 357, 358, 359, 360, 363, 393, 401, 449, 450, 451, 452, 453, 454, 455, 456, 788, 796, 797, 798, 803>

[662] H. Cohen and H. W. Lenstra, Jr., Primality testing and Jacobi sums, *Math. Comp.* 42 (1984) 297–330. <347, 363>

[663] S. Cohen and H. Niederreiter, editors, *Finite Fields and Applications*, volume 233 of *London Mathematical Society Lecture Note Series*, Cambridge, 1996. Cambridge University Press. <31>

[664] S. D. Cohen, The distribution of irreducible polynomials in several indeterminates over a finite field, *Proc. Edinburgh Math. Soc., Ser. II* 16 (1968/1969) 1–17. <81, 82, 85>

[665] S. D. Cohen, Further arithmetical functions in finite fields, *Proc. Edinburgh Math. Soc., Ser. II* 16 (1968/1969) 349–363. <368, 374>

[666] S. D. Cohen, On irreducible polynomials of certain types in finite fields, *Proc. Cambridge Philos. Soc.* 66 (1969) 335–344. <57, 58, 59, 60, 66>

[667] S. D. Cohen, The distribution of polynomials over finite fields, *Acta Arith.* 17 (1970) 255–271. <72, 73, 217, 230, 234, 236, 240>

[668] S. D. Cohen, Some arithmetical functions in finite fields, *Glasgow Math. J.* 11 (1970) 21–36. <80, 83, 85>

[669] S. D. Cohen, Uniform distribution of polynomials over finite fields, *J. London Math. Soc., 2nd Ser.* 6 (1972) 93–102. <77, 79>

[670] S. D. Cohen, The values of a polynomial over a finite field, *Glasgow Math. J.* 14 (1973) 205–208. <368, 374>

[671] S. D. Cohen, The irreducibility of compositions of linear polynomials over a finite field, *Compos. Math.* 47 (1982) 149–152. <60, 63, 66, 70>

[672] S. D. Cohen, The reducibility theorem for linearised polynomials over finite fields, *Bull. Austral. Math. Soc.* 40 (1989) 407–412. <66, 70>

[673] S. D. Cohen, Windmill polynomials over fields of characteristic two, *Monatsh. Math.* 107 (1989) 291–301. <69, 70, 89, 90>

[674] S. D. Cohen, Exceptional polynomials and the reducibility of substitution polynomials, *Enseign. Math., IIe Ser.* 36 (1990) 53–65. <237, 240>

[675] S. D. Cohen, Primitive elements and polynomials with arbitrary trace, *Discrete Math.* 83 (1990) 1–7. <92, 95>

[676] S. D. Cohen, Proof of a conjecture of Chowla and Zassenhaus on permutation polynomials, *Canad. Math. Bull.* 33 (1990) 230–234. <228, 230>

[677] S. D. Cohen, Permutation polynomials and primitive permutation groups, *Arch. Math. (Basel)* 57 (1991) 417–423. <218, 230>

[678] S. D. Cohen, The explicit construction of irreducible polynomials over finite fields, *Des. Codes Cryptogr.* 2 (1992) 169–174. <60, 64, 65, 66, 286, 290>

[679] S. D. Cohen, Dickson polynomials of the second kind that are permutations, *Canad. J. Math.* 46 (1994) 225–238. <226, 230>

[680] S. D. Cohen, Dickson permutations, In *Number-Theoretic and Algebraic Methods in Computer Science*, 29–51, World Sci. Publ., River Edge, NJ, 1995. <226, 230>

[681] S. D. Cohen, Permutation group theory and permutation polynomials, In *Algebras and Combinatorics*, 133–146, Springer, Singapore, 1999. <216, 230>

[682] S. D. Cohen, Gauss sums and a sieve for generators of Galois fields, *Publ. Math. Debrecen* 56 (2000) 293–312. <89, 90, 92, 94, 95>

[683] S. D. Cohen, Kloosterman sums and primitive elements in Galois fields, *Acta Arith.* 94 (2000) 173–201. <93, 95>

[684] S. D. Cohen, Primitive polynomials over small fields, In *Finite Fields and Applications*, volume 2948 of *Lecture Notes in Comput. Sci.*, 197–214, Springer, Berlin, 2004. <93, 95>

[685] S. D. Cohen, Explicit theorems on generator polynomials, *Finite Fields Appl.* 11 (2005) 337–357. <60, 64, 65, 66, 77, 79>

[686] S. D. Cohen, Primitive polynomials with a prescribed coefficient, *Finite Fields Appl.* 12 (2006) 425–491. <92, 95>

[687] S. D. Cohen, Primitive cubics and quartics with zero trace and prescribed norm, *Finite Fields Appl.* 18 (2012) 1156–1168. <92>

[688] S. D. Cohen and M. D. Fried, Lenstra's proof of the Carlitz-Wan conjecture on exceptional polynomials: an elementary version, *Finite Fields Appl.* 1 (1995) 372–375. <218, 230, 238, 239, 240>

[689] S. D. Cohen and M. J. Ganley, Commutative semifields, two-dimensional over their middle nuclei, *J. Algebra* 75 (1982) 373–385. <276, 277, 278, 282>

[690] S. D. Cohen and D. Hachenberger, Primitive normal bases with prescribed trace, *Appl. Algebra Engrg. Comm. Comput.* 9 (1999) 383–403. <89, 90, 116>

[691] S. D. Cohen and D. Hachenberger, Primitivity, freeness, norm and trace, *Discrete Math.* 214 (2000) 135–144. <92, 94, 95>

[692] S. D. Cohen and S. Huczynska, Primitive free quartics with specified norm and trace, *Acta Arith.* 109 (2003) 359–385. <89, 90, 92, 94, 95, 116>

[693] S. D. Cohen and S. Huczynska, The primitive normal basis theorem—without a computer, *J. London Math. Soc., 2nd Ser.* 67 (2003) 41–56. <93, 95>

[694] S. D. Cohen and S. Huczynska, The strong primitive normal basis theorem, *Acta Arith.* 143 (2010) 299–332. <95, 116>

[695] S. D. Cohen and C. King, The three fixed coefficient primitive polynomial theorem, *JP J. Algebra Number Theory Appl.* 4 (2004) 79–87. <93, 95>

[696] S. D. Cohen and R. W. Matthews, A class of exceptional polynomials, *Trans. Amer. Math. Soc.* 345 (1994) 897–909. <239, 240, 301, 302>

[697] S. D. Cohen and R. W. Matthews, Exceptional polynomials over finite fields, *Finite Fields Appl.* 1 (1995) 261–277. <239, 240>

[698] S. D. Cohen and D. Mills, Primitive polynomials with first and second coefficients prescribed, *Finite Fields Appl.* 9 (2003) 334–350. <93, 95>

[699] S. D. Cohen, G. L. Mullen, and P. J.-S. Shiue, The difference between permutation polynomials over finite fields, *Proc. Amer. Math. Soc.* 123 (1995) 2011–2015. <228, 230>

[700] S. D. Cohen and M. Prešern, Primitive finite field elements with prescribed trace, *Southeast Asian Bull. Math.* 29 (2005) 283–300. <92, 95>

[701] S. D. Cohen and M. Prešern, Primitive polynomials with prescribed second coefficient, *Glasgow Math. J.* 48 (2006) 281–307. <92, 95>

[702] S. D. Cohen and M. Prešern, The Hansen-Mullen primitive conjecture: completion of proof, In *Number Theory and Polynomials*, volume 352 of *London Math. Soc. Lecture Note Ser.*, 89–120, Cambridge Univ. Press, Cambridge, 2008. <92, 95>

[703] R. M. Cohn, *Difference Algebra*, Interscience Publishers John Wiley & Sons, New York-London-Sydney, 1965. <238, 240>

[704] C. J. Colbourn, Covering arrays from cyclotomy, *Des. Codes Cryptogr.* 55 (2010) 201–219. <610, 611, 619>

[705] C. J. Colbourn, Covering arrays and hash families, In *Information Security and Related Combinatorics*, NATO Peace and Information Security, 99–136, IOS Press, 2011. <610, 619>

[706] C. J. Colbourn and J. H. Dinitz, editors, *Handbook of Combinatorial Designs*, Discrete Mathematics and its Applications. Chapman & Hall/CRC, Boca Raton, FL, second edition, 2007. <31, 317, 324, 550, 551, 554, 555, 556, 563, 572, 573, 574, 596, 599, 600, 607, 616, 617, 619>

[707] C. J. Colbourn and A. C. H. Ling, Linear hash families and forbidden configurations, *Des. Codes Cryptogr.* 52 (2009) 25–55. <613, 619>

[708] C. J. Colbourn and A. Rosa, *Triple Systems*, Oxford Mathematical Monographs. The Clarendon Press, Oxford University Press, New York, 1999. <590, 594, 599>

[709] G. E. Collins, Computing multiplicative inverses in GF(p), *Math. Comp.* 23 (1969) 197–200. <359, 363>

[710] G. E. Collins, Lecture notes on arithmetic algorithms, 1980, University of Wisconsin. <359, 363>

[711] A. Commeine and I. Semaev, An algorithm to solve the discrete logarithm problem with the number field sieve, In *Public Key Cryptography—PKC 2006*, volume 3958 of *Lecture Notes in Comput. Sci.*, 174–190, Springer, Berlin, 2006. <399, 401>

[712] "Computer Algebra Group, University of Sydney," Magma Computational Algebra System, `http://magma.maths.usyd.edu.au/magma`, as viewed in July, 2012. <33, 48, 49>

[713] A. Conflitti, On elements of high order in finite fields, In *Cryptography and Computational Number Theory*, volume 20 of *Progr. Comput. Sci. Appl. Logic*, 11–14, Birkhäuser, Basel, 2001. <98, 99>

[714] K. Conrad, Jacobi sums and Stickelberger's congruence, *Enseign. Math,. IIe Ser.* 41 (1995) 141–153. <152, 161>

[715] K. Conrad, On Weil's proof of the bound for Kloosterman sums, *J. Number Theory* 97 (2002) 439–446. <154, 155, 161>

[716] S. Contini and I. E. Shparlinski, On Stern's attack against secret truncated linear congruential generators, volume 3574 of *Lecture Notes in Comput. Sci.*, 52–60, Springer, Berlin, 2005. <338, 344>

[717] D. Coppersmith, Fast evaluation of logarithms in fields of characteristic two, *IEEE Trans. Inform. Theory* 30 (1984) 587–594. <348, 363, 370, 374, 398, 401>

[718] D. Coppersmith, Solving linear equations over GF(2): block Lanczos algorithm, *Linear Algebra Appl.* 192 (1993) 33–60. <400, 401, 534, 535>

[719] D. Coppersmith, Solving homogeneous linear equations over GF(2) via block Wiedemann algorithm, *Math. Comp.* 62 (1994) 333–350. <400, 401, 534, 535>

[720] D. Coppersmith, Rectangular matrix multiplication revisited, *J. Complexity* 13 (1997)

42–49. <521, 535>

[721] D. Coppersmith, A. M. Odlyzko, and R. Schroeppel, Discrete logarithms in GF(p), *Algorithmica* 1 (1986) 1–15. <399, 401>

[722] D. Coppersmith, J. Stern, and S. Vaudenay, The security of the birational permutation signature schemes, *J. Cryptology* 10 (1997) 207–221. <769, 775, 779, 783>

[723] D. Coppersmith and S. Winograd, Matrix multiplication via arithmetic progressions, *J. Symbolic Comput.* 9 (1990) 251–280. <382, 521, 535>

[724] T. H. Cormen, C. E. Leiserson, R. L. Rivest, and C. Stein, *Introduction to Algorithms*, MIT Press, Cambridge, MA, second edition, 2001. <46, 49>

[725] C. J. Corrada-Bravo and I. Rubio, Deterministic interleavers for turbo codes with random-like performance and simple implementation, In *Proc. Third International Symposium on Turbo Codes*, Brest, France, 2003. <726, 727>

[726] R. S. Coulter, The classification of planar monomials over fields of prime square order, *Proc. Amer. Math. Soc.* 134 (2006) 3373–3378. <281, 282>

[727] R. S. Coulter and M. Henderson, The compositional inverse of a class of permutation polynomials over a finite field, *Bull. Austral. Math. Soc.* 65 (2002) 521–526. <229, 230>

[728] R. S. Coulter and M. Henderson, Commutative presemifields and semifields, *Adv. Math.* 217 (2008) 282–304. <277, 278, 281, 282>

[729] R. S. Coulter, M. Henderson, and P. Kosick, Planar polynomials for commutative semifields with specified nuclei, *Des. Codes Cryptogr.* 44 (2007) 275–286. <276, 278, 282>

[730] R. S. Coulter, M. Henderson, and R. Matthews, A note on constructing permutation polynomials, *Finite Fields Appl.* 15 (2009) 553–557. <224, 225, 230>

[731] R. S. Coulter and F. Lazebnik, On the classification of planar monomials over fields of square order, *Finite Fields Appl.* 18 (2012) 316–336. <281, 282>

[732] R. S. Coulter and R. W. Matthews, Planar functions and planes of Lenz-Barlotti class II, *Des. Codes Cryptogr.* 10 (1997) 167–184. <271, 273, 279, 280, 282>

[733] R. S. Coulter and R. W. Matthews, On the permutation behaviour of Dickson polynomials of the second kind, *Finite Fields Appl.* 8 (2002) 519–530. <226, 230>

[734] R. S. Coulter and R. W. Matthews, On the number of distinct values of a class of functions over a finite field, *Finite Fields Appl.* 17 (2011) 220–224. <236, 281, 282>

[735] N. T. Courtois, Fast algebraic attacks on stream ciphers with linear feedback, In *Advances in Cryptology—CRYPTO 2003*, volume 2729 of *Lecture Notes in Comput. Sci.*, 176–194, Springer, Berlin, 2003. <248, 252>

[736] N. T. Courtois, Algebraic attacks over GF(2^k), application to HFE Challenge 2 and Sflash-v2, In *Public Key Cryptography—PKC 2004*, volume 2947 of *Lecture Notes in Comput. Sci.*, 201–217, Springer, Berlin, 2004. <782, 783>

[737] N. T. Courtois, M. Daum, and P. Felke, On the security of HFE, HFEv- and Quartz, In *Public Key Cryptography—PKC 2003*, volume 2567 of *Lecture Notes in Comput. Sci.*, 337–350, Springer, Berlin, 2002. <770, 780, 783>

[738] N. T. Courtois, L. Goubin, W. Meier, and J.-D. Tacier, Solving underdefined systems of multivariate quadratic equations, In *PubKeyCrypt 2002*, volume 2274 of *Lecture Notes in Comput. Sci.*, 211–227, Springer, Berlin, 2002. <781, 783>

[739] N. T. Courtois, L. Goubin, and J. Patarin, *SFLASH: Primitive specification (second revised version)*, 2002, `https://www.cosic.esat.kuleuven.be/nessie`, Submissions, Sflash, 11 pages. <773, 783>

[740] N. T. Courtois, A. Klimov, J. Patarin, and A. Shamir, Efficient algorithms for solving overdefined systems of multivariate polynomial equations, In *Advances in Cryptology—EUROCRYPT 2000*, volume 1807 of *Lecture Notes in Comput. Sci.*, 392–407, Springer, Berlin, 2000. <780, 782, 783>

[741] N. T. Courtois and W. Meier, Algebraic attacks on stream ciphers with linear feedback, In *Advances in Cryptology—EUROCRYPT 2003*, volume 2656 of *Lecture Notes in Comput. Sci.*, 345–359, Springer, Berlin, 2003. <248, 252>

[742] N. T. Courtois and J. Patarin, About the XL algorithm over GF(2), In *Topics in Cryptology—CT-RSA 2003*, volume 2612 of *Lecture Notes in Comput. Sci.*, 141–157, Springer, Berlin, 2003. <782, 783>

[743] N. T. Courtois and J. Pieprzyk, Cryptanalysis of block ciphers with overdefined systems of equations, In *Advances in Cryptology—ASIACRYPT 2002*, volume 2501 of *Lecture Notes in Comput. Sci.*, 267–287, Springer, Berlin, 2002. <782, 783>

[744] J.-M. Couveignes and T. Henocq, Action of modular correspondences around CM points, In *Algorithmic number theory (Sydney, 2002)*, volume 2369 of *Lecture Notes in Comput. Sci.*, 234–243, Springer, Berlin, 2002. <787, 796>

[745] J.-M. Couveignes and J.-G. Kammerer, The geometry of flex tangents to a cubic curve and its parameterizations, *J. Symbolic Comput.* 47 (2012) 266–281. <796>

[746] J.-M. Couveignes and R. Lercier, Elliptic periods for finite fields, *Finite Fields Appl.* 15 (2009) 1–22. <121, 122, 123, 128>

[747] J.-M. Couveignes and R. Lercier, Fast construction of irreducible polynomials over finite fields, *Israel Journal of Mathematics* 194 (2013) 77–105. <378, 380>

[748] D. Cox, J. Little, and D. O'Shea, *Ideals, Varieties, and Algorithms*, Undergraduate Texts in Mathematics. Springer, New York, third edition, 2007. <783, 826, 834>

[749] D. A. Cox, *Galois Theory*, Pure and Applied Mathematics (New York). Wiley-Interscience, John Wiley & Sons, Hoboken, NJ, 2004. <3, 9, 11>

[750] R. Crandall, Method and apparatus for public key exchange in a cryptographic system, United States Patent 5, 159, 632, Date: Oct. 27th 1992. <353, 363>

[751] R. Crandall and C. Pomerance, *Prime Numbers*, Springer, New York, second edition, 2005, A computational perspective. <346, 354, 355, 362, 363, 496, 500>

[752] R. M. Crew, Etale p-covers in characteristic p, *Compositio Math.* 52 (1984) 31–45. <487, 488>

[753] H. S. Cronie and S. B. Korada, Lossless source coding with polar codes, In *Proc. (ISIT) Symp. IEEE Int Information Theory*, 904–908, 2010. <739>

[754] E. Croot, Sums of the form $1/x_1^k + \cdots + 1/x_n^k$ modulo a prime, *Integers* 4 (2004) A20, 6. <213>

[755] S. Crozier, J. Lodge, P. Guinand, and A. Hunt, Performance of turbo codes with relative prime and golden interleaving strategies, In *Proc. of the Sixth International Mobile Satellite Conference (IMSC '99)*, 268–275, Ottawa, Ontario, Canada, 1999. <726, 727>

[756] C. Culbert and G. L. Ebert, Circle geometry and three-dimensional subregular translation planes, *Innov. Incidence Geom.* 1 (2005) 3–18. <568, 574>

[757] T. W. Cusick, Value sets of some polynomials over finite fields GF(2^{2m}), *SIAM J. Comput.* 27 (1998) 120–131 (electronic). <235, 236>

[758] T. W. Cusick, Polynomials over base 2 finite fields with evenly distributed values, *Finite Fields Appl.* 11 (2005) 278–291. <235, 236>

[759] T. W. Cusick, C. Ding, and A. Renvall, *Stream Ciphers and Number Theory*, volume 66

of *North-Holland Mathematical Library*, Elsevier Science B.V., Amsterdam, revised edition, 2004. <31, 326, 328, 333, 334, 336>

[760] T. W. Cusick and P. Müller, Wan's bound for value sets of polynomials, In *Finite Fields and Applications*, volume 233 of *London Math. Soc. Lecture Note Ser.*, 69–72, Cambridge Univ. Press, Cambridge, 1996. <233, 235, 236>

[761] S. Czapor, K. Geddes, and G. Labahn, *Algorithms for Computer Algebra*, Kluwer Academic Publishers, 1992. <31, 382, 392>

[762] J. Daemen and V. Rijmen, *The Design of Rijndael: AES – the Advanced Encryption Standard*, Springer-Verlag, 2002. <31, 751, 760, 761, 763>

[763] Z. Dai, Multi-continued fraction algorithms and their applications to sequences, In *Sequences and Their Applications—SETA 2006*, volume 4086 of *Lecture Notes in Comput. Sci.*, 17–33, Springer, Berlin, 2006. <329, 336>

[764] Z. Dai and X. Feng, Classification and counting on multi-continued fractions and its application to multi-sequences, *Sci. China, Ser. F* 50 (2007) 351–358. <329, 336>

[765] Z. Dai, K. Wang, and D. Ye, Multi-continued fraction algorithm on multi-formal Laurent series, *Acta Arith.* 122 (2006) 1–16. <329, 336>

[766] Z. Dai and J. Yang, Multi-continued fraction algorithm and generalized B-M algorithm over $\mathbb{F}_q$, *Finite Fields Appl.* 12 (2006) 379–402. <329, 336>

[767] F. Daneshgaran and M. Mondin, Design of interleavers for turbo codes: iterative interleaver growth algorithms of polynomial complexity, *IEEE Trans. Inform. Theory* 45 (1999) 1845–1859. <726, 727>

[768] A. Danilevsky, The numerical solution of the secular equation, *Matem. Sbornik* 44 (1937) 169–171, In Russian. <375, 380>

[769] G. Darbi, Sulla riducibilità delle equazioni aldebriche, *Ann. Mat. Pura Appl.* 4 (1927) 185–208. <62, 66>

[770] H. Darmon and J.-F. Mestre, Courbes hyperelliptiques à multiplications réelles et une construction de Shih, *Canad. Math. Bull.* 43 (2000) 304–311. <239, 240>

[771] P. Das, The number of permutation polynomials of a given degree over a finite field, *Finite Fields Appl.* 8 (2002) 478–490. <219, 230>

[772] P. Das, The number of polynomials of a given degree over a finite field with value sets of a given cardinality, *Finite Fields Appl.* 9 (2003) 168–174. <235, 236>

[773] P. Das, Value sets of polynomials and the Cauchy Davenport theorem, *Finite Fields Appl.* 10 (2004) 113–122. <235, 236>

[774] P. Das and G. L. Mullen, Value sets of polynomials over finite fields, In *Finite Fields with Applications to Coding Theory, Cryptography and Related Areas*, 80–85, Springer, Berlin, 2002. <234, 236>

[775] H. Davenport, Bases for finite fields, *J. London Math. Soc., 2nd Ser.* 43 (1968) 21–39. <115, 116, 137, 138>

[776] H. Davenport and D. J. Lewis, Character sums and primitive roots in finite fields, *Rend. Circ. Mat. Palermo, 2nd Ser.* 12 (1963) 129–136. <181, 185>

[777] H. Davenport and D. J. Lewis, Notes on congruences. I, *Quart. J. Math. Oxford, 2nd Ser.* 14 (1963) 51–60. <238, 240, 293, 302>

[778] J. H. Davenport, Y. Siret, and É. Tournier, *Calcul Formel : Systèmes et Algorithmes de Manipulations Algébriques.*, Masson, Paris, France, 1987. <382, 392>

[779] J. H. Davenport and B. M. Trager, Factorization over finitely generated fields, In *SYMSAC'81: Proceedings of the Fourth ACM Symposium on Symbolic and Al-*

gebraic Computation, 200–205, ACM, New York, 1981. <387, 392>

[780] G. Davidoff, P. Sarnak, and A. Valette, *Elementary Number Theory, Group Theory, and Ramanujan Graphs*, volume 55 of *London Mathematical Society Student Texts*, Cambridge University Press, Cambridge, 2003. <652, 653, 658>

[781] J. A. Davis, Difference sets in abelian 2-groups, *J. Combin. Theory, Ser. A* 57 (1991) 262–286. <604, 607>

[782] J. A. Davis and J. Jedwab, A unifying construction for difference sets, *J. Combin. Theory, Ser. A* 80 (1997) 13–78. <605, 607>

[783] J. A. Davis and J. Jedwab, Peak-to-mean power control in OFDM, Golay complementary sequences, and Reed-Muller codes, *IEEE Trans. Inform. Theory* 45 (1999) 2397–2417. <843, 844, 849>

[784] E. Dawson and L. Simpson, Analysis and design issues for synchronous stream ciphers, In *Coding Theory and Cryptology*, volume 1 of *Lect. Notes Ser. Inst. Math. Sci. Natl. Univ. Singap.*, 49–90, World Sci. Publ., River Edge, NJ, 2002. <326, 336>

[785] J. De Beule and L. Storme, *Current Research Topics in Galois Geometry*, Nova Academic Publishers, Inc., New York, 2012. <31, 573, 574>

[786] P. de la Harpe and A. Musitelli, Expanding graphs, Ramanujan graphs, and 1-factor perturbations, *Bull. Belg. Math. Soc. Simon Stevin* 13 (2006) 673–680. <657, 658>

[787] M. J. de Resmini, D. Ghinelli, and D. Jungnickel, Arcs and ovals from abelian groups, *Des. Codes Cryptogr.* 26 (2002) 213–228. <278, 280, 282>

[788] M. J. de Resmini and N. Hamilton, Hyperovals and unitals in Figueroa planes, *European J. Combin.* 19 (1998) 215–220. <571, 574>

[789] B. de Smit and H. W. Lenstra, Standard models for finite fields, in preparation. <401, 402, 404>

[790] B. de Smit and H. W. Lenstra, Standard models for finite fields: the definition, 2008, `http://www.math.leidenuniv.nl/~desmit/papers/standard_models.pdf`, [Online]. <402, 404>

[791] T. Decker and P. Wocjan, Efficient quantum algorithm for hidden quadratic and cubic polynomial function graphs, preprint available, `http://arxiv.org/abs/quant-ph/0703195`, 2007. <840, 841>

[792] W. Decker, G.-M. Greuel, G. Pfister, and H. Schönemann, SINGULAR 3-1-5 — A computer algebra system for polynomial computations, 2012, http://www.singular.uni-kl.de. <48, 49>

[793] R. Dedekind, Abriss einer theorie der höheren congruenzen in bezug auf einen reellen primzahl-modulus, *J. Reine Angew. Math.* 54 (1857) 1–26. <9, 11>

[794] P. Deligne, Les constantes des équations fonctionnelles des fonctions L, In *Modular Functions of One Variable, II (Proc. Internat. Summer School, Univ. Antwerp*, 501–597. Lecture Notes in Math., Vol. 349, Springer, Berlin, 1973. <478, 479>

[795] P. Deligne, La conjecture de Weil. I, *Inst. Hautes Études Sci. Publ. Math.* (1974) 273–307. <163, 169, 195, 197, 201, 470, 472, 473, 476, 479>

[796] P. Deligne, *Applications de la Formule des Traces aux Sommes Trigonometriques*, in *Cohomologie Étale*, volume 569 of *Lecture Notes in Mathematics*, Springer-Verlag, Berlin, 1977. <169>

[797] P. Deligne, *Cohomologie Étale*, volume 569 of *Lecture Notes in Mathematics*, Springer-Verlag, Berlin, 1977. <31, 470, 471, 474, 475, 477, 479>

[798] P. Deligne, La conjecture de Weil. II, *Inst. Hautes Études Sci. Publ. Math.* (1980) 137–252. <472, 475, 476, 479, 488>

[799] P. Deligne and N. Katz, *Groupes de Monodromie en Géométrie Algébrique. II*, Lecture Notes in Mathematics, Vol. 340. Springer-Verlag, Berlin, 1973. <165, 167, 169, 486, 488>

[800] P. Delsarte, An algebraic approach to the association schemes of coding theory, *Philips Res. Rep. Suppl.* (1973) vi+97. <251, 252>

[801] P. Delsarte, Four fundamental parameters of a code and their combinatorial significance, *Information and Control* 23 (1973) 407–438. <631, 642, 663, 664, 665, 673, 691, 703>

[802] P. Delsarte, On subfield subcodes of modified Reed-Solomon codes, *IEEE Trans. Inform. Theory* IT-21 (1975) 575–576. <668, 669, 684, 703>

[803] P. Delsarte, Bilinear forms over a finite field, with applications to coding theory, *J. Combin. Theory, Ser. A* 25 (1978) 226–241. <846, 849>

[804] P. Delsarte and J.-M. Goethals, Alternating bilinear forms over $GF(q)$, *J. Combin. Theory, Ser. A* 19 (1975) 26–50. <702, 703>

[805] P. Delsarte, J.-M. Goethals, and F. J. MacWilliams, On generalized Reed-Muller codes and their relatives, *Information and Control* 16 (1970) 403–442. <686, 703>

[806] P. Delsarte and V. I. Levenshtein, Association schemes and coding theory, *IEEE Trans. Inform. Theory* 44 (1998) 2477–2504. <691, 703>

[807] P. Dembowski, *Finite Geometries*, Ergebnisse der Mathematik und ihrer Grenzgebiete, Band 44. Springer-Verlag, Berlin, 1968. <28, 31, 274, 278, 564, 567, 574, 588, 589, 591, 599>

[808] P. Dembowski and T. G. Ostrom, Planes of order n with collineation groups of order n^2, *Math. Z.* 103 (1968) 239–258. <254, 261, 279, 280, 281, 282>

[809] U. Dempwolff, Automorphisms and equivalence of bent functions and of difference sets in elementary abelian 2-groups, *Comm. Algebra* 34 (2006) 1077–1131. <273>

[810] U. Dempwolff, Semifield planes of order 81, *J. Geom.* 89 (2008) 1–16. <275, 278>

[811] U. Dempwolff and M. Röder, On finite projective planes defined by planar monomials, *Innov. Incidence Geom.* 4 (2006) 103–108. <279, 280, 282>

[812] J. Denef and F. Loeser, Weights of exponential sums, intersection cohomology, and Newton polyhedra, *Invent. Math.* 106 (1991) 275–294. <165, 169, 196, 201, 476, 479>

[813] J. Denef and F. Loeser, Character sums associated to finite Coxeter groups, *Trans. Amer. Math. Soc.* 350 (1998) 5047–5066. <146, 161>

[814] J. Denef and F. Loeser, Definable sets, motives and p-adic integrals, *J. Amer. Math. Soc.* 14 (2001) 429–469. <302>

[815] J. Denef and F. Vercauteren, An extension of Kedlaya's algorithm to Artin-Schreier curves in characteristic 2, In *Algorithmic Number Theory*, volume 2369 of *Lecture Notes in Comput. Sci.*, 308–323, Springer, Berlin, 2002. <454, 456>

[816] J. Denef and F. Vercauteren, Counting points on C_{ab} curves using Monsky-Washnitzer cohomology, *Finite Fields Appl.* 12 (2006) 78–102. <491>

[817] J. Denef and F. Vercauteren, An extension of Kedlaya's algorithm to hyperelliptic curves in characteristic 2, *J. Cryptology* 19 (2006) 1–25. <454, 456, 491>

[818] J. Dénes and A. D. Keedwell, *Latin Squares and Their Applications*, Academic Press, New York, 1974. <555, 556>

[819] J. Dénes and A. D. Keedwell, *Latin Squares*, volume 46 of *Annals of Discrete Mathematics*, North-Holland Publishing Co., Amsterdam, 1991. <31>

[820] R. H. F. Denniston, Some maximal arcs in finite projective planes, *J. Combin. Theory* 6 (1969) 317–319. <572, 574>

[821] R. H. F. Denniston, Uniqueness of the inverse plane of order 5, *Manuscripta Math.* 8 (1973) 11–19. <589>

[822] R. H. F. Denniston, Uniqueness of the inversive plane of order 7, *Manuscripta Math.* 8 (1973) 21–26. <589>

[823] J.-M. Deshouillers, G. Effinger, H. te Riele, and D. Zinoviev, A complete Vinogradov 3-primes theorem under the Riemann hypothesis, *Electron. Res. Announc. Amer. Math. Soc.* 3 (1997) 99–104. <497, 500>

[824] M. Deuring, Galoissche Theorie und Darstellungstheorie, *Math. Ann.* 107 (1933) 140–144. <110, 116>

[825] M. Deuring, Die Typen der Multiplikatorenringe elliptischer Funktionenkörper, *Abh. Math. Sem. Hansischen Univ.* 14 (1941) 197–272. <431, 440>

[826] M. Dewar, L. Moura, D. Panario, B. Stevens, and Q. Wang, Division of trinomials by pentanomials and orthogonal arrays, *Des. Codes Cryptogr.* 45 (2007) 1–17. <640, 641, 642>

[827] M. Dewar and D. Panario, Linear transformation shift registers, *IEEE Trans. Inform. Theory* 49 (2003) 2047–2052. <70>

[828] M. Dewar and D. Panario, Mutual irreducibility of certain polynomials, In *Finite Fields and Applications*, volume 2948 of *Lecture Notes in Comput. Sci.*, 59–68, Springer, Berlin, 2004. <70>

[829] J.-F. Dhem, *Design of an Efficient Public Key Cryptographic Library for RISC-Based Smart Cards*, PhD thesis, Faculté des sciences appliquées, Laboratoire de microélectronique, Université catholique de Louvain-la-Neuve, Belgique, 1998, available at `http://users.belgacom.net/dhem/these/index.html`. <355, 363>

[830] A. Díaz and E. Kaltofen, FoxBox a system for manipulating symbolic objects in black box representation, In *ISSAC '98: Proceedings of the 1998 International Symposium on Symbolic and Algebraic Computation*, 30–37, 1998. <392>

[831] J. W. Di Paola, On minimum blocking coalitions in small projective plane games, *SIAM J. Appl. Math.* 17 (1969) 378–392. <560, 563>

[832] P. Diaconis and R. Graham, Products of universal cycles, In E. D. Demaine, M. L. Demaine, and T. Rodgers, editors, *A Lifetime of Puzzles*, 35–55, A. K. Peters Ltd., Wellesley, MA, 2008. <632, 642>

[833] P. Diaconis and R. Graham, *Magical Mathematics: The Mathematical Ideas that Animate Great Magic Tricks*, Princeton University Press, 2011. <632, 642>

[834] P. Diaconis and M. Shahshahani, Generating a random permutation with random transpositions, *Z. Wahrsch. Verw. Gebiete* 57 (1981) 159–179. <651, 652, 658>

[835] J. Dick, Walsh spaces containing smooth functions and quasi-Monte Carlo rules of arbitrary high order, *SIAM J. Numer. Anal.* 46 (2008) 1519–1553. <623, 628, 630>

[836] J. Dick, P. Kritzer, G. Leobacher, and F. Pillichshammer, Constructions of general polynomial lattice rules based on the weighted star discrepancy, *Finite Fields Appl.* 13 (2007) 1045–1070. <624, 630>

[837] J. Dick and H. Niederreiter, On the exact t-value of Niederreiter and Sobol' sequences, *J. Complexity* 24 (2008) 572–581. <628, 630>

[838] J. Dick and H. Niederreiter, Duality for digital sequences, *J. Complexity* 25 (2009) 406–414. <627, 628, 630>

[839] J. Dick and F. Pillichshammer, *Digital Nets and Sequences: Discrepancy Theory and*

Quasi-Monte Carlo Integration, Cambridge University Press, Cambridge, 2010. <619, 623, 624, 625, 628, 630>

[840] L. E. Dickson, The analytic representation of substitutions on a power of a prime number of letters with a discussion of the linear group, *Ann. of Math.* 11 (1896/97) 65–120. <216, 230, 237, 240>

[841] L. E. Dickson, Higher irreducible congruences, *Bull. Amer. Math. Soc.* 3 (1897) 381–389. <62, 63, 66>

[842] L. E. Dickson, A class of groups in an arbitrary realm connected with the configuration of the 27 lines on a cubic surface, *Quart. J. Pure Appl. Math.* 33 (1901) 145–173. <11>

[843] L. E. Dickson, Theory of linear groups in an arbitrary field, *Trans. Amer. Math. Soc.* 2 (1901) 363–394. <11>

[844] L. E. Dickson, A new system of simple groups, *Math. Ann.* 60 (1905) 137–150. <11>

[845] L. E. Dickson, On finite algebras, In *Gesellschaften der Wissenschaften zu Göttingen*, 358–393, 1905. <276, 278>

[846] L. E. Dickson, Criteria for the irreducibility of functions in a finite field, *Bull. Amer. Math. Soc.* 13 (1906) 1–8. <67, 70>

[847] L. E. Dickson, On commutative linear algebras in which division is always uniquely possible, *Trans. Amer. Math. Soc.* 7 (1906) 514–522. <276, 278, 282>

[848] L. E. Dickson, A class of groups in an arbitrary realm connected with the configuration of the 27 lines on a cubic surface (second paper), *Quart. J. Pure Appl. Math.* 39 (1908) 205–209. <11>

[849] L. E. Dickson, A fundamental system of invariants of the general modular linear group with a solution of the form problem, *Trans. Amer. Math. Soc.* 12 (1911) 75–98. <62, 63, 66>

[850] L. E. Dickson, *Linear Groups: With an Exposition of the Galois Field Theory*, with an introduction by W. Magnus. Dover Publications Inc., New York, 1958. <3, 11, 31, 61, 62, 63, 66, 70, 71, 72, 73>

[851] L. E. Dickson, *History of the Theory of Numbers. Vol. I: Divisibility and Primality*, Chelsea Publishing Co., New York, 1966. <3, 11>

[852] C. Diem, The GHS attack in odd characteristic, *Journal of the Ramanujan Mathematical Society* 18 (2003) 1–32. <786, 796, 810, 811>

[853] C. Diem, The XL-algorithm and a conjecture from commutative algebra, In *Advances in Cryptology—ASIACRYPT 2004*, volume 3329 of *Lecture Notes in Comput. Sci.*, 323–337, Springer, Berlin, 2004. <783>

[854] C. Diem, An index calculus algorithm for plane curves of small degree, In *Algorithmic Number Theory*, volume 4076 of *Lecture Notes in Comput. Sci.*, 543–557, Springer, Berlin, 2006. <798, 803, 808, 811>

[855] C. Diem and J. Scholten, Cover Attacks – A report for the AREHCC project, 2003. <802, 803>

[856] C. Diem and E. Thomé, Index calculus in class groups of non-hyperelliptic curves of genus three, *J. Cryptology* 21 (2008) 593–611. <798, 803, 808, 809, 811>

[857] J. Dieudonné, *Sur les Groupes Classiques*, Actualités Sci. Ind., 1040 (Publ. Inst. Math. Univ. Strasbourg Nou. Sér. 1 (1945)). Hermann et Cie., Paris, 1948. <513, 514, 515, 516, 517, 518, 519, 520>

[858] J. A. Dieudonné, *La Géométrie des Groupes Classiques*, Springer-Verlag, Berlin, 1971, IIe ed. <512, 516, 520>

[859] W. Diffie and M. E. Hellman, New directions in cryptography, *IEEE Trans. Inform. Theory* IT-22 (1976) 644–654. <183, 185, 746>

[860] W. Diffie and M. E. Hellman, New directions in cryptography, In *Secure Communications and Asymmetric Cryptosystems*, volume 69 of *AAAS Sel. Sympos. Ser.*, 143–180, Westview, Boulder, CO, 1982. <765, 783>

[861] J. F. Dillon, *Elementary Hadamard Difference-Sets*, ProQuest LLC, Ann Arbor, MI, 1974, Thesis (Ph.D.)–University of Maryland, College Park. <265, 267, 268, 269, 271, 273>

[862] J. F. Dillon, Multiplicative difference sets via additive characters, *Des. Codes Cryptogr.* 17 (1999) 225–235. <239, 240, 260, 261, 602, 607>

[863] J. F. Dillon, Geometry, codes and difference sets: exceptional connections, In *Codes and Designs*, volume 10 of *Ohio State Univ. Math. Res. Inst. Publ.*, 73–85, de Gruyter, Berlin, 2002. <239, 240, 260, 261>

[864] J. F. Dillon and H. Dobbertin, New cyclic difference sets with Singer parameters, *Finite Fields Appl.* 10 (2004) 342–389. <239, 240, 260, 261, 268, 273, 319, 324, 602, 607, 755, 756, 763>

[865] J. F. Dillon and G. McGuire, Near bent functions on a hyperplane, *Finite Fields Appl.* 14 (2008) 715–720. <310>

[866] E. Dimitrova, L. D. García-Puente, F. Hinkelmann, A. S. Jarrah, R. Laubenbacher, B. Stigler, M. Stillman, and P. Vera-Licona, Polynome, Available at `http://polymath.vbi.vt.edu/polynome/`, 2010. <832, 834>

[867] E. Dimitrova, L. D. García-Puente, F. Hinkelmann, A. S. Jarrah, R. Laubenbacher, B. Stigler, M. Stillman, and P. Vera-Licona, Parameter estimation for Boolean models of biological networks, *Theoret. Comput. Sci.* 412 (2011) 2816–2826. <831, 834>

[868] E. S. Dimitrova, A. S. Jarrah, R. Laubenbacher, and B. Stigler, A Gröbner fan method for biochemical network modeling, In *ISSAC 2007*, 122–126, ACM, New York, 2007. <831, 834>

[869] C. Ding, T. Helleseth, and H. Niederreiter, editors, *Sequences and Their Applications*, Springer Series in Discrete Mathematics and Theoretical Computer Science, London, 1999. Springer-Verlag London Ltd. <31, 32>

[870] C. Ding, D. Pei, and A. Salomaa, *Chinese Remainder Theorem: Applications in Computing, Coding, Cryptography,* World Scientific Publishing Co. Inc., River Edge, NJ, 1996. <229, 230>

[871] C. Ding, Z. Wang, and Q. Xiang, Skew Hadamard difference sets from the Ree-Tits slice symplectic spreads in $\mathrm{PG}(3, 3^{2h+1})$, *J. Combin. Theory, Ser. A* 114 (2007) 867–887. <229, 230, 281, 282, 604, 607>

[872] C. Ding, Q. Xiang, J. Yuan, and P. Yuan, Explicit classes of permutation polynomials of $\mathbb{F}_{3^{3m}}$, *Sci. China, Ser. A* 52 (2009) 639–647. <226, 230>

[873] C. Ding, G. Xiao, and W. Shan, *The Stability Theory of Stream Ciphers*, volume 561 of *Lecture Notes in Comput. Sci.*, Springer-Verlag, Berlin, 1991. <325, 326, 329, 336>

[874] C. Ding and J. Yuan, A family of skew Hadamard difference sets, *J. Combin. Theory, Ser. A* 113 (2006) 1526–1535. <229, 230, 279, 282, 604, 607>

[875] C. S. Ding, H. Niederreiter, and C. P. Xing, Some new codes from algebraic curves, *IEEE Trans. Inform. Theory* 46 (2000) 2638–2642. <707, 712>

[876] J. Ding, A new variant of the Matsumoto-Imai cryptosystem through perturbation, In *Public Key Cryptography—PKC 2004*, volume 2947 of *Lecture Notes in Comput.*

Sci., 305–318, Springer, Berlin, 2004. <773, 783>

[877] J. Ding, Mutants and its impact on polynomial solving strategies and algorithms, Privately distributed research note, University of Cincinnati and Technical University of Darmstadt, 2006. <782, 783>

[878] J. Ding, Inverting square systems algebraically is exponential, Cryptology ePrint Archive, Report 2011/275, 2011, `http://eprint.iacr.org/`. <770, 776, 782, 783>

[879] J. Ding, J. Buchmann, M. S. E. Mohamed, W. S. A. M. Mohamed, and R.-P. Weinmann, Mutant XL, First International Conference on Symbolic Computation and Cryptography, preprint available `http://www.cdc.informatik.tu-darmstadt.de/reports/reports/MutantXL_Algorithm.pdf`, 2008. <782, 783>

[880] J. Ding, V. Dubois, B.-Y. Yang, O. C.-H. Chen, and C.-M. Cheng, Could SFLASH be repaired?, In *Automata, Languages and Programming. Part II*, volume 5126 of *Lecture Notes in Comput. Sci.*, 691–701, Springer, Berlin, 2008. <773, 774, 779, 783>

[881] J. Ding and J. E. Gower, Inoculating multivariate schemes against differential attacks, In *Public Key Cryptography—PKC 2006*, volume 3958 of *Lecture Notes in Comput. Sci.*, 290–301, Springer, Berlin, 2006. <775, 778, 783>

[882] J. Ding, J. E. Gower, and D. S. Schmidt, *Multivariate Public Key Cryptosystems*, volume 25 of *Advances in Information Security*, Springer, New York, 2006. <764, 783>

[883] J. Ding and T. Hodges, Cryptanalysis of an implementation scheme of the tamed transformation method cryptosystem, *J. Algebra Appl.* 3 (2004) 273–282. <775, 776, 783>

[884] J. Ding and T. Hodges, Inverting the HFE systems is quasi-polynomial for all fields, In *Advances in Cryptology—CRYPTO 2011*, volume 6841 of *Lecture Notes in Comput. Sci.*, 724–742, Springer, Berlin, 2011. <770, 776, 782, 783>

[885] J. Ding, L. Hu, X. Nie, J. Li, and J. Wagner, High order linearization equation (HOLE) attack on multivariate public key cryptosystems, In *Public key cryptography—PKC 2007*, volume 4450 of *Lecture Notes in Comput. Sci.*, 233–248, Springer, Berlin, 2007. <777, 783>

[886] J. Ding and D. Schmidt, A common defect of the TTM cryptosystem, In *Proceedings of the Technical Track of the ACNS'03*, 68–78. ICISA Press, 2003, http://eprint.iacr.org/2003/085. <775, 776, 783>

[887] J. Ding and D. Schmidt, The new implementation schemes of the TTM cryptosystem are not secure, In *Coding, Cryptography and Combinatorics*, volume 23 of *Progr. Comput. Sci. Appl. Logic*, 113–127, Birkhäuser, Basel, 2004. <775, 776, 777, 783>

[888] J. Ding and D. Schmidt, Cryptanalysis of HFEv and internal perturbation of HFE, In *Public Key Cryptography—PKC 2005*, volume 3386 of *Lecture Notes in Comput. Sci.*, 288–301, Springer, Berlin, 2005. <774, 783>

[889] J. Ding and D. Schmidt, Rainbow, a new multivariable polynomial signature scheme, In *Conference on Applied Cryptography and Network Security—ACNS 2005*, volume 3531 of *Lecture Notes Comput. Sci.*, 164–175, Springer, 2005. <772, 783>

[890] J. Ding, D. Schmidt, and F. Werner, Algebraic attack on HFE revisited, In *Information Security and Cryptology—ISC 2008*, volume 5352 of *Lecture Notes in Comput. Sci.*, Springer, 2007. <770, 776, 783>

[891] J. Ding, D. Schmidt, and Z. Yin, Cryptanalysis of the new TTS scheme in ches 2004,

Int. J. Inf. Sec. 5 (2006) 231–240. <772, 783>

[892] J. Ding, C. Wolf, and B.-Y. Yang, l-invertible cycles for $\mathcal{M}$ultivariate $\mathcal{Q}$uadratic ($\mathcal{MQ}$) public key cryptography, In *Public Key Cryptography—PKC 2007*, volume 4450 of *Lecture Notes in Comput. Sci.*, 266–281, Springer, Berlin, 2007. <775, 783>

[893] J. Ding and B.-Y. Yang, Multivariates polynomials for hashing, In *Information security and cryptology*, volume 4990 of *Lecture Notes in Comput. Sci.*, 358–371, Springer, Berlin, 2008. <783>

[894] J. Ding and B.-Y. Yang, Multivariate public key cryptography, In *Post-quantum cryptography*, 193–241, Springer, Berlin, 2009. <764, 783>

[895] J. Ding, B.-Y. Yang, C.-H. O. Chen, M.-S. Chen, and C.-M. Cheng, New differential-algebraic attacks and reparametrization of rainbow, In *Applied Cryptography and Network Security*, volume 5037 of *Lecture Notes in Comput. Sci.*, 242–257, Springer, 2008. <772, 773, 779, 781, 783>

[896] J. Ding and Z. Yin, Cryptanalysis of TTS and Tame–like signature schemes, In *Third International Workshop on Applied Public Key Infrastructures*, 2004. <775, 783>

[897] J. H. Dinitz, New lower bounds for the number of pairwise orthogonal symmetric Latin squares, In *Proceedings of the Tenth Southeastern Conference on Combinatorics, Graph Theory and Computing*, Congress. Numer., XXIII–XXIV, 393–398, Winnipeg, Man., 1979. <614, 619>

[898] J. H. Dinitz and D. R. Stinson, The construction and uses of frames, *Ars Combin.* 10 (1980) 31–53. <615, 616, 619>

[899] J. H. Dinitz and D. R. Stinson, Room squares and related designs, In *Contemporary Design Theory*, Wiley-Intersci. Ser. Discrete Math. Optim., 137–204, Wiley, New York, 1992. <615, 616, 619>

[900] J. H. Dinitz and G. S. Warrington, The spectra of certain classes of Room frames: the last cases, *Electron. J. Combin.* 17 (2010) Research Paper 74, 13 pages. <616, 619>

[901] J. D. Dixon and D. Panario, The degree of the splitting field of a random polynomial over a finite field, *Electron. J. Combin.* 11 (2004) Research Paper 70, 10 pp. <372, 373, 374>

[902] V. Dmytrenko, F. Lazebnik, and J. Williford, On monomial graphs of girth eight, *Finite Fields Appl.* 13 (2007) 828–842. <229, 230>

[903] H. Dobbertin, Almost perfect nonlinear power functions on GF(2^n): the Niho case, *Inform. and Comput.* 151 (1999) 57–72. <229, 230, 261>

[904] H. Dobbertin, Almost perfect nonlinear power functions on GF(2^n): the Welch case, *IEEE Trans. Inform. Theory* 45 (1999) 1271–1275. <229, 230, 261>

[905] H. Dobbertin, Kasami power functions, permutation polynomials and cyclic difference sets, In *Difference Sets, Sequences and Their Correlation Properties*, volume 542 of *NATO Adv. Sci. Inst. Ser. C Math. Phys. Sci.*, 133–158, Kluwer Acad. Publ., Dordrecht, 1999. <239, 240, 603, 607>

[906] H. Dobbertin, Almost perfect nonlinear power functions on GF(2^n): a new case for n divisible by 5, In *Finite Fields and Applications*, 113–121, Springer, Berlin, 2001. <227, 230, 261>

[907] H. Dobbertin, G. Leander, A. Canteaut, C. Carlet, P. Felke, and P. Gaborit, Construction of bent functions via Niho power functions, *J. Combin. Theory, Ser. A* 113 (2006) 779–798. <268, 273>

[908] H. Dobbertin, D. Mills, E. N. Müller, A. Pott, and W. Willems, APN functions in odd characteristic, *Discrete Math.* 267 (2003) 95–112. <256, 261>

[909] C. Doche, Redundant trinomials for finite fields of characteristic 2, In *Australasian Conference on Information Security and Privacy - ACISP 2005*, volume 3574 of *Lecture Notes in Comput. Sci.*, 122–133, Springer, Berlin, 2005. <351, 353, 363>

[910] G. Dolinar, A. E. Guterman, B. Kuzma, and M. Orel, On the Polya permanent problem over finite fields, *European J. Combin.* 32 (2011) 116–132. <509, 510>

[911] S. Dolinar and D. Divsalar, Weight distribution of turbo codes using random and nonrandom permutations, TDA Progress Report 42-122, JPL, 1995. <726, 727>

[912] G. Dorfer and H. Maharaj, Generalized AG codes and generalized duality, *Finite Fields Appl.* 9 (2003) 194–210. <708, 712>

[913] G. Dorfer, W. Meidl, and A. Winterhof, Counting functions and expected values for the lattice profile at n, *Finite Fields Appl.* 10 (2004) 636–652. <335, 336>

[914] G. Dorfer and A. Winterhof, Lattice structure and linear complexity profile of nonlinear pseudorandom number generators, *Appl. Algebra Engrg. Comm. Comput.* 13 (2003) 499–508. <335, 336>

[915] J. M. Dover, A family of non-Buekenhout unitals in the Hall planes, In *Mostly Finite Geometries*, volume 190 of *Lecture Notes in Pure and Appl. Math.*, 197–205, Dekker, New York, 1997. <571, 574>

[916] K. Drakakis, A review of Costas arrays, *J. Appl. Math.* (2006) Art. ID 26385, 32. <609, 619>

[917] K. Drakakis, R. Gow, and G. McGuire, APN permutations on $\mathbb{Z}_n$ and Costas arrays, *Discrete Appl. Math.* 157 (2009) 3320–3326. <229, 230>

[918] K. Drakakis, F. Iorio, and S. Rickard, The enumeration of Costas arrays of order 28 and its consequences, *Adv. Math. Commun.* 5 (2011) 69–86. <609, 619>

[919] V. G. Drinfeld, Elliptic modules, *Mat. Sb. (New Ser.)* 94(136) (1974) 594–627, 656. <535, 538, 539, 544, 546>

[920] V. G. Drinfeld, Elliptic modules. II, *Mat. Sb. (New Ser.)* 102(144) (1977) 182–194, 325. <535, 538, 539, 546>

[921] M. Drmota and D. Panario, A rigorous proof of the Waterloo algorithm for the discrete logarithm problem, *Des. Codes Cryptogr.* 26 (2002) 229–241. <370, 374>

[922] M. Drmota and R. F. Tichy, *Sequences, Discrepancies and Applications*, volume 1651 of *Lecture Notes in Mathematics*, Springer-Verlag, Berlin, 1997. <31, 32, 174, 185>

[923] G. Drolet, A new representation of elements of finite fields $GF(2^m)$ yielding small complexity arithmetic circuits, *IEEE Trans. Comput.* 47 (1998) 938–946. <822, 823>

[924] V. Dubois, P.-A. Fouque, A. Shamir, and J. Stern, Practical cryptanalysis of SFLASH, In *Advances in Cryptology—CRYPTO 2007*, volume 4622 of *Lecture Notes in Comput. Sci.*, 1–12, Springer, Berlin, 2007. <773, 774, 779, 783>

[925] V. Dubois, P.-A. Fouque, and J. Stern, Cryptanalysis of SFLASH with slightly modified parameters, In *Advances in Cryptology—EUROCRYPT 2007*, volume 4515 of *Lecture Notes in Comput. Sci.*, 264–275, Springer, Berlin, 2007. <773, 774, 778, 783>

[926] V. Dubois and N. Gama, The degree of regularity of HFE systems, In *Advances in Cryptology—ASIACRYPT 2010*, volume 6477 of *Lecture Notes in Comput. Sci.*, 557–576, Springer, Berlin, 2010. <782, 783>

[927] J. Dubrois and J.-G. Dumas, Efficient polynomial time algorithms computing industrial-strength primitive roots, *Inform. Process. Lett.* 97 (2006) 41–45. <349, 363>

[928] W. Duke, On multiple Salié sums, *Proc. Amer. Math. Soc.* 114 (1992) 623–625. <155, 161>

[929] J.-G. Dumas, Q-adic transform revisited, In *Proceedings of the 2008 International Symposium on Symbolic and Algebraic Computation*, 63–69, ACM, New York, 2008. <522, 523, 535>

[930] J.-G. Dumas, L. Fousse, and B. Salvy, Simultaneous modular reduction and Kronecker substitution for small finite fields, *J. Symbolic Comput.* 46 (2011) 823–840. <351, 363, 523, 535>

[931] J.-G. Dumas, T. Gautier, M. Giesbrecht, P. Giorgi, B. Hovinen, E. Kaltofen, B. D. Saunders, W. J. Turner, and G. Villard, LinBox: A generic library for exact linear algebra, In A. M. Cohen, X.-S. Gao, and N. Takayama, editors, *ICMS'2002, Proceedings of the 2002 International Congress of Mathematical Software*, 40–50, World Scientific Pub., 2002. <521, 535>

[932] J.-G. Dumas, T. Gautier, and C. Pernet, Finite field linear algebra subroutines, In *Proceedings of the 2002 International Symposium on Symbolic and Algebraic Computation*, 63–74, ACM, New York, 2002. <523, 535>

[933] J.-G. Dumas, P. Giorgi, and C. Pernet, Dense linear algebra over word-size prime fields: the FFLAS and FFPACK packages, *ACM Trans. Math. Software* 35 (2008) Art. 19, 35. <351, 363, 523, 524, 535>

[934] J.-G. Dumas, C. Pernet, and Z. Wan, Efficient computation of the characteristic polynomial, In *ISSAC'05*, 140–147, ACM, New York, 2005. <529, 535>

[935] J.-G. Dumas and G. Villard, Computing the rank of sparse matrices over finite fields, In V. G. Ganzha, E. W. Mayr, and E. V. Vorozhtsov, editors, *CASC 2002, Proceedings of the Fifth International Workshop on Computer Algebra in Scientific Computing*, 47–62. Technische Universität München, Germany, 2002. <530, 531, 532, 534, 535>

[936] I. I. Dumer, Concatenated codes and their multilevel generalizations, In *Handbook of Coding Theory, Vol. I, II*, 1911–1988, North-Holland, Amsterdam, 1998. <702, 703>

[937] A. Duran, B. Saunders, and Z. Wan, Hybrid algorithms for rank of sparse matrices, In R. Mathias and H. Woerdeman, editors, *SIAM Conference on Applied Linear Algebra*, 2003. <534, 535>

[938] I. Duursma and K.-H. Mak, On lower bounds for the Ihara constants A(2) and A(3), *arXiv:1102.4127v2[math.NT]* (2011). <463, 464, 469>

[939] P. F. Duvall and J. C. Mortick, Decimation of periodic sequences, *SIAM J. Appl. Math.* 21 (1971) 367–372. <313, 317>

[940] B. Dwork, On the rationality of the zeta function of an algebraic variety, *Amer. J. Math.* 82 (1960) 631–648. <163, 169, 479, 488>

[941] B. Dwork, p-adic cycles, *Inst. Hautes Études Sci. Publ. Math.* 37 (1969) 27–115. <479, 488>

[942] B. Dwork, Bessel functions as p-adic functions of the argument, *Duke Math. J.* 41 (1974) 711–738. <479, 488>

[943] B. M. Dwork, On the zeta function of a hypersurface III, *Ann. of Math., 2nd Ser.* 83 (1966) 457–519. <302>

[944] W. Eberly, Black box Frobenius decompositions over small fields, In *Proceedings of the 2000 International Symposium on Symbolic and Algebraic Computation*, 106–113, ACM, New York, 2000. <529, 530, 535>

[945] W. Eberly, Early termination over small fields, In *Proceedings of the 2003 Interna-*

tional Symposium on Symbolic and Algebraic Computation, ISSAC '03, 80–87, ACM, New York, NY, USA, 2003. <530, 535>

[946] W. Eberly, M. Giesbrecht, P. Giorgi, A. Storjohann, and G. Villard, Faster inversion and other black box matrix computations using efficient block projections, In *ISSAC 2007*, 143–150, ACM, New York, 2007. <535>

[947] W. Eberly and E. Kaltofen, On randomized Lanczos algorithms, In *Proceedings of the 1997 International Symposium on Symbolic and Algebraic Computation*, 176–183, ACM, New York, 1997. <531, 535>

[948] G. L. Ebert, Partitioning projective geometries into caps, *Canad. J. Math.* 37 (1985) 1163–1175. <568, 574>

[949] G. L. Ebert, Nests, covers, and translation planes, *Ars Combin.* 25 (1988) 213–233. <567, 574>

[950] G. L. Ebert, Spreads admitting regular elliptic covers, *European J. Combin.* 10 (1989) 319–330. <567, 574>

[951] G. L. Ebert, Partitioning problems and flag-transitive planes, *Rend. Circ. Mat. Palermo Ser. II Suppl.* (1998) 27–44, Combinatorics '98 (Mondello). <569, 574>

[952] G. L. Ebert, G. Marino, O. Polverino, and R. Trombetti, Infinite families of new semifields, *Combinatorica* 29 (2009) 637–663. <276, 278>

[953] Y. Edel, G. Kyureghyan, and A. Pott, A new APN function which is not equivalent to a power mapping, *IEEE Trans. Inform. Theory* 52 (2006) 744–747. <257, 261>

[954] Y. Edel and A. Pott, A new almost perfect nonlinear function which is not quadratic, *Adv. Math. Commun.* 3 (2009) 59–81. <256, 261>

[955] G. A. Edgar and C. Miller, Borel subrings of the reals, *Proc. Amer. Math. Soc.* 131 (2003) 1121–1129. <186, 192>

[956] H. M. Edwards, A normal form for elliptic curves, *Bull. Amer. Math. Soc. (New Ser.)* 44 (2007) 393–422 (electronic). <441, 446>

[957] G. Effinger, A Goldbach theorem for polynomials of low degree over odd finite fields, *Acta Arith.* 42 (1983) 329–365. <498, 500>

[958] G. Effinger, A Goldbach 3-primes theorem for polynomials of low degree over finite fields of characteristic 2, *J. Number Theory* 29 (1988) 345–363. <498, 500>

[959] G. Effinger, Toward a complete twin primes theorem for polynomials over finite fields, In *Finite Fields and Applications*, volume 461 of *Contemp. Math.*, 103–110, Amer. Math. Soc., Providence, 2008. <496, 500>

[960] G. Effinger and D. Hayes, A complete solution to the polynomial 3-primes problem, *Bull. Amer. Math. Soc.* 24 (1991) 363–369. <498, 500>

[961] G. Effinger and D. R. Hayes, *Additive Number Theory of Polynomials over a Finite Field*, Oxford Mathematical Monographs. Oxford University Press, New York, 1991. <31, 32, 497, 498, 499, 500>

[962] G. Effinger, K. Hicks, and G. L. Mullen, Twin irreducible polynomials over finite fields, In *Finite Fields with Applications to Coding Theory, Cryptography and Related Areas*, 94–111, Springer, Berlin, 2002. <496, 500>

[963] G. Effinger, K. Hicks, and G. L. Mullen, Integers and polynomials: comparing the close cousins $\mathbb{Z}$ and $\mathbb{F}_q[x]$, *Math. Intelligencer* 27 (2005) 26–34. <494, 500>

[964] M. Einsiedler and T. Ward, *Ergodic Theory with a View Towards Number Theory*, volume 259 of *Graduate Texts in Mathematics*, Springer-Verlag London Ltd., London, 2011. <337, 338, 344>

[965] T. ElGamal, A public key cryptosystem and a signature scheme based on discrete logarithms, *IEEE Trans. Inform. Theory* 31 (1985) 469–472. <746, 750>

[966] M. Elia and M. Leone, On the inherent space complexity of fast parallel multipliers for $GF(2^m)$, *IEEE Trans. Comput.* 51 (2002) 346–351. <820, 821, 823>

[967] S. Eliahou, M. Kervaire, and B. Saffari, On Golay polynomial pairs, *Adv. in Appl. Math.* 12 (1991) 235–292. <843, 849>

[968] N. D. Elkies, The existence of infinitely many supersingular primes for every elliptic curve over $\mathbb{Q}$, *Invent. Math.* 89 (1987) 561–567. <438, 440>

[969] N. D. Elkies, Distribution of supersingular primes, *Astérisque* 198-200 (1991) 127–132. <438, 440>

[970] N. D. Elkies, Elliptic and modular curves over finite fields and related computational issues, In *Computational Perspectives on Number Theory*, volume 7 of *AMS/IP Stud. Adv. Math.*, 21–76, Amer. Math. Soc., Providence, RI, 1998. <787, 796>

[971] N. D. Elkies, Explicit modular towers, *Proceedings of the Thirty Fifth Allerton Conference on Communication, Control and Computing* (1998) 23–32. <464, 469>

[972] N. D. Elkies, Explicit towers of Drinfeld modular curves, In *European Congress of Mathematics, Vol. II*, volume 202 of *Progr. Math.*, 189–198, Birkhäuser, Basel, 2001. <464, 469>

[973] N. D. Elkies, E. W. Howe, A. Kresch, B. Poonen, J. L. Wetherell, and M. E. Zieve, Curves of every genus with many points. II. Asymptotically good families, *Duke Math. J.* 122 (2004) 399–422. <463>

[974] W. Ellison, Waring's problem, *Amer. Math. Monthly* 78 (1971) 10–36. <499, 500>

[975] B. Elspas, The theory of autonomous linear sequential networks, In *Linear Sequential Switching Circuits*, 21–61, Holden-Day, San Francisco, Calif., 1965. <834>

[976] H. Enderling, M. Chaplain, and P. Hahnfeldt, Quantitative modeling of tumor dynamics and radiotherapy, *Acta Biotheoretica* 58 (2010) 341–353. <831, 834>

[977] A. Enge, Computing discrete logarithms in high-genus hyperelliptic Jacobians in provably subexponential time, *Math. Comp.* 71 (2002) 729–742. <455, 456>

[978] A. Enge, The complexity of class polynomial computation via floating point approximations, *Math. Comp.* 78 (2009) 1089–1107. <787, 796>

[979] A. Enge, Computing modular polynomials in quasi-linear time, *Math. Comp.* 78 (2009) 1809–1824. <788, 796>

[980] A. Enge and P. Gaudry, A general framework for subexponential discrete logarithm algorithms, *Acta Arith.* 102 (2002) 83–103. <455, 456, 808, 811>

[981] B.-G. Englert and Y. Aharonov, The mean king's problem: prime degrees of freedom, *Phys. Lett. A* 284 (2001) 1–5. <835, 841>

[982] S. S. Erdem, T. Yanık, and Ç. K. Koç, Polynomial basis multiplication over $\mathrm{GF}(2^m)$, *Acta Appl. Math.* 93 (2006) 33–55. <109>

[983] P. Erdős and P. Turán, On some problems of a statistical group-theory I, *Z. Wahrscheinlichkeitstheorie und Verw. Gebiete* 4 (1965) 175–186 (1965). <373, 374>

[984] P. Erdős and P. Turán, On some problems of a statistical group-theory II, *Acta math. Acad. Sci. Hungar.* 18 (1967) 151–163. <373, 374>

[985] P. Erdős and P. Turán, On some problems of a statistical group-theory III, *Acta Math. Acad. Sci. Hungar.* 18 (1967) 309–320. <373, 374>

[986] P. Erdős and P. Turán, On some problems of a statistical group-theory IV, *Acta*

Math. Acad. Sci. Hungar 19 (1968) 413–435. <373, 374>

[987] S. Erickson, M. J. Jacobson, Jr., N. Shang, S. Shen, and A. Stein, Explicit formulas for real hyperelliptic curves of genus 2 in affine representation, In *Arithmetic of Finite Fields*, volume 4547 of *Lecture Notes in Comput. Sci.*, 202–218, Springer, Berlin, 2007. <452, 456>

[988] S. Erickson, M. J. Jacobson, Jr., and A. Stein, Explicit formulas for real hyperelliptic curves of genus 2 in affine representation, *Adv. Math. Commun.* 5 (2011) 623–666. <451, 452, 456>

[989] eSTREAM Project, `http://www.ecrypt.eu.org/stream/`. <752, 754, 755, 763>

[990] eSTREAM Project, The eStream report: end of phase II, `http://www.ecrypt.eu.org/stream/PhaseIIreport.pdf`. <758, 763>

[991] O. Etesami and M. A. Shokrollahi, Raptor codes on binary memoryless symmetric channels, *IEEE Trans. Inform. Theory* 52 (2006) 2033–2051. <734>

[992] J. Ethier and G. L. Mullen, Strong forms of orthogonality for sets of frequency hypercubes, Preprint, 2011. <554, 556>

[993] J. Ethier and G. L. Mullen, Strong forms of orthogonality for sets of hypercubes, *Discrete Math.* 312 (2012) 2050–2061. <554, 556>

[994] ETSI, Digital video broadcasting (dvb); second generation framing structure, channel coding and modulation systems for broadcasting, interactive services, news gathering and other broadband satellite applications (dvb-s2): Etsi en 302 307, 2009. <713, 719>

[995] ETSI, Digital video broadcasting (dvb); implementation guidelines for a second generation digital terrestrial television broadcasting system (dvb-t2): Etsi ts 102 831, 2010. <713, 719>

[996] T. Etzion, A. Trachtenberg, and A. Vardy, Which codes have cycle-free Tanner graphs?, *IEEE Trans. Inform. Theory* 45 (1999) 2173–2181. <718, 719>

[997] A. B. Evans, Maximal sets of mutually orthogonal Latin squares. II, *European J. Combin.* 13 (1992) 345–350. <228, 230>

[998] A. B. Evans, *Orthomorphism Graphs of Groups*, volume 1535 of *Lecture Notes in Mathematics*, Springer-Verlag, Berlin, 1992. <171, 185, 228, 230>

[999] R. Evans, Residuacity of primes, *Rocky Mountain J. Math.* 19 (1989) 1069–1081. <141, 161>

[1000] R. Evans, Character sums as orthogonal eigenfunctions of adjacency operators for Cayley graphs, In *Finite Fields: Theory, Applications, and Algorithms*, volume 168 of *Contemp. Math.*, 33–50, Amer. Math. Soc., Providence, RI, 1994. <155, 161>

[1001] R. Evans, Congruences for Jacobi sums, *J. Number Theory* 71 (1998) 109–120. <147, 161>

[1002] R. Evans, Gauss sums and Kloosterman sums over residue rings of algebraic integers, *Trans. Amer. Math. Soc.* 353 (2001) 4429–4445. <160, 161>

[1003] R. Evans, Gauss sums of orders six and twelve, *Canad. Math. Bull.* 44 (2001) 22–26. <150, 151, 161>

[1004] R. Evans, Twisted hyper-Kloosterman sums over finite rings of integers, In *Number Theory for the Millennium I*, 429–448, A. K. Peters, Natick, MA, 2002. <155, 160, 161>

[1005] R. Evans, Hypergeometric ${}_3F_2(1/4)$ evaluations over finite fields and Hecke eigenforms, *Proc. Amer. Math. Soc.* 138 (2010) 517–531. <158, 161>

[1006] R. Evans, Seventh power moments of Kloosterman sums, *Israel J. Math.* 175 (2010) 349–362. <158, 161>

[1007] R. Evans and J. Greene, Clausen's theorem and hypergeometric functions over finite fields, *Finite Fields Appl.* 15 (2009) 97–109. <146, 161>

[1008] R. Evans and J. Greene, Evaluations of hypergeometric functions over finite fields, *Hiroshima Math. J.* 39 (2009) 217–235. <146, 161>

[1009] R. Evans, H. D. L. Hollmann, C. Krattenthaler, and Q. Xiang, Gauss sums, Jacobi sums, and p-ranks of cyclic difference sets, *J. Combin. Theory, Ser. A* 87 (1999) 74–119. <152, 161>

[1010] R. J. Evans, Identities for products of Gauss sums over finite fields, *Enseign. Math., IIe Ser.* 27 (1981) 197–209 (1982). <146, 161>

[1011] R. J. Evans, Pure Gauss sums over finite fields, *Mathematika* 28 (1981) 239–248 (1982). <145, 161>

[1012] R. J. Evans, Period polynomials for generalized cyclotomic periods, *Manuscripta Math.* 40 (1982) 217–243. <160, 161>

[1013] R. J. Evans, Character sum analogues of constant term identities for root systems, *Israel J. Math.* 46 (1983) 189–196. <146, 161>

[1014] R. J. Evans, The evaluation of Selberg character sums, *Enseign. Math., IIe Ser.* 37 (1991) 235–248. <146, 161>

[1015] R. J. Evans, Selberg-Jack character sums of dimension 2, *J. Number Theory* 54 (1995) 1–11. <146, 161>

[1016] R. J. Evans, J. Greene, and H. Niederreiter, Linearized polynomials and permutation polynomials of finite fields, *Michigan Math. J.* 39 (1992) 405–413. <228, 230>

[1017] S. A. Evdokimov, Efficient factorization of polynomials over finite fields and the generalized Riemann hypothesis, Translation of Zapiski Nauchnyck Seminarov Leningradskgo Otdeleniya Mat. Inst. V.A. Steklova Akad. Nauk SSSR (LOMI), volume 176, 1989, 104–117. <379, 380>

[1018] S. A. Evdokimov, Faktorizatsiya razreshimogo mnogochlena nad konechnym polem i Obobshchennaya Gipoteza Rimana, *Zapiski Nauchnyck Seminarov Leningradskgo Otdeleniya Mat. Inst. V.A. Steklova Akad. Nauk SSSR (LOMI)* 176 (1989) 104–117, With English abstract, S. A. Evdokimov, Factoring a solvable polynomial over a finite field and the Generalized Riemann Hypothesis. <381, 382>

[1019] S. A. Evdokimov, Factorization of polynomials over finite fields in subexponential time under GRH, In *Algorithmic Number Theory, First International Symposium, ANTS-I*, volume 877 of *Lecture Notes in Comput. Sci.*, 209–219, Springer, Berlin, 1994. <381, 382>

[1020] G. Everest and T. Ward, *Heights of Polynomials and Entropy in Algebraic Dynamics*, Universitext. Springer-Verlag London Ltd., London, 1999. <337, 338, 344>

[1021] J.-H. Evertse, Linear equations with unknowns from a multiplicative group whose solutions lie in a small number of subspaces, *Indag. Math. (New Ser.)* 15 (2004) 347–355. <302>

[1022] C. Faber and G. van der Geer, Complete subvarieties of moduli spaces and the Prym map, *J. Reine Angew. Math.* 573 (2004) 117–137. <486, 488>

[1023] C. C. Faith, Extensions of normal bases and completely basic fields, *Trans. Amer. Math. Soc.* 85 (1957) 406–427. <129, 130, 138>

[1024] G. Faltings, Finiteness theorems for abelian varieties over number fields, In *Arithmetic Geometry*, 9–27, Springer, New York, 1986. <433, 440>

[1025] H. Fan and Y. Dai, Fast bit-parallel $GF(2^n)$ multiplier for all trinomials, *IEEE Trans.*

Comput. 54 (2005) 485–490. <822, 823>

[1026] H. Fan and M. A. Hasan, A new approach to subquadratic space complexity parallel multipliers for extended binary fields, *IEEE Trans. Comput.* 56 (2007) 224–233. <813, 817, 823>

[1027] H. Fan and M. A. Hasan, Subquadratic computational complexity schemes for extended binary field multiplication using optimal normal bases, *IEEE Trans. Comput.* 56 (2007) 1435–1437. <819, 820, 821, 823>

[1028] H. Fan, J. Sun, M. Gu, and K. Lam, Overlap-free Karatsuba-Ofman polynomial multiplication algorithms, *IET Information Security* 4 (2010) 8–14. <813, 814, 823>

[1029] S. Fan, Primitive normal polynomials with the last half coefficients prescribed, *Finite Fields Appl.* 15 (2009) 604–614. <94, 95, 116>

[1030] S. Fan and W. Han, Character sums over Galois rings and primitive polynomials over finite fields, *Finite Fields Appl.* 10 (2004) 36–52. <92, 93, 95>

[1031] S. Fan and W. Han, p-adic formal series and Cohen's problem, *Glasgow Math. J.* 46 (2004) 47–61. <91, 92, 95>

[1032] S. Fan and W. Han, p-adic formal series and primitive polynomials over finite fields, *Proc. Amer. Math. Soc.* 132 (2004) 15–31. <92, 93, 95>

[1033] S. Fan and W. Han, Primitive polynomial with three coefficients prescribed, *Finite Fields Appl.* 10 (2004) 506–521. <93, 95>

[1034] S. Fan, W. Han, and K. Feng, Primitive normal polynomials with multiple coefficients prescribed: an asymptotic result, *Finite Fields Appl.* 13 (2007) 1029–1044. <91, 94, 95, 116>

[1035] S. Fan, W. Han, K. Feng, and X. Zhang, Primitive normal polynomials with the first two coefficients prescribed: a revised p-adic method, *Finite Fields Appl.* 13 (2007) 577–604. <94, 95, 116>

[1036] S. Fan and X. Wang, Primitive normal polynomials with a prescribed coefficient, *Finite Fields Appl.* 15 (2009) 682–730. <94, 95, 116>

[1037] S. Fan and X. Wang, Primitive normal polynomials with the specified last two coefficients, *Discrete Math.* 309 (2009) 4502–4513. <92, 94, 95>

[1038] X. Fan, T. Wollinger, and Y. Wang, Efficient doubling on genus 3 curves over binary fields, In *Topics in Cryptology CT-RSA 2006*, volume 3860 of *Lecture Notes in Comput. Sci.*, 64–81, Springer, Berlin, 2006. <797, 803>

[1039] R. R. Farashahi, Hashing into Hessian curves, In *Progress in Cryptology—AFRICACRYPT 2011*, volume 6737 of *Lecture Notes in Comput. Sci.*, 278–289, Springer, Berlin, 2011. <796>

[1040] J.-C. Faugére, A new efficient algorithm for computing Gröbner bases (F_4), *J. Pure Appl. Algebra* 139 (1999) 61–88. <781, 783>

[1041] J.-C. Faugère, A new efficient algorithm for computing Gröbner bases without reduction to zero (F_5), In *Proceedings of the 2002 International Symposium on Symbolic and Algebraic Computation*, 75–83, ACM, New York, 2002. <781, 783>

[1042] J.-C. Faugère and A. Joux, Algebraic cryptanalysis of hidden field equation (HFE) cryptosystems using Gröbner bases, In *Advances in Cryptology—CRYPTO 2003*, volume 2729 of *Lecture Notes in Comput. Sci.*, 44–60, Springer, Berlin, 2003. <770, 780, 781, 783>

[1043] J.-C. Faugère and S. Lachartre, Parallel Gaussian elimination for Gröbner bases computations in finite fields, In M. M. Maza and J.-L. Roch, editors, *PASCO 2010, Proceedings of the Fourth International Workshop on Parallel Symbolic*

Computation, 89–97, ACM, New York, 2010. <534, 535>

[1044] J.-C. Faugère and L. Perret, Polynomial equivalence problems: algorithmic and theoretical aspects, In *Advances in Cryptology—EUROCRYPT 2006*, volume 4004 of *Lecture Notes in Comput. Sci.*, 30–47, Springer, Berlin, 2006. <767, 783>

[1045] H. Faure, Discrépance de suites associées à un système de numération (en dimension s), *Acta Arith.* 41 (1982) 337–351. <628, 630>

[1046] K. Fazel and S. Kaiser, *Multi-Carrier and Spread Spectrum Systems from OFDM and MC-CDMA to LTE and WiMAX*, John Wiley & Sons, Ltd, West Sussex, second edition, 2008. <713, 719>

[1047] S. Feisel, J. von zur Gathen, and M. A. Shokrollahi, Normal bases via general Gauss periods, *Math. Comp.* 68 (1999) 271–290. <126, 128>

[1048] H. Feistel, W. Notz, and J. Smith, Some cryptographic techniques for machine-to-machine data communications, *Proceedings of the IEEE* 63 (1975) 1545–1554. <743, 750>

[1049] H. Fell and W. Diffie, Analysis of a public key approach based on polynomial substitution, In *Advances in Cryptology—CRYPTO '85*, volume 218 of *Lecture Notes in Comput. Sci.*, 340–349, Springer, Berlin, 1986. <765, 768, 783>

[1050] G. Fellegara, Gli ovaloidi in uno spazio tridimensionale di Galois di ordine 8, *Atti Accad. Naz. Lincei Rend. Cl. Sci. Fis. Mat. Nat.* 32 (1962) 170–176. <588, 589>

[1051] B. Felszeghy, On the solvability of some special equations over finite fields, *Publ. Math. Debrecen* 68 (2006) 15–23. <210, 213>

[1052] G. Feng, A VLSI architecture for fast inversion in $GF(2^m)$, *IEEE Trans. Comput.* 38 (1989) 1383–1386. <818, 823>

[1053] G. L. Feng and K. K. Tzeng, A generalization of the Berlekamp-Massey algorithm for multisequence shift-register synthesis with applications to decoding cyclic codes, *IEEE Trans. Inform. Theory* 37 (1991) 1274–1287. <329, 336>

[1054] K. Feng, Quantum error-correcting codes, In *Coding Theory and Cryptology*, volume 1 of *Lect. Notes Ser. Inst. Math. Sci. Natl. Univ. Singap.*, 91–142, World Sci. Publ., River Edge, NJ, 2002. <839, 841>

[1055] K. Feng and J. Luo, Value distributions of exponential sums from perfect nonlinear functions and their applications, *IEEE Trans. Inform. Theory* 53 (2007) 3035–3041. <272, 273>

[1056] K. Feng and J. Luo, Weight distribution of some reducible cyclic codes, *Finite Fields Appl.* 14 (2008) 390–409. <205, 206>

[1057] K. Feng, H. Niederreiter, and C. Xing, editors, *Coding, Cryptography and Combinatorics*, volume 23 of *Progress in Computer Science and Applied Logic*, Birkhäuser Verlag, Basel, 2004. <31, 32>

[1058] T. Feng, B. Wen, Q. Xiang, and J. Yin, Partial difference sets from quadratic forms and p-ary weakly regular bent functions, arXiv:1002.2797v2, 2011. <265, 273>

[1059] X. Feng and Z. Dai, Expected value of the linear complexity of two-dimensional binary sequences, In *Sequences and Their Applications—SETA 2004*, volume 3486 of *Lecture Notes in Comput. Sci.*, 113–128, Springer, Berlin, 2005. <330, 336>

[1060] S. T. J. Fenn, M. Benaissa, and D. Taylor, $GF(2^m)$ multiplication and division over the dual basis, *IEEE Trans. Comput.* 45 (1996) 319–327. <822, 823>

[1061] N. Fernando and X.-D. Hou, A piecewise construction of permutation polynomials over finite fields, *Finite Fields Appl.* 18 (2012) 1184–1194. <221, 226, 230>

[1062] F. Fiedler, K. H. Leung, and Q. Xiang, On Mathon's construction of maximal arcs in Desarguesian planes, *Adv. Geom.* (2003) S119–S139. <572, 574>

[1063] J. P. Fillmore and M. L. Marx, Linear recursive sequences, *SIAM Rev.* 10 (1968) 342–353. <312, 317>

[1064] N. J. Fine and I. N. Herstein, The probability that a matrix be nilpotent, *Illinois J. Math.* 2 (1958) 499–504. <502, 510>

[1065] FIPS 180-3, Secure hash standard (SHS), Federal Information Processing Standards Publication 180-3, National Institute of Standards and Technology, 2008. <745, 750>

[1066] FIPS 186-2, Digital signature standard, Federal Information Processing Standards Publication 186-2, 2000, available at `http://csrc.nist.gov`. <353, 363>

[1067] FIPS 186-3, Digital signature standard (DSS), Federal Information Processing Standards Publication 186-3, National Institute of Standards and Technology, 2009. <746, 750>

[1068] FIPS 46-3, Data encryption standard (DES), Federal Information Processing Standards Publication 46-3, National Institute of Standards and Technology, 1999. <744, 750>

[1069] S. Fischer and W. Meier, Algebraic immunity of s-boxes and augmented functions, In *Proceedings of Fast Software Encryption 2007*, volume 4593 of *Lecture Notes in Comput. Sci.*, 366–381, 2007. <248, 252>

[1070] S. D. Fisher, Classroom notes: matrices over a finite field, *Amer. Math. Monthly* 73 (1966) 639–641. <501, 510>

[1071] R. W. Fitzgerald, A characterization of primitive polynomials over finite fields, *Finite Fields Appl.* 9 (2003) 117–121. <88, 90>

[1072] R. W. Fitzgerald, Highly degenerate quadratic forms over finite fields of characteristic 2, *Finite Fields Appl.* 11 (2005) 165–181. <205, 206>

[1073] R. W. Fitzgerald, Highly degenerate quadratic forms over $\mathbb{F}_2$, *Finite Fields Appl.* 13 (2007) 778–792. <204, 206>

[1074] R. W. Fitzgerald, Invariants of trace forms over finite fields of characteristic 2, *Finite Fields Appl.* 15 (2009) 261–275. <205, 206>

[1075] R. W. Fitzgerald, Trace forms over finite fields of characteristic 2 with prescribed invariants, *Finite Fields Appl.* 15 (2009) 69–81. <204, 205, 206>

[1076] R. W. Fitzgerald and J. L. Yucas, Irreducible polynomials over GF(2) with three prescribed coefficients, *Finite Fields Appl.* 9 (2003) 286–299. <56, 59, 79>

[1077] R. W. Fitzgerald and J. L. Yucas, Pencils of quadratic forms over finite fields, *Discrete Math.* 283 (2004) 71–79. <206>

[1078] R. W. Fitzgerald and J. L. Yucas, Sums of Gauss sums and weights of irreducible codes, *Finite Fields Appl.* 11 (2005) 89–110. <141, 161>

[1079] R. W. Fitzgerald and J. L. Yucas, Generalized reciprocals, factors of Dickson polynomials and generalized cyclotomic polynomials over finite fields, *Finite Fields Appl.* 13 (2007) 492–515. <285, 286, 287, 290>

[1080] P. Flajolet, X. Gourdon, and D. Panario, The complete analysis of a polynomial factorization algorithm over finite fields, *J. Algorithms* 40 (2001) 37–81. <365, 368, 369, 373, 374, 382>

[1081] P. Flajolet and A. Odlyzko, Singularity analysis of generating functions, *SIAM J. Discrete Math.* 3 (1990) 216–240. <366, 374>

[1082] P. Flajolet and A. M. Odlyzko, Random mapping statistics, In *Advances in cryptology—EUROCRYPT '89 (Houthalen, 1989)*, volume 434 of *Lecture Notes in Comput. Sci.*, 329–354, Springer, Berlin, 1990. <754, 763>

[1083] P. Flajolet and R. Sedgewick, *Analytic Combinatorics*, Cambridge University Press, Cambridge, 2009. <365, 367, 374>

[1084] P. Flajolet and M. Soria, Gaussian limiting distributions for the number of components in combinatorial structures, *J. Combin. Theory, Ser. A* 53 (1990) 165–182. <367, 368, 374>

[1085] P. Flajolet and M. Soria, General combinatorial schemas: Gaussian limit distributions and exponential tails, *Discrete Math.* 114 (1993) 159–180. <368, 374>

[1086] J. J. Flynn, *Near-Exceptionality over Finite Fields*, PhD dissertation, University of California, Berkeley, Department of Mathematics, 2001. <233, 236>

[1087] S. Fomin and A. Zelevinsky, The Laurent phenomenon, *Adv. in Appl. Math.* 28 (2002) 119–144. <337, 344>

[1088] K. Fong, D. Hankerson, J. López, and A. Menezes, Field inversion and point halving revisited, *IEEE Trans. Comput.* 53 (2003) 1047–1059. <363>

[1089] F. Fontein, Groups from cyclic infrastructures and Pohlig-Hellman in certain infrastructures, *Adv. Math. Commun.* 2 (2008) 293–307. <456>

[1090] G. D. Forney, Jr., On decoding BCH codes, *IEEE Trans. Inform. Theory* IT-11 (1965) 549–557. <694, 702, 703>

[1091] G. D. Forney, Jr., *Concatenated Codes*, M.I.T. Press, Cambridge, MA, 1966. <720, 727>

[1092] G. D. Forney, Jr., Generalized minimum distance decoding, *IEEE Trans. Inform. Theory* IT-12 (1966) 125–131. <697, 702, 703>

[1093] G. D. Forney, Jr., N. J. A. Sloane, and M. D. Trott, The Nordstrom-Robinson code is the binary image of the octacode, In *Coding and Quantization*, volume 14 of *DIMACS Ser. Discrete Math. Theoret. Comput. Sci.*, 19–26, Amer. Math. Soc., Providence, RI, 1993. <701, 703>

[1094] P.-A. Fouque, L. Granboulan, and J. Stern, Differential cryptanalysis for multivariate schemes, In *Advances in Cryptology—EUROCRYPT 2005*, volume 3494 of *Lecture Notes in Comput. Sci.*, 341–353, Springer, Berlin, 2005. <773, 777, 778, 783>

[1095] P.-A. Fouque, G. Macario-Rat, L. Perret, and J. Stern, Total break of the l-IC signature scheme, In *Public Key Cryptography—PKC 2008*, volume 4939 of *Lecture Notes in Comput. Sci.*, 1–17, Springer, Berlin, 2008. <774, 783>

[1096] H. M. Fredricksen, A. W. Hales, and M. M. Sweet, A generalization of Swan's theorem, *Math. Comp.* 46 (1986) 321–331. <69, 70>

[1097] D. Freeman, P. Stevenhagen, and M. Streng, Abelian varieties with prescribed embedding degree, In *Algorithmic Number Theory*, volume 5011 of *Lecture Notes in Comput. Sci.*, 60–73, Springer, Berlin, 2008. <811>

[1098] D. M. Freeman, Converting pairing-based cryptosystems from composite-order groups to prime-order groups, In *Advances in Cryptology—EUROCRYPT 2010*, volume 6110 of *Lecture Notes in Comput. Sci.*, 44–61, Springer-Verlag, Berlin, 2010. <793, 796>

[1099] J. W. Freeman, Reguli and pseudoreguli in PG(3, s^2), *Geom. Dedicata* 9 (1980) 267–280. <570, 574>

[1100] T. S. Freeman, G. Imirzian, E. Kaltofen, and Lakshman Yagati, DAGWOOD: A system for manipulating polynomials given by straight-line programs, *ACM Trans. Math. Software* 14 (1988) 218–240. <391, 392>

[1101] D. Freemann, M. Scott, and E. Teske, A taxonomy of pairing-friendly elliptic curves, *Journal of Cryptology* 23 (2010) 224–280. <793, 795, 796>

[1102] Free Software Foundation, GNU Multiple Precision library, version 5.0.5, 2012, available at `http://gmplib.org/`. <346, 355, 363>

[1103] G. Frei, The unpublished section eight: on the way to function fields over a finite field, In *The Shaping of Arithmetic After C. F. Gauss's Disquisitiones Arithmeticae*, 159–198, Berlin: Springer, 2007. <6>

[1104] G. Frey, Applications of arithmetical geometry to cryptographic constructions, In D. Jungnickel and H. Niederreiter, editors, *Finite Fields and Applications*, 128–161, Springer-Verlag, Berlin, 2001. <786, 796, 809, 811>

[1105] G. Frey and T. Lange, Varieties over special fields, In *Handbook of Elliptic and Hyperelliptic Curve Cryptography*, Discrete Math. Appl., 87–113, Chapman & Hall/CRC, Boca Raton, FL, 2006. <31, 32, 456>

[1106] G. Frey, M. Müller, and H.-G. Rück, The Tate pairing and the discrete logarithm applied to elliptic curve cryptosystems, *IEEE Trans. Inform. Theory* 45 (1999) 1717–1719. <455, 456>

[1107] G. Frey, M. Perret, and H. Stichtenoth, On the different of abelian extensions of global fields, In *Coding Theory and Algebraic Geometry*, volume 1518 of *Lecture Notes in Math.*, 26–32, Springer, Berlin, 1992. <468, 469>

[1108] G. Frey and H.-G. Rück, A remark concerning m-divisibility and the discrete logarithm problem in the divisor class group of curves, *Math. Comp.* 62 (1994) 865–874. <793, 796, 810, 811>

[1109] M. Fried, On a conjecture of Schur, *Michigan Math. J.* 17 (1970) 41–55. <228, 230, 239, 240, 284, 290, 293, 294, 302>

[1110] M. Fried, The field of definition of function fields and a problem in the reducibility of polynomials in two variables, *Illinois J. Math.* 17 (1973) 128–146. <301, 302>

[1111] M. Fried, On a theorem of Ritt and related Diophantine problems, *J. Reine Angew. Math.* 264 (1973) 40–55. <295, 302>

[1112] M. Fried, On a theorem of MacCluer, *Acta Arith.* 25 (1973/74) 121–126. <292, 293, 302>

[1113] M. Fried, On Hilbert's irreducibility theorem, *J. Number Theory* 6 (1974) 211–231. <294, 296, 300, 302>

[1114] M. Fried, Fields of definition of function fields and Hurwitz families—groups as Galois groups, *Comm. Algebra* 5 (1977) 17–82. <292, 302>

[1115] M. Fried, Galois groups and complex multiplication, *Trans. Amer. Math. Soc.* 235 (1978) 141–163. <239, 240, 299, 300, 302>

[1116] M. Fried and R. Lidl, On Dickson polynomials and Rédei functions, In *Contributions to General Algebra, 5*, 139–149, Hölder-Pichler-Tempsky, Vienna, 1987. <284, 290>

[1117] M. Fried and G. Sacerdote, Solving Diophantine problems over all residue class fields of a number field and all finite fields, *Ann. of Math., 2nd Ser.* 104 (1976) 203–233. <301, 302>

[1118] M. D. Fried, The place of exceptional covers among all Diophantine relations, *Finite Fields Appl.* 11 (2005) 367–433. <292, 293, 294, 296, 297, 298, 299, 300, 301, 302>

[1119] M. D. Fried, Variables separated equations: Strikingly different roles for the branch cycle lemma and the finite simple group classification, *Science China Mathematics* 55 (2012) 1–69. <293, 297, 301, 302>

[1120] M. D. Fried, R. Guralnick, and J. Saxl, Schur covers and Carlitz's conjecture, *Israel J. Math.* 82 (1993) 157–225. <218, 230, 237, 240, 294, 301, 302>

[1121] M. D. Fried and M. Jarden, *Field Arithmetic*, volume 11 of *Ergebnisse der Mathematik und ihrer Grenzgebiete (3) [Results in Mathematics and Related Areas (3)]*, Springer-Verlag, Berlin, 1986. <31, 32, 295, 300, 301, 302>

[1122] M. D. Fried and M. Jarden, *Field Arithmetic*, volume 11 of *Ergebnisse der Mathematik und ihrer Grenzgebiete. 3. Folge. A Series of Modern Surveys in Mathematics [Results in Mathematics and Related Areas, 3rd Series. A Series of Modern Surveys in Mathematics]*, Springer-Verlag, Berlin, third edition, 2008, Revised by Jarden. <31, 32, 238, 240>

[1123] M. D. Fried and R. E. MacRae, On curves with separated variables, *Math. Ann.* 180 (1969) 220–226. <300, 302>

[1124] M. D. Fried and R. E. MacRae, On the invariance of chains of fields, *Illinois J. Math.* 13 (1969) 165–171. <294, 302>

[1125] E. Friedman and L. C. Washington, On the distribution of divisor class groups of curves over a finite field, In *Théorie des Nombres*, 227–239, de Gruyter, Berlin, 1989. <450, 456>

[1126] J. Friedman, Some geometric aspects of graphs and their eigenfunctions, *Duke Math. J.* 69 (1993) 487–525. <646, 647, 656, 658>

[1127] J. Friedman, A Proof of Alon's Second Eigenvalue Conjecture and Related Problems, *Mem. Amer. Math. Soc.* 195 (2008). <657, 658>

[1128] J. Friedman, R. Murty, and J.-P. Tillich, Spectral estimates for abelian Cayley graphs, *J. Combin. Theory, Ser. B* 96 (2006) 111–121. <651, 658>

[1129] C. Friesen, A special case of Cohen-Lenstra heuristics in function fields, In *Number Theory*, volume 19 of *CRM Proc. Lecture Notes*, 99–105, Amer. Math. Soc., Providence, RI, 1999. <450, 456>

[1130] C. Friesen, Class group frequencies of real quadratic function fields: the degree 4 case, *Math. Comp.* 69 (2000) 1213–1228. <450, 456>

[1131] C. Friesen, Bounds for frequencies of class groups of real quadratic genus 1 function fields, *Acta Arith.* 96 (2001) 313–331. <450, 456>

[1132] S. Frisch, When are weak permutation polynomials strong?, *Finite Fields Appl.* 1 (1995) 437–439. <232>

[1133] D. Fu and J. Solinas, IKE and IKEv2 authentication using the elliptic curve digital signature algorithm (ECDSA), RFC 4754, Internet Engineering Task Force, 2007, `http://www.ietf.org/rfc/rfc4754.txt`. <785, 796>

[1134] F.-W. Fu, H. Niederreiter, and F. Özbudak, On the joint linear complexity of linear recurring multisequences, In *Coding and Cryptology*, volume 4 of *Ser. Coding Theory Cryptol.*, 125–142, World Sci. Publ., Hackensack, NJ, 2008. <325, 330, 331, 336>

[1135] F.-W. Fu, H. Niederreiter, and F. Özbudak, Joint linear complexity of arbitrary multisequences consisting of linear recurring sequences, *Finite Fields Appl.* 15 (2009) 475–496. <328, 331, 336>

[1136] F.-W. Fu, H. Niederreiter, and F. Özbudak, Joint linear complexity of multisequences consisting of linear recurring sequences, *Cryptogr. Commun.* 1 (2009) 3–29. <330, 331, 336>

[1137] F.-W. Fu, H. Niederreiter, and M. Su, The expectation and variance of the joint linear complexity of random periodic multisequences, *J. Complexity* 21 (2005) 804–822. <331, 336>

[1138] L. Fu, Weights of twisted exponential sums, *Math. Z.* 262 (2009) 449–472. <168, 169>

[1139] L. Fu and C. Liu, Equidistribution of Gauss sums and Kloosterman sums, *Math. Z.* 249 (2005) 269–281. <140, 161>

[1140] L. Fu and D. Wan, Moment L-functions, partial L-functions and partial exponential sums, *Math. Ann.* 328 (2004) 193–228. <169, 198, 201>

[1141] L. Fu and D. Wan, Mirror congruence for rational points on Calabi-Yau varieties, *Asian J. Math.* 10 (2006) 1–10. <200, 201>

[1142] C. A. Fuchs, On the quantumness of a Hilbert space, *Quantum Inf. Comput.* 4 (2004) 467–478. <836, 841>

[1143] C. A. Fuchs and M. Sasaki, Squeezing quantum information through a classical channel: measuring the "quantumness" of a set of quantum states, *Quantum Inf. Comput.* 3 (2003) 377–404. <836, 841>

[1144] R. Fuhrmann, A. Garcia, and F. Torres, On maximal curves, *J. Number Theory* 67 (1997) 29–51. <461, 463>

[1145] R. Fuhrmann and F. Torres, The genus of curves over finite fields with many rational points, *Manuscripta Math.* 89 (1996) 103–106. <461, 463>

[1146] R. Fuji-Hara, K. Momihara, and M. Yamada, Perfect difference systems of sets and Jacobi sums, *Discrete Math.* 309 (2009) 3954–3961. <143, 161>

[1147] W. Fulton, *Algebraic Curves*, Advanced Book Classics. Addison-Wesley Publishing Company Advanced Book Program, Redwood City, CA, 1989. <406, 421, 422>

[1148] M. Fürer, Fast integer multiplication, In *Proceedings of the Thirty-ninth Annual ACM Symposium on Theory of Computing,* San Diego, California, USA, 57–66. ACM, 2007, Preprint available at `http://www.cse.psu.edu/~furer/Papers/mult.pdf`. <380, 382>

[1149] È. M. Gabidulin, Theory of codes with maximum rank distance, *Problemy Peredachi Informatsii* 21 (1985) 3–16. <846, 849>

[1150] A. Gács, A remark on blocking sets of almost Rédei type, *J. Geom.* 60 (1997) 65–73. <560, 563>

[1151] A. Gács, On a generalization of Rédei's theorem, *Combinatorica* 23 (2003) 585–598. <559, 563>

[1152] A. Gács, L. Lovász, and T. Szőnyi, Directions in AG$(2, p^2)$, *Innov. Incidence Geom.* 6/7 (2007/08) 189–201. <559, 563>

[1153] A. Gács, P. Sziklai, and T. Szőnyi, Two remarks on blocking sets and nuclei in planes of prime order, *Des. Codes Cryptogr.* 10 (1997) 29–39. <560, 563>

[1154] S. D. Galbraith, Supersingular curves in cryptography, In *Advances in Cryptology—ASIACRYPT 2001*, volume 2248 of *Lecture Notes in Comput. Sci.*, 495–513, Springer, Berlin, 2001. <454, 456>

[1155] S. D. Galbraith, M. Harrison, and D. J. Mireles Morales, Efficient hyperelliptic arithmetic using balanced representation for divisors, In *Algorithmic Number Theory,* volume 5011 of *Lecture Notes in Comput. Sci.*, 342–356, Springer, Berlin, 2008. <451, 452, 456>

[1156] S. D. Galbraith, F. Hess, and N. P. Smart, Extending the GHS Weil descent attack, In L. Knudsen, editor, *Advances in Cryptology—EUROCRYPT 2002*, volume 2332 of *Lecture Notes in Comput. Sci.*, 29–44, Springer-Verlag, Berlin, 2002. <786, 796>

[1157] S. D. Galbraith, F. Hess, and F. Vercauteren, Hyperelliptic pairings, In *Pairing-Based Cryptography—Pairing 2007*, volume 4575 of *Lecture Notes in Comput. Sci.*, 108–131, Springer, Berlin, 2007. <803>

[1158] S. D. Galbraith, J. F. McKee, and P. C. Valença, Ordinary abelian varieties having small embedding degree, *Finite Fields Appl.* 13 (2007) 800–814. <811>

[1159] S. D. Galbraith and K. G. Paterson, editors, *Pairing-based cryptography—Pairing 2008*, volume 5209 of *Lecture Notes in Comput. Sci.*, Berlin, 2008. Springer. <788, 796>

[1160] S. D. Galbraith and N. P. Smart, A cryptographic application of Weil descent, In M. Walker, editor, *Cryptography and Coding*, volume 1746 of *Lecture Notes in Comput. Sci.*, 191–200, Springer-Verlag, Berlin, 1999. <786, 796>

[1161] Z. Galil, R. Kannan, and E. Szemerédi, On nontrivial separators for k-page graphs and simulations by nondeterministic one-tape Turing machines, *J. Comput. System Sci.* 38 (1989) 134–149. <184, 185>

[1162] R. G. Gallager, Low-density parity-check codes, *IRE Trans.* IT-8 (1962) 21–28. <713, 719>

[1163] R. G. Gallager, *Low-Density Parity-Check Codes*, MIT Press, Cambridge, MA, 1963. <713, 714, 719>

[1164] R. G. Gallager, A simple derivation of the coding theorem and some applications, *IEEE Trans. Inform. Theory* IT-11 (1965) 3–18. <660, 661, 703>

[1165] R. Gallant, R. Lambert, and S. Vanstone, Improving the parallelized Pollard lambda search on binary anomalous curves, *Math. Comp.* 69 (2000) 1699–1705. <786, 796>

[1166] L. H. Gallardo and O. Rahavandrainy, Unitary perfect polynomials over $\mathbb{F}_4$ with less than five prime factors, *Funct. Approx. Comment. Math.* 45 (2011) 67–78. <500>

[1167] L. H. Gallardo and L. N. Vaserstein, The strict Waring problem for polynomial rings, *J. Number Theory* 128 (2008) 2963–2972. <499, 500>

[1168] É. Galois, Sur la théorie des nombres, *Bulletin des Sciences Mathématiques XIII* (1830) 428–435, Reprinted in Écrits et Mémoires Matheématiques d'Évariste Galois, 112-128. <4, 6, 11, 377, 380>

[1169] R. A. Games and A. H. Chan, A fast algorithm for determining the complexity of a binary sequence with period 2^n, *IEEE Trans. Inform. Theory* 29 (1983) 144–146. <329, 336>

[1170] M. J. Ganley, Central weak nucleus semifields, *European J. Combin.* 2 (1981) 339–347. <276, 278, 282>

[1171] F. R. Gantmacher, *The Theory of Matrices. Vol. 1*, AMS Chelsea Publishing, Providence, RI, 1998. <529, 535>

[1172] S. Gao, *Normal Bases over Finite Fields*, PhD thesis, University of Waterloo, Canada, 1993. <60, 63, 64, 66, 102, 109, 112, 116, 119, 128>

[1173] S. Gao, Elements of provable high orders in finite fields, *Proc. Amer. Math. Soc.* 127 (1999) 1615–1623. <98, 99, 341>

[1174] S. Gao, Abelian groups, Gauss periods, and normal bases, *Finite Fields Appl.* 7 (2001) 149–164. <126, 127, 128>

[1175] S. Gao, Absolute irreducibility of polynomials via Newton polytopes, *J. Algebra* 237 (2001) 501–520. <388, 392>

[1176] S. Gao, On the deterministic complexity of factoring polynomials, *J. of Symbolic Comput.* 31 (2001) 19–36. <381, 382>

[1177] S. Gao, Factoring multivariate polynomials via partial differential equations, *Math. Comp.* 72 (2003) 801–822. <386, 387, 392>

[1178] S. Gao and J. von zur Gathen, Berlekamp's and Niederreiter's polynomial factorization algorithms, In *Finite fields: theory, applications, and algorithms*, volume 168 of *Contemp. Math.*, 101–116, Amer. Math. Soc., Providence, RI, 1994. <381, 382>

[1179] S. Gao, J. von zur Gathen, and D. Panario, Gauss periods: orders and cryptographical applications, *Math. Comp.* 67 (1998) 343–352, With microfiche supplement. <125, 128>

[1180] S. Gao, J. von zur Gathen, D. Panario, and V. Shoup, Algorithms for exponentiation in finite fields, *J. Symbolic Comput.* 29 (2000) 879–889. <120, 125, 126, 128, 358, 363, 821, 822, 823>

[1181] S. Gao, J. Howell, and D. Panario, Irreducible polynomials of given forms, In *Finite Fields: Theory, Applications, and Algorithms*, volume 225 of *Contemp. Math.*, 43–54, Amer. Math. Soc., Providence, RI, 1999. <89, 90, 348, 363>

[1182] S. Gao, E. Kaltofen, and A. G. B. Lauder, Deterministic distinct degree factorization for polynomials over finite fields, *J. Symbolic Comput.* 38 (2004) 1461–1470. <387, 392>

[1183] S. Gao and A. G. B. Lauder, Hensel lifting and bivariate polynomial factorisation over finite fields, *Math. Comp.* 71 (2002) 1663–1676. <385, 392>

[1184] S. Gao and H. W. Lenstra, Jr., Optimal normal bases, *Des. Codes Cryptogr.* 2 (1992) 315–323. <118, 128>

[1185] S. Gao and T. Mateer, Additive fast Fourier transforms over finite fields, *IEEE Trans. Inform. Theory* 56 (2010) 6265–6272. <113, 116>

[1186] S. Gao and D. Panario, Density of normal elements, *Finite Fields Appl.* 3 (1997) 141–150. <113, 116>

[1187] S. Gao and D. Panario, Tests and constructions of irreducible polynomials over finite fields, In *Foundations of Computational Mathematics*, 346–361, Springer, Berlin, 1997. <347, 348, 363, 371, 374, 376, 377, 380>

[1188] S. Gao and S. A. Vanstone, On orders of optimal normal basis generators, *Math. Comp.* 64 (1995) 1227–1233. <125, 128, 820, 823>

[1189] Z. Gao and D. Panario, Degree distribution of the greatest common divisor of polynomials over $\mathbb{F}_q$, *Random Structures Algorithms* 29 (2006) 26–37. <371, 372, 374>

[1190] Z. Gao and L. B. Richmond, Central and local limit theorems applied to asymptotic enumeration. IV. Multivariate generating functions, *J. Comput. Appl. Math.* 41 (1992) 177–186. <368, 374>

[1191] M. Z. Garaev, Double exponential sums related to Diffie-Hellman distributions, *Int. Math. Res. Not.* (2005) 1005–1014. <183, 184, 185>

[1192] M. Z. Garaev, An explicit sum-product estimate in $\mathbb{F}_p$, *Int. Math. Res. Not. IMRN* (2007) Art. ID rnm035, 11. <187, 192>

[1193] M. Z. Garaev, A quantified version of Bourgain's sum-product estimate in $\mathbb{F}_p$ for subsets of incomparable sizes, *Electron. J. Combin.* 15 (2008) Research paper 58, 8. <187, 192>

[1194] M. Z. Garaev, The sum-product estimate for large subsets of prime fields, *Proc. Amer. Math. Soc.* 136 (2008) 2735–2739. <186, 192>

[1195] M. Z. Garaev, Sums and products of sets and estimates for rational trigonometric sums in fields of prime order, *Uspekhi Mat. Nauk* 65 (2010) 5–66. <187, 189, 192>

[1196] M. Z. Garaev and V. C. Garcia, Waring type congruences involving factorials modulo a prime, *Arch. Math. (Basel)* 88 (2007) 35–41. <213>

[1197] M. Z. Garaev, F. Luca, and I. E. Shparlinski, Waring problem with factorials, *Bull.*

Austral. Math. Soc. 71 (2005) 259–264. <213>

[1198] M. Z. Garaev, F. Luca, I. E. Shparlinski, and A. Winterhof, On the lower bound of the linear complexity over $\mathbb{F}_p$ of Sidelnikov sequences, *IEEE Trans. Inform. Theory* 52 (2006) 3299–3304. <334, 336>

[1199] A. Garcia, M. Q. Kawakita, and S. Miura, On certain subcovers of the Hermitian curve, *Comm. Algebra* 34 (2006) 973–982. <208, 213>

[1200] A. Garcia and H. Stichtenoth, A tower of Artin-Schreier extensions of function fields attaining the Drinfeld-Vlăduţ bound, *Invent. Math.* 121 (1995) 211–222. <464, 467, 469>

[1201] A. Garcia and H. Stichtenoth, On the asymptotic behaviour of some towers of function fields over finite fields, *J. Number Theory* 61 (1996) 248–273. <467, 469>

[1202] A. Garcia and H. Stichtenoth, On the Galois closure of towers, In *Recent Trends in Coding Theory and its Applications*, volume 41 of *AMS/IP Stud. Adv. Math.*, 83–92, Amer. Math. Soc., Providence, RI, 2007. <468, 469>

[1203] A. Garcia, H. Stichtenoth, A. Bassa, and P. Beelen, Towers of function fields over non-prime finite fields, 2012, arXiv:1202.5922v1. <463, 468, 469>

[1204] A. Garcia, H. Stichtenoth, and H.-G. Rück, On tame towers over finite fields, *J. Reine Angew. Math.* 557 (2003) 53–80. <467, 469>

[1205] A. Garcia, H. Stichtenoth, and C.-P. Xing, On subfields of the Hermitian function field, *Compositio Math.* 120 (2000) 137–170. <461, 463>

[1206] A. Garcia and J. F. Voloch, Fermat curves over finite fields, *J. Number Theory* 30 (1988) 345–356. <212, 213>

[1207] M. García-Armas, S. R. Ghorpade, and S. Ram, Relatively prime polynomials and nonsingular Hankel matrices over finite fields, *J. Combin. Theory, Ser. A* 118 (2011) 819–828. <508, 509, 510>

[1208] F. Gardeyn, A Galois criterion for good reduction of τ-sheaves, *J. Number Theory* 97 (2002) 447–471. <541, 546>

[1209] T. Garefalakis, Irreducible polynomials with consecutive zero coefficients, *Finite Fields Appl.* 14 (2008) 201–208. <77, 79>

[1210] T. Garefalakis, Self-reciprocal irreducible polynomials with prescribed coefficients, *Finite Fields Appl.* 17 (2011) 183–193. <78, 79>

[1211] T. Garefalakis and G. Kapetanakis, On the Hansen–Mullen conjecture for self-reciprocal irreducible polynomials, *Finite Fields Appl.* 18 (2012) 832–841. <78, 79>

[1212] T. Garefalakis and D. Panario, The index calculus method using non-smooth polynomials, *Math. Comp.* 70 (2001) 1253–1264. <370, 374>

[1213] T. Garefalakis and D. Panario, Polynomials over finite fields free from large and small degree irreducible factors, *J. Algorithms* 44 (2002) 98–120. <370, 374>

[1214] M. R. Garey and D. S. Johnson, *Computers and Intractability: A Guide to the Theory of NP-completeness*, W. H. Freeman and Co., San Francisco, Calif., 1979. <781, 783>

[1215] G. Garg, T. Helleseth, and P. V. Kumar, Recent advances in low-correlation sequences, In V. Tarokh, editor, *New Directions in Wireless Communications Research*, chapter 3, 63–92, Springer-Verlag, Berlin, 2009. <317, 323, 324>

[1216] J. von zur Gathen, Factoring sparse multivariate polynomials, In *Twenty Fourth Annual IEEE Symposium on Foundations of Computer Science*, 172–179, Los Alamitos, CA, USA, 1983. <391, 392>

[1217] J. von zur Gathen, Hensel and Newton methods in valuation rings, *Math. Comp.* 42 (1984) 637–661. <385, 392>

[1218] J. von zur Gathen, Irreducibility of multivariate polynomials, *J. Comput. System Sci.* 31 (1985) 225–264. <387, 390, 392>

[1219] J. von zur Gathen, Irreducible polynomials over finite fields, In *Proc. Sixth Conf. Foundations of Software Technology and Theoretical Computer Science*, volume 241 of *Springer Lecture Notes in Comput. Sci.*, 252–262, Delhi, India, 1986. <379, 380>

[1220] J. von zur Gathen, Factoring polynomials and primitive elements for special primes, *Theoretical Computer Science* 52 (1987) 77–89. <381, 382>

[1221] J. von zur Gathen, Tests for permutation polynomials, *SIAM J. Comput.* 20 (1991) 591–602. <217, 230>

[1222] J. von zur Gathen, Values of polynomials over finite fields, *Bull. Austral. Math. Soc.* 43 (1991) 141–146. <235, 236, 238, 240>

[1223] J. von zur Gathen, Irreducible trinomials over finite fields, *Math. Comp.* 72 (2003) 1987–2000. <69, 70, 348, 363>

[1224] J. von zur Gathen, Counting decomposable multivariate polynomials, *Appl. Algebra Engrg. Comm. Comput.* 22 (2011) 165–185. <83, 84, 85>

[1225] J. von zur Gathen and J. Gerhard, Arithmetic and factorization of polynomials over $\mathbb{F}_2$, Technical Report tr-rsfb-96-018, University of Paderborn, Germany, 1996, 43 pages. <382>

[1226] J. von zur Gathen and J. Gerhard, Polynomial factorization over $\mathbb{F}_2$, *Math. Comp.* 71 (2002) 1677–1698. <369, 374, 381, 382>

[1227] J. von zur Gathen and J. Gerhard, *Modern Computer Algebra*, Cambridge University Press, Cambridge, New York, Melbourne, second edition, 2003. <31, 32, 84, 85, 125, 126, 128, 346, 363, 376, 378, 380, 381, 382, 385, 387, 392>

[1228] J. von zur Gathen, J. L. Imaña, and Ç. K. Koç, editors, *Arithmetic of Finite Fields*, volume 5130 of *Lecture Notes in Comput. Sci.*, Springer, Berlin, 2008. <31, 32>

[1229] J. von zur Gathen and E. Kaltofen, Factoring multivariate polynomials over finite fields, *Math. Comp.* 45 (1985) 251–261. <387, 392>

[1230] J. von zur Gathen and E. Kaltofen, Factoring sparse multivariate polynomials, *J. Comput. System Sci.* 31 (1985) 265–287. <390, 391, 392>

[1231] J. von zur Gathen, M. Karpinski, and I. E. Shparlinski, Counting curves and their projections, *Comput. Complexity* 6 (1996/97) 64–99. <489, 491>

[1232] J. von zur Gathen and M. Nöcker, Exponentiation in finite fields: Theory and practice, In *Applied Algebra, Algebraic Algorithms and Error-Correcting Codes, Twelfth International Symposium AAECC-12*, volume 1255 of *Lecture Notes in Comput. Sci.*, 88–113, Springer, 1997. <356, 363>

[1233] J. von zur Gathen and M. Nöcker, Polynomial and normal bases for finite fields, *J. Cryptology* 18 (2005) 337–355. <73, 348, 353, 363>

[1234] J. von zur Gathen and D. Panario, Factoring polynomials over finite fields: a survey, *J. Symbolic Comput.* 31 (2001) 3–17. <381, 382>

[1235] J. von zur Gathen, D. Panario, and B. Richmond, Interval partitions and polynomial factorization, *Algorithmica* 63 (2012) 363–397. <369, 374, 382>

[1236] J. von zur Gathen and F. Pappalardi, Density estimates related to Gauss periods, In *Cryptography and Computational Number Theory*, volume 20 of *Progr. Comput. Sci. Appl. Logic*, 33–41, Birkhäuser, Basel, 2001. <120, 128>

[1237] J. von zur Gathen and G. Seroussi, Boolean circuits versus arithmetic circuits, *Information and Computation* 91 (1991) 142–154. <381, 382>

[1238] J. von zur Gathen, M. A. Shokrollahi, and J. Shokrollahi, Efficient multiplication using type 2 optimal normal bases, In *Arithmetic of Finite Fields*, volume 4547 of *Lecture Notes in Comput. Sci.*, 55–68, Springer, Berlin, 2007. <127, 128, 821, 822, 823>

[1239] J. von zur Gathen and V. Shoup, Computing Frobenius maps and factoring polynomials, *Computational Complexity* 2 (1992) 187–224. <369, 374, 376, 380, 381, 382>

[1240] J. von zur Gathen and I. E. Shparlinski, Orders of Gauss periods in finite fields, In *Algorithms and Computations*, volume 1004 of *Lecture Notes in Comput. Sci.*, 208–215, Springer, Berlin, 1995. <125, 128>

[1241] J. von zur Gathen and I. E. Shparlinski, Orders of Gauss periods in finite fields, *Appl. Algebra Engrg. Comm. Comput.* 9 (1998) 15–24. <98, 99>

[1242] J. von zur Gathen and I. E. Shparlinski, Constructing elements of large order in finite fields, In *Applied Algebra, Algebraic Algorithms and Error-Correcting Codes*, volume 1719 of *Lecture Notes in Comput. Sci.*, 404–409, Springer, Berlin, 1999. <99>

[1243] J. von zur Gathen and I. E. Shparlinski, Gauss periods in finite fields, In *Finite Fields and Applications*, 162–177, Springer, Berlin, 2001. <98, 99>

[1244] J. von zur Gathen, A. Viola, and K. Ziegler, Counting reducible, powerful, and relatively irreducible multivariate polynomials over finite fields, *SIAM J. Discrete Math.*, 27 (2013) 855–891. <84, 85>

[1245] P. Gaudry, An algorithm for solving the discrete log problem on hyperelliptic curves, In *Advances in Cryptology—EUROCRYPT 2000*, volume 1807 of *Lecture Notes in Comput. Sci.*, 19–34, Springer, Berlin, 2000. <455, 456, 798, 802, 803>

[1246] P. Gaudry, Fast genus 2 arithmetic based on theta functions, *Journal of Mathematical Cryptology* 1 (2007) 243–265. <797, 803>

[1247] P. Gaudry, Index calculus for abelian varieties of small dimension and the elliptic curve discrete logarithm problem, *J. Symbolic Comput.* 44 (2009) 1690–1702. <786, 796, 802, 803>

[1248] P. Gaudry and N. Gürel, Counting points in medium characteristic using Kedlaya's algorithm, *Experiment. Math.* 12 (2003) 395–402. <491, 788>

[1249] P. Gaudry and R. Harley, Counting points on hyperelliptic curves over finite fields, In *Algorithmic Number Theory*, volume 1838 of *Lecture Notes in Comput. Sci.*, 313–332, Springer, Berlin, 2000. <454, 456>

[1250] P. Gaudry, F. Hess, and N. P. Smart, Constructive and destructive facets of Weil descent on elliptic curves, *Journal of Cryptology* 15 (2002) 19–46. <786, 796, 809, 810, 811>

[1251] P. Gaudry and F. Morain, Fast algorithms for computing the eigenvalue in the Schoof–Elkies–Atkin algorithm, In J.-G. Dumas, editor, *Proceedings of the 2006 International Symposium on Symbolic and Algebraic Computations—ISSAC MMVI*, 109–115, ACM, New York, 2006. <788, 796>

[1252] P. Gaudry and É. Schost, Construction of secure random curves of genus 2 over prime fields, In *Advances in Cryptology—EUROCRYPT 2004*, volume 3027 of *Lecture Notes in Comput. Sci.*, 239–256, Springer, Berlin, 2004. <803>

[1253] P. Gaudry and É. Schost, Genus 2 point counting over prime fields, *J. Symbolic Comput.* 47 (2012) 368–400. <803>

[1254] P. Gaudry, B. A. Smith, and D. R. Kohel, Counting points on genus 2 curves with real multiplication, In *Advances in Cryptology—ASIACRYPT 2011*, volume 7073 of *Lecture Notes in Comput. Sci.*, 504–519, Springer, Berlin, 2011. <803>

[1255] P. Gaudry and E. Thomé, MPFQ – A finite field library Release 1.0-rc3, 2010, available at `http://mpfq.gforge.inria.fr`. <346, 363>

[1256] P. Gaudry, E. Thomé, N. Thériault, and C. Diem, A double large prime variation for small genus hyperelliptic index calculus, *Math. Comp.* 76 (2007) 475–492. <455, 456, 798, 799, 803>

[1257] C. F. Gauss, *Mathematical Diary.* Original manuscript in Latin: Handschriftenabteilung Niedersächsische Staats- und Universitätsbibliothek Göttingen, Cod. Ms. Gauss Math. 48 Cim., English commented transl. by J. Gray, A commentary on Gauss's Mathematical Diary, 1796–1814, with an English translation. *Exposition. Math.*, 2:97–130, 1984. <6, 11>

[1258] C. F. Gauss, *Disquisitiones Arithmeticae*, Lipsiae: G. Fleischer, 1801, English transl. A. A. Clarke. New Haven: Yale University Press, 1966. <4, 11>

[1259] C. F. Gauss, *Werke,* ed. Königliche Gesellschaft der Wissenschaften zu Göttingen, vol. II*, Höhere Arithmetik.*, Göttingen: Universitäts-Druckerei, 1863. <4, 5, 11>

[1260] D. Gavinsky, M. Rötteler, and J. Roland, Quantum algorithm for the Boolean hidden shift problem, In *Proceedings of the Seventeenth Annual International Computing and Combinatorics Conference (COCCON'11)*, volume 6842 of *Lecture Notes in Comput. Sci.*, 158–167, Springer, 2011. <840, 841>

[1261] D. Gay and W. Vélez, On the degree of the splitting field of an irreducible binomial, *Pacific J. Math.* 78 (1978) 117–120. <62, 66>

[1262] G. Ge and L. Zhu, Authentication perpendicular arrays $\mathrm{APA}_1(2,5,v)$, *J. Combin. Des.* 4 (1996) 365–375. <612, 619>

[1263] W. Geiselmann, *Algebraische Algorithmenentwicklung am Beispiel der Arithmetik in endlichen Körpern*, PhD thesis, Universit at Karlsruhe, 1992. <39, 42, 49, 123, 128>

[1264] W. Geiselmann and D. Gollmann, Duality and normal basis multiplication, In *Cryptography and Coding, III*, volume 45 of *Inst. Math. Appl. Conf. Ser. (New Ser.)*, 187–195, Oxford Univ. Press, New York, 1993. <39, 49>

[1265] W. Geiselmann and D. Gollmann, Self-dual bases in $\mathbf{F}_{q^n}$, *Des. Codes Cryptogr.* 3 (1993) 333–345. <103, 109>

[1266] W. Geiselmann, W. Meier, and R. Steinwandt, An attack on the isomorphisms of polynomials problem with one secret, *Int. Journal of Information Security* 2 (2003) 59–64. <767, 783>

[1267] E.-U. Gekeler, On the coefficients of Drinfeld modular forms, *Invent. Math.* 93 (1988) 667–700. <545, 546>

[1268] M. Genma, M. Mishima, and M. Jimbo, Cyclic resolvability of cyclic Steiner 2-designs, *J. Combin. Des.* 5 (1997) 177–187. <594, 599>

[1269] S. R. Ghorpade, S. U. Hasan, and M. Kumari, Primitive polynomials, Singer cycles and word-oriented linear feedback shift registers, *Des. Codes Cryptogr.* 58 (2011) 123–134. <502, 510>

[1270] S. R. Ghorpade and G. Lachaud, Étale cohomology, Lefschetz theorems and number of points of singular varieties over finite fields, *Mosc. Math. J.* 2 (2002) 589–631. <195, 201>

[1271] P. Gianni and B. Trager, Square-free algorithms in positive characteristic, *Appl. Alg. Eng. Comm. Comp.* 7 (1996) 1–14. <384, 392>

[1272] K. S. Gibbins, M. J. Hoffman, and W. K. Wootters, Discrete phase space based on finite fields, *Phys. Rev. A* 70 (2004) 062101. <836, 841>

[1273] M. Giesbrecht, A. Lobo, and B. D. Saunders, Certifying inconsistency of sparse linear systems, In *Proceedings of the 1998 International Symposium on Symbolic and Algebraic Computation*, 113–119, ACM, New York, 1998. <531, 532, 535>

[1274] V. Gillot, Bounds for exponential sums over finite fields, *Finite Fields Appl.* 1 (1995) 421–436. <207, 213>

[1275] P. Giorgi, C.-P. Jeannerod, and G. Villard, On the complexity of polynomial matrix computations, In *Proceedings of the 2003 International Symposium on Symbolic and Algebraic Computation*, 135–142, ACM, New York, 2003. <535>

[1276] D. Giry and J.-J. Quisquater, Bluekrypt cryptographic key length recommendation, 2011, v26.0, April 18, `http://www.keylength.com/`. <784, 796>

[1277] A. Giulietti, L. van der Perre, and M. Strum, Parallel turbo coding interleavers: avoiding collisions in access to storage elements, *IEE Electronics Letters* 38 (2002) 232–234. <726, 727>

[1278] M. Giulietti, J. W. P. Hirschfeld, G. Korchmáros, and F. Torres, Curves covered by the Hermitian curve, *Finite Fields Appl.* 12 (2006) 539–564. <208, 213>

[1279] M. Giulietti and G. Korchmáros, A new family of maximal curves over a finite field, *Math. Ann.* 343 (2009) 229–245. <462, 463>

[1280] M. Giulietti, G. Korchmáros, and F. Torres, Quotient curves of the Suzuki curve, *Acta Arith.* 122 (2006) 245–274. <461, 463>

[1281] D. Glass and R. Pries, Hyperelliptic curves with prescribed p-torsion, *Manuscripta Math.* 117 (2005) 299–317. <486, 488>

[1282] A. Glibichuk and M. Rudnev, On additive properties of product sets in an arbitrary finite field, *J. Anal. Math.* 108 (2009) 159–170. <192, 212, 213>

[1283] A. A. Glibichuk, Sums of powers of subsets of an arbitrary finite field, *Izv. RAN. Ser. Mat.* 75 (2011) 35–68. <175, 185>

[1284] A. A. Glibichuk and S. V. Konyagin, Additive properties of product sets in fields of prime order, In *Additive Combinatorics*, volume 43 of *CRM Proc. Lecture Notes*, 279–286, Amer. Math. Soc., Providence, RI, 2007. <187, 192>

[1285] D. Gligoroski, S. Markovski, and S. J. Knapskog, Multivariate quadratic trapdoor functions based on multivariate quadratic quasigroup, In *Proceedings of The American Conference on Applied Mathematics—MATH08*, Cambridge, Massachusetts, USA, 2008. <776, 783>

[1286] D. Gluck, A note on permutation polynomials and finite geometries, *Discrete Math.* 80 (1990) 97–100. <281, 282>

[1287] C. Godsil and G. Royle, *Algebraic Graph Theory*, volume 207 of *Graduate Texts in Mathematics*, Springer-Verlag, New York, 2001. <645, 656, 658>

[1288] J.-M. Goethals, Nonlinear codes defined by quadratic forms over GF(2), *Information and Control* 31 (1976) 43–74. <702, 703>

[1289] J.-M. Goethals and J. J. Seidel, Orthogonal matrices with zero diagonal, *Canad. J. Math.* 19 (1967) 1001–1010. <610, 619>

[1290] W. M. Y. Goh and E. Schmutz, The expected order of a random permutation, *Bull. London Math. Soc.* 23 (1991) 34–42. <373, 374>

[1291] J. S. Golan, *Semirings and their Applications*, Kluwer Academic Publishers, Dordrecht, 1999. <28, 32>

[1292] M. Golay, Notes on digital coding, *Proc. IRE* 37 (1949) 657. <683, 702, 703>

[1293] M. Golay, Static multislit spectrometry and its application to the panoramic display of infrared spectra, *J. Opt. Soc. Amer.* 41 (1951) 468–472. <843, 849>

[1294] R. Gold, Characteristic linear sequences and their coset functions, *SIAM J. Appl. Math.* 14 (1966) 980–985. <313, 317>

[1295] R. Gold, Maximal recursive sequences with 3-valued recursive crosscorrelation functions, *IEEE Trans. Inform. Theory* 14 (1968) 154–156. <227, 230, 268, 273, 320, 324>

[1296] D. M. Goldschmidt, *Algebraic Functions and Projective Curves*, volume 215 of *Graduate Texts in Mathematics*, Springer-Verlag, New York, 2003. <406, 422>

[1297] C. Goldstein, N. Schappacher, and J. Schwermer, editors, *The Shaping of Arithmetic After C. F. Gauss's Disquisitiones Arithmeticae*, Springer, Berlin, 2007. <5, 11>

[1298] D. Gollmann, *Design of Algorithms in Cryptography. (Algorithmenentwurf in der Kryptographie.)*, Aspekte Komplexer Systeme. 1. Mannheim: B.I. Wissenschaftsverlag. viii, 158 p. 68.00; öS 531.00; sFr 68.00 /hc , 1994. <103, 105, 109>

[1299] F. Göloğlu, G. McGuire, and R. Moloney, Binary Kloosterman sums using Stickelberger's theorem and the Gross-Koblitz formula, *Acta Arith.* 148 (2011) 269–279. <154, 161>

[1300] S. W. Golomb, *Shift Register Sequences*, With portions co-authored by Lloyd R. Welch, Richard M. Goldstein, and Alfred W. Hales. Holden-Day Inc., San Francisco, Calif., 1967. <70, 312, 317>

[1301] S. W. Golomb, Algebraic constructions for Costas arrays, *J. Combin. Theory, Ser. A* 37 (1984) 13–21. <608, 619>

[1302] S. W. Golomb, Periodic binary sequences: solved and unsolved problems, In *Sequences, Subsequences, and Consequences*, volume 4893 of *Lecture Notes in Comput. Sci.*, 1–8, Springer, Berlin, 2007. <96, 97>

[1303] S. W. Golomb and G. Gong, *Signal Design for Good Correlation: For Wireless Communication, Cryptography, and Radar*, Cambridge University Press, Cambridge, 2005. <31, 32, 172, 185, 243, 252, 317, 324, 603, 607, 756, 763>

[1304] S. W. Golomb and G. Gong, The status of Costas arrays, *IEEE Trans. Inform. Theory* 53 (2007) 4260–4265. <608, 619>

[1305] S. W. Golomb and O. Moreno, On periodicity properties of Costas arrays and a conjecture on permutation polynomials, *IEEE Trans. Inform. Theory* 42 (1996) 2252–2253. <229, 230>

[1306] S. W. Golomb, M. G. Parker, A. Pott, and A. Winterhof, editors, *Sequences and Their Applications*, volume 5203 of *Lecture Notes in Comput. Sci.*, Springer, Berlin, 2008. <31, 32>

[1307] J. Gómez-Calderón, On the cardinality of value set of polynomials with coefficients in a finite field, *Proc. Japan Acad., Ser. A Math. Sci.* 68 (1992) 338–340. <235, 236>

[1308] J. Gómez-Calderón and D. J. Madden, Polynomials with small value set over finite fields, *J. Number Theory* 28 (1988) 167–188. <233, 236>

[1309] D. Gómez-Pérez, J. Gutierrez, and Á. Ibeas, Attacking the Pollard generator, *IEEE Trans. Inform. Theory* 52 (2006) 5518–5523. <338, 344>

[1310] D. Gómez-Pérez and A. P. Nicolás, An estimate on the number of stable quadratic polynomials, *Finite Fields Appl.* 16 (2010) 401–405. <178, 185, 342, 343, 344>

[1311] D. Gómez-Pérez, A. P. Nicolás, A. Ostafe, and D. Sadornil, Stable polynomials over finite fields, preprint available, `http://arxiv.org/abs/1206.4979`, 2011. <343, 344>

[1312] D. Gómez-Pérez, A. Ostafe, and I. E. Shparlinski, Algebraic entropy, automorphisms and sparsity of algebraic dynamical systems and pseudorandom number generator, Preprint, 2012. <339, 340, 344>

[1313] D. Gómez-Pérez, A. Ostafe, and I. E. Shparlinski, On irreducible divisors of iterated polynomials, *Revista Matem. Iberoamer.* (2012), to appear. <337, 342, 343, 344>

[1314] D. Gómez-Pérez and A. Winterhof, Waring's problem in finite fields with Dickson polynomials, In *Finite Fields: Theory and Applications*, volume 518 of *Contemp. Math.*, 185–192, Amer. Math. Soc., Providence, RI, 2010. <213>

[1315] G. Gong, T. Helleseth, H. Hu, and A. Kholosha, On the dual of certain ternary weakly regular bent functions, *IEEE Trans. Inform. Theory* 58 (2012) 2237–2243. <271, 273>

[1316] G. Gong, T. Helleseth, H.-Y. Song, and K. Yang, editors, *Sequences and Their Applications*, volume 4086 of *Lecture Notes in Comput. Sci.*, Springer, Berlin, 2006. <31, 32>

[1317] G. Gong and A. M. Youssef, Cryptographic properties of the Welch-Gong transformation sequence generators, *IEEE Trans. Inform. Theory* 48 (2002) 2837–2846. <756, 763>

[1318] P. Gopalan, V. Guruswami, and R. J. Lipton, Algorithms for modular counting of roots of multivariate polynomials, In *LATIN 2006: Theoretical Informatics*, volume 3887 of *Lecture Notes in Comput. Sci.*, 544–555, Springer, Berlin, 2006. <489, 491>

[1319] V. D. Goppa, A new class of linear correcting codes, *Problemy Peredači Informacii* 6 (1970) 24–30. <684, 702, 703>

[1320] V. D. Goppa, Rational representation of codes and (L, g)-codes, *Problemy Peredači Informacii* 7 (1971) 41–49. <684, 702, 703>

[1321] V. D. Goppa, Codes that are associated with divisors (Russian), *Problemy Peredači Informacii* 13 (1977) 33–39. <703, 712>

[1322] V. D. Goppa, Codes on algebraic curves (Russian), *Dokl. Akad. Nauk SSSR* 259 (1981) 1289–1290. <703, 712>

[1323] V. D. Goppa, Algebraic-geometric codes (Russian), *Izv. Akad. Nauk SSSR Ser. Mat.* 46 (1982) 762–781. <703, 712>

[1324] B. Gordon, W. H. Mills, and L. R. Welch, Some new difference sets, *Canad. J. Math.* 14 (1962) 614–625. <318, 324, 602, 607>

[1325] D. M. Gordon, Discrete logarithms in GF(p) using the number field sieve, *SIAM J. Discrete Math.* 6 (1993) 124–138. <399, 401>

[1326] D. M. Gordon, The prime power conjecture is true for $n < 2,000,000$, *Electron. J. Combin.* 1 (1994) Research Paper 6, approx. 7 pp. <602, 607>

[1327] D. M. Gordon and K. S. McCurley, Massively parallel computation of discrete logarithms, In *Proceedings of the Twelfth Annual International Cryptology Conference on Advances in Cryptology - CRYPTO '92*, 312–323, Springer-Verlag, London, UK, 1993. <348, 363>

[1328] J. A. Gordon, Very simple method to find the minimum polynomial of an arbitrary nonzero element of a finite field, *Electron. Lett.* 12 (1976) 663–664. <350, 363>

[1329] D. Gorenstein and N. Zierler, A class of error-correcting codes in p^m symbols, *J. Soc. Indust. Appl. Math.* 9 (1961) 207–214. <678, 692, 702, 703>

[1330] M. Goresky and A. Klapper, Arithmetic crosscorrelations of feedback with carry shift register sequences, *IEEE Trans. Inform. Theory* 43 (1997) 1342–1345. <336>

[1331] M. Goresky and A. M. Klapper, Fibonacci and Galois representations of feedback-with-carry shift registers, *IEEE Trans. Inform. Theory* 48 (2002) 2826–2836. <336>

[1332] D. Goss, π-adic Eisenstein series for function fields, *Compositio Math.* 41 (1980) 3–38. <545, 546>

[1333] D. Goss, *Basic Structures of Function Field Arithmetic*, volume 35 of *Ergebnisse der Mathematik und ihrer Grenzgebiete*, Springer-Verlag, Berlin, 1996. <31, 32, 535, 536, 538, 541, 542, 543, 546>

[1334] D. Goss, Applications of non-Archimedean integration to the L-series of τ-sheaves, *J. Number Theory* 110 (2005) 83–113. <542, 546>

[1335] D. Goss, ζ-phenomenology, In *Noncommutative Geometry, Arithmetic, and Related Topics: Proceedings of the Twenty-First Meeting of the Japan-U.S. Mathematics Institute*, The Johns Hopkins University Press, Baltimore, MD, 2011. <543, 546>

[1336] K. Goto and R. van de Geijn, High-performance implementation of the level-3 BLAS, *ACM Trans. Math. Software* 35 (2009) Art. 4, 14. <523, 535>

[1337] D. Gottesman, Class of quantum error-correcting codes saturating the quantum Hamming bound, *Phys. Rev. A, 3rd Ser.* 54 (1996) 1862–1868. <835, 838, 841>

[1338] D. Gottesman, Fault-tolerant quantum computation with higher-dimensional systems, In *Quantum Computing & Quantum Communications; First NASA International Conference (QCQC'98)*, volume 1509 of *Lecture Notes in Comput. Sci.*, 302–313, Springer, 1998. <838, 841>

[1339] L. Goubin and N. T. Courtois, Cryptanalysis of the TTM cryptosystem, In *Advances in Cryptology—ASIACRYPT 2000*, volume 1976 of *Lecture Notes in Comput. Sci.*, 44–57, Springer, Berlin, 2000. <769, 773, 774, 779, 783>

[1340] A. Gouget and J. Patarin, Probabilistic multivariate cryptography, In *VIETCRYPT*, volume 4341 of *Lecture Notes in Comput. Sci.*, 1–18, Springer, Berlin, 2006. <770, 783>

[1341] X. Gourdon, *Combinatoire, Algorithmique et Géométrie des Polynômes*, PhD dissertation, École Polytechnique, 1996. <369, 374>

[1342] X. Gourdon, Largest component in random combinatorial structures, *Discrete Math.* 180 (1998) 185–209. <374>

[1343] P. Goutet, An explicit factorisation of the zeta functions of Dwork hypersurfaces, *Acta Arith.* 144 (2010) 241–261. <141, 161>

[1344] P. Goutet, On the zeta function of a family of quintics, *J. Number Theory* 130 (2010) 478–492. <141, 161>

[1345] P. Goutet, Isotypic decomposition of the cohomology and factorization of the zeta functions of dwork hypersurfaces, *Finite Fields Appl.* 17 (2011) 113–137. <472, 479>

[1346] R. Gow and J. Sheekey, On primitive elements in finite semifields, *Finite Fields Appl.* 17 (2011) 194–204. <278>

[1347] W. T. Gowers, A new proof of Szemerédi's theorem, *Geom. Funct. Anal.* 11 (2001) 465–588. <188, 192>

[1348] B. Grammaticos, R. G. Halburd, A. Ramani, and C.-M. Viallet, How to detect the integrability of discrete systems, *J. Phys. A* 42 (2009) 454002, 30. <337, 344>

[1349] L. Granboulan, A. Joux, and J. Stern, Inverting HFE is quasipolynomial, In *Advances in Cryptology—CRYPTO 2006*, volume 4117 of *Lecture Notes in Comput. Sci.*,

345–356, Springer, Berlin, 2006. <782, 783>

[1350] A. Granville, S. Li, and S. Qi, On the number of solution of the equation $\sum_{i=1}^{n} x_i/d_i \equiv 0 \pmod 1$, and of diagonal equations in finite fields, *Sichuan Daxue Xuebao* 32 (1995) 243–248. <209, 213>

[1351] M. Grassl, T. Beth, and M. Rötteler, On optimal quantum codes, *International Journal of Quantum Information* 2 (2004) 55–64, See also arXiv preprint quant-ph/0312164. <838, 841>

[1352] M. Grassl, W. Geiselmann, and T. Beth, Quantum Reed-Solomon codes, In *Proceedings of the Thirteenth Conference on Applied Algebra, Algebraic Algorithms and Error-Correcting Codes (AAECC'13)*, volume 1719 of *Lecture Notes in Comput. Sci.*, 231–244, Springer, Berlin, 1999. <838, 841>

[1353] M. Grassl, M. Rötteler, and T. Beth, Efficient quantum circuits for non-qubit quantum error-correcting codes, *International Journal of Foundations of Computer Science* 14 (2003) 757–775. <838, 841>

[1354] R. M. Gray, Toeplitz and circulant matrices: a review, Technical report, Stanford University, 2001. <507, 510>

[1355] D. R. Grayson and M. E. Stillman, Macaulay2, a software system for research in algebraic geometry, Available at http://www.math.uiuc.edu/Macaulay2/, 1992. <47, 49, 832, 834>

[1356] M. Greig, Some balanced incomplete block design constructions, In *Proceedings of the Twenty-first Southeastern Conference on Combinatorics, Graph Theory, and Computing*, volume 77, 121–134, 1990. <594, 599>

[1357] M. Greig, Some group divisible design constructions, *J. Combin. Math. Combin. Comput.* 27 (1998) 33–52. <597, 599>

[1358] F. Griffin, H. Niederreiter, and I. E. Shparlinski, On the distribution of nonlinear recursive congruential pseudorandom numbers of higher orders, In *Applied Algebra, Algebraic Algorithms and Error-Correcting Codes*, volume 1719 of *Lecture Notes in Comput. Sci.*, 87–93, Springer, Berlin, 1999. <338, 340, 344>

[1359] F. Griffin and I. E. Shparlinski, On the linear complexity profile of the power generator, *IEEE Trans. Inform. Theory* 46 (2000) 2159–2162. <333, 336>

[1360] D. Gross, Hudson's theorem for finite-dimensional quantum systems, *J. Math. Phys* 47 (2006) 122107. <836, 841>

[1361] J. Guajardo and C. Paar, Itoh-Tsujii inversion in standard basis and its application in cryptography and codes, *Des. Codes Cryptogr.* 25 (2002) 207–216. <818, 823>

[1362] K. C. Gupta and S. Maitra, Multiples of primitive polynomials over GF(2), In *Progress in Cryptology—INDOCRYPT 2001*, volume 2247 of *Lecture Notes in Comput. Sci.*, 62–72, Springer, Berlin, 2001. <635, 642>

[1363] S. Gurak, Gauss and Eisenstein sums of order twelve, *Canad. Math. Bull.* 46 (2003) 344–355. <151, 161>

[1364] S. Gurak, Gauss sums for prime powers in p-adic fields, *Acta Arith.* 142 (2010) 11–39. <160, 161>

[1365] S. Gurak, Jacobi sums and irreducible polynomials with prescribed trace and restricted norm, In *Finite Fields: Theory and Applications*, volume 518 of *Contemp. Math.*, 193–208, Amer. Math. Soc., Providence, RI, 2010. <143, 161>

[1366] S. J. Gurak, Kloosterman sums for prime powers in p-adic fields, *J. Théor. Nombres Bordeaux* 21 (2009) 175–201. <160, 161>

[1367] R. Guralnick and D. Wan, Bounds for fixed point free elements in a transitive group and applications to curves over finite fields, *Israel J. Math.* 101 (1997) 255–287.

<233, 236>

[1368] R. M. Guralnick, Rational maps and images of rational points of curves over finite fields, *Irish Math. Soc. Bull.* (2003) 71–95. <233, 236, 240>

[1369] R. M. Guralnick and P. Müller, Exceptional polynomials of affine type, *J. Algebra* 194 (1997) 429–454. <238, 239, 240>

[1370] R. M. Guralnick, P. Müller, and J. Saxl, The rational function analogue of a question of Schur and exceptionality of permutation representations, *Mem. Amer. Math. Soc.* 162 (2003) viii+79. <239, 240, 299, 300, 302>

[1371] R. M. Guralnick, P. Müller, and M. E. Zieve, Exceptional polynomials of affine type, revisited, Preprint, 1999. <238, 240>

[1372] R. M. Guralnick, J. Rosenberg, and M. E. Zieve, A new family of exceptional polynomials in characteristic two, *Ann. of Math., 2nd Ser.* 172 (2010) 1361–1390. <237, 240>

[1373] R. M. Guralnick, T. J. Tucker, and M. E. Zieve, Exceptional covers and bijections on rational points, *Int. Math. Res. Not. IMRN* (2007) Art. ID rnm004, 20. <239, 240>

[1374] R. M. Guralnick and M. E. Zieve, Polynomials with PSL(2) monodromy, *Ann. of Math., 2nd Ser.* 172 (2010) 1315–1359. <237, 240>

[1375] V. Guruswami and A. C. Patthak, Correlated algebraic-geometric codes: improved list decoding over bounded alphabets, *Math. Comp.* 77 (2008) 447–473. <705, 712>

[1376] V. Guruswami and A. Rudra, Limits to list decoding Reed-Solomon codes, *IEEE Trans. Inform. Theory* 52 (2006) 3642–3649. <699, 703>

[1377] V. Guruswami and M. Sudan, Improved decoding of Reed-Solomon and algebraic-geometry codes, *IEEE Trans. Inform. Theory* 45 (1999) 1757–1767. <699, 703>

[1378] F. G. Gustavson, Analysis of the Berlekamp-Massey linear feedback shift-register synthesis algorithm, *IBM J. Res. Develop.* 20 (1976) 204–212. <329, 336>

[1379] J. Gutierrez and D. Gómez-Pérez, Iterations of multivariate polynomials and discrepancy of pseudorandom numbers, In *Applied Algebra, Algebraic Algorithms and Error-Correcting Codes*, volume 2227 of *Lecture Notes in Comput. Sci.*, 192–199, Springer, Berlin, 2001. <338, 340, 344>

[1380] J. Gutierrez and Á. Ibeas, Inferring sequences produced by a linear congruential generator on elliptic curves missing high-order bits, *Des. Codes Cryptogr.* 45 (2007) 199–212. <338, 344>

[1381] J. Gutierrez and I. E. Shparlinski, Expansion of orbits of some dynamical systems over finite fields, *Bull. Aust. Math. Soc.* 82 (2010) 232–239. <344>

[1382] J. Gutierrez, I. E. Shparlinski, and A. Winterhof, On the linear and nonlinear complexity profile of nonlinear pseudorandom number-generators, *IEEE Trans. Inform. Theory* 49 (2003) 60–64. <332, 333, 336>

[1383] C. Guyot, K. Kaveh, and V. M. Patankar, Explicit algorithm for the arithmetic on the hyperelliptic Jacobians of genus 3, *Journal of the Ramanujan Mathematical Society* 19 (2004) 75–115. <799, 803>

[1384] K. Gyarmati and A. Sárközy, Equations in finite fields with restricted solution sets I: Character sums, *Acta Math. Hungar.* 118 (2008) 129–148. <184, 185>

[1385] K. Gyarmati and A. Sárközy, Equations in finite fields with restricted solution sets II: Algebraic equations, *Acta Math. Hungar.* 119 (2008) 259–280. <184, 185>

[1386] D. Hachenberger, On completely free elements in finite fields, *Des. Codes Cryptogr.* 4 (1994) 129–143. <129, 138>

[1387] D. Hachenberger, Explicit iterative constructions of normal bases and completely free elements in finite fields, *Finite Fields Appl.* 2 (1996) 1–20. <129, 138>

[1388] D. Hachenberger, Normal bases and completely free elements in prime power extensions over finite fields, *Finite Fields Appl.* 2 (1996) 21–34. <129, 138>

[1389] D. Hachenberger, *Finite Fields: Normal Bases and Completely Free Elements*, The Kluwer International Series in Engineering and Computer Science, 390. Kluwer Academic Publishers, Boston, MA, 1997. <31, 32, 129, 130, 131, 132, 133, 134, 135, 136, 138>

[1390] D. Hachenberger, A decomposition theory for cyclotomic modules under the complete point of view, *J. Algebra* 237 (2001) 470–486. <131, 132, 133, 134, 138>

[1391] D. Hachenberger, Primitive complete normal bases for regular extensions, *Glasgow Math. J.* 43 (2001) 383–398. <95, 134, 135, 137, 138>

[1392] D. Hachenberger, Universal generators for primary closures of Galois fields, In *Finite Fields and Applications*, 208–223, Springer, Berlin, 2001. <137, 138>

[1393] D. Hachenberger, Generators for primary closures of Galois fields, *Finite Fields Appl.* 9 (2003) 122–128. <137, 138>

[1394] D. Hachenberger, Primitive complete normal bases: existence in certain 2-power extensions and lower bounds, *Discrete Math.* 310 (2010) 3246–3250. <95, 137, 138>

[1395] D. Hachenberger, Primitive complete normal bases for regular extensions II: the exceptional case, *unpublished* (2012). <137, 138>

[1396] D. Hachenberger, H. Niederreiter, and C. P. Xing, Function-field codes, *Appl. Algebra Engrg. Comm. Comput.* 19 (2008) 201–211. <708, 712>

[1397] C. D. Haessig, *L*-functions of symmetric powers of cubic exponential sums, *J. Reine Angew. Math.* 631 (2009) 1–57. <479, 488>

[1398] J. Hagenauer, E. Offer, and L. Papke, Iterative decoding of binary block and convolutional codes, *IEEE Trans. Inform. Theory* 42 (1996) 429–445. <725, 727>

[1399] A. W. Hales and D. W. Newhart, Swan's theorem for binary tetranomials, *Finite Fields Appl.* 12 (2006) 301–311. <68, 70>

[1400] L. Hales and S. Hallgren, An improved quantum Fourier transform algorithm and applications, In *Forty First Annual Symposium on Foundations of Computer Science*, 515–525, IEEE Comput. Soc. Press, Los Alamitos, CA, 2000. <839, 841>

[1401] T. R. Halford, A. J. Grant, and K. M. Chugg, Which codes have 4-cycle-free Tanner graphs?, *IEEE Trans. Inform. Theory* 52 (2006) 4219–4223. <718, 719>

[1402] C. Hall, *L*-functions of twisted Legendre curves, *J. Number Theory* 119 (2006) 128–147. <496, 500>

[1403] J. I. Hall, On the order of Hall triple systems, *J. Combin. Theory, Ser. A* 29 (1980) 261–262. <611, 619>

[1404] M. Hall, Jr., Automorphisms of Steiner triple systems, *IBM J. Res. Develop* 4 (1960) 460–472. <611, 619>

[1405] K. H. Ham and G. L. Mullen, Distribution of irreducible polynomials of small degrees over finite fields, *Math. Comp.* 67 (1998) 337–341. <75, 79>

[1406] N. Hamilton and R. Mathon, More maximal arcs in Desarguesian projective planes and their geometric structure, *Adv. Geom.* 3 (2003) 251–261. <572, 574>

[1407] N. Hamilton and R. Mathon, On the spectrum of non-Denniston maximal arcs in $PG(2, 2^h)$, *European J. Combin.* 25 (2004) 415–421. <572, 574>

[1408] R. W. Hamming, Error detecting and error correcting codes, *Bell System Tech. J.* 29 (1950) 147–160. <684, 702, 703>

[1409] A. R. Hammons, Jr., P. V. Kumar, A. R. Calderbank, N. J. A. Sloane, and P. Solé, The $\mathbf{Z}_4$-linearity of Kerdock, Preparata, Goethals, and related codes, *IEEE Trans. Inform. Theory* 40 (1994) 301–319. <28, 29, 31, 251, 252, 273, 699, 701, 702, 703>

[1410] D.-G. Han, D. Choi, and H. Kim, Improved computation of square roots in specific finite fields, *IEEE Trans. Comput.* 58 (2009) 188–196. <360, 363>

[1411] W. Han, The distribution of the coefficients of primitive polynomials over finite fields, In *Cryptography and Computational Number Theory*, volume 20 of *Progr. Comput. Sci. Appl. Logic*, 43–57, Birkhäuser, Basel, 2001. <93, 95>

[1412] W. B. Han, The coefficients of primitive polynomials over finite fields, *Math. Comp.* 65 (1996) 331–340. <93, 95>

[1413] D. Hankerson, A. Menezes, and S. Vanstone, *Guide to Elliptic Curve Cryptography*, Springer Professional Computing. Springer-Verlag, New York, 2004. <31, 32, 49, 346, 353, 354, 356, 363, 455, 456>

[1414] J. P. Hansen and J. P. Pedersen, Automorphism groups of Ree type, Deligne-Lusztig curves and function fields, *J. Reine Angew. Math.* 440 (1993) 99–109. <461, 463>

[1415] S. H. Hansen, Error-correcting codes from higher-dimensional varieties, *Finite Fields Appl.* 7 (2001) 530–552. <705, 712>

[1416] T. Hansen and G. L. Mullen, Primitive polynomials over finite fields, *Math. Comp.* 59 (1992) 639–643, S47–S50. <75, 79, 88, 90, 92, 95, 97, 349, 363>

[1417] B. Hanson, D. Panario, and D. Thomson, Swan-like results for binomials and trinomials over finite fields of odd characteristic, *Des. Codes Cryptogr.* 61 (2011) 273–283. <69, 70>

[1418] G. Hardy and E. Wright, *An Introduction to the Theory of Numbers*, Oxford University Press, Oxford, 2008. <495, 500>

[1419] G. H. Hardy and J. E. Littlewood, Some problems of 'Partitio Numerorum'; IV: The singular series in Waring's problem and the value of the number G(k), *Math. Z.* 12 (1922) 161–188. <498, 500>

[1420] G. H. Hardy and J. E. Littlewood, Some problems of 'Partitio Numerorum'; III: On the expression of a number as a sum of primes, *Acta Math.* 44 (1923) 1–70. <497, 500>

[1421] R. Harley, Fast arithmetic on genus two curves, Preprint, 2000. <797, 803>

[1422] R. Harley, Asymptotically optimal p-adic point-counting, 2002, Posting to the Number Theory List, available at `http://listserv.nodak.edu/cgi-bin/wa.exe?A2=ind0212&L=NMBRTHRY&P=R1277`. <788, 796>

[1423] N. V. Harrach and C. Mengyán, Minimal blocking sets in $\mathrm{PG}(2, q)$ arising from a generalized construction of Megyesi, *Innov. Incidence Geom.* 6/7 (2007/08) 211–226. <559, 563>

[1424] D. Hart, A. Iosevich, and J. Solymosi, Sum-product estimates in finite fields via Kloosterman sums, *Int. Math. Res. Not. IMRN* (2007) Art. ID rnm007, 14. <186, 192>

[1425] D. Hart, L. Li, and C. Yen Shen, Fourier analysis and expanding phenomena in finite fields, preprint available, `http://arxiv.org/abs/0909.5471`, 2009. <188, 192>

[1426] W. B. Hart et al., Fast Library for Number Theory (Version 2.2), available at `http://www.flintlib.org`. <47, 49, 346, 351, 363>

[1427] R. Hartshorne, *Algebraic Geometry*, Springer-Verlag, New York, 1977, Graduate Texts in Mathematics, No. 52. <291, 295, 302, 421, 422, 469, 479>

[1428] D. Harvey, Faster arithmetic for number-theoretic transforms, Slides presented at FLINT/Sage Days 35, Warwick, 2011. <354, 363>

[1429] D. Harvey, Kedlaya's algorithm in larger characteristic, *Int. Math. Res. Not. IMRN* (2007) Art. ID rnm095, 29. <490, 491, 788, 796>

[1430] D. Harvey, Faster polynomial multiplication via multipoint Kronecker substitution, *J. Symbolic Comput.* 44 (2009) 1502–1510. <351, 363>

[1431] M. A. Hasan, Shift-register synthesis for multiplicative inversion over $GF(2^m)$, In *Proc. IEEE International Symposium on Information Theory*, 49, 1995. <816, 823>

[1432] M. A. Hasan, On matrix-vector product based sub-quadratic arithmetic complexity schemes for field multiplication, In *Proc. SPIE 6697, 669702*, 2007. <817, 823>

[1433] M. A. Hasan and V. K. Bhargava, Bit-serial systolic divider and multiplier for finite fields $GF(2^m)$, *IEEE Trans. Comput.* 41 (1992) 972–980. <817, 823>

[1434] M. A. Hasan and V. K. Bhargava, Division and bit-serial multiplication over $GF(q^m)$, *IEE Proceedings-E, Computers and Digital Techniques* 139 (1992) 230–236. <816, 817, 823>

[1435] M. A. Hasan and V. K. Bhargava, Architecture for low complexity rate-adaptive Reed-Solomon encoder, *IEEE Trans. Comput.* 44 (1995) 938–942. <822, 823>

[1436] M. A. Hasan and T. Helleseth, editors, *Arithmetic of Finite Fields*, volume 6087 of *Lecture Notes in Comput. Sci.*, Springer, Berlin, 2010. <31, 32>

[1437] M. A. Hasan, A. H. Namin, and C. Negre, New complexity results for field multiplication using optimal normal bases and block recombination, Technical Report cacr2010-19, University of Waterloo, Waterloo, 2010. <821, 823>

[1438] M. A. Hasan and C. Negre, Low space complexity multiplication over binary fields with Dickson polynomial representation, *IEEE Trans. Comput.* 60 (2011) 602–607. <822, 823>

[1439] M. A. Hasan, M. Z. Wang, and V. K. Bhargava, A modified Massey-Omura parallel multiplier for a class of finite fields, *IEEE Trans. Comput.* 42 (1993) 1278–1280. <820, 823>

[1440] S. Hasegawa and T. Kaneko, An attacking method for a public key cryptosystem based on the difficulty of solving a system of non-linear equations, In *Proc. Tenth Symposium on Information Theory and Its Applications*, JA5–3, 1987. <768, 783>

[1441] K. Hashimoto, Zeta functions of finite graphs and representations of p-adic groups, In *Automorphic Forms and Geometry of Arithmetic Varieties*, volume 15 of *Adv. Stud. Pure Math.*, 211–280, Academic Press, Boston, MA, 1989. <658>

[1442] S. H. Hassani, S. B. Korada, and R. Urbanke, The compound capacity of polar codes, In *The Forty Seventh Annual Allerton Conference on Communication, Control, and Computing*, 16–21, 2009. <739>

[1443] H. Hasse, Theorie der relativ-zyklischen algebraischen funktionenkrper, insbesondere bei endlichen konstantkrper, *J. Reine Angew. Math.* 172 (1934) 37–54. <162>

[1444] J. Håstad and M. Näslund, The security of all RSA and discrete log bits, *J. ACM* 51 (2004) 187–230. <394, 401>

[1445] P. Hawkes and G. G. Rose, Exploiting multiples of the connection polynomial in word-oriented stream ciphers, In *Advances in Cryptology—ASIACRYPT 2000*, volume 1976 of *Lecture Notes in Comput. Sci.*, 303–316, Springer, Berlin, 2000.

<326, 336>

[1446] P. Hawkes and G. G. Rose, Rewriting variables: the complexity of fast algebraic attacks on stream ciphers, In *Advances in Cryptology—CRYPTO 2004*, volume 3152 of *Lecture Notes in Comput. Sci.*, 390–406, Springer, Berlin, 2004. <248, 252>

[1447] D. R. Hayes, The distribution of irreducibles in GF$[q, x]$, *Trans. Amer. Math. Soc.* 117 (1965) 101–127. <74, 79, 495, 500>

[1448] D. R. Hayes, The expression of a polynomial as a sum of three irreducibles, *Acta Arith.* 11 (1966) 461–488. <497, 500>

[1449] D. R. Hayes, Explicit class field theory for rational function fields, *Trans. Amer. Math. Soc.* 189 (1974) 77–91. <538, 546>

[1450] D. R. Hayes, Explicit class field theory in global function fields, In *Studies in Algebra and Number Theory*, volume 6 of *Adv. in Math. Suppl. Stud.*, 173–217, Academic Press, New York, 1979. <538, 546>

[1451] D. R. Hayes, A brief introduction to Drinfeld modules, In *The Arithmetic of Function Fields*, volume 2 of *Ohio State Univ. Math. Res. Inst. Publ.*, 1–32, de Gruyter, Berlin, 1992. <535, 546>

[1452] L. S. Heath and N. A. Loehr, New algorithms for generating Conway polynomials over finite fields, *J. Symbolic Comput.* 38 (2004) 1003–1024. <402, 404>

[1453] D. R. Heath-Brown, Arithmetic applications of Kloosterman sums, *Nieuw Arch. Wiskd. (5)* 1 (2000) 380–384. <154, 161>

[1454] D. R. Heath-Brown and S. Konyagin, New bounds for Gauss sums derived from kth powers, and for Heilbronn's exponential sum, *Q. J. Math.* 51 (2000) 221–235. <141, 161, 173, 176, 185>

[1455] D. R. Heath-Brown and S. J. Patterson, The distribution of Kummer sums at prime arguments, *J. Reine Angew. Math.* 310 (1979) 111–130. <150, 161>

[1456] A. Hedayat, D. Raghavarao, and E. Seiden, Further contributions to the theory of F-squares design, *Ann. Statist.* 3 (1975) 712–716. <553, 556>

[1457] A. S. Hedayat, N. J. A. Sloane, and J. Stufken, *Orthogonal Arrays, Theory and Applications*, Springer Series in Statistics. Springer-Verlag, New York, 1999. <610, 619, 631, 642>

[1458] A. Hefez, On the value sets of special polynomials over finite fields, *Finite Fields Appl.* 2 (1996) 337–347. <235, 236>

[1459] L. Heffter, Ueber Tripelsysteme, *Math. Ann.* 49 (1897) 101–112. <591, 599>

[1460] H. Heilbronn, Lecture Notes on Additive Number Theory mod p, *California Institute of Technology* (1964). <212, 213>

[1461] R. Heindl, *New Directions in Multivariate Public Key Cryptography*, PhD dissertation, Clemson University, 2009, `http://etd.lib.clemson.edu/documents/1247508584/`. <776, 783>

[1462] J. Heintz and M. Sieveking, Absolute primality of polynomials is decidable in random polynomial time in the number of variables, In *Automata, Languages and Programming*, volume 115 of *Lecture Notes in Comput. Sci.*, 16–28, Springer-Verlag, 1981. <387, 392>

[1463] H. A. Helfgott, Growth and generation in $\mathrm{SL}_2(\mathbb{Z}/p\mathbb{Z})$, *Ann. of Math., 2nd Ser.* 167 (2008) 601–623. <191, 192>

[1464] H. A. Helfgott, Growth in $\mathrm{SL}_3(\mathbb{Z}/p\mathbb{Z})$, *J. Eur. Math. Soc. (JEMS)* 13 (2011) 761–851. <191, 192>

[1465] H. A. Helfgott and M. Rudnev, An explicit incidence theorem in $\mathbb{F}_p$, *Mathematika* 57 (2011) 135–145. <191, 192>

[1466] H. A. Helfgott and A. Seress, On the diameter of permutation groups, *arXiv:1109.3550.* <191, 192>

[1467] T. Helleseth, Some results about the cross-correlation function between two maximal linear sequences, *Discrete Math.* 16 (1976) 209–232. <260, 261, 320, 324>

[1468] T. Helleseth, On the covering radius of cyclic linear codes and arithmetic codes, *Discrete Appl. Math.* 11 (1985) 157–173. <176, 185>

[1469] T. Helleseth, H. D. L. Hollmann, A. Kholosha, Z. Wang, and Q. Xiang, Proofs of two conjectures on ternary weakly regular bent functions, *IEEE Trans. Inform. Theory* 55 (2009) 5272–5283. <271, 273>

[1470] T. Helleseth and A. Kholosha, Monomial and quadratic bent functions over the finite fields of odd characteristic, *IEEE Trans. Inform. Theory* 52 (2006) 2018–2032. <266, 270, 271, 273>

[1471] T. Helleseth and A. Kholosha, On the dual of monomial quadratic p-ary bent functions, In *Sequences, Subsequences, and Consequences*, volume 4893 of *Lecture Notes in Comput. Sci.*, 50–61, Springer, Berlin, 2007. <270, 273>

[1472] T. Helleseth and A. Kholosha, New binomial bent functions over the finite fields of odd characteristic, *IEEE Trans. Inform. Theory* 56 (2010) 4646–4652. <271, 273>

[1473] T. Helleseth and A. Kholosha, Crosscorrelation of m-sequences, exponential sums, bent functions and Jacobsthal sums, *Cryptogr. Commun.* 3 (2011) 281–291. <271, 273>

[1474] T. Helleseth, A. Kholosha, and S. Mesnager, Niho bent functions and Subiaco hyperovals, In M. Lavrauw, G. L. Mullen, S. Nikova, D. Panario, and L. Storme, editors, *Theory and Applications of Finite Fields*, volume 579 of *Contemp. Math.*, 91–101, Providence, Rhode Island, 2012, American Mathematical Society. <268, 270, 273>

[1475] T. Helleseth and P. V. Kumar, Sequences with low correlation, In *Handbook of Coding Theory, Vol. I, II*, 1765–1853, North-Holland, Amsterdam, 1998. <206, 265, 273, 317, 319, 320, 321, 322, 323, 324>

[1476] T. Helleseth and P. V. Kumar, Pseudonoise sequences, In J. D. Gibson, editor, *The Mobile Communications Handbook*, The electrical engineering handbook series, chapter 8, CRC Press, London, second edition, 1999. <317, 324>

[1477] T. Helleseth, P. V. Kumar, and H. Martinsen, A new family of ternary sequences with ideal two-level autocorrelation function, *Des. Codes Cryptogr.* 23 (2001) 157–166. <603, 607>

[1478] T. Helleseth, P. V. Kumar, and K. Yang, editors, *Sequences and Their Applications*, Discrete Mathematics and Theoretical Computer Science, Springer-Verlag, London, 2002. <31, 32>

[1479] T. Helleseth, C. Rong, and D. Sandberg, New families of almost perfect nonlinear power mappings, *IEEE Trans. Inform. Theory* 45 (1999) 474–485. <227, 230, 256, 261>

[1480] T. Helleseth, D. Sarwate, H.-Y. Song, and K. Yang, editors, *Sequences and Their Applications*, volume 3486 of *Lecture Notes in Comput. Sci.*, Springer, Berlin, 2005. <31, 32>

[1481] T. Helleseth and V. Zinoviev, New Kloosterman sums identities over $\mathbb{F}_{2^m}$ for all m, *Finite Fields Appl.* 9 (2003) 187–193. <226, 230>

[1482] M. Henderson, A note on the permutation behaviour of the Dickson polynomials of the second kind, *Bull. Austral. Math. Soc.* 56 (1997) 499–505. <226, 230>

[1483] M. Henderson and R. Matthews, Permutation properties of Chebychev polynomials of the second kind over a finite field, *Finite Fields Appl.* 1 (1995) 115–125. <226, 230>

[1484] M. Henderson and R. Matthews, Dickson polynomials of the second kind which are permutation polynomials over a finite field, *New Zealand J. Math.* 27 (1998) 227–244. <226, 230>

[1485] B. Hendrickson and E. Rothberg, Improving the run time and quality of nested dissection ordering, *SIAM J. Sci. Comput.* 20 (1998) 468–489. <532, 535>

[1486] K. Hensel, Über die Darstellung der Zahlen eines Gattungsbereiches für einen beliebigen Primdivisor, *J. Reine Angew. Math.* 103 (1888) 230–237. <110, 112, 116>

[1487] I. R. Hentzel and I. F. Rúa, Primitivity of finite semifields with 64 and 81 elements, *Internat. J. Algebra Comput.* 17 (2007) 1411–1429. <278>

[1488] C. Hering, Eine nicht-desarguessche zweifach transitive affine Ebene der Ordnung 27, *Abh. Math. Sem. Univ. Hamburg* 34 (1969/1970) 203–208. <569, 574>

[1489] J. R. Heringa, H. W. J. Blöte, and A. Compagner, New primitive trinomials of Mersenne-exponent degrees for random-number generation, *Internat. J. Modern Phys. C* 3 (1992) 561–564. <96, 97>

[1490] F. Hernando and G. McGuire, Proof of a conjecture on the sequence of exceptional numbers, classifying cyclic codes and APN functions, *Journal of Algebra* 343 (2011) 78–92. <260, 261>

[1491] M. Herrmann and G. Leander, A practical key recovery attack on Basic $\mathcal{TCH}o$, In *Public Key Cryptography—PKC 2009*, volume 5443 of *Lecture Notes in Comput. Sci.*, 411–424, Springer, Berlin, 2009. <630, 642>

[1492] F. Hess, Pairing lattices, In S. D. Galbraith and K. Paterson, editors, *Pairing-Based Cryptography—Pairing 2008*, volume 5209 of *Lecture Notes in Comput. Sci.*, 18–38, Springer-Verlag, Berlin, 2008. <791, 796>

[1493] F. Hess and I. E. Shparlinski, On the linear complexity and multidimensional distribution of congruential generators over elliptic curves, *Des. Codes Cryptogr.* 35 (2005) 111–117. <334, 336>

[1494] F. Hess, N. P. Smart, and F. Vercauteren, The eta pairing revisited, *IEEE Trans. Inform. Theory* 52 (2006) 4595–4602. <791, 792, 796>

[1495] F. Heß, A. Stein, S. Stein, and M. Lochter, The magic of elliptic curves and public-key cryptography, *Jahresber. Dtsch. Math.-Ver.* 114 (2012) 59–88. <809, 811>

[1496] A. E. Heydtmann, Sudan-decoding generalized geometric Goppa codes, *Finite Fields Appl.* 9 (2003) 267–285. <708, 712>

[1497] K. Hicks, G. Mullen, J. Yucas, and R. Zavislak, A polynomial analogue of the $3n+1$ problem, *Amer. Math. Monthly* 115 (2008) 615–622. <500>

[1498] J. Hietarinta and C. Viallet, Searching for integrable lattice maps using factorization, *J. Phys. A* 40 (2007) 12629–12643. <337, 338, 344>

[1499] D. Hilbert, Ueber die Irreducibilität ganzer rationaler Functionen mit ganzzahligen Coefficienten, *JFRAM* 110 (1892) 104–129. <386, 392>

[1500] D. Hilbert, Beweis fur die darstellbarkeit der ganzen kahlen durch eine feste anzahl nter potenzen (waringsches problem), *Math. Ann.* 67 (1909) 281–300. <498, 500>

[1501] A. Hildebrand and G. Tenenbaum, Integers without large prime factors, *J. Théor.*

Nombres Bordeaux 5 (1993) 411–484. <399, 401>

[1502] F. Hinkelmann, M. Brandon, B. Guang, R. McNeill, A. Veliz-Cuba, G. Blekherman, and R. Laubenbacher, Adam: Analysis of analysis of dynamic algebraic models, Available at http:/adam.vbi.vt.edu/, 2010. <827, 830, 832, 834>

[1503] F. Hinkelmann and A. S. Jarrah, Inferring biologically relevant models: Nested canalyzing functions, *ISRN Biomathematics* 2012 (2012) Article ID 613174, 7 pages. <834>

[1504] F. Hinkelmann and R. Laubenbacher, Boolean models of bistable biological systems, *Discrete Contin. Dyn. Syst. Ser. S* 4 (2011) 1443–1456. <827, 834>

[1505] F. Hinkelmann, D. Murrugarra, A. Jarrah, and R. Laubenbacher, A mathematical framework for agent based models of complex biological networks, *Bulletin of Mathematical Biology* (2010) 1–20, 10.1007/s11538-010-9582-8. <829, 831, 834>

[1506] Y. Hiramine, A conjecture on affine planes of prime order, *J. Combin. Theory, Ser. A* 52 (1989) 44–50. <281, 282>

[1507] Y. Hiramine, On planar functions, *J. Algebra* 133 (1990) 103–110. <282>

[1508] Y. Hiramine, M. Matsumoto, and T. Oyama, On some extension of 1-spread sets, *Osaka J. Math.* 24 (1987) 123–137. <276, 278, 567, 574>

[1509] J. W. P. Hirschfeld, *Finite Projective Spaces of Three Dimensions*, Oxford Mathematical Monographs, The Clarendon Press, Oxford University Press, New York, 1985. <31, 32, 579, 587, 588, 589>

[1510] J. W. P. Hirschfeld, *Projective Geometries over Finite Fields*, Oxford Mathematical Monographs, The Clarendon Press, Oxford University Press, New York, second edition, 1998. <31, 32, 564, 570, 572, 574, 575, 576, 577, 578, 579, 580, 581, 582, 583, 584, 585, 589>

[1511] J. W. P. Hirschfeld, G. Korchmáros, and F. Torres, *Algebraic Curves over a Finite Field*, Princeton Series in Applied Mathematics, Princeton University Press, Princeton, NJ, 2008. <31, 32, 406, 422, 461, 463, 584, 586, 587, 589>

[1512] J. W. P. Hirschfeld and L. Storme, The packing problem in statistics, coding theory and finite projective spaces, *J. Statist. Plann. Inference* 72 (1998) 355–380. <584, 585, 589>

[1513] J. W. P. Hirschfeld and L. Storme, The packing problem in statistics, coding theory and finite projective spaces: update 2001, In *Finite Geometries*, volume 3 of *Dev. Math.*, 201–246, Kluwer Acad. Publ., Dordrecht, 2001. <573, 574, 584, 585, 589>

[1514] J. W. P. Hirschfeld, L. Storme, J. A. Thas, and J. F. Voloch, A characterization of Hermitian curves, *J. Geom.* 41 (1991) 72–78. <208, 213>

[1515] J. W. P. Hirschfeld and J. A. Thas, *General Galois Geometries*, Oxford Mathematical Monographs, The Clarendon Press, Oxford University Press, New York, 1991. <31, 32, 566, 574, 584, 585, 586, 589>

[1516] A. Hocquenghem, Codes correcteurs d'erreurs, *Chiffres* 2 (1959) 147–156. <678, 702, 703>

[1517] J. H. Hodges, The matrix equation $X^2 - I = 0$ over a finite field, *Amer. Math. Monthly* 65 (1958) 518–520. <502, 510>

[1518] M. van Hoeij, Factoring polynomials and the knapsack problem, *J. Number Theory* 95 (2002) 167–189. <385, 392>

[1519] J. van der Hoeven and G. Lecerf, On the bit-complexity of sparse polynomial and series multiplication, *J. Symbolic Comput.* 50 (2013) 227–254. <382, 387, 392>

[1520] E. Hof and S. Shamai, Secrecy-achieving polar-coding for binary-input memoryless symmetric wire-tap channels, preprint, 2010. <739>

[1521] J. Hoffstein, J. Pipher, and J. H. Silverman, *An Introduction to Mathematical Cryptography*, Undergraduate Texts in Mathematics, Springer, New York, 2008. <31, 32, 750>

[1522] T. Høholdt, Personal communication, 2011. <683, 703>

[1523] T. Høholdt and H. E. Jensen, Determination of the merit factor of Legendre sequences, *IEEE Trans. Inform. Theory* 34 (1988) 161–164. <323, 324, 842, 849>

[1524] T. Høholdt and R. Pellikaan, On the decoding of algebraic-geometric codes, *IEEE Trans. Inform. Theory* 41 (1995) 1589–1614. <705, 712>

[1525] T. Høholdt, J. H. van Lint, and R. Pellikaan, Algebraic geometry codes, In *Handbook of Coding Theory, Vol. I, II*, 871–961, North-Holland, Amsterdam, 1998. <702, 703>

[1526] H. D. L. Hollmann and Q. Xiang, A proof of the Welch and Niho conjectures on cross-correlations of binary m-sequences, *Finite Fields Appl.* 7 (2001) 253–286. <261>

[1527] H. D. L. Hollmann and Q. Xiang, A class of permutation polynomials of $\mathbf{F}_{2^m}$ related to Dickson polynomials, *Finite Fields Appl.* 11 (2005) 111–122. <227, 230>

[1528] S. Hong, Newton polygons of L functions associated with exponential sums of polynomials of degree four over finite fields, *Finite Fields Appl.* 7 (2001) 205–237. <484, 488>

[1529] S. Hong, Newton polygons for L-functions of exponential sums of polynomials of degree six over finite fields, *J. Number Theory* 97 (2002) 368–396. <484, 488>

[1530] I. Honkala and A. Tietäväinen, Codes and number theory, In *Handbook of Coding Theory, Vol. II*, 1141–1194, North-Holland, Amsterdam, 1998. <702, 703>

[1531] C. Hooley, On Artin's conjecture, *J. Reine Angew. Math.* 225 (1967) 209–220. <72, 73>

[1532] C. Hooley, On exponential sums and certain of their applications, In *Number Theory Days*, volume 56 of *London Math. Soc. Lecture Note Ser.*, 92–122, Cambridge Univ. Press, Cambridge, 1982. <164, 169>

[1533] C. Hooley, On the number of points on a complete intersection over a finite field, *J. Number Theory* 38 (1991) 338–358. <195, 201>

[1534] S. Hoory, A lower bound on the spectral radius of the universal cover of a graph, *J. Combin. Theory, Ser. B* 93 (2005) 33–43. <648, 658>

[1535] S. Hoory, N. Linial, and A. Wigderson, Expander graphs and their applications, *Bull. Amer. Math. Soc. (New Ser.)* 43 (2006) 439–561. <642, 643, 647, 649, 658>

[1536] K. J. Horadam, *Hadamard Matrices and Their Applications*, Princeton University Press, Princeton, NJ, 2007. <170, 185>

[1537] J. D. Horton, Orthogonal starters in finite abelian groups, *Discrete Math.* 79 (1990) 265–278. <614, 619>

[1538] A. Hoshi, Explicit lifts of quintic Jacobi sums and period polynomials for $\mathbb{F}_q$, *Proc. Japan Acad., Ser. A, Math. Sci.* 82 (2006) 87–92. <141, 149, 161>

[1539] X.-D. Hou, p-ary and q-ary versions of certain results about bent functions and resilient functions, *Finite Fields Appl.* 10 (2004) 566–582. <264, 273>

[1540] X.-D. Hou, A note on the proof of a theorem of Katz, *Finite Fields Appl.* 11 (2005) 316–319. <199, 201, 213>

[1541] X.-D. Hou, Affinity of permutations of $\mathbb{F}_2^n$, *Discrete Appl. Math.* 154 (2006) 313–325. <229, 230, 256, 261>

[1542] X.-D. Hou, Two classes of permutation polynomials over finite fields, *J. Combin.*

Theory, Ser. A 118 (2011) 448–454. <227, 230>

[1543] X.-D. Hou, Classification of self dual quadratic bent functions, *Des. Codes Cryptogr.* 63 (2012) 183–198. <263, 273>

[1544] X.-D. Hou, A new approach to permutation polynomials over finite fields, *Finite Fields Appl.* 18 (2012) 492–521. <227, 230>

[1545] X.-D. Hou and T. Ly, Necessary conditions for reversed Dickson polynomials to be permutational, *Finite Fields Appl.* 16 (2010) 436–448. <227, 230>

[1546] X.-D. Hou and G. L. Mullen, Number of irreducible polynomials and pairs of relatively prime polynomials in several variables over finite fields, *Finite Fields Appl.* 15 (2009) 304–331. <80, 81, 82, 83, 85>

[1547] X.-D. Hou, G. L. Mullen, J. A. Sellers, and J. L. Yucas, Reversed Dickson polynomials over finite fields, *Finite Fields Appl.* 15 (2009) 748–773. <227, 230, 260, 261>

[1548] X.-D. Hou and C. Sze, On certain diagonal equations over finite fields, *Finite Fields Appl.* 15 (2009) 633–643. <208, 213>

[1549] E. Howe and K. Lauter, Improved upper bounds for the number of points on curves over finite fields, *Ann. Inst. Fourier (Grenoble)* 53 (2003) 1677–1737. <460, 463>

[1550] E. Howe, K. Lauter, C. Ritzenthaler, and G. van der Geer, manYPoints - table of curves with many points, *http://www.manypoints.org/*. <460, 463>

[1551] C.-N. Hsu, The distribution of irreducible polynomials in $\mathbf{F}_q[t]$, *J. Number Theory* 61 (1996) 85–96. <76, 79>

[1552] M.-D. A. Huang, Factorization of polynomials over finite fields and factorization of primes in algebraic number fields, In *Proceedings of the Sixteenth Annual ACM Symposium on Theory of Computing,* Washington DC, 175–182, ACM Press, 1984. <381, 382>

[1553] M.-D. A. Huang, Riemann hypothesis and finding roots over finite fields, In *Proceedings of the Seventeenth Annual ACM Symposium on Theory of Computing,* Providence RI, 121–130, ACM Press, 1985. <381, 382>

[1554] X. Huang and V. Y. Pan, Fast rectangular matrix multiplication and applications, *Journal of Complexity* 14 (1998) 257–299. <380, 382>

[1555] H. Hubrechts, Point counting in families of hyperelliptic curves in characteristic 2, *LMS J. Comput. Math.* 10 (2007) 207–234. <454, 456>

[1556] H. Hubrechts, Point counting in families of hyperelliptic curves, *Found. Comput. Math.* 8 (2008) 137–169. <454, 456, 491>

[1557] S. Huczynska and S. D. Cohen, Primitive free cubics with specified norm and trace, *Trans. Amer. Math. Soc.* 355 (2003) 3099–3116 (electronic). <89, 90, 92, 94, 95>

[1558] W. C. Huffman and V. Pless, *Fundamentals of Error-Correcting Codes*, Cambridge University Press, Cambridge, 2003. <31, 32, 661, 663, 672, 674, 677, 678, 681, 682, 703>

[1559] D. R. Hughes, On t-designs and groups, *Amer. J. Math.* 87 (1965) 761–778. <597, 599>

[1560] D. R. Hughes and F. C. Piper, *Projective Planes*, volume 6 of *Graduate Texts in Mathematics*, Springer-Verlag, New York, 1973. <28, 31, 32, 274, 278, 563, 574, 579, 589>

[1561] T. W. Hungerford, *Algebra*, volume 73 of *Graduate Texts in Mathematics*, Springer-Verlag, New York-Berlin, 1980, Reprint of the 1974 original. <80, 85>

[1562] N. E. Hurt, Exponential sums and coding theory: a review, *Acta Appl. Math.* 46

(1997) 49–91. <154, 161>

[1563] D. Husemöller, *Elliptic Curves*, volume 111 of *Graduate Texts in Mathematics*, Springer-Verlag, New York, second edition, 2004. <31, 32, 422, 440>

[1564] H.-K. Hwang, A Poisson $*$ negative binomial convolution law for random polynomials over finite fields, *Random Structures Algorithms* 13 (1998) 17–47. <368, 374>

[1565] J. Y. Hyun, H. Lee, and Y. Lee, MacWilliams duality and a Gleason-type theorem on self-dual bent functions, *Des. Codes Cryptogr.* 63 (2012) 295–304. <263, 273>

[1566] T. Icart, How to hash into elliptic curves, In S. Halevi, editor, *Advances in Cryptology—CRYPTO 2009*, volume 5677 of *Lecture Notes in Comput. Sci.*, 303–316, Springer-Verlag, Berlin, 2009. <796>

[1567] IEEE, Standard specifications for public key cryptography, Standard P1363-2000, Institute of Electrical and Electronics Engineering, 2000, Draft D13 available at `http://grouper.ieee.org/groups/1363/P1363/draft.html`. <68, 70, 785, 796>

[1568] Y. Ihara, On discrete subgroups of the two by two projective linear group over $\mathfrak{p}$-adic fields, *J. Math. Soc. Japan* 18 (1966) 219–235. <658>

[1569] Y. Ihara, Some remarks on the number of rational points of algebraic curves over finite fields, *J. Fac. Sci. Univ. Tokyo Sect. IA Math.* 28 (1981) 721–724. <460, 462, 463, 464, 469>

[1570] L. Illusie, *Cohomologie l-adique et Fonctions L*, volume 589 of *Lecture Notes in Mathematics*, Springer-Verlag, Berlin, 1977, Séminaire de Géometrie Algébrique du Bois-Marie 1965–1966 (SGA 5), Edité par Luc Illusie. <32, 470, 471, 474, 478, 479>

[1571] L. Illusie, Ordinarité des intersections complètes générales, In *The Grothendieck Festschrift, Vol. II*, volume 87 of *Progr. Math.*, 376–405, Birkhäuser Boston, Boston, MA, 1990. <483, 488>

[1572] L. Illusie, Crystalline cohomology, In *Motives*, volume 55 of *Proc. Sympos. Pure Math.*, 43–70, Amer. Math. Soc., Providence, RI, 1994. <479>

[1573] K. Imamura, On self-complementary bases of $GF(q^n)$ over GF(q)., *Trans. IECE Japan, E* 66 (1983) 717–721. <103, 104, 109>

[1574] K. Imamura and M. Morii, Two classes of finite fields which have no self-complementary normal bases, In *1985 IEEE International Symposium on Information Theory Proceedings (ISIT)*, Brighton, England, 1985. <114, 116>

[1575] K. Ireland and M. Rosen, *A Classical Introduction to Modern Number Theory*, volume 84 of *Graduate Texts in Mathematics*, Springer-Verlag, New York, second edition, 1990. <206, 213>

[1576] T. Itoh and S. Tsujii, A fast algorithm for computing multiplicative inverses in $\mathrm{GF}(2^m)$ using normal bases, *Inform. and Comput.* 78 (1988) 171–177. <360, 363, 818, 823>

[1577] I. D. Ivanović, Geometrical description of quantal state determination, *J. Phys. A* 14 (1981) 3241–3245. <835, 841>

[1578] G. Ivanyos, M. Karpinski, L. Rónyai, and N. Saxena, Trading GRH for algebra: Algorithms for factoring polynomials and related structures, *Math. Comp.* 81 (2012) 493–531. <381, 382>

[1579] G. Ivanyos, M. Karpinski, and N. Saxena, Schemes for deterministic polynomial factoring, In *Proceedings of the 2009 International Symposium on Symbolic and Algebraic Computation*, ISSAC '09, 191–198, ACM, New York, NY, USA, 2009. <381, 382>

[1580] H. Iwaniec, *Topics in Classical Automorphic Forms*, volume 17 of *Graduate Studies in Mathematics*, American Mathematical Society, Providence, RI, 1997. <157, 161>

[1581] H. Iwaniec and E. Kowalski, *Analytic Number Theory*, volume 53 of *American Mathematical Society Colloquium Publications*, American Mathematical Society, Providence, RI, 2004. <140, 154, 156, 157, 160, 161, 181, 185>

[1582] F. Izadi, C_{ab} curves and arithmetic on their Jacobians, In *Algebraic Curves and Cryptography*, volume 58 of *Fields Inst. Commun.*, 99–118, Amer. Math. Soc., Providence, RI, 2010. <807, 811>

[1583] F. Jacob and J. Monod, Genetic regulatory mechanisms in the synthesis of proteins, *Journal of Molecular Biology* 3 (1961) 318–356. <827, 834>

[1584] C. G. J. Jacobi, Uber die kreistheilung und ihre anwendung auf die zahlentheorie, *Gesammelte Werke* 6 (1846) 254–274. <27, 32>

[1585] M. Jacobson, A. Menezes, and A. Stein, Solving elliptic curve discrete logarithm problems using Weil descent, *J. Ramanujan Math. Soc.* 16 (2001) 231–260. <810, 811>

[1586] M. Jacobson, Jr., A. Menezes, and A. Stein, Hyperelliptic curves and cryptography, In *High Primes and Misdemeanours: Lectures in Honour of the 60th Birthday of Hugh Cowie Williams*, volume 41 of *Fields Inst. Commun.*, 255–282, Amer. Math. Soc., Providence, RI, 2004. <452, 455, 456, 809, 811>

[1587] M. Jacobson, Jr., R. Scheidler, and A. Stein, Cryptographic aspects of real hyperelliptic curves, *Tatra Mt. Math. Publ.* 47 (2010) 31–65. <448, 452, 453, 456>

[1588] M. J. Jacobson, Jr., R. Scheidler, and A. Stein, Fast arithmetic on hyperelliptic curves via continued fraction expansions, In *Advances in Coding Theory and Cryptography*, volume 3 of *Ser. Coding Theory Cryptol.*, 200–243, World Sci. Publ., Hackensack, NJ, 2007. <451, 452, 456>

[1589] N. Jacobson, *Basic Algebra. I*, W. H. Freeman and Company, New York, 2nd edition, 1985. <510, 511, 513, 514, 517, 518, 520>

[1590] R. Jain, Error characteristics of fiber distributed data interface (FDDI), *IEEE Transactions on Communications* 38 (1990) 1244–1252. <638, 639, 642>

[1591] K. Jambunathan, On choice of connection-polynomials for LFSR-based stream ciphers, In *Progress in Cryptology—INDOCRYPT 2000*, volume 1977 of *Lecture Notes in Comput. Sci.*, 9–18, Springer, Berlin, 2000. <633, 634, 635, 642>

[1592] N. S. James and R. Lidl, Permutation polynomials on matrices, *Linear Algebra Appl.* 96 (1987) 181–190. <226, 230>

[1593] G. J. Janusz, Separable algebras over commutative rings, *Trans. Amer. Math. Soc.* 122 (1966) 461–479. <28>

[1594] H. Janwa, G. M. McGuire, and R. M. Wilson, Double-error-correcting cyclic codes and absolutely irreducible polynomials over GF(2), *J. Algebra* 178 (1995) 665–676. <260, 261>

[1595] H. Janwa and R. M. Wilson, Hyperplane sections of Fermat varieties in $\mathbf{P}^3$ in char. 2 and some applications to cyclic codes, In *Applied Algebra, Algebraic Algorithms and Error-Correcting Codes*, volume 673 of *Lecture Notes in Comput. Sci.*, 180–194, Springer, Berlin, 1993. <258, 260, 261>

[1596] A. S. Jarrah and R. Laubenbacher, On the algebraic geometry of polynomial dynamical systems, In *Emerging Applications of Algebraic Geometry*, volume 149 of *IMA Vol. Math. Appl.*, 109–123, Springer, New York, 2009. <337, 344>

[1597] A. S. Jarrah, R. Laubenbacher, B. Stigler, and M. Stillman, Reverse-engineering of polynomial dynamical systems, *Advances in Applied Mathematics* 39 (2007) 477–489. <831, 834>

[1598] A. S. Jarrah, R. Laubenbacher, and A. Veliz-Cuba, The dynamics of conjunctive and disjunctive Boolean network models, *Bull. Math. Biol.* 72 (2010) 1425–1447. <834>

[1599] A. S. Jarrah, B. Raposa, and R. Laubenbacher, Nested canalyzing, unate cascade, and polynomial functions, *Phys. D* 233 (2007) 167–174. <833, 834>

[1600] C.-P. Jeannerod and C. Mouilleron, Computing specified generators of structured matrix inverses, In W. Koepf, editor, *Proceedings of the International Symposium of Symbolic and Algebraic Computation—ISSAC 2010*, 281–288. ACM, 2010. <533, 535>

[1601] T. Jebelean, An algorithm for exact division, *J. Symbolic Comput.* 15 (1993) 169–180. <359, 363>

[1602] T. Jebelean, Improving the multiprecision Euclidean algorithm, In *Design and Implementation of Symbolic Computation Systems - DISCO 1993*, volume 722 of *Lecture Notes in Comput. Sci.*, 45–58, Springer, Berlin, 1993. <359, 363>

[1603] J. Jedwab, What can be used instead of a Barker sequence?, In *Finite Fields and Applications*, volume 461 of *Contemp. Math.*, 153–178, Amer. Math. Soc., Providence, RI, 2008. <841, 849>

[1604] J. Jedwab, D. J. Katz, and K.-U. Schmidt, Advances in the merit factor problem for binary sequences, arXiv:1205.0626v1, 2012. <842, 849>

[1605] J. Jedwab, D. J. Katz, and K.-U. Schmidt, Littlewood polynomials with small L^4 norm, arXiv:1205.0260v1, 2012. <323, 324, 842, 849>

[1606] E. Jensen and M. R. Murty, Artin's conjecture for polynomials over finite fields, In *Number Theory*, Trends in Mathematics, 167–181, Birkhauser, Basel, 2000. <497, 500>

[1607] J. M. Jensen, H. E. Jensen, and T. Høholdt, The merit factor of binary sequences related to difference sets, *IEEE Trans. Inform. Theory* 37 (1991) 617–626. <842, 849>

[1608] V. Jha and N. L. Johnson, An analog of the Albert-Knuth theorem on the orders of finite semifields, and a complete solution to Cofman's subplane problem, *Algebras Groups Geom.* 6 (1989) 1–35. <276, 278>

[1609] V. Jha and N. L. Johnson, Nests of reguli and flocks of quadratic cones, *Simon Stevin* 63 (1989) 311–338. <568, 574>

[1610] X. Jiang, J. Ding, and L. Hu, Kipnis-Shamir attack on HFE revisited, In *Information Security and Cryptology*, volume 4990 of *Lecture Notes in Comput. Sci.*, 399–411, Springer, Berlin, 2008. <780, 783>

[1611] N. L. Johnson, Projective planes of prime order p that admit collineation groups of order p^2, *J. Geom.* 30 (1987) 49–68. <281, 282>

[1612] N. L. Johnson, Nest replaceable translation planes, *J. Geom.* 36 (1989) 49–62. <568, 574>

[1613] N. L. Johnson, V. Jha, and M. Biliotti, *Handbook of Finite Translation Planes*, volume 289 of *Pure and Applied Mathematics*, Chapman & Hall/CRC, Boca Raton, FL, 2007. <565, 574>

[1614] N. L. Johnson and R. Pomareda, André planes and nests of reguli, *Geom. Dedicata* 31 (1989) 245–260. <568, 574>

[1615] N. L. Johnson and R. Pomareda, Mixed nests, *J. Geom.* 56 (1996) 59–86. <568, 574>

[1616] S. C. Johnson, Sparse polynomial arithmetic, *ACM SIGSAM Bull.* 8 (1974) 63–71.

<382, 392>

[1617] J.-R. Joly, Équations et variétés algébriques sur un corps fini, *Enseignement Math., IIe Ser.* 19 (1973) 1–117. <206, 213>

[1618] R. Jones, Iterated Galois towers, their associated martingales, and the p-adic Mandelbrot set, *Compos. Math.* 143 (2007) 1108–1126. <337, 342, 344>

[1619] R. Jones, The density of prime divisors in the arithmetic dynamics of quadratic polynomials, *J. Lond. Math. Soc., 2nd Ser.* 78 (2008) 523–544. <337, 342, 344>

[1620] R. Jones and N. Boston, Settled polynomials over finite fields, *Proc. Amer. Math. Soc.* 140 (2012) 1849–1863. <178, 185, 342, 344>

[1621] C. Jordan, *Traité des Substitutions et des Équations Algébriques*, Paris: Gauthier-Villars, 1870. <10, 11>

[1622] H. F. Jordan and D. C. M. Wood, On the distribution of sums of successive bits of shift-register sequences, *IEEE Trans. Computers* C-22 (1973) 400–408. <630, 642>

[1623] J.-P. Jouanolou, *Théorèmes de Bertini et Applications*, volume 42 of *Progress in Mathematics*, Birkhäuser Boston, 1983. <387, 392>

[1624] A. Joux, A one round protocol for tripartite Diffie-Hellman, *J. Cryptology* 17 (2004) 263–276. <748, 750>

[1625] A. Joux, Discrete logarithms in GF(2^{607}) and GF(2^{613}), mailing list announcement, `https://listserv.nodak.edu/cgi-bin/wa.exe?A2=ind0509&L=NMBRTHRY&P=R1490&D=0&I=-3&T=0`, 2005. <400, 401>

[1626] A. Joux and R. Lercier, The function field sieve is quite special, In *Algorithmic Number Theory*, volume 2369 of *Lecture Notes in Comput. Sci.*, 431–445, Springer, Berlin, 2002. <399, 401>

[1627] A. Joux and R. Lercier, Improvements to the general number field sieve for discrete logarithms in prime fields: A comparison with the Gaussian integer method, *Math. Comp.* 72 (2003) 953–967. <399, 401>

[1628] A. Joux and R. Lercier, The function field sieve in the medium prime case, In *Advances in Cryptology—EUROCRYPT 2006*, volume 4004 of *Lecture Notes in Comput. Sci.*, 254–270, Springer, Berlin, 2006. <399, 401>

[1629] A. Joux and V. Vitse, Cover and decomposition index calculus on elliptic curves made practical—Application to a seemingly secure curve over $\mathbb{F}_{p^6}$, In *Advances in Cryptology—EUROCRYPT 2012*, volume 7237 of *Lecture Notes Comput. Sci.*, 9–26, 2011. <786, 796>

[1630] M. Joye, A. Miyaji, and A. Otsuka, editors, *Pairing-Based Cryptography — Pairing 2010*, volume 6487 of *Lecture Notes in Comput. Sci.*, Springer-Verlag, Berlin, 2010. <788, 796>

[1631] D. Jungnickel, *Finite Fields: Structure and Arithmetics*, Bibliographisches Institut, Mannheim, 1993. <13, 31, 32, 39, 49, 101, 103, 105, 106, 109, 113, 116, 117, 123, 128, 170, 183, 185, 325, 327, 328, 329, 336, 503, 504, 505, 506, 510>

[1632] D. Jungnickel, Trace-orthogonal normal bases, *Discrete Appl. Math.* 47 (1993) 233–249. <123, 124, 128>

[1633] D. Jungnickel, T. Beth, and W. Geiselmann, A note on orthogonal circulant matrices over finite fields, *Arch. Math. (Basel)* 62 (1994) 126–133. <506, 510>

[1634] D. Jungnickel and M. J. de Resmini, Another case of the prime power conjecture for finite projective planes, *Adv. Geom.* 2 (2002) 215–218. <573, 574>

[1635] D. Jungnickel, A. J. Menezes, and S. A. Vanstone, On the number of self-dual bases of GF(q^m) over GF(q), *Proc. Amer. Math. Soc.* 109 (1990) 23–29. <104, 109,

115, 116>

[1636] D. Jungnickel and H. Niederreiter, editors, *Finite Fields and Applications*, Springer-Verlag, Berlin, 2001. <31, 32>

[1637] D. Jungnickel and S. A. Vanstone, On primitive polynomials over finite fields, *J. Algebra* 124 (1989) 337–353. <92, 95>

[1638] J. Justesen, A class of constructive asymptotically good algebraic codes, *IEEE Trans. Inform. Theory* IT-18 (1972) 652–656. <689, 702, 703>

[1639] J. Justesen and T. Høholdt, *A Course in Error-Correcting Codes*, EMS Textbooks in Mathematics. European Mathematical Society (EMS), Zürich, 2004. <661, 703>

[1640] V. Kabanets and R. Impagliazzo, Derandomizing polynomial identity tests means proving circuit lower bounds, *Comput. Complexity* 13 (2004) 1–46. <391, 392>

[1641] N. Kahale, Isoperimetric inequalities and eigenvalues, *SIAM J. Discrete Math.* 10 (1997) 30–40. <645, 658>

[1642] T. Kaida, S. Uehara, and K. Imamura, An algorithm for the k-error linear complexity of sequences over GF(p^m) with period p^n, p a prime, *Inform. and Comput.* 151 (1999) 134–147. <329, 336>

[1643] T. Kailath, S. Y. Kung, and M. Morf, Displacement ranks of a matrix, *Bull. Amer. Math. Soc. (New Ser.)* 1 (1979) 769–773. <532, 535>

[1644] B. S. Kaliski, Jr., The Montgomery inverse and its applications, *IEEE Trans. on Computers* 44 (1995) 1064–1065. <360, 363>

[1645] E. Kaltofen, A polynomial reduction from multivariate to bivariate integral polynomial factorization, In *Proceedings of the Fourteenth Symposium on Theory of Computing*, 261–266, ACM, 1982. <387, 392>

[1646] E. Kaltofen, A polynomial-time reduction from bivariate to univariate integral polynomial factorization, In *Proc. Twenty Third Annual Symp. Foundations of Comp. Sci.*, 57–64. IEEE, 1982. <381, 382, 387, 392>

[1647] E. Kaltofen, Effective Hilbert irreducibility, *Information and Control* 66 (1985) 123–137. <387, 392>

[1648] E. Kaltofen, Fast parallel absolute irreducibility testing, *J. Symbolic Comput.* 1 (1985) 57–67. <387, 392>

[1649] E. Kaltofen, Sparse Hensel lifting, In *Proceedings of EUROCAL '85, Vol. 2,* volume 204 of *Lecture Notes in Comput. Sci.*, 4–17, Springer-Verlag, 1985. <387, 391, 392>

[1650] E. Kaltofen, Uniform closure properties of p-computable functions, In *Proc. Eighteenth Annual ACM Symp. Theory Comput.*, 330–337, 1986, also published as part of [1652] and [1653]. <390, 391, 392>

[1651] E. Kaltofen, Deterministic irreducibility testing of polynomials over large finite fields, *J. Symbolic Comput.* 4 (1987) 77–82. <387, 392>

[1652] E. Kaltofen, Greatest common divisors of polynomials given by straight-line programs, *J. ACM* 35 (1988) 231–264. <936>

[1653] E. Kaltofen, Factorization of polynomials given by straight-line programs, In S. Micali, editor, *Randomness and Computation*, volume 5 of *Advances in Computing Research*, 375–412, JAI Press Inc., Greenwhich, Connecticut, 1989. <390, 391, 392, 936>

[1654] E. Kaltofen, Polynomial factorization 1982-1986, In D. V. Chudnovsky and R. D. Jenks, editors, *Computers in Mathematics*, volume 125 of *Lecture Notes in Pure and Applied Mathematics*, 285–309, Marcel Dekker, New York, NY, 1990. <387,

392>

[1655] E. Kaltofen, Polynomial factorization 1987-1991, In *Proc. LATIN '92*, volume 583 of *Lecture Notes in Comput. Sci.*, 294–313, Springer-Verlag, 1992. <381, 382, 387, 392>

[1656] E. Kaltofen, Asymptotically fast solution of Toeplitz-like singular linear systems, In *Proceedings of the International Symposium on Symbolic and Algebraic Computation*, ISSAC '94, 297–304, ACM, New York, NY, USA, 1994. <533, 535>

[1657] E. Kaltofen, Analysis of Coppersmith's block Wiedemann algorithm for the parallel solution of sparse linear systems, *Math. Comp.* 64 (1995) 777–806. <534, 535>

[1658] E. Kaltofen, Effective Noether irreducibility forms and applications, *J. Comput. System Sci.* 50 (1995) 274–295. <386, 392>

[1659] E. Kaltofen, Polynomial factorization: a success story, In *ISSAC '03: Proceedings of the 2003 International Symposium on Symbolic and Algebraic Computation*, 3–4, ACM, 2003. <387, 392>

[1660] E. Kaltofen and P. Koiran, On the complexity of factoring bivariate supersparse (lacunary) polynomials, In *ISSAC '05: Proceedings of the 2005 International Symposium on Symbolic and Algebraic Computation*, 208–215, ACM, 2005. <390, 392>

[1661] E. Kaltofen and P. Koiran, Finding small degree factors of multivariate supersparse (lacunary) polynomials over algebraic number fields, In *ISSAC '06: Proceedings of the 2006 International Symposium on Symbolic and Algebraic Computation*, 162–168, ACM, 2006. <390, 392>

[1662] E. Kaltofen and W. Lee, Early termination in sparse interpolation algorithms, *J. Symbolic Comput.* 36 (2003) 365–400. <392>

[1663] E. Kaltofen and A. Lobo, Factoring high-degree polynomials by the black box Berlekamp algorithm, Technical report, Department of Computer Science, Rensselaer Polytechnic Institute, 1994. <381, 382>

[1664] E. Kaltofen and A. Lobo, Distributed matrix-free solution of large sparse linear systems over finite fields, *Algorithmica* 24 (1999) 331–348. <530, 534, 535>

[1665] E. Kaltofen and V. Pan, Parallel solution of Toeplitz and Toeplitz-like linear systems over fields of small positive characteristic, In *First International Symposium on Parallel Symbolic Computation—PASCO '94*, volume 5 of *Lecture Notes Ser. Comput.*, 225–233, World Sci. Publ., River Edge, NJ, 1994. <509, 510>

[1666] E. Kaltofen and B. D. Saunders, On Wiedemann's method of solving sparse linear systems, In *Applied Algebra, Algebraic Algorithms and Error-Correcting Codes*, volume 539 of *Lecture Notes in Comput. Sci.*, 29–38, Springer, Berlin, 1991. <531, 533, 535>

[1667] E. Kaltofen and V. Shoup, Subquadratic-time factoring of polynomials over finite fields, *Math. Comp.* 67 (1998) 1179–1197. <369, 374, 376, 380, 381, 382>

[1668] E. Kaltofen and B. Trager, Computing with polynomials given by black boxes for their evaluations: Greatest common divisors, factorization, separation of numerators and denominators, In *Proc. Twenty Ninth Annual Symp. Foundations of Comp. Sci.*, 296–305, 1988. <391, 392>

[1669] E. Kaltofen and B. Trager, Computing with polynomials given by black boxes for their evaluations: Greatest common divisors, factorization, separation of numerators and denominators, *J. Symbolic Comput.* 9 (1990) 301–320. <391, 392, 530, 535>

[1670] E. Kaltofen and G. Villard, On the complexity of computing determinants, *Comput. Complexity* 13 (2004) 91–130. <390, 392, 535>

[1671] N. Kamiya, On multisequence shift register synthesis and generalized-minimum-distance decoding of Reed-Solomon codes, *Finite Fields Appl.* 1 (1995) 440–457. <329, 336>

[1672] J.-G. Kammerer, R. Lercier, and G. Renault, Encoding points on hyperelliptic curves over finite fields in deterministic polynomial time, In M. Joye, A. Miyaji, and A. Otsuka, editors, *Pairing-Based Cryptography—Pairing 2010*, volume 6487 of *Lecture Notes in Comput. Sci.*, 278–297, Springer-Verlag, Berlin, 2010. <796>

[1673] W. M. Kantor, An exponential number of generalized Kerdock codes, *Inform. and Control* 53 (1982) 74–80. <251, 252>

[1674] W. M. Kantor, Two families of flag-transitive affine planes, *Geom. Dedicata* 41 (1992) 191–200. <568, 569, 574>

[1675] W. M. Kantor, 2-transitive and flag-transitive designs, In *Coding Theory, Design Theory, Group Theory*, Wiley-Intersci. Publ., 13–30, Wiley, New York, 1993. <569, 574>

[1676] W. M. Kantor, Note on GMW designs, *European J. Combin.* 22 (2001) 63–69. <603, 607>

[1677] W. M. Kantor, Commutative semifields and symplectic spreads, *J. Algebra* 270 (2003) 96–114. <278>

[1678] W. M. Kantor, Finite semifields, In *Finite Geometries, Groups, and Computation*, 103–114, de Gruyter, Berlin, 2006. <278>

[1679] W. M. Kantor, HMO-planes, *Adv. Geom.* 9 (2009) 31–43. <276, 278>

[1680] W. M. Kantor and R. A. Liebler, Semifields arising from irreducible semilinear transformations, *J. Aust. Math. Soc.* 85 (2008) 333–339. <276, 278>

[1681] W. M. Kantor and C. Suetake, A note on some flag-transitive affine planes, *J. Combin. Theory, Ser. A* 65 (1994) 307–310. <568, 569, 574>

[1682] W. M. Kantor and M. E. Williams, Symplectic semifield planes and $\mathbb{Z}_4$-linear codes, *Trans. Amer. Math. Soc.* 356 (2004) 895–938. <276, 278>

[1683] A. A. Karatsuba, The complexity of computations, *Trudy Mat. Inst. Steklov.* 211 (1995) 186–202. <355, 363>

[1684] A. A. Karatsuba and Y. Ofman, Multiplication of multiplace numbers on automata, *Dokl. Acad. Nauk SSSR* 145 (1962) 293–294, English translation in Soviet Physics-Doklady 7, 595–596,1963. <354, 355, 363, 382, 813, 814, 823>

[1685] M. Karzand and E. Telatar, Polar codes for q-ary source coding, In *Proc. IEEE International Symposium on Information Theory*, 909–912, 2010. <739>

[1686] M. Kasahara and R. Sakai, A construction of public key cryptosystem for realizing ciphtertext of size 100 bit and digital signature scheme, *IEICE Trans. Fundamentals* E87-A (2004) 102–109. <772, 783>

[1687] M. Kasahara and R. Sakai, A construction of public-key cryptosystem based on singular simultaneous equations, *IEICE Transactions on Fundamentals of Electronics, Communications and Computer Sciences* E88-A (2005) 74–80. <772, 783>

[1688] T. Kasami, Weight distributions of Bose-Chaudhuri-Hocquenghem codes, In *Combinatorial Mathematics and its Applications*, 335–357, Univ. North Carolina Press, Chapel Hill, N.C., 1969. <258, 261, 321, 324>

[1689] T. Kasami, The weight enumerators for several classes of subcodes of the 2nd order binary Reed-Muller codes, *Information and Control* 18 (1971) 369–394. <258, 259, 261>

[1690] T. Kasami, The weight enumerators for several classes of subcodes of the second order binary Reed-Muller codes, *Inform. and Control* 18 (1971) 369–394. <227, 230>

[1691] T. Kasami, S. Lin, and W. W. Peterson, Generalized Reed-Muller codes, *Electron. Commun. Japan* 51 (1968) 96–104. <686, 688, 689, 703>

[1692] T. Kasami, S. Lin, and W. W. Peterson, Polynomial codes, *IEEE Trans. Inform. Theory* IT-14 (1968) 807–814. <688, 689, 703>

[1693] M. Kassabov, A. Lubotzky, and N. Nikolov, Finite simple groups as expanders, *Proc. Natl. Acad. Sci. USA* 103 (2006) 6116–6119. <652, 658>

[1694] J. Katz and Y. Lindell, *Introduction to Modern Cryptography*, Chapman & Hall/CRC Cryptography and Network Security. Chapman & Hall/CRC, Boca Raton, FL, 2008. <31, 32, 750>

[1695] N. Katz and R. Livné, Sommes de Kloosterman et courbes elliptiques universelles en caractéristiques 2 et 3, *C. R. Acad. Sci. Paris, Sér. I, Math.* 309 (1989) 723–726. <270, 273>

[1696] N. H. Katz and C.-Y. Shen, Garaev's inequality in finite fields not of prime order, *Online J. Anal. Comb.* (2008) Art. 3, 6. <187, 192>

[1697] N. H. Katz and C.-Y. Shen, A slight improvement to Garaev's sum product estimate, *Proc. Amer. Math. Soc.* 136 (2008) 2499–2504. <192>

[1698] N. M. Katz, On a theorem of Ax, *Amer. J. Math.* 93 (1971) 485–499. <199, 201, 210, 213>

[1699] N. M. Katz, Slope filtration of F-crystals, In *Journées de Géométrie Algébrique de Rennes, Vol. I*, volume 63 of *Astérisque*, 113–163, Soc. Math. France, Paris, 1979. <483, 488>

[1700] N. M. Katz, *Sommes Exponentielles*, volume 79 of *Astérisque*, Société Mathématique de France, Paris, 1980. <156, 161, 164, 169>

[1701] N. M. Katz, *Gauss Sums, Kloosterman Sums, and Monodromy Groups*, volume 116 of *Annals of Mathematics Studies*, Princeton University Press, Princeton, NJ, 1988. <31, 32, 140, 154, 155, 156, 159, 161>

[1702] N. M. Katz, An estimate for character sums, *J. Amer. Math. Soc.* 2 (1989) 197–200. <168, 169, 183, 184, 185>

[1703] N. M. Katz, Affine cohomological transforms, perversity, and monodromy, *J. Amer. Math. Soc.* 6 (1993) 149–222. <167, 169>

[1704] N. M. Katz, Estimates for "singular" exponential sums, *Internat. Math. Res. Notices* (1999) 875–899. <165, 169, 197, 201, 340, 344>

[1705] N. M. Katz, Frobenius-Schur indicator and the ubiquity of Brock-Granville quadratic excess, *Finite Fields Appl.* 7 (2001) 45–69. <198, 201>

[1706] N. M. Katz, Sums of Betti numbers in arbitrary characteristic, *Finite Fields Appl.* 7 (2001) 29–44. <473, 474, 476, 479>

[1707] N. M. Katz, Estimates for nonsingular multiplicative character sums, *Int. Math. Res. Not.* 7 (2002) 333–349. <166, 169, 199, 201>

[1708] N. M. Katz, *Moments, Monodromy, and Perversity: a Diophantine Perspective*, volume 159 of *Annals of Mathematics Studies*, Princeton University Press, Princeton, NJ, 2005. <193, 201>

[1709] N. M. Katz, Estimates for nonsingular mixed character sums, *Int. Math. Res. Not. IMRN* (2007) Art. ID rnm069, 19. <168, 169>

[1710] N. M. Katz, Another look at the Dwork family, In *Algebra, Arithmetic, and Geometry: in Honor of Yu. I. Manin. Vol. II*, volume 270 of *Progr. Math.*, 89–126, Birkhäuser Boston Inc., Boston, MA, 2009. <472, 479>

[1711] N. M. Katz, *Convolution and Equidistribution: Sato-Tate Theorems for Finite-Field*

Mellin Transforms, Annals of Mathematics Studies. Princeton University Press, Princeton, NJ, 2012. <144, 157, 161>

[1712] N. M. Katz and G. Laumon, Transformation de Fourier et majoration de sommes exponentielles, *Inst. Hautes Études Sci. Publ. Math.* (1985) 361–418. <167, 169>

[1713] N. M. Katz and Z. Zheng, On the uniform distribution of Gauss sums and Jacobi sums, In *Analytic Number Theory, Vol. 2*, volume 139 of *Progr. Math.*, 537–558, Birkhäuser Boston, Boston, MA, 1996. <140, 144, 161>

[1714] S. Kauffman, C. Peterson, B. Samuelsson, and C. Troein, Genetic networks with canalyzing Boolean rules are always stable, *Proceedings of the National Academy of Sciences of the United States of America* 101 (2004) 17102–17107. <832, 834>

[1715] S. A. Kauffman, Metabolic stability and epigenesis in randomly constructed genetic nets, *Journal of Theoretical Biology* 22 (1969) 437–467. <829, 834>

[1716] N. Kayal, Recognizing permutation functions in polynomial time, *ECCC* TR05-008 (2005). <217, 230, 392>

[1717] W. F. Ke and H. Kiechle, On the solutions of the equation $x^m + y^m - z^m = 1$ in a finite field, *Proc. Amer. Math. Soc.* 123 (1995) 1331–1339. <208, 213>

[1718] K. S. Kedlaya, Counting points on hyperelliptic curves using Monsky-Washnitzer cohomology, *J. Ramanujan Math. Soc.* 16 (2001) 323–338. <454, 456, 491>

[1719] K. S. Kedlaya, Errata for: "Counting points on hyperelliptic curves using Monsky-Washnitzer cohomology" [J. Ramanujan Math. Soc. **16** (2001), no. 4, 323–338], *J. Ramanujan Math. Soc.* 18 (2003) 417–418. <454, 456>

[1720] K. S. Kedlaya, Computing zeta functions via p-adic cohomology, In *Algorithmic Number Theory*, volume 3076 of *Lecture Notes in Comput. Sci.*, 1–17, Springer, Berlin, 2004. <491>

[1721] K. S. Kedlaya and C. Umans, Fast modular composition in any characteristic, In *Forty ninth Annual IEEE Symposium on Foundations of Computer Science*, 146–155, IEEE Computer Society, 2008. <350, 358, 363>

[1722] K. S. Kedlaya and C. Umans, Fast polynomial factorization and modular composition, *SIAM J. Comput.* 40 (2011) 1767–1802. <358, 363, 376, 377, 378, 380, 381, 382>

[1723] W. Keller-Gehrig, Fast algorithms for the characteristic polynomial, *Theoret. Comput. Sci.* 36 (1985) 309–317. <529, 535>

[1724] C. A. Kelley and D. Sridhara, Pseudocodewords of Tanner graphs, *IEEE Trans. Inform. Theory* 53 (2007) 4013–4038. <718, 719>

[1725] D. Kelmer, Distribution of twisted Kloosterman sums modulo prime powers, *Int. J. Number Theory* 6 (2010) 271–280. <156, 161>

[1726] H. Kempfert, On the factorization of polynomials, *Journal of Number Theory* 1 (1969) 116–120. <381, 382>

[1727] O. Kempthorne, A simple approach to confounding and fractional replication in factorial experiments, *Biometrika* 34 (1947) 255–272. <631, 642>

[1728] A. M. Kerdock, A class of low-rate nonlinear binary codes, *Information and Control* 20 (1972) 182–187; ibid. 21 (1972), 395. <251, 252, 701, 703>

[1729] A. Ketkar, A. Klappenecker, S. Kumar, and P. K. Sarvepalli, Nonbinary stabilizer codes over finite fields, *IEEE Trans. Inform. Theory* 52 (2006) 4892–4914. <838, 841>

[1730] H. Kharaghani and B. Tayfeh-Rezaie, A Hadamard matrix of order 428, *J. Combin. Des.* 13 (2005) 435–440. <170, 185>

[1731] K. Khoo, G. Gong, and D. R. Stinson, New family of Gold-like sequences, *IEEE*

Intern. Symp. Inform. Theory 2 (2002) 181. <206>

[1732] D. S. Kim, Codes associated with special linear groups and power moments of multi-dimensional Kloosterman sums, *Ann. Mat. Pura Appl. Ser. IV* 190 (2011) 61–76. <154, 161>

[1733] J. H. Kim, Codes associated with $\mathrm{Sp}(4, q)$ and even-power moments of Kloosterman sums, *Bull. Aust. Math. Soc.* 79 (2009) 427–435. <157, 161>

[1734] J. H. Kim, R. Montenegro, Y. Peres, and P. Tetali, A birthday paradox for Markov chains with an optimal bound for collision in the Pollard rho algorithm for discrete logarithm, *Ann. Appl. Probab.* 20 (2010) 495–521. <397, 401>

[1735] R. Kim and W. Koepf, Parity of the number of irreducible factors for composite polynomials, *Finite Fields Appl.* 16 (2010) 137–143. <69, 70>

[1736] S.-H. Kim and J.-S. No, New families of binary sequences with low correlation, *IEEE Trans. Inform. Theory* 49 (2003) 3059–3065. <206>

[1737] A. Kipnis, J. Patarin, and L. Goubin, Unbalanced oil and vinegar signature schemes, In *Advances in Cryptology—EUROCRYPT '99*, volume 1592 of *Lecture Notes in Comput. Sci.*, 206–222, Springer, Berlin, 1999. <770, 781, 783>

[1738] A. Kipnis and A. Shamir, Cryptanalysis of the oil and vinegar signature scheme, In *Advances in Cryptology—CRYPTO '98*, volume 1462 of *Lecture Notes in Comput. Sci.*, 257–266, Springer, Berlin, 1998. <780, 783>

[1739] A. Kipnis and A. Shamir, Cryptanalysis of the HFE public key cryptosystem by relinearization, In *Advances in Cryptology—CRYPTO '99*, volume 1666 of *Lecture Notes in Comput. Sci.*, 19–30, Springer, Berlin, 1999. <390, 392, 780, 783>

[1740] T. Kiran and B. S. Rajan, Optimal rate-diversity tradeoff STBCs from codes over arbitrary finite fields, In *IEEE Int. Conf. Commun.*, 453–457, 2005. <848, 849>

[1741] T. P. Kirkman, On a problem in combinations, *Cambridge and Dublin Math. J.* 2 (1847) 191–204. <590, 599>

[1742] A. Y. Kitaev, A. H. Shen, and M. N. Vyalyi, *Classical and Quantum Computation*, volume 47 of *Graduate Studies in Mathematics*, American Mathematical Society, Providence, RI, 2002. <835, 841>

[1743] A. Klappenecker and M. Rötteler, Constructions of mutually unbiased bases, In *Finite Fields and Applications*, volume 2948 of *Lecture Notes in Comput. Sci.*, 137–144, Springer, Berlin, 2004. <835, 841>

[1744] A. Klappenecker, M. Rötteler, I. E. Shparlinski, and A. Winterhof, On approximately symmetric informationally complete positive operator-valued measures and related systems of quantum states, *J. Math. Phys.* 46 (2005) 082104, 17. <837, 841>

[1745] A. Klapper, Cross-correlations of geometric sequences in characteristic two, *Des. Codes Cryptogr.* 3 (1993) 347–377. <203, 204, 206>

[1746] A. Klapper, Cross-correlations of quadratic form sequences in odd characteristic, *Des. Codes Cryptogr.* 11 (1997) 289–305. <203, 204, 206>

[1747] A. Klapper, A. H. Chan, and M. Goresky, Cross-correlations of linearly and quadratically related geometric sequences and GMW sequences, *Discrete Appl. Math.* 46 (1993) 1–20. <204, 206>

[1748] A. Klapper and M. Goresky, Feedback shift registers, 2-adic span, and combiners with memory, *J. Cryptology* 10 (1997) 111–147. <335, 336>

[1749] A. M. Klapper, Expected π-adic security measures of sequences, *IEEE Trans. Inform. Theory* 56 (2010) 2486–2501. <336>

[1750] S. L. Kleiman, Bertini and his two fundamental theorems, *Rend. Circ. Mat. Palermo*

Ser. II Suppl. 55 (1998) 9–37. <387, 392>

[1751] E. Kleinfeld, Techniques for enumerating Veblen-Wedderburn systems, *J. Assoc. Comput. Mach.* 7 (1960) 330–337. <275, 278>

[1752] T. Kleinjung, Discrete logarithms in GF(p) – 160 digits, mailing list announcement, `https://listserv.nodak.edu/cgi-bin/wa.exe?A2=ind0702&L=NMBRTHRY&P=R45&D=0&I=-3&T=0`, 2007. <400, 401>

[1753] T. Kleinjung, K. Aoki, J. Franke, A. K. Lenstra, E. Thomé, J. W. Bos, P. Gaudry, A. Kruppa, P. L. Montgomery, D. A. Osvik, H. te Riele, A. Timofeev, and P. Zimmermann, Factorization of a 768-bit RSA modulus, In *Advances in Cryptology—CRYPTO 2010*, volume 6223 of *Lecture Notes in Comput. Sci.*, 333–350, Springer, Berlin, 2010. <400, 401>

[1754] R. Kloosterman, The zeta function of monomial deformations of Fermat hypersurfaces, *Algebra Number Theory* 1 (2007) 421–450. <479, 488>

[1755] A. A. Klyachko, Monodromy groups of polynomial mappings, In *Studies in Number Theory*, volume 6, 82–91, Izdat. Saratov. Univ., Saratov, 1975. <237, 240>

[1756] A. W. Knapp, *Elliptic Curves*, volume 40 of *Mathematical Notes*, Princeton University Press, Princeton, NJ, 1992. <31, 32, 422, 440>

[1757] M. P. Knapp, Diagonal equations of different degrees over p-adic fields, *Acta Arith.* 126 (2007) 139–154. <213>

[1758] N. Knarr and M. Stroppel, Polarities and unitals in the Coulter-Matthews planes, *Des. Codes Cryptogr.* 55 (2010) 9–18. <280, 282>

[1759] E. Knill and R. Laflamme, Theory of quantum error-correcting codes, *Phys. Rev. A* 55 (1997) 900–911. <837, 841>

[1760] A. Knopfmacher and J. Knopfmacher, Counting polynomials with a given number of zeros in a finite field, *Linear and Multilinear Algebra* 26 (1990) 287–292. <368, 374>

[1761] A. Knopfmacher and J. Knopfmacher, Counting irreducible factors of polynomials over a finite field, *Discrete Math.* 112 (1993) 103–118. <368, 374>

[1762] A. Knopfmacher, J. Knopfmacher, and R. Warlimont, Lengths of factorizations for polynomials over a finite field, In *Finite Fields: Theory, Applications, and Algorithms*, volume 168 of *Contemp. Math.*, 185–206, Amer. Math. Soc., Providence, RI, 1994. <368, 374>

[1763] A. Knopfmacher and R. Warlimont, Distinct degree factorizations for polynomials over a finite field, *Trans. Amer. Math. Soc.* 347 (1995) 2235–2243. <368, 374>

[1764] D. E. Knuth, Finite semifields and projective planes, *J. Algebra* 2 (1965) 182–217. <274, 276, 278>

[1765] D. E. Knuth, *The Art of Computer Programming. Vol. 2: Seminumerical Algorithms*, Addison-Wesley Publishing Co., Reading, Mass.-London-Don Mills, Ontario, 1969. <365, 367, 374, 397, 401>

[1766] D. E. Knuth, The analysis of algorithms, In *Actes du Congrès International des Mathématiciens (Nice, 1970), Tome 3*, 269–274, Gauthier-Villars, Paris, 1971. <359, 363>

[1767] D. E. Knuth, *The Art of Computer Programming. Vol. 2, Seminumerical Algorithms*, Addison-Wesley Publishing Company, Reading, MA, second edition, 1981, Addison-Wesley Series in Computer Science and Information Processing. <357, 363>

[1768] D. E. Knuth, *The Art of Computer Programming. Vol. 2, Seminumerical algorithms*, Addison-Wesley Publishing Company, Reading, MA, third edition, 1997,

Addison-Wesley Series in Computer Science and Information Processing. <346, 354, 355, 363>

[1769] N. Koblitz, p-adic variation of the zeta-function over families of varieties defined over finite fields, *Compositio Math.* 31 (1975) 119–218. <486, 488>

[1770] N. Koblitz, *p-adic Numbers, p-adic Analysis, and Zeta-Functions*, volume 58 of *Graduate Texts in Mathematics*, Springer-Verlag, New York, 1977. <480, 488>

[1771] N. Koblitz, Elliptic curve cryptosystems, *Math. Comp.* 48 (1987) 203–209. <746, 750, 784, 796>

[1772] N. Koblitz, Hyperelliptic cryptosystems, *J. Cryptology* 1 (1989) 139–150. <746, 750, 797, 803>

[1773] N. Koblitz, *Introduction to Elliptic Curves and Modular Forms*, volume 97 of *Graduate Texts in Mathematics*, Springer-Verlag, New York, second edition, 1993. <31, 32, 422, 440>

[1774] N. Koblitz, *Algebraic Aspects of Cryptography*, volume 3 of *Algorithms and Computation in Mathematics*, Springer-Verlag, Berlin, 1998. <31, 32, 451, 456>

[1775] Ç. K. Koç and T. Acar, Montgomery multiplication in GF(2^k), *Des. Codes Cryptogr.* 14 (1998) 57–69. <353, 363, 822, 823>

[1776] Ç. K. Koç and B. Sunar, Low-complexity bit-parallel canonical and normal basis multipliers for a class of finite fields, *IEEE Trans. Comput.* 47 (1998) 353–356. <820, 823>

[1777] W. Koepf and R. Kim, The parity of the number of irreducible factors for some pentanomials, *Finite Fields Appl.* 15 (2009) 585–603. <35, 49, 69, 70>

[1778] R. Koetter, W.-C. W. Li, P. O. Vontobel, and J. L. Walker, Characterizations of pseudo-codewords of (low-density) parity-check codes, *Adv. Math.* 213 (2007) 205–229. <718, 719>

[1779] D. R. Kohel, *Endomorphism Rings of Elliptic Curves over Finite Fields*, ProQuest LLC, Ann Arbor, MI, 1996, Thesis (Ph.D.) University of California, Berkeley. <122, 128>

[1780] J. F. Koksma, *Some Theorems on Diophantine Inequalities*, Scriptum no. 5. Math. Centrum Amsterdam, 1950. <340, 344>

[1781] F. Kong, Z. Cai, J. Yu, and D. Li, Improved generalized Atkin algorithm for computing square roots in finite fields, *Inform. Process. Lett.* 98 (2006) 1–5. <360, 363>

[1782] K. Kononen, More exact solutions to Waring's problem for finite fields, *Acta Arith.* 145 (2010) 209–212. <212, 213>

[1783] K. Kononen, M. Moisio, M. Rinta-Aho, and K. Väänänen, Irreducible polynomials with prescribed trace and restricted norm, *JP J. Algebra Number Theory Appl.* 11 (2008) 223–248. <55, 79, 143, 161>

[1784] K. Kononen, M. Rinta-Aho, and K. Väänänen, On the degree of a Kloosterman sum as an algebraic integer, preprint available, `http://arxiv.org/abs/1107.0188v1`, 2011. <154, 161>

[1785] K. P. Kononen, M. J. Rinta-aho, and K. O. Väänänen, On integer values of Kloosterman sums, *IEEE Trans. Inform. Theory* 56 (2010) 4011–4013. <154, 156, 161, 270, 273>

[1786] S. Konyagin, T. Lange, and I. E. Shparlinski, Linear complexity of the discrete logarithm, *Des. Codes Cryptogr.* 28 (2003) 135–146. <333, 336>

[1787] S. Konyagin and F. Pappalardi, Enumerating permutation polynomials over finite fields by degree, *Finite Fields Appl.* 8 (2002) 548–553. <219, 230>

[1788] S. Konyagin and F. Pappalardi, Enumerating permutation polynomials over finite fields by degree. II, *Finite Fields Appl.* 12 (2006) 26–37. <219, 230>

[1789] S. V. Konyagin, Estimates for Gaussian sums and Waring's problem modulo a prime, *Trudy Mat. Inst. Steklov.* 198 (1992) 111–124. <175, 176, 185, 212, 213>

[1790] S. V. Konyagin, Estimates for trigonometric sums over subgroups and for Gauss sums, In *IV International Conference "Modern Problems of Number Theory and its Applications": Current Problems, Part III (Russian)*, 86–114, Mosk. Gos. Univ. im. Lomonosova, Mekh.-Mat. Fak., Moscow, 2002. <141, 161>

[1791] S. V. Konyagin, Estimates for character sums in finite fields, *Mat. Zametki* 88 (2010) 529–542. <183, 185>

[1792] P. Koopman, 32-bit cyclic redundancy codes for internet applications, In *Proc. of Int'l Conf. Dependable Systems and Networks*, 459–468, 2002. <638, 639, 642>

[1793] P. Koopman and T. Chakravarty, Cyclic redundancy code (CRC) polynomial selection for embedded networks, In *Proc. Dependable Systems and Networks*, 145–154, 2004. <635, 642>

[1794] S. B. Korada, *Polar Codes for Channel and Source Coding*, PhD thesis, EPFL, 2009. <739>

[1795] S. B. Korada, A. Montanari, E. Telatar, and R. Urbanke, An empirical scaling law for polar codes, In *2010 IEEE International Symposium on Information Theory*, 884–888. IEEE, 2010. <739>

[1796] S. B. Korada, E. Sasoglu, and R. Urbanke, Polar codes: Characterization of exponent, bounds, and constructions, preprint available, `http://arxiv.com/abs/0901.0536`, 2009. <739>

[1797] S. B. Korada and R. Urbanke, Polar codes are optimal for lossy source coding, *IEEE Trans. Inform. Theory* 56 (2010) 1751–1768. <739>

[1798] G. Korchmáros and T. Szőnyi, Fermat curves over finite fields and cyclic subsets in high-dimensional projective spaces, *Finite Fields Appl.* 5 (1999) 206–217. <209, 213>

[1799] P. Kosick, *Commutative Semifields of Odd Order and Planar Dembowski-Ostrom Polynomials*, PhD thesis, Department of Mathematical Sciences, University of Delaware, USA, 2010. <282>

[1800] R. Kötter and F. R. Kschischang, Coding for errors and erasures in random network coding, *IEEE Trans. Inform. Theory* 54 (2008) 3579–3591. <848, 849>

[1801] Y. Kou, S. Lin, and M. P. C. Fossorier, Low-density parity-check codes based on finite geometries: a rediscovery and new results, *IEEE Trans. Inform. Theory* 47 (2001) 2711–2736. <718, 719>

[1802] A. G. Kouchnirenko, Polyèdres de Newton et nombres de Milnor, *Invent. Math.* 32 (1976) 1–31. <482, 488>

[1803] R. G. Kraemer, Proof of a conjecture on Hadamard 2-groups, *J. Combin. Theory, Ser. A* 63 (1993) 1–10. <604, 607>

[1804] C. Kramp, *Éléments d'arithmétique universelle*, Cologne: Th. F. Thiriart, 1808. <5, 11>

[1805] W. Krandick and T. Jebelean, Bidirectional exact integer division, *J. Symbolic Comput.* 21 (1996) 441–455. <359, 363>

[1806] R. A. Kristiansen and M. G. Parker, Binary sequences with merit factor > 6.3, *IEEE Trans. Inform. Theory* 50 (2004) 3385–3389. <323, 324>

[1807] M. Krivelevich and B. Sudakov, Pseudo-random graphs, In *More Sets, Graphs and Numbers*, volume 15 of *Bolyai Soc. Math. Stud.*, 199–262, Springer, Berlin, 2006.

<646, 650, 658>

[1808] W. Krull, Algebraische Theorie der Ringe. II, *Math. Ann.* 91 (1924) 1–46. <28>

[1809] D. S. Kubert and S. Lichtenbaum, Jacobi-sum Hecke characters and Gauss-sum identities, *Compositio Math.* 48 (1983) 55–87. <146, 161>

[1810] R. Kubota, Waring's problem for $\mathbb{F}_q[x]$, *Dissertationes Math. (Rozprawy Mat.)* 117 (1974) 60pp. <499, 500>

[1811] T. Kumada, H. Leeb, Y. Kurita, and M. Matsumoto, New primitive t-nomials ($t = 3, 5$) over GF(2) whose degree is a Mersenne exponent, *Math. Comp.* 69 (2000) 811–814. <96, 97>

[1812] P. V. Kumar, R. A. Scholtz, and L. R. Welch, Generalized bent functions and their properties, *J. Combin. Theory, Ser. A* 40 (1985) 90–107. <263, 264, 265, 267, 273>

[1813] V. A. Kurbatov and N. G. Starkov, The analytic representation of permutations, *Sverdlovsk. Gos. Ped. Inst. Učen. Zap.* 31 (1965) 151–158. <217, 230>

[1814] H. Kurzweil, M. Seidl, and J. B. Huber, Reduced-complexity collaborative decoding of interleaved Reed-Solomon and Gabidulin codes, preprint available, `http://arxiv.com/abs/1102.3126`, 2011. <739>

[1815] E. N. Kuz'min, Irreducible polynomials over a finite field and an analogue of Gauss sums over a field of characteristic 2, *Sibirsk. Mat. Zh.* 32 (1991) 100–108, 205. <56, 78, 79>

[1816] G. M. Kyureghyan, Crooked maps in $\mathbb{F}_{2^n}$, *Finite Fields Appl.* 13 (2007) 713–726. <259, 261>

[1817] G. M. Kyureghyan, Constructing permutations of finite fields via linear translators, *J. Combin. Theory, Ser. A* 118 (2011) 1052–1061. <225, 229, 230>

[1818] G. M. Kyureghyan and A. Pott, Some theorems on planar mappings, In *Arithmetic of Finite Fields*, volume 5130 of *Lecture Notes in Comput. Sci.*, 117–122, Springer, Berlin, 2008. <281, 282>

[1819] M. K. Kyuregyan, On the theory of the reducibility of polynomials over finite fields, *Akad. Nauk Armyan. SSR Dokl.* 86 (1988) 17–22. <60, 64, 66>

[1820] M. K. Kyuregyan, Recurrent methods of constructing irreducible polynomials over GF(2^s) (Russian), *J. Inform. Process. Cybernet EIK* 27 (1991) 357–372. <60, 61, 63, 64, 66>

[1821] M. K. Kyuregyan, Recurrent methods for constructing irreducible polynomials over GF(2^s), *Finite Fields Appl.* 8 (2002) 52–68. <60, 63, 64, 66, 286, 290>

[1822] M. K. Kyuregyan, Recurrent methods for constructing irreducible polynomials over $\mathbb{F}_q$ of odd characteristics, *Finite Fields Appl.* 9 (2003) 39–58. <60, 61, 64, 65, 66>

[1823] M. K. Kyuregyan, Iterated constructions of irreducible polynomials over finite fields with linearly independent roots, *Finite Fields Appl.* 10 (2004) 323–341. <60, 63, 66>

[1824] M. K. Kyuregyan, Recurrent methods for constructing irreducible polynomials over $\mathbb{F}_q$ of odd characteristics. II, *Finite Fields Appl.* 12 (2006) 357–378. <60, 64, 65, 66>

[1825] M. K. Kyuregyan and G. M. Kyureghyan, Irreducible compositions of polynomials over finite fields, *Des. Codes Cryptogr.* 61 (2011) 301–314. <60, 66>

[1826] G. Lachaud, Sommes d'Eisenstein et nombre de points de certaines courbes algébriques sur les corps finis, *C. R. Acad. Sci. Paris, Sér. I, Math.* 305 (1987) 729–732. <461, 463>

[1827] G. Lachaud, The parameters of projective Reed-Muller codes, *Discrete Math.* 81 (1990) 217–221. <687, 703>

[1828] G. Lachaud and J. Wolfmann, The weights of the orthogonals of the extended quadratic binary Goppa codes, *IEEE Trans. Inform. Theory* 36 (1990) 686–692. <260, 261, 271, 273>

[1829] L. Lafforgue, Chtoucas de Drinfeld et correspondance de Langlands, *Invent. Math.* 147 (2002) 1–241. <545, 546>

[1830] L. Lafforgue, Chtoucas de Drinfeld, formule des traces d'Arthur-Selberg et correspondance de Langlands, In *Proceedings of the International Congress of Mathematicians, Vol. I*, 383–400, Higher Ed. Press, Beijing, 2002. <545, 546>

[1831] J. C. Lagarias, Pseudorandom number generators in cryptography and number theory, In *Cryptology and Computational Number Theory*, volume 42 of *Proc. Sympos. Appl. Math.*, 115–143, Amer. Math. Soc., Providence, RI, 1990. <338, 344>

[1832] Y. Laigle-Chapuy, A note on a class of quadratic permutations over $\mathbb{F}_{2^n}$, In *Applied Algebra, Algebraic Algorithms and Error-Correcting Codes*, volume 4851 of *Lecture Notes in Comput. Sci.*, 130–137, Springer, Berlin, 2007. <224, 230>

[1833] Y. Laigle-Chapuy, Permutation polynomials and applications to coding theory, *Finite Fields Appl.* 13 (2007) 58–70. <218, 223, 230>

[1834] D. Laksov, Linear recurring sequences over finite fields, *Math. Scand.* 16 (1965) 181–196. <635, 642>

[1835] C. Lam, M. Aagaard, and G. Gong, Hardware implementations of multi-output Welch-Gong ciphers, 2011, Technical Report, University of Waterloo, CACR 2011-01. <754, 758, 763>

[1836] C. W. H. Lam, G. Kolesova, and L. Thiel, A computer search for finite projective planes of order 9, *Discrete Math.* 92 (1991) 187–195. <564, 574>

[1837] C. W. H. Lam, L. Thiel, and S. Swiercz, The nonexistence of finite projective planes of order 10, *Canad. J. Math.* 41 (1989) 1117–1123. <564, 574>

[1838] T. Y. Lam and K. H. Leung, Vanishing sums of mth roots of unity in finite fields, *Finite Fields Appl.* 2 (1996) 422–438. <211, 213>

[1839] B. LaMacchia and A. Odlyzko, Solving large sparse linear systems over finite fields, In *Advances in Cryptology—CRYPTO '90*, volume 537 of *Lecture Notes in Comput. Sci.*, 109–133, Springer, Berlin, 1991. <399, 401, 534, 535>

[1840] R. Lambert, *Computational Aspects of Discrete Logarithms*, PhD thesis, University of Waterloo, Ontario, Canada, 1996. <530, 535>

[1841] L. Lan, L. Zeng, Y. Y. Tai, L. Chen, S. Lin, and K. Abdel-Ghaffar, Construction of quasi-cyclic LDPC codes for AWGN and binary erasure channels: a finite field approach, *IEEE Trans. Inform. Theory* 53 (2007) 2429–2458. <718, 719>

[1842] E. S. Lander, *Symmetric Designs: an Algebraic Approach*, volume 74 of *London Mathematical Society Lecture Note Series*, Cambridge University Press, Cambridge, 1983. <273, 600, 603, 606, 607>

[1843] S. Lang, *Elliptic Curves: Diophantine Analysis*, volume 231 of *Grundlehren der Mathematischen Wissenschaften [Fundamental Principles of Mathematical Sciences]*, Springer-Verlag, Berlin, 1978. <31, 32, 422, 440>

[1844] S. Lang, *Abelian Varieties*, Springer-Verlag, New York, 1983. <164, 169>

[1845] S. Lang, *Elliptic Functions*, volume 112 of *Graduate Texts in Mathematics*, Springer-Verlag, New York, second edition, 1987. <31, 32, 422, 440>

[1846] S. Lang, *Algebra*, volume 211 of *Graduate Texts in Mathematics*, Springer-Verlag, New York, third edition, 2002. <415, 422, 536, 546>

[1847] S. Lang and J. Tate (eds.), *The Collected Papers of Emil Artin*, Addison–Wesley Publishing Co., Inc., Reading, Mass.-London, 1965. <72>

[1848] S. Lang and H. Trotter, *Frobenius Distributions in* GL_2*-Extensions*, volume 504 of *Lecture Notes in Mathematics*, Springer-Verlag, Berlin, 1976. <32, 438, 440>

[1849] S. Lang and A. Weil, Number of points of varieties in finite fields, *Amer. J. Math.* 76 (1954) 819–827. <194, 201>

[1850] T. Lange, *Efficient Arithmetic on Hyperelliptic Curves*, PhD thesis, Universität Gesamthochschule Essen, 2001. <801, 802, 803>

[1851] T. Lange, Trace zero subvariety for cryptosystems, *Journal of the Ramanujan Mathematical Society* 19 (2004) 15–33. <802, 803>

[1852] T. Lange, Arithmetic on binary genus 2 curves suitable for small devices, In *Proceedings ECRYPT Workshop on RFID and Lightweight Crypto*, 2005. <798, 803>

[1853] T. Lange, Formulae for arithmetic on genus 2 hyperelliptic curves, *Appl. Algebra Eng. Commun. Comput.* 15 (2005) 295–328. <798, 803>

[1854] T. Lange and M. Stevens, Efficient doubling on genus two curves over binary fields, In *Eleventh International Workshop on Selected Areas in Cryptography*, volume 3357 of *Lecture Notes in Comput. Sci.*, 170–181, Springer, Berlin, 2004. <797, 803>

[1855] P. Langevin, Covering radius of $\mathrm{RM}(1, 9)$ in $\mathrm{RM}(3, 9)$, In *Eurocode '90*, volume 514 of *Lecture Notes in Comput. Sci.*, 51–59, Springer, Berlin, 1991. <245, 252>

[1856] P. Langevin and G. Leander, Monomial bent functions and Stickelberger's theorem, *Finite Fields Appl.* 14 (2008) 727–742. <268, 273>

[1857] V. Laohakosol and U. Pintoptang, A modification of Fitzgerald's characterization of primitive polynomials over a finite field, *Finite Fields Appl.* 14 (2008) 85–91. <88, 90>

[1858] G. Larcher and H. Niederreiter, Generalized (t, s)-sequences, Kronecker-type sequences, and diophantine approximations of formal Laurent series, *Trans. Amer. Math. Soc.* 347 (1995) 2051–2073. <626, 630>

[1859] R. Laubenbacher, A. Jarrah, H. Mortveit, and S. S. Ravi, *Encyclopedia of Complexity and System Science*, chapter A Mathematical Foundation for Agent-Based Computer Simulation, Springer Verlag, New York, 2009. <829, 834>

[1860] R. Laubenbacher and B. Stigler, A computational algebra approach to the reverse engineering of gene regulatory networks, *Journal of Theoretical Biology* 229 (2004) 523–537. <337, 344, 831, 834>

[1861] A. G. B. Lauder, Computing zeta functions of Kummer curves via multiplicative characters, *Found. Comput. Math.* 3 (2003) 273–295. <491>

[1862] A. G. B. Lauder, Counting solutions to equations in many variables over finite fields, *Found. Comput. Math.* 4 (2004) 221–267. <490, 491>

[1863] A. G. B. Lauder, Deformation theory and the computation of zeta functions, *Proc. London Math. Soc., 3rd Ser.* 88 (2004) 565–602. <490, 491>

[1864] A. G. B. Lauder and K. G. Paterson, Computing the error linear complexity spectrum of a binary sequence of period 2^n, *IEEE Trans. Inform. Theory* 49 (2003) 273–280. <329, 336>

[1865] A. G. B. Lauder and D. Wan, Computing zeta functions of Artin-Schreier curves over finite fields. II, *J. Complexity* 20 (2004) 331–349. <454, 456, 491>

[1866] A. G. B. Lauder and D. Wan, Counting points on varieties over finite fields of small characteristic, In *Algorithmic Number Theory: Lattices, Number Fields, Curves and Cryptography*, volume 44 of *Math. Sci. Res. Inst. Publ.*, 579–612, Cambridge Univ. Press, Cambridge, 2008. <454, 456, 489, 490, 491>

[1867] G. Laumon, Majorations de sommes trigonométriques (d'après P. Deligne et N. Katz), In *The Euler-Poincaré characteristic*, volume 83 of *Astérisque*, 221–258, Soc. Math. France, Paris, 1981. <169>

[1868] G. Laumon, Transformation de Fourier, constantes d'équations fonctionnelles et conjecture de Weil, *Inst. Hautes Études Sci. Publ. Math.* 65 (1987) 131–210. <478, 479>

[1869] G. Laumon, Exponential sums and l-adic cohomology: a survey, *Israel J. Math.* 120 (2000) 225–257. <169>

[1870] M. Lavrauw, G. L. Mullen, S. Nikova, D. Panario, and L. Storme, editors, *Proc. Tenth International Conference on Finite Fields and Applications*, volume 579 of *Contemp. Math.*, American Mathematical Society, Providence, RI, 2012. <31>

[1871] M. Lavrauw and O. Polverino, Finite semifields, In L. Storme and J. D. Buele, editors, *Current Research Topics in Galois Geometry*, chapter 6, Nova Publishers, 2011. <274, 278>

[1872] M. Lavrauw, L. Storme, and G. Van de Voorde, A proof of the linearity conjecture for k-blocking sets in $\mathrm{PG}(n, p^3)$, p prime, *J. Combin. Theory, Ser. A* 118 (2011) 808–818. <561, 563>

[1873] K. M. Lawrence, A combinatorial characterization of (t, m, s)-nets in base b, *J. Combin. Des.* 4 (1996) 275–293. <620, 630>

[1874] K. M. Lawrence, A. Mahalanabis, G. L. Mullen, and W. C. Schmid, Construction of digital (t, m, s)-nets from linear codes, In *Finite Fields and Applications*, volume 233 of *London Math. Soc. Lecture Note Ser.*, 189–208, Cambridge University Press, Cambridge, 1996. <625, 630>

[1875] C. F. Laywine and G. L. Mullen, *Discrete Mathematics using Latin Squares*, Wiley-Interscience Series in Discrete Mathematics and Optimization. John Wiley & Sons Inc., New York, 1998. <31, 32, 551, 555, 556>

[1876] C. F. Laywine, G. L. Mullen, and G. Whittle, d-dimensional hypercubes and the Euler and MacNeish conjectures, *Monatsh. Math.* 119 (1995) 223–238. <552, 553, 556>

[1877] D. Lazard, Gröbner bases, Gaussian elimination and resolution of systems of algebraic equations, In *Computer Algebra*, volume 162 of *Lecture Notes in Comput. Sci.*, 146–156, Springer, Berlin, 1983. <781, 783>

[1878] G. Leander and A. Kholosha, Bent functions with 2^r Niho exponents, *IEEE Trans. Inform. Theory* 52 (2006) 5529–5532. <268, 273>

[1879] N. G. Leander, Monomial bent functions, *IEEE Trans. Inform. Theory* 52 (2006) 738–743. <268, 271, 273>

[1880] G. Lecerf, Sharp precision in Hensel lifting for bivariate polynomial factorization, *Math. Comp.* 75 (2006) 921–933. <385, 392>

[1881] G. Lecerf, Improved dense multivariate polynomial factorization algorithms, *J. Symbolic Comput.* 42 (2007) 477–494. <386, 392>

[1882] G. Lecerf, Fast separable factorization and applications, *Appl. Alg. Eng. Comm. Comp.* 19 (2008) 135–160. <383, 384, 392>

[1883] G. Lecerf, New recombination algorithms for bivariate polynomial factorization based on Hensel lifting, *Appl. Alg. Eng. Comm. Comp.* 21 (2010) 151–176. <385, 392>

[1884] C. Lee and C. Chang, Low-complexity linear array multiplier for normal basis of type-II, In *Proc. IEEE International Conf. Multimedia and Expo*, 1515–1518,

2004. <821, 823>

[1885] C. Lee and C. W. Chiou, Scalable Gaussian normal basis multipliers over $GF(2^m)$ using Hankel matrix-vector representation, *Journal of Signal Processing Systems* 69 (2012) 197–211. <822, 823>

[1886] D. B. Leep and C. C. Yeomans, The number of points on a singular curve over a finite field, *Arch. Math. (Basel)* 63 (1994) 420–426. <238, 240>

[1887] A. M. Legendre, Recherches d'analyse indeterminee, *Memoires Acad. Sci. Paris* (1785) 465–559. <181, 185>

[1888] D. H. Lehmer, Euclid's Algorithm for Large Numbers, *Amer. Math. Monthly* 45 (1938) 227–233. <359, 363>

[1889] A. Lempel and H. Greenberger, Families of sequences with optimal Hamming correlation properties, *IEEE Trans. Inform. Theory* IT-20 (1974) 90–94. <845, 846, 849>

[1890] A. Lempel and M. J. Weinberger, Self-complementary normal bases in finite fields, *SIAM J. Discrete Math.* 1 (1988) 193–198. <114, 116>

[1891] D. Lenskoi, On the arithmetic of polynomials over a finite field (Russian), *Volz. Mat. Sb.* 4 (1966) 155–159. <494, 500>

[1892] A. K. Lenstra, Factorization of polynomials, In *Computational Methods in Number Theory, Part I*, volume 154 of *Math. Centre Tracts*, 169–198, Math. Centrum, Amsterdam, 1982. <404>

[1893] A. K. Lenstra, H. W. Lenstra, Jr., and L. Lovász, Factoring polynomials with rational coefficients, *Math. Ann.* 261 (1982) 515–534. <387, 392>

[1894] A. K. Lenstra and E. R. Verheul, Selecting cryptographic key sizes, In H. Imai and Y. Zheng, editors, *Public Key Cryptography—Third International Workshop on Practice and Theory in Public Key Cryptosystems PKC 2000*, volume 1751 of *Lecture Notes in Comput. Sci.*, 446–465, Springer-Verlag, Berlin, 2000. <784, 796>

[1895] H. W. Lenstra, Finding isomorphisms between finite fields, *Math. Comp.* 56 (1991) 329–347. <348, 363, 401, 402, 404>

[1896] H. W. Lenstra, Jr., A normal basis theorem for infinite Galois extensions, *Nederl. Akad. Wetensch. Indag. Math.* 47 (1985) 221–228. <130, 138>

[1897] H. W. Lenstra, Jr., Finding small degree factors of lacunary polynomials, In K. Győry, H. Iwaniec, and J. Urbanowicz, editors, *Number Theory in Progress: Proc. Internat. Conf. Number Theory*, 267–276, Berlin, 1999, de Gruyter. <390, 392>

[1898] H. W. Lenstra, Jr., Exceptional covers, October 1999, MSRI lecture, available at `http://www.msri.org/realvideo/ln/msri/1999/cgt/lenstra/1/index.html`. Accessed April 12, 2012. <238, 239, 240>

[1899] H. W. Lenstra, Jr. and R. J. Schoof, Primitive normal bases for finite fields, *Math. Comp.* 48 (1987) 217–231. <93, 95, 115, 116, 137, 138>

[1900] H. W. Lenstra, Jr. and M. Zieve, A family of exceptional polynomials in characteristic three, In *Finite Fields and Applications*, volume 233 of *London Math. Soc. Lecture Note Ser.*, 209–218, Cambridge Univ. Press, Cambridge, 1996. <240>

[1901] J. S. Leon, J. M. Masley, and V. Pless, Duadic codes, *IEEE Trans. Inform. Theory* 30 (1984) 709–714. <682, 703>

[1902] M. Leone, A new low complexity parallel multiplier for a class of finite fields, In *Proc. Cryptographic Hardware and Embedded Systems (CHES)*, volume 2162 of *Lecture Notes Comput. Sci.*, 160–170. Springer, 2001. <821, 823>

[1903] R. Lercier, *Algorithmique des Courbes Elliptiques dans les Corps Finis*, PhD thesis, École Polytechnique, 1997, In French, available at `http://perso.univ-rennes1.fr/reynald.lercier/file/Ler97a.pdf`. <359, 363>

[1904] R. Lercier and D. Lubicz, Counting points on elliptic curves over finite fields of small characteristic in quasi quadratic time, In E. Biham, editor, *Advances in Cryptology—EUROCRYPT 2003*, volume 2656 of *Lecture Notes in Comput. Sci.*, 360–373, Springer-Verlag, Berlin, 2003. <788, 796>

[1905] R. Lercier and D. Lubicz, A quasi quadratic time algorithm for hyperelliptic curve point counting, *Ramanujan J.* 12 (2006) 399–423. <454, 456, 491>

[1906] C. Leroux, I. Tal, A. Vardy, and W. J. Gross, Hardware architectures for successive cancellation decoding of polar codes, In *2011 IEEE International Conference on Acoustics, Speech and Signal Processing—ICASSP*, 1665–1668, 2011. <739>

[1907] K. H. Leung, S. L. Ma, and B. Schmidt, Nonexistence of abelian difference sets: Lander's conjecture for prime power orders, *Trans. Amer. Math. Soc.* 356 (2004) 4343–4358. <603, 607>

[1908] K. H. Leung, S. L. Ma, and B. Schmidt, New Hadamard matrices of order $4p^2$ obtained from Jacobi sums of order 16, *J. Combin. Theory, Ser. A* 113 (2006) 822–838. <149, 161>

[1909] K. H. Leung, S. L. Ma, and B. Schmidt, On Lander's conjecture for difference sets whose order is a power of 2 or 3, *Des. Codes Cryptogr.* 56 (2010) 79–84. <603, 607>

[1910] K. H. Leung and B. Schmidt, The field descent method, *Des. Codes Cryptogr.* 36 (2005) 171–188. <604, 607>

[1911] V. Levenshtein, Application of Hadamard matrices to a problem of coding theory, *Problemy Kibernetiki* 5 (1961) 123–136. <170, 185>

[1912] A. Levin, *Difference Algebra*, volume 8 of *Algebra and Applications*, Springer, New York, 2008. <238, 240>

[1913] A. B. Levin, Difference algebra, In *Handbook of Algebra*, volume 4, 241–334, Elsevier/North-Holland, Amsterdam, 2006. <238, 240>

[1914] F. Levy-dit Vehel and L. Perret, Polynomial equivalence problems and applications to multivariate cryptosystems, In *Progress in Cryptology—INDOCRYPT 2003*, volume 2904 of *Lecture Notes in Comput. Sci.*, 235–251, Springer, Berlin, 2003. <767, 783>

[1915] H. Li and H. J. Zhu, Zeta functions of totally ramified p-covers of the projective line, *Rend. Sem. Mat. Univ. Padova* 113 (2005) 203–225. <485, 488>

[1916] J. Li, D. B. Chandler, and Q. Xiang, Permutation polynomials of degree 6 or 7 over finite fields of characteristic 2, *Finite Fields Appl.* 16 (2010) 406–419. <216, 230>

[1917] J. Li and D. Wan, On the subset sum problem over finite fields, *Finite Fields Appl.* 14 (2008) 911–929. <213>

[1918] J. Li and D. Wan, Counting subset sums of finite abelian groups, *J. Combin. Theory, Ser. A* 119 (2012) 170–182. <213>

[1919] K.-Z. Li and F. Oort, *Moduli of Supersingular Abelian Varieties*, volume 1680 of *Lecture Notes in Mathematics*, Springer-Verlag, Berlin, 1998. <486, 488>

[1920] L. Li and O. Roche-Newton, An improved sum-product estimate for general finite fields, *SIAM J. Discrete Math.* 25 (2011) 1285–1296. <188>

[1921] W.-C. Li, Character sums and abelian Ramanujan graphs, *J. Number Theory* 41 (1992) 199–217. <650, 658>

[1922] W.-C. Li, *Number Theory with Applications*, volume 7 of *Series on University Math-*

ematics, World Scientific Publishing Co. Inc., River Edge, NJ, 1996. <31, 32, 643, 650, 651, 658>

[1923] W.-C. Li, On negative eigenvalues of regular graphs, *C. R. Acad. Sci. Paris, Sér. I, Math.* 333 (2001) 907–912. <647, 658>

[1924] W.-C. Li, Recent developments in automorphic forms and applications, In *Number Theory for the Millennium II*, 331–354, A. K. Peters, Natick, MA, 2002. <643, 658>

[1925] W.-C. Li, Ramanujan hypergraphs, *Geom. Funct. Anal.* 14 (2004) 380–399. <648, 658>

[1926] W.-C. Li, Zeta functions in combinatorics and number theory, In *Fourth International Congress of Chinese Mathematicians*, volume 48 of *AMS/IP Stud. Adv. Math.*, 351–366, Amer. Math. Soc., Providence, RI, 2010. <658>

[1927] W.-C. Li and P. Solé, Spectra of regular graphs and hypergraphs and orthogonal polynomials, *European J. Combin.* 17 (1996) 461–477. <648, 658>

[1928] Y. Li, S. Ling, H. Niederreiter, H. Wang, C. Xing, and S. Zhang, editors, *Coding and Cryptology*, volume 4 of *Series on Coding Theory and Cryptology*. World Scientific Publishing Co. Pte. Ltd., Hackensack, NJ, 2008. <31, 32>

[1929] Y. Li and M. Wang, On EA-equivalence of certain permutations to power mappings, *Des. Codes Cryptogr.* 58 (2011) 259–269. <226, 230>

[1930] Q. Liao and K. Feng, On the complexity of the normal bases via prime Gauss period over finite fields, *J. Syst. Sci. Complex.* 22 (2009) 395–406. <125, 128>

[1931] Q. Liao and L. You, Low complexity of a class of normal bases over finite fields, *Finite Fields Appl.* 17 (2011) 1–14. <119, 128>

[1932] Y. S. Liaw, More Z-cyclic Room squares, *Ars Combin.* 52 (1999) 228–238. <616, 619>

[1933] R. Lidl and G. L. Mullen, When does a polynomial over a finite field permute the elements of the field?, *Amer. Math. Monthly* 95 (1988) 243–246. <216, 230>

[1934] R. Lidl and G. L. Mullen, Cycle structure of Dickson permutation polynomials, *Math. J. Okayama Univ.* 33 (1991) 1–11. <228, 230>

[1935] R. Lidl and G. L. Mullen, When does a polynomial over a finite field permute the elements of the field?, II, *Amer. Math. Monthly* 100 (1993) 71–74. <216, 217, 230>

[1936] R. Lidl, G. L. Mullen, and G. Turnwald, *Dickson Polynomials*, volume 65 of *Pitman Monographs and Surveys in Pure and Applied Mathematics*, Longman Scientific & Technical, Harlow, 1993. <31, 32, 230, 239, 240, 283, 289, 290, 294, 298, 302, 333, 336>

[1937] R. Lidl and H. Niederreiter, On orthogonal systems and permutation polynomials in several variables, *Acta Arith.* 22 (1972/73) 257–265. <231, 232>

[1938] R. Lidl and H. Niederreiter, *Introduction to Finite Fields and Their Applications*, Cambridge University Press, Cambridge, revised edition, 1994. <13, 31, 32, 71, 73, 312, 317, 393, 401>

[1939] R. Lidl and H. Niederreiter, *Finite Fields*, volume 20 of *Encyclopedia of Mathematics and its Applications*, Cambridge University Press, Cambridge, second edition, 1997. <3, 11, 13, 24, 27, 31, 32, 37, 49, 60, 61, 62, 66, 70, 71, 73, 87, 90, 171, 173, 175, 179, 183, 185, 202, 206, 207, 209, 213, 215, 216, 217, 228, 230, 232, 235, 236, 238, 240, 253, 261, 283, 287, 290, 318, 324, 325, 327, 336, 349, 363, 365, 374, 377, 380, 393, 394, 401, 510>

[1940] R. Lidl and C. Wells, Chebychev polynomials in several variables, *J. Reine Angew.*

Math. 255 (1972) 104–111. <231, 232>

[1941] C. H. Lim and P. J. Lee, More flexible exponentiation with precomputation, In *Advances in Cryptology—CRYPTO '94*, volume 839 of *Lecture Notes in Comput. Sci.*, 95–107, Springer, Berlin, 1994. <357, 363>

[1942] S. Lin, On a class of cyclic codes, In *Error Correcting Codes*, 131–148, John Wiley, New York, 1968. <689, 696, 702, 703>

[1943] S. Lin and D. Costello, *Error Control Coding*, Prentice-Hall, Saddle River, NJ, second edition, 2004. <31, 32, 661, 692, 703>

[1944] J. Lindholm, An analysis of the pseudo-randomness properties of subsequences of long m-sequences, *IEEE Trans. Inform. Theory* 14 (1968) 569–576. <630, 642>

[1945] S. Ling and C. Xing, *Coding Theory: A First Course*, Cambridge University Press, Cambridge, 2004. <31, 32, 661, 675, 680, 685, 703>

[1946] P. Lisoněk and M. Moisio, On zeros of Kloosterman sums, *Des. Codes Cryptogr.* 59 (2011) 223–230. <154, 161>

[1947] C. Liu, Twisted higher moments of Kloosterman sums, *Proc. Amer. Math. Soc.* 130 (2002) 1887–1892. <158, 161>

[1948] C. Liu, The L-functions of twisted Witt extensions, *J. Number Theory* 125 (2007) 267–284. <483, 488>

[1949] C. Liu and D. Wan, T-adic exponential sums over finite fields, *Algebra Number Theory* 3 (2009) 489–509. <483, 488>

[1950] C. Liu and D. Wei, The L-functions of Witt coverings, *Math. Z.* 255 (2007) 95–115. <483, 488>

[1951] M. Liu, Y. Zhang, and D. Lin, Perfect algebraic immune functions, In *Advances in cryptology—ASIACRYPT 2012 (Beijing, China)*, volume 7658 of *Lecture Notes in Comput. Sci.*, Springer, Berlin, 172–189, 2012. <250, 252>

[1952] P. Loidreau, On the factorization of trinomials over $\mathbb{F}_3$, INRIA rapport de recherche 3918, 2000. <69, 70>

[1953] A. F. Long, Jr., Classification of irreducible factorable polynomials over a finite field, *Acta Arith.* 12 (1967) 301–313. <63, 66>

[1954] A. F. Long, Jr., Factorization of irreducible polynomials over a finite field with the substitution $x^{p^r} - x$ for x, *Duke Math. J* 40 (1973) 63–76. <60, 63, 66>

[1955] A. F. Long, Jr., Factorization of irreducible polynomials over a finite field with the substitution $x^{p^r} - x$ for x, *Acta Arith.* 25 (1973/74) 65–80. <60, 63, 66>

[1956] A. F. Long, Jr., A theorem on factorable irreducible polynomials in several variables over a finite field with the substitution $x_i^{p^r} - x_i$ for x_i, *Math. Nachr.* 63 (1974) 123–130. <63, 66>

[1957] A. F. Long, Jr. and T. P. Vaughan, Factorization of $q(h(t)(x))$ over a finite field, where $q(x)$ is irreducible and $h(x)$ is linear II, *Linear Algebra Appl.* 11 (1975) 53–72. <63, 66>

[1958] A. F. Long, Jr. and T. P. Vaughan, Factorization of $q(h(t)(x))$ over a finite field, where $q(x)$ is irreducible and $h(x)$ is linear I, *Linear Algebra Appl.* 13 (1976) 207–221. <63, 66>

[1959] D. Lorenzini, *An Invitation to Arithmetic Geometry*, volume 9 of *Graduate Studies in Mathematics*, American Mathematical Society, Providence, RI, 1996. <459, 463>

[1960] S. R. Louboutin, Efficient computation of root numbers and class numbers of parametrized families of real abelian number fields, *Math. Comp.* 76 (2007)

455–473. <143, 161>

[1961] L. Lovász and A. Schrijver, Remarks on a theorem of Rédei, *Studia Sci. Math. Hungar.* 16 (1983) 449–454. <559, 563>

[1962] R. Lowe and D. Zelinsky, Which Galois fields are pure extensions?, *Math. Student* 21 (1953) 37–41. <61, 66>

[1963] H.-F. Lu and P. V. Kumar, Rate-diversity tradeoff of space-time codes with fixed alphabet and optimal constructions of PSK modulation, *IEEE Trans. Inform. Theory* 49 (2003) 2747–2751. <847, 849>

[1964] H.-F. Lu and P. V. Kumar, A unified construction of space-time codes with optimal rate-diversity tradeoff, *IEEE Trans. Inform. Theory* 51 (2005) 1709–1730. <847, 849>

[1965] Y. Lu and L. Zhu, On the existence of triplewhist tournaments TWh(v), *J. Combin. Des.* 5 (1997) 249–256. <618, 619>

[1966] F. Lübeck, Conway polynomials for finite fields, 2008, `http://www.math.rwth-aachen.de:8001/~Frank.Luebeck/data/ConwayPol`. <402, 404>

[1967] A. Lubotzky, *Discrete Groups, Expanding Graphs and Invariant Measures*, volume 125 of *Progress in Mathematics*, Birkhäuser Verlag, Basel, 1994. <643, 648, 649, 658>

[1968] A. Lubotzky, Expander Graphs in Pure and Applied Mathematics, *ArXiv e-prints* (2011). <643, 649, 652, 658>

[1969] A. Lubotzky, R. Phillips, and P. Sarnak, Ramanujan graphs, *Combinatorica* 8 (1988) 261–277. <652, 653, 654, 655, 658>

[1970] A. Lubotzky and B. Weiss, Groups and expanders, In *Expanding Graphs*, volume 10 of *DIMACS Ser. Discrete Math. Theoret. Comput. Sci.*, 95–109, Amer. Math. Soc., Providence, RI, 1993. <651, 658>

[1971] M. G. Luby, LT codes, In *Proc. Forty Third Ann. IEEE Symp. Foundations of Comput. Sci.*, 271–280, 2002. <731, 732, 733, 734>

[1972] M. G. Luby, M. Mitzenmacher, and M. A. Shokrollahi, Analysis of random processes via And-Or tree evaluation, In *Proceedings of the Ninth Annual ACM-SIAM Symposium on Discrete Algorithms*, 364–373, ACM, New York, 1998. <728, 729, 730, 734>

[1973] M. G. Luby, M. Mitzenmacher, M. A. Shokrollahi, and D. Spielman, Analysis of low density codes and improved designs using irregular graphs, In *Proceedings of the Thirtieth Annual ACM Symposium on the Theory of Computing*, 249–258, ACM, New York, 1998. <728, 734>

[1974] M. G. Luby, M. Mitzenmacher, M. A. Shokrollahi, D. Spielman, and V. Stenman, Practical loss-resilient codes, In *Proceedings of the Twenty Ninth Annual ACM Symposium on the Theory of Computing*, 150–159, ACM, New York, 1997. <728, 730, 734>

[1975] M. G. Luby, M. Mitzenmacher, M. A. Shokrollahi, and D. A. Spielman, Efficient erasure correcting codes, *IEEE Trans. Inform. Theory* 47 (2001) 569–584. <728, 729, 730, 734>

[1976] M. G. Luby, M. Mitzenmacher, M. A. Shokrollahi, and D. A. Spielman, Improved low-density parity-check codes using irregular graphs, *IEEE Trans. Inform. Theory* 47 (2001) 585–598. <729, 734>

[1977] G. Lunardon, Normal spreads, *Geom. Dedicata* 75 (1999) 245–261. <561, 563>

[1978] G. Lunardon and O. Polverino, Blocking sets of size $q^t + q^{t-1} + 1$, *J. Combin. Theory, Ser. A* 90 (2000) 148–158. <561, 563>

[1979] H. Lüneburg, Über projektive Ebenen, in denen jede Fahne von einer nicht-trivialen Elation invariant gelassen wird, *Abh. Math. Sem. Univ. Hamburg* 29 (1965) 37–76. <569, 574>

[1980] H. Lüneburg, *Translation Planes*, Springer-Verlag, Berlin, 1980. <566, 574>

[1981] J. Luo and K. Feng, On the weight distributions of two classes of cyclic codes, *IEEE Trans. Inform. Theory* 54 (2008) 5332–5344. <205, 206>

[1982] Y. Luo, Q. Chai, G. Gong, and X. Lai, Wg-7, a lightweight stream cipher with good cryptographic properties, In *Proceedings of IEEE Global Communications Conference (GLOBECOM'10)*, 2010. <758, 763>

[1983] K. Ma and J. von zur Gathen, The computational complexity of recognizing permutation functions, *Comput. Complexity* 5 (1995) 76–97. <217, 230, 392>

[1984] K. Ma and J. von zur Gathen, Tests for permutation functions, *Finite Fields Appl.* 1 (1995) 31–56. <217, 230>

[1985] F. S. Macaulay, *The Algebraic Theory of Modular Systems*, Cambridge Mathematical Library, Cambridge University Press, Cambridge, 1994. <781, 783>

[1986] C. R. MacCluer, On a conjecture of Davenport and Lewis concerning exceptional polynomials, *Acta Arith* 12 (1966/1967) 289–299. <293, 302>

[1987] D. J. C. MacKay, Good error-correcting codes based on very sparse matrices, *IEEE Trans. Inform. Theory* 45 (1999) 399–431. <713, 715, 716, 719>

[1988] H. F. MacNeish, Euler squares, *Ann. of Math., 2nd Ser.* 23 (1922) 221–227. <552, 556>

[1989] F. J. MacWilliams, Orthogonal matrices over finite fields, *Amer. Math. Monthly* 76 (1969) 152–164. <505, 507, 510>

[1990] F. J. MacWilliams, Orthogonal circulant matrices over finite fields, and how to find them., *J. Combin. Theory, Ser. A* 10 (1971) 1–17. <115, 116, 506, 510>

[1991] F. J. MacWilliams and N. J. A. Sloane, *The Theory of Error-Correcting Codes. I*, North-Holland Mathematical Library, Vol. 16, North-Holland Publishing Co., Amsterdam, 1977. <31, 32, 179, 185, 205, 206, 259, 261, 266, 273, 586, 589, 661, 684, 691, 701, 703>

[1992] F. J. MacWilliams and N. J. A. Sloane, *The Theory of Error-Correcting Codes. II*, North-Holland Mathematical Library, Vol. 16, North-Holland Publishing Co., Amsterdam, 1977. <243, 244, 245, 246, 251, 252>

[1993] H. Mahdavifar and A. Vardy, Achieving the secrecy capacity of wiretap channels using polar codes, *IEEE Trans. Inform. Theory* 57 (2011) 6428–6443. <739>

[1994] S. Maitra, K. C. Gupta, and A. Venkateswarlu, Results on multiples of primitive polynomials and their products over GF(2), *Theoret. Comput. Sci.* 341 (2005) 311–343. <96, 97, 634, 635, 642>

[1995] C. Malvenuto and F. Pappalardi, Enumerating permutation polynomials I: Permutations with non-maximal degree, *Finite Fields Appl.* 8 (2002) 531–547. <220, 230>

[1996] C. Malvenuto and F. Pappalardi, Enumerating permutation polynomials II: k-cycles with minimal degree, *Finite Fields Appl.* 10 (2004) 72–96. <220, 230>

[1997] C. Malvenuto and F. Pappalardi, Corrigendum to: "Enumerating permutation polynomials I: Permutations with non-maximal degree" [Finite Fields Appl. 8 (2002), no. 4, 531–547], *Finite Fields Appl.* 13 (2007) 171–174. <220, 230>

[1998] F. Manganiello, E. Gorla, and J. Rosenthal, Spread codes and spread decoding in network coding, In *Proc. Int. Symp. Inform. Theory*, 881–885, 2008. <848, 849>

[1999] J. I. Manin, The Hasse-Witt matrix of an algebraic curve, *Izv. Akad. Nauk SSSR Ser. Mat.* 25 (1961) 153–172. <486, 488>

[2000] Y. Mansury, M. Kimura, J. Lobo, and T. S. Deisboeck, Emerging patterns in tumor systems: Simulating the dynamics of multicellular clusters with an agent-based spatial agglomeration model, *Journal of Theoretical Biology* 219 (2002) 343–370. <831, 834>

[2001] I. Mantin, *Analysis of the Stream Cipher RC4*, Master's dissertation, The Weizmann Institute of Science, Rehovot, 76100, Israel, 2001. <751, 753, 763>

[2002] Maplesoft^{TM}, Maplesoft - Technical Computing Software for Engineers, Mathematicians, Scientists, Intructors and Students, version 16.01, `http://www.maplesoft.com/`, as viewed in July 2012. <48, 49, 346, 363>

[2003] J. E. Marcos, Specific permutation polynomials over finite fields, *Finite Fields Appl.* 17 (2011) 105–112. <221, 224, 225, 230>

[2004] D. A. Marcus, *Number Fields*, Universitext, Springer-Verlag, New York, 1977. <847, 849>

[2005] G. A. Margulis, Explicit constructions of expanders, *Problemy Peredači Informacii* 9 (1973) 71–80. <649, 658>

[2006] G. A. Margulis, Explicit group-theoretic constructions of combinatorial schemes and their applications in the construction of expanders and concentrators, *Problemy Peredachi Informatsii* 24 (1988) 51–60. <653, 655, 658>

[2007] W. J. Martin and D. R. Stinson, A generalized Rao bound for ordered orthogonal arrays and (t, m, s)-nets, *Canad. Math. Bull.* 42 (1999) 359–370. <621, 630>

[2008] W. J. Martin and D. R. Stinson, Association schemes for ordered orthogonal arrays and (T, M, S)-nets, *Canad. J. Math.* 51 (1999) 326–346. <621, 630>

[2009] W. J. Martin and T. I. Visentin, A dual Plotkin bound for (T, M, S)-nets, *IEEE Trans. Inform. Theory* 53 (2007) 411–415. <621, 630>

[2010] J. L. Massey, *Threshold Decoding*, Massachusetts Institute of Technology, Research Laboratory of Electronics, Tech. Rep. 410, Cambridge, Mass., 1963. <696, 703>

[2011] J. L. Massey, Shift-register synthesis and BCH decoding, *IEEE Trans. Inform. Theory* IT-15 (1969) 122–127. <246, 252, 314, 317, 325, 336, 350, 363, 693, 702, 703>

[2012] J. L. Massey and J. K. Omura, Computational methods and apparatus for finite field arithmetic, *US Patent No. 4,587,627, to OMNET Assoc., Sunnyvale CA, Washington, D.C.: Patent and Trademark Office* (1986). <818, 820, 823>

[2013] J. L. Massey and S. Serconek, Linear complexity of periodic sequences: a general theory, In *Advances in Cryptology—CRYPTO '96*, volume 1109 of *Lecture Notes in Comput. Sci.*, 358–371, Springer, Berlin, 1996. <308, 310, 328, 336>

[2014] E. D. Mastrovito, VLSI designs for multiplication over finite field $GF(2^m)$, In *Proc. Sixth International Conference on Applied Algebra, Algebraic Algorithms and Error-Correcting Codes (AAECC-6)*, 297–309, 1988. <816, 823>

[2015] A. M. Masuda, L. Moura, D. Panario, and D. Thomson, Low complexity normal elements over finite fields of characteristic two, *IEEE Trans. Comput.* 57 (2008) 990–1001. <38, 39, 49, 117, 123, 128>

[2016] A. M. Masuda and D. Panario, Sequences of consecutive smooth polynomials over a finite field, *Proc. Amer. Math. Soc.* 135 (2007) 1271–1277. <500>

[2017] A. M. Masuda and D. Panario, *Tópicos de Corpos Finitos com Aplicações em Criptografia e Teoria de Códigos*, Publicações Matemáticas do IMPA. [IMPA Mathematical Publications]. Instituto Nacional de Matemática Pura e Aplicada (IMPA), Rio de Janeiro, 2007, 26o Colóquio Brasileiro de Matemática. [26th

Brazilian Mathematics Colloquium]. <13, 31, 32>

[2018] A. M. Masuda, D. Panario, and Q. Wang, The number of permutation binomials over $\mathbb{F}_{4p+1}$ where p and $4p+1$ are primes, *Electron. J. Combin.* 13 (2006) Research Paper 65, 15 pp. <217, 218, 223, 230>

[2019] A. M. Masuda and M. E. Zieve, Nonexistence of permutation binomials of certain shapes, *Electron. J. Combin.* 14 (2007) Note 12, 5 pp. <218, 230>

[2020] A. M. Masuda and M. E. Zieve, Permutation binomials over finite fields, *Trans. Amer. Math. Soc.* 361 (2009) 4169–4180. <218, 223, 230>

[2021] E. Mathieu, Nombre de valeurs que peut acquérir une fonction quand on y permute ses variables de toutes les manières possibles, *J. Math.* 5 (1860) 9–42. <10, 11>

[2022] E. Mathieu, Mémoire sur l'étude des fonctions de plusieurs quantités, sur la manière de les former et sur les substitutions qui les laissent invariables, *J. Math. Pures Appl.* 6 (1861) 241–323. <63, 66>

[2023] R. Mathon, Symmetric conference matrices of order pq^2+1, *Canad. J. Math.* 30 (1978) 321–331. <610, 619>

[2024] R. Mathon, New maximal arcs in Desarguesian planes, *J. Combin. Theory, Ser. A* 97 (2002) 353–368. <572, 574>

[2025] R. Mathon and G. F. Royle, The translation planes of order 49, *Des. Codes Cryptogr.* 5 (1995) 57–72. <275, 278>

[2026] M. Matsui, Linear cryptoanalysis method for DES cipher, In *EUROCRYPT*, volume 765 of *Lecture Notes in Comput. Sci.*, 386–397, Springer, Berlin, 1993. <253, 261>

[2027] R. Matsumoto, Improvement of Ashikhmin-Litsyn-Tsfasman bound for quantum codes, *IEEE Trans. Inform. Theory* 48 (2002) 2122–2124. <838, 841>

[2028] T. Matsumoto and H. Imai, Public quadratic polynomial-tuples for efficient signature-verification and message-encryption, In *Advances in Cryptology—EUROCRYPT '88*, volume 330 of *Lecture Notes in Comput. Sci.*, 419–453, Springer, Berlin, 1988. <765, 769, 783>

[2029] T. Matsumoto, H. Imai, H. Harashima, and H. Miyakawa, A cryptographically useful theorem on the connection between uni and multivariate polynomials, *Transactions of the IECE of Japan* 68 (1985) 139–146. <765, 768, 783>

[2030] S. Mattarei, On a bound of Garcia and Voloch for the number of points of a Fermat curve over a prime field, *Finite Fields Appl.* 13 (2007) 773–777. <212, 213>

[2031] R. Matthews, Permutation polynomials over algebraic number fields, *J. Number Theory* 18 (1984) 249–260. <302>

[2032] R. Matthews, Some results on permutation polynomials over finite fields, *Appl. Algebra Engrg. Comm. Comput.* 3 (1992) 63–65. <231, 232>

[2033] R. Matthews, Permutation properties of the polynomials $1+x+\cdots+x^k$ over a finite field, *Proc. Amer. Math. Soc.* 120 (1994) 47–51. <223, 224, 230>

[2034] R. W. Matthews, *Permutation Polynomials in One and Several Variables*, PhD thesis, University of Tasmania, Hobart, Tasmania, Australia, 1982. <226, 230, 273, 278>

[2035] H. F. Mattson and G. Solomon, A new treatment of Bose-Chaudhuri codes, *J. Soc. Indust. Appl. Math.* 9 (1961) 654–669. <678, 702, 703>

[2036] C. Mauduit, H. Niederreiter, and A. Sárközy, On pseudorandom $[0,1)$ and binary sequences, *Publ. Math. Debrecen* 71 (2007) 305–324. <336>

[2037] C. Mauduit and A. Sárközy, On finite pseudorandom binary sequences. I. Measure of pseudorandomness, the Legendre symbol, *Acta Arith.* 82 (1997) 365–377. <182, 185, 335, 336, 842, 849>

[2038] U. M. Maurer, Fast generation of prime numbers and secure public-key cryptographic parameters, *J. Cryptology* 8 (1995) 123–155. <347, 363>

[2039] U. M. Maurer and S. Wolf, The Diffie-Hellman protocol: towards a quarter-century of public key cryptography, *Des. Codes Cryptogr.* 19 (2000) 147–171. <746, 750>

[2040] J. P. May, D. Saunders, and Z. Wan, Efficient matrix rank computation with application to the study of strongly regular graphs, In *ISSAC 2007*, 277–284, ACM, New York, 2007. <534, 535>

[2041] B. Mazur, Frobenius and the Hodge filtration (estimates), *Ann. of Math., 2nd Ser.* 98 (1973) 58–95. <481, 488>

[2042] O. D. Mbodj, Quadratic Gauss sums, *Finite Fields Appl.* 4 (1998) 347–361. <150, 161>

[2043] K. McCann and K. S. Williams, The distribution of the residues of a quartic polynomial, *Glasgow Math. J.* 8 (1967) 67–88. <234, 236>

[2044] K. S. McCurley, Cryptographic key distribution and computation in class groups, In *Number Theory and Applications*, volume 265 of *NATO Adv. Sci. Inst. Ser. C Math. Phys. Sci.*, 459–479, Kluwer Acad. Publ., Dordrecht, 1989. <394, 401>

[2045] B. R. McDonald, *Finite Rings with Identity*, Pure and Applied Mathematics, Vol. 28. Marcel Dekker Inc., New York, 1974. <28, 29, 30>

[2046] R. J. McEliece, Table of polynomials of period e over GF(p), *Math. Comp.* 23 (1969) C1–C6. <62, 66>

[2047] R. J. McEliece, *The Theory of Information and Coding*, Addison-Wesley Publishing Co., Reading, Mass.-London-Amsterdam, 1977. <661, 684, 685, 694, 703>

[2048] R. J. McEliece, A public-key cryptosystem based on algebraic coding theory, DSN progress report #42-44, Jet Propulsion Laboratory, Pasadena, California, 1978. <749, 750>

[2049] R. J. McEliece, *Finite Fields for Computer Scientists and Engineers*, The Kluwer International Series in Engineering and Computer Science, 23. Kluwer Academic Publishers, Boston, MA, 1987. <13, 31, 32>

[2050] R. J. McEliece, E. R. Rodemich, H. Rumsey, Jr., and L. R. Welch, New upper bounds on the rate of a code via the Delsarte-MacWilliams inequalities, *IEEE Trans. Inform. Theory* IT-23 (1977) 157–166. <673, 674, 703>

[2051] R. L. McFarland, A family of difference sets in non-cyclic groups, *J. Combin. Theory, Ser. A* 15 (1973) 1–10. <267, 273, 605, 607>

[2052] G. McGuire, G. L. Mullen, D. Panario, and I. E. Shparlinski, editors, *Finite Fields: Theory and Applications*, volume 518 of *Contemp. Math.*, American Mathematical Society, Providence, RI, 2010. <31, 32>

[2053] B. D. McKay and I. M. Wanless, On the number of Latin squares, *Ann. Comb.* 9 (2005) 335–344. <550, 556>

[2054] H. McKean and V. Moll, *Elliptic Curves: Function Theory, Geometry, Arithmetic*, Cambridge University Press, Cambridge, 1997. <31, 32, 422, 440>

[2055] P. K. Meher, Systolic and super-systolic multipliers for finite field $GF(2^m)$ based on irreducible trinomials, *IEEE Transactions on Circuits and Systems I: Regular Papers* 55 (2008) 1031–1040. <815, 823>

[2056] W. Meidl, Linear complexity and k-error linear complexity for p^n-periodic sequences, In *Coding, Cryptography and Combinatorics*, volume 23 of *Progr. Comput. Sci. Appl. Logic*, 227–235, Birkhäuser, Basel, 2004. <329, 336>

[2057] W. Meidl, Reducing the calculation of the linear complexity of $u2^v$-periodic binary sequences to Games-Chan algorithm, *Des. Codes Cryptogr.* 46 (2008) 57–65. <329, 336>

[2058] W. Meidl and H. Niederreiter, Counting functions and expected values for the k-error linear complexity, *Finite Fields Appl.* 8 (2002) 142–154. <331, 336>

[2059] W. Meidl and H. Niederreiter, Linear complexity, k-error linear complexity, and the discrete Fourier transform, *J. Complexity* 18 (2002) 87–103. <331, 336>

[2060] W. Meidl and H. Niederreiter, On the expected value of the linear complexity and the k-error linear complexity of periodic sequences, *IEEE Trans. Inform. Theory* 48 (2002) 2817–2825. <331, 336>

[2061] W. Meidl and H. Niederreiter, The expected value of the joint linear complexity of periodic multisequences, *J. Complexity* 19 (2003) 61–72. <328, 331, 336>

[2062] W. Meidl and H. Niederreiter, Periodic sequences with maximal linear complexity and large k-error linear complexity, *Appl. Algebra Engrg. Comm. Comput.* 14 (2003) 273–286. <331, 336>

[2063] W. Meidl, H. Niederreiter, and A. Venkateswarlu, Error linear complexity measures for multisequences, *J. Complexity* 23 (2007) 169–192. <331, 336>

[2064] W. Meidl and F. Özbudak, Linear complexity over $\mathbb{F}_q$ and over $\mathbb{F}_{q^m}$ for linear recurring sequences, *Finite Fields Appl.* 15 (2009) 110–124. <325, 336>

[2065] W. Meidl and A. Winterhof, Lower bounds on the linear complexity of the discrete logarithm in finite fields, *IEEE Trans. Inform. Theory* 47 (2001) 2807–2811. <333, 336>

[2066] W. Meidl and A. Winterhof, Linear complexity and polynomial degree of a function over a finite field, In *Finite Fields with Applications to Coding Theory, Cryptography and Related Areas*, 229–238, Springer, Berlin, 2002. <329, 336>

[2067] W. Meidl and A. Winterhof, On the linear complexity profile of explicit nonlinear pseudorandom numbers, *Inform. Process. Lett.* 85 (2003) 13–18. <332, 336>

[2068] W. Meidl and A. Winterhof, On the autocorrelation of cyclotomic generators, In *Finite Fields and Applications*, volume 2948 of *Lecture Notes in Comput. Sci.*, 1–11, Springer, Berlin, 2004. <172, 185>

[2069] W. Meidl and A. Winterhof, On the linear complexity profile of some new explicit inversive pseudorandom numbers, *J. Complexity* 20 (2004) 350–355. <332, 336>

[2070] W. Meidl and A. Winterhof, On the joint linear complexity profile of explicit inversive multisequences, *J. Complexity* 21 (2005) 324–336. <332, 336>

[2071] W. Meidl and A. Winterhof, Some notes on the linear complexity of Sidel′nikov-Lempel-Cohn-Eastman sequences, *Des. Codes Cryptogr.* 38 (2006) 159–178. <334, 336>

[2072] W. Meidl and A. Winterhof, On the linear complexity profile of nonlinear congruential pseudorandom number generators with Rédei functions, *Finite Fields Appl.* 13 (2007) 628–634. <333, 336>

[2073] W. Meier, E. Pasalic, and C. Carlet, Algebraic attacks and decomposition of Boolean functions, In *Advances in cryptology—EUROCRYPT 2004*, volume 3027 of *Lecture Notes in Comput. Sci.*, 474–491, Springer, Berlin, 2004. <247, 252>

[2074] W. Meier and O. Staffelbach, Fast correlation attacks on stream ciphers, In D. Barstow, W. Brauer, P. Brinch Hansen, D. Gries, D. Luckham, C. Moler, A. Pnueli, G. Seegmüller, J. Stoer, N. Wirth, and C. Günther, editors, *Advances in Cryptology—EUROCRYPT'88*, volume 330 of *Lecture Notes in Comput. Sci.*, 301–314, Springer, Berlin, 1988. <630, 642>

[2075] W. Meier and O. Staffelbach, Fast correlation attacks on certain stream ciphers, *J. Cryptology* 1 (1989) 159–176. <247, 252>

[2076] A. Menezes, *Elliptic Curve Public Key Cryptosystems*, The Kluwer International Series in Engineering and Computer Science, 234. Kluwer Academic Publishers, Boston, MA, 1993. <31, 32>

[2077] A. Menezes, I. Blake, X.-H. Gao, R. Mullin, S. Vanstone, and T. Yaghoobian, *Applications of Finite Fields*, The Springer International Series in Engineering and Computer Science, Vol. 199, Springer, 1993. <13, 31, 32, 60, 61, 63, 64, 65, 66, 71, 73, 113, 116>

[2078] A. Menezes and M. Qu, Analysis of the Weil descent attack of Gaudry, Hess and Smart, In *Topics in Cryptology—CT-RSA 2001*, volume 2020 of *Lecture Notes in Comput. Sci.*, 308–318, Springer, Berlin, 2001. <810, 811>

[2079] A. J. Menezes, T. Okamoto, and S. A. Vanstone, Reducing elliptic curve logarithms to logarithms in a finite field, *IEEE Trans. Inform. Theory* 39 (1993) 1639–1646. <439, 440, 793, 796>

[2080] A. J. Menezes, P. C. van Oorschot, and S. A. Vanstone, *Handbook of Applied Cryptography*, CRC Press Series on Discrete Mathematics and its Applications, CRC Press, Boca Raton, FL, 1997. <31, 32, 46, 49, 346, 347, 351, 352, 363, 393, 401, 750, 758, 763, 785, 796, 822, 823>

[2081] G. Menichetti, On a Kaplansky conjecture concerning three-dimensional division algebras over a finite field, *J. Algebra* 47 (1977) 400–410. <277, 278>

[2082] G. Menichetti, n-dimensional algebras over a field with a cyclic extension of degree n, *Geom. Dedicata* 63 (1996) 69–94. <277, 278>

[2083] P. Merkey and E. Posner, Optimum cyclic redundancy codes for noisy channels, *IEEE Trans. Inform. Theory* 30 (1984) 865–867. <633, 635, 638, 639, 642>

[2084] "Mersenne Research, Inc.," GIMPS Home, `http://www.mersenne.org`, as viewed in July, 2012. <46, 49>

[2085] S. Mesnager, Improving the lower bound on the higher order nonlinearity of Boolean functions with prescribed algebraic immunity, *IEEE Trans. Inform. Theory* 54 (2008) 3656–3662. <248, 252>

[2086] S. Mesnager, A new class of bent and hyper-bent Boolean functions in polynomial forms, *Des. Codes Cryptogr.* 59 (2011) 265–279. <154, 161>

[2087] J.-F. Mestre, Construction des courbes de genre 2 à partir de leurs modules, *Progress in Mathematics* 94 (1991) 313–334. <803>

[2088] J.-F. Mestre, Lettre adressée à Gaudry et Harley, `http://www.math.jussieu.fr/~mestre/lettreGaudryHarley.ps`, 2000. <788, 796>

[2089] J.-F. Mestre, Algorithmes pur compter des point de courbes en petite charactéristique et en petit genres, Available at `http://www.math.jussieu.fr/~mestre/`, 2002. <454, 456>

[2090] P. Meyer, Eine Charakterisierung vollständig regulärer, abelscher Erweiterungen, *Abh. Math. Sem. Univ. Hamburg* 68 (1998) 199–223. <129, 130, 138>

[2091] H. Meyn, On the construction of irreducible self-reciprocal polynomials over finite fields, *Appl. Algebra Engrg. Comm. Comput.* 1 (1990) 43–53. <57, 59, 60, 61, 64, 66, 286, 290>

[2092] P. Michel, Some recent applications of Kloostermania, In *Physics and Number Theory*, volume 10 of *IRMA Lect. Math. Theor. Phys.*, 225–251, Eur. Math. Soc., Zürich, 2006. <156, 161>

[2093] T. Migler, K. E. Morrison, and M. Ogle, How much does a matrix of rank k weigh?,

Math. Mag. 79 (2006) 262–271. <501, 510>

[2094] M. Mignotte and C. Schnorr, Calcul des racines d-ièmes dans un corps fini, *Comptes Rendus de l'Académie des Sciences Paris* 290 (1988) 205–206. <381, 382>

[2095] P. Mihăilescu, Fast generation of provable primes using search in arithmetic progressions, In *Advances in Cryptology—CRYPTO '94*, volume 839 of *Lecture Notes in Comput. Sci.*, 282–293, Springer, Berlin, 1994. <347, 363>

[2096] P. Mihăilescu, Optimal Galois field bases which are not normal, 1997, Presented at the Workshop on Fast Software Encryption in Haifa. <353, 363>

[2097] P. Mihăilescu, Medium Galois Fields, their Bases and Arithmetic, 2000, `http://grouper.ieee.org/groups/1363/P1363a/NumThAlgs.html`. <358, 363>

[2098] P. Mihăilescu, F. Morain, and E. Schost, Computing the eigenvalue in the Schoof–Elkies–Atkin algorithm using abelian lifts, In C. W. Brown, editor, *Proceedings of the 2007 International Symposium on Symbolic and Algebraic Computation—ISSAC 2007*, 285–292, ACM, New York, 2007. <788, 796>

[2099] G. L. Miller, Riemann's hypothesis and tests for primality, *J. Comput. System Sci.* 13 (1976) 300–317. <346, 363>

[2100] R. L. Miller, Necklaces, symmetries and self-reciprocal polynomials, *Discrete Math.* 22 (1978) 25–33. <286, 290>

[2101] S. J. Miller and M. R. Murty, Effective equidistribution and the Sato-Tate law for families of elliptic curves, *J. Number Theory* 131 (2011) 25–44. <430, 440>

[2102] V. S. Miller, Use of elliptic curves in cryptography, In *Advances in Cryptology—CRYPTO '85*, volume 218 of *Lecture Notes in Comput. Sci.*, 417–426, Springer, Berlin, 1986. <746, 750, 784, 796>

[2103] D. Mills, Factorizations of root-based polynomial compositions, *Discrete Math.* 240 (2001) 161–173. <67, 70>

[2104] D. Mills, Existence of primitive polynomials with three coefficients prescribed, *JP J. Algebra Number Theory Appl.* 4 (2004) 1–22. <93, 95>

[2105] W. H. Mills, Polynomials with minimal value sets, *Pacific J. Math.* 14 (1964) 225–241. <233, 236>

[2106] W. H. Mills and R. C. Mullin, Coverings and packings, In *Contemporary Design Theory*, Wiley-Intersci. Ser. Discrete Math. Optim., 371–399, Wiley, New York, 1992. <598, 599>

[2107] J. S. Milne, *Elliptic Curves*, BookSurge Publishers, Charleston, SC, 2006. <31, 32, 422, 440>

[2108] R. Mines, F. Richman, and W. Ruitenburg, *A course in constructive algebra*, Universitext. Springer-Verlag, New York, 1988. <383, 392>

[2109] M. Minzlaff, Computing zeta functions of superelliptic curves in larger characteristic, *Math. Comput. Sci.* 3 (2010) 209–224. <490, 491>

[2110] Y. Miyamoto, H. Doi, K. Matsuo, J. Chao, and S. Tsuji, A fast addition algorithm of genus two hyperelliptic curve, In *Proc. of SCIS2002, IEICE Japan*, 497–502, 2002, in Japanese. <798, 803>

[2111] R. T. Moenck, Another polynomial homomorphism, *Acta Informat.* 6 (1976) 153–169. <351, 363>

[2112] R. T. Moenck, On the efficiency of algorithms for polynomial factoring, *Math. Comp.* 31 (1977) 235–250. <381, 382>

[2113] T. Moh, A public key system with signature and master key functions, *Comm. Algebra*

27 (1999) 2207–2222. <774, 783>

[2114] M. S. E. Mohamed, D. Cabarcas, J. Ding, J. Buchmann, and S. Bulygin, MXL$_3$: an efficient algorithm for computing Gröbner bases of zero-dimensional ideals, In *Information Security and Cryptology—ICISC 2009*, volume 5984 of *Lecture Notes in Comput. Sci.*, 87–100, Springer, Berlin, 2010. <782, 783>

[2115] M. S. E. Mohamed, J. Ding, J. Buchmann, and F. Werner, Algebraic attack on the MQQ public key cryptosystem, In *Eighth International Conference on Cryptology and Network Security*, volume 5888 of *Lecture Notes in Comput. Sci.*, 392–401, Springer, Berlin, 2009. <776, 783>

[2116] M. S. E. Mohamed, W. S. A. E. Mohamed, J. Ding, and J. Buchmann, MXL2: Solving polynomial equations over GF(2) using an improved mutant strategy, In *PQCrypto*, volume 5299 of *Lecture Notes in Comput. Sci.*, 203–215. Springer, 2008. <782, 783>

[2117] B. Mohar, Isoperimetric numbers of graphs, *J. Combin. Theory, Ser. B* 47 (1989) 274–291. <649, 658>

[2118] B. Mohar, A strengthening and a multipartite generalization of the Alon-Boppana-Serre theorem, *Proc. Amer. Math. Soc.* 138 (2010) 3899–3909. <647, 648, 658>

[2119] M. Moisio, The moments of a Kloosterman sum and the weight distribution of a Zetterberg-type binary cyclic code, *IEEE Trans. Inform. Theory* 53 (2007) 843–847. <157, 161>

[2120] M. Moisio, On the number of rational points on some families of Fermat curves over finite fields, *Finite Fields Appl.* 13 (2007) 546–562. <208, 213>

[2121] M. Moisio, Kloosterman sums, elliptic curves, and irreducible polynomials with prescribed trace and norm, *Acta Arith.* 132 (2008) 329–350. <75, 79, 270, 273>

[2122] M. Moisio, On the moments of Kloosterman sums and fibre products of Kloosterman curves, *Finite Fields Appl.* 14 (2008) 515–531. <157, 161>

[2123] M. Moisio and K. Ranto, Elliptic curves and explicit enumeration of irreducible polynomials with two coefficients prescribed, *Finite Fields Appl.* 14 (2008) 798–815. <79>

[2124] M. Moisio, K. Ranto, M. Rinta-Aho, and K. Väänänen, On the weight distribution of cyclic codes with one or two zeros, *Adv. Appl. Discrete Math.* 3 (2009) 125–150. <154, 161>

[2125] M. Moisio and D. Wan, On Katz's bound for the number of elements with given trace and norm, *J. Reine Angew. Math.* 638 (2010) 69–74. <196, 201>

[2126] F. Möller, Exceptional polynomials with 2-transitive affine monodromy groups, *Finite Fields Appl.* 18 (2012) 445–457. <238, 240>

[2127] R. A. Mollin and C. Small, On permutation polynomials over finite fields, *Internat. J. Math. Math. Sci.* 10 (1987) 535–543. <223, 230>

[2128] R. Moloney, *Divisibility Properties of Kloosterman Sums and Division Polynomials for Edwards Curves*, PhD dissertation, University College Dublin, College of Engineering, Mathematical and Physical Sciences, 2011. <154, 161>

[2129] M. Monagan and R. Pearce, Polynomial division using dynamic arrays, heaps, and packed exponent vectors, In *Proc. of CASC 2007*, 295–315. Springer-Verlag, 2007. <382, 392>

[2130] M. Monagan and R. Pearce, Parallel sparse polynomial multiplication using heaps, In *ISSAC '09: Proceedings of the 2009 International Symposium on Symbolic and Algebraic Computation*, 263–270, ACM, New York, NY, USA, 2009. <382, 392>

[2131] M. Monagan and R. Pearce, Sparse polynomial multiplication and division in Maple

14, *ACM Communications in Computer Algebra* 44 (2010) 183–220. <382, 392>

[2132] P. L. Montgomery, Modular multiplication without trial division, *Math. Comp.* 44 (1985) 519–521. <352, 360, 363, 822, 823>

[2133] P. L. Montgomery, Speeding the Pollard and elliptic curve methods of factorization, *Math. Comp.* 48 (1987) 243–264. <359, 363, 441, 443, 446>

[2134] P. L. Montgomery, A block Lanczos algorithm for finding dependencies over GF(2), In *Advances in Cryptology—EUROCRYPT '95*, volume 921 of *Lecture Notes in Comput. Sci.*, 106–120, Springer, Berlin, 1995. <400, 401>

[2135] J. W. Moon and L. Moser, On the correlation function of random binary sequences, *SIAM J. Appl. Math.* 16 (1968) 340–343. <842, 849>

[2136] T. Moon, *Error Correction Coding: Mathematical Methods and Algorithms*, John Wiley and Sons, Hoboken, NJ, 2005. <661, 692, 703>

[2137] Y. Moon, J. B. Lee, and Y. H. Park, Counting formula for solutions of diagonal equations, *Bull. Korean Math. Soc.* 37 (2000) 803–810. <213>

[2138] C. Moore, D. Rockmore, A. Russell, and L. J. Schulman, The hidden subgroup problem in affine groups: basis selection in Fourier sampling, preprint available, `http://arxiv.org/abs/quant-ph/0211124`, 2003. <840, 841>

[2139] E. H. Moore, A doubly-infinite system of simple groups, In E. H. Moore *et al.*, editor, *Mathematical Papers Read at the International Mathematics Congress Held in Connection with the World's Columbian Exposition*, 208–242, New York: Macmillan & Co., 1896. <3, 10, 11>

[2140] E. H. Moore, A two-fold generalization of Fermat's theorem, *Bull. Amer. Math. Soc.* 2 (1896) 189–199. <509, 510>

[2141] F. Morain, Implementing the asymptotically fast version of the elliptic curve primality proving algorithm, *Math. Comp.* 76 (2007) 493–505. <347, 363>

[2142] D. J. M. Morales, An analysis of the infrastructure in real function fields, preprint available, `http://eprint.iacr.org/2008/299`, 2008. <456>

[2143] L. J. Mordell, On a sum analogous to a Gauss sum, *Quart. J. Math.* 3 (1932) 161–162. <190, 192>

[2144] C. Moreno, *Algebraic Curves over Finite Fields*, volume 97 of *Cambridge Tracts in Mathematics*, Cambridge University Press, Cambridge, 1991. <32, 208, 213>

[2145] O. Moreno, Discriminants and the irreducibility of a class of polynomials in a finite field of arbitrary characteristic, *J. Number Theory* 28 (1988) 62–65. <67, 70>

[2146] O. Moreno and F. N. Castro, Divisibility properties for covering radius of certain cyclic codes, *IEEE Trans. Inform. Theory* 49 (2003) 3299–3303. <213>

[2147] O. Moreno and F. N. Castro, On the covering radius of certain cyclic codes, In *Applied Algebra, Algebraic Algorithms and Error-Correcting Codes*, volume 2643 of *Lecture Notes in Comput. Sci.*, 129–138, Springer, Berlin, 2003. <213>

[2148] O. Moreno and F. N. Castro, On the calculation and estimation of Waring number for finite fields, In *Arithmetic, Geometry and Coding Theory (AGCT 2003)*, volume 11 of *Sémin. Congr.*, 29–40, Soc. Math. France, Paris, 2005. <211, 213>

[2149] O. Moreno and F. N. Castro, Optimal divisibility for certain diagonal equations over finite fields, *J. Ramanujan Math. Soc.* 23 (2008) 43–61. <210, 212, 213>

[2150] O. Moreno, F. N. Castro, and H. F. Mattson, Jr., Correction to: "Divisibility properties for covering radius of certain cyclic codes" [IEEE Trans. Inform. Theory 49 (2003), 3299–3303] by Moreno and Castro, *IEEE Trans. Inform. Theory* 52 (2006) 1798–1799. <213>

[2151] O. Moreno and C. J. Moreno, Improvements of the Chevalley-Warning and the Ax-Katz theorems, *Amer. J. Math.* 117 (1995) 241–244. <200, 201, 207, 210, 213, 481, 488>

[2152] O. Moreno and I. Rubio, Cyclic decomposition of monomial permutations, *Congr. Numer.* 73 (1990) 147–158. <229, 230>

[2153] O. Moreno, K. W. Shum, F. N. Castro, and P. V. Kumar, Tight bounds for Chevalley-Warning-Ax-Katz type estimates, with improved applications, *Proc. London Math. Soc., 3rd Ser.* 88 (2004) 545–564. <481, 488>

[2154] M. Morf, Doubling algorithms for Toeplitz and related equations, In *Proc. 1980 Int'l Conf. Acoustics Speech and Signal Processing*, 954–959, Denver, Colo., 1980. <533, 535>

[2155] I. H. Morgan, Construction of complete sets of mutually equiorthogonal frequency hypercubes, *Discrete Math.* 186 (1998) 237–251. <554, 556>

[2156] I. H. Morgan and G. L. Mullen, Primitive normal polynomials over finite fields, *Math. Comp.* 63 (1994) 759–765, S19–S23. <88, 90>

[2157] I. H. Morgan and G. L. Mullen, Completely normal primitive basis generators of finite fields, *Utilitas Math.* 49 (1996) 21–43. <89, 90, 95, 137, 138>

[2158] I. H. Morgan, G. L. Mullen, and M. Živković, Almost weakly self-dual bases for finite fields, *Appl. Algebra Engrg. Comm. Comput.* 8 (1997) 25–31. <89, 90, 107, 109>

[2159] J. P. Morgan, Nested designs, In *Design and Analysis of Experiments*, volume 13 of *Handbook of Statist.*, 939–976, North-Holland, Amsterdam, 1996. <595, 599>

[2160] M. Morgenstern, Existence and explicit constructions of $q + 1$ regular Ramanujan graphs for every prime power q, *J. Combin. Theory, Ser. B* 62 (1994) 44–62. <655, 658>

[2161] R. Mori and T. Tanaka, Performance and construction of polar codes on symmetric binary-input memoryless channels, In *Proc. IEEE Int. Symp. Information Theory ISIT 2009*, 1496–1500, 2009. <735, 738, 739>

[2162] R. Mori and T. Tanaka, Performance of polar codes with the construction using density evolution, *IEEE Comm. Letters* 13 (2009) 519–521. <739>

[2163] R. Mori and T. Tanaka, Non-binary polar codes using Reed-Solomon codes and algebraic geometry codes, In *Proc. Information Theory Workshop*, 1–5, 2010. <739>

[2164] M. Morii and M. Kasahara, Generalized key-equation of remainder decoding algorithm for Reed-Solomon codes, *IEEE Trans. Inform. Theory* 38 (1992) 1801–1807. <695, 703>

[2165] B. Morlaye, Équations diagonales non homogènes sur un corps fini, *C. R. Acad. Sci. Paris, Sér. A-B* 272 (1971) A1545–A1548. <208, 213>

[2166] K. E. Morrison, Integer sequences and matrices over finite fields, *J. Integer Seq.* 9 (2006) Article 06.2.1, 28 pp. <502, 510>

[2167] E. Mortenson, Modularity of a certain Calabi-Yau threefold and combinatorial congruences, *Ramanujan J.* 11 (2006) 5–39. <141, 161>

[2168] M. J. Mossinghoff, Wieferich pairs and Barker sequences, *Des. Codes Cryptogr.* 53 (2009) 149–163. <604, 607>

[2169] C. Mulcahy, Card colm, Mathematical Association of America Online, `http://www.maa.org/columns/colm/cardcolm.html`. <632, 642>

[2170] T. Mulders and A. Storjohann, Rational solutions of singular linear systems, In *Proceedings of the 2000 International Symposium on Symbolic and Algebraic Computation*, 242–249, ACM, New York, 2000. <527, 535>

[2171] G. Mullen and H. Stevens, Polynomial functions (mod m), *Acta Math. Hungar.* 44 (1984) 237–241. <229, 230>

[2172] G. L. Mullen, Permutation polynomials in several variables over finite fields, *Acta Arith.* 31 (1976) 107–111. <230, 232>

[2173] G. L. Mullen, Polynomial representation of complete sets of mutually orthogonal frequency squares of prime power order, *Discrete Math.* 69 (1988) 79–84. <553, 556>

[2174] G. L. Mullen, Permutation polynomials and nonsingular feedback shift registers over finite fields, *IEEE Trans. Inform. Theory* 35 (1989) 900–902. <231, 232>

[2175] G. L. Mullen, Dickson polynomials over finite fields, *Adv. in Math. (China)* 20 (1991) 24–32. <226, 230>

[2176] G. L. Mullen, Permutation polynomials over finite fields, In *Finite Fields, Coding Theory, and Advances in Communications and Computing*, volume 141 of *Lecture Notes in Pure and Appl. Math.*, 131–151, Dekker, New York, 1993. <216, 217, 230>

[2177] G. L. Mullen, A candidate for the "next Fermat problem," *Math. Intelligencer* 17 (1995) 18–22. <551, 556>

[2178] G. L. Mullen, Permutation polynomials: a matrix analogue of Schur's conjecture and a survey of recent results, *Finite Fields Appl.* 1 (1995) 242–258. <216, 228, 230>

[2179] G. L. Mullen and C. Mummert, *Finite Fields and Applications*, volume 41 of *Student Mathematical Library*, American Mathematical Society, Providence, RI, 2007. <13, 31, 32>

[2180] G. L. Mullen and D. Panario, Handbook of Finite Fields (Web resource), `http://www.crcpress.com/product/isbn/9781439873786`, as viewed in November, 2012. <32, 49>

[2181] G. L. Mullen, D. Panario, and I. E. Shparlinski, editors, *Finite Fields and Applications*, volume 461 of *Contemp. Math.*, American Mathematical Society, Providence, RI, 2008. <31, 32>

[2182] G. L. Mullen, A. Poli, and H. Stichtenoth, editors, *Finite Fields and Applications*, volume 2948 of *Lecture Notes in Comput. Sci.*, Springer-Verlag, Berlin, 2004. <31, 32>

[2183] G. L. Mullen and W. C. Schmid, An equivalence between (t, m, s)-nets and strongly orthogonal hypercubes, *J. Combin. Theory, Ser. A* 76 (1996) 164–174. <620, 630>

[2184] G. L. Mullen and P. J.-S. Shiue, editors, *Finite Fields, Coding Theory, and Advances in Communications and Computing*, volume 141 of *Lecture Notes in Pure and Applied Mathematics*, Marcel Dekker Inc., New York, 1993. <31, 32>

[2185] G. L. Mullen and P. J.-S. Shiue, editors, *Finite Fields: Theory, Applications, and Algorithms*, volume 168 of *Contemp. Math.*, American Mathematical Society, Providence, RI, 1994. <31, 32>

[2186] G. L. Mullen and I. E. Shparlinski, Open problems and conjectures in finite fields, In *Finite Fields and Applications*, volume 233 of *London Math. Soc. Lecture Note Ser.*, 243–268, Cambridge Univ. Press, Cambridge, 1996. <73, 88, 89, 90, 96, 97>

[2187] G. L. Mullen, H. Stichtenoth, and H. Tapia-Recillas, editors, *Finite Fields with Applications to Coding Theory, Cryptography and Related Areas*, Springer-Verlag, Berlin, 2002. <31, 32>

[2188] G. L. Mullen, D. Wan, and Q. Wang, Value sets of polynomials maps over finite fields,

Quarterly J. of Math., to appear, 2013. <232, 236>

[2189] G. L. Mullen and D. White, A polynomial representation for logarithms in GF(q), *Acta Arith.* 47 (1986) 255–261. <396, 401>

[2190] P. Müller, New examples of exceptional polynomials, In *Finite Fields: Theory, Applications, and Algorithms*, volume 168 of *Contemp. Math.*, 245–249, Amer. Math. Soc., Providence, RI, 1994. <240>

[2191] P. Müller, Primitive monodromy groups of polynomials, In *Recent Developments in the Inverse Galois Problem*, volume 186 of *Contemp. Math.*, 385–401, Amer. Math. Soc., Providence, RI, 1995. <301, 302>

[2192] P. Müller, A Weil-bound free proof of Schur's conjecture, *Finite Fields Appl.* 3 (1997) 25–32. <228, 230, 237, 239, 240>

[2193] P. Müller, Arithmetically exceptional functions and elliptic curves, In *Aspects of Galois Theory*, volume 256 of *London Math. Soc. Lecture Note Ser.*, 180–201, Cambridge Univ. Press, Cambridge, 1999. <239, 240>

[2194] S. Müller, On the computation of square roots in finite fields, *Des. Codes Cryptogr.* 31 (2004) 301–312. <360, 363>

[2195] V. Müller, Fast multiplication on elliptic curves over small fields of characteristic two, *Journal of Cryptology* 11 (1998) 219–234. <801, 803>

[2196] R. Mullin, I. Onyszchuk, S. Vanstone, and R. Wilson, Optimal normal bases in $GF(p^n)$, *Discrete Appl. Math.* 22 (1988/89) 149–161. <818, 819, 823>

[2197] R. C. Mullin and G. L. Mullen, editors, *Finite Fields: Theory, Applications, and Algorithms*, volume 225 of *Contemp. Math.*, American Mathematical Society, Providence, RI, 1999. <31, 32>

[2198] R. C. Mullin and E. Nemeth, An existence theorem for room squares, *Canad. Math. Bull.* 12 (1969) 493–497. <614, 619>

[2199] R. C. Mullin, I. M. Onyszchuk, S. A. Vanstone, and R. M. Wilson, Optimal normal bases in GF(p^n), *Discrete Appl. Math.* 22 (1988/89) 149–161. <117, 128>

[2200] R. C. Mullin, J. L. Yucas, and G. L. Mullen, A generalized counting and factoring method for polynomials over finite fields, *J. Combin. Math. Combin. Comput.* 72 (2010) 121–143. <57, 58, 59>

[2201] D. Mumford, An algebro-geometric construction of commuting operators and of solutions to the Toda lattice equation, Korteweg deVries equation and related nonlinear equation, In *Proceedings of the International Symposium on Algebraic Geometry*, 115–153, Kinokuniya Book Store, Tokyo, 1978. <545, 546>

[2202] D. Mumford, *Algebraic Geometry. I*, Classics in Mathematics. Springer-Verlag, Berlin, 1995. <387, 392>

[2203] D. Mumford, *The Red Book of Varieties and Schemes*, volume 1358 of *Lecture Notes in Mathematics*, Springer-Verlag, Berlin, expanded edition, 1999. <291, 292, 297, 302>

[2204] A. Munemasa, Orthogonal arrays, primitive trinomials, and shift-register sequences, *Finite Fields Appl.* 4 (1998) 252–260. <90, 631, 638, 642>

[2205] A. Muratović-Ribić, A note on the coefficients of inverse polynomials, *Finite Fields Appl.* 13 (2007) 977–980. <229, 230>

[2206] A. Muratović-Ribić, Inverse of some classes of permutation binomials, *J. Concr. Appl. Math.* 7 (2009) 47–53. <229, 230>

[2207] M. R. Murty, *Problems in Analytic Number Theory*, volume 206 of *Graduate Texts in Mathematics*, Springer-Verlag, New York, 2001. <650, 658>

[2208] M. R. Murty, Ramanujan graphs, *J. Ramanujan Math. Soc.* 18 (2003) 33–52. <643, 651, 658>

[2209] D. R. Musser, Multivariate polynomial factorization, *J. Assoc. Comput. Mach.* 22 (1975) 291–308. <385, 392>

[2210] M. Muzychuk, On Skew Hadamard difference sets, arXiv:1012.2089v1, 2010. <604, 607>

[2211] K.-i. Nagao, Improving group law algorithms for Jacobians of hyperelliptic curves, In *Proceedings of the Fourth International Symposium of Algorithmic Number Theory— ANTS-IV*, volume 1838 of *Lecture Notes in Comput. Sci.*, 439–448, Springer, Berlin, 2000. <799, 803>

[2212] K.-i. Nagao, Index calculus attack for Jacobian of hyperelliptic curves of small genus using two large primes, *Japan J. Indust. Appl. Math.* 24 (2007) 289–305. <798, 799, 803>

[2213] M. Nagata, *On Automorphism Group of* $k[x, y]$, Kinokuniya Book-Store Co. Ltd., Department of Mathematics, Kyoto University, Lectures in Mathematics, No. 5, Tokyo, 1972. <768, 783>

[2214] S. Najib, Une généralisation de l'inégalité de Stein-Lorenzini, *J. Algebra* 292 (2005) 566–573. <83, 85>

[2215] A. Naldi, D. Thieffry, and C. Chaouiya, Decision diagrams for the representation and analysis of logical models of genetic networks, In *CMSB'07: Proceedings of the 2007 International Conference on Computational Methods in Systems Biology*, 233–247, Springer-Verlag, Berlin, Heidelberg, 2007. <829, 834>

[2216] A. H. Namin, H. Wu, and M. Ahmadi, A new finite field multiplier using redundant representation, *IEEE Trans. Comput.* 57 (2008) 716–720. <822, 823>

[2217] Y. Nawaz and G. Gong, The WG stream cipher, 2005, preprint available at `http://www.cacr.math.uwaterloo.ca/techreports/2005/cacr2005-15.pdf`. <751, 755, 756, 757, 763>

[2218] M. Nazarathy, S. Newton, R. Giffard, D. Moberly, F. Sischka, W. Trutna, Jr., and S. Foster, Real-time long range complementary correlation optical time domain reflectometer, *IEEE J. Lightwave Technology* 7 (1989) 24–38. <843, 849>

[2219] V. I. Nechaev, On the complexity of a deterministic algorithm for a discrete logarithm, *Mat. Zametki* 55 (1994) 91–101, 189. <394, 401>

[2220] NESSIE: New European Schemes for Signatures, Integrity, and Encryption. Information Society Technologies programme of the European commission (IST-1999-12324), `http://www.cryptonessie.org/`. <773, 783>

[2221] J. C. Néto, A. F. Tenca, and W. V. Ruggiero, A parallel k-partition method to perform Montgomery multiplication, In *Proc. ASAP-2011*, 251–254, 2011. <822, 823>

[2222] E. Netto, Zur Theorie der Tripelsysteme, *Math. Ann.* 42 (1893) 143–152. <590, 599>

[2223] P. M. Neumann, *The Mathematical Writings of Evariste Galois*, European Mathematical Society, Zurich, 2011. <14, 32>

[2224] T. Neumann, *Bent Functions*, PhD thesis, Department of Mathematics, University of Kaiserslautern, Germany, 2006. <273>

[2225] D. K. Nguyen and B. Schmidt, Fast computation of Gauss sums and resolution of the root of unity ambiguity, *Acta Arith.* 140 (2009) 205–232. <141, 161>

[2226] X. Nie, L. Hu, J. Li, C. Updegrove, and J. Ding, Breaking a new instance of ttm cryptosystems., In *ACNS*, volume 3989 of *Lecture Notes in Comput. Sci.*, 210–225, Springer, Berlin, 2006. <775, 783>

[2227] H. Niederreiter, Permutation polynomials in several variables over finite fields, *Proc.*

Japan Acad. 46 (1970) 1001–1005. <231, 232>

[2228] H. Niederreiter, Orthogonal systems of polynomials in finite fields, *Proc. Amer. Math. Soc.* 28 (1971) 415–422. <230, 231, 232>

[2229] H. Niederreiter, Permutation polynomials in several variables, *Acta Sci. Math. (Szeged)* 33 (1972) 53–58. <231, 232>

[2230] H. Niederreiter, On the distribution of pseudo-random numbers generated by the linear congruential method. II, *Math. Comp.* 28 (1974) 1117–1132. <174, 185>

[2231] H. Niederreiter, On the cycle structure of linear recurring sequences, *Math. Scand.* 38 (1976) 53–77. <316, 317>

[2232] H. Niederreiter, Weights of cyclic codes, *Information and Control* 34 (1977) 130–140. <317>

[2233] H. Niederreiter, Distribution properties of feedback shift register sequences, *Problems Control Inform. Theory* 15 (1986) 19–34. <316, 317>

[2234] H. Niederreiter, Low-discrepancy point sets, *Monatsh. Math.* 102 (1986) 155–167. <622, 630>

[2235] H. Niederreiter, Continued fractions for formal power series, pseudorandom numbers, and linear complexity of sequences, In *Contributions to General Algebra, 5*, 221–233, Hölder-Pichler-Tempsky, Vienna, 1987. <330, 336>

[2236] H. Niederreiter, Point sets and sequences with small discrepancy, *Monatsh. Math.* 104 (1987) 273–337. <620, 621, 622, 625, 626, 628, 630>

[2237] H. Niederreiter, A simple and general approach to the decimation of feedback shift-register sequences, *Problems Control Inform. Theory* 17 (1988) 327–331. <313, 317>

[2238] H. Niederreiter, Low-discrepancy and low-dispersion sequences, *J. Number Theory* 30 (1988) 51–70. <626, 627, 628, 630>

[2239] H. Niederreiter, The probabilistic theory of linear complexity, In *Advances in Cryptology—EUROCRYPT '88*, volume 330 of *Lecture Notes in Comput. Sci.*, 191–209, Springer, Berlin, 1988. <329, 330, 336>

[2240] H. Niederreiter, Sequences with almost perfect linear complexity profile, In *Advances in Cryptology—EUROCRYPT '87*, volume 304 of *Lecture Notes in Comput. Sci.*, 37–51, Springer, Berlin, 1988. <327, 329, 330, 336>

[2241] H. Niederreiter, Some new cryptosystems based on feedback shift register sequences, *Math. J. Okayama Univ.* 30 (1988) 121–149. <316, 317>

[2242] H. Niederreiter, A combinatorial approach to probabilistic results on the linear-complexity profile of random sequences, *J. Cryptology* 2 (1990) 105–112. <330, 336>

[2243] H. Niederreiter, Keystream sequences with a good linear complexity profile for every starting point, In *Advances in Cryptology—EUROCRYPT '89*, volume 434 of *Lecture Notes in Comput. Sci.*, 523–532, Springer, Berlin, 1990. <330, 336>

[2244] H. Niederreiter, A short proof for explicit formulas for discrete logarithms in finite fields, *Appl. Algebra Engrg. Comm. Comput.* 1 (1990) 55–57. <396, 401>

[2245] H. Niederreiter, The distribution of values of Kloosterman sums, *Arch. Math. (Basel)* 56 (1991) 270–277. <156, 161>

[2246] H. Niederreiter, The linear complexity profile and the jump complexity of keystream sequences, In *Advances in Cryptology—EUROCRYPT '90*, volume 473 of *Lecture Notes in Comput. Sci.*, 174–188, Springer, Berlin, 1991. <329, 330, 336>

[2247] H. Niederreiter, Low-discrepancy point sets obtained by digital constructions over finite fields, *Czechoslovak Math. J.* 42 (1992) 143–166. <623, 630>

[2248] H. Niederreiter, *Random Number Generation and Quasi-Monte Carlo Methods*, volume 63 of *CBMS-NSF Regional Conference Series in Applied Mathematics*, Society for Industrial and Applied Mathematics (SIAM), Philadelphia, PA, 1992. <174, 185, 317, 619, 621, 627, 630>

[2249] H. Niederreiter, A new efficient factorization algorithm for polynomials over small finite fields, *Appl. Algebra Engrg. Comm. Comput.* 4 (1993) 81–87. <314, 317, 381, 382>

[2250] H. Niederreiter, Factoring polynomials over finite fields using differential equations and normal bases, *Math. Comp.* 62 (1994) 819–830. <381, 382>

[2251] H. Niederreiter, Constructions of (t, m, s)-nets, In *Monte Carlo and Quasi-Monte Carlo Methods*, 70–85, Springer-Verlag, Berlin, 2000. <625, 630>

[2252] H. Niederreiter, editor, *Coding Theory and Cryptology*, volume 1 of *Lecture Notes Series, Institute for Mathematical Sciences, National University of Singapore*, World Scientific Publishing Co. Inc., River Edge, NJ, 2002. <31, 32>

[2253] H. Niederreiter, Linear complexity and related complexity measures for sequences, In *Progress in Cryptology—INDOCRYPT 2003*, volume 2904 of *Lecture Notes in Comput. Sci.*, 1–17, Springer, Berlin, 2003. <330, 336>

[2254] H. Niederreiter, Periodic sequences with large k-error linear complexity, *IEEE Trans. Inform. Theory* 49 (2003) 501–505. <331, 336>

[2255] H. Niederreiter, Digital nets and coding theory, In *Coding, Cryptography and Combinatorics*, volume 23 of *Progr. Comput. Sci. Appl. Logic*, 247–257, Birkhäuser, Basel, 2004. <624, 630>

[2256] H. Niederreiter, Constructions of (t, m, s)-nets and (t, s)-sequences, *Finite Fields Appl.* 11 (2005) 578–600. <625, 630>

[2257] H. Niederreiter, The probabilistic theory of the joint linear complexity of multisequences, In *Sequences and Their Applications—SETA 2006*, volume 4086 of *Lecture Notes in Comput. Sci.*, 5–16, Springer, Berlin, 2006. <330, 336>

[2258] H. Niederreiter, Nets, (t, s)-sequences, and codes, In *Monte Carlo and Quasi-Monte Carlo Methods*, 83–100, Springer-Verlag, Berlin, 2008. <623, 628, 630>

[2259] H. Niederreiter, Quasi-Monte Carlo methods, In *Encyclopedia of Quantitative Finance*, 1460–1472, John Wiley and Sons, Chichester, 2010. <619, 630>

[2260] H. Niederreiter and R. Göttfert, Factorization of polynomials over finite fields and characteristic sequences, *J. Symbolic Comput.* 16 (1993) 401–412. <314, 317>

[2261] H. Niederreiter and F. Özbudak, Constructions of digital nets using global function fields, *Acta Arith.* 105 (2002) 279–302. <624, 630>

[2262] H. Niederreiter and F. Özbudak, Constructive asymptotic codes with an improvement on the Tsfasman-Vlăduţ-Zink and Xing bounds, In *Coding, Cryptography and Combinatorics*, volume 23 of *Progr. Comput. Sci. Appl. Logic*, 259–275, Birkhäuser, Basel, 2004. <712>

[2263] H. Niederreiter and F. Özbudak, Matrix-product constructions of digital nets, *Finite Fields Appl.* 10 (2004) 464–479. <625, 630>

[2264] H. Niederreiter and F. Özbudak, Further improvements on asymptotic bounds for codes using distinguished divisors, *Finite Fields Appl.* 13 (2007) 423–443. <712>

[2265] H. Niederreiter and F. Özbudak, Improved asymptotic bounds for codes using distinguished divisors of global function fields, *SIAM J. Discrete Math.* 21 (2007) 865–899. <712>

[2266] H. Niederreiter and F. Özbudak, Low-discrepancy sequences using duality and global function fields, *Acta Arith.* 130 (2007) 79–97. <629, 630>

[2267] H. Niederreiter and G. Pirsic, Duality for digital nets and its applications, *Acta Arith.* 97 (2001) 173–182. <623, 630>

[2268] H. Niederreiter and K. H. Robinson, Complete mappings of finite fields, *J. Austral. Math. Soc., Ser. A* 33 (1982) 197–212. <228, 230>

[2269] H. Niederreiter and I. E. Shparlinski, On the distribution and lattice structure of nonlinear congruential pseudorandom numbers, *Finite Fields Appl.* 5 (1999) 246–253. <340, 344>

[2270] H. Niederreiter and I. E. Shparlinski, On the distribution of inversive congruential pseudorandom numbers in parts of the period, *Math. Comp.* 70 (2001) 1569–1574. <180, 185, 340, 344>

[2271] H. Niederreiter and I. E. Shparlinski, Dynamical systems generated by rational functions, In *Applied Algebra, Algebraic Algorithms and Error-Correcting Codes*, volume 2643 of *Lecture Notes in Comput. Sci.*, 6–17, Springer, Berlin, 2003. <337, 338, 344>

[2272] H. Niederreiter and I. E. Shparlinski, On the distribution of power residues and primitive elements in some nonlinear recurring sequences, *Bull. London Math. Soc.* 35 (2003) 522–528. <341, 344>

[2273] H. Niederreiter and I. E. Shparlinski, Periodic sequences with maximal linear complexity and almost maximal k-error linear complexity, In *Cryptography and Coding*, volume 2898 of *Lecture Notes in Comput. Sci.*, 183–189, Springer, Berlin, 2003. <331, 336>

[2274] H. Niederreiter and A. Venkateswarlu, Periodic multisequences with large error linear complexity, *Des. Codes Cryptogr.* 49 (2008) 33–45. <331, 336>

[2275] H. Niederreiter and L.-P. Wang, Proof of a conjecture on the joint linear complexity profile of multisequences, In *Progress in Cryptology—INDOCRYPT 2005*, volume 3797 of *Lecture Notes in Comput. Sci.*, 13–22, Springer, Berlin, 2005. <330, 336>

[2276] H. Niederreiter and L.-P. Wang, The asymptotic behavior of the joint linear complexity profile of multisequences, *Monatsh. Math.* 150 (2007) 141–155. <330, 336>

[2277] H. Niederreiter and A. Winterhof, Multiplicative character sums for nonlinear recurring sequences, *Acta Arith.* 111 (2004) 299–305. <341, 344>

[2278] H. Niederreiter and A. Winterhof, Cyclotomic R-orthomorphisms of finite fields, *Discrete Math.* 295 (2005) 161–171. <171, 185, 221, 228, 230>

[2279] H. Niederreiter and A. Winterhof, Exponential sums for nonlinear recurring sequences, *Finite Fields Appl.* 14 (2008) 59–64. <340, 344>

[2280] H. Niederreiter and C. Xing, *Rational Points on Curves over Finite Fields: Theory and Applications*, volume 285 of *London Mathematical Society Lecture Note Series*, Cambridge University Press, Cambridge, 2001. <32, 406, 422, 460, 463, 464, 468, 469, 545, 546, 708, 710, 712>

[2281] H. Niederreiter and C. Xing, *Algebraic Geometry in Coding Theory and Cryptography*, Princeton University Press, Princeton, NJ, 2009. <31, 32, 406, 421, 422, 705, 707, 712>

[2282] H. Niederreiter and C. P. Xing, Low-discrepancy sequences and global function fields with many rational places, *Finite Fields Appl.* 2 (1996) 241–273. <629, 630>

[2283] H. Niederreiter and C. P. Xing, Quasirandom points and global function fields, In *Finite Fields and Applications*, volume 233 of *London Math. Soc. Lecture Note Ser.*, 269–296, Cambridge University Press, Cambridge, 1996. <626, 630>

[2284] H. Niederreiter and C. P. Xing, Towers of global function fields with asymptotically many rational places and an improvement on the Gilbert-Varshamov bound, *Math. Nachr.* 195 (1998) 171–186. <711, 712>

[2285] H. Niederreiter, C. P. Xing, and K. Y. Lam, A new construction of algebraic-geometry codes, *Appl. Algebra Engrg. Comm. Comput.* 9 (1999) 373–381. <705, 706, 712>

[2286] M. A. Nielsen and I. L. Chuang, *Quantum Computation and Quantum Information*, Cambridge University Press, Cambridge, 2000. <835, 841>

[2287] Y. Niho, *Multi-Valued Cross-Correlation Functions Between Two Maximal Linear Recursive Sequences*, PhD thesis, Univ. Southern California, 1972. <261>

[2288] Y. Niitsuma, Counting points of the curve $y^2 = x^{12} + a$ over a finite field, *Tokyo J. Math.* 31 (2008) 59–94. <149, 161>

[2289] A. Nilli, On the second eigenvalue of a graph, *Discrete Math.* 91 (1991) 207–210. <646, 658>

[2290] A. Nilli, Tight estimates for eigenvalues of regular graphs, *Electron. J. Combin.* 11 (2004) Note 9, 4 pp. <646, 647, 658>

[2291] A. Nimbalker, T. K. Blankenship, B. Classon, T. E. Fuja, and D. J. Costello, Jr., Contention-free interleavers, In *Proc. 2004 IEEE International Symposium on Information Theory*, 54, Chicago, IL, 2004. <726, 727>

[2292] NIST, Digital signature standard (DSS), Federal Information Processing Standards Publication 186-3, National Institute of Standards and Technology, 2009. <785, 787, 796>

[2293] I. Niven, Fermat's theorem for matrices, *Duke Math. J.* 15 (1948) 823–826. <501, 510>

[2294] J.-S. No, S. W. Golomb, G. Gong, H.-K. Lee, and P. Gaal, Binary pseudorandom sequences of period $2^n - 1$ with ideal autocorrelation, *IEEE Trans. Inform. Theory* 44 (1998) 814–817. <755, 756, 763>

[2295] J. S. No and P. V. Kumar, A new family of binary pseudorandom sequences having optimal periodic correlation properties and large linear span, *IEEE Trans. Inform. Theory* IT-35 (1989) 371–379. <321, 324>

[2296] W. Nöbauer, On the length of cycles of polynomial permutations, In *Contributions to General Algebra, 3*, 265–274, Hölder-Pichler-Tempsky, Vienna, 1985. <229, 230>

[2297] E. Noether, Normalbasis bei Körpen ohne höhere Verzweigung, *J. Reine Angew. Math.* 167 (1932) 147–152. <110, 116>

[2298] A. W. Nordstrom and J. P. Robinson, An optimum nonlinear code, *Information and Control* 11 (1967) 613–616. <701, 702, 703>

[2299] M. Noro and K. Yokoyama, Yet another practical implementation of polynomial factorization over finite fields, In *ISSAC '02: Proceedings of the 2002 International Symposium on Symbolic and Algebraic Computation*, 200–206, ACM, 2002. <387, 392>

[2300] A. Nowicki, W. Secomski, J. Litniewski, I. Trots, and P. A. Lewin, On the application of signal compression using Golay's codes sequences in ultrasonic diagnostic, *Arch. Acoustics* 28 (2003) 313–324. <843, 849>

[2301] M. Nüsken and M. Ziegler, Fast multipoint evaluation of bivariate polynomials, In *Twelfth Annual European Symposium on Algorithms (ESA)*, volume 3221 of *Lecture Notes in Comput. Sci.*, 544–555, Springer, Berlin, 2004. <381, 382>

[2302] K. Nyberg, Perfect nonlinear S-boxes, In *Advances in Cryptology—EUROCRYPT '91*, volume 547 of *Lecture Notes in Comput. Sci.*, 378–386, Springer, Berlin,

1991. <254, 261, 272, 273>

[2303] K. Nyberg, Differentially uniform mappings for cryptography, In *Advances in Cryptology—EUROCRYPT '93*, volume 765 of *Lecture Notes in Comput. Sci.*, 55–64, Springer, Berlin, 1994. <256, 259, 261>

[2304] K. Nyberg and L. R. Knudsen, Provable security against differential cryptanalysis, In *Advances in Cryptology—CRYPTO '92*, volume 740 of *Lecture Notes in Comput. Sci.*, 566–574, Springer, Berlin, 1993. <254, 255, 261>

[2305] L. O'Connor, An analysis of exponentiation based on formal languages, In *Advances in Cryptology—EUROCRYPT '99*, volume 1592 of *Lecture Notes in Comput. Sci.*, 375–388, Springer, Berlin, 1999. <357, 363>

[2306] A. M. Odlyzko, Discrete logarithms in finite fields and their cryptographic significance, In *Advances in Cryptology*, volume 209 of *Lecture Notes in Comput. Sci.*, 224–314, Springer, Berlin, 1985. <348, 363, 365, 370, 374, 393, 398, 399, 401>

[2307] A. M. Odlyzko, Asymptotic enumeration methods, In *Handbook of Combinatorics, Vol. 2*, 1063–1229, Elsevier, Amsterdam, 1995. <367, 374>

[2308] A. M. Odlyzko, Discrete logarithms: the past and the future, *Des. Codes Cryptogr.* 19 (2000) 129–145. <393, 401>

[2309] C. O'Eigeartaigh and M. Scott, Pairing calculation on supersingular genus 2 curves, In *Selected Areas in Cryptography, 13th International Workshop, SAC 2006, Montreal, Canada, August 17-18, 2006 Revised Selected Papers*, volume 4356 of *Lecture Notes in Comput. Sci.*, 302–316, Springer, Berlin, 2007. <803>

[2310] A. P. Ogg, Abelian curves of small conductor, *J. Reine Angew. Math.* 226 (1967) 204–215. <300, 302>

[2311] A. P. Ogg, Rational points of finite order on elliptic curves, *Invent. Math.* 12 (1971) 105–111. <299, 302>

[2312] E. Okamoto and K. Nakamura, Evaluation of public key cryptosystems proposed recently, In *Proc. 1986 Symposium of Cryptography and Information Security*, volume D1, 1986. <768, 783>

[2313] C. M. O'Keefe and T. Penttila, Ovoids of $PG(3, 16)$ are elliptic quadrics, *J. Geom.* 38 (1990) 95–106. <588, 589>

[2314] C. M. O'Keefe and T. Penttila, Ovoids of $PG(3, 16)$ are elliptic quadrics II, *J. Geom.* 44 (1992) 140–159. <588, 589>

[2315] C. M. O'Keefe, T. Penttila, and G. F. Royle, Classification of ovoids in $PG(3, 32)$, *J. Geom.* 50 (1994) 143–150. <588, 589>

[2316] M. Olofsson, *VLSI Aspects on Inversion in Finite Fields*, PhD thesis, Department of Electrical Engineering, Linkopings Universitet, 2002. <816, 823>

[2317] J. D. Olsen, R. A. Scholtz, and L. R. Welch, Bent-function sequences, *IEEE Trans. Inform. Theory* 28 (1982) 858–864. <252>

[2318] B. Omidi Koma, D. Panario, and Q. Wang, The number of irreducible polynomials of degree n over $\mathbb{F}_q$ with given trace and constant terms, *Discrete Math.* 310 (2010) 1282–1292. <75, 76, 79>

[2319] R. Omrani, O. Moreno, and P. V. Kumar, Improved Johnson bounds for optical orthogonal codes with $\lambda > 1$ and some optimal constructions, In *Proc. Int. Symp. Inform. Theory*, 259–263, 2005. <844, 849>

[2320] H. Ong, C. Schnorr, and A. Shamir, Signatures through approximate representations by quadratic forms, In *Advances in Cryptology—CRYPTO 1983*, 117–131. Plenum Publ., 1984. <765, 767, 783>

[2321] H. Ong, C.-P. Schnorr, and A. Shamir, Efficient signature schemes based on polynomial equations, In *Advances in Cryptology*, volume 196 of *Lecture Notes in Comput. Sci.*, 37–46, Springer, Berlin, 1985. <767, 783>

[2322] F. Oort, Moduli of abelian varieties and Newton polygons, *C. R. Acad. Sci. Paris, Sér. I, Math.* 312 (1991) 385–389. <486, 488>

[2323] O. Ore, Über die reduzibilität von algebraischen gleichungen, *Skrifter Norsk Vid. Akad. Oslo* (1923). <62, 66>

[2324] O. Ore, Contributions to the theory of finite fields, *Trans. Amer. Math. Soc.* 36 (1934) 243–274. <60, 63, 66, 70, 71, 73, 112, 116>

[2325] B. Ors, L. Batina, B. Preneel, and J. Vandewalle, Hardware implementation of an elliptic curve processor over $GF(p)$ with Montgomery modular multiplier, *International Journal of Embedded Systems* 3 (2008) 229–240. <822, 823>

[2326] A. Ostafe, Multivariate permutation polynomial systems and nonlinear pseudorandom number generators, *Finite Fields Appl.* 16 (2010) 144–154. <232, 341, 342, 344>

[2327] A. Ostafe, Pseudorandom vector sequences derived from triangular polynomial systems with constant multipliers, In *Arithmetic of Finite Fields*, volume 6087 of *Lecture Notes in Comput. Sci.*, 62–72, Springer, Berlin, 2010. <341, 342, 344>

[2328] A. Ostafe, Pseudorandom vector sequences of maximal period generated by triangular polynomial dynamical systems, *Des. Codes Cryptogr.* 63 (2012) 59–72. <339, 342, 344>

[2329] A. Ostafe, E. Pelican, and I. E. Shparlinski, On pseudorandom numbers from multivariate polynomial systems, *Finite Fields Appl.* 16 (2010) 320–328. <338, 340, 344>

[2330] A. Ostafe and I. E. Shparlinski, On the degree growth in some polynomial dynamical systems and nonlinear pseudorandom number generators, *Math. Comp.* 79 (2010) 501–511. <339, 341, 342, 344>

[2331] A. Ostafe and I. E. Shparlinski, On the length of critical orbits of stable quadratic polynomials, *Proc. Amer. Math. Soc.* 138 (2010) 2653–2656. <178, 185, 342, 343, 344>

[2332] A. Ostafe and I. E. Shparlinski, Pseudorandom numbers and hash functions from iterations of multivariate polynomials, *Cryptogr. Commun.* 2 (2010) 49–67. <339, 341, 342, 344>

[2333] A. Ostafe and I. E. Shparlinski, On the Waring problem with Dickson polynomials in finite fields, *Proc. Amer. Math. Soc.* 139 (2011) 3815–3820. <192, 213>

[2334] A. Ostafe and I. E. Shparlinski, Degree growth, linear independence and periods of a class of rational dynamical systems, In *Proc. Conf. on Arithmetic, Geometry, Cryptography and Coding Theory*, volume 574 of *Contemp. Math.*, 131–144, Amer. Math. Soc., Providence, RI, 2012. <339, 341, 342, 344>

[2335] A. Ostafe and I. E. Shparlinski, On the power generator and its multivariate analogue, *J. Complexity* 28 (2012) 238–249. <339, 340, 344>

[2336] A. Ostafe, I. E. Shparlinski, and A. Winterhof, On the generalized joint linear complexity profile of a class of nonlinear pseudorandom multisequences, *Adv. Math. Commun.* 4 (2010) 369–379. <341, 342, 344>

[2337] A. Ostafe, I. E. Shparlinski, and A. Winterhof, Multiplicative character sums of a class of nonlinear recurrence vector sequences, *Int. J. Number Theory* 7 (2011) 1557–1571. <341, 342, 344>

[2338] A. M. Ostrowski, Über die Bedeutung der Theorie der konvexen Polyeder für die formale Algebra, *Jahresber. Deutsch. Math.-Verein.* 30 (1921) 98–99. <388, 392, 973>

[2339] A. M. Ostrowski, On the significance of the theory of convex polyhedra for formal algebra, *ACM SIGSAM Bull.* 33 (1999) 5, Translated from [2338]. <388, 392>

[2340] P. Oswald and M. A. Shokrollahi, Capacity-achieving sequences for the erasure channel, *IEEE Trans. Inform. Theory* 48 (2002) 3017–3028. <729, 734>

[2341] F. Özbudak, On maximal curves and linearized permutation polynomials over finite fields, *J. Pure Appl. Algebra* 162 (2001) 87–102. <239, 240>

[2342] C. Paar, A new architecture for a parallel finite field multiplier with low complexity based on composite fields, *IEEE Trans. Comput.* 45 (1996) 856–861. <813, 814, 823>

[2343] L. J. Paige, Neofields, *Duke Math. J.* 16 (1949) 39–60. <28, 32>

[2344] R. Paley, On orthogonal matrices., *J. Math. Phys., Mass. Inst. Techn.* 12 (1933) 311–320. <170, 185>

[2345] R. E. A. C. Paley, On orthogonal matrices, *J. Math. Phys* 12 (1933) 311–320. <609, 619>

[2346] V. Y. Pan, *Structured Matrices and Polynomials: Unified Superfast Algorithms*, Birkhäuser Boston Inc., Boston, MA, 2001. <532, 535>

[2347] D. Panario, What do random polynomials over fiuite fields look likc?, In *Finite Fields and Applications*, volume 2948 of *Lecture Notes in Comput. Sci.*, 89–108, Springer, Berlin, 2004. <365, 374>

[2348] D. Panario, X. Gourdon, and P. Flajolet, An analytic approach to smooth polynomials over finite fields, In *Algorithmic Number Theory*, volume 1423 of *Lecture Notes in Comput. Sci.*, 226–236, Springer, Berlin, 1998. <399, 401>

[2349] D. Panario, B. Pittel, B. Richmond, and A. Viola, Analysis of Rabin's irreducibility test for polynomials over finite fields, *Random Structures Algorithms* 19 (2001) 525–551. <369, 374, 376, 380>

[2350] D. Panario and B. Richmond, Analysis of Ben-Or's polynomial irreducibility test, *Random Structures Algorithms* 13 (1998) 439–456. <369, 374, 377, 378, 380>

[2351] D. Panario and B. Richmond, Exact largest and smallest size of components, *Algorithmica* 31 (2001) 413–432. <371, 374>

[2352] D. Panario and B. Richmond, Smallest components in decomposable structures: exp-log class, *Algorithmica* 29 (2001) 205–226. <371, 374>

[2353] D. Panario, A. Sakzad, B. Stevens, and Q. Wang, Two new measures for permutations: ambiguity and deficiency, *IEEE Trans. Inform. Theory* 57 (2011) 7648–7657. <229, 230>

[2354] D. Panario, O. Sosnovski, B. Stevens, and Q. Wang, Divisibility of polynomials over finite fields and combinatorial applications, *Des. Codes Cryptogr.* 63 (2012) 425–445. <631, 641, 642>

[2355] D. Panario, B. Stevens, and Q. Wang, Ambiguity and deficiency in Costas arrays and APN permutations, In *LATIN 2010: Theoretical Informatics*, volume 6034 of *Lecture Notes in Comput. Sci.*, 397–406, Dekker, New York, 2010. <229, 230>

[2356] D. Panario and D. Thomson, Efficient pth root computations in finite fields of characteristic p, *Des. Codes Cryptogr.* 50 (2009) 351–358. <37, 49, 71, 73>

[2357] D. Panario and A. Viola, Analysis of Rabin's polynomial irreducibility test, In *LATIN'98: theoretical informatics (Campinas, 1998)*, volume 1380 of *Lecture Notes in Comput. Sci.*, 1–10, Springer, Berlin, 1998. <376, 380>

[2358] G. Panella, Caratterizzazione delle quadriche di uno spazio (tridimensionale) lineare sopra un corpo finito, *Boll. Un. Mat. Ital. Ser. III* 10 (1955) 507–513. <588, 589>

[2359] Y. H. Park and J. B. Lee, Permutation polynomials and group permutation polynomials, *Bull. Austral. Math. Soc.* 63 (2001) 67–74. <221, 230>

[2360] K. R. Parthasarathy, *Quantum Computation, Quantum Error Correcting Codes and Information Theory*, Published for the Tata Institute of Fundamental Research, Mumbai, 2006. <835, 841>

[2361] F. Parvaresh and A. Vardy, Correcting errors beyond the Guruswami-Sudan radius in polynomial time, In *Proceedings of the Forty Sixth Annual IEEE Symposium on Foundations of Computer Science*, 285–294, 2005. <699, 703>

[2362] E. Pasalic, On cryptographically significant mappings over GF(2^n), In *Arithmetic of Finite Fields*, volume 5130 of *Lecture Notes in Comput. Sci.*, 189–204, Springer, Berlin, 2008. <226, 230>

[2363] E. Pasalic and P. Charpin, Some results concerning cryptographically significant mappings over GF(2^n), *Des. Codes Cryptogr.* 57 (2010) 257–269. <226, 230>

[2364] J. Patarin, Cryptanalysis of the Matsumoto and Imai public key scheme of Eurocrypt '88, In *Advances in Cryptology—CRYPTO '95*, volume 963 of *Lecture Notes in Comput. Sci.*, 248–261, Springer, Berlin, 1995. <770, 776, 777, 783>

[2365] J. Patarin, Asymmetric cryptography with a hidden monomial and a candidate algorithm for $\simeq$ 64 bits asymmetric signatures, In *Advances in Cryptology—CRYPTO '96*, volume 1109 of *Lecture Notes in Comput. Sci.*, 45–60, Springer, Berlin, 1996. <767, 783>

[2366] J. Patarin, Hidden Field Equations (HFE) and Isomorphisms of Polynomials (IP): two new families of asymmetric algorithms, In *Eurocrypt'96*, volume 1070 of *Lecture Notes in Comput. Sci.*, 33–48, Springer, Berlin, 1996. <767, 783>

[2367] J. Patarin, The oil and vinegar signature scheme, 1997, Dagstuhl Workshop on Cryptography. <770, 783>

[2368] J. Patarin, N. T. Courtois, and L. Goubin, FLASH, a fast multivariate signature algorithm, In *Topics in Cryptology—CT-RSA 2001*, volume 2020 of *Lecture Notes in Comput. Sci.*, 298–307, Springer, Berlin, 2001. <773, 783>

[2369] J. Patarin, N. T. Courtois, and L. Goubin, QUARTZ, 128-bit long digital signatures, In *Topics in Cryptology—CT-RSA 2001*, volume 2020 of *Lecture Notes in Comput. Sci.*, 282–297, Springer, Berlin, 2001. <770, 783>

[2370] J. Patarin, L. Goubin, and N. T. Courtois, C^*_{-+} and HM: Variations around two schemes of T. Matsumoto and H. Imai, In *Asiacrypt'98*, volume 1514, 35–49, Springer, Berlin, 1998. <772, 773, 778, 783>

[2371] J. Patarin, L. Goubin, and N. T. Courtois, Improved algorithms for Isomorphisms of Polynomials, In *Eurocrypt'98*, volume 1403, 184–200, Berlin, 1998, Springer. <767, 783>

[2372] K. G. Paterson, Applications of exponential sums in communications theory, In *Cryptography and Coding*, volume 1746 of *Lecture Notes in Comput. Sci.*, 1–24, Springer, Berlin, 1999. <179, 180, 185>

[2373] S. Paulus and H.-G. Rück, Real and imaginary quadratic representations of hyperelliptic function fields, *Math. Comp.* 68 (1999) 1233–1241. <448, 451, 452, 456>

[2374] S. E. Payne, Spreads, flocks, and generalized quadrangles, *J. Geom.* 33 (1988) 113–128. <567, 574>

[2375] F. Pellarin, Values of certain l-series in positive characteristic, *Ann. of Math., 2nd Ser.* 176 (2012) 2055–2093. <545, 546>

[2376] A. Pellet, Sur les fonctions irréducibles suivant un module premier, *C.R. Acad. Sci. Paris* 93 (1881) 1065–1066. <60, 66>

[2377] A. Pellet, Sur les fonctions réduites suivant un module premier, *Bull. Soc. Math. France* 17 (1889) 156–167. <63, 66>

[2378] A. E. Pellet, Sur les fonctions irréductibles suivant un module premier et une fonction modulaire., *C. R. Acad. Sci. Paris.* 70 (1870) 328–330. <62, 63, 66, 71, 73>

[2379] A. E. Pellet, Sur la décomposition d'une fonction entière en facteurs irréductibles suivant un module premier., *C. R. Acad. Sci. Paris.* 86 (1878) 1071–1072. <66, 67, 70>

[2380] R. Pellikaan, B.-Z. Shen, and G. J. M. van Wee, Which linear codes are algebraic-geometric?, *IEEE Trans. Inform. Theory* 37 (1991) 583–602. <710, 712>

[2381] J. Pelzl, T. Wollinger, and C. Paar, High performance arithmetic for hyperelliptic curve cryptosystems of genus two, *Information Technology: Coding and Computing (ITCC)* 2 (2004) 513–517. <799, 803>

[2382] T. Penttila and G. F. Royle, Sets of type (m, n) in the affine and projective planes of order nine, *Des. Codes Cryptogr.* 6 (1995) 229–245. <571, 574>

[2383] T. Penttila and B. Williams, Ovoids of parabolic spaces, *Geom. Dedicata* 82 (2000) 1–19. <282>

[2384] G. I. Perel'muter, Estimate of a sum along an algebraic curve, *Mat. Zametki* 5 (1969) 373–380. <168, 169>

[2385] S. Perlis, Normal bases of cyclic fields of prime-power degree, *Duke Math J.* 9 (1942) 507–517. <112, 116>

[2386] C. Pernet and A. Storjohann, Faster algorithms for the characteristic polynomial, In *ISSAC 2007*, 307–314, ACM, New York, 2007. <529, 535>

[2387] L. Perret, A fast cryptanalysis of the isomorphism of polynomials with one secret problem, In *Advances in Cryptology—EUROCRYPT 2005*, volume 3494 of *Lecture Notes in Comput. Sci.*, 354–370, Springer, Berlin, 2005. <767, 783>

[2388] O. Perron, Bemerkungen über die Verteilung der quadratischen Reste, *Math. Z.* 56 (1952) 122–130. <319, 324>

[2389] W. W. Peterson, *Error-Correcting Codes*, The M.I.T. Press, Cambridge, Mass., 1961. <661, 674, 692, 703>

[2390] W. W. Peterson and E. J. Weldon, Jr., *Error-Correcting Codes*, The M.I.T. Press, Cambridge, Mass., second edition, 1972. <316, 317, 661, 673, 674, 681, 686, 688, 692, 693, 696, 697, 703>

[2391] K. Petr, Über die Reduzibilität eines Polynoms mit ganzzahligen Koeffizienten nach einem Primzahlmodul, *Časopis pro pěstování matematiky a fysiky* 66 (1937) 85–94. <375, 380, 381, 382>

[2392] E. Petterson, Über die Irreduzibilität ganzzahliger Polynome nach einem Primzahlmodul, *J. Reine Angew. Math.* 175 (1936) 209–220. <60, 62, 66>

[2393] D. Pierce and M. J. Kallaher, A note on planar functions and their planes, *Bull. Inst. Combin. Appl.* 42 (2004) 53–75. <279, 282>

[2394] J. Pila, Frobenius maps of abelian varieties and finding roots of unity in finite fields, *Math. Comp.* 55 (1990) 745–763. <490, 491, 803>

[2395] A. Pincin, Bases for finite fields and a canonical decomposition for a normal basis generator, *Comm. Algebra* 17 (1989) 1337–1352. <111, 116>

[2396] N. Pippenger, On the evaluation of powers and monomials, *SIAM J. Comput.* 9 (1980) 230–250. <357, 363>

[2397] F. Piroi and A. Winterhof, Quantum period reconstruction of binary sequences, In *Applied Algebra, Algebraic Algorithms and Error-Correcting Codes*, volume 3857 of *Lecture Notes in Comput. Sci.*, 60–67, Springer, Berlin, 2006. <839, 841>

[2398] G. Pirsic, J. Dick, and F. Pillichshammer, Cyclic digital nets, hyperplane nets, and multivariate integration in Sobolev spaces, *SIAM J. Numer. Anal.* 44 (2006) 385–411. <624, 630>

[2399] N. L. Pitcher, *Efficient Point-Counting on Genus-2 Hyperelliptic Curves*, ProQuest LLC, Ann Arbor, MI, 2009, Thesis (Ph.D.)–University of Illinois at Chicago. <454, 456>

[2400] D. A. Plaisted, New NP-hard and NP-complete polynomial and integer divisibility problems, *Theoret. Comput. Sci.* 13 (1984) 125–138. <390, 392>

[2401] M. Planat, H. C. Rosu, and S. Perrine, A survey of finite algebraic geometrical structures underlying mutually unbiased quantum measurements, *Found. Phys.* 36 (2006) 1662–1680. <835, 841>

[2402] V. Pless, Q-codes, *J. Combin. Theory, Ser. A* 43 (1986) 258–276. <682, 703>

[2403] V. Pless, Duadic codes and generalizations, In *Eurocode '92*, volume 339 of *CISM Courses and Lectures*, 3–15, Springer, Vienna, 1993. <682, 703>

[2404] V. Pless, *Introduction to the Theory of Error-Correcting Codes*, Wiley-Interscience Series in Discrete Mathematics and Optimization. John Wiley & Sons Inc., New York, third edition, 1998. <31, 32>

[2405] V. S. Pless, W. C. Huffman, and R. A. Brualdi, editors, *Handbook of Coding Theory. Vol. I, II*, North-Holland, Amsterdam, 1998. <31, 32, 661, 683, 690, 691, 703>

[2406] S. C. Pohlig and M. E. Hellman, An improved algorithm for computing logarithms over $GF(p)$ and its cryptographic significance, *IEEE Trans. Inform. Theory* 24 (1978) 106–110. <395, 401, 800, 802, 803>

[2407] L. Poinsot, Réflexions sur les principes fondamentaux de la théorie des nombres, *Journal de mathématiques pures et appliquées* 10 (1845) 1–101. <71, 73>

[2408] P. Polito and O. Polverino, Linear blocking sets in $\mathrm{PG}(2, q^4)$, *Australas. J. Combin.* 26 (2002) 41–48. <561, 563>

[2409] P. Pollack, An explicit approach to hypothesis H for polynomials over a finite field, In *Anatomy of Integers*, volume 46 of *CRM Proc. Lecture Notes*, 259–273, Amer. Math. Soc., Providence, 2008. <496, 500>

[2410] P. Pollack, A polynomial analogue of the twin primes conjecture, *Proc. Amer. Math. Soc.* 136 (2008) 3775–3784. <496, 500>

[2411] P. Pollack, Simultaneous prime specializations of polynomials over finite fields, *Proc. Lond. Math. Soc.* 97 (2008) 545–567. <496, 500>

[2412] P. Pollack, Revisiting Gauss's analogue of the prime number theorem for polynomials over finite fields, *Finite Fields Appl.* 16 (2010) 290–299. <494, 500>

[2413] J. M. Pollard, Monte Carlo methods for index computation (mod p), *Math. Comp.* 32 (1978) 918–924. <397, 401, 746, 750>

[2414] J. M. Pollard, Kangaroos, Monopoly and discrete logarithms, *J. Cryptology* 13 (2000) 437–447. <397, 401>

[2415] J. M. Pollard and C.-P. Schnorr, An efficient solution of the congruence $x^2 + ky^2 = m \pmod{n}$, *IEEE Trans. Inform. Theory* 33 (1987) 702–709. <768, 783>

[2416] O. Polverino, Small minimal blocking sets and complete k-arcs in $\mathrm{PG}(2, p^3)$, *Discrete Math.* 208/209 (1999) 469–476. <562, 563>

[2417] O. Polverino, Small blocking sets in $\mathrm{PG}(2, p^3)$, *Des. Codes Cryptogr.* 20 (2000) 319–324. <561, 562, 563>

[2418] O. Polverino and L. Storme, Small minimal blocking sets in $\mathrm{PG}(2, q^3)$, *European J.*

Combin. 23 (2002) 83–92. <562, 563>

[2419] B. Poonen, Local height functions and the Mordell-Weil theorem for Drinfeld modules, *Compositio Math.* 97 (1995) 349–368. <540, 546>

[2420] A. D. Porto, F. Guida, and E. Montolivo, Fast algorithm for finding primitive polynomials over $GF(q)$, *Electron. Lett.* 28 (1992) 118–120. <350, 363>

[2421] A. G. Postnikov, *Ergodic Problems in the Theory of Congruences and of Diophantine Approximations*, Proceedings of the Steklov Institute of Mathematics, No. 82 (1966), American Mathematical Society, Providence, RI, 1967. <337, 344>

[2422] A. Pott, On the complexity of normal bases, *Bull. Inst. Combin. Appl.* 4 (1992) 51–52. <124, 128>

[2423] A. Pott, *Finite Geometry and Character Theory*, volume 1601 of *Lecture Notes in Mathematics*, Springer-Verlag, Berlin, 1995. <265, 268, 273>

[2424] A. Pott, Y. Tan, T. Feng, and S. Ling, Association schemes arising from bent functions, *Des. Codes Cryptogr.* 59 (2011) 319–331. <265, 273>

[2425] Y. Zhou and A. Pott, A new family of semifields with 2 parameters. *Advances in Mathematics* 234 (2013) 43–60. <282>

[2426] B. Preneel et al., NESSIE security report, Technical Report D20-v2, New European Schemes for Signatures, Integrity, and Encryption, 2003. <784, 796>

[2427] F. P. Preparata, A class of optimum nonlinear double-error-correcting codes, *Information and Control* 13 (1968) 378–400. <701, 702, 703>

[2428] N. Presman, O. Shapira, and S. Litsyn, Binary polar code kernels from code decompositions, 2011, preprint available, `http://arxiv.org/abs/1101.0764`. <739>

[2429] N. Presman, O. Shapira, and S. Litsyn, Polar codes with mixed kernels, In *Proceedings of the 2011 Symposium on Information Theory*, 6–10, 2011. <739>

[2430] R. Pries and H. J. Zhu, p-rank stratification of Artin-Schreier curves, *Ann. Inst. Fourier*, 62 (2012) 707–726. <487, 488>

[2431] M. Ptashne, *A Genetic Switch: Phage Lambda and Higher Organisms*, Blackwell Publishers, 1992. <830, 834>

[2432] S. Qi, On diagonal equations over finite fields, *Finite Fields Appl.* 3 (1997) 175–179. <208, 213>

[2433] G. Quenell, Spectral diameter estimates for k-regular graphs, *Adv. Math.* 106 (1994) 122–148. <645, 658>

[2434] M. Rabin, Probabilistic algorithms in finite fields, *SIAM Journal on Computing* 9 (1980) 273–280. <376, 380>

[2435] M. O. Rabin, Probabilistic algorithm for testing primality, *J. Number Theory* 12 (1980) 128–138. <346, 363>

[2436] R. Raghavendran, Finite associative rings, *Compositio Math.* 21 (1969) 195–229. <28>

[2437] J. Rajsski and J. Tyszer, Primitive polynomials over GF(2) of degree up to 660 with uniformly distributed coefficients, *J. Elect. Testing* 19 (2003) 645–657. <96, 97>

[2438] M. Ram Murty, *Introduction to p-adic analytic number theory*, volume 27 of *AMS/IP Studies in Advanced Mathematics*, American Mathematical Society, Providence, RI, 2002. <806, 811>

[2439] J. Ray and P. Koopman, Efficient high Hamming distance CRCs for embedded networks, In *Dependable Systems and Networks DSN*, 3–12, 2006. <635, 642>

[2440] D. K. Ray-Chaudhuri and R. M. Wilson, Solution of Kirkman's schoolgirl problem, In *Proc. Sympos. Pure Math., Vol. XIX: Combinatorics*, 187–203, Amer. Math.

Soc., Providence, RI, 1971. <592, 599>

[2441] C. Rebeiro, S. S. Roy, D. S. Reddy, and D. Mukhopadhyay, Revisiting the Itoh-Tsujii inversion algorithm for FPGA platforms, *IEEE Transactions on VLSI Systems* 19 (2011) 1508–1512. <818, 823>

[2442] L. Rédei, A short proof of a theorem of Št. Schwarz concerning finite fields, *Časopis Pěst. Mat. Fys.* 75 (1950) 211–212. <62, 66>

[2443] L. Rédei, *Algebra. Vol 1*, International Series of Monographs in Pure and Applied Mathematics. Pergamon Press, Oxford, 1967. <61, 66>

[2444] L. Rédei, *Lückenhafte Polynome über endlichen Körpern*, Birkhäuser Verlag, Basel, 1970, Lehrbücher und Monographien aus dem Gebiete der exakten Wissenschaften, Mathematische Reihe, Band 42. <556, 563>

[2445] L. Rédei, *Lacunary Polynomials over Finite Fields*, North-Holland Publishing Co., Amsterdam, 1973. <31, 32, 556, 557, 559, 563>

[2446] R. Ree, Proof of a conjecture of S. Chowla, *J. Number Theory* 3 (1971) 210–212. <72, 73>

[2447] I. S. Reed and G. Solomon, Polynomial codes over certain finite fields, *J. Soc. Indust. Appl. Math.* 8 (1960) 300–304. <679, 702, 703>

[2448] O. Reingold, S. Vadhan, and A. Wigderson, Entropy waves, the zig-zag graph product, and new constant-degree expanders, *Ann. of Math., 2nd Ser.* 155 (2002) 157–187. <655, 658>

[2449] D. Ren, Q. Sun, and P. Yuan, Number of zeros of diagonal polynomials over finite fields, *Finite Fields Appl.* 7 (2001) 197–204. <209, 213>

[2450] J. M. Renes, R. Blume-Kohout, A. J. Scott, and C. M. Caves, Symmetric informationally complete quantum measurements, *J. Math. Phys.* 45 (2004) 2171–2180. <836, 837, 841>

[2451] A. Reyhani-Masoleh and M. A. Hasan, A new construction of Massey-Omura parallel multiplier over $GF(2^m)$, *IEEE Trans. Comput.* 51 (2002) 511–520. <819, 820, 821, 823>

[2452] A. Reyhani-Masoleh and M. A. Hasan, Efficient multiplication beyond optimal normal bases, *IEEE Trans. Comput.* 52 (2003) 428–439. <819, 820, 823>

[2453] A. Reyhani-Masoleh and M. A. Hasan, Low complexity bit parallel polynomial basis multiplication over $GF(2^m)$, *IEEE Trans. Comput.* 53 (2004) 945–959. <822, 823>

[2454] G. Rhin, Répartition modulo 1 dans un corps de séries formelles sur un corps fini, *Dissertationes Math. (Rozprawy Mat.)* 95 (1972) 75. <76, 79>

[2455] T. Richardson and R. Urbanke, *Modern Coding Theory*, Cambridge University Press, Cambridge, 2008. <714, 716, 717, 719>

[2456] T. J. Richardson, Error floors of LDPC codes, In *Proceedings of the Forty-First Annual Allerton Conference on Communication, Control and Computing*, (2003) 1426–1435. <719>

[2457] T. J. Richardson, M. A. Shokrollahi, and R. L. Urbanke, Design of capacity-approaching irregular low-density parity-check codes, *IEEE Trans. Inform. Theory* 47 (2001) 619–637. <713, 719, 729, 730, 734>

[2458] T. J. Richardson and R. L. Urbanke, The capacity of low-density parity-check codes under message-passing decoding, *IEEE Trans. Inform. Theory* 47 (2001) 599–618. <718, 719, 734>

[2459] C. Ritzenthaler, Optimal curves of genus 1, 2 and 3, In *Actes de la Conférence "Théorie des Nombres et Applications,"* Publ. Math. Besançon Algèbre Théorie

Nr., 99–117, Presses Univ. Franche-Comté, Besançon, 2011. <460, 463>

[2460] R. L. Rivest, Permutation polynomials modulo 2^w, *Finite Fields Appl.* 7 (2001) 287–292. <229, 230>

[2461] R. L. Rivest, M. J. B. Robshaw, R. Sidney, and Y. L. Yin, The RC6 block cipher, 1998. <751, 759, 760, 763>

[2462] R. L. Rivest, A. Shamir, and L. Adleman, A method for obtaining digital signatures and public-key cryptosystems, *Comm. ACM* 21 (1978) 120–126. <744, 750>

[2463] A. M. Robert, The Gross-Koblitz formula revisited, *Rend. Sem. Mat. Univ. Padova* 105 (2001) 157–170. <153, 161>

[2464] J. A. G. Roberts and F. Vivaldi, A combinatorial model for reversible rational maps over finite fields, *Nonlinearity* 22 (2009) 1965–1982. <337, 344>

[2465] F. Rodríguez-Henríquez and Ç. K. Koç, Parallel multipliers based on special irreducible pentanomials, *IEEE Trans. Comput.* 52 (2003) 1535–1542. <822, 823>

[2466] F. Rodríguez-Henríquez, G. Morales-Luna, N. A. Saqib, and N. C. Cortés, Parallel Itoh-Tsujii multiplicative inversion algorithm for a special class of trinomials, *Des. Codes Cryptogr.* 45 (2007) 19–37. <818, 823>

[2467] M. Roitman, On Zsigmondy primes, *Proc. Amer. Math. Soc.* 125 (1997) 1913–1919. <76, 79>

[2468] A. Rojas-León, Estimates for singular multiplicative character sums, *Int. Math. Res. Not.* (2005) 1221–1234. <166, 167, 169, 199, 201>

[2469] A. Rojas-León, Purity of exponential sums on $\mathbb{A}^n$, *Compos. Math.* 142 (2006) 295–306. <163, 169>

[2470] A. Rojas-León, Rationality of trace and norm L-functions, *Duke Math. J.* 161 (2012) 1751–1795. <199, 201>

[2471] A. Rojas-León and D. Wan, Moment zeta functions for toric Calabi-Yau hypersurfaces, *Commun. Number Theory Phys.* 1 (2007) 539–578. <196, 198, 201>

[2472] A. Rojas-León and D. Wan, Improvements of the Weil bound for Artin-Schreier curves, *Math. Ann.* 351 (2011) 417–442. <197, 201>

[2473] S. Roman, *Field Theory*, volume 158 of *Graduate Texts in Mathematics*, Springer, New York, second edition, 2006. <349, 363>

[2474] S. Rønjom and T. Helleseth, A new attack on the filter generator, *IEEE Trans. Inform. Theory* 53 (2007) 1752–1758. <246, 252>

[2475] C. Ronse, *Feedback Shift Registers*, volume 169 of *Lecture Notes in Comput. Sci.*, Springer, Berlin, 1984. <312, 317>

[2476] L. Rónyai, Factoring polynomials over finite fields, In *Proceedings of the 28th Annual IEEE Symposium on Foundations of Computer Science,* Los Angeles CA, 132–137, IEEE Computer Society Press, Washington DC, 1987. <381, 382>

[2477] L. Rónyai, Factoring polynomials over finite fields, *Journal of Algorithms* 9 (1988) 391–400. <381, 382>

[2478] L. Rónyai, Factoring polynomials modulo special primes, *Combinatorica* 9 (1989) 199–206. <381, 382>

[2479] L. Rónyai, Galois groups and factoring polynomials over finite fields, *SIAM Journal on Discrete Mathematics* 5 (1992) 345–365. <381, 382>

[2480] L. Rónyai and Á. Szántó, Prime-field-complete functions and factoring polynomials over finite fields, *Computers and Artificial Intelligence* 15 (1996) 571–577. <381, 382>

[2481] L. Rónyai and T. Szőnyi, Planar functions over finite fields, *Combinatorica* 9 (1989)

315–320. <280, 281, 282>

[2482] L. A. Rosati, Unitals in Hughes planes, *Geom. Dedicata* 27 (1988) 295–299. <571, 574>

[2483] M. Y. Rosenbloom and M. A. Tsfasman, Codes for the m-metric, *Problems Inform. Transmission* 33 (1997) 45–52. <622, 630>

[2484] R. Roth, *Introduction to Coding Theory*, Cambridge University Press, Cambridge, 2006. <661, 680, 684, 692, 703>

[2485] R. M. Roth, Maximum-rank array codes and their application to crisscross error correction, *IEEE Trans. Inform. Theory* 37 (1991) 328–336. <846, 849>

[2486] O. S. Rothaus, On "bent" functions, *J. Combin. Theory, Ser. A* 20 (1976) 300–305. <263, 264, 265, 266, 273>

[2487] M. Rötteler, Quantum algorithms to solve the hidden shift problem for quadratics and for functions of large Gowers norm, In *Mathematical Foundations of Computer Science 2009*, volume 5734 of *Lecture Notes in Comput. Sci.*, 663–674, Springer, Berlin, 2009. <840, 841>

[2488] M. Rötteler, Quantum algorithms for highly non-linear Boolean functions, In *Proceedings of the Twenty First Annual ACM-SIAM Symposium on Discrete Algorithms (SODA)*, 448–457, 2010. <840, 841>

[2489] S. S. Roy, C. Rebeiro, and D. Mukhopadhyay, Generalized high speed Itoh-Tsujii multiplicative inversion architecture for FPGAs, *Integration* 45 (2012) 307–315. <818, 823>

[2490] I. F. Rúa, Primitive and non primitive finite semifields, *Comm. Algebra* 32 (2004) 793–803. <278>

[2491] I. F. Rúa, E. F. Combarro, and J. Ranilla, Determination of division algebras with 243 elements, *Finite Fields Appl.* 18 (2012) 1148–1155. <275, 278>

[2492] I. F. Rúa, E. F. Combarro, and J. Ranilla, Classification of semifields of order 64, *J. Algebra* 322 (2009) 4011–4029. <275, 278>

[2493] K. Rubin and A. Silverberg, Supersingular abelian varieties in cryptology, In *Advances in Cryptology—CRYPTO 2002*, volume 2442 of *Lecture Notes in Comput. Sci.*, 336–353, Springer, Berlin, 2002. <454, 456>

[2494] K. Rubin and A. Silverberg, Using abelian varieties to improve pairing-based cryptography, *J. Cryptology* 22 (2009) 330–364. <811>

[2495] I. M. Rubio and C. J. Corrada-Bravo, Cyclic decomposition of permutations of finite fields obtained using monomials, In *Finite Fields and Applications*, volume 2948 of *Lecture Notes in Comput. Sci.*, 254–261, Springer, Berlin, 2004. <229, 230, 727>

[2496] I. M. Rubio, G. L. Mullen, C. Corrada, and F. N. Castro, Dickson permutation polynomials that decompose in cycles of the same length, In *Finite Fields and Applications*, volume 461 of *Contemp. Math.*, 229–239, Amer. Math. Soc., Providence, RI, 2008. <229, 230, 727>

[2497] H.-G. Rück, A note on elliptic curves over finite fields, *Math. Comp.* 49 (1987) 301–304. <431, 440, 793, 796>

[2498] H.-G. Rück, On the discrete logarithm in the divisor class group of curves, *Math. Comp.* 68 (1999) 805–806. <455, 456>

[2499] H.-G. Rück and H. Stichtenoth, A characterization of Hermitian function fields over finite fields, *J. Reine Angew. Math.* 457 (1994) 185–188. <208, 213, 461, 463>

[2500] M. Rudnev, An improved sum-product inequality in fields of prime order, *Int. Math. Res. Notices* 2012 (2012) 3693–3705. <187, 192>

[2501] A. Rudra, Limits to list decoding of random codes, *IEEE Trans. Inform. Theory* IT-57 (2011) 1398–1408. <699, 703>

[2502] R. A. Rueppel, *Analysis and Design of Stream Ciphers*, Communications and Control Engineering Series. Springer-Verlag, Berlin, 1986. <325, 329, 330, 336>

[2503] R. A. Rueppel, Stream ciphers, In *Contemporary Cryptology*, 65–134, IEEE, New York, 1992. <325, 328, 336>

[2504] W. M. Ruppert, Reduzibilität ebener Kurven, *J. Reine Angew. Math.* 369 (1986) 167–191. <386, 387, 392>

[2505] W. M. Ruppert, Reducibility of polynomials $f(x, y)$ modulo p, *J. Number Theory* 77 (1999) 62–70. <386, 392>

[2506] J. J. Rushanan, *Topics in Integral Matrices and Abelian Group Codes: Generalized Q-Codes*, ProQuest LLC, Ann Arbor, MI, 1986, Thesis (Ph.D.)–California Institute of Technology. <682, 703>

[2507] F. Ruskey, The Object Server Home Page (COS), `http://theory.cs.uvic.ca`, as viewed in July 2012. <46, 49>

[2508] F. Ruskey, C. R. Miers, and J. Sawada, The number of irreducible polynomials and Lyndon words with given trace, *SIAM J. Discrete Math.* 14 (2001) 240–245. <54, 59>

[2509] A. Russell and I. E. Shparlinski, Classical and quantum function reconstruction via character evaluation, *J. Complexity* 20 (2004) 404–422. <840, 841>

[2510] I. Z. Ruzsa, Essential components, *Proc. London Math. Soc., 3rd Ser.* 54 (1987) 38–56. <184, 185>

[2511] W. E. Ryan and S. Lin, *Channel Codes: Classical and Modern*, Cambridge University Press, Cambridge, 2009. <661, 703>

[2512] A. Sackmann, M. Heiner, and I. Koch, Application of petri net based analysis techniques to signal transduction pathways, *BMC Bioinformatics* 7 (2006) 482. <829, 834>

[2513] H. R. Sadjadpour, N. J. A. Sloane, M. Salehi, and G. Nebe, Interleaver design for turbo codes, *IEEE J. Select. Areas Commun.* 19 (2001) 831–837. <630, 634, 642, 726, 727>

[2514] J. Saez-Rodriguez, L. G. Alexopoulos, J. Epperlein, R. Samaga, D. A. Lauffenburger, S. Klamt, and P. K. Sorger, Discrete logic modelling as a means to link protein signalling networks with functional analysis of mammalian signal transduction, *Molecular Systems Biology* 5:331 (2009). <825, 834>

[2515] O. Sahin, H. Frohlich, C. Lobke, U. Korf, S. Burmester, M. Majety, J. Mattern, I. Schupp, C. Chaouiya, D. Thieffry, A. Poustka, S. Wiemann, T. Beissbarth, and D. Arlt, Modeling erbb receptor-regulated g1/s transition to find novel targets for de novo trastuzumab resistance, *BMC Systems Biology* 3 (2009) 1. <825, 834>

[2516] S. Sakata, n-dimensional Berlekamp-Massey algorithm for multiple arrays and construction of multivariate polynomials with preassigned zeros, In *Applied Algebra, Algebraic Algorithms and Error-Correcting Codes*, volume 357 of *Lecture Notes in Comput. Sci.*, 356–376, Springer, Berlin, 1989. <329, 336>

[2517] S. Sakata, Extension of the Berlekamp-Massey algorithm to N dimensions, *Inform. and Comput.* 84 (1990) 207–239. <329, 336>

[2518] A. Sakzad, M.-R. Sadeghi, and D. Panario, Codes with girth 8 Tanner graph representation, *Des. Codes Cryptogr.* 57 (2010) 71–81. <718, 719>

[2519] A. Sakzad, M. R. Sadeghi, and D. Panario, Cycle structure of permutation functions over finite fields and their applications, *Adv. Math. Commun.* 6 (2012) 347–361. <229, 230, 727>

[2520] A. Sălăgean, On the computation of the linear complexity and the k-error linear complexity of binary sequences with period a power of two, *IEEE Trans. Inform. Theory* 51 (2005) 1145–1150. <329, 336>

[2521] R. Sandler, The collineation groups of some finite projective planes, *Portugal. Math.* 21 (1962) 189–199. <276, 278>

[2522] P. Sarkar and S. Maitra, Nonlinearity bounds and constructions of resilient Boolean functions, In *Advances in cryptology—CRYPTO 2000 (Santa Barbara, CA)*, volume 1880 of *Lecture Notes in Comput. Sci.*, 515–532, Springer, Berlin, 2000. <247, 252>

[2523] P. Sarnak, *Some Applications of Modular Forms*, volume 99 of *Cambridge Tracts in Mathematics*, Cambridge University Press, Cambridge, 1990. <644, 658>

[2524] P. Sarnak, Kloosterman, quadratic forms and modular forms, *Nieuw Arch. Wiskd.* 1 (2000) 385–389. <154, 161>

[2525] P. Sarnak, What is ... an expander?, *Notices Amer. Math. Soc.* 51 (2004) 762–763. <643, 646, 658>

[2526] D. Sarwate and M. Pursley, Crosscorrelation properties of pseudorandom and related sequences, *Proceedings of the IEEE* 68 (1980) 593–619. <317, 324>

[2527] D. V. Sarwate, An upper bound on the aperiodic autocorrelation function for a maximal-length sequence, *IEEE Trans. Inform. Theory* 30 (1984) 685–687. <842, 849>

[2528] T. Sasaki, T. Saito, and T. Hilano, Analysis of approximate factorization algorithm. I, *Japan J. Indust. Appl. Math.* 9 (1992) 351–368. <385, 392>

[2529] T. Sasaki and M. Sasaki, A unified method for multivariate polynomial factorizations, *Japan J. Indust. Appl. Math.* 10 (1993) 21–39. <385, 392>

[2530] T. Sasaki, M. Suzuki, M. Kolář, and M. Sasaki, Approximate factorization of multivariate polynomials and absolute irreducibility testing, *Japan J. Indust. Appl. Math.* 8 (1991) 357–375. <385, 392>

[2531] E. Sasoglu, E. Telatar, and E. Arikan, Polarization for arbitrary discrete memoryless channels, preprint available, `http://arxiv.org/abs/0908.0302`, 2009. <739>

[2532] E. Sasoglu, E. Telatar, and E. Yeh, Polar codes for the two-user binary-input multiple-access channel, In *Proc. IEEE Information Theory Workshop (ITW)*, 1–5, 2010. <739>

[2533] T. Satoh, The canonical lift of an ordinary elliptic curve over a finite field and its point counting, *J. Ramanujan Math. Soc.* 15 (2000) 247–270. <491, 788, 796>

[2534] T. Satoh, Generating genus two hyperelliptic curves over large characteristic finite fields, In *Advances in Cryptology - EUROCRYPT 2009*, volume 5479 of *Lecture Notes in Comput. Sci.*, 536–553, Springer, Berlin, 2009. <803>

[2535] T. Satoh and K. Araki, Fermat quotients and the polynomial time discrete log algorithm for anomalous elliptic curves, *Comment. Math. Univ. St. Paul.* 47 (1998) 81–92. <440, 784>

[2536] E. Savas and Ç. K. Koç, The Montgomery modular inverse—revisited, *IEEE Trans. Comput.* 49 (2000) 763–766. <360, 363>

[2537] A. Scheerhorn, Trace and norm-compatible extensions of finite fields, *Appl. Algebra Engrg. Comm. Comput.* 3 (1992) 199–209. <130, 138>

[2538] A. Scheerhorn, Iterated constructions of normal bases over finite fields, In *Finite Fields: Theory, Applications, and Algorithms*, volume 168 of *Contemp. Math.*,

309–325, Amer. Math. Soc., Providence, RI, 1994. <130, 138, 286, 290>

[2539] A. Scheerhorn, Dickson polynomials and completely normal elements over finite fields, In *Applications of Finite Fields*, volume 59 of *Inst. Math. Appl. Conf. Ser. (New Ser.)*, 47–55, Oxford Univ. Press, New York, 1996. <138>

[2540] A. Scheerhorn, Dickson polynomials, completely normal polynomials and the cyclic module structure of specific extensions of finite fields, *Des. Codes Cryptogr.* 9 (1996) 193–202. <138>

[2541] D. M. Schinianakis, A. P. Fournaris, H. E. Michail, A. P. Kakarountas, and T. Stouraitis, An RNS implementation of an F_p elliptic curve point multiplier, *IEEE Transactions on Circuits and Systems I: Regular Papers* 56 (2009) 1202–1213. <822, 823>

[2542] A. Schinzel, *Polynomials with Special Regard to Reducibility*, volume 77 of *Encyclopedia of Mathematics and its Applications*, Cambridge University Press, 2000. <386, 392>

[2543] O. Schirokauer, The special function field sieve, *SIAM J. Discrete Math.* 16 (2002) 81–98. <399, 401>

[2544] O. Schirokauer, The impact of the number field sieve on the discrete logarithm problem in finite fields, In *Algorithmic Number Theory: Lattices, Number Fields, Curves and Cryptography*, volume 44 of *Math. Sci. Res. Inst. Publ.*, 397–420, Cambridge Univ. Press, Cambridge, 2008. <399, 401>

[2545] O. Schirokauer, The number field sieve for integers of low weight, *Math. Comp.* 79 (2010) 583–602. <399, 401>

[2546] B. Schmidt, *Characters and Cyclotomic Fields in Finite Geometry*, volume 1797 of *Lecture Notes in Mathematics*, Springer-Verlag, Berlin, 2002. <600, 601, 607>

[2547] K. Schmidt, *Dynamical Systems of Algebraic Origin*, volume 128 of *Progress in Mathematics*, Birkhäuser Verlag, Basel, 1995. <337, 344>

[2548] W. M. Schmidt, *Equations over Finite Fields. An Elementary Approach*, Lecture Notes in Mathematics, Vol. 536. Springer-Verlag, Berlin, 1976. <31, 32, 176, 185, 193, 194, 199, 201>

[2549] W. M. Schmidt, Construction and estimation of bases in function fields, *J. Number Theory* 39 (1991) 181–224. <330, 336>

[2550] T. Schoen and I. Shkredov, Additive properties of multiplicative subgroups of $\mathbb{F}_p$, *Quart. J. Math.* 63 (2012) 713–822. <212, 213>

[2551] J. Scholten and H. J. Zhu, Families of supersingular curves in characteristic 2, *Math. Res. Lett.* 9 (2002) 639–650. <487, 488>

[2552] J. Scholten and H. J. Zhu, Hyperelliptic curves in characteristic 2, *Int. Math. Res. Not.* (2002) 905–917. <484, 487, 488, 799, 803>

[2553] J. Scholten and H. J. Zhu, Slope estimates of Artin-Schreier curves, *Compositio Math.* 137 (2003) 275–292. <484, 488>

[2554] R. A. Scholtz, The spread spectrum concept, *IEEE Trans. Commun.* COM-25 (1977) 748–755. <842, 845, 849>

[2555] R. A. Scholtz and L. R. Welch, GMW sequences, *IEEE Trans. Inform. Theory* 30 (1984) 548–553. <318, 324>

[2556] T. Schönemann, Grundzüge einer allgemeinen theorie der höheren congruenzen, deren modul eine reele primzahl ist, *J. Reine Agnew. Math.* 31 (1845) 269–325. <9, 11>

[2557] A. Schönhage, Schnelle berechnung von kettenbruchentwicklungen, *Acta Inf.* 1 (1971) 139–144. <359, 363>

[2558] A. Schönhage, Schnelle Multiplikation von Polynomen über Körpern der Charakteristik 2, *Acta Informatica* 7 (1977) 395–398. <382>

[2559] A. Schönhage and V. Strassen, Schnelle Multiplikation grosser Zahlen, *Computing (Arch. Elektron. Rechnen)* 7 (1971) 281–292. <355, 358, 363, 382>

[2560] R. Schoof, Elliptic curves over finite fields and the computation of square roots mod p, *Math. Comp.* 44 (1985) 483–494. <490, 491, 787, 796>

[2561] R. Schoof, Algebraic curves over $\mathbb{F}_2$ with many rational points, *J. Number Theory* 41 (1992) 6–14. <464, 469>

[2562] B. Schumacher and M. D. Westmoreland, Modal quantum theory, In *QPL 2010, 7th Workshop on Quantum Physics and Logic*, 145–149, 2010. <840, 841>

[2563] I. Schur, Über den Zusammenhang zwischen einem Problem der Zahlentheorie und einem Satz über algebraische Funktionen, *S.-B. Preuss. Akad. Wiss. Phys.-Math. Klasse* (1923) 123–134. <239, 240>

[2564] I. Schur, Zur theorie der einfach transitiven permutationgruppen, *S.-B. Preuss. Akad. Wiss. Phys.-Math. Klasse* (1933) 598–623. <239, 240>

[2565] R. Schürer, A new lower bound on the t-parameter of (t, s)-sequences, In *Monte Carlo and Quasi-Monte Carlo Methods*, 623–632, Springer-Verlag, Berlin, 2008. <626, 630>

[2566] M. P. Schützenberger, A non-existence theorem for an infinite family of symmetrical block designs, *Ann. Eugenics* 14 (1949) 286–287. <600, 607>

[2567] Š. Schwarz, Contribution à la reclictibilité des polynômes dans la théorie des congruences, *Věstnik Knálovskè české spol. nauk.* (1939) 1–7. <375, 380>

[2568] Š. Schwarz, A contribution to the reducibility of binomial congruences (Slovak), *Časopis Pěst. Mat. Fys.* 71 (1946) 21–31. <61, 62, 66>

[2569] Š. Schwarz, On the reducibility of binomial congruences and on the bound of the least integer belonging to a given exponent mod p, *Časopis Pěst. Mat. Fys.* 74 (1949) 1–16. <62, 66>

[2570] Š. Schwarz, On the reducibility of polynomials over a finite field, *Quart. J. Math. Oxford* 2 (1956) 110–124. <375, 380>

[2571] Š. Schwarz, On a class of polynomials over a finite field (Russian), *Mat.-Fyz. Časopis. Slovensk. Akad.* 10 (1960) 68–80. <63, 66>

[2572] J. Schwinger, Unitary operator bases, *Proc. Nat. Acad. Sci. U.S.A.* 46 (1960) 570–579. <835, 841>

[2573] M. Scott, Optimal irreducible polynomials for GF(2^m) arithmetic, In *Software Performance Enhancement for Encryption and Decryption (SPEED 2007)*, 2007, Available online (July 2011) `http://www.hyperelliptic.org/SPEED/start07.html`. <32, 49, 68, 70, 353, 363>

[2574] E. J. Scourfield, On ideals free of large prime factors, *J. Théor. Nombres Bordeaux* 16 (2004) 733–772. <399, 401>

[2575] B. Segre, Ovals in a finite projective plane, *Canad. J. Math.* 7 (1955) 414–416. <584, 589>

[2576] B. Segre, On complete caps and ovaloids in three-dimensional Galois spaces of characteristic two, *Acta Arith.* 5 (1959) 315–332 (1959). <588, 589>

[2577] B. Segre, Introduction to Galois geometries, *Atti Accad. Naz. Lincei Mem. Cl. Sci. Fis. Mat. Natur. Sez. I Ser. XIII* 8 (1967) 133–236. <584, 589>

[2578] G. E. Séguin, Low complexity normal bases for $F_{2^{mn}}$, *Discrete Appl. Math.* 28 (1990) 309–312. <119, 128>

[2579] E. S. Selmer, *Linear Recurrence Relations over Finite Fields*, University of Bergen, Bergen (Norway), 1966. <312, 317>

[2580] I. Semaev, Construction of polynomials, irreducible over a finite field, with linearly independent roots, *Mat. Sbornik* 135 (1988) 520–532, In Russian; English translation in Math. USSR-Sbornik, 63:507-519, 1989. <111, 116, 119, 128, 379, 380>

[2581] I. A. Semaev, Evaluation of discrete logarithms in a group of p-torsion points of an elliptic curve in characteristic p, *Math. Comp.* 67 (1998) 353–356. <440, 784>

[2582] G. Seroussi, Table of low-weight binary irreducible polynomials, Technical Report HP-98-135, Computer Systems Laboratory, Hewlett Packard, 1998. <33, 35, 49, 348, 363>

[2583] G. Seroussi and A. Lempel, Factorization of symmetric matrices and trace-orthogonal bases in finite fields, *SIAM J. Comput.* 9 (1980) 758–767. <103, 109>

[2584] G. Seroussi and A. Lempel, On symmetric representations of finite fields, *SIAM J. Algebraic Discrete Methods* 4 (1983) 14–21. <503, 504, 510>

[2585] J.-P. Serre, Géométrie algébrique et géométrie analytique, *Ann. Inst. Fourier, Grenoble* 6 (1955–1956) 1–42. <537, 546>

[2586] J.-P. Serre, *Abelian l-adic Representations and Elliptic Curves*, McGill University lecture notes written with the collaboration of Willem Kuyk and John Labute. W. A. Benjamin, Inc., New York-Amsterdam, 1968. <299, 300, 302>

[2587] J.-P. Serre, Propriétés galoisiennes des points d'ordre fini des courbes elliptiques, *Invent. Math.* 15 (1972) 259–331. <300, 302>

[2588] J.-P. Serre, *A Course in Arithmetic*, volume 7 of *Graduate Texts in Mathematics*, Springer-Verlag, New York, 1973. <29>

[2589] J.-P. Serre, Majorations de sommes exponentielles, In *Journées Arithmétiques de Caen*, 111–126, Astérisque No. 41–42, Soc. Math. France, Paris, 1977. <169, 650, 658>

[2590] J.-P. Serre, Quelques applications du théorème de densité de Chebotarev, *Inst. Hautes Études Sci. Publ. Math.* (1981) 323–401. <300, 302, 438, 440>

[2591] J.-P. Serre, Nombres de points des courbes algébriques sur $\mathbf{F}_q$, In *Seminar on Number Theory*, Exp. No. 22, 8, Univ. Bordeaux I, Talence, 1983. <459, 460, 463>

[2592] J.-P. Serre, Sur le nombre des points rationnels d'une courbe algébrique sur un corps fini, *C. R. Acad. Sci. Paris, Sér. I, Math.* 296 (1983) 397–402. <459, 463, 464, 469>

[2593] J.-P. Serre, Quel est le nombre maximum de points rationnels que peut avoir une courbe algébrique de genre g sur un corps fini?, *Annuaire du Collége de France* 84 (1984) 397–402. <461, 463>

[2594] J.-P. Serre, Répartition asymptotique des valeurs propres de l'opérateur de Hecke T_p, *J. Amer. Math. Soc.* 10 (1997) 75–102. <647, 658>

[2595] J.-P. Serre, On a theorem of Jordan, *Bull. Amer. Math. Soc. (New Ser.)* 40 (2003) 429–440. <300, 302>

[2596] J.-A. Serret, *Cours d'algèbre supérieure*, Paris: Bachelier, 1849, Second ed. Paris: Mallet-Bachelier, 1854. Third. ed. Paris: Gauthier-Villars, 1866. <8, 11>

[2597] J. A. Serret, Mémoire sur la théorie des congruences suivant un module premier et suivant une fonction modularie irréductible, *Mém. Acad. Sci., Inst. de France* 1 (1866) 617–688. <60, 62, 66>

[2598] J. A. Serret, Détermination des fonctions entières irréductibles, suivant un module premier, dans le cas où le degré est égal au module, *J. Math. Pures Appl.* 18 (1873) 301–304. <63, 66>

[2599] J. A. Serret, Sur les fonctions entières irréductibles, suivant un module premier, dans le cas où le degré est une puissance du module, *J. Math. Pures Appl.* 18 (1873) 437–451. <63, 66>

[2600] J.-A. Serret, *Cours d'Algèbre Supérieure. Tome I*, Les Grands Classiques Gauthier-Villars. [Gauthier-Villars Great Classics]. Éditions Jacques Gabay, Sceaux, 1992, Reprint of the fourth (1877) edition. <60, 61, 62, 66, 71, 73, 381, 382>

[2601] H. Shacham and B. Waters, editors, *Pairing-Based Cryptography — Pairing 2009*, volume 5671 of *Lecture Notes in Comput. Sci.*, Springer-Verlag, Berlin, 2009. <788, 796>

[2602] I. R. Shafarevich, *Basic Algebraic Geometry 1: Varieties in Projective Space*, Springer-Verlag, second edition, 1994. <386, 392>

[2603] R. Shaheen and A. Winterhof, Permutations of finite fields for check digit systems, *Des. Codes Cryptogr.* 57 (2010) 361–371. <229, 230>

[2604] A. Shallue and C. E. van de Woestijne, Construction of rational points on elliptic curves over finite fields, In F. Hess, S. Pauli, and M. Pohst, editors, *Algorithmic Number Theory—ANTS-VII*, volume 4076 of *Lecture Notes in Comput. Sci.*, 510–524, Springer-Verlag, Berlin, 2006. <796>

[2605] C. J. Shallue and I. M. Wanless, Permutation polynomials and orthomorphism polynomials of degree six, preprint, 2012. <217, 230>

[2606] A. Shamir, Efficient signature schemes based on birational permutations, In *Crypto*, volume 773 of *Lecture Notes in Comput. Sci.*, 1–12, Springer, Berlin, 1993. <768, 772, 775, 783>

[2607] D. Shanks, Class number, a theory of factorization, and genera, In *1969 Number Theory Institute*, 415–440, Amer. Math. Soc., Providence, RI, 1971. <396, 401>

[2608] C. E. Shannon, A mathematical theory of communication, *Bell System Tech. J.* 27 (1948) 379–423, 623–656. <661, 684, 703, 726, 727>

[2609] C. E. Shannon, Communication theory of secrecy systems, *Bell System Tech. J.* 28 (1949) 656–715. <743, 750>

[2610] R. T. Sharifi, On norm residue symbols and conductors, *J. Number Theory* 86 (2001) 196–209. <146, 161>

[2611] J. T. Sheats, The Riemann hypothesis for the Goss zeta function for $\mathbb{F}_q[t]$, *J. Number Theory* 71 (1998) 121–157. <544, 546>

[2612] G. B. Sherwood, S. S. Martirosyan, and C. J. Colbourn, Covering arrays of higher strength from permutation vectors, *J. Combin. Des.* 14 (2006) 202–213. <610, 619>

[2613] I. P. Shestakov and U. U. Umirbaev, The Nagata automorphism is wild, *Proc. Natl. Acad. Sci. USA* 100 (2003) 12561–12563. <768, 783>

[2614] G. Shimura and Y. Taniyama, *Complex Multiplication of Abelian Varieties and its Applications to Number Theory*, volume 6 of *Publications of the Mathematical Society of Japan*, The Mathematical Society of Japan, Tokyo, 1961. <299, 302>

[2615] K. Shiratani and M. Yamada, On rationality of Jacobi sums, *Colloq. Math.* 73 (1997) 251–260. <146, 161>

[2616] S. G. Shiva and P. Allard, A few useful details about a known technique for factoring $1 + X^{2q-1}$, *IEEE Trans. Inform. Theory* IT-16 (1970) 234–235. <62, 66>

[2617] I. Shmulevich, E. R. Dougherty, S. Kim, and W. Zhang, Probabilistic Boolean networks: a rule-based uncertainty model for gene regulatory networks, *Bioinformatics* 18 (2002) 261–274. <829, 834>

[2618] M. A. Shokrollahi, New sequences of linear time erasure codes approaching the channel capacity, In *Applied Algebra, Algebraic Algorithms and Error-Correcting Codes*, volume 1719 of *Lecture Notes in Comput. Sci.*, 65–76, Springer, Berlin, 1999. <729, 730, 734>

[2619] M. A. Shokrollahi, Codes and graphs, In *STACS 2000*, volume 1770 of *Lecture Notes Comput. Sci.*, 1–12, Springer, 2000. <728, 730, 734>

[2620] M. A. Shokrollahi, Capacity-achieving sequences, In *Codes, Systems, and Graphical Models*, volume 123 of *IMA Vol. Math. Appl.*, 153–166, Springer, New York, 2001. <729, 730, 734>

[2621] M. A. Shokrollahi, Raptor codes, *IEEE Trans. Inform. Theory* 52 (2006) 2551–2567. <732, 733, 734>

[2622] M. A. Shokrollahi and M. Luby, Raptor codes, In *Foundations and Trends in Communications and Information Theory*, volume 6, 213–322, NOW Publishers, 2009. <734>

[2623] P. W. Shor, Polynomial-time algorithms for prime factorization and discrete logarithms on a quantum computer, *SIAM J. Comput.* 26 (1997) 1484–1509. <395, 401, 749, 750, 839, 841>

[2624] V. Shoup, *Removing Randomness From Computational Number Theory*, PhD thesis, University of Wisconsin, Madison, 1989. <379, 380, 381, 382>

[2625] V. Shoup, New algorithms for finding irreducible polynomials over finite fields, *Math. Comp.* 54 (1990) 435–447. <378, 379, 380>

[2626] V. Shoup, On the deterministic complexity of factoring polynomials over finite fields, *Inform. Process. Lett.* 33 (1990) 261–267. <182, 185>

[2627] V. Shoup, Searching for primitive roots in finite fields, In *Symposium on the Theory of Computing – STOC 1990*, 546–554, ACM, 1990. <349, 363>

[2628] V. Shoup, Searching for primitive roots in finite fields, *Math. Comp.* 58 (1992) 369–380. <349, 363>

[2629] V. Shoup, Fast construction of irreducible polynomials over finite fields, *J. Symbolic Comput.* 17 (1994) 371–391. <378, 380>

[2630] V. Shoup, Lower bounds for discrete logarithms and related problems, In *Advances in Cryptology—EUROCRYPT '97*, volume 1233 of *Lecture Notes in Comput. Sci.*, 256–266, Springer, Berlin, 1997. <394, 401>

[2631] V. Shoup, Efficient computation of minimal polynomials in algebraic extensions of finite fields, In *Proceedings of the 1999 International Symposium on Symbolic and Algebraic Computation*, 53–58, ACM, New York, 1999. <350, 363>

[2632] V. Shoup, *A Computational Introduction to Number Theory and Algebra*, Cambridge University Press, Cambridge, second edition, 2009. <31, 32, 346, 347, 349, 363>

[2633] V. Shoup, NTL: A Library for doing Number Theory, version 5.5.2, 2009, available at `http://www.shoup.net/ntl`. <32, 47, 49, 346, 357, 363>

[2634] I. E. Shparlinski, On primitive polynomials, *Problemy Peredachi Informatsii* 23 (1987) 100–103. <72, 73, 96, 97>

[2635] I. E. Shparlinski, Distribution of values of recurrent sequences, *Problems Inform. Transmission* 25 (1989) 120–125. <316, 317>

[2636] I. E. Shparlinski, Some problems in the theory of finite fields, *Uspekhi Mat. Nauk* 46 (1991) 165–200, 240. <380>

[2637] I. E. Shparlinski, *Computational and Algorithmic Problems in Finite Fields*, volume 88 of *Mathematics and its Applications (Soviet Series)*, Kluwer Academic Publishers Group, Dordrecht, 1992. <31, 32, 381, 382>

[2638] I. E. Shparlinski, A deterministic test for permutation polynomials, *Comput. Complexity* 2 (1992) 129–132. <217, 230>

[2639] I. E. Shparlinski, Finding irreducible and primitive polynomials, *Applicable Algebra in Engineering, Communication and Computing* 4 (1993) 263–268. <380>

[2640] I. E. Shparlinski, On finding primitive roots in finite fields, *Theoret. Comput. Sci.* 157 (1996) 273–275. <349, 363>

[2641] I. E. Shparlinski, *Finite Fields: Theory and Computation*, volume 477 of *Mathematics and its Applications*, Kluwer Academic Publishers, Dordrecht, 1999. <31, 32, 381, 382>

[2642] I. E. Shparlinski, On the linear complexity of the power generator, *Des. Codes Cryptogr.* 23 (2001) 5–10. <333, 336>

[2643] I. E. Shparlinski, On a question of Erdős and Graham, *Arch. Math. (Basel)* 78 (2002) 445–448. <213>

[2644] I. E. Shparlinski, *Cryptographic Applications of Analytic Number Theory: Complexity Lower Bounds and Pseudorandomness*, volume 22 of *Progress in Computer Science and Applied Logic*, Birkhäuser Verlag, Basel, 2003. <31, 32, 176, 185, 333, 336>

[2645] I. E. Shparlinski, Bounds of Gauss sums in finite fields, *Proc. Amer. Math. Soc.* 132 (2004) 2817–2824. <141, 161>

[2646] I. E. Shparlinski, On the number of zero trace elements in polynomial bases for $\mathbb{F}_{2^n}$, *Rev. Mat. Complut.* 18 (2005) 177–180. <107, 109>

[2647] I. E. Shparlinski, Playing "hide-and-seek" with numbers: the hidden number problem, lattices and exponential sums, In *Public-Key Cryptography*, volume 62 of *Proc. Sympos. Appl. Math.*, 153–177, Amer. Math. Soc., Providence, RI, 2005. <176, 185>

[2648] I. E. Shparlinski, On some dynamical systems in finite fields and residue rings, *Discrete Contin. Dyn. Syst.* 17 (2007) 901–917. <337, 344>

[2649] I. E. Shparlinski, On the distribution of angles of the Salié sums, *Bull. Austral. Math. Soc.* 75 (2007) 221–227. <157, 161>

[2650] I. E. Shparlinski, On the distribution of Kloosterman sums, *Proc. Amer. Math. Soc.* 136 (2008) 419–425. <156, 161>

[2651] I. E. Shparlinski, On the exponential sum-product problem, *Indag. Math. (New Ser.)* 19 (2008) 325–331. <188, 192>

[2652] I. E. Shparlinski, On the distribution of arguments of Gauss sums, *Kodai Math. J.* 32 (2009) 172–177. <140, 161>

[2653] I. E. Shparlinski, On the average distribution of pseudorandom numbers generated by nonlinear permutations, *Math. Comp.* 80 (2011) 1053–1061. <341, 344>

[2654] I. E. Shparlinski, On the distribution of irreducible trinomials, *Canad. Math. Bull.* 54 (2011) 748–756. <88, 90>

[2655] I. E. Shparlinski and A. Winterhof, Noisy interpolation of sparse polynomials in finite fields, *Appl. Algebra Engrg. Comm. Comput.* 16 (2005) 307–317. <176, 185>

[2656] I. E. Shparlinski and A. Winterhof, Constructions of approximately mutually unbiased bases, In *LATIN 2006: Theoretical Informatics*, volume 3887 of *Lecture Notes in Comput. Sci.*, 793–799, Springer, Berlin, 2006. <835, 841>

[2657] I. E. Shparlinski and A. Winterhof, Quantum period reconstruction of approximate sequences, *Inform. Process. Lett.* 103 (2007) 211–215. <839, 841>

[2658] F. Shuqin and H. Wenbao, Primitive polynomials over finite fields of characteristic two, *Appl. Algebra Engrg. Comm. Comput.* 14 (2004) 381–395. <93, 95>

[2659] V. M. Sidel′nikov, Some k-valued pseudo-random sequences and nearly equidistant codes, *Problemy Peredači Informacii* 5 (1969) 16–22. <319, 324>

[2660] V. M. Sidel′nikov, On mutual correlation of sequences, *Soviet Math. Dokl.* 12 (1971) 197–201. <320, 324>

[2661] V. M. Sidel′nikov, On the cross correlation of sequences, *Problemy Kibernet.* (1971) 15–42. <321, 324>

[2662] V. M. Sidel'nikov, Estimates for the number of appearances of elements on an interval of a recurrent sequence over a finite field, *Discrete Math. Appl.* 2 (1992) 473–481. <316, 317>

[2663] T. Siegenthaler, Correlation-immunity of nonlinear combining functions for cryptographic applications, *IEEE Trans. Inform. Theory* 30 (1984) 776–780. <247, 252>

[2664] M. Sieveking, An algorithm for division of powerseries, *Computing* 10 (1972) 153–156. <380, 382>

[2665] D. Silva and F. R. Kschischang, Universal secure network coding via rank-metric codes, *IEEE Trans. Inform. Theory* 57 (2011) 1124–1135. <121, 128>

[2666] D. Silva, F. R. Kschischang, and R. Kötter, A rank-metric approach to error control in random network coding, *IEEE Trans. Inform. Theory* 54 (2008) 3951–3967. <848, 849>

[2667] J. H. Silverman, *Advanced Topics in the Arithmetic of Elliptic Curves*, volume 151 of *Graduate Texts in Mathematics*, Springer-Verlag, New York, 1994. <31, 32, 422, 440>

[2668] J. H. Silverman, *The Arithmetic of Dynamical Systems*, volume 241 of *Graduate Texts in Mathematics*, Springer, New York, 2007. <337, 338, 344>

[2669] J. H. Silverman, Variation of periods modulo p in arithmetic dynamics, *New York J. Math.* 14 (2008) 601–616. <342, 344>

[2670] J. H. Silverman, *The Arithmetic of Elliptic Curves*, volume 106 of *Graduate Texts in Mathematics*, Springer-Verlag, New York, second edition, 2009. <31, 32, 422, 423, 424, 425, 428, 429, 430, 431, 432, 433, 434, 435, 436, 437, 438, 439, 440>

[2671] J. H. Silverman, A survey of local and global pairings on elliptic curves and abelian varieties, In *Pairing-Based Cryptography (PAIRING 2010)*, volume 6478 of *Lecture Notes in Comput. Sci.*, 377–396, Springer, Berlin, 2010. <435, 440>

[2672] J. H. Silverman and J. Tate, *Rational Points on Elliptic Curves*, Undergraduate Texts in Mathematics, Springer-Verlag, New York, 1992. <31, 32, 422, 440>

[2673] M. K. Simon, J. K. Omura, R. A. Scholtz, and B. K. Levitt, *Spread Spectrum Communications Handbook*, McGraw-Hill, Inc., 1994. <317, 324>

[2674] J. Singer, A theorem in finite projective geometry and some applications to number theory, *Trans. Amer. Math. Soc.* 43 (1938) 377–385. <601, 607>

[2675] M. Skałba, Points on elliptic curves over finite fields, *Acta Arithmetica* 117 (2005) 293–301. <796>

[2676] C. Small, Solution of Waring's problem mod n, *Amer. Math. Monthly* 84 (1977) 356–359. <211, 213>

[2677] C. Small, Sums of powers in large finite fields, *Proc. Amer. Math. Soc.* 65 (1977) 35–36. <211, 213>

[2678] C. Small, Waring's problem mod n, *Amer. Math. Monthly* 84 (1977) 12–25. <211, 213>

[2679] C. Small, Diagonal equations over large finite fields, *Canad. J. Math.* 36 (1984) 249–262. <207, 209, 213>

[2680] C. Small, Permutation binomials, *Internat. J. Math. Math. Sci.* 13 (1990) 337–342. <223, 230>

[2681] C. Small, *Arithmetic of Finite Fields*, volume 148 of *Monographs and Textbooks in Pure and Applied Mathematics*, Marcel Dekker Inc., New York, 1991. <31, 32, 206, 207, 213, 223, 230>

[2682] N. P. Smart, The discrete logarithm problem on elliptic curves of trace one, *J. Cryptology* 12 (1999) 193–196. <440, 784>

[2683] N. P. Smart, The exact security of ECIES in the generic group model, In B. Honary, editor, *Cryptography and Coding*, volume 2260 of *Lecture Notes in Comput. Sci.*, 73–84, Springer-Verlag, Berlin, 2001. <785, 796>

[2684] N. Smart (ed.), ECRYPT II yearly report on algorithms and keysizes (2009-2010), Technical Report D.SPA.13, European Network of Excellence in Cryptology II, 2010. <784, 793, 796>

[2685] B. Smeets, The linear complexity profile and experimental results on a randomness test of sequences over the field $\mathbb{F}_q$, presented at IEEE Int. Symp. on Information Theory 1988, June 19–24. <329, 330, 336>

[2686] B. Smeets and W. Chambers, Windmill generators: a generalization and an observation of how many there are, In *Advances in Cryptology—EUROCRYPT'88*, volume 330 of *Lecture Notes in Comput. Sci.*, 325–330, Springer, Berlin, 1988. <69, 70>

[2687] M. H. M. Smid, Duadic codes, *IEEE Trans. Inform. Theory* 33 (1987) 432–433. <682, 703>

[2688] B. A. Smith, Isogenies and the discrete logarithm problem in Jacobians of genus 3 hyperelliptic curves, *J. Cryptology* 22 (2009) 505–529. <798, 803>

[2689] S. L. Snover, *The Uniqueness of the Nordstrom-Robinson and the Golay Binary Codes*, ProQuest LLC, Ann Arbor, MI, 1973, Thesis (Ph.D.)–Michigan State University. <701, 703>

[2690] I. M. Sobol', Distribution of points in a cube and approximate evaluation of integrals (Russian), *Ž. Vyčisl. Mat. i Mat. Fiz.* 7 (1967) 784–802. <620, 625, 628, 630>

[2691] M. Sodestrand, W. Jenkins, G. A. Jullien, and F. J. Taylor, *Residue Number System Arithmetic: Modern Applications in Digital Signal Processing*, IEEE Press, 1986. <822, 823>

[2692] P. Solé, A quaternary cyclic code, and a family of quadriphase sequences with low correlation properties, In *Coding Theory and Applications*, volume 388 of *Lecture Notes in Comput. Sci.*, 193–201, Springer, New York, 1989. <322, 324>

[2693] J. A. Solinas, Generalized Mersenne numbers, Combinatorics and Optimization Research Report CORR 99-39, University of Waterloo, 1999, available at `http://www.cacr.math.uwaterloo.ca/techreports/1999/corr99-39.ps`. <353, 363>

[2694] R. Solovay and V. Strassen, A fast Monte-Carlo test for primality, *SIAM J. Comput.* 6 (1977) 84–85. <346, 363, 381, 382>

[2695] L. Song and K. K. Parhi, Low-energy digit-serial/parallel finite field multipliers, *The Journal of VLSI Signal Processing* 19 (1998) 149–166. <815, 823>

[2696] A. B. Sørensen, Projective Reed-Muller codes, *IEEE Trans. Inform. Theory* 37 (1991) 1567–1576. <687, 703>

[2697] K. W. Spackman, Simultaneous solutions to diagonal equations over finite fields, *J.*

Number Theory 11 (1979) 100–115. <213>

[2698] S. Sperber, On the p-adic theory of exponential sums, *Amer. J. Math.* 108 (1986) 255–296. <169, 210, 213, 481, 488>

[2699] W. Stahnke, Primitive binary polynomials, *Math. Comp.* 27 (1973) 977–980. <95, 97>

[2700] M. Stamp and C. F. Martin, An algorithm for the k-error linear complexity of binary sequences with period 2^n, *IEEE Trans. Inform. Theory* 39 (1993) 1398–1401. <326, 329, 336>

[2701] H. M. Stark and A. A. Terras, Zeta functions of finite graphs and coverings, *Adv. Math.* 121 (1996) 124–165. <658>

[2702] A. Steane, Multiple-particle interference and quantum error correction, *Proc. Roy. Soc. London Ser. A* 452 (1996) 2551–2577. <838, 841>

[2703] A. Steel, Conquering inseparability: primary decomposition and multivariate factorization over algebraic function fields of positive characteristic, *J. Symbolic Comput.* 40 (2005) 1053–1075. <387, 392>

[2704] L. J. Steggles, R. Banks, O. Shaw, and A. Wipat, Qualitatively modelling and analysing genetic regulatory networks: a Petri net approach, *Bioinformatics* 23 (2007) 336–343. <829, 834>

[2705] D. Stehlé and P. Zimmermann, A binary recursive GCD algorithm, In *Algorithmic Number Theory*, volume 3076 of *Lecture Notes in Comput. Sci.*, 411–425, Springer, Berlin, 2004. <359, 363>

[2706] A. Stein, Sharp upper bounds for arithmetic in hyperelliptic function fields, *J. Ramanujan Math. Soc.* 16 (2001) 119–203. <452, 456>

[2707] A. Stein, Explicit infrastructure for real quadratic function fields and real hyperelliptic curves, *Glas. Mat. Ser. III* 44(64) (2009) 89–126. <452, 456>

[2708] A. Stein and E. Teske, Optimized baby step–giant step methods, *J. Ramanujan Math. Soc.* 20 (2005) 27–58. <396, 401>

[2709] W. A. Stein et al., Sage Mathematics Software (Version 5.3), The Sage Development Team. Available at `http://www.sagemath.org/`, 2012. <48, 49, 346, 363>

[2710] S. A. Stepanov, On the number of polynomials of a given form that are irreducible over a finite field, *Mat. Zametki* 41 (1987) 289–295, 456. <77, 79>

[2711] S. A. Stepanov, *Arithmetic of Algebraic Curves*, Monographs in Contemporary Mathematics, Consultants Bureau, New York, 1994. <31, 32, 177, 185>

[2712] J. Stern, D. Pointcheval, J. Malone-Lee, and N. Smart, Flaws in applying proof methodologies to signature schemes, In M. Yung, editor, *Advances in Cryptology—CRYPTO 2002*, volume 2442 of *Lecture Notes in Comput. Sci.*, 93–110, Springer-Verlag, Berlin, 2002. <785, 796>

[2713] H. Stichtenoth, Transitive and self-dual codes attaining the Tsfasman-Vlăduţ-Zink bound, *IEEE Trans. Inform. Theory* 52 (2006) 2218–2224. <468, 469>

[2714] H. Stichtenoth, *Algebraic Function Fields and Codes*, volume 254 of *Graduate Texts in Mathematics*, Springer-Verlag, Berlin, second edition, 2009. <31, 32, 208, 213, 406, 407, 409, 411, 412, 413, 414, 415, 416, 417, 418, 419, 420, 422, 457, 458, 459, 460, 461, 462, 463, 464, 465, 466, 467, 469, 705, 712>

[2715] H. Stichtenoth and C. P. Xing, Excellent nonlinear codes from algebraic function fields, *IEEE Trans. Inform. Theory* 51 (2005) 4044–4046. <712>

[2716] L. Stickelberger, On a new property of the discriminants of algebraic number fields. (Ueber eine neue Eigenschaft der Discriminanten algebraischer Zahlkörper.), *Verh. d. intern. Math.-Congr.* 1 (1897) 182–193. <66, 67, 70>

[2717] B. Stigler, Polynomial dynamical systems in systems biology, In *Modeling and Simulation of Biological Networks*, volume 64 of *Proc. Sympos. Appl. Math.*, 53–84, Amer. Math. Soc., Providence, RI, 2007. <337, 344>

[2718] D. R. Stinson, On bit-serial multiplication and dual bases in GF(2^m), *IEEE Trans. Inform. Theory* 37 (1991) 1733–1736. <107, 109>

[2719] D. R. Stinson, *Combinatorial Designs: Constructions and Analysis*, Springer-Verlag, New York, 2004. <31, 32, 599, 619>

[2720] D. R. Stinson, *Cryptography: Theory and Practice*, Discrete Mathematics and its Applications. Chapman & Hall/CRC, Boca Raton, FL, third edition, 2006. <31, 32, 393, 401, 750>

[2721] D. R. Stinson, R. Wei, and L. Zhu, New constructions for perfect hash families and related structures using combinatorial designs and codes, *J. Combin. Des.* 8 (2000) 189–200. <613, 619>

[2722] K.-O. Stöhr and J. F. Voloch, Weierstrass points and curves over finite fields, *Proc. London Math. Soc., 3rd Ser.* 52 (1986) 1–19. <462, 463>

[2723] T. Stoll, Complete decomposition of Dickson-type polynomials and related Diophantine equations, *J. Number Theory* 128 (2008) 1157–1181. <288, 290>

[2724] R. Stong, The average order of a permutation, *Electron. J. Combin.* 5 (1998) Research Paper 41, 6 pp. <373, 374>

[2725] T. Storer, *Cyclotomy and Difference Sets*, volume 2 of *Lectures in Advanced Mathematics*, Markham Publishing Co., Chicago, IL, 1967. <603, 604, 607>

[2726] A. Storjohann, Deterministic computation of the Frobenius form (extended abstract), In *Forty Second IEEE Symposium on Foundations of Computer Science*, 368–377, IEEE Computer Soc., Los Alamitos, CA, 2001. <529, 535>

[2727] A. Storjohann and G. Villard, Algorithms for similarity transforms, Technical report, Rhine Workshop on Computer Algebra, 2000. <529, 535>

[2728] A. J. Stothers, *On the Complexity of Matrix Multiplication*, PhD thesis, University of Edinburgh, 2010. <521, 535>

[2729] W. W. Stothers, On permutation polynomials whose difference is linear, *Glasgow Math. J.* 32 (1990) 165–171. <228, 230>

[2730] D. R. Stoutemyer, Which polynomial representation is best?, In *Proceedings of the 1984 MACSYMA Users' Conference*, 221–243, 1984. <382, 392>

[2731] V. Strassen, Gaussian elimination is not optimal, *Numer. Math.* 13 (1969) 354–356. <358, 363>

[2732] V. Strassen, Evaluation of rational functions, In R. E. Miller and J. W. Thatcher, editors, *Complexity of Computer Computations*, Plenum Press, 1972. <381, 382>

[2733] V. Strassen, Vermeidung von Divisionen, *J. Reine Angew. Math.* 264 (1973) 182–202. <390, 392>

[2734] V. Strassen, Algebraische berechnungskomplexität, In *Perspectives in Mathematics, Anniversary of Oberwolfach 1984*, 509–550, Birkhäuser Verlag, Basel, 1984. <381, 382>

[2735] M. Streng, Computing Igusa class polynomials, preprint available, `http://arxiv.org/abs/0903.4766`, 2012. <803>

[2736] S. J. Suchower, Subfield permutation polynomials and orthogonal subfield systems in finite fields, *Acta Arith.* 54 (1990) 307–315. <231, 232>

[2737] S. J. Suchower, Polynomial representations of complete sets of frequency hyperrectangles with prime power dimensions, *J. Combin. Theory, Ser. A* 62 (1993) 46–65.

<554, 556>

[2738] B. Sudakov, E. Szemerédi, and V. H. Vu, On a question of Erdős and Moser, *Duke Math. J.* 129 (2005) 129–155. <188, 192>

[2739] M. Sudan, Decoding of Reed Solomon codes beyond the error-correction bound, *J. Complexity* 13 (1997) 180–193. <698, 699, 703>

[2740] M. Sugita, M. Kawazoe, and H. Imai, Gröbner basis based cryptanalysis of sha-1, Cryptology ePrint Archive, Report 2006/098, 2006, `http://eprint.iacr.org/`. <783>

[2741] Y. Sugiyama, M. Kasahara, S. Hirasawa, and T. Namekawa, A method for solving key equation for decoding Goppa codes, *Information and Control* 27 (1975) 87–99. <694, 703>

[2742] J. Sun and O. Y. Takeshita, Interleavers for turbo codes using permutation polynomials over integer rings, *IEEE Trans. Inform. Theory* 51 (2005) 101–119. <229, 230, 725, 726, 727>

[2743] Q. Sun, The number of solutions of certain diagonal equations over finite fields, *Sichuan Daxue Xuebao* 34 (1997) 395–398. <209, 213>

[2744] Q. Sun and D. Q. Wan, On the solvability of the equation $\sum_{i=1}^{n} x_i/d_i \equiv 0 \pmod 1$ and its application, *Proc. Amer. Math. Soc.* 100 (1987) 220–224. <208, 209, 213>

[2745] Q. Sun and D. Q. Wan, On the Diophantine equation $\sum_{i=1}^{n} x_i/d_i \equiv 0 \pmod 1$, *Proc. Amer. Math. Soc.* 112 (1991) 25–29. <209, 213>

[2746] Z.-W. Sun, On value sets of polynomials over a field, *Finite Fields Appl.* 14 (2008) 470–481. <211, 213, 235, 236>

[2747] B. Sunar, A generalized method for constructing subquadratic complexity $GF(2^k)$ multipliers, *IEEE Trans. Comput.* 53 (2004) 1097–1105. <814, 823>

[2748] B. Sunar, A Euclidean algorithm for normal bases, *Acta Appl. Math.* 93 (2006) 57–74. <360, 363>

[2749] B. Sunar and Ç. K. Koç, Mastrovito multiplier for all trinomials, *IEEE Trans. Comput.* 48 (1999) 522–527. <822, 823>

[2750] B. Sunar and Ç. K. Koç, An efficient optimal normal basis type II multiplier, *IEEE Trans. Comput.* 50 (2001) 83–87. <821, 823>

[2751] A. V. Sutherland, Genus 1 point-counting record modulo a 5000+ digit prime, 2010, Posting to the Number Theory List, `http://listserv.nodak.edu/cgi-bin/wa.exe?A2=ind1007&L=nmbrthry&T=0&F=&S=&P=287`. <788, 796>

[2752] A. V. Sutherland, On the evaluation of modular polynomials, ArXiv 1202.3985v3, to appear in the proceedings of the Tenth Algorithmic Number Theory Symposium ANTS-X, 2012. <788, 796>

[2753] R. G. Swan, Factorization of polynomials over finite fields, *Pacific J. Math.* 12 (1962) 1099–1106. <34, 49, 67, 68, 70, 96, 97>

[2754] N. Szabo and R. I. Tanaka, *Residue Arithmetic and its Application to Computer Technology*, McGraw-Hill, 1967. <352, 363>

[2755] P. Sziklai, On small blocking sets and their linearity, *J. Combin. Theory, Ser. A* 115 (2008) 1167–1182. <561, 563>

[2756] T. Szőnyi, On the number of directions determined by a set of points in an affine Galois plane, *J. Combin. Theory, Ser. A* 74 (1996) 141–146. <558, 563>

[2757] T. Szőnyi, Blocking sets in Desarguesian affine and projective planes, *Finite Fields*

Appl. 3 (1997) 187–202. <561, 563>

[2758] T. Szőnyi, Around Rédei's theorem, *Discrete Math.* 208/209 (1999) 557–575. <563>

[2759] P. Szüsz, On a problem in the theory of uniform distribution, *Comptes Rendus Premier Congrès Hongrois* (1952) 461–472, (in Hungarian). <340, 344>

[2760] L. Taelman, Special L-values of t-motives: a conjecture, *Int. Math. Res. Not. IMRN* (2009) 2957–2977. <540, 546>

[2761] L. Taelman, A Dirichlet unit theorem for Drinfeld modules, *Math. Ann.* 348 (2010) 899–907. <540, 546>

[2762] L. Taelman, The Carlitz shtuka, *J. Number Theory* 131 (2011) 410–418. <541, 546>

[2763] L. Taelman, A Herbrand-Ribet theorem for function fields, *Inventiones Mathematicae* 188 (2012) 253–275. <542, 546>

[2764] L. Taelman, Special L-values of Drinfeld modules, *Ann. of Math., 2nd Ser.* 175 (2012) 369–391. <542, 546>

[2765] Y. Taguchi, The Tate conjecture for t-motives, *Proc. Amer. Math. Soc.* 123 (1995) 3285–3287. <540, 546>

[2766] T. Takagi, T. Okamoto, E. Okamoto, and T. Okamoto, editors, *Pairing-Based Cryptography — Pairing 2007*, volume 4575 of *Lecture Notes in Comput. Sci.* Springer-Verlag, Berlin, 2007. <788, 796>

[2767] T. Takahashi, Good reduction of elliptic modules, *J. Math. Soc. Japan* 34 (1982) 475–487. <539, 546>

[2768] O. Y. Takeshita, On maximum contention-free interleavers and permutation polynomials over integer rings, *IEEE Trans. Inform. Theory* 52 (2006) 1249–1253. <725, 727>

[2769] O. Y. Takeshita, Permutation polynomial interleavers: an algebraic-geometric perspective, *IEEE Trans. Inform. Theory* 53 (2007) 2116–2132. <229, 230>

[2770] O. Y. Takeshita and D. J. Costello, Jr., On deterministic linear interleavers for turbo-codes, In *Proc. 35th Annual Allerton Conference on Communication, Control, and Computing*, 711–712, 1997. <726, 727>

[2771] O. Y. Takeshita and D. J. Costello, Jr., New deterministic interleaver designs for turbo codes, *IEEE Trans. Inform. Theory* 46 (2000) 1988–2006. <726, 727>

[2772] I. Tal and A. Vardy, List decoding of polar codes, In *Proc. IEEE Int. Symp. Information Theory (ISIT)*, 1–5, 2011. <739>

[2773] A. Tamagawa, The Tate conjecture and the semisimplicity conjecture for t-modules, *Sūrikaisekikenkyūsho Kōkyūroku* (1995) 89–94. <540, 546>

[2774] L. Tan and W.-F. Qi, On the k-error linear complexity of l-sequences, *Finite Fields Appl.* 16 (2010) 420–435. <336>

[2775] Y. Tan, A. Pott, and T. Feng, Strongly regular graphs associated with ternary bent functions, *J. Combin. Theory, Ser. A* 117 (2010) 668–682. <265, 273>

[2776] T. Tanaka and R. Mori, Refined rate of channel polarization, preprint available, `http://arxiv.org/abs/1001.2067`, 2010. <739>

[2777] D. Tang, C. Carlet, and X. Tang, Highly Nonlinear Boolean Functions with Optimal Algebraic Immunity and Good Behavior Against Fast Algebraic Attacks. *IEEE Trans. Inform. Theory* 59 (2013) 653–664. <250, 252>

[2778] X. Tang, P. Fan, and S. Matsufuji, Lower bound on correlation of spreading sequence set with low or zero correlation, *Electronics Letters* 36 (2000) 551–552. <323, 324>

[2779] R. M. Tanner, A recursive approach to low complexity codes, *IEEE Trans. Inform.*

Theory 27 (1981) 533–547. <714, 719>

[2780] T. Tao, *Structure and Randomness*, American Mathematical Society, Providence, RI, 2008. <310>

[2781] T. Tao and V. Vu, *Additive Combinatorics*, volume 105 of *Cambridge Studies in Advanced Mathematics*, Cambridge University Press, Cambridge, 2006. <31, 32, 188, 192>

[2782] V. Tarokh, N. Seshadri, and A. R. Calderbank, Space-time codes for high data rate wireless communication: performance criterion and code construction, *IEEE Trans. Inform. Theory* 44 (1998) 744–765. <847, 849>

[2783] J. Tate, Endomorphisms of abelian varieties over finite fields, *Invent. Math.* 2 (1966) 134–144. <431, 433, 440, 459, 463, 806, 811>

[2784] J. Tate, The arithmetic of elliptic curves, *Invent. Math.* 23 (1974) 179–206. <422, 440>

[2785] W. Tautz, J. Top, and A. Verberkmoes, Explicit hyperelliptic curves with real multiplication and permutation polynomials, *Canad. J. Math.* 43 (1991) 1055–1064. <239, 240>

[2786] D. E. Taylor, *The Geometry of the Classical Groups*, volume 9 of *Sigma Series in Pure Mathematics*, Heldermann Verlag, Berlin, 1992. <513, 515, 518, 519, 520>

[2787] A. F. Tenca and Ç. K. Koç, A scalable architecture for Montgomery multiplication, In *Proc. Cryptographic Hardware and Embedded Systems (CHES)*, volume 1717 of *Lecture Notes Comput. Sci.*, 94–108, Springer, Berlin, 1999. <822, 823>

[2788] G. Tenenbaum, *Introduction to Analytic and Probabilistic Number Theory*, volume 46 of *Cambridge Studies in Advanced Mathematics*, Cambridge University Press, Cambridge, 1995. <369, 371, 374>

[2789] A. Terras, *Fourier Analysis on Finite Groups and Applications*, volume 43 of *London Mathematical Society Student Texts*, Cambridge University Press, Cambridge, 1999. <305, 307, 308, 310, 644, 658>

[2790] E. Teske, Square-root algorithms for the discrete logarithm problem (a survey), In *Public-Key Cryptography and Computational Number Theory*, 283–301, de Gruyter, Berlin, 2001. <397, 401>

[2791] F. Thaine, On Gaussian periods that are rational integers, *Michigan Math. J.* 50 (2002) 313–337. <141, 161>

[2792] D. Thakur, Multizeta in function field arithmetic, In *Proceedings of Banff Workshop*, European Mathematical Society (EMS), Zürich. <542, 544, 546>

[2793] D. S. Thakur, *Function Field Arithmetic*, World Scientific Publishing Co. Inc., River Edge, NJ, 2004. <31, 32, 535, 546>

[2794] J. A. Thas, Normal rational curves and k-arcs in Galois spaces, *Rend. Mat. (6)* 1 (1968) 331–334. <585, 589>

[2795] J. A. Thas, The affine plane $\mathrm{AG}(2, q)$, q odd, has a unique one point extension, *Invent. Math.* 118 (1994) 133–139. <589>

[2796] "The GAP Group," GAP system for computational discrete algebra, `http://www.gap-system.org`, as viewed in July, 2012. <47, 49>

[2797] "The GNU Project," The GNU MP Bignum library, `http://www.gmplib.org/`, as viewed in July, 2012. <32, 47, 49>

[2798] "The Magma Group," Computational algebra system for algebra, number theory and geometry, version 2.18-11, `http:magma.maths.usyd.edu.au/magma/`, as viewed in July, 2012. <346, 363, 782, 783>

[2799] "The Mathworks Inc." MATLAB - The Language of Technical Computing, `http://www.mathworks.com/products/matlab/`, as viewed in July 2012. <48, 49>

[2800] "The OEIS Foundation Inc." The on-line encyclopedia of integer sequences™(OEIS™), `http://www.oeis.org`, as viewed in July, 2012. <46, 49>

[2801] "The PARI Group," PARI/GP Development Center, http://pari.math.u-bordeaux.fr/, as viewed in July, 2012. <47, 49>

[2802] N. Thériault, Index calculus attack for hyperelliptic curves of small genus, In *Advances in Cryptology—ASIACRYPT 2003*, volume 2894 of *Lecture Notes in Comput. Sci.*, 75–92, Springer, Berlin, 2003. <455, 456, 798, 803>

[2803] J. J. Thomas, J. M. Keller, and G. N. Larsen, The calculation of multiplicative inverses over GF(P) efficiently where P is a Mersenne prime, *IEEE Trans. Comput.* 35 (1986) 478–482. <360, 363>

[2804] E. Thomé, Fast computation of linear generators for matrix sequences and application to the block Wiedemann algorithm, In *Proceedings of the 2001 International Symposium on Symbolic and Algebraic Computation*, 323–331, ACM, New York, 2001. <535>

[2805] T. M. Thompson, *From Error-Correcting Codes Through Sphere Packings to Simple Groups*, volume 21 of *Carus Mathematical Monographs*, Mathematical Association of America, Washington, DC, 1983. <691, 703>

[2806] T. Tian and W. F. Qi, Primitive normal element and its inverse in finite fields, *Acta Math. Sinica (Chin. Ser.)* 49 (2006) 657–668. <116>

[2807] T. Tian and W.-F. Qi, Typical primitive polynomials over integer residue rings, *Finite Fields Appl.* 15 (2009) 796–807. <90>

[2808] A. Tietäväinen, On systems of linear and quadratic equations in finite fields, *Ann. Acad. Sci. Fenn. Ser. A I No.* 382 (1965) 5. <210, 213>

[2809] A. Tietäväinen, On diagonal forms over finite fields, *Ann. Univ. Turku. Ser. A I No.* 118 (1968) 10. <207, 213>

[2810] A. Tietäväinen, On the nonexistence of perfect codes over finite fields, *SIAM J. Appl. Math.* 24 (1973) 88–96. <672, 684, 703>

[2811] A. Tietäväinen, A short proof for the nonexistence of unknown perfect codes over GF(q), $q > 2$, *Ann. Acad. Sci. Fenn. Ser. A I* (1974) 6. <672, 702, 703>

[2812] R. A. H. Toledo, Linear finite dynamical systems, *Comm. Algebra* 33 (2005) 2977–2989. <834>

[2813] A. Tonelli, Bemerkung über die Auflösung quadratischer Congruenzen, *Nachrichten von der Königl. Gesellschaft der Wissenschaften und der Georg-Augusts-Universität zu Göttingen* (1891) 344–346. <796>

[2814] A. Topuzoğlu and A. Winterhof, Pseudorandom sequences, In *Topics in Geometry, Coding Theory and Cryptography*, volume 6 of *Algebr. Appl.*, 135–166, Springer, Dordrecht, 2007. <337, 338, 344>

[2815] Á. Tóth, On the evaluation of Salié sums, *Proc. Amer. Math. Soc.* 133 (2005) 643–645. <160, 161>

[2816] J. Tromp, L. Zhang, and Y. Zhao, Small weight bases for Hamming codes, In *Computing and Combinatorics*, volume 959 of *Lecture Notes in Comput. Sci.*, 235–243, Springer, Berlin, 1995. <89, 90>

[2817] T. T. Truong, Degree complexity of a family of birational maps. II. Exceptional cases, *Math. Phys. Anal. Geom.* 12 (2009) 157–180. <338, 344>

[2818] B. Tsaban and U. Vishne, Efficient linear feedback shift registers with maximal period,

Finite Fields Appl. 8 (2002) 256–267. <70>

[2819] M. Tsfasman, S. Vlăduţ, and D. Nogin, *Algebraic Geometric Codes: Basic Notions*, volume 139 of *Mathematical Surveys and Monographs*, American Mathematical Society, Providence, RI, 2007. <31, 32>

[2820] M. A. Tsfasman and S. G. Vlăduţ, *Algebraic-Geometric Codes*, volume 58 of *Mathematics and its Applications (Soviet Series)*, Kluwer Academic Publishers Group, Dordrecht, 1991. <31, 32, 710, 712>

[2821] M. A. Tsfasman, S. G. Vlăduţ, and T. Zink, Modular curves, Shimura curves, and Goppa codes, better than Varshamov-Gilbert bound, *Math. Nachr.* 109 (1982) 21–28. <462, 463, 464, 469, 711, 712>

[2822] S. Tsujii, A. Fujioka, and T. Itoh, Generalization of the public key cryptosystem based on the difficulty of solving a system of non-linear equations, In *Proc. Tenth Symposium on Information Theory and Its Applications*, JA5–3, 1987. <765, 768, 783>

[2823] S. Tsujii, K. Kurosawa, T. Itoh, A. Fujioka, and T. Matsumoto, A public key cryptosystem based on the difficulty of solving a system of nonlinear equations, *ICICE Transactions (D) J69-D* 12 (1986) 1963–1970. <765, 768, 783>

[2824] W. J. Turner, *Black Box Linear Algebra with the Linbox Library*, PhD thesis, 2002. <531, 535>

[2825] G. Turnwald, Permutation polynomials of binomial type, In *Contributions to General Algebra, 6*, 281–286, Hölder-Pichler-Tempsky, Vienna, 1988. <218, 223, 230>

[2826] G. Turnwald, A new criterion for permutation polynomials, *Finite Fields Appl.* 1 (1995) 64–82. <217, 228, 230, 233, 235, 236>

[2827] G. Turnwald, On Schur's conjecture, *J. Austral. Math. Soc., Ser. A* 58 (1995) 312–357. <228, 230, 239, 240>

[2828] R. Turyn and J. Storer, On binary sequences, *Proc. Amer. Math. Soc.* 12 (1961) 394–399. <322, 324>

[2829] R. J. Turyn, The linear generation of Legendre sequence, *J. Soc. Indust. Appl. Math.* 12 (1964) 115–116. <333, 336>

[2830] R. J. Turyn, Character sums and difference sets, *Pacific J. Math.* 15 (1965) 319–346. <604, 606, 607>

[2831] R. J. Turyn, An infinite class of Williamson matrices, *J. Combin. Theory, Ser. A* 12 (1972) 319–321. <610, 619>

[2832] R. J. Turyn, Hadamard matrices, Baumert-Hall units, four-symbol sequences, pulse compression, and surface wave encodings, *J. Combin. Theory, Ser. A* 16 (1974) 313–333. <843, 849>

[2833] S. Uchiyama, Note on the mean value of $V(f)$. II, *Proc. Japan Acad.* 31 (1955) 321–323. <368, 374>

[2834] S. Uchiyama, Sur les polynômes irréductibles dans un corps fini. II, *Proc. Japan Acad.* 31 (1955) 267–269. <74, 79>

[2835] D. Ulmer, Jacobi sums, Fermat Jacobians, and ranks of abelian varieties over towers of function fields, *Math. Res. Lett.* 14 (2007) 453–467, `http://people.math.gatech.edu/~ulmer/research/papers/2007c-correction.pdf`. <146, 161>

[2836] C. Umans, Fast polynomial factorization and modular composition in small characteristic, In *STOC'08*, 481–490, ACM, New York, 2008. <358, 363>

[2837] A. Valette, Graphes de Ramanujan et applications, *Astérisque* (1997) Exp. No. 829, 4, 247–276, Séminaire Bourbaki, Vol. 1996/97. <643, 658>

[2838] E. R. van Dam and D. Fon-Der-Flaass, Codes, graphs, and schemes from nonlinear functions, *European J. Combin.* 24 (2003) 85–98. <259, 261>

[2839] E. R. van Dam and W. H. Haemers, Eigenvalues and the diameter of graphs, *Linear and Multilinear Algebra* 39 (1995) 33–44. <645, 658>

[2840] W. van Dam, S. Hallgren, and L. Ip, Quantum algorithms for some hidden shift problems, In *Proceedings of the Fourteenth Annual ACM-SIAM Symposium on Discrete Algorithms (Baltimore, MD, 2003)*, 489–498, ACM, New York, 2003. <840, 841>

[2841] W. van Dam and I. E. Shparlinski, Classical and quantum algorithms for exponential congruences, In *Theory of Quantum Computation, Communication, and Cryptography*, volume 5106 of *Lecture Notes in Comput. Sci.*, 1–10, Springer, Berlin, 2008. <840, 841>

[2842] G. van der Geer and M. van der Vlugt, Reed-Muller codes and supersingular curves. I, *Compositio Math.* 84 (1992) 333–367. <487, 488>

[2843] G. van der Geer and M. van der Vlugt, On the existence of supersingular curves of given genus, *J. Reine Angew. Math.* 458 (1995) 53–61. <486, 488>

[2844] G. van der Geer and M. van der Vlugt, Quadratic forms, generalized Hamming weights of codes and curves with many points, *J. Number Theory* 59 (1996) 20–36. <205, 206>

[2845] G. van der Geer and M. van der Vlugt, An asymptotically good tower of curves over the field with eight elements, *Bull. London Math. Soc.* 34 (2002) 291–300. <467, 469>

[2846] G. van der Geer and M. van der Vlugt, Tables of curves with many points, 2009, `http://www.science.uva.nl/~geer/tables-mathcomp21.pdf`. <460, 463>

[2847] M. van der Put, A note on p-adic uniformization, *Nederl. Akad. Wetensch. Indag. Math.* 49 (1987) 313–318. <544, 546>

[2848] B. L. van der Waerden, *A History of Algebra: From al-Khwārizmī to Emmy Noether*, Springer-Verlag, Berlin, 1985. <3, 11>

[2849] J. H. van Lint, *Introduction to Coding Theory*, volume 86 of *Graduate Texts in Mathematics*, Springer-Verlag, Berlin, third edition, 1999. <31, 32, 586, 589, 661, 663, 668, 670, 671, 672, 673, 674, 683, 684, 685, 686, 689, 690, 703>

[2850] J. H. van Lint and A. Schrijver, Construction of strongly regular graphs, two-weight codes and partial geometries by finite fields, *Combinatorica* 1 (1981) 63–73. <617, 619>

[2851] J. H. van Lint and R. M. Wilson, *A Course in Combinatorics*, Cambridge University Press, Cambridge, 1992. <31, 32, 609, 619>

[2852] P. C. van Oorschot and M. J. Wiener, Parallel collision search with cryptanalytic applications, *J. Cryptology* 12 (1999) 1–28. <397, 401, 746, 750>

[2853] T. van Trung and S. Martirosyan, New constructions for IPP codes, *Des. Codes Cryptogr.* 35 (2005) 227–239. <613, 619>

[2854] P. van Wamelen, New explicit multiplicative relations between Gauss sums, *Int. J. Number Theory* 3 (2007) 275–292. <146, 161>

[2855] R. Varshamov, Estimate of the number of signals in error correcting codes, *Dokl. Akad. Nauk. SSSR* 117 (1957) 739–741. <671, 702, 703>

[2856] R. Varshamov, A general method of synthesizing irreducible polynomials over Galois fields, *Soviet Math. Dokl.* 29 (1984) 334–336. <379, 380>

[2857] R. R. Varshamov, A certain linear operator in a Galois field and its applications (Russian), *Studia, Sci. Math. Hunger.* 8 (1973) 5–19. <63, 66>

[2858] R. R. Varshamov, Operator substitutions in a Galois field, and their applications (Russian), *Dokl. Akad. Nauk SSSR;* 211 (1973) 768–771. <63, 66>

[2859] R. R. Varshamov, A general method of synthesis for irreducible polynomials over Galois fields, *Dokl. Akad. Nauk SSSR* 275 (1984) 1041–1044. <63, 64, 66, 113, 116>

[2860] R. R. Varshamov and G. Ananiashvili, The theory of polynomial reducibility in finite fields (Russian), In *Abstract and Structual Theory of the Construction of Swithching Circuits (Russian) (M.A. Gavrilov, ed.)*, 134–138, Nauka, Moscow, 1966. <60, 66>

[2861] R. R. Varshamov and G. Garakov, On the theory of self-dual polynomials over a Galois field (Russian), *Bull. Math. Soc. Sci. Math. R. S. Roumania (New Ser.)* 13 (1969) 403–415. <61, 66>

[2862] R. C. Vaughan and T. D. Wooley, Waring's problem: a survey, In *Number Theory for the Millennium*, volume III, 301–340, A. K. Peters, Natick, MA, 2002. <499, 500>

[2863] T. Vaughan, Polynomials and linear transformations over finite fields, *J. Reine Angew. Math.* 267 (1974) 179–206. <62, 66>

[2864] A. Veliz-Cuba, A. S. Jarrah, and R. Laubenbacher, Polynomial algebra of discrete models in systems biology, *Bioinformatics* 26 (2010) 1637–1643. <826, 829, 830, 831, 834>

[2865] J. Vélu, Isogénies entre courbes elliptiques, *C. R. Acad. Sci. Paris, Sér. A-B* 273 (1971) A238–A241. <122, 128>

[2866] A. Venkateswarlu and H. Niederreiter, Improved results on periodic multisequences with large error linear complexity, *Finite Fields Appl.* 16 (2010) 463–476. <331, 336>

[2867] F. Vercauteren, Computing zeta functions of hyperelliptic curves over finite fields of characteristic 2, In *Advances in Cryptology—CRYPTO 2002*, volume 2442 of *Lecture Notes in Comput. Sci.*, 369–384, Springer, Berlin, 2002. <454, 456>

[2868] F. Vercauteren, Optimal pairings, *IEEE Trans. Inform. Theory* 56 (2010) 455–461. <791, 796>

[2869] E. R. Verheul, Evidence that XTR is more secure than supersingular elliptic curve cryptosystems, *Journal of Cryptology* 17 (2004) 277–296. <790, 796>

[2870] C.-M. Viallet, Algebraic dynamics and algebraic entropy, *Int. J. Geom. Methods Mod. Phys.* 5 (2008) 1373–1391. <337, 338, 344>

[2871] C. M. Viallet, Integrable lattice maps: Q_V, a rational version of Q_4, *Glasgow Math. J.* 51 (2009) 157–163. <337, 338, 344>

[2872] G. D. Villa Salvador, *Topics in the Theory of Algebraic Function Fields*, Mathematics: Theory & Applications. Birkhäuser Boston Inc., Boston, MA, 2006. <406, 422>

[2873] G. Villard, Further analysis of Coppersmith's block Wiedemann algorithm for the solution of sparse linear systems, In *Proceedings of the 1997 International Symposium on Symbolic and Algebraic Computation*, ISSAC '97, 32–39, ACM, New York, NY, USA, 1997. <534, 535>

[2874] G. Villard, Computing the Frobenius normal form of a sparse matrix, In *Computer Algebra in Scientific Computing*, 395–407, Springer, Berlin, 2000. <531, 535>

[2875] G. Villard, *Algorithmique en Algèbre Linéaire Exacte*, Mémoire d'habilitation, Université Claude Bernard Lyon 1, 2003. <531, 535>

[2876] L. A. Vinh, The Szemerédi-Trotter type theorem and the sum-product estimate in finite fields, *Eur. J. Combinatorics* 32 (2011) 1177–1181. <191, 192>

[2877] I. M. Vinogradov, Representation of an odd number as a sum of three primes, *Comptes Rendus (Doklady)* 15 (1937) 191–294. <497, 500>

[2878] U. Vishne, Factorization of trinomials over Galois fields of characteristic 2, *Finite Fields Appl.* 3 (1997) 370–377. <68, 70>

[2879] A. J. Viterbi, Error bounds for convolutional codes and an asymptotically optimum decoding algorithm, *IEEE Trans. Inform. Theory* 13 (1967) 260–269. <721, 727>

[2880] A. J. Viterbi, *CDMA: Principles of Spread Spectrum Communication*, Addison-Wesley, Reading, MA, 1995. <317>

[2881] S. G. Vlèduts and Y. I. Manin, Linear codes and modular curves, In *Current Problems in Mathematics, Vol. 25*, Itogi Nauki i Tekhniki, 209–257, Akad. Nauk SSSR Vsesoyuz. Inst. Nauchn. i Tekhn. Inform., Moscow, 1984. <545, 546>

[2882] S. G. Vlăduţ and V. G. Drinfeld, The number of points of an algebraic curve, *Funktsional. Anal. i Prilozhen.* 17 (1983) 68–69. <462, 463>

[2883] J. F. Voloch, Diagonal equations over function fields, *Bol. Soc. Brasil. Mat.* 16 (1985) 29–39. <213>

[2884] J. F. Voloch, On the order of points on curves over finite fields, *Integers* 7 (2007) A49, 4. <98, 99>

[2885] J. F. Voloch, Symmetric cryptography and algebraic curves, In *Algebraic Geometry and its Applications*, volume 5 of *Ser. Number Theory Appl.*, 135–141, World Sci. Publ., Hackensack, NJ, 2008. <255, 261>

[2886] J. F. Voloch, Elements of high order on finite fields from elliptic curves, *Bull. Aust. Math. Soc.* 81 (2010) 425–429. <99>

[2887] C. H. Waddington, Canalisation of development and the inheritance of acquired characters, *Nature* 150 (1942) 563–564. <832, 834>

[2888] L. I. Wade, Certain quantities transcendental over $GF(p^n, x)$, *Duke Math. J.* 8 (1941) 701–720. <546>

[2889] A. Wagner, On finite affine line transitive planes, *Math. Z.* 87 (1965) 1–11. <568, 574>

[2890] S. S. Wagstaff, Jr., The Cunningham project, `http://homes.cerias.purdue.edu/~ssw/cun/`, as viewed in July, 2012. <46, 49>

[2891] R. J. Walker, Determination of division algebras with 32 elements, In *Proc. Sympos. Appl. Math., Vol. XV*, 83–85, Amer. Math. Soc., Providence, RI, 1963. <275, 278>

[2892] D. Wan, On the Riemann hypothesis for the characteristic p zeta function, *J. Number Theory* 58 (1996) 196–212. <544, 546>

[2893] D. Wan, Generators and irreducible polynomials over finite fields, *Math. Comp.* 66 (1997) 1195–1212. <75, 79, 168, 169>

[2894] D. Wan, Computing zeta functions over finite fields, In *Finite Fields: Theory, Applications, and Algorithms*, volume 225 of *Contemp. Math.*, 131–141, Amer. Math. Soc., Providence, RI, 1999. <490, 491>

[2895] D. Wan, Dwork's conjecture on unit root zeta functions, *Ann. of Math., 2nd Ser.* 150 (1999) 867–927. <479, 488>

[2896] D. Wan, Higher rank case of Dwork's conjecture, *J. Amer. Math. Soc.* 13 (2000) 807–852. <479, 488>

[2897] D. Wan, Rank one case of Dwork's conjecture, *J. Amer. Math. Soc.* 13 (2000) 853–908. <479, 488>

[2898] D. Wan, Rationality of partial zeta functions, *Indag. Math. (New Ser.)* 14 (2003) 285–292. <198, 201>

[2899] D. Wan, Variation of p-adic Newton polygons for L-functions of exponential sums, *Asian J. Math.* 8 (2004) 427–471. <484, 485, 488>

[2900] D. Wan, Mirror symmetry for zeta functions, In *Mirror Symmetry V*, volume 38 of *AMS/IP Stud. Adv. Math.*, 159–184, Amer. Math. Soc., Providence, RI, 2006. <196, 200, 201>

[2901] D. Wan, Algorithmic theory of zeta functions over finite fields, In *Algorithmic Number Theory: Lattices, Number Fields, Curves and Cryptography*, volume 44 of *Math. Sci. Res. Inst. Publ.*, 551–578, Cambridge Univ. Press, Cambridge, 2008. <491>

[2902] D. Wan, Lectures on zeta functions over finite fields, In *Higher-Dimensional Geometry over Finite Fields*, volume 16 of *NATO Sci. Peace Secur. Ser. D Inf. Commun. Secur.*, 244–268, IOS, Amsterdam, 2008. <193, 196, 201>

[2903] D. Wan, Modular counting of rational points over finite fields, *Found. Comput. Math.* 8 (2008) 597–605. <489, 491>

[2904] D. Q. Wan, On a problem of Niederreiter and Robinson about finite fields, *J. Austral. Math. Soc., Ser. A* 41 (1986) 336–338. <228, 230>

[2905] D. Q. Wan, Permutation polynomials over finite fields, *Acta Math. Sinica (New Ser.)* 3 (1987) 1–5. <218, 223, 230>

[2906] D. Q. Wan, Zeros of diagonal equations over finite fields, *Proc. Amer. Math. Soc.* 103 (1988) 1049–1052. <209, 210, 213>

[2907] D. Q. Wan, An elementary proof of a theorem of Katz, *Amer. J. Math.* 111 (1989) 1–8. <199, 201>

[2908] D. Q. Wan, Permutation polynomials and resolution of singularities over finite fields, *Proc. Amer. Math. Soc.* 110 (1990) 303–309. <218, 230>

[2909] D. Q. Wan, A generalization of the Carlitz conjecture, In *Finite Fields, Coding Theory, and Advances in Communications and Computing*, volume 141 of *Lecture Notes in Pure and Appl. Math.*, 431–432, Dekker, New York, 1993. <218, 230>

[2910] D. Q. Wan, Newton polygons of zeta functions and L functions, *Ann. of Math., 2nd Ser.* 137 (1993) 249–293. <483, 484, 488>

[2911] D. Q. Wan, A p-adic lifting lemma and its applications to permutation polynomials, In *Finite Fields, Coding Theory, and Advances in Communications and Computing*, volume 141 of *Lecture Notes in Pure and Appl. Math.*, 209–216, Dekker, New York, 1993. <217, 230, 233, 236>

[2912] D. Q. Wan, A classification conjecture about certain permutation polynomials, In *Finite Fields: Theory, Applications and Algorithms*, volume 168 of *Contemporary Math.*, 401–402, American Mathematical Society, Providence, RI, 1994. <228, 230>

[2913] D. Q. Wan, Permutation binomials over finite fields, *Acta Math. Sinica (New Ser.)* 10 (1994) 30–35. <218, 223, 230>

[2914] D. Q. Wan, A Chevalley-Warning approach to p-adic estimates of character sums, *Proc. Amer. Math. Soc.* 123 (1995) 45–54. <199, 201>

[2915] D. Q. Wan, Minimal polynomials and distinctness of Kloosterman sums, *Finite Fields Appl.* 1 (1995) 189–203. <154, 161>

[2916] D. Q. Wan and R. Lidl, Permutation polynomials of the form $x^r f(x^{(q-1)/d})$ and their group structure, *Monatsh. Math.* 112 (1991) 149–163. <221, 230>

[2917] D. Q. Wan, G. L. Mullen, and P. J.-S. Shiue, Erratum: "The number of permutation polynomials of the form $f(x) + cx$ over a finite field", *Proc. Edinburgh Math.*

Soc., Ser. II 38 (1995) 133. <555, 556>

[2918] D. Q. Wan, G. L. Mullen, and P. J.-S. Shiue, The number of permutation polynomials of the form $f(x) + cx$ over a finite field, *Proc. Edinburgh Math. Soc., Ser. II* 38 (1995) 133–149. <228, 230, 555, 556>

[2919] D. Q. Wan, P. J.-S. Shiue, and C. S. Chen, Value sets of polynomials over finite fields, *Proc. Amer. Math. Soc.* 119 (1993) 711–717. <217, 230, 234, 236>

[2920] Z.-X. Wan, *Geometry of Classical Groups over Finite Fields*, Science Press, Beijing, second edition, 2002. <31, 32, 512, 514, 516, 519, 520>

[2921] Z.-X. Wan, *Lectures on Finite Fields and Galois Rings*, World Scientific Publishing Co. Inc., River Edge, NJ, 2003. <13, 28, 29, 31, 32>

[2922] Z.-X. Wan, A shorter proof for an explicit formula for discrete logarithms in finite fields, *Discrete Math.* 308 (2008) 4914–4915. <396, 401>

[2923] Z.-X. Wan, *Finite Fields and Galois Rings*, World Scientific Publishing Co. Inc., Singapore, 2012. <31, 32>

[2924] Z.-X. Wan and K. Zhou, On the complexity of the dual basis of a type I optimal normal basis, *Finite Fields Appl.* 13 (2007) 411–417. <125, 128>

[2925] C. C. Wang, An algorithm to design finite field multipliers using a self-dual normal basis, *IEEE Trans. Comput.* 38 (1989) 1457–1460. <39, 49>

[2926] C. C. Wang, T. K. Truong, H. M. Shao, L. J. Deutsch, J. K. Omura, and I. S. Reed, VLSI architectures for computing multiplications and inverses in $GF(2^m)$, *IEEE Trans. Comput.* 34 (1985) 709–716. <819, 820, 823>

[2927] L. Wang, On permutation polynomials, *Finite Fields Appl.* 8 (2002) 311–322. <222, 230>

[2928] L. Wang and Y. Zhu, $F[x]$-lattice basis reduction algorithm and multisequence synthesis, *Sci. China, Ser. F* 44 (2001) 321–328. <330, 336>

[2929] L.-C. Wang and F.-H. Chang, Tractable rational map cryptosystem (version 2), `http://eprint.iacr.org/2004/046`, ver. 20040221:212731. <775, 783>

[2930] L.-C. Wang and F.-H. Chang, Tractable rational map cryptosystem (version 4), `http://eprint.iacr.org/2004/046`, ver. 20060203:065450. <775, 783>

[2931] L.-C. Wang, Y.-H. Hu, F. Lai, C.-Y. Chou, and B.-Y. Yang, Tractable rational map signature, In *Public Key Cryptography—PKC 2005*, volume 3386 of *Lecture Notes in Comput. Sci.*, 244–257, Springer, Berlin, 2005. <772, 783>

[2932] L.-C. Wang, B.-Y. Yang, Y.-H. Hu, and F. Lai, A "medium-field" multivariate public-key encryption scheme, In *Topics in Cryptology—CT-RSA 2006*, volume 3860 of *Lecture Notes in Comput. Sci.*, 132–149, Springer, Berlin, 2006. <775, 783>

[2933] L.-P. Wang and H. Niederreiter, Enumeration results on the joint linear complexity of multisequences, *Finite Fields Appl.* 12 (2006) 613–637. <330, 336>

[2934] L.-P. Wang, Y.-F. Zhu, and D.-Y. Pei, On the lattice basis reduction multisequence synthesis algorithm, *IEEE Trans. Inform. Theory* 50 (2004) 2905–2910. <330, 336>

[2935] M. Wang, Linear complexity profiles and continued fractions, In *Advances in Cryptology—EUROCRYPT '89*, volume 434 of *Lecture Notes in Comput. Sci.*, 571–585, Springer, Berlin, 1990. <329, 336>

[2936] M. Wang and I. F. Blake, Bit serial multiplication in finite fields, *SIAM J. Discrete Math.* 3 (1990) 140–148. <105, 109>

[2937] M. Z. Wang, Linear complexity profiles and jump complexity, *Inform. Process. Lett.* 61 (1997) 165–168. <329, 336>

[2938] P. S. Wang, An improved multivariate polynomial factoring algorithm, *Math. Comp.* 32 (1978) 1215–1231. <385, 392>

[2939] P. S. Wang and L. P. Rothschild, Factoring multivariate polynomials over the integers, *Math. Comp.* 29 (1975) 935–950. <385, 392>

[2940] Q. Wang, Cyclotomic mapping permutation polynomials over finite fields, In *Sequences, Subsequences, and Consequences*, volume 4893 of *Lecture Notes in Comput. Sci.*, 119–128, Springer, Berlin, 2007. <221, 222, 230>

[2941] Q. Wang, On inverse permutation polynomials, *Finite Fields Appl.* 15 (2009) 207–213. <229, 230>

[2942] Q. Wang, On generalized Lucas sequences, In *Combinatorics and graphs*, volume 531 of *Contemp. Math.*, 127–141, Amer. Math. Soc., Providence, RI, 2010. <222, 229, 230>

[2943] Q. Wang, Cyclotomy and permutation polynomials of large indices, arXiv:1209.3828v1, 2012. <221, 230>

[2944] Q. Wang, K. Wang, and Z. Dai, Implementation of multi-continued fraction algorithm and application to multi-sequence linear synthesis, In *Sequences and Their Applications—SETA 2006*, volume 4086 of *Lecture Notes in Comput. Sci.*, 248–258, Springer, Berlin, 2006. <329, 336>

[2945] Q. Wang and J. L. Yucas, Dickson polynomials over finite fields, *Finite Fields Appl.* 18 (2012) 814–831. <284, 285, 287, 288, 289, 290>

[2946] Y. Wang, Linear complexity versus pseudorandomness: on Beth and Dai's result, In *Advances in Cryptology—ASIACRYPT'99*, volume 1716 of *Lecture Notes in Comput. Sci.*, 288–298, Springer, Berlin, 1999. <334, 336>

[2947] K. L. Wantz, A new class of unitals in the Hughes plane, *Geom. Dedicata* 70 (1998) 125–138. <571, 574>

[2948] L. C. Washington, *Introduction to Cyclotomic Fields*, volume 83 of *Graduate Texts in Mathematics*, Springer-Verlag, New York, second edition, 1997. <143, 152, 161>

[2949] L. C. Washington, *Elliptic Curves*, Discrete Mathematics and its Applications. Chapman & Hall/CRC, Boca Raton, FL, 2003. <31, 32>

[2950] L. C. Washington, *Elliptic Curves*, Discrete Mathematics and Its Applications. Chapman & Hall/CRC, Boca Raton, FL, second edition, 2008. <31, 32, 422, 440>

[2951] A. Wassermann, Konstruktion von Normalbasen, *Bayreuth. Math. Schr.* (1990) 155–164. <120, 128>

[2952] A. Wassermann, Zur Arithmetik in endlichen Körpern, *Bayreuth. Math. Schr.* (1993) 147–251, Dissertation, Universität Bayreuth, Bayreuth, 1992. <120, 128>

[2953] Y. Watanabe, N. Takagi, and K. Takagi, A VLSI algorithm for division in $GF(2^m)$ based on extended binary GCD algorithm, *IEICE Transactions on Fundamentals of Electronics, Communications and Computer Sciences* E85-A (2002) 994–999. <816, 823>

[2954] W. C. Waterhouse, Abelian varieties over finite fields, *Ann. Sci. École Norm. Sup. IVe Ser.* 2 (1969) 521–560. <430, 431, 436, 440>

[2955] W. A. Webb, Waring's problem in GF$[q, x]$, *Acta Arith.* 22 (1973) 207–220. <499, 500>

[2956] C. Wei and Q. Sun, The least integer represented by $\sum_{i=1}^{n} x_i/d_i$ and its application, *Acta Math. Sinica (Chin. Ser.)* 49 (2006) 1021–1026. <209, 213>

[2957] Q. Wei and Q. Zhang, On strong orthogonal systems and weak permutation polynomials over finite commutative rings, *Finite Fields Appl.* 13 (2007) 113–120.

<232>

[2958] S. Wei, G. Chen, and G. Xiao, A fast algorithm for determining the linear complexity of periodic sequences, In *Information Security and Cryptology*, volume 3822 of *Lecture Notes in Comput. Sci.*, 202–209, Springer, Berlin, 2005. <329, 336>

[2959] S. Wei, G. Xiao, and Z. Chen, A fast algorithm for determining the linear complexity of a binary sequence with period $2^n p^m$, *Sci. China, Ser. F* 44 (2001) 453–460. <329, 336>

[2960] S. Wei, G. Xiao, and Z. Chen, A fast algorithm for determining the minimal polynomial of a sequence with period $2p^n$ over GF(q), *IEEE Trans. Inform. Theory* 48 (2002) 2754–2758. <329, 336>

[2961] A. Weil, On some exponential sums, *Proc. Nat. Acad. Sci. U. S. A.* 34 (1948) 204–207. <162, 169>

[2962] A. Weil, *Sur les Courbes Algébriques et les Variétés qui s'en dÉduisent*, Actualités Sci. Ind., no. 1041; Publ. Inst. Math. Univ. Strasbourg 7 (1945). Hermann et Cie., Paris, 1948. <162, 169, 497, 500>

[2963] A. Weimerskirch and C. Paar, Generalizations of the Karatsuba algorithm for efficient implementations, 2006, preprint available, `http://eprint.iacr.org/2006/224`. <813, 823>

[2964] L. Welch, Lower bounds on the maximum cross correlation of signals, *IEEE Trans. Inform. Theory* 20 (1974) 397–399. <320, 324>

[2965] L. R. Welch and E. R. Berlekamp, Error Correction for Algebraic Block Codes, U. S. Patent 4,633,470 (1986). <695, 703>

[2966] E. J. Weldon, Jr., Euclidean geometry cyclic codes, In *Combinatorial Mathematics and its Applications*, 377–387, Univ. North Carolina Press, Chapel Hill, N.C., 1969. <689, 696, 703>

[2967] C. Wells, The degrees of permutation polynomials over finite fields, *J. Combin. Theory* 7 (1969) 49–55. <219, 220, 230>

[2968] A. Wells Jr., A polynomial form for logarithms modulo a prime, *IEEE Trans. Inf. Theory* 30 (1984) 845–846. <396, 401>

[2969] G. P. Wene, On the multiplicative structure of finite division rings, *Aequationes Math.* 41 (1991) 222–233. <278>

[2970] A. Weng, *Konstruktion kryptographisch geeigneter Kurven mit komplexer Multiplikation*, PhD thesis, Universität Gesamthochschule Essen, 2001. <803>

[2971] A. Weng, Constructing hyperelliptic curves of genus 2 suitable for cryptography, *Math. Comput.* 72 (2003) 435–458. <803>

[2972] G. Weng, W. Qiu, Z. Wang, and Q. Xiang, Pseudo-Paley graphs and skew Hadamard difference sets from presemifields, *Des. Codes Cryptogr.* 44 (2007) 49–62. <282>

[2973] G. Weng and X. Zeng, Further results on planar DO functions and commutative semifields, *Des. Codes Cryptog.* 63 (2012) 413–423. <282>

[2974] R. C. Whaley, A. Petitet, and J. J. Dongarra, Automated empirical optimizations of software and the ATLAS project, *Parallel Computing* 27 (2001) 3–35. <523, 535>

[2975] A. L. Whiteman, An infinite family of Hadamard matrices of Williamson type, *J. Combin. Theory, Ser. A* 14 (1973) 334–340. <610, 619>

[2976] D. H. Wiedemann, Solving sparse linear equations over finite fields, *IEEE Trans. Inform. Theory* 32 (1986) 54–62. <350, 363, 399, 401, 530, 535>

[2977] D. Wiedermann, An iterated quadratic extension of GF(2), *Fibonacci Quart.* 26

(1988) 290–295. <64, 66>

[2978] M. J. Wiener and R. J. Zuccherato, Faster attacks on elliptic curve cryptosystems, In S. Tavares and H. Meijer, editors, *Selected Areas in Cryptography—SAC '98*, volume 1556 of *Lecture Notes in Comput. Sci.*, 190–100, Springer-Verlag, Berlin, 1999. <786, 796>

[2979] M. L. H. Willems and J. A. Thas, A note on the existence of special Laguerre i-structures and optimal codes, *European J. Combin.* 4 (1983) 93–96. <587, 589>

[2980] M. Willett, Matrix fields over GF(q), *Duke Math. J.* 40 (1973) 701–704. <503, 510>

[2981] K. S. Williams, On general polynomials, *Canad. Math. Bull.* 10 (1967) 579–583. <234, 236>

[2982] K. S. Williams, On exceptional polynomials, *Canad. Math. Bull.* 11 (1968) 279–282. <233, 236>

[2983] K. S. Williams, Polynomials with irreducible factors of specified degree, *Canad. Math. Bull.* 12 (1969) 221–223. <368, 374>

[2984] V. V. Williams, Breaking the Coppersmith–Winograd barrier, 2011, preprint available at `http:cs.berkeley.edu/~virgi/matrixmult.pdf`. <358, 363, 382>

[2985] V. V. Williams, Multiplying matrices faster than Coppersmith-Winograd, In *Proceedings of the Forty Fourth Symposium on Theory of Computing*, STOC '12, 887–898, ACM, New York, NY, USA, 2012. <358, 363, 375, 380, 521, 535>

[2986] R. M. Wilson, Cyclotomy and difference families in elementary abelian groups, *J. Number Theory* 4 (1972) 17–47. <594, 599>

[2987] S. Winograd, On multiplication of 2×2 matrices, *Linear Algebra and Appl.* 4 (1971) 381–388. <524, 535>

[2988] S. Winograd, *Arithmetic Complexity of Computations*, SIAM, 1980. <814, 817, 823>

[2989] A. Winterhof, On the distribution of powers in finite fields, *Finite Fields Appl.* 4 (1998) 43–54. <181, 185>

[2990] A. Winterhof, On Waring's problem in finite fields, *Acta Arith.* 87 (1998) 171–177. <175, 185, 212, 213>

[2991] A. Winterhof, Incomplete additive character sums and applications, In *Finite Fields and Applications*, 462–474, Springer, Berlin, 2001. <181, 185>

[2992] A. Winterhof, A note on Waring's problem in finite fields, *Acta Arith.* 96 (2001) 365–368. <213>

[2993] A. Winterhof, Some estimates for character sums and applications, *Des. Codes Cryptogr.* 22 (2001) 123–131. <181, 185>

[2994] A. Winterhof, A note on the linear complexity profile of the discrete logarithm in finite fields, In *Coding, Cryptography and Combinatorics*, volume 23 of *Progr. Comput. Sci. Appl. Logic*, 359–367, Birkhäuser, Basel, 2004. <333, 336>

[2995] A. Winterhof and C. van de Woestijne, Exact solutions to Waring's problem for finite fields, *Acta Arith.* 141 (2010) 171–190. <212, 213>

[2996] E. Wirsing, Thin essential components, In *Topics in Number Theory*, 429–442. Colloq. Math. Soc. János Bolyai, Vol. 13, North-Holland, Amsterdam, 1976. <184, 185>

[2997] E. Witt, Über steinersche systeme, *Abh. Math. Sem. Univ. Hamburg* 12 (1938) 265–275. <589>

[2998] C. Wolf, A. Braeken, and B. Preneel, Efficient cryptanalysis of RSE(2)PKC and RSSE(2)PKC, In *SCN 2004*, volume 3352 of *Lecture Notes in Comput. Sci.*, 294–309, Springer, Berlin, 2004. <772, 783>

[2999] J. K. Wolf, Adding two information symbols to certain nonbinary BCH codes and some applications, *Bell System Tech. J.* 48 (1969) 2405–2424. <681, 703>

[3000] J. Wolfmann, Formes quadratiques et codes à deux poids, *C. R. Acad. Sci. Paris, Sér. A-B* 281 (1975) Aii, A533–A535. <205, 206>

[3001] J. Wolfmann, The number of solutions of certain diagonal equations over finite fields, *J. Number Theory* 42 (1992) 247–257. <208, 213>

[3002] J. Wolfmann, New results on diagonal equations over finite fields from cyclic codes, In *Finite Fields: Theory, Applications, and Algorithms*, volume 168 of *Contemp. Math.*, 387–395, Amer. Math. Soc., Providence, RI, 1994. <213>

[3003] J. Wolfmann, Some systems of diagonal equations over finite fields, *Finite Fields Appl.* 4 (1998) 29–37. <213>

[3004] "Wolfram Research", Wolfram Research: Mathematica, technical and scientific software, version 8.04, `http://www.wolfram.com/`, as viewed in July, 2012. <48, 49, 346, 363>

[3005] W. K. Wootters and B. D. Fields, Optimal state-determination by mutually unbiased measurements, *Ann. Physics* 191 (1989) 363–381. <835, 841>

[3006] H. Wu, Low complexity bit-parallel finite field arithmetic using polynomial basis, In *Proc. Cryptographic Hardware and Embedded Systems (CHES)*, volume 1717 of *Lecture Notes Comput. Sci.*, 280–291, Springer, Berlin, 1999. <814, 823>

[3007] H. Wu, Bit-parallel finite field multiplier and squarer using polynomial basis, *IEEE Trans. Comput.* 51 (2002) 750–758. <815, 823>

[3008] H. Wu, M. A. Hasan, and I. F. Blake, New low-complexity bit-parallel finite field multipliers using weakly dual bases, *IEEE Trans. Comput.* 47 (1998) 1223–1234. <822, 823>

[3009] H. Wu, M. A. Hasan, I. F. Blake, and S. Gao, Finite field multiplier using redundant representation, *IEEE Trans. Comput.* 51 (2002) 1306–1316. <351, 363, 822, 823>

[3010] M. Wu, X. Yang, and C. Chan, A dynamic analysis of irs-pkr signaling in liver cells: A discrete modeling approach, *PLoS ONE* 4 (2009) e8040. <825, 834>

[3011] P.-C. Wu, Random number generation with primitive pentanomials, *ACM Trans. Modeling and Computer Simulation* 11 (2001) 346–351. <96, 97>

[3012] Q. Xiang, Maximally nonlinear functions and bent functions, *Des. Codes Cryptogr.* 17 (1999) 211–218. <239, 240>

[3013] G. Xiao and S. Wei, Fast algorithms for determining the linear complexity of period sequences., In *Progress in Cryptology—INDOCRYPT 2002*, volume 2551 of *Lecture Notes in Comput. Sci.*, 12–21, Springer, Berlin, 2002. <329, 336>

[3014] G. Xiao, S. Wei, K. Y. Lam, and K. Imamura, A fast algorithm for determining the linear complexity of a sequence with period p^n over GF(q), *IEEE Trans. Inform. Theory* 46 (2000) 2203–2206. <329, 336>

[3015] G. Z. Xiao and J. L. Massey, A spectral characterization of correlation-immune combining functions, *IEEE Trans. Inform. Theory* 34 (1988) 569–571. <247, 252>

[3016] C. Xing and Y. Ding, Multisequences with large linear and k-error linear complexity from Hermitian function fields, *IEEE Trans. Inform. Theory* 55 (2009) 3858–3863. <331, 336>

[3017] C. P. Xing, Goppa geometric codes achieving the Gilbert-Varshamov bound, *IEEE Trans. Inform. Theory* 51 (2005) 259–264. <711, 712>

[3018] C. P. Xing and H. Niederreiter, A construction of low-discrepancy sequences using global function fields, *Acta Arith.* 73 (1995) 87–102. <629, 630>

[3019] C. P. Xing, H. Niederreiter, and K. Y. Lam, A generalization of algebraic-geometry codes, *IEEE Trans. Inform. Theory* 45 (1999) 2498–2501. <707, 712>

[3020] C. P. Xing and S. L. Yeo, New linear codes and algebraic function fields over finite fields, *IEEE Trans. Inform. Theory* 53 (2007) 4822–4825. <707, 712>

[3021] T. Yan, The geobucket data structure for polynomials, *J. Symbolic Comput.* 25 (1998) 285–293. <382, 392>

[3022] B.-Y. Yang and J.-M. Chen, All in the XL family: theory and practice, In *Information Security and Cryptology—ICISC 2004*, volume 3506 of *Lecture Notes in Comput. Sci.*, 67–86, Springer, Berlin, 2005. <782, 783>

[3023] B.-Y. Yang and J.-M. Chen, Building secure tame-like multivariate public-key cryptosystems: The new TTS, In *ACISP 2005*, volume 3574 of *Lecture Notes in Comput. Sci.*, 518–531, Springer, Berlin, 2005. <772, 773, 779, 780, 783>

[3024] B.-Y. Yang, J.-M. Chen, and Y.-H. Chen, TTS: High-speed signatures on a low-cost smart card, In *Proceedings of the Sixth International Workshop on Cryptographic Hardware and Embedded Systems (CHES)*, volume 3156 of *Lecture Notes in Comput. Sci.*, 371–385, Springer, Berlin, 2004. <772, 783>

[3025] J. Yang and Z. Dai, Linear complexity of periodically repeated random sequences, *Acta Math. Sinica (New Ser.)* 11 (1995) 1–7. <331, 336>

[3026] J. Yang, S. X. Luo, and K. Q. Feng, Gauss sum of index 4. II. Non-cyclic case, *Acta Math. Sin. (Engl. Ser.)* 22 (2006) 833–844. <150, 161>

[3027] J. Yang and L. Xia, Complete solving of explicit evaluation of Gauss sums in the index 2 case, *Sci. China Math.* 53 (2010) 2525–2542. <150, 161>

[3028] R. Yang, Newton polygons of L-functions of polynomials of the form $x^d + \lambda x$, *Finite Fields Appl.* 9 (2003) 59–88. <484, 488>

[3029] S. M. Yang and L. L. Qi, On improved asymptotic bounds for codes from global function fields, *Des. Codes Cryptogr.* 53 (2009) 33–43. <712>

[3030] T. Yanik, E. Savas, and C. Koc, Incomplete reduction in modular arithmetic, *Computers and Digital Techniques, IEE Proceedings* - 149 (2002) 46 –52. <351, 363>

[3031] M. Yannakakis, Computing the minimum fill-in is NP-complete, *SIAM J. Algebraic Discrete Methods* 2 (1981) 77–79. <532, 535>

[3032] A. C. C. Yao, On the evaluation of powers, *SIAM J. Comput.* 5 (1976) 100–103. <357, 363>

[3033] C. K. Yap, *Fundamental Problems of Algorithmic Algebra*, Oxford University Press, New York, 2000. <359, 363>

[3034] Y. Ye, A hyper-Kloosterman sum identity, *Sci. China, Ser. A* 41 (1998) 1158–1162. <155, 161>

[3035] C. S. Yeh, I. S. Reed, and T. K. Troung, Systolic multipliers for finite fields $GF(2^m)$, *IEEE Trans. Comput.* 33 (1984) 357–360. <815, 823>

[3036] B. Young and D. Panario, Low complexity normal bases in $\mathbb{F}_{2^n}$, *Finite Fields Appl.* 10 (2004) 53–64. <38, 49, 119, 128>

[3037] H. P. Young, Affine triple systems and matroid designs, *Math. Z.* 132 (1973) 343–359. <611, 619>

[3038] A. M. Youssef and G. Gong, Hyper-bent functions, In *Advances in Cryptology—EUROCRYPT 2001*, volume 2045 of *Lecture Notes in Comput. Sci.*, 406–419, Springer, Berlin, 2001. <264, 273>

[3039] J. Yu, Transcendence and Drinfel'd modules, *Invent. Math.* 83 (1986) 507–517. <546>

[3040] J. Yu, On periods and quasi-periods of Drinfel'd modules, *Compositio Math.* 74 (1990)

235–245. <546>

[3041] J.-D. Yu, Variation of the unit root along the Dwork family of Calabi-Yau varieties, *Math. Ann.* 343 (2009) 53–78. <479, 488>

[3042] J. Yuan, C. Carlet, and C. Ding, The weight distribution of a class of linear codes from perfect nonlinear functions, *IEEE Trans. Inform. Theory* 52 (2006) 712–717. <272, 273>

[3043] J. Yuan and C. Ding, Four classes of permutation polynomials of $\mathbb{F}_{2^m}$, *Finite Fields Appl.* 13 (2007) 869–876. <226, 230>

[3044] J. Yuan, C. Ding, H. Wang, and J. Pieprzyk, Permutation polynomials of the form $(x^p - x + \delta)^s + L(x)$, *Finite Fields Appl.* 14 (2008) 482–493. <226, 230>

[3045] P. Yuan, More explicit classes of permutation polynomials of $\mathbb{F}_{3^{3m}}$, *Finite Fields Appl.* 16 (2010) 88–95. <226, 230>

[3046] P. Yuan and C. Ding, Permutation polynomials over finite fields from a powerful lemma, *Finite Fields Appl.* 17 (2011) 560–574. <221, 224, 226, 230>

[3047] P. Yuan and X. Zeng, A note on linear permutation polynomials, *Finite Fields Appl.* 17 (2011) 488–491. <216, 230>

[3048] J. L. Yucas, Irreducible polynomials over finite fields with prescribed trace / prescribed constant term, *Finite Fields Appl.* 12 (2006) 211–221. <54, 59>

[3049] J. L. Yucas and G. L. Mullen, Irreducible polynomials over GF(2) with prescribed coefficients, *Discrete Math.* 274 (2004) 265–279. <55, 56, 59, 79>

[3050] J. L. Yucas and G. L. Mullen, Self-reciprocal irreducible polynomials over finite fields, *Des. Codes Cryptogr.* 33 (2004) 275–281. <56, 59>

[3051] D. Y. Y. Yun, Fast algorithm for rational function integration, In B. Gilchrist, editor, *Information Processing 77—Proceedings of the IFIP Congress 77*, 493–498, North-Holland, Amsterdam, 1977. <381, 382>

[3052] H. Zassenhaus, On Hensel factorization I, *J. Number Theory* 1 (1969) 291–311. <385, 392>

[3053] H. Zassenhaus, Polynomial time factoring of integral polynomials, *ACM SIGSAM Bull.* 15 (1981) 6–7. <387, 392>

[3054] G. Zeng, Y. Yang, W. Han, and S. Fan, Reducible polynomial over $\mathbb{F}_2$ constructed by trinomial σ-lfsr, In *Information Security and Cryptology*, volume 5487 of *Lecture Notes in Comput. Sci.*, 192–200, Springer, Berlin, 2009. <70>

[3055] L. Zeng, L. Lan, Y. Y. Tai, S. Song, S. Lin, and K. Abdel-Ghaffar, Constructions of nonbinary quasi-cyclic LDPC codes: a finite field approach, *IEEE Trans. Communications* 56 (2008) 545–554. <718, 719>

[3056] X. Zeng, C. Carlet, J. Shan, and L. Hu, More balanced Boolean functions with optimal algebraic immunity and good nonlinearity and resistance to fast algebraic attacks, *IEEE Trans. Inform. Theory* 57 (2011) 6310–6320. <250, 252>

[3057] X. Zeng, X. Zhu, and L. Hu, Two new permutation polynomials with the form $(x^{2^k} + x + \delta)^s + x$ over $\mathbb{F}_{2^n}$, *Appl. Algebra Engrg. Comm. Comput.* 21 (2010) 145–150. <226, 230>

[3058] Z. Zha and L. Hu, Two classes of permutation polynomials over finite fields, *Finite Fields Appl.* 18 (2012) 781–790. <221, 226, 230>

[3059] Z. Zha, G. M. Kyureghyan, and X. Wang, Perfect nonlinear binomials and their semifields, *Finite Fields Appl.* 15 (2009) 125–133. <282>

[3060] Z. Zha and X. Wang, New families of perfect nonlinear polynomial functions, *J. Algebra* 322 (2009) 3912–3918. <282>

[3061] L. Zhang, Q. Huang, S. Lin, K. Abdel-Ghaffar, and I. Blake, Quasi-cyclic LDPC codes: an algebraic construction, rank analysis, and codes on Latin squares, *IEEE Trans. Communications* 58 (2010) 3126–3139. <718, 719>

[3062] Q. Zhang, Polynomial functions and permutation polynomials over some finite commutative rings, *J. Number Theory* 105 (2004) 192–202. <229, 230, 232>

[3063] Z. Zhao and X. Cao, A note on the reducibility of binary affine polynomials, *Des. Codes Cryptogr.* 57 (2010) 83–90. <69, 70>

[3064] G. Zhou and H. Michalik, Comments on "A new architecture for a parallel finite field multiplier with low complexity based on composite field," *IEEE Trans. Comput.* 59 (2010) 1007–1008. <814, 823>

[3065] K. Zhou, A remark on linear permutation polynomials, *Finite Fields Appl.* 14 (2008) 532–536. <216, 230>

[3066] G. Zhu and D. Wan, An asymptotic formula for counting subset sums over subgroups of finite fields, *Finite Fields Appl.* 18 (2012) 192–209. <213>

[3067] H. J. Zhu, p-adic variation of L functions of one variable exponential sums. I, *Amer. J. Math.* 125 (2003) 669–690. <485, 488>

[3068] H. J. Zhu, Asymptotic variation of L functions of one-variable exponential sums, *J. Reine Angew. Math.* 572 (2004) 219–233. <485, 487, 488>

[3069] H. J. Zhu, L-functions of exponential sums over one-dimensional affinoids: Newton over Hodge, *Int. Math. Res. Not.* (2004) 1529–1550. <483, 484, 488>

[3070] N. Zierler, Linear recurring sequences, *J. Soc. Indust. Appl. Math.* 7 (1959) 31–48. <312, 317>

[3071] N. Zierler, Primitive trinomials whose degree is a Mersenne exponent, *Information and Control* 15 (1969) 67–69. <96, 97>

[3072] N. Zierler and W. H. Mills, Products of linear recurring sequences, *J. Algebra* 27 (1973) 147–157. <313, 317>

[3073] M. Zieve, Bivariate factorizations via Galois theory, with application to exceptional polynomials, *J. Algebra* 210 (1998) 670–689. <239, 240>

[3074] M. E. Zieve, On a theorem of Carlitz, arXiv:0810.2834, 2008. <238, 240>

[3075] M. E. Zieve, Some families of permutation polynomials over finite fields, *Int. J. Number Theory* 4 (2008) 851–857. <223, 224, 230>

[3076] M. E. Zieve, On some permutation polynomials over $\mathbb{F}_q$ of the form $x^r h(x^{(q-1)/d})$, *Proc. Amer. Math. Soc.* 137 (2009) 2209–2216. <221, 223, 230>

[3077] M. E. Zieve, Classes of permutation polynomials based on cyclotomy and an additive analogue, In *Additive number theory*, 355–361, Springer, New York, 2010. <221, 224, 230>

[3078] P. Zimmermann, Avoiding adjustments in modular computations, 2012, preprint available at `http:www.loria.fr/~zimmerma/papers/norm.pdf`. <354, 363>

[3079] T. Zink, Degeneration of Shimura surfaces and a problem in coding theory, In *Fundamentals of Computation Theory*, volume 199 of *Lecture Notes in Comput. Sci.*, 503–511, Springer, Berlin, 1985. <463>

[3080] R. Zippel, Probabilistic algorithms for sparse polynomials, In *EUROSAM '79: Proceedings of the International Symposium on Symbolic and Algebraic Computation*, volume 72 of *Lecture Notes in Comput. Sci.*, 216–226, Springer-Verlag, Berlin, 1979. <391, 392>

[3081] R. Zippel, Newton's iteration and the sparse Hensel algorithm (Extended Abstract), In *SYMSAC '81: Proceedings of the fourth ACM Symposium on Symbolic and*

Algebraic Computation, 68–72, ACM, New York, 1981. <391, 392>

[3082] Z. Zlatev, *Computational Methods for General Sparse Matrices*, volume 65 of *Mathematics and its Applications*, Kluwer Academic Publishers Group, Dordrecht, 1991. <532, 535>

[3083] K. Zsigmondy, Über die Anzahl derjenigen ganzen ganzzahligen Functionen $n$ten Grades von x, welche in Bezug auf einen gegebenen Primzahlmodul eine vorgeschriebene Anzahl von Wurzeln besitzen., *Sitzungsber. Wien Abt II* 103 (1894) 135–144. <364, 368, 374>

[3084] D. Zywina, Explicit class field theory for global function fields, preprint, 2011. <538, 546>

Index